The Collected Papers of R. H. Bing

R. H. BING
1914–1986

The Collected Papers of R. H. Bing

Volume 1

Sukhjit Singh, Steve Armentrout, and Robert J. Daverman, Editors

AMERICAN MATHEMATICAL SOCIETY
PROVIDENCE, RHODE ISLAND

1980 *Mathematics Subject Classification* (1985 *Revision*).
Primary 57-06, 54-06, 26-06.

Library of Congress Cataloging-in-Publication Data

Bing, R. H.
[Works. 1988]
The collected papers of R. H. Bing/Sukhjit Singh, Steve Armentrout,
and Robert J. Daverman, editors.
p. cm.
Includes bibliographies.
ISBN 0-8218-0117-1 (v. 1: alk. paper)
1. Manifolds (Mathematics) 2. Bing, R. H. I. Singh, Sukhjit
II. Armentrout, Steve.
III. Daverman, Robert J.
IV. Title.
QA613.B45 1988 88-14445
516′.07–dc19 CIP

Printed in the United States of America

The paper used in this book is acid-free and falls within the guidelines
established to ensure permanence and durability. ♾

TABLE OF CONTENTS

III. Collected Papers of R. H. Bing

I. General Topology

A. Metrization

B. Retracts and Related Topics

B. Cuttings

C. Planar Embeddings

III. Continua

A. Pseudo Arc and Related Topics

B. Locally Connected Continua: Partitions and Decompositions

C. Homogeneity

D. Others

IV. Topological Classifications

A. 2-Sphere

B. 3-Space and 3-Sphere

C. Arcs and Simple Closed Curves

D. Hilbert Space

V. Miscellaneous Topics

E. Fundamental Group Computations

F. Others

VI. Topology of 3-Manifolds

A. Decompositions

C. Approximating and Embedding Surfaces

D. Recognizing Tameness and Related Topics

E. Topics Related to the Poincaré Conjecture

EDITORIAL PREFACE

These volumes bring together all of R. H. Bing's published mathematical papers; however, they include neither his book on the topology of 3-manifolds nor the two collections of research papers he helped edit.

Biographical data appears here in a remarkable document (R. H. Bing: A Study of His Life) put together as a first person narrative based on information, anecdotes, and comments which Bing taped and subsequent transcriptions which he helped modify and, ultimately, approved. See S. Singh's Introduction to the Study for details of how this came to be.

Similarly, the classification of papers by subject matter is significant in that it also was constructed by Bing himself, at the instigation of and in conjunction with S. Singh.

A somewhat nonstandard item included here is a list of abstracts Bing published. Most of the results announced eventually found their way into print in more formal form, frequently with their titles unchanged. One justification for their reproduction is the belief shared by the editors that these abstract dates help accurately pinpoint exactly when certain of Bing's ideas matured to an important state of fruition.

We wish to thank everyone who helped, especially those mathematicians who contributed many missing reprints—the list is too long to mention everyone by name. Mary Lane deserves our special thanks for her part in bringing this project to a successful completion. Finally, we acknowledge our debt to Mary (Mrs. R. H.) Bing for her enthusiastic support of this project.

The Collected Papers of R. H. Bing

I. R. H. Bing: An Introduction

R. H. BING: A STUDY OF HIS LIFE

S. SINGH

0. INTRODUCTION

It was during a visit with R. H. Bing (affectionately called "RH" by his friends and close associates) while he was hospitalized for two weeks in the summer of 1984 that the idea of doing a Bing biography occurred to me. As R. H. greeted me from his bed, he appeared weak but in good spirits, his eyes shining with anticipation of some good conversation about mathematics. Throughout our talk, R. H. would tell me of an interesting event in his life accompanied by an anecdote from his childhood or student days. After such an enlightening visit, it entered my mind how historically crucial it was for the mathematical community to have a written record of R. H.'s life.

R. H. regained his health and strength and returned to his office on the campus of The University of Texas at Austin. It was here that I approached him and solicited his help with this project. My intent was to write his biography under his guidance not only for authenticity but with the hope of capturing the "Bing persona." R. H. reluctantly agreed but became more enthusiastic as the project progressed.

My first goal was to collect biographical data through a series of informal interviews I had with R. H. several times each week in which he answered questions concerning the various stages of his life. However, I soon began to realize that my note-taking alone could not serve as a sole resource for accurate biographical data, much less convey the colorful colloquialisms of Bing. It then occurred to me that the use of audiotapes would be an excellent medium for gathering information.

After a considerable amount of persuasion, R. H. agreed to do a tape, or a sequence of tapes, following an outline of topics which I provided. However, R. H. proceeded only under the condition that the tape be used only to facilitate the writing of the biography and that it not be made public.

R. H. was quite concerned that his random comments might be misunderstood or that he might easily say inaccurate things during the flow of a taping session.

The biographical sketch you are about to read is a culmination of years of separate and joint efforts by R. H. and myself at rearranging, editing, adding, and deleting material from the original transcription of this tape. Originally, I began writing the Bing biography in third person based on this tape, but soon discarded this approach. It occurred to me that the first person narrative style manifest in the tape was more suitable in exemplifying the "Bing persona" as well as his unique conversational style.

I started by carefully rearranging, editing, and polishing a transcription of the tape. This draft served as the starting point. Subsequently, R. H. and I both jointly and separately kept adding to and deleting from this draft. At one point, we agreed that the project was finished, and I left a copy of the final version with him at his home in Austin. Sure enough, R. H. returned this copy to me with some additional changes only one month before his death. I have incorporated these revisions.

So, welcome to a rare insight into the special character and spirit of R. H. Bing. The strength, love, and discipline he received and learned formed the underlying core of his commitment to mathematics. Although many contributed to his lifelong commitment to mathematics, the most notable are his mother Lula May, his teacher R. L. Moore, and his wife Mary.

1. FAMILY BACKGROUND AND CHILDHOOD

My father was reared near Waller, Texas. Grandfather Bing died at a relatively young age, and I never met him. However, I did visit the family farm when I was very young. He and his brother had come from Canterbury, England. His brother, not satisfied with life in this country, returned to England. My father kept in contact with the Canterbury Bings until World War I. When I visited England, I tried to find more about the Canterbury Bings and was told that they had been prominent members of the community and had owned a bottling company. A company there was still called the Bing Bottling Company and manufactured a product called "Bing Soda"; however, it was not owned by the heirs of the original Bings.

My grandfather Bing was a relatively well-educated farmer for his time. Record keeping by farmers was a bit unusual at the turn of the century, but I remember seeing books that he kept showing the price that he paid for farm animals and implements and the price he got in return for crops that he sold.

My maternal grandfather was named Thompson. He had been a bandleader and conducted music at various revivals. This caused him to move about quite a bit. My mother was born as the family moved through Emporia, Kansas. When I visited Emporia several years ago, I looked at newspapers of that time to see if her birth had been recorded, but it had not. Both of my mother's parents died while she was still in high school. This left my mother with a younger sister and brother to be reared by kinfolks in Leonard, Texas.

My mother was quite eager to get out and make a living for herself. She did temporary teaching while still in high school (at age 15). As soon as she finished high school, she took examinations to qualify for a temporary teacher's certificate. My mother told me about how she went about looking for a job. Taking a pen in hand, she closed her eyes, moved her hand over a map of the state of Texas and put the pen down on the map. She did this three times and wrote to the three schools picked in this fashion. My mother received three job offers and accepted the first, located in Oakwood, Texas. This turned out to be the one paying the least, but since she had accepted it earlier, she felt a strong commitment to take the job. Soon after moving to Oakwood she sent for her brother and sister to come and live with her.

My mother contracted a disease during her high school years which caused the loss of one eye, but she was still quite popular and successful as a primary teacher in Oakwood. During her first year, however, she thought herself as nearly being fired when she introduced phonics. But, her enthusiastic persuasion caused the students to read beyond their parents' expectations and she became very respected. After several years of teaching in Oakwood, she married my father who was then superintendent of the Oakwood School District.

Since my father died when I was five years old, I do not have a good memory of him. However, my mother had great respect and affection for my father and much of what I know about him is probably something I learned from her. My mother stopped teaching after marriage. She and my father had two children of which I was the older. I also have a sister. Having a sister has been a valuable experience for me, although at the time my sister did not appreciate it—especially when I encouraged her to have a make-believe doll funeral.

I recall that I loved my father very much and have pictures of him putting me on a horse and giving me a ride. I do remember him telling me stories. During his final illness, I was taken to a neighbor's home and told of his death. I did not comprehend this, and on returning home, looked for him in the barn, under the house, and other places where I thought he might be hiding. Two stories about my Dad come to mind. Some of this I recall and some I remember only because I was told about it later.

During those days it was the responsibility of the superintendent of the schools to see that order was kept in the classes. The women teachers would send students that had been giving them unusual trouble to the superintendent for punishment. Corporal punishment was much more prevalent in those days than it is today and on one occasion my father resorted to whipping a boy for some bad offense. Several days later this boy, along with his father and two of his larger brothers, waylaid my father on the way home from school, intent on beating him up. My mother saw this and came running with sticks and stones. The men relented and left with a warning that they would get my father later.

My mother was very apprehensive for my father's safety and suggested he order a gun to protect himself. My father pretended not to be concerned and told my mother not to worry. However, my mother went ahead and ordered a pistol from the hardware store; not letting my mother know, my father also ordered a shotgun. Hence, we soon had both a pistol and a shotgun. As far as I know, these were never used during my father's lifetime. But I have heard that my mother carried the pistol in her purse and said she would shoot any man that attacked my father.

Several years after my father's death, my mother had occasion to use her pistol which led to her recognition as a marksman of some merit by the local gun club. Hogs belonging to someone down the street frequently broke into our garden. We repaired our fence and ran the hogs off, but a day or so later they would get out and come and root up the garden again. My mother decided that she would try to frighten the hogs by firing shots near them to keep them away. She got out the pistol and asked me to chase the hogs out; her accuracy, however, was not too good. As I chased out a hog and she shot the pistol, the hog fell over on his back and started kicking. Soon it was dead. My mother called the meat market and asked if they would get the hog because she thought it would be useful. The meat market, however, would not take the hog because they did not know its owner. The daughter of one of the neighbors brought over a butcher knife and offered to help butcher the hog but the plan was not pursued.

Soon the story was all over town of how my mother had killed a hog—what a crackshot she was with a gun. The owner of the hog never came forward because he was ashamed that his hog had destroyed my mother's garden. As the hog began to stink, my mother had someone put a rope around the carcass and drag it off to where dead cattle were left. The men of the town had great fun telling this story. They had a gun club that was going away to engage in a contest, and were making my mother captain because she was so accurate at shooting. I am not sure of the significance of this story, but thought it quite interesting because it made such a deep impression on my mother. I do not remember my parents purchasing these guns, but I do remember the kicking hog.

Another story about my father, perhaps more pertinent, involved the local schools for the whites and the schools for the blacks. Although these schools were separate in those days, they were in the same district and hence my father was superintendent of both. He knew how to order books and do other things that were useful to the black schools. I recall that one day, when I was about 10, I met several black youths. They asked me if I was the son of Superintendent Bing. I said that I was. They said, "He was a very good man, he was always fair to us. You can never be as good a man as your father was." This made an impression on me.

My father was successful as a teacher and the Oakwood schools turned out many outstanding people during his tenure; one became president of Chrysler Motor Company. Several years before his death, however, my father decided to leave teaching and take up farming. He acquired several farms and a store and was quite successful at managing them. My mother said that my father was a workaholic. As a matter of fact, work may have led to his early death. One day he had some plowing to do and it began to rain. My mother insisted that he come in out of the rain but he said that he had just a bit more to do and continued plowing. As a result of this, he took influenza or pneumonia or some other disease of that time and died in 1920 leaving my mother to take care of two children—the older being just five.

The following story about my mother reveals something of her character. She was a strong disciplinarian and made it a practice of confronting students who had done something wrong and having them confess what they had done before revealing what she knew. This worked well most of the time, but I remember one case in which it did not.

I was playing in the backyard and a neighbor told my mother of something that the neighbor thought that I had done. My mother called me into the house and asked me what I had done wrong. I did not recall having done anything wrong, so my mother gave me a few paddles and told me to go off and try to recall what it was. I still was unable to decide what she wanted, so I received another paddle. This happened several times and finally my mother thought she would give a hint. She said, "What were you saying in the backyard?" Then I remembered that I had been pounding on a stick, driving it into the ground with a hammer and saying, "Go down, go down, go down." The neighbor had heard me and thought I was cursing and saying, "goddamn, goddamn, goddamn," and this is what was told to my mother. I remember seeing the remorse on my mother's face when she found that she had punished me needlessly. She went off and cried and tried to make up to me for the harm that she had done. This impressed me in two ways. She was interested that I did not use profanity, and she was very concerned when she found that she had wronged me. I long forgot the punishment but not these two things.

My mother felt that it was an excellent idea for mothers to train their children at an early age. Long before starting school, I was able to read and do arithmetic and regarded these things as great fun. I think I owe a great deal to my mother's early training for my interest and success in school.

When I started school, I had already memorized all of the primary books and found school very, very easy. During several months of the first year, I was ill, but by the second year I was put in the third grade. Although I had to work during my first few years to stay up with the class, by the time I had gotten to junior high school, I was able to lead my class. One thing that bothered me was my being younger than my classmates and hence at a disadvantage when competing in athletics with boys who were older.

I credit my mother with much of my success in mathematics. Before starting school, I learned that mathematics was fun. In the seventh grade, I entered the county number sense contest. My mother was coach of the team and she taught shortcuts enabling one to do computations quickly. We learned to approximate answers to harder problems. I later learned that my partner and I made the highest grade in the state in the number sense contest that year—more than likely due to my mother's coaching.

When I took geometry, my mother was quite interested in how I went about proving theorems. She taught me that the purpose of geometry was to discover proofs rather than to memorize them. I was never pushed in my pursuit of mathematics and wondered whether I would have liked it nearly so well if I had been offered a speeded-up course such as we give some of our better students today. Today, as well as then, I am grateful that I had the time to think, to contemplate and to work out mathematics on my own rather than to be pushed ahead to learn proofs that were provided by others.

2. COLLEGE DAYS

Several years after my father's death, our funds had been depleted to the extent that my mother returned to teaching to earn a living. (The pay was only $6.50 per month.) Yet, there was never a question as to whether or not my sister and I would go to college. The question was only how it could be afforded. I recall that one of our friends wanted my mother to borrow enough money to send me to Princeton. He offered to help in fundraising; however, my mother hesitated about this obligation and thought it wiser for me to attend a school that I could afford.

Two of my favorite high school teachers had finished their college education at Southwest Texas State Teacher's College, or Southwest Texas State University as it is now called, and arranged for me to get a job at San Marcos in the college cafeteria. This is the way that I supported myself during college. I was quite frugal; my mother and I figured it had cost me less than $300 to get my bachelor's degree, most of which was for tuition. I did, however, complete college in $2\frac{1}{2}$ years—eager to finish in a hurry so funds could be used to send my sister to school. She also went to Southwest Texas State Teacher's College.

Several years after my sister and I had gotten our degrees from Southwest Texas State Teacher's College, and I had even received a masters and Ph.D. from The University of Texas, my mother, by going to school in the summers, also received a bachelor's degree from Southwest Texas State Teacher's College in San Marcos.

Young R. H. (approximate age 2–3 years; 1916–1917) on a horse with assistance from his father, Rupert Henry.

R. H. (6 years old) with sister Percy May (5 years old) Oakwood, Texas, 1920.

R. H. age 12, Oakwood, Texas, 1926.

R. H. with mother Lula May, circa 1935.

A portrait of R. H. (taken by a studio at the University of Wisconsin) circa 1953.

R. H. pointing to a poster describing Alexander's Horned Sphere during a lecture, circa 1955.

R. H. with wife Mary at Lake Wisconsin, Summer House, Wisconsin, circa 1973.

R. H. (right) with O. Kueker (left) during the Southwest Texas State University Topology Conference, 1984. This conference was held to honor R. H.'s 70th birthday. Mr. Kueker made and presented a model (shown) of the first 3-stages of Bing's Dogbone decomposition.

3. STUDENT OF R. L. MOORE: GRADUATE SCHOOL

After I finished college, I had teaching positions in three Texas high schools. Although these $4\frac{1}{2}$ years of teaching did not add to my mathematical progress, it taught me how to get along with people. While teaching in high school, I went back in the summers and took courses at The University of Texas. It was there that I met Professor R. L. Moore. R. L. Moore appeared to have a very low opinion of courses in pedagogy or education, and I think he suspected I would not be very good since I had spent so long teaching in high school. I was soon able to relieve him of this notion, being able to prove theorems that his students, who had not taken time off to teach in high school, could not prove.

R. L. Moore was quite serious about his teaching, as he was about other things. When anyone talked to him, he looked at the person with penetrating eyes. I was careful to try to not say anything unless it was accurate. He was interested in what a person could do himself. He placed books off limits for his students when these books were dealing with things on which they could work themselves. As a matter of fact, in his course on topology he placed his own Colloquium book off limits.

Moore taught not only in the classroom but in the halls and elsewhere. I remember my landlord telling me how interested R. L. Moore was in my work and how Moore tried to explain to him some of the things I was doing. My landlord did not understand any of this. But this made a profound impression on me that Moore was so interested in my work that he would try to explain some of it to my landlord.

Mrs. Moore was a real asset to Dr. Moore. I remember Mrs. Moore telephoning my wife, Mary, and telling her that she realized what a sacrifice my wife was making in order to make it possible for me to spend so much time on mathematics. I believe the word "sacrifice" came as a surprise to Mary who thought running a household was a 50-50 proposition. But being told by Mrs. Moore that a wife might enhance her husband's career by letting him spend full time on it did have an impression. Mary has always been a big help in entertaining students, managing the family, and providing a pleasant environment.

In Moore's classes, he taught by indirection. I recall Moore having one of his ex-students visit his class. This ex-student had not produced as a research mathematician and after he left, Moore said, "How can I stand having a man like that visit my class? He had a real opportunity and he did not take advantage of it!" The students got the message.

Moore did not like for people to have sloppy appearances. He expected them to sit up straight in their chairs and to behave like ladies and gentlemen. Once he showed a picture of students at the University of Chicago who were being quite sloven and his remarks were, "Do you think the University of

Chicago is that bad? Do the students there think so little of themselves as to behave in such an undignified manner?" Again, we got the message.

One of the teachers at The University of Texas had a wife who would come to the building after hours to pick up her husband and blow on the horn to advise him that she was there. Moore thought it was inappropriate to make noise around a building where research was being done. He would ask his students, "What do you think of a man who would permit his wife to come to the building and honk the horn outside?" Again, we got the message.

Moore was very proud of his students who had succeeded in research. He told us stories of how Ray Wilder had been able to prove a theorem in a short period of time. How Gordon Whyburn had had successes. How Burton Jones had been able to do such and such. How one of his students, Miss Miller, had succeeded where others had failed in proving a certain theorem. Again, we got the message.

Dick Anderson was a student at The University of Texas who got his degree several years after I did. He and I frequently spent part of the noon hour playing chess. Moore did not approve of this and told me that I should not waste Anderson's time which could be better spent on mathematics. I also got the message that Moore thought I could make better use of my time.

On another occasion in the halls someone had asked me who had won the football game the preceding Saturday and whether or not I knew the score. I remembered who had won the game, but I was unable to provide the score. Moore's remark was, "At least, thank goodness for that!"

I will always appreciate the good instruction I received under R. L. Moore. It was to my liking because I enjoyed working out mathematics for myself and this is what he encouraged. However, I did not like the close supervision he gave my thesis. When he had me write things, he wanted it in his own way rather than in mine. I felt very glad when my thesis was finished, for I now felt I had the author's prerogative of saying things the way I wanted to say them. Moore felt that if it was said correctly, it didn't really matter whether or not it was easily understood because it was the reader's responsibility to dig it out. My feelings were that we should say things, often repeatedly, in an effort to make it more understandable. Moore's conciseness in written presentation did not always carry over to his oral discussions since he took care to make them clear.

4. EARLY RESEARCH AT TEXAS

After receiving my degree from Moore, I visited him many times. We never discussed mathematics together. In one case, shortly after I received my degree from him, I proved one of the big theorems that had been baffling quite a few people; namely the Kline sphere characterization. After writing this up, I submitted it as an abstract to the American Mathematical Society and also to a journal without ever having shown or discussed it with Moore. When it became known that I had announced the result, we were flooded with phone calls, telegrams, and letters—mostly addressed to Moore—asking if I had gotten the result and if it had been checked.

Moore's reply was to turn these letters over to me to be answered because he had not seen the proof. Yet, he did not ask me for the proof. I think that he thought that by asking me, he would show lack of confidence in my ability to check the proof myself. I also did not show him the proof because I felt that this might show that I did not have self-confidence and needed someone else to help me check the result.

At Chicago I announced the result. Others had announced partial solutions to the result, and before I got on the program I was offered additional time so that I could better explain my proof. I was only able, however, to give the presentation that I had planned. I do recall Paul Erdös asking me at that meeting whether I had had a first-class mathematician check the proof. I rather resented this because I thought that it was a first-class mathematician who had provided the proof.

My thesis dealt with simple plane webs and I wrote several papers on this subject. I learned, however, that the mathematical community was not very interested in webs. There was much more interest in the Kline sphere characterization; at least in knowing that it was true. I found that in the case of hard theorems, the mathematical community is quite often interested in knowing whether the result is true. But many mathematicians will not go through the work of checking to understand the proof.

I spent two years at The University of Texas after getting my Ph.D. There were other interesting results that I got while I was there; one of the most interesting dealt with the pseudo-arc. Ed Moise had written his thesis about the pseudo-arc and showed that the pseudo-arc had the interesting property that it was homeomorphic to each of its nondegenerate subcontinua. That is the reason he called it the pseudo-arc. It turned out later that other mathematicians had described the pseudo-arc earlier by other methods. I was quite interested in Moise's description, but wondered if a different description might not show that the pseudo-arc was homogeneous and, indeed, found that this was true.

When Moise was still a student and I already had my Ph.D., I was reluctant to announce the result that the pseudo-arc was homogeneous and thought that Moise should at least have a chance to make such an announcement if he were close to discovering it. I asked him if he had considered the matter as to whether or not the pseudo-arc was homogeneous (knowing all the time that it was). He said that he had checked this and found indeed that it was not. So several months later, I announced that a pseudo-arc is homogeneous. I found that while some people were interested in the pseudo-arc, many were not because the pseudo-arc did not have the nice geometric properties of differentiability or linearity of many geometric objects. I decided that mathematicians were much more interested in manifolds than they were in exotic objects.

5. WISCONSIN YEARS AND MATHEMATICS OF THIS TIME

Two years after taking my Ph.D. at Texas, I went to the University of Wisconsin where I had a very successful career. After two years at Wisconsin, I took a leave of absence and spent a year as a visitor at the University of Virginia. I had as an office mate, Ed Floyd, and learned quite a bit about manifolds from him. I taught a class there in which the members were Gordon Whyburn, Ed Floyd, Vic Klee, Truman Botts, and Bob Williams. This was a very interesting class and we studied current literature. I obtained several very good results during the year I was at the University of Virginia.

Floyd told me that people were interested in knowing whether there was an involution of a 3-sphere onto itself different from ordinary involutions. There was a suspicion that probably the union of two solid Alexander horned spheres sewed together along their boundary gave S^3 and that an involution that interchanged these two solid spheres (or Alexander balls) would be an involution inequivalent to a standard involution. By the use of strings, rubber bands, and other methods, I was able to show that the union of two solid horned spheres sewed together along their boundaries with the identity homeomorphism was indeed a 3-sphere.

I found that many people were quite interested in this result. Since a proof was not very hard, others also learned the proof. It is only this year (circa 1984) that I have discovered a new proof which gives other results along this line. Mike Freedman and Richard Skora have made use of these new results.

The pseudo-arc is an indecomposable continuum. It is not the sum of two proper subcontinua. In fact, it is hereditarily indecomposable since each of its proper subcontinua has this property. A student from the University of Virginia, John Kelley, who is now at Berkeley, had shown that if there is a two-dimensional hereditarily indecomposable continuum, then there is an indecomposable continuum of dimension three, one of dimension four, and in fact, for each positive integer n, a hereditarily indecomposable continuum of dimension n. However, it was not known whether there was a hereditarily indecomposable continuum of dimension two.

I remember spending many hours at the University of Virginia trying to solve the problem of whether or not there is a two-dimensional hereditarily indecomposable continuum. As a matter of fact, late one evening, while I was still at my desk thinking about this problem, Vic Klee came by. I remember telling him, "I think I have this problem about solved. If I could put all of the ideas together that I have in mind, I believe that I would have a solution. I am going to sit here and work on this until my mind is able to put these things together and provide a solution." And, indeed, I succeeded; in fact, I was able to provide a proof quite different from that of Kelley's that there are hereditarily indecomposable continua of all dimensions. The example constructed was an exotic continuum and not so many people were interested in it as they were before the problem had been solved.

I have noted that many other results that are of considerable interest before they are proved meet similar fates. Dick Anderson and I gave an elementary proof that there is a homeomorphism between Hilbert space and the cartesian product of a countable number of lines. The question of whether or not there is such a homeomorphism had been of interest not only to topologists but also to many analysts since Hilbert space was in their arena. However, after the result was proved, interest waned, and many who had previously seemed deeply interested, now were willing to accept the result without caring to know why it is true.

A somewhat similar reaction met the discovery that all 3-manifolds can be triangulated. Many topologists now accept the result without knowing a proof that it is true. A complicating factor is that there are no known elementary proofs. I have great respect for those that understand the basis of their own work.

I have long been interested in metrization problems. When I was a graduate student at The University of Texas, I worked on the problem as to whether or not every normal Moore space is metrizable. This question had been called to Moise's attention when he was visiting a meeting of the American Mathematical Society in New York. I noted that an affirmative answer could be given by using a result of F. B. Jones. Jones, however, did not remember how his original proof went and began to suspect that maybe it was false. Others took up the hunt for the solution. On different days we heard the rumor that various Moore students had either proved or disproved the theorem. In fact, Moore himself let it be known that he had a solution. An examination of his archives in the Humanities Research Center at The University of Texas showed that his write up had a gap.

I found many results about normal Moore spaces and finally decided to publish a paper telling what I knew, even though I had not gotten a complete result. For example, I published a paper which showed that a Moore space was metrizable if it was collection-wise normal. For those wanting to know what this means I remark that a Moore space is a regular Hausdorff space which has a countable family of open coverings $G_1, G_2, \ldots$ such as that if p is a point and N is a neighborhood of p, then there is an integer n, such that each element of G_n that contains p lies in N. A space is collection-wise normal if for each discrete collection of closed sets, there is a collection of mutually exclusive open sets such that each of the elements in the discrete collection lies in one and only one element of the collection of open sets.

My paper contained the result that a Moore space is metrizable if it is collection-wise normal. It also contained a necessary and sufficient condition telling when a Hausdorff space is metrizable. It turned out later that both Smirnoff from Russia and Nagata from Japan had gotten similar results at about the same time. This suggests that when it is time for a new result to be born, and that if one person does not get it, another person is likely to. No research mathematician is indispensible for progress in mathematical research.

A square has a convex metric; namely, if p and q are two points of a square, there is another point of the square halfway between them. People have long been interested in the question as to whether or not each compact connected, locally connected metric space has a convex metric. Karl Menger had been very interested in this problem. Gustav Beer had what some people thought was a solution; however, others found fault with his proof. He submitted his paper to a journal many times and each time it came back with suggestions for corrections. Menger told me that the decision to publish the paper was made so that it would have a wider range of critics. I examined this paper and whether or not it had errors, I did not know. But I did see that Gustav Beer was considering only the problem where the space under consideration was one-dimensional. After studying Beer's paper, I decided that indeed a locally connected metric continuum of any dimension could be partitioned and this partition could be used to show that the space had a convex metric. Moise obtained a related result at about the same time but his proof had a minor error.

Although for many years I was quite interested in the normal Moore space problem, I lost some of my interest when Cohen proved that the continuum hypothesis could not be proved on the basis of the Zermelo-Fraenkel axioms. It was later proved that the normal Moore space problem could not be solved on the basis of these. It seems like there were two points of view. One was that something could be considered as true if it could not be shown to be false even if in a larger space it could be proved false. Another point of view is that a thing can be considered as false unless it can be proved to be true in the space under consideration. The notion of forcing is related to these ideas.

While we might find something that at present we do not recognize as an axiom, the above does show that on the basis of the axioms that we do recognize, we are unable to prove all theorems in mathematics. In other words, mathematics cannot be based on axioms (as we know them) alone.

I will now tell of an idea I got as a boy that helped me in later research. I was assigned the task of drawing a map. One way was to take a map, put it on the window, place a piece of paper over it and then trace. I had more difficulty, however, when I was asked to draw a map of a different size. However, I found that if I divided the page containing the original map into rectangles and then similarly divided the paper on which I was to draw the map, then I could make a reasonable copy of the map without resorting to the use of the window. In fact, even if the paper on which the drawing was made was of a different size from the original map and if one put the corresponding parts into the corresponding rectangles, then one got a reasonable copy of the original map, especially if the rectangles were small. A mosaic copy of a picture looks somewhat like the original.

I noted that a person could take a map whose boundary lines were curved (or even squiggly) and approximate it with a map whose boundary lines were polygonal; then if one did not examine the changed map too carefully, it resembled the original. This is the idea behind the fact that if one looks at a mosaic from a distance, it looks like an object which is not polygonal. Also, if one looks at a TV set, one does not see the dots, but a picture.

Let me discuss how I used this notion many years later. In the 1950's, I learned that Moise had proved that if h is a homeomorphism of a 3-cell into Euclidean 3-space, then this homeomorphism could be approximated by a piecewise linear homeomorphism. This result has many applications, and I felt that it was very important. Reading Moise's papers proved very difficult; in fact, I have heard that others had difficulty in reading these papers, even the referee whose name was Hu. Hu is said to have met with Moise frequently in order to have Moise explain the proofs to him. There is somewhat of a related joke. Some people make the remark that Hu (who) refereed Moise's papers. It sounds as though they were asking a question, if "Hu" was mistaken to mean "who". "Hu refereed Moise's paper. Who? Hu! Who? Hu!"

The first step in Moise's proof seems to show that if h is an embedding of the cartesian product of a 2-sphere S^2 and an interval $[0,1]$ into R^3, then there is a polyhedral 2-sphere in this image which separates the two boundary components of $h(S^2 \times [0,1])$. His proof goes something as follows. Let X be the closure of the bounded component of $R^3 - h(S \times \{\frac{1}{2}\})$. The first Vietoris homology of $Bd(X)$ is trivial. In a sequence of steps Moise changed X to a polyhedral 3-manifold M with boundary so that $Bd(M)$ has trivial first Vietoris homology and separates $h(S \times \{0\})$ from $h(S \times \{1\})$ in R^3. Then $Bd(M)$ is the required 2-sphere. Using this as a start, he showed that any homeomorphism of a 3-cell into R^3 could be approximated by a piecewise linear homeomorphism.

As I wrestled with Moise's paper, I decided that there might be a different way of proving the same result and recalled the techniques I had learned for copying a map when I was a boy. Hence, I decided to triangulate R^3 and use the 3-simplexes as I had used the rectangles. I regarded $h(S^2 \times \{\frac{1}{2}\})$ as the boundaries of the states and produced a polyhedral 2-sphere approximating $h(S^2 \times \{\frac{1}{2}\})$ by aping the methods I used as a boy to copy a map. With this step I got the same results that Moise did and was able to make the same applications that he did; namely, that a homeomorphism of a 3-cell into R^3 can be approximated by a piecewise linear homeomorphism and that any 3-manifold can be triangulated. Later I was able to prove the side approximation theorem and got even more results. Many papers resulted from the side approximation theorem. I suspect that my results on the approximation theorem and the side approximation theorem even have a larger impact on 3-dimensional topology than any other of the results I have obtained in this area.

6. COLLABORATION WITH OTHERS

Let me make some remarks about collaboration with other mathematicians. First, I would like to talk about collaborating with one of my own students, Jim Kister. I mentioned to him as I have to my other students that it was the responsibility of my students to keep me informed of what was going on as I became older. Kister and I were visitors at the Institute for Advanced Study at the same time and we discussed the embedding of objects in Euclidean 3-space. I was pleased to learn that Kister had learned many things since I had him as a student, and we put some of these notions into a joint paper.

Another person with whom I wrote a joint paper was Ed Floyd. When I was visiting the University of Virginia and had Floyd as an office mate, I discussed with him partitioning of sets and how the notion of partitioning might be used to duplicate some of the things we knew about manifolds. Now, the triangulation of a manifold is a special example of a partitioning of a 3-manifold. Floyd knew much more about manifolds than I did and as a result of some observations of his, we were able to use the notion of partitionings and expand them.

Burton Jones and I also wrote a joint paper. Actually, this paper did not result from joint work, but resulted from the fact that both Jones and I had gotten some results about the pseudo-arc of the same nature. It seemed inappropriate for Jones and me to write separate papers covering essentially the same topic. We then decided to write a joint paper with each of us writing one part of it based on work we had done independently.

Two joint papers were written with Karol Borsuk. One of these resulted from a discussion we had in Warsaw, Poland, and the other from discussions we had when Borsuk was a visitor at the University of Wisconsin. I was amazed by the amount of knowledge that Borsuk had about 3-space and the dedication with which he worked at mathematics. I learned a great deal from Borsuk—not only about mathematics but what devotion to mathematics really means.

Still another person with whom I wrote a joint paper was Joe Martin. We were both working on the Poincaré conjecture and had frequent discussions about it. Most of the work in our joint papers was done independently.

I have received much satisfaction in writing joint papers with Dick Anderson, Steve Armentrout, Woody Bledsoe, Karol Borsuk, Mort Curtis, Ed Floyd, Burton Jones, Jim Kister, Andrzej Kirkor, Vic Klee, Joe Martin, Dan Mauldin, Mike Starbird, and others. Mathematics has been produced and friendships strengthened. But, much of the joint work was created when partners were working separately. In general, I think better results are obtained by those who stand on their own feet and work alone.

7. TEACHING METHODS AND RELATED PHILOSOPHY

Let me make a few remarks about training students and how my methods of training are similar as well as different from R. L. Moore's. We both try to train students by having them *do* mathematics. Moore, however, would have his students work on things in which he was interested—whether or not these were of great interest in the mathematical community. I have tried to lead my students to work on things that were of general interest to other mathematicians so that if they were to go to any one of the leading mathematical centers in the country, they would find people with kindred interests.

My method agreed with Moore's in that I tried to get students to discover their own thesis topics rather than assign these topics to them. As we study papers I frequently ask them, "Why did not the person writing this paper prove more? What would happen if one changed the hypothesis? Are there examples that would show that the theorem is false if something is dropped?" If a person studies a paper in this frame of mind, most good papers will lead to other research.

Also, like Moore, I tried to make all of the students in my classes participate in proving theorems. Those inept at proving theorems can at least have the experience of trying and encountering difficulties before seeing the proofs of others. They can learn (and perhaps appreciate) definitions. However, I give greater praise to those who can make original contributions themselves, who can discover examples on their own and unearth new attacks or new theorems. I recall one unusually successful class of 40 students I taught at the University of Wisconsin. The class was supposed to meet 3 times a week. I lectured to the students for one of the periods and divided the class into the other 2 periods, letting half of them come one day to give their results and the other half come another day for their results. Several members of this class went on to get Ph.D.'s and to produce even more significant results. I can't help but think that this technique I used encouraged them.

Another innovation I have used in teaching seminars, particularly when studying new results, was to have the seminar meet for about 2 weeks and on the last day only those people who could bring a conjecture related to what had come before were permitted to attend. The idea being that those who were not able to come with a conjecture had not been thinking enough about what was going on to merit being permitted to attend the final conjecture session.

I have a basic feeling that mathematics is fun and should be fun for the participant. It does few people any good to force feed them into learning things that they do not enjoy. They are unlikely to make much use of the material they learned—only having to do it in order to make a grade or meet a requirement.

At lower levels students will not be able to prove many theorems, but they should be able to show what they have learned. We should be interested in what they are doing and praise them for their progress. I do not have much sympathy for programs where we try to give a strong dose of mathematics to students who find mental activity distasteful. What I do have great sympathy for are those who do not find mathematics easy. If I were to limit my friends to those who are good in mathematics, I would lose many of those that I have. I believe we should strive to have students enjoy mathematics whether they are gifted in it or not.

Let me make a few remarks about teaching. I think that as we lecture to people, we should try to put ourselves in the position of the audience and try to see how its members view what we are saying. I like to be able to look into the eyes of my audience so I can gauge listeners' interest. There is no need to lecture to ourselves and go off and leave students. It seems that too many teachers are lecturing to themselves.

Not just in lecturing but in talking to people about mathematics and other things, one should try to look at things from their point of view. I've heard that a person does not really understand another's problems until they have stood in that person's shoes. I recall a word of wisdom given to me by a cleaning woman working at my home. She was a member of a minority. I had heard another member of that minority say something that sounded very ridiculous—something that I thought even reasonable members of that minority would not accept, and discussed this with her. She said this, "You should listen to what people feel rather than to what they say." This is good advice.

Department of Mathematics
Southwest Texas State University
San Marcos, TX 78666

A CHRONOLOGY OF R. H. BING

(October 20, 1914–April 28, 1986)

Education

1935	B.S.	Southwest Texas State University
1938	M.Ed.	The University of Texas, Austin
1945	Ph.D.	The University of Texas, Austin

Professional: Employment

1943–1947	Instructor, Assistant Professor The University of Texas, Austin
1947–1973	Assistant Professor, Associate Professor, Professor The University of Wisconsin, Madison
1949–1950	Visiting Faculty Member The University of Virginia
1957–1958	Visiting Member, The Institute for Advanced Study
1958–1960	Chairman, Mathematics Department The University of Wisconsin, Madison
1962–1963	Visiting Member, The Institute for Advanced Study
1967	Visiting Member, The Institute for Advanced Study

1973–1985	Professor The University of Texas, Austin
1975–1977	Chairman, Department of Mathematics The University of Texas, Austin

Professional: Awards and Service

1952–1954	Council, American Mathematical Society
1952–1953	Chairman, Wisconsin Section Mathematical Association of America
1952–1953	Visiting Lecturer, Mathematical Association of America
1955	Director, Summer Institute on Set Theoretic Topology
1958–1960	Council, American Mathematical Society
1961	Hedrick Lecturer, Mathematical Association of America
1961–1962	Visiting Lecturer, Mathematical Association of America
1963–1964	President, Mathematical Association of America
1965–1986	Member, National Academy of Sciences
1966	U. S. Delegate, International Mathematical Union
1966–1967	Chairman, Conference Board of Mathematical Sciences
1966–1967	President's Committee, National Medal of Science
1967–1969	Chairman, Division of Mathematics National Research Council
1968–1975	Member, National Science Board

1970	Colloquium Lecturer, American Mathematical Society
1970–1973	Chairman, Mathematics Section National Academy of Sciences
1974	Distinguished Service to Mathematics Award from Mathematical Association of America
1974–1976	President's Committee, National Medal of Science
1977–1978	President, American Mathematical Society
1977–1980	Council, National Academy of Sciences
1977–1980	Governing Board, National Research Council
1978	U.S. Delegate, International Mathematical Union

PH.D. STUDENTS OF R. H. BING

1. Cohen, Herman J., *Concerning homogeneous plane continua*, Univ. of Wisconsin, 1949.

2. Thomas, Garth H.M., *Simultaneous partitionings of two sets*, Univ. of Wisconsin, 1952.

3. Sanderson, Donald E., *Isotopic deformations in 3-manifolds*, Univ. of Wisconsin, 1953.

4. Goblirsch, Richard P., *Approximating the area of a surface with the area of a nearby polyhedral one*, Univ. of Wisconsin, 1956.

5. Brown, Morton, *Continuous collections of higher dimensional hereditarily indecomposable continua*, Univ. of Wisconsin, 1958.

6. Lehner, Guydo R., *Extending homeomorphisms on the pseudo-arc*, Univ. of Wisconsin, 1958.

7. Kister, James M., *Isotopies in manifolds*, Univ. of Wisconsin, 1959.

8. Rosen, Ronald H., *Imbedding of decompositions of 3-space*, Univ. of Wisconsin, 1959.

9. McMillan, Daniel R., *On homologically trivial 3-manifolds*, Univ. of Wisconsin, 1960.

10. Casler, Burtis G., *On the sum of two solid Alexander horned spheres*, Univ. of Wisconsin, 1962.

11. Gillman, David S., *Piercing 2-spheres in* E^3, Univ. of Wisconsin, 1962.

12. Hempel, John P., *A surface in* S^3 *is tame if it can be deformed into each complementary domain*, Univ. of Wisconsin, 1962.

13. Glaser, Leslie C., *Contractible complexes in S^n*, Univ. of Wisconsin, 1964.

14. Henderson, David W., *Extensions of Dehn's lemma and the loop theorem*, Univ. of Wisconsin, 1964.

15. Hosay, Norman, *Characterizations of tame sets in E^3*, Univ. of Wisconsin, 1964.

16. Price, Thomas M., *Upper semicontinuous decompositions of E^3*, Univ. of Wisconsin, 1964.

17. Cobb, John I., *Locally tame embeddings mostly in the trivial range*, Univ. of Wisconsin, 1966.

18. Craggs, Robert F., *Small ambient isotopies of a 3-manifold which transform one embedding of a polyhedron into another*, Univ. of Wisconsin, 1966.

19. Dancis, Jerome, *Some nice embeddings of k-complexes and k-manifolds into n-manifolds $n \geq 2k+2$*, Univ. of Wisconsin, 1966.

20. McAllister, Byron L., *A refinement of the Whyburn cyclic element theory*, Univ. of Wisconsin, 1966.

21. Daverman, Robert J., *Locally fenced 2-spheres in S^3*, Univ. of Wisconsin, 1967.

22. Jones, Stephen L., *Concerning collections filling Euclidean n-space*, Univ. of Wisconsin, 1967.

23. Wright, Thomas P., *On stable homeomorphisms of Euclidean n-space*, Univ. of Wisconsin, 1967.

24. Yohe, James M., *Hereditarily infinite dimensional spaces*, Univ. of Wisconsin, 1967.

25. Jensen, Richard A., *Cross sectionally connected spheres*, Univ. of Wisconsin, 1969.

26. Fearnley, Lawrence, *Hereditarily indecomposable circularly chainable continua*, Univ. of London, 1970.

27. Gerlach, Jacob H., *Toroidal decompositions of E^3 which give E^3*, Univ. of Wisconsin, 1970.

28. Olinick, Michael, *Reflexive compact maps of Euclidean spaces and the monotone mapping problem*, Univ. of Wisconsin, 1970.

29. Webster, Dallas E., *Alternate methods in handle-straightening theory*, Univ. of Wisconsin, 1970.

30. Gerlach, Mary A., *Some fibered cellular decompositions of E^3 give E^3*, Univ. of Wisconsin, 1971.

31. Jones, Ralph, *Triangulated open n-manifolds are unions of n open n-cells*, Univ. of Wisconsin, 1971.

32. Shilepsky, Arnold C., *Homogeneity and extension properties of embeddings of S^1 in E^3*, Univ. of Wisconsin, 1971.

33. Crary, Fred D., *Some new engulfing theorems*, Univ. of Wisconsin, 1973.

34. Lewis, Ira W., *Stable homeomorphisms of the pseudo-arc*, The University of Texas at Austin, 1977.

35. Richter, Gary H., *Some properties of patched 2-spheres*, The University of Texas at Austin, 1978.

R H BING*

October 20, 1914 – April 28, 1986

R H Bing, President of the AMS in 1977 and 1978, died at age 71 at his home in his study in Austin, Texas. A major contributor to geometric topology, especially with his work on 3-manifolds, he saw many of his basic ideas and techniques used by others in recent major breakthroughs in higher dimensional manifolds. He was the retired "Mildred Caldwell and Baine Perkins Kerr Centennial Professor Emeritus in Mathematics" at The University of Texas at Austin. From 1947 to 1973 he was on the faculty at the University of Wisconsin, where he was "Rudolph E. Langer Research Professor of Mathematics" during the last nine of these years.

R H Bing was a long time member of the National Academy of Sciences and a former member of the National Science Board. He was also President of the Mathematical Association of America (MAA) in 1963 and 1964, and is one of only four people who have, since the forties, served both the Society and the Association as President.

He was former Chairman of the Conference Board of the Mathematical Sciences, former Chairman of the Mathematics Section of the National Academy of Sciences and of the Division of Mathematical Sciences of the National Research Council, and former Chairman of Section A and Vice-President of the American Association for the Advancement of Science. He served on countless committees for many organizations. In 1974 he received the MAA award for Distinguished Service to Mathematics. Few mathematicians in our lifetime have served our community so well and with such distinction.

He was justly proud of his collection of thirty-five doctoral students at Wisconsin and Texas. Many are in important academic positions throughout the country, and many are active in current research. Participation with him in a seminar or class was a lively, rewarding, and stimulating experience. He had great influence on many young mathematicians who came into contact with him at Wisconsin or at Texas, at the Institute for Advanced Study or at the University of Virginia where he held visiting positions, or in his many travels in this country and abroad. He presented invited lectures at over two hundred colleges and universities with locations in nearly every state of the Union and in seventeen foreign countries. He was the Hedrick Lecturer for the MAA in 1961 and the Colloquium Lecturer for the AMS in 1970. His book on the geometric topology of 3-manifolds was recently published by the AMS.

R H, as he was known to many of us, had, in an old southern tradition, no given names to go with his initials. On occasion, he was required to list his name as R (only) H (only) Bing: hence, his sometime nickname *Ronly Honly Bing.* He was born in Oakwood, Texas, where his father was superintendent of schools. His mother, who was widowed when R H was five, taught school in Texas while rearing R H and his sister. At an early age, he was instilled with a love of learning, of doing, of competing, and of achieving. To R H, mathematics was like a big game in which he competed mightily against the unrevealed geometric truths of mathematics. He loved that game among all others and played it very hard. He won some and he lost some. But in the final reckoning he had won a truly impressive number.

Bing graduated from high school as one of the state-wide winners in what is known as the University Interscholastic League Contests in Texas, a series of contests by grade level on number sense, estimation, and mental arithmetic—eighty problems in ten minutes.

After high school, he went to Southwest Texas State Teachers College at San Marcos (now Southwest Texas State University), graduating at twenty in 1935. As an undergraduate, R H prepared to be a high school mathematics teacher. So he became one, along with becoming a football and track coach at Palestine, Texas. (Can anyone hear some of his students, or their parents, complaining because the football coach was their geometry teacher?) Bing may well be the only high school football coach who ever made it to the National Academy of Sciences in mathematics.

In 1938, while working toward a master's degree in mathematics education at the University of Texas, he took a summer geometry course from R. L. Moore and decided (correctly) that if some of those other people in the course could be doctoral students in mathematics, then so could he. There is some doubt as to whether Moore was particularly impressed with Bing during that first summer. There is no doubt that Moore was very impressed after Bing started serious graduate work in mathematics in 1942 and then solved, along with several other problems, the Kline sphere characterization problem by the time he completed his doctoral degree in 1945. Bing became Moore's standard of comparison for future students from then on. He stayed on as a young faculty member at Texas until 1947 when he went to the University of Wisconsin to begin his long, distinguished career there.

The authors first knew R H, his wife Mary, and their oldest child Robert when we both got out of the Navy and came back to Texas to continue graduate work after the war. And we have known him well and admired him and his mathematics ever since. As a person, R H was friendly, open, cheerful, and enthusiastic. He was also very hard working, often rising at five in the morning and getting right at his research. Our paths have crossed his many times and we are the richer for it.

In his last few years R H had suffered from cancer and other declines in health, but neither his vitality nor his zest for life and mathematics had deteriorated. Those of us who knew him well thought it particularly fitting that he died "at his desk with his boots on." That is the way he would have wanted it.

R H was a dedicated family man. He and his devoted wife Mary had four children: Robert, Susan, Gay, and Mary Pat. The first three are now married and have given the Bings six grandchildren, the youngest just a few months old. The family has always been close and supportive. It has been our pleasure to know them well over the years. R H and Mary were also dedicated to their activities with the Presbyterian Church where R H served as an elder.

While a more detailed account of his mathematical work is expected to be published later, a few comments are in order now.

Bing made perhaps his most lasting research contributions to topology during the decade of the fifties. When Moise (also a graduate student at Texas before and after the war) triangulated 3-manifolds and proved the Hauptvermutung at the beginning of that decade, Bing became intensely interested in the subject and gave alternative proofs of these important results based on his polyhedral approximation theorem for spheres. This led to his involvement in the study of 3-manifolds, especially their pathology, throughout the remainder of his life. In the mid-fifties, he developed two of his most far-reaching results—his side approximation theorem for 2-spheres in Euclidean 3-space and an example, now generally known as the "dogbone space," of a cellular decomposition of Euclidean 3-space that does not yield a manifold. (Shortly thereafter, he showed that the Cartesian product of this nonmanifold with a line is Euclidean four-space.) The side approximation theorem led to extensive work, during the sixties and early seventies, on tame and wild embeddings of 2-manifolds in 3-manifolds and on some generalizations to higher dimensions. In the paper describing the dogbone space, together with another important paper in which he showed that the 3-sphere S^3 results from canonically sewing two solid Alexander horned spheres together along their boundaries, he developed what has since become known as "Bing Shrinking." This procedure has become important in showing whether a cell-like decomposition of a manifold yields a manifold.

In the mid-seventies, Edwards and Cannon showed that various phenomena of the type studied and identified by Bing could be generalized as sufficient conditions to characterize manifolds among all cell-like images of manifolds. With his shrinking criteria and other results, Bing somehow had introduced key phenomena in the pathology of higher dimensional manifolds while studying 3-manifolds. "Bing Shrinking" was at the heart of Torunczyk's characterizations of Q-manifolds and ℓ_2-manifolds among ANR's.

More recently, it was used by Freedman in his remarkable proof of the four-dimensional Poincaré conjecture in the topological category, a key step in opening up four-dimensional topology.

Of those mathematicians closely identified with Texas who have had a great impact on topology and mathematics, two stand out: R. L. Moore in the first half of this century and R H Bing in the second.

Mathematics and mankind are the better for the life of R H Bing.

R. D. Anderson, Louisiana State University

C. E. Burgess, University of Utah

ABSTRACTS BY R. H. BING

Conformal minimal varieties, Bull. Amer. Math. Soc. **49** (1943), 536.

On generalized convex functions (with E. F. Beckenbach), Bull. Amer. Math. Soc. **50** (1944), 348–349.

Concerning the vertex mean-value theorem of convex polynomials (with E. F. Beckenbach), Bull. Amer. Math. Soc. **50** (1944), 667.

Concerning simple plane webs, Bull. Amer. Math. Soc. **51** (1945), 525.

Generalizations of two theorems of Janiszewski, Bull. Amer. Math. Soc. **51** (1945), 666.

Collections cutting the plane, Bull. Amer. Math. Soc. **51** (1945), 887.

Solution of a problem of J. R. Kline, Bull. Amer. Math. Soc. **52** (1946), 73.

Generalization of a theorem of Janiszewski, Bull. Amer. Math. Soc. **52** (1946), 242–243.

Sets cutting the plane, Bull. Amer. Math. Soc. **52** (1946), 243.

Converse linearity conditions, Bull. Amer. Math. Soc. **52** (1946), 423.

Skew sets, Bull. Amer. Math. Soc. **52** (1946), 829.

Solution of a problem proposed by R. L. Wilder, Bull. Amer. Math. Soc. **53** (1947), 70.

Some characterizations of a simple closed curve, Bull. Amer. Math. Soc. **53** (1947), 70.

A homogeneous indecomposable plane continuum, Bull. Amer. Math. Soc. **53** (1947), 502–503.

Extending a metric, Bull. Amer. Math. Soc. **53** (1947), 503.

Some characterizations of arcs and simple closed curves, Bull. Amer. Math. Soc. **54** (1948), 76.

A convex metric for a locally connected continuum, Bull. Amer. Math. Soc. **54** (1948), 483–484.

Partitioning a set, Bull. Amer. Math. Soc. **54** (1948), 1082.

Complements of continuous curves, Bull. Amer. Math. Soc. **55** (1949), 58.

Metrization of topological spaces, Bull. Amer. Math. Soc. **55** (1949), 713.

A normal space which is not collectionwise normal, Bull. Amer. Math. Soc. **55** (1949), 1078–1079.

A necessary and sufficient condition that a topological space be metrizable, Bull. Amer. Math. Soc. **56** (1950), 53.

A characterization of 3-*space by partitionings*, Bull. Amer. Math. Soc. **56** (1950), 183.

Chained hereditarily indecomposable continua are homeomorphic, Bull. Amer. Math. Soc. **56** (1950), 255.

Hereditarily indecomposable continua, Bull. Amer. Math. Soc. **56** (1950), 329.

Coverings with connected intersections (with E. E. Floyd), Bull. Amer. Math. Soc. **56** (1950), 350.

A homeomorphism between the 3-*sphere and the sum of two solid horned spheres*, Bull. Amer. Math. Soc. **57** (1951), 293.

Concerning snake-like continua, Bull. Amer. Math. Soc. **57** (1951), 293.

Convexifying continuous curves so as to obtain many unique segments, Bull. Amer. Math. Soc. **57** (1951), 487.

Locally tame surfaces are tame, Bull. Amer. Math. Soc. **59** (1953), 177.

Decompositions of Euclidean spaces, Bull. Amer. Math. Soc. **60** (1954), 172.

Monotone upper semicontinuous decompositions of E^3, Bull. Amer. Math. Soc. **60** (1954), 383.

Some monotone decompositions of a cube, Bull. Amer. Math. Soc. **60** (1954), 384.

Another homogeneous plane continuum, Bull. Amer. Math. Soc. **60** (1954), 559.

An upper semicontinuous decomposition of E^3 into points and tame arcs whose decomposition space is not E^3, Bull. Amer. Math. Soc. **61** (1955), 322.

Approximating wild surfaces with polyhedral ones, Bull. Amer. Math. Soc. **61** (1955), 565.

A simple closed curve which pierces no disk, Bull. Amer. Math. Soc. **62** (1956), 47.

Periodic homeomorphisms of E^3 which are not orthogonal, Bull. Amer. Math. Soc. **62** (1956), 47.

An alternative proof that 3-manifolds can be triangulated, Bull. Amer. Math. Soc. **62** (1956), 179.

A peculiar decomposition of E^3, Bull. Amer. Math. Soc. **62** (1956), 365.

A necessary and sufficient condition that a 3-manifold be S^3, Bull. Amer. Math. Soc. **62** (1956), 605.

Conditions under which monotone decompositions of E^3 are simply connected, Bull. Amer. Math. Soc. **63** (1957), 143.

Pseudo-arc is only nondegenerate homogeneous chainable compact continuum, Bull. Amer. Math. Soc. **63** (1957), 198.

Conditions under which a surface in E^3 is tame, Bull. Amer. Math. Soc. **63** (1957), 403–404.

E^3 does not contain uncountably many mutually exclusive wild surfaces, Bull. Amer. Math. Soc. **63** (1957), 404.

A surface S is tame in E^3 if $E^3 - S$ is locally simply connected at each point of S, Notices Amer. Math. Soc. **5** (1958), 180–181.

The cartesian product of a certain nonmanifold and a line is E^4, Notices Amer. Math. Soc. **5** (1958), 244.

A 2-manifold-with-boundary in E^3 is tame if its complement is $1 - ULC$, Notices Amer. Math. Soc. **5** (1958), 365.

Each homogeneous plane continuum that contains an arc separates the plane, Notices Amer. Math. Soc. **5** (1958), 696–697.

Each homogeneous plane continuum that contains an arc is a simple closed curve, Notices Amer. Math. Soc. **6** (1959), 255.

Each 2-sphere in E^3 contains a tame arc, Notices Amer. Math. Soc. **6** (1959), 405.

Each disk in each 3-manifold is pierced by tame arc, Notices Amer. Math. Soc. **6** (1959), 510.

Pushing a 2-sphere into its complement, Notices Amer. Math. Soc. **6** (1959), 838.

A 3-cell is the only object whose cartesian product with an arc is a 4-cell, Notices Amer. Math. Soc. **7** (1960), 68.

A wild surface each of whose arcs is tame, Notices Amer. Math. Soc. **7** (1960), 220.

Point-like decompositions of E^3, Notices Amer. Math. Soc. **7** (1960), 946–947.

Embedding circle like continua in the plane, Notices Amer. Math. Soc. **8** (1961), 56.

Almost cartesian product neighborhoods of 2-spheres, Notices Amer. Math. Soc. **8** (1961), 162.

Improving the intersection of lines with surfaces, Notices Amer. Math. Soc. **9** (1962), 108.

An arc is tame in 3-space if and only if it is strongly cellular (with A. Kirkor), Notices Amer. Math. Soc. **9** (1962), 325.

Every simple closed curve in E^3 is unknotted in E^4 (with V. L. Klee), Notices Amer. Math. Soc. **10** (1963), 132.

Uncountably many fixed point-free periodic actions on a 3-sphere (with J. M. Kister), Notices Amer. Math. Soc. **10** (1963), 295.

Improving the side approximation theorem, Notices Amer. Math. Soc. **10** (1963), 453.

Stable homeomorphisms on E^5 can be approximated by piecewise linear ones, Notices Amer. Math. Soc. **10** (1963), 666.

Some remarks concerning topologically homogeneous spaces (with K. Borsuk), Notices Amer. Math. Soc. **11** (1964), 237.

Radial engulfing, Notices Amer. Math. Soc. **12** (1965), 229.

Cubes with knotted holes (with J. M. Martin), Notices Amer. Math. Soc. **16** (1969), 300.

A solution of the monotone mapping problem, Notices Amer. Math. Soc. **16** (1969), 854.

Linear isotopies (with Michael Starbird), Notices Amer. Math. Soc. **22** (1975), A–201.

Linear isotopies (with Michael Starbird), Notices Amer. Math. Soc. **22** (1975), A–575.

II. Classifications of Works of R. H. Bing

PUBLICATIONS OF R. H. BING: CLASSIFIED BY YEAR

1943

Conformal minimal varieties (with E. F. Beckenbach). *Duke Math. J.* **10** (1943), 637–640. (page) 755.

1945

On generalized convex functions (with E. F. Beckenbach). *Trans. Amer. Math. Soc.* **58** (1945), 220–230. 759.

Collections filling up a simple plane web. *Bull. Amer. Math. Soc.* **51** (1945), 674–679. 281.

Generalizations of two theorems of Janiszewski. *Bull. Amer. Math. Soc.* **51** (1945), 954–960. 305.

1946

Generalizations of two theorems of Janiszewski. II. *Bull. Amer. Math. Soc.* **52** (1946), 478–480. 313.

Converse linearity conditions. *Amer. J. Math.* **68** (1946), 309–318. 771.

1947

1948

1949

1950

Coverings with connected intersections (with E. E. Floyd). *Trans. Amer. Math. Soc.* **69** (1950), 387–391. 411.

1951

A characterization of 3-space by partitionings. *Trans. Amer. Math. Soc.* **70** (1951), 15–27. 559.

Metrization of topological spaces. *Canad. J. Math.* **3** (1951), 175–186. 95.

An equilateral distance. *Amer. Math. Monthly* **58** (1951), 380–383. 107.

Concerning hereditarily indecomposable continua. *Pacific J. Math.* **1** (1951), 43–51. 363.

Higher-dimensional hereditarily indecomposable continua. *Trans. Amer. Math. Soc.* **71** (1951), 267–273. 373.

Snake-like continua. *Duke Math. J.* **18** (1951), 653–663. 381.

1952

A homeomorphism between the 3-sphere and the sum of two solid horned spheres. *Ann. of Math.* **56** (1952), 354–362. 1161.

Partitioning continuous curves. *Bull. Amer. Math. Soc.* **58** (1952), 536–556. 417.

The place of topology in a teacher training program. *Symposium on Teacher Education in Mathematics*, Madison, 1952, 4 pages. *

* This paper is not included in these volumes.

1953

A convex metric with unique segments. *Proc. Amer. Math. Soc.* **4** (1953), 167–174. 111.

Examples and counterexamples. *Pi Mu Epsilon J.* **1** (1953), 311–317. 827.

A connected countable Hausdorff space. *Proc. Amer. Math. Soc.* **4** (1953), 474. 119.

Review of *General Topology* by W. Sierpiński. *Bull. Amer. Math. Soc.* **59** (1953), 410. 185.

1954

Locally tame sets are tame. *Ann. of Math.* **59** (1954), 145–158. 1337.

1955

Partially continuous decompositions. *Proc. Amer. Math. Soc.* **6** (1955), 124–133. 439.

Some monotone decompositions of a cube. *Ann. of Math.* **61** (1955), 279–288. 895.

What topology is here to stay? In "Summary of Lectures and Seminars," Summer Institute on Set Theoretic Topology, Madison (1955), Revised (1958), 25–27, University of Wisconsin. 905.

Decompositions of E^3 into points and tame arcs. In "Summary of Lectures and Seminars," Summer Institute on Set Theoretic Topology, Madison (1955), Revised (1958), 41–48, University of Wisconsin. 909.

Approximating surfaces with polyhedral ones. In "Summary of Lectures and Seminars," Summer Institute on Set Theoretic Topology, Madison (1955), Revised (1958), 49–53, University of Wisconsin. 1189.

The pseudo-arc. In "Summary of Lectures and Seminars," Summer Institute on Set Theoretic Topology, Madison (1955), Revised (1958), 72–75, University of Wisconsin. 393.

1956

A simple closed curve that pierces no disk. *J. Math. Pures Appl.* **35** (1956), 337–343. 1351.

1957

Point set topology. Chapter X of *Insights into Modern Mathematics*, Yearbook of National Council of Teachers of Mathematics, 1957. *

Upper semicontinuous decompositions of E^3. *Ann. of Math.* **65** (1957), 363–374. 917.

A decomposition of E^3 into points and tame arcs such that the decomposition space is topologically different from E^3. *Ann. of Math.* **65** (1957), 484–500. 929.

Approximating surfaces with polyhedral ones. *Ann. of Math.* **65** (1957), 456–483. 1195.

1958

The cartesian product of a certain nonmanifold and a line is E^4. *Bull. Amer. Math. Soc.* **64** (1958), 82–84. 1587.

* This paper is not included in these volumes.

Necessary and sufficient conditions that a 3-manifold be S^3. *Ann. of Math.* **68** (1958), 17–37. (See also correction, *Ann. of Math.* **77** (1963), 210.) 573.

1959

An alternative proof that 3-manifolds can be triangulated. *Ann. of Math.* **69** (1959), 37–65. 1555.

Another homogeneous plane continuum (with F. B. Jones). *Trans. Amer. Math. Soc.* **90** (1959), 171–192. 451.

Each homogeneous nondegenerate chainable continuum is a pseudo-arc. *Proc. Amer. Math. Soc.* **10** (1959), 345–346. 397.

Conditions under which a surface in E^3 is tame. *Fund. Math.* **47** (1959), 105–139. 1359.

The cartesian product of a certain nonmanifold and a line is E^4. *Ann. of Math.* **70** (1959), 399–412. 1591.

Geometry, Mathematics for High School (text). School Mathematics Study Group, 1959, joint author. *

1960

Imbedding decompositions of E^3 in E^4 (with M. L. Curtis). *Proc. Amer. Math. Soc.* **11** (1960), 149–155. 947.

A simple closed curve is the only homogeneous bounded plane continuum that contains an arc. *Canad. J. Math.* **12** (1960), 209–230. 473.

Elementary point set topology. Number 8 of the Herbert Ellsworth Slaught Memorial Papers, *Amer. Math. Monthly* **67** (1960), Part II, i–iv, 1–58. 187.

* This paper is not included in these volumes.

Set theory. *McGraw-Hill Encyclopedia of Science and Technology*, 1960, 205–206, McGraw-Hill, New York (reprinted 1977). 247.

1961

A set is a 3 cell if its cartesian product with an arc is a 4 cell. *Proc. Amer. Math. Soc.* **12** (1961), 13–19. 1605.

A wild surface each of whose arcs is tame. *Duke Math. J.* **28** (1961), 1–15. 1395.

Tame Cantor sets in E^3. *Pacific J. Math.* **11** (1961), 435–446. 1411.

On finiteness of the number of mappings carrying a planar nonconvex polygon into a convex one (with N. D. Kazarinoff and written in Russian by V. G. Boltyanski). *Matematicheskoe Prosveshchenie*, No. 8 (1961). 845.

On finiteness of the number of mappings carrying a planar nonconvex polygon into a convex one (with N. D. Kazarinoff and written in Russian by V. G. Boltyanski, translated by H. H. McFaden). *Matematicheskoe Prosveshchenie*, No. 8 (1961). 849.

A surface is tame if its complement is 1–ULC. *Trans. Amer. Math. Soc.* **101** (1961), 294–305. 1423.

1962

Applications of the side approximation theorem for surfaces. *General Topology and its Relation to Modern Analysis and Algebra* (Proceedings of the Symposium, Prague, 1961), 91–95, Academic Press, New York-London, Publ. House Czech. Acad. Sci., Prague, 1962. 1223.

Embedding circle-like continua in the plane. *Canad. J. Math.* **14** (1962), 113–128. 329.

1963

1964

(T. L. Saaty, editor), 93–128, John Wiley & Sons, New York, 1964. 1489.

A 3-dimensional absolute retract which does not contain any disk (with K. Borsuk). *Fund. Math.* **54** (1964), 159–175. 149.

Inequivalent families of periodic homeomorphisms of E^3. *Ann. of Math.* **80** (1964), 78–93. 1171.

The simple connectivity of the sum of two disks. *Pacific J. Math.* **14** (1964), 439–455. 783.

Taming complexes in hyperplanes (with J. M. Kister). *Duke Math. J.* **31** (1964), 491–511. 731.

An arc is tame in 3-space if and only if it is strongly cellular (with A. Kirkor). *Fund. Math.* **55** (1964), 175–180. 1453.

1965

Some remarks concerning topologically homogeneous spaces (with K. Borsuk). *Ann. of Math.* **81** (1965), 100–111. 167.

A translation of the normal Moore space conjecture. *Proc. Amer. Math. Soc.* **16** (1965), 612–619. 121.

Improving the side approximation theorem. *Trans. Amer. Math. Soc.* **116** (1965), 511–525. 1305.

Computing the fundamental group of the complements of curves. Technical Report 2, Washington State Univ., 1965, 28 pages. 801.

1966

Mapping a 3-sphere onto a homotopy 3-sphere. *Topology Seminar Wisconsin 1965* (R. H. Bing and R. J. Bean, editors),

Ann. of Math. Studies, No. 60, 89–99, Princeton University Press, Princeton, N.J., 1966. 1525.

1967

A hereditarily infinite dimensional space. *General Topology and its Relations to Modern Analysis and Algebra* II (Proceedings of the Second Prague Topological Symposium, 1966), 56–62, Academia, Prague, 1967. 497.

Challenging conjectures. *Amer. Math. Monthly* **74** (1967), 56–64. 835.

A toroidal decomposition of E^3 (with S. Armentrout). *Fund. Math.* **60** (1967), 81–87. 997.

Improving the intersections of lines and surfaces. *Michigan Math. J.* **14** (1967), 155–159. 1459.

1968

A complete elementary proof that Hilbert space is homeomorphic to the countable infinite product of lines (with R. D. Anderson). *Bull. Amer. Math. Soc.* **74** (1968), 771–792. 609.

Radial engulfing. *Conference on the Topology of Manifolds 1967* (J. G. Hocking, editor), 1–18, Prindle, Weber & Schmidt, Boston, 1968. 675.

1969

Retractions onto ANR's. *Proc. Amer. Math. Soc.* **21** (1969), 618–620. 179.

Point-set topology. *The Mathematical Sciences Essays for COSRIMS*, (COSRIMS, editors) 1969, 209–216, M.I.T. Press, Cambridge, Massachusetts. 249.

Extending monotone decompositions of 3-manifolds. *Proceedings of Auburn Topology Conference* (W. R. R. Transue, editor), 7–35, Auburn, Alabama, 1969. 1005.

The elusive fixed point property. *Amer. Math. Monthly* **76** (1969), 119–132. 505.

The elusive fixed point property. Sugaku **21** (1969), 203–210. (Japanese translation from Amer. Math. Monthly **76** (1969), 119–132.) 519.

1970

The monotone mapping problem. *Topology of Manifolds* (Proceedings of the University of Georgia Topology of Manifolds Institute, 1969) (J. C. Cantrell and C. H. Edwards, Jr., editors), 99–115, Markham Publ. Co., Chicago, 1970. 1055.

Extending monotone decompositions of 3-manifolds. *Trans. Amer. Math. Soc.* **149** (1970), 351–369. 1035.

Topology of 3-manifolds. Notes distributed by American Mathematical Society in conjunction with Colloquium Lectures at the University of Wyoming, Laramie, Wyoming, August 24–28, 1970, Seventy-fifth summer meeting of the American Mathematical Society, 30 pages. 1615.

Models for S^3. Visiting Scholars' Lectures, Texas Tech. University Mathematics Series No. 9, 1–31, Texas Tech. Press, Lubbock, 1970–1971. 853.

1971

Monotone images of E^3 (with J. M. Martin). *Proceedings of Conference on Monotone Mappings and Open Mappings*, 1970 (L. F. McAuley, editor), 55–77, SUNY at Binghamton, N.Y., 1971. 1073.

Cubes with knotted holes (with J. M. Martin). *Trans. Amer. Math. Soc.* **155** (1971), 217–231. 1537.

1973

Award for distinguished service to Professor Raymond L. Wilder. *Amer. Math. Monthly* **80** (1973), 117–119. 871.

1974

An unusual map of a 3-cell onto itself. *Topology Conference* (R. F. Dickman, Jr., and P. Fletcher, editors), V.P.I. (1973), Lecture Notes in Math. No. 375, 15–33, Springer, Berlin-New York, 1974. 1097.

Sets generated by rectangles (with W. W. Bledsoe and R. D. Mauldin). *Pacific J. Math.* **51** (1974), 27–36. 875.

Topologia conjuntista. *Universitas Enciclopedia Tematica*, **X** (1974), 126–132. *

Topology, General. *Encyclopaedia Britannica* (1974 Edition) **18**, 509–514. 257.

Pulling back feelers. *Symposia Mathematica*, **XVI**, 245–266, Academic Press, New York-London, 1974. 1465.

---

*This paper is not included in these volumes.

1975

Vertical general position. *Geometric Topology* (L. C. Glaser and T. B. Rushing, editors), Lecture Notes in Math. No. 438, 16–41, Springer, Berlin-New York, 1975. 693.

1978

Linear isotopies in E^2 (with M. Starbird). *Trans. Amer. Math. Soc.* **237** (1978), 205–222. 635.

Super triangulations (with M. Starbird). *Pacific J. Math.* **74** (1978), 307–325. 653.

Award for distinguished service to Professor R. D. Anderson. *Amer. Math. Monthly* **85** (1978), 73–74. 885.

1979

A decomposition of S^3 with a null sequence of cellular arcs (with M. Starbird). *Geometric Topology* (J. C. Cantrell, editor), 3–21, Academic Press, New York-London, 1979. 1117.

1980

Sets in R^3 with the two-disk property. *Proceedings of the International Conference on Geometric Topology*, 43, Polish Scientific Publishers, Warszawa, 1980. 1645.

1981

Metrization problems. *General Topology and Modern Analysis* (L. F. McAuley and M. M. Rao, editors), 3–16, Academic Press, New York-London, 1981. 129.

1983

The Geometric Topology of 3-Manifolds. *Amer. Math. Soc. Colloq. Publ.*, Vol. 40, 1–238, Amer. Math. Soc., Providence, Rhode Island, 1983. *

1986

Decompositions that destroy simple connectivity. *Illinois J. Math.* **30** (1986), 527–535. 1137.

Wild surfaces have some nice properties. *Michigan Math. J.* **33** (1986), 403–415. 1321.

Preprint

1984

Shrinking without lengthening. *Topology*, (to appear). 1147.

* This paper is not included in these volumes.

PUBLICATIONS OF R. H. BING: CLASSIFIED BY SUBJECT MATTER*

I. General Topology

A. Metrization

Extending a metric. *Duke Math. J.* **14** (1947), 511–519. 77.

A convex metric for a locally connected continuum. *Bull. Amer. Math. Soc.* **55** (1949), 812–819. 87.

Partitioning a set. *Bull. Amer. Math. Soc.* **55** (1949), 1101–1110. 417.

Metrization of topological spaces. *Canad. J. Math.* **3** (1951), 175–186. 95.

An equilateral distance. *Amer. Math. Monthly* **58** (1951), 380–383. 107.

A convex metric with unique segments. *Proc. Amer. Math. Soc.* **4** (1953), 167–174. 111.

A translation of the normal Moore space conjecture. *Proc. Amer. Math. Soc.* **16** (1965), 612–619. 121.

* Classes are not necessarily disjoint, i.e., an article may appear in several classes. This classification was done by R. H. Bing and S. Singh.

B. AR's and ANR's

C. Homogeneity

D. Others

II. Planar Sets

A. Planar Webs

B. Cuttings

C. Planar Embeddings

D. Others

III. Topological Classifications

A. Of 2-Sphere

B. Of S^3

C. Of Pseudo Arc

D. Of Arcs and Simple Closed Curves

E. Of Hilbert Space

IV. Continua

A. Pseudo Arc

B. Locally Connected Continua: Partitions and Decompositions

V. 3-Manifolds

A. Decompositions

Imbedding decompositions of E^3 in E^4 (with M. L. Curtis). *Proc. Amer. Math. Soc.* **11** (1960), 149–155. 947.

Decompositions of E^3. *Topology of 3-Manifolds and Related Topics* (M. K. Fort, Jr., editor), 5–21, Prentice-Hall, Englewood Cliffs, N.J., 1962. 955.

Point-like decompositions of E^3. *Fund. Math.* **50** (1962), 431–453. 973.

Mapping a 3-sphere onto a homotopy 3-sphere. *Topology Seminar Wisconsin 1965* (R. H. Bing and R. J. Bean, editors), Ann. of Math. Studies, No. 60, 89–99, Princeton University Press, Princeton, N.J., 1966. 1525.

A toroidal decomposition of E^3 (with S. Armentrout). *Fund. Math.* **60** (1967), 81–87. 997.

Extending monotone decompositions of 3-manifolds. *Proceedings of Auburn Topology Conference* (W. W. Transue, editor), 7–35, Auburn, Alabama, 1969. 1005.

The monotone mapping problem. *Topology of Manifolds* (Proceedings of the University of Georgia Topology of Manifolds Institute, 1969) (J. C. Cantrell and C. H. Edwards, Jr., editors), 99–115, Markham Publ. Co., Chicago, 1970. 1054.

Extending monotone decompositions of 3-manifolds. *Trans. Amer. Math. Soc.* **149** (1970), 351–369. 1035.

Monotone images of E^3 (with J. M. Martin). *Proceedings of Conference on Monotone Mappings and Open Mappings*, 1970 (L. F. McAuley, editor), 55–77, SUNY at Binghamton, N.Y., 1971. 1073.

An unusual map of a 3-cell onto itself. *Topology Conference* (R. F. Dickman, Jr., and P. Fletcher, editors), V.P.I. (1973), Lecture Notes in Math. No. 375, 15–33, Springer, Berlin-New York, 1974. 1097.

A decomposition of S^3 with a null sequence of cellular arcs (with M. Starbird). In *Geometric Topology* (J. C. Cantrell, editor), 3–21, Academic Press, New York-London, 1979. 1117.

B. Maps

C. Approximating Surfaces

Algebra (Proceedings of the Symposium, Prague, 1961), 91–95, Academic Press, New York-London, Publ. House Czech. Acad. Sci., Prague, 1962. 1223.

Approximating surfaces from the side. *Ann. of Math.* **77** (1963), 145–192. 1229.

Each disk in E^3 contains a tame arc. *Amer. J. Math.* **84** (1962), 583–590. 1435.

Embedding surfaces in 3-manifolds. *Proceedings International Congr. Mathematicians* (Stockholm, 1962), 457–458, Inst. Mittag-Leffler, Djursholm, 1963. 1277.

Pushing a 2-sphere into its complement. *Michigan Math. J.* **11** (1964), 33–45. 1279.

Improving the side approximation theorem. *Trans. Amer. Math. Soc.* **116** (1965), 511–525. 1305.

Pulling back feelers. *Symposia Mathematica* **XVI**, 245–266, Academic Press, New York-London, 1974. 1465.

D. Recognizing Tameness

Locally tame sets are tame. *Ann. of Math.* **59** (1954), 145–158. 1337.

Conditions under which a surface in E^3 is tame. *Fund. Math.* **47** (1959), 105–139. 1359.

A wild surface each of whose arcs is tame. *Duke Math. J.* **28** (1961), 1–15. 1395.

Tame Cantor sets in E^3. *Pacific J. Math.* **11** (1961), 435–446. 1411.

A surface is tame if its complement is 1−ULC. *Trans. Amer. Math. Soc.* **101** (1961), 294–305. 1423.

Each disk in E^3 contains a tame arc. *Amer. J. Math.* **84** (1962), 583–590. 1435.

Each disk in E^3 is pierced by a tame arc. *Amer. J. Math.* **84** (1962), 591–599. 1443.

Spheres in E^3. *Amer. Math. Monthly* **71** (1964), 353–364. 1293.

An arc is tame in 3-space if and only if it is strongly cellular (with A. Kirkor). *Fund. Math.* **55** (1964), 175–180. 1453.

Improving the intersections of lines and surfaces. *Michigan Math. J.* **14** (1967), 155–159. 1459.

Wild surfaces have some nice properties. *Michigan Math. J.* **33** (1986), 403–415. 1321.

E. Triangulation

Locally tame sets are tame. *Ann. of Math.* **59** (1954), 145–158. 1337.

An alternative proof that 3-manifolds can be triangulated. *Ann. of Math.* **69** (1959), 37–65. 1555.

F. Poincaré Conjecture

Necessary and sufficient conditions that a 3-manifold be S^3. *Ann. of Math.* **68** (1958), 17–37. (See also correction, *Ann. of Math.* **77** (1963), 210.) 573.

Some aspects of the topology of 3-manifolds related to the Poincaré conjecture. *Lectures on Modern Mathematics* **II** (T. L. Saaty, editor), 93–128, John Wiley & Sons, New York, 1964. 1489.

Cubes with knotted holes (with J. M. Martin). *Trans. Amer. Math. Soc.* **155** (1971), 217–231. 1537.

VI. Miscellaneous Topics

A. Linear Isotopies

Linear isotopies in E^2 (with M. Starbird). *Trans. Amer. Math. Soc.* **237** (1978), 205–222. 635.

Super triangulations (with M. Starbird). *Pacific J. Math.* **74** (1978), 307–325. 653.

B. Periodic Homeomorphisms

A homeomorphism between the 3-sphere and the sum of two solid horned spheres. *Ann. of Math.* **56** (1952), 354–362. 1161.

Decompositions of E^3 into points and tame arcs. In "Summary of Lectures and Seminars," Summer Institute on Set Theoretic Topology, Madison (1955), Revised (1958), 41–48, University of Wisconsin. 909.

Inequivalent families of periodic homeomorphisms of E^3. *Ann. of Math.* **80** (1964), 78–93. 1171.

C. Functions

On generalized convex functions (with E. F. Beckenbach). *Trans. Amer. Math. Soc.* **58** (1945), 220–230. 759.

Converse linearity conditions. *Amer. J. Math.* **68** (1946), 309–318. 771.

D. Conformal Varieties

Conformal minimal varieties (with E. F. Beckenbach). *Duke Math. J.* **10** (1943), 637–640. 755.

VII. Fundamental Group

The simple connectivity of the sum of two disks. *Pacific J. Math.* **14** (1964), 439–455. 783.

Computing the fundamental group of the complements of curves. Technical Report 2, Washington State Univ., 1965, 28 pages. 801.

VIII. Higher Dimensional Manifolds

A. Engulfing

B. Cartesian Factors

C. Tameness

IX. Interesting Examples

A. Related to 3-Manifolds

A decomposition of E^3 into points and tame arcs such that the decomposition space is topologically different from E^3. *Ann. of Math.* **65** (1957), 484–500. 929.

A wild surface each of whose arcs is tame. *Duke Math. J.* **28** (1961), 1–15. 1395.

Tame Cantor sets in E^3. *Pacific J. Math.* **11** (1961), 435–446. 1411.

Decompositions of E^3. *Topology of* 3-*Manifolds and Related Topics* (M. K. Fort, Jr., editor), 5–21, Prentice-Hall, Englewood Cliffs, N.J., 1962. 955.

A toroidal decomposition of E^3 (with S. Armentrout). *Fund. Math.* **60** (1967), 81–87. 997.

A 3-dimensional absolute retract which does not contain any disk (with K. Borsuk). *Fund. Math.* **54** (1964), 159–175. 149.

B. Others

Solution of a problem of R. L. Wilder. *Amer. J. Math.* **70** (1948), 95–98. 823.

Higher-dimensional hereditarily indecomposable continua. *Trans. Amer. Math. Soc.* **71** (1951), 267–273. 373.

Examples and counterexamples. *Pi Mu Epsilon J.* **1** (1953), 311–317. 827.

A connected countable Hausdorff space. *Proc. Amer. Math. Soc.* **4** (1953), 474. 119.

Computing the fundamental group of the complements of curves. Technical Report 2, Washington State Univ., 1965, 28 pages. 801.

A hereditarily infinite dimensional space. *General Topology and its Relations to Modern Analysis and Algebra* II (Proceedings of the Second Prague Topological Symposium, 1966), 56–62, Academia, Prague, 1967. 497.

X. Introductory & Expository Articles, Monographs, & Books

* This paper is not included in these volumes.

August 24–28, 1970, Seventy-fifth summer meeting of the American Mathematical Society, 30 pages.* 1615.

Models for S^3. Visiting Scholars' Lectures, Texas Tech. University Mathematics Series No. 9, 1–31, Texas Tech. Press, Lubbock, 1970–1971. 853.

Topology, General. *Encyclopaedia Britannica* (1974 Edition) **18**, 509–514. 257.

Topologia conjuntista. *Universitas Enciclopedia Tematica*, **X** (1974), 126–132. **

The Geometric Topology of 3-Manifolds. *Amer. Math. Soc. Colloq. Publ.*, Vol. 40, 1–238, Amer. Math. Soc., Providence, Rhode Island, 1983. * **

Point-set topology. *The Mathematical Sciences Essays for COSRIMS*, (COSRIMS, editors) 1969, 209–216, M.I.T. Press, Cambridge, Massachusetts. 249.

Sets in R^3 with the two-disk property. *Proceedings of the International Conference on Geometric Topology*, 43, Polish Scientific Publishers, Warszawa, 1980. 1645.

XI. Awards

Award for distinguished service to Professor Raymond L. Wilder. *Amer. Math. Monthly* **80** (1973), 117–119. 871.

Award for distinguished service to Professor R. D. Anderson. *Amer. Math. Monthly* **85** (1978), 73–74. 885.

---

* This reference may be included in all classes related to 3-manifolds.
** This paper is not included in these volumes.

XII. Books Edited

Topology Seminar, Wisconsin, 1965 (R. H. Bing and R. J. Bean, editors), *Annals of Math. Studies*, No. 60, 1–246, Princeton University Press, Princeton, N.J., 1966. *

Continua, Decompositions, Manifolds (R. H. Bing, W. T. Eaton, and M. P. Starbird, editors), *Proceedings of Texas Topology Symposium*, 1980, 1–267, University of Texas, Austin, 1983. *

* This paper is not included in these volumes.

WORKS NOT INCLUDED IN THESE VOLUMES

The place of topology in a teacher training program. *Symposium on Teacher Education in Mathematics*, Madison, 1952, 4 pages.

Point set topology. Chapter X of *Insights into Modern Mathematics*, Yearbook of National Council of Teachers of Mathematics, 1957.

Geometry, Mathematics for High School (text). School of Mathematics Study Group, 1959, joint author.

Topologia conjuntista. *Universitas Enciclopedia Tematica*, X (1974), 126–132.

The Geometric Topology of 3-Manifolds. *Amer. Math. Soc. Colloq. Publ.*, Vol. 40, 1–238, Amer. Math. Soc., Providence, Rhode Island, 1983.

III. Collected Papers of R. H. Bing

I. General Topology

A. Metrization

EXTENDING A METRIC

BY R. H. BING

In this paper we shall prove a covering theorem (Theorem 1) and apply it to get three metrization theorems (Theorems 2, 3, 5).

A topological space is metric provided that for each point P and each point Q, there exists a number $d(P, Q)$ such that

(a) $d(P, Q) = 0$ if and only if $P = Q$,

(b) $d(P, Q) = d(Q, P)$ (symmetry),

(c) $d(P, Q) \leq d(P, R) + d(R, Q)$ for each point R (triangle condition), and

(d) the function $d(x, y)$ preserves limit points.

By (d) we mean that P is a limit point of the set M if and only if for each positive number ϵ, there is a point Q of $M - M \cdot P$ such that $d(P, Q)$ is less than ϵ. The number $d(P, Q)$ is called the distance between P and Q. The distance function $d(x, y)$ of a space is called the metric of the space.

The distance between two points is a positive number. This may be seen by letting P equal Q in (c) and applying (a) and (b). Lindenbaum [7] has shown that we may omit (b) and substitute

(c′) $d(P, Q) \leq d(P, R) + d(Q, R)$

for (c). Then (b) and (c) follow from (a) and (c′). Birkhoff [3; 466] uses the inequality

(c″) $d(P, Q) \leq d(Q, R) + d(R, P)$

in place of (c′) and shows that (b) follows from (a) and (c″).

A covering theorem. The theorem in this section may be useful in studying a topological space some of whose properties are defined by means of a sequence of collections of sets covering the space. A space satisfying R. L. Moore's Axiom 1 [8] is such a space. In particular, we shall show that the theorem may be used in solving three metrization problems.

A collection G of point sets is *coherent* provided that each proper subcollection G' of G contains an element which intersects an element of $G - G'$.

THEOREM 1. *Suppose that r is a positive integer and $H_1, H_2, \cdots$ are collections of sets such that each pair of points that can be covered by a coherent collection of r or fewer elements of H_{i+1} can be covered by an element of H_i. If P and Q are two points whose sum cannot be covered by any element of H_s but which can be covered by a coherent collection of sets $h_1, h_2, \cdots$, and h_n belonging to $H_{\alpha(1)}, H_{\alpha(2)}, \cdots$, and $H_{\alpha(n)}$ respectively, then*

$$2(1/r^{\alpha(1)} + 1/r^{\alpha(2)} + \cdots + 1/r^{\alpha(n)}) > 1/r^s. \tag{1}$$

Received January 14, 1947; presented to the American Mathematical Society April 26, 1947.

Proof. We may suppose that h_1 contains P, that h_i intersects h_j if and only if i is equal to $j - 1$, j, or $j + 1$, and that h_n contains Q. There is an ordered subcollection W of $h_1, h_2, \cdots, h_n$ such that the end elements of W contain P and Q respectively and such that adjacent elements of W intersect each other. Now (1) is true for the collection $h_1, h_2, \cdots, h_n$ if it is true for this subcollection W. The ordered collection $h_1, h_2, \cdots, h_n$ is called a chain of elements from P to Q.

We shall show that

$$1/r^{\alpha(1)} + 2/r^{\alpha(2)} + \cdots + 2/r^{\alpha(n-1)} + 1/r^{\alpha(n)} > 1/r^s, \tag{2}$$

thereby proving (1). We shall prove (2) by induction. If n is 1, then $\alpha(1) < s$ and (2) is true.

Assume that (2) is not true. Let k be an integer such that (2) is true for all pairs of points P, Q and all integers n less than k, but is not true for n equal to k. Suppose that $h_1, h_2, \cdots, h_k$ is a chain of elements from P to Q, that m is an integer such that no element of H_m contains $P + Q$, and that (2) does not hold for k and m equal to n and s respectively.

Denote some point common to both h_i and h_{i+1} by $P(i)$ and denote Q by $P(n)$. Let $j_1, j_2, \cdots, j_r$ be integers such that $j_{i+1} > j_i$, some element of H_{m+1} contains both P and $P(j_1 - 1)$ but no element of H_{m+1} contains both P and $P(j_1)$, some element of H_{m+1} contains both $P(j_1 - 1)$ and $P(j_2 - 1)$ but no element of it contains both $P(j_1 - 1)$ and $P(j_2)$, $\cdots$, and some element of H_{m+1} contains both $P(j_{r-1} - 1)$ and $P(j_r - 1)$ but no element of it contains both $P(j_{r-1} - 1)$ and $P(j_r)$. It is not possible for any element of H_{m+1} to contain both $P(j_t - 1)$ $(t < r)$ and Q, because if it did, then some coherent collection of no more than r elements of H_{m+1} would cover $P + Q$, and accordingly, some element of H_m would cover $P + Q$. Also, $j_1 > 1$ or else $\alpha(1) \leq s$ and (2) holds.

Since (2) is true for n equal to any integer which is less than k, we have that

$$\begin{aligned} &1/r^{\alpha(1)} + 2/r^{\alpha(2)} + \cdots + 2/r^{\alpha(j_1-1)} + 1/r^{\alpha(j_1)} > 1/r^{m+1}, \\ &1/r^{\alpha(j_1)} + 2/r^{\alpha(j_1+1)} + \cdots + 2/r^{\alpha(j_2-1)} + 1/r^{\alpha(j_2)} > 1/r^{m+1}, \\ &\cdots \quad \cdots \quad \cdots \quad \cdots \quad \cdots \quad \cdots \quad \cdots, \\ &1/r^{\alpha(j_{r-1})} + 2/r^{\alpha(j_{r-1}+1)} + \cdots + 2/r^{\alpha(j_r-1)} + 1/r^{\alpha(j_r)} > 1/r^{m+1}. \end{aligned} \tag{3}$$

Adding these inequalities, we obtain (2).

As is demonstrated by the inequality (2), the left member of inequality (1) is unnecessarily large. One might wonder how much could be omitted from this member. By an argument similar to that given above, it can be shown that *if h_1 and h_n contain P and Q respectively and $n > 1$, then the sum of some $n - 1 - R$ terms of* $1/r^{\alpha(1)}$, $1/r^{\alpha(2)}$, $1/r^{\alpha(2)}$, $1/r^{\alpha(3)}$, $1/r^{\alpha(3)}$, $\cdots$, $1/r^{\alpha(n-1)}$, $1/r^{\alpha(n-1)}$, $1/r^{\alpha(n)}$ *is greater than* $1/r^s$, *where R is the remainder obtained on dividing $n - 2$ by $r - 1$.* However, for each triple of integers n, $r \geq 2$, and s, an example can be given of a coherent collection of sets $h_1, h_2, \cdots, h_n$ satisfying the conditions previously given but such that the sum of each $n - 2 - R$ elements of $1/r^{\alpha(1)}, 1/r^{\alpha(2)}, 1/r^{\alpha(2)}, \cdots, 1/r^{\alpha(n-1)}, 1/r^{\alpha(n-1)}, 1/r^{\alpha(n)}$ is less than $1/r^s$.

The general metrization problem. The general metrization problem is to characterize topologically a space which can be assigned a metric. As an application of Theorem 1, we shall give such a characterization. Theorem 2 resembles a well-known result given by Alexandroff and Urysohn. This result states [1] that a *topological space is metrizable provided there exists a sequence* $G_1, G_2, \cdots$ *such that*

(a) *for each integer i,* G_i *is a collection of domains covering space,*

(b) *if D is a domain and P is a point in D, there is an integer n such that every element of* G_n *containing P is a subset of D, and*

(c) *each pair of intersecting elements of* G_{i+1} *is a subset of some element of* G_i.

Kindred theorems have been treated by Niemytzki [9], Chittenden [5], Wilson [12], Frink [6], and Vickery [11].

In their proof of the above result, Alexandroff and Urysohn made use of results of Chittenden [4] concerning the equivalence of distance and uniformly regular écart. Frink [6] proves an equivalent theorem much more directly by using a theorem somewhat like Theorem 1 of this paper. While the following theorem may be proved by either of these methods, the proof that we give resembles that given by Frink.

THEOREM 2. *A topological space is metrizable provided there exists a sequence* $H_1, H_2, \cdots$ *such that*

(a) *for each integer i,* H_i *is a collection of sets covering the space,*

(b) *a point P is a limit point of the set M if and only if for each integer n, some element of* H_n *contains P and intersects* $M - M \cdot P$, *and*

(c) *each pair of points that is covered by the sum of a pair of intersecting elements of* H_{i+1} *can be covered by an element of* H_i.

Proof. We shall not conclude from (c) that (c′) *each pair of points that is covered by either an element of* H_{i+1} *or the sum of a pair of intersecting elements of* H_{i+1} *can be covered by an element of* H_i because it may be that H_{i+1} consists of a collection of mutually exclusive sets; in this case, there would not be a pair of intersecting elements of H_{i+1} and there might be an element of H_{i+1} containing two points which could not be covered by any element of H_i.

However, hypothesis (c′) may be substituted for (c). Let H'_i be the collection of all sets h such that h is either a point or a pair of points which can be covered by an element of H_1, by an element of $H_2, \cdots$, and also by an element of H_i. The sequence of collections $H'_1, H'_2, \cdots$ has the properties indicated in (a), (b), and (c′). We shall suppose that $H_1, H_2, \cdots$ satisfies conditions (a), (b), and (c′).

If P is a point and Q is a point, we define $d(P, Q)$ as the minimum of 1 and the greatest lower bound of the collection of sums of the type $1/2^{\alpha(1)} + 1/2^{\alpha(2)} +$

$\cdots + 1/2^{\alpha(n)}$ where $h_1, h_2, \cdots, h_n$ are the elements of a coherent collection of sets covering $P + Q$ and h_i is an element of $H_{\alpha(i)}$.

We shall show that the metric defined by $d(x, y)$ satisfies the triangle condition. If P, Q, and R are points, U is a coherent collection of sets covering $P + R$, and V is a coherent collection of sets covering $R + Q$, then the coherent collection $U + V$ covers $P + Q$. Hence, $d(P, Q) \leq d(P, R) + d(R, Q)$.

If M is a set not having P as a limit point, there is an integer s such that no element of H_s containing P intersects $M - M \cdot P$. Then by Theorem 1, we have that if Q is a point of $M - M \cdot P$, $d(P, Q) \geq 1/2^{s+1}$. Also, the sum of the elements of H_n which contain P intersects every set having P as a limit point. Hence, a set M has P as a limit point only if for each positive number ϵ, M contains a point Q different from P such that $d(P, Q) < \epsilon$.

Since $d(P, Q) = 0$ only if $P = Q$ and the metric defined by $d(P, Q)$ is symmetric, satisfies the triangle condition, and preserves limit points, it is a metric for the space.

Almost convex metrics. A metric defined by $d(x, y)$ is *convex* provided that for each point P and each point Q there is a point R such that $d(P, R) = d(R, Q) = d(P, Q)/2$. It is *almost convex* [2] provided that for each point P, each point Q, and each positive number ϵ, there is a point R such that $| d(P, R) - d(P, Q)/2 | < \epsilon$ and $| d(R, Q) - d(P, Q)/2 | < \epsilon$.

It has been shown by Urysohn [10] that a normal space has a metric if it is completely separable. However, the method used for defining a metric is such that no point is defined to be between any other two points. Hence, no metric defined by this method is convex. In fact, none is even almost convex.

The metric defined in the following theorem has the interesting property that if a connected space has an almost convex metric $d(x, y)$, then the elements of $H_1, H_2, \cdots$ may be so chosen that the metric defined is $d(x, y)$. . We select the elements of H_i to be the set of all pairs of points whose distance apart is less than $1/2^i$. Hence, the methods we use can give the ordinary metric to a Euclidean space. Theorem 3 can be compared with a result by Aronszajn [2; Theorem 2].

THEOREM 3. *A necessary and sufficient condition that a connected topological space have an almost convex metric is that there exist a sequence* $H_1, H_2, \cdots$ *satisfying conditions* (a), (b), *and* (c) *mentioned in Theorem* 2 *and*

(d) *each pair of points that can be covered by an element of* H_i *can be covered by the sum of two intersecting elements of* H_{i+1}.

If the space is a point, the result is trivial. Hence, we only consider the case in which it is nondegenerate.

Necessity. If the space has an almost convex metric defined by $d(x, y)$, we define H_i to be the set of all pairs of points whose distance apart is less than $1/2^i$. We note that the sequence $H_1, H_2, \cdots$ satisfies conditions (a), (b), (c), and (d).

Sufficiency. From hypothesis (d), we find that H_n $(n > 1)$ contains more than one element. Since space is connected, each element of H_n ($n > 1$) intersects another element of H_n. Hence, condition (c′) follows from hypotheses (c) and (d) and the fact that space is connected.

Using condition (b) and the fact that space is connected, we find that for each pair of points P and Q, there is a finite coherent collection of elements of H_1 which covers $P + Q$. The distance $d(P, Q)$ is defined to be the greatest lower bound of the collection of all sums of the type $1/2^{\alpha(1)} + 1/2^{\alpha(2)} + \cdots + 1/2^{\alpha(n)}$, where $h_1, h_2, \cdots$, and h_n are the elements of a coherent collection covering $P + Q$ and h_i is an element of $H_{\alpha(i)}$. As is shown in Theorem 2, $d(x, y)$ defines a metric. We shall show that this metric is almost convex.

We shall show that if P and Q are two points and ϵ is a positive number, then there is a point R such that $| d(P, R) - d(P, Q)/2 | < \epsilon$ and $| d(R, Q) - d(P, Q)/2 | < \epsilon$. Let $h_1, h_2, \cdots, h_n$ be the elements of a coherent collection covering $P + Q$ such that h_i is an element of $H_{\alpha(i)}$ and $1/2^{\alpha(1)} + 1/2^{\alpha(2)} + \cdots + 1/2^{\alpha(n)} < d(P, Q) + \epsilon$. We shall suppose that h_1 contains P, h_i intersects h_{i+1}, and h_n contains Q.

Using condition (d), we find that each element of H_i in $h_1, h_2, \cdots, h_n$ may be replaced by an intersecting pair of elements of H_{i+1} so that the resulting collection is a coherent collection containing $P + Q$. The resulting sum is the same. This procedure may be continued so as to obtain a coherent collection of elements $h'_1, h'_2, \cdots, h'_m$ covering $P + Q$ such that each h'_i is an element of H_j, $m/2^j < d(P, Q) + \epsilon$, and $1/2^j < \epsilon/2$. Suppose that h'_1 contains P, h'_i intersects h'_{i+1}, and h'_m contains Q. Let R be a point of h'_k where $m/2 \leq k \leq (m + 1)/2$. We find that

$$\begin{aligned} d(P, R) &\geq d(P, Q)/2 + d(P, Q)/2 - d(Q, R) \\ &> d(P, Q)/2 + m/2^{j+1} - \epsilon/2 - (m + 2)/2^{j+1} \\ &= d(P, Q)/2 - \epsilon/2 - 1/2^j > d(P, Q)/2 - \epsilon. \end{aligned}$$

Also,

$$\begin{aligned} d(P, R) &\leq (m + 1)/2^{j+1} = (m/2^{j+1} - \epsilon/2) + \epsilon/2 + 1/2^{j+1} \\ &< d(P, Q)/2 + \epsilon/2 + \epsilon/2. \end{aligned}$$

Similarly, it may be found that $| d(R, Q) - d(P, Q)/2 | < \epsilon$.

We could have given Theorem 3 in the following stronger (for the sufficiency case) but more complicated form:

THEOREM 3′. *Suppose that $N(1)$, $N(2)$, $\cdots$ is a sequence of positive integers infinitely many of which are greater than* 1. *A necessary and sufficient condition that a connected space have an almost convex metric is that there exist a sequence $H_1, H_2, \cdots$ satisfying conditions* (a) *and* (b) *of Theorem* 2 *and such that*

(c) *each pair of points that can be covered by a coherent collection of $N(i)$ elements of H_{i+1} can be covered by an element of H_i and*

(d) *each pair of points that can be covered by an element of* H_i *can be covered by a coherent collection of* $N(i)$ *elements of* H_{i+1}.

In proving the necessity of this theorem, the elements of H_i could be defined as the set of all pairs of points whose distance apart is less than $1/[N(i)N(i-1) \cdots N(1)]$. For the sufficiency case, the distance between two points could be defined to be the greatest lower bound of sums of the type $1/[N(\alpha_1)N(\alpha_1 - 1) \cdots N(1)] + 1/[N(\alpha_2)N(\alpha_2 - 1) \cdots N(1)] + \cdots + 1/[N(\alpha_n)N(\alpha_n - 1) \cdots N(1)]$, where $h_1, h_2, \cdots, h_n$ is a coherent collection of sets covering the two points and h_i is an element of $H_{\alpha(i)}$.

Extending metrics. In this last section, we shall consider the problem of defining a metric for a space that will preserve a given metric that is defined on a closed subset of that space.

The following result may be too simple to be dignified by calling it a theorem. It is included so that it can be used in the proof of Theorem 5.

THEOREM 4. *Suppose H and K are two closed sets and* $d(x, y; H)$ *and* $d(x, y; K)$ *are metrics on H and K respectively, such that if P and Q are points of* $H \cdot K$, *then* $d(P, Q; H) = d(P, Q; K)$. *Then there is a metric* $d(x, y)$ *on* $H + K$ *that preserves* $d(x, y; H)$ *on H and* $d(x, y; K)$ *on K.*

Proof. We shall suppose that $H \cdot K$ exists. If $P + Q$ belongs to H [or K], we define $d(P, Q)$ to be $d(P, Q; H)$ [or $d(P, Q; K)$]. If P belongs to H and Q to K, we define $d(P, Q)$ to be the greatest lower bound of $d(P, R; H) + d(R, Q; K)$ where R is a point of $H \cdot K$. We shall show that $d(x, y)$ satisfies the triangle condition on $H + K$.

Assume that there are three points P, R, and Q such that

$$d(P, Q) - \theta = d(P, R) + d(R, Q) \qquad (\theta > 0).$$

Then two of the points P, R, Q belong to one of the sets H, K and the other of the three points belongs to the other set. If $P + R$ belongs to H and Q is a point of K, there is a point A of $H \cdot K$ such that

$$\begin{aligned} d(P, R) + d(R, Q) &> d(P, R; H) + d(R, A; H) + d(A, Q; K) - \theta \\ &\geq d(P, A; H) + d(A, Q; K) - \theta \geq d(P, Q) - \theta. \end{aligned}$$

Also, if $P + Q$ belongs to H and R is a point of K, there is a point A of $H \cdot K$ and a point B of $H \cdot K$ such that

$$\begin{aligned} d(P, R) + d(R, Q) &> d(P, A; H) + d(A, R; K) - \theta/2 \\ &\qquad + d(R, B; K) + d(B, Q; H) - \theta/2 \\ &\geq d(P, A; H) + d(A, B; K) + d(B, Q; H) - \theta \\ &= d(P, A; H) + d(A, B; H) + d(B, Q; H) - \theta \\ &\geq d(P, Q; H) - \theta = d(P, Q) - \theta. \end{aligned}$$

Hence, the assumption that $d(x, y)$ does not satisfy the triangle condition is false.

THEOREM 5. *Suppose that* K *is a closed subset of a metrizable space* S *and that* $d(x, y; K)$ *is a metric on* K. *Then there is a metric* $d(x, y)$ *on* S *that preserves the metric* $d(x, y; K)$ *on* K; *that is, if* P *and* Q *are points of* K, *then* $d(P, Q) = d(P, Q; K)$.

If $d(x, y; K)$ *is bounded.* First, we shall consider the case in which the diameter of K under $d(x, y; K)$ is finite. For convenience, we shall suppose that this diameter is less than $\frac{1}{8}$.

In showing the existence of a metric $d(x, y)$ on S, we shall make use of a metric $d'(x, y)$ on S which has the property that the distance between two points of K under $d'(x, y)$ is greater than the distance between the same two points under $d(x, y; K)$.

Let $d''(x, y)$ be a metric for S such that the diameter of S under $d''(x, y)$ is less than $\frac{1}{2}$. Let G_i be the collection of all domains in S whose diameter under $d''(x, y)$ is less than $1/2^i$. Denote by G_i' the collection obtained by omitting from G_i all of its domains that contain two points of K whose distance apart under $d(x, y; K)$ is greater than or equal to $1/2^{i+2}$.

Let $H_1, H_2, \cdots$ be a sequence such that H_1 is G_1', H_2 is a subcollection of G_2' covering S such that each element of H_2 is a subset of an element of H_1 and each pair of intersecting elements of H_2 is covered by an element of H_1, and in general, H_{n+1} is a subcollection of G_{n+1}' covering S such that each element of H_{n+1} is a subset of an element of H_n and each pair of intersecting elements of H_{n+1} is covered by an element of H_n. Once H_n is obtained, we can obtain H_{n+1} as follows. For each point P, select an element h_p of H_n containing P and let D_p be an element of G_{n+1}' containing P such that the diameter under $d''(x, y)$ of D_p is less than one-half the distance under $d''(x, y)$ from P to the complement of h_p. Let H_{n+1} be a collection of such D_p's covering space.

We observe that the sequence $H_1, H_2, \cdots$ satisfies the conditions mentioned in Theorem 2 and that S is an element of H_1. Hence, we may use the method given in Theorem 2 to define a metric $d'(x, y)$ on S. We shall show that if P and Q are two points of K, then $d'(P, Q) > d(P, Q; K)$.

Let j be a positive integer such that $1/2^{j+1} \leq d(P, Q; K) < 1/2^j$. Since $P + Q$ does not belong to any element of G_{j-1}', it does not belong to any element of H_{j-1}. If $h_1, h_2, \cdots, h_n$ is a coherent collection of domains covering $P + Q$, where h_i is an element of $H_{\alpha(i)}$, we find from Theorem 1 that $1/2^{\alpha(1)} + 1/2^{\alpha(2)} + \cdots + 1/2^{\alpha(n)} > 1/2^j$. Hence, $d'(P, Q) \geq 1/2^j > d(P, Q; K)$.

If P is a point and Q is a point, we define $d(P, Q)$ to be the minimum of $d'(P, Q)$ and the greatest lower bound of $d'(P, A) + d(A, B; K) + d'(B, Q)$, where A is a point of K and so is B.

If P, Q, A, and B are points of K, then

$$\begin{aligned} d'(P, A) + d(A, B; K) + d'(B, Q) &\geq d(P, A; K) \\ + d(A, B; K) + d(B, Q; K) &\geq d(P, Q; K). \end{aligned}$$

Treating P and Q as A and B, we find that $d(P, Q) = d(P, Q; K)$. Hence, $d(x, y)$ preserves $d(x, y; K)$ on K. We shall now show that $d(x, y)$ is a metric for S.

Clearly, $d(P, Q) = d(Q, P) \geq 0$, the equality holding only if $P = Q$. Also, $d(x, y)$ preserves limit points in the space S. We shall show that $d(x, y)$ satisfies the triangle condition, thereby showing that it is a metric. This demonstrates that Theorem 4 is true if $d(x, y; K)$ is bounded.

Assume that the triangle condition is not satisfied and that there are three points P, R, and Q such that

$$d(P, Q) - \theta = d(P, R) + d(R, Q) \qquad (\theta > 0). \tag{4}$$

If $d(P, R) \neq d'(P, R)$ and $d(R, Q) \neq d'(R, Q)$, then there exists a point A, a point B, a point C, and a point D all of K so that

$$d(P, R) + \theta/2 > d'(P, A) + d(A, B; K) + d'(B, R)$$

and

$$d(R, Q) + \theta/2 > d'(R, C) + d(C, D; K) + d'(D, Q).$$

Adding the above two inequalities, we obtain

$$\begin{aligned} d(P, R) &+ d(R, Q) + \theta \\ &> d'(P, A) + d(A, B; K) + d'(B, R) + d'(R, C) \\ &\qquad + d(C, D; K) + d'(D, Q) \\ &\geq d'(P, A) + d(A, B; K) + d'(B, C) + d(C, D; K) + d'(D, Q) \\ &\geq d'(P, A) + d(A, D; K) + d'(D, Q) \geq d(P, Q). \end{aligned} \tag{5}$$

Now (5) contradicts (4). Similarly, we find that if $d(P, R) = d'(P, R)$ or $d(R, Q) = d'(R, Q)$, the triangle inequality is also satisfied.

If $d(x, y; K)$ is not bounded. Let P be a point of K and $D_1, D_2, \cdots$ be a collection of domains in S such that D_1 does not contain P, D_i contains all points Q of K such that $d(P, Q; K) > i$, D_{i+1} is a subset of D_i, and the common part of $D_1, D_2, \cdots$ does not exist. For convenience, we shall denote $S - D_i$ by E_i. We note that E_i is a closed subset of S and that if P and Q are points of $K \cdot E_i$, then $d(P, Q; K) \leq 2i$.

Since $d(x, y; K)$ is bounded over $K \cdot E_1$, there is a bounded metric $d(x, y; E_1)$ on E_1 that preserves $d(x, y; K)$ on $K \cdot E_1$. By Theorem 4, there is a bounded metric $d(x, y; E_1 + K \cdot E_2)$ on $E_1 + K \cdot E_2$ that preserves $d(x, y; E_1)$ on E_1 and $d(x, y; K)$ on $K \cdot E_2$. Furthermore, there is a bounded metric $d(x, y; E_2)$ on E_2 that preserves $d(x, y; E_1 + K \cdot E_2)$ on $E_1 + K \cdot E_2$. Continuing this process, we find that there are bounded metrics $d(x, y; E_3)$, $d(x, y; E_4)$, $\cdots$ such that $d(x, y; E_i)$ preserves $d(x, y; K)$ on $K \cdot E_i$ and preserves $d(x, y; E_{i-1})$ on E_{i-1}.

If P and Q are points of E_i, then $d(P, Q; E_i) = d(P, Q; E_j)$ for $j > i$. If E_i contains P and Q, define $d(P, Q)$ to be $d(P, Q; E_i)$. Since for each integer i, $d(x, y; E_i)$ preserves $d(x, y; K)$ on $K \cdot E_i$, $d(x, y)$ preserves $d(x, y; K)$ on K.

References

1. P. Alexandroff and P. Urysohn, *Une condition nécessaire et suffisante pour qu'une classe (L) soit une classe (B)*, Comptes Rendus Hebdomadaires des Séances de l'Académie des Sciences, vol. 177(1923), pp. 1274–1276.

2. N. Aronszajn, *Quelques remarques sur les relations entre les notions d'écart régulier et de distance*, Bulletin of the American Mathematical Society, vol. 44(1938), pp. 653–657.

3. G. Birkhoff, *Metric foundations of geometry. I*, Transactions of the American Mathematical Society, vol. 55(1944), pp. 465–492.

4. E. W. Chittenden, *On the equivalence of écart and voisinage*, Transactions of the American Mathematical Society, vol. 18(1917), pp. 161–166.

5. E. W. Chittenden, *On the metrization problem and related problems in the theory of abstract sets*, Bulletin of the American Mathematical Society, vol. 33(1927), pp. 13–34.

6. A. H. Frink, *Distance functions and the metrization problem*, Bulletin of the American Mathematical Society, vol. 43(1937), pp. 133–142.

7. A. Lindenbaum, *Contributions à l'étude de l'espace métrique. I*, Fundamenta Mathematicae, vol. 8(1926), pp. 209–222.

8. R. L. Moore, *Foundations of Point Set Theory*, American Mathematical Society Colloquium Publications, vol. 13, New York, 1932.

9. V. W. Niemytzki, *On the "third axiom of metric space"*, Transactions of the American Mathematical Society, vol. 29(1927), pp. 507–513.

10. P. Urysohn, *Zum Metrisationsproblem*, Mathematische Annalen, vol. 94(1925), pp. 309–315.

11. C. W. Vickery, *Axioms for Moore spaces and metric spaces*, Bulletin of the American Mathematical Society, vol. 46(1940), pp. 560–564.

12. W. A. Wilson, *On semi-metric spaces*, American Journal of Mathematics, vol. 53(1931), pp. 361–373.

The University of Texas.

A CONVEX METRIC FOR A LOCALLY CONNECTED CONTINUUM

R. H. BING

A topological space is metrizable if there is a distance function $D(x, y)$ such that if x, y, z are points, then

$$D(x, y) \geqq 0, \text{ the equality holding only if } x = y, \tag{1}$$

$$D(x, y) = D(y, x) \text{ (symmetry)}, \tag{2}$$

$$D(x, y) + D(y, z) \geqq D(x, z) \text{ (triangle condition)}, \tag{3}$$

$$D(x, y) \text{ preserves limit points.} \tag{4}$$

By (4) we mean that x is a limit point of the set T if and only if for each positive number ϵ there is a point of T at a positive distance from x of less than ϵ. We say that the metric $D(x, y)$ is convex if for each pair of points x, y there is a point u such that

$$D(x, u) = D(u, y) = D(x, y)/2. \tag{5}$$

A subset M of a topological space S is said to have a convex metric (even though S may have no metric) if the subspace M of S has a convex metric.

It is known [5][1] that a compact continuum is locally connected if it has a convex metric. The question has been raised [5] as to whether or not a compact locally connected continuum M can be assigned a convex metric. Menger showed [5] that M is convexifiable if it possesses a metric D such that for each point p of M and each positive number ϵ there is an open subset R of M containing p such that each point of R can be joined in M to p by a rectifiable arc of length (under D) less than ϵ. Kuratowski and Whyburn proved [4] that M has a convex metric if each of its cyclic elements does. Beer considered [1] the case where M is one-dimensional. Harrold found [3] M to be convexifiable if it has the additional property of being a plane continuum with only a finite number of complementary domains.

We shall show that if M_1 and M_2 are two intersecting compact continua with convex metrics D_1 and D_2 respectively, then there is a convex metric D_3 on M_1+M_2 that preserves D_1 on M_1 (Theorem 1). Using this result, we show that any compact n-dimensional locally connected continuum has a convex metric (Theorem 6). We do not

---

Presented to the Society, February 28, 1948; received by the editors June 21, 1948.

[1] Numbers in brackets refer to the references cited at the end of the paper.

answer the question: *Does each compact locally connected continuum have a convex metric?*[2]

In my paper *Extending a metric* it is shown [2, Theorem 5] that if K is a closed subset of the metrizable space S and D_1 is a metric on K, then there is a metric D_2 on S that preserves D_1 on K. The following result is a modification of that result.

THEOREM 1. *If M_1 and M_2 are two intersecting compact continua with convex metrics D_1 and D_2 respectively, there is a convex metric D_3 on M_1+M_2 that preserves D_1 on M_1.*

PROOF. Let $G(x)$ be the least upper bound of $D_1(p, q)$ for all points p, q of $M_1 \cdot M_2$ such that $D_2(p, q) \leqq x$. Then

$$G[D_2(p, q)] \geqq D_1(p, q) \qquad (p,\ q \text{ elements of } M_1 \cdot M_2). \tag{6}$$

Since $G(x)$ is a monotone nondecreasing function of x that approaches zero as x approaches zero, there is a function $F(x)$ $(x>0)$ such that $F(x)$ approaches zero as x approaches zero, $F(x) \geqq G(x)$, and the derivative of $F(x)$ with respect to x is a continuous monotone nonincreasing function greater than one.

If C is an arc which lies except for possibly its end points in $M_2 - M_1 \cdot M_2$ and $\int_C F'[D_2(p, M_1)]ds$ exists, where $F'(x)$ represents the derivative of $F(x)$ with respect to x, p is a variable point of C, s is the length along C under D_2, and $D_2(p, M_1)$ is the greatest lower bound of $D_2(p, q)$ for all points q of M_1, we define the length $L_0(C)$ of C under D_0 to be $L_0(C) = \int_C F'[D_2(p, M_1)]ds$. We note that $L_0(C)$ is not defined for all arcs C. However, if r and q are two points which lie on an arc C with such a length $L_0(C)$, then we call $D_0(r, q)$ the greatest lower bound of $L_0(C)$ for all such arcs C from r to q. Since $F'(x) > 1$, $D_0(r, q) \geqq D_2(p, q)$.

If r and q are two points of M_1 and $D_0(r, q)$ is defined, then

$$D_0(r, q) \geqq D_1(r, q) \tag{7}$$

because for each positive number ϵ there is a curve C from r to q whose interior lies in $M_1 - M_1 \cdot M_2$ such that

$$\begin{aligned} D_0(r, q) + \epsilon &> \int_C F'[D_2(p, M_1)]ds \geqq \int_C F'(s)ds \\ &= F(\text{length } C \text{ under } D_2) \geqq G[D_2(r, q)] \geqq D_1(r, q). \end{aligned} \tag{8}$$

The first "$\geqq$" relationship in (8) follows from the facts that $F'(x)$ is a

[2] Since this paper was submitted, both E. E. Moise [8] and the author [9], working independently, have answered this question in the affirmative.

nonincreasing function and $s \geqq D_2(p, M_1)$ since the end points of C lie on M_1; the second "$\geqq$" results from the facts that $F(x) \geqq G(x)$, length C under $D_2 \geqq D_2(r, q)$, and $G(x)$ is a monotone nondecreasing function; the third "$\geqq$" is a consequence of (6).

If q is a point of $M_2 - M_1 \cdot M_2$ and C is a shortest arc from q to M_1 under D_2, then

$$L_0(C) = \int_C F'[D_2(p, M_1)]ds = \int_C F'(s)ds = F[D_2(q, M_1)].$$

Hence, if $p_1, p_2, \cdots$ is a sequence of points of $M_2 - M_1 \cdot M_2$ such that $D_2(p_i, M_1)$ approaches zero as i increases without limit, then $D_0(p_i, M_1)$ approaches zero as i increases without limit.

If p and q are two points of M_1, we define $D_3(p, q)$ to be $D_1(p, q)$; if p is a point of M_1 and q is a point of $M_2 - M_1 \cdot M_2$, we define $D_3(p, q)$ to be the greatest lower bound of $D_1(p, a) + D_0(a, q)$ for all points a of M_1; if both p and q are points of $M_2 - M_1 \cdot M_2$, we define $D_3(p, q)$ to be the minimum of $D_0(p, q)$ and the greatest lower bound of $D_0(p, a) + D_1(a, b) + D_0(b, q)$, where a is a point of M_1 and so is b. If $p = q$, we define $D_3(p, q)$ to be equal to zero.

The above definition of $D_3(p, q)$ is equivalent to defining $D_3(p, q)$ to be the greatest lower bound of the lengths of all arcs C from p to q where length in M_1 is measured under D_1 and length in $M_2 - M_1 \cdot M_2$ is measured under D_0. It follows from (7) that we need only consider those arcs C which intersect M_1 in a connected piece if at all.

Now the function $D_3(p, q)$ may be shown to satisfy conditions (1), (2), (3), (4), and (5). Hence, it is a convex metric for $M_1 + M_2$ that preserves D_1 on M_1.

THEOREM 2. *If D_1 and D_2 are convex metrics for the intersecting compact continua M_1 and M_2 respectively and $D_2 \geqq D_1$ on $M_1 \cdot M_2$, then for each positive number ϵ there is a convex metric D_3 for $M_1 + M_2$ such that $D_3 = D_1$ on M_1, $D_3 \leqq D_2$ on M_2, and the diameter under D_3 of each component of $M_2 - M_1 \cdot M_2$ is less than ϵ plus twice the diameter under D_1 of the boundary with respect to $M_1 + M_2$ of this component.*

PROOF. Define $E(p, q)$ to be the greatest lower bound of all sums of the type $f(p, p_1) + f(p_1, p_2) + \cdots + f(p_n, q)$ where adjacent points of $p, p_1, \cdots, p_n, q$ belong to the same one of the continua M_1, M_2 and $f(p_i, p_j)$ is $D_1(p_i, p_j)$ or $D_2(p_i, p_j)$ according as $p_i + p_j$ is or is not a subset of M_1. Since $D_2 \geqq D_1$ on $M_1 \cdot M_2$, n need be no larger than 2 for $E(p, q)$ to attain this greatest lower bound. The convex metric E on $M_1 + M_2$ preserves D_1 on M_1 and $E \leqq D_2$ on M_2.

Let X be the set of all points p of $M_2 - M_1 \cdot M_2$ such that the distance from p to M_1 under E is greater than one-half the diameter under E of the boundary with respect to M_1+M_2 of the component of $M_2-M_1\cdot M_2$ containing p.

Let n be a positive number so small that ϵ/n is greater than twice the diameter of M_2 under E. If C is a rectifiable (under E) arc in M_1+M_2 from p to q, define $L(C)$ to be the greatest lower bound of all sums of the type $f(p, p_1)+f(p_2, p_3)+ \cdots +f(p_k, q)$ where $p_1, p_2, \cdots, p_k$ are points of C and $f(p_i, p_j)$ is either n times or 1 times the length under E of the subarc of C from p_i to p_j according as this subarc is or is not a subset of X. If $D_3(p, q)$ is defined to be the greatest lower bound of all such values $L(C)$, D_3 satisfies the conditions of Theorem 2.

THEOREM 3. *Suppose M_2 is a compact continuum with a convex metric, M_2 lies in a complete locally connected space S with a metric D, each component of $S-M_2$ is of diameter under D of less than θ and M_1 is a subcontinuum of M_2 with a convex metric D_1 such that*

$$D_1(p, q) \leqq D(p, q) \qquad \text{if } D_1(p, q) > \theta \ (p, q \text{ elements of } M_1). \tag{9}$$

For each positive number ϵ there is a continuum M_3 containing M_2 and a convex metric D_3 for M_3 such that D_3 preserves D_1 on M_1 and the boundary of each component of $S-M_3$ is of diameter less than $\theta+\epsilon$ under D_3.

PROOF. By Theorem 1, there is a convex metric D_2 for M_2 that preserves D_1 on M_1. Let n be an integer so large that

$$nD_2(p, q) > D(p, q) \qquad \text{if } D(p, q) > \epsilon/8 \ (p, q \text{ elements of } M_2). \tag{10}$$

Let X denote the collection of all pairs of points (x, y) such that both x and y are points of the boundary of the same component of $S-M_2$ and

$$nD_2(x, y) > \theta + \epsilon. \tag{11}$$

There is a finite subcollection X' of X such that for each element (x, y) of X there is an element (x', y') of X' such that both x' and y' are accessible from the same component of $S-M_2$ and

$$nD_2(x, x') + nD_2(y', y) < \epsilon/2. \tag{12}$$

Let $C_1, C_2, \cdots, C_j$ be a finite collection of components of $S-M_2$ irreducible with respect to the property that for each element of X' there is an integer i less than or equal to j such that both points of this element of X' are accessible from C_i.

There is a dendron T_i $(i=1, 2, \cdots, j)$ such that T_i lies except for its ends in C_i and the sum of the ends of T_i is a finite subset Y_i of the boundary of C_i such that Y_i contains all points of the sum of the elements of X' that are accessible from C_i and for each point q of the boundary of C_i there is a point r of Y_i such that $nD_2(r, q) < \epsilon/2$. Let $D(T_i)$ be a convex metric for T_i such that if p and q are two end points of T_i, then

$$\theta + \epsilon/4 < D(T_i; p, q) < \theta + \epsilon/2. \tag{13}$$

Let E be a metric for $M_2 + \sum T_i = M_3$ such that the distance between two points of M_3 under E is the greatest lower bound of the lengths of arcs containing them where length is measured by nD_2 in M_2 and by $D(T_i)$ in T_i. If (x, y) is an element of X, it follows from (12) and (13) that

$$E(x, y) < \theta + \epsilon. \tag{14}$$

The diameter of the boundary of each component of $S - M_3$ is less than $\theta + \epsilon$ under E, for suppose p and q are two points of this boundary; if $p + q$ is a subset of M_2, $E(p, q) < \theta + \epsilon$ by (11) and (14); if neither p nor q is a point of M_2, both belong to some T_i whose diameter is less than $\theta + \epsilon/2$; if p is an interior point of a T_i which does not contain q, there is an end point r of T_i such that $nD_2(r, q) < \epsilon/2$ and then $E(p, q) \leqq D(T_i; p, r) + nD_2(r, q) < \theta + \epsilon$.

We shall show that if r and s are two points of M_1, then $D_1(r, s) \leqq E(r, s)$. Suppose this is not the case and that rs is an arc from r to s in M_3 whose length is less than $D_1(r, s)$ under E. If $D_1(r, s) \leqq \theta$, then rs is a subset of M_2 alone and

$$E(r, s) = nD_2(r, s) \geqq D_2(r, s) = D_1(r, s). \tag{15}$$

Suppose rs is not a subset of M_2 and $p_1p_2, p_3p_4, \cdots, p_{2j-1}p_{2j}$ are the subarcs of rs which lie except for their end points in $M_3 - M_2$ where $p_{2i-1}p_{2i}$ precedes $p_{2i+1}p_{2i+2}$ on rs in the order from r to s. Let $Z(t)$ be 0 or t according as t is less than $\epsilon/8$ or not. If $D_1(r, s) > \theta$, it follows from (9), (3), and (10) that

$$\begin{aligned} D_1(r, s) &\leqq D(r, s) \leqq D(r, p_1) + D(p_1, p_2) + \cdots + D(p_{2j}, s) \\ &< D(r, p_1) + \theta + D(p_2, p_3) + \cdots + \theta + D(p_{2j}, s) \\ &< Z[D(r, p_1)] + Z[D(p_2, p_3)] + \cdots + Z[D(p_{2j}, s)] \\ &\quad + j\theta + (j+1)\epsilon/8 \\ &\leqq nD_2(r, p_1) + nD_2(p_2, p_3) + \cdots + nD_2(p_{2j}, s) \\ &\quad + j(\theta + \epsilon/4) \\ &\leqq \text{length } rs \text{ under } E. \end{aligned} \tag{16}$$

It follows from (15) and (16) that $E(r, s) \geqq D_1(r, s)$.

It follows from Theorem 2 that there is a convex metric D_3 for M_3 such that $D_3 = D_1$ on M_1 and $D_3 \leqq E$. The boundary of each component of $S - M_3$ is of diameter less than $\theta + \epsilon$ under D_3 because it is under E.

THEOREM 4. *Suppose M is a compact locally connected continuum such that if p is a point contained in an open subset R_1 of M, there is an open subset R_2 of R_1 containing p such that the boundary of R_2 with respect to M is a subset of a subcontinuum of M with a convex metric. Then M has a convex metric.*

PROOF. Let F be a metric for M. We shall show that there is a collection of continua $M_1, M_2, \cdots$ in M and a collection of metrics $D_1, D_2, \cdots$ such that:

(a) M_{i+1} contains M_i.

(b) D_i is a convex metric for M_i.

(c) D_{i+1} preserves D_i on M_i.

(d) Under F, each component of $M - M_i$ is of diameter less than $1/4^i$.

(e) Under D_i, the boundary (with respect to M) of each component of $M - M_i$ is of diameter less than $1/4^i$.

(f) Under D_{i+1}, the common part of the M_{i+1} and each component of $M - M_i$ is of diameter less than $3/4^i$.

First, we show that if ϵ is a positive number, there is a subcontinuum W of M with a convex metric such that each component of $M - W$ is of diameter less than ϵ under F. By the Heine-Borel Theorem, we find that there is a finite collection G of subcontinua of M such that each element of G has a convex metric, for each point p of M there is an element g of G such that p belongs either to g or to a component of $M - g$ of diameter under F of less than ϵ, and the sum W of the elements of G is a continuum. Each component of $M - W$ is of diameter less than ϵ under F and it follows from Theorem 1 that W has a convex metric.

Let M_1 be a subcontinuum of M with a convex metric E such that each component of $M - M_1$ is of diameter less than $1/4$ under F. A suitable multiple of E gives a metric D_1 for M_1 which will satisfy conditions (b) and (e).

Let n be an integer so large that

$$nF(p, q) > D_1(p, q) \quad \text{if} \quad D_1(p, q) > 1/17 \quad (p, q \text{ elements of } M_1).$$

There is a continuum W in M containing M_1 such that W has a convex metric and the diameter of each component of $M - W$ is less than

$1/17$ under nF. By Theorem 3, there is a continuum M_2 with a convex metric E such that M_2 contains W, E preserves D_1 on M_1, and the diameter under E of the boundary with respect to M of each component of $M-M_1$ is less than $1/4^2$. Applying Theorem 2 we find that M_2 has a convex metric D_2 such that M_2 and D_2 satisfy conditions (a), (b), (c), (d), (e), and (f).

Similarly, there exist continua $M_3, M_4, \cdots$ and convex metrics $D_3, D_4, \cdots$ satisfying conditions (a), (b), (c), (d), (e), and (f). Let D be a function of the pairs of points of M such that $D(p, q)$ is the lower limit of $D_1(p_1, q_1), D_2(p_2, q_2), \cdots$ where $p_1, p_2, \cdots$ and $q_1, q_2, \cdots$ are sequences of points converging to p and q respectively and p_i+q_i is a subset of M_i. We shall show that D preserves limit points in M.

Let ϵ be a positive number and n an integer so large that $3/4^n+3/4^{n+1}+\cdots<\epsilon/4$. If p and q are points of the same component of $M-M_n$, then $D(p, q)<\epsilon/2$ by (f). If p is a point of M_n, let R be the set of all points r of M_n such that $D_n(p, r)<\epsilon/2$. The sum V of R and all components of $M-M_n$ that have a point of R on their boundaries is an open subset of M containing p and if q is a point of V, then $D(p, q)<\epsilon$. Hence, the set of points q such that $D(p, q)>\epsilon$ is not a limit point of p.

If V is an open subset of M containing p, we shall show that there is a positive number ϵ such that $D(p, M-V)\geqq\epsilon$. Let R be an open subset of V containing p and n an integer such that $F(R, M-V)>3/4^n$. There is a positive number ϵ such that $D_n(r, s)>\epsilon$ if $F(r, s)>1/4^n$. Since each component of $M-M_n$ is of diameter less than $1/4^n$ under F, each arc in M from R to $M-V$ contains points r and s of M_n such that $F(r, s)>1/4^n$. Hence, if k is an integer bigger than n, $D_k(R\cdot M_k, [M-V]\cdot M_k)>\epsilon$. Hence, $D(p, M-V)\geqq\epsilon$.

We have shown that D satisfies conditions (1), (2), and (4). As each D_i satisfies conditions (3) and (5) and D is the limit of $D_1, D_2, \cdots$, then D satisfies these conditions. Hence, it is a convex metric for M.

THEOREM 5. *If M is an n-dimensional locally connected compact continuum and ϵ is a positive number, there is a locally connected continuum W in M such that each component of $M-W$ is of diameter less than ϵ, W is $(n-1)$-dimensional if $n>1$, and W is a dendron (acyclic continuous curve) if $n=1$.*

PROOF. An application of the Heine-Borel Theorem gives that there is an $(n-1)$-dimensional closed subset H of M such that each component of $M-H$ is of diameter less than ϵ. If $n-1>0$, H is contained in an $(n-1)$-dimensional locally connected subcontinuum of M [6, Theorem 1]. If $n-1=0$, a dendron in M contains H [7, Theorem 1].

THEOREM 6. *Each n-dimensional compact locally connected continuum has a convex metric.*

PROOF. If $n=1$, Theorem 6 follows from Theorems 4 and 5 and the fact that a dendron has a convex metric. If $n>1$, Theorem 6 follows from Theorems 4 and 5 and induction on n.

DEFINITION. A set S is said to be finite-dimensional if for each point p of S and each open subset R of S containing p there is an integer n and an open subset R' of R containing p such that the boundary of R' with respect to S is n-dimensional.

The following result may be established by using Theorems 4, 5, and 6.

THEOREM 7. *Each finite-dimensional compact locally connected continuum has a convex metric.*

REFERENCES

1. Gustav Beer, *Beweis des Satzes, dass jede im kleinen zusammenhängende Kurve konvex metrisiert werden kann*, Fund. Math. vol. 31 (1938) pp. 281–320.

2. R. H. Bing, *Extending a metric*, Duke Math. J. vol. 14 (1947) pp. 511–519.

3. O. G. Harrold, Jr., *Concerning the convexification of continuous curves*, Amer. J. Math. vol. 61 (1939) pp. 210–216.

4. C. Kuratowski and G. T. Whyburn, *Sur les éléments cycliques et leurs applications*, Fund. Math. vol. 16 (1930) pp. 305–331.

5. K. Menger, *Untersuchungen über allgemeine Metrik*, Math. Ann. vol. 100 (1928) pp. 75–163.

6. R. L. Wilder, *On the imbedding of subsets of a metric space in Jordan continua*, Fund. Math. vol. 19 (1932) pp. 45–64.

7. Leo Zippin, *A study of continuous curves and their relation to the Janiszewski-Mullikin Theorem*, Trans. Amer. Math. Soc. vol. 31 (1929) pp. 744–770.

8. E. E. Moise, *Grille decomposition and convexification theorems for compact locally connected continua*, to appear in Bull. Amer. Math. Soc.

9. R. H. Bing, *Partitioning a set*, to appear in Bull. Amer. Math. Soc.

UNIVERSITY OF WISCONSIN

METRIZATION OF TOPOLOGICAL SPACES

R. H. BING

A single valued function $D(x, y)$ is a *metric* for a topological space provided that for points x, y, z of the space:

1. $D(x, y) \geqslant 0$, the equality holding if and only if $x = y$,
2. $D(x, y) = D(y, x)$ (*symmetry*),
3. $D(x, y) + D(y, z) \geqslant D(x, z)$ (*triangle inequality*),
4. x belongs to the closure of the set M if and only if $D(x, m)$ (m element of M) is not bounded from 0 (preserves limit points).

A function $D(x, y)$ is a metric for a point set R of a topological space S if it is a metric for R when R is considered as a subspace of S. A topological space or point set that can be assigned a metric is called metrizable.

If a topological space has a metric, this metric may be useful in studying the space. Determining which topological spaces can be assigned metrics leads to interesting and important problems. For example, see [3].

A regular[1] topological space is metrizable if it has a countable basis[2] [7 and 8]. However, it is not necessary that a space be separable[3] in order to be metrizable. Theorem 3 gives a necessary and sufficient condition that a space be metrizable by using a condition more general than perfect separability.

Alexandroff and Urysohn showed [1] that a necessary and sufficient condition that a topological space be metrizable is that there exist a sequence of open coverings $G_1, G_2, \ldots$ such that (a) G_{i+1} is a refinement[4] of G_i, (b) the sum of each pair of intersecting elements of G_{i+1} is a subset of an element of G_i, and (c) for each point p and each open set D containing p there is an integer n such that every element of G_n containing p is a subset of D. We call a sequence of open coverings satisfying condition (c) a development. A developable space is a topological space that has a development. In section 2 we study conditions under which developable spaces can be assigned metrics.

The results of this paper hold in a topological space as defined by Whyburn in [10] or in a Hausdorff space.

Received November 26, 1949.

[1]A topological space is regular if for each open set D and each point p in D there is an open set containing p whose closure lies in D.

[2]A basis for a topological space is a collection G of open sets such that each open set is the sum of a subcollection of G. In [7 and 8] a space with a countable basis was said to satisfy the second axiom of countability. More recently such spaces have been called perfectly separable.

[3]A space is separable if it has a countable dense subset.

[4]The collection G is a refinement of the collection H if each element of G is a subset of an element of H.

1. Screenable spaces. We shall use the following definitions:

Discrete. A collection of point sets is *discrete* if the closures of these point sets are mutually exclusive and any subcollection of these closures has a closed sum.

Screenable. A space is *screenable* if for each open covering H of the space, there is a sequence $H_1, H_2, \ldots$ such that H_i is a collection of mutually exclusive domains and $\sum H_i$ is a covering of the space which is a refinement of H. A space is *strongly screenable* if there exist such H_i's which are discrete collections.

Perfectly screenable. A space is *perfectly screenable* if there exists a sequence $G_1, G_2, \ldots$ such that G_i is a discrete collection of domains and for each domain D and each point p in D there is an integer $n(p, D)$ such that $G_{n(p,D)}$ contains a domain which lies in D and contains p.

Collectionwise normal. A space is *collectionwise normal* if for each discrete collection X of point sets, there is a collection Y of mutually exclusive domains covering X^* such that no element of Y intersects two elements of X. We use X^* to denote the sum of the elements in X.

The following result follows from the definitions of perfectly screenable and strongly screenable.

THEOREM 1. *A perfectly screenable space is strongly screenable.*

The following example shows that a developable space may not be screenable.

EXAMPLE A. *A locally connected*[5] *separable Moore space*[6] *S such that no space homeomorphic with the closure of any open set in S is either normal or screenable.* The points of S are the points of the plane and the open sets of S are given in terms of a development $G_1, G_2, \ldots$ which is described as follows. Let $L_1, L_2, \ldots$ be a sequence of horizontal lines whose sum is dense in the plane. Either of the following types of sets is an element of G_i: (a) the interior of a circle with diameter less than $1/i$ which does not intersect $L_1 + L_2 + \ldots + L_i$, (b) $p + I_1 + I_2$ where p is a point of some L_j and I_1, I_2 are interiors of circles of diameter less than $1/2i$ which are tangent to L_j at p on opposite sides of L_j and such that $I_1 + I_2$ does not intersect $L_1 + L_2 + \ldots + L_i$.

That S is locally connected follows from the fact that vertical lines are connected and horizontal lines other than $L_1, L_2, \ldots$ are connected. The elements of each G_i are connected. Since the plane is separable and any set dense in the plane is dense in S, S is separable. The sequence $G_1, G_2, \ldots$ satisfies the conditions of Axiom 1 of [5], so S is a Moore space.

Let S' be a space homeomorphic with the closure of an open set E in S and K be an interval in E that is a subset of $L_1 + L_2 + \ldots$. If K_1 and K_2 are

[5]A topological space is locally connected if it has a basis such that the elements of the basis are connected. A set is connected if it cannot be expressed as the sum of two non-null sets such that neither contains a point of the closure of the other.

[6]A space satisfying the first three parts of Axiom 1 of [5] is called a Moore space. It is a regular developable space.

subsets of S' corresponding to the points of K with rational and irrational abscissas respectively, K_1 and K_2 are two mutually exclusive closed point sets. That S' is not normal follows from the fact that there do not exist two mutually exclusive domains containing K_1 and K_2 respectively.

If H is an open covering of S' such that no two points of $K_1 + K_2$ belong to the same element of H, any open covering of S' that refines H contains uncountably many elements. Since S' is separable, it does not contain an uncountable collection of mutually exclusive domains. Hence, S' is not screenable because there is not a sequence $H_1, H_2, \ldots$ such that H_1 is a collection of mutually exclusive domains and $\sum H_i$ is a covering of S' that refines H.

The proof of the following theorem may be compared with one given by Tychonoff [7] to show that any regular perfectly separable topological space is normal.

THEOREM 2. *A regular strongly screenable space is collectionwise normal.*

Proof. Suppose $\{A_\alpha\}$ is a discrete collection of closed sets. Let K be a collection of open sets covering the space such that the closure of no element of K intersects two elements of $\{A_\alpha\}$. Since the space is strongly screenable, there is a sequence $H_1, H_2, \ldots$ such that H_i is a discrete collection of domains and $\sum H_i$ is a covering of the space which is a refinement of K.

Let $U_{i\beta}$ be the sum of the elements H_i that intersect A_β and $V_{i\beta}$ be the sum of those intersecting elements of $\{A_\alpha\}$ other than A_β. If $D_\beta = U_{1\beta} + (U_{2\beta} - \overline{V}_{1\beta}) + (U_{3\beta} - [\overline{V}_{1\beta} + \overline{V}_{2\beta}]) + \ldots + (U_{i\beta} - \sum_{j=1}^{i-1} \overline{V}_{j\beta}) + \ldots$, then $\{D_\alpha\}$ is a collection of mutually exclusive domains covering $\sum A_\alpha$ such that no element of $\{D_\alpha\}$ intersects two elements of $\{A_\alpha\}$.

It cannot be concluded that a strongly screenable space is normal for a perfectly separable space may not even be regular. Example B shows us that we cannot conclude that a regular screenable space is normal.

EXAMPLE B. *A screenable, point-wise paracompact,*[7] *nonparacompact, nonnormal Moore space with an open covering H such that the star*[8] *of each point with respect to H is metrizable.*

Points are of three types: (a) elements of a countable sequence of points $p_1, p_2, \ldots$; (b) elements of the collection of all continuous functions $f_\alpha(x)$ $(0 < x < 1)$ such that $\frac{1}{2} < f_\alpha(x) < 1$; and (c) ordered triples (p_i, t, f_α) where p_i is a point of type (a), f_α is one of type (b), and t is a positive number less than one.

The open sets of this space are defined by a development $G_1, G_2, \ldots$ which is described as follows. The elements of G_n are of three sorts: p_j $(j = 1, 2, \ldots)$

[7]A topological space is point-wise paracompact if for each open covering H there is an open covering H' such that H' refines H and no point lies in infinitely many elements of H'. It is paracompact if for each open covering H there are open coverings H' and H'' such that H' refines H and no element of H'' intersects infinitely many elements of H'.

[8]The star of a point set A with respect to an open covering H is the sum of the element of of H that intersect A.

plus all points (p_j, t, f_α) $(t < 1/n)$; (p_j, t_0, f_α) plus all points (p_j, t, f_α) $(|t - t_0| < 1/n)$; an element f_α of type (b) plus all points (p_j, t, f_α) where $1 - t < 1/n$ if $j \leqslant n$ and $1 - t < [1 + nf_\alpha(1/j)]/(n + 1)$ if $j > n$.

It is convenient to think of the space as being the collection of points of type (a) plus the collection of points of type (b) plus open unit intervals joining points of type (a) to points of type (b). An element of G_n is either (i) an element of type (a) plus all points that can be joined to it by intervals of lengths less than $1/n$, or (ii) a point of an open interval plus the collection of all points of the open interval that are nearer than $1/n$ to the point or (iii) a point f of type (c) plus all points that can be joined to it by an interval of length less than $x(j)$ where p is on the interval from f_α to p_j and $x(j) = 1/n$ if $j \leqslant n$ and $x(j) = [1 + nf_\alpha(1/j)]/(n + 1)$ if $j > n$.

The above space is screenable because for any open covering K of it, there is an open covering $K_1 + K_2$ of it which is a refinement of K and such that K_1 is a collection of mutually exclusive domains covering all points of type (a) and a dense set of those of type (c), while K_2 is another such collection covering all points of type (c) not covered by K_1. The space is point-wise paracompact because for each open covering K of it there is an open covering $K_1 + K_2$ such that no point is covered by more than two elements of $K_1 + K_2$.

The space is not normal because for each domain D containing the collection of all points of type (a), there is a point of type (b) which is a limit point of D. To find such a point, let n_i be an integer such that an element of G_{n_i} contains p_i and lies in D. If f_β is a point of type (b) which satisfies $f_\beta(1/i) > 1 - 1/n_i$ $(i = 1, 2, \ldots)$, it is a point of the closure of D. Since the collection of points of type (a) and the collection of points of type (b) are closed sets, and there do not exist mutually exclusive open sets containing these two collections, the space is not normal. Since it is not normal, it is not paracompact.

To show that there is an open covering H such that the star of any point with respect to H is metrizable, let H be any G_i and M be the star of some point with respect to H. There is a connected open set M' that contains M such that M' contains only one point p_j of type (a) and only one point f_α of type (b). If R is the component of $M' - p_j$ containing f_α, $R + p_j$ has a metric $D_1(x, y)$ because $R + p_j$ is homeomorphic with a regular, perfectly separable space. Furthermore, $M' - R$ has a metric $D_2(x, y)$ described as follows: $D_2[p_j, (p_j, t, f_\alpha)] = t$; $D_2[(p_j, t_1, f_\alpha), (p_j, t_2, f_\beta)]$ is $|t_1 - t_2|$ or $t_1 + t_2$ according as f_α is or is not f_β. There is a metric $D(x, y)$ for M' where $D(x, y) = D_1(x, y)$ if x and y are points of R, $D(x, y) = D_2(x, y)$ if x and y are points of $M' - R$, and $D(x, y) = D_1(x, p_j) + D_2(p_j, y)$ if x and y belong to R and $M' - R$ respectively.

The following theorem may be compared with the result of Urysohn [8] which states that a normal perfectly separable topological space is metrizable.

THEOREM 3. *A necessary and sufficient condition that a regular topological space be metrizable is that it be perfectly screenable.*

Proof of sufficiency. Suppose $H_1, H_2, \ldots$ is a sequence of discrete collections of domains such that for each domain D and each point p in D there is an integer $n(p, D)$ such that an element of $H_{n(p,D)}$ lies in D and contains p.

Let K_{ij} be the sum of the elements of H_j whose closures lie in an element of H_i. Since the space is normal, there is a continuous transformation F_{ij} of space into the real numbers between 0 and 1 such that the image of K_{ij} under F_{ij} is 1 and the image of the complement of H^*_i is 0 [9]. For points x, y of the space we define the distance between them to be

$$D(x, y) = \sum\sum \frac{|F_{ij}(x) + R_{ij}(x, y)F_{ij}(y)|}{2^{i+j}}$$

where $R_{ij}(x, y)$ is -1 or 1 according as y does or does not belong to an element of H_i that contains x. It may be found that $D(x, y)$ satisfies the conditions for a metric.

Proof of necessity. A. H. Stone has shown [6] that for each metric space and each positive number ϵ, there is a sequence $R_{\epsilon 1}, R_{\epsilon 2}, \ldots$ such that $\sum R_{\epsilon i}$ covers the space and $R_{\epsilon i}$ is a discrete collection of closed sets each of diameter less than ϵ. A proof of this is also found in Theorem 9 of the present paper. For each element r of $R_{\epsilon i}$ let D_r be a domain covering r such that D_r is of diameter less than ϵ and each point of D_r is more than twice as close to r as to any other element of $R_{\epsilon i}$. If $H_{\epsilon i}$ denotes the collection of all such open sets D_r, $\{H_{\epsilon i}; \epsilon = 1, \frac{1}{2}, \frac{1}{3}, \ldots$ and $i = 1, 2, \ldots\}$ becomes on suitable ordering a countable sequence of discrete collections of open sets which insures that a metric space is perfectly screenable.

In the following modification of Theorem 3 we dispense with the supposition that the elements of H_i are mutually exclusive. E. E. Floyd suggested that such a modification might be possible.

THEOREM 4. *A regular topological space S is metrizable if and only if there is a sequence $G_1, G_2, \ldots$ such that*

(a) *G_i is a collection of open subsets of S such that the sum of the closures of any subcollection of G_i is closed and*

(b) *if p is a point and D is an open set containing p there is an integer $n(p, D)$ such that an element of $G_{n(p,D)}$ contains p and each element of $G_{n(p,D)}$ containing p lies in D.*

Proof. Since a metric space is perfectly screenable, it contains a sequence $G_1, G_2, \ldots$ satisfying the conditions of the theorem. We complete the proof of Theorem 4 by showing that any regular topological space admitting such a sequence is perfectly screenable.

First, we show that any open subset D of S is strongly screenable. Let $H = (h_1, h_2, \ldots, h_\alpha, \ldots)$ be a well ordered collection of open sets whose sum is D. Let $V_{\alpha i}$ be the sum of the elements of G_i whose closures lie in h_α. If $U_{\alpha ij}$ denotes the sum of the elements of G_j whose closures lie in $V_{\alpha i}$ but do not intersect $\sum_{\beta<\alpha}\overline{V}_{\beta i}$ then $W_{ij} = \{U_{\gamma ij}; \gamma = 1, 2, \ldots, \alpha, \ldots\}$ is a discrete collection of open sets which is a refinement of H. To see that $\sum\sum W_{ij}$ covers D, let p be a point of D and h_β be the first element of H containing p. Then p belongs to some $V_{\beta k}$ but does not belong to $\sum_{\alpha<\beta}\overline{V}_{\alpha k}$. Then for some integer m, p lies in an element of G_m whose closure lies in $V_{\beta k}$ but does not intersect $\sum_{\alpha<\beta}\overline{V}_{\alpha k}$ and p is a point of $U_{\beta k m}$.

For each positive integer k let $X_{k1}, X_{k2}, \ldots$ be a sequence of discrete collections of open sets such that each X_{ki} is a refinement of G_k and $\sum_{i=1}^{\infty} X_{ki}$ covers G^*_k. That S is perfectly screenable follows from the fact that the elements of $\{X_{ki}; i, k = 1, 2, \ldots\}$ may be ordered in a sequence fulfilling the conditions to be satisfied by the sequence $G_1, G_2, \ldots$ mentioned in the definition of a perfectly screenable space.

We find from Example C that condition (b) of Theorem 4 could not be weakened to

(b′) *if p is a point and D is an open set containing p, there is an integer $n(p,D)$ such that an element of $G_n(p, D)$ contains p and lies in D.*

Example C. *A regular strongly screenable topological space not satisfying the first axiom of countability.*[9] Points are the points of the plane. A neighbourhood is either (a) an open interval of a line through the origin such that this interval does not contain the origin or (b) the sum of a collection of open intervals each of which contains the origin and such that each line through the origin contains one of these open intervals.

The space is strongly screenable because for each open covering H of it there are two discrete collections H_1, and H_2 such that $H_1 + H_2$ is an open covering of the space which refines H. The space does not satisfiy the first axiom of countability because for each sequence of neighbourhoods of the origin there is a neighbourhood of the origin that does not contain any of these neighbourhoods. Let G_1 be the collection of all neighbourhoods N of type (b) containing the origin such that if p is a point of the boundary of N, then for some integer n, p is at a distance $1/n$ from the origin in the plane. Then collections $G_2, G_3, \ldots$ may be chosen to satisfy conditions (a) and (b′). However, the space is not metrizable.

2. Developable spaces. For each developable topological space there is a sequence $G_1, G_2, \ldots$ such that (a) G_i is a covering of the space with open sets, (b) G_{i+1} is a refinement of G_i, and (c) for each domain and each point p in D there is an integer $n(p, D)$ such that each element of $G_{n(p,D)}$ which contains p lies in D. Condition (b) is not necessary in defining a developable space because if there is a sequence satisfying conditions (a) and (c), there is one satisfying conditions (a), (b), and (c). In fact, in [5] the condition is imposed that G_{i+1} is a subcollection of G_i. Regular developable topological spaces have been studied extensively because a Moore space is such a space.

[9] A topological space satisfies the first axiom of countability at a point p if there is a countable collection G of neighbourhoods of p such that any neighbourhood of p is a subset of an element of G.

As seen from Examples A and D, not all developable spaces are screenable and not all screenable spaces are developable.

EXAMPLE D. *A regular, separable, strongly screenable space that is not perfectly screenable or developable.* Points belong to the x-axis and neighbourhoods are closed intervals minus their right hand end points.

The space is separable because each set of points dense on the x-axis is dense in the space. If H is an open covering of it, there is an open covering H' which refines H such that no two elements of H' intersect each other. The space is strongly screenable because each such open covering H' is a discrete collection.

The space is not perfectly separable because for each countable collection G of neighbourhoods, there is a point p that does not belong to the left end of any element of G and any neighbourhood of p with a left end at p is not the sum of a subcollection of G. Since the space is separable but not perfectly separable, it is not metrizable. It follows from Theorem 3 that it is not perfectly screenable and from Theorem 5 that it is not developable.

THEOREM 5. *A separable screenable developable space is perfectly separable.*

Proof. No separable space contains uncountably many mutually exclusive domains. Hence if H is an open covering of a separable screenable space, there is a countable open covering H' which refines H. If $G_1, G_2, \ldots$ is a development of a space S and G'_i is a countable open covering that refines G_i, then $\sum G'_i$ is a countable basis for S.

A similar argument shows that a separable perfectly screenable space is perfectly separable.

THEOREM 6. *A strongly screenable developable space is perfectly screenable.*

Proof. Let $G_1, G_2, \ldots$ be a development of the space. Since the space is strongly screenable, for each positive integer i there is a sequence $H_{i1}, H_{i2}, \ldots$ such that H_{ij} is a discrete collection of domains and $\sum_{j=1}^{\infty} H_{ij}$ covers the space and is a refinement of G_i. Then $\{H_{ij}; i, j = 1, 2, \ldots\}$ is a countable collection insuring that the space is perfectly screenable.

THEOREM 7. *A regular developable space (Moore space) is metrizable if it is strongly screenable.*

Theorem 7 follows from Theorems 3 and 6. That Theorem 7 cannot be altered by assuming screenability instead of strong screenability may be seen from Example B.

THEOREM 8. *A screenable Moore space is metrizable if it is normal.*

Proof. This result will follow from Theorem 7 if it is shown that a screenable normal developable space is strongly screenable.

Let H be a collection of mutually exclusive open sets, W be the complement of H^*, and $G_1, G_2, \ldots$ be a development of the space. Denote by X_i the sum of all points p such that no element of G_i containing p intersects W. Since the space is normal, there is a domain D containing X_i such that $\bar{D}$ does not intersect W. If H_i is the collection of all domains h such that h is the common part of D and an element of H, then H_i is a discrete collection of domains. Since for each collection H of mutually exclusive domains there is a collection $\sum H_i$ covering H such that H_i is a discrete collection of domains which is a refinement of H, a normal developable space is strongly screenable if it is screenable.

THEOREM 9. *For each open covering H of a developable space there is a sequence $X_1, X_2, \ldots$ such that X_i is a discrete collection of closed sets which is a refinement of both X_{i+1} and H while $\sum X_i$ covers the space.*

Proof. Suppose W is a well ordering of H and $G_1, G_2, \ldots$ is a development of the space such that G_{i+1} is a refinement of G_i. For each element h of H, let $x(h, i)$ denote the sum of all points p such that no element of H that contains p precedes h in W and each element of G_i containing p is a subset of h. If X_i denotes the collection of all such sets $x(h, i)$, X_i is a discrete collection because no element of G_i intersects two elements of X_i. If p is a point and $h(p)$ is the first element of H and W containing p, then for some integer i, $[h(p), i]$ contains p. Hence $\sum X_i$ covers the space.

THEOREM 10. *A Moore space is metrizable if it is collectionwise normal.*

Proof. If it is shown that a collectionwise normal Moore space is screenable Theorem 10 will follow from Theorem 8.

For each open covering H of the space, it follows from Theorem 9 that there is a sequence $X_1, X_2, \ldots$ such that X_i is a discrete collection of closed sets and $\sum X_i$ is a covering of the space which is a refinement of H. Collectionwise normality insures that there is a collection Y_i of mutually exclusive open sets covering X^*_i such that no element of Y_i intersects two elements of X_i but each is a subset of an element of H. Then $Y_1, Y_2, \ldots$ is a sequence such that $\sum Y_i$ is a refinement of H covering the space and Y_i is a collection of mutually exclusive domains. Hence, the space is screenable.

Question. Is there a normal Moore space which is neither screenable nor collectionwise normal? If this question could be answered, it could be determined whether or not each normal Moore space is metrizable. F. B. Jones showed [4] that such a space is metrizable if it is separable and $\aleph_1 = C$. Hence, the space mentioned in Example E is metrizable if $\aleph_1 = C$.

EXAMPLE E. *A separable normal Moore space.* The points of the space are the points of the plane which lie above the x-axis and the points of a subset X of the x-axis such that each subset of X is the common part of X and a G_δ set in the plane. The elements of G_i are of two sorts: (a) the interior of a circle of radius less then $1/i$ which lies above the x-axis and (b) $p + I$ where p is a point of X and I is the interior of a circle of diameter less than $1/i$ which is tangent to the x-axis at p from above.

If X is countable, the above space is metrizable because it is perfectly separable. The set X cannot have the power of the continuum because each subset Y with the power of the continuum in a separable metric space S contains a subset Y' which is not the common part of Y and a G_δ set in S.

In my paper [2] the following theorem is proved.

THEOREM 11. *A topological space is metrizable provided there exists a sequence $H_1, H_2, \ldots$ such that*

(a) *for each integer i, H_i is a collection of sets covering space.*

(b) *a point p is a point of the closure of the set M if and only if for each integer n, some element of H_n contains p and intersects M, and*

(c′) *each pair of points that is covered by either an element of H_{i+1} or the sum of a pair of intersecting elements of H_{i+1} can be covered by an element of H_i.*

In [2] it was falsely stated that (c′) could be replaced by (c) *each pair of points that is covered by the sum of a pair of intersecting elements of H_{i+1} can be covered by an element of H_i.* I am indebted to Dick Wick Hall for calling my attention to the fact that this replacement is not possible. This paper was being studied in one of his classes and L. K. Meals, a member of that class, pointed out that if S is a nonmetrizable space with a development $G_1, G_2, \ldots$ such that G_{i+1} is a refinement of G_i (as in Example A), then if H_i is G_i or S according as i is odd or even, then $H_1, H_2, \ldots$ satisfies conditions (a), (b), and (c).

3. Collectionwise normality. In a developable space, either full normality[10] or collectionwise normality implies metrizability. However, in general, collectionwise normality is weaker than full normality as is shown in the following theorem.

THEOREM 12. *Full normality implies collectionwise normality but not conversely.*

Proof. Let W be a discrete collection of closed sets and H be an open covering of the space such that no element of H intersects two elements of W. If the space is fully normal, there is an open covering H' of the space such that for each point p, the sum of the elements of H' containing p is a subset of an element of H. For each element w of W let D_w be the sum of the elements of H' intersecting w. If w_1 and w_2 are different elements of W, D_{w_1} does not intersect D_{w_2} in a point p or else an element of H containing p intersects both w_1 and w_2. Then the collection of all such domains D_w is a collection of mutually exclusive domains covering W such that no one of these domains intersects two elements of W.

The space described below shows that collectionwise normality does not imply full normality.

[10]A space is fully normal if for each open covering H of the space there is an open covering H' of the space such that the star of each point with respect to H' is a subset of an element of H. Stone showed [6] that the notions of full normality and paracompactness are equivalent for a topological space.

EXAMPLE F. *A collectionwise normal space which is not fully normal.* Points are the elements of an uncountable well ordered collection W such that no element of W is preceded by uncountably many elements of it. A neighbourhood is either the first element of W or the collection of all points that lie between two nonadjacent elements of W.

This space is collectionwise normal because it is normal and does not contain an infinite discrete collection of points. The space is not fully normal because if H is any collection of open sets covering the space, there is a point p such that the star of p with respect to H contains all points that follow p. There is an open covering H such that no elements of H contains all the points that follow some point in it.

THEOREM 13. *Suppose H is an open covering of a collectionwise normal space S and $H_1, H_2, \ldots$ is a sequence such that H_i is a discrete collection of closed sets and $\sum H_i$ is a refinement of H which covers S. Then there is an open covering G of S such that G is a refinement of H and for each point p of an element of H_i there is a domain D containing p such that not more than i elements of G intersect D.*

Since a metric space is developable and collectionwise normal, Theorems 9 and 13 imply that a metric space is paracompact [6]. Theorem 13 would not be true if the hypothesis that $\sum H_i$ covers S were omitted.

Proof of Theorem 13. Since S is collectionwise normal, there is a discrete collection Y_i of open sets covering the sum of the elements of H_i such that Y_i is a refinement of H and each element of Y_i intersects just one element of H_i. As S is normal, there is an open set D_i containing the sum of the elements of H_i such that Y_i covers $\bar{D}_i$. Each element of Y_1 is an element of G and if y is an element of Y_{i+1} not covered by $\bar{D}_1+\bar{D}_2+ \ldots +\bar{D}_i$, $y-(\bar{D}_1+\bar{D}_2 \ldots +\bar{D}_i)$ is an element of G.

THEOREM 14. *Collectionwise normality implies normality but normality does not imply collectionwise normality.*

Proof. Since a collection consisting of two mutually exclusive closed sets is a discrete collection, collectionwise normality implies normality. We shall show that the space described in Example G is normal but not collectionwise normal.

EXAMPLE G. *A normal topological space that is not collectionwise normal.* Let P be an uncountable set, Q the set of all subsets of P, and F the set of all functions f on Q having only 1 and 0 as values. To each element p of P associate the function f_p whose value $f_p(q)$ on q is 1 or 0 according as p belongs to q or not. Let F_p be the set of all such functions f_p. The set F is topologized as follows. Any point f in $F-F_p$ is declared to be a neighbourhood of itself. Given a point f_p in F_p and a finite subset r of Q we define the r neighbourhood of f_p to be the set of all f such that $f(q) = f_p(q)$ whenever q belongs to r.

To show the space F thus topologized is normal consider two mutually exclusive closed subsets H_1 and H_2 of it. Let $A_k(k = 1, 2)$ be the set of points common to H_k and F_p and let q_k be the associated set in P consisting of all p for which f_p belongs to A_k. We suppose that neither A_1 nor A_2 is null because if $A_1 = 0$, H_1 and $F - H_1$ are mutually exclusive domains containing H_1 and H_2 respectively. The set D_k of all f in F such that $f(q_k) = 1$ and $f(q_j) = 0$ $(j \neq k)$ is then an open set containing A_k. Moreover no point in F is common both to D_1 and D_2. Therefore the sets $(D_1 - H_2) + (H_1 - A_1)$ and $(D_2 - H_1) + (H_2 - A_2)$ are mutually exclusive open sets containing H_1 and H_2 respectively.

We now show that F is not collectionwise normal. The subset $F_p = \{f_p\}$ of F is a discrete collection of points. However, there does not exist a collection of mutually exclusive neighbourhoods $\{D_p\}$ such that D_p is a neighbourhood of f_p. For suppose, to the contrary, that there were such a collection. Let D_p be the r_p neighbourhood of f_p. Since r_p is a finite subset of Q and P is uncountable there is an integer n and an uncountable subset W of P such that r_p has exactly n elements for every p in W. For any two elements a and b of W the sets r_a and r_b have an element in common, else D_a and D_b would intersect. Hence there is an element q_1 of Q and an uncountable subset W'_1 of W such that q_1 belongs to r_p for every p in W'_1. Moreover there is a t_1 with value 1 or 0 and an uncountable subset W_1 of W'_1 such that $f_p(q_1) = t_1$ for every p in W_1. Similarly there is an element q_2 of Q different from q_1, a t_2 with value 1 or 0, and an uncountable subset W_2 of W_1 such that q_2 belongs to r_2 and $f_p(q_2) = t_2$ for every p in W_2. Continuing recursively in this fashion we get q_k, t_k, W_k for $k = 1, 2, \ldots, n$. Let r be the set consisting of $q_1, q_2, \ldots, q_n$ and D the set of all f with $f(q_k) = t_k$ for $k = 1, 2, \ldots, n$. Then $r_p = r$ and $D_p = D$ for all p in W_n in contradiction to the choice of D_p as mutually exclusive.

One might wonder if Example G could be modified so as to obtain a normal developable space which is not metrizable. A developable space could be obtained by introducing more neighbourhoods into the space F. However, a difficulty might arise in introducing enough neighbourhoods to make the resulting space developable but not enough to make it collectionwise normal.

Another method of modifying Example G is to replace points by closed sets. No non-isolated point of the space F is the intersection of a countable number of neighbourhoods. In Example H, we modify Example G by replacing the points of F by closed sets so as to get a space in which all closed sets are inner limiting (G_δ) sets.[11]

EXAMPLE H. *A normal topological space that is not collectionwise normal and in which each closed set is an inner limiting* (G_δ) *set.* We define P, Q, and F_p as in Example G but let the points of F be functions f defined on Q such

[11]A set is an inner limiting set or G_δ set if it is the intersection of a countable number of open sets.

that $f(x)$ is a non-negative integer. Each point of $F - F_p$ is a neighbourhood. For each finite subset r of Q, each element f_p of F_p, and each positive integer n, f_p plus all points f of F such that $f(x) > n$ (x element of Q) and $f(x') = f_p(x')$ mod 2 (x' element of r) is a neighbourhood.

References

[1] P. Alexandroff and P. Urysohn, *Une condition nécessaire et suffisante pour qu'une classe* (L) *soit une classe* (B), Comptes Rendus Hebdomadaires des Séances de l'Académie des Sciences, vol. 177 (1923), 1274-1276.

[2] R. H. Bing, *Extending a Metric*, Duke Math. J., vol. 14 (1947), 511-519.

[3] E. W. Chittenden, *On the metrization problem and related problems in the theory of abstract sets*, Bull. Amer. Math. Soc., vol. 33 (1927), 13-34.

[4] F. B. Jones, *Concerning normal and completely normal spaces*, Bull. Amer. Math. Soc., vol. 43 (1937), 671-679.

[5] R. L. Moore, *Foundations of Point Set Theory*, Amer. Math. Soc. Colloquium Publications, vol. 13, New York, 1932.

[6] A. H. Stone, *Paracompactness and product spaces*, Bull. Amer. Math. Soc., vol. 54 (1948), 977-982.

[7] A. Tychonoff, *Über einen Metrisationssatz von P. Urysohn*, Math. Ann., vol. 95 (1926), 139-142.

[8] P. Urysohn, *Zum Metrisationsproblem*, Math. Ann., vol. 94 (1925), 309-315.

[9] ———, *Über die Mächtigkeit der zusammenhängenden Mengen*, Math. Ann., vol. 94 (1925), 262-295.

[10] G. T. Whyburn, *Analytic Topology*, Amer. Math. Soc. Colloquium Publications, vol. 28, New York, 1942.

University of Virginia and
University of Wisconsin

AN EQUILATERAL DISTANCE*

R. H. BING, University of Wisconsin

1. Introduction. A function $D(p, q)$ on the pairs of points of a topological space is called a distance for the space provided

(1) $D(p, q) \geqq 0$, the equality holding if and only if $p=q$,
(2) $D(p, q)=D(q, p)$, (symmetry),
(3) $D(p, q) \leqq D(p, r)+D(r, q)$, (triangle inequality), and
(4) $D(p, q)$ preserves limit points.

By (4) we mean that p is a limit point of the set M if and only if 0 is a limit of the set of numbers $D(p, x)$, where x is a point of M.

Examples of distance functions for the plane are

$$D_1(p, q) = \sqrt{(x_p - x_q)^2 + (y_p - y_q)^2}, \qquad D_2(p, q) = | x_p - x_q | + | y_p - y_q | ,$$

and

$$D_3(p, q) = \max (| x_p - x_q | , | y_p - y_q |).$$

Furthermore, if D is any distance function, then each of the following is a distance:

$$D_4 = cD, \quad D_5 = \sqrt{D}, \quad D_6 = \min (1, D), \quad \text{and} \quad D_7 = D/(1 + D).$$

While any function $F_1(p, q)$ satisfying conditions (2), (3), and (4) in a space without isolated points also satisfies (1),

$F_2(p, q)=2|x_p-x_q|+x_p-x_q+|y_p-y_q|$ does not satisfy (2),
$F_3(p, q)=(x_p-x_q)^2+(y_p-y_q)^2$ does not satisfy (3), and
$F_4(p, q)$ equals 0 or 1, according as p is or is not q, does not satisfy (4).

Although general distances may be quite different from the Euclidean distance, the interiors of spheres are open sets, and spaces with distance functions have a type of uniformity that facilitates their study. Some types of distances are more like the Euclidean distance than others. For example, a distance function $D(p, q)$ is convex provided:

(5) For each pair of points p, q there is a point r different from p and q such that $D(p, q)=D(p, r)+D(r, q)$.

We call $D(p, q)$ an equilateral distance if it satisfies conditions (1), (2), (3), (4), and

(6) For each pair of points p, q there is a point r such that $D(p, q)=D(p, r) =D(r, q)$.

It may be noticed that the point r is not between p and q but forms an equilateral triple with them.

* Excerpt from an address "An equilateral distance" before the Wisconsin Section of the Association, held at Lawrence College, Appleton, Wisconsin, on May 14, 1949.

A misprint influenced the starting of this paper. The author became interested in equilateral triples as a result of looking at Gustav Beer's paper *Beweis des Satzes, dass jede im kleinen zusammenhängende Kurve convex metrisiert werden kann*. A typographical error [Fundamenta Mathematicae, vol. 31 (1938), p. 281] in the introduction of that paper caused a distance to be called convex if it satisfied (6) instead of (5). However, the "+" was intended instead of the "=" because the paper considered the assigning a convex metric to the one-dimensional image of a straight line interval under a continuous transformation. One might wonder if this continuum might be assigned a distance satisfying (6). It did not seem intuitively obvious that even an interval could be assigned such a distance. The ruler that is described below was constructed to show that it could be assigned one.

2. Equilateral triples in the plane. The ordinary distance for Euclidean 2-space (the plane) is equilateral because each pair of points lies on the vertices of an equilateral triangle. However, this is not true of Euclidean 1-space (the line).

THEOREM 1. *If M is a bounded plane set with more than 3 points, some pair of points of M does not belong to an equilateral triple in M under the metric $D(p, q) = \sqrt{(x_p-x_q)^2+(y_p-y_q)^2}$.*

Proof. If each pair of points of M belongs to an equilateral triple, then each pair of points of $\overline{M}$ (M plus its limit points) belongs to an equilateral triple in $\overline{M}$, for suppose that p and q are points of $\overline{M}$. Let $p_1, p_2, \cdots$ and $q_1, q_2, \cdots$ be sequences of points of M converging to p and q, respectively. Suppose r_i is a point of M such that p_i, q_i, r_i is an equilateral triple. Some subsequence of $r_1, r_2, \cdots$ converges to a point r of $\overline{M}$, and p, q, r is an equilateral triple.

For some pair of points a, b of $\overline{M}$, $D(a, b)$ is the diameter of $\overline{M}$ (that is, no pair of points of $\overline{M}$ are farther apart than a and b). Suppose c is a point of $\overline{M}$ such that a, b, c is an equilateral triple.

We now show that there is a point d of $\overline{M}$ outside the triangle abc. Suppose d' is a point of $\overline{M}$ other than a, b, or c, and suppose that z is the one of a, b, c that is farthest from d'. If d' is not outside the triangle abc, then the point of $\overline{M}$ that forms an equilateral triple with z and d' is outside.

Without loss of generality, we suppose that the interval ad intersects the interval bc at f and that b and the point e of $\overline{M}$ which forms an equilateral triple with a and d are on opposite sides of the line through a and d. By the laws of cosines and sines,

$$
\begin{aligned}
D^2(e, b) &= D^2(a, b) + D^2(a, e) - 2D(a, b)D(a, e)\cos\angle eab \\
&> D^2(a, b) + D(a, e)[D(a, f) - 2D(a, b)\cos\angle eab] \\
&= D^2(a, b) + D(a, e)D(a, b)[\sin\angle abf \\
&\qquad - 2\cos\angle eab\sin\angle afb]/\sin\angle afb.
\end{aligned} \tag{7}
$$

Since $\angle eab$ and $\angle afb$ are supplementary, the expression in the bracket of the fourth member of (7) is equal to $\sin(\pi/3)-\sin(2\angle eab)$, which is positive. Hence, the assumption that each pair of points of M belongs to an equilateral triple has led us to the contradiction that e and b of $\overline{M}$ are farther apart than any two points of $\overline{M}$.

3. An equilateral ruler. The ruler portrayed in the figure that follows is the graph of a function $f(x)$ satisfying

(8) for each x, either $f(x)=f(2x)$ or $f(x)=f(x/2)$, and

$$\left\{\begin{matrix} f(a) \leqq f(b)+f(c) \\ f(b) \leqq f(a)+f(c) \\ f(c) \leqq f(a)+f(b) \end{matrix}\right\} \quad \text{provided} \quad \left\{\begin{matrix} 0 \leqq a \leqq b+c \\ 0 \leqq b \leqq a+c \\ 0 \leqq c \leqq a+b \end{matrix}\right\}. \tag{9}$$

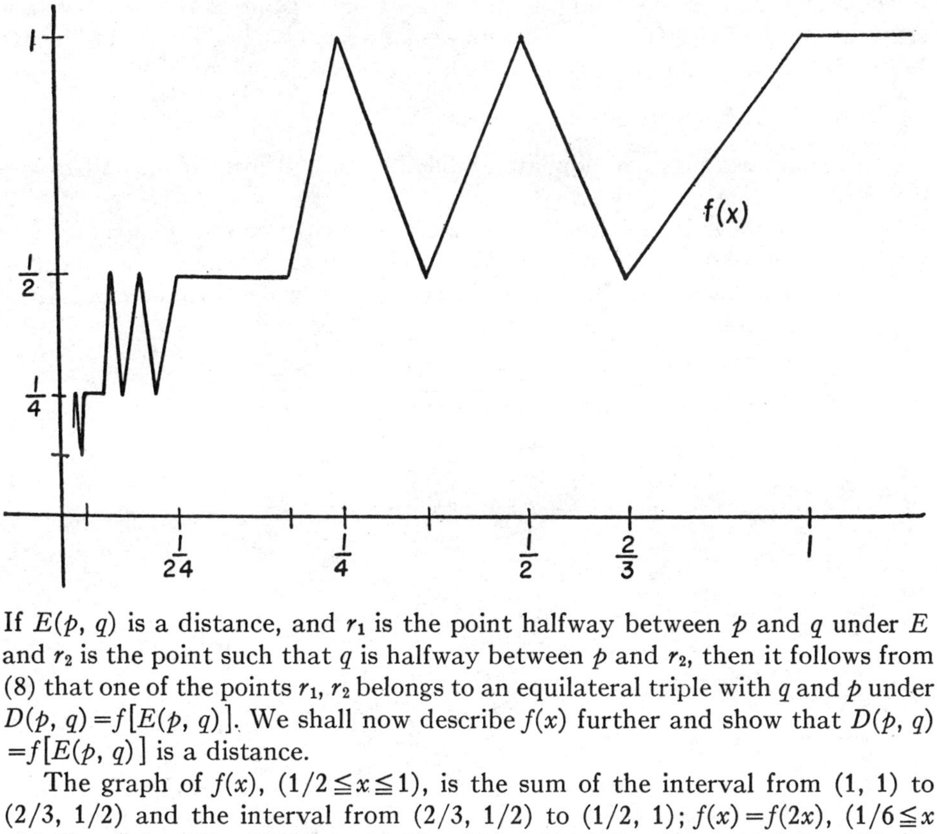

If $E(p, q)$ is a distance, and r_1 is the point halfway between p and q under E and r_2 is the point such that q is halfway between p and r_2, then it follows from (8) that one of the points r_1, r_2 belongs to an equilateral triple with q and p under $D(p, q)=f[E(p, q)]$. We shall now describe $f(x)$ further and show that $D(p, q)=f[E(p, q)]$ is a distance.

The graph of $f(x)$, $(1/2\leqq x\leqq 1)$, is the sum of the interval from (1, 1) to (2/3, 1/2) and the interval from (2/3, 1/2) to (1/2, 1); $f(x)=f(2x)$, $(1/6\leqq x\leqq 1/2)$; $f(x)=1/2$, $(1/24\leqq x\leqq 1/6)$; $f(x)=f(24x)/2$, $(0<x\leqq 1/24)$; $f(0)=0$; $f(x)=1$, $(x\geqq 1)$. It may be noted that

$$f(a)/a\geqq 12^i\cdot 9 \text{ if } 0<a\leqq 1/24^{i+1}, \tag{10}$$

and

$$|f(c)-f(b)|/(c-b) \leqq 12^i \cdot 6 \text{ if } 1/24^{i+1} \leqq b < c. \tag{11}$$

It is necessary that $f(x)$ satisfy condition (9) or else $D(p, q)=f[E(p, q)]$ does not satisfy condition (3) and is not a distance. Suppose $0<a\leqq b\leqq c$ and $1/24^{i+1} \leqq b \leqq 1/24^i$. Then $f(a)\leqq f(b)+f(c)$ because $f(a)$ is not more than twice as large as either $f(b)$ or $f(c)$. If $a\geqq 1/24^{i+1}$, then (9) holds because $c\leqq 2/24^i$ and each of the numbers $f(a), f(b), f(c)$ lies between $1/2^{i+1}$ and $1/2^i$, inclusive. If $a<1/24^{i+1}$, then it follows from (10) and (11) that $f(a)>|f(b)-f(c)|\cdot[a/(c-b)]$. Since $a/(c-b)\geqq 1$, $f(a)>f(b)-f(c)$ and $f(a)>f(c)-f(a)$. Hence, condition (9) holds and $D(p, q)$ satisfies condition (3).

Since $f(x)$ is continuous, $f(0)=0$, and $f(x)>0$ if $x>0$, $D(p, q)$ satisfies (1) and (4). It is symmetric because $E(p, q)$ is symmetric. Hence, $D(p, q)$ is a metric topological equivalent to $E(p, q)$.

To find the distance between p and q with this ruler, place it so that p falls at the origin and q is on the positive x-axis. Then $D(p, q)$ is the height of the graph above q. If this ruler is used to measure a straight line interval of length 7/4 or more, an equilateral distance is obtained. Hence,

THEOREM 2. *Each arc has an equilateral distance.*

This ruler also gives an equilateral distance to a Hilbert space and to any Euclidean space.

If n is a positive integer and bc is an interval, a ruler can be constructed in a similar fashion that will cause each pair of points of bc to belong to a collection of n points of bc such that the distance between any two of them is the distance between any other two.

A CONVEX METRIC WITH UNIQUE SEGMENTS

R. H. BING

1. Introduction. If $D(x,y)$ is a convex metric for a continuous curve M (compact, locally connected, metric continuum), then for each pair of points p, q of M there is an arc pq in M from p to q such that pq is isometric to a straight line interval. We shall call it a segment.

Each continuous curve has a convex metric [1; 3; 5; 6; 2]. However, if $D(x, y)$ is a convex metric for M, there may be two segments from p to q. If $M=S_2$ is the surface of a sphere, $D(x, y)$ is the size of the central angle subtended by x and y, and p, q are diametrically opposite points, then there are many segments from p to q. In fact, we show in §4 that if $D(x, y)$ is any convex metric whatsoever for S_2, each point of S_2 belongs to a pair of points which are not joined by a unique segment.

There is a dense subset W of S_2 such that no two points of W are diametrically opposite. If $D(x, y)$ is the previously mentioned convex metric for S_2, then each pair of points of W are joined by one and only one segment. We shall show that for any continuous curve there is such a dense subset and such a convex metric.

THEOREM. *Each continuous curve M has a dense subset W and a convex metric $D(x, y)$ such that each pair of points of W belongs to a unique segment.*

We shall prove this result in much the same manner that it was shown that any continuous curve can be convexified. The metric is

Presented to the Society September 7, 1951; received by the editors March 21, 1952.

an adaption and modification of that described in Theorem 4 of [5]. We consider a decreasing sequence of core partitionings of M, assign sizes to the pieces of the partitionings, and define the distance between two points in terms of the sizes of the chains joining them. The convex metric $D(x, y)$ is described in §2 and the dense set W is given in §3.

We recall the following definitions. A *partitioning* of M is a finite collection of mutually exclusive connected open subsets of M whose sum is dense in M. A sequence of partitionings is decreasing if for each integer i, G_{i+1} is a refinement of G_i (each element of G_{i+1} lies in an element of G_i) and the mesh of G_i (maximum of diameters of elements of G_i) approaches 0 as i increases without limit.

A partitioning is *regular* if each of its pieces is the interior of its closure. If the partitioning H is a refinement of the regular partitioning K, the elements of H which have a boundary point in common with a boundary of an element of K are called *border elements*. Other elements are interior elements.

We call H a *core refinement* of K if each is regular, H is a refinement of K, each border element of H is adjacent to an interior element, and the sum of the interior elements of H in each element of K has a connected closure.

2. **A convex metric for** M. In this section we describe a convex metric $D(x, y)$ for M. We find in §3 that there is a dense subset W of M such that under $D(x, y)$, each pair of points of W belong to a unique segment.

Description of convex metric $D(x, y)$. It was shown in [3] that M has a decreasing sequence $G_1, G_2, \cdots$ of regular partitionings such that each is a core refinement of the preceding and each element of G_{i+1} is of diameter less than one-third the distance between any two nonadjacent elements of G_i.

Suppose the elements of the partitioning G_i are ordered $g_{i1}, g_{i2}, \cdots, g_{in_i}$. We assign sizes to the elements of G_i as follows. The element g_{1j} of G_1 is assigned a size $1/4^j$. Each interior element g_{2j} of G_2 is assigned a size $1/4^{j+n_1}$ where n_1 is the number of elements in G_1. If g_{2j} is a border element of G_2, its size is one-half the size of the element of G_1 containing it plus $1/4^{j+n_1}$. In general, the size of g_{ij} is the first or second of the following expressions according as g_{ij} is an interior of a border element of G_i.

$$\text{Size } (g_{ij}) = \begin{cases} 1/4^{j+n_1+n_2+\cdots+n_{j-1}} \text{ or} \\ 1/4^{j+n_1+n_2+\cdots+n_{j-1}} + 1/2 \text{ size of element of} \\ G_{i-1} \text{ containing } g_{ij}. \end{cases}$$

Had we not been interested in building a convex metric with unique segments, we would have made the following simplification in the definition of the sizes of the elements of G_i: each interior element of G_i would be given a size $1/(2^i \cdot n_i)$ and each border element of G_i would be given a size of one-half the size of the element of G_{i-1} containing it.

If g is an element of G_i, we denote its size by $S_i(g)$. If p is a point of g, we shall also denote $S_i(g)$ by $S_i(p)$. If G is a collection of elements of G_i, the sum of the sizes of the elements of G is denoted by $S_i(G)$.

Suppose K is a continuum which is the closure of the sum of a subcollection of G_i. The sum of the sizes of the elements of G_i in K is called the ith size of K and is denoted by $S_i(K)$.

If p and q are two points of M, we define $E_i(p, q)$ to be the minimum of all numbers $S_i(K)$ where K is a continuum with ith size which contains $p+q$. We may regard $E_i(p, q)$ as the ith approximation to the distance between p and q. Then $D(p, q) = \lim E_i(p, q)$.

Existence of $\lim E_i(p, q)$. Suppose K is a continuum containing $p+q$ and having an ith size of $E_i(p, q)$. It is the closure of the sum of a chain of elements of G_i such that p and q lie in the closures of the end links of this chain. There is a continuum K' in K containing $p+q$ such that K' has an $(i+1)$st size but does not contain three border elements of G_{i+1} in the same element of G_i. Then $S_{i+1}(K') \leqq S_i(K) + 1/4^{1+n_1+\cdots+n_i} + \cdots + 1/4^{n_1+n_2+\cdots+n_{i+1}}$. Therefore

$$(1)\quad E_{i+1}(p, q) + 1/(3 \cdot 4^{n_1+n_2+\cdots+n_{i+1}}) \leqq E_i(p, q) + 1/(3 \cdot 4^{n_1+n_2+\cdots+n_i}).$$

Now inequality (1) implies that $\{E_i(p, q) + 1/(3 \cdot 4^{n_1+n_2+\cdots+n_i})\}$ is a monotone nonincreasing sequence of positive numbers. Hence $\lim E_i(p, q)$ exists.

In fact, $E_i(p, q)$ converges to $D(x, y)$ uniformly since it can be shown that $|D(p, q) - E_i(p, q)| < 2/2^i$.

That $D(x, y)$ is a convex metric for M follows by an argument similar to that used in [3] to show that a continuous curve has a convex metric.

3. **Unique segments in** M. In this section we show that if M is given the convex metric $D(x, y)$ described in the last section, there is a dense subset W of M such that each pair of points of W belongs to a unique segment.

Description of set W. Let $G_1, G_2, \cdots$ be the decreasing sequence of core partitionings of M described in the last section. Let C_i be the sum of the core elements of G_i and W_i be the intersection of C_i, C_{i+1},

$C_{i+2}, \cdots$. Then W_i is a closed set that intersects each element of G_{i-1}. We shall show that each pair of points of $W=W_2+W_3+\cdots$ belongs to a unique segment in M.

A unique shortest ith chain from p to q. If g and g' are two elements of G_i, the denominators of $S_i(g)$ and $S_i(g')$ will lie between $4^{1+n_1+\cdots+n_{i-1}}$ and $4^{n_1+n_2+\cdots+n_i}$ inclusive but they will not be equal. Therefore, if F_1 and F_2 are two subcollections of G_i,

$$(2) \qquad |S_i(F_1) - S_i(F_2)| \geqq 1/4^{n_1+n_2+\cdots+n_i}.$$

Hence, for each pair of points p, q there is a unique continuum K containing $p+q$ such that $S_i(K)=E_i(p, q)$.

Finding a continuum H of ith size near a continuum K of $(i+1)$st size. If K is a continuum with an $(i+1)$st size and g_1, g_2 are two elements of G_i intersecting K, there is a continuum H with an ith size such that H contains g_1+g_2, H is adjacent to each element of G_i that is adjacent to K, and

$$(3) \qquad S_i(H) < S_{i+1}(K) + S_i(g_1)/2 + S_i(g_2)/2.$$

We obtain the continuum H in the following fashion. Let G' be the collection of border elements of G_{i+1} that lie in K and G'' be the collection of all elements g of G' such that g lies in neither g_1, g_2, nor any element of G_i containing two elements of G'. Then G'' is the sum of two mutually exclusive subcollections G_1'', G_2'' such that each element of G'' is adjacent to an element of $(G'-G'')+G_1''$ and also to an element of $(G'-G'')+G_2''$. Let G_k'' be the one of G_1'', G_2'' such that the sum of the $(i+1)$st sizes of its elements is the smaller. Then H is the closure of the sum of all elements g of G_i such that g contains an element of $(G'-G'')+G_k''$.

Since each element of G' is adjacent to H and the mesh of G_{i+1} is less than the distance between nonadjacent elements of G_i, each element of G_i that is adjacent to K is also adjacent to H.

Let F' be the collection of elements of G_i which contain an element of $G'-G''$ and F'' be the collection of those that contain an element of G_k''. Since $S_i(F'')<S_{i+1}(G'')$ and $S_i(F')<S_{i+1}(G'-G'')+S_i(g_1)+S_i(g_2)$, inequality (3) follows.

Unless the collection of elements of G_i that irreducibly cover the sum of the elements of G_1'' is the same as those that irreducibly cover the sum of the elements of G_2'', it follows from inequality (2) that $S_i(F'')+1/(2\cdot 4^{n_1+n_2+\cdots+n_i})<S_{i+1}(G'')$. Hence, unless H contains K, we obtain the following strong form of inequality (3):

$$(4) \quad S_i(H) + 1/(2\cdot 4^{n_1+n_2+\cdots+n_i}) < S_{i+1}(K) + S_i(g_1)/2 + S_i(g_2)/2.$$

Shortest $(i+1)$*st chain lies in shortest ith chain.* Suppose p and q are two points of W_i and H, K are the two continua such that each contains $p+q$, $S_i(H)=E_i(p, q)$, and $S_{i+1}(K)=E_{i+1}(p, q)$. We shall show that K lies in H.

There is a continuum K' in H containing $p+q$ such that $S_{i+1}(K') \leqq S_i(H)-S_i(p)/2-S_i(q)/2+1/4^{1+n_1+\cdots+n_i}+ \cdots +1/4^{n_1+n_2+\cdots+n_i+1}$. Therefore

$$(5)\quad \begin{aligned} S_{i+1}(K) \leqq S_i(H) &- S_i(p)/2 - S_i(q)/2 \\ &+ 1/(3\cdot 4^{n_1+n_2+\cdots+n_i}) - 1/(3\cdot 4^{n_1+n_2+\cdots+n_{i+1}}). \end{aligned}$$

Now it follows from inequality (3) that there is a continuum H' containing $p+q$ such that $S_i(H')<S_{i+1}(K)+S_i(p)/2+S_i(q)/2$. This and inequality (5) implies that $S_i(H')-S_i(H)<1/4^{n_1+n_2+\cdots+n_i}$ and inequality (2) shows that $H'=H$.

Unless K lies in $H'=H$, it follows from inequality (4) that $S_i(H) +1/(2\cdot 4^{n_1+n_2+\cdots+n_i})<S_{i+1}(K)+S_i(p)/2+S_i(q)/2$. This is impossible because of inequality (5). Therefore K is a subset of H.

The unique segment. Suppose p and q are two points of W and K_i is the continuum containing $p+q$ such that $S_i(K_i)=E_i(p, q)$. If m is an integer such that p and q belong to core elements of G_n if $n\geqq m$, we find that $K_m\supset K_{m+1}\supset K_{m+2}\supset \cdots$. In fact, since the chain of closures of elements of G_{m+i+1} whose sum is K_{m+i+1} runs straight through the chain of closures of elements of G_{m+i} whose sum is K_{m+i} without doubling back and zigzagging, $K_m\cdot K_{m+1}\cdot \cdots$ is an arc pq from p to q. We show that pq is a unique segment from p to q by showing that any point not on pq is not between p and q.

Suppose r is a point between p and q but not on pq. Let H_i be the sum of the closures of all elements of G_i adjacent to K_i. Then there is an integer t greater than m such that r does not belong to H_t. Furthermore, for each positive number ϵ there is a positive integer s and a continuum R_s containing $p+r+q$ such that $S_{t+s}(R_s)<D(p, q) +\epsilon$. We show that there is no such point r by showing that there is no integer s for $\epsilon=2/(3\cdot 4^{n_1+n_2+\cdots+n_t})$.

The continuum R_s is adjacent to $M-H_t$. By inequality (3) there is a continuum R_{s-1} containing $p+q$ such that R_{s-1} is adjacent to $M-H_t$ and

$$\begin{aligned} S_{t+s-1}(R_{s-1}) &< S_{t+s}(R_s) + S_{t+s-1}(p)/2 + S_{t+s-1}(q)/2 \\ &< D(p, q) + \epsilon + S_{t+s-1}(p)/2 + S_{t+s-1}(q)/2. \end{aligned}$$

Furthermore, there is a continuum R_{s-2} containing $p+q$ such that R_{s-2} is adjacent to $M-H_t$ and

$$S_{t+s-2}(R_{s-2}) < S_{t+s-1}(R_{s-1}) + S_{t+s-2}(p)/2 + S_{t+s-2}(q)/2$$
$$< D(p, q) + \epsilon + S_{t+s-1}(p)/2 + S_{t+s-1}(q)/2$$
$$+ S_{t+s-2}(p)/2 + S_{t+s-2}(q)/2.$$

Continuing in this manner we find that there is a continuum R_0 containing $p+q$ such that R_0 is adjacent to $M-H_t$ and

$$S_t(R_0) < D(p, q) + \epsilon$$
$$+ S_t(p)/2 + S_{t+1}(p)/2 + \cdots + S_{t+s-1}(p)/2 \tag{6}$$
$$+ S_t(q)/2 + S_{t+1}(q)/2 + \cdots + S_{t+s-1}(q)/2.$$

Since K_t is not adjacent to $M-H_t$, $R_0 \neq K_t$ and it follows from inequality (2) that

$$S_t(K_t) + 1/4^{n_1+n_2+\cdots+n_t} \leqq S_t(R_0). \tag{7}$$

It follows from inequality (5) that

$$D(p, q) \leqq E_t(p, q) - S_t(p)/2 - S_{t+1}(p)/2 - \cdots$$
$$- S_t(q)/2 - S_{t+1}(q)/2 - \cdots + 1/(3\cdot 4^{n_1+n_2+\cdots+n_t}). \tag{8}$$

But inequalities (7), (6), and (8) imply that

$$1/4^{n_1+n_2+\cdots+n_t} < \epsilon - S_{t+s}(p)/2 - S_{t+s+1}(p)/2 - \cdots$$
$$- S_{t+s}(q)/2 - S_{t+s+1}(q)/2 - \cdots$$
$$+ 1/(3\cdot 4^{n_1+n_2+\cdots+n_t}).$$

Hence, if we take $\epsilon = 2/(3\cdot 4^{n_1+n_2+\cdots+n_t})$, we find that the supposition that there is a point r between p and q which is not on pq leads to a contradiction.

4. Examples and questions. Even though a continuous curve has a convex metric, this metric may be quite different from a Euclidean metric. For example, if S is a horizontally based square plus its interior in the plane and $D(p, q)$ is defined to be the sum of the absolute values of the difference of the ordinates and the difference of the abscissas of p and q, each point of S lies on some segment between opposite vertices of S. It is of interest to get convex metrics that resemble Euclidean metrics.

Continuous curves each of whose segments is unique. If I^n is a cube plus its interior in Euclidean n-space, its Euclidean metric causes each pair of its points to belong to a unique segment. Also, if M is a continuous curve each of whose nondegenerate cyclic elements is topologically a cube of some dimension, M has such a convex metric. Hence, each unicoherent plane continuous curve has such a metric.

If C is a cone whose base is a figure eight, C may be regarded as the sum of two closed discs that are sewn together along a radius of each. This shows that I^2 is not the only continuous curve which is 2-dimensional at each of its points and has a convex metric under which all segments are unique.

Suppose M is a continuous curve with a convex metric under which each pair of points belongs to a unique segment. Then for each point p of M there is a continuous transformation $F(m, t)$ (m element of M, t element of $[0, 1]$) of the cartesian product of M and a straight line interval onto M such that $F(m, 1)=m$, $F(m, 0)=p$, $F(F(m, t_1), t_2) = F(m, t_1, t_2)$. By letting $t_1=0$ we find that the last equation implies that $F(p, t)=p$. We can define $F(m, t)$ as the point which divides the segment from p to m in the same ratio that t divides the interval from 0 to 1.

The interior of each sphere of M is contractible to its center.

Questions. Does the existence of such a transformation $F(m, t)$ guarantee that M can be assigned a convex metric under which segments are unique?

Not only would it be interesting to know a topological characterization of continuous curves which have convex metrics with only unique segments but it would also be interesting to know which continuous curves have the following type of convex metric. If two segments in Euclidean space intersect in more than one point, their sum is a segment. If an n-dimensional continuous curve has such a convex metric, is it necessarily topologically equivalent to an n-cube I^n?

The surface of a sphere has a convex metric under which nearby pairs of points belong to a unique segment. One might wonder if the local structure of a continuous curve determines whether or not it has such a convex metric—that is, does a continuous curve M have such a metric if for each point p of M there is an open subset N_p of M containing p such that N_p is topologically equivalent to a set with such a metric?

Continuous curves with nonunique segments. The surface of a sphere S^n in Euclidean $(n+1)$-space has the property that if p is one of its points there is a point q of it such that there is not a unique segment from p to q. We know this because S^n is not contractible—that is, it cannot be shrunk to a point [4]. Similarly, a continuous curve which is not unicoherent is not contractible [7] and therefore does not have a unique segment between each pair of its points.

Not each contractible continuous curve has a convex metric with only unique segments. Suppose that in the plane J_i $(i=1, 2, \cdots)$ is a circle with radius $1/i$ and center at $(1/i, 0)$. If C is a cone with

$J_1+J_2+\cdots$ as a base, it is contractible but it does not have a convex metric with only unique segments because there is no contractible neighborhood not containing the vertex about the point (0, 0) of the base.

REFERENCES

1. R. H. Bing, *Partitioning a set*, Bull. Amer. Math. Soc. vol. 55 (1949) pp. 1101–1110.

2. ———, *Complementary domains of continuous curves*, Fund. Math. vol. 36 (1949) pp. 306–318.

3. ———, *Partitioning continuous curves*, Bull. Amer. Math. Soc. vol. 58 (1952) pp. 536–556.

4. W. Hurewicz and H. Wallman, *Dimension theory*, Princeton University Press, 1948.

5. E. E. Moise, *Grille decomposition and convexification theorems for compact metric locally connected continua*, Bull. Amer. Math. Soc. vol. 55 (1949) pp. 1111–1121.

6. ———, *A note of correction*, Proceedings of American Mathematical Society vol. 2 (1951) p. 838.

7. G. T. Whyburn, *Analytic topology*, Amer. Math. Soc. Colloquium Publications, vol. 28, New York, 1942.

THE UNIVERSITY OF WISCONSIN

A CONNECTED COUNTABLE HAUSDORFF SPACE

R. H. BING

In Math. Ann. vol. 94 (1925) pp. 262–295, Urysohn gave an example of a connected Hausdorff space with only countably many points. Here is another.

EXAMPLE 1. The points of the space are the rational points in the plane on or above the x-axis. If (a, b) is such a point and $\epsilon > 0$, $(a, b)+\{(r, 0) \mid \text{either } |r-(a+b/3^{1/2})| < \epsilon \text{ or } |r-(a-b/3^{1/2})| < \epsilon\}$ is a neighborhood.

To construct geometrically a neighborhood with center at (a, b), consider an equilateral triangle with base on the x-axis and apex at (a, b). (If $b=0$, regard (a, b) as the triangle.) Then (a, b) plus all rational points on the x-axis whose distances from a base vertex of the triangle are less than ϵ is an ϵ-neighborhood with center at (a, b).

This space satisfies the Hausdorff axioms and has the property that for each pair of neighborhoods, there is a point common to their closures. Hence, the space is connected.

Although this space has a countable basis, it is not regular and hence not metric. Its dimension depends on the definition of dimension used. In the Menger-Urysohn sense (the dimension is defined inductively in terms of boundaries of open sets) the space is one-dimensional, and in the Lebesgue sense (the dimension is defined in terms of orders of coverings) it is infinite-dimensional.

EXAMPLE 2. We may enlarge our description to get a connected countable Hausdorff space of any positive dimension (even infinite) in the Menger-Urysohn sense. We give one of dimension two.

The points of the space are the rational points of Euclidean 3-space whose second and third coordinates are non-negative. If $(a, b, 0)$ is a point, an ϵ-neighborhood with center at $(a, b, 0)$ is $(a, b, 0)$ $+\{(r, 0, 0) \mid \text{either } |r-(a+b/3^{1/2})| < \epsilon \text{ or } |r-(a-b/3^{1/2})| < \epsilon\}$. If (a, b, c), $c \neq 0$, is a point, an ϵ-neighborhood with center at (a, b, c) is the sum of (a, b, c) and all ϵ-neighborhoods with centers at points $(a, r, 0)$ where either $|r-(b+c/2^{1/2})| < \epsilon$ or $|r-(b-c/2^{1/2})| < \epsilon$.

THE UNIVERSITY OF WISCONSIN

Received by the editors August 15, 1952.

A TRANSLATION OF THE NORMAL MOORE SPACE CONJECTURE[1]

R. H. BING

1. **Introduction.** The purpose of this note is to translate an unsolved problem in topology into a nontopological setting so that it can be considered by a wider audience. Those interested in the foundations of set theory might find the new formulation more to their liking. A quickening of interest in logic generated by Paul Cohen's result that the Continuum Hypothesis is independent of the axioms of set theory (Zermelo-Fraenkel or Godel-Bernays) together with the Axiom of Choice suggests that this is an opportune time to consider this reformulation.

Let X denote a set of points, R the cartesian product $X\times X$, and L the diagonal of R consisting of all points $(x, x)\in X\times X$. It may be convenient to think of R as a unit square in the Euclidean plane with diagonal L from (0, 0) to (1, 1) as shown in Figure 1. However, we do not insist that the cardinality of X be that of the continuum.

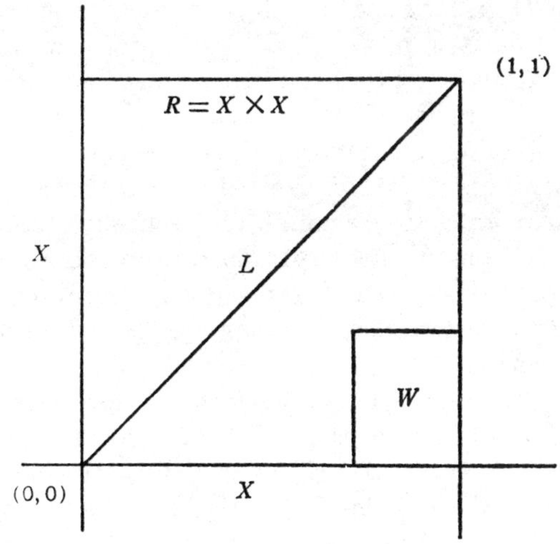

The horizontal projection $h(x, y)=(y, y)$ and the vertical projection $v(x, y)=(x, x)$ sends R onto L. We shall be concerned with sets W as shown in Figure 1 such that $h(W)\cap v(W)=0$. Although W may be as shown, it may be more complicated (such as the set of points

Received by the editors June 5, 1964.

[1] Work on this paper supported by the National Science Foundation under grant GP-2244.

of R with rational abscissas and irrational ordinates, or worse).

Let $f: R-L \rightarrow \{0, 1, 2, \cdots\}$ be a transformation (not necessarily continuous) of $R-L$ into the nonnegative integers. Do the following possible properties of f imply each other?

(a) There is a transformation $F: X \rightarrow \{0, 1, 2, \cdots\}$ such that $\max[F(x), F(y)] > f(x, y)$ for each $(x, y) \in R-L$.

(b) For each subset W of R with $h(W) \cap v(W) = 0$, there is a transformation $F_W: X \rightarrow \{0, 1, 2, \cdots\}$ such that $\max[F_W(x), F_W(y)] > f(x, y)$ for each $(x, y) \in W$.

It is clear that Condition (a) implies Condition (b) but whether or not Condition (b) implies Condition (a) is not obvious. Whether or not there is such an implication is related to the following conjecture.

Normal Moore space conjecture. Each normal Moore space is metrizable.

A Moore space is one which satisfies the first three parts of Axiom 1 of [6]. Hence a Moore space is a regular Hausdorff space S with a sequence of open coverings $H_1, H_2, \cdots$ such that for each point p, $\{\text{Star}(p, H_1), \text{Star}(p, H_2), \cdots\}$ is a countable basis for p. The sequence of open coverings $H_1, H_2, \cdots$ is called a development. Sometimes additional conditions are imposed such as that each H_{i+1} is a refinement of H_i (or even a subcollection of H_i). Although these extra conditions are not essential, we suppose for convenience that H_{i+1} refines H_i.

An alternate way of characterizing a Moore space is to say that it is a regular Hausdorff space such that for each point p of it, there is a countable basis $U(p, 1), U(p, 2), \cdots$ for p such that for each open set V containing a point q, there is an integer $n(q, V)$ such that $q \in U(r, n(q, V))$ implies $U(r, n(q, V)) \subset V$.

Efforts to show that the normal Moore space conjecture is true have been unsuccessful. See for example, [**1, 2, 3, 4, 5, 7, 8**]. Similarly, efforts to construct counterexamples have been frustrating. As may be noted by the tone of this paper, the author does not believe that the conjecture is true. We suggest where a counterexample may lie and show that if there are counterexamples of certain sorts, there are perhaps simpler ones.

A collection of closed sets is *discrete* if the closed sets are mutually exclusive and any subset of the collection has a closed sum. A space is *collectionwise normal* with respect to a discrete collection $\{K_\alpha\}$ of closed sets if for each element $K_\alpha \in \{K_\alpha\}$ there is an open set O_α containing K_α such that all the sets O_α are mutually exclusive.

Counterexample of Type D. It is known [**1**] that if there is a counterexample S to the normal Moore space conjecture, then S contains a discrete collection of closed sets such that S is not collectionwise normal with respect to the collection. We say that S is a *counterexample of Type* D if it has the additional property that it contains a discrete collection of points with respect to which it is not collectionwise normal.

Question. Does the existence of a counterexample to the normal Moore space conjecture imply the existence of one of Type D? By using a result of Worrell [7], Traylor [2] has shown that there is a counterexample of Type D if there is a locally compact counterexample. Theorem 3 shows that each separable counterexample is of Type D.

2. The nonimplication gives a counterexample.

THEOREM 1. *If there is an* X, R, L, $f(x, y)$ *satisfying Condition* (b) *but not Condition* (a), *then there is a counterexample of Type* D.

PROOF. The points of the counterexample are the points of R. Each point of $R-L$ is isolated. For each point $(x, x)\in L$ and each integer n,

$$N((x, x), n) = \{(x, x)\} \cup \{(y, z) \in R \mid f(y, z) > n \text{ and } x \text{ is either } y \text{ or } z\}.$$

Note that $N((x, x), n)$ lies on the sum of the horizontal and vertical segments in R through (x, x). Elements of the open covering H_i of the development $H_1, H_2, \cdots$ are either the one point neighborhoods containing only one point of $R-L$ or neighborhoods of the sort $N((x, x), i)$.

To see that the space is normal, consider two mutually exclusive closed sets A, B in R. Let W_1 be the set of all points (x, y) of R such that $(x, x)\in A$ and $(y, y)\in B$. Also, let $W_2=\{(x, y)\mid (x, x)\in B, (y, y)\in A\}$. Note that $h(W_1)\cap v(W_1)=0=h(W_2)\cap v(W_2)$. Let F_{W_1}, F_{W_2} be the functions promised by Condition (b). Let D_A be the open set consisting of the sum of all $N((a, a), n)$'s with $(a, a)\in A$ and $n=F_{W_1}(a)+F_{W_2}(a)$. Also, let D_B be the sum of all $N((b, b), m)$'s such that $(b, b)\in B$ and $m=F_{W_1}(b)+F_{W_2}(b)$. The only possible point of intersection of $N((a, a), n)$ and $N((b, b), m)$ is either the point (a, b) or the point (b, a). However, $(a, b)\in W_1$ and since $\max[m, n] \geqq \max[F_{W_1}(a), F_{W_1}(b)]>f(a, b)$, (a, b) does not belong to both $N((a, a), n)$ and $N((b, b), m)$. Similarly, we find that $(b, a)\in W_2$ does not belong to both. Then $O_A=(D_A-B)\cup A$ and $O_B=(D_B-A)\cup B$ are mutually exclusive open subsets of R containing A and B respectively.

Finally we show that the topology we have given R makes it a counterexample of Type D unless $f(x, y)$ satisfies Condition (a). Assume R is collectionwise normal with respect to the collection of points in L and let $\{U_x\}$ be a collection of mutually exclusive open sets covering L such that U_x is the only element of $\{U_x\}$ containing $(x, x) \in L$. Let $F(x)$ be a positive integer such that $N((x, x), F(x)) \subset U_x$. Then $F(x)$ shows that Condition (a) is satisfied. If $(x, y) \in R-L$, then $(x, y) \notin N((x, x), F(x)) \cap N((y, y), F(y))$ and (x, y) fails to belong to one of them, say $N((x, x), F(x))$. But then $F(x) \geqq f(x, y)$ and $\max[F(x), F(y)] \geqq f(x, y)$.

The following theorem shows that if Condition (b) fails to imply Condition (a), then X must be uncountable.

THEOREM 2. *Each normal Moore space is collectionwise normal with respect to countable discrete collections of closed sets.*

PROOF. Let $A_1, A_2, \cdots$ be a countable discrete collection of closed sets in a normal Moore space S. For each integer i, let D_i and D_i' be mutually exclusive open sets containing A_i and $\bigcup_{j \neq i} A_j$ respectively. Then $D_1, D_1' \cap D_2, D_1' \cap D_2' \cap D_3, \cdots$ are mutually exclusive collections of open sets showing that S is collectionwise normal with respect to $A_1, A_2, \cdots$.

Question. Let $\aleph'$ be the least cardinal such that some normal Moore space contains a discrete collection X of closed sets such that S is not collectionwise normal with respect to X and cardinality of X is $\aleph'$. What is the size of $\aleph'$? The proof of Theorem 2 shows that if $\aleph'' < \aleph'$, then $\aleph'$ is not the limit of $\aleph''$ cardinals all less than $\aleph'$.

THEOREM 3. *Each separable counterexample to the normal Moore space conjecture is of Type* D.

PROOF. It follows from [1] that each counterexample contains a discrete collection of closed sets with respect to which the counterexample is not collectionwise normal. Selecting one point from each element of this collection, one finds from Theorem 2 that the counterexample contains an uncountable discrete collection of points. The space is not collectionwise normal with respect to this collection since no separable space is collectionwise normal with respect to any uncountable collection.

3. A converse of Theorem 1.

THEOREM 4. *If S is a counterexample of Type* D *and X is a discrete collection of points with respect to which S is not collectionwise normal, then there is a function $f(x, y)$ showing that Condition* (b) *does not imply Condition* (a).

PROOF. Assume S is a counterexample of Type D, X is a discrete collection of points with respect to which S is not collectionwise normal, and $H_1, H_2, \cdots$ is a development of S. For each $(x, y)\in R-L$, let $f(x, y)=0$ if $\text{Star}(x, H_1)\cap\text{Star}(y, H_1)=0$ and otherwise let $f(x, y)=n$ where n is the largest integer for which $\text{Star}(x, H_n)\cap\text{Star}(y, H_n)\neq 0$. Note that $\text{Star}(x, H_{f(x,y)+1})\cap\text{Star}(y, H_{f(x,y)+1})=0$. We show that $f(x, y)$ satisfies Condition (b) but not Condition (a).

Let W be a subset of R such that $h(W)\cap v(W)=0$. Let $A=\{x\in X \mid (x, y)\in W\}$ and $B=X-A$. Since S is normal, there are mutually exclusive open sets D_A, D_B in S containing A and B respectively.

Let F_W be a transformation of X into the positive integers such that $\text{Star}(x, H_{F_W(x)})$ lies in one of D_A, D_B. For each point $(x, y)\in W$, $h(x, y)$, $v(x, y)$ belong to different ones of A, B and $N(x, H_{F_W(x)})$, $N(y, H_{F_W(y)})$ lie in different ones of D_A, D_B. Hence $N(x, H_{F_W(x)})\cap N)y, H_{F_W(y)})=0$ and for $m=\max[F_W(x), F_W(y)]$, $N(x, H_m)\cap N(y, H_m)=0$. Hence $f(x, y)<m$ and $\max[F_W(x), F_W(y)]>f(x, y)$. Hence Condition (b) is satisfied.

We show that Condition (a) is not satisfied by showing that if it were, then S would be collectionwise normal with respect to X. Assume there is an F showing that Condition (a) is satisfied. Let X_i be the set of all points $x\in X$ such that $F(x)=i$. It follows from Theorem 2 that there is a collection of mutually exclusive open sets $D_1, D_2, \cdots$ in S containing $X_1, X_2, \cdots$ respectively. For each point $x\in X_i$, let $U_x=D_i\cap\text{Star}(x, H_{F(x)})$. We show that the assumption that there is an F leads to the contradiction that $\{U_x\}$ is a collection of mutually exclusive open sets showing that S is collectionwise normal with respect to X.

Assume $U_{x_1}\cap U_{x_2}\neq 0$ for some pair of different points x_1, x_2 of X. Then x_1, x_2 belong to the same X_i since $D_i\cap D_j=0$ if $i\neq j$. Suppose $x_1, x_2\in X_{n_1}$. Then $F(x_1)=F(x_2)=n_0\geqq f(x_1, x_2)+1$. Since

$$U_{x_i}\subset\text{Star}(x_i, H_{F(x_i)})\subset\text{Star}(x_i, H_{f(x_1,x_2)+1}),$$

$$U_{x_1}\cap U_{x_2}\subset\text{Star}(x_1, H_{f(x_1,x_2)+1})\cap\text{Star}(x_2, H_{f(x_1,x_2)+1})=0.$$

The assumption that Condition (a) is satisfied led to the contradiction that $U_{x_1}\cap U_{x_2}\neq 0=U_{x_1}\cap U_{x_2}$.

We can state Theorems 1 and 4 together.

THEOREM 5. *A necessary and sufficient condition that there be a counterexample of Type* D *is that there be an X, R, L, $f(x, y)$ satisfying Condition* (b) *but not Condition* (a).

4. **Changing counterexamples.** Suppose one took a counterexample S' of Type D, used Theorem 4 to get an X, R, L and $f(x, y)$ satis-

fying Condition (b) but not Condition (a) and finally used the X, R, L, $f(x, y)$ to get a counterexample S'' of Type D as shown in Theorem 1. The resulting S'' might be simpler that S'. How is S'' related to S'?

One finds that S'' is obtained from S' by isolating points of $S'-X$, throwing certain of these isolated points away, combining others, and splitting others. Let us do it in slow motion.

In the following we suppose that S' is a counterexample of Type D and X is a discrete set of points with respect to which S' is not collectionwise normal. A *partition* $[X_\alpha, X_\alpha']$ of X is a separation of X into two mutually exclusive subsets.

EXAMPLE 1. Let X be an uncountable set, F be the set of all functions $f: X \rightarrow \{0, 1, 2, \cdots\}$, G is the set of bounded elements of F, and G' be a subset of G such that

(c) For each $f \in F$, there is a $g \in G'$ such that $g > f$ on two points of X.

(d) For each partition $[X_\alpha, X_\alpha']$ of X, there is an $f_\alpha \in F$ such that for each $g \in G'$, either $f_\alpha > g$ for each value in X_α or for each value in X_α'.

Our description of a G' satisfying the conditions of Example 1 is dependent on the assumption that there is a counterexample S' of Type D with a discrete set X with respect to which S' is not collectionwise normal. Before describing a G' we point out how the existence of such a G' (no matter how obtained) insures the existence of a counterexample S_1 of Type D.

Let S_1 be the topological space whose points are the elements of $X \cup G'$. Each element of G' is isolated. For each $x \in X$ and each positive integer n, $N(x, n) = \{x\} \cup \{g \in G' \mid g(x) \geqq n\}$ is a neighborhood in the nth element of the development of S_1.

No matter how the set G' is obtained, the resulting S_1 is a counterexample of Type D. Condition (c) implies that S_1 is not collectionwise normal with respect to X and Condition (d) implies that S_1 is normal.

If S' is a normal Moore space with a discrete collection X of points with respect to which S' is not collectionwise normal, the collection G' may be obtained as follows. For each $p \in S'-X$, associate the function $g_p \in G$ such that $g_p(x)$ = largest integer n such that $p \in \text{Star}(x, H_n)$. Then G' is the set of elements of G that are associated with some point of $S'-X$.

Note that in changing from S' to S_1, we isolated the points of $S'-X$ and combined the ones of these isolated points that were associated with the same element of G'.

We note that S_1 does not have more points than S'. We also note that if $C(Z)$ denotes the cardinality of set Z, then $C(S_1) \leqq 2^{C(X)}$.

EXAMPLE 2. This example is the same as Example 1 with the following additional restriction.

(e) Each $g \in G'$ is positive on precisely two values of X.

The G' of Example 2 is obtained from the G' of Example 1 by first throwing away all the elements of G' of Example 1 which are positive on no more than one element of X and then for each other $g' \in G'$ of Example 1, replace this g' with as many elements of G' of Example 2 as there are pairs of points of X on which g' is positive.

If S_2 represents the resulting counterexample of Type D, we obtained S_2 from S_1 by throwing away certain isolated points of S_1, splitting others, and combining certain of the split pieces. Note that after the discarding, splitting, and combining, we have not increased the cardinality since $C(S_2) = C(X) \leqq C(S_1)$.

EXAMPLE 3. This example is the same as Example 2 with the following additional restriction.

(f) If $g \in G'$, $g(x_1) > 0$, $g(x_2) > 0$, then $g(x_1) = g(x_2)$.

If $g' \in G'$ of Example 2 and x, y are different points of X on which g' is positive, then the element g'' of G' of Example 3 corresponding to g' satisfies $g''(x) = g''(y) = \min[g'(x), g'(y)]$ and $g'' = 0$ on $X - (\{x\} \cup \{y\})$.

At first glance it might appear that in reducing the elements of G' we lost Condition (c) but it is not lost. Consider an $f \in F$. To find an element of the new G' that dominates it on two points, we let $X_i = f^{-1}(i)$ and note from the proof of Theorem 4 that there is an integer j such that S_2 is not collectionwise normal with respect to X_j. For some two points x, y of X_j, $\text{Star}(x, H_{j+1}) \cap \text{Star}(y, H_{j+1}) \neq 0$ in S_2. Hence there is a $g \in G'$ of Example 2 such that $g(x) > j$ and $g(y) > j$. The associated element of G' of Example 3 dominates f on x and on y.

If S_3 is the space resulting from Example 3, S_3 is obtained from S_2 by removing certain points from certain neighborhoods and then combining certain of these removed points. When the topology of a space is altered by making the neighborhoods smaller, there is a possibility that a non-collectionwise normal space may become collectionwise normal. However, the proof of Theorem 4 shows that this did not happen.

The function $f(x, y)$ on $R - L$ described in the introduction may be obtained from the G' of Example 3 by letting $f(x, y)$ be the maximal value that any element of G' takes on both x and y. The resulting $f(x, y)$ will satisfy Condition (b) but not Condition (a) if G' satisfies Conditions (c), (d), (e), (f).

The following question is a backhanded version of one that we have asked earlier.

Question. Suppose X is an uncountable set, F is the set of all maps of X into $\{0, 1, 2, \cdots\}$, G is the set of elements of F that are positive on precisely two values of X. Suppose G' is a subcollection of G such that the following holds.

(g) For each $f \in F$, there is a $g \in G'$ such that $g > f$ on the two values of X on which g is positive.

Does Condition (g) imply that G' would necessarily have the following property?

(h) For some partition $[X_\alpha, X_\alpha']$ of X and each $f_\beta \in F$ there is a $g \in G'$ such that $g > f$ for one value of X_α and also for one value of X_α'.

References

1. R. H. Bing, *Metrization of topological spaces*, Canad. J. Math. **8** (1951), 653–663.

2. B. Fitzpatrick, *Metrizability in normal Moore spaces*, Pacific J. Math. (to appear).

3. B. Fitzpatrick and D. R. Traylor, *Two theorems on metrizability of Moore spaces*, Pacific J. Math. (to appear).

4. R. W. Heath, *Screenability, pointwise paracompactness and metrization of Moore spaces*, Canad. J. Math. **16** (1964), 763–770.

5. F. B. Jones, *Concerning normal and completely normal spaces*, Bull. Amer. Math. Soc. **43** (1937), 671–677.

6. R. L. Moore, *Foundations of point set theory*, Amer. Math. Soc. Colloq. Publ. Vol. 13, Amer. Math. Soc., Providence, R. I., 1962.

7. J. M. Worrell, *Concerning upper semi-continuous collections of mutually exclusive closed and compact point sets*, Abstract 590-44, Notices Amer. Math. Soc. **9** (1962), 204.

8. J. N. Younglove, *Concerning metric subspaces of nonmetric spaces*, Fund. Math. **48** (1959), 15–25.

The University of Wisconsin

METRIZATION PROBLEMS*

R. H. Bing

University of Texas, Austin

I. USING EXAMPLES

Burton Jones is the first mathematician with whom I collaborated after receiving my Ph.D. Although R. D. Anderson, E. E. Moise, C. E. Burgess, Mary Ellen Estill (Rudin), Eldon Dyer, Billy Jo Ball were at Texas at this time, they were students and R. L. Moore insisted that his students develop independent work habits. Hence, after receiving my Ph.D., I did not discuss research with these students. However, Jones had received the Ph.D. several years earlier and was just returning from Cambridge where he had done work on underwater sound related to war work. I did not feel restrained in discussing mathematical research with him.

An unsolved problem of considerable interest to me was the question Jones had asked in 1937 [J_1]---is a normal Moore space metrizable? To understand how we worked at unsolved problems, it is necessary to know our modus operandi. Our first approach in attacking a problem was to look for a counterexample. If no one of our vast store of examples worked, we would try modifying known examples to discover a counterexample. It was my gut-reaction (and still is) that there is a real counterexample to the normal Moore space conjecture but it may be more complicated than anything we have examined. I soon learned that Jones had examples in his repetoire that were missing from mine.

*Work on this paper was supported by NSF Grant MCS-790-4709.

ISBN 0-12-481820-X

II. JONES' TIN-CAN-SPACE T

One of these amazing examples is Jones' tin-can-space, which I call T. Jones found the example in the 1930's, showed it to me in the 1940's and wrote up a version of it in the 1960's [J_2]. Points of T are tin cans placed on horizontal shelves. There were ω_1 of these horizontal shelves placed above each other in a best well ordered fashion (each shelf had at most a countable number of other shelves below it). There were only a countable number of tin cans on each level so T had $\aleph_0 \times \aleph_1 = \aleph_1$ points (tin cans).

The bottom level (shelf) L_0 had only one tin can and the can was assigned a size of 0. The next level L_1 had a countable number of disjoint tin cans, and each of these was above the can on the bottom row. These cans were assigned sizes $\frac{1}{2}, \frac{1}{4}, \frac{1}{8}, \frac{1}{16}, \ldots$. There were $\aleph_0$ tin cans on the next level L_2 above each can on L_1 and they were assigned sizes $\frac{1}{2}, \frac{1}{4}, \frac{1}{8}, \ldots$, etc. Inductively cans are also placed on levels $L_2, L_3, \ldots$ and assigned sizes.

Figure 1 shows cans on levels L_0, L_1, L_2 with certain sizes indicated. It was too crowded to show all cans on higher levels, but we note that for levels $L_1, L_2, \ldots, L_\alpha, \ldots$ ($\alpha < \omega_0$), the following properties hold.

1. There are only a countable number of cans on L_i ($0 < i < \omega_0$).
2. If L_α and L_β are levels with L_α below L_β, n is a positive integer, and p is a can on L_α, then there is a stack of cans such that the stack is based on p, has its top in L_β, and the sum of the sizes of the cans in the stack above p (size of p is ignored) is $1/2^n$.

Only a countable number of cans are put on the shelf L_{ω_0} and this causes some difficulty. Stacks of cans lead up to a Cantor set of positions for cans at the L_{ω_0} level, but we ignore most of them and use only a countable number of them---being guided by Properties 1 and 2 of the

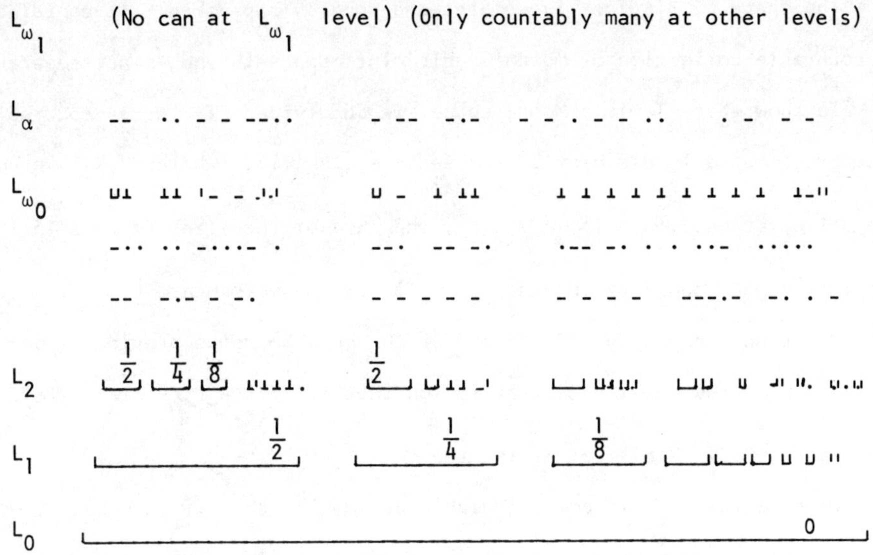

Fig. 1

previous paragraph. This is done as follows. There are only a countable number of cans in the union of all levels below L_{ω_0}. For each such can p on a lower level and each positive integer n, we pick a stack of cans based on p and reaching up to the L_{ω_0} level, so that the sum of the sizes of the cans in the stack above p is $1/2^n$. We then put a can p on the L_{ω_0} level at the top of the stack. We do this for each lower can and each n, but even so---put only a countable number of cans on the L_{ω_0} level. However, conditions 1 and 2 prevail even through level L_{ω_0}. Cans on level L_{ω_0} are assigned a size of 0.

The procedure is continued inductively to get cans on each level L_α with $\alpha < \omega_1$. Pages 116-117 of [J_2] gives details of a variation of the inductive procedure.

For a topology of T, let each can on a nonlimit level be isolated. If p is a can in some limit level, it is the top of some stack. The set of cans in such a stack is a neighborhood.

The space T is locally compact and locally separable. It contains an uncountable collection of mutually disjoint open sets and is not separable.

To show that T is a Moore space, we subdivided T into a countable number of mutually disjoint closed sets $A_{r_1}, A_{r_2}, \ldots$ where r_i is a non-negative rational, and if $p \in A_{r_i}$, the sume of the sizes of cans in the maximal stack topped by p is r_i. To get a development $G_1, G_2, \ldots$, we would let an element $g \in G$ be one of the neighborhoods previously described with the additional restriction that if $p \in A_{r_1} \cup A_{r_2} \cup \cdots \cup A_{r_i}$, and $p \in g \in G_i$, this p is the top can of g.

We show that T is not metrizable by showing that if it were, it would be the union of a monotone increasing sequence of countable sets. With this objective in mind, we suppose T has a bounded metric with diameter less than 1 and for each $p \in T$ let

$$\varepsilon(p) = \mathrm{LUB}\{x \mid x \text{ neighborhood of } p \text{ is countable}\} .$$

For subsets A, B, C of T let

$$f_1(A) = \bigcup_{a \in A} N(a, \varepsilon(a)/2) ,$$

$f_2(B) = \{q$, some element of B lies at same or a higher level than $q\}$, and

$f_3(C) =$ union of $f_2(C)$ and next level.

If X is the one point set which is bottom of T, consider the monotone increasing sequence of countable sets

$$X, f_3 f_2 f_1(X), (f_3 f_2 f_1)^2(X), \ldots .$$

Note that f_3 pushes us past isolated levels and $f_2 f_1$ makes short order of limit ones.

If T were normal, it would be an example showing that the normal Moore space conjecture is false. In his 1965 paper, Jones said he had not yet discovered whether T was normal or not. We discuss the normality of T

in the final section of this paper, but mention that some of what we know about T was learned as a result of the study of it as a special Aronszajn tree. See [F and R].

III. EARLY PROOFS

Not all questions are solved with counterexamples. One of the best known metrization results is the one stating that a second axiom regular Hausdorff space is metrizable. Here the proof consists of constructing a distance function. I recall my disappointment when I saw the traditional argument. While it gave a distance function, the distance function constructed had little resemblance to ordinary distance. If a straight line interval were metrized by such a technique, there would be scant hope that the resulting metrized arc could be isometrically embedded in any Euclidean space. It embeds isometrically in a Hilbert cube.

I resolved to try for a more realistic distance function if I had the opportunity. Such an opportunity came when I attacked the following problem.

> Suppose a regular Hausdorff space H has a development $G_1, G_2, \ldots$ such that if two points are covered by a coherent collection of r elements of G_{i+1}, the pair lies in one element of G_i. Describe a metric for H.

I wondered as to the most realistic way to describe a reasonable distance. It would simplify the situation if the elements of G_{i+1} were the open sets with diameters $1/r^i$. Then I would assign a size of 1 to elements of G_1 and a size of $1/r^i$ to elements of G_{i+1}. The size of a chain would be the sum of the sizes of its links and the distance from p to q would be defined as the greatest lower bound of sizes of all chains joining p to q where links of the chains are elements of $G_1 \cup G_2 \cup \cdots$ As shown by Theorem 1 of [B_1], this greatest lower bound is not 0 if

$p \neq q$, so the resulting limit provides a distance---even in the general case.

This method of taking least upper bounds on sizes of chains not only gave a reasonable distance for simple continua, but also was the basis for a method I used later to show that each compact, locally connected metric continuum has a convex metric [B_3].

The following theorem of Bing, Nagata, and Smirnov [B_2N,Sm] has attracted some attention. Bing's version states:

THEOREM. *A regular Hausdorff space* H *is metrizable if and only if there is a sequence* $G_1, G_2, \ldots$ *such that*

each G_i *is a discrete collection of open sets in* H *, and*

$G_1 \cup G_2 \cup \cdots$ *is a basis for* H .

The "only if" part of the proof was facilitated by a result of Stone [St] that a metric space is paracompact, and the "if" part was provided by a construction described below.

Proof. To get a first approximation to a distance, convolute H so that part of it not covered by G_1 lies in a horizontal plane and points in elements of G_1 lie on hills of heights less than or equal to $1/2$. If p and q are two points,

$d_1(p,q)$ = height p if q is on the bottom level,

$d_1(p,q)$ = |height p - height q| if p,q are on the same hill,

$d_1(p,q)$ = (height p + height q) if p and q are on different hills.

Next, turn to G_2 but get d_2 by deciding that the heights of the hills are less than $1/4$. The procedure is continued and the final distance is defined as $d(p,q) = \Sigma d_i(p,q)$. Note that instead of embedding H in a Hilbert cube as was done in the case where H was second countable, it was embedded in the cartesian product of cones.

It is not our purpose here to show that a collectionwise normal Moore space is metrizable $[B_2]$ but we remark that in getting a metric for it, we may use the "hills approach" $[B_3]$.

IV. RISE AND FALL

We once had great expectations that set theory could provide answers to the questions of general topology. Was the continuum hypothesis true? Did topological spaces that are known to have one set of properties necessarily have another set? Is there a Souslin space? Are all normal Moore spaces metrizable?

The axiomatic method is very popular with mathematicians. Most mathematicians have had a pleasant experience in using the axiomatic approach in synthetic geometry. Would it not be wonderful to have a definite set of axioms and procedures and be able to get answers with precision? Perhaps the rise in expectations for the usefulness of the axiomatic set theory approach was based on hope.

Minor problems began to develop early. Russell's paradox showed that the rules of sets applied to some collections did not apply to others. To bypass this problem, these troublesome collections were called classes instead of sets. One might wonder if this is the beginning of a trend to change definitions or concepts. We discuss other variations later. By and large, mathematicians were glad that axiomatic set theory had been rescued from Russell's paradox. If you cannot trust the axioms, whom can you trust?

The concept of the "nonexcluded middle" caused some confusion. In the 1940's and 1950's certain logicians were using certain axioms of set theory to disprove accepted theorems of topology. For example, L. E. J. Brouwer showed on the basis of axioms of set theory used by many logicians of that day that the Jordan curve theorem was false. On another occasion he showed on the same basis that Brouwer's fixed point theorem is false. I am relieved that many of my present day colleagues assure me that these approaches are no longer fashionable. It is not nice to fool Mother Nature!

We still have a sprinkling of people who do not believe in infinite sets. They cite the computer (which knows nothing of continuity, but uses tables rather than graphs) as evidence that this is the wave of the future. They may use axioms to support their points of view. However, this point of view seems at variance with traditional mathematics and the "real world" as most mathematicians envisage it. Presumably, the axioms of set theory suited for studying general topology are not of this sort.

Mathematicians speak of the set of real numbers (or of *the reals*) rather than some set of reals. This suggests the underlying assumption that there is only one set of reals. If one class in some school is studying the reals and another class somewhere else is studying them also, the two classes are studying the same thing. However, we are told that in axiomatic set theory that there are many sets of reals [C]. To distinguish the real reals from such an axiomatic set of reals, I shall call the latter a set of axiomatic reals. I asked the innocent question as to whether there was always a homeomorphism between any two sets of axiomatic reals, but it was suggested that this was not a proper question. Is the cardinality of two different sets of axiomatic reals the same? Is the cardinality of the reals the same as the cardinality of a set of axiomatic reals? Some might duck the questions by saying that they have no meaning.

However, there is a suspicion that a set of axiomatic reals has fewer numbers than the reals, since in a certain sense it is only a countable collection of numbers [C]. However, when I am told that the cardinality of a set is not an intrinsic property of the set but instead is concerned about a shortage of algorithms, I get the uneasy feeling that someone is tinkering with basic traditional concepts.

The terms uncountable, countable, infinite, finite have different meanings to different people. These concepts are used to indicate how many elements a set has. Would one call a drove of hornets infinite or uncountable just because it is impracticable (or impossible) to look inside their nest? I hope not. An uncountable set has more elements than there are integers. To many this would mean that there is no one-to-one correspondence between the uncountable set and the positive integers. There is none whatsoever. It is not enough that there exists none known to man or beasts or even in our list of algorithms in the model. There must be none known to omnipotent powers looking from outer space. There is no such 1-1 correspondence.

Others call an infinite set uncountable if there is no algorithm *in the model* for exhibiting a 1-1 correspondence between the set and the integers. The two concepts differ and there is confusion in using the same term to denote both. For example, on pages 19-20 of [C], Cohen remarks, "This paradox that a countable model can contain an uncountable set is explained by noting that to say that a set in uncountable merely asserts the nonexistence of a one-to-one mapping of the set with the set of integers. The "uncountable" set in M actually has only countably many members in M, but there is no one-to-one correspondence within M of this set with the set of integers." The algorithm approach is the one used in axiomatic set theory and it leads to situations where countable models have uncountable sets [C]. It leads to the statement of results which appear ridiculous if examined from other viewpoints. There is no meeting of minds and we hear, "Why don't you drop your definitions and accept mine?" "I got there first." "Yours makes no sense." "How can sets of measure 0 contain sets of positive measure?" In any case, there is confusion.

What is the beef? One is that if a mathematician uses traditional concepts in asking a question, the axiomatic set theorist may ignore the concepts intended by the questioner, replace them with related concepts, answer the changed questions, and then announce results with verbage that makes it appear that the original question has been answered. For example, if a traditional mathematician asks if the continuum hypothesis is true, he should not get an answer starting with "It is consistent with...," but instead might be told that "Axiomatic set theory is not equipped to cope with such questions since it contains no place for the traditional concepts of the reals, c, uncountable, $\aleph_1$."

Axiomatic set theory is a strong vital subject in its own right, and despite the fact that it tends to mislead general topologists by shifting the meaning of terms, it provides considerable help to general topologists. It warns them what problems to avoid. For example, axiomatic Martin's axiom plus the negation of the axiomatic continuum hypothesis $(MA + \neg CH)$ implies that the Normal Moore Space conjecture is false [F]. To show that the Normal Moore Space conjecture is true in the real world, one would be faced with a formidable task---one with such scant promise of success that most mathematicians would pass it by to try something more promising.

V. PARTIAL SOLUTIONS TO NORMAL MOORE SPACE CONJECTURE

Although considerable effort has been spent in seeking a solution, a real solution eludes us. We list certain partial solutions.

Jones showed [J_1] that if $2^{\aleph_0} < 2^{\aleph_1}$, then each separable normal Moore space is metrizable. This tends to case doubt on the existence of a separable counterexample since few would be prepared to argue that $2^{\aleph_0} = 2^{\aleph_1}$.

Traylor showed [T_r] that each metacompact separable, normal Moore space is metrizable.

Reed and Zenor got the very interesting result [RZ] that a normal Moore space is metrizable if it is both locally compact and locally connected.

Bing showed [B_2] that a normal Moore space is metrizable if and only if it is collectionwise normal.

Fleissner showed [R, p. 46] that if each set is constructible (V = L) then each normal locally compact Moore space is metrizable. Although Godel showed that (V = L) is consistent with the usual axioms of set theory, many mathematicians feel that (V = L) does not represent the real situation.

Recall that a Q set is an uncountable subset X of the reals such that if $Y \subset X$, then Y is the intersection of X with a G_δ subset of the reals. There is a question about the existence of Q sets, but Bing has shown [B_2] that if there is one, there is a separable normal Moore space that is not metrizable. Conversely, Heath has shown [H] that if there is a separable, normal, nonmetrizable Moore space, then there is a Q-set.

It is to be hoped that the arguments for the above results are real and not dependent on a variable view of the reals, an algorithmic view of cardinality, the principle of constructibility, nor its big brother forcing. However, axiomatic set theory may make its presence felt unexpectedly.

At first glance it might appear that T was defined without axiomatic restrictions. A closer examination shows that $\aleph_1$ was used. Was this traditional, intrinsic $\aleph_1$ or was it the axiomatic, algorithmic $\aleph_1$? It would be interesting to know which of the metrization results we give hold in real spaces.

Let us consider the separable version of the Normal Moore Space conjecture. Is each normal separable Moore space metrizable? Pick your own poison.

(a) There is a model for set theory in which each separable normal Moore space is metrizable.

(b) There is another model containing a separable normal Moore space which is not metrizable.

First we justify (a). Jones [J_1] has shown that a separable normal Moore space is metrizable if $2^{\aleph_0} < 2^{\aleph_1}$. Perhaps it is not true that

$2^{\aleph_0} < 2^{\aleph_1}$ ---but no matter! Play the axiomatic game and do not be confused by seeking the real truths. Godel has shown that there is a model for set theory where the continuum hypothesis holds. In this model

$$2^{\aleph_0} = c < 2^c = 2^{\aleph_1} .$$

Now for (b). Bing $[B_2]$ has shown how to construct a separable normal, nonmetrizable Moore space if there is a Q set.

Tall reports [T] that Silver discovered that a model described by Soloway in 1967 contained a set of reals with a Q set.

VI. NORMALITY OF T

If one could get a real proof that T is normal, one would have a counterexample to the Normal Moore Space conjecture. However, Fleissner shows in §4 of [F] that T is normal. Why is there no celebration?

Fleissner's definitions and methods are axiomatic. He shows that it is consistent with the axioms, definitions, and methods of axiomatic set theory that T be normal. However, it is also consistent that other results hold ---assumptions like the continuum hypothesis, Martin's axiom, and constructibility that fit into the game of axiomatics, but which are not "self-evident-truths" with regard to the traditional reals. In showing that T was normal, Fleissner used restrictions that Martin axiom holds but the continuum hypothesis does not.

By using other restrictions, Fleissner showed in §3 of [F] that T is not normal.

Both [T] and [R] contain excellent accounts about developments. In summary, known results are partial or "iffy." Where do we go from here?

One might abandon T as a possible counterexample and look in other directions for a counterexample.

Another possibility is to select a reasonable set of axiomatic restrictions and solve the conjecture on the basis of them. It may be difficult for some researchers to select a reasonable set of restrictions since their experience conditions them to accept axiomatic reasoning as a game in which all allowable moves are equally acceptable.

The Normal Moore Space conjecture has been very stimulating. It has been an area in which the findings of those in axiomatic set theory and in general topology have bolstered each other. Recently, most of the exciting results have come from set theorists and we in general topology are envious, interested, and grateful. Keep the good results coming. They add to our understanding of the problem. We need all the help we can get.

REFERENCES

[B_1] Bing, R. H., "Extending a metric," *Duke Math J.* *14*(1947), 511-519.

[B_2] ______, "Metrization of topological spaces," *Canadian J. Math* *3*(1951), 175-186.

[B_3] ______, "Partitioning a set," *Bull. Amer. Math. Soc.* *55*(1949), 1101-1110.

[C] Cohen, Paul J., *Set Theory and the Continuum Hypothesis*, W. A. Benjamin, Inc., Massachusetts, 1966.

[F] Fleissner, "When is Jones' space normal?" *Proc. Amer. Math. Soc.* *50* (1975), 375-378.

[H] Heath, R. W., "Screenability, pointwise paracompactness and metrization of Moore spaces," *Canadian J. Math.* *16*(1964), 764-770.

[J_1] Jones, F. B., "Concerning normal and completely normal spaces," *Bull. Amer. Math. Soc.* *43*(1937), 671-677.

[J_2] ______, "Remarks on the normal Moore space metrization problem," in: *Topology Seminar, Wisconsin, 1965*, R. H. Bing and R. J. Bean, eds. Ann. of Math Studies No. 60, Princeton Univ. Press, New Jersey, 1966, 115-119.

[R] Rudin, Mary Ellen, *Lectures on Set Theoretic Topology*, *CBMS Regional Conf.*, E. E. Grace, ed. Sem. in Math. No. 23, Amer. Math. Soc., Rhode Island, 1975.

[RZ] Reed, G. M. and P. L. Zenor, "Metrization of Moore spaces and generalized manifolds," *Fund. Math.* *91*(1976), 203-210.

[T] Tall, Franklin D., "The normal Moore space problem," in: *Topological Structures II*, Mathematical Centre Tracts No. 116, 1979, 243-261.

[T_r] Traylor, R. D., "On normality, pointwise paracompactness, and the metrization question," *Topology Conference, Arizona State Univ.*, Tempe, 1968, 286-289.

[N] Nagata, J.-I., "On a necessary and sufficient condition for metrizability," *J. Inst. Polytech. Osaka City Univ. Ser. A. Math. 1*(1950), 93-100.

[Sm] Smirnov, Ju. M., "A necessary and sufficient condition for metrizability of a topological space," *Dokl. Akad. Nauk. SSSR 77*(1951), 197-200.

[St] Stone, A. H., "Paracompactness and product spaces," *Bull. Amer. Math. Soc. 54*(1948), 977-982.

B. Retracts and Related Topics

RETRACTIONS ONTO SPHERES

R. H. BING, The University of Wisconsin and The Institute for Advanced Study

Suppose that S is a 2-sphere in E^3. It is known (Theorem 8-88 of [1]) that S separates E^3 into exactly two components and is the common boundary of each. We call the unbounded component Ext S and the bounded component Int S.

In case S is round, it is obvious that there is a retraction of $S+\text{Ext } S$ onto S. One which comes quickly to mind sends a point p of $S+\text{Ext } S$ to the point where the segment from p to the center of Int S intersects S. If the sphere is wild, however, as the Alexander horned sphere illustrated on page 176 of [1], it is not so obvious that there is such a retraction. It is the purpose of this note to show that there is.

If X is a subset of Y, a retraction r of Y onto X is a continuous transformation of Y onto X such that r is the identity on X. A set X is called an AR (absolute retract) if whenever X is embedded as a closed set in a metric space Z, there is a retraction of Z onto X. It is called an ANR (absolute neighborhood retract) if whenever it is embedded as a closed set in a metric space Z, there is an open subset U of Z containing X and a retraction of U onto X. It is known (Theorem 2-36 of [1]) that S is an ANR.

Although Theorems 2–4 may be too deep for the beginner, Theorem 1 should be accessible to those acquainted with such topology texts as [1].

THEOREM 1. *If S is a 2-sphere embedded in E^3, there is a retraction of $S+\text{Ext } S$ onto S.*

Proof. We would like to provide an elementary proof that effectively describes the retraction so that one can visualize the retraction and have some idea as to where points are sent.

Let N be a neighborhood of S such that there is a retraction r of $\overline{N}$ onto S. That there is such a retraction follows from the fact that S is an ANR. Perhaps this is the weak link in our effort to describe effectively the retraction but if one considers how one proves that S is an ANR from the Tietze extension theorem, one may be able to see where r sends points. In fact, since we are working in E^3 rather than in an abstract normal space, perhaps the reader can concoct an easier proof of the version of the Tietze extension theorem we need than that given in Section 2-7 of [1]. Such an easy proof is given in [3].

We suppose that $\overline{N}$ is a bounded polyhedron and that $\text{Ext } S-\overline{N}$ is connected. If this is not already the case, we could adjust the N considered in the previous paragraph to bring it about by boring holes in that N and whittling it down to size. We make no effort to keep $\overline{N}$ from having handles and in general it will.

Let C be a large cube whose interior contains $\overline{N}$. There is a retraction of $E^3-\text{Int } C$ onto $\text{Bd } C$ so if we can find a retraction of $C-\text{Int } S$ onto S, the truth of Theorem 1 follows.

Consider a triangulation of $C-N$. Let $s_1, s_2, \cdots, s_n$ be the closed 3-simplexes (tetrahedra) of this triangulation. We suppose that the s's are ordered so that s_n has a 2-face on $\text{Bd } C$, s_{n-1} has a 2-face on $\text{Bd } C+s_n$, s_{n-2} has a 2-face on $\text{Bd } C+s_n+s_{n-1}, \cdots$, and s_1 has a face on $\text{Bd } C+s_n+s_{n-1}+\cdots+s_2$. Note that if N_0 denotes $\overline{N}$ and N_i denotes $N_0+s_1+s_2+\cdots+s_i$, then $\text{Ext } S-N_i$ is connected.

Since $\text{Ext } S-N_{i-1}$ is connected, s_i has a 2-face which does not belong to N_{i-1}. Let f_i be a particular such face of s_i, and g_i be a retraction of s_i onto $\text{Bd } s_i-\text{Int} f_i$.

We now extend r to a retraction taking N_1 onto S. We do this starting at low dimensional faces of s_1 and proceeding one at a time until only f_1 is left. That we can do this follows from Lemma 6-42 of [1]. Denote the extended retraction by r'. Once r has been extended to take Bd $s_i -$ Int f_i onto S, we define $r_1 = r$ on N_0, $r_1 = r'g_1$ on s_1. We continue extending r in this fashion so that r_2 takes N_2 into S, r_3 takes N_3 onto S, $\cdots$, and r_n takes N_n onto S.

We note that we did not use any properties of 2-spheres in E^3 that are not equally true for $(n-1)$-spheres in E^n so we have the following version of Theorem 1.

THEOREM 2. *For each $(n-1)$-sphere S in E^n and each point p of* Int S, *there is a retraction of $E^n - \{p\}$ onto S.*

Theorem 2 can be generalized. Daniel R. McMillan made the interesting observation that we did not use the fact that the space into which S was embedded was E^n and that we could have used a triangulated n-manifold-with-boundary M instead of E^n. The component U of $M - S$, whose closure is to be retracted onto S, would need to have a point removed before retracting unless $U \cdot \text{Bd } M \neq 0$ or $\overline{U}$ is noncompact. By working a bit harder we could even remove the hypothesis that M can be triangulated. In place of M we could even use a pseudo-manifold.

A set is called i-connected if each continuous transformation of the boundary of an $(i+1)$-cell into X can be extended to take the $(i+1)$-cell into X. Note that we did not use in the proof of Theorem 2 the fact that S is an $(n-1)$-sphere but merely that it is an i-connected $(i=0, 1, \cdots, n-2)$ ANR. Hence we have the following

THEOREM 3. *Suppose that X is a closed subset of a compact n-manifold M such that X is an i-connected $(i=0, 1, \cdots, n-2)$ ANR. Then for each point p of a component U of $M - X$ there is a retraction of $U + X - \{p\}$ onto X.*

Proof. Let N be an open subset of M containing X such that there is a retraction r_0 of $\overline{N}$ onto X and $B_1, B_2, \cdots, B_m$ be a finite number of topological n-cells in U such that $U \subset \overline{N} + B_1 + B_2 + \cdots + B_m$.

Let g_1 be a map of Bd B_1 into X that agrees with r_0 on $\overline{N} \cdot \text{Bd } B_1$. To prove that there is such a g_1 we argue as follows: Regard X as embedded in a Hilbert cube H. Since X is an ANR there is an open subset N_1 of H containing X and a retraction g' of N_1 onto X. It follows from the Tietze extension Theorem that there is a map g'' of Bd B_1 into H that agrees with r_0 on $\overline{N} \cdot \text{Bd } B_1$. Let T be a triangulation of Bd B_1 of such small mesh that each simplex of T that intersects $\overline{N} \cdot \text{Bd } B_1$ lies in $g''^{-1}(N_1)$. Let $g_1 = g'g''$ on each simplex of T in $g''^{-1}(N_1)$. Then g_1 may be extended to the other simplexes of T one at a time by starting at those of low dimension and using the fact that X is i-connected $(i=0, 1, \cdots, n-2)$.

Let p_1 be a point of Int B_1 and g_1' be a retraction of $B_1-\{p_1\}$ onto Bd B_1. Let r_1 be the retraction of $(X+U\cdot\overline{N}+B_1)-\{p_1\}$ onto X which is $g_1' g_1$ on $B_1-\{p_1\}$ and r_0 elsewhere.

Continuing as suggested in the last paragraph, we let p_i be a point of Int B_i and get a retraction r_2 of $(X+U\cdot\overline{N}+B_1+B_2)-(\{p_1\}+\{p_2\})$ onto $X, \cdots$, and a retraction r_m of

$$(X+U\cdot\overline{N}+B_1+B_2+\cdots+B_m)-(\{p_1\}+\{p_2\}+\cdots+\{p_m\})$$
$$=X+U-(\{p_1\}+\{p_2\}+\cdots+\{p_m\})$$

onto X.

Let $q_1, q_2, \cdots, q_j$ be a sequence of points in U such that q_j is the p of the statement of Theorem 3, each p_i appears in the sequence, and each adjacent pair q_i, q_{i+1} lies on the interior of an n-simplex K_i in U. By modifying r_m in Int K_1 we obtain a retraction r_{m+1} of $X+U-(\{q_2\}+\{q_3\}+\cdots+\{q_j\})$ onto X. Eliminating the q's one at a time and changing the r's we arrive at a retraction r_{m+j-1} of

$$X+U-\{q_j\}=X+U-\{p\} \text{ onto } X.$$

We can avoid using the hypothesis that M is compact by using a locally finite collection of B's instead of a finite collection. If X is not compact we would embed the one-point-compactification of X in H to get the g's. In fact, if $\overline{U}$ is not compact, we would not even need to remove a point p from U before retracting. However, we shall not pursue this further.

If S is a wild 2-sphere in E^3, $S+\text{Int } S$ is called a *crumpled cube.* M. K. Fort, Jr. showed me a proof using the Side Approximation Theorem that a crumpled cube is an AR. The result generalizes as follows.

THEOREM 4. *If S is an $(n-1)$-sphere in E^n, $S+\text{Int } S$ is an AR.*

This result follows from the fact that $S+\text{Int } S$ is the retract of a closed n-cell as shown in Theorem 2, an n-cell is an AR (Theorem 2-34 of [1]), and any retract of an AR is an AR.

An alternate proof of Theorem 4 can be obtained by studying the homotopy connectedness properties of $S+\text{Int } S$ and applying the results on page 289 of [2].

Research supported by grants NSF-G21514 and NSF-G11665.

References

1. John G. Hocking and Gail S. Young, Topology, Addison-Wesley, Reading, Mass., 1961.
2. Kasimir Kuratowski, Topologie, vol. 2, 3rd ed., Warsaw, 1961.
3. R. H. Bing, Spheres in E^3, this MONTHLY, 71 (1964) 353–364.

FUNDAMENTA
MATHEMATICAE
LIV (1964)

A 3-dimensional absolute retract which does not contain any disk

by

R. H. Bing (Madison) and **K. Borsuk** (Warszawa) *

Some examples of 2-dimensional AR-spaces which do not contain any disks were discovered long ago (see [4], [5], [6]). However, the question as to whether or not there exists a 3-dimensional AR-space which does not contain any disk remained open. The aim of the present note is to give an affirmative answer to this question.

1. Loops and their linking. By a map $f\colon X\to Y$ of a space X into a space Y we mean a single-valued continuous function from X to Y. In particular, if $X=C$ is a simple closed curve, then a map

$$f\colon C\to Y$$

is called a *loop* in the space Y. Two loops $f_0, f_1\colon C\to Y$ are said to be *homotopic in* Y if there exists a map φ of the Cartesian product $C\times I$ of C by the unit interval $I\colon 0\leqslant t\leqslant 1$ into Y such that

$$f_0(x)=\varphi(x,0) \text{ and } f_1(x)=\varphi(x,1) \quad \text{for every} \quad x\in C.$$

In particular, the loop f_0 is said to be *homotopic to zero in* Y (symbolically: $f_0\simeq 0$ in Y) provided f_0 is homotopic in Y to a loop f_1 mapping C onto a single point of Y. If C lies in a plane H, then the homotopy $f_0\simeq 0$ in Y is equivalent to the extendability of the map f_0 to a map f of the disk $D\subset H$ bounded by C into Y.

Let us consider two loops in the Euclidean 3-space E^3:

$$f_1\colon C_1\to E^3, \quad f_2\colon C_2\to E^3,$$

and let us suppose that they are disjoint, i.e. $f_1(C_1)\cap f_2(C_2)=0$. If orientations of C_1, C_2 and E^3 are given, then an integer $\lambda(f_1,f_2)$ called the *linking coefficient* of f_1 and f_2 is defined (see, for instance, Lefschetz [11], p. 124). It describes the algebraic number of times each of loops f_1 and f_2 twists around the other. For our aims it suffices to consider only the absolute value of this coefficient

$$\bar\lambda(f_1,f_2)=|\lambda(f_1,f_2)|,$$

* The first author is supported by NSF grants NSF-G21514 and NSF-G11665.

which does not depend on the choice of orientations of C_1, C_2 and E^3. We shall apply the following properties of the coefficients $\bar{\lambda}$:

(1) $\bar{\lambda}(f_1, f_2) = \bar{\lambda}(f_2, f_1)$.

(2) *If h is a homeomorphism of E^2 into itself, then*

$$\bar{\lambda}(f_1, f_2) = \bar{\lambda}(hf_1, hf_2)\,.$$

(3) *If $0 \leqslant \varepsilon < \varrho\big(f_1(x_1), f_2(x_2)\big)$ for every $x_1 \in C_1$, $x_2 \in C_2$, then for every two maps $g_1\colon f_1(C_1) \to E^3$, $g_2\colon f_2(C_2) \to E^3$ such that $\varrho\big(g_\nu(y), y\big) \leqslant \frac{1}{2}\varepsilon$ for every $y \in f_\nu(C_\nu)$ and $\nu = 1, 2$ one has $\bar{\lambda}(f_1, f_2) = \bar{\lambda}(g_1 f_1, g_2 f_2)$.*

If $f_1\colon C_1 \to E^3$ and $f_1'\colon C_1' \to E^3$ are homeomorphisms of simple closed curves C_1, C_1' onto the same simple closed curve $C \subset E^3$, then for every loop $f_2\colon C_2 \to E^3$ with $f_2(C_2) \subset E^3 - C$, the coefficients $\bar{\lambda}(f_1, f_2)$ and $\bar{\lambda}(f_1', f_2)$ are equal. Let us denote the common value of them by $\bar{\lambda}(C, f_2)$ or by $\bar{\lambda}(f_2, C)$. The vanishing of the number $\bar{\lambda}(C, f_2)$ implies that each of C, $f_2(C_2)$ bounds in the complement of the other.

The corresponding homotopy linking is not symmetric, for of two polygonal simple closed curves in E^3, it may be that the first is homotopic to zero in the complement of the second but the second is not homotopic to zero in the complement of the first. Since the first homology group is obtained by abelianizing the first homotopy group (fundamental group), it follows that if C is a simple closed curve in E^3, then a loop f lying in $E^3 - C$ is not homotopic to zero in $E^3 - C$ if $\bar{\lambda}(C, f) > 0$. However, if C is an unknotted simple closed curve as would lie in a plane, then the first homotopy group of $E^3 - C$ is infinite cyclic and hence already abelian so f is homotopic to zero in $E^3 - C$ if and only if $\bar{\lambda}(C, f) = 0$. In this paper we make use of the following

(4) *If C is a simple closed curve in a plane in E^3, then a loop f lying in $E^3 - C$ is homotopic to zero in $E^3 - C$ if and only if $\bar{\lambda}(C, f) = 0$.*

If both disjoint loops $f_1\colon C_1 \to E^3$ and $f_2\colon C_2 \to E^3$ are homeomorphisms of C_1 onto C_1' and of C_2 onto C_2', then the coefficient $\bar{\lambda}(f_1, f_2)$ depends only on C_1' and C_2'. We shall denote it by $\bar{\lambda}(C_1', C_2')$. If $\bar{\lambda}(C_1', C_2') \neq 0$, then the simple closed curves C_1', C_2' are said to be *homologically linked*. It can be shown that:

(5) *Given:*

1° *A disk D_1 with boundary C_1 and a disk $D_2 \subset D_1$ with boundary C_2,*

2° *a simple closed curve $C \subset E^3$,*

3° *a map $f\colon D_1 \to E^3$ such that $f\big((D_1 - D_2) \cup C_2\big) \subset E^3 - C$,*

then the loops $f_1 = f/C_1$, $f_2 = f/C_2$ satisfy the equation $\bar{\lambda}(f_1, C) = \bar{\lambda}(f_2, C)$.

(6) *Given:*

1° *A simple closed curve $C \subset E^3$ decomposed into the sum of two simple arcs J_1, J_2 with common ends a and b,*

2° *a plane* $H \subset E^3$ *decomposing* E^3 *between* a *and* b,

3° *a simple closed curve* $C' \subset H$ *decomposing* H *between the sets* $J_1 \cap H$ *and* $J_2 \cap H$,

then $\bar{\lambda}(J_1 \cup J_2, C') \neq 0$.

2. Anchor ring and its core. Now let us suppose that C_0 is a circle with radius r_0 lying in a plane $H \subset E^3$. Let D_0 be the disk bounded (in H) by C_0 and let r be a positive number less than r_0. The set

$$A_0 = \{x | x \in E^3,\ \varrho(x, C_0) \leqslant r\}$$

is an anchor ring with the *core* (center line) C_0. The numbers r and r_0 will be said to be the *inner* and the *outer radius* of A_0 and the number $\vartheta = r/r_0$ is called the *profile* of A_0. Evidently two anchor rings are similar if and only if they have the same profile. Let B_0 denote the boundary of A_0. A circle $C_1 \subset E^3 - A_0$ will be said to be *transversal* to the anchor ring A_0 provided $\bar{\lambda}(C_0, C_1) = 1$. We observe:

(7) *Given in* E^3 *an anchor ring* A_0 *with core* C_0 *and a circle* C_1 *transversal to* A_0. *Then there exists a neighborhood* U *of the boundary* B_0 *of* A_0 *such that every loop* f *lying in* U *is homotopic to zero in the set* U *if and only if* $\bar{\lambda}(f, C_0) = \bar{\lambda}(f, C_1) = 0$.

Evidently, if the number $\varepsilon > 0$ is sufficiently small then the set

$$U = \{x | x \in E^3,\ \varrho(x, B_0) < \varepsilon\}$$

satisfies condition (7).

3. Chain which substitutes for an anchor ring. Let A_0 be an anchor ring with outer radius r_0, inner radius r, core C_0, and boundary B_0. Let n be a natural number greater than 3 and let $a_1, a_2, \ldots, a_n$ (where the indices are understood mod n) be vertices of a regular n-polygon inscribed on C_0 in their natural order. Evidently

$$\varrho(a_0, a_1) = \varrho(a_i, a_{i+1}) = 2r_0 \sin\frac{\pi}{n}$$

is small if n is large. Set

$$s_0 = \frac{1}{3}\varrho(a_0, a_1) = \frac{2}{3} r_0 \sin\frac{\pi}{n}.$$

Let a'_k denote the center of the segment $a_k a_{k+1}$, D_k denote the geometrical disk with center a_k, radius s_0, and lying in the plane of the circle C_0. Let D'_k denote the geometrical disk with center a'_k, radius s_0, and lying in the plane perpendicular to the segment joining a'_k with the center of C_0. Let C_k denote the boundary of D_k and C'_k the boundary of D'_k. One sees easily that:

1° $C_k \cap C'_j = 0$ for every k and j.

2° $C_k \cap C_j = C'_k \cap C'_k = 0$ if $k \neq j \pmod{n}$.

3° $\bar{\lambda}(C_k, C_j) = \bar{\lambda}(C'_k, C'_j) = 0$ if $k \neq j \pmod{n}$.

4° $\bar{\lambda}(C_k, C'_j) = 1$ if $0 \leqslant k-j \leqslant 1 \pmod{n}$, and $\bar{\lambda}(C_k, C'_j) = 0$ in all other cases.

5° If x, y are points belonging to two different circles of the collection $C_1, C'_1, C_2, C'_2, \ldots, C_n, C'_n$, then $\varrho(x, y) > \frac{1}{2} s_0$.

Now let us suppose that $\vartheta = r/r_0 \leqslant \frac{1}{4}$, that ε is a positive number less than $2r$, and that the natural number n is so large that $\varrho(a_0, a_1) < \varepsilon$. Denote by A_k the anchor ring with core C_k and inner radius $s = \vartheta s_0$, and by A'_k the anchor ring with core C'_k and with the same inner radius s. It follows that A_k and A'_k are similar to A_0 and their outer radius is s_0. Their diameters are equal to $2(s+s_0) = 2(1+\vartheta)s_0 < 2(1+\frac{1}{4})\frac{1}{3}\varepsilon < \varepsilon$. It follows from 5°, that all anchor rings $A_1, A'_1, A_2, A'_2, \ldots, A_n, A'_n$ are disjoint. Moreover it is clear that for every point $x \in \bigcup_{k=1}^{n} (A_k \cup A'_k)$

$$\varphi(x, C_0) < s_0 + \vartheta s_0 \leqslant \frac{5}{4} s_0 < \frac{1}{2} \varrho(a_0, a_1) < \frac{\varepsilon}{2} < r.$$

We conclude that all anchor rings A_i, A'_i are contained in the interior of the anchor ring A_0 and also in the ε-neighborhood of its core C_0.

The set

$$M^1 = \bigcup_{k=1}^{n} (A_k \cup A'_k) \subset A_0 \tag{8}$$

will be said to be *chain substituting for the anchor ring* A_0, and the anchor rings A_k, A'_k, $k = 1, 2, \ldots, n$, will be said to be *links* of this chain. Thus the chain M^1 has $2n$ links $A_1, A'_1, A_2, A'_2, \ldots, A_n, A'_n$; we shall denote these links (in the same order) by $L^1_1, L^1_2, \ldots, L^1_{2n}$, where the lower indices are understood mod $2n$. It follows by our construction that the cores of two links L^1_i and L^1_j are linked if and only if $|i-j| = 1 \pmod{2n}$. Thus we have:

(9) *For every anchor ring A_0 with core C_0 and profile $\vartheta \leqslant \frac{1}{4}$ and for every $\varepsilon > 0$ there exists an integer $n > 3$ such that in the interior of A there exists a chain M^1 with $2n$ links $L^1_1, L^1_2, \ldots, L^1_{2n}$, these links being congruent anchor rings with profiles ϑ, diameters $\leqslant \varepsilon$, and each lying in the ε-neighborhood of C_0. If C^1_i denotes the core of the link L^1_i, then $\bar{\lambda}(C^1_i, C^1_j) = 1$ if $|i-j| = 1 \pmod{2n}$ and $\bar{\lambda}(C^1_i, C^1_j) = 0$ if $|i-j| > 1 \pmod{2n}$.*

The set

$$N^1 = \bigcup_{i=1}^{2n} C^1_i = \bigcup_{k=1}^{n} (C_k \cup C'_k)$$

will be called the *core of the chain* M^1.

It is known (see, for instance, [8]) that a loop f lying in $E^3 - A_0$ is homotopic to zero in $E^3 - M^1$ if and only if it is homotopic to zero in

$E^3 - A_0$, that is if $\bar{\lambda}(f, C_0) = 0$. This fact may be also formulated as follows: Let D be a disk with the boundary C.

(10) *A loop $f\colon C \to E^3 - A_0$ can be extended to a map $f'\colon D \to E^3 - N^1$ if and only if $\bar{\lambda}(f, C_0) = 0$.*

Now let us prove (with the same notation):

(11) *If $f\colon C \to E^3 - A_0$ and $\bar{\lambda}(f, C_0) > 0$, then for each extension $f'\colon D \to E^3$ of f and for every $\varepsilon > 0$ there exist a link L_i^1 and two disjoint subdisks D_1, D_2 of D with boundaries C_1' and C_2' such that each partial map $\bar{f}_\nu = f'/C_\nu'$, $\nu = 1, 2$, is a lop in $A_0 - M^1$ lying in the ε-neighborhood of link L_i^1 and homologically linked with the core of this link.*

Proof. By the construction of M^1, we infer that the core C_{i+1}^1 of the link L_{i+1}^1 is transversal to the anchor ring L_i^1. It follows by (7) that there exists an open neighborhood U_i of the boundary B_i^1 of L_i^1 contained in A_0 and such that a loop g lying in U_i is homotopic to zero in U_i if and only if $\bar{\lambda}(g, C_\nu^1) = 0$ for $\nu = i, i+1$. Since the sets $B_1^1, B_2^1, \ldots, B_{2n}^1$ are disjoint, we may assume that the neighborhoods $U_1, U_2, \ldots, U_{2n}$ are mutually disjoint and also disjoint from the core N^1 of M^1, and that each U_i lies in the ε-neighborhood of L_i^1. Now let us write:

$$F_i = f'^{-1}(L_i^1), \quad G_i = f'^{-1}(U_i) \quad \text{for every} \quad i = 1, 2, \ldots, 2n.$$

The sets $G_1, G_2, \ldots, G_{2n}$ are open and mutually disjoint while F_i is a compact subset of G_i. It follows that G_i contains an open subset G_i^0 such that

$$F_i \subset G_i^0 \subset G_i \quad \text{for} \quad i = 1, 2, \ldots, 2n$$

and the boundary Z_i of G_i^0 is the union of a finite number of disjoint simple closed curves $C_{i,1}^0, C_{i,2}^0, \ldots, C_{i,m_i}^0$. Each of the partial functions

$$g_{i,j} = f'/C_{i,j}^0, \quad i = 1, 2, \ldots, 2n, \ j = 1, 2, \ldots, m_i,$$

is a loop in U_i. If $\bar{\lambda}(g_{i,j}, C_\nu^1) = 0$ for $\nu = i, i+1$, then $g_{i,j}$ is homotopic to zero in U_i by (7). Consider the component E of $D - \bigcup C_{i,j}^0$ that contains C. Each boundary component of $\bar{E}$ other than C is a $C_{i,j}^0$. If for each such $C_{i,j}^0$, $\bar{\lambda}(g_{i,j}, C_\nu^1) = 0$ for $\nu = i, i+1$, we find that the map $f'/C = f$ can be extended to map of D into $E^3 - N^1$ — namely, the extension is f' on E and takes each component of $D - E$ into an appropriate U_i. Since the existence of such an extension is contrary to (10) we conclude that for at least one $C_{i,j}^0$ and $\nu = i$ or $i+1$ it is $\bar{\lambda}(g_{i,j}, C_\nu') > 0$.

Thus we have shown that there exists a loop $g_{i,j}$ such that for some index ν we have $\bar{\lambda}(g_{i,j}, C_\nu^1) > 0$. Now, fixing the index ν, we can find a maximal C_{i_0,j_0}^0 in the sense that there are indices i_0, j_0 such that $\bar{\lambda}(g_{i_0,j_0}, C_\nu^1)$ is positive but for every other pair of indices i, j satisfying the condition $\bar{\lambda}(g_{i,j}, C_\nu^1) > 0$, the disk $D_{i,j}$, bounded in D by the curve $C_{i,j}^0$, does not contain C_{i_0,j_0}^0 on its interior.

Now let us consider the domain $V \subset D$ bounded by the disjoint simple closed curves C and $C^0_{i_0,j_0}$. Consider the various $C^0_{i,j}$'s in V. For some such $C^0_{i,j}$, $\bar{\lambda}(g_{i,j}, C^1_\nu) > 0$ or else we could obtain a map $g\colon \bar{V} \to E^3 - C^1_\nu$ such that $g/C = f'$, $g/C^0_{i_0,j_0} = f'/C^0_{i_0,j_0}$. But since $\bar{\lambda}(g/C, C^1_\nu) = 0$, we arrive at the contradiction from (5) that $\bar{\lambda}(g_{i_0,j_0}, C^1_\nu) = 0$. Consequently there is a $C^0_{i,j}$ in V such that $\bar{\lambda}(g_{i,j}, C^1_\nu) > 0$. It follows from the maximality of $C^0_{i_0,j_0}$ that this $C^0_{i,j}$ bounds a disk in V. Consequently the proof of (11) is complete.

As an easy consequence of (11) we get:

(11′) *If C is the boundary of a disk D and if the loop $f\colon C \to E^3 - A_0$ satisfies the condition $\bar{\lambda}(f, C_0) > 0$, then for every extension $f'\colon D \to E^3$ of f there exists a disk $D' \subset D$ with boundary C' such that the loop f'/C' is homologically linked with the core of one of the links of the chain M^1.*

4. Wreath substituting for an anchor ring. Suppose we are given an anchor ring A_0 with outer radius 1, profile $\vartheta \leqslant \frac{1}{4}$, core C_0, and chain M^1 substituting for A_0 having $2n$ links $L^1_1, L^1_2, \ldots, L^1_{2n}$ similar to A_0, with outer radii equal to $\frac{2}{3}\sin(\pi/n)$. Let us construct a decreasing sequence of compacta

$$M^1 \supset M^2 \supset \ldots \supset M^k \supset \ldots$$

in the following manner: Suppose that for a natural number k we have a set M^k such that

(12_k) *M^k is the union of $(2n)^k$ disjoint anchor rings L^k_i $\left(i = 1, 2, \ldots, (2n)^k\right)$ with the outer radii equal to $(\frac{2}{3}\sin\pi/n)^k$ and with profile ϑ.*

Let φ_i denote the similarity mapping A_0 onto L^k_i. Then φ_i maps the links L^1_j $(j = 1, 2, \ldots, 2n)$ of the chain M^1 onto disjoint anchor rings similar to A_0 and contained in the interior of L^k_i. The inner radii of these anchor rings are equal to $(\frac{2}{3}\sin\pi/n)^{n+k}$. If we set

$$M^{k+1} = \bigcup_{i=1}^{(2n)^k} \varphi_i(M^1),$$

then we get a set M^{k+1} satisfying the condition (12_{k+1}). The anchor rings L^k_i $\left(i = 1, 2, \ldots, (2n)^k\right)$ are components of M^k and their diameters converge to zero when $k \to \infty$. It follows that the set

$$M = \bigcap_{k=1}^{\infty} M^k$$

is a 0-dimensional compactum, actually a set of Antoine [3]. Evidently each of the sets

$$M_i = M \cap L^1_i, \quad i = 1, 2, \ldots, 2n,$$

is similar to M, and consequently M_i is a set of Antoine lying in the interior of the anchor ring L_i^1. It is known [13] that there exists in the interior of L_i^1 a simple arc J_i containing M_i. The set

$$W = \bigcup_{i=1}^{2n} J_i$$

will be said to be a *wreath* substituting for the anchor ring A_0, and the arcs J_i (components of W) will be said to be *links of the wreath* W. We infer the following from (9):

(13) *For every anchor ring A_0 with core C_0 and profile $\vartheta \leqslant \frac{1}{4}$ and for every $\varepsilon > 0$ there exists in the interior of A_0 a wreath W substituting for A_0, lying in the ε-neighborhood of C_0 and such that the diameters of its links are less than ε.*

Let us observe that (11′) implies that if a simple closed curve C is the boundary of a disk D and if the loop $f\colon C \to E^3 - A_0$ is homologically linked with C_0, then for every continuous extension $f'\colon D \to E^3$ of f and for every $k = 1, 2, \ldots$ there exists a disk $D_0^k \subset D$ with the boundary C_k^0 such that the core C' of one of the components of M^k is homologically linked with the loop f'/C_k^0. It follows that $f'(D) \cap M^k \neq 0$, for every $k = 1, 2, \ldots$, and consequently $f'(D) \cap M \neq 0$.

Moreover, by virtue of (11), there exist two disjoint disks $D_1, D_2 \subset D$ with boundaries C_1', C_2', and link L_i^1 of the chain M^1 such that the core of L_i^1 is homologically linked with each of the loops $f_\nu' = f'/C_\nu'$, $\nu = 1, 2$. Moreover, we can assume that the loops f_1', f_2' lie in $A_0 - M^1$ and in the ε-neighborhood of the link L_i^1. It follows that both sets $f'(D_1)$ and $f'(D_2)$ intersect the set M_i, and consequently also the arc $J_i \supset M_i$. Hence, we have the following:

(14) *Let C be the boundary of a disk D and let $f\colon C \to E^3 - A_0$ with $\bar{\lambda}(f, C_0) > 0$. Then for every continuous extension $f'\colon D \to E^3$ of f and for every $\varepsilon > 0$ there exist two disjoint disks $D_1^0, D_2^0 \subset D$ with boundaries C_1^0, C_2^0 and a link L_i^1 of M^1 such that both loops f'/C_1^0, f'/C_2^0 lie in $A_0 - M^1$ in the ε-neighborhood of L_i^1 and that $f'(D_1^0) \cap L_i^1 \cap W \neq 0 \neq f'(D_2^0) \cap L_i^1 \cap W$.*

In particular it follows (under the hypotheses of (14)) that

$$f'(D) \cap W \neq 0 . \tag{15}$$

5. Broken anchor ring and its substituting wreath. By a *topological anchor ring* we understand the image $A = h(A_0)$ of A_0 by a homeomorphism h. Then the set $h(C_0)$ will be said to be the *core* of A and the sets $h(M^1)$ and $h(W)$ will be said to be the *chain* and the *wreath* substituting for the topological anchor ring A. The links of $h(M^1)$ and of $h(W)$ are defined as components of these sets.

It follows from (13):

(16) *For every topological anchor ring A with core C and for every $\varepsilon > 0$ there exists a chain and a wreath substituting for A, lying in the ε-neighborhood of C and having links with diameters less than ε.*

In particular, if $\hat{C}_0$ is a simple polygon (i.e. a polygonal simple closed curve) lying in a plane $H \subset E^3$ and η is a positive number sufficiently small, then the set

$$\hat{A}_0 = \{x | x \in E^3, \ \varrho(x, C_0) \leqslant \eta\}$$

is a topological anchor ring. More exactly, there exists a homeomorphism h mapping the space E^3 onto itself so that $h(A_0) = \hat{A}_0$ and $h(C_0) = \hat{C}_0$. The set $\hat{A}_0$ will be called a *broken anchor ring* with core $\hat{C}_0$ and *inner radius* η. The sets

$$\hat{M}^1 = h(M^1) \quad \text{and} \quad \hat{W} = h(W)$$

will be said to be the *chain* and the *wreath substituting for the broken anchor ring* $\hat{A}_0$.

The following is a consequence of (2) and (11):

(17) *If C is the boundary of a disk D and $f\colon C \to E^3 - \hat{A}_0$ is a loop with $\bar{\lambda}(f, \hat{C}_0) > 0$ then, for every continuous extension $f'\colon D \to E^3$ of f and for every $\varepsilon > 0$, there exist in the interior of D two disjoint disks D_1, D_2 with boundaries C_1', C_2' and a link of the chain $\hat{M}^1$ such that the loops $\bar{f}_\nu = f'/C_\nu'$, $\nu = 1, 2$ lie in $\hat{A}_0 - \hat{M}^1$ in the ε-neighborhood of the link of $\hat{M}^1$ and they are both homologically linked with the core of this link.*

6. Dense sequences of chords. Let Q be a unit ball in E^3 with boundary S. A segment $K \subset E^3$ will be said to be a *chord* of S provided both its ends belong to S. A sequence $\{K_\nu\}$ of chords of S will be said to be *dense* on S provided for each open subset $G \neq 0$ of S there exists an index ν such that both ends of the chord K_ν belong to G. Let us show:

(18) *There exists a sequence of the disjoint chords $\{K_\nu\}$ dense on S and such that the diameters of K_ν converge to 0.*

Proof. Let $\{a_n\}$ be a sequence of points of S dense in S and such that $a_n \neq a_m$ for $n \neq m$. We define the sequence $\{K_\nu\}$ by the induction:

1. Let a_1' be a point of $S - \{a_n\}$ such that $\varrho(a_1, a_1') \leqslant 1$. We set $K_1 = \overline{a_1 a_1'}$.

2. Suppose that for $\nu = 1, 2, \ldots, m$ we have $K_\nu = \overline{a_\nu a_\nu'}$, where $a_\nu' \in S - \{a_n\}$, $\varrho(a_\nu, a_\nu') \leqslant 1/\nu$ and $K_\nu \cap K_\mu = 0$ for $\mu, \nu = 1, 2, \ldots, m$. Then we see at once that there exists a point $a_{m+1}' \in S - \{a_n\}$ such that $\varrho(a_{m+1}, a_{m+1}') \leqslant 1/(m+1)$ and that the chord $K_{m+1} = \overline{a_{m+1} a_{m+1}'}$ is disjoint from $K_1 \cup K_2 \cup \ldots \cup K_m$.

Thus we get a sequence $\{K_\nu\}$ of disjoint chords with diameters convergent to zero. If x is a point of an open subset G of S, then there exists an $\varepsilon > 0$ such that all points $y \in S$ with $\varrho(x, y) < \varepsilon$ belong to G. Since the sequence $\{a_n\}$ is dense in G, there exists an index $n > 2/\varepsilon$ such that $\varrho(a_n, x) < \varepsilon/2$. Then $\varrho(a_n', x) \leqslant \varrho(a_n', a_n) + \varrho(a_n, x) < \varepsilon$ and consequently both ends of $K_n = \overline{a_n a_n'}$ belong to G.

7. A sequence of broken anchor rings dense in Q. Let $\{\hat{A}_\nu\}$ be a sequence of broken anchor rings lying in the interior of the unit ball Q. We say that $\{\hat{A}_\nu\}$ is *dense* in Q if for every simple closed curve $C \subset Q - S$ there exists an index ν such that $\hat{A}_\nu \subset Q - C$ and the core $\hat{C}_\nu$ of $\hat{A}_\nu$ is homologically linked with C.

Now let us show:

(19) *There exists a sequence $\{A_\nu'\}$ of broken anchor rings lying in $Q - S$ and dense in Q.*

Proof. We apply the following elementary facts about the topology of the plane:

1° If X and Y are disjoint compact subsets of E^2 which lie in a region $G \subset E^2$ but do not cut it, then there exists a simple polygon $\hat{C} \subset G - X - Y$ which separates G between X and Y. Moreover, we can assume that all vertices of $\hat{C}$ have rational coordinates.

Since the set of all simple polygons in E^2 with rational coordinates of vertices is countable, we infer:

2° For every plane region G there exists a countable family of simple polygons lying in G such that every two disjoint compact subsets of G which do not cut G, may be separated by a simple polygon belonging to this family.

Applying 1° and 2° let us show:

3° There exists in $Q - S$ a countable family $\mathfrak{C}$ of plane simple polygons such that for every simple closed curve $C \subset Q - S$, there exists a $\hat{C} \in \mathfrak{C}$ such that $\overline{L}(C, \hat{C}) > 0$.

In order to prove 3°, consider the family $\mathfrak{H}$ of all planes $H \subset E^3$ which intersect Q and pass through three non collinear points with rational coefficients. Manifestly $\mathfrak{H}$ is countable. By 2°, for every $H \in \mathfrak{H}$ there exists a countable family $\mathfrak{C}_H$ of simple polygons lying in the region $G_H = H \cap (Q - S)$ and such that every two disjoint compact subsets of G which do not cut G, can be separated by a simple polygon $\hat{C} \in \mathfrak{C}_H$. Evidently the family

$$\mathfrak{C} = \bigcup_{H \in \mathfrak{H}} \mathfrak{C}_H$$

is also countable.

Now let us consider a simple closed curve $C \subset Q - S$ and two distinct points $a, b \in C$. Then C is the union of two simple arcs J_1 and J_2 with common ends a and b. Consider a plane $H \in \mathfrak{H}$ cutting E^3 between a and b. Then there exists a simple polygon $\hat{C} \in \mathfrak{C}_H$ separating G between the sets $J_1 \cap H$ and $J_2 \cap H$. It follows by (6) that $\bar{\lambda}(C, \hat{C}) > 0$. Thus 3° is proved.

Now let us order the polygons of the family $\mathfrak{C}$ in a sequence $\{C_i'\}$ such that each polygon $\hat{C} \in \mathfrak{C}$ appears in $\{C_i'\}$ an infinite number of times. Denote by A_ν' a broken anchor ring with core C_ν' and with radius $\eta_\nu < 1/\nu$. Evidently η_ν can be chosen so small that $A_\nu' \subset Q - S$. The proof of proposition (19) will be complete, if we shall show that for every simple closed curve $C \subset Q - S$ there exists an index ν such that $A_\nu' \subset Q - C$ and $\bar{\lambda}(C, C_\nu') > 0$.

In order to prove it, let us observe that, by the construction of the sequence $\{C_i'\}$ there exists an index i_0 such that $\bar{\lambda}(C, C_{i_0}') > 0$. Let ε denote the distance between C and C_{i_0}'. Since C_{i_0}' appears in the sequence $\{C_i'\}$ an infinite number of times, there exists an index i such that

$$C_i' = C_{i_0}' \quad \text{and} \quad \frac{1}{i} < \varepsilon .$$

Then the polygon C_i' satisfies the condition $\bar{\lambda}(C_i', C) > 0$. Moreover, if $x \in A_i$ then $\varrho(x, C_i') < 1/i < \varepsilon$, and consequently $A_i \subset Q - C$. Thus (19) is proved.

Let us observe that (19) implies:

(20) *For every neighborhood U_x of each point $x \in Q$ there exists an index ν such that $C_\nu' \cap U_x \neq 0$.*

In fact, there exists in U_x a disk $D \subset Q - S$. By (19), there exists an index ν such that the simple polygon C_ν' is linked with the boundary C of D. It follows $C_\nu' \cap D \neq 0$ and consequently also $C_\nu' \cap U_x \neq 0$.

Now let us prove:

(21) *There exists a sequence $\{A_i\}$ of broken anchor rings in $Q - S - \bigcup_{\nu=1}^{\infty} K_\nu$ and dense in Q such that for each $i = 1, 2, \ldots$ the inner radius of A_i is less than $1/i$ and there exists in A_i a wreath W_i substituting for A_i and such that $W_i \cap W_j = 0$ for $i \neq j$ and the diameters of the links of W_i are less than $1/i$.*

Proof. Consider the sequence $\{A_\nu'\}$ of broken anchor rings of (19). If we tried to prove (21) by shrinking the inner radii of the broken anchor rings of $\{A_\nu'\}$ and putting wreaths in these shrunken broken anchor rings, we would have to exercise care to insure that the wreaths did not intersect each other or the segments in $\{K_\nu\}$. We modify the cores of the broken anchor rings of $\{A_\nu'\}$ before shrinking them.

Let C_1' be the core of A_1'. At most a countable number of planes parallel to the plane containing C_1' contains a segment of $\{K_\nu\}$. We adjust C_1' to obtain a polygonal simple closed curve C_1 in the interior of A_1' such that C_1 lies in a plane, C_1 misses $\bigcup_{\nu=1}^{\infty} K_\nu$, and there is a map of an annulus into the interior of A_1' that takes the two boundary components of the annulus homeomorphically onto C_1' and C_1. We can obtain C_1 by first translating C_1' to a nearby plane H not containing any element of $\{K_\nu\}$ and then adjusting the image of C_1' in H so that it misses the 0-dimensional set consisting of the intersection of H with the segments of $\{K_\nu\}$. It follows from (5) that if C is a simple closed curve in $Q-A_1'$, then $\bar{\lambda}(C, C_1') = \bar{\lambda}(C, C_1)$.

Let A_1 be a broken anchor ring with core C_1 and inner radius less than 1 and so small that $A_1 \subset A_1'$ and $A_1 \cap \bigcup_{\nu=1}^{\infty} K_\nu = 0$. It follows from (16) that there exists a wreath W_1 substituting for the anchor ring A_1 such that the diameters of its links are less than 1.

Suppose the broken anchor rings $A_1, A_2, \ldots, A_k$ have been chosen so that for $i = 1, 2, \ldots, k$, $A_i \subset A_i' - \bigcup_{\nu=1}^{\infty} K_\nu$, the inner radius of A_i is less than $1/i$, each simple closed curve in $Q-A_i'$ that homologically links the core of A_i' also homologically links the core of A_i. Suppose that $W_1, W_2, \ldots, W_k$ have been chosen so that they substitute for the anchor rings $A_1, A_2, \ldots, A_k$ respectively, are mutually disjoint, and the diameter of the links of W_i are less than $1/i$. Suppose the core C_{i+1}' of A_{i+1}' lies in a plane H. Note that at most a countable number of planes parallel to H contain a non degenerate continuum that lies in $\bigcup_{i=1}^{k} W_i$ or any of $\{K_\nu\}$. Let C_{i+1} be a polygonal simple closed curve in the interior of A_{i+1} such that C_{i+1} lies in a plane, C_{i+1} misses $\bigcup_{\nu=1}^{\infty} K_\nu$, C_{i+1} misses each W_1, $W_2, \ldots, W_k$ and there is a map of an annulus into the interior of A_{k+1}' which takes the two boundary components of the annulus into C_{i+1}' and C_{i+1}. Then A_{i+1} is a broken anchor ring with core C_{i+1}' and inner radius less than $1/i+1$ and so small that $A_{i+1} \subset A_{i+1}'$, A_{i+1} misses each W_1, $W_2, \ldots, W_i$, and A_{i+1} misses $\bigcup_{\nu=1}^{\infty} K_\nu$. Then W_{i+1} is a wreath in A_{i+1} substituting for A_{i+1} such that the diameters of the links of W_{i+1} have diameters less than $1/i+1$. If we continue defining A_j's and W_j's in this way, the sequences $\{A_\nu\}$, $\{W_\nu\}$ satisfy (21).

It follows from a modified version of (20) applied to the broken anchor rings $\{A_\nu\}$ of (21) that for every neighborhood U_x of each point $x \in Q$, there are infinitely many indices i such that $A_i \cap U_x \neq 0$. Since each point of A_i lies within the distance $1/i$ of W_i we conclude.

(22) *Each open subset of Q contains a link of an element of $\{W_\nu\}$.*

8. The construction of the space Q^*. Now let us consider in the ball Q a sequence of chords $\{K_\nu\}$ satisfying (18), and sequences of anchor rings $\{A_\nu\}$ and wreaths $\{W_\nu\}$ satisfying (21). The chords K_ν

and the links of the wreaths W_ν constitute a countable family $\mathfrak{A}$ of disjoint arcs with diameters converging to zero. It follows that the decomposition $\mathfrak{M}$ of Q into arcs of the family $\mathfrak{A}$ and the individual points of Q not on one of these arcs is upper semicontinuous. Let Q^* denote the hyperspace of this decomposition and let

$$\varphi: Q \to Q^*$$

be the continuous map whose inverse sets $\varphi^{-1}(x)$, $x \in Q^*$, are exactly the elements of the decomposition $\mathfrak{M}$. It follows by a theorem of Lelek [12] that Q^* is an AR-set. In Section 9 we prove that $\dim Q^* = 3$ and in Section 11 we prove that Q^* does not contain any disk.

9. Dimension Q^*. Let us denote by P the subset of Q consisting of all individual points of the decomposition $\mathfrak{M}$. Then φ maps the set P homeomorphically onto a subset P^* of Q^*, and it maps the set $N = Q - P$ onto the countable set $N^* = \varphi(N) = Q^* - P^*$.

Since the set $N \cap S$ is countable (it consists of the ends of chords K_ν), there exists a simple closed curve $C \subset S - N$. Then S is the union of two disks D_1 and D_2 having C as their common boundary. Since φ is a homeomorphism on C, $C^* = \varphi(C)$ is a simple closed curve and this curve is homotopic to zero in each of the sets $D_1^* = \varphi(D_1)$ and $D_2^* = \varphi(D_2)$. Moreover we have:

$$C^* \subset D_1^* \cap D_2^* \subset C^* \cup N^*,$$

where N^* is a countable set. It follows that the set $S^* = D_1^* \cup D_2^* = \varphi(S)$ contains a 2-dimensional true cycle which is not homologous to zero in S^*, but it is homologous to zero in Q^* (because Q^* is an AR). Consequently (see [1], also p. 151 of [10]) we obtain:

$$\dim Q^* \geqslant 3 . \tag{23}$$

On the other hand, let us observe that (22) implies that the set $N \subset \bigcup_{\nu=1}^{\infty} W_\nu$ is dense in Q and consequently (see p. 44 of [10]) the dimension of the set $P = Q - N \subset Q$ is less than or equal to 2. It follows that the dimension of the set P^*, homeomorphic to P, is less than or equal to 2, and since $Q^* = P^* \cup N^*$, where N^* is countable, we conclude (see p. 32 of [10]) that $\dim Q^* \leqslant 3$. By virtue of (23) we get:

$$\dim Q^* = 3 . \tag{24}$$

10. Lemma on extension of maps. We now prove the following:

LEMMA. *Suppose we are given a closed* AR*-set Y_0 lying in a compact metric* ANR*-space Y. For every neighborhood V of Y_0 in Y there exists a neighborhood U of Y_0 in Y such that every map f_0 of a closed subset X_0 of an arbitrary metric space X into U can be extended to a map f: $X \to V$.*

Proof. We can suppose that Y is a subset of the Hilbert cube Q^ω, consisting of all points $z = (z_1, z_2, \ldots)$ of the Hilbert space E^ω satisfying the condition $0 \leqslant z_i \leqslant 1/i$. Since Y is an ANR, there exists a neighborhood G of Y in E^ω and a retraction

$$r\colon G \to Y$$

Evidently, there exists for G and r a positive number ε such that:

(25) *If $z \in E^\omega$ and $\varrho(z, Y) \leqslant \varepsilon$ then $z \in G$,*

(26) *If $z \in E^\omega$ and $\varrho(z, Y_0) \leqslant \varepsilon$ then $r(z) \in V$.*

Since Y_0 is an AR, there exists a retraction $r_0\colon E^\omega \to Y_0$. Then there is an open neighborhood U of Y_0 such that:

(27) *If $y \in U$ then $\varrho\big(y, r_0(y)\big) \leqslant \varepsilon$.*

Consider now a map $f_0\colon X_0 \to U$. Then $r_0 f_0\colon X_0 \to Y_0$ and consequently $r_0 f_0$ can be extended to a map $\alpha\colon X \to Y_0$. Since the values of f_0 and of $r_0 f_0$ lie in the Hilbert cube Q^ω, the formula

$$\beta_0(x) = f_0(x) - r_0 f_0(x) \quad \text{for every} \quad x \in X_0$$

defines a continuous function with values belonging (by (27)) to the set Z_ε consisting of all points $z = (z_1, z_2, \ldots) \in E^\omega$ such that $|z_i| \leqslant 2/i$ and $\varrho(z, 0) \leqslant \varepsilon$. Evidently Z_ε is an AR-set and consequently β_0 can be extended to a map

$$\beta\colon X \to Z_\varepsilon .$$

Now let us observe that the values of the map α lie in $Y_0 \subset Y$ and that $\varphi\big(\alpha(x) + \beta(x), \alpha(x)\big) = \varrho\big(\beta(x), 0\big) \leqslant \varepsilon$ for every $x \in X$. It follows by (25) that

$$\alpha(x) + \beta(x) \in G \quad \text{for every} \quad x \in X ,$$

and consequently the formula

$$f(x) = r\big(\alpha(x) + \beta(x)\big)$$

defines a continuous function. Since $\alpha(x) \in Y_0$ and $\beta(x) \in Z_\varepsilon$, we infer by (26) that the values of f belong to V. Hence

$$f\colon X \to V .$$

Moreover, if $x \in X_0$, then $f(x) = r\big(\alpha(x) + \beta(x)\big) = r\big(r_0 f_0(x) + f_0(x) - r_0 f_0(x)\big) = r f_0(x) = f_0(x)$, because $f_0(x) \in V \subset Y$. Thus f is an extension of f_0 and the lemma is proved.

11. Q^* does not contain any disk. Let us suppose that D^* is a disk lying in Q^*. Since the set $N^* \subset Q^*$ is countable, the disk D^* contains a subdisk with the boundary lying in the set $P^* = Q^* - N^*$. Consequently we may assume, without loss of generality, that the boundary C^* of D^* lies in P^*. Then $C = \varphi^{-1}(C^*)$ is a simple closed curve in Q and φ maps C onto C^* homeomorphically.

Now we shall consider two cases. First, let us suppose that $D^* \subset S^*$. Then S is the union of two disks D_1 and D_2 with C as their common boundary. Since the set N^* is countable, we infer that the set $D^*-C^*-N^*$ is arcwise connected, and consequently the set $\varphi^{-1}(D^*-C^*-N^*)$ is contained in one of the regions D_1-C and D_2-C. We can assume that $\varphi^{-1}(D^*-C^*-N^*) \subset D_1-C$. We now show that $D^*=\varphi(D_1)$ by showing that the assumption that there is a point p in $D_1-\varphi^{-1}(D^*)$ leads to an absurdity. First, we note that if D^* is given a definite metric, then for each positive integer n there is a simple closed curve J_n in D^*-N of diameter less than $1/n$ such that $\varphi^{-1}(J_n)$ separates p from D_2-C in S. We obtain J_n by considering a triangulation of D^* of mesh less than $1/n$ such that each 1-simplex of the triangulation misses N^* and find that the boundary of one 2-simplex of the triangulation serves as J_n. If $J_{n_1}, J_{n_2}, \ldots$ is a converging subsequence of $J_1, J_2, \ldots$, the limit of $\varphi^{-1}(J_{n_1}), \varphi^{-1}(J_{n_2}), \ldots$ separates p from D_2-C in S. However, this is impossible since this limit contains at most two points. Hence, we conclude that $D^*=\varphi(D_1)$. Now let us consider a chord K_ν of $\{K_\nu\}$ of (18) with ends a and b belonging to D_1. The set $D_1 \cap N$ is countable. Consequently there exists a simple arc J with ends a and b and with interior lying in D_1-C-N. Then $\varphi(J)$ is a simple closed curve lying in D^*-C^* and since D_1-J is arcwise connected, we infer that $D^*-\varphi(J)$ is connected. This contradiction disposes of the first case.

We now consider the second case, where D^* is not contained in S^*. It follows that D^* contains a subdisk lying in Q^*-S^*. Consequently we can assume that the given disk D^* lies in Q^*-S^*, and its boundary C^* lies in P^*. Then $C=\varphi^{-1}(C^*)$ is a simple closed curve lying in $Q-S$ and, by our construction, there exists in $Q-S$ a broken anchor ring $A \in \{A_\nu\}$ of (21) such that $C \subset Q-A$ and that C is linked with the core of A. Let $W \subset N$ be the wreath substituting for A. If $D^* \subset Q^*-N^*$, then $\varphi^{-1}(D^*)$ would be a disk in $Q-S-N \subset Q-W$ which is impossible by (15). Consequently every disk $D^* \subset Q^*-S^*$ intersects the countable set N^*-S^*. Let us order the points of $D^* \cap N^*$ into a sequence $\{a_n\}$ where $a_n \neq a_k$ for $n \neq k$. We can assume that $a_1, a_2, \ldots, a_m$ are the points of N^*-S^* of the form $a_i=\varphi(J_i)$, where J_i is a link of the wreath W. Let M be the chain substituting for A and $\hat{A}_i$ $(i=1,2,\ldots,m)$ be the link of M containing J_i. Let us observe that $\hat{A}_i \cap \hat{A}_j=0$ for $i \neq j$.

Now let us consider a positive number ε so small that

(28) $\varrho(x,y) > 2\varepsilon$ for every $x \in J_i$ and $y \in Q-\hat{A}_i$, for $i=1,2,\ldots,m$,

and let V_i denote the ε-neighborhood of J_i in Q. Then

(29) $\varrho(x,y) > \varepsilon$ for every $x \in V_i$ and $y \in Q-\hat{A}_i$, for $i=1,2,\ldots,m$.

By the lemma of Section 10, there exists a neighborhood $U_i \subset V_i$ of the arc J_i such that every map of a closed subset X_0 of a space X

into U_i can be extended to a map of X into V_i. Now let us consider, for $i = 1, 2, \ldots, m$ a disk $D_i^* \subset D^*$ containing in its interior the point a_i and such that its boundary C_i^* lies in $D^* - N^*$ and that $\varphi^{-1}(D^*) \subset U_i$. Since $U_i \subset V_i \subset \hat{A}_i$, we infer that

$$(30) \qquad D_i^* \cap D_j^* = 0 \quad \text{for} \quad i, j = 1, 2, \ldots, m, \ i \neq j.$$

It follows by (28) that, for $i = 1, 2, \ldots, m$:

(31) *The distance of every point of $\varphi^{-1}(D_i^*)$ from the boundary B_i of $\hat{A}_i$ is greater than ε.*

Moreover, by the definition of U_i, we infer that, for $i = 1, 2, \ldots, m$:

(32) *There exists a map ψ_i: $D_i^* \to V_i$ such that $\psi_i(x) = \varphi^{-1}(x)$ for every $x \in C_i^*$.*

Now suppose that, for some $k \geqslant m$, we have already defined the disks $D_1^*, D_2^*, \ldots, D_k^* \subset D^*$ such that $a_\nu \in D_\nu^*$ and that the boundary C_ν^* of D_ν^* lies in $D^* - N^*$ for $\nu = 1, 2, \ldots, k$. Suppose also that we have defined the neighborhoods $V_1, V_2, \ldots, V_k$ in Q of sets $\varphi^{-1}(a_1)$, $\varphi^{-1}(a_2), \ldots$, $\varphi^{-1}(a_k)$ respectively and also the maps $\psi_1, \psi_2, \ldots, \psi_k$ satisfying (32). Consider the point a_{k+1}. If $a_{k+1} \in \bigcup_{i=1}^{k} D_i^*$, then we choose an index $j \leqslant k$ such that $a_{k+1} \in D_j^*$ and we set $D_{k+1}^* = D_j^*$, $V_{k+1} = V_j$ and $\psi_{k+1} = \psi_j$. If, however, $a_{k+1} \in D^* - \bigcup_{i=1}^{k} D_i^*$, then we set $V_{k+1} = Q - W$ and we define D_{k+1}^* as a disk lying in $D^* - \bigcup_{i=1}^{k} D_i^*$ such that its boundary C_{k+1}^* lies in $D^* - N^*$ and that it is a neighborhood (in D^*) of the point a_{k+1} so small that the partial map φ^{-1}/C_{k+1}^* can be extended to a map ψ_{k+1}: $D_{k+1}^* \to V_{k+1}$, for which the diameter of the set $\psi_{k+1}(D_{k+1}^*)$ is less than twice the diameter of $\varphi^{-1}(a_{k+1})$. By the lemma of Section 10 this is always possible.

Thus we have a sequence of disks $\{D_n^*\}$ and a sequence of maps $\{\psi_n\}$ such that:

(33) *If $n \neq n'$ then either $D_n^* \cap D_{n'}^* = 0$, or $D_n^* = D_{n'}^*$ and $\psi_n = \psi_{n'}$.*

(34) *ψ_n is an extension of φ^{-1}/C_n^*.*

(35) *For every $\varepsilon < 0$ there exists only a finite number of disks D_n^* such that the diameters of $\psi_n(D_n^*)$ are $\geqslant \varepsilon$.*

(36) *If a_n does not belong to $D_1^* \cup D_2^* \cup \ldots \cup D_m^*$ then $\psi_n(D^*) \subset Q - W$.*

Now let us set

$$(37) \qquad g(x) = \begin{cases} \varphi^{-1}(x) & \text{for every} \quad x \in D^* - \bigcup_{i=1}^{\infty} D_i^*, \\ \psi_i(x) & \text{for every} \quad x \in D_i^*, \ i = 1, 2, \ldots \end{cases}$$

It follows from (33) that g is a function g: $D^* \to Q$ and from (34) and (35) that this function is continuous. Moreover, we infer from (36) that

$$(38) \qquad g^{-1}(J_i) \subset D_i^* \quad \text{for} \quad i = 1, 2, \ldots, m.$$

Applying (17), we infer that there exists an index $i_0 \leqslant m$ such that the disk D^* contains two disjoint disks D' and D'' with boundaries C' and C'' such that the loops $g|C'$ and $g|C''$ are linked with the core C_{i_0} of a link $\hat{A}_{i_0}$ of the chain M and that $g(C')$ and $g(C'')$ are subsets of the ε-neighborhood of the boundary B_{i_0} of the link $\hat{A}_{i_0}$.

It follows by (15) that if J_{i_0} is the link of the wreath W lying in A_{i_0} then

$$g(D') \cap J_{i_0} \neq 0 \neq g(D'') \cap J_{i_0}.$$

By virtue of (38), we conclude that

$$D' \cap D^*_{i_0} \neq 0 \neq D'' \cap D^*_{i_0}.$$

Since the disks D' and D'' are disjoint, we infer that the disk $D^*_{i_0}$ intersects the boundary of D', and consequently $D^*_{i_0} \cap C' \neq 0$, hence also $g(D^*_{i_0}) \cap g(C') \neq 0$. But this is impossible, by (28) and (29), because the points of $g(D^*_{i_0})$ belong to the ε-neighborhood of J_{i_0} and the points of $g(C') = \varphi^{-1}(C')$ belong to the ε-neighborhood of B_{i_0}.

Thus the supposition that Q^* contains a disk leads to a contradiction.

Remark. It follows by our construction that the hyperspace X of the decomposition of the set $Q-S$ whose non degenerate elements coincide with the links of the wreaths W_ν, $\nu = 1, 2, \ldots$, does not contain any disk. By a generalization of the theorem of Andrews and Curtis [2] obtained recently by D. S. Gillman and J. M. Martin [9] the Cartesian product of E^1 by the decomposition-space of E^n into points and arcs, with only a countable number of arcs, is homeomorphic with E^{n+1}. It follows that $E^1 \times X$ is topologically E^4 and thus we get a 3-dimensional Cartesian divisor of E^4 which does not contain any disk.

12. Problems. The following questions remain open:

1° Is it true that the space Q^* (constructed in Section 8) does not contain any 2-dimensional AR? More exactly, is it true that the first Betti number of every 2-dimensional subset of Q^* is infinite?

2° Does there exist an AR-space of an arbitrarily given dimension which does not contain any disk?

3° Does there exist a Cartesian divisor of a Euclidean cube such that its dimension is greater than or equal to 2 but it does not contain any disk?

References

[1] P. Alexandroff, *Dimensionstheorie*, Math. Annalen 106 (1932), pp. 161-238.

[2] J. J. Andrews and M. L. Curtis, *n-space modulo an arc*, Annals of Math. 75 (1962), pp. 1-7.

[3] L. Antoine, *Sur l'homéomorphie de figures et de leurs voisinages*, Journ. de Math. 86 (1921), pp. 221-325.

[4] K. Borsuk, *Sur les rétractes*, Fund. Math. 17 (1931), pp. 152-170.

[5] — *On an irreducible 2-dimensional absolute retract*, ibidem 37 (1950), pp. 137-160.

[6] — *On a family of 2-dimensional AR-sets*, ibidem 51 (1962), pp. 283-297.

[7] — *Theorie der Retrakte, und ihre Problems*, Izwiestija na matiematiczeskaja institut 4 (1959), pp. 45-54.

[8] R. P. Coelho, *On the groups of certain linkages*, Portugaliae Mathematica 6 (1947), pp. 57-65.

[9] D. S. Gillman and J. M. Martin, *Countable decompositions of E^3 into point and pointlike arcs*, Amer. Math. Soc. Notices 10 (1963), pp. 74-75.

[10] W. Hurewicz and H. Wallman, *Dimension theory*, Princeton (1941), pp. 1-165.

[11] S. Lefschetz, *Algebraic topology*, Amer. Math. Soc. Coll. Publ. 27 (1942), pp. 1-389.

[12] A. Lelek, *Some mappings of ANR-sets*, Fund. Math. 52 (1963), pp. 225-233.

[13] G. T. Whyburn, *Concerning the proposition that every closed, compact, and totally disconnected set of points is a subset of an arc*, ibidem 18 (1932), pp. 47-60.

Reçu par la Rédaction le 16. 1. 1963

Some Remarks Concerning Topologically Homogeneous Spaces

By R. H. BING and K. BORSUK

1. Homogeneous and locally homogeneous spaces

A space X is said to be *homogeneous* (topologically) provided that, for every two points $x_1, x_2 \in X$, there exists a homeomorphism h mapping X onto itself and satisfying the condition $h(x_1) = x_2$. Evidently this condition implies that X is *locally homogeneous* [8]; that is, for every two points $x_1, x_2 \in X$, there exists a homeomorphism h mapping a neighborhood U_1 of x_1 onto a neighborhood $U_2 = h(U_1)$ of x_2 and satisfying the condition $h(x_1) = x_2$. Some authors use the term *microhomogeneous* introduced by D. van Dantzig [7], instead of the term *locally homogeneous.*

Linear spaces and connected manifolds are two important classes of homogeneous spaces. Other examples of compact homogeneous spaces are the Cantor's discontinuum, the universal curve of Sierpinski [2], the pseudo-arc [3], the solenoids of van Dantzig [8], and the Hilbert cube [10]. The problem if only manifolds among the connected ANR-spaces of a finite dimension are homogeneous, however, still remains open. In the present note we make a contribution to this problem. It may be noted that some analogous questions, involving a little stronger notion of homogeneity, were studied by D. Montgomery [11].

2. Infinite cycles in metric spaces

We will consider only metric spaces. It will be convenient to recall some homological notions concerning these spaces.

By an *n-dimensional ε-simplex* in a metric space X, we mean a system $\sigma = (x_0, x_1, \cdots, x_n)$ consisting of $n + 1$ points (vertices) of X (not necessarily different) in a definite order, satisfying the condition $\rho(x_i, x_j) \leqq \varepsilon$ for $i, j = 0, 1, \cdots, n$. If $i_0, i_1, \cdots, i_n$ is a permutation of indices $0, 1, \cdots, n$, then the simplex $\sigma' = (x_{i_0}, x_{i_1}, \cdots, x_{i_n})$ is considered as equal to σ if the permutation is even, and as equal to $=\sigma$, if the permutation is odd. The finite linear form

$$\kappa - a_1\sigma_1 + a_2\sigma_2 + \cdots + a_k\sigma_k ,$$

where σ_i are n-dimensional ε-simplexes in X, and the coefficients a_i are elements of an abelian group $\mathfrak{A}$ is said to be an *n-dimensional ε-chain in X over* $\mathfrak{A}$. As usual, one introduces the operation ∂ of the boundary, assigning to every n-

The first author is supported by NSF grant GP-2244.

dimensional ε-chain κ in X over $\mathfrak{A}$ an $(n-1)$-dimensional ε-chain $\partial\kappa$ in X over $\mathfrak{A}$. The ε-chains with vanishing boundary are said to be *ε-cycles in X.*

By an *n-dimensional infinite chain* $\boldsymbol{\kappa}$ in X we understand a sequence $\boldsymbol{\kappa} = \{\kappa_i\}$ such that:

1. κ_i is an n-dimensional ε_i-chain in X with coefficients belonging to an abelian group $\mathfrak{A}_i$ (depending on i) and with $\lim_{i\to\infty} \varepsilon_i = 0$.

2. There exists a compact subset X_0 of X such that the vertices of all chains κ_i belong to X_0. The compactum X_0 is said to be a *carrier* of the infinite chain $\boldsymbol{\kappa}$.

If for two n-dimensional infinite chains $\boldsymbol{\kappa} = \{\kappa_i\}$ and $\boldsymbol{\kappa}' = \{\kappa_i'\}$, the coefficients of κ_i and of κ_i' belong to the same group $\mathfrak{A}_i$, for every $i = 1, 2, \cdots$, then the sequence $\{\kappa_i \pm \kappa_i'\}$ is an n-dimensional infinite chain, which we denote by $\boldsymbol{\kappa} \pm \boldsymbol{\kappa}'$.

An infinite chain $\boldsymbol{\gamma} = \{\gamma_i\}$ for which every chain γ_i is an n-dimensional cycle in X is said to be an *n-dimensional infinite cycle in X.* Two n-dimensional infinite cycles $\boldsymbol{\gamma} = \{\gamma_i\}$ and $\boldsymbol{\gamma}' = \{\gamma_i'\}$ are said to be *homologous in X* (in symbols: $\boldsymbol{\gamma} \sim \boldsymbol{\gamma}'$ in X) provided for every $i = 1, 2, \cdots$ the groups of coefficients for γ_i and for γ_i' are the same, and there exists in X an infinite $(n+1)$-dimensional chain $\boldsymbol{\kappa} = \{\kappa_i\}$ such that $\partial\kappa_i = \gamma_i - \gamma_i'$ for $i = 1, 2, \cdots$. Then we write $\partial\boldsymbol{\kappa} = \boldsymbol{\gamma} - \boldsymbol{\gamma}'$.

In particular, $\boldsymbol{\gamma} \sim 0$ in X denotes that there exists in X an infinite $(n+1)$-dimensional chain $\boldsymbol{\kappa} = \{\kappa_i\}$ such that $\partial\kappa_i = \gamma_i$ for $i = 1, 2, \cdots$. An infinite cycle $\boldsymbol{\gamma}$ in X is said to be *essential* if there exists a carrier $X_0 \subset X$ of it such that $\boldsymbol{\gamma}$ is not homologous to zero in X_0.

If some n-dimensional infinite cycle in X is not homologous to 0 in X, then X is said to be *cyclic in dimension n.* If every n-dimensional infinite cycle in X is homologous to zero in X, then X is said to be *acyclic in dimension n.*

Let f be a continuous map of space X into another space Y. Assigning to each simplex $\sigma = (x_0, x_1, \cdots, x_n)$ of X the simplex $f(\sigma) = (f(x_0), f(x_1), \cdots, f(x_n))$ of Y, and to every chain $\kappa = a_1\sigma_1 + a_2\sigma_2 + \cdots + a_k\sigma_k$ of X the chain $f(\kappa) = a_1 f(\sigma_1) + a_2 f(\sigma_2) + \cdots + a_k f(\sigma_k)$ of Y, we get a correspondence by which to every infinite n-dimensional chain $\boldsymbol{\kappa} = \{\kappa_i\}$ in X corresponds an infinite n-dimensional chain $f(\boldsymbol{\kappa}) = \{f(\kappa_i)\}$ in Y. It is clear that by this correspondence to infinite cycles in X correspond infinite cycles in Y, and to homologous infinite cycles — homologous infinite cycles.

We assume as known the following propositions:

1. If some n-dimensional essential infinite cycle in X is homologous to 0 in X, then $\dim X > n$ [1].

2. If X is the union of two closed subsets X_1, X_2 and if an n-dimensional infinite cycle $\boldsymbol{\gamma}$ lying in $X_1 \cap X_2$ is not homologous to zero in the set $X_1 \cap X_2$, but it is homologous to zero in each of the sets X_1 and X_2, then there exists in X an $(n+1)$-dimensional infinite cycle $\boldsymbol{\gamma}'$ not homologous to 0 in X. Actually, if $\boldsymbol{\gamma} = \partial\boldsymbol{\kappa}_1 = \partial\boldsymbol{\kappa}_2$, where $\boldsymbol{\kappa}_j$ is an infinite $(n+1)$-dimensional chain in X_j, for $j = 1, 2$, then it suffices to set $\boldsymbol{\gamma}' = \boldsymbol{\kappa}_1 - \boldsymbol{\kappa}_2$ in order to obtain an $(n+1)$-dimensional infinite cycle not homologous to zero in X [1 and 4]. This proposition is quoted sometimes as the *Phragmen-Brouwer theorem.*

3. Let h_0 and h_1 denote the homeomorphisms of a space A onto subsets $A \times (0)$ and $A \times (1)$ of the cartesian product $A \times \langle 0, 1\rangle$ given by the formulas:

$$h_0(x) = (x, 0)\ , \qquad h_1(x) = (x, 1) \qquad\qquad \text{for every point } x \in A.$$

If $\boldsymbol{\kappa}$ is an infinite chain in A and if the infinite cycle $\partial\boldsymbol{\kappa}$ lies in a subset B of A, then there exists in the set $B \times \langle 0, 1\rangle$ an infinite chain $\boldsymbol{\lambda}$ such that

$$\partial\boldsymbol{\lambda} = \partial h_0(\boldsymbol{\kappa}) - \partial h_1(\boldsymbol{\kappa})\ .$$

Moreover, the infinite cycle $h_0(\boldsymbol{\kappa}) - h_1(\boldsymbol{\kappa}) - \boldsymbol{\lambda}$ is homologous to 0 in the set $A \times \langle 0, 1\rangle$ [1].

3. Locally contractible spaces

A space X is said to be *locally contractible* if, for every point $a \in X$ and for every neighborhood U of a, there exists a neighborhood V of a and a continuous map f: $V \times \langle 0,1\rangle \rightarrow U$ such that $f(x, 0) = x$ and $f(x, 1) = a$ for every point $x \in V$. By a *neighborhood of a point,* we mean an arbitrary set (not necessarily either open or closed) which contains the point on its interior.

Let us recall the following property of locally contractible metric spaces [5]:

4. For every continuous map f of a finite-dimensional compactum A into a locally contractible metric space X, and for every $\varepsilon > 0$, there exists an $\eta > 0$ such that for every continuous map f': $A \rightarrow X$ satisfying the condition $\rho(f, f') < \eta$ there exists a continuous map φ: $A \times \langle 0, 1\rangle \rightarrow X$ such that $\varphi(x, 0) = f(x)$, $\varphi(x, 1) = f'(x)$ and $\rho(\varphi(x, t), f(x)) < \varepsilon$ for every point $x \in X$ and $0 \leqq t \leqq 1$.

For brevity, the locally contractible, separable, metric spaces will be called lcsm-*spaces.* In particular, every compact ANR-space is an lcsm-space.

Instead of working with locally contractible spaces, we could just as well deal with those whose local homology is trivial.

4. Membranes and their maps into lcsm-spaces

Let B be a closed subset of a compactum A. If there exists in B an infinite cycle $\boldsymbol{\gamma}$ homologous to zero in A, but not homologous to zero in any subset of

A which does not contain $A - B$, then A will be said to be a *membrane for* γ *spanned on* B. Applying the Brouwer reduction theorem [9, p. 161] one finds that

5. If an infinite cycle γ lying in a compactum $B \subset X$ is homologous to zero in X, then there exists a membrane $A \subset X$ for γ spanned on B.

Moreover, one sees that

6. If A is a membrane for an n-dimensional infinite cycle γ spanned on a compactum $B \subset A$ and if $G \neq 0$ is an open subset of $A - B$, then there exists in the set $B' = \bar{G} \cap (A - G)$ an n-dimensional infinite cycle γ' such that $\gamma \sim \gamma'$ in $A - G$, and the set $A' = \bar{G}$ is a membrane for γ' spanned on B'.

Applying the Phragmen-Brouwer theorem, we infer easily that

7. If A_1 and A_2 are two distinct membranes for an infinite $(n - 1)$-dimensional cycle γ spanned on a compactum $B \subset A_1 \cap A_2$, then the set $A_1 \cup A_2$ is cyclic in dimension n.

Let A be a membrane for an infinite cycle γ spanned on a compactum B, and let f be a continuous map of A into a space X. A point $x_0 \in X$ is said to be *essential for the map* f (with respect to γ and B) if $x_0 \in f(A) - f(B)$, and if the infinite cycle $f(\gamma)$ is not homologous to 0 in the set $f(A) - \{x_0\}$.

THEOREM 4.1. *Let A be a membrane for an infinite $(n-1)$-dimensional cycle γ spanned on a compactum B, and let X_0 be a subset of the second category (of Baire) of an* lcsm-*space X. Suppose for every point $x \in X_0$ there exists a continuous map f: $A \to X$ satisfying two conditions*:

(a) *x is essential for f (with respect to γ and B).*

(b) *x does not belong to the interior of the set $f(A)$ (in the space X). Then* $\dim X > n$.

PROOF. Let us suppose that $\dim X \leqq n$, and let φ denote the subset of the functional space X^A consisting of all maps f such that for a point $x = x_f \in X_0$ the conditions (a) and (b) are satisfied. Since the space X is separable, the functional space X^A is also separable, and we infer that there exists in φ a sequence of maps $\{f_k\}$ dense in φ.

Let $\{G_i\}$ be a sequence of open sets in $A - B$ such that for each compact set $C \subset A - B$ and for each neighborhood U of C (in $A - B$) there is an index j such that $C \subset G_j \subset \bar{G}_j \subset U$. Following 6 above, we find that there exists in the set $\bar{G}_j - G_j$ an $(n - 1)$-dimensional infinite cycle γ_j such that $\gamma \sim \gamma_j$ in $A - G_j$ and that $\bar{G}_j$ is a membrane for γ_j spanned on the set $\bar{G}_j - G_j$. Since $\gamma_j \sim 0$ in $\bar{G}_j$, the infinite cycle $f_i(\gamma_j)$ is homologous to zero in the set $f_i(\bar{G}_j)$ for every $i = 1, 2, \cdots$. It follows that there exists for $f_i(\gamma_j)$ a membrane $M_{i,j} \subset f_i(\bar{G}_j)$ spanned on the set $f_i(\bar{G}_j - G_j)$. Let us consider the boundary $F_{i,j}$

of the membrane $M_{i,j}$ in X, that is the set

$$F_{i,j} = M_{i,j} \cap \overline{X - M_{i,j}} \,. \tag{1}$$

Since none of the sets $F_{i,j}$ contains a non-empty open subset of X, the set $\bigcup_{i,j=1}^{\infty} F_{i,j}$ is of the first category in X, and we infer that there exists a point

$$x_0 \in X_0 - \bigcup_{i,j=1}^{\infty} F_{i,j} \,. \tag{2}$$

By the local contractibility of X, there is a neighborhood V_0 of the point x_0 contractible in X.

The hypothesis of the theorem says that there is a map $f_0 \in \varphi$ such that the point x_0 is essential for f_0 (with respect to γ and B). Then $x_0 \in f_0(A) - f_0(B)$. Since the set $f_0^{-1}(V_0 \cap f_0(A - B))$ is a neighborhood in $(A - B)$ of the compact set $f_0^{-1}(x_0) \subset A - B$, we infer that there is an index j_0 such that

$$f_0^{-1}(x_0) \subset G_{j_0} \subset \bar{G}_{j_0} \subset f_0^{-1}(V_0 \cap f_0(A - B)) \,. \tag{3}$$

Moreover, for every $\varepsilon > 0$, there exists an index i_0 such that the distance between f_0 and f_{i_0} is less than ε. Since $f_0(\bar{G}_{j_0} - G_{j_0}) \subset V_0 - \{x_0\}$, and since X is locally contractible and $\dim X \leq n$, we infer from 4 of § 3 that there is an index i_0 such that $f_{i_0}(\bar{G}_{j_0}) \subset V_0$, and there is a map

$$g\colon (\bar{G}_{j_0} - G_{j_0}) \times \langle 0, 1 \rangle \longrightarrow V_0 - \{x_0\} \tag{4}$$

such that $g(x, 0) = f_0(x)$ and $g(x, 1) = f_{i_0}(x)$ for every $x \in \bar{G}_{j_0} - G_{j_0}$. It follows that the compactum

$$D = g[(\bar{G}_{j_0} - G_{j_0}) \times \langle 0, 1 \rangle] \subset V_0 - \{x_0\}$$

satisfies the condition

$$f_0(\gamma_{j_0}) \sim f_{i_0}(\gamma_{j_0}) \qquad \text{in } D. \tag{5}$$

We now consider two cases:

Case 1. $x_0 \in X - M_{i_0, j_0}$. Since $f_{i_0}(\gamma_{j_0}) \sim 0$ in M_{i_0, j_0}, we infer by (5) that

$$f_0(\gamma_{j_0}) \sim 0 \qquad \text{in } M_{i_0, j_0} \cup D \,.$$

It follows that there exists for the infinite cycle $f_0(\gamma_{j_0})$ a membrane M spanned on the set $f_0(\bar{G}_{j_0} - G_{j_0})$ and contained in the set $M_{i_0, j_0} \cup D$. Then the point x_0 does not belong to M. On the other hand,

$$f_0(\gamma_{j_0}) \sim 0 \qquad \text{in } f_0(\bar{G}_{j_0}) \,,$$

and since the point x_0 is essential for f_0, there exists in the set $f_0(\bar{G}_{j_0}) \subset V_0$ a membrane M' for $f_0(\gamma_{j_0})$ spanned on the set $f_0(\bar{G}_{j_0} - G_{j_0})$. Let us observe that this membrane contains the point x_0, because x_0 is essential for f_0 (with respect to γ and B) and $f_0(\gamma) \sim f_0(\gamma_{j_0})$ in $f_0(A - G_{j_0}) \subset A - \{x_0\}$. Thus we have shown that there exist in the set V_0 two distinct membranes M and M' for the infinite

$(n-1)$-dimensional cycle $f_0(\gamma_{j_0})$ not homologous to 0 in $f_0(\bar{G}_{j_0}-G_{j_0})$, spanned on the same set $f_0(\bar{G}_{j_0}-G_{j_0})$. We infer by 7 of § 4, that the set $M\cup M'\subset V_0$ carries an n-dimensional infinite cycle γ' such that γ' is not homologous to zero in $M\cup M'$. Since V_0 is contractible in X, we infer that $\gamma'\sim 0$ in X. It follows by 1 of § 2 that $\dim X>n$.

Case 2. $x_0\in M_{i_0,j_0}$. In this case the infinite cycle $f_{i_0}(\gamma_{j_0})$ is homologous to 0 in the set $f_0(\bar{G}_{j_0})\cup D\subset V_0$, and we infer that there is a membrane $M'\subset f_0(\bar{G}_{j_0})\cup D$ for the infinite cycle $f_{i_0}(\gamma_{j_0})$ spanned on the set $f_{i_0}(\bar{G}_{j_0}-G_{j_0})$. Since the point x_0 belongs neither to the interior of the set $f_0(\bar{G}_{j_0})\subset f_0(A)$ nor to the compact set D, we infer that x_0 does not belong to the interior (in the space X) of the membrane M'. On the other hand, the point x_0 belongs in our case to M_{i_0,j_0}, but it does not belong to the set-theoretical boundary F_{i_0,j_0} (in the space X) of the membrane M_{i_0,j_0}. It follows that the membrane M_{i_0,j_0} for the infinite cycle $f_{i_0}(\gamma_{j_0})$ (which is not homologous to zero in the set $f_{i_0}(\bar{G}_{j_0}-G_{i_0})$), spanned on the set $f_{i_0}(\bar{G}_{j_0}-G_{j_0})$ contains x_0 in its interior, and consequently it is distinct from the membrane M'. Thus we have in the set V_0 two distinct membranes M_{i_0,j_0} and M' for the infinite cycle $f_{i_0}(\gamma_{j_0})$, spanned on the same set $f_{i_0}(\bar{G}_{j_0}-G_{j_0})$, in which $f_{i_0}(\gamma_{j_0})$ is not homologous to zero. As in the *Case* 1, we conclude that $\dim X>n$.

Thus the proof of Theorem 4.1 is concluded.

5. Membranes in lcsm-spaces

Let us formulate some corollaries, which follow by Theorem 4.

COROLLARY 5.1. *Let A be a membrane for an $(n-1)$-dimensional infinite cycle γ spanned on a set B, and let X_0 be a subset of the second category (of Baire) in an* lcsm-*space X. If for every point $x\in X_0$ there exists a homeomorphism h_x of A onto a subset $h_x(A)$ of X such that $x\in h_x(A-B)$, but $h_x(A)$ is not a neighborhood of x (in space X), then* $\dim X>n$.

Since the euclidean n-ball (closed) Q^n is a membrane for an $(n-1)$-dimensional infinite cycle spanned on the boundary $\dot{Q}^n$ of this ball, we get the following

COROLLARY 5.2. *If X_0 is a subset of the second category (of Baire) of an* lcsm-*space X such that for every point $x\in X_0$ there exists in X an n-ball Q_x such that $x\in Q_x-\dot{Q}_x$, but Q_x is not a neighborhood of x in the space X, then* $\dim X>n$.

By an *n-dimensional umbrella* we understand the union of an n-ball Q and of a simple arc L such that the set $Q\cap L$ consists of only one point a lying in the set $Q-\dot{Q}$ and being an endpoint of L. The point a will be called the *center* of this umbrella.

It is evident that if an n-ball Q is a subset of a locally contractible space (in particular, of an lcsm-space), and if Q does not contain any open (non-empty) subset of X, then there exists in X a simple arc L such that $Q \cup L$ is an n-dimensional umbrella.

COROLLARY 5.3 ("Umbrella Theorem"). *In an n-dimensional* lcsm-*space X the set of all centers of n-dimensional umbrellas contained in X is of the first category (of Baire).*

PROOF. Let us suppose that the subset X_0 of X consisting of all centers of n-dimensional umbrellas lying in X is of the second category in X. Then there exists, for every point $x \in X_0$, an n-dimensional umbrella $T_x = Q_x \cup L_x \subset X$ with center x. Then the point x belongs to the set $Q_x - \dot{Q}_x$, but Q_x is not a neighborhood of x in the space X. We infer by Corollary 5.2 that $\dim X > n$, contrary to our hypothesis.

REMARK. Another *umbrella theorem*, concerning families of disjoint n-dimensional umbrellas in the euclidean $(n+1)$-space, is given by Gail S. Young [13] as a generalization of Moore's Theorem on simple triods.

6. Locally homogeneous lcsm-spaces and the manifolds

Now let us prove the following

THEOREM 6.1. *An n-dimensional connected* lcsm-*space of the second category is an n-dimensional manifold if and only if it is locally homogeneous and it contains an n-ball.*

PROOF. The necessity of the condition is obvious. Now let us assume that X is a locally homogeneous n-dimensional lcsm-space of the second category which contains an n-ball Q_0. If a point $x_0 \in Q_0 - \dot{Q}_0$ belongs to the closure of the set $X - Q_0$, then one sees at once (by the local contractibility of X) that there exists in X an n-dimensional umbrella. Since X is locally homogeneous, we infer that every point x of X is the center of an n-dimensional umbrella contained in X. But this is impossible by the *umbrella theorem*, because $\dim X = n$. It follows that Q_0 is a neighborhood in the space X of each point $x_0 \in Q_0 - \dot{Q}_0$. Since X is locally homogeneous, we infer that every point $x \in X$ has a neighborhood in X homeomorphic to the euclidean n-space. Since X is connected, we conclude that X is an n-dimensional manifold.

REMARK. We do not know whether every n-dimensional, connected, locally homogeneous lcsm-space is an n-dimensional manifold. In the sequel we will prove it only for locally compact spaces of dimensions $n = 0, 1, 2$. However, some remarkable results concerming characterizations of generalized manifolds (in the sense of R. L. Wilder [12]) obtained some years ago by P. E. Conner and E. E. Floyd [7] suggests that each such is a generalized manifold. As yet this problem remains open.

7. Separation by points

We shall use the following theorem in the next section in establishing Corollary 8.2.

THEOREM 7.1. *Suppose X is a locally connected separable metric space of dimension more than* 1. *Then some point of X fails to locally separate X.*

PROOF. With no loss of generality we suppose that X is connected.

Let $\{x_i\}$ be a sequence of points dense in X. For each pair of distinct points x_j, x_k of $\{x_i\}$, let $U(j, k)$ be a connected open subset of X containing $(x_j) \cup (x_k)$ and such that for any connected open subset U' of X containing $(x_j) \cup (x_k)$, diameter of $U(j, k)$ is less than twice diameter of U'. Let

$$L(j, k) = \{y : y = x_j, x_k, \text{ or } y \text{ serarates } x_j \text{ from } x_k \text{ in } U(j, k)\} .$$

It may be shown that $L(x_j, x_k)$ is homeomorphic to a closed subset of an arc. If $x_j = x_k$, we let $L(j, k) = (x_j)$ Since X has dimension greater than 1, there is a point x_0 of the set $X - \bigcup_{j\,k=1}^{\infty} L(j, k)$. We show that x_0 does not locally separate X.

Assume that U is a connected neighborhood of x_0 such that $U - (x_0)$ is the union of mutually separated non-empty sets U_1, U_2. Let ε be a positive number such that $\rho(x_0, X - U) > 3\varepsilon$ and V be a connected neighborhood of x_0 of diameter less than ε. There are points x_r, x_s of $\{x_i\}$ in $V \cap U_1$, $V \cap U_2$ respectively and $L(r, s) \subset U$. Then $x_0 \in L(r, s)$. The assumption that x_0 locally separates X led to the contradiction that the point x_0 of the set $X - \bigcup_{j,k=1}^{\infty} L(j, k)$ was a point of $L(r, s)$.

COROLLARY 7.2. *No point separates a connected, locally connected, locally homogeneous separable metric space of dimension more than* 1.

Example 7.3. Theorem 7.1 is not true without the assumption that X is separable. Consider an example obtained by erecting a perpendicular line at each point of a plane P. The distance between points on different perpendiculars is taken to be the sum of the distance between the feet of the perpendiculars and the distance of the points to P. Similarly, one can get an example to show that Corollary 7.2 is not true without the restriction on separability. To get such an example, one could start with a 2-sphere, put a tangent 2-sphere on at each point of it, put on more tangent 2-spheres at points of these secondary 2-spheres, and continue to add 2-spheres.

8. Disconnection of locally homogeneous spaces by compacta

Now let us prove the following

THEOREM 8.1. *Let X be a metric separable space satisfying the following conditions*:

1. $\dim X = n$.
2. *X is locally compact.*
3. *X is locally contractible.*
4. *X is locally homogeneous.*

Then every compact subset B of X cyclic in dimension $n-1$ and contractible in a proper subset of X separates X.

PROOF. Let γ be an $(n-1)$-dimensional infinite cycle in B which is not homologous to zero in B. Since B is contractible in a proper subset X' of X, the infinite cycle γ is homologous to zero in X'. Then 5 of § 4 implies that there exists for γ a membrane $A \subset X'$ spanned on B.

In order to prove our theorem, it suffices to show that B separates X between every point of the set $A - B$ and every point of the set $X - A \supset X - X' \neq \emptyset$. Otherwise there exists a point

$$a \in (A - B) \cap \overline{X - A}\ .$$

Let us denote, for every point $x \in X$ and for every $\varepsilon > 0$, by $Q(x, \varepsilon)$ the closed ball in X with center x and radius ε; that is,

$$Q(X, \varepsilon) = \{x' \in X; \rho(x, x') \leqq \varepsilon\}\ .$$

It follows from Condition 3 that there exists an $\varepsilon_1 > 0$ such that for every $0 < \varepsilon \leqq \varepsilon_1$ the ball $Q(a, \varepsilon)$ is compact. Let ε_0 be a positive number less than ε_1 and also less than the distance $\rho(a, B)$. It follows by 6 of § 4 that for every $0 < \varepsilon \leqq \varepsilon_0$ the compactum

$$B(\varepsilon) = Q(a, \varepsilon) \cap \overline{A - Q(a, \varepsilon)}$$

contains an $(n-1)$-dimensional infinite cycle $\gamma(\varepsilon)$ such that the set

$$A(\varepsilon) = A \cap Q(a, \varepsilon)$$

is a membrane for the cycle $\gamma(\varepsilon)$ spanned on the set $B(\varepsilon)$ and that $\gamma(\varepsilon)$ is homologous to γ in the set $A - (A(\varepsilon) - B(\varepsilon))$.

Let us observe that the set

$$\big(A(\varepsilon) - B(\varepsilon)\big) \cap \overline{X - A(\varepsilon)} = \bigcup_{\nu=1}^{\infty} \left(A\left(\frac{\nu\varepsilon}{\nu + 1}\right) \cap \overline{X - A(\varepsilon)}\right)$$

is of the first category (of Baire) in the ball $Q(a, \varepsilon)$, because

$$A\left(\frac{\nu\varepsilon}{\nu+1}\right) \cap \overline{X - A(\varepsilon)} = A\left(\frac{\nu\varepsilon}{\nu+1}\right) \cap \overline{Q(a, \varepsilon) - A(\varepsilon)}$$

$$\subset A\left(\frac{\nu\varepsilon}{\nu+1}\right) \cap \overline{Q(a, \varepsilon) - A\left(\frac{\nu\varepsilon}{\nu+1}\right)}$$

is a compact subset of $Q(a, \varepsilon)$ and its interior is empty.

Since ε_0 is less than the distance $\rho(a, B)$, the set $B(\varepsilon)$ is contained in $A-B$ for every $0 < \varepsilon \leqq \varepsilon_0$ and, as we have already noticed, there exists in the set $B(\varepsilon)$ an $(n-1)$-dimensional infinite cycle $\boldsymbol{\gamma}(\varepsilon)$ homologous to $\boldsymbol{\gamma}$ in the set $A-(A(\varepsilon)-B(\varepsilon))$ and such that $A(\varepsilon)$ is a membrane for $\boldsymbol{\gamma}(\varepsilon)$ spanned on $B(\varepsilon)$. In particular, it is so for $\varepsilon = \varepsilon_0/k$ for every $k = 1, 2, \cdots$. Since the space consisting of all homeomorphisms h of the compactum $Q(a, \varepsilon)$ into X, such that $h(Q(a, \varepsilon))$ is a neighborhoood of the point $h(a)$ in X is separable, there exists, for every $k = 1, 2, \cdots$, a sequence $h_{k,1}, h_{k,2}, \cdots$ of homeomorphisms

$$h_{k,i}\colon Q(a, \varepsilon_0/k) \longrightarrow X$$

dense in this space.

As we have seen, the set $(A(\varepsilon_0/k) - B(\varepsilon_0/k)) \cap \overline{X - A(\varepsilon_0/k)}$ is of the first category (of Baire) in $Q(a, \varepsilon_0/k)$ and we infer that the set

$$h_{k,i}[(A(\varepsilon_0/k) - B(\varepsilon_0/k)) \cap \overline{X - A(\varepsilon_0/k)}]$$

is of the first category (of Baire) in $h_{k,i}(Q(a, \varepsilon_0))$, and consequently also in X. It follows that there exists a point

$$a_0 \in X - \bigcup_{k,i=1}^{\infty} h_{k,i}[(A(\varepsilon_0/k) - B(\varepsilon_0/k)) \cap \overline{X - A(\varepsilon_0/k)}]\,.$$

Since X is locally contractible, there exists a positive number η_0 such that the ball $Q(a_0, \eta_0)$ is a compactum contractible in X. Since X is locally homogeneous, there exists a natural number k_0 and a homeomorphism h_0 mapping $Q(a, \varepsilon_0/k_0)$ onto a closed neighborhood U_0 of the point a_0 such that

$$h_0(a) = a_0 \quad \text{and} \quad U_0 = h_0(Q(a, \varepsilon_0/k_0)) \subset Q\left(a_0, \frac{1}{2}\eta_0\right).$$

Since h_0 is a homeomorphism, the point $a_0 = h_0(a)$ does not belong to $h_0(B(\varepsilon_0/k_0))$ and consequently there exists a positive number η_1 less than $\rho(a_0, h_0(B(\varepsilon_0/k_0)))$ and also less than $\eta_0/2$.

It follows, by the definition of the sequence $h_{k_0,i}$ and by the hypothesis that X is locally contractible, that there exists an index i_0 such that the set $h_{k_0,i_0}(Q(a, \varepsilon_0/k_0))$ is contained in the set $Q(a_0, \eta_0)$ and that the infinite cycle $h_0(\boldsymbol{\gamma}(\varepsilon_0/k_0))$ is homologous to the infinite cycle $h_{k_0,i_0}(\boldsymbol{\gamma}(\varepsilon_0/k_0))$ in the compactum

$$C = \{x \in X\colon \rho(x, h_0(B(\varepsilon_0/k_0))) \leqq \eta_1\} \subset Q(a_0, \eta_0) - \{a_0\}\,.$$

We infer that the $(n-1)$-dimensional infinite cycle $h_0(\boldsymbol{\gamma}(\varepsilon_0/k_0))$ is homologous to 0 in each of the closed subsets $h_0(A(\varepsilon_0/k_0))$ and $h_{k_0,i_0}(A(\varepsilon_0/k_0)) \cup C$ of the compactum $Q(a_0, \eta_0)$.

We now finish the proof of Theorem 8.1 in much the same way that we finished the proof of Theorem 4.1 by breaking the proof into cases.

Case 1. $a_0 \notin h_{k_0,i_0}(A(\varepsilon_0/k_0))$. In this case there is a membrane M for the infinite cycle $h_0(\boldsymbol{\gamma}(\varepsilon_0/k_0))$ spanned on the set $h_0(B(\varepsilon_0/k_0))$ such that $M \subset C \cup h_{k_0,i_0}(A(\varepsilon_0/k_0))$. The membrane M is different from $h_0(A(\varepsilon_0/k_0))$ since $a_0 \in h_0(A(\varepsilon_0/k_0))$ but $a_0 \notin M$. It follows from 7 of § 4 that $M \cup h_{k_0,i_0}(A(\varepsilon_0/k_0))$ carries an n-dimensional cycle $\boldsymbol{\gamma}'$ such that $\boldsymbol{\gamma}'$ is not homologous to zero on $M \cup h_{k_0,i_0}(A(\varepsilon_0/k_0))$. Since $M \cup h_{k_0,i_0}(A(\varepsilon_0/k_0)) \subset Q(a_0, \eta_0)$ is contractible in X, then $\boldsymbol{\gamma}' \sim 0$ in X and it follows from 1 of § 2 that dim $X > n$ in this case.

Case 2. $a_0 \in h_{k_0,i_0}(A(\varepsilon_0/k_0))$. In this case $h_{k_0,i_0}(A(\varepsilon_0/k_0))$ is a neighborhood of a_0 since $a_0 \notin h_{k_0,i_0}[A(\varepsilon_0/k_0) - B(\varepsilon_0/k_0) - \overline{X - A(\varepsilon_0/k_0)}]$ and $a_0 \notin h_{k_0,i_0}(B(\varepsilon_0/k_0))$. There is a membrane M' for the infinite cycle $h_{k_0,i_0}(\boldsymbol{\gamma}(\varepsilon_0/k_0))$ spanned on $h_{k_0,i_0}(B(\varepsilon_0/k_0))$ such that $M' \subset C \cup h_0(A(\varepsilon_0/k_0))$. Then $M' \neq h_{k_0,i_0}(A(\varepsilon_0/k_0))$ since M' contains no neighborhood of a_0 but $h_{k_0,i_0}(A(\varepsilon_0/k_0))$ does. Then $M' \cup h_{k_0,i_0}(A(\varepsilon_0/k_0))$ carries an n-dimensional infinite cycle that is not homologous to zero in $M' \cup h_{k_0,i_0}(A(\varepsilon_0/k_0))$ but which is homologous to zero in X. Again it follows from 1 of § 2 that dim $X > n$.

COROLLARY 8.2. *For dimensions $n = 0, 1, 2$ every connected, n-dimensional, locally compact, locally contractible and locally homogeneous space X is a manifold.*

If $n = 0$, the space X consists of only one point. If $n = 1$, X contains a simple arc. Since X is locally compact, it is of the second category. By Theorem 6.1, X is a manifold. In the case $n = 2$, X is a locally compact and locally contractible space which is (by Theorem 7.1) not locally separated by any of its points. Moreover, it follows by Theorem 8.1 that for each point x of X there is a neighborhood U of x such that every simple closed curve in U separates X. As was shown by Gail Young [14], X is a manifold.

COROLLARY 8.3. *The 2-dimensional compact manifolds are the same as 2-dimensional locally homogeneous compact* ANR*-spaces.*

In particular it follows

COROLLARY 8.4. *No 2-dimensional compact* AR*-space is locally homogeneous.*

Problem 1. Let X be a compactum satisfying Conditions 1–4 of the last theorem. Is it true that:

(1) X is cyclic in dimension n,

(2) no compact subset of X, acyclic in dimension $n-1$, separates X?

Problem 2. Does there exist a locally homogeneous compact 3-dimensional AR-space?

UNIVERSITY OF WISCONSIN
POLISH ACADEMY OF SCIENCE

REFERENCES

1. P. S. ALEXANDROFF, *Dimensionstheorie. Ein Beitrag zur Geometrie der abgeschlossenen Mengen*, Math. Ann., 106 (1932), 161-238.
2. R. D. ANDERSON, *A characterization of the universal curve and a proof of its homogeneity*, Ann. of Math., 67 (1958), 313-324.
3. R. H. BING, *A homogeneous indecomposable plane continuum*, Duke Math. J., 15 (1948), 729-742.
4. K. BORSUK, *Über sphäroidale und h-sphäroidale Räume*, Recueil Math. Moscou, 1 (43) (1936), 543-560.
5. ———, ————, Bull. de l'Ac. Pol. des Sc.
6. H. J. COHEN, *Some results concerning homogeneous plane continua*, Duke Math. J., 18 (1951), 467-474.
7. P. E. CONNER and E. E. FLOYD, *A characterisation of general manifolds*, Michigan Math. J., 6 (1959) 33-43.
8. D. VAN DANTZIG, *Über topologisch homogene Kontinua*, Fund. Math., 15 (1930), 102-125.
9. W. HUREWICZ and H. WALLMAN, Dimension Theory, Princeton 1941.
10. O. H. KELLER, *Die Homöomorphie der kompakten konvexen Mengen im Hilbertschen Raum*, Math. Ann., 105 (1931), 748-758.
11. D. MONTGOMERY, *Locally homogeneous, spaces*, Ann. of Math., 51 (1950), 261-271.
12. R. L. WILDER, Topology of Manifolds, Amer. Math. Soc. Colloquium Publications, Vol. 32, 1949.
13. GAIL S. YOUNG, *A generalization of Moore's theorem on simple triods*, Bull. Amer. Math. Soc., 50 (1944), 714.
14. ———, *Characterization of 2-manifolds*, Duke Math. J., 14 (1947). 979-990.

(Received April 13, 1964)

RETRACTIONS ONTO ANR'S[1]

R. H. BING

The following theorem is proved in [1].

THEOREM 3 OF [1]. *Suppose X is a closed subset of a compact n-manifold M such that X is an i-connected ($i=1, 2, \cdots, n-2$) ANR. Then for each point p of a component U of $M-X$ there is a retraction of $U+X-\{p\}$ onto X.*

It was stated that Theorem 3 could be generalized by dropping the condition that M is compact and that if $\overline{U}$ is not compact, it is not even necessary to remove a point from U. The proof of this extended result is elementary but was not included in [1] since it was thought to be beyond the interests of most readers of the Monthly. However, the extended result has applications [2] so it was decided to publish a proof.

RETRACTION THEOREM. *Suppose M is an n-manifold and X is an i-connected ($i=0, 1, 2, \cdots, n-2$) closed subset of M that is an ANR. For each component U of $M-X$ whose closure is not compact, there is a retraction of $U+X$ onto X.*

PROOF. Let N be an open subset of M containing X such that there is a retraction of $\overline{N}$ onto X.

Let $B_1, B_2, \cdots$ be a locally finite collection of topological n-balls in U such that $U \subset \overline{N} + \sum B_i$. Let $p_i \in \text{Int } B_i - \sum \text{Bd } B_j$ and $P = \sum \{p_i\}$. Also, let $N' = \overline{N} - \sum \text{Int } B_i$, and $B_i' = B_i - \sum_{i+1}^{\infty} \text{Int } B_j$.

Let r_0 be a retraction of N' onto X. The retraction r_0 of N' onto X can be extended inductively to a retraction r_i of $N' + \sum_1^i B_j' - P$ onto X. Once r_i is defined, r_{i+1} is extended as follows. Let f_{i+1} be a retraction of $B_{i+1} - P$ onto Bd B_{i+1} and g_{i+1} be a map of Bd B_{i+1} into X that agrees with r_i on Bd $B_{i+1} \cap (N' + \sum_1^i B_j')$. The connectedness of X is used in showing this extension is possible. Then on $B_{i+1}' - P$, $r_{i+1} = g_{i+1} \circ f_{i+1}$. The limit of the r_i's provide a retraction r of $X+U-P$ onto X.

Let $R_1, R_2, \cdots$ be a locally finite collection of mutually exclusive topological rays in U such that each R_i is closed in M and p_i is the end of R_i. One way to get such a collection is to consider an expanding bull's-eye sequence $C_1, C_2, \cdots$ of compact sets as described later

Received by the editors August 24, 1968.

[1] Research supported by NSF Grant GP-8501.

with the additional condition that each Bd $C_i \cap P = 0$. Then the R_i's may be chosen to move monotonically out through the C_j's so that if $p_i \notin C_j$, then $R_i \cap C_j = 0$ and if $p_i \in C_j$, then $R_i \cap C_j$ is an arc from p_i to a point of Bd C_j at which Bd C_j is locally an $n-1$ manifold.

In each neighborhood of each R_i there is a closed subset E_i of M such that E_i is homeomorphic to closed Euclidean half space and $p_i \in \text{Int } E_i$. To obtain E_i we can consider the discrete sequence $p_i = p_1^i, p_2^i, \cdots$ of points of R_i such that each p_{j+1}^i is further out R_i than p_j^i and each arc $p_j^i p_{j+1}^i$ of R_i lies in a ball in U. Then E_i results from putting a small ball about $p_i = p_1^i$ so that the ball misses $R_i - p_1^i p_2^i$, pushing a feeler from the ball so that the feeler contains p_2^i and is very near $p_1^i p_2^i$ (so near in fact that it misses $R_i - p_1^i p_3^i$), elongating the feeler until it contains p_3^i with the elongation being very close to $p_2^i p_3^i, \cdots$. Hence using the R_i's as guides to pull a boundary point of balls about the ends of the R_i's out to infinity, one finds that there is a locally finite collection $E_1, E_i, \cdots$ of mutually exclusive sets in U such that each E_i is closed in M, each E_i is homeomorphic to closed Euclidean half space, and $p_i \in \text{Int } E_i$. By retracting each of the E_i's onto its boundary, we find that there is a retraction r' of $X+U$ onto $X+U-\sum \text{Int } E_i$.

A retraction promised by the retraction theorem is given by $r \circ r'$.

The retraction theorem applies to manifolds with boundaries. If U intersects Bd M, we do not need to assume that its closure is not compact. It also applies to pseudo manifold and other spaces but we shall not push it to full generality. However, we do describe a sequence of compact sets whose existence is of interest for its own sake.

EXPANDING BULL'S-EYE SEQUENCE. *If M is an n-manifold and X is a closed subset of M that is an ANR, then there is a sequence of compact sets $C_1, C_2, \cdots$ such that*

(1) $\sum C_i = M$,

(2) $C_i \subset \text{Int} C_{i+1}$,

(3) Bd $C_i - X$ *is locally an $n-1$ manifold except at a set of dimension $n-2$, and*

(4) *each component of $(M-X)-C_i$ which is not a component of $M-X$ has a noncompact closure.*

It is easy to get a sequence $C_1, C_2, \cdots$ satisfying conditions (1) and (2) above and such that each C_i is the union of a finite number of balls. These C_i's would then satisfy condition (3). If C_1 does not satisfy condition (4) it could be enlarged by adding the closure of the union of all components of $(M-X)-C_1$ that intersect C_2 and have compact closures. The following lemma implies that this enlargement is compact, where in applying the lemma we are considering $X-C_1$

as an ANR in the manifold $M-C_1$. By enlarging the other C_i's, a sequence satisfying conditions (1), (2), (3), (4) can be obtained.

LEMMA. *Suppose M is a manifold and X is a closed subset of M that is an ANR. Then no compact subset of M intersects infinitely many components of $M-X$.*

PROOF. Assume the lemma is false and p_0 is a point in the compact set such that each neighborhood of p_0 intersects infinitely many components of $M-X$. Suppose the metric of M is locally Euclidean near p_0 and B, B', B'' are balls centered at p_0 with radii 3ϵ, 2ϵ, ϵ, respectively, and having the ordinary Euclidean metric. Let r be a retraction of a neighborhood N of X onto X which does not move any point more than ϵ and V be a component of $(M-X)$ such that $B'\cap V\subset N$ and there is a point $p_1\in V\cap B''$. We are led to the contradiction that there is a map $f: B'\rightarrow \text{Bd } B'$ such that $f|\text{Bd } B'$ is homotopic to the identity on Bd B'. (Of course, there is no such map f since there is no retraction of a ball onto its boundary.) To define the impossible map f of B' onto Bd B' we would let r_1 be the projection from p_1 of $B-\{p_1\}$ onto Bd B' and define $f=r_1$ on $B'-V$ and $f=r_1\circ r$ on $B'\cap V$.

REFERENCES

1. R. H. Bing, *Retractions onto spheres*, Amer. Math. Monthly **71** (1964), 482–484.

2. Robert F. Brown, *A fixed point theorem for open Q-acyclic n-manifolds*, (to appear).

UNIVERSITY OF WISCONSIN

C. Others

General topology. By W. Sierpiński. Trans. by C. C. Krieger. (Mathematical Expositions, No. 7.) University of Toronto Press, 1952. 12+290 pp. $6.00.

This book retains many of the characteristics that brought popularity to the author's earlier book, *Introduction to general topology*. However, it has a different axiomatic treatment and is much enlarged so that it could be classified as more than a revision. Readers who have objected to the corruption of the vocabulary and notation of topology by those who have undertaken to rewrite it may find it refreshing to find the way in which such expressions as compact, bicompact, basis, $+$, $\sum$, $\prod$ are used.

Chapter I treats Fréchet (V) spaces, that is, spaces in which each point is associated with one or more subsets called neighborhoods but these neighborhoods need not satisfy any additional conditions. Topological spaces are studied in Chapter II when it is supposed that the neighborhoods satisfy additional conditions. It is of interest that in the cartesian product of topological spaces, a neighborhood is defined to be the cartesian product of neighborhoods. Topological spaces with a countable basis are considered in Chapter III, while Chapter IV treats Hausdorff spaces with such a basis. Chapter V deals with the properties normality and regularity.

An extended treatment of metric spaces is given in Chapters VI and VII. A large part of the study of topology is devoted to the important metric spaces and Sierpiński devotes over one-half of his book to them. In the study of complete metric spaces in the last chapter, particular attention is given to various kinds of their subsets, such as those that are F_σ, G_δ, Borel, or analytic sets. The treatment is more extensive than that in the corresponding chapter of the *Introduction*. The Appendix is much the same as in the *Introduction*.

It is logically satisfying to give a sequence of axioms and definitions and prove theorems on the basis of these without appealing to the intuition. However, such a sequence of axioms may omit some of the spaces that are of interest to topologists. Nevertheless, this book deals with many of the spaces usually studied. A feature that makes it especially suitable as a text is the collection of problems and theorems left as exercises in each chapter.

R. H. BING

ELEMENTARY POINT SET TOPOLOGY

By

R. H. BING

Professor of Mathematics

The University of Wisconsin

The Eighth

HERBERT ELLSWORTH SLAUGHT

MEMORIAL PAPER

Published as a supplement to the American Mathematical Monthly

Volume 67 Number 7

August-September, 1960

TABLE OF CONTENTS

1. INTRODUCTION

Topology as an offshoot of geometry. In many respects, topology may be considered an offshoot of geometry. It is a study of properties that are somewhat geometric. Just as it would be difficult to define geometry to a person unacquainted with such terms as points, lines, triangles, and planes, it is difficult to give an understandable definition of topology without using some of the terminology of topology. We delay giving a definition of topology until later.

A person may approach the study of topology in somewhat the same spirit that a student begins the study of synthetic Euclidean plane geometry. The student brings into his geometry class a rich background of information about the Euclidean plane. There is a listing of certain properties of the plane. This list is called a set of *postulates* or *axioms*. An effort is made to make the list so complete that the theorems about the plane can be proved on the basis of the postulates alone without making any recourse whatever to our intuitive knowledge about the plane. We strive for efficiency in our list of postulates by eliminating any postulate that can easily be proved from the remaining postulates.

The study of topology frequently starts with a set of postulates or axioms. In a later section we list some axioms that can start us on a study of topology. In plane geometry we use the plane as a model to suggest the postulates. The plane also acts as a model to suggest our beginning axioms for topology. However, a line also makes a good model for these axioms. There are other models as suggested in a later section *Some examples of topological spaces.*

In geometry we try to give explicit definitions of the terms we use. This is in addition to the descriptive discussion accompanied by examples that convey additional information. Explicit *definitions* describe the thing being defined in terms of things whose meaning is already assumed to be understood. This is different from the definitions as given in dictionaries where a "mesh-type" of definition is given and there is no clear understanding as to which term is defined first, but enough synonyms are included so that a person who knows the meaning of many other words is likely to have enough information to decipher the meaning of the word under consideration. In mathematics, definitions are given in a "tree-like" fashion where a term is defined by using previously accepted terms. We try to avoid the "circular" lines of definitions found in dictionaries.

The basic terms that we start with are not defined in terms of previously defined terms. These are our *undefined terms.* If there is to be a beginning, there must be some term used first. Perhaps a more compelling reason for having undefined terms is that although we may have an intuitive feeling as to the meaning of a basic term under consideration, any meaningful formulation of its meaning involves more complicated terms. For example, the fuzzy definition "A point is that which has position but no dimension" may be criticized on the basis that "position" and "dimension" were used. A similar criticism may be made of the definition "A point is the intersection of two nonparallel lines." In topology we use "point" as an undefined term. In the last section of this article we mention another advantage of undefined terms in topology—namely, if a term is left undefined, any meaning can be assigned to it and as long as the axioms are still satisfied, the theorems we have proved on the basis of the axioms hold true with this meaning. This results in richer interpretations of our theorems.

The set of all points under consideration is called *space.* If we are studying plane geometry, we take the Euclidean plane E^2 to be space. In studying solid geometry, we take Euclidean 3-space E^3 to be space. In topology it is customary to take a set of axioms that will hold in many different spaces and the theorems we prove will hold true in each of the spaces. We suppose that the reader started this chapter with a knowledge of the Euclidean line E^1, the Euclidean plane E^2, and Euclidean 3-space E^3. One section of this article is devoted to a listing of various other spaces.

In geometry we prove theorems on the basis of some axioms, definitions, and undefined terms. We do the same thing in topology—although we use different axioms, undefined terms, and definitions. This schematic diagram shows the axiomatic approach.

$$\left.\begin{array}{l}\text{Undefined terms}\\ \quad\text{(point, neighborhood, . . .)}\\ \text{Definitions}\\ \quad\text{(limit point, closed, topologically equivalent, . . .)}\\ \text{Axioms}\end{array}\right\}\Rightarrow\text{Theorems}$$

Before considering a set of axioms for topology, we shall indulge in an intuitive treatment of several aspects of topology. Our treatment contains some rubber-sheet geometry that may be of interest in spite of the fact that it does not add much to our knowledge of topology. We hope that by mentioning the axiomatic approach in this introduction and by returning to it after a brief sojourn to enrich our topological background in which some topological oddities are mentioned, we leave the impression that the true spirit of topology lies in proving theorems rather than in the oddities.

To point out things that are studied in topology but not in geometry we consider the simple closed curves shown in Figure 1. Geometrically they are

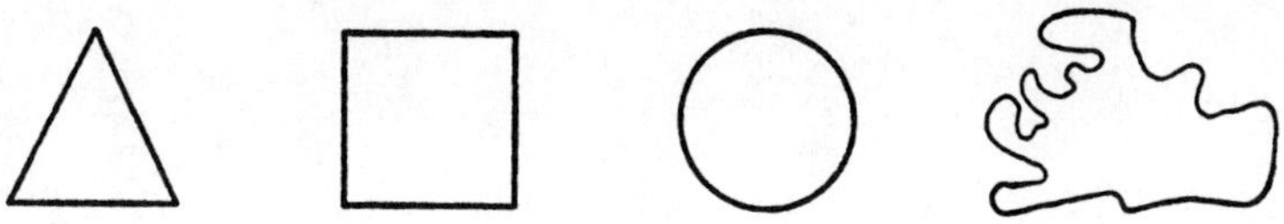

FIG. 1

different. However they are topologically equivalent. In a course in topology we study properties common to such figures. We list three topological properties of a simple closed curve.

1. A simple closed curve is connected. (We give a definition later of "connected" but our intuitive notion of "being in one piece" may suffice here.)

2. If a point is removed from a simple closed curve, the remaining part is connected. (Although a person could not physically remove a point from an object, we can do the operation mentally.)

3. If two points are removed from a simple closed curve, the remaining part is not connected but is in two pieces. (These pieces are called "components" of the remainder.)

Exercise 1. If three points are removed from a simple closed curve, how many components does the remainder have? Into how many pieces does n points separate a simple closed curve?

In some respects there are as many, if not more, simple closed curves in the plane of the fourth type shown in Figure 1 as there are triangles, rectangles, and circles. In fact, if there were some way of selecting a simple closed curve at random from the plane, one might consider it odd indeed if the one chosen even contained a straight line segment or the arc of a circle. It does not require much imagination to see how a person who was seeking to learn about all kinds of plane figures would be led to a study of simple closed curves.

Exercises are scattered throughout the article so that the reader who wishes to do more than skim can exercise. Answers are found at the end of the article. Much of this article resembles the chapter entitled *Point Set Topology* from *Insights into Modern Mathematics*, 23rd yearbook of the National Council of Teachers of Mathematics, and is used with permission.

Perhaps you have already felt a need for better explanations for the meaning of such terms as simple closed curve, topologically equivalent, connected, components, separate. It is hoped that the following sections will partially fill this need. The list of terms found at the end of this article may help the reader in locating definitions.

2. TOPOLOGICAL EQUIVALENCE

We mentioned in the first section of this article that a triangle and a "wiggly" simple closed curve are alike in some respects. We say that they are topologically equivalent. A circle, a square, an ellipse, and a triangle are all called simple closed curves.

Topological equivalence. Two sets A, B are topologically equivalent if there is a 1-1 correspondence between them that is continuous both ways.

The curved arc A and the straight line segment B shown in Figure 2 are topologically equivalent. Although there are many 1-1 correspondences that could be chosen, the projection of A onto B provides a convenient one—that is, a point p of A corresponds to the point $f(p)$ of B directly beneath it.

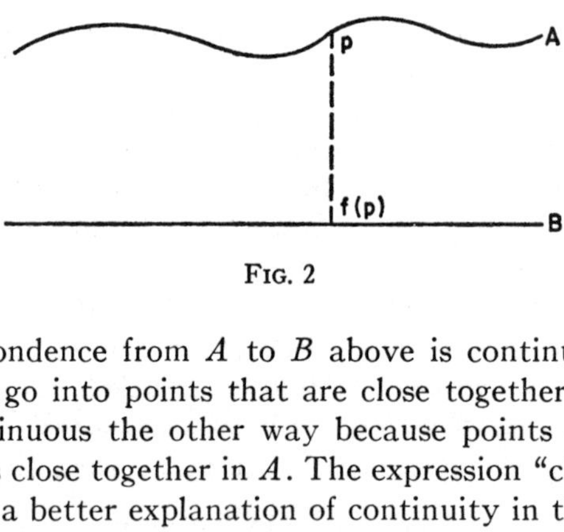

FIG. 2

The 1-1 correspondence from A to B above is continuous because points close together in A go into points that are close together in B. The 1-1 correspondence is continuous the other way because points close together in B correspond to points close together in A. The expression "close together" is not precise and we give a better explanation of continuity in the section *Transformations.*

A 1-1 correspondence of a set A onto a set B that is continuous both ways is called a *homeomorphism* of A onto B. The projection of A onto B is an example of a homeomorphism of A onto B. A homeomorphism of one set onto another set is a special type of transformation and will be discussed further in later sections.

Two sets are *topologically equivalent* if there is a homeomorphism of one onto the other. Accordingly, instead of saying that the two sets are topologically

4

equivalent, we may say that they are *homeomorphic.* Each set is topologically equivalent or homeomorphic to itself because each point of the set may be made to correspond to itself.

It is possible to get a homeomorphism of a short segment I_1 onto a long segment I_2. (See Fig. 3.) The homeomorphism this time is given by a projection from a point. This homeomorphism shows that, in a certain sense, there are as many points on a short segment as on a long segment.

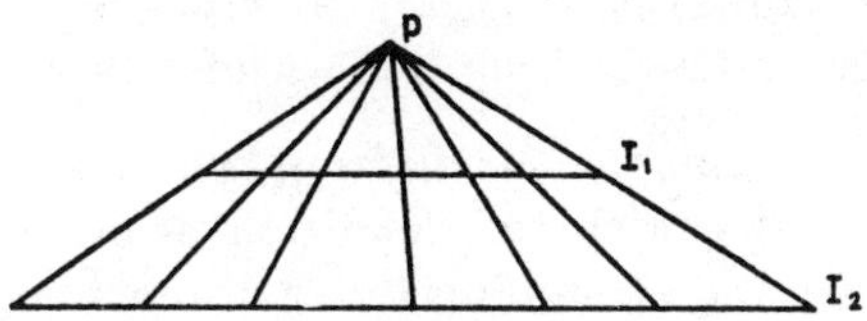

FIG. 3

We mention some other objects that are topologically equivalent or homeomorphic—the surface of a tetrahedron and the surface of a ball; a teacup and a doughnut; the x-axis and the graph of $y=x^2$; the half-open segment* $(0, 1]$ and the graph of $y=1/x$ $(0<x\leqq 1)$. (We suppose that a segment contains its endpoints unless we state otherwise.)

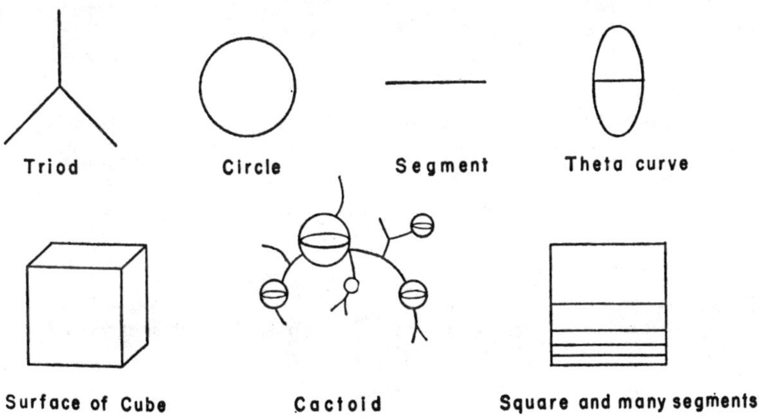

FIG. 4

In Figure 4 we show some objects no two of which are homeomorphic. For example, the circle is not topologically equivalent to the segment because no point of it could be made to correspond (in the proper way) to the end of the segment.

A misconception about topological equivalence. One might suppose that if two sets are topologically equivalent it is possible to deform one onto the other

* By "half-open segment" is meant a segment which contains only one of its endpoints. For example, the half-open segment $(0, 1]$ contains the point 1 but not 0.

by "pulling and stretching but without breaking and tearing." We give some counterexamples to show that this is not always possible.

In E^3 (Euclidean 3-space—the familiar three-dimensional space of everyday experience) consider two sets each consisting of two tangent spheres; in the first set the spheres are tangent externally and in the second set one sphere lies on the interior (except for the point of tangency) of the other. The two sets are topologically equivalent because there is a 1-1 correspondence of the proper sort between the two sets. However, it is not possible to deform one set onto the other in E^3 by "pulling and stretching but without breaking and tearing."

In the next section *Examples of surfaces*, we give other counterexamples to the faulty "pulling and stretching" definition. If one figure can be deformed onto another by pulling and stretching but without breaking and tearing, the figures are topologically equivalent; the converse of this statement is not true. We repeat that two sets are topologically equivalent if there is a 1-1 correspondence between them that is continuous both ways. The definition makes no mention of pulling stretching, and deforming.

Exercise 2. Which pairs of the objects in Figure 5 are topologically equivalent?

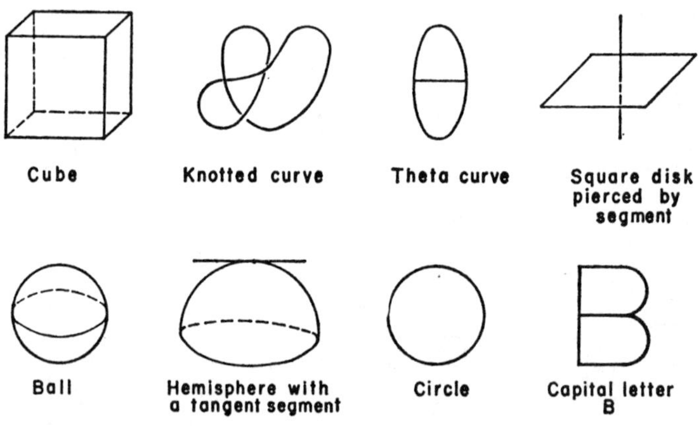

FIG. 5

Exercise 3. Figure 6 shows the capital letters as they may appear without serifs. Group the letters so that letters in the same group are topologically equivalent. Is the grouping different if the letters have width?

FIG. 6

Exercise 4. In setting up a 1-1 correspondence between the objects in Figures 7a and 7b to show they are topologically equivalent there are four of the points c, d, e, f, g, h that would necessarily correspond to particular points in Figure 7b, there is one of the six points for which there would be two choices of points in Figure 7b, and there would be another of the six for which there would be many choices. Which points correspond to points c, d, e, f, g, h?

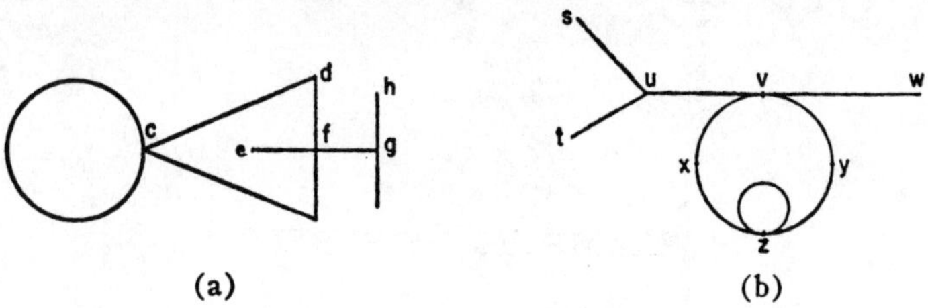

FIG. 7

3. EXAMPLES OF SURFACES

In this section we give examples of surfaces, some of which are familiar. These examples provide more counterexamples to the faulty "pulling and stretching" definition of topologically equivalent discussed in the preceding section.

Bands. Consider rectangular bands as shown in Figure 8. The arrows in Figure 8a indicate that the side AB of the rectangle is to be "sewed" to CD

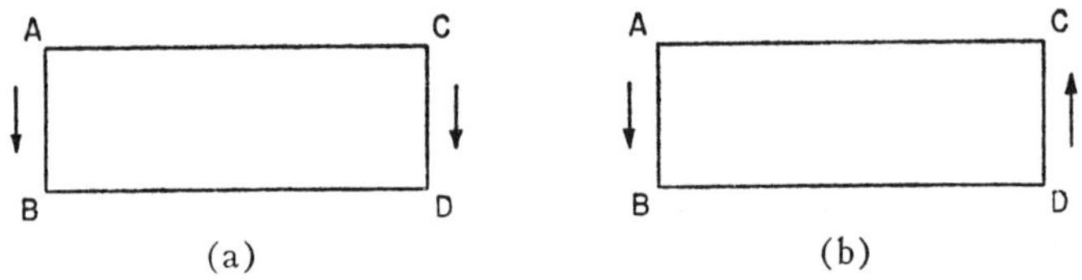

FIG. 8

with A coinciding with C and B with D. Perhaps a cylinder results as shown in Figure 9a but there is no assurance that before the sewing is done, several twists are not made as, say, in Figure 9b, where a full twist is first made. Although the bands in Figure 9a and 9b are topologically equivalent, neither can be obtained from the other by "pulling and stretching without breaking and tearing."

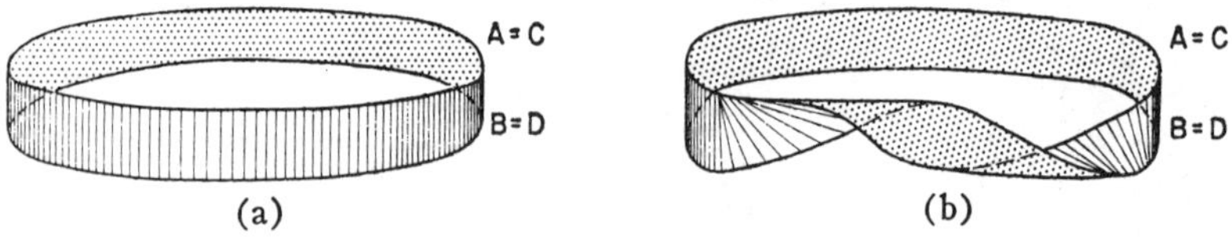

FIG. 9

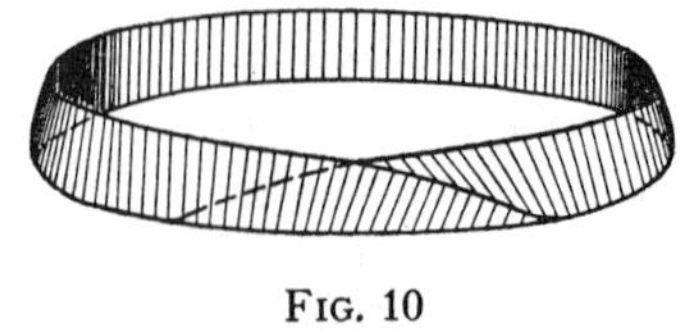

FIG. 10

8

The arrows in Figure 8b indicate that AB is attached to CD with A coinciding with D and B with C. The resulting band is called a Moebius band. The Moebius band shown in Figure 10 could not be deformed into one with more twists even though the two would be topologically equivalent.

The Moebius band is an example of a one-sided surface. If a fly were to walk around the band an odd number of times, he would be on the "other side" (if there were two sides). Machinery belts are sometimes made into Moebius bands so that they will not wear out on only one side and remain good on the other—they only have one side.

If Moebius bands are split, the results may be unexpected. If it is split down the middle, there results a two-sided band with twists. If this twisted two sided band is split down the middle, there results linking twisted two-sided bands.

The Moebius band has been the butt of jokes and stories. We list several.

1. There are stories of people being given the job of painting one side of a Moebius band and removing all the paint off the other side. See "Alexander Botts and the Moebius Strip," a story by Wm. H. Upson in the December 22, 1945, issue of the *Saturday Evening Post.*

2. One lecturer told the story of how a certain witch doctor kept a supply of bands in the hut in which he gave advice to couples wondering if they were suited to each other. If he wished to prophesy a break-up in their proposed union, he would split an untwisted two-sided band and the two pieces would fall apart. If he wished to prophesy that they would stay together, he would split a two-sided band with a full twist. If he wished to prophesy a really solid marriage, he would split a Moebius band where even after splitting it remained one band. Questioned after his lecture as to the practical value of bands, the lecturer replied, "You might use them if you wanted to be a witch doctor."

3. A three-dimensional poster was used to advertise a certain topology lecture. The poster showed a burlesque dancer whose figure was partially covered by a Moebius band or Moebius strip. The caption was "Would you like to see a Moebius strip? Come to the Auditorium at 8 o'clock tonight."

4. The July, 1949, issue of *Ford Times* carried a story by Wm. H. Upson entitled "Paul Bunyan Versus the Conveyor Belt," telling how Paul Bunyan lengthened a belt in the shape of a Moebius band by splitting it down the middle. Bunyan and his mechanic Fordson won two bets from Loud Mouth Johnson by knowing properties of bands.

5. "No-sided Professor" is the title of an article by Martin Gardner which appeared in the January, 1947 issue of *Esquire.* It tells of the mishaps of a research mathematician who, along with one of his colleagues, disappeared into one of the higher-dimensional spaces while experimenting with odd surfaces.

We mention a more practical use of bands. Suppose a person wishes to study the function $z=y\cos(\pi x/5)$ defined in the strip between the parallel lines $y=1$ and $y=-1$ in the plane. If he divides the strip into rectangles by taking vertical segments on the lines $x=0$, $x=\pm 10$, $x=\pm 20 \cdots$, it may be shown that z takes on the same values on one rectangle as on any other rectangle. Hence the function need be studied on only one rectangle. Also if two points on vertical sides of one of the rectangles have the same y coordinates, z assumes the same values on these two points since

$$y \cos (\pi x/5) = y \cos [\pi(x + 10)/5].$$

Hence we may consider the opposite vertical sides of the rectangle joined to make a cylinder. (See Fig. 11.)

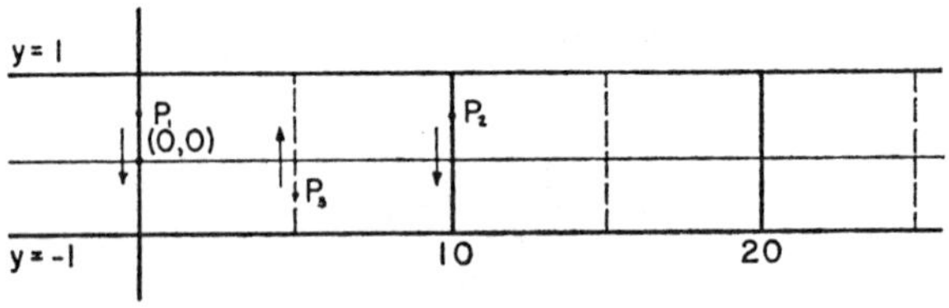

FIG. 11

If one divides the strip into rectangles by taking vertical segments every 5 units rather than every 10 units, two points on opposite vertical sides of one of these smaller rectangles have corresponding z values assigned to them if they are symmetric with respect to the center of the rectangle since

$$y \cos (\pi x/5) = - y \cos [\pi(x + 5)/5].$$

Exercise 5. If the vertical sides of this smaller rectangle are joined so that z takes on the same value on joined points, what figure results?

The large and small rectangles with the two vertical sides joined so that joined points take on the same z values showed why Figure 8a and 8b are useful representations of cylindrical bands and Moebius bands.

Spheres, tori, Klein bottles, and projective planes. If a square as shown in Figure 12a were sewn up as indicated by the arrows, AB would be sewn to AD

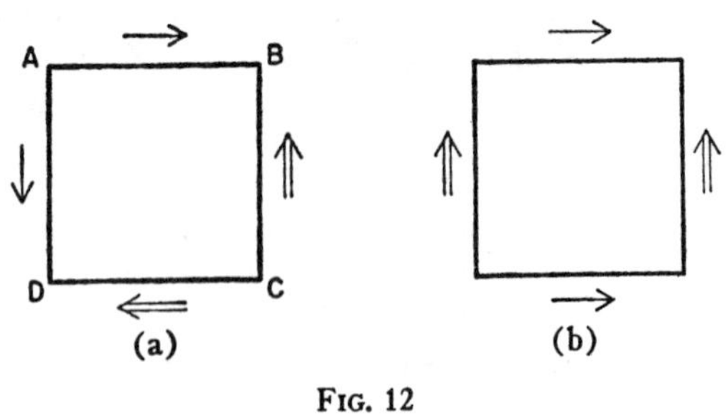

FIG. 12

and CB to CD. There would result a figure that could be distorted into the surface of a ball. A set topologically equivalent to the surface of a ball is called a 2-sphere.

If only the top and bottom of the square shown in Figure 12b were attached, there would result a cylinder. In order to make the rest of the sewing possible in Euclidean 3-space E^3, we suppose that there was no twisting before the first sewing was done. Now suppose the right and left sides are joined as indicated. There are a number of ways that this may be done, three of which are as indicated in Figure 13.

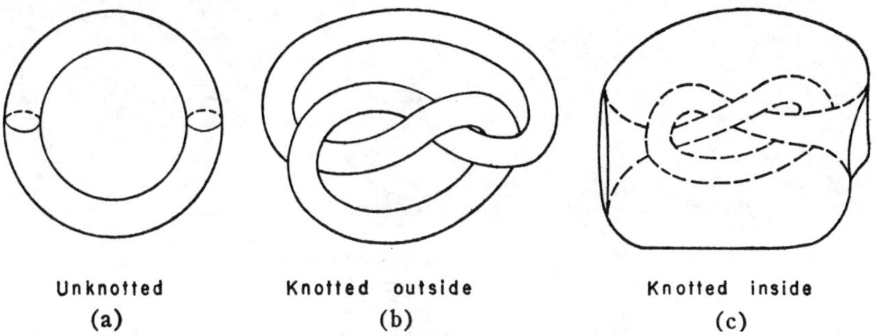

FIG. 13

In Figure 13a, there is no "funny business" and an ordinary *torus* results. In 13b, the cylinder was stretched, a knot was tied in it, and then the final sewing was done. The third is the hardest of the three to visualize. Not only is a knot tied in the cylinder but one end of the cylinder is pulled entirely over the knot before the final sewing is done. If a hole were bored in a cubical block of wood, the surface of the remaining wood would probably be like Figure 13a after certain deformation. If a worm bored a knotted hole in a block of wood, the surface of the remaining wood might resemble Figure 13c. The three surfaces shown in Figure 13 are topologically equivalent to each other but not one of them can be stretched to make it fit on any other one of them. We thus have another counterexample to the faulty pulling and stretching definition.

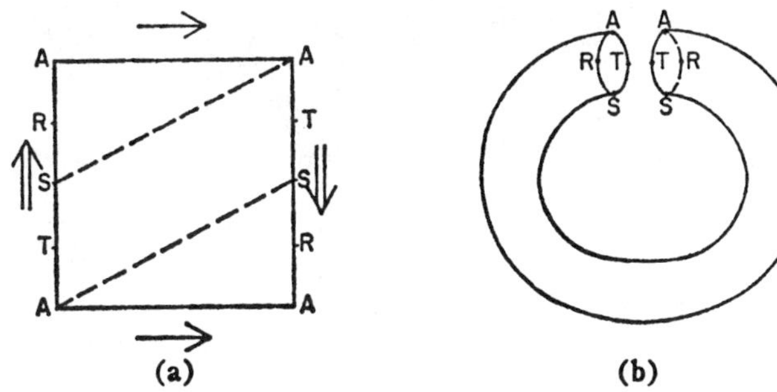

FIG. 14

Not all the objects studied in topology can actually be constructed in E^3. Figure 14 shows such an object. When the first sewing is done a cylinder results as shown in Figure 14b. However, the sewing cannot be completed since the front left arc ARS is to be sewed onto the back right arc ARS and the back left arc ATS is to be sewed onto the right front arc ATS. However we can study objects that we cannot actually construct. If the joining is abstractly made, the resulting surface is called a Klein bottle. If such a surface could be constructed in E^3, it would be a one-sided surface and would have no inside nor outside. However, no such bottle exists in E^3.

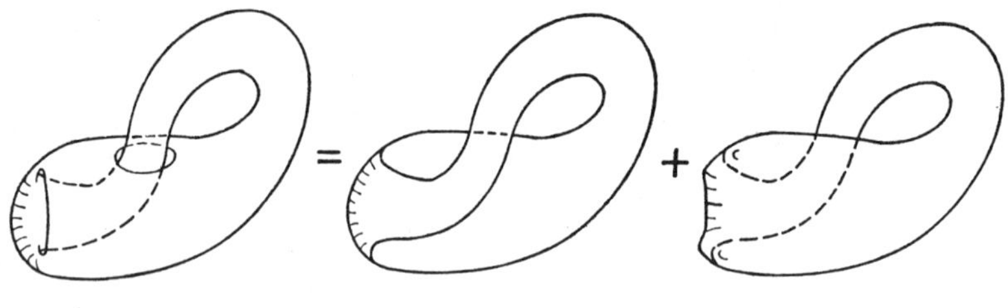

(a) (b)

FIG. 15

Glass models of Klein bottles appear much as shown in Figure 15a. The left end of a glass tube is run inside the tube, the ends of the tube are joined together on opposite sides of the middle of the tube, and the viewer is to use his imagination and suppose that they miss the middle of the tube where they join. The tube may be split down the middle to make two Moebius bands as shown in Figure 15b. The cut is indicated in Figure 14a by dotted lines.

If a sphere is subdivided into spherical triangles, an order may be chosen on the boundary of each of these triangular disks so that if two of them share an edge in common, the order on this edge is different when viewed from the different disks. For example, we may order the boundaries of the disks clockwise when viewed from the outside of the sphere. In Figure 16, the orientation of

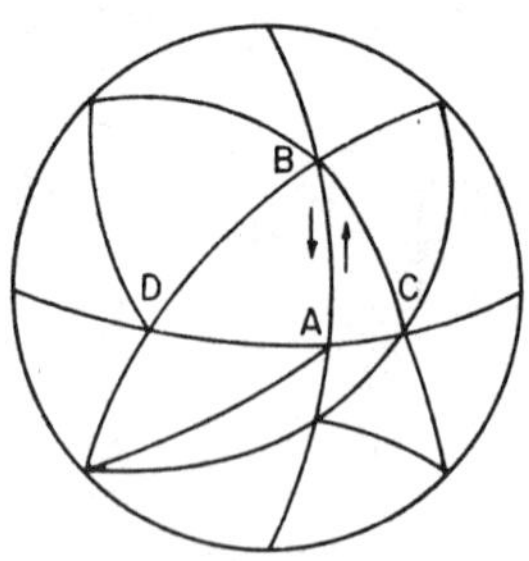

FIG. 16

segment AB is from A to B when considered from spherical triangle ABC but it is from B to A when considered from disk ABD. Since such an orientation may be chosen, we say that a sphere is *orientable*. Since the rectangle in Figure 14a can be subdivided into small triangles, a Klein bottle can be triangulated into curvilinear disks. However, it is not orientable.

Exercise 6. Suppose the square of Figure 14a is subdivided into 9 squares, each of these squares is divided into two isosceles right triangles, and a clockwise orientation is chosen on the boundary of each triangle. When the sewing is done to get a Klein bottle, where will there be two curvilinear triangles sharing an edge so that the order on this edge is the same from both triangles?

Suppose a top is spinning on the surface of the earth and the top is moved in a path so that it returns to its original position. If the direction of spinning does not change, the top will be spinning in the same direction when it returns. Suppose that instead of having the top reaching out from the surface, we consider a direction of turning (say clockwise) about each point. It is possible to select this direction of turning or orientation so that as a point moves on the surface of a ball the orientation does not change.

The condition is somewhat different on a Klein bottle. Suppose an orientation about a point is chosen and the point is moved the long way around the Klein bottle but so that it returns to its original position. If the orientation is

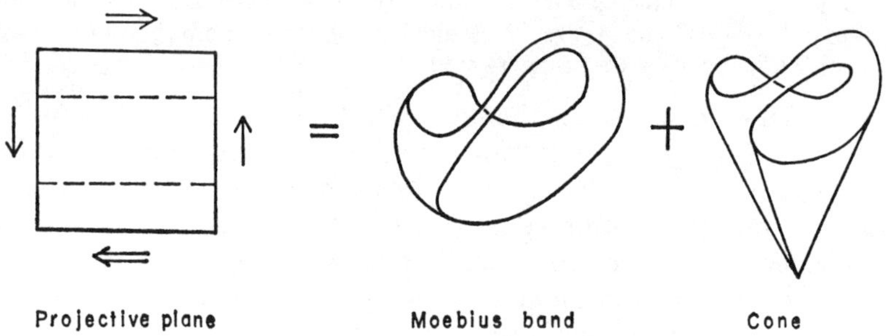

(a) (b)

FIG. 17

not changed as the point moves, the orientation will be different at the end from at the beginning. The Klein bottle is not orientable.

The object shown in Figure 17 is a *projective plane*. It is the sum of a Moebius band and a disk. The fitting seems best if the disk is shown as a cone. If it is cut along the dotted lines shown in Figure 17a, the middle portion makes a Moebius band and the outside pieces make a disk. A projective plane is not orientable.

A projective plane is sometimes defined as a circular disk with diametrically opposite points of the boundary connected. The three-dimensional counterpart of a projective plane is projective three space which may be thought of as a solid ball with diametrically opposite points of the boundary joined.

The projective plane is studied in projective geometry. Yet not everyone who studies projective geometry knows of any geometric model for the projective plane.

Exercise 7. Consider a circular disk with a circular hole in the middle as shown in Figure 18. If diametrically opposite points of the outer boundary are joined, what figure results?

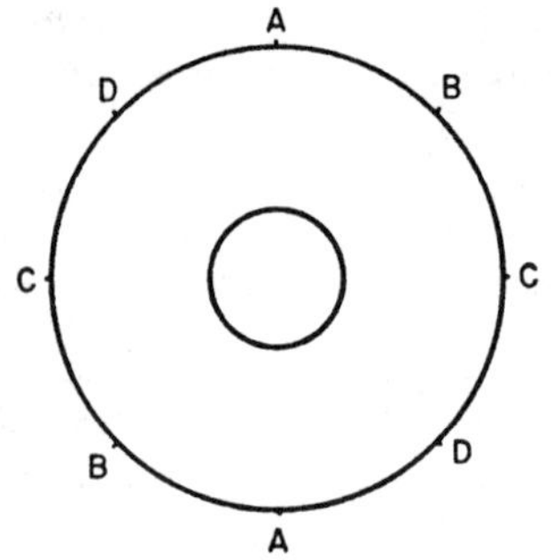

FIG. 18

Exercise 8. Suppose the function $z=\cos x \cos y$ is defined on the horizontally based square in the plane with opposite vertices (0, 0) and (π, π). What figure is formed if opposite vertical sides are sewed together so that z takes on the same value on joined points? If the top and bottom are also joined in this fashion, what figure is formed?

Exercise 9. Describe a function $z=f(x, y)$ that can be defined on a horizontally based square with opposite vertices (0, 0) and (1, 1) so that studying z on the square is equivalent to studying it on a Klein bottle after points of the boundary are joined.

The *complement* of a figure is the set of points not in the figure. (Complement of A in space S is $S-A$.) For example, on the surface of the earth the surface of the oceans, seas, lakes, etc., is the complement of the surface of the land. In the plane the complement of a circle is the sum of its interior and its exterior. The complement of a circle in E^3 would be all in one piece.

We now show that even though two figures in the same space are topologically equivalent, their complements may be topologically different.

In Figure 13, each of the surfaces has an interior and an exterior. It could be shown that the interior of the first is topologically equivalent to the interior of the second but not to the interior of the third. A small creature who knew about topological equivalence but not about straightness, length, etc., could not tell the difference between the inside of the first tube and the interior of the second. However, the exterior of the first tube in Figure 13 is like the exterior of the third but not like the exterior of the second. No two of the three figures have complements that are topologically equivalent.

The knotted curve shown in Figure 5 is topologically equivalent to a circle but their complements are topologically different.

Examples and counterexamples. Although examples are not the main forte of topology, they do serve a useful purpose. A wealth of examples helps our intuition in picking out certain theorems that are likely to be true and labeling others as false.

Examples may be used as counterexamples to certain conjectures and show that it is useless to try to prove them. A counterexample is a convincing way of showing that a "theorem" is, in fact, false. We know in this way that each of the following conjectures is a false statement.

FALSE STATEMENT A. *If two objects in E^3 are topologically equivalent, it is possible to deform one onto the other with pulling and stretching without breaking and tearing.*

Exercise 10. Name 3 pairs of topologically equivalent objects that serve as counterexamples to False Statement A.

FALSE STATEMENT B. *Each surface is orientable.*

Exercise 11. Name three counterexamples to False Statement B.

FALSE STATEMENT C. *Each surface is two-sided.*

Exercise 12. Name a counterexample to False Statement C.

FALSE STATEMENT D. *If two surfaces in E^3 are topologically equivalent, their complements are topologically equivalent also.*

Exercise 13. Give a counterexample to False Statement D.

4. LIMIT POINTS

The notion of a limit is a very important one in mathematics. A student may do well in arithmetic, algebra, and even geometry without understanding limits, but he must learn this concept in order to go far in mathematics. Although limits are used extensively in calculus, some students do the mechanical parts of this subject without gaining an insight into the concept of limits. Limits are used so extensively in point set topology that it is inconceivable that a person could make much headway here without learning a considerable amount about limits. Learning about limit points is a good starting point.

We use the notion of a limit to tell us whether a point is "infinitely close" to a set—whether it "sticks onto" a set. In such a nice space as the Euclidean plane we use the idea of distance to define limit points. In the more general spaces such as those discussed in the section *Some examples of abstract spaces* we use neighborhoods to define limit points.

NEIGHBORHOOD. *A neighborhood in the Euclidean plane E^2 is the interior of a circle. If a point lies on the interior of a circle, this interior is called a neighborhood of the point.*

Each point has many neighborhoods,—some small and some large, some with the point in the middle and some asymmetric with respect to the point. Each neighborhood is a neighborhood of many points—in fact, of each point in it.

A neighborhood in Euclidean 3-space E^3 is the interior of a sphere; in Euclidean 1-space E^1 (or a line) it is an open segment (segment without its end points)—in fact, in any space with a distance, it is the interior of a generalized sphere. In more general spaces such as those discussed in the section *Some examples of abstract spaces*, there may be no distance function and the neighborhoods are certain designated point sets that satisfy certain conditions or axioms for these abstract spaces. A distance function is defined in the section *Some examples of topological spaces* but our intuitive notion of how far things are apart will suffice here.

NEIGHBORHOOD DEFINITION OF LIMIT POINT. *The point p is a limit point of the set X if each neighborhood of p contains a point of X other than p.*

(Note that a limit point of X is not necessarily a point of X.)

DISTANCE DEFINITION OF LIMIT POINT. *The point p is a limit point of the set X if for each positive number ϵ there is a point x of X such that $0<\rho(p, x)<\epsilon$.*

(We use $\rho(p, x)$ to denote the distance from p to x.) This definition is not applicable in abstract spaces without distance functions.

Exercise 14. Suppose that, in the plane, X is the set of all points satisfying the condition $0<x^2+y^2<1$ or $x=y=1$. Which of the following points are points or limit points of X? $p=(0, 0)$, $q=(0, 1)$, $r=(1, 1)$, $s=(2, 0)$, $t=(0, \frac{1}{2})$.

Exercise 15. In an abstract space whose points are the points of the plane and whose neighborhoods are horizontal lines, which of the points p, q, r, s, t are limit points of the set X of Exercise 14?

Sets. A *point set* is a collection, each of whose members (or elements) is a point. The collection of all points with a particular property R is designated by $\{p \mid p$ has property $R\}$. A point set with only one point p is designated by $\{p\}$. We use the following terminology in discussing sets.

SUBSET. *We say that A is a subset of B if each element of A is an element of B.*

We denote this by $A \subset B$. If p is an element of A, we write $p \in A$. We may say that p belongs to A.

SUM OR UNION. *The sum of A and B (denoted by $A+B$ or $A \cup B$) is the set of all elements in either A or B—that is $A+B=\{p \mid p \in A$ or $p \in B\}$.*

INTERSECTION, PRODUCT, OR COMMON PART. *The intersection of A and B* (denoted by $A \cdot B$, $A \cap B$, *or* AB) *is the set of all elements in both A and B—that is,* $A \cdot B=\{p \mid p \in A$ and $p \in B\}$.

If there is no element which is in both A and B, we say that A does not intersect B and may write $A \cdot B=0$.

Some people use the artificial convention that a collection need have no elements. This may lead to complications in having to make exceptions for the null set, as this artificial thing is called. For example, instead of saying that two parallel lines do not intersect, some of those who use this "null set" convention say that two parallel lines meet in the "empty set." Instead of saying that each bounded collection of numbers has a least upper bound, they say that each nonnull bounded collection of numbers has a least upper bound. The definition of "connected" as given in the section *Properties preserved by mappings* would contain an extra condition if the "null set" convention were used. On the other hand, the convention may have some advantage in the algebra of sets (Boolean algebra) for then two sets always have a product and a difference. We believe the disadvantages of introducing the null set in our treatment of point set topology outweighs the advantages so we do not use it here. Anytime we say that X is a set, we imply that there is at least one element in X.

Not only do we add and multiply two sets but we do the same for three sets, four sets . . . or even an infinite collection of sets. Suppose $\{A_\alpha\}$ is a collection (perhaps infinite) of point sets and A_α is a typical set in $\{A_\alpha\}$. The sum (or union) of the elements of $\{A_\alpha\}$ is the set of all points p such that each belongs to some A_α. We use $\sum A_\alpha$ or $\cup A_\alpha$ to denote this sum. Then $\sum A_\alpha = \{p \mid p \in \text{some } A_\alpha\}$. The product, intersection, or common part of the elements of $\{A_\alpha\}$ may be written $\prod A_\alpha$ or $\cap A_\alpha$. Then $\prod A_\alpha = \{p \mid p \in \text{each } A_\alpha\}$. If there is no point common to all elements of $\{A_\alpha\}$ we say that this intersection does exist and write $\prod A_\alpha = 0$.

DIFFERENCE. *We use $A-B$ to denote the collection of elements of A that do not belong to B—that is, $A-B = \{p \mid p \in A$ and $p \notin B\}$. If $A \subset B$, we write $A-B=0$.*

An example. If a person were to squirt some black ink on the plane, the set A of points in the dark spot is an example of a point set. Suppose a set B is determined by squirting some red ink on the plane. Then $A+B$ designates the set of points covered by ink, $A \cdot B$ designates the set covered by both kinds of ink, and $A-B$ designates those covered by black but not red ink. (See Fig. 19.)

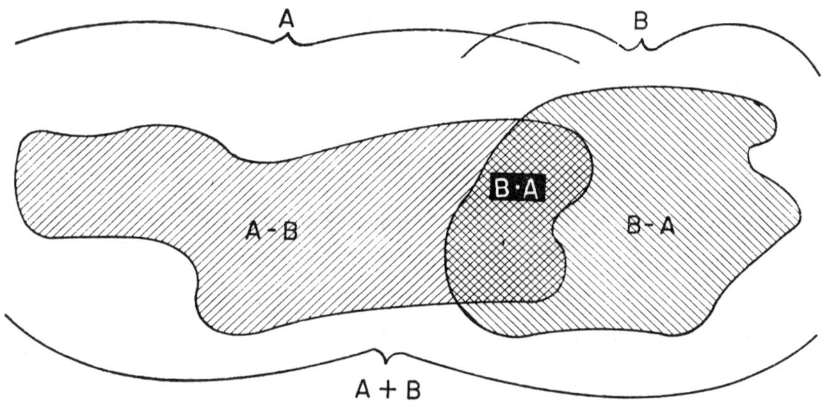

FIG. 19

OPEN SET OR DOMAIN. *A point set D is an open set if for each point p of D there is a neighborhood of p that lies in D. An open set is called a domain. The intersection of a set X and an open set is called an open subset of X.*

(Note that an open subset of X may not be an open set.)

An open set is the sum of neighborhoods, although they may be infinite in number. For the Euclidean plane, the interior of a square is a domain but a disk (circle plus its interior) is not.

Exercise 16. In showing that the sum of two open sets A, B is open, one shows that if p is a point of $A+B$, p belongs in a neighborhood which lies in $A+B$. There are two cases in the proof—either p belongs to A or it does not. If it does not belong to A, show that there is a neighborhood N of p in $A+B$ containing p.

Exercise 17. Suppose S is the interior of a square in the plane. For each point p of S describe a radius r_p such that if $V(p, r_p)$ is the interior of a circle with center at p and radius r_p, then $\sum V(p, r_p) = S$.

CLOSED SET. *A closed set is one which contains all its limit points. Each of its limit points belongs to it. A line is closed in the plane.*

The *closure* of a set X is the sum of X and all its limit points. This closure is designated by $\overline{X}$. The closure of the set X of Exercise 14 is the sum of a disk and $\{r\}$. The closure of the set X of Exercise 15 is the set of all points above the line $y = -1$ and on or below the line $y = 1$.

Exercise 18. For the Euclidean plane give an example of (a) a closed set, (b) an open set, (c) a set neither open nor closed, and (d) a set that is both open and closed.

Exercise 19. What would need to be proved to show that the closure of each set is closed?

Exercise 20. If, in the plane, G is the graph of $y = \sin 1/x$ $(0 < x < 1/\pi)$, what is $\overline{G}$?

It is possible to prove each of the following theorems from the definitions alone without even knowing what a neighborhood looks like:

1. *The sum of two open sets is open.*
2. *The product of two closed sets is closed.*
3. *The complement of a closed set is open.*
4. *The complement of an open set is closed.*
5. *The closure of each set is closed.*

The reader may gain experience by proving these from the definitions. Exercises 16, 21, 22, 23, 24 point in the direction of these proofs. The first two theorems may be generalized to say that the sum of any collection (perhaps infinite) of open sets is open and the product of any collection of closed sets is closed.

Exercise 21. In showing that the product of two closed sets A, B is closed, one considers a limit point p of $A \cdot B$ and shows that p is a point of $A \cdot B$. Show that if p is a limit point of $A \cdot B$, it is a point of A.

Exercise 22. If A is a closed set and p is a point of its complement, show that the assumption that there is no neighborhood of p that misses A leads to the contradiction that $p \in A$.

Exercise 23. If A is an open set and A' is its complement, show that if $p \notin A'$, then p lies in a neighborhood N that misses A'.

Exercise 24. Show that if A is a set and $p \notin \bar{A}$, then p is not a limit point of $\bar{A}$.

5. TRANSFORMATIONS

Ordinarily, a person regards a transformation as a change. A seed becomes a tree, baby becomes an adult, wood turns to smoke and ashes, a bomb changes to fragments, a wad of gum is squeezed, one object is transformed into another.

In mathematics a person may come to regard a transformation as a rule or law associating points of one set X with points of another set Y. If f denotes the transformation and $x \in X$, we use $f(x)$ to denote the point of Y associated with x. (See Fig. 20.)

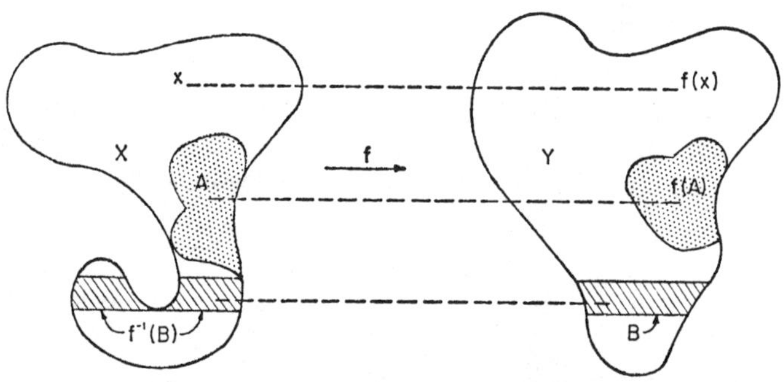

FIG. 20

We call $f(x)$ the *image* of x under f. If A is a subset of X, $f(A)$ denotes the collection of images of points of A. If B is a subset of Y, $f^{-1}(B)$ is the set of all points of X that go under f into a point of B; $f^{-1}(B)$ is called the *inverse image* of B.

A useful and more sophisticated approach is to define a transformation of X into Y to be a collection of ordered pairs that satisfy certain conditions. For example, if one considers the equation $y = x^2$, the corresponding transformation might be regarded as the set of all ordered pairs of the form (a, a^2). The pairs would be the coordinates of points on the parabola. A definition of a transformation based on ordered pairs may seem somewhat artificial and perhaps not exactly describing the intuitive notion that we have in mind, but it does have the advantage of being a precise definition given in terms of more elementary concepts.

ORDERED PAIRS DEFINITION OF TRANSFORMATION. *A transformation of X into Y is a collection of ordered pairs (x, y) such that $x \in X$, $y \in Y$, and each element of X is the first element of one and only one of the ordered pairs.*

FUNCTION DEFINITION OF TRANSFORMATION. *We may regard a transformation of X into Y as a single-valued function that assigns one and only one element of Y to each element of X.*

This definition leaves it to those who define function to decide if it is a change, a rule, a collection of ordered pairs, or something else.

If each element of Y is the image of an element of X under f, we say that f takes X *onto* Y instead of merely into Y. Hence, of the two transformations of the reals into the reals given by $y=x^2$ and $y=x^3$, the second is an onto transformation but the other one is not.

If no element of Y is the image of two elements of X, we say that f is *one-to-one* or 1-1. Hence $y=\sin x$ is 1-1 for $0<x<\frac{1}{2}\pi$ but not for $0<x<\pi$.

Intuitively, we say that a transformation f is continuous if points "close together" in X go into points "close together" in Y. The notion "close together" is not precise so we give the following definition:

NEIGHBORHOOD DEFINITION OF CONTINUOUS. *A transformation f of X into Y is continuous at the point x of X if for each neighborhood U of $f(x)$ there is a neighborhood V of x such that $f(V \cdot X) \subset U$.* (See Fig. 21.)

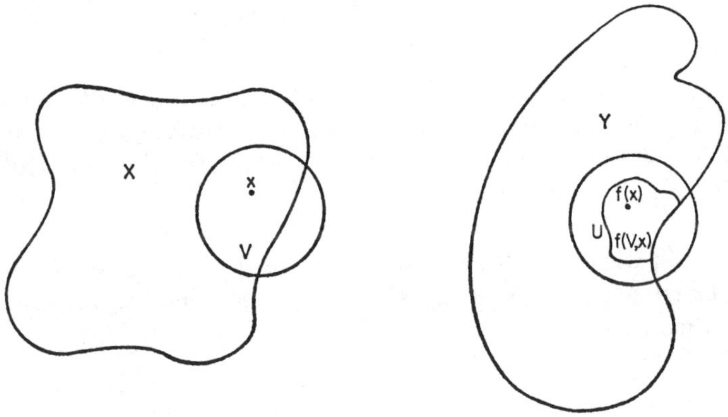

FIG. 21

The size of V in the above definition is a function of x and U—if U is taken to be small, V may need to be chosen small. However, Example 8 below shows that V may not need to be small even if U is.

If the set X on which f is defined is all of space, the definition may be simplified by replacing "$f(V \cdot X) \subset U$" by "$f(V) \subset U$." We could make this change anyway if we called V an open subset of X containing x rather than a neighborhood of x.

The set on which a function is defined is called the *domain of definition* of the function. Note that a domain of definition may not be an open set or a domain as defined earlier.

A *function is continuous* if it is continuous at each point of its domain of definition X. A continuous function is called a *map*.

For metric spaces (Euclidean spaces and other spaces with distance functions), we may use the following definition:

$\epsilon-\delta$ DEFINITION OF CONTINUITY. *The transformation f of X into Y is continuous at the point x of X if for each positive number ϵ there is a positive number δ such that for each point x' of X within a distance δ of x, $f(x')$ is within ϵ of $f(x)$.*

Let us consider some examples of transformations of the reals into the reals described by equations.

1. $f(x)=x^3-1$.
2. $f(x)=x^3-x$.
3. $f(x)=x^2$.
4. $f(x)=\sin x$.
5. $f(x)=\text{Arc tan } x$.
6. $f(x)=0$ if x is irrational or 0, $f(p/q)=1/q$ if $p/q\neq 0$ is rational and in lowest terms.
7. $f(x)=x$ or x^3 according as x is algebraic or nonalgebraic.
8. $f(x)=1-x^2$, 0, or x^2-1 according as $x<-1$, $-1\leqq x\leqq 1$, $x>1$.

Exercise 25. Which of the above examples is (a) onto, (b) 1-1, (c) continuous, (d) continuous at $x=0$?

Here is an example of a homeomorphism f of the interior of a circle of radius 1 onto the plane. The center c of the circle goes into itself and if p is another point of the interior of the circle, $f(p)$ is on the ray from c through p and $\rho(c, f(p))$ is $\rho(c, p)/(1-\rho(c, p))$.

Now let us consider a transformation of one segment AC onto a segment $C'B'$ as shown in Figure 22. The points between A and B go into points of $A'B'$ through parallel projections and the points of BC go into points of $B'C'$ by projections through a point.

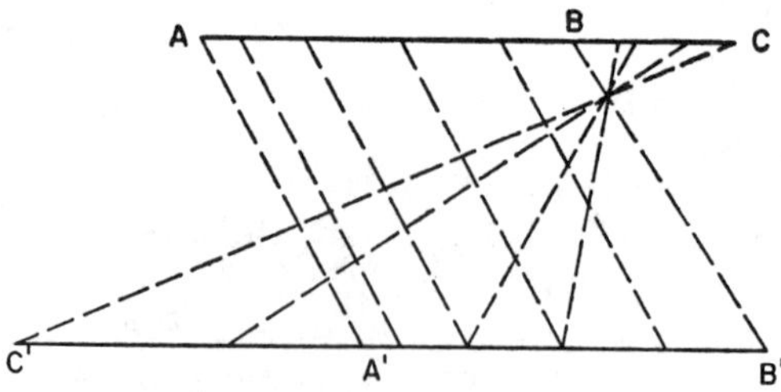

FIG. 22

If P is a point of the segment AB, the ratio in which P divides AB is the quotient obtained by dividing the distance from A to P by the distance from P to B. The map from AB to $A'B'$ is called *linear* if for each point P of AB, P' divides $A'B'$ in the same ratio that P divides AB. We note that while the map shown in Figure 22 is linear on AB and is linear on BC, it is not linear on AC. We make use of linear maps in the section *The Jordan curve theorem*.

This transformation indicated by Figure 22 may be represented by the graph shown in Figure 23. The interval AC is placed on the x-axis and the interval $C'B'$ on the y-axis. To find the image of a point of AC, project the point vertically onto the graph and then project this point horizontally onto the y-axis. The equation in this case is $f(x)=x+2$ or $17-4x$ according as $1\leqq x\leqq 3$ or $3\leqq x\leqq 4$.

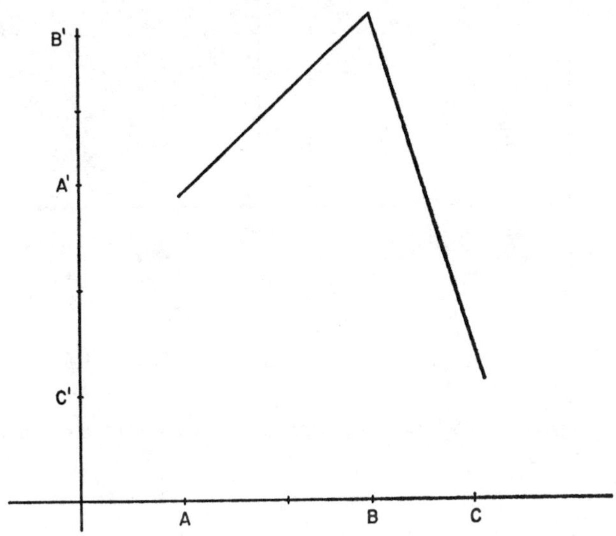

FIG. 23

If X is a square plus its interior and Y is a segment, a transformation of X into Y may be represented by a graph in 3-space which may be a surface. However, if Y were also a square plus its interior, it would require a space of four dimensions to draw the graph—two dimensions for X and two others for Y.

The $\epsilon-\delta$ definitions of continuity of functions given in analytics and calculus texts are puzzling to most students when they are encountered for the first time. In some cases, even an expert could not tell from the definitions given whether or not a function such as $f(x)=\sqrt{(1-x^2)}$, $-1\leqq x\leqq 1$, is continuous at $x=1$ since mention of the domain of definition is sometimes omitted—in other cases the function would be called discontinuous because it is not defined for all values of x. From the definitions we have given, the corresponding transformation would be continuous.

Here is a geometric definition of the continuity of a graph which may be more meaningful to many students than the corresponding $\epsilon - \delta$ definition. It is understood that the graph is in the plane.

DEFINITION OF CONTINUITY OF A GRAPH. *The graph G is continuous at the point p of G if for each pair of horizontal lines, one above and one below p, there is a pair of vertical lines, one to the right and one to the left of p, such that each point of G between the vertical lines is between the horizontal lines.* (See Fig. 24.)

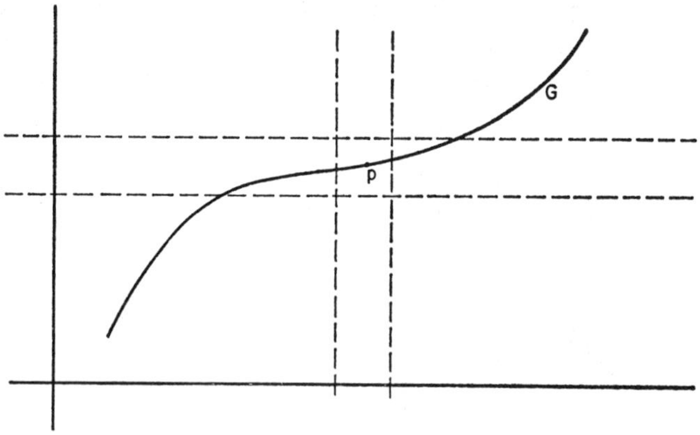

FIG. 24

Exercise 26. What in the above definition corresponds in the $\epsilon - \delta$ definition to (a) x, (b) $f(x)$, (c) ϵ, (d) δ?

Exercise 27. Under what conditions would a graph in 3-space be continuous at one of its points p?

The definition of topologically equivalent given in the section *Topological equivalence* may have been unsatisfactory since no precise definition of continuous had been given. This flaw can now be removed.

A 1-1 transformation f of A onto B is said to be *reversibly continuous* if f^{-1} is continuous at each point of B.

A 1-1 transformation f of A onto B is called a *homeomorphism* of A onto B if it is continuous and reversibly continuous.

Two sets are said to be *topologically equivalent* (or *homeomorphic*) if there is a homeomorphism of one onto the other.

6. PROPERTIES PRESERVED BY MAPPINGS

A continuous transformation is called a *mapping*. In this section we point out that certain properties are preserved by mappings. The following exericse shows that in a certain sense, limit points are preserved.

Exercise 28. Show that if p is a limit point of X and if f is a map of $\bar{X}$ into $Y, f(p)$ is either a point or a limit point of $f(X)$.

Suppose $x_1, x_2, \cdots$, is an infinite sequence of points—we do not suppose that different elements of the sequences are different points. The sequence is said to *converge* to a point x_0 if for each neighborhood U of x_0 there is an integer n such that each point of $x_n, x_{n+1}, \cdots$ lies in U. A proof similar to that used to answer Exercise 28 shows that if f is a mapping of X, X contains $x_0, x_1, x_2, \cdots$, and $x_1, x_2, \cdots$ converges to x_0, then $f(x_1), f(x_2), \cdots$ converges to $f(x_0)$.

The preceding result may remind us of the definition of continuity of a function given in some calculus books. The function $f(x)$ is sometimes said to be continuous at $x=a$ provided the limit, as x approaches a, of $f(x)$ is $f(a)$.

Connectedness is another property that is preserved by mappings. Intuitively, we think that a thing is connected if it is "all in one piece." For example, the disk D shown in Figure 25 is connected but the Cantor set C is not. However, D is the sum of its interior and its boundary, so in a certain sense it is in

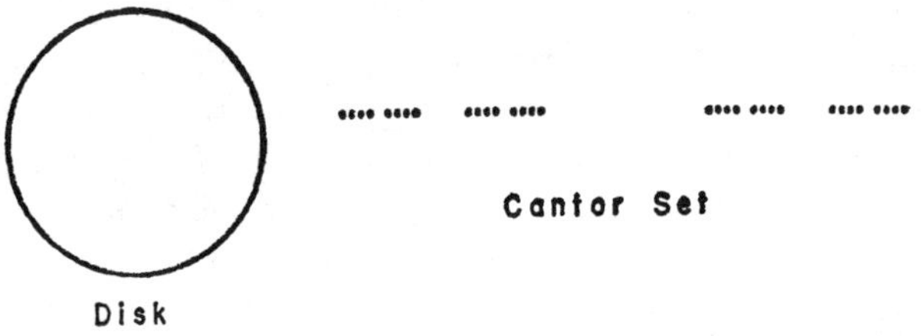

FIG. 25

two pieces. (The *boundary* of M is the intersection of $\overline{M}$ and the closure of the complement of M.) We shall give a more precise definition of connected. The

Cantor set shown in Figure 25 is obtained by starting with a segment, removing the open middle third of it, removing each of the open middle thirds of the remaining pieces, etc.

Two sets are *mutually separated* if neither contains either a point or a limit point of the other. (A, B mutually separated means $\overline{A}\cdot B=A\cdot\overline{B}=0$.) The interior of a disk and its boundary are not mutually separated since a point of the boundary is a limit point of the interior.

A set is *not connected* if it can be expressed as the sum of two mutually separated sets. The Cantor set is not connected because its right-hand half and left-hand half are mutually separated.

DEFINITION OF CONNECTED. *A set is connected if it is not the sum of any two mutually separated sets.*

Exercise 29. Show that if f is a mapping of a connected set X, then $f(X)$ is connected also.

A closed and connected point set is called a continuum. Being a continuum is not preserved under a mapping—we showed in the last section a homeomorphism of the interior of a circle (not a continuum) onto the plane (a continuum). The inverse of this homeomorphism is a mapping of a continuum onto something that is not closed.

A *component* of a set X is a maximal connected subset of X. Hence, the graph of a hyperbola has two components but the graph of a parabola has only one. The complement of a hyperbola in the plane has three components while the complement of a parabola has only two. A continuum has only one component but a Cantor set has infinitely many.

Being open and closed is preserved in a certain sense by the inverses of mappings. In fact, rather than say that a transformation f of X into Y is continuous if it is continuous at each point of X, it is sometimes defined to be continuous if the inverse of each open subset in Y is an open subset of X.

Exercise 30. Show that if f is a continuous transformation of X onto Y, then the inverse of each open subset of Y is an open subset of X.

Exercise 31. Show that the transformation f of X into Y is continuous if the inverse of each open subset of Y is an open subset of X.

Exercises 30, 31 show that a transformation f of X into Y is continuous if and only if the inverse of each open subset of Y is an open subset of X. Similarly it might have been shown that f is continuous if and only if the inverse of each closed subset of Y is a closed subset of X. (A closed subset of X may not be closed but is merely the intersection of X with a closed set.)

As the population of the earth increases, it is inevitable that there be centers of dense population. In contrast, it is felt that if space travel becomes feasible and space is habitable, the population of the universe can increase without us having crowding. Even in the plane, it is possible to have infinitely many points such that no point is within 100 units of any two of them. (Put the points at (0, 0), (300, 0), (600, 0), $\cdots$.) The limited property of the earth we have mentioned reminds us of a topological property called compactness. Compactness is preserved under a mapping if we use the following definition.

DEFINITION OF COMPACTNESS. *A set M is compact if each infinite subset of M has a limit point in M.*

The plane is not compact but, as shown in the next section, a closed segment is. (Some definitions do not impose the condition that the limit point of M need belong to M. If this condition were omitted, the interior of a circle would be compact and Exercise 32 could not be done. Other definitions impose even more stringent conditions.)

Exercise 32. Show that if f is a mapping of X and X is compact, then $f(X)$ is compact also.

A set of tarpaulins is said to cover a football field if each spot of the field is under one of them. Similarly, a collection G of point sets is said to *cover* a point set X if each point of X belongs to an element of G. If each element of G is a domain, G is called an *open covering.*

The collection of all neighborhoods covers the plane but no finite subcollection of them does. However, if a collection of neighborhoods covers the surface of a sphere or a segment, a finite number of them does also. This reminds us of the following topological property that is preserved under a mapping.

DEFINITION OF BICOMPACTNESS. *A set X is bicompact if each open covering G of X contains a finite subcovering—that is, a finite subcollection of G which covers X.*

Exercise 33. Show that bicompactness is preserved by a mapping.

The two properties, compactness and bicompactness, are not equivalent because there are topological spaces that are compact but not bicompact. However, any bicompact space is compact and any compact space with a distance function is bicompact. Therefore, bicompactness is a stronger condition and a person who uses it in the hypothesis of a theorem is not proving as strong a result as one who uses mere compactness instead. Because of lack of agreement on terminology, some people use the term compact to mean bicompact as we have defined it here. Not everyone that mentions the word compact is using the weaker condition.

Exercise 34. Show that bicompactness implies compactness.

A set X is *homogeneous* if for each pair of points p, q of X there is a homeomorphism of X onto itself that takes p onto q. A circle is homogeneous because a rotation will take any point into any other point. A closed segment is not homogeneous because there is no homeomorphism of the segment onto itself that takes an end point onto a non-end point.

A definition of topology. Topology may be defined as a study of properties that remain invariant under homeomorphisms. Such a definition might be meaningless to persons unacquainted with such properties as connectedness, compactness, bicompactness, homogeneity, etc. If one were to give such a definition to an unenlightened gathering, the question might be asked as to what is a homeomorphism. If a definition of "homeomorphism" were given such as appears in this article, the next question might be, what is the meaning of the word "continuous"? Neither the neighborhood nor the $\epsilon-\delta$ definition of continuity is likely to be of much help. Perhaps it is best not to define topology to people who do not already know what it is.

Unsolved problems. A difficulty in discussing unsolved problems in topology is that most of these problems involve complicated concepts meaningful only to the advanced worker. Most people would have neither the technical vocabulary nor the experience to appreciate the discussion. However, here is a problem which we have at least developed the vocabulary to present.

A *simple closed curve* is a set topologically equivalent to a circle. Is a simple closed curve the only homogeneous compact plane continuum other than a point? For many years it was conjectured that the answer was in the affirmative—in fact some papers appeared in scientific journals giving faulty proofs to this effect. However, in 1948, an example was given of a bounded plane continuum which was neither a simple closed curve nor a point but which was homogeneous nevertheless. In 1954 another example was announced. Are there other homogeneous bounded plane continua?

7. THE JORDAN CURVE THEOREM

The interior of circles and triangles appear to be somewhat alike. However, if one has a complicated simple closed curve such as shown in Figure 26, one might have difficulty in deciding if its interior was like the interior of a circle.

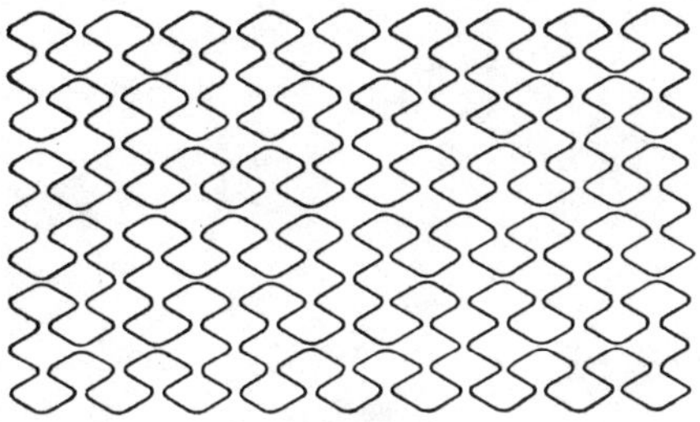

FIG. 26

If the curve had been even more complicated, it might have been impossible to tell by merely looking whether or not it even had an interior.

A subset X of a set Y is said to *separate* Y if Y is connected but $Y-X$ is not connected. It may seem intuitively obvious that a simple closed curve in the plane separates the plane but it is not easy to prove this.

The *Jordan curve theorem* states that the complement of each simple closed curve J in the plane has exactly two components and is the boundary of each of them. One extension of the Jordan curve theorem due to Schoenflies may be stated as follows:

AN EXTENSION OF THE JORDAN CURVE THEOREM. *If J_1 and J_2 are simple closed curves in the plane, there is a homeomorphism of the plane onto itself that takes J_1 onto J_2.*

In plane analytical geometry one studies translations and rotations of the plane onto itself—and perhaps even reflections through a line. These are examples of homeomorphisms of the plane onto itself. Under each of these particular homeomorphisms, a figure would go into a figure congruent to itself. However, the types of homeomorphisms used to show the truth of the above extension of the Jordan curve theorem would frequently change the shapes of objects.

To those who think the theorem needs no proof since it is intuitively obvious, it may come as a surprise that the analogous theorem is not true in 3-space. Although the horned sphere described at the end of this section is topologically like the surface of a ball, there is no homeomorphism of 3-space onto itself that will take one onto the other.

The Jordan curve theorem is one of the more frequently used theorems of plane topology; it is used in such subjects as analysis or complex variables. Although several proofs of this theorem have been given—and some only recently—these proofs are still complicated.

In this section, we only treat the Jordan curve theorem where the simple closed curves involved are uncomplicated. However, if the reader works through even the first of the exercises, he will get some understanding of the meaning of the theorem.

Exercise 35. Describe a homeomorphism of the plane onto itself that takes a given triangle onto a given circle. (See Fig. 27 for a hint as to how to proceed.)

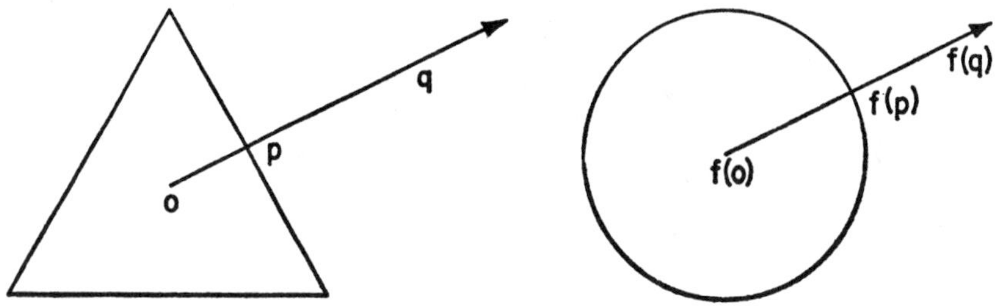

FIG. 27

Exercise 36. Describe a homeomorphism of the plane onto itself that takes a given triangle onto the graph of $\{y-\sqrt{(1-x^2)}\}\{2y-\sqrt{(1-x^2)}\}=0, (-1\leqq x\leqq 1)$.

By using mathematical induction on the number of sides, it can be shown that for any polygon there is a homeomorphism of the plane onto itself that takes the polygon onto a triangle. One without training in topology who obtained an airtight proof of such a theorem could be proud of his accomplishment. Indeed, it is no easy task for the beginner to get even Exercise 37. Some of the methods used in working the following exercise can be used to get the more general result.

Exercise 37. Describe a homeomorphism of the plane onto itself that takes polygon $ABCDEF$ of Figure 28 onto a triangle. (See Fig. 29 for a hint as how to proceed.)

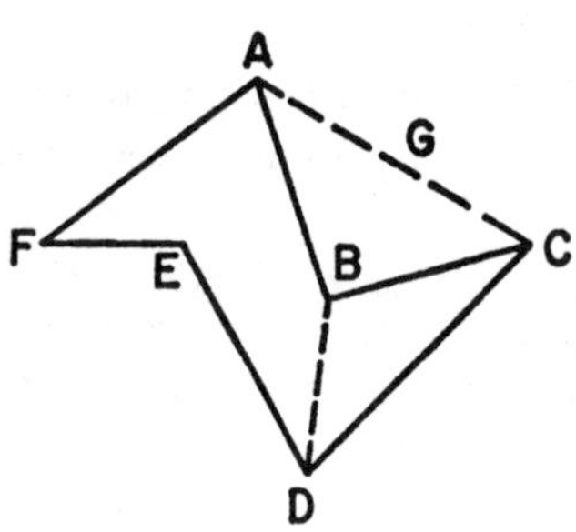

FIG. 28

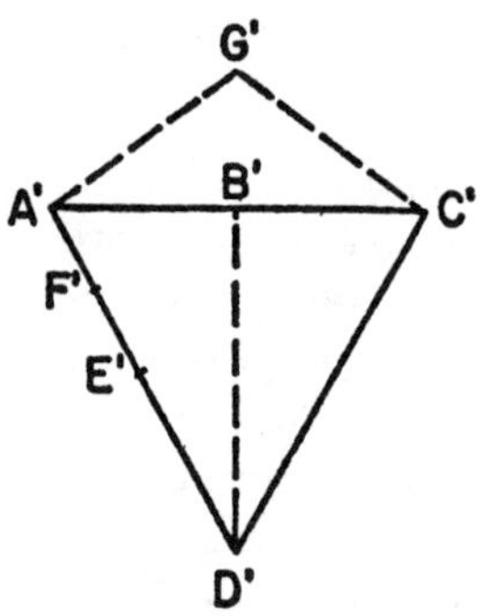

FIG. 29

A set in the plane is *bounded* if it lies on the interior of some large circle. The interior of a triangle in the plane is bounded but its exterior is not.

It is to be noted that we regard a triangle as the sum of its three sides. The complement of a triangle in the plane has two components. If ABC is a plane triangle we use Int ABC to denote the bounded component of the complement of ABC. Either the Jordan curve theorem or the Schoenflies theorem shows that each simple closed curve in the plane has a complement with a bounded component.

The reader who is more interested in getting an overall view of this chapter than in going into details may skip the rest of this section and proceed with the next section *Properties of the real numbers*.

We now give three theorems relating to the Schoenflies theorem. Each is of interest in its own right.

RETRACTION THEOREM. *Suppose P is a polygon and ABC is a triangle such that $P\cdot(ABC+\text{Int } ABC)$ is the sum of two sides AB, BC of ABC. If AC is the third side of ABC, there is a homeomorphism h of the plane onto itself that takes P onto $P-(AB+BC)+AC$.*

Let M be the midpoint of AC and R, S be the points on the line MB as shown in Figure 30 so that P intersects the disk bounded by the polygon $ARCS$

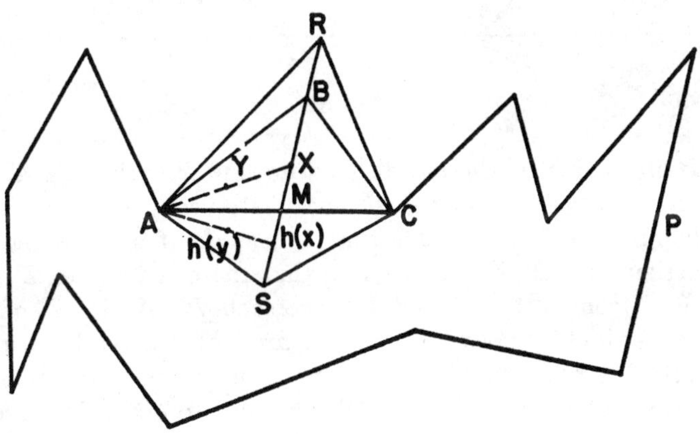

FIG. 30

only in $AB+BC$. The homeomorphism h is the identity outside the disk, takes RB linearly onto RM and BS linearly onto MS.

Exercise 38. If X is a point of RS and Y is a point of Int ARS between A and X, how might h be defined on Y?

SCHOENFLIES THEOREM FOR POLYGONS. *For each polygon P in the plane there is a homeomorphism h of the plane onto itself that takes P onto a triangle.*

SHELLING THEOREM. *If P is a polygon with more than three sides and UV is a side of P, there is a triangle ABC such that AB, BC are sides of P different from UV, and* ($ABC+\text{Int } ABC$) *lies, except for $AB+BC$, on* Int P.

These two theorems may be proved by induction on the number $n(P)$ of sides of P by making use of the retraction theorem. Each holds for $n(P)=3$ or 4 so we assume $n(P)=k$ is the first integer for which there is a polygon P for which one of the two results does not hold.

Let L be a line of support for this P (P intersects L but not both components of the complement of L) such that L contains two points of P. Since the theorems would be easier to prove if P were assumed to be a convex polygon, we suppose that L contains a segment AB that intersects P exactly in its end points. (See Fig. 31.)

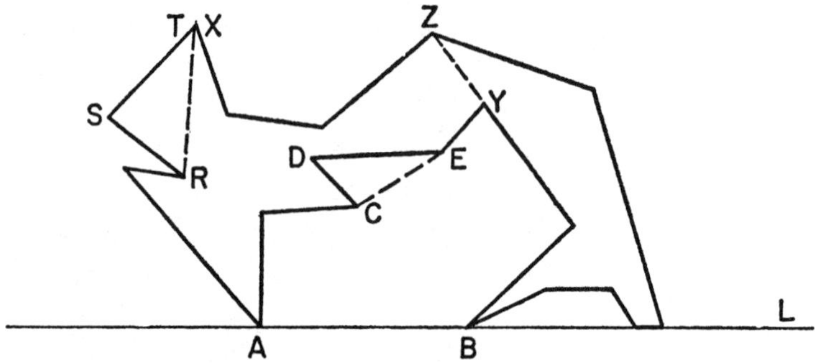

FIG. 31

Let AXB and AYB be the two broken lines in P with ends A and B. With no loss of generality we suppose that X and Y are vertices of P chosen to be at a maximum distance from L and such that X is as far if not farther from L than Y. If X is farther than Y from L, we find that there is a vertex Z of AXB such that Z is farther from L than Y and YZ intersects P only at its end points. If it should happen (which it does not, but we are not yet ready to prove that it does not) that X is the same distance from L as Y, we would select Y so that there is a vertex Z of AXB such that YZ is parallel to L and intersects P only at Y and Z.

We now show that the Schoenflies theorem is satisfied on P. Consider the polygon $AB+AYB$. If it is a triangle, it follows from the retraction theorem that the Schoenflies theorem is satisfied on P. If $AB+AYB$ has more than three sides, the following five exercises indicate why the Schoenflies theorem holds on P.

Exercise 39. Why is the shelling theorem satisfied on $AB+AYB$?

Exercise 40. In applying the shelling theorem to $AB+AYB$ to show that there is a triangle CDE such that CD and DE are sides of P other than UV, and CDE lies, except for $CD+DE$, on Int ($AB+AYB$), what should be used for UV to insure that CD, DE are sides of P?

Exercise 41. Why will not the segment CE of Exercise 40 intersect P at a point other than C and E?

Exercise 42. Name a polygon P' with one fewer sides than P such that the retraction theorem says that there is a homeomorphism of the plane onto itself taking P onto P'.

Exercise 43. If h is a homeomorphism of the plane onto itself taking P onto P' and g is a homeomorphism of the plane onto itself taking P' onto a triangle, name a homeomorphism of the plane onto itself that takes P onto a triangle.

Finally we consider why the shelling theorem is satisfied on P. Let ZAY and ZBY be the two arcs whose sum is P. If UV is a given arc of P, we choose the one of ZAY, ZBY that does not contain it. With no loss of generality we suppose it is ZAY. The shelling theorem is satisfied on polygon $ZAY+YZ$ since this polygon has fewer sides than P. It follows from this shelling theorem that there is a triangle RST such that RS, ST are sides of $ZAY+YZ$ other than YZ and such that $(RST+\text{Int } RST)$ lies, except for $RS+ST$, on $\text{Int }(ZAY+YZ)$. If we show that $\text{Int }(ZAY+YZ)\subset\text{Int } P$, the triangle RST shows that the shelling theorem is satisfied for P. Consider

$$I(P) = \text{Int}(ZAY + YZ) + \text{Int}(ZBY + YZ) + \text{open segment } (YZ).$$

If we show that $\text{Int } P = I(P)$, it will follow that $\text{Int}(ZAY+YZ)\subset\text{Int } P$. Recall that this was the last link in establishing the shelling theorem for P. It follows from the Schoenflies theorem applied to P that the complement of P has only one bounded component. Hence, we show that $I(P)=\text{Int } P$ by showing that $I(P)$ and the complement $E(P)$ of $P+I(P)$ are mutually separated.

Since $P+I(P)$ is closed, $E(P)$ contains no limit point of $I(P)$. We show that $I(P)$ and $E(P)$ are mutually separated by showing that no point of $I(P)$ is a limit point of $E(P)$. Points of the open segment (YZ) are the only ones in question. No point of this open segment is a limit point of $E(P)$ if $\text{Int}(ZAY+YZ)$ contains points on one side of (YZ) and $\text{Int}(ZBY+YZ)$ contains points on the other side. This will be the case if $\text{Int}(ZAY + YZ)$ does not intersect $\text{Int}(ZBY+YZ)$ since each of the open sets has YZ on its boundary. The following two exercises show why these two open sets do not intersect.

Exercise 44. Why does ZBY not intersect $\text{Int }(ZAY+YZ)$? Why does ZAY not intersect $\text{Int }(ZBY+YZ)$?

Exercise 45. Show that there is no point common to $\text{Int }(ZAY+YZ)$ and $\text{Int }(ZBY+YZ)$.

J. W. Alexander described the *horned sphere* illustrated in Figure 32. It is topologically equivalent to the surface of a ball but its exterior is not like the exterior of a ball. Recall that a set topologically equivalent to the surface of a ball is called a 2-sphere but we sometimes use the word sphere alone for brevity.

We might describe the horned sphere as follows. A long cylinder closed at both ends is folded until the ends are near each other and parallel. Then tubes are pushed out each end until they hook as shown. The process is continued by pushing out additional tubes, etc. The resulting set has the property that, although it is topologically equivalent to the surface of a sphere, there is a circle in the exterior of the horned sphere that cannot be shrunk in the exterior of the horned sphere to a point.

Although a horned sphere is topologically equivalent to a sphere, the complement of a horned sphere is not like the complement of a sphere because each simple closed curve in the complement of a sphere can be shrunk to a point without touching the sphere. The simple closed curve J shown in Figure 32 cannot be shrunk to a point in the complement of the horned sphere.

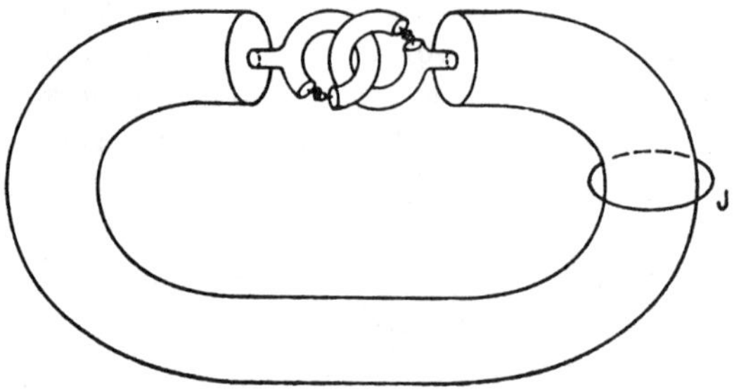

FIG. 32

The four-color problem. There are many unsolved problems in mathematics. It was not known for many years whether or not the Jordan curve theorem was true—though during part of this time many people suspected it to be true. The four color problem is still unsolved. It asks if each map on the surface of a sphere can be colored with four colors if the boundaries of the states are simple closed curves. States whose boundaries meet only at a point are allowed to have the same color, but states whose boundaries meet along an arc are not allowed to have the same color. If some of the states had boundaries that contained part of the graph of $y = \sin 1/x$, we might have a situation such as depicted in Figure 33 so it is essential to impose the condition that the boundaries of the states be simple closed curves.

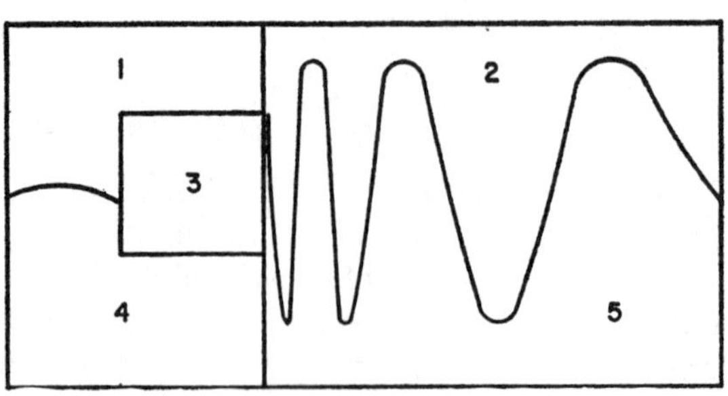

FIG. 33

8. PROPERTIES OF THE REAL NUMBERS

Much of mathematics is devoted to a study of the real numbers. In analytics much use is made of the fact that there is a 1-1 correspondence between the real numbers and the points on a line. Certain of the topological properties of a line can be developed from a study of the real numbers. With the view of getting a set of axioms that will characterize the line, we list certain properties of the set of real numbers E^1.

PROPERTY I. *E^1 is linearly ordered—that is, its elements a, b, c are ordered so that:*

1. *If $a \neq b$, either $a < b$ or $b < a$ (each pair of elements is related);*
2. *$a \not< a$ (no element is less than itself);*
3. *If $a < b$ and $b < c$, then $a < c$ (transitive property).*

When we say that $a = b$, we mean that a is the same as b. Ordinarily we use the plural to denote more than one and, if we say that a, b are elements, we imply that they are distinct.

Any set (whether it is E^1 or not) is said to be *linearly ordered* if it has an order relation satisfying the above conditions.

Exercise 46. Show that if a, b are elements of a linearly ordered set, then not both $a < b$ and $b < a$.

An element a of a linearly ordered set X is said to be a *maximum* of X if for each other element b of X, $b < a$. Also, a is a *minimum* if for each other element b of X, $a < b$.

PROPERTY II. *E^1 has no maximum or minimum.*

Exercise 47. Give an example of a linearly ordered set failing to satisfy Property II.

If A, B are two (nonnull) subsets of a linearly ordered set X, then $[A, B]$ is called a *cut* of X if (1) each point of X belongs to either A or B, and (2) each point of A is less than each point of B.

Exercise 48. (a) Give an example of a cut $[A, B]$ of the reals E^1 so that A has no maximum. (b) Give an example of a cut $[D, E]$ of E^1 so that D has a maximum. (c) In which of the above cases does B or E have a minimum?

PROPERTY III. *E^1 satisfies the Dedekind cut condition—that is, if $[A, B]$ is a cut of E^1, either A has a maximum or B has a minimum.*

Exercise 49. Give an example of set X satisfying Properties I, II, but not III.

Exercise 50. Give an example of a set X different from E^1 satisfying Properties I, II, III.

A set is called *countable* if either it is finite or its elements can be put in a one-to-one correspondence with the positive integers—that is they can be arranged in a sequence. It may be shown that each subset of a countable space is countable.

A set is called *uncountable* if it is not countable. It is known that the rational numbers are countable while the real numbers are uncountable.

a is between b and c if either $b<a<c$ or $c<a<b$.

PROPERTY IV. *E^1 is separable—that is, it contains a countable set of points $p_1, p_2, \cdots$ such that if a, b are any two points of E^1 there is some p_i between a and b.*

Exercise 51. What is a countable set of numbers that shows the real numbers are separable?

Two elements of a linearly ordered set are *adjacent* if there is no element between them.

Exercise 52. Show that no two points of a set satisfying Property IV are adjacent.

If a, b are two points, the set consisting of a, b and all points between a and b is called a *closed interval.* If a, b are adjacent, the corresponding closed interval contains only two points.

Two closed intervals *intersect* if some point belongs to both.

Exercise 53. Show that if G is a collection of closed intervals of the real numbers and no two elements of G intersect, then G is countable.

Example 1. Consider the points of a horizontally based square linearly ordered as follows: $a<b$ means that either a, b are on the same vertical line and b is above a or a lies on a vertical line to the left of b.

Exercise 54. Show that above example does not satisfy Properties II, IV. How may it be modified so as to satisfy Property II?

Example 2. Suppose that each finite ordered set of letters is a *word.* Hence CAT, AAAC, ZZRNT, A, NAACP, KKK, etc., are words. Suppose the words are linearly ordered as would be done in an alphabetical ordering.

Exercise 55. Example 2 satisfies Property I. Show that it has the following properties: a. It has a minimum but no maximum. b. It does not satisfy Property III. c. There is no word preceding CAT which is adjacent to CAT but there is a word following CAT which is adjacent to CAT. d. There are only a countable number of words in the set.

Unsolved problem. Mathematicians have shown that any set L satisfying Properties I, II, III, IV is like the set of reals E^1 in the sense that there is a **1-1** correspondence between the elements of L and the numbers of E^1 that preserves the order relation. It is an unsolved problem as to whether or not any set satisfying Properties I, II, III, IV′, IV″ is like a line where IV′, IV″ are as follows:

PROPERTY IV′. *No two elements of L are adjacent.*

PROPERTY IV″. *Any collection of nonintersecting closed intervals of L is countable.*

This question was raised (in French) in a Polish mathematical journal in 1920 by a Russian mathematician named Souslin. It is called the Souslin problem. Despite many attempts to solve it, it remains unsolved today. One could solve it by either showing that Properties I, II, III, IV′, IV″ imply Property IV or by constructing an example satisfying Properties I, II, III, IV′, IV″ but not Property IV. Of course, one should not do both.

9. A TOPOLOGICAL CHARACTERIZATION OF A LINE

We use the properties of the real numbers considered in the preceding section to give a topological characterization of a line. This characterization is a set of axioms satisfied by the line such that anything that satisfies the axioms is topologically equivalent to a line.

A space is a linear topological space if its points are ordered and its neighborhoods are the sets of the following three sorts—(1) the set of points between two fixed points, (2) the set of points before some fixed point, or (3) the set of points following some fixed point. In studying a line, we only need the first kind of neighborhood but if we study a segment (which is also a linear topological space) the other types of neighborhoods are needed. (Sometimes other kinds of neighborhoods are needed but we do not find this convenient for our present purposes.)

A topological space is said to be *separable* if it contains a countable set of points such that each neighborhood contains one of these points.

The following is a topological characterization of a line L.

AXIOM 1. *L is a linear topological space.*

AXIOM 2. *L has no maximum or minimum.*

AXIOM 3. *L is connected.*

AXIOM 4. *L is separable.*

It may be shown that any space satisfying the above axioms is topologically equivalent to a line. Of course, there is no assurance that this space is "straight."

We recall some examples from the preceding section to see why all these axioms are necessary. To see that Axiom 2 is necessary (*i.e.*, Axioms 1, 3, 4 are not sufficient), consider a segment. To see that Axiom 3 is necessary, one could consider the set obtained by removing one point from a line. To see that Axiom 4 is needed, one could consider the points of the plane between and on two horizontal lines ordered as were the points in Example 1 of the preceding section.

Can Axiom 4 be replaced by the following?

AXIOM 4′. *L does not contain uncountably many mutually exclusive neighborhoods.*

There is no known proof that Axioms 1, 2, 3, 4′ characterize the line. Neither is there a known example of a space satisfying Axioms 1, 2, 3, 4′ that is not topologically equivalent to a line. This again is the Souslin problem.

Since there is 1-1 correspondence between points of a line and the real numbers that preserves order, the above axioms may be used to prove some useful theorems about real numbers. We illustrate with examples.

THEOREM. *A bounded collection of numbers has a least upper bound.*

We shall give a proof of this theorem based on the axioms. First we elaborate on its meaning.

A number x is an *upper bound* for a collection N of numbers if x is greater than or equal to each value of N. The *least upper bound* is the smallest of all the upper bounds. Similarly, a *lower bound* and a *greatest lower bound* are defined. A set of numbers with both an upper and a lower bound is called *bounded.*

Now for a proof of the theorem. Suppose N is a bounded collection of numbers. Let R be the set of all points p of the x-axis such that p is associated with an upper bound of N, and L be the rest of the x-axis. (See Fig. 34.)

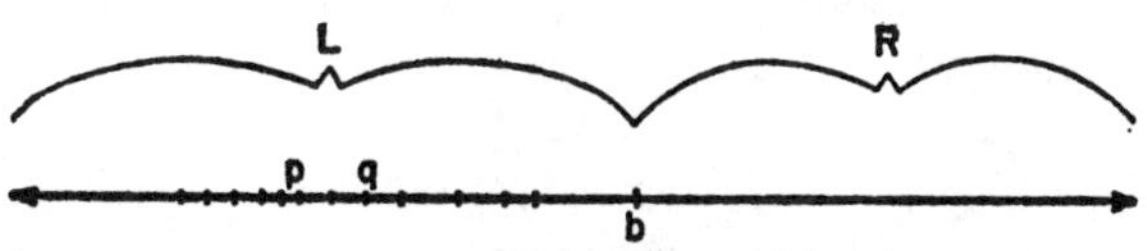

FIG. 34

There is a point in R because N was given to be bounded. There is a point in L because any number less than a number in N is not an upper bound of N.

No point p of L is a limit point of R because if p is in L, there is a point q corresponding to a number of N such that q is to the right of p—no point to the left of q is a point of R.

Axiom 3 tells us that L, R cannot be mutually separated, so some point b of R is a limit point of L. This point will correspond to an upper bound x_b of N since b is in R. Since b is a limit point of L, for each positive number ϵ there is a point of N within ϵ of x_b. Then x_b is the least of all the upper bounds of N.

We now mention some theorems of a related nature frequently used in calculus.

1. *A collection of numbers with a lower bound has a greatest lower bound.*

2. *A bounded monotone increasing sequence of numbers converges to the least upper bound of the sequence.*

3. *The Dedekind cut proposition holds.*

(The proposition, sometimes taken as an alternative axiom, states that if $[A, B]$ is a cut of the reals, either A has a maximum or B has a minimum.)

4. *A bounded infinite collection of numbers has a greatest limit.*

(This limit is sometimes called the upper limit. If the numbers are regarded as points, it will be a limit point that is larger than any other limit point.)

Exercise 56. How would the x-axis be broken into two sets L, R to show that the above Theorem 4 holds?

Exercise 57. Show from the axioms that a segment is connected.

A subset of the plane is said to be bounded if it lies on the interior of some circle. It can be shown that each closed and bounded subset of the plane is bicompact and hence compact. Also, each compact subset of the plane is closed and bounded.

Exercise 58. Show that each compact subset of a line is closed and bounded.

Exercise 59. In showing that each closed and bounded subset of the x-axis is bicompact, one must show that if G is an open covering of a closed and bounded subset S of the x-axis, then a finite number of elements of G cover S. How would the x-axis be broken into two sets L, R to show that the assumption that no finite subcollection of G covers S leads to the contradiction that the x-axis is not connected?

The following theorems used in calculus and concerning a continuous function $f(x)$ defined on a segment $(a \leqq x \leqq b)$ are special cases of more general theorems we have considered.

1. *$f(x)$ is bounded.*

2. *$f(x)$ has a maximum and a minimum.*

3. *If $f(x)$ takes on two values, it takes on each value between them.*

The first of these results follows from the facts that a segment is bicompact (Exercise 59), its image under a continuous transformation is bicompact (Exercise 33) and hence compact (Exercise 34), and a compact set of numbers is closed and bounded (Exercise 58).

The second of these results follows from the further facts that a closed and bounded set of numbers has a least upper bound and a greatest lower bound since it is bounded. It takes on these values since it is closed.

The third result follows from the facts that a segment is connected (Exercise 57) and therefore its image is connected (Exercise 29).

A set X is said to have the *fixed-point property* if for each mapping f of X into itself, for some point p of $X, f(p)=p$. A circle does not have the fixed point property because a rotation can move each of its points. It can be shown that a segment, a disk, and a solid cube do have the fixed point property.

Exercise 60. Show that a segment has the fixed point property.

Unsolved problems. It is conjectured that if C is a bounded plane continuum whose complement is connected in the plane, then C has the fixed-point property. Partial solutions have been given of two types—one of these says that if the continuum has certain properties (such as being like a segment or a disk), it has the fixed-point property; the other type says that if certain restrictions are placed on the mapping, some point is left fixed. In spite of the vast amount of time spent on this problem by some talented workers, the answer still eludes us. Perhaps some student coming along will succeed in verifying the conjecture that has baffled so many of his predecessors—or maybe someone will construct a counterexample showing that the conjecture is false.

An arc is a set topologically equivalent to a segment. If one were asked to name a plane continuum with more than one point which was topologically equivalent to each continuum in it with more than a point, one might think of an arc (See Fig. 4). For many years no one knew of any example other than an arc, although the question was raised in mathematical journals as to whether or not there were any. In 1948 such an example was found. The problem as to whether or not there are still other examples is unsolved.

10. SOME EXAMPLES OF TOPOLOGICAL SPACES

What is a point?

If one is studying the real number system or a real line one might answer that a point is a real number. For analysis, the answer might be—a complex number. In plane analytics, a point is an ordered pair of real numbers—in solid analytics it is an ordered triple of real numbers.

In synthetic geometry and in topology, a point is an undefined term. If we have a collection of elements (which we call points), and certain subcollections of these elements (which we call neighborhoods), we may regard the original collections as a topological space if the neighborhoods satisfy certain conditions called axioms.

In this section we mention some abstract spaces somewhat out of the ordinary. We describe them by telling what the points and neighborhoods of these spaces are.

Euclidean 4-space. Here each point has four coordinates instead of three. (No mention need be made of time. In fact, any student who has worked problems with four variables may consider himself as having worked with 4-space.) A point is an ordered quadruple of numbers and a neighborhood with center at (x_0, y_0, z_0, w_0) and radius ϵ is the set of all points $(x, y, z\ w)$ such that $((x_0-x)^2+(y_0-y)^2+(z_0-z)^2+(w_0-w)^2)^{1/2}<\epsilon$.

Hilbert space. The points of this space are sequences $(a_1, a_2, \cdots)$ of numbers such that $\sum a_n^2$ exists. The distance between two points $(x_1, x_2, \cdots)$, $(y_1, y_2, \cdots)$ is $(\sum(x_n-y_n)^2)^{1/2}$. Neighborhoods are the interiors of generalized spheres.

A function space. A point in this abstract space is a continuous function $y=f(x)(0\leqq x\leqq 1)$. A neighborhood with center at $f_0(x)$ and radius ϵ is the collection of all functions $f(x)$ such that $|f(x)-f_0(x)|<\epsilon$ for each x between 0 and 1 inclusive.

Metric spaces. Many of the spaces we deal with in topology are metric spaces. *Metric spaces* are topological spaces with distance functions $\rho(p, q)$ (called the distance between p and q). *Distance functions* satisfy the following conditions.

1. $\rho(x, y)\geqq 0$, *the equality holding if and only if* $x=y$.
2. $\rho(x, y)=\rho(y, x)$ (*symmetry*).
3. $\rho(x, y)+\rho(y, z)\geqq\rho(x, z)$ (*triangle condition*).
4. $\rho(x, y)$ *preserves limit points.*

This fourth condition means that a point x belongs to the closure of a set M if and only if for each positive number ϵ, there is a point m of M such that $\rho(x, m) < \epsilon$.

Exercise 61. Describe a distance function for the above function space.

We shall now describe a topological space that has no distance function.

Another function space. This is a function space used extensively in studying linear spaces. Again a point in this abstract space is a function $y=f(x)$ $(0 \leqq x \leqq 1)$ but this time we do not suppose that it is continuous. To get a neighborhood in this function space, we select a finite collection of numbers $x_1, \cdots, x_n$ between 0 and 1 inclusive and a positive number ϵ. The neighborhood $N(f_0(x); x_1, \cdots, x_n; \epsilon)$ is the set of all functions $f(x)$ defined for $0 \leqq x \leqq 1$ and such that $|f_0(x_i) - f(x_i)| < \epsilon$ $(i=1, \cdots, n)$. To get another neighborhood, we select another function, another finite set of points, and another positive number.

If p is a limit point of a set X in a metric space, there is a sequence of distinct points of X converging to p.

In the function space we have just described, let $f_0(x)$ be the function such that $f_0(x)=0$ for $0 \leqq x \leqq 1$. Let M be the collection of all functions that take on the value 0 at a finite collection of numbers and on the value 1 everywhere else. Although $f_0(x)$ is a limit point of M, no sequence of points of M converges to $f_0(x)$. This shows that the function space we have described is not a metric space.

Exercise 62. If $f_1(x), f_2(x), \cdots$, is a sequence of points of M in the above function space, what is a neighborhood of $f_0(x)$ that contains no point of $f_1(x), f_2(x), \cdots$?

Exercise 63. If p is a limit point of a set M in a metric space, how might one select a sequence of distinct points of M converging to p?

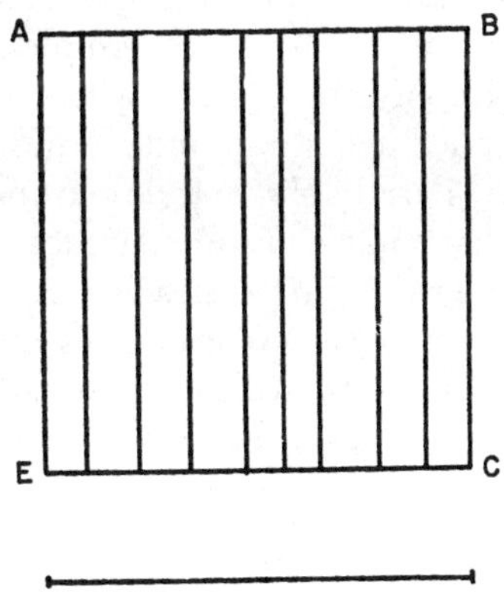

FIG. 35

Decomposition spaces. Suppose a square $ABCE$ plus its interior is decomposed into vertical straight line segments as shown in Figure 35. Let D be the space whose " points" are these vertical segments and whose "neighborhoods" are the sum of such collections of " points" which are open subsets of the square disk. The decomposition space D is topologically equivalent to a segment.

Exercise 64. What "points" of D correspond to the ends of the segment equivalent to D?

If the square disk were decomposed in the manner illustrated in Figure 36, it is possible to define "points" and "neighborhoods" so that the decomposition space is topologically equivalent to the sum of two spherical surfaces and two segments.

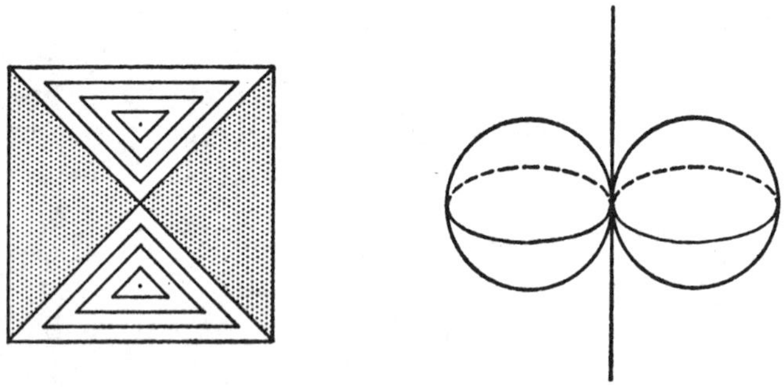

FIG. 36

Exercise 65. Which element of the decomposition space corresponds to the point of tangency of the two spheres? Which elements correspond to the outer ends of the two segments?

These two examples illustrate how topology may be used to study collections of figures rather than ordinary point sets.

The topologist is acquainted with dozens of different spaces—he may invent new spaces as the needs arise. Some of these spaces are much simpler than the ones we have described and others are more complex.

Queer spaces play an important role in topology. A wealth of examples of spaces with certain properties helps our understanding of these properties. It suggests to us that theorems are likely to follow from these properties.

Also, queer spaces are useful as counterexamples. There is no more convincing way of showing that one set of properties do not imply a second set than by exhibiting a space with the first set of properties but without the second set. Of the problems solved during recent years which had been regarded as difficult unsolved problems for many years, a goodly number of them were solved by the use of counterexamples.

11. SOME AXIOMS FOR TOPOLOGY

In geometry we may prove certain theorems on the basis of only part of our axioms—for example, we may omit the parallel axiom and study what theorems can be proved with the remaining set of axioms. While the axioms we then use do not completely describe the plane, we find that we can prove a restricted set of theorems on the basis of them. The beginning axioms we list are satisfied by a line, a plane, and E^3, but do not completely describe any of them. An advantage of working with such a restricted set of axioms is that the theorems we prove on the basis of these axioms will not only hold true on a line, in the plane, and in E^3, but in any space whatsoever that satisfies these axioms.

Suppose we have a collection S (which we call space) of things (which we call points) and certain subcollections of S (which we call neighborhoods). The points and neighborhoods may or may not be like the things we ordinarily think of as points and neighborhoods. We suppose however that they satisfy the following axioms.

AXIOM 1. *If a point p belongs to each of two neighborhoods N_1, N_2, then p lies in a neighborhood N_3 such that $N_3 \subset N_1 \cdot N_2$.*

AXIOM 2. *If p, q are two points, there are neighborhoods N_p, N_q containing p, q, respectively, such that $N_p \cdot N_q = 0$.*

We take "point" and "neighborhood" as undefined terms with the restriction that each neighborhood is a point set. Then without even knowing what these points and neighborhoods are, we can prove certain theorems on the basis of these axioms. Suppose that we have the neighborhood definition of limit point and the definitions of open, closed, sum, intersection, as given earlier. Then the following theorems can be proved.

THEOREM 1. *If p is a limit point of the sum of two sets, it is a limit point of one of them.*

THEOREM 2. *The intersection of two open sets is open.*

THEOREM 3. *The sum of two closed sets is closed.*

THEOREM 4. *A point set with only one point in it is closed.*

The spaces considered in the preceding section *Some examples of topological spaces* satisfied the two axioms we listed. These axioms are called the *Hausdorff axioms* and a space satisfying them is called a *Hausdorff space.* Hence, the four theorems are true in these spaces. Many additional theorems can be proved on the basis of them.

Exercise 66. An equivalent statement of Theorem 1 says that if p is not a limit point of either A or B, it is not a limit point of $A+B$. Prove this alternative statement. Which of the other three theorems is implied by Theorem 1?

Exercise 67. Theorem 2 can be proved from Theorem 3 by resorting to the results that the complements of closed sets are open and the complements of open sets are closed. However, prove it directly from Axiom 1.

Exercise 68. Which of the axioms is used to prove Theorem 4?

Exercise 69. Let $V(p, \epsilon)$ denote the set of all points whose distance from p is less than ϵ. To show that any metric space with the $V(p, \epsilon)$'s as neighborhoods satisfies Axiom 2, we might demonstrate that if p, q are two points, then $V(p, \rho(p, q)/2)\cdot V(q, \rho(p, q)/2)=0$. How do we know that there is no point common to $V(p, \rho(p, q)/2)$ and $V(q, \rho(p, q)/2)$?

Exercise 70. To prove that a metric space satisfies Axiom 1 we might show that if r were a point in $V(p, \epsilon)\cdot V(q, \delta)$, then there is a positive number γ such that $V(r, \gamma)\subset V(p, \epsilon)\cdot V(q, \delta)$. What could we use for γ?

There are many theorems that could not be proved on the basis of Axioms 1 and 2 alone. For example, it could not be proved that a simple closed curve separates the plane. This result holds on the surface of a sphere or in a plane and is a consequence of the Jordan or Schoenflies theorems. However, additional axioms need to be assumed to prove these results from axioms. We frequently introduce additional axioms in studying topology. Such axioms as those given in the section *A topological characterization of a line* might be used. We might add an axiom saying that space is metric. We might axiomatize that neighborhoods are connected, that space is compact, that simple closed curves separate space, etc.

The following two exercises show why the space L of the section *A topological characterization of a line* is a Hausdorff space. If p and q are two points with $p<q$, we use (p, q) to denote the set of all points between p and q.

Exercise 71. To prove that L satisfies Axiom 1 of the present section, we show that if $r\epsilon(p_1, q_1)\cdot(p_2, q_2)$, then there is a (p_3, q_3) such that $r\epsilon(p_3, q_3)\subset(p_1, q_2)\cdot(p_2, q_2)$. What might we take for (p_3, q_3)?

Exercise 72. To show that L satisfies Axiom 2 of this section, we would consider two points p and q of L. Let r be a point between them, and let N_p, N_q be the sets of points preceding and following r respectively. Which of the axioms for L would not be satisfied if there were no point between p and q?

12. DEFINITIONS AND UNDEFINED TERMS

We continue our comparison of topology with geometry. We repeat some of the things we have said about definitions and undefined terms. First we consider undefined terms.

In giving definitions in an order, we find that the first terms to be considered do not have any predecessors. We take these as our undefined terms. In an intuitive sense, these undefined terms have a meaning for us. For example, in geometry we take the terms "set," "point," "line," "plane" as undefined terms. Although we may have an intuitive understanding of the meaning of these terms, we do not define them. If we develop our theorems logically without recourse to our intuitive notions about these terms, our results may have a wide interpretation.

We also have undefined terms in topology. We may have an intuitive feeling as to the meaning of an undefined term, though we may change this intuitive meaning without upsetting the logical structure of our study. In topology we take such words as "point," "neighborhood," "set" as undefined. We also suppose that we have a language at our command and know the meaning of such expressions as "is an element of." As a convenience we use previously understood terms from more elementary branches of mathematics.

Exercise 73. What represents a point in each of the following spaces? (a) The coordinate plane (Euclidean 2-space E^2). (b) Euclidean 3-space E^3. (c) Euclidean 4-space E^4. (d) Hilbert space. (e) The first function space given in the section *Some examples of topological spaces.* (f) The first decomposition space from that section.

In ordinary usage, a neighborhood of us is the set of things near us. There are neighborhoods of many sizes. In topology, our axioms tell us that a neighborhood is a point set but this is not a definition of a neighborhood since there are some point sets that are not neighborhoods. If we were restricting our attention to the plane, we would take a neighborhood to be the interior of a circle. In this case, our neighborhoods vary in size. We say that each point lies in an arbitrarily small neighborhood—meaning, that for each positive number ϵ, there is a neighborhood of the point of diameter less than ϵ. Instead of saying that a set X is arbitrarily close to a point p, we say that each neighborhood containing p contains a point of X. Since we want our axioms to hold in a variety of spaces, we do not insist that neighborhoods be the interiors of circles. However, we use them much as if they were. Perhaps at the start it is convenient to think of a point as a geometric object and a neighborhood as the interior of a circle.

One of the things you may have gained from this article is an increased vocabulary so that you know the meaning of more of the terms of topology. However, definitions given in this article were not always given in the logical order. The four columns in Figure 37 not only give you an opportunity to recall certain of the definitions of topology but shows a possible logical order. The columns are arranged so that the terms in each column (after the first) may be defined by using terms from the preceding columns.

I	II	III	IV
Undefined terms	limit point	closed	
		closure	
		mutually separated	connected
Terms from other		separate	
branches of mathematics		compact	
(circle, ordered pair,	transformation	continuous transformation	reversibly continuous
1-1 correspondence,			
interior of a circle,		inverse	fixed point property
numbers, is greater		onto	
than, etc.)	countable	uncountable	
		separable	
	intersection	Hausdorff space	
Language of "every-day"	open set	open covering	bicompact
vocabulary (contains,	complement		
lies in, is an element of,	bounded plane set		
etc.)	subset		
	sum		
	difference		
	greatest lower bound		
	least upper bound		
	mutually exclusive		
	sequence converges		

FIG. 37

By adding another column we could include the term homeomorphism. The terms homogeneous and topologically equivalent would come in the next column. After we had defined topologically equivalent we could define arc, simple closed curve, and 2-sphere.

The reader may find it a good review of this article to refer to the list of terms following the next section and see how many of the definitions he can recall.

13. TRENDS TOWARD ABSTRACTNESS

One of the aspects of modern mathematics is the trend toward abstractness. This is not only true in topology but also in algebra, analysis, and other branches of mathematics. Let us examine how this trend has affected topology.

In proving a theorem in plane geometry, the student does not use all the properties of the plane but merely uses certain of the axioms. Similarly, in proving a topological theorem in the plane, one might note that his proof depends only on certain properties of the plane. The theorem would then hold true in any space with these essential properties. Let us illustrate this with an example.

THEOREM. *The sum of two closed plane sets is closed.*

This theorem can be proved without mentioning many properties of the plane. There is no need to mention the x-axis, y-axis, or straight lines. The theorem can be proved by using the definitions of closed sets, sum, limit point, along with the fact that if N_1, N_2 are two neighborhoods containing the same point, there is a neighborhood N_3 of the point in $N_1 \cdot N_2$ (Axiom 2 of the preceding section).

Exercise 74. Prove the above theorem on the basis of the definitions and the fact that if a point of the plane lies in each of two neighborhoods $N_1 \cdot N_2$, there is a neighborhood N_3 of the point in $N_1 \cdot N_2$.

As pointed out in the preceding section, we can prove the theorem about the sum of two closed sets being closed in any of the spaces mentioned in the section *Some examples of topological spaces* since all of these spaces satisfy Axiom 2 of the preceding section. One method of studying topology is to consider what theorems can be proved in a space with certain properties. One may not even know what space is under consideration as long as it is known that the space has the properties used in proving the theorem. Naturally the mathematician is interested in getting the maximum in the way of results or conclusions from a minimum in the way of hypotheses and axioms.

If one is able to prove that certain theorems are true on the basis of certain properties of a space, at some time in the future a use may be found for a particular space with these properties, and the theorems proved before the space was invented will be known to hold.

An advantage (other than giving the imagination and mind ample working space) in dealing with abstract spaces is that our theorems tell us more than things about ordinary points and also reveal truths about collections of other sorts—for example, about collections of functions, such as described in Examples 3 and 4 in the section *Some examples of topological spaces.*

There are other abstract spaces which have for points other sorts of unexpected things—closed sets, continua, matrices.

Types of unsolved problems. We finish this article by mentioning some types of unsolved problems.

As we have indicated, some theorems go about as follows: Each space with a certain set of properties has some other property. Quite a few of the many mathematical papers published each year are devoted to proving this kind of theorem. We shall not discuss here the properties of interest but they include such things as being compact, bicompact, and having distance functions.

A more concrete type of theorem tells us that certain theorems (such as the Jordan curve theorem) hold in the plane or in some other familiar space. During the past forty years we have made much headway in proving theorems for the plane. In more recent years some headway has been made at hacking away at some of the complicated theorems in 3-space.

Intermediate between the abstract and concrete types of theorems are the characterization theorems. Show that a well-known space (such as a line, plane, or 3-space) satisfies a certain minimum set of conditions—then show that any space that satisfies these same conditions must be topologically equivalent to the well-known space.

We have some good topological characterizations of the line. However, one of the interesting unsolved problems (Souslin problem) is in this area. It gives a certain set of conditions satisfied by the line and asks if any space which satisfies them must be homeomorphic with a line.

Topological characterizations of the plane have been discovered during the past 40 years. An unsolved problem for many years in this area may be stated as follows: Is the connected compact metric space S topologically equivalent to the surface of a ball if it has more than one point and satisfies the following conditions?

1. *The neighborhoods of S are connected.*
2. *The complement of no simple closed curve is connected.*
3. *The complement of each pair of points is connected.*

The answer was given in the affirmative in 1946. The resulting characterization of a 2-sphere is called the Kline sphere characterization.

Although topological characterizations of 3-space have been given, these are still complicated. In view of the recent spurt toward increasing our knowledge of the topological properties of 3-space, there is hope for improvements here.

It is difficult to specify the types of unsolved problems that are of most interest to research topologists. Researchers vary in their interests and the field is wide. Things in which one worker is intensely interested may not even concern someone else at the moment, because he may be too busy working on something else. Although there are many unsolved problems in topology, those of interest are not likely to have easy solutions. However, we have talented young people coming along each year who are taking their places as ingenious researchers and making contributions that help push the frontier of mathematics farther ahead.

LIST OF TOPOLOGICAL TERMS

The following list showing where topological terms are defined in this article may be of convenience as a reference.

ANSWERS TO EXERCISES

1. Three. n.

2. Cube and ball, knotted curve and circle, theta curve and capital letter **B**, square disk pierced by segment and hemisphere with a tangent segment.

3. (A, R), (B), $(C, G, I, J, L, M, N, S, U, V, W, Z)$, (D, O), (E, F, T, Y), (H, K), (P, Q), (X). Yes—if the letters have width, each letter is equivalent to a disk or a disk with one or two holes in it.

4. The following pairs correspond: (c, z), (e, w), (f, v), (g, u). Either s or t might correspond to h. Any point on the circle $vxzy$ other than v and z might correspond to d.

5. A Moebius band.

6. The triangle with the upper left vertical edge and the one with the lower right vertical edge; or the triangle with the lower left vertical edge and the one with the upper right vertical edge; or the triangles with the leftmost and rightmost middle vertical sides.

7. A Moebius band.

8. A Moebius band. A projective plane.

9. Any function $f(x, y)$ satisfying the conditions $f(x, 0)=f(x, 1)$ and $f(0, y)=f(1, 1-y)$ will suffice. Examples of such functions are $f(x, y)=\cos 2\pi x \cos 2\pi y$ and $f(x, y)=x(x-1)(x-3)+y(y-1)(x+y)(x+y+1)^2(x+y-2)$.

10. A disk pierced by a segment and a hemisphere with a tangent segment, two homeomorphic bands with a different number of twists, some pair of the tori shown in Figure 13.

11. A Moebius band, a projective plane, a Klein bottle.

12. A Moebius band. We avoided mentioning a projective plane and a Klein bottle because they do not lie in E^3 and the sides of a surface refer to parts of the space in which they lie.

13. Any pair of the tori shown in Figure 13.

14. r and t are points of X; p, q, t are limit points.

15. p, q, s, t are limit points of X.

16. If $p \notin A$, $p \in B$ by the definition of $A+B$. Since B is open, p belongs to a neighborhood N in B by the definition of open. Then N lies in $A+B$.

17. Let r_p be the distance from p to the boundary of S. It is also permissible for it to be a fractional part of this distance.

18. (a) A line, (b) the interior of a square, (c) a segment without its end points, (d) the plane.

19. Show that each limit point of the closure belongs to the closure.

20. $\bar{G}$ is the sum of G, the segment from $(0, 1)$ to $(0, -1)$ and the set whose only element is the point $(1/\pi, 0)$.

21. Since p is a limit point of $A \cdot B$, each neighborhood of p contains a point of $A \cdot B-\{p\}$ by the definition of limit point. Then each neighborhood of p contains a point of $A-\{p\}$ since each point of $A \cdot B$ is a point of A. Then p is a limit point of A by the definition of limit point. Since A is closed, $p \in A$.

22. If each neighborhood of p contains a point of A, it contains a point of $A-\{p\}$ since $A=A-\{p\}$. Then p would be a limit point of A by the definition of limit point and $p \in A$ since A is closed.

23. If $p \notin A'$, $p \in A$ by the definition of A. Then p lies in a neighborhood N in A since A is open. No point of N lies in A' since each of its points lies in A.

24. Since $p \notin \bar{A}$, there is a neighborhood N of p that contains no point of A. Then N contains no point of $\bar{A}$ and p is not a limit point of $\bar{A}$.

25. (a) 1, 2, 7, 8. (b) 1, 5, 7. (c) 1, 2, 3, 4, 5, 8. (d) All.

26. (a) Projection of p on x-axis. (b) Projection of p on y-axis. (c) Distance from p to horizontal lines if they are equidistant from p. (d) Distance from p to vertical lines if they are equidistant from p.

27. If for each pair of horizontal planes, one above and one below p, there is a neighborhood of p such that each point of the graph in, above, or below the neighborhood is between the horizontal planes.

28. We show that each neighborhood U of $f(p)$ contains a point of $f(X)$. This insures that $f(p)$ is either a point or a limit point of $f(X)$. Let V be a neighborhood of p such that $f(V\cdot X)\subset U$. Since p is a limit point of X, V contains a point x of X and $f(x)$ is in U.

29. If $f(X)$ were the sum of two mutually separated (nonnull) sets A, B, then X would be the sum of the sets $f^{-1}(A)$, $f^{-1}(B)$. These two sets are mutually exclusive because f is single valued, and each is an open subset of X by Exercise 30. Hence if $f(X)$ were the sum of two mutually separated (nonnull) sets A, B, X would be the sum of two mutually separated (nonnull) sets $f^{-1}(A)$, $f^{-1}(B)$.

30. Suppose p is a point of $f^{-1}(D)$, where D is an open subset of Y. If U is a neighborhood of $f(p)$ with $U\cdot Y\subset D$, the continuity of f implies that there is a neighborhood V of p such that $f(V\cdot X)\subset U$. Then $V\cdot X\subset f^{-1}(D)$. The sum of all such V's for all points p of $f^{-1}(D)$ is an open set W and $W\cdot X=f^{-1}(D)$ is an open subset of X.

31. If U is an open subset of Y containing $f(p)$, we have that $f^{-1}(U)$ is an open subset of X, *i.e.*, $f^{-1}(U)=D\cdot X$, where D is open. We can use a neighborhood V of p in D to show that the definition of continuity is satisfied at p. Since f is continuous at each point p of X, it is continuous.

32. We show that each infinite collection $y_1, y_2, \cdots$ of points ($y_i\neq y_j$ if $i<j$) of $f(X)$ has a limit point in $f(X)$. Let x_i be a point of $f^{-1}(y_i)$. Then $x_1, x_2, \cdots$ has a limit point x_0 in X. Let K be infinite set formed by subtracting $f^{-1}f(x_0)$ from $x_0, x_1, x_2, \cdots$. Since x_0 is a limit point of K but $f(x_0)$ is not a point of $f(K)$ then, by Exercise 28, $f(x_0)$ is a limit point of $f(K)$ and hence of $y_1, y_2 \cdots$.

33. If G is a collection of domains covering $f(X)$ the inverse of each of these is an open subset of X by Exercise 30. If X is bicompact, a finite number of these inverses cover X and the collection of corresponding elements of G covers $f(X)$.

34. We show that if K is an infinite collection of points in a bicompact space X, then K has a limit point in X. For each point p of K let G_p be the sum of p and the complement of $\overline{K}$. No finite subcollection of the G_p's covers X so either the G_p's do not cover X or some G_p is not open. If some point q of X does not lie in any G_p, this point q is a limit point of K because it belongs to $\overline{K}$ but not to K. If some G_p is not open the corresponding point p is a limit point of K because if it were not, G_p would be the sum of two open sets—the complement of $\overline{K}$ and a neighborhood containing p but no other point of K.

35. We use functional notation to describe the homeomorphism—that is, the point corresponding to x is called $f(x)$. Let o be a point of the interior of the triangle and $f(o)$ be the center of the circle. For each point p on the triangle, let $f(p)$ be the point on the circle such that the ray from o through p points in the same direction as the ray from $f(o)$ through $f(p)$. Each point q of the first ray corresponds to the point $f(q)$ of the second ray such that $f(q)$ divides $f(o)$, $f(p)$ in the same ratio (internally or externally) that q divides o, p. (See Fig. 27.)

36. First we describe a homeomorphism g which takes the plane onto itself and the circle with equation $x^2+y^2=1$ onto the given graph. If p is a point of the interior of the circle, $g(p)$ is on the same vertical line with p and divides the points where the vertical line hits the graph in the same ratio that p divides the points where the line hits the circle. If p is beneath the circle, $g(p)$ is on the same vertical line with p and the same distance below the graph as p is below the circle. If p is a point of the plane that is neither interior to, below, or on the lower semicircle of the circle, $g(p)=p$.

Suppose f is a homeomorphism such as given in Exercise 35 of the plane onto itself that takes the given triangle onto the circle with equation $x^2+y^2=1$. Then gf is the required homeomorphism —to find the point that corresponds to p, find the point $f(p)$ that corresponds to p under the homeomorphism f and then find the point $gf(p)$ which corresponds to this point $f(p)$ under g.

37. Construct segment AGC and segment BD as shown in Figure 28. Also construct the "kite-shaped" figure shown in Figure 29. We describe the homeomorphism f of the polygon $ABCDEF$ onto the triangle $A'C'D'$ by steps.

(i) Let f be a homeomorphism that takes AB linearly onto $A'B'$—if p is a point of AB, $f(p)$ divides A', B' in the same ratio that p divides A, B. Extend f to take BC, CD, DE, EF, FA, AG, GC, BD linearly onto $B'C'$, $C'D'$, $D'E'$, $E'F'$. $F'A'$, $A'G'$, $G'C'$, $B'D'$, respectively.

(ii) We now extend f so that it takes the interior of $ABCG$ onto the interior of $A'B'C'G'$. Let

p and $f(p)$ be points of the interiors respectively of these two triangles. If x is a point other than p of the interior of $ABCG$, extend the segment from p through x until it intersects triangle $ABCG$ at a point y. Now $f(y)$ has already been defined in the first step. The point $f(x)$ is the point that divides $f(p)$, $f(y)$ in the same ratio that x divides p, y.

(iii) Similarly extend f to take the interior of triangle BCD onto the interior of triangle $B'C'D'$.

(iv) To extend f to the interior of $ABDEF$, take a point p of this interior such that no ray from p intersects $ABDEF$ in two points (we might choose p on the line which bisects angle E) and let $f(p)$ be any point on the interior of $A'B'D'E'F'$. For each point x other than p of the interior of $ABDEF$, let y be the point where the ray from p through x intersects $ABDEF$ and $f(p)$ be the point that divides $f(p)$, $f(y)$ in the same ratio that x divides p, y.

(v) To extend the homeomorphism f to the exterior of $AGCDEF$, consider the same points p, $f(p)$ used in Step (iv). For each point x of the exterior of $AGCDEF$ let y be the point where the segment from p to x hits $AGCDEF$. Then $f(x)$ is the point that divides $f(p)$, $f(y)$ externally in the same ratio that x divides p, y.

38. So that $h(Y)$ lies on the segment from A to $h(X)$ and divides this segment in the same ratio that Y divides AX.

39. This follows by mathematical induction on $n(P)$ since $AB+AYB$ has fewer sides than P.

40. Let $UV=AB$.

41. Since X lies in Ext $(AB+AYB)$ (X is at a maximum distance from L) AXB lies except for its ends in this exterior. Then CE does not intersect AXB except possibly at C or E. Also, CE was chosen so as not to intersect AYB except at C and E since it intersects $AB+AYB$ only there.

42. $P-(CD+DE)+CE$.

43. gh.

44. B lies in Ext $(ZAY+YZ)$ as does each non-endpoint of ZBY. A lies in Ext $(ZBY+YZ)$ as does each non-endpoint of ZAY.

45. If there were such a point P, consider a segment PQ that misses YZ and has only the point Q on ZAY. We would have a contradiction of the first or second part of Exercise 44 according as PQ does or does not intersect ZBY.

46. If $a<b$ and $b<a$, then $a<a$ by the transitive property. This contradicts the fact that no element is less than itself.

47. The numbers between 0 and 1 inclusive.

48. (a) Let A be the set of negative numbers and B the others. (b) Let D be the set of non-positive numbers and E the others. (c) B has a minimum but E does not.

49. The set of rational numbers or the reals with 0 removed.

50. The integers. Another interesting example is the set of points in the plane on or between two horizontal lines ordered so that $p<q$ if p is below or to the left of q.

51. The rationals.

52. If some two points a, b were adjacent, there would be no point of the countable set $p_1, p_2, \cdots$ between them since there is no point whatsoever between them.

53. Let each of the intervals correspond to a rational number in it. This sets up a 1–1 correspondence between the intervals and a subset of the rationals. The set of rationals is countable so each subset of it is also. Hence, the collection of intervals is countable.

54. Example 1 does not satisfy Property II because the point of the square in the upper right hand corner is a maximum and the point in the lower left hand corner is a minimum. If these two points were removed, the remainder would satisfy Property II. Each vertical segment in the square is a closed interval in the linearly ordered set. The square contains uncountably many such vertical segments, no two of which intersect. Example 1 could not satisfy Property IV because it can be shown by techniques similar to those used in solving Exercise 53 that no space satisfying Property IV can have uncountably many mutually exclusive closed intervals.

55. a. Example 2 has the word A as a minimum. For each given word we can obtain a bigger word by adding another letter.

b. To check Property III, consider an expression such as CCC $\cdots$ (there are infinitely many c's in this expression). It is not a word since only finite ordered sets of letters are words. If A is the set of words preceding CCC $\cdots$ and B is the set following, the cut $[A, B]$ does not satisfy the Dedekind cut condition.

c. The word CATA follows CAT and is adjacent to CAT. However, if a word precedes CAT, we obtain a new word between it and CAT by adding an A to the end of this original word.

d. There are 26 one-letter words, $(26)^2$ two-letter words, $(26)^3$ three-letter words, $\cdots$. We put these words in a 1-1 correspondence with the positive integers as follows: The one-letter words correspond to the integers between 1 and 26 inclusive, the two-letter words correspond to the integers between 27 and $26+(26)^2$ inclusively, the three-letter words correspond to the integers between $27+(26)^2$ and $26+(26)^2+(26)^3$ inclusively, $\cdots$.

56. Let L be the set of points of the x-axis associated with numbers n that have infinitely many numbers of the collection which are bigger than n while R is the rest of the x-axis.

57. Suppose pq is the segment with left end p and right end q, L_p is the part of the line to the left of p and L_q is the part of the line to the right of q. The segment pq is connected because the assumption that it is the sum of two mutually separated sets A, B leads to the contradiction that the line is not connected since it is the sum of two mutually separated sets A', B'; A' is A, $A+L_p$, $A+L_q$, $A+L_p+L_q$ according as A contains neither p nor q, A contains p but not q, A contains q but not p, A contains both p and q; B' is defined in a similar fashion.

58. Suppose that S is a compact (nonnull) subset of the line. Denote a fixed point of S by s_0. If S were not bounded, for each integer i there would be a point s_i in S which would not belong to the neighborhood with center at s_0 and radius i. This is impossible since S is compact but $s_1, s_2, \cdots$ would have no limit point. If S were not closed there would be a point x_0 not of S which is a limit point of S. For each integer i there would be a point s_i of S in the neighborhood with center at x_0 and radius $1/_i$. This is impossible since S is compact and $s_1, s_2, \cdots$ would not have a limit point in S.

59. Let L be the collection of all points p of the x-axis such that a finite subcollection of G covers each point of S to the left of p. Let R be the rest of the x-axis.

60. Suppose f is a mapping of a horizontal segment S into itself, L is the set of points of S that move to the right under f, and R is the set of points of S that move to the left. Unless f leaves some point fixed, the left end of S belongs to L and the right end belongs to R. If a point p of S moves to the right, it follows from the continuity of f that there is a neighborhood of p such that each point of S in the neighborhood moves to the right. Hence, L is an open subset of S. Also, R is an open subset of S. Since S is connected (Exercise 56), it is not the sum of two mutually separated sets L and R. Any point of $S-(L+R)$ is left fixed under f.

61. $\rho(f(x), g(x))$ is the maximum value of $|f(x)-g(x)|$.

62. $N(f_0(x); y; \frac{1}{2})$ where y is a value such that $1=f_1(y)=f_2(y)= \cdots$.

63. For each integer i let m_i be a point of M whose distance from p is less than $1/i$ and such that m_i is different from any of $m_1, \cdots, m_{i-1}$. Then $m_1, m_2, \cdots$ converges to p.

64. The segments AE and BC.

65. The sum of the perimeter of the square and its diagonals. The centers of the upper and lower nested collections of triangles.

66. If p is not a limit point of A, it lies in a neighborhood N_1 which contains no point of $A-\{p\}$; also, if p is not a limit point of B it lies in a neighborhood N_2 which contains no point of $B-\{p\}$. It follows from Axiom 1 that p lies in a neighborhood $N_3 \subset N_1 \cdot N_2$. Then N_3 contains no point of $A+B-\{p\}$ and p is not a limit point of $A+B$. Theorem 3 is implied by Theorem 1.

67. Let p be a point in the intersection of two open sets U_1, U_2. We show that p lies in a neighborhood $N \subset U_1 \cdot U_2$. It follows from the definition of open sets that p lies in a neighborhood N_1 with $N_1 \subset U_1$ and p lies in a neighborhood N_2 with $N_2 \subset U_2$. It follows from Axiom 1 that p lies in a neighborhood $N_3 \subset N_1 \cdot N_2$. Then $N=N_3 \subset U_1 \cdot U_2$.

68. Axiom 2.

69. If r were a point common to $V(p, \rho(p, q)/2)$ and $V(q, \rho(p, q)/2)$, $\rho(p, r)+\rho(r, q)<\rho(p,q)$. This violates the triangle condition for a metric space.

70. Let γ be the minimum of $\epsilon-\rho(p, r)$ and $\delta-\rho(q, r)$.

71. Let p_3 be the larger of p_1, p_2 and q_3 be the smaller of q_1, q_2.

72. L would not be connected. Axiom 2.

73. (a) An ordered pair of numbers. (b) An ordered triple of numbers. (c) An ordered quadruple of numbers. (d) A sequence of numbers the sum of whose squares exists (is a number). (e) A continuous function $y=f(x)$ defined for $0\leqq x\leqq 1$. (f) A vertical segment.

74. To show that the sum $A+B$ of two closed sets A, B is closed, we need to show that each limit point of $A+B$ belongs to $A+B$.

Suppose p is not a point of $A+B$. Then it is not a point of A (from the definition of sum). Since A is given to be closed, p is not a limit point of A (definition of closed). Hence, there is a neighborhood N_1 of p which contains no point of A (definition of limit point). Likewise, p is not a point or limit point of B and there is a neighborhood N_2 of p which contains no point of B.

There is a neighborhood N_3 of p which lies in both N_1 and N_2. (We cannot justify this from the definitions.) No point of A is in N_3 since no point of A is in N_2—also, no point of B is in N_3. Hence N_3 contains no point of $A+B$ since it contains no point of A and no point of $A+B$. Hence, $A+B$ is closed.

References

Elementary articles

1. R. H. Bing, Examples and counterexamples, Pi Mu Epsilon J., vol. 1, 1953, pp. 311–317.
2. ———, Point set topology, Chapter X of Insights into Modern Mathematics, Twenty-third Yearbook of the National Council of Teachers of Mathematics.
3. R. Courant and H. Robbins, What is Mathematics? New York, 1941, Chapter 5.
4. Martin Gardner, Topology and magic, Scripta Math., vol. 17, 1951, pp. 75–83.
5. ———, Topology: A strange new mathematics, Science World, vol. 1, 1957, pp. 7–9.
6. Hans Hahn, Geometry and intuition, Scientific American, vol. 190, 1954, pp. 84–91.
7. J. G. Hocking, Topology, Astounding Science Fiction, vol. 53, 1954, pp. 96–110.
8. B. W. Jones, Elementary Concepts of Mathematics, New York, 1947, Chapter 9.
9. Bruce E. Meserve, Topology for secondary schools, The Mathematics Teacher, vol. 46, 1953, pp. 465–474.
10. ———, Fundamental Concepts of Geometry, Cambridge, 1955, Chapter 9.
11. C. S. Ogilvy, Through the Mathescope, New York, 1956, Chapter 11.
12. A. W. Tucker and H. S. Bailey, Jr., Topology, Scientific American, vol. 182, 1950, pp. 18–24.
13. D. A. Johnson and W. H. Glenn, Topology—the Rubber Sheet Geometry, St. Louis, 1960.

Intermediate articles

14. D. W. Hall and D. E. Spencer, Elementary Topology, New York, 1955.
15. Solomon Lefschetz, Introduction to Topology, Princeton, 1949.
16. M. A. Newman, Elements of the Topology of Plane Sets of Points, Cambridge, 1951.
17. R. L. Wilder, Some unsolved problems in topology, this MONTHLY, vol. 44, 1937, pp. 61–70.

Graduate-level articles

18. W. Hurewicz and H. Wallman, Dimension Theory, Princeton, 1948.
19. J. L. Kelley, General Topology, New York, 1955.
20. S. Lefschetz, Algebraic Topology, Amer. Math. Soc. Colloquium Pub., vol. 27, 1942.
21. R. L. Moore, Foundations of Point Set Theory, Amer. Math. Soc. Colloquium Pub., vol. 13, 1932.
22. E. M. Patterson, Topology, New York, 1956.
23. Waclaw Sierpinski, General Topology, (Tr. by C. C. Krieger), Toronto, 1952.
24. G. T. Whyburn, Analytic Topology, Amer. Math. Soc. Colloquium Pub., vol. 28, 1942.
25. R. L. Wilder, Topology of Manifolds, Amer. Math. Soc. Colloquium Pub., vol. 32, 1949.

Reprinted by permission from

McGRAW-HILL ENCYCLOPEDIA OF SCIENCE AND TECHNOLOGY

SET THEORY

By

R. H. Bing

Set theory

A mathematical term referring to the study of collections or sets. Consider a collection of objects (such as points, dishes, equations, chemicals, numbers, or curves). This set may be denoted by some symbol, such as X. It is useful to know properties that the set X has, irrespective of what the elements of X are. The cardinality of X is such a property.

Cardinality of sets. Two sets A and B are said to have the same cardinal, written $C(A) = C(B)$, provided there is a one-to-one correspondence between the elements of A and the elements of B. For finite sets this notion coincides with the phrase "A has the same number of elements as B." However, for infinite sets the above definition yields some interesting consequences. For example, let A denote the set of integers and B the set of odd integers. The function $f(n) = 2n - 1$ shows that $C(A) = C(B)$. Hence, an infinite set may have the same cardinal as a part or subset of itself.

Subset. A is called a subset of B if each element of A is an element of B, and it is expressed as $A \subset B$. The collection of odd integers is a subset of the reals. Also, each set is a subset of itself.

The continuum hypothesis. An infinite set is called uncountable if it cannot be put in a 1-1 correspondence with the positive integers. Here is one of the unsolved problems of set theory. It is of particular interest since so many mathematicians have tried unsuccessfully to solve it. If X is an uncountable subset of the reals R, is $C(X)$ equal to $C(R)$? The conjecture that the answer is in the affirmative is called the continuum hypothesis.

Comparing cardinality of sets. One says that $C(A) \leqq C(B)$ if there is a 1-1 correspondence between the elements of A and a subset of the elements of B. One useful theorem that can be proved states that any two sets A, B are comparable; that is, either $C(A) \leqq C(B)$ or $C(B) \leqq C(A)$ (possibly both). Another theorem states that if $C(A) \leqq C(B)$ and $C(B) \leqq C(A)$, then $C(A) = C(B)$. Each of these results may be proved by using well orderings of A and B.

Ordering. An ordering is one way of setting up a 1-1 correspondence between two sets of the same cardinality. A relation $<$ is an order relation for a set X if it satisfies the following conditions:

1. If x_1, x_2 are two elements of X, either $x_1 < x_2$ or $x_2 < x_1$ (any two elements are related).
2. $x_1 \nless x_1$ (no element is less than itself).
3. If $x_1 < x_2$ and $x_2 < x_3$, then $x_1 < x_3$ (the order relation is transitive).

The following are examples of ordered sets: a horizontal line where $<$ means "is to the left of"; the reals where $<$ means "is less than"; the collection of words in the dictionary where ordering is alphabetical.

Well ordering. An ordering of a set is called a well ordering if it satisfies the additional condition:

4. Each subset Y of X has a first element; that is, there is an element y_0 of Y such that if y' is another element of Y, $y_0 < y'$.

Used above was the natural convention that each set contains at least one element and not the artificial convention that there is an "empty" set. Some authors use an "empty set" but its introduction is not necessary.

The natural ordering of the positive integers is a well ordering, but neither the natural ordering of the integers nor of the reals is a well ordering. A well ordering for the integers is 0, 1, 2, . . . , -1, -2, ... Since a well ordering of the reals cannot be written down, one might guess that there is none. This guess is shown to be false by the theorem that states that any set X has a well ordering. In proving this, one considers the collection Z of all subsets of X, selects a point $x_\alpha = f(z_\alpha)$ from each element z_α of Z, and well orders X so that if S_α is the set of all elements that precedes x_α, then $x_\alpha = f(Z - S_\alpha)$.

Some of the theorems proved by well ordering are so strange that their truths do not seem intuitively obvious. Well ordering is also used to construct pathological examples which serve as counterexamples to various conjectures. These counterexamples are useful since they show that the conjectures are false and it is useless to try to prove them.

Formation of sets. One approved method of forming a set is to consider a property P possessed by certain elements of a given set X. The set of elements of X having property P may be considered as a set Y. The expression $p \in X$ is used to denote the fact that p is an element of X. Then $Y = \{p/p \in X$ and p has property $P\}$. Another approved method is to consider the set Z of all subsets of a given set X. It may be shown in this case that $C(X) < C(Z)$.

Paradoxically, it is not permissible to regard the collection of all sets as a set. If such a collection X were called a set, and Z were used to denote the set of all subsets of X, one would arrive at the absurdity that $C(X) < C(Z)$.

Operations with sets. In set theory, one is interested not only in the properties of sets but also in operations involving sets: addition, subtraction, multiplication, and mapping.

Sum or union. The sum of A and B ($A + B$ or $A \cup B$) is the set of all elements in either A or B, that is, $A + B = \{p/p \in A$ or $p \in B\}$.

Intersection, product, or common part. The intersection of A and B ($A \cdot B$, $A \cap B$, or AB) is the set of all elements in both A and B, that is, $A \cdot B = \{p/p \in A$ and $p \in B\}$. If there is no element which is in both A and B, one says that A does not intersect B and writes $A \cdot B = 0$.

Difference. The expression $A - B$ is used to denote the collection of elements of A that do not belong to B, that is, $A - B = \{p/p \in A$ and $p \notin B\}$. If $A \subset B$, it is expressed as $A - B = 0$.

An example. If a person were to squirt some black ink on a plane, the set A of points in the dark spot would be an example of a point set. Suppose a set B is determined by squirting some red ink on the plane. Then $A + B$ designates the set of points covered by ink, $A \cdot B$ designates the set covered by both kinds of ink, and $A - B$ designates those covered by black but not red ink (Fig. 1).

Boolean algebra. By using the previous notation, it follows that the sets of Fig. 1 satisfy some of the familiar laws of algebra as

$$A + B = B + A$$
$$A \cdot (B + C) = A \cdot B + A \cdot C$$

However, other identities are not so familiar:

$$X - (A + B) = (X - A) \cdot (X - B)$$
$$X - A \cdot B = (X - A) + (X - B)$$

See BOOLEAN ALGEBRA.

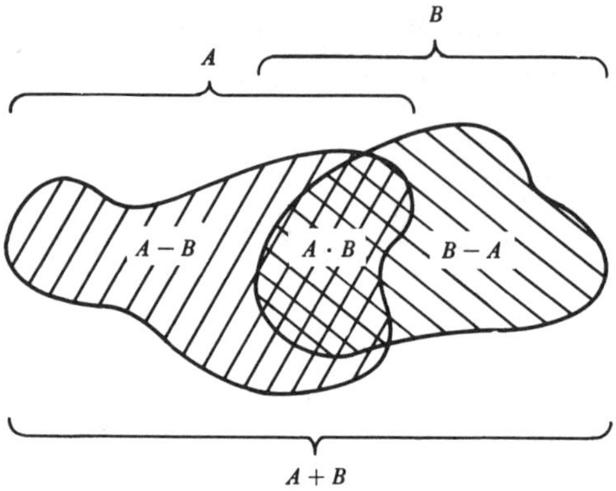

Fig. 1. Sets of point on a plane.

Transformations. A transformation of a set X into a set Y is a function that assigns a point of Y to each point of X. Such a transformation is also called a mapping. The transformation shown in Fig. 2 is the vertical projection of a set X onto a segment Y. The point assigned to X under a transformation f is called the image of x and denoted by $f(x)$. Also, the set of all points x mapped into a particular point y of Y is called the inverse of y and denoted by $f^{-1}(y)$. The inverse of y shown in Fig. 2 is the sum of three segments. The equation $f(x) = x^2$ represents a transformation that takes each real number into its square, for example, 2 to 4, −5 to 25, and so on. Rotations, congruences, and similarities are examples of transformations from geometry. However, in general, a transformation may change both the size and shape of an object.

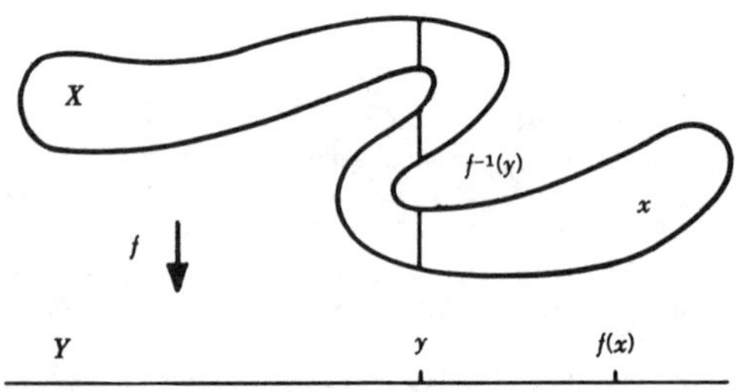

Fig. 2. Vertical projection of a set X onto a segment Y.

Topology. Topology is one of the branches of mathematics that makes extensive use of set theory. Here, not only does one have sets of points for consideration but also collections of interiors of spheres or neighborhoods. These neighborhoods enable one to study limit points and the continuity of transformations. *See* CONFORMAL MAPPING; RING THEORY; TOPOLOGY. [R.H.BI.]

Bibliography: E. Kamke, *Theory of Sets*, trans. from 2d German ed., 1950.

Point-Set Topology

R. H. Bing

Topology is a relatively new branch of mathematics. Those who took graduate training in mathematics thirty years ago did not have the opportunity to take a course in topology at many schools. Others had the opportunity but passed it by, thinking topology was one of those "newfangled" things that was not here to stay. In that respect it was like the automobile.

One who is introduced to topology through popular lectures and entertaining expository articles may get the impression that topology is recreational mathematics. If he were to take a course in topology, expecting it to consist of cutting out pretty figures and stretching rubber sheets, he would be in for a rude awakening. If he pursued the subject further, however, he might be delighted to find that it is rich in substance and beauty. Of the references given at the end of this essay, Newman's book is the most pertinent to the material covered here. Some advanced texts on other subjects have a chapter treating the point-set topology used in their subjects. Some of these are very readable, for example Chapter 1 of Munroe.

In many respects, topology may be considered as an offshoot of geometry. From Euclid we have certain axioms for geometry. On the basis of some undefined terms, definitions, and axioms, certain theorems are proved. So it is with many treatments of topology. The undefined terms, definitions, and axioms that are used are different from those employed in geometry, but the methods are similar.

Topological Equivalence

In studies of the foundations of geometry, one may consider modifying certain of the axioms and seeing how this affects the resulting theorems. Around the end of the nineteenth century, Hilbert studied the effects of changing certain of Euclid's axioms. Veblen made further studies in this respect. One fundamental idea from plane geometry is the notion of betweenness. We use it to develop the notion of straightness. If this notion were modified in a certain way, the two objects shown in Figure 1 would

FIGURE 1

be alike. The first is a triangle and the second is a more general closed curve. In topology, they are considered alike, and we say they are *topologically equivalent.* The concept of topological equivalence corresponds in a way to congruence in Euclidean geometry.

Two sets, A and B, are called topologically equivalent if there is a one-to-one correspondence between them that is continuous both ways, as suggested in the following example. The curved arc A and the straight line segment B shown in Figure 2 are topologically equivalent. Although

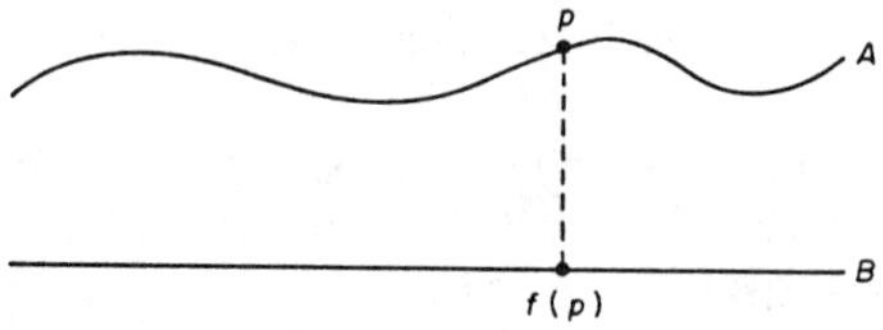

FIGURE 2

there are many one-to-one correspondences that could be chosen, the vertical projection of A onto B provides a convenient one; that is, a point p of A corresponds to the point $f(p)$ of B directly beneath it. This particular one-to-one correspondence from A to B is continuous because points close together in A go into points that are close together in B. The correspondence is also continuous the other way because points close together in B correspond to points close together in A. The expression "close together" is not precise, but this at least gives an intuitive explanation of continuity.

A one-to-one correspondence of a set A onto a set B that is continuous both ways is called a *homeomorphism* of A onto B. The projection of A onto B is an example of a homeomorphism of A onto B. Two sets are topologically equivalent if there is a homeomorphism of one onto the other. Accordingly, instead of saying that the two sets are topologically equivalent, we may say that they are homeomorphic.

It is possible to get a homeomorphism of a short segment I_1 onto a long segment I_2 (see Figure 3). The homeomorphism this time is given by a

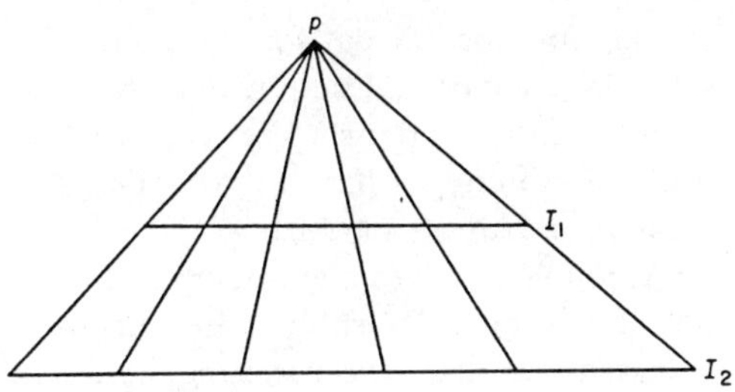

FIGURE 3

projection from a point. This homeomorphism shows that, in a certain sense, there are as many points on a short segment as on a long segment.

In Figure 4 we show some objects no two of which are homeomorphic. For example, the θ-curve is not topologically equivalent to the segment

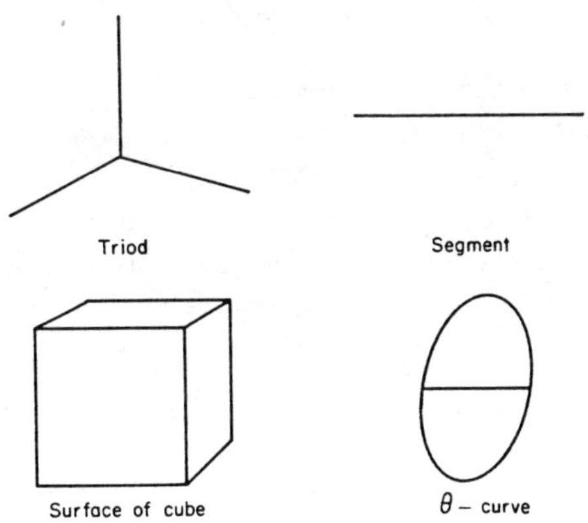

FIGURE 4

because no point of it could be made to correspond (in the proper way) to the end of the segment.

Pulling and Stretching Not Always Possible

Topology has been wrongly called "rubber sheet" geometry, owing to the misconception that, if two sets are topologically equivalent, it is always possible to deform one onto the other by "pulling and stretching but without breaking and tearing." If one figure can be deformed into another simply by pulling and stretching, the figures are topologically equivalent. But the converse of this statement is not true, as the following examples will show.

In E^3 (Euclidean 3-space, the familiar three-dimensional space of everyday experience), consider two sets each consisting of two tangent spheres; in the first set the spheres are tangent externally, and in the second set one sphere lies on the interior (except for the point of tangency) of the other. The two sets are topologically equivalent because there is a one-to-one correspondence of the proper sort between the two sets. However, it is not possible to deform one set onto the other in E^3 by "pulling and stretching but without breaking and tearing."

Another counterexample involves a cylinder or tube formed by sewing two opposite sides of a rectangular rubber sheet together. If the ends of the cylinder are sewed together, the resulting figure may resemble an inner tube. However, if a knot is tied in the cylinder before its ends are joined, the exterior view of the inner tube will be changed but its interior structure will not be. If after the knot is tied and before the final sewing

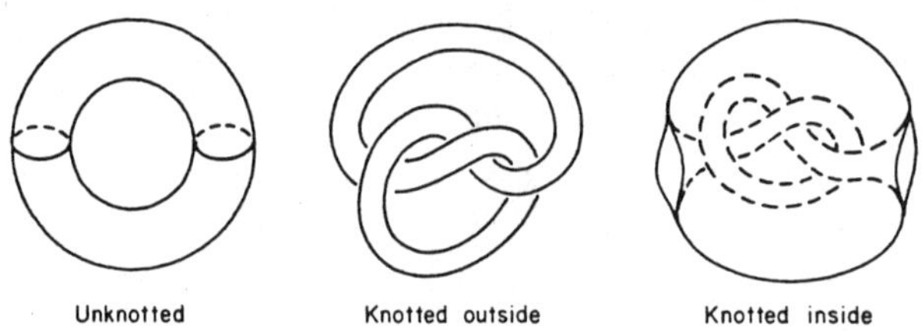

FIGURE 5

is done, one end of the cylinder is stretched and pulled over the knot before the ends are connected, another figure results. The three surfaces shown in Figure 5 are topologically equivalent but no one of them can simply be stretched to make it fit on any other one.

Complements of Figures

The "complement" of a geometric figure is the set of points not in the figure. For example, on the surface of the earth, the surface of the oceans, seas, lakes, and other bodies of water is the complement of the surface of the land. In the plane, the complement of a circle consists of two pieces: its interior and its exterior. The complement of a circle in E^3 would be all in one piece.

Even though two figures in the same space are topologically equivalent, their complements may be topologically different. In Figure 5, each of the surfaces has an interior and an exterior. It could be shown that the interior of the first is topologically equivalent to the interior of the second but not to the interior of the third. A small creature who knew about topological equivalence but not about straightness, length, and so forth, could not tell the difference between the inside of the first tube and the inside of the second. However, the exterior of the first tube in Figure 5 is like the exterior of the third but not like the exterior of the second. No two of the three figures have complements that are topologically equivalent.

J. W. Alexander described the "horned sphere" illustrated in Figure 6. It is topologically equivalent to the surface of a sphere, but its exterior is not like the exterior of a sphere.

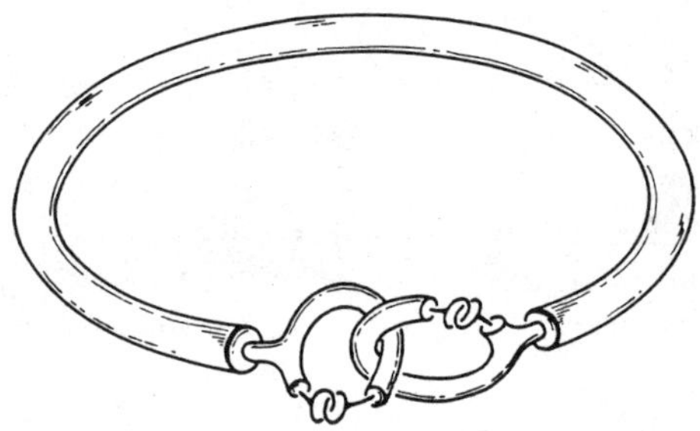

FIGURE 6

We might describe the horned sphere as follows. A long cylinder closed at both ends is folded until the ends are near each other and parallel. Then tubes are pushed out each end until they hook as shown. The process is continued by pushing out additional tubes, and so on. The resulting set has the property that, although it is topologically equivalent to the surface of a sphere, there is a circle in the exterior of the horned sphere that cannot be shrunk to a point without hitting the horned sphere. Thus, the complement of a horned sphere is not like the complement of a round sphere because each simple closed curve in the complement of a round sphere can be shrunk to a point without touching the sphere. (A "simple" curve in this context means a curve that does not cross itself.)

The interiors of circles and triangles appear to be alike. However, if one has a complicated simple closed curve such as shown in Figure 7, he might have difficulty in deciding if its interior is like the interior of a circle. If the curve were even more complicated, it might be impossible to tell merely by looking whether it even has an interior.

The Jordan curve theorem states that the complement of each simple closed curve J in the plane consists of two separate connected pieces and the curve is the boundary of each of them. One extension of the Jordan curve theorem due to Schoenflies may be stated as follows: If J_1 and J_2 are simple closed curves in the plane, there is a homeomorphism of the plane onto itself that takes J_1 onto J_2.

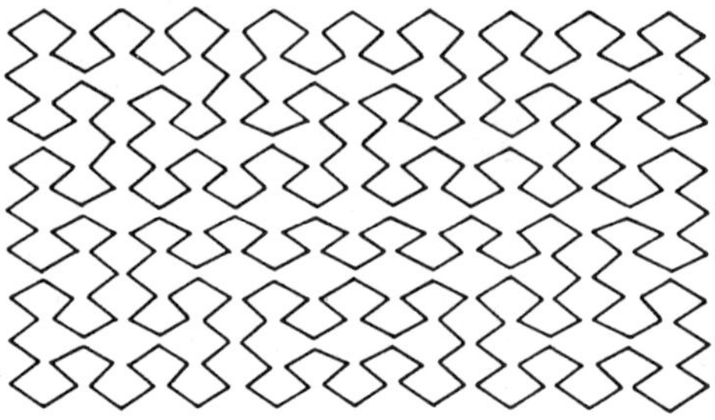

FIGURE 7

In plane analytical geometry, one studies translations and rotations of the plane onto itself — and perhaps even reflections through a line. These are examples of homeomorphisms of the plane onto itself. Under each of these particular homeomorphisms, any Figure F would go into a figure congruent to F. But the types of homeomorphisms used to show the truth of the foregoing extension of the Jordan curve theorem would frequently change the shapes of objects.

To those who think the theorem needs no proof since it is intuitively obvious, it may come as a surprise that the analogous theorem is not true in 3-space. Although the surface of a horned sphere described in previous paragraphs is topologically like the surface of a sphere (just as a simple closed curve is topologically like a circle), there is no homeomorphism of 3-space onto itself that will take the surface of a horned sphere onto the surface of a sphere.

The Jordan curve theorem is one of the more frequently used theorems of plane topology; it is used in such subjects as analysis or complex variables. It was not known for many years whether or not the Jordan curve theorem was true, though during most of this time many people suspected it to be true. Although several proofs of this theorem have now been given (and some only recently), these proofs are still complicated.

Some Unsolved Problems

There are many unsolved problems in topology, such as the famous four-color problem. Suppose a map is drawn on the surface of a sphere in such a way as to divide up this surface into regions, each with a boundary that is a simple closed curve. The four-color problem then asks if each such map on the surface of a sphere can be colored by use of no more than four colors. Regions whose boundaries meet only at a point are allowed to have the same color, but regions with a common boundary line are not allowed to have the same color.

Quite a few unsolved topology problems involve what are known as topological characterizations. Suppose we show first that a well-known space (such as a line, plane, or 3-space) satisfies a certain set of topological conditions and then that any space satisfying these same conditions must be topologically equivalent to the well-known space. We then say that these conditions constitute a *topological characterization* of the well-known space.

We have some good topological characterizations of the line. However, one of the interesting unsolved problems, called Souslin's problem, is in this area. It gives a certain set of conditions satisfied by the line and asks whether any space that satisfies them must be topologically equivalent to a line.

Several topological characterizations of the plane have been discovered during the past forty years, and a problem that was for many years unsolved asked whether or not a certain set of conditions constituted a topological characterization of the surface of a sphere.[1] The answer was finally given in the affirmative in 1946 (Kline sphere characterization).

Although topological characterizations of 3-space have been given, these are still complicated. Problems in this area are discussed by Bing in the reference given below. In view of the recent spurt toward increasing our knowledge of topological properties of 3-space, there is hope for improvements here.

BIBLIOGRAPHY

H. M. A. Newman, *Elements of the Topology of Plane Sets of Points*, Cambridge University Press, Cambridge, England, 1951.

M. E. Monroe, *Introduction to Measure and Integration*, Addison-Wesley, Cambridge, Mass., 1953.

R. H. Bing, "Some Aspects of the Topology of 3-Manifolds Related to the Poincaré Conjecture," *Lectures on Modern Mathematics*, Vol. II, T. L. Saaty, Ed., Wiley, New York, 1964.

BIOGRAPHICAL NOTE

R. H. Bing is a research professor at the University of Wisconsin and a leader of the active group of topologists working there. Wisconsin's topology program attracts many capable graduate students, and the lively topology seminars cause a flow of research topologists to spend their summers and research leaves at Wisconsin.

[1] For those who might be curious as to the precise form such a problem might take, even though they may not be able to understand all the technical terms, we give the complete statement: Is the connected compact metric space S topologically equivalent to the surface of a sphere if it has more than one point and satisfies the following conditions?

1. The neighborhoods of S are connected.
2. The complement of no simple closed curve is connected.
3. The complement of each pair of points is connected.

Dr. Bing's research interests have been chiefly in geometric topology. He did his graduate work at the University of Texas and, for a time, devoted much of his study to the topology of the plane, continuous curves, and abstract spaces. More recently, he has been interested in higher dimensional Euclidean spaces. He feels that the study of Euclidean spaces is applied in that it reveals properties of the real space in which we live.

Topology, General

by

R. H. Bing

An offshoot of geometry, topology was originated during the 19th century. A simple (but not entirely accurate) definition has often been used: Topology is the study of those properties that an object retains under deformation, specifically, bending, stretching and squeezing, but not breaking or tearing. Thus a triangle is topologically equivalent to a circle but not to a straight line segment. Similarly, a solid cube made of modelling clay could be deformed into a ball by kneading. It could not, however, be molded into a solid torus (ring) unless a hole were bored through it or two surfaces were joined together. A solid cube is therefore not topologically equivalent to a finger ring.

More precisely, if there are given two geometric objects or sets of points, and if some two-way transformation (or operation or mapping) takes each point p of either set into one and only one point p' of the other, and if the transformation is continuous in the sense (which can be made precise) that points close to p become points close to p', then the transformation is called a homeomorphism and the two sets are said to be topologically equivalent. Topology is, then, the study of properties that remain invariant under homeomorphisms.

This definition is intended to make it clear that the deformation concept has certain limitations. If two figures are given in Euclidean two-dimensional space, called $\mathfrak{R}^2$—that is, the space of ordinary plane geometry—and if one of them consists of a circle tangent internally to a larger circle, and the other consists of two externally tangent circles, then a homeomorphism exists that transforms one figure into the other, and therefore the two figures are topologically equivalent. One figure cannot, however, be changed to the other by distortion in $\mathfrak{R}^2$. It is possible to turn one of the circles through 180° around the common tangent line as axis, thus carrying it into three-dimensional space $\mathfrak{R}^3$, and effect the deformation. The extra dimension may or may not be available, depending on the conditions of the problem. An internally tangent sphere in $\mathfrak{R}^3$ could be continuously deformed to bring it to a position of external tangency by a rotation in hypothetical four-dimensional space $\mathfrak{R}^4$, which might present no difficulty mathematically but would be impossible to achieve or even visualize in a physical application. The mathematical context may also prevent the use of an additional dimension: In any case, the deformation concept is not used or needed in defining topology.

Examples of problems, methods, and applications of topology. *Euler's theorem.* An example of a topological invariant is provided by Euler's classic theorem on polyhedra. A three-dimensional polyhedron without holes or handles may be considered (it is topologically a ball). If the numbers of faces, edges, and vertices on its surface are F, E, and V, respectively, then $F - E + V = 2$. Thus a cube, with six faces, 12 edges, and eight

vertices, satisfies the property; and so do all polyhedral balls, however complicated and irregular. Furthermore, if a point is removed from a face of the surface of the polyhedral ball and the remainder is suitably mapped onto a planar figure, the property still holds, being topologically invariant; and in fact the theorem is most readily proved with the aid of such a mapping.

Though topology deals with some very generalized sets in abstract spaces that do not resemble point sets of the "real world," there are applications of topology to physics and the other sciences.

The fixed-point theorem. The fact that every direction is south to a person standing at the North Pole is a defect of the system of latitude and longitude. Because the polar regions have become more important places than they once were, it might be convenient to adopt a coordinate grid on which there would be no such singular point anywhere on Earth. On the other hand, such a grid is impossible. It is also true that at any instant there is always at least one place on the Earth's surface where the wind is not blowing. These two facts, seemingly very different from each other, are both direct consequences of the fixed-point theorem, which states that every homeomorphism of a round spherical surface onto itself either leaves at least one point fixed or sends some point into its diametrically opposite. A three-dimensional ball also has the fixed-point property: If one continuously stirs a pot of glue, no matter how long and thoroughly, and even induces slippage along the sides of the pot, so that each particle is moved about, and then stops, there will always be some particle of the glue that has returned to its starting point.

The Möbius band. A driving belt may wear because of friction over the wheels. If such a belt is given a 180° twist before the ends are sewn together, it may last longer because it wears equally on both sides. In fact, it has only one side and one edge. Such a belt is a model of a Möbius band (see MATHEMATICS AS A CALCULATORY SCIENCE). Cutting a Möbius band down the middle gives it a second edge and returns it to a single two-sided surface (with twists).

Jordan's theorem. A simple closed curve in the plane does not cross itself. (A figure-8 curve is not simple.) Jordan's theorem states, in effect, that every simple closed curve divides the plane into two compartments, one inside the curve and one outside it; and that it is impossible to pass continuously from one to the other without crossing the curve. This seemingly trivial and self-evident state of affairs has many important consequences.

The classical "utility problem" or "cranky neighbour problem" asks whether it is possible for three houses (A, B, and C) each to be connected to three utility plants, 1, 2, and 3, without having any of the lines cross. The negative answer is derived from topological considerations. It must be recognized at the outset that it makes no essential difference how the houses and plants are oriented. The houses and utilities may therefore be considered as points on a simple closed curve. If the third house C is now placed inside the curve, it can be satisfactorily connected to 1 and 2; but then the inside will be compartmented, and no matter where 3 is placed, it will be cut off in terms of Jordan's theorem from one of A, B, or C. The same argument holds if C is placed outside.

This method of solution finds serious application in the theory of stamped electrical circuits. The lines of such stamped circuits cannot be permitted to cross. The question of what noncrossing circuits are possible is an extension of the problem of the three utilities.

The role of topology in mathematics. The applications of topology to mathematics itself are much deeper and more far-reaching than the above examples would indicate. Most of those who work in topology are not searching for immediate applications. They study it because of the challenge it offers and because they wish to learn more about the properties of real or abstract spaces. They wish to know whether one set of properties implies another set of properties. They wish to know what is true and what is possible. Topological notions and methods underlie much of modern mathematics, and the topological approach has illuminated and clarified very basic structural concepts in diverse branches of the subject.

In the same way that the Euclidean plane satisfies certain axioms or postulates, it can be shown that certain abstract spaces have definite properties without examining these spaces individually. By approaching topology from this abstract point of view, it is possible to use its methods to study things other than collections of points. Collections of entities that are of concern in analysis or algebra or collections of geometric objects can be treated as spaces and the elements in them as points.

Much use is made of the notion of open sets (subsets of a space that satisfy certain axioms) in studying topological spaces. A point p is said to be a limit point of a set Y in a topological space X, for example, if each open set in X that contains p contains a point of Y other than p. In determining whether or not p is a limit point of Y, it is of no concern whether p belongs to Y (which is stated symbolically as $p \in Y$). In physical terms, this means that p is infinitely close, in a certain sense that can be made precise, to other points of Y. The set of points that are either points or limit points of Y is called the closure of Y and is designated by $\bar{Y}$. A set is called closed if it contains each of its limit points.

The closure of a set

Examples are useful in topology; they help a mathematician to discover theorems that are likely to be true and aid him in labelling others as false. Giving a counter-example is a convincing way of showing that a "proof" is wrong or that a conjecture is false. Examples of some useful topological spaces are described below, some of them being quite different from those studied in geometry.

Building topological spaces. Certain topological spaces are of historical interest because they have been widely used in the past. Others are studied because they illustrate an interesting property. On many occasions the mathematician "manufactures" a topological space to meet his immediate needs.

A topological space X is built up of a collection of sets $\{U\alpha\}$ called a basis. This collection of sets together with all their possible unions are called the open sets of the topology. In the plane, for example, each open set is the union of interiors of circles. The set of interiors of circles, therefore, is a basis for the plane. The set of interiors of rectangles is another basis. In determining whether a point p is a limit point of a set X, it makes no difference whether each interior of a circle that contains p or each interior of a rectangle that contains p contains a point on X other than p. For the line, one basis is the collection of open intervals. For Euclidean three-dimensional space, one basis is the collection of interiors of balls, and another is the collection of interiors of rectangular solids. If a distance is defined for a topological space, the basis is usually taken as the interiors of balls in which the interior of the ball with centre at p and radius r is the set of all points that have distance from p less than r.

Well-known topological spaces

A look at several well-known topological spaces may indicate how topological spaces are constructed and suggest which topological spaces have been considered important.

Euclidean n-dimensional space. Points of this space are n-tuples of numbers $(x_1, x_2, \cdots, x_n)$. Basis elements are the interiors of balls in which the distance between $p = (x_1, x_2, \cdots, x_n)$ and $q = (y_1, y_2, \cdots, y_n)$ is the square root of $(y_1 - x_1)^2 + (y_2 - x_2)^2 + \cdots (y_n - x_n)^2$.

Hilbert space. Points of real Hilbert space are sequences of numbers $x_1, x_2, \cdots$ such that the infinite sum Σx_i^2 is finite. Elements of the basis are the interiors of balls in which the distance between $p = (x_1, x_2, \cdots)$ and $q = (y_1, y_2, \cdots)$ is the square root of $\Sigma(y_i - x_i)^2$.

Discrete spaces. For a collection of objects that is to be regarded as a topological space with its elements as points, it is sufficient to describe the subsets of the collection that are to be called open sets. One possibility is to define all subsets as open, even the one-point sets. The resulting topology is called the discrete topology. It is not

very useful from the topological point of view because a discrete space has no limit points.

Linear spaces. If the elements of a collection are ordered, which means that for any two elements one precedes the other, this ordering can be used to give the collection a linear topology. Basis elements are of three sorts: the set of points between some two points, the set of points before some point, and the set of points after some point. The line and interval are two conspicuous examples of linear spaces. Another example is given below.

A lexicographically ordered square. If I^2 is a square in the plane with its sides parallel to the axes, then each point p of I^2 may be designated by coordinates (x, y), and it is possible to write $p = (x, y)$. The points of I^2 are so ordered that if $p_1 = (x_1, y_1)$ and $p_2 = (x_2, y_2)$, then p_1 precedes p_2 if either $x_1 < x_2$ or $x_1 = x_2$, and $y_1 < y_2$. In this ordering the lower left-hand corner is the first point and the upper right corner is the last. Such points of I^2 are said to be the points of the lexicographically ordered square because the order is analogous to that found in a dictionary. The topology is linear. While the lexicographical square shares some properties with an interval, it differs in that it contains an uncountable collection of disjoint open sets.

Function spaces. If A and B are topological spaces, a function f that maps A into B, which is stated symbolically as $f{:}A \to B$, is a rule that assigns an element of B to each element of A. The point b of B associated with the point a of A is designated by $f(a)$; b is called the image of a. If $\{f_\alpha\}$ is a collection of such functions, it is sometimes convenient to regard this collection as a topological space the points of which are the functions. A collection of subsets of the collection must be selected to serve as open sets. There are many ways of doing this; the description of two of them follows.

Topologies for function space

In the first topology given for $\{f_\alpha\}$, the open sets are made to be large by restricting the basis elements with respect to only a finite number of points of A. A basis element is determined by a particular function belonging to the space—that is, an $f_{\alpha_0} \in \{f_\alpha\}$—a finite number of points $a_1, a_2, \cdots, a_n$ of A; and a finite collection of open sets $U_1, U_2, \cdots, U_n$ in B, such that $f_{\alpha_0}(a_i) \in U_i$ for $i =$

$1, 2, \cdots, n$. The basis element determined by f_{α_0}, these a_i's and these U_i's, is the set of all f's in $\{f_\alpha\}$ such that $f(a_i) \in U_i = 1, 2, \cdots, n$. Other basis elements are determined by other elements of $\{f_\alpha\}$, other finite subsets of A, and other finite collections of open sets in B.

In the second topology given for $\{f_\alpha\}$ it is supposed that B is the interval [0, 1] of the real line $\mathfrak{R}^1$. The distance between each pair of elements f_β, f_γ of $\{f_\alpha\}$ is defined to be the least upper bound of $|f_\beta(x) - f_\gamma(x)|$, x varying over A. Basis elements are open balls.

Something can be learned from the fact that different topologies can be chosen for $\{f_\alpha\}$. A topology can be chosen to suit certain needs, and the choice of topology is usually made with a definite object in mind. In dealing with differentiable functions, for example, it may be advantageous to define distance so that for two functions to be close, their derivatives must also be close.

For the function spaces just described, no use was made of the topology of A; it might just as well have been discrete. The function spaces are related to Cartesian-product spaces, which are covered below.

Cartesian-product spaces. Starting with a pair of topological spaces X_1, X_2, the Cartesian product of X_1 and X_2 (written $X_1 \times X_2$) may be constructed as follows: The points of $X_1 \times X_2$ are ordered pairs (a, b) in which $a \in X_1$ and $b \in X_2$. Basis elements of $X_1 \times X_2$ are sets of the form $U_1 \times U_2$ in which U_1 is a basis element of X_1; U_2 is a basis element of X_2; and $U_1 \times U_2$ denotes the set of all points (x_1, x_2) of $X_1 \times X_2$ in which $x_1 \in U_1$, $x_2 \in U_2$. In a certain sense, the plane $\mathfrak{R}^2$ is the Cartesian product of two lines. Also $\mathfrak{R}^{m+n}$ is the Cartesian product of $\mathfrak{R}^m$ and $\mathfrak{R}^n$.

Cartesian products of topological spaces

Instead of the Cartesian product of just two spaces, the Cartesian product of a collection $\{X_\alpha\}$ of topological spaces can be studied. In this case a point in the Cartesian-product space is a collection $\{p_\alpha\}$ such that $p_\alpha \in X_\alpha$. The collections $\{p_\alpha\}$ and $\{X_\alpha\}$ have like subscripts. To get a basis element, a collection $\{U_\alpha\}$ is considered in which U_α is an open set in X_α and all but a finite number of the U_α's are the whole of the corresponding X_α. The basis element corresponding to U_α is the set of all points $\{q_\alpha\}$ of the Cartesian-product space such that $q_\alpha \in U_\alpha$.

Topological spaces. The collection of open sets of a topological space is axiomatically required to have the following properties:

1. Each point in the space is an element of (belongs to) some open set.
2. The union of any collection of open sets is an open set.
3. If two open sets share a common point, their intersection is open.

It may be shown that $\mathfrak{R}^n$, Hilbert space, discrete spaces, linear spaces, function spaces, and Cartesian-product spaces are all topological spaces. The basis elements were chosen to make this true. Had they been indiscriminately chosen, a topological space might not have resulted. The line would not be a topological space, for example, if its basis elements were defined as closed intervals. The third requirement would not be satisfied.

An unusual topology of the line. The points of an unusual topology called "E^1-bad" are the points of a horizontal line. Basis elements are half-open intervals open on the right—that is, sets of the form $[p, q)$ in which p, q are points of the horizontal line with p to the left of q, and $[p, q)$ is the set consisting of p and all points strictly between p and q. While the space E^1-bad has the same points as a line, its topology is different. It does not have the linear topology. No point of E^1-bad is a limit point of the set of points to the left of it. It will be shown below that E^1-bad differs from $\mathfrak{R}^1$ in other respects.

Topological properties. Although in what follows mention is made of such topological properties as connectedness, separability, being homeomorphic with another space, regularity, normality, and compactness, these are only a sampling of many such topological properties, many others of which can be found in textbooks on topology. Homeomorphisms preserve all topological properties, but it may be shown that mappings preserve some of them; for example, connectedness, separability, and compactness.

Mappings. Suppose A, B are topological spaces and $f{:}A \to B$ is a function that maps A into B. If $p \in A$, f is continuous at p if for each open set U in B containing $f(p)$ there is an open set V in A containing p such that f sends each point of V into U. This property of f is expressed imprecisely by saying that points close to p are sent close to $f(p)$. The function f is continuous if it is continuous at each point of A. Continuous functions are called mappings. Topologists are concerned with what properties are preserved by mappings. They are interested in whether or not mappings can be extended. They classify certain mappings as being differentiable, piecewise linear, or homeomorphic.

Homeomorphisms. Two topological spaces are called topologically equivalent, or homeomorphic, if there is a one-to-one correspondence between them that is continuous both ways. The vertical projection shown in Figure 1 sets up such a one-to-one correspondence between the straight interval X and the curved interval Y. If A and B

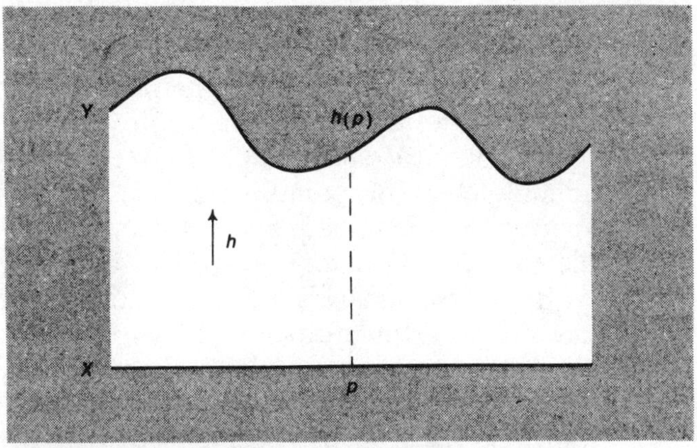

Figure 1: Vertical projection showing a one-to-one correspondence between the straight interval *X* and the curved interval *Y* (see text).

512 Topology, General

Topological equivalence

are topologically equivalent, there is a function $h: A \rightarrow B$ such that h is continuous, h is onto (each point of B is the image of a point of A), h is one to one, and the inverse function, h^{-1}, is continuous. Thus h is called a homeomorphism.

Topological properties are properties such that if one topological space has the property and a second space is topologically equivalent to the first, then the second has the property also. Properties having to do with size or straightness are not topological properties. Various topological properties will be examined in what follows.

Suppose Y is a point set in a topological space X. Regard Y as a subspace of X the open sets of which are the intersection of Y with open sets in X. The set Y has a topological property if it has the property when regarded as a subspace. A function on Y is called continuous if it is continuous when Y is regarded as a subspace.

Connectedness. A topological space X is connected if each map of it into the two-point space $\{0, 1\}$ takes all of X into one of 0 or 1. This is a complicated positive way of saying what is usually stated in a simple negative way. A space is connected if it is not the sum of two (non-null) sets neither of which contains either a point or a limit point of the other. Intuitively, this means that the space is all in one piece. It may be shown that $\mathfrak{R}^n$, Hilbert space, and the lexicographically ordered square are connected but that E^1-bad is not.

The Suslin problem. It is important to know what set of properties characterizes a line. It is known that a line is a connected linear space with no first or last point. The lexicographically ordered square with its end points removed, however, also has these properties.

Separability

A topological space is called separable if it contains a countable set of points such that each open set in the space contains some point of the countable set. It is possible to prove that any separable, connected linear space with no first or last point is topologically equivalent to a line. The Suslin problem asks whether the conditions of separability can be replaced by the condition that the space does not contain an uncountable collection of disjoint open sets. It has been shown that within the framework of the ordinary axioms of set theory, a proof cannot be constructed to show that this replacement is possible. This limitation results partially because these axioms permit the assumption that a set does not exist unless an algorithm is available for describing it and permits the labelling of some subsets of a countable model as uncountable merely because the function is missing from the model being used to match the set one to one with the integers.

Spaces like Hilbert space. Is Hilbert space homeomorphic with $\mathfrak{R}^1 \times \mathfrak{R}^1 \times \cdots$, the Cartesian product of countably many real lines? This question has been of interest, but the answer was hard to obtain. If there were a homeomorphism, it would not be direct by associating the point $(a_1, a_2, \cdots)$ in $\mathfrak{R}^1 \times \mathfrak{R}^1 \times \cdots$ with the point $(a_1, a_2, \cdots)$ in Hilbert space. The point all coordinates of which are 1 lies in $\mathfrak{R}^1 \times \mathfrak{R}^1 \times \cdots$ but not in Hilbert space. The set of points with precisely one coordinate 1 and all others 0 has a limit point in $\mathfrak{R}^1 \times \mathfrak{R}^1 \times \cdots$ but no limit point in Hilbert space. It was shown in 1966 by the U.S. mathematician R.D. Anderson, however, that, contrary to earlier claims, there is indeed a homeomorphism between Hilbert space and $\mathfrak{R}^1 \times \mathfrak{R}^1 \times \cdots$. Investigations are under way to determine what other spaces are homeomorphic with Hilbert space.

Hausdorff spaces. A topological space is called Hausdorff (after the German mathematician Felix Hausdorff) if it satisfies the following condition: For each two points, p, q there are disjoint open sets $U(p)$, $U(q)$ containing p and q, respectively. Being Hausdorff is a topological property. Not all topological spaces are Hausdorff, but most of those that are seriously studied, including all those mentioned so far in this article, are Hausdorff. It is possible to build a non-Hausdorff space by adding an ideal point to an interval as follows.

Non-Hausdorff spaces

A non-Hausdorff interval with two right ends. Let pq be a closed interval with the ordinary linear topology. Basis elements are either open intervals or half-open intervals that contain either p or q. An interval is constructed with two right ends by adding another right end q' to pq as follows: A point of this enlarged space is either q' or a point of pq. A basis element is either a basis element of the original interval pq or the union of the one-point set $\{q'\}$ and an open interval of pq with q as an end. The enlarged space is not Hausdorff because there are no disjoint open sets containing q, q'.

Other separation properties. The Hausdorff property is one of the separation properties. Two sets A, B can be separated if these two sets lie in disjoint open sets. A space is Hausdorff, therefore, if each two points can be separated. A space is regular if each point can be separated from a closed set not containing it. A space is normal if each pair of disjoint closed sets can be separated.

Sometimes a topological space has to be constructed that has one topological property but not another. To build a space that is Hausdorff but not regular, the topology of a line can be changed. Rather than using ordinary open intervals as the basis elements, they can be the set of points consisting of the centre and irrational points of open intervals. While no basis element then has more than one rational point, any closed set containing it has many rational points.

The plane is normal: If A, B are disjoint closed sets, the set of points closer to A than to B is an open set (as is the set of points closer to B than to A). The upper half-plane space described below, however, is regular but not normal.

The upper half-plane space. The points of this space are the points of the plane on or above the x-axis. The basis elements are of two sorts: the interiors of circles above the x-axis, and the union of interiors of circles tangent to the x-axis from above together with the point of tangency (see Figure 2). Physically this space may be regarded as points of an upward flow in which the rate of flow is one unit per second at the x-axis and less above. If p is a point and $r > 0$, the open set generated by p and r

A space regarded as points of a flow

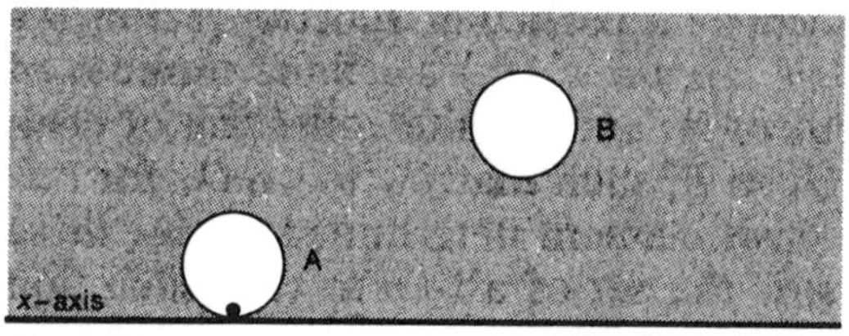

Figure 2: *Two sorts of basis elements for the upper plane space.*
(A) The interiors of circles tangent to the x-axis from above the point of tangency and (B) the interiors of circles above the x-axis (see text).

can be regarded as the set of points to which a person could swim from p in less time than from r if he swims at the rate of one unit per second. The current would prevent him from swimming between any two points of the x-axis. No basis element contains two points of the x-axis in the upper half-plane space.

To see that the upper half-plane space is not normal, one could assign coordinates to the x-axis and let A be the set of points with rational coordinates and B the set of those with irrational coordinates. A rather involved argument shows that it is impossible to separate A from B.

Cartesian products of normal spaces. It can be seen that the Cartesian product of regular spaces is regular. Until the 1970s, one of the unsolved questions in topology asked whether the Cartesian product of a normal space and a straight-line interval is normal. In 1970 the mathematician Mary Ellen Rudin of the United States constructed an example (using box products) of a normal Hausdorff space whose Cartesian product with an interval is not normal.

The first axiom of countability. This axiom is modelled after the fact that if p is any point in the plane and U is an open set containing p, then there is an integer n such that the open ball with centre at p and radius $1/n$ lies in U. A topological space X satisfies the first axiom of countability provided the following condition holds:

For each point p in X there is a sequence of open sets $U_1, U_2, \cdots$ each containing p, such that if U is any open

set whatsoever containing p, then there is an integer n such that U_n is contained in U, symbolically, $U_n \subset U$. Each of $\mathfrak{R}^n$, Hilbert space, discrete spaces, the lexicographically ordered square, E^1-bad, and the upper half-plane space satisfy the axiom. Some Cartesian-product spaces do not satisfy the axiom at any point whatsoever. To get another Hausdorff space not satisfying this first axiom of countability the topology of the line $\mathfrak{R}^1$ can be changed as follows. The basis elements are defined to be of two sorts: any one point set other than the origin, or the complement of a finite set of points. Then the first axiom of countability is not satisfied at the origin.

First and second axioms of countability

The second axiom of countability. A topological space satisfies the second axiom of countability if it has a countable basis. The plane has a countable basis (interiors of circles with radii and centres represented by rational numbers). Neither the lexicographically ordered square, E^1-bad, nor the upper half-plane space does. One very useful theorem in topology is that any space that is regular and that satisfies the second axiom of countability is normal.

Compactness. A topological space is called compact in the limit-point sense (called countably compact by some or merely compact by others) if each infinite set of points has a limit point.

A collection $\{U_\alpha\}$ of open sets is called a covering of a topological space X if each point of X lies in some U_α. A space is called compact in the covering sense (sometimes called bicompact or even merely compact) if each open covering contains a finite number of elements that cover. It may be shown that any space that is compact in the covering sense is compact in the limit-point sense; but the converse is not necessarily true. For a space to be compact in the covering sense, it is necessary and sufficient that each infinite subset Y have a limit point p such that for each open set U containing p, $U \cap Y$ has as many points as Y. For some important spaces (such as metrizable spaces; see below *Moore spaces*), the two kinds of compactness are equivalent so that there is no confusion in omitting qualifications.

There are examples of two topological spaces such that each is compact in the limit-point sense but their

Cartesian product is not. (These spaces could not satisfy the first axiom of countability.) The Tikhonov theorem, however, states that if $\{X_\alpha\}$ is any collection (perhaps even an uncountable collection) of topological spaces, the Cartesian product of the X_α's is compact in the covering sense if and only if each X_α is.

The Tikhonov theorem

Cardinality of compact spaces. The Tikhonov theorem shows that the number of points in a compact topological space can be arbitrarily large. If one forms spaces by taking big Cartesian products, however, the first axiom of countability is not satisfied. Until the late 1960s, one of the unsolved questions in topology asked how many points there could be in a first-axiom topological space that was compact in the covering sense. In 1969 the Russian mathematician A.V. Arkhangelsky showed that no such space could have more points than there are real numbers.

Some kinds of spaces have been more widely studied than others. Metrizable spaces, decomposition spaces, Moore spaces, and uniform spaces, all to be covered below, are Hausdorff spaces with a particularly rich history.

Metrizable spaces. A topological space X is metrizable if one can assign a suitable distance $D(x, y)$ to each pair of its points x, y. This distance function is required to satisfy the following properties:

1. $D(x, y) \geq 0$, the equality holding if and only if $x = y$;
2. $D(x, y) = D(y, x)$ (symmetry condition);
3. $D(x, y) + D(y, z) \geq D(x, z)$ (triangle condition); and
4. D preserves the topology of X.

The fourth condition means that the open balls under the distance function form a basis for X.

One of the standard theorems in topology states: A necessary and sufficient condition that a separable, regular Hausdorff space be metrizable is that it have a countable basis. This theorem permits the supposition that any regular Hausdorff space satisfying the second axiom of countability is metrizable and shows, for example, that certain decomposition spaces are metrizable (even though it would be hard to assign them a metric) and that all theorems that hold in metrizable spaces are true in these spaces.

Not all metrizable spaces are separable. In the late 1940s, working independently in the Soviet Union, Japan, and the United States, three topologists—Y.M. Smirnov,

J. Nagata, and R.H. Bing, respectively—came up with closely related sets of necessary and sufficient conditions that a Hausdorff space be metrizable. This work was probably triggered by that of another topologist, A.H. Stone, working in England.

Conditions that a Hausdorff space be metrizable

A topological space to which a distance function has already been assigned is called a metric space. Two spaces may be topologically equivalent even though they have been assigned quite different metrics. (The whole plane and the interior of a circle in the plane are examples of topologically equivalent spaces the ordinary metrics of which are quite different.) Any metric, however, is a handy tool to use in studying a space (though some may be better than others), so that it is of interest to determine which topological spaces are metrizable.

Decomposition spaces. Suppose a topological space X is decomposed into a disjoint collection G of closed sets, each compact in the limit-point sense such that if $g \in G$, and U is an open set in X containing g, then there is an open set U' in U containing g such that U' is the union of elements of G. Such a decomposition is called an upper semi-continuous decomposition. Breaking a square into vertical intervals gives an example of such a decomposition. One can build a topological space by regarding the elements of G as points and choosing certain subcollections of G as open sets. The usual selection defines a set of (generalized) points to be open in the decomposition space if and only if the union of the closed sets identified with these points is open in X. In the case of the decomposition of the square mentioned above, the decomposition space would be homeomorphic to an interval. In general, if X is regular, the decomposition space is regular, and if X satisfies the second axiom of countability, then the decomposition space does also. Hence it can be assigned a metric even though this metric might not be readily determined.

The study of collections of closed sets as points of a decomposition space is yet another illustration of how topology can be used to study collections other than ordinary point sets.

Collections other than point sets

One of the interesting properties of decomposition space is that if an upper semi-continuous decomposition of the plane is such that the elements of the decomposition are connected and do not separate the plane, then the decomposition space is topologically equivalent to the plane. Efforts to extend this result to higher dimensions have resulted in some interesting research and some unsolved problems.

Moore spaces. Robert Lee Moore introduced a kind of space that has been widely studied. Moore spaces are more restrictive than first-axiom regular Hausdorff spaces and more general than metrizable spaces. They share many of the properties of metrizable spaces, and many theorems that hold in metrizable spaces can be shown to hold in Moore spaces.

If X is a topological space, $x \in X$, and G is an open covering of X, Star (x, G) denotes the union of the elements of G containing x. A Moore space may be defined as a regular Hausdorff space X with a sequence of open coverings, $G_1, G_2, \cdots$ such that if $p \in X$ and U is an open set containing p, then there is an integer n (a function of p and U) such that Star $(p, G_n) \subset U$. It may be noted that a metric space is a Moore space because G_i can be taken to be the set of open sets with diameters less than $1/i$. The upper half-plane space, however, is a Moore space, but it is not normal and hence not metrizable.

For many years topologists tried to decide whether all normal Moore spaces are necessarily metrizable. It is now known that some of the same considerations that

prevent one from obtaining an affirmative answer to the Suslin problem by using the ordinary axioms of set theory make it impossible to show by using these axioms that even all separable normal Moore spaces are metrizable.

Uniform spaces. With each covering of a space X there may be associated a subset U of $X \times X$, in which U is the set of all points $(x, y) \in X \times X$, such that x and y lie in the same element of G. The set U is called a neighbourhood of the diagonal because it contains all points of $X \times X$ the two coordinates of which are equal. A collection of neighbourhoods of the diagonal is called a uniformity if it satisfies a certain set of requirements. These requirements, not listed here, were fashioned so as to be useful in studying functions defined on X. Any space for which a uniformity has been selected is called a uniform space.

Neighbourhoods of the diagonal

Some mathematicians are more interested in studying the continuous functions that can be defined on a space than they are in studying the space itself. Uniform spaces are particularly interesting to them because by using the uniformity they can consider functions on the uniform spaces that share many of the properties of uniformly continuous functions on metric spaces.

Problems of current interest. Many of the problems of research interest in topology are concerned with manifolds and involve an interplay between the methods of general topology and those of algebraic topology (see TOPOLOGY, ALGEBRAIC). An n-manifold $\mathfrak{M}^n$ is a separable metric space such that each point of $\mathfrak{M}^n$ lies in an open set homeomorphic with $\mathfrak{R}^n$. Some interesting results obtained after 1950 and some open questions are given below without defining all terms.

Planar fixed-point problem. The following has been called the most interesting unsolved problem in plane topology: Does each compact continuum that is the intersection of a decreasing sequence of disks have the fixed-point property?

Polyhedral Schoenflies problem. It was shown in 1960 by Morton Brown of the United States that an $(n - 1)$-sphere in $\mathfrak{R}^n$ is tame under certain conditions. A set in $\mathfrak{R}^n$ is tame if there is a homeomorphism of $\mathfrak{R}^n$ onto itself that takes the set onto a polyhedron. The polyhedral Schoenflies question asks whether each polyhedral $(n - 1)$-sphere in $\mathfrak{R}^n$ bounds a piecewise linear n-ball. The answer is known to be in the affirmative if $n = 1, 2, 3$ but is unsolved in higher dimensions.

Triangulation problem. Approximating homeomorphisms with piecewise linear homeomorphisms led to the result that all three-dimensional manifolds can be triangulated, or divided into three-dimensional triangles (or tetrahedra). It was announced in 1969, however, that not all n-manifolds ($n \geq 6$) have a combinatorial triangulation. Which (if any) higher dimensional manifolds can be triangulated?

Triangulation

Free-surface problem. Is a two-dimensional sphere S^2 in $\mathfrak{R}^3$ tame if for each $\varepsilon > 0$, there are ε-maps f_1, f_2 of S^2 into opposite components of $\mathfrak{R}^3 - S^2$? It was shown in 1961 by the U.S. mathematician R.H. Bing that the answer is in the affirmative if f_1 and f_2 are homeomorphisms. It was shown that S^2 is tame if each component of $\mathfrak{R}^3 - S^2$ is locally 1-connected.

Approximating surfaces. It was shown in 1959 by R.H. Bing that if $\mathfrak{C}^2$ is a two-dimensional complex and h is a homeomorphism of $\mathfrak{C}^2$ into a combinatorial three-dimensional manifold $\mathfrak{M}^3$, then for each $\varepsilon > 0$ there is a piecewise linear homeomorphism h' of $\mathfrak{S}^2$ into $\mathfrak{M}^3$ such that h' is within ε of h. If $\mathfrak{S}^2$ is a connected two-dimensional manifold and $\mathfrak{M}^3$ is connected, it was shown that h' can be a side approximation in that if U is a compact of $\mathfrak{M}^3 - h(\mathfrak{S}^2)$, h' could be selected so that each component of $h'(\mathfrak{S}^2) - U$ is of diameter less than ε. To what extent do these results extend to higher dimensions?

Poincaré conjecture. The axioms of plane geometry give a geometric characterization of the plane. It is known that a connected compact two-dimensional manifold is topologically a two-dimensional sphere if each topological circle in it separates it. It is natural to seek a nice characterization of a three-dimensional sphere. A conjecture rich in history is that a compact connected three-dimensional manifold $\mathfrak{M}^3$ is topologically a three-dimensional sphere if each map of the boundary of a disk into $\mathfrak{M}^3$ can be extended to send the whole disk into $\mathfrak{M}^3$. A variation of this conjecture that turned out to be false was proposed by the French mathematician Henri Poincaré in 1900. Numerous unsuccessful attempts have been made to answer the question. It is known that $\mathfrak{M}^3$ is a three-dimensional sphere if each simple closed curve in $\mathfrak{M}^3$ lies in a topological three-dimensional ball in $\mathfrak{M}^3$. Solutions by C.D. Papakyriakopoulos of the United States to Dehn's lemma and the loop theorem have facilitated the study of three-dimensional manifolds. Although the Poincaré problem still remains unsolved, it is of interest that in 1960 two U.S. mathematicians Stephen Smale and John Stallings gave independent solutions of a higher dimensional version of the Poincaré conjecture in $\mathfrak{M}^n$, $n \geq 5$.

Characterization of a three-dimensional sphere

BIBLIOGRAPHY. Among elementary treatments of topology with material understandable to those at the advanced high-school level are: B.H. ARNOLD *Intuitive Concepts in Elementary Topology* (1962); R.H. BING, *Elementary Point Set Topology* (1960); RICHARD COURANT and HERBERT ROBBINS, *What Is Mathematics?* (1941); and L.A. STEEN and J.A. SEEBACH, JR., *Counterexamples in Topology* (1970). Although Arnold's approach is intuitive, some theorems are proved. Bing gives some topological examples and discusses the axiomatic approach to topology. Courant and Robbins' book discusses aspects of several branches of mathematics, and chapter 5 is devoted to interesting theorems and problems in topology. The book of Steen and Seebach contains examples prepared by teams of students working under the authors' direction.

Intermediate-level books suitable for those with the background of a course in calculus include: J.D. BAUM, *Elements of Point Set Topology* (1964); M.C. GEMIGNANI, *Elementary Topology* (1967); ROBERT H. KASRIEL, *Undergraduate Topology* (1971); KAZIMIERZ KURATOWSKI, *Wstęp do teorii mnogści i topologii* (Eng. trans., *Introduction to Set Theory and Topology*, 1962); and M.H. NEWMAN, *Elements of the Topology of Plane Sets of Points*, 2nd ed. (1951). The first three of these are especially suitable as texts for an undergraduate course in topology. Newman's book gives an elementary treatment of many of the topological properties of the Euclidean plane. WACLAW SIERPINSKI, *Topologja ogólna* (1928; Eng. trans., *General Topology*, 2nd ed., 1956), was the text that introduced many graduate students to the study of topology. JOHN L. KELLEY, *General Topology* (1955), followed as a widely used graduate text. It treated general topology in a way particularly useful to those working in analysis. Although DICK WICK HALL and GUILFORD L. SPENCER, *Elementary Topology* (1955), is labelled elementary, it has a careful treatment of some of the difficult properties of the plane not easy to understand. J.G. HOCKING and G.S.YOUNG, *Topology* (1961), has been widely used as a graduate text and not only treats general topology but also introduces many geometric ideas useful in the study of manifolds. More recent texts include JAMES DUGUNDJI, *Topology* (1965); RYSZARD ENGELKING, *Zarys topologii ogólnej* (1965; Eng. trans., *Outline of General Topology*, 1966); S.T. HU, *Introduction to General Topology* (1966); KAZIMIERZ KURATOWSKI, *Topologie*, 4th ed., vol. 1 (1958; Eng. trans., *Topology*, new ed., vol. 1, 1966); and STEPHEN WILLARD, *General Topology* (1970).

(R.H.Bi.)

II. Planar Sets

A. Planar Webs

COLLECTIONS FILLING UP A SIMPLE PLANE WEB

R. H. BING

A web has been defined [2][1] by R. L. Moore. A compact plane continuum W is said [1] to be a simple plane web if there exist an upper semicontinuous collection G of mutually exclusive continua filling up W and another such collection H also filling up W such that (1) G is a dendron with respect to its elements and so is H and (2) if g and h are elements of G and H respectively, the common part of g and h exists and is totally disconnected. It has been shown [1] that we have an equivalent definition if we substitute for (1) in the above definition the condition that (1′) G is a dendron with respect to its elements and H is an arc with respect to its elements. The present paper shows that an equivalent definition is obtained if condition (1) is omitted.

This paper gives conditions under which the collections filling up a simple plane web are *non-equicontinuous*. A collection of mutually exclusive continua is said to be *equicontinuous* if for every positive number ϵ, there is a positive number δ_ϵ such that if P and Q are points of an element g of G at a distance apart of less than δ_ϵ, there is an arc from P to Q in g of diameter less than ϵ. If a collection is not equicontinuous, it is said to be *non-equicontinuous*. If there exists a point P belonging to the limiting set of G such that every subcollection of G whose limiting set contains P is non-equicontinuous, then G is said to be *hereditarily non-equicontinuous* at P.

I wish to thank Professor R. L. Moore for his valuable suggestions in the preparation of this paper.

A simple plane web is a continuous curve [3]. In establishing some of the following theorems, we shall make use of the fact [1, Theorem 1] that a necessary and sufficient condition that a continuous curve be a simple plane web is that it remain connected and locally connected on the omission of any countable set of points.

THEOREM 1. *A necessary and sufficient condition that a compact plane continuum W be a simple plane web is that there exist an upper semicontinuous collection G of mutually exclusive continua filling up W and another such collection H filling up W such that if g and h are elements of G and H respectively, the common part of g and h exists and is totally disconnected.*

Presented to the Society, September 17, 1945; received by the editors June 25, 1945.

[1] Numbers in brackets refer to the references cited at the end of the paper.

PROOF. The necessity of the above condition follows from the definition of a simple plane web. To demonstrate its sufficiency, we shall show that if Z is a countable subset of W, $W-Z$ is connected and locally connected. The argument is similar to that used in proving the necessity of Theorem 1 of [1].

Let M be the sum of all elements of G that do not intersect Z and let N be the sum of all elements of H that do not intersect Z. As $M+N$ is connected and dense in W, $W-Z$ is connected.

Assume that $W-Z$ is not locally connected. Let d be a domain containing a point P of $W-Z$ such that any domain d' containing P contains points of an infinitude of components of $(W-Z)\cdot d$. An infinitude of such components contain points of $M\cdot d'$. Let C_1, C_2, C_3 be circles in d such that C_1 encloses P, C_2 encloses C_1, and C_3 encloses C_2 but no element of G or H. Then C_1 encloses a sequence of points $P_1, P_2, \cdots$ of M such that (1) no two of them belong to the same component of $(W-Z)\cdot d$ and (2) if g_n is the component containing P_n of the common part of d and the element of G containing P_n, then $g_1, g_2, \cdots$ has a sequential limiting set. There exists a sequence of points $Q_1, Q_2, \cdots$ of N converging to a point of C_2 such that Q_n is a point of a continuum in g_m $(m \geqq n)$ irreducible from C_1 to C_3. Let h_n be the component containing Q_n of the common part of d and the element of H containing Q_n. It follows [2, pp. 391–392] that some nondegenerate continuum F is a subset of the common part of the limiting set of $g_1, g_2, \cdots$ and the limiting set of $h_1, h_2, \cdots$. As G and H are upper semicontinuous collections, F is a subset of the common part of an element of G and an element of H. This is contrary to the hypothesis of this theorem.

THEOREM 2. *If x is the closure of a connected set which is the sum of a countable number of points and complementary domains of a simple plane web W, the common part of W and x is a continuous curve.*[2]

PROOF. It has been shown [1, Theorem 6] that x is locally connected. As W is a continuous curve, for any positive number ϵ there are only a finite number of complementary domains of W of diameter more than ϵ. Since $W\cdot x$ is equal to x minus the sum of a set of complementary domains of W, then $W\cdot x$ is a continuous curve.

[2] Indeed, $W\cdot x$ may be shown to be *hereditarily* a continuous curve. Furthermore, it may be shown that in order that the compact point set K should be the boundary of the closure of a connected point set which is the sum of a countable number of points and complementary domains of some simple plane web W it is necessary and sufficient that K should be a continuous curve such that (1) every nondegenerate cyclic element of K is a simple closed curve, (2) K does not have uncountably many cut points, and (3) no cyclic element of K encloses a point of K.

THEOREM 3. *If x is the closure of a connected set which is the sum of a countable number of points and complementary domains of a simple plane web W and R is a nondegenerate connected subset of the common part of W and x, then R is arcwise connected.*[3]

PROOF. Let A and B be two points of R and let AB be an arc in $W\cdot x$. If AB contains a point P not of R, there is an arc AOB in $W\cdot x-P$. There is a simple closed curve which is the sum of arcs $A'PB'$ and $A'QB'$ of AB and AOB respectively. It has been shown [1, Theorem 8] that each point of $W\cdot x$ is a limit point of the connected set $W-W\cdot x$. Therefore, no subset of $W\cdot x$ contains a simple closed curve plus an arc having only its end points on the simple closed curve. Using this fact and the fact that there is an arc from A to B in $W\cdot x$ minus a finite number of points not belonging to R, we find that $A'QB'$ is a subset of R. By replacing $A'PB'$ and other subarcs of AB by arcs which are subsets of R, we may obtain an arc from A to B in R.

THEOREM 4. *If x is the closure of a connected set which is the sum of a countable number of points and complementary domains of a simple plane web W and AB is an arc lying in the common part of W and x, then AB contains an arc lying on the boundary of some complementary domain of W.*

PROOF. Since AB is [1, Theorem 7] the sum of a countable number of points and closures of individual complementary domains of W, the boundary of some one of these domains contains a subarc of AB.

THEOREM 5. *Let G be an upper semicontinuous collection of mutually exclusive continua filling up a simple plane web W and let H be another such collection filling up W such that* (1) *H is a dendron with respect to its elements and* (2) *if g and h are elements of G and H respectively, the common part of g and h exists and is totally disconnected. Let g_1 and g_2 be two elements of G belonging to x_1 and x_2 respectively where x_i* $(i=1, 2)$ *is the closure of a connected set which is the sum of a countable number of points and complementary domains of W. Then g_1+g_2 is a subset of the boundary of a complementary domain of W.*

PROOF. By Theorems 3 and 4 there is a complementary domain d of W such that g_1 and the boundary of d have an arc in common. The boundary of d intersects two elements h_1 and h_2 of H and as g_2 intersects each of them, each element of H that separates h_1 from h_2 in W also disconnects both g_2 and $W\cdot x_2$.

[3] In fact, R is *locally* arcwise connected.

Let Z be a countable set of points such that Z plus the sum of a set of complementary domains of W is a connected set whose closure is x_2. Then each element of H that separates h_1 from h_2 intersects Z or contains a point of x_2 belonging to the boundary of d. As Z is countable and as there are uncountably many elements of H each of which separates h_1 from h_2 in W, the boundary of d contains uncountably many points of g_2.

Assume that g_i ($i=1$ or 2) is not a subset of the boundary of d. By Theorems 3 and 4 there is a complementary domain d' of W other than d such that some arc is a subset of the common part of g_i and the boundary of d'. Both g_1 and g_2 intersect the boundary of d' and by Theorem 3 there are two mutually exclusive arcs in $W \cdot x_1$ and $W \cdot x_2$ respectively such that each has an end point on the boundary of d and an end point on the boundary of d'. This is impossible, for no subset of $W \cdot x_1 + W \cdot x_2$ plus the boundaries of d and d' is a simple closed curve plus an arc having only its end points on the simple closed curve.

THEOREM 6. *Suppose W is a simple plane web with two complementary domains. Suppose that G is an upper semicontinuous collection of mutually exclusive continua filling up W, that H is another such collection also filling up W, that G is a dendron with respect to its elements and so is H and finally that if g and h are elements of G and H respectively, then the common part of g and h exists and is totally disconnected. Then there is an element of G which does not separate W and which is not a subset of the closure of any connected set which is the sum of a countable number of points and complementary domains of W.*

PROOF. Those elements of G that do not separate W are called *end elements* of G. Assume that each end element of G is a subset of the closure of a connected set which is the sum of a countable number of points and complementary domains of W. Then by Theorem 5, if g_1 and g_2 are end elements of G, there is a complementary domain d of W whose boundary contains g_1 and g_2.

Suppose that g_3 is an end element of G other than g_1 or g_2. As the boundaries of two complementary domains of a simple plane web have no more than a point in common, it follows by Theorem 5 that g_3 is a subset of the boundary of d. However, if it were a subset of the boundary of d, no two elements of H could intersect each of the sets g_1, g_2, and g_3.

Suppose that g_1 and g_2 are the only end elements of G. If d' is a complementary domain of W other than d, its boundary is a subset of an element of H, for no three mutually exclusive continua in W could intersect the boundary of d' as well as both g_1 and g_2. Likewise, the boundary of d' is a subset of an element of G, for no three mutually exclusive continua in W could intersect both the boundary of d and the boundary of d'. This is impossible as no element of G intersects an element of H in a nondegenerate continuum.

THEOREM 7. *Let W be a simple plane web with two complementary domains. Let G be an upper semicontinuous collection of mutually exclusive continua filling up W and let H be another such collection filling up W such that* (1) *G is a dendron with respect to its elements and so is H and* (2) *if g and h are elements of G and H respectively, the common part of g and h exists and is totally disconnected. Then the collection G is non-equicontinuous.*

PROOF. Let X be a set such that x is an element of X only if x is maximal with respect to being the closure of a connected set which is the sum of a countable number of points and complementary domains of W. By Theorem 6 there is an end element g_e of G which is not a subset of an element of X. There is [1, Theorem 8] an arc AB in W which contains no point of an element of X and which has only its end points in common with g_e.

Let d be one of the complementary domains of g_e+AB which has AB as a subset of its boundary. If F is a point of W in d and E is a point of W belonging to neither d, AB, nor g_e, then there is an arc EF in W which contains no point of AB. Let C be the first point of the closure of d on EF in the order from E to F. If ρ is one-half the distance from C to AB, we shall show that for any positive number ϵ there exist points P and Q belonging to an element g of G such that the distance from P to Q is less than ϵ and the diameter of any subcontinuum of g containing P and Q is of diameter more than ρ.

There exist points P_1 and P_2 such that each can be joined to C by an arc in W of diameter less than the minimum of ρ and $\epsilon/2$ and such that P_1 belongs to d and P_2 belongs to neither d nor g_e. As g_e is an end element of G, there is an element g of G which separates g_e from P_1+P_2 in W. There exist on g points P and Q at a distance from C of less than the minimum of ρ and $\epsilon/2$ such that P belongs to d and Q does not. Every subcontinuum of g that contains P and Q intersects AB and therefore is of diameter more than ρ.

THEOREM 8. *Let W be a simple plane web with two complementary domains. Let G be an upper semicontinuous collection of mutually exclusive continua filling up W and let H be another such collection filling up W such that* (1) *G is an arc with respect to its elements,* (2) *H is a* *dendron with respect to its elements, and* (3) *if g and h are elements of G and H respectively, the common part of g and h exists and is totally disconnected. Then W contains a point at which G is hereditarily non-equicontinuous.*

PROOF. Obtain g_e, AB, C, ρ, and g as in Theorem 7. Of every countable sequence of different elements of G having a subset of g as a limiting set, all but a finite number separate g from g_e. Hence G is hereditarily non-equicontinuous at C.

REFERENCES

1. Bing, R. H., *Concerning simple plane webs*, submitted to Trans. Amer. Math. Soc.

2. Moore, R. L., *Concerning continua which have dendratomic subsets*, Proc. Nat. Acad. Sci. U. S. A. vol. 29 (1943) pp. 384–389.

3. ———, *Concerning webs in the plane*, Proc. Nat. Acad. Sci. U. S. A. vol. 29 (1943) pp. 389–393.

THE UNIVERSITY OF TEXAS

CONCERNING SIMPLE PLANE WEBS

BY

R. H. BING

A compact continuum W is said [1](1) to be a simple web if there exists an upper semi-continuous collection G of mutually exclusive continua filling up W and another such collection H also filling up W such that (1) G is a dendron with respect to its elements and so is H and (2) if g and h are elements of G and H, respectively, the common part of g and h exists and is totally disconnected. Hence, a simple web is a web(2).

Examples of simple plane webs. Although a simple web is not necessarily a subset of the plane, this paper will deal only with those of this type. A square plus its interior is a simple web. We may consider the elements of G to be intervals parallel to one pair of sides of the square and the elements of H to be intervals parallel to the other pair of sides. We see from Theorem 1 of this paper that if $C_1, C_2, \cdots$ are circles no two of which intersect each other and such that C_1 incloses C_i $(i=2, 3, \cdots)$, then C_1 plus its interior minus the sum of the interiors of $C_2, C_3, \cdots$ is a simple web.

A square plus its interior plus an interval intersecting both the interior and the exterior of the square is not a simple web. Hence, a plane web is not necessarily a simple web. By the use of Theorem 1 of this paper, we find that the continuous curve described on page 273 of [7] which is left connected but not locally connected on the omission of some countable subset is not a simple web.

This paper gives necessary and sufficient conditions that a compact plane continuum be a simple plane web. Professor R. L. Moore suggested the problem of finding such conditions. Much credit is due him for the development of this paper.

It has been shown [6] that every simple plane web is a continuous curve. In the present paper, the following theorem will be established.

THEOREM 1. *In order that a compact plane continuous curve be a simple plane web, it is necessary and sufficient that it remain connected and locally connected on the omission of any countable subset.*

Presented to the Society, September 17, 1945; received by the editors May 12, 1945 and, in revised form, December 24, 1945.

(1) Numbers in brackets refer to the references cited at the end of the paper.

(2) A web has been defined by Moore [3] to be a compact continuum M for which "there exist two upper semicontinuous collections H_1 and H_2 of mutually exclusive continua such that (1) each of these collections fills up M, (2) each of them is a dendron with respect to its elements, and (3) there exists an uncountable subcollection W of the collection H_1 such that no element of W is a subset of any element of H_2."

Proof of necessity. The following argument is similar to that found in the proof of Theorem 1 of [1]. Suppose that G and H are collections satisfying with respect to the simple plane web W all(3) of the conditions described in the preceding definition. Suppose that K is a countable subset of W. If W_Z $(Z=G, H)$ denotes the sum of all of the continua of the collection Z that do not intersect K, then any two points of W_G+W_H lie on a continuum which is the sum of three continua of the collection $G+H$ that do not intersect K. As W_G+W_H is dense in W, $W-K$ is connected.

If $W-K$ is not locally connected at P, there exist a domain(4) d containing P and a sequence of points of $W-K$ converging to P such that each belongs to a different component of $(W-K)\cdot d$. Let C_1, C_2, C_3 denote three circles such that C_1 incloses P, C_2 incloses C_1, C_3 incloses C_2 but no element of G or H and C_3 plus its interior is a subset of d. Now C_1 incloses a sequence of points $P_1, P_2, \cdots$ of W_G no two of which belong to the same component of $(W-K)\cdot d$ such that if g_i is the component containing P_i of the common part of d and the element of G containing P_i, then $g_1, g_2, \cdots$ has a sequential limiting set F. There exists a sequence $Q_1, Q_2, \cdots$ in W_H converging to a point of C_2 such that Q_n is a point of a subcontinuum of g_m $(m\geqq n)$ irreducible from C_1 to C_3. If h_i denotes the component containing Q_i of the common part of d and the element of H containing Q_i, then some nondegenerate subcontinuum of F belongs (see [6, Theorem 2]) to the limiting set of $h_1, h_2, \cdots$. However, as G and H are upper semi-continuous collections, this nondegenerate continuum is both a subset of an element of G and a subset of an element of H. This is contrary to the hypothesis that the common part of an element of G and an element of H is totally disconnected.

This establishes the necessity of the condition of Theorem 1. A number of preliminary theorems will be proved before the establishment of its sufficiency. When the term "simple plane web" is used in the following theorems, it is to be understood that we refer to a compact plane continuum that remains connected and locally connected on the omission of any countable subset.

THEOREM 2. *Each complementary domain of a simple plane web is bounded by a simple closed curve but no two boundaries of its individual complementary domains have more than one point in common.*

Proof. The outer boundary of a bounded complementary domain of a continuous curve is a simple closed curve [4]. As a simple plane web has no cut points, the boundary of one of its complementary domains is also the outer boundary of this domain. If the boundaries of two such complementary

(3) However, in the argument no use is made of the condition that G and H are dendrons with respect to their elements.

(4) A domain is an open set not necessarily connected. However, a complementary domain is connected [7, pp. 11 and 153].

domains intersected in a pair of points, this pair of points would separate the simple plane web.

THEOREM 3. *If the connected set x is the sum of a countable number of points and complementary domains of the simple plane web W, then each point of $W\cdot\bar{x}$ is a limit point of $W-W\cdot\bar{x}$.*

Proof. Suppose that d is a domain containing a point P of $W\cdot\bar{x}$. As $W-W\cdot x$ is a connected and locally connected inner limiting set, the common part of it and d contains [8] an arc AQB. There exists in W an arc ACB containing no point of the countable point set $W\cdot x+Q$. The point set $AQB+ACB$ contains a simple closed curve J which contains an interval $A'QB'$ of AQB. There exists in W an arc α from Q to $J-A'QB'$ which contains no point of $W\cdot x+A'+B'$. The arc α contains an arc $Q'OL$ having in common with J only its end points Q' and L which belong to $A'QB'$ and $J-A'QB'$ respectively. Let $Q'A'L$ and $Q'B'L$ be arcs lying on J with points as indicated. Then $Q'A'L+Q'B'L+Q'OL$ divides space into three mutually exclusive domains only one of which contains the connected set x. Now one of the segments(5) $\langle Q'A'L\rangle$, $\langle Q'B'L\rangle$ and $\langle Q'OL\rangle$ contains no point of $\bar{x}$ and each of them contains a point of d. Therefore d contains a point of $W-W\cdot\bar{x}$. Since any domain containing a point of $W\cdot\bar{x}$ contains a point of $W-W\cdot\bar{x}$, then each point of $W\cdot\bar{x}$ is a limit point of $W-W\cdot\bar{x}$.

THEOREM 4. *Under the hypotheses of Theorem 3, the closure of x does not separate W.*

Proof. If P is a point of the plane S, then $S-(x+P)$ is a connected and locally connected inner limiting set. Hence [8] it is arcwise connected. Furthermore [2, Theorem 12], $S-\bar{x}$ is connected.

Suppose that $W-W\cdot\bar{x}$ is the sum of two mutually separated sets S_1 and S_2. There exists in $S-\bar{x}$ an arc AB from a point A of S_1 to a point B of S_2. Let P denote the last point of the closed point set $S_1\cdot AB$ in the order from A to B on AB. There is a point Q following P on AB such that the interval PQ of AB contains no point of S_2. Let P' denote the first point in the order from Q to B that the interval QB of AB has in common with β, the boundary of d, the complementary domain of W that contains Q. Since the closure of $x+d$ does not separate S, one arc of β from P to P' contains no point of $\bar{x}$. But P belongs to S_1 and P' belongs to S_2. Thus, the supposition that $W-W\cdot\bar{x}$ is not connected leads to a contradiction.

THEOREM 5. *If R is a connected point set consisting of a plane domain plus a totally disconnected point set and the complement of R is a locally connected inner limiting set, then $\bar{R}$ is a continuous curve.*

Proof. If $\bar{R}$ is not locally connected, there exist a domain d, a sequence

(5) By $\langle\alpha\rangle$ is meant the arc α minus its end points.

of components $c_1, c_2, \cdots$ of $\bar{d}\cdot\bar{R}$, and a sequence of arcs $\alpha_1, \alpha_2, \cdots$ such that (1) for each positive integer j, α_j is a subset of $\bar{d}-\bar{d}\cdot\bar{R}$; (2) if h, i, j are positive integers such that $h<i<j$, then α_i separates c_h from c_j in d and c_i separates α_h from α_j in d; (3) d intersects c, the sequential limiting set of $c_1, c_2, \cdots$. The sequence $\alpha_1, \alpha_2, \cdots$ converges to c.

As no point of c is a point of a domain lying in R and some nondegenerate subcontinuum c' of c is a subset of d, the complement of R contains two points P_1 and P_2 of c'. There exist domains d_1 and d_2 containing P_1 and P_2 respectively such that d_1 and d_2 are mutually exclusive subsets of d. As the complement of R is locally arcwise connected at P_1 and P_2, there exists a positive integer j such that each of the point sets $d_1-d_1\cdot R$ and $d_2-d_2\cdot R$ contains an arc intersecting both α_j and α_{j+2}. Therefore, there is a simple closed curve J lying in $d-d\cdot R$ and intersecting α_j and α_{j+2}. As R is connected and contains no point of J, it does not intersect both the interior and the exterior of J. Therefore, no component of $\bar{d}\cdot\bar{R}$ separates α_j from α_{j+2} in d and $\bar{R}$ is locally connected.

THEOREM 6. *Under the hypotheses of Theorem 3, $\bar{x}$ is a continuous curve each of whose degenerate cyclic elements is a point of W in x and each of whose nondegenerate cyclic elements is the closure of a complementary domain of W in x.*

Proof. As the complement of x is a locally connected inner limiting set, it follows from Theorem 5 that $\bar{x}$ is a continuous curve. If P is a cut point of $\bar{x}$, it is a cut point of x and therefore must be a point of W in x.

If J is a simple closed curve in $\bar{x}$, we shall show that it is a subset of the closure of a complementary domain of W in x. Since $\bar{x}$ does not separate space (see Theorem 4), one of the complementary domains of J is a subset of $\bar{x}$. By Theorem 3, we have that J separates no two points of $\bar{x}-x$ from each other. Therefore, one of the complementary domains of J is a subset of x and is therefore a subset of a complementary domain of W in x.

THEOREM 7. *Suppose W is a simple plane web, β is its outer boundary and X is the collection of all nondegenerate continua x such that x is maximal with respect to being the closure of a connected set which is the sum of a countable number of points and bounded complementary domains of W. If A and B are points of W, there is an arc AB in W such that AB minus its end points contains no point of either β or any element of X.*

Proof. Suppose that A belongs to x_A of X and that B belongs to neither β nor any element of X. As $W-W\cdot x_A+A$ is connected, A is accessible [10, Theorem 1] from $W-W\cdot x_A$. Since the continua of the collection X are mutually exclusive and no one of them separates the plane, then x_A plus the complement of the sum of the elements of X is a connected and locally connected inner limiting set which if regarded as space satisfies Axioms 1–5 of [7]. In this space, there exists an arc AB having no point except A in common with

$x_A+\beta$. The arc AB is a subset of W.

THEOREM 8. *Under the hypotheses of Theorem 7, if P is a point belonging to both β and an element x_P of X, Q is another point of β and c is a component of x_P-P, then there exists in W a simple closed curve J inclosing c but no point of x_P-c and containing P and Q but no other point either of β or of an element of X.*

Proof. Let X' denote the set of all continua of the collection X except x_P and x_Q where x_Q denotes the element of X containing Q if there is one but denotes Q if there is not. Let Z denote the collection whose elements are the continua of X' and the points of the plane that belong to no element of X'. The collection Z is upper semi-continuous and no one of its elements separates the plane. It follows [5] that the space Σ_1 whose "points" are elements of Z is topologically equivalent to the plane. If Σ_2 denotes the space whose "points" are the "points" of Σ_1 other than the elements of X' not intersecting β, then it is topologically equivalent to a plane minus a countable number of points. If β' denotes the "point" set whose points are the points of β that belong to no element of X' and the continua of X' that intersect β, then in the space Σ_2, β' is a simple closed curve and x_P and x_Q are continuous curves each lying except for one point wholly in the interior of β'. In this space, there exists an arc EQ from a point E of c to Q such that $x_P+x_Q+\beta'$ intersects EQ only at $E+Q$. Furthermore [7, Theorem 12, p. 217], there exists a simple closed curve intersecting $x_P+x_Q+\beta'$ only at $P+Q$ and inclosing $c+EQ-Q$ but no point of x_P-c.

Description of dendrons G and F. In this section, W will designate a compact plane continuous curve which is left connected and locally connected on the omission of any countable point set. Also, X will denote the set of all nondegenerate continua x such that x is maximal with respect to being the closure of a connected set which is the sum of a countable number of points and bounded complementary domains of W.

As a step toward proving the sufficiency of Theorem 1, we shall describe an upper semi-continuous collection G of mutually exclusive continua filling up W such that G is a dendron with respect to its elements. At the same time, we shall describe a dendron F in W such that each element of G intersects F in only one point.

As the description of G is somewhat complicated, we shall describe it by steps. First, we consider a transformation T throwing a square plus a subset of its interior into W. Next, we describe an upper semi-continuous collection of mutually exclusive continua which is an arc with respect to its elements and which covers a part of the inverse of W under transformation T. We cover more of this inverse by another upper semi-continuous collection of mutually exclusive continua. The process is continued until we have an upper semi-continuous collection of mutually exclusive continua filling up the inverse of W under T such that this collection is a dendron with respect to its elements. The image of the elements of this collection with respect to T gives the elements of G.

Transformation T. Let $P_1P_2P_3$ and $P_1P_4P_3$ be two arcs whose sum is the boundary β of the unbounded complementary domain of W and whose ends are not points of X^*(6). There exist in W arcs $P_1P_5P_3$ and $P_1P_6P_3$ which except for their end points are mutually exclusive and contain no point of $X^*+\beta$. Let R be the interior of a square with diagonals V_1V_3 and V_2V_4 and let V_5, V_6 be the midpoints of V_1V_4, V_2V_3 respectively.

There exists a continuous transformation T of $\overline{R}$ into $W+X^*$ such that $T(V_1V_5) = T(V_4V_5) = P_1P_5$, $T(V_2V_6) = T(V_3V_6) = P_3P_6$, $T(V_5V_2) = P_5P_3$, $T(V_4V_6) = P_1P_6$ and each point of $W+X^*-P_1P_5-P_6P_3$ is the image of only one point of $\overline{R}-V_1V_5V_4-V_2V_6V_3$ under the transformation T.

Let M denote the set of all points P of $\overline{R}$ such that the image of P under T is a point of W. We note that M is a compact continuous curve which is left connected and locally connected on the omission of any countable subset. Let Y be the collection of all continua y such that y is the inverse under T of an element of X. We note that y is maximal with respect to being the closure of a connected set which is the sum of a countable number of points and bounded complementary domains of M.

The elements of Y are ordered $y_1, y_2, \cdots$. By Theorem 7 and the fact that M is separable, there exists a countable point set $z_1, z_2, \cdots$ dense in M and such that for each positive integer i, z_i belongs to M but not to Y^* or the boundary of R. The bounded complementary domains of M are ordered $d_1, d_2, \cdots$.

We shall describe an upper semi-continuous collection G' of mutually exclusive continua filling up M such that (1) G' is a dendron with respect to its elements, (2) no element of G' contains more than two points of Y^*, (3) each element of G' intersects V_5V_2 and V_4V_6, and (4) each element of G' is a continuum of condensation of M. Also, we shall describe a dendron F' in M such that each element of G' intersects F' in only one point.

First step. If some element of Y has a point P in common with V_1V_2 or V_4V_3 and P is nearer the midpoint than either end point of the one of V_1V_2, V_4V_3 on which it lies, let i be the least positive integer n such that y_n is such an element of Y and let P be the point that y_i has in common with $V_1V_2+V_4V_3$. If there is no such i, let P be the midpoint of V_1V_2. Let P_1 and P_2 be the projections of P on V_1V_2 and V_4V_3 respectively. Let k be the least positive integer n, if there is one, such that y_n has no point in common with the boundary of R. There are two arcs P_1AP_2 and P_1DP_2 lying in M, intersecting Y^* plus the boundary of R only at P_1 and P_2 and forming a simple closed curve J such that (1) J incloses z_1 and y_k, (2) the interior of J together with P_1 and P_2 contains each element of Y (if any) that contains

(6) By X^* is meant the set of all points P such that P is a point of an element of X.

P_1 or P_2, and (3) P_1AP_2 separates $\langle P_1DP_2\rangle$ from V_1V_4 in M.

We shall consider the case in which P_1 and P_2 belong to elements $y(P_1)$ and $y(P_2)$ of Y respectively. Let $\langle P_1BP_2\rangle$ and $\langle P_1CP_2\rangle$ be mutually exclusive open arcs in M which are inclosed by J, which contain no point of Y^*, and which are such that P_1BP_2 separates $\langle P_1CP_2\rangle$ from V_1V_4 in M, the simple closed curve formed by $P_1AP_2+P_1BP_2$ incloses $y(P_1)-P_1$, the simple closed curve formed by $P_1BP_2+P_1CP_2$ incloses $y(P_2)-P_2$, and P_2C of P_1CP_2 intersects V_5V_2.

If $y(P_1)-P_1$ is not connected, let $c_1, c_2, \cdots$ denote its components. Let i, j be the least positive integers m, n respectively such that z_m, y_n are inclosed by $P_1AP_2+P_1BP_2$. There is a simple closed curve J_1 in M containing P_1 and P_2 and inclosing c_1, z_i, y_j but no point of $y(P_1)-c_1$ which is such that $J_1-P_1-P_2$ contains no point of Y^* and is inclosed by $P_1AP_2+P_1BP_2$. Let O represent the simple closed curve which is a subset of $P_1AP_2+P_1BP_2+J_1$ and which incloses c_2 but no point of J_1. Let i, j be the least positive integers m, n such that z_m, y_n are inclosed by O. There is a simple closed curve J_2 in M containing P_1 and P_2 and inclosing c_2, z_i, y_j but no point of $Y(P_1)-c_2$ which is such that $J_2-P_1-P_2$ contains no point of Y^* and is inclosed by O. This process is continued until for each c_i there is a unique J_i inclosing it.

The interior of $P_1AP_2+P_1BP_2$ is topologically equivalent to the bounded complementary domain of $P_1BP_2+P_2C+P_1DP_2$. Treating $P_2BP_1+P_1DP_2$ as P_1AP_2 and P_2C+CP_2 as P_1BP_2, we consider curves $J_1', J_2', \cdots$ inclosing the components of $y(P_2)-P_2$ in a manner similar to that in which we considered simple closed curves inclosing the components of $y(P_1)-P_1$.

Let g_1' denote the closure of $J+P_1BP_2+P_2C+J_1+J_2+ \cdots +J_1'+J_2' + \cdots$. The continuum g_1' is a subset of an element of G'. It is to be noted that g_1' is a continuum of condensation of M intersecting both V_5V_2 and V_4V_6 and if C is the common part of M and a bounded complementary domain of g_1', then (1) C intersects the interiors of triangles $V_1V_2V_5$ and $V_4V_3V_6$, (2) $\bar{C}$ is a simple plane web, and (3) no two elements of Y intersect both C and the boundary of C with respect to M.

Second step. Let i be the least positive integer n such that y_n intersects one of the intervals V_1P_1 and V_4P_2 in a point P nearer the midpoint than either end point of that interval. If there is no such i, let P be the midpoint of V_1P_1. Let P_3, P_4 be the projections of P on V_1P_1 and V_4P_2 respectively. By treating $V_1P_1+P_1AP_2+P_2V_4+V_4V_1$ as the boundary of R and following a procedure similar to that used in finding g_1', we obtain g_2'. By treating $P_1V_2+V_2V_3+V_3P_2+P_2DP_1$ as the boundary of R, we get g_3'. This process is continued until every point of M belongs to either the closure of the sum of all the g_n''s or to a bounded complementary domain of one of them.

If U is a component of M minus the common part of M and the sum of the g_n''s together with their bounded complementary domains and U is closed, U will have a point in common with V_1V_2 and a point in common with V_4V_3

and is to be considered as a g'. If U is not closed, it will not have a point in common with either V_1V_2 or V_4V_3 and, as every two g_n''s are separated in M by a third g_n', there is only one g_n' which contains a limit point of U. If there are such U's, g_n' plus all U's which have a point of g_n' as a limit point is a g'. Each g' described is a subset of an element of G'.

If G_1' is the set of maximal g''s thus described, G_1' is an arc with respect to its elements, each point of $G_1'^*$ is a limit point of $M'-G_1'^*$, no element of G_1' has two points in common with an element of Y and each element of G_1' has both a point in common with V_1V_2 and a point in common with V_4V_3. There exists a dendron F_1' in M such that $F_1'\cdot G_1'^*$ is V_4V_3, $F_1'-V_4V_3$ intersects neither V_4V_6 nor any element of Y and if C is a component of $M-G_1'^*$, then $F_1'\cdot C$ is an arc minus an end point of diameter less than the diameter of V_1V_2 and if K is a connected subset of C that intersects both this arc and an element of Y that intersects the boundary of C with respect to M, then K intersects both V_5V_2 and V_4V_6. Let D_1 be a domain containing $F_1'-V_4V_3$ such that each point of D_1 is nearer F_1' than $G_1'^*+V_4V_6$. Then $\bar{D}_1$ intersects $G_1'^*$ only in V_4V_3.

Third step. If C is a component of $M-G_1'^*$, C is the common part of M and a bounded complementary domain of some element of G_1'. We now consider the breaking up of $\bar{C}$ into elements of G'. We note that C intersects the interiors of the triangles $V_1V_2V_5$ and $V_4V_3V_6$. If β denotes the boundary of C with respect to M, then $C+\beta$ is a simple plane web and no more than one element of Y intersects both C and β.

Let y_a be such an element of Y intersecting β at P_1 and let c denote the closure of the common part of y_a and C. Let Q_1Q_2 be an arc such that $Q_1Q_2-Q_1$ is $F_1'\cdot C$. Let i be the least positive integer n such that d_n is a subset of c. There is a point P_2 of the boundary of d_i such that $c+\beta-P_2$ is connected. There is a subcontinuum w of c such that (1) w contains P_1 and P_2, (2) if a point of a complementary domain of M belongs to w, the complementary domain is a subset of w, and (3) no subcontinuum of w has properties (1) and (2). There is an upper semi-continuous collection E of mutually exclusive continua filling up w such that E is an arc from P_1 to P_2 with respect to its elements and such that every element of E is either a point or an arc which has only its end points in common with the boundary of c. Designate the arc of elements of E by $\xi(P_1P_2)$. Each connected subset of C that intersects $\xi(P_1P_2)$ and Q_1Q_2 also intersects V_5V_2 and V_4V_6. To adjust the argument to the case where no element of Y has a point in common with both C and β, let $\xi(P_1P_2)$ be an arc in $C+P_1$ from a point P_1 of β to a point P_2 of C such that $\xi(P_1P_2)$ contains no point of Y^* and any connected subset of C intersecting both $\xi(P_1P_2)$ and Q_1Q_2 also intersects both V_5V_2 and V_4V_6.

Let j, k be the least positive integers m, n respectively such that z_m, y_n are subsets of C. There is a simple closed curve J lying in C, inclosing y_k, intersecting $\xi(P_1P_2)+Q_1Q_2$ only at P_2+Q_2 and such that Q_1Q_2+J contains z_j.

We note that J plus the common part of M and the interior of J is a simple plane web intersecting both V_5V_2 and V_4V_6.

A breaking up of the portion of $\overline{C}$ not in the interior of J into elements of G' will now be considered. Considering the elements of $\xi(P_1P_2)$ as points, $\xi(P_1P_2)$ as V_1V_2, β as V_1V_4, $\xi(P_1P_2)$ as V_4V_3, J as V_2V_3, and Q_1Q_2 as V_5V_6, we divide the portion of $\overline{C}$ not in the interior of J in a manner somewhat analogous to that described in the second step. We divide the portion of $\overline{C}$ corresponding to $V_1V_2V_6V_5$ in the manner in which $V_1V_2V_3V_4$ was divided in the second step and we divide the portion of $\overline{C}$ corresponding to $V_5V_6V_3V_4$ in the same fashion, care being taken that each pair of g''s thus defined that intersect on Q_1Q_2 also intersect the same elements of $\xi(P_1P_2)$.

The closures of all components of $M-G_1'^*$ are broken up just as was $\overline{C}$. If a g' as defined in this step intersects another g' defined in either this or the second step, the sum of the two is a g'. Let G_2' denote the set of maximal g''s defined. We note that the boundary of d_i of y_a is a subset of $G_2'^*$. Also, each element of G_2' intersects V_5V_2 and V_4V_6 and so does each component of $M-G_2'^*$.

There exists a dendron F_2' in M such that $F_2' \cdot G_2'^*$ is F_1', $F_2'-F_1'$ is a subset of D_1, $F_2'-F_1'$ intersects no element of Y, and if C is a component of $M-G_2'^*$, then $F_2' \cdot C$ is an arc minus one end point of diameter less than one-half the diameter of V_1V_2. Let D_2 be a domain containing $F_2'-F_1'$ such that each point of D_2 is nearer F_2' than $G_2'^*\,\overline{D}_2$ and is a subset of D_1+F_1'.

Fourth step. We shall now describe the breaking up of a component C of $M-G_2'^*$. The boundary of C with respect to M is a simple closed curve J. If an element of Y intersects both C and J, break up $\overline{C}$ as described in the third step. Otherwise, let j be the least integer equal to n such that y_n is inclosed by J. Let P be a point of y_j such that y_j-P is connected. Let $Q_1Q_2-Q_1$ be the common part of F_2' and the interior of J. There is an arc Q_1P in M which except for its end points intersects neither $J+Q_1Q_2$ nor any element of Y and which is such that any connected subset of the interior of J intersecting Q_1Q_2 and y_j but not Q_1P intersects both V_5V_2 and V_4V_6. Treating Q_1P+J, Q_1Q_2, and y_j as β, Q_1Q_2, and c respectively were treated in the third step, we break up the common part of M and the interior of J in a manner analogous to that in which C was broken up as considered in the third step.

Let G_3' be the set of all g''s considered after all components of $M-G_2'^*$ are broken up as described above. We define a dendron F_3' in D_2 containing F_2' and satisfying conditions analogous to those satisfied by F_2'. In like manner, we consider G_4', G_5', $\cdots$ and F_4', F_5', $\cdots$. It is to be noted that (1) for each positive integer j, there is an n such that z_j and the boundary of d_j belongs to $G_n'^*$, (2) for no n do three points of Y^* belong to the same element of G_n', (3) each element of G_n' intersects V_5V_2 and V_4V_6, and (4) each component of $M-G_n'^*$ intersects V_5V_2 and V_4V_6.

Definition of dendron G. Let G' be the set of all continua such that g'

is an element of G' only if either (1) for some positive integer n, g' belongs to G_n' and G_{n+1}' or (2) g' is a component of $M-(G_1'^*+G_2'^*+\cdots)$. With respect to its elements, G' is a dendron. Each element of G' is a continuum intersecting both V_5V_2 and V_4V_6. Each element of G' is a continuum of condensation of M. No element of G' intersects an element of Y in more than two points.

We define G to be the collection of all point sets g such that g is the image of an element of G' under T. We note that G is an upper semi-continuous collection of mutually exclusive continua filling up W such that G is a dendron with respect to its elements.

Definition of dendron F. We define F' to be the limiting set of F_1', F_2', $\cdots$. We note that it is a dendron that intersects each element of G' in only one point. We define F to be the image of F' under transformation T. We note that F is a dendron that intersects each element of G in only one point and the outer boundary of W in only an arc.

Description of arc H. We shall describe an upper semi-continuous collection H of mutually exclusive continua filling up W such that H is an arc with respect to its elements and such that if g and h are elements of G and H respectively, then $g\cdot h$ is totally disconnected. First, we shall establish four additional theorems.

THEOREM 9. *If R is the interior of a square and E is a collection of point sets covering $\overline{R}$ such that for each point P and each domain D there is a point P' of D which belongs to no element of E that contains P, then for any positive ϵ and any two points A and B of $\overline{R}$ there is in $R+A+B$ an arc ACB such that* (1) *for any two points Q_1 and Q_2 of ACB*(7), $\cos(Q_1Q_2, AB)>1-\epsilon$ *and* (2) *no element of E contains a subarc of ACB.*

Proof. There is a point P_1 such that(8) $d(A, P_1)=d(P_1, B)$, the intervals AP_1 and P_1B are subsets of $R+A+B$, and $\cos(AP_1, AB)>1-\epsilon$. There exists a quadrilateral Q lying in $R+A+B$ and inclosing $AP_1+P_1B-(A+B)$. Within Q there exist (1) a point P_2 not belonging to any element of E that contains A and such that $\max[d(A, P_2);\ d(P_2, P_1)]<(2/3)d(A, P_1)$, $\cos(AP_2, AB)>1-\epsilon$ and $\cos(P_2P_1, AB)>1-\epsilon$ and (2) a point P_3 not belonging to any element of E that contains P_1 and such that $\max[d(P_1, P_3); d(P_3, B)]<(2/3)d(P_1, B)$, $\cos(P_1P_3, AB)>1-\epsilon$ and $\cos(P_3B, AB)>1-\epsilon$. Likewise we get within Q a point P_4 not belonging to any element of E containing A and such that $\max[d(A, P_4); d(P_4, P_2)]<(2/3)d(A, P_2)$ and $\cos(AP_4, AB)>1-\epsilon$ and $\cos(P_4P_2, AB)>1-\epsilon$. This process is continued.

Let L denote the closure of the sum of all such P_n's. Since L is closed and compact and as its projection on the line AB is the interval AB and each line

(7) By $\cos(AB, CD)$ is meant the cosine of the smaller angle between the lines AB and CD. If AB and CD are coincident or parallel, $\cos(AB, CD)=1$.

(8) By $d(A, B)$ is meant the distance between A and B.

perpendicular to this interval contains only one point of L, then L is an arc from A to B.

As each of the numbers cos (P_2P_1, AB) and cos (P_1P_3, AB) is greater than $1-\epsilon$ and as the projection of P_1 on AB is between the projections of P_2 and P_3, it follows that cos $(P_2P_3, AB)>1-\epsilon$. By induction, we get for different integers j and k that cos $(P_jP_k, AB)>1-\epsilon$. Hence if Q_1 and Q_2 are points of the continuum L, cos $(Q_1Q_2, AB)>1-\epsilon$.

Each subarc of L contains a P_j and a P_k such that P_j does not belong to any element of E containing P_k. Hence, no subarc of L is a subset of an element of E.

THEOREM 10. *Let R be the interior of a square with diagonals V_1V_3 and V_2V_4; let E be a collection of mutually exclusive point sets filling up $\overline{R}$ such that* (1) *no element of E contains a nondegenerate connected subset of $V_1V_4+V_2V_3$ and* (2) *no element of E is dense in a domain; then there exists an upper semicontinuous collection H of mutually exclusive arcs filling up $\overline{R}$ such that H is an arc with respect to its elements, each element of H has one end point on V_1V_2 and one on V_4V_3, each point of $V_1V_2+V_4V_3$ is an end point of an element of H, and no element of E contains a subarc of an element of H.*

Proof. With the help of the preceding theorem, it will be shown that there exists a set of mutually exclusive arcs $\alpha_1, \alpha_2, \cdots$ such that (1) each of them has one end point on V_1V_2 and the other on V_4V_3 and has no arc in common with an element of E and (2) the components of $\overline{R}-(\alpha_1+\alpha_2+\cdots)$ are arcs having one end point on V_1V_2 and one on V_4V_3 and, furthermore, no such component has an arc in common with an element of E.

We shall assume that $\overline{R}-R$ is a unit square. Consider points Q_1, Q_2, Q_3, Q_4, Q_5, Q_6 in the orders $V_1Q_1Q_2Q_3V_4$ and $V_2Q_4Q_5Q_6V_3$ and such that Q_1, Q_2, Q_3 and Q_4, Q_5, Q_6 divide V_1V_4 and V_2V_3 respectively into four equal parts. Choose a positive number ϵ less than 1. Let P_0 be the midpoint of V_1V_2 and P_1, P_2 be points of V_1V_2 at a distance from P_0 equal to $\epsilon/6$ and in the order $V_1P_1P_0P_2V_2$. Let P_3 and P_4 denote points on V_1V_2 at a distance from P_0 equal to $2\epsilon/6$ and in the order $V_1P_3P_0P_4V_2$. Similarly, consider points $P_5, P_6, \cdots, P_{2j-1}, P_{2j}$ where P_{2j-1} and P_{2j} are at a distance from P_0 equal to $j\epsilon/6$ and from V_1 and V_2 respectively of not more than $\epsilon/6$ and are in the order $V_1P_{2j-1}P_0P_{2j}V_2$. Let P_{2j+1} be the midpoint of V_1P_{2j-1} and in general let $P_{2j+2k+1}$ be the midpoint of $V_1P_{2j+2k-1}$ for all positive integers k. Also $P_{2j+2k+2}$ is the midpoint of $P_{2j+2k}V_2$. Let $P_{i,1}$, $P_{i,2}$, $P_{i,3}$, $P_{i,4}$ be the projections of P_i on Q_1Q_4, Q_2Q_5, Q_3Q_6, V_4V_3 respectively.

It is assumed that each of the following arcs has the properties that (1) no subarc of it is a subset of E and (2) if A and B are two points of it, cos $(AB, V_1V_4)>1/(1+\epsilon^2)^{1/2}$. There exist in P_0+P_2+R mutually exclusive arcs $P_0P_{0,1}$ and $P_2P_{2,1}$ satisfying the above two conditions and such that no arc from P_0P_2 to $P_{0,1}P_{2,1}$ in $P_0P_2+P_{0,1}P_{2,1}$ plus the interior of $P_0P_2+P_2P_{2,1}$

$+P_{2,1}P_{0,1}+P_{0,1}P_0$ is a subset of an element of E. In the case where the elements of E contain arcs, this may be seen by constructing $P_0P_{0,1}$ and $P_2P_{2,1}$ so that two elements of E contain arcs from $P_0P_{0,1}$ to $P_2P_{2,1}$ in $P_0P_{0,1}+P_2P_{2,1}$ plus the interior of $P_0P_2+P_2P_{2,1}+P_{2,1}P_{0,1}+P_{0,1}P_0$. Also, there exist in R mutually exclusive arcs $P_{0,2}P_{0,3}$ and $P_{2,2}P_{2,3}$ which satisfy the above two conditions and such that no arc from $P_{0,2}P_{2,2}$ to $P_{0,3}P_{2,3}$ in $P_{0,2}P_{2,2}+P_{0,3}P_{2,3}$ plus the interior of $P_{0,2}P_{2,2}+P_{2,2}P_{2,3}+P_{2,3}P_{0,3}+P_{0,3}P_{0,2}$ is a subset of an element of E. Furthermore, there are mutually exclusive arcs $P_{0,1}P_{0,2}$ and $P_{1,1}P_{1,2}$ in R satisfying the above two conditions and such that no arc from $P_{1,1}P_{0,1}$ to $P_{1,2}P_{0,2}$ in $P_{1,1}P_{0,1}+P_{1,2}P_{0,2}$ plus the interior of $P_{1,1}P_{0,1}+P_{0,1}P_{0,2}+P_{0,2}P_{1,2}+P_{1,2}P_{1,1}$ is a subset of an element of E. Arcs satisfying corresponding conditions join $P_{1,3}$, $P_{0,3}$ to $P_{1,4}$, $P_{0,4}$ respectively. Furthermore, arcs which intersect no arc previously described and which satisfy conditions analogous to those given join P_3, P_1 to $P_{3,1}$, $P_{1,1}$; $P_{3,2}$, $P_{1,2}$ to $P_{3,3}$, $P_{1,3}$; $P_{2,1}$, $P_{4,1}$ to $P_{2,2}$, $P_{4,2}$; $P_{2,3}$, $P_{4,3}$ to $P_{2,4}$, $P_{4,4}$; $P_{5,1}$, $P_{3,1}$ to $P_{5,2}$, $P_{3,2}$; $\cdots$ in such a manner that if α_i designates $P_iP_{i,1}+P_{i,1}P_{i,2}+P_{i,2}P_{i,3}+P_{i,3}P_{i,4}$, then V_1V_4, V_2V_3 are the sequential limiting sets of $\alpha_1, \alpha_3, \cdots$ and $\alpha_0, \alpha_2, \cdots$ respectively.

Let H_1 be the set consisting of V_1V_4, V_2V_3, $\alpha_0, \alpha_1, \cdots$. If h_1 and h_2 are elements of H_1 which are not separated from each other in $\overline{R}$ by any element of H_1 and C is set of all points of $\overline{R}$ between h_1 and h_2, no arc from Q_2Q_5 to $V_1V_2+V_4V_3$ in $Q_2Q_5+V_1V_2+V_4V_3$ plus the bounded complementary domain of $V_1V_2+V_4V_3+h_1+h_2$ is a subset of an element of E.

Designate $V_1V_2\cdot\overline{C}$ and $V_4V_3\cdot\overline{C}$ by U_1U_2 and U_4U_3 respectively, where the points U_1, U_2, U_3, U_4 are in the orders $V_1U_1U_2V_2$ and $V_4U_4U_3V_3$. Treating U_1U_2 and U_4U_3 as V_1V_2 and V_4V_3 were treated, we break up $\overline{C}$ in a manner somewhat analogous to that in which $\overline{R}$ was divided except that instead of three lines Q_1Q_4, Q_2Q_5, Q_3Q_6, we use seven lines parallel to U_1U_2 that divide a segment from U_1U_2 to U_4U_3 into eight equal parts. The division of all such C's gives us H_2. This division of C is possible, for no use was made above of the fact that V_1V_4 and V_2V_3 were straight lines. In fact, it can be shown that if (1) P_1P_2 and $P_1'P_2'$ are parallel intervals, (2) P_1P_1' and P_2P_2' are mutually exclusive arcs such that if A and B are either two points of P_1P_1' or two points of P_2P_2', then $\sin(AB, P_1P_2)>1/(1+\epsilon^2)^{1/2}$, (3) Q_1, Q_2, Q_1', Q_2' are points in the orders $P_1Q_1Q_2P_2$ and $P_1'Q_1'Q_2'P_2'$ and $\min[\sin(Q_1Q_2', P_1P_2); \sin(Q_2Q_1', P_1P_2)]>1/(1+\epsilon^2)^{1/2}$, and (4) E is a collection of mutually exclusive point sets filling up $P_1P_2+P_2P_2'+P_2'P_1'+P_1'P_1$ plus its interior and such that no element of E is dense in a domain, then there exist in its interior plus the points Q_1, Q_2, Q_1' and Q_2' two mutually exclusive arcs Q_1Q_1' and Q_2Q_2' such that no arc from Q_1Q_2 to $Q_1'Q_2'$ in $Q_1Q_2+Q_2Q_2'+Q_2'Q_1'+Q_1'Q_1$ plus its interior is a subset of an element of E and such that if C and D are two points of Q_1Q_1' or two points of Q_2Q_2', then $\sin(CD, P_1P_2)>1/(1+\epsilon^2)^{1/2}$.

No element of E contains an arc in $\overline{R}-(H_1^*+H_2^*)$ whose projection of V_1V_4 is of length 3/8. With the use of fifteen lines parallel to V_1V_2, H_3 is defined in such a way that no element of E contains an arc in $\overline{R}-(H_1^*+H_2^*+H_3^*)$ whose projection on V_1V_4 is of length 3/16. If a suitable pattern is used in defining $H_4, H_5, \cdots$ and H is the set such that h is an element of H only if it is an element of some H_n or a component of $\overline{R}-(H_1^*+H_2^*+\cdots)$, then H satisfies the conditions of our theorem.

Our method actually gives the following result.

THEOREM 11. *Let R be the interior of a square with diagonals V_1V_3 and V_4V_2; let E be a collection of point sets covering $\overline{R}$ such that no element of E contains a nondegenerate connected subset of $V_1V_4+V_2V_3$ and if D_1 and D_2 are two domains, there exist nondegenerate connected subsets f_1 and f_2 of D_1 and D_2 respectively such that no element of E intersects both f_1 and f_2. Then there exists an upper semi-continuous collection H of mutually exclusive arcs filling up $\overline{R}$ such that H is an arc with respect to its elements, each element of H has one end point on V_1V_2 and one on V_4V_3, each point of $V_1V_2+V_4V_3$ is an end point of an element of H and no element of E contains a subarc of an element of H.*

THEOREM 12. *Let R be the interior of a square with diagonals V_1V_3 and V_2V_4; let F be a dendron in $\overline{R}$ intersecting the boundary of R only in V_4V_3; let E be a collection of point sets covering $\overline{R}$ such that no element of E contains a nondegenerate connected subset of $V_1V_4+V_2V_3$ and if D_1 and D_2 are two domains, there exist nondegenerate connected subsets f_1 and f_2 in D_1 and D_2 respectively such that no element of E intersects both f_1 and f_2. There exists an upper semi-continuous collection H of mutually exclusive continua filling up $\overline{R}$ such that H is a dendron with respect to its elements, V_1V_4 is a subset of an element of H and so is V_2V_3, no element of E contains a nondegenerate subcontinuum of an element of H and each element of H intersects F in only one point and V_1V_2 in only a totally disconnected point set.*

Indication of proof. We construct dendron H so that each element of H intersects F in only one point in a manner analogous to that in which we constructed G' so that it intersected F' in only one point. We use the methods of Theorem 10 to insure that no element of H intersects an element of E in a nondegenerate continuum.

Definition of arc H. We shall now define an upper semi-continuous collection H of mutually exclusive continua filling up W such that H is an arc with respect to its elements, each element of H intersects each element of G, and the intersection of an element of G and an element of H is a totally disconnected point set.

As the points of $M-M\cdot Y^*$ and the elements of Y form an upper semi-continuous collection of mutually exclusive continua filling up $\overline{R}$, it follows [5] that there is a continuous transformation T' carrying $M+Y^*$ into $\overline{R}$ such that (1) T' carries each point of $V_5V_2+V_4V_6$ plus the portion of the boundary of R not belonging to Y^* into itself, (2) T' carries each point of $M-M\cdot Y^*$

into a point, (3) T' carries each element of Y into a point, and (4) each point of $\bar{R}$ is either the image of only one point of $M-M\cdot Y^*$ or the image of only one element of Y. If G'' is the set of all $T'(g')$'s where g' is an element of G', the elements of G'' intersect V_5V_2 and V_4V_6. By the methods of Theorem 11, it can be shown that there is an upper semi-continuous collection H'' of mutually exclusive continua filling up $\bar{R}$ such that (1) H'' is with respect to its elements an arc with ends at V_1 and V_3, (2) every element of H'' other than V_1 and V_2 is an arc having only one point on $V_1V_2+V_2V_3$ and only one point on $V_1V_4+V_4V_3$, these points being its end points, (3) if g'' is an element of G'' intersecting an element h'' of H'', $g''\cdot h''$ is totally disconnected, and (4) one element of H'' contains V_5+V_2 and another contains V_4+V_6 and each of these elements lies except for its end points on the interior of the parallelogram $V_5V_2V_6V_4$.

If H' is a set such that h' is an element of H' only if it is a subcontinuum of M which is maximal with respect to $T'(h')$ being an element of H'', then H' is a collection of mutually exclusive continua filling up M. Furthermore, it is an arc with respect to its elements. As no element of G' has more than two points in common with the boundary of an element of Y, if g' is an element of G' having a point in common with an element h' of H', $g'\cdot h'$ is totally disconnected.

Let H be a set such that h is an element of H only if it is a continuum which is maximal with respect to being the image under T of the sum of two or three elements of H'. Then H is the required collection of mutually exclusive continua filling up W. The sufficiency of Theorem 1 is demonstrated.

Theorem 1 may also be stated as follows:

THEOREM 1. *A necessary and sufficient condition that a compact plane continuum W remain connected and locally connected on the omission of any countable subset is that there exist two continuous monotone transformations T_1 and T_2 such that both $T_1(W)$ and $T_2(W)$ are nondegenerate dendrons and if a and b are elements of $T_1(W)$ and $T_2(W)$ respectively, then $T_1^{-1}(a)\cdot T_2^{-1}(b)$ exists and is totally disconnected*(9).

Alternate definition of a simple plane web. Actually, we have shown that the following is an equivalent definition to that previously given.

Definition. A compact plane continuum W is a simple plane web provided there exist an upper semi-continuous collection G of mutually exclusive continua filling up W and another such collection H also filling up W such that (1) G is a dendron with respect to its elements and H is an arc with respect to its elements and (2) if g and h are elements of G and H respectively, then $g\cdot h$ exists and is totally disconnected. It has been shown [1] that condition (1) is unnecessary.

(9) G. T. Whyburn and some of his students have worked with monotone mappings onto dendrons. See [9] for references.

However, we cannot substitute for (1) in the above definition the condition (1′) that G is an arc with respect to its elements and so is H. Consider the plane continuum W obtained by subtracting from a circle C_0 plus its interior the interiors of five mutually exclusive circles C_1, C_2, C_3, C_4, C_5, each of which is tangent to C_0 internally. Now C_i is not both a subset of an element of G and also a subset of an element of H. Assume that C_j $(j=1, 2, 3)$ is not a subset of an element of G. Then there are uncountably many elements of G that intersect C_j. It may be seen that some element g_j of G separates a point P_j of C_j from $C_0+C_j+\cdots+C_5-C_j$ in W and intersects C_i $(i=0, 1, \cdots, 5)$ only if i is equal to j. No one of the elements g_1, g_2, g_3 separates the other two from each other in W. Hence, G is not an arc with respect to its elements.

THEOREM 13. *If W is a simple plane web, there exist a dendron F in W, an arc α in F, an upper semi-continuous collection G of mutually exclusive continua filling up W and another such collection H also filling up W such that* (1) *G is a dendron with respect to its elements and H is an arc with respect to its elements,* (2) *if g and h are elements of G and H respectively, $g\cdot h$ exists and is totally disconnected,* (3) *each element of G intersects F in only one point, and* (4) *each element of H intersects α in only one point.*

Proof. Consider the dendron F and the upper semi-continuous collections G and H already defined. If α is the common part of F and the outer boundary of W, then F, α, G, and H satisfy the conditions of our theorem.

THEOREM 14. *If W is a simple plane web, there exist a dendron F in W, an upper semi-continuous collection G of mutually exclusive continua filling up W and another collection H also filling up W such that* (1) *G is a dendron with respect to its elements and so is H,* (2) *if g and h are elements of G and H respectively, $g\cdot h$ exists and is totally disconnected,* (3) *each element of G intersects F in only one point and so does each element of H.*

Proof. The dendron F and the upper semi-continuous collection G are the same as those previously described.

We shall use the same symbols as those used in our definition of the arc H. Let F'' be the image of F' under T'. If h is the element of H'' containing V_4+V_6, let H_1'' denote the set of all continua h'' such that h'' is either V_1, h, or an element of H'' between V_1 and h. Using the methods of Theorem 12, it can be shown that there exists an upper semi-continuous collection H_2'' of mutually exclusive continua filling up $h+(\overline{R}-H_1''^*)$ such that (1) H_2'' is a dendron with respect to its elements, (2) h is a subset of an element of H_2'', (3) each element of H_2'' intersects F'' in only one point and V_6V_3 in a totally disconnected set, and (4) if g'' is an element of G'' intersecting an element h'' of H_2'', $g''\cdot h''$ is totally disconnected. We define H from $H_1''+H_2''$ in the same manner in which the arc H was defined from H''.

Theorem 14 may also be stated as follows.

THEOREM 14. *Suppose that W is a compact plane continuum and that T_1, T_2 are continuous monotone transformations of W such that if a and b are elements of $T_1(W)$ and $T_2(W)$ respectively, then $T_1^{-1}(a)$ and $T_1^{-2}(b)$ are nondegenerate sets whose common part exists and is totally disconnected. Then there exist monotone retractions R_1 and R_2 of W such that $R_1(W)$ is a dendron, $R_2(W)$ is the same dendron and if a and b are elements of this dendron, $R_1^{-1}(a) \cdot R_2^{-1}(b)$ exists and is totally disconnected.*

REFERENCES

1. R. H. Bing, *Collections filling up a simple plane web*, Bull. Amer. Math. Soc. vol. 51 (1945) pp. 674–679.

2. S. Eilenberg, *Sur les transformations d'espaces métriques en circonférence*, Fund. Math. vol. 24 (1935) pp. 160–176.

3. R. L. Moore, *Concerning continua which have dendratomic subsets*, Proc. Nat. Acad. Sci. U.S.A. vol. 29 (1943) pp. 384–389.

4. ———, *Concerning continuous curves in the plane*, Math. Zeit. vol. 15 (1922) pp. 254–260.

5. ———, *Concerning upper semi-continuous collections of continua*, Trans. Amer. Math. Soc. vol. 27 (1925) pp. 416–428.

6. ———, *Concerning webs in the plane*, Proc. Nat. Acad. Sci. U.S.A. vol. 29 (1943) pp. 389–393.

7. ———, *Foundations of point set theory*, Amer. Math. Soc. Colloquium Publications, vol. 13, New York, 1932.

8. ———, Bull. Amer. Math. Soc. vol. 33 (1927) p. 141 and vol. 35 (1929) p. 776.

9. G. T. Whyburn, *Analytic topology*, Amer. Math. Soc. Colloquium Publications, vol. 28, New York, 1942.

10. R. L. Wilder, *Concerning continuous curves*, Fund. Math. vol. 7 (1925) pp. 340–377.

THE UNIVERSITY OF TEXAS

B. Cuttings

GENERALIZATIONS OF TWO THEOREMS OF JANISZEWSKI

R. H. BING

Janiszewski proved [2][1] the following for the plane.

THEOREM A. *The sum of two compact continua does not cut the point A from the point B provided neither cuts A from B and their common part is connected or does not exist.*

THEOREM B. *The sum of two compact continua cuts the plane provided their common part is not connected.*

Many generalizations and modifications of these theorems appear in the literature (for example, see [1, 3, 4, 6, 8, 9, 10, 11, 12]). This paper generalizes the two theorems by considering, instead of continua, sets that are neither open nor closed. All sets referred to in this paper are in the plane.

DEFINITION. The set R cuts the point A from the point B in the set W provided some continuum in W contains A and B and R intersects neither A nor B but it intersects every continuum in W containing A and B. If R cuts A from B in the plane, we simply say that it cuts A from B. Thus we use *cut* in the sense that *coupe* was used in Fundamenta Mathematicae (for example, see [1, p. 75, and 10, p. 15]) and not in the sense used by some writers to mean separate. If a closed set cuts two points from each other in the plane, it separates them from each other; this is not true for more general sets.

Generalizations of Theorem A. We shall make use of the following known result [12, p. 129, and 4, pp. 35–36].

THEOREM 1. *If neither of the closed point sets H and K separates the point A from the point B, the common part of H and K is connected or does not exist and the part of H in the complement of K is compact, then the sum of H and K does not separate A from B.*

THEOREM 2. *If neither of the point sets H and K separates the point A from the point B, each of the sets is closed in their sum,[2] one of the sets is compact[3] and their common part is a continuum or does not exist, then the sum of H and K does not separate A from B.*

Presented to the Society, September 17, 1945; received by the editors July 24, 1945.

[1] Numbers in brackets refer to the references cited at the end of the paper.

[2] The set H is closed in the set S if $\overline{H} \cdot S$ is equal to H.

[3] In Theorems 2, 5, 6, 7, 8, 9, 10 instead of assuming that one of the sets is compact, we can assume that the part of it in the complement of the other is compact.

PROOF. Assume that $H+K$ separates A from B. Then $H+K$ contains a continuum M separating A from B [8, Theorem 73, p. 150, and Theorem 24, p. 194]. However, neither of the closed point sets $M\cdot H+H\cdot K$ and $M\cdot K+H\cdot K$ separates A from B because neither H nor K does. Hence, by Theorem 1 their sum does not. But their sum contains M.

Example. Theorem 2 would not be true if instead of requiring that $H\cdot K$ be a continuum we only required that it be connected. Let $H\cdot K$ be the sum of the interval from (0, 0) to (−1, 2), the interval from (−1, 2) to (1, 1) and the image of

$$y = \sin(\pi/2x), \qquad 0 < x \leqq 1;$$

let $H-H\cdot K$ be the image of

$$0 \leqq y < \sin(\pi/2x), \qquad 0 < x \leqq 1;$$

let $K-H\cdot K$ be the image of

$$0 \geqq y > \sin(\pi/2x), \qquad 0 < x \leqq 1.$$

THEOREM 3. *If the sum of the simple closed curve J and the arc PQ separates the point A from the point B then their sum contains a simple closed curve J' such that J' minus the common part of J' and PQ is a connected subset of J.*

PROOF. There is a subset M of J irreducible with respect to $M+PQ$ separating A from B. Also, $M+PQ$ contains a continuum J' which is irreducible with respect to separating A from B. This continuum is a simple closed curve containing M. If M were the sum of two mutually separated point sets M_1 and M_2, neither of the closed sets M_1+PQ and M_2+PQ would separate A from B and their sum would not. As M is irreducible with respect to $M+PQ$ separating A from B, it contains no point of PQ and is therefore equal to $J'-J'\cdot PQ$.

THEOREM 4. *If the points A and B belong to the same component of the complement of the connected domain D and J is a simple closed curve separating A from B then J contains a continuum which contains no point of D and which cuts A from B in the complement of D.*[4]

PROOF. Assume that J intersects D; the theorem is obviously true if it does not. Let $O_1, O_2, \cdots$ be a sequence of points dense in $D\cdot J$. There is an arc O_1O_2 in D and by Theorem 3 there is a simple closed curve J' separating A from B and such that $J'-J'\cdot O_1O_2$ is an open

[4] By Theorem 7 it follows that J contains a continuum cutting A from B in the complement of D and intersecting $\overline{D}$ in no more than two points.

arc (P_1Q_1) of J containing neither O_1 nor O_2. If P_1 is not O_3 there is an arc P_1O_3 in D and there is a simple closed curve J'' separating A from B such that $J''-J''\cdot(O_1O_2+P_1O_3)$ is an open arc (P_2Q_2) of (P_1Q_1). This process is continued to get an infinitude of (P_nQ_n)'s where (P_nQ_n) is a subset of $(P_{n-1}Q_{n-1})$ containing neither $O_1, O_2, \cdots,$ nor O_{n+1} but separating A from B in the complement of D.

The common part M of $(P_1Q_1), (P_2Q_2), \cdots$ is the same as the common part of $P_1Q_1, P_2Q_2, \cdots$ and is therefore a continuum. As no point of D belongs to each P_nQ_n, M does not intersect D. If C is a continuum containing A and B but no point of D, then for each positive integer n it contains a point of P_nQ_n and therefore contains a point of M.

Example. Theorem 4 would not be true if the word "cuts" were replaced by "separates." Consider the domain D which is the image of

$$\sin(\pi/x) < y < 1 + \sin(\pi/x), \qquad 0 < x < 1,$$

and a circle J with center at (0, 0) and radius 1. Now J contains no continuum which does not intersect D but which separates (0, 0) from (2, 0) in the complement of D.

THEOREM 5. *The sum of two domains does not separate the point A from the point B provided neither of the domains separates A from B, one of them is compact and their common part is connected or does not exist.*

PROOF. First, assume that H and K are two compact domains satisfying the above hypotheses whose sum separates A from B. Then $H+K$ contains a simple closed curve [8, Theorem 9, p. 184] separating A from B. By Theorem 4, this simple closed curve contains a continuum M which contains no point of $H\cdot K$ but which cuts A from B in the complement of $H\cdot K$. As M contains no point of $H\cdot K$, it is a subset of either H or K, say H. Then the domain H cuts A from B and therefore separates A from B.

The theorem may be proved in the case in which only one of H, K is compact by regarding the plane as a sphere minus one point.

THEOREM 6. *Suppose that H and K are sets such that one is compact, neither cuts the point A from the point B, their common part is connected and on the subtraction of their common part from each, the remainders are mutually separated. Then the sum of H and K does not cut A from B.*[5]

[5] Using other methods, Eilenberg showed [1, Theorem 8, p. 78] that the sum of the two compact sets H and K does not cut A from B provided neither cuts A from B, each is open or closed in $H+K$ and $H\cdot K$ is connected.

PROOF. Suppose that C_H and C_K are continua such that each contains $A+B$, C_H does not intersect H and C_K does not intersect K. There exist domains D_0, D_1, D_2 such that D_0 is connected and covers $H \cdot K$ but no point of C_H+C_K, D_1 covers H but no point of C_H, D_2 covers K but no point of C_K, $D_1-D_0 \cdot D_1$ and $D_2-D_0 \cdot D_2$ are mutually separated and one of D_1, D_2 is compact. Considering D_0+D_1 and D_0+D_2 as the domains of Theorem 5, we find that $D_0+D_1+D_2$ does not cut A from B. Neither does $H+K$ which is a subset of $D_0+D_1+D_2$.

THEOREM 7. *If H is a compact closed set cutting the point A from the point B in the complement of the connected set K, then H contains a subset H' irreducible with respect to being a closed set cutting A from B in the complement of K. Also, H' is a continuum that is not separated by any subset of the closure of K.*

PROOF. Let $D_1, D_2, \cdots$ be a sequence of domains such that if D is a domain containing a point P of H, there exists an integer i such that D_i is a subset of D containing P. Let H_1 be a closed subset of H such that (1) H_1 cuts A from B in the complement of K and (2) H_1 intersects D_1 only if every closed subset of H satisfying (1) intersects D_1. In general, let H_n be a closed subset of H_{n-1} such that (1) H_n cuts A from B in the complement of K and (2) H intersects D_n only if every closed subset of H_{n-1} satisfying (1) intersects D_n.

The common part H' of $H_1, H_2, \cdots$ is a closed subset of H. It cuts A from B in the complement of K for each continuum in the complement of K containing A and B intersects each of $H_1, H_2, \cdots$. If H' contained a proper closed subset M cutting A from B in the complement of K, then there would be an integer i such that D_i contained a point P of H' but no point of M. Then H_i would not contain P.

Assume that $\overline{K}$ contains a subset R such that $H'-R$ is the sum of two mutually separated point sets E_1 and E_2. Then there exist mutually exclusive domains F_1 and F_2 containing E_1 and E_2 respectively. However, as neither $K+(H'-H' \cdot F_1)$ nor $K+(H'-H' \cdot F_2)$ cuts A from B, then by Theorem 6, $K+H'$ could not.

Generalizations of Theorem B. We shall now consider some generalizations of Theorem B.

DEFINITION. The *junction* [5, p. 206] of the sets H and K is the common part of H and the closure of K plus the common part of K and the closure of H, that is $H \cdot \overline{K}+\overline{H} \cdot K$.

THEOREM 8. *If H and K are connected sets one of which is compact* *and such that H is continuumwise connected and the junction of H and K is the sum of two mutually separated sets each of which contains a point of H, then the sum of H and K cuts the plane.*[6]

PROOF. Let the junction M of H and K be the sum of two mutually separated point sets M_A and M_B which contain points A and B respectively of H. There exists [7, Corollary 2, p. 221] a compact continuum J irreducible with respect to separating A from B and such that if the components of $\overline{M}_A \cdot \overline{M}_B$ are regarded as points, J is a simple closed curve (or a point) containing no point of $\overline{M} - \overline{M}_A \cdot \overline{M}_B$. Let T be the set of all continua t such that t is either a point of J not belonging to $\overline{M}$ or a component of $\overline{M}_A \cdot \overline{M}_B$.

Let C be a continuum in H containing A and B. No point of $C \cdot J$ is a limit point of K, for J contains no point of M. Using an argument similar to that employed by Kuratowski [7, Theorem 11, pp. 222–223] we find that T contains a finite number of elements $t_1, t_2, \cdots, t_n$ none of which intersect $C+K$ and each of which belongs both to the boundary of a component of $J-(t_1+\cdots+t_n)$ which contains a point of C but no point of K and also to the boundary of a component of $J-(t_1+\cdots+t_n)$ which contains a point of K but no point of C. Let W_C be the sum of those components containing a point of C and let W_K be the sum of those containing a point of K.

Assume that $C+K$ does not cut between any two elements of $t_1, t_2, \cdots, t_n$. Then there is a continuum V containing $t_1+t_2+\cdots+t_n$ but no point of $C+K$. As neither the continuum $V+W_C$ nor the continuum $V+W_H$ separates A from B, then by Theorem 1 their sum does not. However, their sum contains J.

Let i and j be integers such that $C+K$ cuts t_i from t_j and let α_i and α_j be maximal connected subsets of J containing t_i and t_j respectively but no point belonging to an element of T containing a point of $C+K$. Now $C+K$ cuts every point of α_i from every point of α_j. But α_m $(m=i, j)$ contains a point P_m not belonging to H for J contains no point of M. Therefore $H+K$ cuts P_i from P_j.

Example. Theorem 8 would not be true if we did not require that H be continuumwise connected. Let H be the sum of the intervals having end points at $(-3, 0)$ and $(-3, 1)$, at $(-3, 0)$ and $(-1, 0)$, at $(1,0)$ and $(2, 0)$ and at $(-3, 1/2^n)$ and $(2, 1/2^n)$ where $n=1, 2, \cdots$. Let K be the sum of the intervals having end points at $(3, 0)$ and $(3, 1)$, at $(1, 0)$ and $(3, 0)$, at $(-2, 0)$ and $(-1, 0)$ and at $(-2, 1/3^n)$ and $(3, 1/3^n)$.

[6] Nicodym proved this theorem in the case where H is a bounded continuum [10, Theorem 2, p. 20].

THEOREM 9. *If H and K are connected sets one of which is compact, the common part of the closure of H and the closure of K is the sum of two mutually separated point sets each of which contains a point of the junction of H and K, then $H+K$ cuts the plane.*

PROOF. There exists a simple closed curve containing no point of $\overline{H}\cdot\overline{K}$ and separating some two points of the junction of H and K. Using an argument similar to that used in Theorem 8 we find that this simple closed curve contains two points which $H+K$ cuts from each other.

THEOREM 10. *The sum of two continuumwise connected sets separates the plane provided one of them is compact and their common part exists but does not belong to any continuumwise connected subset of their junction.*

PROOF. Suppose that H and K are sets satisfying the hypotheses of Theorem 10 and that the complement of $H+K$ is connected. Neither the complement of $H+\overline{H}\cdot K$ nor the complement of $K+\overline{K}\cdot H$ cuts between any two points of $H\cdot K$ and, by Theorem 6, the complement of $\overline{H}\cdot K+H\cdot\overline{K}$ does not.

This type of argument gives the following result.

THEOREM 11. *The sum of two connected domains one of which is compact separates the plane provided their common part is not connected.*

REFERENCES

1. S. Eilenberg, *Transformations continues en circonférence et la topologie du plan*, Fund. Math. vol. 26 (1936) pp. 61–112.

2. Z. Janiszewski, *Sur les coupures du plan faites par les continus* (in Polish), Prace matematyczno-fizyczne vol. 26 (1913) pp. 11–63.

3. F. B. Jones, *Certain consequences of the Jordan Curve Theorem*, Amer. J. Math. vol. 63 (1941) pp. 531–544.

4. B. Knaster and C. Kuratowski, *Sur les continus non-bornés*, Fund. Math. vol. 5 (1924) pp. 23–58.

5. ———, *Sur les ensembles connexes*, Fund. Math. vol. 2 (1921) pp. 206–255.

6. C. Kuratowski and S. Straszewicz, *Généralisation d'un théorème de Janiszewski*, Fund. Math. vol. 12 (1928) pp. 152–157.

7. C. Kuratowski, *Sur la séparation d'ensembles situés sur le plan*, Fund. Math. vol. 12 (1928) pp. 214–239.

8. R. L. Moore, *Foundations of point set theory*, Amer. Math. Soc. Colloquium Publications, vol. 13, New York, 1932.

9. Anna M. Mullikin, *Certain theorems relating to plane connected sets*, Trans. Amer. Math. Soc. vol. 24 (1922) pp. 144–162.

10. S. Nikodym, *Sur les coupures du plan faites par les ensembles connexes et les continus*, Fund. Math. vol. 7 (1925) pp. 15–23.

11. S. Straszewicz, *Über die Zerschneidung der Ebene durch abgeschlossene Mengen*, Fund. Math. vol. 7 (1925) pp. 159–187.

12. ———, *Über eine Verallgemeinerung des Jordan'schen Kurvensatzes*, Fund. Math. vol. 4 (1923) pp. 128–135.

THE UNIVERSITY OF TEXAS

GENERALIZATIONS OF TWO THEOREMS OF JANISZEWSKI. II

R. H. BING

The purpose of this note is to strengthen Theorems 5 and 6 of [1][1] and to make corrections regarding assumptions of compactness in that paper. The following theorems hold in the plane.

THEOREM 1. *If neither of the domains D_1, D_2 separates the point A from the point B, the boundary of D_1 is compact and the common part of D_2 and each component of D_1 is connected or does not exist, then D_1+D_2 does not separate A from B.*

PROOF. Assume that D_1+D_2 separates A from B. Considering there to be a point P at infinity, we find that D_1+D_2+P contains a simple closed curve J separating A from B. Let d_2 be a component of D_2 intersecting J. We find [1, Theorem 4] that $J-J\cdot d_2$ contains a continuum M cutting A from B in the complement of d_2 and such that any open arc of J containing M separates A from B in the complement of d_2. Let d_1 be a component of D_1 covering a point of M on the boundary of d_2. Now d_1 covers M or else it would intersect two components of D_2. But by Theorem 5 of [1], d_1+d_2 does not separate A from B.

Instead of assuming that the boundary of D_1 is compact, we could assume that the part of D_1 in the complement of D_2 is compact.

THEOREM 2. *If neither of the domains D_1, D_2 cuts the point A from the point B, the boundary of D_1 is compact and the common part of D_2 and each component of D_1 is connected or does not exist, then D_1+D_2 does not cut A from B.*

PROOF. Let C_i $(i=1, 2)$ be the component of the complement of D_i containing $A+B$, let D_i' be the complement of C_i and let D_2'' be the sum of all components of D_2' that are not covered by D_1'. Neither D_1' nor D_2'' separates the plane. The boundary of D_1' is a subset of the boundary of D_1 and is therefore compact. If d' is a component of D_1', we shall show that $d'\cdot D_2''$ is connected or does not exist. It will follow from Theorem 1 that $D_1'+D_2''$ does not separate the plane. Hence, its complement is a continuum containing $A+B$ and its subset D_1+D_2 does not cut A from B.

Presented to the Society, February 23, 1946; received by the editors January 7, 1946.

[1] Number in brackets refers to the reference cited at the end of the paper.

Assume that $d' \cdot D_2''$ contains two components c_1 and c_2. There exist an arc in d' from a point of c_1 to a point of c_2 and a simple closed curve J in D_1 separating this arc from the boundary of d'. Let d be the component of D_1 containing J and let P_iQ_i be an arc in J irreducible from a point P_i of $J \cdot c_i$ to the boundary of c_i. Since Q_i is a point of C_2, P_iQ_i must contain a point of D_2. Then both c_1 and c_2 contain points of $d \cdot D_2$. But it is contrary to a hypothesis of this theorem that $d \cdot D_2$ not be connected. Hence, $d' \cdot D_2''$ is connected or does not exist.

Example. Theorem 2 would not be true if instead of assuming that the boundary of D_1 is compact, we assume that the part of D_1 in the complement of D_2 is compact. Let $D_1 \cdot D_2$ be the set of all points having positive ordinates less than 1 other than those on the lines joining $(1, 1/n)$ to $(n, 1/n)$, $(n, 1/n)$ to $(1, 1)$, $(-1, 1/n)$ to $(-n, 1/n)$ and $(-n, 1/n)$ to $(-1, 1)$ for $n=2, 3, \cdots$; let D_i $(i=1, 2)$ be the sum of $D_1 \cdot D_2$ and the interior of a unit circle with center at $([-1]^i, 1)$. Neither D_1 nor D_2 cuts $(0, 0)$ from $(0, 1)$ but their sum does.

THEOREM 3. *Suppose that neither of the sets H, K cuts the point A from the point B, that the boundary of H is compact, that the junction of H and K is equal to $H \cdot K$ and that H is the sum of a collection of mutually exclusive sets no one of which contains either a limit point of the sum of the others or two components of $H \cdot K$. Then $H+K$ does not cut A from B.*

PROOF. We note that H is contained by a domain D, no component of which contains two components of $H \cdot K$. Let C_H and C_K be two continua in the complements of H and K respectively such that each contains $A+B$. Let D_0 be a subdomain of D with a compact boundary such that D_0 contains $H \cdot K$ but no point of $C_H + C_K$ and each component of D_0 contains a point of $H \cdot K$. There exist domains D_1 and D_2 such that D_1 is a subset of D having a compact boundary and containing $H - H \cdot D_0$ but no point of C_H, D_2 contains $K - K \cdot D_0$ but no point of C_K and $D_1 \cdot D_2$ is a subset of D_0. Considering $D_0 + D_1$ and $D_0 + D_2$ as the domains of Theorem 2, we find that $D_0 + D_1 + D_2$ does not cut A from B. Hence, its subset $H+K$ does not.

THEOREM 4. *If H is a compact closed set cutting the point A from the point B in the complement of the connected set K, then H contains a subset H' irreducible with respect to being a closed set cutting A from B in the complement of K. If K is compact, H' is a continuum that is not separated by any subset of the closure of K.*

The proof is as given in Theorem 7 of [1]. If K is not compact, H' need not be a continuum as is shown by the following example.

Let H be the sum of the points (1, 1) and $(-1, 1)$; let K be the common part of domains D_1 and D_2 described in the example in Theorem 2; let A and B be the points (0, 0) and (0, 1).

Corrections to [1]. The example given in Theorem 2 shows that 6, 7, 10 should have been omitted from the third footnote of [1]. As pointed out in Theorem 4, it is necessary to suppose that K is compact in Theorem 7 of [1]. Accordingly, D must be assumed compact in the fourth footnote of [1].

REFERENCE

1. R. H. Bing, *Generalizations of two theorems of Janiszewski*, Bull. Amer. Math. Soc. vol. 51 (1945) pp. 954–960.

THE UNIVERSITY OF TEXAS

ANNALS OF MATHEMATICS
Vol. 47, No. 3, July, 1946

SETS CUTTING THE PLANE*

BY R. H. BING

(Received August 31, 1945)
(Revised January 24, 1946)

DEFINITIONS. Suppose R is a set containing neither of the points A or B. If there is a continuum in the set W containing $A + B$ but each such continuum contains a point of R, R is said to *cut* A from B in W. If there is an arc from A to B in W but each such arc contains a point of R, R is said to *disrupt* A from B in W. If R does *not cut* (or *not disrupt*) A from B in W, there is a continuum (or an arc) in $W - W \cdot R$ containing $A + B$. If a set cuts or disrupts A from B in the plane, we simply say that it cuts or disrupts A from B. Although the results of this paper are stated only for the plane, they hold on a sphere and on a cactoid.

THEOREM 1. *Suppose G is a finite collection of connected sets, the elements of G have a point in common, all but at most two elements of G are bounded and the sum of no two elements of G cuts A from B. Then the sum of the elements of G does not separate A from B. If the elements of G are bounded, their sum does not cut A from B.*

This is a generalization of Kuratowski's "theorem of three continua" [5; see also 2, Theorem 9, p. 78].

PROOF. Designate the elements of G by $g_1, g_2, \cdots, g_n$ where $g_3, \cdots, g_n$ are bounded. Let $C_{ij}(C_{ij} \equiv C_{ji})$ be a continuum containing $A + B$ but no point of $g_i + g_j$ and let D_i be a connected domain containing g_i but no point of $C_{i1} + C_{i2} + \cdots + C_{in}$ and such that D_i is bounded if g_i is. If $D_1 + D_2 + \cdots + D_n$ does not separate A from B, then neither does $g_1 + g_2 + \cdots + g_n$.

If $D_1 + D_2 + \cdots + D_n$ separates A from B, it contains a continuum J separating A from B such that J is either a simple closed curve or an open curve [4, Corollary 3, p. 221]. We shall assume that J is bounded, for if it is an open curve, there exists an open curve J' which is a subset of $D_1 + D_2$ such that $J - J \cdot J'$ is bounded; as J' does not cut A from B, some bounded subset of $J + J'$ does. Let H be the set of all elements h such that h is a component of the common part of J and some D_i. By the Borel theorem, a finite subcollection H' of H covers J. A finite collection K of arcs covers J such that each element of K is a subset of an element of H'. Let P be a point common to all elements of G and let α_i be an arc in D_i containing P and each element of K which is a subset of D_i. As the sum of no two elements of $\alpha_1, \alpha_2, \cdots, \alpha_n$ separates A from B, an extension of "the theorem of three continua" gives that their sum does not. But their sum covers J.

If a bounded domain cuts A from B, it separates A from B. Hence, the sum

* Presented to the American Mathematical Society, February 23, 1946.

of the elements of G does not cut A from B if this sum is bounded. That it is necessary to suppose that each element of G is bounded in order to conclude that this sum does not cut A from B is shown by the following example. Let g_1 be the set of all points with positive ordinates less than 1 other than points on the lines joining $(1,1/n)$ to $(n,1/n)$, $(n,1/n)$ to $(1,1)$, $(-1,1/n)$ to $(-n,1/n)$ and $(-n,1/n)$ to $(-1,1)$ for $n = 2,3, \cdots$; let g_2 be the line from $(-1,1)$ to $(0,1/2)$; let g_3 be the line from $(0,1/2)$ to $(1,1)$. The sum of no two of the sets g_1, g_2, g_3 cuts $(0,0)$ from $(0,1)$ but the sum of the three sets does.

THEOREM 2. *If G is a finite collection of connected sets which have a point in common and the sum of no two elements of G disrupts the point A from the point B, then the sum of the elements of G does not disrupt A from B.*

PROOF. Let the elements of G be $g_1, g_2, \cdots, g_n$, let α_{ij} ($\alpha_{ij} \equiv \alpha_{ji}$) be an arc from A to B in the complement of $g_i + g_j$ and let D_i be the complementary domain of $\alpha_{i1} + \alpha_{i2} + \cdots + \alpha_{in}$ containing g_i. Since each complementary domain of $D_1 + D_2 + \cdots + D_n$ is a continuous curve, $D_1 + D_2 + \cdots + D_n$ cuts A from B if it disrupts A from B. However, the type of argument used in Theorem 1 shows that $D_1 + D_2 + \cdots + D_n$ does not cut A from B.

THEOREM 3. *If K is a bounded connected point set not cutting the point A from the point B, then there exists a continuum C containing $A + B$ and such that if $C \cdot \bar{K}$ exists, it belongs to every continuum in the complement of K containing $A + B$.*

This theorem has been proved [3, Theorem 12] by Eilenberg by other methods.

PROOF. Let T denote the set of points common to all continua in the complement of K containing $A + B$. We note that T is a closed subset of $\bar{K} - K + A + B$. For each point P of $\bar{K} - \bar{K} \cdot T$, let D_P denote a connected domain containing P such that the diameter of D_P is less than one-half the distance from P to some continuum in the complement of $K + P$ which contains $A + B$. Let $D_1, D_2, \cdots$ be a sequence of D_P's whose sum covers $\bar{K} - \bar{K} \cdot T$. Since the sum of $K + D_i$ and $K + D_j$ does not cut A from B [1, Theorem 6], it follows from Theorem 1 that $K + D_1 + D_2 + \cdots + D_i$ does not. Hence, some component of the complement of $D_1 + D_2 + \cdots$ contains $A + B$.

THEOREM 4. *If the connected set K does not disrupt the point A from the point B, then there exists an arc AB such that if a point belongs to $AB \cdot \bar{K}$, it belongs to every continuum in the complement of K that contains $A + B$.*

PROOF. Let AOB be an arc containing no point of K and let T be the common part of all the continua in the complement of K containing $A + B$. It is to be noted that T is a closed subset of $\bar{K} - K + A + B$. By Theorem 3, there is a continuum C containing $A + B$ but no point of $\bar{K} - \bar{K} \cdot T$. To show that there is an arc AB containing no point of K that does not belong to T, it is sufficient to demonstrate that if E amd F are points of $AOB \cdot C$ such that there is no point of C between them on AOB and if D is a domain containing the open arc $EF - (E + F)$ of AOB, then there is an open arc from E to F in D which contains no point of $\bar{K}$.

Since $EF + C$ contains no point of the connected set K, only one of its complimentary domains contains a point of $\bar{K}$. But there are two complementary domains of $EF + C$ each of which has EF on its boundary. Let D' be one of these two domains which contains no point of $\bar{K}$. There is an arc from E to F in $E + F + D \cdot D'$. This arc does not intersect $\bar{K}$ except possibly at E or F.

THEOREM 5. *If K is a bounded connected set not cutting the point A from the point B and T is the common part of all continua in the complement of K containing $A + B$, then in the space whose "points" are the components of T and the points of the complement of T, there is an arc (or a point) containing $A + B$ but no point of $\bar{K}$ that does not belong to T.*

PROOF. We shall suppose that T has more than one component. There is, by Theorem 3, a compact continuum M' in the complement of $\bar{K} - \bar{K} \cdot T$ that contains $A + B$. Let M be a subcontinuum of M' irreducible from A to B. We shall show that if t is a component of T containing neither A nor B, then t separates A from B in M.

Let C_A and C_B denote the maximal continuumwise connected subsets of $M - t$ containing A and B respectively. On the assumption that some point P of $M - t$ belongs to both $\bar{C}_A$ and $\bar{C}_B$, we shall show that there is a continuum in the complement of $K + t$ containing $A + B$.

Let $D_1, D_2, \cdots$ and $D_1', D_2', \cdots$ be sequences of domains closing down on t and P respectively. Let C_i be the component of $C_A - C_A \cdot D_i$ containing A. Suppose that for no integer i does P belong to C_i. Since $C_A = C_1 + C_2 + \cdots$, there is an integer n_1 such that C_{n1} intersects D_1'. As the continuum C_{n_1} does not contain P, for some integer m, D_m' contains no point of C_{n_1}. There is an integer n_2 such that C_{n_2} intersects D_m'. Then $C_{n_2} - C_{n_1}$ contains a connected subset intersecting both D_2 and D_2'. In general, there exists an integer n_i such that $C_{n_i} - C_{n_{i-1}}$ contains a connected subset intersecting both D_i and D_i'. Let L be the limiting set of $C_{n_1}, C_{n_2} - C_{n_1}, C_{n_3} - C_{n_2}, \cdots$. Some continuum L' in $L - L \cdot t$ contains P and a point Q of the complement of T. There is an integer i such that a continuum N in the complement of $\bar{K}$ contains both Q and a point of C_i. Then $C_i + N + L'$ is a continuum containing $A + P$ but no point of $K + t$. Likewise, there is a continuum containing $B + P$ but no point of $K + t$. That there is a continuum containing $A + B$ but no point of $K + t$ is contrary to the definitions of t and T. Therefore, no point of $M - t$ belongs to both $\bar{C}_A$ and $\bar{C}_B$. As $\bar{C}_A + t + \bar{C}_B = M$, it follows that t separates A from B in M.

Let $D_1, D_2, \cdots$ be a sequence of domains closing down on M and let $E_1, E_2, \cdots$ be the complementary domains of $\bar{K}$ containing points of M. Only two components t_1 and t_2 of T are continuumwise accessible from E_i. Regarding t_1 and t_2 as points, there is an arc α_i from t_1 to t_2 in $E_i \cdot D_i + t_1 + t_2$. In the space whose "points" are the components of T and the points of the complement of T, $T + \alpha_1 + \alpha_2 + \cdots$ is an arc from A to B.

Using Theorem 5, we obtain the following:

THEOREM 6. *Suppose that K is a bounded set with a connected complement,* *that W is a closed set plus a countable set of points and that no component of W cuts the point A from the point B in K; then W does not cut A from B in K.*

THE UNIVERSITY OF TEXAS.

REFERENCES

1. R. H. BING, *Generalizations of two theorems of Janiszewski*, Bull. Amer. Math. Soc., vol. 51 (1945), pp. 954–960.
2. S. EILENBERG, *Transformations continues en circonférence et la topologie du plan*, Fund. Math., vol. 26 (1936), pp. 61–112.
3. ———, *Sur les transformations d'espaces métriques en circonférence*, Fund. Math., vol. 24 (1935), pp. 160–176.
4. C. KURATOWSKI, *Sur la séparation d'ensembles situés sur le plan*, Fund. Math., vol. 12 (1928), pp. 214–239.
5. ———, *Théorème sur trois continus*, Monatschefte für Mathematik und Physik, vol. 36 (1929), pp. 77–80.

C. Planar Embeddings

SKEW SETS.*

By R. H. Bing.

A skew set is one which is not topologically equivalent to any subset of the surface of a sphere. Kuratowski proved [1] that a compact skew continuous curve which contains only a finite number of simple closed curves contains one of the two following types of curves:

Type 1. A skew curve S_1 is of type 1 provided there exist six distinct points P_1, P_2, P_3, Q_1, Q_2, Q_3 such that S_1 is the sum of nine arcs $P_1Q_1, P_1Q_2, \cdots, P_3Q_3$ with end points as indicated and with the common part of two of these arcs that intersect each other being an end point of each.

Type 2. A skew curve S_2 is of type 2 provided there exist five distinct points P_1, P_2, P_3, P_4, P_5 such that S_2 is the sum of ten arcs $P_1P_2, P_1P_3 \cdots, P_4P_5$ with end points as indicated and with the common part of two of these arcs that intersect each other being an end point of each.

Claytor showed [2] that a compact cyclic continuous curve is skew only if it contains one of these two types of curves. It would be interesting to get a simple characterization of a general planar set. The study of skew sets may lead in that direction.

In this paper we shall consider the following two types of skew sets which are generalizations of the types of skew curves studied by Kuratowski:

Skew set of type 1. The set S_1 is a skew set of type 1 provided there exist six distinct points P_1, P_2, P_3, Q_1, Q_2, Q_3 such that S_1 is the sum of nine connected sets $M(P_1Q_1), M(P_1Q_2), \cdots, M(P_3Q_3)$ where $M(P_iQ_j)$ contains $P_i + Q_j$ and contains a limit point of $M(P_rQ_s)$ only if $P_i + Q_j$ intersects $P_r + Q_s$.

Skew set of type 2. The set S_2 is a skew set of type 2 provided there exist five points P_1, P_2, P_3, P_4, P_5 such that S_2 is the sum of ten connected sets $M(P_1P_2), M(P_1P_3), \cdots, M(P_4P_5)$ where $M(P_iP_j)$ contains $P_i + P_j$ and contains a limit point of $M(P_rP_s)$ only if $P_i + P_j$ intersects $P_r + P_s$.

* Received July 16, 1946; Presented to the American Mathematical Society, August 22, 1946.

The justification for calling these skew sets follows from Theorems 3 and 4.

Because of their lengths, the proofs of the following two theorems are omitted.

THEOREM 1. *Suppose that $P_1, P_2, P_3, Q_1, Q_2, Q_3$ are six distinct points and that $P_1Q_1, P_1Q_2, \cdots, P_3Q_3$ are nine arcs with end points as indicated and such that two of these arcs intersect each other only if they have an end point in common. Then the sum of these arcs is not homeomorphic to any plane set.*

THEOREM 2. *Suppose that P_1, P_2, P_3, P_4, P_5 are five distinct points such that $P_1P_2, P_1P_3, \cdots, P_4P_5$ are ten arcs with end points as indicated and such that two of these arcs intersect each other only if they have an end point in common. Then the sum of these arcs is not homeomorphic to any plane set.*

The space S considered in the next two theorems has the property that any pair of points of a component of an open set in S can be joined by an arc lying in that open set.

THEOREM 3. *If S contains a skew set of type 1, it contains a skew curve of type 1.*

Proof. Let $M(P_1Q_1), M(P_1Q_2), \cdots, M(P_3Q_3)$ be sets in S satisfying the conditions mentioned in the definition of a skew set of type 1. Let $D(P_1Q_1)$ be an open set containing $M(P_1Q_1)$ but no point of $M(P_2Q_2) + M(P_2Q_3) + M(P_3Q_3)$ and let P_1Q_1 be an arc from P_1 to Q_1 in $D(P_1Q_1)$. If we replace $M(P_1Q_1)$ by the arc P_1Q_1 in the sequence $M(P_1Q_1), M(P_1Q_2), \cdots, M(P_3Q_3)$, the sum of the resulting sequence is a skew set of type 1. Similarly, we may replace $M(P_1Q_2)$ by an arc P_1Q_2, $M(P_1Q_3)$ by an arc $P_1Q_3, \cdots$, and $M(P_3Q_3)$ by an arc P_3Q_3 in such a way that two of the arcs $P_1Q_1, P_1Q_2, \cdots, P_3Q_3$ intersect each other only if they have an end point in common.

We shall now show that $P_1Q_1 + P_1Q_2 + \cdots + P_3Q_3$ contains nine arcs $P_1O_{11}Q_1, P_1O_{12}Q_2, \cdots, P_3O_{33}Q_3$ with end points as indicated such that two of these arcs intersect each other only if they have an end point in common and the common part of two of these arcs that intersect each other has only a finite number of components.

Let P_1Q_1 be $P_1O_{11}Q_1$. We shall determine $P_1O_{12}Q_2$ as follows. Since $P_1O_{11}Q_1 \cdot P_1Q_2$ is a closed and compact set, it can be covered by a finite set of components of $P_1O_{11}Q_1 - P_1O_{11}Q_1 \cdot (P_2Q_1 + P_3Q_1)$. Let $P_1O_{12}Q_2$ be an arc from P_1 to Q_2 in P_1Q_2 plus the sum of these components such that $P_1O_{11}Q_1 \cdot P_1O_{12}Q_2$ has only a finite number of components.

Now $P_1O_{11}Q_1 \cdot P_1Q_3$ can be covered by a finite set T_1 of components of $P_1O_{11}Q_1 - P_1O_{11}Q_1 \cdot (P_2Q_1 + P_3Q_1)$ and $P_1O_{12}Q_2 \cdot P_1Q_3$ can be covered by a finite set T_2 of components of $P_1O_{12}Q_2 - P_1O_{12}Q_2 \cdot (P_2Q_2 + P_3Q_2)$. Let $P_1O_{13}Q_3$ be an arc from P_1 to Q_3 in P_1Q_3 plus the sum of the elements of $T_1 + T_2$ such that each of the sets $P_1O_{11}Q_1 \cdot P_1O_{13}Q_3$ and $P_1O_{12}Q_2 \cdot P_1O_{13}Q_3$ has only a finite number of components.

This process is continued. To get $P_3O_{33}Q_3$, we let T_{ij} be a finite collection of components of $P_iO_{ij}Q_j - P_iO_{ij}Q_j \cdot (P_1O_{11}Q_1 + P_1O_{12}Q_2 + P_2O_{21}Q_1 + P_2O_{22}Q_2)$ covering $P_iO_{ij}Q_j \cdot P_3Q_3$ and let $P_3O_{33}Q_3$ be an arc from P_3 to Q_3 in P_3Q_3 plus the sum of the elements of $T_{13} + T_{23} + T_{31} + T_{32}$ such that the common part of $P_3O_{33}Q_3$ and each of the sets $P_1O_{13}Q_3$, $P_2O_{23}Q_3$, $P_3O_{31}Q_1$, $P_3O_{32}Q_2$ has only a finite number of components. Denote the collection of arcs $P_1O_{11}Q_1, P_1O_{12}Q_2, \cdots, P_3O_{33}Q_3$ by G and the sum of the elements of G by G^*. We note that two elements of G intersect each other only if they have an end point in common and the common part of two intersecting elements of G has only a finite number of components.

Some subset of G^* is irreducible with respect to containing six distinct points $P'_1, P'_2, P'_3, Q'_1, Q'_2, Q'_3$ and nine arcs $P'_1Q'_1, P'_1Q'_2, \cdots, P'_3Q'_3$ such that two of these arcs intersect each other only if they have an end point in common. For convenience, suppose that G^* is such a subset of itself.

No point of G^* is a point of order five or more. If there were such a point P', it would belong to some three elements of G and the point P' could be used instead of the common end point of the three arcs of G containing P'. Then G^* would not be irreducible in the sense supposed. Likewise, we find that G^* contains no point of order four that belongs to three elements of G.

Since the common part of no two elements of G has infinitely many components, G^* does not contain infinitely many simple closed curves. We know by Theorem 1 that G^* is a skew set and by the previously mentioned theorem of Kuratowski that it contains either a skew curve of type 1 or a skew curve of type 2.

Assume that G^* contains a skew curve of type 2. Let $A_1, A_2, \cdots, A_5$ be five distinct points and $A_1A_2, A_1A_3, \cdots, A_4A_5$ be ten arcs in G^* such that the common part of two of these arcs that intersect each other is an end point of each. If some component of $G^* - (A_1 + A_2 + \cdots + A_5)$ has three of the five points $A_1, A_2, \cdots, A_5$ as limit points, a careful examination of G^* shows that it contains a skew set of type 1. Compare this notion with a theorem by Hall [3]. Assume that no such component contains three of these points as limit points.

Let X be a component of $G-(A_1+A_2+\cdots+A_5)$ that contains P_1 and has A_i and A_j as limit points. If BA_i and CA_i are arcs in $X+A_i$ each of which is a subset of an element of G, then BA_i and CA_i do not belong to different elements of G or else A_i is a point of order four or more belonging to three elements of G.

There are arcs from P_1 to Q_1 in each of the sets $P_1O_{11}Q_1$, $P_1O_{12}Q_2 + P_2O_{22}Q_2 + P_2O_{21}Q_1$, and $P_1O_{13}Q_3 + P_3O_{33}Q_3 + P_3O_{31}Q_1$ but no two of these sets contains an arc lying in $X+A_i$ and containing A_i; also no two of these sets contains an arc lying in $X+A_j$ and containing A_j. Therefore Q_1 is a point of X. Likewise, we find that each of the six points P_1, P_2, P_3, Q_1, Q_2, Q_3 is a point of X. But G^*-X is not a subset of one arc. Hence, G^* contains a skew curve of type 1.

The following theorem may be established in a like manner.

THEOREM 4. *If S contains a skew set of type 2, it contains either a skew curve of type 1 or a skew curve of type 2.*

Example. That we cannot conclude under the hypotheses of Theorem 4 that S contains a skew curve of type 2 may be seen from the following example showing that a skew curve of type 1 contains a skew set of type 2. Consider the skew curve which is the sum of nine arcs $P_1Q_1, P_1Q_2, \cdots, P_3Q_3$ satisfying conditions given in the definition of a skew curve of type 1. Let P_1Q_1 be the sum of the two arcs P_1R and RQ_1. The sum of the arcs $RQ_1+P_2Q_1$, $RQ_1+P_3Q_1$, $P_1R+P_1Q_2$, $P_1R+P_1Q_3$, $P_2Q_1+P_3Q_1$, P_2Q_2, P_2Q_3, P_3Q_2, P_3Q_3, and $P_1Q_2+P_1Q_3$ is a skew set of type 2. However, it contains no skew curve of type 2.

THEOREM 5. *Suppose that $\alpha_1, \alpha_2, \cdots, \alpha_n$ are arcs in the plane such that two of these arcs intersect only if they have an end point in common. There exists a collection of arcs $\beta_1, \beta_2, \cdots, \beta_n$ in the plane such that β_i $(i=1,2,\cdots,n)$ has the same end points α_i and contains no non end point of β_j $(j\neq i)$.*

Proof. Let γ_i $(i=1,2,\cdots,n)$ be an arc (not necessarily in the plane) having the same end points as α_i and such that γ_i contains no non end point of γ_j $(j \neq i)$. We shall show that $\gamma_1 + \gamma_2 + \cdots + \gamma_n$ is topologically equivalent to a plane set. The truth of Theorem 5 will follow as a consequence.

If $\gamma_1 + \gamma_2 + \cdots + \gamma_n$ is not homeomorphic with a plane set, it contains a skew curve of either type 1 or type 2. Assume that it contains nine arcs $P_1Q_1, P_1Q_2, \cdots, P_3Q_3$ satisfying conditions given in the definition of a skew curve of type 1. Then P_iQ_j is the sum of a subcollection of the arcs $\gamma_1, \gamma_2, \cdots, \gamma_n$; suppose that it is the sum of $\gamma_r, \cdots$, and γ_s. Let $M(P_iQ_j)$ be the sum of the arcs $\alpha_r, \cdots$, and α_s. Then the sum of $M(P_1Q_1), M(P_1Q_2)$, $\cdots$, and $M(P_3Q_3)$ is a skew set of type 1. But the plane contains no skew set of type 1. Also, the assumption that $\gamma_1 + \gamma_2 + \cdots + \gamma_n$ contains a skew curve of type 2 leads to the contradiction that $\alpha_1 + \alpha_2 + \cdots + \alpha_n$ contains a skew set of type 2.

Question. It would be interesting to know if the following statement is true: If $\alpha_1, \alpha_2, \cdots, \alpha_n$ are arcs two of which intersect only if they have an end point in common, then there exist arcs $\beta_1, \beta_2, \cdots, \beta_n$ such that two of these arcs intersect only if they have an end point in common, the common part of two of these arcs that intersect is connected, and β_i $(i = 1, 2, \cdots, n)$ is an arc in $\alpha_1 + \alpha_2 + \cdots + \alpha_n$ having the same end points as α_i.

THEOREM 6. *No plane set G contains a collection of five mutually separated points sets such that the closure of the sum of any pair of these five sets is the closure of a connected subset of G which is open in G.*

Proof. Assume that the theorem is false. Let S_1, S_2, S_3, S_4, S_5 be five mutually separated sets in the plane set G such that the closure of the sum of any pair of these five sets is the closure of a connected subset of G which is open in G. Denote a point of S_i $(i = 1, 2, 3, 4, 5)$ by P_i. Let T_{ij} $(i = 1, 2, 3, 4, 5;\ j = 1, 2, 3, 4, 5)$ be a connected subset of G which is open in G and such that the closure of T_{ij} is the closure of $S_i + S_j$. If $M(P_iP_j)$ is the sum of P_i, P_j, and T_{ij}, the sum of the sequence $M(P_1P_2)$, $M(P_1P_3)$, $\cdots, M(P_4P_5)$ is a skew set of type 2. But the plane contains no skew set of type 2 because it contains no skew curve of either type 1 or type 2.

Problem. The following problem could be posed to a person not acquainted with terms in topology. Consider a nation composed of more than four states. Suppose that there are five political parties, that each state favors some one of these parties, and that each party is favored by some state. Would it be possible for each pair of the parties to hold a joint convention at some place so that each of the states favoring these two parties could send a delegate to the convention so that this delegate could remain in the states but not cross the boundary of any "unfriendly" state, that is one favoring a party other than one of the two holding the joint convention.

We see by Theorem 6 that the answer is in the negative. We let the sum of the states be G and the sum of the interiors of the states favoring a particular party be one of the five sets.

THE UNIVERSITY OF TEXAS.

REFERENCES

1. C. Kuratowski, "Sur le problème des courbes gauches en topologie," *Fundamenta Mathematicae*, vol. 15 (1930), pp. 271-283.
2. S. Claytor, "Topological immersion of Peanian continua in a spherical surface," *Annals of Mathematics*, vol. 35 (1934), pp. 809-835.
3. D. W. Hall, "A note on primitive skew curves," *Bulletin of the American Mathematical Society*, vol. 49 (1943), pp. 935-936.

EMBEDDING CIRCLE-LIKE CONTINUA IN THE PLANE

R. H. BING

A finite sequence of open sets $L_1, L_2, \ldots, L_n$ is called a linear chain if each L_i intersects only the L's adjacent to it in the sequence. The finite sequence is a circular chain if we also insist that the first and last links intersect each other. The 1-skeleton of the covering is an arc for a linear chain and a simple closed curve for a linear chain.

A compact metric continuum X is called snake-like if for each $\epsilon > 0$, X can be covered by a linear chain of mesh less than ϵ. Likewise, X is called circle-like if for each $\epsilon > 0$, X can be irreducibly covered with a circular chain of mesh less than ϵ. This definition is more restrictive than that given in **(3**, p. 210**)** for there a pseudo-arc is not circle-like but here it is. The present usage is in keeping with definitions of Burgess.

When a person thinks of a linear chain in the plane, he is likely to think of a finite sequence of open disks. However, the definition of a chain does not insist that the links be open disks or even that they be connected. Hence, a person who thinks only of chains with connected links is considering only a special sort of embedding as pointed out by Example 3 of **(1)**. Theorems 10 and 11 show that in the category sense, most plane continua can be covered by chains whose links are small open disks. Theorems 5, 6, and 7 show that even for chainable continua which cannot be covered by chains with small connected links, it is the embedding that is special rather than the topology of the embedded continua.

1. Circling. We use $C(L_1, L_n, \ldots, L_2)$ to denote the chain C with links $L_1, L_2, \ldots, L_n$. In dealing with circular chains, we use L_{i-1} as the link preceding L_i and in case $i = 1$, we interpret L_{i-1} to mean L_n. If $i = n$, the link L_{i+1} following L_n is L_1. To facilitate this convention, we understand L_0 to be another name for L_n.

Number of times C_{i+1} circles C_i. Suppose the circular chain C_{i+1} $(A_1, A_2, \ldots, A_n)$ is a refinement of the circular chain $C_i(B_1, B_2, \ldots, B_m)$. Let $f'(A_k)$ be the subscript of one of the elements of C_i containing A_k. In some cases there are two choices for $f'(A_k)$ but a definite one of them is selected. We note that $f'(A_k), f'(A_{k+1})$ are adjacent or one is m and the other is 1. Also, $f'(A_1), f'(A_n)$ are adjacent in this sense.

Let f be a map of the integers $(0, 1, 2, \ldots, n)$ into the integers defined as follows.

Received December 12, 1960. Work on this paper was supported by contract NSF–G11665.

$$f(0) = f'(A_n),$$

$$f(i+1) = \begin{cases} f(i) - 1 \text{ if } f'(A_{i+1}) \text{ precedes } f'(A_i), \\ f(i) \text{ if } f'(A_{i+1}) = f'(A_i), \\ f(i) + 1 \text{ if } f'(A_{i+1}) \text{ follows } f'(A_i). \end{cases}$$

Thus it is true that $f(i) \equiv f'(A_i) \bmod m$ and $f(0) = f'(A_n) \equiv f(n) \bmod m$. Recall that we are using the convention that $f'(A_{i+1})$ precedes $f'(A_i)$ if $f'(A_{i+1}) = m$ and $f'(A_i) = 1$ while $f'(A_{i+1})$ follows $f'(A_i)$ if $f'(A_{i+1}) = 1$ and $f'(A_i) = m$—so in a sense we sometimes treat m as 0. We note that $f'(A_i)$ and $f(i)$ may differ by a multiple of m as may $f(0)$ and $f(n)$.

The number of times that C_{i+1} circles C_i is defined to be

$$|f(n) - f(0)|/m.$$

We note that this number is invariant under taking different elements of C_{i+1}, C_i as the first element or in ordering the elements in a counter fashion. Nor does it matter which of the two choices for $f'(A_k)$ we made when we had a choice.

THEOREM 1. *If C_2 circles C_1 x times and C_3 circles C_2 y times, then C_3 circles C_1 xy times.*

Proof. Suppose $(A_1, A_2, \ldots, A_m)$, $(B_1, B_2, \ldots, B_n)$, and $(D_1, D_2, \ldots, D_r)$ are the links of C_3, C_2, C_1 respectively and f, g, h are the functions determining how many times C_3 circles C_2, C_2 circles C_1, and C_3 circles C_1. We suppose that the first links of C_2 and C_1 are selected so that $f(0) = g(0) = 0$ and that the choice of $h'(A_i)$ is such that

$$h'(A_i) = g'(B_{f'(A_i)}).$$

We note that g is only defined on the integers between 0 and n inclusive. For convenience we suppose $g(n) \geqslant g(0)$. Suppose that g is extended to all the integers so that

$$g(i + kn) = kxr + g(i).$$

It follows by induction that

$$h(i) = g(f(i)).$$

Hence

$$|h(m) - h(0)| = |g(f(m)) - 0| = g(\pm yn)| = |\pm xyr| = xyr.$$

THEOREM 2. *Suppose X is a compact continuum in a locally connected metric space and ϵ is a positive number such that X cannot be covered by any ϵ-chain. If C_1, C_2 are two circular ϵ chains covering X such that C_2 is a refinement of C_1, then there is a chain C_3 circling C_1 a positive number of times such that each link of C_3 is a component of a link of C_2.*

Note that we do not assert that C_3 covers X.

Proof. Let $D_1, D_2, \ldots, D_m$ be a finite collection of components of links of C_2 such that the collection covers X. Suppose the D's are ordered so that D_{i+1} intersects $\sum_{j=1}{}^{i} D_j$.

Let $f'(D_k)$ be a subscript of a link of C_1 containing D_k and f be a map of the integers $(1, 2, \ldots, m)$ into the integers defined as follows:

$$f(1) = f'(D_1).$$

After $f(1), f(2), \ldots, f(i-1)$ have been defined, pick a $j < i$ such that $D_i \cdot D_j \neq 0$ and define

$$f(i) = \left\{ \begin{array}{l} f(j) - 1 \text{ if } f'(D_i) \text{ precedes } f'(D_j), \\ f(j) \text{ if } f'(D_i) = f'(D_j), \\ f(j) + 1 \text{ if } f'(D_i) \text{ follows } f'(D_j) \end{array} \right\}.$$

Case 1. If f has the property that $D_i \cdot D_k \neq 0$ implies that $|f(i) - f(k)| \leqslant 1$, X can be covered by a linear ϵ-chain each of whose links is the sum of D's. The link containing D_j is the sum of all D_i's such that $f(j) = f(i)$. In this case the definition of $f(i)$ was independent of the D_j picked that intersects D_i.

Case 2. Suppose there are two elements D_i, D_j such that $D_i \cdot D_j \neq 0$ but $|f(i) - f(j)| > 1$. Let n be a minimal integer such that there is an ordered subcollection $(E_1, E_2, \ldots, E_n)$ of the D's such that E_{i+1} intersects $\sum_{j=1}{}^{i} E_j$ and there is a g defined on the E's in a fashion similar to that in which f was defined on the D's that is not independent of the $j < i$ selected such that $E_i \cdot E_j \neq 0$. We shall show that a reordering of the E's gives a circular chain which circles in C_1. A reordering is needed since E_3 may intersect E_1 instead of E_2.

The reordering is accomplished as follows. Let $F_1 = E_1$, $F_2 = E_2$, F_3 be one of the remaining E's intersecting F_2, F_4 one of the remaining E's intersecting $F_3, \ldots$. If we come to a place where we cannot select another F, we have used up our E's or else our collection of E's was not minimal with respect to there being no independent g. Also, F_n (the last F) intersects an F_r such that for g defined on the F's, $|g(n) - g(r)| > 1$. Furthermore, $r = 1$ or we could delete $F_1, F_2, \ldots, F_{r-1}$ from our minimal collection. If F_n intersects any F_j $(j \neq 1, n-1, n)$, then $F_{j+1}, \ldots, F_{n-1}$ could have been deleted from our minimal collection. Hence F_n intersects only F_1 and F_{n-1}. A similar argument shows that each F intersects only the F's adjacent to it in $F_1, F_2, \ldots, F_n, F_1$.

THEOREM 3. *Suppose ϵ is a positive number, X is a circle-like continuum in the plane that cannot be covered by a linear ϵ-chain, and C_1 is a circular ϵ-chain in the plane covering X. Then there is a positive number n such that no chain C_2 covering X circles C_1 more than n times.*

Proof. Let T be a triangulation of the plane such that each closed 2-simplex of T that intersects X lies in a link of $C_1(L_1, L_2, \ldots, L_m)$. For each link L_i of C_1, let S_i be the collection of closed 2-simplexes of T that intersect X and lie in L_i. Let K be the collection of 1-simplexes common to an element of S_1 and an element of $S_2 - S_1$. The number n promised by the theorem is the number of elements in K.

Assume some chain C_2 covers X and circles C_1 more than n times. We suppose that the links of C_2 are "cut down to size" so that each point of a link of C_2 lies in a 2-simplex of T that intersects X. It follows from Theorem 2 that there is a chain C_3 with connected links which circles C_2 a positive number of times and from Theorem 1 that such a C_3 would circle C_1 more than n times. We are not interested in whether or not C_3 covers X.

Let $(D_1, D_2, \ldots, D_r)$ be the links of C_3 and f be a map of the integers $(0, 1, 2, \ldots, r)$ into the integers showing that C_3 circles C_1 more than n times. Then

D_r lies in the $f(0)$th link of C_1,
D_i lies in the $(f(i) \bmod m)$th link of C_1,
$|f(i) - f(i+1)| \leqslant 1$, and
$|f(r) - f(0)| = n'm$ where n' is an integer larger than n.

Let $A_1A_2, A_2A_3, \ldots, A_rA_1$ be a collection of polygonal arcs such that

$$A_iA_{i+1} \subset D_i, \qquad A_rA_1 \subset D_r,$$

$A_1A_2 + A_2A_3 + \ldots + A_rA_1$ is a simple closed curve J, and
J does not contain an endpoint of an element of K.

We suppose that no vertex of J lies on an element of K and no straight line segment in J intersects two elements of K.

Let $P_1, P_2, \ldots, P_s$ be the vertices of J ordered in a natural fashion on J. We suppose that each A_i is a vertex of J, that $P_1 = A_1$, and the first P's are vertices of A_1A_2. Let g be a map of the integers $(0, 1, 2, \ldots, s)$ into the integers such that

$g(j) = f(i)$ if P_j is a point of $(A_iA_{i+1} - A_{i+1})$,
$g(j) = f(r)$ if P_j is a point of $(A_rA_1 - A_1)$,
$g(0) \equiv g(s) \bmod m$ and is adjacent to $g(1)$.

Then

$$|g(s) - g(0)| = |f(r) - f(0)| = n'm.$$

We now adjust g on certain integers. If P_j lies in an element of S_1, then it lies in D_1 but perhaps also in either D_2 or D_r. We wish to treat it as though it were only in D_1. If P_j is in an element of $S_2 - S_1$, we treat it as though it did not lie in D_1. So as not to disturb the adjusted g on 0 and s, we suppose that P_1 lies in an element of S_1. We define the adjusted g as follows.

$g'(j) = g(j)$ unless P_j lies in an element of $S_1 + S_2$,
$g'(j) = f(1)$ if P_j lies in an element of S_1,
$g'(j) = f(2)$ if P_j lies in an element of $S_2 - S_1$.

We note that

$$|g'(i+1)| - g'(i)| \leqslant |,$$
$$|g'(s) - g'(0)| = n'm.$$

Let i be an integer such that one of $g'(i)$, $g'(i+1)$ is 1 mod m and the other is 2 mod m. The corresponding segment P_iP_{i+1} of J crosses an element of K. The crossing is from an element of S_1 to an element of $S_2 - S_1$ if $g'(i+1) > g'(i)$ and from an element of $S_2 - S_1$ to an element of S_1 if $g'(i+1) < g'(i)$. There must be n' more crossings one way than the other since $|g'(s) - g'(0)| = n'm$. We show that this is impossible since K has only n elements and on no one of these elements can the number of crossings one way exceed by more than 1 the number of crossings in the other direction.

Suppose K_i is an element of K and these crossings are ordered on K_i rather than on J. Since as one moves along K_i he alternately moves in and out of the disk bounded by J as he passes these crossings, J crosses K_i from different directions for two adjacent crossings (adjacent on K_i). Hence, the number of crossings of K_i by J in one direction does not exceed by more than 1 the number of crossings in the other direction and it is false that $n' > n$. The contradiction arose from the false assumption that there is a chain C_2 in X which circles C_1 more than n times.

Question. Theorem 3 gives no clue as to the minimum size of n. Suppose C_1 has i links and no link of C_1 has more than j components. Perhaps the minimum n is a function of i and j. An early conjecture that $n = j$ turned out to be false on considering a circle J in the plane covered by a circular chain of twelve links $D_1, D_2, \ldots, D_{12}$ such that each D_i is the interior of a round disk. One notes that J circles the chain consisting of $(D_1 + D_5 + D_9$, $D_2 + D_6 + D_{10}$, $D_3 + D_7 + D_{11}$, $D_4 + D_8 + D_{12})$ three times and each link has three components. However, it is possible to join D_1 to D_5 by a path in the exterior of J, D_6 to D_{10} by a path in the exterior of J, D_3 to D_7 by a path on the interior of J and D_8 to D_{12} by a path on the interior of J so as to get a chain of four links $(D_1 + D_5 + D_9 +$ path from D_5 to $D_9)$, $(D_2 + D_6 + D_{10}$ $+$ path from D_6 to $D_{10})$, $(D_3 + D_7 + D_{11} +$ path from D_3 to $D_7)$, $(D_4 + D_8$ $+ D_{12} +$ path from D_8 to $D_{12})$ each of which has only two components and J circles the chain three times. Here we spoke of a curve rather than a chain circling but no confusion should result.

There are several proofs of the result that ordinary solenoids cannot be embedded in the plane (the circle is a degenerate solenoid which can be embedded). One of these follows from Bing's result (3) that the simple closed curve is the only homogeneous bounded plane continuum that contains an arc. Another follows from the results that for a decomposition of E^3 whose only non-degenerate element is an ordinary solenoid, the decomposition space is not simply connected but for a decomposition of E^3 whose only non-degenerate element is a plane continuum, the decomposition space is simply connected (0). However, Theorem 3 gives a more straightforward proof of this result so we state it as a corollary.

COROLLARY. *The circle is the only solenoid which can be embedded in the plane.*

2. Plane embeddings. Theorem 3 with its corollary showed that certain circle-like continua cannot be embedded in the plane. In this section we show what sorts of circle-like continua can be embedded in the plane.

THEOREM 4. *Suppose X is a circle-like continuum and $C_1, C_2, \ldots,$ is a sequence of circular chains covering X such that mesh C_i approaches 0 as i increases without limit and C_{i+1} circles C_i exactly once. Then there is a homeomorphism h of X into the plane such that for each positive number ϵ, $h(X)$ can be covered by a circular chain each of whose links is the interior of a round disk of diameter less than ϵ.*

Proof. By dropping out certain elements from the chains $C_1, C_2, \ldots,$ one can get a sequence whose meshes converge to 0 very fast. Hence, we suppose with no loss of generality that C_{i+1} has such small mesh that each link L of C_{i+1} is contained in a link L' of C_i such that the distance from L to the boundary of L' is more than twice the mesh of C_{i+1}.

Suppose C_i has n_i links. Let f_i be a map of the integers $(0, 1, 2, \ldots, n_{i+1})$ into the integers such that

the n_{i+1}th link of C_{i+1} lies in the $f_i(0)$th link of C_i,
the jth link of C_{i+1} lies in the $(f_i(j) \bmod n_i)$th link of C_i,
$|f_i(j) - f_i(j+1)| \leqslant 1$,
$|f_i(n_{i+1}) - f_i(0)| = n_i$,
one of $f_i(j-1), f_i(j+1)$ is equal to $f_i(j)$.

That it is possible to place the last restriction on f_i follows from the fact that each link of C_{i+1} was required to lie way inside a link of C_i.

Let J_1 be a polygonal simple closed curve in the plane which is the sum of n_1 arcs

$$A_1A_2, A_2A_3, \ldots, A_{n_1}A_1$$

such that each of these arcs is of length less than 1 and if two of the arcs intersect each other, the intersection is an endpoint of each. A first approximation to h takes X onto J_1 by taking the part of X in the ith link of C_1 onto the arc A_iA_{i+1}.

Let $F_1(L_1, L_2, \ldots, L_{n_1})$ be a circular chain of mesh less than 1 covering J_1 such that L_i covers A_iA_{i+1}, L_{n_1} covers $A_{n_1}A_1$, and the sum of the links of F_1 is the sum of the elements of a circular chain each of whose elements is the interior of a round disk of diameter less than 1. Note that as we build the F_i's in the plane to match the C_i's, we shall not insist that the links of the F_i's be circular disks but rather that the sum of these links be the sum of the links of a circular chain each of whose links is the interior of a round disk.

We want to build a chain F_2 of mesh less than $\frac{1}{2}$ such that F_2 has the same number of links as C_2, the jth link of F_2 lies in the $(f_1(j) \bmod n_1)$th link of F_1, and the sum of links of F_2 is the sum of the elements of a circular chain each of whose elements is the interior of a round disk of diameter less than $\frac{1}{2}$. To facilitate the construction of such an F_2 we build a simple closed curve J_2 very close to J_1 that follows J_1 around in the same fashion that C_2 follows C_1. There is a problem in showing that there is such a J_2 since one might suppose that if P is a path near J_1, that follows J_1 around in a particular way, then P would have to cross itself if its two ends were to be joined. The following two paragraphs show that as long as C_2 circles C_1 exactly once, P need not cross itself and there is a J_2.

If g is a map of the interval $(0, 2\pi)$ into the positive reals such that $g(2\pi) - g(0) = 2\pi$ and ϵ is a small positive number, then

$$h(1, \theta) = (1 - \epsilon(g(\theta) - \theta), g(\theta))$$

is a homeomorphism of the unit circle onto a simple closed curve where the equation is given in polar co-ordinates since if $g(\theta_1) - \theta_1 = g(\theta_2) - \theta_2$ and $g(\theta_1) = g(\theta_2)$, then $\theta_1 = \theta_2$.

Let J_2 be a polygonal simple closed curve very close to J_1 which is the sum of the arcs $B_1B_2, B_2B_3, \ldots, B_{n_2}B_1$ each of which is of length less than $\frac{1}{2}$ such that $B_iB_{i+1} \subset L_j$ if $f_1(i) \equiv j \bmod n_1$. In getting the B_1B_{i+1}'s to be of diameter less than $\frac{1}{2}$ we make use of the conditions that the A_jA_{j+1}'s have length less than 1 and one of $f_1(k-1)$, $f_1(k+1)$ is equal to $f_1(k)$. To see that J_2 need not cross itself we regard J_2 as a circle (rather than a polygon) with A_i the point $(1, 2\pi i/n_2)$ in polar co-ordinates. Let g be defined on the values $2\pi j/n_2$ $(j = 0, 1, \ldots, n_2)$ by $g(2\pi j/n_2) = 2\pi f_1(j)n_1$. Then extend g linearly to all of $[0, 2\pi]$. Suppose $f_1(i-1) \neq f_1(i) = f_1(i+1) = \ldots = f_1(i+k) \neq f_1(i+k+1)$. Then $B_iB_{i+1} + B_{i+2}B_{i+3} + \ldots + B_{i+k}B_{i+k+1}$ is the graph of $h(1, [2\pi a/n_2, 2\pi b/n_2])$ where h is the homeomorphism given in the preceding paragraph, a is i or $i-1$ according as $f_1(i-1)$ is less than or greater than $f_1(i)$ and b is $i+k+1$ or $i+k$ according as $f_1(i+k+1)$ is greater than or less than $f_1(i+k)$.

Let F_2 be a circular chain covering J_2 such that the ith link of F_2 contains B_iB_{i+1}, the closure of the ith link of F_2 lies in the $(f_1(i) \bmod n_1)$th link of F_1, each link of F_2 is of diameter less than $\frac{1}{2}$, and the sum of the elements of F_2 is the sum of the links of a circular chain each of whose links is a round disk of diameter less than $\frac{1}{2}$.

We continue in this fashion to define chains $F_3, F_4, \ldots$ such that

F_i is mesh less than 2^{1-i},
F_i has n_1 links,
the closure of the jth links of F_{i+1} lies in the $(f_i(j) \bmod n_i)$th link of F_i,
the sum of the links of F_i is the sum of the links of a circular chain each of whose links is the interior of a round disk of diameter less than 2^{1-i}.

The homeomorphism h is defined as follows. For each point p of X let $R_i(p)$ denote the sum of the closures of all links of F_i whose subscripts differ mod n_i from the subscript of a link of C_i containing p by no more than 1. Then $R_i(p)$ is the sum of the closures of three or four links of F_i according as p belongs to one or two links of C_i. Then $R_1(p)$, $R_2(p)$, . . . , is a decreasing sequence of closed sets converging to the point $h(p)$. The map h defined may be shown to be a homeomorphism.

Had we wanted to insist that the links of the circular chain be interiors of square disks, interiors of rectangular disks, interiors of triangular disks, or even annuli instead of interiors of round disks, this would have been no problem since the methods used in the proof of Theorem 4 could have been used to get a continuum so embedded that it could have been covered with circular chains whose links are of the required sort. In Theorem 4 of **(1)** we embedded a snake-like continuum in the plane so that it could be covered by linear chains whose links were the interiors of rectangles. The methods of Theorem 4 of the present paper can be used to give the following variation of that result.

THEOREM 5. *Each snake-like continuum can be embedded in the plane in such a way that for each* $\epsilon > 0$ *there is a linear chain covering the embedded snake-like continuum such that each link of the chain is the interior of a round disk of diameter less than* ϵ.

To construct such an embedding of a snake-like continuum X one would consider a sequence of linear chains $C_1, C_2, \ldots$, covering X and with meshes converging to 0 so fast that each collection of five consecutive links of C_{i+1} lies in a link of C_i.

Suppose C_i has n_i links. Let $A_1A_2 \ldots A_{n_1}A_{n_1+1}$ be a polygonal arc in the plane such that each A_iA_{i+1} is a segment of length less than 1. Let

$$F_1(L_1, L_2, \ldots, L_{n_1})$$

be a linear chain of mesh less than 1 covering $A_1A_2 \ldots A_{n_1+1}$ such that L_i contains A_iA_{i+1} and the sum of the links of F_1 is the sum of the elements of a linear chain each of whose elements is the interior of a round disk of diameter less than 1.

Let $B_1B_2 \ldots B_{n_2+1}$ be a polygonal arc very close to $A_1A_2 \ldots A_{n_1+1}$ such that each B_iB_{i+1} is a segment of length less than $\frac{1}{2}$ and lies in a link L_j of F_1 such that the ith link of C_2 lies in the jth link of C_1. Let F_2 be a linear chain of mesh less than $\frac{1}{2}$ such that this ith link of F_2 contains B_iB_{i+1} and its closure lies in the jth link of F_1 while the sum of the links of F_2 is the sum of the links of a linear chain each of whose elements is the interior or a round disk of diameter less than $\frac{1}{2}$.

Continuing as in the proof of Theorem 4 we get chains $F_3, F_4, \ldots$. Let X' be the intersection of the sum of the links of F_1, the sum of the links of $F_2, \ldots$. Then X' is the homeomorphic image of X that can for each positive number ϵ be covered by a linear chain each of whose links is the interior of a round disk of diameter less than ϵ.

The following result may also be proved in a similar fashion.

THEOREM 6. *If X is an indecomposable snake-like continuum, then there is a homeomorphism h of X into the plane such that for each positive number ϵ there is a linear chain C irreducibly covering $h(X)$ such that each link of C is the interior of a round disk of diameter less than ϵ and the sum of the first and last disks of C lies in a round disk of diameter less than ϵ such that this disk does not intersect any other links of C other than the second and the next to the last.*

It follows as a corollary of the above result that each indecomposable snake-like continuum is also circle-lke. In fact, the class of snake-like continua that are also circle-like is exactly the class which are also indecomposable.

The following theorem which follows as a consequence of Theorems 3, 4, and 6 answers a question raised by Burgess in **(5)**.

THEOREM 7. *Each circle-like continuum which can be embedded in the plane can be embedded in such a way that for each positive number ϵ, the embedded continuum can be irreducibly covered by a circular chain each of whose links is a round disk of diameter less than ϵ.*

THEOREM 8. *Each circle-like continuum can be embedded in the cartesian product of a triod and an arc.*

Proof. The proof of this theorem is similar to the proof of Theorem 4. The only difference is that if we have a sequence of circular chains defining the circle-like continuum, we cannot suppose that the chains do not circle each other more than once. We use a fin sticking up out of one of the links as a cross-over place.

Let $C_1, C_2, \ldots,$ be a sequence of circular chains covering the circle-like continuum X where each link L of C_{i+1} is contained in a link L' of C_i such that the mesh of C_{i+1} is less than the distance from L to the boundary of L'. We suppose that C_i has n_i links.

Let D be a disk in the $z = 0$ plane and E be a disk in the $x= 0$ plane such that $D \cdot E$ is an arc spanning D and $D + E$ is homeomorphic to the cartesian product of an arc and a triod.

Let J_1 be a polygonal simple closed curve in D which is the sum of n_1 arcs $A_1A_2, A_2A_3, \ldots, A_{n_1}A_1$ such that each of these arcs is of diameter less than 1, A_1A_2 crosses the straight line segment $D \cdot E$, and if two of the arcs intersect each other, the intersection is an endpoint of each. A first approximation of a homeomorphism taking X into $D + E$ takes X onto J_1 by taking the part of X in the ith link of C_1 onto the arc A_iA_{i+1}.

Let $F_1(L_1, L_2, \ldots, L_{n_1})$ be a circular chain of mesh less than 1 covering J_1 such that the L_1 is an open subset of $D + E$ covering A_1A_2. We can throw away part of certain links of F_1. However, we keep all of L_1. Let $L_i' = L_i \cdot D$ if $i > 1$ and $L_1' = L_1$. It may be convenient to think of L_1' as a disk with a fin, L_i' $(i > 1)$ as a disk, and $\sum L_i'$ as an open annular ring with a fin.

We build F_2 in $\sum L_i'$ as was done in Theorem 4, using $L_1' \cdot E$ as a cross-over place in case C_2 circles C_1 more than once. Each link of F_2 except one is then reduced to a disk and this link has a fin that is used as a cross-over for building F_3 in case C_3 circles C_2 more than once. We continue to build circular chains $F_3, F_4, \ldots$, as done in the proof of Theorem 4. The limiting set of the sum of the links of F_1, the sum of the links of $F_2, \ldots$, is a circle-like continuum in $D + E$ homeomorphic to X.

Since separating the plane into a certain number of pieces is a topological property for closed sets and the continuum $h(X)$ described in Theorem 4 separates the plane into exactly two pieces, but no snake-like continuum separates the plane, we have the following result which shows that part of the hypothesis of a result announced in **(6)** is not needed.

THEOREM 9. *For each circle-like plane continuum X which is not snake-like, the complement of X in the plane has exactly two components and X is the boundary of each.*

3. Categories of continua in the plane. The space of compact continua of a metric space S is the metric space $C(S)$ whose points are the compact continua of S, where the distance between two elements C_1, C_2 of $C(S)$ is the Hausdorff distance between them in S, namely,

$$\text{Least upper bound } \rho(x, C_i)(x \in C_1 + C_2, i = 1, 2).$$

The metric space $C(E^n)$ is a complete metric space.

A subset of a complete metric space is of the second category if the subset contains a dense G_δ (inner limiting set) subset of the space. The complement of a set of the second category is of the first category.

We say that most compact continua of E^n have a certain property if in the metric space $C(E^n)$, the set of elements with the property is associated with a subset of $C(E^n)$ of the second category. It is shown in **(2)** that most compact continua in E^n ($n \geqslant 2$) are pseudo-arcs. We now show that most of these can be covered by chains each of whose links is the interior of a small round disk.

THEOREM 10. *Most bounded continua in the plane are pseudo-arcs which for each $\epsilon > 0$ can be covered with a linear chain each of whose links is the interior of a round disk of diameter less than ϵ.*

Proof. The proof is the same as that given in **(2)**. Let F_i be the collection of all compact continua C in E^2 such that C cannot be covered by a linear chain each of whose links is the interior of a disk of diameter less than $1/i$. If $C_1, C_2, \ldots$, is a sequence of elements of F_i converging to a compact continuum C_0, C_0 is an element of F_i because if a chain of a certain sort covers C_0, it also covers some C_j. Hence, F_i is closed in $C(E^2)$.

The collection of all simple polygonal arcs is dense in $C(E^2)$. To get a polygonal arc within ϵ of a bounded continuum X one could let U be a neighbourhood of X such that each point of U is within ϵ of X, let $p_1, p_2, \ldots, p_n$ be a finite set of points of X which is ϵ dense in X, and take any polygonal arc in U that contains the p's. Since each of these broken lines can be covered by a chain each of whose links is the interior of a small circular disk, $C(E^2)-\sum F_i$ is a dense G_δ subset of $C(E^2)$.

As pointed out in **(2)**, the set of all compact continua in E^2 which are not hereditarily indecomposable is the sum of a countable number of nowhere dense closed sets. Hence, the set of elements of $C(E^2)$ which are not pesudo-arcs that can be covered by chains whose links are the interiors of small round disks is of the first category.

The following theorem can be proved in a similar fashion.

THEOREM 11. *Suppose U is the interior of a circle in E^2 and W is the collection of all compact continua C in E^2 such that U lies in a bounded component of $E^2 - C$. Then most elements of W are circle-like continua C such that*

each proper subcontinuum of C is a pseudo-arc and

for each $\epsilon > 0$, C can be covered by a circular chain each of whose links is the interior of a round disk of diameter less than ϵ.

Questions. If X_1, X_2 are two non-degenerate hereditarily indecomposable linearly chainable compact continua, then X_1, X_2 are pseudo-arcs (and hence are topologically equivalent) and each is homogeneous. Suppose Y_1, Y_2 are hereditarily indecomposable circularly chainable compact planar continua neither of which is linearly chainable. Are they topologically equivalent? Is each homogeneous? See the question raised about Example 2 in **(2)**.

4. Circle-like continua in 2-manifolds. We show in this section that each circle-like continuum that can be embedded in a 2-manifold can be embedded in the plane. Of course it cannot be concluded that an open subset of the 2-manifold can be embedded along with the circle-like continuum as can be seen by considering the centre simple closed curve in a Moebius band. However, we have the following result.

THEOREM 12. *If a circle-like continuum X lies in a 2-manifold M, either it lies in an open subset of M topologically equivalent to a subset of the plane or it lies in an open subset of M topologically equivalent to a subset of a Moebius band.*

Proof. Let T be a triangulation of M such that no vertex of T lies on X. We suppose that M is metrized so that each 1-simplex of T is a unit segment, each 2-simplex of T is an equilateral triangle, and the distance between two points is the length of some simple polygonal arc joining the two points. (We assume M to be connected.)

Let 2ϵ be a positive number such that no vertex of T is within 2ϵ of X. About each vertext v of T remove from M the closed disk whose centre is v

and whose radius is ϵ. Let M' be the remainder of M after all of these disks are removed.

Let $C(L_1, L_2, \ldots, L_n)$ be a circular chain of mesh less than $\epsilon/2$ irreducibly covering X. We are supposing that M is the total space. Hence, the links of C are open subsets of M'. If the sum of two adjacent links intersects two 2-simplexes of T, the two 2-simplexes share a common edge.

The open subset of M promised by the Theorem is $\sum L_i$, the sum of the links of C. In order to embed $\sum L_i$ in a set of the proper sort, we consider a covering space of M' into which $\sum L_i$ can be lifted.

Covering spaces for M'. Suppose S is a unit equilateral triangular disk from which has been removed each point whose distance from a vertex is less than or equal to ϵ. We call S an ϵ-blunted triangle. We note that the intersection of M' and a 2-simplex of T is an ϵ-blunted triangle. An ϵ-blunted triangle is a 2-manifold-with-boundary with three boundary components.

Suppose N is a 2-manifold with boundary which is the sum of a locally finite collection of ϵ-blunted triangles such that if two of these ϵ-blunted triangles are adjacent, their intersection is a common boundary component of each. If to each boundary component of N is added an ϵ-blunted triangle, there results a new 2-manifold-with-boundary N' with twice as many boundary components as N such that Int N' is homeomorphic with Int N. We suppose that no two of the added ϵ-blunted triangles are adjacent to each other. We say that N grew to N' in one step. If in turn we permit N' to grow a step, the resulting set to grow a step, . . . , the result after a countably infinite number of steps is a 2-manifold without boundary which is homeomorphic with Int N.

Suppose N is a 2-manifold-with-boundary with a decomposition into ϵ-blunted triangles and g is a map of N into M' such that

g is a local homeomorphism and
g is an isometry in taking each ϵ-blunted triangle in the decomposition of N onto the intersection of M' and a 2-simplex of T.

If N grows in one step to N', there is a unique extension of g so that the new g has the required properties for g and takes N' into M'. In fact, if N^∞ is the result of N after a countably infinite number of steps of growth, N^∞ is a 2-manifold without boundary homeomorphic with Int N and g can be uniquely extended to take N^∞ onto M' so that g is a local homeomorphism and g is an isometry in taking each ϵ-blunted triangle in the decomposition of N^∞ into M'.

The group G. We now consider a certain group G that will be useful in building an open set to contain $\sum L_i$. Suppose the 2-simplexes of T are ordered $D_1, D_2, \ldots$, and the 1-simplexes of T are ordered $E_1, E_2, \ldots$. Consider the free group G with generators $g_1, g_2, \ldots$. If an arc crosses E_i in going from D_j to D_{j+k}, the crossing generates the element g_i of G. If an arc crosses E_i in going from D_{j+k} to D_j, the crossing corresponds to g_i^{-1}. If a path from D_j

goes to E_i and then back into D_j without crossing into D_{j+k}, the element $g_i g_i^{-1}$ is generated. If a path from D_{j+k} goes to E_i and then back into D_{j+k} without crossing into D_j, the element $g_i^{-1}g_i$ is generated. We note that the group G is not the fundamental group of M' since it contains elements not generated by any loop. However, certain loops in M' do correspond to elements in the group G.

Let G' be the collection of words that can be made with letters $g_1, g_1^{-1}, g_2, g_2^{-1}, \ldots$. We use words of G' to represent loops rather than for purposes of multiplication. A word of G' is a finite ordered set whose elements are in $g_1, g_1^{-1}, g_2, g_2^{-1}, g_3, \ldots$, and an element of G is an equivalence class of words of G'. Two words are different as elements of G' if they have different spellings even though they may be in the same equivalence class corresponding to an element of G.

If a link L_i of C lies in a 2-simplex of T, let a_i be the centre of this 2-simplex. If L_i intersects two 2-simplexes of T, let a_i be the centre of the 1-simplex common to these two 2-simplexes. Consider the closed curve (perhaps singular) consisting of the segments from a_1 to a_2, a_2 to $a_3, \ldots$, and a_n to a_1. It generates a word W of G' as it crosses or touches the various E_i's.

Suppose $W_1, W_2, \ldots, W_m$ is a sequence of words of G' such that:

$W = W_1$.

W_{i+1} is obtained from W_i by cancelling two letters in W_i such that one of these letters is the inverse of the other and the two letters are either adjacent or they are at opposite ends of W_i.

W_m cannot be further reduced.

The word W_1 corresponds to the closed curve $a_1a_2 \ldots a_na_1$ and W_{i+1} corresponds to a closed curve (perhaps singular) in M' which is homotopic to the preceding in M'. We consider two cases.

Case 1. *W_m has no letters.* We suppose that W_{m-1} is $g_1g_1^{-1}$ and that E_1 is the common edge of D_1 and D_2. Let N^1 be a 2-manifold with boundary which is the sum of two ϵ-blunted triangles joined along a common boundary component of each and g be a homeomorphism of N^1 onto $M' \cdot (D_1 + D_2)$ which is an isometry in taking each ϵ-blunted triangle of N^1 onto the intersection of M' and one of D_1, D_2. Since g is a homeomorphism, there is a closed curve in N^1 that is taken by g onto the closed curve associated with W_{m-1}.

The closed curve associated with W_{m-2} is the same as the closed curve associated with W_{m-1} except that an arc in the first closed curve which crosses an E_i and then crosses back (or merely touches and then backs away) is replaced by an arc which does not touch E_i. Hence, if N^2 is the result of N^1 after a one-step growth, there is a closed curve in N^2 that maps under g onto the closed curve associated with W_{m-2}. By continuing this procedure, one finds that if N^{m-1} is the result of N^1 after $m-2$ steps of growth, there is a closed curve $b_1b_2 \ldots b_nb_1$ in N^{m-1} that maps by the extended g onto $a_1a_2 \ldots a_na_1$.

The homeomorphism h of $\sum L_i$ into Int N^{m-1} is described as follows. If a_1 is the centre of D_j, $h = g^{-1}$ in taking L_i into the sum of the two ϵ-blunted triangles of N^{m-1} having b_i on their common edge. We note that h takes $\sum L_i$ homeomorphically into Int N^{m-1} and Int N^{m-1} is topologically equivalent to the plane since Int N^1 is.

Case 2. $W_m = x_1x_2 \ldots x_r$. Let $E(x_i)$ be the 1-simplex of T associated with x_i or x_i^{-1}. Since there is a closed curve in M' associated with W_m, $E(x_i)$ and $E(x_{i+1})$ are different edges of the same 2-simplex of T as are $E(x_1)$ and $E(x_r)$.

Let $A_1, A_2, \ldots, A_r$ be r ϵ-blunted triangles joined in a fashion to be described. Let h be an isometry of A_i onto the intersection of M' and the 2-simplex of T having $E(x_i)$ and $E(x_{i+1})$ as edges (h takes A_r onto the intersection of M' and the 2-simplex of T having $E(x_1)$ and $E(x_r)$ as edges). Suppose that A_i is joined to A_{i+1} along an edge of each so that h is the same on this edge. Also, A_1 and A_r are joined in a similar fashion. We note that A_{i+2} and A_i are not joined to A_{i+1} along the same edge since $E(x_i)$, $E(x_{i+1})$ are different edges of the same simplex of T. We note that if N is the sum of these A_i's, Int N is either the interior of an annulus ring or the interior of a Moebius band.

By the same argument as that used in Case 1, it follows that $\sum L_i$ can be embedded in the result after N is grown through several steps. Hence, either $\sum L_i$ can be embedded in an annulus ring (and hence in the plane) or it can be embedded in a Moebius band.

The methods used in the proof of Theorem 12 give the following result.

THEOREM 13. *Each tree-like continuum on a 2-manifold M lies on an open subset of M which can be embedded in the plane.*

Since each neighbourhood of a planar tree-like continuum contains a disk containing the continuum, Thorem 13 actually implies that some disk in M contains the tree-like continuum. This was proved in Lemma 2 of (7) by other methods.

THEOREM 14. *Each circle-like continuum that can be embedded in a 2-manifold can be embedded in the plane.*

Proof. We suppose that $C_1, C_2, \ldots,$ is a sequence of circular chains covering X such that mesh C_i approaches 0 as i increases without limit and C_{i+1} is a refinement of C_i. Since a circle-like continuum X can be embedded in the plane if it is snake-like, we suppose that each C_{i+1} circles C_i a positive number of times. Since it follows from Theorem 4 that X can be embedded in the plane if with only a finite number of exceptions C_{i+1} circles C_i exactly once, we do not consider this case. In view of Theorem 1 we suppose with no loss of generality that each C_{i+1} circles C_i more than twice. We finish the proof of Theorem 14 by showing that this case cannot occur.

Suppose $C_1(L_1, L_2, \ldots, L_n)$ is the chain described in the proof of Theorem 12 such that either $\sum L_i$ can be embedded in the plane or $\sum L_i$ can be embedded in a Moebius band so that C_1 circles the band exactly once. If C_1 circles a Moebius band exactly once, then C_2 circles the band more than twice and it follows from Theorem 2 that there is a chain with connected links in C_2 that circles C_1 more than twice. Hence, there is a simple closed curve in the Moebius band that circles the band more than twice. But no simple closed curve circles a Moebius band more than twice.

COROLLARY. *The circle is the only solenoid that can be embedded in a 2-manifold.*

THEOREM 15. *If X is a circle-like continuum on an orientable 2-manifold M, then X lies on an annulus in M.*

Proof. We showed in Theorem 12 that X lies in an open subset U of M such that U can either be embedded in the plane or U can be embedded in a Moebius band.

Suppose U can be embedded in the plane and M' is a connected 2-manifold-with-boundary in U such that Int M' contains X. Then M' is a punctured disk. Since $E^2 - X$ has one or two components according as X is snake-like or not, it is possible to remove canals from M' so that X lies in the remaining punctured disk which has at most two boundary components. If the remainder has two boundary components, it is an annulus. If it has only one boundary component, it is a disk. An annulus in the disk contains X.

Finally we consider the case where no open subset of M which contains X can be embedded in the plane. We show that this case cannot occur. It follows from Theorem 13 that if it did, X would not be snake-like.

Let $C_1, C_2, \ldots,$ be a decreasing sequence of circular chains covering X such that each C_{i+1} circles in C_i exactly once, each link of each C_i is an open subset of M, and the sum of the links of C_1 can be embedded in a Moebius band as described in Theorem 12 with C_1 circling this Moebius band exactly once.

Suppose that C_i has n_i links and each link of C_{i+1} lies in a component of a link of C_i. Let f_i $(i = 1, 2)$ be a map of $(0, 1, 2, \ldots, n_{i+1})$ into the integers such that the $f_i(j)$th link of C_i is the first link of C_i with a component that contains the jth link of C_{i+1} and $f_i(0) = f_i(n_{i+1})$. Since C_{i+1} circles C_i once,

$$\sum_{j=0}^{n_i+1} (f_i(j+1) - f_i(j)) = \pm n_i$$

where $(n_i - 1)$, $(1 - n_i)$ are considered as -1, 1 in computing this sum. Since C_3 circles C_1 once, we also have that

$$\sum_{j=0}^{n_3} (f_1f_2(j+1) - f_1f_2(j)) = \pm n_1.$$

Let U_j be the component of the $f_2(j)$th link of C_2 containing the jth link of C_3. Suppose U_r and U_s intersect with $r < s$. Then either

$$f_1f_2(r) - f_1f_2(s) + \sum_{j=r}^{s-1} (f_1f_2(j+1) - f_1f_2(j))$$

or

$$f_1f_2(s) - f_1f_2(r) + \sum_{j=0}^{r-1} (f_1f_2(j+1) - f_1f_2(j)) + \sum_{j=s}^{n_3} (f_1f_2(j+1) - f_1f_2(j))$$

is an odd multiple of n_1.

By a continuation of the elimination of the U's, one can obtain a circular chain of the U's that circles C_1 an odd number of times. Then a simple closed curve J in the sum of these U's circles the Moebius band an odd number of times. This odd number cannot be more than one since no simple closed curve in a Moebius band circles the Moebius band more than twice. Any open subset of the Moebius band that contains J also contains a Moebius band. Hence, the sum of the links of C_1 contains a Moebius band which is contrary to the hypothesis that M is an orientable manifold.

COROLLARY. *If X is a circle-like continuum in an orientable connected 2-manifold M, then $M - X$ has at most two components.*

REFERENCES

1. R. H. Bing, *Snake-like continua*, Duke Math. J., *18* (1951), 653–663.
2. ——— *Concerning hereditarily indecomposable continua*, Pacific J. Math., *1* (1951), 43–51.
3. ——— *A simple closed curve is the only homogeneous bounded plane continuum that contains an arc*, Can. J. Math., *12* (1960), 209–230.
4. ——— *Conditions under which monotone decompositions of E^3 are simply connected*, Abstract 325, Bull. Amer. Math. Soc., *63* (1957), 143.
5. C. E. Burgess, *Homogeneous continua which are almost chainable*, Can. J. Math., *13* (1961), 519–528.
6. ——— *Homogeneous continua which are circularly chainable and contain an arc*, Abstract 545-16, Amer. Math. Soc. Notices, *5* (1958), 229.
7. R. H. Roberts and N. E. Steenrod, *Monotone transformations of two-dimensional manifolds* Ann. Math., *39* (1938), 851-863.

University of Wisconsin

III. Continua

A. Pseudo Arc and Related Topics

A HOMOGENEOUS INDECOMPOSABLE PLANE CONTINUUM

By R. H. Bing

In this paper we shall answer the following question raised [2] by Knaster and Kuratowski: *If a nondegenerate bounded plane continuum is homogeneous, is it necessarily a simple closed curve?*

Mazurkiewicz gave [3] a partial solution to the above question by showing that the simple closed curve is the only homogeneous nondegenerate bounded locally connected plane continuum.

Waraszkiewicz announced [5] that the answer to the above question was in the affirmative. Choquet went even further in announcing that every homogeneous bounded closed plane set belongs to one of the following types: (1) a finite number of points; (2) a totally disconnected perfect set; (3 and 4) a set homeomorphic to the sum of a collection of concentric circles such that the common part of this sum and a line through the center of the circles is either a finite set or a totally disconnected perfect set. However, the findings of this paper do not hold with these announcements.

In this paper we shall give an example of a homogeneous nondegenerate bounded plane continuum which is not a simple closed curve.

1. **Definitions.** We give the following definitions of some terms that we shall use.

Homogeneous. A set S is homogeneous if, for each pair of its points P_1 and P_2, there is a homeomorphism that carries S into itself and P_1 into P_2. This is somewhat of a "point" condition and it might be well to call S pointwise homogeneous if these conditions are satisfied. However, we shall follow the usual terminology and call it homogeneous. Theorem 15 shows that the continuum we describe has an even stronger type of homogeneity.

Chain. A chain $D = [d_1, d_2, \cdots, d_n]$ is a collection of domains $d_1, d_2, \cdots, d_n$ such that d_i intersects d_j if and only if i is equal to $j - 1$, j, or $j + 1$. The element d_i is the i-th link of the chain D and d_1 and d_n are its end links. Links which are not end links are called interior links. Two links are adjacent if they intersect. If P and Q are points belonging to d_1 and d_n respectively but to no other link of the chain, then $D = [d_1, d_2, \cdots, d_n]$ is called a chain from P to Q. We have not supposed that the links of a chain are necessarily connected.

Contain. The chain D contains the chain E if each link of E is a subset of a link of D.

Subchain. If E is a chain each of whose links is a link of the chain D, then E is a subchain of D. If the end links of E are the i-th and j-th links of D, we may denote E by $D(i, j)$.

Received December 9, 1947; presented to the American Mathematical Society April 26, 1947.

Crooked. The chain $E = [e_1, e_2, \cdots, e_n]$ is crooked in the chain $D = [d_1, d_2, \cdots, d_m]$ provided D contains E and E is such that if $E(i, j)$ is one of its subchains with e_i and e_j intersecting d_h and d_k respectively, and $|h - k|$ being greater than 2, then $E(i, j)$ is the sum of three chains $E(i, r)$, $E(r, s)$, and $E(s, j)$ such that $(s - r)(j - i)$ is positive and e_r and e_s are subsets of links of $D(h, k)$ adjacent to d_k and d_h respectively. Compare this notion of being crooked with that of being very crooked as given by Moise [4].

Consolidation. The chain E is a consolidation of the chain D if each link of E is the sum of a subcollection of D and D is contained in E.

Pattern. If $(x_1, y_1), (x_2, y_2), \cdots, (x_n, y_n)$ is a collection of ordered pairs of positive integers, the chain D follows the pattern $(x_1, y_1), (x_2, y_2), \cdots, (x_n, y_n)$ in the chain E if the x_i-th link $(i = 1, 2, \cdots, n)$ of D is a subset of the y_i-th link of E.

Using the above terms, we shall describe a continuum M. In Theorem 12, we show that two continua are homeomorphic with each other if each is defined in the same manner as M, and in Theorems 10 and 13 we show that each such continuum is indecomposable and homogeneous. The continuum M may lie in the plane. It is topologically equivalent to the continuum which Moise showed [4] to be homeomorphic to each of its nondegenerate subcontinua. I am indebted to Moise for some of the ideas in this paper because I attended a lecture at which he presented [4], and some of the results of this paper are modifications of those in [4].

Description of M. Let P and Q be two points of a compact metric space. Let $D_1, D_2, \cdots$ be a sequence of chains from P to Q such that for each positive integer i, (1) D_{i+1} is crooked in D_i, (2) no link of D_i has a diameter

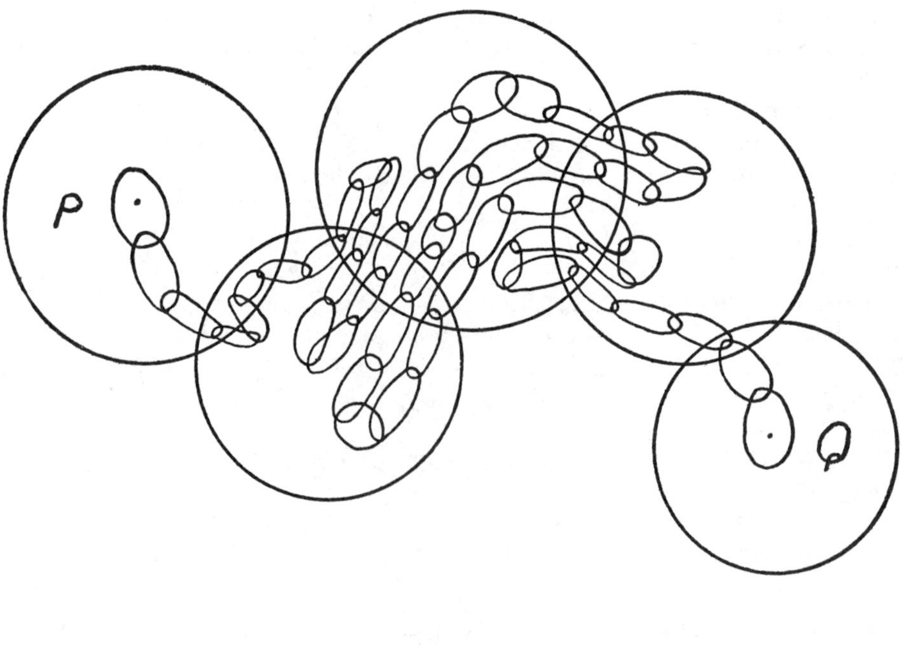

of more than $1/i$, and (3) the closure of each link of D_{i+1} is a subset of a link of D_i. See the figure. Then M is the common part of $D_1^*, D_2^*, \cdots$, where D_i^* denotes the sum of the links of D_i.

Before proving that M is homogeneous, we shall prove some preliminary theorems.

2. **Properties of crooked chains.** In this section we shall prove some theorems that we shall use in studying the properties of the continuum M.

THEOREM 1. *If D, E, and F are chains such that D is a consolidation of E and F is crooked in E, then F is crooked in D.*

Proof. Suppose f_h and f_k are links of F intersecting links d_r and d_s respectively of D, $|r - s| > 2$, and no interior link of $F(h, k)$ intersects either d_r or d_s. Then no link of $F(h, k)$ intersects any link of D not belonging to $D(r, s)$. Let e_m and e_n be links of E contained in d_r and d_s such that e_m and e_n intersect f_h and f_k respectively.

As each link of $E(m, n)$ intersects a link of $F(h, k)$ and each interior link of $F(h, k)$ is a subset of an interior link of $D(r, s)$, $E(m, n)$ is contained in $D(r, s)$. Since the link of $E(m, n)$ adjacent to e_m intersects an interior link of $F(h, k)$, it is not contained in d_r. Hence, the link of $E(m, n)$ adjacent to e_m is in the link of $D(r, s)$ adjacent to d_r. Also, the link of $E(m, n)$ adjacent to e_n is in the link of $D(r, s)$ adjacent to d_s.

As F is crooked in E, $F(h, k)$ is the sum of three chains $F(h, u)$, $F(u, v)$, and $F(v, k)$ such that $(k - h)(v - u)$ is positive and f_u and f_v lie in the links of $E(m, n)$ adjacent to e_n and e_m respectively. Then f_u and f_v are in links of $D(r, s)$ adjacent to d_s and d_r respectively. Hence, F is crooked in D.

Theorem 1 would not be true if we weakened its hypothesis to the extent of supposing that D contains E instead of supposing that D is a consolidation of E.

THEOREM 2. *If D, E, and F are chains such that E contains F and is crooked in D, then F is crooked in D.*

Proof. Suppose f_h and f_k are links of F intersecting links d_r and d_s respectively of D and $|r - s| > 2$. Let e_m and e_n be elements of E containing links f_h and f_k respectively. Since E is crooked in D, $E(m, n)$ is the sum of three chains $E(m, u)$, $E(u, v)$, and $E(v, n)$ such that $(n - m)(v - u)$ is positive and e_u and e_v are in links of $D(r, s)$ adjacent to d_s and d_r respectively.

Since e_u is between e_m and e_n in $E(m, n)$, there is a link f_x of $F(h, k)$ in e_u. Also, there is a link f_y of $F(x, k)$ in e_v. Then $F(h, k)$ is the sum of three chains $F(h, x)$, $F(x, y)$, and $f(y, k)$ such that $(k - h)(y - x)$ is positive and f_x and f_y are in links of $D(r, s)$ adjacent to d_s and d_r respectively.

COROLLARY. *If D, E, and F are chains such that E is crooked in D and is a consolidation of F, then F is crooked in D.*

THEOREM 3. *If the chain D is crooked in the chain* $E = [e_1, e_2, \cdots, e_m]$ *and* d_i *is a particular link of D, then there is a chain F such that F is a consolidation of D,* d_i *is contained in only the first link of F, each link of F is a subset of the sum of two adjacent links of E, and any link of F containing an end link of D which intersects* e_1 *or* e_m *is a subset of* $e_1 + e_2$ *or* $e_{m-1} + e_m$.

Proof. We shall prove the theorem by induction on m. It is rather easy to see that the theorem holds for m equal to either 1, 2, 3, or 4. In these cases, we let one link of F be the sum of the links of D contained in the sum of the first two links of E, and the other link (if any) of F be the sum of the other links of D.

If d_i is a subset of $e_1 + e_2$, the theorem is easily established irrespective of the value of m since F may be defined so that its first link is the sum of the links of D contained in $e_1 + e_2$, its second link is the sum of the links of D contained in e_3 but not in e_2, $\cdots$, and its last link is the sum of the links of D contained in $e_{m-1} + e_m$. Likewise, we find that the theorem is satisfied if d_i is a subset of $e_{m-1} + e_m$.

Assume that the theorem is false. Let r be an integer such that the theorem holds for m less than r but does not hold for m equal to r.

Special Case. Let us consider the case in which no interior link of D intersects either e_1 or e_r. For convenience, suppose that the end links d_1 and d_n of D intersect e_1 and e_r respectively. Since D is crooked in E, D is the sum of three chains $D(1, h)$, $D(h, k)$, and $D(k, n)$ such that $1 < h < k < n$, d_h is a subset of e_{r-1}, and d_k is a subset of e_2.

Suppose first that d_i is a link of $D(1, h)$. Since $E(1, r-1)$ contains $D(1, h)$ and has fewer than r links, there is a chain H such that H is a consolidation of $D(1, h)$, only the first link of H contains d_i, each link of H is a subset of the sum of two adjacent links of $E(1, r-1)$, any link of H containing d_1 is a subset of either $e_1 + e_2$ or $e_{r-2} + e_{r-1}$, and any link of H containing d_h is a subset of either $e_1 + e_2$ or $e_{r-2} + e_{r-1}$.

Let h_u be the first link of H that contains d_h. Since $r > 4$, d_h is not a subset of $e_1 + e_2$. Hence h_u is a subset of $e_{r-2} + e_{r-1}$. The chain F is selected as follows: $H(1, u-1)$ is a subchain of F; $D(k+1, n)$ is a subchain of F; if D' denotes the links of D contained in neither $H(1, u-1)$ nor $D(k+1, n)$, the sum of the links of D' in $e_1 + e_2$ is a link of F, the sum of the links of D' in e_3 but not in e_2 is another link of F, $\cdots$, and the sum of the links of D' in $e_{r-2} + e_{r-1}$ is another link of F. The chain F is the required chain.

By a similar line of argument, we find that this special case of our theorem holds if d_i is a link of either $D(h, k)$ or $D(k, n)$.

General Case. Let $D(h, k)$ be a subchain of D maximal with respect to $D(h, k)$ containing d_i and no interior link of $D(h, k)$ intersecting either e_1 or e_r. There is a chain H such that H is a consolidation of $D(h, k)$, only the first link of H contains d_i, each link of H is a subset of the sum of two adjacent links of E, and any link of F containing an end link of $D(h, k)$ which intersects e_1 or

e_r is a subset of $e_1 + e_2$ or $e_{r-1} + e_r$. If no link of $D(h, k)$ intersects e_1, or if no link of $D(h, k)$ intersects e_r, the existence of H follows from the induction on m; if one link of $D(h, k)$ intersects e_1 and another intersects e_r, the existence of H follows from the special case of the theorem previously discussed.

Let h_u be the first link of H that intersects $e_1 + e_r$. For convenience, suppose that h_u intersects e_1. Then it is a subset of $e_1 + e_2$. We define F as follows: $H(1, u - 1)$ is a subchain of F; if D' denotes the links of D not in $H(1, u - 1)$, the sum of the elements of D' in $e_1 + e_2$ is a link of F, the sum of the links of D' in e_3 but not in e_2 is another link, $\cdots$, and the sum of the links of D' in $e_{r-1} + e_r$ is another link. The chain F satisfies the prescribed conditions. This completes the proof of Theorem 3.

THEOREM 4. *If $D = [d_1, d_2, \cdots, d_n]$ is a chain crooked in the chain $E = [e_1, e_2, \cdots, e_m]$ and $D(r, s)$ is a subchain of D such that a link of $D(r, s)$ intersects e_1 and a link of $D(r, s)$ intersects e_m, then there is a chain F such that F is a consolidation of D, each element of F is a subset of the sum of two adjacent links of E, d_r is contained in only the first link of F, and d_s is contained in only the last.*

Proof. Let $D(h, k)$ be a subchain of $D(r, s)$ such that d_h intersects e_1 and d_k intersects e_m. For convenience, suppose that $h < k$ and that d_r and d_s are links of $D(1, h)$ and $D(k, n)$ respectively.

By Theorem 3, there is a chain H which is a consolidation of $D(1, h)$ such that each link of H is a subset of the sum of two adjacent elements of E, d_r is contained in only the first link of H, and any link of H containing d_h is a subset of either $e_1 + e_2$ or $e_{m-1} + e_m$; also, there is a chain G such that G is a consolidation of $D(k, n)$, each link of G is a subset of the sum of two adjacent elements of E, d_s is contained in only the first link of G, and any link of G containing d_k is a subset of $e_1 + e_2$ or $e_{m-1} + e_m$.

Let h_u and g_v be the first elements of H and G containing d_h and d_k respectively. Since the case of our theorem where $m \leq 4$ is easily established, we suppose that h_u and g_v are subsets of $e_1 + e_2$ and $e_{m-1} + e_m$ respectively. The required chain F is defined as follows: The first links of F are the links of $H(1, u - 1)$; the last links of F are the links of $G(v - 1, 1)$; if D' denotes the links of D in neither $H(1, u - 1)$ nor $G(v - 1, 1)$, the sum of the links of D' in $e_1 + e_2$ is a link of F, the sum of the links of D' in e_3 but not in e_2 is another link, $\cdots$, and the sum of the links of D' in $e_{m-1} + e_m$ is another link of F.

THEOREM 5. *Suppose $(1, x_1), (2, x_2), \cdots, (n, x_n)$ is a collection of ordered pairs of positive integers such that $h = x_1 \leq x_i \leq x_n = k$ and $|x_i - x_{i+1}| \leq 1$ for $i = 1, 2, \cdots, n - 1$. Suppose $D_1, D_2, \cdots$ is a sequence of chains from P to Q such that for each positive integer i, D_{i+1} is crooked in D_i, the closure of each link of D_{i+1} is a compact subset of a link of D_i, and no link of D_i has a diameter of more than $1/i$. Denote the r-th link of D_i by $d(i)_r$. Suppose that the subchain $D_2(u, v)$ of D_2 is contained in the subchain $D_1(h, k)$ of D_1 and the closures of $d(2)_u$ and $d(2)_v$ are mutually exclusive subsets of $d(1)_h$ and $d(1)_k$ respectively. Then for* *each integer w there is an integer j greater than w and a chain* $E = [e_1, e_2, \cdots, e_n]$ *following the pattern* $(1, x_1), (2, x_2), \cdots, (n, x_n)$ *in* D_1 *such that E is a consolidation of the links of* D_j *contained in* $D_2(u, v)$ *and no interior link of E intersects* $d(2)_u + d(2)_v$.

Proof. This theorem may be proved by induction on n.

Since the closure of each link of D_2 is a compact closed subset of a link of D_1, there is a positive number ϵ such that any set which is of diameter less than ϵ and which intersects a link of D_2 is a subset of a link of D_1. Using this result, the fact that the closures of $d(2)_u$ and $d(2)_v$ are subsets of $d(1)_h$ and $d(1)_k$ respectively, and the hypothesis that no link of D_i is of diameter more than $1/i$, we find that there is an integer m greater than w such that each five-linked subchain of D_m contained in $D_2(u, v)$ is such that the closure of the sum of this subchain is a subset of a link of $D_1(h, k)$, every five-linked subchain of D_m, one of whose links intersect $d(2)_u$, is contained in $d(1)_h$, and every five-linked subchain of D_m, one of whose links intersects $d(2)_v$, is contained in $d(1)_k$.

Since $| x_i - x_{i+1} | \leq 1$, n is no less than $k - h + 1$. If $n = k - h + 1$, the theorem holds, for let j be any integer greater than m. Then the sum of all links of D_j in the common part of $d(1)_h$ and $D_2^*(u, v)$ may be taken as e_1, the sum of all links of D_j in the common part of $d(2)_{h+1}$ and $D_2^*(u, v)$ but not in $d(1)_h$ may be taken as $e_2, \cdots$, and the sum of all links of D_j in the common part of $d(2)_k$ and $D_2^*(u, v)$ may be taken as e_n.

Assume that the theorem is not true. Suppose n is a fixed positive integer such that the theorem does not hold for this value of n but does hold for all smaller positive integral values and suppose that h and k are positive integers such that $n > k + 1 - h$ and such that the theorem does not hold for these particular values of n, h, and k.

Let us consider the case in which $x_1 = x_2$. Because of the induction on n, there is an integer s greater than m and a chain $F = [f_1, f_2, \cdots, f_{n-1}]$ such that F is a consolidation of the links of D_s in $D_2(u, v)$, F follows the pattern $(1, x_2), (2, x_3), \cdots, (n - 1, x_n)$ in D_1, only the first link of F contains any point of $d(2)_u$, and only the last link of F contains any point of $d(2)_v$. Since no link of D_i has a diameter of more than $1/i$ and the closure of $d(2)_u$ is a subset of $d(1)_h$, a positive integer j and a chain E can be found satisfying the prescribed conditions. Although we omit the details of the determination of j and $E = [e_1, e_2, \cdots, e_n)$, we note that they can be chosen so that e_1 is a subset of $f_1 + d(2)_u$, e_2 is a subset of $f_1 + f_2$, e_3 is a subset of $f_3, \cdots$, and e_n is a subset of $f_{n-1} + d(2)_v$. Hence, $x_1 \neq x_2$, or else the assumption that the theorem does not hold for the values assigned to n, h, and k is false.

Now, let us consider the case in which there are integers r and t such that $1 < t < r < n$, $x_1 = x_r < x_t$, and $x_i \leq x_t$ for $i = 1, 2, \cdots, r$.

Suppose that a subchain $D_{m+1}(p, q)$ of D_{m+1} is contained in $D_2(u, v)$ and that its end links $d(m + 1)_p$ and $d(m + 1)_q$ intersect $d(2)_u$ and $d(1)_{x_t+1}$ respectively. Because of the way that the closures of subchains of D_m are subsets of the links of D_1 and since D_{m+1} is crooked in D_m, it may be found that $D_{m+1}(p, q)$ is the sum of three chains, $D_{m+1}(p, y)$, $D_{m+1}(y, z)$, and $D_{m+1}(z, q)$, such that $(q - p)(z - y)$ is positive. $D_{m+1}(p, z)$ is a subset of $D_1(h, x_t)$, but no link of it intersects $d(2)_v$, the closure of $d(m + 1)_y$ is a subset of $d(1)_{x_t}$, and the closure of $d(m + 1)_z$ is a subset of $d(1)_h$.

Hence, the collection of links of D_{m+1} in $D_2(u, v)$ is the sum of five subcollections R, U, S, V, and T such that R contains each link of D_{m+1} in $d(2)_u$, the closure of U^* is a subset of $d(1)_{x_t}$, the closure of V^* is a subset of $d(1)_h$, T contains each link of D_{m+1} in $d(2)_v$, no link of $R + U$ intersects any link of $V + T$, and each link of $R + U + S + V$ is a subset of $D_1^*(h, x_t)$.

By using induction on n, we find that there is an integer $j(R + U)$ greater than m and a chain F such that F is a consolidation of the links of $D_{j(R+U)}$ in the common part of $D_2^*(u, v)$ and $(R + U)^*$, F follows the pattern $(1, x_1)$, $(2, x_2)$, $\cdots$, (t, x_t) in D_1, and only the first and last links of F contain points of $d(2)_u$ and U^* respectively. Also, there is an integer $j(U + S + V)$ greater than m and a chain G such that G is a consolidation of the links of $D_{j(U+S+V)}$ in the common part of $D_2^*(u, v)$ and $(U + S + V)^*$, G follows the pattern $(1, x_t)$, $(2, x_{t+1})$, $\cdots$, $(r + 1 - t, x_r)$ in D_1, and only the first and last links of G contain points of U^* and V^* respectively. Furthermore, there is an integer $j(V + T)$ greater than m and a chain H such that H is a consolidation of the links of $D_{j(V+T)}$ in the common part of $D_2^*(u, v)$ and $(V + T)^*$, H follows the pattern $(1, x_r)$, $(2, x_{r+1})$, $\cdots$, $(n + 1 - r, x_n)$ in D_1, only the first link of H contains any point of V^*, and only the last link of H contains any point of $d(2)_v$.

Let j be the largest of the integers $j(R + U)$, $j(U + S + V)$, and $j(V + T)$. A chain E satisfying the conditions of our theorem may be defined as follows: e_1 is the sum of all links of D_j in $D_2(u, v)$ that are contained in either $d(2)_u$ or the first link of F; e_i $(i = 2, 3, \cdots, t - 1)$ is the sum of all links of D_j in the i-th link of F; e_t is the sum of all links of D_j in the sum of the last link of F and the first link of G; $\cdots$; and e_n is the sum of all links of D_j in $D_2(u, v)$ that are contained in either $d(2)_r$ or the last link of H. Hence, the theorem is satisfied if for some integer i greater than 1, $x_i = x_1$.

Assume that for each integer i greater than 1, $x_i \neq x_1$. Let W be the set of all links of $D_2(u, v)$ that are not in $d(1)_h$ and W' be the set of all elements of W that intersect $d(1)_h$. The closure of each link of W' is a subset of $d(1)_{h+1}$. Because of the induction on n, there is an integer j greater than m and a chain $F = [f_1, f_2, \cdots, f_{n-1}]$ such that F is a consolidation of the links of D_j in W, F follows the pattern $(1, x_2)$, $(2, x_3)$, $\cdots$, $(n - 1, x_n)$ in D_1, f_1 is the only element of F intersecting W', and f_{n-1} is the only link of F intersecting $d(2)_k$. A chain E satisfying the prescribed conditions may be obtained as follows: e_1 is the sum of all elements of D_j in the common part of $D_2^*(u, v)$ and $d(1)_h$, e_2 is f_1, $\cdots$, and e_n is f_{n-1}.

As an immediate consequence of Theorem 5, we have the following result which we shall use in the next section.

THEOREM 6. *Suppose* $(1, x_1)$, $(2, x_2)$, $\cdots$, (n, x_n) *is a collection of ordered pairs of positive integers such that* $1 = x_1 \leq x_i \leq x_n$ *and* $| x_i - x_{i+1} | \leq 1$ *for* $i = 1, 2, \cdots, n-1$. *Suppose* $D_1, D_2, \cdots$ *is a sequence of chains from P to Q such that* D_1 *has* x_n *links and for each positive integer i,* D_{i+1} *is crooked in* D_i, *the closure of each link of* D_{i+1} *is a compact subset of a link of* D_i, *and no link of* D_i *has a diameter of more than* $1/i$. *Then there is an integer j and a chain E from P to Q such that E is a consolidation of* D_j *and E follows the pattern* $(1, x_1)$, $(2, x_2)$, $\cdots$, (n, x_n) *in* D_1.

THEOREM 7. *Suppose the chain D is crooked in the chain* $E = [e_1, e_2, \cdots, e_n]$ $(n > 2)$, *one link of D is in* e_1, *and one link of D is in* e_n. *Then there is a chain F which is a consolidation of D and is such that F has its end links in* e_1 *and* e_n, *and F is crooked in E, but no proper consolidation of F has these properties. Furthermore, F follows the pattern* $(1, 1)$, $(2, 2)$, $(3, 3)$, $(4, 2)$, $(5, 3)$, $(6, 4)$, $\cdots$, $[i, y(i)]$, $\cdots$, $[z(n), n]$ *in E where* $z(n)$ *is the number of links of F,* $z(3) = 3$, $z(4) = 6$, $z(n) = 2z(n-1) + z(n-2) - 2$, *and* $y(i) = 1 + y(1 + | z(r-1) + z(r-2) - 1 - i |)$ *where* $z(r-1) \leq i \leq z(r)$ *for some integer r.*

An alternate method of defining the pattern $P(n)$ of F in E is as follows: $P(3) = (1, 1), (2, 2), (3, 3)$; $P(4) = (1, 1), (2, 2), (3, 3), (4, 2), (5, 3), (6, 4)$; if the chains $F[z(n-1), z(n-1) + z(n-2) - 1]$, $F[z(n-1) + z(n-2) - 1, z(n)]$, $E[n-1, 2]$, $E[2, n]$ are denoted by F', F'', E', E'' respectively, where the ordering of links is such that $f_{z(n-1)+z(n-2)-1}$ is both the last link of F' and the first link of F'' and e_2 is both the last link of E' and the first link of E'', then $F[1, z(n-1)]$ follows the pattern $P(n-1)$ in $E[1, n-1]$, F' follows the pattern $P(n-2)$ in E', and F'' follows the pattern $P(n-1)$ in E''.

Proof. This theorem may be proved by induction on n. It is easy to verify for the cases in which n is equal to 3 or 4. Hence, suppose that the theorem holds for all integers n greater than 2 and less than m. We shall show that it holds for n equal to m.

Define the sum of all links of D in e_1 to be f_1 and the sum of all links of D in e_m to be $f_{z(m)}$. Each element of D in e_2 which intersects e_1 but does not lie in e_1 is a subset of f_2, and each element of D in e_{m-1} which intersects e_m but does not lie in e_m is a subset of $f_{z(m)-1}$. Other links of D will be placed in appropriate links of F as described below.

Let $D(r, s)$ be a subchain of D such that d_r intersects e_1 and d_s intersects e_m but no interior link of $D(r, s)$ intersects $e_1 + e_m$. Then $D(r, s)$ is the sum of three chains $D(r, j)$, $D(j, k)$, $D(k, s)$ such that $(r - s)(j - k) > 0$, d_k is in e_2, and d_j is in e_{m-1}. Because of the induction on n, some consolidation F_1 of $D(r, j)$ plus a link of D in e_1 is crooked in E and follows the pattern $(1, 1)$, $(2, 2)$, $\cdots$, $[i, y(i)]$, $\cdots$, $[z(m-1), m-1]$ in E. We note that F_1 may not be a chain because its first link may not intersect its second. It is a chain in other respects. Also, some consolidation F_3 of $D(k, s)$ plus a link of D in e_m is crooked in E and follows the pattern $(1, 2)$, $(2, 3)$, $\cdots$, $[i, y(i) + 1]$, $\cdots$, $[z(m-1), m]$ in E. Furthermore, some consolidation F_2 of $D(j, k)$ is crooked in E and follows the pattern $(1, m - 1), (2, m - 2), \cdots, [i, m - y(i)], \cdots,$ $[z(m - 2), 2]$ in E. The i-th link of F_1 belongs to the i-th link of F, the i-th link of F_3 belongs to the $[z(m - 1) + z(m - 2) + (i - 2)]$-th link of F, and the i-th link of F_2 belongs to the $[z(m - 1) + i - 1]$-th link of F.

Other elements of D may be placed in appropriate elements of F so as to form a chain F satisfying the prescribed conditions.

THEOREM 8. *Suppose the chain D_n $(n = 1, 2)$ is crooked in the chain E, each end link of E contains a link of D_n, and no proper consolidation of D_n has these properties. Then D_1 follows a pattern in E that D_2 follows in E, and the chain whose links are the sums of corresponding links of D_1 and D_2 is crooked in E.*

Proof. Both D_1 and D_2 follow the pattern in E that F follows in E as given in Theorem 7. Hence, both follow the same pattern. That the chain whose links are the sum of the corresponding links D_1 and D_2 is crooked in E may be proved by induction on the number of links in E.

THEOREM 9. *Suppose the chain $D = [d_1, d_2, \cdots, d_n]$ is crooked in the chain $E = [e_1, e_2, \cdots, e_m]$, d_1 is in e_1, d_n is in e_m, d_r is in e_m, d_s is in e_1 $(r < s)$. Then for each integer t less than s there is a consolidation F of D such that the first link of F contains d_1 and the last link contains d_n, each link of F is a subset of the sum of two adjacent links of E, and any subchain of F whose end links contain d_t and d_n respectively contains $D(r, n)$.*

Proof. By Theorems 7 and 8 there are chains G_1 and G_2 which are consolidations of $D(1, r)$ and $D(r, s)$ respectively such that the first links of G_1 and G_2 are subsets of e_1 and contain d_1 and d_s respectively, the last links of G_1 and G_2 are subsets of e_m and both contain d_r, and the chain $H = [h_1, h_2, \cdots, h_u]$ whose links are the sums of corresponding links of G_1 and G_2 is crooked in E.

Let h_v be the first link of H that contains d_t. By Theorem 4 there is a consolidation K of $H + D(s + 1, n)$ such that the first link of K contains h_v, the last link contains d_n, and each link of K is a subset of the sum of two adjacent links of E. The first $v - 1$ links of F are the links of $G_1(1, v - 1)$, the v-th link of F is the sum of the links of D that lie in the first link of K but not in any link of $G_1(1, v - 1)$, and each other link of F is the sum of links of D that lie in a link of K but do not intersect a link of $G(1, v - 1)$.

3. **Properties of the continuum** M. The chains $D_1, D_2, \cdots$ mentioned in this section refer to those used in defining M.

THEOREM 10. *Each subcontinuum of M is indecomposable.*

Proof. Assume that the subcontinuum M' of M is the sum of two proper subcontinua H and K of M'. There are points P and Q of M' and an integer j such that the distance from P to H is greater than $2/j$ and the distance from Q to K is greater than $2/j$.

Let $D_j(h, k)$ $(h < k)$ and $D_{j+1}(u, v)$ be subchains of D_j and D_{j+1} respectively, such that P and Q belong to end links of each of these subchains. Suppose P is a point of $d(j)_h$. Since M' is a continuum, each link of $D_j(h, k) + D_{j+1}(u, v)$ contains a point of M'; in particular, $d(j)_{h+1}$ contains a point of K but none of H and $d(j)_{k-1}$ contains a point of H but none of K.

As D_{j+1} is crooked in D_j, $D_{j+1}(u, v)$ contains three links $d(j+1)_r$, $d(j+1)_s$, and $d(j+1)_t$ such that $r < s < t$, $d(j+1)_s$ is a subset of $d(j)_{h+1}$, and $d(j+1)_r + d(j+1)_t$ is a subset of $d(j)_{k-1}$. But H is not a continuum since both $d(j+1)_r$ and $d(j+1)_t$ contain points of H but $d(j+1)_s$ does not. Hence, the assumption that a nondegenerate subcontinuum of M is the sum of two proper subcontinua is false.

THEOREM 11. *Suppose M_n $(n = 1, 2)$ is a compact closed set; ϵ_1, ϵ_2, $\cdots$ is a sequence of positive numbers with a finite sum; and $X_{n,1}$, $X_{n,2}$, $\cdots$ is a sequence of well-ordered collections of domains such that for each positive integer i, (1) $X_{n,i}$ covers M_n, (2) each element of $X_{n,i}$ intersects M_n, (3) no element of $X_{n,i}$ has a diameter of more than ϵ_i, and (4) if the j-th element of $X_{n,i+1}$ intersects the k-th element of $X_{n,i}$, then the distance between the j-th element of $X_{m,i+1}$ $(m = 1, 2)$ and the k-th element of $X_{m,i}$ is less than ϵ_i. Then there is a homeomorphism T carrying M_1 into M_2.*

The generality obtained by supposing that $X_{n,i}$ $(n = 1, 2)$ has more than a finite number of elements is only superficial because if some r elements of $X_{1,i}$ cover M_1 and some s elements of $X_{2,i}$ cover M_2, we need consider no more than $r + s$ elements of $X_{1,i}$ and $X_{2,i}$. However, in the proof to follow, we make no reference to the number of elements in $X_{n,i}$ but only use the notion of a $1 - 1$ correspondence between the elements of $X_{1,i}$ and $X_{2,i}$. In the application of this theorem (Theorem 12), $X_{n,i}$ has only a finite number of elements.

Proof. We shall define the image $T(P)$ of a point P of M_1 under the homeomorphism T.

Let $x(n, i)_k$ denote the k-th element of $X_{n,i}$, $y(n, i)_k$ denote the set of all points Q such that the distance from Q to $x(n, i)_k$ is less than $\epsilon_i + 2(\epsilon_{i+1} + \epsilon_{i+2} + \cdots)$, and $Y_{n,i}$ denote the well-ordered collection whose k-th element is $y(n, i)_k$.

Suppose that $x(n, i)_r$ intersects $x(n, i+1)_s$. Since the distance between $x(m, i)_r$ and $x(m, i+1)_s$ is less than ϵ_i and the diameter of $x(m, i+1)_s$ is no more than ϵ_{i+1}, the closure of $y(m, i+1)_s$ is a subset of $y(m, i)_r$. In general, if $i < j$ and $x(n, i)_r$ intersects $x(n, j)_s$, the closure of $y(m, j)_s$ is a subset of $y(m, i)_r$.

If $x(1, 1)_a$, $x(1, 2)_b$, $\cdots$ is a sequence of domains containing P, we define $T(P)$ to be the common part of $y(2, 1)_a$, $y(2, 2)_b$, $\cdots$. Since the closure of $y(2, i+1)_u$ is a subset of $y(2, i)_v$ if $x(1, i+1)_u$ intersects $x(1, i)_v$, the same point $T(P)$ is defined no matter what sequence $x(1, 1)_p$, $x(1, 2)_q$, $\cdots$ of domains containing P is selected.

Suppose $T(P)$ is a point of M_2 and D is a domain containing $T(P)$. There is an integer j such that any element of $Y_{2,j}$ containing $T(P)$ is a subset of D. If $x(1, j)_r$ is an element of $X_{1,j}$ containing P, the image under T of the common part of M_1 and $x(1, j)_r$ is a subset of $y(2, j)_r$. But $y(2, j)_r$ contains $T(P)$ and is a subset of D. Hence, T is a continuous transformation.

We shall show that $T(M_1) = M_2$ by showing that each domain D containing a point P of M_2 contains a point of $T(M_1)$. Let j be an integer such that every element of $Y_{2,j}$ containing P is a subset of D. Suppose $x(2, j)_r$ is an element of $X_{2,j}$ containing P. The image under T of the common part of M_1 and $x(1, j)_r$ is a subset of $y(2, j)_r$ and $y(2, j)_r$ is a subset of D. Hence, D contains a point of $T(M_1)$. Since M_1 is closed and compact, this implies that P is a point of $T(M_1)$.

Finally, we show that T is reversible. Assume it is not reversible. Then there are two points P_1 and P_2 of M_1 such that $T(P_1) = T(P_2)$. There is an integer k such that no element of $Y_{1,k}$ intersects both P_1 and P_2. Let $x(2, k)_r$ be an element of $X_{2,k}$ containing $T(P_1) = T(P_2)$. There is an integer j greater than k such that every element of $Y_{2,j}$ containing $T(P_1) = T(P_2)$ is a subset of $x(2, k)_r$. Let $x(1, j)_u$ and $x(1, j)_v$ be elements of $X_{1,j}$ containing P_1 and P_2 respectively. Since both $y(2, j)_u$ and $y(2, j)_v$ contain $T(P_1) = T(P_2)$, both $x(2, j)_u$ and $x(2, j)_v$ are subsets of $x(2, k)_r$. Then $x(1, j)_u$ and $x(1, j)_v$ are subsets of $y(1, k)_r$. But no element of $Y_{1,k}$ contains P_1 and P_2.

Since the transformation T is a continuous and reversible transformation of a compact closed set into a compact closed set, it is a homeomorphism.

THEOREM 12. *Suppose M_n $(n = 1, 2)$ is a compact continuum; P_n and Q_n are points of M_n; $\epsilon_1, \epsilon_2, \cdots$ is a sequence of positive numbers having zero as a limit; and $D_{n,1}, D_{n,2}, \cdots$ is a sequence of chains from P_n to Q_n such that for each positive integer i, (1) $D_{n,i+1}$ is crooked in $D_{n,i}$, (2) the closure of each link of $D_{n,i+1}$ is a subset of a link of $D_{n,i}$, (3) no link of $D_{n,i}$ has a diameter of more than ϵ_i, and (4) M_n is the common part of the sum of the links of $D_{n,1}$, the sum of the links of $D_{n,2}, \cdots$. Then there is a homeomorphism carrying M_1 into M_2 that carries P_1 and Q_1 into P_2 and Q_2 respectively.*

Proof. We shall select a sequence $X_{n,1}, X_{n,2}, \cdots$ of collections of domains covering M_n and satisfying the conditions of Theorem 11.

Let $X_{1,1}$ be a $D_{1,i}$ such that no link of $X_{1,1}$ has a diameter of more than 1/2. Let $X_{2,1}$ be a chain from P_2 to Q_2 which has the same number of links as $X_{1,1}$ and which is a consolidation of some $D_{2,i}$.

Since $X_{2,1}$ is a consolidation of some $D_{2,i}$, there is an integer k such that $D_{2,k}$ is contained in $X_{2,1}$ and no element of $D_{2,k}$ has a diameter of more than 1/2. Let $X_{2,2}$ be such a $D_{2,k}$. By Theorem 6, there is an integer j and a chain $X_{1,2}$ from P_1 to Q_1 such that $X_{1,2}$ is a consolidation of $D_{1,j}$ and follows a pattern in $X_{1,1}$ that $X_{2,2}$ follows in $X_{2,1}$.

Let $X_{1,3}$ be a $D_{1,i}$ such that $X_{1,3}$ is contained in $X_{1,2}$ and no link of $X_{1,3}$ is of diameter more than 1/4. By Theorem 6, there is an integer j and a chain

$X_{2,3}$ from P_2 to Q_2 such that $X_{2,3}$ is a consolidation of $D_{2,i}$ and follows a pattern in $X_{2,2}$ that $X_{1,3}$ follows in $X_{1,2}$.

This process described in the last two paragraphs may be continued to get a sequence $X_{n,1}$, $X_{n,2}$, $\cdots$ of collections of domains such that for each positive integer i, (1) $X_{n,i}$ covers M_n, (2) each element of $X_{n,i}$ intersects M_n, (3) no element of $X_{n,2i}$ has a diameter of more than $1/2^i$, (4) $X_{n,i+1}$ is a chain from P_n to Q_n that follows a pattern in $X_{n,i}$ that $X_{m,i+1}$ follows in $X_{m,i}$. Then it follows from Theorem 11 that there is a homeomorphism carrying M_1 into M_2. The homeomorphism described in Theorem 11 carries P_1 and Q_1 into P_2 and Q_2 respectively.

THEOREM 13. *The continuum M is homogeneous.*

Proof. We find from Theorem 10 that M is indecomposable. Let R and S be points belonging to different composants of M. Then for each integer j, there is an integer k greater than j such that the subchain of D_k from R to S has a link that intersects the first link of D_j and has a link that intersects the last link of D_j. Assume that this is not true and let $D_i(R, S)$ be the subchain of D_i from R to S. The limiting set of $D^*_{j+1}(R, S)$, $D^*_{j+2}(R, S)$, $\cdots$ is a continuum containing R and S. Since R and S belong to different composants of M, this continuum is M. Hence, there is an integer k greater than j such that $D_k(R, S)$ intersects both end links of D_j.

Using Theorem 4, we find that there is an integer h and a chain E_1 from R to S such that E_1 is a consolidation of D_h and no element of E_1 has a diameter of more than 1. Let j be an integer greater than h such that no element of D_j is of diameter more than 1/2. By Theorem 1, D_j is crooked in E_1. There is an integer k such that the closure of the sum of each pair of intersecting links of D_k is a subset of a link of D_j. By Theorem 4, there is a chain E_2 from R to S such that E_2 is a consolidation of D_k and E_2 is contained in D_j. By Theorem 2, E_2 is crooked in E_1.

Continuing this process, we get a sequence E_1, E_2, $\cdots$ of chains from R to S such that for each integer i, (1) E_{i+1} is crooked in E_i, (2) the closure of each element of E_{i+1} is a subset of an element of E_i, (3) no element of E_i has a diameter of more than $1/i$, and (4) E_i is a consolidation of some D_j.

Suppose that P_1 and P_2 are two points of M. Let Q_n ($n = 1, 2$) be a point belonging to a composant of M not containing P_n. From the above result, we find that there is a sequence $Y_{n,1}$, $Y_{n,2}$, $\cdots$ of chains from P_n to Q_n such that for each positive integer i, (1) $Y_{n,i+1}$ is crooked in $Y_{n,i}$, (2) the closure of each link of $Y_{n,i+1}$ is a subset of a link of $Y_{n,i}$, (3) no link of $Y_{n,i}$ has a diameter of more than $1/i$, and (4) $Y_{n,i}$ is a consolidation of some D_j. Then by Theorem 12, there is a homeomorphism carrying M into itself that carries P_1 into P_2. Therefore M is homogeneous.

THEOREM 14. *Suppose (P_i, Q_i, R_i) $(i = 1, 2)$ is a triple of points of the continuum M such that P_i and R_i belong to the same composant of M and Q_i*

belongs to a different composant. Then there is a homeomorphism T carrying P_1 into P_2, Q_1 into Q_2, R_1 into R_2, and M into itself.

Proof. For convenience we shall denote an element of D_j which contains R by $d(j)_R$ and a subchain of D_j from R to Q by $D_j(R, Q)$. Let W_i be a subcontinuum of M irreducible from P_i to R_i.

Let $X_{i,1}, X_{i,2}, \cdots$ be a sequence of chains from P_i to Q_i such that (1) $X_{i,n}^*$ covers M, (2) $X_{i,n+1}$ is crooked in $X_{i,n}$, (3) each element of $X_{i,n}$ is of diameter less than $1/n$, and (4) $X_{i,n}(P_i, R_i)$ covers W_i. That there exist such sequences follows as a consequence of Theorem 9.

Denote $X_{1,2}$ by $Y_{1,1}$. Let $Y_{2,1}$ be a chain from P_2 to Q_2 such that $Y_{2,1}$ is a consolidation of some $X_{2,j}$, $Y_{2,1}$ has the same number of links as $Y_{1,1}$, $Y_{2,1}(P_2, R_2)$ has the same number of links as $Y_{1,1}(P_1, R_1)$, and $Y_{2,1}^*(P_2, R_2)$ covers W_2. Let $Y_{2,2}$ be some $X_{2,j}$ contained in $Y_{2,1}$ such that each link of $Y_{2,2}$ is of diameter less than $1/2$.

There is an integer j and a point A of M such that $X_{1,j}$ is contained in $Y_{1,1}$, $X_{1,j}(P_1, A)$ is contained in $Y_{1,1}(P_1, R_1)$, the closure of $x(1, j)_A$ is a subset of $y(1, 1)_{P_1}$, $x(1, j)_{R_1}$ is an element of $X_{1,j}(P_1, A)$ and its closure is a subset of $y(1, 1)_{R_1}$, and each element of $X_{1,j}$ is of diameter less than $1/4$. By Theorem 5 there is an integer k and chains E_1, E_2, and E_3 such that E_1 is a consolidation of the links of $X_{1,k}$ in $X_{1,j}(P_1, R_1)$ and follows a pattern in $Y_{1,1}$ that $Y_{2,2}(P_2, R_2)$ follows in $Y_{2,1}$, E_2 is a consolidation of the links of $X_{1,k}$ in $X_{1,j}(A, R_2)$ and follows this same pattern in $Y_{1,1}$, and E_3 is a consolidation of the links of $X_{1,k}$ in $X_{1,j}(A, Q_1)$ and follows a pattern in $Y_{1,1}$ that $Y_{2,2}$ follows in $Y_{2,1}$. Let $Y_{1,2}$ be a chain whose n-th link is the sum of the n-th links of E_1, E_2, and E_3. Then $Y_{1,2}$ follows the same pattern in $Y_{1,1}$ that $Y_{2,2}$ follows in $Y_{2,1}$, R_1 and R_2 belong to corresponding links of $Y_{1,2}$ and $Y_{2,2}$, and W_1 is covered by $Y_{1,2}(P_1, R_1)$.

Similarly, we obtain chains $Y_{1,3}, Y_{2,3}, Y_{2,4}, Y_{1,4}, Y_{1,5}, \cdots$ such that (1) $Y_{i,n+1}$ is a chain from P_i to Q_i that follows a pattern in $Y_{i,n}$ that $Y_{j,n+1}$ follows in $Y_{j,n}$ $(i, j = 1, 2)$, (2) R_1 and R_2 belong to corresponding links of $Y_{1,n}$ and $Y_{2,n}$, and (3) the diameter of each link of $Y_{i,2n}$ is of diameter less than $1/2^n$. It follows from Theorem 11 that there is a homeomorphism T carrying P_1 into P_2, Q_1 into Q_2, R_1 into R_2, and M into itself.

THEOREM 15. *Suppose W_i $(i = 1, 2)$ is a nondegenerate subcontinuum of M, P_i is a point of W_i and Q_i is a point of a composant not containing W_i of M_i. Then there is a homeomorphism carrying M into itself, and W_1, P_1, Q_1 into W_2, P_2, Q_2 respectively.*

Proof. As W_i is indecomposable (Theorem 10), there is a point R_i belonging to a different composant of W_i from P_i. But R_i and P_i belong to the same composant of M. According to Theorem 14, there is a homeomorphism carrying M into itself and P_1, R_1, Q_1 into P_2, R_2, Q_2 respectively. This homeomorphism also carries W_1 into W_2 because for each pair of points A, B of M

there is only one subcontinuum of M irreducible between them: namely, the limiting set of $X_1^*(A, B)$, $X_2^*(A, B)$, $\cdots$ where $X_i(A, B)$ is the subchain of X_i from A to B.

References

1. Gustave Choquet, *Prolongements d'homéomorphies. Ensembles topologiquement nommables. Caractérisation topologique individuelle des ensembles fermés totalement discontinus*, Comptes Rendus Hebdomadaires des Séances de l'Académie des Sciences, vol. 219(1944), pp. 542–544.
2. B. Knaster and C. Kuratowski, Problème 2, Fundamenta Mathematicae, vol. 1(1920), p. 223.
3. Stefan Mazurkiewicz, *Sur les continus homogènes*, Fundamenta Mathematicae, vol. 5 (1924), pp. 137–146.
4. E. E. Moise, *An indecomposable plane continuum which is homeomorphic to each of its nondegenerate subcontinua*, Transactions of the American Mathematical Society, vol. 63 (1948), pp. 581–594.
5. Z. Waraszkiewicz, *Sur les courbes planes topologiquement homogènes*, Comptes Rendus Hebdomadaires des Séances de l'Académie des Sciences, vol. 204(1937), pp. 1388–1390.

The University of Wisconsin.

CONCERNING HEREDITARILY INDECOMPOSABLE CONTINUA

R. H. BING

1. Introduction. A continuum is indecomposable if it is not the sum of two proper subcontinua. It is hereditarily indecomposable if each of its subcontinua is indecomposable. In [3] Knaster gave an example of a hereditarily indecomposable continuum which was not a point. In this paper we study some properties of the Knaster example and describe some other hereditarily indecomposable continua.

2. Chained hereditarily indecomposable continua are homeomorphic. The hereditarily indecomposable continuum given [3] by Knaster was a plane continuum which was described in terms of covering bands. For each positive number ϵ, it could be covered by an ϵ-chain. Moise used [5] a hereditarily indecomposable continuum to exhibit a continuum which was topologically equivalent to each of its nondegenerate subcontinua. He called it a pseudo-arc and noted that it was similar (if not in fact topologically equivalent) to Knaster's example. It could be chained. Bing used [2] such a continuum as an example of a homogeneous plane continuum. Anderson showed [1] that the plane is the sum of a continuous collection of such continua. Theorem 1 reveals that all of these continua are topologically equivalent.

We follow the definitions used in [2]. In particular, we recall the following. A *chain* $D = [d_1, d_2, \cdots, d_n]$ is a collection of open sets $d_1, d_2, \cdots, d_n$ such that d_i intersects d_j if and only if i is equal to $j-1, j$, or $j+1$. If the links are of diameter less than ϵ, the chain is called an ϵ-*chain*. We do not suppose that the links of a chain are necessarily connected.

If the chain $E = [e_1, e_2, \cdots, e_n]$ is a refinement of the chain $D = [d_1, d_2, \cdots, d_m]$, E is called *crooked* in D provided that if $k - h > 2$ and e_i and e_j are links of E in links d_h and d_k of D, respectively, then there are links e_r and e_s of E in links d_{k-1} and d_{h+1}, respectively, such that either $i > r > s > j$ or $i < r < s < j$.

EXAMPLE 1. *The pseudo-arc.* The following description of a chained hereditarily indecomposable continuum appeared in [2] and is much like one given

Received June 1, 1950.

Pacific J. Math. 1 (1951), 43-51.

earlier in [5]. In the plane let D_1, D_2, $\cdots$ be a sequence of chains between the distinct points p and q such that for each positive integer i, (a) D_{i+1} is crooked in D_i, (b) no link of D_i has a diameter of more than $1/i$, and (c) the closure of each link of D_{i+1} is a compact subset of a link of D_i. The common part of the sum of the links of D_1, the sum of the links of D_2, $\cdots$ is a pseudo-arc. That it is hereditarily indecomposable is shown in [2] and [5].

A continuum can be chained if for each positive number ϵ, the continuum can be covered by an ϵ-chain. A *composant* of a continuum W is a set H such that for some point p of W, H is the sum of all proper subcontinua of W containing p. We recall that a nondegenerate indecomposable compact continuum has uncountably many composants and no two of its composants intersect each other. The following result holds in a metric space.

THEOREM 1. *The compact nondegenerate hereditarily indecomposable continua M and M' are homeomorphic if each can be chained.*

In fact, if p and q are points of different composants of M while p' and q' are points of different composants of M', there is a homeomorphism carrying M into M', p into p', and q into q'.

Proof. Since M can be chained, there is a sequence C_1, C_2, $\cdots$ such that C_i is a $1/i$-chain covering M, each element of C_i intersects M, and C_{i+1} is a refinement of C_i.

First, we show that there is an integer n_2 so large that C_{n_2} is crooked in $C_1 = [c_{1,1}, c_{1,2}, \cdots, c_{1,t_1}]$. If this were not true there would be elements $c_{1,h}$ and $c_{1,k}$ of C_1 such that $k - h > 2$ and for infinitely many integers m, $C_m = [c_{m,1}, c_{m,2}, \cdots, c_{m,t_m}]$ would have two links $c_{m,i}$ and $c_{m,j}$ in $c_{1,h}$ and $c_{1,k}$ respectively such that if $c_{m,r}$ is in $c_{1,k-1}$ and between $c_{m,i}$ and $c_{m,j}$, then there is not a link of C_m in $c_{1,h+1}$ which is between $c_{m,r}$ and $c_{m,j}$. Denote by W_m the sum of $c_{m,i}$, $c_{m,r}$, and the elements of C_m between them, where we suppose that no element of C_m in $c_{1,k-1}$ is between $c_{m,i}$ and $c_{m,r}$. Let V_m be the sum of $c_{m,r}$, $c_{m,j}$, and the elements of C_m between them.

Let a_1, a_2, $\cdots$ be an increasing sequence of integers such that both W_{a_1}, W_{a_2}, $\cdots$ and V_{a_1}, V_{a_2}, $\cdots$ converge. But the limiting set W of W_{a_1}, W_{a_2}, $\cdots$ is a continuum which intersects $\bar{c}_{1,h}$ but not $\bar{c}_{1,k}$. Also, the limiting set V of V_{a_1}, V_{a_2}, $\cdots$ is a continuum which intersects $\bar{c}_{1,k}$ but not $\bar{c}_{1,h}$. Hence the assumption that there is no integer n_2 such that C_{n_2} is crooked in C_1 has led to the contradiction that the hereditarily indecomposable continuum M contains the decomposable continuum $W + V$.

Hence, there is a subsequence $C_{n_1}, C_{n_2}, C_{n_3}, \cdots$ of $C_1, C_2, C_3, \cdots$ such that $C_{n_{i+1}}$ is crooked in C_{n_i}.

Let p and q be points belonging to different composants of M. Then for each integer j, there is an integer k greater than j such that the subchain of C_{n_k} from p to q has a link that intersects the first link of C_{n_j} and has a link that intersects the last link of C_{n_j}. To see that this is so, let W_i be the sum of the links of the subchain of C_{n_i} from p to q. Since the limiting set of each subsequence $W_1, W_2, \cdots$ is a continuum in M containing $p + q$, each such limiting set is M. Hence, some $W_k (k > j)$ intersects both the first and last links of C_{n_k}, and the subchain of C_{n_k} corresponding to W_k has links intersecting the first and last links of C_{n_j}.

We find from Theorem 4 of [2] that there is a chain E_j such that the first link of E_j contains p, the last contains q, E_j is a consolidation of C_{n_j}, while each link of E_j lies in the sum of two adjacent links of C_{n_j} and is therefore of diameter less than $2/j$. Hence, there is no loss of generality in supposing that each of the chains $C_1, C_2, \cdots$ is from p to q. Therefore, there is a sequence $D_1, D_2, \cdots$ of chains from p to q such that for each positive integer i, (a) D_{i+1} is crooked in D_i, (b) the closure of each link of D_{i+1} is a subset of a link of D_i, (c) no link of D_i has a diameter of more than $1/i$, and (d) M is the common part of the sum of the links of D_1, the sum of the links of $D_2, \cdots$.

Similarly, we find that if p' and q' are points of different composants of M', there is a sequence $D_1', D_2', \cdots$ of chains from p' to q' such that (a) D_{i+1}' is crooked in D_i', (b) the closure of each element of D_{i+1}' is a subset of an element of D_i', (c) no element of D_i' has a diameter of more than $1/i$, and (d) D_i covers M'. Theorem 12 of [2] shows that there is a homeomorphism that carries M into M', p into p', and q into q'.

The preceding theorem shows that M is homogeneous and homeomorphic with each of its nondegenerate subcontinua. It also reveals that the continua studied by Knaster [3], Moise [5], Bing [2], and Anderson [1] are all topologically equivalent and are pseudo-arcs.

QUESTION. It would be interesting to know if each nondegenerate bounded hereditarily indecomposable plane continuum which does not separate the plane is homeomorphic to M. This question would be answered in the affirmative if it were shown that each bounded atriodic plane continuum which does not separate the plane can be chained (see Section 6, below).

3. Most continua are pseudo-arcs. Mazurkiewicz showed [4] that the continua contained in a circle plus its interior which were not hereditarily indecomposable were of the first category. We go even further to show that those which are not pseudo-arcs are of the first category.

The *space of compact continua* in the metric space S is the metric space $C(S)$ whose points are the compact continua of S, where the distance between two elements g_1, g_2 of $C(S)$ is the Hausdorff distance between them in S, namely L.U.B. (distance from x to g_i in S), $x \in g_1 + g_2$; $i = 1,2$. By saying that *most compact continua of S have a certain property*, we mean that there is a dense inner limiting (G_δ) subset W of $C(S)$ such that each element of W has the property (when regarded as a continuum in S). The collection of continua of S with this property is said to be of the second category.

The following theorem holds in a space S which is either a Hilbert space or a Euclidean n-space ($n > 1$).

THEOREM 2. *Most compact continua are pseudo-arcs.*

Proof. Let F_i be the collection of all compact continua f in S such that f can not be covered by a $1/i$-chain. If $f_1, f_2, \cdots$ is a sequence of elements of F_i converging to a compact continuum f_0, f_0 is an element of F_i because if a $1/i$-chain covers f_0, it covers some f_j. Hence, F_i is closed in $C(S)$.

Let G_i be the set of all compact continua K such that K contains a subcontinuum K' which is the sum of two continua K_1 and K_2 such that K_1 contains a point at a distance (in S) of $1/i$ or more from K_2, while K_2 contains a point at a distance of $1/i$ or more from K_1. Then G_i is a closed subset of $C(S)$. Furthermore, the collection of points of S is closed in $C(S)$.

The collection of all pseudo-arcs is dense in $C(S)$. For suppose that g is an element of $C(S)$ and ϵ is a positive number. There is a broken line ab whose distance (in $C(S)$) from g is less than $\epsilon/2$. Let D be an $\epsilon/2$-chain from a to b covering ab such that each element of D is the interior of a sphere. There is a pseudo-arc h containing $a + b$ which is covered by D. The distance from g to h is less than ϵ in $C(S)$.

Each element of $C(S)$ not belonging to ΣF_i can be chained and each element of $C(S)$ not belonging to ΣG_i is hereditarily indecomposable. Let W be the set of all elements g of $C(S)$ such that g is not a point of S, g is not an element of any F_i, and g is not an element of any G_i. By Theorem 1, each element of W is topologically equivalent to M. However, W is a dense inner limiting subset of $C(S)$.

4. Decomposition of a pseudo-arc. Like a simple closed curve, the pseudo-arc is homogeneous; and like an arc it is homeomorphic with each of its nondegenerate subcontinua. We now show that like both a simple closed curve and an arc, it is topologically equivalent to itself under a nontrivial monotone decomposition.

THEOREM 3. *If an upper semi-continuous collection of continua fills a chained compact continuum, the resulting decomposition space is chained.*

Proof. Suppose G is an upper semi-continuous collection of continua filling the chained continuum M and that the resulting decomposition space M' is given a metric. We show that for each positive number ϵ, an ϵ-chain in M' covers M'.

Let δ be a positive number so small that if two elements of G are no farther apart in M than δ, then the points in M' corresponding to them are no farther apart than $\epsilon/5$ in M'. Let $[c(1), c(2), \cdots, c(n)]$ be a δ-chain covering M. Let $n_1 = 1$, $n_2, n_3, \cdots, n_j = n$ be a monotone increasing sequence of integers such that an element of G intersects $c(n_i)$ and $c(n_{i+1})$, but none intersects $c(n_i)$ and $c(n_{i+1}+1)$. Denote by $D(i,j)$ the open subset of M' consisting of those points corresponding to elements of G which are covered by $c(i) + c(i+1) + \cdots + c(j)$. Then

$$D(n_1, n_5),\quad D(n_4, n_8),\quad D(n_7, n_{11}), \cdots,\quad D(n_{3k+1}, n_j),$$

$$j - 5 \leq 3k + 1 < j - 3$$

is an ϵ-chain covering M'.

The following result follows from Theorems 1 and 3.

THEOREM 4. *If M is a pseudo-arc and G is an upper semi-continuous collection of proper subcontinua of M filling M, the resulting decomposition space is topologically equivalent to M.*

5. Other types of hereditarily indecomposable plane continua. The pseudo-arc can be imbedded in the plane. We now show that there are nondegenerate hereditarily indecomposable plane continua which are not topologically equivalent to the pseudo-arc. In fact, there are as many topologically different hereditarily indecomposable plane continua as there are plane continua.

A method which differs from the one below of showing that there are topologically different nondegenerate plane continua is to consider in 3-space the intersection of a plane with a hereditarily indecomposable continuum which separates 3-space. That there are such continua in 3-space will be shown in my paper, *Higher dimensional hereditarily indecomposable continua,* which proves that there are hereditarily indecomposable continua of all dimensions.

EXAMPLE 2. *A hereditarily indecomposable continuum that separates the plane.* A circular chain differs from a chain in that its first and last links intersects each other. In the plane let $D_1, D_2, \cdots$ be a sequence of circular chains such that (a) each element of D_i is the interior of a circle of diameter less than $1/i$, (b) the closure of each element of D_{i+1} lies in an element of D_i, (c) the sum of the elements of D_i is topologically equivalent to the interior of an annulus ring, (d) each complementary domain of the sum of the elements of D_{i+1} contains a complementary domain of the sum of the elements of D_i, and (e) if E_i is a proper subchain of D_i and E_{i+1} is a subchain of D_{i+1} contained in E_i, then E_{i+1} is crooked in E_i. Condition (c) is superfluous since it follows from condition (a). The required continuum M is the intersection of the sum of the elements of D_1, the sum of the elements of $D_2, \cdots$.

Suppose that a chain $D_i = [d_1, d_2, \cdots, d_n]$ satisfying the preceding conditions has been found. To see that there is a chain D_{i+1} satisfying the required conditions, it might be convenient first to consider a chain $D_i' = [d_1', d_2', \cdots, d_{3n}']$ in D_i which follows the following pattern: $d_1', d_{n+1}', d_{2n+1}'$ are subsets of d_1; d_2', d_{n+2}', d_{3n}' are subsets of d_2; $d_3', d_{n+3}', d_{3n-1}'$ are subsets of d_3; $\cdots$; d_n', d_{2n}', d_{2n+1}' are subsets of d_n. Roughly speaking, D_i' goes through D_i twice in one direction and once in the opposite direction. Then the circular chain D_{i+1} satisfying the required conditions is the sum of two chains one of which is crooked in $[d_1', d_2', \cdots, d_{2n+1}']$ and the other of which is crooked in

$$[d'_{2n+1}, d'_{2n+2}, \cdots, d'_{3n}, d'_i].$$

That M separates the plane follows from conditions (c) and (d). We show that it is hereditarily indecomposable by showing that each of its proper subcontinua is indecomposable.

Suppose M' is a proper subcontinuum of M. Let m be an integer so large that some element of D_m does not intersect M'. If $E_k (k \geq m)$ is the collection of elements of D_k which intersect M', then $E_m, E_{m+1}, \cdots$ is a sequence of chains such that E_{k+1} is crooked in E_k. If M' were the sum of two proper subcontinua H and K, there would be a point p of H; a point q of K, and an integer w such that the distances from p to K and from q to H are each greater than $2/w$. Suppose e_i and e_j are elements of E_{w+1} containing p and q respectively. Since E_{w+1} is crooked in E_w, there are elements e_r and e_s in E_{w+1} such that e_r separates e_i from e_s in E_{w+1}, each point of e_s is nearer than $2/w$ to p, and each point of e_r is nearer to q than $2/w$. Then H would not be connected because it has a point in e_s, a point in

e_i, but none in e_r.

QUESTIONS. Theorem 1 shows that if M_1 and M_2 are two nondegenerate chained hereditarily indecomposable continua, they are topologically equivalent and each is homogeneous. Suppose W_1 and W_2 are two continua, each defined as described in Example 2. It would be interesting to know if W_1 would necessarily be homeomorphic to W_2. Also, is any such continuum homogeneous?

EXAMPLE 3. *Another hereditarily indecomposable continuum.* The following example of a hereditarily indecomposable plane continuum is somewhat of a combination of Examples 1 and 2.

If W is a nondegenerate indecomposable plane continuum, it contains a nondegenerate subcontinuum H such that no point of H is accessible from the complement of W. To see that this is true, consider two parallel lines L_1 and L_2, each of which separates W. Let G be an uncountable collection of mutually exclusive subcontinua of W each irreducible from L_1 to L_2, and let K be the sum of L_1, L_2, and the closure of the sum of the elements of G. If D is a complementary domain of K between L_1 and L_2, its closure does not intersect three elements of G. Hence, some element of G is not accessible from any complementary domain of K between L_1 and L_2. If H is a subcontinuum of this element of G which does not intersect $L_1 + L_2$, no point of H is accessible from the complement of W.

If W_0 is a nondegenerate hereditarily indecomposable plane continuum, there is a point p of W_0 such that if W' is a nondegenerate subcontinuum of W_0 containing p, then p is not accessible from the complement of W'. To find such a point, let $W_1, W_2, \cdots$ be a sequence of continua such that W_i is a subcontinuum of W_{i-1}, no point of W_{i+1} is accessible from the complement of W_i, and W_i is of diameter less than $1/i$. Then $W_0 \cdot W_1 \cdot W_2 \cdot \cdots$ is a point p; and if W' is a nondegenerate subcontinuum of W_0 containing p, it contains one of the W_i's. Hence, p is not accessible from the complement of W'.

Let M_1 be a pseudo-arc in the plane and M_2 be a hereditarily indecomposable plane continuum as described in Example 2. Let p be a point of M_1 such that p is not accessible from the complement of any nondegenerate subcontinuum of M_1. By a theorem of R. L. Moore, there is a continuous transformation T of the plane into itself such that $T^{-1}(p) = (M_2$ plus its interior) and the inverse of each other point is a point. We show that $M_3 = (T^{-1}(M_1)$ minus the interior of $M_2)$ is hereditarily indecomposable. That M_2 and M_3 are not homeomorphic follows from the fact that M_2 is irreducible with respect to separating the plane but M_3 is not.

If W is a subcontinuum of M_3 intersecting M_2 and containing a point of $M_3 - M_2$, it contains M_2. For suppose that it does not contain some point of M_2. Then there is an arc α from the exterior of M_2 to M_2 that does not intersect W. But $T(\alpha)$ would be an arc, revealing that p is accessible from the complement of $T(W)$. This is contrary to the definition of p.

Suppose that some subcontinuum M' of M_3 is the sum of two proper subcontinua H and K. Since M_2 is hereditarily indecomposable, M' is not a subset of M_2. Since M_1 is hereditarily indecomposable, we may suppose that $T(H) = T(M')$. But H would equal M' because T is one-one on the exterior of M_2 and H contains M_2 if it intersects it.

A variation of the method used in obtaining Example 3 may be used to get other topologically different hereditarily indecomposable plane continua. Instead of replacing a point of M_1 by a continuum homeomorphic to M_2, we can replace each of several points of M_1 by such a continuum.

THEOREM 5. *There are as many topologically different hereditarily indecomposable bounded plane continua as there are real numbers.*

Proof. Suppose $n_1, n_2, \cdots$ is a monotone increasing sequence of positive integers. The collection of such sequences has the power of the continuum. For each such sequence we describe a hereditarily indecomposable plane continuum such that no two of these continua are topologically equivalent.

The hereditarily indecomposable continuum associated with $n_1, n_2, \cdots$ will have one composant containing exactly n_1 continua each topologically equivalent to M_2 of Example 2, another composant containing exactly n_2 continua each topologically equivalent to M_2, $\cdots$; no other composant contains a continuum topologically equivalent to M_2.

Let M_1 be a pseudo-arc in the plane. Suppose

$$p_{1,1}, p_{1,2}, \cdots, p_{1,n_1}, p_{2,1}, \cdots$$

is a converging sequence of different points of M_1 such that $p_{i,j}$ is not accessible from the complement of any nondegenerate subcontinuum of M_1 containing it, and $p_{i,j}$ belongs to the composant containing $p_{r,s}$ if and only if $i = r$.

Suppose $M_{1,1}, M_{1,2}, \cdots, M_{1,n_1}, M_{2,1}, \cdots$ is a sequence of mutually exclusive continua in the plane all topologically equivalent to M_2 of Example 2 such that the sequence converges to a point and each of the continua lies in the exterior of each of the others. There is a continuous transformation T of the plane into itself such that $T^{-1}(p_{i,j})$ is ($M_{i,j}$ plus its interior), and $T^{-1}(q)$ is a point if q is not a $p_{i,j}$. The argument used in Example 3 shows that $M_3 = (T^{-1}(M_1)$ minus the sum of the interiors of the $M_{i,j}$'s) is a hereditarily indecomposable continuum. Furthermore, one composant of M_3 contains exactly n_1 mutually exclusive subcontinua, each topologically equivalent to M_2, another composant contains exactly n_2 such subcontinua, $\cdots$, while the other composants of M_3 contain no such continua.

6. Added in proof. R. D. Anderson answered the question at the end of Section 2 in the negative by announcing at the February, 1951, meeting of the American Mathematical Society in New York that there are nondegenerate bounded hereditarily indecomposable plane continua other than pseudo-arcs which do not separate the plane.

References

1. R. D. Anderson, *On collections of pseudo-arcs*, Abstract 337t, Bull. Amer. Math. Soc. 56 (1950), 350.

2. R. H. Bing, *A homogeneous indecomposable plane continuum*, Duke Math. J. 15 (1948), 729-742.

3. B. Knaster, *Un continu dont tout sous-continu est indécomposable*, Fund. Math. 3 (1922), 247-286.

4. S. Mazurkiewicz, *Sur les continus absolument indécomposable*, Fund. Math. 16 (1930), 151-159.

5. E. E. Moise, *An indecomposable plane continuum which is homeomorphic to each of its nondegenerate subcontinua*, Trans. Amer. Math. Soc. 63 (1948), 581-594.

UNIVERSITY OF VIRGINIA AND UNIVERSITY OF WISCONSIN

HIGHER-DIMENSIONAL HEREDITARILY INDECOMPOSABLE CONTINUA

BY

R. H. BING

1. **Introduction and definitions.** A continuum is *indecomposable* if it is not the sum of two proper subcontinua. It is *hereditarily indecomposable* if each of its subcontinua is indecomposable. Knaster gave [5](1) an example of a 1-dimensional hereditarily indecomposable continuum. Moise described [6] a plane hereditarily indecomposable continuum (it was topologically equivalent [2] to the one described by Knaster) which was topologically equivalent to each of its nondegenerate subcontinua. I showed [1] that the same continuum was homogeneous. Recently, I gave [2] examples of many topologically different hereditarily indecomposable continua. However, each of these continua was 1-dimensional. A point is a 0-dimensional hereditarily indecomposable continuum. (In this paper we define a continuum to be a compact closed connected set.)

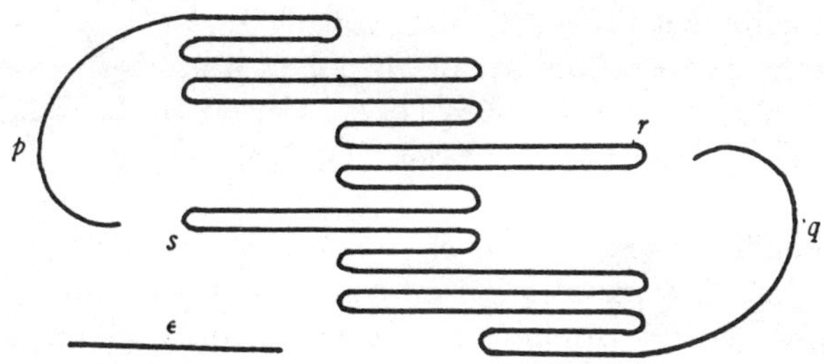

FIG. 1

Kelly has shown [4] that if there is a hereditarily indecomposable continuum of dimension more than one, there is one of infinite dimension. In this paper we show that there are hereditarily indecomposable continua of all dimensions. In describing these continua, we make use of the following term.

ε-crooked. An arc is *ε-crooked* if for each pair of its points p and q there are points r and s between p and q on the arc such that r lies between p and s, $\rho(p, s)<\epsilon$, and $\rho(r, q)<\epsilon$ where $\rho(x, y)$ denotes the distance between x and y. See Fig. 1. It may be noted that if $\epsilon'>\epsilon$, an arc is ϵ'-crooked if it is ε-crooked. An arc intersecting sets H and K is called ε-crooked with respect to them if each of its subarcs pq with ends p and q on H and K respectively contains

Presented to the International Congress of Mathematicians, September 6, 1950; received by the editors February 21, 1951.

(1) Numbers in brackets refer to the references cited at the end of the paper.

points r and s such that r lies between p and s on pq, $\rho(s, H) < \epsilon$, and $\rho(r, K) < \epsilon$.

We use E_n to denote an *n-dimensional Euclidean space* and I_ω to denote a *Hilbert cube*. The points of a Hilbert cube are countable sequences $(x_1, x_2, \cdots)$ such that $|x_i| \leqq 1/i$ and the distance between two points is given by $\rho(x, y) = \sum |x_i - y_i|/2^i$. We make use of the facts that I_ω is compact and any separable metrizable space can be topologically imbedded in I_ω [3, p. 64]. Throughout this paper we restrict ourselves to metric spaces.

2. **Description of a 2-dimensional hereditarily indecomposable continuum.** We shall show that if b and c are two points of E_3, there is a sequence $S_1, S_2, \cdots$ of bounded connected domains satisfying the following conditions:

(1) S_i contains $\bar{S}_{i+1}$.

(2) S_i separates b from c.

(3) $E_3 - S_i$ has only two components and no point of S_i is farther than $1/i$ from either of them.

(4) Each arc in S_i is $1/i$-crooked.

The intersection S_0 of $S_1, S_2, \cdots$ is a 2-dimensional hereditarily indecomposable continuum which has exactly two complementary domains and which is irreducible with respect to separating E_3.

We may take S_1 to be the set of all points p of E_3 such that minimum $(1/4, \rho(b, c)/3) < \rho(b, p) <$ minimum $(1/2, 2\rho(b, c)/3)$. That $S_2, S_3, \cdots$ may be chosen so as to satisfy conditions (1), (2), (3), and (4) follows from Theorem 2.

Since, by condition (1), $S_0 = \bar{S}_1 \cdot \bar{S}_2 \cdot \cdots$, it is a continuum. We find from condition (2) that S_0 separates b from c and from condition (3) that S_0 has only two complementary domains and is irreducible with respect to separating E_3. That S_0 is 2-dimensional follows from the fact that it is irreducible with respect to separating E_3.

Condition (4) insures that S_0 is hereditarily indecomposable. To see that this is true, assume that H and K are two intersecting continua in S_0 such that neither H nor K contains the other. There would be a positive integer n and points p and q of H and K respectively such that $\rho(p, K) > 1/n$ and $\rho(q, H) > 1/n$. Let O_H and O_K be connected open subsets of S_n containing H and K respectively such that $\rho(p, O_K) > 1/n$ and $\rho(q, O_H) > 1/n$. Let pxq be an arc from p to q in $O_H + O_K$ such that O_H and O_K contain px and xq respectively. Since $\rho(p, xq) > 1/n$ and $\rho(px, q) > 1/n$, pxq is not $1/n$-crooked. The assumption that S_0 is not hereditarily indecomposable has led to the contradiction that an arc in S_n is not $1/n$-crooked.

3. **Crooked domains.** The following two theorems hold in either E_n or in I_ω.

THEOREM 1. *Suppose D is a bounded domain that separates the point b from the point c, H and K are two point sets, and ϵ is a positive number. Then there* *is a domain E in D that separates b from c such that each arc in E from a point of H to a point of K is ε-crooked with respect to H and K.*

Proof. Let $K'=\{x|x \text{ element of } K,\ \rho(x, H)\geqq\epsilon\}$ where the symbol $\{x|\mathrm{P}\}$ is used to denote the collection of all points x such that x satisfies condition P. We suppose that neither K' nor H is empty because if either is, we can let E be D.

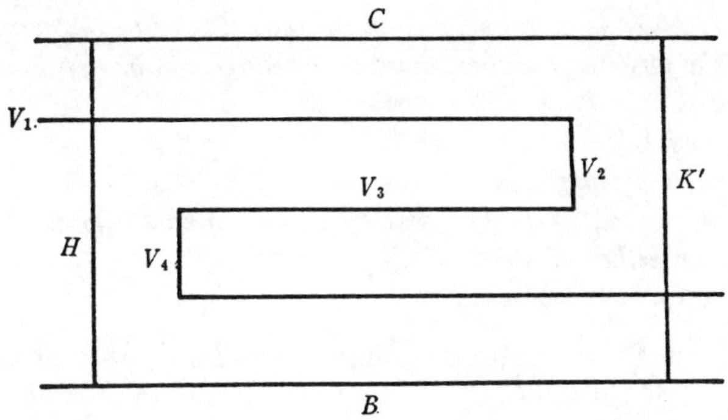

FIG. 2

Suppose the complement of D is the sum of the mutually exclusive closed sets B and C which contain b and c respectively, and that δ is a positive number less than the distance from any point of B to any point of C. We define V to be the sum of the following five sets $V_1, V_2, \cdots, V_5$.

$$V_1 = \{x \mid \rho(x, B) = 3\delta/4,\ \rho(x, K') \geqq \epsilon/2\}.$$
$$V_2 = \{x \mid \delta/2 \leqq \rho(x, B) \leqq 3\delta/4,\ \rho(x, K') = \epsilon/2\}.$$
$$V_3 = \{x \mid \rho(x, B) = \delta/2,\ \rho(x, H + K') \geqq \epsilon/2\}.$$
$$V_4 = \{x \mid \delta/4 \leqq \rho(x, B) \leqq \delta/2,\ \rho(x, H) = \epsilon/2\}.$$
$$V_5 = \{x \mid \rho(x, B) = \delta/4,\ \rho(x, H) \geqq \epsilon/2\}.$$

We find that V separates B from C (see Fig. 2) because it contains the boundary of the closed set

$$\{x \mid \rho(x, B) \leqq \delta/4\} + \{x \mid \delta/4 \leqq \rho(x, B) \leqq \delta/2,\ \rho(x, H) \leqq \epsilon/2\} + \{x \mid \delta/2 \leqq \rho(x, B) \leqq 3\delta/4,\ \rho(x, K') \geqq \epsilon/2\}$$

which contains B but no point of C.

Let $E, E_1, E_2, \cdots, E_5$ be γ neighborhoods of $V, V_1, V_2, \cdots, V_5$ respectively where γ is the minimum of $\epsilon/4$ and $\delta/16$. We note that $\bar{E}$ lies in D. Also, E separates b from c because it contains V. No point of E_4 is as far as ϵ from H and no point of E_2 is as far as ϵ from K.

We shall show that if pq is an arc in E from a point p of H to a point q of

K, there are points r and s on pq between p and q such that r is between p and s, $\rho(s, H) < \epsilon$, and $\rho(r, K) < \epsilon$. If q is nearer than ϵ to H, r and s may be found near q. If q is a point of K', q belongs to E_5 and p belongs to E_1. In Fig. 2, E would be a small neighborhood of V. Since E_2 separates $E \cdot H$ from E_4 and E_4 separates E_2 from $E \cdot K'$, there are points r and s of pq belonging to E_2 and E_4 respectively such that r is between p and s on pq. Hence, any arc in E from a point of H to a point of K is ϵ-crooked with respect to H and K.

THEOREM 2. *If D is a bounded domain that separates the point b from the point c and ϵ is a positive number, there is a connected domain E satisfying the following conditions:*

(1) *D contains $\overline{E}$.*

(2) *E separates b from c.*

(3) *The complement of E has exactly 2 components and no point of E is farther than ϵ from either of them.*

(4) *Each arc in E is ϵ-crooked.*

Proof. Suppose G is a finite covering of $D = D_0$ by sets of diameter less than $\epsilon/2$. Let $(g_1, g_1'), (g_2, g_2'), \cdots, (g_n, g_n')$ be the finite collection of pairs of elements of G. By repeated applications of Theorem 1 we find that there is a domain D_i $(i = 1, 2, \cdots, n)$ such that $\overline{D}_i$ lies in D_{i-1}, D_i separates b from c, each arc in D_i from g_i to g_i' is $\epsilon/2$-crooked with respect to g_i and g_i'. The domain D_n satisfies conditions (1), (2), and (4). Since E_n and I_ω are unicoherent, one component of D_n separates b from c. By subtracting the sum of a collection of arcs from this component we may obtain a connected domain E satisfying conditions (1), (2), (3), and (4).

4. **Higher-dimensional hereditarily indecomposable continua.** Using the same scheme used in §2 to define S_0 in E_3, we can define hereditarily indecomposable continua in E_n and in I_ω. The following result holds in either E_n or I_ω.

THEOREM 3. *If H and K are two mutually exclusive continua, there is a hereditarily indecomposable continuum which has exactly two complementary domains and which is irreducible with respect to separating H from K.*

Using Theorem 3, we can get the following result which shows that there are hereditarily indecomposable continua of all dimensions.

THEOREM 4. *There are infinite-dimensional hereditarily indecomposable continua in I_ω and there are n-dimensional hereditarily indecomposable continua in E_{n+1}.*

THEOREM 5. *Each $n+1$-dimensional continuum contains an n-dimensional hereditarily indecomposable continuum.*

Proof. If M is an $n+1$-dimensional continuum, it contains two mutually exclusive closed subsets B and C such that no closed subset of M of dimension less than n separates B from C [3, p. 26]. We find from Theorem 6 of the next section that there is a closed subset H of M which separates B from C such that each component of H is hereditarily indecomposable. One component C of H is not of dimension less than n [3, p. 94]. There is an n-dimensional subcontinuum of C and this subcontinuum is hereditarily indecomposable because C is.

5. Separating with collections of hereditarily indecomposable continua. In this section we consider separations by closed point sets such that the components of these closed point sets are hereditarily indecomposable.

THEOREM 6. *If B and C are two mutually exclusive closed subsets of a continuum M, there is a closed point set H in M separating B from C in M such that each component of H is hereditarily indecomposable.*

Proof. There is a continuous transformation T of M into I_ω such that $T(B)$ is a point, $T(C)$ is another point, and T is 1-1 on $M-(B+C)$. We find from Theorem 3 that there is a hereditarily indecomposable continuum K in I_ω which separates $T(B)$ from $T(C)$. Then $H=T^{-1}[K\cdot T(M)]$.

The following result can be obtained in a similar fashion.

THEOREM 7. *If B and C are two mutually exclusive bounded closed subsets of E_n, there is a closed bounded point set H separating B from C such that each component of H is hereditarily indecomposable.*

THEOREM 8. *If p and q are two points of E_n, there is a bounded closed point set H in E_n separating p from q such that each continuum in E_n containing $p+q$ contains a component of H.*

Before proving Theorem 8 we state the following interesting corollary of it.

COROLLARY. *For each pair of points p, q of E_n there is a bounded closed set H in $E_n-(p+q)$ such that each arc from p to q in E_n contains a degenerate component of H.*

Proof of Theorem 8. Suppose E_n is a plane in E_{n+1}. Then H is the intersection of E_n with a hereditarily indecomposable continuum K in E_{n+1} which separates p from q in E_{n+1}.

We shall show that the assumption that there is a continuum C in E_n which contains $p+q$ but no component of H leads to the contradiction that K contains a decomposable continuum.

Suppose there is such a continuum C in E_n. We now show that there is a unicoherent locally connected continuum M in E_{n+1} such that $M\cdot E_n=C$. Let P be a compact locally connected continuum in E_n containing C and r be a point whose distance from E_n is greater than the diameter of P. We define $M=\{x\mid x$ lies on closed interval between r and a point $p(x)$ of P, $\rho[x, p(x)]$

$\geqq \rho[p(x), C]\}$.

If M' is the cone with vertex r and base P, we may think of M as the image of M' under a homeomorphism that moves toward r those points of the cone M' which are not between r and C. Since M' is a cone whose base is a locally connected continuum, M' is locally connected and unicoherent. Hence, M is also.

Since M is a locally connected continuum and $M \cdot K$ separates p from q in M, some closed subset S_1 of $M \cdot K$ is irreducible with respect to separating p from q. Since M is unicoherent, S_1 is connected. In order for S_1 to separate p from q in M it must intersect C but must not be a subset of E_n. Let s be a point of $S_1 \cdot E_n$ and S_2 be the component of H that contains s. Now S_2 does not lie in S_1 because C contains no component of H and S_1 does not lie in S_2 because it is not a subset of E_n. Hence the assumption that there is a continuum C in E_n which contains $p+q$ but no component of H has led to the contradiction that the hereditarily indecomposable continuum K contains the decomposable continuum S_1+S_2.

THEOREM 9. *For each pair of points p, q of a continuum M there is a closed set H in $M-(p+q)$ such that each subcontinuum of M containing $p+q$ contains a component of H.*

Proof. The proof is similar to that used in Theorem 8. We suppose M lies in a hyperplane in I_ω and let H be the intersection of M with a hereditarily indecomposable continuum in I_ω that separates p from q.

6. **Homogeneity.** A set M is homogeneous if for each pair of its points p, q there is a homeomorphism of M into itself that carries p into q. I used [1] a pseudo-arc [6] (a nondegenerate hereditarily indecomposable continuum that can be chained [2]) to show that the simple closed curve is not the only type of nondegenerate homogeneous plane continuum. However, we find from the following result that if n is an integer greater than one, no n-dimensional hereditarily indecomposable continuum is homogeneous.

THEOREM 10. *If M is an n-dimensional hereditarily indecomposable continuum, there is a point p such that each nondegenerate subcontinuum of M containing p is n-dimensional.*

Proof. Suppose M is the sum of a finite number of closed sets each of diameter less than 1. One of the closed sets had dimension n and one of its components has dimension n. Hence, M has an n-dimensional subcontinuum M_1 of diameter less than 1. Similarly, M_1 has an n-dimensional subcontinuum M_2 of diameter less than 1/2, M_2 has an n-dimensional subcontinuum M_3 of diameter less than 1/3, $\cdots$. If $p = M_1 \cdot M_2 \cdot \cdots$, each nondegenerate subcontinuum C of M containing p is n-dimensional because it contains one of the M_i's. If M_i does not contain C, C contains M_i because $C+M_i$ is hereditarily indecomposable.

References

1. R. H. Bing, *A homomogeneous indecomposable plane continuum*, Duke Math. J. vol. 15 (1948) pp. 729–742.

2. ———, *Concerning hereditarily indecomposable continua*, Pacific Coast Mathematical Journal vol. 1 (1951) pp. 43–51.

3. W. Hurewicz and H. Wallman, *Dimension theory*, Princeton University Press, 1948.

4. J. L. Kelly, *The hyperspaces of a continuum*, Trans. Amer. Math. Soc. vol. 52 (1942) pp. 22–36.

5. B. Knaster, *Un continu dont tout sous-continu est indécomposable*, Fund. Math. vol. 3 (1922) pp. 247–286.

6. E. E. Moise, *An indecomposable plane continuum which is homeomorphic to each of its nondegenerate subcontinua*, Trans. Amer. Math. Soc. vol. 63 (1948) pp. 581–594.

University of Wisconsin,
Madison, Wis.

SNAKE-LIKE CONTINUA

By R. H. Bing

1. **Introduction.** A *chain* is a finite collection of open sets $d_1, d_2, \cdots, d_j$ such that d_i intersects d_j if and only if $i = j - 1, j$, or $j + 1$. We do not impose the condition that the d_i's are connected. If the links of a chain are of diameter less than ϵ, the chain is called an *ϵ-chain.* A compact continuum is called *snake-like* if for each positive number ϵ it can be covered by an ϵ-chain. The term snake-like was suggested to me by Gustav Choquet.

An arc is snake-like while a triod is not. Also, a pseudo-arc [7], [2], [3] is snake-like. If A is the closure of the graph of $y = \sin 1/x$ $(0 < x \leq 1)$, B is the closure of the graph of $y = \sin 1/x$ $(-1 \leq x < 0)$, C is the graph of $y = 1$ $(0 \leq x \leq 1)$, and D is the graph of $y = 0$ $(0 \leq x \leq 1)$, then each of the sets $A, B, C, D, A + B$, and $B + C$ is snake-like. However, $B + D$ is not snake-like because it contains a triod.

A continuum M is called a triod [11] if it contains a subcontinuum N such that $M - N$ is the sum of three mutually separated sets. A continuum which contains no triod is atriodic [8]. Each snake-like continuum is atriodic.

A *circular chain* has more than two links and the first and last links intersect. A finite coherent collection G of open sets is called a *tree chain* if no three of the open sets have a point in common and no subcollection of G is a circular chain. (A collection G is *coherent* if for each proper subcollection G' of it, an element of G' intersects an element of $G - G'$.) A compact continuum is called *tree-like* if for each positive number ϵ there is a tree chain covering it such that each element of the tree chain is of diameter less than ϵ.

A snake-like continuum has a type of linearity that facilitates its study. For example, it has been shown [5] that a snake-like continuum has the fixed point property; that is, if T is a continuous transformation of a snake-like continuum into itself, some point is left fixed. However, it is not known whether or not each bounded plane continuum which does not separate the plane has this property. In fact, it would be of interest to know if each tree-like plane continuum has the fixed point property.

The following three results were proved in [3].

Theorem 1. *If M and N are two nondegenerate snake-like hereditarily indecomposable continua, they are homeomorphic.*

A continuum is indecomposable if it is not the sum of two proper subcontinua. A continuum each of whose subcontinua is indecomposable is called *hereditarily indecomposable.* A continuum homeomorphic with M is called a *pseudo-arc.*

Received April 20, 1951.

THEOREM 2. *In a Euclidean space or Hilbert space "most" compact continua are snake-like.*

By the above statement we mean that if C is the metric space whose points are the compact continua in the Euclidean or Hilbert space such that the distance between c_1 and c_2 is the Hausdorff distance $[\rho(c_1, c_2) =$ least upper bound $\rho(x, c_i)$ $(x \varepsilon c_1 + c_2, i = 1, 2)]$ between them, then the collection of points in C corresponding to snake-like continua contains a dense G_δ subset of C.

THEOREM 3. *Being snake-like is preserved under a monotone decomposition.*

By Theorem 3 we mean that if T is a continuous transformation of a snake-like continuum M onto a continuum N such that the inverse under T of each point of N is connected, then N is snake-like.

2. **Imbedding the continua in the plane.** We find that snake-like continua are essentially planar.

THEOREM 4. *Each snake-like continuum is homeomorphic with a plane continuum.*

Proof. Let M be a snake-like continuum. Since it is compact, there is a $\frac{1}{2}$-chain C_1 covering it and a positive number ϵ such that each pair of nonadjacent elements of C_1 are farther apart than ϵ and each element of C_1 intersects M. There is a positive number δ such that each δ-chain covering M is a refinement of C_1, if each of its elements intersects M. We recall that one chain is a refinement of another if each link of it lies in a link of the other.

Let C_2 be a γ-chain ($\gamma =$ minimum $\frac{1}{4}, \delta, \epsilon/8$) covering M such that each element of C_2 intersects M and each pair of nonintersecting elements of C_2 are a positive distance apart. Since $\gamma \leq \delta$, C_2 is a refinement of C_1. Similarly, we find that there is a sequence of chains $C_1, C_2, \cdots$ such that C_i covers M but no proper subchain of it does, C_i is a $\frac{1}{2}^i$-chain, C_{i+1} is a refinement of C_i, and no subchain of C_{i+1} of less than nine links intersects two nonadjacent links of C_i.

We say that the chain $(d_1, d_2, \cdots, d_n)$ follows the pattern $(1, a_1), (2, a_2), \cdots, (n, a_n)$ in the chain E if d_i lies in the a_i-th link of E. It can be shown that in the plane there is a sequence of chains $D_1, D_2, \cdots$ such that D_i is a $\frac{1}{2}^i$-chain whose elements are interiors of rectangles, D_{i+1} follows a pattern in D_i that C_{i+1} follows in C_i, and each element of D_i contains the closure of an element of D_{i+1} while the closure of each element of D_{i+1} lies in an element of D_i. It follows from Theorem 11 of [2] that M is homeomorphic with $D_1^* \cdot D_2^* \cdot \cdots$, where D_i^* denotes the sum of the elements of D_i.

The following example shows that Theorem 4 would not have been true if we had substituted "tree-like" for "snake-like".

EXAMPLE 1. *A skew tree-like continuum.* If G is the closure of the graph in polar coordinates of $\rho = 1/\theta + |\sin(3\theta/2)|^\theta$ $(\pi \leq \theta < \infty)$ it is the sum of a simple triod and a curve spiraling down on it. Suppose A is a straight line interval in 3-space that intersects G only at the point $(\pi, 1)$. Then $G + A$ is a tree-like continuum but it is not homeomorphic with any plane continuum.

EXAMPLE 2. *A skew continuum covered by circular chains.* The solenoid is defined as the intersection of a sequence of solid tori in 3-space such that each is wrapped around twice in the preceding. For each positive number ϵ it can be covered by a circular chain whose links are of diameter less than ϵ. However, it cannot be imbedded in the plane.

THEOREM 5. *If M is a snake-like continuum, there is an uncountable collection W of mutually exclusive continua in the plane such that each element of W is homeomorphic with M.*

Proof. The planar continuum we described in Theorem 4 which was homeomorphic with M had the property that for each positive number ϵ, it is covered by an ϵ-chain whose elements are the interiors of rectangles. Hence, it follows from Theorem 1 of [10] that there is an uncountable collection of mutually exclusive plane continua each homeomorphic with M.

Two plane sets are *equivalent* if there is a homeomorphism of the plane into itself that carries one into the other. It cannot be concluded that each pair of homeomorphic snake-like plane continua are equivalent.

EXAMPLE 3. *Homeomorphic snake-like plane continua which are not equivalent.* Let $f_1(x)$ be the function whose graph is the sum of (a) the graph A_1 of $y = 2 \sin 3x/2$ $(0 \le x \le 2\pi/3)$, (b) the graph A_2 of $y = 3 \sin 6x$ $(2\pi/3 \le x \le 5\pi/6)$, (c) the graph A_3 of $y = -\cos 3x$ $(\pi/2 \le x \le 5\pi/6)$, and (d) the set $B_1 + B_2 + B_3$, where B_i is symmetric to A_i with respect to the point $(\pi/2, 0)$. Then $f_1(x)$ is a single valued function for some values, triple valued for others, and double valued for two values. The closure M_1 of the graph of $y = f_1(\pi/x \bmod \pi)$ $(0 < x \le 1)$ is a snake-like continuum but it does not have the property that for each positive number ϵ there is an ϵ-chain covering it whose links are connected.

Let $f_2(x)$ be the single valued function whose graph is the sum of A_1, A_2, the graph A_4 of $y = \cos 3x$ $(5\pi/6 \le x \le 7\pi/6)$ and the set $C_1 + C_2 + C_4$, where C_i is symmetric to A_i with respect to the point $(7\pi/6, 0)$. The closure M_2 of the graph of $y = f_2(\pi/x \bmod 7\pi/3)$ $(0 < x \le 3/7)$ is homeomorphic with M_1 but M_2 can be covered by ϵ-chains with connected links.

There does not exist an uncountable collection G of mutually exclusive continua such that for each element g of G there is a homeomorphism of the plane into itself that carries M_1 into g. This property of M_1 is shared with the continuum of the next example.

EXAMPLE 4. *A snake-like indecomposable plane continuum which is not equivalent to uncountably many mutually exclusive plane continua.* The continuum described by Roberts in Example 2 of [10] is such a continuum.

EXAMPLE 5. *An uncountable collection of mutually exclusive tree-like plane continua.* This example is due to Anderson. He has given [1] an example of a hereditarily indecomposable tree-like plane continuum M such that no nondegenerate subcontinuum of it is snake-like. Hence, by taking one nondegenerate continuum from each composant of M we get an uncountable collection of mutually exclusive continua no one of which is snake-like. Each such subcontinuum is tree-like because each subcontinuum of a tree-like continuum is tree-like.

QUESTION. Suppose S is an atriodic bounded plane continuum which does not separate the plane. It would be interesting to know if M has uncountably many mutually exclusive homeomorphic images in the plane. This question resembles the one asked by R. L. Moore [10]. Theorem 5 with Corollary 2 of §4 answers the question in the affirmative if S contains no nondegenerate indecomposable continuum.

3. Tree-like continua.

We pointed out in the last section that it is not always possible to cover a snake-like continuum by an ϵ-chain with connected links. However, we find that the corresponding result does not hold for tree-like continua.

THEOREM 6. *Each bounded plane continuum which does not contain a continuum which separates the plane is tree-like.*

Proof. Let M be a bounded plane continuum which does not contain a separating continuum and ϵ be a positive number. Since M is not dense in any open set, there is a finite ϵ-collection $q_1, q_2, \cdots, q_n$ of mutually exclusive interiors of quadrilaterals such that no point of M is on the vertex of any of the quadrilaterals and $\bar{q}_1, \bar{q}_2, \cdots, \bar{q}_n$ covers M. There is an ϵ collection $d_1, d_2, \cdots, d_n$ of interiors of simple closed curves such that d_i contains $M \cdot \bar{q}_i$ and no point belongs to three of the d_i's.

Since M does not separate the plane, there is an unbounded continuum W which does not intersect M which is such that the complement of W is the interior d_0 of a simple closed curve and $d_1, d_2, \cdots, d_n$ covers d_0. Let H be the collection of all interiors h of simple closed curves such that h is a component of the intersection of d_0 and one of the d_i's ($i = 1, 2, \cdots, n$). Since M is bicompact, there is a finite subcollection H' of H which covers M.

We now show that H' is a tree-chain. No point belongs to three elements of H' because none belongs to three elements of $d_1, d_2, \cdots, d_n$. Suppose H' contained a circular chain $h_1, h_2, \cdots, h_m$ ($m \geq 3$). Let $p_1p_2 \cdots p_mp_1$ be a simple closed curve such that p_1p_2 lies in h_1, p_2p_3 lies in h_2, $\cdots$, and p_mp_1 lies in h_m. There is a point q on the interior of this simple closed curve which is a boundary point of h_1, a boundary point of h_2, and a boundary point of $h_1 \cdot h_2$. Since W does not intersect $p_1p_2 \cdots p_mp_1$, an element h of H contains q and h intersects $h_1 \cdot h_2$. The assumption that H' contains a circular chain has led to the contradiction that there is a point common to three elements of H.

In the proof of Theorem 6 we establish the following result.

THEOREM 7. *If M is a tree-like plane continuum and ϵ is a positive number, there is a tree-chain covering M whose links are the interiors of simple closed curves of diameters less than ϵ.*

4. Snake-like continua that are hereditarily decomposable.

A continuum is *decomposable* if it can be expressed as the sum of two proper subcontinua. A continuum is *unicoherent* provided that if it is the sum of two subcontinua H and K, then $H \cdot K$ is connected. A continuum is called *hereditarily decomposable* (*hereditarily unicoherent*) if each nondegenerate subcontinuum of it is decomposable (unicoherent).

In this section we suppose that S is a nondegenerate compact continuum which is atriodic, hereditarily unicoherent, and hereditarily decomposable. In Theorem 11 we show that S is snake-like.

For convenience we suppose that S is all of the space. We find that S has the following properties:

1. *The intersection of any collection of subcontinua of S is a continuum.* This follows from the fact that S is hereditarily unicoherent.

2. *If H and K are continua in S, $S - H$ does not have more than two components and $S - (H + K)$ does not have more than three components.* This follows from the fact that S is atriodic.

3. *If each pair of the continua H_1, H_2, H_3 intersect, one of them is a subset of the sum of the other two.* The sets H_1, H_2, and H_3 have a point in common because their sum is unicoherent and is the sum of the two continua $H_1 + H_2$ and H_3 whose intersection is the sum of $H_1 \cdot H_3$ and $H_2 \cdot H_3$. It follows from condition (1) that $H_0 = H_1 \cdot H_2 + H_2 \cdot H_3 + H_3 \cdot H_1$ is connected. We find from condition (2) that $H_1 + H_2 + H_3 - H_0$ does not have more than two components. Therefore, one of the sets H_1, H_2, H_3 is a subset of H_0.

4. *If each of three mutually exclusive continua H_1, H_2, H_3 contains interior points, one of the continua separates the other two from each other in S.* (A set A *contains* a set B *interiorly* if A contains an open set which contains B. A point which is contained interiorly is called an *interior point.*) If the condition were not satisfied there would be a continuum K_i which contains H_i and intersects each of the three continua H_1, H_2, H_3 but does not contain interior points of any of them except H_i. Condition (3) is violated by K_1, K_2, K_3.

5. *If each of two mutually exclusive continua H_1, H_2 contains interior points, there is a continuum with interior points which separates H_1 from H_2.* If C is the component of $S - (H_1 + H_2)$ with limit points on both H_1 and H_2, $\bar{C}$ is the sum of three continua K_1, K_2, K_3 each of which contains an interior point not in either of the others. If one of the K_i's does not intersect $H_1 + H_2$, it separates H_1 from H_2 because of condition (4). But if H_1 intersects K_1, H_2 intersects each of K_2 and K_3, and K_1 intersects K_2, then the three continua $H_1 + K_1 + K_2$, $H_2 + K_1 + K_2$, $\bar{C}$ violate condition (3).

6. *If H_1 and H_2 are continua with interior points such that H_1 contains H_2 interiorly, there is a continuum H_3 such that H_1 contains H_3 interiorly and H_3*

contains H_2 interiorly. If C is a component of $S - H_1$, we find from condition (5) that there is a continuum K that separates $\bar{C}$ from H_2. Let $H(C)$ be the sum of K and the component of $S - K$ containing H_2. If $S - H_1$ has only one component, $H_3 = H(C)$; if $S - H_1$ has two components C_1 and C_2, $H_3 = H(C_1) \cdot H(C_2)$; if $H_1 = S$, $H_3 = H_1$.

The following result can be compared with Theorem 6 of [9], and Theorem 2.2 of [6].

THEOREM 8. *There is an upper semi-continuous collection G of mutually exclusive continua filling the continuum S such that* (a) *G is an arc with respect to its elements,* (b) *no element of G contains interior points, and* (c) *if r and s are points of different end elements of G, S is irreducible from r to s.*

Proof. For each point p let g_p denote the intersection of all continua which contain interiorly a continuum that contains p interiorly. We find from condition (1) that g_p is a continuum. The elements of G are the g_p's.

We now show that if g_p contains q, then $g_p = g_q$. Assume that g_q does not contain a point p' of g_p. Then there is a continuum H_1 in $S - p'$ which contains interiorly a continuum H_2 which contains q interiorly. Repeated applications of condition (6) shows that there are continua H_3 and H_4 such that H_1 contains H_3 interiorly, H_3 contains H_4 interiorly, and H_4 contains H_2 interiorly.

Then H_4 does not contain p or else H_3 contains p interiorly, H_1 contains H_3 interiorly, and g_p does not contain p' which is not in H_1.

Let K_i ($i = 2, 4$) be the closure of the component of $S - H_i$ containing p. But K_2 contains K_4 interiorly and K_4 contains p interiorly. But g_p lies in K_2 and K_2 does not contain q. Hence, the assumption that there is a point p' in g_p which is not in g_q has led to a contradiction. Similarly, we find that each point in g_q is in g_p.

If the sequence of points $p_1, p_2, \cdots$ converges to p, the superior limit of $g_{p_1}, g_{p_1}, \cdots$ is a subset of g_p. Hence, G is an upper semi-continuous collection.

Using conditions (2) and (4) we find that if g_1, g_2, g_3 are three elements of G, one of them separates the other two from each other. Hence G is an arc with respect to its elements.

Suppose the element g of G contains an open set. Since S contains no decomposable continuum, g is the sum of three continua such that each contains a point interiorly that belongs to neither of the others. It follows from condition (3) that some pair of these three continua do not intersect. Using conditions (5) and (6) we find that these two continua do not belong to the same element of G. Hence, no element of G contains an open set.

Suppose r and s are points of different end elements of G and M is a continuum containing $r + s$. If M were not S, there would be three elements of G which would not lie in M. But the sum of M and these three elements would be a triod. Hence, S is irreducible from r to s.

It may be noted that in Theorem 8 we might have supposed that S contained no indecomposable continuum with interior points instead of supposing it to be hereditarily decomposable.

We shall be interested in points p of a compact continuum M that satisfy the following condition:

(A) Each nondegenerate subcontinuum of M containing p is irreducible from p to some other point.

A point satisfying condition A is a special kind of terminal point as defined by Miller [6; 190]. The following result may be compared with Theorem 3.4 of [6].

THEOREM 9. *If G is the upper semi-continuous collection mentioned in Theorem 8, each end element of G contains a point p satisfying condition* A *for M equal to S.*

Proof. Let H be the collection of all proper subcontinua h of S such that if h is a proper subcontinuum of a subcontinuum S' of S, there is a point p of S' such that S' is irreducible from h to p. Then the end elements g_1 and g_2 of G are elements of H. The property of being an element of H is inductive because for each monotone decreasing sequence $h_1, h_2, \cdots$ of elements of H, Πh_i is an element of H. Hence, in each of g_1 and g_2 there is an element of H which is irreducible with respect to being an element of H and each irreducible element of H is a point.

THEOREM 10. *If D is a chain covering a compact continuum M and p is a point of M satisfying condition* A, *there is a chain $E = (e_1, e_2, \cdots, e_n)$ covering M such that E is a refinement of D and $e_1 - \bar{e}_2$ contains p.*

Proof. Assume the theorem is false. Then there is a subcontinuum M' of M containing p such that M' is irreducible with respect to the theorem being false for this fixed chain D. For convenience we suppose that $M = M'$.

Suppose each element of D intersects M. Then M contains a proper subcontinuum M'' that contains p and intersects each element of $D = (d_1, d_2, \cdots, d_m)$. Since M'' is a proper subcontinuum of M, there is a chain $F = (f_1, f_2, \cdots, f_j)$ covering M'' such that F is a refinement of D, $f_1 - \bar{f}_2$ contains p, and no chain with fewer links has these properties.

The last link of F intersects $M'' \cdot (d_1 + d_m)$ or else a chain with fewer links than F with the required properties covers M''. Suppose d_1 is the one of d_1, d_m that $M'' \cdot f_j$ intersects and N is an open set such that N intersects M'' and $\bar{N}$ lies in $d_1 \cdot f_j - p$.

Since p satisfies condition A, the component of $M - N$ containing p is a subset of M'' and is covered by F. Then $M - N$ is the sum of two mutually separated sets H and K such that H contains p and is covered by F.

Using the normality of M, we find that there are mutually exclusive open sets O_H and O_K containing H and K respectively. Then $E = (O_H \cdot f_1, O_H \cdot f_2, \cdots, O_H \cdot (f_{j-1} - \bar{N}), O_H \cdot f_j, (d_1 - \bar{O}_H) + N, O_K \cdot d_2, \cdots, O_K \cdot d_m)$ is a chain covering M satisfying the conditions of Theorem 10.

THEOREM 11. *Necessary and sufficient conditions that a compact hereditarily decomposable continuum be snake-like are that it be atriodic and hereditarily unicoherent.*

Proof. Since it is easy to see that a snake-like continuum is atriodic and hereditarily unicoherent, we establish Theorem 11 by showing that S is snake-like.

Suppose ϵ is fixed. If Theorem 11 were not true, there would be a subcontinuum S' of S which would be irreducible with respect to not being coverable by any ϵ-chain. For convenience we suppose that $S = S'$.

If G is the upper semi-continuous collection mentioned in Theorem 8, there is an uncountable collection G' of G such that if g_1 and g_2 are the end elements of G and g' is an element of G', there is a sequence of elements of G converging to g' such that infinitely many elements of this sequence separate g' from g_1 and infinitely many separate g' from g_2.

Let g' be a fixed element of G'. If g' is nondegenerate, let p' be a point satisfying condition A for M equal to g'; otherwise, let p' be the only point in g'. Denote by H the sum of g' and all elements of G that it separates from g_2 and by K the sum of g' and all elements of G that it separates from g_1. Then $H + K = S$ and $H \cdot K = g'$. Also, p' satisfies condition A for M equal to either H or K.

Since H and K are proper subcontinua of S, they are covered by ϵ-chains. We find by Theorem 10 that there are ϵ-chains $h_1, h_2, \cdots, h_m$ and $k_1, k_2, \cdots, k_n$ covering H and K respectively such that p' is a point of both $h_m - \bar{h}_{m-1}$ and $k_1 - \bar{k}_2$.

Let N be a neighborhood of p' in $(h_m - \bar{h}_{m-1}) \cdot (k_1 - \bar{k}_2)$. Then $S - N$ is the sum of two mutually separated sets H' and K' such that K' contains $K - N$ and $k_1, k_2, \cdots, k_n$ covers K'. An ϵ-chain covering S is $H' \cdot h_1, H' \cdot h_2, \cdots, H' \cdot h_{m-1}, (H' \cdot h_m) + N, K' \cdot k_1 + N, K' \cdot k_2, \cdots, K' \cdot k_n$.

The following two results are corollaries of Theorem 11.

COROLLARY 1. *A necessary and sufficient condition that a hereditarily decomposable tree-like continuum be snake-like is that it be atriodic.*

COROLLARY 2. *Each atriodic hereditarily decomposable bounded plane continuum which does not separate the plane is snake-like.*

5. **End points of snake-like continua.** Suppose the point p of the compact continuum M satisfies one of the following conditions.

(A) *Each nondegenerate subcontinuum of M containing p is irreducible from p to some other point.*

(B) *If each of two subcontinua of M contains p, one of the subcontinua contains the other.*

(C) *For each positive number ϵ there is an ϵ-chain covering M such that only the first link of the chain contains p.*

We call p an *end point* of M if it satisfies condition C. In Theorem 13 we find that a point is an end point of a snake-like continuum M if it satisfies either condition A or B.

THEOREM 12. *Conditions* A *and* B *are equivalent.*

Proof. If H and K are two continua such that neither of them contains the other and each of them contains p, $H + K$ is not irreducible from p to any point. Hence, condition A implies condition B.

Suppose M' is a nondegenerate subcontinuum of M containing p. By Theorem 5 of [4] there is a point q of $M' - p$ such that the sum of all continua in $M' - q$ containing p is dense in M'. Condition B implies condition A because if p satisfies condition B, M' is irreducible from p to q. If H were a proper subcontinuum of M' containing $p + q$, there would be a continuum K in $M - q$ such that K contains p but is not a subset of H. But neither H nor K would contain the other.

THEOREM 13. *A necessary and sufficient condition that a point p be an end point of a snake-like continuum M is that it satisfy either condition* A *or condition* B.

Proof. The sufficiency follows from Theorems 10, 11, and 12.

To see that the condition is necessary, suppose H and K are two subcontinua of M such that each contains p and each contains a point that is at a distance of ϵ from the other. Let $d_1, d_2, \cdots, d_n$ be an ϵ-chain covering M such that $d_1 - d_2$ contains p. Let d_i be the last element of this chain that contains a point of $H + K$. If d_i contains a point of K, each element of $d_1, d_2, \cdots, d_i$ contains a point of K and no point of H is farther than ϵ from H.

We say that a chain $d_1, d_2, \cdots, d_n$ is a chain from the point p to the point q provided $d_1 - d_2$ contains p and $d_n - d_{n-1}$ contains q. The points p and q of the snake-like continuum M are called *opposite end points* of M if for each positive number ϵ there is an ϵ-chain from p to q covering M.

We find from Theorems 8, 9, and 13 that each hereditarily decomposable snake-like continuum contains two end points between which it is irreducible. The following result shows that such a continuum has opposite end points.

THEOREM 14. *If the snake-like continuum M is irreducible between two of its end points, these end points are opposite end points of it.*

Proof. Suppose M is irreducible between its end points p and q. Let $d_1, d_2, \cdots, d_n$ be an ϵ-chain covering M such that d_1 intersects M and q lies in $d_n - d_{n-1}$. Since M is irreducible from p to q, there is a positive number δ such that if E is a δ-chain covering M, there is a link of E in d_1 which is between the links of E containing p and the links of E containing q. Let $F = (f_1, f_2, \cdots, f_m)$ be a δ-chain covering M such that $f_1 - f_2$ contains p, f_j $(j > 1)$ lies in d_1, and q belongs to $N = f_j + f_{j+1} + \cdots + f_m$. Then $f_1, f_2, \cdots, f_{j-1}, N \cdot d_1, N \cdot d_2, \cdots, N \cdot d_n$ is an ϵ-chain from p to q covering M.

THEOREM 15. *Each nondegenerate snake-like continuum contains a snake-like continuum with opposite end points.*

Proof. If M is a nondegenerate snake-like continuum, there is a sequence C_1, C_2, $\cdots$ of chains such that C_1 has more than two links, C_i is a $1/i$-chain, each link of C_i intersects M while C_i covers M, each link of C_i contains the closure of a link of C_{i+1}, and the closure of each link of C_{i+1} lies in a link of C_i. Let $D_1 = C_1$, D_2 be a subchain of C_2 such that each end link of D_1 contains the closure of an end link of D_2, D_3 be a subchain of C_3 such that D_3 is a refinement of D_2 and each end link of D_2 contains the closure of an end link of D_3, $\cdots$. Then $D_1^* \cdot D_2^* \cdot \cdots$ is a subcontinuum of M with opposite end points.

EXAMPLE 6. *A snake-like continuum without opposite end points.* Let C be the Cantor set in the plane which is obtained by deleting the middle open third of the interval from (0, 0) to (1, 0), deleting the middle open thirds of the remaining intervals, $\cdots$. Let M_0 be the sum of all semicircles in the upper half plane with both end points on C which are symmetric with respect to the line $x = \frac{1}{2}$. Let M_i $(i = 1, 2, \cdots)$ be the sum of all semicircles in the lower half plane with both end points on C which are symmetric with respect to the line $x = 5/(3^i \cdot 2)$. Then $M = M_0 + M_1 + M_2 + \cdots$ is a snake-like continuum with only one end point, namely the origin.

EXAMPLE 7. *A snake-like continuum without any end points.* This continuum is the sum of M of Example 6 and the continuum which is symmetric to it with respect to the origin.

THEOREM 16. *A necessary and sufficient condition that a nondegenerate compact continuum be a pseudo-arc* [7], [2], [3] *is that each point of it be an end point of it.*

Proof of necessity. It was shown in Theorem 1 of [3] that if p and q are points of different composants of a pseudo-arc, they are opposite ends of it.

Proof of sufficiency. A snake-like continuum M is a pseudo-arc if it is hereditarily indecomposable. If M is not hereditarily indecomposable, it contains two intersecting continua H and K neither of which contains the other. Under these conditions, no point of $H \cdot K$ is an end point of M.

REFERENCES

1. R. D. ANDERSON, Abstract 236, Bulletin of the American Mathematical Society, vol. 57(1951), p. 185.
2. R. H. BING, *A homogeneous indecomposable plane continuum*, this Journal, vol. 15(1948), pp. 729–742.
3. R. H. BING, *Concerning hereditarily indecomposable continua*, Pacific Coast Journal of Mathematics, vol. 1(1951), pp. 43–51.
4. R. H. BING, *Some characterizations of arcs and simple closed curves*, American Journal of Mathematics, vol. 70(1948), pp. 497–506.
5. O H. HAMILTON, *A fixed point theorem for pseudo-arcs and certain other metric continua*, Proceedings of the American Mathematical Society, vol. 2(1951), pp. 173–174.
6. HARLAN C. MILLER, *On unicoherent continua*, Transactions of the American Mathematical Society, vol. 69(1950), pp. 179–194.

7. EDWIN E. MOISE, *An indecomposable plane continuum which is homeomorphic to each of its nondegenerate subcontinua*, Transactions of the American Mathematical Society, vol. 63(1948), pp. 581–594.
8. R. L. MOORE, *Concerning triodic continua in the plane*, Fundamenta Mathematicae, vol. 13(1929), pp. 261–263.
9. R. L. MOORE, *Concerning continua which have dendratomic subsets*, Proceedings of the National Academy of Sciences, vol. 29(1943), pp. 384–389.
10. J. H. ROBERTS, *Concerning atriodic continua*, Monatshefte für Mathematik und Physik, vol. 37(1930), pp. 223–230.
11. R. H. SORGENFREY, *Concerning triodic continua*, American Journal of Mathematics, vol. 66(1944), pp. 439–460.

University of Wisconsin.

The Pseudo-Arc

R. H. Bing

A continuum that has been used extensively as a counter example to explode several conjectures is the pseudo-arc.

Knaster described [**5**] a particular continuum in 1922 to show that there was a continuum each of whose subcontinua was indecomposable.

Moise used different methods [**6**] in 1948 to describe a continuum which was topologically equivalent to each of its nondegenerate subcontinua. Since it had this property in common with an arc, he called it a pseudo-arc.

Bing showed [**3**] in 1951 that the continua that Knaster and Moise had described by different methods were topologically equivalent. In fact, any two nondegenerate snake-like hereditarily indecomposable continua are topologically equivalent [**3**].

Any continuum which is topologically equivalent to the continuum described by Knaster and Moise is a pseudo-arc—any continuum which differs from it by the slightest whit is not. Each snake-like hereditarily indecomposable continuum with more than a point is a pseudo-arc.

We frequently regard the bounded continua in a metric space as points and take the distance between two of these ideal points (continua) to be the Hausdorff distance. We say that most continua have a certain property if there is a dense G_δ subset of them in the ideal space with the property. The collection of continua with the property is said to be of the first category; the set of those without it is of the second category. In this sense, it was discovered [**3**] that most continua in E^n ($n > 1$) are pseudo-arcs. The set of all bounded continua (as arcs, disks, triods, spheres, points, etc.) that differ from a pseudo-arc in the least is of the second category. This being the case, if a person is dealing with a continuum whose shape is unknown (as a physicist using one of the obsolete treatments of the atom), it might be more reasonable to suspect it to be in the shape of a pseudo-arc than to suspect it to be spherical.

We list some properties of the pseudo-arc.

1. It is homogeneous. [**2**, **7**]
2. It is topologically equivalent to each of its nondegenerate subcontinua.
3. It is characterized by the fact that it is a nondegenerate snake-like hereditarily indecomposable continuum. [**3**]

4. It has the fixed point property. [**4**]

5. Each nondegenerate monotone continuous image of it is topologically equivalent to it. [**3**]

6. There is a continuous collection of pseudo-arcs filling the plane. [**1**]

7. Most continua in E^n $(n > 1)$ are pseudo-arcs. [**3**]

Although homogeneity is discussed elsewhere, we include a chronological account of some interesting occurrences regarding it here:

1920 The following question is raised: Is the simple closed curve the only nondegenerate bounded homogeneous plane continuum?
1922 Knaster describes the pseudo-arc.
1924 Mazurkiewicz gives a partial affirmative answer (puts on additional conditions).
1951 Cohen gets a better partial affirmative answer (imposes fewer additional conditions).
1937 A false affirmative solution is given.
1944 The false affirmative solution is extended.
1948 Moise describes the pseudo-arc.
1948 Bing shows the pseudo-arc is homogeneous.
1949 Moise shows the pseudo-arc is homogeneous.
1953 Two papers appear defending the false solutions.
1954 Bing and Jones, working independently, simultaneously discover another homogeneous plane continuum.

Future ?

Here is a condensed description of the pseudo-arc described in [**2**] or [**3**]. Suppose p, q are two points and $D_1 = (d_1^1, d_2^1, \ldots, d_n^1)$ is a chain from p to q. Suppose $D_2 = (d_1^2, d_2^2, \ldots, d_m^2)$ is a chain from p to q such that D_2 is crooked in D_1 (it zig-zags back and forth so much in D_1 that if $k - h > 2$ and d_i^2, d_j^2 are links of D_2 in links d_h^1, d_k^1 respectively of D_1, then there are links d_r^2, d_s^2 of D_2 in d_{k-1}^1, d_{h+1}^1 respectively such that either $i < r < s < j$ or $i > r > s > j$). Continuing, we define a sequence of chains $D_1, D_2, D_3, \ldots,$ from p to q such that (a) D_{i+1} is crooked in D_i, (b) as i increases without limit, mesh D_i approaches 0, and (c) the closure of each link of D_{i+1} is a compact subset of a link of D_i. The common part of the sum of the links of D_1, the sum of the links of $D_2, \ldots,$ is a pseudo-arc. It can be shown that any two continua obtained in this fashion are topologically equivalent.

Questions: Here are some questions that remain unanswered.

1. Since most snake-like continua are pseudo-arcs, a person might suspect that if a snake-like continuum were described in a random fashion, a pseudo-arc would result. Hence, suppose D_1 is a chain from p to q with n links. Let the pattern [**2**] that D_{i+1} follows in D_i be determined by the flip of a coin as follows: d_1^{i+1}, d_2^{i+1} lie in D_1^i; if d_{2k}^{i+1} lies in d_j^i and d_j^i is not the last link of D_i, a coin is tossed; if it falls heads, d_{2k+1}^{i+1}, d_{2k+2}^{i+1} fall in d_{j+1}^i; if it falls tails, d_{2k+1}^{i+1}, d_{2k+2}^{i+1} lie in d_{j-1}^i or d_j^i according as $i > 1$ or $i = 1$. When some d_{2h}^{i+1} is found to lie

in the last link of D_i, the process is completed and D_{i+1} is defined. What is the probability that for some integer m, D_m will be crooked in D_1? If the probability is 100%, then it appears that if we constructed a snake-like continuum in such a random fashion, the chance of getting a pseudo-arc is excellent. Of course, it might be that at each stage D_{i+1} would run straight through D_i and we would end up with an arc, but the possibility seems remote.

2. What characterizes all continuous images of a pseudo-arc? (A take-off on the fact that all images of an arc are continuous curves.)

3. Is the identity of the only homeomorphism that leaves *each composant* of a pseudo-arc fixed? Answer the same question for some open set.

4. Suppose K, K' are two subsets of a pseudo-arc and h is a homeomorphism of K onto K'. Under what conditions can h be extended to a homeomorphism taking the pseudo-arc onto itself? Can it (for example) if for each 0-dimensional closed subset W of K, h on W can be extended to take the pseudo-arc onto itself?

References

1. R. D. Anderson, (sic)

2. R. H. Bing, *A homogeneous indecomposable plane continuum*, Duke Journal, vol. 15, 1948, pp. 729–742.

3. R. H. Bing, *Concerning hereditarily indecomposable continua*, Pacific Coast Journal of Math., vol. 1, 1951, pp. 43–51.

4. O. H. Hamilton, *A fixed point theorem for pseudo-arcs and certain other metric continua*, Proc. Am. Math. Soc., vol. 2, 1951, pp. 173–174.

5. B. Knaster, *Un continu dont tout sous-continu est indécomposable*, Fund. Math., vol. 3, 1922, pp. 247–286.

6. E. E. Moise, *An indecomposable plane continuum which is homeomorphic to each of its nondegenerate subcontinua*, Trans. Am. Math. Soc., vol. 63, 1948, pp. 581–594.

7. E. E. Moise, *A note on the pseudo-arc*, Trans. Am. Math. Soc., vol. 63, 1949, pp. 57–58.

EACH HOMOGENEOUS NONDEGENERATE CHAINABLE CONTINUUM IS A PSEUDO-ARC

R. H. BING[1]

The endeavor to find all homogeneous plane continua continues. The simple closed curve and the point are obvious examples. The discovery of the pseudo-arc [1; 6] should have exploded (but did not) the conjectures that there are no others. A history of the problem with a discussion of various false starts is given in [4]. Finding the circle of pseudo-arcs [4] raised the number of known examples to four. Are there others as yet undiscovered? Jones showed [5] that each one which does not separate the plane is indecomposable. The theorem in this paper narrows the field for search still further.

We recall the following definitions:

A set X is *homogeneous* if for each pair of points p, q of X there is a homeomorphism of X onto itself that takes p onto q.

A continuum is *nondegenerate* if it contains more than one point.

An *ϵ-chain* is a finite ordered collection $d_1, d_2, \cdots, d_n$ of open sets, each of diameter less than ϵ, such that d_i intersects d_j if and only if i and j are adjacent integers.

A *snakelike* or *chainable* continuum is a compact metric continuum M such that for each positive number ϵ, M can be covered by an ϵ-chain.

A point p is an *end point* of a snakelike continuum M if for each positive number ϵ there is an ϵ-chain covering M such that the first link of the chain contains p.

A continuum is *indecomposable* if it is not the sum of two proper subcontinua. It is *hereditarily indecomposable* if each subcontinuum of it is indecomposable.

A *pseudo-arc* is a nondegenerate, hereditarily indecomposable, chainable continuum. Any two such continua are homeomorphic [2].

THEOREM. *Each homogeneous, nondegenerate, chainable continuum is a pseudo-arc.*

PROOF. First we show that M has an end point p. For each integer n, let q_n be a point of M such that a $1/n$-chain covers M and an end link of this chain contains q_n. Some subsequence of $q_1, q_2, \cdots$ converges to a point q. Then the point q of M has the following property:

Presented to the Society, February 23, 1957; received by the editors October 16, 1958.

Work on this paper was supported by Contract NSF G1219.

Property of q. For each neighborhood N of q and each positive number ϵ there is an ϵ-chain covering M one of whose end links intersects M and lies in N. It follows from the homogeneity of M and the fact that each homeomorphism of M onto itself is uniformly continuous that each point of M has the *Property of q*.

Let d_1 be an end link of a 1-chain covering M such that d_1 contains a point p_1 of M. Since p_1 has the *Property of q*, there is an end link d_2 of a 1/2-chain covering M such that d_1 contains $\bar{d}_2$ and d_2 contains a point p_2 of M. Also, there is an end link d_3 of a 1/3-chain covering M such that d_2 contains $\bar{d}_3$ and d_3 contains a point p_3 of M. Similarly, we obtain $d_4, d_5, \cdots$. Then the point p which is the intersection of $d_1, d_2, \cdots$ is an end point of M.

Finally we show that M is hereditarily indecomposable. Assume M contains a continuum H which is the sum of two proper subcontinua H', H''. Let p be a point of $H' \cdot H''$. Then it follows from the homogeneity of M that p is an end point of M. However, as noted in [**3**], this would imply that one of H', H'' contains the other and this is impossible since each is a proper subcontinuum of their sum.

References

1. R. H. Bing, *A homogeneous indecomposable plane continuum*, Duke Math. J. vol. 15 (1948) pp. 729–742.

2. ———, *Concerning hereditarily indecomposable continua*, Pacific J. Math. vol. 1 (1951) pp. 43–51.

3. ———, *Snake-like continua*, Duke Math. J. vol. 18 (1951) pp. 653–663.

4. R. H. Bing and F. B. Jones, *Another homogeneous plane continuum*, Trans. Amer. Math. Soc. vol. 90 (1959) pp. 171–192.

5. F. B. Jones, *Certain homogeneous unicoherent indecomposable continua*, Proc. Amer. Math. Soc. vol. 2 (1951) pp. 855–859.

6. E. E. Moise, *A note on the pseudo-arc*, Trans. Amer. Math. Soc. vol. 64 (1949) pp. 57–58.

University of Wisconsin

B. Locally Connected Continua: Partitions and Decompositions

PARTITIONING A SET

R. H. BING

1. Introduction. Suppose space has a metric $D(x, y)$.

Under certain conditions in calculus we define $\int_b^a f(x)dx$ to be equal to $\lim \sum f(\xi_i)\Delta x_i$. The dividing of the interval from a to b into a finite number of pieces with lengths Δx_i is a partitioning of the interval from a to b. Similarly, partitioning may be used in integrating over an arbitrary set. The partitioning may provide a basis for a measure on the range of the function which is being integrated. We shall define partitioning on sets for which there is no accepted unit of measure such as length, area, or volume.

The notion of partitioning may be used to assign a convex metric to certain sets. See Theorems 7, 8, 9, and 10. Theorem 8 provides an answer to a question raised by Menger in [3].[1] See [1] for a discussion of and references to partial solutions of this problem. Further applications of partitioning are treated in my paper *Complements of continuous curves*, appearing in Fund. Math.

Most of the results of this paper (including Theorem 7 but not Theorems 1 and 8) were contained in an address I gave before the Mathematics Club at the University of Chicago in May, 1948. After finishing Theorems 1 and 8 in August, I found that Moise, working independently, had obtained a solution [4] to the above mentioned problem by Menger six weeks earlier.

Partitioning. A set M can be partitioned if for each positive number ϵ there is a finite collection G of connected mutually exclusive open subsets of M such that each element of G is of diameter less than ϵ and the sum of the elements of G is dense in M. We shall say that G is an ϵ-partitioning of M.

Refinement. If G and H are two partitionings of M, we say that G is a refinement of H if each element of G is a subset of an element of H.

Property S. A set M has property S if for each positive number ϵ, M is the sum of a finite number of connected subsets such that the diameter of each of these subsets is less than ϵ. This property is discussed in [5, 6, 7, 8].

Convex metric. A metric E for a set M is convex if for each pair of points x, y of M there is a point z of M such that $E(x, z) = E(y, z)$

Presented to the Society, September 10, 1948; received by the editors September 2, 1948.

[1] Numbers in brackets refer to the bibliography at the end of the paper.

$=E(x, y)/2$. The metric F for M is almost convex [2] if for each pair of points x, y of M and each positive number ϵ there is a point z of M such that

$$|F(x, z) - F(x, y)/2| + |F(y, z) - F(x, y)/2| < \epsilon.$$

2. **Partitionable sets.** Theorem 1 characterizes sets that can be partitioned. First we consider two lemmas that will be used.

LEMMA 1. *For each connected set M with property S there is a compact locally connected continuum H and a homeomorphism T of M into a dense subset H' of H such that the diameter of each connected subset X of M is the same as the diameter of $T(X)$ and for each connected open subset R of H, $T^{-1}(R \cdot H')$ is connected.*

PROOF. The continuum H is the complete enclosure of the relative distance space for M. See [8; pp. 154–158].

LEMMA 2. *If M is a connected set with property* S *and H and K are subsets of M at a positive distance from each other, then there is a finite collection of connected, mutually exclusive open subsets of M such that the sum of this collection is dense in M, no element of this collection intersects both H and K, and each element of this collection that intersects K has property* S.

PROOF. We shall prove the lemma for the case where M is closed and compact. It will follow from Lemma 1 for the general case.

Let H' and K' be two sets at a positive distance from each other, containing H and K respectively, such that each point of M belongs to an arc of diameter less than 1/2 that intersects $H'+K'$. Denote by H_1 (or K_1) the set of all points of M that belong to a connected subset of M which intersects H' (or K') and is of diameter less than $D(H', K')/3$. We note that H_1 and K_1 have only a finite number of components.

Let W be a finite collection of points of M such that each point of M belongs to an arc in M of diameter less than 1/4 that intersects W. There is a finite collection A of arcs such that each element of A lies in $M-\overline{K}_1$ and intersects H_1 while A^*, the sum of the elements of A, contains each point of W which belongs to a component of $M-\overline{K}_1$ which intersects H_1. Let B be a finite collection of arcs such that A^*+B^* contains W and each element of B is of diameter less than 1/2, lies in $M-(\overline{H}_1+A^*)$, and intersects K_1.

Denote by H_2 (or K_2) the set of all points that belong to a connected subset of M which intersects H_1+A^* (or K_1+B^*) and is of diameter less than $D(H_1+A^*, K_1+B^*)/3$. We note that each point of K_2 belongs to a connected subset of K_2 which intersects K_1 and is of diameter less than $1/2+1/6$.

Similarly, there exist open subsets $H_3, K_3, H_4, K_4, \cdots$ of M such that (1) H_i is at a positive distance from K_i, (2) H_{i+1} and K_{i+1} contain $\overline{H}_i$ and $\overline{K}_i$ respectively, (3) each component of H_{i+1} contains a component of H_i, (4) each point of K_{i+1} belongs to a connected subset of K_{i+1} that intersects K_i and is of diameter less than $1/2^i+1/(3\cdot 2^i)$, and (5) each point of M belongs to an arc of diameter less than $1/2^i$ that intersects H_i+K_i. Now $\sum(H_i+K_i)$ is dense in M and has only a finite number of components while $\sum K_i$ has property S.

THEOREM 1. *A necessary and sufficient condition that a set M can be partitioned is that it have property* S.

PROOF OF SUFFICIENCY. We show that M can be partitioned by showing that each of its components C can be partitioned.

Let $(p_1, p_2, \cdots, p_n)$ be a finite collection of points of C such that each point of M is at a distance from $\sum p_i$ of less than $\epsilon/4$. Let H_i be the set of all points of C that are at a distance from p_i of less than $\epsilon/4$ and K_i be the set of those that are at a distance from p_i of more than $\epsilon/2$.

By Lemma 2 there are two mutually exclusive open subsets U_1 and V_1 of C such that U_1+V_1 is dense in C, U_1 has a finite number of components and contains H_1, and V_1 has property S and contains K_1. Also, there are two open subsets U_2 and V_2 of V_1 such that U_2+V_2 is dense in V_1, U_2 contains $V_1\cdot H_2$ and has a finite number of components, while V_2 contains $V_1\cdot K_2$ and has property S. Similarly, we define sets $U_3, V_3, U_4, V_4, \cdots, U_{n-1}, V_{n-1}$. The components of $U_1+U_2+\cdots+U_{n-1}+V_{n-1}$ are finite in number, each is of diameter less than ϵ, and their sum is dense in M.

3. **Refinements of partitionings.** We shall be interested in partitioning a set in such a way that each of the pieces into which it is partitioned can be partitioned. These pieces must have property S if this is to be done.

THEOREM 2. *If H and K are two connected subsets of the connected partitionable set M and ϵ is a positive number such that any connected subset of M intersecting both H and K is of diameter more than ϵ, then there are two connected, mutually exclusive open subsets U and V of M containing H and K respectively such that $U+V$ has property* S *and is dense in M.*

PROOF. The general case will follow from Lemma 1 if we prove the case where M is a compact continuum. Hence, we suppose it is one.

For each integer i, let G_i be a $1/i$-partitioning of M. Let H' and K' be mutually exclusive subcontinua of M containing H and K respectively such that each element of G_2 intersects $H'+K'$. Let U_1 (or V_1) be the set of all points that belong to a connected subset of M of diameter less than $D(H', K')/3$ that intersects H' (or K').

Let A and B be finite collections of arcs such that (1) each element of A intersects U_1 while each element of B intersects V_1, (2) each element of $A+B$ is a subset of an element of G_2, (3) no element of A intersects either $\overline{V}_1$ or an element of B, while no element of B intersects $\overline{U}_1$, and (4) each element of G_4 contains a point of A^*+B^*.

Let U_2 (or V_2) be the set of all points that belong to a connected subset of M of diameter less than $D(U_1+A^*, V_1+B^*)/3$ that intersects U_1 (or V_1). We note that each point of U_2 (or V_2) belongs to a connected subset of U_2 (or V_2) of diameter less than $1/2+D(U_1+A^*, V_1+B^*)/3<1/2+1/6$ that intersects U_1 (or V_1).

Similarly, we obtain connected open subsets U_3, V_3, U_4, V_4, $\cdots$ such that (1) U_{i+1} contains $\overline{U}_i$ while V_{i+1} contains $\overline{V}_i$, (2) U_i+V_i intersects each element of G_{2^i}, and (3) each point of U_{i+1} (or V_{i+1}) belongs to a connected subset of U_{i+1} (or V_{i+1}) which is of diameter less than $1/2^i+1/(3\cdot 2^i)$ and intersects U_i (or V_i). Then $U=\sum U_i$ and $V=\sum V_i$ are the required sets.

Each of the following three theorems follows from repeated applications of the theorem preceding it.

THEOREM 3. *If H is a finite collection of connected subsets of the connected partitionable set M and ϵ is a positive number such that no connected subset of M of diameter less than ϵ intersects two elements of H, then there is a finite collection U of connected, mutually exclusive open subsets of M such that each element of U has property* S *and contains one and only one element of H, each element of H is contained in an element of U, and the sum of the elements of U is dense in M.*

THEOREM 4. *If M is partitionable, then for each positive number ϵ there is an ϵ-partitioning G of M such that each element of G has property* S.

THEOREM 5. *If M is a partitionable set, there is a sequence $G_1, G_2, \cdots$ such that G_i is a $1/i$-partitioning of M and G_{i+1} is a refinement of G_i.*

THEOREM 6. *Suppose space has a complete metric and R is a connected domain with property* S *and a nondegenerate boundary B. Then R contains a connected set E with property* S *such that $\overline{E}=E+B$, but no connected subset of E has this property. Also, E is topologically* *equivalent to a subset of a dendron, and any nondegenerate continuum in E is a dendron with a finite number of ends.*

PROOF. Let $G_1, G_2, \cdots$ be a sequence such that G_i is a $1/i$-partitioning of R and G_{i+1} is a refinement of G_i. Denote the collection of closures of elements of G_i by H_i. Let H_1' be a subset of H_1 irreducible with respect to covering B and H_{i+1}' be a subcollection of H_{i+1} such that each element of H_{i+1}' is a subset of an element of H_i', and H_{i+1}' is irreducible with respect to covering B.

Let m_1 be an integer so large that H_{m_1}' has more than one element. There is a dendron T_1 in R which intersects each element of H_{m_1}' but which contains no subcontinuum doing so. Let m_2 be an integer so large that $D(T_1, B) > 3/m_2$. There is a dendron T_2 such that (1) T_2 contains T_1, (2) T_2 intersects each element of H_{m_2}', (3) each component of $T_2 - T_1$ is a subset of an element of G_{m_1}, and (4) no subcontinuum of T_2 has these properties. Similarly, we define $T_3, T_4, \cdots$. Then $E = \sum T_i$.

4. **Convexification of partitionable sets.** We can assign a convex metric to certain partitionable sets.

THEOREM 7. *If M is a compact partitionable continuum, it can be assigned a convex metric.*

FIRST PROOF OF THEOREM 7. We shall define a sequence $E_1(x, y)$, $E_2(x, y), \cdots$ of functions on the pairs of points of M such that $E(x, y) = \lim E_i(x, y)$ is a convex metric for M.

Let $G_1, G_2, \cdots$ be a sequence such that G_i is a $1/i$-partitioning of M and G_{i+1} is a refinement of G_i. There is a dendron T_1 in M such that T_1 intersects each element of G_1, and there is a convex metric F_1 for T_1 such that the diameter of T_1 under F_1 is less than $1/4$. Let U_1 be the set of all elements u such that u is a component of $g - g \cdot T_1$ for some element g of G_1. Denote by V_1 the set of all continua v such that v is either a point of T_1 or the closure of an element of U_1.

If v and w are two elements of V_1, we define $E_1(v, w)$ to be the greatest lower bound of all numbers of the sort $r/2 + s$, where some continuum is the sum of v, w, r other elements of V_1 and some arcs in T_1 whose total length under F_1 is s.

Let m_2 be an integer so large that $D(v, w) > 3/m_2$ if v and w are elements of V_1 such that $E_1(v, w) > 1/16$. In each element u_i of U_1 that contains an element of G_{m_2} there is a dendron $T_{2,i}$ which intersects each element of G_{m_2} that intersects u_i. It may be noted that there are only a finite number of such u_i's. Let $F_{2,i}$ be a convex metric for $T_{2,i}$ such that the diameter of $T_{2,i}$ under $F_{2,i}$ is less than $1/16$.

Let U_2 be the collection of all components of the sort $g-g\cdot(T_1+\sum T_{2,i})$, where g denotes an element of G_{m_2}. Let V_2 designate the set of all continua v such that v is either a point of $T_1+\sum T_{2,i}$ or the closure of an element of U_2.

If v and w are two elements of V_2, we define $E_2(v, w)$ to be the greatest lower bound of all numbers of the type $r/4+s$, where some continuum is the sum of v, w, r other elements of V_2 and a collection of arcs in $T_1+\sum T_{2,i}$ whose total length under their respective F's is s.

Similarly, we define m_3, G_{m_3}, $T_{3,j}$, $F_{3,j}$, U_3, V_3, $E_3(v, w)$, m_4, G_{m_4}, $T_{4,j}$, $F_{4,j}$, U_4, V_4, $E_4(v, w)$, $\cdots$. In general, m_{i+1} is an integer so large that

$$(1)\quad D(v, w) > 3/m_{i+1} \quad \text{if} \quad E_i(v, w) > 1/4^{i+1} \quad (v, w \text{ elements of } V_i).$$

Consider an element $u_{i,n}$ of U_i that contains an element of $G_{m_{i+1}}$. Let $K_{i,n}$ denote the collection of all continua k such that k is the closure of a component of the common part of $u_{i,n}$ and an element of $G_{m_{i+1}}$. Denote by $K'_{i,n}$ the finite collection of all elements k of $K_{i,n}$ such that k intersects an element of V_i which does not intersect any component of $T_1+\sum T_{2,j}+\cdots+\sum T_{i,j}$ which intersects k. There is a dendron $T_{i+1,n}$ in $u_{i,n}$ such that $T_{i+1,n}$ intersects each element of $K'_{i,n}$ and each element of $G_{m_{i+1}}$ which intersects $u_{i,n}$. Let $F_{i+1,n}$ be a convex metric for $T_{i+1,n}$ such that

$$(2)\qquad \text{diameter } T_{i+1,n} \text{ under } F_{i+1,n} < 1/4^{i+1}.$$

Denote by U_{i+1} the collection of all components of sets of the sort $g-g\cdot(T_1+\sum T_{2,j}+\cdots+\sum T_{i+1,j})$, where g denotes an arbitrary element of $G_{m_{i+1}}$. The elements of V_{i+1} are the points of $T_1+\sum T_{2,j}+\cdots+\sum T_{i+1,j}$ and the closures of the elements of U_{i+1}.

Let

$$E_{i+1}(v, w) = \text{greatest lower bound } (r/2^{i+1}+s) \quad (v, w \text{ elements of } V_{i+1}),$$

where some continuum is the sum of v, w, r other elements of V_{i+1} and a collection of arcs in $T_1+\sum T_{2,j}+\cdots+\sum T_{i+1,j}$ whose lengths under their respective F's total s. It may be noted that if v intersects w, $E_i(v, w)=0$.

If p and q are two points of M, we define $E_i(p, q)$ to be the greatest lower bound of all numbers $E_i(v, w)$, where v and w are elements of V_i containing p and q respectively. We shall show that $\lim E_i(p, q) = E(p, q)$ exists and $E(x, y)$ is a convex metric for M.

Using (1) we find that if some continuum is the sum of the elements v and w of V_i and n elements of V_{i+1}, then $E_i(v, w) < n/2^{i+1}$. From this and the fact that two elements of V_{i+1} are subsets of the same element of V_i if they intersect the same arc in $\sum T_{i+1,j}$, we find that if v and w are two elements of V_{i+1}, then there are elements v' and w' of V_i containing v and w respectively such that $E_i(v', w') \leqq E_{i+1}(v, w)$. Hence, $E_i(p, q)$ is a monotone nondecreasing function of i.

If a component C of $T_1 + \sum T_{2,j} + \cdots + \sum T_{i,j}$ intersects the sum of two elements v, w of V_i which intersect each other, it follows from (1) and (2) that under its respective metric

$$\text{diameter } C \cdot (v + w) < 1/4^i. \tag{3}$$

If v and w are two elements of V_{i+1} which are subsets of the same element of V_i, then

$$E_{i+1}(v, w) < 1/4^{i+1} + \delta(1/2^{i+1} + 1/4^i) \tag{4}$$

where δ is equal to 2, 1, or 0 according as neither, one, or both of the elements v, w are nondegenerate subsets of elements of $K'_{i,j}$ for some integer j.

Using (3) and (4) we find that if v and w are elements of V_{i+1} which are subsets of the sum of a coherent collection of n elements of V_i, then

$$E_{i+1}(v, w) < n(1/2^i + 1/4^i + 1/4^{i+1}) + 1/4^i < (n/2^i)(1 + 3/2^i). \tag{5}$$

Using (5) we find that

$$E_{i+1}(p, q) < (1 + 3/2^i)\left[E_i(p, q) + 2/2^i\right]. \tag{6}$$

Repeated applications of (6) give that

$$E_{i+k}(p, q) < (1 + 3/2^i)(1 + 3/2^{i+1}) \cdots (1 + 3/2^{i+k-1})\left[E_i(p, q) + 2/2^i + 2/2^{i+1} + \cdots + 2/2^{i+k-1}\right].$$

Since

$$(1 + 3/2^i)(1 + 3/2^{i+1}) \cdots < \left[1 + 3/(2^i - 6)\right]\left[1 + 3/(2^{i+1} - 6)\right] \cdots = 2^i/(2^i - 6) \qquad (i > 3),$$

then

$$E_i(p, q) \leqq E(p, q) < \left[2^i/(2^i - 6)\right]\left[E_i(p, q) + 4/2^i\right] \qquad (i > 3). \tag{7}$$

That $\lim E_i(p, q) = E(p, q)$ exists follows from the fact that $E_1(p, q)$, $E_2(p, q)$, $\cdots$ is a bounded monotone sequence.

That $E(x, y)$ satisfies the triangle condition follows from the fact that for each triple of points p, q, r, $E_i(p, q) \leqq E_i(p, r) + E_i(r, q) + 1/2^i$. Since $E_i(x, y)$ is symmetric, $E(x, y)$ is also; that is, $E(x, y) = E(y, x)$. If R is an open subset of M containing the point p, there is an integer n so large that $D(p, M-R)>3/m_n$. Since $E_n(p, M-R)$ is positive, $E(p, M-R)$ is also. If p and q belong to the closure of the same element of G_{m_n+1}, then by (4), $E_n(p, q)<2/2^n$, and by (7), $E(p, q) <[2^n/(2^n-6)](2/2^n+4/2^n)$ if $n>3$. Hence $E(p, q_1)$, $E(p, q_2)$, $\cdots$ has zero as a limit if and only if $q_1, q_2, \cdots$ converges to p. Therefore $E(x, y)$ is a metric for M that preserves its topology.

That $E(x, y)$ is almost convex follows from the fact that for each pair of points p, q and each integer n there is a point r such that $|E_n(p, r)-E_n(p, q)/2|+|E_n(r, q)-E_n(p, q)/2|\leqq 1/2^n$. Since M is compact, $E(x, y)$ is a convex metric for M.

ALTERNATE PROOF OF THEOREM 7. The proof of Theorem 7 can be simplified by making use of Theorem 6 of this paper and Theorem 4 of [1] which states that *a compact locally connected continuum M has a convex metric if for each point p contained in an open subset R_1 of M, there is an open subset R_2 of R_1 containing p such that the boundary of R_2 with respect to M is a subset of a subcontinuum of M with a convex metric*. Hence, we need to show only that $\bar{E}$ of Theorem 6 has a convex metric.

Now E is the sum of a dendron T_1 and a finite collection U_1 of connected, mutually exclusive open subsets of E such that the common part of T_1 and the closure of an element of U_1 is an end point of T_1. There is a convex metric F_1 for T_1 such that the distance between any two of its end points is between 1/2 and 1. Let V_1 be the collection of all continua v such that v is either a point of T_1 or the closure of an element of U_1. If v and w are elements of V_1, define $E_1(v, w)$ to be the greatest lower bound of all numbers of the type $r/2+s$ where some continuum is the sum of v, w, r other elements of V_1 and arcs in T_1 whose lengths under F_1 total s.

Let ϵ_2 be a number so small that if v and w are two elements of V_1 which do not intersect, then $D(v, w)>3\epsilon_2$ and $D(T_1, \bar{E}-E)>3\epsilon_2$. Now E is the sum of a dendron T_2 and a finite collection U_2 of connected, mutually exclusive open subsets of E such that each element of U_2 is of diameter less than ϵ_2, and the common part of T_2 and the closure of an element of U_2 is an end point of T_2. Let F_2 be a convex metric for T_2 that preserves F_1 on T_1 and such that any arc in T_2 is of length between $1/2^2$ and 1/2 if the end points of this arc are also end points of the closure of a component of T_2-T_1. Denote by V_2 the collection of all elements v such that v is either a point of T_2 or the closure of an element of U_2. If v and w are elements of V_2, let $E_2(v, w)$ be the greatest lower bound of numbers of the type $r/4+s$, where some continuum is the sum of v, w, r other elements of V_2 and arcs in T_2 whose lengths under F_2 total s.

This process is continued to get a sequence of functions E_3, E_4, $\cdots$. In general, ϵ_{i+1} is a number so small that if v and w are any two elements of V_i that do not intersect each other, then $D(v, w)>3\epsilon_{i+1}$ and $D(T_i, \overline{E}-E)>3\epsilon_{i+1}$. There is a dendron T_{i+1} and a finite collection U_{i+1} of connected, mutually exclusive open subsets of E such that each element of U_{i+1} is of diameter less than ϵ_{i+1}, and the common part of T_{i+1} and the closure of an element of U_{i+1} is an end point of T_{i+1}. There is a convex metric F_{i+1} for T_{i+1} that preserves F_i on T_i such that any arc in T_{i+1} is of length, under F_{i+1}, of between $1/2^{i+1}$ and $1/2^i$ if the end points of this arc are end points of the closure of the same component of $T_{i+1}-T_i$. Use V_{i+1} to denote the collection of points of T_{i+1} and the closures of elements of U_{i+1}. For elements v and w of V_{i+1}, define $E_{i+1}(v, w)$ to be the greatest lower bound of $r/2^{i+1}+s$, where some continuum is the sum of v, w, r other elements of V_{i+1} and arcs in T_{i+1} whose lengths under F_{i+1} total s.

Define $E_i(p, q)$ to be the greatest lower bound of $E_i(v, w)$ where p and q are points of elements v and w respectively of V_i. The argument to show that $\lim E_i(p, q)$ exists and determines a convex metric for $\overline{E}$ is similar to and somewhat easier than that used in the first proof.

Applying Theorem 1 to Theorem 7 we obtain the following.

THEOREM 8. *Each compact, locally connected continuum has a convex metric.*

The application of Lemma 1 and Theorem 1 to Theorem 7 gives Theorems 9 and 10.

THEOREM 9. *Each connected partitionable set has an almost convex metric.*

THEOREM 10. *Each connected set with property* S *has an almost convex metric.*

REFERENCES

1. R. H. Bing, *A convex metric for a locally connected continuum*, Bull. Amer. Math. Soc. vol. 55 (1949) pp. 812–819.

2. ———, *Extending a metric*, Duke Math. J. vol. 14 (1947) pp. 511–519.

3. K. Menger, *Untersuchungen über allgemeine Metrik*, Math. Ann. vol. 100 (1928) pp. 75–163.

4. E. E. Moise, *Grille decomposition and convexification theorems for compact locally connected continua*, Bull. Amer. Math. Soc. vol. 55 (1949) pp. 1111–1121.

5. R. L. Moore, *Concerning connectedness im kleinen and a related property*, Fund. Math. vol. 3 (1922) pp. 232–237.

6. W. Sierpinski, *Sur une condition piur qu'un continu soit une courbe jordanienne,* Fund. Math. vol. 1 (1920) pp. 44–60.

7. G. T. Whyburn, *Concerning S-regions in locally connected continua*, Fund. Math. vol. 20 (1933) pp. 131–139.

8. ———, *Analytic topology*, Amer. Math. Soc. Colloquium Publications, vol. 28, New York, 1942.

UNIVERSITY OF WISCONSIN

COVERINGS WITH CONNECTED INTERSECTIONS

BY

R. H. BING AND E. E. FLOYD

If G is a collection of subsets of a set, then a *subintersection* of G is a non-null set which is the common part of the elements of a subcollection of G.

Suppose that a space X is a compact, locally connected, metric continuum. We show that X has a countable basis whose subintersections are connected and uniformly locally connected. In fact, there is a basis for X with the additional property that the collection of closures of elements of this basis is a family of continuous curves such that each subintersection of this family is a continuous curve. This extends a result of Anderson [1](1) showing that there is a sequence $G_1, G_2, \cdots$ such that G_i is a finite $1/i$-collection of continuous curves covering X and the subintersections of $\sum G_i$ are locally connected.

The notion of partitioning [2, 3, 4] will be used in proving these results. A *partitioning* of X is a finite collection of mutually exclusive connected domains whose sum is dense in X. The partitioning U is a *brick partitioning* if each of its elements is uniformly locally connected and equal to the interior of its closure while the interior of the closure of the sum of two adjacent elements of U is connected and uniformly locally connected. If each element of U is of diameter less than ϵ, U is an *ϵ-paritioning.* In general, if each element of a collection is of diameter less than ϵ, the collection is called an *ϵ-collection.*

The brick partitioning V is a *core refinement* of the brick partitioning U if (a) V is a refinement of U, (b) for each pair of adjacent element u', u'' of U there is a pair of adjacent element v', v'' of V in u' and u'' respectively such that $\bar{v}'+\bar{v}''$ is a subset of the interior of $\bar{u}'+\bar{u}''$, and (c) for each element u of U, the elements of V in u may be ordered $v_0, v_1, \cdots, v_n$ such that $\bar{v}_0$ intersects each $\bar{v}_i$ while $\bar{v}_i$ intersects the boundary of u if and only if $i>0$. We call v_0 a *core element* and $v_1, v_2, \cdots, v_n$ *border elements.*

If B is a subset of X and G is a collection of subsets of X, we use $S(B, G)$ to denote the interior of the closure of the sum of the elements of G which have limit points on $\overline{B}$.

We shall use the following result which was proved in [3].

THEOREM 1. *For each brick partitioning U of X and each positive number ϵ, there is a brick ϵ-partitioning V of X which refines U.*

Although the following result is a corollary of Theorem 6, it is given here since its proof is much simpler than that of Theorem 6.

Presented to the Society, April 28, 1950; received by the editors March 13, 1950.

(1) Numbers in brackets refer to the references cited at the end of the paper.

THEOREM 2. *The space X has a countable basis whose subintersections are connected and uniformly locally connected.*

Proof. Let $U_1, U_2, \cdots$ be a decreasing sequence of brick partitionings of X (U_{i+1} refines U_i and the maximum of the diameters of elements of U_i approaches 0 with $1/i$). The basis is the collection of sets $S(p, U_i)$ for $p \in X$ and $i = 1, 2, \cdots$.

If $G = [g]$ is an infinite subcollection of this basis, Πg is either empty or a single point. The conclusion follows in this case. If G is finite, there is a largest integer n such that some $S(p, U_n)$ is an element of G. Any two elements of U_n in Πg are adjacent; moreover, the interior of the closure of their sum is connected, uniformly locally connected, and contained in Πg. Since Πg is the interior of the sum of the closures of adjacent pairs of elements of U_n, it is connected and uniformly locally connected.

THEOREM 3. *For each brick partitioning U of X and each positive number ϵ there is a brick partitioning V of X such that V is a core refinement of U and each border element of V is of diameter less than ϵ.*

Proof. Since U is a brick partitioning of X, there is a positive number ϵ' less than ϵ such that the common boundary of each pair of adjacent elements of U contains a point that is farther than ϵ' from any other element of U. By Theorem 1 there is a brick ϵ'-partitioning U' of X which refines U. For each element u of U, let T_u be a dendron in u which intersects each element of U' in u. There is a brick $(\delta/2)$-partitioning U'' of X which refines U' where δ is less than the distance between any T_u and the corresponding $X-u$. The core element v_0 of V in u is the set which is maximal with respect to being a connected domain containing T_u and being the interior of the closure of the sum of some elements of U'' whose closures lie in u. The border elements of V in u are components of the intersection of elements of U' with $u-\bar{v}_0$.

THEOREM 4. *For each ordering $u_1, u_2, \cdots, u_n$ of the elements of the brick partitioning U of X and each positive number ϵ there is a brick partitioning V of X such that*

(1) *V is a core refinement of U;*

(2) *each border element of V has a diameter of less than ϵ;*

(3) *if v_1, v_2 are adjacent elements of V, one of the sets $S(v_1, U)$, $S(v_2, U)$ contains the other;*

(4) *if v_1, v_2 are adjacent elements of V and the element of U containing v_1 precedes in $u_1, u_2, \cdots, u_n$ the element of U containing v_2, then $S(v_1, U)$ contains $S(v_2, U)$.*

Proof. We first show that there is a brick partitioning V satisfying (1), (2), and (3). Suppose N is a fixed positive integer. Denote by Lemma N the result obtained by replacing in Theorem 4 conditions (3) and (4) by

(3') *if* v_1, v_2 *are adjacent elements of* V, *one of the sets* $S(v_1, U)$, $S(v_2, U)$ *contains the other if each contains* N *or more elements of* U.

Lemma N holds if N is greater than the number of elements in U. We show that it holds for $N=M$ if it holds for $N=M+1$. Induction then establishes Lemma N; Lemma N for $N=1$ is Theorem 4 with condition (4) deleted.

Let U' be a brick partitioning of X satisfying the conditions of Lemma N for $N=M+1$. Define A to be the set of all points p such that $S(p, U)$ contains at least $M+1$ elements of U. We define a brick partitioning U'' of X whose elements are of two types; (a) each element of U' in $W=S(A, U')$ is an element of U''; (b) if u is an element of U, $u-\overline{W}$ is an element of U''. We note that U'' is a refinement of U and a consolidation of U'. However, it may not be a core refinement of U.

There is a positive number δ_1 so small that if B is a subset of $X-W$ of diameter less than δ_1, $S(B, U)$ does not contain $M+1$ elements of U.

Let δ_2 be the minimum of the distances between nonadjacent elements of U'. We note that if B is a subset of X of diameter less than δ_2 and u' is an element of U' with a limit point on B, then $S(u', U)$ contains $S(B, U)$.

We now describe a brick partitioning V which insures that Lemma N holds for $N=M$. Let V' be a brick partitioning of X which is a core refinement of U'' and such that each border element of V' is of diameter less than min $(\delta_1/2, \delta_2)$. If u'' is an element of U'' of type (b) in an element u of U, the core element of V in u is the interior of the closure of the sum of the elements of V' in u'' whose closures lie in u. The other elements of V are the elements of U' in W and the elements of V' which are not in W and whose closures do not lie in any element of U. We find that V is a core refinement of U.

We now show that the elements of V satisfy conditions (3'). Suppose v_1 and v_2 are two adjacent elements of V such that each of $S(v_1, U)$ and $S(v_2, U)$ contains M or more elements of U. We may suppose that neither v_1 nor v_2 is a core element, for if v_1 is a core element, $S(v_2, U)$ contains $S(v_1, U)$. Hence, if v_i is not in W, it may be supposed to be of diameter less than either $\delta_1/2$ or δ_2.

If both v_1 and v_2 are subsets of W, they are elements of U' and condition (3') holds for them because each of the sets $S(v_1, U)$, $S(v_2, U)$ contains $M+1$ elements of U.

If v_1 is a subset of W and v_2 is not, then v_1 is an element of U'. Since the diameter of $\bar{v}_2$ is less than δ_2, $S(v_1, U)$ contains $S(v_2, U)$.

If neither v_1 nor v_2 is a subset of W, v_1+v_2 is of diameter less than δ_1. Then $S(v_1+v_2, U)$ does not contain $M+1$ elements of U. Hence $S(v_1, U)=S(v_2, U)$.

By induction we find that Lemma N holds for all values N. Since it holds for $N=1$, there is a sequence $U=V_0, V_1, \cdots, V_n$ of brick partitionings of X such that V_{i+1} satisfies (1) and (3) where U is V_i and the diameters of the border elements of V_{i+1} are less than the distance between any nonadjacent elements of V_i and less than the ϵ mentioned in the statement of Theorem 4.

Consider the core partitioning V of U where the core element of V in u_i

is the interior of the closure of the sum of the elements of V_i whose closures lie in u_i. The border elements of V in u_i are the elements of V_i in u_i which are not in this core.

If v_1 and v_2 are two adjacent elements of V in the same element u_i of V, one of the sets $S(v_1, U)$, $S(v_2, U)$ contains the other because if neither v_1 nor v_2 is a core element, then both are elements of V_i. If v_1 and v_2 are adjacent elements, v_1 is in u_i, v_2 is in u_j, and $j>i$, then $S(v_1, U)$ contains $S(v_2, U)$ because the diameter of v_2 is less than the distance between any two nonadjacent elements of V_i. Hence, V satisfies conditions (1), (2), (3), and (4).

THEOREM 5. *Suppose U is a brick partitioning of X and G is a collection of open sets satisfying the following conditions:*

(a) *each element of G is the interior of the closure of the sum of the elements of a subcollection of U;*

(b) *the subintersections of G are connected and uniformly locally connected;*

(c) *For each subcollection of G, the intersection of the closures of the elements of the subcollection is the closure of the intersection of the elements of the subcollection.*

Then for each positive number ϵ there are a brick partitioning V of X and an ϵ-covering H of X such that $G+H$ satisfies the above conditions (a), (b), *and* (c) *with V substituted for U in condition* (a).

Proof. Let U' be a brick $(\epsilon/2)$-partitioning of X that refines U. We note the G satisfies condition (a) with U' substituted for U.

Since U' has only a finite number of elements, G has only a finite number. Let $u_1, u_2, \cdots, u_n$ be an ordering of the elements of U' such that if $i<j$, u_j intersects as many elements of G as u_i does. It follows from condition (c) that if $\bar{u}_i$ intersects $\bar{u}_j$, then each element of G containing u_i also contains u_j.

Let δ be a positive number so small that if B is a subset of X of diameter less than δ, then there exists a point p of X such that $S(p, U')$ contains $S(B, U')$. Suppose V is a core refinement of U' such that V satisfies conditions (3) and (4) of Theorem 4 and the border elements of V are of diameter less than $\delta/2$. Let $v_1, v_2, \cdots, v_m$ be an ordering of the elements of V such that v_i precedes v_j provided either (1) v_i lies in an element of U' which precedes the element of U' containing v_j in the ordering $u_1, u_2, \cdots, u_n$ or (2) v_i and v_j lie in the same element of U' and $S(v_i, U')$ contains more elements of U' than $S(v_j, U')$ does. We note that if $\bar{v}_i$ intersects $\bar{v}_j$ and $i<j$, then each element of G containing v_i also contains v_j.

For each point p, define $h(p)$ to be the interior of the closure of the sum of all elements of V whose closures lies in $S(p, U')$. Let H be the collection of all such sets $h(p)$. We prove that H is an ϵ-covering of X and that $G+H$ satisfies conditions (a), (b), and (c) with V substituted for U in condition (a).

To prove that H is a covering, consider a point q. Since each border element of V is of diameter less than $\delta/2$, there exists a point p such that $S(p, U')$

contains $S[S(q, V), U']$. Then $h(p)$ contains q. As the elements of U' are of diameter less than $\epsilon/2$, each $S(p, U')$ is of diameter less than ϵ and the elements of H are of diameter less than ϵ.

We next verify condition (b). Let J be a subcollection of $G+H$ and π be the intersection of the elements of J. Since π is the interior of the closure of the sum of the elements of a subcollection of V, it is uniformly locally connected. If J contains no element of H, then π is connected by hypothesis. Suppose J contains an element $h(p)$ of H. Let v_i, v_j be elements of V contained in π and u_r, u_s be the elements of U' containing v_i and v_j respectively. Then $\bar{u}_r \cdot \bar{u}_s$ is not null. Hence π contains v_i, v_j, the core elements of V in u_r and u_s, and the border elements of V whose boundaries intersect the closure of no elements of U' except u_r and u_s. Then π is connected.

Finally, we check condition (c). If $\bar{v}_i$ intersects $\bar{v}_j$ and $i<j$, $S(v_i, U')$ contains $S(v_j, U')$. Hence $\bar{v}_j$ lies in $S(v_i, U')$ and if $h(p)$ contains v_i, it also contains v_j. Hence each element of $G+H$ containing v_i also contains v_j. If p is a point of the intersection of the closures of the elements of J, the last element of $v_1, v_2, \cdots, v_m$ having p on its closure is in each element of J. Hence $\bar{\pi}$ contains p and is the intersection of the closures of the elements of J.

THEOREM 6. *The space X has a countable basis G whose subintersections are connected and uniformly locally connected and such that if G' is a subcollection of G, then the intersection of the closures of the elements of G' is the closure of the intersection of the elements of G'.*

Proof. Let G_0 be the covering of X whose only element is X itself, and U_0 be the brick partitioning of X whose only element is X itself. Repeated applications of Theorem 5 give a sequence $G_0, G_1, \cdots$ of coverings of X such that G_i $(i\geqq 1)$ is a $(1/i)$-covering and such that $G_1+G_2+\cdots+G_i$ satisfies conditions (b) and (c). Define G to be $\sum G_i$ and the theorem follows. The following result is a consequence of Theorem 6.

THEOREM 7. *For each positive integer i, X is the sum of a finite $(1/i)$-collection G_i of continuous curves such that each subintersection of $\sum G_i$ is a continuous curve.*

REFERENCES

1. R. D. Anderson, *Concerning upper semi-continuous collections of continua*, Trans. Amer. Math. Soc. vol. 67 (1949) pp. 451–460.

2. R. H. Bing, *Partitioning a set*, Bull. Amer. Math. Soc. vol. 55 (1949) pp. 1101–1110.

3. ———, *Complements of continuous curves*, Fund. Math. vol. 36 (1950) pp. 303–318.

4. E. E. Moise, *Grille decomposition and convexification theorems for compact locally connected continua*, Bull. Amer. Math. Soc. vol. 55 (1949) pp. 1111–1121.

UNIVERSITY OF WISCONSIN,
MADISON, WIS.
UNIVERSITY OF VIRGINIA,
CHARLOTTESVILLE, VA.

PARTITIONING CONTINUOUS CURVES

R. H. BING

I. Introduction

1. **The continuous curve.** By a continuous curve we mean a compact, locally connected, metric continuum. For definitions of locally connected, continuum, and other terms used in this paper, see either [**18**] or [**21**]. Those who like to visualize topology concretely may wish to think of a continuous curve as a chunk out of Euclidean 3-space—one that is connected (all in one piece), one that is bounded (lies on the interior of a sphere), and one that is locally connected (nearby points belong to small connected subsets). A wad of paper, an irregularly shaped rock, or the earth itself may be considered as examples. However, our remarks about continuous curves will apply equally well to those in Euclidean spaces of all dimensions and to those in a Hilbert space.

When Jordan first introduced the term continuous curve, he defined it analytically to be the image (in the plane) of a straight line interval under a continuous transformation. It was not until over twenty years later that it was discovered that any compact locally connected metric continuum was the image of a straight line interval under a continuous transformation and conversely. This interesting and unusual discovery adds spice to the study of mathematics [**24**, p. 12]. Another interesting aspect of this discovery is that it was made independently by two mathematicians, Hahn and Mazurkiewicz. Since Peano had shown earlier that a square plus its interior is the image of a straight line interval, a continuous curve is sometimes called a Peano continuum.

In this discussion we shall be interested in the continuous curve itself and not the continuous transformation of an interval. Hence, we use the definition in the first paragraph rather than the analytic one. In this discussion we shall be interested in the structure of a continuous curve.

2. **Examples of continuous curves.** A straight line interval, a square plus its interior in the plane, and a cube plus its interior in 3-space are examples of continuous curves. In fact, any closed n-cell or n-simplex is an example. One can get a less familiar continuous

An address delivered before the Minneapolis meeting of the Society, September 6, 1951, by invitation of the Committee to Select Hour Speakers for Annual and Summer Meetings; received by the editors March 5, 1952.

curve by taking a continuum which is the sum of a finite number of dissimilar continuous curves.

If M is any compact point set whatever, it is possible to obtain a continuous curve by adding to M a countable collection of arcs. We use this fact in constructing the following two examples of continuous curves which we shall use.

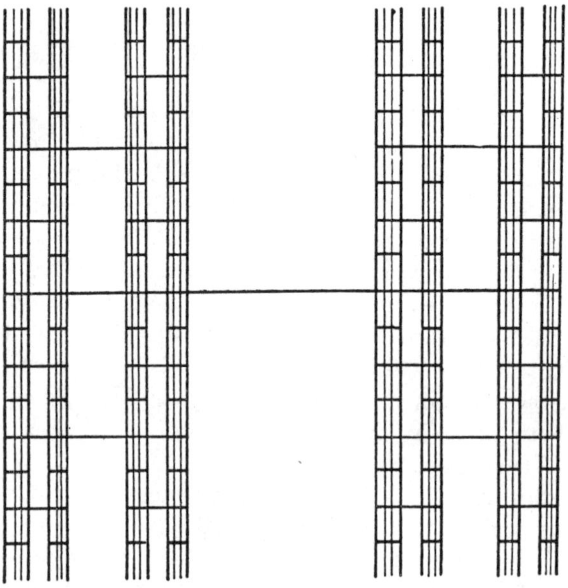

FIG. 1

EXAMPLE A. Let M be the cartesian product of a straight line interval and a Cantor set. Form a continuous curve by adding to M a decreasing sequence of horizontal straight line intervals as indicated in Figure 1. These intervals extend from (0, 1/2) to (1, 1/2), and (0, 1/4) to (1/3, 1/4), from (2/3, 1/4) to (1, 1/4), from (0, 3/4) to (1/3, 3/4), $\cdots$.

EXAMPLE B. It is shown in [5] that there is a bounded continuum S in Euclidean 3-space that is irreducible with respect to separating 3-space and has the property that if two of its subcontinua intersect, one is a subset of the other. The continuum S is 2-dimensional. It is called hereditarily indecomposable because no subcontinuum of it is the sum of two proper subcontinua of the subcontinuum. A continuous curve M may be formed by adding to S the sum of a decreasing sequence of mutually exclusive arcs. If K is a subset of M irreducible with respect to separating M, each component of K is indecomposable.

3. **Partitioning.** An interval can be partitioned. This fact is used in defining the integral of a continuous function over an interval. ($\int_a^b f(x)dx = \lim \sum f(\xi_i)\, \Delta x_i$ where ξ_i is a point of and Δx_i is the length of the ith element of the partitioning of $[a, b]$.)

In studying a square plus its interior it is sometimes convenient to partition it into small pieces. For example, in integrating over a square region, we may partition it into small rectangles (if we use rectangular coordinates) or into pieces of other shapes (in polar coordinates for instance).

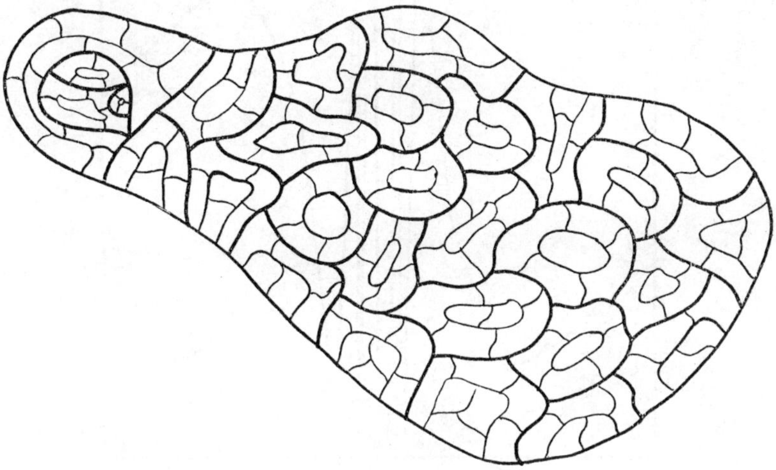

FIG. 2

A partitioning of a continuous curve M is a finite collection G of mutually exclusive connected open subsets of M whose sum is dense in M [6]. If the mesh of G is less than ϵ (each element of G is of diameter less than ϵ), G is called an ϵ-partitioning. The essential difference between a partitioning and a grating [13, p. 279] is that the elements of a partitioning are connected. It is much like a grille decomposition [15].

In defining the integral over a plane domain R ($\int\int_R f(x, y)dS = \lim$ (mesh of partitioning $\to 0$) $\sum f(\xi_i, \eta_i)\Delta S_i$) we do not need to impose the conditions that the elements of the partitioning are either connected, mutually exclusive, or open. However, we shall be using partitionings in connections where these properties are important.

4. **Decreasing sequence of partitionings.** If G and H are two partitionings of the same continuous curve, G is called a refinement of H if each element of G is a subset of an element of H. In Figure 2, the partitioning represented by light lines is a refinement of the one represented by heavy lines. A sequence $G_1, G_2, \cdots$ of partitionings is called a decreasing sequence if, for each positive integer i, G_{i+1} is a refinement of G_i and the mesh of G_i approaches 0 as i increases without limit.

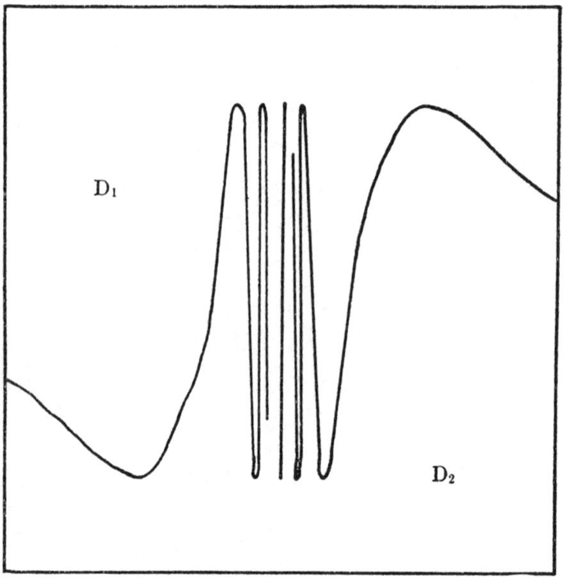

FIG. 3

While it is apparent that there is a decreasing sequence of partitionings of a square plus its interior, it is not so obvious that every continuous curve has such a sequence of partitionings. However, it is true that any continuous curve has such a sequence. This has given us new information about the continuous curve.

5. **Purpose of paper.** It is only within the last three years that it has been known that each continuous curve can be partitioned [**6**; **4**; **15**; **16**]. It is the purpose of this paper to point out some recent developments, to show ways in which a continuous curve can be partitioned, and to point out analogies between the ways in which a square plus its interior can be partitioned and the ways in which an arbitrary continuous curve can be partitioned. Details of some proofs

are included for completeness, but some readers may prefer to skip them.

II. Breaking a continuous curve into two pieces

1. The type of division needed. If one pursues the proper attack, it is not difficult to show that any continuous curve can be partitioned. However, if one gets off on the wrong track, considerable difficulty may be encountered.

Suppose it is desired to partition a continuous curve into pieces of diameters less than a fixed positive number ϵ. A straightforward method of attack is to divide the continuous curve into two pieces, divide each of the two pieces, and continue this procedure until an ϵ-partitioning results. However, unless certain precautions are taken, this method does not succeed.

If $\epsilon=1$, M is a square plus its interior, and at the first stage it is partitioned into two pieces as shown in Figure 3, the above procedure will not lead to a partitioning. The common boundary of D_1 and D_2 is a part of the closure of the graph of $y=\sin 1/x$ and neither D_1 nor D_2 is the sum of a finite number of connected subsets each of diameter less than 1.

A set has property S if, for each positive number ϵ, the set is the sum of a finite number of connected subsets each of diameter less than ϵ. This notion was used by Sierpinski and when R. L. Moore used it later, he called it property S [**19**; **17**; **18**].

A precaution to take in partitioning a continuous curve is to divide it at each step so that the pieces have property S.

We shall be interested in the following result [**6**].

THEOREM 1. *If H and K are two mutually exclusive closed subsets of a continuous curve M, there are two mutually exclusive open subsets D_H and D_K of M containing H and K respectively such that* (a) *each of D_H and D_K has property* S *and* (b) $M=\overline{D}_H+\overline{D}_K$.

Before proving this theorem, we shall review the method of constructing a set with property S.

2. Gradual growing process. This is a method used to form sets with property S. Compare with [**22**; **21**].

A set H is said to ϵ-grow into a set H' in a set M if M contains H', each point of H' belongs to a connected ϵ-subset (subset of diameter less than ϵ) of H' that intersects H, and for some positive number δ, H' contains all connected δ-subsets of M that intersect H.

Let M be a set with property S, H be a subset of M, and ϵ be a positive number. Suppose one wishes to obtain a subset of M with property S which contains H but no point whose distance is more than ϵ from H. Such a set can be obtained by the following gradual growing process. Let $\epsilon_1, \epsilon_2, \cdots$ be a sequence of positive numbers whose sum is no more than ϵ. Let H ϵ_1-grow into a set D_1 in M, D_1 ϵ_2-grow into a set D_2 in $M, \cdots, D_i$ ϵ_{i+1}-grow into a set D_{i+1} in $M, \cdots$. Then the sum D_0 of $D_1, D_2, \cdots$ is an open subset of M, has property S, contains H but no point at a distance more than ϵ from H.

3. **An attempt to divide.** One might try to prove Theorem 1 by letting each of the sets H and K grow gradually in M into mutually exclusive open subsets whose sum is dense in M. Some difficulty may be encountered in trying to keep the two subsets from intersecting and at the same time making their sum dense in M. This difficulty could be overcome if we could answer the following question in the affirmative.

QUESTION. Does there exist a positive integer n such that the following result holds for each continuous curve M, each positive number ϵ, and each pair of mutually exclusive closed subsets H and K of M? *If R is a finite subset of M such that each point of R belongs to an arc in M of diameter less than ϵ that intersects $H+K$, there are two collections A_H and A_K of arcs satisfying the following conditions*:

(a) *Each element of A_H intersects H but not K and each element of A_K intersects K but not H nor any element of A_H.*

(b) *Each element of R belongs to an element of A_H+A_K.*

(c) *Each element of A_H+A_K is of diameter less than $n\epsilon$.*

We cannot let $n=1$ and always get such collections A_H and A_K, for let M be the sum of the interval from $(-10, -1)$ to $(10, 1)$ and the interval from $(-10, 1)$ to $(10, -1)$, $H=\{(-10, 1)\}$, $K=\{(10,1)\}$, $R=\{(-10, -1), (10, -1)\}$, and $\epsilon=11$. Had the answer been in the affirmative for $n=1$, the attempts made in [15] to partition a continuous curve might have succeeded [16].

Since we do not know the answer to the above question, we shall get some other results before proving Theorem 1.

4. **A crude division.** Before proving Theorem 1, it is convenient to prove a weaker theorem.

THEOREM 1′. *Theorem* 1 *is true if condition* (a) *is replaced by the weaker conditions that D_H has property* S *and D_K has only a finite number of components.*

There are two essentially different methods of proving Theorem 1′. One method is due to Whyburn [22, Corollary, p. 137]. He considers an uncountable family of sets with property S such that of any two of the sets, one contains the closure of the other. He shows that one of the sets can be modified so that it still has property S and the complement of its closure has a finite number of components.

An essentially different method is to let H grow gradually and K in a wild fashion to form sets D_H and D_K satisfying the conditions of Theorem 1′ [6]. This method is similar to another attack we shall use later so we include the details.

Details of second method of proof of Theorem 1′. The open subsets D_H and D_K are formed by permitting H and K to grow so that H grows in a convergent fashion. The procedure is as follows.

We use $\rho(p, q)$ to denote the distance between p and q.

Suppose that, at the ith stage, H and K have been enlarged to sets H_i and K_i such that $\rho(H_i, K_i)>0$ and each point of M belongs to an arc in M of diameter less than $1/2^i$ that intersects H_i+K_i. Let R_i be a finite subset of M such that each point of M belongs to an arc in M of diameter less than $1/2^{i+1}$ that intersects R_i. For each point p of R_i let A_p be an arc in $M-\overline{H}_i$ from p to a point of K_i if there is one; if there is no such arc A_p, let B_p be an arc of diameter less than $1/2^i$ in $M-\overline{K}_i$ from p to a point of H_i.

Use H_i' to denote H_i plus the sum of the B_p's, K_i' to denote K_i plus the sum of the A_p's, and δ_i to denote $\rho(H_i', K_i')$. Let $H_{i+1} = \{p \mid p$ lies on an arc of diameter less than $\delta_i/3$ in M that intersects $H_i'\}$ and $K_{i+1}=\{p \mid p$ lies on an arc of diameter less than $\delta_i/3$ in M that intersects $K_i'\}$.

Since H_i has ϵ_i-grown into H_{i+1} for $\epsilon_i=7/(2^{i+1}\cdot 3)$ and the ϵ_i's have a finite sum, $\sum H_i=D_H$ has property S.

Since K_i has only a finite number of components and each component of K_{i+1} intersects K_i, $\sum K_i=D_K$ has only a finite number of components.

Since no point of M is at a distance of more than $1/2^i$ from H_i+K_i, $M=\overline{D}_H+\overline{D}_K$.

5. **Proof of Theorem 1.** By using Theorem 1′, it is shown in Theorem 3′ of the next section that for each positive number ϵ, there is an ϵ-partitioning of M. We use this result in establishing Theorem 1. For each integer i, let G_i be a $1/2^i$-partitioning of M.

The general idea is to let both H and K grow gradually into sets D_H and D_K. Suppose at the ith stage H and K have been enlarged to H_i and K_i so that $\rho(H_i, K_i)>0$ and H_i+K_i intersects each element of G_i. Let $\{p_j\}$ be a finite number of points such that each element of G_{i+1} contains an element of $\{p_j\}$ and each point of $\{p_j\}$ lies in an element of G_i.

If there is an arc in both $M-\overline{H}_i$ and an element of G_i from p_j to a point of K_i, let A_j be such an arc; if there is no such arc, let B_j be an arc in both $M-\overline{K}_i$ and an element of G_i which is irreducible from p_j to $\overline{H}_i$.

Let $\rho(H_i', K_i')=\delta_i$ where H_i' is the sum of H_i and all the B_j's while K_i' is the sum of K_i and all the A_j's.

Suppose $H_{i+1}=\{p \mid$ there is an arc in M of diameter less than $\delta_i/3$ from p to $H_i'\}$ and $K_{i+1}=\{p \mid$ there is an arc in M of diameter less than $\delta_i/3$ from p to $K_i'\}$.

The sets H_i and K_i have ϵ_i-grown into H_{i+1} and K_{i+1} where $\epsilon_i=7/(2^{i+1}\cdot 3)$. Hence, the sets $D_H=\sum H_i$ and $D_K=\sum K_i$ have property S. Since H_i+K_i intersects each element of G_i, $M=\overline{D}_H+\overline{D}_K$.

III. Partitioning

1. Treating S-sets as continuous curves. It is necessary that a set M have property S in order to be partitionable. (A set M is partitionable if for each positive number ϵ, there is an ϵ-partitioning of it.) The converse is true but it is not so obvious. However this converse follows from the facts that a continuous curve is partitionable and that any connected set with property S can be imbedded in a continuous curve so that any partitioning of the continuous curve produces a partitioning of it with no larger mesh.

If G is a partitioning of a continuous curve, the pieces of G are not continuous curves. However, if these pieces have property S, they can be satisfactorily dealt with.

Suppose M is a connected set with property S and $R(p, q)$ is the relative distance for M, that is, $R(p, q)$ is the greatest lower bound of the diameters of all connected subsets of M containing $p+q$. It may be shown that $R(p, q)$ preserves the topology of M and the diameter of any connected set is not increased under $R(p, q)$. Hence, any ϵ-partitioning of M under the metric $R(p, q)$ is an ϵ-partitioning of it under its original metric.

If we add only enough points to M to make it complete under $R(p, q)$, the resulting set M' is a continuous curve. Any ϵ-partitioning of M' induces an ϵ-partitioning of M. In fact we have the following result [6, Lemma 1; 21, pp. 154–158].

THEOREM 2. *If M is a connected set with property* S *with a relative distance metric, M' is the complete enclosure of M, and D is a connected open subset of M', then $M\cdot D$ is a connected open subset of M.*

2. Crude partitioning. Before showing that a continuous curve can be partitioned in various fashions, we show that it can be partitioned in some manner.

THEOREM 3′. *For each positive number ϵ and each continuous curve M, there is an ϵ-partitioning G of M.*

We use the following corollary which results from applying Theorems 1′ and 2 to each component of M.

COROLLARY OF THEOREM 1′. *Suppose that the set M has property* S *and that H and K are two of its subsets such that $\rho(H, K)>0$. Then M contains two mutually exclusive open subsets D_H and D_K containing H and K respectively such that D_H+D_K is dense in M, D_H has property* S, *and D_K has only a finite number of components.*

The general idea in proving Theorem 3′ is to use Theorem 1′ and divide M into two pieces D_H and D_K so that D_H has property S, D_K has only a finite number of components and is of diameter less than ϵ. We apply the corollary of Theorem 1′ to D_H and divide it into two pieces one of which has property S and the other of which has only a finite number of components each of diameter less than ϵ. The procedure is continued so as to get an ϵ-partitioning of M.

DETAILS OF PROOF OF THEOREM 3′. Let $(p_1, p_2, \cdots, p_n)$ be a finite subset of M such that no point of M is at a distance as much as $\epsilon/4$ from $\{p_i\}$. Let K_i be the set of all points of M that are at a distance from p_i of less than $\epsilon/4$ and H_i be the set of those that are at a distance from p_i of more than $\epsilon/3$.

Applying Theorem 1′ to M, we find that there are mutually exclusive subsets D_{H_1} and D_{K_1} of M containing H_1 and K_1 respectively such that $\overline{D}_{H_1}+\overline{D}_{K_1}=M$, D_{H_1} has property S, and D_{K_1} has only a finite number of components. The components of D_{K_1} are of diameters less than ϵ and are elements of G. We divide the components of D_{H_1} further.

If D_{H_1} intersects both H_2 and K_2, we apply the corollary of Theorem 1′ to D_{H_1} and find that there are open subsets D_{H_2} and D_{K_2} of D_{H_1} containing $H_2\cdot D_{H_1}$ and $K_2\cdot D_{H_1}$ respectively such that $D_{H_2}+D_{K_2}$ is dense in D_{H_1}, D_{H_2} has property S, and D_{K_2} has only a finite number of components. The components of D_{K_2} are elements of G. If K_2 does not intersect D_{H_1}, we do not add elements to G at this stage but let $D_{H_2}=D_{H_1}$.

The process is continued. For each positive integer i less than $n-1$ for which D_{H_i} intersects both H_{i+1} and K_{i+1} we apply the corollary to D_{H_i} and get mutually exclusive open subsets $D_{H_{i+1}}$, $D_{K_{i+1}}$ of D_{H_i} containing $H_{i+1}\cdot D_{H_i}$ and $K_{i+1}\cdot D_{H_i}$ respectively such that $D_{H_{i+1}}+D_{K_{i+1}}$ is dense in D_{H_i}, $D_{H_{i+1}}$ has property S, and $D_{K_{i+1}}$ has only a finite number of components. If D_{H_i} does not intersect K_{i+1}, we suppose $D_{K_{i+1}}=0$ and $D_{H{i+1}}=D_{H_i}$.

The elements of G are the components of $D_{K_1}, D_{K_2}, \cdots, D_{K_{n-1}}$, and $D_{H_{n-1}}$.

3. **A more useful partitioning.** If G is an arbitrary partitioning of a continuous curve as described in Theorem 3′, it may not be the member of any decreasing sequence of partitionings because some element of it may not have property S. However, if we use Theorem 1 instead of Theorem 1′, we obtain the following extension of Theorem 3′.

THEOREM 3. *For each continuous curve M and each positive number ϵ there is an ϵ-partitioning G of M such that each element of G has property* S.

A partitioning whose pieces have property S is called an S-partitioning. If its mesh is less than ϵ, it is an ϵ-S-partitioning.

4. **Decreasing sequence of partitionings.** Theorem 2 may be applied to any set with property S to get the following corollary of Theorem 3.

COROLLARY OF THEOREM 3. *For each positive number ϵ and each set M with property* S, *there is an ϵ-*S*-partitioning of M.*

Using the corollary we find that if G is an S-partitioning of a continuous curve and ϵ is a positive number, each piece of G may be ϵ-S-partitioned. Hence, we have the following useful result [6].

THEOREM 4. *Each continuous curve has a decreasing sequence of partitionings.*

IV. CORE REFINEMENTS

In Figure 2, the partitioning represented by light lines is a core refinement of the one represented by heavy lines.

A partitioning is regular if each of its elements is the interior of its closure. This prevents pieces from having unnecessary boundary points.

If the partitioning H is a refinement of the partitioning G, the elements of H which have a boundary point in common with the boundary of an element of G are called border elements of H. Other elements are called interior elements.

The partitioning H is called a core refinement of the partitioning G if the following conditions are satisfied:

(a) Each border element of H is adjacent to an interior element.

(b) For each element g of G, the closure of the sum of the interior elements of H in g is connected.

(c) Both G and H are regular partitionings.

In [8] additional conditions were used in defining a core refinement for brick partitionings but we shall not need them here. We regard the interior of the closure of the sum of the interior elements of H in g as the core of g.

In some of the applications of partitionings, use is made of the following result.

THEOREM 5. *For each continuous curve M there is a decreasing sequence $G_1, G_2, \cdots$ of partitionings of M such that G_{i+1} is a core refinement of G_i.*

DETAILS OF PROOF. We prove this theorem by showing that if G is a regular S-partitioning of M and ϵ is a positive number, there is an ϵ-S-partitioning H which is a core refinement of G.

For each element g of G, let H' be a regular ϵ-S-partitioning of g. There is a dendron T in g that intersects each element of H'. Let H'' be a regular S-partitioning of g that refines H' and has a mesh less than the distance between T and the boundary of g.

Each element h'' of H'' with a boundary point on the boundary of g is a subset of a border element of H. If h' is the element of H' containing h'', the border element h of H containing h'' is the interior of the closure of the sum of all elements k of H'' such that there is an arc in h' from h'' to k that does not intersect the closure of any element of H'' whose closure intersects T. All elements of H'' not in such a border element of H are interior elements of H.

V. THE CONVEXIFICATION PROBLEM

1. The method of solution. A metric is a convex metric for a continuous curve M if for each pair of points of M there is a point of M halfway between them. In 1928 Menger raised [14] the question as to whether or not each continuous curve has a convex metric that preserves its topology. During the following twenty years a number of contributions [1; 3; 9; 10; 12] were made toward the solution of this question but it was not until the notion of partitioning was introduced that the final answer was given [6; 15].

THEOREM 6. *Each continuous curve has a convex metric.*

The fact that each continuous curve has a decreasing sequence $G_1, G_2, \cdots$ of partitionings is used in showing that each continuous curve has a convex metric. Each piece of each G_i is assigned a size and the distance between two points p, q is defined in terms of the sizes of the chains from p to q whose links are elements of the G_i's.

2. **Definition of convex metric.** Let $G_1, G_2, \cdots$ be a sequence of partitionings of M such that G_{i+1} is a core refinement of G_i and the mesh of G_{i+1} is less than one third the distance between any two nonadjacent elements of G_i.

The elements of G_1 are assigned a size of $1/2$. Each border element of G_{i+1} is assigned a size equal to one half the size of the element of G_i containing it. Each interior element of G_{i+1} is given a size equal to $1/(2^{i+1}\cdot N_{i+1})$ where N_{i+1} is the number of elements in G_{i+1}.

If K is a continuum which is equal to the closure of the sum of the elements of G_i, the ith size of K is the sum of the sizes of the elements of G_i in K. We define $E_i(p, q)$ as the minimum of the ith sizes of all continua containing $p+q$. It may be noted that if $E_i(p, q)$ is the ith size of K, then K is the sum of a chain $C_1, C_2, \cdots, C_n$ of continua such that C_1 contains p, C_n contains q, C_j intersects C_k if and only if $|j-k| \leqq 1$ and C_j is the closure of an element of G_i. The convex metric for M is $D(p, q) = \lim E_i\,(p, q)$.

We note that $D(p, p)=0$ and $D(p, q)\geqq 0$. Also $D(p, q)=D(q, p)$ because $E_i(p, q)=E_i(q, p)$.

If p, q, r are three points, $E_i(p, q)+E_i(q, r)\geqq E_i(p, r)$ because if K_{pq} and K_{qr} are continua with ith sizes containing $p+q$ and $q+r$ respectively, the ith size of the continuum $K_{pq}+K_{qr}$ is no more than the sum of the ith sizes of K_{pq} and K_{qr}. Hence, $D(x, y)$ satisfies the triangle condition.

We now show that if p is a limit point of the set A, then $D(p, A)=0$. Since $E_i(p, q)+1/2^{i+1}\geqq E_{i+1}(p, q)$, $D(p, q)\leqq E_i(p, q)+1/2^i$. However, there is a point q of A such that $p+q$ belongs to the closure of the same element of G_i. Then $D(p, q)\leqq 1/2^{i-1}$ and $D(p, a)$ (a element of A) is not bounded from 0.

We now show that if p is not a point of the closed set A, $D(p, A) >0$. Suppose i is an integer such that no element of G_i is adjacent to both an element of G_i whose closure contains p and also to an element of G_i whose closure intersects A. If ϵ is the least of the sizes of the elements of G_i, we shall show that $D(p, A)\geqq\epsilon$.

If K is a continuum containing p, intersecting A, and having an $(i+k)$th size, K contains 2^k elements g of G_{i+k} such that g is a border element of G_{i+k} and lies in border elements of $G_{i+1}, G_{i+2}, \cdots, G_{i+k-1}$. The size of g is as much as $\epsilon/2^k$ and the $(i+k)$th size of K is more than ϵ. Hence, for each point a of A, $E_{i+k}(p, a)>\epsilon$ and $D(p, a)\geqq\epsilon$.

We show that the metric $D(x, y)$ is convex by showing that if p and q are two points, there is a point halfway between them. Let K be a continuum containing $p+q$ of ith size $E_i(p, q)$. Then K is the sum of two subcontinua K_p and K_q containing p and q respectively each of whose ith sizes differ from one half the ith size of K by less than $1/2^i$. There is a point r_i in $K_p \cdot K_q$ such that neither $E_i(p, r_i)$ nor $E_i(q, r_i)$ differs from $E_i(p, q)/2$ by more than $1/2^i$. If r is a limit point of $r_1, r_2, \cdots$, it is halfway between p and q under the metric $D(x, y)$.

Although a solution to the convexification problem has been given by means of partitionings, it would still be interesting to see the answer given by other methods.

3. **Unique segments.** If a circle is given a convex metric, for each point p there is a point q and two points r_1, r_2 such that both r_1 and r_2 are halfway between p and q. There is not a unique segment from p to q.

Although under a convex metric there is not a unique segment between each pair of points of a circle, there is a dense subset of a circle such that each pair of points of this dense subset is joined by a unique segment. In another paper I shall show that each continuous curve has such a convex metric. The result may be stated as follows.

THEOREM 7. *Each continuous curve has a dense subset R and a convex metric D such that each pair of points of R are the end points of a unique segment under D.*

For the surface of a sphere, there is a convex metric and a positive number ϵ such that each pair of points whose distance apart is less than ϵ are end points of a unique segment. It would be interesting to know a topological characterization of such sets.

A closed n-cell in a Euclidean space has a convex metric such that if two segments intersect in more than one point, the sum of these two segments is a segment. It would be interesting to obtain a topological characterization of continuous curves that can be given such a convex metric.

Each continuous curve with a convex metric that gives unique segments can be shrunk to each of its points in a very nice fashion. One might wonder if those continuous curves with such nice convex metrics could be characterized by the fashion in which they can be shrunk to points.

VI. Mapping continuous curves onto intervals

1. **Introduction.** We have seen that if H and K are two mutually exclusive closed subsets of a continuous curve M, M can be divided into two mutually exclusive open subsets D_H and D_K containing H and K respectively such that each of the sets D_H, D_K has property S. We shall now see how uncountably many such divisions can be made.

We map M onto a straight line interval $[0, 1]$ so that H goes into 0, K goes into 1, and the inverse of each nondegenerate connected subset of $[0, 1]$ has property S [**4**]. Then except for a countable number of points x of $[0, 1]$, the collection of components of $T^{-1}[0, x)$ and $T^{-1}(x, 1]$ is an S-partitioning of M. We use $(a, b]$ to denote the closed interval $[a, b]$ minus the point a.

2. **Details of the transformation.** To prove the theorem regarding the mapping of M onto $[0, 1]$ we first prove the following lemma.

LEMMA FOR THEOREM 8. *Suppose C_1 and C_2 are mutually exclusive closed subsets of a continuous curve M, $M-C_i$ $(i=1, 2)$ has property* S, *and ϵ is a positive number such that C_j $(j\neq i)$ can ϵ-grow into $M-C_i$. Then for each positive number δ, M contains mutually exclusive open subsets D_1 and D_2 satisfying the following conditions:*

(a) *C_i can δ-grow into D_i and $(\epsilon+\delta)$-grow into $M-D_j$.*

(b) *$M-(D_1+D_2)$ can δ-grow into $M-(C_1+C_2)$.*

(c) *D_i and $M-D_i$ have property* S.

PROOF OF LEMMA. Let G be a γ-S-partitioning of $M-(C_1+C_2)$ where γ is less than either δ or $\rho(C_1, C_2)/3$.

Let A_i be a finite collection of arcs in $M-C_j$ $(i\neq j)$ such that each of these arcs is of diameter less than ϵ, each intersects C_i, and each element of G intersects both an arc of A_1 and also an arc of A_2.

Let G' be a γ'-S-partitioning of $M-(C_1+C_2)$ which refines G, where γ' is less than the distance between C_i and any arc of A_j $(i\neq j)$. Let R' be the closure of the sum of all elements g of G' such that g either lies in an element of G not adjacent to C_1+C_2 or g lies in an element of G adjacent to C_i and $\bar{g}$ intersects an arc in A_j $(i\neq j)$. Then R is the sum of R' and the closure of all elements g' of G' of the following sort: g' lies in an element g of G which is adjacent to C_i but R' separates g' from C_i in $g+C_i$.

We define D_i to be the interior of the closure of the sum of C_i and all elements of G' which do not lie in R but lie in an element of G adjacent to C_i.

Each point of D_i-C_i belongs to an arc intersecting C_i and lying in both D_i and also in the closure of an element of G. This shows the first part of condition (a) is satisfied. An arc showing that the second part of this condition is satisfied may be obtained by going out an element of A_i and then continuing in the closure of an element of G.

To see that condition (b) is satisfied, note that each point of $M-(C_1+C_2)$ belongs to an arc that intersects R and lies in both $M-(C_1+C_2)$ and also in the closure of an element of G.

Condition (c) is satisfied because G' is an S-partitioning.

THEOREM 8. *If H and K are mutually exclusive closed subsets of the continuous curve M, there is a continuous transformation T of M onto the interval* $[0, 1]$ *such that* (a) $T(H)=0$ *and* $T(K)=1$, (b) *if W is a nondegenerate connected subset of* $[0, 1]$, *$T^{-1}(W)$ has property* S, (d) *if* 0 *belongs to W, each component of $T^{-1}(W)$ intersects H and if* 1 *belongs to W, each component of $T^{-1}(W)$ intersects K.*

We prove this result by repeated applications of the preceding lemma. To get a start we obtain $C_1=T^{-1}(0)$ and $C_2=T^{-1}(1)$. Let G be an ϵ-S-partitioning of M where $\epsilon=\rho(H, K)/3$. Then C_1 is the sum of the closures of all elements of G whose closures intersect H plus all points that this sum separates from K. Similarly, C_2 is the sum of the closures of all elements of G whose closures intersect K plus all points that this sum separates from H.

In our first application of the lemma we use $T^{-1}(0)$, $T^{-1}(1)$, $1/2$ for C_1, C_2, δ. The sets $M-(D_1+D_2)$, D_1, D_2 we obtain are called $T^{-1}(1/2)$, $T^{-1}[0, 1/2)$, $T^{-1}(1/2, 1]$.

In our next application of the lemma we let $T^{-1}(0)$, $T^{-1}[1/2, 1] =T^{-1}(1/2)+T^{-1}(1/2, 1]$, $1/2$, $1/4$ be the C_1, C_2, ϵ, δ of the lemma and denote by $T^{-1}(1/4)$, $T^{-1}[0, 1/4)$, $T^{-1}(1/4, 1]$ the sets $M-(D_1+D_2)$, D_1, D_2 obtained.

In general, suppose that $T^{-1}[0, i/2^j]$ and $T^{-1}[(i+1)/2^j, 1]$ have been defined. We apply the lemma using $T^{-1}[0, i/2^j]$, $T^{-1}[(i+1)/2^j, 1]$, $1/2^j$, $1/2^{j+1}$ for C_1, C_2, ϵ, δ and denote by $T^{-1}[(2i+1)/2^{j+1}]$, $T^{-1}[0, (2i+1)/2^{j+1})$, $T^{-1}((2i+1)/2^{j+1}, 1]$ the sets $M-(D_1+D_2)$, D_1, D_2 obtained.

After $T^{-1}(x)$ has been defined for all numbers of $[0, 1]$ of the form $x=i/2^j$, the definition of T is continued in a continuous fashion—that is, $T^{-1}(x)$ is the intersection of all closed sets $T^{-1}[i/2^j, (i+1)/2^j]$ for which $i/2^j\leqq x\leqq(i+1)/2^j$.

In the next section we need to use the following information about the transformation T.

THEOREM 9. *If T is a continuous transformation mentioned in Theorem* 8, *then except for a countable number of values x* $(0\leqq x\leqq 1)$, *both $T^{-1}[0, x)$ and $T^{-1}(x, 1]$ are uniformly locally connected. Furthermore, if M is the closure of a uniformly locally connected domain D and B is a subset of the boundary of D, except for a countable number of values x, each point of $T^{-1}(x)\cdot B$ is a limit point of both $T^{-1}[0, x)\cdot B$ and $T^{-1}(x, 1]\cdot B$.*

A set D is uniformly locally connected if for each positive number ϵ there is a positive number δ such that each pair of points of D

whose distance apart is no more than δ belongs to a connected subset of D of diameter less than ϵ.

The proof of Theorem 9 is contained in the proof of Theorem 5 of [4].

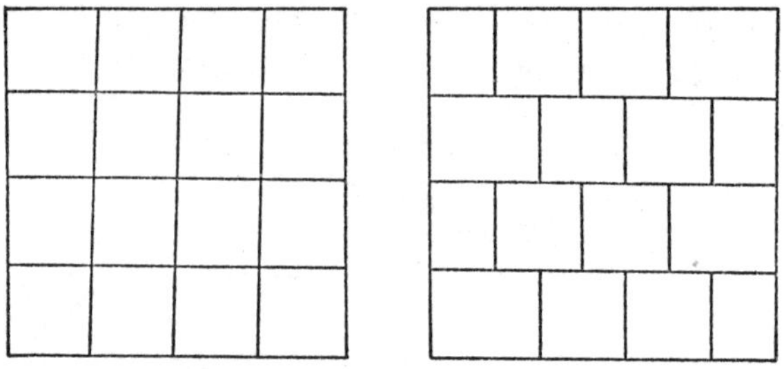

a b

FIG. 4

VII. BRICK PARTITIONING

Figure 4 shows two ways in which a square plus its interior may be partitioned. The partitioning represented in Figure 4b is called a brick partitioning. A partitioning G is a brick partitioning if each of its elements is uniformly locally connected and equal to the interior of its closure while the interior of the sum of the closures of each pair of elements of G is uniformly locally connected. If G is a brick partitioning, not only can the elements of G be subdivided so as to form a partitioning G' which refines G but also adjacent elements of G and part of their common boundary can be added so as to form a partitioning G'' such that G refines G''.

In Figure 2, the left-most piece of the partitioning shown by heavy lines is not uniformly locally connected. Also, the interior of the sum of the closures of the two left-most pieces is not uniformly locally connected. If the three left-most pieces are replaced by the interior of the sum of their closures, a brick partitioning results.

Wilder showed [**24**, Theorem 3.4, p. 79] that if p is a point of a continuous curve M and D is a domain containing p, there is a uniformly locally connected domain E containing p and lying in D. However, to get a brick partitioning, we need to know that there is such a domain E such that $M-\overline{E}$ is uniformly locally connected. We get this result from Theorems 8 and 9.

It would be interesting to know if a modification of the methods used by Wilder would give this result. It appears that Moise had this in mind in [**15**].

Repeated applications of Theorems 8 and 9 show the following result [**4**].

THEOREM 10. *Each continuous curve has a decreasing sequence of brick partitionings.*

The first part of Theorem 9 is used in getting the elements of the partitionings to be uniformly locally connected and the second part is used to get the interior of the sum of the closures of two elements to be uniformly locally connected.

VIII. THE BOUNDARIES OF THE PIECES

1. **Locally connected boundaries.** In partitioning a square plus its interior as shown in Figure 4, the boundaries of the pieces are continuous curves and can be partitioned. If M is a finite complex and G is the collection of simplexes of M which are not part of the boundary of any other simplex, the elements of G have boundaries which are the sums of a finite number of continuous curves. These boundaries can be partitioned. (G may not be a brick partitioning.)

One might wonder if each continuous curve can be partitioned so that the boundaries of the pieces are locally connected. However, Example A (Figure 1) described in §I shows that this is not always possible. If the continuous curve is partitioned into pieces of diameters less than 1, each piece which is not a subset of a horizontal line has a boundary with uncountably many components.

However, in the partitioning of Example A, each component of the boundary of a piece of a partitioning is either a point or an interval and is therefore locally connected. One might wonder if each continuous curve can be partitioned so that the components of the boundaries of the pieces are locally connected. Example B of §I shows that this is not always possible.

It would be interesting to be able to characterize the types of continuous curves that can be partitioned into pieces with nice boundaries.

2. **Dimensionally regular partitionings.** Suppose that M is an n-dimensional continuous curve. It is known that for each positive number ϵ there is an open ϵ-covering H of M such that no point of M is covered by more than $n+1$ elements of H [11].

A partitioning G of the n-dimensional continuous curve M is called dimensionally regular if it is regular (each piece is the interior of its closure) and if the following inequality is satisfied when $j \leqq n+1$.

$$\text{Dimension (intersection of any } j \text{ boundaries)} \leqq n + 1 - j.$$

In another paper I shall prove the following result.

THEOREM 11. *There is a decreasing sequence of dimensionally regular partitionings for each n-dimensional continuous curve.*

It may be noted that the boundaries of the pieces will be of dimension less than n.

IX. APPLICATIONS OF PARTITIONINGS

The study of partitionings throws light on the structure of continuous curves and is of interest aside from its applications. In fact, I regard the result that any continuous curve is partitionable as more interesting and topologically important than any of its applications.

The first use made of the fact that any continuous curve is partitionable was to solve the convexification problem as mentioned in §V.

In studying a continuous curve M, one may consider a decreasing sequence of finite open coverings of M. A decreasing sequence of partitionings might be used instead. For example, in showing that each connected open subset of a continuous curve is arcwise connected, one may use chains whose links are connected open sets to define the arcs [**18**]. Chains whose links are pieces of partitionings could be used instead. This was done in the following application.

1. **The Kline sphere characterization.** A nondegenerate continuous curve M is topologically equivalent to the surface of a sphere if and only if it satisfies the following two conditions:

(a) No pair of points separates M.

(b) Each simple closed curve separates M.

Several mathematicians contributed to the proving of this result ([**25**] and other references in [**7**]). The final link was in showing that conditions (a) and (b) also imply that no arc separates M [**7**]. This was done by showing that if there is an arc which separates M, there is a simple closed curve that does not. The simple closed curve was constructed by means of chains whose links are open sets. By using the notion of brick partitioning, the above result was strengthened [**4**] by showing that condition (b) could be replaced by the following weaker condition:

(b′) Each simple closed curve whose complement has property S separates M.

2. **Basis with connected intersections.** A countable basis for a continuous curve is a countable collection $D_1, D_2, \cdots$ of open subsets such that if p is a point and E is an open set containing p there is an integer $i(p, E)$ such that D_i contains p and lies in E. If S is a square plus its interior, there is a countable basis for S whose elements are

the intersections of S with interiors of circles with rational radii and rational centers. The intersection of any two elements of such a basis is connected. One might wonder if each continuous curve has such a basis.

E. E. Floyd noticed that if G is a brick partitioning of M and H is the collection of all open sets h such that for some point p, h is the interior of the sum of the closures of all elements of G whose closures contain p, then the intersection of any two elements of H is connected. Consequently, Floyd and I showed [8] that any continuous curve has a countable basis such that the intersections of the elements of this basis are connected and uniformly locally connected.

3. **Characterizing 3-space.** Suppose H and K are two sets such that for each positive number ϵ there are ϵ-partitionings G_H and G_K of H and K respectively and a 1-1 correspondence between the elements of G_H and the elements of G_K such that two elements of G_H have a common boundary point if and only if the corresponding elements of G_K have a common boundary point. One might wonder if this implies that H is homeomorphic to K. It does not because if H is a square plus its interior and K is the sum of H and a straight line interval that intersects H in just one point, there are such partitionings G_H and G_K. However, it is possible to use the notion of partitionings to characterize certain sets.

The two-manifold can be partitioned so that the boundary of the elements of the partitionings are simple closed curves. Some 3-manifolds can be partitioned so that the boundaries of the pieces are simple surfaces (topologically equivalent to the surface of a sphere). It is an open question as to whether or not all compact 3-manifolds without boundary have this property. The 3-manifold obtained by adding an ideal point at infinity to ordinary Euclidean 3-space is called a 3-sphere. In fact any set homeomorphic to it is called a 3-sphere—for example, the surface of a sphere in Euclidean 4-space. Using the notion of partitioning, we can get the following characterization of a 3-sphere [2].

A necessary and sufficient condition that a continuous curve M be a 3-sphere is that one of its decreasing sequences of regular partitionings $G_1, G_2, \cdots$ have the following properties:

(a) *The boundary of each element of G_i is a simple surface.*

(b) *If the boundaries of two elements of $\sum G_i$ intersect, this intersection is a 2-cell.*

(c) *The intersection of the boundaries of three elements of G_i is one-dimensional at each of its points.*

(d) *If g is an element of* G_{i-1} ($g = M$ *if* $i = 1$) *the elements of* G_i *may be ordered* $g_1, g_2, \cdots, g_n$ *so that the boundary of* g_j ($j = 1, 2, \cdots, n$) *intersects the boundary of g plus the boundary of* $g_1 + g_2 + \cdots + g_{j-1}$ *in a connected set.*

4. **Simultaneous partitionings.** In a forthcoming paper [**20**] Garth Thomas shows that if the continuous curve M_1 is imbedded in the continuous curve M_2, there is a partitioning G_2 of M_2 such that the intersections of M_1 and the elements of G_2 give a partitioning G_1 of M_1.

X. Questions

We have already raised certain questions in the preceding sections but we mention some more here.

1. If M is locally simply connected, can it be partitioned into simply connected pieces?

2. If M is a chain from H to K where each of the sets H and K contains only one point, is there an open transformation satisfying the conditions of Theorem 8? Whyburn shows such a transformation in [**23**] if M is 1-dimensional. If the dimension of M is finite, is there such a transformation such that $T^{-1}(x)$ is of dimension less than the dimension of M?

3. Is there a decreasing sequence of partitionings $G_1, G_2, \cdots$ of M and a boundary operation carrying subsets of G_i into subsets of G_{i+1} that gives a homology theory?

4. Suppose M_1 and M_2 are two continuous curves such that one can be given a convex metric with unique segments but the other cannot. Is there an essential difference between the types of partitionings that M_1 and M_2 can have?

Many other things could be said concerning partitioning, but there is much to learn. We summarize the results of this paper in one sentence: Any continuous curve can be partitioned in somewhat the same way that a square plus its interior can be partitioned.

References

1. Gustav Beer, *Beweis des Satzes, dass jede im kleinen zusammenhängende Kurve convex metrisiert werden kann*, Fund. Math. vol. 31 (1938) pp. 281–320.

2. R. H. Bing, *A characterization of 3-space by partitionings*, Trans. Amer. Math. Soc. vol. 70 (1951) pp. 15–17.

3. ———, *A convex metric for a locally connected continuum*, Bull. Amer. Math. Soc. vol. 55 (1949) pp. 812–819.

4. ———, *Complementary domains of continuous curves*, Fund. Math. vol. 36 (1949) pp. 306–318.

5. ———, *Higher dimensional hereditarily indecomposable continua*, Trans. Amer. Math. Soc. vol. 71 (1951) pp. 267–273.

6. ———, *Partitioning a set*, Bull. Amer. Math. Soc. vol. 55 (1949) pp. 1101–1110.

7. ———, *The Kline sphere characterization problem*, Bull. Amer. Math. Soc. vol. 52 (1946) pp. 644–653.

8. R. H. Bing and E. E. Floyd, *Coverings with connected intersections*, Trans. Amer. Math. Soc. vol. 69 (1950) pp. 387–391.

9. L. M. Blumenthal, *Distance geometries*, University of Missouri Studies, vol. 13, No. 2, 1938.

10. O. G. Harrold, Jr., *Concerning the convexification of continuous curves*, Amer. J. Math. vol. 61 (1930) pp. 210–216.

11. W. Hurewicz and H. Wallman, *Dimension theory*, Princeton University Press, 1948.

12. C. Kuratowski and G. T. Whyburn, *Sur les éléments cycliques et leurs applications*, Fund. Math. vol. 16 (1930) pp. 305–331.

13. S. Lefschetz, *Algebraic topology*, Amer. Math. Soc. Colloquium Publications, vol. 27, New York, 1942.

14. K. Menger, *Untersuchungen über allgemeine Metrik*, Math. Ann. vol. 100 (1928) pp. 75–163.

15. E. E. Moise, *Grille decomposition and convexification theorems for compact metric locally connected continua*, Bull. Amer. Math. Soc. vol. 55 (1949) pp. 1111–1121.

16. ———, *A note of correction*, Proceedings of the American Mathematical Society vol. 2 (1951) p. 838.

17. R. L. Moore, *Concerning connectedness im kleinen and a related property*, Fund. Math. vol. 3 (1922) pp. 232–237.

18. ———, *Foundations of point set theory*, Amer. Math. Soc. Colloquium Publications, vol. 13, New York, 1932.

19. W. Sierpinski, *Sur une condition pour qu'un continu soit une courbe jordanienne*, Fund. Math. vol. 1 (1920) pp. 44–60.

20. Garth Thomas, *Simultaneous partitionings*, Bull. Amer. Math. Soc. Abstracts 57-4-366 and 57-6-553.

21. G. T. Whyburn, *Analytic topology*, Amer. Math. Soc. Colloquium Publications, vol. 28, New York, 1942.

22. ———, *Concerning S-regions in locally connected continua*, Fund. Math. vol. 20 (1933) pp. 131–139.

23. ———, *The existence of certain transformations*, Duke Math. J. vol. 5 (1939) pp. 647–655.

24. R. L. Wilder, *Topology of manifolds*, Amer. Math. Soc. Colloquium Publications, vol. 32, New York, 1949.

25. L. Zippin, *On continuous curves and the Jordan curve theorem*, Amer. J. Math. vol. 52 (1930) pp. 331–350.

THE UNIVERSITY OF WISCONSIN

PARTIALLY CONTINUOUS DECOMPOSITIONS

R. H. BING

1. Introduction. In this paper we consider collections G of mutually exclusive compact sets whose sum is a subset X of a metric space. (We use the convention that a *compact* set is closed.) We shall be interested in whether or not G is upper or lower semicontinuous. See [8; 9] for treatments of such collections.

The collection G is *upper semicontinuous* (or *lower semicontinuous*) if for each element g of G and each positive number ϵ there is a positive number δ such that if g' is an element of G such that $\rho(g, g') < \delta$, then $g' \subset V(g, \epsilon)$ (or $g \subset V(g', \epsilon)$). Here $\rho(A, B)$ denotes the distance between the sets A and B and $V(A, \epsilon)$ is the set of all points whose distance from A is less than ϵ.

If $\{g_i\}$ is a sequence of elements of G, the *superior limit* (or *inferior limit*) of $\{g_i\}$ is the set of all points p such that each neighborhood of p intersects infinitely many (or all but a finite number of the) elements of $\{g_i\}$. If these two limits are equal, they are called the limit of $\{g_i\}$, and $\{g_i\}$ is said to converge to this limit.

In the case where X is compact, the collection G is upper semicontinuous if for each converging sequence $\{p_i\}$ of points of X, the superior limit of $\{g_i\}$ (g_i contains p_i) lies in an element of G. The collection is lower semicontinuous if for each such sequence, each element that intersects the inferior limit of $\{g_i\}$ lies in it. The collection G is continuous if it is both upper and lower semicontinuous.

The following scheme reminds us of the definitions of upper and lower semicontinuity as given in real variables.

$$\begin{aligned} &\text{Lower semicontinuity} &&\leftrightarrow g_0 \subset \text{limit } \{g_i\}. \\ &\text{Upper semicontinuity} &&\leftrightarrow g_0 \supset \text{limit } \{g_i\}. \\ &\text{Continuity} &&\leftrightarrow g_0 = \text{limit } \{g_i\}. \end{aligned}$$

Here $\{g_i\}$ represents a converging sequence of elements of G and g_0 is an element of G that intersects the limit of this sequence.

If the elements of G are connected, G is called a *monotone* decomposition of X. An element (collection) is *nondegenerate* if it has more than one point (element). We show in Theorem 4 that if G is a lower semicontinuous collection each of whose elements is a nondegenerate continuous curve and X is a closed plane set, then G is also upper semicontinuous.

Presented to the Society, April 30, 1954; received by the editors May 5, 1954.

2. **Decompositions and transformations.** While upper semicontinuous decompositions have been widely studied, the notion of lower semicontinuous decompositions has not been used nearly so widely. We list some relationships between partially continuous decompositions and transformations. For the most part, these statements are well known for upper semicontinuous decompositions and are no more difficult to prove in the lower semicontinuous case.

Suppose f is a transformation (not necessarily continuous) of a compact metric space X onto a set $f(X)$ such that for each point y of $f(X)$, $f^{-1}(y)$ is closed. Then f imposes a decomposition G on X—the elements of G are the inverses of the points of $f(X)$. Here are two tests for the upper or lower semicontinuity of G:

1. *G is upper semicontinuous if and only if, for each closed subset A of X, $f^{-1}f(A)$ is closed.*

2. *G is lower semicontinuous if and only if, for each open subset D of X, $f^{-1}f(D)$ is open.*

Now suppose $f(X)$ has a topology and f is continuous. Then we have the following results:

3. *G is upper semicontinuous if $f(X)$ is metric (even Hausdorff).*

4. *G is lower semicontinuous if $f(X)$ is T_1 and f is open.*

Conversely, suppose G is a decomposition of X. Then there is a continuous transformation g of X determining G and satisfying the following conditions.

5. *$g(X)$ is metric if G is upper semicontinuous.*

6. *$g(X)$ is T_1 and g is open if G is lower semicontinuous.*

We may regard the elements of G as the elements of the space $g(X)$. If U is an open set in X, the collection of elements of G which lie in U is an open set in the metric space $g(X)$ of statement 5; the collection of elements of X which intersect U is an open set in the T_1 space $g(X)$ of statement 6.

3. **Reducing lower semicontinuous collections.** Here we see how to reduce a locally compact continuum with a lower semicontinuous decomposition to a compact continuum with a decomposition which comes closer to being continuous. We define a *continuum* to be a closed connected set which is not necessarily compact.

If g_1, g_2 are two sets, the *Hausdorff distance* $H(g_1, g_2)$ between them is the least upper bound of $\rho(x, g_i)$; $i=1, 2$; $x\in g_1+g_2$.

THEOREM 1. *Suppose G is a monotone lower semicontinuous decomposition of a locally compact continuum X and G has more than one element. For each element g_0 of G and each positive number ϵ there is a compact continuum M such that M is the sum of elements of G, M prop-*

erly contains g_0, and the Hausdorff distance between each pair of elements of G in M is less than ϵ.

PROOF. Regard X as all of space. Since a locally compact connected metric space is perfectly separable, there is a countable basis for X—that is, a countable collection $\{D_i\}$ of open sets such that each open set in X is the sum of elements of $\{D_i\}$.

If there is a continuum in $X-D_1$ that properly contains g_0 and which is the sum of elements of G, let M_1 be such a continuum; otherwise $M_1=X$. If there is a continuum in M_1-D_2 which properly contains g_0 and is the sum of elements of G, let M_2 be such a continuum; otherwise $M_2=M_1$. In general, M_{i+1} is a subcontinuum of M_i which properly contains g_0 and is the sum of elements of G; it misses D_{i+1} if this is possible; otherwise $M_{i+1}=M_i$.

The intersection of $M_1, M_2, \cdots$ is a closed set M' that contains g_0 and is the sum of elements of G. We show that $M'=g_0$.

Let $N(\epsilon)$ be the sum of all elements of G in M' whose distance (not Hausdorff distance) from g_0 is greater than or equal to ϵ. It follows from the lower semicontinuity of G that $N(\epsilon)$ is closed. Since $M'=g_0+N(1)+N(1/2)+N(1/3)+\cdots$, it follows from the Baire category theorem that unless $g_0=M'$, there is an integer j such that $N(1/j)$ contains a nonnull subset E of M' open relative to M'.

Assume g_0 is not a component of M'. Let R be a continuum in $M'-N(1/j)$ that properly contains g_0 and M'' be the closure of the sum of all elements of G that intersect R. Then M'' does not intersect E, but properly contains g_0, and is the sum of elements of G. If p is a point of E, there is an integer r such that D_r contains p and $D_r \cdot M'$ lies in E. But M_r would not contain p because M'' is a continuum in $M'-D_r$ (hence in $M_{r-1}-D_r$) which properly contains g_0 and is the sum of elements of G. Hence, the assumption that g_0 is not a component of M' has led to the contradiction that the point p of M' does not belong to M'.

Since g_0 is compact, lies in an open set D whose closure is compact, and is a component of the intersection of the decreasing sequence $M_1, M_2, \cdots$ of continua, g_0 is this intersection.

Since G is lower semicontinuous there is a positive number δ such that if g is an element of G such that $\rho(g, g_0)<\delta$, then for each point p of g_0, $\rho(p, g)<\epsilon/2$. Since g_0 is the intersection of the continua of $M_1, M_2, \cdots$ and lies in an open set whose closure is compact, there is an integer k such that M_k is compact and for each point p of M_k, $\rho(p, g_0)$ is less than either $\epsilon/2$ or δ. We can let $M=M_k$ because if g is an element of G in M_k, then $H(g, g_0)<\epsilon/2$.

We note that the preceding theorem is not true if we replace lower semicontinuity by upper semicontinuity. The theorem implies that if g_0 is an element of G there is a sequence of different elements of G converging to g_0.

Consider two examples of lower semicontinuous decompositions.

EXAMPLE 1. We describe a lower semicontinuous decomposition G_ϵ of a unit cube into arcs and disks. The decomposition is not upper semicontinuous.

Consider a base B of this cube. If S is a slice of the cube parallel to B and at an irrational distance from B, S is an element of G_ϵ.

If S is a slice of the cube at a distance p/q (in lowest terms) from B, there is a continuous decomposition $G(S, \epsilon/q)$ of S such that each element of $G(S, \epsilon/q)$ is an arc which comes within ϵ/q of each point of S. Each element of $G(S, \epsilon/q)$ is an element of G.

EXAMPLE 2. We may be unable to get M for Theorem 1 on which G is continuous. We give here an example of a lower semicontinuous decomposition G of a Hilbert cube such that if G' is any infinite subcollection of G whose sum is a continuum, G' fails to be upper semicontinuous.

We may regard the Hilbert cube as the cartesian product $C_1 \times C_2 \times C_3 \times \cdots$ where C_i is a unit cube. Frequently a Hilbert cube is considered as the cartesian product of closed intervals I_n where $I_n = [-1/n, 1/n]$ but by adjusting the metric we may suppose $I_n = [0, 1]$. Similarly, we may choose a metric for $C_1 \times C_2 \times C_3 \cdots$ to give it the familiar metric for the Hilbert cube. For each positive number ϵ and each cube C_i let $G(\epsilon, C_i)$ be a particular decomposition of C_i into disks and arcs as described in Example 1. There is a countable collection $W(\epsilon, C_i) = \{w_j\}$ of arcs in $G(\epsilon, C_i)$ such that each continuum in C_i which is the sum of infinitely many elements of $G(\epsilon, C_i)$ contains some element of $W(\epsilon, C_i)$. Let $G'(\epsilon, C_i) = G(\epsilon, C_i) - W(\epsilon, C_i)$.

If $g^1 \in G'(1, C_1)$, $g^1 \times C_2 \times C_3 \times \cdots$ is an element of G. If $w_i^1 \in W(1, C_1)$, consider $G(\epsilon_i^1, C_2)$ where $\epsilon_i^1 = 1/iq$ and w_i^1 is in the slice of C_1 at a distance of p/q from the base of C_1.

If $g^2 \in G'(\epsilon_i^1, C_2)$, $w_i^1 \times g^2 \times C_3 \times \cdots$ is an element of G. If $w_j^2 \in W(\epsilon_i^1, C_2)$, consider $G(\epsilon_j^2, C_3)$ where $\epsilon_j^2 = \epsilon_i^1/jq$ and w_j^2 is in the slice of C_2 at a distance of p/q from the base of C_2. If $g^3 \in G'(\epsilon_j^2, C_3)$, $w_i^1 \times w_j^2 \times g^3 \times C_4 \times \cdots$ is an element of G.

The process is continued. An element of G is either of the form

$$\overset{1}{w}_{n_1} \times \overset{2}{w}_{n_2} \times \overset{3}{w}_{n_3} \times \cdots, \overset{1}{g} \times C_2 \times C_3 \times \cdots, \quad \text{or}$$

$$\overset{1}{w}_{n_1} \times \overset{2}{w}_{n_2} \times \cdots \times \overset{i}{w}_{n_i} \times \overset{i+1}{g} \times C_{i+2} \times \cdots$$

where $w_n^i \in W(\epsilon_{n_{i-1}}^{i-1}, C_i)$, $g^i \in G'(\epsilon_{n_{i-1}}^{i-1}, C_i)$, $\epsilon_{n_i}^i = \epsilon_{n_{i-1}}^{i-1}/n_i q$, and $w_{n_i}^i$ is in the slice of C_i at a distance of p/q from the base of C_i.

Suppose M' is a continuum in the Hilbert cube which is the sum of a subcollection G' of G. First we show that if M' intersects each of $S' \times C_2 \times C_3 \times \cdots$ and $S'' \times C_2 \times C_3 \times \cdots$ where S' and S'' are different slices of C_1 parallel to its base, then G' is not upper semicontinuous. There is a sequence of slices $S_1, S_2, \cdots$ each at an irrational distance from the base of C_1 converging to a slice S_0 at a rational distance from this base and such that each S_i separates S' from S''. Then G' is not upper semicontinuous at any of its elements in $S_0 \times C_2 \times C_3 \times \cdots$.

Suppose M' lies in $S \times C_2 \times C_3 \times \cdots$ but intersects two elements of $G(1, C_1)$ in the slice S. Then there is a sequence $g_1, g_2, \cdots$ of elements of $G'(1, C_1)$ in this slice converging to an element w of $W(1, C_1)$ in this slice and such that M' intersects and contains each $g_i \times C_2 \times C_3 \times \cdots$. But G' is not upper semicontinuous at any element in $w \times C_2 \times C_3 \times \cdots$.

We have found that if G' is continuous, the first coordinates of any two elements of G' are the same. Here $w_{n_i}^i$, g^i, and C_i are regarded as coordinates of an element of G'. A similar argument shows that the other coordinates are also equal. Hence, G' has only one element.

Question. For some continua M (for an arc but not for a Hilbert cube) the following statement is true: For each nondegenerate monotone lower semicontinuous decomposition G of M there is a subcontinuum M' of M which is the sum of an infinite subcollection G' of G such that G' is continuous. It would be interesting to know for what types of continua this statement is true. The statement with "lower" replaced by "upper" is not true for any nondegenerate compact continuum.

If we had not been interested in getting a connected M, we would not have needed to suppose that G is monotone. We would have been able to conclude that the collection of elements of G in M is continuous. Consider the following result.

Theorem 2. *Suppose G has uncountably many elements and is a lower semicontinuous decomposition of a locally compact separable metric space X. There is an uncountable subcollection G' of G such that G' is continuous and the sum of the elements of G' is compact.*

Proof. Since X has a countable covering $\{D_i\}$ by open sets with compact closures and each element of G is covered by a finite subcollection of $\{D_i\}$, there is a finite subcollection of $\{D_i\}$ that covers uncountably many elements of G. Hence, with no loss of generality we suppose that X is compact.

The methods of Theorem 1 reveal that there is a compact set M_0 which is the sum of uncountably many elements of G such that if g_1, g_2 are two elements of G in M_0, then $H(g_1, g_2) < 1$.

Again these methods reveal that M_0 contains two mutually exclusive compact subsets C_1^1, C_2^1 such that each is the sum of uncountably many elements of G and such that if g_1, g_2 are elements of G in C_i $(i=1, 2)$, then $H(g_1, g_2) < 1/2$. Let $M_1 = C_1^1 + C_2^1$.

In general, M_i is the sum of 2^i mutually exclusive compact sets $C_1^i, C_2^i, \cdots, C_{2^i}^i$ such that C_j^i is the sum of uncountably many elements of G and if g_1, g_2 are two elements of G in C_j^i, $H(g_1, g_2) < 1/2^i$. Each C_j^i contains two of the compact sets C_r^{i+1}, C_s^{i+1} in M_{i+1}.

The collection G' of elements of G in $M' = M_0 \cdot M_1 \cdot \cdots$ is continuous. Also, G' has uncountably many elements since it is a Cantor set with respect to its elements as points.

Although this concludes the proof of Theorem 2, it is of interest that if X is compact, and $N(\epsilon)$ is the sum of the elements g of G at which G fails to be upper semicontinuous by ϵ (for each i there is a g_i in G such that $\rho(g, g_i) < 1/i$, $g_i \not\subset V(g, \epsilon)$), then $N(\epsilon)$ is closed and contains no open set. Hence the sum of the elements of G at which G is continuous is a dense G_δ set.

In a bicompact connected T_1 space it does not follow that for each pair of points there is a continuum irreducible between them. Hence, Theorem 3 shows that not each bicompact T_1 space is a space $g(X)$ of the type referred to in statement 6 of the preceding section *Transformations and decompositions*. The following result follows from the same type of argument given in the proof of Theorem 1.

THEOREM 3. *If G is a monotone lower semicontinuous decomposition of a connected compact metric space X and A, B are two nonnull closed subsets of X, there is a continuum M in X which is irreducible with respect to being a continuum which is the sum of elements of G and intersects each of A, B.*

If M_A and M_B denote the sum of the elements of G in M which intersect A and B respectively, then each of $M - M_A$, $M - M_B$, and $M - (M_A + M_B)$ is connected.

4. **Decompositions whose elements are continuous curves.** In this section we shall be interested in a lower semicontinuous decomposition G of a closed planar set such that the elements of G are nondegenerate continuous curves. (A continuous curve is a compact locally connected continuum.) We find that G must be upper semicontinuous also. Example 1 shows that the hypothesis of Theorem 4

must include the condition that the sum of the elements of G is planar.

THEOREM 4. *Suppose G is a lower semicontinuous decomposition of a closed planar set such that each element of G is a nondegenerate continuous curve. Then G is also upper semicontinuous.*

PROOF. Suppose E_1, E_2, E_3 are three point sets. We say that the arc α is *hooked* with respect to E_1, E_2, E_3 if it contains two arcs α_1, α_2 with at most an end point in common such that α_1 intersects all three of E_1, E_2, E_3 and α_2 intersects two of them. Compare this notion of being hooked with the idea of a sequence of chains being hooked as given by Anderson on page 650 of [1].

On the assumption that G is not upper semicontinuous we show that there is a decreasing sequence of compact continua $M_1, M_2, \cdots$ and three decreasing sequences of domains $D_{i1}, D_{i2}, \cdots$ $(i=1, 2, 3)$ such that:

(a) M_j is the sum of uncountably many elements of G.

(b) $M_1 \cdot M_2 \cdot \cdots$ is an element α_0 of G which is an arc.

(c) Each element of G in M_j which is an arc is hooked with respect to D_{1j}, D_{2j}, D_{3j}.

(d) $\overline{D}_{ij+1} \subset D_{ij}$ and $D_{i1} \cdot D_{i2} \cdot \cdots$ is a point p_i with $p_1 \neq p_2 \neq p_3 \neq p_1$.

But since the arc α_0 cannot be hooked with respect to the three points p_1, p_2, p_3 there is an integer j such that α_0 is not hooked with respect to D_{1j}, D_{2j}, D_{3j}. Hence the assumption that G is not upper semicontinuous will have led us to a contradiction.

Suppose g is an element of G at which G is not upper semicontinuous. Then there is a sequence of elements $g_1, g_2, \cdots$ converging to a closed set W which properly contains g. From Theorem 1 there is a compact continuum M in W which properly contains g, is the sum of elements of G, and is such that if g', g'' are two elements of G in M, then the Hausdorff distance $H(g', g'') < 1$.

Let G_M be the collection of elements of G in M. We now establish two properties of G_M that we shall use.

PROPERTY 1. All but a countable number of elements of G_M are arcs. Since the plane does not contain uncountably many triods, all but a countable number of elements of G_M are arcs or simple closed curves. If G_M contained uncountably many simple closed curves, it would contain three such that some one of the three separated the other two from each other in the plane. This is impossible for the sum of the three simple closed curves would belong to the inferior limit of a sequence of plane continua which does not intersect the sum.

PROPERTY 2. G_M is not upper semicontinuous. If G_M were upper semicontinuous, it would not contain infinitely many elements of the sequence $g_1, g_2, \cdots$. Anderson has shown [1] that if G' is a continuous collection of nondegenerate continuous curves filling a compact continuum M', then M' contains the interior of a circle. But M cannot contain the interior of a circle since it is a subset of the inferior limit of the sequence $g_1, g_2, \cdots$ which has infinitely many elements which do not intersect M. Hence, G_M is not upper semicontinuous.

We note that if M' is any subcontinuum of M which is the sum of uncountably many elements of G, then the collection of elements of G in M' has properties 1 and 2.

Let α be an arc in M which is an element of G, and D_{11}, D_{21}, D_{31} be three mutually exclusive domains intersecting α. We show that there is a continuum M_1 in M which is the sum of uncountably many elements of G and such that each arc in M_1 which is an element of G is hooked with respect to D_{11}, D_{21}, D_{31}.

Let $D_1, D_2, \cdots, D_n$ be a chain of open topological disks covering α such that $D_i \cdot D_j$ is connected or null according as i and j are or are not adjacent and each of D_{11}, D_{21}, D_{31} contains a link of the chain which intersects α. We suppose that D_{n_i} intersects α, $D_{n_i} \subset D_{i1}$, and $n_1 < n_2 < n_3$. For convenience we suppose that $D_{n_1}, D_{n_2}, D_{n_3}$ are the interiors of circles with centers on the x-axis and that if L is the part of the x-axis to the left of D_{n_1} and R is the part to the right of D_{n_3}, then $L+R$ does not intersect any link of the chain.

By Theorem 1 there is a continuum M' in M such that M' is covered by $D_1, D_2, \cdots, D_n$, M' contains α, and M' is the sum of an uncountable subcollection G' of G such that each element of G' will intersect each of D_{n_1}, D_{n_3}. Each element will also intersect D_{n_2} because D_{n_2} is between D_{n_1} and D_{n_3}.

In each element g of G', select an arc h_g which is irreducible from $\overline{D}_{n_1}$ to $\overline{D}_{n_3}$ and denote the collection of all such arcs by H. The elements of H have a natural linear order—for example, h' is above h'' if it lies in the complementary domain of $L+D_{n_1}+h''+D_{n_2}+R$ which is unbounded from above.

Since G' is not upper semicontinuous, there is a converging sequence $g_2', g_3', \cdots$ of elements of G' and two elements g_0', g_1' of G' which belong to the inferior limit of $g_2', g_3', \cdots$. Let h_i be the element of H in g_i'. Since some subsequence of $h_2, h_3, \cdots$ is monotone, we suppose that h_{i+1} is above h_i if $i>1$.

In the linear order of H, one of the sets h_0, h_1 is separated from all but a finite number of elements of $h_2, h_3, \cdots$ by some element of $h_0, h_1, h_2, \cdots$. For convenience we suppose that h_j separates h_0 from each element of $h_r, h_{r+1}, \cdots$.

Let N be a connected neighborhood in D_{n_2} which contains a point of h_0 but no point of h_j. Since G' is lower semicontinuous, some element g_k' of $g_r', g_{r+1}', \cdots$ intersects N. Then the component of $g_k' -(D_{n_1}+D_{n_3})$ that contains h_k also intersects $\overline{D}_{n_1}$, D_{n_2}, and $\overline{D}_{n_3}$ but not N, while another component of it intersects N. Now if M_1 is a continuum sufficiently close to g_k' (and there is one sufficiently close by Theorem 1) and the sum of uncountably many elements of G', then each element of G' in M_1 which is an arc is hooked with respect to D_{n_1}, D_{n_2}, and D_{n_3} and hence with respect to D_{11}, D_{21}, D_{31}.

Now that we have obtained D_{11}, D_{21}, D_{31}, and M_1, all we need is an iteration of the preceding argument to get the three sequences $D_{i1}, D_{i2}, \cdots$ $(i=1, 2, 3)$ and the sequence $M_1, M_2, \cdots$. Once M_n, D_{1n}, D_{2n}, D_{3n} are found, we consider an arc α in M_n which is an element of G. Let D_{jn+1} $(j=1, 2, 3)$ be an open set of diameter less than $1/n$ which intersects α and whose closure lies in D_{jn}. We use a $1/n$-chain covering α and show the existence of M_{n+1} by the same methods used to show the existence of M_1. The reason that condition (b) is satisfied is that an arc is the only nondegenerate snake-like continuous curve. (A continuum is snake-like [4] if for each positive number ϵ it can be covered by an ϵ-chain.)

Theorem 4 does not hold if we omit some of the conditions in the hypothesis. We cite some examples to illustrate this.

EXAMPLE 3. If the requirement were not made that the elements of G be nondegenerate, we could have let G be the collection of points on the y-axis from $(0, 0)$ to $(0, 1)$ and the collection of vertical intervals of length 1 with lower ends on the sect $(a, b]$ where $a=(0, 0)$ and $b=(1, 0)$. Then G is lower semicontinuous but not upper semicontinuous.

EXAMPLE 4. It is essential to suppose that the elements of G be continuous curves and not just continua. Let M be a nondegenerate hereditarily indecomposable continuum. Such continua have been described by Knaster [6], Moise [7], and Bing [2; 3]. Kelley has shown [5] that there is a continuous decomposition G' of such a continuum into nondegenerate continua. Let g be an element of G' and G'' be a continuous decomposition of g into nondegenerate continua. Then the collection consisting of elements of G'' and elements of G' other than g is lower semicontinuous but not upper semicontinuous.

REFERENCES

1. R. D. Anderson, *Continuous collections of continuous curves in the plane*, Proc. Amer. Math. Soc. vol. 3 (1952) pp. 647–657.

2. R. H. Bing, *Concerning hereditarily indecomposable continua*, Pacific Journal of Mathematics vol. 1 (1951) pp. 43–51.

3. ———, *Higher-dimensional hereditarily indecomposable continua*, Trans. Amer. Math. Soc. vol. 71 (1951) pp. 267–273.

4. ———, *Snake-like continua*, Duke Math. J. vol. 18 (1951) pp. 653–663.

5. J. L. Kelley, *The hyperspaces of a continuum*, Trans. Amer. Math. Soc. vol. 52 (1942) pp. 22–36.

6. B. Knaster, *Un continu dont tout sous-continu est indécomposable*, Fund. Math. vol. 3 (1922) pp. 247–286.

7. E. E. Moise, *An indecomposable plane continuum which is homeomorphic to each of its nondegenerate subcontinua*, Trans. Amer. Math. Soc. vol. 63 (1948) pp. 581–594.

8. R. L. Moore, *Foundations of point set theory*, Amer. Math. Soc. Colloquium Publications, vol. 13, New York, 1932.

9. G. T. Whyburn, *Analytic topology*, Amer. Math. Soc. Colloquium Publications, vol. 28, New York, 1942.

University of Wisconsin

C. Homogeneity

ANOTHER HOMOGENEOUS PLANE CONTINUUM

BY

R. H. BING AND F. B. JONES

1. **History of problem.** A set X is homogeneous if for each pair of points x, y of X there is a homeomorphism of X onto itself that takes x into y.

In 1920 Knaster and Kuratowski [21] raised the following question: If a nondegenerate bounded plane continuum is homogeneous, is it necessarily a simple closed curve?

In 1922, Knaster [20] described a hereditarily indecomposable plane continuum. It is reported that he suspected that this continuum had other interesting properties and suggested to his students the problem of discovering if this Knaster continuum (as it came to be called) had the property possessed by an arc of being topologically equivalent to each of its nondegenerate subcontinua. This Knaster continuum is homogeneous and furnishes a counterexample to an affirmative answer of the above question, but this was not discovered until 1951.

A partial affirmative solution was given to the question in 1924 when Mazurkiewicz [22] showed that the bounded nondegenerate homogeneous plane continuum is a simple closed curve if it is locally connected. This result was improved in 1951 when Cohen [12] showed that the continuum is a simple closed curve if it either is arcwise connected or contains a simple closed curve.

A false affirmative solution was announced [25] in 1937. (Of course, at the time, it was not known that the solution was false—this only developed eleven years later when a counter-example was given.) This false solution was extended [11] in 1944 when an attempt was made to classify all homogeneous bounded closed plane sets.

That a pseudo-arc is homogeneous was shown first by Bing [3] in 1948 and shortly thereafter by essentially the same methods by Moise [24]. Both of these proofs made use of the description of the pseudo-arc given by Moise [23] to show that a pseudo-arc is topologically equivalent to each of its nondegenerate subcontinua.

In 1951 Bing [4] discovered that the pseudo-arc described by Moise in 1948 is actually topologically equivalent to the continuum Knaster described twenty-six years earlier by different methods. In fact, it was shown that any two nondegenerate snake-like hereditarily indecomposable continua are topologically equivalent. Also, it was shown that in the category sense, most

Presented to the Society, September 2, 1954 by R. H. Bing and by F. B. Jones under the title *On a certain type of homogeneous plane continuum;* received by the editors April 15, 1957.

bounded continua in E^n ($n>1$) are pseudo-arcs. Hence, most bounded plane continua are homogeneous.

In 1953 two papers [18; 19] appeared questioning the homogeneity of the pseudo-arc. The second of the papers made an unsuccessful attempt to correct an error in the first. Inasmuch as these papers received not unfavorable reviews (Mathematical Reviews vol. 15 (1954) p. 146; p. 335) despite the errors in them, it seems desirable to mention that the pseudo-arc has not been abandoned as an example of a homogeneous plane continuum.

In 1954, working independently, Bing and Jones each discovered a homogeneous plane continuum that was neither a simple closed curve nor a pseudo-arc. Neither knew of the others work until the titles of the papers appeared adjacent to each other on the 1954 summer program of the American Mathematical Society [6; 15]. Inasmuch as both had discovered the same example, it was decided to make this a joint paper. The first part of this paper showing that the example—a circle of pseudo-arcs—is homogeneous was written by Bing. The latter part showing that such a circle of pseudo-arcs can be imbedded in the plane was prepared by Jones.

Perhaps the future holds the answer as to whether or not there are other homogeneous bounded plane continua. Jones' result [14] that each bounded homogeneous plane continuum which does not separate the plane is indecomposable may guide our search. Bing has described a continuum (Example 2 of [4]) that is suspected of being homogeneous. However, at the moment, the only nondegenerate bounded plane continua known to be homogeneous are the simple closed curve, the pseudo-arc, and the circle of pseudo-arcs described in this paper. Other pertinent references are found in the bibliography of this paper.

2. **The example.** In this section we describe a homogeneous plane continuum which we call a circle of pseudo-arcs. First we define some terms that we shall use.

A *chain* is a finite collection D of open sets $d_1, d_2, \cdots, d_n$ such that d_i intersects d_j if i, j are adjacent integers and otherwise $\rho(d_i, d_j)$ is positive. We say that i, j are *adjacent* if $|i-j| \leqq 1$—hence i is adjacent to itself. We use ρ to denote the *distance* function. In earlier papers the weaker condition that nonadjacent links did not intersect was used instead of the stronger condition that they were a positive distance apart but in some respects this is less convenient for our present purposes than the stronger condition.

The elements d_1, d_n of the above chain D are called *end links* of D and the other links are called *interior links*. If each link is of diameter less than ϵ, D is called an *ϵ-chain*. The *subchain* of D consisting of $d_r, d_{r+1}, \cdots, d_s$ is denoted by $D(r, s)$.

Suppose $D=d_1, d_2, \cdots, d_n$ is a chain which *covers* a point set X. Then each point of X lies in some link of D. We say that D *properly covers* X if each link of D contains a point of X. Also, D *irreducibly covers* X of each link of D contains a point of X that is not covered by any other link of D. If a chain irreducibly covers a set, then it properly covers the set, but not conversely.

A compact continuum is called *snake-like* [5] if for each positive number ϵ, it can be covered by an ϵ-chain. The *pseudo-arc* described in [23; 3; 4; 7] is an example of a snake-like continuum.

A continuum G is called an *arc of continua* $\{g_x | 0 \leqq x \leqq 1\}$ if there is a map f of G onto $[0, 1]$ such that $f^{-1}(x)$ is the continuum g_x. Then $\{g_x\}$ is an upper semicontinuous collection of continua filling G such that the corresponding decomposition space is an arc. Then g_0 and g_1 are called the *end elements* of $\{g_x\}$. We use $G(a, b)$ to denote the sum of all continua in the collection $\{g_x | a \leqq x \leqq b\}$.

Such a continuum G is a *snake-like arc of continua* $\{g_x | 0 \leqq x \leqq 1\}$ if it is snake-like. If each element of $\{g_x\}$ is a pseudo-arc, G is called a *snake-like arc of pseudo-arcs*. If in addition, the collection $\{g_x\}$ is continuous (f is open), then G is called a *continuous snake-like arc of pseudo-arcs*. In Theorem 10 we show that any two continuous snake-like arcs of pseudo-arcs of homeomorphic—indeed there is a very strong type of homeomorphism between them.

A *circular chain* differs from a regular chain in that the first and last links intersect. A compact nonsnake-like continuum which for each positive number ϵ can be covered by an ϵ-circular chain is called *circle-like*.

The bounded plane continuum which is shown in this paper to be homogeneous is a continuous circle-like circle of pseudo-arcs—which we call for brevity, a *circle of pseudo-arcs*. It is a circle-like continuum M such that there is a continuous decomposition of M into pseudo-arcs such that the decomposition space is a simple closed curve. Hence, there is an open map f of M onto a circle such that the inverse of each point of the circle is a pseudo-arc.

Perhaps certain of the continua used by Anderson in [1; 2] were circles of pseudo-arcs but this was not pointed out there. It follows from Theorem 10 that any two circles of pseudo-arcs are topologically equivalent and that each is homogeneous.

In establishing homeomorphisms between snake-like continua it is frequently convenient to consider sequences of chains covering them and relations between these chains. The chain D_2 refines the chain D_1 if each link of D_2 lies in a link of D_1. A *chain map* H of a chain E into a chain D is a single valued function that assigns a link of D to each link of E such that the images of adjacent links are adjacent—that is $H(e_i)$ and $H(e_{i+1})$ are adjacent links of D. If each link of D is the image of a link of E, we say that H maps E *onto* D. We note that the chain map H does not operate on the points in the links of E but only on the links. Chain maps play much the same role in this paper as *following a pattern* did in [3].

3. Snake-like arcs of continua. In this section we prove some theorems about snake-like continua. A modification of the argument in the following proof shows that each tree-like continuum as defined in [5] is unicoherent but we shall not be concerned with tree-like continua in this paper.

THEOREM 1. *Each snake-like continuum is unicoherent.*

Proof. Suppose a snake-like continuum M is the sum of two continua M_1, M_2. We show that M is unicoherent by showing that each pair of points p, q of $M_1 \cdot M_2$ belongs to a component in $M_1 \cdot M_2$.

Suppose $d_1, d_2, \cdots, d_n$ is a chain covering M, $p \in d_i$, $q \in d_j$, and $i<j$. Since each of M_1, M_2 is connected and contains $p+q$, each intersects each link of $d_i, d_{i+1}, \cdots, d_j$. Hence there is a sequence of points $p=p_i, p_{i+1}, \cdots, p_j = q$ where $p_r \in d_r \cdot M_1$. If $d_1, d_2, \cdots, d_n$ is an ϵ-chain, then $\rho(p_r, M_2) < \epsilon$ and $\rho(p_r, p_{r+1}) < 2\epsilon$.

Since M is snake-like, for each positive number ϵ it can be covered by an ϵ-chain. The results of the preceding paragraph show that there is a sequence $R_1, R_2, \cdots$ such that R_k is a finite number of points $p=p_1^k$, $p_2^k, \cdots, p_t^k = q$ such that $p_r^k \in M_1$, $\rho(p_r^k, M_2) < 1/k$, and $\rho(p_r^k, p_{r+1}^k) < 1/k$.

Some subsequence of $R_1, R_2, \cdots$ converges to a set L. But L is connected, contains $p+q$, and belongs to both M_1 and M_2. Since p and q are arbitrary points of $M_1 \cdot M_2$, we have that $M_1 \cdot M_2$ is connected and M is unicoherent.

THEOREM 2. *Suppose G is a snake-like arc of continua $\{g_x | 0 \leqq x \leqq 1\}$ and $0<a<b<c<d<1$. There is a positive number ϵ such that if D is any ϵ-chain whatever covering G, any link of D intersecting $G(b, c)$ is between any link of D intersecting $G(0, a)$ and any link intersecting $G(d, 1)$.*

Proof. The required number ϵ is any positive number less than each of $\rho(G(0, a), G(b, c))$, $\rho(G(b, c), G(d, 1))$, and $\rho(G(0, b), G(c, 1))/2$.

Let $D = d_1, d_2, \cdots, d_n$ be an ϵ-chain covering G and d_i, d_j, d_k be elements of D intersecting $G(0, a)$, $G(b, c)$, and $G(d, 1)$ respectively where $i<k$. We show that $i<j<k$.

Since each of $G(0, b)$, $G(c, 1)$ is connected, there are subchains D', D'' of D containing d_i and d_k respectively and covering $G(0, b)$, $G(c, 1)$ respectively such that each link of D' intersects $G(0, b)$ and each link of D'' intersects $G(c, 1)$. Since $\rho(G(0, b), G(c, 1)) > 2\epsilon$, some link d_t of D lies between the links of D' and the links of D''. Hence $i<t<k$.

If $j<i<t$, $G(b, c)$ intersects d_i because $G(b, c)$ is connected and intersects each of d_j and d_t. This is impossible because $G(0, a)$ intersects d_i and $\rho(G(0, a), G(b, c)) > \epsilon$. Similarly, it is impossible that $t<k<j$. Hence $i<j<k$.

THEOREM 3. *Suppose G is a snake-like arc of continua $\{g_x | 0 \leqq x \leqq 1\}$. Then for each positive number ϵ there is an ϵ-chain D covering G such that the first link of D intersects g_0 and the last link intersects g_1.*

Proof. Suppose δ is a positive number less than each of $\epsilon/2$ and $\rho(g_0, g_1)/2$.

Since G is snake-like, there is a δ-chain $E=e_1, e_2, \cdots, e_n$ covering G. Suppose e_i, e_j are the first and last members of this chain intersecting g_0 while e_r, e_s are the first and last members of the chain intersecting g_1. Then $E(i, j)$ and $E(r, s)$ are subchains of E properly covering g_0 and g_1 respectively and no link of $E(i, j)$ intersects any link of $E(r, s)$. For convenience we suppose that $i<j<r<s$.

Let a, b be numbers such that $0<a<b<1$, $E(i, j)$ covers $G(0, a)$, and $E(r, s)$ covers $G(b, 1)$. An application of Theorem 2 gives a positive number γ such that if F is a γ-chain covering G, then any link of F intersecting $G(a, b)$ lies between any link of F intersecting $G(0, a/2)$ and any intersecting $G((b+1)/2, 1)$. We put the further restriction on γ that if F is a γ-chain irreducibly covering G, then F refines E, any link of F intersecting $G(0, a)$ lies in a link of $E(i, j)$, and any link of F intersecting $G(b, 1)$ lies in a link of $E(r, s)$.

Let $F=f_1, f_2, \cdots, f_m$ be a γ-chain irreducibly covering G and f_t, f_u be links of F intersecting g_0 and g_1 respectively such that no link of F between f_t and f_u intersects g_0+g_1. For convenience we suppose that $t<u$. Let $F(v, w)$ be the subchain of $F(t, u)$ which is maximal with respect to the property that f_v lies in an end link of $E(i, j)$ and f_w lies in an end link of $E(r, s)$. Then $E(i,j)$ covers each link of $F(1, v)$ and $E(r, s)$ covers each link of $F(w, m)$. For convenience we suppose that f_v lies in e_j and f_w lies in e_r.

Use A to denote the sum of the elements of $F(1, v)$ and B to denote the sum of the elements of $F(v, m)$. Then the elements of D are $A\cdot e_i, A\cdot e_{i+1}, \cdots, A\cdot e_j, f_{v+1}, f_{v+2}, \cdots, f_{w-1}, B\cdot e_r, B\cdot e_{r+1}, \cdots, B\cdot e_s$.

THEOREM 4. *Suppose G is a snake-like arc of continua $\{g_x | 0\leqq x\leqq 1\}$ and p is a point of $G-(g_0+g_1)$. Then each neighborhood of p which does not intersect g_0+g_1 separates g_0 from g_1 in G.*

Proof. Suppose U is a neighborhood of p which does not intersect g_0+g_1 and ϵ is a positive number such that $\epsilon<\rho(p, G-U)$.

By Theorem 3 there is an ϵ-chain $D=d_1, d_2, \cdots, d_n$ covering G such that d_1 intersects g_0 and d_n intersects g_1. Let d_i be an element of D that contains p. Then $G-U$ is the sum of the mutually exclusive closed sets $G\cdot(d_1+d_2+\cdots+d_i)-U$ and $G\cdot(d_i+d_{i+1}+\cdots+d_n)-U$ while the first of these sets contains g_0 and the second contains g_1.

THEOREM 5. *Suppose G is a snake-like arc of indecomposable continua $\{g_x | 0\leqq x\leqq 1\}$ and E is a chain covering G such that E properly covers each g_x. Then if U is an open set such that $U\cdot G\neq 0$, $U\cdot(g_0+g_1)=0$, and $\overline{U}$ lies in a link of E, then $G-U$ may be expressed as the sum of two mutually separated sets A, B such that $g_0\subset A$, $g_1\subset B$, and for each x $(0\leqq x\leqq 1)$, either $A\cdot g_x$ intersects each link of E or $B\cdot g_x$ intersects each link of E.*

Proof. Let $D_1, D_2, \cdots$ be a decreasing sequence of chains covering G such that the first link of each D_i intersects g_0 and the last link intersects g_1.

Theorem 3 shows that there are such chains.

Let d_1 be a link of some D_i such that $d_1 \subset U$ and diameter d_1 is less than 1. Then Theorem 4 implies that $G-d_1$ is the sum of two mutually separated sets A_1, B_1 containing g_0, g_1 respectively. Using the assumption that the theorem is false, we find that there are numbers a_1, b_1 such that $0<a_1<b_1<1$ and for each x $(a_1 \leqq x \leqq b_1)$, neither $A_1 \cdot g_x$ nor $B_1 \cdot g_x$ intersects each link of E.

Let d_2 be a link of some D_i such that $d_2 \subset d_1$, $d_2 \cdot (G(0, a_1)+G(b_1, 1))=0$, and diameter d_2 is less than $1/2$. Then $G-d_2$ is the sum of two mutually separated sets A_2, B_2 containing $G(0, a_1)$ and $G(b_1, 1)$ respectively. If the theorem is false, there are numbers a_2, b_2 such that $a_1<a_2<b_2<b_1$ and for each g_x $(a_2 \leqq x \leqq b_2)$ neither $A_2 \cdot g_x$ nor $B_2 \cdot g_x$ intersects each link of E.

We continue this procedure to get a sequence of open sets $d_1, d_2, d_3, \cdots$, a sequence of numbers $a_1, a_2, a_3, \cdots$ and a sequence of numbers $b_1, b_2, b_3, \cdots$ such that $d_{j+1} \subset d_j$, d_j is of diameter less than $1/j$ and is a link of some D_i, $a_j<a_{j+1}<b_{j+1}<b_j$, and $G-d_j$ is the sum of two mutually separated sets A_j, B_j containing $G(0, a_{j-1})$, $G(b_{j-1}, 1)$ respectively.

Let c be a number such that $0<a_1<a_2< \cdots <c< \cdots <b_2<b_1<1$. We show that the assumption that the theorem is false leads to the contradiction that g_c is decomposable. For some increasing sequence of integers $n(1)$, $n(2), \cdots$, $A_{n(1)} \cdot g_c$, $A_{n(2)} \cdot g_c, \cdots$ converges to a set A_c and $B_{n(1)} \cdot g_c$, $B_{n(2)} \cdot g_c, \cdots$ converges to a set B_c. Now A_c is a continuum because $A_i \cdot g_c$ is not the sum of two sets whose distance apart is more than $1/i$. It is a proper subcontinuum of g_c since no $A_i \cdot g_c$ intersects each link of E. Also, B_c is a proper subcontinuum of g_c. But g_c is decomposable because it is the sum of the two proper subcontinua A_c, B_c.

4. **Chain maps.** In this section we give some theorems about chain maps. The first of these might be labeled a fixed point theorem for chain maps.

THEOREM 6. *Suppose $D=d_1, d_2, \cdots, d_n$ and $E=e_1, e_2, \cdots, e_m$ are chains and H_1, H_2 are two chain maps of D into E such that H_i $(i=1, 2)$ takes a link of D into e_{m-1}. Then for some link d_i of D, the link $H_1(d_i)$ of E is adjacent to the link $H_2(d_i)$.*

Proof. Suppose $H_1(d_1)$ precedes $H_2(d_1)$ in $e_1, e_2, \cdots, e_m$. If each link $H_1(d_j)$ of E precedes the corresponding link $H_2(d_j)$, let d_i be a link of D such that $H_1(d_i)=e_{m-1}$. Then $H_1(d_i)$ is adjacent to $H_2(d_i)$.

If for some link d_j of D, $H_1(d_j)$ does not precede $H_2(d_j)$, let d_i be the first such link of D. Then $H_1(d_i)$ is adjacent to $H_2(d_i)$.

We note that if one chain covers a pseudo-arc, then another chain covering it can be inscribed in the first in a prescribed way.

THEOREM 7. *Suppose $D=d_1, d_2, \cdots, d_n$ is a chain properly covering a pseudo-arc P and H is a chain map of a chain $X=x_1, x_2, \cdots, x_m$ onto D. Then there is a chain $E=e_1, e_2, \cdots, e_m$ properly covering P such that $e_i \subset d_j$ if $H(x_i)=d_j$.*

In fact, if A, B are closed sets in $P\cdot d_1$, $P\cdot d_n$ respectively and $H(x_r)=d_1$, $H(x_s)=d_n$, the chain E may be selected so that $A\subset e_r$, $B\subset e_s$.

Proof. The proof would be slightly easier if $A\subset d_1-\bar{d}_2$, $B\subset d_n-\bar{d}_{n-1}$, $r=1$, $s=m$, x_r is the only element of X that H takes into d_1, and x_s is the only element of X that H takes into d_n. We now prove the theorem in this special case.

Let p_1, p_2 be points of different components of P in d_1-d_2, d_n-d_{n-1} respectively and $D_1, D_2, \cdots$ be a sequence of chains from p_1 to p_2 such that each D_i covers P, D_i is of mesh less than $1/i$, D_{i+1} is crooked in D_i, and $D=D_1$. Then it follows from Theorem 6 of [3] that there is a chain $E=e_1, e_2, \cdots, e_m$ from p_1 to p_2 such that E covers P, $e_i\subset d_j$ if $H(x_i)=d_j$, and for some integer r, each link of E is the sum of links of D_r.

Now that we have shown the theorem is true in the special case we alter D, X, H and obtain D', X', H' so that the special case applies. Then we adjust the E' obtained to get the required E. We suppose with no loss of generality that $r<s$.

The chain D' is obtained as follows. Let U_0, U_1, U_2 be open subsets of d_1, $d_1\cdot d_2$, d_2 respectively such that $P\cdot(d_1+d_2)\subset U_0+U_1+U_2$, $P\cdot(d_1-d_2)+A\subset U_0$, $P\cdot(d_2-d_1)\subset U_2$, $U_0\cdot\overline{U}_2=0$, and $A\cdot\overline{U}_1=0$. Also, let U_{n-1}, U_n, U_{n+1} be open subsets of d_{n-1}, $d_{n-1}\cdot d_n$, d_n respectively such that $P\cdot(d_{n-1}+d_n)\subset U_{n-1}+U_n+U_{n+1}$, $P\cdot(d_{n-1}-d_n)\subset U_{n-1}$, $P\cdot(d_n-d_{n-1})+B\subset U_{n+1}$, $\overline{U}_{n-1}\cdot\overline{U}_{n+1}=0$, and $\overline{U}_n\cdot B=0$. Then $D'=U_0, U_1, U_2, d_3, \cdots, d_{n-2}, U_{n-1}, U_n, U_{n+1}$.

The chain $X'=y', y_r, y_{r-1}, \cdots, y_2, x_1, x_2, \cdots, x_m, z_{m-1}, \cdots, z_s, z'$ is obtained from X by adding r elements on at the front and $m+1-s$ on at the end. Then H' is the chain map of X' onto D' so that $H'(y')=U_0$, $H'(z')=U_{n+1}$, $H'(x_i)=H'(y_i)=H'(z_i)=U_j$ (or d_j) if $H(x_i)=d_j$.

Since the special case applies to D', X', and H', there is a chain $E'=f', f_r, f_{r-1}, \cdots, f_2, e_1', e_2', \cdots, e_m', g_{m-1}, \cdots, g_s, g'$ such that $A\subset f'$, $B\subset g'$, and a link e' of E' lies in a link d of D' provided d is the image under H' of the link of X' corresponding to e'.

The chain $E=e_1, e_2, \cdots, e_m$ satisfying the conclusions of the theorem is obtained by adding together certain elements of E'. The link e_r is the sum of f', f_r, e_r' while e_s is the sum of g', g_s, e_s'. In general e_i is the sum of e_i', f_i, g_i (if there are such f's and g's).

The following theorem concerning the existence of chains that cover a snake-like arc of pseudo-arcs in a prescribed fashion will be of use in showing that certain such continua are homeomorphic.

THEOREM 8. *Suppose P is a snake-like arc of pseudo-arcs $\{p_x\mid 0\leq x\leq 1\}$, $D=d_1, d_2, \cdots, d_n$ is a chain covering P such that each link of D intersects each p_x, and H is a chain map of a chain $Y=y_1, y_2, \cdots, y_m$ onto D. Then there is a chain $E=e_1, e_2, \cdots, e_m$ covering P such that each p_x intersects each link of E and $e_i\subset d_j$ if $H(y_i)=d_j$.*

Proof. First we consider the case where $H(y_1)=d_1$, $H(y_m)=d_n$. Let U_A, U_B be open sets with closures in d_1, d_2 respectively such that U_A, U_B intersect each element of $\{p_x\}$.

It follows from Theorem 7 that for each a $(0\leqq a\leqq 1)$ there is a chain $E(a)=e(a)_1, e(a)_2, \cdots, e(a)_m$ covering p_a such that $\overline{U}_A\cdot p_a\subset e(a)_1$, $\overline{U}_B\cdot p_a\subset e(a)_m$, $e(a)_i\subset d_j$ if $H(y_i)=d_j$.

The upper semicontinuity of the collection $\{p_x\}$ implies that a lies in a connected open subset I_a of $(0\leqq x\leqq 1)$ such that if $x\in I_a$, then $E(a)$ covers p_x, $\overline{U}_A\cdot p_x\subset e(a)_1$, and $\overline{U}_B\cdot p_x\subset e(a)_m$. A finite collection $I_{a_1}, I_{a_2}, \cdots, I_{a_t}$ of such open connected subsets cover $(0\leqq x\leqq 1)$. Let $E(a_1), E(a_2), \cdots, E(a_t)$ be the corresponding $E(a)$'s.

We now show how to form E from $E(a_1), E(a_2), \cdots E(a_t)$. It follows from Theorem 4 that P is covered by open sets $U_1, U_2, \cdots, U_t$ such that $U_i\cdot U_j=U_A+U_B$ if $i\neq j$ and if $p_x\cdot(U_i-(U_A+U_B))\neq 0$, then $x\in I_{a_i}$. Then the chain E whose kth link $e_k=e(a_1)_k\cdot U_1+e(a_2)_k\cdot U_2+\cdots+e(a_t)_k\cdot U_t$ is the required chain.

The purpose of the restriction that $H(y_1)=d_1$, $H(y_m)=d_n$ was to make it apparent that each p_x intersected each link of E. To prove the theorem in the more general case, suppose $H(y_r)=d_1$, $H(y_s)=d_n$. Then alter Y as was done in the proof of Theorem 7 by adding elements onto the first of it and elements onto the last of it so that the resulting chain Y' has ends in d_1, d_n respectively and such that the added elements of Y' can be combined with the original links to get a chain resembling Y. Then end chains of the resulting chain E' may be added onto the center of E' as was done in the proof of Theorem 7 so as to get a chain E satisfying the conditions of the theorem.

In stating the following corollary to Theorem 8, the same symbols are used as appear at the place in the proof of Theorem 10 where it is applied.

COROLLARY TO THEOREM 8. *Suppose $Q(i/k, (i+1)/k)$ is a snake-like arc of pseudo-arcs $\{q_x/i/k\leqq x\leqq(i+1)/k\}$, E is a chain covering $Q(i/k, (i+1)/k)$ such that each q_x intersects each link of E, and H_1 is a chain map of a chain $D(r, s)$ onto E. Then there is a positive number ϵ such that if $F(t, u)$ is an ϵ-chain irreducibly covering $Q(i/k, (i+1)/k)$, then there is a chain map $R_{(2i+1)/2k}$ of $F(t, u)$ onto $D(r, s)$ such that for each link f_i of $F(t, u)$, $f_i\subset H_1R_{(2i+1)/2k}(f_i)$, and if $f_v, f_{v+1}, \cdots, f_w$ is a subchain of $F(t, u)$ irreducibly covering some q_x, then each link of $D(r, s)$ is the image of three consecutive links of $f_v, f_{v+1}, \cdots, f_w$.*

In obtaining this result from Theorem 8, we let the P, $\{p_x|0\leqq x\leqq 1\}$, and D of the theorem be the $Q(i/k, (i+1)/k)$, $\{q_x|i/k\leqq x\leqq(i+1)/k\}$, and E of the corollary, Y be a chain with four times as many links as $D(r, s)$, and H be a chain map that takes links of Y numbered $4i+1$, $4i+2$, $4i+3$, $4i+4$ into the image under H_1 of the ith link of $D(r, s)$. Then ϵ is a number so small that any ϵ-chain properly covering $Q(i/k, (i+1)/k)$ is a refinement of the chain E promised by Theorem 8.

5. **A homeomorphism between pseudo-arcs.** The following theorem shows that if the same ϵ-chain properly covers two pseudo-arcs, there is a homeomorphism of one onto the other that moves no point by more than ϵ. It also shows how a chain map can be approximated with a homeomorphism.

THEOREM 9. *Suppose D, E are chains properly covering pseudo-arcs P, Q and H is a chain map of D onto E such that each link of E is the image of a link of D which contains a point of P not on the closure of any other link of D. Then there is a homeomorphism h of P onto Q such that for each link d_i of D, $h(P\cdot d_i) \subset H(d_i)$.*

The rather awkward condition that each link of E is the image of a link of D that covers a point of P not covered by the closure of any other link of D is necessary because it may be that $D=d_1, d_2, \cdots, d_n$ and $E=e_1, e_2, \cdots, e_n$ have the same number of links, $H(d_i)=e_i$, $P\subset \bar{d}_2+\bar{d}_3+\cdots+\bar{d}_{n-1}$, and $Q\not\subset \bar{e}_2+\bar{e}_3+\cdots+\bar{e}_{n-1}$. The less complicated but more stringent condition that each link of E is the image of an interior link of D might be substituted.

We use two simplifications in this proof so as to make the remaining part of the proof essentially like the proof of Theorem 12 of [3].

FIRST SIMPLIFICATION. We now show that there is no loss of generality in supposing that D and $E=e_1, e_2, \cdots, e_n$ have the same number of links, H takes the ith link of D into the ith link of E, and each end link of D contains a point not on the closure of any other link of D.

Let $D''=d_1'', d_2'', \cdots, d_n''$ be the chain such that d_i'' is the sum of the links of D that go into the ith link of E under H. Then D'' is a chain covering P such that each end link of D'' contains a point not contained on the closure of any other link of D''. Also, if h is a homeomorphism of P onto Q such that $h(P\cdot d_i'')\subset e_i$, then h is the required homeomorphism. Hence we suppose with no loss of generality that $D''=d_1'', d_2'', \cdots, d_n''=d_1, d_2, \cdots, d_n=D$, $H(d_i)=e_i$, $P\cdot(d_1-\bar{d}_2)$ contains a point p_1, and $P\cdot(d_n-\bar{d}_{n-1})$ contains a point p_2 such that p_1, p_2 belong to different composants of P.

SECOND SIMPLIFICATION. We replace chains $D=d_1, d_2, \cdots, d_n$ and $E=e_1, e_2, \cdots, e_n$ left after the first simplification with chains $D'=d_1', d_2', \cdots, d_{4n-3}'$ and $E'=e_1', e_2', \cdots, e_{4n-3}'$ covering P and Q respectively such that $p_1\in(d_1'-\bar{d}_2')$, $p_2\in(d_{4n-3}'-\bar{d}_{4n-4}')$, and if h is any homeomorphism of P onto Q such that $h(P\cdot d_i')\subset \bar{e}_{i-1}'+\bar{e}_i'+\bar{e}_{i+1}'$, then h is the required homeomorphism.

First we suppose that q_1, q_2 are points of different composants of Q in e_1-e_2 and e_n-e_{n-1} respectively. If there are not already such points, points may be deleted from e_2 and e_{n-1} to make the condition satisfied.

Suppose that ϵ is a positive number so small that each subset of Q of diameter less than 5ϵ lies in one link of E. Let F be an ϵ-chain properly covering Q. Then e_1' is the sum of all links of F whose closures intersect e_1-e_2, e_5' is the sum of all links of F whose closures intersect $e_2-(e_1+e_3)$, e_9' is the sum of all links of F whose closures intersect $e_3-(e_2+e_4)$, $\cdots$, and e'_{4n-3} is the sum of all links of F whose closures intersect e_n-e_{n-1}. The links of F with closures in $e_1\cdot e_2$ are combined to form e'_2, e'_3, e'_4; the links of F with closures in $e_2\cdot e_3$ are combined to form e'_6, e'_7, e'_8, $\cdots$, and the links of G with closures in $e_{n-1}\cdot e_n$ are combined to form e'_{4n-6}, e'_{4n-5}, e'_{4n-4}. (See Figure 1.)

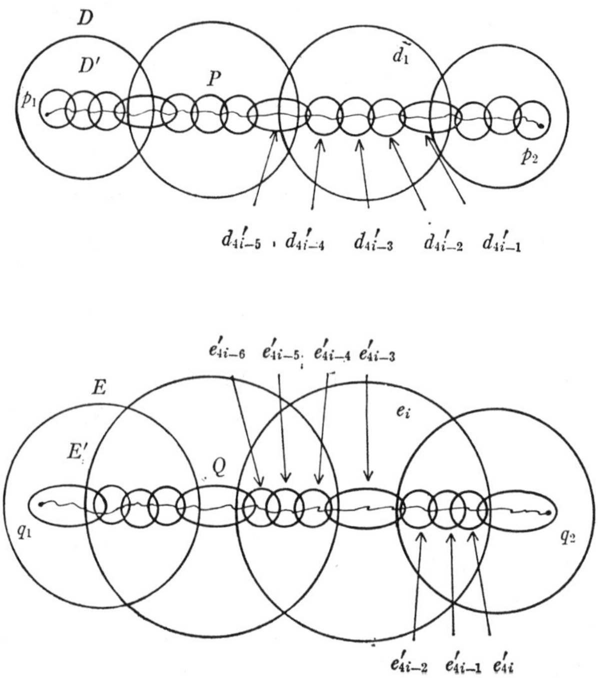

FIG. 1

Let δ be a positive number so small that the distance between any pair of nonadjacent links of $D=d_1, d_2, \cdots, d_n$ is more than 5δ and $\rho(p_1+p_2, d_2+d_{n-1}) >3\delta$. Let G be a δ-chain properly covering P and refining D. Then d'_3 is the sum of the links of G that intersect $d_1\cdot d_2$, d'_7 is the sum of the links of G that intersect $d_2\cdot d_3$, $\cdots$, and d'_{4n-5} is the sum of the links of G that intersect $d_{n-1}\cdot d_n$. Also, d'_1 is the sum of the links of G whose closures contain p_1; d'_2 is the sum of the other links of G in e_1-e_2; d'_4, d'_5, d'_6 are formed by combining links of G in $e_2-(e_1+e_3)$, $\cdots$, d'_{4n-3} is the sum of the links of G whose closures contain p_2; d'_{4n-4} is the sum of the other links of G in e_n-e_{n-1}. See Figure 1.

We now show that if h is a homeomorphism of P onto Q such that $h(P\cdot d'_i)\subset\bar{e}'_{i-1}+\bar{e}'_i+\bar{e}'_{i+1}$, then h is the required homeomorphism because $h(P\cdot d_i)\subset e_i$. The reason is that $h(P\cdot d_i)\subset h(P\cdot(d'_{4i-5}+d'_{4i-4}+d'_{4i-3}+d'_{4i-2} + d'_{4i-1}))\subset \bar{e}'_{4i-6}+\bar{e}'_{4i-5}+\bar{e}'_{4i-4}+\bar{e}'_{4i-3}+\bar{e}'_{4i-2}+\bar{e}'_{4i-1}+\bar{e}'_{4i}\subset\bar{e}_i$.

Completion of proof. It follows from Theorem 13 of [3] and the definition of a pseudo-arc that there is a sequence of chains $F_1, F_2, \cdots$ from p_1 to p_2 such that F_i covers P, F_{i+1} is crooked in F_i, and F_i is of mesh less than $1/i$. Also, there is a sequence of chains $G_1, G_2, \cdots$ from q_1 to q_2 such that G_i covers Q, G_{i+1} is crooked in G_i, and G_i is of mesh less than $1/i$.

In order to define the homeomorphism h we shall obtain a sequence of chains $D_1, D_2, \cdots$ from p_1 to p_2 and a sequence of chains $E_1, E_2, \cdots$ from q_1 to q_2 such that (1) D_i covers P and E_i covers Q; (2) D_i, E_i $(i>2)$ are of mesh less than $1/i$; and (3) $D_i=d_1^i, d_2^i, \cdots, d_{n_i}^i$ and $E_i=e_1^i, e_2^i, \cdots, e_{n_i}^i$ have the same number of links and there is a chain map H_i of E_{i+1} onto E_i and D_{i+1} onto D_i such that $d_j^{i+1}\subset H_i(d_j^{i+1})$, $e_j^{i+1}\subset H_i(e_j^{i+1})$, and if $H_i(d_j^{i+1})=d_k^i$, then $H_i(e_j^{i+1})=e_k^i$.

We set $D_1=D'$ and $E_1=E'$ where D' and E' are the chains obtained in the second simplification.

Let E_2 be any G_i of mesh less than $1/3$ that refines E_i and H_2 be any chain map of E_2 onto E_1 such that $e_j^2\subset H_1(e_j^2)$ for each j. Then by Theorem 6 of [3] we find that there is a chain D_2 from p_1 to p_2 covering P such that for some integer r, each link of D_2 is the sum of links of F_r, D_2 has the same number of links as E_2, and $d_j^2\subset d_k^1=H_1(d_j^2)$ if $H_1(e_j^2)=d_k^1$.

Let D_3 be any F_i of mesh less than $1/4$ that refines D_2 and H_2 be any chain map of D_3 onto D_2 such that $d_j^3\subset H_2(d_j^3)$ for each j. Again it follows from Theorem 6 of [3] that there is a chain E_3 from q_1 to q_2 covering Q such that for some integer r, each link of E_3 is the sum of links of G_r, E_3 has the same number of links as D_3, and $e_j^3\subset e_k^2=H_2(e_j^3)$ if $H_2(d_j^3)=d_k^2$.

The sequences $D_1, D_2, \cdots$ and $E_1, E_2, \cdots$ are obtained by a repetition of this procedure.

Let $d(p, i)$ denote the sum of the elements of D_i containing the point p of P and $e(p, i)$ denote the sum of the corresponding elements of E_i. Then $d(p, i+1)\subset d(p, i)$. Also, $e(p, i+1)\subset e(p, i)$ since if e_j^{i+1} lies in $e(p, i+1)$, d_j^{i+1} contains p, $H_i(d_j^{i+1})=d_k^i$ contains p and lies in $d(p, i)$, and e_k^i lies in $e(p, i)$ and contains e_j^{i+1}. The homeomorphism h is defined so that $h(p)$ is the intersection of the closures of the decreasing sequence of open sets $e(p, 1), e(p, 2), \cdots$. That h is a homeomorphism of P onto Q follows from an argument similar to that contained at the end of the proof of Theorem 10 of this paper or in Theorem 11 of [3].

Finally we show that $h(P\cdot d_i^1)\subset \bar{e}_{i-1}^1+\bar{e}_i^1+\bar{e}_{i+1}^1$. If p is a point of d_i^1, $d(p, 1)\subset d_{i-1}^1+d_i^1+d_{i+1}^1$ and $h(p)\in \bar{e}(p, 1)\subset \bar{e}_{i-1}^1+\bar{e}_i^1+\bar{e}_{i+1}^1$. Hence $h(P\cdot d_i^1)\subset \bar{e}_{i-1}^1+\bar{e}_i^1+\bar{e}_{i+1}^1$.

6. **Homeomorphisms between arcs of pseudo-arcs.** The theorem of this section shows that there is a strong type of topological equivalence between any two continuous snake-like arcs of pseudo-arcs. It is a key theorem to the showing that each circle of pseudo-arcs is homogeneous.

THEOREM 10. *Suppose P, Q are continuous snake-like arcs of pseudo-arcs* $\{p_x\}$, $\{q_x\}$ $(0\leqq x\leqq 1)$. *Then each homeomorphism h that takes the sum of the* *ends of P onto the sum of the ends of Q may be extended to a homeomorphism taking P onto Q. In fact, if* $h(p_0)=q_0$, *the extended homeomorphism h may be chosen so that* $h(p_x)=q_x$.

Proof. Here is an outline of our plan. We shall get a sequence of chains $D_1, D_2, \cdots$ covering P and a sequence of chains $E_1, E_2, \cdots$ covering Q and use these sequences to define our homeomorphism h. Also, we obtain sequences of chain maps $H_1, H_2, \cdots$ and $K_1, K_2, \cdots$ such that H_i takes D_i onto E_i and K_i takes E_{i+1} onto D_i. This is illustrated as follows:

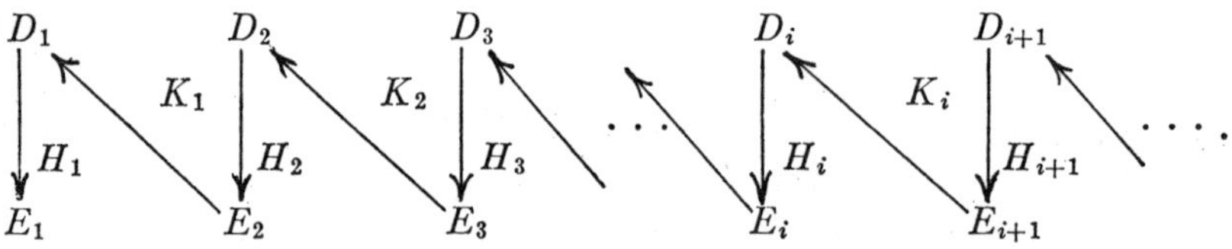

The chain map H_1 will be an approximation to the homeomorphism h we are seeking in that for each point p of P, there is an integer i such that p belongs to the ith link d_i^1 of D_1 and $h(p)$ belongs to the link $H_1(d_i^1)$ of E_1. Also, K_1 will be an approximation to h^{-1} in that for each point q of Q there is an integer i such that q is an element of the ith link e_i^2 of E_2 and $h^{-1}(q)$ is in the link $K_1(e_i^2)$ of D_1. Furthermore, K_1 agrees with H_1^{-1} in that each link of E_2 lies in its image under H_1K_1.

The D's, E's, H's, K's will be defined in the following order: E_1, D_1, H_1, E_2, K_1, D_2, H_2, E_3, K_3, $\cdots$. The meshes of D_i and E_i are less than $1/2^i$. In general, e_j^{i+1} is contained in $H_iK_i(e_j^{i+1})$ where e_j^{i+1} is the jth link of E_{i+1} and d_j^{i+1} is contained in $K_iH_{i+1}(d_j^{i+1})$. (See Figure 2.)

The D's, E's, H's and K's will be chosen so that for each point p of p_x, there is a decreasing sequence of links $d_{n_1}^1, d_{n_2}^2, \cdots$ containing p $(d_{n_i}^i \in D_i)$ such that $K_{i-1}H_i(d_{n_i}^i)=d_{n_{i-1}}^{i-1}$ and $H_1(d_{n_1}^1), H_2(d_{n_2}^2), \cdots$ is a sequence of links $(H_i(d_{n_i}^i)\in E_i)$ whose closures contain a point q of q_x. The homeomorphism h is chosen so that $h(p)=q$. In order to define the D's, E's, H's, and K's properly, we extend the homeomorphism h to certain elements of $\{p_x\}$ as we go.

We shall use $P(a, b)$ to denote the sum of the elements of $\{p_x \mid a \leqq x \leqq b\}$ and $Q(a, b)$ to denote the sum of the elements of $\{q_x \mid a \leqq x \leqq b\}$.

Now for the details of the proof.

Step 1. In this step we define E_1 and extend the homeomorphism h to some more elements of $\{p_x\}$. The step contains three parts.

(a) Consider a chain E_1 irreducibly covering Q and of mesh less than $1/2$.

(b) It follows from the continuity of the collection $\{p_x\}$ and the bicompactness of an arc that there is a positive integer n such that if $0 \leqq b-a \leqq 1/n$, then the subchain of E_1 that irreducibly covers $Q(a, b)$ properly covers each element of $\{q_x/a \leqq x \leqq b\}$.

(c) Extend the homeomorphism h already defined on p_0+p_1 to $p_0+p_{1/n}$

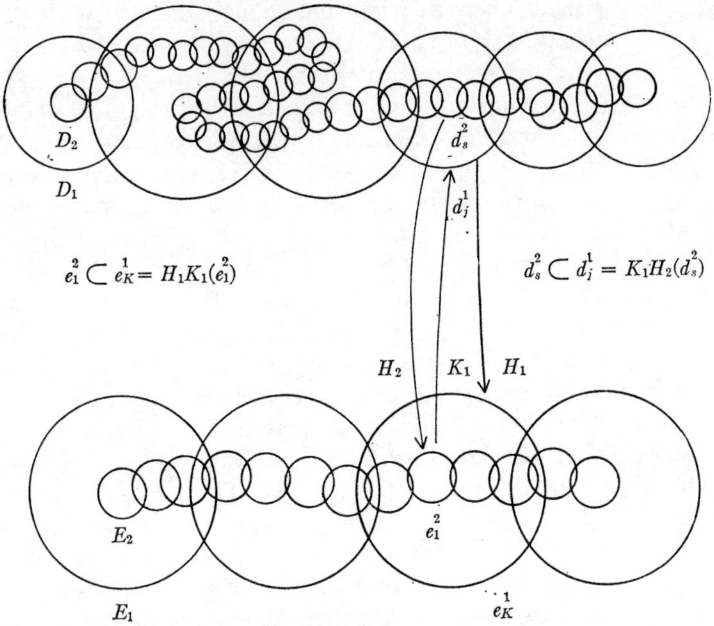

FIG. 2

$+p_{2/n}+ \cdots +p_1$. We do not need to exercise the care in making this extension that we will need to do in later steps. We merely impose the condition that h takes $p_{i/n}$ homeomorphically onto $q_{i/n}$.

Step 2. In this step we define D_1, H_1, and extend the map h still further by defining it on more elements of $\{p_x\}$. We wish the following conditions to be satisfied.

I. If link d_j of D_1 intersects $p_{i/n}$, then $H_1(d_j)$ contains $h(d_j \cdot p_{i/n})$.

II. For each x, H_1 takes any subchain of D_1 that properly covers p_x onto a subchain of E_1 that properly covers q_x.

Parts (a)–(e) of Step 2 are used in getting D_1 and H_1. Parts (f)–(h) show that H_1 and D_1 satisfy the above Conditions I and II. In Part (i) we extend the homeomorphism to other elements of $\{p_x\}$.

(a). In Step 2a we get a preliminary approximation F to the chain D_1. We let F be an ϵ-chain covering P where ϵ is small enough to make the following statements true. (1) $\epsilon<1/2$. (2) No link of F intersects both p_a and p_b if $1/2n \leqq b-a$. (3) The image under h of any subset of $p_0+p_{1/n}+ \cdots +p_1$ of diameter less than 5ϵ lies in an element of E_1. (4) Any subchain of F covering an element of $\{p_x\}$ has at least six times as many links as E_1.

(b) In this part of Step 2 we get approximations $R_0, R_{1/2n}, R_{1/n}, R_{3/2n}, \cdots, R_1$ to H_1.

First we describe $R_{i/n}$ $(i=0, 1, \cdots, n)$. Consider the subchain $f_t, f_{t+1}, \cdots, f_u$ of F irreducibly covering $p_{i/n}$ and the subchain $e_r, e_{r+1}, \cdots, e_s$ of E_1 irreducibly covering $q_{i/n}$. Then $R_{i/n}$ is a chain map of $f_t, f_{t+1}, \cdots, f_u$ onto $e_r, e_{r+1}, \cdots, e_s$ such that $h(f_j \cdot p_{i/n}) \subset R_{i/n}(f_j)$ and for each element e_k of $e_r, e_{r+1}, \cdots, e_s$ there are three consecutive elements of $f_t, f_{t+1}, \cdots, f_u$ that go into e_k under $R_{i/n}$. A precaution to take to insure that each e_k is the image of three consecutive elements of $f_t, f_{t+1}, \cdots, f_u$ is to decide that $R_{i/n}(f_{j-2}) = R_{i/n}(f_{j-1}) = R_{i/n}(f_j) = R_{i/n}(f_{j+1}) = R_{i/n}(f_{j+2}) = e_k$ if e_k is the only element of E_1 such that $h(f_j \cdot p_{i/n}) \subset e_k$. Condition 3 of Step 2a enables us to do this. Unless f_j is the first or last element of F, such an $R_{i/n}$ sends more than three consecutive elements of F into e_k.

Now we describe $R_{(2i+1)/2n}$ $(i=0, 1, \cdots, n-1)$. Let $f_t, f_{t+1}, \cdots, f_u$ and $e_r, e_{r+1}, \cdots, e_s$ be the subchains of F and E_1 that irreducibly cover $P(i/n, (i+1)/n)$ and $Q(i/n, (i+1)/n)$ respectively. It may be that the subscripts t and u mentioned here may differ considerably from the t and u mentioned in the last paragraph but the r and s used here do not differ by more than 1 from those used there. It follows from Step 1b that $e_r, e_{r+1}, \cdots, e_s$ properly covers each element of $\{q_x/i/n \leqq x \leqq (i+1)/n\}$. Then $R_{(2i+1)/2n}$ is any chain map whatever of $f_t, f_{t+1}, \cdots, f_u$ onto $e_r, e_{r+1}, \cdots, e_s$ so long as it is true that for each element e_k of $e_r, e_{r+1}, \cdots, e_s$ and each subchain $f_v, f_{v+1}, \cdots, f_w$ of $f_t, f_{t+1}, \cdots, f_u$ irreducibly covering an element of $\{p_x | i/n \leqq x \leqq (i+1)/n\}$ there are three consecutive elements of $f_v, f_{v+1}, \cdots, f_w$ that go into e_k under $R_{(2i+1)/2n}$. We must be more careful in describing the analogue of $R_{(2i+1)/2n}$ in Step 3 but we can accomplish the result here by letting $R_{(2i+1)/2n}$ send the first three elements of $f_t, f_{t+1}, \cdots, f_u$ into e_r; the next three into $e_{r+1}, \cdots$, the next three into e_s, the next three into $e_{s-1}, \cdots$. Condition 4 of Step 2a assures us that $f_v, f_{v+1}, \cdots, f_w$ has enough elements that such a procedure will cause each element of $e_r, e_{r+1}, \cdots, e_s$ to be the image of some three consecutive members of $f_v, f_{v+1}, \cdots, f_w$ under $R_{(2i+1)/2n}$.

If it were true that all the R's agreed, we could use F and an extension of the R's for D_1 and H_1. However, there is no reason why they must agree so we get a chain D_1 that refines F and a chain map H_1. In obtaining H_1 we shall be influenced by $R_{i/n}$ near $p_{i/n}$ and by $R_{(2i+1)/2n}$ on the part of P between $p_{i/n}$ and $p_{(i+1)/n}$.

(c) We mentioned that we are influenced by $R_{i/n}$ near $p_{i/n}$. We now describe the part P_i of P near $p_{i/n}$ where we were influenced.

Let ϵ be a positive number less than $1/2n$ and so small that if $f_t, f_{t+1}, \cdots, f_u$ is the subchain of F irreducibly covering $p_{i/n}$ and $e_r, e_{r+1}, \cdots, e_s$ is the image of $f_t, f_{t+1}, \cdots, f_u$ under $R_{i/n}$, then $f_t, f_{t+1}, \cdots, f_u$ properly covers each element of $\{p_x | |x-i/n| \leqq \epsilon\}$ and $e_r, e_{r+1}, \cdots, e_s$ properly covers each element of $\{q_x | |x-i/n| \leqq \epsilon\}$. We shall describe P_i so that $P_i \subset P(i/n-\epsilon, i/n+\epsilon)$.

Since each of $R_{i/n}$, $R_{(2i+1)/2n}$, and $R_{(2i-1)/2n}$ takes $f_t, f_{t+1}, \cdots, f_u$ (and possibly more) onto a subchain (possibly a different one) of E_1 that properly covers $q_{i/n}$, it follows from Theorem 6 that there are an f_y and an f_z of f_t,

$f_{t+1}, \cdots, f_u$ such that $R_{(2i+1)/n}(f_y)$, $R_{i/n}(f_y)$ are adjacent and $R_{(2i-1)/2n}(f_z)$, $R_{i/n}(f_z)$ are adjacent.

By Theorem 5 there is an open set U_i in $P(i/n, i/n+\epsilon)\cdot f_y$ such that $\overline{U}_i \subset f_y$ and $P-U_i$ is the sum of two mutually exclusive closed sets A_i, B_i such that $P(0, i/n)\subset A_i$, $P(i/n+\epsilon, 1)\subset B_i$, and for each element p_x of $\{p_x | i/n \leqq x \leqq i/n+\epsilon\}$, there is one of the sets $p_x\cdot A_i$, $p_x\cdot B_i$ that intersects each element of $f_t, f_{t+1}, \cdots, f_u$.

Also, there is an open set V_i in $P(i/n-\epsilon, i/n)$ such that $\overline{V}_i\cdot\overline{U}_i=0$, $\overline{V}_i\subset f_z$, and $P-V_i$ is the sum of two mutually exclusive closed sets A_i', B_i' such that $P(0, 1/n-\epsilon)\subset A_i'$, $P(i/n, 1)\subset B_i'$, for each element p_x of $\{p_x | i/n-\epsilon\leqq x\leqq i/n\}$ there is one of the sets $p_x\cdot A_i'$, $p_x\cdot B_i'$ that intersects each element of $f_t, f_{t+1}, \cdots, f_u$.

Then $P_i=A_i\cdot B_i'$ if $i=1, 2, \cdots, n-1$. If $i=0$, $P_i=A_i$; if $i=n$, $P_i=B_i'$. We note that $p_{i/n}\subset P_i$.

(d) In this step we define the chain D_1. Let D_1 be a chain irreducibly covering P of mesh so small that (1) if an element d of D_1 intersects P_i, it lies in $P_i+U_i+V_i$ and in a link of the subchain of F that irreducibly covers $p_{i/n}$, (2) if d does not intersect any P_i, then it lies in some $P(i/n, (i+1)/n)$ and in a link of the subchain of F that irreducibly covers $P(i/n, (i+1)/n)$, (3) d does not intersect both U_i and V_i but it lies in f_y or f_z according as it intersects U_i or V_i. We note that if d intersects P_i, it lies in an element of F on which $R_{i/n}$ is defined and otherwise it lies in an element of F on which some $R_{(2i+1)/2n}$ is defined.

(e) We now describe the chain map H_1 of D_1 onto E_1. Suppose d is an element of D_1 that lies in f_k of F. If d intersects the U_i mentioned in Step 2c, then we pick f_k to be f_y; then $H_1(d)$ is $R_{i/n}(f_y)$ or $R_{(2i+1)/2n}(f_y)$ according as d does or does not intersect P_i. If d intersects V_i, then we pick f_k to be f_z; $H_1(d)$ is $R_{i/n}(f_z)$ or $R_{(2i-1)/2n}(f_z)$ according as d does or does not intersects P_i. If d intersects P_i but neither U_i nor V_i, pick f_k to be any link of F on which $R_{i/n}$ is defined; then $H_1(d)=R_{i/n}(f_k)$. If d does not intersect any $V_i+P_i+U_i$, it lies in some $P(i/n, (i+1)/n)$ so pick f_k to be some link of F on which $R_{(2i+1)/2n}(f_k)$ is defined; then $H_1(d)=R_{(2i+1)/2n}(f_k)$.

(f) Here we show that H_1 is a chain map—that is $H_1(d_j)$ is adjacent to $H_1(d_{j+1})$ where d_j, d_{j+1} are adjacent elements of D_1. If both d_j and d_{j+1} intersect P_i, then $H_1(d_j)$ is adjacent to $H_1(d_{j+1})$ because $R_{i/n}$ is a chain map. If one intersects P_i and the other does not, either both intersect U_i or both intersect V_i. If both intersect U_i, both lie in f_y and $R_{i/n}(f_y)$ is adjacent to $R_{(2i+1)/2n}(f_y)$; if both intersect V_i, both lie in f_z and $R_{i/n}(f_z)$ is adjacent to $R_{(2i-1)/2n}(f_z)$. If neither d_j nor d_{j+1} intersect any P_i, then $H_1(d_j)$ is adjacent to $H_1(d_{j+1})$ because each $R_{(2i+1)/2n}$ is a chain map.

(g) In Step 2g we show that the chain map H_1 satisfies Condition I mentioned at the beginning of Step 2.

Suppose d_j intersects $p_{i/n}$, $d_j\subset f_k$, and $H_1(d_j)=R_{i/n}(f_k)$. We find from the definition of $R_{i/n}$ in Step 2b that $h(f_k \cdot p_{i/n}) \subset R_{i/n}(f_k)$. Hence $h(d_j \cdot p_{i/n}) \subset R_{i/n}(f_k) = H_1(d_j)$.

(h) Now we turn to Condition II. Suppose that d_j intersects p_x $(i/n \leqq x \leqq (i+1)/n)$. First we show that $H_1(d_j)$ intersects q_x.

There is an element f_k of F containing d_j such that $H_1(d_j)$ is equal to either $R_{i/n}(f_k)$, $R_{(2i+1)/2n}(f_k)$, or $R_{(i+1)/n}(f_k)$. If $H_1(d_j) = R_{i/n}(f_k)$, $H_1(d_j)$ is a link of the subchain of E_1 irreducibly covering $q_{i/n}$ and hence intersects q_x as a result of conditions in Step 1b; if $H_1(d_j) = R_{(i+1)/n}(f_k)$, $H_1(d_j)$ is a link of the subchain of E_1 irreducibly covering $q_{(i+1)/n}$ and hence intersects q_x; if $H_1(d_j) = R_{(2i+1)/2n}(f_k)$, $H_1(d_j)$ is a link of the subchain of E_1 irreducibly covering $Q(i/n, (i+1)/n)$ as noted in Step 2b and hence intersects each element of $\{q_x / i/n \leqq x \leqq (i+1)/n\}$ by Step 1b.

Suppose $d_v, d_{v+1}, \cdots, d_w$ is the subchain of D_1 that irreducibly covers p_x. Now we show that if e_k is an element of the subchain of E_1 irreducibly covering q_x, then there is an element d_j of $d_v, d_{v+1}, \cdots, d_w$ such that $e_k = H_1(d_j)$.

If p_x intersects $P_i + U_i$ it follows from the definition of ϵ in Step 2c that e_k is an element of the subchain $e_r, e_{r+1}, \cdots, e_s$ of E_1 that irreducibly covers $q_{i/n}$.

If $p_x \cdot P_i$ intersects each element of the subchain $f_t, f_{t+1}, \cdots, f_u$ of F that irreducibly covers $p_{i/n}$, consider the three consecutive elements f_{a-1}, f_a, f_{a+1} of $f_t, f_{t+1}, \cdots, f_u$ that go into e_k under $R_{i/n}$ as mentioned in Step 2b. There is a d_j of $d_v, d_{v+1}, \cdots, d_w$ such that $d_j \cdot p_x \cdot P_i$ intersects f_a. Then $H_1(d_j)$ is equal to either $R_{i/n}(f_{a-1})$, $R_{i/n}(f_a)$, or $R_{i/n}(f_{a+1})$ and all three of them are equal to e_k.

In case p_x intersects $P_i + U_i$ but $p_x \cdot P_i$ does not intersect each of $f_t, f_{t+1}, \cdots, f_u$, then $p_x - (U_i + P_i)$ intersects each element of $f_t, f_{t+1}, \cdots, f_u$ by definition of A_i, B_i in Step 2c. Now let f_{a-1}, f_a, f_{a+1} be three consecutive elements of the subchain $f_t, f_{t+1}, \cdots, f_u$ that go into e_k under $R_{(2i+1)/2n}$. There is a d_j of $d_v, d_{v+1}, \cdots, d_w$ such that d_j lies in f_a and intersects $p_x - (U_i + P_i)$. Then $H_1(d_j)$ is either $R_{(2i+1)/2n}(f_{a-1})$, $R_{(2i+1)/2n}(f_a)$, or $R_{(2i+1)/2n}(f_{a+1})$ and all three are equal to e_k.

In case p_x intersects $V_{i+1} + P_{i+1}$, we also find that there is an element d_j of $d_v, d_{v+1}, \cdots, d_w$ such that $e_k = H_1(d_j)$.

In case p_x does not intersect $P_i + U_i + V_{i+1} + P_{i+1}$, we find that e_k is a link of the subchain of E_k that irreducibly covers $Q(i/n, (i+1)/n)$. Let $f_t, f_{t+1}, \cdots, f_u$ be the subchain of F that irreducibly covers p_x and f_{a-1}, f_a, f_{a+1} be the three consecutive elements of $f_t, f_{t+1}, \cdots, f_u$ that go into e_k under $R_{(2i+1)/2n}$. Again we find that there is an element d_j of $d_v, d_{v+1}, \cdots, d_w$ in f_a such that $H_1(d_j) = e_k$.

(i) As in Step 1b we find that there is an integer k such that k is a multiple of n and if $0 \leqq b - a \leqq 1/k$, the subchain of D_1 that irreducibly covers $P(a, b)$ properly covers each element of $\{p_x | a \leqq x \leqq b\}$ and the image of this subchain under H_1 properly covers each element of $\{q_x | a \leqq x \leqq b\}$.

We now extend the homeomorphism h to $p_0, p_{1/k}, p_{2/k}, \cdots, p_1$. This is done so that $h(p_{i/k}) = q_{i/k}$ and such that for each element d_j of the subchain of D_1 that properly covers $p_{i/k}$, $h(p_{i/k} \cdot d_j) \subset H_1(d_j)$. Theorem 9 assures us that h can be extended in this fashion.

Step 3. In this step we define a chain E_2 covering Q, a chain map K_1 of E_2 onto D_1, and extend the map h to additional elements of $\{p_x\}$. The step is much like Step 2. We are getting an approximation to h^{-1} instead of to h.

We show that there is a 1/4-chain E_2 irreducibly covering Q and a chain map K_1 of E_2 onto D_1 such that

I. If link e_i of E_2 intersects $q_{j/k}$, then $h^{-1}(e_i \cdot q_{j/k}) \subset K_1(e_i)$.

II. For each x, K_1 takes any subchain of E_2 properly covering q_x onto a subchain of D_1 properly covering p_x.

III. $e_i \subset H_1 K_1(e_i)$ for each link e_i of E_2.

The thing making Step 3 more complicated than Step 2 is Condition III. The proof that there are E_2 and K_1 satisfying Conditions I and II is similar to that showing that there are D_1 and H_1 satisfying Conditions I and II of Step 2.

Here we show how to alter the arguments of Step 2 to get E_2 and K_1 to satisfy Condition III in addition to I and II.

As in Step 2, we get an approximation F to E_2. The F of Step 3 will be an ϵ-chain irreducibly covering Q where ϵ is small enough to make the following statements true. (1) $\epsilon < 1/4$. (2) No link of F intesects q_a and q_b if $1/2k < b - a$. (3) The image under h^{-1} of any subset of $q_0 + q_{1/k} + \cdots + q_1$ of diameter less than 5ϵ lies in an element of E_1. (4) ϵ is less than the ϵ mentioned in the corollary to Theorem 8 where the $D(r, s)$ there is the subchain of D irreducibly covering $P(i/k, (i+1)/k)$ $(i = 0, 1, \cdots, k-1)$ and E is the image of $D(r, s)$ under H_1.

The chain map $R_{i/k}$ we define here is similar to the $R_{i/n}$'s defined in Step 2b in that $R_{i/k}$ takes the subchain $f_t, f_{t+1}, \cdots, f_u$ of F that irreducibly covers $q_{i/k}$ onto the subchain $d_v, d_{v+1}, \cdots, d_w$ of D_1 that irreducibly covers $p_{i/k}$ such that each element of $d_v, d_{v+1}, \cdots, d_w$ is the image of three consecutive elements of $f_t, f_{t+1}, \cdots, f_u$ and for each j, $h^{-1}(q_{i/k} \cdot f_j) \subset R_{i/k}(f_j)$. Operating on both sides of $h^{-1}(q_{i/k} \cdot f_j) \subset p_{i/k} \cdot R_{i/k}(f_j)$ by h we find that $q_{i/k} \cdot f_j \subset h(p_{i/k} \cdot R_{i/k}(f_j))$. Since $h(p_{i/k} \cdot R_{i/k}(f_j)) \subset H_1 R_{i/k}(f_j)$ by Step 2i, we have that $q_{i/k} \cdot f_j \subset H_1 R_{i/k}(f_j)$.

To complete the analogue of Step 2b, we only need to pick a chain map $R_{(2i+1)/2k}$. It takes the subchain $f_t, f_{i+1}, \cdots, f_u$ of F that irreducibly covers $Q(i/k, (i+1)/k)$ onto the subchain $d_r, d_{r+1}, \cdots, d_s$ of D_1 that irreducibly covers $P(i/k, (i+1)/k)$ so that (1) $f_j \subset H_1 R_{(2i+1)/2k}(f_j)$ and (2) for each x $(i/k \leqq x \leqq (i+1)/k)$, $R_{(2i+1)/2k}$ takes the subchain $f_v, f_{v+1}, \cdots, f_w$ of F irreducibly covering q_x onto $d_r, d_{r+1}, \cdots, d_s$ so that each element of $d_r, d_{r+1}, \cdots, d_s$ is the image of three consecutive links of $f_v, f_{v+1}, \cdots, f_w$. That there is such a map $R_{(2i+1)/2k}$ follows from the fact that the mesh of the chain

F was taken smaller than the ϵ mentioned in the corollary to Theorem 8.

In the analogue to Step 2c we pick ϵ so small that the Q_i, V_i, U_i we obtain will have the property that if f_j is an element of the subchain of F irreducibly covering $q_{i/k}$, then $(V_i+Q_i+U_i)\cdot f_j \subset H_1R_{i/k}(f_j)$. (We already have that $q_{i/k}\cdot f_j \subset H_1R_{i/k}(f_j)$ from the definition of $R_{i/k}$.)

Now the argument in Step 3 proceeds as the argument in Step 2 through parts c, d, e, f, g, h, and i. The purpose in extending the homeomorphism h in part (i) is to help in setting up Step 4. We complete the discussion of Step 3 by showing that the E_2 we obtained in the analogue of Step 2d satisfies Condition III with respect to the K_1 we obtained in the analogue of Step 2e.

Suppose the link e_j of E_2 intersects Q_i (where Q_i is the analogue of P_i in Step 2c), $e_j \subset f_z$, and $K_1(f_z)=R_{i/k}(f_z)$. Then f_z is a link of the subchain of F irreducibly covering $q_{i/k}$ and $e_j \subset V_i+Q_i+U_i$. Since $(V_i+Q_i+U_i)\cdot f_z \subset H_1R_{i/k}(f_z)$, $e_j \subset H_1K_1(e_j)$.

Suppose e_j does not intersect any Q_i. Then there is an f_z such that $e_j \subset f_z$ and $K_1(e_j)=R_{(2i+1)/2k}(f_z)$. But since $f_z \subset H_1R_{(2i+1)/2k}(f_z)$, $e_j \subset H_1K_1(e_j)$.

Steps 4, 5, $\cdots$. These steps are essentially repetitions of Step 3. We have Conditions A, A′, B, B′ satisfied which are modifications of Conditions II and III of Step 3.

In Step 2n we find a chain D_n irreducibly covering P and a chain map H_n of D_n onto E_n such that the following conditions hold.

A. For each x $(0\leqq x\leqq 1)$, H_n takes any subchain of D_n properly covering p_x onto a subchain of E_n properly covering q_x.

B. $d_i \subset K_{n-1}H_n(d_i)$ for each link d_i of D_n.

In Step $2n+1$ we get a chain E_{n+1} irreducibly covering Q and a chain map K_n of E_{n+1} onto D_n satisfying the following conditions.

A′. For each x $(0\leqq x\leqq 1)$, K_n takes any subchain of E_{n+1} properly covering q_x onto a subchain of D_n properly covering p_x.

B′. $e_i \subset H_nK_n(e_i)$ for each link e_i of E_{n+1}.

Although we carry along analogues of Condition I of Step 3, the only part of this that will interest us henceforward is the following.

C. If link d_i of D_n intersects p_0+p_1, then $h(d_i\cdot(p_0+p_1))\subset H_n(d_i)$.

C′. If link e_i of E_{n+1} intersects q_0+q_1, then $h^{-1}(e_i\cdot(q_0+q_1))\subset K_n(e_i)$.

Definition of homeomorphism h. For each point p of P let $d(p, i)$ denote the sum of the elements of D_i containing p and $H_id(p, i)$ denote the sum of the images under H_i of the elements of D_i in $d(p, i)$. We note that $d(p, i)$ is the sum of one or two adjacent links of D_i while $H_id(p, i)$ is the sum of one or two adjacent links of E_i.

Of course, $d(p, i+1)\subset d(p, i)$. We now show that $H_{i+1}d(p, i+1)\subset H_id(p, i)$ by showing that if $e_j^{i+1}=H_{i+1}(d_k^{i+1})$ where d_k^{i+1} is an element of D_{i+1} in $d(p, i+1)$, then $e_j^{i+1}\subset H_id(p, i)$. Since $d_k^{i+1}\subset K_iH_{i+1}(d_k^{i+1})$ by Condition B, the link $K_iH_{i+1}(d_k^{i+1})=K_i(e_j^{i+1})$ of D_i lies in $d(p, i)$. Also, $e_j^{i+1}\subset H_iK_i(e_j^{i+1})$ by Condition B′ and $H_iK_i(e_j^{i+1})=H_i(d_k^{i+1})\subset H_id(p, i)$.

For each point p of P, let $h(p)$ be the intersection of the closures of the decreasing sequence of open sets $H_1(d(p, 1))$, $H_2(d(p, 2))$, $H_3(d(p, 3))$, $\cdots$. The intersection exists because $H_{i+1}d(p, i+1))\subset H_i(d(p, i))$. It is a point because the diameter of the closure of $H_i(d(p, i))$ is less than $2/2^i$.

If q is any point of $d(p, i)$, the diameter of $H_i(d(p, i))+H_i(d(q, i))$ is less than $3/2^i$ so $\rho(h(p), h(q))<3/2^i$. Therefore h is continuous.

If $p\in p_x$, we find from Condition A that $H_i(d(p, i))$ intersects q_x. Hence $h(p)\in q_x$.

We now show that h takes P onto Q. Let q be a point of Q and e_j^i be an element of E_i containing q. There is an element d_k^i of D_i such that $H_i(d_k^i)=e_j^i$. Therefore for some point p of P, $e_j^i\subset H_i(d(p, i))$. Since the diameter of $H_i(d(p, i))$ is less than $2/2^i$, $\rho(q, h(p))<2/2^i$. This shows that $h(P)$ is dense in Q. Since $h(P)$ is closed, it is equal to Q.

The transformation h we have defined agrees with the given homeomorphism h on p_0+p_1 because for each point p of p_0+p_1, $H_i(d(p, i))$ contains $h(p)$ by Condition C.

Finally we show that h is 1-1. Suppose $h(p)=h(q)$. Since the closures of $H_id(p, i)$ and $H_id(q, i)$ intersect and we are dealing in this paper only with chains whose nonadjacent links do not have closures that intersect, then $H_id(p, i)$ intersects $H_id(q, i)$. Since K_{i-1} is a chain map, the set $K_{i-1}H_id(p, i)$ intersects the set $K_{i-1}H_id(q, i)$. But the closures of these two sets contain p and q respectively so $\rho(p, q)<4/2^i$. Therefore $p=q$ if $h(p)=h(q)$.

7. **A circle of pseudo-arcs.** It follows from Theorem 10 that each circle of pseudo-arcs is homogeneous and that any two of them are homeomorphic. In this section we show that the plane contains a circle of pseudo-arcs. But we first describe an analogous upper-semicontinuous collection G of arcs in the plane.

Let W_1 and W_2 be circles in the plane with center at (0, 0) and radii equal to one and two, respectively. We shall define a collection $\{g_x \mid -\pi\leqq x\leqq\pi\}$ of mutually exclusive arcs such that each g_x is a straight line interval which is irreducible from W_1 to W_2 (or which is the sum of two such straight line intervals whose intersection is a common end point), $g_{-\pi}=g_\pi$, and $\sum g_x(-\pi\leqq x\leqq\pi)$ is a circle-like continuum.

Let $g_{-\pi}$ be the sum of the two straight line intervals from $(r=1, \theta=-\pi)$ to $(r=2, \theta=-\pi\pm\pi/12)$. Let $g_\pi=g_{-\pi}$. Let g_0 be the sum of the two straight line intervals from (1, 0) to (2, $\pm\pi/12$). Let $g_{\pi/2}$ be the sum of the straight line intervals from (2, $\pi/2$) to (1, $\pi/2\pm\pi/12$) and let $g_{-\pi/2}$ be the sum of the straight line intervals from (2, $-\pi/2$) to (1, $-\pi/2\pm\pi/12$). Because of their shape, these elements of $\{g_x\}$ will be called V's.

Now let U denote the set of all points of the annulus W_1W_2 not belonging to any V defined thus far. There are a finite number of components of U which intersect both W_1 and W_2 and in each of these components two V's will be defined. Let the arc T_i $(i=1, 2)$ be the closure of the intersection of such a component of U with W_i. One end point of T_i $(i=1, 2)$ is an end point of a V and the other end point of T_i is the vertex of a V. Now construct two disjoint V's in U such that one of these V's has its end points on T_i $(i=1, 2)$, the other has its vertex on T_i and these three points are the quarter points of T_i. The reader should note that on W_1 or W_2 no vertex of a V is adjacent to a vertex of a V and no end point of a V is adjacent to an end point of *another* V.

After the above construction of a pair of V's has been carried out for every component of U which intersects both W_1 and W_2, a new U may be defined and the process continued countably many $(\aleph_0)$ times. Let G denote the collection of all the V's together with each straight line interval which is the limit of a convergent sequence of V's but which is not a subset of a V. By appropriately choosing the subscripts x, $\{g_x| -\pi \leqq x \leqq \pi\}$ is the collection G. It is easy to see that G is an upper-semicontinuous collection of arcs, G is a circle (if its elements are thought of as points) and G^* (the sum of the elements of G) is a circle-like continuum. Something quite similar to the fact that G^* is circle-like has been previously observed by Roberts [25].

The reader will note that G is *not* a *continuous* collection. This is due fundamentally to the fact that in the plane an arc has two sides. Furthermore G^* is not homogeneous for still another reason, namely, some points of G are local separating points (the vertices of the V's) while others are not. But if pseudo-arcs are substituted for the V's, it should be possible to eliminate these two properties from G and G^* respectively while in general keeping G and G^* unchanged in other respects.

As before let W_1 and W_2 be concentric circles in the plane with center at (0, 0) and radii one and two, respectively. Since there are countably many V's of the preceding construction, let them be $V_1, V_2, V_3, \cdots$ in inverse order to the polar angle between a_i and c_i (the end points of V_i). Let b_i be the vertex of V_i and for convenience let b_i belong to W_1 when i is odd but belong to W_2 when i is even.

Now there exists a sequence $D_1, D_2, D_3, \cdots$ of circular chains of simple domains in the plane such that

(1) for each positive integer i, the closure of each element of D_{i+1} is a subset of some element of D_i;

(2) for each i, each element of D_i intersects the annulus W_1W_2 and not both of two intersecting links of D_i intersect W_1+W_2,

(3) if (for each i) δ_i is the maximum diameter of a link of D_i, then $\delta_i \to 0$,

(4) the subscripts of the elements of D_i which intersect W_1 preserve the counterclockwise order on W_1 and the subscripts of these intersecting W_2 preserve the counterclockwise order on W_2,

(5) if a_i, b_i, and c_i are the end points and vertex of V_i, there is a natural number $m(i)$ such that the (shortest) subchain of $D_{m(i)}$ irreducible from a_i to c_i contains b_i, the subchain of $D_{m(i)+1}$ irreducible from a_i to b_i contains c_i,

the subchain of $D_{m(i)+2}$ irreducible from b_i to c_i contains a_i, the subchain of $D_{m(i)+3}$ irreducible from a_i to c_i contains b_i, etc.,

(6) $(W_1+W_2)\cdot \prod D_i^* =$ closure of $\sum(a_i+b_i+c_i)$ as in the preceding example,

(7) for each i, D_i is the sum of finitely many subchains $T_{i1}, T_{i2}, \cdots, T_{in(i)}$ such that (a) $T_{i1}^*, T_{i2}^*, \cdots, T_{in(i)}^*$ is a circular chain, and (b) for each j, $[1 \leqq j \leqq n(i)]$, T_{ij} is either irreducible from W_1 to W_2 or (for some k) irreducible about $a_k+b_k+c_k$, and

(8) if $h<i$, each element of $\{T_{ij}^*\}_{j=1}^{n(i)}$ is a subset of two intersecting elements of $\{T_{hj}^*\}_{j=1}^{n(h)}$ and

(9) if $h<i$ and T_{ij} is a refinement of $T_{hk}+T_{hl}$, T_{ij} is crooked [3] with respect to $T_{hk}+T_{hl}$ where $l=(k+1) \bmod n(h)$.

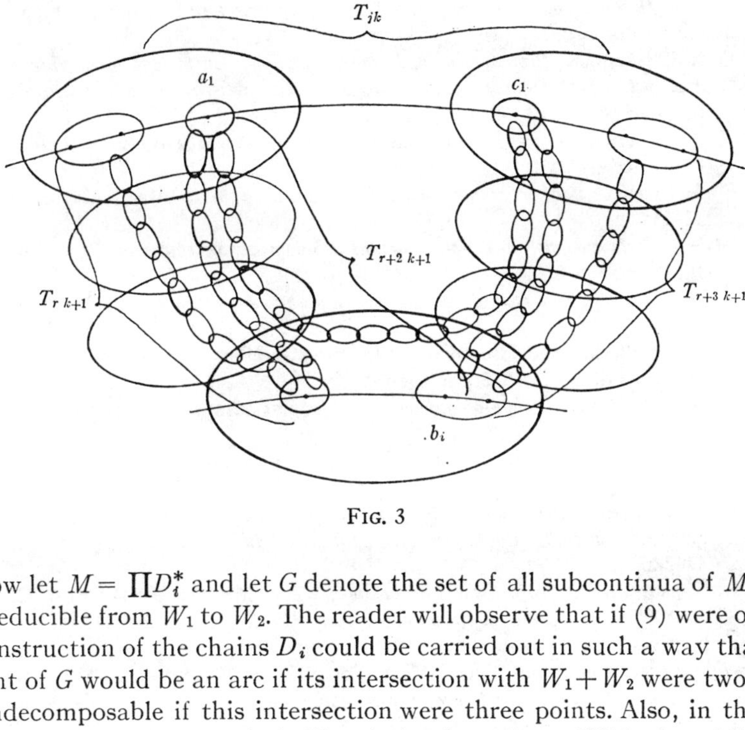

FIG. 3

Now let $M= \prod D_i^*$ and let G denote the set of all subcontinua of M which are irreducible from W_1 to W_2. The reader will observe that if (9) were omitted the construction of the chains D_i could be carried out in such a way that each element of G would be an arc if its intersection with W_1+W_2 were two points and indecomposable if this intersection were three points. Also, in this case M would be a continuous circle-like circle of continua. With the addition of (9) one may see with the help of [3] that each element of G is a pseudo-arc. It follows from Theorem 10 that M is homogeneous.

BIBLIOGRAPHY

1. R. D. Anderson, *On monotone interior mappings in the plane*, Trans. Amer. Math. Soc. vol. 73 (1952) pp. 211–222.

2. ———, *Report on results on continuous collections of pseudo-arcs in the plane and homogeneity of the universal curve*, Summary of Lectures and Seminars, Summer Institute on Set Theoretic Topology, Madison, 1955, pp. 76–77.

3. R. H. Bing, *A homogeneous indecomposable plane continuum*, Duke Math. J. vol. 15 (1948) pp. 729–742.

4. ———, *Concerning hereditarily indecomposable continua*, Pacific J. Math. vol. 1 (1951) pp. 43–51.

5. ———, *Snake-like continua*, Duke Math. J. vol. 18 (1951) pp. 653–663.

6. ———, *Another homogeneous plane continuum*, Bull. Amer. Math. Soc. Abstract 60-6-766.

7. ———, *The pseudo-arc*, Summary of Lectures and Seminars, Summer Institute on Set Theoretic Topology, Madison, 1955, pp. 70–73.

8. C. E. Burgess, *Some theorems on n-homogeneous continua*, Proc. Amer. Math. Soc. vol. 5 (1954) pp. 136–143.

9. ———, *Certain types of homogeneous continua*, Proc. Amer. Math. Soc. vol. 6 (1955) pp. 348–350.

10. ———, *Homogeneous continua*, Summary of Lectures and Seminars, Summer Institute on Set Theoretic Topology, Madison, 1955, pp. 73–76.

11. Gustav Choquet, *Prolongements d'homéomorphies. Ensembles topologiquement nommables, Caractérization topologique individuelle des ensembles fermés totalement discontinus*, C.R. Acad. Sci. Paris, vol. 219 (1944) pp. 542–544.

12. H. J. Cohen, *Some results concerning homogeneous plane continua*, Duke Math. J. vol. 18 (1951) pp. 467–474.

13. F. B. Jones, *A note on homogeneous plane continua*, Bull. Amer. Math. Soc. vol. 55 (1949) pp. 113–114.

14. ———, *Certain homogeneous unicoherent indecomposable continua*, Proc. Amer. Math. Soc. vol. 2 (1951) pp. 855–859.

15. ———, *On a certain type of homogeneous plane continuum*, Bull. Amer. Math. Soc. Abstract 60-6-770.

16. ———, *On homogeneity*, Summary of Lectures and Seminars, Summer Institute on Set Theoretic Topology, Madison, 1955, pp. 66–68.

17. ———, *On a certain type of homogeneous plane continuum*, Proc. Amer. Math. Soc. vol. 6 (1955) pp. 735–740.

18. I. Kapuano, *Sur une proposition de M. Bing*, C. R. Acad. Sci. Paris vol. 236 (1953) pp. 2468–2469.

19. ———, *Sur les continus linéaires*, C. R. Acad. Sci. Paris, vol. 237 (1953) pp. 683–685.

20. B. Knaster, *Un continu dont tout sous-continu est indécomposable*, Fund. Math. vol. 3 (1922) pp. 247–286.

21. B. Knaster and C. Kuratowski, *Problème* 2, Fund. Math. vol. 1 (1920) p. 223.

22. S. Mazurkiewicz, *Sur les continus homogènes*, Fund. Math. vol. 5 (1924) pp. 137–146.

23. E. E. Moise, *An indecomposable plane continuum which is homeomorphic to each of its nondegenerate subcontinua*, Trans. Amer. Math. Soc. vol. 63 (1948) pp. 581–594.

24. ———, *A note on the pseudo-arc*, Trans. Amer. Math. Soc. vol. 64 (1949) pp. 57–58.

25. J. H. Roberts, *Collections filling a plane*, Duke Math. J. vol. 2 (1936) pp. 10–19.

26. Z. Waraszkiewicz, *Sur les courbes planes topologiquement homogenes*, C. R. Acad. Sci. Paris, vol. 204 (1937) pp. 1388–1390.

THE UNIVERSITY OF WISCONSIN,
MADISON, WIS.
THE UNIVERSITY OF NORTH CAROLINA,
CHAPEL HILL, N. C.

A SIMPLE CLOSED CURVE IS THE ONLY HOMOGENEOUS BOUNDED PLANE CONTINUUM THAT CONTAINS AN ARC

R. H. BING

1. Introduction. One of the unsolved problems of plane topology is the following:

Question. What are the homogeneous bounded plane continua?

A search for the answer has been punctuated by some erroneous results. For a history of the problem see **(6)**.

The following examples of bounded homogeneous plane continua are known: a point; a simple closed curve; a pseudo arc **(2, 12)**; and a circle of pseudo arcs **(6)**. Are there others?

The only one of the above examples that contains an arc is a simple closed curve. In this paper we show that there are no other such examples. We list some previous results that point in this direction. Mazurkiewicz showed **(11)** that the simple closed curve is the only non-degenerate homogeneous bounded plane continuum that is locally connected. Cohen showed **(8)** that the simple closed curve is the only homogeneous bounded plane continuum that contains a simple closed curve. Cohen showed **(8)** that the simple closed curve is the only non-degenerate homogeneous bounded plane continuum that is arcwise connected.

In this paper we prove the following theorem:

THEOREM 1. *The simple closed curve is the only homogeneous bounded plane continuum that contains an arc.*

Theorem 1 is proved by listing certain properties possessed by any homogeneous bounded plane continuum that contains an arc but is not a simple closed curve (these properties with their consequences are listed in §§ 2, 3, 5, 6, and 10) and then showing (Theorem 6) that no homogeneous bounded plane continuum could have one of these properties. The proof of Theorem 1 is completed in § 10.

In this paper, all sets are assumed to be metric. For the most part we will deal with planar sets but since some of the results apply to more general metric spaces, we do not suppose that sets discussed are planar unless this is stated. We recall some definitions, related results, and related questions.

A set is *homogeneous* if for each pair of its points p, q there is a homeomorphism of the set onto itself that takes p to q.

An ϵ collection is a collection each of whose elements is of diameter no more than ϵ.

Received April 2, 1959.

An ϵ chain is a finite ordered ϵ collection of open sets $d_1, d_2, \ldots, d_n$ such that d_i intersects d_j if and only if i is adjacent to j.

A compact continuum X is snake-like if for each positive number ϵ, X can be covered by an ϵ chain. It is known **(5)** that the only non-degenerate homogeneous snake-like continuum is a pseudo arc.

It is convenient to associate with any open covering G a 1-complex $C(G)$, called the 1-nerve of G, such that there is a $1 - 1$ correspondence between the elements of G and the vertices of $C(G)$ and two elements of G intersect if and only if the corresponding vertices of $C(G)$ are joined by a 1-simplex in $C(G)$. Note that the 1-nerve of an ϵ chain is topologically an arc.

A compact continuum X is tree-like if for each positive number ϵ there is an ϵ collection G of open sets covering X such that the 1-nerve of G contains no simple closed curve. Each 1-dimensional compact plane continuum that does not separate the plane is tree-like **(3)**.

Question. Is there a homogeneous tree-like continuum that contains an arc?

Jones has shown **(10)** that each homogeneous tree-like compact continuum is indecomposable. Perhaps each is hereditarily indecomposable.

A compact continuum X is circle-like if it is not snake-like but for each positive number ϵ there is an ϵ collection G of open sets irreducibly covering X such that the 1-nerve of G is topologically a circle. A simple closed curve and a circle of pseudo arcs are examples of circle-like homogeneous planar continua. Example 2 of **(4)** is not known to be non-homogeneous. A solenoid is an example of a compact homogeneous continuum that contains an arc. However, the simple closed curve is the only solenoid that is planar.

A solenoid may be defined as the intersection of a sequence of tori $T_1, T_2, \ldots,$ such that T_{i+1} runs smoothly around inside T_i n_i times longitudinally without folding back and T_i has cross diameter of less than $1/i$. The sequence $n_1, n_2, \ldots,$ determines the topology of the solenoid. If it is $1, 1, \ldots,$ after some place, the solenoid is a circle. If it is $2, 2, \ldots,$ the solenoid is the dyadic solenoid.

There is no loss of generality in supposing that each integer in the sequence $n_1, n_2, \ldots,$ used in defining a solenoid is prime, for if n_i is not prime, it may be replaced in the sequence by its prime factors. The order of the elements of the sequence does not affect the topology of the solenoid—that is, if $m_1, m_2, \ldots,$ is a reordering of $n_1, n_2, \ldots,$ the solenoids determined by the sequences are topologically equivalent. Also, the first few terms of the sequence does not affect the topology of the solenoid. Hence, solenoids determined by the sequences of primes $n_1^1, n_2^1, \ldots,$ and $n_1^2, n_2^2, \ldots,$ are topologically equivalent if it is possible to remove a finite number of elements from each so that each prime greater than 1 occurs the same number of times in each of the remainders. Perhaps the converse of this is true.

Another way of describing a solenoid is to consider a unit circle C in the plane with centre at the origin and a sequence of maps $f_1, f_2, \ldots,$ of C onto itself so that in polar co-ordinates

$$f_i(1, \theta) = (1, n_i\theta).$$

This solenoid is the inverse limit of the circles and the f_i's and consists of all points $p_1 \times p_2 \times p_3 \times \ldots$ of the Cartesian product $C \times C \times C \times \ldots$ such that for each i, $p_i = f_i(p_{i+1})$.

We show in Theorem 9 that if a circle-like homogeneous continuum contains an arc, it is a solenoid. Although a solenoid may not be planar, it is locally planar. Anderson has shown **(1)** that the only 1-dimensional locally connected continuum that is not locally planar at any point is the Menger universal curve. It is homogeneous **(1)**.

Question. Are solenoids and the Menger universal curve the only homogeneous 1-dimensional compact continua that contain arcs?*

2. Some elementary properties of M. In §§ 2 and 3 we suppose that M is a homogeneous, non-degenerate, bounded plane continuum that contains an arc but is not a simple closed curve. Our plan is to list enough of the properties of such an assumed M to show that it cannot exist. This section lists some elementary properties of M.

Property 1. *M is not locally connected.* Mazurkiewicz **(11)** showed that the simple closed curve is the only non-degenerate homogeneous bounded locally connected plane continuum.

Property 2. *M is not connected im kleinen at any point.* A set X is connected *im kleinen* at a point x if for each neighbourhood U of x there is a neighbourhood N of x such that $N \cdot X$ lies in the component of $U \cdot X$ containing x. If M were connected *im kleinen* at one point, it would follow from the homogeneity of X that it is connected *im kleinen* at every point. Since a continuum is locally connected at each point if it is connected *im kleinen* at each point, Property 1 implies Property 2.

Property 3. *M contains an open set U with uncountably many components.* Property 3 follows from Property 2 and the following theorem.

THEOREM 2. *If a complete metric space fails to be connected im kleinen at each point of a dense G_δ set, it contains an open set U with uncountably many components.*

Proof. Suppose X is a complete metric space that fails to be connected *im kleinen* at each point of a dense G_δ set Y and Y is the intersection of the open sets $U_1, U_2 \ldots$.

---

*At the 1959 Summer Meeting of the American Mathematical Society J. H. Case presented an abstract announcing another such continuum.

Assume that X fails to contain an open set U with uncountably many components. Let V_1 be an open subset of U_1 of diameter less than $\frac{1}{2}$. It follows from the Baire Category Theorem that V_1 contains an open subset V_2 such that

$$\bar{V}_2 \subset V_1 \cdot U_2,$$
$$\text{diameter } V_2 < 1/2^2, \text{ and}$$
$$V_2 \text{ lies in a component of } V_1.$$

Similarly, there is an open set V_3 satisfying

$$\bar{V}_3 \subset V_2 \cdot U_3,$$
$$\text{diameter } V_3 < 1/2^3,$$
$$V_3 \text{ lies in a component of } V_2.$$

If one continues to get $V_4, V_5, \ldots,$ one finds that X is connected *im kleinen* at $\bar{V}_1 \cdot \bar{V}_2 \ldots.$ This contradicts the fact that $\bar{V}_1 \cdot \bar{V}_2 \ldots$ is a point of Y.

Grace **(9)** has given an example of a compact metric continuum that fails to be locally connected anywhere but which contains no open subset with uncountably many components. This shows that Theorem 2 cannot be weakened by replacing the property of not being connected *im kleinen* with the property of not being locally connected.

Property 4. *M contains no simple triod.* A simple triod is the sum of three arcs such that the intersection of any two of them is the same point p. If M contained a simple triod, it would follow from the homogeneity of M that each component of U of Property 3 would contain a simple triod. This would violate the fact that the plane does not contain uncountably many mutually exclusive simple triods. See Theorem 4.

Property 5. *M contains no simple closed curve.* Cohen showed **(8)** that if a bounded homogeneous plane continuum contains a simple closed curve, it is one.

3. Arc components of M. An arc component of a set X is a subset of X maximal with respect to the property that each pair of points of the subset belongs to an arc in X. In this section we show that the closure of each arc component of the assumed homogeneous bounded plane continuum M is homogeneous. In doing this we find it convenient to work with only certain parts of the arc components. These parts are called rays and are defined as follows.

Suppose p and q are two points of the same arc component of M. The sum of all arcs in M that have p as end point and contain q is called a *ray* starting at p. One may note that this ray differs from an ordinary ray of the plane in that it is neither straight nor closed. However, it has a starting point and is the image of an ordinary ray under a $1-1$ continuous transformation.

Property 6. *Each ray in M is the sum of a countable number of arcs.* Let p be the starting point of a ray R and $\{p_i\}$ be a countable dense subset of R. Then R is the sum of the arcs $pp_1, pp_2 \dots$. If there were a point r of R not in any pp_i, we would consider the arc pr. It follows from the homogeneity of M that r is the interior point of an arc so pr may be extended to an arc ps so that r is contained in the interior of ps. Since M contains no simple triod, each p_i belongs to the arc pr. However, $\{p_i\}$ would not be dense in R since no p_i is near s.

Property 7. *For each point p of an arc component A of M, A is the sum of two rays R_1, R_2 starting at p such that $R_1 \cdot R_2 = p$.* It follows from the homogeneity of M that p is an interior point of an arc ab. It follows from the fact that M contains no simple triod (Property 4) that A is the sum of two rays starting at p and going through a, b respectively. Since M contains no simple closed curve (Property 5), these rays intersect only at p.

Property 8. *M has uncountably many arc components.* If M had only countably many arc components, it would follow from Properties 6 and 7 that M is the sum of a countable collection of arcs. It would then follow from the Baire Category Theorem that one of these arcs contains an open subset of M. The homogeneity of M would then imply that M is a 1-manifold. However, the simple closed curve is the only compact connected 1-manifold.

Property 9. *If R is a ray of M and p is a point of $\bar{R}$, one of the rays starting at p lies in $\bar{R}$.* If neither of the rays starting at p lies in $\bar{R}$, p belongs to an arc ab in M such that neither a nor b is a point of $\bar{R}$. It follows from Property 8 and the homogeneity of M that there is an uncountable family $\{h_\alpha\}$ of homeomorphisms of $(ab + \bar{R})$ into M such that if $\alpha \neq \beta$, $h_\alpha(ab)$, $h_\beta(ab)$ belong to different arc components of M.

It follows from Theorem 3 of § 4 that there is an arc A_0 of the collection $\{h_\alpha(ab)\}$ with two sequences $A_1, A_3, \dots,$ and $A_2, A_4, \dots,$ of arcs of $\{h_\alpha(ab)\}$ converging homeomorphically to A_0 from opposite sides. For convenience we suppose $A_0 = ab$. Some two of the arcs A_{2i}, A_{2i+1} near A_0 would separate some point of $\bar{R}$ from A_0 in $\bar{R}$ and hence two points p, q of R from each other in R. But then the arc pq in R would cross either A_{2i} or A_{2i+1} and violate Property 4.

Property 10. *If $\bar{R}_1$ is the closure of a ray of M, it contains a continuum $\bar{R}$ that is irreducible with respect to being the closure of a ray.* Let $D_1, D_2, \dots,$ be a countable basis of open sets for the plane and $R_1, R_2, \dots,$ be a sequence of rays such that

1. R_{i+1} is a ray in $\bar{R}_i$ missing D_i if any ray in $\bar{R}_i$ misses D_i; $R_{i+1} = R_i$ if each ray in $\bar{R}_i$ intersects D_i.

If p is a point common to the elements of the decreasing sequence $\bar{R}_1, \bar{R}_2, \bar{R}_3, \dots,$ it follows from Property 9 that one of the rays R starting at p lies in infinitely many of the $\bar{R}_i$'s. Hence it lies in $\bar{R}_1 \cdot \bar{R}_2 \cdot \bar{R}_3 \dots$. If R' is a ray in $\bar{R}$, it follows from Condition 1 above that R' intersects each D_i that R intersects. Hence $\bar{R}' = \bar{R}$.

Property 11. *If R is a ray in an arc component A of M, $\bar{R} = \bar{A}$.* Assume p is a point in $A - \bar{R}$. It follows from Property 10 and the homogeneity of M that p is the starting point of a ray R' whose closure is irreducible with respect to being the closure of a ray. Then $\bar{R}'$ does not contain R and there is a point q of $R - \bar{R}'$. Let R'' be a ray starting at q whose closure is irreducible with respect to being the closure of a ray. Since neither of the rays R', R'' contains the other and the starting point q of R'' does not belong to R', the rays R', R'' do not intersect. Either some point of pq belongs to both $\bar{R}'$ and $\bar{R}''$ or some point of pq belongs to neither. We show that in either case, the assumption that $\bar{R} \neq \bar{A}$ has led to a contradiction.

If some point r of the arc pq fails to belong to $\bar{R}' + \bar{R}''$, there is no ray starting at r whose closure is irreducible with respect to being the closure of a ray. This violates Property 10 and the homogeneity of M.

If some point r of pq belongs to both $\bar{R}'$ and $\bar{R}''$, there are two mutually exclusive rays in A each missing r and such that r belongs to the closure of each. This violates the homogeneity of M since there are not two mutually exclusive rays in A each missing q such that q belongs to the closure of each.

Property 12. *If the closures of two arc components of M intersect, the closures are equal.* Suppose A_1, A_2 are two arc components whose closures contain the point p. Let A_p be the arc component of M containing p. It follows from Property 9 that one of the rays starting at p lies in $\bar{A}_1$ and from Property 11 that A_p lies in $\bar{A}_1$. Similarly A_p lies in $\bar{A}_2$. It follows from Property 10, Property 11, and the homogeneity of M that the closure of each arc component of M is irreducible with respect to being the closure of an arc component. Hence, $\bar{A}_p = \bar{A}_1 = \bar{A}_2$.

Property 13. *The closure of each arc component A of M is homogeneous.* We show that $\bar{A}$ is homogeneous by showing that if p is a point of A and q is a point of $\bar{A}$, there is a homeomorphism of $\bar{A}$ onto itself taking p to q. The homogeneity of M implies that there is a homeomorphism h of M onto itself taking p to q. Since such a homeomorphism takes arc components onto arc components, it follows from Property 12 that $h(\bar{A}) = \bar{A}$.

If q and r are points of $\bar{A} - A$ and one wishes a homeomorphism of $\bar{A}$ onto itself taking q onto r, one could use the preceding paragraph to show that there is a homeomorphism h_1 of $\bar{A}$ onto itself taking q to p and a homeomorphism h_2 of $\bar{A}$ onto itself taking p to r. The required homeomorphism is h_2h_1.

4. Collections of arcs in the plane. In this section we digress from our consideration of homogeneity to consider collections of arcs in the plane.

Theorem 3 is used in establishing Properties 9 and 15 but is of interest aside from these applications.

We recall the following notions concerning the abutting of arcs in the plane E^2 Suppose ab, cd, and ef are arcs in E^2 such that $ab \cdot cd = c$ and $ab \cdot ef = e$ are interior points of ab. Then cd and ef are said to *abut on opposite sides* of ab if there is a homeomorphism of E^2 onto itself that takes ab onto a horizontal segment and cd, ef onto vertical segments which lie except for their points of contact with ab on opposite sides of the line containing ab.

A sequence of arcs $A_1, A_2, \ldots,$ is said to *converge homeomorphically* to an arc A_∞ if for each positive number ϵ there is an integer n such that if $n < i$, there is a homeomorphism of A_i onto A_∞ that moves no point more than ϵ.

Suppose ab, cd, ef are arcs such that cd and ef abut on ab from opposite sides. A sequence of arcs $A_1, A_2, \ldots,$ converging homeomorphically to ab is said to *converge homeomorphically from the cd side of ab* if none of the arcs intersect ab and all but possibly a finite number of these arcs intersect cd. Two sequences or arcs converging homeomorphically to ab are said to *converge homeomorphically from opposite sides* if one of the sequences converges from the cd side of ab and the other from the ef side of ab.

THEOREM 3. *If W is an uncountable collection of mutually exclusive arcs in E^2, then there is an element w of W and two sequences of elements of W converging homeomorphically to w from opposite sides.*

This result follows as a corollary of the following result which has a more cumbersome statement.

THEOREM 3′. *Each uncountable collection of mutually exclusive arcs in E^2 has a countable subcollection W' such that each element w_0 of $W - W'$ has the following property:*

For each pair of arcs cd, ef abutting on w_0 from opposite sides and each positive number ϵ there are uncountable subcollections W_1, W_2 of $W - W'$ such that

1. *each element of W_1 intersects cd,*
2. *each element of W_2 intersects ef, and*
3. *for each element w of $W_1 + W_2$ there is a homeomorphism of w onto w_0 that moves no point by more than ϵ.*

Proof of Theorem 3′. Let W' be the collection of all elements w of W with the property that there is an arc cd abutting on w from one side and a positive number ϵ such that no uncountable subcollection W_1 satisfies Conditions 1 and 3 of the statement of Theorem 3′. Theorem 3′ is established by showing that the collection W′ does not have uncountably many elements. Assume W' is uncountable.

For each element w_α of W' let v_α be an arc abutting on w_α from one side and ϵ_α be a positive number such that

4. v_α intersects only a countable number of elements w of W' such that there is a homeomorphism of w_α onto w that moves no point by more than ϵ_α.

Let ϵ' be a positive number so small that for an uncountable subcollection W'' of W', ϵ' will serve as the ϵ_α for each element w_α of W''.

Suppose T is a triod which is the sum of an arc ab and an arc cd abutting on ab from one side. For each element w_α of W'' let h_α be a homeomorphism of $ab + cd = T$ onto $w_\alpha + v_\alpha$ that takes ab onto w_α. Let ρ denote the ordinary distance function for the plane. The homeomorphisms h_α may be regarded as points of a function space metrized as follows:

$$D(h_\alpha, h_\beta) = \max_{t \epsilon T} \rho(h_\alpha(t), h_\beta(t)).$$

Then $\{h_\alpha\}$ is a separable metric space and some element h_0 of it is a limit point of an uncountable order (each neighbourhood of h_0 contains uncountably many points of $\{h_\alpha\}$).

Let H be the set of all elements of $\{h_\alpha\}$ within $\frac{1}{2}\epsilon'$ of h_0 and W''' be the set of all elements of W'' that are images of ab under an element of H. We note that if w_1, w_2 are two elements of W''' then there is a homeomorphism of w_1 onto w_2 that moves no point by more than ϵ'.

For convenience we suppose that $h_0(ab) = w_0$ is the horizontal diameter of a unit circle C with centre at the origin and $h_0(cd) = v_0$ is a vertical radius of C which extends upward. Also, we suppose $\epsilon' < 1$.

Since each element of H is within $\frac{1}{2}\epsilon'$ of h_0, each element of W''' intersects the y-axis. Let p_α be the highest point where w_α intersects this axis. Let p_γ be one of these p_α's which has uncountably many other p_α's above it. But then v_γ intersects all of the w_α's such that p_α lies above p_γ. This contradicts the definition of v_γ given in Condition 4. The assumption that W' was uncountable led to this contradiction.

THEOREM 4. *Suppose $B, B_1, B_2, \ldots,$ is a sequence of mutually exclusive arcs in E^2 such that $B_1, B_3, \ldots,$ and $B_2, B_4, \ldots,$ converge homeomorphically to B from opposite sides. If C is a continuum intersecting each B_i but neither end of B and h is a homeomorphism of $C + B + B_1 + B_2 + \ldots$ into E^2, then $h(B_1), h(B_3), \ldots,$ and $h(B_2), h(B_4), \ldots,$ converges homeomorphically to $h(B)$ from opposite sides.*

Proof. The proof is divided into two steps.

Step 1. *C contains two subcontinua C_1, C_2 such that C_1 intersects all but possibly a finite number of the odd B_i's but no even B_i and C_2 intersects all but possibly a finite number of the even B_i's but no odd B_i.* With no loss of generality we suppose that B is the horizontal interval ab, that each odd B_i intersects the perpendicular bisector of ab at a point above ab and each even B_i intersects this perpendicular bisector at a point below C. The two continua C_1, C_2 that we describe will lie except for their intersections with ab on opposite sides of the line containing ab.

Let ϵ be a positive number so small that neither a nor b lies within ϵ of C. Let K_1, K_2 be circles with centres at a, b respectively with radii equal to

$\frac{1}{2}\epsilon$. Since we can throw away a finite number of B_i's, we suppose that for each B_i there is a homeomorphism of B_i onto B that moves no point by more than $\frac{1}{2}\epsilon$.

Let X_i be an arc of B_i irreducible from K_1 and K_2 and Y_i be the arc from a to b obtained by adding to X_i a radius of K_1 and a radius of K_2. Each of $Y_1, Y_3, \ldots,$ lies except for a, b above ab. We suppose that the ordering is such that Y_{2i+1} is above (except at a, b) Y_{2i+3}.

Let D be the disc bounded by $ab + y_1$. If each arc in D from a to b intersects C, it follows from the unicoherence of D that some component C_1 of $D \cdot C$ separates a from b in D. This continuum C_1 intersects each Y_{2i+1} and is the C_1 promised in Step 1. If there is an arc in D from a to b that misses C, there is such an arc Z which intersects ab only at a, b. Let D' be the disc bounded by $ab + Z$ and X_{2i+1} be an arc on the interior of D'. Any component C_1 of $D' \cdot C$ that intersects X_{2i+1} intersects each X_{2j+1} $(j \geqslant i)$ and this C_1 will serve as the C_1 promised by Step 1. The continuum C_2 is obtained in a similar fashion.

Step 2. *If cd and ef are arcs abutting on* $h(B)$ *from opposite sides and infinitely many of the* $h(B_{2i+1})$'s *intersect cd, all but a finite number of the* $h(B_{2i})$'s *intersect ef.* We suppose with no loss of generality that $h(B) = ab$ is a horizontal segment, cd is a vertical segment pointing upward from ab, and ef is a vertical segment pointing downward from ab.

Let ϵ be a positive number so small that neither a nor b is within ϵ of $cd + ef + h(C_1) + h(C_2)$. Following Step 1, we let K_1, K_2 be circles with centres at a, b respectively and radii equal to $\frac{1}{2}\epsilon$. Since we can disregard any finite collection of the $h(B_i)$'s, we suppose with no loss of generality that there is a homeomorphism of each $h(B_i)$ onto $h(B)$ that moves no point by more than $\frac{1}{2}\epsilon$.

We let X_i be a subarc of $h(B_i)$ irreducible from K_1 to K_2 and Y_i be the arc from a to b obtained by adding to X_i radii of K_1 and K_2 respectively. Since infinitely many of the $h(B_{2i+1})$'s intersect cd, infinitely many of the Y_{2i+1}'s lie above ab (except for a, b).

Suppose Y_1 lies above ab (except for a, b) and let D be the disc bounded by $Y_1 + ab$. Since each point of $\operatorname{Int} D$ is separated from ab by a Y_{2i+1} and each Y_{2i+1} misses C_2, C_2 does not intersect the interior of D. Since no X_{2i} lies interior to D (each intersects C_2), all but a finite number of these X_{2i}'s lie below ab. Hence, all but a finite number of the X_{2i}'s (and hence the $h(B_{2i})$'s) intersect ef.

Since all but a finite number of the $h(B_i)$'s intersect $cd + ef$ we suppose with no loss of generality that infinitely many $h(B_{2i+1})$'s intersect cd. It follows from Step 2 that all but a finite number of the $h(B_{2i})$'s intersect ef and by a repetition of Step 2 that all but a finite number of the $h(B_{2i+1})$'s intersect cd.

It is known that the plane does not contain uncountably many mutually exclusive triods (**3**, Theorem 5, p. 254). In extending this result to higher dimensions it is convenient to think of a simple triod as having a topological 1-simplex as base and having a feeler sticking out from an interior point of this base. The following theorem is a strengthening of this result concerning triods in the plane.

THEOREM 5. *Suppose W is an uncountable collection of simple triods in the plane such that each of these triods has a designated base and feeler. If no two of the bases of the elements of W intersect, some feeler intersects uncountably many bases.*

Proof. If the bases are mutually exclusive, it follows from Theorem 3′ that there is a base b_0 with uncountably many bases arbitrarily close on either side of b_0. The feeler from b_0 would intersect uncountably many of these nearby bases.

5. The reduced continuum M'**.** In this section we return to a study of the assumed homogeneous bounded plane continuum M studied in §§ 2 and 3 which contains an arc but is not a simple closed curve. It follows from Properties 5 and 13 that if there is such an M, there is a continuum M' that is the closure of one of its arc components. We list some properties that such an M' would need to possess in order to show that there is no such M' and hence no M. In §§ 5, 6, 10, we use M' to denote a homogeneous bounded plane continuum one of whose arc components is dense in M' but which is not a simple closed curve.

Property 14. *If C is a non-degenerate subcontinuum of M' that is not an arc, C intersects uncountably many arc components of M'.* This is true by Property 8 if $C = M'$ so we suppose C is a proper subcontinuum of M'. Let p be a point of $M' - C$ and A be the arc component of M' containing p. Since each ray is dense in M', there is a sequence of points $p_1, p_{-1}, p_2, p_{-2}, \ldots$, of $A - C$ such that A is the sum of the arcs $p_i p_{i+1}$ and no two of the $p_i p_{i+1}$'s intersect except possibly at an end point of each. If one considers the intersections of these arcs $p_i p_{i+1}$ with C, one finds that $A \cdot C$ is the sum of a countable collection of mutually exclusive closed sets. Since no continuum is the sum of a countable number more than one of mutually exclusive closed point sets, C intersects uncountably many arc components of M'.

Property 15. *Each non-degenerate proper subcontinuum* of *M' is an arc.* If M' contains a non-degenerate proper subcontinuum C that is not an arc, it follows from Property 14 and the fact that each ray is dense in M' that M' contains an uncountable collection of mutually exclusive arcs each of which intersects C but no one of which has an end on C. It follows from Theorem 3 of § 4 that there is one of these arcs B that has two sequences of arcs $B_1, B_3, \ldots$, and $B_2, B_4, \ldots$, of the arcs converging homeomorphically to B from opposite sides. It follows from Theorem 4 of § 4 that under no homeomorphism h of

$C + B + B_1 + B_2 \ldots$ into the plane is the image of any interior point of B accessible from the complement of $h(C + B + B_1 + B_2 + \ldots)$. This violates the homogeneity of M' since some points of it are accessible from the complement of M'.

Property 16. *M' is indecomposable.* If M' were the sum of two proper subcontinua, it would follow from Property 15 that these subcontinua were arcs. The only homogeneous continuum that is the sum of two arcs is a simple closed curve.

6. Continua each of whose proper subcontinua is an arc. A solenoid is a non-degenerate homogeneous compact continuum each of whose proper subcontinua is an arc. Other examples are not at hand. We note that Property 15 shows that M' has this property. The following question is related to the last two given in § 1.

Question. Are solenoids the only non-degenerate homogeneous compact continua each of whose proper subcontinua is an arc?

Theorems 7 and 9 answer this question in the affirmative for the cases of tree-like and circle-like continua.

In developing the following property, we use merely the fact that each proper subcontinuum of M' is an arc (Property 15) rather than the facts that M' is homogeneous and lies in the plane.

Property 17. *For each positive number ϵ and each arc xy in M' there is an ϵ-chain $d_1, d_2, \ldots, d_n$ covering xy such that x, y belong to d_1, d_n respectively and $M' \cdot Bd \sum d_i \subset \bar{d}_1 + \bar{d}_n$.* Let $e_1, e_2, \ldots, e_n$ be an ϵ-chain covering xy such that x, y belong to e_1, e_n respectively. There are open sets O_1, O_2 in e_1, e_2 respectively such that xy is an arc component of $M' - (O_1 + O_2)$. It follows from Property 15 that xy is a component of $M' - (O_1 + O_2)$.

Since no component of $M' - (O_1 + O_2)$ intersects both xy and $M' - \sum e_i$, then $M' - (O_1 + O_2)$ is contained in two mutually exclusive open sets A, B such that $xy \subset A$, $M' - \sum e_i \subset B$ (see **13**, Theorem 35, p. 21). The link d_i of the chain $d_1, d_2, \ldots, d_n$ is defined to be $e_i \cdot (A + O_1 + O_2)$.

Since

$$M' \cdot \sum \bar{d}_i = M' \cdot A + M' \cdot (B + O_1 + O_2) \cdot \sum \bar{d}_i \subset \sum d_i + (\bar{O}_1 + \bar{O}_2),$$

one finds on subtracting $\sum d_i$ from the ends of the above inequality that

$$M' \cdot Bd \sum d_i \subset \bar{O}_1 + \bar{O}_2 \subset \bar{d}_1 + \bar{d}_n.$$

Property 18. *For each positive number ϵ there is a positive number δ such that if ab is an arc in M' such that $\rho(a, b) < \delta$, then either diameter $ab < \epsilon$ or ab is ϵ dense in M'.* Assume that there is no such δ. Then for each integer i there is an arc a_ib_i in M' such that

$$\rho(a_i, b_i) < 1/i,$$
$$\text{diameter } a_ib_i > \epsilon,$$
$$a_ib_i \text{ is not } \epsilon \text{ dense in } M'.$$

Some subsequence of $a_1b_1, a_2b_2, \ldots,$ converges to a non-degenerate proper subcontinuum. Hence it is an arc. Some subarc of this arc is the limit of a folded sequence of arcs in M' (each in an a_ib_i). The assumption that there is no δ leads to the contradiction of Theorem 6 of the next section.

Property 19. *If a point p of M' is accessible from a component U of $E^2 - M'$, each point of any arc in M' containing p is accessible from the same side that p is.* Let xy be an arc containing the two points p, q on its interior. With no loss of generality we suppose that xy is horizontal, q is between p and y, and rp is a vertical interval lying except for p below xy and in U. We show that q is accessible from U from below.

Let

$$\epsilon = \min \rho(r, p), \qquad \rho(x, p), \qquad \tfrac{1}{2}\rho(q, y),$$

δ be the positive number promised by Property 18, with $\delta < \epsilon$, and D be a δ-chain covering xy and satisfying the conditions of Property 17. We use D^* to denote the sum of the links of D. If q is not accessible from U from below, there is a point s in $M' \cdot D^*$ which is beneath the point q. Let ab be an arc in $M' \cdot D^*$ containing s such that ab lies below xy and each end of ab is in an end link of D. The arc rp prevents either a or b from being in the end link of D containing x so $\rho(a, b) < \delta$. Also, ab is not ϵ dense in M' since it is not near x. However, diameter $ab > \epsilon$ since $\frac{1}{2}\rho(q, y) \geqslant \epsilon$. The assumption that q was not accessible from U from below led to a contradiction of Property 18.

7. Folded sequences of arcs. A solenoid is an example of a homogeneous continuum each of whose proper subcontinua is an arc. The arcs in this continuum seem to run in a parallel fashion and not to "zig-zag" or "fold back." We find from Theorem 6 that such folding is impossible in a homogeneous continuum each of whose proper subcontinua is an arc. No use is made of the fact that the continuum lies in the plane.

Suppose $a_1b_1, a_2b_2, \ldots,$ is a sequence of arcs converging (not necessarily homeomorphically) to an arc xy. The sequence is called a *folded sequence* converging to xy if $a_1, b_1, a_2, b_2, \ldots,$ converges to x.

THEOREM 6. *Suppose X is a homogeneous compact continuum each of whose proper subcontinua is an arc. Then no folded sequence of arcs in X converges to an arc.*

Proof. Assume $a_1b_1, a_2b_2, \ldots,$ is a sequence of arcs converging to an arc xy such that $a_1, b_1, a_2, b_2, \ldots,$ converges to x. If $\epsilon < \frac{1}{2}\rho(x, y)$, y has the following property:

Property (y, ϵ). For each positive integer n there is a $1/n$ chain E such that X intersects each link of E, the distance between y and the first link of D is less than $1/n$, the distance between the end links of E is more than ϵ, and $X \cdot Bd\, E^*$ lies in the closure of the last link of E.

We can obtain such an E as follows.

1. Let D be a $1/n$ chain covering xy such that the first link of D contains y, the last link contains x, and for each other link d, $\bar{d} \cdot X \subset D^*$. The existence of such a D is guaranteed by the proof given in Property 17 of the last section.

2. Let a_ib_i be an arc of the folded sequence converging to xy such that each of a_i, b_i lies in the last link of D, some point of a_ib_i lies in the first link of D, and D covers a_ib_i.

3. Let D' be a chain covering a_ib_i such that D' refines D, the first and last link of D' lie in the last link of D, some link of D' lies in the first link of D, and $X \cdot Bd\, D'^*$ lies in the sum of the closures of the two end links of D'.

Then E can be formed as follows. The jth link of E is the sum of the elements of D' in the jth link of D.

Let $X(\epsilon)$ be the set of all points y of X such that y has Property (y, ϵ). Then $X(\epsilon)$ is closed and it follows from the homogeneity of X that each point of X belongs to some $X(\epsilon)$ for some ϵ. It follows from the Baire Category Theorem that there is an integer m and an open set U such that $U \subset X(1/m)$.

We obtain a sequence of chains $E_1, E_2, \ldots$, such that

1. E_i is a $1/i$ chain such that the distance between its end links is more than $1/m$,

2. $X \cdot Bd\, E_i^*$ lies in the closure of the last link of E_i, and

3. the first link e_1^1 of E_1 lies in U, and the closure of the first link e_1^{i+1} of each E_{i+1} lies in the first link of E_i.

The intersection of the e_1^i's is a point that cannot be the interior point of any arc in X. This contradiction results from the false assumption that there is a folded sequence of arcs in X converging to an arc.

The following result gives an immediate application of Theorem 6. The result is not needed in the proof of Theorem 1 but it can be used in lieu of Property 19 in finishing the proof of Theorem 1 for the case where M does not separate the plane, since each 1-dimensional bounded plane continuum that does not separate the plane is tree-like **(3)**.

THEOREM 7. *There exists no non-degenerate, homogeneous, tree-like continuum each of whose proper subcontinua is an arc.*

Proof. Assume X is a non-degenerate, homogeneous, tree-like continuum each of whose proper subcontinua is an arc. It follows from **(10)** that X is indecomposable. Let D_i be a $1/i$ tree-chain covering X and a_ib_i be an arc in X such that both ends of a_ib_i lie in the same link of D_i and

$$\text{diameter } X/4 - 1/i < \text{diameter } a_ib_i < \text{diameter } X/2.$$

Such an arc $a_i b_i$ may be found by considering an arc of large diameter in X both of whose ends lie in the same link of D_i, reducing this arc by throwing away the part of it in this link and considering one of the larger components of the remainder, reducing the component in a similar fashion, . . . , and stopping this reduction when an arc of the required diameter is found.

Some subsequence of $a_1b_1, a_2b_2, \ldots$, converges to a proper subcontinuum of X. It follows from the hypothesis that this subcontinuum is an arc ab. However, there is a folded sequence of arcs (each in one of the a_ib_i's) converging to a subarc of ab. This contradicts Theorem 6.

8. A nearly homogeneous example. Consider the example Y shown in Figure 1. At a glance it might appear to be homogeneous. The example Y intersects the x axis in a Cantor set and is the sum of semicircles with ends on this Cantor set. Also, the example may be obtained by starting with a punctured disc with three holes and digging canals into the disc from the four complementary domains of the punctured disc.

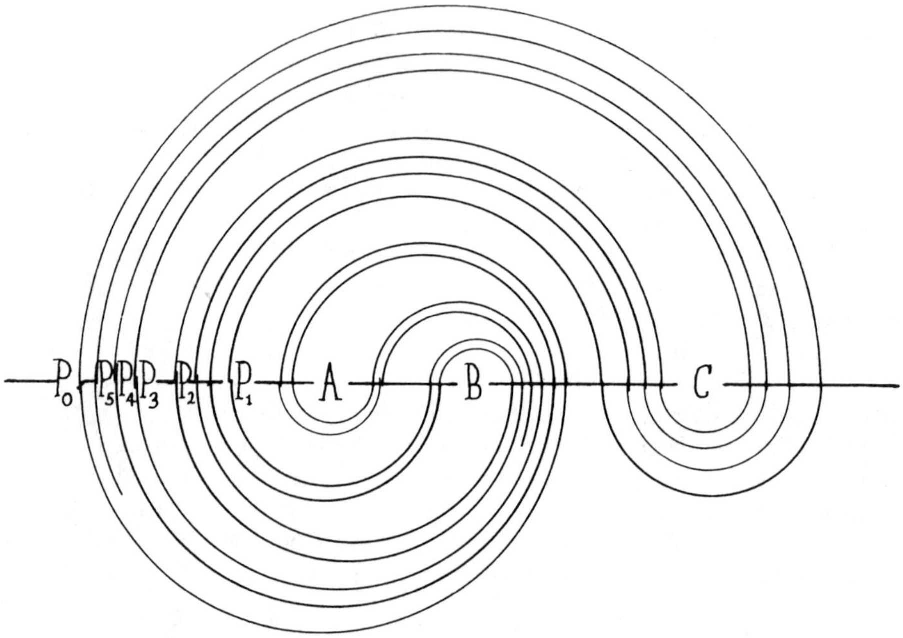

FIGURE 1

The canal from the unbounded complementary domain may be defined in terms of its right bank as follows. Let p_0 be the point furthest to the left of Y and consider the ray R starting at p_0, going along the upper semicircle of X, then along the lower right semicircle, and then down the right bank of the canal leading from the unbounded complementary domain of Y. Let p_1 be the first point on R which is between p_0 and A on the x axis, p_2 be the first point of R that is between p_0 and p_1 on the x axis, and in general, p_{i+1} is the first point of R that is between p_0 and p_i on the x axis.

As viewed from C, p_0p_1 circles C. However, it circles neither A nor B when viewed from A, B respectively. Furthermore, p_0p_2 circles B and C but not A, p_0p_3 circles B, p_0p_4 circles A and B, p_0p_5 circles A, p_0p_6 circles A and C, p_0p_7 circles C.... The other canals from A, B, C run between the canals from the unbounded complementary domains and each canal is dense in Y. Figure 1 does not show the banks of the canals from A, B, C but shows only an arc on the outer bank. This arc does not separate the plane. There are points of X not shown in the figure that are nearer A, B, C than any point on the outer bank. These points keep the complementary domains containing A, B, C from running into each other.

We could write the equation of Y by giving functions f, g such that $(x, f(x))$ are abscissas of the ends of semicircles of Y in the upper half-plane and $(x, g(x))$ are the abscissas of the ends of semicircles of Y in the lower half-plane. However, we shall not do this since we are interested in Y's topological properties rather than its equation.

Example Y is locally homogeneous—that is, for each pair of points p, q of Y there are arbitrarily small homeomorphic open subsets N_p, N_q containing p, q respectively. In fact, the open subsets may be taken to be homeomorphic with the Cartesian product of a Cantor set with an open interval.

Also, example Y is nearly homogeneous—if p and q are points of Y, then for each open subset U of Y containing q there is a homeomorphism of Y onto itself taking p into U. One may see that this is true since each arc component is dense in Y and each arc lies in an open subset homeomorphic with the Cartesian product of a Cantor set and an open interval. See **(7)** for a discussion of various types of homogeneity.

However, M is not homogeneous. If it were, for each positive number ϵ and each point p there would be a homeomorphism h of Y onto itself such that h moves no point by more than ϵ and p, $h(p)$ belong to different composants of Y. (See Theorem 8 of § 9.) Suppose that ϵ is taken to be less than the distance across the canal leading from the unbounded complementary domain at a wide point and p is taken to be the highest point of Y. There is a canal leading from the outside that locally separates p from $h(p)$ in the plane. As p is moved parallel to this canal and in the direction of its wide spot, the canal continues to separate the moving p from the corresponding $h(p)$. However, as the canal widens, it is no longer possible for p to be within ϵ of its image under h.

This intuitive reason of why Y is not homogeneous is refined in § 10 to establish Property 20 and finish the proof of Theorem 1.

9. Homeomorphisms near the identity. In indicating why the nearly homogeneous Example Y of § 8 is not homogeneous, we made use of the fact that if Y were homogeneous there would be a homeomorphism of Y onto itself that does not move any point far but which takes one arc component of Y onto another. We formalize this in the following theorem.

THEOREM 8. *If p is a point of a homogeneous, compact, indecomposable, non-degenerate continuum X, then there is a sequence of homeomorphisms h_1, $h_2, \ldots$, converging to the identity such that no two $h_i^{-1}(p)$'s belong to the same composant of X.*

Proof. Let $\{x_\alpha\}$ be an uncountable collection of points all belonging to different composants of X and h_α be a homeomorphism of X onto itself that takes x_α to p.

If the collection of homeomorphisms $\{h_\alpha\}$ is metrized by the distance function

$$D(h_\alpha, h_\beta) = \max_{x \epsilon X} \rho(h_\alpha(x), h_\beta(x)),$$

the collection $\{h_\alpha\}$ becomes an uncountable subset of a separable metric function space and some sequence $h_1', h_2', \ldots$, of elements of $\{h_\alpha\}$ converges to an element h_0' of $\{h_\alpha\}$. Then

$$h_i = h_i' h_0'^{-1}.$$

Since $x_1, x_2, \ldots$, belong to different composants of X, $h_0'(x_1), h_0'(x_2), \ldots$, also belong to different composants. These are the $h_i^{-1}(p)$'s.

10. M' contains a folded sequence of arcs. In this section we complete the proof of Theorem 1. We showed in §§ 2 and 3 that if there is a homogeneous bounded plane continuum M that contains an arc, there is one such M' each of whose proper subcontinua is an arc. Theorem 6 showed that no such M' contains a folded sequence of arcs converging to an arc. Finally, we show that there is no such M' except a circle, for if there were, it would have the following property.

Property 20. *M' contains a folded sequence of arcs converging to an arc.* Let a_0a_6 be an arc in M' which is accessible from a component of $E^2 - M'$. With no loss of generality we suppose that a_0a_6 is horizontal and $a_1, a_2, \ldots, a_5$ are points of a_0a_6 such that

$$\text{abscissa } a_i = i \quad (i = 0, 1, \ldots, 6).$$

We suppose furthermore that a_0a_6 is accessible from $E^2 - M$ from below. It follows from the methods used in establishing Property 19 that there is a positive number ϵ_1 such that

$$\text{no point of } M' \text{ below } a_0a_6 \text{ is within } \epsilon_1 \text{ of } a_1a_5.$$

Assume M' contains no folded sequence of arcs converging to an arc. Then there is a positive number ϵ_2 such that if D is an ϵ_2-chain covering a_1a_5 with

a_1 in one end link of D and a_5 in the other, then each arc of M' being covered by D and having both ends in the same link of D is of diameter less than $\frac{1}{2}$. Note that $\epsilon_2 \leqslant \frac{1}{2}$. Let D be such an ϵ_2 chain covering a_1a_5 satisfying conditions of Property 17.

Let rs be an arc such that rs lies above a_1a_5; rs misses M'; rs is irreducible from the vertical line containing a_1 to the vertical line containing a_5; the vertical segments ra_1, sa_5 lie in end links of D, and each point between rs and a_1a_5 lies in a link of D. We find that there is such an rs as follows. Cover a_0a_6 by a chain of small mesh satisfying the conditions of Property 17, consider an accessible point of M' above a_3 and in one link of this chain, and note from Property 17 and Theorem 6 that this point lies in an accessible arc in M' slightly above a_0a_6 and with ends near the ends of a_0a_6. It follows from Property 19 that there is an arc in the complement of M' slightly to one side of this first arc. It is this second arc that contains rs.

Let K be the topological disc bounded by a_1a_5, a_1r, rs, and a_5s. We note that if p is a point of $M' \cdot K$ that is above a_2a_4, then the closure of the component of $M' \cdot \operatorname{Int} K$ containing p is an arc irreducible from a_1r to a_5s. If it were not, an arc being covered by D and having diameter more than $\frac{1}{2}$ would have ends in the same link of D.

Let

$$\epsilon_3 = \min(\epsilon_1, \rho(rs, M')).$$

It follows from Property 18 that there is a positive number ϵ_4 such that if ab is an arc in M' with $\rho(a, b) < \epsilon_4$, then either

diameter $ab < \epsilon_3$, or
ab is ϵ_3 dense in M'.

Let A be the arc component of M' containing a_3. It follows from Theorem 8 that there is a homeomorphism h of M' onto itself that moves no point by more than ϵ_4 and which takes a_3 into a point of $M' - A$. Then $h(a_3)$ is a point of K and lies above a_2a_4. Also, $h(a_3)$ lies on an arc in $M' \cdot K$ that is irreducible from a_1r to a_5s.

Since A is dense in M', there is an arc xy in $A \cdot K$ such that xy is irreducible from a_1r to a_5s and xy separates $h(a_3)$ from a_1a_5 in K. By considering points slightly above a_3 we find that xy has the following property.

Special Separating Property. The arc xy separates two points of $K \cdot (M' - A)$ from each other in K such that the first of the points is above a_3 and the other is the image of the first under h.

Let $x_1x_2x_3 \ldots x_{2n}$ be the arc in A such that $x_1x_2 = xy$, $x_{2n-1}x_{2n} = a_1a_5$, and $x_1x_2, x_3x_4, x_5x_6, \ldots, x_{2n-1}x_{2n}$ are the closures of the components of $x_1x_2x_3 \ldots x_{2n} \cdot \operatorname{Int} K$ that are irreducible from a_1r to a_5s. Then x_1x_2 has the special separating property but $x_{2n-1}x_{2n}$ does not.

We now show that if $x_{2i-1}x_{2i}$ has the special separating property, then so does $x_{2i+1}x_{2i+2}$. The resulting contradiction arises as a result of consequences of our false assumption that there is an M' that contains no folded sequence of arcs converging to an arc.

Suppose p, $h(p)$ are points of $K\cdot(M'-A)$ such that p is above a_3 and $x_{2i-1}x_{2i}$ separates p from $h(p)$ in K. For convenience we suppose that $x_{2i+1}x_{2i+2}$ is below $x_{2i-1}x_{2i}$ and $h(p)$ is above $x_{2i-1}x_{2i}$. (Other cases are handled with similar arguments to that given in this case.) Then $x_{2i+1}x_{2i+2}$ separates p from $h(p)$ in K unless p is above $x_{2i+1}x_{2i+2}$, so we suppose p is between $x_{2i-1}x_{2i}$ and $x_{2i+1}x_{2i+2}$. Our proof now breaks down into two cases.

Case 1. *If* x_{2i}, x_{2i+1} *belong to the same one of* a_1r, a_5s, (see Figure 2). Let tu be the closure of the component of $M'\cdot\text{Int } K$ containing p. It is an arc irreducible from a_1r to a_5s. We suppose u belongs to the vertical line through a_1 containing x_{2i} and x_{2i+1}.

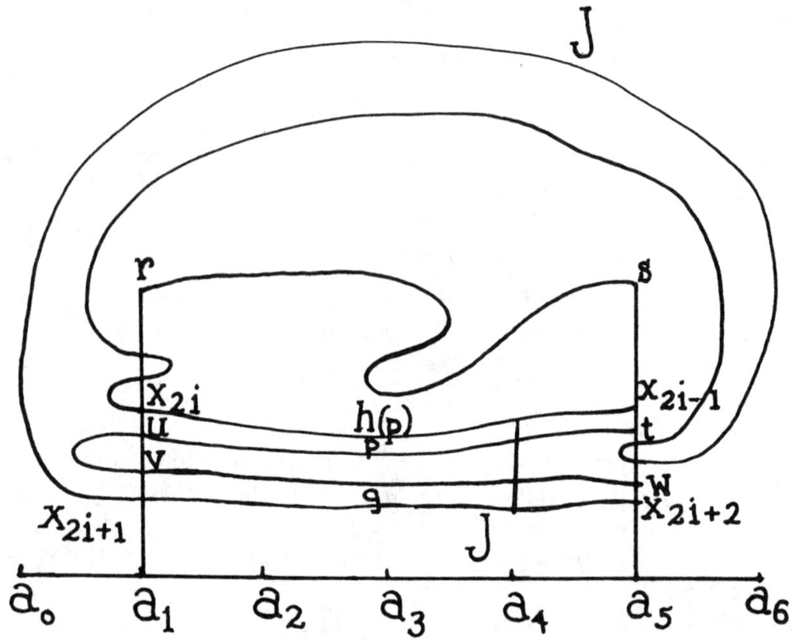

FIGURE 2

Suppose a point moves in an arc in M' through p, past u, and (vw) is the next component of $M'\cdot\text{Int } K$ it meets whose closure vw is an arc irreducible from a_1r to a_5s. Let q be a point of vw directly above a_3. It follows from the Jordan curve theorem that q lies between $x_{2i-1}x_{2i}$ and $x_{2i+1}x_{2i+2}$. Also, q is below tu and $\rho(q, x_{2i-1}x_{2i}) \geqslant \epsilon_4$ or else the arc $tuvw$ contains an arc ab such that

$$\rho(a, b) < \epsilon_4,$$
$$\text{diameter } ab > \epsilon_3, \text{ and}$$
$$\rho(ab, a_4) > \epsilon_3.$$

If q were above tu, we would take $p = a$ and let b be a point of vw between p and $x_{2i-1}x_{2i}$. If q were below tu and $\rho(q, x_{2i-1}x_{2i}) < \epsilon_4$, we would take q to be b and a to be a point of tu between q and $x_{2i-1}x_{2i}$. Since the existence of such an arc ab violates the definition of ϵ_4, we suppose that

$$\rho(q, x_{2i-1}x_{2i}) \geqslant \epsilon_4.$$

We now show that $x_{2i+1}x_{2i+2}$ separates q from $h(q)$ in K. Note that q is above $x_{2i+1}x_{2i+2}$.

Consider the simple closed curve J that is the sum of a vertical interval in K above a_4 and an arc in $x_{2i-1}x_{2i+2}$ that contains $x_{2i}x_{2i+1}$. Note that no point of the arc pq in M is within ϵ_4 of this vertical part of J above a_4. As a point moves from p to q, the image of the point under h does not intersect J. Hence, $h(q)$ is either above $x_{2i-1}x_{2i}$ or below $x_{2i+1}x_{2i+2}$. It is not above $x_{2i-1}x_{2i}$ because $\rho(q, x_{2i-1}x_{2i}) > \epsilon_4$ and $\rho(q, h(q)) < \epsilon_4$. Hence $x_{2i+1}x_{2i+2}$ has the special separation property.

Case 2. *If* x_{2i} *and* x_{2i+1} *belong to different vertical lines, see Figure* 3. We suppose x_{2i}, u, x_{2i+2} belong to a_5s and define vw and q as in Case 1.

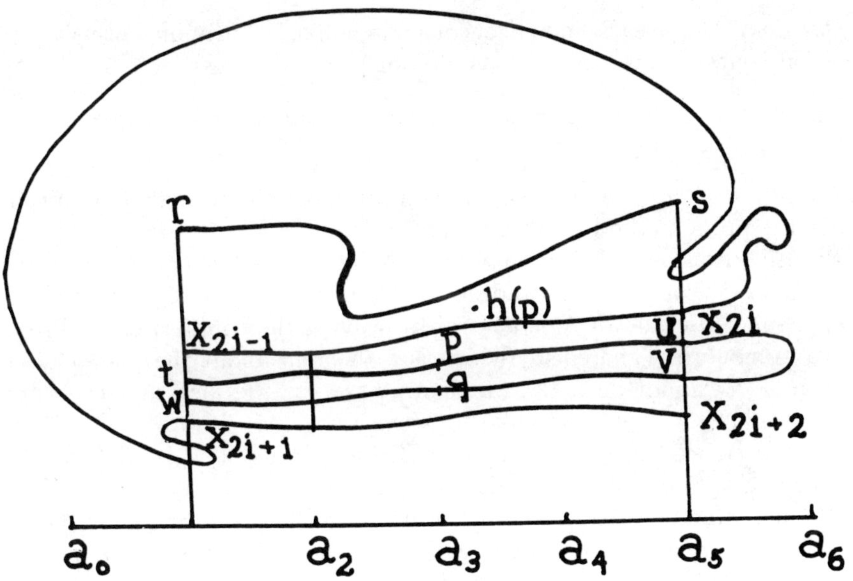

FIGURE 3

If v is below $x_{2i+1}x_{2i+2}$, q is below $x_{2i+1}x_{2i+2}$ and $h(q)$ is above.

If v is above $x_{2i+1}x_{2i+2}$ it is between the points x_{2i} and x_{2i+2}, q is between $x_{2i-1}x_{2i}$ and $x_{2i+1}x_{2i+2}$, and $h(q)$ is above $x_{2i-1}x_{2i}$. Since each of p, q is within ϵ_4 of $x_{2i-1}x_{2i}$, it follows as in Case 1 that $tuvw$ contains an arc ab such that

$\rho(a, b) < \epsilon_4$,
diameter $ab > \epsilon_3$,
$\rho(ab, a_2) > \epsilon_3$.

These conditions violate the definition of ϵ_4.

We note that in establishing Property 20 we used properties of the plane or 2-sphere and not just properties of an arbitrary 2-manifold alone. If this could be by-passed, one might get an affirmative answer to the following question.

Question. Suppose X is a 1-dimensional homogeneous compact continuum that contains an arc and lies on a compact 2-manifold. Is X necessarily a simple closed curve?

11. Circle-like continua and tree-like continua. Solenoids and a circle of pseudo-arcs are the known examples of homogeneous circle-like continua. Since each proper subcontinuum of a circle-like continuum is snake-like and each homogeneous non-degenerate snake-like continuum is a pseudo-arc, one might suspect that the answer to the following is in the affirmative.

Question. Does each homogeneous circle-like continuum other than a solenoid contain a pseudo-arc? We do not provide an answer.

THEOREM 9. *Each homogeneous circle-like continuum that contains an arc is a solenoid.*

Indication of proof. This theorem is much easier to establish than Theorem 1 but the same method of attack may be used.

By using rays as in § 3, it may be shown that the homogeneous circle-like continuum X contains a non-degenerate subcontinuum X' such that each proper subcontinuum of X is an arc. In proving the counterpart of Property 9, we cannot use Theorem 3 (which is a theorem about the plane) to show that $ab + \bar{R}$ cannot lie in X, but instead we use the fact that each proper subcontinuum of X is snake-like to prove this.

We may as well suppose that $X' = X$, for if it is not, it is snake-like, it is a pseudo-arc **(5)**, and it contains no arc.

We finish the indication of proof of Theorem 9 by showing that there is a sequence of circular chains (open coverings whose 1-nerves are simple closed curves) $D_1, D_2, \ldots$, covering X such that

1. D_{i+1} is a refinement of D_i,
2. D_{i+1} circles around $D_i n_i$ times without any folding back, and
3. the mesh of D_{i+1} is less than $1/2^i$ times the distance between any two non-adjacent elements of D_i.

It is then only a matter of getting an open covering of a similar kind of the solenoid which is the intersection of the tori $T_1, T_2, \ldots$, where T_{i+1}

winds about T_i n_i times and use the two coverings to get a homeomorphism of X onto the solenoid. (D_{i+1} is said to fold back in D_i if D_i contains two adjacent links d_x, d_y, and D_{i+1} contains a subchain E such that each link of E lies in either d_x or d_y, each end link of E intersects $d_x - d_y$ but not each link of E lies in d_x.)

Suppose that D_i has already been obtained and it is such that there is a positive number ϵ such that if D' is any circular chain of mesh less than ϵ covering X, then D' refines D_i and circles about D_i without any folding back. We show how to get D_{i+1}, With no loss of generality we suppose that $\epsilon < 1/2^i$ times the distance between any two non-adjacent elements of D_i.

We apply Theorem 6 to show that no folded sequence of arcs in X converges to an arc. Hence, there is a δ such that if ab is an arc of diameter greater than $\epsilon/14$, no δ chain D'' covers ab in such a way that both a, b lie in the same link of D''. We suppose $\delta < \epsilon/14$.

Let D be a δ circular chain covering X with its links ordered $d_1, d_2, \ldots, d_n$. Let $d_{n1} = d_1$; d_{n2} is the first link of D whose distance from d_{n1} is more than $\epsilon/14$; d_{n3} is the next link of D after d_{n2} whose distance from d_{n2} is more than $\epsilon/14 \ldots$. Let d_{n2r-1} or d_{n2r} be the last such link obtained.

The first link of D_{i+1} is the sum of the links between d_{n1} and d_{n4} inclusive; the next link of D_{i+1} is the sum of the links of D between d_{n3} and d_{n6} inclusive; $\ldots$; and the last link of D_{i+1} is the sum of the links between d_{n2r-1} and d_{n2} inclusive (this link contains d_n and d_1). Each link of D_{i+1} other than the last is of diameter less than $4\epsilon/14 + 7\delta$ and the last is of diameter less than $5\epsilon/14 + 9\delta$. In either case, D is of mesh less than ϵ. If D' is a refinement of D_{i+1} of mesh less than δ, D' circles about D_{i+1} without any folding back.

A triodic continuum is the sum of three continua A, B, C such that $A \cdot B = A \cdot C = B \cdot C$ is a proper subcontinuum of each of A, B, C. Theorem 7 did not provide an answer as to whether or not each homogeneous tree-like continuum fails to contain an arc. Our methods do not give this because we fail to prove the counterpart of Properties 4 and 9.

THEOREM 10. *A homogeneous tree-like continuum contains no arc if it contains no triodic continuum.*

To establish Theorem 10 we use the hypothesis that the continuum contains no triodic continuum to establish the counterpart of Property 9. Property 15 then follows and reduces Theorem 10 to Theorem 7.

References

1. R. D. Anderson, *A characterization of the universal curve and a proof of its homogeneity*, Ann. Math., *67* (1958), 313–324.

2. R. H. Bing, *A homogeneous indecomposable plane continuum*, Duke Math. J., *15* (1948), 729–742.

3. ——— *Snake-like continua*, Duke Math. J., *18* (1951), 653–663.

4. ——— *Concerning hereditarily indecomposable continua*, Pac. J. Math., *1* (1951), 43–51.

5. ——— *Each homogeneous chainable continuum is a pseudo arc*, Proc. Amer. Math. Soc., *10* (1959), 345–346.

6. R. H. Bing and F. B. Jones, *Another homogeneous plane continuum*, Trans. Amer. Math. Soc., *90* (1959), 171–192.

7. C. E. Burgess, *Continua and various types of homogeneity*, Trans. Amer. Math. Soc., *88* (1958), 366–374.

8. H. J. Cohen, *Some results concerning homogeneous plane continua*, Duke Math. J., *18* (1951), 467–474.

9. E. E. Grace, *Totally nonconnected im kleinen continua*, Proc. Amer. Math. Soc., *9* (1958), 818–821.

10. F. B. Jones, *Certain homogeneous unicoherent indecomposable continua*, Proc. Amer. Math. Soc., *2* (1951), 855–859.

11. S. Mazurkiewicz, *Sur les continus homogènes*, Fund. Math., *5* (1924), 137–146.

12. E. E. Moise, *A note on the pseudo-arc*, Trans. Amer. Math. Soc., *64* (1949), 57–58.

13. R. L. Moore, *Foundations of point set theory*, Amer. Math. Soc. Coll. Publ., *13* (1932).

University of Wisconsin

D. Others

A HEREDITARILY INFINITE DIMENSIONAL SPACE

R. H. BING

Madison

1. Introduction

In recent papers [1, 2, 3] David Henderson describes examples of infinite dimensional compact metric spaces which contain no 1-dimensional closed subsets. In this paper we modify Henderson's approach slightly to give alternative descriptions of such examples.

The plan for getting an example is to start with a Hilbert cube H. We regard H as the Cartesian product of $I_1 \times I_2 \times \ldots$ where $I_i = [-1/2^i, 1/2^i]$. The metric for H is Euclidean.

Note that H has a countable basis $U_1, U_2, \ldots$ where U_i is of the form $B^n \times \times I_{n+1} \times I_{n+2} \times \ldots$ where $i \geqq n$ and B^n is the intersection of an open round n-ball in E^n with $I_1 \times I_2 \times \ldots \times I_n$ where for the moment we regard $I_1 \times I_2 \times \ldots \times I_n$ as lying in Euclidean n-space E^n.

An example of an infinite dimensional compact metric space with no 1-dimensional closed subset is obtained as the intersection of a countable number of closed subsets $K_1, K_2, \ldots$ of H where the K_i's have two important properties.

The first of these important properties is the following:

Property 1. *Any continuum in K_i from U_i to $H - U_i$ contains a subset of* Bd U_i *of dimension greater than or equal to* 2.

This property insures that $K = K_1 \cap K_2 \cap \ldots$ contains no closed 1-dimensional subset.

The second important property of the K_i's is chosen to insure that K is not 0-dimensional or null. In fact, we choose the K_i's so that K contains a continuum that joins the first pair of opposite faces of H. This pair is $\{-1/2\} \times I_2 \times I_3 \times \ldots$ and $\{1/2\} \times I_2 \times I_3 \times \ldots$. In choosing the property we are guided by a generalization of the following interesting property of a canonical cube C^3. If X is a closed set that separates the front face from the back face of C^3 and Y is a closed set that separates the left face from the right, then $X \cap Y$ contains a continuum joining the top and bottom of C^3. See Proposition A on page 40 of [4].

If X, Y are subsets of H, we say that Y separates X wrt I_i if $X - Y$ is the union of two mutually separated sets one of which contains $X \cap (I_1 \times I_2 \times \ldots \times I_{i-1} \times$

$\times \{-1/2^i\} \times I_{i+1} \times \ldots)$ and the other of which contains $X \cap (I_1 \times I_2 \times \ldots \times \times I_{i-1} \times \{1/2^i\} \times I_{i+1} \times \ldots)$. We say that Y weakly separates X wrt I_i if $X - Y$ is the union of two mutually separated sets (either or both of which may be null) one of which misses $I_1 \times I_2 \times \ldots I_{i-1} \times \{-1/2^i\} \times I_{i+1} \times \ldots$ and the other of which misses $I_1 \times I_2 \times \ldots \times I_{i-1} \times \{1/2^i\} \times I_{i+1} \times \ldots$.

Our remarks about C^3 may be extended to H as follows. If $X_2, X_3, \ldots$ are closed subsets of H such that X_2 weakly separates H wrt I_2, X_3 weakly separates X_2 wrt I_3, X_4 weakly separates X_3 wrt $I_4, \ldots$, then it can be shown that $X_2 \cap X_3 \cap \ldots$ contains a continuum which joins the first pair of opposite faces of H. This result and related ones are given by Theorems 3, 4, 5 in Section 4.

The second important property of the K_i's is the following:

Property 2. *K_i is a closed subset of a subcontinuum R_i of H such that:*

a) *R_i weakly separates H* wrt *I_{2i} and*
b) *K_i weakly separates R_i* wrt *I_{2i+1}.*

It follows from Theorem 5 that $K = K_1 \cap K_2 \cap \ldots$ is not 0-dimensional since it contains a continuum joining the first pair of opposite faces of H.

Theorem 1. *The set $K = K_1 \cap K_2 \cap \ldots$ is an infinite dimensional compact metric space with no 1-dimensional closed subset.*

2. Description of the R_i's and K_i's

First we consider the case where B^n is a round ball in $I_1 \times I_2 \times \ldots \times I_n$. Let R_i be the union of $H - U_i$ and the set of all points p of U_i whose $2i$th coordinate is $1/2^{2i} \sin (1/\varrho(p, \text{Bd } U_i))$. We use ϱ to denote the distance function. Note that R_i weakly separates H wrt I_{2i}.

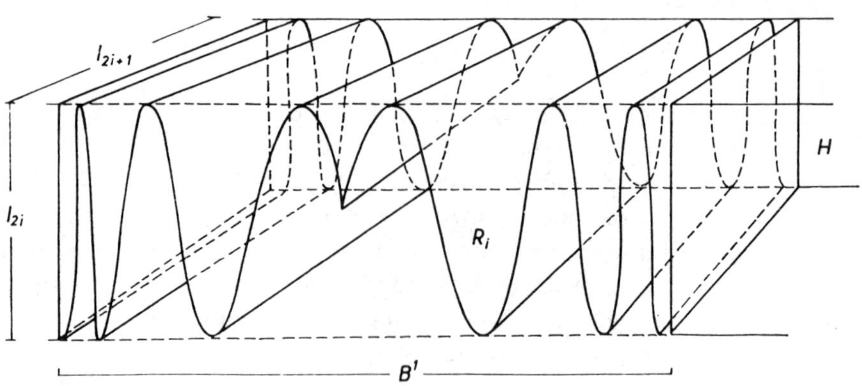

Figure 1.

Figure 1 gives a diagramatic view of R_i in case B^n is 1-dimensional and Figure 2 shows it if B^n is of dimension 2. In Figure 2 we are reminded of a vibrating drum or ripples on a pond where the period becomes short near the boundary but the amplitude remains constant.

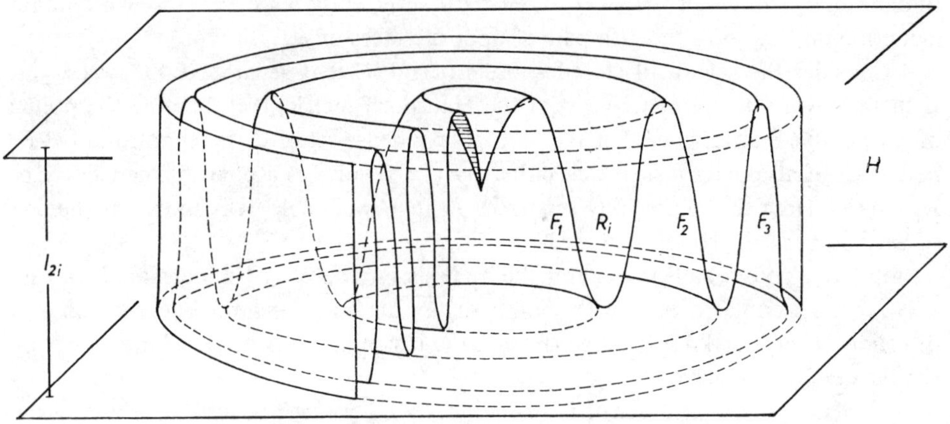

Figure 2.

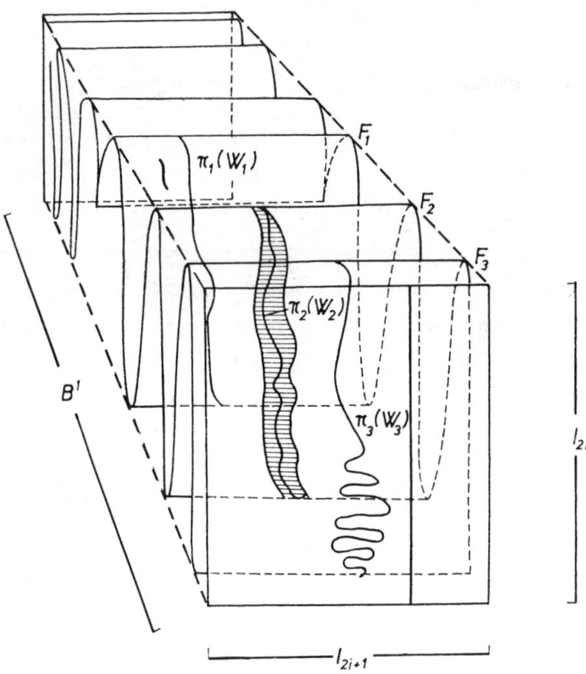

Figure 3.

By considering the variable half periods of $y = \sin 1/x$ we find that $R_i \cap U_i$ contains a countable number of mutually exclusive sets $F_1, F_2, \ldots$ such that there is a homeomorphism π_j of Bd U_i onto F_j such that if $\pi_j(x_1, x_2, \ldots) = (y_1, y_2, \ldots)$, then $y_k = x_k$ if $k > n$ and $(y_1, y_2, \ldots, y_n)$ is between $(x_1, x_2, \ldots, x_n)$ and the center of B^n. Any infinite subsequence of the F_j's will converge to Bd U_i and if C is a continuum in R_i from U_i to $H - U_i$, then for j sufficiently large, $F_j \cap C$ will contain a continuum that joins the $2i$th pair of opposite faces of H.

Consider the set of all closed subsets of Bd U_i that separate Bd U_i wrt I_{2i+1}. If these closed sets are metrized with the Hausdorff metric, they become the points of a separable metric space. Let $W_1, W_2, \ldots$ be a dense set of these separators. Let K_i be the set of all points p such that either $p \in \bigcup \pi_j W_j$ or p is a point of the closure of $R_i - \bigcup F_j$. Note that K_i weakly separates R_i in H wrt I_{2i+1}. It is shown at the end of Section 3 that K_i has Property 1.

Figure 3 shows half of some of the $\pi_j(W_j)$'s in case B^n is 1-dimensional. In case $n = 2$, F_j resembles a pipe with length the $2i$ direction and thickness the $2i + 1$ direction, where $\pi_j(W_j)$ is a set separating the inside lateral surface of the pipe from the outside lateral surface.

In case B^n is not a subset of $I_1 \times I_2 \times \ldots \times I_n$, we let t be a positive number so small that $tB^n \subset I_1 \times I_2 \times \ldots \times I_n$. If R_i' (K_i') is the set like R_i (K_i) we get by using tB^n, then R_i (K_i) is the set of all points $(x_1, x_2, \ldots)$ of H such that $(tx_1, tx_2, \ldots, tx_n, x_{n+1}, \ldots)$ is a point of R_i' (K_i').

3. Preventing 1-dimensionality

How does one prove that a set is of dimension greater than 1? In Lemma 2 of [2], Henderson used the criteria that a set is of dimension greater than 1 if there is an essential map of it onto a square. In this section we use a modification of Lemma 2 proposed by Harry Row.

Theorem 2. *Suppose in a metric space X that A_1, A_2 are mutually exclusive closed sets; B_1, B_2 are mutually exclusive sets; and Y is a compact set such that each subset of X that separates A_1 from A_2 in X intersects Y in a set containing a continuum from B_1 to B_2. Then dimension $Y \geqq 2$. In fact, if W separates A_1 from A_2 in X then there is a point of $W \cap Y$ at which Y is of dimension greater than* 1.

Proof. That dimension $Y \geqq 2$ follows immediately from Proposition B on page 34 of [4] but we include another proof. Let W' be a closed subset of W such that $X - W'$ is the union of two mutually separated sets V_1, V_2 containing A_1, A_2 respectively. Assume Y is of dimension less than or equal to 1 at each point of $Y \cap W'$.

At each point p of $Y \cap W'$, let O_p be an open set such that $p \in O_p$, dimension $(Y \cap \text{Bd } O_p) \leqq 0$, and $\bar{O}_p \cap (A_1 \cup A_2) = \emptyset$. Let $O_1, O_2, \ldots, O_n$ be a finite number of O_p's covering $Y \cap W'$.

Let $W'' = (W' - \bigcup_{i=1}^{n} O_i) \cup \bigcup_{i=1}^{n} \text{Bd } O_i$. Note that W'' separates A_1 from A_2 since $X - W''$ is the union of the mutually separated sets $V_1 - \bigcup_{i=1}^{n} \bar{O}_i$ and $(V_2 \cup \bigcup_{i=1}^{n} \bar{O}_i) - \bigcup_{i=1}^{n} \text{Bd } O_i)$. Since dimension $(W'' \cap Y) \leqq 0$, $(W'' \cap Y)$ does not contain a continuum from B_1 to B_2. This contradiction resulted from the assumption that Y is of dimension less than or equal to 1 at each point of $Y \cap W'$.

Theorem 2 and Property 1. It follows from the above Theorem 2 that if K_i is as described in Section 2 and C is a continuum in K_i from U_i to $H - U_i$, then dimension $(C \cap \text{Bd } U_i) \geqq 2$. To see this, let Bd U_i of Section 2 be the X of Theorem 1, the intersection of Bd U_i with the $2i$th pair of opposite faces of H be B_1, B_2, the intersection of Bd U_i with the $(2i + 1)$st pair of opposite faces of H be A_1, A_2, and $C \cap \text{Bd } U_i$ be Y. To see that if W is a subset of Bd U_i that separates Bd U_i wrt I_{2i+1}, then $W \cap C$ contains a continuum joining the $2i$th pair of opposite faces of H we proceed as follows. Let $W(n_1), W(n_2), \ldots$ be a subsequence of $W_1, W_2, \ldots$ converging to W. For k sufficiently large, there is a continuum $C(n_k)$ in $C \cap \pi_{n_k} W(n_k)$ joining the $2i$th pair of opposite faces of H. Some subsequence of $C(n_1), C(n_2), \ldots$ converges to a continuum in W. This continuum lies in $W \cap C$ and joins the $2i$th pair of opposite faces in H.

4. Essential maps

A map f of a set X onto a cell B is said to be inessential if there is a map $g : X \to \text{Bd } B$ such that $f = g$ on f^{-1} Bd B. If there is no such map g, we say that f is essential. In the following theorem we use I^n to denote an n-cell and I^1 to denote $[0, 1]$.

Theorem 3. *Suppose f is an essential map of a metric space X onto $I^{n+1} = I^n \times I^1$ and π is the projection map of $I^n \times I^1$ onto I^n. If Y is a subset of X that separates $f^{-1}(I^n \times \{0\})$ from $f^{-1}(I^n \times \{1\})$, then $\pi f/Y$ is an essential map of Y onto I^n.*

Proof. Suppose $X - Y$ is the union of mutually separated sets U, V where $f^{-1}(I^n \times \{0\}) \subset U$ and $f^{-1}(I^n \times \{1\}) \subset V$. We suppose Y is closed since by using the fact that X is hereditarily normal, we find that if an arbitrary set in X separates two sets in X, a closed set in the arbitrary set separates the same two sets in X.

Assume $\pi f/Y$ is inessential. Then there is a map $g' : Y \to \text{Bd } I^n$ such that $g' = \pi f$ on $Y \cap (\pi f)^{-1} \text{Bd } I^n$. Let $g : Y \cup f^{-1} \text{Bd } (I^n \times I^1) \to \text{Bd } (I^n \times I^1)$ be such that $g = f$ on $f^{-1} \text{Bd } (I^n \times I^1)$ and for each point $x \in Y$, $g(x) = (g'(x), t)$ where t is the second coordinate of $f(x)$. Let $\bar{g}$ be an extension of g to all of X such that $\bar{g}(U) \subset (I^n \times \{0\}) \cup (\text{Bd } I^n) \times I^1$ and $\bar{g}(V) \subset (I^n \times \{1\}) \cup (\text{Bd } I^n) \times I^1$. Note that $\bar{g}$ is an extension of the map $f/f^{-1} \text{Bd } (I^n \times I^1)$. The assumption that $\pi f/Y$ is inessential led to the contradiction that f is inessential.

Theorem 3 can be extended as follows.

Theorem 4. *If in Theorem 3 we supposed merely that $X - Y$ was the sum of two mutually separated sets one of which missed $f^{-1}(I^n \times \{0\})$ and the other of which missed $f^{-1}(I^n \times \{1\})$, the conclusion still holds that $\pi f/Y$ is an essential map of Y onto I^n.*

Proof. The proof is the same as the proof of Theorem 3 except that instead of being able to require that $g = f$ on $f^{-1}\,\mathrm{Bd}\,(I^n \times I^1)$, we would merely suppose that $g = f$ on $f^{-1}((\mathrm{Bd}\,I^n) \times I^1)$, $g(f^{-1}(I^n \times \{0\})) \subset I^n \times \{0\}$, and $g(f^{-1}(I^n \times \times \{1\})) \subset I^n \times \{1\}$. Since $g/f^{-1}\,\mathrm{Bd}\,(I^n \times I^1) \to \mathrm{Bd}\,(I^n \times I^1)$ is homotopic to f and g can be extended to take X onto $\mathrm{Bd}\,(I^n \times I^1)$, the Homotopy Extension Theorem (see Theorem VI 5 of [4]) says that $f/f^{-1}\,\mathrm{Bd}\,(I^n \times I^1)$ can be so extended.

Theorem 5. *If $X_2, X_3, \ldots$ is a sequence of sets in H such that X_2 weakly separates H* wrt *I_2, X_3 weakly separates X_2* wrt *I_3, X_4 weakly separates X_3* wrt *$I_4, \ldots$ then $X_2 \cap X_3 \cap \ldots$ contains a continuum joining the first pair of opposite faces of H.*

Proof. Let $\pi_{n,j}$ be the projection of H onto $I_1 \times I_{n+1} \times \ldots \times I_{n-1+j}$. It follows from Theorem 4 and induction on n that $\pi_{n,j}/X_2 \cap X_3 \cap \ldots \cap X_n$ is an essential map of $X_2 \cap X_3 \cap \ldots \cap X_n$ onto $I_1 \times I_{n+1} \times \ldots \times I_{n-1+j}$.

Since $\pi_{n,1}/X_2 \cap X_3 \cap \ldots \cap X_n$ takes $X_2 \cap X_3 \cap \ldots \cap X_n$ essentially onto I_1, there is a continuum C_n in $X_2 \cap X_3 \cap \ldots \cap X_n$ joining the first pair of opposite faces of H. Some subsequence of $C_2, C_3, \ldots$ converges to a continuum in $X_2 \cap X_3 \cap \cap \ldots$ and this continuum joins the first pair of opposite faces of H.

5. Variations in the definition of K

The proof of Theorem 3 is slightly easier than that of Theorem 4 so it would have been easier to prove that dimension $(K_1 \cap K_2 \cap \ldots) \geqq 2$ if we had replaced "weakly separates" by "separates" in Property 2. We could do this by taking a new R_i whose points have coordinates the same as the points of the old R_i except that the $2i$th coordinate is divided by 2. To get a new K_i we would divide both the $2i$th and the $(2i + 1)$st coordinates of points of the old K_i by 2. While this variation simplifies the proof, it complicates the construction.

Another variation of description of K_i permits us to avoid using the projection π_j. Instead of letting $W_1, W_2, \ldots$ be a dense set of separators of $\mathrm{Bd}\,U_i$ wrt I_{2i+1} and projecting these separators onto the F_i's to obtain K_i we could have let $W_1', W_2', \ldots$ be a dense set of separators of R_i wrt I_{2i+1} and used $W_i' \cap F_i$ instead of $\pi_i W_i$ in defining K_i. Instead of proving that dimension $(C \cap \mathrm{Bd}\,U_i) \geqq 2$ we would have shown that dimension $C \geqq 2$ but this would have been just as good.

6. Infinite dimensional continuous curves

Of course, $K = K_1 \cap K_2 \cap \ldots$ is not locally connected or it would contain an arc. However it is possible to change any compact metric space to a continuous curve (Peano continuum) by adding to it the union of a null sequence of arcs so that no two of them intersect each other except possibly at an end point of each. Hence such an addition would convert K into an infinite dimensional continuous curve with no 2-dimensional subcontinuum.

By picking a nondegenerate component K', adjoining a null sequence of mutually exclusive arcs to it to convert K' to a continuous curve C and then shrinking the arcs to points, there results an infinite dimensional continuous curve C' with no 2-dimensional subcontinuum. Actually C' results from a decomposition of K' whose nondegenerate elements are point pairs. Also C' has the property that each open subset of it is infinite dimensional.

7. Questions

What conditions imposed on an infinite dimensional compact metric space implies that the space contains a 1-dimensional subcontinuum? Would the triviality of the 1st homology (Čech) imply this? Would the triviality of all the homology groups imply it?

The triviality of the 0-th homotopy implies that a space is arcwise connected. Does some restriction on the global and local homotopy groups of an infinite dimensional compact metric space imply that it contains a 2-dimensional subcontinuum? Need an infinite dimensional compact metric space contain a 2-dimensional closed subset even if it is an absolute retract?

References

[1] *David Henderson:* Finite dimensional subsets of infinite dimensional spaces. Wisconsin Topology Conference, Summer 1965, Annals of Mathematics Studies, Princeton University Press (1966), 141—146.

[2] *David Henderson:* An infinite-dimensional compactum with no positive-dimensional compact subsets — a simpler construction. American Journal of Mathematics *89* (1967), 105—121.

[3] *David Henderson:* Each strongly infinite-dimensional compactum contains a hereditarily infinite-dimensional compact subset. American Journal of Mathematics *89* (1967), 122—123.

[4] *W. Hurewicz and H. Wallman:* Dimension Theory. Princeton University Press, 1941.

UNIVERSITY OF WISCONSIN

THE ELUSIVE FIXED POINT PROPERTY

R. H. BING,* University of Wisconsin

A set X has the *fixed point property* if each map $f: X \rightarrow X$ leaves some point fixed—that is, there is a point $x \in X$ such that $f(x) = x$.

Satisfactory necessary and sufficient conditions have not been found for determining whether or not a set has the fixed point property. The sufficient conditions that have been found are too restrictive to be necessary. On the other hand, many examples have been shown to have the fixed point property—sometimes with the method of proof tailored to the example.

The paper is intended primarily as an expository article to bring together some of the interesting results about fixed points. The only new result is Theorem 15. It is hoped that a graduate student with a beginning course in geometric topology will be able to dig through most of the material. Although such specialized concepts as Betti number, Lefschetz number, Euler characteristic, indecomposable continuum, pseudo arc, contractible, degree of a map, Green's Theorem, Sperner's Lemma, absolute neighborhood retract, cyclic element, nerve of a covering, snake-like continuum, tree-like continuum, orientation preserving homeomorphism, Čech homology, etc. are mentioned when this is natural, the treatment is such that the reader can pass on and understand much of what follows, waiting until later to learn about the implications of these concepts from relevant sources if he so desires. Some theorems are included for historical information or the advanced worker, but it is hoped that many readers can provide or follow proofs of Theorems 2, 3, 6, 12–15; 20–22.

Continua discussed in the paper are compact and metrizable. All polyhedra considered are finite and connected.

* Professor Bing is best known for his research in geometric topology. Since his graduate study and post-doctoral instructorship at the Univ. of Texas, he has been at the Univ. of Wisconsin, where he now is a Rudolph E. Langer Professor. He has been a leader in many professional organizations: AAAS (Vice President and Chairman, Section A), CBMS (Chairman), AMS (Council Member, Vice President), MAA (Section Chairman, Hedrick lecturer, President), NRC (Chairman, Div. Math. Sciences). Prof. Bing is a member of the National Academy of Sciences and currently serves on the National Science Board. *Editor.*

1. Sets with the fixed point property. Perhaps the best-known result about fixed points is the following:

THEOREM 1. *Each n-cell has the fixed point property.*

The result is frequently called the Brouwer Fixed Point Theorem although the work of Brouwer [**7**] was probably preceded by that of Bohl [**4**]. However, even before Brouwer's paper [**7**] appeared, reference had been made to the Brouwer Fixed Point Theorem. (See Hadamard's reference on page 472 of [**23**].) In proving the theorem, Bohl considered differentiable maps and used Green's Theorem to show that equivalent integrals did not match if the n-cell had a fixed point free map into itself. On the other hand, Brouwer proved the theorem by showing that homotopic maps of an $(n-1)$-sphere onto itself had the same degree; hence, there is no retraction of an n-cell onto its boundary; hence, each map of an n-cell into itself is fixed point free. Still another proof (using combinatorial techniques) was supplied by Knaster, Kuratowski, and Mazurkiewicz [**18**]. They used Sperner's Lemma in showing that there is no retraction of an n-cell onto its boundary. Hirsch gives a beautiful elementary proof of this result in [**14**].

A subset Y of a set X is a *retract* of X if there is a map $f: X \to Y$ such that $f|Y =$ identity—that is, f leaves each point of Y fixed. The map f is called a *retraction* of X onto Y. The metric set Y is called an *absolute neighborhood retract* if for each metric space Z containing Y, there is an open set U in Z with $Y \subset U$ and a retraction of U onto Y.

A convenient way of showing that a set Y has the fixed point property is to use the following result.

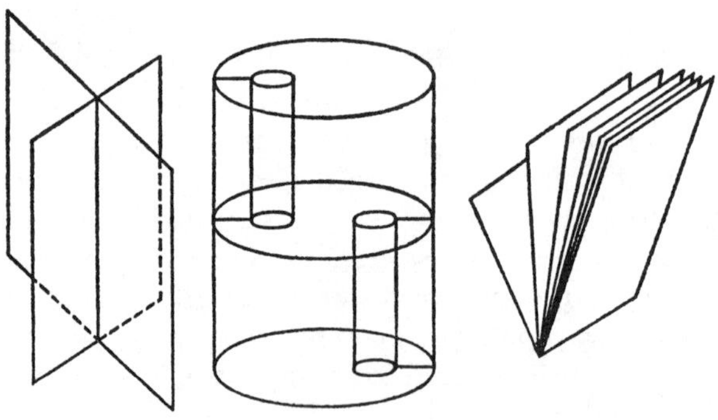

a b c

FIG. 1

THEOREM 2. *If Y is a retract of X and X has the fixed point property, then so does Y.*

For example, Theorem 2 in conjunction with Theorem 1 shows that such objects as represented by Figures 1a, 1b have the fixed point property. Figure 1a represents four pages of a book while Figure 1b represents a house-with-two-rooms [9].

Figure 1c represents infinitely many 2-cells converging to a 2-cell such that any two of the 2-cells share a common edge. Although the object is not a retract of a 3-cell (cube), for each $\epsilon > 0$, there is a retraction of the object onto the union of a finite number of the cells that does not move any point more than ϵ. Hence the object of Figure 1c can be shown to have the fixed point property by using Theorems 1, 2, 3.

THEOREM 3. *If f is a fixed point free map of a compact set X into itself, there is a positive number ϵ such that for each $x \in X$, $\rho(x, f(x)) > \epsilon$, where ρ denotes the distance function. The number ϵ is a function of f but not of x.*

Lefschetz was able to generalize Theorem 1 by defining for each map $f: X \to X$ an integer $L(f)$. This number has since been called the Lefschetz number of f. We shall not define it here but mention that if for each $i > 0$, the ith Betti number of continuum X is 0 (as it is for a cell), then for each $f: X \to X$, $L(f) = 1$.

THEOREM 4. (Lefschetz [20]). *A sufficient condition that an absolute neighborhood retract X have the fixed point property is that for each map $f: X \to X$, $L(f) \neq 0$.*

Theorem 4 can be improved for the case in which X is a compact 1-connected manifold M. Fadell [10] and Fuller [11] have shown that M has the fixed point property if and only if for each map $f: M \to M$, $L(f) \neq 0$.

Any continuous image of an interval [0, 1] is called a *Peano continuum* or *continuous curve*. A *nondegenerate cyclic element* of a continuous curve C is the union of a simple closed curve J in C and all simple closed curves that intersect C in two or more points. Ayres provided us with the following information about continuous curves.

THEOREM 5. (Ayres [1]). *A necessary and sufficient condition that a continuous curve have the fixed point property is that each of its non-degenerate cyclic elements does.*

If X and Y are sets, the *wedge* of X and Y (denoted by $X \vee Y$) is the set obtained by joining X and Y at a point $x \in X$ and $y \in Y$—that is, x is identified with y. The reader should be able to prove the following variation of Theorem 5. A deeper variation is found in [16].

THEOREM 6. *If each of X and Y has the fixed point property, so does $X \vee Y$.*

Suppose P is a triangulated polyhedron of dimension n and $\alpha_m (m = 0, 1, \cdots, n)$ is the number of m simplexes in the triangulation of P. Then $\chi(P) = \alpha_0 - \alpha_1 + \alpha_2 - \cdots + (-1)^n \alpha_n = \sum_0^n (-1)^i \alpha_i$ has been called the Euler characteristic of P. It follows from the Euler-Poincaré formula (see pages 241–243 of [15])that $\chi(P)$ is also equal to $\sum_0^n (-1)^i p_i$ where p_i is the ith Betti number of P. Lopez has described a polyhedron with the following properties.

THEOREM 7. (Lopez [21]). *There is a connected polyhedron X with the fixed point property such that* $\chi(X)$ *is a positive even integer.*

We mention this result since it leads to some interesting anomalies which we mention in Section 2. We do not describe the polyhedron used by Lopez since it is of dimension 8.

QUESTION 1. *Is there a two dimensional polyhedron with the fixed point property which has even Euler characteristic?*

If $G=\{U_1, U_2, \cdots, U_k\}$ is an open covering of a space X, there is associated with the covering a geometric complex $N(G)$ with vertices $v_1, v_2, \cdots, v_k$ such that $\langle v_{i_1}, v_{i_2}, \cdots, v_{i_j}\rangle$ is a simplex of $N(G)$ if and only if $U_{i_1}\cap U_{i_2}\cap \cdots \cap U_{i_j}\neq 0$. The complex $N(G)$ is called the *nerve* of G. If the mesh of G is small, $N(G)$ is a close approximation to X in the sense that there is a map of X into $N(G)$ with small inverses.

A *snake-like* continuum is a compact metric continuum X such that for each $\epsilon>0$, there is an ϵ-covering (one of mesh less than ϵ) G of X such that $N(G)$ is an arc. A *tree-like* continuum has coverings of arbitrarily small meshes such that the nerves of these coverings are trees (contain no simple closed curves). Hamilton has proved the following:

THEOREM 8. (Hamilton [13]). *Each snake-like continuum has the fixed point property.*

We shall not give the proof here but mention that it is a variation of the *dead-end* type proof given in connection with the proof of Theorem 13.

One of the most interesting unsolved problems in geometric topology is the following:

QUESTION 2. *Does each tree-like continuum have the fixed point property?*

Each bounded plane continuum which does not separate the plane is the intersection of a decreasing sequence of topological disks (sometimes called merely disks for short). Question 2 seems related to the following which has been called the most interesting outstanding problem in plane topology.

QUESTION 3. *Does the intersection of each decreasing sequence of disks have the fixed point property?*

We mention several related partial results.

A continuum is called *indecomposable* if it is not the sum of two proper subcontinua. Hence, an arc is decomposable but a pseudo arc is not. (See [22] or [2] for a description of a pseudo arc.)

THEOREM 9. (Hamilton [12]). *If T is a tree-like continuum which contains no indecomposable continuum, then each homeomorphism of T into itself leaves some point fixed.*

THEOREM 10. (Hamilton [12]). *If X is a bounded plane continuum which* *does not separate the plane and whose boundary in the plane contains no indecomposable continuum, then each homeomorphism of X into itself leaves some point fixed.*

THEOREM 11. (Cartwright and Littlewood [8]). *If h is an orientation preserving homeomorphism of the plane onto itself and $X = h(X)$ is a bounded plane continuum which does not separate the plane, then there is a point $x \in X$ such that $h(x) = x$.*

References [5, 16, 24, 25, 27] give other sets of conditions that imply that compact metric continua have the fixed point property. Rather than consider other sets of conditions here, we turn to three instructive examples.

THEOREM 12. *A disk with a spiral about its boundary has the fixed point property.*

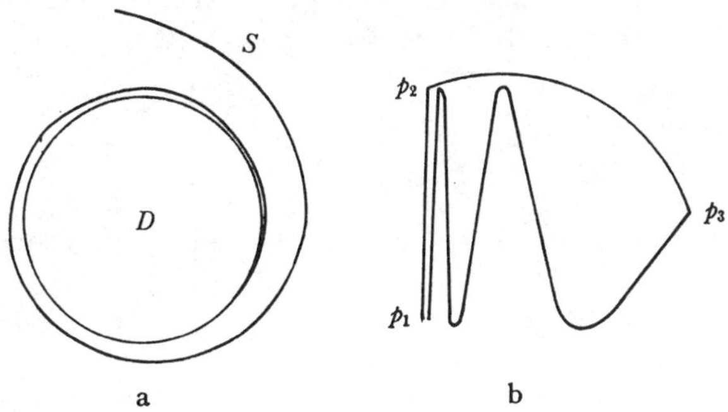

FIG. 2

This object is represented in Figure 2a as a disk D and a spiral S about the boundary of D. If $f:(D \cup S) \to (D \cup S)$, either $f(D) \subset D$ or $f(D) \subset S$ since D is arcwise connected, arcwise connectedness is preserved by maps, and there is no arc in $D \cup S$ from D to S. If $f(D) \subset D$, Theorem 1 implies that f leaves a point fixed. If on the other hand, $f(D) \subset S$, a consideration of f at any point of the boundary of D shows that f takes some point of S into S. Since S is arcwise connected, $f(D \cup S) \subset S$. Since $D \cup S$ is compact, $f(D \cup S)$ is either a point or an arc. If it is a point, the point is left fixed by f. If it is an arc, Theorem 1 implies that some point of the arc is left fixed.

THEOREM 13. *The* sin $1/x$ *circle has the fixed point property.*

This object is represented in Figure 2b as the union of the closure of the graph of $y = \sin 1/x$ $(0 < x \leqq 1/\pi)$ and an arc from $p_2 = (0, 1)$ to $p_3 = (1/\pi, 0)$. The point $(0, -1)$ is labeled p_1. Suppose f is a map of the sin $1/x$ circle into itself.

We use the *dead-end* method to prove that f leaves some point fixed. Consider a dog chasing a rabbit, where a variable point x represents the dog, $f(x)$

the rabbit, and the direction along the arc in the sin $1/x$ circle from x to $f(x)$ the direction of the chase. Let the dog start at p_1. After he moves slightly beyond p_1 toward p_2 in his chase of his image, we note that the rabbit is in front of the dog in the sense that x is between p_1 and $f(x)$ on the arc $[p_1, f(x)]$ of the sin $1/x$ circle. It follows by continuity that even as x moves past p_3 and then back along the sin $1/x$ graph to near $[p_1, p_2]$, $f(x)$ stays in front of x. Each point q_0 of $[p_1, p_2]$ is the limit of a sequence of points $q_1, q_2, \cdots$, of the sin $1/x$ graph. Since $f(q_i)$ is ahead of q_i, then $f(q_1), f(q_2), \cdots$, approaches $[p_1, p_2]$. Hence $f(q_0) \in [p_1, p_2]$. Since $f[p_1, p_2] \subset [p_1, p_2]$, Theorem 1 implies that f leaves some point of $[p_1, p_2]$ fixed.

The following question suggested by Young in [27] seems related to Theorem 13. The continuum X' used in the proof of Theorem 15 of the next section shows the reason for inserting the planar assumption.

QUESTION 4. *Does each arcwise connected compact planar continuum containing no simple closed curve have the fixed point property?*

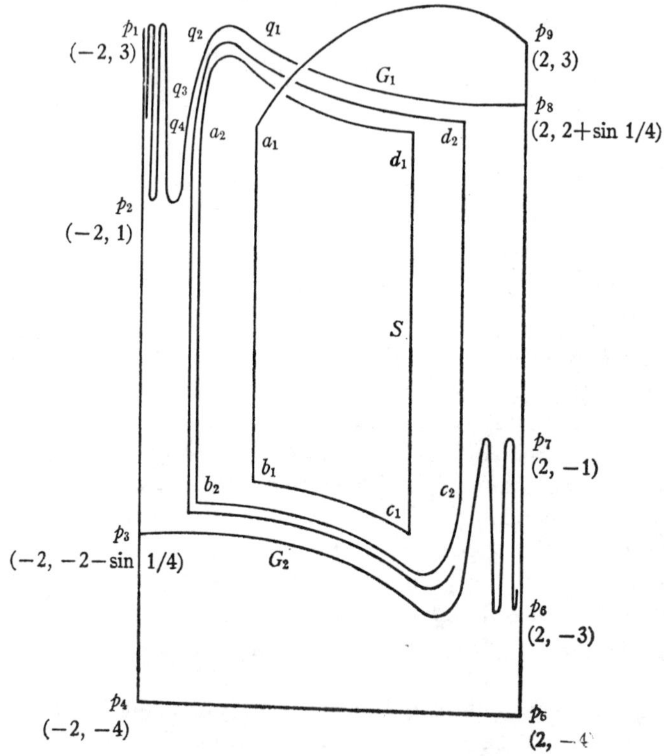

FIG. 3

We consider one more example which plays an important role in the following two sections.

THEOREM 14. *The object represented by Figure 3 has the fixed point property.*

This object (which we call X) is the union of three segments $[p_1, p_4]$, $[p_4, p_5]$, $[p_5, p_9]$, two sin $1/x$ curves G_1, G_2, and a spiral S. We describe them more precisely before showing that X has the fixed point property.

The points $p_1, p_2, p_3, p_4, p_5, p_6, p_7, p_8, p_9$ have coordinates $(-2, 3)$, $(-2, 1)$, $(-2, -2-\sin 1/4)$, $(-2, -4)$, $(2, -4)$, $(2, -3)$, $(2, -1)$ $(2, 2+\sin 1/4)$, $(2, 3)$ as shown in Figure 3.

The sin $1/x$ curve G_1 has equation $y=2+\sin 1/(x+2)(-2<x\leqq 2)$ while G_2 is the reflection of G_1 through the origin $(0, 0)$.

To describe S we select a sequence of points $q_1, q_2, \cdots$ on G_1 such that the length of the arc $[q_i, p_8]$ of G_1 is $2+\sum_1^i 1/j$. Since a harmonic sequence is involved, the points $q_1, q_2, \cdots$ approach the segment $[p_1, p_2]$. On the vertical line through q_i, let a_i and b_i be the points $1/i$ below G_1 and $1/(i+1)$ above G_2 respectively. Let d_i be the reflection of b_i through the origin and c_i the point of the vertical line through d_i that is $1/(i+1)$ units above G_2.

The spiral S starts at p_9, runs to a_1 along a semicircle in a plane normal to the plane containing G_1, follows a straight path to b_1, goes from b_1 to c_1 along an arc parallel to G_2, goes straight to d_1, goes to a_2 along an arc parallel to $G_1, \cdots$. In general, the vertical segments $[a_i, b_i]$ and $[c_i, d_i]$ are in S as are arcs from b_i to c_i parallel to G_2 and arcs from d_i to a_{i+1} parallel to G_1.

To prove that X has the fixed point property, we assume $f: X\rightarrow X$ is fixed point free. The same dead-end method used in the proof of Theorem 13 can be used to show that f leaves some point fixed.

Start the dog at p_3. If $f(p_3)\in[p_1, p_3]$, the dog chases the rabbit to a dead-end along $[p_3, p_1]$ and f leaves some point of $[p_3, p_1]$ fixed. If $f(p_3)\in G_2$, move x along G_1 toward $f(x)$. It is noted that if $x\in G_2-\{p_3\}$, $f(x)$ lies ahead of x in the sense that x lies on the arc in G_2 from p_3 to $f(x)$. An argument like that used in the proof of Theorem 13 shows that f leaves some point of $[p_6, p_7]$ fixed.

If $f(p_3)\notin[p_1, p_3]\cup G_2$, one finds that after the dog leaves p_3 in his chase of the rabbit, x lies between p_3 and $f(x)$ on the arc in X from p_3 to $f(x)$. The chase continues until the dog arrives at the fork p_8. The rabbit is now on either G_1 or $S\cup[p_8, p_9]$. If it is on G_1, we have a situation similar to that in which $f(p_3)\in G_2$ so we suppose $f(p_8)\in S\cup[p_8, p_9]$.

As x moves past p_8, p_9, and into S, it is found that $f(x)$ remains ahead of x in the sense that x lies between p_9 and $f(x)$ on the arc in S from p_9 to $f(x)$. Since the tail end of S converges to $[p_1, p_3]\cup G_2\cup[p_6, p_8]\cup G_1$, if $x_1, x_2, \cdots$ is a sequence of points of S converging to p_8, then $f(x_1), f(x_2), \cdots$ converges to a point of $[p_1,p_3]\cup G_2\cup[p_6, p_8]\cup G_1$. This violates the assumption that $f(p_8)\in S\cup[p_8, p_9]$.

2. Fixed point anomalies. We find that in many ways fixed points do not behave in an expected fashion.

One might think that Theorem 6 would generalize to say that the union along an arc of two sets with the fixed point property would also have the fixed point property. We show that this is not true.

THEOREM 15. *There is a 1-dimensional continuum X with the fixed point property and a disk D such that $D\cap X$ is an arc but $D\cup X$ does not have the fixed point property.*

The continuum X is shown in Figure 3 and D is the rectangular disk with vertices p_5, p_9, p_{10}, p_{11} where $p_{10}=(3, 3)$ and $p_{11}=(3, -4)$. Note that $X\cup D$ retracts onto the continuum

$$X' = (X \cup \mathrm{Bd}\, D) - (p_8, p_9).$$

This continuum X' resembles the continuum which Young [27] showed failed to have the fixed point property. We point out a fixed point free map $f: X'\to X'$.

Restricted to $[p_1, p_3]\cup G_2\cup [p_6, p_8]\cup G_1$, f is a reflection through the origin. Also f takes homeomorphically $[p_3, p_5]$ to $[p_8, p_{11}]$; $[p_6, p_5]$ to $[p_1, p_{11}]$; $[p_5, p_{11}]$ to $[p_{11}, a_1]$; $[p_{11}, a_1]$ to $[a_1, c_1]$, and each $[a_i, a_{i+1}]$ to $[c_i, c_{i+1}]$. The only other precaution to take to see that f is continuous is to be sure that as a point moves far out on S, its image under f is very near its reflection through the origin.

QUESTION 5. *Does $X\times[0, 1]$ have the fixed point property?*

It may be noted that f is a homeomorphism on much of X. This suggests the following question suggested by Young [27].

QUESTION 6. *If C is an arcwise connected continuum which contains no simple closed curve, does each homeomorphism of C into itself leave some point fixed?*

The continuum X of Theorem 15 did not lie in the plane because of the semicircle from p_9 to a_1.

QUESTION 7. *If C is a plane continuum with the fixed point property and D is a disk that intersects C in an arc, must $C\cup D$ have the fixed point property?*

There may be some who feel that the X of Theorem 15 is a pathological example and such unusual things could not happen if the X were required to be polyhedral. This is not an accurate appraisal since Lopez [21] has constructed two continua X, Y with the fixed point property such that $X\vee Y$ has Euler characteristic $=\chi(X\cup Y)$ zero. We shall not describe this example since it is of dimension 17. However, disks can be pasted over the separating points of $X\vee Y$ so as to remove the separating points and it follows from the following result due to Wecken that the union of $X\vee Y$ and the disks does not have the fixed point property.

THEOREM 16. (Wecken [26]). *If K is a connected polyhedron without local separating points and $\chi(K)=0$, then K admits a fixed point free map.*

THEOREM 17. (Lopez [21]). *There is a polyhedron P with the fixed point property and a disk D such that $D\cap P$ is an arc but $D\cup P$ does not have the fixed point property.*

Since the dimension of Lopez's example is 17, it is natural to seek something simpler.

QUESTION 8. *What is the lowest dimension for such a polyhedron P as promised by Theorem* 17?

We shall note in Theorems 18, 23 that cartesian products need not retain the fixed point properties of their factors. However the following is open.

QUESTION 9. *If a compact* 1-*dimension continuum has the fixed point property, does its cartesian product with a segment?*

QUESTION 10. *If a bounded plane continuum has the fixed point property, does its cartesian product with an arc?*

If one takes a polyhedron with a separating point and multiplies the polyhedron with a segment, the cartesian product will have the same Euler characteristic as the polyhedron but will have lost the property of having a separating point. Hence the example of Lopez in conjunction with the result of Wecken shows the following:

THEOREM 18. (Lopez [21]). *There is a polyhedron with the fixed point property such that its cartesian product with a closed* 1-*simplex does not have the fixed point property.*

QUESTION 11. *If P and Q are polyhedra without local separating points but with the fixed point property, must $P \times Q$ have the fixed point property?*

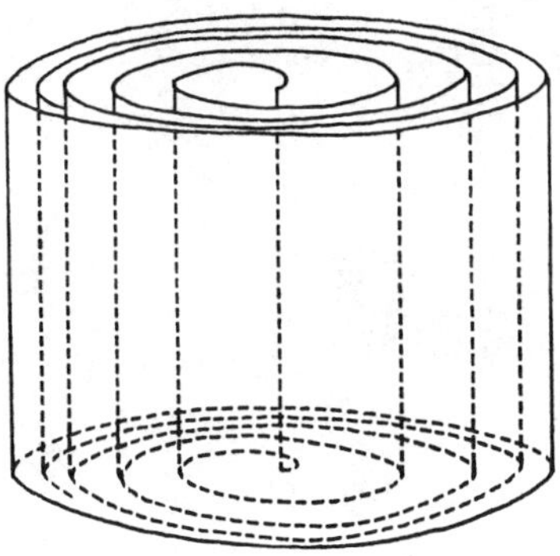

FIG. 4

Polyhedra with trivial ith Betti numbers ($i>0$) have the fixed point property as a result of Theorem 4. Hence compact contractible polyhedra have the fixed point property. In particular, a cone over a polyhedron has the fixed point property. This does not hold in the nonpolyhedral case.

Kinoshita described the example illustrated in Figure 4. It consists of a cylindrical can with a bottom but no top plus the part of the solid cylinder which lies above a spiral on the base converging to the edge of the base. It resembles a can with a spring inside and has been called a can-with-a-roll-of-toilet-paper.

THEOREM 19. (Kinoshita [17]). *The can-with-a-roll-of-toilet-paper has the following properties: it is contractible; it does not have the fixed point property; the cone over it does not have the fixed point property.*

The above properties are treated in [17]. The can-with-a-roll-of-toilet-paper does not answer the following.

QUESTION 12. *Does each homeomorphism of a contractible continuum onto itself leave some point fixed?*

Borsuk [6] gave an example of a 3-dimension continuous curve such that there is a fixed point free homeomorphism of the continuous curve onto itself even though in Čech homology, all p cycles ($p>0$) bound. Bing [3] gave a 2-dimensional example on which the following theorem is based.

THEOREM 20. (Bing [3]). *There is a continuous curve of dimension two which is the intersection of a decreasing sequence of topological cubes and which admits a fixed point free homeomorphism onto itself.*

Consider two cones in Euclidean 3-space each with bases on the cylinder $x^2+y^2=4$ such that the cones have bases in the planes $z=1/2$ and $z=-1/2$ and vertices at $(-2, 0, 1/2)$ and $(2, 0, 1/2)$. Let X denote the image of the union of these two cones if each horizontal plane is rotated about the z axis through an angle of $\tan \pi z$. We do not draw X to scale in Figure 5 but that figure illustrates how the vertex of each cone spirals around the base of the other. For those who want an equation, we suggest the following:

$$\{[x-(2z+1)\cos\tan\pi z]^2+[y-(2z+1)\sin\tan\pi z]^2-(1-2z)^2\}$$
$$\{[x-(2z-1)\cos\tan\pi z]^2+[y-(2z-1)\sin\tan\pi z]^2-(1+2z)^2\}=0.$$

For each θ between 0 and 2π, there is a fixed point free homeomorphism h_θ of X onto itself. Each of the bases of X is rotated through an angle of size θ by h_θ. On other horizontal cross sections, h_θ is a result of a rotation of size θ, about the z-axis, a lift, and a stretch. The lift carries the line of centers of the rotated cross section into the line of centers of the higher cross section. Those wanting an equation can use $([\arctan(\theta+\tan\pi z)]/\pi)-z$ for the size of the lift. Keeping the line of centers fixed, one of the two circles in the rotated and lifted cross section is expanded and the other contracted to make the moved cross section agree with the cross section in the plane to which the lift was made.

It is natural to expect that the cone over a compact continuum would have the fixed point property. That this is true for polyhedra follows from Theorem 4. However, the following theorem due to Knill shows that the result does not hold in general. (Theorem 19 also shows this.)

FIG. 5

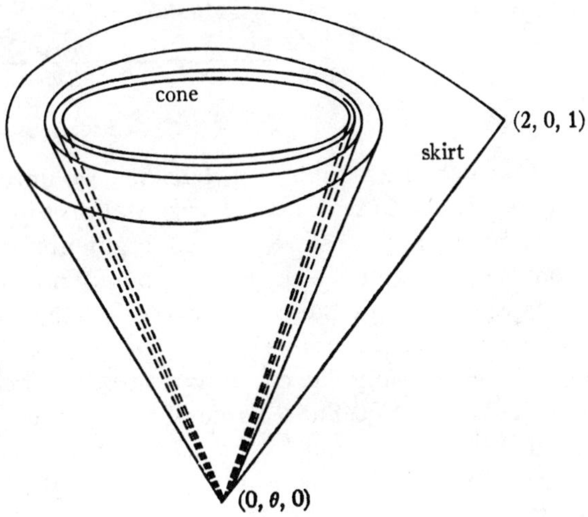

FIG. 6

THEOREM 21. (Knill [**19**]). *The cone over a circle with a spiral as shown in Figure* 6 *does not have the fixed point property.*

Suppose that in cylindrical coordinates (r, θ, z) the circle is defined by $[z=1,\ r=1]$, the spiral by $[z=1,\ r=1+1/(1+\theta),\ \theta \geqq 0]$, and the vertex by $(0, \theta, 0)$. We call the cone over the spiral, the *skirt* of the cone.

The fixed point free map f of the cone-with-a-skirt onto itself that we shall define sends the skirt to the skirt and the cone to a union of the cone and a segment from its vertex.

If $0 \leqq z_0 \leqq 1/3$, the horizontal cross section of the cone at level z_0 goes to the point that divides the segment from $(2, 0, 1)$ to $(0, \theta, 0)$ in the same ratio that z_0 divides 0 to 1/3.

Suppose f_z is the function that sends 0 to 1, 1/3 to 0, 2/3 to 1, 1 to 0 and is linear on each of $[0, 1/3]$, $[1/3, 2/3]$, $[2/3, 1]$, while $f_\theta(\theta, z) = \theta + z - 2/3$. For each point (r, θ, z) of the cone with $1/3 \leqq z \leqq 1$, the fixed point free map is defined by $f(r, \theta, z) \equiv (f_z(z), f_\theta(\theta, z), f_z(z))$. Note that $f_z(z) = z$ only at levels $z = 1/4, 1/2, 3/4$. At levels 1/2 and 3/4 there is rotation and the 1/4 level is not sent onto the cone.

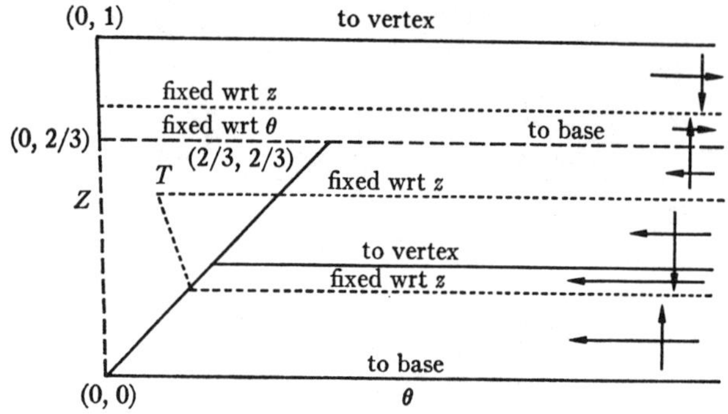

FIG. 7

In order to describe f on the skirt, we unroll it. The unrolled skirt shown in Figure 7 is not to scale since horizontal rays in the half strip represent horizontal cross sections of the skirt, vertical segments in the half strip represent segments from the vertex of the skirt to the base, and the top edge of the half strip represents the vertex of the cone. Arrows on the right of the figure indicate motion in both θ and z directions.

We use rectangular coordinates to describe f on the skirt. We ignore r since $r=z(1+1/(1+\theta))$. The θ value of $f(\theta, z)$, for a point (θ, z) of the strip, is given by 0 if $0 \leqq z \leqq 1/3$ and by max $(0, \theta+z-2/3)$ for $1/3<z\leqq 1$. There is lack of continuity on the horizontal ray to the right from $(1/3, 1/3)$ but this does not matter since this ray will be sent into the vertex of the cone. The θ values of all points of the skirt are changed except those on the vertical segment from $(0, 0)$ to $(0, 2/3)$ and the horizontal ray to the right from $(0, 2/3)$. This invariant set with respect to θ is shown with a dashed line in Figure 7.

We now describe the z values of images of points of the skirt. Except on the triangular disk T with vertices $(0, 0)$, $(0, 2/3)$, $(2/3, 2/3)$, these z values are given by $f_z(z)$ where f_z was the function used to give the z values of images of points of the cone. For each point (θ, z) of T, we define the z value of its image to be 3 max $[|z-1/3|, |\theta-1/3|]$.

Points of the skirt whose z values are invariant under f are shown as dotted. Since the dotted set misses the dashed one, each point of the skirt is moved.

It is interesting to note that the cone over the cone-with-a-skirt does not have the fixed point property either. This result follows from the theorem of Kinoshita [**15**] which states that if a contractible set does not have the fixed point property, then the cone over it does not have the fixed point property either.

One might be misled into thinking that the reason the cone-with-a skirt did not have the fixed point property was that its base did not. This is disproved by the following result.

THEOREM 22. *If $D\cup S$ is a disk with a spiral about its boundary, shown in Figure* 2a, *the cone over $D\cup S$ does not have the fixed point property.*

There is a retraction of this cone onto a cone-with-a-skirt, so it follows from Theorem 21 and 2 that the cone over $D \cup S$ does not have the fixed point property.

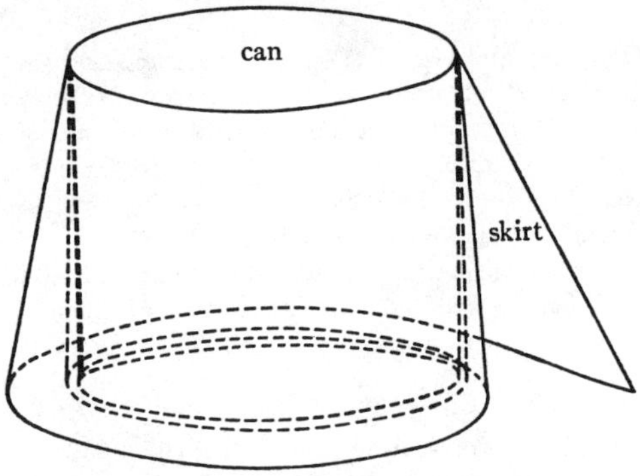

FIG. 8

Let B be a can-with-a-skirt shown in Figure 8. The can is described in cylindrical coordinates as $\{(r, \theta, z) \mid [z=0, r \leqq 1] \text{ or } [r=1, 0 \leqq z \leqq 1]\}$ while the skirt is $\{(r, \theta, z) \mid 0 \leqq z \leqq 1, \theta \geqq 0, r=1+(1-z)/(1+\theta)\}$. Knill has shown that this can-with-a-skirt has the following properties.

THEOREM 23. (Knill [**19**]). *If B is the can-with-a-skirt, then B has the fixed point property; $B \times [0, 1]$ does not have the fixed point property, and there is a 3-cell C (namely, the solid can) such that although each of B and C has the fixed point property and $B \cap C$ is a disk, $B \cup C$ does not have the fixed point property.*

Research on this paper supported by NSF grant GP-3857 and a grant from University of Wisconsin.

References

1. W. L. Ayres, Some generalizations of the Scherer fixed-point theorem, Fund. Math., 16 (1930) 332–336.

2. R. H. Bing, A homogeneous indecomposable plane continuum, Duke Math. J., 15 (1948) 729–742.

3. ———, Challenging conjectures, this MONTHLY, 74 (1967) 56–64.

4. P. Bohl, Über die Bewegung eines mechanischen Systems in der Nähe einer Gleichgewichtslage, J. Reine Angew, Math., 127 (1904) 179–276.

5. K. Borsuk, A theorem on fixed points, Bull. Acad. Polon. Sci. Cl. III, 2 (1954) 17–20.

6. ———, Sur un continu acyclique qui se laisse transformer topologiquement en lui-même sans points invariants, Fund. Math., 24 (1935) 51–58.

7. L. E. J. Brouwer, Über Abbildung von Mannigfaltigkeiten, Math. Ann., 71 (1912) 97–115.

8. M. L. Cartwright and J. E. Littlewood, Some fixed point theorems, Ann. of Math., 54 (1951) 1–37.

9. B. G. Casler, An imbedding theorem for connected 3-manifolds with boundary, Proc. Amer. Math. Soc., 16 (1965) 559–566.

10. E. Fadell, On a coincidence theorem of F. B. Fuller, Pacific J. Math., 15 (1965) 825–834.

11. F. B. Fuller, The homotopy theory of coincidences, Ann. of Math., 59 (1954) 219–226.

12. O. H. Hamilton, Fixed points under transformations of continua which are not connected im kleinen, Trans. Amer. Math. Soc., 44 (1938) 18–24.

13. ———, A fixed point theorem for pseudo-arcs and certain other metric continua, Proc. Amer. Math. Soc., 2 (1951) 173–174.

14. Morris W. Hirsch, A proof of the nonretractibility of a cell onto its boundary, Proc. Amer. Math. Soc., 14 (1963) 364–365.

15. John G. Hocking and Gail S. Young, Topology, Addison-Wesley, Reading, 1961.

16. J. L. Kelley, Fixed sets under homeomorphisms, Duke Math. J., 5 (1939) 535–537.

17. Shin'ichi Kinoshita, On some contractible continua without the fixed-point property, Fund. Math., 40 (1953) 96–98.

18. B. Knaster, C. Kuratowski, S. Mazurkiewicz, Ein Beweis des Fixpunktsatzes für n-dimensionale Simplexe, Fund. Math., 14 (1929) 132–137.

19. Ronald J. Knill, Cones, products and fixed points, Fund. Math., 60 (1967) 35–46.

20. S. Lefschetz, On the fixed point formula, Ann. of Math., 38 (1937) 819–822.

21. William Lopez, An example in the fixed point theory of polyhedra, Bull. Amer. Math. Soc., to appear.

22. E. E. Moise, An indecomposable plane continuum which is homeomorphic to each of its nondegenerate subcontinua, Trans. Amer. Math. Soc., 63 (1948) 581–594.

23. Jules Tannery, Introduction à la Théorie des Fonctions d'une variable, vol. 2, Hermann, Paris, 1910.

24. I. van der Walt, Fixed and Almost Fixed Points, Amsterdam Mathematisch Centrum, Amsterdam, 1963.

25. L. E. Ward, Jr., A fixed point theorem for chained spaces, Pacific J. Math., 9 (1959) 1273–1278.

26. Franz Wecken, Fixpunktklassen III, Mindestzahlen von Fixpunkten, Math. Ann., 118–119 (1941–43) 544–577.

27. G. S. Young, Fixed-point theorems for arcwise connected continua, Proc. Amer. Math. Soc., 11 (1960) 880–884.

不動点の遁走性

(The Elusive Fixed Point Property)

Uinv. of Wisconsin R. H. Bing
早稲田大学 野口 広訳

集合 X が**不動点性** (fixed point property) をもつとは，任意の写像 $f: X \to X$ がある点を不動にする…つまり $f(x)=x$ となる点 $x \in X$ が存在することである．

ある集合が不動点性をもつか否かを決定できる満足な必要十分条件は見い出されていない．いままでに知られている十分条件はどれも必要条件とするには余りにも制限が強い．他方，不動点性をもつ多くの例が見い出されていて…ときにはその例にあうように証明の方法が考え出されることもある．

この論文は不動点に関する興味あるいくつかの結果をひとまとめに紹介するのが目的で，新しい結果としては定理 15 があるのみである．幾何学的なトポロジーの入門課程を終えた大学院学生がなんとか，この記事の大部分を理解してくれるとよいと思う．Betti 数，Lefschetz 数，Euler の特性数，分解不可能な連続体，擬似弧，可縮，写像の次数，Green の定理，Sperner の補題，絶体近傍縮体，巡回元，被覆の脈体，蛇状連続体，木状連続体，向きを保存する位相写像，Čech ホモロジー，等々の特別な概念も必要に応じて用いるが，読者はこうした概念については必要のあるときに適当な本から後で学ぶまで待つことにしてこのまま読み通しても大部分は理解できるようにしてある．歴史的な意味を持つゆえにあるいは進んで研究する人のためにのべた定理もあるが，多くの読者が定理 2, 3, 6, 12–15；の証明を自分で考えまた定理 20–22 の証明を理解することが望ましい．

この論文であつかう連続体はコンパクトであり距離空間である．またすべて多面体は有限であり連結であるとする．

1. 不動点性をもつ集合． 不動点に関して最もよく知られている結果は多分次の定理であろう．

定理 1． n–胞体は不動点性をもつ．

この結果はしばしば Brouwer の不動点定理と呼ばれるが，Brouwer の仕事[7]よりは Bohl[4] の方が先である．Bohl はこの定理の証明に微分可能写像を扱い Green の定理を用いて，もし n–胞体が不動点のない自身への写像をもつと，同値な積分が一致しなくなることを示した．他方，Brouwer は $(n-1)$–球面の自身上へのホモトープな写像は同じ次数をもつことを示し，したがって n–胞体からその境界上への縮小は存在しない；したがって n–胞体の自身への任意の写像は不動点をもつことを示した．その他にも (組合せの手法を用いた) Knaster, Kuratowski および Mazurkiewicz [16] による証明がある．彼等は Sperner の補題を用いて n–胞体からその境界への縮小が存在しないこと示した．Hirsch は [12] でこの結果の見事な初等的証明を与えている．

集合 X の部分集合 Y が X の**縮体** (retract) であるとは，$f|Y=id$–…つまり，f は Y のどの点も不動にする，ような写像 $f: X \to Y$ が存在することである．写像 f は X の Y 上への縮小 (retraction) とよばれる．距離空間 Y が**絶体近傍縮体** (absolute neighbourhood retract) であるとは，Y を含む任意の距離空間 Z に対して Y がその縮体となるような $Y \subset U$ である Z の開集合 U が存在することである．

集合 Y が不動点性をもつことを示すのに次の結果を用いると便利である．

定理 2． 不動点性をもつ X の縮体 Y は不動点性をもつ．

例えば，定理 1 の結果に定理 2 を用いて図 1 a, 1 b は不動点性をもつことが分かる．図 1 a は 4 頁の本を示し，図 1 b は 2 室の家[9]を示している．

図 1 c はどの二つも一つの辺を共通し一つの 2–胞体に収束する無限個の 2–胞体を示している．これは 3–胞体 (立体) の縮体ではないが，どんな $\varepsilon>0$

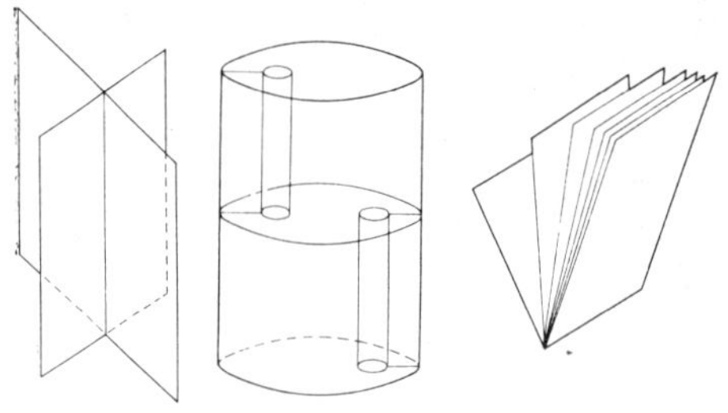

a　　　　b　　　　c

図 1

に対しても有限個の2-胞体の和集合の上へ，どの点も ε 以上動かさないで，縮少する．よって定理 1, 2, 3 を用いて図 1 c の集合は不動点性をもつことが示される．

定理 3.　コンパクトな集合 X の自身への不動点をもたない写像 f に対してどの点 $x\in X$ をとっても $\rho(x, f(x))>\varepsilon$ となるような正数 ε が存在する．ただし ρ は距離関数，数 ε は f の関数であって点 x の関数ではない．

Lefschetz は任意の写像 $f: X\to X$ に整数 $L(f)$ を定義して定理 1 を拡大することに成功した．この数はそれ以来 f の Lefschetz 数とよばれている．ここでこの定義はのべないが，(胞体のように) i 次元 Betti 数が $i>0$ でいつも 0 であると；連続体 X の任意の $f: X\to X$ に対して $L(f)=1$ となる．

定理 4. (Lefschetz[18]).　絶体近傍可縮体 X が不動点性をもつ十分条件は任意の写像 $f: X\to X$ に対して $L(f)\neq 0$ となることである．

[0,1] 区間の連続写像は Peano **連続体**あるいは**連続曲線**とよばれる．連続曲線 C の**非退化巡回元**とは C の中の一つの単純閉曲線 J と，J と二つあるいはそれ以上の点で交るすべての単純閉曲線との和集合のことである．Ayres は連続曲線に関する次の情報を用意した．

定理 5. (Ayres[1]).　連続曲線が不動点性をもつ必要十分条件はその非退化巡回元がいつでも不動点性をもつことである．

集合 X と Y の**楔** (wedge) ($X\vee Y$ で示す) とは X と Y をそれらの 1 点 $x\in X$ と $y\in Y$ で重ねたもの——すなわち，x と y を同視した集合である．すると読者はきっと次のような定理 5 の言い変えを証明できるであろう．[14] にはもっと深刻な言い変えがのべられている．

定理 6.　X と Y が不動点性をもつと $X\vee Y$ は不動点性をもつ．

P が次元 n の三角分割された多面体で a_m ($m=0, 1, \cdots, n$) がその三角分割の m 単体の個数とする．このとき $\chi(P)=a_0-a_1+a_2-\cdots+(-1)^n a_n=\sum_0^n(-1)^i a_i$ を P の Euler 特性数とよぶ．Euler-Poincarè の公式 ([13] の 241-243 参照) によって $\chi(P)$ はまた $\sum_0^n(-1)^i p_i$ に等しい．ここで p_i は P の i-次元 Betti 数である．Lopez は次のような性質をもつ多面体を構成した．

定理 7. (Lopez[19]).　$\chi(X)$ が正の偶整数であり，不動点性をもつ連結多面体が存在する．

この定理から 2 章でのべる興味ある変則的な性質が得られるが，Lopez の多面体は 8 次元であるからここではのべない．

問題 1.　Euler 特性数が偶数である，不動点性をもつ 2 次元多面体は存在するか？

$G=\{U_1, U_2, \cdots, U_k\}$ が空間 X の開被覆であるとその被覆に対応して $v_1, \cdots, v_k$ の頂点を持つ (幾何学的) 複体 $N(G)$ が，$\langle v_{i_1}, v_{i_2}, \cdots, v_{i_j}\rangle$ が $N(G)$ の単体である必要十分条件は $U_{i_1}\cap U_{i_2}\cap\cdots\cap U_{i_j}\neq 0$ であるとして定まる．複体 $N(G)$ は G の**脈体** (nerve) とよばれる．G の直径を小さくするにしたがい X より $N(G)$ への写像で各点の逆像が小さくなるものがあり，こうした意味で $N(G)$ は X の近い近似である．

蛇状連続体 (snake-like continuum) とはすべての $\varepsilon>0$ に対し $N(G)$ が弧となるような ε-被覆 (直径が ε 以下) が存在するコンパクトな距離をもった連続体 X のことである．**木状連続体** (tree-like continuum) はその脈体が木である (単純閉曲線を含まない) ような任意に小さい直径の被覆をもつ．Hamilton は次を証明している．

定理 8. (Hamilton[11]).　蛇状連続体は不動点性をもつ．

ここでは上の証明はしないが，定理 13 の証明で用いる**行き止り** (dead-end) 型の証明法の変型を用いて証明される．幾何学的トポジローでの未解決問題の中で次は最も興味あるものの一つである：

問題 2. 木状連続体は不動点性をもつか？

平面を分離しないような有界な平面連続体は次第に小さくなる位相的円板（単に円板とちぢめてよぶこともある）の列の共通部分である．そこで問題2は平面トポロジーで最高に面白いといわれている次の問題と関連しているようである．

問題 3. 次第に小さくなる円板列の共通部分は不動点性をもつか？

以下でこれらに関連している部分的な結果をいくつかのべよう．

連続体は，二つの真部分連続体の和にならないとき**分解不可能** (indecomposable) という．例えば，弧は分解できるが擬似弧は分解不可能である（擬似弧については [20] あるいは [2] を見よ）．

定理 9. (Hamilton[10]). T が分解不可能な連続体を含まない木状連続体であると，T の自身へのすべての位相写像はある点を不動にする．

定理 10. (Hamilton[10]). X が平面を分離せず，平面における境界が分解不可能な連続体を含まない有界な平面連続体であると，X の自身へのすべての位相写像はある点を不動にする．

定理 11. (Cartwright, Littlewood [8]). h が平面の自身の上への向きを保存する位相写像で $X=h(X)$ が平面を分離しない有界な平面連続体であると，$h(x)=x$ となる点 $x\in X$ が存在する．

文献[5, 14, 21, 23]にコンパクトな距離連続体が不動点性をもつためのいくつもの条件が示されているが，それらの代りにこれから三つの教訓的な例をのべよう．

定理 12. 境界の周りに螺線がまきついている円板は不動点性をもつ．

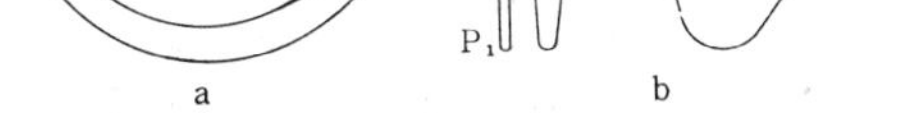

図　2

この集合は図 2a のように円板 D と D の境界にまきついている螺線 S で示される．$f:(D\cup S)\to(D\cup S)$ なる写像は S が弧状連結であり，弧状連結性は写像で保存され，また D から S へいくような弧が $D\cup S$ には存在しないから $f(D)\subset D$ であるか $f(D)\subset S$ でなければならない．もし $f(D)\subset D$ なら定理1で f はある点を不動にする．そうでなければ $f(D)\subset S$ であるから D の境界の任意の点が f で写される状態を調べると f は S のある点を S の中に写すことがわかる．S は弧状連結であるから，$f(D\cup S)\subset S$ である．$D\cup S$ はコンパクトであるから，$f(D\cup S)$ は1点であるか弧であらねばならない．もし1点ならその点が f で不動にされる．もし弧であれば定理1によってその弧のある点が不動にされる．

定理 13. $\sin 1/x$ で示される円周は不動点性をもつ．

この集合は図 2b のように $y=\sin 1/x$ $(0<x\leqq 1/\pi)$ のグラフの閉苞と $p_2=(0,1)$ より $p_3=(1/\pi,0)$ への弧との和集合として示される．点 $(0,-1)$ を p_1 とする．f が円周 $\sin 1/x$ の自身への写像とする．

ここで f が不動点をもつことを示すのに行き止まり (dead-end) 法を用いる．変数の点 x が犬を示し $f(x)$ が兎であるとし，円周 $\sin 1/x$ を x より $f(x)$ に至る弧に沿った方向に犬が兎を追いかけたとする．犬は p_1 から出発したとすると犬が p_1 をこえて p_2 の方向にすこし自分の影を追いかけて動くと，兎は犬の前方にいることがすぐ分かる．すなわち x は円周 $\sin 1/x$ の弧 $[p, f(x)]$ 上の p_1 と $f(x)$ の間にある．連続性によって x が p_3 を通りこしても，そしてさらに $\sin 1/x$ のグラフに沿って $[p_1, p_2]$ の近くに戻ってきても，$f(x)$ は x の前方にある．$[p_1, p_2]$ の各点 q_0 は $\sin 1/x$ のグラフの点列 $q_1, q_2, \cdots$ の極限である：$f(q_i)$ は q_i より前方にあるから，$f(q_1), f(q_2), \cdots$ は $[p_1, p_2]$ に近づく．よって $f(q_0)\in[p_1, p_2]$ である．すなわち $f[p_1, p_2]\subset[p_1, p_2]$ であるから，定理1によって f は $[p_1, p_2]$ のある点を不動にする．

Young が[23]でのべた次の問題は定理13と関連しているようである．次章の定理 15 の証明で用いる連続体 X' がここで平面上にあるという仮定を入れなければならない理由を示すものである.

問題 4. 単純閉曲線を含まない弧状連結なコンパクトな平面連続体は不動点性をもつか?

ここで次の章で重要な役割を果すもう一つの例を与えよう.

定理 14. 図3で示される集合は不動点性をもつ.

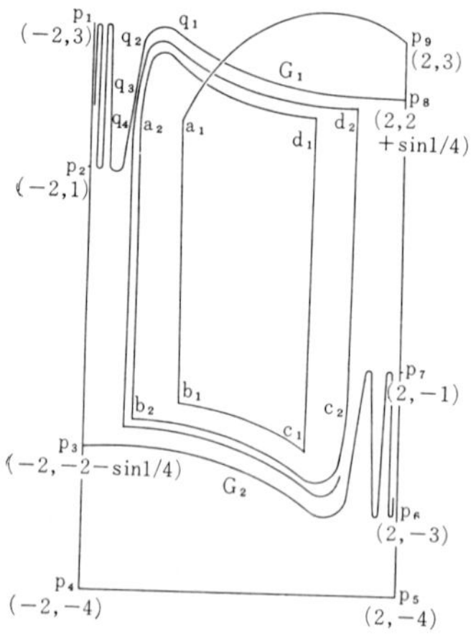

図 3

その集合 (X とよぶ) は三つの線分 $[p_1, p_4]$, $[p_4, p_5]$, $[p_5, p_9]$ と二つの $\sin 1/x$ 曲線 G_1, G_2 と螺線 S の和集合である. X が不動点性をもつことを示す前にもっと正確に X をのべよう.

点 $p_1, p_2, p_3, p_4, p_5, p_6, p_7, p_8, p_9$ は図3に示されるように $(-2, 3)$, $(-2, 1)$, $(-2, -2-\sin 1/4)$, $(-2, -4)$, $(2, -4)$, $(2, -3)$, $(2, -1)$, $(2, 2+\sin 1/4)$, $(2, 3)$ の座標をもつ. $\sin 1/x$ 曲線 G_1 は方程式 $y=2+\sin 1/(x+2)$ $(-2<x\leqq 2)$ でまた G_2 は G_1 の原点 $(0, 0)$ に関する反射像である.

S はまず G_1 上の点列 $q_1, q_2, \cdots$ を G_1 上で弧 $[q_i, p_8]$ の長が $2+\sum_1^i 1/j$ であるようにとる. 調和数列が表われているから, 点 $q_1, q_2, \cdots$ は線分 $[p_1, p_2]$ に近づく. q_i を通る垂直線上に a_i と b_i なる点をそれぞれ G_1 より $1/i$ だけ下にまた G_2 より $1/(i+1)$ だけ上にとる. d_i は b_i の原点に関する反射像で c_i は d_i を通る垂直線上の G_2 より $1/(i+1)$ だけ上にある点である.

螺線 S は p_9 から始まり, G_1 を含む平面に垂直な平面上の半円周に沿って a_1 へ行き, 直線に沿って b_1 へ行き, b_1 より c_1 へは G_2 に平行な弧に沿って行き, 直線に沿って d_1 へ行き, G_1 に平行な弧に沿って a_2 へ行く…. すべての i で垂直線分 $[a_i, b_i]$ と $[c_i, d_i]$ は S に属しまた G_2 に平行な b_i から c_i へ行く弧と G_1 に平行で d_i から a_{i+1} へ行く弧が S に属する.

X が不動点性をもつことを示すのに $f: X \to X$ が不動点なしと仮定しよう. 定理 13 の証明で用いたと同じ行き止まり法を用いて矛盾が起ることを示す.

犬は p_3 から出発する. もし $f(p_3) \in [p_1, p_3]$ であれば, 犬は $[p_3, p_1]$ に沿って兎を行き止りに追いつめるから, f は $[p_3, p_1]$ のある点を不動にする. もし $f(p_3) \in G_2$ であれば x は G_2 に沿って $f(x)$ の方へ動く, このときもし $x \in G_2 - \{p_3\}$ であると $f(x)$ は x の前方にある, つまり x は p_3 より $f(x)$ への G_2 の弧の上にある. 定理13の証明で用いたと同様の議論で f は $[p_6, p_7]$ のある点を不動にすることが示される.

もし $f(p_3) \notin [p_1, p_3] \cup G_2$ であれば犬が p_3 を動き出して兎を追うとき, x は p_3 より $f(x)$ への X の弧上にある. この追いかけっこは犬が p_8 のフォークにつくまで続く. そこで兎は G_1 にいるか $S \cup [p_8, p_9]$ にいることになる. もし G_1 にいると $f(p_3) \in G_2$ のときと類似の状態であるから $f(p_8) \in S \cup [p_8, p_9]$ である場合を調べる. x が p_8, p_9 を通り越して動くと, $f(x)$ は x の前にある, つまり x は p_9 より $f(x)$ への S の弧上の点で p_9 と $f(x)$ の間にある. S の末端は $[p_1, p_3] \cup G_2 \cup [p_6, p_8] \cup G_1$ に収束しているから, もし $x_1, x_2, \cdots$ が p_8 に収束する S の点列とすると $f(x_1), f(x_2), \cdots$ は $[p_1, p_3] \cup G_2 \cup [p_6, p_8] \cup G_1$ の一点に収束する, これは仮定の $f(p_8) \in S \cup [p_8, p_9]$ に反する.

2. 不動点の変則性. いろいろな意味で不動点は思ったようには振りまわらない.

定理6は弧を共有する二つの不動点性を持つ集合の和集合はまた不動点性をもつと拡張されそうに思えるが, これは正しくないことを示そう.

定理 15. D を円板とするとき $D \cap X$ が弧であるのに $D \cup X$ が不動点性をもたないような不動点

性をもつ1次元連続体 X が存在する.

この連続体 X は図3の X で D は頂点 p_5, p_9, p_{10}, p_{11} をもつ矩形円板である，ただし $p_{10}=(3,3)$ で $p_{11}=(3,-4)$ である．まず $X\cup D$ は連続体

$$X'=(X\cup BdD)-(p_8, p_9)$$

の上へ縮小する．この連続体 X' は Young[23]が不動点性をもつことを示すのに失敗した連続体に似ている．不動点のない写像 $f: X'\to X'$ を示そう.

$[p_1, p_3]\cup G_2\cup[p_6, p_8]\cup G_1$ に制限してみると，f は原点のまわりの反射である．また f は $[p_3, p_5]$ を $[p_8, p_{11}]$ の上へ $[p_6, p_5]$ を $[p_1, p_{11}]$ の上へ $[p_5, p_{11}]$ を $[p_{11}, a_1]$ へ $[p_{11}, a_1]$ を $[a_1, c_1]$ へまた，さらに各 $[a_i, a_{i+1}]$ を $[c_i, c_{i+1}]$ へ位相的に写す．こうしてできた写像 f が連続であることを知るためには S から遠く離れている点が f によって原点に関する反射点のごくそばに動かされることを確かめればよい.

問題 5. $X\times[0,1]$ は不動点性をもつか？

ここで f として多くの場合 X の位相写像を考えることがある．このことは Young [23] の次の問題を暗示している.

問題 6. 単純閉曲線を含まない弧状連結連続体 C の自身への位相写像はある点を不動にするか？

定理15の連続体 X は p_9 より a_1 へいく半円周があったから一つの平面上にはない.

問題 7. 不動点性をもつ平面連続体 C と円板 D が弧で交わるとき $C\cup D$ は不動点性をもつか？

定理15の X は病理学的な例でこうした異常な事柄は X が多面体ならば起らないと思う人もいることであろう．しかし Lopez [19] が $X\vee Y$ の Euler の特性数 $=\chi(X\cup Y)=0$ となる不動点性をもつ二つの多面体 X, Y を構成したことから正確な評価とはいえない．この例は次元が17なのでここにはのべないが，$X\vee Y$ の分離点上にいくつかの円板を分離点が分離点でなくなるように貼り合せてみると，次にのべる Wecken の結果によって，$X\vee Y$ と円板等の和集合は不動点性をもたない.

定理 16. (Wecken[19]). K が $\chi(K)=0$ である局所分離点を有しない連結多面体とすると K は不動点のない写像をもつ.

定理 17. (Lopez[19]). D が円板であるとき $D\cap P$ が弧で $D\cup P$ が不動点性をもたないような不動点性をもつ多面体 P が存在する.

Lopez の例は次元が17であるから，もっと単純なものを求めるのは当然である.

問題 8. 定理17をみたすような多面体 P の最少次元はいくつか？

定理18と23で直積の不動点性はその因子の不動点性から得られないことをのべるが，次の事実は不明である.

問題 9. 不動点性をもつ1次元コンパクト連続体と線分の直積は不動点性をもつか？

問題 10. 不動点性をもつ有界な平面連続体と弧との直積は不動点性をもつか？

分離点をもつ多面体をとり線分との直積をつくると，直積は元の多面体と同じ Euler 特性数をもつが分離点をもつという性質を失う．よって Lopez の例と Wecken の結果と合わせると次が示される.

定理 18. (Lopez[19]). 閉じた1次元単体との直積は不動点性をもたないような，自身は不動点をもつ多面体が存在する.

問題 11. P と Q が局所分離点は有しないが不動点性をもつ多面体であるとき $P\times Q$ は不動点性をもつか？

i 次元 Betti 数 $(i>0)$ が0である多面体は定理4の結果から不動点性をもつ．よってコンパクトな可縮多面体は不動点性をもつ．特に任意の多面体の上の錐は不動点性をもつ．この事柄は多面体でないと成り立たない.

樹下は図4で示した例を与えている.

これは底ぶたがあって上ぶたのない円柱状の罐に底の円周に収束する底ぶた上の螺線の上の円柱の1部を加えてできる．これはスプリングが中に入っている罐に似ていて，**巻紙つきの罐**(can-with-a-roll-of-toilet-paper)という.

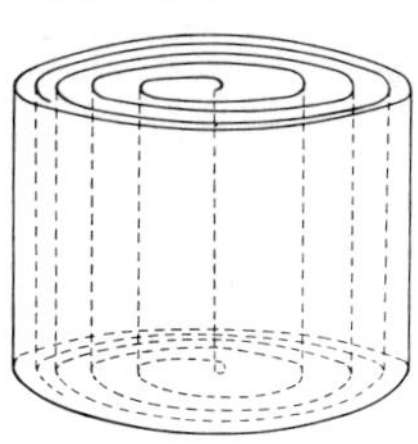

図 4

定理 19. (樹下[15]). 巻紙つきの罐は可縮で；不動点性をもたず；この罐の上の錐は不動点性をもたない.

上の性質は[15]で示されているが，巻紙つきの罐は次の問には答えていない.

問題 12. 可縮な連続体の自身の上への位相写像は不動点をもつか？

Borsuk[6]は3次元の連続曲線でČechのホモロジーですべての $p>0$ で p サイクルが有界であるにもかかわらず自身上への不動点のない位相写像の存在する例を上げている. Bing[3]は次の定理を用いて2次元的な例を作った.

定理 20. (Bing[3]). 自身上への不動点のない位相写像をもち次第に小さくなる位相立体の共通部分となる次元2の連続曲線が存在する.

ユークリッド3-空間で円柱 $x^2+y^2=4$ 上の $z=1/2$ および $z=-1/2$ との交りを底とし $(-2,0,-1/2)$ および $(2,0,1/2)$ を頂点とする二つの錐面を考える. 各水平面を z 軸のまわりに $\tan \pi z$ の角度だけ回転してこの二つの錐面の和集合を写した像を X で示す. 図5は長さを正確に計って X を画いたものではないが各錐面の頂点が他の錐面の底のまわりを巡る有様を示している. 式で書きたい人には次のようになる.

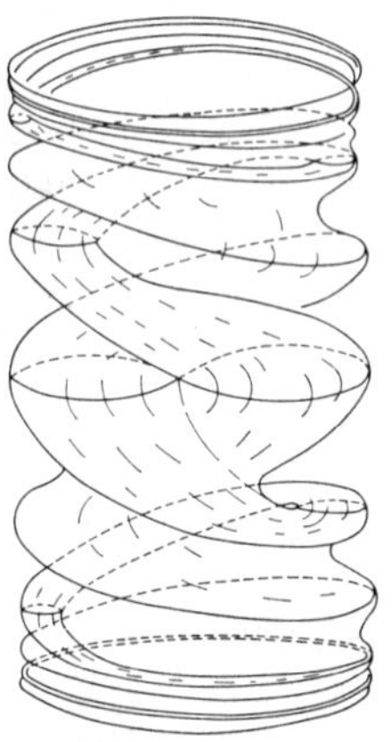

図 5

$$\{[x-(2z+1)\cos \tan \pi z]^2 + [y-(2z+1)\sin \tan \pi z]^2-(1-2z)^2\} \times \{[x-(2z-1)\cos \tan \pi z]^2 + [y-(2z-1)\sin \tan \pi z]^2-(1+2z)^2\}=0$$

0と 2π の間の任意の θ に対して X を自身の上へ写す，不動点のない位相写像 h_θ が存在する. X の底のおのおのは h_θ によって θ の大きさの角だけ回転する. h_θ は他の水平切断部を z-軸の回りで θ の大きさの回転をし，上にもちあげてそしてのばす. 上への持ち上げは回転された切断部の中心線を高い方の切断部の中心線へと移す. 方程式を知りたい人には，$([\arc \tan(\theta+\tan \pi z)]/\pi)-z$ がその持ち上げである. 中心線を固定すると回転されもち上げられた切断部の二つの円周の一つは拡がり他方は縮まって，動いた切断部はもち上げが行なわれた平面での切断部と一致する.

コンパクトな連続体上の錐面は不動点性をもつであろうと思うのは自然である. このことは定理4によって多面体については正しい. しかし次のKnillの定理によって一般には成り立たないことが示される. もちろん定理19からもわかることである.

定理 21. (Knill[17]). 図6で示されるような螺線をもつ円周を底とする錐は不動点性をもたない.

円柱座標 (r,θ,z) を用いて円周は $[z=1,\ r=1]$，螺線は $[z=1,\ r=1+1/(1+\theta),\ \theta\geqq 0]$ で頂点は $(0,\theta,0)$ で定義されるとする. この螺線の上の錐を錐の**スカート** (skirt) という.

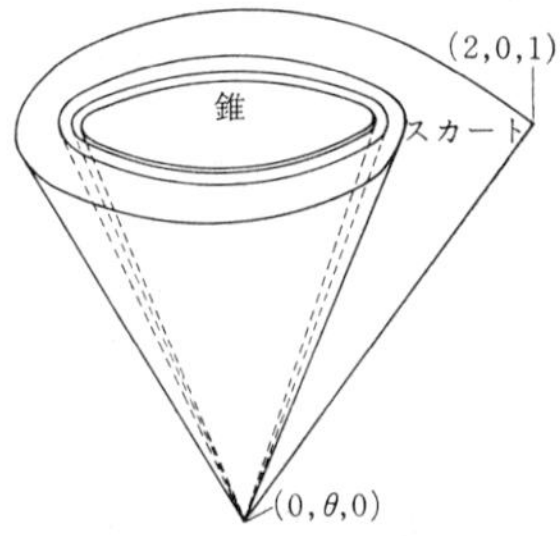

図 6

このスカートつきの錐 (cone-with-a-skirt) の自身の上への不動点を持たない写像 f をこれから定義するが，f はスカートをスカートへそして錐面は錐面とその頂点から出る線分との和集合へ写す.

$0\leqq z_0\leqq 1/3$ であるとき高さ z_0 における錐面の水平切断部は z_0 が0より1/3を分割すると同じ割合で $(2,0,1)$ より $(0,\theta,0)$ の線分を分割する点に写る.

f_z が0を1へ，1/3を0へ，2/3を1へ，1を0へ写す関数で $[0,1/3]$, $[1/3,2/3]$, $[2/3,1]$ のおのおので線型であり他方 $f_\theta(\theta,z)=\theta+z-2/3$ であるとする. $1/3\leqq z\leqq 1$ に対する錐面の各点 (r,θ,z) に対し不動点のない写像は

$$f(r,\theta,z)=(f_z(z),f_\theta(\theta,z),f_z(z))$$

として定義される. ここで $f_z(z)=z$ は $z=1/4$, 1/2, 3/4 のみで起る. 1/2, 3/4 の値では回転が存

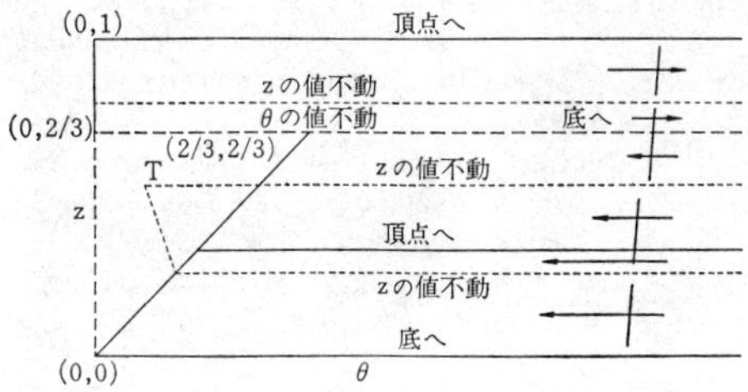

図 7

在し1/4の高さの切断部は錐面上に写されない．

fをスカート上でのべるためにそれをほどいてみる．ほどかれたスカートは図7で示される．この図は正確に測ったものではなく，半分帯にある水平の一方向のみ無限に広がる半直線は錐の水平切断面を表わし，半分帯にある垂直線分はスカートの頂点より底へ行く線分を示し，半分帯の上方の辺は錐の頂点を示している．図で右方の矢印はθとzの両方向の運動を示している．

スカート上でのfを直交座標を用いてのべてみる．$r=z(1+1/(1+\theta))$であるからrを無視する．帯の点(θ, z)に対する$f(\theta, z)$のθの値は$0 \leqq z \leqq 1/3$のとき0で$1/3<z \leqq 1$のとき$\max(0, \theta+z-2/3)$である．$(1/3, 1/3)$より右の水平半直線上で連続性が破れるがこの半直線は錐面の頂点に写されるので問題にならない．スカートのすべての点のθの値は$(0, 0)$より$(0, 2/3)$への水平線分と$(0, 2/3)$より右の水平半直線上の点とを除いてすべて変る．このθに関して不変な集合は図7で破線で示されている．

次にスカートの点の像のzの値を示そう．$(0, 0)$, $(0, 2/3)$, $(2/3, 2/3)$の頂点をもつ三角板T上を除いてzの値は$f_z(z)$で与えられる，ここでf_zは錐面の点のzの値を与えるのに用いた関数であった．Tの各点(θ, z)に対してその像のzの値は$3\max[|z-1/3|, |\theta-1/3|]$であると定義する．

スカートの点でzの値がfで変らないものは図の点線で示した．点線は破線と交らないのでスカートの点はどれも動くことがわかる．

ここでスカートつきの錐面上の錐はまた不動点性を持たないことは注目に値する．この結果は可縮な集合が不動点性をもたなければその上の錐もまた不動点性をもたないという樹下の定理からえられる．

スカートつきの錐が不動点性を有しない理由としてその底が不動点性を有しないからと誤って思うかもしれないが，このことは次の結果から誤りであることが示される．

定理 22. 図2aで示した円板Dとその境界のまわりの螺線Sとの和集合$D \cup S$を底とする錐は不動点性をもたない．

この錐からスカートつきの錐への縮小が存在する，よって定理21と2によって$D \cup S$上の錐は不動点性を有しない．

Bが図8で示したスカートつきの罐であるとするとこの罐は円柱座標で$\{(r, \theta, z) | [z=0, r \leqq 1]$あるいは$[r=1, 0 \leqq z \leqq 1]\}$で表わされる，ここでスカートは$\{(r, \theta, z) | 0 \leqq z \leqq 1,\ \theta \geqq 0,\ r=1+(1-z)/(1+\theta)\}$である．Knillはこのスカートつきの罐は次の性質をもつことを証明した．

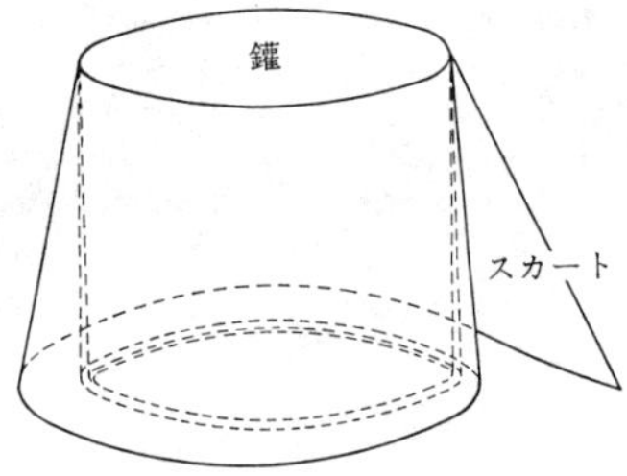

図 8

定理 23. (Knill[17]). Bがスカートつきの罐であると，Bは不動点性をもつ，$B \times [0, 1]$は不動点性をもたない，またBとCは不動点性をもつが$B \cap C$は円板であり$B \cup C$は不動点性をもたないような3-胞体C（つまり，立方罐(solid can)）が存在する．

文献

[1] W. L. Ayres, Some generalizations of the Scherrer fixed-point theorem, Fud. Math., **16** (1930), 332-336.

[2] R. H. Bing, A homogeneous indecomposable plane continuum, Duke Math. J., **15** (1948), 729-742.

[3] ———, Challenging conjectures, Amer. Math. Monthly, **74** (1967), 56-64.

[4] P. Bohl, Über die Bewegung eine mechanischen Systems in der Nähe enier Gleichgewichtslage,

J. Reine Angew. Math., **127** (1904), 179-276.

[5] K. Borsuk, A theorem on fixed points, Bull. Acad. Polon. Sci. Cl. III, **2** (1954), 17-20.

[6] K. Borsuk, Sur un continu acyclique qui se laisse transformer topologiquement en lui même sans points invariants, Fund. Math., **24** (1935), 51-58.

[7] L. E. J. Brouwer, Über Abbildung von Mannigfaltigkeiten, Math. Ann., **71** (1912), 97-115.

[8] M. L. Cartwright and J. E. Littlewood, Some fixed point theorems, Ann., of Math., **54** (1951), 1-37.

[9] B. G. Casler, An imbedding theorem for connected 3-manifolds with boundary, Proc. Amer. Math. Soc., **16** (1965), 559-566.

[10] O. H. Hamilton, Fixed points under transformations of continua which are not connected im kleinen, Trans. Amer. Math. Soc., **44** (1938), 18-24.

[11] ———, A fixed point theorem for pseudo-arcs and certain other metric continua, Proc. Amer. Math. Soc., **2** (1951), 173-174.

[12] Morris W. Hirsch, A proof of the nonretractibility of a cell onto its boundary, Proc. Amer. Math. Soc., **14** (1963), 364-365.

[13] John G. Hocking and Gail S. Young, *Topology*, Addison Wesley, 1961.

[14] J. L. Kelley, Fixed sets under homeomorphisms, Duke Math. J., **5** (1939), 535-537.

[15] Shin'ichi Kinoshita, On some contractible continua without the fixedpoint property, Fund. Math., **40** (1953), 96-98.

[16] B. Knaster, C. Kuratowski, S. Mazurkiewicz, Ein Beweis des Fixpunktsatzes für n-dimensionale Simplexe, Fund. Math. **14** (1929), 132-137.

[17] Ronald J. Knill, Cones, products and fixed points, Fund. Math., **60** (1967), 35-46.

[18] S. Lefschetz, On the fixed point formula, Ann. of Math., **38** (1937), 819-822.

[19] William Lopez, An example in the fixed point theory of polyhedra, Bull. Amer. Math. Soc., to appear.

[20] E. E. Moise, An indecomposable plane continuum which is homeomorphic to each of its nondegenerate subcontinua, Trans. Amer. Math. Soc., **63** (1948), 581-594.

[21] L. E. Ward, Jr., A fixed point theorem for chained spaces, Pacific J. Math., **9** (1959), 1273-1278.

[22] Franz Wecken, Fixpunktklassen III, Mindestzahlen von Fixpunkten, Math. Ann., **118-119** (1941-43), 544-577.

[23] G. S. Young, Fixed-point theorems for arcwise connected continua Proc. Amer. Math. Soc., **11** (1960), 880-884.

IV. Topological Classifications

A. 2-Sphere

THE KLINE SPHERE CHARACTERIZATION PROBLEM

R. H. BING

The object of this paper is to give a solution to the following problem proposed by J. R. Kline: Is a nondegenerate, locally connected, compact continuum which is separated by each of its simple closed curves but by no pair of its points homeomorphic with the surface of a sphere? The answer is in the affirmative.

A solution to the Kline problem gives a characterization of a simple closed surface. Partial solutions of this problem have been made by Hall [1, 2][1] and Jones [3]. Other characterizations of a simple closed surface have been given by Kuratowski [4], Zippin [5, 6], Wilder [7] and Claytor [8]. Previous to the giving of these characterizations, Moore gave [9] two sets of axioms, each set of which characterized a set topologically equivalent to a plane.

DEFINITION. We say that M disrupts X from Y in D if there is an arc from X to Y in D but each such arc contains a point of M.

We shall make use of the following lemma.

LEMMA. *Suppose that space is locally connected and cannot be separated by the omission of any pair of its points, that the boundary of the connected domain D is equal to the sum of the mutually exclusive sets M, N and E, each of which contains a point which is accessible from D, and that D' is a connected subdomain of D such that no point of D either disrupts D' from $E+M$ in $D+E+M$ or disrupts D' from $E+N$ in $D+E+N$. Then there is an open arc from M to N in D that does not disrupt D' from E in $D+E$.*

PROOF. Consider the arc AB in $D+B$ from a point A of D' to a point B of E. Let W_1 be the set of all points P of AB such that there is an open arc from P to E in D that does not intersect some open arc from M to N in D. Assume that the first point R of AB in the order from A to B on the closure of W_1 does not belong to D'.

If R disrupts D' from E in $D+E$, there are an arc from D' to M in $D+M-R$ and an arc from D' to N in $D+N-R$. In the sum of these two arcs plus D' there is an open arc from M to N in D which does not intersect RB. This is contrary to the definition of R. Hence, R does not disrupt D' from E in $D+E$.

Let $A'B'$ be an arc in $D+B'-R$ from a point A' of D' to a point

Presented to the Society, November 23, 1945; received by the editors December 3, 1945.

[1] Numbers in brackets refer to the references cited at the end of the paper.

B' of E. Since D' is connected and does not contain R, we may suppose that A' is A. Let X be the last point of $A'B'$ in the order from A' to B' on AR and let Y be a point of AB between R and B such that RY contains no point of XB'.

Choose a metric for $XB'+XY$. Let P_1 be a point of $W_1 \cdot XY$ such that no point of W_1 at a distance of more than 1 from P_1 is between P_1 and X on XY. Denote by V_1 the set of all points P of XB' such that there are in D an open arc from P to E and an open arc from XP_1 to E such that the sum of these two open arcs does not intersect some open arc from M to N in D. Let Q_1 be a point of V_1 such that no point of V_1 at a distance of more than 1 from Q_1 is between Q_1 and X on XB'.

Denote by W_2 the set of all points P of XY such that there are in D an open arc from P to E and an open arc from XQ_1 to E such that the sum of these open arcs does not intersect some open arc from M to N in D. Let P_2 be a point of $W_2 \cdot XP_1$ such that no point of W_2 at a distance of more than $1/2$ from P_2 is between P_2 and X on XY.

In general, let W_n (or V_n) be the set of all points P of XY (or XB') such that there are an open arc from P to E in D and an open arc from XQ_{n-1} (or XP_n) to E in D such that the sum of these open arcs does not intersect some open arc from M to N in D. Let P_n (or Q_n) be a point of W_n (or V_n) on XP_{n-1} (or XQ_{n-1}) such that no point of W_n (or V_n) at a distance of more than $1/n$ from P_n (or Q_n) is between P_n (or Q_n) and X on XY (or XB').

Let P_0 and Q_0 be the limit points of $P_1, P_2, \cdots$ and $Q_1, Q_2, \cdots$ respectively. We note that for no integer n is P_n between X and P_0 on XY or is Q_n between X and Q_0 on XB'.

Since no pair of points separates space, there is an arc from A to E in space that contains neither P_0 nor Q_0. Let $A''B''$ be a subset of this arc irreducible from $AX+XP_0+XQ_0$ to E. We shall suppose that A'' is a point of $AX+XP_0$ since the argument to follow may be adjusted to take care of the case where it belongs to $AX+XQ_0$.

There is an integer i such that AA'' contains no point of W_i and $A''B''$ intersects neither P_0P_i nor Q_0Q_i. There is an arc $M'N'$ in $D+M'+N'$ from a point M' of M to a point N' of N such that there are open arcs $\langle P'B_1\rangle$ and $\langle Q'B_2\rangle$ in $D-\langle M'N'\rangle$, where P' and Q' are points of P_0P_i and Q_0Q_i respectively and B_j $(j=1, 2)$ is a point of E. Denote by P'' the first point of XP' in the order from X to P' on $P'B_1+Q'B_2$ and denote by Q'' the first point of XQ' in the order from X to Q' on $P'B_1+Q'B_2$. Let $P''B_3$ and $Q''B_4$ be arcs in $P'B_1+Q'B_2$ where B_3 and B_4 are points of B_1+B_2.

Denote the set $AX+XP''+XQ''$ by T. Since no point of AX is a point of W_i, $P''B_3+Q''B_4$ does not intersect AX. Then $P''B_3+Q''B_4$ intersects T only at $P''+Q''$.

Now $M'N'$ intersects T or else there is an open arc from A to E in $T+P''B_3$ and R is not as assumed. Also, $M'N'$ does not intersect T in only one point, for if it did, there would be an open arc from X to E in $T+P''B_3+Q''B_4$ which would not intersect $M'N'$. But X is not a point of W_i.

Let M'' and N'' be the first and last points of $M'N'$ in the order from M' to N' on T. Replace the part of $M'N'$ between M'' and N'' by an arc α from M'' to N'' in T. Now one of the points M'', N'' belongs to XP'', for if neither were a point of XP'', there would be an open arc from X to E in $XP''+P''B_3$ which would not intersect $M'M''+\alpha+N''N'$. This contradicts the fact that X is not a point of W_i. Also, one of M'', N'' is a point of XQ''. Since A'' is not a point of W_i, it is not between P'' and $M''+N''$ on XP''.

Let C be the first point of $A''B''$ in the order from A'' to B'' on $M+M'M''+N+N''N'+E+P''B_3+Q''B_4$. Now C is not a point of $E+P''B_3+Q''B_4$, for if it were, there would be an open arc from A'' to E in $A''C+P''B_3+Q''B_4$ that would not intersect $M'M''+\alpha+N''N'$. But A'' is not a point of W_i.

Assume that C is a point of $M+M'M''$. Consider an arc β from M to N in $A''C+M'M''+T+N''N'$ containing $A''C$. We note that β does not contain M''. Now A'' is not a point of XP'', for if it were, there would be an open arc from A'' to E in $P''B_3+Q''B_4+T$ which would not intersect β. Also, A'' is not a point of AX or else there would be an open arc from X to E in $P''B_3+Q''B_4+T$ which would not intersect β. But X is not a point of W_i. Hence, C is not a point of $M+M'M''$. Likewise, we find that C is not a point of $N+N''N'$.

This establishes the lemma. The assumption that R is not a point of D' leads to the contradiction that the point C of $M+M'M''+N+N''N'+E+P''B_3+Q''B_4$ belongs to neither $E+P''B_3+Q''B_4$, $M+M'M''$ nor $N+N''N'$.

THEOREM. *Suppose that S is a nondegenerate, compact continuous curve such that no pair of points but every simple closed curve of S separates it. Then S is topologically equivalent to the surface of a sphere.*

PROOF. Regard S as space. It is known [6, 10] that S is a simple closed surface if no arc separates it. We shall show that no arc separates S.

Before giving the details of the proof, we shall briefly outline what we intend to do. On the assumption that some arc separates space, we shall get a finite collection H_1 of connected domains such that their sum does not separate space and such that the sum of any two nonintersecting elements of H_1 separates the sum of the elements of H_1. Collections $H_2, H_3, \cdots$ are defined which satisfy corresponding conditions and which are such that the closure of each element of H_{n+1} is a subset of the sum of the elements of H_n. See the figure. There

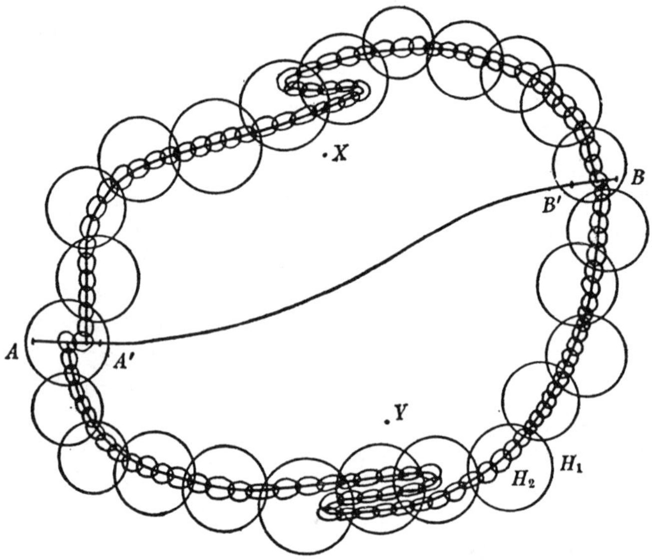

FIG. 1

are actually many more elements in H_1 and H_2 than are shown in the figure. The collections $H_1, H_2, \cdots$ are described in such a way that the common part of their sums is a simple closed curve not separating space. Hence, we shall show that the assumption that some arc separates space leads to the contradiction that some simple closed curve does not. We now consider the details of the proof.

Assume that an arc separates the point X from the point Y. Then there is an arc AB that separates X from Y such that no proper subarc of AB separates X from Y. Let D_X and D_Y be the complementary domains of AB containing X and Y respectively.

Description of collection H_1. Assume that a metric has been chosen for S and let ϵ_1 be a positive number less than one one-hundredth of the distance from A to B. We shall describe a collection H_1 of connected domains. The sum of the elements of H_1 will be denoted by H_1^*.

The collection H_1 of connected domains $h_{1,1}, h_{1,2}, \cdots, h_{1,t}$ $(t>100)$ will satisfy the following conditions:

(1) $h_{1,i}$ intersects $h_{1,j}$ only if i is equal to either $j-1$, j or $j+1$ $(h_{1,t+1}\equiv h_{1,1}$ and $h_{1,1-1}\equiv h_{1,t})$;

(2) $S-H_1^*$ is connected;

(3) some point of $S-H_1^*$ is accessible from $h_{1,i}$;

(4) the diameter of $h_{1,i}$ is less than ϵ_1;

(5) no connected subset of H_1^* that intersects $h_{1,i}$ and $h_{1,i+2}$ is of diameter less than $\epsilon_1/4$.

Denote by $D_1, D_2, \cdots, D_n$ the elements of a finite collection of connected domains covering S such that the diameter of each is less than $\epsilon_1/100$. Suppose that each of the domains $D_1, D_2, \cdots, D_j$ intersects the complement of D_X+D_Y, each of the domains $D_{j+1}, D_{j+2}, \cdots, D_k$ is a subset of D_X and each of the domains $D_{k+1}, D_{k+2}, \cdots, D_n$ is a subset of D_Y.

Let $\alpha_1, \alpha_2, \cdots, \alpha_j$ be a collection of arcs in the complement of D_X+D_Y+A+B such that α_i $(i=1, \cdots, j)$ intersects D_i and AB.

Let A' and B' be points of the arc AB such that AA' and $B'B$ are arcs that do not intersect $\alpha_1+\alpha_2+\cdots+\alpha_j$, each of the arcs AA' and $B'B$ is covered by an element of $D_1, D_2, \cdots, D_n$, some point of $A'B'$ is accessible from $D_Z(Z=X, Y)$ and if D_p is an element of $D_1, D_2, \cdots, D_n$ in D_Z, then no point of D_Z either disrupts D_p from AB' in D_Z+AB' or disrupts D_p from $A'B$ in $D_Z+A'B$.

Considering $D_X, D_{j+1}, \overline{D}_X\cdot(AA'-A'), \overline{D}_X\cdot(B'B-B')$ and $\overline{D}_X\cdot A'B'$ as D, D', M, N and E of the preceding lemma, we find that there is an arc α_{j+1} from $A'B'$ to D_{j+1} in $D_X+A'B'$ that does not disrupt $AA'-A'$ from $B'B-B'$ in $D_X+AB-A'B'$. Let D' be a component of $D_X-D_X\cdot\alpha_{j+1}$ that contains an open arc from a point of $AA'-A'$ to a point of $B'B-B'$. If D_{j+2} is not a subset of D', let α_{j+2} be an arc in $D_X-D'+A'B'$ from $A'B'$ to a point of D_{j+2}. If D_{j+2} is a subset of D', we shall apply the lemma to get an arc α_{j+2} from D_{j+2} to $A'B'$ in $D_X+A'B'$ such that $\alpha_{j+1}+\alpha_{j+2}$ does not disrupt $AA'-A'$ from $B'B-B'$ in $D_X+AB-A'B'$. The procedure is described in the following paragraph.

Let R be a point of D'. Since no point of D_X disrupts D_{j+2} from AB' in D_X+AB', there is an arc β from D_{j+2} to AB' in $D_X+AB'-R$. A subarc of β in $D'+AB'+\alpha_{j+1}-R$ intersects D_{j+2} and $AB'+\alpha_{j+1}$. Hence, R does not disrupt D_{j+2} from $AB'+\alpha_{j+1}$ in $D'+\overline{D}'\cdot(AB'+\alpha_{j+1})$. Also, R does not disrupt D_{j+2} from $A'B+\alpha_{j+1}$ in $D'+\overline{D}'\cdot(A'B+\alpha_{j+1})$. Applying the lemma, we find that there is an arc from D_{j+2} to $A'B'+\alpha_{j+1}$ in $D'+A'B'+\alpha_{j+1}$ that does not disrupt $AA'-A'$ from $B'B-B'$ in $D'+AB-A'B'$. Then there is an arc α_{j+2} in $D_X+A'B'$

from D_{j+2} to $A'B'$ such that $\alpha_{j+1}+\alpha_{j+2}$ does not disrupt $AA'-A'$ from $B'B-B'$ in $D_X+AB-A'B'$.

Likewise, we find that there is an arc α_{j+3} from D_{j+3} to $A'B'$ in $D_X+A'B'$ such that $\alpha_{j+1}+\alpha_{j+2}+\alpha_{j+3}$ does not disrupt $AA'-A'$ from $B'B-B'$ in $D_X+AB-A'B'$. A continuation of this process gives that there are arcs $\alpha_{j+1}, \alpha_{j+2}, \cdots, \alpha_k$ in $D_X+A'B'$ whose sum does not disrupt $AA'-A'$ from $B'B-B'$ in $D_X+AB-A'B'$ and such that $\alpha_p (p=i+1, \cdots, k)$ intersects D_p and $A'B'$. Also, there are arcs $\alpha_{k+1}, \alpha_{k+2}, \cdots, \alpha_n$ in $D_Y+A'B'$ whose sum does not disrupt $AA'-A'$ from $B'B-B'$ in $D_Y+AB-A'B'$ and such that α_p $(p=k+1, \cdots, n)$ intersects D_p and $A'B'$.

Let G be the collection of all domains g such that g is a component of the common part of the complement of $AB+\alpha_1+\alpha_2+\cdots+\alpha_n$ and some domain of $D_1, D_2, \cdots, D_n$. If P is a point of D_i, there is an arc in D_i from P to α_i. Hence, if g is an element of G, some point of $AB+\alpha_1+\alpha_2+\cdots+\alpha_n$ is accessible from g.

Let g_A and g_B be domains of diameters less than $\epsilon_1/100$ that cover $AA'-A'$ and $B'B-B'$ respectively but no point of $A'B'+\alpha_1+\alpha_2+\cdots+\alpha_n$. There exists a finite collection G_X of domains of G such that this collection but no collection of fewer elements of G satisfies the condition that the sum of the elements of G_X is a connected subset of D_X and intersects both g_A and g_B. Denote the elements of G_X by $g_2, g_3, \cdots, g_r$, where g_A intersects g_2, g_i $(i=2, \cdots, r-1)$ intersects g_{i+1} and g_r intersects g_B. Denote g_A by g_1 and g_B by g_{r+1}.

Also, there exists a finite collection G_Y of domains of G such that this collection but no collection of fewer elements of G satisfies the condition that the sum of the elements of G_Y is a connected subset of D_Y and intersects both g_A and g_B. Denote the elements of G_Y by $g_{r+2}, g_{r+3}, \cdots, g_s$ where g $(i_i=r+1, \cdots, s-1)$ intersects g_{i+1} and g_s intersects g_1.

Let E denote $g_1+g_2+\cdots+g_s$ plus all points that it separates from $A'B'$. Each component of the common part of E and an element of G intersects an element of $g_1, g_2, \cdots, g_s$. However, it is to be noted that no such component intersects two g_i's that do not belong to a consecutive set of three domains of $g_1, g_2, \cdots, g_s, g_1, g_2$. Denote by g_i' the sum of g_i and all such components that intersect g_i. We note that g_i' is a diameter less than $\epsilon_1/33$.

If three is a factor of s, denote the sum of the first three elements of $g_1', g_2', \cdots, g_s'$ by h_1, the sum of the next three elements by $h_2, \cdots$ and the sum of the last three elements by h_v. If three is a factor of $s+1$, then $h_1, h_2, \cdots$ and h_v are defined as before except that h_v is the sum of the last four elements of $g_1', g_2', \cdots, g_s'$ instead of the last three. If three is a factor of $s+2$, each of h_{v-1} and h_v is the sum of four elements of $g_1', g_2', \cdots, g_s'$. Since each h_i contains either g_A, g_B or an element g of G which does not intersect g_A+g_B and since a point of $AB+\alpha_1+\alpha_2+\cdots+\alpha_n$ is accessible from g, then a point of $A'B'+\alpha_1+\alpha_2+\cdots+\alpha_n$ is accessible from h_i. Now h_i is of diameter less than $\epsilon_1/8$ and the collection $h_1, h_2, \cdots, h_v$ satisfies conditions analogous to conditions (1), (2) and (3) to be satisfied by $h_{1,1}, h_{1,2}, \cdots, h_{1,t}$.

Let $h_{1,1}$ be the sum of $h_1, h_2, \cdots, h_n$ where some connected subset of $h_1+h_2+\cdots+h_v$ of diameter less than $\epsilon_1/4$ intersects $h_1, h_2, \cdots$ and h_n but no such subset intersects both h_1 and h_{n+1}; let $h_{1,2}$ be the sum of $h_{n+1}, h_{n+2}, \cdots, h_m$ where some connected subset of $h_1+h_2+\cdots+h_v$ of diameter less than $\epsilon_1/4$ intersects $h_{n+1}, h_{n+2}, \cdots$ and h_m but no such subset intersects both h_{n+1} and h_{m+1}; $\cdots$; and let $h_{1,t}$ be the sum of $h_{p+1}, h_{p+2}, \cdots, h_v$ where some connected subset of $h_1+h_2+\cdots+h_v$ of diameter less than $3\epsilon_1/4$ intersects $h_{p+1}, h_{p+2}, \cdots$ and h_v but no subset of $h_1+h_2+\cdots+h_v$ of diameter less than $\epsilon_1/4$ intersects both h_{p+1} and h_1. We use $3\epsilon_1/4$ in the last case instead of $\epsilon_1/4$ in order to insure that no connected set in $h_1+h_2+\cdots+h_v$ of diameter less than $\epsilon_1/4$ intersects both h_{p+1} and h_1. The collection H_1 satisfies conditions (1), (2), (3), (4) and (5).

Description of collection H_2. Choose a positive number ϵ_2 less than one one-hundredth of the diameter of any connected set in H_1^* that intersects $h_{1,i}$ and $h_{1,i+2}$. We shall describe a collection H_2 of connected domains $h_{2,1}, h_{2,2}, \cdots, h_{2,s}$ such that:

(1) $h_{2,i}$ intersects $h_{2,j}$ only if i is equal to either $j-1$, j or $j+1$ ($h_{2,s+1}\equiv h_{2,1}$ and $h_{2,1-1}\equiv h_{2,s}$);

(2) $S-H_2^*$ is connected;

(3) some point of $S-H_2^*$ is accessible from $h_{2,i}$;

(4) the diameter of $h_{2,i}$ is less than ϵ_2;

(5) no connected subset of H_2^* that intersects $h_{2,i}$ and $h_{2,i+2}$ is of diameter less than $\epsilon_2/4$;

(6) each $h_{1,i}$ contains 98 consecutive elements of $h_{2,1}, h_{2,2}, \cdots, h_{2,s}$;

(7) if $H(n; i, j)$ denotes $h_{n,i-100}+\cdots+h_{n,i}+\cdots+h_{n,j}+\cdots+h_{n,j+100}$ and $h_{1,i}$ and $h_{1,j}$ intersect $h_{2,m}$ and $h_{2,k}$ respectively, then either $H(1;i,j)$ covers the closure of $H(2;m,k)$ and $H(1;j,i)$ covers the closure of $H(2;k,m)$ or $H(1;i,j)$ covers the closure of $H(2;k,m)$ and $H(1;j,i)$ covers the closure of $H(2;m,k)$.

Denote by C the component that contains $h_{1,5}+h_{1,6}$ of the common part of $h_{1,2}+h_{1,3}+\cdots+h_{1,9}$ and the complement of the closure of $h_{1,1}+h_{1,10}$. We shall show that if P is a point of $h_{1,5}+h_{1,6}$ and R is a point of $C-P$, then R does not disrupt P from $S-H_1^*$ in $S-H_1^*+C$. Let PQ be an arc in $S-R$ from P to a point Q of $S-H_1^*$. Let Q' be the first point of PQ in the order from P to Q on $S-C$. If PQ' intersects $h_{1,3}$, then there is an arc from $PQ'-Q'$ to $S-H_1^*$ in $S-H_1^*+h_{1,3}$ because a point of $S-H_1^*$ is accessible from $h_{1,3}$. Also, if PQ' intersects $h_{1,8}$, R does not disrupt P from $S-H_1^*$ in $S-H_1^*+C$. If PQ' intersects neither $h_{1,3}$ nor $h_{1,8}$, then Q' is a point of $S-H_1^*$. This demonstrates that R does not disrupt P from $S-H_1^*$ in $S-H_1^*+C$.

Let G be a finite collection of connected domains covering $h_{1,5}+h_{1,6}$ such that each intersects $h_{1,5}+h_{1,6}$ and is of diameter less than $\epsilon_2/1200$. No point of C disrupts an element of G from $S-H_1^*$ in $S-H_1^*+C$. Repeated applications of the preceding lemma give that there is a continuum K in $S-H_1^*+C$ that intersects $A'B'$ and each element of G but does not disrupt $h_{1,1}$ from $h_{1,10}$ in $h_{1,1}+h_{1,2}+\cdots+h_{1,10}$. Let G' be the set of all domains g' such that g' is either the common part of $h_{1,2}+h_{1,3}+\cdots+h_{1,9}$, the complement of K and an element of G, or the common part of the complement of K and h_i for i equal to 2, 3, 4, 7, 8 or 9.

There exists a finite collection G'' of elements of G' such that the sum of the elements of G'' is a connected domain intersecting $h_{1,1}$ and $h_{1,10}$ but the sum of no subcollection of G' having fewer elements than G'' is a connected domain intersecting $h_{1,1}$ and $h_{1,10}$. Assume that g_1 of G'' intersects $h_{1,1}$, g_i $(i=1, \cdots, r-1)$ intersects g_{i+1} and g_r intersects $h_{1,10}$.

There exists a collection $g_1', g_2', \cdots, g_r'$ of connected domains such that g_1' intersects $h_{1,1}$, g_i' intersects g_{i+1}', g_r' intersects $h_{1,10}$ and the closure of g_k' $(k=1, \cdots, r)$ is a subset of g_k.

Let E denote $h_{1,1}+g_1'+\cdots+g_r'+h_{1,10}+\cdots+h_{1,t}$ plus all points that it separates from $A'B'$. Each component of the common part of E and an element of G' intersects one of the domains $h_{1,1}$, $g_1', \cdots, g_r'$, $h_{1,10}$ but no such component intersects two of these domains that do not belong to a consecutive set of three of these domains. Add such components to the ones of $h_{1,1}$, g_i', $h_{1,10}$ that they intersect to form the sets h_1, g_i'', h_{10}. We note that the diameter of each g_i'' not intersecting $h_{1,2}+h_{1,3}+h_{1,4}+h_{1,7}+h_{1,8}+h_{1,9}$ is less than $\epsilon_2/400$.

Consecutive elements of $g_1'', g_2'', \cdots, g_r''$ may be combined by threes and fours in a manner previously described so as to get a collection $g_{1,1}, g_{1,2}, \cdots, g_{1,u}$ such that the collection $h_1, g_{1,1}, \cdots, g_{1,u}, h_{10}, h_{1,11}, \cdots, h_{1,t}$ satisfies conditions analogous to conditions (1), (2) and (3) to be satisfied by $h_{2,1}, h_{2,2}, \cdots, h_{2,s}$. We note that the closure of $g_{1,i}$ $(2\leqq i<u)$ is a subset of $h_{1,1}+h_{1,2}+\cdots+h_{1,10}$.

In a manner similar to that in which $h_{1,1}+h_{1,2}+\cdots+h_{1,10}$ was replaced by $h_1+g_{1,1}+\cdots+g_{1,u}+h_{10}$, we replace $h_{1,11}+\cdots+h_{1,20}$ by $h_{11}+g_{11,1}+\cdots+g_{11,v}+h_{20}, \cdots$ and $h_{1,t-m}+\cdots+h_t$ $(9\leqq m \leqq 18)$ by $h_{t-m}+g_{t-m,1}+\cdots+g_{t-m,w}+h_t$.

Let $g_{i,0}$ be the fourth element of $g_{i,1}, g_{i,2}, \cdots, g_{i,n}$ which follows all of these elements that intersect $h_{1,i+3}$. We note that $g_{i,0}$, the three domains immediately preceding $g_{i,0}$, and the three domains immediately following $g_{i,0}$ are each a subset of $h_{1,i+4}$ of diameter less than $\epsilon_2/100$.

In the manner described above, replace $g_{1,0}+\cdots+h_{10}+h_{11}+\cdots+g_{11,0}$ by $g'_{1,0}+h'_{2,2}+\cdots+h'_{2,r}+g'_{11,0}$; replace $g'_{11,0}+\cdots+g_{21,0}$ by $h'_{2,r+1}+\cdots+h'_{2,v}+g'_{21,0}; \cdots$ and replace $g'_{t-m,0}+\cdots+g'_{1,0}$ by $h'_{2,n}+\cdots+h'_{2,u}+h'_{2,1}$. The closure of $h'_{2,2}+\cdots+h'_{2,r}$ is a subset of $h_{1,4}+\cdots+h_{1,20}, \cdots$ and the closure of $h'_{2,n}+\cdots+h'_{2,u}+h'_{2,1}$ is a subset of $h_{1,t-m+3}+\cdots+h_{1,10}$. Consecutive elements of $h'_{2,1}, h'_{2,2}, \cdots, h'_{2,u}$ may be combined in a manner previously described so as to form a collection H_2 of connected domains $h_{2,1}, h_{2,2}, \cdots, h_{2,s}$ satisfying conditions (1), (2), (3), (4), (5), (6) and (7).

Description of simple closed curve J. For each positive integer i greater than one, we define a collection H_i of connected domains $h_{i,1}, h_{i,2}, \cdots, h_{i,n}$ satisfying conditions analogous to those satisfied by H_2 where ϵ_i is a positive number less than one one-hundredth the diameter of any connected set in H^*_{i-1} intersecting $h_{i-1,j}$ and $h_{i-1,j+2}$. We shall show that the common part J of $H_1^*, H_2^*, \cdots$ is a simple closed curve that does not separate S.

As the closure of H^*_{i+1} is a connected subset of H_i^* (condition 7) and as each $h_{i,j}$ contains an element of H_{i+1} (condition 6), then J is a nondegenerate continuum. This continuum does not separate space because the complement of each H_i^* is connected.

To show that J is a simple closed curve, we shall show that any pair of points P, Q of J separates it. Suppose that h_{i,P_i} and h_{i,Q_i} are elements of $h_{i,1}, \cdots, h_{i,n}$ that contain P and Q respectively. For convenience in notation, we shall assume that it is $H(i; P_i, Q_i)$ that covers the closure of $H(i+1; P_{i+1}, Q_{i+1})$ and that it is $H(i; Q_i, P_i)$ that covers the closure of $H(i+1; Q_{i+1}, P_{i+1})$. If J_{PQ} is the common part of $H(1; P_1, Q_1), H(2; P_2, Q_2), \cdots$, we find that J is the sum of two continua J_{PQ} and J_{QP} which have only P and Q in common.

Hence, the assumption that an arc separates S leads to the conclusion that some simple closed curve does not.

References

1. D. W. Hall, *A partial solution of a problem of J. R. Kline*, Duke Math. J. vol. 9 (1942) pp. 893–901.

2. ———, *A note on primitive skew curves*, Bull. Amer. Math. Soc. vol. 49 (1943) pp. 935–936.

3. F. B. Jones, Bull. Amer. Math. Soc. Abstract 48-11-340.

4. C. Kuratowski, *Une caractérisation topologique de la surface de la sphère*, Fund. Math. vol. 13 (1929) pp. 307–318.

5. L. Zippin, *A study of continuous curves and their relation to the Janiszewski-Mullikin Theorem*, Trans. Amer. Math. Soc. vol. 31 (1929) pp. 744–770.

6. ———, *On continuous curves and the Jordan Curve Theorem*, Amer. J. Math. vol. 52 (1930) pp. 331–350.

7. R. L. Wilder, *A converse of the Jordan-Brouwer separation theorem in three dimensions*, Trans. Amer. Math. Soc. vol. 32 (1930) pp. 632–651.

8. S. Claytor, *Topological immersion of Peanian continua in a spherical surface*, Ann. of Math. vol. 35 (1934) pp. 809–835.

9. R. L. Moore, *Concerning a set of postulates for plane analysis situs*, Trans. Amer. Math. Soc. vol. 20 (1919) pp. 169–178.

10. E. R. van Kampen, *On some characterizations of 2-dimensional manifolds*, Duke Math. J. vol. 1 (1935) pp. 74–93.

The University of Texas

EXTRAIT DE »FUNDAMENTA MATHEMATICAE« T. XXXVI (1949).

Complementary domains of continuous curves [1]).

By

R. H. Bing (Madison, Wisc., U. S. A.).

1. *Introduction.* Suppose that space is metric, compact, connected, and locally connected. It is known that each pair of points can be joined by an arc. If the space is locally topologically equivalent to a subset of the plane, then each complementary domain of this arc has property S. In this paper we show that the arc may be so chosen that its complementary domains have property S even if the space is not locally planar.

Much of this paper is devoted to the development of theorems regarding partitionings. The results regarding the complements of continuous curves are applications of these results. Although finite coverings by open sets could be used to prove these results, once the theorems regarding partitionings are demonstrated, partitionings seem to be a more efficient method of accomplishing them. It is hoped that partitionings will be useful in other connections. As pointed out in sections 4 and 5, problems handled by other methods can sometimes be treated effectively by using partitionings. In fact, partitionings provided a means for showing [2,3] that a compact continuous curve (locally connected continuum) can be convexified. The study of partitionings throws light on the structure of continuous curves and is of interest aside from its applications.

We shall use the following definitions.

Property S. A set M has *property* S if for each positive number ϵ, M is the sum of a finite number of connected subsets each of diameter less than ϵ.

Uniformly locally connected. A set M is *uniformly locally connected* if for each positive number ϵ there is a positive number $\delta(\epsilon)$ such that each pair of points of M at a distance apart of less than $\delta(\epsilon)$ belong to a connected subset of M of diameter less than ϵ.

[1]) Presented to the American Mathematical Society, November 26, 1948.

Distance. We denote the distance between two points p,q by $D(p,q)$. If p and q belong to a connected set M, $E(M;p,q)$ denotes the greatest lower bound of the diameters of all connected subsets of M containing $p+q$. We note that E is a metric for M that preserves its topology if M is locally connected. Also, M is uniformly locally connected under E if M has property S under D.

Partitioning. A finite collection G of mutually exclusive connected open subsets of M is a *partitioning* of M if the sum of the elements of G is dense in M. If each element of G is of diameter less than ϵ, G is an *ϵ-partitioning.* If each element of G has property S, it is an *S partitioning.* (Moise used the expression grille-decomposition instead of partitioning. See [2,3]).

Brick Partitioning. An S partitioning G of M is a *brick partitioning* if:

(a) each domain containing a point of M which is a limit point of each of two elements of G also contains a point of M which is a limit point of each of these same two elements of G but of no other element of G,

(b) each element of G is uniformly locally connected under $E(M;x,y)$,

(c) each boundary point in M of an element of G is a boundary point of another element of G.

Refinement. A collection H is a refinement of the collection G if each element of H is a subset of an element of G.

Partition Chain. A partition chain is a finite collection $C=[c_1,c_2,\dots,c_n]$ of connected and uniformly locally connected domains such that:

(a) $\bar{c}_i$ intersects $\bar{c}_j$ if and only if $i=j-1$, j, or $j+1$,

(b) c_i is regular, that is $c_i=\bar{c}_i-F(\bar{c}_i)$ [$F(A)$ denotes the boundary of A],

(c) $\bar{c}_i\cdot\bar{c}_{i+1}$ contains a point which is not a limit point of $F(\bar{C}^*)$ (C^* denotes the sum of the elements of C).

If the point p belongs to one end element of C and the point q to the other, C is called a chain from p to q. If each element of C is of diameter less than ϵ, C is an ϵ-partition chain.

Circular Partition Chain. The definition is the same as the above except that in (a) we use the convention that $n+1=1$.

Run Straight Through. A partition chain E is said to *run straight through* the partition chain C if E is a refinement of C and E is such that if two elements of it are subsets of the same element of C, then each element of E between them is a subset of this same element of C.

2. *Brick Partitioning.* The elements of a brick partitioning G of space are packed in something like brick; that is, for each pair of elements of G whose boundaries intersect each other, there is a point on their common boundary that is not a boundary point of any other element of G. An advantage of such a partitioning is that two adjacent elements of G may be consolidated into one element so that the resulting partitioning is also a brick partitioning.

In proving the theorems of this section, we make use of Theorem 4 of *Partitioning a Set* [2] which states that for each positive number ϵ any set with property S can be S ϵ-partitioned.

Theorem 1. *Suppose M is a connected domain with property S and B is a subset of the boundary of M. Then for each positive number ϵ there is an S partitioning $[g_0, g_1, \ldots, g_n]$ of M such that the diameter of g_i $(i=1,2,\ldots,n)$ is less than $\epsilon, \bar{g}_0$ intersects $M \cdot \bar{g}_i$, and $E(M+B;B,g_j)\geqslant 0$, the inequality holding if and only if $j=0$.*

Proof. Let H be an S ϵ-partitioning of M and T be a connected subset of M such that T intersects each element of H but $E(M+B;B,T)>0$. For each element h_j of H, let H_j be an S partitioning of h_j such that each element of H_j is of diameter less than $\frac{1}{3}E(M+B;B,T)$. There is a finite collection W_j of connected subsets of h_j such that:

(a) each pair of elements of W_j are at a positive distance from each other under $E(h_j)$,

(b) each element of H_j that is at a distance of zero from $B+T$ under $E(M+B)$ is a subset of an element of W_j,

(c) no element of W_j is at a zero distance from both B and T under $E(M+B)$,

(d) no collection having fewer elements than W_j satisfies conditions (a),(b), and (c).

By Theorem 3 of *Partitioning a Set* [2], there is an S partitioning H_j' of h_j such that each element of H_j' contains exactly one element of W_j. If H_j'' denotes the collection of those elements of H_j' which are at a zero distance from B under $E(M+B)$, then the elements of $\sum H_j''$ are the elements $g_1, g_2, \ldots, g_n$ while $g_0 = M - \sum_{i=1}^{n} \bar{g}_i$. Now g_0 is connected because it contains T and each of its components intersects T. Also, $\bar{g}_0$ intersects $M \cdot \bar{g}_i$ or else the condition (d) above was not satisfied by some W_j.

Theorem 2. *Suppose:*

(a) *R is a connected domain with property S,*

(b) *$B_1, B_2, \ldots, B_n$ is a collection of mutually exclusive closed subsets of the boundary of R,*

(c) *ϵ is a positive number such that for each point p of R there is an arc of diameter less than ϵ in $R+B_i$ from p to B_i.*

Then for each positive number δ there is an S partitioning $g_0, g_{11}, g_{12}, \ldots, g_{1n_1}, g_{21}, \ldots, g_{nn_n}$ of R such that

(a) *the diameter of g_{ij} is less than δ,*

(b) *$\bar{g}_0$ intersects $R \cdot \bar{g}_{ij}$,*

(c) *$\bar{g}_{ij}$ intersects B_i but $\bar{g}_0$ does not,*

(d) *for each point p of g_0 there is an arc of diameter less than $\epsilon + \delta$ from p to B_i in $g_0 + \bar{g}_{i1} + \bar{g}_{i2} + \ldots + \bar{g}_{in_i}$.*

Proof. Suppose that 2δ is less than the distance between two elements of $B_1, B_2, \ldots, B_n$. Let:

(a) $[h_1, h_2, \ldots, h_s]$ be an S δ-partitioning of R,

(b) p_i be a point of h_i,

(c) T be a dendron in R containing $\sum p_i$,

(d) a_{ij} be an arc of diameter less than ϵ in $R + B_i$ from p_j to B_i,

(e) θ be a positive number less than either δ or min $D(T + a_{jk}, B_i)$ $(i \neq j)$. If h_t has a point of B_i on its boundary, we let h_t, θ, and $\bar{h}_t \cdot B_i$ be the M, ϵ, and B of Theorem 1 and obtain a partitioning $g_0, g_1, \ldots, g_n$ satisfying the conditions of that theorem. Then $g_1, g_2, \ldots, g_n$ are elements of $g_{i1}, g_{i2}, \ldots, g_{in_i}$. Similarly, all elements g_{ij} are obtained and g_0 is defined to be $R - R \cdot \sum \sum \bar{g}_{jk}$.

Theorem 3. *Suppose R is a connected set with property S while R_1 and R_2 are two connected subsets of R such that $E(R; R_1, R_2) > 0$. There is a continuous transformation T of R into a straight line interval I such that $T(R_1)$ and $T(R_2)$ are the ends of I while for each connected subset X of I containing an end point of I, $T^{-1}(X)$ is connected and has property S* [2]).

[2]) Using results obtained in the proof of Theorem 5, we find that T can be chosen so that each $T^{-1}(X)$ is uniformly locally connected under $E(R)$.

Proof. For convenience we suppose that R is closed [Lemma 1 of 2]. Let R_1' and R_2' be mutually exclusive closed subcontinua of R such that R_i' contains R_i and $R-(R_1'+R_2')$ has property S. Letting 3δ be the distance from R_1' to R_2' and applying Theorem 1 to each component of $R-(R_1'+R_2')$, we find that there is an S partitioning $[g_1, g_2, \ldots, g_n]$ of R such that (a) g_1 contains R_1', (b) g_2 contains R_2', (c) $D(g_1, g_2)>0$, and (d) $\bar{g}_i$ $(i=3,4,\ldots,n)$ intersects both $\bar{g}_1$ and $\bar{g}_2$. We define $T(\bar{g}_1)$ and $T(\bar{g}_2)$ to be 0 and 1 respectively.

Each component C_j of $R-(\bar{g}_1+\bar{g}_2)$ has property S and by Theorem 1 there is an S partitioning $g_{j0}, \bar{g}_{j1}, \ldots, \bar{g}_{jn}$ of C_j such that the diameter of g_{ji} $(i=1,2,\ldots,n)$ is less than either $1/2$ or $\frac{1}{3}D(g_1, g_2)$, $\bar{g}_{j0}$ intersects $\bar{g}_{ji}$, and $D(\bar{g}_{ji}, \bar{g}_1+\bar{g}_2)\geqslant 0$, the inequality holding if and only if $i=0$. We define $T(\sum \bar{g}_{j0})$ to be $1/2$.

This process is continued as follows: Suppose that $T(X_1)$ and $T(X_2)$ have been defined to be $p/2^q$ and $(p+1)/2^q$ and let C_s be a component of $R-(X_1+X_2)$ whose closure intersects both X_1 and X_2. By Theorem 2 there is an S partitioning $h_{s0}, h_{s11}, \ldots, h_{s1m_1 1}, h_{s21}, \ldots, h_{s2m_2}$ of C_s such that:

(a) the diameter of h_{sij} is less than either $D(X_1, X_2)/3$ or $1/2^{q+1}$,

(b) $\bar{h}_{s0}$ intersects $\bar{h}_{sij}$,

(c) $\bar{h}_{sij}$ intersects X_i but $\bar{h}_{s0}$ does not,

(d) for each point r of h_{s0} there is an arc of diameter less than $1/2^q+1/2^{q+1}$ from r to X_i in $\bar{h}_{s0}+\bar{h}_{si1}+\ldots+\bar{h}_{sim_i}$.

Then $T(\sum \bar{h}_{s0})=(2p+1)/2^{q+1}$.

The continuation of this process defines T except on a subset of R which is not dense in any open subset of R. The transformation is defined by continuity on this set. It may be shown to satisfy the conditions mentioned in the statement of Theorem 3.

Theorem 4. *A necessary and sufficient condition that a domain D with property S not be uniformly locally connected is that there be a point x of $\bar{D}$ and an arc pxq in $D+x$ such that $E(D; px-x, xq-x)>0$.*

Theorem 5. *Suppose R is a connected and uniformly locally connected domain, R_1 and R_2 are two connected subsets of R which are at a positive distance from each other, and $B_1, B_2, \ldots, B_n$ is a collection of subsets of the boundary of R. Then there is a brick partitioning $[h_1, h_2]$ of R such that h_j contains R_j and each domain containing a point $\bar{B}_i\cdot\bar{h}_j$ $(i=1,2,\ldots,n;\ j=1,2)$ also contains a point of $B_i\cdot\bar{h}_j$ which is not a limit point h_k $(k\neq j)$.*

Proof. By Theorem 3 there is an uncountable collection W of S partitionings of R such that:

(a) each element of W contains exactly two elements, one of which contains R_1 and the other which contains R_2,

(b) R is a subset of the sum of these two elements plus their common boundary,

(c) for each pair of elements H, K of W there is an element h of H and element k of K such that $E(R;h,k)>0$.

Using Theorem 4, we find that if W_0' is the set of all elements $[h_1,h_2]$ of W such that not both h_1 and h_2 are uniformly locally connected, W_0' is at most countable. Also, W_i' is not uncountable if it is the set of all elements $[h_1,h_2]$ of W such that there is a point of $\bar{B}_i\cdot\bar{h}_1\cdot\bar{h}_2$ which is not a limit point of both $B_i\cdot(\bar{h}_1-\bar{h}_1\cdot\bar{h}_2)$ and $B_i\cdot(\bar{h}_2-\bar{h}_1\cdot\bar{h}_2)$. Each element of the uncountable collection $W-(W_0'+W_1'+\ldots+W_n')$ satisfies the conditions of the theorem.

Each of the following three theorems follows from repeated applications of the theorem preceding it.

Theorem 6. *Suppose R is a connected and uniformly locally connected domain, W is a finite collection of subsets of R such that the distance between two elements of W is positive, and $B_1,B_2,\ldots,B_n$ is a collection of subsets of the boundary of R. Then there is a brick partitioning $G=[g_1,g_2,\ldots,g_m]$ of R such that,*

(a) *no element of G intersects two elements of W,*

(b) *each domain containing a point of $\bar{g}_j\cdot\bar{B}_k$ $(j=1,2,\ldots,m;$ $k=1,2,\ldots,n)$ also contains a point of B_k which is a limit point of g_j but of no other element of G,*

(c) *each domain containing a point of $\bar{g}_j\cdot\bar{g}_k$ $(j,k=1,2,\ldots,m)$ contains a point of R which is a limit point of both g_j and g_k but of no other element of G.*

Theorem 7. *Suppose R is a connected and uniformly locally connected domain and $B_1,B_2,\ldots,B_n$ is a collection of subsets of the boundary of R. Then for each positive number ϵ there is a brick ϵ-partitioning G of R satisfying conditions* (b) *and* (c) *of Theorem* 6.

Theorem 8. *If M is a set with property S, then there is a sequence $G_1,G_2,\ldots$ such that G_i is a brick $(1/i)$-partitioning of M and G_{i+1} is a refinement of G_i.*

3. *Partition Chains.* The first theorem of this section is an extension of the lemma used extensively in *The Kline Sphere Characterization Problem* [1] and may be proved by the methods used there.

We say that *M disrupts X* from *Y* in *R* if there is an arc from *X* to *Y* in *R* but each such arc contains a point of *M*. An arc minus its end points is called an open arc.

Theorem 9. *Suppose the boundary of the connected domain R is the sum of the mutually exclusive sets M, N, and E, each of which is accessible from R, and U is a connected subdomain of R such that no pair of points of R disrupts U from F(R) in $\bar{R}$ and no point of R disrupts U from either $M+E$ in $R+M+E$ or $N+E$ in $R+N+E$. Then there are open arcs (α) and (β) in R such that α intersects both M and N, β intersects both U and E, but α does not intersect β.*

Theorem 10. *Suppose $C=[c_1,c_2,\ldots,c_n]$ is a partition chain from p to q, B is a closed subset of $\bar{C}^*$, and α is an arc from p to q in $\bar{C}^*-[F(\bar{C}^*)+B]$ such that $\alpha\cdot\bar{c}_i\cdot\bar{c}_{i+1}$ $(i=1,2,\ldots,n-1)$ is a point. Then there is a partition chain $K=[k_1,k_2,\ldots,k_n]$ running straight through C from p to q and such that B does not intersect $\bar{K}^*$ but each component of $\bar{c}_i-\bar{k}_i$ $(i=1,2,\ldots,n)$ has a limit point in B if $c_i\neq k_i$.*

Proof. Let G be a brick partitioning of $\bar{C}^*-F(\bar{C}^*)$ such that G is a refinement of C and each element of G is of diameter less than $D(\alpha,B)/3$. Denote by k_i the maximal connected domain in c_i which contains $\alpha\cdot c_i$ and is such that $\bar{k}_i$ is the sum of the closures of elements of G which do not have a limit point in B. Then $K=[k_1,k_2,\ldots,k_n]$ is the required chain.

Theorem 11. *Suppose that in addition to the hypotheses of Theorem 9, it is given that $[g_1,g_2]$ is a brick partitioning of R such that M does not intersect $\bar{g}_2$ and N does not intersect $\bar{g}_1$. Then there are arcs α and β satisfying the conditions of Theorem 9 and such that $\alpha\cdot\bar{g}_1\cdot\bar{g}_2$ is a point.*

Proof. Using Theorem 9, we find that there are open arcs (α) and (β) in R such that α intersects both M and N but does not intersect β, $\alpha\cdot\bar{g}_1\cdot\bar{g}_2$ is a point, β intersects E, and $\beta+E$ intersects both $\bar{g}_1$ and $\bar{g}_2$. (If $\bar{g}_i$ did not intersect E, in applying Theorem 9 we would let g_i be U of that theorem). Since we find from the method used in Theorem 10 that E may be replaced by a set containing $\beta+E$, there is no loss of generality in supposing that both $\bar{g}_1$ and $\bar{g}_2$ intersect E. Hence, we suppose this.

For convenience we suppose that U intersects g_2. Let (cb_1) be an open arc in g_2 from a point c of U to E. For each set A denote by $W(A)$ the set of all points p of R such that there is an open arc (β_p) from p to E in R and an open arc (α_p) from M to N in R such that $\alpha_p\cdot\bar{g}_1\cdot\bar{g}_2$ is a point and α_p does not intersect β_p+A. Assume that $W(E)$ does not intersect U and let r be the nearest point of the closure of $W(E)\cdot cb_1$ to c in the order from c to b_1 on cb_1.

Now r does not disrupt E from U in $R+E$ or else, by using the fact that r does not disrupt U from $N+E$ in $R+N+E$ or U from $M+E$ in $R+M+E$, it could be shown that there is an open arc (γ) from M to N in R such that $\gamma\cdot\bar{g}_1\cdot\bar{g}_2$ is a point and γ does not intersect rb_1 of cb_1. But then r would not be the nearest point of the closure of $W(E)\cdot cb_1$ to c on cb_1.

There is an open arc (cb_2) in $R-r$ from c to a point of E such that either (cb_2) is a subset of g_2 or $\bar{g}_1\cdot\bar{g}_2\cdot(cb_2)$ is a point while both $g_1\cdot cb_2$ and $g_2\cdot cb_2$ are connected. For convenience we suppose that $cb_1\cdot cb_2$ is an arc cx.

By using the methods of the lemma in *The Kline Sphere Characterization Problem* [1], it may be found that there are:

(a) an arc pxq such that px is a subset of xb_2, xq is a subset of xb_1, and p is a point of g_2 if $xb_2\cdot W(E)$ intersects g_2;

(b) an open arc (mn) in R from a point m of M to a point n of N such that $mn\cdot\bar{g}_1\cdot\bar{g}_2$ is a point and mn contains pxq;

(c) an open arc (pb_3) in R from p to E such that (pb_3) does not intersect mn;

(d) an open arc (qb_4) in R from q to E such that this open arc does not intersect mn;

(e) an open arc (st) in R from a point of s of (pxq) to a point r of $M+N+E$ such that $st\cdot pxq=s$ and s is not a point of $W[(pb_3)+(qb_4)]$.

For convenience suppose that p precedes q on mn in the order from m to n. Let v be the first point of

$$(M+mp)+(N+qn)+(E+pb_3+qb_4)$$

on st in the order from s to t. Now v is not a point of $E+pb_3+qb_4$ or else s is a point of $W[(pb_3)+(pb_4)]$. Also, v is not a point of $N+qn$, for assume that it is. If sv does not intersect $\bar{g}_1$, there is an arc γ

from m to N in $mp+ps+sv+qn-q$ such that $\gamma\cdot\bar{g}_1\cdot\bar{g}_2$ is a point, but this contradicts the fact that s is not a point of $W\cdot[(pb_3)+(qb_4)]$. If sv intersects $\bar{g}_1$ in a point u, it can also be seen that there is an arc γ from M to N in $\bar{g}_1+qn+uv$ of sv such that γ does not intersect $xq+(qb_4)$ and $\gamma\cdot\bar{g}_1\cdot\bar{g}_2$ is a point. But x is not a point of $W(E)$. Similarly, we find that v is not a point of $M+mp$. Hence, the theorem is satisfied by showing that the assumption that U does not contain a point of $W(E)$ leads to a contradiction.

Theorem 12. *Suppose the boundary of the connected and uniformly locally connected domain R is the sum of the mutually exclusive sets M, N, and E while U is a connected subdomain of R such that no point of R disrupts U from either $M+E$ in $R+M+E$ or $N+E$ in $R+N+E$. Then either* (a) *there are open arcs* (α) *and* (β) *in R such that α intersects both M and N, β intersects both U and E, but α does not intersect β, or* (b) *there is a closed set X in $\bar{R}$ such that each component of X intersects E, $X+E$ does not disrupt M from N in $\bar{R}$, but there are a pair of points p_1, p_2 of R such that p_1+p_2 disrupts U from $F(R)$ in $\bar{R}$ and $X+E+p_i$ disrupts M from N in $\bar{R}$.*

Proof. Suppose that condition (a) is not satisfied. There is a positive number δ so small that if a subset of R disrupts U from $F(R)$ in $\bar{R}$, then this subset has a diameter of more than 3δ. By Theorem 7 there is a brick δ-partitioning G of R satisfying conditions (b) and (c) of that theorem with B_1, B_2, and B_3 replaced by M, N, and E.

Let A be a finite collection of arcs in $\bar{R}$ such that A^*+E does not disrupt M from N, each element of A intersects E, and if g is an element of G which does not intersect A^*, then each arc from g to E in $\bar{R}$ disrupts M from N in $\bar{R}-(A^*+E)$. Let α be an arc irreducible from M to N in $\bar{R}-(A^*+E)$. There is a brick partitioning H of R such that H is a refinement of G, each element of H is of diameter less than $D(\alpha,A^*)/3$, and H satisfies conditions (b) and (c) of Theorem 7 with B_1, B_2, and B_3 replaced by M, N, and E.

Let W_1 be the set of all connected sets w such that w is maximal with respect to being the common part of R and a domain whose closure is the closure of the sum of a collection of elements of G which do not intersect A^*. It may be seen that no element of W_1 has a limit point in $E+A^*$. Let W_2 be the set of all connected sets w such that w is maximal with respect to being the common part of an element of G which does intersect A^* and a domain whose closure does not intersect A^* but is the closure of the sum of a subcollection of H. There is a partition chain $C=[c_1,c_2,\dots,c_m]$ such that the elements of C are elements of W_1+W_2, one end link of C has a point of M on its boundary and the other end link has a point of N on its boundary, and no chain with fewer elements than C has these properties. It may be noted that each element of H which is not in an element of W_1+W_2, has a limit point on A^*.

Let X be the closure of the sum of A^* and all elements of H which are not subsets of either an element of W_1 or of an element of C. Now each component of X intersects E and $R-R\cdot X=R'$ is a connected domain which contains each element of C and each element of W_1. Let Y be the sum of all subdomains y of R' such that each arc from y to E in $R+E$ disrupts M from N in $\bar{R}-(X+E)$. Denote the component of Y containing U by U'. It follows from Theorem 9 that some pair of points p_1, p_2 of R' disrupts U' from $M+N+X+E$ in $\bar{R}'$. We shall show that each of these points disrupts M from N in $\bar{R}-(X+E)$, thereby completing the proof of the theorem.

Now p_1 and p_2 are on the closures of elements of c_i and c_j $(i<j)$ of C which are not elements of W_1. These elements are not adjacent because they are of diameter less than δ. However, $j=i+2$ or else C could be replaced by a chain with fewer links. Now c_{i+1} is an element of W_1 and c_i and c_{i+2} are not.

If V denotes the collection of all points q of R' such that q separates M from N in $\bar{R}-(X+E)$, both p_1 and p_2 belong to V, for suppose that one does not. Then there is a simple closed curve J in $\bar{R}-(X+E)$ containing p_1+p_2 and two points r, s of $V+M+N$. For convenience, suppose r and s are points of V. Since $r+s$ is not the boundary of U', an arc in $R+E-(r+s)$ intersects both J and E. But then one arc of J from r to s would not intersect any element of W_1. This contradicts the fact that each arc in J from r to s contains a points of W_1 because one of the points r, s belongs to $\sum_{n=1}^{i}\bar{c}_n$ and the other belongs to $\sum_{n=i+2}^{m}\bar{c}_n$.

By using a type of argument similar to the preceding and by applying Theorem 11, we can obtain the following extension of Theorem 12.

Theorem 13. *Suppose that in addition to the hypotheses of Theorem* 12, *it is given that* $[g_1, g_2]$ *is a brick partitioning of R such that M does not intersect* $\bar{g}_2$ *and N does not intersect* $\bar{g}_1$. *Then either* (a) *there are open arcs* (α) *and* (β) *in R such that* α *intersects both M and N,* $\alpha\cdot\bar{g}_1\cdot\bar{g}_2$ *is a point,* β *intersects both U and E, but does not intersect* α *or* (b) *there are a closed set X in* $\bar{R}$ *such that each component of X intersects E, an arc* γ *in* $\bar{R}-(X+E)$ *from M to N such that* $\gamma\cdot\bar{g}_1\cdot\bar{g}_2$ *is a point, and a pair of points* p_1, p_2 *of R such that* p_1+p_2 *disrupts U from* $M+N+E$ *in* $\bar{R}$ *and* $X+E+p_i$ *disrupts M from N in* $\bar{R}$.

Theorem 14. *Suppose C is a partition chain from p to q such that no pair of points of* $\bar{C}^*$ *separates space and each element of C has a limit point on* $F(\bar{C}^*)$. *Then for each positive number* ϵ, *there is an* ϵ*-partition chain E from p to q such that:*

(a) *E runs straight through C,*

(b) $\bar{E}^*$ *does not intersect* $F(\bar{C}^*)$,

(c) *each element of E has a limit point on* $F(\bar{E}^*)$,

(d) $\bar{C}^*-\bar{E}^*$ *has property S and each point of it is joined to* $F(\bar{C}^*)$ *by an arc which does not intersect* $\bar{E}^*$ *but which is a subset of the closure of the sum of eleven elements of C.*

Part (d) of the above conclusion can be strengthened by using a smaller number than eleven, but eleven is used because this simplifies the proof and in the applications of Theorem 14 in the next sections, reducing the number does not improve the results.

Proof. Let G be a brick ϵ-partitioning of $\bar{C}^*-F(\bar{C}^*)$ such that G is a refinement of $C=[c_1, c_2, \dots, c_n]$, one element of G contains p, and another contains q.

By repeated applications of Theorem 11, it may be found that there is a finite collection A of arcs such that each element of A intersects $F(\bar{C}^*)$, each element of A is in the sum of the closures of eleven elements of C, each element of G intersects an element of A, and there is an arc α from p to q in $\bar{C}^*-[A^*+F(\bar{C}^*)]$ such that $\alpha\cdot\bar{c}_i\cdot\bar{c}_{i+1}$ $(i=1,2,\dots,n-1)$ is a point. Before establishing the existence of such a set A, we proceed to show how the theorem can be completed if there is such a collection.

Denote the distance from α to $F(\bar{C}^*)+A^*$ by. 3δ. Let be H a brick δ-partitioning of $\bar{C}^*-F(\bar{C}^*)$ which is a refinement of G. Denote by X the closure of the sum of all elements of H which have a point of $F(\bar{C}^*)+A^*$ on their boundaries. Let R be the set of all elements r such that r is a component of $g-X$ for some element g of G. From the elements of R which intersect a, form a chain E which runs straight through C and from p to q. The chain E satisfies the conditions of the theorem.

A method of verifying that A exists is now outlined. For convenience we suppose that n is of the form $6i+1$. It follows from repeated application of Theorem 11 that there is a finite collection A' of arcs such that no element of A' intersects $c_1+c_7+c_{13}+\dots+c_n$ but each intersects $F(\bar{C}^*)$, A'^* intersects each element of G in c_r $(r=3,4,5,9,10,11,15,\dots,n-2)$, and there is an arc from p to q in $\bar{C}^*-[F(\bar{C}^*)+A'^*]$ such that this arc intersects the common boundary of two adjacent elements of C in exactly one point. Using Theorem 10, we obtain a chain $K=[k_1,k_2,\dots,k_n]$ such that $\bar{K}^*$ does not intersect A'^*, k_s is a subset of c_s, $k_t=c_t$ $(t=1,7,13,\dots,n)$, $F(\bar{K}^*)$ is accessible from each element of K, and each point of $\bar{C}^*-\bar{K}^*$ belongs to an arc in $\bar{C}^*-\bar{K}^*$ which intersects $F(\bar{C}^*)$.

By using Theorem 11 again we find that there is a finite collection A'' of arcs such that no element of A'' intersects $k_4+k_{10}+k_{16}+\dots+k_{n-3}$ but each intersects $F(\bar{K}^*)$, A''^* intersects each element of G not intersected by A'^*, and there is an arc from p to q in $\bar{K}^*-[F(\bar{K}^*)+A''^*]$ such that this arc intersects the common boundary of two adjacent elements of K in exactly one point. Then $A=A'+A''$.

Had we used Theorem 13 instead of Theorem 11 in the preceding argument, we could have obtained the following.

Theorem 15. *Suppose C is a partition chain from p to q such that no point of $\bar{C}^*$ separates space,* (a) *those elements of C which are not open arcs have a point of $F(\bar{C}^*)$ on their boundaries, and* (b) *each pair of complementary domains of $\bar{C}^*$ are separated by a pair of points each of which is either p, q, or a point of $\bar{C}^*$ which separates p from q in $\bar{C}^*$.*

Then for each positive number ϵ there is an ϵ-partition chain E from p to q satisfying conditions (a) *and* (b) *above with E substituted for C and such that* (c) *E runs straight through C,* (d) *$\bar{E}^*$ does not intersect $F(\bar{C}^*)$, and* (e) *if a complementary domain D of $\bar{E}^*$ intersects $F(\bar{C}^*)$, then each point of $D\cdot\bar{C}^*$ belongs to an arc in D which intersects $F(\bar{C}^*)$ and is a subset of the closure of the sum of eleven elements of C.*

The methods of Theorem 14 yield the following result for circular partition chains.

Theorem 16. *Suppose C is a circular partition chain such that no pair of points of $\bar{C}^*$ separ ates space and each element of C has a limit point on $F(\bar{C}^*)$. Then for each positive number ϵ there is a circular ϵ-partition chain E satisfying conditions* (b), (c), *and* (d) *of Theorem* 14 *and such that E is a refinement of C, each element of C contains an element of E, and if two elements of E are subsets of the same element of C, a subchain of E containing them lies in this same element of E.*

4. *Complements of Arcs and Dendrons.* The theorems of this section are applications of Theorems 14 and 15. There are several known methods [for example, see Whyburn's *Analytic Topology*, 4] of showing that each pair of points of a compact locally connected continuum belongs to an arc in this continuum. Defining an arc by means of a sequence of partition chains gives another method of doing this.

Theorem 17. *If p, q, and r are three points of the compact locally connected continuum M which is not separated by any pair of its points, then there is an arc from p to q in M such that the complement of this arc is connected, has property S and contains r.*

Proof. By Theorem 5 there is a brick partitioning $[C_1, k]$ of M such that C_1 contains $p+q$ and k contains r. There is a regular (1/2)-partition chain C_2 in C_1 from p to q and satisfying conditions like those given in Theorem 14. In C_2 there is a regular (1/4)-partition chain C_3 satisfying like conditions. Similarly, there are chains $C_4, C_5, \ldots$ The common part of $\bar{C}_1, \bar{C}_2^*, \bar{C}_3^*, \ldots$ is an arc from p to q whose complement is connected, has property S, and contains r.

Theorem 18. *If p and q are two points of the compact locally connected continuum M, there is an arc from p to q such that each complementary domain of this arc has property S and each such pair of complementary domains is separated in M by a pair of points. Furthermore, for each positive number ϵ there are only a finite number of these complementary domains of diameter more than ϵ.*

Proof. The proof is similar to that given for the previous theorem except that we apply Theorem 15 instead of Theorem 14.

Theorem 19. *Suppose W is a totally disconnected closed subset of the compact locally connected continuum M. Then there is a dendron (acyclic continuous curve) T in M containing W such that each component of $M-T$ has property S and for each positive number ϵ there are no more than a finite number of such components of diameter more than ϵ.*

Proof. Let G_1 be a brick 1/2-partitioning of M and H_1 be the collection of elements of G_1 whose closures contain points of W. By the preceding theorem, there is a finite collection A_1 of arcs in M such that for each positive number ϵ and each element a of A_1, $M-a$ has no more than a finite number of components of diameter more than ϵ, each of these components has property S, $\bar{H}_1^*+A_1^*=M_1$ is connected, but if B is a proper closed subset of A_1^*, $\bar{H}_1^*+B$ is not connected.

Let G_2 be a brick 1/4-partitioning of M_1 and H_2 be the elements of G_2 whose closures intersect W. There is a finite collection A_2 of arcs in M_1 satisfying conditions like those satisfied by the collection A_1. We denote $\bar{H}_2^*+A_2^*$ by M_2. Similarly, we define $M_3, M_4, \ldots$ The dendron T is the common part of $M_1, M_2, \ldots$

It would be interesting to know the answer to the following.

Question. For each closed set W whose complement has property S, does there exist a countable collection A of arcs such that $W+A^*$ is a continuous curve each of whose complementary domains has property S?

5. *A Sphere Characterization.* J. R. Kline conjectured that a compact locally connected nondegenerate continuum was a simple surface (set topologically equivalent to the surface of a sphere) if it was separated by no pair of its points but by each simple closed curve in it. In the following theorem I use much the same methods as I used in proving this conjecture was true. However, partitionings are used here instead of open coverings.

Theorem 20. *If M is a compact locally connected nondegenerate continuum which is not separated by any pair of its points, then either M is a simple surface or there is a simple closed curve J in M such that $M-J$ is connected and has property S.*

Proof. Suppose M is not a simple surface. Zippin has shown [5] that some arc pq in M separates two points from each other in M. Without loss of generality we suppose that pq separates x from y but that no proper subarc of pq separates x from y. Let D_x, and D_y be the components of $M-pq$ containing x and y respectively [3]).

[3]) By methods more complicated than those used here, I have shown that there is a simple closed curve J satisfying the conditions of this theorem, containing $p+q$, and such that one component of $J-(p+q)$ lies except for a totally disconnected set in D_x and the other component lies except for a totally disconnected set in D_y.

Let ϵ_1 and ϵ_2 be positive numbers so small that (a) no pair of subsets of M each of diameter less than ϵ_1 separates two points of M from each other if these points are further from the subsets than $D(p,q)/8$ and (b) if β is a subarc of pq containing an end point of pq, then β is of diameter less than $\epsilon_1/3$ if its end points are closer together than ϵ_2. Let G be a brick ϵ_2-partitioning of M such that p and q belong to the elements g_p and g_q of G.

There is an arc α_z $(z=x,y)$ in D_z which is irreducible from $\bar{g}_p$ to $\bar{g}_q$. Let H be a brick partitioning of M such that H is a refinement of G and no element of H is of diameter more than $D(pq, \alpha_x+\alpha_y)/3$. Denote by R the sum of all closures r of elements of H such that r either intersects pq or is a subset of $\bar{g}_p+\bar{g}_q$. Now R is the sum of three mutually exclusive connected sets R_1, U, and R_3 such that R_1 is closed and contains $\bar{g}_p$, R_3 is closed and contains $\bar{g}_q$, while U is an open subset of R whose boundary in R is a subset of $\bar{g}_p+\bar{g}_q$.

Since both R_1 and R_3 are of diameter less than ϵ_1, each component of $M-R$ that has both a boundary point on R_1 and a boundary point on R_3 is accessible from U. Let C_2 and C_4 be the components of $M-R$ containing (α_x) and (α_y) respectively, E be the sum of $\bar{U}$ and the closures of all components of $M-(R+C_2+C_4)$ such that these components have a point of U on their boundaries, C_i $(i=1,3)$ be a regular connected domain whose closure is the sum of R_i and the closures of all components of $M-R$ which have a boundary point on R_i but none on U. Then $C_1=[c_1,c_2,c_3,c_4]$ is a circular partition chain such that $M-\bar{C}_1^*$ is a connected domain which has property S and whose closure is E.

Theorem 16 gives that there is a sequence $C_1, C_2, \ldots$ of circular partition chains such that (a) C_{i+1} is a refinement of C_i and each element of C_i contains an element of C_{i+1}, (b) each element of C $(i=2,3,\ldots)$ is of diameter less than $1/2^i$, (c) $\bar{C}_{i+1}^*$ does not intersect $F(\bar{C}_i^*)$, (d) each element of C_i is accessible from the closure $M-\bar{C}_i^*$, (e) if two elements of C_{i+1} are subsets of the same element of C_i, then a subchain of C_{i+1} between these two elements lies in this same element of C_i and (f) $M-\bar{C}_i^*$ has property S while each point of $\bar{C}_i^*-\bar{C}_{i+1}^*$ is joined to the boundary of $\bar{C}_i^*$ by an arc in $M-\bar{C}_{i+1}^*$ which is a subset of the closure of the sum of eleven elements of C_i.

The common part of $\bar{C}_1^*, \bar{C}_2^*, \ldots$ is a simple closed curve J and $M-J$ is connected and has property S.

References.

[1] Bing R. H., *The Kline sphere characterization problem*, Bull. Amer. Math. Soc. **25** (1946), pp. 644-653.

[2] — *Partitioning a set* (accepted by Bulletin of the American Mathematical Society).

[3] Moise E. E., *Grille decomposition and convexification theorems for compact metric locally connected continua* (accepted by Bulletin of the American Mathematical Society).

[4] Whyburn G. T., *Analytic Topology*, Colloquium Amer. Math. Soc. **28** (1942), New York.

[5] Zippin L., *On continuous curves and the Jordan Curve Theorem*, Amer. Journ. Math. **52** (1930), pp. 331-350.

University of Wisconsin.

B. 3-Space and 3-Sphere

A CHARACTERIZATION OF 3-SPACE BY PARTITIONINGS

BY

R. H. BING

While it is apparent that any Euclidean n-sphere can be partitioned, it was shown only recently that any compact locally connected metric continuum is partitionable(1). In this paper it is shown how a 3-sphere can be characterized in terms of its partitionings.

Definitions and notation. We designate *Euclidean n-space* by E^n. The *unit n-sphere with center at the origin* in E^{n+1} is designated by S^n. A continuum is called a *simple closed curve, simple surface* (or *2-sphere*), or *3-sphere* according as it is topologically equivalent to S^1, S^2, or S^3. While a 3-sphere may be regarded as the set of points in E^4 with coordinates (x, y, z, w) satisfying $x^2+y^2+z^2+w^2=1$, we prefer to think of it as E^3 plus a point added in such a fashion that the exterior of a cube is topologically equivalent to its interior.

A simple surface C in E^3 is called *tame* if there is a homeomorphism of E^3 into itself that carries C into S^2. If there is no such homeomorphism, C is called *wild*.

We shall suppose that *space* S is metric, compact, locally connected, and connected.

A *partitioning* of S is a collection of mutually exclusive open sets whose sum is dense in S. A sequence $G_1, G_2, \cdots$ of partitionings is a *decreasing sequence of partitionings* if G_{i+1} is a refinement of G_i and the maximum of the diameters of the elements of G_i approaches 0 as i increases without limit. A partitioning is *regular* if each of its elements is the interior of the closure of this element.

A regular partitioning G is a *brick* partitioning if each element of G is uniformly locally connected and the interior of the closure of the sum of each pair of elements of G is uniformly locally connected. We know that *S has a decreasing sequence of brick partitionings*(2).

The *boundary* of a set A will be denoted by $F(A)$.

Characterization of a 3-sphere. A 3-sphere may be partitioned in many ways. The boundaries of some elements of a partitioning may be simple surfaces, the boundaries of others may be tori, and those of some may not even

---

Presented to the Society, December 27, 1949; received by the editors, January 31, 1950.

(1) R. H. Bing, *Partitioning a set*, Bull. Amer. Math. Soc. vol. 55 (1949) pp. 1101–1110, and E. E. Moise, *Grille decomposition and convexification theorems for compact continua*, Bull. Amer. Math. Soc. vol. 55 (1949) pp. 1111–1121.

(2) R. H. Bing, *Complements of continuous curves*, Fund. Math. vol. 36 (1950) pp. 303–318.

be locally simply connected. However, a 3-sphere has a partitioning in which all the elements of the partitioning have simple surfaces for boundaries. The following theorem gives a characterization of a 3-sphere in terms of the manner in which it may be partitioned.

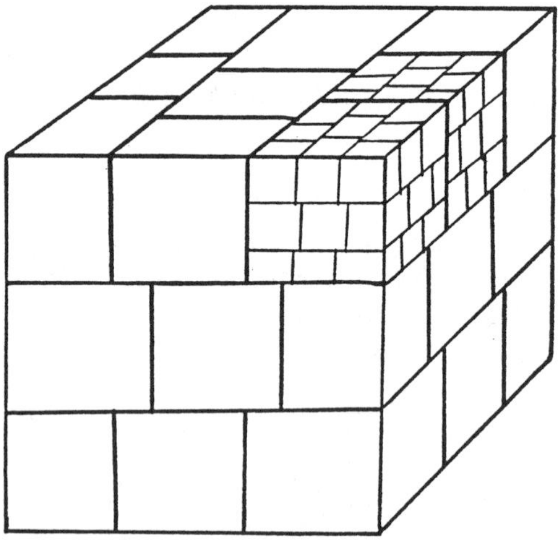

FIG. 1

THEOREM 1. *A necessary and sufficient condition that S be a 3-sphere is that one of its decreasing sequences of regular partitionings $G_1, G_2, \cdots$ have the following properties:*

(1.1) *The boundary of each element of G_i is a simple surface.*

(1.2) *If the boundaries of two elements of $\sum G_i$ intersect, this intersection is a 2-cell.*

(1.3) *The intersection of the boundaries of three elements of G_i is one-dimensional at each of its points.*

(1.4) *If g is an element of G_{i-1} ($g=S$ if $i=1$) the elements G_i in g may be ordered $g_1, g_2, \cdots, g_n$ so that $F(g_j)$ $[j=1$ (if $i>1$), $2, \cdots, n]$ intersects $F(g)+F(g_1)+\cdots+F(g_{j-1})$ in a connected set.*

To see that the condition is necessary, one may consider the ways in which a cube may be filled with oriented rectangular solids all the same height. See Figure 1. Before showing that the condition is sufficient, we develop some other theorems to use. In Theorem 3 we show that there is a sequence of partitionings of S^3 that corresponds to $G_1, G_2, \cdots$. Using Theorem 5 we show further that there is such a sequence of partitionings of S^3 which is decreasing. The proof of the sufficiency in Theorem 1 follows Theorem 6.

The following is a combinatorial result. Although it will not be used in proving Theorem 1, it may be of interest for its own sake.

THEOREM 2. *If C is a finite complex which is a 3-manifold without boundary, then C is a 3-sphere if the 3-simplexes of C may be ordered $c_1, c_2, \cdots, c_n$ such that c_j $(j=2, \cdots, n)$ intersects $c_1+c_2+\cdots+c_{j-1}$ in a connected set.*

Proof. Consider c_i to be a regular tetrahedron whose edges are of length 1. Let c_{ij} $(j=1, 2, 3, 4)$ be a regular tetrahedron in c_i such that c_{ij} has a vertex

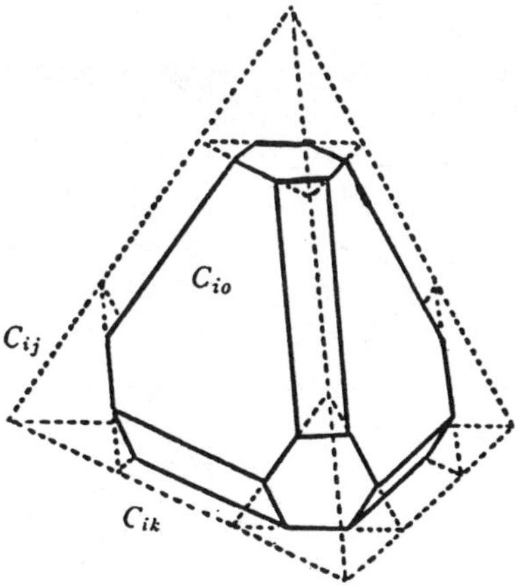

FIG. 2

in common with a vertex of c_i and the edges of c_{ij} are of length 1/3. Let c_{ik} $(k=5, 6, \cdots, 10)$ be a 5-sided polyhedron in c_i such that two faces of c_{ik} are triangles which lie in faces of different c_{ij}'s, two other faces are trapezoids of heights 1/9 which lie in faces of c_i, and the fifth face is a rectangle whose interior lies in the interior of c_i. Let c_{i0} be the 14-sided polyhedron equal to the closure of $c_i-(c_{i1}+c_{i2}+\cdots+c_{i10})$. See Figure 2.

Denote by G the regular partitioning of C where the closures of the elements of G are of three types: (1) a c_{i0}; (2) if p is a vertex of a 3-simplex in C, the sum of all c_{ij}'s containing p; and (3) if e is an edge of a 3-simplex in C, the sum of all c_{ik}'s having the middle third of e as an edge.

Each element of G is an open 3-cell whose boundary is a simple surface. To establish that elements of types 2 and 3 have these properties, one uses the hypothesis that C is a 3-manifold(3). If the boundaries of two elements of G intersect, the intersection is a 2-cell. If the boundaries of three elements of G intersect, the intersection is an arc. Furthermore, the elements of G may be ordered $g_1, g_2, \cdots, g_m$ so that $F(g_j)$ intersects $F(g_1)+F(g_2)+\cdots+F(g_{j-1})$ in a connected set. To get such an ordering, suppose that g' precedes g'' if the first element of $c_{1\,0}, c_{1\,1}, \cdots, c_{1\,10}, c_{2\,0}, \cdots, c_{n\,10}$ which is in $\bar{g}'+\bar{g}''$ is in $\bar{g}'$. Hence it follows from Theorem 3 that C is a 3-sphere.

THEOREM 3. *Suppose T is a homeomorphism carrying a simple surface F_0*

(3) H. Seifert and W. Threlfall, *Lehrbuch der Topologie*, New York, 1947, Theorem 1, p. 205.

in S into a tame simple surface C in E^3, *and G is a regular partitioning of S satisfying the following conditions*:

(3.1) *The boundary of each element of G is a simple surface.*

(3.2) *If the boundaries of two elements of G intersect, their intersection is the sum of a finite number of mutually separated 2-cells.*

(3.3) *The intersection of the boundaries of three elements of G is one-dimensional at each of its points.*

(3.4) F_0 *is the boundary of the closure of the sum of an ordered subcollection* $[g_1, g_2, \cdots, g_n]$ *of G such that* $F(g_j)$ $(j=1, 2, \cdots, n)$ *intersects* $F_0+F(g_1)+\cdots+F(g_{j-1})$ *in a connected set* .

Then there is a partitioning $[h_0, h_1, \cdots, h_n]$ *of* E^3 *and a homeomorphism* T' *of the boundary (in S) of* $g_1+g_2+\cdots+g_n$ *into the boundary (in* E^3*) of* $h_1+h_2+\cdots+h_n$ *such that* h_0 *is the exterior of C,* $F(h_i)$ *is a tame simple surface,* $T=T'$ *on* F_0, *and* $T'[F(g_i)]=F(h_i)$.

Proof. We shall use induction on n. If $n=1$, the theorem is true because we can let h_1 be the interior of C and T' be T.

If p is an interior point of a 2-cell K (interior with respect to K) which lies in the intersection of the boundaries of two elements g_i, g_j of G, p is not on the boundary of any other element g_r of G or else $F(g_i)\cdot F(g_r)$ would contain a 2-cell intersecting p. But this is impossible since both K and $F(g_i)\cdot F(g_r)$ lie on the simple surface $F(g_i)$, and $F(g_r)\cdot K$ is prevented by condition (3.3) from being two-dimensional. Hence, the common part of the boundaries of three elements of G is a subset of the boundaries of the sum of the 2-cells which is the common part of some two of them. Hence, condition (3.3) may be replaced by the following condition.

(3.3)′ *if the boundaries of three elements of G intersect, their intersection is the sum of a finite number of arcs.*

We shall show that $\bar{g}_1+\bar{g}_2+\cdots+\bar{g}_n-F_0$ is connected. If it were the sum of two mutually separated sets H and K, F_0 would be the boundary of $H+K$ and each of the sets $\overline{H}, \overline{K}$ would intersect F_0. Then $\overline{H}\cdot\overline{K}$ would contain a point p of F_0. There would be elements g_i and g_j of $[g_1, g_2, \cdots, g_n]$ in H and K respectively such that $\bar{g}_i\cdot\bar{g}_j$ contains p. It follows from (3.3) that $F_0\cdot F(g_i)\cdot F(g_j)$ is one-dimensional and from (3.2) that $F(g_i)\cdot F(g_j)$ contains a 2-cell. Hence H and K could not be mutually separated because the points of $\bar{g}_i\cdot\bar{g}_j$ not in F_0 would be limit points of H and K.

We now show that if $n\geqq 2$, then $F_0\cdot F(g_1)$ is topologically equivalent to a bounded subset of the plane whose boundary is the sum of a finite number of mutually exclusive circles. Since $F(g_1)$ is a simple surface which contains a point not of F_0, $F_0\cdot F(g_1)$ is a proper subset of F_0. From condition (3.4) we find that $F_0\cdot F(g_1)$ is connected. Furthermore, it follows from condition (3.2) that $F_0\cdot F(g_1)$ is the sum of a finite number of 2-cells each of which is a component of the intersection of $F(g_1)$ and the boundary of an element of G not in $[g_1, g_2, \cdots, g_n]$. As a result of condition (3.3′), if two of these 2-cells intersect, their common part is the sum of a finite number of arcs. Hence, $F_0 \cdot F(g_1)$ is topologically equivalent to a bounded subset of the plane whose boundary is the sum of a finite collection of mutually exclusive circles.

Let $J_1, J_2, \cdots, J_t$ be the components of the boundary with respect to F_0 of $F_0 \cdot F(g_1)$, I_i $(i=1, 2, \cdots, t)$ be the component of $F_0 - J_i$ containing no point of $F(g_1)$, and K_i be the component of $F(g_1) - J_i$ containing no point of F_0. The boundary of $\bar{g}_2 + \bar{g}_3 + \cdots + \bar{g}_n$ is $\sum(I_i + J_i + K_i)$ as can be shown from conditions (3.1), (3.2), and (3.3). However, no component of $\bar{g}_2 + \bar{g}_3 + \cdots + \bar{g}_n$ intersects two elements of $\{I_i + J_i + K_i\}$. If this were not true there would be an integer j such that $\bar{g}_2 + \bar{g}_3 + \cdots + \bar{g}_j$ contains a continuum intersecting two elements of $\{I_i + J_i + K_i\}$ but $\bar{g}_2 + \bar{g}_3 + \cdots + \bar{g}_{j-1}$ contains no such continuum. However, $F(g_j)$ would not intersect $F_0 + F(g_1) + \cdots + F(g_{j-1})$ in a connected set as required by condition (3.4). Hence, each component of $\bar{g}_2 + \bar{g}_3 + \cdots + \bar{g}_n$ has one of the simple surfaces of $\{I_i + J_i + K_i\}$ for a boundary.

If H_i $(i=1, 2)$ is a collection of t mutually exclusive simple closed curves on a simple surface such that no element of H_i separates any other two elements of H_i from each other on the surface, there is a homeomorphism of the surface into itself that carries the elements of H_1 into the elements of H_2. Also, for each homeomorphism T'' of the surface of a cube in E^3 into itself, there is a homeomorphism of E^3 into itself that preserves T'' on the surface of the cube. Hence, if T is the transformation mentioned in the statement of Theorem 3, without loss of generality we may suppose that C is the surface of a cube and the $T(J_1), T(J_2), \cdots, T(J_t)$ are circles $C_1, C_2, \cdots, C_t$ all lying in one base of C. Let W_i $(i=1, 2, \cdots, t)$ be the hemispherical surface which contains C_i and lies in C plus its interior. There is a homeomorphism T_i of K_i into W_i such that if p is a point of J_i, $T(p) = T_i(p)$. Suppose $T'(p) = T(p)$ if p is a point of F_0 and $T'(p) = T_i(p)$ if p is a point of K_i. Then T' is a homeomorphism carrying $F_0 + F(g_1)$ into $C + \sum W_i$.

Each bounded complementary domain of $C + \sum W_i$ is a 3-cell and the boundary of each of these domains is tame and the image under T' of either $F(g_1)$ or the boundary of a component of $\bar{g}_2 + \bar{g}_3 + \cdots + \bar{g}_n$. By induction on n, the complementary domains of $C + \sum W_i$ can be partitioned and the homeomorphism T' extended on the boundaries of these partitionings so that a partitioning $[h_0, h_1, \cdots, h_n]$ and a homeomorphism T' are obtained which satisfy the conditions of the theorem.

THEOREM 4. *Suppose C is a tame simple surface in E^3 and $G_1, G_2, \cdots$ is a sequence of partitionings of E^3 satisfying the following conditions:*

(4.1) *The boundary of each element of G_i is a tame simple surface.*

(4.2) *If g is an element of G_i whose closure intersects C, $F(g) \cdot C$ is connected and does not separate $F(g)$.*

(4.3) *Each G_{i+1} is a refinement of G_i.*

(4.4) *One element of* G_i *is the exterior of* C.

(4.5) *For each positive number* ϵ *there is a positive integer* $n(\epsilon)$ *such that the closure of no element of* G_n *other than the exterior of* C *contains two points which belong to the boundary of the sum of the elements of* G_1 *and are farther apart than* ϵ.

Let K *be a closed set and* R *be a closed proper subset of* C *such that if* g *is an element of* G_1 *other than the exterior of* C, *then* $F(g)\cdot C$ *does not intersect both* R *and* K. *Then there are an integer* m *and a homomeomorphism* T *of* E^3 *into itself satisfying the following conditions*:

(4.6) *Each point of the boundary of an element of* G_1 *is invariant under* T.

(4.7) *If* g *is an element of* G_m *other than the exterior of* C, $T(\bar{g})$ *does not intersect both* R *and* K.

Proof. Let $g_1, g_2, \cdots, g_j$ be the elements of G_1 that lie on the interior of C and whose boundaries intersect R. Let J_i be a simple closed curve on $F(g_i)$ such that one complementary domain I_i of J_i in $F(g_i)$ contains $F(g_i)\cdot R$ but I_i+J_i does not contain a point of K. It follows from condition (4.5) that there is an integer m such that if g is an element of G_m other than the exterior of C, the intersection of $\bar{g}$ with the boundary of the sum of the elements of G_1 is a subset of $I_1+I_2+\cdots+I_j$ if it intersects R.

There is a homeomorphism T_i carrying $\bar{g}_i$ into itself that leaves each point of $F(g_i)$ invariant and such that if g is an element of G_m in g_i and $\bar{g}$ intersects R, then $T_i(\bar{g})$ does not intersect K. (To see that this is so, think of $\bar{g}_i$ as a cube plus its interior whose base is I_i+J_i. Then T_i is a homeomorphism that moves each point of g_i toward the base.) Then the required transformation T satisfying conditions (4.6) and (4.7) is one that is the identity except on $g_1+g_2+\cdots+g_j$ and is T_i on g_i.

Theorem 4 could be improved by replacing G_i by G_1 in conditions (4.1), (4.2), and (4.3) but the stronger conditions as stated are needed in Theorem 5.

THEOREM 5. *Suppose that* C *is a tame simple surface in* E^3 *and* $G_1, G_2, \cdots$ *is a sequence of partitionings of* E^3 *satisfying conditions* (4.1), (4.2), (4.3), *and* (4.4) *as well as the following*:

(5.1) *For each positive integer* j *and each positive number* ϵ *there is a positive integer* $n(j, \epsilon)$ *such that the closure of no element of* G_n *other than the exterior of* C *contains two points of the boundary of the sum of the elements of* G_j *which are farther apart than* ϵ.

Then for each positive number δ, *there is an integer* m *and a homeomorphism* T *of* E^3 *into itself such that* T *leaves each point of the exterior of* C *invariant and* T *carries each other element of* G_m *into a set of diameter less than* δ.

We could not substitute for condition (4.2) one like condition (3.2) of Theorem 3.

If we tried to make such a substitution, we could not take care of the case where two mutually exclusive 2-cells D_1 and D_2 on C are "hooked together" by two elements of G_1, by four elements of G_2, by eight elements of $G_3, \cdots$ as illustrated in Figure 3. For the sake of clarity, only the elements of G_1 which "hook" D_1 and D_2 together are shown in the diagram but if g_1 and g_2 are these two elements, one may see how the other elements of G_1 can be placed by considering a homeomorphism carrying C plus its interior, $\bar{g}_1$, and $\bar{g}_2$ into the oriented rectangular solids with opposite vertices at (0, 0, 0) and (1, 1, 1), (2/5, 0, 1/5) and (3/5, 1, 2/5), and (2/5, 0, 3/5) and (3/5 ,1, 4/5) respectively.

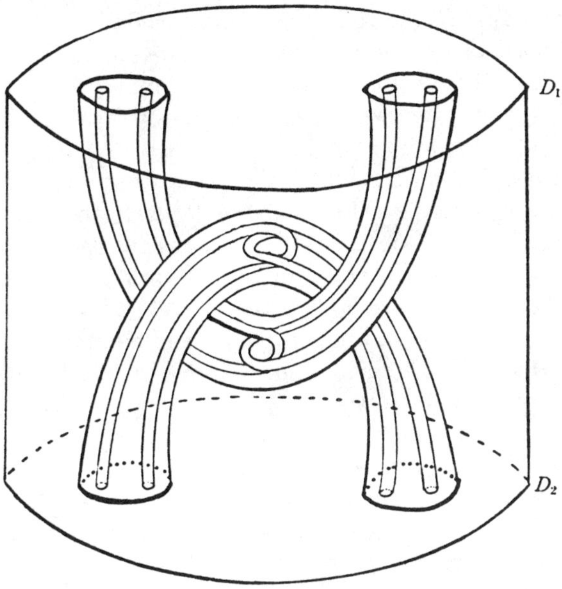

FIG. 3

Proof of Theorem 5. For convenience we suppose that C is the surface of a cube oriented with the x, y, and z axes. (If it is not one, there is a homeomorphism T_0 of E^3 into itself that carries it into one. If the elements of G_i are replaced by their images, conditions (4.1), (4.2), (4.3), (4.4), and (5.1) still hold. Also, there is a positive number δ' such that if p and q are two points of $T_0(C+\text{its interior})$ whose distance apart is less than δ', then $T_0^{-1}(p)$ and $T_0^{-1}(q)$ are no farther apart than δ.)

Let n be an integer so large that if g is an element of G_n other than the exterior of C, $F(g)$ does not intersect C in two points that are farther apart than $\delta/12$. Let S_1 be a single-valued surface defined above the base of C such that each point of $S_1 \cdot C$ is $\delta/4$ units above the base of C.

We find from Theorem 4 that there is an integer n_2 and a homeomorphism T_2 of E^3 into itself such that (1) T_2 leaves invariant each point of the boundary of the sum of the elements of G_n and (2) if g is a bounded element of G_{n_2}, $T_2(\bar{g})$ does not intersect both S_1 and a point of C whose distance is as much as $2\delta/4$ from the base of C. In this application of Theorem 4 we let K and R in the statement of Theorem 4 be S_1 and the points of C which are no nearer the base of C than $2\delta/4$. There is a single-valued surface S_2 above the base of C and S_1 such that (1) each point of $S_2 \cdot C$ is $2\delta/4$ units above the base of C and (2) S_2 does not intersect the image under T_2 of an element of G_{n_2} that intersects S_1. To ensure that (2) will be satisfied, we need only take S_2 sufficiently close to the portion of C that lies above S_2.

In a second application of Theorem 4 we let C, G_1, K, and R in the statement of Theorem 4 be C, the collection of images under T_2 of the elements of G_{n_2}, S_2, and the set of points of C which are not nearer the base of C than $3\delta/4$. We find that there is an integer n_3 and a homeomorphism T_3 of E^3 into itself such that (1) T_3 preserves T_2 on the sum of the boundaries of the elements of G_{n_2}, and (2) if g is a bounded element of G_{n_3}, $T_3(\bar{g})$ does not intersect both S_2 and a point of C whose distance is as much as $3\delta/4$ from the base of C. There is a single-valued surface S_3 between S_2 and the top of C such that (1) each point of $S_3 \cdot C$ is $3\delta/4$ units above the base of C and (2) the image under T_3 of no element of G_{n_3} intersects both S_2 and S_3.

Similarly there are mutually exclusive surfaces $S_3, S_4, \cdots, S_j$ ($j\delta/4 <$ height of $C \leqq (j+1)\delta/4$), an integer $n_j = n_z$, and a homeomorphism T_j of E^3 into itself such that (1) S_i is single-valued above the base of C, (2) each point of $S_i \cdot C$ is $i\delta/4$ units above the base of C, (3) T_j leaves invariant each boundary point of each element of G_n, and (4) no image under T_j of an element of G_{n_z} intersects two of the surfaces $S_1, S_2, \cdots, S_j$.

There is a homeomorphism T' of E^3 into itself which leaves each point of C invariant, which carries $S_1, S_2, \cdots, S_j$ into surfaces parallel to the base of C and such that the image of a point under T' has the same x and y values as the point has. Then the transformation $T'T_j = T_z$ is one which (1) leaves invariant each point of C, (2) does not move any point of the boundary of the sum of the elements of G_n in the x or y directions, and (3) if p and q are two points of a bounded element of G_{n_z}, the z coordinates of $T_z(p)$ and $T_z(q)$ do not differ by more than $\delta/2$.

Similarly, we find that there is a homeomorphism T_y of E^3 into itself and an integer n_y greater than n_z such that (1) T_yT_z leaves invariant each point of C plus its exterior, (2) if p is a point of the boundary of an element of G_{n_z}, $T_yT_z(p)$ has the same x and z coordinates as $T_z(p)$, (3) if p and q are two points of the same bounded element of G_{n_y} the y coordinates of $T_yT_z(p)$ and $T_yT_z(q)$ do not differ by more than $\delta/2$. Also, there is an integer n_x greater than n_y and a homeomorphism T_x of E^3 into itself such that (1) $T_xT_yT_z$ leaves invariant each point of C plus its interior, (2) if p is a point of the boundary of an element of G_{n_y}, $T_xT_yT_z(p)$ has the same y and z coordinates as $T_yT_z(p)$, and (3) if p and q are two points of the same bounded element of G_{n_x}, then the x coordinates of $T_xT_yT_z(p)$ and $T_xT_yT_z(q)$ do not differ by more than $\delta/2$.

If p and q are two points of a bounded element of G_{n_x}, the y coordinates of $T_xT_yT_z(p)$ and $T_xT_yT_z(q)$ do not differ by more than $\delta/2$ because p and q belong to an element of G_{n_y}, the y coordinates of $T_yT_z(p)$ and $T_yT_z(q)$ do not differ by as much as $\delta/2$, and T_x does not alter the length in the y direction of an element of G_{n_y}. Furthermore, we find that if p and q are two points of a bounded element of G_{n_x}, the z coordinates of $T_xT_yT_z(p)$ and $T_xT_yT_z(q)$ do not differ by more than $\delta/2$. Hence, the image under $T_xT_yT_z$ of each bounded element of G_{n_x} has a diameter of less than δ. Then n_x is the integer and $T_xT_yT_z$ the transformation required in Theorem 5.

THEOREM 6. *The compact continua M_1 and M_2 are homeomorphic if there is a decreasing sequence of partitionings $G_{i1}, G_{i2}, \cdots$ $(i=1, 2)$ for M_i and a* 1-1 *correspondence between the elements of G_{1j} and G_{2j} such that* (1) *two elements of G_{1j} have a boundary point in common if and only if the corresponding elements of G_{2j} have a boundary point in common and* (2) *corresponding elements of G_{1j+1} and G_{2j+1} are subsets of corresponding elements of G_{1j} and G_{2j}.*

Proof of sufficiency in Theorem 1. Suppose that the ordering of G_1 mentioned in condition (1.4) is $g_1, g_2, \cdots, g_n$. We first show that $\bar{g}_1$ is topologically equivalent to a 3-cell.

Let C be the surface of a cube in E^3 and T_1 be a homeomorphism carrying $F(g_1)$ into C. We find from Theorem 3 that there is a sequence of partitionings $H_1, H_2, \cdots$ of E^3 and a sequence of homeomorphisms $T_1, T_2, \cdots$ satisfying the following:

(1.5) The boundary of each element of H_i is a tame simple surface.

(1.6) H_{i+1} is a refinement of H_i.

(1.7) One element of H_i is the exterior of C.

(1.8) There is a 1-1 correspondence between the bounded elements of H_i and the elements of G_i in g_1 such that

(a) two bounded elements of H_i are adjacent if and only if the corresponding element of G_i in g_1 are adjacent and

(b) corresponding elements of H_{i+1} and G_{i+1} are subsets of corresponding elements of H_i and G_i.

(1.9) T_i is a homeomorphism of the boundary of the sum of the elements of G_i in g_1 into the boundary of the sum of the elements of H_i such that T_i preserves T_{i-1} on the boundary of the sum of the elements of G_{i-1} in g_1 and T_i carries the boundary of an element of G_i in g_1 into the boundary of the corresponding element of H_i.

Although the diameters of the bounded elements of H_i may not approach 0 as i increases without limit, $H_1, H_2, \cdots$ will satisfy a condition like (5.1) because $G_1, G_2, \cdots$ is a decreasing sequence of partitionings and T_{i+n} preserves T_i on the sum of the boundaries of the elements of G_i in $\bar{g}_1$.

We find from Theorem 5 that there is an integer n_1 and a homeomorphism F_1 of E^3 into itself such that F_1 leaves each point of the exterior of C invariant and F_1 carries each bounded element of H_{n_1} into a set of diameter less than 1/2. We let K_1 be the partitioning of E^3 whose elements are the images of the elements of H_{n_1} under F_1.

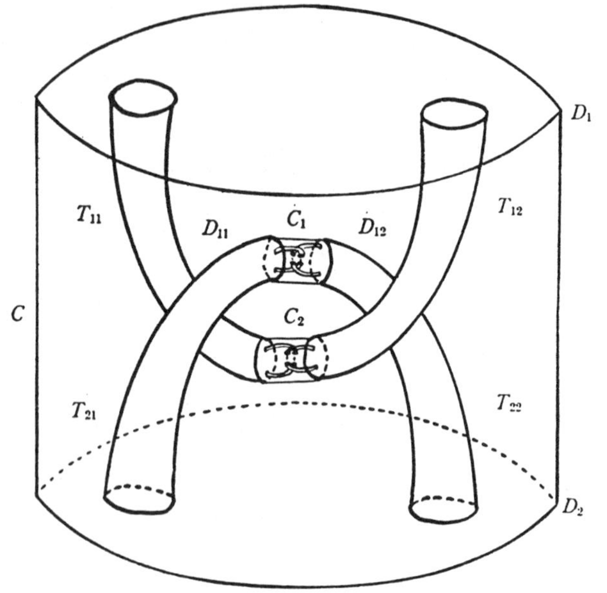

FIG. 4

An application of Theorem 5 to the elements of K_1 reveals that there is an integer n_2 and a homeomorphism F_2 of E^3 into itself such that (1) if p is a boundary point of an element of K_1, p is invariant under F_2, and (2) if h is a bounded element of H_{n_2}, the diameter of the image of h under F_2F_1 is less than 1/4. We denote the collection of images of elements of H_{n_2} under F_2F_1 by K_2. If T_{n_1} is the transformation mentioned in (1.9), we note that $F_1T_{n_1}$ is a homeomorphism of the boundary of the sum of the elements of G_{n_1} in g_1 into the boundary of the sum of the elements of K_1 and that $F_2F_1T_{n_2}$ is a homeomorphism of the boundary of the sum of the elements of G_{n_2} in g_1 into the boundary of the sum of the elements of K_2. Furthermore, $F_2F_1T_{n_2}$ preserves $F_1T_{n_1}$ on these boundaries.

Similarly, we find that there are integers $n_3, n_4, \cdots$ and partitionings $K_3, K_4, \cdots$ of E^3 satisfying conditions like conditions (1.5), (1.6), (1.7), and (1.8) satisfied by $H_3, H_4, \cdots$ and such that each bounded element of K_i is of diameter less than $1/2^i$.

We find from Theorem 6 that these conditions imply that $\bar{g}_1$ and C plus its interior are homeomorphic. A similar line of argument shows that $\bar{g}_2 + \bar{g}_3 + \cdots + \bar{g}_n$ is also homeomorphic to C plus its interior. Since each of these correspondences relates $F(g_1)$ to C, S is topologicallyequivalent to S^3.

Possible strengthening of Theorem 1. It may be noted that in neither Theorem 3 nor in Theorem 5 did we use the full strength of condition (1.2). Hence, at the expense of conciseness, a strengthening of the sufficiency part of Theorem 1 could be made here. For purposes of application in Theorem 1, condition (3.2) deals with $F(g')\cdot F(g'')$ where g', g'' are elements of G_{i+1} in the same element of G_i; condition (4.2) deals with $F(g')\cdot F(g'')$ where g', g'' are elements of different G_i's and g' contains g''.

Consider the right cylinder C with bases D_1 and D_2. The interiors of two mutually exclusive discs in D_i $(i=1, 2)$ are replaced by the surfaces of tubes T_{i1} and T_{i2} and discs D_{i1} and D_{i2} as shown in Figure 4 where D_{i1} and D_{i2} are the bases of a right circular cylinder C_i and $D_1+T_{11}+C_1+T_{12}$ is "hooked" to $D_2+T_{21}+C_2+T_{12}$ as shown. Discs in the bases of the cylinders C_1 and C_2 are replaced by surfaces of tubes and discs as before. If the process is continued and the limit taken, a wild simple surface M is obtained. If g is an element of G_1 and condition (1.2) is replaced by one like (3.2), M might be the image of $F(g)$ in a homeomorphism of S into S^3 if S is topologically equivalent to S^3 and the elements of G_2, G_3, $\cdots$ in g are fitted together as suggested after Theorem 4 in Figure 3. In fact, if M_1 and M_2 are two Alexander "horned spheres"(4) in S^3, T is a homeomorphism of M_1 into M_2, and U_i $(i=1, 2)$ is one of the complementary domains of M_i, then S may be topologically equivalent to $U_1+M_1+U_2$ where it is understood that a point p of M_1 is a limit point of a subset V of U_2 if and only if $T(p)$ is a limit point of V(5).

Condition (1.3) prevents S from being the sum of three 3-cells such that the common part of any two is the boundary of each. However, Theorem 1 would be somewhat stronger if the expression "1-dimensional" could be replaced by "0-dimensional or 1-dimensional."

Condition (1.4) rules out many 3-manifolds (as well as some spaces that are not 3-manifolds). It imposes a type of sequential unicoherence. However, unicoherence alone does not appear to be strong enough because projective 3-space has a decreasing sequence of regular partitionings satisfying conditions (1.1), (1.2), and (1.3). One might wonder if simple connectedness might be substituted for condition (1.4).

Since E^3 is topologically equivalent to S^3 minus a point, the following is immediately obtained from Theorem 1.

THEOREM 7. *A metric space S' is topologically equivalent to E^3 provided it has a sequence of regular partitionings G_1, G_2, $\cdots$ satisfying conditions* (1.1), (1.2), (1.3), *and* (1.4) *as well as the following: G_{i+1} is a refinement of G_i; exactly one element of G_i fails to have a compact closure; for each point p there is an*

(4) J. W. Alexander, *An example of a simply connected surface bounding a region which is not simply connected*, Proc. Nat. Acad. Sci. U.S.A. vol. 10 (1924) pp. 8–10.

(5) R. L. Wilder, *Topology of manifolds*, Amer. Math. Soc. Colloquium Publications, vol. 32, New York, 1949, Problem 4.6, p. 383.

integer $n(p)$ such that p does not belong to the closure of this exceptional element of $G_{n(p)}$; the diameters of the elements of G_i other than this exceptional element approach 0 *as i increases without limit.*

Characterization of a simple closed curve. Since a 1-manifold without boundary is a simple closed curve, a simple closed curve is characterized by its local properties. Hence the following characterization of a simple closed curve is shorter than the corresponding characterizations of a simple surface and a 3-sphere.

THEOREM 8. *A necessary and sufficient condition that S be a simple closed curve is that one of its decreasing sequences of regular partitionings $G_1, G_2, \cdots$ have the following properties:*

(8.1) *The boundary of each element of G_i is a pair of distinct points.*

(8.2) *No three elements of G_i have a boundary point in common.*

In fact, each decreasing sequence of partitionings $H_1, H_2, \cdots$ of a simple closed curve has properties (8.1) and (8.2) if H_1 has more than one element. Condition (8.2) could be replaced by the condition that the partitionings are brick partitionings.

Characterization of a simple surface. When a simple closed curve is partitioned into more than two elements, the boundary of each element of the partitioning is a pair of points. However, when a simple surface is partitioned there is a bigger variety of possible boundaries. They may either be simple closed curves or the sum of a finite number of simple closed curves.

THEOREM 9. *A necessary and sufficient condition that S be a simple surface is that one of its decreasing sequences of regular partitionings $G_1, G_2, \cdots$ have the following properties:*

(9.1) *The boundary of each element of G_i is a simple closed curve.*

(9.2) *The intersection of the boundaries of 3 elements of G_i contains no arc.*

(9.3) *If g is an element of G_{i-1} ($g=S$ if $i=1$) the elements of G_i in g may be ordered $g_1, g_2, \cdots, g_n$ so that $F(g_j)$ [$j=1$ (if $i>1$), $2, \cdots, n$] intersects $F(g)+G(g_1)+ \cdots +F(g_{j-1})$ in a nondegenerate connected set.*

Indication of proof. Conditions (9.1), (9.2), and (9.3) correspond to conditions (1.1), (1.3), and (1.4) of Theorem 1. The word "nondegenerate" in condition (9.3) replaces the equivalent of condition (1.2) of Theorem 1. This condition cannot be deleted entirely or else S could be the sum of two simple surfaces with only one point in common.

To prove Theorem 9, prove a theorem resembling Theorem 3 showing that there are a sequence of partitionings $H_1, H_2, \cdots$ of S^2 such that H_{i+1} refines H_i and a sequence of homeomorphisms $T_1, T_2, \cdots$ such that T_i carries the boundary of the sum of the elements of G_i into the boundary of the sum of the elements of H_i and T_{i+1} preserves T_i. Next prove a result like Theorem 5 showing that there is no loss of generality in supposing that the maximum of the diameters of the elements of H_i approaches 0 as i increases without limit. Finally, use Theorem 6 to establish a homeomorphism between S and S^2.

The following result throws some light on the role of condition (9.3). It follows from the fact that for continua H, K in S^2, $S^2-(H+K)$ is not connected if $H\cdot K$ is not connected.

THEOREM 10. *Suppose* G_1 *and* G_2 *are two partitionings of* S^2, G_2 *refines* G_1, *the boundaries of the elements of* G_1 *and* G_2 *are connected, and* g *is either* S^2 *or an element of* G_1 *whose boundary is nondegenerate. Then the elements of* G_2 *in* g *may be ordered* $g_1, g_2, \cdots, g_n$ *such that* $F(g_j)$ *intersects* $F(g)+F(g_1)+\cdots+F(g_{j-1})$ *in a nondegenerate connected set.*

It would be interesting to know what the corresponding result is for S^3. In particular, if $G_1, G_2, \cdots$ are partitionings of S^3 satisfying conditions (1.1), (1.2), and (1.3), would they automatically satisfy condition (1.4)?

UNIVERSITY OF VIRGINIA,
CHARLOTTESVILLE, VA.
UNIVERSITY OF WISCONSIN,
MADISON, WIS.

ANNALS OF MATHEMATICS
Vol. 68, No. 1, July, 1958
Printed in Japan

NECESSARY AND SUFFICIENT CONDITIONS A 3-MANIFOLD BE S^3

BY R. H. BING*

(Received January 31, 1957)

1. Introduction

A 3-*manifold* is a metric space each point of which has a neighborhood topologically like E^3 (euclidean 3-space). Examples of 3-manifolds are E^3, projective 3-space, an open subset of E^3, the surface of a sphere in E^4 (euclidean 4-space), and spaces obtained by taking the cartesian product of a circle or line with a 2-manifold (such as a projective plane, a 2-sphere, a 2-sphere with handles and crosscuts). Hence, we regard a 3-manifold as a concrete object rather than an abstraction of combinatorially equivalent systems of symbols.

The surface of a sphere S^3 in E^4 is topologically equivalent to the space obtained by adding a point at infinity to E^3, and deciding that the sum of this ideal point and the exterior of an ordinary sphere is a neighborhood of this ideal point. It is frequently convenient to think of S^3 as the sum of two topological cubes sewed together along their boundaries. We note that S^3 is a compact, connected, simply connected 3-manifold.

A simple closed curve J in a set X *can be shrunk to a point* in X if there is a map f of a disk E into X such that f takes Bd E homeomorphically onto J. If we regard E as $\{(x, y)/x^2 + y^2 \leqq 1\}$, one may consider J as having been shrunk to $f(\{(x, y)/x^2 + y^2 = 1 - t\})$ at time $t(0 \leqq t \leqq 1)$. Also, if f is a map of a circle into X, we say that f *can be shrunk to a point* if the map f can be extended to take a disk bounded by the circle into X. A set is said to be *simply connected* if each map of a circle into it can be shrunk to a point in it—the set is *simply connected in the weak sense* if each simple closed curve in it can be shrunk to a point in it.

The properties of being simply connected, and being simply connected in the weak sense, are equivalent on a 3-manifold. Since we shall be dealing primarily with 3-manifolds in this paper, we shall mainly use this weak form of simple connectedness. Relations between the two types of simple connectedness will be considered in another paper [5].

A compact, connected 2-manifold is topologically S^2 (surface of a sphere in E^3), if each simple closed curve in it can be shrunk to a point in it.

* Work on this paper was supported by Contract NSF-G479

The *Poincaré conjecture* is that a compact, connected 3-manifold M is topologically S^3, if it is simply connected. The conjecture has not been proved, and I suspect that perhaps being simply connected is not enough alone to insure that M is topologically S^3. In Theorem 1, we give a condition that is strong enough. The example used in Section 2 suggests a reason for imposing the condition.

Poincaré announced in [10] that two compact combinatorial n-manifolds were topologically equivalent if they had the same homological characters, but disproved this in [11] by constructing a three dimensional manifold which was not simply connected, even though all its 1 and 2 cycles bounded. J.H.C. Whitehead gave a faulty proof of the Poincaré conjecture in [13], but pointed out his error shortly thereafter in [14].

THEOREM 1. *A compact, connected 3-manifold M is topologically S^3 if each simple closed curve in M lies in a topological cube in M.*

The proof of Theorem 1 follows from direct applications of Lemmas 4, 5, 7 developed in Sections 4, 5.

Theorem 2 of Section 7 uses the condition of simple closed curves lying in cubes to determine whether an open subset of E^3 is topologically E^3. Other characterizations of S^3 are given by Theorems 3 and 4 of Section 7.

With a slight variation, the property mentioned in the hypothesis of Theorem 1 was used by Fox and Artin. In fact, they suggested [6, p. 987] that, it seemed to them, a decisive point for settling the Poincaré conjecture would be the existence or non-existence of a compact simply connected 3-manifold M with a simple closed curve W, such that no homeomorphism of M onto itself can transform W into a sufficiently small neighborhood of any point. Theorem 1 supports their conjecture because if, for each simple closed curve W in M, there is a homeomorphism of M onto itself taking M into a sufficiently small neighborhood, the hypothesis of Theorem 1 is satisfied. An example is given at the end of Section 4 to show that a topological cube in a 3-manifold may contain a simple closed curve J, for which there is no homeomorphism of the 3-manifold onto itself taking J into a small neighborhood. Hence, the property of Theorem 1 is essentially, though not identically, that used by Fox and Artin.

2. A simply connected open subset of E^3

Examples are known [6, 9, 15] of bounded open subsets of E^3 that are topologically different from E^3 even though they are connected, simply connected, and have connected boundaries. We shall examine such an open set U and note a property wherein it differs from E^3. The example is equivalent to the one given in [9, 15] but the property we note is the one considered in the statement of Theorem 1, - namely, that some simple closed curve fails to lie in a topological cube.

Consider the torus T imbedded in the torus T' in E^3 as shown in Figure 1. It may be noted that T can be untwisted and taken into T'—more precisely, there is a homeomorphism h of E^3 onto itself that takes T onto T', and such that h is the identity exterior to some sphere. Then $h(T') = h^2(T)$ is a torus whose interior contains T'. The open set U we shall examine is the sum of the tori T, $h(T)$, $h^2(T)$, $h^3(T)$, $\cdots$. Being an open subset of E^3, U is a 3-manifold. Each compact set in U lies in some $h^n(T)$.

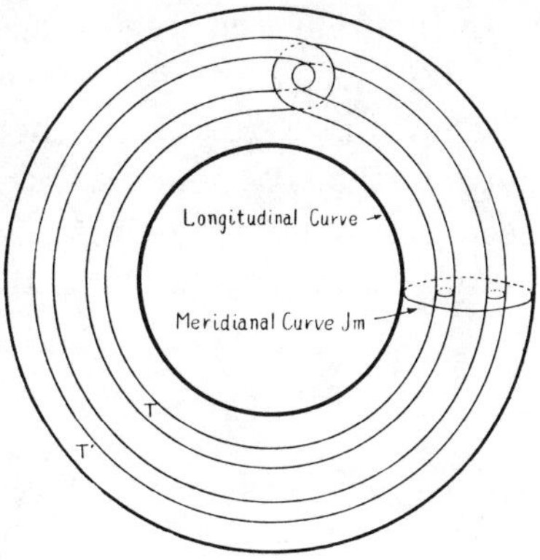

Fig. 1.

Each simple closed curve in T can be shrunk to a point in T'. The open set U is simply connected because each simple closed curve in U lies in some $h^n(T)$, and can be shrunk to a point in $h^{n+1}(T)$.

An *n-manifold with boundary* is a metric space A, such that each point of A has a neighborhood whose closure is topologically an n-dimensional cube. We use Bd A to denote the set of all points of A that do not have neighborhoods homeomorphic with E^n, and call Bd A the *boundary of A*. For example, Bd A is a pair of points, a simple closed curve, or a 2-sphere according as A is an arc, a topological disk, or a cube. We use Int A (*interior of A*) to denote A-Bd A. In the case of a torus A, Bd A is a 2-sphere with a handle.

We show that U is topologically different from E^3 by proving that a simple closed curve on Bd T that circles T longitudinally does not lie on the interior of a topological cube in U. We do this by showing that each topological cube whose interior contains J, also contains a simple closed curve on Bd $h(T)$ that circles $h(T)$ longitudinally. It will then follow that for each positive integer n, the cube contains a simple closed curve in Bd $h^n(T)$ that circles $h^n(T)$ longitudinally. Hence the cube could not lie in U.

Let J be a simple closed curve on Bd T that circles T once longitudinally, but not at all meridianally. Suppose C is a cube whose interior contains J. We show that Int C contains a simple closed curve that circles Bd T' longitudinally once, and meridianally not at all.

Shrink C slightly to get a cube C' such that $J \subset C' \subset \text{Int } C$. Let X be a 2-manifold with boundary in Bd T' such that $C' \cdot \text{Bd } T' \subset X \subset \text{Int } C$. we show that X contains a simple closed curve that circles T' once longitudinally, and no times meridianally.

In the following five paragraphs we show that no simple closed curve in Bd $T' - X$ is isotopic in Bd T' to the meridianal simple closed curve J_m shown in Figure 1. If there were such a simple closed curve we could suppose that it bounds a disk D as shown in Figure 2 and that J intersects D in just two points p, q as shown in Figure 2. (A homeomorphism may be necessary to bring about such a convenient position).

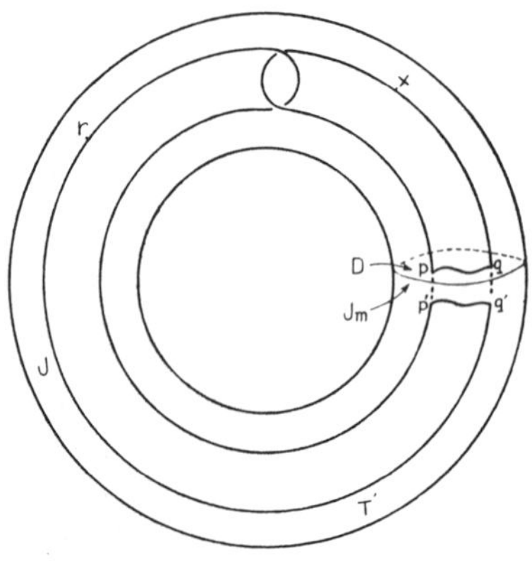

Fig. 2.

The disk D separates in C' the point x from the point y shown in Figure 2, because if not, there would be a simple closed curve K in C', such that K contains xqy and $K \cdot D = q$. But then K would link Bd D---this is impossible since $K \subset C'$ and Bd $D \subset \text{Ext } C'$. An elementary discussion of linking is found in [3].

Some component V of $D \cdot \text{Int } C$ contains both p and q. This is because C' is unicoherent. Since $D \cdot C'$ separates x from y in C', some component of it does—this component, V', contains both p and q, and $V' \subset V$.

Let pq be an arc from p to q in V. Let p', q' be points of J slightly below p, q as shown in Figure 2, and $p'q'$ be an arc slightly below pq and in C. Then the simple closed curves $pq + pxq$ and $p'q' + p'yq'$ link, and neither can be shrunk to a point in the complement of the other.

Let f be a map of a disk E into C that takes Bd E homeomorphically onto $pq + pxq$. Let R be the component of $C - D$ containing $p'q' + p'yq'$. There is a map g of E into $C + D$, such that $f = g$ on $E - f^{-1}(R)$ and $gf^{-1}(R) \subset D$. Then g shows that $pq + pxq$ can be shrunk to a point in the complement of $p'q' + p'yq'$. This contradiction arose from the false assumption that Bd $T' - X$ contains a simple closed curve that circles T' meridianally but not longitudinally.

Since Bd $T' - X$ does not contain a simple closed curve isotopic on Bd T' to J_m, X contains a simple closed curve that circles T' longitudinally. It follows from Lemma 1, that Int C contains a simple closed curve on Bd T' that circles T' longitudinally, but not meridianally.

Now E^3 has the property that each simple closed curve in it lies on the interior of a topological cube in it. Since U does not have this property, it is topologically different from E^3. We shall show in Theorem 2, that each bounded, connected, open subset of E^3 is topologically E^3, if it has a connected boundary and this property. In fact, we do not need to use the property in all its strength. Instead of supposing that all simple closed curves of the open set lie interior to topological cubes, we only suppose that polyhedral ones do. Also, we do not need to know that these polyhedral simple closed curves lie interior to topological cubes, but merely that they lie in topological cubes. To weaken the condition in this last respect makes the proof of the theorem more difficult (see Lemma 5), but the resulting theorem is stronger because the hypothesis is weaker.

LEMMA 1. *If C is a cube whose interior contains a simple closed curve J_1 on* Bd T' *that circles T' both longitudinally and meridianally, then each simple closed curve on* Bd T' *is isotopic on* Bd T' *to a simple closed curve in* Int $C \cdot$ Bd T'.

PROOF. Shrink C slightly to get a cube C' such that $J_1 \subset C' \subset$ Int C. Let X be a 2-manifold with boundary in Bd T' such that Bd $T' \cdot C' \subset X \subset$ Int C. Denote the component of X containing J_1 by Y. If each component of Bd $T' - Y$ is an open disk, each simple closed curve on Bd T' is isotopic on Bd T' to a simple closed curve on Y. We now show that if J_2 is a component of Bd Y, then J_2 bounds a disk in Bd T'.

If J_2 does not bound a disk in Bd T', it circles T' the same number of times both longitudinally and meridianally as does J_1, since $J_1 \cdot J_2 = 0$. Also, J_1 is homotopic to a point in the complement of J_2 since $J_1 \subset C'$, $C' \cdot J_2 = 0$. We finish the proof of the lemma by showing that the following is true : *If J_1, J_2 are mutually exclusive simple closed curves on* Bd T *such that each circles T' both longitudinally and meridianally* (*perhaps several times*), *then neither is homotopic to a point in the complement of the other*.

There is an isotopy of Bd T' onto itself that straightens out J_1 and J_2 so that they go smoothly around T' in a parallel fashion. We suppose that J_2 goes around T' meridianally m times and longitudinally n times. Then perspectively J_1 crosses under J_2 mn times and always in the same direction. If the fundamental group of $E^3 - J_2$ is abelianized to form the infinite cyclic group (the one-dinensional homology group) with generator a, J_1 corresponds to either the element a^{mn} or a^{-mn} in this group, so J_1 cannot be shrunk to a point in $E^3 - J_2$.

3. Punctured cubes

Suppose $C, C_1, C_2, \cdots, C_n$ is a finite collection of cubes in E^3 such that

$C_i \cdot C_j = 0$ if $i \neq j$ and $C_i \subset \text{Int } C$. Then $C - \Sigma \text{ Int } C_i$ is a cube with holes in it, and any set homeomorphic to $C - \Sigma \text{ Int } C_i$ is called a punctured cube. We shall make use of punctured cubes in showing that certain manifolds are topologically S^3 and that certain others are topologically E^3.

The boundary of a punctured cube is the sum of a finite number of topological 2-spheres, and two such punctured cubes are homeomorphic if their boundaries have the same number of components.

If B_1, B_2 are two components of the boundary of a punctured cube H, there is a homeomorphism of H onto itself that takes B_1 onto B_2. In fact, if A_1, A_2 are annuli on Bd H, there is a homeomorphism of H onto itself taking A_1 onto A_2. This may be used to establish the following lemma.

LEMMA 2. *If H_1, H_2 are two punctured cubes such that $H_1 \cdot H_2$ is an annulus on the boundary of each, then $H_1 + H_2$ is a punctured cube.*

It follows from a result of Moise [8, Theorem 2] that if H is a polyhedral punctured cube in a triangulated 3-manifold, then there is a piecewise linear homeomorphism of H into E^3. Alexander showed [1] that a polyhedral 2-sphere in E^3 bounds a polyhedral cube in E^3. These results may be used to give the following lemma.

LEMMA 3. *If D is a polyhedral disk in a polyhedral punctured cube H such that $D \cdot \text{Bd } H = \text{Bd } D$, then H is the sum of two polyhedral punctured cubes H_1, H_2, such that $H_1 \cdot H_2 = D$.*

4. Punctured cubes in 3-manifolds

In this section we suppose that M is a connected 3-manifold and give (Lemma 4) conditions under which M is topologically S^3. First we consider some properties of M.

Moise has shown [8] that M can be triangulated. Bing has given [3, 4] an alternative proof of this result. Hence, we suppose there is a *triangulation* T of M. We regard T as a locally finite collection of tetrahedra filling M, such that if two of them intersect, the intersection is either a vertex, an edge, or a face of each. The sum of the edges is called the 1-*skeleton* R_1 of T. The sum of the vertices is called the 0-*skeleton* R_0, and the sum of the faces is called the 2-*skeleton* R_2.

The triangulation T imposes a locally affine structure on M. In each tetrahedron of T, it is meaningful to consider straight line intervals, triangles, and tetrahedra. A subset K of M is called a *polyhedron* if it is the sum of a finite number of such intervals, triangles, tetrahedra, and points. (These intervals, triangles, tetrahedra, points may lie in different elements of T).

A subset X of M is said to be locally polyhedral at a point p of X if there is a neighborhood N of p in M, such that $\bar{N}\cdot X$ is a polyhedron. A map f of a subset X of M into E^3 (or M) is locally piecewise linear at a point p of X, if there is a neighborhood N of p in M and a triangulation T' of $\bar{N}$ compatible with T, such that f is linear on $X\cdot t$ for each tetrahedron t of T'.

The following lemma brings us a bit closer to the proof of Theorem 1.

LEMMA 4. *A triangulated* 3-*manifold* M *is topologically* S^3 *if there is a polyhedral punctured cube* H *in* M, *such that* Int H *contains the* 1-*skeleton of the triangulation.*

PROOF. We note that M is compact since its 1-skeleton lies in a compact subset H of M.

First we consider the case where $R_2 \subset \text{Int } H$. If Bd H has just one component, Bd H lies on the interior of an element t_i of T. Since each polyhedral 2-sphere in E^3 bounds a polyhedral cube in E^3, Bd H bounds a polyhedral cube C in t_i [1]. Then $M = H + C$, and M is topologically S^3. If Bd H has several components, these may be eliminated one at a time by filling in components of $M - H$.

If $R_2\cdot \text{Bd } H \neq 0$, we suppose that Bd H and R_2 are in general position in the sense that, if p is a point of $R_2\cdot\text{Bd } H$, there is a neighborhood N of p, and a homeomorphism of N into E^3 that takes $N\cdot R_2$ onto the xy plane, and $N\cdot\text{Bd } H$ onto the xz plane. If Bd H is not already in such a position, a slight adjustment of H is enough to cause the adjusted H to be better located. Then $R_2\cdot\text{Bd } H$ is the sum of a finite number of polygonal simple closed curves. The proof of the lemma is finished by using induction on the number of these simple closed curves.

Suppose t', t'' are two elements of T such that their common face contains a simple closed curve of $R_2\cdot\text{Bd } H$. Now we select a small simple closed curve of $t'\cdot t''$, in the sense that we consider a simple curve J in $R_2\cdot\text{Bd } H$, such that one component I of $t'\cdot t'' - J$ contains no point of $R_1 + \text{Bd } H$.

If I does not belong to Int H, there is a polyhedral cube C in $t' + t''$ such that $I \subset \text{Int } C$, Bd $C\cdot$Bd H is an annulus containing J on its interior and $R_2\cdot\text{Bd } C = J$. Then it follows Lemma 2, that $H + C$ is a punctured cube. But $H + C$ contains R_1 and $R_2\cdot\text{Bd }(H + C)$ has one less component than $R_2\cdot\text{Bd } H$.

If I does belong to Int H, Lemma 3 reveals that H is the sum of two polyhedral punctured cubes H_1, H_2, such that $H_1\cdot H_2 = I + J$. Since R_1 is connected, one of Int H_1, Int H_2 contains R_1 (say H_1). If H_1 is pushed away from $t'\cdot t''$ in a neighborhood of $I + J$, there results a punctured cube H' containing R_1, such that Bd H', R_2 are in general position, and $R_2 \cdot$Bd H' has fewer components than $R_2 \cdot$Bd H.

LEMMA 5. *If a polyhedral simple closed curve J lies in a topological cube in M, then it lies on the interior of a polyhedral cube in M.*

PROOF. Suppose H is a punctured cube containing J. We enumerate the steps in the proof.

1. First we shrink H slightly to get a punctured cube H' in H containing J, such that Bd H' is locally polyhedral at each point of Bd $H' - J$. The details are as follows.

Let h be a homeomorphism of H into E^3, such that h(Bd H) is the sum of a finite number of boundaries of cubes. It follows from [8], that there is a homeomorphism g of H onto $h(H)$, such that g is locally piecewise linear at each point of $H -$ Bd H. Shrink $h(H)$ slightly near h(Bd $H - J$) so as to obtain a punctured cube G, such that $J \subset G \subset J +$ Int H, and Bd G is locally polyhedral except at points of $J \cdot$Bd $h(H)$. Then $H' = g^{-1}(G)$ is a punctured cube in M, such that $J \subset H'$ and Bd H' is locally polyhedral at each point of Bd $H' - J$.

In Steps 2 to 7, we consider the case where J intersects Int H'.

2. Suppose A is an arc of J, such that A contains $J \cdot$Bd H'.

Polyhedral cubes about A can be found so as to get a family of polyhedral cubes $\{C_t, 0 \leqq t < 1\}$, such that if $x < y$, $A \subset C_x \subset$ Int C_y and Bd $C_x \cdot J$ is two points.

3. In this step we shrink H' to get a polyhedral punctured cube H'', such that $H'' + C_0$ contains H'. This punctured cube H'' is obtained by shrinking the punctured cube G of Step 1 near Bd $G \cdot h(J)$, and applying g^{-1} to the resulting shrunken punctured cube. We note that Bd $C_t \cdot$Bd $H' =$ Bd $C_t \cdot$Bd H'' $(0 < t < 1)$.

4. In this step we select an element C_s $(0 < s < 1)$ of $\{C_t\}$ so that Bd C_s and Bd H'' are in general position in the sense that if p is a point of Bd $C_s \cdot$Bd H'', there is a neighborhood N of p in M, and a homeomorphism of N onto E^3 that takes $N \cdot$Bd C_s and $N \cdot$Bd H'' onto the xy and xz planes, respectively. There is such a C_s because, for not more than a countable number of elements C_x of $\{C_t\}$, can Bd C_x fail to be in general position with respect to Bd H''. The cube C_s has the convenient property that Bd $C_s \cdot$Bd H'' is the sum of a finite collection of mutually exclusive polygonal simple closed curves.

5. Suppose axb and ayb are the two arcs of J, such that $axb \subset C_s$ and $ayb - (a + b) \subset M - C_s$. In this step we point out that no one of the simple closed curves in Bd $C_s \cdot$Bd H'' separates a from b in Bd C_s.

Since J can be shrunk to a point in H', there is a map f of a disk E

into H', so that f takes Bd E homeomorphically onto J, and $f(\text{Int } E) \subset$ Int H'. Then it will be evident that no simple closed curve of Bd $C_s \cdot$ Bd H'' separates a from b in Bd C_s, when we show that a and b belong to the same component of Bd $C_s \cdot f(E)$. This component belongs to $a + b + f(\text{Int} E)$, hence to Int H', and hence to Int H''.

Suppose that a and b belong to different components of Bd $C_s \cdot f(E)$. Then a simple closed curve K in Bd $C_s - f(E)$ separates a from b in Bd C_s. There is a retraction r of $f(E) \cdot C_s$ onto axb, such that $r(f(E) \cdot \text{Bd } C_s) = a + b$. The retraction r takes all of Bd $\cdot C_s \cdot f(E)$ in one component of Bd $C_s - K$ into a, and the rest of Bd $C_s \cdot f(E)$ into b. The map r can be extended to take all of $f(E)$ onto J, so that ayb is taken homeomorphically onto itself. But the map $f^{-1}rf$ violates the proposition that there is no retraction of a disk onto its boundary.

6. If Bd $H'' \cdot$ Bd C_s is one simple closed curve, there is a polyhedral disk E in $C_s \cdot$ Bd H'', such that Bd $E =$ Bd $H'' \cdot$ Bd C_s. By using a simplification of Lemma 1, where " disk " is used instead of " annulus ", we find that $H'' + C_s$ is a polyhedral punctured cube. By cutting polyhedral tubes in $H'' + C_s$, a polyhedral cube may be obtained whose interior contains J.

7. If Bd $H'' \cdot$ Bd C_s has several simple closed curves, these simple closed curves are removed one at a time. Our method is suggested by the proof of Lemma 4. If E is a disk in Bd C_s bounded by a simple closed curve in Bd $H'' \cdot$ Bd C_s, such that Int $E \cdot$ Bd $H'' = 0$, we add a polyhedral cube to H'' if $H'' \cdot$ Int $E = 0$ and cut H'' in two pieces with a disk slightly away from E if $E \subset H''$, keeping the piece that contains ayb.

8. Finally we consider the case where J does not intersect Int H'. In this case it bounds a disk D on Bd H' which is locally polyhedral at each point of Int D. In fact it follows from the methods of [7] that J bounds a polyhedral disk D' in M. This disk D' in turn lies on the interior of a polyhedral disk D''. If D'' is thickened as suggested in [7], a polyhedral cube is obtained whose interior contains D', and hence J.

EXAMPLE. It cannot be concluded that if a nonpolyhedral simple closed curve J lies in a topological cube C in M, then J lies on the interior of a polyhedral cube in M. Consider a Cantor set X on the interior of a torus T such that no simple closed curve that circles Bd T meridianally can be shrunk to a point in $T - X$. Antoine described such a Cantor set X [2]. There is a topological cube C (solid horned sphere in this case) on Int T, such that $X \subset$ Bd C. Let J be a simple closed curve on Bd C containing X. Then J lies in a topological cube C in Int T but no polyhedral cube in Int T contains J on its interior.

5. Building cubes about finite graphs

A finite graph is the sum of a finite number of arcs such that if two of them intersect, their intersection is an end point of each. In this section we suppose that M is triangulated. The 1-skeleton of M is a finite graph if M is compact. We shall be interested in getting a polyhedral cube in M whose interior contains this 1-skeleton, so as to be able to apply Lemma 4 to get Theorem 1. This is done in Lemma 7. First we prove a lemma to help us there.

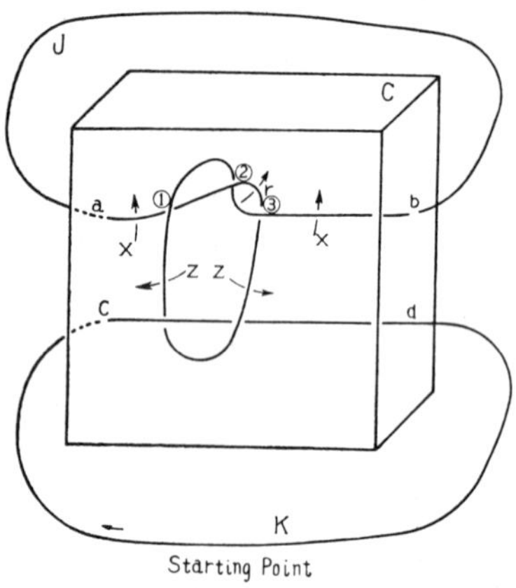

Fig. 3.

LEMMA 6. *Suppose C is a cube and ab, cd are the two arcs as shown in Figure 3. Then cd is not homotopic in $C - ab$ to any arc in* Bd C.

PROOF. Let J be the simple closed curve containing the arc ab as shown in Figure 3. The fundamental group G of $E^3 - J$ has generators x, y, z, and three relations

1. $x = z^{-1}yz$,
2. $z = y^{-1}xy$,
3. $y = x^{-1}zx$.

The generator x may be regarded as a path starting at the eye, going to the tail of the arrow marked x, going under J along the arrow, and then returning from the point of the arrow to the eye. The generators y and z are regarded in similar fashions. The three relations result from the three perspective crossings of J marked 1, 2, 3.

The element of the fundamental group G of $E^3 - J$ corresponding to the simple closed curve K is z.

It may be noted from Relation 3, that x and z may be used as generators rather than all of x, y, z. When y is eliminated as a generator, Relations 1, 2, 3 reduce to

$$4. \ xzx = zxz .$$

Assume cd is homotopic in $C - ab$ to an arc cxd on Bd C. This arc cxd in turn is homotopic on Bd $C - (a + b)$ to an arc cyd on the left, bottom, and right faces of Bd C. The element of the fundamental group G of $E^3 - J$ corresponding to $K' = (K - cd) + cyd$ is x^n. Since K and K' are homotopic in $E^3 - J$, $z = x^n$.

We show that this is untrue by giving an example of a group G' with generators x', z' satisfying Relation 4, (and others) where $z' \neq x'^n$ for any integer n. The matrices

$$x' = \begin{pmatrix} 1 & 0 \\ 0 & -1 \end{pmatrix}, \quad z' = \begin{pmatrix} -1/2 & -3/4 \\ -1 & 1/2 \end{pmatrix}$$

determine such a group G'. A homomorphism carrying x, z into x', z', respectively, takes G onto G'. Since $z' \neq x'^n$ for any integer n, $z \neq x^n$.

A permutation group G'' with generators

$$x'' = (12)\,(3)$$
$$z'' = (13)\,(2)$$

could have been used instead of the matrix group G' to show that $z \neq x^n$.

LEMMA 7. *If each polyhedral simple closed curve in M lies in a topological cube, each polyhedral finite graph lies on the interior of a polyhedral cube.*

PROOF OF SPECIAL CASE. First, we consider the case of a connected finite polyhedral graph G, such that each point of G is of order 2 or of order 4---that is, at each point G is locally topologically like a line, or locally like the sum of two intersecting lines at their point of intersection. We give the argument in 7 steps.

1. Let $p_1 p_2, \cdots, p_n$ be the points of order 4 of G and $C_1, C_2, \cdots, C_n$ be mutually exclusive polyhedral cubes in M such that $p_j \subset \text{Int } C_j$ and $G \cdot \text{Bd } C_j$ is 4 points a, b, c, d.

2. Let J be a polyhedral simple closed curve in $G + \Sigma\, C_i$ which contains $G - \Sigma \text{ Int } C_i$ and such that $J \cdot C_j$ ia $ab + cd$ as shown in Figure 3.

3. It follows from Lemma 5, that there is a polyhedral punctured cube H, such that $J \subset \text{Int } H$. We suppose that H is adjusted so that $\text{Bd } H$, $\Sigma \text{ Bd } C_i$ are in general position. Then $\text{Bd } H \cdot \text{Bd } C_j$ is either null or the sum of a finite number of polygonal simple closed curves. In order to use induction on the number of components of $\text{Bd } H \cdot \Sigma \text{ Bd } C_i$ in the proof of the special case of the theorem, we suppose that for no polyhedral punctured cube H' with $J \subset \text{Int } H'$ and $\text{Bd } H'$, $\Sigma \text{ Bd } C_i$ in general position, does $\text{Bd } H' \cdot \Sigma \text{ Bd } C_i$ have fewer components than does $\text{Bd } H \cdot \Sigma \text{ Bd } C_i$.

4. If $\text{Bd } H \cdot \Sigma \text{ Bd } C_i = 0$, we can fill in the holes of H in the various C_i's so as to obtain a polyhedral punctured cube whose interior contains $J + \Sigma\, C_i$. The boundary of each of these holes is a polyhedral 2-sphere in a cube, bounds a topological cube by [1], so the sum of H and this cube is a punctured cube. The interior of this polyhedral punctured cube contains G, and some polyhedral cube in it contains G on its interior.

5. We finish the proof of the special case of the theorem in the next

three steps by showing that Bd $H \cdot \Sigma$ Bd $C_i = 0$. In the present step we show that Bd $H \cdot$Bd C_j does not contain a simple closed curve that bounds a disk D in Bd C_j such that $D \cdot J = 0$. If it does, there is such a disk D, such that Int $D \cdot$Bd $H = 0$, and we consider the two cases--- Int $D \subset H$ or Int $D \cdot H = 0$.

If Int $D \subset H$, D separates H into two polyhedral punctured cubes by Lemma 3, only one of these H'' contains J, and if H'' is shoved to one side of Bd C_j near D, there results a polyhedral punctured cube H', such that Bd $H' \cdot \Sigma$ Bd C_i has fewer components than Bd $H \cdot \Sigma$ Bd C_i.

If Int $D \cdot H = 0$, we add to H a polyhedral cube obtained by thickening D and find from Lemma 2, that we can get a polyhedral punctured cube H', such that Bd $H' \cdot \Sigma C_i$ has one less component than Bd $H \cdot \Sigma C_i$.

6. Suppose D is a disk in Bd C_j such that $D \cdot$Bd $H =$ Bd D. Then by Step 5, Int D contains one of a, b, c, d but not all of them. Since D separates H (Lemma 3) and J minus a point is connected, D contains exactly two of a, b, c, d. Then there is another disk D' in Bd C_j, such that $D' \cdot$Bd $H =$ Bd D' and D' contains the other two of a, b, c, d. We show in this step that it is impossible that $a + b$ belong to either D or D'.

Suppose $c + d \subset D$ and $a + b \subset D'$. Since D separates $cd - (c + d)$ from ab in H, the component X of $C_j \cdot H$ containing cd does not intersect Int $D' + ab$. Let cxd be an arc in D, and f be a map of a disk E into H that takes Bd E homeomorphically onto $cd + cxd$. Let g be a map of E into C_j, such that $g = f$ on $f^{-1}(X)$ and $g(E - f^{-1}(X)) \subset$ Bd $C_j -$ Int D'. Then g reveals a homotopy in $C_j - ab$ of cd onto cxd. But Lemma 6 shows that this is impossible. This shows that $a + b$ cannot belong to either D or D'.

7. We now consider the case where the D and D' of Step 6 contain $a + c$ and $b + d$, respectively. For convenience we suppose that C_j is a real cube as shown in Figure 3, $a + c$ and $b + d$ are in opposite faces F, F' of C_j, $F \subset$ Int D, $F' \subset$ Int D' and cd is a straight line interval from the center of F to the center of F'. Let N be a polyhedral cube in $C_j \cdot H$ containing cd such that Bd $N \cdot$ Bd $C_j = F + F'$ and any straight line perpendicular to cd and through a point of $cd - (c + d)$ intersects Bd N in exactly two points. Keeping the two faces F, F' fixed and moving the points of $C_j - (ab + F + F')$ directly away from cd in a line perpendicular to cd, N may be taken onto C_j and H onto a polyhedral punctured cube H' so that Int H' contains J and $C_j \subset$ Int H'. Then Bd $H' \cdot \Sigma$ Bd C_i has fewer components than Bd $H \cdot \Sigma$ Bd C_i.

A similar demonstration shows that it is not true that $a + d \subset D$ and $b + c \subset D'$. Hence Bd $H \cdot \Sigma$ Bd $C_i = 0$, and Step 4 applies.

GENERAL CASE. Since some finite graph containing G has the properties of being connected, and each point being of an even order, we suppose that G has these properties. To see that the theorem holds in this case, we consider cubes containing the points of order greater than 4, shift some arcs of G near these boints, so that G is changed to a finite graph G', each point of which is of order 2 or 4, and finally adjust the cube covering G' to get one covering G.

6. Open subsets of E^3

The results about punctured cubes may be used to generalize Theorem 1 to the noncompact case, and get the following result.

THEOREM 2. *A bounded connected open subset U of E^3 is topologically E^3 if the boundary of U is connected aud each polyhedral simple closed curve in U lies in a topological cube in U.*

PROOF. We find from Lemma 7, that each polyhedral finite graph in U lies on the interior of a polyhedral cube in U. Then U is topologically E^3 if it is the sum of a countable collection of polyhedral cubes $C_1, C_2, \cdots$, such that $C_i \subset \text{Int } C_{i+1}$. That U is the sum of such cubes follows from the following lemma, which is proved by " filling holes ", as was done in the proof of Lemma 4.

LEMMA 8. *Suppose U is an open subset of E^3, T is a triangulation of U, and X is a compact subset of U, such that there is a topological cube that contains the sum of the edges of each tetrahedron of T that intersects X. Then there is a polyhedrnl punctured cube in U whose interior contains X.*

In Theorem 2, we could dispense with the condition that U is bounded if we insisted that its boundary be unbounded.

7. Cubes with handles

A solution of the Poincaré conjecture is elusive. Theorem 1 provides a partial solution. In this section, we raise four questions. Answers to the first three of these would lead to other partial solutions.

QUESTION. The example given in Section 2 shows that Theorem 2 would not be true if we replaced the hypothesis that *each polyhedral simple closed curve in U lies in a topological cube in U*, with the condition that *each polyhedral simple closed curve J in U lies in a torus T in U, such that J can be shrunk to a point in T*. Could such a replacement be made in Theorem 1? That is---is a compact, connected 3-manifold M topologically S^3, if each simple closed curve J in M lies in a torus T in M, such that J can be shrunk to a point in T?

A torus is a cube with a handle. A cube with n handles is the sum of $n+1$ topological cubes $C_0, C_1, \cdots, C_n$ such that $C_0 \cdot C_i = \text{Bd } C_0 \cdot \text{Bd } C_i$ is the sum of two mutually exclusive disks, $C_0 + C_i$ is a torus, and $C_i \cdot C_j = 0$ $(i \neq j, i \neq 0 \neq j)$.

Each connected oriented compact 3-manifold M is the sum of two homeomorphic cubes with handles M_1, M_2, such that $M_1 \cdot M_2 = \text{Bd } M_1 = \text{Bd } M_2$. To get M_1, M_2, consider a triangulation of M and let M_1 be the sum of tubes about the 1-skeleton on the triangulation and M_2 be the closure of $M - M_1$ [12, p. 219].

Tying cubes with handles together provides us with means of describing 3-manifolds with various properties. We illustrate with an example.

EXAMPLE. The Kline sphere characterization says that a compact continuous curve is topologically S^2, if it is separated by each of its simple closed curves, but by no pair of its points. One might hope that this result could be extended to say that a compact connected triangulated 3-manifold is topologically S^3, if it is separated by each polyhedral 2-manifold X. It has been known for many years that this hope could not be realized. The following easy example shows why.

Consider the 3-manifold M which is the sum of two solid tori M_1, M_2 tied together along their boundaries, such that if J, J' are the simple closed curves on $\text{Bd } M_1 = \text{Bd } M_2$ circling M_1 meridianally and longitudinally, respectively, then J circles M_2 three times longitudinally, and once meridianally; while J' circles M_2 twice longitudinally, and once meridianally. Such 3-manifolds as M have been described differently [12, p. 210] and called lens spaces--- M is the lens space (3,1). We suppose that M is triangulated so as to make $\text{Bd } M_1 = \text{Bd } M_2$ a polyhedral surface.

We shall find from the proof of Theorem 3, that M is not simply connected because J circles M_2 three times longitudinally, rather than once. (Actually, its fundamental group is cyclic of order three, but we shall not need this). Hence, M is topologically different from S^3.

We now point out why it is that each polyhedral simple closed curve K in M bounds a polyhedral 2-manifold in M. First, K is homologous (using as coefficient group the cyclic group of order two) in M to a 1-cycle K' on $\text{Bd } M_1 = \text{Bd } M_2$, and K' is homologous on $\text{Bd } M_1 = \text{Bd } M_2$ to either 0, J, J', or $J + J'$. But J bounds a disk in M_1, and J' bounds a Möbius strip in M_2. Hence, K bounds a 2-complex X. A slight adjustment in X causes it to be a 2-manifold except on a finite number of simple closed curves. But it can then be altered in a tubular neighborhood of these simple closed curves to get a polyhedral 2-manifold bounded by K.

Assume some polyhedral 2-manifold Y in M fails to separate M. Then there is a polyhedral simple closed curve J in M that intersects Y at only one point p, pierces it there, and this point p does not belong to the 1-skeleton of Y. The preceding paragraph shows that K bounds a polyhedral 2-manifold X. With respect to a triangulation of M whose 2-skeleton intersects Y only in a finite graph missing p, X may be adjusted to a 2-manifold X' with boundary K such that the 0-skeleton of X' misses Y, the 0-skeleton of Y misses X', and the 1-skeletons of X' and Y do not intersect. But then $X \cdot Y$ is a finite graph that is of odd order at p, and of even order elsewhere. This absurdity resulted from the false assumption that Y does not separate M.

Perhaps there is a counter-example that shows that the Poincaré conjecture is untrue. If so, this example might be discovered by tying cubes with handles together. The following theorem shows that if we are to obtain a counter-example in this fashion, the cubes with handles used would have to contain more than one handle.

THEOREM 3. *A simply connected 3-manifold M is topologically S^3, if it is the sum of two* tori, M_1, M_2, *such that* $M_1 \cdot M_2 = \text{Bd } M_1 = \text{Bd } M_2$.

Although the preceding theorem is known, we include a brief proof for completeness.

PROOF. Suppose D_1 is a flat disk in M_1 such that $M_1 - (D_1 + \text{Bd } M_1)$ is topologically an open 3-cell. Let D_2 be a corresponding disk in M_2. Suppose Bd D_1 circles M_2 i times longitudinally, and j times meridianally. We show that $i = \pm 1$.

Let K_1 be a simble closed curve on Bd M_2 that circles M_2 once longitudinally, but not at all meridianally. Since K_1 can be shrunk to a point in M, and $M - (D_1 + D_2 + \text{Bd } M_2)$ is the sum of two open 3-cells, K_1 can be shrunk to a point in $D_1 + D_1 + \text{Bd } M_2$.

Let G be the fundamental group of Bd M_2 with two generators x, y where x is associated with a loop circling M_2 longitudinally in a fixed direction, and y is associated with a loop circling meridianally. The group may be shown to be abelian, so the simple closed curve Bd D_1 of Bd M_2 is associated with the element $x^i y^j$. We suppose x is chosen so that i is positive.

Suppose K_1 is shrunk to a point in Bd $M_2 + D_1 + D_2$. Then there is a finite number of loops $K_1, K_2, \cdots, K_n$ on Bd M_2, such that K_n is a point, and for each r, K_r is shrunk to K_{r+1} in Bd $M_2 + D_s$ ($s = 1$ or 2), and no point of Int D_s is used twice in the shrinkings. Then the element of G associated with K_r differs from the element associated with K_{r+1} by either $x^i y^j$ or $(x^i y^j)^{-1}$ if $s = 1$, and by either y or y^{-1} if $s = 2$. But K_1 is associated with x, and K_n with 1, and x cannot be changed to 1 by multiplying by $(x^i y^j)^u$ and y^v unless $i = 1$. Hence $i = 1$.

Suppose M_2 is sliced meridianally, given an integral number of twists, and then sewed back together so that Bd D_1 circles the adjusted M_2 once longitudinally, but not at all meridianally.

A cube, resulting from thickening D_1, is sliced from M_1 (leaving a cube C_1 as the closure of remainder of M_1), and sewed onto the adjusted M_2 (changing it to a cube C_2). Then M is the sum of the two cubes C_1, C_2 sewed together along their common boundary and is therefore topologically S^3.

Theorem 3 can sometimes be used to show that a 3-manifold is topologically S^3, even if it is the sum of two solid spheres with handles M_1, M_2, where M_i has more than one handle. It may be possible to remove the handles from M_1 one at a time and add them to M_2, so that the closures of the resulting sets remain solid spheres with handles. We illustrate this with an example.

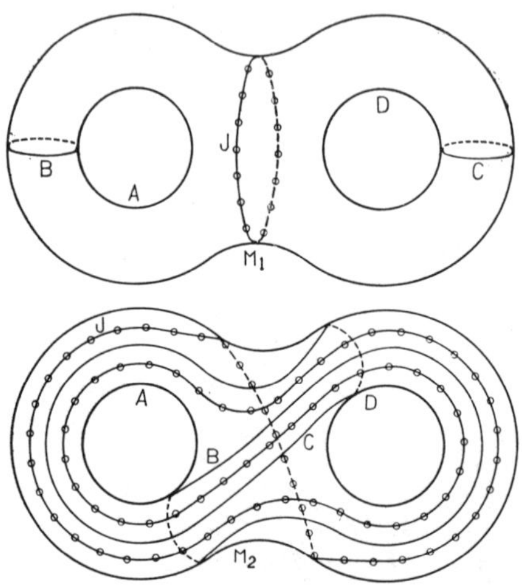

Fig. 4

EXAMPLE. Let M_1, M_2 be the solid spheres with two handles shown in Figure 4, and suppose that they are joined together along their boundaries, so that the simple closed curve A, B, J, C, D of one go into the corresponding simple closed curves of the other.

At first glance, it might appear that $M_1 + M_2$ is not simply connected because neither A nor D can be shrunk to a point in either M_1 or M_2. However, A is homotopic to C in M_2, and C can be shrunk to a point in M_1. Also, D is homotopic to B in M_2 which, in turn, is homotopic to 1 in M_1. Hence $M_1 + M_2$ is simply connected.

We show that $M_1 + M_2$ is topologically S^3 by using Theorem 3. First we change M_1 to a torus by removing the topological cube resulting from thickening the horizontal disk bounded by B. That M_2 plus this topological cube is a torus may be seen in the following three steps. (1) Slice the left handle of M_2 (leaving a torus); (2) add on the thickened disk removed from M_1 (leaving a topological cube); (3) reunite the part severed at Step 1 (leaving a torus). The left handle of M_2 would run through the right hole in M_2, if the reunion suggested in Step 3 were physically made.

MODIFICATION OF PRECEDING EXAMPLE. In the preceding example the simple closed curve B circled the right handle of M_2 once, and the left not at all, while C circled the left handle once, but the right not at all. Suppose that, instead, B had circled the right handle of M_2 i times, and the left handle j times, while C had circled the right handle r times, and the left s times. If $|is - rj| = 1$, each polyhedral simple closed curve in $M_1 + M_2$ bounds a 2-complex in $M_1 + M_2$.

QUESTION. Suppose M_i $(i = 1, 2)$ is a cube with two handles. Is a simply connected 3-manifold topologically S^3 if it is the sum of M_1 and M_2, such that $M_1 \cdot M_2 = \text{Bd } M_1 = \text{Bd } M_2$?

EXAMPLE. In simplifying the sum of two cubes with handles, one may try to remove a handle from one and add it to the other. However, unless care is exercised, the resulting sum may not be a cube with handles, because Lemma 2 is not true if we replace punctured cubes in the statement of that lemma with cubes with handles. Figure 5 shows that the

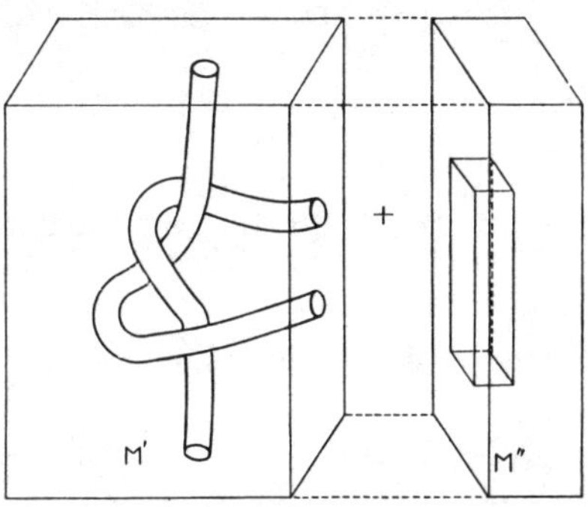

Fig. 5

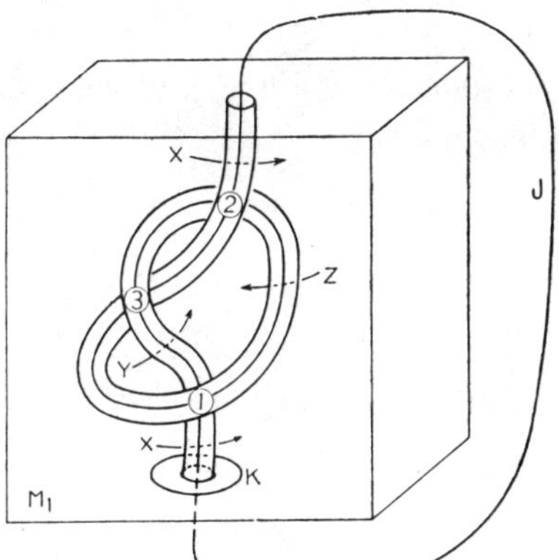

Fig. 6

sum of a cube with two handles M', and a topological cube M'', may be a cube with a knotted hole in it as shown in Figure 6, even though the intersection of M' and M'' is an annulus.

LEMMA. 9. *Suppose M_1 is the solid shown in Figure 6, and D is a disk (not necessarily in E^3), such that $D \cdot M_1 =$ Bd D is a simple closed curve on* Bd M_1 *that goes through the hole in M_1 i times, and circles this hole j times. Then $M_1 + D$ is simply connected if and only if $i = 0$, $j = \pm 1$.*

PROOF. Let J be the simple closed curve going through the center of the hole in M_1, as shown in Figure 6. The fundamental group of $E^3 - J$ (with base point on Bd D) is the same as the fundamental group of M_1, since there is a retraction of $E^3 - J$ onto M_1. The fundamental group of

$E^3 - J$ has generators x, y, z and relations

$$(1)\ x = zyz^{-1}\ ,$$
$$(2)\ y = xzx^{-1}\ ,$$
$$(3)\ z = yxy^{-1}\ .$$

The fundamental group of $M_1 + D$ satisfies the additional relation

$$(4)\ (zxy)^i x^j = 1\ .$$

Any other relationship is a consequence of these four.

If $i = 0$, $j = \pm 1$, Relation (4) implies that $x = 1$, Relation (3) implies that $z = 1$, and Relation (2) implies $y = 1$. Hence $M_1 + D$ is simply connected if $i = 0$, $j = \pm 1$.

Unless $3i + j = \pm 1$, we can let $x = y = z$, and find a nontrivial group G' satisfying Relations (1), (2), (3), (4), as well as others. A homomorphism of the fundamental group of $M_1 + D$ onto G' shows that, if $M_1 + D$ is simply connected, then $3i + j = \pm 1$. Hence we suppose that $j = 1 - 3i$.

If we use Relation (3) to eliminate z, we find that Relations (1), (2), (3) telescope into

$$(5)\ xyx = yxy$$

and Relation (4) becomes (4′) $(yxy^{-1}xy)^i = x^{3i-1}$.

Using Relation 5, we find that x and $yxy^{-1}xy$ commute, and Relation (4′) simplifies to (4″) $(xy)^{3i} = x^{6i-1}$.

Substituting w for xy we find that Relations (4″) and (5) become

$$(6)\ w^{3i} = x^{6i-1}$$
$$(7)\ (xw)^2 = w^3\ .$$

The group G_1, from the following examples, shows that Relations (6) and (7) are not enough to make the fundamental group of $M_1 + D$ trivial if $i \neq 0$. Hence $M_1 + D$ is not simply connected unless $i = 0$, $j = \pm 1$.

EXAMPLE OF GROUPS. The fundamental group of a cube with a tubular hole can be represented as a group with generators and relations. This approach was used in proving Lemmas 6 and 9. There is no algorism for telling which " words " written with generators correspond to the identity element—in fact, at times, we cannot be sure whether or not all " words " correspond to the identity element and the group is trivial. Sometimes we can show that a group is nontrivial by mapping it homomorphically onto a group known to be nondegenerate. We mention three nondegenerate groups used for this purpose.

Group G_1. For each positive integer $k > 0$ we describe a nontrivial group G_1 with generators x, w satisfying

$$x^k = w^3 = (xw)^2 = 1 .$$

The group G_1 satisfies other relations, but we are not interested in them here, since G_1 is used in Lemma 9, to show that the above relations do not imply that G_1 is trivial.

If $k = 2$, we let G_3 be a subgroup of the permutation group on three elements where

$$x = (12)(3) ,$$
$$w = (321) .$$

If $k = 3n$, we let

$$x = (123 \cdots k)$$
$$w = (k\ 21)(k-1\ 43) \cdots (2n+1\ 2n\ 2n-1) .$$

If $k = 3n + 1$, we replace the expression $(2n + 1\ 2n\ 2n - 1)$ in w by $(2n+ 2\ 2n\ 2n - 1)(2n + 1)$.

If $k = 2n + 2$, we replace it by $(2n + 3\ 2n + 1\ 2n - 1)(2n + 2)(2n)$.

Group G_2. We describe a group G_2 with generators x, y satisfying

$$xyx = yxy ,$$
$$xy^{3i} \neq x^{4i} (i \neq 0) .$$

Consider the infinite permutation group G_2 on elements

$$\cdots, a_{-2}, a_{-1}, a_0, a_1, a_2, \cdots, b_{-1}, b_0, b_1, \cdots, c_{-1}, c_0, c_1, \cdots$$

where

x moves a_n to b_n, b_n to a_{n+1}, and leaves c_n fixed,

and y moves c_n to b_n, b_n to c_{n+1}, and leaves a_n fixed .

Then each of xyx and yxy

moves a_n to c_{n+1}, b_n to b_{n+1}, and c_n to a_{n+1} .

We note that $(xy)^{3i}$

moves a_n to a_{n+2i}, b_n to b_{n+2i}, and c_n to c_{n+2i}

while x^{6i}

moves a_n to a_{n+3i}, b_n to b_{n+3i}, and leaves c_n fixed .

Group G_3. The matrix group G_3 with generators

$$x = \begin{pmatrix} 0 & 0 & 1 & 0 \\ 0 & 1 & 0 & 0 \\ 1 & 0 & 0 & 0 \\ 1 & 0 & 0 & 1 \end{pmatrix} , \quad y = \begin{pmatrix} 1 & 0 & 0 & 1 \\ 0 & 0 & 1 & 0 \\ 0 & 1 & 0 & 0 \\ 0 & 1 & 0 & 1 \end{pmatrix}$$

also satisfies the relation and inequality mentioned for group G_2. Either of groups G_2, G_3 can be used to show that *a closed curve on* Bd M_1 *of Figure* 6, *can be shrunk to a point on* Bd M_1, *if it can be shrunk to a point in* M_1. Both of the groups G_2, G_3 were suggested to me by J. B. Kruskal. I was originally using a more cumbersome group for this purpose.

THEOREM 4. *Suppose a 3-manifold M is the sum of the solid M_1 shown in Figure* 6, *and a torus M_2, such that* $M_1 \cdot M_2 =$ Bd $M_1 =$ Bd M_2. *Then M is topologically S^3 if it is simply connected.*

PROOF. Let E be a tame disk in M_2, such that Bd E circles M_2 meridianally and Int $E \subset$ Int M_2. Then $M_1 + E$ is simply connected and it follows from Lemma 9, that Bd E circles the hole in M_1 once, but does not go through the hole. Hence E may be thickened to form a topological cube C, such that both $M_1 + C$ and the closure of $M_2 - C$ are topological cubes. The sum M of $M_1 + C$ and the closure of $M_2 - C$ is topologically S^3.

QUESTION. Suppose M is a simply connected 3-manifold that is the sum of a torus and a cube with a tubular hole. Theorem 3 shows that M is topologically S^3 if the hole is straight, and Theorem 4 shows that M is topologically S^3 if it is knotted in a certain way. It has been pointed out to me by C. D. Papakyriakopoulos that, by using a combination of the methods of homology and homotopy, it can be shown that M is topologically S^3 no matter how the hole is knotted. If a simply connected 3-manifold is formed by subtracting a torus (perhaps knotted) from Int M_1, and then adding two tori to the resulting set, is the resulting 3-manifold topologically S^3?

QUESTION. Which compact, connected orientable 3-manifolds can be obtained from S^3 as follows : Remove a finite collection of mutually exclusive (but perhaps knotted and linking) polyhedral tori $T_1, T_2, \cdots, T_n$ from S^3, and sew them back. Handles can be formed by removing handles and sewing them back in a different fashion. The question may be stated in an alternate way as follows : Which compact orientable 3-manifolds M contain a finite collection of mutually exclusive polyhedral tori $T_1, T_2, \cdots, T_n$, such that $M - \Sigma T_i$ can be imbedded in E^3? If it could be shown that each simply connected compact manifold belongs to this class, this result might lead to a solution of the Poincaré hypothesis.

BIBLIOGRAPHY

1. J. W. ALEXANDER, *On the subdivision of 3-space by a polyhedron*, Proc. Nat. Acad. Sci. U.S.A., 10 (1924), 6–8.

2. LOUIS ANTOINE, *Sur l'homeomorphie de deux figures et de leurs voisinages*, J. Math. Pures Appl., 86 (1921), 221–325.
3. R. H. BING, Approximating surfaces with polyhedral ones, Ann. of Math., 65 (1957), 454–483.
4. ———, *An alternate proof that 3-manifolds can be triangulated*, Abstract 270, Bull. Amer. Math. Soc., 62 (1956), 179–180.
5. ———, Conditions under which monotone decompositions of E^3 are simply connected, abstract 325, Bull. Amer. Math. Soc. 63 (1957), p. 143.
6. RALPH H. FOX and EMIL ARTIN, *Some wild cells and spheres in three-dimensional space*, Ann. of Math. 49 (1948), 979–990.
7. O. G. HARROLD, Jr., Some consequence of the approximation theorem of Bing, Summary of Lectures and Seminars Summer of Institute on Set Theoretic Topology, Madison, 1955,pp. 53–56.
8. E. E. MOISE, *Affine structures in 3-manifolds*, V. *The triangulation theorem and Hauptvermutung*, Ann. of Math., 56 (1952), 96–114.
9. M. H. A. NEWMAN and J. H. C. Whitehead, *On the group of a certain linkage*, Quart. J. of Math, (Oxford), 8 (1937), 14–21.
10. H. POINCARÉ, *Second Complément a l'Analysis Situs*, Proc. London Math. Soc., 32 (1900), 277–308.
11. ———, *Cinquième Complément a l'Analysis Situs*, Rend. Circ. Mat. Palermo, 18 (1904), 45–110.
12. H. SEIFERT and W. THRELFALL, Lehrbuch der Topologie, Chelsea, 1947.
13. J. H. C. WHITEHEAD, *Certain theorems about three-dimensional manifords*, I, Quart. J. of Math. (Oxfoad), 5 (1934), 308–320.
14. ———, *Three-dimensional manifolds* (corrigendum), *Quart. J. of Math.* (Oxford), 6 (1935), 80.
15. ———, *A certain open manifold whose group is unity*, Quart, J. of Math. (Oxford), 6 (1935), 268–279.

ANNALS OF MATHEMATICS
Vol. 77, No. 1, January, 1963
Printed in Japan

Correction to

NECESSARY AND SUFFICIENT CONDITIONS THAT A 3-MANIFOLD BE S^3.

BY R. H. BING

These Annals, Vol. 68 (1958), 17-37
(Received April 19, 1962)

Can one obtain a counter-example to the Poincaré conjecture by cutting a knotted torus from S^3 and sewing it back in differently?

It was shown in [1] that one could not obtain a counter-example if the center of the solid torus is a trefoil knot. John Hempel has generalized this result to show [2] that no counter-example is obtained if the center is any torus knot whatever.

The knot with four crossings is not a torus knot so is a logical candidate for examination.

It is stated on page 36 of [1] that it had been "pointed out to me by C.D. Papakyriakopoulos that, by using a combination of the methods of homology and homotopy, it can be shown that" no counter-example is obtained, no matter how the hole is knotted. However, if the proof was as Bing recalls it, the proof was in error.

Tominaga attempted to generalize this result still further in [3] by showing that two compact triangulated 3-manifolds M_1, M_2 are topologically equivalent if they contain polyhedral tori T_1, T_2 such that $T_i (i = 1, 2)$ can be homotopically shrunk to a point in M_i and $M_1 - \text{Int } T_1$ is topologically equivalent to $M_2 - \text{Int } T_2$. The last line of the proof of this is not convincing. It states that, since there is a homeomorphism of $M_1 - \text{Int } T_1$ onto $M_2 - \text{Int } T_2$ that takes a longitudinal curve on T_1 onto a longitudinal curve on T_2, it follows from "the standard method" that M_1 is topologically M_2.

UNIVERSITY OF WISCONSIN

REFERENCES

1. R. H. BING, *Necessary and sufficient conditions that a 3-manifold be S^3*, Ann. of Math. 68 (1958), 17-37.
2. JOHN HEMPEL, *A simply connected 3-monifold is S^3 if it is the sum of a soild torus and the complement of a torus knot*, Abstract 594-27, Notices of A.M.S., 9 (1962), 466.
3. AKIRA TOMINAGA, *A unique imbedding of a torus homotopic to 0 in a* 3-manifold, J. Sci. Hiroshima University, 25 (1961), 1-2.

C. Arcs and Simple Closed Curves

SOME CHARACTERIZATIONS OF ARCS AND SIMPLE CLOSED CURVES.*

By R. H. Bing.

1. Introduction. This paper uses the notion of *cuttings* to give some characterizations of arcs and simple closed curves. We make a distinction between *cutting* and *separating* as given in the following definitions:

Definitions. The set R *separates* A from B in the connected set M if $M - M \cdot R$ is the sum of two mutually separated sets containing A and B respectively. If M contains a continuum intersecting both A and B and R is a set containing neither A nor B but intersecting each continuum in M intersecting both A and B, we say that R cuts A from B in M. If a set separates (or cuts) two points from each other in M, we say that it separates (or cuts) M.

A *continuum* is a closed and connected set. It is *irreducible from* H to K if it intersects H and K but contains no proper subcontinuum doing so. It is *irreducible about* R if it contains R but none of its proper subcontinua does.

The theorems of this paper hold in a metric space. In fact, they hold in a Moore space, that is in a space satisfying Axiom 0 and the first three parts of Axiom 1 as given in R. L. Moore's *Foundations of Point Set Theory.*

J. R. Kline has shown [3] that for the plane, a nondegenerate continuum is a simple closed curve if it remains connected on the omission of any connected subset. This characterization is similar to the definition by R. L. Moore which defines [5] a simple closed curve to be a nondegenerate compact continuum which is separated by each pair of its points. However, the separating of a continuum by a pair of its points is not the same as the cutting of a continuum by a pair of its points. Every pair of points of a compact indecomposable continuum cuts it but no pair of points separates it. Therefore, the above mentioned characterizations by Kline and Moore differ from Theorem 11 of this paper which states that a nondegenerate compact continuum is a simple closed curve if it is cut by no one but by each pair of its points.

C. Kuratowski has shown [4] that if no subcontinuum of a nondegenerate bounded continuum C cuts it, then C is a simple closed curve. One of the results (Theorem 10) of this paper proves that instead of assuming that no

* Received September 25, 1947; Presented to the American Mathematical Society, November 28, 1947.

subcontinuum of C cuts C, we need only assume that no point cuts C and no subcontinuum of C separates it.

If a point separates a continuum, it separates two open subsets of the continuum from each other. Hence, to assume that a point cuts a continuum between two open subsets is a weaker hypothesis than to suppose that it separates the continuum. Some known results may be strengthened by replacing the notion of *separation* by the idea of *cutting between two open subsets.* For example, Corollary 2 is stronger than the known result that every compact continuum contains two points neither of which separates it and Theorem 9 is stronger than the known result that a compact continuum is a simple closed curve if every pair of its points separates it.

2. Cuttings. The five theorems of this section do not depend on each other. Most of them are used in later sections.

THEOREM 1. *Suppose that a, b, and c are three points of the continuum M which is irreducible from a to b. A necessary and sufficient condition that c separate a from b in M is that for each point p of $M - c$ there be a continuum in $M - p$ containing c and intersecting $a + b$.*

Proof. We shall prove only the sufficiency case of this theorem.

Suppose that for each point p of $M - c$ there is a continuum in $M - p$ containing c and intersecting $a + b$. Let X be the set of all points p such that some continuum in $M - p$ contains $c + b$ and let Y be the set of all points p such that some continuum in $M - p$ contains $c + a$. Each point of $M - c$ belongs to $X + Y$. We note that X contains a and Y contains b. Since M is irreducible from a to b, no point belongs to both X and Y.

For each point p of X there is a continuum in $M - p$ containing $c + b$ and there is a domain containing p but no point of the continuum. Each point of M in this domain is a point of X. Hence, X is open in M. Also, Y is open in M. Then X and Y are mutually separated sets. Since c separates X from Y, it separates a from b.

Example. Theorem 1 would not be true if instead of assuming that for each point p of $M - c$ there is a continuum in $M - p$ containing c and intersecting $a + b$, we only assumed that no point of $M - (a + b + c)$ cuts c from both a and b in M. To demonstrate this, let M be the closure of the graph in polar coordinates of $\rho = 1 + 1/\theta$ $(\theta \geq \pi)$; let a, b, and c be the points whose polar coordinates are $(1 + (1/\pi), \pi)$, $(1, \pi)$, and $(1, 0)$.

We shall make use of the following definition:

DEFINITION. A set M is aposyndetic [1] at the point p if p belongs to M and for each point x of M distinct from p, there exists an open subset of M which contains p and belongs to a connected and relatively closed subset of M lying in $M - x$.

THEOREM 2. *The compact continuum M which is not separated by any of its subcontinua is locally connected at the point p if it is aposyndetic at p.*

Proof. Let D be an open subset of M containing p. For each point q of $M - p$ there is a continuum $C(q)$ lying in $M - q$ and containing an open subset of M which contains p. Denote $M - C(q)$ by $D(q)$. Since $M - D$ is compact, there are a finite number of points $q_1, q_2, \cdots, q_n$ such that $D(q_1) + D(q_2) + \cdots + D(q_n)$ covers $M - D$.

Since no continuum separates M, $D(q_i)$ is connected and its closure $\bar{D}(q_i)$ is a continuum in $M - p$. We shall show that p does not cut $\bar{D}(q_i)$ from $\bar{D}(q_j)$ in M. Assume that this is not the case. Then $M - [\bar{D}(q_i) + \bar{D}(q_j)]$ is not a connected set, for if it were, the closure of this connected set would be a continuum in M separating q_i from q_j. Let $M - [\bar{D}(q_i) + \bar{D}(q_j)]$ be the sum of the mutually separated sets X and Y where Y contains p. Since neither $\bar{D}(q_i)$ nor $\bar{D}(q_j)$ separates M, $\bar{D}(q_i) + X + \bar{D}(q_j)$ is a continuum. Hence, p does not cut $\bar{D}(q_i)$ from $\bar{D}(q_j)$.

Since p does not cut any two of the continua $\bar{D}(q_1), \bar{D}(q_2), \cdots, \bar{D}(q_n)$ from each other in M, there is a continuum E in $M - p$ containing the sum of $\bar{D}(q_1), \bar{D}(q_2), \cdots$, and $\bar{D}(Q_n)$. The complement of E is a connected open subset of D containing p. This demonstrates that M is locally connected at p.

THEOREM 3. *If D and E are mutually exclusive open subsets of the compact continuum M and R is a subset of $M - (D + E)$ which cuts D from E in M, then the sum of the components of $M - (D + E)$ which intersect $\bar{R}$ separates D from E in M.*

Proof. Let $E_1, E_2, \cdots$ and $D_1, D_2, \cdots$ be sequences of open subsets of M such that $\Sigma E_i = E$, $\Sigma D_i = D$, D_{i+1} contains $\bar{D}_i$, and E_{i+1} contains $\bar{E}_i$. Then M is the sum of three mutually exclusive sets X_i, Y_i, and Z_i where Y_i is the sum of all components of $M - (D_i + E_i)$ intersecting $\bar{R}$, X_i is the sum of D_i and all components of $M - (D_i + E_i)$ that intersect D but not $\bar{R}$, and Z_i is the sum of E_i and all components of $M - (D_i + E_i)$ that intersect E but not $\bar{R}$. Now ΠY_i is the sum of the components of $M - (D + E)$ that intersect $\bar{R}$ and $M - \Pi Y_i$ is the sum of the mutually separated sets ΣX_i and ΣZ_i which contain D and E respectively.

COROLLARY 1. *If a point cuts between the open subsets D and E of the compact continuum M, then a continuum in $M-(D+E)$ separates D from E in M.*

THEOREM 4. *If D and E are mutually exclusive open subsets of the compact continuum M and R is a subset of $M-(D+E)$ which cuts $\bar{D}$ from $\bar{E}$ in M, then the sum of the components of $M-(D+E)$ which intersect R separates D from E in M.*

Proof. Let N be the sum of the components of $M-(D+E)$ which intersect R and X be the set of all points p of $M-N$ such that some continuum in $M-N$ intersects both $\bar{D}$ and p. We shall show that $M-N$ is the sum of two mutually separated sets one of which is X.

Now $\bar{X}$ is a subset of $X+N$; for let $p_1, p_2, \cdots$ be a sequence of points of X converging to a point p and denote the component of $M-(D+E)$ containing p_i by C_i. The limiting set C of $C_1, C_2, \cdots$ is a continuum intersecting both p and $\bar{D}$. Now C is a subset of N or X according as it does or does not intersect N. Likewise, the closure of $M-(N+X)$ is a subset of $M-X$. Therefore N separates X from $M-(N+X)$ in M and hence separates D from E.

THEOREM 5. *For each proper subset R of the compact continuum M there is a point p of $M-R$ such that the sum of all continua that lie in $M-p$ and intersect R is dense in M.*

Proof. Let N be a subcontinuum of M irreducible about R. First consider the case where $M=N$. Let p be any point of $N-R$. Assume that N contains an open subset E of itself such that each subcontinuum of N intersecting both $\bar{E}$ and R contains p. Then each component of $N-E$ which contains a point of R also contains p. Then there is a continuum in $N-E$ containing R contrary to the assumption that N is irreducible about R.

Next, consider the case where $M \neq N$. Since M is completely separable, there is a sequence of open subsets $D_1, D_2, \cdots$ of $M-N$ such that if D is an open subset of $M-N$ containing a point q, then there is an integer i such that D_i contains q and is a subset of D.

If an open subset of a proper subcontinuum of M contains $N+D_1$, let N_2 be such a subcontinuum; otherwise, let N_2 be N. If some proper subcontinuum of M has an open subset which contains N_2+D_2, let N_3 be such a subcontinuum; otherwise, let N_3 be N_2. Similarly, we obtain $N_4, N_5, \cdots$. If ΣN_i is dense in M, a point of $M-\Sigma N_i$ is the required point p.

Suppose that ΣN_i is not dense in M. For convenience, suppose that $\Sigma N_i = N$. Then N does not belong to any open subset of any proper subcontinuum of M.

Let E_i be the component of $M - D_i$ containing N and F_i be the closure of $M - E_i$. Since $E_i \cdot F_i$ does not contain any open subset of M, the Theorem of Baire assures us that some point p of $M - N$ does not belong to $\Sigma E_i \cdot F_i$. If $F_i + N$ were the sum of two mutually separated sets X and Y where X contains N, N would belong to an open subset of the continuum $X + E_i$. Hence, $F_i + N$ is a continuum. Let G_i be the one of the continua E_i or $F_i + N$ that does not contain p. Although ΣG_i is dense in M, it does not contain p.

COROLLARY 2. *Each nondegenerate compact continuum contains two points neither of which cuts between two open subsets of the continuum.*

3. Characterizations of arcs. Using the facts that a compact continuum M is an arc between its points a and b if every point of $M - (a + b)$ separates a from b in M, we have the following theorem as an application of Theorem 1.

THEOREM 6. *A necessary and sufficient condition that the compact continuum M which is irreducible from the point a to the point b be an arc from a to b is that for each pair of points p, q of M there is a continuum in $M - p$ containing q and intersecting $a + b$.*

It is to be noted that p is allowed to be either a or b. As is shown by the example following Theorem 1, the result would not be true if we restricted the range of p to $M - (a + b)$.

THEOREM 7. *The compact continuum M is an arc between its points a and b provided each pair of points of $M - (a + b)$ cuts a from b in M and cuts an open subset of M from $a + b$ in M.*

Proof. Since M is irreducible from a to b, a does not cut M or else some two points which it cuts from b would not cut an open subset of M from b. Also, b does not cut M. As neither a nor b cuts M, we have by Theorem 6 that M is an arc if no point of $M - (a + b)$ cuts any point from $a + b$ in M. We shall show that the assumption that $M - (a + b)$ contains points p and q such that p cuts q from $a + b$ leads to the contradiction that $p + q$ does not cut any open subset of M from $a + b$.

Suppose that $p+q$ cuts the open subset D of M from $a+b$. Let ab be an arc from a to b such that $ab \cdot M = a+b$. The arc ab is not necessarily in the space containing M. By Theorem 3 there are continua H and K in $M-D$ such that H contains p, K contains q, X contains ab, and $M+ab$ is the sum of two closed proper subsets X and Y such that $X \cdot Y = H+K$.

Let F and G be the components of $M \cdot X$ containing a and b respectively. Each intersects $H+K$ but not both intersect the same one of H, K or else some pair of points of Y does not cut a from b in M. But p does not cut q from a or from b according as K intersects F or G.

THEOREM 8. *The compact continuum M is an arc between its points a and b provided each pair of points of $M-(a+b)$ cuts an open subset of M from an open subset of M containing $a+b$.*

Proof. Assume first that some subcontinuum N of M is irreducible from a to b and is not an arc. Then by the preceding theorem, $N-(a+b)$ contains two points p and q whose sum does not cut any open subset of N from $a+b$ in N. However, $p+q$ cuts between two open subsets D and E of M in M where E does not intersect N and D contains $a+b$.

By Theorem 3, there are two continua H and K in $M-(D+E)$ whose sum separates D from E in M and such that H contains p and K contains q. Suppose that $M-(H+K)$ is the sum of two mutually separated sets $S(D)$ and $S(E)$ which contain D and E respectively. By Theorem 5 there is a point r of $S(E)-S(E) \cdot N$ that does not cut any open subset of $S(E)$ from $H+K+N$ in M.

If $S(D)$ is not a subset of N, by Theorem 5 there is a point t of $S(D)-S(D) \cdot N$ that does not cut any open subset of $S(D)$ from $H+K+N$ in M. Then $r+t$ does not cut any open subset of M from N and hence does not cut any open subset of M from $a+b$.

We shall show that in case N contains $S(D)$, there is a point t of $S(D)$ that does not cut either H or K from any domain containing $a+b$ in $N+H+K$. Then $r+t$ will not cut any open subset of M from an open subset of M containing $a+b$.

If N were indecomposable, we could use any point t in $S(D)$ belonging to a composant of N not intersecting $a+b+p+q$. Hence, suppose that it is decomposable. Then either a or b does not cut p from $a+b$ in N. Also, either a or b does not cut q from $a+b$. If a (or b) cuts neither p nor q from b (or a), then there is a domain containing a (or b) such that any point of $S(D)-a$ (or $S(D)-b$) in this domain will serve as t. Hence, suppose that a cuts p from b and b cuts q from a in N.

Let $D_1, D_2, \cdots$ be a sequence of open subsets of N closing down on a where D_1 does not intersect $p+q$ and let C_{p_i} and C_{q_i} be the components of $N-D_i$ containing p and q respectively. C_{p_i} does not intersect C_{q_i} for any integer i because there is an integer j such that C_{q_j} contains b. Let F be an open subset of $S(D)$ containing a but no point of C_{q_j}. Since the limiting set of $C_{q_1}, C_{q_2}, \cdots$ is n, there is an integer k such that C_{q_k} intersects F. Then no point t of $F \cdot C_{q_k}$ cuts q from b and p from any domain containing a in N.

We have shown that the assumption that there is a subcontinuum of M irreducible from a to b which is not an arc leads to the contradiction that some pair of points r, t of $M-(a+b)$ does not cut an open subset of M from an open subset of M containing $a+b$. Hence, every subcontinuum of M irreducible from a to b is an arc. Let ab be one such arc. We shall show that if M is not ab, then there are two points of $M-(a+b)$ whose sum does not cut any open subset of M from any open subset of M containing $a+b$.

Let X be the collection of all continua x such that x is the closure of a maximal continuumwise connected subset of $M-ab$. The sum of a countable subcollection of X is dense in $M-ab$. If this countable subcollection failed to cover two points of $M-ab$, this pair of points would not cut an open subset of M from ab in M.

Since a countable subcollection of X covers $M-ab$, some element x_1 of X contains an open subset of $M-ab$. First, consider the case where x_1 covers $M-ab$. If x_1 contains ab, no pair of points of $ab-(a+b)$ would cut an open subset of M from an open subset of M containing $a+b$ for, if so, each open subset of M would intersect the maximal continuumwise connected subset of $M-ab$ whose closure is x_1. If r is a point of $ab-(a+b)$ not belonging to x_1 and t is a point of $M-ab$ not cutting any open subset of M from ab (Theorem 5) then $r+t$ would not cut any open subset of M from $a+b$ in M.

If x_1 does not cover $M-ab$, there is another element x_2 of X which contains an open subset of $M-ab$. Let C_i $(i=1,2)$ be the closure of the sum of ab and all elements of X other than x_i. By Theorem 5 there is a point p_i of $M-C_i$ that does not cut any open subset of M from C_i in M. Any subcontinuum of M irreducible from p_i to ab is a subset of x_i. Then p_i+p_2 do not cut any open subset of M from ab in M. Hence, M is ab.

4. Characterizations of a simple closed curve. We shall now use cuttings of continua to give some characterizations of a simple closed curve.

THEOREM 9. *The compact continuum M is a simple closed curve if each pair of its points cuts between two open subsets of M.*

Proof. No subcontinuum N of M separates M because if $M - N$ were the sum of two mutually separated sets X and Y, then by Theorem 5 there would be points p and q in X and Y respectively such that continuumwise connected subsets of $N + X - p$ and $N + Y - q$ would contain N and be dense in $N + X$ and $N + Y$ respectively. Then $p + q$ would not separate any two open subsets of M from each other.

If the points p and q cut between two open subsets, then by Theorem 3 there are subcontinua H and K of M such that $M - (H + K)$ is the sum of two mutually separated sets X and Y. Since no subcontinuum of M separates it, $\bar{X}$ is ireducible from H to K or else the sum of H, K, and a proper subcontinuum of $\bar{X}$ is a continuum separating M. By Theorem 7 we find that every point of X separates H from K in $H + X + K$. Then some two points separate M. Reapplying Theorem 7, we find that M is the sum of two arcs between these points and is in fact a simple closed curve.

THEOREM 10. *A nondegenerate compact continuum is a simple closed curve if it is neither cut by any point nor separated by any one of its subcontinua.*[1]

Proof. Let M be such a continuum. It has been shown [2] by F. B. Jones that a compact continuum which is not cut by any of its points is aposyndetic on a dense subset. Using this result and Theorem 2, we find that M is locally connected at two of its points a, b.

Let $D_1, D_2, \cdots$ and $E_1, E_2, \cdots$ be sequences of connected domains closing down on a and b respectively such that D_1 does not intersect E_1. Now $M - (D_i + E_i)$ is the sum of two mutually separated sets X_i and Y_i, for if it were connected. it would be a continuum separating a from b in M. No proper subcontinuum of $\bar{D}_i + X_i + \bar{E}_i$ contains both D_i and E_i or else a subcontinuum of M separates a point of X_i from Y_i. Also $\bar{Y}_i$ is irreducible from $\bar{D}_i$ to $\bar{E}_i$. Suppose that X_{i+1} intersects X_i. Then X_i is a subset of X_{i+1}. Now M is the sum of two continua $a + \Sigma X_i + b$ and $a + \Sigma Y_i + b$ and it follows from Theorem 6 that each of these is an arc from a to b.

[1] By modifying our arguments we could have strengthened Theorems 2, 6, 10, 11, and 12 by weakening the hypotheses of these theorems by supposing the sets to be locally peripherally compact instead of compact.

THEOREM 11. *A nondegenerate compact continuum is a simple closed curve if it is cut by no one but by each pair of its points.*

Proof. Suppose that M is such a continuum. It will follow from Theorem 10 that M is a simple closed curve if we can show that none of its subcontinua separates it.

Assume that M contains a continuum H such that $M - H$ is the sum of two mutually separated sets X and Y containing points p and q respectively. We shall show that this assumption leads to the contradiction that $p + q$ does not cut any point of M from H in M.

Assume that $p + q$ cuts the points a of X from H in M. Since no point cuts M, there is a continuum C in $M - p$ containing $a + H$. Now $C - C \cdot Y$ is a closed set in $M - (p + q)$ containing $a + H$. Also, it is a continuum, for if it were the sum of the mutually separated sets R and S where S contains H, then C would be the sum of the mutually separated sets R and $S + C \cdot Y$. Also, $p + q$ does not cut any point of Y from H and therefore does not cut M.

THEOREM 12. *A nondegenerate compact connected space which is cut by no one of its points is a simple closed curve if each of its proper subcontinua is cut by each interior point of this subcontinuum.*

Proof. Assume that the complement of the continuum H is the sum of two mutually separated sets X and Y and let p be a point of X. Then p does not cut the continuum $H + X$ because it does not cut the space. Hence, no continuum separates the space and Theorem 12 follows from Theorem 10.

THEOREM 13. *The nondegenerate compact continuum M which is cut by no one of its points is a simple closed curve provided that for each three points a, b, p of M there is a point q such that $p + q$ cuts a from b in M.*

Proof. Again we show that no subcontinuum H of M separates it. Assume that $M - H$ is the sum of two mutually separated sets X and Y. Let a and b be two points of a continuum K in X and let p be a point of Y. If there were a point q such that $p + q$ cuts a from b, then q belongs to K. But there is a continuum in $M - q$ containing $a + H$ and another continuum in $M - q$ containing $b + H$. The sum of these two continua is a continuum in $M - (p + q)$ containing $a + b$. Hence, no subcontinuum separates M.

Example. We give an example to show that a compact continuum which is cut by no point need not be a simple closed curve even though for each pair of its points there is another pair of points cutting between these two in the continuum. Let the continuum be the sum of the graph of $y = \sin(1/x)$ $(0 < x \leqq 4)$ and the square with opposite vertices at $(0, 2)$ and $(4, -2)$.

THE UNIVERSITY OF WISCONSIN.

REFERENCES.

1. F. B. Jones, "Aposyndetic continua and certain boundary problems," *American Journal of Mathematics*, vol. 63 (1941), pp. 545-553.

2. ———, "Concerning non-aposyndetic continua," *American Journal of Mathematics*, vol. 70 (1948), pp. 403-413.

3. J. R. Kline, "Closed connected sets which remain connected upon the removal of certain connected subsets," *Fundamenta Mathematicae*, vol. 5 (1924), pp. 3-10.

4. C. Kuratowski, "Contribution à l'étude de continus de Jordan," *Fundamenta Mathematicae*, vol. 5 (1924), pp. 112-127.

5. R. L. Moore, "Concerning simple continuous curves," *Transactions of the American Mathematical Society*, vol. 21 (1920), pp. 333-347.

D. Hilbert Space

A COMPLETE ELEMENTARY PROOF THAT HILBERT SPACE IS HOMEOMORPHIC TO THE COUNTABLE INFINITE PRODUCT OF LINES[1]

BY R. D. ANDERSON AND R. H. BING

1. **Introduction.** In this paper we give a complete and self-contained proof that real Hilbert space, l_2, is homeomorphic to the countable infinite product of lines, s; symbolically $l_2 \sim s$. We assume only that the reader understands the material of a first year course in topology: for example, elementary notions of complete metric spaces, product spaces, continuity, homeomorphisms, and the Tietze Extension Theorem. While the treatment is elementary, the arguments are not necessarily simple. The reader should be prepared to draw figures and verify continuity statements.

Among those who read an earlier version of this paper we particularly thank Frederic Ancel, William Barit, William Jaco, Dallas Webster, and James West who made very useful comments.

We define l_2 as the space of square summable sequences of reals—that is, $l_2 = \{(x_i)_{i>0} \mid x_i$ is a real number and $\sum x_i^2 < \infty\}$. For $x, y \in l_2$, the distance between $x = (x_i)_{i>0}$ and $y = (y_i)_{i>0}$ is defined by $d(x, y) = [\sum (x_i - y_i)^2]^{1/2}$. It is easy to verify that under such a distance function, l_2 is a complete metric space.

We define s as the space of all sequences $(x_i)_{i>0}$ of reals with the usual product topology. Thus $s = \prod_{i>0} R_i$, where for each i, R_i is the space of reals. A basis for the topology of s is the collection of sets of the form $\prod_{i>0} U_i$, where U_i is an open subset of R_i and for all but a finite number of the i's, $U_i = R_i$.

Since for each positive integer i, the real line R_i is homeomorphic with the open interval $(-1, 1) = I_i^0$, we may alternatively write $s = \prod_{i>0} I_i^0$. We find this formulation convenient in §3.

At times (§§7 and 9) we need to regard s as a complete metric space. A convenient complete metric to use is $d(x, y) = \sum (\min [1/2^i, |x_i - y_i|])$. An alternate complete metric is the traditional Fréchet metric $d(x, y) = \sum |x_i - y_i| / [2^i(1 + |x_i - y_i|)]$. It is immaterial, for our purposes, which we use, so let us suppose for simplicity that we use the first.

---

[1] The preparation of this expository article was supported by the National Science Foundation under grant GN 530 to the American Mathematical Society. The research leading to the paper was partially supported by NSF grant GP-6867 and GP-3857. Submitted by invitation of the editors; received by the editors April 19, 1968.

The question as to whether l_2 is homeomorphic to s was raised by Fréchet [10] in his book, *Les éspaces abstraits* (1928), and also by Banach [5] in his book, *Théorie des opérations linéaires* (1932). In an added footnote in [5], Banach erroneously attributed to Mazur a negative answer to the question. More recently, Kadec, Bessaga, Pelczynski, and others have studied homeomorphisms between various infinite dimensional Banach and Fréchet spaces. The work up to about 1964 is summarized by Bessaga in [6]. In the direction of proving that $l_2 \sim s$, Bessaga and Pelczynski [9] had proved that $l_2 \sim (l_2 \times s)$ and that l_2 was homeomorphic to the countable product of copies of itself. Kadec [11], [12] proved that all separable infinite dimensional Banach spaces are homeomorphic. In 1966, Anderson [1] proved that $l_2 \sim s$. This was done by using results in [2] together with the result that $l_2 \sim (l_2 \times s)$. Based on an earlier Bessaga and Pelczynski result cited in [6], the results of Kadec [11] and Anderson [1] established that all separable infinite dimensional Fréchet spaces are homeomorphic to each other.

In the present paper, the result that $l_2 \sim s$ appears in §3 as Theorem 1.1, but its proof makes use of lemmas and theorems from later sections. The procedure is summarized in §10. §11 lists open questions.

2. **Further definitions and notation.** We let $X \setminus Y$ denote the set of all elements of X not in Y.

Let Z denote the set of positive integers. For $\alpha \subset Z$, we let $\alpha' = Z \setminus \alpha$. We let $s^\alpha = \prod_{i \in \alpha} R_i$ and l_2^α denote the space of sequences $(x_i)_{i \in \alpha}$ of reals for which $\sum_{i \in \alpha} x_i^2 < \infty$ with $d(x, y) = (\sum_{i \in \alpha} (x_i - y_i)^2)^{1/2}$.

The following two propositions are well known and easy to prove.

LEMMA 2.1. *If α is a nonempty subset of Z for which α' is nonempty, then $s \sim s^\alpha \times s^{\alpha'}$. If α is infinite, then $s_2^\alpha \sim s$.*

LEMMA 2.2. *If α is a nonempty subset of Z for which α' is nonempty, then $l_2 \sim l_2^\alpha \times l_2^{\alpha'}$. If α is infinite, then $l_2 \sim l_2^\alpha$.*

We use the convention that if α is an infinite subset of Z, then propositions concerning l_2 and s are considered as applicable to l_2^α and s^α.

If $\alpha \subset Z$, let τ_α denote the projection of l_2 or s onto l_2^α or s^α. If $\alpha = \{i\}$, we may write τ_α as τ_i.

A set K in l_2 or s is *deficient in the ith direction* if $\tau_i(K)$ consists of a single element. A set K in l_2 or s is *infinitely deficient* if for some infinite set $\alpha \subset Z$, $\tau_i(K)$ consists of a single element for each $i \in \alpha$. In this case we also say that K is *deficient with respect to α.*

We use the term map to denote a continuous function. For f a map of X into Y and $K \subset X$, the notation $f|K$ denotes the map f

restricted to the domain K. We use Id to denote the identity homeomorphism.

Let G be an open covering of a space X, Y be a subset of X, and f be a map of Y into X. We say that *f is limited by G* if for each $p \in Y$, there exists a $g_p \in G$ such that $p \in g_p$ and $f(p) \in g_p$. The covering G provides a measure as to how close f is to the identity. If X is a metric space, the *mesh* of G is the least upper bound of the diameters of the elements of G.

Suppose $f_1, f_2, \cdots$ is a sequence of maps such that the limit of $f_1, f_2 \circ f_1, \cdots$ exists. We denote this limit by $\lim(f_i \circ \cdots \circ f_2 \circ f_1)_{i>0} = L\prod_{i>0} f_i$ and call it the infinite left product of the f_i.

In §5 we define *invertible isotopies* and adopt notation to be used in working with isotopies. In §7 we define an *invertibly continuous family of invertible isotopies.*

We use S_1 to denote the set of all points of l_2 whose distance from the origin is 1—that is, $S_1 = \{x \in l_2 \mid \sum x_i^2 = 1\}$.

We use E^i to denote $\{x \in l_2 \mid x_j = 0 \text{ if } j > i\}$. Then $\cup E^i$ is the union of all finite dimensional coordinate planes of l_2. Sometimes $\cup E^i$ is called infinite dimensional Euclidean space. We note that $l_2 \backslash \cup E^i = \{x \in l_2 \mid x_i \neq 0 \text{ for infinitely many } i\}$.

3. **A homeomorphism of l_2 onto s.** Throughout this section we regard $s = \prod_{i>0} I_i^0$ and let $C_i = \{y \in s \mid \sum y_j^2 \leq i \text{ and for each } j > 0, |y_j| \leq 1 - 1/i\}$. Note that C_i is compact since it is a closed subset of the product of closed intervals.

The main result of the paper is the following.

THEOREM 3.1. *$l_2 \sim s$.*

The plan for showing that $l_2 \sim s$ is to exhibit homeomorphisms through middle spaces as follows:

$$l_2 \backsim l_2 \backslash \cup E^i \backsim S_1 \cap (l_2 \backslash \cup E^i) \backsim s \backslash \cup C_i \backsim s.$$

The four homeomorphisms are guaranteed by Corollary 9.3, Lemma 3.3, Lemma 3.2, and Corollary 9.4 respectively.

The topology of the first three spaces above is determined by the topology of l_2 while the topology of the last two is determined by the topology of s, so we may start by considering the bridge between l_2 and s, namely the homeomorphism of $S_1 \cap (l_2 \backslash \cup E^i)$ onto $s \backslash \cup C_i$. This bridge homeomorphism suggests why we chose to define a homeomorphism on $S_1 \cap (l_2 \backslash \cup E^i)$ instead of on l_2.

LEMMA 3.2. *$S_1 \cap (l_2 \backslash \cup E^i) \sim s \backslash \cup C_i$.*

PROOF. Throughout this proof we denote $S_1 \cap (l_2 \setminus \bigcup E^i)$ by S_1^*. Hence the coordinates of points of both S_1^* and s lie in $(-1, 1)$.

There are many 1-1 functions taking S_1^* onto a dense subset of s. Since most of the coordinates of each point of S_1^* are close to 0, one might hope to divide each of these coordinates by a suitable divisor and obtain the corresponding coordinate of the corresponding point of s. Let us consider such an elementary function $h: S_1^* \to s$ where we let

$$x = (x_1, x_2, \cdots) \in S_1^*, \quad \text{and} \quad h(x) = (y_1, y_2, \cdots) \in s.$$

We note that $x_1 \in (-1, 1)$ and $y_1 \in (-1, 1)$ so we set $y_1 = x_1$.

Once x_1 is fixed, x_2 has the limited domain $(-(1-x_1^2)^{1/2}, (1-x_1^2)^{1/2})$ while y_2 can still be anywhere in $(-1, 1)$. We take the domain of x_2 linearly onto the domain of y_2 and thus define

$$y_2 = x_2/(1 - x_1^2)^{1/2}.$$

Continuing in this fashion, we define

$$y_i = x_i \Big/ \left(1 - \sum_1^{i-1} x_j^2\right)^{1/2}.$$

One may note that the function h we have defined is 1-1 and coordinate wise continuous. Since s is a product space, coordinatewise continuity into s implies continuity. Hence h is continuous and thus is a map.

One reason for working with S_1^* rather than S_1 is that we would have had difficulties trying to avoid dividing by 0.

In order to study h^{-1}, we may solve for $(x_1, x_2, \cdots)$ in terms of $(y_1, y_2, \cdots)$. One finds that $h^{-1} = g \mid h(S_1^*)$ where $g: s \to l_2$ is defined by

$$x_i = y_i(1 - y_1^2)^{1/2}(1 - y_2^2)^{1/2} \cdots (1 - y_{i-1}^2)^{1/2}.$$

Also, g is 1-1 and coordinatewise continuous. Coordinatewise convergence in l_2 does not imply convergence as can be seen by considering the sequence of points such that the ith member of the sequence has all coordinates 0 except the ith which is 1/2. However, it is well known and easy to prove that coordinatewise continuity into S_1 does imply continuity. Therefore $h^{-1} = g \mid h(S_1^*)$ is continuous and h is a homeomorphism. (Incidentally, it may be shown that g defined on all of s is not a homeomorphism.) Another reason for defining the bridge homeomorphism on S_1^* instead of on l_2 is that functions into S_1 are continuous if they are coordinatewise continuous.

The remaining part of the proof of Lemma 3.2 is to show that $s\backslash h(S_1^*) = \bigcup C_i$. This equality is no accident since the definition of C_i was chosen only after the definition of h.

One finds from inductive use of the equation describing the x_i's in terms of the y_i's that

$$\sum_1^i x_j^2 = 1 - (1 - y_1^2)(1 - y_2^2) \cdots (1 - y_i^2).$$

Hence, if $y \in s$, the sum of the squares of the first i coordinates of $g(y)$ is less than 1. In fact, one finds that

$$g(s) = \left\{ x \in l_2 \mid \text{ for each } i, \sum_1^i x_j^2 < 1 \right\}$$

since for each $x \in l_2$ such that for each i, $\sum_1^i x_j^2 < 1$, one can inductively define y_i coordinates so that $g(y) = x$. In particular, $S_1^* \subset g(s)$ and

$$s\backslash h(S_1^*) = g^{-1} \left\{ x \in l_2 \mid \sum_1^\infty x_i^2 < 1 \right\}.$$

To prove that $h(S_1^*) = s\backslash \bigcup C_i$, we need only show that if $y \in h(S_1^*)$, $\sum_1^\infty y_i^2 = \infty$ and if $y \in s\backslash h(S_1^*)$, $\sum_1^\infty y_i^2$ is finite.

If $y \in s\backslash h(S_1^*)$ and $g(y) = x$, then $\sum_1^\infty x_i^2 < 1$. Since

$$y_i = x_i \Big/ \left(1 - \sum_1^{i-1} x_j^2\right)^{1/2}, \quad \text{then } y_i^2 \leqq x_i^2 \Big/ \left(1 - \sum_1^\infty x_j^2\right)$$

$$\text{and } \sum y_i^2 \leqq \sum x_i^2 \Big/ \left(1 - \sum_1^\infty x_j^2\right).$$

Hence, $\sum_1^\infty y_i^2$ is finite.

If $y \in h(S_1^*)$, we use a plan suggested by Robert Connelly and show that $\sum_1^\infty y_i^2 = \infty$ by demonstrating that for each integer n, $\sum_n^\infty y_i^2 > 1$. This is because

$$\sum_n^\infty y_i^2 = \frac{x_n^2}{\sum_n^\infty x_j^2} + \frac{x_{n+1}^2}{\sum_{n+1}^\infty x_j^2} + \cdots > \frac{x_n^2 + x_{n+1}^2 + \cdots}{\sum_n^\infty x_j^2} = 1.$$

LEMMA 3.3. $l_2 \backslash \bigcup E^i \sim S_1 \cap (l_2 \backslash \bigcup E^i)$.

PROOF. The required homeomorphism is $h_1 \circ h_2$ where h_2 is the homeomorphism which takes each point $(x_1, x_2, \cdots)$ to $(1, x_1, x_2, \cdots)$ and for each $x \in h_2(l_2)$, $h_1(x)$ is the point of S_1 between x and $(-1, 0, 0, \cdots)$.

4. Convergence of homeomorphisms. It is trivial to verify that the product (or composition) of any two homeomorphisms of a space X onto itself is a homeomorphism of X onto itself. Therefore, any finite product of such homeomorphisms is a homeomorphism of X onto itself. However, infinite left products, that is products of the form $\cdots \circ h_i \circ \cdots \circ h_2 \circ h_1$, may or may not converge so as to define a homeomorphism of X onto itself or even of a subset of X onto X.

Equivalently, we may consider the possible convergence of the sequence $h_1, h_2 \circ h_1, \cdots$ of homeomorphisms of X onto itself. We may denote this sequence by $(h_i \circ \cdots \circ h_2 \circ h_1)_{i>0}$. Frequently h_i is close to the identity so that members of the sequence are products of the previous element of the sequence by a homeomorphism near the identity.

There are two basic types of procedures which we shall use to establish that sequences of the form $(g_i)_{i>0} = (h_i \circ \cdots \circ h_2 \circ h_1)_{i>0}$ converge to desired homeomorphisms of some subset Y of X onto X.

Convergence Procedure I. A simple form of this procedure is represented by the following trivial but useful theorem.

THEOREM 4.1. *The sequence $(g_i)_{i>0}$ of homeomorphisms of X onto X converges to a homeomorphism of the subset Y of X onto X provided that*

(1) *for each $p \in Y$ there exists a neighborhood U of p in X and an integer $n(U)$ such that for each $n > n(U)$, $g_n | U = g_{n(U)} | U$ and*

(2) *for each $q \in X$ there exists a neighborhood V of q and an integer $n(V)$ such that for each $n > n(V)$, $g_n^{-1} | V = g_{n(V)}^{-1} | V$ and $g_n^{-1}(V) \subset Y$.*

Observe that if for each $i > 0$, $g_i = h_i \circ \cdots \circ h_2 \circ h_1$, then the condition that $g_n | U = g_{n(U)} | U$ becomes the condition that h_n restricted to $h_{n(U)} \circ \cdots \circ h_2 \circ h_1(U)$ is the identity restricted to $h_{n(U)} \circ \cdots \circ h_2 \circ h_1(U)$. Similarly, the condition $g_n^{-1} | V = g_{n(V)}^{-1} | V$ becomes $h_n^{-1} | V = \mathrm{Id} | V$.

Convergence Procedure II. In some instances it does not seem feasible to construct a desired homeomorphism by a sequential process using Convergence Procedure I. To handle such cases we introduce another procedure. While a simpler version of this convergence procedure has been applied to compact spaces, as far as the authors know, the formulation we give for complete metric spaces is new.

If G is an open covering of a space X and h is a homeomorphism of X, we use $h(G)$ to denote the collection of images of elements of G. Suppose $(h_i)_{i>0}$ is a sequence of homeomorphisms of a complete metric space X onto itself. We say that $(h_i)_{i>0}$ satisfies the *inductive convergence criterion* if there is a sequence $(G_i)_{i>0}$ of open coverings of X such that for each positive integer i,

mesh $G_i < 1/2^i$,
mesh$(h_i \circ \cdots \circ h_2 \circ h_1)^{-1}(G_i) < 1/2^i$, and
h_{i+1} is limited by G_i.

THEOREM 4.2. *If $(h_i)_{i>0}$ is a sequence of homeomorphisms of a complete metric space X onto itself which satisfies the inductive convergence criterion, then $(g_i)_{i>0} = (h_i \circ \cdots \circ h_2 \circ h_1)_{i>0}$ converges to a homeomorphism of X onto itself.*

PROOF. For each $p \in X$, $(g_i(p))_{i>0}$ is a Cauchy sequence since $g_i(p)$ and $h_{i+1} \circ g_i(p) = g_{i+1}(p)$ lie in some one element of G_i and mesh $G_i < 1/2^i$. This implies that the limit $(g_i)_{i>0}$ is uniquely defined. Since each g is continuous and $d(g_i(p), g_{i+1}(p)) < 1/2^i$ for all p, the limit $(g_i)_{i>0}$ is continuous.

Also, for any $q \in X$, q and $h_{i+1}^{-1}(q)$ lie in some one element of G_i. Since mesh $g_i^{-1}(G_i) < 1/2^i$, the distance between $g_i^{-1}(q)$ and $g_i^{-1} \circ h_{i+1}^{-1}(q) = g_{i+1}^{-1}(q)$ is less than $1/2^i$. Hence $(g_i^{-1}(q))_{i>0}$ converges. Indeed $(g_i^{-1})_{i>0}$ converges to a continuous function and thus $(g_i)_{i>0}$ converges to a homeomorphism of X onto itself.

The following is a variation of Theorem 4.2 that we shall use. If A, B are sets we use $d(A, B)$ to denote the greatest lower bound of $d(a, b)$ where $a \in A$, $b \in B$.

THEOREM 4.3. *Suppose $(K_i)_{i>0}$ is a sequence of closed sets in a complete metric space X and for each positive integer i, h_i is a homeomorphism of $X \backslash h_{i-1} \circ \cdots \circ h_2 \circ h_1 \circ \mathrm{Id}(K_i \backslash \bigcup_1^{i-1} K_j)$ onto X. Let G_i be an open covering of $X \backslash \bigcup_1^{i-1} K_j$ such that* mesh $G_i < 1/2^i$, mesh $h_i \circ \cdots \circ h_2 \circ h_1(G_i) < 1/2^i$, *and if $g \in G_i$,* diameter $g < d(g, \bigcup_1^{i-1} K_j)/2^i$. *Then $(h_i \circ \cdots \circ h_2 \circ h_1)_{i>0}$ converges to a homeomorphism of $X \backslash \bigcup K_i$ onto X if each h_{i+1} is limited by $h_i \circ \cdots \circ h_2 \circ h_1(G_i)$.*

The proof is similar to the proof of Theorem 4.2. The condition that diameter $g < d(g, \bigcup_1^{i-1} K_j)/2^i$ is used to show that $(h_1^{-1} \circ h_2^{-1} \circ \cdots \circ h_i^{-1}(q))_{i>0}$ converges to a point of $X \backslash \bigcup K_j$ rather than to a point of $\bigcup K_j$.

5. **Isotopies.** Several times in this paper we shall be concerned with isotopies, which we define for our purposes as follows. For a metric space X, an *isotopy* H of X onto X is a continuous 1-parameter family of homeomorphisms H_t $(0 \leq t \leq 1)$ of X onto X such that $H_0 = \mathrm{Id}$; it is required that H be simultaneously continuous in t and X rather than that homeomorphisms be near each other.

One can think of an isotopy as a motion of X onto itself starting with the identity H_0, ending with H_1, and using t as a time variable.

Although it is required that an isotopy H be continuous in both t and X, it is not ordinarily required that $\{H_t^{-1}\}_{0\leq t\leq 1}$ be continuous in this sense. If it is, we call the isotopy *invertible*. In this paper we shall only use isotopies that are invertible.

For compact spaces, all isotopies are invertible, but this is not true for spaces in general. A noninvertible isotopy H of l_2 onto itself can be defined as follows. Let f_i be a map of $[0, 1]$ onto $[1/i, 1]$ such that $f_i[0, 1/i+1]=f_i[1/i, 1]=1$. For each $(x_1, x_2, \cdots)\in l_2$, let $H_t(x_1, x_2, \cdots)=(x_1\cdot f_1(t), x_2\cdot f_2(t), \cdots)$. Although H_0^{-1} is continuous in l_2 and $H_t^{-1}(0, 0, \cdots)$ is continuous in t, H^{-1} is not continuous in t and l_2 at $t=0$, $(0, 0, \cdots)\in l_2$.

For $K\subset X$, an *invertible isotopy H pushing K off X is a* 1-parameter family of homeomorphisms H_t $(0\leq t\leq 1)$ onto X such that

$$H_0=\mathrm{Id},$$
$$H_t(X)=X \quad \text{for } (0\leq t<1),$$
$$H_1(X\backslash K)=X,$$

and

$$H \text{ and } H^{-1} \text{ are continuous in } t \text{ and } X.$$

Those who prefer to think of an isotopy as a map H of $X\times[0, 1]$ onto X that is an onto homeomorphism at every level may regard this invertible isotopy pushing K off X as a map from $(X\times[0,1])\backslash(K\times\{1\})$ onto X satisfying the appropriate conditions.

If K is a single point p, we say that H is an *invertible isotopy pushing p off X.*

In Lemma 7.1 and Step 1 of Lemma 8.2 we exhibit invertible isotopies pushing points off s and a copy $l_2(1)$ of l_2. The motions in the two cases are rather different. The first pushes a point off s in some one direction while the second pushes a point off $l_2(1)$ by increasing its norm (distance from the origin) using infinitely many directions. In both instances, the isotopies pushing individual points off the spaces are the key devices in pushing certain countable unions of closed sets of infinite deficiency off s and $l_2(1)$.

For any isotopy H on a space X and numbers a, b $(0\leq a<b\leq 1)$, we let $H[a, b]$ denote the isotopy which acts on the interval $[a, b]$ as H acts on $[0, 1]$. Specifically,

$$H[a, b]_t = \begin{cases} \mathrm{Id} & \text{for } 0 \leq t \leq a, \\ H_1 & \text{for } b \leq t \leq 1, \\ H_{(t-a)/(b-a)} & \text{for } a \leq t \leq b. \end{cases}$$

Observe that for H and F isotopies on X, a motion which is essentially H followed by F may be represented as the isotopy defined for each t $(0 \leqq t \leqq 1)$ and any a, b, c $(0 \leqq a < b < c \leqq 1)$ as $F[b, c]_t \circ H[a, b]_t$.

The following proposition is a variation of Theorem 4.1.

LEMMA 5.1. *Suppose that for each positive integer i, H^i is an invertible isotopy of X onto X and $p_0 \in X$. Then*

$$H = \left\{ L \prod_{i>0} H^i[1 - (1/i), 1 - (1/i + 1)]_t \right\}$$

is an invertible isotopy pushing p_0 off X provided that

(1) *for each $p \in X \backslash \{p_0\}$, there is a neighborhood U of p in X and an integer $n(U)$ such that for each $n > n(U)$, $H^n_t = \mathrm{Id}$ on $H^{n(U)}_1 \circ \cdots \circ H^2_1 \circ H^1_1(U)$ and*

(2) *for each $q \in X$, there is a neighborhood V of q in X and an integer $n(V)$ such that for each $n > n(V)$, $(H^n_t)^{-1} = \mathrm{Id}$ on V and $(H^{n(V)}_1 \circ \cdots \circ H^2_1 \circ H^1_1)^{-1}(V) \subset X \backslash \{p_0\}$.*

PROOF. For any t $(0 \leqq t < 1)$, let i be such that $1 - (1/i) > t$. Then $H^i[1 - (1/i), 1 - (1/i+1)]_t$ is the identity. Hence H_t may be considered as a finite product of homeomorphisms of X onto itself. Hence, the continuity of H and H^{-1} for values of t in $(0 \leqq t < 1)$ follows from the continuity of the finite factors to be considered.

For $x_0 \in X \backslash \{p_0\}$ and $t = 1$, the continuity of H at $(x_0, 1)$ follows from Condition 1 in the statement of the lemma while if $x \in X$, the continuity of H^{-1} at $(x, 1)$ follows from Condition 2.

6. Straightening sets in s.

There are two related ways of showing that if a countable collection of compact sets are pushed out of s, the remainder is homeomorphic with s. In one of these procedures, s is regarded as the pseudo interior of the Hilbert cube. This approach was used in [2]. The procedure used in [2] offers an alternative (and perhaps easier) method of proving Corollary 9.4. However, in the present paper we adopt a procedure that parallels our treatment of the proof that $l_2 \sim l_2 \backslash \cup E^i$. In §§ 6 and 7 we regard s as $\prod_{i>0} R_i$ where R_i denotes a copy of the real line. In this section we show how to move sets into such nice positions that they can be pushed out of s with a procedure to be described in §9.

LEMMA 6.1. *For any compact subset C of s there exists a homeomorphism h of s onto s such that $h(C)$ projects onto a single point of R_1—that is, $\tau_1 h(C)$ is a single element.*

PROOF. The proof is in two steps. In the first step we adjust C so that its image intersects any line parallel to the R_1-axis in at most one point. In the second step we move points in the 1st direction so that the final image of C lies in a hyperspace of s perpendicular to the R_1-axis.

Step 1. For each $i>1$, let f_i be a homeomorphism of $R_1\times R_i$ onto itself such that (1) f_i does not change the 1st coordinate of any point and (2) for some closed rectangular region D_i in $R_1\times R_i$ with $\tau_{\{1,i\}}(C)\subset D_i$, $f_i(D_i)$ is a closed parallelogram region intersecting no line parallel to the R_1-axis in a set of diameter greater than $1/2^i$.

Let f be the homeomorphism of s onto itself such that $f(x_1, x_2, \cdots) = (x_1, y_2, y_3, \cdots)$ where y_i is the 2nd coordinate of $f_i(x_1, x_i)$. Each point of s has a unique image in s and a unique inverse image in s determined coordinatewise. Continuity follows from the continuity of coordinate functions determined by the f_i's.

For any $i>0$, we know from the definition of f_i that $f(C)$ cannot intersect any line parallel to the R_1-axis in a set of diameter more than $1/2^i$. Therefore $f(C)$ intersects any such line in at most a single point.

Step 2. Let $s(0)$ be the set of points in s whose 1st coordinate is 0 and τ^* denote the projection function of s onto $s(0)$. Then $\tau^*|f(C)$ is a homeomorphism into $s(0)$.

Let $\phi : \tau^*(f(C))\to R_1$ be defined by $\phi=\tau_1\tau^{*-1}|\tau^*(f(C))$. By the Tietze Extension Theorem there is a map $\Phi : s(0)\to R_1$ such that $\Phi|\tau^*(f(C))=\phi$.

Let h^* be the homeomorphism of s onto itself such that for each point $p\in s(0)$ and line L_p through p parallel to the R_1-axis, $h^*|L_p$ is a translation of $-\Phi(p)$ units in the 1st direction. The homeomorphism promised by Lemma 6.1 is $h=h^*\circ f$. Note that $h(C)\subset s(0)$ and thus $\tau_1(h(C))=0$.

THEOREM 6.2. *For each collection* $\{C_i\}_{i>0}$ *of compact subsets of* s *there is a homeomorphism* g *of* s *onto* s *such that each* $g(C_i)$ *is infinitely deficient.*

PROOF. Let $\alpha_1, \alpha_2, \cdots$ be disjoint infinite subsets of Z such that $\bigcup\alpha_i=Z$. Let s be written as $\prod_{i>0} s^{\alpha_i}$ where $s^{\alpha_i}=\prod_{j\in\alpha_i} R_j$.

Let $\theta: Z\to Z$ be such that for each $i>0$, $\theta^{-1}(i)$ is infinite. We regard each s^{α_i} as a copy of s and learn from Lemma 6.1 that there is a homeomorphism g_i of s^{α_i} onto itself such that $g_i(\tau_{\alpha_i}(C_{\theta(i)}))$ is deficient with respect to the first element of α_i.

Let g be the homeomorphism of $s=\prod_{i>0} s^{\alpha_i}$ onto itself defined coordinatewise as g_i on s^{α_i}. From the definitions of θ and of the g_i's it follows that for each $i>0$, $g(C_i)$ is infinitely deficient.

7. Isotopies pushing points off s. We shall need to consider a 2-parameter $(0<r\leqq 1,\ 0\leqq t\leqq 1)$ family of homeomorphisms ${}_rH_t$ such that for r fixed, ${}_rH$ is an isotopy pushing the origin off s. We call such a family an *invertibly continuous family of invertible isotopies* if both ${}_rH_t$ and ${}_rH_t^{-1}$ are continuous in r, t, and s. We shall want each ${}_rH_t$ to be the identity outside the r-neighborhood of the origin in s. We are supposing that s has the complete metric defined by $d(x, y) = \sum(\min(1/2^i, |x_i - y_i|))$.

We first describe an invertible isotopy pushing the origin off s and then modify the isotopy to obtain invertibly continuous families of invertible isotopies each pushing the origin off s.

LEMMA 7.1. *There is an invertible isotopy H pushing the origin p_0 off s.*

PROOF. We shall push p_0 off s along the positive ray of the x_1-axis. For convenience, we regard each finite product of lines as having a Euclidean metric.

For each positive integer i, let F^i be an invertible isotopy on $R_1\times R_{i+1}$ such that (1) F_t^i leaves each point whose 1st coordinate is less than $i-1-1/2^{i-1}$ fixed while (2) F_1^i carries the vertical ray from $(i-1, -1/2^i)$ through $(i-1, 1)$ isometrically onto the horizontal ray from $(i-1/2^i, 0)$ through $(i+1, 0)$. The isotopy F^i can result from a "limited rotation" about $(i-1, 0)$ (not all points of the plane being rotated) followed by a "limited translation."

Let $a\in R_2\times\cdots\times R_i$, $b\in R_1\times R_{i+1}$, and $c\in\prod_{j>i+1}R_j$. For any $i>1$ and any $a\in R_2\times\cdots\times R_i$ let $\phi_i(a)=\max\{0, 1-2^{i+1}\cdot d(0, a)\}$ where $d(0, a)$ denotes the distance from a to the origin of Euclidean $(i-1)$-space $R_2\times\cdots\times R_i$. To avoid a special argument, we suppose that ϕ_1 is the constant 1. Let H^i be an isotopy on s defined for each t $(0\leqq t\leqq 1)$ as

$$H_t^i(a, b, c) = (a, F^i_{t\cdot\phi_i(a)}(b), c).$$

When we verify that $H^1, H^2, \cdots$ satisfies Conditions 1 and 2 of the statement of Lemma 5.1, it will follow from that lemma that

$$H_t = \left\{L\prod_{i>0} H^i[1-1/i, 1-1/i+1]_t\right\}$$

is an invertible isotopy pushing p_0 off s. Let h^i denote $H_1^i\circ\cdots\circ H_1^2\circ H_1^1$.

(1) To see that the H^i's satisfy Condition 1, one need only consider the geometry of the successive motions. Specifically, we let $p\in s\backslash p_0$ with $p=(0, \cdots, a_j, a_{j+1}, \cdots)$ where a_j is the first nonzero coordinate of p. Let $h^j(p)=(b_1, b_2, \cdots, b_j, b_{j+1}, a_{j+2}, \cdots)$ and $h^{j+1}(p)$

$=(c_1, b_2, \cdots, b_{j+1}, c_{j+2}, a_{j+3}, \cdots)$. If $j>1$, each of $b_2, \cdots, b_j$ is zero but we shall not use this information. If any one of $b_2, \cdots, b_{j+1}, c_{j+2}$ is nonzero, there is a neighborhood U of p and an integer k such that $\phi_k=0$ on the projection of $h^k(U)$ into $R_2\times \cdots \times R_k$. Hence for $k'\geqq k$, $H_t^{k'}=\mathrm{Id}$ on $h^k(U)$. If each of $b_2, \cdots, b_{j+1}$ is zero, then $b_1\neq j$ since $h^j(0, \cdots, 0, a_{j+2}, \cdots)=(j, 0, \cdots, 0, a_{j+2}, \cdots)$. If $b_1\neq j$ and $c_{j+2}=0$, $c_1<j+1-1/2^{j+1}$ since F_1^{j+1} takes the ray from $(j, -1/2^{j+1})$ through $(j, 1)$ onto the ray from $(j+1-1/2^{j+1}, 0)$ through $(j+2, 0)$. But if $c_1<j+1-1/2^{j+1}$, there is a neighborhood U of p such that the first coordinate of each point of $h^{j+1}(U)$ is less than $j+1-1/2^{j+1}$. Then for $k\geqq j+1$, $H_t^k=\mathrm{Id}$ on $h^{j+1}(U)$.

(2) To see that the H^i's satisfy Condition 2, it is only necessary to notice that H_t^{i+1} does not move a point unless its first coordinate is greater than $i-1/2^i$ and the first coordinate of $h^i(p_0)$ is i.

The following lemma is included as a stepping stone toward Lemma 7.3.

LEMMA 7.2. *Suppose s is expressed as $s=s^\alpha\times s^{\alpha'}$ as suggested in Lemma 2.1 where s^α, $s^{\alpha'}$ are copies of s with origins p_0^α, $p_0^{\alpha'}$. Then if H is an invertible isotopy pushing p_0^α off s^α and ϕ_r is a continuous 1-parameter $(0<r\leqq 1)$ family of maps of $s^{\alpha'}$ into $[0, 1]$ such that for each r, $\phi_r^{-1}(1)=p_0^{\alpha'}$, then*

$$ {}_rH_t(p, q) = (H_{t\cdot\phi_r(q)}(p), q) $$

defines an invertibly continuous 1-parameter $(0<r\leqq 1)$ family of invertible isotopies each pushing the origin off s.

The proof is left to the reader.

LEMMA 7.3. *There is an invertibly continuous 1-parameter $(0<r\leqq 1)$ family of invertible isotopies ${}_rH$ each pushing the origin p_0 off s such that ${}_rH_t$ is the identity outside the r-neighborhood of p_0.*

PROOF. For any r $(0<r\leqq 1)$, ${}_rH$ is to be an isotopy whose "action" pushes the origin off a certain subspace s_0 of s and is "phased out" in a neighborhood U of the origin in the complementary subspace s_0' (i.e., $s=s_0\times s_0'$). The two "factors" which determine the size of the domain of support of the isotopy ${}_rH$ are (1) the diameter of the subspace s_0 and (2) the diameter of the neighborhood U of the origin in s_0'. We shrink the former, as r tends to zero, by a process of swapping axes and the latter by a reduction process introduced after each swap of axes.

Let α be an infinite subset of Z such that α' is infinite and $1,2,3\in\alpha'$. The distances in s^α and $s^{\alpha'}$ are given by $d(x, y)$

$=\sum_{i\in\alpha}(\min(1/2^i, |x_i-y_i|))$ and $d(x, y)=\sum_{j\in\alpha'}(\min(1/2^i, |x_j-y_j|))$ respectively. Note that the diameter of s^α is less than $1/2^3$.

Let H be an isotopy pushing the origin off s^α as guaranteed by Lemma 7.1.

Let ϕ be a map of $s^{\alpha'}$ onto $[0, 1]$ such that $\phi^{-1}(1)=p_0^{\alpha'}$ and $\phi=0$ for all points outside a 1/8-neighborhood of $p_0^{\alpha'}$. We could have specified that $\phi(q)=\max[0,\ 1-8\sum_{i\in\alpha'}(\min\ 1/2^i,\ |x_i|)]$ for $q=(x_{n_1}, x_{n_2}, \cdots)$ but we shall not be concerned with the particular description of ϕ. In fact, it is perhaps best not to define ϕ in terms of the distance from $p_0^{\alpha'}$ since this distance changes when coordinates are interchanged. We define ${}_1H$ so that for each $(p, q)\in s^\alpha\times s^{\alpha'}$ (with $(p, q)\neq p_0$ for $t=1$)

$$ {}_1H_t(p, q) = (H_{t\cdot\phi(q)}(p), q). $$

Note that ${}_1H_t(p, q)=(p, q)$ if $d((p, q), p_0)\geqq 1/4$. This is because $1, 2, 3\in\alpha'$ and $d((p, q), p_0)\geqq 1/4$ implies $d(q, p_0^{\alpha'})\geqq 1/8$ and $\phi(q)=0$.

As r changes from 1 to 1/2, we define ${}_rH$ by modifying ${}_1H$ by swapping axes and reducing ϕ. We describe each of these operations separately.

Swapping axes. Our purpose of swapping axes is to move the small positive integers out of α and hence reduce the diameter of s^α. We recall that 1, 2, 3 are in α'. Let j be the least integer in α and k be an integer in α' larger than j. Let α_1' be the subset of Z obtained from α' by replacing k by j and α_1 be the subset obtained from α by replacing j by k.

The action of the isotopy ${}_1H$ is defined on s^α and is phased out in a neighborhood of the origin in $s^{\alpha'}$. We gradually transfer the action over $3/4\leqq r\leqq 1$ so that the action of ${}_{3/4}H$ will be defined on s^{α_1} and be phased out in a neighborhood of the origin in $s^{\alpha_1'}$. The following formulas describe a suitable version of this process.

For $0\leqq\lambda\leqq 1$, let F_λ be the rotation of s such that if $F_\lambda(x_1, x_2, \cdots)=(y_1, y_2, \cdots)$, then $y_i=x_i$ unless $i\in\{j, k\}$ and (y_j, y_k) is the image of (x_j, x_k) under a clockwise rotation of the plane $R_j\times R_k$ by $\lambda\cdot\pi/2$ about the origin. Let $f(r)$ be the linear function that sends 1 to 0 and 3/4 to 1. Then for $3/4\leqq r\leqq 1$,

$$ {}_rH_t = F_{f(r)}^{-1}\circ {}_1H_t\circ F_{f(r)}. $$

We note that if $d(x, p_0)\geqq 1/2$, then ${}_rH_t(x)=x$ if $3/4\leqq r\leqq 1$. This is because for such x $d(F_{f(r)}(x), p_0)\geqq 1/2-1/2^j\geqq 1/4$, ${}_1H_t\circ F_{f(r)}(x)=F_{f(r)}(x)$ and $F_{f(r)}^{-1}\circ {}_1H_t\circ F_{f(r)}(x)=x$.

Reducing ϕ. Here we define ${}_rH$ for $1/2\leqq r\leqq 3/4$. Let $x=(p', q')$

where $p' \in s^{\alpha_1}$, $q' \in s^{\alpha_1'}$. From the definitions of ${}_1H$ and F_λ and specifically from the fact that the rotation of $R_j \times R_k$ which determines F_1 merely interchanged vertical and horizontal lines in $R_j \times R_k$, it follows that

$$ {}_{3/4}H_t(p', q') = (H_t'._{\phi_{3/4}(q')}(p'), q') $$

where $\phi(3/4)$ is defined below and where H_t' is $T^{-1} \circ H_t \circ T$ applied to s^{α_1} instead of s^{α} with T being a permutation of a finite number of coordinates. The reason for introducing T is that k may not be the least integer in α_1 and in this case we need to shift k to the first element of α_1 and to move all the others in α_1 which are less than k into their successors in α_1. Also, $\phi_{3/4}$ is a map of $s^{\alpha_1'}$ onto $[0, 1]$ such that $\phi_{3/4}^{-1}(1) = p_{01}^{\alpha'}$ and $\phi_{3/4} = 0$ except in a small neighborhood of the origin $p_{01}^{\alpha}{}'$. In fact, $\phi_{3/4} = \phi \circ T'$ where T' is the map from $s^{\alpha_1'}$ onto $s^{\alpha'}$ which merely moves the jth coordinate of a point to the kth coordinate and changes its sign. Note that if $d(x, p_0) \geqq 1/2$, then $\phi_{3/4}(q') = 0$ as $d(q', p_{01}^{\alpha'}) \geqq 1/2 - 1/16$ and $d(T'(q'), p_0^{\alpha'}) \geqq 1/2 - 1/16 - 1/32 \geqq 1/8$ which implies that $\phi_{3/4}(q') = \phi(T'(q')) = 0$.

Let $\phi_{1/2}$ be a map of $s^{\alpha_1'}$ onto $[0, 1]$ so that $\phi_{1/2} \leqq \phi_{3/4}$, $\phi_{1/2}^{-1}(1) = p_{01}^{\alpha'}$, and $\phi_{1/2} = 0$ except in the $1/2^4$-neighborhood of $p_{01}^{\alpha'}$. For each $q' \in s^{\alpha_2'}$ let, $\phi_r(q')$ divide $\phi_{1/2}(q')$, $\phi_{3/4}(q')$ in the same ratio that r divides $1/2$, $3/4$ with $\phi_r(q') = \phi_{1/2}(q')$ if $\phi_{1/2}(q') = \phi_{3/4}(q')$. Then for $1/2 \leqq r \leqq 3/4$, define

$$ {}_rH_t(p', q') = (H'_{t \cdot \phi_r(q')}(p'), q'). $$

As r moved from 1 to 1/2, we defined ${}_rH_t$ from ${}_1H_t$ by swapping axes and reducing ϕ. By similarly swapping axes so as to replace the least element of α_1 with a larger element of α_1' and by further reducing ϕ, we define ${}_rH_t$ $(1/4 \leqq r \leqq 1/2)$ from ${}_{1/2}H_t$ in an analogous fashion. Similarly, we define ${}_rH_t$ as r shrinks to $1/2^3$, to $1/2^4$, $\cdots$.

8. **Isotopies pushing points off l_2.** We prove lemmas about pushing points off l_2 similar to those we proved about pushing points out of s in the last section. We recall that S_1 is the set of all points of l_2 at a distance of 1 from the origin.

LEMMA 8.1. *There is an invertible isotopy F pushing a point off S_1.*

PROOF. Let $l_2(1)$ be the set of all points of l_2 whose first coordinate is 1. For each positive integer i, let p_i denote the point of l_2 whose first i coordinates are each 1 and whose other coordinates are 0. We prove Lemma 8.1 in two steps.

Step 1. In this step we exhibit a particular invertible isotopy H

on $l_2(1)$ pushing p_1 out of $l_2(1)$. The plan is to push p_1 to p_2, then to p_3, then to $p_4, \cdots$, and hence off l_2. We take care that no point other than p_1 undergoes more than a finite number of these pushes.

For each integer $i \geqq 2$, let U_i be the neighborhood of p_i in $l_2(1)$ such that

$$U_i = \left\{(1, x_2, \cdots) \in l_2(1) \mid \sum_2^i (1 - x_j)^2 < 1/i^2\right\}.$$

Let H^1 be the isotopy on $l_2(1)$ such that H_t^1 is a translation of t units in the x_2 direction. For each integer $i \geqq 2$, H^i is the isotopy on $l_2(1)$ such that $H_t^i(p)$ is the point resulting from translating p by $i \cdot t \cdot d(p, (l_2(1) \setminus U_i))$ in the x_{i+1} direction. Then

$$H_t = L \prod_{i>0} H^i[1 - 1/i, 1 - 1/(i+1)]_t.$$

It may be shown with the use of Theorem 4.1 and Lemma 5.1 that H is an invertible isotopy pushing p_1 out of $l_2(1)$.

Step 2. In this step we describe an invertible isotopy F pushing p_1 off S_1. Let S_1^+ be the set of all points of S_1 with positive 1st coordinate and ρ be the projection of S_1^+ onto $l_2(1)$ from the origin. The invertible isotopy F promised by Lemma 8.1 may be defined as

$$\begin{aligned} F_t(p) &= \rho^{-1} \circ H_t \circ \rho(p) \quad \text{if } p \in S_1^+, \\ &= p \quad \text{if } p \in S_1 \setminus S_1^+. \end{aligned}$$

To complete the proof we need only verify the continuity of F and F^{-1} at points $p \in S_1$ whose 1st coordinate is 0. Suppose the ith coordinate of p is not 0 and δ is one third the absolute value of this coordinate. A calculation shows that if U is the δ-neighborhood of p in S_1 and $q \in \rho(U \cap S_1^+)$, then the absolute value of the ith coordinate of q is greater than 2. For $k \geqq i$, H_t^k and $(H_t^k)^{-1}$ do not move q. Hence, to verify the continuity of F and F^{-1}, we only need to check the effect of a finite number of $\rho^{-1} \circ H_t^i \circ \rho$'s near p; in fact we only need to check one of them since the composition of a finite collection of continuous functions is continuous.

A computation shows that if $p = (0, x_2, x_3, \cdots) \in S_1$ and $q = (y_1, y_2, \cdots) \in S_1^+$ within δ of p, then for some $0 \leqq \theta \leqq 1$,

$$\rho^{-1} \circ (H_t^{j-1})^{-1} \circ \rho(q) = (k \cdot y_1, k \cdot y_2, \cdots, k(y_j + \theta \cdot y_1), \cdots)$$

where $k = (1 + 2 \cdot y_j \cdot y_1 \cdot \theta + y_1^2 \cdot \theta^2)^{-1/2}$. If δ is small, then y_1 is small and k is near 1. Hence coordinatewise, $\rho^{-1} \circ (H_t^j)^{-1} \circ \rho(q)$ is near p. Coordinatewise continuity in S_1 implies continuity there so F is continuous at p. Similarly, F^{-1} is continuous at p also.

LEMMA 8.2. *There is an invertible isotopy H pushing the origin p_0 off l_2 such that each H_t is the identity outside the 1-neighborhood of p_0.*

PROOF. For each point $p \neq p_0$, we let a_p denote the projection of p from p_0 on S_1.

Let ϕ be the map of the nonnegative reals onto $[0, 1]$ such that $\phi(x)=0$ if $x \geqq 1$, $\phi(0)=0$, $\phi(1/2)=1$, and ϕ is linear on each of $[0, 1/2]$ and $[1/2, 1]$.

The isotopy H promised by Lemma 8.2 is defined by $H_t = H^2[1/2, 1]_t \circ H^1[0, 1/2]_t$ where H^1 and H^2 are isotopies defined as follows. The isotopy H^1 is the identity outside the 1-neighborhood of p_0 and $H^1_1(p_0) = (1/2, 0, 0, \cdots)$. For each t $(0 \leqq t \leqq 1)$, $H^2_t(p)$ is at the same distance from p_0 as p and the projection of $H^2_t(p)$ from p_0 on S_1 is $F_{t \cdot \phi(d(p,p_0))}(a_p)$ where F is the isotopy of Lemma 8.1.

LEMMA 8.3. *There is an invertibly continuous 1-parameter $(0<r\leqq 1)$ family of invertible isotopies $_rH$ each pushing the origin p_0 off l_2 such that for any t $(0 \leqq t \leqq 1)$, $_rH_t$ is the identity outside the r-neighborhood of p_0.*

PROOF. The required isotopy is defined by

$$_rH_t = r \circ H_t \circ 1/r$$

where H is the isotopy of Lemma 8.2 and $r(p)$ is the point whose coordinates are obtained by multiplying those of p by r.

9. **Eliminating sets with small pushes.** Can a small push send a mathematician out of society without leaving a vacancy? We shall be interested in conditions under which a set can be pushed out of a space with a small push.

LEMMA 9.1. *Let X denote l_2 or s, G be an open covering of X, α be an infinite subset of Z, and K be a closed subset of X which is deficient with respect to α. Then for each open set U containing K there exists a homeomorphism h of $X \backslash K$ onto X such that h is limited by G, h is the identity except on U, and $\tau_i(x) = \tau_i(h(x))$ for each $i \in \alpha'$.*

PROOF. We suppose $\alpha' \neq \emptyset$ and write $X = X^\alpha \times X^{\alpha'}$ as suggested by Lemmas 2.1 and 2.2. With no loss of generality we suppose that $\tau_\alpha(K) = \{p_0^\alpha\}$ where p_0^α is the origin in X^α. Let $\tau_{\alpha'}(K) = K' \subset X^{\alpha'}$. See Figure 1.

We need a function r from $X^{\alpha'}$ into $(0, 1]$ to measure how far each

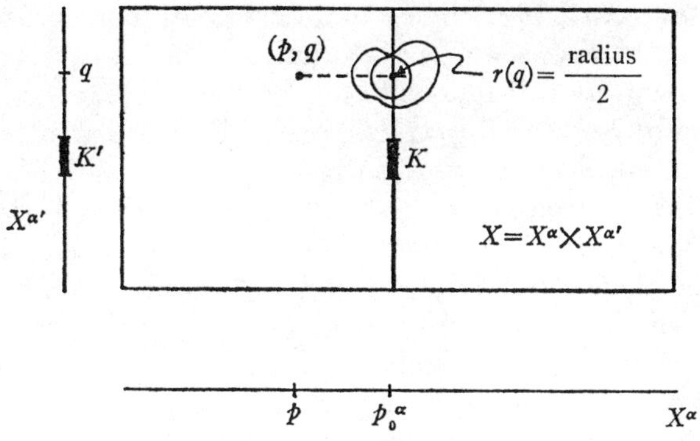

FIGURE 1

point (p_0^α, q) lies inside elements of G. Let G' be an open covering of X of mesh less than 1 such that each element of G' lies in an element of G and each element of G' intersecting K lies in U. Letting $V_\epsilon(x)$ denote the ϵ-neighborhood of x in X, we define

$$r(q) = \sup \{\epsilon \mid V_{2\epsilon}(p_0^\alpha, q) \text{ lies in some element of } G'\}.$$

Note that for each $q \in X^{\alpha'}$, $r(q) > 0$.

Next, we need a function t from $X^{\alpha'}$ into $[0, 1]$ to measure the degree of closeness of points of $X^{\alpha'}$ to K'. Let

$$A = \{x \in X^{\alpha'} \mid V_{r(x)}(p_0^\alpha, x) \not\subset U\}$$

and

$$t(q) = d(q, A)/(d(q, A) + d(q, K')).$$

(We supposed $A \neq \emptyset$. If $A \neq \emptyset$, define $t(q) = 1$.)

Note that $\bar{A} \cap K' = \emptyset$, $t(K') = 1$, and $t(A) = 0$.

By Lemma 8.3 for l_2 (and thus for l_2^α) and by Lemma 7.3 for s (and thus for s^α), there exists an invertibly continuous family ${}_rH$ $(0 < r \leqq 1)$ of invertible isotopies each pushing p_0^α off X^α and such that for each t $(0 \leqq t \leqq 1)$, ${}_rH_t$ is the identity outside the r-neighborhood of p_0^α in X^α.

We shall verify that the desired homeomorphism h of our lemma is

$$h(p, q) = ({}_{r(q)}H_{t(q)}(p), q).$$

Since $t(q) = 1$ for $q \in K'$, we note that for such q, ${}_{r(q)}H_{t(q)}$ is a homeomorphism of $(X^\alpha \setminus \{p_0\}) \times \{q\}$ onto $X^\alpha \times \{q\}$. Also, since $t(q) < 1$ for $q \in X^{\alpha'} \setminus K'$, we note that for such q, ${}_{r(q)}H_{t(q)}$ is a homeomorphism of

$X^\alpha \times \{q\}$ onto $X^\alpha \times \{q\}$. Since r and t are continuous functions on X and ${}_rH$ is a continuous family of isotopies, then h is continuous. Since ${}_rH$ is an invertibly continuous family of invertible isotopies and h^{-1} is described by $h^{-1}(p) = ({}_{r(q)}H_{t(q)}^{-1}(p), q)$, h^{-1} is also continuous. Hence, h is a homeomorphism of $X \backslash K$ onto X.

Since q is the second coordinate of $h(p, q)$, $\tau_i(x) = \tau_i(h(x))$ for each $i \in \alpha'$. We note that h is the identity outside U since ${}_rH_t$ is the identity outside the r neighborhood of p_0^α and $V_{r(q)}(p_0^\alpha, q) \subset U$ if $t(q) \neq 0$.

It only remains to observe that h is limited by G. Since ${}_rH_t$ is the identity except on the r-neighborhood of p_0^α in X^α, either $h(p, q) = (p, q)$ or each of p and ${}_{r(q)}H_{t(q)}(p)$ lie within $r(q)$ of p_0^α. In this latter case, the definition of $r(q)$ insures that both (p, q) and $h(p, q)$ lie in an element of G containing (p_0, q).

We are now in a position to prove a theorem having two important corollaries.

THEOREM 9.2. *Let X denote either l_2 or s, U be an open set in X, and $\{K_i\}_{i>0}$ be a countable collection of closed sets of X which lie in U, with each K_i having infinite deficiency. Then there exists a homeomorphism h of $X \backslash \bigcup K_i$ onto X such that $h | X \backslash U$ is the identity.*

PROOF. By a standard sequential process augmenting one finite set at a time, it is routine to show that there exists a collection $\{\alpha_i\}_{i>0}$ of disjoint infinite subsets of Z such that K_i is deficient with respect to α_i. Using Lemma 9.1 to assert the existence of h_i, we can inductively define a sequence of homeomorphisms $(h_i)_{i>0}$ and a sequence of coverings $(G_i)_{i>0}$ satisfying the hypotheses of Theorem 4.3 (specifying that h_i does not change the j-coordinate of any point unless $j \in \alpha_i$) so that $L \prod_{i>0} h_i$ is a homeomorphism of $X \backslash \bigcup K_i$ onto X.

COROLLARY 9.3. *$l_2 \backslash \bigcup E^i \sim l_2$.*

COROLLARY 9.4. *If $\{C_i\}_{i>0}$ is a countable family of compact subsets of s, then $s \backslash \bigcup C_i$ is homeomorphic to s.*

PROOF. It follows from Theorem 6.2 that there is a homeomorphism h of s onto itself such that each $h(C_i)$ is infinitely deficient. Then Corollary 9.4 follows from Theorem 9.2.

The following corollary may be useful in studying spaces which are locally like l_2 or s.

COROLLARY 9.5. *If U is an open subset of l_2 and $K_1, K_2, \cdots$ are compact subsets in U, then there is a homeomorphism h of $l_2 \backslash \bigcup K_i$ onto l_2 such that $h | l_2 \backslash U$ is the identity.*

PROOF. For convenience, we regard l_2 as s. It follows from Theorem 6.2 that there is a homeomorphism g of s onto itself such that each $g(K_i)$ is infinitely deficient. We learn from Theorem 9.2 that there is a homeomorphism h' of $s\backslash \bigcup g(K_i)$ onto s such that $h'|s\backslash g(U)$ is the identity. Then $h = g^{-1} \circ h \circ g$.

10. Summary of the procedure. In the foregoing we have established the existence of a homeomorphism between l_2 and s (Theorem 3.1). Our method of attack was to use several intermediate spaces and homeomorphisms as follows:

$$l_2 \overset{A}{\sim} l_2\backslash \bigcup E^i \overset{B}{\sim} S_1 \cap (l_2\backslash \bigcup E^i) \overset{C}{\sim} s\backslash \bigcup C_i \overset{D}{\sim} s.$$

Of the four homeomorphisms, B was easily established in Lemma 3.3 and C was established in Lemma 3.2 by exhibiting explicit formulas. The bulk of the paper was concerned with apparatus for establishing A and D which were stated as Corollaries 9.3 and 9.4. §4 identified criteria for the convergence of sequences of homeomorphisms and §5 introduced isotopy procedures to be used later.

§§6 and 7 were concerned with properties of s while §8 was concerned with properties of l_2 analogous to those of §7 for s. In this treatment, it appears as if l_2 is a simpler space to handle than is s, but in fact the procedure was designed for l_2 and then adapted to s since we were unable to adapt to l_2 a more natural simpler procedure for s.

Finally, in §9, the results of §§7 and 8 pushing individual points off s and l_2 were used to push $\bigcup C_i$ off s and $\bigcup E^i$ off l_2.

11. Some open questions related to s or l_2. Various topological properties of s or l_2 have been studied by Klee [13], [14], [15], Anderson [1], [2], [3], [4] and others. In various survey papers to appear in the Proceedings of the Symposium on Infinite Dimensional Topology (Annals of Mathematics Study), such properties are discussed. Here we list a number of open questions suggested conversationally or in print by one or more of several authors including Bessaga, Pelczynski, Klee, Fort, Borsuk, Eells, Palais, Henderson, Corson and others.

In this section we restrict ourselves to questions directly related to s or l_2 (and thus to separable metric spaces). See Bessaga [6] for many questions concerning homeomorphisms between various topological linear spaces.

Questions Concerning the Hilbert Cube as a Compactification of s. Let the Hilbert cube, I^∞, be written as $\prod_{i>0} I_i$ where for each $i>0$, I_i is the closed interval $[-1, 1]$. As in §3, we may regard $s=\prod_{i>0} I_i^0$ where for each $i>0$, I_i^0 is the open interval $(-1, 1)$. Then $s\subset I^\infty$ and I^∞ is a compactification of s. Indeed, both s and $I^\infty\backslash s$ are dense in I^∞. A *β^*-homeomorphism* is a homeomorphism h of I^∞ onto itself such that $h(s)=s$ and a *β-homeomorphism* is a homeomorphism h of I^∞ onto itself such that $h(s)\supset s$. A closed set K in a space X is said to have *Property Z* if for every nonnull homotopically trivial open set U in X, $U\backslash K$ is nonnull and homotopically trivial.

(1) Let $K\subset I^\infty$. What are necessary and sufficient conditions on K (or on $I^\infty\backslash K$) in order that there exist a homeomorphism of I^∞ onto itself carrying K onto $B(I^\infty)$? Anderson has recently shown that if $K\supset B(I^\infty)$ then in order that such a homeomorphism exist it is necessary and sufficient that K be the countable union of closed sets with Property Z.

(2) A special case of (1) is the following. Let $s\subset M\subset I^\infty$. What are necessary and sufficient conditions on M in order for there to exist a β-homeomorphism h such that $h(s)=M$? In particular, is it necessary and sufficient that (a) $I^\infty\backslash M$ be the countable union of compact sets and (b) each compact subset of M have Property Z in M? (Conditions (a) and (b) are clearly necessary.)

(3) A different version of question (2) would ask simply for necessary and sufficient conditions that s and M be homeomorphic.

(4) Let s_f and I_f^∞ be the sets of all points of s and I^∞ with only finitely many nonzero coordinates. Characterize the subsets K of s (or I^∞) for which there exist homeomorphisms of s (or I^∞) onto itself carrying K onto s_f (or I_f^∞).

(5) Regarding $s\subset I^\infty$ it is easy to show that there exist homeomorphisms of s onto itself that cannot be extended to β^*-homeomorphisms, but the following question seems more interesting. For any homeomorphism h of s onto itself does there exist a homeomorphism f of s onto itself such that $f^{-1}hf$ can be extended to a β^*-homeomorphism?

Questions on Infinite Products. (6) Is every product of a compact absolute retract by I^∞ homeomorphic to I^∞?

(7) Is every countable infinite product of compact absolute retracts homeomorphic to I^∞?

A set is a topologically complete absolute retract if it admits a complete metric and is a retract of every metric space in which it is embedded as a closed set.

(8) Is every product of a topologically complete absolute retract by s homeomorphic to s?

(9) Is every countable infinite product of topologically complete

absolute retracts (with infinitely many noncompact factors) homeomorphic to s?

It is reasonable to weaken questions (6) to (8) by replacing "absolute retract" by "contractible finite complex" or a similar condition. In an as yet unpublished paper, Anderson has shown that any countable infinite product of dendrons is homeomorphic to I^∞. (A dendron is an acyclic locally connected one-dimensional continuum.) He feels it likely that the methods of the proof of this proposition can be modified to show that any countable infinite product of contractible finite complexes is homeomorphic to I^∞.

(10) Is every product of I^∞ by a finite connected complex homeomorphic to a product of I^∞ by some manifold-with-boundary?

(11) Is every product of s by a finite connected complex homeomorphic to a product of s by some manifold-with-boundary?

Questions on Topological Banach Manifolds. A number of authors have studied the differential topology of so-called Banach manifolds. For our purposes, we consider a *topological Banach manifold* (T.B.M.) to be a connected separable metric space locally like s—that is with an open covering of sets each homeomorphic to s. (In effect we are specifying that our Banach manifolds are modeled on some separable infinite dimensional Banach space.)

(12) Can every T.B.M. be imbedded in s as an open set in s?

(13) If a T.B.M. can be imbedded in s as an open subset of s, then the T.B.M. has a local compactification as a metric space admitting a cover by open sets homeomorphic to open subsets of I^∞. Furthermore the T.B.M. intersects such open sets in I^∞ in the manner induced by the open imbedding. As a weaker version of the imbedding problem of (12), we can ask if such a local compactification of each T.B.M. is possible.

(14) Is every contractible T.B.M. homeomorphic to s? What about the special case where the T.B.M. is homeomorphic to an open subset of s?

(15) More generally than in (14), are every two T.B.M.'s that are of the same homotopy type, homeomorphic to each other?

Questions on Unions of Two Sets. (16) Let $M=H\cup K$. If H, K, and $H\cap K$ are all homeomorphic to I^∞, is $M\sim I^\infty$?

(17) Let $M=H\cup K$. If H and K are both closed or both open and if H, K, and $H\cap K$ are all homeomorphic to s, is $M\sim s$?

(18) Let $M=H\cup K$. If H, K, and M are all homeomorphic to I^∞, is $H\cap K\sim I^\infty$?

(19) Let $M=H\cup K$. If H and K are both closed or both open and if H, K, and M are all homeomorphic to s, is $H\cap K\sim s$?

Questions on Spaces of Closed Subsets. (20) Is the space of all closed subsets (with the Hausdorff metric) of an n-cell (or of I^∞) homeomorphic to I^∞? This is not even known for $n=1$.

(21) Is the space of all subcontinua of an n-cell ($n>1$) or of I^∞ homeomorphic to I^∞?

Questions on Spaces of Homeomorphisms. (22) Is s homeomorphic to the space of all homeomorphisms of a disk onto itself which are the identity on the boundary of the disk? It is known that s is homeomorphic to the space of orientation preserving homeomorphisms of an interval onto itself.

(23) For any geometric n-sphere or n-cube X and any sufficiently small $\epsilon>0$, is s homeomorphic to the space of all homeomorphisms of X onto itself which are within ϵ of the identity? The answer is almost certainly *yes* for $n=1$.

(24) Is s homeomorphic to the space of all homeomorphisms of I^∞ onto itself?

REFERENCES

1. R. D. Anderson, *Hilbert space is homeomorphic to the countable infinite product of lines*, Bull. Amer. Math. Soc. **72** (1966), 515–519.

2. ———, *Topological properties of the Hilbert cube and the infinite product of open intervals*, Trans. Amer. Math. Soc. **126** (1967), 200–216.

3. ———, *On topological infinite deficiency*, Michigan Math. J. **14** (1967), 365–383.

4. ———, *On a theorem of Klee*, Proc. Amer. Math. Soc. **17** (1966), 1401–1404.

5. S. Banach, *Théorie des opérations linéaires*, Monografie Mat. PWN, Warsaw, 1932.

6. C. Bessaga, *On topological classification of complete linear metric spaces*, Fund. Math. **56** (1965), 251–288.

7. C. Bessaga and V. L. Klee, *Two topological properties of topological linear spaces*, Israel J. Math. **2** (1964), 211–220.

8. ———, *Every non-normable Fréchet space is homeomorphic with all its closed convex bodies*, Math. Ann. **163** (1966), 161–166.

9. C. Bessaga and A. Pelczynski, *Some remarks on homeomorphisms of F-spaces*, Bull. Acad. Polon. Sci. Sér. Sci. Math. Astronom. Phys. **10** (1962), 265–270.

10. M. Fréchet, *Les espaces abstraits*, Gauthier-Villars, Paris, 1928.

11. M. I. Kadec, *On topological equivalence of separable Banach spaces*, Dokl. Akad. Nauk SSSR **167** (1966), 23–25 = Soviet Math. Dokl. **7** (1966), 319–322.

12. ———, *Proof of the topological equivalence of all separable infinite-dimensional Banach spaces*, Functional Anal. and Appl. **1** (1967), 53–62.

13. V. L. Klee, *Convex bodies and periodic homeomorphisms in Hilbert space*, Trans. Amer. Math. Soc. **74** (1953), 10–43.

14. ———, *Some topological properties of convex sets*, Trans. Amer. Math. Soc. **78** (1955), 30–45.

15. ———, *A note on topological properties of normed linear spaces*, Proc. Amer. Math. Soc. **7** (1956), 673–674.

LOUISIANA STATE UNIVERSITY, BATON ROUGE AND
UNIVERSITY OF WISCONSIN, MADISON

V. Miscellaneous Topics

A. Linear Isotopies

TRANSACTIONS OF THE
AMERICAN MATHEMATICAL SOCIETY
Volume 237, March 1978

LINEAR ISOTOPIES IN E^2

BY

R. H. BING AND MICHAEL STARBIRD[1]

ABSTRACT. This paper deals with continuous families of linear embeddings (called linear isotopies) of finite complexes in the Euclidean plane E^2. Suppose f and g are two linear embeddings of a finite complex P with triangulation T into a simply connected open subset U of E^2 so that there is an orientation preserving homeomorphism H of E^2 to itself with $H \circ f = g$. It is shown that there is a continuous family of embeddings h_t: $P \to U$ ($t \in [0, 1]$) so that $h_0 = f$, $h_1 = g$, and for each t, h_t is linear with respect to T.

It is also shown that if P is a PL star-like disk in E^2 with a triangulation T which has no spanning edges and f is a homeomorphism of P which is the identity on Bd P and is linear with respect to T, then there is a continuous family of homeomorphisms h_t: $P \to P$ ($t \in [0, 1]$) such that $h_0 = \text{id}$, $h_1 = f$, and for each t, h_t is linear with respect to T. An example shows the necessity of the "star-like" requirement. A consequence of this last theorem is a linear isotopy version of the Alexander isotopy theorem–namely, if f and g are two PL embeddings of a disk P into E^2 so that $f|\text{Bd } P = g|\text{Bd } P$, then there is a linear isotopy with respect to some triangulation of P which starts at f, ends at g, and leaves the boundary fixed throughout.

1. Introduction. This paper deals with finite complexes linearly embedded in the Euclidean plane E^2 and continuous families of such linear embeddings which are called linear isotopies.

Suppose (P, T) is a compact triangulated PL complex in a simply connected open subset U of E^2 and g is a linear homeomorphism of P into U that can be extended to an orientation preserving homeomorphism of E^2 onto itself. It is shown in Theorem 7.2 that there is a linear isotopy h_t ($t \in [0, 1]$) of (P, T) into U such that $h_0 = \text{id}$ and $h_1 = g$. Each h_t is linear on each simplex of T.

Theorem 7.2 follows from the result (Theorem 7.1) that if (D, T) is a triangulated disk in E^2, then there is a linear isotopy h_t ($t \in [0, 1]$) of (D, T) into E^2 such that $h_0 = \text{id}$, $h_1(D)$ is a convex disk, and h_t is decreasing in the sense that if $0 \leqslant s \leqslant t \leqslant 1$, $h_t(D) \subset h_s(D)$.

Suppose (P, T) is a complex linearly embedded in E^2 and g is a linear

Presented to the Society, January 23, 1975 under the title *Linear isotopies*; received by the editors December 1, 1975 and, in revised form, November 3, 1976.

AMS (MOS) subject classifications (1970). Primary 57A05, 57C35, 50B99.

Key words and phrases. Linear isotopy, push, linear embedding.

[1]Research partially supported by the National Science Foundation.

homeomorphism of (P, T) into E^2 that is fixed on a certain subcomplex P' of P. Conditions are investigated under which there is a linear isotopy h_t ($t \in [0, 1]$) of (P, T) into E^2 such that

$h_0 = \text{id}$, each h_t is fixed on P', and $h_1 = g$.

It is shown in Theorem 4.1 that if (P, T) is a star-like disk without spanning edges and $P' = \text{Bd } P$, then for each g there is such an h_t ($t \in [0, 1]$). However, Example 4.2 shows that the result does not always hold if some 1-simplex of T is permitted to span P.

§2 treats the question of extending the linear embedding into E^2 of a subcomplex C^1 of a finite complex C to a linear embedding of all of C into E^2. Although we do not get necessary and sufficient conditions that such an embedding can be extended, we get some partial solutions (mostly in the case where C is a triangulated disk and $C^1 = \text{Bd } D$). Our methods give a new proof of Fary's theorem which states that the extension is possible if C is a 1-complex which can be topologically embedded in E^2 and $C^1 = \varnothing$.

A linear isotopy version of the Alexander Isotopy Theorem for PL disks in E^2 appears in §5, namely, if P is a PL disk in E^2 and f is a PL homeomorphism of P fixed on Bd P, then there is a triangulation T of P and a linear isotopy $h_t\colon (P, T) \to E^2$ ($t \in [0, 1]$) with $h_0 = \text{id}$, $h_1 = f$, and for each t, $h_t|\text{Bd } P = \text{id}$.

§3 contains the definitions of a linear isotopy and the related notion of a push. It is shown that a linear isotopy of a finite complex in E^n can be extended to a linear isotopy of all of E^n.

§§4 and 5 contain the theorems about linear isotopies of disks with fixed boundaries.

Linear degeneration is defined in §6. This concept is a variation of a collapse and is used to prove the theorems in §7 mentioned in the second paragraph of the introduction.

In future papers Starbird will make use of some variations of the theorems in this paper to investigate questions about linear isotopies in E^3.

2. Linear embeddings. In this section we investigate the question of extending the linear embedding of a part of a complex into E^2 to a linear embedding of the entire complex.

DEFINITION. Let C be a complex with a triangulation T. Then $h\colon C \to E^n$ is a *linear embedding* of C into E^n if h is an embedding which is linear on each simplex in T.

Notation. When we wish to emphasize a particular triangulation on a complex, we may write (C, T) to stand for a complex C with a triangulation T. We use St(v) and Lk(v) to denote the star and link respectively of a vertex v in a complex C.

DEFINITION. A disk D in E^2 is *star-like* if and only if there exists a point s

in E^2 (necessarily in Int D) such that for every point x in Bd D, the straight segment xs meets Bd D only at x. We say that such a point s *can see* every point of Bd D.

Note that if D is a star-like polygonal disk in E^2, the set of points from which all of Bd D can be seen is a convex, open subset of Int D.

THEOREM 2.1. *Let P be a polygonal disk with a triangulation T which contains no spanning edge, i.e., if v and w are nonadjacent vertices on* Bd P, *then vw is not a simplex in the triangulation T. Then if h*: Bd $P \to E^2$ *is a linear embedding of* Bd P *such that h*(Bd P) *bounds a star-like disk in E^2, then there exists a linear embedding H: $P \to E^2$ shich extends h.*

PROOF. The proof is by induction on n, the number of interior vertices in the triangulation T.

If $n = 0$, P is a 2-simplex and there is nothing to do.

Assume that the theorem is true for triangulations with k interior vertices where $k < n$. Suppose T has n interior vertices. Let v be an interior vertex of T so that St(v) contains an edge on Bd P. Let s be a point in E^2 which can see every point of h(Bd P). We will have H map v to s. For every vertex $w \in$ (Bd $P \cap$ St(v)), $H(vw)$ will be the straight segment from $h(w)$ to s. These segments divide the interior of h(Bd P) into star-like subdisks each of which meets the requirements of the theorem. Complete the proof by applying induction in each subdisk.

The following theorem is a corollary of Theorem 2.1.

THEOREM 2.2. *Let P be a polygonal disk with a triangulation T and let h*: Bd $P \to E^2$ *be a linear embedding such that h*(Bd P) *bounds a convex disk D and for each spanning* 1*-simplex vw of P, $h(v)$ and $h(w)$ are the ends of a straight spanning segment of D. Then there exists a linear embedding H: $P \to E^2$ which extends h.*

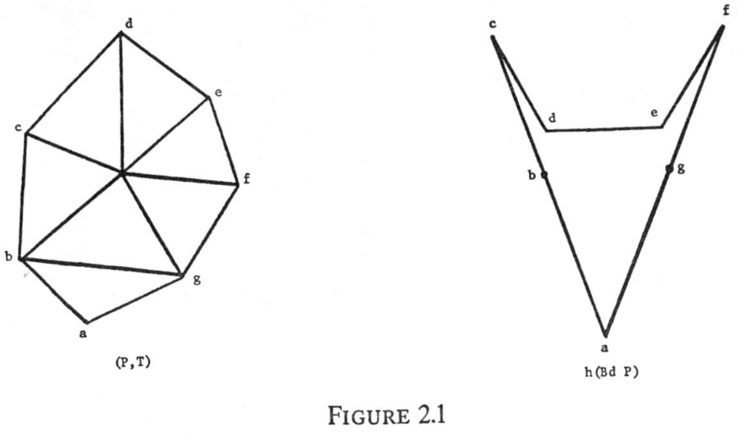

FIGURE 2.1

EXAMPLE 2.1. It is to be noted that Theorem 2.2 is false if instead of supposing that D is convex we suppose instead that it is merely star-like as can be seen by Figure 2.1.

THEOREM 2.3 (FARY [4]). *Let Γ be a finite graph with no loops or multiple edges. If Γ can be topologically embedded in E^2, then it can be linearly embedded in E^2.*

PROOF. Incorporate the embedded Γ into the 1-skeleton of a topological triangulation of a disk in E^2 with no spanning edges. Re-embed the boundary of the disk linearly so that it bounds a star-like disk in E^2 and apply Theorem 2.1. Then Γ will be embedded in the 1-skeleton of a linearly embedded disk and its edges will therefore be straight.

Question 2.1. If C^1 is a subcomplex of a complex C, under what conditions can a linear embedding of C^1 into E^2 be extended to a linear embedding of C? Theorem 2.3 provides an answer if C is a finite 1-complex and $C^1 = \emptyset$. Theorem 2.1 only gives a sufficient condition even in the case where C is a triangulated disk without spanning edges and $C^1 = \text{Bd } C$. In the case where C is a triangulated disk and $C^1 = \text{Bd } C$, might a necessary and sufficient condition be that the embedding could be extended to send each acyclic 1-dimensional subcomplex of C into the disk in E^2 bounded by image of C^1.

Question 2.2. There are 2-complexes which can be topologically embedded in E^3, but not linearly embedded. Suppose a 2-complex can be topologically embedded in E^4, can it then be linearly embedded in E^4? What higher dimensional analogs of the Fary Theorem are true?

3. Linear isotopies and the Fundamental Extension Theorem. In this section we define linear isotopies and pushes of complexes in E^n and prove that a linear isotopy of a finite complex in E^n can be extended to a linear isotopy of all of E^n. This is the only section in which theorems about E^n for n greater than two are proved. Whitehead proved that any finite complex linearly embedded in E^n can be incorporated into a triangulation of E^n without subdividing the complex. This result is given as Theorem 3.1 below and we include a proof for completeness.

DEFINITION. Let C be a complex with a triangulation T. A *linear isotopy* of C is a continuous family of embeddings h_t: $C \to E^n$ ($t \in [0, 1]$) so that for every $t \in [0, 1]$, h_t is a linear embedding of C in E^n. If we wish to emphasize the triangulation involved, we will write that h_t ($t \in [0, 1]$) is a linear isotopy of (C, T). A linear isotopy h_t: $C \to E^n$ ($t \in [0, 1]$) is called a *simple push* if it is fixed on each vertex except one. A finite sequence of such simple pushes, performed one after the preceding, is called a *push*. It may be noted that if there is a linear isotopy taking one embedding h of a finite complex C into E^n onto another embedding g, then *$h(C)$ can be pushed to $g(C)$*.

THEOREM 3.1 (WHITEHEAD [8, THEOREM 5]). *Suppose* (C, T_c) *is a triangulated finite complex linearly embedded in* E^n. *Then there is a triangulation* T *of* E^n *such that each simplex of* T_c *is a simplex of* T.

PROOF. The proof is by induction on the number of simplexes in T_c. The theorem is true if T_c has only one simplex. Assume T_c has n simplexes and the theorem is true for complexes with fewer. Let σ be a simplex of T_c that is not a face of any other simplex of T_c and let T' be a triangulation of E^n such that

(1) each simplex of T_c other than σ is an element of T'.

We also suppose that if t' is a simplex of T' that contains an element of T_c that is not a face of σ, then

(2) $t' \cap \operatorname{Int} \sigma = \varnothing$.

If condition (2) above is not satisfied by the first T' selected, some subdivision of this T' leads to one satisfying both conditions (1) and (2).

Let T'' be a triangulation of E^n such that

σ is a simplex of T'' and $t' \cap t''$ is either empty or a face of σ where $t'' \in T''$ containing σ as a face and $t' \in T'$ containing as a face an element of T_c which is not a face of σ.

Let S' be the convex cellular subdivision of E^n whose elements are the intersections of elements of T' and T''. Note that each proper face of σ is an element of S'. By subdividing S' we get a triangulation S of E^n such that

each proper face of σ is an element of S and S is a subdivision of both T' and T''.

Let

$$U = (C - \sigma) \cup (\cup \operatorname{Int} t_i') \quad \text{for } t_i' \in T' \text{ with } t_i' \cap (C - \sigma) \neq \varnothing,$$

$$V = \cup \operatorname{Int} t_i'' \quad \text{for } t_i'' \in T'' \text{ having } \sigma \text{ as a face.}$$

Then U, V are disjoint open sets and each of $\bar{U}$, $\bar{V}$, $E^n - (U \cup V)$ is the union of elements of S.

We start defining T by putting each simplex of S which is in $E^n - (U \cup V)$ in T. The proper faces of σ are now in T.

Now consider $\bar{U}$. It is the union of simplexes of T' and Bd U is already triangulated by T. To get other simplexes of T in $\bar{U}$ we subdivide the simplexes of T' in $\bar{U}$ one at a time proceeding upward through dimensions. If $t' \in T'$ in $\bar{U}$ and triangulation T on Bd t' has already been defined, we let t' be an element of T if T did not properly divide Bd t', but if T properly divided Bd t' we obtain simplexes of T in t' by coning over Bd t' from the baricenter of t'. Note that each simplex of T_c other than σ has now been placed in T.

The simplexes of T in $\bar{V}$ are obtained by subdividing the elements of T'' in V in a fashion similar to that in which the elements of T' in $\bar{U}$ were subdivided as described in the preceding paragraph.

In the proof of Theorem 3.1 we regarded (C, T_c) as a finite complex so that we could use induction. However, once the result is proved for the finite case, it can be extended to the infinite case as follows.

THEOREM 3.2. *Suppose (C, T_c) is a triangulated complex (perhaps infinite) linearly embedded as a closed subset of E^n. Then there is a triangulation T of E^n such that each simplex of T_c is a simplex of T.*

PROOF. The theorem is proved by enlarging C until the enlargement contains all of E^n.

Let B_1 be the closed n-ball with center at the origin and radius 1 while U_1 is a bounded open set containing B_1 such that each element of T_c that intersects B_1 lies in U_1. By Theorem 3.1 there is a triangulation T_1 of E^n such that each simplex of T_c that intersects $\bar{U}_1$ is a simplex in T_1. We suppose that T_1 is adjusted so that each simplex of T_1 that intersects B_1 lies in U_1. Let C_1 be the union of C and the simplexes of T_1 that intersect B_1. A simplex in the triangulation T_{c_1} of C_1 is either a simplex of T_c or the face of a simplex of T_1 that intersects B_1.

In a similar fashion we get a sequence of triangulated complexes (C_1, T_{c_1}), (C_2, T_{c_2}), ... such that $T_{c_i} \subset T_{c_{i+1}}$ and C_i contains the ball with center at the origin and radius i. A simplex is an element of the required triangulation T if and only if it belongs to some T_{c_i}.

THEOREM 3.3 (THE FUNDAMENTAL EXTENSION THEOREM). *Let (C, T_c) be a triangulated finite complex linearly embedded in E^n and let $h_t\colon C \to E^n$ $(t \in [0, 1])$ be a linear isotopy of C such that $h_0 =$ identity. Then there is a triangulation T of E^n with $T_c \subset T$ and a linear isotopy $H_t\colon E^n \to E^n$ $(t \in [0, 1])$ with compact support such that $H_0 = \mathrm{id}$ and $H_t|C = h_t$.*

PROOF. For every $s \in [0, 1]$, let T_s be a triangulation of E^n which contains $h_s(C, T_c)$ as a subcomplex. Theorem 3.1 asserts that there is such a T_s. For each T_s there is an $\varepsilon_s > 0$ and a linear isotopy

$$H_t^s\colon E^n \to E^n \qquad (t \in [s - \varepsilon_s, s + \varepsilon_s] \cap [0, 1])$$

that moves the vertices of T_s which are in $h_s(C)$ according to h_t, leaves all other vertices of T_s fixed, and extends linearly to the simplexes of T_s.

We can find two finite collections of points in $[0, 1]$, $\{t_i\}$ $(i = 0, 1, \ldots, m)$ with $0 = t_0 < t_1 < \cdots < t_m = 1$ and $\{s_i\}$ $(i = 1, 2, \ldots, m)$ so that $H_t^{s_i}$ is defined for $t \in [t_{i-1}, t_i]$, that is, $[t_{i-1}, t_i] \subset [s_i - \varepsilon_{s_i}, s_i + \varepsilon_{s_i}]$.

Now $H_t\colon E^n \to E^n$ $(t \in [0, 1])$ is defined to be

$$H_t^{s_1}(H_{t_0}^{s_1})^{-1} \quad \text{if } t \in [t_0, t_1],$$

$$H_t^{s_2}(H_{t_1}^{s_2})^{-1}H_{t_1}^{s_1}(H_{t_0}^{s_1})^{-1} = H_t^{s_2}(H_{t_1}^{s_2})^{-1}H_{t_1} \quad \text{if } t \in [t_1, t_2],$$

$$\cdots$$

$$H_t^{s_{i+1}}(H_{t_i}^{s_{i+1}})^{-1}H_{t_i} \quad \text{if } t \in [t_i, t_{i+1}],$$

One can obtain T as follows. Let $T_0 = H_{t_0}^{s_1}(T_{s_1})$. Then T_0 is a triangulation of E^n with $T_c \subset T_0$ and H_t is linear on T_0 for t in $[0, t_1]$. Also $H_{t_1}^{-1}$ is linear on $H_{t_1}(T_0)$. Let T_1' be a triangulation of E^n which is a refinement of both $H_{t_1}(T_0)$ and $H_{t_1}^{s_2}(T_{s_2})$ and has $H_{t_1}(T_c)$ as subcomplex and let $T_1 = H_{t_1}^{-1}(T_1')$. Also let T_2' be a triangulation of E^n which is a refinement of each of $H_{t_2}(T_1)$ and $H_{t_2}^{s_3}(T_{s_3})$ and has $H_{t_2}(T_c)$ as a subcomplex and let $T_2 = H_{t_2}^{-1}(T_2')$. Continuing in this fashion one gets $T_3, T_4, \ldots, T_{m-1}$. The required triangulation T is T_{m-1}.

Question 3.1. Is Theorem 3.3 true if we drop the finite condition, permit (C, T) to be an infinite complex linearly embedded as a closed subset of E^n, and do not require compact support?

4. Linear isotopies of disks with fixed boundaries. Let f and g be two linear embeddings of a triangulated disk (P, T) into E^2 so that $f|\text{Bd } P = g|\text{Bd } P$. In this section we investigate conditions under which there is a linear isotopy or push from f to g which leaves the boundary fixed. Cairns proved the following theorem in this vein.

CAIRNS' THEOREM [2]. *Let (P, T) be a triangulated disk linearly embedded in E^2 where T has only* 3 1-*simplexes in* Bd P *and let f be a linear homeomorphism of (P, T) which is the identity on the boundary. Then there is a push of (P, T) from the identity to f which keeps* Bd P *fixed.*

In this section we prove several variations of Cairns' Theorem. The first variation (Theorem 4.1) removes the hypothesis that T has only 3 1-simplexes in the boundary and replaces it by the hypotheses that P is star-like and T has no spanning 1-simplex. Examples show that neither the "star-like" hypothesis nor the "no spanning 1-simplex" hypothesis can be removed.

THEOREM 4.1. *Let (P, T) be a triangulated, star-like disk linearly embedded in E^2 where T contains no spanning* 1-*simplex. Let g: $P \to E^2$ be a linear homeomorphism of (P, T) which is the identity on* Bd P. *Then there is a push taking (P, T) to $(P, g(T))$ which leaves* Bd P *fixed throughout.*

PROOF. The proof is by induction on n, the number of interior vertices in the triangulation T. The proof bears a resemblance to the proof of Theorem 2.1. The scheme of the proof is to take an interior vertex v in T whose star contains an edge on Bd P and push P, leaving Bd P fixed, by h_t^1 ($t \in [0, 1]$)

so that $h_1^1(v)$ can see all of Bd P. Then we push $(P, g(T))$, leaving Bd P fixed, by h_t^2 $(t \in [0, 1])$ so that $h_1^1(v) = h_1^2(g(v))$. For every edge vw in T where $w \in \text{Bd } P$, $h_1^1(vw) = h_1^2 g(vw)$. As in Theorem 2.1, we then apply induction in each of the subdisks into which the $h_1^1(vw)$'s divide P to finish the proof. What we need to show is that there is a push of (P, T) taking v to a point which can see all of Bd P and similarly a push of $(P, g(T))$ taking $g(v)$ to the same point. The following two lemmas are helpful in constructing these pushes.

LEMMA 4.2. *Let (P, T) be a triangulated star-like disk linearly embedded in E^2. Let $v_0v_1v_2$ be a simplex of T such that v_0 is an interior vertex and there is a point $s \in \text{Int } v_0v_1v_2$ which can see every point of* Bd P. *Then for each neighborhood N of s there is a push of (P, T) that moves v_0 into N, leaves* Bd P *fixed, and is such that any vertex that is moved at all is moved in a straight line toward s.*

PROOF. We first describe a continuous family of linear maps f_t: $(P, T) \to P$ $(t \in [0, 1])$ which (except for not being homeomorphisms) satisfy the conclusion of the lemma. Then we pick a t_0 near 1 and approximate f_t $(t \in [0, t_0])$ by a suitable linear isotopy.

The family f_t $(t \in [0, 1])$ will be described by specifying each map f_t on the vertices of T and then extending linearly to P.

The map f_t $(t \in [0, 1])$ is described as follows. It begins by moving v_0 directly toward s leaving all other vertices fixed. We suppose the motion is linear on v_0. We will extend this map by coning from v_0 to $\text{Lk}(v_0)$. Either this extension is a homeomorphism for every position of v_0 as it moves to s or it is not. If it is, the lemma is proved. If not, let t_1 be the first moment at which the cone map is not a homeomorphism.

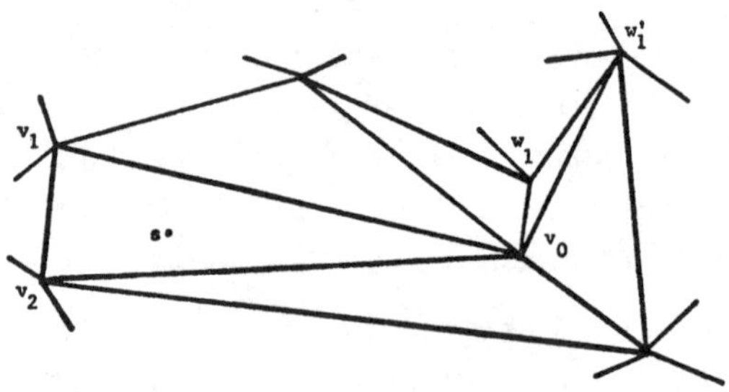

FIGURE 4.1

If $t_1 < 1$, let w_1 be a vertex which is in the way. See Figure 4.1. (If there is more than one such vertex, move each of them as described for w_1 in what follows.) That is, w_1 lies on the segment between $f_{t_1}(v_0)$ and w_1', a vertex in $\text{Lk}(v_0)$. (Note that w_1 is also a vertex in $\text{Lk}(v_0)$.) The map f_{t_1} smashes the simplex $v_0w_1w_1'$ and we say that w_1 is killed. From now on f_t on w_1 will be defined as follows: $f_t(w_1)$ ($t \in [t_1, 1]$) will be the point on the intersection of the segment w_1s and the segment $f_t(v_0)f_t(w_1')$.

We continue defining f_t by moving v_0 and w_1 toward s as described and leaving all other vertices fixed and extending linearly. Let $t_2 \in [t_1, 1]$ be the first point in $[t_1, 1]$ at which the map $f_{t_2}f_{t_1}^{-1}$ is not a homeomorphism of $f_{t_1}(P)$. If $t_2 < 1$, we find a vertex w_2 analogous to w_1 above and a 1-simplex $w_2'w_2''$ so that $w_2 \in f_{t_2}(w_2'w_2'')$. We say that w_2 is killed at time t_2. We define the map f_t ($t \in [t_2, 1]$) on w_2, as before, to be the point of intersection of the segment w_2s and $f_t(w_2')f_t(w_2'')$. Continue until v_0 has been moved to s.

Let $w_1, w_2, \ldots, w_n$ be the vertices that were killed and t_i ($i = 1, 2, \ldots, n$) be the time at which w_i was killed. For convenience we suppose $0 < t_1 < t_2 < \cdots < t_n < 1$. Let t_0 be an element of $(t_n, 1)$ such that $f_{t_0}(v_0) \in N$. We show how to adjust f_t ($t \in [0, t_0]$) so that the adjustment is a linear isotopy.

We note that w_n was killed when $w_n'w_n''$ was moved up to it from the non-s side. If w_n had been moved toward s slightly before $w_n'w_n''$ arrived, the killing would have been delayed. In fact, if w_n had been continuously moved toward s and slightly ahead of $f_t(w_n)$, it would not have been killed in the time interval $[0, t_0]$. (The interval $[0, t_0]$ was used instead of $[0, 1]$ because difficulty would have been encountered at $t = 1$.) Similarly w_{n-1} could have been moved continuously ahead of $f_t(w_{n-1})$ to prevent its being killed in the time interval $[0, t_0]$. Moving backward through the w_i's we find that all of the w_i's could have been moved slightly ahead of their images under f_t so as to have a linear isotopy rather than a 1-parameter family of maps.

Note that every vertex that moved, moved directly toward s, a point which could see all of Bd P. Any vertex w_i which was killed was on $f_{t_i}(w_i'w_i'')$. Therefore, w_i originally was in the triangle $s\,w_i'w_i''$. Hence w_i was an interior vertex since the triangle $s\,w_i'w_i''$ is contained in P. Note that Bd P remained fixed in f_t ($t \in [0, t_0]$) and also remained fixed by the linear isotopy approximation. Replacing the linear isotopy by a push completes the proof of this lemma.

LEMMA 4.3. *Let (P, T) be a star-like disk with no spanning* 1-*simplex linearly embedded in* E^2. *Let $s \in$* Int P *be a point which can see* Bd P *and let N be a neighborhood of s. Then there is a push of P such that* Bd P *is left fixed and each interior vertex is taken into N.*

PROOF. We suppose N is the interior of a circle each of whose points can see all of Bd P. Let v be an interior vertex belonging to a 2-simplex of T whose interior intersects N. Use Lemma 4.2 to move v into N. Iterated use of Lemma 4.2 carries the other interior vertices into N.

We use Lemmas 4.2 and 4.3 to finish the proof of Theorem 4.1. Let s be a point that can see Bd P. Let v be an interior vertex of T which contains an edge uw of Bd P in its star. Using Lemma 4.3 on (P, T) move v to a point v'

very close to point s and using it on $(P, g(T))$ move $g(v)$ to a point v'' very close to s. They should be so close that there is a point $s' \in uwv' \cap uwv''$ so that s' can see Bd P. Now apply Lemma 4.2 to move v' to s' (by aiming for a point beyond s' on ray $v's'$) and v'' to s'. Apply induction as described in the beginning of the proof of Theorem 4.1 and the theorem is proved.

We list two corollaries of Theorem 4.1. The first is a variation of a theorem of Ho [6].

COROLLARY 4.4. *Let (P, T) be a triangulated, convex disk linearly embedded in E^2. Let $g\colon P \to P$ be a linear homeomorphism of (P, T) which is the identity on* Bd P. *Then there is a push of (P, T) taking (P, T) to $(P, g(T))$ which leaves* Bd P *fixed throughout.*

COROLLARY 4.5 (CAIRNS). *Let (B, T) be a triangulated complex. Let g_0 and g_1 be two linear embeddings of (B, T) into E^2 so that there is an orientation-preserving homeomorphism f of E^2 with $f \circ g_0 = g_1$. Then there is a push h_t $(t \in [0, 1])$ of (B, T) so that $h_0 = g_0$ and $h_1 = g_1$.*

PROOF. Incorporate $g_0(B)$ and $g_1(B)$ into equivalent triangulations of a convex PL disk and apply Corollary 4.4.

In a recent paper [7], an example is presented to show that the analog in E^3 of Corollary 4.5 is false.

We now give an example to show that the hypothesis of star-likeness in Theorem 4.1 is necessary.

EXAMPLE 4.1. This is an example of a polygonal disk (P, T) linearly embedded in E^2 and a linear homeomorphism $g\colon P \to P$ with $g|\text{Bd } P$ the identity such that there is no linear isotopy of (P, T) taking the identity map to g while leaving Bd P fixed.

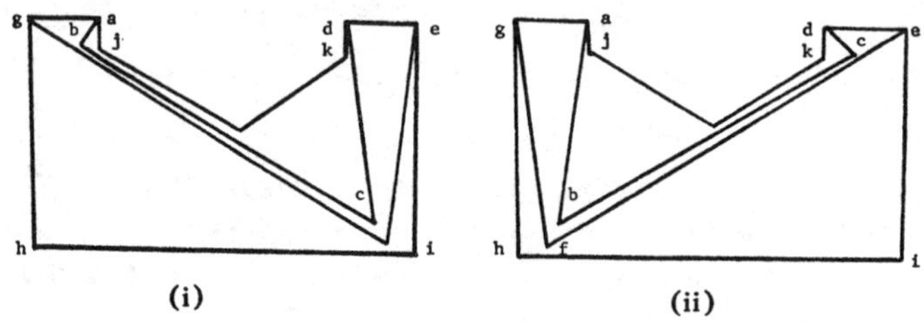

FIGURE 4.2

The (i) and (ii) parts of Figure 4.2 show two embeddings of two spanning arcs inside a polygonal disk P. Although it is possible to isotope one embedding of these arcs to the other keeping Bd P fixed, it is impossible to linearly isotope one to the other keeping Bd P fixed since at the moment when the image of the segment bc is horizontal it is impossible to linearly embed *efg*. Since these spanning arcs in the two embeddings can be incorporated into equivalent triangulations of P, the example shows that Theorem 4.1 is false without the star-like requirements.

EXAMPLE 4.2. A slight modification of Example 4.1 shows the necessity of the "no spanning 1-simplex" hypothesis in Theorem 4.1. Just add a rectangular disk to the bottom of the disk P of Example 4.1 to make the resulting disk P' star-like but keep the bottom of P as a spanning 1-simplex of P'.

5. The Alexander Isotopy Theorem. The standard proof [1] of the Alexander Isotopy Theorem shows that for every PL homeomorphism f of a PL planar disk P to itself which fixes the boundary there is a PL isotopy h_t: $P \to P$ ($t \in [0, 1]$) so that $h_0 = \text{id}$, $h_1 = f$, and for each $t \in [0, 1]$, $h_t|\text{Bd } P = \text{id}$. It is not necessarily true that for each t, h_t is a linear homeomorphism of P with respect to the same triangulation of P. In fact, the standard proof of Alexander's theorem makes use of a process which squeezes badness out of existence in a limit. This type of argument cannot yield a linear isotopy. Here we show that the PL isotopy h_t *can* be chosen so that there is one triangulation T with respect to which each h_t is a linear homeomorphism. In a future paper Starbird will prove the analogous theorem for 3-cells in E^3.

LEMMA 5.1. *Let P be a PL planar disk in E^2 and r and s be two PL embeddings of an arc A into P so that $r(A) \cap \text{Bd } P = s(A) \cap \text{Bd } P = r(\text{Bd } A) = s(\text{Bd } A) = \{p, q\}$. Then there is a triangulation T_1 of A and a linear isotopy h_t: $(A, T_1) \to E^2$ ($t \in [0, 1]$) so that $h_0 = r$, $h_1 = s$, and for each t, $h_t(\text{Bd } A) = h_t(A) \cap \text{Bd } P = \{p, q\}$.*

PROOF. Find a spanning PL arc B in P so that $B \cap \text{Bd } P = \text{Bd } B \cap \text{Bd } P = B \cap r(A) = B \cap s(A) = \{p, q\}$. One can use the disks bounded by $B \cup r(A)$ and $B \cup s(A)$ respectively to guide a push of $r(A)$ to B and a push of B to $s(A)$.

THEOREM 5.2. *Let P be a PL disk PL embedded in E^2 and f be a PL homeomorphism of P onto itself which leaves* Bd P *fixed. Then there is a triangulation T and a linear isotopy h_t: $(P, T) \to E^2$ ($t \in [0, 1]$) so that $h_0 = \text{id}$, $h_1 = f$, and, for each $t \in [0, 1]$, $h_t|\text{Bd } P = \text{id}$.*

PROOF. Let T' be a triangulation of P with respect to which f is linear. We first find a subdivision T of T' and a linear isotopy g_t: $(P, T) \to E^2$ ($t \in [0, 1]$) with the properties that $g_0 = \text{id}$, for each t in $[0, 1]$, $g_t|\text{Bd } P = \text{id}$, and $g_1|(\text{1-skeleton of } T') = f|(\text{1-skeleton of } T')$. The linear isotopy g_t is obtained as follows. Let $\{\sigma_1, \sigma_2, \ldots, \sigma_n\}$ be a shelling of the triangulation T'. Use Lemma 5.1 to push Bd σ_1 to $f(\text{Bd } \sigma_1)$ leaving Bd P fixed and extend this push to a push of all of P using Theorem 3.3. Now Bd σ_2 is partly where it should be under f and partly subdivided and in the wrong place. Now push Bd σ_2 to where it should be under f using Lemma 5.1 and Theorem 3.3 as above leaving Bd P and all of the moved σ_1 fixed. Continue in this manner until the linear isotopy g_t is produced.

Next note that for each 2-simplex σ_i of T', $g_1|\sigma_i$ and $f|\sigma_i$ are two linear homeomorphisms of $(\sigma_i, T|\sigma_i)$ such that $g_1|\text{Bd } \sigma_i = f|\text{Bd } \sigma_i$. By Corollary 4.4 there is a linear isotopy taking g_1 to f which leaves the boundary of σ_1 fixed. Performing g_t ($t \in [0, 1]$) followed by these linear isotopies on each σ_i yields a desired linear isotopy of P from id to f.

Question 5.1. If P is a polygonal disk in E^2, is there an $\varepsilon > 0$ such that for any triangulation T of P of mesh less than ε and any linear homeomorphism h of (P, T) into E^2 fixed on Bd P, it is possible to push $(h(P), h(T))$ back to (h, T) while leaving Bd P fixed?

6. Linear degeneration. In §7 we improve Corollary 4.5 by restricting the linear isotopy in the conclusion to take place in a restricted part of the plane. (See Theorems 7.1, 7.2, and 7.3.) The proofs of those theorems make use of the concept of linear degeneration which is a method of transforming a triangulated complex (C, T) similar to a collapse. However, we are interested in the linear structure of C, so each transformation will be linear with respect to the given triangulation T. Theorem 6.1 will be used in the next section.

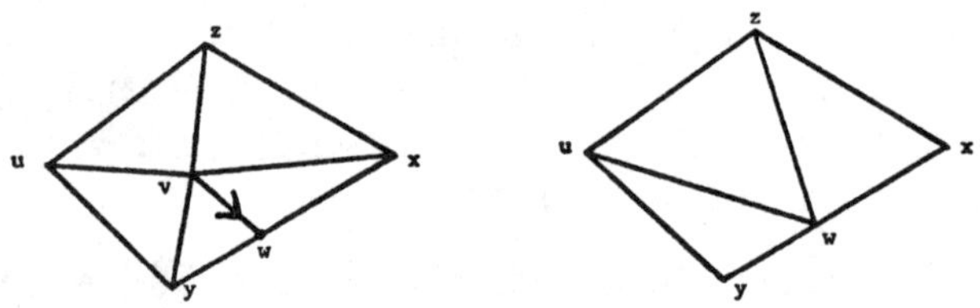

ILLUSTRATION 6.1

A linear degeneration is shown in Illustration 6.1 and its formal definition is given below. It consists of moving the vertex v to a vertex w in its link so that at each moment before the end, the map is a linear homeomorphism whereas at the last moment v was mapped to w and the 2-simplexes vxw and vyw were mapped simplicially to wx and wy respectively. Note that if the vertices u and z did not exist, this movement would have resulted in a collapse. Moving v to y would not be a linear degeneration because the final map would not be a linear embedding of the simplex vxw.

DEFINITION. Let (C, T) be a triangulated complex linearly embedded in E^2 and let vw be a 1-simplex of C. A continuous family of maps $f_t\colon C \to E^2$ ($t \in [0, 1]$) obtained by moving v along vw to w and coning out to Lk(v) while leaving $C - \text{St}(v)$ fixed is a *simple linear degeneration* provided

(1) f_t is monotone nonincreasing in the sense that if $s \leqslant t$, $f_t(C) \subseteq f_s(C)$,

(2) for t in $[0, 1)$, f_t is a linear embedding with respect to T, and

(3) f_1 is a linear embedding of each 2-simplex of T which does not have vw as an edge.

Then $f_1(C)$ is regarded as a new complex linearly embedded in E^2 and having one fewer vertex than C. If a finite sequence of simple linear degenerations transforms a complex C to a complex C^1, then C *linearly degenerates* to C^1 and the composition of the simple linear degenerations is a *linear degeneration.*

The main tools used in this section are lemmas about the existence of low order vertices in triangulations of disks. Ho [5, Lemma 2.17] noted that the Euler characteristic formula $V - E + F = 2$ can be used to verify the following.

6-LEMMA. *Let T be a triangulation of a disk D, V_I be the number of interior vertices of T, V_B be the number of boundary vertices, and for each vertex v, $o(v)$ be the order of v (number of* 1*-simplexes ending at v). Then*

$$\left(6V_I - \sum_{v_i \in \operatorname{Int} D} o(v_i)\right) + \left(4V_B - \sum_{v_j \in \operatorname{Bd} D} o(v_j)\right) = 6.$$

The 6-Lemma implies the following.

5-3 LEMMA. *If T is a triangulation of a disk, then either some interior vertex of T has order less than or equal to* 5 *or there are three boundary vertices each of which has order less than or equal to* 3.

We find the 5-3 Lemma useful when it is used in conjunction with the following fact about 3, 4, and 5 sided polygons.

STAR-LIKE LEMMA. *Let J be a polygonal simple closed curve in E^2 with* 3, 4, *or* 5 *sides. Then J bounds a star-like disk D and has a vertex v such that for each point w of J not on an edge of J containing v, vw spans D. Also, if a point s of* Int D *can see every point of J, then every point on the open segment (sv) can see all of J.*

We use the 5-3 Lemma and the Star-like Lemma to prove a theorem similar to a shelling theorem.

THEOREM 6.1. *Let P be a polygonal disk in E^2 having a triangulation T. Then (P, T) linearly degenerates to a* 2*-simplex in such a way that at each stage the image of P is a disk. In fact, if vw is a boundary* 1*-simplex of T, we can pick the degeneration so that vw remains fixed during the degeneration and is an edge of the final* 2*-simplex.*

PROOF. The proof is by induction on the number n of 2-simplexes in T. If $n = 1$, there is nothing to do. Assume the theorem for k less than n and let T have n 2-simplexes.

If T has a spanning 1-simplex st, then st divides P into two disks D_1 and D_2 where $vw \subset D_2$. Induction implies that D_1 degenerates to a 2-simplex having st as a side and a move of the opposite vertex of this 2-simplex produces a degeneration to D_2. Induction implies that D_2 in turn degenerates to a 2-simplex having vw as an edge. Hence, we suppose that T has no spanning 1-simplex. Hence no boundary vertex is of order 2 (unless P is a 2-simplex).

Case 1. Suppose there is an interior vertex s of order less than 6. The Star-like Lemma implies that there is a vertex t in $\mathrm{Lk}(s)$ so that s can be moved to t to produce a simple linear degeneration. The image of Bd P is left fixed under this degeneration and induction implies that the linear degeneration of P to a 2-simplex can be completed.

Case 2. Suppose there is no interior vertex of order less than 6. It follows from the 5-3 Lemma that there is a boundary vertex s other than v or w which is of order 3. One can obtain a decreasing linear degeneration by moving s. Note that vw remained fixed during this move.

EXAMPLE 6.1. While we could strengthen Theorem 6.1 so as to leave any given 1-simplex of T fixed or any 2-simplex that intersects Bd P fixed, the example illustrated by Figure 6.1 shows that we could not hope to keep a given interior 2-simplex fixed. No linear degeneration of P to a 2-simplex leaves the center 2-simplex fixed.

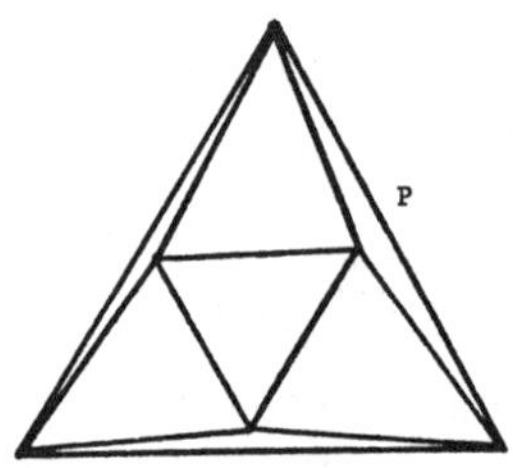

FIGURE 6.1

We next investigate the possibilities of degenerating a triangulated disk P to complexes other than a 2-simplex. A necessary condition for linearly degenerating a complex K to a complex L is that there exist a linear map of K onto L. In the next theorem we investigate the possibility of linearly degenerating a disk to an arc. We make use of the Shortest Arc Property defined below.

DEFINITION. Let P be a disk with triangulation T and A be an arc which is the union of 1-simplexes of T. Then A has the *Shortest Arc Property* if and only if no edge path in T that joins the ends of A contains fewer 1-simplexes of T than A does.

Note that if there is a linear retraction from a triangulated disk (P, T) to a boundary arc A, then A must have the Shortest Arc Property.

THEOREM 6.2. *Suppose* (P, T) *is a triangulated disk and* A *is an arc in the 1-skeleton of* T *with its ends on* Bd P. *Then the following are equivalent*:

(1) *The disk* (P, T) *linearly degenerates to* A *leaving* A *fixed.*

(2) *There is a linear retraction of* T *to* A.

(3) *The arc* A *has the Shortest Arc Property.*

PROOF. Our plan is to show that

$$\mathrm{LR} \Rightarrow \mathrm{SAP} \Rightarrow \mathrm{LD} \Rightarrow \mathrm{LR}$$

where LR means T linearly retracts to A, SAP means that A has the Shortest Arc Property in P, and LD means that T linearly degenerates to A. Since the first and third of these implications are trivial, we concentrate on the second. This implication is demonstrated by using induction on the number n of 2-simplexes in T. In the cases where T has only one or two 2-simplexes, SAP $\Rightarrow$ LD. We suppose T has n 2-simplexes and that the implication follows for triangulations with fewer.

Our plan is to linearly degenerate the closure of each component of $P - A$ onto the part of A on its boundary so we suppose with no loss of generality that $A \subset \mathrm{Bd}\, P$.

Case 1. Suppose there is an edge path E joining two vertices v_0' and v_1' of A such that $E \not\subset \mathrm{Bd}\, P$ and E has the same number of 1-simplexes as the number in A between v_0' and v_1'. Then there is an edge path B between vertices v_0 and v_1 of A with the above properties and such that $B \cap A = \{v_0, v_1\}$. Then B separates P. Let C be the closure of a component of $P - B$ missing the part of A between v_0 and v_1. We show in the next paragraph that $(B \cup A) \cap C$ has the Shortest Arc Property with respect to C.

If $(B \cup A) \cap C$ did not have the Shortest Arc Property with respect to C, there would be a too-short arc B' in C from w_0 to w_1 such that $B' \cap [(B \cup A) \cap C] = \{w_0, w_1\}$. This is impossible because there is no place for w_0 and w_1 to lie. They cannot both be on A or else A does not have the Shortest Arc Property. They cannot be both on B or else we could replace a part of B by B' and get a shorter arc than B between v_0 and v_1. One cannot belong to A and the other to B for if w_0 were in A, w_1 in B, and v_0 between w_0 and w_1 on $(B \cup A) \cap C$, the union of B' and the part of B from w_1 to v_1 would be shorter than the arc on A from w_0 to v_1. Hence there is no place for w_0 and w_1 and $(B \cup A) \cap C$ has the Shortest Arc Property in C.

Since C has fewer than n 2-simplexes, it follows by induction that C linearly degenerates to $(B \cup A) \cap C$. Note that $B \cap C$ is a spanning arc of P and the closure of $P - C$ is a disk D which has fewer than n 2-simplexes. Induction implies that there is a linear degeneration of D onto $D \cap A$. The composition of these two linear degenerations is the required one for Case 1.

Case 2. Suppose there is no edge path E as supposed in Case 1. The 5-3

Lemma implies that there is a vertex v_0 of T such that v_0 is an interior vertex of order less than 6 or v_0 is a boundary point of order less than 4 which is not an end of A.

We note that v_0 does not lie on A. If it did, it could not be of order 2 since A has the Shortest Arc Property and it could not be of order 3 or else Case 1 would apply.

If $v_0 \in \operatorname{Bd} P - A$, the linear degeneration can be started by moving v_0. Induction can be used to show that it can be completed.

If v_0 is an interior vertex, it follows from the Star-like Lemma that we can start the linear degeneration by moving v_0 to a vertex w_0 of $\operatorname{Lk}(v_0)$. This degeneration reduced the number of 2-simplexes and induction implies that the linear degeneration can be completed if A has the Shortest Arc Property with respect to the new triangulation. However, it does have this property since if there were a too-short arc B' in the new triangulation, it would pass through w_0 and the inverse of B' would contain an arc B through v_0 showing that Case 1 applied.

One can easily construct an example showing the necessity in Theorem 6.2 of requiring that the ends of A be on Bd P. See Figure 6.2.

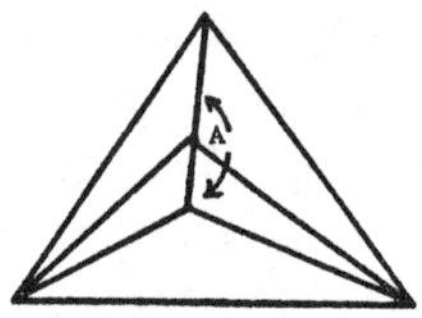

FIGURE 6.2

7. Linear isotopies in open subsets of E^2. In this section we show that only a part of E^2 need be used in pushing a complex from one linear embedding to another. The following theorems are variations of Cairn's result listed above as Corollary 4.5.

THEOREM 7.1. *Let (P, T) be a triangulated disk linearly embedded in E^2. Then there is a linear isotopy h_t: $P \to P$ $(t \in [0, 1])$ of (P, T) such that $h_0 =$ identity, $h_1(P)$ is convex, and $h_t(P)$ $(t \in [0, 1])$ is monotonically decreasing.*

PROOF. By Theorem 6.1, P can be linearly degenerated to a 2-simplex of T. This degeneration can be approximated by a monotonically nonincreasing push of (P, T) which pushes P to a convex set. This fact can be proved by induction on n where the inductive statement is that if (P, T) is a triangulated disk in E^2 with no three collinear vertices, then an n step linear degeneration of P followed by a nonincreasing push taking P to a convex disk can be replaced by a nonincreasing push of P to a convex disk.

The push so obtained can be replaced by a monotonically decreasing linear isotopy to prove Theorem 7.1.

THEOREM 7.2. *Let (P, T) be a triangulated disk, U be a connected open subset of E^2, and f_0 and f_1 be linear embeddings of P in U such that there is an orientation preserving homeomorphism G of E^2 so that $G \circ f_0 = f_1$. Then there is a push h_t: $P \to U$ $(t \in [0, 1])$ of (P, T) into U so that $h_0 = f_0$ and $h_1 = f_1$.*

PROOF. Using Theorem 7.1, $f_0(P)$ can be linearly isotoped to a convex set which can in turn be linearly isotoped to a small convex set. Likewise $f_1(P)$ can be linearly isotoped to a small set. These linear isotopies can be replaced by pushes. We can then push the shrunk $f_0(P)$ into a convex open set which also contains the shrunk $f_1(P)$. Now Corollary 4.4 can be applied to finish the proof as was done in Corollary 4.5.

THEOREM 7.3. *Let (C, T) be a triangulated complex, D be an open disk in E^2, and f_0 and f_1 be two linear embeddings of C into D such that there is an orientation preserving homeomorphism G of E^2 so that $G \circ f_0 = f_1$. Then there is a push h_t: $C \to D$ $(t \in [0, 1])$ of (C, T) into D so that $h_0 = f_0$ and $h_1 = f_1$.*

PROOF. Incorporate $f_0(C)$ and $f_1(C)$ into equivalently triangulated disks contained in D and apply Theorem 7.2.

EXAMPLE 7.1. It should be noted that the hypothesis that D be a disk in Theorem 7.3 is necessary as shown by Figure 7.1.

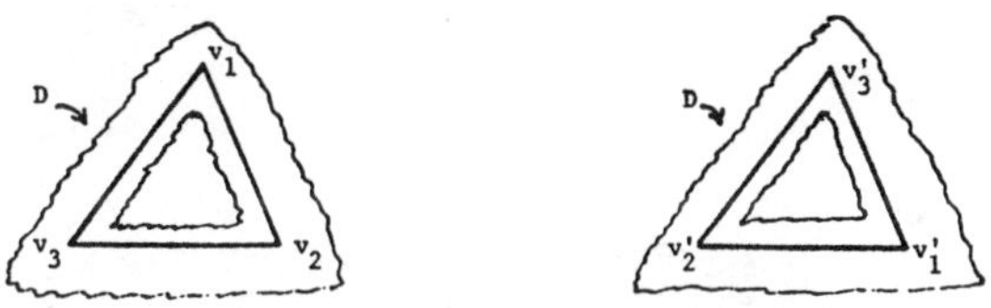

FIGURE 7.1

Question 7.1. In §4 and this section, Cairns' Theorems were modified by keeping part of a complex fixed during a push and by restricting the part of the plane in which the push takes place. Another possible avenue of inquiry is suggested in the following question: Let $\{A_i\}_{i=1}^n$ be a collection of arcs. Let f_0, f_1: $\bigcup A_i \to E^2$ be maps of the disjoint union of the A_i's into E^2 so that $f_0, f_1|A_i$ is linear for each i and so that there is an orientation preserving homeomorphism G of E^2 so that $G \circ f_0 = f_1$. Then is there an isotopy h_t: $E^2 \to E^2$ $(t \in [0, 1])$ so that $h_0 = \text{id}$, for each i, $h_1(f_0(A_i)) = f_1(A_i)$, and for each $t \in [0, 1]$ and each i, $h_t|f_0(A_i)$ is a linear homeomorphism?

REFERENCES

1. J. W. Alexander, *On the deformation of an n-cell*, Proc. Nat. Acad. Sci. USA **9** (1923), 406–407.

2. S. S. Cairns, *Isotopic deformations of geodesic complexes on the 2-sphere and the plane*, Ann. of Math. (2) **45** (1944), 207–217. MR **5**, 273.

3. ______, *Deformations of plane rectilinear complexes*, Amer. Math. Monthly **51** (1944), 247–252. MR **5**, 273.

4. I Fáry, *On straight line representation of planar graphs*, Acta. Univ. Szeged Sect. Sci. Math. **11** (1948), 229–233. MR **10**, 136.

5. C. W. Ho, *On certain homotopy properties of some spaces of linear and piecewise linear homeomorphisms*. I, Trans. Amer. Math. Soc. **181** (1973), 213–233. MR **48** #1252.

6. ______, *Deforming P. L. homeomorphisms on a convex 2-disk*, Bull. Amer. Math. Soc. **81** (1975), 726–728. MR **51** #11529.

7. M. Starbird, *A complex which cannot be pushed around in E^3*, Proc. Amer. Math. Soc. **63** (1977), 363–367.

8. J. H. C. Whitehead, *On subdivisions of complexes*, Proc. Cambridge Philos. Soc. **31** (1935), 69–75.

Department of Mathematics, University of Texas, Austin, Texas 78712

PACIFIC JOURNAL OF MATHEMATICS
Vol. 74, No. 2, 1978

SUPER TRIANGULATIONS

R. H. BING AND MICHAEL STARBIRD

This paper concerns itself with continuous families of linear embeddings of triangulated complexes into E^2. In [2] Cairns showed that if f and g are two linear embeddings of a triangulated complex (C, T) into E^2 so that there is an orientation preserving homeomorphism k of E^2 with $k \circ f = g$, then there is a continuous family of linear embeddings h_t: $(C, T) \to E^2 (t \in [[0, 1])$ so that $h_0 = f$ and $h_1 = g$. In this paper we prove various relative versions of this result when C is an arc, a θ-curve, or a disk.

Introduction. To appreciate where the results in this paper fit into the literature, it is useful to be aware of the following examples which were described in [1, Example 4.1].

EXAMPLE 1. This example is a triangulated 1-complex (C, S) linearly embedded in E^2 consisting of a simple closed curve J with two disjoint spanning arcs in its interior. The complex C is homeomorphic to ⦶. There is a homeomorphism $g: E^2 \to E^2$ fixed on J such that $f = g|C$ is linear with respect to S but there is no linear isotopy $h_t: (C, S) \to E^2 (t \in [0, 1])$ with $h_0 = id$ and $h_1 = f$ which keeps J fixed.

EXAMPLE 2. Example 1 can be modified by incorporating (C, S) into the 1-skeleton of a triangulated disk (P, T) with boundary J to produce an example of a disk with properties similar to those of (C, S). Namely, the triangulated disk (P, T) is linearly embedded in E^2 and admits a linear homoemorphism k fixed on $\operatorname{Bd} P$ for which there is no linear isotopy $h_t: (P, T) \to E^2 (t \in [0, 1])$ with $h_0 = \mathrm{id}$ and $h_1 = k$ which leaves the boundary fixed throughout.

It is known that no such example can be found where P is convex [1, Corollary 4.4] nor could P be star-like if T has no spanning edge [1, Theorem 4.1].

In this paper it is shown (Theorem 2.4) that no 1-complex homeomorphic to a θ-curve can have the properties of Example 1. Then in Theorem 4.2 it is proved that Example 2 can not retain its properties under all subdivisions. In fact each triangulation T of a disk P has a subdivision T' which is a super triangulation of P. A super triangulation T' of a disk P is one which is as flexible as possible. Namely, any linear embedding of $\operatorname{Bd} P$ into E^2 extends to a linear embedding of (P, T') and for any two linear homeomorphisms f, g of (P, T') into E^2 with $f | \operatorname{Bd} P = g | \operatorname{Bd} P$, there is a linear isotopy

$h_t: (P, T') \to E^2(t \in [0, 1])$ with $h_0 = f$ and $h_1 = g$ which agrees with f and g on Bd P throughout. (The formal definition of super triangulation appears in §3.)

DEFINITION. Let (C, T) be a finite complex C with triangulation T. A *linear embedding* of C (or (C, T)) into E^n is an embedding which is linear on each simplex of T. A *linear isotopy* $h_t: (C, T) \to E^n(t \in [0, 1])$ is a continuous family of linear embeddings of (C, T) into E^n. A *simple push* is a linear isotopy which is fixed except on the open star of one vertex. A *push* is a sequence of simple pushes each one performed after the preceding one.

2. Linear isotopies of arcs and θ-curves. This section begins with a theorem which states that two triangulated arcs linearly embedded in E^2 with the same endpoints can be linearly isotoped to a common arc in E^2 keeping the endpoints fixed where the movement takes place only in the closure of the unbounded component of the complement of the union of the two arcs.

THEOREM 2.1. *Let (A, T_A) and (B, T_B) be two triangulated arcs linearly embedded in E^2 which share common endpoints v and w. Then there are linear isotopies $g_t: (A, T_A) \to E^2(t \in [0,1])$ and $h_t: (B, T_B) \to E^2(t \in [0, 1])$ such that*

(1) $g_0 = \text{id}$ *and* $h_0 = \text{id}$,

(2) $g_1(A) = h_1(B)$ *as subsets of* E^2,

(3) *for all* $t \in [0, 1]$ *both* h_t *and* g_t *leave both* v *and* w *fixed, and*

(4) *for* $0 \leqq s \leqq u \leqq 1$, $g_u(A) \cup h_u(B)$ *misses the unbounded component of* $E^2 - (g_s(A) \cup h_s(B))$.

The proof of Theorem 2.1 uses the following lemma to reduce bends.

LEMMA 2.2. *Let (A, T_A) be a triangulated arc linearly embedded in E^2 from v to w, a be a vertex of A, and x be another point of A such that*

(1) *the segment ax meets A only at its ends and*

(2) *the disk D_{ax} bounded by $ax \cup A_{ax}$ (where A_{ax} denotes the subarc of A from a to x) contains neither v nor w in its interior. Then there is a push $h_t: (A, T_A) \to E^2(t \in [0, 1])$ so that*

(1) $h = \text{id}$,

(2) *each h_t is fixed on each of a, v, and w,*

(3) *for $0 \leqq s < u \leqq 1$, $h_u(A) \subset (A - A_{ax}) \cup D^s_{ax}$ where D^s_{ax} is the disk bounded by $ax \cup h_s(A)_{ax}$, and*

(4) $h_1(A) = (A - A_{ax}) \cup ax$.

Proof of lemma. Hypotheses 1 and 2 imply that $D_{ax} \cap A = A_{ax}$. If A_{ax} contains only one vertex a_j of A other than a, push a_j straight to x, shortening the edge of A containing x. If x were a vertex of A and the open arc (A_{ax}) contains only one vertex a_j, a_j could be pushed to the center of ax. If the open arc (A_{ax}) contains more than one vertex of A, triangulate D_{ax} without adding interior vertices. Find a shelling of the triangulation which leaves the 2-simplex containing ax until last. Let the shelling guide a desired push of A.

The proof of Theorem 2.1 uses the following lemma to reduce the number of components of $A \cap B$.

LEMMA 2.3. *Suppose that (A, T_A) and (B, T_B) are triangulated arcs linearly embedded in E^2 from v to w such that each of v and w is accessible from the unbounded component of $E^2 - (A \cup B)$. Then there is a linear isotopy $h_t: (A, T_A) \to E^2 (t \in [0, 1])$ such that*

(1) $h_0 = \text{id}$,

(2) *each h_t is fixed on v and w,*

(8) *for $0 \leq s \leq u \leq 1$, $h_u(A)$ misses the unbounded component of $E^2 - (B \cup h_s(A))$, and*

(4) *for each edge b_jb_{j+1} of B, $b_jb_{j+1} \cap h_1(A)$ is either empty, connected, or consists of exactly two points x and y so that x and y are interior points of two consecutive edges of A.*

Proof of lemma. Let b_jb_{j+1} be an edge of B.

Case 1. If $b_jb_{j+1} \cap A$ contains a vertex of A, then by repeated applications of Lemma 2.2, A can be pushed to make $b_jb_{j+1} \cap A$ connected.

Case 2. Suppose that $(b_jb_{j+1} \cap A) = \{x_1, x_2, \cdots, x_r\}$ where each x_i is an interior point of an edge of A and $x_1 < x_2 < x_3 < \cdots < x_r$ on b_jb_{j+1}.

If, for some i, $A_{x_ix_{i+1}}$ contains two or more vertices of A, triangulate $D_{x_ix_{i+1}}$ without adding interior vertices. As in the proof of Lemma 2.2, follow a shelling that leaves the 2-simplex containing x_ix_{i+1} until last to push A to $(A - A_{x_ix_{i+1}}) \cup x_ix_{i+1}$. Now apply Case 1.

Suppose each $A_{x_ix_{i+1}}$ contains one vertex of A and $b_jb_{j+1} \cap A$ contains at least three points x_1, x_2, and x_3. Let a_i and a_{i+1} be respectively the vertex of A between x_1 and x_2 and the vertex of A between x_2 and x_3. By moving a_i toward x_1 and a_{i+1} toward x_3, pivoting about x_2, a_i and a_{i+1} can be linearly isotoped down to x_1 and x_3 respectively and Case 1 can again be applied to make the image of A intersect b_jb_{j+1} in a connected set.

The movements we used did not complicate the intersection of the image of A and the other edges of B so we can finish Lemma 2.3 by considering the edges of B one at a time.

EXAMPLE 6.1. We note that Lemma 2.3 cannot be strengthened to require a push rather than a linear isotopy as demonstrated by Figure 2.1.

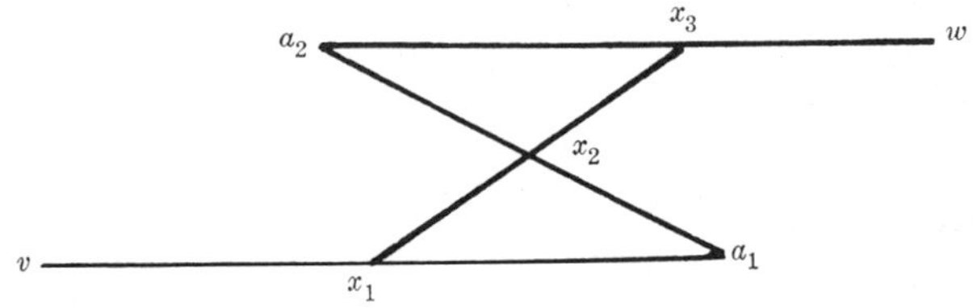

FIGURE 2.1

Note that no interior vertex of A or B can be pushed at all without moving A or B partially into the unbounded component of $E^2 - (A \cup B)$.

Proof of Theorem 2.1. The proof uses induction on n, the sum of the number of bends in A and the number in B. If $n = 0$, A and B coincide already.

We assume the theorem for k less than n and suppose that the sum of the bends in A and B is n. Let $\{a_i\}_{i=0}^m$ and $\{b_i\}_{i=0}^p$ be vertices of A and B in order so that $a_0 = b_0 = v$, $a_m = b_p = w$, and $(m - 1) + (p - 1) = n$.

Case 1. Suppose v or w (say v) is not accessible from the unbounded component of $E^2 - (A \cap B)$. In this case find a point x on A or B (say A) such that $vx \cap a_1a_2 = \varnothing$, $vx \cap b_1b_2 = \varnothing$, and vx meets $A \cup B$ only at its ends.

If $w \notin \operatorname{Int} D_{vx}$, apply Lemma 2.2 to push A onto $vx \cup A_{xw}$ thereby reducing the number of bends in A.

If $w \in \operatorname{Int} D_{vx}$, find a point y on $\operatorname{Bd} D_{vx}$ so that $wy \cap vx = \varnothing$ and $wy \cap a_{m-2}a_{m-1} = \varnothing$. Let y' be the nearest point of $wy \cap A_{va_{m-2}}$ to w. Since $(A_{xw} \cup wy') \subset D_{vx}$, then $D_{wy'}$ is a subset of D_{vx} and hence does not intersect vx. Since w is not in $\operatorname{Int} D_{wy'}$ either, apply Lemma 2.2 to push A to $(A - A_{y'w}) \cup y'w$ and thereby reduce the number of bends in A. One can use the facts that the movement occurs in $D_{wy'}$, $D'_{wy'} \subset D_{vx}$, and D_{vx} misses the unbounded component of $E^2 - (A_{vx} \cup B)$ to show that Condition 4 of Theorem 2.1 is satisfied.

Case 2. Suppose v and w are both accessible from the unbounded component of $E^2 - A \cup B$. By isotoping A according to Lemma 2.3 and then changing the roles of A and B and isotoping B according to the lemma, we can linearly isotope A and B to a position where the conclusion of Lemma 2.3 is satisfied by both A and B. If this adjustment reduced the number of bends in $A \cup B$, then Theorem 2.1 follows by induction. We assume that it did not and proceed to consider five subcases of Case 2.

Subcase 2a. Suppose $a_1 = b_1$. We apply induction to the two arcs A_{a_1w} and B_{b_1w}.

Subcase 2b. Suppose $va_1 \subset vb_1$ and $a_1a_2 \cap B = \{a_1\}$. Then push a_1 toward b_1, pivoting on a_2, until the moved a_1a_2 hits a point of $A_{a_3w} \cup B_{b_1w}$. Let a_1' be the position of a_1 at that moment. If $a_1'a_2 \cap A_{a_3w}$ contains a vertex a_j, push A to $va_1 \cup a_1a_j \cup A_{a_jw}$ by an application of Lemma 2.2 and thereby reduce the number of bends in A.

If $a_1'a_2$ contains a vertex b_j of B_{b_2w}, push B to $va_1 \cup a_1b_j \cup B_{b_jw}$. This push moves the vertex b_1 to the point a_1 and, therefore, throws us into Subcase 2a which was already considered.

If $a_1' = b_1$ and $a_1'a_2 \cap (A_{a_3w} \cup B_{b_2w}) = \emptyset$, then push a_1 to b_1 which puts us into Subcase 2a.

Subcase 2c. Suppose $va_1 \subset vb_1$ and $a_1a_2 \cap B \neq \{a_1\}$. By Lemma 2.3, $a_1a_2 \cap B = \{a_1\} \cup \{y\}$ where y is an interior point of b_1b_2.

Several things could happen. First, if b_1b_2 intersects A in a point z other than y, then z is an interior point of a_2a_3 by conclusion 4 of Lemma 2.3. Furthermore $z \in yb_2$. This is true because A_{a_2w} cannot intersect the triangle a_1b_1w since such an intersection would violate either Lemma 2.3 or the fact that w is accessible from the unbounded component of $E^2 - (A \cup B)$. We move a_1 to b_1 and a_2 to z, pivoting about y. This procedure takes us to Subcase 2a which has already been considered.

Second, if b_1b_2 misses A_{a_2w}, it may be that the segment b_1b_2 can be extended slightly beyond b_2 without meeting the unbounded domain of $E^2 - (A \cup B)$. In this case, we let x be the first point at which this extension of b_1b_2 intersects $A \cup B$. If $x \in B$, apply Lemma 2.2 to push B to $(B - B_{b_2x}) \cup b_2x$ and thereby reduce the bends in B. If x is a vertex of A push A to $(A - A_{yx}) \cup yx$ and then to $(A - A_{vy}) \cup vb_1 \cup b_1y$ to reduce the number of bends in A. If x is not a vertex of A but $x \in a_2a_3$, move a_2 to x and a_1 to b_1 while pivoting about y. This puts us in Subcase 2a. If $x \in A_{a_3w}$, triangulate disk bounded by $A_{yx} \cup yx$ without adding interior vertices. As in the proof of Lemma 2.2, find a shelling of this disk which leaves the 2-simplex containing yx until last. Let this shelling guide a linear isotopy of A onto the set $va_1 \cup a_1y \cup yx \cup A_{xw}$. Now a_1 can be moved to b_1 putting us in Case 2a.

Third, and last, if $b_1b_2 \cap A_{a_2w} = \varnothing$ and b_1b_2 cannot be extended beyond b_2 without intersecting the unbounded component of $E^2 - (A \cup B)$ then move b_2 toward b_3 as you move b_1 toward a_1, pivoting about y. If the pivoting segment b_1b_2 never meets a point of A_{a_3w} nor B_{b_3w} while b_1 is moved onto a_1, then this linear isotopy puts us in Subcase 2a. If not, then the first moment at which b_1b_2 meets A_{a_3w} or B_{b_3w} locates a vertex a_j or $b_j (j \geqq 3)$ such that the segment ya_j or yb_j satisfies the hypotheses of Lemma 2.2. An application of Lemma 2.2 would then reduce the bends in A or B.

Subcase 2d. Suppose $vb_1 \subset va_1$. This case is identical to Subcase 2b and 2c with the roles of A and B exchanged.

Subcase 2e. Suppose $va_1 \cap B = v$,

Perhaps the segment from v through a_1 can be extended beyond a_1 without going into the unbounded domain of $E^2 - (A \cup B)$. If it can, extend it until the extension hits a point x of $A \cup B$. If $x \in A$, use Lemma 2.2 to push A to $vx \cup A_{xw}$, reducing the number of bends in A. If $x \in B$, use Lemma 2.2 to push B to $vx \cup B_{xw}$. This either reduces the number of bends in B or carries us to a previous case.

If va_1 cannot be extended as considered in the last paragraph, examine the segment vx as x moves from a_1 to a_2 and find the first x_0 at which vx_0 meets $B \cup A_{a_2w}$ in a point other than v.

If vx_0 meets A_{a_2w}, let a_j be the point of $vx_0 \cap A_{a_2w}$ nearest v. Lemma 2.2 implies that A can be pushed to $va_j \cup A_{a_jw}$, thereby reducing the number of bends in A.

If vx_0 misses A_{a_1w}, use Lemma 2.2 to push A to $vx_0 \cup A_{x_0w}$. The first edge vx_0 of $vx_0 \cup A_{x_0w}$ contains a point of B other than v and we are in a previous subcase.

The following theorem is used in the proof of Theorem 4.2.

THEOREM 2.4. *Let $(J \cup A, T)$ be a triangulated θ-curve linearly embedded in E^2 so that A is a spanning arc of the disk bounded by the simple closed curve J. Let k be a homeomorphism of E^2 such that $k|J = \mathrm{id}$ and $k|A$ is a linear embedding of $(A, T|A)$. Then there is a linear isotopy $h_t: (J \cup A, T) \to E^2 (t \in [0, 1])$ such that $h_0 = \mathrm{id}$, $h_1 = k|J \cup A$, and for each t in $[0, 1]$, $h_t|J = \mathrm{id}$.*

Proof. Let $(A, T|A)$ and $(k(A), k(T|A))$ be the two arcs in the hypothesis of Theorem 2.1. Let $f_t': (A, T|A) \to E^2 (t \in [0, 1])$ and $g_t: (k(A), k(T|A)) \to E^2 (t \in [0, 1])$ be linear isotopies satisfying the conclusion of Theorem 2.1. Each of these linear isotopies can be extended to J by the identity. We do so and abuse the notation slightly by letting f'_t and g_t now denote those linear isotopies of $(J \cup A, T)$ and $(k(J \cup A), k(T))$ respectively.

Now $f'_1(A) = g_1 \circ k(A)$ as sets; however, it may not be the case that $f'_1 | A$ equals $g_1 \circ k | A$. (See Figure 2.2.) In order to rectify this situation find a linear isotopy $f''_t: (J \cup A, T) \to E^2 (t \in [0, 1])$ such that $f''_0 = f'_1$, $f''_1 | A = g_1 \circ k | A$, and for each t in $[0, 1]$, $f''_t | J = \text{id}$. The linear isotopy f''_t simply moves the vertices of $f'_1(A)$ until they are in the position to which they are mapped by $k \circ g_1$. Note that for any $\varepsilon > 0$ f''_t can be chosen so that for each t in $[0, 1]$, $f''_t(A)$ lies in the ε-neighborhood of $f'_1(A)$; however, as illustrated in Figure 2.2, it may not be possible to have $f''_t(A) = f'_1(A)$ for each t.

A linear isotopy h_t satisfying the conclusion of Theorem 2.4 can now be obtained by performing three linear isotopies in succession. First perform $f'_t(t \in [0, 1])$, second perform $f''_t(t \in [0, 1])$, and finally perform $g_{1-t} \circ k(t \in [0, 1])$.

Note that for any $\varepsilon > 0$, h_t could be chosen so that for t, $h_t(A)$ almost misses the unbounded component C of $E^2 - (A \cup k(A))$ where "almost" means that $h_t(A)$ misses C except for an ε-neighborhood of $f'_1(A)$.

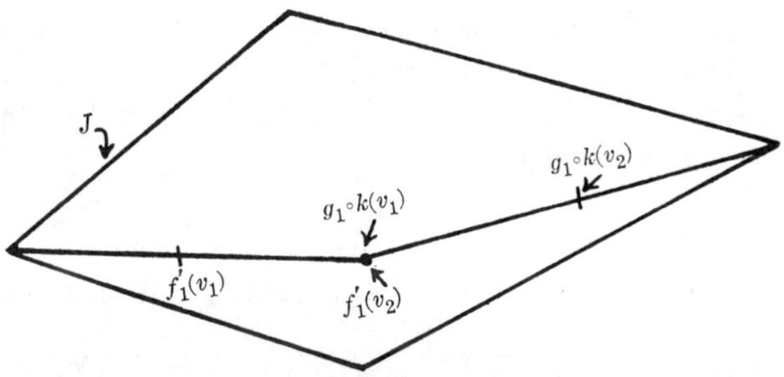

FIGURE 2.2

3. Super triangulations. A triangulation T of a disk P is *super* if and only if it has the following three properties.

(1) Every linear embedding of Bd P in E^2 can be extended to a linear embedding of (P, T).

(2) If f and g are two linear embeddings of (P, T) which agree on Bd P, then there is a linear isotopy $h_t: (P, T) \to E^2 (t \in [0, 1])$ such that $h_0 = f$, $h_1 = g$, and for all $t \in [0, 1]$, $h_t | \text{Bd}\, P = f | \text{Bd}\, P = g | \text{Bd}\, P$.

(3) If h_0 and h_1 are two linear embeddings of (P, T) into E^2 and f_t is a linear isotopy of Bd P into E^2 from $h_0 | \text{Bd}\, P$ to $h_1 | \text{Bd}\, P$, then f_t can be extended to a linear isotopy of P from h_0 to h_1.

It may be noted that Properties 1 and 2 imply Property 3. To see that this is true one could use Property 1 and the compactness of $[0, 1]$ to cover $[0, 1]$ with subintervals $[t_0, t_1], [t_1, t_2], \cdots, [t_{n-1}, t_n]$ such that for each i there is a linear isotopy $h_t^i\colon (P, T) \to E^2$, $t \in [t_i, t_{i+1}]$ with $h_t^i|\operatorname{Bd} P = f_t|\operatorname{Bd} P$. Although $h_{t_{i+1}}^i$ and $h_{t_{i+1}}^{i+1}$ need not agree on interior vertices of T, Property 2 can be used to adjust them so they do.

In this section we produce (in Theorem 3.4) for each integer $n (n \geqq 3)$ a super triangulation T_n of a disk P_n which has n 1-simplexes on the boundary. For $n = 3$, T_3 could be chosen to be a single 2-simplex. For $n = 4$ or 5, T_n could be chosen to be obtained by coning from an interior point to $\operatorname{Bd} P_n$. This T_n is super since any linear embedding of $\operatorname{Bd} P_n (n = 4, 5)$ would bound a star-like disk. To produce T_6, we add an annulus A to P_5 so that A has five 1-simplexes in the boundary component which is $\operatorname{Bd} P_5$, but six in the other. This annulus is given a specific, simple triangulation. It is shown in Theorem 3.3 that if one begins with a super triangulation of a disk with n sides and enlarges the disk and triangulation by the addition of an annulus which is triangulated as specified in Theorem 3.3, then the new triangulation of the new disk is also super. Thus to produce T_7 another annulus is added on to P_6. This process is continued to produce triangulations T_n where each T_n has a bull's-eye pattern.

Theorems 3.1 and 3.2 have the same general form as Theorem 3.3 except where different triangulations of the added annulus are considered. They are included in this section because their proofs contain techniques used in the proof of Theorem 3.3. They are used explicitly in the next section.

DEFINITION. Let J be a PL simple closed curve in E^2. A point x in $\operatorname{Int} J$ *can see* J if and only if for each point y in J the segment xy meets J only at y.

THEOREM 3.1. *Suppose (P, T) is a triangulated disk, A is a subcomplex of T which is an annulus containing* $\operatorname{Bd} P$, *the closure of $P - A$ is a disk D, and A has the following triangulation. Namely, A is the union of n 4-sided disks $v_i v_{i+1} w_{i+1} w_i (i = 1, 2, \cdots, n$, counting is* mod n) *where for each i, $v_i \in \operatorname{Bd} P$ and $w_i \in \operatorname{Bd} D$ and the 2-simplexes of T in A are precisely those of the form $v_i w_i v_{i+1}$ or $w_i v_{i+1} w_{i+1}$. Then T super if T is restricted to D is super.*

Proof. Let h be a linear embedding of $\operatorname{Bd} P$ in E^2 with $h(v_i) = v_i'$. To show that (P, T) has Property 1 we can pick the image of w_i to be w_i', a point in $\operatorname{Int} h(\operatorname{Bd} P)$ near v_i' on the bisection of angle $v_{i-1}' v_i' v_{i+1}'$. This linearly embeds A. The fact that T restricted to D is super ensures that the embedding can be extended.

Next, we show that (P, T) has Property 2. For convenience we suppose that f is the identity and $g(w_i) = w_i'$ is as described in the preceding paragraph. Our plan is to push each w_i to w_i' and use the fact that D has Property 3 to show that (P, T) has Property 2.

Special Case. If $v_iv_{i+1}w_{i+1}w_i$ is convex we could push w_i to a point near v_i along w_iv_i and then push it to a point near v_i so that if w_i'' is the new w_i, both $v_{i-1}v_iw_i''w_{i-1}$ and $v_iv_{i+1}w_{i+1}w_i''$ are convex. Similarly, w_{i-1} can be pushed to w_{i-1}'', a point near v_{i-1} so that both $v_{i-2}v_{i-1}w_{i-1}''w_{i-2}$ and $v_{i-1}v_iw_i''w_{i-1}''$ are convex. Continuing back through the w_i's (and counting mod n) each w_i can be pushed to w_i'' a point near v_i so that each quadrilateral $v_jv_{j+1}w_{j+1}''w_j''$ is convex. Now we can push each w_i'' onto the bisector of $v_{i-1}v_iv_{i+1}$ and then to w_i'.

General Case. Finally we show that some w_{i+1} can be pushed to w_{i+1}''' so that the resulting $v_iv_{i+1}w_{i+1}'''w_i$ is convex. To do this, we pick an i so that w_iw_{i+2} is a spanning arc of D. (If $n = 3$ there is no spanning arc and we use w_1w_3 for w_iw_{i+2}.) We note that w_{i+1} can see Bd $v_{i+1}v_{i+2}w_{i+2}w_i$ so it can be pushed in a straight line to a point w_{i+1}''' near the side w_iv_{i+1} so that w_{i+1}''' can see Bd $v_iv_{i+1}v_{i+2}w_{i+2}w_i$. The resulting $v_iv_{i+1}w_{i+1}'''w_i$ is convex and we proceed as in the Special Case.

THEOREM 3.2. *Suppose (P, T) is a triangulated disk and A is a subcomplex of (P, T) such that A is an annulus containing* Bd P, *the closure of $P - A$ is a disk D, and A has the following triangulation. Namely, A is the union of n 4-sided disks $v_iv_{i+1}w_{i+1}w_i$ ($i = 1, 2, \cdots, n$, counting is* mod n) *where for each j, $v_j \in$* Bd P *and $w_j \in$* Bd D, *and T restricted to A is determined by coning over the boundary of each of these 4-sided disks from an interior point.* (*See Figure* 3.1.) *Then T is super if T restricted to D is super.*

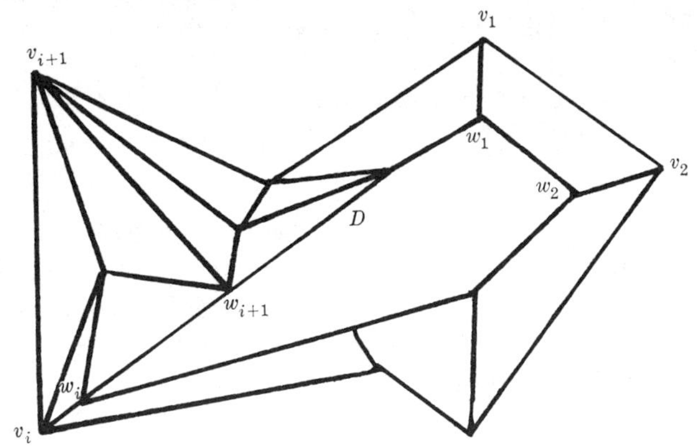

FIGURE 3.1

Proof. To show that (P, T) has Property 1 we let h be a linear embedding of Bd P in the plane and place the image of w_i near $h(v_i)$

in Int $h(\mathrm{Bd}\,P)$ and on the bisector of angle $h(v_{i-1})h(v_i)h(v_{i+1})$. The linear embedding of P can be completed since T restricted to D is super.

To show that (P, T) has Property 2 we suppose that f is the identity and $g(w_i) = w_i'$ is near v_i and on the bisector of angle $v_{i-1}v_iv_{i+1}$. We wish to push w_i to w_i'.

Let X_i be an interior diagonal of the disk $v_iv_{i+1}w_{i+1}w_i$. If the disk is convex, there are two choices of X_i, but if it is concave, there is only one choice.

If the X's can be chosen so that some w_j is not the end of any X, push this w_j along w_jv_j to a point near v_j. Then push it to a point w_j'' near v_j so that $v_{j-1}v_jw_j''w_{j-1}$ and $v_jv_{j+1}w_{j+1}w_j''$ are convex. Now their diagonals can be chosen in two ways. We pick them to contain w_j'' and find that there is another w_k with no X containing it. It in turn is moved into a position w_k'' analogous to w_j'' above. Continuing, we find that if we can get started and can move one w_i to a good position, the others can be pushed into position also. Once each w_j has been moved to a point w_j'' near v_j so that each quadrilateral $v_jv_{j+1}w_{j+1}''w_j''$ is convex, each w_j'' is pushed to a point on the bisector of angle $v_{j-1}v_jv_{j+1}$. Finally, each w_j is moved along the bisector to w_j'. As the boundary of each quadrilateral disk $v_iv_{i+1}w_{i+1}v_i$ is moved it is a minor matter to move the vertex that is on the interior of the quadrilateral.

If the X's are chosen so that each w_j is the end of an X then each w_j is the end of precisely one X since there are the same number of X's as w's. By considering a new triangulation T' of A whose 2-simplexes are those into which the X's divide the quadrilaterals, we note that the new triangulation of A makes it satisfy the hypothesis of Theorem 3.1. The result then follows from Theorem 3.1.

The next theorem is a generalization of Theorem 3.2 in which we allow the annulus to have a slightly different triangulation.

THEOREM 3.3. *Suppose* (P, T), A, D, *are as in Theorem* 3.2 *except that* $\mathrm{Bd}\,P$ *has an extra vertex* v_{n+1}, *counting is* $\mathrm{mod}\,n$ *in subscripting the* w'*s but* $\mathrm{mod}\,n+1$ *in subscripting the* v'*s and* T *restricted to* A *has an additional* 2-*simplex* $v_{n+1}v_1w_1$. *Then* T *is super if* T *restricted to* D *is super.*

Figure 3.2 gives a schematic view of the quadrilaterals in A. It does not show the vertices of T in the interiors of these quadrilaterals since these interior vertices can be dragged along as the boundaries of the quadrilaterals are moved.

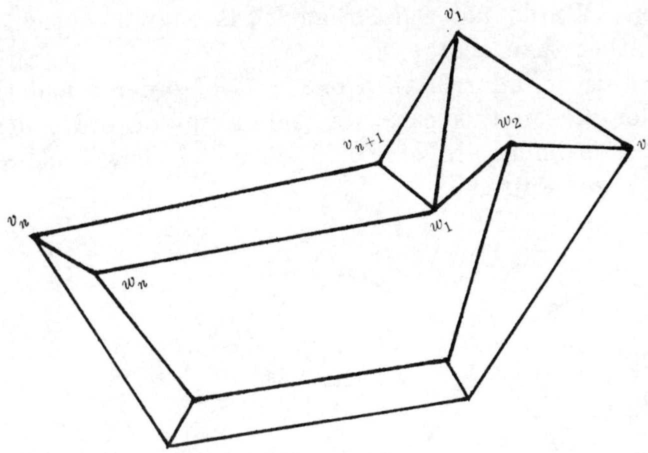

FIGURE 3.2

Proof that (P, T) *has Property* 1. Let H be a linear embedding of Bd P in E^2. For convenience we denote $h(v_i)$ by v_i and the disk bounded by $h(\text{Bd}\, P)$ by P. We cannot hope to put the images of the w's near the corresponding v's if $v_n v_1$ is not a spaninng arc of P.

Let k be the largest of $1, 2, \cdots, n+1$ for which $v_{k-1}v_{k+1}$ is a spanning arc of P. The image of w_i is denoted by w_i' and is located as follows: w_i' is in Int P near v_i and on the bisector of angle $v_{i-1}v_iv_{i+1}$ if $i = 1, 2, \cdots, k-1$; w_i' is near v_{i+1} and on the bisector of angle $v_iv_{i+1}v_{i+2}$ if $i = k, \cdots, n$.

Proof that (P, T) *has Property* 2. We suppose f is the identity map and $g(w_i) = w_i'$ is as described in the preceding paragraph. The cases where $n = 3, 4$ need a slightly different approach so we do not include them.

We start by retriangulating A as was done in the proof of Theorem 3.2. This is done by removing the vertex inside each quadrilateral $v_iv_{i+1}w_{i+1}w_i$ and using a diagonal X_i to divide the quadrilateral into two 2-simplexes. If the quadrilateral is convex there two choices for X but if it is concave there is only one choice. Note that with the new triangulation T' of A, the sum of the orders of the vertices on Bd D is $4n + 1$.

The proof is now broken into steps with Steps a, b, and c being used to push certain w's near their corresponding v's, Steps d and e to push all w's near Bd P, and Step f to push the w's to the w''s.

Step a. *Pushing one* w. If some $w_i (i \neq 1)$ is not on any X, it is of order 3 in T' and we push w_i along w_iv_i to a point near v_i and then to a point w_i'' near v_i so that both $v_{i-1}v_iw_i''w_{i-1}$ and $v_iv_{i+1}w_{i+1}w_i''$

are convex. We do not claim that w''_i is near w'_i since w'_i may be near v_{i+1} rather than near v_i.

If no w_i is of order 3, then one w is of order 5 and the others are of order 4. In this case we pick a w_j of order 4 such that $w_{j-1}w_{j+1}$ is a spanning arc of D. Figure 3.3 shows one possibility where $j \neq 1$ and dotted

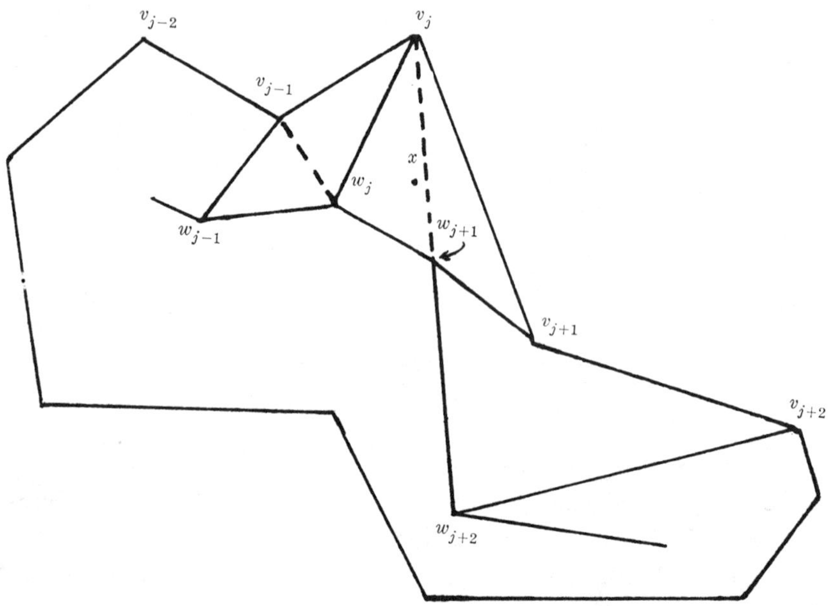

FIGURE 3.3

X's lean one way. For any spanning segment $(w_{j+1}v_j)$ of a 5-sided planar disk $(w_{j-1}w_{j+1}v_{j+1}v_jv_{j-1})$ one end or the other is in the closure of the points which can see the whole boundary. Let x be a point which can see the boundaries of both $w_{j-1}w_{j+1}v_jv_{j-1}$ and $w_{j-1}w_{j+1}v_{j+1}v_jv_{j-1}$ and push w_j to x. For convenience suppose that $x = w_j$. Note that $w_jw_{j+1}v_{j+1}v_j$ is convex and its triangulation can be changed by replacing X_j by the other diagonal. This makes w_j of order 5 in the new triangulation and causes some other w_i to be of order 3. While the above argument is based on Figure 3.3 the argument is similar if the X's lean the other way or if $j = 1$.

Since the new triangulation made w_i of order 3, it can be pushed near v_i as previously described. We say that w_i was crushed. Since each of the quadrilaterals containing this crushed w_i is convex, we suppose that the triangulation of the adjusted A is such that the crushed w_i is of order 5.

Step b. *Pushing another* w. By considering another spanning segment w_iw_{i+2} of D, it can be shown that we can move other w's (leaving the one fixed considered in Step a) and get a new triangulation of the resulting A such that another w_j is of order 3 in the new triangulation. This w_j is pushed to a point w_j'' near v_j so that the quadrilaterals of A containing w_j'' are both convex.

Step c. *Pushing a string of* w's. We continue pushing the w_i's $(i \neq 1)$ to points near their corresponding v_i's as long as this can be done. It can be shown that if both w_i and w_j have been crushed with $i \neq j$ then all w's between them have been crushed. Hence a string of adjacent w's have been crushed and the string has at least two crushed w's. Figure 3.4 shows the situation. The diagonals in the quadrilaterals with a crushed vertex are not shown since they are convex and the X's can lean either way. The quadrilaterals which do not have a crushed vertex are concave and have their X's as drawn. (Figure 3.4 is schematic and shows them as convex rather than as concave.) Let w_f and w_l be the first and last w's respectively that are crushed.

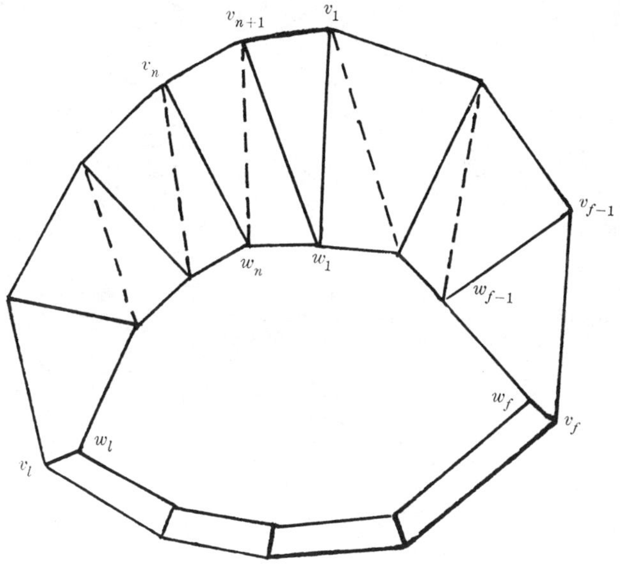

FIGURE 3.4 (Schematic)

Step d. *Leaning* w's. Let $w_{j-1}w_{j+1}$ be a spanning segment of D. Since w_j can be adjusted and the triangulation adjusted to make w_j of order 5, we note that $j \in \{f-1, f, \cdots, l, l+1\}$. We suppose j was selected to be minimal. It is not l or $l+1$.

If $j = f-1$, we consider the 5-sided disk $w_{f-2}w_fv_fv_{f-1}v_{f-2}$ and note that w_{f-1} can either be pushed to a point near w_f or to a point near v_{f-1}. Since we are assuming that $v_{f-1}w_{f-1}w_{f-2}v_{f-2}$ is not convex, we know that w_{f-1} cannot be pushed to a point near v_{f-1}. Therefore we push w_{f-1} to a point near w_f and say that w_{f-1} leans forward.

After w_{f-1} is moved to a point near w_f, w_f is moved to a point near v_{f+1}, w_{f+1} to a point near v_{f+2}, $\cdots$, w_n to a point near v_{n+1}, w_1 to a point near v_1, $\cdots$, and w_{f-2} to a point near v_{f-2}. In this case we have pushed all the w's near the v's but not necessarily to the corresponding v''s.

Similarly, if $j = f + 1, \cdots, l - 1$ we can lean certain w's forward and send others near their corresponding v's.

If $j = f$, we are faced with a different situation. See Figure 3.5. Although $w_{f-1}w_{f+1}$ is a spanning arc of D, $v_{f-1}v_{f+1}$ need not be a spanning arc of P. However, we push w_f along w_fw_{f+1} to a point

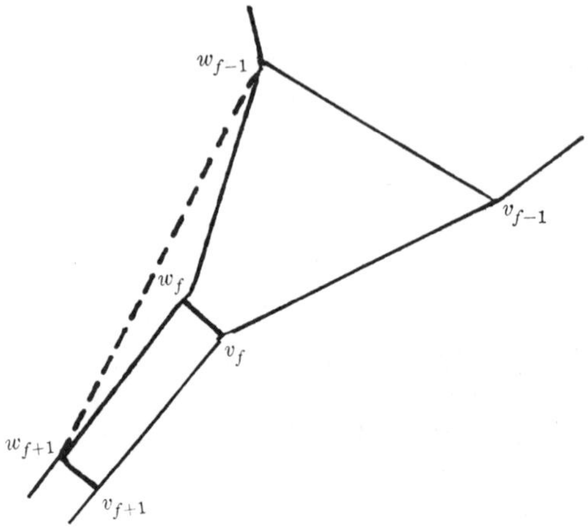

FIGURE 3.5

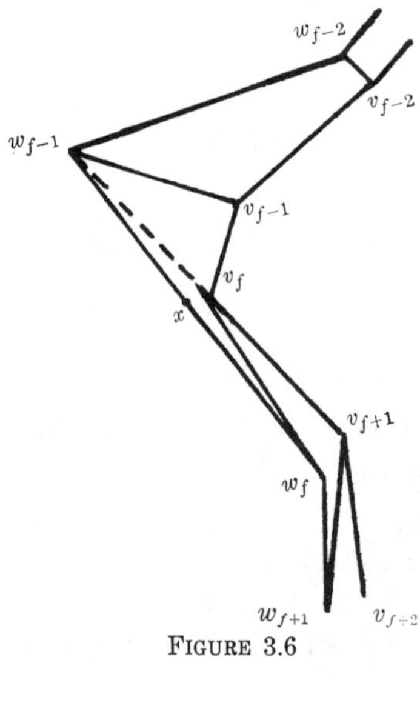

FIGURE 3.6

near w_{f+1}, which is close to v_{f+1}. This move may destroy the convexity of $v_f w_f w_{f-1} v_{f-1}$. However we push w_{f+1} to a point near $v_{f+2}, \cdots, w_n$ to a point near v_{n+1}, w_1 to a point near $v_1, \cdots$, and w_{f-2} to a point near v_{f-2}. Figure 3.6 shows the difficulty of proceeding to push w_{f-1} down to a point near v_{f-1}.

Step e. *Pushing the last* w *close to* Bd P. If $w_f w_{f-1} v_{f-1} v_f$ were convex we could push w_{f-1} to a point near v_{f-1}, but if it is concave, v_f may block the edge $w_{f-1} w_f$. We suppose this is the case and push w_{f-1} toward v_{f-1} but stop before $w_{f-1} w_f$ hits v_f. Now there is a point x on $w_{f-1} w_f$ very close to v_f so that x can see v_{f-1} and v_{f+1}.

Let $w_{j-1} w_{j+1}$ be a spanning arc of the new D such that $j \neq f-2$ or $f-1$. If $j = f$, consider $w_{f+1} v_{f+1} v_f w_{f-1}$ and push w_f to x. Now w_f is near v_f and w_{f-1} can be pushed along $w_{f-1} v_{f-1}$ until it is near v_{f-1}.

If $j = f+1, \cdots, n$, we push w_j to a point near v_j, then w_{j-1} to a point near $v_{j-1}, \cdots, w_f$ to a point near v_f and w_{f-1} to a point near v_{f-1}. If $j = 1, 2, \cdots, f-3$ we push w_j to a point near w_{j+1}, then w_{j+1} to a point near $v_{j+2}, \cdots$, and finally w_{f-1} to a point near v_f. It is to be noted that each w_i is now near either v_i or v_{i+1}, that a string (perhaps null) of w's lean forward and the rest are near their corresponding v's.

Step f. *The final moves.* We recall that k is the largest of $1, 2, \cdots, n+1$ for which $v_{k-1} v_k$ is a spanning arc of P. Hence it is not 1 or 2.

If there is no w near v_k, the w's are now in their standard position.

Suppose some w is near v_k. This w is either w_{k-1} or w_k. We let v_r be the v with no w near it.

We now show that w_k is not near v_k. If it were, $w_1, w_2, \cdots, w_k$ would be near $v_1, v_2, \cdots, v_k$ respectively and $r = k+1, k+2, \cdots$, or $n+1$. But then there would be a spanning arc $v_{r-1} v_r$ of P and this violates the definition of k.

If w_{k-1} is near v_k, we push w_{k-1} to a point near v_{k-1}, w_{k-2} to a point near $v_{k-2}, \cdots$, and w_r to a point near v_r. Each w is now near the correct v and only a small adjustment is necessary to move each vertex w to standard position.

THEOREM 3.4. *For each integer* $n \geqq 3$, *there is a triangulated disk* (P_n, T_n) *such that* P_n *has* n *sides and* T_n *is super.*

Proof. For $n = 4$ or 5 we can produce T_n by coning over Bd P_n.

Once a (P_{n-1}, T_{n-1}) is obtained, one can construct a (P_n, T_n) by putting an annulus A as described in Theorem 3.3 about P_{n-1}.

Question 3.1. It may be noted that for $n > 4$, the T_n we described has $2n^n - n - 40$ 2-simplexes. How low could one go?

Question 3.2. Let T be a triangulation of a disk which satisfies Property 1 of a super triangulation. Is T super?

4. Super subdivisions. In this section we prove (Theorem 4.2) that every triangulation of a disk has a subdivision which is super and does not subdivide the boundary. Theorem 4.1 is the principal tool used in the proof of Theorem 4.2 and is of the same type as Theorems 3.1, 3.2, and 3.3.

THEOREM 4.1. *Suppose (P, T) is a triangulated disk and A is a subcomplex of (P, T) such that A is an annulus containing* Bd P, *the closure of $P - A$ is a disk D, and A has the following triangulation. Namely, A is the union of n 4-sided disks $v_i v_{i+1} w_{i+1,1} w_{i,k_i}$ $(i = 1, 2, \cdots, n$, counting is* mod n, *and for each j and k, $v_j \in$* Bd P *and $w_{j,k} \in$* Bd D*) together with 2-simplexes $v_i w_{i,j} w_{i,j+1}$ $(j = 1, 2, \cdots, k_i - 1)$ where T restricted to A contains the 2-simplexes $v_i w_{i,j} w_{i,j+1}$ together with those determined by coning over the boundary of each 4-sided disk from an interior point. (See Figure 4.1.)*

Then T is super if T restricted to D is super.

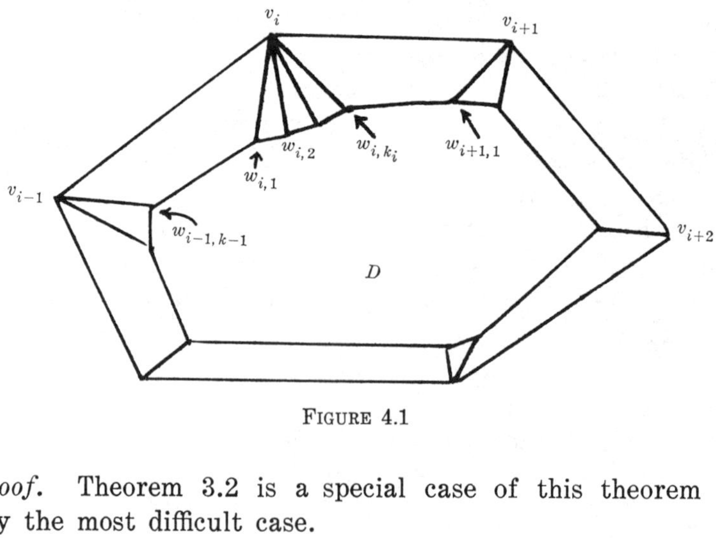

FIGURE 4.1

Proof. Theorem 3.2 is a special case of this theorem and is actually the most difficult case.

The proof that (P, T) has Property 1 is essentially identical to that in Theorem 3.2 so we do not do it.

If $k_i = 1$ for each i, then (P, T) satisfies the hypotheses of Theorem 3.2 and, therefore, T is super.

We assume that there is a j for which $k_j \neq 1$ and proceed to prove that T has Property 2. We assume that f is the identity and that for each i and j, $g(w_{i,j})$ is a point near v_i and near the bisector of angle $v_{i-1}v_iv_{i+1}$.

For each j where $1 < j < k_i$, push $w_{i,j}$ straight along $v_iw_{i,j}$ to a point near v_i.

In each quadrilateral $v_iv_{i+1}w_{i+1,1}w_{i,k_i}$ draw a diagonal X_i. We have added n diagonals. Since for some j, $k_j \neq 1$, there is a vertex $w_{i,1}$ or w_{i,k_i} which is not met by any X. Suppose $w_{i,1}$ is not met. (The other case is analogous.) That vertex $w_{i,1}$ can be pushed straight along $v_iw_{i,1}$ to a point near v_i and then to a point $w'_{i,1}$ so that $v_{i-1}v_iw'_{i,1}w_{i-1,k_{i-1}}$ is convex. Replace its diagonal by the one that contains $w'_{i,1}$. This change guarantees the existence of another vertex $w_{j,1}$ or w_{j,k_j} which is not met by any diagonal. This vertex can now be pushed toward v_j as was done before. Continuing in this fashion, all the vertices $w_{i,j}$ can be pushed near their corresponding vertices v_i. An additional slight adjustment will bring each $w_{i,j}$ to the desired location $g(w_{i,j})$.

THEOREM 4.2. *Every triangulation T of a disk P has a super subdivision which does not subdivide the boundary.*

Proof. The proof is by induction on n, the number of interior vertices of T.

Case $n = 0$. Suppose T is a triangulation of a disk P which contains no interior vertices. Suppose $\mathrm{Bd}\, P$ has k sides.

Let $\{A_i\}_{i=1}^{k-3}$ be the 1-simplexes of T which hit $\mathrm{Int}\, P$. Subdivide each A_i by adding k interior vertices. The super subdivision T' of T is now obtained by examining each 2-simplex σ of T, noting that $\mathrm{Bd}\, \sigma$ has been subdivided and giving σ a super triangulation without further subdivision of $\mathrm{Bd}\, \sigma$ using Theorem 3.4.

We now claim that T' is super. First notice that any linear embedding of $\mathrm{Bd}\, P$ can be extended to a linear embedding of the A_i's since they each have so many bends that they can be laid along the embedded $\mathrm{Bd}\, P$. Since each subdisk has a super triangulation, this embedding of $\mathrm{Bd}\, P \cup (\bigcup_{i=1}^{k-3} A_i)$ can be extended over each subdisk into which the A_i's divide P.

Next we show that T' has Property 2. Suppose $g, h\colon (P, T') \to E^2$ are two linear embeddings which agree on $\mathrm{Bd}\, P$. The plan is to push both $g(A_i)$ and $h(A_i)$ to a common arc for each i and then use the properties of the super triangulations of each subdisk of P into which the A_i's divide P to complete the proof.

Suppose $A_1 = v_1v_3$ is a 1-simplex of T belonging to a shellable 2-simplex $v_1v_2v_3$ of T. Use Theorem 2.4 to push $g(A_1)$ to an arc $g'(A_1)$ with a push that leaves $g(\text{Bd}\, P \cup (\bigcup_{i=2}^{k-3} A_i))$ fixed and makes $g'(A_1)$ lie smoothly near $g(v_1v_2 \cup v_2v_3)$ so that $g'(A_1) \cap h(v_1v_2v_3) = g(v_1) \cup g(v_3)$.

Next use Theorem 2.4 to push $h(A_1)$ to the arc $h'(A_1)$ with a push that leaves $h(\text{Bd}\, P \cup (\bigcup_{i=2}^{k-3} A_i))$ fixed and makes $h'(A_1) = g'(A_1)$. Following a shelling of T and repeating the above process of moving $g(A_i)$ first and then moving $h(A_i)$ to agree, the $g(A_i)$'s can be made to agree with the $h(A_i)$'s.

Using the fact that each subdisk into which the A_i's divide P has a super triangulation, these pushes of arcs can be extended to make the pushed g agree with the pushed h.

The inductive step. Suppose T has n interior vertices and the theorem is true for triangulations with fewer than n interior vertices.

Let w be an interior vertex of T so that $\text{Lk}\,(w) \cap \text{Bd}\, P$ contains a vertex v. Let A be a tight annular neighborhood of $\text{Bd}\, P$ which contains no interior vertices of P. Let $D = \text{Cl}\,(P - A)$. Let $z = vw \cap \text{Bd}\, D$. Let z' and z'' be two points on $\text{Bd}\, D$ on either side of z and very close to z. Let A^+ be the larger annulus whose inner boundary component contains $z'w$ and $z''w$ rather than $z'z$ and $z''z$. Find a triangulation $T(A^+)$ of A^+ which is a subdivision of T and makes A^+ into an annulus as described in Theorem 4.1. Now $\text{Cl}\,(P - A^+)$ can be given a triangulation $T(\text{Cl}\,(P - A^+))$ which is a subdivision of T, which has no additional interior vertices, and so that $T(A^+) \,|\, \text{Bd}\,(\text{Cl}\,(P - A^+))$ is a subcomplex. By induction $T(\text{Cl}\,(P - A^+))$ has a super subdivision T' which does not subdivide $\text{Bd}\,(\text{Cl}\,(P - A^+))$. By Theorem 4.1, $T' \cup T(A^+)$ is a super triangulation of P.

The following result is an immediate corollary of Theorem 4.2. It appears with a different proof in [1, Theorem 5.2].

COROLLARY 4.3. *Let f be a* PL *homeomorphism of a* PL *disk P in E^2 which is fixed on* Bd P. *Then there is a triangulation T of P and a push of (P, T) which takes the identity to f and leaves* Bd P *fixed throughout.*

Question 4.1. Let T be a super triangulation of a disk P. Is every subdivision of T which does not subdivide $\text{Bd}\, P$ also super?

Note. Example 2 in the introduction can be constructed with only three interior vertices. No such example could have only one interior vertex. In a preprint of this paper we posed the question of whether such an example could be constructed with only two interior vertices. C. W. Ho has recently answered this question in the negative by proving that the space of all linear homeomorphisms of an n-cell (C, T) which agree on Bd C and where T has only two interior vertices is a contractible space given the compact-open topology [3, p. 2].

REFERENCES

1. R. H. Bing and M. Starbird, *Linear isotopies in* E^2, (to appear in Trans. Amer. Math. Soc.).
2. S. S. Cairns, *Isotopic deformations of geodesic complexes on the 2-sphere and plane,* Ann. of Math., **45** (1944), 207–217.
3. C. W. Ho, *On the space of the linear homeomorphisms of an n-cell with two interior vertices,* (preprint).

Received April 13, 1977. This work was supported by NSF Grant MCS 76-07242.

UNIVERSITY OF TEXAS
AUSTIN, TX 78712

B. Engulfing

Radial Engulfing

R. H. BING†

The Engulfing Theorem has been one of the most useful discoveries to piecewise linear topology during the past decade. The theorem was stated by Stallings in Theorem 3.1 of [7] as follows.

ENGULFING THEOREM (*Stallings*). Let M^n be a piecewise linear manifold; U an open subset of M^n, P a polyhedron of M^n of dimension r. Let (M^n, U) be r-connected; let $P \cap (M^n - U)$ be compact; let $r \leq n - 3$. Then there is a compact $E \subset M^n$, and there is a piecewise linear homeomorphism $h\colon M^n \to M^n$, such that $P \subset h(U)$ and

$$h|(M^n - E) = \text{Id}|(M^n - E).$$

Throughout this paper we shall use M^n to denote a connected piecewise linear n manifold (not necessarily compact) without boundary,

† Address given at Conference on Topology of Manifolds, Michigan State University, March 22, 1967. Research on this paper supported by NSF grant GP-3857.

U is an open set in M^n, P is an $(n - 3)$-dimensional polyhedron (not necessarily compact), $Q \subset P$ is a subpolyhedron in U; R is the closure $P - Q$, and r is the dimension of R. Although we speak briefly of the case where R is noncompact, we shall regard it as a finite polyhedron unless we state otherwise. We use Id to denote the identity map.

In proving the Engulfing Theorem, Stallings considered a subcomplex Q of P already in U and such that R is compact. He then let U act as an amoeba and sent out feelers to engulf the 0-simplexes of R, one at a time, all the time keeping Q covered. Once a simplex was engulfed, it was added to Q so that none of the previously considered simplexes was uncovered. No effort was made to prevent the feelers from knotting. After the 0-simplexes of R were covered, the engulfing was extended to the 1-simplexes, one at a time. In this case it was not so much like sending out feelers but more like sliding the new U sidewise along a singular disk bounded by the 1-simplex to be engulfed and an arc in the extended U joining the ends of the simplex.

In case the feelers reaching out to these ends twist badly, it is the low dimension that saves the proof since there is no actual knotting of low-dimensional polyhedral spheres in high-dimensional Euclidean spaces.

As more and more of R is engulfed, care must be taken not to uncover any of Q or any essential part of R already covered. As simplexes of higher and higher dimensions are engulfed, the task becomes increasingly difficult since shadows leading out to singularities must be identified and engulfed to prevent uncovering something that has already been considered and added to Q. In fact, when U is finally required to reach out and engulf some of the r-dimensional simplexes of R, in the case $r = n - 3$, the amoeba may be required to regurgitate some of the $(r + 1)$-dimensional path it has previously eaten so as to be able to hold on to Q and eat its way forward.

In this paper we give a slightly different treatment of the Engulfing Theorem and show how this different approach yields additional conclusions that are sometimes useful. Instead of engulfing the simplexes of R one at a time, we consider a homotopy $H^1: R \times [0, 1] \to M^n$ so that $H_0^1 = \text{Id}$, $H_1^1(P) \subset U$, and $H_t^1 = \text{Id}$ on Q. We approximate H^1 with a piecewise linear homotopy and in moving U are guided by paths in this approximation. We have the same problems as Stallings (maybe even worse ones) in dealing with singularities that prevent something that has been previously treated from being uncovered, but we build some lower-dimensional shadows which if engulfed first enable us to proceed without difficulty. A description of this method of engulfing is given in Sections 2, 4, and 5.

There are many modifications of the Engulfing Theorem. Perhaps there is no strongest possible version, so the researcher who hopes to make substantial use of the theorem should understand various methods used in proving it rather than learn many stated versions of it. We point out several modifications of the theorem that the engulfing technique we describe in Sections 2, 4, and 5 is well suited to treat. Certain of these modifications are more useful in certain applications than the originally stated version of the Engulfing Theorem. The Modifications 1–4 are treated in this paper, while Modification 5 will be treated elsewhere.

MODIFICATION 1. Instead of supposing that P is of dimension r, it would have been sufficient to suppose that dimension $P \leq n - 3$ and dimension $R = r$. Also, the homeomorphism could have been chosen so that $h = \text{Id}$ on Q. These improvements can be achieved by a slight adjustment in the original argument given by Stallings in [7]. His inductive proposition only needs to be changed to include them in it. The engulfing technique described in Sections 2, 4, and 5 is such as to incorporate these improvements also.

MODIFICATION 2. The engulfing by Stallings tended not to control the tracks of points, either in diameter or in location. In certain applications it is convenient to have the engulfing push out in prescribed directions. For example, Connell needed to make such a modification when he used a version (Lemma 2 of [4]) of the Engulfing Theorem to show that stable homeomorphisms of E^n $(n \geq 7)$ onto itself can be approximated by piecewise linear ones. Connel permitted $M^n = E^n$ to have a curvilinear combinatorial triangulation and let U be an open ball under the ordinary structure of E^n. He wanted a piecewise linear homeomorphism (under the curvilinear triangulation) that would shove points of U almost straight out (under the ordinary structure of E^n) from the center of U so as to engulf an $(n - 4)$-dimensional polyhedron (in the curvilinear triangulation). Engulfing Theorem D given in Section 5 shows that he could have engulfed an $(n - 3)$-dimensional polyhedron, so as pointed out on page 337 of [4] and in [1], stable homeomorphisms of E^n $(n \geq 5)$ onto itself can be approximated by piecewise linear ones. In Sections 2 and 5, we discuss engulfings that keep tracks near certain preassigned sets.

MODIFICATION 3. The statement of the Engulfing Theorem did not say how far points would move under h nor how far they would move from U. It is sometimes convenient to have bounds for these movements. For example, in [4] Connell did not want to move points that were near his U (open hollow ball) too far away from U during a pushing out engulfing. Engulfing Theorem A″ gives a bound on the sizes of tracks. Engulfing Theorem E puts a bound on how far points are moved from U and R. Engulfing Theorems E′ and F put a limit on the diameters of tracks.

MODIFICATION 4. Sometimes a person wants U to engulf an r-dimensional polyhedron P^r even if (M^n, U) is not r-connected. In case there is a homotopy pulling P^r into U this can be accomplished even if (M^n, U) is only $(2r + 2 - n)$-connected as pointed out in Section 4.

In [2] Cernavskii showed that if $k < (2n/3) - 1$, every homeomorphism $g: E^n \to E^n$ is representable as a composition $g = g_k g_n$ of two homeomorphisms, where g_n is stable and g_k is the identity on a preassigned k-simplex. The fundamental lemma used in proving this result required that in $M^n = E^n - T$, an open set $U = h(Q_\iota)$ engulf a k-dimensional set $P^k = B^k - B^{k-1}$ even though (M^n, U) is only $(n - k - 1)$-connected. (The second halves of the equalities refer to Cernavskii's notation.) However, Cernavskii notes that there is a homotopy pulling P^k into U. Hence, the engulfing is possible even if (M^n, U) is only $(2k + 2 - n)$-connected. This happens if $2k + 2 - n \leq n - k - 1$ or if $k \leq (2n/3) - 1$. The discrepancy between $k < (2n/3) - 1$ used by Cernavskii and the $k \leq (2n/3) - 1$ we obtained perhaps resulted from Cernavskii using less than an optimal form of the engulfing theorem. This does not seem to matter since Cernavskii has since [3] changed his restriction to $k \leq n - 3$ as a consequence of using some results of Homma [6].

MODIFICATION 5. It is sometimes useful to be able to engulf infinite polyhedra. Indeed, Cernavskii in [2] used a version of the Engulfing Theorem that enabled him to engulf such infinite polyhedra, although for simplicity, we neglected to point this out in the preceding paragraph. When R is not compact, one cannot require that E be compact but can sometimes use a substitute. Certain extra requirements must be placed on M^n; U, and P, for even in $M^n = E^3$, one would not expect to find a homeomorphism that would cause an open ball U to engulf $P^0 = \{(i, 0) \mid i = 1, 2, 3, \ldots\}$. One might require that (M^n, U) be uniformly locally r-connected as well as r-connected and that for each $\epsilon > 0$, the part of P outside the open ϵ-neighborhood of U is compact. We leave it to another paper to give a full treatment of the engulfing of infinite polyhedra but only mention in passing that the techniques we describe in Sections 2 and 5 are often particularly well suited to cope with such situations.

Although we have been discussing engulfing as introduced by Stallings, we should call attention to the important version of the Engulfing Theorem proved by Zeeman. Using exponents to denote the dimensions of complexes, Zeeman gave the following version of the Engulfing Theorem as Lemma 38 in [9]. It is mentioned as Lemma 2 in [8].

ENGULFING THEOREM (*Zeeman*). If X^x is homotopic to a point in M^n, then there exist subspaces $Y^y, Z^z \subset M^n$ such that $X^x \subset Y^y \searrow Z^z$, $y \leq x + 1$ and $z \leq 2x - m + 2$.

1. DEFINITIONS

We use the terms *piecewise linear n-manifold* and *combinatorial n-manifold* synonymously. Each has a triangulation such that the closed star of each vertex has a subdivision that is isomorphic to a rectilinear triangulation of a canonical n-simplex.

Although we do not emphasize this, the *engulfing isotopies* $H: M^n \times [0, 1] \to M^n$ that we use can be taken to be the result of a finite number of *small pushes*—that is, there is a finite sequence of numbers $0 = t_0 < t_1 < \cdots < t_m = 1$ such that not only is each H_t a piecewise linear homeomorphism of M^n onto itself but also for each i there is a combinatorial n-cell C_i in M^n such that $H: M^n \times [t_i, t_{i+1}]$ is a piecewise linear map such that if $x \notin C_i$, $H(x \times [t_i, t_{i+1}]) = x$.

By saying that (M^n, U) is *r-connected* we mean that if $0 \leq i \leq r$ and f is a map of an i-cell I^i into M^n which takes $Bd\ I^i$ into U, there is a homotopy fixed on $Bd\ I^i$ that pulls f into U. Note that we use $Bd\ X$ to denote the combinatorial boundary of X. In the case $i = 0$, I^0 is a point, $Bd\ I^0 = \varnothing$, so the 0-connectedness of (M^n, U) would merely mean that $U \neq \varnothing$ (since M^n is connected).

If f is a map of K into Y, the *singularities* of f are the points of K which lie in the closure (in K) of the set of points $x \in K$ such that $x \neq f^{-1}f(x)$.

If $H: K \times [0, 1] \to Y$ is a homotopy (or isotopy), the *track* of a point $x \in K$ is the set $H(x \times [0, 1])$.

If K and Y lie in the same metric space, and $f: K \to Y$, we say that f is an ϵ *map* (or ϵ *homeomorphism* if f is a homeomorphism) if the distance between no point and its image under f is more than ϵ. Also, if $\epsilon > 0$, the ϵ-*neighborhood* of a set K in a metric space Y is the set of all points in Y whose distance from K is less than ϵ.

If K is a triangulated set, the r-*skeleton* of K is the union of all simplexes of the triangulation whose dimensions are less than or equal to r.

If K is a triangulated set, a *cylindrical triangulation* of $K \times [0, 1]$ is a triangulation T compatible with the triangulation of K and the linear structure of $[0, 1]$ and such that the vertical projection of $K \times [0, 1]$ onto $K \times 0$ takes each simplex of T linearly onto a simplex of T. A *vertical segment* in $K \times [0, 1]$ is a set of the form $x \times [0, 1]$ where $x \in K$.

A *polyhedron* in a triangulated set K is a closed subset of K which is the union of simplexes under some subdivision of the triangulation of K. We do not insist that it be compact. For example, the x axis is a polyhedron in the plane. If $L_1 \subset L_2$ and each is a polyhedron in K, we call L_1 a *subpolyhedron* of L_2.

2. RADIAL ENGULFING IN CODIMENSION 4

Let $\{A_\alpha\}$ be a collection of sets in M^n. We define what is meant by the statement that finite r-complexes in M^n can be pulled into U along $\{A_\alpha\}$. This means that if P^k is any finite k-dimensional polyhedron whatsoever in M^n and Q' is a subpolyhedron of P^k that lies in U, then there is a homotopy $H: P^k \times [0, 1] \to M^n$ such that $H_0 = \text{Id}$, $H_1(P^k) \subset U$, $H_t = \text{Id}$ on Q', and for each point $x \in P^k$, $H(x \times [0, 1])$ lies in an element of $\{A_\alpha\}$. For example, if $M^n = E^n -$ origin, U is open ball with center at origin and with origin removed, and $\{A_\alpha\}$ is the collection of all open rays leading from the origin, then for each $i \leq n$, finite i-complexes in M^n can be pulled into U along $\{A_\alpha\}$.

ENGULFING THEOREM A. Suppose $r \leq n - 4$ and $\{A_\alpha\}$ is a collection of sets such that finite r-complexes in M^n can be pulled into U along $\{A_\alpha\}$. Then for each $\epsilon > 0$ there is an engulfing isotopy $H: M^n \times [0, 1] \to M^n$ such that $H^0 = \text{Id}$, $H_t = \text{Id}$ on Q, $P \subset H_1(U)$, and for each $x \in M^n$, there are $r + 1$ elements of $\{A_\alpha\}$ such that the track $H(x \times [0, 1])$ lies in the ϵ-neighborhood of the sum of these $r + 1$ elements.

Proof: The proof is by induction on r. In the case $r = 0$, mutually exclusive polygonal arcs are run from the points of R to U so that these polygonal arcs lie near elements of $\{A_\alpha\}$. Then U is pushed out near these polygonal arcs. Suppose the theorem is true for nonnegative integers less than r. We then show that it is true for r if $r \leq n - 4$.

Let $H^1: R \times [0, 1] \to M^n$ be a homotopy pulling R into U along $\{A_\alpha\}$ in the sense that $H_0^1 = \text{Id}$, $H_1^1(R) \subset U$, $H_t^1 = \text{Id}$ on $Q \cap R$, and for each $x \in R$, $H^1(x \times [0, 1])$ lies in an element of $\{A_\alpha\}$.

Let H^2 be an approximation of H^1 and T be a triangulation of $R \times [0, 1]$ such that H^2 is a linear homeomorphism on each simplex of T, each $H^2(x \times [0, 1])$ lies in an $\epsilon/4$ neighborhood of some element of $\{A_\alpha\}$, and $H^2([(Q \cap R) \times (0, 1)] \cup [R \times 1]) \subset U$. We suppose the triangulation T is extended to $Q \times 0$ and $H_0^2 = \text{Id}$ on $P = P \times 0$.

Let H^3 be a piecewise linear approximation of H^2 such that

- $H_0^3 = \text{Id}$,
- $H^3[(R \times 1) \cup ([Q \cap R] \times [0, 1])] \subset U$,
- each $H^3(x \times [0, 1])$ lies in an $\epsilon/2$ neighborhood of some element of $\{A_\alpha\}$,
- H^3 is a piecewise linear (not necessarily linear) homeomorphism on each simplex of T, and
- H^3 is in general position with respect to T in the sense that if $\mathring{\sigma}^i, \mathring{\sigma}^j$ are different open simplexes of T with dimensions i, j respectively, then dimension $H^3(\mathring{\sigma}^i) \cap H^3(\mathring{\sigma}^j) \leq i + j - n$.

That there is such a general position approximation to H^2 follows from Corollary IIa of [5]. To get H^3, we consider the simplexes of T not in $P \times 0$ one at a time and those of lowest dimension first. The image under H^2 of each of these simplexes lies in the interior of a piecewise linear n-cell in M^n so we can regard our action as taking place in E^n. By considering these piecewise linear simplexes one at a time (with those of low dimension being considered first) and giving each a piecewise linear push so as not to move the preceding simplexes and so as to leave the map a piecewise linear homeomorphism on each simplex, the general position map H^3 is achieved.

Let X denote the set of singularities of H^3 and note that

$$\dim X \leq 2(r + 1) - n \leq r - 2.$$

Of course, some singularities of H^3 may result from the intersections of Q with $H^3(R \times [0, 1])$ while others may result from the intersections of images under H^3 of simplexes of T in $R \times [0, 1]$.

Let $L(X)$ be the union of all vertical segments in $R \times [0, 1]$ through points of X. Then $L(X)$ is an example of a *shadow* which will be treated in more detail in the next section. Shadows are so simple in codimension 4 that they need no special attention. Note that

$$\dim L(X) \leq r - 1.$$

In applying induction, we let $Q \cup H^3[(R \times 1) \cup ((Q \cap R) \times [0, 1])]$ take the place of Q and $H^3(L(X))$ take the place of R. It follows by induction that there is an engulfing isotopy $H\colon M^n \times [0, \frac{1}{2}] \to M^n$ such that $H_0 = \text{Id}$, $Q \cup H^3[(R \times 1) \cup ((Q \cap R) \times [0, 1]) \cup L(X)] \subset H_{1/2}(U)$, $H_t = \text{Id}$ on $Q \cup H^3([R \times 1] \cup [(Q \cap R) \times (0, 1)])$, and for each $x \in M^n$, there are r elements of $\{A_\alpha\}$ such that $H(x \times [0, \frac{1}{2}])$ lies in the ϵ neighborhood of the union of these r elements.

To finish the engulfing isotopy it only remains to extend $H\colon M^n \times [0, \frac{1}{2}]$ to $M^n \times [\frac{1}{2}, 1]$. That this extension is possible follows from Engulfing Theorem B.

ENGULFING THEOREM B. Suppose

K is a finite polyhedron,
T is a triangulation of $K \times [0, 1]$,
$g\colon K \times [0, 1] \to M^n$ is a map which is a piecewise linear homeomorphism on each simplex of T,
L = subpolyhedron of $K \times [0, 1]$ containing all singularities of g and which is the union of vertical segments of $K \times [0, 1]$,
C = closed set in M^n such that
$C \cap g(K \times [0, 1]) = g(K \times 1) \cup g(L)$.
U = open set in M^n containing $g(K \times 1) \cup g(L)$.

Then for each $\epsilon > 0$ there is an engulfing isotopy $H\colon M^n \times [0, 1] \to M^n$ such that

$H_0 = \text{Id}$,
$g(K \times [0, 1]) \subset U$,
$H_t = \text{Id}$ on C, and
for each $x \in M^n$, either $H(x \times [0, 1])$ is a point or there is a point $y \in K$ such that $H(x \times [0, 1])$ lies in the ϵ neighborhood of $g(y \times [0, 1])$.

Proof: We used K instead of R in stating Engulfing Theorem B since it is true irrespective of the dimension of K. When we applied the theorem in the preceding proof we were regarding K as R and C as $Q \cup H^3([(Q \cap R) \times (0, 1)] \cup [R \times 1] \cup L(X))$.

Let L' be a polyhedron in $K \times [0, 1]$ which is the union of vertical segments in $K \times [0, 1]$ and such that $g(L') \subset U$ and L contains no limit points of $K \times [0, 1] - L'$. If we add $g(L')$ to C and replace $K \times [0, 1]$ by the closure of $K \times [0, 1] - L'$, we see that we can assume with no loss of generality that g has no singularities. Hence, we suppose $g\colon K \times [0, 1] \to M^n$ is a homeomorphism and $K \times 0 = K$.

Let δ be a positive number such that any connected set which lies in the δ neighborhood of the sum of two $g(x \times [0, 1])$'s actually lies in the ϵ-neighborhood of one of them.

Let T^1 be a subdivision of T such that L is the union of simplexes of T^1 and g is a linear homeomorphism on each simplex of T^1 that takes it onto a set in M^n with diameter less than δ. Let T^2 be a subdivision of T^1 that is cylindrical—that is, the vertical projection of $K \times [0, 1]$ onto $K \times 0$ sends each simplex of T^2 linearly onto a simplex of T^2.

We now prove the theorem by induction on the dimension of K. If this dimension is 0, we push out in M^n along mutually exclusive polygonal arcs in $g(K \times [0, 1])$. If $\dim K = k > 0$, we let K' be the $(k - 1)$-skeleton of T^2 restricted to $K \times 0 = K$ and let $\mathring{\sigma}_1, \mathring{\sigma}_2, \ldots, \mathring{\sigma}_m$ be the open k simplexes of T^2 in $K \times 0 = K$ which miss L'. Let $V_1, V_2, \ldots, V_m$ be mutually exclusive open sets in M^n such that $V_i \cap g(K \times [0, 1]) = g(\mathring{\sigma}_i \times [0, 1])$, $V_i \cap C = 0$, and V_i lies in the δ neighborhood of $g(x_i \times [0, 1])$, where x_i is the barycenter of $\mathring{\sigma}_i$.

It follows by induction on k that there is an engulfing isotopy $H\colon M^n \times [0, \frac{1}{2}] \to M^n$ such that $H_0 = \mathrm{Id}$, $H^t = \mathrm{Id}$ on C, $g(K' \times [0, 1]) \subset H_{1/2}(U)$ and if $x \in M^n$, either $H(x \times [0, \frac{1}{2}])$ is a point or there is a point $y \in K$ such that $H(x \times [0, \frac{1}{2}])$ lies in a δ neighborhood of $g(y \times [0, 1])$. The engulfing $H\colon M^n \times [0, \frac{1}{2}] \to M^n$ is extended to $H\colon M^n \times [\frac{1}{2}, 1] \to M^n$ by pushing down through the V_i's.

ENGULFING THEOREM A′. Suppose f is a map of M^n into a metric space Y such that for each open covering $\{G_\alpha\}$ of Y, finite r-complexes ($r \leq n - 4$) in M^n can be pulled into U along $\{f^{-1}(G_\alpha)\}$. Then for each open covering $\{G'_\beta\}$ of Y there is an engulfing isotopy $H\colon M^n \times [0, 1] \to M^n$ such that $H_0 = \mathrm{Id}$, $H_t = \mathrm{Id}$ on Q, $P \subset H_1(U)$, and for each $x \equiv M^n$, $H(x \times [0, 1])$ lies in some $f^{-1}(G'_\beta)$.

Proof: Let $\{G''_\alpha\}$ be an open covering of Y such that any connected set that lies in the sum of $r + 1$ elements of $\{G''_\alpha\}$ lies in one element of $\{G'_\beta\}$. It follows from the techniques of the proof of Engulfing Theorem A that there is an engulfing isotopy $H\colon M^n \times [0, 1] \to M^n$ such that $H_0 = \text{Id}$, $H_t = \text{Id}$ on Q, $P \subset H_1(U)$, and for each $x \in M^n$, $H(x \times [0, 1])$ lies in the sum of $r + 1$ elements of $\{G''_\alpha\}$. Then H is the required isotopy.

ENGULFING THEOREM A″. Suppose that $g_i(\epsilon)$ is a function such that the i-complexes in M^n that lie within ϵ of U can be pulled into U along $\{f^{-1}(G_\alpha)\}$ so that no point is moved farther than $g_i(\epsilon)$. Then if P lies in the ϵ neighborhood of U, an engulfing can be found satisfying the conclusions of Engulfing Theorem A′ and such that the diameter of the track of each point is less than $g_r(\epsilon) + g_{r-1}[g_r(\epsilon)] + \cdots + g_0(g_1[\cdots g_r(\epsilon)\cdots])$.

Proof: The proof is by induction on r. We can pick H^3 so that for $x \in R$, diameter $H^3(x \times [0, 1]) < g_r(\epsilon)$. Hence the last part of the isotopy need add no more than $g_r(\epsilon)$ to the diameter of the track of any point.

Since L lies in the $g_r(\epsilon)$ neighborhood of U and $\dim L \leq r - 1$ it follows by induction that the tracks of points during the first half of the engulfing (where L is engulfed) need have diameters of no more than $g_{r-1}[g_r(\epsilon)] + g_{r-2}(g_{r-1}[g_r(\epsilon)]) + \cdots + g_0(g_1[\cdots g_r(\epsilon)\cdots])$. The sum of these two terms gives the expression in the conclusion of the theorem.

3. SOPHISTICATED SHADOWS IN CODIMENSION 3

Let us consider how we might extend Engulfing Theorem A (or A′) by permitting $\dim R = r \leq n - 3$. The description of an engulfing isotopy could be started as before by defining H^1, H^2, T, H^3, and X. However we should not define a shadow $L(X)$ as before because in the case $r = n - 3$, the dimension of this shadow would be as big as the dimension of R and our induction would be blocked. Hence, we describe a more sophisticated shadow L. This is done by considering a subset X_1 of X and homeomorphism g of $(Q \times 0) \cup (R \times [0, 1])$ onto itself. It turns out that the L we describe will be a shadow for H^3g rather than for H^3.

Since H^3 is a homeomorphism on each simplex of T, no singularity results from considering simplexes of T one at a time. Let X_1 be the set of singularities of H^3 resulting from considering pairs of simplexes of T such that at least one member of this pair has dimension less than $n - 2$. Note that even if $r = n - 3$,

$$\dim X_1 \leq (n - 3) + (r + 1) - n = r - 2.$$

Our real goal in Engulfing Theorem A was to engulf P but we actually ended up with all of $Q \cup H^3(R \times [0, 1])$ engulfed. In working in codimension 3 we shall content ourselves with engulfing the $n - 3$ skeleton of T, merely using the $(n - 2)$-simplexes in moving across from one face of the simplex to another face and then in succeeding stages of the engulfing to being willing to uncover interior points of this $(n - 2)$-simplex which do not belong to the shadow. We permit the amoeba to regurgitate part of the path it has previously eaten in order to move ahead to the important task of working its way out to P. It is for this reason that we can use X_1 as a partial substitute for X.

A polyhedron $L \subset R \times [0, 1]$ is called a *shadow* for H^3 if

$\dim L \leq r - 1$,
L is the sum of vertical segments of $R \times [0, 1]$,
$X_1 \subset L$, and
if $x, y \in R \times [0, 1]$ with $H^3(x) = H^3(y)$, then $x \in L$ if and only if $y \in L$.

In engulfing P, the shadow is first engulfed and this shadow is kept engulfed during the remaining part of the engulfing isotopy. Things might have gone wrong if we had omitted the last requirement in the definition of a shadow. As we pushed across an $(n - 2)$-simplex $H^3(\sigma_1)$ containing $H^3(x)$ we might have uncovered part of $H^3(L)$ through an $(n - 2)$-simplex $H^3(\sigma_2)$.

Let L_1 be the union of all vertical segments in $R \times [0, 1]$ through points of X_1. It satisfies all of the requirements of a shadow except possibly the last one. However, we may need to modify H^3 before getting a shadow satisfying the last condition. The following theorem tells that such a modification is possible.

SHADOW BUILDING THEOREM. Suppose T is a triangulation of $(Q \times 0) \cup (R \times [0, 1])$ and $H^3: (Q \times 0) \cup (R \times [0, 1]) \rightarrow M^n$ such that

$H^3(x \times 0) = x$,
H^3 is a piecewise linear homeomorphism on each simplex of T,

H^3 is in general position with respect to T in the sense that if $\mathring{\sigma}^i, \mathring{\sigma}^j$ are different open simplexes of T with dimensions i, j respectively, then $\dim H^3(\mathring{\sigma}^i) \cap H^3(\mathring{\sigma}^j) \leq i + j - n$;

X is the set of singularities of H^3; X_1 is the set of singularities of H^3 resulting from considering pairs of elements of T at least one of which is of dimension less than $n - 2$; and K is the $n - 3$ skeleton of T. Then for each $\epsilon > 0$ there is a piecewise linear homeomorphism g of $(Q \times 0) \cup (R \times [0, 1])$ onto itself such that $g = \text{Id}$ on $K \cup X_1$ and there is a shadow L for $H^3 g$.

Proof: Unless $r = n - 3$, we can let $g = \text{Id}$ and L be the union of all vertical segments through X. Hence, we suppose $r = n - 3$.

The first approximation of L is L_1. It contains X_1 and is of dimension less than or equal to $r - 1$. Unfortunately it may contain many points of $X - X_1$ so we shall move some of these off L_1 before getting our second approximation L_2.

Since $\dim L_1 \leq n - 4$ and $\dim X \leq n - 4$, there is a piecewise linear homeomorphism g_1 of $(Q \times 0) \cup (R \times [0, 1])$ onto itself that is the identity of $K \cup X_1$ such that g_1 moves points very little and $\dim L_1 \cap g_1(X - X_1) \leq n - 8$. Let X_2 be the set of all points x of $g_1(X)$ such that for some point $y \in L_1 \cap g_1(X)$, $H^3 g_1(x) = H^3 g_1(y)$. Note that $\dim (X_2 - X_1) \leq n - 8$. Let L_2 be the union of all vertical segments in $R \times [0, 1]$ through points of X_2. Note that $L_1 \subset L_2$ and $\dim (L_2 - L_1) \leq n - 7$.

Let g_2 be a piecewise linear homeomorphism of $(Q \times 0) \cup (R \times [0, 1])$ onto itself that equals g_1 on $K \cup L_1 \cup X_2$ and such that $\dim (L_2 - L_1) \cap g_2(X - X_2) \leq n - 11$. Let L_3 be the union of all vertical segments in $R \times [0, 1]$ through points of X_3. Note that $\dim (L_3 - L_2) \leq n - 10$.

We continue defining X_i's, L_i's, and g_i's until we have reduced the dimension of $X_{j+1} - X_j$ to a negative. Then $g = g_{j+1}^{-1}$ and $L = L_j$. If we have taken care that the g_i's are sufficiently close to the identity, we will have g is an ϵ-homeomorphism.

4. REDUCING CONNECTEDNESS REQUIREMENTS

In this section we do not give any results that could not have been obtained just as well by Stalling's method of engulfing by considering simplexes one at a time. However, we follow the technique described in the proof of Engulfing Theorem A where we push out along directions determined by vertical segments in $R \times [0, 1]$ after first covering images of a shadow. The main difference is that we build a shadow once and continue to use it rather than build a new shadow each time we get ready to move across a simplex. We shall ignore $\{A_\alpha\}$ in this section and return to consideration of radial engulfing in the next.

ENGULFING THEOREM C. If (M^n, U) is $(r-1)$-connected and there is a homotopy fixed on Q and pulling R into U, then there is an engulfing isotopy fixed on Q that causes U to engulf P.

Proof: The proof is by induction on r. As in the proof of Engulfing Theorem A, let H^1 be a homotopy pulling R into U, H^2 be an approximation to H^1, and T be a triangulation of $(Q \times 0) \cup (R \times [0, 1])$ such that H^2 is a linear homeomorphism on each simplex of T. It will be convenient for our argument that T restricted to $R \times [0, 1]$ be a cylindrical triangulation, so we suppose this. We then find a general position approximation H^3 of H^2 satisfying the same requirements as the H^3 of the proof of Engulfing Theorem A except that we drop the requirement concerning $\{A_\alpha\}$ and add a requirement that there is a shadow L for H^3. The Shadow Building Theorem insures that there is such an H^3 with a shadow L.

Let $\sigma_1, \sigma_2, \ldots, \sigma_{j-2}$ be the r-simplexes of T in $R = R \times 0$; K be the $r-2$ skeleton of T restricted to R; and K' be the $r-1$ skeleton. Since (M^n, U) is $(r-1)$-connected and $\dim (K \times [0, 1]) \cup L \leq r-1$, there is a homotopy fixed on Q pulling $H^3(K \times [0, 1] \cup L)$ into U. It follows by induction on r that there is an engulfing isotopy $H: M^n \times [0, 1/j] \rightarrow M^n$ such that

$$H_0 = \mathrm{Id},$$
$$H_t = \mathrm{Id} \text{ on } Q \cup H^3((R \times 1) \cup [(Q \cap R) \times (0, 1)])$$

and

$$H^3[K \times (0, 1) \cup L] \subset H_{1/j}(U).$$

One notes that H^3 is a homeomorphism on $K' \times [0, 1] - L$ so one can apply Engulfing Theorem B to extend $H: M^n \times [0, 1/j] \rightarrow M^n$ to $H: M^n \times [1/j, 2/j] \rightarrow M^n$ so that $H^3(K' \times [0, 1]) \subset H_{2/j}(U)$ and for $1/j \leq t \leq 2/j$, $H_t = H_{1/j}$ on $Q \cup H^3([(Q \cap R) \times (0, 1)] \cup (R \times 1) \cup [K \times (0, 1)] \cup L)$.

Now consider σ_1 and all the $(r + 1)$-simplexes of T above it. Suppose there are j_1 of these. By starting at the top and working his way down one finds in j_1 applications of Engulfing Theorem B that $H: M^n \times [0, 2/j] \to M^n$ can be extended to an isotopy $H: M^n \times [2/j, 3/j] \to M^n$ such that

$$H^3(\text{all } r\text{-simplexes of } T \text{ above } \sigma_1) \subset H_{3/j}(U)$$

and for $2/j \leq t \leq 3/j$, $H_t = H_{2/j}$ on $Q \cup H^3([(Q \cap R) \times (0, 1)] \cup (R \times 1) \cup [K' \times (0, 1)] \cup L)$. In each application of Engulfing Theorem B, we cover up an $(r + 1)$-simplex of T^2 but in working down to those beneath it we do not necessarily keep it covered since L was not required to contain all singularities of H^3.

By working the way down the cylinders above $\sigma_2, \sigma_3, \ldots, \sigma_{j-2}$ one at a time, one finds, on repeated applications of Engulfing Theorem B, that H can be extended to an isotopy $H: M^n \times [0, 1] \to M^n$ such that

$$H_0 = \text{Id, and}$$
$$H^3(r\text{-skeleton of } T) \subset H_1(U).$$

This H is the required engulfing.

One may note that the only place the $(r - 1)$-connectedness of (M^n, U) was used was in showing that the $H^3(L)$ could be homotopically pulled into U. If the shadow L were of lower dimension, even weaker connectedness could have been assumed. If P (as well as R) is of dimension less than or equal to r, then the argument used in the Shadow Building Theorem shows that the dimension of L could have been made as small as $(2r + 2 - n)$. Hence, we have the following modified version of Engulfing Theorem C.

ENGULFING THEOREM C′. If P is r-dimensional $(r \leq n - 3)$, (M^n, U) is $(2r + 2 - n)$-connected, and there is a homotopy fixed on Q and pulling R into U, then there is an engulfing isotopy fixed on Q that causes U to engulf P.

5. RADIAL ENGULFING IN CODIMENSION 3

We now return to a topic started in Section 2.

ENGULFING THEOREM D. Suppose f is a map of M^n onto a finite dimensional metric space Y such that for each open covering

$\{G_\alpha\}$ of Y, finite r-complexes ($r \leq n - 3$) in M^n can be pulled into U along $\{f^{-1}(G_\alpha)\}$. Then for each open covering $\{G'_\beta\}$ of Y there is an engulfing isotopy H: $M^n \times [0, 1] \to M^n$ such that $H_0 = \text{Id}$, $H_t = \text{Id}$ on Q, $P \subset H_1(U)$, and for each $x \in M^n$, there is a $G'_\gamma \in \{G'_\beta\}$ such that $H(x \times [0, 1])$ lies in $f^{-1}(G'_\gamma)$.

Proof: Let $k = \dim Y$ and $\{G''_\alpha\}$ be an open covering of Y such that any connected subset of Y that lies in the sum of $k + 2$ elements of $\{G''_\alpha\}$ lies in one element of $\{G'_\beta\}$.

For $i = 1, 2, \ldots, k + 1$ let $\{G^i_{\alpha_i}\}$ be a discrete collection of open sets in Y such that $\{G^i_{\alpha_i}\}$ refines $\{G''_\alpha\}$ and $\{G'''_\alpha\} = \{G^1_{\alpha_1}\} \cup \{G^2_{\alpha_2}\} \cup \cdots \cup \{G^{k+1}_{\alpha_{k+1}}\}$ covers Y. That there is such a collection follows from the k-dimensionality of Y.

With the preliminaries over, the proof follows the pattern of the proof of Engulfing Theorem C where we define H^1, H^2, T, and H^3 but with the added precaution that the mesh of T is so fine and H^3 is so close to H^1 that if σ is an element of T in $R = R \times 0$, then $H^3(\sigma \times [0, 1])$ lies in an element of $f^{-1}\{G'''_\alpha\}$.

We then define K and K' as before but take special care in ordering the r-simplexes $\sigma_1, \sigma_2, \ldots, \sigma_{j-2}$ of T in $R = R \times 0$. This is done so that there are $k + 2$ integers $1 = k_1 < k_2 < \cdots < k_{k+2} = j - 2$ so that if $k_i < m \leq k_{i+1}$, then $H^3(\sigma_m \times [0, 1])$ lies in an element of $\{f^{-1}(G^i_{\alpha_i})\}$. If one then continues as in the proof of Engulfing Theorem C, the resulting engulfing isotopy is the required one for Engulfing Theorem D.

6. CURBING TRACKS

Our technique has been to engulf the shadow L and then move points which are close to $H^3(R \times [0, 1])$. We can get an upper bound on the diameter of the track of a point under an isotopy by adding up the amount that each of these two stages contribute. If our homotopies pulling R into U and pulling L into U take place near U, we can pick an engulfing isotopy that is fixed except near U. In general, it is the engulfing of part of L that sends us farthest from U.

Let $g_i(\epsilon)$ be an increasing function such that if P^i is a finite i-complex in the ϵ-neighborhood of U (in M^n) and Q' is a subcomplex of P^i which is contained in U, then there is a homotopy fixed on Q' and pulling P^i

into U so that the diameters of tracks are less than $g_i(\epsilon)$. In Engulfing Theorem F we suppose that this pulling is along $\{f^{-1}(G_\alpha)\}$.

ENGULFING THEOREM E. If (M^n, U) is r-connected ($r \leq n - 3$) and P lies in the ϵ-neighborhood of U, then U can engulf P with an isotopy that is fixed on Q and outside the intersections of the $g_0(g_1(\cdots g_r(\epsilon)\cdots))$ neighborhoods of U and R.

Proof: The proof is by induction on r. Note that H^3 can be chosen so that $H^3(R \times [0, 1])$ lies within the intersection of the $[g_r(\epsilon)]$-neighborhood of U and the $[g_r(\epsilon)]$-neighborhood of R. Hence, L lies in this intersection. The last part of the isotopy can be taken to be fixed except near here.

Since $\dim L \leq r - 1$, it follows by induction that the engulfing of it need move no point outside the intersection of the $g_0[g_1(\cdots g_{r-1}(\delta)\cdots)]$-neighborhood of U where $\delta \leq g_r(\epsilon)$.

The following engulfing theorem is an immediate consequence of Engulfing Theorems D and E.

ENGULFING THEOREM E′. Suppose that in Engulfing Theorem D, $g'(\epsilon)$ is a function such that any connected subset of an $f^{-1}(G'_\beta)$ which lies within the intersection of the ϵ neighborhoods of U and R is of diameter less than $g'(\epsilon)$. Then if each point of R lies in the ϵ-neighborhood of U, the engulfing isotopy of Engulfing Theorem D can be such that the tracks of points have diameters less than $g'[g_0(g_1[\cdots g_r(\epsilon)\cdots])]$.

If we look at the isotopy closer we can get a bound that is more delicate in cases where g' is large.

ENGULFING THEOREM F. Under the hypotheses of Engulfing Theorem E′ there is a required isotopy such that the diameters of tracks are less than

$$g'[g_r(\epsilon)] + g_{r-1}[g_r(\epsilon)] + \cdots + g_0[g_1(\cdots g_r(\epsilon)\cdots)].$$

Proof: The term $g'[g_r(\epsilon)]$ determines a bound on the diameter of the part of track generated by the last portion of the isotopy when movement was near $H^3(R \times [0, 1])$ and hence within $g_r(\epsilon)$ of R and within $g_r(\epsilon)$ of U. Since these parts of tracks were restricted to $f^{-1}(G'_\beta)$'s, the first term of the expression in the statement of Engulfing Theorem F gives a bound on the movement of points during the last portion of the isotopy. The first portion of the isotopy was used to engulf the shadow L and since L is of dimension less than $n - 3$, Engulfing Theorem A″ provides us with the remaining terms of the expression.

7. QUESTIONS

a. Is Engulfing Theorem D true if we do not suppose that Y is finite dimensional?

b. Does Engulfing Theorem A hold for $r \leq n - 3$?

c. Under what conditions can we suppose that U can engulf finite $(n-2)$-complexes? For example—could it if M^n is compact, U is $(n-2)$-connected, and (M^n, U) is $(n-2)$-connected? An affirmative answer would imply the Poincaré conjecture. The answer is not affirmative if one drops the condition that U is $(n-2)$-connected as can be seen by letting $M^n = S^2 \times S^1$, D be a disk on S^2, and U be the interior of a knotted solid torus in the interior of $D \times S^1$ and passing around the hole in $D \times S^1$.

d. Under what conditions can we engulf $(n-1)$-complexes? As a sample of a pertinent specific question we suggest the following: If D is a canonical $(n-1)$-cell in E^n, U is an open set containing Int D, E is a locally flat $(n-1)$-cell with $Bd\, D = Bd\, E$, is there an isotopy $H: E^n \times [0, 1] \to E^n$ such that $H_0 = \text{Id}$, $H_t = \text{Id}$ on $Bd\, D$, and $E \subset H_1(U)$? One can get a homotopy fixed on $Bd\, D$ pulling E into U. It would require engulfing of infinite complexes as discussed in Modification 5 early in this chapter to produce the engulfing if it could be obtained at all.

Discussion of Bing's paper raised the following questions:

1. What are necessary and sufficient conditions on an n-manifold M^n without boundary in order that it have the property that each Cantor set in M^n lies in an open n-cell in M^n?

2. What obstructions are there to engulfing in codimensions other than those in the known engulfing theorems?

3. Can engulfing be better characterized by specifying Bing's collection $\{A_\alpha\}$ in some particular way?

4. If M^n is a compact n-manifold in which each properly embedded topological polyhedron lies in an open n-cell, is $M^n = S^n$?

5. Is the union of an $(n-1)$-simplex and a flat $(n-1)$-cell in E^n (which meet only on their common boundary) a flat $(n-1)$-sphere?

6. In the language of Stalling's engulfing theorem, let $\tilde{M}^n$ be a covering space of M^n. Then does $\tilde{U}$ engulf P in codimension 2?

REFERENCES

[1] R. H. Bing, Stable homeomorphisms on E^5 can be approximated by piecewise linear ones, Abstract 607–16, Notices of Amer. Math. Soc., 10 (1963), 666.

[2] A. V. Cernavskii, Isotopy of elements and spheres in n-dimensional space for $k < \frac{2}{3}n - 1$, Dokl. Akad. Nauk SSSR, 158 (1964), 62–65 or Soviet Math. Dokl., 5 (1964), 1194–1197.

[3] ———, Singular points of topological imbeddings of manifolds and the union of locally flat cells, Dokl. Akad. Nauk. SSSR, 167 (1966), 528–530 or Soviet Math. Dokl., 7 (1966), 433–436.

[4] E. H. Connell, Approximating stable homeomorphisms by piecewise linear ones, Annals of Math., 78 (1963), 326–338.

[5] David W. Henderson, Relative general position, (to appear).

[6] Tatsuo Homma, Piecewise linear approximation of embeddings of manifolds, Florida State University (1965), Theorem 2.1.

[7] John Stallings, The piecewise linear structure of Euclidean space, Proc. of the Cambridge Phil. Soc., 58 (1962), 481–488.

[8] E. C. Zeeman, The Poincaré conjecture for $n \geq 5$, Topology of 3-Manifolds and Related Topics, Prentice-Hall (1962), 198–204.

[9] ———, Seminar on Combinatorial Topology, Institut des Hautes Études Scientifiques (1963), Chapter 7.

THE UNIVERSITY OF WISCONSIN

VERTICAL GENERAL POSITION

BY

R. H. BING

1. INTRODUCTION. Suppose f_1 is a map of a finite geometric k-complex $|K|$ into a geometric PL m-manifold M^m without boundary, $m \geq k$. We often need to approximate f_1 with a map f that is nicer in some sense than f_1. Ordinary general position may give us some (or even all) of the niceness we need for certain purposes. In other cases, we need more.

We use d to denote the distance function. For f to be an approximation to f_1, we mean that we have preselected a positive ε and then picked f so that for each point $p \in |K|$, $d(f_1(p), f(p)) \leq \varepsilon$. If $|K|$ has been noncompact, we might have used a positive continuous function $\varepsilon(x)$ defined on M rather than a constant ε. Note that we let $M^m = M$ and use the exponent only when we wish to emphasize the dimension. All of the manifolds considered in this paper are without boundary.

If σ_i is a simplex, we use $\mathring{\sigma}_i$ to denote the corresponding open simplex; i.e., what is left of σ_i after all its proper faces are removed. We follow the convention that if σ_i is a 0-simplex, then $\mathring{\sigma}_i = \sigma_i$.

A preliminary requirement of a general position approximation f is that there be a subdivision K' of K such that f is a PL homeomorphism on each simplex of K'. An additional requirement for f to be some kind of general position map is that a slight change in it will not make the set of singularities in $|K| = |K'|$ any simpler. Wh.t is meant by simpler may vary according to ones needs.

Perhaps the simplest such requirement for general position is that if $\mathring{\sigma}_i$, $\mathring{\sigma}_j$ are different open simplexes of K', then

$$\dim(f((\mathring{\sigma}_i) \cap f(\mathring{\sigma}_j)) \leq \dim \sigma_i + \dim \sigma_j - m = m - (\text{codim } \sigma_i + \text{codim } \sigma_j) .$$

To keep the middle expression from reading maximum $(-1, \dim \sigma_i + \dim \sigma_j - m)$ we use the convention, not that the dimension of the empty set is -1, but rather that $X = 0$ if $\dim X < 0$. The codimension of σ is $\dim M - \dim \sigma$.

An advantage of using such a simple kind of general position as the above is that it is easy to get an approximation that satisfies it. For example, if f_1 were a map of $|K|$ into Euclidean m-space R^m, then to get f one would only first take a fine subdivision K' of K. Next, pick f to be a close approximation of f_1 on the vertices of K' and such that the images of these vertices are distinct, no three of them lie in the same line, no four in the same 2-plane,..., and no $m + 1$ in the same $(m-1)$-dimensional hyperplane. Then f would be extended to a linear homeomorphism on each simplex of K'.

Because of its simplicity, the type of general position mentioned above is perhaps preferable if it suffices. However, there are times when we are concerned with not only considering the images of pairs of simplexes of K' but also triples, quadruples,... . Under these conditions we require that if $\sigma_1, \sigma_2, \dots, \sigma_j$ are j different simplex of K', then

$$\dim \cap f(\overset{\circ}{\sigma}_i) \leq \Sigma \dim \sigma_i - (j-1)m = m - \Sigma \operatorname{codim} \sigma_i .$$

If f is a map of $|K|$ into M, the singularity set S_f of f is the closure of the set of all points $p \in |K|$ such that $p \neq f^{-1}f(p)$. A moments reflection shows that by no means does the above restriction make the set of singularities as simple as possible. For example, if it should happen that for some point $p \in M$, $f^{-1}(p)$ is a set of three collinear points (or two parallel segments) in a simplex of K', we could certainly modify the map to change this. We might send a 2-sphere into R^3 with a general position map such that the set of singularities is a finite set of points. These could certainly be eliminated. It is not our purpose to make the approximating map as simple as possible in all ways but only in certain ways which meet our needs at the moment.

If K is a complex, we call $|K|$ a polyhedron. We do not insist that it be compact but if P is a polyhedron in M, we also mean that P is a closed subset of M. Although K provides $|K|$ with a triangulation, there are other rectilinear triangulations. If f is a map of a noncompact P, we shall insist that f be closed so that the image of closed sets will be closed.

A map f of a polyhedron P with a triangulation T into a PL manifold M^m is a GENERAL POSITION MAP with respect to T if

f is a closed map,

f is a PL flat homeomorphism on each simplex of T, and for different simplexes $\sigma_1, \sigma_2, \dots, \sigma_j$ of T,

$$\dim \cap f(\overset{\circ}{\sigma}_i) \leq m - \Sigma \operatorname{codim} \sigma_i .$$

A PL homeomorphism h of a simplex Δ into a PL manifold M^m is FLAT if $f(\Delta)$ lies on the interior of a PL m-cell B^m in M and there is a PL homeomorphism g of B onto a canonical m-cell I^m such that gh takes Δ linearly into a simplex in I^m. In the above definition of general position, some leave out the restriction that f be flat on the simplexes of T but we include it since it is sometimes useful and the following important theorem is just as easy to prove with it included.

THEOREM 1.1. *Suppose M^m is a PL m-manifold, P is a polyhedron, Q is a subpolyhedron of P, T' is a triangulation of P such that Q is the union of simplexes of T', f_1 is a closed map of P into M such that f_1 restricted to Q is a general position map with respect to T' restricted to Q, and $\varepsilon(x)$ is a positive continuous function on M. Then there is a triangulation T of P and a general position map f of P with respect to T into M such that*

T restricted to Q is the same as T' restricted to Q,

$f = f_1$ on Q,

$d(f(x), f_1(x)) < \varepsilon(f(x))$ for each $x \in |P|$.

The proof of Theorem 1.1 is standard and we do not provide complete details here. Some of the details are included in the proof of Lemma 2.6. The overall plan is to extend the triangulation of T' restricted to Q to a triangulation T of P without subdividing elements of T' in Q and then approximate f_1 with a map f_2 that extends f_1 on Q and is such that f_2 is a PL flat homeomorphism on each simplex of T. The simplexes of T not in Q are ordered $\sigma_1, \sigma_2, \ldots$ such that each proper face of σ_i either procedes σ_i in the ordering or lies in Q. Using Lemma 2.3 of the next section we get a sequence of homeomorphisms $g_1, g_2, \ldots$ such that g_i is close to g_{i-1} and equal to g_{i-1} except on simplexes of T having σ_i as a face, g_i is PL flat on each simplex of T, and if $\Delta_1, \Delta_2, \ldots, \Delta_n$ are different simplexes of T each of which either lies in Q or in $\sigma_1, \sigma_2, \ldots, \sigma_i$, then

$$\dim \cap g_i(\mathring{\Delta}_j) \leq \Sigma \dim \Delta_j - (n-1)m .$$

The homeomorphism g_i is obtained by regarding $g_{i-1}(\sigma_i)$ as a simplex in a PL n-cell and pushing at its barycenter.

Let us consider the case where f_1 is a closed map of the cartesian product $P \times I$ rather than a map of P. For each point $p \in P$, we call $p \times I$ a vertical segment. We may want the singularities S_{f_1} of f_1 in $P \times I$ to be nice with respect to vertical segments.

Suppose P is a polyhedron, X is a subpolyhedron of $P \times I$, T is a triangulation of $P \times I$, and T(X) is a triangulation of X such that each simplex of T(X) lies in a simplex of T. If $\Delta \varepsilon T(X)$, we use $\sigma(\Delta)$ to denote the simplex of T such that $\mathring{\Delta} \subset \mathring{\sigma}(\Delta)$. Also, ρ denotes the projection of $P \times I$ onto $P \times 0$. We say that the projection ρ is VERTICALLY NICE with respect to T and T(X) if for each collection of different simplexes $\Delta_1, \Delta_2, \ldots, \Delta_j$ of $T(X)$

$$\dim \cap \rho(\mathring{\Delta}_i) \leq \sum_{i=1}^{j} \dim (\Delta_i) - \sum_{i=2}^{j} \dim \rho\sigma(\Delta_i) .$$

If we designate $\dim \rho\sigma(\Delta_i) - \dim (\Delta_i)$ by $\operatorname{codim} \Delta_i$, the right side of the above becomes $\dim \rho\sigma(\Delta_1) - \Sigma \operatorname{codim} \Delta_i$. Since the ordering of the Δ's was arbitrary we could even get

$$\dim \cap \rho(\mathring{\Delta}_i) \leq \min (\dim \rho\sigma(\Delta_i)) - \Sigma \operatorname{codim} \Delta_i .$$

Rather than use an arbitrary rectilinear triangulation of $P \times I$, it is sometimes convenient to use a cylindrical triangulation where simplexes are stacked vertically. A rectilinear triangulation T of $P \times I$ is called a CYLINDRICAL TRIANGULATION if the projection of $P \times I$ onto $P \times 0$ takes each simplex of T onto a simplex of T.

Suppose M^m is a PL m-manifold, P^r is a r-dimensional polyhedron ($r \leq m-3$), $\varepsilon(x)$ is a positive continuous function defined on M^m and f_1 is a closed map of $P \times I$ onto M such that f_1 restricted to $P \times 0$ is a general position map with respect to some rectilinear triangulation of $P \times 0$. We say that f is a VERTICAL GENERAL POSITION ε-APPROXIMATION of f_1 if there is a cylindrical triangulation T of $P \times I$ and a rectilinear triangulation $T(S_f)$ of the singularity set S_f of f such that

f is a general position map with respect to T,

the projection of S_f onto $P \times 0$ is vertically nice with respect to T and $T(S_f)$,

$d(f_1(x), f(x)) < \varepsilon(f(x))$ for each $x \in P \times I$.

A vertical general position map resembles a fibered general position map as treated by Millett in [8].

In Section 2 we prove the following vertical general approximation theorem.

THEOREM 2.1. *If M^m is a PL m-manifold, P^r is an r-dimensional polyhedron with $r \leq m - 3$, $\varepsilon(x)$ is a positive continuous function defined on M and f_1 is a closed map of $P \times I$ into M, that is a general position map with respect to some triangulation of $P \times 0$, then f_1 has a vertical general position ε-approximation f that agrees with f_1 on $P \times 0$.*

We shall make use of the vertical general position approximation Theorem 2.1 in Section 3 to build a sophisticated shadow. This sophisticated shadow in turn will be used in Section 4 to prove the following infinite engulfing theorem.

THEOREM 4.1. *Suppose that*

M^m is a PL manifold without boundary,

U is an open subset of M,

P is a polyhedron (not necessarily compact) in M with $\dim P \leq m - 3$,

Q is a subpolyhedron of P with $Q \subset U$,

$R^r = Cl\,(P - Q)$ is a polyhedron (not necessarily compact),

$\{A_\alpha\}$ is a collection of sets in M such that r-complexes in M can be pulled into U along $\{A_\alpha\}$.

Then for each continuous positive function $\varepsilon(x)$ on M there exists an engulfing isotopy $H : M \times I \to M$ such that

$H_0 = Id$,

$H_t = Id$ on a neighborhood of Q,

$H_1(U) \supset P$, and

for each point $p \in M$ that is moved by H, then there are s elements of $\{A_\alpha\}$ such that $H(p \times I)$ lies in the $\varepsilon(x)$-neighborhood of the union of these s elements where

$$s = \begin{cases} 1 & \text{if } r = 0 \\ 1 & \text{if } 1 = r \leq m - 4 \\ r & \text{if } 2 \leq r \leq m - 4 \\ 2 & \text{if } 1 = r = m - 3, \text{ and} \\ r + 1 & \text{if } 2 \leq r = m - 3 \,. \end{cases}$$

Suppose M is a PL manifold, U, V are open subsets of M, and $\{A_\alpha\}$ is a collection of subsets of M. We say that K-COMPLEXES IN V CAN BE PULLED INTO U ALONG $\{A_\alpha\}$ if for each k-complex K with $|K|$ a closed subset of M in V and each subcomplex L of K with $|L| \subset U$, there is a closed homotopy $H : |K| \times I \to M$ such that $H_0 = Id$, $H_t = Id$ on $|L|$, and for each $p \in K$, $H(p \times I)$ lies in an element of $\{A_\alpha\}$. We sometimes work with the case where $V = M$ and then merely say that K-COMPLEXES CAN BE PULLED INTO U ALONG $\{A_\alpha\}$. If $H : K \times I \to M$ is a closed homotopy (or isotopy) the track of a point $x \in k$ is the set $H(x \times I)$.

Theorem 4.1 speaks of an ENGULFING ISOTOPY $H_t (0 \leq t \leq 1)$. In the case where R is noncompact this means that for each compact set $C_0 \subset M$ there is a partitioning $0 = t_0 < t_1 < .. < t_n = 1$ of I and a finite collection of polyhedral n-cells $C_1, C_2, \ldots, C_n$ in M such that

$H_t (t_{i-1} \leq t \leq t_i)$ restricted to C_i is a PL isotopy and

$H_t = H_{t-1}$ on $\bigcup_{j=0}^{n} C_j - \text{Int}\, C_i$ for $t_{i-1} \leq t \leq t_i$.

In case R is compact we have

$H_t = H_i$ on $M^m - \text{Int}\, C_i$ for $t_{i-1} \leq t \leq t_i$.

Theorem 4.1 speaks of the $\varepsilon(x)$-neighborhood of certain sets where $\varepsilon(x)$ is a positive function on M rather than a constant. It is understood that the $\varepsilon(x)$-

neighborhood of a set A in M is the set of all points $p \in M$ such for some point $q \in A$, $d(p,q) < \varepsilon(q)$.

The proof of Theorem 4.1 is given in Section 4 but we indicate here why our version of Theorem 4.1 is better than some others. For the case where R is compact the proof is much like that given by Crary in [5] but we find it easier to build sophisticated shadows to take care of the two codimension 3 cases since we have the vertical general position approximation Theorem 2.1 at our disposal. Crary was able to steal a dimension over that used by others by starting the induction at the case where $r = 1$ rather than at $r = 0$. The length of our proof of Theorem 4.1 is mainly due to the treatment of the case where R is noncompact. An easy treatment for the reader who is only interested in the compact case could consist of following Crary for the cases where $r = 0$, or $1 \leq r \leq m - 4$, then using the vertical general position approximation Theorem 2.1 to build sophisticated shadows as suggested in Theorem 3.1 and then returning to Crary for cases where $1 \leq r \leq n - 3$.

Other versions of Theorem 4.1 have been treated in [2,4,5,6,9,10,11,12,13,14]. In Theorem A of [2], Bing treated the compact codimension 4 case but since he started induction at $r = 0$ instead of at $r = 1$ he only concluded that each track lies near $r + 1$ elements of $\{A_\alpha\}$ rather than near r of them. Theorems D, E, F, of [2] considered the compact codimension 3 case but were restricting tracks by the use of a decomposition of M rather than $\{A_\alpha\}$'s.

Seebeck [10] stated a compact codimension 3 version of Theorem 4.1 but let $s = r + 3$. In [6], Edwards and Glaser stated a compact codimension 3 version of Theorem 4.1 but let $s = r + 2$. Perrin Wright wrestled with the compact codimension 3 case in [12, 13].

Crary treated the noncompact case in Theorem 4.4 and 4.5 of [5]. Also, Rushing gave a treatment of infinite radial engulfing in [9].

Lois Broussard read a preliminary version of this paper and corrections and improvements suggested by her were included in this version.

2. Vertical General Positions.

We consider some lemmas to be used in the proof of Theorem 2.1. The following is a crucial lemma which suggests how one can adjust a simplex (by jiggling its vertices) so as to simplify its intersection with another simplex.

Lemma 2.2. *Suppose in Euclidean n-space* R^n *that*

σ^r *is an r-simplex,*

σ^s *is an s-simplex,*

v_0 *is a particular vertex of* σ^s, *and*

Y *is the union of the vertices of* σ^r *and the vertices of* σ^s *other than* v_0. *Then if* v_0 *does not lie in any hyperplane of* R^n *determined by fewer than* $n + 1$ *points of* Y,

$$dim(\sigma^r \cap \mathring{\sigma}^s) \leq r + s - n .$$

PROOF. Assume Δ^k $(k = r + s + 1 - n)$ is a k-simplex in $\sigma^r \cap \mathring{\sigma}^s$. Then σ^s lies in a plane determined by the $k + 1$ vertices of Δ^k and $s - k$ vertices of σ^s in Y. Then v_0 lies in the plane determined by the vertices of σ^r and (s-k) vertices of σ^s in Y (or by $r + 1 + s - k$ elements of Y). But $r + 1 + s - k = n$ and v_0 does not lie in any plane determined by fewer than $n + 1$ points of Y.

Repeated applications of Lemma 2.2 gives the following result.

LEMMA 2.3. *Suppose*

σ *is a simplex in some Euclidean space,*

X *is a polyhedron in* σ, *with a triangulation* T_x,

Δ *is a simplex in* σ,

v *is a particular vertex of* Δ *that does not lie in any hyperplane determined by fewer than* $(1 + dim\ \sigma)$ *points in the union of the set of vertices of* T_x *and the vertices of* Δ *other than* v.

Then

$$dim(\mathring{\Delta} \cap X) \leq dim\ \Delta + dim\ X - dim\ \sigma.$$

Lemma 2.3 can be used to verify the following useful result.

LEMMA 2.4. *Suppose*

σ *is an* n*-simplex,*

X *is a geometric* m*-complex in* σ,

T *is a rectilinear triangulation of* σ *such that* X *is the union of simplexes of* T *and each simplex of* T *in* X *that intersects* $\mathring{\sigma}$ *has a vertex in* $\mathring{\sigma}$,

$\{v_i\}$ *is the set of vertices of* T *in* $X \cap \mathring{\sigma}$,

U_i *is a neighborhood of* v_i *for each* $v_i \in \{v_i\}$, *and*

V *is a geometric* r*-complex in* σ.

Then there are neighborhoods $\{U_i'\}$ *with* $U_i' \subset U_i$ *such that if* h *is a map of* σ *into itself such that*

h *is linear on each simplex of* T,

h *is fixed on vertices of* T *not in* $\{v_i\}$, *and*

$h(v_i) \in U_i'$,

then h *is a* PL *homeomorphism of* σ *onto itself such that*

$$dim(V \cap \mathring{\sigma} \cap h(X)) \leq m + r - n .$$

PROOF. The elements of $\{U_i\}$ are assumed to be so small that for any selection of $\{U_i'\}$, h is a homeomorphism.

Express V as the union of a finite number of simplexes and let Y be the set of all points v such that v is either a vertex of one of these simplexes or a vertex of T not in $\{v_i\}$. Suppose $\sigma \subset R^n$. Pick

v_1' to be a point of U_1 not on any hyperplane determined by fewer than $n+1$ points of Y,

v_2' to be a point of U_2 not on any hyperplane determined by fewer than $n+1$ points of $Y \cup \{v_1'\}$,

v_k' to be a point in U_k not on any hyperplane determined by fewer than $n+1$ points of $Y \cup \{v_1'\} \cup \{v_2'\} \cup \dots \cup \{v_{k-1}'\}$.

If v_k is the last element of $\{v_i\}$, the required U_i''s are sufficiently small neighborhoods of the v_i''s.

THEOREM 2.5. *Suppose*

K *is a finite geometric* k*-complex,*

T *is a rectilinear triangulation of* $K \times I$,

ρ *is the projection map of* $K \times I$ *onto* $K \times 0$,

T_1 *is a subdivision of* T,

X *is a polyhedron in* $K \times I$ *such that*

> X *is the union of simplexes of* T_1 *and if* s *is simplex of* T_1 *in* X *and* s *lies in a simplex* σ *of* T, *then* $\dim s < \dim \rho\sigma$,

T_2 *is the barycentric subdivision of* T_1, *and*

$\varepsilon > 0$.

Then there is a homeomorphism g *of* $K \times I$ *onto itself such that*

$d(x, g(x)) < \varepsilon$ *for each* $x \in K \times I$,

g *is linear on each simplex of* T_2,

$g(\sigma) = \sigma$ *for each simplex* σ *in* T,

ρg *is a homeomorphism on each simplex of* T_1 *in* X,

and if $\Delta_1, \Delta_2, \dots, \Delta_j$ *are different simplexes of* T_2 *in* X *and* $\sigma(\Delta_i)$ *is the simplex in* T *with* $\mathring{\Delta}_i \subset \mathring{\sigma}(\Delta_i)$, *then*

$$\dim \bigcap_{i=1}^{j} \rho g(\mathring{\Delta}_i) \leq \sum_{i=1}^{j} \dim \Delta_i - \sum_{i=2}^{j} \dim \rho\sigma(\Delta_i) .$$

PROOF. The term $\sum_{i=2}^{j} \dim \rho\sigma(\Delta_i)$ corresponds to the term $(j-1)n$ in Theorem 1.1.

We first take care of the requirement that ρg is a homeomorphism on each simplex of T_1 in X. Let Y be the set of vertices of T_1 in X and $s_1, s_2, \dots, s_k$ be an ordering of the simplex of T_1 in X. For each $x \in K \times I$, use $\sigma(x)$ to denote the element of T such that $x \in \mathring{\sigma}(x)$. Let $g_1, g_2, \dots, g_k$ be a sequence of homeomorphisms of $K \times I$ onto itself such that

each g_i is linear on each simplex of T_1,

ρg_i is a homeomorphism of each element of $s_1, s_2, \dots, s_i$,

$g_i = g_{i-1}$ on each point of Y other than possibly one vertex v of s_i, and for this v, $g_i(v)$ lies near v in $\mathring{\sigma}(v)$.

To find a vertex v of s_i to adjust to make ρg_i be a homeomorphism on s_i, we pick one such that $\rho\sigma(v)$ does not lie in the planar subspace of $\rho\sigma(s_i)$ determined by $\rho g_{i-1}(s_i)$. (If $\dim \rho\sigma(s_i) = \dim \rho g_{i-1}(s_i)$ we do not need to adjust a v since ρg_{i-1} is already a homeomorphism on s_i.) We adjust v so little that ρg_i remains a homeomorphism on each simplex of $s_1, s_2, \ldots, s_{i-1}$.

Note that ρg_k is a homeomorphism on each simplex of T_1. We suppose that the $g_i(v)$'s are picked so close to the v's that g_k is an $\varepsilon/2$-approximation of the identity.

Let $v_{u+1}, v_{u+2}, \ldots v_k$ be the vertices of T_1 in X (note that we started counting at $u+1$ rather than 1 for notatonal convenience). We suppose with no loss of generality that no two $g_k(v)$'s have the same projection. Let $v_{k+1}, v_{k+2}, \ldots, v_m$ be the vertices of T_2 in X which are not vertices of T_1. We suppose $v_{k+1}, v_{k+2}, \ldots, v_m$ are ordered lexicographically primarily by $\dim \rho\sigma(v_i)$ and secondarily by $\dim \sigma(v_i)$; hence, either $\dim \rho\sigma(v_i) < \dim \rho\sigma(v_{i+1})$ or $\dim \rho\sigma(v_i) = \dim \rho\sigma(v_{i+1})$ and $\dim \sigma(v_i) \leq \dim \sigma(v_{i+1})$. Let X_i be the set of simplexes of T_2 in X that have all their vertices in $v_{u+1}, v_{u+2}, \ldots, v_i$. Note that v_{k+1} is the barycenter of a 1-simplex of T_1 and if v_{r+1} is the barycenter of a simplex s of T_1 then each proper face of s is a simplex of X_r.

The required homeomorphism g is the last term of a sequence of homeomorphisms $g_{k+1}, g_{k+2}, \ldots, g_m$ such that

each g_i is linear on each simplex of T_2,

g_i agrees with g_{i-1} on each vertex of T_2 other than v_i,

if $\Delta_1, \Delta_2, \ldots, \Delta_j$ are different elements of X_r, then

$$\dim \cap \rho g_r(\mathring{\Delta}_i) \leq (\Sigma \dim \Delta_i - \Sigma \dim \rho\sigma(\Delta_i)) + \dim \rho\sigma(\Delta_1) .$$

By taking the $g_i(v_i)$'s sufficiently close to the v_i's, we insure that $g_m = g$ is an ε-homeomorphism and that ρg is a homeomorphism on each simplex of T_1 in X.

Suppose we have defined g_r with the required properties and wish to define g_{r+1}. We can relocate v_{r+1}. We use $\mathrm{Cl}(U)$ to denote the closure of U. So as to be able to use hyperplanes, we suppose that $K \times 0$ lies in some high dimensional Euclidean space R. Let Y_r be a finite point set in R such that if $\Delta_1, \Delta_2, \ldots, \Delta_j$ is a collection of different elements of X_r, then $\mathrm{Cl}(\cap \rho g_r(\mathring{\Delta}_i))$ has a triangulation with vertices in Y_r. Since X_r has only a finite number of elements, there is a single Y_r that serves for all combinations of elements of X_r. We pick $g_{r+1}(v_{r+1})$ in $\mathring{\sigma}(v_{r+1})$ such that $\rho g_{r+1}(v_{r+1})$ does not lie in any hyperplane determined by fewer than $(1 + \dim \rho\sigma(v_{r+1}))$ points of Y_r.

To show that g_{r+1} has the required properties, let $\Delta_1, \Delta_2, \ldots, \Delta_j$ be a collection of different elements of X_{r+1}. Since $g_{r+1} = g_r$ on each element of X_r and g_r is vertically nice, we suppose at least one Δ_i lies in $X_{r+1} - X_r$. Since there is one element of T_1 that contains all elements of $X_{r+1} - X_r$ and g_{r+1} is a homeomorphism of this element, we suppose only one Δ_i lies in $X_{r+1} - X_r$ - otherwise, $\cap \rho g_{r+1}(\mathring{\Delta}_i) = 0$. The lexicographical ordering of the v_i's shows that there is no loss of generality in supposing that this element is Δ_j. Lemma 2.3 insures that

$$\dim (\rho g_{r+1}(\mathring{\Delta}_j) \cap \mathrm{Cl}(\bigcap_{i=1}^{j-1} \rho g_r(\mathring{\Delta}_i)) \leq \dim (\Delta_j) +$$

$$\dim (\bigcap_{i=1}^{j-1} \rho g_r(\mathring{\Delta}_i)) - \dim \rho\sigma(\Delta_j) .$$

This implies that

$$\dim \bigcap_{i=1}^{j} \rho g_{r+1}(\mathring{\Delta}_i) \leq \dim (\Delta_j) + (\sum_{i=1}^{j-1} \dim \Delta_i - \sum_{i=2}^{j-1} \dim \rho\sigma(\Delta_i)) -$$

$$\dim \rho\sigma(\Delta_j) = \sum_{i=1}^{j} \dim \Delta_i - \sum_{i=2}^{j} \rho\sigma(\Delta_i) .$$

QUESTION. Is Theorem 2.5 true if we do not introduce T_2 but conclude that g is linear on each simplex of T_1.

LEMMA 2.6. *Suppose*

K is a geometric k-complex,

T_0 is a triangulation of $K \times I$,

M is a PL manifold, and

f_0 is a PL map of $K \times I$ into M such that f_0 is flat on each simplex of T_0 and f_0 is general position on $K \times 0$ with respect to T_0 restricted to $K \times 0$.

Then for each continuous position function $\varepsilon(x)$ defined on $K \times I$ there is a cylindrical triangulation T of $K \times I$ and a map f of $K \times I$ into M such that

T is a subdivision of T_0,

f is general position with respect to T,

$f = f_0$ on $K \times 0$, and

f is an ε-approximation of f_0.

PROOF. Let T_1 be a cylindrical triangulation of $K \times I$ that refines T_0. To give us room to push vertices of T_1 in $K \times 0$ about we suppose that if Δ, σ are elements of T_1, T_0 respectively in $K \times 0$ with $\mathring{\Delta} \subset \mathring{\sigma}$, then Δ has a vertex in $\mathring{\sigma}$. We are not to conclude that f_0 is a general position map with respect to T_1 since we may have introduced some low dimensional simplexes that intersect S_{f_0}. In fact, even on $K \times 0$ where $f = f_0$, we cannot be sure that f_0 is a general position map with respect to T_1 restricted to $K \times 0$. We modify T_1 in the next paragraph to cure this situation.

We suppose with no loss of generality that the vertical 1-simplexes of T_1 that intersect $K \times 0$ are so short that each lies except for its bottom point in an open simplex of T_0. It follows from repeated applications of Lemma 2.4 that the vertices of T_1 in $K \times 0$ can be moved slightly so that there is a triangulation T_2 of $K \times 0$ and a homeomorphism h of $K \times 0$ onto itself such that

f_0 is a general position map with respect to T_2,

h takes each simplex T_2 linearly onto a simplex to T_1 in $K \times 0$,

h takes each simples of T_0 in $K \times 0$ onto itself, and

for each vertex v of T_2, $h(v)$ is close to v.

(The meaning of "closeness" will become clearer as we proceed.) Let H be the homeomorphism of $K \times I$ onto itself defined by $H(p \times t) = (h(p) \times t)$. The required triangulation T for Lemma 2.6 has for its simplexes the image of the simplexes of T_1 under H^{-1}. We note that f_0 is a general position map of $K \times 0$ under T restricted to $K \times 0$. We modify f_0 in the following paragraph so that the modified map f_1 is flat on each simplex of T.

For each simplex Δ of T whose vertices miss $K \times 0$, we let $f_1 = f_0 H$ on Δ. Since $H(\Delta)$ is a simplex in an element of T_0 and f_0 is flat on each element of T_0, f_1 is flat on Δ. If Δ is a simplex of T some of whose vertices lie in $K \times 0$, we let H_Δ be the linear map of Δ such that if v is a vertex of Δ, $H_\Delta(v)$ is v or $H(v)$ according as $v \in K \times 0$ or $v \notin K \times 0$. Recall that vertical 1-simplexes of T_1 that intersect $K \times 0$ were short. We suppose that h is so close to the identity that H_Δ takes Δ homeomorphically into a simplex of T_0. Then $f_1 = f_0 H_\Delta$ on Δ. We suppose that h is so close to the identity that f_1 is an $\varepsilon/2$-approximation of f_0.

Finally, we modify f_1 to get a map f that is general position with respect to T. Let $\Delta_1, \Delta_2, \ldots$ be an ordering of the simplexes of T not in $K \times 0$ so that each proper face of each Δ_i either lies in $K \times 0$ or precedes Δ_i in the ordering. We seek a PL approximation f of f_1 so that f is general position with respect to T. One can obtain T as the limit of a sequence of homeomorphism maps $g_0, g_1, \ldots$ such that

$g_0 = f_1$,

$g_i = g_{i-1}$ except on simplexes of T having Δ_i as a face, and

if $\Delta_{n_1}, \Delta_{n_2}, \ldots, \Delta_{n_i}$, are different simplexes of T such that each Δ_{n_j} either lies in $K \times 0$ or is an element of $\Delta_1, \Delta_2, \ldots, \Delta_i$, then $\dim \cap g_i(\mathring{\Delta}_{n_j}) \leq \Sigma \dim \Delta_{n_j} - (n_i - 1)\, m$

To get g_i, we would suppose that $g_{i-1}(\Delta_i)$ lies in a simplex in an n-cell in M and pushes at its barycenter so that the image of this barycenter misses certain hyperplanes as suggested in Lemma 2.3. During the push, $g_{i-1}(\Delta_1 \cup \Delta_2 \cup \ldots \cup \Delta_{i-1})$ is not moved and the push is so small that $f = \lim g_i$ is an ε-approximation of f_0.

PROOF OF THEOREM 2.1. It follows from Theorem 1.1 that there is a triangulation T_0 of $P \times I$ and a PL $\varepsilon/3$-approximation f_2 of f_1 such that f_2 agrees with f_1 on $P \times 0$ and f_2 is general position with respect to T_0. It follows from Lemma 2.6 that there is a cylindrical triangulation T of $P \times I$ and a $2\varepsilon/3$-approximation f_3 of f_2 such that f_3 agrees with f_1 on $P \times 0$ and f_3 is general position with respect to T.

Let T_1 be a subdivision of T such that S_{f_3} is the union of simplexes of T_1, T_2 be the barycentric subdivision of T_1, and $\delta(x)$ be a positive continuous function defined on $P \times I$ such that if h is any homeomorphism of $P \times I$ that δ-approximates the identity, then f_3h^{-1} is an ε-approximation of f_1.

We apply Theorem 2.5 letting P, T, T_1, S_{f_3}, T_2, $\delta(x)$ replace the K, T, T_1, X, T_2, ε in the statement of that theorem and conclude that there is a map g such that g is vertically nice. Although Theorem 2.5 was only stated for a finite geometric complex P, an examination of its proof shows that it holds for the noncompact P of Theorem 2.1 with $\delta(x)$ replacing ε.

The required map f is f_3g^{-1}. The singularity set of $S_f = g(X)$ and a simplex in the required triangulation $T(S_f)$ is the image under g of a simplex of T_2 in X.

3. BUILDING SHADOWS.

Shadows can be used to prove engulfing theorems. The ease with which one proves theorems is sometimes determined by where he starts. Certain shadows were described in [2,5] but the existence of these could have been verified much more easily if the Vertical General Position Theorem 2.1 had been known.

Suppose M^m is a PL m-manifold, P^r $(r \leq m-3)$ is a geometric r-complex, T is a triangulation of $P \times I$, f is a closed general position map with respect to T of $P \times I$ into M, and S_f is the set of singularities of f in $P \times I$. If $r < m - 3$ then $\dim S_f \leq r - 2$. The union $L(S_f)$ of all vertical segments in $P \times I$ through points of S_f is called a SIMPLE shadow of f. Note that if $r < m - 3$, $\dim L(S_f) < r$.

If $r = m - 3$, one can not conclude that $\dim S_f \leq r - 2$ nor that $\dim L(S_f) \leq r - 1$. In this case we define a set L to be a SOPHISTICATED SHADOW for f if

$\dim L \leq r - 1$,

L is the union of vertical segments of $P \times I$,

L contains each point $x \in S_f$ such that $f^{-1}f(x)$ intersects the geometric r-skeleton of T, and

$f^{-1}f(L) = L$.

While not all general position maps of $P \times I$ into M have sophisticated shadows, all vertical general position maps do.

THEOREM 3.1. *Suppose M^m is an m-manifold, P^r $(r = m - 3)$ is a geometric* *r-complex, and f is a vertical general position map of* $P \times I$ *into* M *with respect to some triangulation* T *of* $P \times I$ *and some triangulation* $T(S_f)$ *of* S_f. *Then* f *has a sophisticated shadow* L *such that no vertical segment in* $(P \times I) - L$ *contains two points of* S_f.

PROOF. If S_1 is the set of all points $x \in S_f$ such that $f^{-1}f(x)$ intersects the r-skeleton of T, the general position of f implies that $\dim S_2 \leq r - 2$.

If S_2 is the closure of the set of all points $x \in S_f$ such that $f^{-1}f(x)$ has more than two points, the general position of f implies that $\dim S_2 \leq r - 3$.

If S_3 is the singularity set of ρ in S_f, it follows from the vertical niceness of ρ that $\dim S_3 \leq r - 2$.

Let S be the set of all points $x \in S_f$ such that $f^{-1}f(x) \in S_1 \cup S_2 \cup S_3$. Then $\dim S \leq r - 2$. A sophisticated shadow L for Theorem 3.1 is the union of all vertical segments in $P \times I$ through points of S.

4. INFINITE RADIAL ENGULFING. In this section we provide a proof of Theorem 4.1. The proof follows the methods used to prove Radial Engulfing Theorems A, D of [2] with certain improvements suggested by Crary in Theorem 3.1 of [5].

The proof of Theorem 4.1 makes several uses of the following trivial lemma.

LEMMA 4.2. *Suppose* $G_0, G_1, \ldots$ *is a sequence of coverings of a topological space* M *such that each* G_i *refines* G_{i-1} *and the union of any pair of intersecting elements of* G_i *lie in an element of* G_{i-1}. *If* g_i $(i = 1,2,\ldots,j)$ *is an element of* G_i *with* $g_i \cap g_{i+1} \neq 0$, *then there is an element* $g_0 \in G_0$ *such that*

$$\bigcup_{i=1}^{j} g_i \subset g_0 .$$

PROOF. Starting at the right and moving backwards, we find that

$g_j \cup g_{j-1}$ lies in an element g'_{j-2} of G_{j-2},

$g_j \cup g_{j-1} \cup g_{j-2} \subset g'_{j-2} \cup g_{j-2} \subset g'_{j-3}$ for some $g'_{j-3} \in G_{j-3}$,

. . .

$g_j \cup g_{j-1} \cup \ldots \cup g_1 \subset g_0$ for some $g_0 \in G_0$.

Our proof of Theorem 4.1 will also make use of the following variation of Theorem B of [2].

THEOREM 4.3. *Suppose*

M is a PL manifold,

K is a geometric k-complex (possibly infinite),

f is a closed PL map of $K \times I$ into M,

L is a closed subset of $K \times I$ that is the union of vertical segments of $K \times I$,

$S_f \subset L$,

C is a closed set in M such that $C \cap f(K \times I) = f((K \times 1) \cup L)$

U is an open subset of M that contains $f((K \times 1) \cup L)$, and

G_0' is a collection of open sets in $M - f(L)$ such that the image under f of each vertical segment in $(K \times I) - L$ lies in an element of G_0'.

Then there is an engulfing isotopy H_t $(0 \leq t \leq 1)$ of M onto itself such that

$H_0 = Id$,

$H_t = Id$ in a neighborhood of C,

$f(K \times I) \subset H_1(U)$,

if H moves a point, the track of the point lies in an element of G_0'.

PROOF. We suppose with no loss of generality that f is a homeomorphism and L is a polyhedron in $K \times I$. To see that the theorem easily reduces to this special case we note that we can increase L to a polyhedron L' in $K \times I$ such that L' is the union of vertical segments in $K \times I$, $f(L') \subset U$, and L contains no limit point of $(K \times I) - L'$. Enlarge C to make it contain $f(L')$. Replace $K \times I$ by the closure of $(K \times I) - L'$. Replace L by $L' \cap Cl((K \times I) - L')$. Reduce the elements of G_0 by removing the part of $f(L')$ in them. Then f is a homeomorphism on the new $K \times I$ and any engulfing isotopy for the new things, will be an engulfing isotopy for the new things, will be an engulfing isotopy for the old.

Let T be a cylindrical triangulation of $K \times I$ such that L is the union of simplexes of T and f is a PL flat homeomorphism on each simplex of T. Let T_K be a triangulation of K whose simplexes are the projections of simplexes of T. Use T_K^i to denote the geometric i-skeleton of T_k. Our plan for the required homotopy H is first to push U down so that it covers $f(T_K^0 \times I)$, then so that it covers $f(T_K^1 \times I)$, ... , and finally so that it covers $f(K \times I)$. To guide our pushes, we define certain open covers of $M - f(L)$.

Let G be a decomposition of $M - f(L)$ whose nondegenerate elements are the images under f of vertical segments in $(K \times I) - L$. Since G is an upper semicontinuous decomposition, the resulting decomposition space is metrizable. We suppose the elements of G_0' are unions of elements of G. Let G_0 be the open covering of $M - f(L)$ consisting of elements of G_0' and the single open set $M - (f(K \times I))$. Let $G_0, G_1, \ldots$ be a sequence open covers of $M - f(L)$ such that

each G_i refines G_{i-1},

if two elements of G_i intersect, their union lies in an element of G_{i-1}, and

each G_i is the union of elements of G.

During the (i+1) stage of our engulfing isotopy when we have already engulfed

$f(T_K^{i-1} \times I)$, we will engulf $f(T_K^i \times I)$ so that nondegenerate tracks miss a neighborhood of C and will stay in elements of G_{i+1} that intersect $f(K \times I)$. Lemma 4.2 will insure that nondegenerate tracks will lie in elements of G_0'.

Since it is easy to engulf $f(T_K^0 \times I)$ so that nondegenerate tracks lie in elements of G_1 we omit details of this part of our engulfing isotopy. We will suppose that $f(T_K^{i-1} \times I) \subset H_{i/(i+1)}(U)$ and show how to define H_t $(i/(i+1) \leq t \leq (i+1)/i+2))$.

Let $\sigma_1, \sigma_2, \ldots$ be the i-simplexes in T_K such that $\sigma_i \times I$ is not contained in L. For each σ_j, let Δ_j be an i-simplex in $\mathring{\sigma}_j$ such that

$$f((\sigma_j - \mathring{\Delta}_j) \times I) \subset H_{i/(i+1)}(U) .$$

Let $N_1, N_2, \ldots$ be a locally finite (in M) collection of disjoint open subsets of $M - f(L)$ such that $f(\Delta_j \times I) \subset N_j$. By regarding $f((T_K^i - \Delta_j) \times I)$ as a part of C, it follows from Theorem B of [2] that there is an engulfing isotopy H_t^j $((i/(i+1) \leq t \leq (i+1)/(i+2))$ such that $H_t^j = H_{i/(i+1)}$ on $(M - N_j) \cup f((T_K^i - \Delta_j) \times I)$ and a neighborhood of C, $f(\Delta_j \times I) \subset H^j_{(i+1)/(i+2)}(U)$, and nondegenerate tracks lie in elements of G_{i+1}. The required isotopy $H_t(i/(i+1) \leq t \leq (i+1)/(i+2))$ is the composition of the various H_t^j's.

PROOF OF THEOREM 4.1. Suppose M, U, P, Q, R, $\{A_\alpha\}$, $\varepsilon(x)$ are as given in the statement of Theorem 4.1 and $H'(P \times I)$ is a closed homotopy pulling P into U along $\{A_\alpha\}$ as guaranteed by the hypothesis of the theorem. It follows from Theorems 1.1 and 2.1 that there is a cylindrical triangulation T of $P \times I$ and an approximation f of H' that is general position with respect to T and vertical general position with respect to T and some triangulation of S_f. Also f agrees with H' on $P \times 0$, $f((Q \cap R) \times I) \cup f(R \times 1) \subset U$, and each $f(p \times I)$ lies in the $\varepsilon(x)$-neighborhood of $H'(p \times I)$. We do not need vertical general position in the first three cases of the proof but include it since it is used in the last two cases.

So that we can dispense with $\varepsilon(x)$ for the rest of the proof, we let G_0' be a collection of open sets in M such that each $f(p \times I)$ lies in an element of G_0' and each G_0' lies in the $\varepsilon(x)$-neighborhood of some $H'(q \times I)$ and hence in the $\varepsilon(x)$-neighborhood of some A_α.

CASE 1. $0 = r \leq m - 3$. This is the easy case in which R is a discrete set of points and $f(R \times I)$ is a locally finite collection of disjoint polygonal arcs in $M - Q$. We consider a locally finite collection of neighborhoods of these arcs such that the neighborhoods have disjoint closures, the closures miss Q, and each lies in an element of G_0'. The required engulfing isotopy is the identity outside the union of the neighborhoods and in each causes U to engulf the component of

$f(R \times I)$ in the neighborhood.

CASE 2. $1 = r \leq m - 4$. As in Case 1, f is a homeomorphism on $(Q \times 0) \cup (R \times I)$. The engulfing isotopy is guaranteed by Theorem 4.3 where we let $R \times I$ and $f((Q \times 0) \cup ((Q \cap R) \times I) \cup (R \times 1)$ be the $K \times I$ and C of that theorem. The elements of G_0' are reduced so as to miss $f((Q \cap R) \times I)$.

CASE 3. $2 \leq r \leq m - 4$. We hold m fixed and prove this case by induction on r. Suppose the theorem is known to be true in this case for smaller values of r. Case 2 handled the case where $r = 1$.

For values of r greater than 1, f may have singularities but on $(Q \times 0) \cup (R \times I)$

$$\dim S_f \leq (m-3) + (r+1) - m = r - 2 .$$

The simple shadow $L(S_f)$ consisting of the union of the vertical segments in $R \times I$ through points of S_f is such that

$$\dim L(S_f) \leq r - 1 .$$

It follows from induction that there is an engulfing isotopy H_t $(0 \leq t \leq \frac{1}{2})$ such that

(i) $H_0 = \mathrm{Id}$,

(ii) $H_t = \mathrm{Id}$ on a neighborhood of $Q \cup f((R \cap Q) \times I) \cup (R \times 1))$, $f(L(S_f)) \subset H_{\frac{1}{2}}(U)$, and

if H moves a point of M, its track lies in the $\varepsilon(x)$-neighborhood of the union of $r - 1$ elements of $\{A_\alpha\}$.

All of the engulfing isotopies H we use henceforth in Cases 3, 4, and 5 satisfy Conditions (i) and (ii) above so we do not repeat them.

Theorem 4.3 implies that there is an isotopy H_t $(\frac{1}{2} \leq t \leq 1)$ such that

$H_t = H_{\frac{1}{2}}$ on a neighborhood of $f(L(S_f))$,

$f(R \times I) \subset H_1(U)$,

each nondegenerate track of a point during $\frac{1}{2} \leq t \leq 1$ lies in an element of G_0'.

In applying Theorem 4.3 we let R, $L(S_f) \cup ((R \cap Q) \times I)$, and $Q \cup f(L(S_f) \cup ((R \cap Q) \times I) \cup (R \times 1))$ be the K, L and C of that theorem.

CASE 4. $1 = r = m - 3$. In this case, $m = 4$ and

$$\dim S_f \leq (r+1) + (r+1) - m = 0 .$$

We shall not use a shadow since it might be of dimension one and as hard to engulf as R itself.

It may be seen that S_f does not intersect any 1-simplex of T and consists of a discrete collection of isolated points of the interiors of 2-simplexes of T. Each of these points is paired with another point of S_f with the same image under f. The vertical general position quality implies that no two points of S_f lie on the same vertical segment of $R \times I$.

If $p \times I$ and $q \times I$ are different vertical segments of $R \times I$ whose images under f intersect each other, $f((p \times I) \cup (q \times I))$ is a polygonal figure X. Let G be the decomposition of M each of whose nondegenerate elements is either one of those figure X's or the image (under f) of a vertical segment of $R \times I$ whose image does not intersect the image of any other vertical segment. The decomposition G is upper semicontinuous and the corresponding decomposition space is metrizable. We shall use G to describe certain open sets in M to guide our pushes.

If $f((p \times I) \cup (q \times I)) \in G$, it lies in the union of two elements of G_0'. If $f(p \times I) \in G$, it lies in one element of G_0'. Let G_0 be an open covering of M such that each element of G_0 is the union of elements of G, $M - f(R \times I)$ is one element of G_0, and each element of G_0 that intersects $f(R \times I)$ lies in an element of G_0' or in the union of two such elements. Let G_1 be an open covering of M such that

G_1 refines G_0,

each element of G_1 is the union of elements of G, and

if two elements of G_1 intersect each other, their union lies in an element of G_0.

The metrizability of the decomposition space is used to get such a G_1.

Let T_R be the triangulation of R whose simplexes are the projections of the simplexes of T. Let T' be a refinement of T_R such that if σ is a 0-simplex of T',

$$S_f \cap (\sigma \times I) = 0;$$

if σ is a 1-simplex of T',

$(\sigma \times I)$ contains at most one element of S_f and

$f(\sigma \times I)$ lies in an element of G_1;

if σ_1, σ_2 are two 1-simplexes of T' with $(\sigma_1 \times I)$, $(\sigma_2 \times I)$ containing associated points of S_f

$f((\sigma_1 \times I) \cup (\sigma_2 \times I))$ lies in an element of G_1.

Let R' be the union of all simplexes σ of T' such that $S_f \cap (\sigma \times I) = 0$. It follows from Theorem 4.3 that there is an engulfing isotopy H_t $(0 \leq t \leq 1/3)$ such that

$$f(R' \times I) \subset H_{1/3}(U) .$$

For each pair of different 1-simplexes σ', σ'' of T' with $f(\mathring{\sigma}' \times I) \cap f(\mathring{\sigma}'' \times I) \neq 0$, let Δ', Δ'' be simplexes in $\mathring{\sigma}'$, $\mathring{\sigma}''$ respectively such that $f((\sigma' - \mathring{\Delta}') \cup (\sigma'' - \mathring{\Delta}'')) \times I) \subset H_{1/3}(U)$ and $f((\mathring{\Delta}' \cup \mathring{\Delta}'') \times I)$ contains the figure X element of G in $f((\sigma' \cup \sigma'') \times I)$. We let $w_1, w_2, \ldots$ be the collection of all such $f((\Delta' \cup \Delta'') \times I)$'s.

Let $N_1, N_2, \ldots$ be a locally finite collection of disjoint open sets in M such that $w_i \subset N_i$ and N_i lies in an element of G_1. By applying Theorem 4.3 twice to each N_i we find that there are engulfing isotopies H_t $(1/3 \leq t \leq 2/3)$ and H_t $(2/3 \leq t \leq 1)$ such that

each $H_t = H_{1/3}$ on $f(R' \times I)$,
each nondegenerate track during $1/3 \leq t \leq 1$ lies in some N_i,
each $f(\Delta' \times I)$ lies in $H_{2/3}(U)$,
each $f(\Delta' \times 0)$ and each $f(\Delta'' \times I)$ lies in $H_1(U)$.

During the last third of the complete engulfing isotopy, sets near the $f((\Delta' \times I) \cap S_f)$ may get uncovered but this does not matter since it was our purpose to engulf $f(R \times 0)$ and not $f(R \times I)$.

CASE 5. $2 \leq r = m - 3$. The proof of this case uses techniques of Cases 3 and 4 as well as some others.

Let L be a sophisticated shadow for the vertical general position map f as described in Section 3. It follows from Case 3 that there is an engulfing isotopy H_t $(0 \leq t \leq \frac{1}{2})$ such that

$f(L) \subset H_{\frac{1}{2}}(U)$ and
each nondegenerate track lies in the $\varepsilon(x)$-neighborhood of the union of $(r - 1)$ elements of $\{A_\alpha\}$.

Let T_R be the triangulation of R whose simplexes are the projections of simplexes of T. Let $\sigma_1, \sigma_2, \ldots$ be an ordering of the simplexes of T_R such that each proper face of each σ_i precedes σ_i in the ordering. Our plan for finishing the required isotopy is to push U down the $f(\sigma_i \times I)$'s one at a time. Our aim is not to engulf $f(R \times I)$ but rather $f(T^r)$ where T^r is the geometric r-skeleton of T. We suppose the metric for M is such that the $1/i$ neighborhood of each $f(\sigma_i \times I)$ has a compact closure. We denote this neighborhood by N_i.

Two vertical segments of $(R \times I) - L$ are paired if their images under f intersect. Let G be the decomposition of $M - f(L)$ whose nondegenerate elements are either figure X's which are the images under f of union of two paired vertical segments in $(R \times I) - L$ or a polygonal arc which is the image of a vertical segment in $(R \times I) - L$ that is not paired with any other.

We shall describe an open cover G_0 of $M - f(L)$ by associating with each element of G an open set containing it. The difference between the elements of this cover G_0 and those of the G_0 of Case 4 is that they satisfy the following additional restrictions:

they are open subsets of $M - f(L)$ rather than open sets in M,

if $f(p \times I)$ is an element of G, the associated element of G_0 lies in N_i if $p \in \mathring{\sigma}_i$, and

if $f((p \times I) \cup (q \times I)) \in G$ where $p \in \mathring{\sigma}_i$, $q \in \mathring{\sigma}_j$, the associated element of G_0 lies in $N_i \cup N_j$.

These last restrictions are made so that the partial engulfing H_t $(0 \leq t < 1)$ to be described will extend properly at $t = 1$.

Let $G_0, G_1, G_2, \ldots$ be a sequence of open coverings of $M - f(L)$ such that for each positive integer i

G_i is a refinement of G_{i-1},

if two elements of G_i intersect, their union lies in an element of G_{i-1}, and

each element of G_i is the union of elements of G.

Since $f(\sigma_1 \times I)$ is of dimension 1, there is no challenge in extending $H_{1/2}$ to $(1/2 \leq t \leq 2/3)$ so that

$H_t = H_{1/2}$ on $f(L)$,

each nondegenerate track during $(1/2 \leq t \leq 2/3)$ lies in the intersection of N_1 with an element of G_1 that intersects $f(\sigma_1 \times I)$, and

$f(T^r \cap (\sigma_1 \times I)) \subset H_{2/3}(U)$.

Suppose we have defined H_t $(0 \leq t \leq i/(i+1))$ so that

$$f(T^r \cap (\bigcup_{j=1}^{i-1} \sigma_j \times I)) \cup f(L) \subset H_{i/(i+1)}(U)$$

and wish to extend H_t to $(0 \leq t \leq (i+1)/(i+2))$ so that

$$f(T^r \cap (\bigcup_{j=1}^{i} \sigma_j \times I)) \cup f(L) \subset H_{(i+1)/(i+2)}(U) \quad \text{and}$$

each nondegenerate track during $i/(i+1) < t < (i+1)/(i+2)$ lies in the intersection of N_i with an element of G_i that intersects $f(\sigma_i \times I)$.

If $\dim \sigma_i < r-1$, we use Theorem 4.3 to get this extension since $(\sigma_i \times I) - L$ contains no points of S_f but if $\dim \sigma_i = m - 3$, we give the matter special attention in the next paragraph.

In case $\dim \sigma_i = m - 3$, consider the simplexes of T above $\sigma_i \times 0$ of dimension $m - 2$. Suppose there are n_i of these. We start at the top and work our way down in n_i applications of Theorem 4.3 taking care that in the kth application, if a point is shoved about, its track lies in an element of G_{i+k} that intersects $f(\sigma_i \times I)$. Lemma 4.2 implies that if the track of a point is nondegenerate in any one of the n_i moves, its entire track in the totality of the n_i moves lies in an element of G_i that intersects $f(\sigma_i \times I)$. In each of these n_i applications of Theorem 4.3 we cause the image of U to cover the image under f of the bottom $(m-2)$-simplex of T being considered but in working down to those beneath it or in later $f(\sigma_j \times I)$'s we may uncover a part of the interior of the image of this $(m-2)$-simplex since the simplex may contain points of S_f that get in our way at later stages.

We extend H_t $(0 \leq t < 1)$ to $t = 1$ by showing that if $p_0 \in M$, there is a neighborhood N_0 of p_0 and an integer n_0 such that $H_t = H_{n_0/(n_0+1)}$ on N_0 for $n_0/(n_0+1) \leq t < 1$. To get this N_0 we note that $\{N_i\}$ is a locally finite collection and there is a neighborhood N_0 of p_0 that intersects only a finite number of N_i's. Note that if Star Y denotes the union of the elements of $\{N_i\}$ intersecting Y, then each of Star N_0, $\text{Star}^2\, N_0 = \text{Star}\,(\text{Star}\, N_0)$, has a compact closure. Since Star N_0 intersects only a finite number of N_i's, there is an integer n_0 such that if $j \geq n_0$, then $N_j \cap \text{Star}^2\, N_0 = 0$. Since each partial track $H(x \times [0,1))$ lies in an element of G_0 and each element of G_0 that intersects $f(R-L) \times I$ lies in the union of two intersecting elements of $\{N_i\}$, $H(N_0 \times [0,1)) \subset \text{Star}^2\, N_0$. Since $H_t(x) = H_{j/(j+1)}(x)$ for $j/(j+1) \leq t \leq (j+1)/(j+2)$ unless $H_{j/(j+1)}(x) \in N_j$, H_t stops shoving points of N_0 about for $n_0/(n_0+1) \leq t < 1$.

The restriction in Theorem 4.1 that tracks lie near the union of s elements of $\{A_\alpha\}$ may not be very restrictive since some A_α's may be quite large. If P were U, the proof of Theorem 4.1 suggests that we could keep tracks near the parts of the A_α's close to U and need not use the parts that are far away. That it may be clumsy to state as much as we have actually proved is indicated by the following.

THEOREM 4.4. *Suppose M, U, P, R, r, $\varepsilon(x)$ are as in Theorem 4.1; U_r, U_{r-1}, ..., U_0 are open sets in M such that*

$$U \subset U \cup P \subset U_r \subset U_{r-1} \subset \cdots \subset U_0;$$

and $\{A_\alpha^0\}$, $\{A_\alpha^1\}$, ..., $\{A_\alpha^r\}$ are collections of subsets of U_0, U_1, ..., U_r respectively such that

P can be pulled into U along $\{A_\alpha^r\}$ keeping Q fixed, and
j complexes in U_{j+1} can be pulled into U along $\{A_\alpha^j\}$ for $0 \leq j \leq r - 1$.

Then there is a closed engulfing isotopy H_t $(0 \leq t \leq 1)$ *such that*

$H_0 = Id$,

$H_t = Id$ *in a neighborhood of* Q,

$P \subset H_1(U)$, *and*

if $x \in M$ *and* $H(x \times I)$ *is nondegenerate, it lies in the* ε*-neighborhood of certain sets as indicated:*

some A^0_α *if* $0 = r \leq m - 3$,

some A^1_α *if* $1 = r \leq m - 4$,

union of some A^1_α, *some* A^2_α, ... , *and some* A^r_α *if* $2 \leq r \leq m - 4$,

union of two A^1_α*'s if* $1 = r = m - 3$, *and*

union of two A^r_α*'s, one* A^{r-1}_α, ... , *and one* A^1_α *if* $2 \leq r \leq m - 3$.

In the above theorem, the A^r_α's guide the homotopy pulling P into U, the A^{r-1}_α's guide the homotopy pulling the associated shadow into U, etc.

To keep our isotopies from oscillating too much at infinity, we may find it convenient to consider a PL manifold M^m as an open subset of a larger metric space $M \cup C$. Under this condition, we require that a polyhedron (or geometric complex) in M be relatively closed in M (but not necessarily in $M \cup C$). If $\{A_\alpha\}$ is a collection of subsets of M, we say that k-COMPLEXES IN M CAN BE PULLED INTO U ALONG $\{A_\alpha\}$ WITHOUT DISTURBING C if for each geometric k complex K in M and each geometric subcomplex L with $L \subset U$, there is a closed homotopy $f_t: K \cup C$ into $M \cup C$ such that $f_0 = Id$, $f_t = Id$ on $L \cup C$, $K \subset f_1(U)$, and each nondegenerate track lies in an element of $\{A_\alpha\}$. The methods of Theorem 4.1 yield the following result.

THEOREM 4.5. *Suppose* M^n, U, P, Q, $\varepsilon(x)$, $\{A_\alpha\}$ *are as given in Theorem 4.1, that* M *is an open subset of the metric space* $M \cup C$ *and k-complexes can be pulled into* M *into* U *along* $\{A_\alpha\}$ *without disturbing* C. *Then there is a closed engulfing isotopy* H_t $(0 \leq t \leq 1)$ *as given in Theorem 4.1 such that in addition to satisfying the conditions of Theorem 4.1,* H *can be extended to* $M \cup C$ *so that each* H_t *is the identity on* C.

There are other variations of Theorem 4.1. If $\dim Q < m - 3$, we could have gotten better restrictions on tracks. However, rather than knowing the statements of many theorems, it seems better to understand methods of proof so that one can prove theorems as he needs them rather than being reduced to memorizing many variations.

QUESTION. How could one state the full strength of what was proved in Theorem 4.1? Also, could one simplify tracks? For example, in Cases 1 and 2 of Theorem 4.1, it seems that not only does each nondegenerate track lie in an ε-neighborhood of some $\{A_\alpha\}$ but that there is an ε-homeomorphism of the track onto some arc in the $\{A_\alpha\}$. In Case 4 one can analyze oscillation of tracks near elements of G. While one might possibly even describe the oscillation in Case 3, it becomes harder for Case 5. How much can one restrict the zigzagging of tracks?

5. Applications of Radial Engulfing. Connell [4] used radial engulfing to show that if h is a stable homeomorphism of R^n $(n \geq 7)$ onto itself, then h can be approximated by a PL homeomorphism. Bing [1] noted that Connell's key lemma held for $n \geq 5$. In stating this key lemma for approximating stable homeomorphisms with a PL one, we let $O(r)$ denote the open r-neighborhood of the origin in R^n and r_1, r_2, ε be positive numbers such that $0 < r_1 < r_1 + \varepsilon < r_2 - \varepsilon < r_2$.

Key Lemma 5.1. *For each exotic curvilinear PL triangulation T of R^n $(n \geq 5)$ and each $\theta > 0$, there is a homeomorphism $h : R^n \to R^n$ such that*

$h = Id$ on $\overline{O}(r_1) \cup (R^n - O(r_2))$,

$O(r_2 - \varepsilon) \subset h(O(r_1 + \varepsilon))$,

for each point $p \in R^n$ other than the origin, the segment from p to $h(p)$ subtends an angle of less than θ when viewed from the origin, and

h is PL with respect to T.

Proof. The theorem is proved with a push-pull operation.

Let T' be a triangulation of $O(r_2) - \overline{O}(r_1)$ compatible with the structure of T and such that the simplexes of T' subtend small angles from the origin. Let T'' be the algebraic complex consisting of all n-simplexes (and their faces) in T' that intersect $\overline{O}(r_2 - \varepsilon/2) - O(r_1 + \varepsilon/2)$, P^{n-3} be the union of the (n-3)-simplexes of T'', and P^2 be the geometric dual 2-skeleton of T''. We let $\{A_\alpha\}$ be the set of all intersections of $O(r_2) - \overline{O}(r_1 + \varepsilon/2)$ with rays from the origin. Regarding $O(r_2) - \overline{O}(r_1 + \varepsilon/2)$ as space, we note that (n-3)-complexes can be pulled into $O(r_1 + \varepsilon) - \overline{O}(r_1 + \varepsilon/2)$ along $\{A_\alpha\}$.

The definition of an engulfing isotopy required that the isotopy have compact support if R is compact. It follows from Theorem 4.1 that there is an engulfing isotopy H_t $(0 \leq t \leq \frac{1}{2})$ such that

$H_t = Id$ on $\overline{O}(r_1 + \varepsilon/2) \cup (R^n - O(r_2))$,

$P^{n-3} \subset H_{\frac{1}{2}}(O(r_1 + \varepsilon) - \overline{O}(r_1))$,

nondegenerate tracks lie near elements of $\{A_\alpha\}$, and H_1 is PL with respect to T'.

The homeomorphism $H_{\frac{1}{2}}$ is the first half of our push. Before describing the other half, we describe the pull.

An argument similar to that used in the preceding paragraph shows that there is an engulfing isotopy G_t $(0 \leq t \leq 1)$ such that

$G_t = \text{Id}$ on $(R^n - O(r_2 - \varepsilon/2)) \cup \overline{O}(r_1)$,

$P^2 \subset G_1(O(r_2) - \overline{O}(r_2 - \varepsilon))$,

nondegenerate tracks lie near rays from the origin, and

G_1 is PL with respect to T'.

The pull operation is G_1^{-1}.

Note that if σ is an n-simplex of T'' and $x \in \sigma$, then for some $p_1 \in P^{n-3}$, $p_2 \in P^2$, x lies on the segment from p_1 to p_2. One can get a homeomorphism g of p_1p_2 onto itself such that g is fixed on the ends of p_1p_2 and pushes points toward p_2 such that

$$p_1p_2 \subset G_1(O(r_2) - \overline{O}(r_2 - \varepsilon)) \cup g(p_1p_2 \cap H_{\frac{1}{2}}(O(r_1 + \varepsilon) - \overline{O}(r_1)) .$$

This suggests that H_t $(0 \leq t \leq \frac{1}{2})$ can be extended to $0 \leq t \leq 1$ so that

$R^n = H_1(O(r_1 + \varepsilon)) \cup G_1(R^n - \overline{O}(r_2 - \varepsilon))$,

H_1 is PL with respect to T,

$H_tH_{\frac{1}{2}}^{-1}$ is invariant on each simplex of T' for $\frac{1}{2} \leq t \leq 1$, and

$H_tH_{\frac{1}{2}}^{-1}$ is fixed on $\overline{O}(r_1 + \varepsilon/2) \cup (R^n - O(r_2))$ and on each simplex of T' contained in $H_{\frac{1}{2}}(O(r_1 + \varepsilon)) \cup G_1(R^n - \overline{O}(r_2 - \varepsilon))$.

The required homeomorphism is $G_1^{-1}H_1$. Since

$$R^n = G_1^{-1}(R^n) = G_1^{-1}(H_1(O(r_1 + \varepsilon)) \cup G_1(R^n - \overline{O}(r_2 - \varepsilon)),$$

$\overline{O}(r_2 - \varepsilon) \subset G_1^{-1}H_1(O(r_1 + \varepsilon))$.

Connell used the Key Lemma 5.1 to show that for R^n's in which the Key Lemma applies, then if h is a homeomorphism of O(1) onto R^n given in polar coordinates by $h(\rho,\theta) = (\rho/(1-\rho),\theta)$ and T_1, T_2 are arbitrary curvilinear PL triangulations of R^n, then h can be approximated by homeomorphism h_1, h_2 that are PL with respect to T_1, T_2 respectively. If T_1, T_2 are compatible on O(1), then for suitably chosen h_1, h_2, $h_2h_1^{-1}$ is an approximation of the identity which is PL with respect to taking R^n under T_1 onto R^n under T_2. The ability to change structures if the structures agreed near the origin produces [4] the following result.

THEOREM 5.2. *Each stable homeomorphism of* R^n $(n \geq 5)$ *onto itself can be approximated by a PL homeomorphism.*

Kirby [7] added interest to the above result when he showed that any orientation preserving homeomorphism of R^n $(n \geq 5)$ (or of S^n) onto itself is stable. It seems that Kirby's methods may be more complicated than Connell's.

Before Kirby proved his more general result, Cernavskii [3] used infinite engulfing to show that if $k < (2n/3) - 1$ and $n \geq 4$, then any homeomorphism of R^n onto itself is the composition of a stable homeomorphism and a homeomorphism that is fixed on some rectilinear k-cell. Bing noted [2] that a better engulfing theorem would have enabled Cernavskii to improve what he called his Fundamental Lemma and conclude the following.

FUNDAMENTAL LEMMA 5.3. *Suppose in* R^n $(n \geq 4)$ *that* B^k $(k \leq (2n/3) - 1)$ *is a rectilinear k-simplex,* V *is an open subset of* R^n *containing* B^k, B^{k-1} *is one of the* $(k-1)$*-faces of* B^k, *and* R^{k-1} *is the hyperplane of* R^n *containing* B^{k-1}. *If* h *is a homeomorphism of* B^k *into* $(R^n - R^{k-1}) \cup B^{k-1}$ *such that* h *is fixed on* B^{k-1} *and* U *is an open set containing* $h(B^k - B^{k-1})$, *then there is an isotopy* H_t $(0 \leq t \leq 1)$ *of* R^n *onto itself such that*

$H_0 = Id$,
H_t *is fixed on* R^{k-1}, *and outside* V, *while*
$B^k - B^{k-1} \subset H_1(U)$.

We wish to ignore V. To this end we suppose that ε is a positive number such that the ε-neighborhood of B^{k-1} lies in V. With no loss of generality we suppose that both B^k and $h(B^k)$ lie in this ε-neighborhood. (If this is not already the case we pick a k-simplex B' in B^k with B^{k-1} as a face such that $B' \cup h(B')$ lies in the ε-neighborhood and prove the theorem for this B'.) The image of U under the end of the isotopy could then be pushed out in V to make it cover the original $B^k - B^{k-1}$. Henceforth, we ignore V since all our homotopies and isotopies will be fixed far away from B^{k-1}.

The tracks of our isotopy will be guided by various homotopies rather than by elements of a collection $\{A_\alpha\}$ so we suppose that $R^n - R^{k-1}$ is the only element of $\{A_\alpha\}$.

We now describe a homotopy F_t $(0 \leq t \leq 1)$ of B^k into R^n such that

$F_0 = Id$,
$F_t = Id$ on B^{k-1},
$F_1 = h$, and
$F_t(B^k - B^{k-1}) \cap R^{k-1} = 0$.

To get such an F one could regard $R^n - R^{k-1}$ as $R^{k-1} \times (R^{n-k+1}$ - origin) and then as $R^{k-1} \times S^{n-k} \times (0,\infty)$. Hence, each point of $R^n - R^{k-1}$ has three coordinates in $R^{k-1} \times S^{n-k} \times (0,\infty)$. Suppose $p \in B^k - B^{k-1}$ has coordinates (a_p, b_p, c_p) and $h(p)$ has coordinates (a_p', b_p', c_p'). Then $F(p \times I)$ is defined so that the first coordinate of $F_t(p)$ is $(1-t)a_p + t(a_p')$ in R^{k-1}, and the last coordinate is $(1-t)c_p + t(c_p')$ in $(0,\infty)$. It is possible to define F_t on the second coordinate since any two maps of $B^k - B^{k-1}$ onto S^{n-k} are homotopic. The map F can be extended continuously to B^{k-1} so that each F_t is the identity there.

Let T be a cylindrical triangulation of $(B^k - B^{k-1}) \times I$ and F^1 be a vertical general position map with respect to T of $(B^k - B^{k-1}) \times I$ into $R^n - R^{k-1}$ such that

$F_0^1 = \mathrm{Id}$,
$F_1^1(B^k - B^{k-1}) \subset U$, and
$d(F_t(x), F_t^1(x)) < d(F_t(x), R^{k-1})$.

If L is the sophisticated shadow of F^1,

$\dim L \leq 2k + 2 - n \leq n - k - 1.$

Since $R^n - R^{k-1}$ has the same connectivity as S^{n-k}, it is i-connected and i-ULC for $i \leq n - k - 1$. Hence (n-k-1)-complexes in $R^n - R^{k-1}$ can be pulled into U without disturbing R^{k-1}. A check for $n = 4, 5, \ldots$ shows that $k \leq n - 3$.

It follows from the proofs of Cases 1, 2, 3 of Theorem 4.1 that there is an engulfing isotopy H_t $(0 \leq t \leq \frac{1}{2})$ of $R^n - R^{k-1}$ such that

$H_0 = \mathrm{Id}$,
$H_t = \mathrm{Id}$ on $F_1^1(B^{k-1})$ and outside ε-neighborhood of B^{k-1},
$F^1(L \times I) \subset H_{\frac{1}{2}}(U)$, and
H_t can be extended to R^n to be the identity on R^{k-1}.

It follows from the proofs of Cases 4 and 5 of Theorem 4.1 that H_t $(0 \leq t \leq \frac{1}{2})$ can be extended to $0 \leq t \leq 1$ so that it satisfies the above conditions as well as the additional one that

$$B^k - B^{k-1} = F_0^1(B^k - B^{k-1}) \subset H_1(U) .$$

UNIVERSITY OF TEXAS
AUSTIN, TEXAS

REFERENCES

1. R. H. Bing, *Stable homeomorphisms on E^5 can be approximated by piecewise linear ones,* Abstract 607-16, Notices of the Amer. Math. Soc., 10, 1963, 666.

2. __________, *Radial engulfing,* Conference on Topology of Manifolds, 1967 (John G. Hocking, ed.), 1-18, Prindle, Weber, and Schmidt, Boston, 1968.

3. A. V. Cernavskii, *Isotopy of elements and spheres in n-dimensional space for $k < 2/3\ n-1$,* Dokl. Akad. Nauk. SSSR, 1958 (1964), 62-65, Soviet. Math. Dokl., 5 (1964), 1194-97.

4. E. H. Connell, *Approximating stable homeomorphisms with piecewise linear ones,* Ann. of Math., 78 (1963), 326-338.

5. Fred D. Crary, *Some new engulfing theorems,* Thesis, University of Wisconsin-Madison, 1973.

6. R. D. Edwards and L. C. Glaser, *A method for shrinking decompositions of certain manifolds,* Trans. Amer. Math. Soc. 165 (1972), 45-56.

7. Robion C. Kirby, *Stable homeomorphisms and the annulus conjecture,* Ann. of Math., (2) vol. 89 (1969), 575-582.

8. K. P. Millett, *Homotopy groups of automorphism spaces,* Geometrical Topology, The Proceedings of the 1974 Park City Conference, (L. C. Glaser and T. B. Rushing, ed.), Notes in Mathematics Series, Springer Verlag.

9. T. Benny Rushing, *Infinite engulfing,* (Mimeographed notes), 1972. [See a. Remark 4.14.3 of "Topological Embeddings", Academic Press, 1973; b. A summation of results on infinite engulfing, Proc. Univ. of Oklahoma Topology Conference, 1972; and c. Zentralblatt 246 Abstract 57007.]

10. C. L. Seebeck III, *Collaring an (n-1)-manifold in an n-manifold,* Trans. Amer. Math. Soc., 148 (1970), 63-68.

11. J. R. Stallings, *The piecewise-linear structure of Euclidean space,* Proc. Cambridge Philos. Soc. 58 (1962), 481-488.

12. Perrin Wright, *A uniform generalized Schoenflies theorem,* Ann. of Math. 89 (1969), 292-304.

13. ____________, *Radial engulfing in codimension three,* Duke Math. J., 39 (1971), 295-298.

14. E. C. Zeeman, *Seminar on Combinatorial Topology,* Institut des Hautes Etudes Scientifiques, 1963, Lemma 38.

C. Tameness

EVERY SIMPLE CLOSED CURVE IN E^3 IS UNKNOTTED IN E^4

R. H. BING† *and* V. L. KLEE‡

1. *Introduction.*

The *standard embedding* of E^n in E^{n+j} takes the point $(x^1, x^2, \ldots, x^n)$ of E^n onto the point $(x^1, x^2, \ldots, x^n, 0, \ldots, 0)$ of E^{n+j}; with reference to this embedding, we may regard E^n as *contained in* E^{n+j}. The *standard* $(n-1)$-*sphere* in E^n is the set $\Sigma^{n-1} = \{x \in E^n : \|x\| = 1\}$, and for $0 \leqslant k \leqslant n-1$ the *standard* k-*sphere* is the set $\Sigma^k = \Sigma^{n-1} \cap E^{k+1}$. A k-sphere S^k in E^n is said to be *unknotted* in E^n provided there is a homeomorphism of E^n onto itself which takes S^k onto Σ^k; otherwise S^k is *knotted* in E^n. All spheres are unknotted in E^2, and for $n \geqslant 3$ various conditions on the embedding of S^k in E^n restrict the values of k for which S^k can be knotted in E^n. (See Zeeman [**12**] for a survey of such results.) But when the embedding is merely assumed to be topological, E^n contains knotted k-spheres for all $n \geqslant 3$ and $1 \leqslant k \leqslant n-1$. Even here it is intuitively clear (and also true) that for sufficiently large j, every S^k in E^n is unknotted in the containing E^{n+j}. This suggests the following problem.

1.1. PROBLEM. *For $n \geqslant 3$ and $1 \leqslant k \leqslant n-1$, determine the smallest integer $j = J(n, k)$ such that every k-sphere in E^n is unknotted in E^{n+j}.*

The literature contains two results which are related to this problem. Here is the first one [**10**].

1.2. THEOREM. *Suppose a Banach space E is the direct sum of its closed linear subspaces E' and E'', X is a closed subset of E', and h is a homeomorphism of X onto a closed subset of E''. Then h can be extended to a homeomorphism of E onto E.*

1.3. COROLLARY. *Suppose X is a compact subset of E^n and X can be topologically embedded in E^j. Then every homeomorphism of X into E^n can be extended to a homeomorphism of E^{n+j} onto E^{n+j}.*

Proof. Let $E'' = \{x \in E^{n+j} : x^1 = \ldots = x^n = 0\}$. By hypothesis, there exists a homeomorphism f of X into E''. Consider an arbitrary homeomorphism g of X into E^n. By 1.2, the mappings f and fg^{-1} can be extended to homeomorphisms u and v of E^{n+j} onto itself, and $v^{-1}u$ is the desired extension of g.

Received 9 November, 1962.

† On leave from the University of Wisconsin. Research supported in part by the National Science Foundation, U.S.A. (NSF–G21514).

‡ Research supported in part by the National Science Foundation, U.S.A. (NSF–GP–378).

[JOURNAL LONDON MATH. SOC., 39 (1964), 86–94]

Since S^k can be embedded in E^{k+1}, it follows from 1.3 that

$$J(n, k) \leqslant k+1.$$

A sphere S^k in E^n is *locally unknotted* provided that for each point w of S^k there exists a neighbourhood W of w in E^n and a homeomorphism of W onto E^n which takes $S^k \cap W$ onto E^k. The second result related to 1.1 is as follows.

1.4. THEOREM (Stallings [**11**]). *Suppose $n \geqslant 5$ and S^k is a knotted k-sphere in E^n which is locally unknotted. Then k is equal to $n-2$ or $n-1$.*

1.5. COROLLARY. *For $n \geqslant 4$ and $k = 1$ as well as for $n \geqslant 3$ and $2 \leqslant k \leqslant n-1$, it is true that $J(n, k) \leqslant k$; that is, each k-sphere in E^n is unknotted in E^{n+k}.*

Proof. The sphere S^k is locally unknotted in E^{n+k} by 1.2 and hence is unknotted in E^{n+k} by 1.4.

Probably the bound in 1.5 can be improved, and it is conceivable even that $J(n, k) = 1$. Note that in all the interesting cases but one, 1.5 yields a better bound than does 1.3 (k instead of $k+1$). However, in the simplest case, where $k = 1$ and $n = 3$, the best that can be wrung from either 1.3 or 1.5 is that $J(3, 1) \leqslant 2$; *i.e.*, every simple closed curve in E^3 is unknotted in E^5. The main purpose of this note is to show that in fact $J(3, 1) = 1$; *i.e.*, *every simple closed curve in E^3 is unknotted in E^4*. It seems surprising that the literature contains no proof of this fact, especially since it has long been known that every *polyhedral* 1-sphere in E^n is unknotted in E^{n+1}. However, several of our colleagues have mentioned the problem recently so it seems worth while to publish the elementary proof. We show also that every topological 1-complex in E^n is tame in E^{n+1}. The proofs are contained in §§2–3 below, while §4 discusses some unsolved problems.

(While this paper was being written, Gluck [**6**] was proving independently that every locally unknotted 1-sphere in E^4 is in fact unknotted. His proof depends on a recent result of Homma [**8**]. As an unknotting theorem, Gluck's result is considerably stronger than ours, However, our method is more elementary and applies also to the taming of 1-complexes. More recently, Bing and Kister [**2**] have extended the method to show that for $k \leqslant n-2$, every topological k-complex in E^n is tame in E^{n+k}.)

2. *Unknotting a* 1-*sphere.*

For convenience in discussing the cartesian product $M \times T$ of a manifold M and a set T of real numbers, we call each set $\{m\} \times T$ *vertical* and each set $M \times \{t\}$ a *horizontal slice* (for $m \in M$, $t \in T$). If M is a manifold with boundary B, each point of $(M \sim B) \times \{t\}$ is called an *interior point* of the slice $M \times \{t\}$. When $M \times \{1\}$ is shrunk to a point to change

$M \times [0, 1]$ into a cone over $M \times \{0\}$, we call the image of each slice a *concentric layer* and the image of each vertical segment $\{m\} \times [0, 1]$ a *radius*.

2.1. PROPOSITION. *Suppose the subset C of E^n is the union of arcs $A_1, A_2, \ldots, A_r$ such that for $i \neq j$, each point of $A_i \cap A_j$ is an endpoint of both A_i and A_j. Then for each $\epsilon > 0$, there is a homeomorphism h of $E^{n+1} = E^n \times E^1$ onto itself such that*

h moves no point more than ϵ,

h is the identity outside an ϵ-neighbourhood of C,

h preserves vertical lines, and

no arc $h(A_i)$ has more than one point in common with any horizontal slice.

Proof. Let $v_1, v_2, \ldots$ be the endpoints of the arcs. On an arc A_i having endpoints v_j and v_k, let f be a homeomorphism of A_i onto the interval $[\epsilon/4j, \epsilon/4k] \subset E^1$ with $f(v_j) = \epsilon/4j$ and $f(v_k) = \epsilon/4k$. As thus defined, f is a continuous mapping of C into $[0, \epsilon/4]$ and hence can be extended to a continuous mapping of E^n into $[0, \epsilon/4]$ such that $f = 0$ outside the $\epsilon/2$-neighbourhood of C in E^n. For each $x \in E^n$, let h be the identity on the rays $\{x\} \times]-\infty, -\epsilon/2]$ and $\{x\} \times [\epsilon/2, \infty[$, and let h take the segments $\{x\} \times [-\epsilon/2, 0]$ and $\{x\} \times [0, \epsilon/2]$ affinely onto the segments $\{x\} \times [-\epsilon/2, f(x)]$ and $\{x\} \times [f(x), \epsilon/2]$. Then h has all the desired properties.

2.2. PROPOSITION. *Suppose M is an n-manifold with boundary B and $A_1, A_2, \ldots, A_r$ are mutually exclusive arcs in $M \times [\alpha, \beta]$ such that each arc intersects each horizontal slice in exactly one point, this being an interior point of the slice. Then there is a homeomorphism h of $M \times [\alpha, \beta]$ onto itself such that*

h is the identity on $(M \times \{\alpha\}) \cup (B \times [\alpha, \beta])$

h preserves horizontal slices, and

each set $h(A_i)$ is vertical.

The same conclusion holds when $[\alpha, \beta]$ is replaced by $[0, \infty[$ and the arcs A_i are replaced by topological rays R_i.

Proof. We first consider the case of $[\alpha, \beta]$, choosing

$$\alpha = t_0 < t_1 < \ldots < t_k = \beta$$

such that for $0 \leqslant j \leqslant k$ there are mutually exclusive n-cells $N_{1j}, N_{2j}, \ldots N_{rj}$ in M for which $A_i \cap (M \times [t_{j-1}, t_j]) \subset (\operatorname{Int} N_{ij}) \times [t_{j-1}, t_j]$. Each n-cell N_{ij} is homeomorphic with an n-simplex, and working in terms of some particular homeomorphism we may regard N_{ij} as a cone over its boundary from each interior point. For $\alpha \leqslant t \leqslant \beta$, let the point $a_i(t)$ be such that

$\big(a_i(t), t\big) \in A_i \cap (M \times \{t\})$. Let h_{ij} be a homeomorphism of $N_{ij} \times [\alpha, \beta]$ onto itself defined as follows:

h_{ij} is the identity on $N_{ij} \times [\alpha, t_{j-1}]$;

for each point x of $\operatorname{Bd} N_{ij}$, h_{ij} takes the "segment" in $M \times \{t\}$ from $\big(a_i(t), t\big)$ to (x, t) "affinely" onto the "segment" from $\big(a_i(t_{j-1}), t\big)$ to (x, t) when $t_{j-1} \leqslant t \leqslant t_j$; and

for $t_j \leqslant t \leqslant \beta$, h_{ij} takes the "segment" from $\big(a_i(t_j), t\big)$ to (x, t) "affinely" onto the "segment" from $\big(a_i(t_{j-1}), t\big)$ to (x, t).

Let h_j be the homeomorphism of $M \times [\alpha, \beta]$ onto itself that is equal to h_{ij} on each set $N_{ij} \times [\alpha, \beta]$ and is the identity elsewhere. Note that h_j takes vertical segments in $M \times [t_j, \beta]$ onto vertical segments. The desired homeomorphism h is $h_1 h_2 \dots h_k$ where we first apply h_k, then $h_{k-1}, \dots,$ and finally h_1. This "backing down" process, starting at the top and working down, takes care of the arcs A_i in $M \times [\alpha, \beta]$. It does not apply to the rays R_i in $M \times [0, \infty[$, for $[0, \infty[$ has no maximum, but for this case we will work upward by using the construction just completed.

By the result for $M \times [\alpha, \beta]$, there is a homeomorphism g_1 of $M \times [0, 1]$ onto itself that is the identity on $(M \times \{0\}) \cup (B \times [0, 1])$ and takes each arc $R_i \cap (M \times [0, 1])$ onto a vertical segment. Let g_1 be extended to $M \times [0, \infty[$ so as to take vertical rays of $M \times [1, \infty[$ onto vertical rays. By a second application of this procedure, there is a homeomorphism g_2 of $M \times [1, 2]$ onto itself that is the identity on $(M \times \{1\}) \cup (B \times [0, 2])$ and takes each arc $g_1\big(R_i \cap (M \times [1, 2])\big)$ onto a vertical segment; g_2 is extended to the rest of $M \times [0, \infty[$ so as to be the identity on $M \times [0, 1]$ and to take vertical rays in $M \times [2, \infty[$ onto vertical rays. Continuing in similar fashion, we define g_n as a homeomorphism of $M \times [0, \infty[$ onto itself which takes each arc $g_{n-1} \cdots g_2 g_1(R_i) \cap (M \times [n-1, n])$ onto a vertical segment, is the identity on $M \times [0, n-1]$, etc. The homeomorphism required by the theorem is equal to g_n on $M \times [0, n]$.

2.3. PROPOSITION. *Suppose K is a cone with vertex p over an n-manifold with boundary B. Suppose $A_1, A_2, \dots, A_r$ are arcs in K such that $A_i \cap A_j = \{p\}$ for $i \neq j$ and each arc intersects each concentric layer in exactly one point, this being an interior point of the layer. Then there is a homeomorphism of K onto itself that preserves layers, is the identity on both M and the cone over B, and takes each arc A_i onto a radius.*

Proof. This is immediate from 2.2, since we may regard K as the one-point compactification of $M \times [0, \infty[$.

2.4. THEOREM. *Every 1-sphere in E^n is unknotted in E^{n+1}.*

Proof. Let S be a 1-sphere in E^n, S being the union of two arcs A and A' which intersect only at their common endpoints v and w. By 2.1 there is a homeomorphism h_1 of E^{n+1} onto itself such that neither of the arcs $h_1(A)$ and $h_1(A')$ has more than one point in common with any horizontal slice. We may assume without loss of generality that

$$h_1(v) \in E^n \times \{-1\},\ h_1(w) \in E^n \times \{1\}, \text{ and } h_1(S) \subset E^n \times [-1, 1].$$

For each $t \in [-1, 1]$, let $a(t)$ be the point of E^n for which $(a(t), t) \in h_1(A)$; then define $a(t) = a(-1)$ for $t \leqslant -1$ and $a(t) = a(1)$ for $t \geqslant 1$. Let h_2 be the homeomorphism of E^{n+1} onto itself which translates all points of the slice $E^n \times \{t\}$ by means of the vector $(-a(t), t)$; *i.e.*, $h_2(x, t) = (x - a(t), t)$. Then $h_2 h_1(A)$ is the vertical segment $\theta \times [-1, 1]$ where θ is the origin of E^n. (See the figures opposite.)

For each $t \in [-1, 1]$, let $a'(t)$ be the point of E^n for which

$$(a'(t), t) \in h_2 h_1(A');$$

then $a'(-1) = \theta = a'(1)$. Let h_3 be a homeomorphism of E^{n+1} onto itself which is the identity on $E^n \times]-\infty, -1]$ and on $E^n \times [-1, \infty[$ and which, for $t \in [-1, 1]$, contracts the slice $E^n \times \{t\}$ toward the point (θ, t) so that the distance from $h_3(a'(t), t)$ to (θ, t) is less than $1 - |t|$. For example, we could define $h_3(x, t) = (\mu_t(\|x\|)x, t)$, where $\mu_t(r) = 1$ when $|t| \geqslant 1$ and also when $\|a'(t)\| \leqslant (1 - |t|)/2$, while for $|t| \leqslant 1$ and $\|a'(t)\| > (1 - |t|)/2$ we have

$$\mu_t(r) = \frac{1 - |t|}{2|a'(t)|} \quad \text{when } 0 \leqslant r \leqslant \|a'(t)\|$$

and

$$\mu_t(r) = \frac{r - \left(\|a'(t)\| - \dfrac{1 - |t|}{2}\right)}{r} \quad \text{when } r \leqslant \|a'(t)\|.$$

Then h_3 preserves horizontal slices, does not move the points of $h_2 h_1(A)$, and brings the points of $h_2 h_1(A')$ inside the cones from $(\theta, -1)$ and $(\theta, 1)$ over the n-cell $M = \{(x, 0) : \|x\| \leqslant 1\}$. Finally, 2.3 guarantees the existence of a homeomorphism h_4 of E^{n+1} onto itself that is the identity outside these cones and also on the segment $h_3 h_2 h_1(A)$, and takes the arc $h_3 h_2 h_1(A)$ onto the union of two radii. (See the figures opposite.) But then the set $h_4 h_3 h_2 h_1(S)$ is actually a triangle in E^{n+1}, so of course can be taken onto the standard 1-sphere Σ^1 by a fifth homeomorphism of E^{n+1} onto itself. Thus S is unknotted in E^{n+1}.

3. *Taming a 1-complex.*

A subset X of E^n is said to be *polyhedral* provided X is the union of a finite number of geometrical simplexes. And X is *tame* in E^n provided

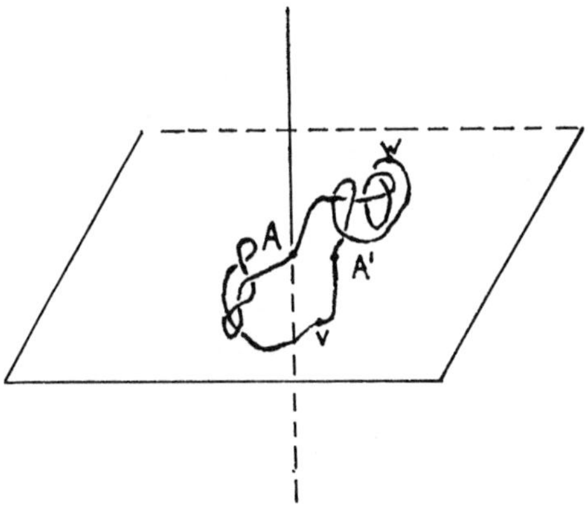
A
A'
v
w

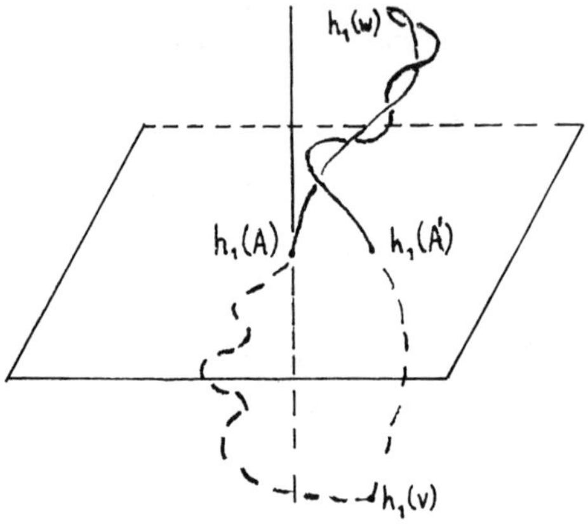
h1(w)
h1(A)
h1(A')
h1(v)

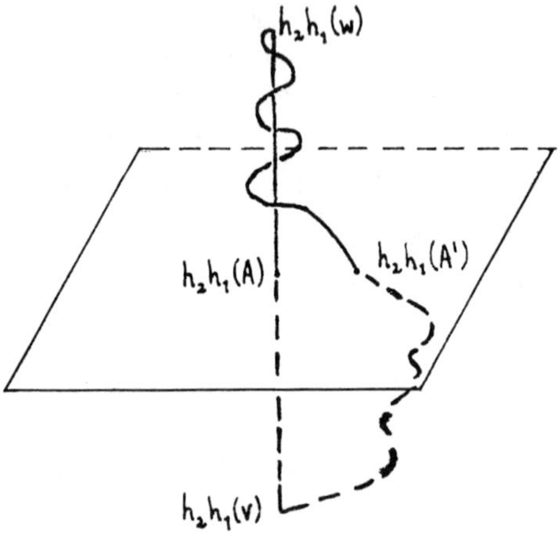
h2h1(w)
h2h1(A)
h2h1(A')
h2h1(v)

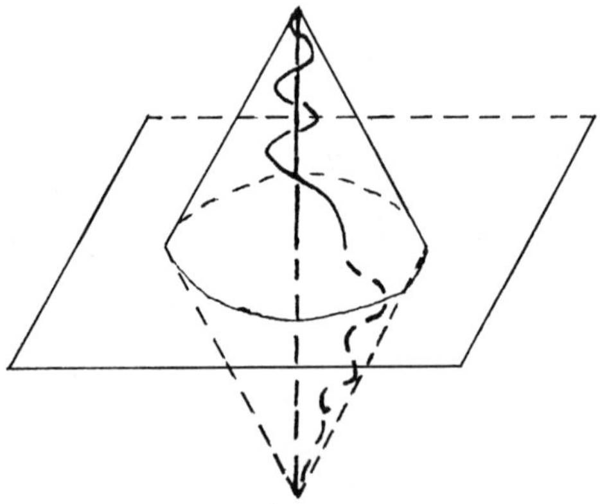

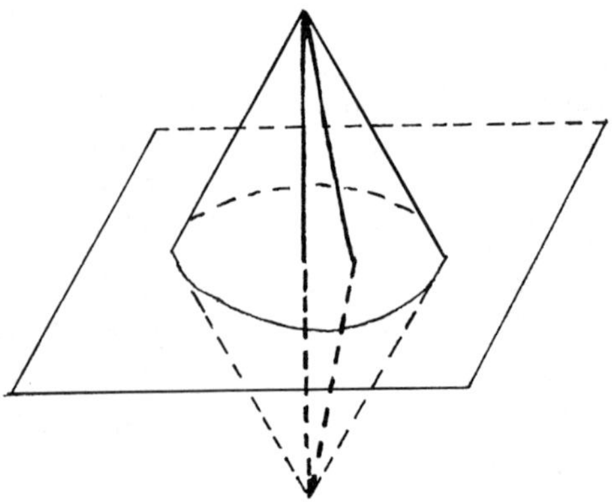

X can be taken onto a polyhedral set by some homeomorphism of E^n onto itself. In §2 we showed explicitly that every 1-sphere in E^n is unknotted in E^{n+1}. However, it would have been enough to show that every 1-sphere in E^n is tame in E^{n+1}, for it has long been known that every 1-sphere is unknotted in E^2 and that a tame 1-sphere can be knotted in E^k only for $k=3$.

A *topological* 1-*complex* is a set C which is the union of a finite number of arcs $A_1, A_2, \ldots, A_r$ such that for $i \neq j$, $A_i \cap A_j$ is empty or consists of a single common endpoint of A_i and A_j. In the present section we show that each topological 1-complex C in E^n is tame in E^{n+1}, and that in fact the taming can be effected by means of a homeomorphism which keeps C in E^n and does not move any point very far. When $n \leqslant 2$ it is already known that such a homeomorphism can be obtained in E^n itself without using the additional freedom provided by E^{n+1}. Thus in Theorem 3.2 below, we treat only the case $n \geqslant 3$.

3.1. PROPOSITION. *Suppose A is an arc in $E^{n+1} = E^n \times E^1$ such that no horizontal slice intersects A in more than one point. Then for each $\epsilon > 0$ there is a homeomorphism h of E^{n+1} onto itself such that*

$h(A)$ is polyhedral,

h moves no point more than ϵ,

h is the identity outside the ϵ-neighbourhood of A,

h preserves horizontal slices, and

h is the identity on each horizontal slice missing A.

Proof. If the third condition were omitted, we could define h by the same shoving process used for h_2 in the proof of 2.4. As it is, we require a localized version of that process which is similar to the first part of the proof of 2.2. Suppose $t_0 < t_1 < \ldots < t_k$, the ends of A lie in $E^n \times \{t_0\}$ and $E^n \times \{t_k\}$, and for $0 \leqslant i \leqslant k$ there is in E^n a Euclidean ball N_i of radius $\epsilon/2$ such that

$$A \cap (E^n \times [t_{i-1}, t_i]) \subset (\operatorname{Int} N_i) \times [t_{i-1}, t_i].$$

For $t_{i-1} \leqslant t \leqslant t_i$, h is the identity on $(E^n \sim N_i) \times \{t\}$. On A, h is defined so as to keep each point in its original horizontal slice and to take the arc A onto the polygonal arc from $A \cap (E^n \times \{t_0\})$ to $A \cap (E^n \times \{t_1\})$ to $A \cap (E^n \times \{t_2\})$ to $\ldots$ to $A \cap (E^n \cap \{t_k\})$. Finally, for each $x \in \operatorname{Bd} N_i$ and $t \in [t_{i-1}, t_i]$, h takes the segment from $A \cap (E^n \cap \{t\})$ to (x, t) affinely onto the segment from $h\big(A \cap (E^n \times \{t\})\big)$ to (x, t).

3.2. THEOREM. *For each topological 1-complex C in E^n and each $\epsilon > 0$ there is a homeomorphism h of $E^{n+1} = E^n \times E^1$ onto itself such that*

$h(C)$ is polyhedral in E^n,

h moves no point more than ϵ, and

h is the identity outside the ϵ-neighbourhood of C.

Proof. The homeomorphism h will be obtained as a composition $h = h_3 h_2 h_1$, where the construction of h_3 depends on the assumption that $n \geqslant 3$. The homeomorphism h_1 is that described in 2.1, where h_1 and $\epsilon/4$ in our present construction correspond to the h and ϵ of 2.1. (Thus h_1 moves no point more than $\epsilon/4$, etc.) The homeomorphism h_2 takes $h_1(C)$ onto a polyhedron. It is obtained by using 2.3 to adjust $h_1(C)$ in a neighbourhood of its vertices and then using 3.1 to make the adjustments elsewhere. The homeomorphism $h_2 h_1$ moves no point more than $\epsilon/2$ and is the identity outside the $\epsilon/2$-neighbourhood of C.

Let L_0 denote the line $\{\theta\} \times E^1$ and let $Y_1, Y_2, \ldots, Y_s$ be straight line segments whose union is the polyhedron $h_2 h_1(C)$. For each i and j there is a 3-dimensional linear subspace Z_{ij} of E^{n+1} such that Z_{ij} includes the origin $(\theta, 0)$ and some translate of Z_{ij} contains $Y_i \cup Y_j$. Since $n \geqslant 3$, the union U of all the Z_{ij}'s with $E^n \times \{0\}$ is not equal to E^{n+1}, and in fact there must be lines L through $(\theta, 0)$ in E^{n+1}, arbitrarily close to L_0, for which $L \cap U = \{(\theta, 0)\}$. Each line parallel to such an L has exactly one point in common with $E^n \times \{0\}$ and at most one point in common with the set $h_2 h_1(C)$. Thus by choosing L sufficiently close to L_0 and regarding E^{n+1} as the linear sum of $E^n \times \{0\}$ and L, we can produce a homeomorphism h_3 of E^{n+1} onto itself which projects $h_2 h_1(C)$ linearly into $E^n \times \{0\}$, moves no point more than $\epsilon/4$, and is the identity outside the ϵ-neighbourhood of C. Then the homeomorphism $h = h_3 h_2 h_1$ has all the desired properties.

4. *Related problems.*

From §§1–2 and some classical results of topology (Jordan curve theorem, existence of knotted S^k in E^n for $n \geqslant 3$) we know that $J(2, 1) = 0$, while for $n \geqslant 3$, $J(n, 1) = 1$ and $1 \leqslant J(n, k) \leqslant k$ when $2 \leqslant k \leqslant n-1$. Probably the inequality can be improved, and perhaps even $J(n, k) = 1$ whenever $n \geqslant 3$ and $1 \leqslant k \leqslant n-1$. This is part of the more general problem of determining the *n-level* $L_n(X)$ of a compact subset X of E^n, where $L_n(X)$ is defined as the smallest integer j such that every homeomorphism of X into E^n can be extended to a homeomorphism of E^{n+j} onto itself. Obviously $J(n, k) = L_n(S^k)$. By 1.3, $L_n(X) \leqslant n$ for all compact $X \subset E^n$. It would be of interest to determine $L_n(X)$ for various individual subsets X of E^n (spheres, cells, manifolds, etc.), and also to determine the numbers $\sup\{L_n(X) : X \in \mathbf{X}\}$ for various important classes $\mathbf{X}$ of subsets of E^n (compact sets, continua, locally connected continua, acyclic continua, etc.). (For a result on topological complexes see Bing and Kister [**2**].)

Some problems of this sort have been studied by Choquet [**3, 4, 5**], who notes [**3**] that for $n \geqslant 3$ and compact $X \subset E^n$, $L_n(X) = 0$ if and only if X is countable. He also states [**3**] that if X is a compact subset of E^2 which separates E^2, then $L_n(X) = 0$ if and only if X is a simple closed curve. These results express a limitation on the topological homogeneity properties of E^n. Another statement of Choquet [**5**] is that $L_2(X) \leqslant 1$ for every compact $X \subset E^2$. His papers [**3**, **4**, **5**] contain several other results which are related to the problem of determining $L_n(X)$ for various X, but apparently the proofs have never been published. For the more modest goal of proving that $L_2(X) \leqslant 1$ for every locally connected continuum X in E^2, some results of Adkisson-MacLane [**1**] and Homma [**7**] might be useful.

The material of this and the preceding sections can be modified so as to treat isotopies rather than merely homeomorphisms. An *isotopy* of E^n is a one-parameter continuous family ζ of homeomorphisms of E^n onto itself; that is, ζ is a continuous mapping of $E^n \times [0, 1]$ onto E^n such that for each $t \in [0, 1]$, the mapping ζ_t which takes x onto $\zeta(x, t)$ is a homeomorphism of E^n onto itself. Two homeomorphisms h_0 and h_1 of E^n onto itself are said to be *isotopic* provided there is an isotopy ζ with $\zeta_0 = h_0$ and $\zeta_1 = h_1$. The unknotting and taming processes described in §2 can all be carried out by means of homeomorphisms which are isotopic to the identity; this is almost obvious from the way in which they were defined. For a compact subset X of E^n, the *isotopy level* $I_n(X)$ may be defined as the smallest integer j such that every homeomorphism of X into E^n can be extended to a homeomorphism of E^{n+j} onto itself which is isotopic to the identity mapping of E^{n+j}. From the stronger form of 1.2 established in [**10**] it follows that always $I_n(X) \leqslant n$. Ivanov [**9**] outlines a proof of an isotopy theorem related to the first result of Choquet stated above.

References.

1. V. W. Adkisson and Saunders MacLane, "Extending maps of plane Peano continua," *Duke Math. J.*, 6 (1940), 216–228.
2. R. H. Bing and J. M. Kister, "Taming complexes in hyperplanes", to appear.
3. Gustave Choquet, "Prolongement d'homéomorphies", *C. R. Acad. Sci. Paris*, 206 (1938), 634–636.
4. ———, "Homéomorphies", *ibid.*, 210 (1940), 129–131.
5. ———, "Prolongements d'homéomorphies. Ensembles topologiquement nommables. Caractérisation topologique individuelle des ensembles fermés totalement discontinues", *ibid.*, 219 (1944), 542–544.
6. Herman Gluck, "Unknotting S^1 in S^4", *Bull. American Math. Soc.*, 69 (1963), 91–94.
7. Tatsuo Homma, "An extension of the Jordan curve theorem", *Yokohama Math. J.* 1 (1953), 125–129.
8. ———, "On the imbedding of polyhedra in manifolds", *ibid.*, 10 (1962), 5–10.
9. V. A. Ivanov, "Isotopy of compacta in Euclidean spaces" (Russian), *Dokl. Akad. Nauk SSSR*, 71 (1950), 1021–1022.
10. V. L. Klee, Jr., "Some topological properties of convex sets", *Trans. American Math. Soc.*, 78 (1955), 30–45.
11. John R. Stallings, "The topological unknottedness of certain spheres", (Outlines of one-hour and half-hour addresses), *International Congress of Mathematicians*, Stockholm, 1962, 2.40–2.41.
12. E. C. Zeeman, "Isotopies and knots in manifolds", *Topology of 3-manifolds and related topics* (Prentice-Hall, Englewood Cliffs (New Jersey), 1962, 187–193).

Institute for Advanced Study, Princeton, N.J., U.S.A.

University of Washington, Seattle 5, Washington, U.S.A.

TAMING COMPLEXES IN HYPERPLANES

By R. H. Bing and J. M. Kister

1. Introduction. In this paper we investigate conditions that suffice for an imbedding of a complex in a Euclidean space to be tame (cf. §§2, 7 for definitions). We also consider the more general question of when two imbeddings are equivalent. We show, for example, with a dimension restriction that if the image under an imbedding of a k-complex is contained in a hyperplane of codimension k, then the imbedding is tame. More precisely we prove:

THEOREM 1.1. *Let K be a finite k-dimensional complex and h an imbedding of K into an n-plane E^n in E^{n+k}, where $k + 2 \leq n$. Let ϵ be any positive number Then there exists an isotopy $g_t(t \in I)$ of E^{n+k} onto itself such that*

1) *g_0 is the identity,*
2) *g_1h is piecewise linear,*
3) *g_t is the identity outside an ϵ-neighborhood of $h(K)$ for each t in I,*
4) *each point of E^{n+k} moves along a polygonal path under $g_t(t \in I)$ having length less than ϵ.*

The case $k = 1$ of this theorem was substantially proved in [1].

In Corollary 5.6 we establish the corresponding result showing any two such imbeddings into hyperplanes are equivalent (and under an economical isotopy). In doing this we give a new proof (Theorem 5.5) of a result of Gugenheim [5] which leads to a sharpened form of his Theorem 5.

The first result of the type we prove in Corollary 5.6 is due to Klee [9] who showed that if a compact set K can be imbedded in E^l, then any two imbeddings of K into hyperplanes of codimension l are equivalent. In Theorem 6.2 we improve Klee's result for certain sets, not necessarily complexes, which leads to stronger results in one sense (we can drop the requirement that $k + 2 \leq n$) than Corollary 5.6 for 2-manifolds (Corollary 6.3) and k-spheres with handles (Corollary 6.4). In [10] Stallings establishes the equivalence of locally flat imbeddings of k-spheres in E^n, $n \geq k + 3$, which, when combined with Klee's result, would provide an alternate proof of Corollary 6.4 in the special case of a k-sphere.

Homma [6] and Gluck [3] have shown that the tameness of an imbedding in a combinatorial manifold is guaranteed if the imbedding is well-behaved in a neighborhood of the image. Gluck, for example, has shown that a locally tame imbedding is tame in the dimension range we are considering. (The authors have just recently learned that C. A. Greathouse [4] has also obtained

Received May 20, 1963. The first author was supported by grants NSF-G21514 and NSF-G11665; the second by grant NSF G24156 and a grant from the Institute for Advanced Study.

this result.) For a more exact statement see Theorem 7.1. This result together with Theorem 1.1 gives us a generalization of Theorem 1.1 to imbeddings in combinatorial manifolds.

Since the proof of Theorem 1.1 is rather long, we give a short outline. The proof depends on five lemmas which we summarize and relate to the theorem before getting into the full details.

Suppose we start with a topological k-complex K in a hyperplane E^n in E^{n+k}. By a deformation of E^{n+k}, using a modified technique of Klee, K is "freed" from the hyperplane and put into a form so that Lemma 3.3 is applicable. Using the concept of an admissible family (cf. §2) we can produce isotopies (Lemma 3.1 and 3.2) which by Lemma 3.3 deform the freed complex to one whose topological (open) simplexes are locally polyhedral—almost what is required. Call this image K'.

Lemma 3.4 shows how to deform the topological r-skeleton of K' (where $r < k$) onto a polyhedron, when the $(r - 1)$-skeleton is already polyhedral. This step relies mainly on general position arguments as well as the isotopies produced by Lemma 3.2.

Then Lemma 3.4 is used inductively on the dimension of the skeletons until the $(k - 1)$-skeleton has been made polyhedral. Finally Lemma 3.5 shows how to handle the topological simplexes of highest dimension and is the finishing step in making K polyhedral.

2. **Definitions.** A simplex, denoted by ρ, σ, τ etc., will mean an open simplex in this paper and $\bar{\rho}$, $\bar{\sigma}$, $\bar{\tau}$ etc., the corresponding closed simplex. If A and B are subsets of a Euclidean space, $A * B$ indicates the set obtained by taking the union of all line segments from A to B and in most uses in this paper coincides with the abstract join. We denote the barycenter of σ by $\beta(\sigma)$ and the line segment in Euclidean space E^m joining the points a and b by $[a, b]$.

If K is a k-complex, a *prismatic triangulation* of $K \times I$ means that there is an ordering of the vertices of K, $v_1, v_2, \cdots$, such that all s-simplexes of $K \times I$, $0 \le s \le k + 1$, have vertices $(v_{i_1}, 1), (v_{i_2}, 1), \cdots, (v_{i_j}, 1), (v_{i_j}, 0), \cdots, (v_{i_s}, 0)$, for some choice of integers $i_1 < i_2 < \cdots < i_s$ where $v_{i_1}, v_{i_2}, \cdots, v_{i_s}$ span a simplex in K, and for some j, $1 \le j \le s$.

A *polyhedron* or *polyhedral set* P is a subset of a geometric complex K which is a subcomplex of some (linear) subdivision of K. A set P in a complex K is *locally polyhedral* at a point x in P if there is a neighborhood N of x in K such that $\bar{N} \cap P$ is a polyhedron. A topological imbedding h of a complex L in a complex K is *tame* if there is a homeomorphism g of K onto itself such that gh is piecewise linear. Two imbeddings h_1, h_2 of a set into a space X are *equivalent* if there is a homeomorphism g of X onto itself such that $gh_1 = h_2$. Hence an imbedding is tame if and only if it is equivalent to a piecewise linear imbedding.

E^m can be regarded, of course, as a geometric complex by taking any rectilinear triangulation; however, it is equivalent to say a compact set is polyhedral in E^m if it is the union of a finite number of closed linear simplexes relative to the vector space structure of E^m, leaving unexpressed any specific triangulation of E^m. Similarly a map f of a complex K into E^m is piecewise linear if f maps each simplex in some subdivision of K linearly into E^m.

We shall regard $E^n \times 0$ and $0 \times E^k$ in $E^n \times E^k = E^{n+k}$ as E^n and E^k respectively. Let S be a set in E^k and $A : S \to (E^{n+k})^I$, a map into the space of paths in E^{n+k}, provided with the compact-open topology. Then for each z in S, $A(z)$ is a map of $I = [0, 1]$ into E^{n+k}. Let $A_z = A(z)(I)$, a set in E^{n+k}, and let $A_z^0 = A(z)(0)$ and $A_z^1 = A(z)(1)$. $\pi : E^{n+k} \to E^k$ is the projection. We shall say $\{A_z \mid z \in S\}$ is an *admissible family* for S if for each $z \in S$,

1) A_z is a polygon arc with A_z^0 and A_z^1 as end-points *or* A_z is a single-point (degenerate).
2) $A_z \subset \pi^{-1}(z)$.
3) If A_z is not degenerate and B_z is the straight line segment joining A_z^0 to A_z^1 then each $(n-1)$-plane orthogonal to B_z in $\pi^{-1}(z)$ intersects A_z in just one point.

Note that the particular parametrization of each A_z given by A is not a part of the concept of an admissible family. We require only that some continuous parametrization exists. However, from condition 3) a preferred parametrization emerges, namely: Let $A(z)(t)$ be the point of A_z on the $(n-1)$-plane in $\pi^{-1}(z)$ orthogonal to B_z which divides B_z in the ratio t to $1-t$. Subsequently whenever we refer to a parametrization we shall mean the preferred one.

An *arc* is the homeomorphic image of I whereas a *path* is merely the continuous image of I. In some of our lemmas we require that points move along arcs but in others we settle for paths. It would be interesting to know if we could improve some of these theorems (say Theorem 1.1 or Lemma 5.1) to make points move along arcs there also.

3. Five lemmas.

LEMMA 3.1. *Let σ be a linear simplex in E^k and $\{A_z \mid z \in \bar{\sigma}\}$ an admissible family for $\bar{\sigma}$ such that for $z \in$ Bd σ, A_z is degenerate. Let V be an open set in E^{n+k} containing every element in $\{A_z \mid z \in \sigma\}$. Then $\{A_z \mid z \in \bar{\sigma}\}$ can be extended to an admissible family for E^k, $\{A_x \mid x \in E^k\}$, so that*

1) *Given any x in E^k there is a $z \in \sigma$ such that the length of A_x is no greater than the length of A_z.*
2) *If A_x is non-degenerate, then $A_x \subset V$.*

Proof. Let $\epsilon(z)$ be a positive continuous function defined on σ such that for any $z \in \sigma$ the $\epsilon(z)$-neighborhood of A_z is contained in V. We can choose such a function so that $\epsilon(z) \to 0$ as $z \to$ Bd σ.

Now use $\epsilon(z)$ to define a Cartesian product neighborhood N of σ in E^k as follows. If $z \in \sigma$, where σ is an r-simplex, let D_z be the open $(k-r)$-disk in the plane orthogonal to σ in E^k at z, having radius $\epsilon(z)$ and center at z. $N = \bigcup_{z \in \sigma} D_z$.

Now we define an admissible family for $\bar{N}$ which extends $\{A_z \mid z \in \bar{\sigma}\}$. If $x \in \bar{N} - \bar{\sigma}$, then $x = tz + (1 - t)y$ where $z \in \sigma$ and $y \in \text{Bd } D_z$. Then $A(z)[0, t]$ is a subarc of A_z having as one end-point A_z^0. Define A_x to be the translation of $A(z)[0, t]$ in E^{n+k} which takes z to x. Since this translation moves no point more then $\epsilon(z)$ and the $\epsilon(z)$-neighborhood of A_z is contained in V, then $A_x \subset V$. Now extend $\{A_x \mid x \in \bar{N}\}$ to E^k by using Tietze's extension theorem on $x \to A_x$ viewed as a map from Bd N into E^n. Then for $x \notin N$, A_x is degenerate and the extension is defined.

LEMMA 3.2. *Let $\{A_z \mid z \in E^k\}$ be an admissible family for E^k, and let $\epsilon(z)$ be a non-negative map of E^k into the reals such that $\epsilon(z)$ is positive whenever A_z is non-degenerate. Then there is an isotopy $\psi_t(t \in I)$ of E^{n+k} onto itself such that*

1) *ψ_0 is the identity.*
2) *$\psi_1(A_z^0) = A_z^1$ for each $z \in E^k$.*
3) *ψ_t is invariant on each $\pi^{-1}(z)$.*
4) *Each point of $\pi^{-1}(z)$ that is moved is moved along a polygonal arc of length no more than the length of A_z.*
5) *No point of $\pi^{-1}(z)$ outside the $\epsilon(z)$-neighborhood of A_z is moved at all.*

Proof. For each z we define a homeomorphism h^z of $\pi^{-1}(z)$ onto itself taking A_z onto B_z, the segment joining A_z^0 to A_z^1. If A_z is degenerate, let h^z be the identity; otherwise, since each $(n - 1)$-plane P intersecting and orthogonal to B_z, say at y_P, intersects A_z in one point, say x_P, we can define such a transformation by requiring $h^z \mid P$ to be the translation of P taking x_P into y_P. Do this for each P orthogonal to the line determined by B_z, where translation by zero is used if P fails to intersect B_z. Since $\{A_z\}$ is admissible, it is easy to verify that if h is defined on E^{n+k} by $h \mid \pi^{-1}(z) = h^z$, then h is a homeomorphism.

Let $\delta(z)$ be a non-negative continuous function positive for each $z \in E^k$ such that A_z is non-degenerate. Again in $\pi^{-1}(z)$, we define an isotopy $\theta_t^z(t \in I)$ taking A_z^0 to A_z^1 along B_z, moving no point along a path longer than the length of B_z, and fixing every point outside a $\delta(z)$-neighborhood of B_z. Let a_z be the point on the line determined by B_z such that A_z^0 is between a_z and A_z^1, and a_z lies $\delta(z)$ units from A_z^0. Let b_z be the point on this line $\delta(z)$ away from A_z^1 such that A_z^1 is between b_z and A_z^0. Let P be an $(n - 1)$-plane in $\pi^{-1}(z)$ orthogonal to B_z at the mid-point β and let γ be a sphere of radius $\delta(z)$ in P centered at β. Then $[a_z, b_z] * \gamma$ is a closed neighborhood of B_z. Define θ_t^z to be the identity on γ, to take $[a_z, A_z^0]$ linearly onto $[a_z, (1 - t)A_z^0 + tA_z^1]$ and $[A_z^0, b_z]$ linearly onto $[(1 - t)A_z^0 + tA_z^1, b_z]$. The definition of θ_t^z is completed by having it be the join map on $[a_z, b_z] * \gamma$ and the identity everywhere else. θ_t^z satisfies the required properties. Define θ_t on E^{n+k} by $\theta_t \mid \pi^{-1}(z) = \theta_t^z$. Because the functions $z \to A_z^i$, $i = 0, 1$ and $\delta(z)$ are continuous, $\theta_t(t \in I)$ is an isotopy of E^{n+k}.

Now define $\psi_t = h^{-1}\theta_t h$. Clearly 1), 2), 3) and 4) are satisfied, and property 5) follows if $\delta(z)$ is chosen small enough.

LEMMA 3.3. *Let $\bar{\sigma}$ be a closed linear simplex in E^k and $f : \bar{\sigma} \to E^{n+k}$, an im-*

bedding, such that $\pi f(x) = x$ *for each* $x \in \bar{\sigma}$. *Let* $\epsilon(x)$ *be a continuous real function defined on* E^k *which is positive on* σ. *Let* C *be an open convex set in* E^k *containing* σ, *and* $\epsilon > 0$. *Then there is an isotopy* $\psi_t (t \in I)$ *of* E^{n+k} *onto itself such that*

1) ψ_0 *is the identity.*
2) $\psi_1 f(\sigma)$ *is locally polyhedral at each point.*
3) ψ_t *is invariant on each* $\pi^{-1}(z)$.
4) $\psi_t \mid \pi^{-1}(z)$ *is the identity if* $z \notin C$.
5) *If* $x \in \bar{\sigma}$, *then* $\psi_t \mid \pi^{-1}(x)$ *is the identity outside the* $\epsilon(x)$*-neighborhood of* $f(x)$.
6) ψ_t *is the identity outside an* ϵ*-neighborhood of* $f(\sigma)$.
7) *Each point of* E^{n+k} *moves along a polygonal arc having length less than* ϵ.

Proof. Triangulate the open r-simplex σ rectilinearly and let $\bar{\rho}$ be one of the infinitely many (if $r > 0$) closed r-simplexes occurring in this triangulation. Suppose the vertices of $\bar{\rho}$ are $w_0, w_1, \cdots, w_r$. Define $u_i = f(w_i)$ and $\bar{\rho}'$ to be the closed r-simplex spanned by $u_0, u_1, \cdots, u_r$. We suppose the triangulation of σ is so fine that for each such $\bar{\rho}$ the diameter of both $\bar{\rho}'$ and $f(\bar{\rho})$ is less than the minimum of $\epsilon/3$ and $\frac{1}{3}$ min $\epsilon(x)$, for $x \in \bar{\rho}$. For each $x \in \bar{\rho}$ define $A_x^0 = f(x)$, $A_x^1 = \pi^{-1}(x) \cap \bar{\rho}'$ and A_x to be the line segment (possibly degenerate) joining A_x^0 and A_x^1. A_x has length less than $\frac{2}{3}\epsilon(x)$ and less than $\frac{2}{3}\epsilon$. By parametrizing each A_x from 0 to 1 in a linear fashion we get an admissible family $\{A_x \mid x \in \bar{\sigma}\}$ for $\bar{\sigma}$ by letting A_x be the degenerate segment $f(x)$ for $x \in \text{Bd}\,\sigma$.

By Lemma 3.1 letting $V = \pi^{-1}(C)$ we can extend this to an admissible family for E^k such that A_z is degenerate if $z \notin C$ and each A_z has length less than $\frac{2}{3}\epsilon$. By Lemma 3.2, letting the "$\epsilon(z)$" there be an extension of $\frac{1}{3}\epsilon(x)$ on $\bar{\sigma}$ to E^k which vanishes off C and is bounded by $\frac{1}{3}\epsilon$, we get an isotopy ψ_t taking each $f(\bar{\rho})$ onto the corresponding $\bar{\rho}'$, and satisfying Lemma 3.3.

LEMMA 3.4. *Let* $\bar{\sigma}$ *be a closed linear r-simplex whose interior* σ *is contained in an open convex subset* C *in* $E^k \subset E^{n+k}$ *where* $r < k < n - 1$; *let* $\overline{\sigma_1'}, \cdots, \overline{\sigma_m'}$ *be a collection of topological r-simplexes in* E^{n+k} *whose interiors are disjoint and let* $\overline{\sigma_1''}, \cdots, \overline{\sigma_m''}$ *be a collection of linear topological r-simplexes in* E^{n+k} *whose interiors are also disjoint. Furthermore, suppose that the projection map* $\pi : E^{n+k} \to E^k$ *takes* $\overline{\sigma_i'}$ *and* $\overline{\sigma_i''}$ *in a 1-1 fashion onto* $\bar{\sigma}$, $i = 1, 2, \cdots, m$, *and that* $\text{Bd}\,\overline{\sigma_1'} = \text{Bd}\,\overline{\sigma_i''}$. *Finally, assume the distance between* $\pi^{-1}(z) \cap \overline{\sigma_i'}$ *and* $\pi^{-1}(z) \cap \overline{\sigma_i''}$ *is less than* ϵ *for all* z *in* $\bar{\sigma}$ *and* $i = 1, 2, \cdots, m$.

Then there is an isotopy $\varphi_t (t \in I)$ *on* E^{n+k} *such that*

1) φ_0 *is the identity.*
2) $\varphi_1(\sigma_i') = \sigma_i''$, $i = 1, \cdots, m$.
3) φ_t *is invariant on* $\pi^{-1}(z)$, *for each* $z \in E^k$.
4) φ_t *is the identity on* $\pi^{-1}(z)$, *if* $z \notin C$.
5) φ_t *is the identity outside an* ϵ*-neighborhood of* $\bigcup_{i=1}^m \overline{\sigma_i'}$.
6) *Each point of* E^{n+k} *moves along a polygonal path of length less than* ϵ.

Proof. By making use of Lemma 3.3 we can get a small isotopy on E^{n+k} moving points only in a small neighborhood of each σ_i' (and leaving Bd σ_i' fixed) taking each σ_i' onto a locally polyhedral set. Hence we may assume without loss of generality that each σ_i' is locally polyhedral to begin with. Furthermore we may suppose that σ is triangulated rectilinearly such that given an r-simplex $\bar{\rho}$ in this triangulation, then for each i there is a linear simplex $\bar{\rho}_i'$ in σ_i' with $\pi(\bar{\rho}_i') = \bar{\rho}$. Let $\bar{\rho}_i'' = \pi^{-1}(\bar{\rho}) \cap \sigma_i''$.

We want to adjust the vertices occurring in each such $\bar{\rho}_i'$ and $\bar{\rho}_i''$ to put them in a type of general position. Specifically, for a fixed $\bar{\rho}$ in σ, let its vertices be $w_0, \cdots, w_r$. Define $u_{ij} = \pi^{-1}(w_j) \cap \bar{\rho}_i'$ and $v_{ij} = \pi^{-1}(w_j) \cap \bar{\rho}_i''$. Let $S_{ij} = \{u_{i0}, u_{i1}, \cdots, u_{ij}, v_{ij}, v_{ij+1}, \cdots, v_{ir}\}$. If we denote by $\langle S\rangle$ the closed convex hull of a set S in E^{n+k}, then we want to adjust each u_{ij} and v_{ij} slightly, keeping them in $\pi^{-1}(w_j)$, so that

a) Each $\langle S_{ij}\rangle$ is an $(r+1)$-simplex.
b) $\langle S_{ij}\rangle \cap \langle S_{i'j'}\rangle = \phi$ for $i \neq i'$.

To see that we can do this we look at all sets of the form $S_{ij} \cup S_{i'j'}$ where $i \neq i'; i = 1, 2, \cdots, m; j = 0, 1, \cdots, r$. Then $S_{ij} \cup S_{i'j'}$ has $2r + 4$ points (not all necessarily distinct). We want to adjust points so that $\dim \langle S_{ij} \cup S_{i'j'}\rangle = 2r + 3$ (which is less than $n + k$). We show how this can be done for $S_{10} \cup S_{20}$.

By adjusting u_{ij} and v_{ij}, $i = 1, 2$, $j = 0, \cdots, r$, by small amounts in $\pi^{-1}(w_j)$ we maximize the dimension of $\langle S_{10} \cup S_{20}\rangle$. If this dimension is not $2r + 3$, then some point, say u_{10}, lies in the plane L determined by the other points in $S_{10} \cup S_{20}$, and L is at most $(2r + 2)$-dimensional. If L contains $\pi^{-1}(w_0)$, then L would be $(n + r)$-dimensional, since $\pi(L)$, which contains $\bar{\rho}$, is r-dimensional. But $n + r > 2r + 2$, hence L cannot contain $\pi^{-1}(w_0)$, so we can adjust u_{10} by an arbitrarily small amount keeping it in $\pi^{-1}(w_0)$ and thereby increase the dimension of $\langle S_{10} \cup S_{20}\rangle$. We conclude that $\langle S_{10} \cup S_{20}\rangle$ was already $(2r + 3)$-dimensional. Also sufficiently small further adjustments of these vertices will not reduce the dimension so we can repeat the procedure on the other sets of this type.

Properties a) and b) follow from the fact that $\langle S_{ij} \cup S_{i'j'}\rangle$ is $(2r + 3)$-dimensional. Furthermore the union C_i of all $\langle S_{ij}\rangle$, $j = 0, 1, \cdots r$ is a complex which is isomorphic to a prismatic triangulation of $\bar{\rho} \times I$. Note that $C_i \cap C_{i'} = \phi$ for $i \neq i'$.

Since for each $\bar{\rho}_1'$ in σ_1' the adjusted vertices u_{1j}, $j = 0, \cdots r$ can be chosen, with the desired properties, arbitrarily close to the originals, then using Lemma 3.1 and Lemma 3.2 in the same manner as in the proof of Lemma 3.3 there is a small isotopy invariant on each $\pi^{-1}(z)$ taking each $\bar{\rho}_1'$ onto the r-simplex spanned by the new vertices u_{1j}, $j = 0, \cdots, r$ and moving points only very near to $\bar{\rho}_1'$. Similarly there is an arbitrarily small isotopy taking each $\bar{\rho}_1''$ onto the r-simplex spanned by the adjusted vertices v_{1j}, $j = 0, \cdots, r$; and there are small isotopies taking each $\bar{\rho}_i'$ and $\bar{\rho}_i''$, $i = 2, \cdots, m$ into general position. Hence we assume that we had this general position to begin with.

For a given $\bar{\rho}$ in σ and for a given i and j we see that $\pi^{-1}(z) \cap \langle S_{ij}\rangle$ is a line segment (or point) for each $z \in \bar{\rho}$. Such a segment is parallel to the segment from u_{ij} to v_{ij} and the ends of the segment lie on Bd $\langle S_{ij}\rangle$. Furthermore, if C_i is the cylinder $\bigcup_{j=0}^{r} \langle S_{ij}\rangle$ constructed above, then for each $z \in \bar{\rho}$, $\pi^{-1}(z) \cap C_i$ is a polygonal arc $A_{z,i}$ having one end point $A^0_{z,i}$ on $\bar{\rho}'_i$ and the other end point $A^1_{z,i}$ on $\bar{\rho}''_i$ such that $A_{z,i}$ is the image of $z \times I$ under the linear map from the primatic triangulation of $\bar{\rho} \times I$ onto C_i. By defining $A_{z,i}$ to be the point $\pi^{-1}(z) \cap \bar{\sigma}'_i$, for $z \in$ Bd σ, we obtain a family $\{A_{z,i} \mid z \in \bar{\sigma}\}$ which will be admissible providing property 3) for admissible families is satisfied.

To obtain property 3) it might be necessary to subdivide the triangulation that we have of σ so that the diameter of each $\bar{\rho}'_i$ and $\bar{\rho}''_i$ is small compared to the shortest distance between them. It might be necessary again to alter with a small isotopy each $\overline{\sigma'_i}$ and $\overline{\sigma''_i}$ putting the new vertices in their interiors into general position, this time keeping the diameter of each of the associated $\bar{\rho}'_i$'s and $\bar{\rho}''_i$'s small compared to the distance between them. Again we will assume this has been done. This means that the segment $[\beta(\bar{\rho}'_i), \beta(\bar{\rho}''_i)]$ is nearly parallel to $[u_{ij}, v_{ij}]$, and $[u_{ij}, v_{ij}]$ is parallel to $\pi^{-1}(z) \cap \langle S_{ij}\rangle$ (if not degenerate) for each $z \in \bar{\rho}$. Hence each straight segment in $A_{z,i}$, for $z \in \bar{\rho}$, is nearly parallel to $[\beta(\bar{\rho}'_i), \beta(\bar{\rho}''_i)]$, and we can therefore conclude property 3) is established and for each $i = 1, \cdots, m\{A_{z,i} \mid z \in \bar{\sigma}\}$ is an admissible family. Furthermore, since $[\beta(\bar{\rho}'_i), \beta(\bar{\rho}''_i)]$ has length less than ϵ, by hypothesis, we can assume the length of each $A_{z,i}$ has length less than ϵ and therefore lies in an ϵ-neighborhood of $\overline{\sigma'_i}$.

Now let V_i be an open set in E^{n+k} contained in the ϵ-neighborhood of $\overline{\sigma'_i}$ containing each non-degenerate $A_{z,i}$ and such that $\pi(V_i) \subset C$, $i = 1, 2, \cdots, m$ and $V_i \cap V_{i'} = \phi$, $i \neq i'$. Use Lemma 3.1 to extend $\{A_{z,i} \mid z \in \bar{\sigma}\}$ to an admissible family $\{A_{z,i} \mid z \in E^k\}$ so that $A_{z,i} \not\subset V_i$ implies $A_{z,i}$ is a point, such that each non-degenerate $A_{z,i}$ has length less than ϵ and lies in an ϵ-neighborhood of $\overline{\sigma'_i}$. Then $A_{z,i}$ is disjoint from $A_{z,i'}$ whenever either is non-degenerate and $i \neq i'$, and $\notin C$ implies $A_{z,i}$ is degenerate.

Next find a continuous non-negative function $\epsilon(z)$ defined on E^k such that

i) $\epsilon(z)$ is positive whenever $A_{z,i}$ is non-degenerate for some i and 0 otherwise.

ii) The $\epsilon(z)$-neighborhood of $A_{z,i}$ is contained in V_i if $\epsilon(z) > 0$.

For each i we can employ Lemma 3.2 on $\{A_{z,i} \mid z \in E^k\}$ for the given $\epsilon(z)$. We obtain an isotopy $\psi^i_t(t \in I)$ of E^{n+k} taking each $A^0_{z,i}$ onto $A^1_{z,i}$, hence $\overline{\sigma'_i}$ onto $\overline{\sigma''_i}$. Because of property ii) no point in E^{n+k} is moved by more than one of the isotopies ψ^i_t, $i = 1, 2, \cdots, m$; hence if we define $\varphi_t = \psi^m_t * \cdots * \psi^2_t * \psi^1_t$, for $t \in I$, then φ_t satisfies conditions 1) to 6) of Lemma 3.4 and we are done.

LEMMA 3.5. *Let $\bar{\sigma}$, $\overline{\sigma'_i}$, $\overline{\sigma''_i}$ and ϵ be as in Lemma 3.4, with the exception that that $r = k < n - 1$. Then there is an isotopy $\varphi_t(t \in I)$ of E^{n+k} satisfying*

1) *φ_0 is the identity.*

2) *$\varphi_1(\overline{\sigma'_i})$ is polyhedral, $i = 1, \cdots, m$.*

3) *φ_t is invariant on $\pi^{-1}(z)$, for each $z \in E^k$.*

4) *φ_t is the identity on $\pi^{-1}(z)$, if $z \notin \sigma$.*

5) φ_t *is the identity outside an* ϵ*-neighborhood of* $\bigcup_{i=1}^m \overline{\sigma_i'}$.
6) *Each point of* E^{n+k} *moves along a polygonal path of length less than* ϵ.

Proof. The proof is very similar to that of Lemma 3.4; however, because of dimension considerations our general position will necessarily be different. As in Lemma 3.4 we can assume without loss of generality that each $\overline{\sigma_i'}$ is locally polyhedral at each point of its interior, and in fact that there is a triangulation of σ so that if $\bar{\rho}$ is a linear r-simplex in this triangulation then $\bar{\rho}_i' = \pi^{-1}(\bar{\rho}) \cap \sigma_i'$ is a *linear* r-simplex, $i = 1, \cdots, m$. Further we can assume that $\bar{\rho}_i'$ is disjoint from $\bar{\rho}_i'' = \pi^{-1}(\bar{\rho}) \cap \sigma_i''$ and the diameters of each are small compared to the distance between them. This we can do by adjusting the vertices as before to get the simplexes disjoint which is possible since $2r < n + k$.

Finally we adjust the vertices of the triangulation of σ_i' so that if $\bar{\rho}$ is an r-simplex in the triangulation of σ with vertices $w_0, \cdots, w_r$ and $u_{ij}, v_{ij}, i = 1, \cdots, m, j = 0, \cdots r$, $\bar{\rho}_i', \bar{\rho}_i''$ and S_{ij} are all as defined in the proof of Lemma 3.4, then $\langle S_{ij}\rangle \cap \bar{\rho}_{i'}' = \phi$ and $\langle S_{ij}\rangle \cap \bar{\rho}_{i'}'' = \phi$, if $i \neq i'$, and each $\langle S_{ij}\rangle$ is an $(r + 1)$-simplex. To do this we consider sets of the form $T_{iji'}' = S_{ij} \cup \{u_{i'0}, \cdots, u_{i'r}\}$ and $T_{iji'}'' = S_{ij} \cup \{v_{i'0}, \cdots, v_{i'r}\}, i \neq i'$, and the S_{ij} themselves. Maximize the dimension of each $\langle T_{iji'}'\rangle$, $\langle T_{iji'}''\rangle$ and $\langle S_{ij}\rangle$ by adjusting very slightly the points u_{ij} and v_{ij} in $\pi^{-1}(w_j)$. The claim is the dimension of $\langle T_{iji'}'\rangle$ and $\langle T_{iji'}''\rangle$ is then $2r + 2$ and that of $\langle S_{ij}\rangle$ is $r + 1$. Suppose not, say dim $\langle T_{102}'\rangle < 2r + 2$. Then some point, say u_{10} is in the hyperplane L determined by the rest of the points in T_{102}' . L has dimension at most $2r + 1 = 2k + 1 < n + k$. But L cannot contain $\pi^{-1}(w_0)$, for otherwise $L = E^{n+k}$, since $\pi(L)$ contains $\bar{\rho}$, hence E^k. Therefore we can rechoose u_{10} outside L and increase the dimension of $\langle T_{102}'\rangle$, a contradiction. This shows the dimension of $\langle T_{iji'}'\rangle$, $i \neq i'$, can be taken to be $2r + 2$, hence $\langle S_{ij}\rangle \cap \bar{\rho}_{i'}' = \phi$ whenever $i \neq i'$. Similarly the dimension of $\langle S_{ij}\rangle$ may be taken to be $r + 1$. Since the adjustments of the vertices were arbitrarily small, we may assume without loss of generality, as we did in the proof of Lemma 3.4, that we already had this kind of general position of the vertices and proceed to the proof in this special case.

As before, let $C_i = \bigcup_{j=0}^r \langle S_{ij}\rangle$, a complex which is isomorphic to a prism triangulation of $\bar{\rho} \times I$ with $\bar{\rho} \times 0$ corresponding to $\bar{\rho}_i'$ and $\bar{\rho} \times 1$ corresponding to $\bar{\rho}_i''$. Then C_i fails to intersect $\bar{\rho}_{i'}'$ and $\bar{\rho}_{i'}''$, for $i \neq i'$. If we successively used C_1 , then C_2 , etc. to define isotopies taking $\bar{\rho}_1'$ onto $\bar{\rho}_1''$, $\bar{\rho}_2'$ onto $\bar{\rho}_2''$, etc., there would be no guarantee that conditions 5) and 6) of the Lemma would be met, due to the fact that C_i and $C_{i'}$ may not be disjoint.

We avoid this difficulty (and get the weaker condition 2)) by only restoring a neighborhood of Bd σ_i'' as follows. As before for $z \in \bar{\rho} \subset \sigma$ let $A_{z,i}$ be the polygonal arc $\pi^{-1}(z) \cap C_i$, with $A_{z,i}^0$ the end-point on $\bar{\rho}_i'$ and $A_{z,i}$ the end-point on $\bar{\rho}_i''$. For $z \in E^k - \sigma$ let $A_{z,i}$ be the single point in the intersection of the k-plane containing $\overline{\sigma_i''}$ with $\pi^{-1}(z)$. For each i, $\{A_{z,i} \mid z \in E^k\}$ is an admissible family.

Let τ be a concentric k-simplex whose closure lies in σ. If τ is large enough, $A_{z,i}$, for each $z \varepsilon E^k - \tau$, $i = 1, 2, \cdots, m$, will have length less than $\epsilon/2m$. Extend $\{A_{z,i} \mid z \varepsilon E^k - \tau\}$ to $\{A_{z,i} \mid z \varepsilon E^k\}$ as follows. Let ω be a k-simplex whose closure lies in and is concentric with τ. Let z be in $\bar{\tau} - \omega$ and let x and y be the intersection of the line joining z and $\beta(\tau)$ with Bd τ and Bd ω respectively. Then $z = tx + (1 - t)y$ and define $A'_{z,i}$ to be $A_{x,i}[0, t]$ a subarc of $A_{x,i}$ having $A^0_{x,i}$ as one end-point. Define $A_{z,i}$ to be $A'_{z,i}$ translated so that $A^0_{x,i}$ is taken into $A^0_{z,i} = \pi^{-1}(z) \cap \sigma'_i$. For $z \varepsilon$ Bd ω, each $A_{z,i}$ is a point in σ'_i, hence defining $A_{z,i}$, for z in ω, to be $\pi^{-1}(z) \cap \sigma'_i$ extends $\{A_{z,i} \mid z \varepsilon E^k - \tau\}$ to an admissible family $\{A_{z,i} \mid z \varepsilon E^k\}$ with no arc as long as $\epsilon/2m$. If ω is large enough each $A_{z,i}$ is in an $\epsilon/2$-neighborhood of Bd σ'_i. Define $\epsilon(z)$ on E^k, vanishing outside σ, positive but less than $\epsilon/2m$ on σ and such that for each z in $\bar{\sigma} - \tau$ and each i and i', with $i \neq i'$, the $\epsilon(z)$-neighborhood of $A_{z,i}$ misses both $\overline{\sigma'_{i'}}$ and $\overline{\sigma''_{i'}}$. Let $\psi^i_t(t \varepsilon I)$ be the isotopy guaranteed by Lemma 3.2 for $\{A_{z,i} \mid z \varepsilon E^k\}$ and $\epsilon(z)$.

Then $\psi^i_t(t \varepsilon I)$ moves no point along a path of length as much as ϵ/m and ψ^i_1 takes $\pi^{-1}(\bar{\sigma} - \tau) \cap \bar{\sigma}'_i$, a neighborhood of Bd σ'_i in $\bar{\sigma}'_i$, onto $\pi^{-1}(\bar{\sigma} - \tau) \cap \overline{\sigma''_i}$, a polyhedral neighborhood of Bd σ''_i in $\overline{\sigma''_i}$. Furthermore ψ^i_t leaves $\overline{\sigma'_{i'}}$ and $\overline{\sigma''_{i'}}$ fixed for $i \neq i'$ and moves no point outside an ϵ-neighborhood of $\sigma_{i'}$. Define $\psi_t = \psi^m_t * \cdots * \psi^2_t * \psi^1_t$ for $t \varepsilon I$. Then $\psi_t(t \varepsilon I)$ satisfies all the properties in the Lemma except 2). $\psi_1(\overline{\sigma'_i})$ is polyhedral only on a neighborhood of Bd $\overline{\sigma'_i}$, namely in $\pi^{-1}(\bar{\sigma} - \tau)$. However, $\pi^{-1}(\bar{\tau}) \cap \psi_1(\overline{\sigma'_i})$ is disjoint from $\pi^{-1}(\bar{\tau}) \cap \psi_1(\overline{\sigma'_{i'}})$, for $i \neq i'$, and we can use the method employed in the proof of Lemma 3.3 to make each $\psi_1(\overline{\sigma'_i})$ polyhedral in $\pi^{-1}(\bar{\tau})$ by a small isotopy $\psi'_t(t \varepsilon I)$ keeping fixed points in $\pi^{-1}(E^k - \tau)$. Then let $\varphi_t = \psi'_t * \psi_t(t \varepsilon I)$ and we are done.

4. Proof of Theorem 1.1. Since $n + k > 2k + 1$, we can get a piecewise linear imbedding $p : K \to E^{n+k}$ such that for each $x \varepsilon K$ the distance between $p(x)$ and $h(x)$ is less than δ for any preassigned positive δ. Let δ be small relative to ϵ—just how small will be specified later. Furthermore we can assume the vertices of $p(K)$ are in general position so that if $\bar{\sigma}$ is a closed simplex in $p(K)$, $\pi \mid \bar{\sigma}$ is 1-1. It will be convenient to think of the geometric representation $p(K)$ as K itself and let $f = hp^{-1}$. Then f moves no point as much as δ and $f(K)$ is the set which we make polyhedral with a small isotopy, using K as a model toward which we pull $f(K)$. By taking a subdivision of K we may assume that any two open simplexes σ and τ in K have disjoint images under π or else have the same images. This determines a complex structure for $\pi(K)$. Note that $\pi(K)$ is contained in the δ-neighborhood of the origin in E^k.

Next we apply a technique due to Klee [9] and used in an analogous way in [1]. We want to lift each point x of $f(K)$ into the n-plane through $f^{-1}(x)$ orthogonal to E^k. Note that $x + \pi f^{-1}(x)$ lies in this plane also, hence we consider the line segment A_x having x as one end-point A^0_x and $x + \pi f^{-1}(x)$ as the other end-point A_x^1. This defines an admissible family $\{A_x \mid x \varepsilon f(K)\}$ (with n and k interchanged) which we want to extend to one for E^n keeping the length of all segments less than δ, and degenerate outside a δ-neighborhood N of $f(K)$. This we do by applying Tietze's Extension Theorem to the map $x \to \pi f^{-1}(x)$ on $f(K)$ and $x \to 0$ outside N. Let φ be an extension to all of E^n taking images only in a δ-neighborhood of 0 in E^k. For each z in E^n define A_z^0 to be z, A_z^1 to be $z + \varphi(z)$ and A_z to be the segment joining A_z^0 and A_z^1. Then $\{A_z \mid z \varepsilon E^n\}$ is admissible and using Lemma 3.2, with $\epsilon(z) = \delta$, we get an isotopy $g_t(t \varepsilon [0, \frac{1}{2}])$, where $g_0 = 1$, $g_{\frac{1}{2}}(A_z^0) = A_z^1$, hence $\pi g_{\frac{1}{2}} f(y) = \pi(y)$ for each y in K, no point is moved along a path of length as much as δ, and finally no point outside a 2δ-neighborhood of $f(K)$ is moved at all.

The procedure now is to use Lemma 3.4 successively to restore the skeletons of K in dimensions $0, 1, \cdots, k-1$ and then to make the resulting deformed $f(K)$ polyhedral using Lemma 3.5. To be more precise, let $f_0 = g_{\frac{1}{2}} f$. The distance $d(x, f_0(x)) < \delta$ for each $x \varepsilon K$. Then for each 0-simplex $\bar{\sigma}$ in $\pi(K)$ there will be finitely many 0-simplexes $\overline{\sigma''}$ in K such that $\pi(\overline{\sigma''}) = \bar{\sigma}$. Let these be $\overline{\sigma_1''}, \cdots, \overline{\sigma_m''}$. Let $\overline{\sigma_i'} = f_0(\overline{\sigma_i''})$. Choose for each 0-simplex σ a convex neighborhood C_σ in E^k so that if τ is another 0-simplex and C_τ is its associated neighborhood, then $C_\sigma \cap C_\tau = \phi$. By Lemma 3.4 there is an isotopy $\psi_t^\sigma(t \varepsilon I)$ taking each $\overline{\sigma_i'}$ onto $\overline{\sigma_i''}$, moving no point outside a δ-neighborhood of $f_0(K)$ (hence a 3δ-neighborhood of $f(K)$) and moving points along paths having length less than δ. Furthermore, if we define ψ_t to be the composition of all ψ_t^σ, for σ a 0-simplex in $\pi(K)$, then ψ_t has these same properties. Define $g_t(t \varepsilon [\frac{1}{2}, \frac{1}{2} + 1/2(k+1)])$ by $g_t = \psi_s g_{\frac{1}{2}}$ where $s = 2(k+1)(t - \frac{1}{2})$ and $f_1 = g_a f = \psi_1 f_0$ where $a = \frac{1}{2} + 1/2(k+1)$. Then $\pi f_1(x) = \pi(x)$ and $d(x, f_1(x)) < 2\delta$, for each x in K, and $f_1 \mid K^0 = 1$. Here K^r denotes the r-skeleton of K. Also $g_t(t \varepsilon [0, \frac{1}{2} + 1/2(k+1)])$ moves no point along a path of length as much as 2δ.

Now let r be an integer less than k and suppose that $g_t(t \varepsilon [0, \frac{1}{2} + r/2(k+1)])$ has been defined, (1) moving only those points in a $2^{r+1}\delta$-neighborhood of $f(K)$, and (2) moving points along paths of length less than $2^r\delta$. Let $f_r = g_q f$ where $q = \frac{1}{2} + r/2(k+1)$ and suppose further that (5) $f_r \mid K^{r-1} = 1$, (3) $\pi f_r = \pi \mid K$, and (4) $d(x, f_r(x)) < 2^r\delta$, for each x in K. Use Lemma 3.4 for extending the isotopy to $g_t(t \varepsilon [0, \frac{1}{2} + (r+1)/[2(k+1)]])$, as we did above for $r = 0$, so that (1)–(5) are satisfied for r replaced by $r+1$.

Now we suppose $g_t(t \varepsilon [0, \frac{1}{2} + r/2(k+1)])$ and f_r have been defined satisfying (1)–(5) for $r = k$. This time use Lemma 3.5 to complete the isotopy $g_t(t \varepsilon I)$, satisfying (1)–(4) for $r = k+1$ where we define $f_{k+1} = g_1 f$, and $f_{k+1}(K)$ is polyhedral. Choosing $\delta = \epsilon/2^{k+2}$ finishes the proof of Theorem 1.1.

5. Equivalence of imbeddings.

LEMMA 5.1. *Let K be a k-complex and let $K \times I$ have a prismatic triangulation. Suppose $\epsilon > 0$ and f is an imbedding of $K \times I$ into E^n such that f is* *linear on each simplex in* $K \times I$ *and* $f(v \times I)$ *has length less than* ϵ *for each vertex* v *in* K. *Then there is an isotopy* $h_t(t \in I)$ *of* E^n *onto itself such that*

1) h_0 *is the identity*
2) $h_1 f(x, 0) = f(x, 1)$ *for each* x *in* K
3) *no point of* E^n *moves along a path of length as much as* ϵ.

Proof. Let $v_1, v_2, \cdots, v_m$ be an ordering of the vertices of K which determine the given prismatic triangulation on $K \times I$. For convenience denote $f(v_i, 0)$ and $f(v_i, 1)$ by u_i and w_i respectively, and let K_0 be $f(K \times 0)$. The isotopy $h_t(t \in I)$ will be constructed by taking the u's onto the w's one at a time.

The isotopy is broken into m pieces—one for each vertex of K. Each h_t will be linear on each simplex of K_0. In general $h_{i/m}(u_j)$ is w_j or u_j depending on whether $j \leq i$ or $j > i$. Hence $h_{1/m}$ moves u_1 to w_1, $h_{2/m}$ moves u_2 to w_2, etc. Since each h_t is linear on each simplex of K_0, this completely describes each $h_{i/m}(K_0)$. We denote $h_{i/m}(K_0)$ by $K_{i/m}$.

Let L be the link of a vertex v_i of K. Denote by L_i the complex $h_{(i-1)/m} f(L \times 0)$. Then L_i is the link of u_i in $K_{(i-1)/m}$.

For each simplex σ in L_1, the join $[u_1, w_1] * \sigma$ intersects K_0 only in $u_1 * \sigma$ since $[u_1, w_1] * \sigma$ is the image under f of a simplex of $K \times I$. Then $[u_1, w_1] * L_1$ intersects K_0 only in $u_1 * L_1$. In general $[u_i, w_i] * L_i$ intersects $K_{(i-1)/m}$ only in $u_i * L_i$.

The first part $h_t(0 \leq t \leq 1/m)$. Let N_1 be a convex open set containing $[u_1, w_1]$ such that $N_1 * L_1$ intersects K_0 exactly in $u_1 * L_1$. That there is such an N_1 follows from the fact that K_0 is a polyhedron and $[u_1, w_1] * L_1$ intersects K_0 exactly in $u_1 * L_1$ as pointed out in the preceding paragraph. Then $N_1 * L_1$ is starlike with respect to each point of N_1. We suppose further that N_1 lies within a $1/z$-neighborhood of $[u_1, w_1]$ where z is a positive integer to be specified later when we consider the lengths of paths.

As t moves from 0 to $1/m$, let h_t move u_1 linearly to w_1. For each t, $h_t(0 \leq t \leq 1/m)$ is fixed on $E^n - (N_1 * L_1)$ and h_t is linear on each segment from u_1 to Bd $(N_1 * L_1)$. Note that if $h_t(0 \leq t \leq 1/m)$ moves a point at all it moves it in a straight path parallel to the vector from u_1 to w_1.

Note that $h_t(0 \leq t \leq 1/m)$ does not move any points of K_0 except possibly those in $u_1 * L_1$. If p is a point in a simplex of K_0 in $u_1 * L_1$ and p has barycentric coordinates $(x_1, x_2, \cdots, x_j)$ with $\Sigma x_i = 1$ and x_1 associated with the vertex u_1, the length of the path through which p is moved under $h_t(0 \leq t \leq 1/m)$ is x_1 times the distance from u_1 to w_1.

The next parts of h_t. In a similar fashion, the other u_i's are pushed one at a time over to the corresponding w_i's. In general, let N_{i+1} be a convex neighborhood of $[u_{i+1}, w_{i+1}]$ so that $N_{i+1} * L_{i+1}$ intersects $K_{i/m}$ in precisely $u_{i+1} * L_{i+1}$ and each point of N_{i+1} lies within $1/z$ of $[u_{i+1}, w_{i+1}]$. Choose $g_t(i/m \leq t \leq (i+1)/m)$ to be an isotopy such that

$g_{i/m}$ = identity,
g_t = identity on $E^n - (N_{i+1} * L_{i+1})$,

$g_t(i/m \leq t \leq (i+1)/m)$ moves u_{i+1} straight to w_{i+1}, and
for each t, g_t is linear on each segment from u_{i+1} to Bd $(N_{i+1} * L_{i+1})$.

The isotopy $h_t(i/m \leq t \leq (i+1)/m)$ is defined to be $g_t h_{i/m}$.

Lengths of certain paths. We denote the distance from p to q by $D(p, q)$. Note that if p is a point of K_0 whose barycentric coordinate with respect to u_{i+1} is x_{i+1}, the length of the path swept out by p under $h_t(i/m \leq t \leq (i+1)/m)$ is $x_{i+1} D(u_{i+1}, w_{i+1})$. If u_{i+1} is not a vertex of the simplex of K_0 containing p, $h_t(p) = h_{i/m}(p)$ under this part of our isotopy.

Suppose p is a point of the simplex of K_0 with vertices $u_i, u_j, \cdots, u_k$ and the barycentric coordinates of p with respect to these vertices are $(x_i, x_j, \cdots, x_k)$. Then the path swept out by p under $h_t(0 \leq t \leq 1)$ is swept out entirely during $((i-1)/m \leq t \leq i/m)$, $((j-1)/m \leq t \leq j/m)$, $\cdots$, $((k-1)/m \leq t \leq k/m)$. The length of the path swept out is $x_i D(w_i, u_i) + x_j D(w_j, u_j) + \cdots + x_k D(w_k, u_k) \leq (x_i + x_j + \cdots + x_k)$ maximum $(D(w_i, u_i), D(w_j, u_j), \cdots, D(w_k, u_k)) < \epsilon$.

Lengths of other paths. Although points of K_0 are not moved along paths with lengths as much as ϵ, there may be points of E^n that sweep out long paths unless we restrict the N's to be small. For example, if N_1 and N_2 were taken unnecessarily large and w_1 was near u_2, a point near u_1 might be moved close to w_1 under $h_{1/m}$ and then near w_2 under $h_{2/m}$—thereby sweeping out a path of length greater than ϵ. The obvious way to prevent this is to insist that each point of N_i is very near $[w_i, u_i]$—within $1/z$ of it for some large integer z.

Assume that no matter how large we make z, there is a point q of E^n that sweeps out a path of length greater than or equal to ϵ. For each integer j let $h_{j,t}$ be an isotopy such as we have described with the restriction that $j \leq z$. Assume that q_j is a point of E^n such that q_j sweeps out a path of length greater than or equal to ϵ under $h_{j,t}$. Then by selecting subsequences m times from both the sequences of $h_{i,t}$'s and q_i's one finds that there are sequences of $h_{i,t}$'s and q_i's such that for $j = 0, 1, 2, \cdots, m$, the sequence $h_{1,j/m}(q_1)$, $h_{2,j/m}(q_2)$, $h_{3,j/m}(q_3)$, $\cdots$ converges to a point p_j. We show the existence of a suitable z by showing that the assumption that there is none leads to the contradiction that $D(p_0, p_1) + D(p_1, p_2) + \cdots + D(p_{m-1}, p_m) < \epsilon$.

We note that p_0 belongs to $f(K \times I)$. If it did not, then for j sufficiently large $h_{j,t}$ would not move p_0. In fact, points near p_0 are not moved until we move the first u_i in the simplex of K_0 containing p_0.

For convenience we suppose that u_1 is a vertex of the simplex of $f(K \times I)$ containing p_0. Let p_0' be the point of K_0 in this simplex which is on the line through p_0 parallel to $[w_1, u_1]$. The path swept out by p_0' under $h_{j,t}$ is the same irrespective of the value of j and is of length less than ϵ. However, the broken line $p_0 p_1 p_2 \cdots p_m$ is a subset of the path swept out by p_0'. Hence $D(p_0, p_1) + D(p_2, p_2) + \cdots + D(p_{m-1}, p_m) < \epsilon$.

LEMMA 5.2. *Suppose K is a k-complex and $K \times I$ is triangulated prismatically. Let f be a map of $K \times I$ into E^n which is linear on each simplex in*

$K \times I$. *Suppose a and b are numbers, with* $0 \leq a < b \leq 1$, *and* $\epsilon > 0$, *so that* $f \mid K \times [a, b]$ *is an imbedding and* $f(v_i \times [a, b])$ *has length less than* ϵ *for each vertex* v_i *in* K. *Then there is an isotopy* $h_t (t \varepsilon I)$ *of* E^n *onto itself such that*

1) h_0 *is the identity*
2) $h_1 f(p, a) = f(p, b)$ *for each point* p *in* K
3) *no point of* E^n *moves along a path of length as much as* ϵ.

Proof. In proving this lemma one is tempted to seek a certain kind of triangulation of $K \times [a, b]$ that will enable one to apply the preceding lemma. However, in the case $k = 2$, $n = 3$, there is no triangulation of $K \times [a, b]$ such that each simplex lies in a simplex of $K \times I$ and has a vertical edge. Hence we use a different approach.

Let $v_1, v_2, \cdots, v_m$ be an ordering of the vertices of K which determine the given prismatic triangulation on $K \times I$.

Let $w_i = f(v_i, a)$, $u_i = f(v_i, b)$, $K_0 = f(K \times a)$, and T be the cellular subdivision of $f(K \times [a, b])$ such that C is an element of T if and only if C is the image under f of the intersection of $K \times [a, b]$ with a simplex in $K \times I$.

Let C_i be the sum of all elements of T which have $[w_i, u_i]$ as an edge. Let A_i be a collection of all maximal segments in C_i parallel to $[w_i, u_i]$. We regard these segments as oriented in the direction from w_i to u_i and refer to the sum of the lower ends of all such segments as the bottom of C_i and the sum of all upper ends as the top of C_i. Our isotopy is broken into m pieces—one for each C_i.

The first part $h_t (0 \leq t \leq 1/m)$. The only points of K_0 that move under this part of the isotopy are those on the bottom of C_1 and $h_{1/m}$ carries these bottom points of C_1 to the corresponding top points of C_1. Any point of E^n that moves at all, moves parallel to $[w_1, u_1]$ and in the direction from w_1 to u_1. No point whose distance from C_1 is as much as $1/z$ moves where z is a large integer to be determined later. To get such an isotopy one might first enlarge the collection A_1 to a continuous collection B_1 such that each element of B_1 is either a point or a segment parallel to $[w_1, u_1]$, each line parallel to $[w_1, u_1]$ contains one and only one element of B_1, no nondegenerate element of B_1 intersects K_0 except those in A_1, and each point of a nondegenerate element of B_1 lies within $1/(2z)$ of C_1. Let $\epsilon(b)$ be a continuous map of B_1 into $[0, 1/(2z)]$ such that $\epsilon(b) = 0$ if and only if b is a degenerate element of B_1 and if b^+ denotes the segment obtained by extending b a distance of $\epsilon(b)$ in both directions, $b^+ - b$ misses K_0. The isotopy is fixed except on the b^+'s and moves points along them so that the lower end of each b is moved to its top.

The next parts of h_t. We define $h_t = g_t h_{1/m} (1/m \leq t \leq 2/m)$ where g_t is as follows: g_0 is the identity; the only parts of $h_{1/m}(K_0)$ that move under g_t are those on the bottom of C_2 and $g_{2/m}$ carries these bottom points to the corresponding top points of C_2; any point of E^n that moves at all, moves parallel to $[w_2, u_2]$ and in the direction from w_2 to u_2; no point is moved whose distance from C_2 is as much as $1/z$.

We continue defining $h_t(0 \leq t \leq t)$ in this fashion until K_0 has been moved onto $f(K \times b)$. In general $h_t = g_t h_{(i-1)/m}((i-1)/m \leq t \leq i/m)$ where $g_{(i-1)/m}$ is the identity, $g_{i/m}$ pushes the bottom of C_i to the top, and g_t otherwise behaves in a decent fashion as suggested above.

The length of paths. Suppose $K \times I$ is metrized by giving K some metric, [0, 1] the usual metric, and deciding that $D((k_1, t_1), (k_2, t_2)) = [D(k_1, k_2)^2 + (t_1 - t_2)^2]^{\frac{1}{2}}$. For each point p of K, the length of $p \times [a, b]$ is $b - a$. The path swept out by $f(p, a)$ under the isotopy h_t we have defined is $f(p \times [a, b])$. We show that this path is of length less than ϵ by showing that the length of $f(p \times [a, b])$ is less than ϵ.

Consider a simplex σ of $K \times I$ having $v_i \times I$ as vertical edge. The length of $v_i \times [a, b]$ is $b - a$ while the length of $f(v_i \times [a, b]) = [w_i, u_i]$ is less than ϵ. Hence, if L is a segment in $\sigma \cap (K \times [a, b])$ parallel to $[w_i, u_i]$, length $f(L)/\text{length } L = \text{length } [w_i, u_i]/(b - a) < \epsilon/(b - a)$. Since $p \times [a, b]$ is the sum of "vertical" segments in certain elements of T and f of one of these segments is of length less than $\epsilon/(b - a)$ times the length of the segment, length $f(p \times [a, b]) < (\epsilon/(b - a)) \cdot \text{length } (p \times [a, b]) = (\epsilon/(b - a))(b - a) = \epsilon$.

The proof that for z sufficiently large, the lengths of other paths not starting in K_0 are less than ϵ follows from an argument similar to that used in the proof of Lemma 5.1.

DEFINITIONS. The point p *divides the sets* A, B *in* E^n *in the ratio* a *to* b if p is in $A * B$, the join of A and B, and divides the segment from A to B on which it lies in the ratio a to b. In E^n, a set of points X is in *ratio changing general position* with respect to a set of points Y if the following two conditions are satisfied.

1. No point p of X lies in an r hyperplane $(r \leq n - 1)$ determined by $r + 1$ points of $X \cup Y - \{p\}$.

2. If σ_1, σ_2 are disjoint closed simplexes with vertices in X and τ_1, τ_2 are disjoint closed simplexes with vertices in Y with $\dim \sigma_1 + \dim \sigma_2 + \dim \tau_1 + \dim \tau_2 \leq n - 2$, then either $(\sigma_1 * \tau_1) \cap (\sigma_2 * \tau_2) = \phi$ or the point of intersection of these two joins divides σ_1, τ_1 in a different ratio from the ratio in which it divides σ_2, τ_2.

LEMMA 5.3. *Suppose that* X, Y *are finite sets of points in* E^n *such that* X *is in ratio changing general position with respect to* Y. *Then each open set* U *contains an open set* N *such that for each point* p *of* N, $X \cup \{p\}$ *is in ratio changing general position with respect to* Y.

Proof. To satisfy the first restriction imposed by the ratio changing general position requirement one only needs to select an open N_0 in U to miss a finite number of r hyperplanes determined by the various $(r + 1)$ tuples of points of $X \cup Y$.

To satisfy the second restriction we consider disjoint closed simplexes σ_1, σ_2 with vertices in X and disjoint closed simplexes τ_1, τ_2 with vertices in Y such that $\dim \sigma_1 + \dim \sigma_2 + \dim \tau_1 + \dim \tau_2 \leq n - 3$. Let p_1 be any point in N_0 and let σ_1' be the simplex $p_1 * \sigma_1$.

If $\sigma_1' * \tau_1$ misses $\sigma_2 * \tau_2$, choose N_1 to be a neighborhood of p_1 so small that $N_1 \subset N_0$ and for each point p in N_1, $(p * \sigma_1) * \tau_1$ misses $\sigma_2 * \tau_2$.

If $\sigma_1' * \tau_1$ intersects $\sigma_2 * \tau_2$ in a point q but q does not divide σ_1', τ_1 in the same ratio that it divides σ_2, τ_2, then we also pick N_1 to be a small neighborhood of p_1.

If q divides σ_1', τ_1 in the same ratio that it divides σ_2, τ_2, consider the simplex τ_1' consisting of all points p such that q divides some segment from p to a point of τ_1 in the same ratio that q divides σ_2, τ_2. Let p_2 be a point of N_0 which is in the r-plane determined by $\sigma_1' * \tau_1$ but which is not in the plane determined by σ_1 and τ_1'. Then p_2 serves the role of p_1 in the preceding two paragraphs and we can get a small N_1 containing it.

A finite number of applications of this method to all possible σ_i's and τ_i's, $i = 1, 2$, where each N_j is chosen to lie in the preceding, leads to an N satisfying the conditions of the lemma.

LEMMA 5.4. *Let K be a rectilinear k-complex in E^n and let $v_1, v_2, \cdots, v_m$ be the vertices of K. Then for each $\epsilon > 0$ and each finite set Y there is an isotopy $h_t(t \in I)$ of E^n onto itself such that*

1) *$h_0 =$ identity,*
2) *$\{h_1(v_1), h_1(v_2), \cdots, h_1(v_m)\}$ is in ratio changing general position with respect to Y,*
3) *h_1 is linear on each simplex of K,*
4) *each point of E^n moves along a polygonal path of length less than ϵ.*
5) *each h_t is the identity outside an ϵ-neighborhood of K.*

Proof. Our isotopy is broken into pieces—one for each v_i. For convenience in notation in describing the ith piece we suppose that $v_1, v_2, \cdots, v_{i-1}$ are already in ratio changing general position with respect to Y and $h_{(i-1)/m}$ is the identity.

Let L be the link of v_i in K and U be a small convex neighborhood of v_i in E^n such that $U * L$ intersects K precisely in $v_i * L$. Lemma 5.3 implies that there is a point v_i' in U very close to v_i so that $\{v_1, v_2, \cdots, v_{i-1}, v_i'\}$ is in ratio preserving general position with respect to Y. Then $h_t(i - 1/m \leq t \leq i/m)$ is the identity outside $U * L$, h_t moves v_i linearly to v_i' as t moves from $(i-1)/m$ to i/m, and for each t, h_t is linear on each segment from v_i to Bd $(U * L)$. Note that paths swept out are short—no more than $D(v_i, v_i')$.

THEOREM 5.5. *Suppose K_1, K_2 are two rectilinear k-complexes in E^n, $n \geq 2k + 2$, and h is a piecewise linear homeomorphism of K_1 onto K_2 that does not move any point as far as ϵ. Then there is an isotopy $h_t(t \in I)$ of E^n onto itself such that*

1) *$h_0 =$ identity,*
2) *$h_1 = h$ on K_1,*

3) *each* h_t *is the identity outside an* ϵ*-neighborhood of* K_1 ,

4) *each point of* E^n *moves along a polygonal path of length less than* ϵ.

Proof. Regard K_1 as triangulated so finely that h is linear on each simplex of K_1 . Let $v_1, v_2, \cdots, v_m$ be the vertices. Using Lemma 5.4 we suppose with no loss of generality that $\{v_1, v_2, \cdots, v_m\}$ is in ratio changing general position with respect to $\{h(v_1), h(v_2), \cdots, h(v_m)\}$.

In proving the theorem, one might be tempted to use the methods of Lemma 5.1 and move the v_i's to the $h(v_i)$'s. If this method had worked, we could have used ordinary general position arguments and not resorted to the more complicated ratio changing general position. However, this method fails. Suppose σ_1 , σ_2 are two disjoint closed k-simplexes of K_1 having v_1 , v_2 respectively as vertices. If $h(v_1) * \sigma_1$ intersects $h(v_2) * \sigma_2$ in a point q near both $h(v_1)$ and v_2 , then the first piece of an isotopy like that used in Lemma 5.1 would move a point near σ_1 to q and the second piece would move q near $h(v_2)$—thereby sweeping out a path of length perhaps longer than ϵ. Gugenheim in [5] was not interested in moving points only short distances but Homma in [6] needed to be careful about this. Hence, when Homma says that Gugenheim's proof induces the more general theorem, details are called for such as may be supplied by the argument which follows.

Let $K_1 \times I$ denote the complex determined by a prismatic triangulation relative to the ordering $v_1, v_2, \cdots, v_m$. Let f be a map of $K_1 \times I$ in E^n defined by $f(v_i, 0) = v_i$ and $f(v_i, 1) = h(v_i)$, and f is linear on each simplex in $K_1 \times I$. It is not necessarily true that f is a homeomorphism since there may be two simplexes whose images intersect. However, since $v_1, v_2, \cdots, v_m$ are in ratio changing general position with respect to their images under h, there is a sequence $0 = t_0 < t_1 < \cdots < t_j = 1$ such that f is a homeomorphism on each $K_1 \times [t_{i-1}, t_i]$. Hence the theorem follows from j applications of Lemma 5.2.

The following corollary is an immediate consequence of Theorems 1.1 and 5.5.

COROLLARY 5.6. *Let* K *be a finite* k*-complex and for* $i = 1, 2$ *let* $h_i : K \to E^{n+k}$ *be an imbedding of* K *into an* n*-plane,* $k + 2 \leq n$. *Suppose* $h_2 h_1^{-1}$ *moves no point as far as* ϵ. *Then there is an isotopy* $g_t (t \in I)$ *of* E^{n+k} *onto itself such that*

1) g_0 *is the identity*

2) $g_1 h_1 = h_2$

3) *each* g_t *is the identity outside an* ϵ*-neighborhood of* $h_1(K)$

4) *each point of* E^{n+k} *moves along a polygonal path of length less than* ϵ.

6. Equivalence of imbeddings of special sets.

LEMMA 6.1. *Each map* f *of a subset* L *of* E^k *into* S^n, $n \geq k$, *is inessential.*

Proof. The proof is in two halves. The first half describes a homotopy $f_t (t \in I)$ of L into S^n so that $f_0 = f$ and f_1 can be extended to an open subset

U of E^k. The second half shows that the extended f_1 is inessential on U, and hence on L.

Let $\{U_i\}$ be a locally finite collection of open subsets of E^k so that $\{U_i\}$ covers L and each $f(\bar{U}_i \cap L)$ lies in an n-ball in S^n. The first half of the proof has a countable number of steps—one for each U_i.

The first step of the first half. For convenience we suppose that $f(\bar{U}_1 \cap L)$ lies in a disk D of S^n with linear coordinates. Define $f_{\frac{1}{2}} = f_0$ on $L - U_1$. In the metric space $U_1 \cup (L \cap \text{Bd } U_1)$, the set $L \cap \text{Bd } U_1$ is closed. It follows from the Tietze Extension Theorem that there is a map $f_{\frac{1}{2}}$ of $U_1 \cup (L \cap \text{Bd } U_1)$ into D which is an extension of $f_{\frac{1}{2}}$ on $L \cap \text{Bd } U_1$. For each point p of $L - U_1$ we define $f_t(p) = f_0(p) (0 \leq t \leq \frac{1}{2})$ and for $p \in L \cap U_1$ we define $f_t(p) = 2tf_0(p) + (1 - 2t)f_{\frac{1}{2}}(p)$.

The second step in the first half. We suppose $f_{\frac{1}{2}}$ is extended to $L \cup U_1$ and treat $L \cup U_1$ in this second step like we treated L in the first step. Since $f_{\frac{1}{2}}$ may not take $\bar{U}_2 \cap (L \cup U_1)$ into a disk in S^n, we reduce U_2 to an open set V_2 in U_2 so that $L \cap (U_1 \cup U_2) = L \cap (U_1 \cup V_2)$ and $f_{\frac{1}{2}}(\bar{V}_2 \cap (L \cup U_1)$ lies in a disk in S^n. Following the techniques of the preceding step, we obtain a homotopy $f_t (\frac{1}{2} \leq t \leq \frac{3}{4})$ of $L \cup U_1$ into S^n such that $f_t = f_{\frac{1}{2}}$ on $(L \cup U_1) - V_2$ and $f_{\frac{3}{4}}$ can be extended to V_2 to take $L \cup U_1 \cup V_2$ into S^n.

The other steps in the first half. We continue defining the homotopy f_t and extending it to V's so that for each positive integer i,

$$L \cap (U_1 \cup U_2 \cup \cdots \cup U_i) = L \cap (U_1 \cup V_2 \cup \cdots \cup V_i),$$

$f_{(2^i-1)/2^i}$ is defined on $L \cup U_1 \cup V_2 \cup \cdots \cup V_i$,

$$f_{(2^i-1)/2^i} = f_{(2^{i-1}-1)/2^{i-1}} \quad \text{on} \quad (L \cup U_1 \cup V_2 \cup \cdots \cup V_{i-1}) - V_i .$$

Since the V_i's form a locally finite collection, $\text{limit}_{t \to 1} f_t = f_1$ is defined and continuous on the open set $U_1 \cup V_2 \cup V_3 \cup \cdots = U$ which contains L.

The second half. We show that the map f_1 of U into S^n is inessential by showing that there is a map F of $U \times I$ into S^n such that $F(p, 0) = f_1(p)$ and $F(U \times 1)$ is a point of S^n. Give U a triangulation and order the closed k-simplexes of the triangulation $\sigma_1, \sigma_2, \cdots$ so that each σ_i shares a $(k - 1)$-face with some σ_j which follows it. To get such an ordering, one could first express U as the sum of an ascending sequence of finite complexes $K_1, K_2, \cdots$ so that each component of $U - K_i$ contains infinitely many simplexes of the triangulation. Order the k-simplexes of U by taking first those of K_1, then those of $K_2 - K_1, \cdots$. To order those in $K_{i+1} - K_i$, take a preliminary ordering by starting at Bd K_{i+1} and digging in. Then reverse the preliminary ordering.

We suppose F is defined on $(U \times 0) \cup (U \times 1)$. We extend it to take $\sigma_1 \times [0, 1]$ into S^n, then $\sigma_2 \times [0, 1], \cdots$. Once F has been extended to take $(\sigma_1 \cup \sigma_2 \cup \cdots \cup \sigma_{i-1}) \times [0, 1]$ into S^n, we extend F to $\sigma_i \times [0, 1]$ as follows. Let ρ_i be a $(k - 1)$-face that σ_i shares with a σ that follows it and W_i be the union of the other $(k - 1)$-faces of σ_i. Since dimension $W_i \times [0, 1] \leq n$, it follows from the corollary on page 84 of [8] that F can be extended to take

$W_i \times [0, 1]$ into S^n. Then F can be extended to $\sigma_i \times [0, 1]$ since $\sigma_i \times [0, 1]$ is topologically equivalent to the cartesian product of a segment with the k-cell $(\sigma_i \times 0) \cup (\sigma_i \times 1) \cup (W_i \times [0, 1])$.

THEOREM 6.2. *Let π be the projection map of E^{k+1} onto a k-plane E^k. Suppose M is the union of two of its compact subsets M_1 and M_2 and suppose there is some imbedding $i : M \to E^{k+1}$ satisfying*

1) *$\pi i(M_1) \supset \pi i(M_2)$ and*
2) *$\pi i \mid M_j$ is an imbedding, $j = 1, 2$.*

If h_1 and h_2 are any two imbeddings of M into n-planes in E^{n+k}, then there is a homeomorphsim g of E^{n+k} onto itself such that $h_2 = gh_1$.

Proof. First consider the case $n > k$. Using the hypothesis we produce a "standard" imbedding h of M into E^{n+k} and prove the existence of g_j, a homeomorphism of E^{n+k} onto itself such that $h = g_j h_j (j = 1, 2)$. Regard E^{n+k} as $E^{n-1} < E^1 \times E^k$ and let D denote the distance function in M. We have $\pi i : M \to E^k$ by hypothesis. Define $h : M \to E^{n+k}$ by

$$h(x) = \begin{cases} (0, 0, \pi i(x)) & \text{for } x \text{ in } M_1 \\ (0, D(x, M_1), \pi i(x)) & \text{for } x \text{ in } M_2 . \end{cases}$$

It is easy to see that h is an imbedding.

We may assume without loss of generality that $h_1(M) \subset E^n = E^{n-1} \times E^1 \times 0$. Define $h_3 : M \to E^{n+k} = E^n \times E^k$ by $h_3(x) = (h_1(x), \pi i(x))$. The construction of g_1 takes place in four steps, the first of which is to take h_1 onto h_3 with a global homeomorphism f_1. More precisely, for each z in $h_1(M)$ consider the line segment (possibly degenerate) A_z in E^{n+k} having z as one end-point A_z^0 and $h_3 h_1^{-1}(z)$ as the other end-point A_z^1. The set of all such A_z is an admissible family. Using Lemma 3.2 we get a homeomorphism f_1 of E^{n+k} onto itself (the end of the isotopy) such that $f_1(A_z^0) = A_z^1$ for each z in $h_1(M)$ and f_1 is invariant on each k-plane orthogonal to E^n. Thus $f_1 h_1(x) = h_3(x)$ for each x in M.

The next step takes $h_3(x)$ onto $h(x)$ for all x in M_1. This time for each z in $h_3(M_1)$ let A_z be the segment joining $A_z^0 = z$ to $A_z^1 = hh_3^{-1}(z)$. As before, this will be an admissible family (for $h(M_1) \subset E^k$) and there will then be a homeomorphism f_2 of E^{n+k} invariant on n-plane orthogonal to E^k (i.e. the E^k-coordinate is unchanged by f_2), and such that $f_2(A_z^0) = A_z^1$ for all z in $h_3(M_1)$. Hence $f_2 h_3(x) = h(x)$ for all x in M_1 and we would be done if $M_1 = M_2$. If we denote by μ the projection map of E^{n+k} onto E^k, then we also have $\mu f_2 h_3(x) = \pi i(x) = \mu h(x)$ for all x in M_2. Let $h_4 = f_2 h_3$. We have to adjust the E^n-coordinate of the images of M_2 under h_4, leaving fixed the images of M_1.

Let ν be the projection map of E^{n+k} onto E^n. Note that $\nu\, h_4(x) = 0$ for all x in M_1, hence by condition (1) $\nu\, h_4(x) \leq 0$ for all x in $M_2 - M_1$. Let r be the radial projection of $E^n - \{0\}$ onto the unit sphere S^{n-1}. Then by Lemma 6.1, since $n > k$, $r\nu[\mu \mid h_4(M_2 - M_1)]^{-1}\colon \mu h_4(M_2 - M_1) \to S^{n-1}$ is inessential. In fact this map can be extended to an inessential map $\theta : U \to S^{n-1}$, where U is a relatively open subset of $0 \times E^k = E^k$ containing $\mu h_4(M_2 - M_1)$. Consider the map $p : SO(n) \to S^{n-1}$, defined by $p(T) = T(w)$, where T is a rotation of E^n and w is the point (0, 1) in $E^n = E^{n-1} \times E^1$. Then p has the bundle property, hence there is a homotopy $\varphi_t (t \in I)$ of U into $SO(n)$ such that $p\varphi_0 = \theta$ and $\varphi_1(U) = e$, the identity element (cf. [7; 62–65]).

In the third step we use φ_t to define a homeomorphism f_3 of E^{n+k} onto itself such that $f_3 \mid h_4(M_1) = 1$ and $r\nu f_3 h_4(x) = w$ for all x in $M_2 - M_1$. Let $\psi(z)$ be the positive-valued function defined on $\mu h_4(M_2 - M_1)$ by letting $\psi(z)$ be the distance from $\nu h_4(x)$ to the origin, where $\mu h_4(x) = z$. Extend $\psi(z)$ to Bd U by taking $\psi(z) = 0$ for every z in Bd U. Then $\psi(z)$ is continuous and can be extended again to a continuous non-negative function on $0 \times E^k$ which vanishes off U. We define f_3 by describing it on each n-plane $\mu^{-1}(z) = E^n \times z$ for z in E^k. For z not in U, $f_3 \mid \mu^{-1}(z)$ is the identity. For z in U f_3 acts on the solid ball of radius $\psi(z)$ in $E^n \times z$ by the rotation $[\varphi_0(z)]^{-1}$ on the first factor; on an $(n - 1)$-sphere of radius $(1 + t)\cdot\psi(z)$, where $t \in I$, f_3 acts as $[\varphi_t(z)]^{-1}$; and outside the ball of radius $2\psi(z)$, f_3 is the identity. Then $f_3 \mid \mu^{-1}(z)$ is a homeomorphism of $\mu^{-1}(z)$ onto itself, and if z is in $\mu h_4(M_2)$ and x_j is in M_j with $\mu h_4(x_j) = z$, $j = 1, 2$, then $f_3 h_4(x_1) = h_4(x_1) = (0, 0, z)$ in $E^{n+k} = E^{n-1} \times E^1 \times E^k$, and $f_3 h_4(x_2) = (0, \psi(z), z)$. It is easy to check that f_3 is a homeomorphism on E^{n+k}, so we will be done as soon as the middle coordinate of $h_5 = f_3 h_4 : M \to E^{n-1} \times E^1 \times E^k$ is altered on images of M_2.

We perform this last step by defining a homeomorphism f_4 of E^{n+k} as follows. Let x_2 be in M_2 with $\mu h_5(x_2) = z$ in E^k, and define $\psi'(z) = D(x_2, M_1)$. Then ψ' is continuous on $\mu h_5(M_2)$ and can be extended to a continuous non-negative function on E^k, denoted also by ψ', such that $\psi'(z) = 0$ if and only if $\psi(z) = 0$. For $z \in E^k$ let ξ_z be the homeomorphism on E^1 taking the interval (possibly degenerate) $[0, \psi(z)]$ linearly onto $[0, \psi'(z)]$, $[\psi(z), \infty)$ onto $[\psi'(z), \infty)$ by a translation, and leaving $(-\infty, 0]$ fixed. Then $f_4(x, y, z) = (x, \xi_z(y), z)$ defines a homeomorphism of $E^{n-1} \times E^1 \times E^k = E^{n+k}$ such that $f_4 h_5 = h$ and taking $g_1 = f_4 f_3 f_2 f_1$ finishes the proof for the case $n > k$.

For $n \leq k$, let a k-plane containing $h_1(M)$ be E^k and let $M_1 = M_2 = M$. Then $h_1 = i$ satisfies the hypothesis and we would be done at the end of the second step by taking $g_1 = f_2 f_1$.

Note: The homeomorphism g in the statement of Theorem 6.2 can be taken so as to leave fixed all points outside some compact set, hence we can assume g is isotopic to 1 and conclude that there is a deformation taking h_1 onto h_2. Perhaps Theorem 6.2 can also be adapted to show there is a small isotopy taking M onto a polyhedron if M is a combinatorial manifold or complex and $n = k + 1$ or k.

COROLLARY 6.3. *If M is any 2-manifold, then any two imbeddings of M into n-plane in E^{n+2} are equivalent.*

Proof. If $n = 3$, then M is a 2-sphere with handles hence admits an imbedding i into E^3 satisfying the hypothesis of Theorem 6.2. If $n > 3$, we use the Corollary 5.6.

COROLLARY 6.4. *Any two imbeddings of a k-sphere with handles in an n-plane in* E^{n+k} *are equivalent.*

Question. If M^3 is a 3-manifold which can be imbedded in E^4 does M^3 admit an imbedding i in E^4 satisfying the hypothesis of Theorem 6.2?

7. **Imbeddings in manifolds.** Let $h : K^k \rightarrow M^n$ be a topological imbedding of a k-complex K in a combinatorial n-manifold M. Then h is said to be *locally tame* if for each point x of $h(K)$ there is a combinatorial n-ball N topologically imbedded in M^n containing x in its interior such that $h \mid h^{-1}(N \cap h(K))$ is piecewise linear into the combinatorial structure of N (not necessarily compatible with that of M). We say h is *locally l-flat* if for each point x of $h(K)$ there is a neighborhood N of x and a homeomorphism g of $E^l \times E^{n-l}$ onto N such that $N \cap h(K) \subset g(0 \times E^{n-l})$. If K^k is a triangulated k-manifold, then h is *locally flat* if it is locally $(n - k)$-flat.

Note that with the definitions of local tameness and local flatness that we have used, an embedding might conceivably be locally flat but not locally tame since local flatness does not require a piecewise linear embedding of $h^{-1}(N \cap h(K))$ into the combinatorial structure of N.

In the following theorems we need not insist that the complexes be considered as finite. As pointed out by Gluck in [2], Homma's results apply equally well to noncompact complexes.

THEOREM 7.1 (Gluck [3] and Greathouse [4]). *If* $n \geq 2k + 2$, *every locally tame imbedding of a k-complex in a combinatorial n-manifold is tame.*

So, for example, one has the following corollary:

COROLLARY 7.2.([2], [9]). *Every locally flat imbedding of a triangulated k-manifold in a combinatorial n-manifold,* $n \geq 2k + 2$, *is tame.*

It is only by use of Theorem 1.1 that we get local flatness to imply local tameness for the dimensions we need. How this is done follows from the proof of the following generalization of Corollary 7.2.

THEOREM 7.3. *If h is a locally k-flat imbedding of a k-complex K into a combinatorial n-manifold M and* $n \geq 2k + 2$, *then h is tame.*

Proof. By using Theorem 7.1 we merely have to verify that h is locally tame. Let x be in $h(K)$ and let N be a neighborhood of x and g a homeomorphism of $E^k \times E^{n-k}$ onto N such that $N \cap h(K) \subset g(0 \times E^{n-k})$. Subdivide K so that $h^{-1}(x)$ is a vertex and the closed star V of $h^{-1}(x)$ in this subdivision is mapped by h into N. Then $g^{-1}h \mid V$ is an imbedding of a complex having dimension at most k into an $(n - k)$-plane in E^n with $n - k \geq k + 2$, hence by Theorem 1.1 there is a homeomorphism g_1 of E^n onto itself such that

$g_1g^{-1}h \mid V$ is piecewise linear. Triangulate E^n and subdivide V so that $g_1g^{-1}h \mid V$ is simplicial and so that the closed star B of the vertex $g_1g^{-1}(x)$ in E^n intersects $g_1g^{-1}(h(K) \cap N)$ only in $g_1g^{-1}h(V)$. Then $C = gg_1^{-1}(B)$ is a combinatorial ball topologically imbedded in M and $h^{-1}(C \cap h(K)) \subset V$ is piecewise linearly imbedded by h in the combinatorial structure of C.

Questions. Is an imbedding of a 2-sphere or 2-disk in a 3-plane in E^4 tame? (Added in proof. Since this paper was written, each of Norman Hosay and David Gillman has shown that each 2-sphere in a 3-plane in E^4 is tame in E^4.) Is an imbedding of a 2-complex in a 3-plane in E^5 tame? It is not clear that an affirmative answer to either of these questions would imply that any two such imbeddings would be equivalent.

References

1. R. H. Bing and V. L. Klee, *Every simple closed curve in E^3 is unknotted in E^4*, Proceedings of the London Mathematical Society, to appear.
2. H. Gluck, *Unknotting S^1 in S^4*, Bulletin of the American Mathematical Society, vol. 69(1963), pp. 91–94.
3. H. Gluck, *Imbeddings in the trivial range*, Bulletin of the American Mathematical Society, vol. 69(1963), pp. 824–831.
4. C. A. Greathouse, *Locally flat, locally tame, and tame embeddings*, Bulletin of the American Mathematical Society, vol. 69(1963), pp. 820–823.
5. V. K. Gugenheim, *Piecewise linear isotopy and embedding of elements and spheres* I, Proceedings of the London Mathematical Society, vol. 3(1953), pp. 29–53.
6. T. Homma, *On the imbedding of polyhedra in manifolds*, Yokohama Mathematical Journal, vol. 10(1962), pp. 5–10.
7. S. -T. Hu, *Homotopy Theory*, New York, 1959.
8. W. Hurewicz and H. Wallman, *Dimension Theory*, Princeton, N. J., 1948.
9. V. L. Klee, *Some topological properties of convex sets*, Transactions of the American Mathematical Society, vol. 78(1955), pp. 30–45.
10. J. Stallings. *On topologically unknotted spheres*, Annals of Mathematics, vol. 77(1963), pp. 490–503.

University of Wisconsin and Institute for Advanced Study
University of Michigan and Institute for Advanced Study

D. Varieties, Convex Functions, and Converse Linearity

CONFORMAL MINIMAL VARIETIES

By E. F. Beckenbach and R. H. Bing

Let the functions

$$(1)\qquad x^i = x^i(u^1, u^2, \cdots, u^n) \qquad (i = 1, 2, \cdots, m;\ m \geq n \geq 2),$$

defined for $(u^1, u^2, \cdots, u^n)$ in a domain D of Euclidean n-space, be coördinate functions of an n-dimensional variety V_n in Euclidean m-space.

The map of D on V_n given by (1) is conformal if and only if there exists a function $\lambda(u^1, u^2, \cdots, u^n)$ in D such that

$$(2)\qquad \sum_{i=1}^{m} \frac{\partial x^i}{\partial u^j}\frac{\partial x^i}{\partial u^k} \equiv [\lambda(u^1, u^2, \cdots, u^n)]\,\delta_{j,k} \qquad (j, k = 1, 2, \cdots, n),$$

where $\delta_{j,k}$ is the Kronecker delta.

By analogy with the notion of a pair of conjugate harmonic functions in the theory of functions of a complex variable, if the functions (1) satisfy conditions (2) and

$$\sum_{j=1}^{n} \frac{\partial^2 x^i}{\partial u^{j2}} = 0 \qquad (i = 1, 2, \cdots, m),$$

then the functions (1) have been called [1], [3] *a set of conjugate harmonic functions* in D.

An extension of a theorem of Weierstrass gives the following geometric interpretation of the above definitions:

A necessary and sufficient condition that a V_n be a minimal variety in conformal representation is that the coördinate functions form a set of conjugate harmonic functions.

It is well known that any smooth V_2 can be mapped conformally on a domain in the plane and that the most general conformal map of a Euclidean space E_n on itself for $n \geq 3$ is the product of inversions with respect to hyperspheres, rigid motions, and transformations of similitude [2; 375–376]. A geometric characterization of spaces conformally equivalent to Euclidean n-space, for $n \geq 4$, has been given by Haantjes [4].

It follows that not necessarily can all minimal varieties V_n be represented conformally on E_n. Indeed, a simple proof has been given [1] that the only sets of conjugate harmonic functions, for $m = n \geq 3$, are sets of constants and linear functions; and it has been stated [1] that this result holds also for $m \geq n \geq$

Received May 17, 1943; in revised form, July 17, 1943; presented to the American Mathematical Society, May 4, 1943.

3. This latter statement and the details of its proof constitute the following theorem and its proof.

THEOREM 1. *If the functions*

$$x^i = x^i(u^1, u^2, \cdots, u^n) \qquad (i = 1, 2, \cdots, m;\ m \geq n \geq 3) \tag{3}$$

are a set of conjugate harmonic functions, then the functions (3) *necessarily are constants or linear functions.*

This theorem is proved by the use of the following lemma.

LEMMA. *If*

$$\mathfrak{x} \equiv (x^1, x^2, \cdots, x^m)$$

is the image of $(u^1, u^2, \cdots, u^n)$ *and*

$$\mathfrak{x}_{rrs} \equiv \frac{\partial^3 \mathfrak{x}}{\partial u^{r2} \partial u^s} \qquad (r, s = 1, 2, \cdots, n)$$

exists in D, then

$$\begin{aligned}(n-2)\sum_{r,s} \mathfrak{x}_{rs}^2 + n\Big(\sum_r \mathfrak{x}_{rr}\Big)^2 \\ \equiv n \sum_{r \neq s} (\mathfrak{x}_r\mathfrak{x}_s)_{rs} + \sum_{r,s} [\mathfrak{x}_r^2 - \mathfrak{x}_s^2]_{rr} - 2(n-1)\sum_{r,s} \mathfrak{x}_s\mathfrak{x}_{rrs},\end{aligned} \tag{4}$$

where subscripts denote differentiation, vector products are scalar, and r and s are summed from 1 *to n where indicated.*

Proof of Lemma. Differentiating

$$\mathfrak{x}_r\mathfrak{x}_s \equiv \sum_{i=1}^m \frac{\partial x^i}{\partial u^r}\frac{\partial x^i}{\partial u^s} \qquad (r, s = 1, 2, \cdots, n),$$

we obtain

$$(\mathfrak{x}_r\mathfrak{x}_s)_r \equiv \mathfrak{x}_{rr}\mathfrak{x}_s + \mathfrak{x}_r\mathfrak{x}_{rs}$$

and

$$(\mathfrak{x}_r\mathfrak{x}_s)_{rs} \equiv \mathfrak{x}_{rrs}\mathfrak{x}_s + \mathfrak{x}_{rr}\mathfrak{x}_{ss} + \mathfrak{x}_{rs}^2 + \mathfrak{x}_r\mathfrak{x}_{rss}.$$

The first term of the second member of (4) is accordingly

$$n \sum_{r \neq s} (\mathfrak{x}_r\mathfrak{x}_s)_{rs} \equiv n \sum_{r,s} [2\mathfrak{x}_s\mathfrak{x}_{rrs} + \mathfrak{x}_{rr}\mathfrak{x}_{ss} + \mathfrak{x}_{rs}^2] - 2n \sum_r [\mathfrak{x}_r\mathfrak{x}_{rrr} + \mathfrak{x}_{rr}^2]. \tag{5}$$

Similarly, the second term of the second member of (4) is computed to be

$$\sum_{r,s} [\mathfrak{x}_r^2 - \mathfrak{x}_s^2]_{rr} \equiv 2n \sum_r [\mathfrak{x}_r\mathfrak{x}_{rrr} + \mathfrak{x}_{rr}^2] - 2\sum_{r,s} [\mathfrak{x}_s\mathfrak{x}_{rrs} + \mathfrak{x}_{rs}^2]. \tag{6}$$

Now (4) follows as an identity from (5) and (6).

Proof of Theorem 1. Since the functions (3) are harmonic, we have

$$\sum_r \mathfrak{x}_{rr} = 0,$$

so that the last term of each member of (4) is equal to zero. From (2) we get

$$\mathfrak{x}_r\mathfrak{x}_s = 0 \qquad (r \neq s)$$

and

$$\mathfrak{x}_r^2 - \mathfrak{x}_s^2 = 0.$$

Therefore, the first two terms of the second member of (4) are each equal to zero. It follows that, under the hypotheses of Theorem 1, the identity (4) reduces to

$$(n - 2) \sum_{r,s} \mathfrak{x}_{rs}^2 = 0.$$

Since $n \neq 2$, we then have

$$\sum_{r,s} \mathfrak{x}_{rs}^2 \equiv \sum_{i=1}^{n} \sum_{r,s} \left(\frac{\partial^2 x^i}{\partial u^r \partial u^s}\right)^2 = 0. \tag{7}$$

Since the sum of real squares in (7) is equal to zero, each term is equal to zero; that is, all of the second partial derivatives of the functions (3) vanish identically in D.

It follows from the extended law of the mean that the functions (3) must be constants or linear, and the theorem is proved.

We note in passing that the weaker hypotheses, (2) and

$$\sum_r \mathfrak{x}_{rr} = \text{const.}, \tag{8}$$

or even (2) and

$$\sum_{r,s} \mathfrak{x}_s\mathfrak{x}_{rrs} = 0, \tag{9}$$

imply, in accordance with (4), the linearity of the functions (3) for $n \geq 3$. But for $n = 2$, (2) and either (8) or (9) imply only that the functions (1) are harmonic; that is, the variety is a minimal surface in conformal representation.

By Theorem 1, the only sets of conjugate harmonic functions for $n \geq 3$ are sets of constants or linear functions,

$$x^i = a^i + \sum_{j=1}^{n} a_j^i u^j \qquad (i = 1, 2, \cdots, m),$$

where a^i and a_j^i are constants, satisfying

$$\sum_{i=1}^{m} a_j^i a_k^i = c\delta_{j,k} \qquad (j, k = 1, 2, \cdots, n).$$

The map of D on V_n by such functions is accordingly the product of rigid motions, transformations of similitude, and reflections in hyperplanes. The following result is then an immediate consequence of Theorem 1 and the above extended theorem of Weierstrass.

THEOREM 2. *A necessary and sufficient condition that a minimal variety* V_n, *with* $n \geq 3$, *in Euclidean m-space* E_m *can be represented conformally on a domain* D *in Euclidean n-space is that* V_n *be contained in a Euclidean subspace* E_n *of* E_m.

Characteristic properties [1] of minimal varieties V_n in E_m, in conformal representation on domains D in E_n, for $m \geq n$ and for $m = n$, accordingly yield the following result.

THEOREM 3. *A necessary and sufficient condition that the functions*

$$x^i = x^i(u^1, u^2, \cdots, u^n) \qquad (i = 1, 2, \cdots, m; m \geq n \geq 3),$$

defined and continuous in a domain D, *be constant or linear functions,*

$$x^i = a^i + \sum_{j=1}^{n} a^i_j u^j,$$

satisfying

$$\sum_{i=1}^{m} a^i_j a^i_k = c\delta_{j,k},$$

is that for all real values of the parameters c^i *the function*

$$\left[\sum_{i=1}^{m} (x^i + c^i)^2\right]^{\frac{1}{2}(2-n)}$$

be harmonic if $m = n$ *or be subharmonic if* $m \geq n$.

BIBLIOGRAPHY

1. E. F. BECKENBACH, *Conjugate harmonic functions*, this Journal, vol. 10(1943), pp. 335–339.
2. L. BIANCHI, *Lezioni di geometria differenziale*, vol. 1, Pisa, 1902, pp. 375–376.
3. N. CIORANESCO, *Sur les fonctions harmoniques conjuguées*, Bulletin des Sciences Mathématiques, vol. 56(1932), pp. 55–64.
4. J. HAANTJES, *Eine Charakterisierung der konformeuklidischen Räume*, Koninklijke Nederlandsche Akademie van Wetenschappen, Proceedings, vol. 43(1940), pp. 91–94.

THE UNIVERSITY OF TEXAS.

ON GENERALIZED CONVEX FUNCTIONS

BY

E. F. BECKENBACH AND R. H. BING

1. **Introduction.** Let $\{F(x)\}$ be a family of real functions defined in the interval $a<x<b$ and satisfying the following conditions:

(1) Each $F(x)$ is a continuous function of x in $a<x<b$.

(2) There is a unique member of the family which, at arbitrary x_1, x_2 satisfying $a<x_1<x_2<b$, takes on arbitrary values y_1, y_2 respectively.

By $F(x; x_1, y_1; x_2, y_2)$ we shall denote the member of $\{F(x)\}$ which takes on values y_1, y_2 at x_1, x_2, respectively. For a given function $f(x)$ we shall for brevity denote $F[x; x_i, f(x_i); x_j, f(x_j)]$ by $F_{i,j}(x)$. We shall further denote the interval $a<x<b$ by (a, b).

Definition. The real function $f(x)$, defined in (a, b), is a *sub-$F(x)$ function* provided the inequality

$$f\left(\frac{x_1+x_2}{2}\right) \leqq F_{1,2}\left(\frac{x_1+x_2}{2}\right) \tag{3}$$

holds for all x_1, x_2 satisfying $a<x_1<x_2<b$.

If the functions $F(x)$ are linear, then the class of sub-$F(x)$ functions is the class of convex functions. Since not all convex functions are continuous [4](1), it follows that (1), (2) and (3) do not imply that $f(x)$ is continuous. For a discussion of convex functions, see [2, 5, 6, 8].

We note that the above definition of a sub-$F(x)$ function is more general than that previously given [1], in which (3) was replaced by

$$f(x) \leqq F_{1,2}(x) \tag{4}$$

for all x_1, x_2, x satisfying $a<x_1<x<x_2<b$. Functions $f(x)$ satisfying (4) necessarily are continuous (Corollary 4, below); and for continuous functions $f(x)$, conditions (3) and (4) are equivalent (Lemma 3, below).

2. **The family** $\{F(x)\}$. We shall have need for the first two results given in this section. The third is included to show that for some families $\{F(x)\}$ satisfying (1) and (2) the class of sub-$F(x)$ functions is not topologically equivalent to the class of convex functions.

Lemma 1 [1]. *For a given x_0 with $a<x_0<b$, let $F_r(x)$ and $F_s(x)$ be two members of a family $\{F(x)\}$ satisfying* (1) *and* (2), *such that*

$$F_r(x_0) = F_s(x_0), \tag{5}$$

Presented to the Society, April 29, 1944; received by the editors June 12, 1944.

(1) Numbers in brackets refer to the references cited at the end of the paper.

$$F_r(x) \not\equiv F_s(x) \qquad (a < x < b); \tag{6}$$

then $F_r(x) < F_s(x)$ *for all* x *in* (a, b) *on one side of* x_0, *while* $F_r(x) > F_s(x)$ *for all* x *in* (a, b) *on the other side of* x_0.

Proof. By (2), (5) and (6), $F_r(x) \neq F_s(x)$ in (a, b), except at x_0; consequently, by (1), on either side of x_0, one of $F_r(x)$, $F_s(x)$ is greater than the other. Suppose it could be the same one, say $F_s(x)$, which is greater on each side; we shall show that this is impossible.

Let x_1, x_2 satisfy $a < x_1 < x_0 < x_2 < b$, and consider the member $F_t(x)$ of $\{F(x)\}$, determined by $F_t(x) \equiv F[x;\ x_1,\ F_s(x_1);\ x_2,\ F_r(x_2)]$. We have $F_t(x_2) < F_s(x_2)$, so that $F_t(x) < F_s(x)$, $x_1 < x < b$; in particular,

$$F_t(x_0) < F_s(x_0). \tag{7}$$

Similarly,

$$F_t(x_0) > F_r(x_0). \tag{8}$$

Now (7) and (8) contradict (5).

Lemma 2. *Let* x_0 *satisfy* $a < x_0 < b$ *and let* E *be a closed and bounded set of numbers in* (a, x_0) *(or in* (x_0, b)*); then for any real* y_0, M *there is a member* $F^*(x)$ *of* $\{F(x)\}$ *satisfying* $F^*(x_0) = y_0$ *and* $F^*(x) > M$ *for all* x *in* E.

Proof. For each x' in E there is an $h = h(x')$ such that if $I = I(x')$ denotes the interval $(x' - h, x' + h)$, we have

$$F(x;\ x_0,\ y_0;\ x',\ 2M) > M$$

for all x in I. By the Heine-Borel theorem there is a finite set $x_1', x_2', \cdots, x_n'$ of numbers in E such that each number in E is contained in at least one of $I(x_1'), I(x_2'), \cdots, I(x_n')$; therefore there is a finite set

$$R:\quad F_1(x), F_2(x), \cdots, F_n(x),$$

of members of $\{F(x)\}$ such that for each x' in E, we have

$$\max\,[F_1(x'), F_2(x'), \cdots, F_n(x')] > M.$$

By Lemma 1, some member of R satisfies the conditions specified for $F^*(x)$.

Remark [7]. *Not all families* $\{F(x)\}$ *satisfying* (1) *and* (2) *are topologically equivalent to the family* $\{L(x)\}$ *of all nonvertical line segments terminating on* $x = a$, $x = b$.

Proof. Let the family $\{F(x)\}_0$ be determined for $-\infty < x < +\infty$ as follows: A function $F(x)$ is a member of $\{F(x)\}_0$ if and only if its graph in Cartesian coordinates is (i) a line of nonpositive slope or (ii) a continuous broken line, with break on the x-axis, such that the slope is positive for $y < 0$ and half as much for $y > 0$.

The family $\{F(x)\}_0$ satisfies conditions (1) and (2) in any interval (a, b).

Let ABC and $A'B'C'$ be triangles in the half-strip $a<x<b$, $y<0$, such that the lines AA', BB', CC' intersect in a point P of $a<x<b$, $y>0$; the slopes of AA' and BB' are negative while that of CC' is positive; and (see Desargues' Theorem [3]) the corresponding sides of the triangles intersect in collinear points in $a<x<b$, $y<0$.

Let $F_A(x)$, $F_B(x)$, $F_C(x)$ be the members of $\{F(x)\}_0$ whose graphs contain A, A'; B, B'; C, C', respectively. Then P is on $F_A(x)$ and on $F_B(x)$, but not on $F_C(x)$.

Suppose there were a reversible continuous transformation of the strip $a<x<b$ into itself carrying $\{F(x)\}_0$ into $\{L(x)\}$. Since a reversible continuous transformation preserves concurrency of graphs, triangles ABC and $A'B'C'$ are carried into triangles whose corresponding sides intersect in collinear points, so that, by the converse of Desargues' Theorem, the lines joining corresponding vertices are parallel or concurrent. This contradicts the fact that P is on $F_A(x)$ and on $F_B(x)$ but not on $F_C(x)$.

Actually we have shown more, that no plane domain D containing a segment of the x-axis can be mapped topologically on a plane domain D' in such a way that the graphs of the functions $\{F(x)\}_0$ in D are mapped on unbroken lines or line-segments in D'.

3. **Continuous sub-$F(x)$ functions.** We shall use the following lemma.

LEMMA 3. *If $f(x)$ is continuous in (a, b), then a necessary and sufficient condition that $f(x)$ be a sub-$F(x)$ function in (a, b) is that we have*

$$f(x) \leqq F_{1,2}(x) \tag{9}$$

for all x_1, x_2, x satisfying $a<x_1<x<x_2<b$.

Necessity. If (9) does not hold, then there exist x_3, x_4, x_5, with $a<x_3<x_5<x_4<b$, such that

$$f(x_5) > F_{3,4}(x_5). \tag{10}$$

By (10) and the continuity of $f(x)$ and of $F_{3,4}(x)$, there exist x_1, x_2, with $x_3 \leqq x_1 < x_5 < x_2 \leqq x_4$, such that

$$f(x) \geqq F_{3,4}(x) \qquad (x_1 \leqq x \leqq x_2), \tag{11}$$

the sign of equality holding at the end points but not elsewhere; then $F_{3,4}(x) \equiv F_{1,2}(x)$, so that (11) can be written as

$$f(x) > F_{1,2}(x) \qquad (x_1 < x < x_2). \tag{12}$$

Now (12) contradicts (3); hence if (9) does not hold, then $f(x)$ is not a sub-$F(x)$ function.

Sufficiency. If (9) holds, then (3) holds as a special case of (9), and consequently $f(x)$ is a sub-$F(x)$ function.

Our argument gives also the following result.

COROLLARY 1. *A necessary and sufficient condition that a function $f(x)$ continuous in (a, b) be a sub-$F(x)$ function is that for each pair of numbers x_1, x_2 satisfying $a<x_1<x_2<b$, there exist an x_3 satisfying $x_1<x_3<x_2$ such that $f(x_3) \leqq F_{1,2}(x_3)$.*

LEMMA 4. *A necessary and sufficient condition that a function $f(x)$ continuous in (a, b) be a sub-$F(x)$ function is that there exist a set E of numbers dense in (a, b) such that for each x_1, x_2 in E we have*

$$f\left(\frac{x_1 + x_2}{2}\right) \leqq F_{1,2}\left(\frac{x_1 + x_2}{2}\right). \tag{13}$$

Necessity. If $f(x)$ is a sub-$F(x)$ function, then the existence of a set E of numbers dense in (a, b) such that (13) holds for each x_1, x_2 in E follows from the definition of a sub-$F(x)$ function.

Sufficiency. As shown in Lemma 3, (12), if $f(x)$ is not a sub-$F(x)$ function, then there exist x_1, x_2 such that $f(x)>F_{1,2}(x)$ $(x_1<x<x_2)$.

Let x_0 satisfy $a<x_0<x_1$ and let x_3 and x_4 divide (x_1, x_2) into three equal parts, $x_3<x_4$. For each x' such that $x_3 \leqq x' \leqq x_4$ there is a y' and an $h=h(x')$ such that if $I=I(x')$ is the interval $(x'-h, x'+h)$, then

$$f(x) > F[x;\, x_0, F_{1,2}(x_0);\, x', y'] > F_{1,2}(x)$$

for all x in I. A finite number of the intervals I covers (x_3, x_4). By Lemma 1, one of the corresponding functions $F[x;\, x_0,\, F_{1,2}(x_0);\, x',\, y']$, which we shall denote by $F^*(x)$, accordingly satisfies the inequalities

$$f(x) > F^*(x) > F_{1,2}(x) \qquad (x_3 \leqq x \leqq x_4). \tag{14}$$

For any set E of numbers dense in (a, b), there are numbers x_5, x_6 in E such that $x_1<x_5<x_3<x_4<x_6<x_2$, for which

$$f(x_5) < F^*(x_5), \qquad f(x_6) < F^*(x_6). \tag{15}$$

Since we have $x_3<(x_5+x_6)/2<x_4$, we find from (14), (15) and Lemma 1 that

$$f\left(\frac{x_5 + x_6}{2}\right) > F^*\left(\frac{x_5 + x_6}{2}\right) > F_{5,6}\left(\frac{x_5 + x_6}{2}\right).$$

Therefore if $f(x)$ is not a sub-$F(x)$ function, there does not exist a set E of numbers dense in (a, b) for which (13) is satisfied.

Actually our method of proof gives more, as stated in the following corollary.

COROLLARY 2. *A necessary and sufficient condition that a function $f(x)$ continuous in (a, b) be a sub-$F(x)$ function is that there exist a positive number ϵ*

and a set E of numbers dense in (a, b) *such that for each pair of numbers* x_1, x_2 *in E satisfying* $a<x_1<x_2<b$, *there is an* x_3 *in E such that*

$$x_1 + \epsilon(x_2 - x_1) < x_3 < x_2 - \epsilon(x_2 - x_1),$$

for which $f(x_3) \leqq F_{1,2}(x_3)$.

As stated in Corollary 1, if the function $f(x)$ is continuous in (a, b) and if for each pair of numbers x_1, x_2 satisfying $a<x_1<x_2<b$ there is an x_3 satisfying $x_1<x_3<x_2$, such that $f(x_3) \leqq F_{1,2}(x_3)$, then $f(x)$ necessarily is a sub-$F(x)$ function. However, there does exist a function $f(x)$ continuous in (a, b), and a set E dense in (a, b), such that for each pair of numbers x_1, x_2 in E satisfying $a<x_1<x_2<b$, there is an x_3 in E satisfying $x_1<x_3<x_2$, such that $f(x_3) \leqq F_{1,2}(x_3)$, and such that $f(x)$ is not a sub-$F(x)$ function. For let $(a, b)=(0, 1)$, and define $f(x)$ as follows:

$$\begin{aligned} f(x) &= 1/2 && (1/3 \leqq x \leqq 2/3),\\ f(x) &= 1/4 && (1/9 \leqq x \leqq 2/9),\\ f(x) &= 3/4 && (7/9 \leqq x \leqq 8/9),\\ &\cdots\cdots && \cdots\cdots\cdots, \end{aligned}$$

$$f(x) \text{ continuous in } (0, 1).$$

If E is the set of all numbers x satisfying

$$1/3 < x < 2/3, \text{ or } 1/9 < x < 2/9, \text{ or } 7/9 < x < 8/9, \text{ or } \cdots,$$

and $\{F(x)\}$ is the set of all nonvertical straight line segments in the strip $0<x<1$, then $f(x)$ satisfies the conditions set forth above.

THEOREM 1. *If* $f(x)$ *is a sub-*$F(x)$ *function in* (a, b), *then for any fixed* x_1, x_2 *satisfying* $a<x_1<x_2<b$, *and for all* ρ *in* (a, b) *dividing* (x_1, x_2) *rationally (internally or externally), the points* $[\rho, f(\rho)]$ *lie on the graph of a uniquely determined continuous sub-*$F(x)$ *function.*

Proof. For fixed x_1, x_2 with $a<x_1<x_2<b$, let $F_s(x)$ be a member of $\{F(x)\}$ satisfying $F_s(x_1) \geqq f(x_1)$, $F_s(x_2) \geqq f(x_2)$; then for all non-negative integers m, n such that $m/2^n \leqq 1$, we have

$$f\left[x_1 + \frac{m}{2^n}(x_2 - x_1)\right] \leqq F_s\left[x_1 + \frac{m}{2^n}(x_2 - x_1)\right]. \tag{16}$$

For, by Lemma 1, we have

$$F_{1,2}(x) \leqq F_s(x) \qquad (x_1 \leqq x \leqq x_2). \tag{17}$$

From (3) and (17) it follows that we have

$$f\left(x_1 + \frac{x_2 - x_1}{2}\right) \leqq F_{1,2}\left(x_1 + \frac{x_2 - x_1}{2}\right) \leqq F_s\left(x_1 + \frac{x_2 - x_1}{2}\right).$$

The inequality (16) follows by induction.

Next, we shall show that, for the above x_1, x_2, if $F_r(x)$ is a member of $\{F(x)\}$ satisfying

$$F_r(x_1) \geqq f(x_1), \qquad F_r(x_2) \leqq f(x_2),$$

then for all positive integers m, n such that

$$\frac{m}{2^n} \geqq 1, \qquad x_1 + \frac{m}{2^n}(x_2 - x_1) < b,$$

we have

$$f\left[x_1 + \frac{m}{2^n}(x_2 - x_1)\right] \geqq F_r\left[x_1 + \frac{m}{2^n}(x_2 - x_1)\right]. \tag{18}$$

From Lemma 1 we obtain

$$F_r(x) \leqq F_{1,2}(x) \qquad (x_2 \leqq x < b). \tag{19}$$

For any positive integers m, n such that $1 \leqq m/2^n \leqq 2$, there is a non-negative integer m' such that $(m+m')/2^{n+1} = 1$. Since x_2 is the average of the numbers

$$x_1 + \frac{m'}{2^n}(x_2 - x_1), \qquad x_1 + \frac{m}{2^n}(x_2 - x_1),$$

and since from (16) we have

$$f\left[x_1 + \frac{m'}{2^n}(x_2 - x_1)\right] \leqq F_{1,2}\left[x_1 + \frac{m}{2^n}(x_2 - x_1)\right],$$

it follows from (3), (19) and Lemma 1 that if $x_1 + m(x_2 - x_1)/2^n$ is in (a, b), then

$$f\left[x_1 + \frac{m}{2^n}(x_2 - x_1)\right] \geqq F_{1,2}\left[x_1 + \frac{m}{2^n}(x_2 - x_1)\right] \geqq F_r\left[x_1 + \frac{m}{2^n}(x_2 - x_1)\right].$$

By induction we have (18) also for $m/2^n > 2$, provided only that $x_1 + m(x_2 - x_1)/2^n$ is in (a, b).

Likewise, we find that if $F_t(x)$ is a member of $\{F(x)\}$ satisfying

$$F_t(x_1) \leqq f(x_1), \qquad F_t(x_2) \geqq f(x_2),$$

then for all non-negative integers m, n such that

$$x_1 - \frac{m}{2^n}(x_2 - x_1) > a,$$

we have

$$f\left[x_1 - \frac{m}{2^n}(x_2 - x_1)\right] \geqq F_i\left[x_1 - \frac{m}{2^n}(x_2 - x_1)\right]. \tag{20}$$

Let j be a positive integer sufficiently large so that the number

$$x_0 = x_1 - \frac{1}{2^j}(x_2 - x_1)$$

satisfies $a < x_0 < x_1$. By (16) and (18), for all non-negative integers m, n such that $m/2^n \leqq 1$, we have

$$\begin{aligned} F_{1,2}\left[x_1 + \frac{m}{2^n}(x_2 - x_1)\right] &\geqq f\left[x_1 + \frac{m}{2^n}(x_2 - x_1)\right] \\ &\geqq F_{0,1}\left[x_1 + \frac{m}{2^n}(x_2 - x_1)\right]. \end{aligned} \tag{21}$$

For numbers $x = x_1 + (m/2^n)(x_2 - x_1)$, where m, n are non-negative integers such that $m/2^n \leqq 1$, designate by W the set of points $[x, f(x)]$. The boundedness of W follows from (1) and (21). Since the set of abscissas of points of W is dense in (x_1, x_2), either the closure $\overline{W}$ of W is a continuous curve which is the graph of a single-valued function $y = \overline{W}(x)$ continuous in (x_1, x_2), or else there exists an x_3 satisfying $x_1 < x_3 < x_2$, and numbers y_1, y_2 with $y_1 < y_2$, such that the points with coordinates (x_3, y_1) and (x_3, y_2) belong to $\overline{W}$.

Assume that there do exist such x_3, y_1, y_2. Consider members $F^*(x)$ and $F^{**}(x)$ of $\{F(x)\}$ determined by

$$F^*(x_1) = f(x_1), \qquad F^*(x_3) = F^{**}(x_3) = \frac{y_1 + y_2}{2}, \qquad F^{**}(x_2) = f(x_2).$$

Since (x_3, y_2) is a point of $\overline{W}$, it follows from the continuity of $F^*(x)$ and of $F^{**}(x)$, and from (18) and Lemma 1, that there are positive integers j, k, with $j/2^k < 1$, such that the number $x_4 = x_1 + (j/2^k)(x_2 - x_1)$ is distinct from x_3 and satisfies

$$f(x_4) > F^*(x_4), \qquad f(x_4) > F^{**}(x_4). \tag{22}$$

If $x_4 > x_3$, let

$$x_5 = x_1 + \frac{j+1}{2^k}(x_2 - x_1).$$

From (20) we obtain $f(x) \geqq F_{4,5}(x)$ for all x less than x_4 satisfying

$$x = x_1 + \frac{m}{2^n}(x_2 - x_1),$$

where m, n are positive integers. Since the point (x_3, y_1) of $\overline{W}$ has ordinate $y_1 < F^{**}(x_3)$, it follows from the continuity of $F^{**}(x)$ and of $F_{4,5}(x)$ that for some number x_6 satisfying $x_3 < x_6 < x_4$, we have

$$F^{**}(x_6) = F_{4,5}(x_6). \tag{23}$$

But from (18) and the definition of $F^{**}(x)$, we obtain $F^{**}(x_2) \geqq F_{4,5}(x_2)$, so that there is a number $x_7 > x_4$ such that

$$F^{**}(x_7) = F_{4,5}(x_7). \tag{24}$$

Now (22), (23) and (24) contradict (2), whence the continuity of $\overline{W}(x)$ in the interval $x_1 < x < x_2$ is proved on the assumption that $x_4 > x_3$.

Considering $F^*(x)$ instead of $F^{**}(x)$, we can modify our argument so as to prove $\overline{W}(x)$ continuous in the interval $x_1 < x < x_2$ on the assumption that $x_4 < x_3$.

By considering suitable x_1', x_2', with $a < x_1' < x_1 < x_2 < x_2' < b$, we see that the range of $\overline{W}(x)$ can be extended so that $\overline{W}(x)$ remains continuous in the closed interval $x_1 \leqq x \leqq x_2$.

We shall show that for any positive rational number $p/q < 1$ we have

$$f(x) = \overline{W}(x) \qquad \left[x = x_1 + \frac{p}{q}(x_2 - x_1)\right]. \tag{25}$$

Assume that there is a rational number p_0/q_0 with $0 < p_0/q_0 < 1$, such that

$$f(x_0) \neq \overline{W}(x_0) \qquad \left[x_0 = x_1 + \frac{p_0}{q_0}(x_2 - x_1)\right]. \tag{26}$$

By the method of the last paragraph, there is a function $G(x)$ continuous in the interval $x_1 \leqq x \leqq x_0$ such that for all non-negative integers m, n with $m/2^n \leqq 1$ we have

$$f(x) = G(x) \qquad \left[x = x_1 + \frac{m}{2^n}(x_0 - x_1)\right]. \tag{27}$$

From the continuity of $\overline{W}(x)$ and of $G(x)$, and from (26) and (27), it follows that there is a positive h such that

$$G(x) \neq \overline{W}(x) \qquad (x_0 - h < x \leqq x_0). \tag{28}$$

There exist positive integers m, n, m', n' such that

$$\frac{m}{2^n} < 1, \qquad \frac{m}{2^n}\left(\frac{p_0}{q_0}\right) = \frac{m'}{2^{n'}}, \qquad \frac{m}{2^n}(x_0 - x_1) > x_0 - x_1 - h.$$

But for

$$x_8 = x_1 + \frac{m'}{2^{n'}}(x_2 - x_1) = x_1 + \frac{m}{2^n}(x_0 - x_1),$$

we have

$$x_0 - h < x_1 + \frac{m}{2^n}(x_0 - x_1) = x_8 < x_0 \tag{29}$$

and

$$G(x_8) = f(x_8) = \overline{W}(x_8). \tag{30}$$

Since (29) and (30) contradict (28), (25) follows.

Since $f(x)$ is a sub-$F(x)$ function and since $\overline{W}(x)$ is continuous in the interval $x_1 < x < x_2$, it follows from (25) and Lemma 4 that $\overline{W}(x)$ is a sub-$F(x)$ function in (x_1, x_2).

By considering suitable x_1', x_2', with $a < x_1' < x_1 < x_2 < x_2' < b$, we see that the range of $\overline{W}(x)$ can be extended so that $\overline{W}(x)$ remains continuous in (a, b) and satisfies the conditions of Theorem 1.

4. **Bounded sub-$F(x)$ functions.** We shall use the theorem proved in this section in establishing a more general result in §5.

LEMMA 5. *If $f(x)$ is a sub-$F(x)$ function in the interval (a, b), and is bounded from above in an interval (x_1, x_2) with $a < x_1 < x_2 < b$, then:*

(i) For any x_3 with $a < x_3 < x_1$ there is an $F^*(x)$ such that $F^*(x_3) = f(x_3)$ and such that $F^*(x) \geqq f(x)$ for all x satisfying $x_3 \leqq x \leqq x_2$.

(ii) For any x_4 with $x_2 < x_4 < b$ there is an $F^{**}(x)$ such that $F^{**}(x_4) = f(x_4)$ and such that $F^{**}(x) \geqq f(x)$ for all x satisfying $x_1 \leqq x \leqq x_4$.

Proof. By Lemma 2 there is an $F^*(x)$ satisfying $F^*(x_3) = f(x_3)$, such that $F^*(x) > f(x)$ for all x satisfying $x_1 \leqq x \leqq x_2$. For any x' in (x_3, x_1) there is an x'' in (x_1, x_2) such that x' divides (x_3, x'') rationally. Therefore, by Theorem 1 and Lemmas 1 and 3, we have $F^*(x) \geqq f(x)$ for all x satisfying $x_3 \leqq x \leqq x_2$.

In like manner, it can be shown that there is an $F^{**}(x)$ satisfying the second part of Lemma 5.

The following corollary follows from Lemma 5 and the continuity of the members of $\{F(x)\}$.

COROLLARY 3. *If $f(x)$ is a sub-$F(x)$ function in the interval (a, b), and is bounded from above in some subinterval of (a, b), then $f(x)$ is bounded from above in every closed subinterval of (a, b).*

THEOREM 2. *If $f(x)$ is a sub-$F(x)$ function in the interval (a, b), and is bounded from above in some subinterval of (a, b), then $f(x)$ is continuous in (a, b).*

Proof. It will be shown that $f(x)$ is continuous at an arbitrary point x_0 of (a, b).

If I: $x_1 \leqq x \leqq x_2$ is a closed subinterval of (a, x_0), then by Corollary 3, $f(x)$ has an upper bound in I. By Lemma 5, there is an $F^*(x)$ such that

$$F^*(x_0) = f(x_0), \qquad F^*(x) \geqq f(x) \qquad (x_1 \leqq x \leqq x_0). \tag{31}$$

For any x' in (x_0, b), there is an x'' in (x_1, x_0) such that x_0 divides (x'', x') rationally. By Theorem 1 and Lemma 3 we have

$$F^*(x) \leqq f(x \qquad (x_0 \leqq x < b). \tag{32}$$

Similarly, there is an x_3 in (x_0, b) and an $F^{**}(x)$ satisfying

$$F^{**}(x_0) = f(x_0), \qquad F^{**}(x) \geqq f(x) \qquad (x_0 \leqq x \leqq x_3), \tag{33}$$

and

$$F^{**}(x) \leqq f(x) \qquad (a < x \leqq x_0). \tag{34}$$

From (31)–(34) and the continuity of the members of $\{F(x)\}$, the continuity of $f(x)$ at x_0 follows. Since x_0 is arbitrary, $a<x_0<b$, the continuity of $f(x)$ in (a, b) is established.

Corollary 4. *If $f(x)$ is defined in (a, b) and satisfies $f(x) \leqq F_{1,2}(x)$ for all x_1, x_2, x with $a<x_1<x<x_2<b$, then $f(x)$ is continuous in (a, b).*

There is a direct proof [1] of Corollary 4 which is simpler than the above proof of Theorem ?.

5. **Measurable sub-$F(x)$ functions.** We shall use Theorem 2 and the following lemma to obtain a result which is more general than Theorem 2.

Lemma 6. *If $f(x)$ is a sub-$F(x)$ function in (a, b) and is bounded from above on a set E of positive measure in (a, b), then $f(x)$ is bounded from above in some subinterval of (a, b).*

Proof. There is a closed subinterval $x_1 \leqq x \leqq x_2$ of (a, b) such that the measure of the part of E in (x_1, x_2) is greater than $2(x_2-x_1)/3$. Let x_3 and x_4, with $x_3<x_4$, divide (x_1, x_2) into three equal parts, and let x' be any value satisfying $x_3 \leqq x' \leqq x_4$. Since the measure of the part of E in (x_1, x_2) is greater than $2(x_2-x_1)/3$, it follows that the measure of the part of E in

$$[x' - (x_4 - x_3),\ x' + (x_4 - x_3)]$$

is greater than x_4-x_3; consequently, there is a $\delta(x')$ less than x_4-x_3 such that $x'-\delta(x')$ and $x'+\delta(x')$ both belong to E.

By assumption there is an M such that we have $f(x)<M$ for all x in E. By Lemma 2, there is an $F^*(x)$ such that $F^*(x)>M$ for all x satisfying $x_1 \leqq x \leqq x_2$. From Lemma 1 and the fact that $f(x)$ is a sub-$F(x)$ function, it follows that

$$F^*(x) > F[x;\, x - \delta(x), f(x - \delta(x));\, x + \delta(x), f(x + \delta(x))] \geqq f(x) \qquad (x_3 \leqq x \leqq x_4).$$

Since $F^*(x)$ is continuous, we conclude that $f(x)$ is bounded, $x_3 \leqq x \leqq x_4$.

THEOREM 3. *If $f(x)$ is a sub-$F(x)$ function in (a, b), and is bounded from above on a set E of positive interior measure in (a, b), then $f(x)$ is continuous in (a, b).*

Proof. Let E_0 be a closed subset of E having positive measure. By Lemma 6 there is a subinterval of (a, b) on which $f(x)$ is bounded from above, so that the continuity of $f(x)$ in (a, b) follows from Theorem 2.

REFERENCES

1. E. F. Beckenbach, *Generalized convex functions*, Bull. Amer. Math. Soc. vol. 43 (1937) pp. 363–371.

2. H. Blumberg, *On convex functions*, Trans. Amer. Math. Soc. vol. 20 (1919) pp. 40–44.

3. W. C. Graustein, *Introduction to higher geometry*, New York, 1930.

4. G. Hamel, *Eine Basis aller Zahlen und die unstetigen Lösungen der Funktionalgleichung* $f(x+y)=f(x)+f(y)$, Math. Ann. vol. 60 (1905) pp. 459–462.

5. G. H. Hardy, J. E. Littlewood, G. Pólya, *Inequalities*, Cambridge, England, 1934.

6. J. L. W. V. Jensen, *Sur les fonctions convexes et les inégalités entre les valeurs moyennes*, Acta Math. vol. 30 (1906) pp. 175–193.

7. F. R. Moulton, *A simple non-desarguesian plane geometry*, Trans. Amer. Math. Soc. vol. 3 (1902) pp. 192–195.

8. W. Sierpiński, *Sur les fonctions convexes mesurables*, Fund. Math. vol. 1 (1920) pp. 125–129.

THE UNIVERSITY OF TEXAS

CONVERSE LINEARITY CONDITIONS.*

By R. H. BING.

1. Introduction. A point is a *between-point* of the plane set M if it is between two points of M; it is a *middle-point* of M if it bisects some interval having its end points on M. The function $f(x)$ satisfies the *between-point condition* on the set E of x's if for each x_0 of E, the point of the graph of $f(x)$ having x_0 as an abscissa is a between-point of the graph of $f(x)$; $f(x)$ satisfies the *middle-point condition* on E if for each x_0 of E, the point of the graph of $f(x)$ having x_0 as an abscissa is a middle-point of the graph of $f(x)$.

Examples of functions satisfying the middle-point condition on their ranges are

$$f(x) = mx + k \ (a < x < b) \quad \text{and} \quad f(x) = \sin x \ (-\infty < x < \infty).$$

It has been shown [2, p. 253; also see 1] that a continuous function $f(x)$ $(a \leq x \leq b)$ is linear if it satisfies the middle-point condition for values of x between a and b. Theorem 1 is a modification of this result.

THEOREM 1. *The continuous function $f(x)$ $(a \leq x \leq b)$ is linear if it satisfies the between-point condition for values of x between a and b.*

Proof. Letting $F(x)$ denote the function whose graph is the interval having end points at $[a, f(a)]$ and $[b, f(b)]$, we consider

$$u(x) = f(x) - F(x) \qquad (a \leq x \leq b). \tag{1}$$

We note that $u(x)$ satisfies the between-point condition for values of x between a and b and that $u(a) = u(b) = 0$. As $u(x)$ is continuous and has a closed range, it takes on a least upper bound on a closed subset E of its range. If x_0 is the maximum element of E and if $u(x_0) \neq 0$, then $u(x)$ does not satisfy the between-point condition at x_0. Therefore

$$u(x) \leq 0 \qquad (a \leq x \leq b). \tag{2}$$

Considering the greatest lower bound of $u(x)$, we find that

$$u(x) \geq 0 \qquad (a \leq x \leq b). \tag{3}$$

* Received February 11, 1946. Presented to the American Mathematical Society, April 27, 1946.

From (2) and (3) we have that

$$u(x) = 0 \qquad (a \leq x \leq b) \tag{4}$$

and from (1) and (4) that $f(x)$ is linear.

THEOREM 2. *The continuous function* $f(x)$ $(a < x < b)$ *is linear if* $\int_a^b f(x)dx$ *exists and for each* x_0 *between* a *and* b *there are an* $h = h(x_0)$ *and a* $k = k(x_0)$ *such that*

$$k^2[hf(x_0) - \int_{x_0-h}^{x_0} f(x)dx] = h^2[\int_{x_0}^{x_0+k} f(x)dx - kf(x_0)] \tag{5}$$
$$(a \leq x_0 - h < x_0 < x_0 + k \leq b).$$

Proof. First, we shall show that there is a continuous function $f'(x)$ $(a \leq x \leq b)$ such that $f'(x) \equiv f(x)$ $(a < x < b)$.

Assume that (b, y_1) and (b, y_2) are limit points of the graph of $f(x)$. Suppose that c and d are values such that $a < c < d < b$. Since $\int_a^b f(x)dx$ exists, $\int_x^d f(y)dy$ $(a \leq x \leq d)$ is continuous and takes on a maximum M and a minimum m. Let $L(x)$ and $R(x)$ $(a \leq x \leq b)$ be continuous linear functions such that $L(b) = R(b) = (y_1 + y_2)/2$, $L(x) > f(x) > R(x)$. $(c < x < d)$, $L(x) > 0 > R(x)$ $(a < x < c)$, $\int_c^d L(x)dx > M$ and $m > \int_c^d R(x)dx$. Since $L(x) > f(x)$ $(c < x < d)$ and $\int_x^d L(y)dy > M$ $(a \leq x \leq c)$, it follows that

$$\int_x^d [f(y) - L(y)]dy < 0 \qquad (a \leq x < d). \tag{6}$$

Let r and s be values between d and b such that $r < s$, $f(r) > L(r)$ and $f(s) < R(s)$. Denote by x' the largest value for which $f(x) - L(x)$ $(d \leq x \leq s)$ assumes a maximum. Assume that there are values h and k such that (5) is satisfied for x' equal to x_0. We note that if $f(x)$ is linear, then (5) holds for all h and k such that $a \leq x_0 - h < x_0 < x_0 + k \leq b$. Hence,

$$k^2\{h[f(x') - L(x')] - \int_{x'-h}^{x'} [f(x) - L(x)]dx\} \tag{7}$$
$$= h^2\{\int_{x'}^{x'+k} [f(x) - L(x)]dx - k[f(x') - L(x')]\}.$$

From (6) and the fact that $f(x') - L(x')$ is a positive maximum for $f(x) - L(x)$ $(d \leqq x \leqq s)$, it follows that the first member of (7) is not negative. Therefore,

$$\int_{x'}^{x'+k} [f(x) - L(x)]dx \geqq k[f(x') - L(x')]. \tag{8}$$

Since x' is the largest value for which $f(x) - L(x)$ $(d \leqq x \leqq s)$ assumes a maximum, it follows from (8) that $x' + k > s$ and $\int_s^{x'+k} f(x)dx > 0$.

Let E be the set of all values x such that $s \leqq x \leqq b$ and $\int_s^x [f(y) - L(y)]dy \geqq 0$. By the methods of the last paragraph we find that b is either a point or a limit point of E. Since $\int_a^b f(x)dx$ exists, it follows that $\int_s^b [f(x) - L(x)]dx \geqq 0$. A similar argument leads to the contradiction that $\int_s^b [f(x) - R(x)]dx \leqq 0$.

Assume that $f(x) \to \infty$ as $x \to b$. There is a continuous linear function $G(x)$ such that $f(x) > G(x)$ $(c < x < d)$, $G(x) < 0$ $(a < x < d)$, $m > \int_c^d G(x)dx$ and there is a value t between d and b for which $f(t) < G(t)$. Let x' be the largest value for which $f(x) - G(x)$ $(d \leqq x < b)$ assumes its minimum. But there are not an h and a k such that (5) is satisfied for x' equal to x_0. Likewise, the assumption that $f(x) \to -\infty$ as $x \to b$ leads to a similar contradiction. Hence, $f(x)$ may be defined so as to be continuous at b. Also, it may be defined so as to be continuous at a. Therefore, there is a continuous function $f'(x)$ $(a \leqq x \leqq b)$ such that $f'(x) \equiv f(x)$ $(a < x < b)$.

Let $F(x)$ $(a \leqq x \leqq b)$ be a continuous linear function such that $f'(a) = F(a)$ and $f'(b) = F(b)$. Assume that $f'(x) - F(x)$ is positive for some value between a and b. Let x' be the largest value for which $f'(x) - F(x)$ takes on its maximum. But there are not an h and a k such that (5) is satisfied for x' equal to x_0. Also, the assumption that $f'(x) - F(x)$ is negative for some value between a and b leads to a contradiction.

COROLLARY. *The continuous function $f(x)$ $(a < x < b)$ is linear if $\int_a^b f(x)dx$ exists and if for each x_0 between a and b there is an $h = h(x_0)$ such that*

$$f(x_0) = (1/2h) \int_{x_0-h}^{x_0+h} f(x)dx \qquad (a \leqq x_0 - h < x_0 + h \leqq b).$$

2. Middle-points of graphs. We shall use Theorem 3 in the next section. Theorem 4 is included because it is of interest for itself.

THEOREM 3. *If (x_0, y_1) and (x_0, y_2) are middle-points of the graph of the continuous function $f(x)$ $(a < x < b)$ and if $y_1 < y_0 < y_2$, then (x_0, y_0) is a middle-point of the graph of $f(x)$.*

Proof. Let $u_1(x)$, $u_0(x)$ and $u_2(x)$ be functions whose graphs are the images of the graph of $f(x)$ under a rotation of an angle of π radians about (x_0, y_1), (x_0, y_0) and (x_0, y_2) respectively. As $y_1 < y_0 < y_2$, we have

$$(9) \qquad u_1(x) < u_0(x) < u_2(x) \qquad (2x_0 - b < x < 2x_0 - a).$$

Since (x_0, y_1) and (x_0, y_2) are middle-points of the graph of $f(x)$, there are values x_1, x_2 such that for $i = 1, 2$ we have

$$(10) \qquad u_i(x_i) = f(x_i) \qquad [x_0 < x_i < \min(b, 2x_0 - a)].$$

From (9) and (10) it follows that

$$(11) \qquad u_0(x_1) > f(x_1) \quad \text{and} \quad u_0(x_2) < f(x_2).$$

From (11) and the continuity of $f(x)$ and $u_0(x)$, it follows that there is a number x_3 between x_1 and x_2 such that $u_0(x_3) = f(x_3)$. Then (x_0, y_0) bisects the interval from $[x_3, f(x_3)]$ to $[2x_0 - x_3, f(2x_0 - x_3)]$ and is therefore a middle-point of the graph of $f(x)$.

THEOREM 4. *If the continuous function $f(x)$ $(a < x < b)$ satisfies the middle-point condition on its range and for no value c between a and b is either $f(x)$ $(a < x < c)$ or $f(x)$ $(c < x < b)$ linear, then the set of all middle-points of the graph of $f(x)$ consists of the sum of two connected domains plus a subset of their boundaries.*

Proof. The existence of a continuous function $f(x)$ $(a < x < b)$ satisfying the middle-point condition and being nonlinear on every connected subset of its range will be shown in the next section.

First, it will be shown that if (x_0, y_1) and (x_0, y_2) are middle-points of the graph of $f(x)$ and if $y_1 < y_0 < y_2$, then there is a positive number ϵ such that if $|\delta| < \epsilon$, $(x_0 + \delta, y_0)$ is a middle-point of the graph of $f(x)$. Employing the notation and the methods used in the proof of Theorem 3, we have that there are an x_1 and an x_2 such that

$$u_0(x_1) > f(x_1) \qquad [x_0 < x_1 < \min(b, 2x_0 - a)]$$

and

$$u_0(x_2) < f(x_2) \qquad [x_0 < x_2 < \min(b, 2x_0 - a)].$$

From the continuity of $u_0(x)$ and $f(x)$, we find that there is a positive number ϵ such that if $|2\delta| < \epsilon$, then

$$u_0(x_1) > f(x_1 + 2\delta) \qquad [x_0 < x_1 + 2\delta < \min(b, 2x_0 - a)]$$

and

$$u_0(x_2) < f(x_2 + 2\delta) \qquad [x_0 < x_2 + 2\delta < \min(b, 2x_0 - a)].$$

From the continuity of $u_0(x)$ and $f(x)$, there is an x_3 between x_1 and x_2 such that $u_0(x_3) = f(x_3 + 2\delta)$. Then $(x_0 + \delta, y_0)$ is a midpoint of the interval from $[2x_0 - x_3, f(2x_0 - x_3)]$ to $[x_3 + 2\delta, f(x_3 + 2\delta)]$.

There are positive numbers ϵ_1 and ϵ_2 such that if $|\delta| < \epsilon_i$ $(i = 1, 2)$, then $[x_0 + \delta, (y_0 + y_i)/2]$ is a middle-point of the graph of $f(x)$. Applying Theorem 3, we find that a connected domain contains (x_0, y_0) which contains only middle-points of the graph of $f(x)$.

We shall show that if x_0 is a number between a and $(a + b)/2$, then there is more than one middle-point of the graph of $f(x)$ with an abscissa equal to x_0. Assume that (x_0, y) is a middle-point of the graph of $f(x)$ only if $y = f(x_0)$. If h is a positive number such that $a < x_0 - h$, then the midpoint of the segment from $[x_0 - h, f(x_0 - h)]$ to $[x_0 + h, f(x_0 + h)]$ is a middle-point of the graph of $f(x)$ having an abscissa equal to x_0. The ordinate of this midpoint is $f(x_0)$ because (x_0, y) is a middle-point of the graph of $f(x)$ only if $y = f(x_0)$. Hence, $f(x)$ $(a < x < 2x_0 - a)$ is symmetric with respect to $[x_0, f(x_0)]$. As $f(x)$ is continuous at $2x_0 - a$, there is a continuous function $g(x)$ $(a \leq x \leq 2x_0 - a)$ such that $g(x) = f(x)$ for values of x between a and $2x_0 - a$. Since $g(x)$ is not linear, we have by Theorem 1 that there is a number x_1 between a and $2x_0 - a$ such that $[x_1, g(x_1)]$ is not a middle-point of the graph of $g(x)$. Since $g(x)$ is symmetric with respect to $[x_0, g(x_0)]$, we may assume that $x_1 < x_0$. Then $[x_1, f(x_1)]$ is not a middle-point of the graph of $f(x)$. Hence, the assumption that there is only one middle-point of the graph of $f(x)$ with an abscissa equal to x_0 leads to the contradiction that $f(x)$ does not satisfy the middle-point condition at x_1.

If x_0 is a value between a and $(a + b)/2$, then the nondegenerate connected set of all middle-points of the graph of $f(x)$ with abscissas equal to x_0 is covered by two points plus a domain consisting of middle-points of the graph of $f(x)$. Then the set of middle-points of the graph of $f(x)$ with abscissas less than $(a + b)/2$ consists of a simply connected domain D plus a subset of its boundary. Also, the set of middle-points of the graph of $f(x)$ with abscissas greater than $(a + b)/2$ consists of a simply connected domain D' plus a subset of its boundary. The set of middle-points of the graph of $f(x)$ with abscissas equal to $(a + b)/2$ is a subset of both the boundary of D and the boundary of D'.

3. Functions satisfying the middle-point condition. An example [2, pp. 253-255] has been given of a continuous nonlinear function $f(x)$ $(a < x < b)$ which satisfies the middle-point condition on its range. The graph of this function is the sum of a countable number of straight line intervals. One might wonder whether a continuous function $f(x)$ $(a < x < b)$ which satisfies the middle-point condition on its range could be nonlinear on each subinterval of its range.

THEOREM 5. *There exists a bounded function $F(x)$ $(a \leq x < b)$ having a derivative everywhere on its range, satisfying the middle-point condition for values of x between a and b and being nonlinear on each subinterval of its range.*

Proof. Designate by $a_0, a_1, a_2, \cdots, a_n, \cdots$ the values 1, 3/4, 9/16, $\cdots$, $(3/4)^n, \cdots$ and by $a_{-1}, a_{-2}, \cdots, a_{-n}, \cdots$ the values 5/4, 23/16, $\cdots$, $2-(3/4)^n, \cdots$. Let

(12) $$F(x) = (-1)^n x^2 \qquad [(4a_{n+1}+a_n)/5 \leq x \leq a_n].$$

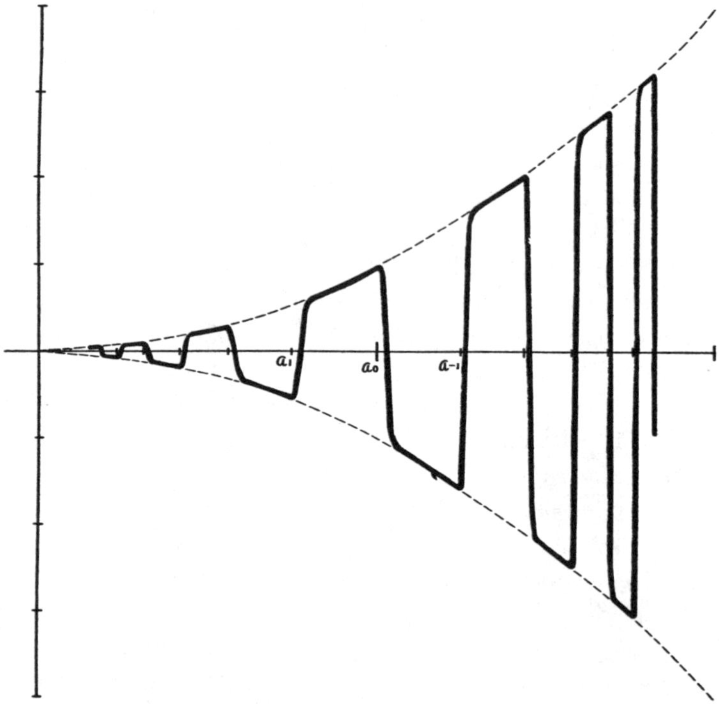

For other x such that $0 \leq x \leq 2$, $F(x)$ is defined so as to be nonlinear on each subinterval, so as to have a derivative at every point of its range and so that $|F(x)| \leq x^2$. See the figure.

We shall show that for each x_0 between 0 and 2, there are middle-points (x_0, y_1) and (x_0, y_2) of the graph of $F(x)$ such that $y_1 > x_0^2$ and $y_2 < -x_0^2$. An application of Theorem 3 will give Theorem 5.

Let W_n be the set of points on the x-axis whose abscissas x satisfy $(4a_{n+1}+a_n)/5 \leq x \leq a_n$. We shall show that if x_0 is a number between 0 and 2, then $(x_0, 0)$ is both a middle-point of $\sum W_{2n}$ and a middle-point of $\sum W_{2n+1}$.

Suppose that $(x_0, 0)$ is not a middle-point of any W_{2n+1}. There is a number j such that $(x_0, 0)$ is between a point of W_{2j+1} and a point of W_{2j-1}. A computation shows that if $j > 0$, then $(x_0, 0)$ bisects an interval having its end points on $W_{2j-1} + W_{2j-2}$ and $\sum_{j}^{\infty} W_{2n+1}$ respectively; if $j < 0$, then $(x_0, 0)$ bisects an interval having its end points on $W_{2j+3} + W_{2j+1}$ and $\sum_{-\infty}^{j} W_{2n-1}$ respectively; if $j = 0$ and $x_0 \leq 1$, then $(x_0, 0)$ bisects an interval from W_{-1} to $W_1 + W_3$; if $j = 0$ and $1 + (7/160)(3/4)^{2m+1} \leq x_0 \leq 1 + (7/32)(3/4)^{2m+1}$ for $m = 0, 1, \cdots$, then $(x_0, 0)$ bisects an interval from W_{2m+1} to W_{-2m-3}. Hence, each point of the x-axis between $(0, 0)$ and $(2, 0)$ is a middle-point of $\sum W_{2n+1}$. Likewise, each such point is a middle-point of $\sum W_{2n}$.

There are points $(x_1, 0)$ and $(x_2, 0)$ of $\sum W_{2n+1}$ such that $x_0 = (x_1 + x_2)/2$. We have by (12) that $[x_0, -(x_1^2 + x_2^2)/2]$ is a middle-point of the graph of $F(x)$. But $y_2 = -(x_1^2 + x_2^2)/2 < -x_0^2$. Likewise, there is a point (x_0, y_1) which is a middle-point of the graph of $f(x)$ and such that $y_1 > x_0^2$. We have by Theorem 3 that $F(x)$ satisfies the middle-point condition for values between 0 and 2.

THEOREM 6. *If each between-point of the graph of $f(x)$ $(a < x < b)$ is a middle-point of this graph and if $f(x)$ $(a \leq x < b)$ has a derivative at $x = a$, then $f(x)$ is linear.*

Proof. There is a continuous linear function $L(x)$ $(a \leq x < b)$ whose graph is tangent to the graph of $f(x)$ at $[a, f(a)]$. Assume that there is an x_0 between a and b such that $f(x_0) \neq L(x_0)$. Let $M(x)$ be the function whose graph is the interval joining $[a, f(a)]$ and $[x_0, f(x_0)]$. As the graph of $L(x)$ is tangent to the graph of $f(x)$ at $[a, f(a)]$, there is an x_1 such that $|L(x) - f(x)| < |M(x) - f(x)|$ for values of x between a and x_1. Then $[(2a + x_1)/3, M([2a + x_1]/3)]$ is a limit point of between-points of the graph of $f(x)$ $(a < x < b)$ but it is not a limit point of middle-points of this graph.

THEOREM 7. *Suppose that AB is an interval from the point A to the point B. There exist two totally disconnected closed subsets H and K of AB such that $H \cdot K = A + B$ and such that each point of $AB - (A + B)$ is both a middle-point of $H - (A + B)$ and a middle-point of $K - (A + B)$.*

Proof. First, we shall describe a totally disconnected closed subset $S(I)$ of an interval I. Let E_1 be the segment consisting of the middle one-fifth of I. Then $I - E_1$ is the sum of two intervals. Let E_2 be the sum of two segments each of which is the middle one-tenth of one of these intervals. In general, let E_{i+1} be the sum of all segments s such that s is the middle $1/(5\cdot 2^i)$ of a component of $I-(E_1+E_2+\cdots+E_i)$. The desired point set $S(I)$ is $I-(E_1+E_2+\cdots)$.

We shall note some properties of $S(I)$ that make it useful in proving Theorem 7. It is closed and totally disconnected. We shall show that if PQ is a subinterval of I such that P is an end point of I, then

(13) $$\text{measure } S(I)\cdot PQ \geq 1/2 \text{ length } PQ.$$

Since $E_1+E_2+\cdots$ is dense in I, (13) is true if it holds for each point Q of $E_1+E_2+\cdots$. Assume that (13) is false. Let $\bar{E}_j$ be the first element of $\bar{E}_1, \bar{E}_2, \cdots$ containing a point Q such that (13) does not hold. For convenience, we shall assume that the length of I is 1. Then the measure of $E_1+E_2+\cdots$ is $1/5+1/(5\cdot 2)+\cdots=2/5$ and the measure of $S(I)$ is $1-2/5=3/5$. The length of each component of E_i is $1/(5\cdot 4^{i-1})$ and the length of each component of $I-(E_1+E_2+\cdots+E_i)$ is

$$(1/2^i)\{1-[1/5+1/(5\cdot 2)+\cdots+1/(5\cdot 2^{i-1})]\}$$
$$=(3\cdot 2^i+2)/(5\cdot 4^i).$$

The measure of the common part of $S(I)$ and a component of $I-(E_1+E_2+\cdots+E_i)$ is $(1/2^i)(3/5)$. If Q is a point of $\bar{E}_1$, the measure of $S(I)\cdot PQ$ is $3/10$ and the length of PQ is no more than $3/5$. Therefore, $\bar{E}_1$ is not $\bar{E}_j$. If Q' is the last point of $P+\bar{E}_1+\bar{E}_2+\cdots+\bar{E}_{j-1}$ on PQ in the order from P to Q, then

$$\text{measure } S(I)\cdot PQ' \geq 1/2 \text{ length } PQ'.$$

But,

$$\text{measure } S(I)\cdot QQ' = (1/2^i)(3/5)$$

and

$$\text{length } QQ' \leq (3\cdot 2^i+2)/(5\cdot 4^i)+1/(5\cdot 4^{i-1})$$
$$=(3\cdot 2^i+6)/(5\cdot 4^i).$$

Then,

$$\text{measure } S(I)\cdot QQ' \geq 2^i/(2^i+2) \text{ length } QQ'.$$

Hence, (13) holds for all Q in I. Using a similar line of thought, we find that if UV is a subarc of I not intersecting $E_1+E_2+\cdots+E_i$ and U is a point of $\bar{E}_i$, then

(14) $$\text{measure } S(I)\cdot UV \geq 2/3 \text{ length } UV.$$

Each middle-point of I is a middle-point of $S(I)$. To see this, let I' and $S(I')$ be the images of I and $S(I)$ under a rotation of π radians about a middle-point R of I. If PQ is the common part of I and I', we have by (13) that both the measure of $S(I) \cdot PQ$ and the measure of $S(I') \cdot PQ$ is as much as 1/2 length PQ. Hence $S(I)$ intersects $S(I')$ and R is a middle-point of $S(I)$.

If K is an interval, let $S(K)$ denote the image of $S(I)$ under a similarity transformation of I into K. It will be shown that if I and K are two intervals that intersect in an interval L and neither I nor K is as much as five times as long as the other, then $S(I)$ intersects $S(K)$ on a set of positive measure. It will follow that if M and N are subintervals of a straight line and neither M nor N is as much as five times as long as the other, then each middle-point of $M + N$ is a middle-point of $S(M) + S(N)$.

If neither I nor K is a subset of the other, then one end point of L is an end point of I and the other end point of L is an end point of K. By (13) we have that

$$\text{measure } S(I) \cdot L \geqq 1/2 \text{ length } L$$

and that

$$\text{measure } S(K) \cdot L \geqq 1/2 \text{ length } L.$$

As $S(I) \cdot L$ and $S(K) \cdot L$ are closed, they intersect on a set of positive measure.

If K is a subset of I, let E_i be the first element of $E_1, E_2, \cdots$ intersecting K. Since

$$\text{length } E_i \leqq 1/5 \text{ length } I < \text{length } K,$$

K is not a subset of E_i. Let L' be a maximal subinterval of L containing no point of E_i. By (14) we have that

$$\text{measure } S(I) \cdot L' \geqq 2/3 \text{ length } L'. \tag{15}$$

But one end point of L' is an end point of L and therefore of K. By (13) we have that

$$\text{measure } S(K) \cdot L' \geqq 1/2 \text{ length } L'. \tag{16}$$

Now (15) and (16) give that $S(I)$ and $S(K)$ intersect on a set of positive measure.

Suppose that AB is the interval from $(0, 0)$ to $(2, 0)$. As in Theorem 5, designate by $a_0, a_1, \cdots, a_n, \cdots$ the values $1, 3/4, \cdots, (3/4)^n, \cdots$ and by $a_{-1}, a_{-2}, \cdots, a_{-n}, \cdots$ the values $5/4, 23/16, \cdots, 2 - (3/4)^n, \cdots$. Let M_n be the set of points $(x, 0)$ such that $(7a_{n+1} + a_n)/8 \leqq x \leqq a_n$, let $H - (A + B) = \sum S(M_{2n})$ and let $K - (A + B) = \sum S(M_{2n+1})$.

If $(x_0, 0)$ is between two points of M_{2n+1}, it is a middle-point of $S(M_{2n+1})$. If $(x_0, 0)$ is a point of $AB - (A + B)$ but is not between any two points of any M_{2n+1}, there is a j such that $(x_0, 0)$ is between a point of M_{2j+1} and a point of M_{2j-1}. A computation shows that if $j < 0$, then $(x_0, 0)$ is a middle-point of $M_{2j+1} + M_{2j-1} + M_{2j-3}$ and is therefore a middle-point of $S(M_{2j+1}) + S(M_{2j-1}) + S(M_{2j-3})$; if $j > 0$, then $(x_0, 0)$ is a middle-point of $S(M_{2j+3}) + S(M_{2j+1}) + S(M_{2j-1})$; if $j = 0$ and $x_0 \leq 1$, then $(x_0, 0)$ is a middle-point of $S(M_3) + S(M_1) + S(M_{-1})$; if $j = 0$ and $1 + (7/256)(3/4)^{2m+1} \leq x \leq 1 + (7/32)(3/4)^{2m+1}$ for $m = 0, 1, \cdots$, then $(x_0, 0)$ is a middle-point of $S(M_{2m+1}) + S(M_{-2m-3})$. Hence, each point of $AB - (A + B)$ is a middle-point of $K - (A + B)$. Likewise, it can be shown that each point of $AB - (A + B)$ is a middle-point of $H - (A + B)$.

THEOREM 8. *There exists a bounded function $F(x)$ $(a < x < b)$ satisfying the middle-point condition and having a derivative on its range such that $F(x)$ is nonlinear on each subinterval of its range and every between-point of the graph of $F(x)$ is a middle-point of this graph.*

Proof. The function

$$f(x) = \frac{1}{x - x^2} \sin \frac{1}{x - x^2} \qquad (0 < x < 1)$$

satisfies all of the conditions of this theorem except the one of boundedness.

Let AB be the interval from $(0, 0)$ to $(1, 0)$ and let H and K be totally disconnected closed subsets of AB such that $H \cdot K = A + B$ and each point of $AB - (A + B)$ is both a middle-point of $H - (A + B)$ and a middle-point of $K - (A + B)$. We define $F(x)$ $(0 < x < 1)$ to be a function having a derivative on its range and such that $|F(x)| \leq x$, the projection on the x-axis of the points of the graph of $F(x)$ for which $F(x) = x$ is the set $H - (A + B)$, the projection on the x-axis of the points of the graph of $F(x)$ for which $F(x) = -x$ is the set $K - (A - B)$ and $F(x)$ is nonlinear on each subinterval. Since each point between $(0, 0)$ and $(1, 1)$ is a middle-point of the graph of $F(x)$ and each point between $(0, 0)$ and $(1, -1)$ is a middle-point of the graph of $F(x)$, it follows by Theorem 3 that every between-point of the graph of $F(x)$ is a middle-point of this graph.

THE UNIVERSITY OF TEXAS.

REFERENCES.

1. E. F. Beckenbach, "On a characteristic property of linear functions," *Bulletin of the American Mathematical Society*, vol. 51 (1945), pp. 923-930.

2. R. Courant and D. Hilbert, *Methoden der mathematischen Physik*, vol. II, Berlin, 1937.

E. Fundamental Group Computations

THE SIMPLE CONNECTIVITY OF THE SUM OF TWO DISKS

R. H. BING

1. Introduction. The following question was called to the author's attention several years ago by Eldon Dyer.

Question. Is the sum of two disks simply connected if their intersection is connected?

Later, the author saw a communication in which an erroneous proof was given that Example 1 of the present paper is not simply connected. We show in § 2 that Example 1 is simply connected. However, we give some examples (Examples 2, 3, 4) in §§ 3, 4, 5 that are not simply connected.

A topological characterization is given in § 4 of intersections that will prevent closed curves which finitely oscillate between two disks from being shrunk. If the intersection is snake-like or arcwise connected, such finitely oscillating curves can always be shrunk but there are examples in which infinitely oscillating curves cannot. It is the topology of the intersection which prevents the sum of two disks from being simply connected rather than the embeddings of the intersection in the disks as shown in §§ 4 and 5. In fact, as pointed out in § 6, much of what we have learned about the sums of disks applies to the sums of continuous curves.

We use Example 1 in § 7 to construct a peculiar group and show that a certain relation kills it.

All sets treated in this paper are metric.

Let I^n denote an *n-cell* and *Bd* I^n its *boundary*. A set A is *n-connected* if each map (continuous transformation) f of *Bd* I^{n+1} into A can be extended to map I^{n+1} into A. We say that f (*Bd* I^{n+1}) *can be shrunk to a point* if the map can be extended. A set is called an *ε-set* if its diameter is less than or equal to ε. A set A is *n-ULC* if for each $\varepsilon > 0$ there is a $\delta > 0$ such that each map of *Bd* I^{n+1} onto a δ-subset of A can be shrunk to a point on an ε-subset. A compact continuum is called a *continuous curve* if it is 0-*ULC*. A set is *simply connected* if it is 1-connected. It is *uniformly locally simply connected* if it is 1-*ULC*. We shall not treat higher types of connectivity in this paper.

We find it convenient to consider an abstract disk D rather than the square I^2. A map of *Bd D* is a *closed curve*. If h is a homeomorphism, h(*BdD*) is a *simple closed curve*.

Received March 6, 1963. Research supported by NSF grants NSF-G21514 and NSF-G11665.

We shall use cylindrical coordinates (ρ, θ, z) to describe examples in E^3 (Euclidean 3 space). If no z coordinate is given, it is understood that $z = 0$. When we use D alone without subscripts it is understood that we mean the unit disk $(\rho \leqq 1)$ in the $z = 0$ plane.

Let f be a map of $Bd\ D$ into E^2 so that the ρ value of each point of $f(Bd\ D)$ is positive. Let $k(\theta)$ be a map of the reals into the reals such that $k(\theta)$ mod 2π is the θ value of $f(1, \theta)$. We say that f *circles the origin n times* if $k(2\pi) - k(0) = 2\pi n$.

2. A false example. Let a, b be fixed numbers with $0 < a < b < 1$ and K_1 be a spiral connecting the circles $\rho = a$ and $\rho = b$ as shown in Figure 1 and given by the following formula.

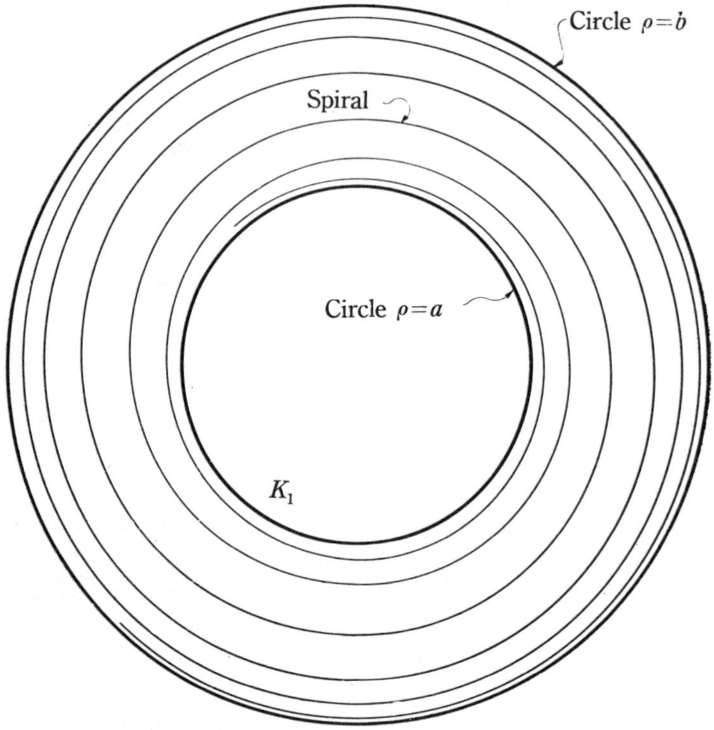

Figure 1

$$K_1 = \{(\rho, \theta)/\rho = a, b, \text{ or } (b + ae^\theta)/(1 + e^\theta)\} .$$

M. K. Fort showed [3] that any bounded plane continuum which has K_1 as a continuous image separates the plane.

EXAMPLE 1. Let D_1 be a disk in E^3 defined by

$$D_1 = \{(\rho, \theta, z)/\rho \leqq 1, z = \text{distance from } (\rho, \theta, 0) \text{ to } K_1\} .$$

Let D_2 be the reflection of D_1 through the $z = 0$ plane. Then $D_1 + D_2$ is the sum of two disks whose intersection is the connected set K_1.

THEOREM 1. *Example 1 is simply connected.*

Proof. Let f be a map of $Bd\ D$ into $D_1 + D_2$. We show that $D_1 + D_2$ is simply connected by showing that f can be extended to map D into $D_1 + D_2$.

With no loss of generality we suppose that the ρ value of each point of $f(Bd\ D) \geqq a$.

Special case. (The θ value of each point is fixed under f and the ρ value of each point of $f(Bd\ D) < b$.) In this special case we start by extending f to the circle $\rho = a$ by insisting that f is fixed on this circle.

For each point $q = (1, \theta_q)$ of $Bd\ D$ such that $f(q) \in K_1$, let S_q be the spiral from q about the circle $\rho = a$ described by the formulas $\rho \leqq 1$, $\rho = (2 - a + ae^{(\theta-\theta_q)})/(1 + e^{(\theta-\theta_q)})$, $\theta_q \leqq \theta$. Let f be extended to map S_q into K_1 so that f preserves the θ value of each point of S_q. This extension is made for each such spiral S_q for each point q of $Bd\ D$ such that $f(q) \in K_1$. Note that we have mapped a closed subset of D into K_1 and each component of $D - f^{-1}(K_1)$ other than the interior of $\rho = a$ intersects $Bd\ D$ in an open arc.

Let g_1 be the map of $f^{-1}(K_1 + D_1 \cdot f(Bd\ D)) = f^{-1}(D_1)$ into D_1 given by extended f. Then g_1 can be extended to take D into D_1. For convenience we also call this extended map g_1. Similarly there is a map g_2 of D into D_2 such that $g_2 = f$ on $f^{-1}(K_1 + D_2 \cdot f(Bd\ D)) = f^{-1}(D_2)$. Let g be the map of D into $D_1 + D_2$ given by g_1 on each component of $D - f^{-1}(K_1)$ which has an arc which goes into D_1 under f and $g = g_2$ on the rest of D.

Less special case. (f circles the origin once and the ρ value of each point of $f(Bd\ D)$ is less than b.) We show that there is a homotopy $h_t (0 \leqq t \leqq 1)$ of $Bd\ D$ into $D_1 + D_2$ such that $h_0 = f$, h_1 preserves the θ value of each point of $Bd\ D$ while the ρ value of each point of $h_1(Bd\ D)$ is less than b. The less special case then follows from the special case.

Let $k(\theta)$ be the function that shows that f circles the origin once. For convenience we suppose that $k(0) = 0$, $k(2\pi) = 2\pi$. Let $k_t(\theta) = t\theta + (1 - t)k(\theta)$, $(0 \leqq t \leqq 1)$. As t goes from 0 to 1, $k_t(\theta)$ goes from $k(\theta)$ to θ. For each point $p = (1, \theta_p)$ of $f^{-1}(K_1)$ we define $h_t(p)$ as a point in K_1 so that the θ value of $h_t(p)$ is $k_t(\theta_p)$. The ρ value of $h_t(p)$ is uniquely determined since the three arc components of K_1 are 1-manifolds almost normal to lines through the origin.

The homotopy h_t on $f^{-1}(K_1)$ is extended to $Bd\ D$ so that $h_t(p) \in D_i$ $(i=1, 2)$ if $f(p) \in D_i$, h_1 preserves the θ value of points of BdD and the value of each point of $h_1(Bd\ D)$ is less than b.

The following version of the less special case follows by a similar argument.

Alternative less special case. (f circles the origin once and the value of each point of $f(Bd\ D)$ is greater than a.)

General case. We suppose that $f(Bd\ D)$ intersects the spiral of K_1 in at least three points. Subdivide $Bd\ D$ into arcs $x_1x_2, x_2x_3, \cdots, x_nx_1 (n \geqq 3)$ so that no $f(x_ix_{i+1})$ (addition on subscripts is mod n so that $x_nx_{n+1} = x_nx_1$) intersects both circles in K_1 but each $f(x_i)$ is on the spiral of K_1. Let $x_iz_ix_{i+1}$ be the chord in D from x_i to x_{i+1}.

Extend the map f of $Bd\ D$ into $D_1 + D_2$ to map the chord $x_iz_ix_{i+1}$ into D_1 so that $f(x_iz_ix_i)$ misses the circles in K_1 and f on $x_ix_{i+1} + x_iz_ix_{i+1}$ circles the origin once. It follows from applications of the less special case and its alternative form that we can extend f to take the interiors of the $(x_ix_{i+1} + x_iz_ix_{i+1})$'s into $D_1 + D_2$. We can then extend f to the disk in D bounded by the chords into D_1.

3. **A true example.** Let C be a Cantor set on the numbers between 1/2 and 1. Let K_2 be the set in the plane consisting of the sum of circles in the plane with centers at the origin and radii in C and spirals joining adjacent circles as shown in Figure 2 and given by the following formula.

$$K_2 = \{(\rho, \theta)/\rho \in C \text{ or } \rho = (b + ae^{\theta})/(1 + e^{\theta})\}$$

where a, b are adjacent numbers of C with $a < b$.

EXAMPLE 2. Let E_1 be the disk in E^3 defined by

$$E_1 = \{(\rho, \theta, z)/\rho \leqq 1, z = \text{distance from } (\rho, \theta, 0) \text{ to } K_2\}\ .$$

Let E_2 be the reflection of E_1 through the plane $z = 0$. Then $E_1 + E_2$ is the sum of two disks whose sum is the connected set K_2.

Before proving that Example 2 is not simply connected we investigate an interesting property of K_2. M. K. Fort, Jr. showed [3] that any compact continuum in the plane separates the plane if it maps onto K_1. We modify his argument slightly to show the following.

THEOREM 2. *If f maps a closed bounded connected subset of the*

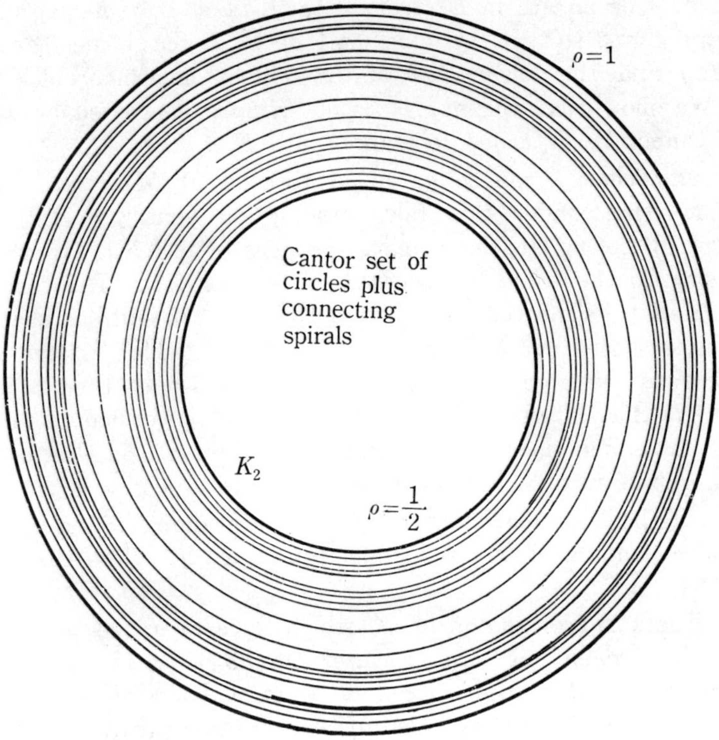

Figure 2

plane onto K_2 *then for each circle* J *in* K_2, *each component of* $f^{-1}(J)$ *separates the plane.*

Proof. Let S^1 be the circle $\rho = 1$ and define g: $K_2 \to S^1$ by $g(\rho, \theta) = (1, \theta)$. It is easy to verify that (K_2, S^1, g) is a locally trivial fiber space with totally disconnected fibers.

Suppose X is a component of $f^{-1}(J)$ that does not separate the plane. There is a homotopy pulling the map gf of X into S^1 to a constant map. Since S^1 is an ANR there is a neighborhood N of X such that the map gf of $\bar{N} \cdot f^{-1}(K_2)$ into S^1 is homotopic to a constant map. Take N so close to X that $\bar{N}$ does not cover $f^{-1}(K_2)$.

Let Y be a continuum in $f^{-1}(K_2)$ irreducible from $E^2 - N$ to X. Note that $Y \subset \bar{N}$, $Y \cdot X \neq 0$, and $Y \not\subset X$. It follows from the lemma on page 542 of [3] that $f(Y)$ is contained in an arc component of K_2. This violates the condition that the arc component of K_2 containing J does not intersect $K_2 - J$.

THEOREM 3. *Example* 2 *is not simply connected.*

Proof. Let x and y be points on the inner and outer circles in

K_2 and xz_iy be an arc in E_i from x to y. Let f be a map of $Bd\ D$ onto $xz_1y + xz_2y$ so that the upper half of $Bd\ D$ goes homeomorphically onto xz_1y and the lower half of $Bd\ D$ goes homeomorphically onto xz_2y. We show that Example 2 is not simply connected by showing that f cannot be extended to map D into $E_1 + E_2$.

Assume that f can be extended to send all of D into $E_1 + E_2$. We show that under this false assumption that $p = (1, 0, 0)$ and $q = (1, \pi, 0)$ belong to the same component of $f^{-1}(K_2)$. If they did not belong to the same component, it follows from Theorem 14 on page 171 of [6] (Theorem 10 on page 185 of 1932 edition) that there is a simple closed curve J in the plane $z = 0$ which misses $f^{-1}(K_2)$ and separates p from q in this plane. There would then be an arc A in $J \cdot D$ that intersects both the upper and lower halves of $Bd\ D$. This is impossible since f takes the upper half of $Bd\ D$ into E_1 and the lower half into E_2 but no point of A into $E_1 \cdot E_2 = K_2$.

Let Y be the component of $f^{-1}(K_2)$ containing p and q. Let Z be a subcontinuum of Y irreducible from p to q. Note that f maps Z onto K_2.

If F is a subcontinuum of Z which separates the plane E^2, no bounded component of $E^2 - F$ intersects Z since Z is irreducible from p to q and neither p nor q is in a bounded component of $E^2 - F$. Hence Z does not contain uncountably many mutually exclusive subcontinua each of which separates E^2. This contradicts Theorem 2 which says that for each circle J in K_2, $Z \cdot f^{-1}(J)$ separates E^2.

4. Finitely oscillating curves. A map of a simple closed curve J into the sum of two disks has only *finite oscillation* with respect to the two disks if J is the sum of a finite number of arcs such that the image of each lies in one of the disks. In some examples (Examples 3, 4, 5 to follow) finitely oscillating curves can be shrunk to points but some others cannot. The proof of Theorem 3 showed that Example 2 contained a finitely oscillating curve which could not be shrunk to a point.

We shall show that whether or not all finitely oscillating curves in the sum of two disks can be shrunk to points in the sum is dependent on whether or not the intersection has a certain extremal inverse property. A set X has the *extremal inverse property* with respect to its points p, q if there is a continuum Z in disk D with points p', q' on $Bd\ D$ and a map of Z into X that takes p', q' to p, q respectively.

Let K_3 be the sum of a triod T and a spiral S about T as shown in Figure 3 and given by the following equations.

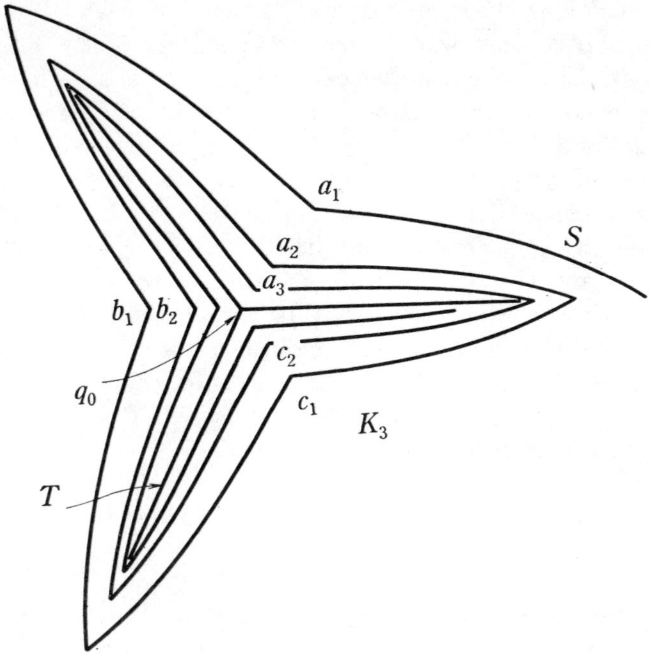

Figure 3

$$T = \{(\rho, \theta)/\rho \leqq 1, \theta = 0, 2\pi/3, \text{ or } 4\pi/3\},$$
$$S = \{(\rho, \theta)/\rho = |\cos 3\theta/2|^{\theta} + 1/\theta, \theta \geqq 2\pi\}.$$

EXAMPLE 3. Let D_1, D_2 be two disks whose intersection is K_3.

THEOREM 4. *K_3 has the extremal inverse property with respect to each pair of its points.*

Proof. We consider only the case where $p \in T$ and $q \in S$. Let S' be another spiral about T which misses S, f be a retraction of $S' + T$ onto T, and p' be a point of S' that maps onto p under f. Extend f to the identity on $S + T$. There is a disk containing $T + S + S'$ which has p' and q on its boundary.

THEOREM 5. *Each snake-like continuum has the extremal inverse property with respect to each pair of its points.*

Proof. Apply the following result to a subcontinuum of the snake-like continuum irreducible between the two points under consideration.

THEOREM 6. *Each snake-like continuum is the image of a pseudo-arc.*

Proof. This theorem has been proved by each of Fearnley [2], Lelek [4], and Mioduszewski [5] but we include a slightly different proof.

Let $D_1, D_2, \cdots$ be a sequence such that D_i is a $1/i$ chain properly covering snake-like continuum X and such that D_{i+1} is a refinement of D_i. It follows from Theorem 7 of [1] that if P is a pseudo-arc there is a sequence of proper open coverings $E_1, E_2, \cdots$ of P such that E_i has the same number of links as D_i and for the jth link of D_{i+1} there is an integer $n(i, j)$ such that the jth link of D_{i+1} lies in the $n(i, j)$th link of D_i and the jth link of E_{i+1} lies in the $n(i, j)$th link of E_i.

For each point p of P let $e(p, i)$ be the sum of the links of E_i containing p and $d(p, i)$ be the sum of the corresponding links of D_i. Note that $e(p, i+1) \subset e(p, i)$ and $d(p, i+1) \subset d(p, i)$. For each point p of P let $f(p)$ be the intersection of the closures of $d(p, i)$'s. Then f is a continuous transformation of P onto X.

THEOREM 7. *If a set has the extremal inverse property, so does each of its continuous images.*

THEOREM 8. *Each arcwise connected set has the extremal inverse property with respect to each pair of its points.*

Note that the following theorem applies to simply connected and uniformly locally simply connected continuous curves as well as merely to disks.

THEOREM 9. *Let A_1, A_2 be two compact sets each of which is* 0-*connected,* 1-*connected,* 0-ULC, *and* 1-ULC. *A necessary and sufficient condition that each finitely oscillating curve with respect to A_1, A_2 can be shrunk to a point in $A_1 + A_2$ is that $A_1 \cdot A_2$ has the extremal inverse property with respect to each pair of its points.*

Proof. If $A_1 \cdot A_2$ does not have the extremal inverse property with respect to point x, y of $A_1 \cdot A_2$, let f be a map of $Bd\, D$ into $A_1 + A_2$ such that the upper half of $Bd\, D$ goes into a path in A_1 from x to y and the lower half of $Bd\, D$ goes into a path in A_2 from x to y. It follows from the proof of Theorem 3 that $f(Bd\, D)$ cannot be shrunk to a point in $A_1 + A_2$.

To prove the sufficiency case consider a map f of $Bd\, D$ into $A_1 + A_2$ so that $Bd\, D$ is the sum of arcs $x_1x_2, x_2x_3, \cdots, x_nx_1$ so that each $f(x_i)$ lies on $A_1 \cdot A_2$ and each $f(x_ix_{i+1})$ lies in one of A_1, A_2. Just as we used chords in the general case of the proof of Theorem 1, we consider continua $Z_1, Z_2, \cdots, Z_n$ in D so that Z_i contains x_i and x_{i+1} and f can be extended to take $Z_1 + Z_2 + \cdots + Z_n$ into $A_1 \cdot A_2$. Then as in the proof of the General Case of the proof of Theorem 1 we extend f to take the components of $D - (Z_1 + Z_2 + \cdots + Z_n)$ which intersect $Bd\, D$ into the appropriate one of A_1, A_2 and then extend f to take the rest of D into A_1. To extend f to take D into A_i for example, one would add a null sequence of arcs in D to $Z_1 + Z_2 + \cdots + Z_n$ to get a set Z_0 so that $D - Z_0$ is a null sequence of open disks, use the fact that A_i is 0-connected and 0-ULC to extend f to Z_0, and finally use the fact that A_i is 1-connected and 1-ULC to extend f to take D into A_i.

THEOREM 10. *Suppose D_1, D_2 are two disks whose intersection is a continuum X and axb is an arc in D_1 that intersects X only at a and b. If f is a map of $Bd\, D$ into $D_1 + D_2$ such that f takes the upper half of $Bd\, D$ onto axb and the lower half into D_2, then $f(Bd\, D)$ can be shrunk to a point in $D_1 + D_2$.*

Proof. Since X has the extremal inverse property with respect to a and b as shown by its embedding in D_1, there is a continuum X' in D intersecting the inverses under f of a and b such that f may be extended to $X' + Bd\, D$. Then f is extended to take the part of D in component of $D - X'$ that contains upper arc of $Bd\, D$ into D_1 and the rest of D into D_2.

Question. The question suggests itself as to which continua have the extremal inverse property with respect to each pair of their points. Example 2 does not have it. Example 1 and 3 do. So do Examples 4 and 5 to be given in the next section. Perhaps Example 2 is unnecessarily complicated as an example of a continuum without the extremal inverse property in that it separates the plane into infinitely many pieces. Perhaps there is an example that does not separate the plane. Does each three branched tree-like plane continuum have the extremal inverse property with respect to each pair of its points? (A compact continuum is a three branched tree-like continuum if it is not snake-like but for each positive number ε it has an ε-cover whose 1-nerve is a triod.)

5. Infinite oscillation. We use rectilinear coordinates to define the two sets shown in Figures 4, 5.

$$K_4 = \{(x, y)/(x = 0,\ -2 \leqq y \leqq 2),\ (y = 1 + \sin 1/x,\ 0 < x \leqq 1)\,, \text{ or } (y = -1 + \sin 1/x,\ -1 \leqq x < 0)\}\,.$$

K_5 = sum of points of K_3 on or to the left of the vertical line through $(1, 0)$ plus the interval from $(1, 0)$ to $(1, 1)$.

Theorems 5 and 8 show that K_4 and K_5 have the extremal

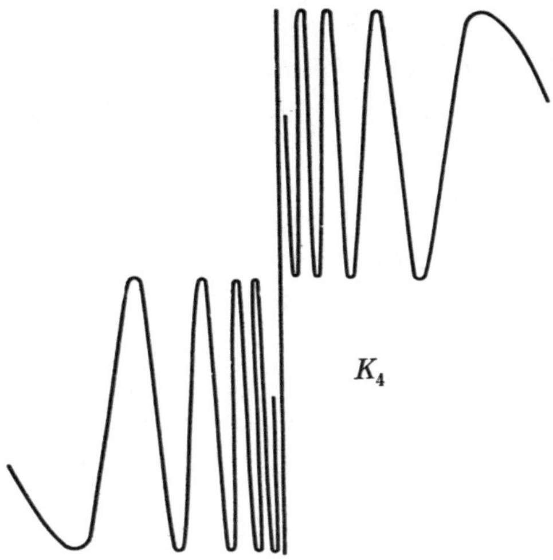

Figure 4

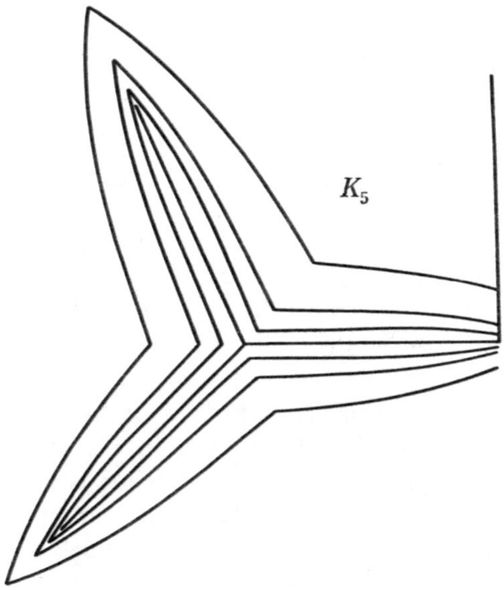

Figure 5

inverse property with respect to each pair of their points. It follows from Theorem 9 that finitely oscillating curves in the following two examples can be shrunk to points in the examples.

EXAMPLE 4. Two disks sewed together along K_4.

EXAMPLE 5. Two disks sewed together along K_5.

THEOREM 11. *Examples* 3, 4, 5 *are not simply connected.*

We only prove the first third of this theorem since the other parts are analogous. We suppose that the disks D_1, D_2 of Example 3 are obtained by pushing parts of a circular disk in the $z = 0$ plane up and down respectively as done in Examples 1 and 2. Theorem 13 shows that there is no loss of generality in supposing this. The disks would be larger than those in Examples 1 and 2 since K_3 is larger than those disks.

Proof that Example 3 *is not simply connected.* Let $a_1, a_2, \cdots$ be the points of K_3 on the open ray $\theta = \pi/3$ ordered inversely according to their distances from the origin q_0. Let $b_1, b_2, \cdots$ be the corresponding points of K_3 on the open ray $\theta = \pi$ and $c_1, c_2, \cdots$ be the corresponding points on the ray $\theta = 5\pi/3$.

Let p_i be the point of *Bd D* whose θ value is $1/i$. Let p_0 be the point of *Bd D* whose θ value is 0. Use p_ip_j to denote the arc on *Bd D* in a clockwise direction from p_i to p_j.

Let f be a map of *Bd D* into $D_1 + D_2$ satisfying the following conditions.

$$
\begin{aligned}
&f(p_j) = q_0 \text{ (origin) } (j = 0, 2, 4, 6, \cdots) , \\
&f(p_{6i-5}) = a_i , \\
&f(p_{6i-3}) = b_i , \\
&f(p_{6i-1}) = c_i , \\
&f(p_{2i-2}p_{2i-1}) \subset D_1 , \\
&f(p_{2i-1}p_{2i}) \subset D_2 .
\end{aligned}
$$

The θ value on each $f(p_ip_{i+1})$ is a constant and f takes the θ values of p_ip_{i+1} linearly onto the ρ values of $f(p_ip_{i+1})$.

Assume f can be extended to take D into $D_1 + D_2$. We call the extended map f. In this extension we suppose that no component of $f^{-1}(q_0)$ separates the $z = 0$ plane. (If a component X did separate the plane, we could modify f to map X plus each of its bounded complementary domains in $z = 0$ into q_0.)

Let F be the part of $D_1 + D_2$ whose ρ value is less than or equal to 1. A finite number of spanning arcs in D separates p_0 from $f^{-1}(D_1 + D_2 - F)$ in D so that no one of the arcs intersects $f^{-1}(q_0)$. Hence, we can cut down the disk D to a disk E such that E agrees with D in a neighborhood of $p_0, f(E) \subset F$, each point of $Bd\, E \cdot f^{-1}(q_0)$ lies on *Bd D*, and E agrees with D in a neighborhood of each such point. Let Z be the closure of $Bd\, E - Bd\, D$. Note that $q_0 \notin f(Z)$. We shall obtain a contradiction to the assumption that f can be extended to take D into $D_1 + D_2$ by showing that the map f of *Bd E* into F cannot be extended to take E into F.

With no loss of generality we suppose that $p_0, p_1, p_2, \cdots$ all belong to *Bd E*. Note that no component of $E \cdot f^{-1}(K_3)$ intersects two p_i's unless perhaps they both have even subscripts since no two of $a_1, a_2, \cdots, b_1, b_2, \cdots, c_1, c_2, \cdots$ belong to the same component of $K_3 \cdot F$. For i odd, the component Y_i of $E \cdot f^{-1}(K_3)$ containing p_i separates $Bd\, E$ since $f(Bd\, E)$ crosses from D_1 to D_2 at $f(p_i)$. Since $Bd\, D$ intersects Y_i in at most a finite number of points and $f(Bd\, D)$ does not cross from D_1 to D_2 at the image of any of these points other than p_i, Y_i must intersect Z. Let q_i be a point of $Y_i \cdot Z$.

Note that since $a_1, a_2, \cdots, b_1, b_2, \cdots, c_1, c_2, \cdots$ belong to different components of $F \cdot K_3$, $Y_i \neq Y_j$ if i, j are different odd positive integers. Let q_∞ be a limit point of $q_1, q_3, q_5, \cdots$. Since for i sufficiently large Y_{i+2} separates Y_i from p_0 in E, $q_1, q_3, \cdots$ converges to q_∞ from the clockwise side. Since q_∞ is a limit point of each of ΣY_{6i-5}, ΣY_{6i-3}, ΣY_{6i-1}, then $f(q_\infty)$ is a limit point of each of $f(\Sigma Y_{6i-5})$, $f(\Sigma Y_{6i-3})$, $f(\Sigma Y_{6i-1})$. The only point common to the closures of these sets is the point q_0, so $f(q_\infty) = q_0$. However, $f(q_\infty) \neq q_0$ since $q_\infty \in Z$ and $q_0 \notin f(Z)$.

THEOREM 12. *The sum of two disks is simply connected if their intersection is connected and locally arcwise connected.*

Proof. Let f be a map of $Bd\, D$ into the sum of two disks D_1, D_2 such that $D_1 \cdot D_2$ is a continuous curve. For each arc ab of $Bd\, D$ which intersects $f^{-1}(D_1 \cdot D_2)$ only in its end points, extend f to map the chord acb of D into an arc in $D_1 \cdot D_2$ such that the diameter of $f(acb)$ is no more than twice the diameter of any other arc in $D_1 \cdot D_2$ from $f(a)$ to $f(b)$. Let f_i be a mapping of D into D_i that agrees with f on the part of D going into D_i under f. Then the extended f is f_1 on the components of D minus the chords which contain a point of $Bd\, D$ that f sends into $D_1 - D_2$ and is f_2 on the rest of D.

THEOREM 13. *The topology of the intersection of two disks determines whether or not their sum is simply connected.*

Proof. Suppose D_1, D_2, E_1, E_2 are disks and h is a homeomorphism of $D_1 \cdot D_2$ onto $E_1 \cdot E_2$. Let D be a circular disk and f a map of $Bd\, D$ into $D_1 + D_2$. We assume that $E_1 + E_2$ is simply connected and show that this assumption implies that f can be extended to map D into $D_1 + D_2$. We assume there are at least three points of $Bd\, D$ that f sends into $D_1 \cdot D_2$

Let g be a map of $f^{-1}(D_1 \cdot D_2 \cdot f(Bd\, D))$ into $E_1 \cdot E_2$ given by $g = hf$. For each arc ab of $Bd\, D$ which intersects $f^{-1}(D_1 \cdot D_2 \cdot f(Bd\, D))$ only in its end points, extend g to map the chord acb of D onto an arc in E_i if $f(ab) \subset D_i$ with the restriction that the diameter of

$g(acb)$ is not more than twice the diameter of any other arc in E_i from $g(a)$ to $g(b)$. Let E be the subdisk of D such that g has been defined to map $Bd\, E$ into $E_1 + E_2$.

Since $E_1 + E_2$ is simply connected, we extend g to map E into $E_1 + E_2$. Call the extension g. Consider $g^{-1}(E_1 \cdot E_2 \cdot g(E)) = X$. No two points of $Bd\, D$ can be joined by an arc in $D - X$ unless the points go into the same one of D_1, D_2 under f.

Define f on X to be $h^{-1}g$. Let f_i be the extended f restricted to $f^{-1}(D_i \cdot f(X + Bd\, D))$. Extend f_i to map D into D_i and call the extended map f_i. The extended map f is f_1 on each component of $D - X$ which has points of $Bd\, D$ which are sent by f into D_1 and is f_2 on the rest of D.

6. **Adding continuous curves.** What we have learned about the sum of disks partially applies to the sums of other continua. If the intersection of two disks is so bad as to make the sum of the disks not simply connected, it is bad enough to keep any two continuous curves whatever with the same intersection from being simply connected. The following example illustrated in Figure 6 shows that the converse is not true.

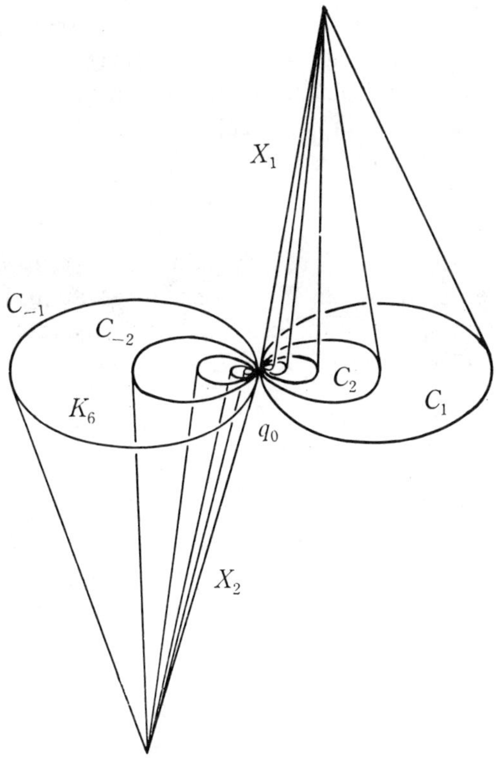

Figure 6

EXAMPLE 6. Let C_i $(i = 1, 2, 3, \cdots)$ be the circle in the x, y plane with equation $(x - 1/i)^2 + y^2 = (1/i)^2$. Denote the origin by q_0. Let $K_6 = C_1 + C_2 + \cdots + q_0 + C_{-1} + C_{-2} + \cdots$. The cone X_1 over $C_1 + C_2 + \cdots + q_0$ from a point above the xy plane is simply connected as is the cone X_2 over $q_0 + C_{-1} + C_{-2} + \cdots$ from a point below the xy plane. Although $X_1 \cdot X_2$ is a point, $X_1 + X_2$ is not simply connected.

THEOREM 14. *Suppose D_1, D_2 are two disks and F_1, F_2 are two continuous curves such that $D_1 \cdot D_2$ is homeomorphic with $F_1 \cdot F_2$. Then $D_1 + D_2$ is simply connected if $F_1 + F_2$ is.*

Proof. The proof is the same as the proof of Theorem 13 except that g maps $Bd\,E$ into $F_1 + F_2$ instead of into $E_1 + E_2$.

THEOREM 15. *Suppose G_1, G_2, G_3, G_4 are four simply connected and uniformly locally simply connected continuous such that $G_1 \cdot G_2$ is topologically equivalent to $G_3 \cdot G_4$. Then the fundamental group of $G_1 + G_2$ is isomorphic to the fundamental group of $G_3 + G_4$.*

Proof. Whether or not a loop in $G_1 + G_2$ can be shrunk to a point depends on how it crosses back and forth between G_1 and G_2. Suppose h is a homeomorphism of $G_1 \cdot G_2$ onto $G_3 \cdot G_4$ and x_0 is a point of $G_1 \cdot G_2$ that acts as a starting point of loops in $G_1 + G_2$ to determine the fundamental group of $G_1 + G_2$. We use $h(x_0)$ as a starting point for the loops in $G_3 + G_4$ to determine the fundamental group of $G_3 + G_4$.

Let $\{f\}$ be an element of the fundamental group of $G_1 + G_2$. It is an equivalence class of maps of the interval $[0, 1]$ into $G_1 + G_2$ such that the ends of $[0, 1]$ go into x_0. Let f be an element of $\{f\}$. Let f' be a map of $[0, 1]$ into $G_3 + G_4$ such that

$$f'(x) = hf(x) \text{ if } f(x) \in G_1 \cdot G_2 ,$$
$$f'(x) \in G_{i+2} \text{ if } f(x) \in G_i .$$

Although these two conditions do not precisely define f', any two maps satisfying this condition are homotopic in $G_3 + G_4$. The element of the fundamental group of $G_3 + G_4$ corresponding to the element $\{f\}$ of the fundamental group of $G_1 + G_2$ is the equivalence class of loops containing f'.

Question. The preceding theorem suggests a topological invariant of compact closed sets. Two sets A, B are alike in a certain sense provided the sum of two Hilbert cubes sewed together along A have the same fundamental group as the sum of two Hilbert cubes sewed together along B. Is there a simpler characterization of this property?

7. An interesting group. One might attempt to compute the fundamental group of Example 1 by cutting it into two pieces with a vertical plane through the origin, fatten each piece to make them intersect in an open subset of their sum, find the fundamental group of each piece, and then apply Van Kampen's theorem to get the fundamental group of Example 1. We ignore the fattening since, being equivalent to taking the slice slightly to one side of the origin, it does not change the fundamental group of the pieces.

Each piece can be folded like a fan and deformed onto a set topologically equivalent to a set K_7 shown in Figure 7 and defined as follows.

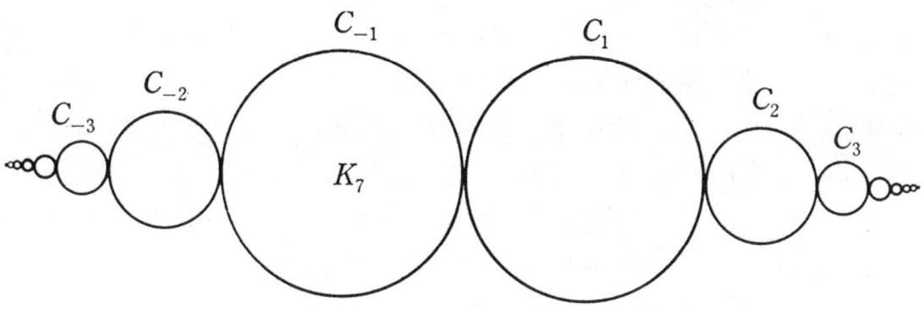

Fignre 7

$$K_7 = \text{closure of } (C_1 + C_2 + \cdots + C_{-1} + C_{-2} + \cdots)$$

where C_i is the circle in the xy plane with $[(i-1)/i, \ i/(i+1)]$ as diameter and C_{-i} is the circle with $[(-(i-1)/i, \ -i/(i+1)]$ as diameter. (We used $[a, b]$ to denote the interval on the x axis from a to b.) The fundamental group of each piece into which we cut Example 1 is the same as the fundamental group of K_7.

Consider the origin as the starting point of loops in K_7 to determine its fundamental group $G(K_7)$. Then a loop is a map of the interval $[0, 1]$ into K_7 that sends the ends of the interval to the origin and an element of $G(K_7)$ is an equivalence class of loops. We can associate words with loops. If a loop goes across the top semicircle of C_i from left to right we write i; if it goes across this semicircle from right to left we write $\bar{i}$ and say i inverse. We call i and $\bar{i}$ letters and say that the letter is positive or negative according as i is positive or negative. Since we are permitting C_i's with negative subscripts, i inverse differs from $-i$. The inverse of $\bar{i}$ is i. A loop then corresponds to an ordered collection of letters (called a word) with the following restrictions.

a. No letter appears in any word more than a finite number of times.

b. There is not infinite oscillation between positive and negative letters.

Let us consider what words are equated if two pieces into which we divided Example 1 are joined together again. If a loop is slid from one piece to the other until it comes back to the first in one direction, each letter i (or $\bar{i}$) in it has been changed to $i+1$ (or $\overline{i+1}$) and if the loop is slid in the other direction, these are replaced by $i-1$ (or $\overline{i-1}$). Since we skipped 0 in putting subscripts on the C_i's we suppose $-1+1=1$ and $1-1=-1$. When we replace each i or $\bar{i}$ in a word W by $i+1$ or $\overline{i+1}$, we have produced a right shift and call the new word $R(W)$. We note that if $W_1 = R(W_2)$, then W_2 may be obtained from W_1 by a left shift and say $W_2 = \bar{R}(W_1)$.

Let us change the group G {equivalence classes of W_α's} by also putting words in the same equivalence class if they are equivalent after a shift. This shifting operation is to be permitted in equating words only a finite number of times as opposed to cancellation which was permitted infinitely often. We call the resulting group G {equivalence classes of W_α's/$R(W_\alpha) = W_\alpha$}. Since the fundamental group of Example 1 is trivial, it follows that this group is trivial.

The inverse of a word is obtained by reversing the order of the letters and replacing each letter with its inverse. If in a word there appears two adjacent subwords which are inverses of each other, the word obtained by canceling the subwords belongs to the same equivalence class with the original word. Infinite cancellation is permitted so that for example $(1, \bar{1}, 2, \bar{2}, 3, \bar{3}, \cdots)$ is equivalent to the trivial word.

Two words are equivalent if and only if they can be cancelled down to a common word. (We could have given more extensive rules but they boil down to this.) To multiply two words, we write one after the other. If $\{W_\alpha\}$ denotes the collection of words and G {equivalence classes of W_α's} denotes the group of equivalence classes of words, then

$$G(K_7) = G\{\text{equivalence classes of } W_\alpha\text{'s}\} .$$

To show algebraically that G {equivalence classes of W_α's/$R(W_\alpha) = W_\alpha$} is trivial, consider a word W. Since we did not permit infinite oscillation between positive and negative letters of W, we can express W as $W_1 W_2 \cdots W_n$ where each W_i has either all positive or all negative letters. We show that W is trivial by showing that each W_i is. We consider only the case where W_i consists of positive letters since the other case is analogous.

Consider $X = \bar{W}_i R(\bar{W}_i) R^2(\bar{W}_i) R^3(\bar{W}_i) \cdots$. It is a word since it only contains positive letters and none appears more than a finite number of times. Then

$$W_i = W_i X \bar{X} = R(\bar{W}_i) R^2(\bar{W}_i) \cdots \bar{X} = R(X)\bar{X} = X\bar{X} = 1 .$$

One might wonder what would have happened if we had not imposed the condition that there is not infinite oscillation between the positive and negative letters in words. This would have been equivalent to the fundamental group of K_7 after the sum of the bottom simicircles were shrunk to a point. Even after a shift, it seems that the group is not killed. After the shift we would have the fundamental group of Example 1 if the annulus in D_2 outside $\rho = a$ is shrunk to the circle $\rho = a$.

REFERENCES

1. R. H. Bing and F. B. Jones, *Another homogeneous plane continuum,* Trans. Amer. Math. Soc., **90** (1959), 171-192.

2. L. Fearnley, *Continuous mappings of the pseudo-arc,* Notices of the American Mathematical Society Abstract 564-171, **6** (1959), 841.

3. M. K. Fort, Jr., *Images of plane continua,* Amer. J. Math., **81** (1959), 541-546.

4. A. Lelek, *On weakly chainable continua,* Fundamenta Mathematicae, **51** (1962), 271-282.

5. J. Mioduszewski, *A functional conception of snake-like continua,* Fundamenta Mathematicae, **51** (1962), 179-189.

6. R. L. Moore, *Foundations of Point Set Theory,* Amer. Math. Soc. Colloq. Publ., vol. 13, revised edition, 1962.

THE UNIVERSITY OF WISCONSIN
THE INSTITUTE FOR ADVANCED STUDY

Computing the Fundamental Group of the Complements of Curves*

R. H. Bing

We discuss an algorithm for computing the fundamental group of the complements of finite graphs in E^3. Other treatments are found in R. H. Crowell and R. H. Fox, *Introduction to knot theory*, Ginn (**1963**); K. Reidemeister, *Knotentheorie*, Chelsea (**1948**); and R. H. Fox, "A quick trip through knot theory," *Topology of* 3-*manifolds*, Prentice Hall (**1962**). Before discussing the algorithm, we give some preliminary definitions.

1. Paths and loops. A *path* in a topological space X is a map f of the segment $[0, 1]$ into X. By a map, we mean a continuous transformation. The images $f(0)$ and $f(1)$ are called the beginning and end points of the path.

MULTIPLICATION OF PATHS. If f_1, f_2 are paths in X such that the end of f_1 is the beginning of f_2, the product of f_1 and f_2 (written $f_1 * f_2$) is the path

$$\begin{aligned} f_1 * f_2(t) &= f_1(2t), & 0 \le t \le \tfrac{1}{2} \\ &= f_2(2t-1), & \tfrac{1}{2} \le t \le 1. \end{aligned}$$

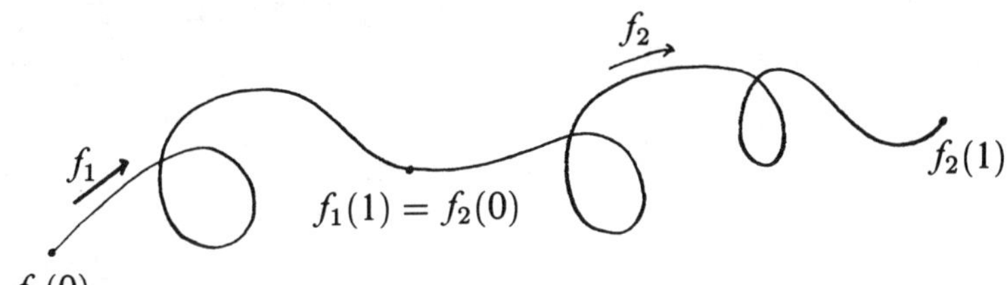

FIGURE 1.

In moving along the path $f_1 * f_2$ as suggested in Figure 1, one moves along the path f_1 at double speed until the end is reached and then moves along f_2 at

*This paper embodies some of the material presented in lectures given by Professor Bing while he was Edwin W. Rice Visiting Professor of Mathematics at Washington State University in April, 1965.

double speed. It is to be noted that $f_1 * f_2$ is not defined unless the end of f_1 is the beginning of f_2.

INVERSE. If f is a path, we call

$$f^{-1}(t) = f(1-t), \qquad 0 \le t \le 1$$

the inverse of f. As indicated in Figure 2, we are moving along f in the opposite direction.

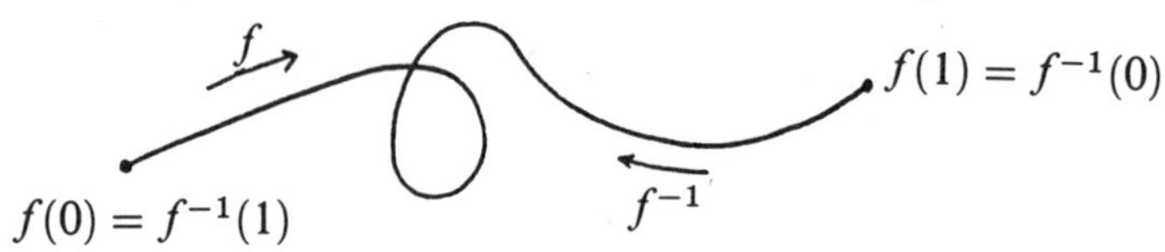

FIGURE 2.

HOMOTOPY OF PATHS. If f_1, f_2 are two paths with the same beginnings and ends we say that f_1 is homotopic to f_2 with ends fixed, (written $f_1 \sim f_2$) if there is a homotopy F_t: $[0,1] \to X$ $(0 \le t \le 1)$ such that $F_0 = f_1$, $F_1 = f_2$, $F_t(0) = f_1(0) = f_2(0)$, $F_t(1) = f_1(1) = f_2(1)$. It may be convenient to regard the homotopy as a continuous shifting of the path f_1 until it becomes the path f_2 as suggested in Figure 3. We may regard F as mapping the unit planar square with the opposite vertices $(0,0)$, $(1,1)$ into X such that $F(x,y) = F_y(x)$. Note that F is constant on each vertical side of the square. On the bottom of the square, F is determined by f_1 and on the top, it is determined by f_2.

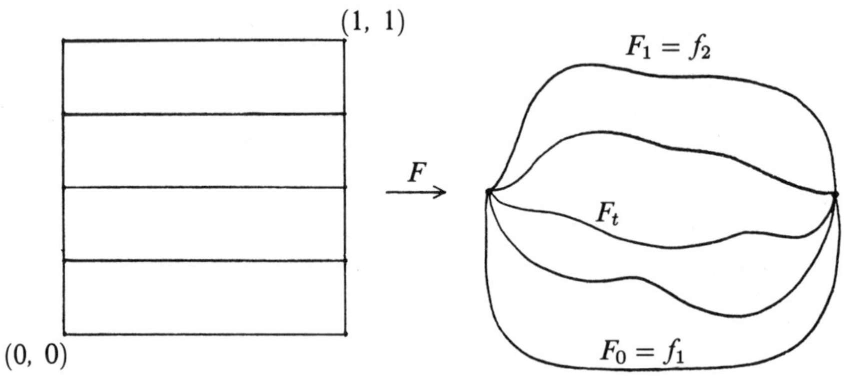

FIGURE 3.

EQUIVALENCE CLASSES OF PATHS. The above homotopy breaks the paths in X into equivalence classes. We use $[f]$ to denote the set of all paths g in X such that $f \sim g$. To see that $f \sim g$ implies $g \sim f$ one need only reflect the unit square across a horizontal segment through its center. To visualize the fact that $f_1 \sim f_2$ and $f_2 \sim f_3$ implies $f_1 \sim f_3$ one might consider the maps F and G on the unit squares showing that $f_1 \sim f_2$ and $f_2 \sim f_3$ respectively, squash each of these squares to one half their normal heights, and then set the second squashed square on top of the first. The homotopy H showing that $f_1 \sim f_3$ is given by

$$\begin{aligned} H_t &= F_{2t}, & 0 \le t \le \tfrac{1}{2} \\ &= G_{2t-1}, & \tfrac{1}{2} \le t \le 1. \end{aligned}$$

If $[f_1] \cap [f_2] \neq 0$, $[f_1] = [f_2]$.

ASSOCIATIVITY. Suppose f_1, f_2, f_3 are paths such that the end of f_i is the beginning of f_{i+1}, $i = 1, 2$. Then $(f_1 * f_2) * f_3$ and $f_1 * (f_2 * f_3)$ are defined, although they may be different as paths. Under $(f_1 * f_2) * f_3$ one considers moving along f_1 at four times normal speed, along f_2 at four times normal speed, and along f_3 at double speed. The rates of speed on f_1 and f_3 are interchanged for $f_1 * (f_2 * f_3)$. If F is defined so as to be constant on each of the slanting segments shown in Figure 4, there results a homotopy showing that $f_1 * (f_2 * f_3) \sim (f_1 * f_2) * f_3$.

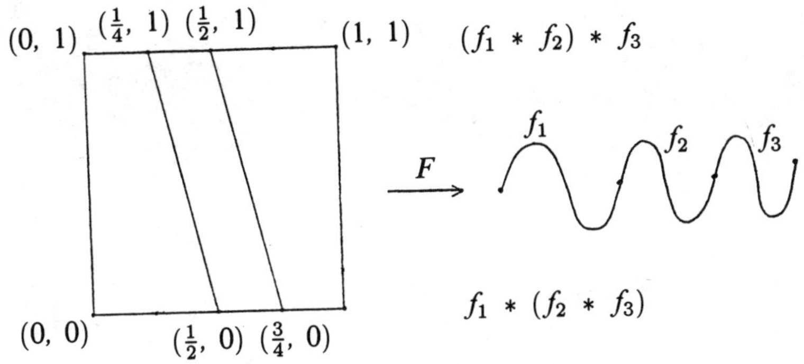

FIGURE 4.

LOOPS. Select a point $x_0 \in X$ and designate it as a *base point.* A *loop* in X based at x_0 is a path in X that has this base point as both a beginning and an end point. See Figure 5. Note that any two loops based at x_0 can be multiplied. This puts us in a position to use equivalence classes of loops based at x_0 as elements of a group.

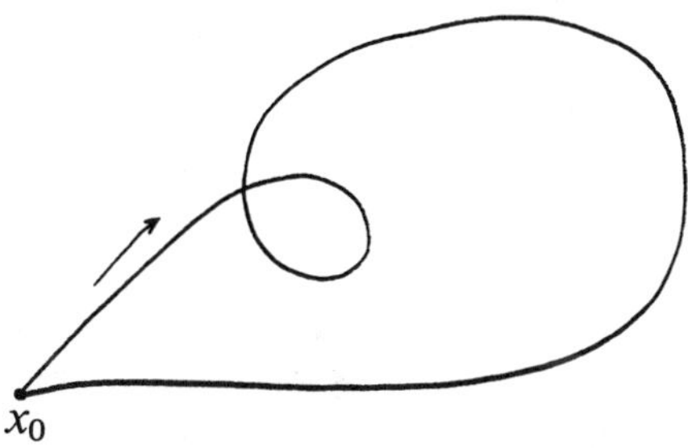

FIGURE 5.

2. The fundamental group. Let x_0 be a base point in a topological space X. We use $\pi_1(X, x_0)$ to denote the *fundamental group* of X based at x_0. An element of this fundamental group is an equivalence class of loops in X based at x_0. If $[f] \in \pi_1(X, x_0)$, $[f]^{-1}$ is the equivalence class of loops containing f^{-1}. To multiply two elements $[f]$, $[g]$ of $\pi_1(X, x_0)$ we select any two elements $f \in [f]$, $g \in [g]$ and let $[f]*[g] = [f*g]$. It can be shown that this multiplication is independent of the elements chosen in the class. It turns out that the identity element of $\pi_1(X, x_0)$ is the equivalence class containing the constant map $f[0, 1] = x_0$.

It may be shown that if there is a path f in X from x_0 to x_1, then $\pi_1(X, x_1)$ is isomorphic to $\pi_1(X, x_0)$. To get such an isomorphism one could let the loop g based at x_1 correspond to the loop $f * g * f^{-1}$ based at x_0 as shown in Figure 6. This correspondence between loops is extended to a correspondence between equivalence classes of loops.

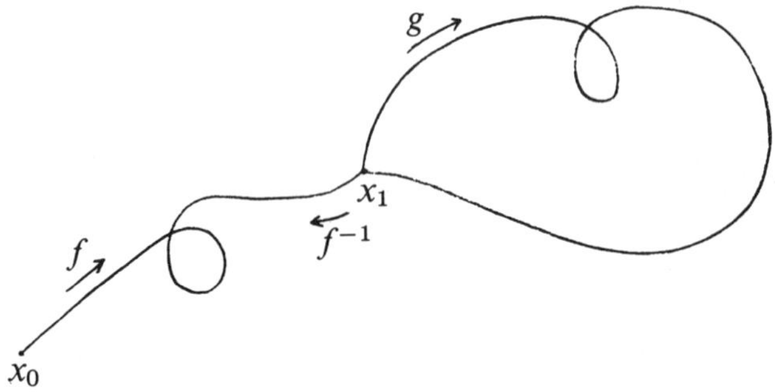

FIGURE 6.

When dealing with a pathwise connected space X, it is customary to ignore the base point and use $\pi_1(X)$ to designate $\pi_1(X, x_0)$. We shall frequently do this in the following sections.

3. Graphs. A *finite graph* in E^3 is the sum of a finite number of straight segments such that if two of these segments intersect, the intersection is an end point of each. It is a geometric 1-complex.

A plane P in E^3 is called in *regular position* with respect to a finite graph G if only a finite number of lines normal to P intersect G in more than one point and each of these exceptional normal lines intersects G in precisely two points which are interior points of the segments used to describe G. We wish to alter situations as shown in the left side of Figure 7 so that they become as shown in the right side.

If a plane P is in regular position with respect to G, any plane parallel to P is in regular position with respect to G. If one considers the unit vectors from the origin which are normal to planes which are not in regular position with respect to G, these vectors intersect the unit sphere with center at the origin in at most a 1-dimensional set. Hence, there are many vectors such that any plane normal to this vector are in regular position with respect to G.

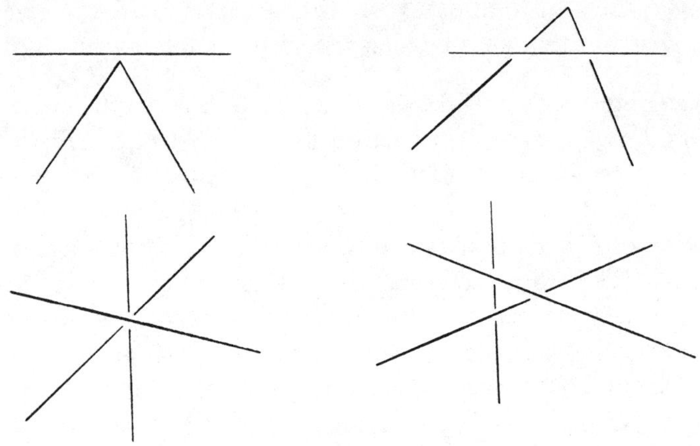

FIGURE 7.
(Projections of G into P)

Consider a *normal projection* G' of G into the plane P which is in regular position with respect to G. For convenience we regard P as horizontal so that each point of G is on a vertical line with its corresponding point in G'. If one segment of G contains a point p which is beneath a second segment we say that p is an *undercrossing point* and the first segment *passes beneath* the second. In drawing G', it is convenient to show a gap in the lower segment if one passes beneath the other. In the graph G (or its projection G') shown in Figure 8 there are five undercrossing points, three branch points and one end point.

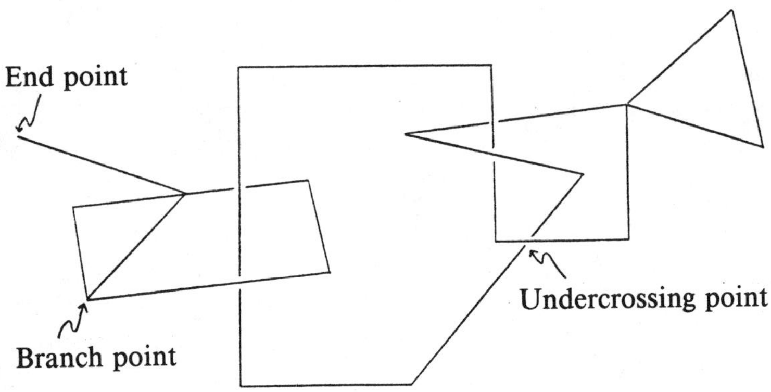

FIGURE 8.

Although for technicalities in giving proofs, one may give lip service to the use of polyhedral graphs, in practical situations one usually draws smooth graphs. The graph shown in Figure 8 is shown as a smooth graph in Figure 9. In working with smooth graphs one insists that if one smooth segment of G passes underneath another, there is actual crossing in G' and not mere tangency.

There is a theory for studying the fundamental groups of the complements of wild graphs but that goes beyond the scope of this paper.

4. Associating words with loops. Suppose G is a graph (either polygonal or smooth) in E^3 with a nice projection as shown in either Figure 8 or 9. Let $(v_1, v_2, \ldots, v_m)$ be the points of G that are end points, branch points, or undercrossing points. Let $A_1, A_2, \ldots, A_n$ be the collection of open arcs and simple closed curves of G such that no point of any A_i is a v_j but each boundary point of A_i is a v_j. The closure of each A_i is either an arc in G or a simple closed curve in G which intersects at most one v_j. To each A_i we associate an arrow going under A_i and designate it with a letter a_i as shown in Figure 9. It does not matter in which direction the arrows (short oriented vectors) point but it is customary to think of moving along the A_i's of G in some direction and for the arrows to go under from the left side of A_i to the right. In most drawings, there is no need to label the v_i's and the A_i's since it is only the a_i's that are used.

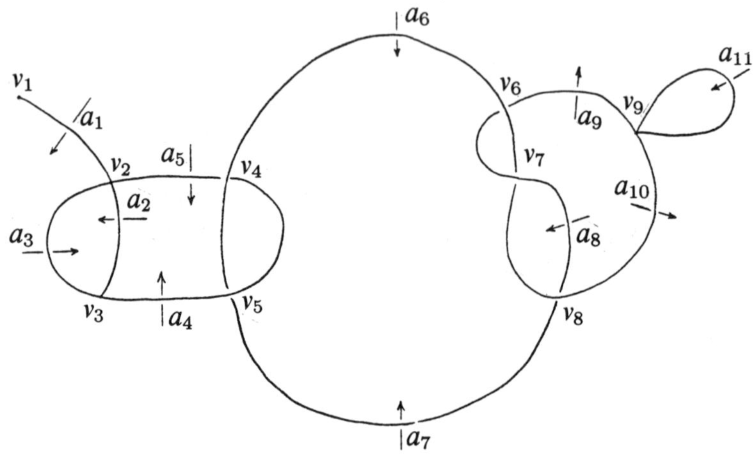

FIGURE 9.

Suppose x_0 is a point of $E^3 - G$ selected as a base point and f is a loop based at x_0 taking $[0, 1]$ into $E^3 - G$. We suppose that f is normal with respect to G in the sense that $[0, 1]$ can be broken into a finite number of segments $[b_0, b_1], [b_1, b_2], \ldots, [b_{r-1}, b_r]$ $(0 < b_0 < b_1 < \cdots < b_r = 1)$ such that f is a homeomorphism on each $[b_{i-1}, b_i]$ and either no point of $f[b_{i-1}, b_i]$ is beneath G or exactly one point of $[b_{i-1}, b_i]$ is beneath G and f crosses under a segment of G at this point but not under an undercrossing point of G. Although not each loop of $[f]$ has such nice properties, there are many elements of $[f]$ that are piecewise linear and many of the piecewise linear elements of $[f]$ are normal with respect to G.

To each of the r pieces of $[0, 1]$ we associate either a blank or a letter. If $f[b_{i-1}, b_i]$ does not cross under G, we associate a blank. If $f[b_{i-1}, b_i]$ crosses under A_j in the direction of the arrow a_j, we associate the letter a_j; if it crosses under A_j in the opposite direction, we associate a_j^{-1}.

To the loop f we associate a word made up of letters a_1, a_1^{-1}, $a_2, a_2^{-1}, \ldots$, a_n, a_n^{-1} written as follows: We ignore blanks and first write down the letter associated with $f[b_0, b_1]$, then the letter associated with $f[b_1, b_2], \ldots$, and finally the letter associated with $f[b_{r-1}, b_r]$. In case f crosses under G we obtain a word with a finite number of letters. The word associated with the loop shown in Figure 10 is $a_1^{-1}a_6a_7^{-1}a_7a_{10}$. It does not contain the letters a_{10}^{-1} and a_8^{-1} since f went over G at these crossings rather than under.

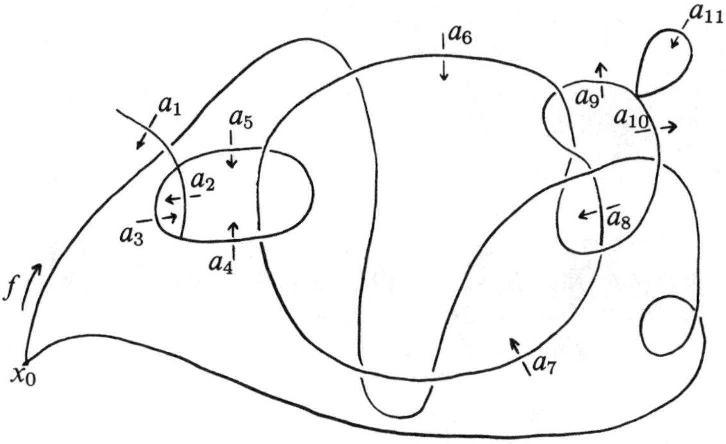

FIGURE 10.

Let $W(f)$ be the word associated with f.

Suppose f_1, f_2 are two loops associated with the same word $W(f_1) = W(f_2) = x_1x_2\cdots x_s$. It may be shown in this case that $f_1 \sim f_2$ so f_1, f_2 belong to the same element of $\pi_1(E^3 - G)$. We show this in the next two paragraphs.

Let f_3 be a loop homotopic to f_1 so that f_3 crosses under G at the same places that f_2 does. Figure 11 shows how to modify f_1 on $[b_{i-1}, b_i]$ to obtain f_3 on $[b_{i-1}, b_i]$. Note that $f_3[b_{i-1}, b_i]$ first runs along $f_1[b_{i-1}, b_i]$, then goes up near A_j, then runs along parallel to A_j, then drops down to $f_2[0, 1]$ and follows it under A_j, then moves up near A_j, and runs parallel to it, and finally returns to $f_1[b_{i-1}, b_i]$ and proceeds along it. All of this is realized by a homotopy that is invariant except near the inverses of points where f_1 goes under G.

Let $y_1, y_2, \ldots, y_s$ and $y_1', y_2', \ldots, y_s'$ be the points of $[0, 1]$ such that $f_2(y_i) = f_3(y_i')$ is under G. Let h be a homeomorphism of $[0, 1]$ onto itself that takes $0, y_1, y_2, \ldots, y_s, 1$ onto $0, y_1', y_2', \ldots, y_s', 1$ respectively. Then $f_3h \sim f_3$ on $f_3[0, 1]$. Since $f_3h[y_{i-1}, y_i]$ has the same ends as $f_2[y_{i-1}, y_i]$ and each runs above G except at its end points, there is a homotopy of f_3h to f_2 in $E^3 - G$ that leaves each of $[0, y_1, \ldots, y_s, 1]$ fixed. Since $f_1 \sim f_3 \sim f_3h \sim f_2$, $[f_1] = [f_2]$ is the same element of $\pi_1(E^3 - G)$.

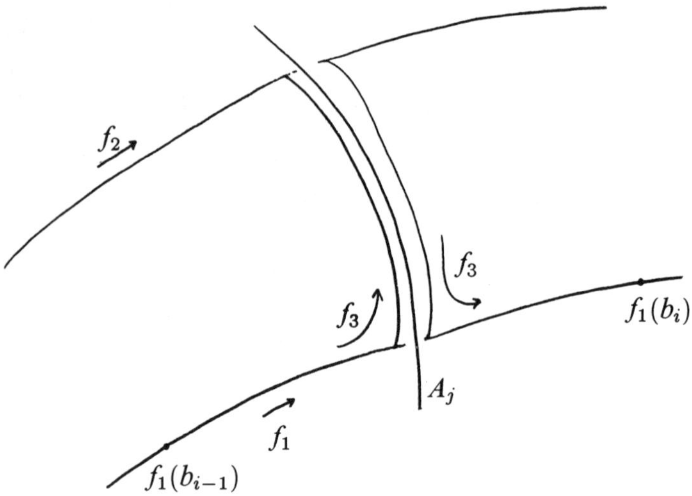

FIGURE 11.

Although we have shown that $f_1 \sim f_2$ if $W(f_1) = W(f_2)$, one should not conclude the converse that $f_1 \sim f_2$ implies $W(f_1) = W(f_2)$. Such is not the case. It may be that f_1 ducks under A_i as shown in Figure 12 and picks up the letters $a_i a_i^{-1}$. In fact two loops f_1, f_2 are homotopic if $W(f_2)$ can be obtained

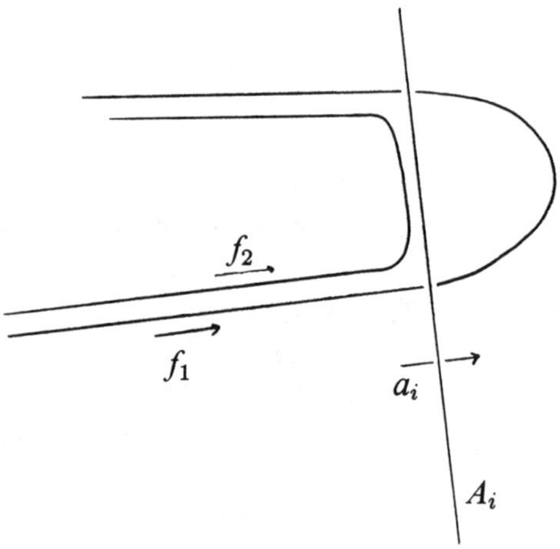

FIGURE 12.

by either cancelling or inserting the adjacent letters $a_i a_i^{-1}$ or $a_i^{-1} a_i$ anyplace in $W(f_1)$. In the next section we point out other conditions under which different words are associated with homotopic loops.

5. Relations. We shall consider operations by which $W(f_1)$ can be changed to $W(f_2)$ if $f_1 \sim f_2$ in $E^3 - G$. To do this we consider a relation at each end point, branch point, or undercrossing point of G. A relation (designated by $r_i = 1$) is a word r_i such that it is permitted to cancel or insert either r_i or r_i^{-1} anywhere in any word. (If $r_i = x_1 x_2 \cdots x_s$, $r_i^{-1} = x_s^{-1} x_{s-1}^{-1} \cdots x_1^{-1}$.)

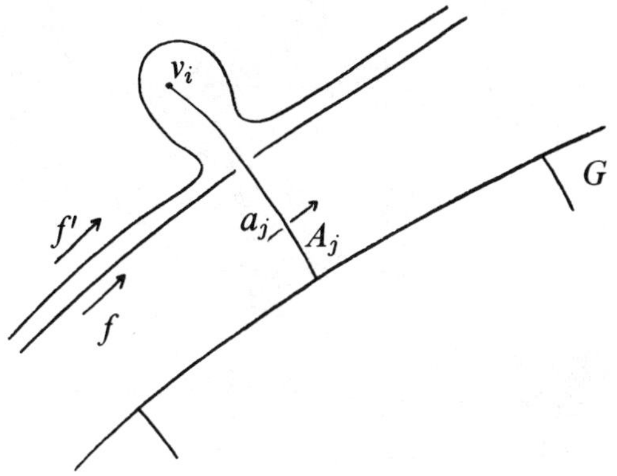

FIGURE 13.

END POINT RELATION. If A_j reaches out to v_i as shown in Figure 13, the loop f going under A_j is homotopic to the loop f' which agrees with f generally except that it goes around the v_i end of A_j instead of crossing under. The relation $r_i = 1$ corresponding to v_i is $a_j = 1$.

BRANCH POINT RELATION. If a loop f goes under the A's reaching out from a branch point as shown in Figure 14, it is homotopic to the loop f' partially shown. The relation $r_i = 1$ in this case is $a_s^{-1} a_k a_j a_r^{-1} = 1$.

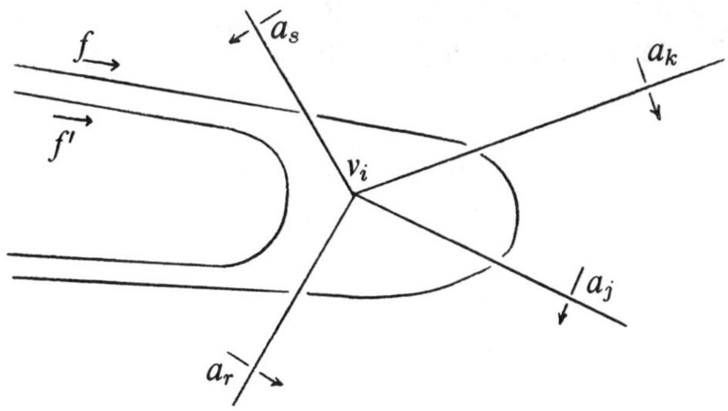

FIGURE 14.

UNDERCROSSING RELATION. Suppose f goes around an undercrossing as shown in Figure 15. Then $f' \sim f$. The relation $r_i = 1$ is $a_j a_s a_j^{-1} a_r^{-1} = 1$.

Suppose we associate a relation $r_i = 1$ with each end point, branch point, and undercrossing point of G. Then $f_1 \sim f_2$ in $E^3 - G$ if $W(f_1)$ can be changed to $W(f_2)$ by a finite sequence of the following operations:

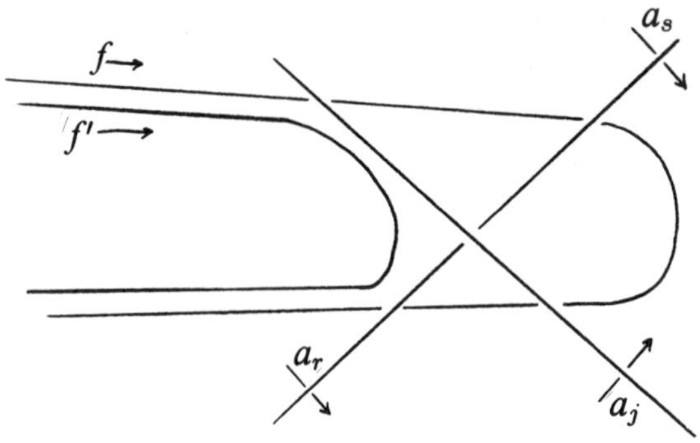

FIGURE 15.

1. Insert xx^{-1} or its inverse anywhere.
2. Cancel xx^{-1} or its inverse anywhere.
3. Insert any r_i or its inverse anywhere.
4. Cancel any r_i or its inverse anywhere.

We shall show in the next section that the converse of the preceding paragraph is true in that if $f_1 \sim f_2$ in $E^3 - G$, then $W(f_1)$ can be changed to $W(f_2)$ with operations of these types.

We did not impose definite rules as to how to get the r_i's. For example, one person looking at Figure 15 might have written $a_j a_s a_j^{-1} a_r^{-1} = 1$ and another might have written $a_s a_j^{-1} a_r^{-1} a_j = 1$. However, this doesn't matter since $a_s a_j^{-1} a_r^{-1} a_j = a_j^{-1} a_j a_s a_j^{-1} a_r^{-1} a_j = a_j^{-1} a_j = 1$. To cancel out $a_s a_j^{-1} a_r^{-1} a_j$ one might first insert $a_j^{-1} a_j$ before a_s, then cancel out the middle four letters and then the remaining two. We do not need to choose any particular r_i while looking at Figure 14 or 15. In fact, frequently, instead of reading a relation like $a_j a_s a_j^{-1} a_r^{-1} = 1$ from Figure 15, it might be simpler to use either $a_r = a_j a_s a_j^{-1}$ or $a_j a_s = a_r a_j$.

6. Changing words. In this section we show that if $f_1 \sim f_2$ in $E^3 - G$ then $W(f_1)$ can be changed to $W(f_2)$ with a finite number of operations of the following four types:

1. Insert xx^{-1} or its inverse anywhere.
2. Cancel xx^{-1} or its inverse anywhere.
3. Insert any r_i or its inverse anywhere.
4. Cancel any r_i or its inverse anywhere.

It simplifies matters to regard things as polyhedral so we suppose that G is a finite graph, P is a horizontal plane in regular position with respect to G, ρ is the vertical projection into P, $G' = \rho(G)$, and f_1, f_2 are piecewise linear loops normal with respect to G such that $f_1 \sim f_2$. No vertex of $f_i[0, 1]$ is on $\rho^{-1}(G')$, and no point of $f_i[0, 1]$ intersects any vertical line through either a vertex of G or an undercrossing point of G. Since $f_1 \sim f_2$ there is a map F' of the unit square S of E^2 with opposite vertices $(0,0)$ and $(1,1)$ into $E^3 - G$ such that $F'(x,0) = f_1(x)$, $F'(x,1) = f_2(x)$, $F'(0,y) = F'(1,y) = x_0$.

Let T be a triangulation of S of small mesh. We shall delay until later discussing the size of the simplexes of T. The 2-simplexes of T are ordered $\Delta_1^2, \Delta_2^2, \ldots, \Delta_q^2$ so that Δ_1^2 shares an edge with the bottom edge of S, Δ_q^2 shares an edge with the top and there is a polygonal arc L_i in S from its left edge to its right that separates $(\operatorname{Int}\Delta_1^2 \cup \operatorname{Int}\Delta_2^2 \cup \cdots \cup \operatorname{Int}\Delta_i^2)$ from $(\operatorname{Int}\Delta_{i+1}^2 \cup \operatorname{Int}\Delta_{i+2}^2 \cup \cdots \cup \operatorname{Int}\Delta_q^2)$ in S. We use L_0 to denote the bottom of S and L_q to denote its top. For any triangulation of S there is such a sequence of L's, but rather than prove this it might be more convenient to take a triangulation like that shown in Figure 16 where there is clearly such an ordering.

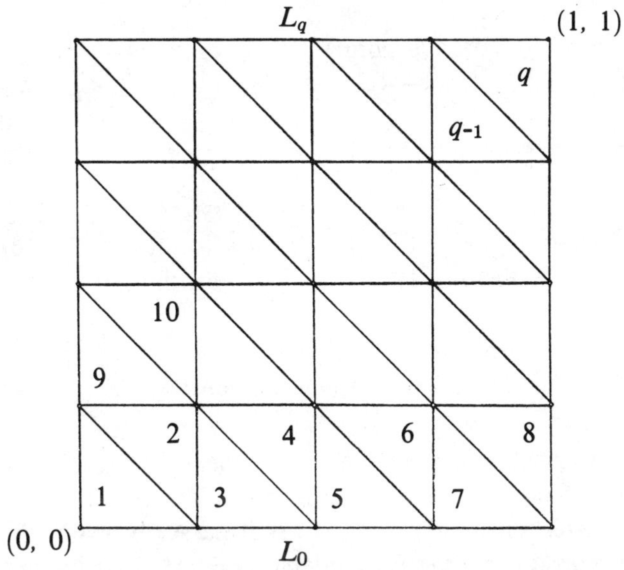

FIGURE 16.

Let F be a map of S into E^3 such that $F|\operatorname{Bd}S = F'|\operatorname{Bd}S$, F is linear on each simplex of T, and for each vertex v of T, $F(v)$ is close to $F'(v)$. We suppose the mesh of T is so small and the $F(v)$'s are so close to the $F'(v)$'s that $F(S) \subset E^3 - G$ and for no 2-simplex Δ_i^2 of T does $\rho F(\Delta_i^2)$ contain the vertical projection of two vertices of G. We also suppose the $F(v)$'s are so chosen that either $G' \cap \rho F(\Delta_i^2) = 0$ or $G' \cap \rho F(\Delta_i^2)$ looks as shown in either Figures 17, 18, 19, or 20.

In all of the figures, G' crosses $\operatorname{Bd}\rho F(\Delta_i^2)$ at points where they intersect. In Figure 17, $\operatorname{Bd}\rho F(\Delta_i^2)$ intersects G' in two points and $\rho F(\Delta_i^2)$ contains the projection of no end point, branch point, or undercrossing point of G. In Figures 18, 19 and 20, $\rho F(\Delta_i^2)$ contains the projection of an end point, branch point, and undercrossing point respectively and arcs emanating from them but no other parts of G' except as shown.

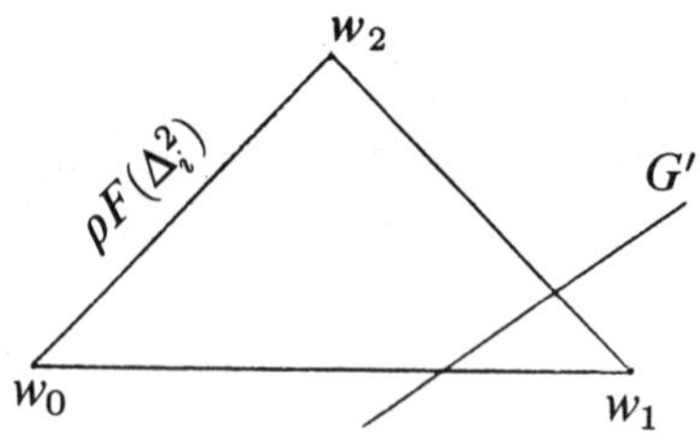

FIGURE 17.

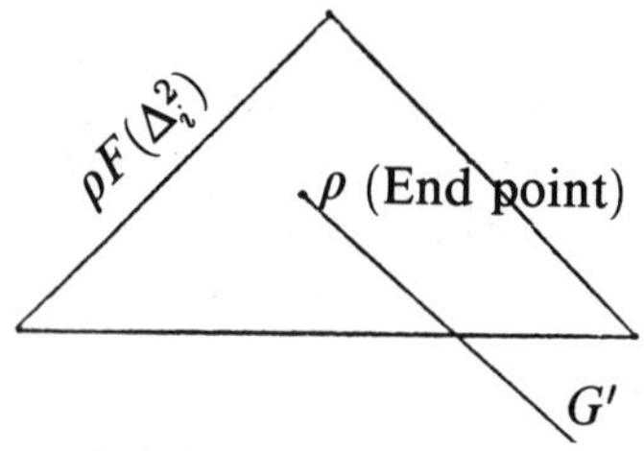

FIGURE 18.

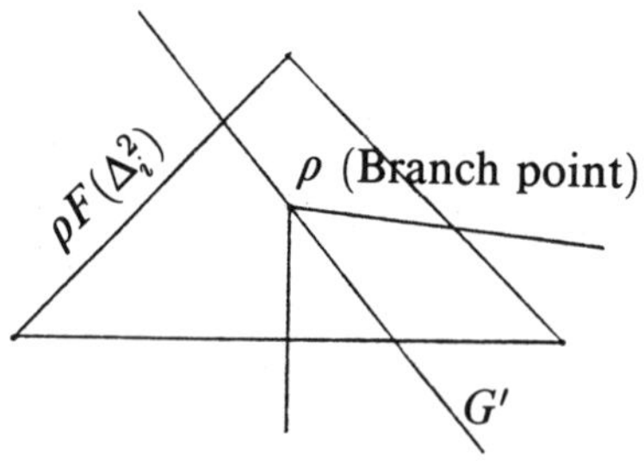

FIGURE 19.

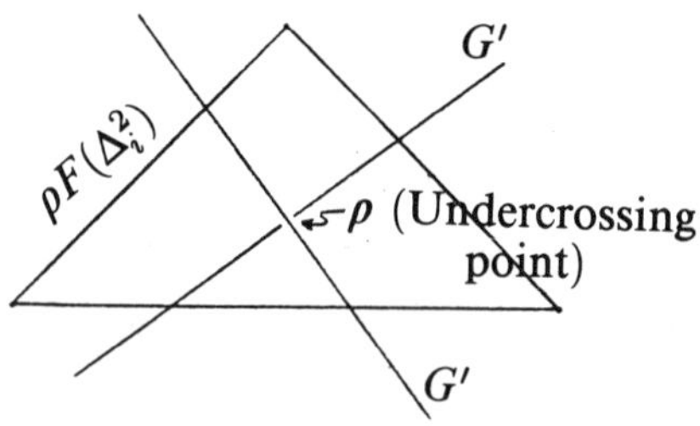

FIGURE 20.

If h_i is a homeomorphism of $[0, 1]$ onto L_i with $h_i(0)$ on the left side of S, Fh_i is a loop in $E^3 - G$. Let $W_i = W(Fh_i)$. Then $W_0 = W(f_1)$ and $W_q = W(f_2)$.

If $\rho F(\Delta_i^2)$ misses G', $W_{i-1} = W_i$.

If $\rho F(\Delta_i^2)$ intersects G' as in Figure 17, then since $F(\Delta_i^2) \cap G = 0$, either $\text{Bd}\, F(\Delta_i^2)$ passes beneath G at both crossing points or it crosses above at both points. If it passes above, $W_{i-1} = W_i$. If it passes beneath, there are several possibilities according as how L_{i-1} intersects $\text{Bd}\,\Delta_i^2$. If it intersects it in w_0w_1, w_1w_2, $w_0w_2 \cup w_1w_2$, or $w_2w_0 \cup w_0w_1$, $W_{i-1} = W_i$. If L_{i-1} intersects $\text{Bd}\,\Delta_{i-1}^2$ in w_0w_2, W_{i-1} can be changed to W_i by an operation of Type 1—insert an $a_ia_i^{-1}$ or its inverse somewhere. If L_{i-1} intersects $\text{Bd}\,\Delta_{i-1}^2$ in $w_0w_1 \cup w_1w_2$, W_{i-1} can be changed to W_i by an operation of Type 2.

If $\rho F(\Delta_i^2)$ intersects G' as shown in Figure 18, then $W_{i-1} = W_i$ if $\text{Bd}\, F(\Delta_i^2)$ is above G and if $\text{Bd}\, F(\Delta_i^2)$ is below, W_{i-1} can be changed to W_i by an operation of Type 3 or 4, where the relation used is determined at the end point which projects into $\text{Int}\, \rho F(\Delta_i^2)$.

If $\rho F(\Delta_i^2)$ intersects G' as shown in Figure 19, since $F(\Delta_i^2)$ misses G, then $\text{Bd}\, F(\Delta_i^2)$ is above G or it is below it at each crossing point. If it is above, $W_{i-1} = W_i$ and if it is below, W_{i-1} can be changed to W_i by using a finite number of operations of Types 1 and 2 and one of Type 3 or 4 determined by the relation determined at the branch point which projects into $\text{Int}\, \rho F(\Delta_i^2)$. The reason for using a finite number of relations of Types 1 and 2 is that r_i might be something like $a_ja_ka_la_m = 1$ and W_i might be obtained from W_{i-1} by replacing

$a_j a_k a_l$ by a_m^{-1}. This could be done by inserting $a_m a_m^{-1}$ in W_{i-1} and canceling out $a_j a_k a_l a_m$.

If $\rho F(\Delta_i^2)$ intersects G' as shown in Figure 20, then either $\operatorname{Bd} F(\Delta_i^2)$ is above G, below G, or above one of the segments shown and below the other. In case $\operatorname{Bd} F(\Delta_i^2)$ is above both, $W_{i-1} = W_i$. If it is above only one of the segments, either $W_{i-1} = W_i$ or W_{i-1} can be changed to W_i by an operation of Type 1 or 2 as discussed in connection with Figure 17. If it is below both segments, W_{i-1} changes to W_i by a finite number of operations of Types 1 and 2 and one operation of Type 3 or 4.

By shifting from L_0 to L_1 to L_2 to $\cdots$ to L_q, one finds that if $f_1 \sim f_2$ in $E^3 - G$, then $W(f_1)$ can be changed to $W(f_2)$ with a finite number of operations of Types 1, 2, 3, and 4.

7. Presentation of groups. We use $H = \{a_1, a_2, \dots\}$ to denote the free group with generators $a_1, a_2, \dots$. An element of H is an equivalence class of words (finite ordered set of letters) where the letters of the words are the generators $a_1, a_2, \dots$ and their inverses $a_1^{-1}, a_2^{-1}, \dots$. Two words belong to the same equivalence class if and only if one can be changed to the other with a finite number of operations where each operation consists of either the insertion of two adjacent letters xx^{-1} or $x^{-1}x$ somewhere in the word, or the cancellation of such a two letter combination. The identity element is the class which contains the word with no letters.

Suppose $H' = \{a_1, a_2, \cdots / r_1 = r_2 = \cdots = 1\}$ is a group and two words belong to the same equivalence class if one can be changed to the other with a finite number of operations of one of the following types:

1. Insert xx^{-1} or its inverse somewhere.
2. Cancel xx^{-1} or its inverse somewhere.
3. Insert an r_i or its inverse somewhere.
4. Cancel an r_i or its inverse somewhere.

An alternative way to look at $H' = \{a_1, a_2, \cdots / r_1, r_2, \cdots = 1\}$ is to let $H' = H/K$ where K is the smallest normal subgroup of H which has elements containing the r_i's.

We find from Sections 5 and 6 of this paper that $\pi_1(E^3 - G) = \{a_1, a_2, \cdots, a_n / r_1 = r_2 = \cdots = r_m = 1\}$ where the a_i's are the letters attached to the arrows going under arcs in G and the r_i's are the relations obtained at the end points, branch points and undercrossing points of G.

If G_1, G_2 are two finite graphs in E^3 such that there is a homeomorphism h of E^3 onto E^3 such that $h(G_1) = G_2$, then $\pi_1(E^3 - G_1) = \pi_1(E^3 - G_2)$. However one should not conclude the converse that if $\pi_1(E^3 - G_1) = \pi_1(E^3 - G_2)$ then there is such a homeomorphism h. Leaving a dangling feeler off a graph does not alter the fundamental group of its complement. In fact, there are smooth simple closed curves (tame knots) in E^3 such that the fundamental groups of their complements are the same even though there is no space homeomorphism taking one onto the other. However, the difference of the fundamental groups of their complements does imply the lack of a space homeomorphism.

8. Short cuts. We mention several simplifications that one discovers as he deals with the fundamental group of the complement of a curve in E^3. The first three involve changing the graph and the last three involve simplifying the presentation of the group.

a. SIMPLIFYING PROJECTIONS. If the regular projection of a curve G_1 looked like G_1' as shown in the left half of Figure 21, its complement would have the same fundamental group as the complement of curve G_2 obtained from G_1 by a homeomorphism of E^3 onto itself. The projection G_2' of G_2 shown in the right half of Figure 21 is simpler.

b. SHRINKING ARCS. If a smooth arc in E^3 is shrunk to a point, the resulting decomposition space is topologically E^3. If one considers an arc, in the sense of Section 4, of G_2 of the right side of Figure 21 and shrinks it to a point to obtain G_3 whose projection is shown in the left side of Figure 22, $E^3 - G_2$ is topologically equivalent to $E^3 - G_3$ and $\pi_1(E^3 - G_2) = \pi_1(E^3 - G_3)$.

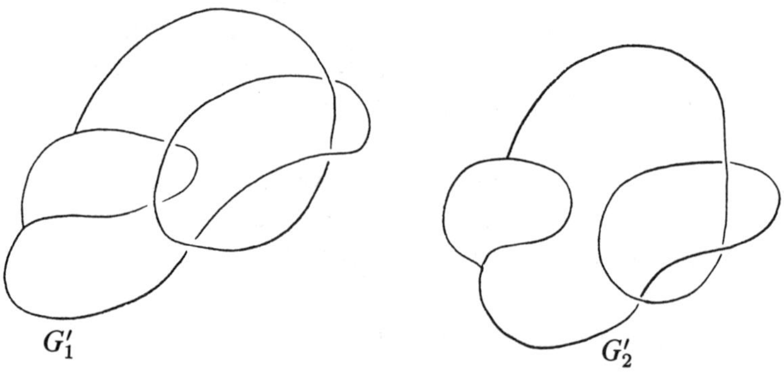

FIGURE 21.

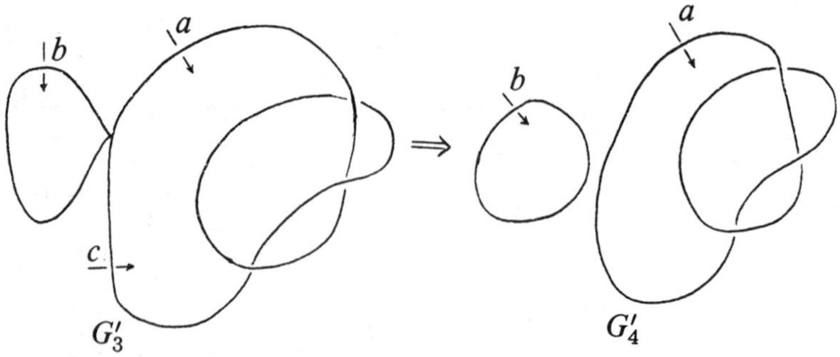

FIGURE 22.

Shrinking arcs enables us to eliminate such dangling arcs from G as shown in Figure 13.

c. PULLING CURVES APART. Note that the relation $bb^{-1}ca^{-1} = 1$ obtained at the branch point of G_3 merely tells us that $a = c$. Since no other relation involves b, $\pi_1(E^3 - G_3) = \pi_1(E^3 - G_4)$, where the projection G_4' of G_4 is shown in the right half of Figure 22. In general it is permissable to pull a curve G apart at a vertex v if the projection of v is a separating point of the regular projection G' of G. To compute $\pi_1(E^3 - G_1)$, it would be a short cut to compute $\pi_1(E^3 - G_4)$.

d. DROPPING A RELATION. Instead of pulling a path f under a particular branch point or an undercrossing point as suggested in Figure 14 or 15, it is possible to get an equivalent homotopy by pulling the path f to the path f' by moving under the rest of G. Hence any particular relation r_i of $\pi_1(E^3 - G) = \{a_1, a_2, \ldots, a_n / r_1 = r_2 = \cdots = r_m = 1\}$ is a consequence of the others and there is no loss in dropping it from the presentation $\{a_1, a_2, \ldots, a_n / r_1 = r_2 = \cdots = r_m = 1\}$.

e. EQUATING GENERATORS. Suppose G is a curve whose regular projection G' is shown in Figure 23. If one followed the suggestions made earlier, one would assign different letters to the vectors a_i shown. However, one finds by either moving under the part of G on the left, or on the right, that a path going under the lower arc is homotopic to one going under the upper arc. If one writes down the presentation of $\pi_1(E^3 - G)$ as suggested in the preceding section, it follows from the algebra that $a_1 = a_2$. It is a short cut to designate each of a_1, a_2 by the same letter, say a, without going through the algebra.

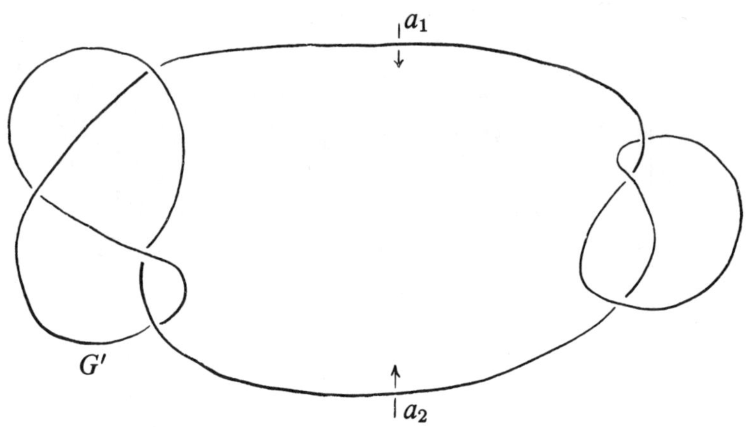

FIGURE 23.

f. FURTHER ELIMINATION OF GENERATORS AND RELATIONS. Consider a graph G whose projection G' is shown in Figure 24. Instead of giving the vector c a name, it would be a short cut to call it $b^{-1}ab$. It is convenient not to assign separate letters to vectors associated with arcs of G that do not pass above

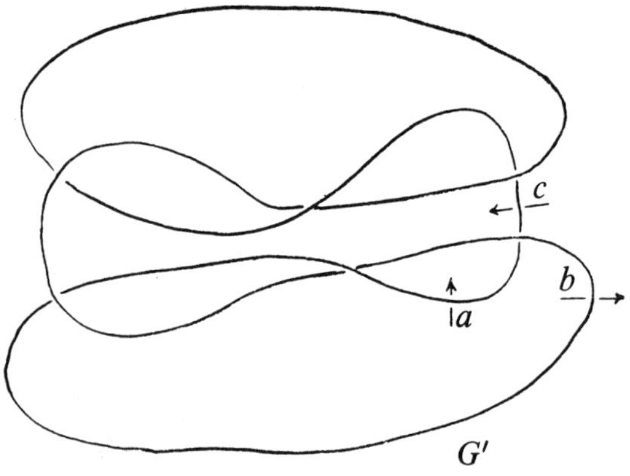

FIGURE 24.

points of G, since these can be assigned words instead of letters. By using $b^{-1}ab$ instead of c, one also eliminates the relation $b^{-1}abc = 1$.

As a consequence of this observation, one obtains the proof of the following theorem:

THEOREM. *Suppose G is a simple closed curve which is the union of n spanning vertical segments of a cube C and n arcs on the boundary of C. Then $\pi_1(E^3 - G)$ has a presentation with n generators and $n-1$ relations.*

In Figure 25 we show the cube C with the front face smaller than the back and the curve G missing the back face. The generators associated with the arcs on the boundary of C are the only ones needed.

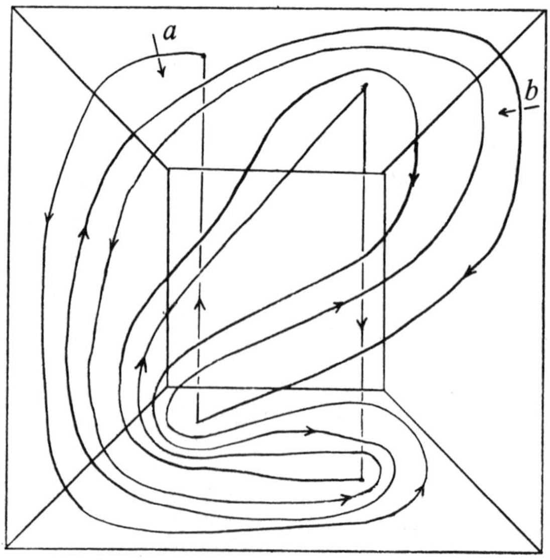

FIGURE 25.

Also by assigning the last generator to be named a word rather than a letter, one gets the following result:

THEOREM. *Suppose G is a smooth curve in E^3 with n undercrossing points. If $n \geq 2$, then $\pi_1(E^3 - G)$ has a presentation with $n-1$ generators and $n-2$ relations.*

9. Why compute fundamental groups?. Some mathematicians study the fundamental groups of the complements of curves for their own sake and because they find the study interesting. We point out some applications the study has to the topology of 3-manifolds.

Many years ago Antoine described a Cantor set C (called Antoine's necklace) such that there is no homeomorphism of E^3 onto itself taking C onto the ordinary Cantor set. It was many years after Antoine described this Cantor set that it was proved to be wild. To show there was no homeomorphism of E^3 onto itself taking C onto an ordinary Cantor set, one needed to show that if T is a solid torus, as shown in Figure 26, and $J_1, J_2, \ldots, J_n$ is a chain of linked simple closed curves, as shown, then a closed curve in $E^3 - T$ could not be shrunk to a point in $E^3 - \bigcup J_i$ if it could not be shrunk to a point in $E^3 - T$. This was done by computing $\pi_1(E^3 - \bigcup J_i)$.

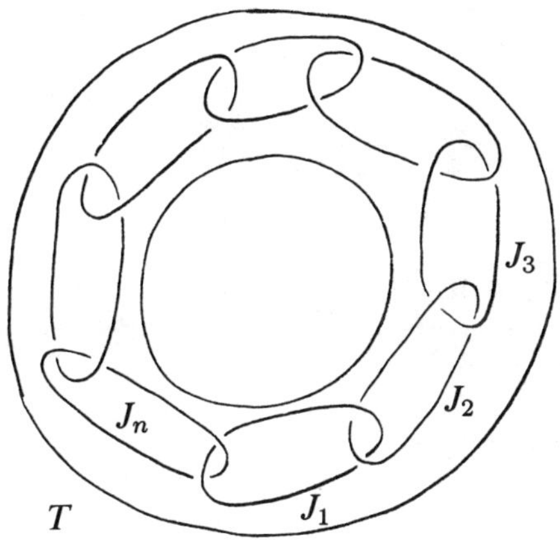

FIGURE 26.

After Antoine described a wild Cantor set, Alexander described a wild sphere such that the fundamental group of its exterior was not trivial. To show that the fundamental group of its exterior is nontrivial, one shows that if X is the curve in the solid torus T as shown in Figure 27, then any closed curve in $E^3 - T$ that cannot be shrunk to a point in $E^3 - T$, cannot be shrunk in $E^3 - X$.

It is an unsolved question (called the Poincaré conjecture) as to whether or not each compact connected 3-manifold whose fundamental group is trivial is a topological 3-sphere. In trying to construct counterexamples, one might remove from S^3 (regard S^3 as E^3 compactified by the addition of the point at

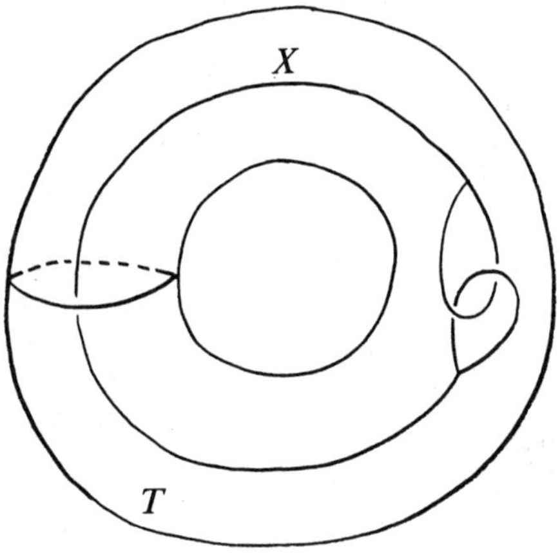

FIGURE 27.

infinity) a tubular neighborhood of a smooth curve in S^3 and sew this tubular neighborhood back in differently. It is known that if the Poincaré conjecture is false, a counterexample can be obtained in this way*. The fundamental group of the complement of the tubular neighborhood is the same as the fundamental group of the complement of the curve.

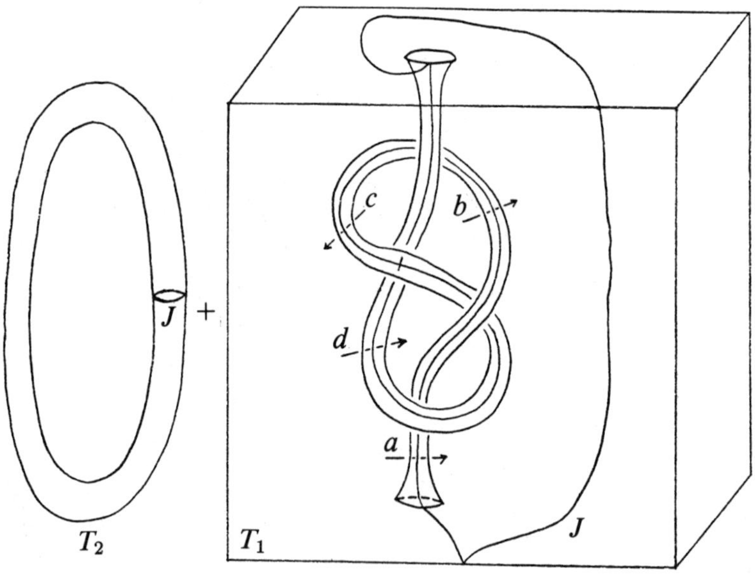

FIGURE 28.

*Editors' note: Whether a counterexample can be so obtained is part of the question whether all knots have property P, which was still unknown in early 1988. See Bing and Martin "Cubes with knotted holes", 1537–1551, in Collected Papers of R. H. Bing for more about property P.

Suppose one considered a cube with a knotted hole bored in it as shown in Figure 28. One computes the fundamental group of T_1 to be $\{a, b, c, d / dbd^{-1}a^{-1} = cdc^{-1}a^{-1} = cbd^{-1}b^{-1} = aba^{-1}c^{-1} = 1\}$. If one sews the solid torus T_2 to T_1 with a homeomorphism of the boundary of T_2 onto the boundary of T_1 so that the meridian simple closed curve J on the boundary of T_2 goes onto the curve J shown on the boundary of T_1, the fundamental group of the union is $\{a, b, c, d / dbd^{-1}a^{-1} = cbd^{-1}b^{-1} = aba^{-1}c^{-1} = da^{-1}bc^{-1}a = 1\}$. Is this group trivial?

In general, one would like to know if there is a cube with a peculiar knotted hole so that one could obtain a 3-manifold with a trivial fundamental group by sewing on a solid torus in an unusual way. What about obtaining a 3-manifold by boring two knotted holes in a cube and sewing on the tubular neighborhood of a figure eight?

UNIVERSITY OF WISCONSIN

F. Others

SOLUTION OF A PROBLEM OF R. L. WILDER.*

By R. H. Bing.

The following problem was proposed by R. L. Wilder in "Concerning Simple Continuous Curves and Related Point Sets," *American Journal of Mathematics*, vol. 53 (1931), pp. 39-55: *Suppose that M is a connected im kleinen set which is the sum of two sets M_1 and M_2 such that each is irreducibly connected from the point A to the point B and such that the common part of M_1 and M_2 is A and B. Is M necessarily a simple closed curve?* Wilder showed that M is a simple closed curve if it is a locally compact continuum. This note gives an example to show that M need not be a simple closed curve if this extra requirement is not placed on it.

We shall prove the following for the plane: *Suppose that S is a square plus its interior. There exists a collection G of point sets filling up S such that each element of G is irreducibly connected from a point A to a point B, the common part of two elements of G is $A + B$, and the sum of two elements of G is a locally connected subset of S which is dense in S.*

Let A and B be opposite vertices of S and let E and F be the other vertices. Denote by $[X]$ the collection of all straight line intervals EP where P is a point of $AF + FB — (A + B)$. For each element X of $[X]$, let $L(X)$ denote the collection of all straight line intervals which have their end points on the boundary of S and which are parallel to X. We note that with respect to its elements, each $L(X)$ is an open arc from A to B. Let $[L]$ be the set of all such collections $L(X)$.

There exists a well ordered sequence $W(L)$ such that the elements of $W(L)$ are the elements of $[L]$ and such that if L is an element of $[L]$, then the set of all elements of $[L]$ which precede L in $W(L)$ has a power less than that of $[L]$. We note that $[L]$ has the power of the continuum. We shall denote the element of $W(L)$ whose ordinal number is α by L_α.

Let $W(S)$ be a well ordering of the points of S.

Denote the collection of all subcontinua of S by $[C]$. Points as well as nondegenerate continua are included in $[C]$. Let $W(C)$ be a well ordered sequence such that the elements of $W(C)$ are the elements of $[C]$ and such that if C is a subcontinuum of S, then the collection of all elements of $[C]$

* Received March 25, 1947; Presented to the American Mathematical Society, November 30, 1946.

that precede C in $W(C)$ does not have the power of $[C]$. Since $[C]$ has the power of the continuum, this subcollection has a power less than that of the continuum. We shall denote the element of $W(C)$ whose ordinal number is α by C_α.

Before beginning the assigning of points of S to the elements of G, we shall briefly outline what we intend to do. We shall describe collections $M_1, M_2, \cdots, M_\alpha, \cdots$ such that each of these collections is irreducibly connected from A to B. These collections will be the elements of G. To insure that M_β $(\beta = 1, 2, \cdots, \alpha, \cdots)$ will be irreducible from A to B, it will be defined so as to intersect every element of $[C]$ which is not a subset of an element of L_β and so as to intersect each element of L_β in only one point.

Our procedure will be to sort out the points of S into the collections $M_1, M_2, \cdots, M_\alpha, \cdots$. With the exception of some skips, the points of S are sorted out to the elements of G in the following order. First, a point of C_1 is assigned to M_1; next, a point of C_2 is assigned to M_1 and another point of C_2 is assigned to M_2; $\cdots$; then a point of C_α is assigned to M_1, another point of C_α is assigned to $M_2, \cdots$, and another point of C_α is assigned to M_α; $\cdots$. Since only one point of S is assigned at each step, this assigning will require uncountably many steps.

Now for the details of the description of the elements of G. The points A and B are assigned to each of the collections of G. No other point is assigned to any two elements of G.

First, consider C_1. If it intersects $A + B$, no point of it is assigned to an element of G at this stage. If it does not intersect $A + B$, the first point of $W(S)$ in C_1 is assigned to M_1.

Next, consider C_2. No point of it is assigned to M_1 at this stage if either C_2 contains a point already assigned to M_1 or C_2 is a subset of an element of L_1 that contains a previously assigned point of M_1. However, if C_2 is not such a set, then to M_1 is assigned the first point of $W(S)$ that is in the common part of C_2 and the sum of the elements of L_1 containing no previously assigned point of M_1.

Now, if C_2 intersects neither A nor B and is not a point previously assigned to M_1, then to M_2 is assigned the first point of $W(S)$ in the subset of C_2 that has not been assigned previously to M_1. Otherwise, no point of C_2 is assigned to M_2 at this stage.

This process of assigning points of S to elements of G is continued as follows. Suppose that at some stage, U is the collection of points of S that have not been assigned previously to elements of G. Let V be the collection of all elements C_α of $W(C)$ for which there is an element C_β of $W(C)$ such that C_β is equal to C_α or precedes C_α in $W(C)$, no point of C_α has been assigned to M_β, and C_α is not a subset of an element of L_β that contains a point which has been assigned to M_β. If C_α is a point of U, it is an element of V since no point would be assigned to M_α before C_α were assigned to an element of G. For convenience, suppose that C_α is the first element of V in $W(C)$. Then a point of C_α is assigned to an element of G at this stage as is described below.

Let R be the collection of all elements L_β of $W(L)$ such that no point of C_α has been assigned to M_β and C_α is not a subset of an element of L_β which contains a point which has been assigned to M_β. Suppose L_β is the first element of R in $W(L)$. Then to M_β is assigned the first point of $W(S)$ in the common part of $C_\alpha \cdot U$ and the sum of the elements of L_β containing no previously assigned points of M_β. If β is the same ordinal number as α, this is the first point to be assigned to M_α.

Suppose that at some stage, Y is the set of points of S that have been assigned previously to elements of G in this process and P is a point of $S - Y$. Then P is an element C_γ of $W(C)$. Now Y is not of the power of the continuum since none of the sets $M_1 \cdot Y, M_2 \cdot Y, \cdots, M_\gamma \cdot Y$ has the power of the continuum.

Each point of S was assigned to some element of G in this process for suppose that U is the collection of all points of S that were not assigned to G even though the process were continued. But as is seen above, a point of U is assigned to an element of G.

No point of $S - (A + B)$ was assigned to two elements of G. Hence, S is the sum of the elements of G and the common part of two elements of G is $A + B$.

We shall now show that each element E of L_α contains a point of M_α. Since E contains a collection of nondegenerate subcontinua such that this collection is of the power of the continuum, there is a nondegenerate subcontinuum C_γ of E such that C_γ follows C_α in $W(C)$. Each point of $S - (A + B)$ was assigned to an element of G while considering an element of $W(C)$. Let Z be the set of all points of S that are assigned to elements of G before considering any element of $W(C)$ that follows C_γ in $W(C)$. Since none of the collections $M_1 \cdot Z, M_2 \cdot Z, \cdots, M_\gamma \cdot Z$ has the power of the continuum, Z does not. Then at some stage, an element of $W(C)$ was considered that followed C_γ in $W(C)$. But this element would not have been considered unless a point of the element of L_α containing C_γ had been assigned to M_α. Hence, each element of L_α contains a point of M_α. However, no two points of an element of L_α are assigned to any element of M_α.

Each subcontinuum C of S which is not a subset of any element of L_a contains a point of M_a because C contains a collection R of subcontinua no one of which is a subset of an element of L_a and such that the collection R has the power of the continuum. Hence, some element of R follows C_a in $W(C)$ and a point of it is assigned to M_a.

We have shown that M_a contains a point of each element of L_a and that each subcontinuum of S which is not a subset of an element of L_a contains a point of M_a. Hence, M_a is connected. If it were the sum of two mutually separated sets, some continuum in its complement and in S would separate some two points of it in S. But each subcontinuum of S that separates S is either an element of L_a or intersects two elements of L_a. Since M_a contains only one point of each element of L_a and since with respect to its elements, L_a is an open arc from A to B, M_γ is irreducibly connected from A to B.

We shall show that if M_a and M_β are two elements of G, then $M_a + M_\beta$ is locally connected. Each nondegenerate subcontinuum of S intersects either two elements of L_a or two elements of L_β and hence contains a point of $M_a + M_\beta$. If D is a domain such that $D \cdot S$ is connected, then $D \cdot (M_a + M_\beta)$ is connected, for if it were the sum of two mutually separated sets, then some continuum R in the complement of $D \cdot (M_a + M_\beta)$ would separate some two points of $D \cdot (M_a + M_\beta)$ from each other in the plane. But some nondegenerate subcontinuum of R would be a subset of $D \cdot S$ and would contain a point of $D \cdot (M_a + M_\beta)$. Hence, $M_a + M_\beta$ is locally connected. Not only is it locally connected but it has the additional property that if P is a point, then $M_a + M_\beta + P$ is locally connected. Hence, the sum of two elements of G has property S.

Question. It can be shown that the sum of no two elements of G is an inner limiting (G_δ) set. It would be interesting to know whether or not it can be concluded that M is a simple closed curve if it is a connected im kleinen inner limiting set which is the sum of two sets M_1 and M_2 such that each is irreducibly connected from the point A to the point B and such that the common part of M_1 and M_2 is A and B.

UNIVERSITY OF TEXAS.

EXAMPLES AND COUNTEREXAMPLES

R. H. Bing, University of Wisconsin

It is not intended that each reader assimilate all the topology in the following expository article. However, it is hoped that he will finish with the impression that mathematics is a living and growing subject and that although we have learned many interesting facts about the structure of things, there is much yet to learn.

Topological Equivalence. Two figures are topologically equivalent if there is a 1-1 correspondence between the points of the two figures such that points "close together" in one figure correspond to points "close together" in the other. For example, a square, a triangle, and a circle are topologically equivalent.

One might suppose that if two sets are topologically equivalent, it is possible to deform one into the other by "pulling and stretching but without breaking or tearing". However, in E^3 (Euclidean 3-space) consider two sets each consisting of two tangent spheres; in the first set the spheres are tangent externally and in the second set one sphere lies on the interior (except for the point of tangency) of the other. The two sets are topologically equivalent because there is a 1-1 correspondence of the proper sort between the two sets. However, it is not possible to deform one set into the other in E^3 by "pulling and stretching but without breaking or tearing".

Let us consider another counterexample to the "pulling and stretching" definition. One can form a cylinder or tube by sewing two opposite sides of a rectangular rubber sheet together. If the ends of the cylinder are sewn together, the resulting figure may resemble an inner tube. However, if a knot is tied in the cylinder before its ends are united, the exterior view of the inner tube will be changed but its

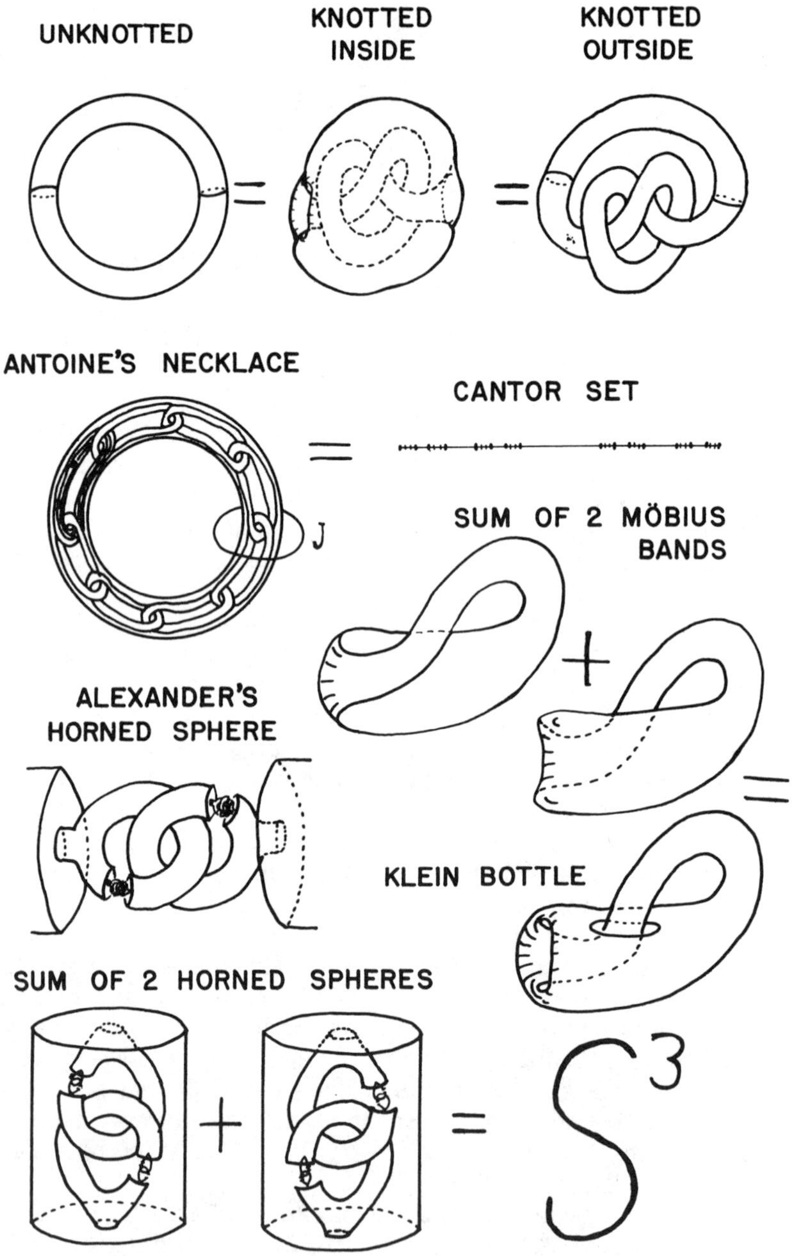
UNKNOTTED
KNOTTED INSIDE
KNOTTED OUTSIDE
ANTOINE'S NECKLACE
CANTOR SET
J
SUM OF 2 MÖBIUS BANDS
ALEXANDER'S HORNED SPHERE
KLEIN BOTTLE
SUM OF 2 HORNED SPHERES
S3

interior structure will not be. If after the knot is tied and before the final sewing is done, one end of the cylinder is stretched and pulled over the knot before the ends are connected, another figure results. The three surfaces shown at the top of the frontispiece are topologically equivalent but not one of them can be stretched to make it fit on any other one of them. The interior of the unknotted surface is topologically equivalent to the outside of the one knotted inside, while the interior of the unknotted surface is like the interior of the one knotted outside.

The complement of a given figure is the set of points not in the figure. For example, on the surface of the earth, the surface of the oceans, seas, lakes, etc., is the complement of the surface of the land. In the plane, the complement of a circle is the sum of the interior and the exterior of the circle. The complement of a circle in E^3 would be connected. We shall give further examples of sets which are topologically equivalent but whose complements are not. They emphasize the contention that two sets may be topologically equivalent even though neither may be pulled to make it fit on the other.

Antoine's Necklace. Suppose M_1 is a solid torus. We will carve from it a necklace as shown on the frontispiece.

Replace M_1 by a closed chain of solid tori as shown. A person who has whittled a chain from a stick will understand how a part of M_1 may be removed so as to leave a closed chain. We call this chain M_2.

Replace each link of M_2 by a closed chain to get M_3. To avoid complications in the figure, we only show this done in one link. Actually each link of M_3 is replaced by a closed chain.

The process is continued a countable number of steps. The points remaining after infinitely many steps is called Antoine's necklace and may be shown (we do not do it) to have the following properties:

(1) Antoine's necklace has uncountably many points.
(2) The biggest connected piece of Antoine's necklace is a point.

(3) If J is the circle illustrated, the circle cannot be shrunk to a point without touching Antoine's necklace.
(4) Each simple surface (set topologically equivalent to a sphere) that has a point of Antoine's necklace on its interior and a point of Antoine's necklace on its exterior also contains a point of Antoine's necklace.

Antoine's necklace gets its name because it was first described by Louis Antoine and, like a set of beads, it is made up of small pieces. From properties (3) and (4) we may see that if it were hung around one's neck, not a point of it would break off and fall to the floor. Although the set is a thing of beauty, what is it good for except possibly to give to one's wife? However, in the last section of this article we discuss some of the uses to which topologists put queer examples.

Cantor Set. A Cantor set as shown on the frontispiece is obtained by starting with a straight line interval, removing its open middle third, removing the open middle thirds of each of the remaining pieces, removing the open middle thirds of the remaining pieces, etc.

A Cantor set can be shown to be topologically equivalent to Antoine's necklace (we do not show it). It has properties (1) and (2) but not property (4). Hence, E^3 cannot be pulled and stretched until Antoine's necklace fits onto a Cantor set.

We call any set topologically equivalent to a Cantor set a topological Cantor set. Antoine's necklace is a topological Cantor set.

It is possible to construct many other interesting topological Cantor sets. There is one in a square plus its interior such that each line intersecting the square intersects the topological Cantor set. There is one in E^3 that would cast a dense shade. There are others in E^3 with positive "volumes". These sets are of interest in studying surface area. Since each topological Cantor set in E^3 lies on a surface with "small area" (in a certain sense), it is possible to obtain surfaces whose "areas" are less than their "volumes".

Sum of Two Möbius Bands. A Möbius band is sometimes called a one-sided band. If a belt connecting pulleys is twisted like a Möbius band, it will wear on both sides. There is the story of the workman who was assigned to the task of painting one side of such a belt and removing all the paint from the other side.

The boundary of a Möbius band is a simple closed curve (set topologically equivalent to a circle). If the boundaries of two Möbius bands are sewn together as shown on the frontispiece, there results a Klein bottle. It has been called a bottle with no inside and no outside.

If one tries to construct a Klein bottle physically by sewing together Möbius bands as shown, a kink develops just before the sewing is finished, making it impossible to complete the construction in E^3. There does not exist a Klein bottle in E^3. However, a topologist does not limit his study to only those figures occurring in E^3. The study of abstract spaces is very interesting.

In traveling around a circle in one direction in the plane, one either goes in a clockwise or a counterclockwise direction. Also, it is possible to assign an orientation to small simple closed curves on a sphere. However, if a cyclone follows a closed path on a Klein bottle and comes back to its starting point, its direction of spinning about the point may have changed, even though it never changed its direction of spinning as it moved.

Another example of a nonorientable surface is the projective plane. It may be obtained by sewing a circle plus its interior to a Möbius band along the boundary of each. Physically, this sewing could not be done in E^3 but it could be done in E^4 and some other spaces.

Horned Sphere. J. W. Alexander described the horned sphere illustrated on the frontispiece. It is topologically equivalent to the surface of a sphere but its exterior is not topologically equivalent to the exterior of a sphere. We might describe the horned sphere as follows. A long cylinder closed at both ends is folded until the ends are near each other and parallel. (In the figure we show only the ends of the closed cylinder.) Then tubes are pushed out of each end until they almost hook as shown. The process is continued by pushing out additional tubes, pushing out additional tubes, etc. The resulting set has the property that although it is topologically equivalent to the surface of a sphere, there is a circle in the exterior of the horned sphere that cannot be shrunk in the exterior of the horned sphere to a point.

A set is simply connected if each simple closed curve (set topologically equivalent to a circle) in it can be shrunk to a point in it. It is uniformly locally simply connected if each "small" simple closed curve in it can be shrunk to a point on a "small" piece of it. The complement of a straight line interval in E^3 is simply connected but not uniformly locally simply connected. A torus is uniformly locally simply connected but not simply connected. Neither the complement in E^3 of Antoine's necklace nor of the horned sphere is either simply connected or uniformly locally so. However, both the complement in E^3 of a Cantor set and a sphere is simply connected and also uniformly locally so.

A simple surface is tame if there is a homeomorphism (1-1 correspondence of the sort previously mentioned) of E^3 onto itself that takes the simple surface onto the surface of a sphere. The surface of an ellipsoid is tame but a horned sphere is not. However, at certain points the horned sphere is smooth and fits into E^3 in a nice fashion. We call it locally tame here. It has been recently discovered that a simple surface is tame if it is locally tame at each point.

Here is an unanswered question. Suppose S is a simple surface in E^3 such that the complement of S is both simply connected and uniformly locally simply connected. Is S tame? Considering the rate at which we are learning new things about E^3, I would conjecture that we shall find the answer before many years.

Sum of Two Solid Horned Spheres. In the last figure on the frontispiece we show two horned spheres. The tubes, instead of being pushed out, were pushed in. The interiors of the horned spheres are not simply connected. Each of the horned spheres plus its interior is called a solid horned sphere.

One of the horned spheres is the reflection of the other through a plane halfway between them. Suppose the two solid horned spheres are sewn together along the points on their boundaries made to correspond under this reflection. To what is the resulting set topologically equivalent?

If a sphere plus its interior is sewn to another sphere plus its interior along the spheres, there results a set topologically equivalent to E^3 with a point added "at infinity". This set may be shown to be topologically equivalent to S^3 (the surface of a sphere in Euclidean 4-space).

Is the set obtained by sewing the two horned spheres together in the above prescribed way topologically equivalent to S^3? The question went unanswered from 1930 to 1951. Some mathematicians conjectured that it was but others thought not. When it was discovered in 1951 that the sum was topologically equivalent to S^3, this discovery exploded a conjecture concerning periodic transformation. (See False Statement H in next section.)

We do not yet know the answer to the following question. If the boundaries of the horned spheres were sewn together in some other fashion (we do not prescribe that a point be joined to its reflection), is the sum topologically equivalent to S^3?

Usefulness of Examples. If we have examples of the things we are studying, these examples may give us a better understanding of our subject. A wealth of examples helps our intuition in picking out certain theorems that are likely to be true and in labeling others as false.

A counterexample is a convincing way of showing that a theorem is false. One of the chief roles of examples is to show that certain statements are false and that it is futile to try to prove them. We illustrate this by mentioning some false statements. Consider the following.

False Statement A. If two sets in E^3 are topologically equivalent and one has the property that a simple surface in its complement separates two points of it from each other, the other set has this property also.

False Statement B. Each surface is orientable.

False Statement C. If two sets in E^3 are topologically equivalent, their complements are also.

False Statement D. The complement of a simple surface is simply connected.

Although the statements may sound very pleasing, a topologist acquainted with Antoine's necklace and a Cantor set would not make (A), one acquainted with a Klein bottle would not make (B), one acquainted with a horned sphere would not make (D), and one acquainted with either Antoine's necklace or a horned sphere would not make (C).

Perhaps it has been many years since any competent topologist considered as true any of the preceding false statements. However, we now mention some, each of which was conjectured as true for many years and each of which was exploded during the past five years. We use the terminology of topology and do not expect the reader to grasp the meaning of each statement.

False Statement E. The arc is the only nondegenerate plane continuum topologically equivalent to each of its nondegenerate subcontinua.

False Statement F. Each nondegenerate indecomposable continuum is 1-dimensional.

False Statement G. No set in the plane is a minimal dispersion set.

False Statement H. Each mapping of period two of E^3 onto itself is topologically orthogonal.

We could mention many other conjectures that have been exploded by counterexamples but instead mention some which still persist.

Conjecture A. Each normal Moore space is metrizable.

Conjecture B. A normal Hausdorff space is countably paracompact.

Conjecture C. A tree-like plane continuum has the fixed point property.

Conjecture D. A connected linear Haudorff space is separable if each uncountable subset of it contains a limit point of it.

The reader may not have understood each of the preceding statements. Nevertheless he should have gotten the point that examples may be very useful in solving difficult problems.

CHALLENGING CONJECTURES

R. H. BING, University of Wisconsin

1. Importance of research to teaching. Mathematics is an alive and growing subject. It is best taught by teachers who are mathematically alive and mathematically growing.

The editorial in the June 18, 1965 issue of *Science* was entitled "What Are Professors For?" It says that the professor's primary activities should be teaching and research. Concerning the teaching portion of his duties the editorial has this to say, "If transfer of information were enough, society could dispense with most professors and education could be almost completely mechanized. The professor's most important role is to provide various forms of guidance for the students. As a result of his university experience, a student should be motivated to pursue continuing scholarship throughout his life."

Motivated to pursue continuing scholarship! If the teacher himself does not pursue continuing scholarship, how effective is he in encouraging his students to do so?

If the teacher pretends to be interested in getting solutions to difficult problems, pretends to be interested in proving challenging theorems, pretends to be seeking knowledge, pretends to have an inquiring mind—how many students are dumb enough to be fooled by the fakery if the teacher is merely pretending but in fact is resting on his background?

The effective teacher is the one who moves ahead. He should not be like the little boy who asked his mother, "Why do I always go around in circles?" and gets the reply "Shut up or I'll nail your other foot to the floor." The teacher should move ahead on two feet—both in teaching and in acquiring mathematical proficiency.

There has been much publicity given recently to the emphasis on research to the extent that teaching is neglected. The layman is disturbed by this. A recent article in the *Reader's Digest* was entitled "Is There a Teacher on the Faculty?" University and college people are less frightened because they know that most researchers do teach and are all the more effective as teachers because of their research experience. We occasionally hear of fellowship bums and those who shun teaching to get contracts. My experience suggests that there is relatively little of this abuse. If it were widespread it would be most unfortunate since those working at the frontier can be and frequently are our most effective teachers. Mathematicians as a whole are an energetic, intelligent, dedicated lot who try for original results themselves but who also see teaching as a way of promoting mathematics.

Let us look at the other side of the coin. It is most unfortunate for the teaching profession when the professor goes to sleep.

The slogan "Publish or Perish" might be interpreted to suggest that the teacher who does not make progress in contributing to knowledge by discovering new mathematical truths or finding new uses of mathematics loses his effectiveness as a mathematics teacher. Recent institutes for college teachers have turned up examples of teachers who have not kept up and have practically perished mathematically.

One of the important innovations of the last decade has been the massive fellowship and grant programs instituted to stimulate more research and better teaching. Governmental agencies such as the National Science Foundation, Office of Naval Research, Air Force Office of Scientific Research, U. S. Army Research Office, evidencing the nation's concern that the United States obtain and maintain a position of scientific leadership in the world, have developed fellowship and grant programs. Public-spirited private foundations such as the Sloan Foundation, the Ford Foundation, the Research Corporation, the Carnegie Corporation, the Rockefeller Foundation, the Guggenheim Foundation, I.B.M., Bell Telephone, U. S. Steel, and others have made valuable contributions.

Many of these programs were aimed at promoting more and better research but a side effect has been that we are getting better teaching. Of course, even those programs aimed at better teaching will eventually result in more and better research. Thanks in a large measure to these fellowship and grant programs and to the leadership and hard work of many of you, I believe that we are providing students in the United States today with the best mathematical training that any group has ever had in our history.

There is room for improvement and we should continue to move ahead. In particular I hope we find better means of keeping the younger faculty of our smaller colleges active. Experience has shown that it is far easier to keep working people alive than it is to resurrect the dead.

Although doing research and finding new applications of mathematics are perhaps the most interesting ways of working at mathematics, other means of maintaining mathematical vigor include keeping acquainted with new developments in mathematics by reading mathematical articles, attending seminars, participating in the development of new curricula and teaching techniques, solving mathematical puzzles and problems, and coaching mathematics teams. The two publications *Mathematics Magazine* and *The American Mathematical Monthly* put out by the Mathematical Association of America can be of help in pursuing some of these activities.

In the better schools, teaching and research move forward hand in hand. That mathematics is an alive and growing subject makes it interesting from the standpoint of both teaching and research. Research acts as a leaven—as a catalytic agent that improves the whole loaf.

Let us look at some challenging conjectures that have stimulated much mathematical thought. We shall take a brief look at each of the following:

Continuum hypothesis.
Normal Moore Space conjecture.
Fixed point conjecture for cellular plane sets.
Poincaré conjecture.
Annulus conjecture.

2. Continuum hypothesis. Perhaps the oldest of these conjectures and the one to which most thought has been given is the continuum hypothesis—which is that C is the first uncountable cardinal.

I would like to discover whether one could prove on the basis of a reasonable set of assumptions that for each uncountable subset X of the Euclidean line E^1 there is a 1-1 correspondence between the points of X and the points of E^1, or whether there is some uncountable subset of E^1 for which there is no such correspondence.

There have been some very interesting theorems by Paul Cohen and others showing that these questions cannot be answered in certain restricted logical systems. This reminds us a bit of the situation in which Euclid and his associates were unable to prove certain theorems without the Parallel Postulate. However, these logical systems which have been shown to be inadequate to handle the continuum hypothesis have countable models and axiomatic systems based on these logical systems do not purport to describe adequately the working model of the real number system as I and many other mathematicians envision it. When I ask if these things can be proved on the basis of a reasonable set of assumptions, I do not know what these might be but would hope to include only things that would be viewed by many as "self evident truths" and not use assumptions as far fetched as the continuum hypothesis itself.

For example, I would be quite willing to have as one of my reasonable assumptions the assertion that any uncountable subset of E^1 is intrinsically uncountable—it is uncountable, not merely because one does not have a handy effective way of setting up a 1-1 correspondence between it and the integers, but rather, it has so many points that even with additional consistent assumptions, there is no such 1-1 correspondence. Not everyone would agree to this assumption, but enough do to meet my criterion of reasonableness.

Also, I would regard the "integers" as a well defined set on which most mathematicians agree even without axiomatization, rather than things satisfying certain noncategorical systems. Similarly, I regard the "reals" and E^1 as fixed, invariant, agreed upon sets. These viewpoints do not lend themselves well to systems with countable models where any uncountable set is at best conditionally uncountable, rather than intrinsically uncountable. It would be interesting to know if one steps outside these abstract "reals" with countable models and into the real reals, whether or not the continuum hypothesis is true.

3. Normal Moore space conjectures. The normal Moore space conjecture perhaps stands with one foot in Foundations and the other in Topology. It suggests that a certain set of conditions imposed on a topological space is enough to insure that the space is metrizable.

The notation of a distance is important in geometry, analysis, and topology—but its meaning in analysis and topology is slightly more abstract than its meaning in geometry. We note that to each pair of points x, y of a Euclidean space there is associated a number $D(x, y)$ called the distance between x and y, satisfying the following conditions:

1. $D(x, y) \geqq 0$, the equality holding if and only if $x=y$.
2. $D(x, y)=D(y, x)$ (symmetric condition).
3. $D(x, y)+D(y, z) \geqq D(x, z)$ (triangle inequality).
4. D preserves limit points. The last condition means that p is a point of the closure of a set X if and only if $D(p, X)=0$.

These conditions alone do not furnish us with all the properties of a Euclidean distance, any more than a set of axioms for the reals based on a countable model of set theory gives a complete picture of the reals as mentioned in the discussion of the continuum hypothesis. However, the properties are powerful enough to be useful in proving many theorems and in getting some counterexamples.

A topological space with a distance function has a sequence of open coverings such that the meshes of the open coverings approach zero. Such sequences of open coverings are very useful in studying topological spaces since they provide us with means of proving theorems, building arcs, approximating the space with polyhedra, and computing certain algebraic properties of the space.

In his book, *Foundations of Point Set Theory*, R. L. Moore takes as axiomatic the condition that a space has a decreasing sequence $G_1, G_2, \cdots$ of open coverings. He does not assume a distance function so cannot discuss the meshes of the coverings, but uses a substitute. He does this by supposing that if N is a neighborhood of a point p, there is a covering G_i of the sequence of coverings such that each open set in G_i which contains p lies in N. A regular Hausdorff space with such a decreasing sequence of open coverings is called a *Moore space*.

A Moore space has many of the properties of a metric space and many theorems about metric spaces can be proved equally well in such a space. There are examples known of Moore spaces which do not have a distance function since they are not normal—that is, they contain two mutually exclusive closed sets which do not lie in mutually exclusive open sets. Such an example is given by the following flow space.

Suppose water is flowing across a square at a varying rate—say varying

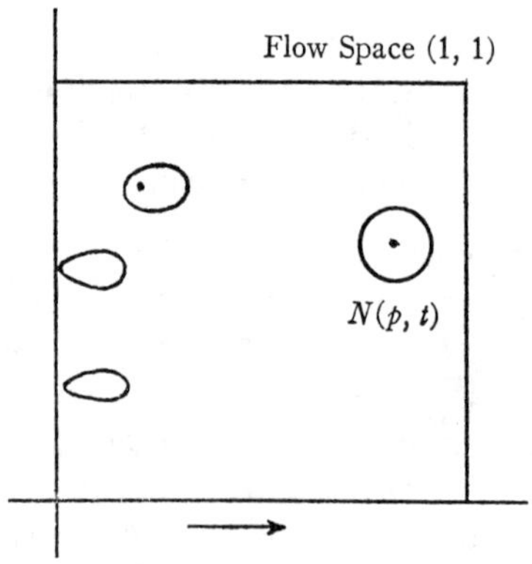

FIG. 1. (rate $=1-x$)

linearly from one foot per second on the left to zero on the right. Points of the flow space are points of the square and for each point p of the square and each $t>0$, $N(p, t)$ is the set of all points of the square to which one could swim from p in time less than t if one swims at a rate of one foot per second. The $N(p, t)$ neighborhoods on the right of the square are nearly round while those on the left are oblong as shown in Figure 1.

No two points of the left side of the square belong to the same $N(p, t)$. It may be shown that the square whose basis of neighborhoods are the $N(p, t)$'s is a Moore space. It is not normal because if one takes as A and B, the rational and irrational points on the left side of the square, these are mutually exclusive closed sets which do not lie in mutually exclusive open sets.

Many research hours have been spent in trying to decide if each normal Moore space is metrizable. Partial solutions have been given. Just as the continuum hypothesis has been translated into equivalent problems, the normal Moore space conjecture has been translated. In the August 1965 issue of the *Proc. Amer. Math. Soc.*, I show relations between the conjecture and the existence of certain functions from abstract sets into the positive integers. Because of such considerations, the normal Moore space conjecture lies partially in Foundations and partially in Topology.

4. Fixed point conjecture for cellular plane sets. A straight line segment has the fixed point property in the sense that if f is a map of I^1 into I^1, then some point of I^1 goes onto itself under f. To see why this is true, one can consider I^1 as horizontal and let R be the set of points of I^1 that move to the right under f and L be those which move to the left. If there is no fixed point, the left point of I^1 is in R and the right point in L. It may be shown that each of L and R is open in I^1. Since I^1 is connected, it is not the sum of two non-null mutually exclusive open sets. Hence there is a point of I^1 in neither R nor L. This point is left fixed under f.

It may also be shown that a square has the fixed point property. If one were to pick up a piece of paper, wad it up, and drop it back on the spot where it lay, there would be some point directly above where it once was.

Cells of all dimensions are known to have the fixed point property.

A set is cell-like if it is the intersection of a decreasing sequence of topological cells. There are many properties that are inherited in the sense that a set has the property if it is the intersection of a decreasing sequence of sets with the property. Being closed and being compact are such properties. People once wondered if the fixed point property was such an inherited property. They now know that it is not.

However, it is still unknown whether or not a set X in the plane has the fixed point property if it is the interesection of a decreasing sequence of topological disks. I have received papers to referee in which erroneous proofs were given. I saw a preprint of another paper giving an erroneous counterexample. Once I attended a lecture in which the speaker was giving a proof and as a result of a question from the audience during his talk, he stopped his talk because the question uncovered a flaw. It is a tantalizing question because it seems that an answer should be available.

In 1948 Borsuk discovered an example of a set which did not have the fixed point property even though it was the intersection of a decreasing sequence of topological cubes. Here is another example.

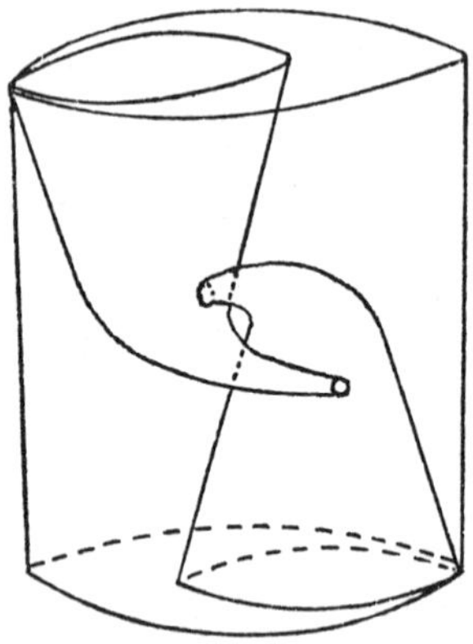

FIG. 2

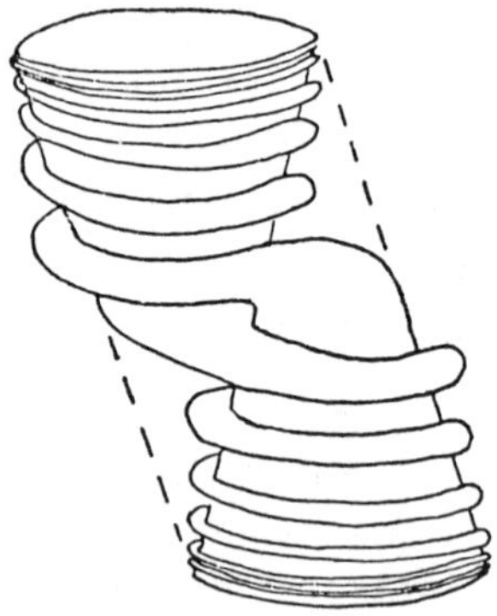

FIG. 3

Start with a solid cylinder as shown in Figure 2. Bore a hole from the top, a hole from the bottom, and whittle away at the sides. Finally, have left a set like that shown in Figure 3. The horizontal cross sections of the ends are circles and other horizontal cross sections are the unions of tangent circles. The set of points of tangency spirals about the bottom circles and about the top circle. It is what one would obtain by taking two cones, one right side up and the other upside down and sewing them together so that the vertex of one spirals around the base of the other as shown in Figure 4. To get a map that leaves no point fixed,

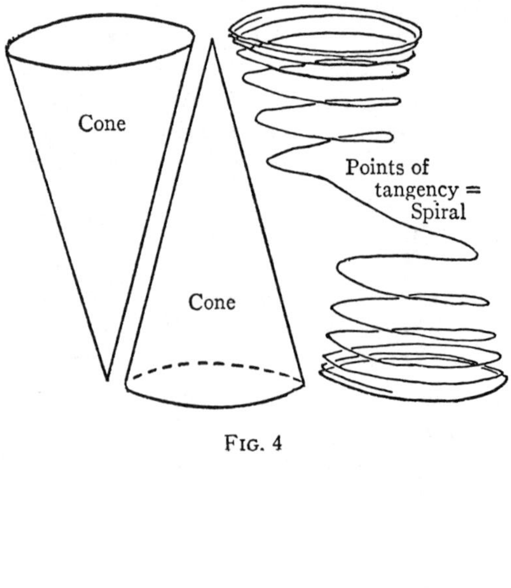

FIG. 4

one merely needs to insist that the top circle goes onto itself under a rotation, the bottom circle goes onto itself under a rotation, and each other horizontal cross section goes onto a slightly higher cross section.

5. The Poincaré conjecture. In high school geometry we work with a set of axioms that characterize the plane. In this spirit it is of interest to consider sets of conditions such that Euclidean 3-space satisfies these conditions and any topological space which satisfies the conditions is topologically equivalent to Euclidean 3-space. The Poincaré conjecture gives a set of conditions that appear to some to be enough to insure that any topological space satisfying them would be topologically like closed 3-space—that is Euclidean 3-space with a point added at infinity to make it a 3-sphere.

The Poincaré conjecture perhaps arose as a result of a mistake. In 1900 Poincaré announced that any two n-manifolds were topologically alike if they had the same homological properties. He perhaps indicated a proof, but it was in error since he showed in a 1904 paper that there is a 3-manifold with the same homological properties as a 3-sphere but which is topologically different.

Homotopy properties are stronger than homological properties but in 1919 Alexander showed that there are lens spaces with the same homotopy properties that are combinatorially different. As a result of some more recent work of Moise we now know that these lens spaces are topologically different.

There have been many false solutions to the Poincaré conjecture—most of them never reached print. One by Whitehead appeared in 1934, but he made a retraction in 1935. I gave a partial solution in 1958 when I showed that a stronger condition than that stipulated in the Poincaré conjecture was enough to insure that the space actually was topologically a 3-sphere. For a while it looked like I might withdraw my paper since it was rumored in 1957 that both Moise and Papakyriakopoulos had complete solutions. Naturally, I did not wish to publish a partial solution if others got the whole thing. Both of their proofs sprung leaks.

In 1957 Poenaru wrote a long paper trying to give a solution but an error was discovered in his proof by the referee of a journal and it did not reach print. Koseki published a long proof in 1958 but the proof is not clear at one crucial spot.

Last summer, it appeared that each of Haken and Poenaru might have solutions. Although each tried novel methods and got interesting results, neither actually got a solution to the Poincaré conjecture.

The precise statement of the Poincaré conjecture is that a compact connected 3-manifold is topologically a 3-sphere if it is simply connected. We shall not discuss it further, but those interested in its history and various partial solutions of it might find of interest the chapter entitled "Some aspects of the topology of 3-manifolds related to the Poincaré conjecture" in volume II of *Lectures on Modern Mathematics* (Wiley, New York, 1964). Any set homeomorphic to an annulus is called an annulus.

6. The Annulus Conjecture. The closure of the part of the plane between two concentric circles is called an annulus. Generalizing this notion we say that the closure of the part of E^n between two concentric $(n-1)$-spheres is an n-dimensional annulus.

Suppose S is a smooth $(n-1)$-sphere in E^n and h is a homeomorphism of E^n onto itself such that $h(S)$ lies in the bounded component of E^n-S. Although $h(S)$ may not be smoothly embedded, it has a Cartesian product neighborhood since h is a space homeomorphism. Is the closure of the part of E^n between S and $h(S)$ an annulus? The answer is yes if $n=1$, 2, or 3. These are precisely the dimensions in which we can triangulate manifolds. If we knew how to prove the annulus conjecture for $n\geqq 4$, this would be useful in studying higher dimensional manifolds.

The annulus conjecture has been shown to be equivalent to other hard conjectures. False solutions to it have been proposed and submitted for publication, but so far as I know, no solutions (false or otherwise) have appeared in print.

7. Mathematics moves forward. The conjectures we have discussed are difficult and it is only infrequently that problems as difficult as these are solved. In fact, perhaps as many will be settled by counterexamples as by proofs. I have a feeling that most topology theorems that are true are true because of some simple underlying geometric principle. If there is such a principle we may write a clumsy complicated proof at first but once the underlying principle is understood, the proof clears up. If there is no such easy underlying geometric principle, the chances are that the conjecture is false.

Although the above conjectures may be hard to solve, I believe it is fruitful to ponder them. Good results are often found when one is looking for something else. With more space I could point out many interesting results that were obtained in unsuccessful attempts to answer the preceding conjectures.

Today, new results are being proved at an increasing rate but these proven theorems reveal many interesting things yet to try. As evidence of the amount of research being done I mention that in 1964 over 1200 abstracts were presented at various meetings of the American Mathematical Society.

Of course, only a fractional part of the mathematical research being done in the world is done in the United States. *Mathematical Reviews* which summarizes papers from all over the world reviewed 7000 papers during the first six months of 1965. Estimating the average length of a research paper to be 10 pages (some are much longer and some shorter) this comes to 140,000 pages of mathematics a year—more pages than many students read in an entire college career. I would not suggest that you read much of it. It contains few expository articles or textbooks. Most of the review articles are concerned with careful proofs, new methods of doing things, new applications of mathematics, reasons why certain things are true.

One nice thing about working at mathematics is that you do not need to know everything about everything. You should have a solid background, and if you are to work successfully at the frontier you need to know almost everything about something.

For those students who follow mathematics as a career, I hope we can get more of them to remain mathematically active. To train them properly we need teachers with mathematical vigor who will remain learners.

I have heard that when learning ceases, rigor mortis sets in. Whether or not this is true, I feel that our schools today are turning out too many finished people—people whose learning is essentially finished when they graduate.

In the United States today there is a need for scholarly people: those who will remain students, scholars, learners, thinkers after college—teachers who can devise and teach a different and better course than they were ever taught.

The progress of our country depends in a large measure on active people. Our most effective teachers are the sort of people who subscribe to professional journals, participate in mathematical activities, read, think, ponder, plan, forge ahead. They work at mathematics whether it be at solving problems in the *Monthly* or doing profound mathematical research.

This is a good day to be a mathematician and teacher. Things are moving. There is excitement and controversy. Changes are being made. The rewards are better than ever before.

But on the other hand, more is expected of us. The high school teacher is supposed to teach not only the facts of mathematics but also an understanding of why these facts are true. It is hoped that the college teacher will impart not merely facts and understanding but also a glimpse of the charm and challenge of mathematics.

I wish you success in leading your better students to see the beauty of mathematics, to give them a taste of the type of mathematics that grips a mathematician's interest, tugs at his imagination, and makes him willing and eager to work long and hard to get a result that challenges him.

We realize that most students will not be research mathematicians, most will not reach the frontier—but they appreciate mathematics more if they know that the frontier is there, know that it is not inaccessible, get an occasional glimpse of it, and realize that it is being pushed back—sometimes steadily but often spasmodically.

Mathematics is not merely a subject of the past, but it is an alive and growing subject. It is not just a subject for the exceptional genius but it is a subject with a challenge for all of us.

О конечности числа отражений, переводящих плоский невыпуклый многоугольник в выпуклый

Р. Г. Бинг и Н. Д. Казаринов
R. H. Bing and N. D. Kazarinoff
Мадисон, Висконсин, США

Пусть P — простой невыпуклый n-угольник (рис. 1), AB — сторона выпуклой оболочки H многоугольника P(1), имеющая с P только две общие точки A и B. Часть $H - P$, прилежащую к AB (на рис. 1 заштрихована), отобразим симметрично от AB; в результате из исходного многоугольника путем замены в нем соответствующей части периметра ее зеркальным отображением, получим новый многоугольник $s(P)$, где s — примененная к P *операция отражения*. Ясно, что $s(P)$ есть k-угольник $(n-2 \le k \le n)$, стороны которого равны соответствующим сторонам P. Рассмотрим последовательность многоугольников $\{r_m(P)\}$, где $r_0(P) = P$, $r_m(P) = r_m(r_{m-1}(P))$ $(m > 0)$, и все r_m при $m > 0$ суть произвольно выбранные операции отражения s.

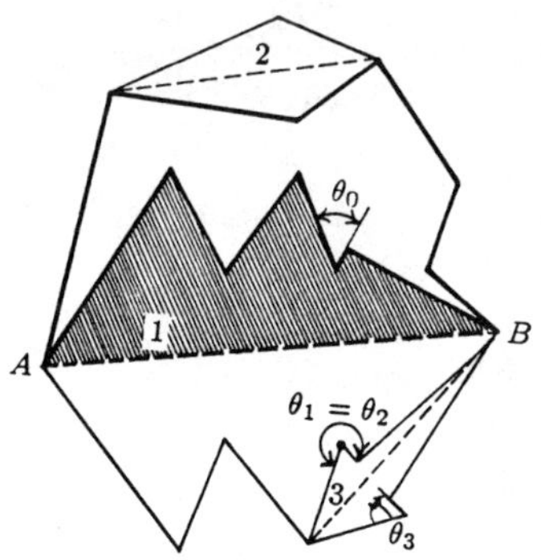

Рисунок 1

(1) Выпуклой оболочкой многоугольника называется наименьший выпуклый многоугольник, содержащий данный. (*Все примечания к настоящей заметке — редакционные.*)

ТЕОРЕМА. *Последовательность* $\{r_m(P)\}$ *конечна, и ее последний член есть выпуклыи* k*-угольник* $(k \leq n)$. [2]

ДОКАЗАТЕЛЬСТВО. Пусть V_0 — произволная вершина P и $V_m = r_m(V_{m-1})$ $(m \geq 1)$. Допустим, что последовательность $\{r_m(P)\}$ бесконечна.

1°. Докажем сначала, что $V_m \to V$ при $m \to \infty$; отсюда будет следовать, что последовательность $\{r_m(P)\}$ точечно сходится и имеет пределом некоторый предельный k-угольник, причем $k \leq n$. Для доказательства заметим, что если E, F, G — три внутренние точки H, не лежащие на одной прямой, то три последовательности расстояний $\{EV_m\}$, $\{FV_m\}$, $\{GV_m\}$ являются ограниченными последовательностями неубывающих положительных чисел.[3] Пусть $R_E = \lim_{m\to\infty} EV_m$; аналогично определяются R_F и R_G. Три окружности с центрами в точках E, F и G и с радиусами R_E, R_F, R_G должны иметь по крайней мере одну общую точку, а поскольку точки E, F, G не лежат на одной прямой, то они имеют единственную общую точку V. Это доказывает 1°.

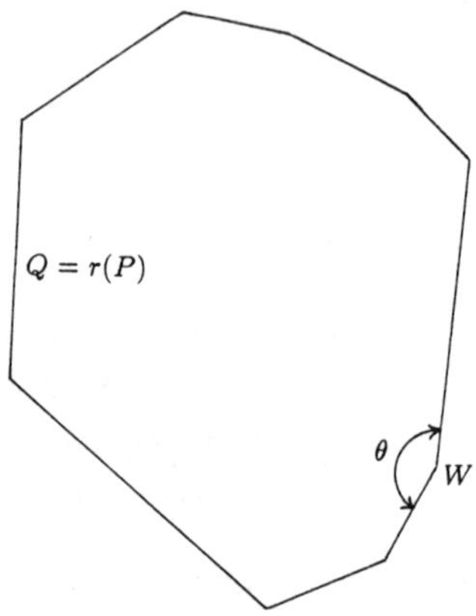

Рисунок 2

Пусть теперь W — любая вершина выпуклой оболочки Q предельного многоугольника $r(P)$ (рис. 2).

2°. Докажем, что любая вершина $r(P)$ получена в результате *конечного* числа N_W отражений. Чтобы доказать это, будем рассуждать следующим образом. Пусть θ_m есть угол многоугольника $r_m(P)$, соответствующий углу θ многоугольника $r(P)$ при вершине W. Мы знаем из 1°, что $\lim_{m\to\infty} \theta_m = \theta$. Кроме того, $\theta < \pi$. Но если W_m не совпадает с W_{m+1} для некоторого m и

---

[2] Эта теорема позволяет свести решение ряда экстремальных задач для простых многоугольников к (более простым!) задачам касающимся выпуклых многоугольников. Несколько иное доказательство той же теоремы см. Ю. Г. Решетняк, Об одном приеме превращения невыпуклой ломаной в выпуклую, Успехи матем. наук **12**, № 3(75), 1957 стр. 189–191.

[3] Ограниченность следует, например, из того, что все $r_m(P)$, как многоугольники одного и того же периметра p, имеют диаметр, меньший $p/2$, и, кроме того, имеют общую часть с H.

если $\theta_m < \pi$, то $\theta_{m+1} = 2\pi - \theta_m > \pi$, а если $\theta_m > \pi$, то $\theta_{m+1} = 2\pi - \theta_m < \pi$ (см. рис. 1). Следовательно, $\lim_{m\to\infty} \theta_m = \theta < \pi$ только в том случае, если все θ_m одинаковы, начиная с некоторого m, т.е. тогда и только тогда, когда W получится в результате конечного числа отражений.

3°. Пусть теперь $N = \max N_W$ для всех вершин W многоугольника Q. Тогда $r_N(P) = Q = r(P)$.(4)

Это очевидно! Доказательство завершено.

ПРЕДПОЛОЖЕНИЕ. При фиксированном n для всех n-угольников и при любом вюборе операций обращения r_m число N ограничено величиной $2n$.

ЗАМЕЧАНИЕ. Многоугольник P не обязательно должен быть простым.

Доказательство этой теоремы, данное Б. Секефальви-Надем [см. B. Sz.-Nagy, Amer. Math. Monthly **46**, 1939, стр. 176–177], неверно.

---

(4) Q совпадает с $r_N(P)$, так как в противном случае существовала бы операция отражения, выводящая $r_N(P)$ за пределы Q, и Q не могло бы быть выпуклой оболочкой $r(P)$.

On Finiteness of the Number of Mappings Carrying a Planar Nonconvex Polygon into a Convex One

R. H. Bing and N. D. Kazarinoff

Let P be a simple nonconvex n-gon (Figure 1), and AB a side of the convex hull[1] H of P, with AB having only the two points A and B in common with P. We map the part of $H - P$ adjacent to AB (shaded in Figure 1) symmetrically with respect to AB; as a result we get from the original polygon a new polygon $s(P)$ by replacing the corresponding part of the perimeter by its mirror reflection, where s is a *reflection operation* applied to P. It is clear that $s(P)$ is a k-gon $(n-2 \leq k \leq n)$ whose sides are equal to the corresponding sides of P. We consider the sequence $\{r_m(P)\}$ of polygons, where $r_0(P) = P$, $r_m(P) = r_m(r_{m-1}(P))$ $(m > 0)$, and all the r_m for $m > 0$ are arbitrarily chosen reflection operations s.

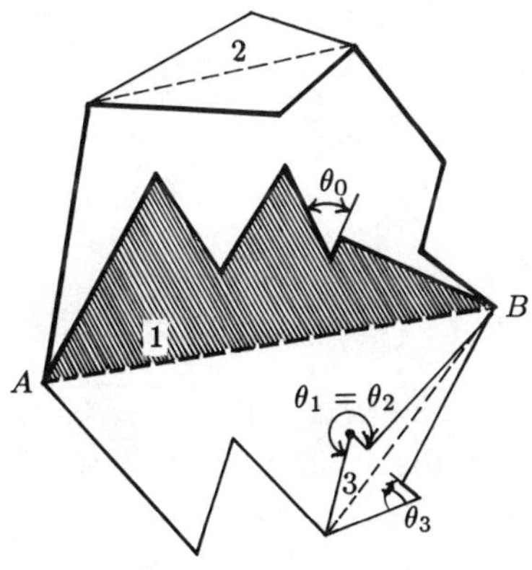

FIGURE 1

THEOREM. *The sequence $\{r_m(P)\}$ is finite, and its last term is a convex k-gon $(k \leq n)$.*[2]

1980 *Mathematics Subject Classification* (1985 *Revision*). Primary 52A10, 52A25.

[1]The convex hull of a polygon is defined to be the smallest convex polygon containing the given one. (All footnotes in this note are due to the editors.)

[2]This theorem enables one to reduce the solution of a number of extremal problems for simple polygons to (simpler!) problems about convex polygons. For a somewhat different proof of the same theorem see Yu. G. Reshetnyak, *On a method for turning a nonconvex polygonal curve into a convex one*, Uspekhi Mat. Nauk **12** (1957), no. 3(75), 189–191.

PROOF. Let V_0 be an arbitrary vertex of P and let $V_m = r_m(V_{m-1})$ $(m \geq 1)$. Assume that the sequence $\{r_m(P)\}$ is infinite.

1°. We first prove that $V_m \to V$ as $m \to \infty$; this implies that the sequence $\{r_m(P)\}$ converges pointwise and has some limiting k-gon with $k \leq n$ as a limit. For a proof, note that if E, F, and G are three interior points of H not on a single line, then the three sequences $\{EV_m\}$, $\{FV_m\}$, and $\{GV_m\}$ of distances are bounded sequences of nondecreasing positive numbers.[3] Let $R_E = \lim_{m\to\infty} EV_m$; R_F and R_G are defined similarly. The three circles about E, F, and G with radii R_E, R_F, and R_G must have at least one common point, and since E, F, and G are not collinear, the circles have a unique common point V. This proves 1°.

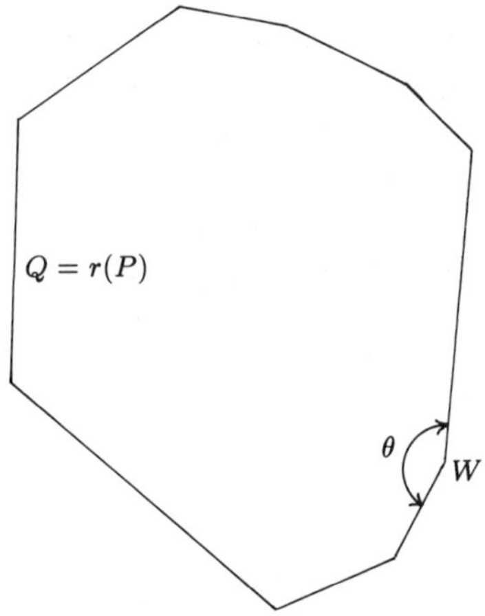

FIGURE 2

Suppose now that W is any vertex of the convex hull Q of the limiting polygon $r(P)$ (Figure 2).

2°. We prove that any vertex of $r(P)$ is obtained as a result of a *finite* number N_W of reflections. To prove this we argue as follows. Let θ_m be the angle of the polygon $r_m(P)$ corresponding to the angle θ of the polygon $r(P)$ at the vertex W. We know from 1° that $\lim_{m\to\infty} \theta_m = \theta$. Moreover, $\theta < \pi$. But if W_m does not coincide with W_{m+1} for some m, and if $\theta_m < \pi$, then $\theta_{m+1} = 2\pi - \theta_m > \pi$, while if $\theta_m > \pi$, then $\theta_{m+1} = 2\pi - \theta_m < \pi$ (see Figure 1). Consequently, $\lim_{m\to\infty} \theta_m = \theta < \pi$ only in the case when all the θ_m are the same beginning with some m, i.e., if and only if W is obtained as a result of finitely many reflections.

3°. Now let $N = \max N_W$ over all vertices W of the polygon Q. Then $r_N(P) = Q = r(P)$.[4]

[3]The boundedness follows, for example, from the fact that all the $r_m(P)$, as polygons with the same perimeter p, have diameter less than $p/2$, and, moreover, have a part in common with H.

[4]Q coincides with $r_N(P)$, for otherwise there would exist a reflection operation carrying $r_N(P)$ beyond the limits of Q, and Q could not be the convex hull of $r(P)$.

This is obvious! The proof is complete.

ASSUMPTION. For fixed n the number N is bounded by $2n$ for all n-gons and for any choice of the reflection operations r_m.

REMARK. The polygon P need not be a simple polygon.

The proof of this theorem given by B. Sz.-Nagy in Amer. Math. Monthly **46** (1939), 176–177, is not valid.

Translated by H. H. McFaden

Models for S^3

R. H. Bing

1. Spheres in general. Spheres are among the objects studied in geometric topology. It is convenient for us to have models of objects that we are studying since these models give a better intuitive understanding of our subject. Models may be of help to us in conjecturing theorems and building counter examples.

We may define an *n-sphere* as an object homeomorphic with the boundary of an $(n+1)$-ball. The *canonical n-sphere* is the set of points in Euclidean $(n+1)$-space R^{n+1} satisfying the equation

$$x_1^2 + x_2^2 + \cdots + x_{n+1}^2 = 1.$$

We use S^n to denote this canonical n-sphere. Any object topologically equivalent to this canonical n-sphere is called an n-sphere and on occasions we find it convenient to denote one of these other n-spheres by S^n. If B^{n+1} is an $(n+1)$-*ball*, Bd denotes *boundary*, and $=_t$ means *is topologically equivalent to*, then $\text{Bd}\, B^{n+1} =_t S^n$.

The above definition has more intuitive meaning in low dimensions. For example, one sees that

$$S^0 =_t \text{ pair of points,}$$
$$S^1 =_t \text{ circle } =_t \text{ simple closed curve,}$$
$$S^2 =_t \text{ boundary of a round ball.}$$

Also, the boundaries of ellipsoids, rectangular solids, and tetrahedra are examples of 2-spheres. The boundary of a rough rock may be a more exotic example. It is more difficult to visualize examples of higher dimensional spheres. There are no such readily available examples in Euclidean 3-space R^3, the space in which we presume we live.

A preliminary version of this paper was read by David G. Wright, Elizabeth Powell, and Ralph Jones. They made suggestions for improving its clarity.

2. Geometric models of S^3. Describing S^3 with an equation might be considered an algebraic approach. In this section we describe ten models of S^3 under subheadings $a, b, \ldots, j$. Some of the descriptions are related. The first nine of these descriptions are more geometric than the tenth. Even if the reader does

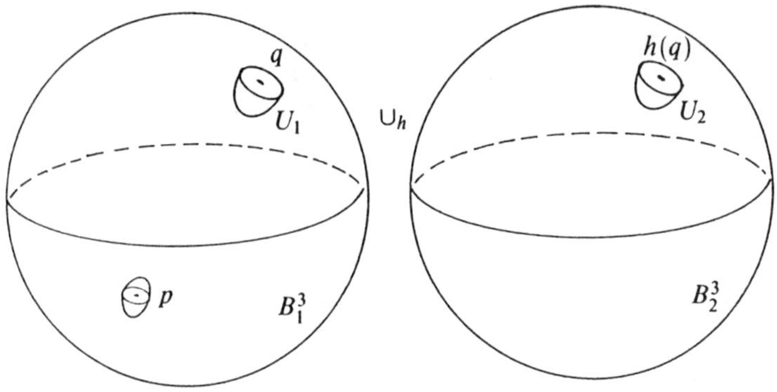

Figure 2a

not understand some model, he can likely proceed to the next since in most cases the description of a model does not depend on the preceding ones.

a. S^3 *as the union of balls.* One may get a model of S^3 as the union of two balls. Before describing the model, we first consider some lower dimensional aspects of such a model.

One may decompose a circle into two semicircles. Hence, if one takes two disjoint arcs and uses a homeomorphism to sew the end points of one arc onto the end points of the other, one has an example of a 1-sphere. One can get a model of a 2-sphere by considering two disks D_1, D_2 and joining them along their boundaries. If h is the homeomorphism of $\operatorname{Bd} D_1$ onto $\operatorname{Bd} D_2$ such that each point p of $\operatorname{Bd} D_1$ is fitted to the point $h(p)$ of $\operatorname{Bd} D_2$, we may write $S^2 =_t D_1 \cup_h D_2$.

Similarly, one may obtain a model of S^3 by sewing two balls B_1^3, B_2^3 together with a homeomorphism h of $\operatorname{Bd} B_1^3$ onto $\operatorname{Bd} B_2^3$. The sewing is abstract rather than physical since it cannot be realized in R^3. See Figure 2a. We write $S^3 =_t B_1^3 \cup_h B_2^3$.

The topology of $B_1^3 \cup_h B_2^3$ is as follows. If Int denotes *interior* and $p \in \operatorname{Int} B_1^3$, then any open set in $\operatorname{Int} B_1^3$ containing p is a neighborhood of p in $B_1^3 \cup_h B_2^3$. (We regard a neighborhood of p as an open set containing p unless we specify otherwise.) If $q \in \operatorname{Bd} B_1^3$, then $\{q\} \cup \{h(q)\}$ is considered a point in $B_1^3 \cup_h B_2^3$. To get a neighborhood of such a $\{q\} \cup \{h(q)\}$, one considers an open set U_1 in B_1^3 containing q and an open set U_2 in B_2^3 containing $h(q)$ such that $h(U_1 \cap \operatorname{Bd} B_1^3) = U_2 \cap \operatorname{Bd} B_2^3$. The associated neighborhood of $\{q\} \cup \{h(q)\}$ is

$$U_1 \cup_{h/(U_1 \cap \operatorname{Bd} B_1^3)} U_2.$$

See Figure 2a.

b. S^3 *as the one point compactification of* R^3. Before describing S^3 as the one-point compactification of R^3, we consider a 2-dimensional version of this model.

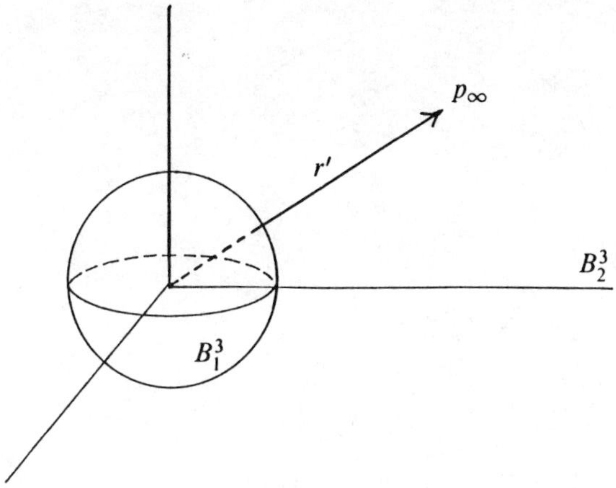

Figure 2b

Suppose S^2 is a round 2-sphere tangent to the plane R^2 at a point p. Let q be the point of S^2 diametrically opposite to p. Consider the projection from q of $S^2 - \{q\}$ onto R^2. If $x \in S^2 - \{q\}$, its image is the point where the line joining q and x passes through R^2. Hence $S^2 - \{q\} =_t R^2$. This suggests that the one-point compactification of R^2 is a model for S^2—the point of compactification added to R^2 corresponds to q.

The points of the 1-point compactification of R^3 are the points of R^3 together with an ideal point which we call the point at infinity. Let us use p_∞ to denote this ideal abstract point. The canonical neighborhoods in this enlarged R^3 are of two sorts: (i) the interior of a ball in R^3; (ii) the union of $\{p_\infty\}$ and the complement in R^3 of a closed 3-ball with center at the origin—the larger the 3-ball, the smaller the neighborhood of p_∞.

Actually, this model of S^3 is related to the one given in the preceding Section 2a. Suppose B_1^3 is the ball in R^3 with its center at the origin and radius 1. Then $\{p_\infty\} \cup (R^3 - \operatorname{Int} B_1^3)$ may be considered as a topological ball B_2^3 with center at p_∞. If r is a ray reaching out from the origin and r' is the part of this ray outside $\operatorname{Int} B_1^3$, then $\{p_\infty\} \cup r'$ can be considered a radius of B_2^3. Then the 1-point compactification of R^3 is the union of B_1^3 and B_2^3 sewn together with the identity map of $\operatorname{Bd} B_1^3$ onto $\operatorname{Bd} B_2^3$. See Figure 2b.

c. *S^3 as the union of two solid tori.* Consider two solid tori T_1^3, T_2^3 linked as shown in Figure 2c$_1$. Each of these tori is topologically like a doughnut. Can the two tori be blown up until their boundaries fit? If this could be done, their union would be a model of S^3. There is a bit of a problem in getting one of them to swallow the point at infinity. Perhaps one could visualize this expansion better after considering a more precise abstract approach. We describe one below.

Let M_i be a meridional circle on $\operatorname{Bd} T_i^3$ (M_i bounds a disk cross section D_i of T_i^3 but no disk in $\operatorname{Bd} T_i^3$) and L_i be a longitudinal circle on $\operatorname{Bd} T_i^3$ (L_i bounds

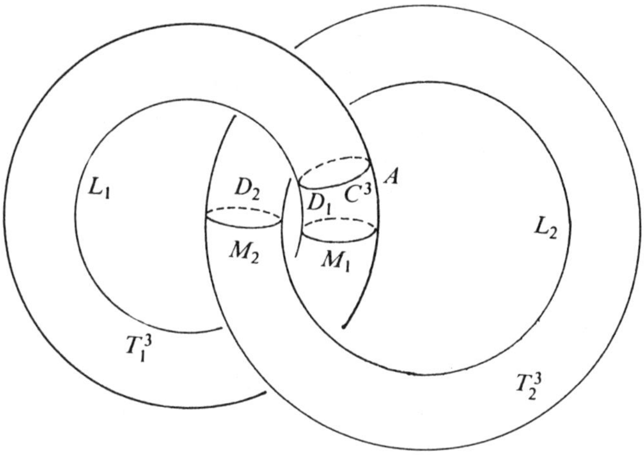

Figure $2c_1$

a disk in $R^3 - \operatorname{Int} T_i^3$ but not on $\operatorname{Bd} T_i^3$) as shown in Figure $2c_1$. Let h be a homeomorphism of $\operatorname{Bd} T_1^3$ onto $\operatorname{Bd} T_2^3$ that takes M_1 onto L_2 and L_1 onto M_2. Then

$$S^3 =_t T_1^3 \cup_h T_2^3.$$

The topology of $T_1^3 \cup_h T_2^3$ is as described in Section 2a.

This model of S^3 is a disguised version of the union of two balls given in Section 2a. If one thickens the disk D_1 to get a pillbox (solid cylinder) C^3 whose lateral surface is an annulus A going meridionally around $\operatorname{Bd} T_1^3$, the closure of $T_1^3 - C^3$ is a 3-ball B_1^3, i.e.

$$B_1^3 = \overline{T_1^3 - C^3}.$$

Also,

$$B_2^3 = C^3 \cup_{h/A} T_2^3$$

is a 3-ball and

$$S^3 =_t T_1^3 \cup_h T_2^3 =_t B_1^3 \cup_{h'} B_2^3$$

where

$$h' = \left\{ \begin{matrix} h \quad \text{on } \operatorname{Bd} T_1^3 - A \\ \text{Identity on bases of } C^3 \end{matrix} \right\}.$$

Before leaving this particular model of S^3, we consider other 3-manifolds obtained by sewing two solid tori together with a homeomorphism between their boundaries. These 3-manifolds are called toroidal manifolds and denoted by $T(i, j)$. To get the toroidal manifold $T(i, j)$ one considers a simple closed curve $J_{i,j}$ on $\operatorname{Bd} T_2^3$ that goes around the solid torus i times meridionally and j times longitudinally and a homeomorphism $h_{i,j}$ of $\operatorname{Bd} T_1^3$ onto $\operatorname{Bd} T_2^3$ that takes M_1 onto $J_{i,j}$. Then

$$T(i, j) =_t T_1^3 \cup_{h_{i,j}} T_2^3.$$

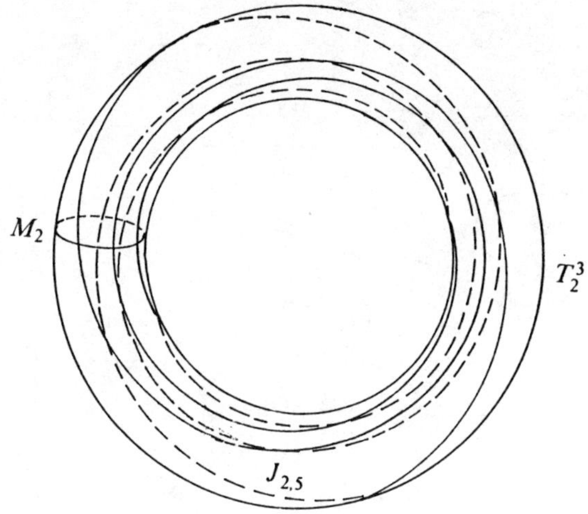

Figure $2c_2$

For there to be an $h_{i,j}$, not both i and j can be 0 and they can have no common divisor bigger than 1. Note that

$$S^3 =_t T(i,1), \text{ and}$$
$$S^2 \times S^1 =_t T(1,0).$$

Figure $2c_2$ shows $J_{2,5}$. Note that as a part of $J_{2,5}$ circles T_2^3 once longitudinally, it goes around 2/5 of a turn meridionally. By the time it has gone around five times longitudinally, it is back to the starting point and has made $5 \times (2/5) = 2$ turns meridionally.

Instead of drawing $J_{2,5}$ on Bd T_2^3, it is sometimes convenient to use a diagrammatic view of Bd T_2^3 as shown in Figure $2c_3$.

Working the following exercises may add to one's understanding of toroidal manifolds.

Exercise $2c_1$. If i, j are relatively prime integers, why is there a simple closed curve $J_{i,j}$ on Bd T that goes around i times longitudinally and j times meridionally?

Exercise $2c_2$. If i, j are not relatively prime, why is there no such simple closed curve $J_{i,j}$?

Exercise $2c_3$. Why is there no $T(2,0)$?

Exercise $2c_4$. Why is $T(1,0)$ homeomorphic with $S^2 \times S^1$?

Exercise $2c_5$. Why is $T(i,1)$ homeomorphic to S^3?

Exercise $2c_6$. If T_1^3, T_2^3, M_1, $J_{i,j}$ are as described, why is there a homeomorphism $h_{i,j}$ of Bd T_1^3 onto Bd T_2^3 taking M_1 onto $J_{i,j}$?

Exercise $2c_7$. If $h'_{i,j}$ is another such homeomorphism, why is it that

$$T_1^3 \cup_{h_{i,j}} T_2^3 =_t T_1^3 \cup_{h'_{i,j}} T_2^3?$$

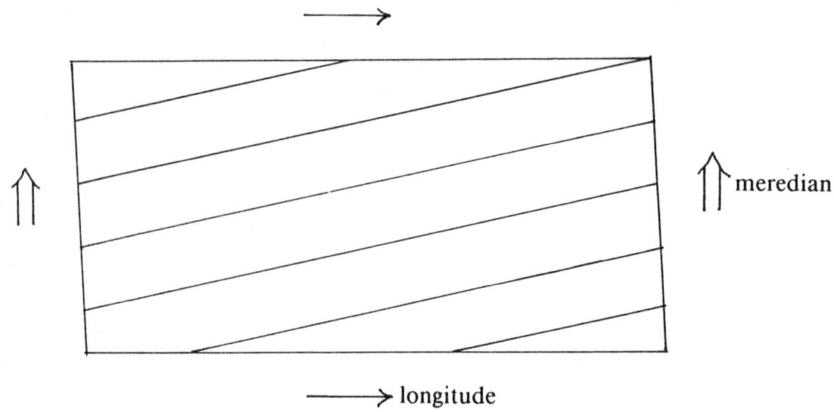

Figure $2c_3$

d. S^3 *as a lens space.* Let B^3 be the unit ball with center at the origin in R^3, i/j be a rational number in lowest terms, and $L(i, j)$ be the space obtained by identifying the upper half of $\mathrm{Bd}\, B^3$ with the lower half of $\mathrm{Bd}\, B^3$ by the following rule h. If $p \in$ upper half of $\mathrm{Bd}\, B^3$, $h(p)$ is obtained by a rotation about the z-axis through i/j complete revolutions ($2i\pi/j$ radians) and a reflection through the xy-plane. If p is a point of the upper half of $\mathrm{Bd}\, B^3$ and $h(p)$ is its associated point in the lower half of $\mathrm{Bd}\, B^3$ as shown in Figure 2d, a neighborhood of $\{p\} \cup \{h(p)\}$ is as described in Section 2a. Each point on the equator is associated with $j - 1$ other points. A neighborhood of such a j-tuple of points resembles an orange with j sections where each of the sections corresponds to a part of B^3 about one of the points. The points of $\mathrm{Int}\, B^3$ have their ordinary neighborhoods in $\mathrm{Int}\, B^3$. It may be noted that $S^3 =_t L(0, 1)$.

If one wanted $L(i, j)$ to have a differential structure near the image of the equator, one would want B^3 in the shape of a lens where the dihedral angle at the equator is $2\pi/j$ radians. If N is a closed ε-neighborhood of the equator, then under the sewing N goes into a torus T_2^3 in $L(i, j)$, the equator covers the center line of T_2^3 j times, and each cross-section of N becomes a sector $(1/j)$ of a cross-section of T_2^3. In Figure 2d we show the dihedral angle as $72° = 2\pi/5$ radians. Each boundary component of the annulus $N \cap \mathrm{Bd}\, B^3$ goes onto the same simple closed curve $J_{2,5}$ in $\mathrm{Bd}\, T_2^3$ and near an arc of the center line of T_2^3, the image of the annulus $N \cap BdB^3$ resembles 5 sheets with a common edge. Also, the closure of $B^3 - N$ becomes a solid torus T_1^3 such that the boundary components of the annulus $N \cap \mathrm{Bd}\, B^3$ correspond to a meridional simple closed curve M_1 of T_1^3. Hence $L(2, 5) =_t T(2, 5)$.

e. *Heegaard splittings of* S^3. Suppose M_1^3, M_2^3 are cubes with n handles as shown in Figure $2e_1$, $J_1, J_2, \ldots, J_n$ are simple closed curves on $\mathrm{Bd}\, M_1^3$ that go meridionally around the handles of M_1^3 as shown and $K_1, K_2, \ldots, K_n$ are

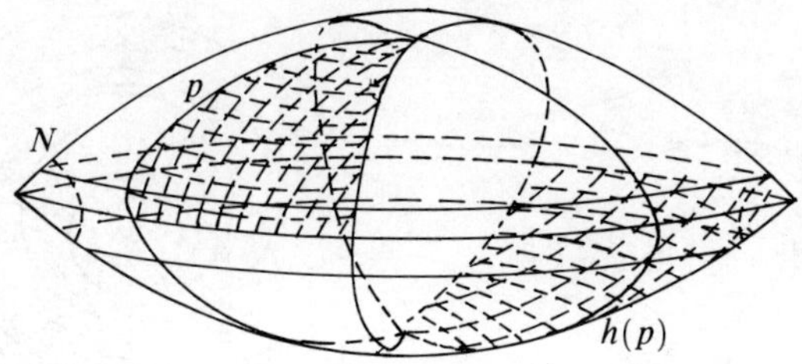

Figure 2d

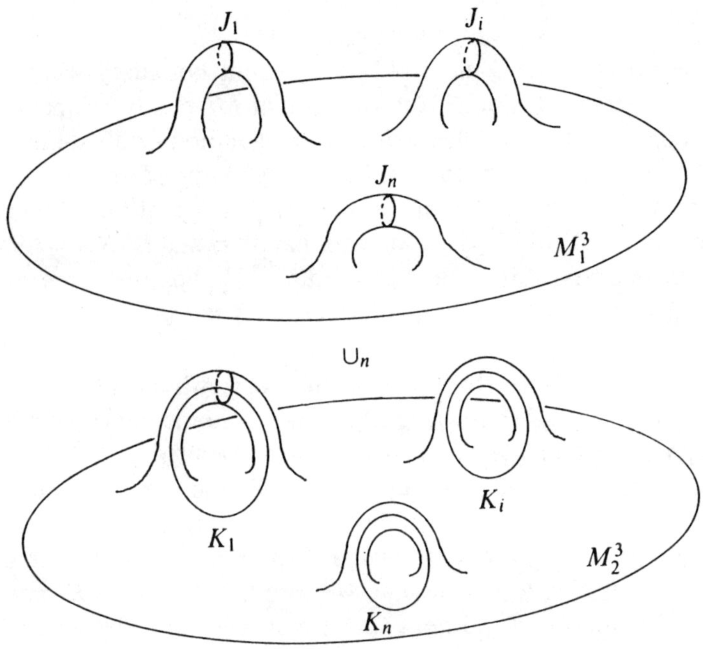

Figure $2e_1$

simple closed curves that go longitudinally around the handles of M_2^3 as shown. Let h be a homeomorphism of Bd M_1^3 onto Bd M_2^3 that takes each J_i onto the corresponding K_i. It makes a good exercise to prove there is such an h but the clue is that if Bd M_1^3 is split along the J_i's (or Bd M_2^3 along the K_i's), it becomes a disk with holes. Then

$$S^3 =_t M_1^3 \cup_h M_2^3.$$

To see that this is a model of S^3 one only needs to consider pillboxes in M_1^3 obtained by thickening disks bounded by the J_i's. Then M_1^3 is changed to a 3-ball if one removes these pillboxes and closes the remainder. Also, M_2^3 is changed to a 3-ball if these pillboxes are added to it.

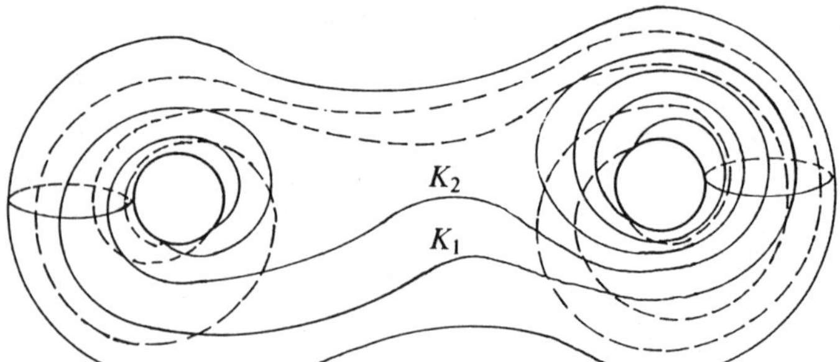

Figure 2e$_2$

Any connected compact 3-manifold M^3 without boundary whatsoever can be obtained in a similar fashion. It is known that M^3 can be triangulated. The union of the edges of the 3-simplexes of this triangulation is called the 1-skeleton of the triangulation. If N_1^3 is a closed tubular neighborhood of this 1-skeleton, it is a cube with handles. It can be shown that if N_1^3 is appropriately chosen, then the complement N_2^3 of Int N_1^3 is a cube with handles; that is, $N_2^3 = M^3 - \text{Int } N_1^3$ is a cube with handles. Since Bd N_1^3 = Bd N_2^3, N_2^3 has the same number of handles as N_1^3. Then $M^3 = N_1^3 \cup_h N_2^3$ where h is the identity map of Bd N_1^3 onto Bd N_2^3.

If M^3 is a compact 3-manifold without boundary which is the union of two cubes with handles M_1^3, M_2^3 sewn together with a homeomorphism of Bd M_1^3 onto Bd M_2^3, we say that (M_1^3, M_2^3) is a Heegaard splitting of M^3. There are many Heegaard splittings of S^3 other than the simple one we have described. For example, if each M_i^3 has two handles, J_1, J_2 go around the handles as shown in Figure 2e$_1$, and K_1, K_2 are as shown in Figure 2e$_2$, it is with considerable difficulty that one is able to decide that $M^3 =_t S^3$. In this case K_1 goes around the left handle 3 times and the right four times while K_2 goes around the left handle 2 times and the right 3 times.

Question. Suppose M^3 is a compact 3-manifold without boundary which is simply connected. (A space X is *simply connected* if each continuous function sending the boundary of a disk into X can be extended to send the disk continuously into X.) It is known that M^3 is topologically S^3 if it is a simply connected toroidal manifold. Hence M^3 is a 3-sphere if it has a Heegaard splitting (M_1^3, M_2^3) where each M_i^3 is of genus 1 (has one handle). It is not known if it is a 3-sphere if each M_i^3 has genus 2. This is an unsolved case of the well-known Poincaré conjecture.

f. *Folding up the sides of a cube.* Consider a cube $a_1a_2a_3a_4a_1'a_2'a_3'a_4'$ as shown in Figure 2f. Suppose the top base is subdivided as shown and each side face $a_ia_{i+1}a_i'a_{i+1}'$ is folded up and over and made to fit homeomorphically onto $a_ia_{i+1}a_i''a_{i+1}''$. Also the bottom base is folded around and sewn homeomorphically onto $a_1''a_2''a_3''a_4''$. This gives a model of S^3. This space is related to the lens

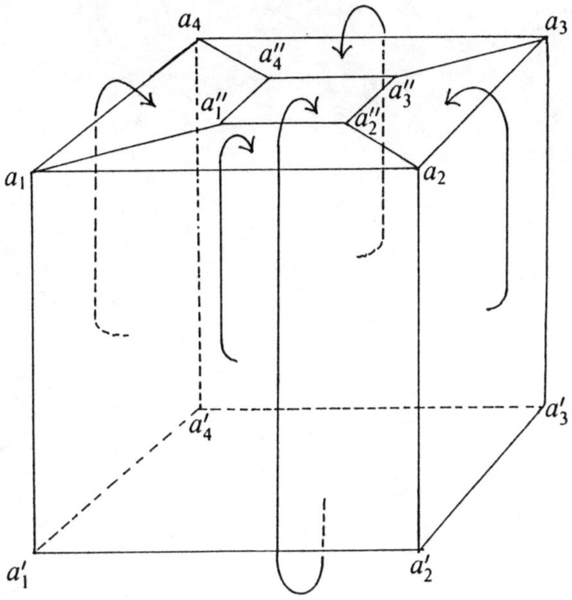

Figure 2f

space $L(0, 1)$ where the top base of the cube corresponds to the top of Bd B^3. We give this example here in hopes that its introduction will lead to a better understanding of the 1-1 maps discussed in Sections 4a, 4b, 5.

g. *Shrinking the boundary of a ball to a point.* If one takes the boundary of a disk and pulls the boundary of the disk to a point (much as one pulls the boundary of a sack together with a draw string), one has a model of S^2. If the disk were in a plane the shrinking could be done in 3-space containing the plane but not in the plane. Similarly, one gets a model of S^3 if one takes a round ball B^3 and draws its boundary to a point. This could not be done in R^3.

Rather than pulling Bd B_3^3 to a point (which is physically impossible in R^3), one could require that the points of our space (points being undefined objects anyway) be the ordinary points of the interior of the ball together with the ideal points (which is the boundary of the ball). It is clear what neighborhoods would be.

h. *The interior of a ball and one point on its boundary.* Rather then identifying Bd B^3 to a point as in g, one could select a point $p \in \text{Bd}\, B^3$ and let our model be Int $B^3 \cup \{p\}$. Rather than use ordinary neighborhoods of p we decide that a canonical neighborhood of p is the union of $\{p\}$ and Int $B^3 - B_\varepsilon^3$ where B_ε^3 is a ball concentric with B^3 and of radius ε less than the radius of B^3.

i. *S^3 as the join of two simple closed curves.* It is known that S^{m+n+1} is topologically the join of S^m and S^n. Hence S^3 is topologically equivalent to the join of two simple closed curves.

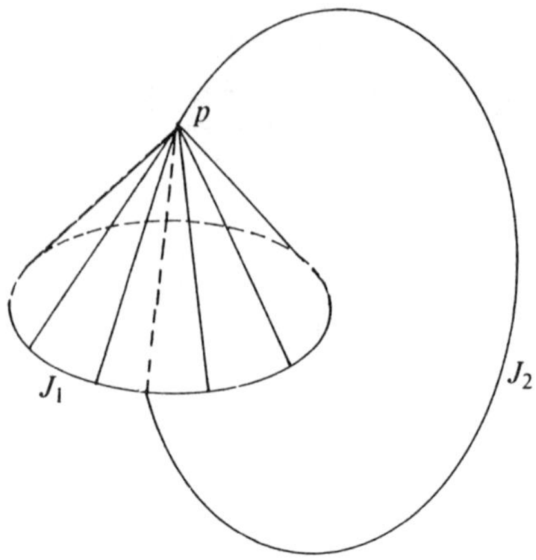

Figure 2i

If A, B are conveniently located in some Euclidean space R^i, we can regard the join of A and B as the union of all straight line segments from points of A to points of B. However, to have this simple model of the join it is usually required that A, B be disjoint compact subsets of R^i and if two segments from A to B in R^i intersect each other, then the intersection of the segments is an end point of each. If this situation does not hold, either imagine that the segments do not intersect, re-embed A and B to bring about a better situation, or use an abstract definition of join.

An easy join to visualize is that of A and B where A is a segment of the z-axis centered at the origin and B is a circle in the xy-plane centered at the origin. In this case, the join is the union of two solid cones with a common base.

Figure 2i shows J_1 and J_2 linked. Note that if $p \in J_2$, then the join of J_1 and $\{p\}$ is a disk (pictured as a cone). Hence, the join of J_1 and J_2 resembles the union of a circle of cones, all with a common boundary.

j. S^3 *as a third symmetric product.* Before describing this model, we discuss a two-dimensional analogue. Consider the one and two point sets of a circle S^1. If this collection of sets is topologized with a Hausdorff distance, what is the resulting object? Recall that if A, B are two sets, the Hausdorff distance between A and B is defined as

$$\text{Max}\left\{\underset{a\in A}{\text{LUB}}\, d(a,B),\ \underset{b\in B}{\text{LUB}}\, d(A,b)\right\}$$

where Max, LUB, d, denote *maximum, least upper bound, distance*, respectively. One might think at first that the space is topologically equivalent to $S^1 \times S^1$ but a closer examination shows that one point sets and diametrically opposite pairs introduce quirks. One can show that the space is actually a Moebius band.

In contrast, if one gives a Hausdorff distance to the one, two, and three point sets of S^1, the resulting object is a 3-sphere. We do not prove that here but remark that this is a peculiar way of getting a model of S^3.

3. 1-1 maps onto S^3. A map is a continuous function. A homeomorphism has the property that both h and h^{-1} are 1-1 and continuous. Not each 1-1 map is a homeomorphism for it is not required that f^{-1} be continuous. For example, there is a 1-1 map of a half open interval onto a circle.

If there is a 1-1 map of X onto S^3, in a certain sense we may regard X as a model of S^3. The points of this model are the points of X. We need to identify which subsets of X are to be regarded as neighborhoods in S^3. The usefulness of the model is determined somewhat by how easy it is to identify these neighborhood sets. In Section 2h where $S^3 =_t \operatorname{Int} B^3 \cup \{p\}$ under a certain topology for $\operatorname{Int} B^3 \cup \{p\}$, the open neighborhoods were rather easy to identify.

We describe S^3 as the 1-1 image of various familiar sets and leave it to the reader to determine how these can be converted into models of S^3. The first of our theorems is suggested by Section 2h.

THEOREM 3a. *There is a* 1-1 *map from the union of the interior of a ball and a* 1*-point set on its boundary onto* S^3.

Before stating the next theorem, we consider the consequence of shrinking certain simple closed curves on the boundary of a solid torus to points. Suppose $A^2 \times [0, 1]$ as the cartesian product of an annulus A^2 and the segment $[0, 1]$ as shown in Figure 3b. If J is the smaller boundary component of A^2 and each $J \times t$ is shrunk to a point, then $A^2 \times [0, 1]$ becomes a solid cylinder whose axis comes from $J \times [0, 1]$ as shown on the right side of Figure 3b. We obtain the same object that we would have obtained by sewing a pillbox in the hole of $A^2 \times [0, 1]$ along $J \times [0, 1]$.

Suppose we remove most of the boundary of $A^2 \times [0, 1]$ and leave only the open arc $p_0 \times (0, 1)$ for some point $p_0 \in J$. Then there is a 1-1 map of this object onto the interior of a cylinder which sends $p_0 \times (0, 1)$ onto the open axis of this cylinder. If we had used $p_0 \times [0, 1)$ instead of $p_0 \times (0, 1)$, there would be a one point set on the boundary of this cylinder. An application of Theorem 3a yields the following result.

THEOREM 3b. *There is a* 1-1 *map onto* S^3 *from the union of an open solid torus and a half open arc on its boundary.*

If instead of sewing a pillbox into the hole of $A^2 \times [0, 1]$ along $J \times [0, 1]$, we sewed it along some annulus on $\operatorname{Bd}(A^2 \times [0, 1])$ (perhaps one which circled $A^2 \times [0, 1]$ many times longitudinally and meridionally), we could have gotten another toroidal manifold minus an open ball rather than just a cylinder (S^3-open ball). Hence, we have the following extension of Theorem 3b.

THEOREM 3b′. *For each toroidal manifold* $T(i, j)$ *there is a* 1-1 *map onto* $T(i, j)$ *from the union of an open solid torus and a half open arc on its boundary.*

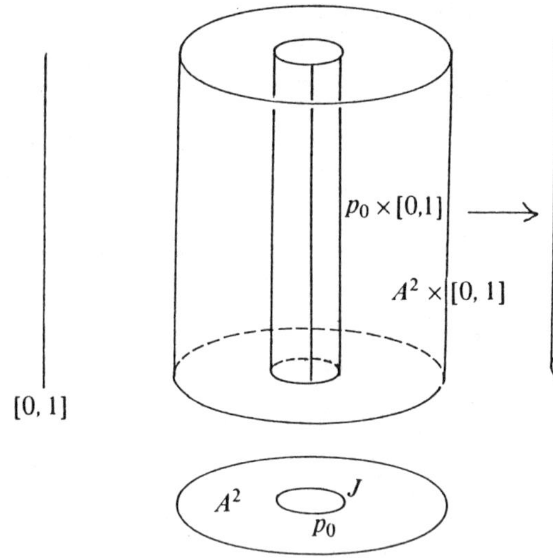

Figure 3b

A cube with two handles is called a solid double torus. It is sometimes called a handlebody of genus 2.

THEOREM 3c. *There is a* 1-1 *map onto* S^3 *from the union of an open solid double torus and an open arc on its boundary.*

To prove Theorem 3c one could regard the open arc as the union of an open arc and a disjoint half open arc. The open arc can be used to remove one of the handles and bring us into a position to apply Theorem 3b.

Using ideas from Section 2e, we state the following extension of Theorem 3c.

THEOREM 3c′. *If* M^3 *is any connected compact* 3-*manifold without boundary whatsoever, then* M^3 *is the image under a* 1-1 *map of the union of some open cube with handles and disjoint open arcs on its boundary* (*one fewer open arcs than handles*).

4. 1-1 **maps of** R^3_+ **onto** S^3. We use R^3_+ to denote 3-dimensional Euclidean half space—the set of all points of R^3 on or above the xy-plane. We describe several 1-1 maps of R^3_+ onto S^3. We summarize the situation as follows.

THEOREM 4. *There are many* 1-1 *maps of* R^3_+ *onto* S^3.

The reader might find it instructive to decide what the image of the boundary of R^3_+ looks like in each of the five maps we describe. Three of the five images are alike.

Let I^3 be a cube and I^2 be its top base. Then $\operatorname{Int} I^3 \cup \operatorname{Int} I^2$ is a model for R^3_+.

a. *Laying down cones.* Suppose four cones are pushed up from I^2 as shown in the left side of Figure 4a. The base of each of these cones is tear shaped so that the base intersects Bd I^2 only at the corner points. Each lateral radius of a cone from a corner point has the length of an edge of the cube. Then the union of I^3 and the four solid cones becomes a model of R^3_+ if the bottom and four lateral faces of I^3 are removed. Let h be a homeomorphism of R^3_+ onto this object.

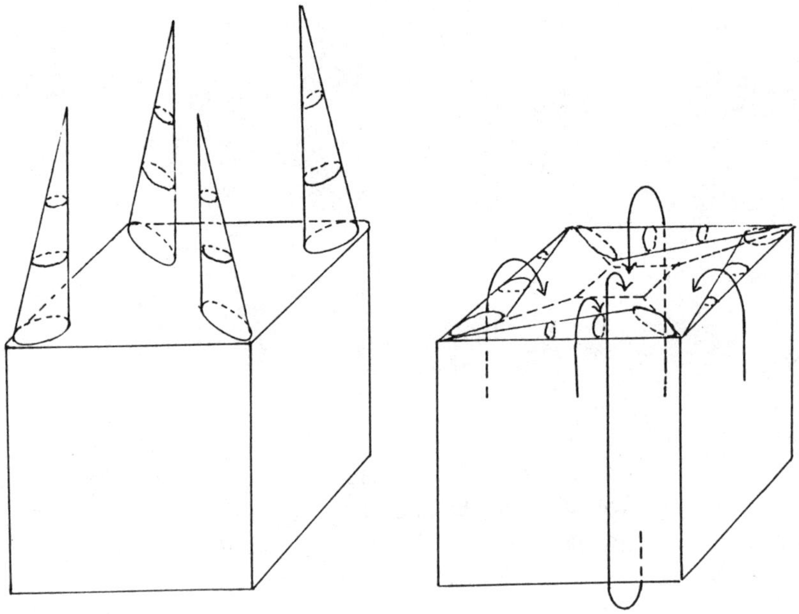

Figure 4a

Suppose that the four cones are pushed over as shown in the right side of Figure 4a. Each base of each of the cones is left fixed and its lateral radius from the corner goes onto a side of I^2. Otherwise the tilted and twisted cone does not intersect I^2. Let f_1 be a map that pushes over the cones. Then $f_1/h(R^3_+)$ is 1-1 since $\text{Bd}\, I^2 \cap h(R^3_+) = 0$. However, $\text{Bd}\, I^2 \subset f_1 h(R^3_+)$ because Bd I^2 is the images of the lateral radii from the corners. Our purpose of pushing over the cones was to get Bd I^2 in the image.

The top of $f_1 h(R^3_+)$ is no longer flat but we subdivide it as shown by the dotted lines in somewhat the same fashion that the top of the cube was subdivided in Section 2f. Let f_2 be a map that folds the sides and bottom of $f_1 h(R^3_+)$ out, up, and over to make it fit onto the top. The folding is similar to that done in §2f except that the lateral sides are made to fit around the tilted cones and pushed into the inside crevices between the tilted cones and I^2. Then $f_2 f_1 h$ is a 1-1 map of R^3_+ onto S^3.

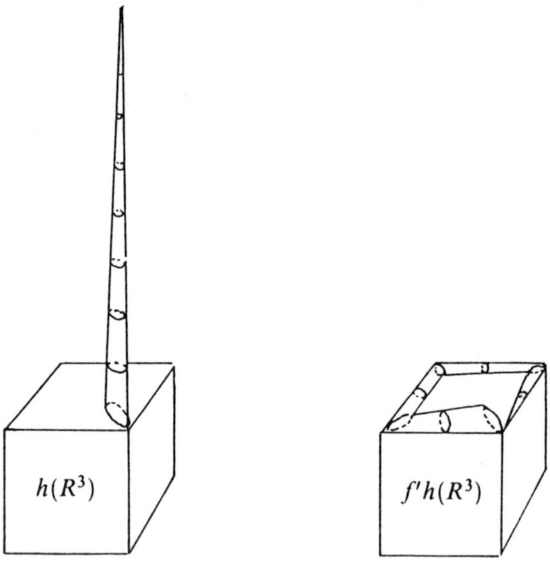

Figure 4b

b. *Laying down one cone.* Consider the case where there is only one cone raised on the top of I^2 as shown in the left side of Figure 4b but where the lateral radius of this cone from the corner of I^2 has length equal to the perimeter of I^2. Let h be the homeomorphism of R^3_+ onto this figure (with the lateral sides and bottom gone) and f_1 be the map that pushes the cone over as shown in the right side of Figure 4b. If R^2 is the xy-plane, $f_1 h(R^2)$ is a dunce's cap. Note that $f_1 h(R^2)$ is compact even though R^2 is not.

Let f_2 be the map that pushes the lateral and bottom sides of I^3 out, up, and over to make it fit onto the top of $f_1 h(R^3)$. The resulting object is a model of S^3 that was obtained as the image of R^3_+ under the 1-1 map $f_2 f_1 h$.

c. *Factoring through a solid torus.* Suppose only one cone is erected in the back right corner of I^2 and that the lateral radius of this cone from the corner is as long as a side. Suppose this cone is folded forward as indicated in Figure 4c. Instead of folding the sides and bottom of the solid out, up, and over as done in Sections 4a and 4b, only the bottom is pulled around to be made to fit on the top as shown in Figure 4c. Hence there is a 1-1 map of R^3_+ onto an open solid torus with a half open arc on its boundary. Theorem 3b assures us that this map can be composed with a 1-1 map to send R^3 onto S^3.

Theorem 3b′ shows us that we can get a 1-1 map of R^3_+ onto any toroidal manifold whatsoever.

d. *Factoring through a double torus.* Let A be a spanning segment of I^2 that subdivides it into two congruent rectangles D_1, D_2 as shown in Figure 4d. Rather than lay down cones, suppose two open tubes are pulled around from

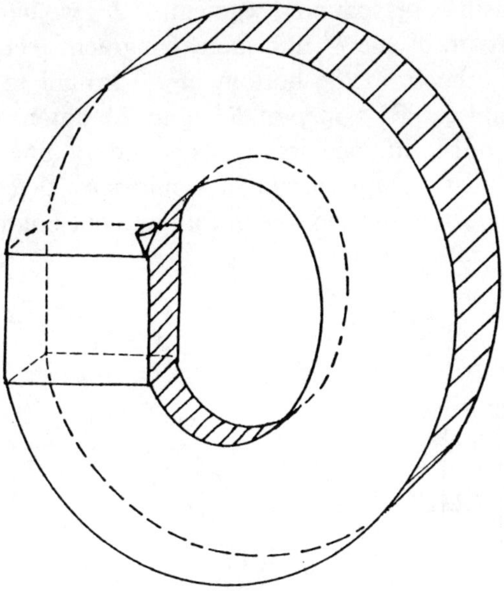

Figure 4c

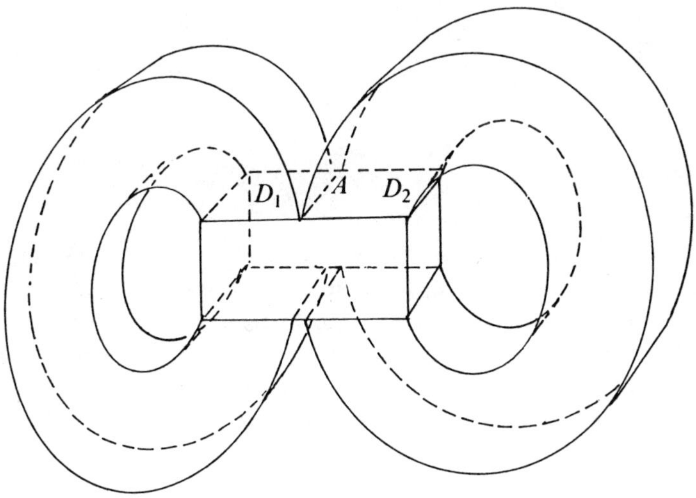

Figure 4d

the bottom of I^3 to be made fit onto Int D_1, Int D_2, respectively. The only part of Int I^2 left exposed is the open arc $\mathring{A}$. There is a 1-1 map of R^3_+ onto the union of an open solid torus and an open arc on its boundary. Theorem 3c assures us that this map can be composed with another to get a 1-1 map of R^3_+ onto S^3.

If instead of using one spanning segment of I^2, we had used many disjoint spanning segments to divide I^2 into many congruent rectangles and if we had pushed an open tube from the bottom of I^3 around to each of these open rectangles we could get a 1-1 map of R^3_+ onto the union of an open cube with handles and many disjoint open arcs on its boundary (one fewer open arcs than handles). An application of Theorem 3c′ would show that there is a 1-1 map of R^3_+ onto any compact connected 3-manifold without boundary.

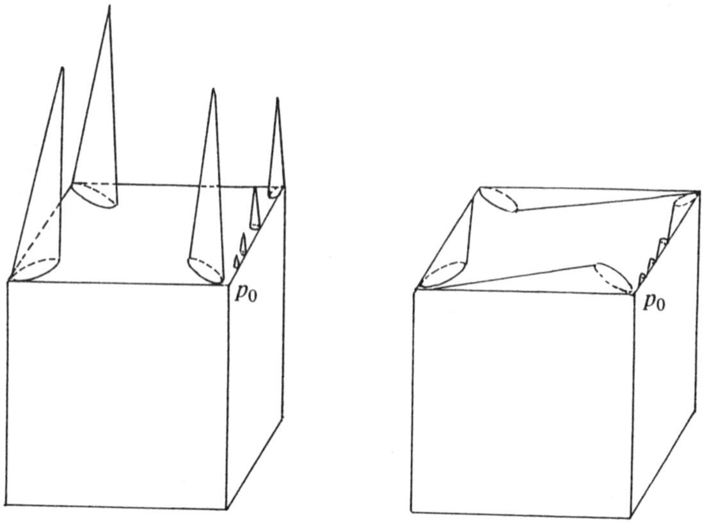

Figure 5

e. *Using third symmetric products.* We describe one last 1-1 map of R^3_+ onto S^3. We recall from Section 2j that the collection of one, two, and three point sets of S^1 topologized with the Hausdorff distance is a model of S^3. There is a natural 1-1 map of a half open interval onto S^1. Hence, if we could show that collection of one, two, and three point sets of $(0, 1]$ topologized with the Hausdorff distance is topologically R^3_+, then we would have still another way of getting a 1-1 map of R^3_+ onto S^3.

It is easier to discover the object equivalent to the collection of one, two, and three point sets of $[0, 1]$ topologized with the Hausdorff distance that (sic) it is to find the corresponding object for S^1. For each triple of numbers x_1, x_2, x_3 we can order them by size with the smallest first as a, b, c and associate this triple with the point of R^3 with cylindrical coordinates (r, θ, z) where $z = a$, $r = c - a$, and $\theta = 2\pi(c - b)/(c - a)$. Each single number a is associated with the point with coordinates $(0, 0, a)$ and each pair (a, b) with $a < b$ is associated with the point $(b - a, 0, a)$. Hence, the object equivalent to the one, two, and three point sets of $(0, 1]$ topologized with the Hausdorff distance is topologically a solid cone with its base gone. This is a model of R^3_+.

5. A 1-1 map of R^3_+ onto R^3. The invariance of domain theorem shows that there is no 1-1 map of R^3 onto R^3_+. It may be somewhat surprising that there is a 1-1 map of R^3_+ onto R^3. This was one of the key theorems used in giving a negative answer to the following monotone mapping question for R^3 whose answer was only recently discovered—does each map of R^3 onto R^3 in which point inverses are compact continua have the property that inverses of compact sets are compact?

The methods used in Sections 4a, 4b to get a 1-1 map of R^3_+ onto S^3 can be modified to get a 1-1 map of R^3_+ onto R^3. When one pushed over the tall cone in Section 4b or the four small ones in Section 4a, $\mathrm{Bd}\, I^2$ got covered by the images of radii. However, suppose one considers a sequence of cones getting shorter and shorter as shown in the left side of Figure 5 so that if they are pushed over as shown in the right side of Figure 5, the corner point p_0 is not covered. If the sides and bottom of the resulting solid are pushed out, up, and over as before, there results a 1-1 map of R^3_+ onto S^3 minus a one point set. Hence, there is a 1-1 map of R^3_+ onto R^3.

THE UNIVERSITY OF WISCONSIN

Reprinted from the AMERICAN MATHEMATICAL MONTHLY
Vol. 80, No. 2, February, 1973
pp. 117–119

AWARD FOR DISTINGUISHED SERVICE TO PROFESSOR RAYMOND L. WILDER

This year's Award for Distinguished Service goes to a man with wide interests. He is known for his contributions to mathematics and logic, his fondness for anthropology and the history of the development of mathematical concepts, and his service through excellence in teaching and able leadership of mathematical organizations.

Raymond L. Wilder was born in Palmer, Massachusetts in 1896. His versatility showed itself early. He was quite good at the piano, and the local movie house hired him to accompany their silent movies.

While Ray liked mathematics, logic, history, and anthropology, he did not see a ready market for a knowledge of these subjects and decided to study actuarial mathematics. Brown and Texas were leading universities in this area at that time. Wilder pursued his study of actuarial mathematics through the bachelor's and master's degree at Brown University, and then went to the University of Texas to continue these studies.

Wilder wanted a better acquaintance with pure mathematics, and asked to enroll in one of Professor R. L. Moore's classes in topology. Moore at first refused him admission, since Wilder's interest in topology was only secondary. Moore remarked that he doubted that Wilder would like the rigors of proving theorems, and perhaps would not be much good at it even if he did. However, Wilder persisted that he was really interested in pure mathematics and answered some of Moore's questions (such as, "What is an axiom?") well enough that Moore relented and let him enroll. For a while he was ignored by Moore, but as Wilder was able to prove some difficult theorems, Moore's enthusiasm began to grow. Wilder continued his actuarial studies, but when Moore learned that Wilder had solved a problem that had baffled J. R. Kline and others, he suggested that Wilder write it up for a thesis promptly. The deadline for taking language and qualifying exams had already passed, but Moore cut through red tape and arranged for Wilder to take his exams after the deadline. Wilder took his Ph. D. in topology that very same year and gave up his actuarial studies.

Wilder spent two years at Ohio State University after finishing at Texas. He then went to the University of Michigan and worked his way through the ranks to become their first Research Professor of Mathematics. Professor Wilder was very influential in helping build the University of Michigan into the giant center of excellence in mathematics that it is today. One year he was their Henry Russel Lecturer. This is the University of Michigan's highest award, presented for outstanding scholarship and to a person who has reached full maturity in his scholarly work.

Professor Wilder's students report that he is a great teacher. Not that his lectures were always polished (which they were not) but, more important, they were stimulating and challenging. Word got around that Wilder not only taught mathematics, but about mathematics. His course in Foundations was widely discussed, even outside mathematical circles, and the course became one of the most popular on campus. The general student learned the history of mathematical concepts and that mathematics was an important part of our culture. Each found in the course something related to his own interests. The course appealed to a diversified audience.

Professor Wilder taught all students in his classes, and not just the elite. He regarded an ordinary student reaching to the limits of his capacity as more exciting than an excellent student just coasting along. While he would not try to make a purse out of a sow's ear or turn each student into a mathematics major, he challenged each student according to that student's ability, and encouraged the student to develop his own talents. Some students who had not intended to major in mathematics changed their majors when they became impressed with the cultural importance of mathematics and were fired with the excitement of discovery. Wilder stimulated superior undergraduates to become mathematicians by admitting them to his advanced courses and letting them talk, prove theorems, and learn by doing. The article on the Distinguished Service Award to Ed Begle four years ago reported that Begle received his predoctoral training at the University of Michigan, where under the influence of Ray Wilder he became interested in topology.

Professor Wilder is superb in leading others to do research. He directed the theses of twenty-five students, including Leon Cohen, Paul Swingle, Sam Kaplan, Morton Curtis, Alice Dickinson, Joe Shoenfield, Tom Brahana, Frank Raymond, and Kyung Kwun. He influenced the work of students who wrote their theses with others, including Norman Steenrod (whose first paper came from work he did in Wilder's class) and Stephen Smale (whose first two papers were connected with Wilder's seminars and questions raised by Wilder). Wilder's seminars were lively affairs that had a stimulating impact on research. They influenced not only the work of graduate students, but also the research of Wilder himself and that of his many colleagues who attended.

One of the things that makes Professor Wilder so successful as a teacher is the high quality of his research. His earlier research dealt with properties of the plane and continuous curves, but he broadened his interests to higher dimensions and along with his students developed the important notion of generalized manifolds. He wrote over 70 research papers and three books. His colloquium volume is a profound treatment of the topology of generalized manifolds; his second book dealt with foundations; the third is concerned with the evolution of mathematical concepts. The breadth of Wilder's scholarship is evidenced by the fact that although most of his papers deal with geometric topology, he addressed the International Congress of Mathematicians on "The Cultural Basis of Mathematics," his Gibbs lecture was entitled "Trends and Social Implications of Research," and one of his latest papers appears in a medical journal.

Another quality that makes Ray Wilder a great teacher is his interest in people. The humanitarian aspect of his character is one of his finest assets. He likes people and they like him. Students sought his council. He became to many a personal friend and one whom they never forgot. His colleagues report that students from long ago have dropped by to visit with Ray and enjoy the memories of his gentle introduction to mathematics, culture, history, and humanity that they recall with so much relish. Ray would remember them and usually prove it with an anecdote. Wilder's students have a warm place in their hearts for him and his talented and charming wife, Una.

A dramatic shift in the position of the United States in mathematical affairs occurred just prior to and during World War II. Immigration of leading foreign mathematicians played a big part in this. Professor Wilder was instrumental in influencing Michigan and the federal establishment to find places for some of these talented immigrants. His motivation was both humanitarian and scientific. Reactions such as "surely there must be an American just as well qualified" had to be overcome. Wilder's efforts in getting Sammie Eilenberg to join Michigan's staff brought great mathematical rewards. Our country today is much stronger mathematically because of the efforts of Wilder and others during these trying times.

Professor Wilder has devoted himself to working with mathematical organizations as well as to teaching and research. He has been an advisor to the Mathematics Division of the Air Force Office of Scientific Research, the NRC Fulbright Committee, the Michigan Mathematical Journal, and to many educational groups. He has been active on committees of both MAA and AMS, and has served as president of both organizations. He continues to work through these organizations for the improvement of teaching, the promotion of research, and the well-being of people.

Professor Wilder contends that mathematics is one of the most important cultural components of every modern society. He feels that a knowledge of mathematics and its methods should be a part of the intellectual and cultural background of each well-trained person; whether he be a teacher, businessman, legislator, public servant, or housewife. People well trained in mathematics are often able to analyze in a unique fashion. Wilder believes that the need for mathematically trained people is increasing — but in areas not traditionally associated with mathematics — positions in government, industry, economics, and particularly in managerial positions. New courses and new teaching techniques will be needed to give these sorts of people the special kind of mathematical training that will appeal to them and be most useful to them. We need more Wilder-type teaching.

R. H. Bing

PACIFIC JOURNAL OF MATHEMATICS
Vol. 51, No. 1, 1974

SETS GENERATED BY RECTANGLES

R. H. BING, W. W. BLEDSOE,
AND R. D. MAULDIN

For any family F of sets, let $\mathscr{B}(F)$ denote the smallest σ-algebra containing F. Throughout this paper X denotes a set and $\mathscr{R}$ the family of sets of the form $A \times B$, for $A \subseteqq X$ and $B \subseteqq X$. It is of interest to find conditions under which the following holds:

(1) Each subset of $X \times X$ is a member of $\mathscr{B}(\mathscr{R})$

The interesting case is when

$$\omega_1 < \text{Card}\, X \leqq c \, ,$$

since results for other cases are known.

It is shown in Theorem 9 **that** (1) **is equivalent to**

(2) There is a countable ordinal α such that each subset of $X \times X$ can be generated from $\mathscr{R}$ is α Baire process steps .

It is also shown that the two-dimensional statements (1) **and** (2) **are equivalent to the one-dimensional statement**

(3) There is a countable ordinal α such that for each family H of subsets of X with Card $H =$ Card X, there is a countable family G such that each member of H can be generated from G in α steps .

It is shown in Theorem 5 **that the continuum hypothesis (CH) is equivalent to certain statements about rectangles of the form** (1) **and** (2) **with** $\alpha = 2$.

Rao [7, 8] and Kunen [2] have shown that

THEOREM 1. *If* Card $X \leqq \omega_1$ (*the first uncountable cardinal*) *then* (1) *is true and if* Card $X > c$ *then* (1) *is false.*

The question of whether (1) is true (without the requirement Card $X \leqq \omega_1$) was raised by Johnson [1] and earlier by Erdös, Ulam, and others (see [8], p. 197). The arguments in Kunen's thesis actually showed that if Card $X \leqq \omega_1$ then

(4) Each subset of $X \times X$ can be generated from $\mathscr{R}$ in 2 steps (i.e., each subset is a member of $\mathscr{R}_{\sigma\delta}$. See definitions in § 2.) .

In Theorem 5 we generalize Theorem 1 and Kunen's result (4), and give a new characterization of CH by showing it to be equivalent to certain statements about rectangles of the form (1) and (4).

If CH is assumed the α appearing in statements (2) and (3) above is 2 (see Theorem 10). This raises the intriguing (but unanswered) question of whether α must *always* be 2 if (1) holds and CH is false.

It would also be interesting to know whether statements (1), (2), and (3) are equivalent to statement (5) below. Clearly (3) imples (5).

(5) If H is a family of subsets of X with Card $H =$ Card X, then there is a countable family G for which $H \subseteqq \mathscr{B}(G)$.

The equivalence of (1) and (2) means for example, (assuming CH), that there is a countable family G from which all real Borel sets (or analytic sets, or projective sets) can be generated in *two* steps (i.e., Borel sets $\subseteqq G_{\sigma\delta}$). This is remarkable in view of the well known result [4, 8] that if G is a countable basis for the real topology, then the Borel sets cannot be generated from G in less than ω_1 steps.

As a generalization of this well known result we show in Theorem 12 that any countable family G which is closed to complementation and which generates the Borel sets (i.e., Borel sets $\subseteqq \mathscr{B}(G)$) must have order ω_1. That is

$$\mathscr{B}(G) \nsubseteq G_\alpha$$

for any countable ordinal α. Thus, even though G might generate the Borel sets in α steps (or 2 steps if CH is assumed), the process, nevertheless, continues to produce new members of $\mathscr{B}(G)$ until we reach G_{ω_1}.

We would like to point out in conjunction with our characterization of CH that Kunen [2] has proved that if Martin's Axiom A holds (see [6]) and Card $X \leqq c$ then (4) holds. He also proved that if $\omega_1 <$ Card $X \leqq c$ then (1) is independent of ZFC (Zermo-Frankel Axioms + the Axiom of Choice) together with the negation of CH.

2. Notation and definitions.

If G is any family of sets, let G_0 be the family G, and for each ordinal α, $\alpha > 0$, let G_α be the family of all countable unions (intersections) of sets in $\bigcup_{\gamma<\alpha} G_\gamma$, if α is odd (even). Limit ordinals will be considered even. (Compare Kuratowski [3].) Thus we have

$$G_0 = G,\ G_1 = G_\sigma,\ G_2 = G_{\sigma\delta},\ G_3 = G_{\sigma\delta\sigma},\ \cdots,\ G_\alpha,\ \cdots .$$

Also $G_\alpha \subseteqq G_{\alpha+1}$ for each ordinal α and $G_{\omega_1} = G_{\omega_1+1}$, where ω_1 is the first uncountable ordinal. If $\alpha > 0$, then the family G_α is closed under countable unions (intersections) if α is odd (even).

We define the *order of* G to be the first ordinal α, $\alpha > 0$, such that $G_{\alpha+1} = G_\alpha$.

For each $A \subseteqq X$ (or $A \subseteqq X \times X$), let A' be the complement of A with respect to X (or $X \times X$), and for each family G of subsets of X (or $X \times X$) let $\mathscr{C}(G)$ be the family of complements of G. Note that if $\mathscr{C}(G) \subseteqq G$, or even if $\mathscr{C}(G) \subseteqq G_{\omega_1}$, then the family G_{ω_1} is the family $\mathscr{B}(G)$, the σ-algebra generated by G. Thus, since

$$(A \times B)' = A \times B' \cup A' \times X \in \mathscr{R}_1 ,$$

it follows that $\mathscr{R}_{\omega_1} = \mathscr{B}(\mathscr{R})$.

If G is a family of subsets of X, let $VG = \{A \times B\colon A \subseteqq X, B \in G\}$, and let $HG = \{A \times B\colon A \in G, B \subseteqq X\}$.

If $Z \subseteqq X \times X$ and $x \in X$, let Z_x denote the vertical section of Z at x, $Z_x = \{y\colon (x, y) \in Z\}$.

3. Results. The following lemma is easily proved by transfinite induction.

LEMMA 2. *If $1 \leqq \alpha < \omega_1$ and $A \in G_\alpha$, then there is a set B in G_1 such that $A \subseteqq B$.*

THEOREM 3. *If G is a countable family of subsets of X, $Z \subseteqq X \times X$, and $0 < \alpha < \omega_1$, then $Z \in (VG)_\alpha$ if and only if $Z_x \in G_\alpha$ for each $x \in$ domain Z.*

Proof. By considering the natural projections of the sets involved on the second coordinate axis, it is easily seen that

$$\text{if } Z \in (VG)_\alpha, \text{ then } Z_x \in G_\alpha \text{ for each } x \in \text{domain } Z .$$

Now suppose that $Z_x \in G_\alpha$, for each $x \in$ domain Z, and let $G = \{\theta_1, \theta_2, \theta_3, \cdots\}$. We complete the proof by transfinite induction on α.

Case 1. $\alpha = 1$.

For each n, let $A_n = \{x \in \text{domain } Z\colon \theta_n \subseteqq Z_x\}$, and let $Z_n = A_n \times \theta_n$. Then $Z_n \in VG$, for each n, and

$$Z = \bigcup_{n=1}^{\infty} Z_n \in (VG)_1 .$$

Now suppose $1 < \alpha < \omega_1$, and that the theorem holds for every γ, $0 < \gamma < \alpha$.

Case 2. α is even.

Let $\{\gamma_n\}_{n=1}^{\infty}$ be a sequence of odd ordinals less than α such that each odd ordinal less than α appears infinitely often in $\{\gamma_i\}_{n=1}^{\infty}$. For each $x \in$ domain Z, let

$$D_1(x),\ D_2(x),\ D_3(x),\ \cdots$$

be a sequence such that $D_i(x) \in G_{\gamma_i}$ for each i, and

$$Z_x = \bigcap_{i=1}^{\infty} D_i(x)\ .$$

This can be done in view of Lemma 2. For each i, let

$$Z^i = \bigcup_{x \in \text{domain } Z} \{x\} \times D_i(x)\ .$$

First note that $Z = \bigcap_{i=1}^{\infty} Z^i$. Also each nonempty section $(Z^i)_x$ of Z^i is equal to $D_i(x) \in G_{\gamma_i}$. Hence, by the induction hypothesis, $Z^i \in (VG)_{\gamma_i}$, for each i, and therefore

$$Z = \bigcap_{i=1}^{\infty} Z^i \in (VG)_\alpha\ ,$$

by the definition of the family $(VG)_\alpha$.

Case 3. α is odd and greater than 1.

For each $x \in$ domain Z, let $\{D_i(x)\}_{i=1}^{\infty}$ be a sequence of members of $G_{\alpha-1}$ for which $Z_x = \bigcup_{i=1}^{\infty} D_i(x)$, and let $Z^i = \bigcup_{x \in \text{domain } (Z)} \{x\} \times D_i(x)$, for each i.

Again it follows that $Z^i \in G_{\alpha-1}$, for each i, and

$$Z = \bigcup_{i=1}^{\infty} Z^i \in (VG)_\alpha\ .$$

COROLLARY 4. *If* $Z \subseteqq X \times X$ *is the graph of a function then* $Z \in \mathscr{R}_2 \subseteqq \mathscr{B}(\mathscr{R})$.

Proof. Let G be a countable basis for the real topology and note that, for each $x \in X$, Z_x is a singleton and hence $Z_x \in G_2$. Thus by Theorem 3, $Z \in (VG)_2 \subseteqq \mathscr{R}_2 \subseteqq \mathscr{B}(\mathscr{R})$. Also see [7].

THEOREM 5. *Let X be the real numbers and let G be a countable base for the usual topology on X. The following three statements are equivalent*:

(1) CH *holds*

(2) *if* $Z \subseteqq X \times X$, *then* $Z = A \cap B$, *where* $A \in (VG)_2$ *and* $B \in (HG)_2$ *and*

(3) *if* $E \subseteqq X \times X$, *then* $E = C \cup D$, *where* $C \in \mathscr{B}(VG)$ *and* $D \in \mathscr{B}(HG)$.

Proof. First, assume CH and suppose $Z \subseteq X \times X$. As is well known [7], the complement of Z is the union of two sets H and K such that each vertical section of H is countable and each horizontal section of K is countable.

Let A be the complement of H and let B be the complement of K. Then each vertical section of A is a G_2 set and by Theorem 3, $A \in (VG)_2$. Similarly, $B \in (HG)_2$. Of course, $Z = A \cap B$.

Since $A \in (VG)_2 \subseteq \mathscr{R}_2$ and $B \in (HG)_2 \subseteq \mathscr{R}_2$ and $\mathscr{R}_2$ is closed under finite intersections, $Z \in \mathscr{R}_2$. Thus, if CH holds, then the order of $\mathscr{R}$ is $\leqq 2$. Since the graph of the identity function, $f(x) = x$, is not in $\mathscr{R}_1$, it follows that the order of $\mathscr{R}$ is 2.

Now, suppose statement 2 holds and $E \subseteq X \times X$. Then, the complement of E can be expressed as the intersection of sets A and B with $A \in (VG)_2$ and $B \in (HG)_2$. It follows that $A' \in (VG)_3 \subseteq \mathscr{B}(VG)$ and $B' \in (HG)_3 \subseteq \mathscr{B}(HG)$. Thus, E is the union of two sets C and D, where $C \in \mathscr{B}(VG)$ and $D \in \mathscr{B}(HG)$.

Finally, assume statement 3 holds. Let T be a totally imperfect subset of X of cardinality c. The existence of such a set can be proven without assuming CH [3, p. 514]. Let $E = T \times T$ and let $E = C \cup D$, with $C \in \mathscr{B}(VG)$ and $D \in \mathscr{B}(HG)$. Then each vertical section of C is a subset of T which is a Borel set. Since an uncountable Borel set contains a perfect set and T contains no perfect set, we have that each vertical section of C is countable. Similarly, each horizontal section of D is countable. But, as is well known [10] this implies CH.

This completes the proof of Theorem 5.

The following two lemmas are well known.

LEMMA 6. *If F is a family of sets, α is a countable ordinal, and $A \in F_\alpha$, then there is a countable subfamily J of F for which $A \in J_\alpha$.*

LEMMA 7. *If F is a family of sets, $\mathscr{C}(F) \subseteq F$, and $A \in \mathscr{B}(F)$ then there is a countable subfamily J of F and a countable ordinal α for which $A \in J_\alpha$.*

THEOREM 8. (a) *The following two statements are equivalent:*

(i) *For each subset Z of $X \times X$ there is a countable ordinal α such that $Z \in \mathscr{R}_\alpha$.*

(ii) *If H is a family of subsets of X and* Card $H =$ Card X, *then there is a countable family G of subsets of X and a countable ordinal α for which $H \subseteq G_\alpha$.*

(b) *If α is a countable ordinal, the following two statements are equivalent:*

(i) *Each subset of $X \times X$ is a member of $\mathscr{R}_\alpha$.*

(ii) *If H is a family of subsets of X and* Card $H =$ Card X *then there is a countable family G of subsets of X for which $H \subseteqq G_\alpha$.*

Proof. The proof of part (b) is similar to that of part (a) which is given below.

First suppose (i) holds, and suppose that H satisfies the hypotheses of (ii). Define the subset $Z \subseteqq X \times X$ by letting each member of H be a vertical section of Z. More precisely, let f be a 1–1 function from X to H and let

$$Z = \bigcup_{x \in X} \{x\} \times f(x) .$$

By (i) there is a countable ordinal α such that $Z \in \mathscr{R}_\alpha$ and hence by Lemma 6, there is a countable subfamily J of $\mathscr{R}$ for which $Z \in J_\alpha$. Let

$$G = \{B\colon A \times B \in J\} ,$$

note that $Z \in (VG)_\alpha$ and use Theorem 3 to conclude that $H \subseteqq G_\alpha$.

Now suppose (ii) holds, and that $Z \subseteqq X \times X$. Let H be the family of vertical sections of Z, and use (ii) to secure a countable family G and a countable ordinal α for which $H \subseteqq G_\alpha$. Thus $Z_x \in G_\alpha$ for each $x \in$ domain Z and by Theorem 3

$$Z \in (VG)_\alpha \subseteqq \mathscr{R}_\alpha .$$

THEOREM 9. *The following four statements are equivalent:*

(i) *Each subset of $X \times X$ is a member of $\mathscr{B}(\mathscr{R})$.*

(ii) *If H is a family of subsets of X and* Card $H =$ Card X *then there is a countable family G and a countable ordinal α for which $H \subseteqq G_\alpha$.*

(iii) *There is a countable ordinal α such that, for each family H of subsets of X with* Card $H =$ Card X, *there is a countable family G for which $H \subseteqq G_\alpha$.*

(iv) *There is a countable ordinal $\alpha \geqq 2$ such that each subset of $X \times X$ is a member of $\mathscr{R}_\alpha$.*

Proof. Statements (i) and (ii) are equivalent by Lemma 7 and Theorem 8a. Clearly (iii) implies (ii) and (iv) implies (i). Also by Theorem 8b it follows that (iii) implies (iv). α cannot be equal to 1 in (iv) because by (i) the identity function $f(x) = x$ is not in $\mathscr{R}_1$.

We complete the proof by showing that (ii) implies (iii). Since this result is immediate if X is countable we will assume that Card $X \geqq \omega_1$.

Suppose that (ii) holds and that (iii) does not. Then for each $\alpha < \omega_1$, there is a family $H(\alpha)$ of subsets of X for which Card $H(\alpha) =$

Card X and

(1) for each countable G, $H(\alpha) \nsubseteq G_\alpha$.

Let $H' = \bigcup_{\alpha<\omega_1} H(\alpha)$. Thus Card H' = Card X and hence by (ii) there is a countable family G' and a countable ordinal α' for which $H' \subseteqq G'_{\alpha'}$. But then $H(\alpha') \subseteqq H' \subseteqq G'_{\alpha'}$ in contradiction of (1).

Therefore (ii) implies (iii).

In part (ii) above the family G can be chosen so that G_{ω_1} is closed to complementation (i.e., is a σ-algebra).

In view of condition (ii) of Theorem 9, it is interesting to note that R. Mansfield has shown that if G is a countable family of Lebesgue measurable sets, then $B(G)$ does not contain all analytic sets [5].

As was mentioned in the introduction it would be interesting to know whether the formula "$H \subseteqq G_\alpha$" in Theorem 9 could be replaced by $H \subseteqq \mathscr{B}(G)$. We do not know the answer to this question.

THEOREM 10. *If* CH *holds,* Card $X = c$, H *is a family of subsets of* X, *and* Card $H = c$, *then there is a countable family* G *for which* $H \subseteqq G_2$.

Proof. By Theorem 5 each subset Z of $X \times X$ is a member of $\mathscr{R}_2$. The desired conclusion now follows from Theorem 8b.

4. Generating Borel sets. Let R be the set of reals, and let H be the family of all Borel subsets of R. This family has cardinality c. Suppose G is a countable family of subsets of R such that $H \subseteqq G_{\omega_1}$ and G_{ω_1} is closed to complementation. The next two theorems show that, even if the family G generates all the Borel sets at an early stage, the order of G is ω_1. This is a generalization of the well known result [4, 9] that if G is a countable basis for the real topology then G has order ω_1. Our proof which is a usual "diagonal" type argument, parallels somewhat Lebesgue's proof of that result [3, p. 368].

Let $G = \{V_1, V_2, V_3, \cdots\}$, let N be the set of irrational numbers between 0 and 1 and let K be the family $\{\theta_1, \theta_2, \theta_3, \cdots\}$ of all intersections of the members of G with N,

$$\theta_i = V_i \cap N .$$

It will be shown that the order of K is ω_1. It then follows that the order of G is ω_1.

For each $z \in N$, let $(z_1, z_2, z_3, \cdots)$ be the sequence of integers appearing in the continued fraction expansion of z. This defines a reversible transformation from N onto the set of all sequences of positive integers. Let

$$(*)\qquad \begin{aligned} z^1 &= (z_1, z_3, z_5, \cdots) \qquad \text{(odd indices)} \\ z^2 &= (z_2, z_6, z_{10}, \cdots) \\ z^3 &= (z_4, z_{12}, z_{20}, \cdots) \\ &\vdots \\ z^n &= (z_{2^n-1}, z_{3\cdot 2^n-1}, z_{5\cdot 2^n-1}, \cdots) \\ &\vdots \end{aligned}$$

This defines a homeomorphism between N and $N^{\aleph_0}$ (see Kuratowski [3], p. 369). Also note that if f is a continuous function from N into N, then the functions f_n from N into the space of positive integers are continuous, where

$$f(z) = (f_1(z), f_2(z), f_3(z), \cdots)\,, \quad \text{or} \quad (f_n(z) = f(z)_n)\,.$$

Recall that $K = \{\theta_1, \theta_2, \theta_3, \cdots\}$. The family K_α which appears in Theorem 11 is defined in § 2.

THEOREM 11. *For each countable ordinal α, $\alpha > 0$, there is a function U_α from N onto K_α such that if f is a continuous function from N into N, then the set*

$$A_f = \{z\colon z \in U_\alpha(f(z))\}$$

is in $\mathscr{B}(K)$.

Proof. Let $U_1(z) = \bigcup_{n=1}^{\infty} \theta_{z_n}$, for each $z \in N$. Clearly U_1 maps N onto K_1.

Let f be a continuous function from N onto N.

We have

$$\begin{aligned} A_f &= \{z\colon z \in U_1(f(z))\} \\ &= \left\{z\colon z \in \bigcup_{n=1}^{\infty} \theta_{f_n(z)}\right\} \\ &= \bigcup_{n=1}^{\infty} \{z\colon z \in \theta_{f_n(z)}\}\,. \end{aligned}$$

For each n,

$$\{z\colon z \in \theta_{f_n(z)}\} = \bigcup_{i=1}^{\infty} \{J_{n_i} \cap \theta_i\}$$

where $J_{n_i} = \{z\colon f_n(z) = i\}$. Since each f_n is continuous it follows that each J_{n_i} is open and therefore the set A_f belongs to G_{ω_1}.

Suppose $1 < \alpha < \omega_1$ and suppose that the function U_γ has been defined for each ordinal γ with $1 \leqq \gamma < \alpha$. (Induction hypothesis.)

If α is odd, let

$$U_\alpha(z) = \bigcup_{n=1}^{\infty} U_{\alpha-1}(z^n), \quad \text{for} \quad z \in N .$$

Clearly U_α maps N onto K_α.

If α is even, let $\{\gamma_n\}_{n=1}^{\infty}$ be a sequence of odd ordinals less than α such that each odd ordinal less than α appears infinitely often in $\{\gamma_i\}_{n=1}^{\infty}$ and let

$$U_\alpha(z) = \bigcap_{n=1}^{\infty} U_{\gamma_n}(z^n) , \quad \text{for} \quad z \in N .$$

If $A \in K_\alpha$ (α is still even), then

$$A = \bigcap_{n=1}^{\infty} D_n ,$$

where $D_n \in K_{\gamma_n}$, for each n. For each n, let y_n be a point of N such that

$$D_n = U_{\gamma_n}(y_n) .$$

And let z be the point mapped by the transformation described by (*) to the point $(y_1, y_2, y_3, \cdots)$ of $N^{\aleph_0}$. Thus

$$U_\alpha(z) = A$$

and U_α maps N onto K_α.

This completes the definition of the functions U_α. Now let f be a continuous function from N into N. It will be shown that if α is even the set

$$A_f = \{z: z \in U_\alpha(f(z))\}$$

is in G_{ω_1}. The argument for the case α is odd is similar.

We have

$$A_f = \left\{z: z \in \bigcap_{n=1}^{\infty} U_{\gamma_n}((f(z))^n)\right\}$$
$$= \bigcap_{n=1}^{\infty} \{z: z \in U_{\gamma_n}((f(z))^n)\} .$$

But, for each n, the function $z \to (f(z))^n$, being the composition of two continuous functions, is a continuous function from N to N.

Thus by the induction hypothesis, the sets $\{z: z \in U_{\gamma_n}((f(z))^n)\}$ are in the family G_{ω_1}. Therefore $A_f \in G_{\omega_1}$.

THEOREM 12. *If G is a countable family of subsets of real numbers with $\mathscr{C}(G) \subseteq G$, and each Borel set is a member of $\mathscr{B}(G)$ then G has order ω_1.*

Proof. Let α be any countable ordinal, and let

$$I_\alpha = \{z: z \notin U_\alpha(z)\}.$$

Suppose $I_\alpha \in K_\alpha$, and let $U_\alpha(z) = I_\alpha$. If $z \in I_\alpha$ then $z \in U_\alpha(z)$. But this contradicts the definition of I_α. If $z \notin I_\alpha$, then $z \in U_\alpha(z) = I_\alpha$, $z \in I_\alpha$. This contradiction shows that $I_\alpha \notin K_\alpha$.

Since $\mathscr{B}(G) = G_{\omega_1}$ (because $\mathscr{C}(G) \subseteqq G$), and $I'_\alpha = \{z: z \in U_\alpha(z)\} \in G_{\omega_1}$ by Theorem 11, it follows that $I_\alpha \in G_{\omega_1} - G_\alpha$. Thus $G_\alpha \neq G_{\omega_1}$, and hence G has order ω_1 [3, p. 371].

REFERENCES

1. J. A. Johnson, Amer. Math. Monthly, **79** (1972), 307.
2. Kenneth Kunen, *Inaccessibility Properties of Cardinals*, Ph. D. Thesis, Department of Mathematics, Stanford University, August, 1968.
3. C. Kuratowski, *Topology I*, Academic Press, New York, 1966.
4. H. Lebesgue, *Sur les fonctions representable analytiquement*, J. de Math., (6) **1** (1905), 139-216.
5. D. A. Martin and R. M. Solovay, *Internal Cohen extensions*, Annals of Math. Logic, **2** (1970), 143-148.
6. B. V. Rao, *On discrete Borel spaces and projective sets*, Bull. Amer. Math. Soc., **75** (1969), 614.
7. ———, *On discrete Borel spaces*, Acta Math. Acad. Scien. Hungaricae, **22** (1971), 197.
8. W. Sierpinski, *Sur l'existence de diverses classes d'ensembles*, Fund. Math., **14** (1929), 82-91.
9. ———, *Hypothèse du Continu*, Warsaw, 1934, 9.

Received November 14, 1972 and in revised form July 12, 1973. This work was partially supported by NSF Grant GJ-32269.

UNIVERSITY OF WISCONSIN
UNIVERSITY OF TEXAS
AND
UNIVERSITY OF FLORIDA

Reprinted from the AMERICAN MATHEMATICAL MONTHLY
Vol. 85, No. 2, February 1978
pp. 73–74

AWARD FOR DISTINGUISHED SERVICE TO PROFESSOR R. D. ANDERSON

Richard Davis Anderson has been one of the most active leaders of our time in promoting mathematics. He has worked through SMSG and CUPM to improve the mathematics curriculum at the secondary and collegiate levels. As an individual teacher and through committees he has promoted strong undergraduate and graduate programs in mathematics. His analyses of the job market have alerted us to the consequences of a heavy Ph.D. production. His emphasis of the long range demographic factors in academic employment has helped the mathematical community and many of its members adjust to the changed conditions of the seventies. His success at research, his interest in people, and his participation in worldwide conferences have made him an effective international spokesman for our national mathematical community.

He is known professionally as R. D. Anderson, but his friends call him Dick. Dick obtained his bachelor's degree from Minnesota in two years. At a meeting of the Association and the Society in the summer following his graduation he was recruited by Professor R. L. Moore to come to Texas. He arrived in Texas at the age of 19 to begin graduate work.

Anderson's work as a graduate student was interrupted by World War II. He served in the Navy as an officer on a submarine in the Atlantic and later as an engineering officer on two naval auxiliaries in the Pacific. With the war over in 1945, Dr. Moore immediately contacted his star student about returning to The University of Texas to resume his study.

We know Anderson as a fierce competitor whether it be at debate, tennis, chess, or mathematics. Competition had started early for Dick since he had a twin brother, Jack. However, Jack went into Chemistry and left Mathematics to Dick. Those doing graduate work in Topology at Texas during Anderson's time included Gail Young, R. H. Bing, Ed Moise, Ed Burgess and Mary Ellen Estill (later to become Mrs. Rudin)—not a slouch in the crowd. (Even as Anderson was completing his doctorate, Eldon Dyer, Mary Elizabeth Hamstrom and Billy Joe Ball were beginning to prove good theorems.)

Anderson was the only one of these who could nearly hold his own in heated debates with Ed Moise. I sometimes played chess with Anderson during our lunch hour at Texas, until Dr. Moore discouraged it. Dr. Moore asked me if I was not concerned that this recreation would rob Anderson of time that he might better spend on proving theorems. The thought had not occurred to me since my concern was in checkmating his king—which I was able to do only too infrequently.

After receiving his Ph.D., Anderson went through the ranks of Instructor, Assistant Professor and Associate Professor at Pennsylvania. During this period he spent two years at the Institute for Advanced Study. Eleven years after receiving his Ph.D. he had been named a Boyd Professor at LSU.

At LSU Anderson was very effective in setting up a program in Infinite Dimensional Topology. He and his students—including such stalwarts as Raymond Wong, Jim West and Tom Chapman—did fundamental work in this area which attracted many to attend LSU sponsored conferences.

Because of the depth of his mathematics and the charm of his personality Anderson is welcomed abroad. He has spent two sabbaticals in Holland and has made many visits to Eastern Europe. His rapport with foreign mathematicians is no accident. For example, in preparation for his first visit to Russia, he studied the Russian language for a considerable time. At last year's St. Louis meeting he wrote down a list from the top of his head of the names of about 25 Russian topologists (probably with the correct spelling) who should have copies of Chapman's book on Q-manifolds and arranged for them to be sent.

Anderson's influence has been boosted by the vigorous support of his wife Jeanette. They have made friends all over the world and are very effective ambassadors. Jeanette carries on correspondence with friends far and wide. Anderson's true spirit cannot be fully appreciated until one understands the concern of the Anderson family for others.

Anderson has been on 25 Association committees dealing with such diverse topics as finances, employment, membership, developing colleges, visiting lecturers, consultants bureau, the undergraduate curriculum, and cooperation with other organizations. He led many of these committees and in particular chaired the important and very busy Committee on the Undergraduate Program in Mathematics (CUPM). Currently he is on the AMS–MAA–SIAM Joint Projects Committee in Mathematics.

Anderson does not restrict his contributions to MAA work. He was a vice-president of the Society and led its Committee on Employment and Educational Policy to study the changing job situation. Anderson is Chairman of the Board of Trustees of the Conference Board. For NSF he has been on committees giving advice on science education and minority programs. He earlier chaired the NSF Mathematics Advisory Panel. For GRE he is a member of the Panel on Graduate Record Examinations in Mathematics. For NRC he is a member of the Advisory Committee of the Office of Mathematical Sciences and was formerly on the Soviet and Eastern European Exchange Program Advisory Committee.

Anderson is a man who has served the mathematical community well. The Award for Distinguished Service in Mathematics is given for past performance, but in the case of a man as vigorous as Anderson we can look for additional contributions in the years that lie ahead.

R. H. BING

ABCDEFGHIJ — 898

The Collected Papers of R. H. Bing

R. H. BING
1914–1986

The Collected Papers of R. H. Bing

Volume 2

Sukhjit Singh, Steve Armentrout, and Robert J. Daverman, Editors

AMERICAN MATHEMATICAL SOCIETY
PROVIDENCE, RHODE ISLAND

1980 *Mathematics Subject Classification* (1985 *Revision*).
Primary 57-06, 54-06, 26-06.

Library of Congress Cataloging-in-Publication Data

Bing, R. H.
[Works. 1988]
The collected papers of R. H. Bing/Sukhjit Singh, Steve Armentrout, and Robert J. Daverman, editors.
p. cm.
Includes bibliographies.
ISBN 0-8218-0117-1 (v. 1: alk. paper)
1. Manifolds (Mathematics) 2. Bing, R. H. I. Singh, Sukhjit
II. Armentrout, Steve.
III. Daverman, Robert J.
IV. Title.
QA613.B45 1988 88-14445
516′.07–dc19 CIP

The paper used in this book is acid-free and falls within the guidelines
established to ensure permanence and durability. ♾

TABLE OF CONTENTS

I. R. H. Bing: An Introduction

II. Classifications of Works of R. H. Bing

III. Collected Papers of R. H. Bing

I. General Topology

A. Metrization

B. Retracts and Related Topics

B. Cuttings

C. Planar Embeddings

III. Continua

A. Pseudo Arc and Related Topics

B. Locally Connected Continua: Partitions and Decompositions

C. Homogeneity

D. Others

IV. Topological Classifications

A. 2-Sphere

B. 3-Space and 3-Sphere

C. Arcs and Simple Closed Curves

D. Hilbert Space

V. Miscellaneous Topics

VI. Topology of 3-Manifolds

A. Decompositions

B. Periodic Homeomorphisms

C. Approximating and Embedding Surfaces

D. Recognizing Tameness and Related Topics

E. Topics Related to the Poincaré Conjecture

VI. Topology of 3-Manifolds

A. Decompositions

Annals of Mathematics
Vol. 61, No. 2, March, 1955
Printed in U.S.A.

SOME MONOTONE DECOMPOSITIONS OF A CUBE*

By R. H. Bing

(Received March 23, 1954)

1. Introduction

In this paper we suppose that M is a sphere plus its interior in E^3, H' is a monotone upper semicontinuous decomposition of Bd M, and H is the decomposition of M whose elements are the elements of H' and the points of the interior of M. We shall be interested in determining the topology of the decomposition space H.

We let f be the map (continuous transformation) associated with the upper semicontinuous decomposition H. Our interest (in terms of mappings) will be to determine the topology of $f(M)$ where f is 1–1 on the interior of M and monotone on Bd M.

Characterizing a 3-sphere. In studying the decomposition space H (or $f(M)$), we need a characterization of S^3. We use the following one taken from [1].

A necessary and sufficient condition that a continuous curve S be a 3-sphere is that one of its decreasing sequences of regular partitionings $G_1, G_2, \cdots$ have the following properties:

(1) *The boundary of each element of G_i is a simple surface.*

(2) *If the boundaries of two elements of $\sum G_i$ intersect, this intersection is a closed 2-cell.*

(3) *The intersection of the boundaries of three elements of G_i is one-dimensional at each of its points.*

(4) *If g is an element of G_{i-1} ($g = S$ if $i = 1$), the elements of G_i in g may be ordered $g_1, g_2, \cdots, g_n$ so that* Bd g_j ($j = 1$ (if $i > 1$), $2, \cdots, n$) *intersects* Bd g + Bd $g_1 + \cdots +$ Bd g_{j-1} *in a connected set.*

The above characterization of S^3 leaves much to be desired. One of its disadvantages is that it is difficult to show that a continuous curve has a decreasing sequence of regular partitionings satisfying the prescribed conditions. It is to be hoped that as the study of topology progresses, a more satisfactory topological characterization of S^3 will be found. However, in the present study we were able to use the above characterization.

Definitions. We use E^n to denote Euclidean n-space and S^n to denote the surface of a sphere in E^{n+1}. Perhaps it is simpler to think of S^n as E^n which has been made compact by the addition of a point. Sets topologically equivalent to S^1, S^2, S^3 are called simple closed curves, simple surfaces, and 3-spheres.

If X is a compact metric space, an upper semicontinuous decomposition G of X is a collection of closed subsets of X such that X is the sum of the elements of G and for each open set U containing an element g of G, there is an open set V containing g such that each element of G which intersects V lies in U. See [6] and [7].

* This work was supported in part by funds from the National Science Foundation under Contract NSF-G479.

The collection G will also be considered as a decomposition space G. We use the same letter to denote the collection as the decomposition space but the elements of the first are point sets in X and the elements of the second are points. If g is an element of the upper semicontinuous decomposition G and U is an open set in X containing g, the sum of all elements of the decomposition space G which are associated with sets that lie in U is a neighborhood of the point g of the decomposition space G. If the elements of G are connected, it is called monotone. Similarly, a map f is monotone if the inverse of each point is connected.

If A is a subset of a space B, we use Bd A to denote the boundary of A in B—that is, Bd $A = \bar{A} \cdot \overline{B - A}$.

A partitioning of a metric space is a finite collection of mutually exclusive connected open sets whose sum is dense in the space. A sequence $G_1, G_2, \cdots$ of partitionings is a decreasing sequence of partitionings if G_{i+1} is a refinement of G_i and the maximum of the diameters of the elements of G_i approaches 0 as i increases without limit. A partitioning is regular if each of its elements is the interior of its closure. It is known that each continuous curve (compact, locally connected, metric continuum) has a decreasing sequence of regular partitionings [2, 3].

Since the image of a continuous curve under a continuous transformation is a continuous curve, the decomposition space associated with any upper semicontinuous decomposition of a solid sphere M is a continuous curve. In the next sections we seek to find what types of continuous curves some of these decomposition spaces are.

2. If H' is a dendron

In the following sections we suppose that M, H', H, and f are as defined in the first two paragraphs of the introduction.

THEOREM 1. *If H' is a dendron, the decomposition space H is topologically equivalent to S^3.*

THEOREM 1'. *If f(Bd M) does not contain a simple closed curve, $f(M)$ is a 3-sphere.*

We intend to show that H is a 3-sphere by showing that it has a decreasing sequence of regular partitionings $G_1, G_2, \cdots$ satisfying conditions 1, 2, 3, 4.

For convenience we suppose that M is a solid sphere with radius 1.

We use $S(\varepsilon)$ to denote the simple surface consisting of all points of M whose distance from Bd M is ε. Also, $I(\varepsilon)$ will be used to denote the points of M whose distance from the center of M is less than $1 - \varepsilon$. Then Bd $I(\varepsilon) = S(\varepsilon)$. When convenient we will regard $I(\varepsilon)$ and $S(\varepsilon)$ as subsets of H rather than of M. We now consider some surfaces in H.

Disks in H. We describe some disks in the decomposition space H that will lie on the boundaries of elements of the partitionings $G_1, G_2, \cdots$.

Let $h_1, h_2, \cdots$ be a sequence of elements of H' such that $h_1 + h_2 + \cdots$ is dense in the decomposition space H' and each h_i separated Bd M (and hence H') into exactly two connected pieces. Let D_i denote one component of Bd $M - h_i$.

Since D_i is topologically equivalent to the interior of a circle, there is a continuous real valued function f_i of $D_i + h_i$ such that $f_i(h_i) = 0$, $0 < f_i(x) \leqq \frac{3}{4}$ if $x \in D_i$, and for each value ε between 0 and $\frac{3}{4}$, $f_i^{-1}(\varepsilon)$ is a simple closed curve.

Let $p(x)$ be the projection transformation from the center of M that takes M minus its center onto Bd M.

The set of all points x of M such that $p(x) \in D_i + h_i$ and $\rho(x, p(x)) = f_i(p(x))$ will be called the graph of f_i. Here $\rho(x, y)$ denotes the distance from x to y in M. We use F_i to denote the elements of H on this graph and $F_i(\varepsilon)$ to denote those elements of F_i whose distance from Bd M is less than or equal to ε. We note that F_i is a simple surface and that $F_i(\varepsilon)$ $(0 < \varepsilon < \frac{3}{4})$ is topologically a closed 2-cell. In the proof of Theorem 3 we refer to $F_i(\varepsilon)$ as a fence.

Consider how the disk $F_i(\varepsilon)$ separates $H - \bar{I}(\varepsilon)$. Use $D_i(\varepsilon)$ and $E_i(\varepsilon)$ to denote the collection of all elements of H in the sets

$$\{x \mid x \in M - \bar{I}(\varepsilon), p(x) \varepsilon D_i, \rho(x, p(x)) < f_i(p(x))\} \qquad \text{and}$$

$$\{x \mid x \in M - \bar{I}(\varepsilon), \text{either } p(x) \notin D_i + h_i \text{ or } \rho(x, p(x)) > f_i(p(x))\}.$$

In our argument to follow we find that both $D_i(\varepsilon)$ and $E_i(\varepsilon) (0 < \varepsilon < \frac{3}{4})$ are topologically equivalent to the interior of a sphere.

Description of G_1. The partitioning G_1 has the following three elements: $I(\frac{1}{2})$, $D_1(\frac{1}{2})$, and $E_1(\frac{1}{2})$. The boundary of $I(\frac{1}{2})$ is the simple surface $S(\frac{1}{2})$, the boundary of $D_1(\frac{1}{2})$ is $F_1(\frac{1}{2})$ plus a disk in $S(\frac{1}{2})$, and the boundary of $E_1(\frac{1}{2})$ is $F_1(\frac{1}{2})$ plus a disk in $S(\frac{1}{2})$.

Description of G_2. Since $\bar{I}(\frac{1}{2})$ is a tame 3-cell in M, we do not bother to describe the elements of G_2 in it. They can be taken so as to satisfy conditions 1, 2, 3, 4 and such that each is of diameter less than $\frac{1}{2}$. More comments concerning the partitioning of a tame 3-cell on the interior of M are found under *Subdividing an interior element of* G_{i-1}.

We now describe the elements $g_1, g_2, \cdots, g_n$ of G_2 in $D_1(\frac{1}{2})$. Consider the first element of $h_1, h_2, \cdots$ that lies in $D_1(\frac{1}{2})$. For convenience we suppose it is h_2. Then there is a positive number ε such that $\varepsilon < \frac{1}{2}$ and $F_2(\varepsilon)$ lies in $D_1(\frac{1}{2})$. Let g_1 be the one of $D_1(\frac{1}{2}) \cdot D_2(\varepsilon)$, $D_1(\frac{1}{2}) \cdot E_2(\varepsilon)$ that has $F_1(\varepsilon)$ on its boundary and g_2 be the other one. See Figure 1.

Now Bd g_1 is the sum of $F_1(\varepsilon)$, $F_2(\varepsilon)$, and an annulus ring in $S(\varepsilon)$ while Bd g_2 is the sum of $F_2(\varepsilon)$ and a disk in $s(\varepsilon)$. Also Bd $g_1 \cdot \bar{E}_1(\frac{1}{2})$ is the disk $F_1(\varepsilon)$ and Bd $g_1 \cdot$ Bd g_2 is the disk $F_2(\varepsilon)$.

We could not let $D_1(\frac{1}{2}) - (\bar{g}_1 + \bar{g}_2)$ be an element of our partitioning G_2 for then the boundary of this element would intersect Bd g_1 in an annulus ring instead of a closed 2-cell as required by condition 2. However it is possible to continue the partitioning of $D_1(\frac{1}{2}) - (\bar{g}_1 + \bar{g}_2)$ so that conditions 1, 2, 3, 4 will be satisfied with respect to the elements of G_2 in $D_1(\frac{1}{2})$. Some comments on how this may be done is found in *Subdividing an interior element of* G_{i-1}.

We now describe the elements of G_2 in $E_1(\frac{1}{2})$. Consider the first element of h_1, $h_2, \cdots$ in $E_1(\frac{1}{2})$. Suppose it is h_3. Let ε' be a positive number such that $\varepsilon' < \varepsilon$ and $F_3(\varepsilon')$ lies in $E_1(\frac{1}{2})$. Then $E_1(\frac{1}{2}) \cdot D_3(\varepsilon')$ and $E_1(\frac{1}{2}) \cdot E_3(\varepsilon')$ are elements of G_2

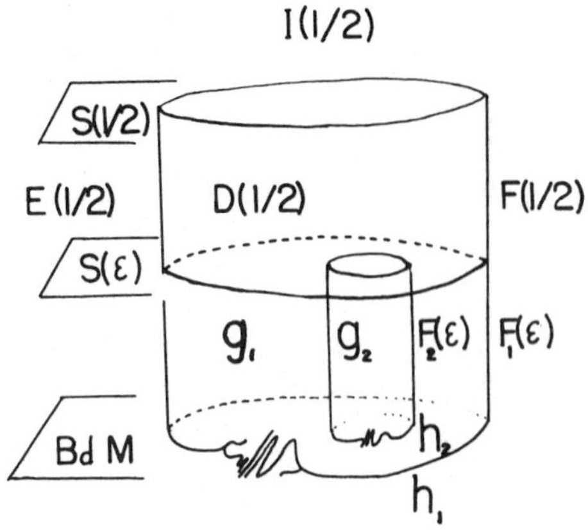

FIG. 1

in $E_1(\frac{1}{2})$ and the other elements may be selected so as to satisfy conditions 1, 2, 3, 4. The partitioning of $E_1(\frac{1}{2})\cdot I(\varepsilon')$ follows a pattern such as outlined in *Subdividing an interior element of* G_{-1}.

Description of G_i. Suppose certain elements of G_{i-1} have been partitioned to get a partitioning G_i' which satisfies conditions 1, 2, 3, 4 and is the start of the partitioning G_i but that g is an element of G_{i-1} which has not been partitioned. We break the description of the elements of G_i in g into two cases.

Subdividing an interior element of G_{i-1}. We wish to subdivide each element g of G_{i-1} which does not intersect H' into pieces with diameters less than $1/i$.

Since the closures of such elements g are tame in M, we suppose that $\bar{g}$ is a cube. The partitioning G_i' imposes a partitioning of Bd g into topological 2-cells. The sum of the boundaries of these 2-cells is denoted by K. Since G_i' satisfies conditions 2 and 3, there is no point of K which is an end point of four mutually exclusive open arcs in K. Hence, we suppose with no loss of generality that K is polygonal and that each straight line interval in K is parallel to but does not lie in an edge of the cube $\bar{g}$.

Let ε be a positive number less than either $1/i$ or the distance between any two vertices of K. Let $P(g)$ be a partitioning of g such as indicated in Figure 1, page 16 of [1] and such that each element of $P(g)$ is of diameter less than ε and the intersection of the boundaries of no two elements of $P(g)$ contains an arc in K. Then the partitioning resulting from replacing g by the elements of $P(g)$ in G_i will satisfy conditions 1, 2, 3, 4.

Subdividing an element of G_{i-1} *which intersects* H'. In this case there is a positive number ε and integers $n_1, n_2, \cdots, n_k$ such that Bd $g = F_{n_1}(\varepsilon) + F_{n_2}(\varepsilon) + \cdots + F_{n_k}(\varepsilon) + A$ where A is a connected subset of $S(\varepsilon)$ which is bounded in $S(\varepsilon)$ by k mutually exclusive simple closed curves. See Figure 2.

Let h_{r_0} be the first element of $h_1, h_2, \cdots$ in $g\cdot H'$. Since Bd $g\cdot H$ is a dendron with ends $h_{n_1}, h_{n_2}, \cdots, h_{n_k}$, there are elements $h_{r_1}, h_{r_2}, \cdots, h_{r_k}$ of $h_1, h_2, \cdots$ in $g\cdot H'$ such that h_{r_j} separates h_{n_j} from the other elements of $h_{r_0}, h_{r_1}, \cdots, h_{r_k}, h_{n_1}, \cdots, h_{n_k}$ in H'.

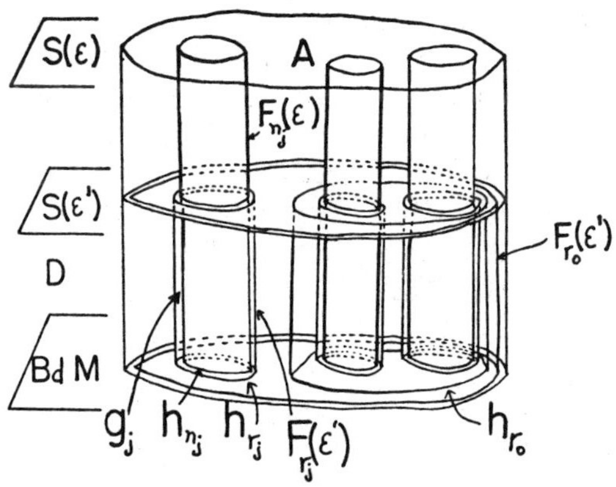

FIG. 2

Let ε' be a positive number less than $1/i$ such that each element of G_i' intersects $I(\varepsilon')$ and $F_{r_0}(\varepsilon')$, $F_{r_1}(\varepsilon')$, $\cdots$, $F_{r_k}(\varepsilon')$ are mutually exclusive sets which lie in g.

Now $S(\varepsilon')$ divides g into two connected pieces: the one in $I(\varepsilon')$ is topologically equivalent to the cartesian product of the interior of A and an open interval; the other, D, has for its boundary a simple surface in H. We now describe the boundaries of the elements of G_i in D.

For each positive integer j less than or equal to k, there is an element g_j' of G_i in D bounded by $F_{n_j}(\varepsilon') + F_{r_j}(\varepsilon')$ plus an annulus ring in $S(\varepsilon')$. See Figure 2. There are two other elements g_{k+1}', g_{k+2}' of G_i in D but we can not precisely define the boundary of each without knowing how h_{r_0} divides h_{r_1}, h_{r_2}, $\cdots$, h_{r_k} in H'. The sum of the boundaries of g_{k+1}' and g_{k+2}' is $F_{r_0}(\varepsilon') + F_{r_1}(\varepsilon') + \cdots + F_{r_k}(\varepsilon')$ plus a connected subset of $S(\epsilon')$ and the intersection of their boundaries is $F_{r_0}(\varepsilon')$.

We mention here the reason for introducing h_{r_1}, h_{r_2}, $\cdots$, h_{r_k}. If we had divided D instead of $D - (\bar{g}_1' + \cdots + \bar{g}_k')$ by $F_{r_0}(\varepsilon')$ we would have obtained elements whose boundaries are simple surfaces but we could not be sure that condition 2 is met because Bd $g \cdot$ Bd g_{k+1}' might not be connected. We note that we did not need all of h_{r_1}, h_{r_2}, $\cdots$, h_{r_k} but only enough of them to divide each pair of h_{n_1}, h_{n_2}, $\cdots$, h_{n_k}.

The elements of G_i in D we have described are not the first elements of G_i in g with respect to condition 4. In fact, they are the last. Since $\bar{g} \cdot \bar{I}(\varepsilon')$ is topologically equivalent to the cartesian product of A and an interval, by modifying the methods used in the preceding case and by starting near A one can get a partitioning g_1, g_2, $\cdots$, g_s of $g \cdot I(\varepsilon')$ into tame open 3-cells such that the partitioning obtained by replacing g by g_1, g_2, $\cdots$, g_s, g_1', $\cdots$, g_{k+2}' in G_i' satisfies conditions 1, 2, 3, 4.

Mesh of partitionings. We shall show that G_1, G_2, $\cdots$ is a decreasing sequence of partitions of H by showing that if R and T are mutually exclusive closed subsets of H, there is a G_i such that no element of G_i intersects both R and T.

A finite number of elements of $h_1, h_2, \cdots$ separate $R \cdot H'$ from $T \cdot H'$ in H'. Then there is an integer j such that the closure of no element of G_j contains both a point of $R \cdot H'$ and a point of $T \cdot H'$. There is a positive integer k such that the closure of no element of G_k contains both a point of $R \cdot (M - I(1/k))$ and a point of $T \cdot (M - I(1/k))$. Also there is an integer n greater than either j or k such that the distance in M between $R \cdot I(1/n)$ and $T \cdot I(1/n)$ is greater than $1/n$. Since the elements in G_{n+1} which intersect $I(1/n)$ are of diameters less than $1/n$, no element of G_{n+1} intersects both R and T.

3. If H' is a simple surface

The following result is due to Floyd and Fort [4]. An outline of the proof is included. The result will be used in the next section.

THEOREM 2. *If H' is topologically equivalent to* Bd M, *there is a homeomorphism of H onto M. In fact there is such a homeomorphism under which each element of H and its image are the same distance from the center of M.*

THEOREM 2'. *If f(Bd M) is a simple surface, $f(M)$ is a solid sphere.*

OUTLINE OF PROOF. What we show is that if g is any monotone map of Bd M onto itself, g may be extended in a 1–1 fashion to the interior of M to map M onto itself. If p is a point of M, $g(p)$ and p will be the same distance from the center of M.

J. W. T. Youngs has shown [8] that for each positive number ε and each monotone map g of Bd M onto itself there is a homeomorphism f_ε of Bd M onto itself such that $\rho(g, f_\varepsilon) < \varepsilon$.

Floyd and Fort showed [4] that for each positive number ε there is a positive number δ such that if f_1, f_2 are homeomorphisms of Bd M onto itself within δ of each other, then there is an ε-isotopy pulling f_1 to f_2.

Using the above mentioned results of Youngs, Floyd, and Fort it may be shown that there is a continuous function $f_t(x)$ ($x \in$ Bd M, $0 \leqq t \leqq 1$) such that $f_1 = g$ and $f_t(0 \leqq t < 1)$ is a homeomorphism of Bd M onto itself.

Suppose the radius of M is 1, its center is m_0, and $p(x)$ is the projection from m_0 of the point x of $M - m_0$ into Bd M. Let $\varepsilon = \rho(x, m_0)$. Then $g(x)$ (under the extension of g) is the point that divides the interval from m_0 to $f_\varepsilon(p(x))$ in the same ratio that x divides the interval from m_0 to $p(x)$.

4. If H' is arbitrary

A nondegenerate cyclic element of a continuous curve Y is a subset of Y which is maximal with respect to the property that each pair of points of the subset belongs to a simple closed curve in Y. A cactoid is a continuous curve each of whose nondegenerate cyclic elements is a simple surface. R. L. Moore has shown [5] that if H' is a monotone upper semicontinuous decomposition of a simple surface the decomposition space H' is a cactoid. Both a dendron and a simple surface are types of cactoids.

THEOREM 3. *H is topologically equivalent to a cactoid C in S^3 plus one of its complementary domains. The cactoid C will have the same number of nondegenerate cyclic elements as H'.*

THEOREM 3'. *$f(M)$ is topologically equivalent to a subset K of S^3. K will have* *the same number of complementary domains as $f(M)$ has simple surfaces and the boundary of each complementary domain of K is a simple surface.*

Plan of attack. The proof of this theorem is somewhat like but a bit more complicated than the proof of Theorem 1. We avoid some of the complications in the present proof by supposing that the proof of Theorem 1 is understood before the present proof is begun.

Now H' has not more than a countable number of nondegenerate cyclic elements. Since the proof is simplified if they are finite in number, we suppose that there are infinitely many and they are ordered $C_1, C_2, \cdots$. We fill up holes in H by tying a countable collection of solid spheres to H along the simple surfaces C_i as described below.

If H is imbedded in the proper sort of space, there are mutually exclusive open sets $X_1, X_2, \cdots$ such that there is a homeomorphism of $X_i + C_i$ onto a cube which takes C_i onto the boundary of this cube, $X_i \cdot H = 0$, and in H (not in Bd M) C_i has the same diameter as X_i.

We finish the proof of Theorem 3 by showing that $H + \sum X_i$ is topologically equivalent to S^3. We do this by getting a decreasing sequence of regular partitionings of $H + \sum X_i$ which satisfies conditions 1, 2, 3, 4.

Partitionings of H'. In the proof of Theorem 1, the partitionings $G_1, G_2, \cdots$ of that proof imposed partitionings on H'. In the description of G_i in that proof, before partitioning an element g of G_{i-1} that intersected H', we first partitioned $g \cdot H'$. For example, when $g \cdot H'$ was bounded in H' by $h_{n_1} + h_{n_2} + \cdots + h_{n_k}$, we used $h_{r_0}, h_{r_1}, \cdots, h_{r_k}$ to divide $g \cdot H'$ into pieces $g_1' \cdot H'$, $g_2' \cdot H'$, $\cdots$, $g_{k+2}' \cdot H'$.

In the present proof, rather than describe the partitionings of H' as we describe the partitionings of H, we first prescribe a decreasing sequence of regular partitionings $G_1', G_2', \cdots$ of H' that will be used later in describing the partitionings of $G_1, G_2, \cdots$ of H.

Let W be the set of points of $\sum C_i$ which separate H'. They are points to be avoided as noted below.

If C_i is a simple surface in H', there is a decreasing sequence of regular partitionings $K_1^i, K_2^i, \cdots$ of C_i such that the boundary of each element of K_j^i is a simple closed curve in $C_i - W$, the intersection of three boundaries of elements of K_j^i does not contain more than a point and if the boundaries of two elements of $K_1^i + K_2^i + \cdots$ intersect, the intersection is an arc. We shall use the partitionings $K_1^i, K_2^i, \cdots$ in describing the partitionings $G_1', G_2', \cdots$ of H'. We note that each K_j^i has three or more elements.

Let W' be the collection of all elements h of H' such that h does not belong to $\sum C_i$ but $H' - h$ has exactly two components. Then $W' + \sum C_i$ is dense in H'.

There is a decreasing sequence of regular partitionings $G_1', G_2', \cdots$ of H' such that if the boundaries of two elements of $\sum G_j'$ intersect, the intersection is either a point of W' or an arc, and if G_k' divided C_i at all, it divided it in the same fashion as some K_j^i does. Then the boundary of an element of G_j' is the sum of a finite number of points of W' and a finite number of simple closed curves in $\sum C_i$.

Building fences. We recall that M is a solid sphere in E^3 with radius 1, that $I(\varepsilon)$ is the interior of a sphere with the same center and radius $1 - \varepsilon$, and $S(\varepsilon) = \text{Bd } I(\varepsilon)$. In the proof of Theorem 1, for each element h_i of the sequence $h_1, h_2, \cdots$ mentioned in that proof, there was a set F_i in H described such that $F_i \cdot S(\varepsilon)$ $(0 < \varepsilon < \frac{3}{4})$ was a simple closed curve in both H and M. We may think of $F_i(\varepsilon)$ as a fence built in M so that its base rests in h_i and the top of the fence extends ε units toward the center of M. We need to build such fences in M to prove the present theorem.

For each element w of W', there is a closed point set $F(w)$ in H, such that $F(w) \cdot S(\frac{3}{4})$ is a point, $F(w) \cdot H' = w$, $F(w) \cdot S(\varepsilon)$ $(0 < \varepsilon < \frac{3}{4})$ is a simple closed curve, and $F(w)$ separates each of H and H' into two pieces. The proof that there is such an $F(w)$ is the same as the proof that there is an F_i in Theorem 1. We use $F(w, \varepsilon)$ to denote the elements of H in $F(w)$ whose distance in M from Bd M is less than or equal to ε.

Now we wish to build fences having for bases, not only elements of W', but also finite graphs in the various nondegenerate cyclic elements of H'. To this end let H_i' be the monotone upper semicontinuous decomposition of Bd M such that each element of $C_i - W$ is an element of H_i' and each component of Bd $M - (C_i - W)$ is an element of H_i'. Then the decomposition space H_i' is a simple surface.

Let H_i be the upper semicontinuous decomposition of M whose elements are the elements of H_i' and the points of the interior of M. Then by Theorem 2 there is a homeomorphism g that takes H_i onto M such that if h is an element of H_i, $g(h)$ and h are the same distance from the center of M in M.

Suppose Y is a finite graph in both H_i' and $C_i - W$. Let $F(Y)$ be the set of all elements h of H_i such that for some element h' of Y, $g(h)$ is on the interval from $g(h')$ to the center of M. We use $F(Y, \varepsilon)$ to denote $F(Y) - I(\varepsilon)$.

We note that $F(Y, \varepsilon)$ is a closed set in H since $F(Y, \varepsilon) \cdot H'$ lies in $C_i - W$. Also $F(Y) \cdot S(\varepsilon)$ $(0 < \varepsilon < \frac{3}{4})$ in either H, M, or H_i is a finite graph topologically equivalent to Y. Furthermore, $F(Y, \varepsilon)$ separates each of $H - I(\varepsilon)$, H', and $S(\varepsilon)$ into the same number of pieces.

For convenience we suppose that $X_i + C_i$ is metrized like a solid sphere with radius r. We let $S_i(\varepsilon)$ be the simple surface consisting of those points of X_i whose distance from C_i is εr.

The fence $F(Y, \varepsilon)$ is extended into X_i as follows. $F'(Y, \varepsilon)$ is the collection of all points of $X_i + C_i$ between Y and the center of $X_i + C_i$ which lies within εr of C_i. We use $F''(Y, \varepsilon)$ to denote $F(Y, \varepsilon) + F'(Y, \varepsilon)$ and $F'''(Y, \varepsilon)$ to denote $F(Y, \varepsilon) + F'(Y, \varepsilon) + S_i(\varepsilon)$.

Description of G_1. Let $Y_1, Y_2, \cdots, Y_n$ be the components of the sum of the boundaries of the elements of the partitioning G_1' of H'. Then Y_j is either a point of W' or a finite graph in some C_i. If it intersects C_i, it does not intersect W but divides C_i into some partitioning K_k^i of C_i. In any case, a fence $F(Y_j)$ has been described for Y_j and for a sufficiently small positive number ε, no two of the fences $F(Y_1, \varepsilon)$, $F(Y_2, \varepsilon)$, $\cdots$, $F(Y_n, \varepsilon)$ intersect.

For convenience we suppose that $F'''(Y_j, \varepsilon) = F(Y_j, \varepsilon)$ if Y_j is an element of W'. Now $S(\varepsilon) + F'''(Y_1, \varepsilon) + F'''(Y_2, \varepsilon) + \cdots + F'''(Y_n, \varepsilon)$ divides $H + \sum X_i$ into many pieces $g_0, g_1, \cdots, g_r, g_{r+1}, \cdots, g_s$, where $g_0 = I(\varepsilon)$, $g_i(1 \leqq i \leqq r)$ intersects H', and $g_i(r < i \leqq s)$ lies in $\sum X_i$. Here r is the number of elements of G_1' and $s - r$ is the number of nondegenerate elements of $Y_1, Y_2, \cdots, Y_n$.

The partitioning $g_0, g_1, \cdots, g_s$ satisfies conditions 1, 2, 3, 4 except that Bd $g_0 = S(\varepsilon)$ may intersect the boundary of some other g_i in a set topologically equivalent to a closed 2-cell minus a finite number of open disks instead of in a closed 2-cell.

However, as noted in *Subdividing an interior element of* G_{i-1} in the proof of Theorem 1, there is a partitioning $g_1', g_2', \cdots, g_j'$ of g_0 so that the partitioning $g_1', g_2', \cdots, g_j', g_1, \cdots, g_s$ of $H + \sum X_i$ satisfies conditions 1, 2, 3, 4.

Description of G_i. Our description here closely resembles the description of G_i given in the proof of Theorem 1. Suppose that certain elements of G_{i-1} have been partitioned to get a partitioning G_i'' which is the start of the partitioning G_i but g is an element of G_{i-1} which has not yet been partitioned.

If g does not intersect H', it is tame in either M or some X_i and as noted in *Subdividing an interior element of* G_{i-1} in the proof of Theorem 1, there is a partitioning $P(g)$ of g into tame pieces of diameter less than $1/i$ so that the partitioning resulting from replacing g by the elements of $P(g)$ in G_i'' satisfies conditions 1, 2, 3, 4.

Now consider the case where g intersects H'. It may be enlightening to consider its boundary. The boundary (in H') of $g \cdot H'$ is the sum of a finite number of simple closed curves $J_1, J_2, \cdots, J_m$ in $\sum C_k - W$ and a finite number of points $w_1, w_2, \cdots, w_n$ in W'. Then there is a positive number ε such that Bd $g = A + F''(J_1, \varepsilon) + A_1 + F''(J_2, \varepsilon) + A_2 + \cdots + F''(J_m, \varepsilon) + A_m + F(w_1, \varepsilon) + \cdots + F(w_n, \varepsilon)$ where A_k is a disk in $F'''(J_k, \varepsilon)$ and A is a connected subset of $S(\varepsilon)$ bounded by $m + n$ mutually exclusive simple closed curves.

The partitioning G_i'' imposes a partitioning on H'. There is an integer j greater than i such that G_j' is a refinement of this partitioning of H' and there is an integer k greater than j such that the closure of no element of G_k' intersects two components of the sum of the boundaries of the elements of G_j'. Suppose the components of the sum of the boundaries of the elements of G_k' in $g \cdot H'$ are $Y_1, Y_2, \cdots, Y_r, w_1, w_2, \cdots, w_s$ where $Y_1, Y_2, \cdots, Y_r$ are finite graphs in $\sum C_t - W$ and $w_1, w_2, \cdots, w_s$ are points of W'.

There is a number ε' less than $1/i$ such that $F''(Y_1, \varepsilon'), F''(Y_2, \varepsilon'), \cdots, F''(Y_r, \varepsilon'), F(w_1, \varepsilon'), \cdots, F(w_s, \varepsilon')$ are mutually exclusive fences which lie in $\bar{g}$ and which do not intersect the closure of any element of G_i'' which does not contain a point of H'.

Now $S(\varepsilon') + F'''(Y_1, \varepsilon') + F'''(Y_2, \varepsilon') + \cdots F'''(Y_r, \varepsilon') + F(w_1, \varepsilon') + \cdots + F(w_s, \varepsilon')$ separates g into a finite number of pieces $g_0, g_1, \cdots, g_m, g_{m+1}, \cdots, g_n$ where $g_0 = g \cdot I(\varepsilon')$, $g_t(1 \leqq t \leqq m)$ intersects H', and

$g_t(m < t \leqq n)$ lies in $\sum X_j$. Each of these except g_0 has a simple surface for a boundary. However, as noted in *Subdividing an element of G_{i-1} which intersects H'* in the proof of Theorem 1, $\bar{g}_0$ is topologically equivalent to the cartesian product of A and a straight line interval and there is a partitioning $g_1', g_2', \cdots, g_j'$ of g_0 such that the partitioning resulting from replacing g_0 by $g_1', g_2', \cdots, g_j', g_1, \cdots, g_n$ in G_i'' satisfies conditions 1, 2, 3, 4.

Mesh of partitionings. The proof that the sequence $G_1, G_2, \cdots$ we have described for $H + \sum X_i$ is a decreasing sequence is similar to that given in the proof of Theorem 1. This completes our proof that $H + \sum X_i$ is topologically equivalent to S^3.

Questions. It would be interesting to know if each arc in the cactoid C we have described is tame. Also, we might look at this study from another direction and suppose that C was a cactoid in S^3. We might wonder what restrictions we would have to put on C in order to conclude that there is a monotone upper semicontinuous decomposition H of M such that the decomposition space H is topologically equivalent to C plus one of its complementary domains.

THE UNIVERSITY OF WISCONSIN

REFERENCES

1. R. H. BING, *A characterization of 3-space by partitionings*, Trans. Amer. Math. Soc., 70 (1951), pp. 15–27.
2. R. H. BING, *Partitioning a set*, Bull. Amer. Math. Soc., 55 (1949), pp. 1101–1110.
3. R. H. BING, *Partitioning continuous curves*, Bull. Amer. Math. Soc., 58 (1952), pp. 536–556.
4. E. E. FLOYD and M. K. FORT, *A characterization theorem for monotone mappings*, Proc. Amer. Math. Soc., 4 (1953), pp. 828–830.
5. R. E. MOORE, *Concerning upper semi-continuous collections*, Monatshefte für Mathematik und Physik, 36 (1929), pp. 81–88.
6. R. E. MOORE, Foundations of point set theory, Amer. Math. Soc. Colloquium Publications, vol. 13, New York, 1932.
7. G. T. WHYBURN, Analytic topology, Amer. Math. Soc. Colloquium Publications, vol. 28, New York, 1942.
8. J. W. T. YOUNGS, *The topological theory of Fréchet surfaces*, Ann. of Math., 45 (1944), pp. 753–785.

What Topology is Here to Stay?

R. H. Bing

Theorems that are stated in a simple and concise fashion are easier to remember and are more likely to survive than statements given in a complicated and involved fashion. Proofs that are understandable make a greater impact than unintelligible ones.

Two aspects of topology of current (and possibly future) interest which were not discussed in previous lectures were considered—the axiomatic approach and decomposition of Euclidean spaces.

A typical problem encountered in an axiomatic treatment is to prove that one set A of conditions implies a second set B. The mathematician tries to minimize the set A, maximize the set B, and prove the results in as general a space as possible. If the set of conditions is easily understood and correlate with other mathematics, and the implications are nontrivial, the problem is likely to be interesting.

The difficulty that the earlier geometers had in trying to define a point led them to take "point" as an undefined term. This actually enlarged the horizon for mathematics because we now consider many things as points in abstract spaces—continua, functions, matrices, sequences, and other things.

Upper semicontinuous decompositions of E^2 were discussed. Examples were given of such decompositions. The plane is a wonderful training ground for graduate students—the problems encountered here develop brain power. However, there has been much study devoted to the plane and simple problems that have not already been attacked are difficult to find in this area. Many of the interesting results about the plane discovered in recent years have dealt with decompositions of the plane and the imbedding of continua in the plane.

On the other hand, E^3 is essentially a virgin forest. For many years the problems were so forbidding that few attacks were successful. Now mathematicians are beginning to venture into the woods. We are developing new tools to cope with the problems encountered there. Simultaneously as we learn about E^3, we learn about collections of objects in E^3, decompositions of E^3. This seems a fertile and fruitful field.

Consider the following result due to R. L. Moore:

> If G is a nondegenerate upper semicontinuous decomposition of S^2 such that each element of G is a continuum which does not separate S^2, the decomposition space is topologically S^2.

The result does not generalize to S^3 directly because the decomposition of S^3 whose only nondegenerate element is a circle is no longer simply connected on the removal of a point. It was pointed out in Abstract 325, Bull. Amer. Math. Soc., vol. 63 (1957) p. 143, that the decomposition space whose only degenerate element is a solenoid is not simply connected.

Consider the effect of imposing additional restrictions on the set G. Suppose the nondegenerate elements of G have one of the following properties:

1. Has a complement like the complement of a point.
2. Is a tame arc.
3. Is a convex body.
4. Is a starlike body.
5. Is a tame arc and these tame arcs are countable.
6. Is a convex continuum.
7. Is an interval.
8. Is an interval oriented with one of the axes.

An example was given in the seminar on 3-dimensional manifolds to show that there are decompositions of S^3 satisfying conditions (1) and (2) but whose decomposition spaces were topologically different from S^3.[1] A forthcoming paper[2] will show that the decomposition space is S^3 if condition (3), (4), or (5) is satisfied. The implication of (6), (7), or (8) is not resolved. However, interest engendered by the institute has led several of its members to believe that a decision may be forthcoming soon. Hamstrom and Dyer have shown (Abstract 544, Bull. Amer. Math. Soc. vol. 62 (1956) p. 398) that the decomposition space is E^3 if each element satisfies (1) and lies in a horizontal plane.

It is possible to let some elements of the decomposition be circles, wild arcs, solid horned spheres, or complicated continua and still get a decomposition space like S^3 if the remaining members of the decomposition are suitably chosen. An example was given where some of the elements of the decomposition were circles and figure-eights and where the decomposition space was S^3.

It might be useful in deciding questions in E^3 if we had a better topological characterization of S^3. One difficulty in getting a characterization of S^3 is that although we are able to construct arcs in continuous curves, we have devised no suitable criteria for deciding when they contain surfaces. Perhaps if we are willing to put in our hypotheses that the continuous curve under consideration contains a family of surfaces satisfying certain conditions (similar to those that polyhedral surfaces satisfy in E^3), we might get a more adequate characterization. The paper Transactions of the American Mathematical Society, vol. 70 1951, pp. 15–27, provides a start in this direction but it leaves much to be desired.

As topology grows, we should grow with it. We should continue to learn new methods of attack and new results. This does not mean that we should flip-flop from one field of mathematics to another, trying to find something easy to do. The river that cuts a deep furrow, that leaves a lasting impression, is one that follows a steady course and does not meander all over the map. However, if we are to go all out to get results that challenge us, with no holds barred, we must be prepared to use results that we are understand from many sources.

Those of us who are to be successful in research must remain students. Our universities are turning out too high a percentage of "finished products"—people whose training is finished when they graduate. No matter how good your background is, no matter what training we have, it is not good enough to keep us up front for long unless we continue to grow. Perhaps it is more important that graduate training provide students with the ability and stimulus to be active researchers than that it give them a broad background.

Footnotes added in 1957

[1] See pages 45–48 of this summary [in *The collected papers of R. H. Bing*, pp. 911–915, Editors], also R. H. Bing, *A decomposition of E^3 into points and tame arcs such that the decomposition space is topologically different from E^3*, Annals of Mathematics, vol. 65 (1957) pp. 484–500 [this publication, pp. 929-945].

[2] R. H. Bing, *Upper semicontinuous decompositions of E^3*, Annals of Mathematics, vol. 65 (1957) pp. 363–374 [this publication, pp. 917–928].

Decompositions of E^3 into Points and Tame Arcs

R. H. Bing

Part I. Decomposition space is the sum of 2 horned spheres. A description was given of the upper semicontinuous decomposition of E^3 into points and tame arcs as described in *A homeomorphism between the* 3*-sphere and the sum of two solid horned spheres*, Annals of Mathematics, vol. 56, 1952, pp. 354–362. Each nondegenerate element of the decomposition was an arc which was symmetric with respect to the xy-plane and did not intersect any horizontal plane in two points. An outline was given of the proof that the decomposition space was topologically E^3.

After a one point compactification of E^3, the set of elements of the decomposition which have a point with a nonnegative z coordinate is a solid horned sphere (as described by J. W. Alexander in *An example of a simply connected surface bounding a region which is not simply connected*, Proc. Nat. Acad. Sci., U.S.A., vol. 10, 1924, pp. 8–10) in the decomposition space—also the set of those with a nonpositive coordinate is a solid horned sphere. This shows that if f is a homeomorphism of a solid horned sphere S onto a solid horned sphere S' and S, S' are sewed together along their boundaries so that a point x of $\mathrm{Bd}(S)$ is sewed to $f(x)$, the resulting set is topologically S^3.

Question. Would the sum of S and S' necessarily be S^3 if we only suppose that f is a homeomorphism of $\mathrm{Bd}\,S$ onto $\mathrm{Bd}\,S'$?

In discussing the above question, it was pointed out that the sum would be equivalent to a decomposition of a one point compactification of E^3 where the nondegenerate elements of the decomposition were arcs that did not intersect any horizontal plane in two points. It was conjectured that the question would be resolved within a few years, either by making use of the methods used in showing that the decomposition space described earlier was S^3 or by the methods to be used in the second part of the lecture to show that a certain decomposition of E^3 into points and tame arcs did not have a decomposition space topologically like E^3.

The preceding decomposition of E^3 into points and tame arcs shows that there is a homeomorphism of period two of E^3 onto itself that is not topologically orthogonal. The homeomorphism (described in the decomposition space) would be a reflection through the xy-plane. Since the set of elements of the decomposition intersecting the xy-plane is the fixed point set, and this set is a wild surface in the decomposition space, the homeomorphism is not topologically either a rotation or a reflection.

Question. Deane Montgomery raised the question as to whether or not there are non-orthogonal homeomorphisms of E^3 onto itself of other periods. He suggested a decomposition of E^3 into points and tame triods that would show that there is such a homeomorphism of period three if it could be shown that the decomposition space is topologically E^3. (The following day it was shown that the decomposition space suggested by Montgomery is topologically E^3. An extension of this example shows that there are non-orthogonal homeomorphisms of E^3 onto itself of all periods. A summary of this argument obtained the following day is included here.)

A periodic transformation of period 3 that is not orthogonal. The transformation is described by using a decomposition of E^3 into points and tame triods as suggested by Deane Montgomery. Here we describe the decomposition.

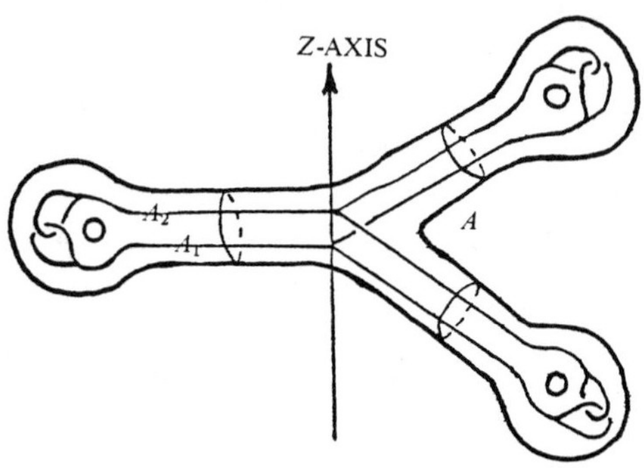

Suppose A is a tame solid as shown which is topologically equivalent to a solid sphere with three handles. It goes onto itself under a rotation of 120° about the z-axis. It may be convenient to think of A as a triod in the xy-plane with center on the z-axis plus circles in the xy-plane attached to the ends of the triods plus all points in E^3 whose distances from the triod and circles are not more than some positive number ε.

Then A_1, A_2 are two tame solids equivalent to A and imbedded in it as shown. The ends are linked as shown and each A_i goes onto itself under a rotation of 120° about the z-axis.

Two continua A_{i1}, A_{i2} are imbedded in A_i $(i = 1, 2)$ as A_1, A_2 were imbedded in A. There are 2^2 A_{ij}'s defined at the second stage. At the third stage, two A's are placed in each of these. The process is continued.

The A's might have been defined so that each component of $A_0 = A \cdot \sum A_i \cdot \sum A_{ij} \cdot \sum A_{ijk} \cdot \cdots$ is a triod that does not intersect any circular cylinder with z-axis as axis in more than 3 points, so we suppose this was done.

Let G be the upper semicontinuous decomposition space whose elements are the components of A_0 and the points of $E^3 - A_0$. We shall indicate that the decomposition space is topologically E^3.

The periodic transformation of the decomposition space onto itself will result from a rotation of 120° about the z-axis. Let Z be the collection of elements of G that intersect the z-axis. Under the rotation, each element of Z goes onto itself. However, there is no homeomorphism that takes Z onto a straight line because each Cantor set on a straight line in E^3 has a simply connected complement but the sum of the nondegenerate elements of G does not have a simply connected complement in E^3.

Reasons why the decomposition space G is topologically E^3. We show that for each positive number ε there is an integer n and a homeomorphism h of A onto itself that is fixed on the boundary of A and takes each of the A's at the nth stage onto a set of diameter less than ε. Then it follows as indicated from the bottom of page 359 to the top of page 361 of *A homeomorphism between the 3-sphere and the sum of two solid horned spheres*, Annals, vol. 56, 1952, that the decomposition space is topologically E^3. The following indication of proof is brief and it is assumed that the reader is acquainted with the paper in the Annals.

There is a homeomorphism h_1 of A onto itself that pulls A_1, A_2 into positions as shown. The cuts are such that any set of diameter as much as ε intersects at least four of them.

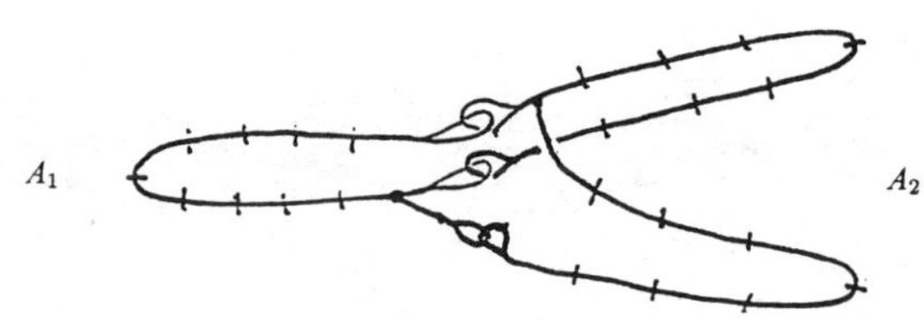

A_1 above is in a convenient position because all cuts are on one end of it. The two continua A_{21}, A_{22} may be placed in A_2 so that the cuts are on one end of each of them.

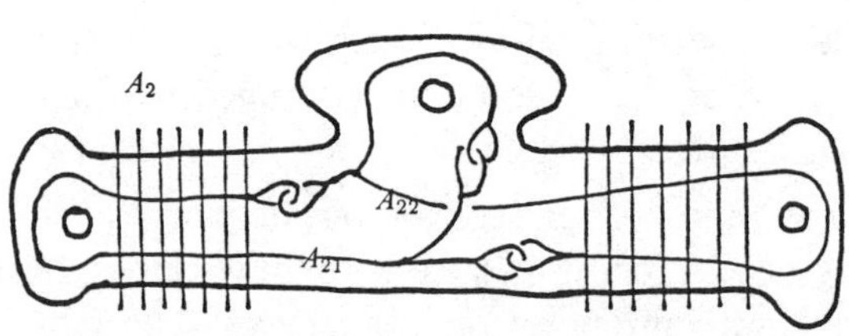

We concern ourselves with the case where only one end of A has more than one cut. Suppose the long end has n cuts. The following diagram indicates how

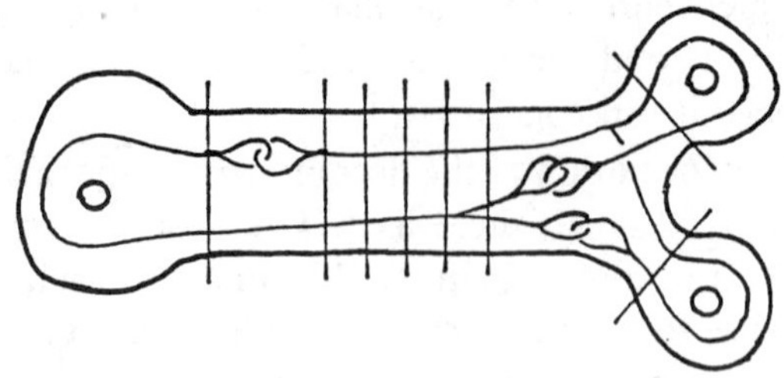

to pull the two A's in it so that they essentially have only $n - 1$ cuts on their long ends and only one cut on each short end.

Hence in $n - 1$ steps, we arrive at A's that can be deformed so that no one of them intersects four cuts. Hence, each of these deformed A's is of diameter less than ε.

Non-orthogonal transformation of period n. The construction of the decomposition space is similar to the above but here instead of a triod we have n intervals branching from a common end. In the proof that the A's can be shrunk, it takes a few more steps to arrive at a position where only one end has more than one cut. Then the reduction is accomplished as indicated.

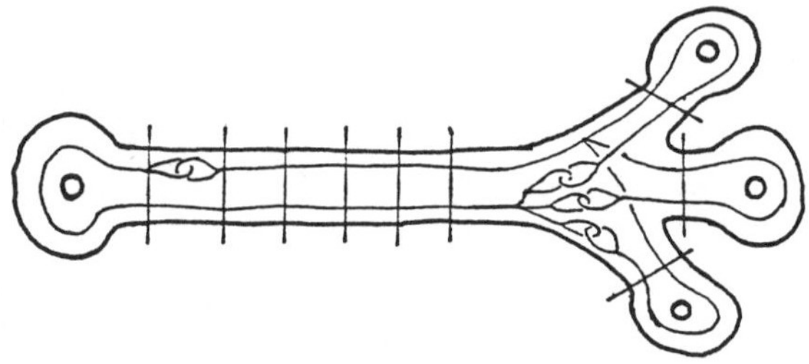

Part II. Decomposition space is not E^3. For some time it had been suspected that the decomposition space discussed in Part I was topologically different from E^3. When it was discovered that this was not so, an effort was made to complicate the decomposition so as to get a decomposition of E^3 into points and tame arcs such that the decomposition space is different from E^3. The following example resulted.

Here we describe an upper semicontinuous decomposition G of E^3 such that the corresponding decomposition space is topologically different from E^3 even though the elements of G are points and tame arcs. The arcs which are elements of G are components of a closed set A_0 which is described as the intersection of a decreasing sequence of closed sets A, $(A_1 + A_2 + A_3 + A_4)$, $\sum A_{ij}$, $\sum A_{ijk}, \ldots$. First we describe A.

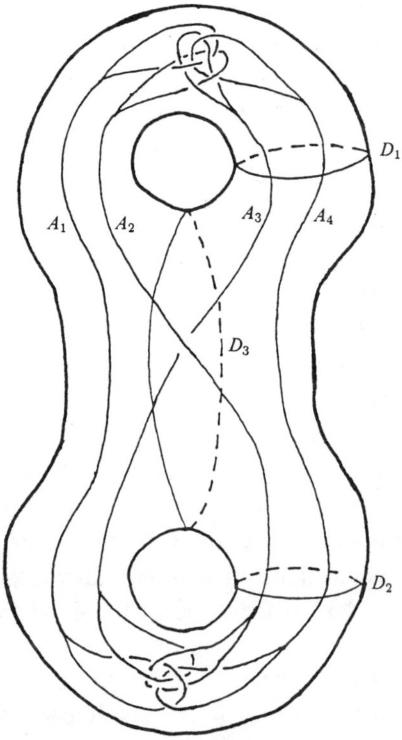

1. Suppose A is a topological solid sphere with two handles as shown in the figure. It is polyhedral. It may be convenient to think of A as an elongated figure eight plus all points whose distances from it are no more than a positive number ε.

2. Suppose A_1, A_2, A_3, A_4 are four polyhedra imbedded in A as shown. Although the A_i's are shown as finite graphs, each is topologically equivalent to A. The upper loops of each pair of A_i's are linked as are the lower loops. We call A_1, A_2, A_3, A_4 the A's defined at the first stage.

3. Four continua A_{i1}, A_{i2}, A_{i3}, A_{i4} are imbedded in A_i as A_1, A_2, A_3, A_4 were imbedded in A. There are now 4^2 A_{ij}'s defined at the second stage of our description.

4. Four continua A_{ij1}, A_{ij2}, A_{ij3}, A_{ij4} are imbedded in A_{ij} in this same manner. Similarly, for each integer n, we define 4 continua in each of the 4^n A's defined at the nth stage.

5. We could have placed the A's so that each component of $A_0 = A \cdot \sum A_i \cdot \sum A_{ij} \cdots$ is an arc that does not intersect any horizontal plane in two points, so we suppose that this was done.

Each element of G is either a point in the complement of A_0 or a component of A_0. Each element of G in A_0 is a tame arc and the collection of such tame arcs corresponds to a Cantor set in the decomposition space.

Reasons why the decomposition space $G \neq E^3$.

1. Suppose D_1, D_2 are planar disks as shown such that D_1 cuts the upper handle of A and D_2 cuts the lower handle. It may be shown that if h is any homeomorphism of A onto itself that is fixed on Bd A, then one of the 4^n A's defined at the nth stage has an image under h that intersects both D_1 and D_2. This shows that there is not an isotopy of E^3 onto itself that is fixed outside A and shrinks the A's to small size.

2. It may be shown that if the decomposition space G were topologically equivalent to E^3, there would be a map f of A onto itself that is fixed on Bd A and such that for each point x of A, $f^{-1}(x)$ is an element of G. This may be done in the following fashion.

Consider the three disks D_1, D_2, D_3 slicing A as shown. The disks may be modified in the decomposition space G slightly at the elements of G in A_0 so that the modified disks D_1', D_2', D_3' are mutually exclusive and tame in G. The two polyhedral cubes in E^3 bounded by Bd $A + D_1 + D_2 + D_3$ are taken into the cubes in decomposition space G bounded by Bd $A + D_1' + D_2' + D_3'$.

3. If there were a map f mentioned in (2) above, there would be an integer n such that the image under f of no one of the 4^n A's defined at the nth stage intersects both D_1 and D_2. The assumption that there is such an f leads to a contradiction with (1) above because we could take h defined in (1) to be f outside the 4^n A's and a homeomorphism inside each of the 4^n A's.

Questions. The following questions were raised in connection with the preceding example.

1. What topological property distinguishes the preceding decomposition space from E^3? It was suggested that it would be interesting to see which of E^3's properties the space has.

2a. Can the above example be modified to make each of its nondegenerate elements a broken line interval? (M. K. Fort, Jr., *A note concerning a decomposition space defined by Bing*, Ann. of Math., vol. 65 (1957) pp. 501–504, subsequently described a decomposition G^1 of E^3 into points and broken line intervals, each broken in only two places, such that the decomposition space G^1 appeared to resemble the decomposition space G.)

2b. Can the above example be modified to make each of its nondegenerate elements a straight line interval?

3. If G is a monotone decomposition of E^3, what restrictions can be placed on the elements of G to insure that the decomposition space is E^3?

4. Suppose H is a monotone decomposition of E^3 onto itself such that the elements of H are points and tame arcs, H' is a subcollection of H whose sum is compact, D is an open set containing the elements of H', and ε is a positive number. Is there necessarily an isotopy of E^3 onto itself that is fixed outside D and which shrinks each element of H' onto a set of diameter less than ε? (Such a shrinking process is used to show that many decomposition spaces are like E^3.)

References

1. For pages 43–44 of this summary: Abstract 157, R. H. Bing, *Periodic homeomorphisms of E^3 which are not orthogonal*, Bull. Amer. Math. Soc., vol. 62 (1956) p. 47.

2. For pages 45–48 of this summary: R. H. Bing, *A decomposition of E^3 into points and tame arcs such that the decomposition space is topologically different from E^3*, Ann. of Math., vol. 65 (1957) pp. 484–500.

ANNALS OF MATHEMATICS
Vol. 65, No. 2, March, 1957
Printed in U.S.A.

UPPER SEMICONTINUOUS DECOMPOSITIONS OF E^3*

BY R. H. BING

(Received March 9, 1956)

In this paper it is shown that monotone upper semicontinuous decompositions of E^3 satisfying certain additional conditions have decomposition spaces which are topologically equivalent to E^3. When these results were first obtained several years ago, we had some misgivings about imposing certain of these conditions since it was not known at that time that they were necessary. There was some suspicion that all of those upper semicontinuous decompositions of E^3 whose elements were compact continua with complements topologically equivalent to the complement of a point might be topologically equivalent to E^3. See questions raised on page 70 of [7], page 710 of [6], page 862 of [5]. However, an example has been given of an upper semicontinuous decomposition of E^3 into points and tame arcs such that the decomposition space is not E^3 [4]. Hence, some extra conditions are required.

Many upper semicontinuous decompositions do satisfy the conditions we impose. For example, an upper semicontinuous decomposition of E^3 whose nondegenerate elements are convex bodies has a decomposition space topologically equivalent to E^3 by Theorem 2. We suppose that a convex body contains an interior point and do not know if the statement would be true with the word "bodies" replaced by "continua."

A collection G of compact sets is upper semicontinuous if for each element g of G and each open set U containing g there is an open set V containing g such that each element of G which intersects V lies in U. The elements of the upper semicontinuous collection G are the points of the decomposition space G. A set U in the decomposition space G is open if the sum of the elements of the upper semicontinuous collection G associated with points in U is open in the original space. If each element of G is connected, G is called monotone.

THEOREM 1. *Suppose G is a monotone upper semicontinuous decomposition of E^3 such that the complement of each element of G is topologically equivalent to the complement of a point, G has only a countable number of nondegenerate elements, and their sum is a G_δ set. Then the decomposition space G is topologically equivalent to E^3.*

The theorem is proved by repeated applications of the following lemma.

LEMMA 1. *For each positive number ε and each open set U containing the sum of the nondegenerate elements of G of Theorem 1, there is a homeomorphism f_ε of E^3 onto itself that shrinks each element of G into a set of diameter less than ε and which is fixed on $E^3 - U$.*

PROOF OF LEMMA. Let $g_1, g_2, \cdots$ be the nondegenerate elements of G and $D_1, D_2, \cdots$ be a sequence of open sets such that $D_1 \cdot D_2 \cdot \cdots = g_1 + g_2 + \cdots$.

* Work on this paper was supported by NSF Contract G1219.

In the decomposition space G, $g_1 + g_1 + \cdots$ is countable and therefore 0-dimensional. Hence we suppose with no loss of generality that each component of U is bounded.

Since the complement of g_1 is topologically equivalent to the complement of a point, there is an open set V containing g_1 such that $V \subset D_1 \cdot U$ and Bd V is locally tame and topologically equivalent to the surface of a sphere We use Bd V to denote the boundary of V. Then Bd V is tame [3] and $\bar{V}$ is topologically a closed 3-cell [1]. Then there is a homeomorphism h_1 of E^3 onto itself that is fixed on $E^3 - V$ which shrinks g_1 to a set of diameter less than ε.

Let G_1 be the collection of images of elements of G under h_1 and W be the sum of the elements of G_1 with diameters greater than or equal to ε. Then $h_1^{-1}(W)$ is closed and it follows from the normality of E^3 that there is an open set U_2 in E^3 such that U_2 is the sum of elements of G, $h_1^{-1}(W) \subset U_2$, and $\bar{U}_2 \subset D_1 \cdot D_2 \cdot U - g_1$. Then h_2 is a homeomorphism that shrinks g_2 to a set of diameter less than ε and which is equal to h_1 on $E^3 - U_2$. We have no assurance that g_2 is an element of $E^3 - U_2$ but in this case its image under h_1 is already of diameter less than ε.

We now describe a homeomorphism h_3 that shrinks g_3 to a set of diameter less than ε. Let G_2 be the collection of images of elements of G under h_2, W_2 be the sum of elements of G_2 with diameters as much as ε, and U_3 be an open set such that U_3 is the sum of elements of G, $h_2^{-1}(W_2) \subset U_3$, and $\bar{U}_3 \subset D_3 \cdot U_2 - g_2$. We select h_3 so that diameter $h_3(g_3) < \varepsilon$ and $h_3 = h_2$ on $E^3 - U_3$.

In a similar fashion we obtain U_4, h_4, U_5, h_5, $\cdots$. The homeomorphism f_ε is the limit of h_1, h_2, $\cdots$. We show that this limit exists and is a homeomorphism.

Each point p of E^3 fails to belong to some $\bar{U}_i$. Since $h_i = h_{i+1} = \cdots$ on $E^3 - \bar{U}_i$, f_ε is defined and continuous at p. Since each h_i is a homeomorphism, no two points of E^3 go into the same point under f_ε.

We finish showing that f_ε is a homeomorphism by showing that each point p of E^3 has an inverse under f_ε and f_ε^{-1} is continuous.

If $p \in E^3 - U$, $f_\varepsilon^{-1}(p) = p$ and $f_\varepsilon^{-1} = h_1^{-1}$ on $h_1(E^3 - \bar{U}_1)$.

If $p \in U$, each $h_i^{-1}(p)$ belongs to the component of U containing p. Since this component is bounded, there is a point q such that some subsequence of $h_1^{-1}(p)$, $h_2^{-1}(p)$, $\cdots$ converges to q. Then for some integer i, $q \in E^3 - \bar{U}_i$ and $f_\varepsilon = h_i = h_{i+1} = \cdots$ on $E^3 - \bar{U}_i$. If p were not equal to $f_\varepsilon(q)$, there would be a neighborhood N of $f_\varepsilon(q)$ in $h_i(E^3 - \bar{U}_i) - p$ and for n greater than i, $h_n^{-1}(p)$ would not lie in $h_i^{-1}(N)$ which is a contradiction. Hence, $f_\varepsilon^{-1}(p) = q$ and $f_\varepsilon^{-1} = h_i^{-1}$ near p.

Proof of Theorem 1. We describe a monotone map f of E^3 onto itself that takes each element of G into a unique point. The monotone map is the limit of a sequence of homeomorphisms f_1, f_2, $\cdots$.

The first element of f_1, f_2, $\cdots$ is a homeomorphism that takes each element of G onto a set of diameter less than 1. Lemma 1 insures that there is such a homeomorphism. Before we describe the others we shall describe some open sets in the decomposition space G.

Since G has only a countable number of nondegenerate elements, their sum is 0-dimensional in the decomposition space. Hence there is a sequence of open sets $V_1, V_2, \cdots$ in the decomposition space G such that V_i contains each element of G associated with a nondegenerate set in E^3 and no component of V_i is of diameter more than $1/i$ in G. We use the fact here that the decomposition space G has a metric.

We also use V_i to denote in E^3 the sum of the elements of G in V_i of G. The components of this V_i are not required to be of diameter less than $1/i$.

Since the diameters in E^3 of the images of each element of G under f_1 is less than 1, it follows from the 0-dimensionality of the decomposition space G that there is an open set U_1 containing the sum of the nondegenerate elements of G such that $U_1 \subset V_1$ and each component of $f_1(U_1)$ is of diameter less than 1. Then $f = f_1$ on $E^3 - U_1$.

By Lemma 1 there is a homeomorphism f_2 of E^3 onto itself that is equal to f_1 on $E^3 - U_1$ and which shrinks each element of G to a set of diameter less than $\frac{1}{2}$. There is an open set U_2 covering the sum of the nondegenerate elements of G such that $U_2 \subset U_1 \cdot V_2$ and each component of $f_2(U_2)$ is of diameter less than $\frac{1}{2}$. Then $f = f_2$ on $E^3 - U_2$.

Similarly there are homeomorphisms $f_3, f_4, \cdots$ and open sets $U_3, U_4, \cdots$ so that f_i shrinks each element of G to a set of diameter less than $1/i$, $f_{i+1} = f_i$ on $E^3 - U_i$, $U_{i+1} \subset U_i \cdot V_{i+1}$, and each component of $f_i(U_i)$ has a diameter less than $1/i$.

We show that $f = \lim f_1, f_2, \cdots$ is continuous. If p belongs to a component D of U_i, each $f_{i+k}(p)$ belongs to $f_i(D)$. Since the diameters of the components of $f_i(U_i)$ are less than $1/i, f_1(p), f_2(p), \cdots$ is a Cauchy sequence and converges to $f(p)$. Also, $\rho(f_i(p), f(p)) \leqq 1/i$. Then g is continuous since $\rho(f_i, f) \leqq 1/i . \rho(x, y)$ $=$ the distance between x and y and $\rho(f_i, f) =$ G.L.B. $\rho(f_i(x), f(x))$.

If p and q belong to the same element of G, $f_i(p) = f_i(q)$ so $f(p) = f(q)$.

If p and q belong to different elements of G, there is an integer i such that the components D_p, D_q of $V_i + p + q$ containing p and q respectively are a positive distance apart. Since $f_{i+k}(p) \in f_i(D_p)$ and $f_{i+k}(q) \in f_i(D_q)$, $f(p) \neq f(q)$.

We finish by showing that for each point p of E^3, $p \in f(E^3)$. Let H be a closed bounded subset of E^3 containing each point whose distance from p is less than or equal to 1. Since f is continuous and $f_1^{-1}(H)$ is closed and bounded, $ff_1^{-1}(H)$ contains p if it has p for a limit point. But it has p as a limit point for $f_k^{-1}(p) \in f_1^{-1}(H)$ because $\rho(f_1, f_k) \leqq 1$ and $\rho(ff_k^{-1}(p), p) \leqq 1/k$ because $\rho(f_k, f) \leqq 1/k$.

QUESTIONS. The only place we used the hypothesis that the sum of the nondegenerate elements of G was a G_δ set was in Lemma 1. Could we dispense with this condition?[1] The place that we used this condition was in showing that the sequence of functions described in the proof of Lemma 1 converged to a homeomorphism. We did this by showing that each point of E^3 failed to belong to some $\bar{U}_i$.

[1] As pointed out in the Remarks at the end of this paper, a negative answer to this question has been given.

In the proof of Lemma 1 we made no attempt to prevent the homeomorphisms $f_1, f_2, \cdots$ from expanding elements of G. Had we devised some method of preventing such an explansion, we might have dispensed with the condition that the sum of the nondegenerate elements of G was a G_δ set.

A modification of the argument for Theorem 1 gives a pseudoisotopy of E^3 onto itself that is fixed on $E^3 - U$ and which shrinks each element of G to a point. This isotopy may expand some elements considerably before shrinking them. Is there such an isotopy which never expands any element to more than twice its original diameter? For each positive number ε, is there one which never expands any element to more than its original diameter plus ε?

Figure 1 shows the difficulty encountered in trying to shrink g_i without expanding one of $g_{i+1}, g_{i+2}, \cdots$. In this case g_i is a thickened spherical surface with a small circular hole in the top and $g_{i+1}, g_{i+2}, \cdots$ are arcs converging to the boundary of this hole. If g_i is gradually shrunk near itself, the boundary of the hole would expand during this shrinking and one of $g_{i+1}, g_{i+2}, \cdots$ might be taken into something larger.

A compact continuum g is starlike if it contains a point p such that the intersection of g and each line through p is an interval. We say that g is starlike with respect to p.

THEOREM 2. *Suppose G is an upper semicontinuous decomposition of E^3 such that G has only a countable number of nondegenerate elements and each is starlike. Then the decomposition space G is topologically equivalent to E^3.*

The proof of this theorem will parallel that of Theorem 1 after a lemma similar to Lemma 1 is proved. However, the proof of Lemma 2 differs from the proof of Lemma 1.

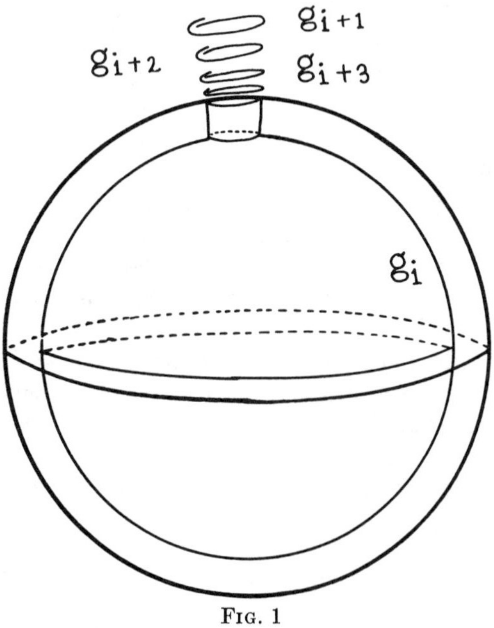

FIG. 1

LEMMA 2. *For each positive number* ε *and each open set* U *containing the sum of the nondegenerate elements of* G *of Theorem* 2, *there is a homeomorphism* f_ε *of* E^3 *onto itself that shrinks each element of* G *into a set of diameter less than* ε *and which is fixed on* $E^3 - U$.

PROOF OF LEMMA. As in the proof of Lemma 1 we suppose that each component of U is bounded. In fact, we suppose that if D is a component of U there is an element of G in D whose diameter is within 1 of the diameter of D.

Our plan is to work on each component D of U which contains elements of G with diameters as much as ε and get a homeomorphism f_D of E^3 onto itself that is fixed on $E^3 - D$ and which takes each element of G in D into a set of diameter less than ε. Since each bounded subset of E^3 intersects only a finite number of such components, the homeomorphism f_ε is a product of the homeomorphisms f_D.

Here is our plan for getting the homeomorphism f_D. Let $g_1, g_2, \cdots$ be the nondegenerate elements of G in D. First we shrink g_1, then $g_2, \cdots$. However, in shrinking these elements we take care that elements which are already small are not permitted to expand too much. Here is where we use the fact that each g_i is starlike. We show that in this sequence of shrinkings, there is a place where each g_i has been shrunk to a set of diameter less than ε. We do not take the limit as we did in Lemma 1.

Our main difficulty will come in devising a scheme that prevents our homeomorphisms which shrink elements of G from expanding others. If at some stage the image of g_i is in the shape of a hollowed out sphere with a small hole connecting the inside and outside and one attempted to contract the image in itself, sets near the small hole would be expanded in the contracting process. This situation is similar to that shown in Figure 1. To handle situations such as this we describe a "size" of images of $g_1, g_2, \cdots$ which is larger than the diameters of these images. We would want some of $g_{i+1}, g_{i+2}, \cdots$ of Figure 1 to have sizes approximating the diameter of g_i.

For each element g_i select a point p_i such that g_i is starlike with respect to p_i. If K is a subset of g_i, let $P(K, t)$ denote the sum of $K \cdot V(p_i, t)$ and the projection of $K - V(p_i, t)$ on Bd $V(p_i, t)$ toward p_i. We use $V(p, \varepsilon)$ to denote $\{q \mid \rho(p, q) < \varepsilon\}$. Then $P(K, t)$ is a set resulting from pulling certain points of K toward p_i. See Figure 2.

We say that $D(K, K') < \delta$ if there is a collection of sets $K_1, K_2, \cdots, K_n$ and positive numbers $t_1, \delta_1, t_2, \delta_2, \cdots, \delta_{n-1}$, such that $K = K_1$, $K' = K_n$, $\sum \delta_i < \delta$, K_i is a subset of an element of G and $K_{i+1} \subset V(P(K_i, t_i), \delta_i)$. As shown in Figure 2, we do not suppose that the K_i's all belong to the same element of G.

If f is a homeomorphism of E^3 onto itself and K is a subset of some g_i, we call the δ-size of K under f the least upper bound of the diameters of all $f(K')$'s where $D(K, K') < \delta$. Since diameter $f(P(K, t))$ may be considerably more than diameter $f(K)$ as previously pointed out (see Figure 1), the δ-size of K under f may be considerably more than diameter $f(K)$. However, the δ-size of K under the identity homeomorphism is not more than $2\delta +$ diameter of K.

We shall describe a sequence of homeomorphisms $f_0, f_1, \cdots$ of E^3 onto itself,

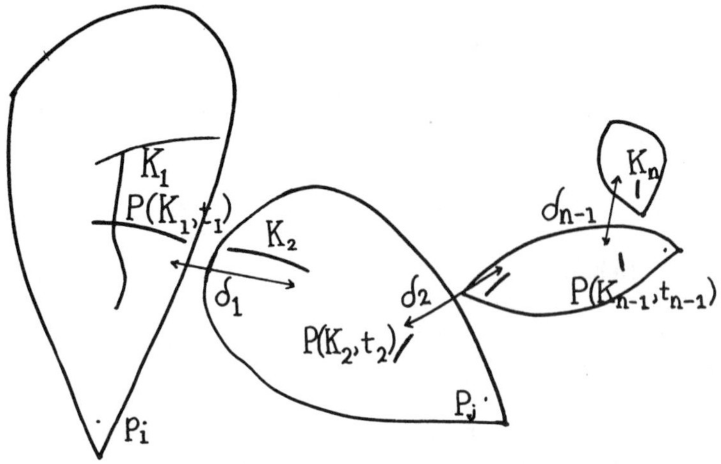

FIG. 2

a sequence of open sets $U_0, U_1, \cdots$ and a sequence of positive number $\varepsilon_0, \varepsilon_1, \cdots$ satisfying the following conditions:

(a) The diameter of g_i under f_i is less than ε.

(b) U_i contains $g_1 + g_2 + \cdots + g_i$ and $f_i = f_{i+1} = \cdots$ on U_i.

(c) The ε_i size of each g_j under $f_i \leqq \varepsilon/2^{i+1} + \varepsilon_{i-1}$-size of g_j under f_{i-1}.

(d) Each f_i is the identity outside D.

We let f_0 be the identity homeomorphism, $\varepsilon_0 = \varepsilon/8$, and $U_0 = 0$.

Before describing the other members of the sequences $f_0, f_1, \cdots, \varepsilon_0, \varepsilon_1, \cdots,$ and $U_0, U_1, \cdots$, we show that if there are such sequences, there is an element of $f_0, f_1, \cdots$ that will serve as f_D.

If for no integer i could we use f_i for f_D, then for each i there would be a g_{n_i} such that $f_i(g_{n_i})$ is of diameter more than ε. There would be an element g_0 of G which is a limit of $g_{n_1} + g_{n_2} + \cdots$. We show that some f_i can be used for f_0 by showing that such a g_0 could be neither degenerate nor nondegenerate.

Now g_0 is not nondegenerate because if it were a $g_k (k = 1, 2, \cdots)$, there would be a neighborhood N of g_k in U_k such that the image under f_k of each element of G in N has a diameter of less than ε. But for i greater than k, g_{n_i} is not in N because $f_k = f_{k+1} = \cdots$ on N by Condition b and diameter $f_i(g_{n_i}) < \varepsilon$.

We obtain the following inequalities from Condition c.

$$\begin{aligned}
\varepsilon_i \text{ size } g_j \text{ under } f_i &< \varepsilon/2^{i+1} + \varepsilon_{i-1}\text{-size } g_j \text{ under } f_{i-1} \\
&< \varepsilon/2^{i+1} + \varepsilon/2^{i} + \varepsilon_{i-2}\text{-size } g_j \text{ under } f_{i-2} \\
&\cdots \\
&< (\varepsilon/2^{i+1} + \varepsilon/2^{i} + \cdots + \varepsilon/2^{2}) + \varepsilon_0\text{-size } g_j \text{ under } f_0 \\
&< \varepsilon/2 + \varepsilon/8\text{-size } g_j \text{ under } f_0 \\
&< 3\varepsilon/4 + \text{diameter } g_j .
\end{aligned}$$

This implies that the diameter of $f_i(g_j) < 3\varepsilon/4 +$ diameter g_j. Hence the diameter of each $g_{n_i} > \varepsilon/4$ and $g_{n_1} + g_{n_2} + \cdots$ cannot have a degenerate element of G as a limit point.

The proof of Lemma 2 is completed by showing the existence of $f_1, f_2, \cdots, \varepsilon_1, \varepsilon_2, \cdots$, and $U_1, U_2, \cdots$. Once f_i has been described, we pick U_i to be an open set containing $g_i + U_{i-1}$ and such that each element of G which intersects $\bar{U}_i$ has an image under f_i of diameter less than ε.

We describe $f_1, f_2, \cdots$ and $\varepsilon_1, \varepsilon_2, \cdots$ inductively by the use of Lemma 3. If g_{i+1} intersects $\bar{U}_i$, we let $f_{i+1} = f_i$ and $\varepsilon_{i+1} = \varepsilon_i$. Otherwise, we let V be an open set containing g_{i+1} but no element of G which intersects $\bar{U}_i$ and use the following result.

LEMMA 3. *Suppose f_i is a homeomorphism of E^3 onto itself, $\varepsilon > 0$, $\varepsilon_i > 0$, and V is an open set containing an element g_{i+1} of G of Theorem 2. Then there is a positive number ε_{i+1} and a homeomorphism f_{i+1} of E^3 onto itself such that the diameter of $f_{i+1}(g_{i+1}) < \varepsilon$, $f_{i+1} = f_i$ on $E^3 - V$, and*

(c) *ε_{i+1}-size of each g_j under $f_{i+1} \leqq \varepsilon/2^{i+2} + \varepsilon_i$-size of g_j under f_i.*

PROOF. Let ε_{i+1} be a positive number so small that $V(g_{i+1}, \varepsilon_{i+1}) \subset V$, $\varepsilon_{i+1} < \varepsilon_i/2$, and each set in V of diameter less than $2\varepsilon_{i+1}$ has an image under f_i of diameter less than $\varepsilon/2^{i+3}$.

Let k be an integer so large that $k\varepsilon_{i+1} >$ diameter of $V(g_{i+1}, \varepsilon_{i+1})$.

We call a bounded open set S ideally starlike with respect to a point p in S if each ray starting at p intersects Bd S in only one point. Let $S_k, S_{k-1}, \cdots, S_1$ be a collection of ideally starlike regions with respect to p_{i+1} such that each contains g_{i+1}, $S_k \subset V(g_{i+1}, \varepsilon_{i+1})$, and each element of G which intersects $\bar{S}_i$ lies in S_{i+1}.

We shall now describe a homeomorphism f of E^3 onto itself that pulls certain points toward p_{i+1}. Then $f_{i+1} = f_i f$. We describe f only on a ray R with one end at p_{i+1}.

Let s_j be the point of Bd S_j on R and r_j be the point of R such that

$$\rho(p_{i+1}, r_j) = j\varepsilon_{+1}.$$

We note that k was taken so large that s_k is between p_{i+1} and r_k.

Let f' be the transformation that takes the interval $p_{i+1}s_1$ linearly onto the interval $p_{i+1}r_1$, the interval s_1s_2 linearly onto the interval $r_1r_2, \cdots$, and the interval $s_{k-1}s_k$ linearly onto the interval $r_{k-1}r_k$. The homeomorphism f is then defined to take $R - p_{i+1}s_k$ identically onto itself and takes the point p of $p_{i+1}s_k$ into the one of p, $f'(p)$ that is nearest p_{i+1}. We note that $f(s_k) = s_k$ because $f'(s_k) = r_k$ is farther than s_k from p_{i+1}.

Assume $f_{i+1} = f_i f$ does not satisfy Condition c. Then there is a g_j and a set K in an element of G such that $D(g_j, K) < \varepsilon_{i+1}$ and

(1) $$\text{diameter } f_{i+1}(K) > \varepsilon/2^{i+2} + \varepsilon_i\text{-size } g_j \text{ under } f_i.$$

We show that there is no such K. We do this by considering two possibilities—either K does not intersect S_{k-1} or K lies in S_k.

If K does not intersect S_{k-1}, f does not move any point of K by more than ε_{i+1} and each point of $f_{i+1}(K) = f_i f(K)$ would lie within $\varepsilon/2^{i+3}$ of a point of $f_i(K)$. Therefore

$$(2) \qquad \text{diameter } f_{i+1}(K) < \varepsilon/2^{i+2} + \text{diameter } f_i(K).$$

Inequalities (1) and (2) imply that

$$(3) \qquad \text{diameter } f_i(K) > \varepsilon_i\text{-size of } g_j \text{ under } f_i .$$

But inequality (3) contradicts the definition of the ε_i-size of g_j under f_i since $D(K, g_j) < \varepsilon_i$.

If K lies in S_k, there is an integer m such that $1 \leqq m \leqq k - 1$ and $K \subset S_{m+1} - S_{m-1}$. (We suppose $S_0 = 0$.) We show that in this second case, Inequality (1) implies that there is a closed set K' in g_{i+1} such that $D(K, K') < \varepsilon_{i+1}$, $D(g_j, K') < \varepsilon_i$, and as in Inequality (3), diameter $f_i(K') > \varepsilon_i$-size of g_j under f_i. This again would violate the definition of the ε_i-size of g_j under f_i.

Let $K' = P(g_{i+1} \cdot V(K, \varepsilon_{i+1}), m\varepsilon_{i+1})$. Then $D(K, K') \leqq \varepsilon_{i+1}$ and $D(g_j, K') \leqq \varepsilon_i$.

We complete the proof of the lemma by showing that each point of $f_{i+1}(K)$ lies within $\varepsilon/2^{i+3}$ of a point of $f_i(K')$. We do this by showing that for each point p of K there is a point q' of K' such that

$$(4) \qquad \rho(f(p), q') < 2\varepsilon_{i+1} .$$

There is a point q of $g_{i+1} \cdot V(p,_i \varepsilon_{+1})$. Denote $P(q, m\varepsilon_{i+1})$ by q'. Then q' is a point of K' and either $q' = q$ or $\rho(p_{i+1}, q') = m\varepsilon_{i+1}$. We show that in either case, Inequality (4) holds.

Consider pqp_{i+1} as a triangle with base pq and apex p_{i+1}. The base of this triangle is of length less than ε_{i+1}.

If $q \neq q'$ and $p \neq f(p)$, then q' and $f(p)$ belong to the lateral sides of triangle pqp_{i+1} and their distances from the apex p_{i+1} do not differ by more than ϵ_{i+1}. Then Inequality (4) holds.

If $q \neq q'$ and $p = f(p)$, $\rho(p_{i+1}, q') = m\varepsilon_{i+1}$ and $(m - 1)\varepsilon_{i+1} \leqq \rho(p_{i+1}, p) \leqq (m + 1)\varepsilon_{i+1}$. Since the distance of p and q' from the apex p_{i+1} do not differ by more than ε_{i+1}, Inequality (4) follows.

If $q = q'$ and $p \neq f(p)$, $\rho(p_{i+1}, q) \leqq m\varepsilon_{i+1}$ and $(m - 1)\varepsilon_{i+1} \leqq \rho(p_{i+1}, f(p)) < \rho(p_{i+1}, p)$. Thus $f(p)$ is not more than ε_{i+1} from the perpendicular projection of q on $p_{i+1}p$ and Inequality (4) follows.

THEOREM 3. *Suppose G is an upper semicontinuous decomposition of E^3 such that G has only a countable number of nondegenerate elements and each is a tame arc. Then the decomposition space is topologically equivalent to E^3.*

PROOF OF THEOREM. The proof follows as in Theorems 1 and 2 if it is shown that for each open covering U of the nondegenerate elements of G and each positive number ε there is a homeomorphism of E^3 onto itself that is fixed on $E^3 - U$ and that takes each element of G into a set of diameter less than ε. This result follows as shown in the proof of Theorem 2 if it is shown that if f_i is a homeomorphism of E^3 onto itself, g_{i+1} is an element of G, and V is an open set containing g_{i+1}, there is a homeomorphism f_{i+1} of E^3 onto itself such that $f_{i+1} = f_i$ on $E^3 - V$, diameter $f_{i+1}(g_{i+1}) < \varepsilon$, and for each element g_j of G, diameter $f_{i+1}(g_j) \leqq \varepsilon/2^{i+2} +$ diameter $f_i(g_j)$. We find from the following lemma that there is such a homeomorphism f_{i+1} which shrinks g_{i+1} but does not permit the other elements of G to expand too much.

LEMMA 4. *If G is an upper semicontinuous decomposition of E^3, $\varepsilon > 0$, g_0 is an element of G which is a tame arc and V is an open set containing g_0, then there is a homeomorphism f of E^3 onto itself such that f is fixed on $E^3 - V$, diameter $f(g_0) < \varepsilon$, and for each element g of G, either diameter $f(g) < \varepsilon$ or $f(g) \subset V(g, \varepsilon)$.*

PROOF OF LEMMA. The proof of the lemma is modeled after the proof of Lemmas 2 and 3. We are able to dispense with the notion of "size" and use diameters instead.

First we consider the case where g_0 is a straight line interval. Let p_0 be one of its ends.

Let k be a positive integer so large that $k\varepsilon/4 >$ diameter $V(g_0, \varepsilon/4)$ and $S_k, S_{k-1}, \cdots, S_1$ be a collection of ideally starlike regions with respect to p_0 such that each S_i contains g_0, $S_k \subset V \cdot V(g_0, \varepsilon/4)$, and each element of G which intersects $\bar{S}_{i-1}$ lies in S_i.

If f moves a point at all, it moves it straight toward p_0. Hence, we only define f on a ray R with its end point at p_0. Let s_j be the point of $R \cdot \text{Bd}\, S_j$ and r_j be the point of R whose distance from p_0 is $j\varepsilon/4$.

Let f' be the map that takes p_0s_1 linearly onto p_0r_1, s_1s_2 linearly onto r_1r_2, $\cdots$, and $s_{k-1}s_k$ linearly onto $r_{k-1}r_k$. Then $f'(s_i) = r_i (i = 1, 2, \cdots, k)$.

Then f is the identity on $R - p_0s_k$. If p is a point of p_0s_k, $f(p)$ is the one of $p, f'(p)$ that is nearest p_0.

Suppose g_j is an element of G belonging to $S_{m+1} - S_{m-1}$. Each point of g_j which is moved at all is moved into $V(p_0, (m+1)\varepsilon/_4) - V(p_0, (m-1)\varepsilon/4$. Since the intersection of this set and $[S_k \subset V(g_0, \varepsilon/4)]$ has a diameter of less than ε, either diameter $f(g_j) < \varepsilon$ or $f(g_j) \subset V(g_j, \varepsilon)$.

If g_0 is not a straight line interval, let h be a homeomorphism of E^3 onto itself that takes g_0 onto a straight line interval. Let δ be a positive number such that each subset of $h(V(g_0, \varepsilon))$ of diameter less than δ has an image under h^{-1} of diameter less than ε. By the first case there is a homeomorphism h' of E^3 onto itself such that h' is fixed on $E^3 - h(V(g_0, \varepsilon) \cdot V)$, diameter $h'h(g_0) < \delta$, and for each element g of G, either diameter $h'h(g) < \delta$ or $h'h(g) \subset V(h(g), \delta)$. Then $f = h^{-1}h'h$.

Suppose K_1, K_2 are solid horned spheres with boundaries as described by Alexander. It was shown in [2] that $K_1 + K_2$ is topologically S^3 if $K_1 \cdot K_2 = \text{Bd}\, K_1 = \text{Bd}\, K_2$ and there is a homeomorphism of K_1 onto K_2 that is fixed at each point of $\text{Bd}\, K_1 = \text{Bd}\, K_2$. The following theorem does not answer the question raised on page 361 of [2] but it does throw a little light on the space formed by sewing two solid horned spheres together along their boundaries.

THEOREM 4. *Suppose the compact continuum M is the sum of two solid horned spheres K_1, K_2 such that $K_1 \cdot K_2 = \text{Bd}\, K_1 = \text{Bd}\, K_2$. Then there is an upper* *semicontinuous decomposition G of E^3 such that each nondegenerate element of G is a tame arc which does not intersect any horizontal plane in two points and M is topologically equivalent to a compactification of the decomposition space G.*

PROOF. Suppose P is a horizontal plane in E^3 and L_1, L_2 are the two sides of P. Let k_0 be a point of $K_1 \cdot K_2$ at which both K_1 and K_2 are locally simply connected. It was shown on pages 356–357 of [2] that there is an upper semicontinuous decomposition G_i of $L_i + P$ such that each nondegenerate element of G_i is an arc which intersects no horizontal plane in two points and the decomposition space G_i is topologically equivalent to $K_i - k_0$.

Each point of $K_1 \cdot K_2 - k_0$ corresponds to an element of G_i intersecting P and we suppose that if $k \in K_1 \cdot K_2 - k_0$, the element of G_1 corresponding to k intersects the element of G_2 corresponding to k. (If this is not already so, it can be arranged by a homeomorphism of $L_1 + P$ onto itself.)

If two elements of $G_1 + G_2$ intersect, their sum is an element of G; each element of $G_1 + G_2$ which does not intersect any other element of $G_1 + G_2$ is an element of G. Then each element of G is either a point, an arc of G_1, an arc of G_2, or an arc which is the sum of an arc in G_1 and an arc in G_2. In any case, each nondegenerate element of G is a tame arc which does not intersect any horizontal plane in two points.

The following theorem shows that if the monotone upper semicontinuous decomposition is sufficiently simple, then the corresponding monotone map can be made simple also.

THEOREM 5. *If G is an upper semicontinuous decomposition of E^3 such that each nondegenerate element of G is a vertical interval then the decomposition space is topologically equivalent to E^3. In fact there is a monotone map f of E^3 onto itself such that f takes each element of G into a unique point, f is fixed on the xy-plane, and $f(p)$ is on the same vertical line with p.*

PROOF. Since f only moves a point vertically, we only give the z coordinate of $f(g)$ for each element g of G.

We alternate between describing a family of surfaces M of elements of G and their images Z in E^3 and describing a family of surfaces Z in E^3 and their inverses M.

Let $M(0)$, $M(1)$, $M(-1)$, $M(2)$, $\cdots$ be a sequence of continuous surfaces in E^3 such that $M(0)$ is the xy-plane, no element of G intersects both $M(i)$ and $M(i+1)$, and $M(i+1)$ is the graph of a single valued function and is at least one unit above $M(i)$.

Let $Z(i)(i = 0, 1, -1, 2, \cdots)$ be the surface $z = i$. If an element g of G intersects $M(i)$, then $f(g) \in Z(i)$.

Let $Z(i + \frac{1}{2})$ be the surface halfway between $Z(i)$ and $Z(i+1)$. (We say that the surface C is halfway between the surfaces A and B if for each vertical line L, $L \cdot C$ is halfway between $L \cdot A$ and $L \cdot B$. Each point of $A \cdot B$ belongs to C.)

There is a continuous surface $M(i + \frac{1}{2})$ between (not necessarily halfway) $M(i)$ and $M(i+1)$ such that no element of G intersects both $M(i + \frac{1}{2})$ and $M(i) + M(i+1)$. If an element g of G intersects $M(i + \frac{1}{2})$, $f(g) \in Z(i + \frac{1}{2})$.

Let $M(i + \frac{1}{4})$ be the surface halfway between $M(i)$ and $M(i + \frac{1}{2})$.

Let $Z(i + \frac{1}{4})$ be a continuous surface between $Z(i)$ and $Z(i + \frac{1}{2})$ inclusive such that if L is a vertical line and g is element of G on L intersecting $M(i + \frac{1}{4})$, then $L \cdot Z(i + \frac{1}{4})$ is either on $Z(i)$, on $Z(i + \frac{1}{2})$, or between them according as g intersects $M(i)$, $M(i + \frac{1}{2})$, or neither.

Now $f(M(i + \frac{1}{4})) = Z(i + \frac{1}{4})$. Similarly $f(M(i + \frac{3}{4})) = Z(i + \frac{3}{4})$.

Let $Z(i + \frac{1}{8})$ be the surface halfway between $Z(i)$ and $Z(i + \frac{1}{4})$. We now describe a surface $M(i + \frac{1}{8})$ such that $M(i + \frac{1}{8}) = Z(i + \frac{1}{8})$.

Now $M(i + \frac{1}{8})$ is between $M(i)$ and $M(i + \frac{1}{4})$ inclusive. It is continuous except at points which belong to an element of G which intersects both $M(i)$ and $M(i + \frac{1}{4})$. If L is a vertical line and g is an element of G on L intersecting $M(i + \frac{1}{8})$, then g intersects $M(i)$, $M(i + \frac{1}{4})$, neither, of both according as $L \cdot Z(i + \frac{1}{8})$ is on $Z(i)$, on $Z(i + \frac{1}{4})$, on neither, or on both.

Here is the reason there is such an $M(i + \frac{1}{8})$. Let $F(i, +)$ be the collection of highest points belonging to elements of G intersecting $M(i)$ and $F(i + \frac{1}{4}, -)$ be the collection of lowest points belonging to elements of G intersecting $M(i + \frac{1}{4})$. Then $F(i, +)$ is upper semicontinuous and $F(i + \frac{1}{4}, -)$ is lower semicontinuous. Also $F(i + \frac{1}{4}, -)$ is above $F(i, +)$ except on elements of g which intersects both $M(i)$ and $M(i + 1)$. Hence there is the surface $M(i + \frac{1}{8})$.

After describing $M(i + \frac{3}{8})$, $Z(i + \frac{3}{8})$, $\cdots$, $Z(i + \frac{7}{8})$, we let $M(i + \frac{1}{16})$ be the surface halfway between $M(i)$ and $M(i + \frac{1}{8})$. It is continuous except on elements of g which intersect $M(i)$ and $M(i + \frac{1}{8})$. Then we describe $Z(i + \frac{1}{16})$. The procedure is continued to define f. It may be shown to be continuous.

Example. Theorem 4 might lead one to suspect that if G is an upper semicontinuous decomposition of E^3 such that each nondegenerate element of G is either an oriented interval, oriented rectangle, or oriented rectangular solid, then there is a transformation $f_t (t = x, y, \text{ or } z)$ such that f_t moves points only in the t direction, flattens out intervals of G parallel to the t axis into points, flattens out rectangles of G parallel to t axis into intervals perpendicular to t axis, and flattens out rectangular solids into rectangles perpendicular to t axis but does not permit images of different elements of G to meet. The following example shows that it is not this easy to shrink elements of some G's.

Let g_0 be a vertical interval in the xy-plane and g_1, g_3, $\cdots$ be a null sequence of horizontal intervals converging to the top point of g_0 from the right such that some point of g_{2i+1} is above some point of g_{2i-1}. Let g_2, g_4, $\cdots$ be a null sequence of horizontal intervals converging to the bottom point of g_0 from the right such that some point of g_{2i} is above some point of g_{2i+2}. It is impossible to flatten g_0 to a point by a transformation f_y that is a homeomorphism on $E^3 - g_0$, moves only points parallel to the y axis, and leaves g_1, g_2, $\cdots$ horizontal. We can also get sequences of intervals h_0, h_1, $\cdots$ and k_0, k_1, $\cdots$ in the other coordinate planes which shows that there is a decomposition G of E^3 into points and a countable number of oriented intervals for which there will not be an f_t with the properties mentioned in the preceding paragraph.

Questions. No doubt the results of this paper can and will be extended. If G is an upper semicontinuous decomposition of E^3 such that G has only a countable number of nondegenerate elements and each of these has a complement topologically like the complement of a point, what additional conditions are needed to insure that the decomposition space is topologically like E^3?

Here are some conditions that might be imposed singly or in combinations on each element g of G: (a) g lies in a horizontal plane. (I have a candidate for a counterexample that suggests that it is not enough to suppose that g lies in a plane.) (b) g contains no nondegenerate indecomposable continuum. (c) g is a continuous curve (even arc, dendron, or topological disk). (d) There is a map of E^3 into itself taking a unit sphere S into g and $E^3 - S$ homeomorphically onto $E^3 - g$. (e) There is an equicontinuous family of maps such as mentioned in (d).

In this paper we gave scant attention to the case where G has uncountably many nondegenerate elements. We do not know in this case, for example, if the decomposition space G is topologically E^3 if each element of G is convex—or even if each is a point or a straight line interval—or for that matter, even if each is a point or an interval oriented with one of the three axes.

REMARKS. After this paper was submitted, Eldon Dyer and Mary-Elizabeth Hamstrom announced at the April, 1956, meeting of the Society in New York City that if G is an upper semicontinuous decomposition of E^3 such that each element of G has a complement like the complement of a point in E^3 and is a bounded continuum in a horizontal plane, then the decomposition space G is topologically E^3. Hence, Condition (a) suggested in the questions is enough to insure that the decomposition space is topologically E^3. Part of Theorem 5 would follow as a trivial corollary of this result.

Also, I announced at the April, 1956, meeting of the Society in Chicago an upper semicontinuous decomposition of E^3 whose decomposition space differs from E^3 even though it contains only a countable number of nondegenerate elements and each is a bounded continuum, lies in a plane, and has a complement like the complement of a point. This shows that in Theorem 1 we cannot dispense with the condition that the sum of the nondegenerate elements of G is a G_δ set.

UNIVERSITY OF WISCONSIN

REFERENCES

1. J. W. ALEXANDER, *On the subdivision of 3-space by a polyhedron*, Proc. Nat. Acad. Sci. U.S.A., vol. 10 (1924), pp. 6–8.
2. R. H. BING, *A homeomorphism between the 3-sphere and the sum of two solid horned spheres*, Ann. of Math., vol. 56 (1952), pp. 354–362.
3. R. H. BING, *Locally tame sets are tame*, Ann. of Math., vol. 59 (1954), pp. 145–158.
4. R. H. BING, *Decompositions of E^3 into points and tame arcs*, Summary of Lectures and Seminars, Summer Institute on Set Theoretic Topology, Madison, 1955, pp. 40–47.
5. J. H. ROBERTS and N. E. STEENROD, *Monotone transformations of two-dimensional manifolds*, Ann. of Math., vol. 39 (1938), pp. 851–862.
6. JAMES F. WARDELL, *Continuous transformations preserving all topological properties*, Amer. J. Math., vol. 58 (1936), pp. 709–726.
7. G. T. WHYBURN, *Structure of continua*, Bull. Amer. Math. Soc., vol. 42 (1936), pp. 49–73.

ANNALS OF MATHEMATICS
Vol. 65, No. 3, May, 1957
Printed in U.S.A.

A DECOMPOSITION OF E^3 INTO POINTS AND TAME ARCS SUCH THAT THE DECOMPOSITION SPACE IS TOPOLOGICALLY DIFFERENT FROM E^3

BY R. H. BING*

(Received April 21, 1956)

1. Introduction

We find that a useful theorem for E^2 does not generalize to E^3. R. L. Moore showed (Theorem 22 of [12]) that if G is an upper semicontinuous decomposition of E^2 such that the elements of G are bounded continua which do not separate E^2, then the decomposition space is topologically equivalent to E^2. One might wonder if a decomposition space G is topologically equivalent to E^3 in the case where G is an upper semicontinuous decomposition of E^3 such that the complement of each element of G is topologically equivalent to the complement of a point in E^3. (See questions raised in [13], [15], [16].) In this paper we show that there are such decompositions of E^3 whose decomposition spaces are not topologically E^3. The decomposition described here is the one given in [2]. That short paper [2] contains the important ideas of this longer paper so the reader may find it convenient to read pages 44–47 of the short summary in order to get an overall view of the general plan of attack before plunging into the details of the present paper.

Although upper semicontinuous collections are defined for more general spaces than metric spaces, we shall only consider them in E^3. A collection G of closed bounded point sets in E^3 is said to be upper semicontinuous if for each element g of G and each open set U in E^3 containing g, there is an open set V containing g such that each element of G that intersects V lies in U—that is (Star of V with respect to G) $\subset U$.

We use the same letter to denote the collection G as to denote the decomposition space. If U is an open set in the original space, the collection of elements of G in U is regarded as an open set U in the decomposition space.

If g is a continuous map of a closed subset X of E^3 onto a metric space Y such that the inverse of each point of Y is a closed bounded set in X, the collection of such inverses is an upper semicontinuous decomposition of X. The decomposition space is topologically equivalent to Y. Conversely, if G is an upper semicontinuous decomposition of X, there is a metric space Y topologically equivalent to the decomposition space G and a map g of X onto Y such that for each point y of Y, $g^{-1}(y)$ is an element of G. Hence, questions involving the topology of decomposition spaces are equivalent to questions involving the topology of image spaces.

* Work on this paper was supported by the National Science Foundation under Contract NSF-C479.

If one has a simple upper semicontinuous decomposition H of E^3 such as one consisting of a finite collection of mutually exclusive straight line intervals and points not on these intervals, not only is the decomposition space topologically equivalent to E^3 but also the nondegenerate elements of G can be shrunk to points in E^3 by a continuous *pseudo isotopy* $f(x, t)$ $(x \in E^3, 0 \leqq t \leqq 1)$ such that

(1) for each fixed t_0 less than 1, $f(x, t_0)$ is a homeomorphism of E^3 onto E^3,

(2) $f(x, 0)$ is the identity—that is $f(x, 0) = x$,

(3) $f(x, 1)$ takes each element of G onto a distinct point of E^3.

In the above case, if U is an open set containing the intervals, there is such a pseudo isotopy $f(x, t)$ that is fixed outside U—that is $f(x, t) = x$ if $x \in E^3 - U$.

The following example was constructed to show that there are decompositions of E^3 whose elements are points and tame arcs such that the arcs cannot be shrunk to points by a pseudo isotopy. After constructing the example it was found that the corresponding decomposition space was topologically different from E^3. The decomposition space is described in a fashion similar to the one used in describing a decomposition space in [3].

After describing the upper semicontinuous decomposition G of E^3 in Section 2, we develop some properties of E^3 in Sections 3, 4, 5, 6 to use in Sections 7 and 8 to show that the decomposition space G is topologically different from E^3. It is believed that these properties of E^3 are of interest aside from their applications in Section 7 and 8 and will find applications elsewhere.

If M is an n-manifold with boundary, we use Bd M to denote the set of points of M with no neighborhood in M topologically equivalent to E^n. Hence, if D is a disk in E^3, Bd D is a simple closed curve.

The Euclidean metric is denoted by ρ.

2. The example

Here we describe an upper semicontinuous decomposition G of E^3 such that the corresponding decomposition space is topologically different from E^3 even though the elements of G are points and tame arcs. The nondegenerate elements of G are the components of a closed set A_0 which is described as the intersection of a decreasing sequence of closed sets A, $\sum A_i$, $\sum A_{ij}$, $\cdots$. First we describe A.

1. Suppose A is a topological solid sphere with two handles as shown in Figure 1. It is polyhedral. It may be convenient to think of A as an elongated figure 8 plus all points whose distances from it are no more than some positive number ε.

2. Suppose A_1, A_2, A_3, A_4 are four polyhedra imbedded in A as shown. Although the A_i's are shown as finite graphs, they are topologically equivalent to A. The upper loops of each pair of A_i's are linked as are the lower loops. We call A_1, A_2, A_3, A_4 the A's defined at the first stage.

3. Four continua A_{i1}, A_{i2}, A_{i3}, A_{i4} are imbedded in A_i as A_1, A_2, A_3, A_4 were imbedded in A. There are 4^2 A_{ij}'s defined at the second stage of our description.

4. Four continua A_{ij1}, A_{ij2}, A_{ij3}, A_{ij4} are imbedded in A_{ij} in this same man-

ner. Similarly, for each integer n, we define 4 continua in each of the 4^n A's defined at the nth stage.

5. We could have placed the A's so that each component of $A_0 = A \cdot \sum A_i \cdot \sum A_{ij} \cdot \sum A_{ijk} \cdot \cdots$ is an arc that does not intersect any horizontal plane in two points so we suppose that this was done. Each element of G is either a point in the complement of A_0 or a component of A_0. Each element of G in A_0 is a tame arc and the collection of such tame arcs corresponds to a Cantor set in the decomposition space.

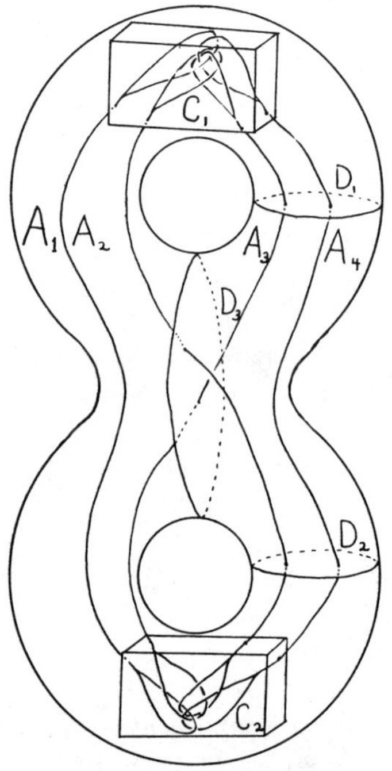

FIG. 1

The idea that leads one to suspect that there is not a pseudo isotopy that shrinks elements of G to points and that is fixed outside A is that if some of the A_i's are shrunk, it would appear that at least one of them would have to be stretched. Then perhaps it could be shown that one of the A_{ij}'s in the stretched A_i could not be shrunk, one of the A_{ijk}'s in it could not be shrunk, etc.

If one is to get a pseudo isotopy $f(x, t)$ fixed outside A and satisfying conditions (1), (2), and (3) mentioned in the introduction, for each positive number ε there would need to be a number t_0 between 0 and 1 such that for each element g of G, $f(g, t_0)$ is of diameter less than ε. Then there would be an integer n such that the image under $f(x, t_0)$ of each of the A's at the nth stage would be of diameter less than ε. In Section 7 we show that there is no integer n and no isotopy fixed outside A that shrinks the A's at the nth stage to sets with diameters less than the distance between the handles of A.

3. Linking curves

In this section, we give the results (Theorems 1, 2, 3) concerning linking that are used in the proof of Theorem 10.

A finite 2-complex X (for our present purposes) is a finite collection of triangular disks (called 2-simplexes). For the sake of simplicity we do not impose the extra condition that if two of the 2-simplexes intersect, the intersection is a vertex or edge of each. The collection of those edges that are edges of exactly an odd number of the triangles of X is called the boundary of the complex X.

If J_1, J_2 are two mutually exclusive polyhedral simple closed curves, J_1 may be regarded at times as a collection of segments or edges. There is a 2-complex X with 2-simplexes $T_1, T_2, \cdots, T_n$ such that J_1 is the boundary of the complex X, no vertex of J_2 lies on any T_i, and no edge of any T_i intersects J_2. (Our simplified definition of 2-complex makes it easier to find such an X.) We say that J_1 *links* J_2 if $\sum N(T_i, J_2)$ is odd where $N(T_i, J_2)$ is the number of points in $T_i \cdot J_2$. It may be shown that whether or not J_1 links J_2 is not a function of the 2-complex X chosen and that J_2 links J_1 if J_1 links J_2. See Section 9 of [5].

Suppose J_1, J_2 are two arbitrary simple closed curves (not necessarily polyhedral) and $\rho(J_1, J_2) > \varepsilon$. Let K_j^i $(i, j = 1, 2)$ be polygonal simple closed curves such that there is a homeomorphism h_j^i of J_j onto K_j^i such that for each x in J_j, $\rho(x, h_j^i(x)) < \varepsilon/2$. Then it follows from the polyhedral version of Theorem 5 that K_1^1 links K_2^1 if and only if K_1^2 links K_2^2. Hence, to determine whether or not two arbitrary disjoint simple closed curves link, we investigate the linking of polygonal simple closed curves near them; that is, J_1 *links* J_2 if and only if K_1^1 links K_2^1.

The following two theorems are proved in Section 9 of [5].

THEOREM 1. *If in E^3 a simple closed curve J does not intersect a certain θ curve, then J links exactly two of the simple closed curves in the θ curve if it links one of them.*

THEOREM 2. *If J_1, J_2, J_3 are three simple closed curves in E^3 such that J_1 links J_2 and J_2 is homotopic to J_3 in the complement of J_1, then J_1 links J_3.*

The preceding two theorems may be generalized by considering linkings of maps rather than linkings of simple closed curves. Suppose f_1, f_2 are two maps of simple closed curves C_1, C_2 into E^3 such that $\rho(f_1(C_1), f_2(C_2)) > \varepsilon$. Consider two homeomorphisms h_1, h_2 of C_1, C_2 into E^3 such that for each point x of C_i, $\rho(f_i(x), h_i(x)) < \varepsilon/2$. The map f_1 *links* the map f_2 if and only if $h_1(C_1)$ links $h_2(C_2)$.

In the proof of Theorem 3 we make use of Theorem 1 as it applies to mappings so we state it as follows.

THEOREM 1′. *Suppose f is the map of a θ curve T into E^3 where f_1, f_2, f_3 are the parts of f operating on the simple closed curves in T. If f_4 is a map of a circle J into E^3 such that $f_4(J) \cdot f(T) = 0$, then f_4 links exactly two of f_1, f_2, f_3 if it links one of them.*

THEOREM 3. *Suppose J_1, J_2 are two linking simple closed curves in an open set U in E^3 such that each of J_1, J_2 can be shrunk to a point in U. For each tame disk D which contains neither J_1 nor J_2 there is a component of $U - D$ that intersects both J_1 and J_2.*

It is essential in the hypothesis of Theorem 3 to suppose that each of J_1, J_2 can be shrunk to a point in U. The theorem is not true with the omission of the condition that J_1 can be shrunk to a point in U as we can see by letting U be the open torus obtained by revolving the interior of the circle $x^2 + (y - 4)^2 = 9$ in the xy plane about the x axis, J_1 be the circle obtained by revolving the point $(0, 4, 0)$ about the x axis, J_2 be the circle $z = 0$, $x^2 + (y - 4)^2 = 4$, and D be a disk obtained by revolving about the line $x = 0$, $y = 4$ the broken line from $(0, 0, 1)$ to $(0, 3, 0)$ to $(0, 0, -1)$ to $(0, 4, 0)$.

PROOF OF THEOREM 3. If J_1 intersects a component C of $D \cdot U$ but no two arcs in J_1 abut on C from opposite sides, it is possible to pull J_1 slightly to one side of C. Then if a component of $U - D$ intersects both J_1 and the adjusted J_1, it would intersect both J_1 and the original. Hence, we suppose with no loss of generality that if either J_1 or J_2 intersects a component of $D \cdot U$, it contains arcs abutting on opposite sides of this component.

Let $C_1, C_2, \cdots, C_n$ be the components of $D \cdot U$ that intersect $J_1 + J_2$. If a component of $(U - D) + C_1 + \cdots + C_n$ intersects both J_1 and J_2, a component of $U - D$ does also because if J_i intersects C_j, it contains arcs abutting on both sides of C_j. We establish the theorem by showing that a component of $(U - D) + C_1 + \cdots + C_n$ intersects both J_1 and J_2.

Since J_i can be shrunk to a point in U, there is a map f_i of a disk E into U such that f_i takes Bd E homeomorphically onto J_i. Let X_i be the set of all points p of E such that each arc from p to Bd E intersects $f_i^{-1}(D - C_1 - \cdots - C_n)$. Then X_i is a closed subset of E and does not intersect Bd E.

Since $f_i(E - X_i)$ belongs to the same component of $(U - D) + C_1 + \cdots + C_n$ as J_i, we shall suppose that $f_1(E - X_1) \cdot f_2(E - X_2) = 0$.

Consider a θ curve in $E - X_1$ such that one simple closed curve in the θ curve is Bd E. By applying Theorem 1 we find that there is simple closed curve J other than Bd E in the θ curve such that f_1 on J links f_2 on Bd E. An oscillating repetition of this procedure shows that there are decreasing sequences E_{11}, E_{12}, E_{12}, $\cdots$ and E_{21}, E_{22}, E_{23}, $\cdots$ of topological disks in E such that Bd $E_{ij} \subset E - X_i$, f_1 on Bd (E_{1i+1}) links f_2 on Bd E_{2i}, f_2 on Bd E_{2i} links f_1 on Bd E_{1i}, and $E_{i1} \cdot E_{i2} \cdot E_{i3} \cdot \cdots$ is either a point or lies in a component of X_i.

Since Bd $E_{ij} \subset E - X_i$, $f_i(\text{Bd } E_{ij})$ belongs to the component of $(U - D) + C_1 + \cdots + C_n$ containing J_i.

Limit f_i (Bd E_{ij}) $= R_i$ is either a point or a subset of

$$D - (C_1 + C_2 + \cdots + C_n).$$

If R_1 is a point not in $D - (C_1 + C_2 + \cdots + C_n)$, this point is also in R_2 and J_1, J_2 belong to the same component of $(U - D) + C_1 + \cdots + C_n$ containing this point.

If R_1 is a subset of $D - (C_1 + C_2 + \cdots + C_n)$, it belongs to one component C_0 of $D \cdot U - (C_1 + C_2 + \cdots + C_n)$. Then R_2 must also lie in C_0 and intersect R_1 at a point p_0 or else it is not true that for each i, f_1 on Bd E_{ij} links f_2 on Bd E_{2j}. Also, it is necessary that the $f_1(E_{ij})$'s converge to R_1 on the same side (call it the "upper" side) of D that the $f_2(E_{2j})$'s converge to R_2 from or else there is no linking. Then the component of $(U - D) + C_1 + \cdots + C_n$ containing a sequence of points converging to p_0 from the "upper" side of D has points of both the $f_1(E_{1j})$'s and the $f_2(E_{2j})$'s and hence of J_1 and J_2.

We make no application of the following theorem in this paper but include it since it is a variation of Theorem 3.

THEOREM 4. *Suppose U is an open subset of E^3, J is a simple closed curve that can be shrunk to a point in U, D is a topological disk, and D' is a component of $D \cdot U$ that misses J. Then no simple closed curve in D' links J.*

PROOF. Assume J' is a simple closed curve in D' that links J. By repeated application of Theorem 1′ to θ curves in D', it may be shown that for each positive number ε there is a simple closed curve J_ε in D' such that J_ε links J and lies within ε of Bd U. This violates the condition that J can be shrunk to a point in U.

QUESTION. If U is an open set, D is a tame disk, and J_1, J_2, J_3 are three simple closed curves each pair of which satisfies the conditions of Theorem 3, it cannot be concluded that there is a component of $U - D$ intersecting each of J_1, J_2, J_3. One may see this by letting U be an open torus.

However, suppose U is an open subset of E^3, C is a polyhedral cube in U, and $J_1, J_2, \cdots, J_n$ are simple closed curves in U such that J_i links $J_j (i \neq j)$ and J_i can be shrunk in $U - (J_1 + \cdots + J_{i-1} + J_{i+1} + \cdots + J_n)$ to a simple closed curve in C. Is it true that if D is a tame disk containing no J_i, then there is a component of $U - D$ intersecting each J_i?

If a ready affirmative answer had been apparent, this result, instead of a combination of Theorems 3 and 5, would have been used to good advantage in Part I of the proof of Theorem 10 and we would have avoided considering the abstract cube X shown in Figure 3.

4. Separating cubes with manifolds

In the proof of Theorem 10 we shall use the fact that there are arcs missing two disks but joining simple closed curves in a cube. Here we give the theorem insuring that there are such arcs missing the disks. We assume that the open set U in question is the interior of a cube but any unicoherent open subset of E^3 would work equally as well.

THEOREM 5. *Suppose U is the interior of a topological cube, Y is a collection of bounded continua in U, and M is a compact 2-manifold with boundary such that for each pair of elements of Y there is a component of $U - M$ intersecting both of these elements. Then there is a component of $U - M$ intersecting each element of Y.*

PROOF. Let y_0 be an element of Y and $U_1, U_2, \cdots, U_n$ be the components of $U - M$ intersecting y_0.

Let Z be a continuum such that

(a) Z is the sum of elements of $\bar{U}_1, \bar{U}_2, \cdots, \bar{U}_n$,

(b) each element of Y intersect some U_i in Z, and

(c) no proper subcontinuum of Z has properties (a) and (b).

When we show that Z contains only one U_i, the truth of Theorem 5 will follow from Condition (b).

If Z contains two elements of $U_1, U_2, \cdots, U_n$, some component M' of $M \cdot U$ separates these two elements from each other in U and Z is the sum of two subcontinua Z_1, Z_2 such that each contains elements of $U_1, U_2, \cdots, U_n$ and $Z_1 \cdot Z_2 = M'$. Because of Condition (c), some element y_1 of Y intersects the sum of the elements of $U_1, U_2, \cdots, U_n$ in Z_1 but misses those in Z_2 and some element y_2 intersects the sum of the elements of $U_1, U_2, \cdots, U_n$ in Z_2 but misses those in Z_1. But then no component of $U - M'$ would intersect both y_1 and y_2.

5. Homotopy on 3-manifolds

Here we give the two results concerning homotopy to be used in the proof of Theorem 10. Although the results are applicable in other spaces than 3-manifolds with boundaries, we only state them in the simple case where they will be used.

THEOREM 6. *Suppose M is a 3-manifold with boundary, D_1, D_2 are two mutually exclusive closed subsets of M, and pxq, pyq are homotopic arcs in M such that $pxq \cdot D_1 = pyq \cdot D_2 = 0$. Then there is an arc pzq homotopic to pxq in M such that $pzq \cdot (D_1 + D_2) = 0$.*

PROOF. Let r, t and s, u be opposite vertices of a square plus its interior S in the plane. Since pxq is homotopic to pyq in M, there is a map f of S into M such that f takes $rs + st$ homeomorphically onto pxq and $ru + ut$ homeomorphically onto pyq.

Since $f^{-1}(D_1)$ and $f^{-1}(D_2)$ are mutually exclusive closed subsets of S neither of which separates r from t, their sum does not. Hence, there is an arc rwt in $S - (f^{-1}(D_1) + f^{-1}(D_2))$. Also there is a homeomorphism h of rwt into an arc pzq in M such that h is so near f on rwt that pzq is homotopic to pxq in M and $pzq \cdot (D_1 + D_2) = 0$.

THEOREM 7. *Suppose M is a 3-manifold with boundary, D_1, D_2 are two mutually exclusive closed subsets of M, and pxq, pyq are homotopic arcs in M such that $px \cdot D_1 = xq \cdot D_2 = py \cdot D_2 = yq \cdot D_1 = 0$. Then there is an arc pzq homotopic to pxq in M such that either $pzq \cdot D_1 = 0$ or $pzq \cdot D_2 = 0$.*

PROOF. Take r, s, t, u, S, f as in the proof of Theorem 6 with the additional condition that $f(rs) = px$, $f(st) = xq$, $f(ru) = py$, $f(ut) = yq$.

If no component of either $f^{-1}(D_1)$ or $f^{-1}(D_2)$ separates r from t, the proof goes as in the proof of Theorem 9.

If a component C of $f^{-1}(D_1)$ separates r from t in S, C intersects both ru and st and there is an arc rwt so near $ru + C + st$ that rwt does not intersect D_1. Near $f(rwt)$ there is an arc pzq homotopic in M to pxq which does not intersect D_2.

Similarly, if a component of $f^{-1}(D_2)$ separates r from t, there is an arc pzq missing D_1 and homotopic in M to pxq.

6. Replacing disks by polyhedral disks

Finally, we give the theorems to be used in the lemma to Theorem 12.

THEOREM 8. *Suppose D is a topological disk in E^3 which is locally polyhedral except on a closed 0-dimension set C on $D -$ Bd D. Then for each open set U containing C there is a polyhedral disk D' in $D + U$ such that* Bd $D' =$ Bd D.

The truth of this theorem follows from Theorem 7 and 8 of [5] but we give an argument much simpler than that used in [5].

PROOF. Cover C with the interiors $U_1, U_2, \cdots, U_n$ of a finite number of cubes such that $U_i \subset U$, $U_i \cdot \text{Bd } D = 0$, and $D \cdot \text{Bd } \bar{U}_i$ does not contain any triod. We can pick such a finite number of U_i's because C is bicompact and D does not contain uncountably many mutually exclusive triods.

Now $D \cdot \text{Bd } \bar{U}_i$ is locally connected except at points of C and, since C is 0-dimensional, each component of $D \cdot \text{Bd } \bar{U}_i$ is a locally connected continuum containing no triod. (If a continuum is not locally connected, it fails to be locally connected at a nondegenerate continuum of points.) Hence each such component is either a point, an arc, or a simple closed curve. For each positive number ε, $D \cdot \text{Bd } \bar{U}_i$ has only a finite number of elements with diameters more than ε. The collection of nondegenerate components of $D \cdot \text{Bd } \bar{U}_i$ forms a null sequence.

Let K be the closure of the component of $D - U_1$ containing Bd D. Then $D - K$ is the sum of a null sequence of topological open disks $V_1, V_2, \cdots$ such that Bd V_i is a simple closed curve in $D \cdot \text{Bd } \bar{U}_1$ and Bd $V_i \cdot$ Bd $V_j = 0$ if $i \neq j$. The open disks $V_1, V_2, \cdots$ in D may be replaced by open disks $V_1', V_2', \cdots$ in U_1 such that $K + V_1' + V_2' + \cdots$ is a topological disk D_1 that is locally polyhedral except possibly at points of $C - U_1$. We suppose that U_2 is adjusted slightly so that $D_1 \cdot \text{Bd } U_2$ contains no triod.

Next we replace other open disks in D_1 by open disks in U_2 to get a topological disk D_2 which is locally polyhedral except at points of $C - (U_1 + U_2)$. Continuing in this fashion, we get a topological disk D_n in $D + U_1 + U_2 + \cdots + U_n$ such that Bd D_n = Bd D and D_n is locally polyhedral except at points of $C - (U_1 + U_2 + \cdots + U_n) = 0$. Then $D_n = D'$ is locally polyhedral and hence polyhedral.

THEOREM 9. *Suppose D_1, D_2 are two polyhedral disks in E^3 such that* Bd $D_1 \cdot D_2$ = Bd $D_2 \cdot D_1 = 0$. *Then for each continuum K that does not intersect $D_1 + D_2$ except possibly in* Bd D_1 + Bd D_2 *there is a polyhedral disk D_3 such that $K \cdot D_3 \subset$* Bd D_3 = Bd D_2, $D_1 \cdot D_3 = 0$, *and D_3 coincides with D_2 in a neighborhood of* Bd D_2.

PROOF. Since there is a piecewise linear homeomorphism of E^3 onto itself throwing D_1 into a plane, we suppose with no loss of generality that it lies in the xy plane. First we alter D_2 slightly near D_1 to get a polyhedral disk D_3' such that $D_3' \cdot K \subset \text{Bd } D_3' = \text{Bd } D_2$, D_3' agrees with D_2 in a neighborhood of Bd D_2, and $D_1 \cdot D_3'$ is the sum of a finite number of mutually exclusive simple closed curves $J_1, J_2, \cdots, J_n$. We obtain D_3' by considering a triangulation T of D_2 of mesh less than $\rho(D_1, \text{Bd } D_2)$ and then moving each vertex of T on D_1 slightly to one side of D_1.

We obtain D_3 from D_3' by removing these simple closed curves. We now indicate how one of them can be removed.

Suppose J_i is one of these simple closed curves such that the interior of J_i in the xy plane does not intersect D_3'. Use Int J_{i1} and Int J_{i3} to denote the components of $D_1 - J_i$ and $D_3' - J_i$ missing Bd D_1 and Bd D_3' respectively. We get rid of the simple closed curve J_i in D_3' by replacing Int J_{i3} by Int J_{i1} in D_3' and then pushing the new part slightly to one side of the xy plane. A repetition of this procedure removes the other simple closed curves and we obtain a polyhedral disk D_3 satisfying the conclusion of Theorem 9.

7. Impossibility of shrinking the A's

Suppose D_1, D_2 are planar disks as shown in Figure 1 such that D_1 cuts the upper handle of A and D_2 cuts the lower handle as shown. We shall show that if h is any homeomorphism of E^3 onto itself that is fixed outside A and n is any positive integer, then one of the 4^n A's defined at the nth stage has an image under h that intersects both D_1 and D_2. This will show the impossibility of shrinking the A's.

Before launching into the rather involved proof of this more general result, let us give a relatively easy indication of proof that one of A_1', A_2', A_3', A_4' intersects both D_1 and D_2 where A_i' is the image of A_i $(i = 1, 2, 3, 4)$ under h and we have the following:

1. $A_1' + A_4'$ intersects both D_1 and D_2.
2. $A_2' + A_3'$ intersects both D_1 and D_2.
3. $A_1' + A_3'$ intersects D_1.
4. $A_2' + A_4'$ intersects D_1.
5. $A_1' + A_2'$ intersects D_2.
6. $A_3' + A_4'$ intersects D_2.

Assume no A_i' intersects both D_1 and D_2. Then if A_1' misses D_1, A_4' intersects D_1 by (1), A_3' intersects D_1 by (3), and one of A_3', A_4' intersects both D_1, D_2 by (6). If A_1' misses D_2, A_4' intersects D_2 by (1), A_2' intersects D_2 by (5), and one of A_2', A_4' intersects both D_1, D_2 by (4).

One might suspect that not each A_{ij} can be shrunk so as to miss one of D_1, D_2, not each A_{ijk} can be so shrunk, etc. In fact, we show this, but the proof is more involved than the simple argument we outlined above.

Mathematical induction may be regarded by some freshmen merely as a tool for showing that various sequences have certain sums. Actually, it is of much use in other branches of mathematics. We use it to show that under each deformation of E^3 that is fixed outside A, one of the 4^n A's defined at the nth stage has an image that intersects each of D_1, D_2.

As in the case in proving many theorems by mathematical induction, care must be used in choosing the condition that we want to hold at each stage. If we pick a condition too weak, as

(a) the image of the set intersects both D_1 and D_2,

the fact that the set $A_{ij\cdots k}$ has the property is no assurance that there is an s $(s = 1, 2, 3, 4)$ such that the set $A_{ij\cdots ks}$ has the property. On the other hand, if an exceedingly strong condition is chosen as

(b) D_1 cuts one loop of its image and D_2 cuts the other loop,

it follows that some $A_{ij\cdots ks}$ has Property (a) if the set $A_{ij\cdots k}$ has Property (b) but it cannot be concluded under this condition that some $A_{ij\cdots ks}$ has Property (b). We seek a condition stronger than (a) but weaker than (b).

We noted that we could regard A as the sum of a linear graph F which is a figure 8 and all points whose distance from F is no more than a positive number

ε. We call F the center of A. In Figure 1 we only showed the centers of A_1, A_2, A_3, A_4 instead of making them look solid. If B is the image of A under a particular homeomorphism h, we call $h(F)$ the center of B—also, the images of the centers of A_1, A_2, A_3, A_4 are called the centers of $h(A_1)$, $h(A_2)$, $h(A_3)$, $h(A_4)$ and the image of the center of any $A_{ij\cdots k}$ is called the center of $h(A_{ij\cdots k})$. The center of B is therefore a function of the particular homeomorphism h of A onto B chosen.

If h is a homeomorphism of A onto itself, the center of A with respect to h may not even be homotopic in A to F. Consider the disks D_1, D_2, D_3 shown in Figure 1 and a homeomorphism h of A onto itself leaving D_1 fixed and interchanging D_2 and D_3. We find that $h(F)$ is not homotopic in A to F because one loop of it links both Bd D_1 = Bd $h(D_1)$ and Bd D_2 = Bd $h(D_3)$. However, we find in the proof of Theorem 11 that $h(F)$ is homotopic to F in A if h is a homeomorphism of A onto itself that is fixed on Bd A.

If F' is a topological figure 8 in A, we say that it is homotopic to the center of A if there is a homotopy or continuous map $f(x, t)$ $(x \in F, 0 \leqq t \leqq 1)$ of $F \times I$ into A such that $f(x, 0) = x$ and $f(x, 1)$ takes F homeomorphically onto F'.

We note that if F' is any topological figure 8 in A which is homoto pic to the center of A, then one loop of F' intersects D_1 and another loop intersects D_2. Then F' has the following property:

Property P. A topological figure 8 has Property P if it contains two points p, q in opposite loops such that any arc from p to q in it intersects both D_1 and D_2.

Also, A has the following property under the identity.

Property Q. An image of A under a homeomorphism h has Property Q if each topological figure 8 in it homotopic to its center has property P.

We show that if h is any homeomorphism of E^3 onto itself that is fixed outside A, then for one of the A's defined at the nth stage, say $A_{ij\cdots k}$, $h(A_{ij\cdots k})$ intersects both D_1 and D_2 because it has Property Q. This result follows either from mathematical induction or from iterated use of the following theorem.

THEOREM 10. *If a continuum B in A is the image of A under a homeomorphism h of E^3 onto itself and B has Property Q, then one of $h(A_1)$, $h(A_2)$, $h(A_3)$, $h(A_4)$ has Property Q.*

PROOF. Suppose neither $h(A_1)$, $h(A_2)$, $h(A_3)$, nor $h(A_4)$ has Property Q. Let F_i be a topological figure 8 in A_i which is homotopic in A_i to the center of A_i but such that $h(F_i)$ does not have Property P. With no loss of generality we suppose that neither loop of F_i lies entirely in $h^{-1}(D_1 + D_2)$. We denote by F_{iu} the loop of F_i homotopic in A_i to the upper loop of the center of A_i.

The proof of this theorem is broken into two large parts. Using the F_i's which we assumed existed, we now show in Part I of the proof that there are four arcs pq_ir_is $(i = 1, 2, 3, 4)$ such that the interiors of these arcs do not intersect each other, $h(pq_is_ir)$ does not intersect both D_1 and D_2, and the sum of the four arcs is homotopic in A to the arcs shown in Figure 2. After showing the existence of these arcs, we finish the proof in Part II by showing that if there were such arcs, there is a topological figure 8 in B which shows that B does not have Property Q.

PART I OF PROOF. At times it is convenient to regard each A_i as the sum of two solid anchor rings (tori) and a solid tube connecting them. The tori are called the loops of A_i.

Let C_1, C_2 be two mutually exclusive polyhedral cubes in A as shown in Figure 1 such that C_1 contains the upper loop of each A_i, C_2 contains the lower loop, Bd $C_i \cdot$ Bd A_j is a simple closed curve, and $A_i \cdot$ Bd C_j is a topological disk that cuts the upper loop of A_i from the lower.

Since the upper loops of the centers of A_i and A_j $(i \neq j)$ link, it follows from Theorem 2 that F_{iu} and F_{ju} link. However, since each of F_{iu} and F_{ju} can be shrunk to a point in $A_i + C_1 + A_j$, it follows from Theorem 3 that there is an arc from F_{iu} to F_{ju} in $A_i + C_1 + A_j$ that misses $h^{-1}(D_1 + D_2)$.

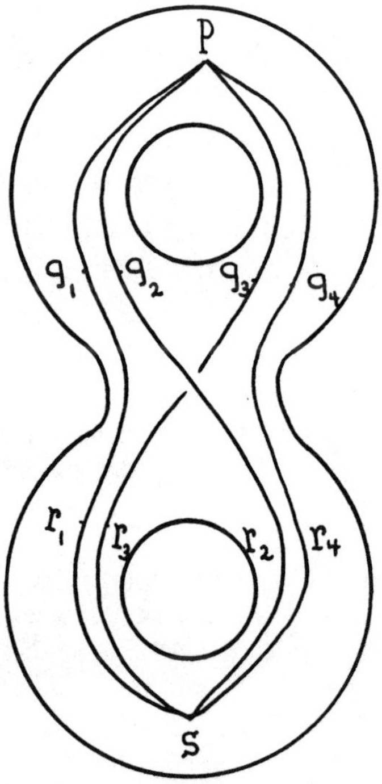

FIG. 2

If there were a topological cube in A containing $C_1 + \sum A_i + C_2$, we could use this cube and apply Theorem 5 to find the pq_is_ir's. Since there is no such cube in A, we use an abstract cube X. In Figure 3 we show C_1, A_1, A_2, A_3, A_4, C_2 as they would look if the lower loops of A_1, A_2, A_3, A_4 were pulled apart and C_2 were shown four times as C'_{21}, C'_{22}, C'_{23}, C'_{24}. It represents a topological cube X such that there is a map f of X onto $C_1 + A_1 + A_2 + A_3 + A_4 + C_2$ that takes C'_1, A'_i, C'_{2i} homeomorphically onto C_1, A_i, and C_2 respectively. Then f would be 1–1 except on $\sum C'_{2i}$ and it would be 4–1 there. We let F'_i be the part of $f^{-1}(F_i)$ in A'_i. Actually, $f^{-1}(F_i)$ intersects each of C'_{21}, C'_{22}, C'_{23}, C'_{24} but we do not wish to consider all of it. Also F'_{iu} is the $f^{-1}(F_{iu})$ in A'_i.

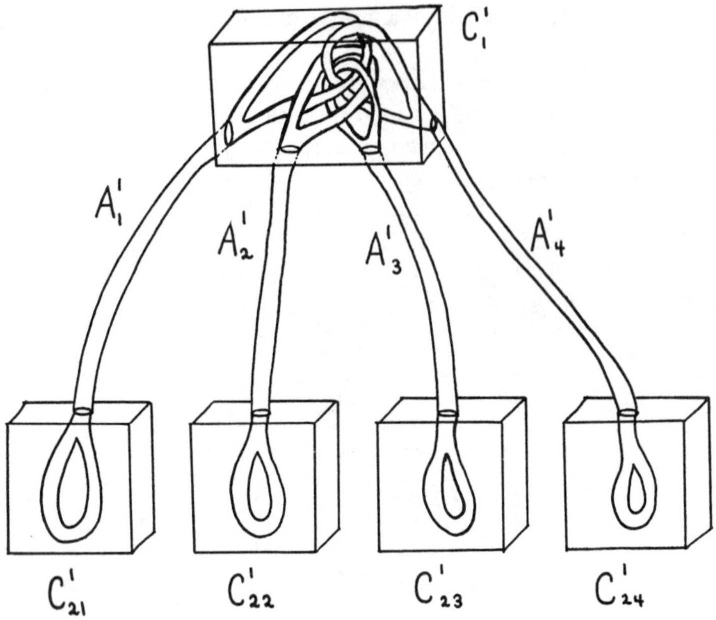

FIG. 3

Let D_i' be $f^{-1}h^{-1}(D_i)$ $(i = 1, 2)$ where D_1, D_2 are the disks shown in Figure 1 cutting the loops of A. It may be that D_i' is skew (cannot be imbedded in the plane) even though $h^{-1}(D_i)$ is not skew. We suppose that f is defined on an open set containing X so that $f^{-1}h^{-1}(D_i) = D_i'$ is a 2-manifold containing a compact 2-manifold with boundary containing $X \cdot D_i'$.

Using Theorem 5 we find that one component of $X - (D_1' + D_2')$ contains points of each of F_{1u}', F_{2u}', F_{3u}', F_{4u}'. This component would need to intersect C_1'. Hence there are four arcs $p'q_i'$ $(i = 1, 2, 3, 4)$ in $X - (D_1' + D_2')$ such that $p'q_i' \cdot p'q_j' = p'$ $(i \neq j)$ is a point of C_1' and $p'q_i' \cdot (\sum F_j')$ is a point q_i' of F_{iu}'.

We note that since X is a cube, if $p'x_iq_i'$ is any arc in $C_1' + A_i'$ from p' to q_i', there is a homotopy in X fixed on p' and q_i' that takes $p'q_i'$ onto $p'x_iq_i'$.

We suppose that $pq_i = f(p'q_i')$ is an arc that intersects $\sum F_j$ only at $q_i \in F_{iu}$ and that $pq_i \cdot pq_j = p$ $(i \neq j)$. It may require a slight adjustment to make these be arcs with these properties since f is 4–1 on $\sum C_{2i}'$, $f(p'q_i')$ may be a more general continuous curve than an arc and $f(p'q_i') \cdot f(p'q_j')$ may contain more than p. However, our main concern is to get pq_i's that miss $D_1 + D_2$ and these can be found near the $f(p'q_i')$'s. If px_iq_i is any arc in $C_1 + A_i$ from p to q_i, there is a homotopy in A of pq_i onto px_iq_i.

A similar argument shows that there are arcs r_is $(i = 1, 2, 3, 4)$ from a point r_i of the lower loop of F_i to a point s of C_2 such that $r_is \cdot r_js = s$ $(i \neq j)$, $\sum r_is \cdot F_j = r_j$, $\sum r_is \cdot \sum pq_j = 0$, $r_is \cdot h^{-1}(D_1 + D_2) = 0$, and r_is is homotopic in A to any arc from r_i to s in $C_2 + A_i$.

Since $h(F_i)$ does not have Property P, there is an arc q_ir_i from q_i to r_i in F_i

which does not intersect both $h^{-1}(D_1)$ and $h^{-1}(D_2)$. Then the arc pq_ir_is is homotopic in A to an arc ps in $C_1 + A_i + C_2$ so the arcs $pq_ir_is(i = 1, 2, 3, 4)$ are homotopic in A to those shown in Figure 2. This completes Part I of the proof of Theorem 10.

PART II OF PROOF. We break Part II of the proof into cases determined by which of $pq_ir_is(i = 1, 2, 3, 4)$ hit $h^{-1}(D_1)$ and which hit $h^{-1}(D_2)$.

Case 1. If $pq_1r_1s + pq_2r_2s + pq_3r_3s = \theta$ misses $h^{-1}(D_1)$, then there is a figure 8 in $A - h^{-1}(D_1)$ near θ that is homotopic to the center of A. (We essentially shrink pq_1r_1s to a point to get the figure 8.) This contradicts the fact that B has property Q.

A similar argument disposes of the cases where any three of $pq_ir_is(i = 1, 2, 3, 4)$ miss the same one of $h^{-1}(D_1)$, $h^{-1}(D_2)$.

Case 2. Suppose $pq_1r_1s + pq_2r_2s$ miss one of $h^{-1}(D_1)$, $h^{-1}(D_2)$ (say $h^{-1}(D_1)$) and $pq_3r_3s + pq_4r_4s$ misses the other (say $h^{-1}(D_2)$). It follows from Theorem 6 that there is a simple closed curve J in $A - h^{-1}(D_1 + D_2)$ which contains s and there is a homotopy in A that is fixed at s and which takes J onto $sr_1q_1p + pq_2r_2s$. Actually, we are using a corollary to Theorem 6 pertaining to simple closed curves because the "arcs" we use are not arcs at all but the simple closed curves $sr_1q_1pq_2r_2s$ and $sr_3q_3pq_4r_4s$.

Then near $J + sr_2q_2p + pq_4r_4s$ there is a topological figure 8 homotopic in A to F such that its lower loop J intersects neither $h^{-1}(D_1)$ not $h^{-1}(D_2)$, the left half of its upper loop is near pq_2r_2s and misses $h^{-1}(D_1)$, while the right half of its upper loop is near pq_4r_4s and misses $h^{-1}(D_2)$. This contradicts the fact that B has Property Q.

A similar argument disposes of the cases where pq_1r_1s, pq_3r_3s miss the same one of $h^{-1}(D_1)$, $h^{-1}(D_2)$ and pq_2r_2s, pq_4r_4s miss the other.

Case 3. Suppose pq_1r_1s, pq_4r_4s miss the same one of $h^{-1}(D_1)$, $h^{-1}(D_2)$ (say $h^{-1}(D_1)$) and pq_2r_2s, pq_3r_3s miss the other one. Then it follows from a corollary to Theorem 7 pertaining to simple closed curves that there is a simple closed curve pzp containing p such that pzp misses one of $h^{-1}(D_1)$, $h^{-1}(D_2)$ and there is a homotopy in A fixed at p that takes $pq_2r_2s + sr_4q_4p$ onto pzp. In applying Theorem 7, we use pq_1r_1s, sr_3q_3p, pq_2r_2s, and sr_4q_4p for the px, xq, py, yp mentioned in the statement of that theorem. If pzp misses $h^{-1}(D_1)$, there is a figure 8 near $pzp + pq_1r_1s + sr_4q_4p$ revealing a contradiction; if pzp misses $h^{-1}(D_2)$, there is such a figure 8 near $pzp + pq_2r_2s + sr_3q_3p$.

THEOREM 11. *If h is any homeomorphism of A onto itself that is fixed on* Bd A, *there is an element g of the decomposition G such that $h(g)$ intersects both D_1 and D_2.*

PROOF. First we show that A under h has Property Q.

The center of A under h is homotopic in A to F because there is a figure 8 on Bd A which is homotopic to F in A. This also follows from a result announced in [14] which would imply that there is an isotopy of E^3 onto itself that is fixed outside A and which pulls h onto the identity.

Let F_u be the upper loop of F and F'_u be its image under a homotopy in A. Then F'_u intersects D_1, for consider a planar disk D slightly larger than D_1 such that $J =$ Bd D lies outside A and links F_u. Then F'_u links J by Theorem 2 and hence intersects D_1 or else J can be shrunk to a point on D without hitting F'_u.

Also, the image under any homotopy in A of the lower loop of F intersects D_2. Therefore A under h has Property Q since the image of F under any homotopy in A has Property P.

It follows from Theorem 10 that $h(A_i)$ has Property Q for i equal to one of 1, 2, 3, 4. A second application of Theorem 10 shows that there is a j such that $h(A_{ij})$ has Property Q. Similarly, we get an A_{ijk}, an A_{ijkl}, $\cdots$ all with Property Q. Then $A_i \cdot A_{ij} \cdot A_{ijk} \cdot \cdots$ is an element g of G and $h(g)$ interesects both D_1 and D_2.

8. The decomposition space G differs from E^3

As a first step in showing that the decomposition space G differs from E^3 we prove the following lemma. We then prove in Theorem 12 that the conditions of the hypothesis of the lemma are not true by showing that its conclusions are false.

LEMMA FOR THEOREM 12. *If the decomposition space G were topologically equivalent to E^3, there would be a map f of A onto itself that is fixed at each point of* Bd A *such that for each point x of A, $f^{-1}(x)$ is an element of G.*

PROOF. Suppose there is a homeomorphism of the decomposition space G onto E^3. Then there is a map f_1 of E^3 onto itself such that for each point x of E^3, $f_1^{-1}(x)$ is an element of G.

Consider the three mutually exclusive planar disks D_1, D_2, D_3 shown in Figure 1. We note that A is the sum of two polyhedral cubes X_1, X_2 where $X_1 \cdot X_2 = D_1 + D_2 + D_3$.

We suppose with no loss of generality that $f_1(\text{Bd } A + D_1 + D_2 + D_3)$ is locally polyhedral except possibly at points of $f_1(A_0)$. (We recall that A_0 is the sum of the nondegenerate elements of the upper semicontinuous collection G.) The following approximation theorem proved by Moise shows that we can do this.

THEOREM 2 OF [11]. *If K, K' are two triangulated 3-manifolds, U is an open set of K, f is a homeomorphism of U into K', and ϕ is a continuous positive function defined on U, then there is a locally piecewise linear homeomorphism f' of U into K' such that $\rho(f(x), f'(x)) < \phi(x)$.*

An alternate proof of this theorem is to be contained in the paper abstracted in [6]. In applying the theorem, we let $K = K' = E^3$, $U = E^3 - A_0$, $\phi(x) = \rho(x, A_0)/2$, $f = f_1$. The homeomorphism f' on U promised by the theorem may be extended to all of E^3 so as to be a map of E^3 onto itself such that for each point x of E^3, $f'^{-1}(x)$ is an element of G and $f'(\text{Bd } A + D_1 + D_2 + D_3)$ is locally polyhedral except possibly at points of $f'(A_0)$.

We find from Theorem 8 that there is a polyhedral disk $E'_i (i = 1, 2, 3)$ in $f_1(D_i + \text{Interior } A)$ such that $\text{Bd } E'_i = \text{Bd } f_1(D_1)$. By applying Theorem 9 three times we find that there are three mutually exclusive polyhedral disks E_1, E_2, E_3 in $f_1(A - \text{Bd } A + D_1 + D_2 + D_3)$ such that $\text{Bd } E_i = f_1(\text{Bd } D_i)$. Then $E_1 + E_2 + E_3 + f_1(\text{Bd } A)$ is polyhedral.

There is a homeomorphism h of $E_1 + E_2 + E_3 + f_1(\text{Bd } A)$ onto $D_1 + D_2 +$

$D_3 + \text{Bd}\, A$ that agrees with f_1^{-1} on Bd A. It follows from [1] that this homeomorphism may be extended to a homeomorphism taking the bounded complementary domain U_1 of $f_1(\text{Bd}\, A \cdot \text{Bd}\, X_1) + E_1 + E_2 + E_3$ onto $X_1 - \text{Bd}\, X_1$ and the bounded complementary domain U_2 of $f_1(\text{Bd}\, A \cdot \text{Bd}\, X_2) + E_1 + E_2 + E_3$ onto $X_2 - \text{Bd}\, X_2$. But $f_1(\text{Bd}\, A) + E_1 + E_2 + E_3 + U_1 + U_2 = f_1(A)$ so there is a homeomorphism h of $f_1(A)$ onto A that agrees with f_1^{-1} on $f_1(\text{Bd}\, A)$.

The map f promised in the lemma is hf_1.

THEOREM 12. *The decomposition space G is topologically different from E^3.*

PROOF. Assume the decomposition space G is topologically equivalent to E^3. The preceding lemma would then say that there is a map f_1 of A onto itself such that f_1 is fixed at each point of Bd A and for each point x of A, $f^{-1}(x)$ is an element of G. There would be an integer n so that no one of the $4^n A_{ij\cdots k}$'s defined at the nth stage would have an image under f_1 that intersects both D_1 and D_2.

Using the preceding false assumption, we describe an impossible homeomorphism h of A onto itself. The homeomorphism h is equal to f_1 on A minus the $4^n A_{ij\cdots k}$'s defined at the nth stage. Using the preceding lemma we find that there is a map $f_{ij\cdots k}$ of $A_{ij\cdots k}$ onto itself that is fixed at each point of Bd $A_{ij\cdots k}$ and such that each $f_{ij\cdots k}^{-1}(x)$ is an element of G. Then $h = f_1 f_{ij\cdots k}^{-1}$ on $A_{ij\cdots k}$. Now h is a homeomorphism of A onto itself that violates Theorem 11. The assumption that led to this contradiction was that the decomposition space G was topologically equivalent to E^3.

THEOREM 13. *The decomposition space G fails to be locally E^3 at each point corresponding to a nondegenerate element of the upper semicontinuous decomposition G.*

PROOF. This proof is a modification of the proof of Theorem 12. If the decomposition space were locally E^3 at a point g of the decomposition space G corresponding to a nondegenerate element in the upper semicontinuous decomposition G of E^3, there would be a neighborhood U of g topologically equivalent to E^3 and an $A_{ij\cdots k}$ such that each element of the upper semicontinuous decomposition G in $A_{ij\cdots k}$ corresponds to a point in U. Following the lemma for Theorem 12 we find that there is a map f of $A_{ij\cdots k}$ onto itself that is identity on Bd A such that for each point p of $A_{ij\cdots k}$, $f^{-1}(p)$ is an element of the upper semicontinuous decomposition G. But it is no more possible for there to be such a map f of $A_{ij\cdots k}$ onto itself than it is for there to be a map f of A onto itself that is fixed on Bd A and such that for each point p of A, $f^{-1}(p)$ is an element of the upper semicontinuous decomposition G. We showed the impossibility of such a latter map f in the proof of Theorem 12.

QUESTIONS. What topological property distinguishes the decomposition space G from E^3? M. L. Curtis and R. L. Wilder showed that the space was a homotopy manifold [8].

M. K. Fort, Jr., [10] modified the upper semicontinuous decomposition G to get a decomposition G' of E^3 equivalent to G such that each nondegenerate element of G' was a broken line interval with exactly two breaks. Is there an upper semicontinuous decomposition of E^3 into points and straight line intervals such that the decomposition space is topologically different from E^3? The answer is not known even if we impose the additional restriction that each of the intervals is parallel to one of the axes.

If G is a monotone upper semicontinuous decomposition of E^3, what restrictions must be placed on the elements of G to insure that the decomposition space is topologically E^3? It is known to be topologically E^3 if it satisfies one of the following sets of conditions.

(1) Each element of G lies in a horizontal plane and has a complement like the complement of a point [9].

(2) G has only a countable number of nondegenerate elements and each is star like [4].

(3) G has only a countable number of nondegenerate elements and each is a tame arc [4].

(4) G has only a countable number of nondegenerate elements, each has a complement like the complement of a point, and the sum of these nondegenerate elements is a G_δ set [4].

An example has recently been constructed [7] to show that one cannot replace in (4) above the restriction that the sum of the nondegenerate is a G_δ set with the restriction that each lies in a plane.

UNIVERSITY OF WISCONSIN

REFERENCES

[1] J. W. ALEXANDER, *On the subdivision of 3-space by a polyhedron*, Proc. Nat. Acad. Sci. U.S.A., vol. 10 (1924), pp. 6–8.

[2] R. H. BING, *Decomposition of E^3 into points and tame arcs*, Summary of Lectures and Seminars, Summer Institute on Set Theoretic Topology, Madison, 1955, pp. 40–47.

[3] R. H. BING, *A homeomorphism between the 3-sphere and the sum of two solid horned spheres*, Ann. of Math., vol. 56 (1952), pp. 354–362.

[4] R. H. BING, *Upper semicontinuous decompositions of E^3*, Ann. of Math., vol. 65 (1957), pp. 363–374.

[5] R. H. BING, *Approximating surfaces with polyhedral ones*, Ann. of Math., vol. 65 (1957), pp. 456–483.

[6] R. H. BING, *An alternate proof that 3-manifolds can be triangulated*, Abstract 270, Bull. Amer. Math. Soc., vol. 62 (1956), pp. 179–180.

[7] R. H. BING, *A peculiar decomposition of E^3*, Abstract 452, Bull. Amer. Math. Soc., vol. 62 (1956), p. 365.

[8] M. L. CURTIS and R. L. WILDER, *The existence of certain types of manifolds*, Abstract 273, Bull. Amer. Math. Soc., vol. 62 (1956), p. 180.

[9] ELDON DYER and MARY-ELIZABETH HAMSTROM, *Certain completely regular mappings*, Abstract 544, Bull. Amer. Math. Soc., vol. 62 (1956), p. 398.

[10] M. K. FORT, JR., *A note concerning a decomposition space defined by Bing*, Ann. of Math., vol. 65 (1957), pp. 501–504.

[11] E. E. MOISE, *Affine structures in 3-manifolds*, V. *The triangulation theorem and Hauptvermutung*, Ann. of Math., vol. 56 (1952), pp. 96–114.

[12] R. L. MOORE, *Concerning upper semicontinuous collections of continua*, Trans. Amer. Math. Soc., vol. 27 (1925), pp. 416–428

[13] J. H. ROBERTS and N. E. STEENROD, *Monotone transformations of two-dimensional manifolds*, Ann. of Math., vol. 39 (1938), pp. 851–862.

[14] D. E. SANDERSON, *An extension of a deformation theorem of Alexander's, Abstract* 576, Bull. Amer. Math. Soc., vol. 61 (1955), p. 323.

[15] JAMES F. WARDWELL, *Continuous transformations preserving all topological properties*, Amer. J. Math., vol. 58 (1936), pp. 709–726.

[16] G. T. WHYBURN, *Structure of continua*, Bull. Amer. Math. Soc., vol. 42 (1936), pp. 49–73.

IMBEDDING DECOMPOSITIONS OF E^3 IN E^4

R. H. BING[1] AND M. L. CURTIS[1]

1. **Introduction.** We consider monotone decompositions of E^3 and study the possibility of imbedding the decomposition spaces in E^4. Attention has recently been called [**1**; **2**] to the fact that certain useful decompositions of E^3 can be imbedded in E^4. It would be of interest to find a simple decomposition of E^3 that could not be so imbedded.

Our first observation is that if the decomposition contains only two nondegenerate elements, then such an imbedding is always possible. However, if the nondegenerate elements of the decomposition consist of nine tamely imbedded (actually planar) circles, appropriately linked, then such an imbedding is not possible. We conjectured that the same was true in case the nondegenerate elements were three planar circles each pair of which was linked, but a neat proof by Goblirsch [**4**] shows our intuition was faulty.

THEOREM 1. *If a monotone decomposition of E^3 contains only two nondegenerate elements, then the decomposition space is imbeddable in E^4.*

PROOF. Let C_1 and C_2 be the nondegenerate elements and consider E^3 to be the x_1, x_2, x_3 plane in E^4. Let P_1 and P_2 be points in E^4 with x_4 coordinates 1 and -1 respectively. Let B_i be the cone over C_i from P_i, $i=1, 2$. Let U_1 and U_2 be disjoint neighborhoods of C_1 and C_2 respectively, and let f_i be a function mapping E^3 onto the unit interval such that the complement of U_i is sent to 0 and precisely all points of C_i are sent to 1, for $i=1, 2$.

We map the E^3 plane into E^4 as follows. A point $x=(x_1, x_2, x_3, x_4)$ in U_i goes to the point on the segment from x to P_i whose fourth coordinate is $(-1)^i f_i(x)$. Points not in $U_1 \cup U_2$ are left fixed. It is clear that the image under this map is just the decomposition space, and it is in E^4. This proves the theorem.

Theorem 1 has the following obvious generalization.

THEOREM 2. *Let $C_1, \cdots, C_n$ be the nondegenerate elements of a monotone decomposition of E^3, and consider E^3 to be a 3-plane in E^4. If there exist points $P_1, \cdots, P_n$ in $E^4 - E^3$ such that the cones B_i (B_i being the*

Presented to the Society, November 28, 1958; received by the editors March 2, 1959.

[1] The work of the first author was supported in part by Grant NSF-G4223, and of the second by NSF-G4211.

cone over C_i from P_i) are all disjoint, then the decomposition space of E^3 can be imbedded in E^4.

As an application we note that if the nondegenerate elements of the decomposition consist of the three "Ballantine Ale" circles, then the decomposition space can be imbedded in E^4.

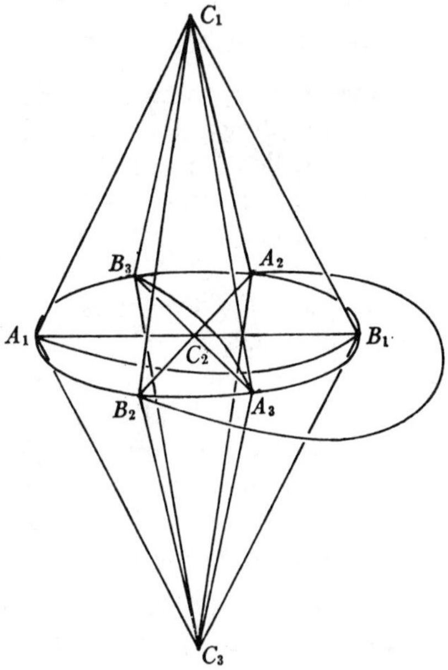

FIG. 1

In order to show that certain decomposition spaces of E^3 cannot be imbedded in E^4 we will show that they contain certain 2-polyhedra which cannot be imbedded in E^4. The example we use is a generalization of the "utility," or "cranky neighbors" puzzle where there are six points A_1, A_2, A_3, B_1, B_2, B_3 in the plane and the problem is to join each of the A's to each of the B's so that no two of the joining arcs intersect except possibly in an end point of each. The generalization of this skew curve of type 1 in the next dimension is the 2-complex P with nine vertices A_1, A_2, A_3, B_1, B_2, B_3, C_1, C_2, C_3 and twenty-seven 2-simplexes of the sort $A_iB_jC_k$. Flores showed [**3**] that this complex cannot be imbedded in E^4. (An independent proof of the fact that P cannot be imbedded in E^4 can be obtained using the work of Shapiro [**5**] as indicated in §4.) In the next section we show that a certain decomposition of E^4 whose nondegenerate elements are nine circles cannot be imbedded in E^4 because it contains the polyhedron P.

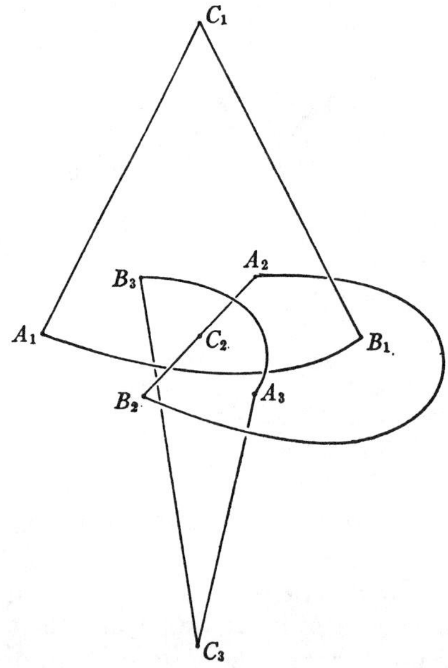

FIG. 2

2. **A construction.** We take nine points in E^3 as shown in Figure 1, and note that all of the "triangles" of the type $A_iB_jC_k$ except three can be filled in in E^3. If all such triangles are added (not in E^3, but say in E^5) the result is the Flores polyhedron P. We fill in all except the three $A_iB_iC_i$ triangles (see Figure 2), and denote the resulting polyhedron by Q.

Let J_i be the circle $A_iB_iC_i$ for $i=1, 2, 3$. The polyhedron R obtained from Q by shrinking the three J_i to points will be in the decomposition space of E^3 resulting from the decomposition having the J_i as the only nondegenerate elements. However, as we show in the next section, R can be imbedded in E^4. Goblirsch's result [5] gives an independent proof of this imbeddability. However, if we put in some more circles to shrink, we can obtain the Flores polyhedron P.

Figure 3 shows the circle $A_1B_1C_1$ with a strip S joined to it. To prevent S intersecting other 2-simplexes we cut small open disks (with boundaries β_1, β_2, β_3) out of the 2-simplexes of $A_3B_3C_1$, $A_2B_1C_2$, $A_1B_3C_2$ respectively. If the four circles β_1, β_2, β_3, β_4 are all shrunk to points, S becomes a disk as does each of the 2-simplexes from which we have removed small open disks. Hence, if we put in four such circles for each of $A_2B_2C_2$ and $A_3B_3C_3$, we get twelve circles, and when they are shrunk the resulting 2-polyhedron is the Flores polyhedron P. Thus we have proved that there exists twelve planar circles in E^3

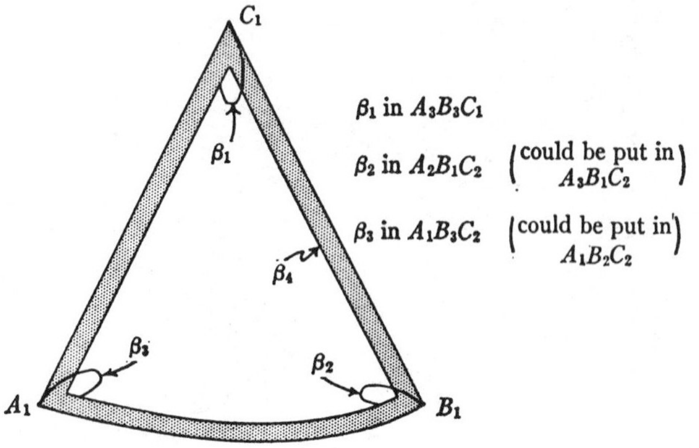

FIG. 3

such that if each is shrunk to a point, the resulting space cannot be imbedded in E^4.

When Ronald H. Rosen saw the preceding example, he suggested the following improvement of it. If we permit the curve β_4 to touch $A_1B_1C_1$ at C_1, S would no longer be an annulus but would be topologically equivalent to a disk with a circular hole in it such that the boundary of the hole is tangent to the boundary of the disk. However, S would become a disk if β_4 were shrunk to a point. An advantage in using the new S rather than the old is that no hole bounded by β_1 need be cut in $A_3B_3C_1$. Hence, this example shows the following.

THEOREM 3. *There exist nine planar circles in* E^3 *such that if each is shrunk to a point, then the resulting space cannot be imbedded in* E^4.

The nine circles referred to in Theorem 3 consist of three linking curves similar to those shown in Figure 2 together with two small circles linking each of these large ones. Two of the larger curves can be joined by a straight line interval so that the eight sets consisting of this sum and the other seven circles satisfy the hypothesis of Theorem 2. Hence, we have the following result.

THEOREM 4. *If either a* 1-*simplex or a* 2-*simplex of the Flores polyhedron* P *is shrunk to a point, the resulting set can be imbedded in* E^4.

3. **Imbedding** R **in** E^4. Returning to Figure 1, we consider the polyhedron R obtained from Q by shrinking J_1, J_2, J_3 to points. Part of the resulting polyhedron is shown in Figure 4 in which we let $V_i = A_i = B_i = C_i$. The part T of R shown in Figure 4 is obtained from R by omitting the six 2-simplexes with vertices V_1, V_2, V_3 and it consists of three 2-spheres joined at poles so as to form a loop.

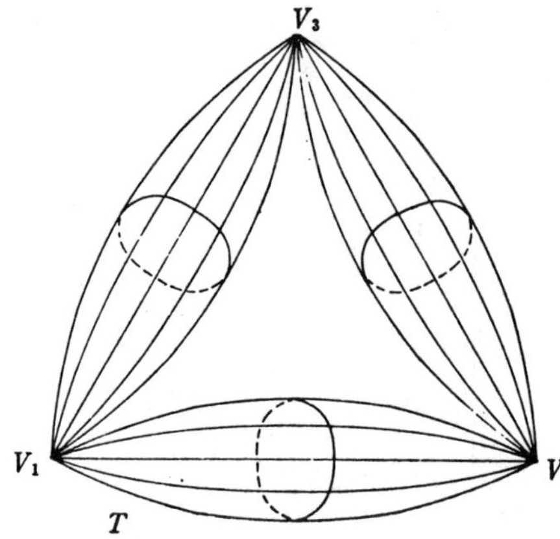

FIG. 4

Each 2-sphere is divided into six 2-cells corresponding to six 2-simplexes in Q.

Now we consider T in E^3 with E^3 in E^4, and we assert that the remaining six simplexes of R can be added to T in E^4. This is demonstrated by using the chart in Figure 5 in which the vertices have been "spread out." The six simplexes to be added to T are indicated by S_1, S_2, S_3, S_4, S_5, S_6 in Figure 5. We see in Figure 5 that S_4, S_2 and S_5 are nonintersecting. This implies that these three 2-simplexes can be added to T in E^3. Of the remaining three, S_1 and S_3 do not intersect. Hence we can add the two simplexes for these in one domain of E^4-E^3. We add S_6 in the other domain and the imbedding is completed.

Questions. We list two questions raised by this paper.

1. If three mutually exclusive bounded continua in E^3 are shrunk to points, can the resulting space be imbedded in E^4? Theorem 1 gives an answer in the case of two continua and Goblirsch's result gives an answer in case the three continua are simple closed curves appropriately linked.

2. What is the minimum integer n such that there are n mutually exclusive circles in E^3 such that if they are shrunk to points, the decomposition space cannot be imbedded in E^4? There is a gap between the integer 9 of Theorem 3 and the integer 3 in Goblirsch's result.

4. Nonimbeddability of P in E^4. Let the vertices of P be placed in general position in E^4. This will define a map $f: P\to E^4$ by extending linearly, and only pairs of images of 2-simplexes can have nonempty intersection. Let $\bar{P}^*$ denote the deleted symmetric product of P [**5**, p. 257]. The cocycle which assigns to the 4-cell $\alpha\times\beta$ of $\bar{P}^*$ the

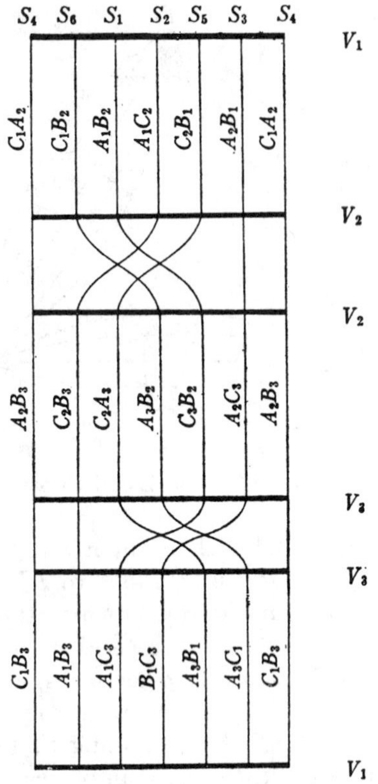

FIG. 5

intersection number of $f(\alpha)$ and $f(\beta)$ is denoted by $m_f(P)$. Its cohomology class is independent of f [5, p. 259]. Hence, if P can be imbedded in E^4, then for any f we must have that $m_f^4(P)$ is cohomologous to zero. We will choose a specific f, assume that $m_f^4(P)$ is a coboundary, and obtain a contradiction.

We place the vertices of P along a G-curve [5, p. 259] in the following order $A_1B_1C_1A_2B_2C_2A_3B_3C_3$ and extend linearly to obtain f. Now the images of two 2-simplexes (having no common vertex) will intersect if and only if their vertices alternate in the above order. If the images of α and β intersect, then $m_f^4(\alpha\times\beta)=1$ and otherwise $m_f^4(\alpha\times\beta)=0$. For P we obtain nine values one and ninety nine zero for the 108 4-cells in $\overline{P}^*$. Suppose C is a 3-cochain such that $\delta C=m_f^4(P)$. Then C can be considered to be an integral linear combination of the 324 3-cells of $\overline{P}^*$, and the equation $\delta C=m_f^4(P)$ becomes a system of 108 equations in 324 unknowns. It is a routine (but tedious) matter to show that these equations have no integral solution, so that P cannot be imbedded in E^4.

BIBLIOGRAPHY

1. R. H. Bing, *The cartesian product of a certain nonmanifold and a line is E^4*, to appear in Ann. of Math. A summary of this paper appears in the Bull. Amer. Math. Soc. vol. 64 (1958) pp. 82–84.

2. M. L. Curtis, *An imbedding theorem*, Duke Math. J. vol. 24 (1957) pp. 349–352.

3. E. Valle Flores, *Über n-dimensionale Komplexe die im R_{2n+1} absolut selbstverschlungen sind*, Ergebnisse eines Mathematischen Kolloquiums, vol. 6, pp. 4–7.

4. R. P. Goblirsch, *On decompositions of 3-space by linkages*, Proc. Amer. Math. Soc. vol. 10 (1959) pp. 728–730.

5. A. Shapiro, *Obstructions to the imbedding of a complex in a euclidean space*, I. *The first obstruction*, Ann. of Math. vol. 66 (1957) pp. 256–269.

THE UNIVERSITY OF WISCONSIN AND
THE UNIVERSITY OF GEORGIA

Decompositions† of E^3

R. H. Bing

1. INTRODUCTION

We investigate the question as to which monotone decompositions of E^3 give E^3. All decompositions are supposed to be monotone and upper semicontinuous as defined by G. T. Whyburn in the preceding lecture. In a sense, we are investigating the possibility of extending R. L. Moore's theorem [20] which states that each decomposition of E^2 into continua

† Work on this paper was supported by contract NSF–G11665.

which fail to separate E^2 yields a decomposition space topologically equivalent to E^2.

2. HUREWICZ EXAMPLE

Hurewicz showed that for each compact metric space X there is a monotone decomposition of E^3 such that X can be embedded in the resulting decomposition space. To see that this is true, consider two Cantor sets C_1, C_2 which lie on opposite edges of a tetrahedron. Let $f_i (i = 1, 2)$ be a continuous map of C_i onto X. For each point x of X let $f^{-1}(x)$ denote the sum of all intervals from points of $f_1^{-1}(x)$ to points of $f_2^{-1}(x)$. Then $f^{-1}(x)$ is connected. The upper semicontinuous decomposition of E^3 whose non-degenerate elements are $f^{-1}(x)$'s is such that the decomposition space contains a set topologically equivalent to X as a subset. We could take X to be the Hilbert cube and show that any separable metric space can be embedded in the resulting decomposition space.

3. CIRCLES AND SOLENOIDS

We give a simpler example to show that not every monotone decomposition of E^3 yields E^3. Consider the decomposition of E^3 whose only non-degenerate element is a circle. Although the decomposition is simply connected, it does not remain so on the removal of the point corresponding to this circle. Hence, the decomposition space is topologically different from E^3.

In the first edition of *Summary of Lectures and Seminars* of the Institute on Point Set Topology [24, p. 26], it is erroneously stated that the decomposition space mentioned in the preceding paragraph was not simply connected. This mistake was clerical and resulted from a careless statement of known results. It did, however, set off an investigation as to what monotone decompositions of E^3 failed to be simply connected. As a result, it was discovered [4] that if G is a monotone decomposition of E^3 whose decomposition space is not simply connected, then there is one element of G which causes this. This element could not be locally connected, could not be the intersection of a decreasing sequence $C_1, C_2, \ldots$ of continua such that C_{i+1} can be shrunk into C_i in a $1/i$ neighborhood of C_i, and would in fact be much like a solenoid. In fact, if G has only one non-degenerate element and this element is a solenoid (other than a circle), the resulting decomposition space is not simply connected.

4. REPAIRING BAD STARTS

One should not conclude, however, that if one element of a decomposition G of E^3 is bad, the resulting decomposition space is not E^3. Suppose one element of G is a circle g_0. How might one complete the decomposition space so that the resulting decomposition space is topologically E^3? In order to make the decomposition space simply connected even after a point is removed, one might have a circle g_1 which links g_0 as another non-degenerate element of G. In order to make the decomposition space simply connected even after two points are removed, one might add figure eights as shown in Figure 1, such that one loop of each of the figure eight's links g_0 and the

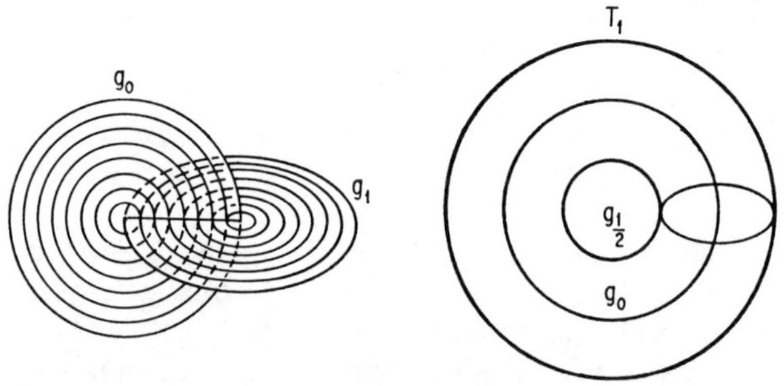

FIGURE 1

other links g_1. The sum of the figure eight's and the two circles is the sum of two perpendicular disks which intersect along a radius of each. Suppose this radius is of length 1, and we call g_t the figure eight which intersects this radius in a point whose distance from g_0 is t. Its two loops have radii t and $1 - t$. The g's are the only non-degenerate elements of G—see [24, p. 27]. In his lecture, Moise pointed out why the decomposition space of G is topologically E^3 but we give another reason.

To see that the decomposition space is E^3, put the point at infinity in E^3 to make it S^3 and regard S^3 as the sum of two solid tori T_1, T_2 where their common boundary is generated by revolving $g_{1/2}$ about the line perpendicular to the plane of g_0 through the center of g_0. As shown in Figure 1, let T_1 be the one of these solid tori with g_0 as the circular axis. Note that each g_t $(0 < t < 1/2)$ lies on a 2-dimensional torus separating this axis from $Bd\ T_1$. The decomposition of one of these 2-dimensional tori whose only non-degenerate element is the corresponding g_t is a 2-sphere. Hence, the decomposition of T_1 yields the sum of a collection of concentric 2-spheres plus their common center (g_0). Hence, T_1 corresponds to a 3-cell. So does T_2. Hence, the decomposition of E^3 whose only non-degenerate elements are g_t's $(0 \leq t \leq 1)$ yields E^3.

It may be shown that, for any finite collection of mutually exclusive non-separating compact continua in E^3, there is a decomposition of E^3 which has these continua as elements and which yields E^3. There will need to be other non-degenerate elements to counteract complications introduced by this finite collection if its elements are not already pointlike.

5. POINTLIKE DECOMPOSITIONS

One might dispense with considering decompositions whose elements are bad and only consider those that are nice. It does little good to suppose that the elements are topologically nice, since there are arcs embedded in E^3 so that their complements are not simply connected. We need to consider the embedding as well as the topological structure of the elements.

A compact continuum C in E^3 is said to be *pointlike* if $E^3 - C$ is topologically equivalent to the complement of a point in E^3. A decomposition of E^3 all of whose elements are pointlike is called a *pointlike* decomposition of E^3. Henceforth, we restrict ourselves to pointlike decompositions of E^3.

In a paper called "Continuous Transformations Preserving all Topological Properties" [23], James F. Wardwell raised the question as to whether each pointlike decomposition of a space yields the space as a decomposition space. One might wonder what mappings preserve all topological properties. A map of an object onto itself does this. The title is catchy.

In contrast to such an attractive title are the non-revealing ones which some mathematicians use. One may conjecture that many mathematicians communicate their results with a care that varies inversely with their chances of being seen. They give details of proof with great care (as they should)—though few readers may actually read these details. But abstracts presented before the society (when the author has at least a skeleton audience—perhaps a larger audience than reads less important papers) are frequently ill prepared, apparently with no purpose of making a meaningful impression. Professor Tomlinson Fort once remarked that if newspaper editors showed the same lack of imagination in titling their articles, the headline above a story about a great calamity might read "Concerning a certain unfortunate event."

6. DOGBONE SPACE

Efforts to prove that even a pointlike decomposition of E^3 yields E^3, if it meets certain conditions, may meet with a counterexample. For example, the dogbone space shown in Figure 2 was discovered when an effort was made to show that the decomposition space was E^3 if each of its non-degenerate elements is a tame arc.

To get the dogbone space, one starts with a double torus T, as shown in Figure 2, and in it considers four double tori T_1, T_2, T_3, T_4 linked as shown. The tori T_i are long and thin and such that any intersection of a horizontal plane with one of them has a small diameter. In each T_i are placed four double tori T_{i1}, T_{i2}, T_{i3}, T_{i4} in this same fashion so that the intersections of T_{ij}'s with horizontal planes are very, very small. The process is continued. Each component of the intersection of T, $\Sigma\, T_i$, $\Sigma\, T_{ij}$, ... is a tame arc which intersects no horizontal plane in more than one point. The decomposition of E^3 whose non-degenerate elements are these arcs does not yield E^3 as shown in [3].

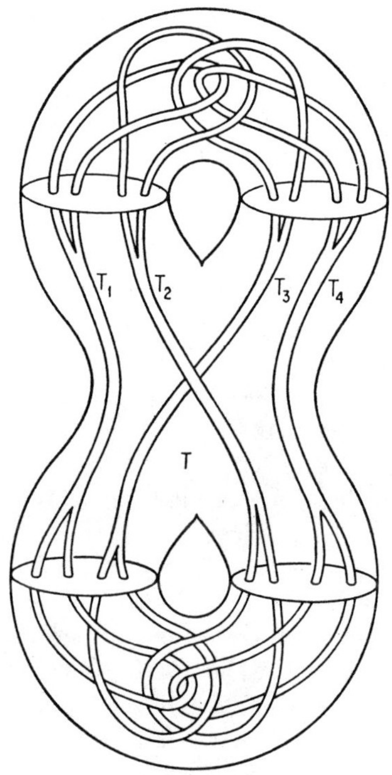

FIGURE 2

To prove that two spaces are topologically different, it is convenient to discover some simple property that one of them has but the other does not. It is less elegant to prove the lack of a homeomorphism. This less elegant method was used in [3] to show that the dogbone space was topologically different from E^3.

I had hoped that others writing about the dogbone space would give a more elegant proof that it was topologically different from E^3. Several have not availed themselves of this opportunity. Fort modified the example to make the non-degenerate elements polygonal arcs [15]. Ball used the example to get two horned spheres (not Alexander horned spheres, however) sewed together on their boundaries so that their sum was topologically different from E^3 [2]. It was shown in [7] that the cartesian product of the dogbone space with a line was topologically E^4. Rosen complicated the dogbone space in [22] to exhibit a totally non-Euclidean space whose cartesian product with a line is E^4. In each of the above treatments, however, the authors referred to [3] for the result that the dogbone space was topologically different from E^3 rather than giving a new proof.

Morton Curtis has some hope of showing that the dogbone space is topologically different from E^3 by showing that certain points of it have no small simply connected neighborhoods. At last report, however, he said that his proof was still "in limbo."

7. SOME DELICATE PROPERTIES OF E^3

Since pointlike decompositions of E^3 may share many of the algebraic properties (homotopy and homological) with E^3, one is driven to some very delicate properties to distinguish some decomposition spaces from E^3. We mention some of these properties that may be useful in this respect.

(a) Each point is separated from a closed set not containing it by a 2-sphere. The decomposition space described in Section 9 was shown in [11] not to share this property with E^3. In that example, each point may be separated from a closed set not containing it by a 2-dimensional torus but not by a 2-sphere.

E^3 has a stronger property: each point can be separated from a closed set not containing it by an uncountable family of mutually exclusive 2-spheres. (The non-degenerate elements of a decomposition tend to make 2-spheres become singular and seemingly mutually exclusive 2-spheres to intersect in the decomposition space. Note Figure 3 which shows how a non-degenerate element causes two disks to intersect. Figure 3 is described in the discussion following Property *e*.)

(b) Each point lies in a small simply connected neighborhood. As we mentioned at the end of the preceding section, Curtis hopes to show that the dogbone space does not share this property with E^3. The property is stronger than the one of being merely 1-ULC (small curves can be shrunk to points on small sets—the sets may not be as small as the curves).

(c) Dehn's lemma. Suppose f is a map of a disk D into a set X such that $Bd\,D$ contains no point of singularity of f. [A point x of D is a point of singularity if f^{-1} is not a homeomorphism on any neighborhood of $f(x)$ in $f(D)$.] We do not suppose that f is piecewise linear—this would have no meaning in many spaces X. If X is a 3-manifold, it follows from the piecewise linear version of Dehn's lemma as proved by Papakyriokopoulos [21] and the polyhedral approximation of surfaces results of [5, 6] that $f(Bd\,D)$ bounds a disk which lies near $f(D)$. One might wonder if in this non-piecewise linear version of Dehn's lemma, one could replace the condition that $Bd\,D$ contains no points of singularity with the weaker condition that f^{-1} is a homeomorphism on $f(Bd\,D)$.

(d) Each 2-sphere with a cartesian product neighborhood bounds a 3-cell. That an all-dimensional version of this holds is shown in recent results by Morton Brown [13], but the result holds in dimension 3 from earlier results of Moise [18] applied to results in the polyhedral case given by Alexander [1]. Even if a 2-sphere has only a one-sided cartesian product neighborhood from the inside, it still follows that it bounds a 3-cell. These facts, used in conjunction with Property *e* mentioned next helps us to chop up certain 3-manifolds-with-boundaries in E^3 and show they are cubes with handles.

(e) Disks which do not intersect each other's boundaries may be pulled apart in the sense that if D_1, D_2, are two disks in a 3-manifold and $D_1 \cdot Bd\, D_2 = D_2 \cdot Bd\, D_1 = 0$, then there are disks D_1', D_2' lying close to $D_1 + D_2$ such that $D_1' \cdot D_2' = 0$ and $Bd\, D_i' = Bd\, D_i$ $(i = 1, 2)$. To see that this is true, suppose that the containing 3-manifold is triangulated [19, 12] and that $D_i (i = 1, 2)$ is approximated by a disk which has the same boundary and which is locally polyhedral mod this boundary [5, 6]. Adjust the approximations so that they are in general position. The polygonal simple closed curves which lie in their intersection may be removed one at a time by adjusting one of the approximating disks.

Property (e) is one which was actually used in [3] to show that the dogbone space was topologically different from E^3 and it has promise of further use. Let us consider it further.

Although we are now restricting ourselves to pointlike decompositions, let us return briefly to the decomposition of E^3 whose only non-degenerate element is a circle C. Consider the solid torus M about C as shown in Figure 3. The simple closed curves J_1, J_2 on $Bd\, M$ correspond to curves in the decomposition space which bound disks in the part of the decomposition space corresponding to M. Try as we may, we cannot prevent these disks from intersecting in the decomposition space.

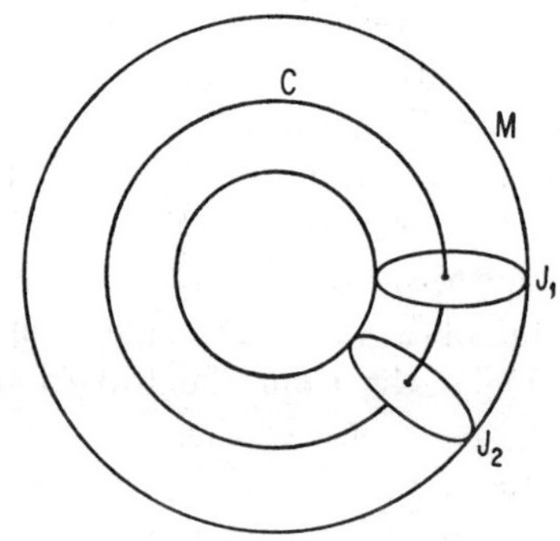

FIGURE 3

Suppose M is a connected bounded 3-manifold-with-boundary in E^3. It follows from [18] that there is no loss of generality in supposing that M (not the closure of $E^3 - M$, however) is polyhedral, so we do this. Then M is a cube with holes [16]. If J is a simple closed curve on $Bd\, M$ that can be shrunk to a point in M, it follows from Dehn's lemma that J bounds a tame disk in $(M - Bd\, M) + J$. If $Bd\, M$ is a 2-dimensional torus and contains a simple closed curve J which can be shrunk to a point in M but not in $Bd\, M$, then M is a solid torus, and each simple closed curve in $Bd\, M - J$ bounds a disk in M. Even if $Bd\, M$ is not a 2-dimensional torus, but $J_1, J_2, \ldots, J_n$ is a collection of mutually exclusive simple closed curves on $Bd\, M$ each of which can be shrunk to a point in M, then $J_1, J_2, \ldots, J_n$ bound mutually exclusive tame disks in $(M - Bd\, M) + \Sigma J_i$. Such techniques show that if $Bd\, M$ is a sphere with handles and simple closed curves circling these handles can be shrunk to a point in M, then M is a cube with handles.

The techniques cited were used in [3] to show that the dogbone space is topologically different from E^3. If it were E^3, the part of it corresponding to T of Figure 2 is topologically equivalent to T. It can then be shown that there is an isotopy of E^3 onto itself fixed outside T which shrinks all the T's at some future stage until no one of them intersects any pair of the four disks shown in Figure 2. It may be shown, however, that the T's are so well linked that if an isotopy shrinks some of them, it stretches others.

8. AN UNUSED EXAMPLE

In planning the example which resulted in the dogbone space, I was prepared to do much more complicated things. I was considering a long cube-with-many-handles T_0 (with n handles rather than the mere 2 of Figure 2) as shown in Figure 4. In it, I intended to put 2^n copies of T_0. Only two of these copies are shown and they are shown as linear although in fact they are solid. These 2^n copies would follow all possible paths from the top to the bottom of T_0, and each pair of them would link between each pair of handles of T_0. The procedure would be continued so that the intersection of the sum of the T's at the various stages would be a Cantor set of tame arcs. There would be no isotopy of E^3 onto itself that would be fixed outside T_0 and pull all of these arcs to such small sizes that no one of them would intersect any two of the disks shown. There would need to be such an isotopy if the decomposition space were topologically E^3.

In proving that the decomposition space of the example above was topologically different from E^3, it was found that more complications were introduced than were needed and that two handles sufficed. Hence, the example has not yet been put to use, since the dogbone example is simpler.

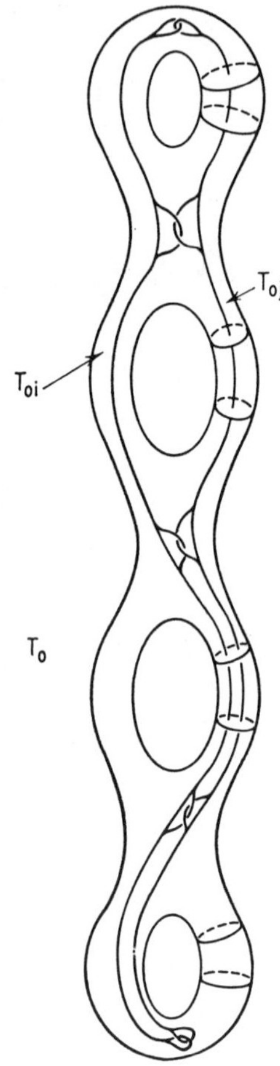

FIGURE 4

9. A DECOMPOSITION WITH COUNTABLY MANY BIG ELEMENTS

It is known that a pointlike decomposition of E^3 yields E^3 if it has only a countable number of non-degenerate elements whose sum is a G_δ set [10]. We now describe a pointlike decomposition of E^3 with only a countable number of non-degenerate elements such that the corresponding space differs from E^3. Each of the non-degenerate elements will lie in either the xy plane or the yz plane and will be pointlike because it will not separate these planes. The decomposition is described in detail in [11].

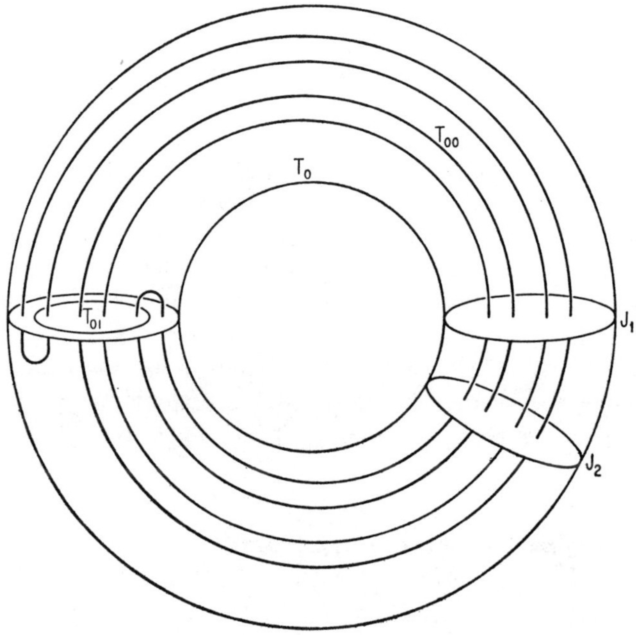

FIGURE 5

Consider a circle in the xy plane and fatten it up to a solid torus T_0 as shown in Figure 5. In the xy plane there is an arc which winds around T_0 twice. The arc is replaced by a folded solid torus T_{00} as shown in Figure 5 and a second circular solid torus T_{01} is used to link the ends of T_{00} so that when T_{00}, T_{01} are regarded as a chain, they go around T_0 twice. The diameter of T_{01} is less than one-half that of T_0, and the center axis of T_{01} lies in the xy plane. Although Figure 5 shows T_{00} and T_{01} as linear, they are indeed solid.

Just as T_{00}, T_{01} were built in T_0, there are two tori T_{0i0}, T_{0i1} considered in each T_{0i} such that their sum circles T_{0i} twice. The diameter of T_{0i1} is less than one-half that of T_{0i}. Similarly we define four T_{0ij0}'s, four T_{0ij1}'s, eight T_{0ijk0}'s, ... Let Y be the intersection of T_0, $\Sigma\, T_{0i}$, $\Sigma\, T_{0ij}$, ... Each component of Y is either a point or an indecomposable continuum. These non-degenerate indecomposable continua in Y were defined in [11] so as to lie in either the xy or xz planes, not to separate these planes, to be the sum of semicircles, and in fact, such that any two of them are homeomorphic under a similarity map. The decomposition space we consider in this section is the one whose non-degenerate elements are the non-degenerate components of Y.

A one-to-one correspondence can be set up between the components of Y and sequences of 0's and 1's which start with a 0. For example, the sequence $0, i, j, k, \ldots$, corresponds to $T_0 \cdot T_{0i} \cdot T_{0ij} \cdot \ldots$. A sequence with infinitely many 1's in it will correspond to a degenerate component of Y, since diameter $T_{0i\ldots jl} <$ (diameter $T_{0i\ldots j})/2$. Hence, Y has only a countable number of non-degenerate components.

Consider disks bounded by the closed curves J_1, J_2 of Figure 5. It can be shown—as is done in [11]—that, for any two disks which they bound, there is one component of Y such that the two disks intersect this same component. This fact can in turn be used to show that the decomposition space cannot be topologically E^3, for it would not satisfy Property e of Section 7.

It was shown in [11] that if p is a point of the decomposition space corresponding to a component of Y and X is a closed set corresponding to $Bd\ T_0$, then there is no 2-sphere in the decomposition space separating p from $Bd\ T_0$.

10. DECOMPOSITIONS INTO SEGMENTS AND POINTS

Let us consider a decomposition of E^3 into points and arcs. It is known that the decomposition space is topologically E^3 if it satisfies any one of the following conditions:

(a) The set of arcs is countable and each arc is tame [10].

(b) Each of the arcs lies in a horizontal plane [14].

(c) The arcs are all straight and point in only a countable number of directions [17].

The following example, H, was constructed in an effort to show that the decomposition of E^3 into points and straight-line segments need not yield E^3.

Although this decomposition looks enough like the decomposition described in [3] to suggest that the decomposition space may be topologically different from E^3, it looks enough like the decomposition in [9] to make one cautious in venturing such a conjecture.

A description of H. Let θ_1 and θ_2 be two horizontal planes in E^3 with θ_1 below θ_2. Each non-degenerate element of H is a straight segment with one end in θ_1 and the other in θ_2. Each horizontal plane that intersects one of the segments will intersect the sum of the segments in a topological Cantor set.

The sum of the non-degenerate elements of H is the intersection of a decreasing sequence of open sets $U_1, U_2, \ldots$. Each U_i is a tubular neighborhood of a finite graph G_i. Figure 6 shows G_1. We describe it as follows:

Let $Q(1)$, $Q(2)$, $Q(3)$, $Q(4)$ be four points in θ_1 and $P(1)$, $P(2)$ be two points in θ_2 such that the line through $P(1)$, $P(2)$ is not parallel to any line through any two of the Q's. We use $[P(i), Q(j)]$ to denote the straight segment from $P(i)$ to $Q(j)$. Then G_1 is the sum of the eight $[P(i), Q(j)]$'s as shown in Figure 6. U_1 is a tubular neighborhood of G_1.

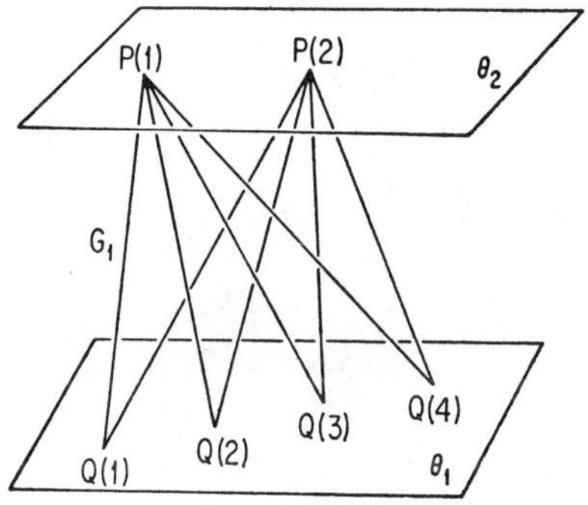

FIGURE 6

We obtain G_2 by replacing each major segment $[P(i), Q(j)]$ in G_1 by a copy of G_1. The four copies of G_1 in G_2 that are near $P(i)$ ($i = 1, 2$) link each other, and the two copies of G_1 in G_2 near $Q(j)$ ($j = 1, 2, 3, 4$) cross in a manner to be described presently. Before describing the eight copies of G_1 in G_2 in detail, we remark that their sum lies in U_1. Just as we obtained G_2 by replacing each major segment in G_1 by a copy of G_1, we obtain G_3 by replacing each major segment of G_2 by a copy of G_1. Continuing in this fashion, we get a sequence of finite graphs $G_1, G_2, G_3, G_4, \ldots$ and a sequence of tubular neighborhoods $U_1, U_2, U_3, U_4, \ldots$ of these graphs such that $G_{i+1} \subset U_{i+1} \subset \bar{U}_{i+1} \subset U_i$. The tubular neighborhoods get progressively so thin that the intersection of the U_i's is the sum of a Cantor set of segments each with one end on θ_1 and the other on θ_2. These segments are the non-degenerate elements of the decomposition H.

The copy of G_1 in G_2 replacing $[P(i), Q(j)]$ is denoted by $G(i, j)$. It is the sum of eight straight segments $[P(i, j, m), Q(i, j, n)]$ ($m = 1, 2$; $n = 1, 2, 3, 4$), where the $P(i, j, m)$'s are points in θ_2 near $P(i)$ and the $Q(i, j, n)$'s are points in θ_1 near $Q(j)$. We shall describe these P's and Q's presently, but first we show how we want $G(i, j)$ and $G(i, k)$ to entwine near $P(i)$.

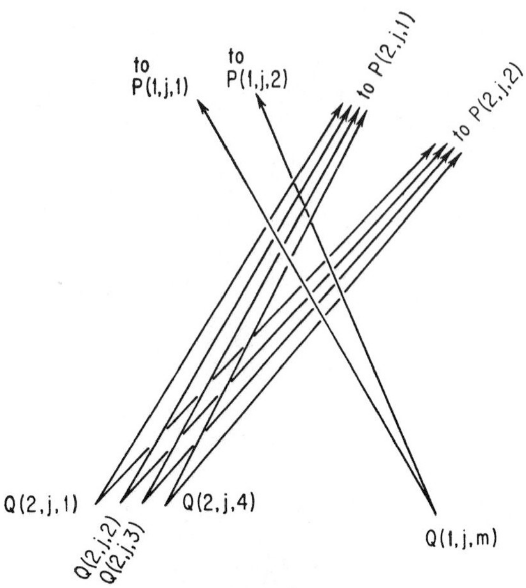

FIGURE 7

Entwining near P(i). Figure 7 shows how certain segments of the $G(1, j)$'s look in perspective near $P(1)$. The general rule is that for $j < k$, $[P(i, j, s), Q(i, j, t)]$ passes between the eye and $[P(i, k, r), Q(i, k, u)]$ for $t \leq u$, and $[P(i, k, r), Q(i, k, u)]$ passes between the eye and $[P(i, j, s), Q(i, j, t)]$ for $u < t$. $P(i, j, 1)$ and $P(i, j, 2)$ are so close together— as are $P(i, k, 1)$ and $P(i, k, 2)$—that the crossings are the same irrespective of whether or not r (or s) is 1 or 2.

Entwining near Q(j). Figure 8 shows how one broken arc in $G(1, j)$ entwines with four broken arcs in $G(2, j)$ near $Q(j)$. Each of the broken arcs in $G(1, j)$ entwines with these four arcs in $G(2, j)$ in this fashion, but we did not show this in Figure 8 to avoid complication. Note that $[P(1, j, 1), Q(1, j, m)]$ passes between the eye and $[P(2, j, 1), Q(2, j, n)]$ $(n = 1, 2, 3, 4)$, $[P(2, j, 1), Q(2, j, n)]$ passes between the eye and $[P(1, j, 2), Q(1, j, m)]$, and $[P(i, j, 2), Q(1, j, m)]$ passes between the eye and $[P(2, j, 2), Q(2, j, n)]$.

It is shown in [11] that we can really get $G(i, j)$'s that entwine as we have suggested.

A thing that makes one suspect that the decomposition space of H is topologically different from E^3 is that if one shrinks some G's, it seems that others will be stretched. It was shown in [11] that if the decomposition space were topologically equivalent to E^3, then there is an isotopy

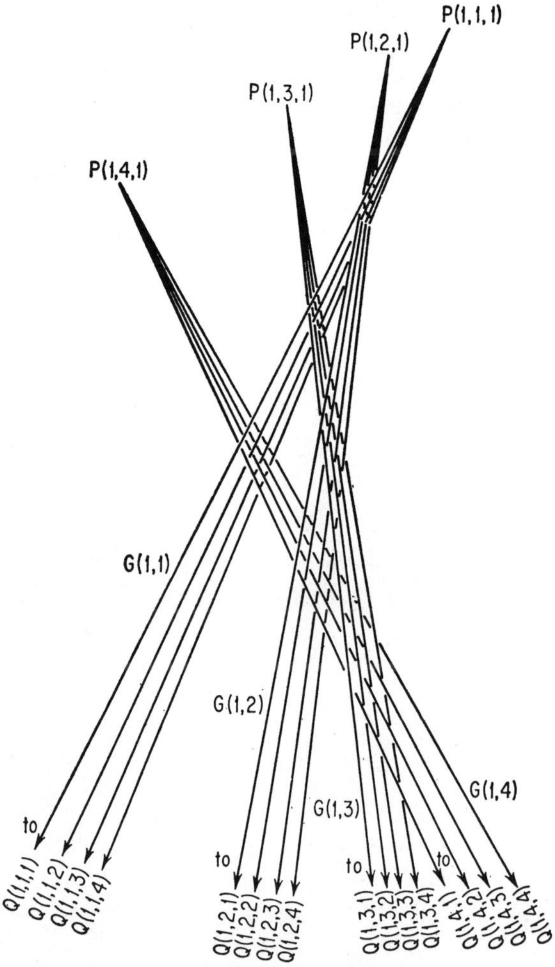

FIGURE 8

fixed outside U_1 which shrinks each G at a later stage to an arbitrarily small size.

11. VERTICALIZING SEGMENTS

Suppose G is a continuous collection of mutually exclusive spanning segments of a unit cube such that each of the segments has one end on the top of the cube and one on the bottom. Is there a homeomorphism of the cube onto itself that takes each of these segments to a vertical position? One is reminded of the game of pick up sticks.

R. H. Fox pointed out an isotopy that works in certain cases. Suppose the bottom base of the cube lies in the xy plane and that, for an element g of G, the intersections of g with the planes $z = 0$ and $z = 1$ are $p(g)$ and $q_0(g)$ respectively, as shown in Figure 9. Let $q_y(g)$ be the intersection of the plane $z = 1$ with the segment from $p(g)$ to a point $t/(1 - t)$ units directly above $q_0(g)$. Let g_t be the segment from $p(q)$ to $q_t(g)$. As t increases from 0 toward 1, g_t approaches a vertical position. No two of the g_t's intersect at a particular time t, since we have merely changed the scale on the z axis.

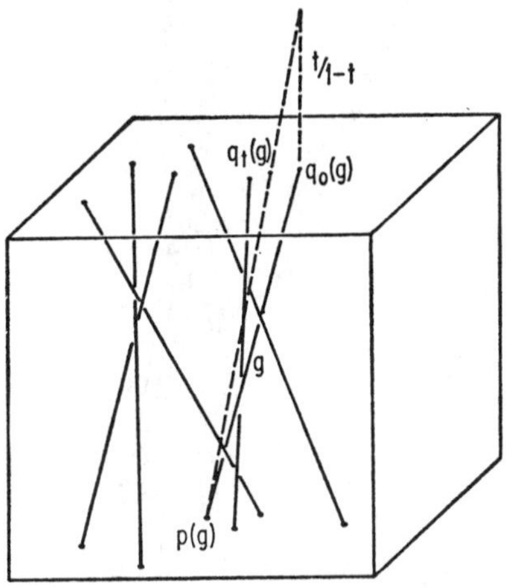

FIGURE 9

If there are only a finite number of spanning segments to be adjusted, the movement we have described can be extended to an isotopy of the cube onto itself. Problems arise, however, when there are infinitely many arcs and the question raised in the first paragraph of this section has a negative answer.

Consider a cube which has as spanning arcs the segments described in Section 10. There is a simple closed curve J on the interior of this cube which misses these segments but which is not homotopic in the cube minus the segments to a curve on the boundary of the cube. If there were a homeomorphism h of the cube onto itself that takes the segments into vertical positions, there would be a homotopy $H_t(0 \leq t \leq 1)$ pulling $h(J)$ onto the base of C such that, in the pulling, each point only moves straight down. Then $h^{-1}H_t h$ violates the condition that J is not homotopic in the cube minus the segments to a curve on the boundary of the cube.

In the decomposition space of Section 10, the set of non-degenerate elements corresponds to a Cantor set with a non-simply connected complement. This alone is not enough to prevent the decomposition space from being E^3, since we have known from the time of Antoine that there were such Cantor sets.

12. *Questions*

(a) Suppose G is a pointlike decomposition of E^3 with only a countable number of non-degenerate elements and each is an arc (perhaps wild). Is the decomposition space topologically E^3? (During the last week of the institute, David Gillman and Joseph Martin produced a strong candidate for a counterexample.)

(b) Suppose G is a decomposition of E^3, each of whose non-degenerate elements is a segment. Is the decomposition space topologically E^3?

(c) Does a pointlike decomposition of E^3 yield E^3 if it yields a 3-manifold?

(d) Does each pointlike decomposition of E^3 yield an absolute retract?

(e) Is the cartesian product of a line with the decomposition space of Section 9 topologically E^4? The space of any pointlike decomposition?

[John Hempel has shown that the cartesian product of a line with the decomposition space of Section 10 is topologically E^4. In October following the institute, Jim Andrews sent me the outline of a proof which suggests that the cartesian product of a line with the decomposition space of Section 9 is E^4.]

(f) Let $G_i (i = 1, 2)$ be a pointlike decomposition of E^3 such that each G_i yields E^3 and if a non-degenerate element of G_1 intersects a non-degenerate element of G_2, the elements are the same. Let G be the decomposition of E^3 whose non-degenerate elements are the non-degenerate elements of G_1 and G_2. Does G yield E^3?

(g) Suppose G is a decomposition of E^3 into points and tame arcs such that, for each element g of G and each closed set X not intersecting g, there is a 2-dimensional torus separating g from X and intersecting no non-degenerate element of G. Does G yield E^3? The techniques used in proving that the sum of two Alexander horned spheres sewed together on their boundaries with the identity mapping was topologically E^3 may apply [9].

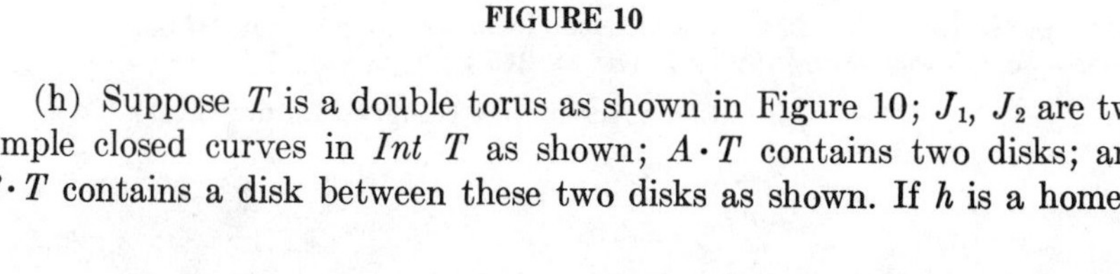

FIGURE 10

(h) Suppose T is a double torus as shown in Figure 10; J_1, J_2 are two simple closed curves in *Int* T as shown; $A \cdot T$ contains two disks; and $B \cdot T$ contains a disk between these two disks as shown. If h is a homeomorphism of E^3 onto itself that is fixed on $E^3 - T$, need one of $h(J_1)$, $h(J_2)$ contain an arc with four points of $A + B$ such that no two of these points which are adjacent on the arc belong to the same one of A and B? If an affirmative answer could be obtained, methods used in getting it might be useful as suggested in [11] in showing that the decomposition space H of Section 10 does not yield E^3.

REFERENCES

1. Alexander, J. W., "On the subdivision of 3-space by a polyhedron," *Proc. Nat. Acad. Sci., U.S.A.*, **10** (1924), 6–8.

2. Ball, B. J., "The sum of two solid horned spheres," *Ann. Math.*, **69** (1959), 253–57.

3. Bing, R. H., "A decomposition of E^3 into points and tame arcs such that the decomposition space is topologically different from E^3," *Ann. Math.*, **65** (1957), 484–500.

4. ———, "Conditions under which monotone decompositions of E^3 are simply connected," *Bull. Amer. Math. Soc.*, Abstract 325, **63** (1957), 143.

5. ———, "Approximating surfaces with polyhedral ones," *Ann. Math.*, **61** (1957), 456–83.

6. ———, "Approximating surfaces from the side," *Ann. Math.* (in press).

7. ———, "The cartesian product of a certain nonmanifold and a line is E^4," *Ann. Math.*, **70** (1959), 399–412.

8. ———, "Locally tame sets are tame," *Ann. Math.*, **59** (1954), 145–58.

9. ———, "A homeomorphism between the 3-sphere and the sum of two solid horned spheres," *Ann. Math.*, **56** (1952), 354–62.

10. ———, "Upper semicontinuous decompositions of E^3," *Ann. Math.*, **65** (1957), 363–74.

11. ———, "Point-like decompositions of E^3," *Fundamenta Mathematicae* (in press).

12. ———, "An alternative proof that 3-manifolds can be triangulated," *Ann. Math.*, **69** (1959), 37–65.

13. Brown, Morton, "A proof of the generalized Schoenflies theorem," *Bull. Amer. Math. Soc.*, **66** (1960), 74–76.

14. Dyer, Eldon and Mary-Elizabeth Hamstrom, "Certain completely regular mappings," *Fundamenta Mathematicae*, **45** (1958), 103–18.

15. Fort, M. K., Jr., "A note concerning a decomposition space defined by Bing," *Ann. Math.*, **65** (1957), 501–04.

16. Fox, R. H., "On the imbedding of polyhedra in 3-space," *Ann. Math.*, **49** (1948) 462–70.

17. McAuley, L. F., "Some upper semicontinuous decompositions of E^3 into E^3," *Ann. Math.*, **73** (1961), 437–57.

18. Moise, E. E., "Affine structures in 3-manifolds, IV: Piecewise linear approximations of homeomorphisms, *Ann. Math.*, **55** (1952), 215–22.

19. ———, V., "The triangulation theorem and Hauptvermutung," *Ann. Math.*, **56** (1952), 96–114.

20. Moore, R. L., "Concerning upper semicontinuous collections of continua," *Trans. Amer. Math. Soc.*, **27** (1925), 416–28.

21. Papakyriokopoulos, "On Dehn's lemma and the asphericity of knots," *Ann. Math.*, **66** (1957), 1–26.

22. Rosen, Ronald, "Imbeddings of decompositions of 3-space," (Unpublished doctoral dissertation, University of Wisconsin, August, 1959).

23. Wardwell, James F., "Continuous transformations preserving all topological properties, *Amer. Jour. Math.*, **58** (1936), 709–26.

24. *Summary of Lectures and Seminars,* Summer Institute on Set Theoretic Topology, University of Wisconsin, Madison, Wisc., 1955.

FUNDAMENTA
MATHEMATICAE
L (1962)

Point-like decompositions of E^3

by

R. H. Bing (Wisconsin) *

1. Introduction. An upper semicontinuous decomposition G of E^n (Euclidean n space) is *point-like* if for each element g of G, E^n-g is topologically equivalent to the complement of a point. We call such an upper semicontinuous decomposition a *point-like decomposition*.

If G is a point-like decomposition of E^2 (or E^1), the resulting decomposition space is topologically E^2 (or E^1) [17]. However, there are examples of point-like decompositions of E^3 whose decomposition spaces are topologically different from E^3. We give another such example in this paper and suggest a decomposition which may be an example.

Suppose that G is a point-like decomposition of E^3 and H is the collection of nondegenerate elements of G. It is known that G yields E^3 if it satisfies any one of the the following conditions:

1) Each element of G is a starlike body (Theorem 2 of [4]).

2) Each element of H lies in a horizontal plane [11].

3) There is a countable number of planes such that each element of H is an interval normal to one of these planes [14].

There are examples ([3], [12]) which show that the union of the following two conditions is not enough to ensure that G yields E^3.

4) Each element of H is a tame arc.

5) The sum of the elements of H is a G_δ set.

There are no published examples which resolve the question as to whether or not either of the following conditions implies that G yields E^3.

6) Each element of H is a straight line interval.

7) H is countable.

In this paper we give an example to show that Condition 7 does not imply that G yields E^3 and another example to cast doubt on the potency of Condition 6.

It is known (Theorems 3, 1 of [4]) that Condition 7 together with any one of Conditions 4, 5, 6 is enough to insure that G yields E^3. Open are

* Work on this paper was supported by contract NSF-G11665.

the effects of imposing such conditions as that H is a family of planar disks or that H is a countable family of arcs (not necessarily tame) or even continuous curves.

2. An indecomposable plane continuum. Each nondegenerate element of the point-like decomposition of E^3 that we describe in Section 3 is a copy of the following plane continuum X which is described as the intersection of a countable number of annuli $A_1, A_2, A_3, \ldots$

The annulus A_1 shown in Figure 1 is bounded by two concentric circles. The annulus A_2 plus its bounded complementary domain lies in A_1 and wraps around A_1 twice with the ends of the "wraps" near each other on the left side of A_1 as shown in Figure 1. The annulus A_3 (only part of which is shown) wraps around A_2 in a similar fashion. Similarly, A_4 wraps around A_3, A_5 around A_4, ... The continuum $X = A_1 \cdot A_2 \cdot A_3 \cdot \ldots$ is the sum of a collection of semicircles.

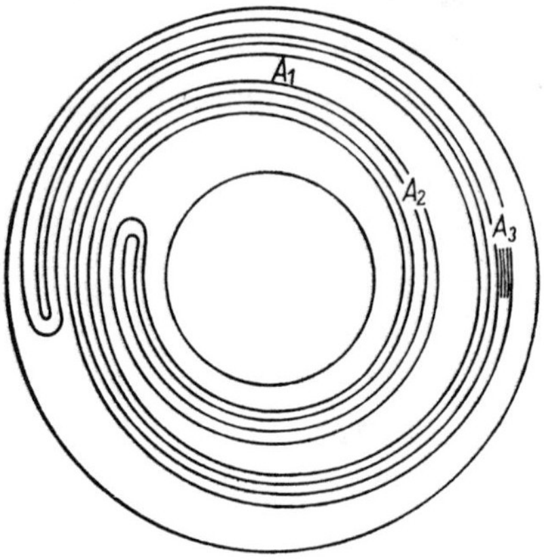

Fig. 1

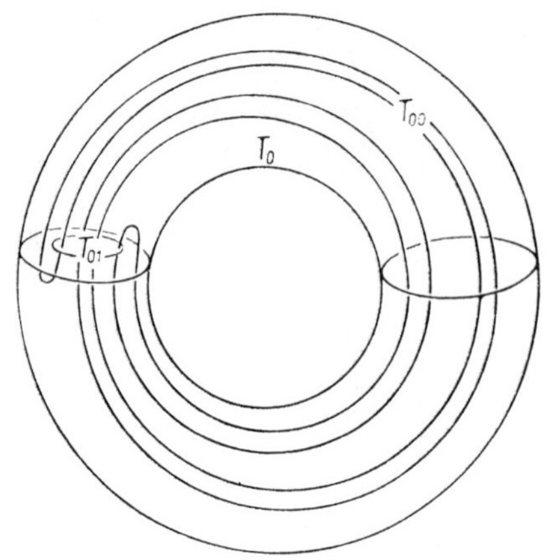

Fig. 2

Note that X is also the intersection of a sequence of topological disks. If X lies in a plane in E^3, it is the intersection of a sequence of tame cubes. Hence, it is point-like in E^3. Also, X may be regarded as the intersection of a sequence of solid tori in E^3. It is in this light that it is viewed in the next section.

3. A decomposition of E^3. In this section we describe a point-like decomposition G of E^3 such that G has only a countable number of nondegenerate elements, each of the nondegenerate elements is a copy of the indecomposable continuum X described in Section 2 and the decomposition space is topologically different from E^3. The proof that the decomposition space is different from E^3 is found in Section 5. The nondegenerate elements of G are described as the intersection of a sequence of tori.

Suppose that T_0 is a round solid torus as shown in Figure 2. Inside T_0 are two solid tori T_{00}, T_{01} linked as shown. These are solid even though they are shown as linear. Their sum essentially goes around T_0 twice. The diameter of T_{00} is comparatively large but the diameter of T_{01} is less than one half that of T_0. The center axis of T_{00} lies in the same plane as the axis of T_0 but the axis of T_{01} lies in a plane perpendicular to this one.

Just as T_{00} and T_{01} are built in T_0, so is a T_{0i0} and a T_{0i1} built in each T_{0i}. Similarly we define four T_{0ij0}'s, four T_{0ij1}'s, eight T_{0ijk0}'s, ... Each component of $Y = T_0 \cdot \sum T_{0i} \cdot \sum T_{0ji} \cdot \sum T_{0ijk} \cdot ...$ is an element of G. Also, each point of $E^3 - Y$ is an element of G.

Note that Y has a Cantor set of components. If $a_1a_2a_3...$ is an infinite decimal where each a_i is 0 or 1, we may let this decimal correspond to the component $T_{0a_1} \cdot T_{0a_1a_2} \cdot T_{0a_1a_2a_3} \cdot ...$ If the decimal has infinitely many 1's, the diameter of the component is 0. Hence, G has only a countable number of nondegenerate elements. We could define G so that each of these nondegenerate components either lies in the xy-plane or the yz-plane and any two of them are congruent under a similarity homeomorphism.

4. The intersection of disks with tori. In this section we consider the intersections of certain disks with the tori used in Section 3 to define Y. Although these tori were drawn as curved, we would get the same set Y if we had used polyhedral tori. We suppose in this section that each of T_0, T_{00}, T_{01} has a polyhedral boundary (is the sum of a finite number of triangular disks). We use $\operatorname{Bd} T$ to denote the boundary of solid torus T and $\operatorname{Int} T$ to denote $T - \operatorname{Bd} T$. For a disk D we use $\operatorname{Bd} D$ to denote its edge and $\operatorname{Int} D$ to denote $D - \operatorname{Bd} D$.

We shall consider the intersections of various polyhedral disks D with T_0, T_{00}, T_{01}. Each of the disks will be in general position with respect to $\operatorname{Bd} T$ ($T = T_0, T_{00}, T_{01}$) in that if p is a point of $D \cdot \operatorname{Bd} T$, there is a neighborhood U of p and a homeomorphism h of U onto E^3 such that $h(U \cdot \operatorname{Bd} T)$ is the $z = 0$ plane and $h(U \cdot D)$ is either the $x = 0$ plane or the $x = 0$, $z \geqslant 0$ halfplane. This ensures that each component of $D \cdot \operatorname{Bd} T$ is a polygonal simple closed curve.

PROPERTY P. *The disk D has Property* P *with respect to T_0 if D is a polyhedral disk in general position with respect to T_0, T_{00}, T_{01} and* $\operatorname{Bd} D$ *is a simple closed curve on* $\operatorname{Bd} T_0$ *that circles* $\operatorname{Bd} T_0$ *meridianally.*

Note that Property P implies that some component of $D \cdot T_0$ is a punctured disk such that the outer boundary component (with respect to $\operatorname{Bd} D$) of the punctured disk circles $\operatorname{Bd} T_0$ meridianally and the other boundary components (if any) bound disks on $\operatorname{Bd} T_0$. To get such a component one could consider the various simple closed curves on $D \cdot \operatorname{Bd} T_0$ that circle $\operatorname{Bd} T_0$ meridianally, consider one of these which bounds a disk D' of D which contains no other one of the simple closed curves, and finally take the component of $D' \cdot T_0$ which contains $\mathrm{Bd}\, D'$.

THEOREM 1. *If D is a polyhedral disk in general position with respect to* $\mathrm{Bd}\, T$ *and J is a component of $D \cdot \mathrm{Bd}\, T$, then J either bounds a disk on* $\mathrm{Bd}\, T$ *or J circles* $\mathrm{Bd}\, T$ *once longitudinally and no times meridianally, or J circles* $\mathrm{Bd}\, T$ *once meridianally and no times longitudinally.*

Proof. Suppose that J does not bound a disk on $\mathrm{Bd}\, T$. Let J' be an inner such simple closed curve on D with respect to this property in the sense that J' is a simple closed curve on $D \cdot \mathrm{Bd}\, T$ which does not bound a disk on $\mathrm{Bd}\, T$ but the subdisk D' of D bounded by J' contains no other such simple closed curve.

The closure of the component of $D' - \mathrm{Bd}\, T$ containing J' is a punctured disk such that the boundary of each hole bounds a disk on $\mathrm{Bd}\, T$. If the holes in this punctured disk are filled by disks near $\mathrm{Bd}\, T$ and then the resulting disk is pushed slightly away from $\mathrm{Bd}\, T$ near these fill-ins, there results a disk D'' which does not intersect $\mathrm{Bd}\, T$ except in its boundary $\mathrm{Bd}\, D'' = \mathrm{Bd}\, D' = J'$. If $\mathrm{Int}\, D'' \subset T$, J' circles $\mathrm{Bd}\, T$ meridianally and if $\mathrm{Int}\, D'' \cdot T = 0$, J' circles $\mathrm{Bd}\, T$ longitudinally. Any other simple closed curve (such as J) on $\mathrm{Bd}\, T - J'$ which does not bound a disk on $\mathrm{Bd}\, T$ would circle $\mathrm{Bd}\, T$ in the same fashion. This is because $\mathrm{Bd}\, T - J'$ is an open annulus and any simple closed curve on an annulus which does not bound a disk on the annulus circles the annulus exactly once.

THEOREM 2. *The fundamental group of $E^3 - (T_{00} + T_{01})$ is a free group on two generators. A simple closed curve that circles* $\mathrm{Bd}\, T_0$ *meridianally cannot be shrunk to a point in $E^3 - (T_{00} + T_{01})$.*

Proof. If we were only going to prove the first half of the theorem we would merely note that there is a homeomorphism of E^3 onto itself that takes T_{00} and T_{01} onto two nonlinking circular solid tori. It is a bit complicated to see where a meridianal simple closed curve on $\mathrm{Bd}\, T_0$ goes under this homeomorphism so we formally compute the fundamental group of $E^3 - (T_{00} + T_{01})$.

The fundamental group of $E^3 - (T_{00} + T_{01})$ can be read from Figure 3. Where the eye is regarded as the starting point, the letters on the arrows represent a path starts at the eye and goes under T_{00} or T_{01} in the direction of the arrow and returns to the starting point, and relations result at perspective crossing points of T_{00} and T_{01}. If one admissible polygonal path is homotopic to another, it can be deformed in a series of jumps so that no jump involves going past more than one perspective crossing. Hence the word corresponding to the first path can be changed by a finite number of the relations into the word representing the other path.

As read from Figure 3 the fundamental group of $E^3 - (T_{00} + T_{01})$ is given by generators a, b, c, d, x, y, z, w and relations $xd = ax$, $yd = dx$,

$bz = yb$, $cz = wc$, $wc = cx$, $xb = ax$, $dx = xc$, $xc = bx$. The relations merely imply that $a = xbx^{-1}$, $c = x^{-1}bx$, $d = b$, $y = bxb^{-1}$, $z = x$, $w = x^{-1}bxb^{-1}x$. Hence the group is generated by b and x with no relations.

As may be seen by examining the simple closed curve J on the right, a simple closed curve circling $\mathrm{Bd}\,T_0$ meridianally may correspond to the element $cb^{-1}da^{-1} = x^{-1}bx^2b^{-1}x^{-1}$. Since this element is non-trivial in the free group, J cannot be shrunk to a point in $E^3-(T_{00}+T_{01})$.

THEOREM 3. *If D is a disk with Property* P *with respect to T_0, then either $D\cdot\mathrm{Bd}\,T_{00}$ contains a simple closed curve that circles $\mathrm{Bd}\,T_{00}$ meridianally or $D\cdot\mathrm{Bd}\,T_{01}$ contains a simple closed curve that circles $\mathrm{Bd}\,T_{01}$ meridianally.*

Proof. First we show that there is no loss of generality in supposing that no simple closed curve on $D\cdot\mathrm{Bd}\,T_{00}$ bounds a disk on $\mathrm{Bd}\,T_{00}$ and no simple closed curve on $D\cdot\mathrm{Bd}\,T_{01}$ bounds a disk on $\mathrm{Bd}\,T_{01}$. We do this by eliminating unwanted simple closed curves by a method used in the proof of Theorem 1. This elimination is accomplished by considering the punctured disk which is the closure of the component which contains $\mathrm{Bd}\,D$ of D minus the sum of the simple closed curves of $D\cdot(\mathrm{Bd}\,T_{00}+\mathrm{Bd}\,T_{01})$ which bound disks on $\mathrm{Bd}\,T_{00}+\mathrm{Bd}\,T_{01}$, then one-at-a-time (starting with a hole bounded by an inner most simple closed curve on $\mathrm{Bd}\,T_{00}+\mathrm{Bd}\,T_{01}$) fill in the holes of this punctered disk with disks on $\mathrm{Bd}\,T_{00}+\mathrm{Bd}\,T_{01}$, and push these fill-ins to one side of $\mathrm{Bd}\,T_{00}+\mathrm{Bd}\,T_{01}$. Hence we suppose that if J is a component of $D\cdot(\mathrm{Bd}\,T_{00}+\mathrm{Bd}\,T_{01})$, then J does not bound a disk on $\mathrm{Bd}\,T_{00}+\mathrm{Bd}\,T_{01}$.

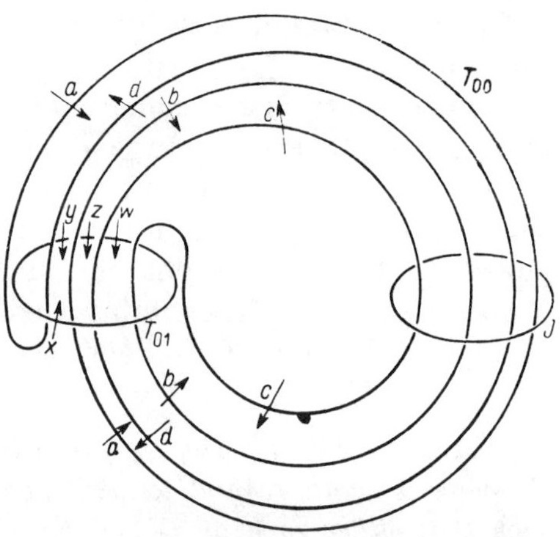

Fig. 3

If a component J of $D\cdot\mathrm{Bd}\,T_{00}$ is a simple closed curve that circles $\mathrm{Bd}\,T_{00}$ longitudinally, then to one side of J there is a simple closed curve K on $\mathrm{Bd}\,T_{00}$ that circles $\mathrm{Bd}\,T_{00}$ longitudinally and which misses D. There is a homeomorphism h of E^3 onto itself that shrinks T_{00} to so near K that h is the identity on $T_{01}+\mathrm{Bd}\,T_0$ and $D\cdot h(T_{00}) = 0$. Hence we suppose that D misses T_{00}.

In a similar fashion, it can be shown that unless Theorem 3 is true we may assume that D misses T_{01} also. However, this would contradict Theorem 2 which says that $\mathrm{Bd}\,D$ cannot be shrunk to a point in $E^3-(T_{00}+T_{01})$.

Theorem 4. *If D is a disk with Property* P *with respect to T_0, then D contains two mutually exclusive subdisks such that each of the subdisks has Property* P *with respect to T_{00} or each of the subdisks has Property* P *with respect to T_{01}.*

Proof. As in the proofs of Theorems 1 and 3 we adjust D so that the adjusted disk contains no simple closed curve that bounds a disk on either $\mathrm{Bd}\,T_{00}$ or $\mathrm{Bd}\,T_{01}$. If we can find two disks satisfying the conclusion of Theorem 4 on the adjusted disk, we can find two on D. For convenience in notation we suppose that the adjusted disk is D. It follows from Theorem 3 that D contains a disk D' that has Property P with respect to either T_{00} or T_{01}. We suppose that D' has Property P with respect to T_{00}. Furthermore, we suppose that D' is maximal in the sense that it is not a proper subdisk of any other subdisk of D that has Property P with respect to T_{00}. Note that although $\mathrm{Bd}\,D$ can be shrunk to a point in the complement of T_{00}, $\mathrm{Bd}\,D'$ cannot be shrunk to a point in the complement of the central curve of T_{00}.

Since $\mathrm{Bd}\,D'$ is not homotopic to $\mathrm{Bd}\,D$ in $(E^3 - \mathrm{Int}\,T_{00})$, $D - D'$ contains a simple closed curve J such that J circles $\mathrm{Bd}\,T_{00}$ meridianally. To see that this is true consider the annulus A in D bounded by $\mathrm{Bd}\,D$ and $\mathrm{Bd}\,D'$. Some point of A must lie in $\mathrm{Int}\,T_{00}$ since $\mathrm{Bd}\,D'$ can be pulled to $\mathrm{Bd}\,D$ on A but not in $E^3 - \mathrm{Int}\,T_{00}$. Since $\mathrm{Bd}\,D$ does not intersect T_{00}, $A \cdot \mathrm{Bd}\,T_{00}$ contains a component J other that $\mathrm{Bd}\,D'$. The disk in D bounded by J is another disk in D that has Property P with respect to T_{00}.

Theorem 5. *If D_1, D_2 are two mutually exclusive disks each with Property* P *with respect to T_0, then D_i $(i = 1, 2)$ contains a subdisk D_i' such that either each of D_1', D_2' has Property* P *with respect to T_{00} or each of D_1', D_2' has Property* P *with respect to T_{01}.*

Proof. We shall show that D_i $(i = 1, 2)$ contains a simple closed curve J_i such that either J_1 and J_2 lie on $\mathrm{Bd}\,T_{00}$ and circle T_{00} meridianally or J_1 and J_2 lie on $\mathrm{Bd}\,T_{01}$ and circle T_{01} meridianally. Then D_i' will be the subdisk of D_i bounded by J_i.

Following an argument similar to that employed in the proofs of Theorems 1 and 3, we assume with no loss of generality that no simple closed curve on D_i bounds any disk on either $\mathrm{Bd}\,T_0$, $\mathrm{Bd}\,T_{00}$, $\mathrm{Bd}\,T_{01}$. Furthermore we suppose that $D_i \cdot \mathrm{Bd}\,T_0 = \mathrm{Bd}\,D_i$. Then $\mathrm{Int}\,D_i \subset \mathrm{Int}\,T_0$.

If D_1 contains a simple closed curve that circles $\mathrm{Bd}\,T_{00}$ longitudinally, there is a simple closed curve K on $\mathrm{Bd}\,T_{00}$ that circles T_{00} longitudinally such that K misses each of D_1 and D_2. There is then a homeomorphism h of E^3 onto itself fixed on $\mathrm{Bd}\,T_0$ and T_{01} that pulls T_{00} so near K that $(D_1 + D_2) \cdot h(T_{00}) = 0$. It then follows from Theorem 3 that each D_1, D_2 contains a simple closed curve that circles $\mathrm{Bd}\,T_{01}$ meridianally. Hence, we suppose that neither D_1 nor D_2 contains any simple closed curve that circles either Bd T_{00} or Bd T_{01} longitudinally.

Since it follows from Theorem 4 that each D_i intersects one of T_{00}, T_{01}, we assume with no loss of generality that D_1 intersects T_{00}. The condition of Theorem 5 is satisfied if D_2 intersects T_{00} so we suppose that D_2 misses T_{00} and intersects T_{01}. Also we suppose that D_1 misses T_{01}. We complete the proof of Theorem 5 by showing that it is impossible that $D_1 \cdot T_{01} = D_2 \cdot T_{00} = 0$.

Consider the universal covering space of T_0. It is represented by Figure 4 where it appears that T_0 has been rolled out to make infinitely many copies of T_{00} and T_{01}. Let T'_{01}, T''_{01}, T'''_{01} be three adjacent copies of T_{01} and T'_{00} be the copy of T_{00} that links T'_{01} and T'''_{01} as shown. It follows from Theorem 3 of [3] that there is an arc A'_1 from T'_{01} to T'_{00} and an arc A'_2

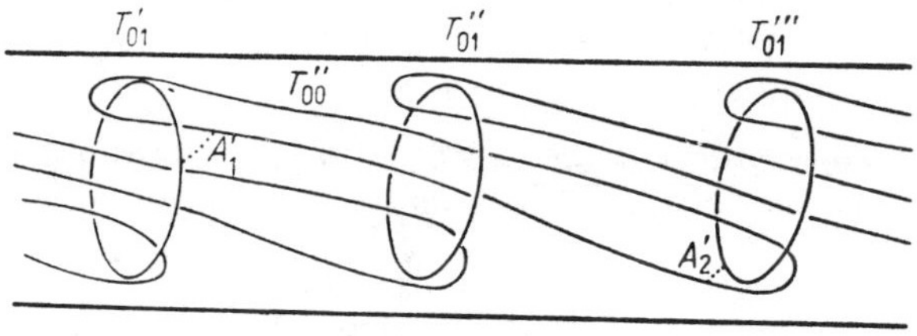

Fig. 4

from T'_{00} to T'''_{01} such that these arcs miss the various copies of D_1 and D_2. With no loss of generality we suppose that the projections of A'_1 and A'_2 into T_0 are mutually exclusive arcs A_1 and A_2.

Let B_1 be an arc in T_{00} from the last point of A_1 to the first point of A_2 and B_2 be an arc in T_{01} from the last point of A_2 to the first point of A_1. We suppose the B's intersect the A's at only the points indicated so that $A_1+B_1+A_2+B_2$ is a simple closed curve that circles T_0 twice. This simple closed curve would be the sum of two arcs, one A_2+B_2 missing D_1 and the other A_1+B_1 missing D_2 if $D_1 \cdot T_{01} = 0 = D_2 \cdot T_{00}$. This is impossible as pointed out in the following paragraph.

Since D_1 and D_2 chop T_0 into two 3-cells C_1, C_2, we may regard an arc going from D_1 to D_2 in C_i as going halfway around T_0. The arc A_1+B_1 would not go as much as once around T_0 since it misses D_2 and the arc A_2+B_2 would go as much as once around T_0 since it misses D_1. This contradicts the fact that the closed curve $A_1+B_1+A_2+B_2$ goes around T_0 exactly twice.

THEOREM 6. *Suppose that S is a polyhedral 2-sphere in* Int T_0 *such that S does not lie in either T_{00} or T_{01}, S is in general position with respect to* Bd T_{00} + Bd T_{01}, *S contains no simple closed curve that bounds a disk in either* Bd T_{00} *or* Bd T_{01}, *and the bounded component of $E^3 - S$ intersects T_{00}.*

Then S contains a disk D such that either D has Property P *with respect to T_{00} or D has Property* P *with respect to T_{01}.*

Proof. If S intersects $\text{Bd}\, T_{00}$, $S \cdot \text{Bd}\, T_{00}$ contains a simple closed curve J such that either J circles $\text{Bd}\, T_{00}$ meridianally (in which case the condition of Theorem 6 is easily established) or circles $\text{Bd}\, T_{00}$ longitudinally. In this latter case there is a simple closed curve K on $\text{Bd}\, T_{00}$ such that K circles $\text{Bd}\, T_{00}$ longitudinally and K lies in the bounded component of $E^3 - S$. There is a homeomorphism h of E^3 onto itself that is fixed on $T_{01} + \text{Bd}\, T_0$ which takes T_{00} so near K that $h(T_{00})$ lies in the bounded component of $E^3 - S$. Hence we suppose that T_{00} lies in the bounded component of $E^3 - S$.

In a similar fashion it is shown that unless Theorem 6 is true, we can assume with no loss of generality that T_{01} lies in the bounded component of $E^3 - S$. However, this would violate Theorem 2 since a simple closed curve circling $\text{Bd}\, T_0$ meridianally can be shrunk to a point in the unbounded component of $E^3 - S$.

5. The decomposition space Γ. We use Γ to denote the decomposition space whose points are the elements of G. There are several ways of showing that the decomposition space Γ is different from E^3.

Perhaps the easiest way to show that Γ is topologically different from E^3 is to follow the methods used in [3]. Using this approach, one would show that if Γ is topologically equivalent to E^3, then there is a pseudo isotopy fixed outside T_0 which shrinks elements of G to points. It would then be shown that this is impossible since any homeomorphism h of E^3 onto itself that is fixed outside T_0 takes some T_{0i} onto a set of large size, some T_{0ij} in this T_{0i} would have an image of large size under h, some T_{0ijk} in it would have large size under h, ...

Although we could carry out the program suggested in the preceding paragraph, we do not suggest that it is easy. On the contrary, one can find from examining the proof in [3] that such an approach can be quite tedious.

We show that Γ is topologically different from E^3 by showing that there are points of Γ without small neighborhoods bounded by 2-spheres. Although this approach is more difficult than the one outlined above, it has the merits that it is different from that used in [3] and gives a handy property to distinguish the decomposition space from E^3. Let f be the map of E^3 onto Γ that takes each element of G onto a unique point of Γ.

THEOREM 7. *If g is a nondegenerate element of G, $f(T_0)$ contains no neighborhood of $f(g)$ whose boundary is a 2-sphere.*

Proof. This theorem is established by applying the results of Section 4. Assume that there is a neighborhood N of $f(g)$ in $f(T_0)$ such that $\text{Bd}\, N$ is a 2-sphere. We show in five steps that this is impossible.

Step 1. *Putting boundaries in convenient positions.* In this paragraph we show that if there is an N, there is one such that $f^{-1}(\mathrm{Bd}\,N)$ is locally polyhedral except possibly at points of Y. Just because $\mathrm{Bd}\,N$ is a 2-sphere in Γ is no reason to suppose that $f^{-1}(\mathrm{Bd}\,N)$ is a 2-sphere in E^3. In fact it may even not be a 2-manifold. However $f^{-1}(\mathrm{Bd}\,N)$ is locally a 2-manifold away from Y and $f^{-1}(\mathrm{Bd}\,N)-Y$ is a 2-manifold in E^3-Y. Note that $f^{-1}(\mathrm{Bd}\,N)$ separates g from $\mathrm{Bd}\,T_0$ in E^3. There is a positive number ε such that if h is a map of $f^{-1}(\mathrm{Bd}\,N)$ into E^3 that moves no point more than ε, then $hf^{-1}(\mathrm{Bd}\,N)$ separates g from $\mathrm{Bd}\,T_0$ (Theorem VI, 10 of [13]). Let $F(x)$ be the real function which is the minimum of ε and one half the distance from x to Y. By regarding $f^{-1}(\mathrm{Bd}\,N)-Y$ as a 2-manifold in E^3-Y it follows from Theorem 7 of [6] that there is a homeomorphism h of $f^{-1}(\mathrm{Bd}\,N)$ into E^3 such that h does not move x by more than $F(x)$ and $hf^{-1}(\mathrm{Bd}\,N)$ is locally polyhedral at $h(x)$ if $F(x)>0$. Note that $E^3-hf^{-1}(\mathrm{Bd}\,N)$ is the sum of two components U_1, U_2 containing g, $\mathrm{Bd}\,T_0$ respectively. Then $f(U_1)$ is an open set bounded by the 2-sphere $fhf^{-1}(\mathrm{Bd}\,N)$. Replacing N by $f(U_1)$ we get an open set in $f(T_0)$ containing $f(g)$ such that $\mathrm{Bd}\,f(U_1)$ is a 2-sphere whose inverse under f is locally polyhedral except possibly on Y. For convenience in notion we suppose that $N=f(U_1)$ and $f^{-1}(\mathrm{Bd}\,N)$ is locally polyhedral except possibly at Y.

As mentioned at the beginning of Section 4 there is no loss of generality in supposing that the boundaries of the T's are polyhedral. Hence, we suppose that each $\mathrm{Bd}\,T$ is polyhedral and in general position with respect to $f^{-1}(\mathrm{Bd}\,N)$.

Step 2. $\mathrm{Bd}\,N$ *contains a disk D such that $f^{-1}(\mathrm{Bd}\,D)$ circles some* $\mathrm{Bd}\,T$ *meridianally.* Since $f(g)$ lies in N, there is a $T_{0a_1a_2\ldots a_n}$ such that $f(T_{0a_1a_2\ldots a_n})$ lies in N. We show that for some $T_{0b_1b_2\ldots b_m}$ $(m\leqslant n)$, $f^{-1}(\mathrm{Bd}\,N)$ contains a simple closed curve that circles $\mathrm{Bd}\,T_{0b_1b_2\ldots b_m}$ meridianally. We do this by applying Theorem 6.

To apply Theorem 6 we must replace $f^{-1}(\mathrm{Bd}\,N)$ by a polyhedral 2-sphere. (We recall that $f^{-1}(\mathrm{Bd}\,N)$ need not even be a 2-sphere.) There is a punctured polyhedral disk A on $f^{-1}(\mathrm{Bd}\,N)$ such that $f^{-1}(\mathrm{Bd}\,N)-A$ lies on $\sum \mathrm{Int}\,T_{0c_1c_2\ldots c_{n+2}}$. The boundary components of A can be shrunk to points in $\sum \mathrm{Int}\,T_{0c_1c_2\ldots c_{n+1}}$. Hence, there results a singular 2-sphere. By the Sphere Theorem [18] there is a polyhedral 2-sphere S in $A+$ $+\sum \mathrm{Int}\,T_{0c_1c_2\ldots c_{n+1}}$ such that $T_{0a_1a_k\ldots a_n}$ lies in the bounded component of E^3-S.

We assume that S does not contain any simple closed curve that bounds a disk on any $\mathrm{Bd}\,T_{0c_1c_2\ldots c_i}$ $(i\leqslant n)$. That we can do this follows from arguments used in the proofs of Theorems 1 and 3. Let $T_{0d_1d_2\ldots d_j}$ be the smallest T containing S. We note that $j<n$. It follows from Theorem 6 that S contains a simple closed curve J such that J circles one of

$T_{0d_1d_2...d_j0}$, $T_{0d_1d_2...d_j1}$ meridianally. This simple closed curve would belong to A and hence to $f^{-1}(\mathrm{Bd}\,N)$.

With no loss of generality we assume that $J \subset \mathrm{Bd}\,T_{00}$.

Step 3. *The disk D of Step 2 contains two mutually exclusive subdisks D_1, E_1 such that $f^{-1}(\mathrm{Bd}\,D_1), f^{-1}(\mathrm{Bd}\,E_1)$ circles the same one of* $\mathrm{Bd}\,T_{000}$, $\mathrm{Bd}\,T_{001}$ *meridianally.* Just as we applied the Sphere Theorem in Step 2, we apply Dehn's Lemma in this step to replace $f^{-1}(D)$ by a disk D' in $f^{-1}(D)+\sum T_{0ijk}$ that contains $f^{-1}(D)-\sum T_{0ij}$. It follows from Theorem 4 that D' contains two mutually exclusive disks D'_1, E'_1 with Property P with respect to the same one of T_{000}, T_{001}. If D_1, E_1 are the subdisks of D which have the same boundaries as $f(D'_1)$, $f(E'_1)$, they are the disks promised by Step 3.

Step 4. *Disks D_1, E_1 contain disks D_2, E_2 respectively such that there is a T_{00ij} such that $f^{-1}(\mathrm{Bd}\,D_2)$, $f^{-1}(\mathrm{Bd}\,E_2)$ each circles* $\mathrm{Bd}\,T_{00ij}$ *meridianally.* This follows from Theorem 5 applied to disks replacing $f^{-1}(D_1)$, $f^{-1}(E_1)$.

Step 5. *D_1 intersects E_1.* It follows from repetitions of Step 4 that there are sequences of disks $D_1, D_2, D_3, \ldots$ and $E_1, E_2, E_3, \ldots$ such that $D_1 \supset D_2 \supset D_3 \supset \ldots, E_1 \supset E_2 \supset E_3 \supset \ldots$, and $f^{-1}(\mathrm{Bd}\,D_i)$, $f^{-1}(\mathrm{Bd}\,E_i)$ circle meridianally the same $\mathrm{Bd}\,T_{00c_1c_2...c_i}$. This implies that D_1 intersects E_1.

The assumption that there is a neighborhood N of $f(g)$ in $f(T_0)$ whose boundary is a 2-sphere led to the contradiction between Steps 3 and 5.

6. A decomposition H of E^3 into points and segments. In this section we describe a decomposition H of E^3 into points and straight segments. Although this decomposition looks enough like the decomposition described in [3] to suggest that the decomposition space may be topologically different from E^3, it looks enough like the decomposition in [7] to make one cautious in venturing such a conjecture.

A descriprion of H. Let Θ_1 and Θ_2 be two horizontal planes in E^3 with Θ_1 below Θ_2. Each nondegenerate element of H is a straight segment with one end in Θ_1 and the other in Θ_2. Each horizontal plane that intersects one of the segments will intersect the sum of the segments in a topological Cantor set.

The sum of the nondegenerate elements of H is the intersection of a decreasing sequence of open sets $U_1, U_2, \ldots$ Each U_i is a tubular neighborhood of a finite graph G_i. Figure 5 shows G_1. We describe it as follows.

Let $Q(1), Q(2), Q(3), Q(4)$ be four points in Θ_1 and $P(1)$, $P(2)$ be two points in Θ_2 such that the line through $P(1), P(2)$ is not parallel to any line through any two of the Q's. We use $[P(i), Q(j)]$ to denote the straight segment from $P(i)$ to $Q(j)$. Then G_1 is the sum of the eight $[P(i), Q(j)]$'s as shown in Figure 5. U_1 is a tubular neighborhood of G_1.

We obtain G_2 by replacing each major segment $[P(i), Q(j)]$ in G_1 by a copy of G_1. The four copies of G_1 in G_2 that are near $P(i)$ $(i = 1, 2)$ link each other and the two copies of G_1 in G_2 near $Q(j)$ $(j = 1, 2, 3, 4)$ cross in a manner to be described presently. Before describing the eight copies of G_1 in G_2 in detail, we remark that their sum lies in U_1. Just as we obtained G_2 by replacing each major segment in G_1 by a copy of G_1, we obtain G_3 by replacing each major segment of G_2 by a copy of G_1. Continuing in this fashion we get a sequence of finite graphs $G_1, G_2, G_3, G_4, \ldots$ and a sequence of tubular neighborhoods $U_1, U_2, U_3, U_4, \ldots$ of these graphs such that $G_{i+1} \subset U_{i+1} \subset \bar{U}_{i+1} \subset U_i$. The tubular neighborhoods get progressively so thin that the intersection of the U_i's is the sum of a Cantor set of segments each with one end on Θ_1 and the other on Θ_2. These segments are the nondegenerate elements of the decomposition H.

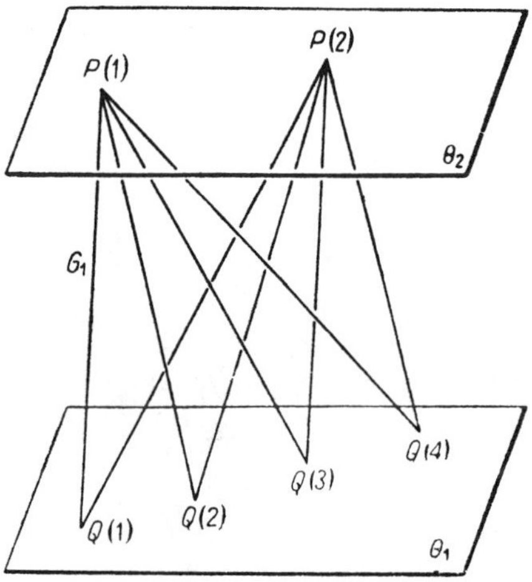

Fig. 5

The copy of G_1 in G_2 replacing $[P(i), Q(j)]$ is denoted by $G(i, j)$. It is the sum of eight straight segments $[P(i, j, m), Q(i, j, n)]$ $(m = 1, 2;$ $n = 1, 2, 3, 4)$ where the $P(i, j, m)$'s are points in Θ_2 near $P(i)$ and the $Q(i, j, n)$'s are points in Θ_1 near $Q(j)$. We shall describe these P's and Q's presently but first show how we want $G(i, j)$ and $G(i, k)$ to entwine near $P(i)$.

Entwining near $P(i)$. Figure 6 shows how certain segments of the $G(1, j)$'s look in perspective near $P(1)$. The general rule is that for $j < k$, $[P(i, j, s), Q(i, j, t)]$ passes between the eye and $[P(i, k, r), Q(i, k, u)]$ for $t \leqslant u$ and $[P(i, k, r), Q(i, k, u)]$ passes between the eye and $[P(i, j, s), Q(i, j, t)]$ for $u < t$. $P(i, j, 1)$ and $P(i, j, 2)$ are so close together (as are $P(i, k, 1)$ and $P(i, k, 2)$) that the crossings are the same irrespective of whether or not r (or s) is 1 or 2.

Entwining near $Q(j)$. Figure 7 shows how one broken arc in $G(1, j)$ entwines with four broken arcs in $G(2, j)$ near $Q(j)$. Each of the broken arcs in $G(1, j)$ entwines with these four arcs in $G(2, j)$ in this fashion but we did not show this in Figure 7 to avoid complication. Note that $[P(1, j, 1), Q(1, j, m)]$ passes between the eye and $[P(2, j, 1), Q(2, j, n)]$ $(n = 1, 2, 3, 4)$, $[P(2, j, 1), Q(2, j, n)]$ passes between the eye and $[P(1, j, 2), Q(1, j, m)]$, and $[P(i, j, 2), Q(1, j, m)]$ passes between the eye and $[P(2, j, 2), Q(2, j, n)]$.

The existence of the $G(i, j)$'s. It remains to be shown that we can really get $G(i, j)$'s that entwine as we have suggested. As a step toward doing this, we replace each $[P(i), Q(j)]$ by a triangular disk $T(i, j)$. The triangles are as shown in Figure 8 and are described as follows.

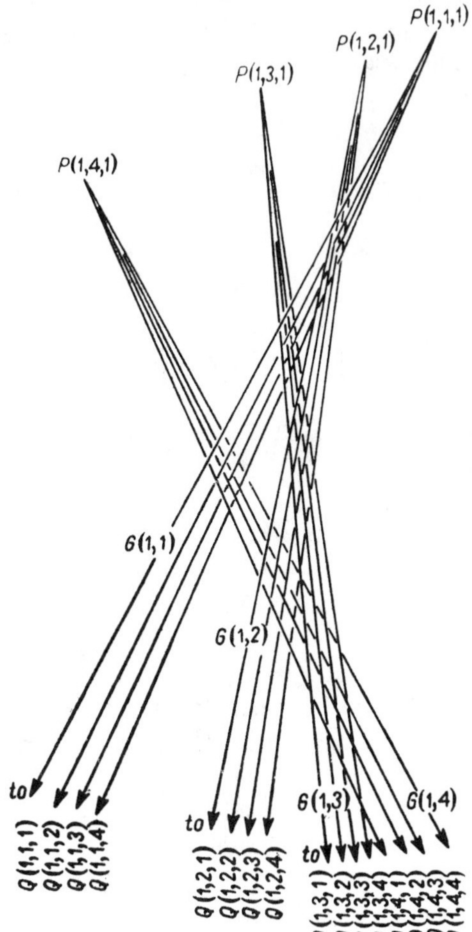

Fig. 6

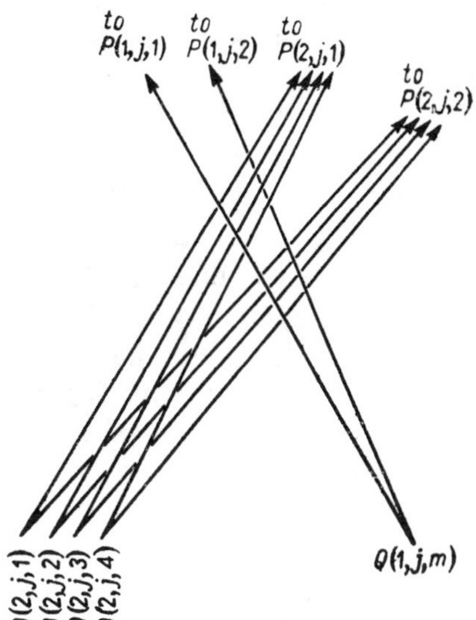

Fig. 7

Through $Q(j)$ consider a short segment $I(Q(j))$ parallel to $[P(1), P(2)]$. The base of $T(i, j)$ is on $I(Q(j))$. There is a line L parallel to $[P(1), P(2)]$ and a segment $I(P(i))$ on L near $P(i)$ such that $T(i, 1) \cdot L = T(i, 2) \cdot L = T(i, 3) \cdot L = T(i, 4) \cdot L = I(P(i))$. The lateral edges of $T(1, j)$ and $T(2, j)$ cross as shown. We suppose that $T(i, j)$ is close to $[P(i), Q(j)]$ and lies in U_1.

In describing the $Q(i, j, k)$'s it is convenient to think of L (and hence the base of $T(i, j)$) as running from left front to right back, with $I(P(2))$

to the right of $I(P(1))$. Let $Q(i, j, 1)$ be the left most point of the base of $T(i, j)$, $Q(i, j, 4)$ the right most point, $Q(i, j, 2)$ a point that divides the base in the ratio 1 to 2 from left to right, and $Q(i, j, 3)$ the point that divides it in the ratio 2 to 1.

The vertices $P(i, j, 1)$ and $P(i, j, 2)$ of $G(i, j)$ in Θ_2 are obtained by adjusting the apex of $T(i, j)$, but before describing this adjustment, let us comment on a restriction that if placed on these P's will ensure that $G(1, j)$ and $G(2, j)$ entwine near $Q(j)$ as shown in Figure 7. If $P(1, j, 1)$, $P(2, j, 1)$ are on the same side of the plane of $T(1, j)+T(2, j)$

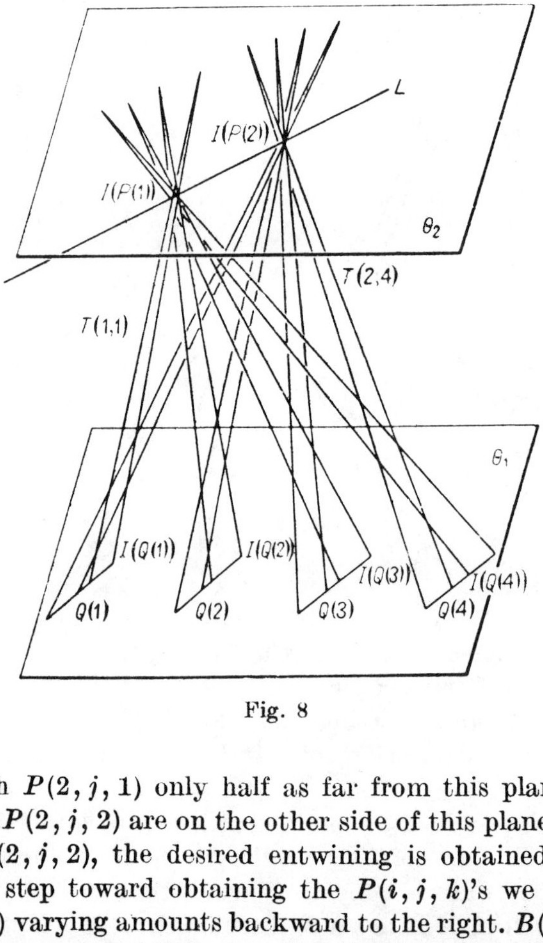

Fig. 8

as the eye with $P(2, j, 1)$ only half as far from this plane as $P(1, j, 1)$ and $P(1, j, 2)$, $P(2, j, 2)$ are on the other side of this plane with $P(1, j, 2)$ nearer than $P(2, j, 2)$, the desired entwining is obtained.

As another step toward obtaining the $P(i, j, k)$'s we move the apex $A(i, j)$ of $T(i, j)$ varying amounts backward to the right. $B(i, j)$ is the point in Θ_2 obtained by moving $A(i, j)$ enough to the right that $[B(i, j), Q(i, j, 1)]\cdot$ $\cdot I(P(i))$ divides $I(P(i))$ in the ratio j to $12-j$ from right to left. If $P(i, j, 1)$ and $P(i, j, 2)$ are obtained from $B(i, j)$ by slight adjustments moving them out of the plane of $T(i, j)$ as suggested in the preceding paragraph, the $G(i, j)$'s entwine near $P(i)$ as suggested by Figure 6.

An intermediate step. Instead of considering eight components of U_2 in U_1, it is more convenient at times to consider two open sets in U_1 as shown in Figure 9 and then consider four components of U_2 in each of these open sets. In Figure 9 we do not emphasize straightness, but the figure is useful in showing the entwining near the Q's as suggested in Figure 7. This intermediate step is useful in Section 8 where we make suggestions as to why the decomposition space of H may be topologically different from E^3.

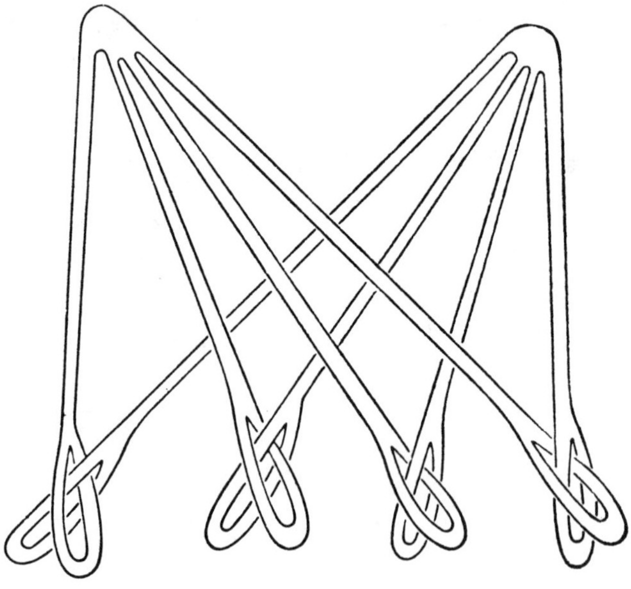

Fig. 9

7. Disks in Υ. We use Υ to denote the decomposition space whose points are the elements of H of the last section and f to be the map taking E^3 onto Υ. For each element y of Υ, $f^{-1}(y)$ is an element of H. We use C to denote the Cantor set in Υ which is the image of the sum of the nondegenerate elements of H. We note that $\Upsilon - C$ is a 3-manifold.

If X, Y are sets in a 3-manifold M, X is locally tame mod Y if for each point p of $X - Y$ there is a neighborhood N of p in M and a homeomorphism h of $\overline{N}$ onto a cube in E^3 such that $h(X \cdot \overline{N})$ is a polyhedron in the cube. Even if M is not a 3-manifold but $M - Y$ is one, we extend the notion to say that X is locally tame mod Y if for each point p of $X - Y$ there is such a neighborhood N of p. Hence we speak of locally tameness of sets in Υ mod C even though Υ may not be a manifold.

As a help toward learning the properties of Υ we consider how a disk in E^3 may be associated with a disk in Υ.

THEOREM 8. *Suppose that D is a disk in Υ such that D is locally tame* mod C *and* $C \cdot \mathrm{Bd}\, D = 0$. *Then for each integer i there is a tame disk D_i in E^3 such that $f(\mathrm{Bd}\, D_i) = \mathrm{Bd}\, D$ and $D_i \subset f^{-1}(D) + U_i$.*

Proof. As a result of Theorem 9 of [9], Theorem 9 of [8], or Theorem 1 of [16], we suppose with no loss of generality that $f^{-1}(D)$ is locally polyhedral mod $f^{-1}(C)$. Since the theorem is true for a particular integer i if is true for a larger one, we suppose that we are proving the theorem for an i so large that $f^{-1}(\mathrm{Bd}\, D) \cdot U_i = 0$.

Let X be a punctured disk in $D - C$ such that the boundary components of X are $\mathrm{Bd}\, D$ and simple closed curves in $f(U_{i+1})$. Then $f^{-1}(X)$ is a punctured disk in E^3. We suppose with no loss of generality that the boundary components of $f^{-1}(X)$ are $f^{-1}(\mathrm{Bd}\, D), J_1, J_2, \ldots, J_n$ where J_j is a polygon in U_{i+1}. Since each simple closed curve in U_{i+1} can be shrunk to a point U_i, J_j bounds a polyhedral singular disk in U_i. Let Y be the sum of $f^{-1}(X)$ and singular disks, one corresponding to each J_j. Note that Y is a polyhedral singular disk all of whose singularities lie in U_i.

It follows from Dehn's lemma as proved by Papakyriokopoulos [18] that for each neighborhood N of the sum of the singularities of Y, there is a nonsingular polyhedral disk D_i in $Y + N$ with the same boundary as D. To ensure that D_i satisfies conditions of Theorem 8, we only need to take N to lie in U_i.

The next three theorems contain the questionable hypothesis that Υ is topologically E^3.

THEOREM 9. *Suppose that* $\mathrm{Bd}\, U_1$ *is polyhedral and* $J_1, J_2, \ldots, J_n$ *is a finite collection of mutually exclusive polygonal simple closed curves on* $\mathrm{Bd}\, U_1$ *each of which can be shrunk to a point in* U_1. *Then if Υ is topologically equivalent to E^3, there is an integer j and a finite collection of mutually exclusive polyhedral disks* $D_1, D_2, \ldots, D_n$ *such that* $\mathrm{Bd}\, D_i = J_i$, $\mathrm{Int}\, D_i \subset U_1$, *and no component of U_j intersects two D_i's.*

Proof. It follows from Theorem 9 of [8], Theorem 9 of [9] or Theorem 1 of [16] that we can suppose that f is piecewise linear except at C. We do this.

Since J_i can be shrunk to a point in U_1, it follows from Dehn's lemma [18] that J_i bounds a disk E_i which lies except for J_i in U_1. Then $f(J_i)$ bound the singular disk $f(E_i)$. Although $f(E_i)$ may not be polyhedral, there is a singular polyhedral disk X_i such that $\mathrm{Bd}\, X_i = f(J_i)$, $\mathrm{Int}\, X_i \subset f(U_1)$, and X_i has no singularities on $\mathrm{Bd}\, X_i$. It follows from Dehn's lemma as proved by Papakyriokopoulos [18] that $\mathrm{Bd}\, X_i$ bounds a polyhedral disk Y_i which lies except for its boundary in $f(U_1)$.

We wish to adjust the Y's so that the resulting disks will not intersect each other. Adjust Y_2 so that it is in general position with respect to Y_1. If Y_1, Y_2 intersect, consider a disk D in Y_1 such that $\mathrm{Int}\, D \cdot Y_2 = 0$ and

$\operatorname{Bd} D \subset Y_2$. Replace the disk in Y_2 bounded by $\operatorname{Bd} D$ by D and then shove the adjusted Y_2 to one side of Y_1 near D. This reduces the number of intersections of Y_1 and Y_2. A repetition of this procedure permits us to adjust Y_2 so that the adjustment misses Y_1. In a similar fashion we fix each of $Y_2, Y_3, \ldots, Y_n$ so that the resulting disks miss Y_1. Then $Y_3, Y_4, \ldots, Y_n$ are adjusted further so that the resulting disks will intersect neither Y_1 nor the new Y_2. In a finite number of such steps we can untangle the Y's from each other so we suppose with no loss of generality that they do not intersect. Note that $f^{-1}(Y_i)$ need not be a disk. Since Y_i is polyhedral in Υ and f is piecewise linear except on C, $f^{-1}(Y_i)$ is locally polyhedral except on C.

Let j be an integer so large that the closure of no component of $f(U_j)$ intersects two of the Y's. We suppose that $f(\operatorname{Bd} U_j)$ is in general position with respect to each Y.

It follows from Theorem 8 that there are tame disks F_i $(i = 1, 2, \ldots, n)$ such that $\operatorname{Bd}(F_i) = f^{-1}(\operatorname{Bd} Y_i) = J_i$, and F_i lies in $f^{-1}(Y_i) + U_j$. The closure of no component of U_j intersects two of the F's. It follows from Theorem 8 of [6] that a slight adjustment in the F's takes them into polyhedral disks that satisfy the conditions of Theorem 9.

We recall that an isotopy of E^3 onto itself is a one parameter family h_t $(0 \leqslant t \leqslant 1)$ of homeomorphisms of E^3 onto E^3. (If we permit h_1 to be a map rather than a homeomorphism we call the family a *pseudoisotopy* as mentioned in the beginning of Section 5.)

THEOREM 10. *If Υ is topologically E^3, then for each integer i and each positive number ε there is an integer j and an isotopy h_t $(0 < t \leqslant 1)$ of E^3 onto itself such that*

$h_0 = I$ (the identity),

$h_t = I$ outside U_i, and

(diameter of each image of a component of U_j under h_1) $< \varepsilon$.

Proof. Let N be a polyhedral tubular neighborhood of G_{i+1} so thin that $N \subset U_{i+1}$ and there is a finite number of mutually exclusive horizontal disks $E_1, E_2, \ldots, E_n$ such that $\operatorname{Bd} E_i \subset \operatorname{Bd} N$, $\operatorname{Int} E_i \subset N$, and the closure of each component of $N - \sum E_a$ is a tame 3-cell of diameter less than $\varepsilon/2$.

Let g_t $(0 \leqslant t \leqslant 1)$ be an isotopy of E^3 onto itself such that $g_0 = I$, $g_t = I$ outside U_i, g_1 is piecewise linear and takes U_{i+1} onto N.

It follows from Theorem 9 that there is an integer j and a finite collection of mutually exclusive polyhedral disks $D_1, D_2, \ldots, D_n$ such that for $a = 1, 2, \ldots, n$, $\operatorname{Bd} D_a = g_1^{-1}(\operatorname{Bd} E_a)$, $\operatorname{Int} D_a \subset U_{i+1}$, and no component of U_j intersects two D_a's.

We now show that there is an isotopy k_t $(0 \leqslant t \leqslant 1)$ of E^3 onto itself such that $k_0 = I$, $k_t = I$ outside U_{i+1}, and k_1 takes D_1 onto $g_1^{-1}(E_1)$. The first part of our isotopy is to free D_1 from $g_1^{-1}(E_2)$. Suppose D_1 and $g_1^{-1}(E_2)$ are in general position and that T is a 2-sphere in their sum which is the sum of a disk in D_1 and a disk in $g_1^{-1}(E_2)$. It follows from Alexander's Sphere Theorem ([1], [15]) that we can suppose with no loss of generality that T is the boundary of a tetrahedron $abcd$ whose base abc is the disk in $T\cdot g_1^{-1}(E_2)$. Since the three lateral faces of a tetrahedron can be pushed down onto the base of a tetrahedron by an isotopy that does not move any point that is far from the solid tetrahedron, we find that there is an isotopy of E^3 onto itself that is fixed outside U_{i+1} and which pushes D_1 onto $\big(D_1-(abd+acd+bcd)\big)+abc$. More details of such an isotopy are found in the proof of Theorem 7.1 of [10]. The isotopy may be continued to push D_1 to one side of abc and thereby reduce the number of components of the intersection of the adjusted D_1 and $g_1^{-1}(E_2)$. A finite number of such reductions changes D_1 so that its image does not intersect $g_1^{-1}(E_2)$. Another sequence of isotopies alters D_1 until it does not intersect any $g_1^{-1}(E_i)$ $(i\neq 1)$. In constructing this sequence of isotopies, care is taken not to alter any simplification accomplished at a preceding stage. We suppose that $k_{1/2}(D_1)$ is a polyhedral disk which lies except for its boundary in the 3-cell C which is the closure of a component of $N-\big(g_1^{-1}(E_2)+g_1^{-1}(E_3)+\ldots+g_1^{-1}(E_n)\big)$. It follows from Alexander's Sphere Theorem that there is a homeomorphism of C onto itself that is fixed on $\operatorname{Bd} C$ which takes $k_{1/2}(D_1)$ onto $g_1^{-1}(E_1)$. It follows from Alexander's Deformation of an n-Cell Theorem [2] that k_t $(0\leqslant t\leqslant \frac{1}{2})$ can be extended to an isotopy k_t $(0\leqslant t\leqslant 1)$ such that $k_s=k_{1/2}$ $(\frac{1}{2}\leqslant s\leqslant 1)$ outside C and $k_1(D_1)=g_1^{-1}(E_1)$.

By repeating the procedure suggested in the preceding paragraph for each D_i, we find that there is an isotopy K_t $(0\leqslant t\leqslant 1)$ of E^3 onto itself such that $K_0=I$, $K_t=I$ outside U_{i+1}, and K_1 takes each D_a onto the corresponding $g_1^{-1}(E_a)$. The required isotopy h_t is g_tK_t.

Let V be a component of U_j. To see that diameter $h_1(V)<\varepsilon$, note that V lies in the sum of a D_a and two adjacent components of $U_{i+1}-\sum D_a$. However, each component of $U_{i+1}-\sum D_a$ goes under h_1 into a component of $N-\sum E_a$ and hence goes into a set of diameter less than $\varepsilon/2$. Hence V would go into a set of diameter less than ε.

THEOREM 11. *If Υ is topologically E^3, then for each integer i there is a pseudo isotopy h_t $(0\leqslant t\leqslant 1)$ of E^3 onto itself such that*

$h_0=I$,

$h_t=I$ *outside* U_i,

h_t *is a homeomorphism for* $0\leqslant t<1$, *and*

$h_1^{-1}(p)$ *is an element of H for each point p of E^3.*

Proof. This result follows from repeated applications of Theorem 10.

It follows from Theorem 10 that there is an integer i_1 and an isotopy h_t $(0 \leqslant t \leqslant \frac{1}{2})$ such that $h_0 = I$, $h_t = I$ outside U_i, and diameter $h_{1/2}(V) < \frac{1}{2}$ for each component V of U_{i_1}.

Let ε_1 be a positive number so small that diameter $h_{1/2}(X) < 1/3$ if X is a subset of U_{i_2} of diameter less than ε_1.

It follows from Theorem 10 that there is an integer i_2 and an isotopy k_t $(\frac{1}{2} \leqslant t \leqslant \frac{3}{4})$ of E^3 onto itself such that

$k_{1/2} = I$,

$k_t = I$ *outside* U_{i_1}, *and*

diameter $k_{3/4}(V) < \varepsilon_1$ *if* V *is a component of* U_{i_1}.

Then h_t $(\frac{1}{2} \leqslant t \leqslant \frac{3}{4})$ is $k_t h_{1/2}$. Note that diameter $h_{3/4}(V) < \frac{1}{3}$ if V is a component of U_{i_2}.

Continuing in this fashion we define h_t $(\frac{3}{4} \leqslant t \leqslant \frac{7}{8})$ so that $h_t = h_{3/4}$ outside U_{i_2} and there is an integer i_3, such that diameter $h_{7/8}(V) < \frac{1}{4}$ if V is a component of U_{i_3}. Also h_t $(\frac{7}{8} \leqslant t \leqslant \frac{15}{16})$ is such that $h_t = h_{7/8}$ outside U_{i_3} and there is an i_4 such that diameter $h_{15/16}(V) < \frac{1}{5}$ if V is a component of U_{i_4}. In similar fashion we define h_t $(\frac{15}{16} \leqslant t \leqslant \frac{31}{32})$, h_t $(\frac{31}{32} \leqslant t \leqslant \frac{63}{64})$, ... The limit of h_t as t approaches 1 is h_1.

8. Why four Q**'s?** Let A, B be two planes between Θ_1 and Θ_2 as shown in Figure 10. We suppose that U_1 is so thin that $\overline{U}_1 \cdot A$ (as is $\overline{U}_1 \cdot B$) is the sum of eight mutually exclusive disks such that each disk contains a point of G_1.

One of the things that leads one to believe that Υ may be topologically different from E^3 is the suspicion that if h is any homeomorphism of E^3 onto itself that is fixed outside U_1, then for each integer n there is a component V of G_n such that $h(V)$ intersects both A and B. If this is shown, it follows from either Theorem 9, 10, or 11 that Υ is topologically different from E^3.

We mention why we use four Q's in describing each component of G_i instead of two or three.

Why not two Q**'s?** If we used only two Q's in describing G_1, there would be a homeomorphism h of E^3 onto itself which is fixed except near G_1 as shown in Figure 11 such that $h[P(i), Q(j)]$ $(i = 1, 2;\ j = 1, 2)$ does not intersect both A and B. Hence there is a homeomorphism h of E^3 onto itself fixed except near G_1 such that for each component V of G_2, $h(V)$ does not intersect both A and B.

Why not three Q**'s?** If there were three Q's in G_1 instead of two, it could be shown that if h is a homeomorphism of E^3 onto itself fixed except near G_1, then for some component V of G_2, $h(V)$ intersects both A and B. However, $h(V)$ might fail, to intersect A and B in as essential a fashion as does G_1. For example, if V is a component of G_2 replacing

$[P(1), Q(1)]$ of G_1 and $h(V)$ intersects A and B as shown in Figure 12, then h can be such for that for no component V' of G_3 near V does $h(V')$ intersect both A and B.

With three Q's there is a very effective entwining of the $G(i, j)$'s near the Q's because, as shown in Figures 7 and 9, if any $Q(i, j, m)$ is moved upward to shorten $[P(1, j, 1), Q(1, j, m)] + [P(1, j, 2), Q(1, j, m)]$ then each $[P(2, j, n), Q(2, j, n)] + [P(2, j, 2), Q(2, j, n)]$ is stretched.

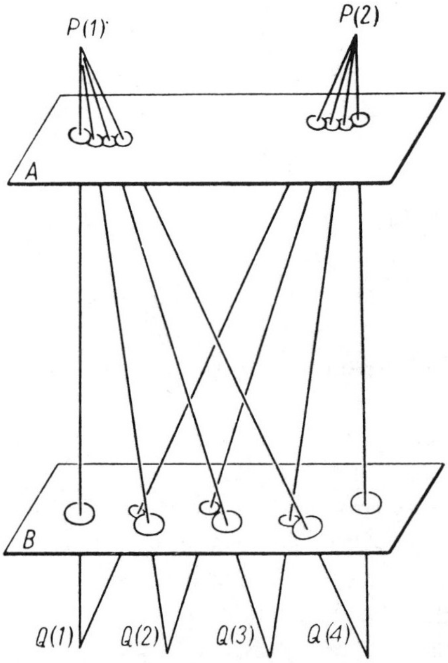

Fig. 10

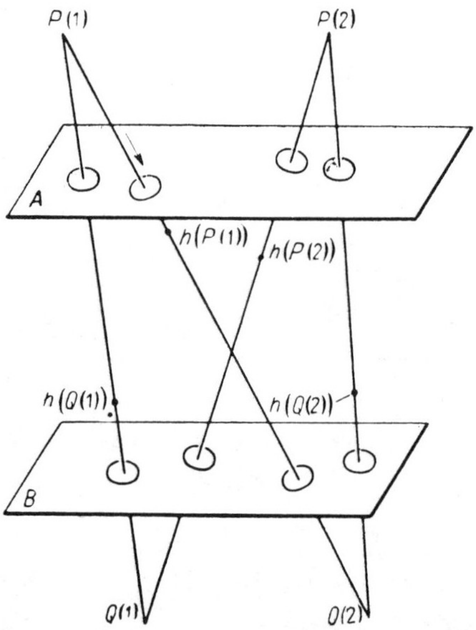

Fig. 11

However, the entwining is not so effective near the P's because, as shown in Figure 6, $P(1, 1, 1)$ may be pulled down to shorten $[P(1, 1, 1), Q(1, 1, 1)] + [P(1, 1, 1), Q(1, 1, 2)]$ without altering any of $[P(1, 2, 1), Q(1, 2, 2)]$, $[P(1, 2, i), Q(1, 2, 3)]$, $[P(1, 3, i), Q(1, 2, 2)]$, or $[P(1, 2, i), Q(1, 2, 3)]$. This is a reason that we use more than three Q's. There is a homeomorphism h such that h pulls $P(1, 1, i)$, $P(1, 2, i)$, and $P(2, 3, i)$ $(i = 1, 2)$ below A to shorten the upper half of $[P(1, 1, i), Q(1, 1, 1)]$, $[P(1, 1, i), Q(1, 1, 2)]$, $[P(1, 2, i), Q(1, 2, 2)]$, $[P(1, 2, i), Q(1, 2, 3)]$, $[P(2, 3, i), Q(2, 3, 2)]$, $[P(2, 3, i), Q(2, 3, 3)]$ and pulls each $Q(i, j, k)$ above B to shorten the lower half of each $[P(2, 1, r), Q(2, 1, s)]$, $[P(2, 2, r), Q(2, 2, s)]$, and $[P(1, 3, r), Q(1, 3, s)]$. Although for each component V of G_2, $h(V)$ intersects each of A and B, each intersects in an innessential fashion as depicted in Figure 12.

Suppose that U is a component of U_i and h is a homeomorphism of E^3 onto itself. We would like to find a definition of an essential intersection of $h(U)$ with A and B such that if $h(U)$ intersects A and B in this effective way, then for some component U' of U_{i+1} in U, $h(U')$ intersects A and B in this essential fashion. We could then use induction, along with Theorem 10, to show that Υ is topologically different from E^3. In the next three paragraphs we make a guess as to what an essential intersection might be.

An arc is said to have oscillation k with respect to A and B if it contains k points of $A+B$ such that no two of these points which are adjacent on the arc belong to the same one of A and B. Note that if an arc has oscillation k, it has oscillation j for any positive integer j less than k. A simple closed curve has oscillation k if it contains an arc of oscillation k with respect to A and B.

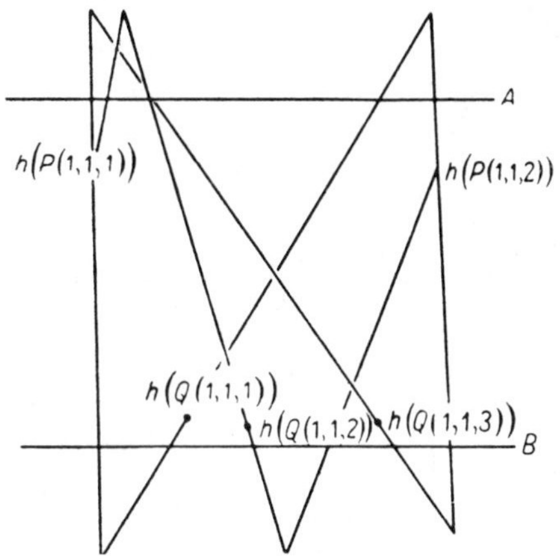

Fig. 12

Property X. An image G' of G_1 has Property X if either some simple closed curve in G' has oscillation 6 or G' contains a θ curve such that each simple closed curve in the θ curve has oscillation 4.

Essential intersection. Suppose that U is a component of U_i and h is a homeomorphism of E^3 onto itself. $h(U)$ has an essential intersection with A and B if for each homeomorphism h' of E^3 onto itself that agrees with h on E^3-U, $h'(U \cdot G_i)$ has Property X.

Suppose that the G's are defined with four Q's as suggested in Section 6 and h is a homeomorphism of E^3 onto itself that is fixed outside U_1. Then $h(U_1)$ intersects A and B is an essential fashion. It follows from Theorem 10 that Υ is topologically different from E^3 if the following conjecture is true.

CONJECTURE. *For each component U of U_i such that $h(U)$ intersects A and B in an essential fashion there is a component U' of $U \cdot U_{i+1}$ such that $h(U')$ intersects A and B in an essential fashion.*

While we do not prove the conjecture, we can prove the following theorem.

THEOREM 12. *If $h(G_1)$ has Property X, then there are an i, j, and k such that $h([P(i), Q(j)]+[P(i), Q(k)])$ has oscillation* 3.

The above theorem is proved by considering where the various points of $A \cdot h(G_1)$ and $B \cdot h(G_1)$ might be.

Here is how one might use Theorem 12 to try to establish the conjecture. Suppose $U = U_1$. Consider the two intermediate open sets in U_1

as depicted in Figure 9, and let W be the one of these containing $[P(i), Q(j)]+[P(i), Q(k)]$ of Theorem 12. Try to show from Theorem 12 that we have a situation something like that depicted in Figure 13 where $A\cdot h(\overline{W})$ contains two disks separating the loops of $h(W)$ associated with $h(Q(i))$ and $h(Q(j))$ while $B\cdot h(\overline{W})$ contains a disk between these two disks as shown. If we were to get a situation exactly like that depicted

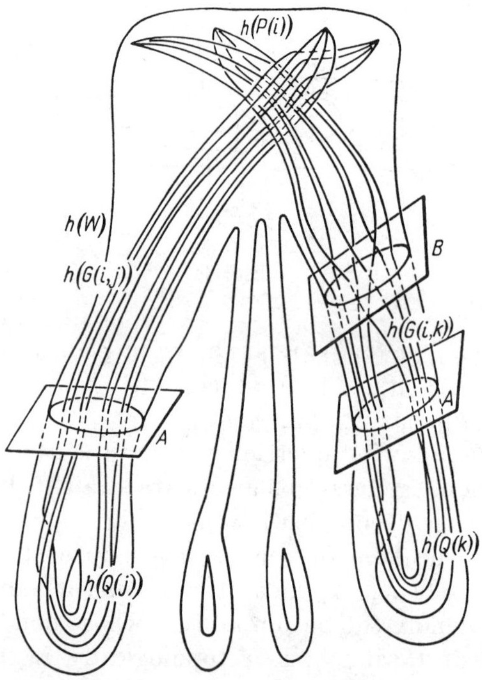

Fig. 13

in Figure 13, it could probably be shown that if h' is a homeomorphism of E^3 onto itself that agrees with h on E^3-W, then one of $h'(G(i, j))$, $h'(G(i, k))$ has Property X. It would be desirable to recognize a situation that is enough like that shown in Figure 13 so that it could be shown that one of $h'(G(i, j))$, $h'(G(i, k))$ has Property X but is general enough so that it can be shown that the situation occurs.

The more Q's we use in defining the G's, the more complicated the decomposition space Υ appears. The reason we stop at four is that we feel that the conjecture may be true.

9. Questions. Is the decomposition space Υ disussed in the preceding three sections topologically different from E^3? Is the conjecture of Section 8 true?

Suppose that T is a double torus as shown in Figure 14; J_1, J_2 are two simple closed curves in Int T as shown; $A \cdot T$ contains two disks; and $B \cdot T$ contains a disk between these two disks as shown. If h is a homeomorphism of E^3 onto itself that is fixed on $E^3 - T$, need one of $h(J_1)$, $h(J_2)$ have oscillation 4 with respect to A and B? An affirmative answer might be obtained by considering a universal covering space of T (see proof of Theorem 5). An affirmative answer might help in

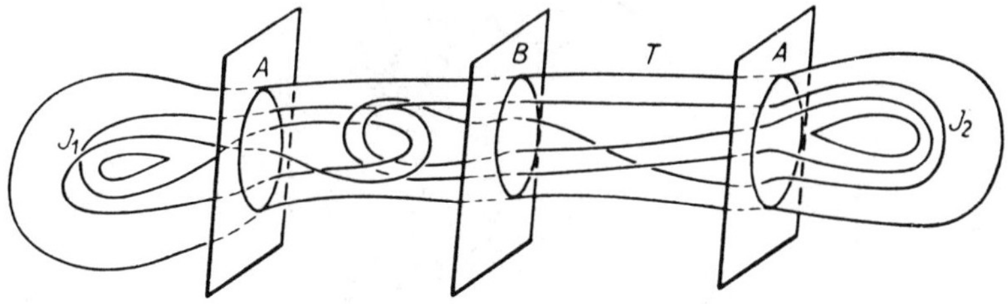

Fig. 14

showing that if $G(i, j)$, $G(i, k)$ are as shown in Figure 13 and h' is a homeomorphism of E^3 onto itself that agrees with h on $E^3 - W$, then one of $h'(G(i, j))$, $h'(G(i, k))$ has Property X.

In each of the following questions we suppose that H' is a decomposition of E^3 whose nondegenerate elements are straight line segments and Υ' denotes the resulting decomposition space.

Suppose that the sum of the nondegenerate elements of H' is the sum of two closed sets X_1, X_2 each of which is the sum of elements of H' and H_i is used to denote the decomposition whose nondegenerate elements are the elements of H' in X_i. Is Υ topologically E^3 if each of Υ_1, Υ_2 is? An answer even in the case where $X_1 \cdot X_2 = 0$ would be interesting. McAuley's result [14] is one phase of considering the case where there are infinitely many X's.

If Υ' is topologically different from E^3, does H' contain a Cantor set of segments such that if H_3 is the decomposition of E^3 whose nondegenerate elements are elements of this Cantor set of segments, then Υ_3 is topologically different from E^3?

For each planar circular disk D and each pair of positive numbers ε, δ let $H(D, \varepsilon, \delta)$ denote the elements of H' with diameters greater than or equal to ε which intersect D within δ of its center and whose directions differ from the normal to D by no more than δ degrees. If H' is topologically different from E^3, does there exist a planer circular disk D and a positive number ε such that for each $\delta > 0$, the decomposition space whose only nondegenerate elements are those of $H(D, \varepsilon, \delta)$ is topologically different from E^3?

Does a point-like decomposition of E^3 yield E^3 if it yields a 3-manifold?

Does each point of the decomposition space Γ discussed in Sections 3-5 have an arbitrarily small simply connected neighborhood?

Does one obtain Euclidean 4-space if one takes the cartesian product of a Euclidean line with either the decomposition space Γ of Sections 3-5 or the decomposition space Υ of Sections 6-8? This question is motivated by [5].

References

[1] J. W. Alexander, *On the subdivision of 3-space by a polyhedron*, Proc. Nat. Acad. Sci. U. S. A. 10 (1924), pp. 6-8.

[2] — *On the deformation of an n-cell*, Proc. Nat. Acad. Sci. U. S. A. 9 (1923), pp. 406-407.

[3] R. H. Bing, *A decomposition of E^3 into points and tame arcs such that the decomposition space is topologically different from E^3*, Ann. of Math. 65 (1957), pp. 484-500.

[4] — *Upper semicontinuous decompositions of E^3*, Ann. of Math. 65 (1957), pp. 363-374.

[5] — *The cartesian product of a certain nonmanifold and a line is E^4*, Ann. of Math. 70 (1959), pp. 399-412.

[6] — *Approximating surfaces with polyhedral ones*, Ann. of Math. 65 (1957), pp. 456-483.

[7] — *A homeomorphism between the 3-sphere and the sum of two solid horned spheres*, Ann. of Math. 56 (1952), pp. 354-362.

[8] — *An alternative proof that 3-manifolds can be triangulated*, Ann. of Math. 69 (1959), pp. 37-65.

[9] — *Locally tame sets are tame*, Ann. of Math. 59 (1954), pp. 145-158.

[10] — *Conditions under which a surface in E^3 is tame*, Fund. Math. 47 (1959), pp. 106-139.

[11] E. Dyer and M. E. Hamstrom, *Completely regular mappings*, Fund. Math. 45 (1958), pp. 103-118.

[12] M. K. Fort, Jr., *A note concerning a decomposition space defined by Bing*, Ann. of Math. 65 (1957), pp. 501-504.

[13] W. Hurewicz and H. Wallman, *Dimension theory*, Princeton 1948.

[14] L. F. McAuley, *Some upper semicontinuous decompositions of E^3 into E^3*, Ann. of Math. 73 (1961), pp. 437-457.

[15] E. E. Moise, *Affine structures in 3-manifolds, II. Positional properties of 2-spheres*, Ann. of Math. 55 (1952), pp. 172-176.

[16] — *Affine structures in 3-manifolds, IV. Piecewise linear approximations of homeomorphisms*, Ann. of Math. 55 (1952), pp. 215-222.

[17] R. L. Moore, *Concerning upper semicontinuous collections of continua*, Trans. Amer. Math. Soc. 27 (1925), pp. 416-428.

[18] C. D. Papakyriakopoulos, *On Dehn's lemma and the asphericity of knots*, Ann. of Math. 66 (1957), pp. 1-26.

THE UNIVERSITY OF WISCONSIN

Reçu par la Rédaction le 15. 3. 1961

FUNDAMENTA
MATHEMATICAE
LX (1967)

A toroidal decomposition of E^3

by

Steve Armentrout (Iowa City, Iowa) and **R. H. Bing** (Madison, Wisc.)*

1. Introduction. An upper semi-continuous decomposition G of E^3 is defined to be a *toroidal* decomposition if and only if the following condition holds: There is a sequence $M_1, M_2, M_3, \ldots$ of compact 3-manifolds-with-boundary in E^3 such that (1) for each i, $M_{i+1} \subset \operatorname{Int} M_i$ and each component of M_i is a solid torus (cube with one handle) and (2) A is a non-degenerate element of G if and only if A is a non-degenerate component of $M_1 \cdot M_2 \cdot M_3 \ldots$.

Bing proved in [3] that the union of two solid Alexander horned spheres, sewed together along their boundaries, was homeomorphic to S^3. A major step in this proof consists of showing that for a certain toroidal decomposition H of E^3 into tame arcs and one-point sets, the decomposition space associated with H is homeomorphic to E^3. Keldyš raised the following question in [7]:

Does every toroidal decomposition of E^3 into tame arcs and one-point sets yield E^3 as its decomposition space?

In this paper, we give a negative answer to this question.

The example of this paper provides another monotone decomposition of E^3 into tame arcs and one-point sets such that the associated decomposition space is topologically distinct from E^3. The first such example given was Bing's dogbone decomposition of E^3 [4].

It was proved by Bing in [5] that there is a toroidal decomposition of E^3 into point-like continua such that the associated decomposition spaces is not homeomorphic to E^3. In this particular example, each non-degenerate element is an indecomposable continuum and hence is not locally connected.

Keldyš has announced in [7] that if G is a toroidal decomposition of E^3 into locally connected point-like continua such that the set of all non-degenerate elements of G is a continuous collection, then the de-

* The first and second authors were supported by the National Science Foundation under grants GP-4508 and GP-3857 respectively.

composition space is homeomorphic to E^3. She also described an example to show that if "locally connected" is deleted in the preceding statement, the resulting proposition is false. Another such example can be obtained by a simple modification of the example of Bing's mentioned in the preceding paragraph.

It follows from a theorem proved by Andrews and Rubin in [1] that if X is the space of the decomposition described in this paper, then $X \times X^1$ is homeomorphic to E^4.

In Section 2 of this paper, we describe the decomposition to be studied and in Section 3, we prove that its decomposition space is not homeomorphic to E^3. In Section 4, we show that each non-degenerate element of the decomposition is a tame arc.

If M is a manifold with boundary, then Int M and Bd M denote the interior and boundary, respectively, of M.

2. The example. Let J_1 be a circle in E^3 of circumference 1 and T_1 be a tubular neighborhood of J_1 of small cross radius.

Let $J_{11}, I_{12}, \dots, J_{1n_1}$ be a chain of simple closed curves as shown in Figure 1 such that each J_{1i} has length 1, adjacent J_{1i}'s are linked as shown with J_{1n_1} linked to J_{11} so that as a chain they go around T_1 more than $n_1/2$ times. Let T_{1i} be a tubular neighborhood of J_{1i} of very small cross radius so that no two of the T_{1i}'s intersect and each lies in T_1.

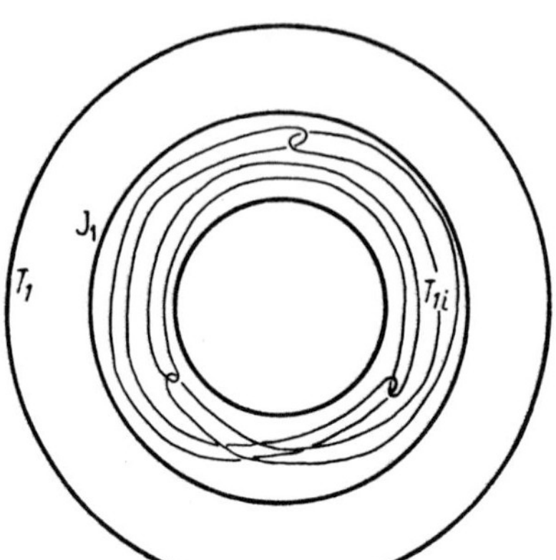

Fig. 1

For simplicity in drawing Figure 1, we showed n_1 as 3 but if the cross radius of T_1 were very small, n_1 would have to be more than 3 since each J_{1i} goes slightly more than half way around T_1 in order for the chain to go more than $n_1/2$ times around T_1. It is possible to find such an n_1 since there are cores of T_1 slightly shorter than J_1 and the J_{1i}'s can be placed close to such a shorter core.

In the interior of each T_{1i}, let $J_{1i1}, J_{1i2}, \ldots, J_{1in_{1i}}$ be a chain of simple closed curves which go around T_{1i} more than $n_{1i}/2$ times and such that each J_{1ij} is of length 1. Also, T_{1ij} is a tubular neighborhood of J_{1ij} of very very small cross radius so that no two of the T_{1ij}'s intersect and each lies in $\operatorname{Int} T_{1i}$. Similarly we define T_{1ijk}'s in the interiors of the T_{1ij}'s more T's in these, and so on.

Let G be the collection whose elements are the components of $T_1 \cdot (\bigcup T_{1i}) \cdot (\bigcup T_{1ij}) \cdot (\bigcup T_{1ijk}) \ldots$. We note that G has a Cantor set of elements. The decomposition of E^3 that we study is the decomposition whose only non-degenerate elements are the non-degenerate elements of G.

3. The decomposition space is topologically distinct from E^3. In this section we show that the space of the decomposition described in Section 2 is not homeomorphic to E^3. We do this by studying homeomorphisms from T_1 onto T_1 that are pointwise fixed on $\operatorname{Bd} T_1$.

Suppose that D_1 and D_2 are disjoint disks in T_1 such that if $r = 1$ or 2, $\operatorname{Bd} D_r$ is a meridional simple closed curve on $\operatorname{Bd} T_1$ and $\operatorname{Int} D_r \subset \operatorname{Int} T_1$. We shall prove that if h is any homeomorphism from T_1 onto T_1, pointwise fixed on $\operatorname{Bd} T_1$, there is an element A of G such that $h[A]$ intersects both D_1 and D_2. From this it follows that the decomposition space is not homeomorphic to E^3, because if it were, there would be, according to Theorem 3 of [2], a homeomorphism f from T_1 onto T_1, pointwise fixed on $\operatorname{Bd} T_1$ and such that if A is any element of G, $f[A]$ intersects at most one of D_1 and D_2.

Suppose that M is a solid torus. By a *homotopy centerline* of M we mean any simple closed curve in M homotopic in M to a core of M.

LEMMA 1. *If M is any solid torus lying in T_1 and m is any positive integer, then the following two statements are equivalent:*

(1) *Each homotopy centerline of M intersects both D_1 and D_2.*

(2) *If J is a simple closed curve in M that goes around M m times, then there is a sequence of distinct points $p_1, q_1, p_2, q_2, \ldots, p_m, q_m$ of J in the order $p_1 q_1 p_2 q_2 \ldots p_m q_m p_1$ on J such that each p_i belongs to D_1 and each q_i belongs to D_2.*

Proof. It is clear that (2) implies (1). Suppose that (1) holds but (2) fails. We may assume that D_1, D_2 and $\operatorname{Bd} M$ are polyhedral and if $r = 1$ or 2, $\operatorname{Bd} M$ and D_r are in general position. Each component of $(D_1 \cup D_2) \cdot \operatorname{Bd} M$ is a simple closed curve. Further, by Theorem 1 of [5], each component of $(D_1 \cup D_2) \cdot \operatorname{Bd} M$ either bounds a disk on $\operatorname{Bd} M$, or circles $\operatorname{Bd} M$ once longitudinally and no times meridionally, or circles $\operatorname{Bd} M$ once meridionally and no times longitudinally.

No component of $(D_1 \cup D_2) \cdot \operatorname{Bd} M$ is a longitudinal simple closed curve on $\operatorname{Bd} M$. If there were such a simple closed curve, then each component of $(D_1 \cup D_2)\cdot\mathrm{Bd}\,M$ would either be longitudinal or bounds a disk on Bd M, and there would be a homotopy centerline of M on Bd M that would miss $D_1 \cup D_2$. This contradicts (1), so no component of $(D_1 \cup D_2)\cdot\mathrm{Bd}\,M$ is longitudinal. Further, some component of $D_1\cdot\mathrm{Bd}\,M$ is meridional on Bd M, or else some homotopy centerline of M misses D_1. Similarly, some component of $D_2\cdot\mathrm{Bd}\,M$ is meridional on Bd M.

Let L_r $(r = 1, 2)$ be a component of $D_r\cdot\mathrm{Bd}\,M$ that is meridional on Bd M and is an innermost simple closed curve on D_r with that property. The component E_r of $D_r\cdot M$ containing L_r is a disk or punctured disk such that each boundary curve of E_r distinct from L_r bounds a disk on Bd M.

By supposition, there is a simple closed curve J in M that goes around M m times but for which there is no sequence of points satisfying (2). There exist disjoint disks E_1' and E_2' such that if $r = 1$ of 2, then $\mathrm{Bd}\,E_r' = L_r$, $\mathrm{Int}\,E_r' \subset \mathrm{Int}\,M$, and $E_r'\cdot J = E_r\cdot J$. We construct E_r' by capping, on Bd M, boundary curves of E_r distinct from L_r and then deforming the caps slightly into Int M.

Since J goes around M m times, it can be shown that there exist m mutually disjoint open arcs $A_1, A_2, \ldots, A_m$ on J such that each $\bar{A}_i$ abuts on different sides of E_1' and each A_i is disjoint from E_1'. Since each such A_i intersects E_2 and the endpoints of $\bar{A}_i$ belong to E_1, there exist points $p_1, q_1, p_2, q_2, \ldots, p_m$ and q_m of J satisfying (2). This is a contradiction, and Lemma 1 is proved.

Lemma 2. *Suppose that h is a homeomorphism from T_1 onto T_1, pointwise fixed on* $\mathrm{Bd}\,T_1$. *Suppose that each homotopy centerline of $h[T_{1ij\ldots k}]$ intersects both D_1 and D_2. Then for some positive integer m, $m \leqslant n_{1ij\ldots k}$, each homotopy centerline of $h[T_{1ij\ldots km}]$ intersects both D_1 and D_2.*

Proof. Suppose that Lemma 2 is false. If $t = 1, 2, \ldots,$ or $n_{1ij\ldots k}$, let L_t be a homotopy centerline of $h[T_{1ij,\ldots kt}]$ that is disjoint from one of D_1, D_2. Observe that $L_1, L_2, \ldots, L_{n_{1ij\ldots k}}$, as a chain, goes around $h[T_{1ij\ldots k}]$ more than $n_{1ij\ldots k}/2$ times.

Let T^* be the universal covering space of $h[T_{1ij\ldots k}]$ and let p be the projection map. Let L_1' be an image under p^{-1} of L_1, L_2' be the image under p^{-1} of L_2 that is linked with $L_1', \ldots, L_{n_{1ij\ldots k}}'$ the image under p^{-1} of $L_{n_{1ij\ldots k}}$ that is linked with $L_{(n_{1ij\ldots k})-1}'$, and L_1'' the image under p^{-1} of L_1 that is linked with $L_{n_{1ij\ldots k}}$. By Theorem 3 of [4], if $t = 1, 2, \ldots,$ or $(n_{1ij,\ldots k})-1$, there is an arc K_t in T^* from a point of L_t' to a point of L_{t+1}' missing $p^{-1}(D_1 \cup D_2)$, and there is such an arc $K_{n_{1ij\ldots k}}$ from a point of $L_{n_{1ij\ldots k}}'$ to a point of L_1''. Near $L_1 \cup \ldots \cup L_{n_{1ij\ldots k}} \cup p[K_1] \cup p[K_2] \cup \ldots \cup p[K_{n_{1ij\ldots k}}]$ there lies a simple closed curve J such that (1) J goes around M more than $n_{1ij\ldots k}/2$ times and (2) J is the union of arcs $a_1, a_2, \ldots, a_{n_{1ij\ldots k}}$ such that these arcs chain J circularly and no one intersects both D_1

and D_2. Since J goes around M more than $n_{1ij...k}/2$ times, this contradicts Lemma 1.

We now show that the decomposition space is not homeomorphic to E^3. Suppose that h is any homeomorphism from T_1 onto T_1, pointwise fixed on $\mathrm{Bd}\, T_1$. It is clear that each homotopy centerline of $h[T_1]$ intersects both D_1 and D_2. By induction and Lemma 2, there is a sequence $1, i, j, k, \ldots$ such that for each t, each homotopy centerline of $h[T_{1ij...t}]$ intersects both D_1 and D_2. Let A be $T_1 \cdot T_{1i} \cdot T_{1ij} \cdot T_{1ijk} \ldots$. Then $h[A]$ intersects both D_1 and D_2.

As we noted in the second paragraph of this section, this is sufficient to show that the decomposition space is not homeomorpic to E^3.

4. Arcs in G are tame. Let A be a non-degenerate element of G. We show in this section that A is a tame arc.

Let $T_1, T_2, T_3, \ldots$ be the decreasing sequence of solid tori such that T_i is the torus at the ith stage in the definition of G which contains A. Let J_i be the core of T_i used in defining T_i in Section 2. We suppose that T_i is fibered into a simple closed curve of mutually exclusive disks such that each of these disks is of diameter less than $1/4^i$, each intersects J_i in precisely one point, and each which intersects T_{i+1} intersects it in precisely two fibers of T_{i+1} except possibly those fibers of T_i that can be joined in T_i to one of the two bends of J_{i+1} by an arc of length less than $1/4^{i+1}$. These exceptional fibers of T_i near the bends of J_{i+1} intersect T_{i+1} in either a point, a disk, the union of two tangent disks, or the union of two mutually exclusive disks which need not be fibers of T_{i+1}.

Let C_i be the union of the disks in the fibering of T_i that intersect A. In case where diameter $A \geqslant 1/4^i$, C_i is a cylinder.

Since A can be covered by the C's, it is snakelike. Since the length of the J's is 1, this snakelike continuum does not oscillate much and is an arc. We show that it is tame by showing that it is locally tame except possibly at its endpoints and has penetration index 1 there.

The fibers of C_{i+1} may not match up with the fibers of C_i near the bends of J_{i+1} so there is no assurance that $C_{i+1} \subset C_i$. Let C_i^+ be the union of all fibers of T_i that can be joined to C_i by an arc in T_i of length less than or equal to $1/4^i$. We find that $C_{i+1}^+ \subset \mathrm{Int}\, C_i^+$. See Figure 2.

It i is large, then except possibly near the ends of C_i, each of the disk in the fibering of C_i intersects C_{i+1}^+ in precisely a fiber of C_{i+1}. We explain why in the next three paragraphs.

In a certain sense, each C_{i+1} is longer than C_i but not much longer. Let $M(C_i) = L(C_i) - 1/4^i$ where $L(C_i)$ is the shortest length of an arc in C_i joining the two bases of C_i. Any arc in C_{i+1} joining the ends of C_{i+1} can be joined in C_i to either end of C_i by an arc in a fiber of C_{i+1}. Hence $L(C_i) < L(C_{i+1}) + 2/4^{i+1}$, and $M(C_1), M(C_2), \ldots$ is a monotonic increasing

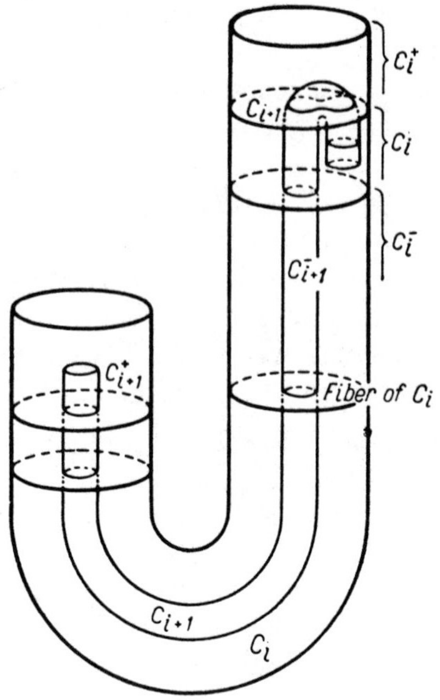

Fig. 2

sequence. Since $\{M(C_i)\}$ is bounded above by 1, it has a limit $M(A)$. Let $\varepsilon_i = M(A) - M(C_i)$. Then $M(C_i) < M(C_{i+j}) < M(C_i) + \varepsilon_i$ and $L(C_{i+j}) < L(C_i) + \varepsilon_i + 1/4^i$.

Each end of C_{i+1} can be joined in C_{i+1} by an arc of length less than $\varepsilon + 6/4^{i+1}$ to a fiber of C_{i+1} that intersects an end of C_i or else an arc in C_{i+1} of length less than $L(C_i)$ intersects the two ends of C_i. Any bend of J_{i+1} in C_i would be within $1/4^{i+1}$ of an end of C_i so each fiber of C_i that fails to intersect C_{i+1} in precisely a fiber of C_{i+1} can be joined to an end of C_i by an arc in C_{i+1} of length less than $\varepsilon_i + 7/4^{i+1}$.

Let C_i^- denote the union of all disks in the fibering of C_i such that each arc in C_i from such a disk to an end of C_i is of length greater than or equal to $\varepsilon_i + 2/4^i$. Then each fiber in C_i^- intersects C_{i+1}^+ only in a fiber of C_{i+1} which in turn is a fiber of C_{i+1}^-.

Some of the C_i^-'s may be disks or null but for convenience we suppose that each is a cylinder. We note that $A \cdot C_i^- = C_i^- \cdot C_{i+1}^- \cdot C_{i+2}^- \ldots$ and hence A intersects each fiber of C_i^- in precisely one point. Hence A is locally tame at each point of $A \cdot \operatorname{Int} C_i$. The only points of A that do not belong to some $\operatorname{Int} C_i$'s are the endpoints of A and the boundaries of the closures of the components of $C_i^+ - C_i^-$ show that A has penetration index 1 at each endpoint. It follows from results of [6] that A is a tame arc.

References

[1] J. J. Andrews and L. Rubin, *Some spaces whose product with E^1 is E^4*, Bull. Amer. Math. Soc. 71 (1965), pp. 675-677.

[2] S. Armentrout, *Decompositions of E^3 with a compact 0-dimensional set of non-degenerate elements*, Trans. Amer. Math. Soc. 123 (1966), pp. 165-177.

[3] R. H. Bing, *A homeomorphism between the 3-sphere and the sum of two solid horned spheres*, Ann. of Math. 56 (1952), pp. 354-362.

[4] — *A decomposition of E^3 into points and tame arcs such that the decomposition space is topologically different from E^3*, Ann. of Math. 65 (1957), pp. 484-500.

[5] — *Point-like decompositions of E^3*, Fund. Math. 50 (1962), pp. 431-453.

[6] O. G. Harrold, Jr., H. C. Griffith, and E. E. Posey, *A characterization of tame curves in 3-space*, Trans. Amer. Math. Soc. 79 (1955), pp. 12-35.

[7] L. V. Keldyš, *Imbedding into E^4 of some monotone images of E^3*, Soviet Math. Dokl. 2 (1961), pp. 9-12.

UNIVERSITY OF IOWA
UNIVERSITY OF WISCONSIN

Reçu par la Rédaction le 17. 1. 1966

Extending Monotone Decompositions of 3-Manifolds

R. H. Bing

1. Introduction. Suppose X is a closed subset of Euclidean 3-space E^3 and f is a compact monotone map of X. We shall be concerned with conditions under which f may be extended to a compact monotone map of E^3 such that the image space is topologically E^3.

The following is typical of some of the theorems we shall prove.

Theorem 7.5. Suppose Y is a closed subset of E^3 such that each component of Y is compact but does not separate E^3. Then there is a compact monotone map of E^3 onto itself such that the components of Y are point inverses, and nondegenerate point inverses not in Y are polyhedral 1-complexes. The image of the union of Y and the nondegenerate point inverses is a closed set whose 1-point compactification is a dendron with a closed set of end points.

A fuller version of these results with a more complete treatment of proofs is included in a paper [2] submitted to Transactions of the American Mathematical Journal. We learn that Ralph J. Bean studies similar situations in a paper [1] to appear in Duke Mathematical Journal.

We find it convenient to state our preliminary results for compact 3-manifolds, so in first sections of this paper we deal with the 3-sphere S^3 rather than E^3. We suppose that E^3 and S^3 have their customary rectilinear structures.

Work on this paper supported by ONR grant NR 043-374. 7

By an _n-manifold_ we mean a separable metric space which is locally like Euclidean n-space. We do not imply without specification that it is either connected, compact, triangulated, or orientable. However, we do imply that it is without boundary. If we want to permit the possibility of a boundary, we call it a manifold with boundary. It is called a _manifold with non null boundary_ if we want to say that it has a boundary for certain.

For a manifold with non null boundary (such as a ball, arc, disk), we use Bd M to denote the boundary of M (set of points which have open neighborhoods homeomorphic to closed Euclidean half space) and Int M to denote the interior of M (set of points with open neighborhoods homeomorphic to Euclidean space).

When we call a manifold or geometric object _triangulated_, we mean that a particular triangulation is assigned to it. If no such assignment has been made but the object is isometric to some geometric complex, we say that the object is _polyhedral_. If the object is topologically equivalent to some geometric complex, we say that it is _triangulable_ or _can be triangulated_. For example, in the plane a polygon is regarded as triangulated and the closed disk bounded by the polygon is polyhedral. If a particular triangulation were assigned to the polygonal disk, it would then be triangulated. A round disk would be neither triangulated nor polyhedral, but it would be triangulable.

Since it is known [4, 6, 11, 12] that any 3-manifold with boundary is triangulable, it does not restrict those being studied to limit our consideration to those that are polyhedral. By using the polyhedral requirement, we are able to specify that certain of our maps are piecewise linear and certain subsets are polyhedral.

A 2-manifold is orientable if it contains no Möbius band; a 3-manifold is orientable if it contains no solid Klein bottle. A solid Klein bottle is a set topologically equivalent to the cartesian product of a Möbius band and a segment.

A closed polyhedral object is called a polyhedron. We do not insist that it be compact. A polyhedral n-cell is a polyhedron that can be given a triangulation compatible with its metric which is isomorphic to some rectilinear triangulation of the ordinary n-cell. A compact 1-dimensional polyhedron is called a polyhedral 1-complex.

We say that there is an elementary geometric collapse of a polyhedron P_1 onto a subpolyhedron P_2 if for some integer m there is a polyhedral m-cell B^m such that $P_1 = P_2 \cup B^m$ and $P_2 \cap B^m$ is a polyhedral (m - 1)-cell on Bd B^m. We write $P_1 \searrow P_2$. We say that P collapses to Q if there is a finite sequence of elementary collapses such that

$P = P_1 \searrow P_2 \searrow \dots \searrow P_j = Q.$

Suppose P is a compact polyhedron in a polyhedral n-manifold with boundary M^n. We say that N^n is a regular neighborhood of P in M^n if N^n is an n-manifold with boundary in M^n, N^n collapses to P, and P lies in an open subset of M^n in N^n. It is known [10, 13, 15] that if N_1^n, N_2^n

9

are regular neighborhoods of P, then there is a piecewise linear homeomorphism of N_1^n onto N_2^n.

We call an i-dimensional polyhedron P^i an i-<u>spine</u> of a compact polyhedral n-manifold with boundary M^n if M^n collapses to P^i and P^i does not collapse onto any proper subpolyhedron.

A collection G of mutually exclusive compact sets in a metric space X is called an *upper semicontinuous collection* if X is the union of elements of G and for each open set U in X containing an element of G, the union of elements of G in U is open. The *decomposition space* X/G has as points the elements of G and as open sets collections of these "points" whose union is open in X. The *decomposition map* of X onto X/G is the one which sends the point $x \in X$ into the "point" $g_X \in X/G$ which contains it (when g_X is regarded as a set in X). The decomposition map may alternatively be called an *upper semicontinuous decomposition* or for brevity merely a *decomposition*. We speak of the decomposition of X as being *into* sets (elements of G) and *onto* X/G. The decomposition is *monotone* if each element of G is connected.

If A is a compact subset of X rather than an upper semicontinuous decomposition of X, we use X/A to denote X/G_A where G_A is the upper semicontinuous decomposition of X whose only nondegenerate element is A.

Closely related to the notion of a monotone upper semicontinuous decomposition is that of a monotone mapping. A continuous transformation is called a _map_ or _mapping_. A map f of X is _monotone_ if the point inverses are compact continua—that is, for each $y \in f(X)$, $f^{-1}(y)$ is compact and connected.

If f is a monotone map of a space X onto a space Y such that the point inverses form an upper semicontinuous collection G, consideration of the 1-1 continuous map of $[0, \infty)$ onto a circle shows that one cannot conclude that Y is homeomorphic to X/G. However, they are homeomorphic if the map f is compact. A mapping is _compact_ if the inverses of compact sets are compact. See [11] for a discussion of relationships between decompositions and mappings.

A map f of a polyhedron P_1 into a polyhedron P_2 is _piecewise linear_ if P_1, P_2 have triangulations T_1, T_2 compatible with the metrics of P_1, P_2 so that f takes each simplex of T_1 linearly _into_ a simplex of P_2. In case P_1 is compact we can say that f takes each simplex of T_2 linearly _onto_ a simplex of T_1 but for noncompact polyhedra, this is too restrictive. If $g(x) = 1/x$ $(x \geq 1)$ there is a piecewise linear map f of $[1, \infty)$ into $[0, 1]$ such that $|f(x) - g(x)| \leq 1/2x$.

Throughout this paper we shall use M^3 to denote a _compact, connected, polyhedral_ 3-_manifold with boundary_. Perhaps it has no boundary, but if this is required it is so specified.

2. Boring holes in M^3. Suppose H^3 is a polyhedral 3-cell in M^3 so that $H^3 \cap \text{Bd } M^3$ is the union of two disjoint disks D_0^2, D_1^2. If we form a new 3-manifold with boundary M_1^3 by removing $\text{Int } H^3 \cup \text{Int } D_0^2 \cup \text{Int } D_1^2$ from M^3, we say that H^3 is a hole bored in M^3, and D_0^2, D_1^2 are the ends of the hole. We say that M_1^3 was obtained by boring a hole in M^3.

It is known that if M^3 has a non null boundary, then it can be changed into a cube with handles (some of the handles will be twisted or nonoriented if M^3 is not orientable) by boring a finite number of holes in M^3. A convenient way to find the holes is to bore them about spanning arcs in the 1-skeleton of some triangulation of M^3. Theorem 2.2 is a variation of this result. In order to prove Theorem 2.2, we first use Theorem 2.1 to build a 2-spine Q^2 which perhaps looks like 3-pages of a book at some places. Casler proved a stronger version of Theorem 2.1 in [9] where he built a Q^2 with a standard structure. We do not need this standard structure. An elementary proof of Theorem 2.1 was included in [1].

Theorem 2.1. If M^3 has a non null boundary, there is a spine Q^2 for M^3 such that if two disks in Q^2 intersect in an interior point of each, their intersection is 2-dimensional.

Theorem 2.2. If M^3 has precisely two boundary components, it is possible to obtain a 3-manifold with boundary M_0^3 from M^3 by boring a finite number of holes in it such that for some polyhedral 2-manifold M^2 there is a piecewise linear homeomorphism of $M^2 \times [0, 1]$ onto M_0^3.

Proof. We shall show that by boring holes in M^3, it can be changed to a polyhedral 3-manifold with boundary M_0^3 such that M_0^3 has two boundary components and M_0^3 collapses to a 2-spine M^2 which is a 2-manifold M^2. The truth of Theorem 2.2 then follows from the facts that M_0^3 is a regular neighborhood of M^2, any two regular neighborhoods of M^2 are piecewise linearly homeomorphic, and there is a regular neighborhood of M^2 piecewise linearly homeomorphic to $M^2 \times [0, 1]$.

It follows from Theorem 2.1 that M^3 has a 2-spine Q^2 such that if two disks in Q^2 meet in an interior point of each, their intersection is 2-dimensional. We suppose Q^2 is such a spine that lies in Int M^3. The 2-manifold M^2 promised by the preceding paragraph is obtained by altering Q^2.

If there is a 1-cell A_1 and three disks D_1^2, D_2^2, D_3^2 in Q^2 such that $D_i^2 \cap D_j^2 = A^1$ $(i \neq j)$, some one of the disks is not a subset of the part of Q^2 irreducible with respect to separating the two boundary components from each other in M^3. Holes can be bored in M^3 and through this disk so as to obtain a 3-manifold with two boundary components which collapses onto a proper subpolyhedron of Q^2.

Let M_1^3 be a 3-manifold with two boundary components obtained by boring holes in M^3 such that M_1^3 contains a spine $Q_1^2 \subset Q^2$ such that no further boring of holes in M_1^3 results in a 3-manifold with two boundary components and a spine which is a proper subset of Q_1^2. Note that Q_1^2 is the union of a 2-manifold M_1^2 and a 1-complex. By adjusting this 1-complex we may suppose that it is the union of a finite number of disjoint arcs A_1^1, $A_2^1, \ldots, A_m^1$ such that each $A_i^1 \cap M_1^2 = \mathrm{Bd}\, A_i^1$.

Boring holes in M_1^3 with the A_i^1's as centers of the holes produces a 3-manifold with boundary M_0^3 such that M_0^3 has two boundary components and a 2-manifold M^2 as a spine where M^2 is obtained from M_1^2 by replacing the A_i^1's and disks on M_1^2 about their ends by annuli.

3. Decomposing M^3 into 2-manifolds and pinched 2-manifolds.

A pinched manifold results from identifying two points of the same manifold. If two objects are joined at a point, their union is called the wedge of the two objects. The wedge of two tangent round 2-spheres is an example of a pinched 2-manifold. If a cube with handles has one of its handles squeezed to a point, the boundary of the resulting object may be an example of a pinched 2-manifold.

Suppose B^2 is a boundary component of M^3. It is known (see for example Lemma 1 of [11]) that there is a piecewise linear homeomorphism h of $B^2 \times [0, 1]$ into M^3 so that for each $x \in B^2$, $h(x \times 0) = x$ and $h(B^2 \times (0, 1]) \subset \mathrm{Int}\, M^3$. We call $h(B^2 \times [0, 1])$ a polyhedral cartesian product neighborhood of B^2.

In the following three theorems we show that certain 3-manifolds can be decomposed into 2-manifolds and pinched 2-manifolds. A map is open if the images of open sets are open.

Theorem 3.1. Suppose M^3 has precisely two boundary components B_0^2, B_1^2. Then there is a monotone piecewise linear open map f of M^3 onto $[0,1]$ such that $f^{-1}(0) = B_0^2$, $f^{-1}(1) = B_1^2$, and for each $t \in [0,1]$, $f^{-1}(t)$ is either a 2-manifold or a pinched 2-manifold. If $[a,b] \subset [0,1]$ and for each $t \in [a,b]$, $f^{-1}(t)$ is a 2-manifold, there is a piecewise linear homeomorphism of $f^{-1}[a,b]$ onto $f^{-1}(a) \times [a,b]$ that takes $f^{-1}(t)$ onto $f^{-1}(a) \times t$.

Proof. Let h_0 be a piecewise linear homeomorphism of $B_0 \times [0, 1/3]$ onto a cartesian product neighborhood of B_0^2 in M^3 so that $h_0(x \times 0) = x$ and let h_1 be a piecewise linear homeomorphism of $B_1^2 \times [2/3, 1]$ onto a cartesian product neighborhood of B_1^2 in M^3 so that $h_1(x \times 1) = x$. We suppose these two cartesian product neighborhoods do not intersect each other. Then $M^3 - h_0(B_0^2 \times [0, 1/3)) - h_1(B_1^2 \times (2/3, 1]) = M_1^3$ is a 3-manifold with precisely two boundary components $h_0(B_0^2 \times 1/3)$, $h_1(B_1^2 \times 2/3)$.

It follows from Theorem 2.2 that it is possible to bore a finite number of mutually exclusive polyhedral holes $H_1^3, H_2^3, \ldots, H_n^3$ in M_1^3 such that each of these holes has both ends on the same boundary component of M_1^3 and for some 2-manifold M^2, there is a piecewise linear homeomorphism h of $M^2 \times [1/3, 2/3]$ onto the closure of $(M_1^3 - \cup H_i^3)$. We suppose $h(M^2 \times 1/3)$, $h(M^2 \times 2/3)$ are the altered $h_0(B_0^2 \times 1/3)$, $h_1(B_1^2 \times 2/3)$, respectively, and define $f(h(M^2 \times t)) = t$, $t \in [1/3, 2/3]$.

15

Let $H_1^3, H_2^3, \ldots, H_m^3$ be the holes bored in M_1^3 from $h_0(B_0^2 \times 2/3)$. For each H_i^3, let C_i^3, $C_i^{3\prime}$ be caps over the ends of H_i^3 so that each C_i^3 is a polyhedral 3-cell in $h_0(B_0^2 \times [0, 1/3])$ such that $C_i^3 \cap h_0(B_0^2 \times 1/3)$ is an end of H_i^3, $\text{Bd}\, C_i^3 \cap h_0(B_0^2 \times 1/6)$ is a point and for each $t \in (1/6, 1/3)$, $\text{Bd}\, C_i^3 \cap h_0(B_0^2 \times t)$ is a simple closed curve; the $C_i^{3\prime}$ are similarly defined. We suppose that no two of the caps intersect each other.

If $t \in [0, 1/3)$ and $p \in h_0(B_0^2 \times t) - \cup \text{Int}\, C_i^3 - \cup \text{Int}\, C_i^{3\prime}$, we define $f(p) = t$.

For each $i = 1, 2, \ldots, m$, consider the polyhedral 3-cell $H_i^3 \cup C_i^3 \cup C_i^{3\prime}$. We note that f has already been defined on its boundary. Triangulate this boundary so that f is linear with respect to the triangulation. Then triangulate $H_i^3 \cup C_i^3 \cup C_i^{3\prime}$ by coning from an interior point p_i.

Let $\varepsilon_1, \varepsilon_2, \ldots, \varepsilon_m$ be different numbers in $(1/6, 1/3)$. Define $f(p_i) = \varepsilon_i$ and extend f linearly to the simplexes of the conical triangulation of $H_i^3 \cup C_i^3 \cup C_i^{3\prime}$. Then $(H_i^3 \cup C_i^3 \cup C_i^{3\prime}) \cap f^{-1}(t)$ is the union of two points if $t = 1/6$, the union of two disjoint disks (cones) if $t \in (1/6, \varepsilon_i)$, the union of two cones with a common vertex p_i if $t = \varepsilon_i$, and an annulus if $t \in (\varepsilon_i, 1/3]$. It is to be noted that f is open at p_i. See Figure 1.

Note that for some neighborhood $(\varepsilon_i - \delta, \varepsilon_i + \delta)$ of ε_i, there is a piecewise linear projection r_i of $f^{-1}[\varepsilon_i - \delta, \varepsilon_i + \delta]$ onto $f^{-1}(\varepsilon_i)$ such that for $t \in [\varepsilon_i - \delta, \varepsilon_i)$, r_i restricted to $f^{-1}(t)$ is one-to-one except onto p_i and the inverse of p_i is a pair of points while if $t \in (\varepsilon_i, \varepsilon_i + \delta)$, r_i

16

restricted to $f^{-1}(t)$ is one-to-one except onto p_i and the inverse of p_i is a simple closed curve.

In a similar fashion we define f on $h_1(B_1^2 \times [2/3, 1])$ and the holes with ends on $h_1(B_1^2 \times 2/3)$.

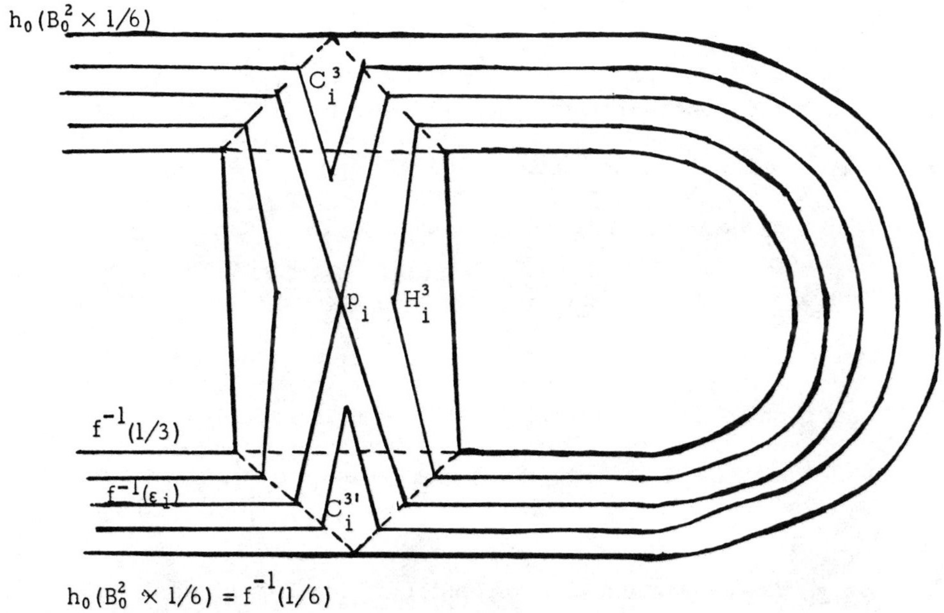

Figure 1

Definitions. A dendron is a compact locally connected metric continuum which contains no simple closed curve. A point p is of order $\leqq n$ if for each neighborhood U of p there is an open set V with $p \in V \subset U$ such that the boundary (frontier) of V has at most n points. A point is of order n if it is of order $\leqq n$ but not of order $\leqq n - 1$. A point of order 1 is an end point of the dendron and a point which is not of order $\leqq 2$ is a branch point. A polyhedral dendron (sometimes called a tree) can have at most a finite number of branch and end points. A dendron with precisely three end points (and hence one branch point) is called a triod.

The following is a variation of Theorem 3.1.

Theorem 3.2. If M^3 has precisely three boundary components, there is a piecewise linear monotone map f of M^3 onto a polyhedral triod T^1 such that each boundary component of M^3 is an inverse of an end point of T^1 and for each $t \in T^1$, $f^{-1}(t)$ is either a 2-manifold or a pinched 2-manifold.

Proof. The plan is to find a polyhedral pinched 2-manifold in M^3 which separates each pair of the three boundary components. Then apply Theorem 3.1 to each piece. Details are found in [2].

In a certain sense, Theorem 3.2 is proved by combining two boundary components of M^3. Repetitions of this procedure permits us to prove the following extension of Theorem 3.2.

Theorem 3.3. If M^3 has more than one boundary component, there is a piecewise linear monotone map f of M^3 onto a tree T^1 such that each point of T^1 is of order $\leqq 3$, the boundary components of M^3 are the inverses of end points of T^1, and for each $t \in T^1$, $f^{-1}(t)$ is either a 2-manifold or a pinched 2-manifold.

Theorem 7.2 of Section 7 gives results about the mapping of M^3 onto a dendron if M^3 has no boundary.

4. Decomposition maps onto 2-spheres. In the last section, we decomposed M^3 into 2-manifolds and pinched 2-manifolds by letting these be the point inverses in a map onto a dendron. In Section 6 we shall map M^3 onto a punctured cube. This is done by considering the point inverses in the map onto a dendron and decomposing M^3 by decomposing each of these inverses. In this section we consider useful decompositions of 2-manifolds and pinched 2-manifolds.

Consider a simplex Δ which is the join of two simplexes Δ_1, Δ_2. Then $\Delta = \Delta_1 \circ \Delta_2$ is the union of segments from Δ_1 to Δ_2. For each $t \in [0,1]$, let A_t be the set of all points such that for some $p_1 \in \Delta_1$, $p_2 \in \Delta_2$, p divides the segment from p_1 to p_2 in the ratio t to $1-t$. We denote p by (p_1, p_2, t) and note that $A_t = \{(p_1, p_2, t) \mid p_1 \in \Delta_1,\ p_2 \in \Delta_2\}$. Then $A_0 = \Delta_1$, $A_1 = \Delta_2$, and the midsection $A_{1/2}$ is homeomorphic to $\Delta_1 \times \Delta_2$. We use A_B to denote $\cup_{t \in B} A_t$.

Let Δ/Δ_1 be the cell obtained from Δ by shrinking Δ_1 to a point. To give Δ/Δ_1 a structure, we regard it as the union of $A_{[1/2, 1]}$ and the cone C over $A_{1/2}$ from a point v_0 where $A_{[1/2,1]} \cap C = A_{1/2}$.

The natural decomposition map f from Δ to Δ/Δ_1 is the identity on $A_{[1/2,1]}$ and for $t \in [0,1/2]$, $f(p_1, p_2, t)$ is the point that divides the segment from v_0 to $(p_1, p_2, 1/2)$ in the ratio 2t to 1 - 2t.

We note that although Δ is a simplex and $f(\Delta) = \Delta/\Delta_1$ is a piecewise linear cell, the map f is not piecewise linear.

If L is a subcomplex of complex K, K/L results by regarding L as a point. To give K/L a cellular structure, we let T' be the first barycentric subdivision of the triangulation of K and note that if Δ is a simplex in T', then either Δ misses L, lies in L, or intersects it in a proper face Δ_L. The natural decomposition map f is such that f(L) is a vertex v_0 of K/L; f is the identity on each simplex of T' that misses L; for each simplex Δ of T' whose intersection with L is a proper face Δ_L of Δ, f is the natural decomposition map from Δ to Δ/Δ_L.

The following result is well known.

Theorem 4.1. If M^2 is a compact connected polyhedral 2-manifold, there is a connected polyhedral 1-complex C^1 in M^2 such that M^2/C^1 is a polyhedral 2-sphere.

To obtain C^1, triangulate M^2 and remove an open 2-simplex Int Δ^2 from M^2. Then C^1 is obtained by collapsing M^2 - Int Δ^2.

The following two results may be proved in a similar fashion.

Theorem 4.2. If N^2 is a compact connected triangulated pinched 2-manifold with no separating point, there is a connected polyhedral 1-complex

C^1 in the 1-skeleton of the triangulation such that N^2/C^1 is a polyhedral 2-sphere and C^1 is of order 2 at the pinch point.

We note that C^1 abuts on the pinch point from both sides in N^2.

Theorem 4.3. If N^2 is a compact connected triangulated pinched 2-manifold which is separated by its pinch point, there is a connected polyhedral 1-complex C^1 in the 1-skeleton of the triangulation such that C^1 abuts on the pinch point from both sides and is of order 2 there while N^2/C^1 is the wedge of two polyhedral 2-spheres.

5. Changing pseudo spines. A pseudo spine for a compact connected polyhedral 2-manifold is a connected polyhedron C in M^2 such that M^2/C is topologically a 2-sphere. To get C one might remove an open 2-simplex from M^2 and collapse the remainder. Note that the pseudo spine differs from a spine for M^2 in that an open simplex was removed before the collapsing. Also, the collapsing was not necessarily continued as much as possible. To emphasize that the pseudo spines we shall be using are 1-dimensional polyhedra, we may call them pseudo 1-spines.

If M^2 is a compact, connected, polyhedral 2-manifold, $M^2 \times [0,1]$ is its cartesian product with an interval, and $C_0^1 \times 0$, $C_1^1 \times 1$ are pseudo 1-spines of $M^2 \times 0$, $M^2 \times 1$, we shall be interested in finding pseudo 1-spines for the intermediate layers.

Theorem 5.1. Suppose M^2 is a compact connected polyhedral 2-manifold and C_0^1, C_1^1 are pseudo 1-spines of M^2. Then C_0^1, C_1^1 can be extended to a 1-parameter family of pseudo 1-spines C_t^1 $(0 \leqq t \leqq 1)$ of M^2 such that there is a monotone map f of $M^2 \times [0,1]$ onto $S^2 \times [0,1]$ with $f(M^2 \times t)$ being equal to $S^2 \times t$ and $C_t^1 \times t$ being the only non-degenerate inverse in $M^2 \times t$.

Proof. Let $D_1^2, D_2^2, \ldots$ be a sequence of polyhedral disks in M^2 such that $D_i^2 \subset \text{Int } D_{i+1}^2$ and $\cup D_i^2 = M^2 - C_0^1$ and $E_1^2, E_2^2, \ldots$ be such a sequence of disks whose union is $M^2 - C_1^1$.

Since E_1^2 is a disk which lies on the interior of a larger disk in M^2 with the larger disk not being a subset of $M^2 - \text{Int } D_1^2$, there is a piecewise linear isotopy fixed except on this larger disk that pushes $M^2 - \text{Int } D_1^2$ off E_1^2. Denote this isotopy by H_t $(0 \leqq t \leqq 1/4)$ where H_0 = identity and $E_1^2 \cap H_{1/4}(M^2 - \text{Int } D_1^2) = 0$. Then $H_{1/4}(D_1^2) - \text{Int } E_1^2$ is an annulus and H_t $(0 \leqq t \leqq 1/4)$ can be extended to a piecewise linear isotopy H_t $(0 \leqq t \leqq 1/2)$ such that $H_{1/2}(D_1^2) = E_1^2$.

The isotopy $(0 \leqq t \leqq 1/2)$ can be extended to a piecewise linear isotopy H_t $(0 \leqq t \leqq 3/4)$ so that $H_{3/4}(D_2^2) = E_2^2$ and for $x \in D_1^2$ and $1/2 \leqq s \leqq 3/4$, $H_s(x) = H_{1/2}(x)$.

Similarly, we extend H_t to H_t $(0 \leqq t < 1)$. In general, H_t is such that $H_{t_i}(D_i) = E_i$ for $t_i = (1 - 1/2^i)$ and $H_s = H_{t_i}$ on D_i for $s \geqq (1 - 1/2^i)$.

For $0 \leqq t < 1$, $C_t^1 = H_t(C_0^1)$.

Let f_0 be a decomposition map of M^2 onto S^2 such that C_0^1 is the only nondegenerate inverse. The map f of $M^2 \times [0, 1)$ is defined by $f(x \times t) = f_0 H_t^{-1}(x) \times t$. It is extended to $M^2 \times 1$ by continuity.

Questions. It would appear that Theorem 5.1 is a special case of some more general theorem. If f is a map of a compact continuum X onto $[0, 1]$ such that the point inverses are all homeomorphic to each other, under what condition can it be concluded that X is homeomorphic to $f^{-1}(0) \times [0, 1]$? If g is a monotone map of $Y \times [0, 1]$ onto Z such that the $g(Y \times t)$'s are homeomorphic but no two intersect, under what condition is Z homeomorphic to $g(Y \times 0) \times [0, 1]$? Is it enough that each $g(Y \times t)$ be a 2-sphere? What if each $Y \times t$ contain at most one nondegenerate point inverse? Something needs to be imposed because if P is a pseudo arc, P' is a nondegenerate proper subcontinuum of $P \times 0$, then $(P \times [0, 1])/P'$ is not homeomorphic to $P \times [0, 1]$ since it contains a triod. However, it is known [3] that $(P \times 0)/P'$ is homeomorphic to P.

It is to be noted that the map f is not piecewise linear and is especially bad at $t = 1$. Can we find such an f that has the nice properties of a natural decomposition map?

6. Decomposing 3-manifolds with boundaries. Suppose $B_0^3, B_1^3, \ldots, B_n^3$ are polyhedral 3-cells such that $B_i^3 \subset \operatorname{Int} B_0^3$ for $i = 1, 2, \ldots, n$ and $B_i^3 \cap B_j^3 = 0$ for $0 < i < j \leqq n$. Then $B_0^3 - \bigcup_{i=1}^n \operatorname{Int} B_i^3$ is a polyhedral punctured cube.

If the boundary components of M^3 are already nicely decomposed with the decomposition space of each being a 2-sphere, we shall be interested in extending the decomposition to M^3 so that the decomposition space is a punctured cube.

Theorem 6.1. Suppose M^3 has precisely two boundary components B_0^2, B_1^2 with pseudo 1-spines C_0^1, C_1^1. Then there is a monotone map g of M^3 onto $S^2 \times [0,1]$ such that $g^{-1}(S^2 \times 0) = B_0^2$, $g^{-1}(S^2 \times 1) = B_1^2$, each $g^{-1}(S^2 \times t)$ is either a polyhedral 2-manifold or polyhedral pinched 2-manifold, C_0^1 and C_1^1 are point inverses, each nondegenerate point inverse is a polyhedral 1-complex, and the image of the union of the non-degenerate point inverses is an arc from $S^2 \times 0$ to $S^2 \times 1$ that intersects each $S^2 \times t$ only once.

Proof. The plan is to get a piecewise linear map of M^3 onto an arc as suggested in Theorem 3.1. There are only a finite number of pinched manifolds in this decomposition and each is provided with a pseudo 1-spine. Theorem 5.1 is used to provide pseudo 1-spines for the intermediate manifolds. Details of construction are found in[2].

Theorem 6.2. If M^3 has only one boundary component B^2 and C^1 is a pseudo 1-spine of B^2, then there is a monotone map g of M^3 onto a polyhedral 3-cell B^3 such that $g^{-1}(\mathrm{Bd}\, B^3) = B^2$, each nondegenerate inverse is a polyhedral 1-complex, C^1 is the only such inverse on B^2, and the image under g of the union of the nondegenerate inverses is a polyhedral arc in B^3.

Proof. A polyhedral open ball is removed from M^3 to change it to a 3-manifold with two boundary components. A polygonal arc is selected to serve as a pseudo 1-spine for the new boundary component. Theorem 6.1 assures us that there is a suitable map of the remainder onto $S^2 \times [0, 1]$. This map may be extended to all of M^3 to get the required map onto B^3.

Theorem 6.3. Suppose M^3 has boundary components $B_1^2, B_2^2, \ldots, B_n^2$ ($n \geqq 2$) with polyhedral pseudo 1-spines $C_1^1, C_2^1, \ldots, C_n^1$. Then there is a monotone map g of M^3 onto a polyhedral punctured cube K^3 with boundary components $E_1^2, E_2^2, \ldots, E_n^2$ such that $g^{-1}(E_i^2) = B_i^2$, C_i^1 is the only nondegenerate inverse on B_i^2, each nondegenerate inverse is a polyhedral 1-complex, and the image under g of the union of the nondegenerate inverses is a tree in K^3 each of whose ends is on Bd K^3.

Proof. Theorem 3.3 provides a monotone map of M^3 onto a tree such that each point inverse is either a 2-manifold or a pinched 2-manifold. Each point inverse that is the wedge of two 2-manifolds is sent by g onto tangent polyhedral 2-spheres and intermediate 2-manifolds and pinched 2-manifolds are treated as in Theorem 6.1. See [2] for details.

7. Extending monotone mappings. In this section we suppose M^3 is without boundary and X is a closed proper subset of M^3 such that if U is a connected open subset of M^3, then either $U - (X \cap U)$ is connected or X intersects the boundary (frontier) of U.

The principal result (Theorem 7.3) of this section is that there is a monotone map of M^3 onto S^3 that has the components of X as point inverses. Before proving this result, we prove some theorems about X.

Theorem 7.1. There is a decreasing sequence $M_1^3, M_2^3, \ldots,$ such that $X = M_1^3 \cap M_2^3 \cap \ldots$ where each M_i^3 is a 3-manifold with boundary which is a polyhedron in M^3, $M_{i+1}^3 \subset \text{Int } M_i^3$, and each component of M_i^3 has a connected boundary.

Proof. Suppose M_i^3 has been constructed. We prove the theorem by showing how to get an M_{i+1}^3 in any open subset V of Int M_i^3 which contains X.

Let T be a triangulation of M^3 of mesh less than one-half the distance from X to $M^3 - V$ and T'' be the second baricentric subdivision of T. Let N^3 be the union of the closed simplexes of T which intersect X and N_1^3 be the union of the closed simplexes of T'' which intersect N^3. Then N_1^3 is a regular neighborhood of N^3 and is hence a polyhedral 3-manifold with boundary [10, 13, 15]. If C^3 is a component of N_1^3, it may have several boundary components, but boring holes in C^3 which miss X reduces the number of boundary components to one. Hence M_{i+1}^3 is obtained from N_1^3 by boring holes.

The following result follows from Theorem 7.1 and repeated applications of Theorem 3.3

Theorem 7.2. There is a dendron D^1 and a monotone map f of M^3

26

onto D^1 such that each point of D^1 is of order $\leqq 3$, D^1 is locally polyhedral mod its set of end points, the components of X are the inverses of the end points of D^1, the inverse of each non-end point is either a polyhedral 2-manifold or a polyhedral pinched 2-manifold, and f is locally piecewise linear mod X.

Theorem 7.3. There is a monotone map g of M^3 onto S^3 such that each component of X is a point inverse, g(X) lies on a straight segment, and each nondegenerate point inverse outside X is a polyhedral 1-complex. In fact, there are also a dendron D^1 in S^3 and a monotone map f of M^3 onto D^1 such that D^1 is locally polyhedral mod its set of end points, each branch point of D^1 is of order 3, X is the inverse under f of the set of end points of D^1, $f = g$ on X and on each nondegenerate point inverse under g, and if x is a non end point of D^1, $f^{-1}(x)$ is either a polyhedral 2-manifold or pinched 2-manifold, $f^{-1}(x)$ contains at most one nondegenerate point inverse under g, and $gf^{-1}(x)$ is either a 2-sphere or the wedge of two 2-spheres.

Proof. The dendron D^1 and the map f are given in Theorem 7.2. The embedding of D^1 in S^3 is described in [2].

Questions. Alden Wright noted in the preceding two theorems the decomposition space was S^3 rather than M^3. He asks under what conditions M^3 can be used as the decomposition space.

Russell McMillan noted that our target space had trivial homotopy. He asks if N^3 and W^3 are 3-manifolds with or without boundary and

$f : N^3 \to W^3$ such that f_* sends $\pi_1(N^3, x_0)$ onto $\pi_1(W^3, f(x_0))$, when is f homotopic to a monotone map. He is willing to assume piecewise linearity.

The unicoherence of S^3 permits us to weaken the restrictions on X (where such a weakening would not be possible for such a 3-manifold as $S^2 \times S^1$). Hence, by ignoring details about f we may state Theorem 7.3 for S^3 as follows.

Theorem 7.4. If Y is a closed subset of S^3, no component of which separates S^3, there is a monotone map of S^3 onto itself such that the components of Y are point inverses, each nondegenerate point inverse not in Y is a polyhedral 1-complex, and the image of the union of Y and these nondegenerate point inverses is a dendron.

We can deal with non-compact 3-manifolds since each connected non compact 3-manifold is the union of a monotone increasing sequence of connected compact 3-manifolds with boundaries. For example, for Euclidean 3-space E^3 we have Theorem 7.5 stated in Section 1.

Questions. In [14], Whyburn raises the question as to whether or not each monotone map of E^3 onto itself need be compact. Bing answers this question in [8] by describing a non-compact monotone map of E^3 onto itself. Could a monotone map be shown to be compact in case each point inverse is either a 1-complex or a component of Y?

It is to be noted that the maps g of Theorems 7.4, 7.5 were not necessarily cellular since some of the 1-complexes which were point inverses

28

might contain cycles. Could one obtain a cellular g if the components of Y were pointlike? In particular, what happens if Y is the union of the tame arcs in the dogbone decomposition described in [5]?

Is there a monotone map g of E^3 onto itself having the tame arcs of the dogbone decomposition as point inverses and such that the image of the nondegenerate inverses is 0-dimensional.

<u>Definitions</u>. Euclidean n-space $(n > 1)$ has only <u>one end</u>. An <u>end</u> of a connected, locally compact, non-compact metric space S may be defined as an equivalence class of sequences $U_1, U_2, \ldots$ where U_i is a non null connected open subset of X with a compact boundary (frontier), $\overline{U_{i+1}} \subset U_i$, and $\cap U_i = 0$. The sequence $U_1, U_2, \ldots$ is in the same equivalence class as $V_1, V_2, \ldots$ if and only if the V_i's satisfy similar conditions, each U_i contains a V_j and each V_i contains a U_j. Hence, S <u>has one end</u> if for each compact set C, the closure of S - C has precisely one non-compact component.

For other non-compact 3-manifolds, our methods give the following result.

<u>Theorem 7.6</u>. Suppose W^3 is a non-compact, connected, polyhedral, 3-manifold without boundary and Y is a closed subset of W^3 such that each component of Y is compact and if U is a connected open subset of W^3, either Y intersects the boundary of U or $U - (Y \cap U)$ is connected. Then there is a straight segment A in S^3, a dendron D^1 which is locally polyhedral mod its set of end points and which intersects A in its set of

end points, a closed subset C^0 of the set of end points of D^1, and a monotone map g of W^3 onto $S^3 - C^0$ such that the components of Y are point inverses of the end points of D^1 not in C^0, the nondegenerate point inverses not in Y are polyhedral 1-complexes in W^3, and the image of the union of Y and the nondegenerate point inverses is $D^1 - C^0$.

The points C^0 correspond to the ends of W^3.

8. Monotone maps that cannot be extended. It would be interesting to know which monotone decompositions of a portion of a 3-manifold can be extended to the whole 3-manifold so that the decomposition space is a 3-manifold.

Suppose C_1, C_2 are two linking round circles in a simplex of S^3 and G is a decomposition of $C_1 \cup C_2$ whose only nondegenerate element is C_2. Is there a monotone upper semicontinuous decomposition of S^3 onto itself such that each element of G is an element of this decomposition? Is there a monotone map g of S^3 onto itself that is a homeomorphism on C_1 and sends C_2 to a point off $g(C_1)$? The example on page 7 of [7] gives an affirmative answer to the second question while Theorem 8.1 gives a negative answer to the first.

Definitions. Suppose U is an open subset of a polyhedral 3-manifold and J^1 is polygonal simple closed curve in U. We say that nJ^1 bounds (mod the integers) in U if J can be regarded as an oriented 1-cycle so that there is some oriented complex C^2 in U such that $\partial C^2 = nJ^1$. By

30

adjusting C^2 to remove certain multiple points we may suppose with no loss of generality that it is locally a 2-manifold off J^1 and that at J^1 it locally resembles n sheets with a common edge.

Notice that in considering complexes, we are using the geometric approach rather than an abstract one and regard a simplex as a geometric object (perhaps oriented), a complex as a geometric object (with perhaps its simplexes oriented), and the boundary of an oriented complex as an oriented complex of one lower dimension.

If B^3 is a polyhedral 3-cell and J_1, J_2 are disjoint simple closed curves in Int B , we say that the (unsigned) linking number of J_1 and J_2 is n if J_1 is homotopic in Int $B^3 - J_2$ to a polyhedral simple closed curve K_1^1, K_1^1 can be oriented so that it bounds an orientable polyhedral 2-manifold M^2 in Int B^3, and J_2 is homotopic in Int $B^3 - K_1^1$ to a polyhedral simple closed curve K_2^1 such that the piercing number of K_2^1 and M^2 is n. It is known that the linking number of J_1 and J_2 is independent of the K_1^1, M^2, K_2^1 selected and is equal to the linking number of J_2 and J_1. We say that J_1 and J_2 link if this linking number is not 0. If the linking number is 0, it is possible to adjust M^2 so as to get an oriented polyhedral 2-manifold with boundary M_1^2 so that Bd $M_1^2 = K_1^1$ and $M_1^2 \cap J_2 = 0$.

The following theorem is included for simplicity. It is a special case of the more complicated Theorem 8.3.

Theorem 8.1. There is no monotone upper semicontinuous decomposition of S^3 cnto itself such that the elements of G are elements of this decomposition.

Proof. Assume the decomposition can be extended. Then there is a monotone map g of S^3 onto itself such that the elements of G are point inverses under g.

Let J be a simple closed curve in $S^3 - g(C_1)$ that cannot be shrunk to a point in $S^3 - g(C_1)$. We show that there is no map g by showing that J can be shrunk to a point in $S^3 - g(C_1)$.

It follows from the monotonicity of g and the local Euclidean structure of S^3 that for each open set U containing J there is a homeomorphism h of J into $g^{-1}(U)$ such that the identity on J is homotopic to gh in U. We suppose that h is selected for an open subset U containing J such that $U \cap g(C_1) = 0$.

Let H_t $(0 \leqq t \leqq 1)$ be a pseudo isotopy of $S^3 - C_1$ into $S^3 - C_1$ such that $H_0 =$ Identity and $H_1(S^3 - C_1) = C_2$. Then a homotopy shrinking J to a point in $S^3 - g(C_1)$ spends the first part of its parameter pulling the identity on J to gh and the last part as gH_th $(0 \leqq t \leqq 1)$.

Example. One cannot weaken Theorem 8.1 by replacing "onto itself" by "onto a subset of itself." If one considers S^3 as the join of circles C_1 and C_2, one sees that there is a monotone decomposition of S^3 onto a disk that homeomorphically sends C_1 to the boundary of the disk and C_2 to the center. Other point inverses are simple curves that link C_1.

The proofs of the following three theorems are found in [2].

Theorem 8.2. Suppose J is a simple closed curve (perhaps wild) in a polyhedral 3-manifold W^3 such that J can be shrunk to a point in W^3. Then for each point $p \in J$ and each neighborhood U of p, there is polyhedral simple closed curve $K^1 \subset U - J$ such that no positive multiple of K^1 bounds in $W^3 - J$.

Theorem 8.3. If J is a simple closed curve in a polyhedral, orientable 3-manifold W^3 and J lies in an open subset U of W^3, there is a polyhedral simple closed curve K^1 in $U - J$ such that no positive multiple of K^1 bounds in $W^3 - J$.

Example. Theorem 8.3 is false if one omits the orientable requirement. Let P^2 be the projective plane, J_0 be a simple closed curve in P^2 such that $P^2 - J_0$ is an open disk. $W^3 = P^2 \times (-1, 1)$, and $J = J_0 \times 0$. If K^1 is any polyhedral simple closed curve in $W^3 - J$, $2K^1$ bounds in $W^3 - J$.

Theorem 8.4. Suppose B^3 is a polyhedral 3-cell in a polyhedral 3-manifold with or without boundary N^3 and C_1, C_2 are linking simple curves in Int B^3. Then if G is the decomposition of $C_1 \cup C_2$ whose only degenerate element is C_2, there is no monotone upper semicontinuous decomposition of N^3 onto a 3-manifold W^3 such that the elements of G are elements of the decomposition.

Question. Would Theorem 8.4 be true if instead of saying that C_1 and C_2 link in Int B^3 we required that they be polygonal simple closed curves in Int B^3 that 1-link there? Recall that two curves 1-link if they do not bound mutually exclusive orientable 2-manifolds. **(See Footnote.)**

THE UNIVERSITY OF WISCONSIN

REFERENCES

1. R. J. Bean, Repairing embeddings and decompositions in S^3, Duke Math. J.

2. R. H. Bing, Extending monotone decompositions of 3-manifolds, submitted to Trans. Amer. Math. Soc.

3. __________, Concerning hereditarily indecomposable continue, Pacific J. Math. 1 (1951), 43-51.

4. __________, Locally tame sets are tame, Ann. of Math. 59 (1954), 145-158.

5. __________, A decomposition of E^3 into points and tame arcs such that the decomposition space is topologically different from E^3, Ann. of Math., 65 (1957), 484-500.

6. __________, An alternative proof that 3-manifolds can be triangulated, Ann. of Math. 69(1959), 37-65.

7. __________, Decompositions of E^3, Topology of 3-Manifolds and Related Topics, Prentice-Hall (1962), 5-21.

8. __________, Solution of the monotone mapping problem, to appear.

9. B. G. Casler, An embedding theorem for connected 3-manifolds with boundary, Proc. Amer. Math. Soc. 16(1965), 559-566.

10. J.F.P. Hudson and E. C. Zeeman, On regular neighborhoods, Proc. London Math. Soc., (3) 14(1964), 719-745.

11. E. E. Moise, Affine structures in 3-manifolds, V. The triangulation theorem and Hauptvermutung, Ann. of Math. 56 (1952), 96-114.

12. __________, Affine Structures in 3-manifolds, VIII. Invariance of the knot-types; local tame embedding, Ann. of Math. 59(1954), 159-170.

13. J. H. C. Whitehead, Simplicial spaces, nuclei, and m-groups, Proc. London Math. Soc., 45 (1939), 243-327.

14. G. T. Whyburn, Decomposition spaces, Topology of 3-Manifolds and Related Topics, Prentice-Hall (1932), 2-3.

15. E. C. Zeeman, Polyhedral n-manifolds: II. Embeddings, Topology of 3-Manifolds and Related Topics, Prentice-Hall (1942), 64-70.

<u>Footnote</u>: Since this paper was presented, Alden Wright gave a negative answer to the question following Theorem 8.4 by showing in Corollary 6 to Theorem 4 of his thesis that if C_1, C_2 are unlinked polyhedral simple closed curve in S^3, then there is a monotone decomposition of S^3 onto itself such that C_2 and points of C_1 are elements of the decomposition. Wright's arguments work even if C_2 is not polyhedral but do not seem to apply if C_1 is wild.

TRANSACTIONS OF THE
AMERICAN MATHEMATICAL SOCIETY
Volume 149, June 1970

EXTENDING MONOTONE DECOMPOSITIONS OF 3-MANIFOLDS

BY
R. H. BING(1)

1. **Introduction.** Suppose X is a closed subset of Euclidean 3-space E^3 and f is a compact monotone map of X. We shall be concerned with conditions under which f may be extended to a compact monotone map of E^3 such that the image space is topologically E^3.

It was our original intention to show that if f has only a countable number of point inverses and none of these separates E^3, then there is a monotone upper semicontinuous decomposition of E^3 onto itself with the point inverses being elements of the decomposition. As we developed tools for proving this, we found that these tools proved more. Since this paper was written we learn that Ralph J. Bean studies similar situations in a paper entitled "Repairing embeddings and decompositions in S^3" to appear in Duke Mathematical Journal.

We find it convenient to state our preliminary results for compact 3-manifolds, so in first sections of this paper we deal with the 3-sphere S^3 rather than E^3. We suppose that E^3 and S^3 have their customary rectilinear structures.

By an *n-manifold* we mean a separable metric space which is locally like Euclidean n-space. We do not imply without specification that it is either connected, compact, triangulated, or orientable. However, we do imply that it is without boundary. If we want to permit the possibility of a boundary, we call it a manifold with boundary. It is called a *manifold with nonnull boundary* if we want to say that it has a boundary for certain.

For a manifold with nonnull boundary (such as a ball, arc, disk), we use Bd M to denote the boundary of M (set of points which have open neighborhoods homeomorphic to closed Euclidean half space) and Int M to denote the interior of M (set of points with open neighborhoods homeomorphic to Euclidean space).

When we call a manifold or geometric object *triangulated*, we mean that a particular triangulation is assigned to it. If no such assignment has been made but the object is isometric to some geometric complex, we say that the object is *polyhedral.* If the object is topologically equivalent to some geometric complex, we say that it is *triangulable* or *can be triangulated.* For example, in the plane, a polygon is regarded as triangulated and the closed disk bounded by the polygon is polyhedral. If a particular triangulation were assigned to the polygonal disk, it would

Received by the editors December 18, 1968.

(1) Paper partially written while a visitor at Australian National University with support from Contract NSF-3857 and University of Wisconsin. Completed under ONR Contract.

then be triangulated. A round disk would be neither triangulated nor polyhedral, but it would be triangulable.

Since it is known ([3], [4], [8], [9]) that any 3-manifold with boundary is triangulable, it does not restrict those being studied to limit our consideration to those that are polyhedral. By using the polyhedral requirement we are able to specify that certain of our maps are piecewise linear and certain subsets are polyhedral.

A 2-manifold is *orientable* if it contains no Möbius band; a 3-manifold is *orientable* if it contains no solid Klein bottle. A solid Klein bottle is a set topologically equivalent to the cartesian product of a Möbius band and a segment.

A closed *polyhedral* object is called a polyhedron. We do not insist that it it be compact. A *polyhedral n-cell* is a polyhedron that can be given a triangulation compatible with its metric which is isomorphic to some rectilinear triangulation of the ordinary n-cell. A compact 1-dimensional polyhedron is called a polyhedral 1-complex.

We say that there is an *elementary geometric collapse* of a polyhedron P_1 onto a subpolyhedron P_2 if for some integer m there is a polyhedral m-cell B^m such that $P_1=P_2 \cup B^m$ and $P_2 \cap B^m$ is a polyhedral $(n-1)$-cell on Bd B^m. We write $P_1 \searrow P_2$. We say that P collapses to Q if there is a finite sequence of elementary collapses such that

$$P = P_1 \searrow P_2 \searrow \cdots \searrow P_j = Q.$$

Suppose P is a compact polyhedron in a polyhedral n-manifold with boundary M^n. We say that N^n is *a regular neighborhood of* P in M^n if N^n is an n-manifold with boundary in M^n, N^n collapses to P, and P lies in an open subset of M^n in N^n. It is known ([7], [10], [12]) that if N_1^n, N_2^n are regular neighborhoods of P, then there is a piecewise linear homeomorphism of N_1^n onto N_2^n.

We call an i-dimensional polyhedron P^i an *i-spine* of a compact polyhedral n-manifold with boundary M^n if M^n collapses to P^i and P^i does not collapse onto any proper subpolyhedron.

A collection G of mutually exclusive compact sets in a metric space X is called an *upper semicontinuous collection* if X is the union of elements of G and for each open set U in X containing an element of G, the union of elements of G in U is open. The *decomposition space* X/G has as points the elements of G and as open sets collections of these "points" whose union is open in X. The *decomposition map* of X onto X/G is the one which sends the point $x \in X$ into the "point" $g_X \in X/G$ which contains it (when g_X is regarded as a set in X). The decomposition map may alternatively be called an *upper semicontinuous decomposition* or for brevity merely a decomposition. We speak of the decomposition of X as being *into* sets (elements of G) and *onto* X/G. The decomposition is *monotone* if each element of G is connected.

If A is a compact subset of X rather than an upper semicontinuous decomposition of X, we use X/A to denote X/G_A where G_A is the upper semicontinuous decomposition of X whose only nondegenerate element is A.

Closely related to the notion of a monotone upper semicontinuous decomposition is that of a monotone mapping. A continuous transformation is called a *map* or *mapping*. A map f of X is *monotone* if the point inverses are compact continua—that is, for each $y \in f(X)$, $f^{-1}(y)$ is compact and connected.

If f is a monotone map of a space X onto a space Y such that the point inverses form an upper semicontinuous collection G, consideration of the 1-1 continuous map of $[0, \infty)$ onto a circle shows that one cannot conclude that Y is homeomorphic to X/G. However, they are homeomorphic if the map f is compact. A mapping is *compact* if the inverses of compact sets are compact. See [11] for a discussion of relationships between decompositions and mappings.

A map f of a polyhedron P_1 into a polyhedron P_2 is *piecewise linear* if P_1, P_2 have triangulations T_1, T_2 compatible with the metrics of P_1, P_2 so that f takes each simplex of T_1 linearly *into* a simplex of P_2. In case P_1 is compact we can say that f takes each simplex of T_2 linearly *onto* a simplex of T_1 but for noncompact polyhedra, this is too restrictive. If $g(x) = 1/x$ $(x \geqq 1)$ there is a piecewise linear map f of $[1, \infty)$ into $[0, 1]$ such that $|f(x) - g(x)| \leqq 1/2x$.

Throughout this paper we shall use M^3 to denote a *compact, connected, polyhedral 3-manifold with boundary*. Perhaps it has no boundary but if this is required, it is so specified.

2. **Boring holes in** M^3. Suppose H^3 is a polyhedral 3-cell in M^3 so that $H^3 \cap \text{Bd } M^3$ is the union of two disjoint disks D_0^2, D_1^2. If we form a new 3-manifold with boundary M_1^3 by removing $\text{Int } H^3 \cup \text{Int } D_0^2 \cup \text{Int } D_1^2$ from M^3, we say that H^3 is a *hole bored* in M^3, and D_0^2, D_1^2 are the *ends* of the hole. We say that M_1^3 *was obtained by boring a hole in* M^3.

It is known that if M^3 has a nonnull boundary, then it can be changed into a cube with handles (some of the handles will be twisted or nonoriented if M^3 is not orientable) by boring a finite number of holes in M^3. A convenient way to find the holes is to bore them about spanning arcs in the 1-skeleton of some triangulation of M^3. Theorem 2.2 is a variation of this result. In order to prove Theorem 2.2 we first use Theorem 2.1 to build a 2-spine Q^2 which perhaps looks like 3-pages of a book at some places. Casler proved a stronger version of Theorem 2.1 in [6] where he built a Q^2 with a standard structure. We do not need this standard structure so for completeneness we include an elementary proof of Theorem 2.1.

THEOREM 2.1. *If M^3 has a nonnull boundary, there is a spine Q^2 for M^3 such that if two disks in Q^2 intersect in an interior point of each, their intersection is 2-dimensional.*

Proof. Let T be a triangulation of M^3. Our intention is to change T to a cellular decomposition T' such that if two disks in the 2-skeleton of T' intersect in an interior point of each, their intersection is 2-dimensional.

Blunt each 3-simplex of T by removing 3-simplexes about each vertex as shown in Figure 2 on p. 17 of [1]. Each of these removed simplexes is similar to the original and has edges one-third as long. For each vertex of T, the union of the blunted ends containing the vertex is a 3-cell which is an element of the cellular subdivision T'.

Consider each of the blunted 3-simplexes. What is left of its edges are trimmed off as shown in Figure 2 on p. 17 of [1]. The union of the prismatic pieces about the middle third of each edge is a 3-cell which is an element of the subdivision T'.

The other 3-dimensional cells in T' are the blunted and trimmed 3-simplexes of T.

The spine Q^2 promised by Theorem 2.1 is obtained by considering the 3-cells of T' and ordering them $B_1^3, B_2^3, \dots, B_m^3$ so that each Bd B_i^3 contains a polyhedral 2-cell B_i^2 which it shares in common with either Bd M^3 or some Bd B_j $(j<i)$. Then

$$M^3 \searrow M^3-\text{Int } B_1^3-\text{Int } B_1^2 \searrow \cdots \searrow M^3-\bigcup_{i=1}^{m} \text{Int } B_i^3-\bigcup_{i=1}^{m} \text{Int } B_i^2 = Q_1^2.$$

Then Q^2 is obtained by collapsing Q_1^2 as much as possible. That M^3 is a regular neighborhood of this reduced Q^2 follows from the result ([7], [10], [12]) that if one polyhedron collapses to a second, each regular neighborhood of the larger polyhedron is a regular neighborhood of the smaller. Likely Q^1 is a 2-spine but its dimension might be less than 2.

THEOREM 2.2. *If M^3 has precisely two boundary components, it is possible to obtain a 3-manifold with boundary M_0^3 from M^3 by boring a finite number of holes in it such that for some polyhedral 2-manifold M^2 there is a piecewise linear homeomorphism of $M^2\times[0, 1]$ onto M_0^3.*

Proof. We shall show that by boring holes in M^3, it can be changed to a polyhedral 3-manifold with boundary M_0^3 such that M_0^3 has two boundary components and M_0^9 collapses to a 2-spine M^2 which is a 2-manifold M^2. The truth of Theorem 2.2 then follows from the facts that M_0^3 is a regular neighborhood of M^2, any two regular neighborhoods of M^2 are piecewise linearly homeomorphic, and there is a regular neighborhood of M^2 piecewise linearly homeomorphic to $M^2\times[0, 1]$.

It follows from Theorem 2.1 that M^3 has a 2-spine Q^2 such that if two disks in Q^2 meet in an interior point of each, their intersection is 2-dimensional. We suppose Q^2 is such a spine that lies in Int M^3. The 2-manifold M^2 promised by the preceding paragraph is obtained by altering Q^2.

If there is a 1-cell A_1 and three disks D_1^2, D_2^2, D_3^2 in Q^2 such that $D_i^2\cap D_j^2=A^1$ $(i\neq j)$, some one of the disks is not a subset of the part of Q^2 irreducible with respect to separating the two boundary components from each other in M^3. Holes can be bored in M^3 and through this disk so as to obtain a 3-manifold with two boundary components which collapses onto a proper subpolyhedron of Q^2.

Let M_1^3 be a 3-manifold with two boundary components obtained by boring holes in M^3 such that M_1^3 contains a spine $Q_1^2\subset Q^2$ such that no further boring of holes in M_1^3 results in a 3-manifold with two boundary components and a spine which is a proper subset of Q_1^2. Note that Q_1^2 is the union of a 2-manifold M_1^2 and a 1-complex. By adjusting this 1-complex we may suppose that it is the union of a finite number of disjoint arcs $A_1^1, A_2^1, \ldots, A_m^1$ such that each $A_i^1 \cap M_1^2 = \text{Bd } A_i^1$.

Boring holes in M_1^3 with the A_i^1's as centers of the holes produces a 3-manifold with boundary M_0^3 such that M_0^3 has two boundary components and a 2-manifold M^2 as a spine where M^2 is obtained from M_1^2 by replacing the A_i^1's and disks on M_1^2 about their ends by annuli.

3. **Decomposing M_3 into 2-manifolds and pinched 2-manifolds.** A *pinched manifold* results from identifying two points of the same manifold. If two objects are joined at a point, their union is called the *wedge* of the two objects. The wedge of two tangent round 2-spheres is an example of a pinched 2-manifold. If a cube with handles has one of its handles squeezed to a point, the boundary of the resulting object may be an example of a pinched 2-manifold.

Suppose B^2 is a boundary component of M^3. It is known (see for example Lemma 1 of [8]) that there is a piecewise linear homeomorphism h of $B^2 \times [0, 1]$ into M^3 so that for each $x \in B^2$,

$$h(x \times 0) = x \quad \text{and} \quad h(B^2 \times (0, 1]) \subset \text{Int } M^3.$$

We call $h(B^2 \times [0, 1])$ a *polyhedral cartesian product neighborhood of* B^2.

In the following three theorems we show that certain 3-manifolds can be decomposed into 2-manifolds and pinched 2-manifolds. A map is *open* if the images of open sets are open.

THEOREM 3.1. *Suppose M^3 has precisely two boundary components B_0^2, B_1^2. Then there is a monotone piecewise linear open map f of M^3 onto $[0, 1]$ such that $f^{-1}(0) = B_0^2$, $f^{-1}(1) = B_1^2$, and for each $t \in [0, 1]$, $f^{-1}(t)$ is either a 2-manifold or a pinched 2-manifold. If $[a, b] \subset [0, 1]$ and for each $t \in [a, b]$, $f^{-1}(t)$ is a 2-manifold, there is a piecewise linear homeomorphism of $f^{-1}[a, b]$ onto $f^{-1}(a) \times [a, b]$ that takes $f^{-1}(t)$ onto $f^{-1}(a) \times t$.*

Proof. Let h_0 be a piecewise linear homeomorphism of $B_0^2 \times [0, 1/3]$ onto a cartesian product neighborhood of B_0^2 in M^3 so that $h_0(x \times 0) = x$ and let h_1 be a piecewise linear homeomorphism of $B_1^2 \times [2/3, 1]$ onto a cartesian product neighborhood of B_1^2 in M^3 so that $h_1(x \times 1) = x$. We suppose these two cartesian product neighborhoods do not intersect each other. Then

$$M^3 - h_0(B_0^2 \times [0, 1/3)) - h_1(B_1^2 \times (2/3, 1]) = M_1^3$$

is a 3-manifold with precisely two boundary components $h_0(B_0^2 \times 1/3)$, $h_1(B_1^2 \times 2/3)$.

It follows from Theorem 2.2 that it is possible to bore a finite number of mutually exclusive polyhedral holes $H_1^3, H_2^3, \ldots, H_n^3$ in M_1^3 such that each of these holes has both ends on the same boundary component of M_1^3 and for some 2-manifold M^2, there is a piecewise linear homeomorphism h of $M^2 \times [1/3, 2/3]$ onto the closure of $(M_1^3 - \bigcup H_i^3)$. We suppose $h(M^2 \times 1/3)$, $h(M^2 \times 2/3)$ are the altered $h_0(B_0^2 \times 1/3)$, $h_1(B_1^2 \times 2/3)$ respectively and define

$$f(h(M^2 \times t)) = t, \qquad t \in [1/3, 2/3].$$

Let $H_1^3, H_2^3, \ldots, H_m^3$ be the holes bored in M_1^3 from $h_0(B_0^2 \times 1/3)$. For each H_i^3, let C_i^3, $C_i^{3\prime}$ be caps over the ends of H_i^3 so that each C_i^3 is a polyhedral 3-cell in $h_0(B_0^2 \times [0, 1/3])$ such that $C_i^3 \cap h_0(B_0^2 \times 1/3)$ is an end of H_i^3, $\text{Bd } C_i^3 \cap h_0(B_0^2 \times 1/6)$ is a point and for each $t \in (1/6, 1/3)$, $\text{Bd } C_i^3 \cap h_0(B_0^2 \times t)$ is a simple closed curve; the $C_i^{3\prime}$ are similarly defined. We suppose that no two of the caps intersect each other.

If $t \in [0, 1/3)$ and $p \in h_0(B_0^2 \times t) - \bigcup \text{Int } C_i^3 - \bigcup \text{Int } C_i^{3\prime}$, we define $f(p) = t$.

For each $i = 1, 2, \ldots, m$, consider the polyhedral 3-cell $H_i^3 \cup C_i^3 \cup C_i^{3\prime}$. We note that f has already been defined on its boundary. Triangulate this boundary so that f is linear with respect to the triangulation. Then triangulate $H_i^3 \cup C_i^3 \cup C_i^{3\prime}$ by coning from an interior point p_i.

Let $\varepsilon_1, \varepsilon_2, \ldots, \varepsilon_m$ be different numbers in $(1/6, 1/3)$. Define $f(p_i) = \varepsilon_i$ and extend f linearly to the simplexes of the conical triangulation of $H_i^3 \cup C_i^3 \cup C_i^{3\prime}$. Then $(H_i^3 \cup C_i^3 \cup C_i^{3\prime}) \cap f^{-1}(t)$ is the union of two points if $t = 1/6$, the union of two disjoint disks (cones) if $t \in (1/6, \varepsilon_i)$, the union of two cones with a common vertex p_i if $t = \varepsilon_i$, and an annulus if $t \in (\varepsilon_i\ 1/3]$. It is to be noted that f is open at p_i.

Note that for some neighborhood $(\varepsilon_i - \delta, \varepsilon_i + \delta)$ of ε_i, there is a piecewise linear projection r_i of $f^{-1}[\varepsilon_i - \delta, \varepsilon_i + \delta]$ onto $f^{-1}(\varepsilon_i)$ such that for $t \in [\varepsilon_i - \delta, \varepsilon_i)$, r_i restricted to $f^{-1}(t)$ is one-to-one except onto p_i and the inverse of p_i is a pair of points while if $t \in (\varepsilon_i, \varepsilon_i + \delta]$, r_i restricted to $f^{-1}(t)$ is one-to-one except onto p_i and the inverse of p_i is a simple closed curve.

In a similar fashion we define f on $h_1(B_1^2 \times [2/3, 1])$ and the holes with ends on $h_1(B_1^2 \times 2/3)$.

DEFINITIONS. A *dendron* is a compact locally connected metric continuum which contains no simple closed curve. A point p is of *order* $\leqq n$ if for each neighborhood U of p there is an open set V with $p \in V \subset U$ such that the boundary (frontier) of V has at most n points. A point is of *order* n if it is of order $\leqq n$ but not of order $\leqq n-1$. A point of order 1 is an *end point* of the dendron and a point which is not of order $\leqq 2$ is a *branch point*. A polyhdral dendron (sometimes called a *tree*) can have at most a finite number of branch and end points. A dendron with precisely three end points (and hence one branch point) is called a *triod*.

The following is a variation of Theorem 3.1.

THEOREM 3.2. *If M^3 has precisely three boundary components, there is a piecewise linear monotone map f of M^3 onto a polyhedral triod T^1 such that each boundary component of M^3 is an inverse of an end point of T^1 and for each $t \in T^1$, $f^{-1}(t)$ is either a 2-manifold or a pinched 2-manifold.*

Proof. Let B_1^2, B_2^2 be two of the boundary components of M^3; h_1 be a piecewise linear homeomorphism of $B_1^2 \times [0, 1]$ onto a cartesian product neighborhood of

B_1^2 in M^3 such that for each $x \in B_1^2$, $h_1(B_1^2 \times 0) = x$; and h_2 be a piecewise linear homeomorphism of $B_2^2 \times [0, 1]$ onto a cartesian product neighborhood of B_2^2 in M^3 such that for each $x \in B_2^2$, $h_2(x \times 0) = x$. We suppose these two cartesian product neighborhoods are disjoint.

Let H^3 be a hole in $M^3 - h_1(B_1^2 \times [0, 1)) - h_2(B_2^2 \times [0, 1))$ with one end on $h_1(B_1^2 \times 1)$ and the other on $h_2(B_2^2 \times 1)$. It follows by the methods of the proof of Theorem 3.1 that there is a monotone piecewise linear map f_1 of

$$H^3 \cup h_1(B_1^2 \times [0, 1]) \cup h_2(B_2^2 \times [0, 1])$$

into a polyhedral triod T_1^1 such that the inverse of the branch point of T_1^1 is the wedge of two 2-manifolds, the inverse of each other point of T_1^1 is a 2-manifold. The inverses of the end points of the triod are $h_1(B_1^2 \times 0)$, $h_2(B_2^2 \times 0)$, and the boundary component of $H^3 \cup h_1(B_1^2 \times [0, 1]) \cup h_1(B_2^2 \times [0, 1])$ intersecting Bd H^3.

It follows from Theorem 3.1 that the piecewise linear map f_1 of

$$H^3 \cup h_1(B_1^2 \times [0, 1]) \cup h_2(B_2^2 \times [0, 1])$$

onto T_1^1 can be extended to a piecewise linear map f of M^3 onto a larger triod T^1 satisfying the conditions of Theorem 3.2. The map f we define is open except on the inverse of the branch point of the triod.

In a certain sense, we proved Theorem 3.2 by combining two boundary components of M^3. Repetitions of this procedure permits us to prove the following extension of Theorem 3.2.

THEOREM 3.3. *If M^3 has more than one boundary component, there is a piecewise linear monotone map f of M^3 onto a tree T^1 such that each point of T^1 is of order* $\leqq 3$, *the boundary components of M^3 are the inverses of end points of T^1, and for each $t \in T^1$, $f^{-1}(t)$ is either a 2-manifold or a pinched 2-manifold.*

Theorem 7.2 of §7 gives results about the mapping of M^3 onto a dendron if M^3 has no boundary.

4. **Decomposition maps onto 2-spheres.** In the last section we decomposed M^3 into 2-manifolds and pinched 2-manifolds by letting these be the point inverses in a map onto a dendron. In §6 we shall map M^3 onto a punctured cube. This is done by considering the point inverses in the map onto a dendron and decomposing M^3 by decomposing each of these inverses. In this section we consider useful decompositions of 2-manifolds and pinched 2-manifolds.

Consider a simplex Δ which is the join of two simplexes Δ_1, Δ_2. Then $\Delta = \Delta_1 \circ \Delta_2$ is the union of segments from Δ_1 to Δ_2. For each $t \in [0, 1]$, let A_t be the set of all points such that for some $p_1 \in \Delta_1$, $p_2 \in \Delta_2$, p divides the segment from p_1 to p_2 in the ratio t to $1-t$. We denote p by (p_1, p_2, t) and note that

$$A_t = \{(p_1, p_2, t) \mid p_1 \in \Delta_1, p_2 \in \Delta_2\}.$$

Then $A_0 = \Delta_1$, $A_1 = \Delta_2$, and the midsection $A_{1/2}$ is homeomorphic to $\Delta_1 \times \Delta_2$. We use A_B to denote $\bigcup_{t \in B} A_t$.

Let Δ/Δ_1 be the cell obtained from Δ by shrinking Δ_1 to a point. To give Δ/Δ_1 a structure we regard it as the union of $A_{[1/2,1]}$ and the cone C over $A_{1/2}$ from a point v_0 where $A_{[1/2,1]} \cap C = A_{1/2}$.

The natural decomposition map f from Δ to Δ/Δ_1 is the identity on $A_{[1/2,1]}$ and for $t \in [0, 1/2]$, $f(p_1, p_2, t)$ is the point that divides the segment from v_0 to $(p_1, p_2, 1/2)$ in the ratio $2t$ to $1-2t$.

We note that although Δ is a simplex and $f(\Delta) = \Delta/\Delta_1$ is a piecewise linear cell, the map f is not piecewise linear.

If L is a subcomplex of complex K, K/L results by regarding L as a point. To give K/L a cellular structure, we let T' be the first barycentric subdivision of the triangulation of K and note that if Δ is a simplex in T', then either Δ misses L, lies in L, or intersects it in a proper face Δ_L. The natural decomposition map f is such that $f(L)$ is a vertex v_0 of K/L; f is the identity on each simplex of T' that misses L; for each simplex Δ of T' whose intersection with L is a proper face Δ_L of Δ, f is the natural decomposition map from Δ to Δ/Δ_L.

The following result is well known.

THEOREM 4.1. *If M^2 is a compact connected polyhedral 2-manifold, there is a connected polyhedral 1-complex C^1 in M^2 such that M^2/C^1 is a polyhedral 2-sphere.*

To obtain C^1, triangulate M^2 and remove an open 2-simplex Int Δ^2 from M^2. Then C^1 is obtained by collapsing $M^2 - \text{Int}\,\Delta^2$.

The following two results may be proved in a similar fashion.

THEOREM 4.2. *If N^2 is a compact connected triangulated pinched 2-manifold with no separating point, there is a connected polyhedral 1-complex C^1 in the 1-skeleton of the triangulation such that N^2/C^1 is a polyhedral 2-sphere and C^1 is of order 2 at the pinch point.*

We note that C^1 abuts on the pinch point from both sides in N^2.

THEOREM 4.3. *If N^2 is a compact connected triangulated pinched 2-manifold which is separated by its pinch point, there is a connected polyhedral 1-complex C^1 in the 1-skeleton of the triangulation such that C^1 abuts on the pinch point from both sides and is of order 2 there while N^2/C^1 is the wedge of two polyhedral 2-spheres.*

5. **Changing pseudo spines.** A *pseudo spine* for a compact connected polyhedral 2-manifold is a connected polyhedron C in M^2 such that M^2/C is topologically a 2-sphere. To get C one might remove an open 2-simplex from M^2 and collapse the remainder. Note that the pseudo spine differs from a spine for M^2 in that an open simplex was removed before the collapsing. Also, the collapsing was not necessarily continued as much as possible. For example, a 1-simplex on S^2 serves as a pseudo spine for it. To emphasize that the pseudo spines we shall be using are 1-dimensional polyhedra, we may call them pseudo 1-spines.

If M^2 is a compact, connected, polyhedral 2-manifold, $M^2 \times [0, 1]$ is its cartesian product with an interval, and $C_0^1 \times 0$, $C_1^1 \times 1$ are pseudo 1-spines of $M^2 \times 0$, $M^2 \times 1$, we shall be interested in finding pseudo 1-spines for the intermediate layers.

THEOREM 5.1. *Suppose M^2 is a compact connected polyhedral 2-manifold and C_0^1, C_1^1 are pseudo 1-spines of M^2. Then C_0^1, C_1^1 can be extended to a 1-parameter family of pseudo 1-spines C_t^1 $(0 \leqq t \leqq 1)$ of M^2 such that there is a monotone map f of $M^2 \times [0, 1]$ onto $S^2 \times [0, 1]$ with $f(M^2 \times t)$ being equal to $S^2 \times t$ and $C_t^1 \times t$ being the only nondegenerate inverse in $M^2 \times t$.*

Proof. Let $D_1^2, D_2^2, \ldots$ be a sequence of polyhedral disks in M^2 such that $D_i^2 \subset \text{Int } D_{i+1}^2$ and $\bigcup D_i^2 = M^2 - C_0^1$ and $E_1^2, E_2^2, \ldots$ be such a sequence of disks whose union is $M^2 - C_1^1$.

Since E_1^2 is a disk which lies on the interior of a larger disk in M^2 with the larger disk not being a subset of $M^2 - \text{Int } D_1^2$, there is a piecewise linear isotopy fixed except on this larger disk that pushes $M^2 - \text{Int } D_1^2$ off E_1^2. Denote this isotopy by H_t $(0 \leqq t \leqq 1/4)$ where $H_0 = \text{identity}$ and $E_1^2 \cap H_{1/4}(M^2 - \text{Int } D_1^2) = 0$. Then $H_{1/4}(D_1^2) - \text{Int } E_1^2$ is an annulus and H_t $(0 \leqq t \leqq 1/4)$ can be extended to a piecewise linear isotopy H_t $(0 \leqq t \leqq 1/2)$ such that $H_{1/2}(D_1^2) = E_1^2$.

The isotopy H_t $(0 \leqq t \leqq 1/2)$ can be extended to a piecewise linear isotopy H_t $(0 \leqq t \leqq 3/4)$ so that $H_{3/4}(D_2^2) = E_2^2$ and for $x \in D_1^2$ and $1/2 \leqq s \leqq 3/4$, $H_s(x) = H_{1/2}(x)$.

Similarly, we extend H_t to H_t $(0 \leqq t < 1)$. In general, H_t is such that $H_{t_i}(D_i) = E$ for $t_i = (1 - 1/2^i)$ and $H_s = H_{t_i}$ on D_i for $s \geqq (1 - 1/2^i)$.

For $0 \leqq t < 1$, $C_t^1 = H_t(C_0^1)$.

Let f_0 be a decomposition map of M^2 onto S^2 such that C_0^1 is the only nondegenerate inverse. The map f of $M^2 \times [0, 1)$ is defined by $f(x \times t) = f_0 H_t^{-1}(x) \times t$. It is extended to $M^2 \times 1$ by continuity.

QUESTIONS. It would appear that Theorem 5.1 is a special case of some more general theorem. If f is a map of a compact continuum X onto $[0, 1]$ such that the point inverses are all homeomorphic to each other, under what condition can it be concluded that X is homeomorphic to $f^{-1}(0) \times [0, 1]$? If g is a monotone map of $Y \times [0, 1]$ onto Z such that the $g(Y \times t)$'s are homeomorphic but no two intersect, under what condition is Z homeomorphic to $g(Y \times 0) \times [0, 1]$? Is it enough that each $f(Y \times t)$ be a 2-sphere? What if each $Y \times t$ contain at most one nondegenerate point inverse? Something needs to be imposed because if P is a pseudo arc, P' is a nondegenerate proper subcontinuum of $P \times 0$, then $(P \times [0, 1])/P'$ is not homeomorphic to $P \times [0, 1]$ since it contains a triod. However it is known [2] that $(P \times 0)/P'$ is homeomorphic to P.

It is to be noted that the map f is not piecewise linear and is especially bad at $t = 1$. Can we find such an f that has the nice properties of a natural decomposition map?

6. **Decomposing 3-manifolds with boundaries.** Suppose $B_0^3, B_1^3, \ldots, B_n^3$ are

polyhedral 3-cells such that $B_i^3 \subset \text{Int } B_0^3$ for $i=1, 2, \ldots, n$ and $B_i^3 \cap B_j^3 = 0$ for $0<i<j\leqq n$. Then $B_0^3 - \bigcup_{i=1}^n \text{Int } B_i^3$ is a *polyhedral punctured cube*.

If the boundary components of M^3 are already nicely decomposed with the decomposition space of each being a 2-sphere, we shall be interested in extending the decomposition to M^3 so that the decomposition space is a punctured cube.

THEOREM 6.1. *Suppose M^3 has precisely two boundary components B_0^2, B_1^2 with pseudo 1-spines C_0^1, C_1^1. Then there is a monotone map g of M^3 onto $S^2\times[0, 1]$ such that $g^{-1}(S^2\times 0)=B_0^2$, $g^{-1}(S^2\times 1)=B_1^2$, each $g^{-1}(S^2\times t)$ is either a polyhedral 2-manifold or polyhedral pinched 2-manifold, C_0^1 and C_1^1 are point inverses, each nondegenerate point inverse is a polyhedral 1-complex, and the image of the union of the nondegenerate point inverses is an arc from $S^2\times 0$ to $S^2\times 1$ that intersects each $S^2\times t$ only once.*

Proof. Let f be a monotone piecewise linear map of M^3 onto $[0, 1]$ as described in Theorem 3.1 such that $f^{-1}(0)=B_0^2$, $f^{-1}(1)=B_1^2$, and each inverse is either a 2-manifold or a pinched 2-manifold. We plan to define g so that $gf^{-1}(t)=S^2\times t$. To get such a g we need to get a 1-complex C_t^1 on each $f^{-1}(t)$ so that $f^{-1}(t)/C_t^1$ is homeomorphic to S^2. We shall first describe the C_t^1's on the pinched inverses, then on the inverses near these pinched inverses, and finally use Theorem 5.1 to fill in the gaps.

Since f is piecewise linear, there are only a finite number of pinched 2-manifolds among the inverses. Let $0=t_1<t_2<\cdots<t_{3n+2}=1$ be such that the pinched inverses are the $f^{-1}(t_{3i})$'s $(i=1, 2, \ldots, n)$. For each $i=1, 2, \ldots, n$, let $C_{t_{3i}}^1$ be a polyhedral 1-complex in $f^{-1}(t_{3i})$ such as guaranteed by Theorem 4.2 so that $f^{-1}(t_{3i})/C_{t_{3i}}^1$ is a 2-sphere.

In the proof of Theorem 3.1 we analyzed the nature of the map f in the neighborhood of the pinched inverses. There is a projection of the $f^{-1}(t)$'s near $f^{-1}(t_{3i})$ onto $f^{-1}(t_{3i})$ so that the projection is one-to-one except onto the pinched point and the inverse of the pinched point under the projection is either a pair of points or a polyhedral simple closed curve. For $t\in[t_{3i-1}, t_{3i+1}]$ let C_t^1 be the inverse of $C_{t_{3i}}^1$ under the projection of $f^{-1}(t)$ onto $f^{-1}(t_{3i})$. There is a map g_i of $f^{-1}[t_{3i-1}, t_{3i+1}]$ onto $S^2\times[t_{3i-1}, t_{3i+1}]$ such that $gf^{-1}(t)=S^2\times t$ and C_t^1 is the only nondegenerate inverse in $g^{-1}(S^2\times t)$.

Although there is a piecewise linear homeomorphism of $f^{-1}[t_{3j-2}, t_{3j-1}]$ for $j=1, 2, \ldots, n+1$ onto $f^{-1}(t_{3j-2})\times[t_{3j-2}, t_{3j-1}]$, there is no assurance that $C_{t_{3j-2}}^1$ and $C_{t_{3j-1}}^1$ line up. However Theorem 5.1 shows that polyhedral C_t^1's can be selected for $t\in(t_{3j-2}, t_{3j-1})$ so as to insure that there is a level preserving monotone map of $f^{-1}[t_{3j-2}, t_{3j-1}]$ onto $S^2\times[t_{3j-2}, t_{3j-1}]$ so that the only nondegenerate inverses are the C_t^1's.

The map g is obtained by patching together the various maps of the $f^{-1}[t_{3i-1}, t_{3i+1}]$ onto the $S^2\times[t_{3i-1}, t_{3i+1}]$ for $i=1, 2, \ldots, n$ and the maps of the $f^{-1}[t_{3j-2}, t_{3j-1}]$ onto the $S^2\times[t_{3j-2}, t_{3j-1}]$ for $j=1, 2, \ldots, n+1$.

THEOREM 6.2. *If M^3 has only one boundary component B^2 and C^1 is a pseudo 1-spine of B^2, then there is a monotone map g of M^3 onto a polyhedral 3-cell B^3 such that $g^{-1}(\text{Bd } B^3)=B^2$, each nondegenerate inverse is a polyhedral 1-complex, C^1 is the only such inverse on B^2, and the image under g of the union of the nondegenerate inverses is a polyhedral arc in B^3.*

Proof. A polyhedral open ball is removed from M^3 to change it to a 3-manifold with two boundary components. A polygonal arc is selected to serve as a pseudo 1-spine for the new boundary component. Theorem 6.1 assures us that there is a suitable map of the remainder onto $S^2\times[0,1]$. This map may be extended to all of M^3 to get the required map onto B^3.

THEOREM 6.3. *Suppose M^3 has boundary components $B_1^2, B_2^2, \ldots, B_n^2$ $(n\geqq 2)$ with polyhedral pseudo 1-spines $C_1^1, C_2^1, \ldots, C_n^1$. Then there is a monotone map g of M^3 onto a polyhedral punctured cube K^3 with boundary components $E_1^2, E_2^2, \ldots, E_n^2$ such that $g^{-1}(E_i^2)=B_i^2$, C_i^1 is the only nondegenerate inverse on B_i^2, each nondegenerate inverse is a polyhedral 1-complex, and the image under g of the union of the nondegenerate inverses is a tree in K^3 each of whose ends is on* Bd K^3.

Proof. Let f be a piecewise linear map such as guaranteed by Theorem 3.3 and described in the proofs of Theorems 3.1 and 3.2 such that $f(M^3)$ is a tree T^1, each branch point of T^1 is of order 3, the inverses of the end points of T^1 are the boundary components of M^3, and for each $x\in T^1$, $f^{-1}(x)$ is either a 2-manifold or a pinched 2-manifold.

Let $f^{-1}(x_1), f^{-1}(x_2), \ldots, f^{-1}(x_{n-2})$ be the pinched 2-manifolds among the inverses which are separated by their pinch points. Each $f^{-1}(x_i)$ separates M^3 into three parts. Assign a 1-complex $C_{x_i}^1$ to $f^{-1}(x_i)$ which satisfies the conditions of Theorem 4.3.

Let $T_1^1, T_2^1, \ldots, T_{n-2}^1$ be mutually exclusive trees in T^1 about the x_i's such that the branch point of T_i^1 is the only point of it that has an inverse which is pinched and no end point of T^1 is in any T_i^1.

For each $x\in T_i^1$, there is a natural projection of $f^{-1}(x)$ into $f^{-1}(x_i)$. From two sides, the projection is one-to-one onto manifolds in $f^{-1}(x_i)$ and from the third side the projection is one-to-one onto $f^{-1}(x_i)$ except onto the pinch point and the inverse of the pinch point is a polyhedral simple closed curve.

For the $f^{-1}(x)$'s that project one-to-one onto a 2-manifold in $f^{-1}(x_i)$, we let C_x^1 be the inverse under the projection of the part of $C_{x_i}^1$ in this 2-manifold.

For the $f^{-1}(x)$'s that project onto all of $f^{-1}(x_i)$, we let C_x^1 be the inverse of $C_{x_i}^1$ under the projection with an open arc removed from the inverse of the pinch point under the projection so that $f^{-1}(x)/C_x^1$ is a 2-sphere rather than the wedge of two 2-spheres. There is a monotone map g_1 of $f^{-1}(T_i^1)$ onto a polyhedral punctured cube K_i^3 with three boundary components such that C_x^1 is the only nondegenerate inverse intersecting $f^{-1}(x)$.

Consider the components of the part of T^1 for which C^1's have not been assigned to the inverses of its points. There are only a finite number of such components and the closure of each is an arc A. Already C^1's have been assigned to the inverses of the ends of A. Theorem 6.1 provides C^1's to the inverses of other points of A as well as a monotone map of $f^{-1}(A)$ into a hollow ball. The map g promised by Theorem 6.3 follows by patching together the monotone maps g_i on the $f^{-1}(T_i^1)$'s and those on the A's.

7. **Extending monotone mappings.** In this section we suppose M^3 is without boundary and X is a closed proper subset of M^3 such that if U is a connected open subset of M^3, then either $U-(X \cap U)$ is connected or X intersects the boundary (frontier) of U.

The principal result (Theorem 7.3) of this section is that there is a monotone map of M^3 onto S^3 that has the components of X as point inverses. Before proving this result, we prove some theorems about X.

THEOREM 7.1. *There is a decreasing sequence $M_1^3, M_2^3, \ldots$ such that $X = M_1^3 \cap M_2^3 \cap \cdots$ where each M_i^3 is a 3-manifold with boundary which is a polyhedron in M^3, $M_{i+1}^3 \subset M_i^3$, and each component of M_i^3 has a connected boundary.*

Proof. Suppose M_i^3 has been constructed. We prove the theorem by showing how to get an M_{i+1}^3 in any open subset V of Int M_i^3 which contains X.

Let T be a triangulation of M^3 of mesh less than one-half the distance from X to $M^3 - V$ and T'' be the second barycentric subdivision of T. Let N^3 be the union of the closed simplexes of T which intersect X and N_1^3 be the union of the closed simplexes of T'' which intersect N^3. Then N_1^3 is a regular neighborhood of N^3 and is hence a polyhedral 3-manifold with boundary ([7], [10], [12]). If C^3 is a component of N_1^3, it may have several boundary components, but boring holes in C^3 which miss X reduces the number of boundary components to one. Hence M_{i+1}^3 is obtained from N_1^3 by boring holes.

The following result follows from Theorem 7.1 and repeated applications of Theorem 3.3.

THEOREM 7.2. *There is a dendron D^1 and a monotone map f of M^3 onto D^1 such that each point of D^1 is of order $\leqq 3$, D^1 is locally polyhedral mod its set of end points, the components of X are the inverses of the endpoints of D^1, the inverse of each non end point is either a polyhedral 2-manifold or a polyhedral pinched 2-manifold, and f is locally piecewise linear mod X.*

THEOREM 7.3. *There is a monotone map g of M^3 onto S^3 such that each component of X is a point inverse, $g(X)$ lies on a straight segment, and each nondegenerate point inverse outside X is a polyhedral 1-complex. In fact, there are also a dendron D^1 in S^3 and a monotone map f of M^3 onto D^1 such that D^1 is locally polyhedral mod its set of end points, each branch point of D^1 is of order 3, X is the inverse under f of the set of end points of D^1, $f=g$ on X and on each nondegenerate point inverse under g,* *and if x is a non end point of* D^1, $f^{-1}(x)$ *is either a polyhedral* 2*-manifold or pinched* 2*-manifold,* $f^{-1}(x)$ *contains at most one nondegenerate point inverse under* g, *and* $gf^{-1}(x)$ *is either a* 2*-sphere or the wedge of two* 2*-spheres.*

Proof. The dendron D^1 and the map f are as given in Theorem 7.2, (except that the embedding of D^1 in S^3 is yet to be defined).

Let $T_1^1, T_2^1, \ldots$ be a monotone increasing sequence of trees in D^1 such that each point of order $\leqq 2$ of T_i^1 is a point of order 2 of T_{i+1}^1, each inverse under f of an end point of T_i^1 is a 2-manifold, and $\bigcup T_i^1$ contains each non end point of D^1.

By Theorem 6.3 there is a monotone map g_1 of $f^{-1}(T_1^1)$ onto a punctured cube K_1^3 such that each nondegenerate point inverse under g_1 is a polyhedral 1-complex, the image of the set of nondegenerate point inverses is a tree $T_{1'}^1$, and for each $x \in T_1^1$, $f^{-1}(x)$ contains exactly one such 1-complex, and $g_1 f^{-1}(x)$ is either a 2-sphere or the wedge of 2-spheres.

Let A be a straight segment in S^3. We suppose that K^3 lies in S^3, each component of $S^3 - K^3$ is of diameter less than $\frac{1}{2}$ and intersects A, $T_{1'}^1 = T_1^1$, $f = g$ on $f^{-1}(T_1^1)$.

Let $T_{2,i}^1$ $(i = 1, 2, \ldots, m)$ be the closures of the components of $T_2^1 - T_1^1$. It follows from Theorem 6.3 that there is a monotone map $g_{2,i}$ of $f^{-1}(T_{2,i}^1)$ onto a punctured cube $K_{2,i}^3$ such that each nondegenerate point inverse under $g_{2,i}$ is a polyhedral 1-complex, the image of the union of these 1-complexes is a tree $T_{2,i'}^1$, and for each $x \in T_{2,i}^1$, $f^{-1}(x)$ contains exactly one such 1-complex, while $g_{2,i} f^{-1}(x)$ is either a 2-sphere or the wedge of two 2-spheres.

We suppose that $K_{2,i}^3$ lies in S^3 so that only one component of $S^3 - K_{2,i}^3$ is of diameter more than 1/4, each component of $S^3 - K_{2,i}^3$ intersects A, and for $x_i = T_1^1 \cap T_{2,i}^1$, g_1, $g_{2,i}$ agree on $f^{-1}(x_i)$ with $g_1 f^{-1}(x_i) = g_{2,i} f^{-1}(x_i) = K_1^3 \cap K_{2,i}^3$, $T_{2,i'}^1 = T_{2,i}^1$, and $f = g$ on $f^{-1}(T_2^1)$.

For simplicity we use g to designate g_1 and the $g_{2,i}$'s. The map g is extended to $T_3^1, T_4^1, \ldots$ so that each component of $S^3 - gf^{-1}(T_i^1)$ is of mesh less than $1/2^i$ and intersects A. We place D^1 so that $f = g$ on $f^{-1}(D^1)$. Then g is extended to M^3 so that if y_0 is an end point of D^1 and y_i is the end point of T_i^1 toward y_0, $gf^{-1}(y_0)$ is the intersection of the small balls in S^3 bounded by the $gf^{-1}(y_i)$'s.

The unicoherence of S^3 permits us to weaken the restrictions on X (where such a weakening would not be possible for such a 3-manifold as $S^2 \times S^1$). Hence, by ignoring details about f we may state Theorem 7.3 for S^3 as follows.

THEOREM 7.4. *If* Y *is a closed subset of* S^3 *no component of which separates* S^3, *there is a monotone map of* S^3 *onto itself such that the components of* Y *are point inverses, each nondegenerate point inverse not in* Y *is a polyhedral* 1*-complex, and the image of the union of* Y *and these nondegenerate point inverses is a dendron.*

We can deal with noncompact 3-manifolds since each connected noncompact 3-manifold is the union of a monotone increasing sequence of connected compact

3-manifolds with boundaries. For example, for Euclidean 3-space E^3 we have the following result.

THEOREM 7.5. *Suppose Y is a closed subset of E^3 such that each component of Y is compact but does not separate E^3. Then there is a compact monotone map of E^3 onto itself such that the components of Y are point inverses, and nondegenerate point inverses not in Y are polyhedral* 1*-complexes. The image of the union of Y and the nondegenerate point inverses is a closed set whose* 1*-point compactification is a dendron with a closed set of end points.*

QUESTIONS. In [11] Whyburn raises the question as to whether or not each monotone map of E^3 onto itself need be compact. Could it be shown to be compact in case each point inverse is either a 1-complex or a component of Y? Would it help to know that the image of the union of the nondegenerate point inverses lies in a closed set whose 1-point compactification is a dendron?

It is to be noted that the maps g of Theorems 7.4, 7.5 were not necessarily cellular since some of the 1-complexes which were point inverses might contain cycles. Could one obtain a cellular g if the components of Y were pointlike? In particular, what happens if Y is the union of the tame arcs in the dogbone decomposition?

DEFINITIONS. Euclidean n-space $(n>1)$ has only *one end*. An *end* of a connected, locally compact, noncompact metric space S may be defined as an equivalence class of sequences $U_1, U_2, \ldots$ where U_i is a nonnull connected open subset of X with a compact boundary (frontier),

$$\overline{U_{i+1}} \subset U_i, \quad \text{and} \quad \bigcap U_i = 0.$$

The sequence $U_1, U_2, \ldots$ is in the same equivalence class as $V_1, V_2, \ldots$ if and only if the V_i's satisfy similar conditions, each U_i contains a V_j and each V_i contains a U_j. Hence, S *has one end* if for each compact set C, the closure of $S-C$ has precisely one noncompact component.

For other noncompact 3-manifolds, our methods give the following result.

THEOREM 7.6. *Suppose W^3 is a noncompact, connected, polyhedral, 3-manifold without boundary and Y is a closed subset of W^3 such that each component of Y is compact and if U is a connected open subset of W^3, either Y intersects the boundary of U or $U-(Y\cap U)$ is connected. Then there is a straight segment in S^3, a dendron D^1 which is locally polyhedral* mod *its set of end points and which intersects A in its set of end points, a closed subset C^0 of the set of end points of D^1, and a monotone map g of W^3 onto S^3-C^0 such that the components of Y are point inverses of the end points of D^1 not in C^0, the nondegenerate point inverses not in Y are polyhedral* 1*-complexes in W^3, and the image of the union of Y and the nondegenerate point inverses is D^1-C^0.*

The points C^0 correspond to the ends of W^3.

8. **Monotone maps that cannot be extended.** It would be interesting to know which monotone decompositions of a portion of a 3-manifold can be extended to the whole 3-manifold so that the decomposition space is a 3-manifold.

Suppose C_1, C_2 are two linking round circles in a simplex of S^3 and G is a decomposition of $C_1 \cup C_2$ whose only nondegenerate element is C_2. Is there a monotone upper semicontinuous decomposition of S^3 onto itself such that each element of G is an element of this decomposition? Is there a monotone map g of S^3 **onto** itself that is a homeomorphism on C_1 and sends C_2 to a point off $g(C_1)$? The example on p. 7 of [5] gives an affirmative answer to the second question while the following theorem gives a negative answer to the first.

The following theorem is included for simplicity. It is a special case of the more complicated Theorem 8.4.

THEOREM 8.1. *There is no monotone upper semicontinuous decomposition of S^3 onto itself such that the elements of G are elements of this decomposition.*

Proof. Assume the decomposition can be extended. Then there is a monotone map g of S^3 onto itself such that the elements of G are point inverses under g.

Let J be a simple closed curve in $S^3 - g(C_1)$ that cannot be shrunk to a point in $S^3 - g(C_1)$. We show that there is no map g by showing that J can be shrunk to a point in $S^3 - g(C_1)$.

It follows from the monotonicity of g and the local Euclidean structure of S^3 that for each open set U containing J there is a homeomorphism h of J into $g^{-1}(U)$ such that the identity on J is homotopic to gh in U. We suppose that h is selected for an open subset U containing J such that $U \cap g(C_1) = 0$.

Let H_t $(0 \leqq t \leqq 1)$ be an isotopy of $S^3 - C_1$ into $S^3 - C_1$ such that $H_0 = \text{Identity}$ and $H_1(S^3 - C_1) = C_2$. Then a homotopy shrinking J to a point in $S^3 - g(C_1)$ spends the first part of its parameter pulling the identity on J to gh and the last part as gH_th $(0 \leqq t \leqq 1)$.

EXAMPLE. One cannot weaken Theorem 8.1 by replacing "onto itself" by "onto a subset of itself". If one considers S^3 as the join of circles C_1 and C_2, one sees that there is a monotone decomposition of S^3 onto a disk that homeomorphically sends C_1 to the boundary of the disk and C_2 to the center. Other point inverses are simple curves that link C_1.

DEFINITIONS. Suppose U is an open subset of a polyhedral 3-manifold and J^1 is polygonal simple closed curve in U. We say that nJ^1 *bounds* (mod the integers) in U if J can be regarded as an oriented 1-cycle so that there is some oriented complex C^2 in U such that $\partial C^2 = nJ^1$. By adjusting C^2 to remove certain multiple points we may suppose with no loss of generality that it is locally a 2-manifold off J^1 and that at J^1 it locally resembles n sheets with a common edge.

Notice that in considering complexes, we are using the geometric approach rather than an abstract one and regard a simplex as a geometric object (perhaps oriented), a complex as a geometric object (with perhaps its simplexes oriented), and the boundary of an oriented complex as an oriented complex of one lower dimension.

If M^2 is a connected polyhedral oriented 2-manifold in an oriented polyhedral 3-manifold, M^2 locally has two sides. If J^1 is a polyhedral simple closed curve that intersects M^2 in only a finite number of points, it pierces it at each of these, and pierces it n more times in one direction than in the other, we call n the (unsigned) *piercing number* of J^1 and M^2. Similarly, if J^1 intersects a 2-complex C^2 in only a finite number of points, none of these are on the 1-skeleton of C^2, and it pierces C^2 at each of these points, we may define the piercing number of J^1 and C^2 in a similar fashion.

THEOREM 8.2. *Suppose J is a simple closed curve (perhaps wild) in a polyhedral 3-manifold W^3 such that J can be shrunk to a point in W^3. Then for each point $p \in J$ and each neighborhood U of p there is polyhedral simple closed curve $K^1 \subset U-J$ such that no positive multiple of K^1 bounds in W^3-J.*

Proof. Let J be the union of two arcs px_0q, px_1q and B_1^3, B_2^3 be polyhedral 3-cells in W^3 such that $p \in \text{Int } B_2^3 \subset B_3^3 \subset (\text{Int } B_1^3 \cap U)$, $q \in W^3 - B_2^3$, and there is a homotopy pulling J into B_1^3 that is the identity on $J \cap B_2^3$.

The 2-sphere Bd B_2^3 is the union of two polyhedral disks B_0^2, B_1^2 such that $B_0^2 \cap B_1^2 = \text{Bd } B_0^2 = \text{Bd } B_1^2$, $px_0q \cap \text{Bd } B_2^3 \subset \text{Int } B_0^2$, and $px_1q \cap \text{Bd } B_2^3 \subset \text{Int } B_1^2$. Then $K^1 = \text{Bd } B_0^2 = \text{Bd } B_1^2$ is the required polyhedral simple closed curve.

Suppose n is positive integer such that nK^1 bounds in W^3-J. There is a polyhedral simple closed curve J_1^1 close to J so that nK^1 bounds in $W^3-J_1^1$, J_1^1 is the union of arcs py_0q, py_1q such that $py_0q \cap \text{Bd } B_2^3 \subset \text{Int } B_0^2$, $py_1q \cap \text{Bd } B_2^3 \subset \text{Int } B_1^2$, and there is a homotopy pulling J_1^1 into B_1^3 that is the identity on $J_1^1 \cap B_2^3$.

We suppose that the homotopy H_t $(0 \leq t \leq 1)$ pulling J_1^1 into B_1^3 is such that there is a finite sequence $0 = t_0 < t_1 < t_2 < \cdots < t_k = 1$ such that each H_{t_i} is a homeomorphism, $H_0 = \text{Identity}$, $H_1(J_1^1) \subset \text{Int } B_1^3$, and $H_{t_i} = H_{t_{i+1}}$ except on an arc A_i^1 of J_1^1 such that $H_{t_i}(A_i^1) \cup H_{t_{i+1}}(A_i^1)$ bounds a polyhedral disk D_i^2 in $W^3 - B_2^3$. Changing a 2-complex in $W^3-J_1^1$ bounded by nK^1 by adding oriented handles near the polyhedral disks D_i^2 produces in each $W^3 - H_{t_i}(J_1^1)$ a 2-complex bounded by nK^1. In particular, nK^1 bounds in $W^3 - H_1(J_1^1)$.

As a result of the homotopy discussed in the last paragraph, we suppose with no loss of generality that $J_1^1 \subset \text{Int } B_1^3$. Let C^2 be an oriented 2-complex in $W^3-J_1^1$ such that $\partial C^2 = nK^1$. Suppose B_1^2 is oriented so that $C^2 + nB_1^2$ is a 2-cycle. (We regard nB_1^2 as n oriented disks all nearly parallel and with the same boundary.) Since J_1^1 misses C^2 and when ordered py_1qy_0p pierces B_1^2 once more in going into B_2^3 than in going out, the piercing number of J_1^1 and $C^2 \cup nB_1^2$ is n, where in lieu of an orientation of W^3 we use an orientation of B_1^3. Let M_1^2 be an oriented polyhedral 2-manifold obtained from $C^2 + nB_1^2$ so that J_1 pierces M_1^2 n more times in one direction than it does in the other. We can talk of the direction of the piercing since $nB_1^2 \subset B_1^3$.

Let M_2^2 be a polyhedral orientable 2-manifold in Int B_1^3 such that Bd $M_2^2 = J_1^1$.

We suppose that M_1^2, M_2^2 are in such general positions that their intersection is a 1-manifold with boundary along which M_1^2 and M_2^2 locally cross.

Let $A_1^1, A_2^1, \ldots, A_k^1$ be the components of $M_1^2 \cap M_2^2$ that are arcs. Remove a tubular neighborhood of A_1^1 from M_2^2 so as to change M_2^2 to an oriented polyhedral 2-manifold with boundary M_3^2. There is no assurance that M_3^2 is connected but we suppose it retains the orientation of M_2^2. Certainly, Bd M_3^2 is not connected. The removal reduces by two the number of points in $M_1^2 \cap$ Bd M_2^2. However, since the piercing was from different sides, this leaves the piercing number of Bd M_3^2 and M_1^2 equal to n. By removing tubular neighborhoods about the others A_i^1's, we arrive at the contradiction that there is a 2-manifold with boundary M_{k+2}^2 such that Bd M_{k+2}^2 misses M_1^2 but its piercing number with M_1^2 is n. Hence, our assumption that nK^1 bounds in $W^3 - J$ is false.

Note that in proving Theorem 8.2 we made two uses of the fact that J can be shrunk to a point in W^3—first, that J can be pulled into the orientable part B_1^3 of W^3 and second, that J bounds. If W^3 is already orientable we do not need to pull J into B_1^3 and hence obtain the following variation of Theorem 8.2.

THEOREM 8.3. *If J is a simple closed curve in a polyhedral, orientable 3-manifold W^3 and J lies in an open subset U of W^3, there is a polyhedral simple closed curve K^1 in $U - J$ such that no positive multiple of K^1 bounds in $W^3 - J$.*

Proof. As in the proof of Theorem 8.2 we express J as the union of two arcs px_0q, px_1q and let B^3 be polyhedral 3-cell such that $p \in \text{Int } B^3 \subset B^3 \subset U$ and $q \in W^3 - B^3$. We express Bd B^3 as the union of two polyhedral disks B_0^2, B_1^2 such that $B_0^2 \cap B_1^2 = \text{Bd } B_0^2 = \text{Bd } B_1^2$, $px_0q \cap \text{Bd } B^3 \subset \text{Int } B_0^2$, and $px_1q \cap \text{Bd } B_2^3 \subset \text{Int } B_1^2$. Then Bd $B_0^2 =$ Bd $B_1^2 = K_1^1$ is a candidate for K^1.

If n is a positive integer such that nK_1^1 bounds in $W^3 - J$, we let C^2 be a 2-complex in $W^3 - J$ such that $\partial C^2 = nK_1^1$. Suppose B_1^2 is oriented so that $C^2 + nB_1^2$ is a 2-cycle.

Let K_2^1 be a polygonal simple closed curve in $U - J$ so oriented and so close to J that K_2^1 misses C^2, K_2^1 intersects B_1^2 in only a finite number of places, pierces it at each of these, and it pierces B_1^2 once more in going into B^3 than in going out. Let M_1^2 be an oriented polyhedral 2-manifold obtained from $C^2 + nB_1^2$ so that K_2^1 pierces M_1^2 n more times in one direction than it does in the other. We shall show that no multiple of K_2^1 bounds in $W^3 - J$—in fact, not even in W^3.

Assume K_2^1 is given an orientation so that mK_2^1 bounds in W^3 and C_2^2 is an oriented complex so that $\partial C_2^2 = mK_2^1$, C_2^2 is locally a 2-manifold off K_2^1, and on K_2^1 it locally resembles m sheets with a common edge. Trim C_2^2 back a bit from K_2^2 to obtain an oriented polyhedral 2-manifold M_2^2 (without singularities) so that the piercing number of Bd M_2^2 and M_1^2 is mn. We do not claim that Bd M_2^2 is connected or even that M_2^2 is connected but we do suppose that M_1^2, M_2^2 are in such general positions that their intersection is a 1-manifold with boundary along which M_1^2 and M_2^2 locally cross.

Let $A_1^1, A_2^1, \ldots, A_j^1$ be the components of $M_1^2 \cap M_2^2$ that are arcs. Remove a tubular neighborhood of A_1^1 from M_2^2 so as to change M_2^2 to an oriented polyhedral 2-manifold with boundary M_3^2. The removal reduces by two the number of points in $M_1^2 \cap \text{Bd } M_2^2$. However, since the piercing was from different sides, this leaves the piercing number of Bd M_3^2 and M_1^2 equal to mn. By removing tubular neighborhoods about the other A_i^1's, we arrive at the contradiction that there is a 2-manifold with boundary M_{j+2}^2 such that Bd M_{j+2}^2 misses M_1^2 but its piercing number with M_1^2 is mn. Hence our assumptions that nK_1^1 bounds in $W^3 - J$ and mK_2^1 in W^3 cannot both be true.

EXAMPLE. Theorem 8.3 is false if one omits the orientable requirement. Let P^2 be the projective plane, J_0 be a simple closed curve in P^2 such that $P^2 - J_0$ is an open disk, $W^3 = P^2 \times (-1, 1)$, and $J = J_0 \times 0$. If K^1 is any polyhedral simple closed curve in $W^3 - J$, $2K^1$ bounds in $W^3 - J$.

DEFINITIONS. If B^3 is a polyhedral 3-cell and J_1, J_2 are disjoint simple closed curves in Int B^3, we say that the (unsigned) *linking number* of J_1 and J_2 is n if J_1 is homotopic in Int $B^3 - J_2$ to a polyhedral simple closed curve K_1^1, K_1^1 can be oriented so that it bounds an orientable polyhedral 2-manifold M^2 in Int B^3, and J_2 is homotopic in Int $B^3 - K_1^1$ to a polyhedral simple closed curve K_2^1 such that the piercing number of K_2^1 and M^2 is n. It is known that the linking number of J_1 and J_2 is independent of the K_1^1, M^2, K_2^1 selected and is equal to the linking number of J_2 and J_1. We say that J_1 and J_2 *link* if this linking number is not 0. If the linking number is 0, it is possible to adjust M^2 so as to get an oriented polyhedral 2-manifold with boundary M_1^2 so that Bd $M_1^2 = K_1^1$ and $M_1^2 \cap J_2 = 0$.

THEOREM 8.4. *Suppose B^3 is a polyhedral* 3-*cell in a polyhedral* 3-*manifold with or without boundary N^3 and C_1, C_2 are linking simple curves in* Int B^3. *Then if G is the decomposition of $C_1 \cup C_2$ whose only nondegenerate element is C_2, there is no monotone upper semicontinuous decomposition of N^3 onto a* 3-*manifold W^3 such that the elements of G are elements of the decomposition.*

Proof. We show that there is no such decomposition by showing that there is no compact monotone map g of N^3 onto W^3 that has elements of G as point inverses. For convenience we suppose W^3 is polyhedral.

Assume there is such a g. Let U be an open subset of W^3 containing $g(C_1)$ such that $g^{-1}(U) \subset \text{Int } B^3$ and let J^1 be a polyhedral simple closed curve in $U - g(C_1)$ such that no multiple of J^1 bounds in $W - g(C_1)$. That there is such a J^1 follows from Theorem 8.2.

Let h be a homeomorphism of J^1 into $g^{-1}(U) - C_1$ such that $h(J^1)$ is a polyhedron, and the identity on J^1 is homotopic to gh in $U - g(C_1)$.

If $h(J^1)$ does not link C_1, there is an orientation for $h(J^1)$ and an oriented 2-complex C^2 in Int $B^3 - C_1$ such that $\partial C^2 = h(J^1)$. Then $g(C^2)$ and a singular annulus joining J^1 and $gh(J^1)$ provides the image of an oriented 2-complex in $M^3 - g(C_1)$ whose boundary is J^1. An approximation to this image gives an oriented

2-complex in $W^3-g(C_1)$ whose boundary is J^1. The case where $h(J^1)$ does not link C_1 leads to the contradiction that J^1 bounds in $W^3-g(C_1)$.

There remains the case where $h(J^1)$ links C_1 in Int B^3. Suppose the linking number is n and that the linking number of C_2 with C_1 is m. Let K_2^1 be a polyhedral simple closed curve homotopic in Int B^3-C_1 to C_2. Then $mh(J^1)$ and nK_2^1 each link C_1 mn times. Hence there are orientations for $h(J^1)$ and K_2^1 and an oriented 2-complex C_2^2 in Int B^3-C_1 such that $\partial C_2^2=mh(J^1)+nK_2^1$. An approximation to m singular annuli each joining J^1 and $gh(J^1)$, $g(C_2^2)$, and the image under g of n singular annuli each joining K_2^1 and C_2 provides us with an oriented 2-complex in $W^3-g(C_1)$ whose boundary is mJ^1. The assumption that there was a decomposition of N^3 into W^3 having elements of G as elements led to the contradiction that no multiple of J^1 bounds in $M^3-g(C_1)$ but mJ^1 bounds there.

QUESTION. Would Theorem 8.4 be true if instead of saying that C_1 and C_2 link in Int B^3 we required that they be polygonal simple closed curves in Int B^3 that 1-link there? Recall that two curves 1-link if they do not bound mutually exclusive orientable 2-manifolds.

REFERENCES

1. R. H. Bing, *A characterization of 3-space by partitionings*, Trans. Amer. Math. Soc. **70** (1951), 15–27. MR **13**, 484.

2. ———, *Concerning hereditarily indecomposable continua*, Pacific J. Math. **1** (1951), 43–51. MR **13**, 265.

3. ———, *Locally tame sets are tame*, Ann. of Math. (2) **59** (1954), 145–158. MR **15**, 816.

4. ———, *An alternative proof that 3-manifolds can be triangulated*, Ann. of Math. (2) **69** (1959), 37–65. MR **20** #7269.

5. ———, "Decompositions of E^3" in *Topology of 3-manifolds and related topics* (Proc. Univ. of Georgia Institute, 1961), Prentice-Hall, Englewood Cliffs, N. J., 1962, pp. 5–21. MR **25** #4501.

6. B. G. Casler, *An imbedding theorem for connected 3-manifolds with boundary*, Proc. Amer. Math. Soc. **16** (1965), 559–566. MR **31** #2730.

7. J. F. P. Hudson and E. C. Zeeman, *On regular neighbourhoods*, Proc. London Math. Soc. (3) **14** (1964), 719–745. MR **29** #4063.

8. E. E. Moise, *Affine structures in 3-manifolds.* V: *The triangulation theorem and Hauptvermutung*, Ann. of Math. (2) **56** (1952), 96–114. MR **14**, 72.

9. ———, *Affine structures in 3-manifolds.* VIII: *Invariance of the knot-types; local tame imbedding*, Ann. of Math. (2) **59** (1954), 159–170. MR **15**, 889.

10. J. H. C. Whitehead, *Simplicial spaces, nuclei, and m-groups*, Proc. London Math. Soc. (2), **45** (1939), 243–327.

11. G. T. Whyburn, "Decomposition spaces" in *Topology of 3-manifolds and related topics* (Proc. Univ. of Georgia Institute, 1961), Prentice-Hall, Englewood Cliffs, N. J., 1962, pp. 2–4. MR **25** #4499.

12. E. C. Zeeman, "Polyhedral n-manifolds: II: Embeddings" in *Topology of 3-manifolds and related topics* (Proc. Univ. of Georgia Institute, 1961), Prentice-Hall, Englewood Cliffs, N. J., 1962, pp. 65–70. MR **28** #1596.

UNIVERSITY OF WISCONSIN
MADISON, WISCONSIN 53706

THE MONOTONE MAPPING PROBLEM

R. H. Bing

1. Introduction

Three related topics are treated in this paper: the monotone mapping problem, monotone decompositions of a 3-cell, and drainage systems for a 3-cell. We pick the title from the first of these three topics since there has been wide interest in the problem.

An earlier version of this paper was read by E. H. Anderson, Ralph Jones, Michael Olinick, and Robin Soloway, and they made useful suggestions for improving it.

Euclidean 3-space is denoted by E^3. The part of E^3 on or above the xy-plane is called E^3_+. The 3-sphere S^3 is the one-point compactification of E^3, or, equivalently, the union of two 3-cells sewed together along their boundaries with a homeomorphism. We use I^3 to denote an oriented 3-cell in E^3 and I^2 to denote its lower base.

A *mapping* (continuous transformation) is *monotone* if each point inverse is a compact connected set. It is *compact* if the inverse of each compact set is compact. The *monotone mapping problem* asks if each monotone map of euclidean n-space onto itself is compact. The answer is trivial if $n = 1$. Whyburn gave an affirmative solution [15] for $n = 2$ and Glaser a negative one [11] for $n \geq 4$. Other partial solutions are included in [5], [6], [7], [13], and [14]. The solution is completed by Theorem 6, which gives a monotone noncompact map of E^3 onto itself. This answers a question raised by Whyburn in [16]. Theorem 15 tells us more about the map.

In proving Theorem 6, use is made of the interesting result (Theorem 2) that there is a monotone decomposition of I^3 onto itself such that each element of the decomposition meets I^2. Other decompositions of this sort are given by Theorems 3, 4, and 12.

Before describing the monotone decomposition of I^3, we find it convenient to bore holes in I^3. Let us first consider a 2-dimensional analogue. Regard the surface of a territory as a 2-dimensional object and its drainage system as holes. The water in the main river comes from smaller streams, these smaller streams branch into still smaller ones, and the branching continues until each point of the watershed is on the closure of the sum of the streams. Some points are at divides, so that for a drop of rain falling there, several paths lead to the sea.

A physical example of a 3-dimensional drainage system may be suggested by the air tube leading into the lungs. Perhaps a tube (with its multiple branchings) is dense in the lung. In the following definition we regard I^3 as the lung and $f(\text{Int } I^2 \cup \text{Int } I^3)$ as the tube.

If f is a map of I^3 onto I^3 which is a homeomorphism on Int I^3 and takes I^2 homeomorphically into I^2, we call $f(\text{Int } I^2 \cup \text{Int } I^3)$ a *drainage system* for I^3. Note that the drainage system is dense in I^3. It follows from invariance of domain that the drainage system misses $f(\text{Bd } I^3 - \text{Int } I^3)$.

The proof of the following theorem is delayed until Section 4 so that we can first give some applications of it in Sections 2 and 3.

Theorem 1. *There is a drainage system* $U = f(\text{Int } I^2 \cup \text{Int } I^3)$ *for* I^3 *and a monotone map* g *of* $I^3 - U$ *onto* I^3 *which is the identity on* $(\text{Bd } I^3) - (\text{Int } I^2)$.

2. A Monotone Decomposition of a Cube onto Itself All of Whose Elements Intersect the Base

The following theorem is an elementary version of Theorem 12 given in Section 6.

Theorem 2. *There is a monotone map* g_1 *of* I^3 *onto itself such that* g_1 *is the identity on* $(\text{Bd } I^3) - (\text{Int } I^2)$ *and* $g_1(I^2) = I^3$.

Proof. Let $U = f(\text{Int } I^2 \cup \text{Int } I^3)$ be the drainage system for I^3 promised by Theorem 1 and g be the associated monotone map of $I^3 - U$ onto I^3. Also, let r be a retraction of I^3 onto Bd I^3 − Int I^2 such that r sends I^2 homeomorphically onto Bd I^3 − Int I^2 and each nondegenerate point inverse is a spanning segment of I^3.

If $p \in I^3 - U$, $f^{-1}(p)$ is a closed subset of Bd I^3 − Int I^2, and unless $f^{-1}(p) \in \text{Bd } I^2$, $r^{-1}f^{-1}(p)$ is the union of segments in I^3 reaching from $f^{-1}(p)$ to Int I^2. Then $fr^{-1}f^{-1}(p)$ is the union of arcs in $U \cup \{p\}$ reaching from p to $f(\text{Int } I^2)$. A point inverse of the g_1 promised by Theorem 2 is the union of $g^{-1}(p)$ and all such arcs reaching down from points of $g^{-1}(p)$ to $f(\text{Int } I^2)$.

For those who prefer a formula, we give $g_1 = gfrf^{-1}$. Since frf^{-1} is the identity on $I^3 - U$ and f^{-1} is one-to-one on U, the map g_1 is well defined. That $gfrf^{-1}$ is continuous follows from the fact (th. VI 3.2 of [8]) that if X_1, X_2, X_3 are topological spaces, f is a closed mapping of X_1 onto X_2, and ϕ is a mapping of X_1 onto X_3, then $\phi \circ f^{-1}$ is continuous if it is well defined.

The following three results are easy consequences of Theorem 2.

Theorem 3. *There is a monotone map g of the tetrahedron $v_0v_1v_2v_3$ onto itself such that g is the identity on the three faces containing v_3, $g(v_0v_1v_2) = v_0v_1v_2v_3$, and $g^{-1}(v_0) = v_0$.*

Proof. We regard the tetrahedron as the union of $\{v_0\}$ and frustums $F_1, F_2, \ldots$, where $F_i \cap F_{i+1}$ is the cross section parallel to $v_1v_2v_3$ whose intersection with v_1v_0 divides it in the ratio i to 1. It follows from Theorem 2 that there is a monotone map g_i of F_i onto itself such that g_i is the identity on all faces of F_i except $F_i \cap v_0v_1v_2$ and g_i takes $F_i \cap v_0v_1v_2$ onto F_i. The required map g is the identity on v_0 and g_i on each F_i.

Theorem 4. *There is a monotone map g of the 3-sphere S^3 onto a tetrahedron T such that for some point $p \in \text{Bd } T$, $g^{-1}(p)$ is degenerate.*

Proof. We regard S^3 as the union of two copies of the tetrahedron $T = v_0v_1v_2v_3$ sewn together along their common boundary with the identity map. The map g is the map g of Theorem 3 applied to each copy. The required point p is v_0.

Theorem 5. *There is a monotone map of E^3 onto E^3_+.*

Proof. This theorem follows from Theorem 4 since E^3 is homeomorphic with $S^3 - g^{-1}(p)$ and E^3_+ is homeomorphic with $v_0v_1v_2v_3 - \{v_0\}$.

QUESTION. Suppose X is a closed subset of I^3 and g is a monotone map of X onto I^3. Under what condition can g be extended to a monotone map of I^3 onto itself? In particular, let $v_0v_1v_2v_3$ be an equilateral tetrahedron, f be a map of v_0v_1 onto the tetrahedron, h be a congruency of v_0v_1 onto v_2v_3, and x_p denote the join of $f^{-1}(p)$ and $hf^{-1}(p)$. If X is the union of all such x_p's, the map which sends x_p to p is a monotone map of X onto the tetrahedron. Can the map be extended to a monotone map of all of $v_0v_1v_2v_3$ onto itself? If this map could be extended in a straightforward manner, we might have an approach to a weaker version of Theorem 2 [deleting the requirement that g is identity on $(\text{Bd } I^3) - (\text{Int } I^2)$] that is easier than the tortuous path through Theorem 1.

3. A Monotone Noncompact Map of E^3 onto Itself

A more specific form of the following theorem is provided by Theorem 15.

Theorem 6. *There is a monotone noncompact map of E^3 onto itself.*

Proof. The required map is $f \circ g$, where g is the monotone map of E^3 onto E^3_+ guaranteed by Theorem 5 and f is the one-to-one noncompact map of E^3_+ onto E^3 guaranteed by the following Theorem 7.

Theorem 7 is merely Glaser's Lemma 2 of [10], but we give an independent proof for completeness. Glaser noted in [11] that Theorem 6 followed from his Lemma 2 of [10] if there is a monotone map of E^3 onto E^3_+. Glaser used a monotone map $f^4_{3,3}$ of S^3 onto I^3 promised by Anderson in [1] to show that there is a monotone map of E^4 onto E^4_+ but since he was unsure if there is a point p of Bd I^3 such that $f^4_{3,3}{}^{-1}(p)$ is pointlike in S^3, he was unable to conclude Theorem 5.

Theorem 7. *There is a one-to-one noncompact map of E^3_+ onto E^3.*

Proof. We regard E^3_+ as Int $I^3 \cup$ Int $I^2 = X_0$. Let p_0p_1 be an edge of I^2 and $D_1, D_2,\ldots$ be a sequence of mutually exclusive square disks on I^2 converging to p_0 such that each has an edge on p_0p_1. Let D_0 be the disk which is the closure of $I^2 - \cup D_i$.

By sewing the top base of I^3 to D_0, one finds that there is a map ϕ' of I^3 onto a cube with a handle H such that ϕ' is one-to-one on X_0 and sends X_0 onto the union of Int H and a part of Bd H which satisfies the hypotheses of the following theorem. The required map is $\phi \circ \phi'$, where ϕ is the map guaranteed by Theorem 8. That ϕ' restricted to Int $I^3 \cup$ Int I^2 is noncompact follows from the fact that if A is an arc in ϕ'(Int $I^3 \cup$ Int D_0) which crosses $\phi'(D_0)$, then $\phi'^{-1}(A)$ is noncompact.

Theorem 8. *Suppose H is a cube with holes; $p \in$* Bd *H; $D_1, D_2,\ldots$ is a sequence of mutually exclusive disks on* Bd *H converging to p; and I_i $(i = 1, 2,\ldots)$ is an arc on* Bd *D_i. Then there is a one-to-one noncompact map ϕ of* Int $H \cup (\cup$ Int $D_i) \cup (\cup$ Int $I_i)$ *onto E^3.*

Proof. Suppose H has k holes. Let $A_1, A_2,\ldots, A_k$ be k mutually exclusive annuli on Bd H such that H is converted to a 3-cell if k pillboxes are attached to H along the A_i's. We suppose that $A_i \cap D_j = 0$ unless $i = j$ and A_i is the union of D_i and a disk D_i' such that $D_i \cap D_i'$ is the union of two arcs, one of which is I_i.

Let $B_1, B_2,\ldots, B_k$ be annuli on Bd H such that the B_i's lie in k disjoint disks on Bd H, $B_i \cap A_j = 0$, $B_i \cap D_j = 0$ unless $j = k + i$, and B_i is the union of D_{k+i} and a disk D'_{k+i} such that $D_{k+i} \cap D'_{k+i}$ is the union of two arcs one of which is I_{k+i} and if A_i is attached to B_i with a homeomorphism that sews D_i to D'_{k+i} and D_{k+i} to D_i' so that I_i misses I_{k+i}, then the sewn up H is orientable.

Suppose all the A_i's are attached to the B_i's in this fashion. Let ϕ_1 be the map of H onto the sewn-up H. Then ϕ_1 is one-to-one and noncompact on Int $H \cup (\cup$ Int $D_i) \cup (\cup$ Int $I_i)$.

It may be noted that $\phi_1(H)$ is still a cube with k holes (in fact, even a cube with handles), and if C is a compact subset of Int H, then $\phi_1(C)$ lies in a 3-cell in Int $\phi_1(H)$. To see this, consider removing k 3-cells from H so that these 3-cells miss C and intersect BdH in disks which contain the B_i's. Use these 3-cells as pillboxes to sew to the reduced H and convert it to a 3-cell. Then reattach the k 3-cells to make a cube with k handles.

We note that $\phi_1(H)$, $\phi_1(p)$, $\phi_1(D_{2k+1})$, and $\phi_1(I_{2k+1})$ satisfy the hypotheses of Theorem 8. By using the first $2k$ of the D_{2k+i}'s, we get a map ϕ_2 of $\phi_1(H)$ onto a cube with k handles such that ϕ_2 is one-to-one on $\phi_1(\text{Int } H \cup \text{Int } D_i) \cup (\cup \text{Int } I_i)$, each compact subset of

$$\phi_2 \circ \phi_1\left(\text{Int } H \cup \left(\bigcup_1^{2k} \left(\text{Int } D_i\right) \cup \left(\bigcup_1^{2k} \text{Int } I_i\right)\right)\right)$$

lies in a 3-cell in Int $\phi_2 \circ \phi_1(H)$, and the requirements of Theorem 8 are still met on the adjusted sets.

Further identifications convert Int $H \cup (\cup \text{Int } D_i) \cup (U\, I_i)$ into a 3-manifold. Since each compact subset of this 3-manifold lies in a topological open 3-cell, it is the union of an increasing sequence of open 3-cells. It follows from [4] that the manifold is topologically E^3.

QUESTION. Theorem 8 carries unnecessary restrictions, but we used them since they were convenient and they were what we needed in applying the theorem in the proof of Theorem 7. We could just as well have replaced $(\cup \text{Int } D_i) \cup (\cup \text{Int } I_i)$ by an open disk on Bd H—or perhaps by many other sorts of 2-manifolds. Perhaps H need not be compact and might have infinitely many holes. What conditions must the union of the interior of a 3-manifold-with-boundary and a portion of its boundary satisfy in order to ensure that there is a one-to-one map of the union onto E^3?

4. A Drainage System for Theorem 1

The drainage system $U = f(\text{Int } I^3 \cup \text{Int } I^2)$ which we shall describe to verify Theorem 1 will resemble a cave eroded from I^3, with the entrance lying in Int I^2 and the cave being dense in I^3. The monotone map g (defined in Section 5) from $I^3 - U$ onto I^3 will have each point inverse lying in a horizontal plane and the image under g of each horizontal cross section of $I^3 - U$ will be a disk. Here is our general plan.

The left side of Figure 1 shows how we decompose the domain space I^3. We regard I^3 as the union of nonoverlapping polyhedral cubes $\{C_i^j | j \geq 1, 1 \leq i \leq n_j\}$ together with a Cantor set lying on $(\text{Bd } I^3) - I^2$. For each positive integer m, the C_i^m's are disjoint cubes, each meeting Bd I^3 in a punctured disk, while $C^m = \cup \{C_i^j | 1 \leq j \leq m, 1 \leq i \leq n_j\}$ is a cube meeting each C_i^{m+1} in a disk. We also suppose that each component of $I^3 - C^m$ has diameter less than $1/2^m$ (thus restricting the diameter of each C_i^{m+1}) and that the closure of each such component is a cube bounded by the union

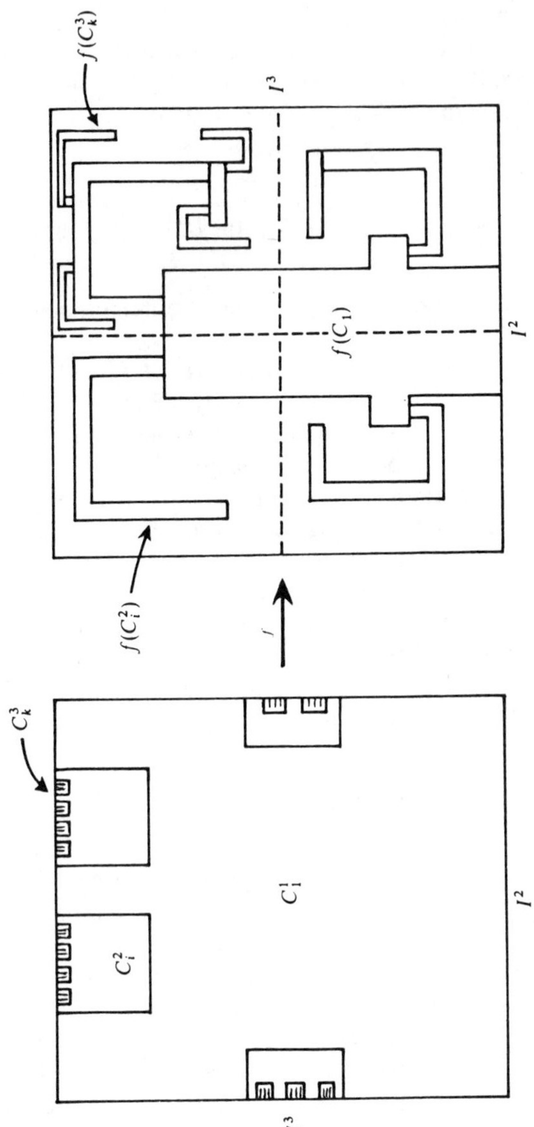

Figure 1

of a disk in Bd I^3 and a disk in some Bd C_i^m. These latter disks will each lie in some Bd C_k^{m+1}.

Let T_1 be a cellular subdivision of I^3 into rectangular solids with faces parallel to the faces of I^3 and such that the mesh of T_1 is less than $\frac{1}{2}$. We define the map f on $C_1^1 = C^1$ to be a homeomorphism into I^3 such that for $m = 1$,

(a) no cell of T_m is separated by $f(C^m)$;

(b) Bd $f(C^m)$ is the union of finitely many rectangular disks, each parallel to a side of I^3;

(c) each component of each horizontal cross section of $f(C^m)$ is a disk;

(d) the interior of each cell of T_m either lies in $f(C^m)$ or contains a horizontal disk in Bd $f(C^m)$; and

(e) (Bd I^3) $\cap f(C^m) = f(I^2) \subset$ Int I^2.

In general, T_{m+1} will be a cellular subdivision of I^3 into rectangular solids such that mesh $T_{m+1} < 1/2^{m+1}$, T_{m+1} refines T_m, and the closure of $I^3 - f(C^m)$ is the union of elements of T_{m+1}. There is a one-to-one correspondence between the n_{m+1} cells of T_m not contained in $f(C^m)$ and the C_i^{m+1}'s. Property (d) allows us to extend f to a homeomorphism of C^{m+1} into I^3 so that $f(C_i^{m+1})$ lies on the interior of the corresponding cell of T_m and properties (a), (b), (c), (d), and (e) continue to hold. The $f(C_i^{m+1})$'s are attached to $f(C^m)$ along horizontal disks in the interior of horizontal disks in Bd $f(C^m)$ − Bd $f(C^{m-1})$ so that components of horizontal cross sections of $f(C^m)$ continue to be components of horizontal cross sections of $f(C^{m+1})$. Since each $f(C_i^{m+1})$ has diameter less than $1/2^m$ and since property (d) implies that $f(C^m)$ is $1/2^m$ dense in I^3, f has a unique extension to a map from I^3 onto I^3.

5. A Monotone Map g for Theorem 1

Consider the drainage system $U = f(\text{Int } I^2 \cup \text{Int } I^3) = f(\text{Int } I^2) \cup f(\cup \text{Int } C^i)$ defined in the last section. Before defining a monotone map of $I^3 - U$ onto I^3 as required by Theorem 1, we define a monotone upper semicontinuous decomposition G of I^3.

Let H be the set of all continua D such that for some C^i, D is a component of a horizontal cross section of $f(C^i)$. The elements of H are disjoint polygonal disks. In each horizontal cross section of I^3, none of these intersect the boundary of this cross section and for each $\varepsilon > 0$, not more than a finite number of disks in the cross section have diameters more than ε.

Let G be the upper semicontinuous decomposition of I^3 consisting of the elements of H and the points of I^3 not in any element of H. Then G induces an upper semicontinuous decomposition of each horizontal cross section of I^3 into points and disks no one of which separates the cross section or intersects its boundary. It follows from [12] that there is a monotone map of the cross section onto a disk such that each point inverse is an element of G and the boundary of the cross section goes onto the boundary of the disk.

Let ϕ be the map which assigns to each element of G its height in the cube. Then the above $\phi^{-1}(t)$ is homeomorphic to a disk for each t, $0 \leq t \leq 1$. It follows from Theorem 7 of [9] that there is a level-preserving monotone map g' of I^3 onto I^3 such that each point inverse is an element of G. Then g' takes Bd I^3 − Int I^2 homeomorphically onto Bd I^3 − Int I^2. Without loss of generality we assume that g' is the identity on Bd I^3 − Int I^2.

The map required by Theorem 1 is $g'|(I^3 - U)$. Let us see why g is a monotone map of $I^3 - U$ onto I^3. If $g'^{-1}(p)$ is a point inverse under g', it is either a point of $I^3 - U$ or a disk D which is a component of a horizontal cross section of some $f(C^i)$. In this latter case, D is also a component of a horizontal cross section of $f(C^{i+1})$, and unless $D \subset \text{Int } I^2$, $D \cap (I^3 - U) = D - f(\text{Int } C^{i+1})$ is a disk or a disk with a finite number of mutually exclusive open polygonal disks removed. If $D \subset \text{Int } I^2$, $D \cap (I^3 - U) = \text{Bd } D$. In any case, $g'^{-1}(p) \cap (I^3 - U) = g^{-1}(p)$ is connected.

QUESTION. The drainage system we described in Section 4 somewhat resembles a portion of a fattened dendron. It seems rather artificial to slice the $f(C^i)$'s horizontally rather than in some other reasonable fashion. The reason we sliced them horizontally was to be able to use the available results of Dyer and Hamstrom [9]. However, it seems that if one had tubes reasonably situated in I^3 and decomposed these tubes into reasonable slices (not necessarily horizontal slices), one should be able to define "reasonable" so as to conclude that there is a monotone map of I^3 onto I^3 whose nondegenerate point inverses are the slices. In particular, suppose D is a dendron, p_0 is an end point of D, R is its set of other ends, and g is a monotone map of I^3 onto D such that g restricted to Bd I^3 is monotone, $g^{-1}(p_0) = I^2$, and the inverse of each point of R is a point of Bd $I^3 - I^2$. Let f be a map of I^3 onto I^3 that is the identity on I^2 and takes each tame 3-cell in $I^3 - g^{-1}(R)$ homeomorphically onto a tame 3-cell in $I^2 \cup \text{Int } I^3$. Is there a monotone map of I^3 onto itself whose nondegenerate point inverses are of the sort $fg^{-1}(p)$, where $p \in D - R$? It follows from [3] that there is an affirmative answer if f sends I^3 homeomorphically onto a tame 3-cell in $I^2 \cup \text{Int } I^3$. One might need to restrict the nondegenerate $fg^{-1}(p)$'s to be nice disks. Ralph Bean advises me that he has results along these lines.

6. Simplifying Decomposition Elements

Recall that in the monotone map g_1 of Theorem 2, each point inverse $g_1^{-1}(p)$ was the union of a point inverse $g^{-1}(p)$ of the map g of Theorem 1 and arcs leading from $g^{-1}(p)$ to $f(\text{Int } I^2)$. Some points of $g^{-1}(p)$ had many arcs leading from them. The complexity of $g_1^{-1}(p)$ is influenced by the nature of $g^{-1}(p)$ and the way that arcs dangle from $g^{-1}(p)$ as determined by the drainage system.

The nondegenerate $g^{-1}(p)$'s of Theorem 1 as described in Section 4 were horizontal polygonal disks with a finite number of mutually exclusive open disks removed. We see that by modifying f we can limit the number of kinds of point inverses of g.

Theorem 9. *There is a drainage system $U = f(\text{Int } I^2 \cup \text{Int } I^3)$ for I^3 and a monotone map g of $I^3 - U$ onto I^3 such that each point inverse of g lies in a horizontal plane and is either a point, a polygonal simple closed curve, or a polygonal θ-curve.*

Proof. We prove this result by pointing out how to modify the drainage system for Theorem 1 described in Section 4. In that description we supposed that $f(C^i)$ was the union of elements of T_{i+1} and that each $\text{Bd}\, f(C^i)$ was the union of rectangular disks each parallel to a side of I^3. These restrictions were not used fully and were included primarily to simplify the presentation.

We regard $f(C^1)$ as a polyhedral tube. It can have slanting sides and assume many positions. Suppose it is adjusted so that each component of each horizontal cross section of $f(C^1)$ is either a point or a horizontal disk, and each component of each horizontal cross section of $\text{Bd}\, f(C^1)$ is either a point, a simple closed curve, a θ-curve, or a horizontal disk, with the additional proviso that the only horizontal disks permitted are $f(I^2)$ and one horizontal disk on the interior of each 3-cell of T_1. These disks are where the $f(C_i^2)$'s fit onto $f(C^1)$. We suppose that $f(C^1)$ does not separate any 3-cell of T_1. We have dropped the requirement that $\text{Bd}\, f(C^i)$ be the union of rectangular disks each parallel to a side of I^3 and are permitting $\text{Bd}\, f(C^i)$ to contain oblique sides.

The T_i's are cellular subdivisions of I^3 into rectangular solids so that mesh $T_i < 1/2^i$, T_{i+1} refines T_i, and $f(C^i)$ does not separate any 3-cell of either T_i or T_{i+1}. We have dropped the restriction that $f(C^i)$ be the union of elements of T_{i+1}. There is a one-to-one correspondence between the C_j^{i+1}'s and the 3-cells of T_i not in $f(C^i)$ while $f(C_j^{i+1})$ lies on the interior of its corresponding element of T_i. The $f(C_j^{i+1})$'s have the property that each component of each horizontal cross section of $f(C_j^{i+1})$ is either a point or a disk, while each component of each horizontal cross section of $\text{Bd}\, f(C_j^{i+1})$ is either a point, a simple closed curve, a θ-curve, or a disk. The only horizontal disks permitted in $\text{Bd}\, f(C_j^{i+1})$ are $f(C_j^{i+1}) \cap f(C^i)$ and one on the interior of each of the 3-cells of T_{i+1} which it intersects and where some $f(C_k^{i+2})$ is attached.

The following theorem permits us to limit the number of arcs dangling down from points of $I^3 - U$ as described in the proof of Theorem 2.

Theorem 10. *There is a drainage system* $U = f(\text{Int } I^2 \cup \text{Int } I^3)$ for I^3 *and a monotone map g of* $I^3 - U$ *onto* I^3 *such that each point inverse of g lies in a horizontal plane and is either a point, a polygonal simple closed curve, or a polygonal θ-curve, while each* $f^{-1}(p)$ *has at most four points (and at most three if* $p \in \text{Bd } I^3$).

Proof. Note that in defining f for Theorem 1 in Section 3, $f^{-1}(p)$ is precisely one point if either $p \in U$ or for each i, $p \notin f(C^i)$ but p is an interior point of a rectangular solid of T_i. It is only the p's that lie in many elements of some T_i that have many inverses.

Theorem 10 is proved by replacing the cellular subdivisions T_i used in the proofs of Theorems 1 and 9 by cellular subdivisions so that no point belongs to more than four elements of any of the cellular subdivisions and no point on Bd I^3 belongs to more than three of them. The illustration on page 16 of [2] suggests such a brick partitioning. It is also required that no vertex of any rectangular solid of any T_i lie on any Bd $f(C^j)$.

We now show that if $f^{-1}(p)$ misses each C^i, then it has at most four points. If it contained five points, there would be an integer i so large that $I^3 - C^i$ would contain five components, each intersecting $f^{-1}(p)$. This is impossible since each component of $I^3 - C^i$ is sent by f into a different rectangular solid of T_i.

A similar argument shows that $f^{-1}(p)$ has at most four points if it intersects some C^i. Another shows that $f^{-1}(p)$ has at most three points if $p \in \text{Bd } I^3$.

Theorem 10 may be extended as follows.

Theorem 11. *There is a drainage system* $U = f(\text{Int } I^2 \cup \text{Int } I^3)$ *for* I^3 *and a monotone map g of* $I^3 - U$ *onto* I^3 *such that each point inverse of g lies in a horizontal plane, each nondegenerate point inverse of g is a simple closed curve or θ-curve,* f^{-1} *is a homeomorphism on each such nondegenerate inverse, and each* $f^{-1}(p)$ *has at most four points (at most three if* $p \in \text{Bd } I^3$).

The description of the drainage system for Theorem 11 requires several modifications over that described in Section 4 and altered in the proofs of Theorems 9 and 10. We delay the description of these modifications until the next section and finish this section by giving some applications of Theorem 11.

Theorem 12. *There is a disk D in* Int I^2 *and a monotone map of* I^3 *onto itself which is the identity on* Bd I^3 − Int I^2 *and whose point inverses are as follows.* Bd D, *annuli each of which lies in* Int I^3 *except for one of its bases which is on* Int D, *sets homeomorphic to the cartesian product of an arc and a θ-curve* *where each of these sets lies in* Int I^3 *except for one of its bases, which is on* Int D; *topological cones over one, two, three, or four points of* Int D, *where the rest of the cone lies in* Int I^3; *and topological cones over one, two, or three points of* Int D, *where the vertex of the cone lies on* Bd $I^3 - I^2$ *and the rest of the cone lies in* Int I^3.

Proof. If the monotone map g_1 as described in the proof of Theorem 2 is applied to the drainage system U of Theorem 11 and its accompanying monotone map g of $I^3 - U$ onto I^3, the resulting monotone map satisfies the requirements of Theorem 12 with $D = f(I^2)$.

Theorem 13. *There is a monotone map of a 3-sphere S^3 onto I^3 whose point inverses are either annuli, sets each homeomorphic to the cartesian product of an arc and a θ-curve, or a topological finite graph. The only finite graphs that need to be used are arcs, simple closed curves, θ-curves, bouquets of two or three simple closed curves, and the suspensions of four point sets.*

Proof. This monotone map results from regarding S^3 as the union of two copies of I^3 sewed together along their boundaries with the identity map. The monotone map from Theorem 12 applied to each copy gives the map required for Theorem 13.

QUESTION. Is there a monotone map of S^3 onto I^3 that would have fewer sorts of inverse images than that given by Theorem 13? Simpler inverse images? We could have altered Theorems 6, 10, 11, 12, and 13 by replacing θ-curves by figure-eights, but this did not appear to be a significant improvement.

Theorem 14. *There is a monotone map of E^3 onto E^3_+ such that each point inverse is either a finite graph, an annulus, or the cartesian product of an arc and a θ-curve. The only finite graphs that are needed are arcs, simple closed curves, θ-curves, bouquets of two circles, bouquets of three circles, or suspensions of four point sets.*

Proof. The map results from carefully applying the proof of Theorem 12 to the frustums in the proof of Theorem 3 and then proceeding as in the proofs of Theorems 4 and 5. We say "carefully" since in considering two adjacent frustums, care is taken that no point of the intersection ever belongs to too many elements of the cellular subdivisions of the anion of the two frustums. These elements of the cellular subdivisions are not real rectangular solids (since we are subdividing frustums) but are reasonable facsimiles.

Theorems 14 and 7 yield the following extension of Theorem 6.

Theorem 15. *There is a monotone noncompact map of E^3 onto itself such that each point inverse is either a finite graph, an annulus, or the cartesian product of an arc and a θ-curve. The only finite graphs that need be used are those mentioned in the statements of Theorems* 13 *and* 14.

7. A Drainage System for Theorem 11

The $f(C^i)$'s used to define the drainage system for Theorem 11 are not polyhedral, but it may be noted that the polyhedral condition was dropped from the statement of Theorem 11. We say that a point $p \in f(C^i)$ *belongs to a suitable horizontal slice of* $f(C^i)$ if the component X_p containing p of the horizontal cross section of $f(C^i)$ is a disk or a point while $X_p \cap \text{Bd}\, f(C^i)$ is either a point, a simple closed curve, or an θ-curve. The points of $f(\text{Bd}\, C^i - I^2)$ will belong to suitable horizontal slices of $f(C^i)$.

The really important difference between the drainage system used for Theorem 11 and those previously described in Section 4 and modified in Theorems 9 and 10 is that if an element Δ of T_{i+1} intersects $f(C^i)$, then the feelers reaching out into Int Δ come from this intersection. The other modifications we make are aimed at making this important modification possible and useful. We now consider other modifications.

A minor modification is that the mesh of T_{i+1} is so small that if $\Delta \in T_{i+1}$, diameter $f^{-1}(\Delta \cap f(C^i)) < 1/2^i$.

Since feelers will be extended from dense subsets of $\text{Bd}\, f(C^i) - f(\text{Int}\, I^2)$, $I^3 - C^i$ will have infinitely many components. The components form a null sequence. The closure of each component is a 3-cell of diameter less than $1/2^i$, bounded by the union of a disk on Bd C^i and a disk in Bd I^3. In fact, $(\text{Bd}\, C^i - \text{Int}\, I^2) \cap \text{Bd}\, I^3$ is a Sierpiński curve.

Before mentioning other modifications, we point out the significance of those we have already given. These can be used to show that if $p \in (I^3 - U) \cap \text{Bd}\, f(C^i)$, then $f^{-1}(p)$ is a point. Let X be the union of the elements of T_{i+1} containing p. Then diameter $f^{-1}(X \cap f(C^i)) < 2/2^i$. Since the components of $f(C^{i+1}) - f(C^i)$ which intersect Int X reach out from $X \cap f(C^i)$ and the components of $C^{i+1} - C^i$ have diameters less than $1/2^i$, diameter $f^{-1}(\text{Int}\, X \cap U) < 4/2^i$. Hence $f^{-1}(p)$ is a point.

We now proceed to give other modifications necessitated by the important one. If Δ_j is a rectangular solid of T_i which does not lie in $f(C^i)$, instead of having one $f(C_m^{i+1})$ in Int Δ_j, we may have several. If Δ_k is a rectangular solid of T_{i+1} which lies in Δ_j and intersects but does not lie in $f(C^i)$, then each feeler $f(C_m^{i+j})$ which intersects Δ_k lies in Int Δ_j and attaches to $f(C^i)$ in a disk in $\Delta_k \cap f(C^i)$. If $\Delta_k \cap f(C^i) = 0$, then only one $f(C_m^{i+1})$ intersects Δ_k. Since we are willing to have several $f(C_m^{i+1})$'s in Δ_j, it is not important that $f(C^i)$ not separate Δ_j or Δ_k.

If a rectangular solid Δ of T_{i+1} intersects $f(C^i)$, then the feelers reaching out into Int Δ come from this intersection. Hence it is important that $f(C^i)$ intersect Int Δ if it intersects Δ. With this in mind, we impose the restraint that if $p \in \text{Bd}\, f(C^i)$ and Δ is a rectangular solid of some T_j with $p \in \Delta$, then

p is a limit point of $f(C^i) \cap \text{Int } \Delta$. After $f(C^i)$ and T_i have been chosen satisfying the restraint, in order to pick T_{i+1} satisfying it, one could use the property that if Δ is a rectangular solid in T_i, then there are only a countable number of cross sections P of Δ parallel to a face of Δ such that $P \cap f(C^i)$ contains a point p and $\Delta - P$ has a component Y such that p is not a limit point of $Y \cap f(C^i)$.

If feelers are to be sent out from $f(C^i)$ on a dense subset of $f(\text{Bd } C^i - \text{Int } I^2)$, it is not possible for $f(\text{Bd } C^i)$ to be a polyhedron if each of these feelers is to be attached along a horizontal disk. We do not define $f(C^i)$ to be polyhedral but rather pimpled, as discussed in Section 8. A definition of pimpled is found in the opening paragraph of that section. We delay until Section 8 showing how a polyhedral $f(C^i)$ can be changed into one with a pimpled boundary but point out in the next five paragraphs the nature of $f(\text{Bd } C^i)$ and the manner in which feelers are attached to $f(C^i)$.

Our $f(C^i)$'s for Theorem 11 share the property with the $f(C^i)$'s of Theorem 10 that each component of a horizontal cross section of $f(C^i)$ is either a disk or a point. Each component of a horizontal cross section of $f(\text{Bd } C^i - \text{Int } I^2)$ is either a point, a simple closed curve, or a θ-curve. The points of $f(\text{Bd } C^i - I^2)$ lie on suitable horizontal slices of $f(C^i)$. Unlike the $f(C^i)$'s of Theorem 10, the point components are dense on $f(\text{Bd } C^i - \text{Int } I^2)$. The simple closed curves and θ-curves are not required to be polyhedral.

We now consider how to attach a feeler to $f(C^i)$. Suppose D is a pimpled disk of $f(\text{Bd } C^i - \text{Int } I^2)$ and it is required to attach a small feeler to $f(C^i)$ along some disk in D maintaining the property that each point of the boundary of the enlarged $f(C^i)$ not on $f(I^2)$ lies on a suitable horizontal slice of the enlarged $f(C^i)$. A dense set of points of D are the tips of pimples with the property that if p is such a point and $\varepsilon > 0$, there is a polyhedral 3-cell $C(p, \varepsilon)$ of diameter less than ε such that $p \in \text{Int } C(p, \varepsilon)$, $f(\text{Bd } C^i) \cap \text{Bd } C(p, \varepsilon)$ lies in a horizontal cross section of I^3, while the extreme horizontal cross sections of $C(p, \varepsilon)$ are points and the others are polyhedral disks. Then the attached feeler is $C(p, \varepsilon) - \text{Int } f(C^i)$.

Note that part of the boundary of the feeler is polyhedral rather than pimpled. If it is necessary to extend the feeler at this stage to make it intersect certain 3-cells of T_{i+1}, this extension can perhaps best be done from the polyhedral portion, maintaining the required horizontal cross-section properties and the restraint. After the feeler has been extended as far as necessary, Theorem 17 assures us that the polyhedral portion of its boundary can be pimpled, maintaining cross-section properties and the restraint.

When we added a feeler, we had a wide choice of tips of pimples about which to build $C(p, \varepsilon)$'s. Suppose as we continue to add feelers, we never attach a feeler so that it intersects the disk along which a previous feeler had been attached.

Note that as we added feelers, we lost the condition used in Theorems 1, 9, and 10 that each component of a horizontal cross section of $f(C^i)$ is a component of a horizontal cross section of $f(C^{i+1})$, since some points near the tip of the pimples had the components containing them enlarged. However, no component is enlarged more than once, and if a component of a horizontal cross section of $f(C^i)$ has diameter more than $1/2^i$, it is never enlarged. The counterpart of the H defined in Section 5 would be the set of all continua D such that for some C^i, D is a component of a horizontal cross section of $f(C^i)$ and diameter $D > 1/2^i$.

8. Pimpled Cubes

Suppose C is a topological cube in I^3. We say that a point $p \in \text{Bd}\ C$ *is the tip of a pimple* if for each $\varepsilon > 0$ there is a polyhedral 3-cell $C(p, \varepsilon)$ in I^3 satisfying

(a) $p \in \text{Int}\ C(p, \varepsilon)$;

(b) each horizontal cross section of Bd $C(p, \varepsilon)$ is a point or a simple closed curve;

(c) Bd $C \cap$ Bd $C(p, \varepsilon)$ is a horizontal simple closed curve; and

(d) each point of $C(p, \varepsilon) \cap$ Bd C belongs to a suitable horizontal slice of C.

We say that a subset D of Bd C is pimpled if each point of D belongs to a suitable horizontal slice of C and a dense set of points of D are the tips of pimples.

Theorem 16. *Suppose C is a topological cube in I^3 and Z_1 is a locally polyhedral open subset of* Bd $C \cap$ Int I^3 *such that no vertical segment lies in Z_1 and each point of Z_1 lies on a suitable horizontal slice of C. Then for each open subset V of I^3 containing a dense subset of Z_1 there is a homeomorphism h of I^3 onto itself such that $h(Z_1)$ is pimpled on $h(C)$, h is the identity on $I^3 - V$, and h is invariant on vertical spanning segments of I^3.*

Proof. We prove Theorem 16 by erecting a countable number of tetrahedral hills on C.

Let $\{A_\alpha\}$ be the collection of all components of horizontal cross sections of Bd C that are simple closed curves. Since Bd C contains only a countable number of mutually exclusive triods, the union of the elements of $\{A_\alpha\}$ contains a dense subset of Z_1.

We will now describe a procedure for building a tetrahedral hill which we shall use infinitely often. Consider a 2-simplex D in Z_1. We suppose with no loss of generality that C is beneath D in the sense that each point of Int D is the upper end of a vertical segment in C. Consider a 3-simplex $v_0v_1v_2v_3$ such that

(a) $v_0v_1v_2v_3 \cap \text{Bd } C = v_0v_1v_2 \subset \text{Int } D$;
(b) v_0v_1 is a subset of an element of $\{A_\alpha\}$;
(c) v_2 is closer to I^2 than v_0 is to I^2; and
(d) v_3 is directly above the barycenter v_4 of $v_0v_1v_2$ and farther from I^2 than v_0 is from I^2.

Consider the homeomorphism g on Bd C which is the identity off Int $v_0v_1v_2$ and is the vertical projection of $v_0v_1v_2$ onto the union of the other three faces of $v_0v_1v_2v_3$. Then each point of $g(Z_1)$ lies in a suitable horizontal slice of $g(C)$ and $g(v_4)$ is the tip of a pimple of $g(C)$. The homeomorphism g is extended to a homeomorphism of I^3 onto itself such that g is invariant on vertical spanning segments of I^3, and g is the identity except near $v_0v_1v_2v_3$ and on vertical segments through points of Int $v_0v_1v_2$.

Let T_1 be a triangulation of $V \cap Z_1$ such that the diameter of each simplex of T_1 is less than the minimum of $\frac{1}{2}$ and the distance of the simplex to $(\text{Bd } C - Z_1) \cup (I^3 - V)$. Although T_1 is locally finite, it has infinitely many simplexes and the diameters of the simplexes approach 0 as the simplexes approach Bd $C - (V \cap Z_1)$.

The open set V is reduced to an open set V_1 such that the components of V_1 are mutually exclusive interiors of suspensions of 2-simplexes of T_1 such that the diameters of the suspensions are the same as the diameters of the corresponding 2-simplexes. The homeomorphism h will be defined as the limit of a sequence of homeomorphisms $h_1, h_2, \ldots$ and each of these is the identity off V_1. The first of these, h_1, is obtained by building a very small tetrahedral hill over a small disk in each open 2-simplex of T_1, being careful that the associated v_0v_1's on different edges of the tetrahedrons belong to different elements of $\{A_\alpha\}$.

Let Z_2 be a dense open subset of $V_1 \cap h_1(Z_1)$ obtained by removing the 1-skeleton of T_1, the tips of the pimples introduced by h_1, and sequences of horizontal triangles in $h_1(Z_1)$ converging to each such tip. Let T_2 be a triangulation of Z_2 such that each simplex of T_2 has diameter less than the minimum of $1/2^2$ and the distance from the simplex to $(I^3 - V_1) \cup (\text{Bd } h_1(C) - Z_2)$ and the inverse under h_1 of each simplex has diameter less than $1/2^2$. We continue the pattern we have started by reducing V_1 to V_2, defining h_2 by raising tetrahedral hills on bases in open 2-simplexes of T_2, reducing Z_2 to Z_3, getting a triangulation T_3 of Z_3, reducing V_2 to V_3, defining $h_3, Z_3, T_3, V_3, h_4, \ldots$. The requirement is made that each component of V_i along with its associated inverse is of diameter less than $1/2^i$ to facilitate the proof that $h_1, h_2, \ldots$ converges to a homeomorphism with the required properties.

The proof of Theorem 11 used the following variation of Theorem 16.

Theorem 17. *Suppose C is a cube in I^3 and Z is a locally polyhedral open subset of* Bd $C \cap$ Int I^3 *such that each point of Z lies on a suitable horizontal slice of C. Then for each open set V containing a dense subset of Z and each $\varepsilon > 0$ there is an ε-homeomorphism h of I^3 onto itself such that $h(Z)$ is pimpled on $h(C)$, h is the identity outside V and each point of* Bd $C - Z$ *is on a suitable horizontal slice of C if and only if it is on a suitable horizontal slice of $h(C)$.*

Proof. The homeomorphism is the composition of two homeomorphisms, the first of which changes vertical portions of Z into locally polyhedral nonvertical portions and the second of which is ensured by Theorem 16.

9. A Gravity-Flow Drainage System

In the drainage systems we have described, a drop of water might need to travel up, down, and sideways in order to get out. We call a drainage system a *gravity-flow system* if for each point p of $I^3 - I^2$ there is in $U \cup \{p\}$ an arc from p to I^2 that is vertically monotonic in the sense that each horizontal cross section of it is a point. In another paper we shall prove the following result.

Theorem 18. *There is a gravity-flow drainage system U for I^3 and a monotone map g of $I^3 - U$ onto I^3 such that each point inverse lies in a horizontal cross section of I^3.*

10. Dimension 2

Theorem 12 has the following analogue for a 2-cell.

Theorem 19. *For each pair of mutually exclusive arcs A_1, A_2 on* Bd I^2, *there is a map of I^2 onto itself which is the identity on A_1 and whose point inverses are as follows:* Bd A_2, *pairs of disjoint arcs each of which has an end in* Int A_2 *and otherwise lies in* Int I^2, *triples of disjoint arcs each of which has an end in* Int A_2 *and otherwise lies in* Int I^2, *topological cones over one, two, or three points of* Int A_2 *where the rest of the cone lies in* Int I^2, *and topological cones over one or two points of* Int A_2 *where the vertex of the cone lies on* Bd $I^2 - A_2$ *and the rest of the cone lies in* Int I^3.

Can one get a nicer map of I^2 onto itself such that each component of each point inverse intersects Bd I^2?

Acknowledgment

Work on this paper was supported by ONR Grant NR-043-374.

References

1. R. D. Anderson, *Open mappings of compact continua*, Proc. Nat. Acad. Sci. U.S. 42 (1956), 347–349.
2. R. H. Bing, *A characterization of* 3-*space by partitionings*, Trans. Amer. Math. Soc. 70 (1951), 15–27.
3. R. H. Bing, *Some monotone decompositions of a cube*, Ann. of Math. 61 (1955), 279–288.
4. Morton Brown, *The monotone union of open n-cells is an open n-cell*, Proc. Amer. Math. Soc. 12 (1961), 812–814.
5. E. H. Connell, *Acyclic maps of* E^n *onto* E^n *are compact*, in *Topology of* 3-*manifolds and Related Topics* (edited by M. K. Fort, Jr.), Prentice-Hall, Englewood Cliffs, N.J., 1962, p. 5.
6. E. H. Connell, *Images of* E^n *under acyclic maps*, Amer. J. Math. 83 (1961), 787–790.
7. A. C. Conner and S. L. Jones, $n - 2$ *monotone maps are compact*, mimeographed notes.
8. James Dugundji, *Topology*, Allyn and Bacon, Boston, 1966.
9. E. Dyer and M. E. Hamstrom, *Completely regular mappings*, Fund. Math. 45 (1957), 103–118.
10. L. C. Glaser, *Dimension lowering monotone non-compact mappings of* E^n, Fund. Math. 58 (1966), 177–181.
11. L. C. Glaser, *Monotone non-compact mappings of* E^r *onto* E^k *for* $r \geq 4$ *and* $k \geq 3$ (to appear).
12. R. L. Moore, *Concerning upper semi-continuous collections of continua*, Trans. Amer. Math. Soc. 27 (1925), 416–428.
13. Michael Olinick, *Factoring monotone maps of* E^n, these proceedings.
14. Jussi Väisälä, *Minimal mappings in Euclidean spaces*, Ann. Acad. Sci. Fenn., Ser. A.I. (1965), 3–22.
15. G. T. Whyburn, *Compactness of certain mappings*, Amer. J. Math. 81 (1959), 306–314.
16. G. T. Whyburn, *Decomposition spaces*, in *Topology of* 3-*Manifolds and Related Topics* (edited by M. K. Fort, Jr.), Prentice-Hall, Englewood Cliffs, N.J., 1962, pp. 2–4.

MONOTONE IMAGES OF E^3

R. H. Bing and Joseph M. Martin

1. Decompositions of a cube whose elements hit the boundary. A continuous function is called a map. A map is monotone if its point inverses are compact and connected. In my address [2] before the 1969 Georgia Conference on the Topology of Manifolds, I showed that there is a monotone map f of a 3-cell I^3 onto itself such that each point inverse intersects Bd I^3 (the boundary of I^3) and some point of Bd I^3 has a degenerate inverse. The map described there was somewhat complicated and I would prefer to have a simpler one. Hence, I seek a map f of I^3 onto itself such that

a) f is monotone,

b) each point inverse intersects Bd I^3,

c) some point of Bd I^3 has a degenerate inverse, and

e) f is easy to describe and understand.

Note that the map described in [2] implies the following result.

Theorem 1.1 *There is a map of I^3 onto itself satisfying conditions* a, b, and c.

While this address was given by Bing, it represents joint work with Martin. Bing's research was supported by an ONR contract and Martin's by NSF grant GP-14396.

It would be interesting to get such a map that is open, one which has many points of Bd I^3 with degenerate inverses, one which has only simple point inverses, or one with a variety of other nice properties. However, since we already have the one described in [2] which satisfies conditions a, b, c above and this one can be used to give some surprising results, our main interest is in getting one which also satisfies condition e (e for easy).

Could one get such a map as follows? Regard a tetrahedron T^3 as the join of two opposite edges I_1, I_2. Let f_0 be a map of the unit interval [0, 1] onto T^3 such that $f_0(0) \varepsilon$ Bd T^3, $f_0^{-1}f_0(0) = 0$, and each point inverse is a one, two, three, or four point set. Let h_i be a homeomorphism of I_i onto [0, 1] and $f_i = f_0 h_i$ be a map of I_i into T^3. Let g be the map of a part of T^3 onto T^3 such that for each point $p \varepsilon T^3$, $g^{-1}(p)$ is the join of $f_1^{-1}(p)$ and $f_2^{-1}(p)$.

Can g be extended to a monotone map sending all of T^3 onto T^3?

Note that if $q = h_i(0)$ is the 0 end of I_1, $g^{-1}g(q)$ is an interval on Bd T^3. If the extension could be made without enlarging such a point inverse, this point inverse could then be shrunk to a point to change the extended g to a map satisfying condition c.

In his lecture before this conference [9], Wilson described a monotone map of I^3 onto I^3 different from that described in [2] and which might be modified to satisfy conditions a, b, c.

2. <u>A monotone map of E^3 onto E^3_+</u>. We use E^3 to denote ordinary Euclidean 3-space and E^3_+ to denote the set of points of E^3 on or above the xy-plane. If $p_0 \in \text{Bd } I^3$, $I^3 - \{p_0\}$ is a model for E^3_+. If I^2 is a base of I^3, $\text{Int } I^3 \cup \text{Int } I^2$ is another model for E^3_+.

Let the 3-sphere S^3 be the union of two 3-balls B^3_1, B^3_2 with a common boundary and h_1, h_2 be homeomorphisms of I^3 onto B^3_1, B^3_2 such that $h_1 = h_2$ on $\text{Bd } I^3$. Then for each $p_0 \in \text{Bd } I^3$, $S^3 - \{h_1(p_0)\} = S^3 - \{h_2(p_0)\}$ is a model for E^3.

Suppose f is a map of I^3 onto itself satisfying conditions a, b, c and $f(p_0)$ is a point of $\text{Bd } I^3$ with a degenerate inverse under f. Let g be the map of $S^3 - \{h_1(p_0)\}$ onto $I^3 - f(p_0)$ described by fh_i^{-1} on $B^3_i - [h_i(p_0)]$. The map is an example of a compact monotone map of E^3 onto E^3_+.

<u>Theorem</u> 2.1. <u>There is a compact monotone map of E^3 onto E^3_+</u>.

3. <u>Monotone maps of E^3 onto other 3-manifolds</u>. By using the results of the preceding section that there is a monotone map of E^3 onto E^3_+ and that $\text{Int } I^3 \cup \text{Int } I^2$ is a model of E^3_+, one finds by sewing I^2 onto the opposite base of I^3 that there is a monotone map of E^3 onto an open solid torus. (The lower portion

of I^3 is stretched and swung about as shown in Figure 6b). If the sewing is changed, one can get a monotone map of E^3 onto an open solid Klein bottle. We shall exhibit monotone mappings of E^3 onto many other 3-manifolds. The following question is suggested.

Which 3-manifolds are the monotone images of E^3?

In another paper Joe Martin and I answer this question by showing that all connected 3-manifolds are the monotone images of E^3. We do this by exhibiting 1-1 maps of E^3_+ onto these 3-manifolds. In fact, for each connected 3-manifold we exhibit several ways of getting a 1-1 map of E^3_+ onto it. It seems that we have more methods than theorems. Certain of these methods are described in the following sections.

One might vary the preceding question to get other questions.

What are the monotone images of E^3?

Could one get another 3-manifold as an image if the point inverses are of dimension less than or equal to 1?[#] If they are polyhedra?

Which 3-manifolds are the monotone images of other familiar sets?

[#]Since this talk was given, Bing constructed a monotone map of E^3 onto E^3 all of whose point inverses are finite graphs.

In connection with restricting point inverses, one who is unfamiliar with the peculiar things that can be done with monotone maps of E^3 onto itself might think that no point inverse of such a monotone map could separate E^3. However, in another paper we get the following rather surprising result.

Theorem. *If X is any bounded continuum in E^3 whatsoever, there is a monotone map of E^3 onto itself having X is a point inverse.*

We shall prove other startling results in that paper but shall not describe them here.

The peculiar monotone maps we have discussed are not compact. (Some compact sets do not have compact inverses.) The monotone mapping of E^3 onto E^3_+ given in Section 2 was compact but the 1-1 map of E^3_+ which was composed with this map was not compact.

G. T. Whyburn [7] showed that each monotone map of E^2 onto itself is compact. Efforts were made to extend this result to E^3 but the efforts failed. The extra dimension of E^3 provided more room into which to fit big point inverses since long continua can bypass each other in E^3 without intersecting. There is no compact monotone map of E^2 onto E^2_+, and even if there were, there would be no 1-1 map of E^2_+ onto E^2. One reason for this last statement is that if h is a 1-1 map of $E^1 = \text{Bd}\, E^2_+$ onto a closed subset of E^2, then $h(E^1)$ separates E^2.

The monotone mapping problem asks if each monotone map of E^n onto itself is compact. There is a trivial affirmative answer for $n = 1$. Whyburn gave an affirmative answer for $n = 2$. Using unpublished results of R. D. Anderson [1], Glaser [4] gave a negative answer for $n \geq 4$. The noncompact maps of E^3 onto E^3 described in [2] and given by Theorem 6.2 finish the last case.

Theorem 3.1. *The monotone mapping problem has an affirmative answer for* $n = 1, 2$ *and a negative answer* for $n = 3, 4, \cdots$.

4. *Using third symmetric products to send* E^3_+ *onto* S^3. The following is a good lunch-time problem. Consider the one and two point sets of a circle S^1. If this collection of sets is topologized with the Hausdorff distance, what is the resulting object? Recall that if A, B are two sets, the Hausdorff distance between A and B is defined as

$$\text{maximum}\left\{\underset{a \in A}{\text{least upper bound}}\ d(a, B),\ \underset{b \in B}{\text{least upper bound}}\ d(A, b)\right\}$$

where d denotes distance. We give the hint that the obvious guess, $S^1 \times S^1$, is a bad guess. In fact, the object is not even a 2-manifold without boundary. See [6] for a discussion of symmetric products.

In contrast, if one gives a Hausdorff distance to the one, two, and three point sets of S^1, the resulting object is 3-sphere S^3. This is a peculiar way of describing S^3. See [6].

It is easier to discover the object equivalent to the collection of one, two, and three points sets of $[0, 1]$ topologized with the Hausdorff metric. For each triple of numbers x_1, x_2, x_3 we order them by size with the smallest first as a, b, c and associate this triple with the point in E^3 with coordinates (a, b, c). Each single number a is associated with the point with coordinates (a, a, a). For each pair of numbers a, b with $a < b$, we associate this pair with the two points of E^3 with coordinates (a, a, b) and (a, b, b). The set of points in E^3 with coordinates (x, y, z) satisfying $0 \leq x \leq y \leq z \leq 1$ is the cone from the origin over the triangle with vertices $(0, 0, 1)$, $(0, 1, 1)$, and $(1, 1, 1)$. The faces of this tetrahedron with edges on the line with equation $x = y = z$ have coordinates (a, a, b) and (a, b, b). The operation of identifying these faces can be realized by folding one of these faces about the common edge until it fits on the other and thereby stretching the tetrahedron into a slanted cone. The folding operation takes the horizontal face of the tetrahedron onto a circular disk with center at $(1, 1, 1)$ where the side of the horizontal face from $(0, 0, 1)$ to $(0, 1, 1)$ is stretched into the circumference of the disk. The vertex of the resulting cone is the origin and its slanted axis is on the line with equation $x = y = z$.

The base of the cone corresponds to the sets containing 1. Hence, the object equivalent to the one, two, and three point sets of [0, 1] topologized by the Hausdorff distance is a topological solid cone.

The collection of one, two, and three point sets of [0, 1) topologized with the Hausdorff distance is equivalent to the above cone minus its base. This is a model for E^3_+. Since there is a 1-1 map of [0, 1) onto S^1, there is a 1-1 map of the one, two, and three point sets of [0, 1) onto those of S^1. Hence, there is a 1-1 map of E^3_+ onto S^3.

Theorem 4.1. *There is a noncompact* 1-1 *map of* E^3_+ *onto* S^3.

If one composes the maps of Theorems 2.1 and 4.1 one obtains the following result.

Theorem 4.2. *There is a noncompact monotone map of* E^3 *onto* S^3.

The above method of regarding the one, two, and three point sets of S_1 as a model of S^3 provides a way of showing the existence of a 1-1 map of E^3_+ onto S^3 without showing the nature of the map. Joe Martin showed me this clever method of mapping E^3_+ onto S^3. If one investigates what Bd E^3_+ goes onto under the above map, a careful analysis shows that its image is the dunce's cap. This shows that the map is related to the one discussed in the next section.

5. Fold-downs which yield S^3. Taking a hint from the preceding paragraph, we consider a model of E^3_+ which is $\text{Int } I^3 \cup \text{Int } I^2$. A solid cone is pushed up as shown in Figure 1a where the length of the radius of the cone is the perimeter of I^2. This solid with its bottom and lateral faces (of the cube) gone is a model of E^3_+.

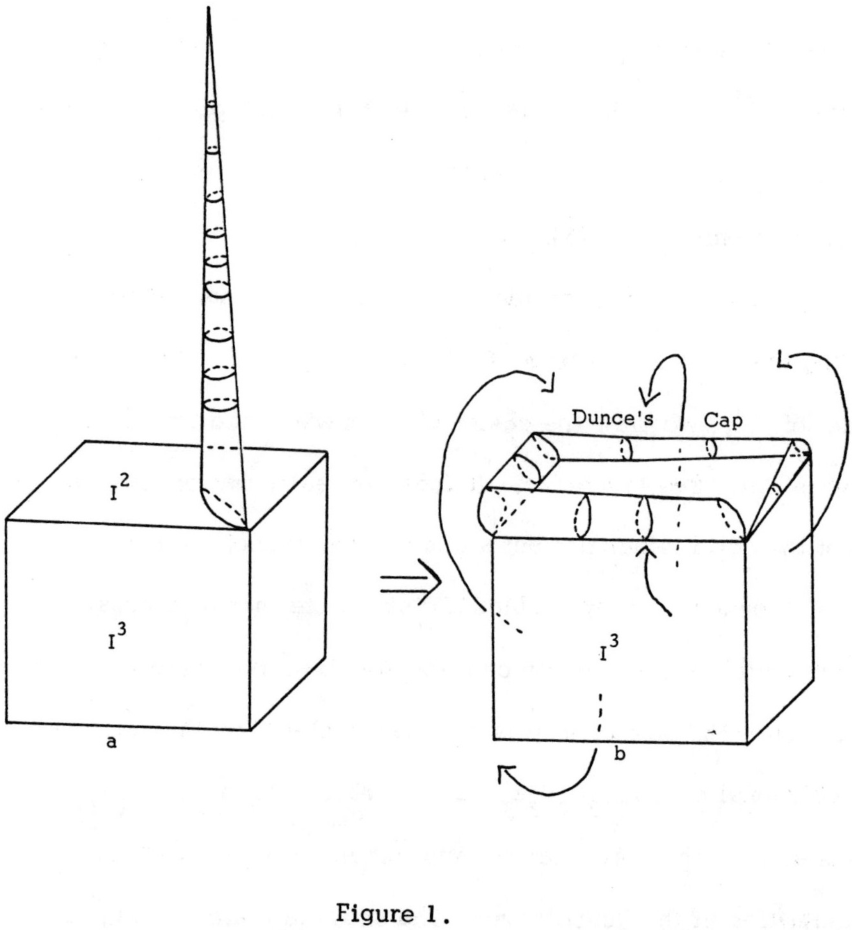

Figure 1.

Suppose the cone is folded over until a radius is joined to the perimeter of I^2 as shown in Figure 1b. The top becomes a dunce's cap (sometimes called a dunce hat).

The laying down of this cone is a 1-1 operation since Bd I^2 was not in the first object. After the cone was pushed over, the top of the resulting solid is compact even though it had not been before.

Suppose the sides and bottom of the solid are folded out, up, and over to make it fit into the top. The resulting object is a model of S^3 that was obtained as the 1-1 image of E^3_+. Glaser draws pictures and gives explanations describing this sewing operation in his thesis [5].

For those who have trouble in visualizing this procedure of folding the sides and bottom of I^3 out, up, and over to form a model of S^3, we draw the easier diagram where the top of the solid is flat. See Figure 2a. It does not really matter that the top of the solid is not flat but a dunce's cap. Looked at from above, a dunce's cap resembles a disk. Note that it is possible to sew a disk into a dunce's cap from above as indicated in Figure 2b. The sewing map is a homeomorphism off the union of the polygonal arc $a_2b_2c_2d_2a_1$ and the rectangle $a_1b_1c_1d_1a_1$ and sends each of the polygonal arc and rectangle onto the set of singularities of the dunce's cap. The dotted arc arc a_1xya_2 is mapped onto the base of the cone that was laid over. The arc

$a_2b_2c_2d_2a_1$ is tucked into the inner portion of the fold of the dunce's cap and the rectangle $a_1b_1c_1d_1a_1$ fits onto the outer part. To visualize the inverse of the sewing map, one could split the dunce's cap along its set of singularities, taking care that the outer portion of the roll is split off and the inner portion is left attached to the central part of the cap.

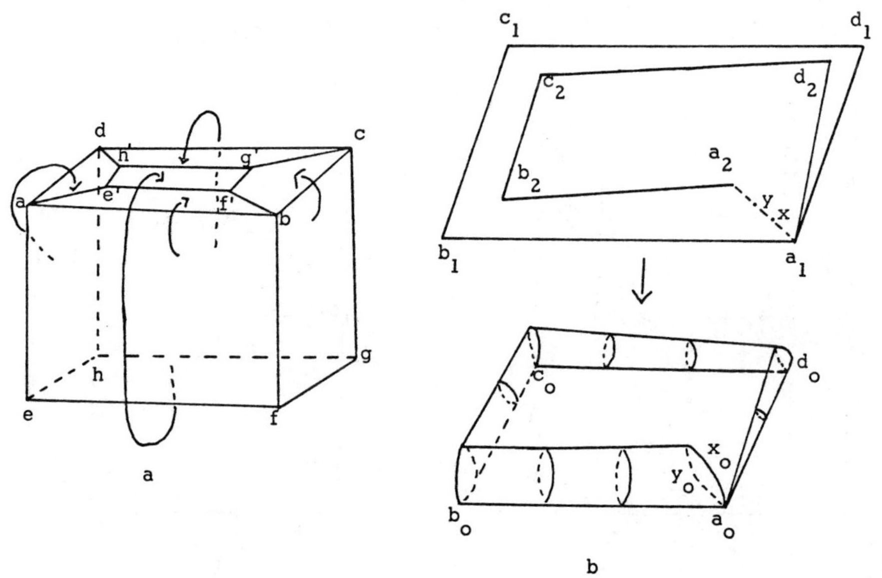

Figure 2.

<u>Theorem</u> 5.1. <u>There is a</u> 1-1 <u>noncompact map of</u> E^3_+ <u>onto</u> S^3 <u>that sends</u> Bd E^3_+ <u>onto a dunce's cap</u>.

The dunce's cap in the above theorem is the set of points of the image at which the inverse of the map is not continuous.

It is difficult to draw a picture of a dunce's cap. Instead of raising one cone on the top of Int $I^3 \cup$ Int I^2, one might raise four shorter cones as shown in Figure 3a where the radii of the cones have lengths equal to the lengths of the sides of I^2. If one lays down the four cones as shown in Figure 3b and folds the sides and bottom of the resulting solid out, up, and over, there results an easier map of E^3_+ onto S^3.

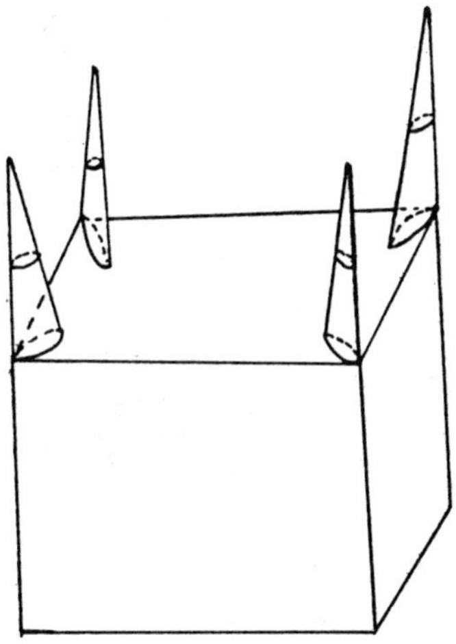

a

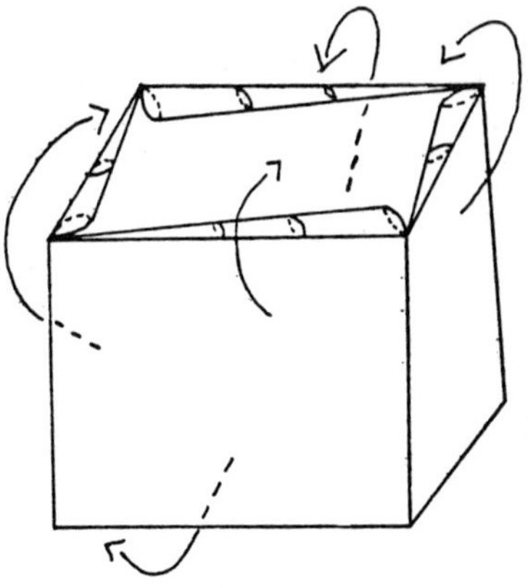

b

Figure 3.

<u>Theorem</u> 5.2. <u>There is a</u> 1-1 <u>noncompact map of</u> E^3_+ <u>onto</u> S^3 <u>such that the image of</u> Bd E^3_+ <u>is a hat with a finite number of peaks laid down as shown in Figure</u> 3b.

6. <u>Laying down cones to get</u> E^3. When one pushed over the tall cone in Figure 1 or the four small ones in Figure 3, the entire perimeter of I^2 got covered by radii. However, suppose one considers a sequence of cones getting shorter and shorter as shown in Figure 4a so that if they are pushed over as shown in Figure 4b, the point p_0 is not covered. If the sides and bottom of the resulting solids are pushed out, up and over, there results a 1-1 map of E^3_+ onto S^3 minus a point. Hence, there is a 1-1 map of E^3_+ onto E^3.

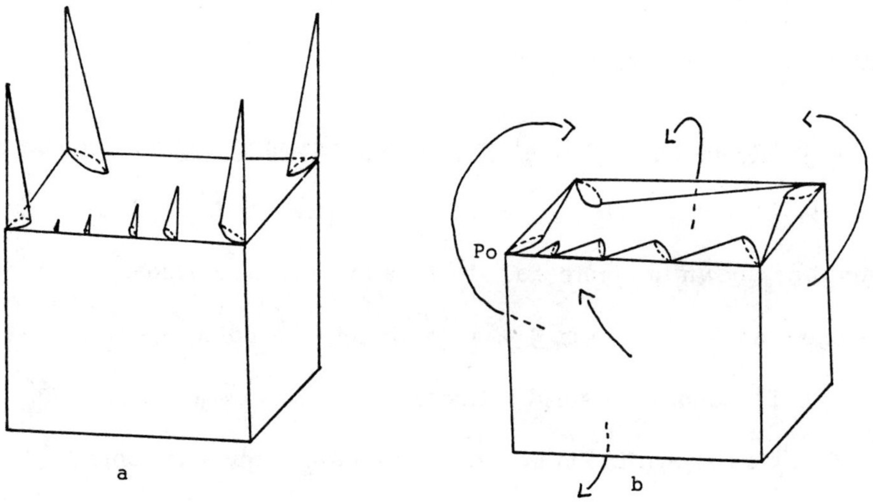

Figure 4.

<u>Theorem</u> 6.1. <u>There is a 1-1 noncompact map of</u> E^3_+ <u>onto</u> E^3.

If one composes the map of Theorem 2.1 with that of Theorem 6.1, the following result is obtained.

Theorem 6.2. *There is a noncompact monotone map of* E^3 *onto* E^3.

L. C. Glaser describes a 1-1 map of E^3_+ onto E^3 in [3]. Another description is given in [2]. Related mappings are considered in [5, 8].

If X is a closed subset of the simple closed curve Bd I^2, we can lay down cones to cover Bd I^2 - X. This shows that the following result holds.

Theorem 6.3. *For each closed subset* X *of a tame arc in* S^3, *there is a* 1-1 *map of* E^3_+ *onto* S^3 - X.

7. *Shrinking concentric simple closed curves to points*. Suppose $A^2 \times [0, 1]$ is the cartesian product of an annulus A^2 and an interval as shown in Figure 5a. If J is the smaller boundary component of A^2 and each $J \times t$ is shrunk to a point, then $A^2 \times [0, 1]$ becomes a solid cylinder whose axis comes from $J \times [0, 1]$ as shown in Figure 5b. We obtained the same object that we would have obtained by sewing a pill box to $A^2 \times [0, 1]$ along $J \times [0, 1]$. (At one time, pills were dispensed in short cylindrical boxes. To sew a pill box onto an object, we select an annulus in the boundary of the object and consider the union of the object and a thickened disk where the identification is realized by a homeomorphism between the sides of the thickened disk and the annulus).

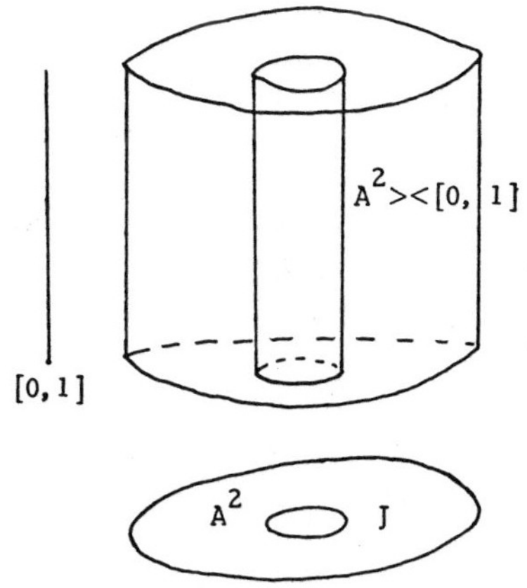

Figure 5.

Suppose we removed most of the boundary of $A^2 \times [0, 1]$ and left only the open arc $p_0 \times (0, 1)$ for some point $p_0 \varepsilon J$. Then there is a 1-1 map of this object onto the interior a cylinder which sends $p_0 \times (0, 1)$ onto the open axis of the cylinder.

We use this technique to describe another 1-1 map of E^3_+ onto E^3. Suppose a row of cones are laid down as shown in Figure 6a. Instead of folding the sides and bottom of the solid out, up and over, we only fold the bottom around as shown in Figure 6b. The model of E^3_+ is sent by a 1-1 map onto an open solid torus with an open arc on its boundary. As noted in the preceding paragraph, there is a 1-1 map of this object onto E^3. We have noted that Theorem 6.1 follows as a corollary of the following result.

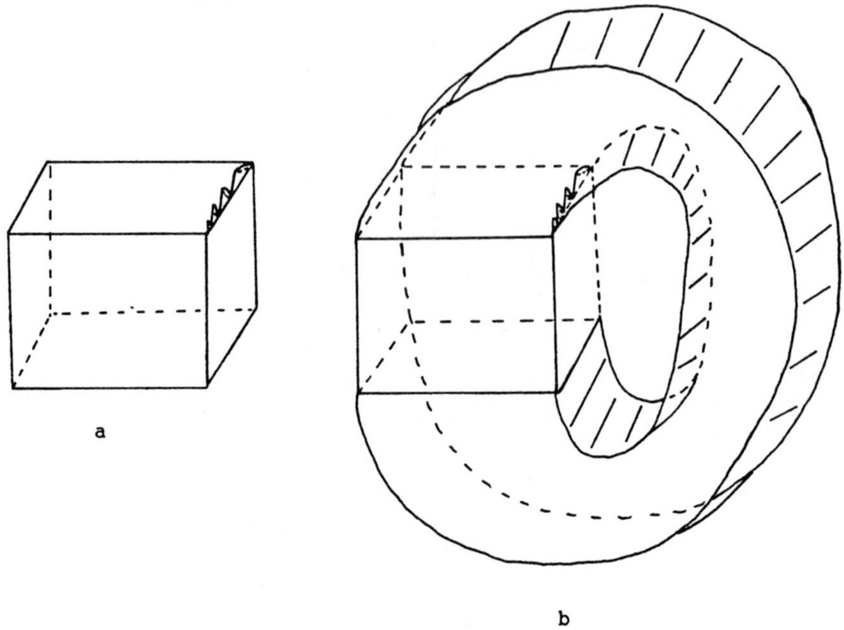

Figure 6.

Theorem 7.1. *There is a* 1-1 *map of* E^3_+ *onto the union of an open solid torus and an open arc on its boundary*.

Corollary 7.2. *Theorem* 6.1.

The open arc of Theorem 7.1 is not only the spanning arc of an annulus that goes around the solid torus longitudinally but it is also a spanning arc of other annuli. If one sews a pill box to a solid torus along an annulus on the boundary of the solid torus that goes around the torus i times meridionally and j times longitudinally, one obtains a toroidal manifold T(i, j) with an open ball removed. Suppose the open arc of Theorem 7.1 is moved about until it is a spanning arc of this annulus that goes around the torus i times meridionally and j times longitudionally. The arc serves as a substitute for the sewing on of a pill box as noted in Figure 5, so the following result follows.

Theorem 7.3. *If p is a point of a toroidal manifold T(i, j) (or a lens space), then there is a 1-1 map of* E^3_+ *onto* $T(i, j) - \{p\}$.

If instead of laying down a row of cones as shown in Figure 5a, one only laid down one cone, then a sewing like that shown in Figure 5b would yield the following result.

Theorem 7.4. *There is a 1-1 map of* E^3_+ *onto the union of an open solid torus and a half open arc on its boundary.*

Using the open part of the arc to plug up a hole, one finds by using the techniques in obtaining Corollary 7.2 that there is a 1-1 map of the image space of Theorem 7.4 onto the union of an open ball and a one point set on its boundary. The one point set can be used to eliminate the boundary so we have the following result.

Corollary 7.5 *to Theorem* 7.4. *Theorem* 4.1.

Corollary 7.6 *to Theorem* 7.4. *Each toroidal manifold is the image of* E^3_+ *under a* 1-1 *map.*

8. *Another mapping onto* S^3. Again we regard Int $I^3 \cup$ Int I^2 as a model for E^3_+. This time two tubes are run from the base of I^3 opposite I^2 and made to fit onto I^2 as shown in Figure 7. Note that the only part of Int I^2 left exposed is an open spanning arc.

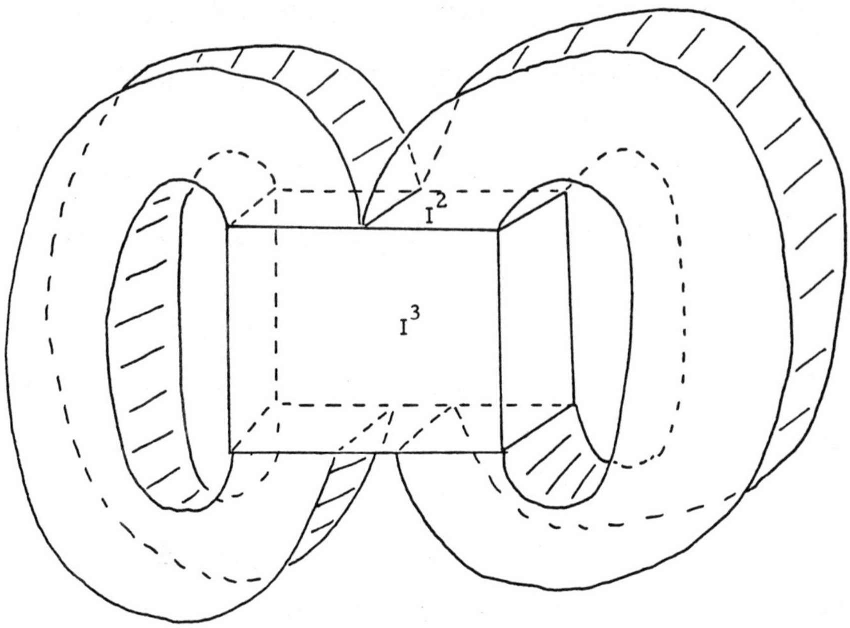

Figure 7.

Theorem 8.1 *There is a* 1-1 *map of* E^3_+ *onto the union of an open solid double torus and an open arc on its boundary.*

Suppose we regard the open arc of Theorem 8.1 as the union of an open arc and a disjoint half open arc. The open arc can be used to remove one of the handles and bring us to an image like that described in Theorem 7.4.

Theorem 8.2. *The 3-sphere S^3 is the image under a 1-1 map of the union of an open solid double torus and an open arc on its boundary.*

9. *Maps onto other manifolds.* In this section, we establish the following result.

Theorem 9.1. *If M^3 is a compact 3-manifold without boundary, then there is a 1-1 map of E^3_+ onto M^3.*

To find a map onto M^3, consider a triangulation of M^3 and a tubular neighborhood N^3 of the 1-skeleton of this triangulation. We suppose that N^3 is so nice that if F^2 is a 2-simplex of the triangulation, then $F^2 - N^3$ is an open disk. This disk can be thickened to a pill box P^3 so that $P^3 \cap N^3$ is an annulus. One can build pill boxes $P^3_1, P^3_2, \cdots, P^3_j$ for the various 2-simplexes of the triangulation so that $N^3 \cup P^3_1 \cup \cdots \cup P_j$ is obtained from M^3 by the removal of an open ball from every 3-simplex of the triangulation.

Let W_4 be the union of Int $(N^3 \cup Q_4^3 \cup \cdots \cup Q_i^3)$ and two disjoint open arcs on its boundary. One of these open arcs serves as a substitute for the hole plugged by Q_3^3 and we find that there is a 1-1 map of W_4 onto W_3. Continuing in this fashion we find that if W_{i+1} is the union of Int N^3 and $(i - 1)$ disjoint open arcs on its boundary, then there is a 1-1 map of W_{i+1} onto M^3.

Since N^3 is a cube with handles (some of which may not be oriented) the truth of Theorem 9.1 then follows from the following result.

Theorem 9.2. *There is a* 1-1 *map of* E_+^3 *onto the union of the interior of a cube with* n *handles and* (n - 1) *disjoint open arcs on its boundary.*

The map for Theorem 9.2 is shown in Figure 8. We have subdivided I^2 into n rectangles and brought tubes from the base of I^3 opposite I^2 to fit into these rectangles. The only part of Int I^2 left exposed is the union of common open edges of pairs of rectangles.

Note that $M^3 - (N^3 \cup P_1^3 \cup \cdots \cup P_j^3)$ is the union of a collection of mutually exclusive open balls or holes. Each P_i^3 connects two of the holes. By not using some of the connecting P^3's it is possible to find a subcollection $Q_1^3, Q_2^3, \cdots, Q_i^3$ of the P^3's so that $M^3 - (N^3 \cup Q_1^3 \cup Q_2^3 \cup \cdots \cup Q_i^3)$ is only one open ball.

Let W_1 be the union of $\mathrm{Int}\,(N^3 \cup Q_1^3 \cup \cdots \cup Q_i^3)$ and a one point set on its boundary. Note that the complement of $N^3 \cup Q_1^3 \cup \cdots \cup Q_i^3$ in M^3 is an open ball. If the corresponding closed ball is regarded as a point, the resulting decomposition space is still topologically like M^3. Since W_1 has only one point on this closed ball, there is a 1-1 map of W_1 onto M^3.

Let W_2 be the union of $\mathrm{Int}\,(N^3 \cup Q_2^3 \cup \cdots \cup Q_i^3)$ and a half open arc on its boundary. Using techniques discussed in Section 7, we can use the open portion of the arc as a substitute for (Q_1 -ends of Q_1) and get a 1-1 map of W_2 onto W_1. Hence, M^3 is the image of W_2 under a 1-1 map.

Let W_3 be the union of $\mathrm{Int}\,(N^3 \cup Q_3^3 \cup \cdots \cup Q_i^3)$ and an open arc on its boundary. We can regard the open arc as the union of an open arc and a disjoint half open arc. The open arc serves as a substitute for the hole plugged by Q_2 and we find that there is a 1-1 map of W_3 onto W_2. Hence M^3 is the image of W_3 under some 1-1 map.

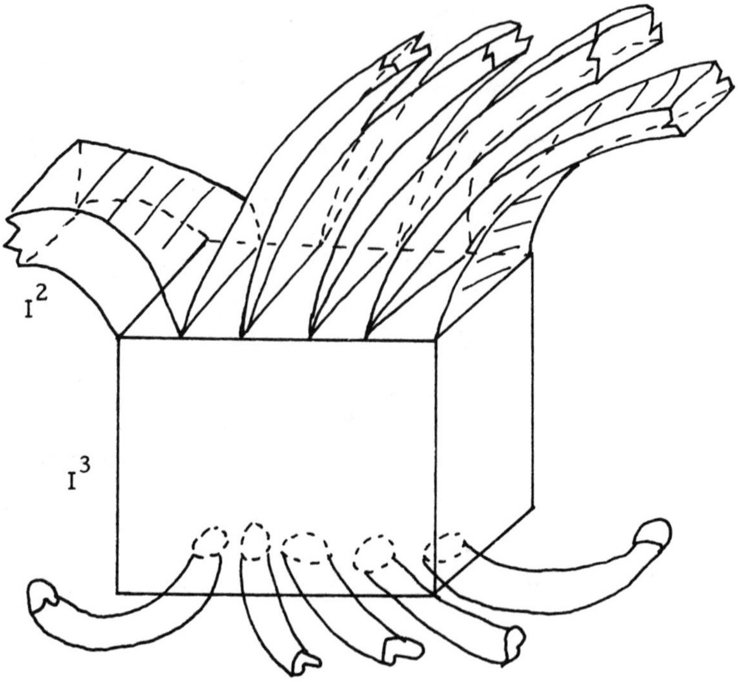

Figure 8.

There are other things of this nature that we can do but we leave discussion of them to another paper.

References

1. R. D. Anderson, Open mappings of compact continua, Proc. Nat. Acad. Sci. U. S. A., 42 (1956), 347-349.

2. R. H. Bing, The monotone mapping problem, Topology of Manifolds Conference at University of Georgia 1969, Markham Publishing Co.

3. L. C. Glaser, Dimension lowering monotone non-compact mappings of E^n, Fund. Math., 58 (1966), 177-181.

4. __________, Monotone non-compact mappings of E^r onto E^k for $r \geq 4$ and $k \geq 3$, to appear.

5. __________, Contractible complexes in S^n, Ph. D. thesis, University of Wisconsin, 1964.

6. R. M. Schori, Symmetric products and the space of closed subsets of n-cells, Topology Conference, Arizona State University, 1967, 258-261.

7. G. T. Whyburn, Compactness of certain mappings, Amer. J. Math., 81 (1959), 306-314.

8. Kenneth Whyburn, A nontopological 1-1 mapping onto E^3, Bull. Amer. Math. Soc., 71 (1965), 533-537.

9. David Wilson, Monotone mappings on manifolds, these Proceedings.

University of Wisconsin.

AN UNUSUAL MAP OF A 3-CELL ONTO ITSELF

by

R. H. Bing

1. Introduction. The 3-cell I^3 is regarded as $I^2 \times I$ where $I = [0,1]$ and $I^2 = I \times I$. The upper and lower bases of I^3 are $I^2 \times I$ and $I^2 \times 0$ respectively. They are referred to as the top and bottom of I^3. A map is a continuous function. A map is monotone if each point inverse is compact and connected. A map is compact if the inverse of each compact set is compact. We use R^3 to denote analytic 3-space regarded as the cartesian product of the real line with itself twice. We use R_+^3 to denote the upper half space of R^3 -- the set of all points in R^3 whose last coordinate is non-negative. The 3-sphere is denoted by S^3. A finite graph is a topological finite geometric 1-complex. It is the union of a finite number of arcs such that if any two of them intersect, the intersection is an end point of each. We say that p is a branch point of order n ($n \geq 3$) if n of these arcs come together at p. A set is degenerate if it has only one element and nondegenerate if it has more than one.

Theorem 1.1 promises a monotone map f of a 3-cell I^3 onto itself with the following properties.

a) f is the identity on the upper base of I^3.

b) Each point inverse intersects the lower base of I^3.

c) The images of the nondegenerate point inverses lie in the union of a countable number of disks.

d) Each nondegenerate point inverse is locally a finite graph except possibly at certain of its end points where it intersects the bottom of I^3.

e) Any simple closed curve in any point inverse lies in the bottom of I^3.

f) The images of the point inverses which are neither arcs nor points lie in the union of a countable number of arcs.

Later we shall refine the map f in Theorem 8.1 so that f has the following additional properties.

g) The set of point inverses with more than one cycle is countable.

h) No point inverse has more than two independent cycles.

i) Each point inverse is either a point or a tame finite graph each of whose branch points is of order 3.

j) The intersection of any point inverse with Bd I^3 has at most four components.

k) There is a countable collection of arcs on the bottom of I^3 such that each point of the bottom of I^3 not on one of these arcs has a degenerate inverse.

In addition to being of interest for its own unusual properties, such a map of I^3 onto itself can be used to generate other interesting maps.

<u>Corollary</u> 1 <u>to Theorem</u> 8.1. There is a monotone map of S^3 onto I^3 such that each nondegenerate inverse is a finite graph and many points on Bd I^3 have degenerate inverses.

To get a map for the corollary, one could regard the 3-sphere S^3 as the union of two 3-balls sewed together with a homeomorphism on their boundaries and use a map like f to send each of the 3-balls onto I^3.

<u>Questions</u>. If one proves the above corollary as suggested above, the point inverses would not have Property h since under the map of the 3-balls into I^3, some point inverse in one of the balls would intersect the boundary of the ball in several components and when this point inverse is united with its mirror image in the other ball, their union would have several independent cycles. Is there a monotone map of S^3 onto I^3 having Properties h and i? If there is, one would expect to be able to find one having Properties c and f also and perhaps one such that for some Cantor set X in S^3, each point inverse is locally polyhedral mod X.

Results in Knoblauch's thesis [6] imply that for each monotone map of one compact 3-manifold without boundary onto another such that some point inverse has nontrivial first homology, then there is one point inverse that requires more than one generator to express its first homology. However, the monotone map of a round 3-ball onto S^3 whose only nondegenerate elements are circles of latitude on the boundary of the ball shows that Knoblauch's result does not hold for 3-manifolds with boundary. Is there a monotone map of I^3 onto itself such that all point inverses intersect Bd I^3 and the nondegenerate inverses are finite graphs with at most one cycle?

Since one obtains a model of R^3 by removing a point from S^3 and a model of R^3_+ by removing a boundary point from I^3, one obtains the following.

Corollary 2 to Theorem 8.1. There is a compact monotone map of R^3 onto R^3_+ such that each point inverse is either a point or a tame finite graph and many points of Bd R^3_+ have degenerate inverses.

It is known [3] that for each compact connected 3-manifold M^3, there is a 1 - 1 map of R^3_+ onto M^3. Such maps could not be compact. However, if one composes the map of the above corollary with such 1 - 1 maps one gets the following interesting corollary.

Corollary 3 to Theorem 8.1. Each compact connected 3-manifold is the image of R^3 under a monotone map whose nondegenerate point inverses are finite graphs.

The monotone mapping question asked if each monotone map of R^n onto itself is compact. The answer is easily seen to be the affirmative if $n = 1$. Whyburn [7] showed that it was in the affirmative if $n = 2$. Glaser [5], using some results promised by Anderson in [1], gave a negative answer for $n \geq 4$ and Bing [2] gave a negative answer for $n \geq 3$. To give the negative answer, Bing considered

$$R^3 \xrightarrow{g_1} R^3_+ \xrightarrow{g_2} R^3$$

where g_1 is a compact monotone map of R^3 onto R^3_+ (as guaranteed by Corollary 2 to Theorem 8.1) and g_2 is a 1 - 1 noncompact map of R^3_+ onto R_3. The map $g_2 g_1$ provides a negative answer to the monotone mapping question for $n = 3$.

It is important to know which maps are compact. If G is an upper semicontinuous decomposition of a space S, the associated decomposition map of S into the decomposition space is a compact map. Hence, the image of a space under a noncompact map may be topologically different from the decomposition space whose elements are the point inverses.

Corollary 4 to Theorem 8.1. There is a monotone noncompact map of R^3 onto itself whose nondegenerate point inverses are finite graphs.

By using the method of laying down cones discussed by Bing and Martin in [3], it is easy to get a 1 - 1 noncompact map $g_2 : R^3_+ \to R^3$. However, the monotone map

$g_1 : R_3 \to R^3_+$ given in [2] was not easy to understand since it was based on a very complicated monotone map g of I^3 onto itself such that all point inverses intersect Bd I^3. Its nondegenerate point inverses were not finite graphs. Many readers did not understand this complicated map and one of the purposes of this paper is to provide a simpler one. The map we describe has features in common with one described by Wilson in [8].

To get a map satisfying Conditions a,b,c,d,e,f,g,h,i,j,k it is convenient to first describe a map satisfying only some of the conditions and then modify this map so that the modification will satisfy the others. As a start, we shall show the following.

Theorem 1.1. There is a monotone map f of I^3 onto I^3 satisfying Conditions a,b,c,d, e,f.

We shall describe the map f for Theorem 1.1 in Sections 2-5. We show in Section 2 that f has Property a, in Section 3 that it has Properties b,c, and d, in Section 5 that it is monotone and has Property e, and in Section 6 that it has Property f. In Sections 7 and 8 we refine f to make it have additional properties.

David Wright read a preliminary version of this paper and made useful suggestions for improving it.

2. A natural map of a disk onto a sphere. Frequently we shall refer to the union of the top and four lateral side of a 3-cell so for brevity we refer to this union as B. (Think of B as a subset of Bd -- but not as bottom.)

As a start of the description of the map f for Theorem 1.1, we turn our attention to how f behaves on $B(I^3)$. We define f here by means of a natural map.

A map g of a disk D onto a 2-sphere S^2 is called natural if g(Bd D) is a point and g takes Int D homeomorphically onto S^2 - g(Bd D). Such natural maps can not be PL even if D and S^2 are polyhedral since if T is a triangulation of D and Δ^2 is a 2-simplex of T with an edge of Bd D, then g could not be linear on Δ^2. However, we could describe a natural map g of D onto S^2 that is locally PL on Int D. Since we do not insist that f be PL anywhere, we do not pursue this problem.

As a start toward the description of the map f promised by Theorem 1.1 we shall use a special natural map of the disk $B(I^3)$ onto Bd I^3. We describe this map f in detail below.

On $I^2 \times I$ (the top of I^3), f is the identity. This insures that the map f satisfies Property a. The map f takes Bd $I^2 \times 0$ onto the center z_0 of $I^2 \times 0$. For $p \in$ Bd I^2, f takes the segment $p \times [\frac{1}{2},1]$ linearly onto the segment $p \times [0,1]$ and the segment $p \times [0,\frac{1}{2}]$ linearly onto the segment from z_0 to $p \times 0$. Hence, if $v_i v_{i+1}$ is a side of I^2, f takes the rectangle $v_i v_{i+1} \times [\frac{1}{2},1]$ linearly onto a lateral face of I^3, and f takes the rectangle $v_i v_{i+1} \times [0,\frac{1}{2}]$ onto a triangle on the bottom of I^3. See Figure 2.1.

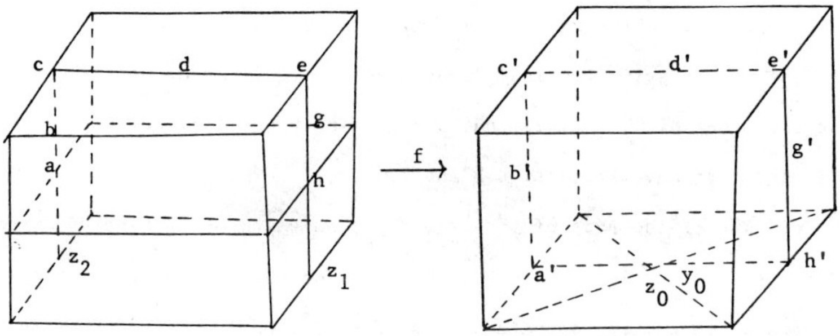

Figure 2.1.

Express the image space I^3 as the union of two 3-cells I_1', I_2' with a common face $I^2 \times \frac{1}{2}$ as shown in the right side of Figure 2.2. The left of Figure 2.2 shows two disjoint small cells I_1, I_2 which have their bottoms in the interior of the bottom of I^3.

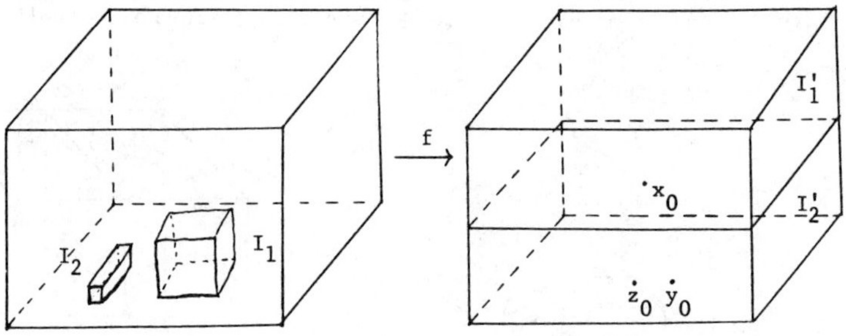

Figure 2.2.

The map f takes $B(I_1)$ onto Bd I_1' in somewhat the same fashion that it took $B(I^3)$ onto Bd I^3 -- it takes the top of I_1 linearly onto the top of I_1' and the boundary of the bottom of I_1 onto the center x_0 of the bottom of I_1'.

If $p \in B(I_1')$, then there are points p_1, p_2 on $B(I^3)$, $B(I_1)$ respectively such that $f(p_1) = f(p_2) = p$. The map f is defined so that it takes the segment from p_1 to p_2 to p. See Figure 4.1 given later. We suppose that I_2 is so low that it does not meet any of these segments.

The map f restricted to $B(I_2)$ is a natural map onto Bd I_2' but instead of taking the boundary of the bottom of I_2 onto the center of the base of I_2', it takes it onto a point y_0 slightly to the right of z_0. This adjustment is made to prevent the inverse of z_0 from having too many independent cycles. We do not need this adjustment for the proof of Theorem 1.1 but are getting ready for Theorem 7.1 in which we shall want f to have Properties h and i.

The closure of the part of $I^3 - (I_1 \cup I_2)$ on which we have not defined f is a solid torus. In Sections 4 and 5 we define f on this solid torus so as to make point inverses connect appropriate points of Bd I^3, Bd I_1, and Bd I_2. We try to define f so that point inverses intersect this solid torus in arcs but find it useful to use some finite graphs with a cycle and one with even a θ-curve. For now, we suppose we have defined f on the closure of $I^3 - (I_1 \cup I_2)$ (even though this is not completed until the end of Section 5).

3. <u>The plan of attack</u>. The map f is defined on part of I_1, I_2 in much the same way that we have defined it on part of I^3. We subdivide I_1' into two 3-cells I_{i1}', I_{i2}' as shown in Figure 3.1.

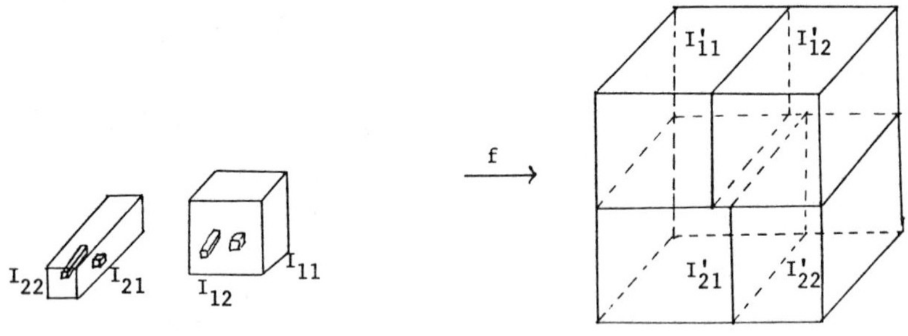

Figure 3.1

Note that the subdivision is staggered so as to get a brick partitioning. No point belongs to $I'_{11} \cap I'_{12} \cap I'_{21} \cap I'_{22}$. We do not need this staggered subdivision for the description of f for Theorem 1.1 but we will need it when we insist that f have Property j. We let I_{i1}, I_{i2} be two disjoint small 3-cells in I_i situated in I_i much as I_1, I_2 were placed in I^3. The map f is a natural map taking $B(I_{ij})$ onto Bd I'_{ij}. The map f is extended to take the rest of the closure of $I_i - (I_{i1} \cup I_{i2})$ into I'_1 except on a set whose closure is a solid torus. The map f is then extended to this solid torus by methods described in Sections 4 and 5.

Similarly we subdivide each I'_{ij} into two cells I'_{ij1}, I'_{ij2} as shown in Figure 3.2 so as to have a brick partitioning of I^3 into eight pieces each about half the size of I^3. We pick I_{ij1}, I_{ij2} as small 3-cells situated in I_{ij} as I_1, I_2 were in I^3 and let f send part of I_{ijk} into I'_{ijk} in much the same way f sent part of I_i into I'_i.

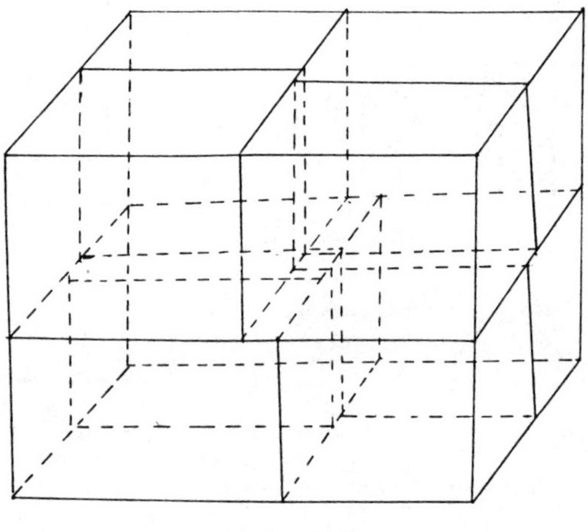

Figure 3.2.

We continue subdividing the $I'_{ij\ldots k}$'s, building $I_{ij\ldots k}$'s, and extending f ad infinitum. Finally we have defined f on all of I^3 except on the Cantor set $X = I^3 \cap (I_1 \cup I_2) \cap (\cup I_{ij}) \cap (\cup I_{ijk}) \cap \ldots$ and f sends $I^3 - X$ onto the set $Y = \text{Bd } I^3 \cup (\text{Bd } I'_1 \cup \text{Bd } I'_2) \cup (\cup \text{Bd } I_{ij}) \cup \ldots$. The map f is extended to X by continuity. Note that Y is the union of countably many disks and each point of $I^3 - Y$ has a one point inverse. Hence, f has Property c. Since we shall define f suitably at each stage, the final map f will have Property d. Since X lies in the bottom of I^3 and $f(\cup I_{ij\ldots k}) = \cup I'_{ij\ldots k} = I^3$, $f(X) = I^3$ and f has Property b.

4. Extending f on a torus. Let A_α be a vertical planar cross section of I^3 through z_0 as shown in Figure 4.1. The segments shown as solid lines are part of point inverses we described in Section 3. We extend the definition of f to the triangular disk $g_\alpha g'_\alpha z_{1\alpha}$ such that the spanning segments of these triangles parallel to $g_\alpha g'_\alpha$ are parts of point inverses. These new partial point inverses are shown dashed. Also, spanning segments in $b_\alpha b'_\alpha z_{2\alpha}$ parallel to $b_\alpha b'_\alpha$ are parts of point inverses.

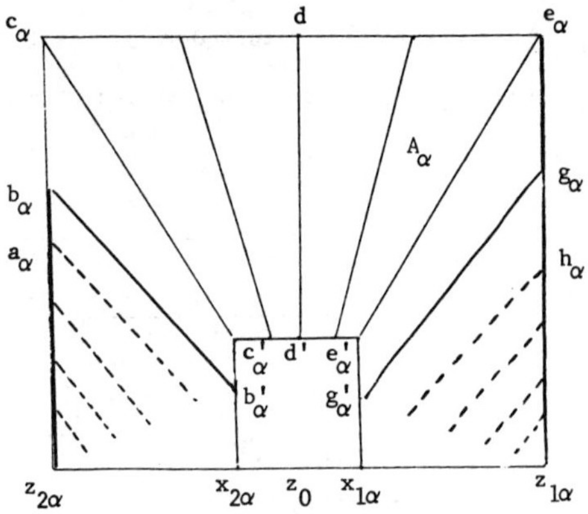

Figure 4.1.

Let T be the closure of the part of $I^3 - (I_1 \cup I_2)$ on which we have not yet defined f. It is a solid torus as shown in Figure 4.2. The map f has been defined so as to take $I_1 \cap T$ onto the top of I'_2 and the union of the four trapezoids on the top of T to the rest of Bd I'_2. The rectangular 3-cell I_2 is shown in T as low and thin.

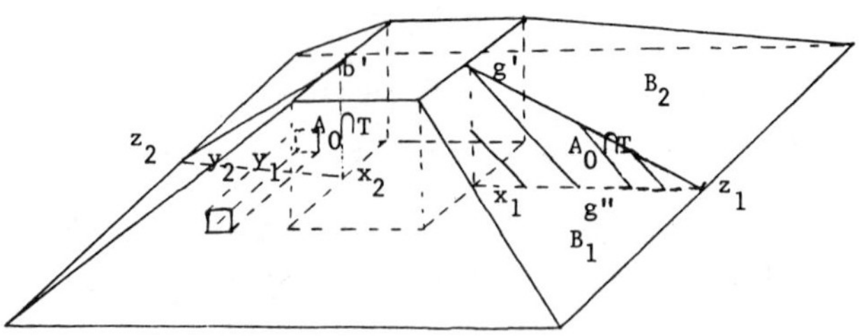

Figure 4.2.

Since 3-balls are easier to work with than solid tori, we divide T into two 3-balls B_1, B_2 by use of A_0, the vertical planar cross section A_α parallel to the front face of I^3. Figure 4.2 shows how A_0 intersects T. We define f on the right triangle $x_1g'z_1$ by deciding that the segment from g' to the center g" of x_1z_1 is a part of a point inverse as is each spanning segment of $x_1g'z_1$ parallel to g'g". Such precision of definition is not important since we merely need to fill the disk $x_1g'z_1$ with $\{x_1\}$, $\{z_1\}$, and a continuous family of spanning arcs from $x_1g'z_1$ to $x_1g''z_1$, but it is easier to make the description with precision than with generality.

Before defining f on the left part of $A_0 \cap T$, we define f on the bottom of T. Figure 4.3 shows the bottom of I_1 as a shaded square and the bottom of I_2 as a shaded rectangle. Partial point inverses are rectangular simple closed curves except for the point inverse containing the boundary of the bottom of I_2 -- this partial point inverse is a θ-curve. We use the previous definition of f on $x_1g''z_1$ to determine where f takes each of these partial point inverses.

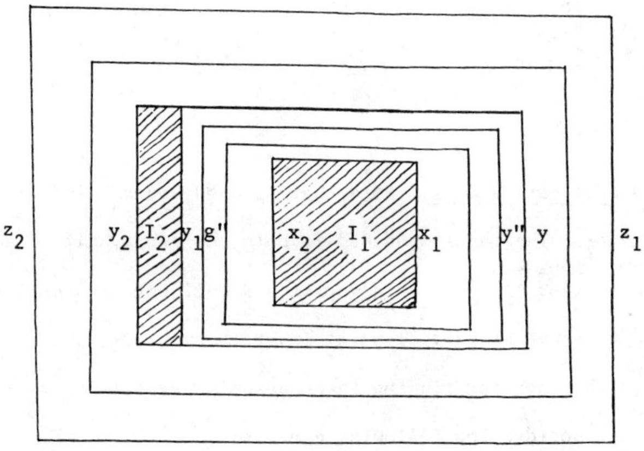

Figure 4.3.

Finally we define f on the left part of $A_0 \cap T$ as shown in Figure 4.4. It has already been defined on the boundary of this disk and a study of this description shows that for each point $p \in A_0 \cap B(I_2)$ there is another point on $y_2z_2b'x_2y_1$ with the same image. We fill this left part of $A_0 \cap T$ with $\{y_1\}, \{y_2\}$, and a continuous collection of disjoint polygonal spanning arcs joining points with the same image. We do not need to exercise great care in defining these polygonal arcs as long as their ends have the same image. Hence, we leave the description vague rather than make a precise description.

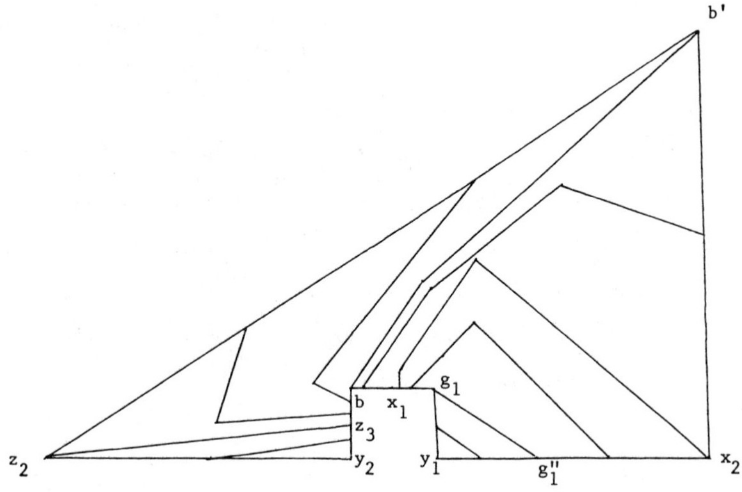

Figure 4.4.

5. <u>A map of a</u> 3-<u>cell onto a disk</u>. Note that A_0 subdivides T into a front 3-cell B_1 and a back 3-cell B_2. The map f sends Bd B_1 onto the front half of Bd I_2' and Bd B_2 onto the back half of Bd I_2'. We use a map of a 3-cell onto a 3-cell to get spanning arcs in the B_i's to serve as partial point inverses.

To get us in the mood for finding these spanning arcs in B_i to serve as partial point inverses, we consider the following exercise.

<u>Exercise</u> 5.1. Suppose B^3 is a round 3-ball and f is a map of Bd B^3 onto a disk D such that f takes each of the upper half D_1 and the lower half D_0 of Bd B^3 homeomorphically onto D. Then f can be extended to take B^3 onto D such that point inverses are either points of the equator of B^3 or tame spanning arcs of B^3.

Solution of Exercise 5.1. Let f_1 = f restricted to B_1 and ρ be the vertical projection of D_1 onto D_0. Let h_t $(0 \le t \le 1)$ be an isotopy of D_0 onto itself such that

$$h_0 \text{ is the identity,}$$
$$h_t \text{ is the identity on Bd } D_0\text{, and}$$
$$h_1 = \rho f_1^{-1} f.$$

If S_t is the curved surface in B_3 which divides each vertical segment from bottom to top in the ratio t to (1 - t) and $p \in \text{Int } D_0$, $f^{-1}f(p)$ is the spanning arc of B^3 that intersects S_t in a point whose vertical projection on D_0 is $h_t(p)$.

Answered question. The point inverses constructed in the solution of Exercise 5.1 may not be polyhedral. In my lecture at Blacksburg I posed the question as to whether or not one could solve the exercise so as to get polyhedral point inverses. (In a lower dimensional version where one uses a round disk instead of a round 3-ball, the point inverses can even be made straight.) I doubted that one could get polyhedral point inverses for all f's but thought it would be very interesting if one could and offered a dinner to the first person in the audience to get a solution. Wlodzirmierz Kuperberg won the dinner by constructing a map f such that $\rho f_1^{-1} f$ twists so badly near a certain Cantor set that it is impossible to get polyhedral point inverses.

Lemma 5.2. Suppose B^3 is a 3-ball and f is a map of Bd B^3 onto a disk D such that

$f^{-1}(p)$ is a nonseparating continuum on Bd B^3 if $p \in$ Bd D,

f takes each of the two components of Bd B^3

$-f^{-1}$(Bd D) homeomorphically onto Int D.

Then the f can be extended to a map of B^3 onto D such that each point inverse not in f^{-1}(Bd D) is a tame spanning arc of B^3.

Proof. It is known [4] that there is a monotone map g of B^3 onto itself such that g is a homeomorphism on $B^3 - f^{-1}$(Bd D) and g takes f^{-1}(Bd D) onto a simple closed curve so that if $p \in f^{-1}$(Bd D), then $g^{-1}g(p) = f^{-1}f(p)$. To prove the lemma one could suppose gf^{-1}Bd D is the equator and let f' be an extension of the map fg^{-1} as promised by Exercise 5.1. For each $p \in$ Bd B^3, $f^{-1}f(p) = g^{-1}f'^{-1}f'g(p)$.

We now return to consideration of defining f on each of the 3-balls B_1, B_2 into which A_0 divided T in Section 4. It may be noted that we have already defined f on each Bd B_i to take it to a disk D_i on Bd I_2' so as to satisfy the hypothesis of Lemma 5.2. Hence Lemma 5.2 assures us that we can extend f to B_i so that each partial point inverse is either an arc on Bd B_i or a tame spanning arc of B_i.

This completes the description of the map f for Theorem 1.1. The map is monotone since each partial point inverses in the closure of $I^3 - (\cup I_{ij...k})$ is connected. Property e holds since each simple closed curve in a point inverse lies in the bottom of some torus.

Had we exercised care, we could have made the partial point inverses in B_i polyhedral. If we did this at each stage, then the final map f would be such that each point inverse is locally polyhedral except at points of $X = I^3 \cap (I_1 \cup I_2) \cap (\cup I_{ij}) \cap \ldots$. However, this would have complicated our description and violated our desire to simplify the description of f.

6. Detecting arcs. In this section we show that f has Property f.

Let W be the closure of some typical $I_{ij...k} - (I_{ij...kl} \cup I_{ij...k2})$. A nondegenerate point inverse $f^{-1}(p)$ is a tame arc if for each W, either $W \cap f^{-1}(p) = 0$ or $W \cap f^{-1}(p)$ is a tame spanning arc of W. Since f is defined on all W's in a fashion similar to the way in which it is defined on $I^3 - (I_1 \cup I_2)$, we examine how point inverses intersect $I^3 - (I_1 \cup I_2)$ and note that the image of the set of point inverses which intersect the closure $I^3 - (I_1 \cup I_2)$ in something other than a tame spanning arc is the union of an arc and a simple closed curve. This insures that f has Property f.

If a point inverse intersects $T \cap$ (bottom I^3), it contains a cycle and is not a spanning arc of the closure of $I^3 - (I_1 \cup I_2)$. The image of this bottom of T is an arc x_0z_0 on Bd I_2' that runs from the center x_0 of the top of I_2' down the right face of I_2', and then to the center z_0 of the bottom of I_2'.

The inverse of each point of the boundary of the top of I_2' intersects the closure of $I^3 - (I_1 \cup I_2)$, not in a spanning arc, but in the union of two spanning arcs with a common end point on Bd I_1. This common end point will be a branch point of the complete point inverse in I^3.

The preceding two paragraphs list all point inverses that intersected the closure of $I^3 - (I_1 \cup I_2)$ in something other than a tame spanning arc. The image of this set is the union of the arc x_0z_0 and the simple closed curve Bd (top of I_2').

7. <u>Refining</u> f. In this section we show that if the f which we defined in Section 2-5 were suitably chosen, then it would have Properties g,h,i,j also.

<u>Theorem</u> 7.1. There is a monotone map f of I^3 onto itself having Properties a,b,c,d,e, f,g,h,i, and j.

Recall that we

> first defined f on the top and four lateral sides of I^3,
>
> then on the closure of $I^3 - (I_1 \cup I_2)$,
>
> then on the closure of $I_1 - (I_{11} \cup I_{12})$,
>
> then on the closure of $I_2 - (I_{21} \cup I_{22})$,
>
> then on the closure of $I_{11} - (I_{111} \cup I_{112})$,
>
> ...
>
> then on the closure of $I_{ij\ldots k} - (I_{ij\ldots k1} \cup I_{ijk\ldots k2})$,
>
> ... .

At each of these stages in the extension of f, we used an xyz. At the first stage we picked z_0 to be the center of the bottom of I^3 (but notice that it could just as well have been any other point of the interior of this bottom -- in fact, any point of Bd I_2' - top I_2'), picked x_0 to be the center of $I_1' \cap I_2'$ (but notice that we could have used any other point of $\mathrm{Int}(I_1' \cap I_2')$) and picked y_0 to be a point on x_0z_0 near z_0 (but we did not specify how near). At later stages we use similar arcs, which we also call xyz's. We prove Theorem 7.1 by picking the xyz's suitably at the various stages.

Note that if p lies in an xyz at some stage, then $f^{-1}(p)$ contains a cycle but if p does not lie in any xyz, $f^{-1}(p)$ contains no cycle. If p is a y at some stage, $f^{-1}(p)$ contains a θ-curve and hence two independent cycles. To count the number of independent cycles in $f^{-1}(p)$ we merely need to know how the various xyz's relate to p.

We now list the precautions to be used in picking the xyz's. It is clear how these xyz's can be picked one at a time to achieve these restrictions so we do not elaborate on the selection procedure.

1. In picking any xyz, the y is chosen so as not to lie on any previous xyz and

xyz is chosen so as not to intersect any previous y (unless this previous y is the z for xyz). This precaution insures that the $f^{-1}(p)$'s that contain θ-curves do not contain other cycles.

2. In picking an xyz, we pick it so that it does not intersect any other xyz in more than a finite number of points and does not intersect the intersection of two previous xyz's. This precaution insures the f has Properties g and h.

Let us consider how branch points might have been introduced other than being part of a cycle introduced when one considered an xyz. The only other way of introducing a branch point when f is defined on the closure of an $I_{ij...k} - (I_{ij...k1} \cup I_{ij...k2})$ is on the $f^{-1}(p)$'s such that $p \in Bd(I'_{ij...k1} \cap I'_{ij...k2})$ and in this case the branch point is not completed until a later stage. Since in the brick partitionings no point belongs to more than two $Bd(I_{ij...k1} \cap I_{ij...k2})$'s, not more than two branch points are introduced in any $f^{-1}(p)$ by this method. Hence, the nondegenerate $f^{-1}(p)$'s are finite graphs. The monotonic behavior of the $I_{ij...k}$'s and the tameness of the preliminary stages of the $f^{-1}(p)$'s insure that f has Property i.

Since no $f^{-1}(p)$ has more than two branch points introduced by considering the $Bd(I_{ij...k1} \cap I_{ij...k2})$'s, no $f^{-1}(p)$ has more than four end points. This insures that f has Property j at the p's whose $f^{-1}(p)$'s do not contain cycles. To make the f have this property at other p's, we take two additional precautions.

3. The xyz's are picked so that if p either lies on two xyz's or is the y of any xyz, then p does not belong to any $Bd(I_{ij...k1} \cap I_{ij...k2})$. This insures that f has Property j at those p's whose $f^{-1}(p)$'s contain two cycles since these $f^{-1}(p)$'s have at most two ends.

4. The xyz's are picked so that no point of any xyz belongs to the intersection of two $Bd(I_{ij...k1} \cap I_{ij...k2})$'s. This insures that the f has Property j on the p whose $f^{-1}(p)$ has only one cycle since these $f^{-1}(p)$'s have at most three ends.

8. Points of Bd I^3 with degenerate inverses. For any f defined according to the instructions in Sections 2-5, the only points of I^3 with degenerate point inverses are those that do not lie on any Bd $I'_{ij...k}$. Hence no f as defined in Sections 2-5 would have Property k.

We prove Theorem 8.1 by modifying the map used to prove Theorems 1.1 and 7.1.

The key to the modification is to take the point of view that if $I'_{ij\ldots k}$ is a 3-cell with its bottom on the bottom of I^3, then the interior of the bottom of $I'_{ij\ldots k}$ belongs to the interior of $I'_{ij\ldots k}$ rather than to its boundary. We define f so that each point of the bottom of I^3 will have a degenerate inverse unless it lies on the lateral side of some $I'_{ij\ldots k}$ -- i.e., on some $B(I'_{ij\ldots k})$.

We now point out how the f used to justify Theorem 8.1 would differ from the one used for Theorems 1.1 and 7.1. We show how changes could be made in each of the preceding sections.

Modification of Section 2. Instead of defining f to be a natural map taking $B(I^3)$ onto Bd I^3, we would define it to be the identity taking $B(I^3)$ onto itself. We are not anxious at early stages to send points to the bottom of I^3.

We define I_1, I_2, I'_1, I'_2 as before. Our new f agrees with the old on $B(I_1)$, but it takes $B(I_2)$ homeomorphically onto $B(I'_2)$. As before, if p_1, p_2 are points of $B(I^3)$, $B(I_1)$ with the same image, f takes the segment from p_1 to p_2 to this image point.

Modification of Section 3. We define the $I'_{ij\ldots k}$'s and $I_{ij\ldots k}$'s as before. If the bottom of $I'_{ij\ldots k}$ does not lie on the bottom of I^3, f is a natural map as before taking $B(I_{ij\ldots k})$ onto Bd $I'_{ij\ldots k}$ but if the bottom of $I'_{ij\ldots k}$ does lie on the bottom of I^3, then f takes $B(I_{ij\ldots k})$ homeomorphically onto $B(I'_{ij\ldots k})$.

Modification of Sections 4 and 5. The description of f on the closure of $I_{ij\ldots k} - (I_{ij\ldots k1} \cup I_{ij\ldots k2})$ breaks into three cases depending on whether or not neither, both, or only the second of $I'_{ij\ldots k1}, I'_{ij\ldots k2}$ has its bottom on the bottom of I^3.

Case 1. The bottom of $I_{ij\ldots k}$ does not lie on the bottom of I^3. This is the complicated case but it is the one whose treatment has already been described in Sections 4 and 5. We make no changes in our original description for this hard case.

Case 2. The bottom of each $I'_{ij\ldots k1}, I'_{ij\ldots k2}$ lies on the bottom of I^3. This is the easy case. So as to prevent point inverses from having branch points of order 4 at certain points of $B(I_{ij\ldots k})$ we first define f in a regular neighborhood of $B(I_{ij\ldots k})$. Shrink $I_{ij\ldots k}$ slightly to get a 3-cell $I''_{ij\ldots k}$ contained in $I_{ij\ldots k}$ so that $(I_{ij\ldots k} - I''_{ij\ldots k}) \cup B(I''_{ij\ldots k})$ is the union of a collection of disjoint striaght segments collinear with the center of the bottom of $I_{ij\ldots k}$ and reaching from $B(I_{ij\ldots k})$ to $B(I''_{ij\ldots k})$. These segments are parts of point inverses.

Extend f to take the spanning disk D of $I''_{ij\ldots k}$ shown in Figure 8.1 onto $I'_{ij\ldots k1} \cap I'_{ij\ldots k2}$. The disk D divides $I''_{ij\ldots k}$ into 3-cells $I''_{ij\ldots k1}, I''_{ij\ldots k2}$ containing $I_{ij\ldots k1}, I_{ij\ldots k2}$ as shown. The part of a point inverse in the closure of $I''_{ij\ldots kn} - I_{ij\ldots kn}$ $(n = 1,2)$ is an arc joining points of $B(I''_{ij\ldots kn})$ and $B(I_{ij\ldots kn})$ with the same image. We show the arcs as straight. Note that point inverses have branch points at points of Bd $D \cap B(I''_{ij\ldots k})$.

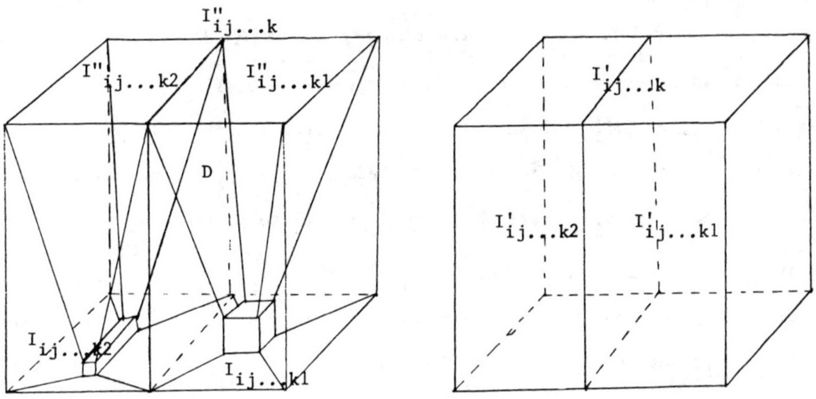

Figure 8.1.

Case 3. This is the half-and-half case where the bottom of $I'_{ij\ldots k2}$ lies on the bottom of I^3 but the bottom of $I'_{ij\ldots k1}$ does not. We define f on $B(I_{ij\ldots k1})$ as before. On the vertical planar cross section A_α, f is defined as shown in Figure 4.1. Now b and g are at the centers of the sides rather than 3/4 the way up. Also, f is similarly defined on the triangular disks $g_\alpha g'_\alpha z_{1\alpha}, b_\alpha b'_\alpha z_{2\alpha}$.

On the bottom of T, point inverses are as shown in Figure 8.2. Some of them are arcs reaching from a point of the boundary of the base of $I_{ij\ldots k2}$ to the corresponding point on the base of $I_{ij\ldots k}$, some of them are rectangular simple closed curves concentric with the boundary of the bottom of $I_{ij\ldots k1}$, and one of them is a rectangular simple closed curve with two feelers.

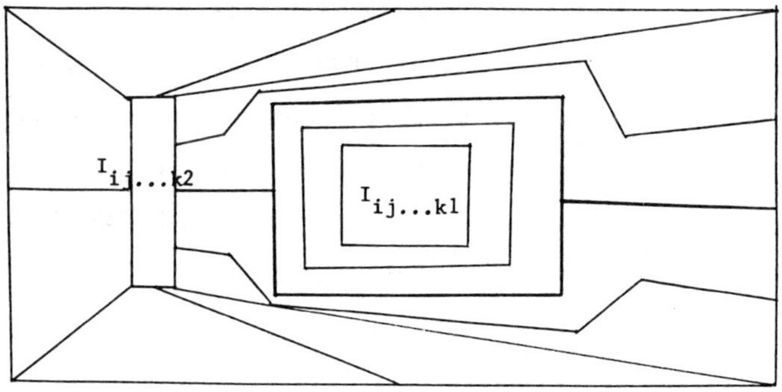

Figure 8.2.

Each of the two disks in $A_0 \cap T$ is filled with spanning arcs joining points with the same image. The description of f is then completed on T as described in Section 5.

To make our map f have Properties g,h,i,j, we would exercise the same precautions as those given in Section 7.

Which is simpler? The description of the maps for Theorems 1.1 and 7.1 were given before that of Theorem 8.1 because it was easier to describe. The description was easier because it made repetitive use of the same step. In the description of the f for Theorem 8.1, some of the steps were simpler but judgement had to be exercised to determine which step to use.

9. A surprising result. D. R. McMillan raised the following question: Is there a monotone map of R^3 onto itself such that some point inverse is a 2-sphere? The first impression is that the question has a negative answer. A 2-sphere separates R^3 but a point does not. Maps preserve connectedness. Many queried about the question anticipated a negative answer. However:

Theorem 9.1. There is a monotone map of R^3 onto itself such that some point inverse is a 2-sphere.

To get a map f verifying the theorem, one could first define a map f of a 3-ball B^3 in R^3 onto a straight segment ab in R^3 such that $f^{-1}(a) = \mathrm{Bd}\, B^3$, $f^{-1}(b) = \text{center } B^3$, and other $f^{-1}(p)$'s are concentric 2-spheres.

One could then consider a monotone map g_1 of R^3 onto itself such that B^3 is the only nondegenerate point inverse. Then one could let g_2 be a monotone map of R^3 onto R^3_+ such that some point of Bd R^3_+ has $g_1(B^3)$ as an inverse. A handy model of R^3_+ is Int $I^3 \cup$ Int(top I^3). One even has a model of R^3_+ if cones are erected as shown in Figure 9.1. Each of the cones has one point of its base which does not lie in the model. We show $g_2g_1(B^3)$ as the vertex of the right front cone.

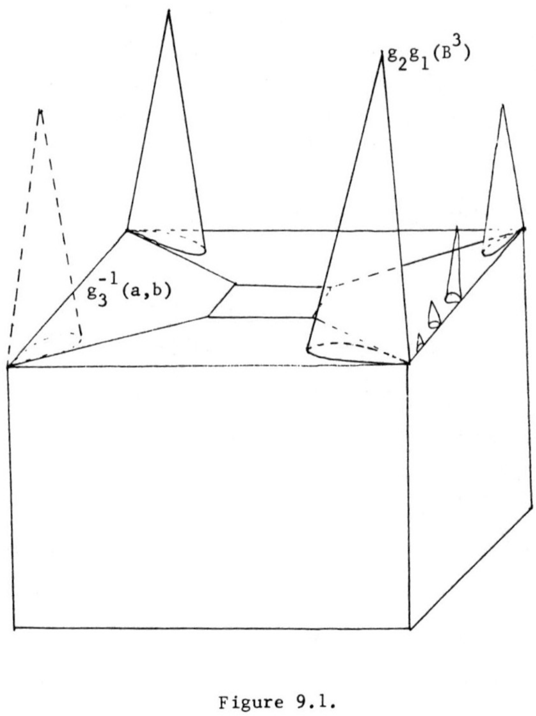

Figure 9.1.

The dotted cone is not a part of the model. If it had been there, all the cones could have been pushed over, and the sides and bottom of I^3 could have been folded over and attached as discussed in [3], so as to obtain a 1 - 1 map of R^3_+ into R^3. With the dotted cone not there, one obtains a 1 - 1 map g_3 of R^3_+ onto R^3 - (ab] such that $g_2g_1(B^3) = g_3^{-1}(a)$. On points not in B^3, the map f is defined by $g_3g_2g_1$.

The surprising Theorem 9.1 caused C. E. Burgess to make the following remark about monotone maps that were not compact.

Metacorollary. Monotone maps that are not compact are no good.

Where did our intuition about the answer to McMillan's question go wrong? Perhaps we assumed that since maps preserved connectedness, they preserved nonconnectedness also. This is not true for noncompact maps -- even those that are monotone.

It is easy to see that Theorem 9.1 has the following extension.

Theorem 9.2. For each compact continuum C in R^3, there is a monotone map of R^3 onto itself that has C as a point inverse.

One might wonder which monotone maps of closed subsets of R^3 into R^3 can be extended to monotone maps of R^3 onto itself, but we do not pursue that here.

BIBLIOGRAPHY

1. R. D. Anderson, Open mappings of compact continua, Proc. Nat. Acad. Sci. U.S.A., 42 (1956), 347-349.

2. R. H. Bing, The monotone mapping problem, Topology of Manifolds, (Edited by J. C. Cantrell and C. H. Edwards, Jr.), 1970, Markham Publishing Co., 99-115.

3. R. H. Bing and Joseph M. Martin, Monotone images of E^3, Proceedings of the Conference on Monotone Mappings and Open Mappings, 1970, State University of Binghamton, 55-77.

4. E. E. Floyd and M. K. Fort, A characterization theorem for monotone mappings, Proc. Amer. Math. Soc., 4 (1953), 828-830.

5. L. C. Glaser, Monotone non-compact mappings of E^r onto E^k for $r \geq 4$ and $k \geq 3$, to appear.

6. T. E. Knoblauch, Decompositions of three-manifolds into toroidal continua, Thesis, University of Wisconsin, 1972.

7. G. T. Whyburn, Compactness of certains mappings, Amer. J. Math., 81 (1959), 306-314.

8. David Wilson, Monotone mappings on manifolds, Proceedings of the Conference on Monotone Mappings and Open Mappings, 1970, State University of Binghamton, 37-54.

A DECOMPOSITION OF S^3 WITH A NULL SEQUENCE OF CELLULAR ARCS

R. H. Bing[1]
Michael Starbird[1]

Department of Mathematics
The University of Texas
Austin, Texas

An example is described showing the following:

THEOREM. There exists a null sequence of disjoint cellular arcs $\{A_i\}$ *in* S^3 *such that the decomposition space* $S^3/\{A_i\}$ *is topologically different from* S^3.

David Gillman and Joe Martin announced a sequence of arcs implying the above theorem many years ago but their example was never published. The example given here is related to that of Gillman and Martin, but differs in details.

The example is modeled after Bing's example of a null sequence of cellular continua $\{G_i\}$ *in* S^3 *where* $S^3/\{G_i\}$ *is not a manifold. Each of the* G_i*'s was a planar indecomposable continuum which is the intersection of a decreasing sequence of solid tori. Each of the* A_i*'s in the present example is the intersection of a decreasing sequence of handle bodies each of which is a long tube with a handle on each end. Instead of having big oscillations, the tubes have small ones which hook like knitting.*

[1]*Research supported in part by NSF Grant MCS76-07242A01.*

ISBN 0-12-158860-2

This example is of current interest in view of recent results discussed elsewhere in these Proceedings, showing that under many decompositions G of S^3, S^3/G *is topologically* S^3. *The present example establishes a bound on what can be proved.*

I. INTRODUCTION

We give an example called the Main Example, which proves the following theorem.

MAIN THEOREM. There exists a null sequence of disjoint cellular arcs $\{A_i\}$ *in* S^3 *such that the decomposition space* $S^3/\{A_i\}$ *is topologically different from* S^3.

A sequence of sets is *null* if for each $\varepsilon > 0$, at most a finite number of them have diameters greater than ε. A set X in S^3 is *cellular* if $X = \bigcap_{i \in \omega_0} B_i$ where for each i, B_i is a 3-cell in S^3 and $B_{i+1} \subset \text{Int } B_i$. We use $S^3/\{A_i\}$ to denote the decomposition space whose only non-degenerate elements are the A_i's.

D. Gillman and J. Martin announced the existence of a sequence of arcs implying the Main Theorem many years ago [8], but their example, which made use of the eyebolt construction employed in [5], was never published. The example given here, which was constructed by Bing with Starbird's aid, is similar to the Gillman-Martin example but differs in details. The decomposition described here has a defining sequence each stage of which is a disjoint collection of solid double tori.

This example is of current interest because of the flurry of recent results [7,9,10,11,12] showing that S^3/G is topologically equivalent to S^3 under various sets of restrictions on G. In particular, in these Proceedings there appears a result [7,10] stating that $S^3/\{A_i\}_{i\in\omega}$ is topologically equivalent to S^3 if each

A_i is a tamely embedded cellular polyhedron. The example in this paper gives a bound on what more can be proved.

The dogbone space [2] gives a decomposition of S^3 into points and tame arcs such that the decomposition space differs from S^3. However, the decomposition in it has a Cantor set of tame arcs rather than a countable number. Bing showed [3, Theorem 3] that if G is an upper semicontinuous decomposition of S^3 with only a countable number of nondegenerate elements $\{G_i\}$ and each G_i is a tame arc, then S^3/G is topologically S^3.

In [6, §2-5], Bing constructed an example of null sequence of disjoint, cellular continua $\{G_i\}$ so that $S^3/\{G_i\}$ is topologically different from S^3. In that example each G_i is an indecomposable continuum. The example here is modeled after that example, but modifications are made to ensure that each non-degenerate element of the present decomposition is a cellular arc rather than an indecomposable continuum.

One feature which the present example shares with Bing's previous example in [6] is the method by which the decomposition space is shown to differ from S^3. That method employs the concept of the two disk property which is defined in the next section. The third section contains theorems which can be used in describing decompositions G of S^3 so that S^3/G is topologically different from S^3. These theorems are used in Section 4 to construct the Main Example. They can also be used in proving the following more general theorem which will appear in a future paper.

THEOREM. Let $\{X_i\}_{i\in\omega_0}$ be a countable collection of nondegenerate continua each of which admits a cellular embedding in S^3. Then there exist embeddings $h_i : X_i \to S^3$ so that for each i, $h_i(X_i)$ is a cellular embedding, $\{h_i(X_i)\}$ is a null sequence of disjoint sets, and $S^3/\{h_i(X_i)\}$ is topologically different from S^3.

Throughout this paper we use PL sets if that is convenient. For example, if we define a cellular set as the intersection of 3-cells, it is assumed that the 3-cells are PL. If D is a meridional disk of a solid torus T, it is assumed that D and T are PL. If D intersects a handlebody H it is understood that H is PL and D, H are in general positon. Of course, the cellular arcs which we describe in the Main Example as an intersection of handlebodies are not PL.

II. THE TWO DISK PROPERTY

To show that a cellular upper semicontinuous decomposition G of S^3 is not shrinkable, it is frequently convenient to construct a defining sequence for G each stage of which is a collection with the two disk property defined below.

Definition. A finite collection M of disjoint closed subsets in the interior of a solid torus T has the *two disk property* if and only if for every pair of disjoint meridional disks D_0, D_1 of T, there is an element of M which intersects both D_0 and D_1. A disk D in T is *meridional* if $D \cap Bd\ T = Bd\ D$ and $Bd\ D$ does not bound a disk in $Bd\ T$.

Example 2.1. The collection M which contains only one element, a core of T, has the two disk property.

Example 2.2. The collection M consisting of the two eyeglasses J_1 and J_2 illustrated in Figure 2.1 has the two disk property. (See Lemma 3.3 for a proof of this fact.)

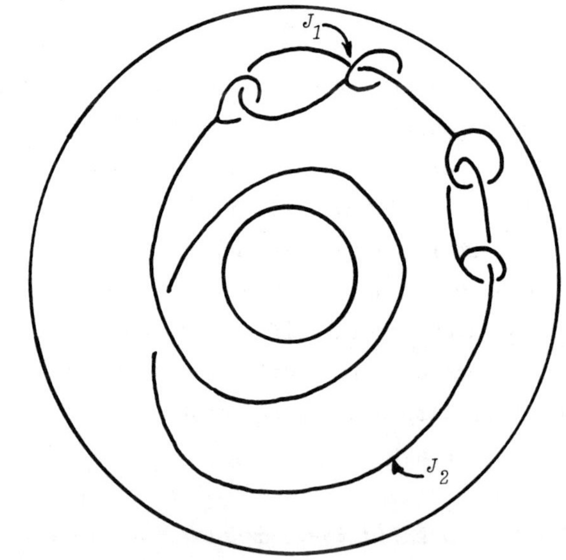

Fig. 2.1.

Definition. A decomposition G of S^3 is *defined by a sequence* $\{M_i\}_{i\in\omega}$ if and only if for each i, M_i is a finite collection of disjoint 3-manifolds with boundary in S^3 so that each element of M_{i+1} is contained in the interior of an element of M_i, and each element of G is either a component of $\bigcap_{i\in\omega_0} (\bigcup M_i)$ or a point of $S^3 - \bigcap_{i\in\omega_0} (\bigcup M_i)$.

THEOREM 2.1. *Let T be a solid torus and G be a cellular decomposition of S^3 defined by the sequence $\{M_i\}_{i\in\omega}$ where each M_i has the two disk property in T. Then S^3/G is not homeomorphic to S^3.*

The properties mentioned in Theorem 2.1 are a favorite way of showing that a decomposition space of S^3 is topologically different from S^3. See for example [2, Lemma for Theorem 12], [6, Section 5] and [1, Theorem 3]. We do not repeat the proof of Theorem 2.1.

III. PRESERVING THE TWO DISK PROPERTY

One method for constructing a defining sequence for an upper semicontinuous decomposition G is to specify how an element at stage i contains the element or elements of stage $i + 1$. In constructing examples of decompositions G for which S^3/G is not homeomorphic to S^3, it is useful to construct a defining sequence for G each stage of which has the two disk property. This section contains theorems which say that if a given collection M has the two disk property then certain modifications of M will yield a new collection which also has the two disk property.

We make an easy start.

LEMMA 3.1. Let T be a solid torus and L be an element of M, a collection with the two disk property. If L^+ is a closed set in T containing L and disjoint from the other sets in M, then the collection M' obtained from M by replacing L by L^+ also has the two disk property.

LEMMA 3.2. Let W be a solid torus and X be a subset of W with n components each of which is a pair of eyeglasses, i.e., two simple closed curves joined by an arc, and so that the chain of eyeglasses is embedded in W as shown in Figure 3.1. Then each meridional disk of W intersects X.

Proof. Let $\{A_i\}_{i=1}^{2n}$ be the planar disks bounded by the simple closed curves in X so that for each i, $X \cap \text{Int } A_i$ is a single point. (See Figure 3.1.) Suppose the lemma is not true. Then there exists a PL, meridional disk D of W so that $D \cap X = \emptyset$, D is in general position with respect to $E = \bigcup_{i=1}^{2n} A_i$ and each component of $D \cap E$ is essential in $E - X$. There is a subdisk

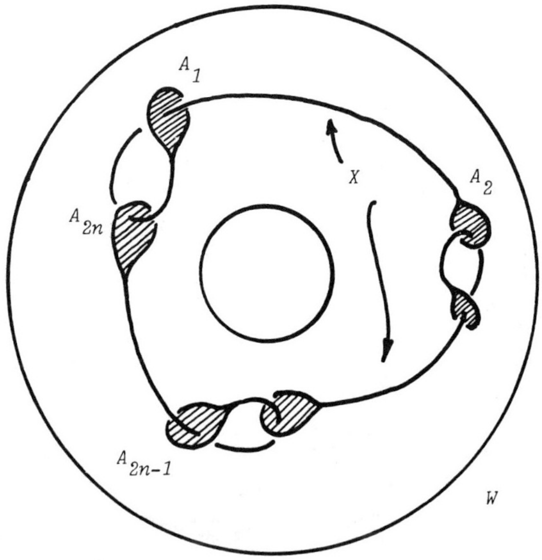

Fig. 3.1.

D' of D such that Int $D' \cap (X \cup E) = \emptyset$ and $Bd\ D' \subset E$. However, this supposed disk D' cannot exist since $Bd\ D' \subset$ Int A_j for some j and is essential in Int $A_j - X$. But each essential curve in Int $A_j - X$ links a simple closed curve in $(E - \text{Int } A_j) \cup X$. Therefore D' cannot exist and the lemma is proved.

Figure 3.2 represents a solid torus W in which two disjoint sets Y and Z are embedded. The set Z is one pair of small eyeglasses while Y is the union of several eyeglasses. The chain of eyeglasses $Y \cup Z$ goes around W twice.

LEMMA 3.3. Let M be a collection of subsets of a torus T so that M has the two disk property. Suppose the solid torus W is an element of M. Then the collection M', obtained from M by replacing W by the two sets Y and Z as described above, still has the two disk property.

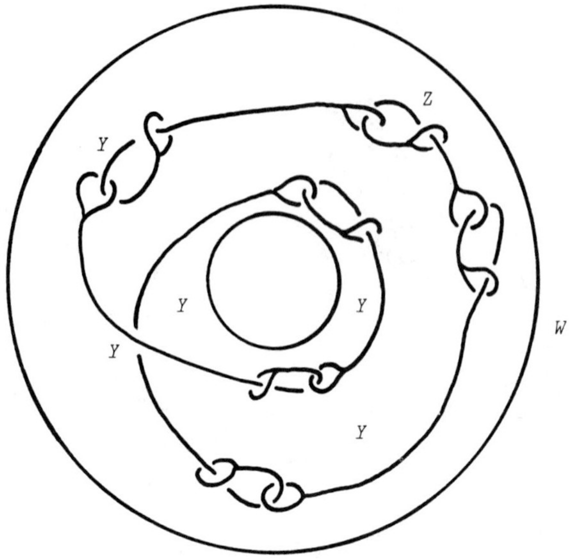

Fig. 3.2.

Proof. Suppose that M' does not have the two disk property. Then there are disjoint, PL, meridional disks D_0, D_1 of T which witness the falsity, are in general position with respect to W, and have no component of intersection with $Bd\ W$ which is trivial on $Bd\ W$. Since M has the two disk property there are subdisks D_0', D_1' of D_0, D_1, respectively, so that $(D_0' \cup D_1') \subset W$, and $(D_0' \cup D_1') \cap Bd\ W = Bd\ D_0' \cup Bd\ D_1'$ -- that is, D_0' and D_1' are meridional disks of W. Consider the double cover T of W. Then one lift $\tilde{Z}$ of Z together with a lift $\tilde{Y}$ of Y encircles T once as in Lemma 3.2. By Lemma 3.2 any meridional disk of τ must intersect $\tilde{Y} \cup \tilde{Z}$. The two lifts of D_0' separate Int τ into two 3-cells B_1, B_2. If $D_0 \cap Z = \emptyset$, then $D_0 \cap Y \neq \emptyset$ and $\tilde{Z}$ is in one of B_1 or B_2. Then the lift of D_1' which intersects the other 3-cell must intersect $\tilde{Y}$ since it must intersect $\tilde{Y} \cup \tilde{Z}$ and it does not intersect $\tilde{Z}$. Therefore D_1 intersects Y and the lemma is proved.

Next we show how to replace a cube with many handles by a torus. For this purpose we need the following lemma whose proof follows from [4, Steps 6 and 7 of proof of Lemma 7]. We include a proof for completeness.

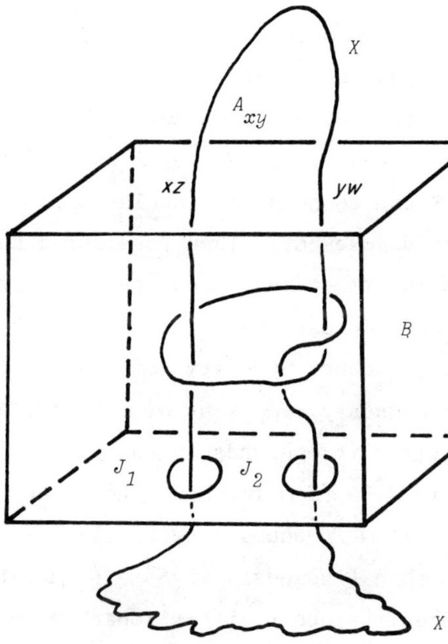

Fig. 3.3.

LEMMA 3.4. Let X be a compact set in S^3, B be a tame 3-cell in S^3 so that $X \cap B$ is the union of two arcs embedded in B as illustrated in Figure 3.3, X contains an arc A_{xy} joining x to y in $S^3 -$ Int B, and neither simple closed curve J_1 nor J_2 is homotopic to a point in $S^3 - (X \cup \text{Int } B)$. Let D be a disk in $S^3 - X$ so that $\text{Bd } D \cap B = \emptyset$, D is in general position with respect to Bd B, and each component of $D \cap \text{Bd } B$ is non-trivial in Bd $B - X$. Then there is a homeomorphism $h : S^3 \to S^3$ fixed outside a small regular neighborhood of B so that $h(D) \cap (X \cup B) = \emptyset$.

Proof. Consider a component K of $D \cap Bd\ B$ which is innermost on D. Let D_K be the subdisk of D bounded by K. Then $D_K \cap (\text{Int } B)=\emptyset$ since each curve in $Bd\ B - X$ which is trivial in $B - X$ is also trivial on $Bd\ B - X$. This fact follows from [4, Lemma 6] in conjunction with the Loop Theorem. Therefore D_K must lie to the outside of Int B. However, K cannot separate x from y on $Bd\ B$ because then K would link a simple closed curve in $A_{xy} \cup \text{Int } B$. The hypothesis on J_1 and J_2 implies that K must separate $Bd\ B$ into two disks, one of which contains x and y and the other of which contains the other two points of $Bd\ B \cap X$.

Next we claim that every component of $Bd\ B \cap D$ is parallel to K in $Bd\ B - X$ in that it separates x and y from the other points of $Bd\ D \cap X$ and cobounds an annulus with K on $Bd\ B$ which misses $Bd\ B \cap X$. Suppose not. Then consider a curve L of $Bd\ B \cap D$ which is innermost on D among those curves not parallel to K. Then L is one boundary component of a disk with holes F in D so that each of the other boundary components of F is parallel to K. Each other boundary component of F bounds a disk in $S^3 - (X \cup \text{Int } B)$ since it cobounds an annulus on $Bd\ B - X$ with K and K bounds D_K in $S^3 - (X \cup \text{Int } B)$. Let G be a singular disk in $[S^3 - (X \cup \text{Int } B)] \cup F$ bounded by L. If F is contained in $S^3 - (X \cup \text{Int } B)$, then L bounds G in $S^3 - (X \cup \text{Int } B)$. This is contrary to hypothesis since L either separates x from y on $Bd\ B$ (and hence links the union of A_{xy} and an arc in B) or else is parallel to one of J_1, J_2. Hence, if there is an F, it is contained in $B - X$. Then L can be pushed into Int B. Use L and G to denote the pushed in L and the correspondingly adjusted G. The points x and y can be joined by an arc C_{xy} in $Bd\ B - G$ and the bottom points of $X \cap Bd\ B$ can be joined by an arc C_{zw} in $Bd\ B - G$. But L links the simple closed curve $J = (X \cap B) \cup C_{xy} \cup C_{zw}$. On the other hand L bounds G which is contained in $S^3 - J$. This contradiction allows us to conclude that all components of $Bd\ B \cap D$ are parallel to K.

There is a homeomorphism $g_1 : S^3 \to S^3$ so that g_1 restricted to the arc xz is pointwise fixed, $g_1(Bd\ B) = Bd\ B$, and $g_1(D) \cap g_1(Bd\ B)$ is contained in the lateral sides of B. There is then a further homeomorphism $g_2 : S^3 \to S^3$ fixed outside D a small neighborhood of B obtained by blowing away from the arc xz The homeomorphism $g = g_1^{-1} \circ g_2 \circ g_1$ establishes the lemma.

LEMMA 3.5. Let T be a solid torus and M be a collection of subsets of T with the two disk property one element of which is a cube with n handles H. Then the collection M', obtained from M by replacing H by a simple closed curve J which is embedded in H as illustrated in Figure 3.4, also has the two disk property.

Proof. Let D_0 and D_1 be disjoint meridional disks of T in general with respect to $\bigcup_{i=1}^{n-1} B_i$ where the B_i's are 3-cells shown in Figure 3.4. Suppose $D_0 \cap J = \emptyset$. Consider $D_0 \cap (\bigcup_{i=1}^{n-1} Bd\ B_i)$. First find new disjoint disks D_0 and D_1 so that the new D_i $(i = 0,1)$ does not intersect any element of M' which the original $D_i (i = 0,1)$ did not and so that each component of $D_0 \cap (\bigcup_{i=1}^{n-1} Bd\ B_i)$ is non-trivial in $\bigcup_{i=1}^{n-1} Bd\ B_i - X$. By Lemma 3.4, D_0 and D_1 can be moved so that the new D_0 misses $\bigcup_{i=1}^{n-1} B_i$. All of this is done without creating new intersections with the other elements of M'. Hence the new D_0 misses a core of H and can be pushed off H. Since M has the two disk property, D_0 and D_1 must intersect some common element of M other than H. Since that element belongs to M', M' has the two disk property.

The next two lemmas show a method by which a cube with many handles can be replaced by a pair of eyeglasses.

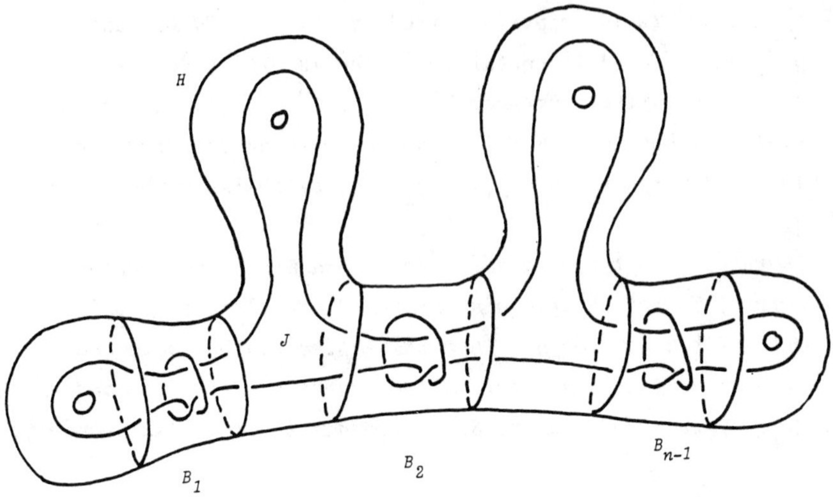

Fig. 3.4.

LEMMA 3.6. Consider the simple closed curve J and eyeglasses L illustrated in Figure 3.5. Then J is not homotopically trivial in $S^3 - L$.

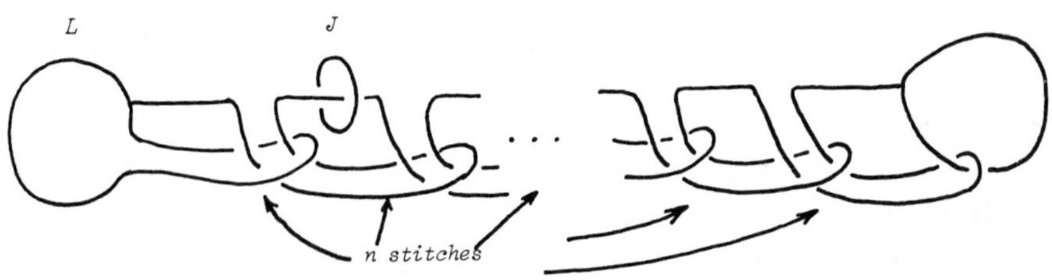

Fig. 3.5.

Proof. The proof is by induction on n, the number of stitches. Note that if J were homotopic to a point in $S^3 - L$, then J would bound a real disk D in $S^3 - L$ by Dehn's Lemma. Consider the case $n = 1$. (See Figure 3.6.)

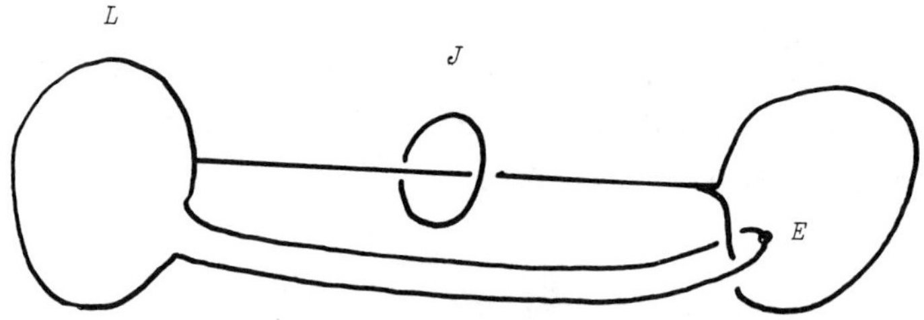

Fig. 3.6.

If J bounds a real disk in D in $S^3 - L$, consider all the general position intersections with the disk E illustrated in Figure 3.6. Trivial curves of $(E \cap D)$ in $E - L$ are removed. Any intersection innermost on D when pushed to one side of E then bounds a disk missing $L \cup E$. This is impossible since any such curve links a simple closed curve in $L \cup E$.

Suppose we have proved the lemma for k stiches and we are given the $k + 1$ case. Consider intersections with disk E as before. As before, an innermost curve of $D \cap E$ when pushed off E cannot be pushed down (see Figure 3.7) because then it would

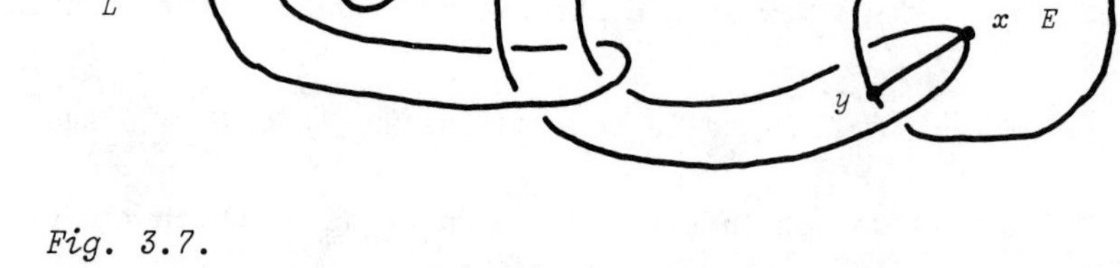

Fig. 3.7.

link *yxwzy*. On the other hand if it is pushed up, the right hand end of the eyeglass could be replaced by the new curve *yxwzy* and induction could be applied to finish the proof.

LEMMA 3.7. Let H be a standardly embedded cube with n handles which is one element of a collection M with the two disk property. Then the collection M', obtained by replacing the element H of M by the eyeglasses G embedded in H as illustrated in Figure 3.8, still has the two disk property.

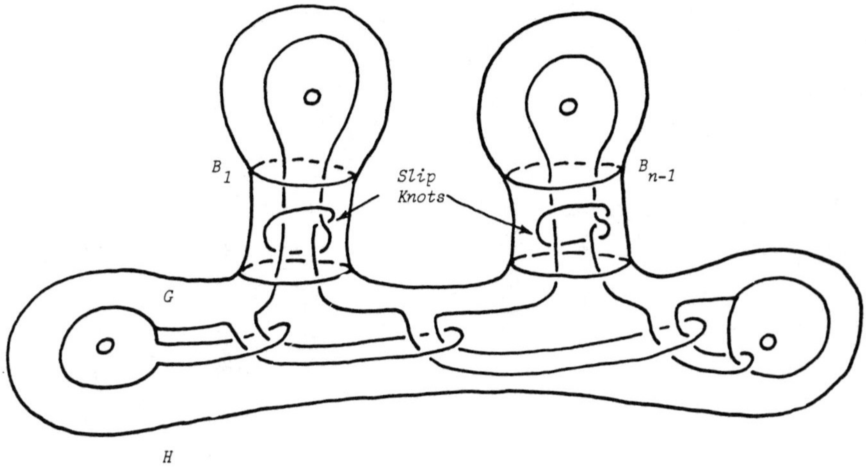

Fig. 3.8.

Proof. Suppose D_0 and D_1 are disjoint meridional disks in the solid torus T, $D_0 \cap G = \emptyset$, and D_0 is in general position with respect to $\bigcup_{i=1}^{n-2} B_i$. New such disks can be found so that each component of $D_0 \cap (\bigcup_{i=1}^{n-2} Bd\ B_i)$ is non-trivial in $\bigcup_{i=1}^{n-2} Bd\ B_i - G$ and the new D_i $(i=0,1)$ do not interest any element of M' which the old ones did not. Note that since H is standardly embedded and those slip knots can be seen to be unknotted in S^3, Lemma 3.6 applies and can be used to show that the hypotheses to Lemma 3.4 are satisfied. Use Lemma 3.4 to find new meridional disks D_0 and D_1 so that $D_0 \cap (G \cup (\bigcup_{i=1}^{n-2} B_i)) = \emptyset$ and each new D_i $(i=0,1)$ does not intersect any element of M' which the old D_i did not. Now D_0 misses a core of H. Hence, since M has the two disk property, D_0 and D_1 must intersect a common element of M other than H. That element is also in M', so M' has the two disk property.

IV. THE MAIN EXAMPLE

There is a null sequence of disjoint cellular arcs $\{A_i\}$ in S^3 such that the decomposition space $S^3/\{A_i\}$ is not homeomorphic to S^3. In fact the decomposition G is describable by a sequence $\{M_i\}_{i \in \omega}$ where each M_i $(i \geq 1)$ consists of 2^i standardly embedded solid double tori. Of course these double tori are tangled up with each other.

The defining sequence $\{M_i\}_{i \in \omega}$ for G is produced inductively. Let M_0 contain a single element, a solid torus T. Let collection M_1 consist of two solid double tori embedded in T as illustrated in Figure 2.1. For each $i \geq 2$, M_i is obtained from M_{i-1} by replacing each solid double torus in M_{i-1} by two solid double tori in such a manner that the following conditions are satisfied.

1. M_i consists of 2^i standardly embedded solid double tori.
2. Each element of M_i is contained in the interior of a 3-cell contained in an element of M_{i-1}.
3. For each element of M_{i-1}, one of the two solid double tori of M_i which it contains has diameter less than $1/i$.
4. Let C be a solid double torus in M_i. Then C is divided into a chain consisting of two solid tori at the ends and 3-cells in between. (See Figure 4.1) Call those ordered links $\{c_i\}_{i=1}^{n_c}$. Then each c_i has diameter less than $1/i$.

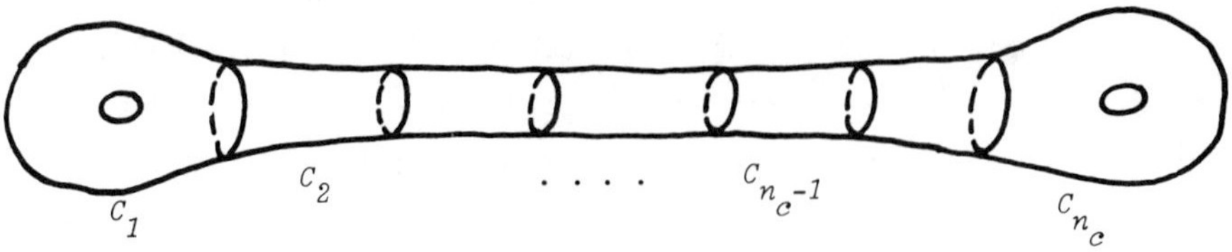

Fig. 4.1.

5. Suppose E is a double torus belonging to M_i which is contained in a double torus C belonging to M_{i-1}. Let $\{e_i\}_{i=1}^k$ and $\{c_j\}_{j=1}^n$ be the links of E and C respectively. Then E runs straight through C in the following sense. If for some i and j, $e_i \cap c_j \neq \emptyset$, then for each $r < i$, $e_r \cap c_{j+2} = \emptyset$.

6. M_i has the two disk property.

PROPOSITION 4.1. If G is a decomposition of E^3 described by a sequence of collections $\{M_i\}_{i\in\omega}$ satisfying the above properties, then G is an upper semi-continuous decomposition of S^3 into points and a null sequence of cellular arcs for which S^3/G is not homeomorphic to S^3.

Proof. Property 2 guarantees that each element of G is cellular. Property 3 implies that the non-degenerate elements of G form a null sequence. Properties 4 and 5 imply that the elements of G are points and arcs. Property 6 implies by Theorem 2.1 that S^3/G is not homeomorphic to S^3.

It remains only to construct the defining sequence $\{M_i\}_{i\in\omega}$ which satisfies Properties 1-6.

Description of the M_i's. Recall that collection M_1 lies in the solid torus T as illustrated in Figure 2.1. One can choose these double tori so that they can be divided into links satisfying Property 4.

Lemma 3.3 together with Lemma 3.1 implies that M_1 has the two disk property.

For each $i \geq 1$, M_{i+1} is obtained from M_i in three steps as follows.

Step 1. Replacing double tori by tori. For each solid double torus H in M_i find a torus W in H embedded as illustrated in Figure 4.2.

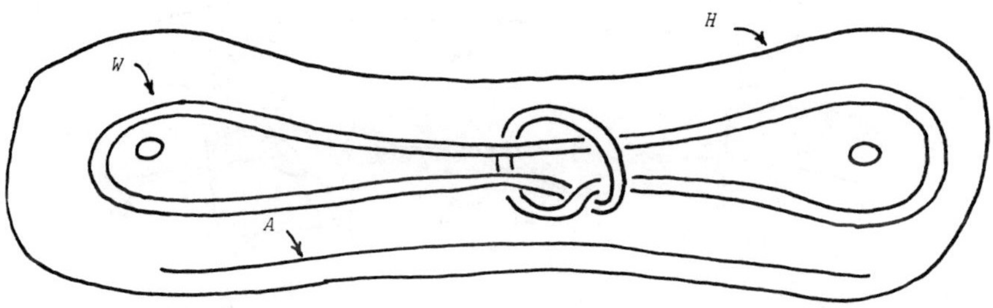

Fig. 4.2.

By Lemma 3.5, the collection M' obtained from M by making the above replacements also has the two disk property.

Step 2. Replacing each torus by a double torus and a cube with many handles. Each element of M_i' is a solid torus W contained in a solid double torus H which is an element of M_i.

Find an arc A in Int $H - W$ so that A intersects each link of H in one subarc of A. In W put a chain of small eyeglasses, one called Z, the others called $\{Y_i\}$, so that each eyeglass intersects at most two links of H and so that the eyeglasses are embedded in W as described in Lemma 3.3. For each eyeglass Y_i in W find an arc E_i joining that eyeglass to A as illustrated in Figure 4.3 so that E_i lies in one link of H, $E_i \cap A$ is one point, and so that the set Y, equal to $A \cup (\cup E_i) \cup (\cup Y_i)$ can be isotoped in H down near A. Therefore Y is contained in a cell contained in H. A regular neighborhood of Y is a standardly embedded cube with handles. By Lemmas 3.3 and 3.1, the collection M''_i, obtained from M'_i by replacing each solid torus W of M'_i by regular neighborhoods of the two objects Y and Z, still has the two disk property.

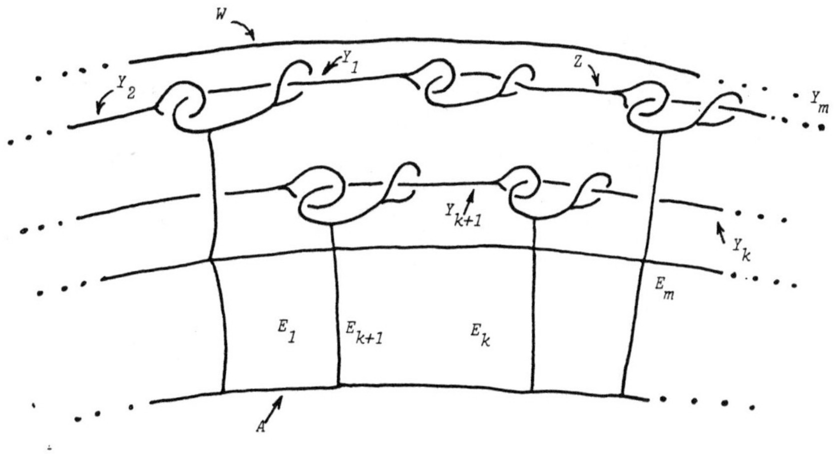

Fig. 4.3.

Step 3. Replacing each cube with many handles by a solid double torus. Use Lemma 3.7 to make this last replacement in such a manner that Property 5 is satisfied. This replacement produces a collection M_{i+1} which satisfies all six Properties so the example is complete.

REFERENCES

1. Armentrout, Steve, Decompositions of E^3 with a compact 0-dimensional set of nondegenerate elements, Trans. Amer. Math. Soc., 123 (1966), 165-177.
2. Bing, R. H., A decomposition of E^3 into points and tame arcs such that the decomposition space is topologically different from E^3, Ann. of Math., 65 (1957), 484-500.
3. ________, Upper semicontinuous decompositions of E^3, Ann. of Math., 65 (1957), 363-374.
4. ________, Necessary and sufficient conditions that a 3-manifold be S^3, Ann. of Math., 68 (1958), 17-37.
5. ________, A wild surface each of whose arcs is tame, Duke Math. J., 28 (1961), 1-16.
6. ________, Point-like decompositions of E^3, Fund. Math., 50 (1962), 431-453.
7. Everett, Dan, Shrinking countable decompositions of E^3 into points and tame cells, (these Proceedings).
8. Gillman, David S., and Martin, Joseph M., Countable decompositions of E^3 into points and point-like arcs, Notices Amer. Math. Soc., 10 (1963), 74.
9. Starbird, Michael, Cell-like 0-dimensional decompositions of E^3, Trans. Amer. Math. Soc. (to appear).
10. Starbird, Michael and Woodruff, Edythe P., Decompositions of E^3 with countably many nondegenerate elements, (these Proceedings).
11. Woodruff, Edythe P., Decomposition spaces having arbitrarily small neighborhoods with 2-sphere boundaries, Trans. Amer. Math. Soc., (to appear).
12. ________, Decomposition of E^3 into cellular sets, (these Proceedings).

ILLINOIS JOURNAL OF MATHEMATICS
Volume 30, Number 4, Winter 1986

DECOMPOSITIONS THAT DESTROY SIMPLE CONNECTIVITY

BY

R.H. BING[†]

We shall be concerned with a monotone decomposition of R^3 with only one nondegenerate decomposition element X. We use g to denote the decomposition map and $g(R^3)$ the decomposition space. Also, D denotes a disk. To determine if $g(R^3)$ is simply connected we shall be concerned with whether maps of Bd D into $g(R^3)$ can be extended to D.

At the Summer Institute on Set Theoretic Topology at Wisconsin in 1955 I gave a talk entitled "What topology is here to stay" in which I envisioned decompositions of R^3 as a very viable area for research. I mentioned R.L. Moore's monotone decomposition theorem [3] for S^2 which states that if G is a nondegenerate upper semicontinuous decomposition of S^2 each of whose elements is a continuum that does not separate S^2, then the decomposition space is S^2. I pointed out that the theorem was false if one replaced S^2 by S^3 and gave as an example the decomposition whose only nondegenerate element is a circle. The earlier version of the Summary of Lectures and Seminars [1] reported on page 26 that the reason I gave that the decomposition space differed from S^3 was that it *is not simply connected*. The second printing of [1] made the correction by replacing the *is not simply connected* part of the statement by *does not remain simply connected* on the removal of some point. It was also claimed there and in [2] that the decomposition space of S^3 (or R^3) whose only nondegenerate element is a solenoid is not simply connected. When I was assembling copies of my publications it was called to my attention that a proof of this claim had not been published. It is the purpose of this paper to fill that gap. Other claims were made in [2] about the simple connectivity of other monotone decompositions (perhaps with many nondegenerate elements) of R^3, but we shall not treat them in this paper.

Richard Skora read an early draft of this paper and made valuable suggestions for improving some proofs.

1. X is a standard solenoid

In this case X is the intersection of smooth unknotted tori $T_1, T_2, \ldots$ where T_{i+1} winds around T_i smoothly more than once, the meridional cross sections

Received July 16, 1985.

[†] Died April 28, 1986.

of T_{i+1} are round planar disks that lie in those of T_i, and the diameters of the meridional cross sections of the T_i's converge to 0 as i increases without limit. Sometimes the restriction of "more than once" is omitted and a circle is permitted to be a solenoid—but we shall not do that in this paper.

THEOREM 1. *If X is a standardly embedded solenoid, $g(R^3)$ is neither simply connected nor locally simply connected.*

Proof. If we seek a map f of Bd D into $g(R^3)$ that cannot be extended to D, we should seek one such that $g(X) \in f(\text{Bd } D)$ because if $g(X) \notin f(\text{Bd } D)$ there is an extension of $g^{-1}f$ on Bd D to take D into R^3. This extension followed by g would extend f to map D into $g(R^3)$.

Let pq be an arc in a meridional cross section of T_1 that intersects X only in its end points where these end points belong to different arc components of X. We show that $g(R^3)$ is not simply connected by showing that a homeomorphism f of Bd D onto $g(pq)$ cannot be extended to map D into $g(R^3)$. Since for each open subset U of $g(R^3)$ containing $g(X)$ there is a pq with $g(pq)$ in U, this will also show that $g(R^3)$ is not locally simply connected. See Figure 1.

Assume f is a homeomorphism of Bd D onto $g(pq)$ and that f can be extended to a map F of D into $g(R^3)$. It would be nice if $F^{-1}(g(X))$ were 0-dimensional, so we adjust F to get a new map F_2 where $F_2^{-1}(g(X))$ is 0-dimensional. First, let F_1 be a map of D into $g(R^3)$ such that $F_1 = F$ on the component of $D - F^{-1}(g(X))$ intersecting Bd D and F_1 takes the rest of D to $g(X)$. Next we let k be a map of D onto itself that is the identity on Bd D and whose point inverses are the components of $F_1^{-1}(g(X))$ and points of $D - F_1^{-1}(g(X))$. Moore's decomposition theorem [3] mentioned earlier is used to get k. Then $F_2 = F_1k^{-1}$. For simplicity we suppose $F = F_2$.

Let a_0b_0 be a spanning arc of D such that $F(a_0b_0)$ misses $g(X)$ and the subdisk D_0 of D bounded by the union of a_0b_0 and the subarc of Bd D from a_0 to b_0 through $f^{-1}(g(X))$ lies in $F^{-1}(g(\text{Int } T_1))$. Since $g^{-1}(F(a_0))$ and $g^{-1}(F(b_0))$ lie in the same meridional cross section of T_1, we can speak of the number of times that $g^{-1}F(a_0b_0)$ winds around T_1.

For some large r let a_rb_r be a spanning subarc of D_0 such that a_r lies on BdD between a_0 and $f^{-1}g(X)$, b_r lies on Bd D between b_0 and $f^{-1}g(X)$, $F(a_rb_r)$ misses $g(X)$, and $F(a_rb_r)$ lies in $g(T_r)$. Let D_r be the subdisk of D_0 bounded by union of a_0b_0, a_rb_r and two subarcs of Bd D. See Figure 1.

Since p and q belong to different arc components of X, for large r, $g^{-1}F(a_rb_r)$ winds around T_1 many times—even more than $g^{-1}F(a_0b_0)$ does. Let $y(r)$ be the number of times that $g^{-1}(F(\text{Bd } D_r))$ winds around T_1. Suppose r is so large that $y(r) > 0$.

Let $z(s)$ be the number of times that T_s winds around T_1. We suppose s is so large that $z(s) > y(r)$ and $g^{-1}F(\text{Bd } D_r)$ misses Bd T_s. We suppose that on D_r near $F^{-1}(g(\text{Bd } T_s))$, F has enough general position so that $D \cap F^{-1}g(\text{Bd } T_s)$ is the union of a finite number of mutually disjoint simply closed

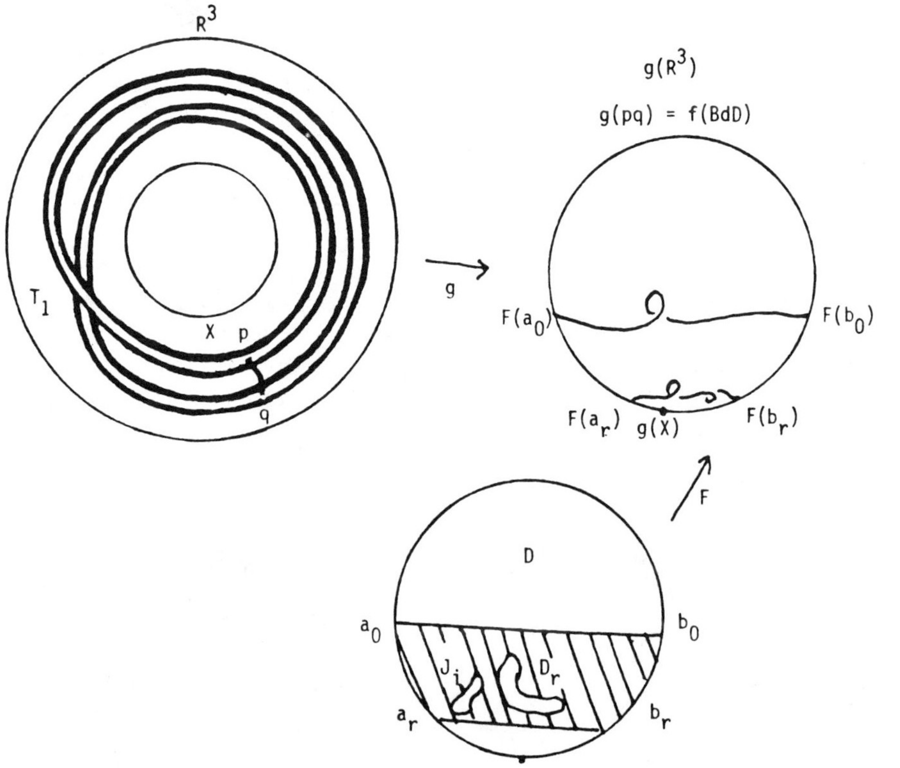

FIG. 1

curves $J_1, J_2, \ldots, J_n$. Since each $g^{-1}F(J_i)$ lies on Bd T_s, it winds around T_1 some integral multiple of $z(s)$.

Let E_r be the finitely holed D_r obtained by deleting from D_r the interiors of the subdisks of D_r bounded by the J_i's. We now come to the contradiction caused by the assumption tha f on Bd D could be extended to F on D. The boundary of each of the holes of E_r winds around T_1 some integral multiple of $z(s)$, but $y(r)$ is not an integral multiple of $z(s)$.

2. X is an embedded solenoid

The complement of an embedded solenoid may be quite different from the complement of a standardly embedded solenoid. We no longer can speak of tori about the embedded solenoid. However, we still find that the decomposition space is not simply connected.

THEOREM 2. *If X is an embedded solenoid, $g(R^3)$ is neither simply connected nor locally simply connected.*

Proof. We use X' to denote the standardly embedded solenoid of Theorem 1, $p'q'$ the arc called pq there, and g' the decomposition map called g there. Let β be a homeomorphism of X onto X' and pq an arc from $\beta^{-1}(p')$ to $\beta^{-1}(q')$ in R^3 that intersects X only at p and q. This is possible since the dimension of X is 1. Extend β to a map of R^3 onto itself that takes pq homeomorphically to $p'q'$. For convenience, call the extension β. It is a map rather than a homeomorphism. We finish the proof of Theorem 2 by showing that $g(pq)$ cannot be shrunk to a point in $g(R^3)$.

Let f be a homeomorphism of Bd D onto $g(pq)$. Assume that f can be extended to map F taking D into $g(R^3)$. This leads to the contradiction that a homeomorphism of Bd D onto $g'(p'q')$ can be extended to a map $g'\beta g^{-1}F$ of D into $g'(R^3)$. See Figure 2. The proof of Theorem 1 showed that no homeomorphism of Bd D onto $g'(p'q')$ could be extended to map D into $g'(R^3)$.

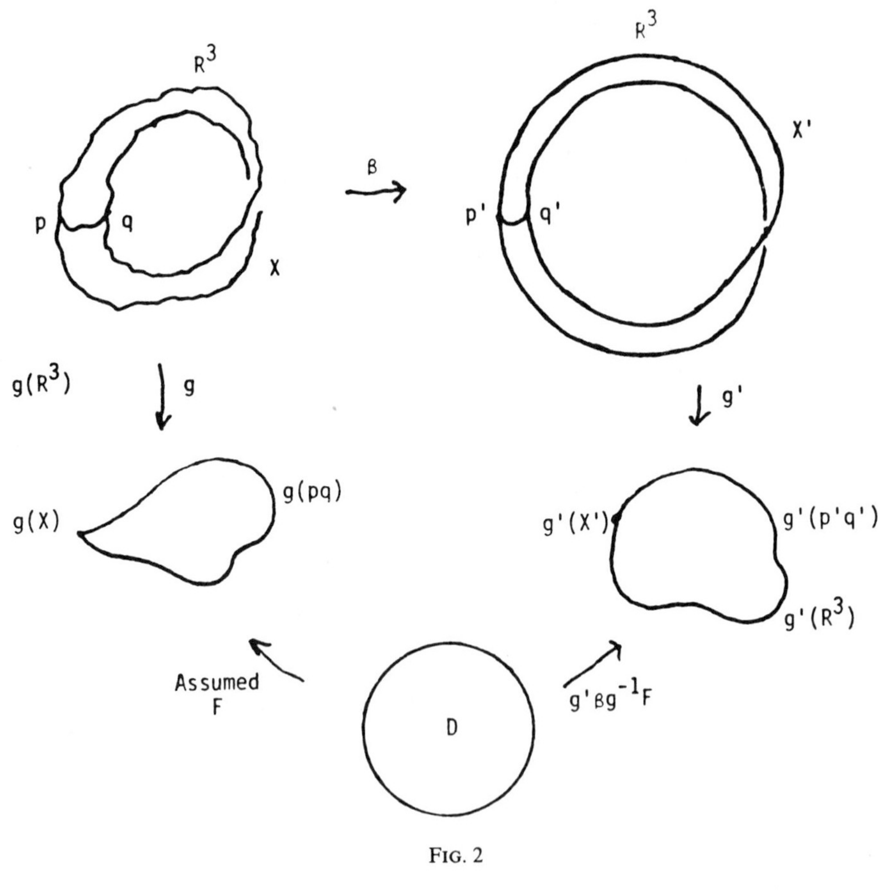

FIG. 2

Since pq could be picked close to X, this also shows that $g(X)$ is not locally simply connected.

3. X is unlike-a-solenoid

We say that a disk D can be converted to a disk with finitely many holes E if there is a finite collection of mutually disjoint disks in Int D and E is obtained from D by removing the interiors of these subdisks. These interiors are called holes in D and E is called a *finitely holed* D. We call D a finitely holed D even if there are no holes and $D = E$.

Let N_i be the $1/i$-neighborhood of X in R^3—that is, the set of points of R^3 whose distance from X is less than $1/i$.

If $f(\text{Bd } D) \subset N_i - X$, we say that *$f$ can be pulled in $N_i - X$ arbitrarily close to X* if for arbitrary large s, f can be extended to take a finitely holed D into $N_i - X$ so that the boundary of each hole is sent into N_s. Note that s is picked before the extension. If a different s had been chosen, we might have needed a different extension. (If f can be extended to map D into N_i, we could have picked an extension independent of s. If f can be extended to map D into $N_i - X$, then technically the definition says that f can be pulled arbitrarily close to X even though $f(D)$ misses X. This is because we call D a finitely holed D.) We say that *X is unlike-a-solenoid* if for each N_i there is an $N_{r(i)}$ such that each map f of Bd D into $N_{r(i)} - X$ can be pulled in $N_i - X$ arbitrarily close to X.

THEOREM 3. *If X is unlike-a-solenoid, then $g(R^3)$ is simply connected and locally simply connected.*

Proof. We first show that $g(R^3)$ is locally simply connected at $g(X)$. We show that if U is a neighborhood in $g(R^3)$ of $g(X)$, there is a neighborhood V of $g(X)$ such that each map f of Bd D into V can be extended to take D into U. We use f as a map of Bd D into V and $g^{-1}f$ to send Bd D into R^3.

To show that $g(R^3)$ is locally simply connected, without loss of generality we pick U to be $g(N_j)$ and V to be $g(N_{r(j)})$ where $r(j)$ is an integer such that any map of Bd D into $N_{r(j)} - X$ can be pulled in $N_j - X$ arbitrarily close to X on a finitely holed D.

We consider the sequence $n_1, n_2, \ldots$ where $n_1 = j$, and $n_{i+1} = r(n_i)$. Although the r's are defined for maps of Bd D into $R^3 - X$, we realize that in the case of the f of Bd D into $V = g(N_{r(j)})$ we wish to extend it to a map from D into $U = g(N_j)$ and this $f(\text{Bd } D)$ may contain $g(X)$.

It may be that $f^{-1}(g(X))$ has several components. We wish to avoid this. With that purpose in mind we suppose D is a round planar disk and let C be the convex hull of $f^{-1}(g(X))$ and partially extend f to send C to $g(X)$. For convenience we call the extension f. Now f is defined except on a collection

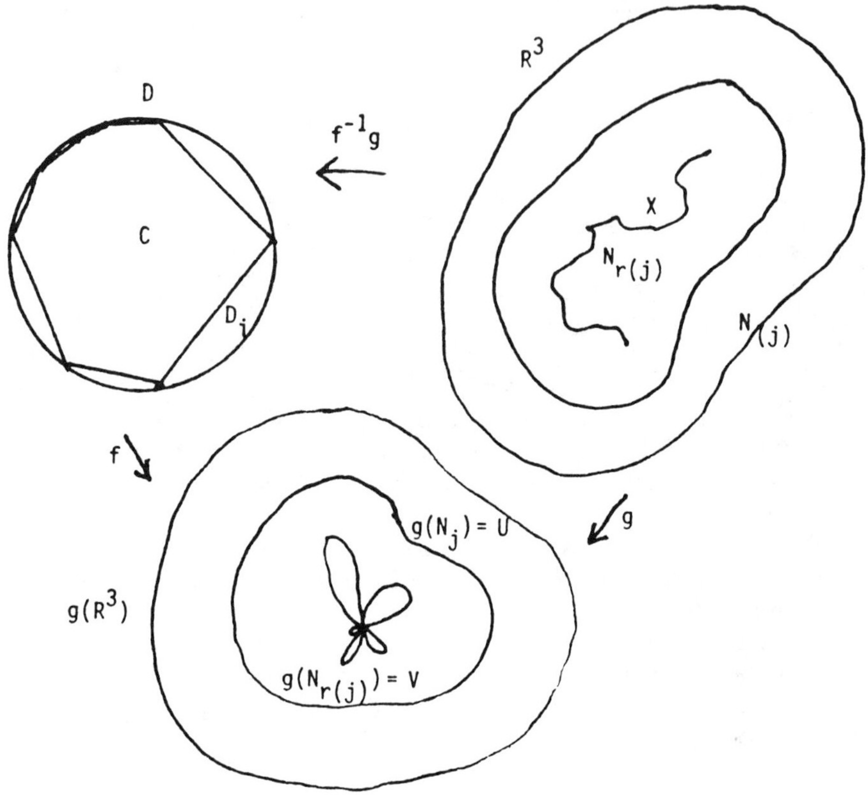

FIG. 3

(possibly infinite) of open disks. We call these disks D_i's. For each of these disks, f sends the straight subarc of its boundary to $g(X)$ and the open curved part into $g(R^3 - X)$. We have simplified the situation so that $f^{-1}g(X)$ intersects each Bd D_i in a connected set. See Figure 3.

Let D_i be one of the subdisks on whose interior f has not been defined. Each Bd $D_i \subset g(N_{n_2})$ and for all but a finite number of these D_i's $f(\text{Bd } D_i) \subset g(N_{n_3})$. For this finite number where $f(\text{Bd } D_i) \not\subset g(N_{n_3})$, we extend f to a part of their interiors.

Let ab be a spanning arc of D_i that cuts D_i into two subdisks D_i', D_i'' where $f(\text{Bd } D_i' - ab)$ misses $g(X)$ and the closure of $f(\text{Bd } D_i'' - ab)$ lies in $g(N_3)$. Extend f to take ab into $g(N_{n_3} - X)$. Extend f further on a finitely holed D_i' to take the finitely holed D_i' into $g(N_{n_1} - X)$ where the boundary of the holes go into $g(N_{n_3})$. To get the extension we use the hypothesis that X is unlike-a-solenoid.

We now find that f is defined except on open disks whose boundaries are sent by f into $g(N_{n_3})$. In fact on all but a finite number of these disks, f sends their boundaries into $g(N_{n_4})$. We extend f into a part of the interiors of this finite collection so that now f is defined except on open disks whose boundaries are sent into $g(N_{n_4})$. Using the hypothesis that X is unlike-a-solenoid we pick the extension on the new part to take this new part into $g(N_{n_2})$.

The extension is extended a countable number of times and finally we define the extension to take the remaining part of D to $g(X)$. We have now shown that $g(R^3)$ is locally simply connected. That it is simply connected follows from the following theorem.

THEOREM 4. *If $g(R^3)$ is locally simply connected, it is simply connected.*

Proof. Let f be a map of Bd D into $g(R^3)$. We show that f is simply connected by showing that f can be extended to map D into $g(R^3)$.

It follows from the local simple connectivity of $g(R^3)$ that there is a neighborhood U of $g(X)$ such that each map of Bd D into U can be extended to map D into $g(R^3)$. Suppose D is a round disk and $A_1, A_2, \ldots, A_n$ are the components of Bd $D - f^{-1}g(X)$ that are not sent into U by f. Let B_i be an open arc on A_i such that each $f(A_i - B_i) \subset U$. Let C_i be a straight arc in D joining the two components of $A_i - B_i$.

Extend f to C_i so that the extension takes C_i into $U - g(X)$. Call the extension f. On the subdisk D_i of D bounded by C_i and a part of A_i, extend $g^{-1}f$ to take each D_i into R^3 and follow this extension by g to extend f to take D_i into $g(R^3)$. The boundary of the remaining part of D is sent by f into U so the local connectivity of $g(R^3)$ shows that f can be extended to the rest of D. This shows that $g(R^3)$ is simply connected.

4. X is solenoid-like

We say that X is solenoid-like if there is a neighborhood N of X in R^3 such that for any neighborhood N' of X there is a map f of Bd D into $N' - X$ which cannot be pulled in $N - X$ on a finitely holed D arbitrarily close to X. One might note if X is solenoid-like, then it is untrue that X is unlike-a-solenoid.

THEOREM 5. *If X is solenoid-like, $g(R^3)$ is not locally simply connected.*

Proof. We show that for some neighborhood $U = g(N)$ of $g(X)$, and each smaller neighborhood $V = g(N')$ of $g(X)$ there is a map of Bd D into V that cannot be extended to map D into U. Here we use N, N', f as in definition of solenoid-like and use gf for the map of Bd D into $V - g(X)$ that cannot be extended to map D into U. If gf could be extended by F to send D into U, $g^{-1}F$ would show that f can be pulled in $N - X$ on a finitely holed D arbitrarily close to X.

Question Recall that X is a continuum in R^3 and $g(R^3)$ is the decomposition space whose only nondegenerate point inverse is X. Is $g(R^3)$ locally simply connected if it is simply connected? If $g(R^3)$ is not locally simply connected, could it be simply connected?

5. Necessary and sufficient conditions

Theorems 3 and 5 provide a necessary and sufficient condition that $g(R^3)$ is not locally simply connected. However the condition is dependent on the embedding of X and does not say that if X' is homeomorphic to X then $g(R^3) = R^3/X$ is locally simply connected if and only if R^3/X' is.

THEOREM 6. *A necessary and sufficient condition that* $g(R^3)$ *not be locally simply connected is that* X *be solenoid-like.*

Proof. The sufficiency is provided by Theorem 5 and the necessity by Theorem 3.

THEOREM 7. *If* X *and* X' *are homeomorphic continua in* R^3 *and* $\dim X = 1$, *then* $g(R^3) = R^3/X$ *fails to be locally simply connected if and only if* $g'(R^3) = R^3/X'$ *does.*

Proof. The proof of Theorem 7 is modelled after that of Theorem 2.

Suppose $g'(R^3)$ is not locally simply connected. Then it is solenoid-like and there are a neighborhood N' of X' and a sequence of mutually disjoint simple closed curves $J_1', J_2', \ldots$ in $N' - X'$ such that J_1' lies in the $1/i$ neighborhood of X' and a map f_i' of Bd D onto J_i' that cannot be pulled in $N' - X'$ arbitrarily close to X' on a finitely-holed D.

Let β be a homeomorphism of X onto X'. We now pick a sequence of simple closed curves $J_1, J_2, \ldots$ in $R^3 - X$ so that β can be extended to a homeomorphism taking $X \cup J_1 \cup J_2 \cup \ldots$ onto $X' \cup J_1' \cup J_2' \cup \ldots$. We assume $J_1, J_2, \ldots, J_{i-1}$ have been found with β extended to them and describe J_i and β on it.

Express J_i' as the union of arcs $a_1'a_2', a_2'a_3', \ldots, a_n'a_{n+1}'$ each of diameter less than $1/i$. Let b_j' be a point of X' in the $1/i$ neighborhood of a_j'. Note that the distance between two adjacent b_j''s is less than $3/i$. Let a_j be a point of

$$R^3 - (X \cup J_1 \cup \cdots \cup J_{i-1})$$

in the $1/i$ neighborhood of $\beta^{-1}(a_j')$. Note that distance between two adjacent a_j's is less than $\varepsilon + 2/i$ where ε is a positive number such that the image under β^{-1} of any $3/i$-subset of X' has diameter less than ε. Hence adjacent a_j's are close if i is large. We suppose the a_j's are distinct and let J_i be a simple closed curve in

$$R^3 - (X \cup J_1 \cup J_2 \cup \cdots \cup J_{i-1})$$

which is the union of arcs $a_1a_2, a_2a_3, \ldots, a_na_{n+1}$ where a_ja_{j+1} lies in the $1/i$ neighborhood of the straight line interval from a_j to a_{j+1}. It is in showing that there are such a_ja_{j+1}'s in $R^3 - (X \cup J_1 \cup J_2 \cup \cdots \cup J_{i-1})$ that we use the fact that dim $X = 1$. The homeomorphism β is extended to J_i so that $\beta(a_j) = a_j'$.

Suppose all J_i's are defined. Call the extended homeomorphism β. Now extend β to a map taking R^3 onto R^3 and call this extension β also. This final extension need not be a homeomorphism. Let f_i be a homeomorphism of Bd D onto J_i such that $\beta f_i = f_i'$.

Assume that $g(R^3) = R^3/X$ is locally simply connected. We prove the theorem by showing that this assumption is false. Local simple connectivity of $g(R^3)$ implies that for a large r, gf_r on Bd D can be extended to a map F_r taking D into $g\beta^{-1}(N')$. Let C be the set of points of D that are carried by $\beta g^{-1}F_r$ onto X' or points of X'. Change D to E, a finitely holed D in $D - C$, so that the boundaries of the holes of E are very close to C. Although $\beta g^{-1}F_r$ need not be a map on D, it is one on E. Also $\beta g^{-1}F_r = f_r'$ on Bd D. Hence $\beta g^{-1}F_r$ on E shows that f_r' on Bd D can be pulled in $N' - X'$ arbitrarily close to X on a finitely holed D. This contradicts the selection of J_r' and f_r'.

REFERENCES

1. *Summary of Lectures and Seminars*, Summer Institute on Set Theoretic Topology, University of Wisconsin, 1955, revised 1958.

2. R.H. BING, *Conditions under which monotone decompositions of E^3 are simply connected*, Bull. Amer. Math. Soc., vol. 63 (1957), 143.

3. R.L. MOORE, *Concerning upper semicontinuous collections of continua*, Trans. Amer. Math. Soc., vol. 27 (1925), pp. 416–428.

THE UNIVERSITY OF TEXAS AT AUSTIN
AUSTIN, TEXAS

Shrinking without lengthening

R.H. Bing

It was shown in [**2**] that the union of two solid Alexander horned balls sewed together along their boundaries with the identity map is a 3-sphere. (Being sewed together with the identity map means that there is a homeomorphism of one of the solid Alexander balls onto the other and the identification (or sewing) is done on their boundaries with this homeomorphism.) This led to the conclusion that there is a homeomorphism of period two (an involution) of a 3-sphere onto itself having as its fixed point set a wild 2-sphere. Such an involution is not equivalent to any PL or differential homeomorphism and is considered wild. (Two homeomorphisms h_1, h_2 are equivalent if there is a homeomorphism h such that $h_1 = h^{-1}h_2h$.) Later it was discovered [**1,3,5**] that there are uncountably many mutually inequivalent wild involutions and indeed there are wild periodic homeomorphisms of S^3 onto itself of all periods.

The original description of the union of the Alexander horned balls considered each to appear as shown in Figure 1.

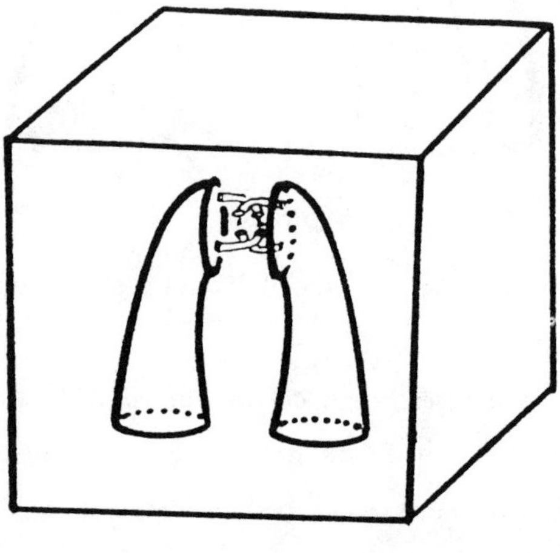

Figure 1.

1

It was noted that this Alexander horned ball is the decomposition space of a 3-cell I^3 as shown in Figure 2 where the nondegenerate elements of the decomposition are tame arcs.

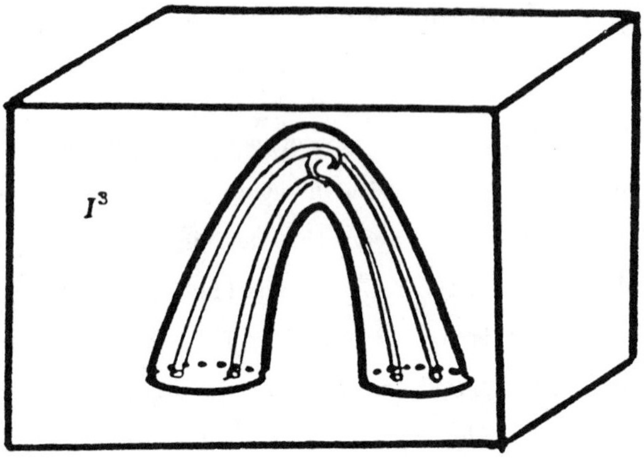

Figure 2.

Each of these arcs is the intersection of a decreasing sequence of folded cylinders $C, C_i, C_{ij}, \ldots$ with bases on the base of I^3 as shown in Figure 3,

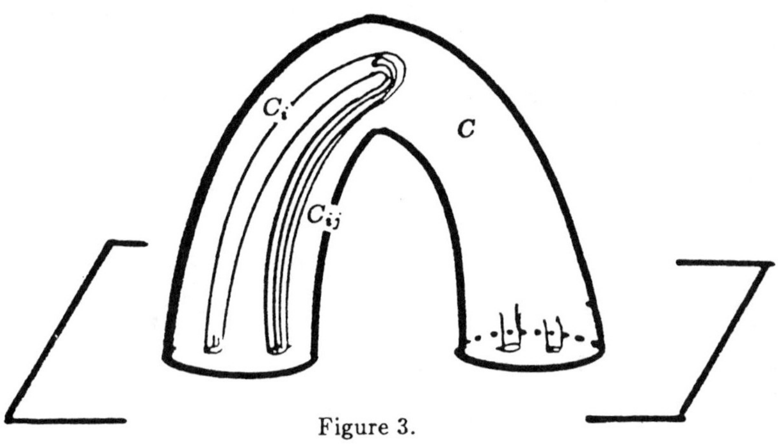

Figure 3.

but each arc has just one end on the base of I^3 and each horizontal cross section of each arc is just one point.

It is easier to visualize the union of two copies of I^3 with their associated decompositions than it is to picture the union of two Alexander horned balls. The union of the decomposition spaces would be a decomposition of $I_1^3 \cup I_2^3 = S^3$ whose

2

nondegenerate elements are tame arcs each of which is the intersection of solid tori $T, T_i, T_{ij}, \ldots$ as shown in Figure 4. We say that $T, T_i, T_{ij}, \ldots$ are tori at the 0th, 1st, 2nd, ... stages respectively.

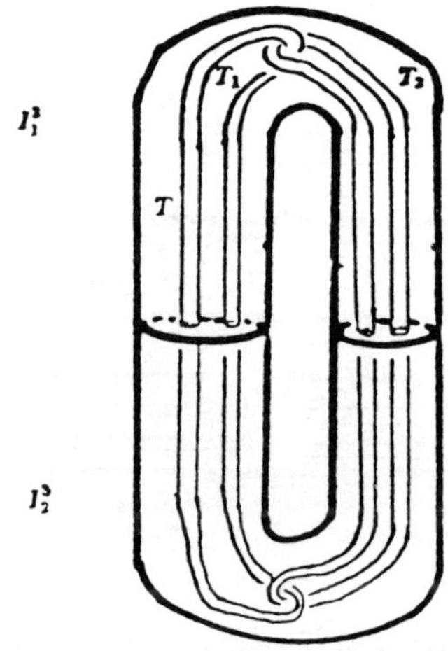

Figure 4.

It was shown in [**2**] that the decomposition space for Figure 4 is a 3-sphere by showing that for each $\epsilon_1 > 0$, there is an integer n_1 such that there is a homeomorphism h_1 of S^3 onto itself fixed outside T such that the images of the solid tori at the n_1th stage have diameters less than ϵ_1. (See Theorem 1 below.) The homeomorphism h_1 was followed by a homeomorphism h_2 that agrees with h_1 outside the solid tori at the n_1th stage but shrinks solid tori at some n_2th stage to images with diameters less than $\epsilon_2 = \epsilon_1/2$. Iterating the procedure gives a decomposition map $(\lim h_1, h_2, \ldots)$ showing that the decomposition space is a 3-sphere and indeed, the union of two Alexander horned balls sewed together on their boundaries with the identity map is a 3-sphere.

In getting the homeomorphisms h_1 to shrink the tori at the n_1th stage, no effort was made to decrease the lengths of their centerlines. Let us consider the case where T is long and thin (so that we can ignore vertical distances) and the length of the centerline of T is 2ϵ. Suppose we want to get a homeomorphism h that is fixed on BdT and takes the inside T's at some future stage to images with diameters less than $\epsilon/4$. Let us consider several ways to proceed.

One such way might be to move T_1 and T_2 about as shown in Figure 5 so that the new T_i's have diameters less than $\epsilon/2$. This method is sometimes called the "standard mistake" because there is no way to move the new T_{i1} and T_{i2} about in the new T_i to significantly reduce their diameters.

Figure 5.

To avoid this apparent trap a different procedure was followed in [**2**]. Rather than get new T_i's at the first step with diameters less than $\epsilon/2$, we contented ourselves to get them with diameter less than 3/4 as shown in Figure 6.

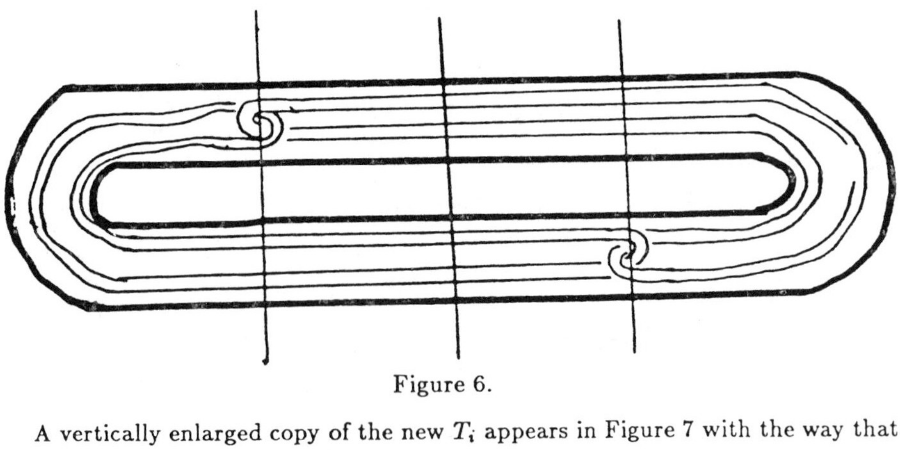

Figure 6.

A vertically enlarged copy of the new T_i appears in Figure 7 with the way that the new T_{i1} and T_{i2} might appear in it. Each of the new T_{i1} and T_{i2} has a diameter less than $\epsilon/2$ (even though the length of the centerline of T_{i1} is more than 2ϵ). It is an easy task to place a new T_{i21} and a new T_{i22} in the new T_{i2} (using the

4

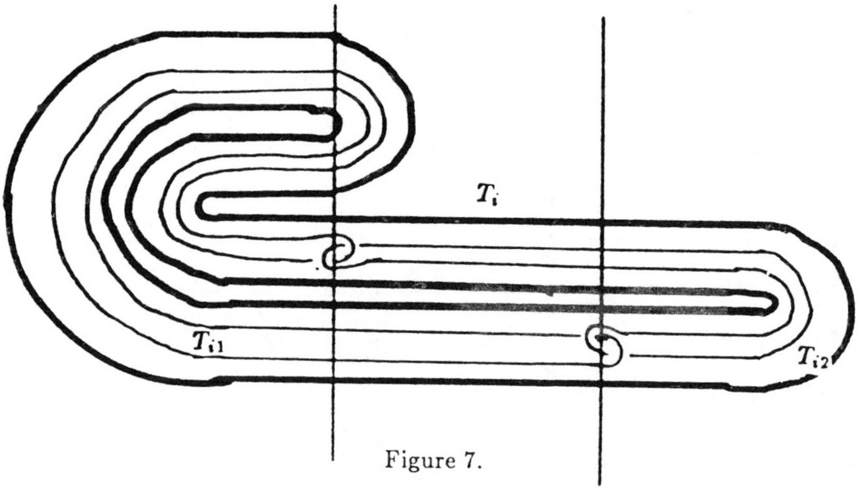

Figure 7.

"standard mistake") to make them have diameters less than $\epsilon/4$ and Figure 8 shows how new T_{i1} and T_{i12} could be placed in the new T_{i1} to make them have diameters less than $\epsilon/4$. (We hesitate to show all of T_{i11} since it is so long.)

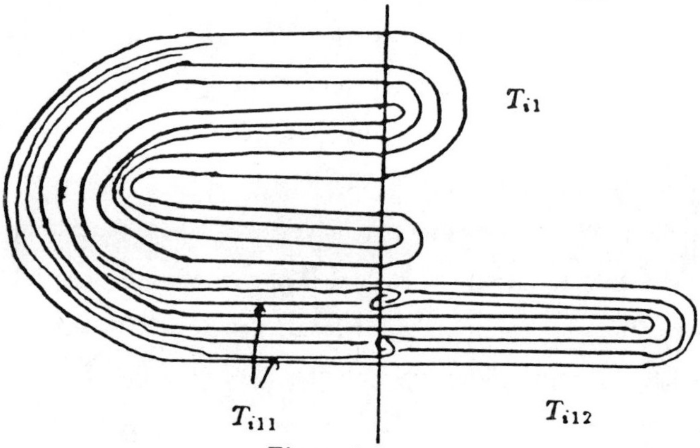

Figure 8.

By using an arbitrary number n in place of 4 and following a similar procedure we can get new T's so that for $m = 1, 2, \ldots, n-1$ the new T's at the mth stage have diameters less than $1 - (m/n)$. Hence, we have the following result.

Theorem 1. *Suppose the length of the centerline of $T \leq 2\epsilon$ and n is a positive integer. Then there is a homeomorphism h of T onto itself that is fixed on $Bd\,D$*

5

such that for each $T_{ij\ldots k}$ at the nth stage

$$\text{diameter } h(T_{ij\ldots k}) < \epsilon/n .$$

A homeomorphism h such as we have described to prove Theorem 1 would have increased the lengths of the images of centerlines of some of the $T_{1ij\ldots k}$'s drastically. Michael Freedman asked if there is a homeomorphism h that does not. We give an affirmative answer in Theorem 2 in the case where n is of the form 2^m. Theorem 2 has had applications in [**4**].

Theorem 2. *Suppose the length of the centerline of $T \leq 2\epsilon$ and m is a nonnegative integer. Then there is a homeomorphism h of T onto itself that is fixed on BdT and such that for each integer m with $0 \leq m \leq n$ and for each $T_{ij\ldots k}$ at the 2^mth stage*

$$\text{diameter of } h(T_{ij\ldots k}) < \epsilon/2^m \text{ and}$$

$$\text{length of centerline of } h(T_{ij\ldots k}) < 2\epsilon .$$

Description of h. By reselecting the centerline of T we can suppose that the reselected centerline S has length less than 2ϵ. Each of the centerlines S_1, S_2 of the adjusted T_1, T_2 runs parallel to one half of S, makes a turn, then runs back parallel to the same half, and finally makes a turn to connect to the start. We say that S_1 and S_2 "hook elbows" at the turns. The lengths of the turns are so short that they will be ignored. We suppose that the halves of S_1 and S_2 are so close and parallel to halves of S that the length of each S_i is about twice the length of one half of S. By cutting corners we could make the lengths of S_1 and S_2 less than the length of S but shall not concern ourselves with this. We note that S_1 and S_2 are each so close to one half of S that their diameters are less than ϵ. Hence, we suppose that the diameters of the new T_i's at the 1st stage are less than $\epsilon = \epsilon/2^0$ where $2^0 = 1$.

Centerlines $S_{ij\ldots k}$ of each new $T_{ij\ldots k}$ will zig-zag along S and be so close and parallel to S that we ignore the lengths at the turns and estimate the length of $S_{ij\ldots k}$ by the sum of the lengths of the projections of its pieces on S. The bends in the zig-zagging of $S_{ij\ldots k}$ are determined by two kinds of turns — those that occur at "linking elbows" and those forced by the fact that the centerlines lie in

6

zig-zagging tubes. These interior turns in the zig-zagging tubes were created by "hooking elbow" turns at previous stages. A "hooking elbow" turn at one stage (as well as an interior turn at that stage) causes an interior turn at the next.

Our goal is to describe new $T_{ij\ldots k}$'s (which for convenience we call $T_{ij\ldots k}$'s) so that for each integer m with $0 \leq m \leq n$, the diameters of the $T_{ij\ldots k}$'s at the 2^mth stage are less than $\epsilon/2^m$.

Each $S_{ij\ldots k}$'s will be the union of two arcs joined at their ends and which otherwise run parallel to each other and to the same zig-zagging arc $P_{ij\ldots k}$. We regard the length of $S_{ij\ldots k}$ to be about twice the length of $P_{ij\ldots k}$ and the diameter of $T_{ij\ldots k}$ to be about the same as the diameter of $P_{ij\ldots k}$. If $P_{ij\ldots k}$ is of length less than ϵ and it bends only at points that divide it into 2^m equal pieces, we suppose that the diameter of $P_{ij\ldots k}$ (and hence that the diameter of $T_{ij\ldots k}$) is less than $\epsilon/2^m$.

We use Figure 9 to show how S_{i1} and S_{i2} lie in T_i. Note that we have used the "standard mistake" but we are only at the 2^1th stage and only require that diameters be less than $\epsilon/2^1$.

Figure 9.

We draw $P_{ij\ldots k}$ as straight or curved depending on which is convenient. We place circles on $P_{ij\ldots k}$ to show where it bends. Figure 10a shows P_i straight (with no circle on it and P_{ij} with one bend (and one circle) and P_{ijk} with two bends (and two circles). Figure 10b shows P_{ijk} with two circles and P_{ijks} with three.

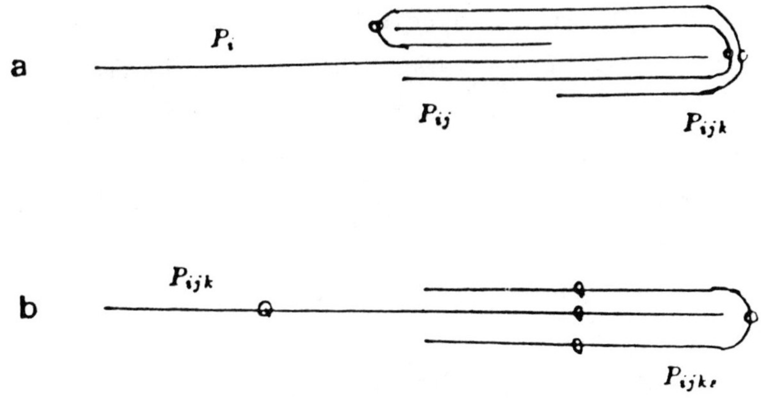

Figure 10.

Note that the bend points of P_{ijk} are at the quarter and three-quarter marks but there is none at its center. The diameter of P_{ijk} is about the same as that of P_{ij} but that of P_{ijks} is only about half as much.

All of the $P_{ij...k}$'s at any stage have their bends at similar places. We denote a typical $P_{ij...k}$'s at the mth stage by Q_m and show it straight but with circles at the bends. Figure 11 shows $Q_1, Q_2, \ldots, Q_8$.

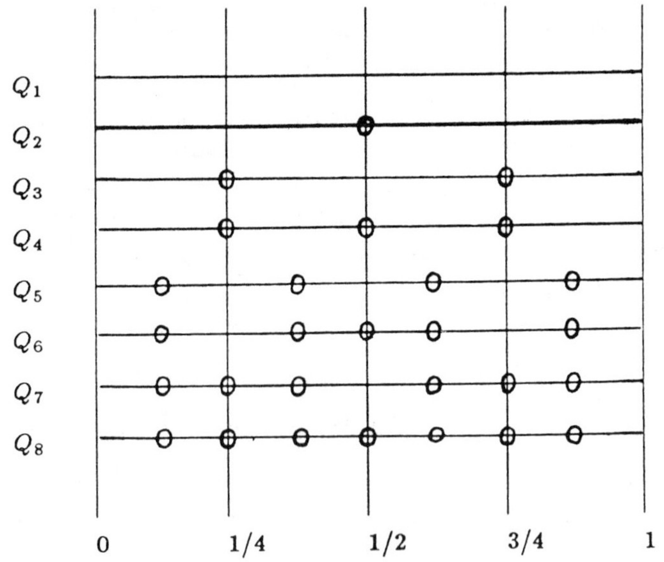

Figure 11.

8

It is to be noticed that if the length of Q_1 is less than ϵ, the length of the pieces between turns of Q_1, Q_2, Q_4, Q_8 are less than ϵ, $\epsilon/2$, $\epsilon/4$, $\epsilon/8$ respectively so the diameter of the T's at the 1st, 2nd, 4th, and 8th stages have diameter less than $\epsilon/2^0$, $\epsilon/2^1$, $\epsilon/2^2$, $\epsilon/2^3$.

We now make certain observations about the Q's that will help us define other Q's and make similar observations about them. Each of the Q's is symmetric about its center, as is each of the halves of a Q. (For considering a half, we ignore any circle at an end of the half.) In general, we will define Q's so that if they are broken into 2^i equal pieces, each of these pieces is alike and is a shrunken version of a previously defined Q. If m is odd, Q_m does not have a circle at its center and Q_{m+1} is formed from Q_m by placing a circle on Q_m at $1/2$. If m is even, Q_{m+1} does not have a circle at its center and each half of it looks like a one-half multiple of Q_r where $r = (m+2)/2$. Once we have obtained Q_i's ($i = 1, 2, \ldots, m$) satisfying these observations, these observations produce a unique Q_{m+1} satisfying them.

Finally, we show how we can slide the T_i's at the $(m+1)$st stage about in those at the mth to get Q's as we have described. Figure 12 shows how to do this if m is odd and Q_m has no circle at its center. The extra circle at the center of Q_{m+1} is where Q_{m+1} goes around the end. We considered each T_i at the mth stage and rotated the inner part of it containing the T_i's at $(m+1)$st stage in it through $\pi/2$ radians.

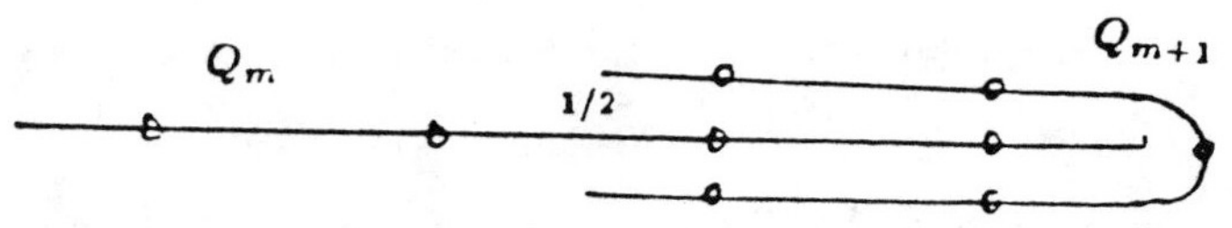

Figure 12.

We note that each half of Q_{m+1} (as does each half of Q_m) looks like a reduced copy of Q_r for $r = (m+2)/2$.

Figure 13 shows how the slide should be done if m is even and the "hooking

9

elbows" occurs at $1/2^t$ where $1/2^t$ is the largest of the numbers $1/2, 1/4, 1/8, \ldots$ denoting a point where Q_m is not circled. Rotation was through $\pi/2^t$. The Q's can be described sequentially.

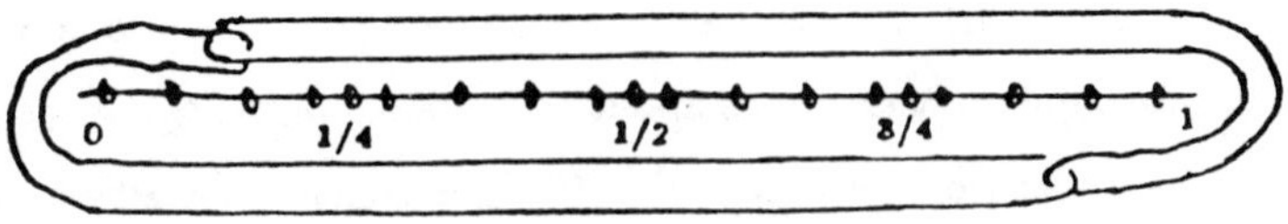

Figure 13.

Actually, if $1/2$ is uncircled, m is odd and we are at the preceding case. If $1/2^t$ is the first such uncircled point of Q_m, then all points of the form $s/2^t$ are uncircled for s odd and circled for s even. The symmetry and equivalences of the pieces of Q_m of size $1/2^t$ shows Q_{m+1} has circles at the required points.

A few philosophical comments seem apppropriate now that Theorem 2 is complete. Figure 11 appears simple but it was not easy to discover. It was actually found by working backwards. If in eight steps one had Q_8, what would Q_7 look like? There was only one possibility. The big gap would have to be in the middle. Once Q_7 were found, what would Q_6 look like? Where would the big gaps be? Actually, more complicated Q's like Q_{32}, Q_{128}, Q_{2^m} were being considered at the time, but Q_8 illustrates the principle just as well and it is easier to work back to Q_1 from Q_8 than from Q_{128}. This suggests several adages.

> *Learning to understand a complicated situation may be aided by considering an easier one. [Q_8 is easier than Q_{2^m}.]*
>
> *In trying to find the steps of a proof it is sometimes useful to work backwards and get the last steps first. [Count 8,7,6,... rather than 1,2,3,... .]*
>
> *In trying to solve a difficult problem, do not overlook the possibility of rephrasing it and looking at the problem from a different point of view. [Decompositions were used to study unions.]*

10

References

1. W.R. Alford, *Uncountably many different involutions of S^3*, Proc. Amer. Math. Soc. **17** (1966), 186-196.

2. R.H. Bing, *A homeomorphism between the 3-sphere and the sum of two solid horned spheres*, Ann. of Math. **56** (1952), 354-362.

3. ———————, *Inequivalent families of periodic homeomorphisms of E^3*, Ann. of Math. **80** (1964), 78-93.

4. Michael H. Freedman and Richard Skora, *Strange actions of groups on spheres*, preprint.

5. Deane Montgomery and Leo Zippin, *Examples of transformation groups*, Proc. Amer. Math. Soc. **5** (1954), 460-465.

Department of Mathematics
The University of Texas at Austin
Austin, Texas 78712

B. Periodic Homeomorphisms

Annals of Mathematics
Vol. 56, No. 2, September, 1952
Printed in U.S.A.

A HOMEOMORPHISM BETWEEN THE 3-SPHERE AND THE SUM OF TWO SOLID HORNED SPHERES*

By R. H. Bing

(Received December 17, 1951)

1. Introduction

We use E^n to denote Euclidean n-space, S^n to denote the surface of a sphere in E^{n+1}, and I^{n+1} to denote the interior of S^n in E^{n+1}.

A set U is *simply connected* if each simple closed curve J in U can be shrunk to a point in U—that is, there is a continuous transformation of $S^1 + I^2$ into U that carries S^1 homeomorphicly into J. Each complementary domain of S^n in E^{n+1} is simply connected.

Alexander described [1] a simple surface K (a set homeomorphic with S^2) in S^3 such that one component of $S^3 - K$ was not simply connected. One complementary domain of K in S^3 is homeomorphic with I^3 but the other is not. If a point p not of K is deleted from S^3, the resulting space is homeomorphic with E^3 and K is associated with a wild simple surface. (A simple surface M in E^3 is tame or wild according as there is or is not a homeomorphism of E^3 into itself that carries M into S^2.) The Alexander horned sphere K shows that certain generalizations of the Jordan Curve Theorem for E^2 cannot be extended to E^3.

Suppose K is an Alexander horned sphere in S^3, D is its complementary domain which is not simply connected, and T is a homeomorphism of $K + D$ into $K' + D'$. Suppose $K + D$ is sewn to $K' + D'$ along K and K' so that the point p of K is identified with the point $T(p)$ of K'. Wilder has raised the question [7; Problem 8 of 4; Problem 4.6 of 8] as to whether or not the resulting set is homeomorphic with S^3. In Section 3 we show that it is.

Professor E. E. Floyd's interest in periodic transformations caused me to study the horned sphere. Aside from showing the existence of queer surfaces in E^3 and S^3, the homeomorphism between S^3 and the sum of two solid horned spheres has some implications regarding periodic transformations. See the last part of Problem 39 proposed by Montgomery in [4].

Antoine has described [2] a Cantor set C in E^3 such that $E^3 - C$ is not simply connected. The set has been called a necklace because it will not fall apart (no simple surface in $E^3 - C$ separates C) although it is 0-dimensional. In Section 5 we describe a Cantor set C whose complement is not simply connected but such that each pair of points of C can be separated by a simple surface in $E^3 - C$.

In giving a characterization of S^3 by using partitionings, I placed restrictions on the elements of the partitionings to cause their boundaries to be tame simple surfaces instead of wild ones. See page 25 of [3]. In the following section we use the same description of a horned sphere as given in [3].

---

* Presented to the Society, April 27, 1951.

2. The horned sphere

The notation used in describing the horned sphere will be used in describing a homeomorphism of the sum of two solid horned spheres into S^3.

Suppose C is a right circular cylinder in E^3 with bases D_1 and D_2. Two mutually exclusive discs in D_i ($i = 1, 2$) are replaced by the surfaces of tubes T_{i1} and T_{i2} and discs D_{i1} and D_{i2} as shown in Figure 1 where D_{i1} and D_{i2} are the bases of a right circular cylinder C_i and $D_1 + T_{11} + C_1 + T_{12}$ is hooked to $D_2 + T_{21} + C_2 + T_{22}$ as shown.

Discs in the bases of the cylinder C_i ($i = 1, 2$) are replaced by the surfaces of tubes T_{i11}, T_{i12}, $T_{i2\ 1}$, T_{i22} and discs D_{i11}, D_{i12}, D_{i21}, D_{i22} as before. The

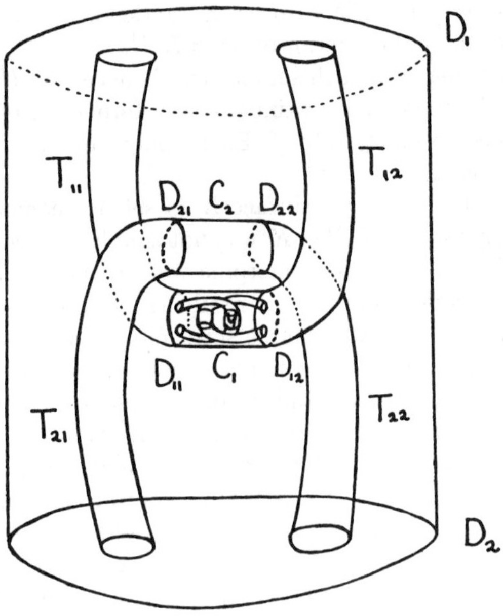

Figure 1

process is continued to get the horned sphere M. We use M_1 to denote the part of M which is the closure of the part of M on the exterior of $C_1 + C_2$, namely $M_1 = C - \sum$ (discs cut from $D_1 + D_2$) $+ \sum T_{ij}$. Likewise, M_i denotes the closure of the part of M on the exterior of $\sum C_{n_1 n_2 \cdots n_i}$. It is topologically equivalent to S^2 minus 2^i open discs. Let M_0 be the Cantor set $M - \sum M_i$.

Although M is homeomorphic to S^2, its interior U is not simply connected. We call this complementary domain of M which is not simply connected the *bad complementary domain* of M. A horned sphere plus its bad complementary domain is called a *solid horned sphere.* The part of U which is on the exterior of $\sum C_{n_1 n_2 \cdots n_i}$ is denoted by U_i.

3. The sum of solid horned spheres

The following result shows the effect of sewing two solid horned spheres together.

THEOREM. *A continuum is homeomorphic with* S^3 *if it is the sum of three mutually exclusive sets* M, U^1, *and* U^2 *such that* (a) *there is a homeomorphism of* $M + U^i$ $(i = 1, 2)$ *into a solid horned sphere that carries* M *into the horned sphere and* (b) *there is a homeomorphism of* $M + U^1$ *into* $M + U^2$ *that leaves each point of* M *fixed.*

PROOF. Suppose p is an ideal point at infinity such that $E^3 + p$ is homeomorphic with S^3. We first describe a decomposition of $E^3 + p$ into arcs and points such that the resulting decomposition space X is homeomorphic with $M + U^1 + U^2$. We complete the proof by showing that this decomposition space is topologically equivalent to S^3.

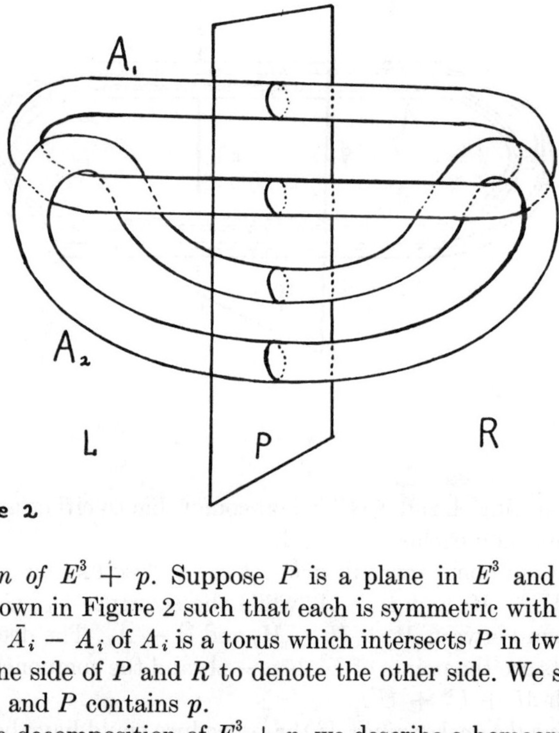

Figure 2

I. *Decomposition of* $E^3 + p$. Suppose P is a plane in E^3 and $\bar{A}_1$, $\bar{A}_2$ are anchor rings as shown in Figure 2 such that each is symmetric with respect to P and the boundary $\bar{A}_i - A_i$ of A_i is a torus which intersects P in two circles. We use L to denote one side of P and R to denote the other side. We shall suppose that each of $\bar{L}$, $\bar{R}$, and P contains p.

As we describe a decomposition of $E^3 + p$, we describe a homeomorphism between $M + U$ and each of two parts of the decomposition space X of $E^3 + p$, namely, the part that intersects $\bar{L}$ and the part that intersects $\bar{R}$. The notation used refers to that used in Section 2.

There is a homeomorphism F_1 of $\bar{U}_1$ into $\bar{L} - (A_1 + A_2)$ such that $F_1(M_1) = P - (A_1 + A_2)$ and $F_1(\bar{U}_1 \cdot C_i) = \bar{L} \cdot (\bar{A}_i - A_i)$. There is a homeomorphism G_1 of $\bar{U}_1$ into $\bar{R} - (A_1 + A_2)$ such that $G_1(x)$ and $F_1(x)$ are symmetric with respect to P. We note that $F_1 = G_1$ on M_1.

Consider A_i ($i = 1, 2$). Put a chain of two anchor rings $\bar{A}_{i1}$ and A_{i2} in it as shown in Figure 3 so that each is symmetric with respect to P. There is a homeomorphism F_2 of $\bar{U}_2$ into $\bar{L} - \sum A_{ij}$ that preserves F_1 on $\bar{U}_1$ such that $F_2(M_2) = P - \sum A_{ij}$ and $F_2(\bar{U}_2 \cdot C_{ij}) = \bar{L} \cdot (\bar{A}_{ij} - A_{ij})$. Then G_2 is defined on $\bar{U}_2$ so that $G_2(x)$ is symmetric to $F_2(x)$ with respect to P.

The process is continued. We get more anchor rings, more F_i's, and more G_i's. We use A_0 to denote $(A_1 + A_2) \cdot (\sum A_{ij}) \cdot (\sum A_{ijk}). \cdots$. The anchor rings could have been defined so that each component of A_0 is an arc which intersects no plane parallel to P in two points. Therefore, we suppose that this was done and that each component of A_0 is such an arc. Furthermore, we shall suppose that the anchor rings are so smoothly shaped that if P' is a plane parallel to P that intersects the interior A of one of the anchor rings, then $A \cdot P'$ is irreducible with

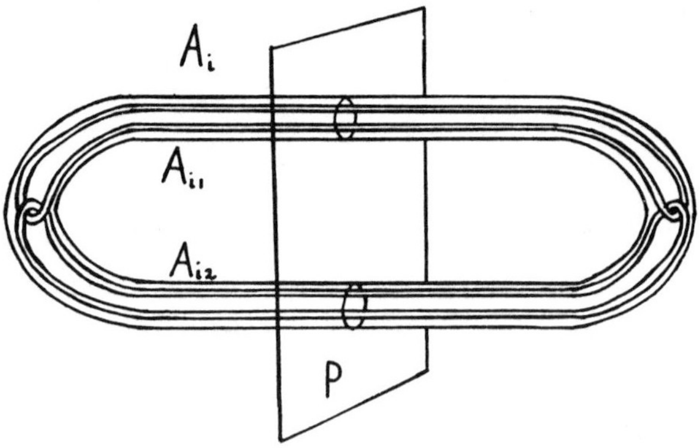

Figure 3

respect to separating A and $A \cdot P'$ is homeomorphic to either an open disc or the sum of two mutually exclusive open discs.

The sequence of homeomorphism $F_1, F_2, \cdots$ describes a homeomorphism between $M + \mathrm{U} - M_0$ and $\bar{L} - A_0$ while the sequence $G_1, G_2, \cdots$ describes a homeomorphism between $M + U - M_0$ and $\bar{R} - A_0$. The decomposition space X whose points are the points of $E^3 + p - A_0$ and the components of A_0 is homeomorphic with $M + U^1 + U^2$.

II. *Shrinking the anchor rings.* If the procedure could have been carried out so that the diameters of the anchor rings at the ith stage converged to 0 with $1/i$, the components of A_0 would be points and the required homeomorphism between $M + U^1 + U^2$ and $E^3 + p$ would be given.

In Figure 4 we show two ways in which a chain of two anchor rings can be imbedded in the same anchor ring. We note that the diameters of the inside anchor rings in Figure 4b are less than those in Figure 4a. This gives a clue as to how to proceed. There is a homeomorphism of $\bar{A}_1$ into itself that leaves each point of

$\bar{A}_1 - A_1$ fixed and carries A_{1i} into a position as shown in Figure 4b. We may think of A_1 being moved in a clockwise direction while its boundary is kept fixed. The part of the plane P in A_1 would be distorted into a surface but no point would be moved farther than the diameter of A_1.

We shall not try to shrink anchor rings at each stage but will follow a procedure that will do it eventually. We shall use the followimg lemma.

LEMMA. *For each positive number ε and each positive integer j there is a positive integer k and a homeomorphism T of E^3 into itself such that each point of $E^3 - \sum A_{n_1 n_2 \cdots n_j}$ is fixed under T and the image of each $A_{n_1 n_2 \cdots n_k}$ under T has a diameter of less than ε.*

4a

4b

Figure 4

PROOF *of Lemma.* For brevity we shall denote $\{A_{n_1 n_2} \cdots {}_{n_j}\}$ by $\{A^j\}$ and the sum of the elements of $\{A^j\}$ by $\sum A^j$.

Since each component of A_0 is an arc which does not intersect any plane parallel to P in two points, for each positive number ε, there is an integer j' and a finite number of planes $P_1, P_2, \cdots, P_n$ all parallel to P such that each connected subset of $\sum A^{j'}$ of diameter more than ε intersects at least two of the planes $P_1, P_2, \cdots, P_n$. For convenience we suppose that $j' = j$.

For each element A of $\{A^j\}$ we shall describe a transformation T that will leave each point of the exterior of A fixed and will carry each element of $\{A^k\}$ in A into a set of diameter less than ε. Since $\{A^j\}$ has only a finite number of elements, a product of these homeomorphisms will be one satisfying the lemma.

If $n = 1$, the diameter of A is less than ε. In this case we let $k = j$ and T be the identity transformation.

If $n = 2$, we let $k = j + 1$ and T carry the elements of $\{A^k\}$ in A into sets as indicated in Figure 5a. The elements of $\{A^k\}$ are represented as simple closed curves.

If $n = 3$, we let $k = j + 2$. The images of the four elements of $\{A^k\}$ in A are placed as indicated in Figure 5b. We may think of a fold being made in the left most anchor ring shown in Figure 5a and two images of the anchor rings of Figure 5a being hooked together.

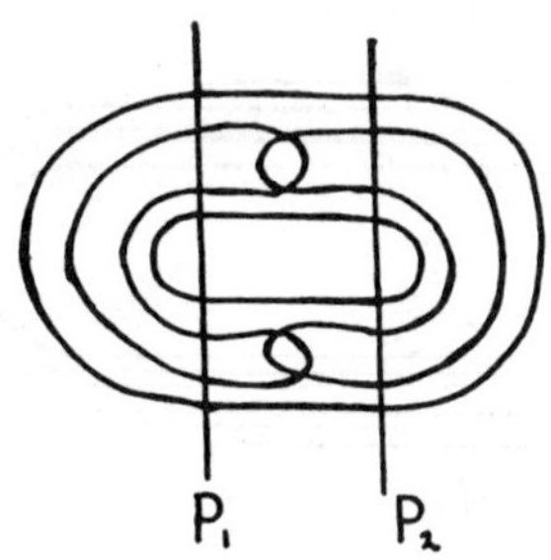

Figure 5a

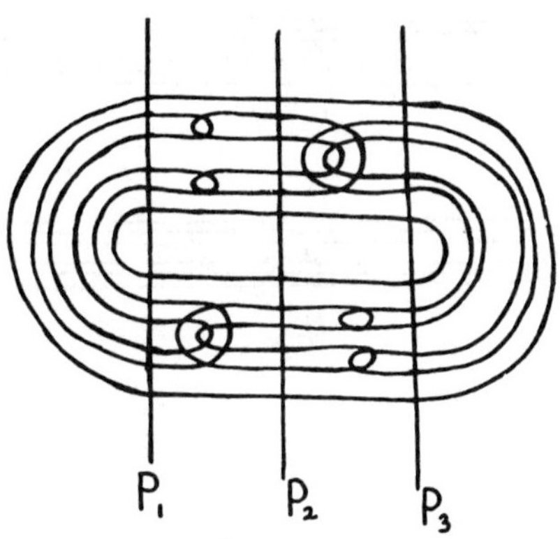

Figure 5b

If $n = 4$, we let $k = j + 3$. We think of a fold being made in the left end of Figure 5b and hooking two images together. See Figure 5c.

If $n = 5$, we let $k = j + 4$. For convenience in Figure 5d, we only show one half of the images of the sixteen elements of $\{A^k\}$ in A. In general, $k = j + n - 1$.

III. *A homeomorphism of* $M + U^1 + U^2$ *into* $E^3 + p$. We find from the preceding lemma that there is an integer j_1 and a homeomorphism T_1 of E^3 into itself that leaves each point of $E^3 - (A_1 + A_2)$ fixed and carries each element of $\{A^{j_1}\}$ into a set of diameter less than 1. There is a positive number ε such that a set is of diameter less than $\frac{1}{2}$ if its image under T_1 is of diameter less than ε. Reapplying the lemma, we find that there is an integer j_2 and a homeomorphism T_2' of E^3 into itself such that T_2' leaves each point of $E^3 - \sum A^{j_1}$ fixed and carries each element of $\{A^{j_2}\}$ into a set of diameter less than ε. Then $T_2 = T_1T_2'$ is a

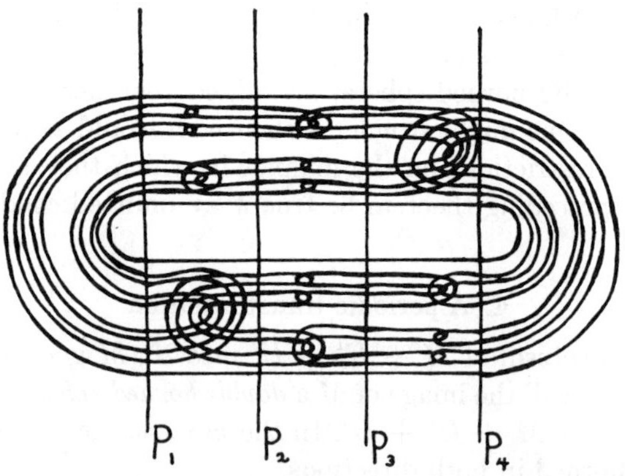

Figure 5c

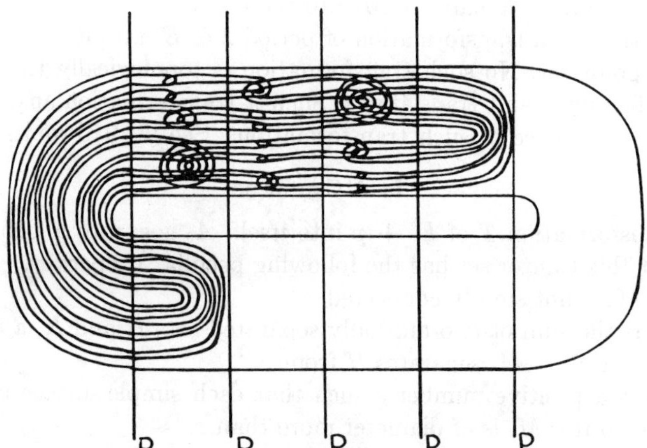

Figure 5d

homeomorphism of E^3 into itself that carries each element of $\{A^{j_i}\}$ $(i = 1, 2)$ into a set of diameter less than $1/i$ and $T_2 = T_1$ on $E^3 - \sum A^{j_i}$.

Likewise, there is an integer j_3 and a homeomorphism T_3 of E^3 into itself such that $T_3 = T_2$ on the exterior of $\sum A^{j_2}$ and the diameter of the image of each element of $\{A^{j_3}\}$ is less than $\frac{1}{3}$. Continuing this procedure, we obtain a sequence of homeomorphisms $T_1, T_2, \cdots$ and a sequence of integers $j_1, j_2, \cdots$ such that $T_{i+1} = T_i$ on the exterior of $\sum A^{ji}$ and the diameter of the image under T_i of each element of $\{A^{ji}\}$ is less than $1/i$.

We note that $T = \lim T_1, T_2, \cdots$ is a homeomorphism that carries the decomposition space X defined in Part I into $E^3 + p$. Hence, $M + U^1 + U^2$ is homeomorphic to $E^3 + p$ and therefore to S^3. If $F = \lim F_1, F_2, \cdots$ and $G = \lim G_1, G_2, \cdots$, the homeomorphism of $M + U^1 + U^2$ into $E^3 + p$ is TF on U^1, is TG on U^2, and carries M into $T(P)$.

Question. If two solid horned spheres are added together in a certain fashion, their sum is like S^3. One might wonder if their sum would necessarily be like S^3 if we placed fewer restrictions on the manner in which they are added. For example, would the preceding theorem be true if we omitted condition *b* from the hypothesis.

4. A periodic transformation

In the homeomorphism of $M + U^1 + U^2$ into S^3, M goes into a set homeomorphic with S^2. We call the image of M a *double horned sphere*. For convenience, we suppose that $S^3 = M + U^1 + U^2$. In the next section we examine the part of M where it is horned in both directions.

There is a periodic transformation of period 2 of S^3 into itself that has for its fixed point set the double horned sphere M. The transformation will carry each of the complementary domains of M into the other.

Similarly, there is a transformation of period 2 of E^3 into itself that has $T(P)$ for its fixed point set. No such transformation is topologically the products of rotations, reflections, and translations. Such a transformation answers speculation as to the existence of such transformations. See [6; 9; Problem 39 of 4].

5. A Cantor set

In the transformation T of $E^3 + p$ into itself, A_0 goes into a Cantor set M_0. We find that this Cantor set has the following peculiar properties.

(1) $E^3 - M_0$ is not simply connected.

(2) If M_0 is the sum of two mutually separated sets H and K, a tame simple surface in $E^3 - M_0$ separates H from K.

(3) There is a positive number ε such that each simple surface in $E^3 - M_0$ that separates M_0 is of diameter more than ε.

If J is a simple closed curve in P that separates $\bar{A}_1 \cdot P$ from $\bar{A}_2 \cdot P$ and j is a positive integer, J cannot be shrunk to a point in $E^3 + p$ without intersecting $\sum \bar{A}^j$. Hence J cannot be shrunk to a point in $E^3 - A_0$ and $E^3 - M_0$ is not simply connected.

It may be noted that there is a tame simple surface in E^3 that separates $\bar{A}_1$ from $\bar{A}_2$. See Figure 2. If K is a tame simple surface having A_i on its interior, there is a tame simple surface on the interior of K that separates $\bar{A}_{i1}$ from $\bar{A}_{i2}$. Hence, if $\bar{A}_{11} + \bar{A}_{12} + \bar{A}_{21} + \bar{A}_{22}$ is expressed as the sum of two mutually separated sets H and K, there is a tame simple surface in E^3 that separates H from K. Continuing in this manner, we find that if $\sum \bar{A}^j$ is expressed as the sum of two mutually separated sets, a tame simple surface in E^3 separates them. Hence, if M_0 is the sum of two mutually separated sets H and K, there is a tame simple surface in E^3 that separates H from K.

Although there are simple surfaces in $E^3 - M_0$ that separate M_0, we are not to conclude that there are any such simple surfaces with small diameters. Suppose J is a simple closed curve in P that separates $\bar{A}_1 \cdot P$ from $\bar{A}_2 \cdot P$. If R is a closed proper subset of M_0, J can be shrunk to a point in $E^3 - R$. Hence, if K is a simple surface in $E^3 - M_0$ that separates M_0 the following theorem shows that K intersects J.

THEOREM. *Suppose R is a subset of E^3, K is a simple surface in $E^3 - R$, D is a complementary domain of K, and J is a simple closed curve in $D + K$ that can be shrunk to a point in $E^3 - R$. Then J can be shrunk to a point in $D + K - R$.*

PROOF. Let T be a continuous transformation of $S^1 + I^2$ such that T carries S^1 homeomorphicly into J and $T(S^1 + I^2) = A$ is contained in $E^3 - R$. There is a continuous transformation F of $T^{-1}[A \cdot (E^3 - D)]$ into K that preserves T on $T^{-1}[A \cdot K]$. [Theorem VI 4, p. 83 of 5.] Then the transformation of $S^1 + I^2$ that is F on $T^{-1}[A \cdot (E^3 - D)]$ and T on $T^{-1}[A \cdot \bar{D}]$ is one showing that J can be shrunk to a point in $\bar{D} - R$.

A corollary of the preceding theorem is the result that the closure of a complementary domain of a simple surface in E^3 is simply connected.

THE UNIVERSITY OF WISCONSIN

REFERENCES

1. J. W. ALEXANDER, *An example of a simply connected surface bounding a region which is not simply connected*, Proc. Nat. Acad. Sci., U. S. A., vol. 10 (1924), pp. 8–10.
2. LOUIS ANTOINE, *Sur l'homeomorphie de deux figures et de leurs voisinages*, J. Math. Pures Appl., vol. 86 (1921), pp. 221–325.
3. R. H. BING, *A characterization of 3-space by partitionings*, Trans. Amer. Math. Soc., vol. 70 (1951), pp. 15–27.
4. SAMUEL EILENBERG, *On the problems of topology*, Ann. of Math., vol. 50 (1949), pp. 247–260.
5. W. HUREWICZ and H. WALLMAN, Dimension Theory, Princeton University Press, 1948.
6. P. A. SMITH, *Transformations of finite period*, Ann. of Math., vol. 39 (1938), pp. 127–164.
7. R. L. WILDER, *The strong symmetrical cut sets of closed euclidean n-space*, Fund. Math., vol. 27 (1930), pp. 135–139.
8. R. L. WILDER, Topology of manifolds, Amer. Math. Soc. Colloquium Publications, vol. 32, New York, 1949.
9. LEO ZIPPIN, Transformation Groups, Lectures in Topology, The University of Michigan Press, Ann Arbor, 1941, pp. 191–221.

Inequivalent Families of Periodic Homeomorphisms of E^{3*}

By R. H. BING

1. Introduction

It is sometimes convenient to regard the 3-sphere S^3 as E^3 (euclidean 3 space) plus a point added at infinity. For example, if one thinks of periodic transformations of S^3 whose fixed point sets are either a 2-sphere, 1-sphere, or 0-sphere, it is convenient to consider in E^3 reflections through planes, rotations about lines, or reflections through points.

A closed set C in E^3 is *tame* if there is a homeomorphism h of E^3 onto itself such that $h(C)$ is a subcomplex of some rectilinear triangulation of E^3. If there is no such homeomorphism, we say that C is *wild*. The set C is *locally tame* at a point p of C if there is a neighborhood N of p and a homeomorphism of N into E^3 that takes $N \cdot C$ onto a tame set. If there is no such neighborhood N of p, we say that C is *locally wild* at p. It is known [4] that a closed set is tame in E^3 if it is locally tame at each of its points. The set of points of C at which it is locally tame is an open subset of C, and the set at which it is *locally wild* is closed. The subset at which it is locally wild is called the *wild subset* of C.

Two homeomorphisms h_1, h_2 of E^3 (or S^3) are *equivalent* if there is a homeomorphism g of E^3 (or S^3) onto itself such that $h_1 = g^{-1}h_2g$.

Smith has shown [17] that for any orientation preserving periodic homeomorphism of S^3 onto itself, the fixed point set is either null, the 1-sphere, or S^3 itself; while for any orientation reversing periodic homeomorphism, the fixed point set is either a 0-sphere or a 2-sphere. From this result it follows that, if h is a periodic homeomorphism of E^3 whose fixed point set is a topological plane P, then h is an involution (homeomorphism of period 2). If P is tame, then h is equivalent to an ordinary reflection through a horizontal plane. This last result follows from the fact that locally tame sets are tame [4] coupled with the theorem of Harrold and Moise which states [9] that, if S is a 2-sphere in S^3 which is locally polyhedral mod one point, then the closure of at least one component of $S^3 - S$ is a topological cube. One might wonder if one could prove that h is equivalent to an ordinary reflection if, instead of supposing that P is tame, one supposes merely that the fundamental group of

* Work supported by grants NSF-G21514 and NSF-G11665.

each component of $E^3 - P$ is finitely generated.

In showing [2] that the sum of two solid Alexander horned spheres sewed together along their boundaries with the identity map is topologically S^3, Bing exhibited an involution of S^3 whose fixed point set is a wild 2-sphere. Wu showed [18] that, for certain groups G, there is a 2-sphere S in S^3 such that the fundamental group of the components of $S^3 - S$ is G, while S is the fixed point set of an involution on S^3. Alford modified Wu's examples to show [1] that there are uncountably many mutually inequivalent such wild involutions on S^3. In the present paper we show:

THEOREM 2.1. *For each closed subset X of E^2 without isolated points there is a reflection of E^3 through a topological plane P such that the wild subset of P is topologically equivalent to X.*

This gives an alternate proof of Alford's result.

COROLLARY 2.2. *There are uncountably many mutually inequivalent involutions of S^3 each of whose fixed point set is a 2-sphere.*

Examples are known [6] of 2-spheres in E^3 which are not locally tame at any point. The following gives another.

COROLLARY 2.3. *There exists a wild 2-sphere S in E^3 not locally tame at any point but such that, if M denotes the crumpled cube in E^3 bounded by S, then S^3 is the sum of two copies of M sewed together on their boundaries with the identity map.*

A *crumpled cube* is a set homeomorphic to the sum of a 2-sphere in E^3 with its bounded complementary domain. Although a periodic homeomorphism of S^3 whose fixed point set is a 2-sphere is not necessarily equivalent to an ordinary reflection, it is a pseudo reflection in the sense that it interchanges two crumpled cubes while leaving their common boundary fixed.

Montgomery and Zippin described [15] an involution of E^3 about a wild line. It was shown in [3] that there are such non-orthogonal rotations of other periods. We extend the result to show the following:

THEOREM 4.1. *For each closed subset X of E^1 without isolated points and for each integer n greater than* 1, *there is a rotation of period n of E^3 about a topological line L such that the wild subset of L is topologically equivalent to X.*

COROLLARY 4.2. *For each integer n greater than* 1, *there are uncountably many mutually inequivalent rotations of E^3 of period n.*

Although the known examples of non-orthogonal homeomorphisms of S^3

of period n (and which leave a simple closed curve J fixed) are not ordinary rotations, they are pseudo rotations about J in the sense that S^3 is the sum of n crumpled cubes $C_1, C_2, \cdots, C_n$ each of which contains J, each is tame mod J, and $h(C_i) = C_{i+1}$. One might wonder if each periodic homeomorphism of S^3 whose fixed point set is a simple closed curve is such a pseudo rotation.

In the case of a periodic homeomorphism of S^3 whose fixed point set is a tame simple closed curve, it is not known whether or not the homeomorphism is necessarily equivalent to a piecewise linear homeomorphism. Perhaps the first problem to attack is the one of deciding whether or not an involution on E^3 is equivalent to a piecewise linear one if its fixed point set is the z-axis. In the case of a periodic piecewise linear homeomorphism on S^3 whose fixed point set is a simple closed curve, it is not known whether or not the fixed simple closed curve could be knotted. Montgomery and Samelson have shown [16] that piecewise linear involutions of S^3 cannot have certain types of knots as fixed point sets. Fox and Kinoshita have enlarged the classes of knots that are outlawed [7, 8, 11]. Moise has shown [14] that a periodic piecewise linear homeomorphism of S^3 whose fixed point set is an unknotted simple closed curve is equivalent to a rotation.

Livesay has shown [12] that any involution of S^3 without fixed points is equivalent to the antipodal map. For any free action, as would be given for example by a homeomorphism of prime period, it can be shown that the homeomorphism is equivalent to a piecewise linear one. However we show that for each composite positive integer n there is a homeomorphism of S^3 onto itself which is of period n, without fixed points, and inequivalent to a piecewise linear homeomorphism. The first example of this sort we are familiar with resulted from a conversation the author had with James M. Kister. In fact we get the following result.

THEOREM 5.1. *For each composite positive integer n there is an uncountable family of mutually inequivalent homeomorphisms of S^3 onto itself such that each is of period n and fixed point free.*

More recently Livesay has shown [13] by using a start made by Hirsch and Smale [10] that any two involutions of S^3 whose fixed point sets are 0-spheres are equivalent to each other. Examples of periodic transformations of E^3 with only one fixed point may be found by considering the product of a reflection through a plane with a rotation about a line normal to the plane. One might wonder if each periodic transformation of E^3 with only one fixed point is the product of a pseudo reflection through a topological plane followed by a pseudo rotation about a topological line. We show the following.

THEOREM 6.1. *For each integer n greater than* 1, *there are uncountably many mutually inequivalent periodic homeomorphisms of* S^3 *onto itself of period* $2n$ *such that the fixed point set of each is a* 0-*sphere.*

2. Pseudo reflections. Proof of Theorem 2.1

Denote the horizontal $z = 0$ plane in E^3 by E^2. Let X be the closed set mentioned in the statement of Theorem 2.1, and $p_1, p_2, \cdots$ be a sequence of points of X which is dense in X.

We obtain an upper semi-continuous decomposition G of E^3 into points and arcs such that each of the arcs is symmetric with respect to E^2, each intersects X, and the sum of the intersections is dense in X. We then show that the corresponding decomposition space is topologically E^3, and use this decomposition to get a reflection satisfying the conclusion of Theorem 2.1. We obtain a sequence $G_1, G_2, \cdots$ of decompositions of E^3 such that an arc belongs to G if and only if it belongs to some G_i.

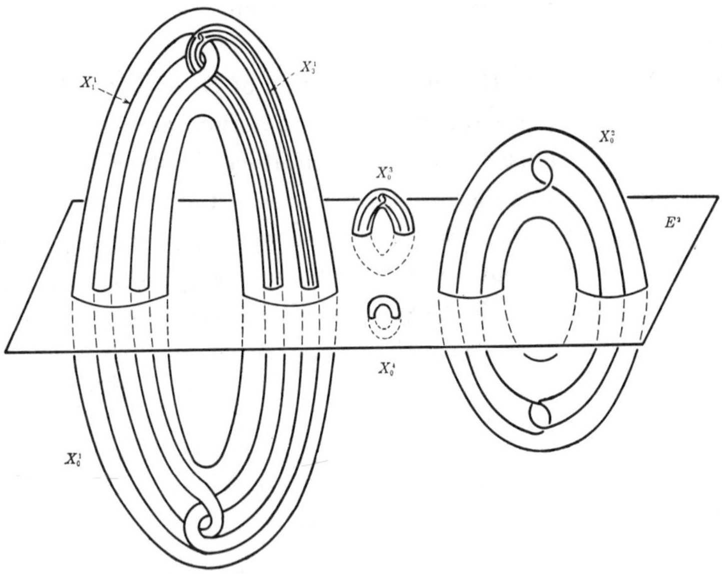

Figure 1.

Description of decomposition space G_1. Let $X_0^1, X_1^1, X_2^1, \cdots$ be a sequence such as described in [2] and as shown in Figure 1 such that

(a.1) diameter $X_0^1 < 1$,

(b.1) $p_1 \in X_i^1$,

(c) X_i^1 is the sum of 2^i mutually exclusive topological anchor rings,

(d) the interior of each component of X_i^1 contains precisely two components of X_{i+1}^1,

(e) each component of X_i^1 is symmetric with respect to E^2,

(f) each horizontal cross section of each component of X_i^1 has diameter less than $1/i$, and

(g) any closed curve in $E^3 - (E^2 + X_i^1)$ that cannot be shrunk to a point in $E^3 - (E^2 + X_i^1)$ cannot be shrunk to a point in $E^3 - (E^2 + X_{i+1}^1)$.

It follows from Theorem 9 of [6] that the X_i^1's we show in Figure 1 satisfy Condition (g). We impose the additional restriction that

(h) each component of $E^2 \cdot \text{Int}\, X_i^1$ intersects X.

Let G_1 be the decomposition of E^3 whose non-degenerate elements are the components of $X_0^1 \cdot X_1^1 \cdot X_2^1 \cdot \cdots$. It follows from Condition (f) that each non-degenerate element of G_1 is a tame arc which does not intersect any horizontal plane in more than one point. Condition (e) implies that each of these arcs is symmetric with respect to E^2. Let H_1 be the sum of the non-degenerate elements of G_1. Condition (h) implies that $H_1 \cdot E^2 \subset X$.

We showed in [2] that the decomposition space of G_1 is topologically E^3; in fact, there is a map f_1 of E^3 onto itself such that

(i) $f_1 = I$ on $E^3 - X_0^1$, and

(j) each $f_1^{-1}(p) \in G_1$ for $p \in E^3$.

We note that $f_1(E^2)$ is a topological plane which is locally tame at a point $f_1(p)$ if $p \in E^2 - H_1$. However, if $p \in H_1$, then arbitrarily close to p there is a simple closed curve in some $E^3 - (E^2 + X_i^1)$ that cannot be shrunk to a point in this $E^3 - (E^2 + X_i^1)$. Condition (g) implies that the closed curve cannot be shrunk to a point in $E^3 - (E^2 + H_1)$. The image of such a simple closed curve under f_1 could not be shrunk to a point in $E^3 - f_1(E^2 + H_1) = E^3 - f_1(E^2)$. Then $f_1(E^2)$ is locally wild at $f_1(p)$ because $E^3 - f_1(E^2)$ is not locally simply connected at $f_1(p)$. Hence, $f_1(H_1)$ is the wild subset of $f_1(E^2)$.

Description of G_2. If $X \subset H_1$, we would not need G_2; but we suppose that some of X is not in H_1. Let q_2 be the first point of $p_2, p_3, \cdots$ not in H_1.

Let ε_1 be a positive number so small that, for each pair of points p, q of E^3

(k.1) $\rho(p, q) < \varepsilon_1 \Rightarrow \rho(f_1(p), f_1(q)) < 1/2$.

We use ρ to indicate the distance function.

Let $X_0^2, X_1^2, X_2^2, \cdots$ be a sequence of topological anchor rings satisfying Conditions (c) to (h) with X_i^2 replacing X_i^1 in the statement of these conditions, and with X_0^2 satisfying

(a.2) diameter $X_0^2 < \varepsilon_1$,

(b.2) $q_2 \in X_i^2$, and

(m.2) $X_0^2 \cdot H_1 = 0$.

Let G_2 be the upper semi-continuous decomposition of E^3 whose non-degenerate elements are the components of $X_0^2 \cdot X_1^2 \cdot X_2^2 \cdot \cdots$, and H_2 be the sum of the non-degenerate elements of G_2. Note that $H_1 \cdot H_2 = 0$ since Condition (m.2) says $X_0^2 \cdot H_1 = 0$.

Let f_2 be the map of E^3 onto itself satisfying Conditions (i), (j) with f_1, X_0^1, G_1 replaced by f_2, X_0^2, G. Let ε_2 be a positive number so small that, for each pair of points p, q of E^3,

(k. 2) $\rho(p, q) < \varepsilon_2 \Rightarrow \rho(f_1 f_2(p), f_1 f_2(q)) < 1/4$.

Description of G_i. We continue defining q's, X's, G's, H's, f's, and ε's. After G_{i-1}, H_{i-1}, f_{i-1}, ε_{i-1} have been defined, we let q_i be the first point of p_i, p_{i+1}, $\cdots$ not in $H_1 + H_2 + \cdots + H_{i-1}$, and pick a sequence X_0^i, X_1^i, X_2^i, $\cdots$ satisfying conditions similar to Conditions (c)–(h) and such that

(a. i) diameter $X_0^i < \varepsilon_{i-1}$.

(b. i) $q_i \in X_j^i$, and

(m.i) $X_0^i \cdot (H_1 + H_2 + \cdots + H_{i-1}) = 0$.

Let G_i be the resulting decomposition, H_i be the sum of the non-degenerate elements of G_i, and f_i be the map of E^3 onto itself satisfying restrictions like Conditions (i), (j). Then ε_i is a positive number so small that, for each pair of points p, q of E^3,

(k. i) $\rho(p, q) < \varepsilon_i \Rightarrow \rho(f_1 f_2 \cdots f_i(p), f_1 f_2 \cdots f_i(q)) < 1/2^i$.

The decomposition G. Let G be the monotone decomposition of E^3 such that a continuum is a non-degenerate element of element of G if and only if it is a non-degenerate element of some G_i. Let H denote $H_1 + H_2 + \cdots$. We note that $\overline{H \cdot E^2} = X$.

We now show that the decomposition space of G is topologically E^3. Let $f = \text{limit}\,(f_1, f_1 f_2, f_1 f_2 f_3, \cdots)$. Conditions (k.i) and (a.i + 1) imply that, for each point p of E^3,

(n) $\rho(f_1 f_2 \cdots f_{i+1}(p), f_1 f_2 \cdots f_i(p)) < 1/2^i$.

Hence $f(p)$ is uniquely defined. Since each $f_1 f_2 \cdots f_i$ is uniformly continuous, Condition (n) implies that f is uniformly continuous. Since each $f_1 f_2 \cdots f_i$ is onto, Condition (n) also implies that f is onto.

We now show that, for each non-degenerate element g of G, $f(g)$ is a point. There is an integer i such that $g \in G_i$. Condition (m) implies that no X_0^j $(j > i)$ intersects g, so each f_j $(j > i)$ is the identity on g. Then $f_i(g)$ is a point and $f(g)$ is the image of this point under $f_1 f_2 \cdots f_{i-1}$.

Let x_1, x_2 be points belonging to different elements of G. We show that $f(x_1) \neq f(x_2)$. Let $F_i = \text{limit}\,(f_i, f_i f_{i+1}, \cdots)$. If $f(x_1) = f(x_2)$, $F_2(x_1)$ and $F_2(x_2)$

belong to the same element of G_1. This element is degenerate since F_2 is the identity on each non-degenerate element of G_1. Similarly $F_3(x_1)$ and $F_3(x_2)$ belong to the same degenerate element of G_2. It follows by induction that for each integer i, $F_{i+1}(x_1)$ and $F_{i+1}(x_2)$ belong to the same degenerate element of G_i. Since limit $(F_2(x_k), F_3(x_k), \cdots) = x_k$ $(k = 1, 2)$, we arrive at the contradiction that $x_1 = x_2$.

Since f is uniformly continuous and 1–1 from the decomposition space of G to E^3, and E^3 is compact with the point added at infinity, the decomposition space of G is topologically E^3.

Note that f is a homeomorphism on E^2 and hence on X. For each point p of $E^2 - X$, f is a homeomorphism in a neighborhood of p. Hence, $f(E^2)$ is locally tame at $f(p)$. However, if $p \in X$, consider a subsequence $q_{n_1}, q_{n_2}, \cdots$ of $q_2, q_3, \cdots$ converging to p. Let J_i be a simple closed curve on Bd $X_0^{n_i}$ which misses E^2 and which circles Bd $X_0^{n_i}$ meridianally. It follows from Condition (g) that J_i cannot be shrunk to a point in $E^3 - (E^2 + H_{n_i})$. Then $f(J_i)$ cannot be shrunk to a point in $E^3 - f(E^2)$. This shows that $E^3 - f(E^2)$ is not locally simply connected at points of $f(X)$. Hence $f(X)$ is the wild subset of $f(E^2)$.

Let r be the reflection of E^3 through E^2. Then frf^{-1} is an involution of E^3 whose fixed point set is a plane $P = f(E^2)$ whose wild set is $f(X)$.

The following is an immediate corollary of Theorem 2.1.

COROLLARY 2.4. *There exists a wild plane P in E^3 such that P is not locally tame at any point, but P is the fixed point set of some involution of E^3.*

By adding the point at infinity to E^3, we obtain the following two corollaries to Theorem 2.1.

COROLLARY 2.5. *For each perfect subset X of a 2-sphere, there is an involution of S^3 whose fixed point set is a topological 2-sphere whose wild set is topologically equivalent to X.*

COROLLARY 2.6. *There are uncountably many mutually inequivalent involutions of S^3, each of whose fixed point sets is a topological 2-sphere.*

We end our discussion of periodic homeomorphisms of E^3 whose fixed point sets are planes by listing the following result we mentioned in the last section.

THEOREM 2.7. *If h is a periodic homeomorphism of E^3 whose fixed point set is a topological plane, then h is an involution. If P is tame, h is equivalent to a reflection through a horizontal plane.*

3. Pseudo rotations about a wild line

The following theorem was proved in [3] but we repeat the proof here for completeness.

THEOREM 3.1. *There is a rotation of period* 3 *of* E^3 *about a wild line.*

PROOF. The proof is given in four steps.

1. *A decomposition* G. The transformation is described by using an upper semi-continuous decomposition G of E^3 into points and tame triods. At the Summer Institute on Set Theoretic Topology at the University of Wisconsin in the summer of 1955, Deane Montgomery raised the question as to whether or not such a decomposition as described here might be topologically E^3. It was subsequently shown in [3] that the decomposition G was indeed topologically E^3.

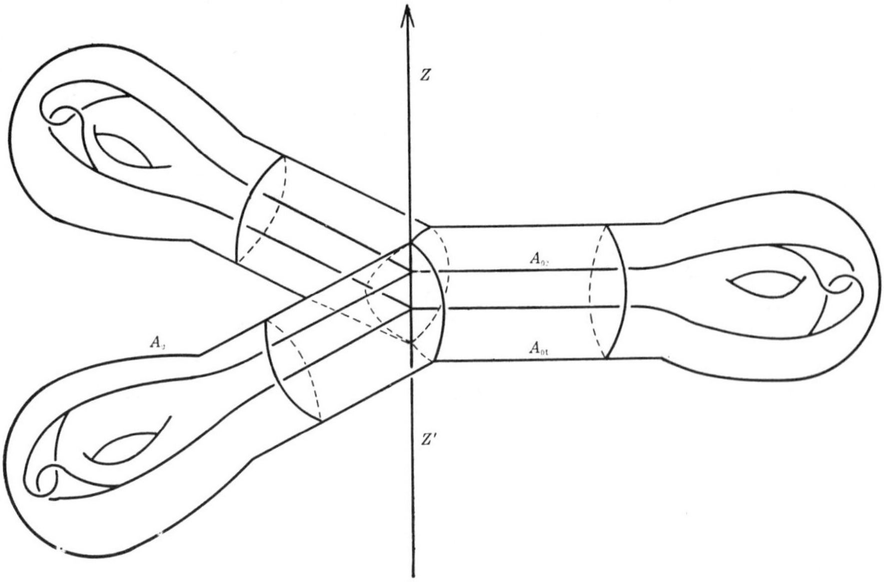

Figure 2.

Suppose A_0 is a tame solid as shown in Figure 2 which is topologically equivalent to a ball with three handles. It goes onto itself under a rotation of 120° about the z-axis Z. We may get such an A_0 as follows. In the xy-plane, consider the triod with center at the origin and with ends at $(1, 0)$, $(-1/2, 3^{1/2}/2)$, $(-1/2, -3^{1/2}/2)$. To this triod add circles with radii 1, and with centers at $(2, 0)$, $(-1, 3^{1/2})$, $(-1, -3^{1/2})$. Then for some positive number ε_0 less than 1, A_0 is the closed ε_0 neighborhood of the sum of the triod and the three circles.

Denote the segment of Z from $(0, 0, -1)$ to $(0, 0, 1)$ by Z'.

Let A_{01}, A_{02} be two tame solids each topologically equivalent to A_0 and imbedded in it as shown in Figure 2. Each A_{0i} is invariant under a rotation of 120° about Z and the ends are linked so that each closed curve in $E^3 - (A_0 + Z')$

that cannot be shrunk to a point in $E^3 - (A_0 + Z')$, cannot be shrunk to a point in $E^3 - (A_{01} + A_{02} + Z')$. Theorem 9 of [6] shows that this last condition holds.

Two continua A_{0i1}, A_{0i2} are imbedded in A_{0i} ($i = 1, 2$) as A_{01}, A_{02} were imbedded in A_0. There are $2^2 A_{0ij}$'s at this second stage. At the third stage, two A's are placed in each of these so there are $2^3 A$'s defined. The process is continued.

The A's might have been defined so that each component of $H = A_0 \cdot (\sum A_{0i}) \cdot (\sum A_{0ij}) \cdot \cdots$ is a triod which is invariant under a rotation of 120° about Z, and which does not intersect any circular cylinder with Z as center in more than three points, so we suppose this was done.

Let G be the upper semi-continuous decomposition of E^3 whose non-degenerate elements are the components of H.

2. *Shrinking the* A's. We show that, for each positive number ε, there is an integer n and a homeomorphism h of E^3 onto itself such that h is fixed on $E^3 - A_0$, and h takes each A at the nth stage onto an ε set. Figures are more effective in giving an easily understood description of h than are equations, so we employ them. We show in Figures 3, 4, and 5 how h acts on A_{01} and A_{02}. The method of shrinking is analogous to that employed in [2] and Figures 4, 5 of that paper may help the reader to understand the argument.

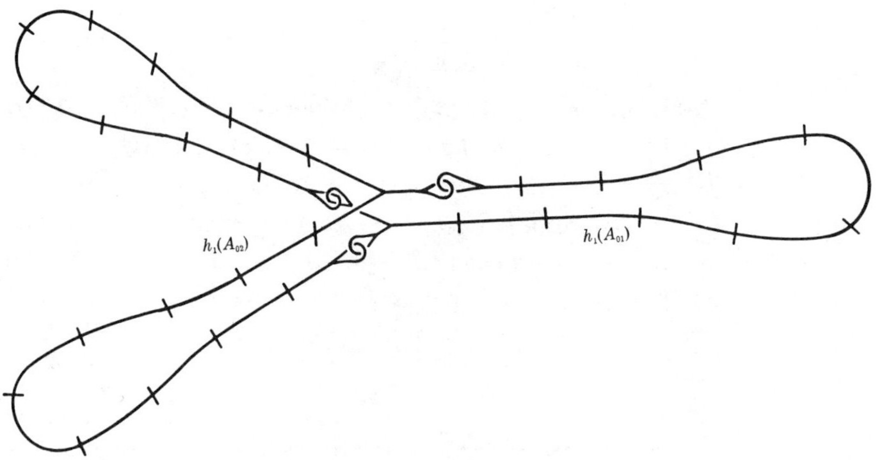

Figure 3.

We define h as the product of several homeomorphisms. The first of these is called h_1 and shrinks the right end of A_{02} and the two left ends of A_{01}. The corresponding linked ends are stretched to bring about a situation as depicted in Figure 3. The slashes indicate planar cuttings and are such that any connected ε subset of $h_1(A_{01} + A_{02})$ intersects at least four of them. Then $h = h_1$ on $E^3 - (A_{01} + A_{02})$.

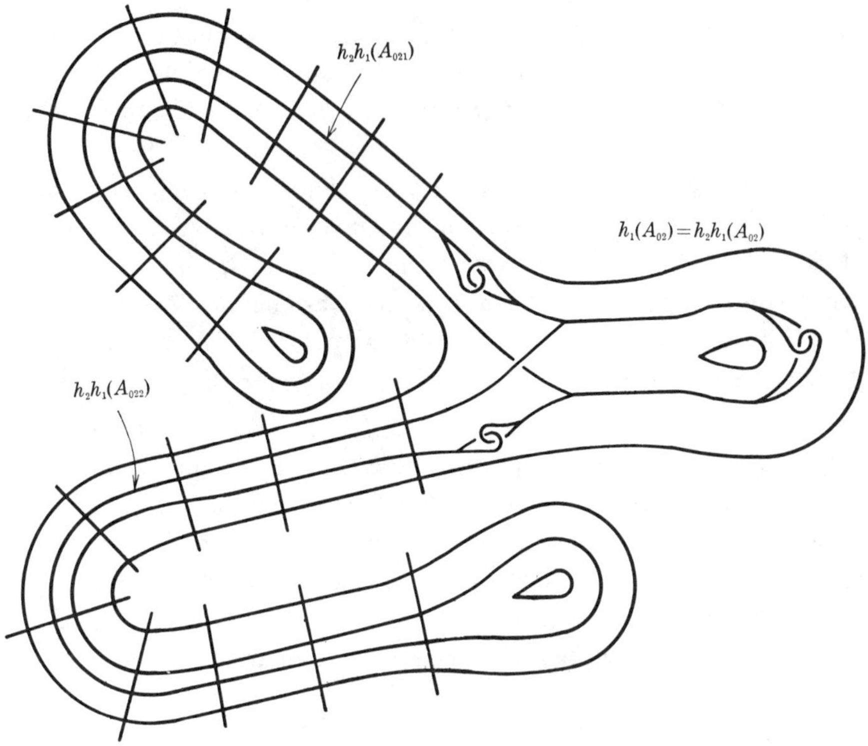

Figure 4.

At first it might appear that it would be easier to define h inside A_{01} than inside A_{02}, since $h_1(A_{01})$ has only one long end while $h_1(A_{02})$ has two. However, $h_1(A_{02})$ is shown enlarged in Figure 4, so that we can see the images of A_{021} and A_{022} inside it after we consider a homeomorphism h_2 of E^3 onto itself that is the identity outside $h_1(A_{02})$, and such that h_2h_1 pulls A_{021} and A_{022} as shown so that each has only one long end. We suppose that $h = h_2h_1$ on $A_{02} - (A_{021} + A_{022})$

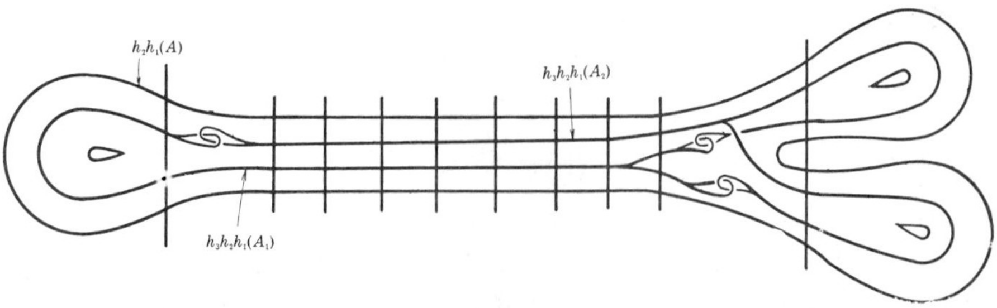

Figure 5.

so we only concern ourselves henceforth with A's with one long end. We use A to denote any one of A_{01}, A_{021}, A_{022}, and shall use A_1, A_2 to denote the two A's in A.

Although we arrived at a situation in Figures 3 and 4 where $h_2h_1(A)$ has no cuttings on its short ends, we show it in Figure 5 with a single slash cutting the two short ends to make our inductive argument easier. Suppose the long end of $h_2h_1(A)$ has $n-1$ cuttings. Figure 5 shows how h_3 acts on $h_2h_1(A_1)$ and $h_2h_1(A_2)$. Note that $h_3h_2h_1(A_1)$ and $h_3h_2h_1(A_2)$ have only one long end. The last cutting of a long end of these can be ignored (as can the last cutting of each short end) since this cutting is already represented by an adjacent one. Any connected ε subset of $h_3h_2h_1(A_1 + A_2)$ still intersects at least 4 of the remaining cuttings.

We have now arrived at a simpler situation in which each long end of $h_3h_2h_1(A_i)$ has $n-2$ cuttings instead of $n-1$, where A_i's represent an A at the second or third stage. For simplicity we suppose it is the third. By a repetition of the above procedure we get an h such that no end of an A at the fourth stage has more than $n-3$ cuttings, none at the fifth stage more than $n-4$ cuttings, $\cdots$, and none at the nth stage has more than 1 cutting. Then the image of an A at the nth stage under h has diameter less than ε since it does not intersect four cuttings.

What we have just shown about A_0 is equally true about the other A's at later stages. For each open set U containing H and each $\varepsilon > 0$, there is a homeomorphism h_ε of E^3 onto itself such that $h_\varepsilon = I$ on $E^3 - U$, and the diameter of each component of $h_\varepsilon(H)$ is less than ε.

3. *The decomposition space of G is topologically* E^3. This follows from an argument like that used on pages 359–361 of [2] to show that the decomposition space of G_1 described in the proof of Theorem 2.1 of the present paper is topologically E^3. It also follows from the following lemma which is proved on pages 364–366 of [5] (though not stated there).

LEMMA 3.2. *Suppose G is a monotone upper semi-continuous decomposition of E^3, and H is the sum of the non-degenerate elements of G. Then the decomposition space of G is topologically E^3 if each component of H is an element of G, and for each open set U containing H and each $\varepsilon > 0$, there is a homeomorphism h_ε of E^3 onto itself which is fixed on $E^3 - U$ and takes each component of H onto an ε set. In fact, for each open set V containing H the homeomorphism between the decomposition G and E^3 may be realized by a map f such that, for each point p of E^3, $f^{-1}(p)$ is an element of G and $f = I$ on $E^3 - V$.*

4. *The image of Z is wild.* Let f be the map showing that the decomposition space of G is topologically E^3. If J is a closed curve in $E^3 - (A_0 + Z')$ that cannot be shrunk to a point in $E^3 - (A_0 + Z')$, it follows from Theorem 9 of [6] that J cannot be shrunk to a point in $E^3 - (Z' + \sum A_{01})$, nor in $E^3 - (Z' + \sum A_{0ij})$, nor in $\cdots$, and hence not in $E^3 - (Z' + H)$. Then $f(J)$ cannot be shrunk to a point in the complement of $f(Z' + H) = f(Z')$. Since f is a homeomorphism on Z, $f(Z')$ is a wild arc whose complement is not simply connected. In fact, $f(H)$ is the wild subset of $f(Z)$.

5. *The rotation.* Let r be a rotation of 120° of E^3 about Z. Then frf^{-1} gives a rotation of period 3 of E^3 about the wild line $f(Z)$.

Theorem 3.1 can be extended as follows.

THEOREM 3.3. *For each integer n_0 greater than* 1, *there is a rotation of period n_0 of E^3 about a wild line.*

PROOF. The decomposition G we use to prove this theorem is like that used to prove Theorem 3.1 except that the A's have n_0 ends instead of 3.

The proof that the decomposition space of G is topologically E^3 is modeled after the corresponding part of the proof of Theorem 3.1. To get a homeomorphism h which is the identity on $E^3 - A_0$, and which shrinks the A's at some later stage, we start by taking a homeomorphism h_1 of E^3 onto itself so that neither $h_1(A_{01})$ nor $h_1(A_{02})$ has more than $(n_0 + 1)/2$ long ends. In a finite number of such steps, we arrive at an $h_i h_{i-1} \cdots h_1$ such that the images of each A at the ith stage has at most one long end. We suppose the long end of $h_i h_{i-1} \cdots h_1(A)$ has $n - 1$ cuttings and that a single cutting suffices for the other ends. Then we reduce the number of cuttings of the long end as we did in the proof of Theorem 3.1 and as indicated in Figure 6.

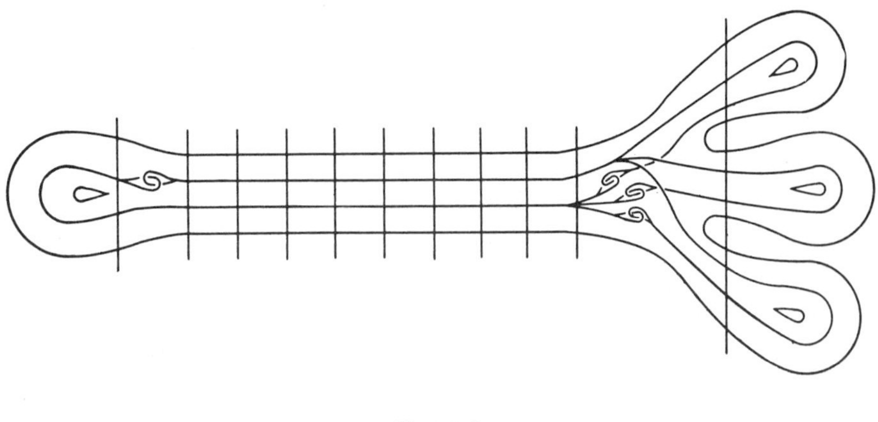

Figure 6.

4. Families of rotations. Proof of Theorem 4.1

Theorem 4.1 follows from Theorem 3.2 in the same fashion that Theorem 2.1 follows from properties of the decomposition of G_1 of Theorem 2.1. Let Z be the z axis, X be a closed subset of Z mentioned in the statement of Theorem 4.1, and $p_1, p_2, \cdots$ be a sequence of points of X dense in X.

Let G_1 be an upper semi-continuous decomposition of E^3 as described in Theorems 3.1 and 3.3 such that p_1 lies in a non-degenerate element of G_1, each non-degenerate element of G_1 intersects X, and each non-degenerate element of G_1 goes onto itself under a rotation of $(360/n)°$ about Z. Let f_1 be the map of E^3 onto itself such that, for each point p of E^3, $f_1^{-1}(p) \in G_1$ and $f_1 = 1$ on $E^3 - A_0$.

If some point of X does not belong to any non-degenerate element of G_1, let q_2 be the first point $p_2, p_3, \cdots$ not belonging. Then ε_1 is taken to be a positive number so small that if p, q are points of E^3 with $\rho(p, q) < \varepsilon_1$, then $\rho(f_1(p), f_1(q)) < 1/2$. A decomposition G_2 is obtained like G_1; but, so that the A_0 used in selecting it contains q_2, is of diameter less than ε_1 and does not intersect any non-degenerate element of G_1. We select G_2 so that each non-degenerate element of G_2 intersects X and one contains q_2.

We continue selecting $G_3, G_4, \cdots$, and let G be the decomposition of E^3 such that g is a non-degenerate element of G if and only if it is a non-degenerate element of some G_i. It is shown that the decomposition space of G is topologically equivalent to E^3 in the same fashion that it is shown that the decomposition space of G of Theorem 2.1 is topologically E^3.

Let f be the uniformly continuous map of E^3 onto itself such that, for each point p of E^3, $f^{-1}(p) \in G$. Then if r is a rotation of E^3 about Z of period n, frf^{-1} gives a rotation of E^3 about the wild line $f(Z)$ whose wild subset is $f(X)$.

Looking at S^3 instead of E^3, we get the following variations of Theorem 4.1.

COROLLARY 4.3. *For each integer n greater than* 1, *and each perfect subset X of a circle, there is a homeomorphism h of S^3 onto itself of period n such that the fixed point set of h is a simple closed curve whose wild subset is topologically equivalent to X.*

COROLLARY 4.4. *For each integer n greater than* 1 *there are uncountably many mutually inequivalent periodic homeomorphisms of S^3 of period n each of whose fixed point sets is a simple closed curve.*

Looking at a cylinder instead of all of E^3, we get the following variation of Theorem 4.1. This variation is used in the next section.

THEOREM 4.5. *Suppose C is the solid cylinder $x^2 + y^2 \leq 1$, $0 \leq z \leq 1$ in*

E^3. *For each integer n greater than* 1, *and each closed subset X of an arc, there is a homeomorphism h of period n of C onto itself such that the fixed point set of h is a spanning arc of C whose wild subset is topologically equivalent to X, but on* Bd C, *h is the ordinary rotation of* $(360/n)°$ *about the z-axis.*

5. Fixed point free maps. Proof of Theorem 5.1

Suppose n_1 is a factor of the n of Theorem 5.1.

We regard S^3 as the sum of two solid tori T_1, T_2. Let h' be the homeomorphism of S^3 onto itself such that h' rotates T_1 through $(360/n_1)°$ longitudinally and $(360/n)°$ meridianally, while h' rotates T_2 through $(360/n)°$ longitudinally and $(360/n_1)°$ meridianally. Each homeomorphism of the family promised by Theorem 5.1 is h' on T_2 but will be a modification of h' on T_1.

Chop T_1 up into n_1 congruent curved cylinders as shown in Figure 7. Suppose r is a wild rotation of C_1 as described in Theorem 4.5 such that on Bd C_1, r is an ordinary rotation of $(360/n)°$ about the axis of T_1 but the fixed point set is a wild arc with a prescribed wild subset. Suppose r is extended to all of T_1 so that, if t is the longitudinal rotation of T_1 by $(360/n_1)°$, then $r = t^i r t^{-i}$ for each integer i.

On T_1, $h = tr$. Note that, although h leaves no point fixed, h^{n_1} leaves a wild simple closed curve fixed. Theorem 5.1 follows from the fact that we

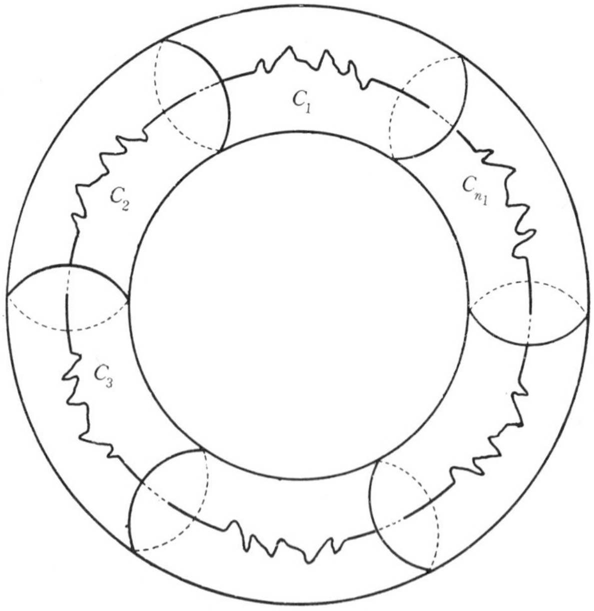

Figure 7.

could take the wild subset of the fixed point set of C_1 under r in uncountably many topologically different ways.

6. 0-spheres as fixed point sets

Consider the following version of Theorem 6.1.

THEOREM 6.2. *For each integer n greater than* 1, *there are uncountably many mutually inequivalent homeomorphisms of E^3 onto itself of period $2n$ such that the fixed point set of each is a point.*

To see that Theorem 6.2 follows from Theorem 4.1, we might consider an uncountable family of closed subsets of Z such that each misses the origin, each is symmetric with respect to the origin, and no two are topologically equivalent. The decomposition we select is such that each non-degenerate element is invariant under a rotation of $(360/2n)^\circ$, and each such element has another element symmetric to it with respect to the xy-plane. If f represents the map of E^3 onto itself that shrinks the elements of the decomposition to points, r_1 denotes the reflection of E^3 through the xy plane, and r_2 denotes a rotation of E^3 about Z of period $2n$, then $fr_2r_1f^{-1}$ gives a periodic homeomorphism of period $2n$ with only the origin fixed. However, $(fr_2r_1f^{-1})^2$ has a wild line of fixed points.

If n is odd, we could use a combination of Theorems 2.1 and 4.1 to get a homeomorphism h of period $2n$ whose fixed point set is a point such that h^2 has a wild line of fixed points and h^n has a wild plane of fixed points.

There is no loss of generality in restricting ourselves to periodic transformations of even period in the above theorem for, as pointed out in [17], any periodic homeomorphism of E^3 which leaves only one point fixed reverses orientation and hence is of even period.

UNIVERSITY OF WISCONSIN AND
INSTITUTE FOR ADVANCED STUDY

REFERENCES

1. W. R. ALFORD, Uncountably many different involutions of the three sphere, Abstract 589-4, Amer. Math. Soc. Notices, 9 (1962), 105.
2. R. H. BING, *A homeomorphism between the* 3-*sphere and the sum of two solid horned spheres*, Ann. of Math., 56 (1952), 354-362.
3. ———, Decompositions of E^3 into points and tame arcs, in Summary of Lectures and Seminars, Summer Institute on Set Theoretic Topology, Madison, 1955, pp. 41-48.
4. ———, *Locally tame sets are tame*, Ann. of Math., 59 (1954), 145-158.
5. ———, *Upper semicontinuous decompositions of* E^3, Ann. of Math., 65 (1957), 363-374.
6. ———, *A wild surface each of whose arcs is tame*, Duke J. Math., 28 (1961), 1-16.
7. R. H. FOX, Knots and periodic transformations, in Topology of 3-Manifolds and Related Topics, Prentice Hall, 1962, pp. 177-182.
8. ———, *On knots whose points are fixed under a periodic transformation of the*

3-*sphere*, Osaka Math. J., 10 (1958), 31-35.
9. O. G. HARROLD, JR. and E. E. MOISE, *Almost locally polyhedral spheres*, Ann. of Math., 57 (1953), 575-578.
10. M. W. HIRSCH and S. SMALE, *On involutions of the* 3-*sphere*, Amer. J. Math., 81 (1959), 893-900.
11. S. KINOSHITA, *On knots and periodic transformations*, Osaka Math. J., 10 (1958), 43-52.
12. G. R. LIVESAY, *Fixed point free involutions on the* 3-*sphere*, Ann. of Math., 72 (1960), 603-611.
13. ———, *Involutions with two fixed points on three-sphere*, Ann. of Math., 78 (1963), 582-593.
14. EDWIN MOISE, *Periodic homeomorphisms of the* 3-*sphere*, Illinois J. Math., 6 (1962), 206-225.
15. DEANE MONTGOMERY and LEO ZIPPIN, *Examples of transformation groups*, Proc. Amer. Math. Soc., 5 (1954), 460-465.
16. DEANE MONTGOMERY and HANS SAMELSON, *A theorem of the fixed points of involutions in* S^3, Canad. J. Math., 7 (1955), 208-220.
17. P. A. SMITH, *Transformations of finite period*: II, Ann. of Math., 40 (1939), 690-711.
18 WU, TA-SUN, On the involutions of the 3-sphere, Abstract 61T-268, Amer. Math. Soc. Notices, 8 (1961), p. 518.

(Received March 28, 1963)

C. Approximating and Embedding Surfaces

Approximating Surfaces With Polyhedral Ones

R. H. Bing

Suppose a person had a large rough lump of glass that he wanted to have polished and cut to resemble a diamond. What instructions could he give to have the work done? If he gave instructions to have rough spots filed off, a workman, blindly following his instructions, might file into tunnels from the other side so that the surface of the lump would no longer be like the surface of a sphere. See the left figure below.

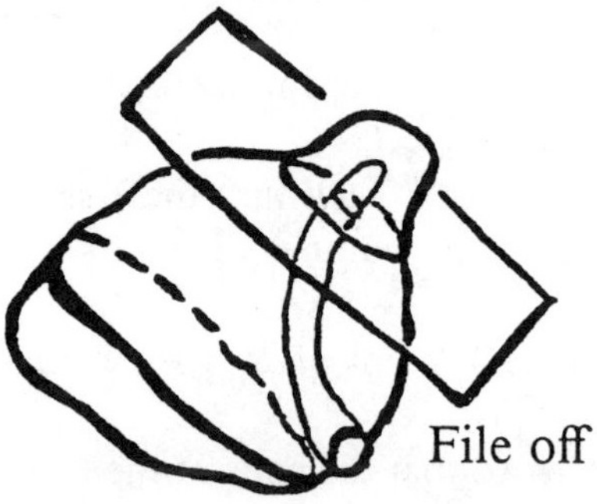

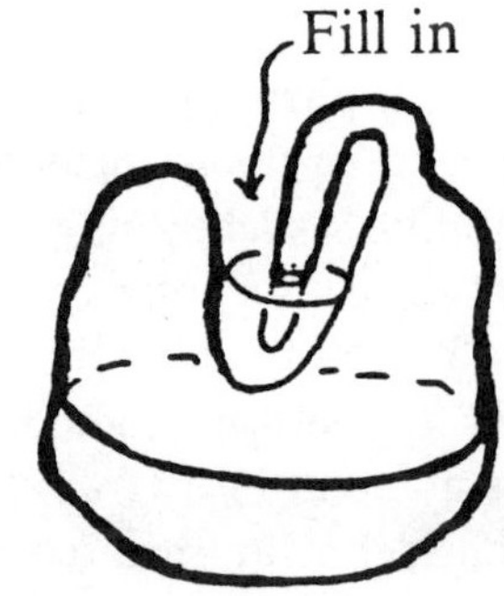

If he ordered that holes and crevices be filled in, a handle might result. See the right figure above. Perhaps it would be easier to use the large rough lump as a model and make a polyhedral solid which approximated it. To get a suitable approximation, it might be well to get a ruling or lattice subdivision of E^3 so as to see where to put our approximation. In the proof of the following result, we use a triangulation of E^3 with small mesh for this purpose.

THEOREM. *If S is a simple surface in E^3 and ε is a positive number, there is a polyhedral simple surface S'' and a homeomorphism h of S onto S'' such that for each point s of S, $h(s)$ is within ε of s.*

The general idea of attack is as follows: Triangulate 3-space. Consider a component of $S \cdot C$ as shown where C is one of the tetrahedra of the triangulation. Replace this component with a polyhedral disk in C as shown. For all such tetrahedra C of the triangulation, a selected collection of the components

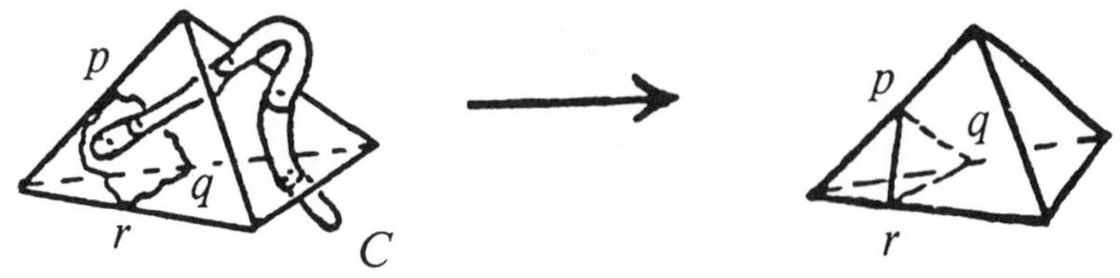

of $C \cdot S$ can be replaced by disks and the disks sewn together so as to form a polyhedral surface approximating S.

Now for a few more details.

1. There is a number δ such that no continuum in S of diameter less than δ divides S into two pieces each of diameter greater than $\varepsilon/2$.

This is impossible!

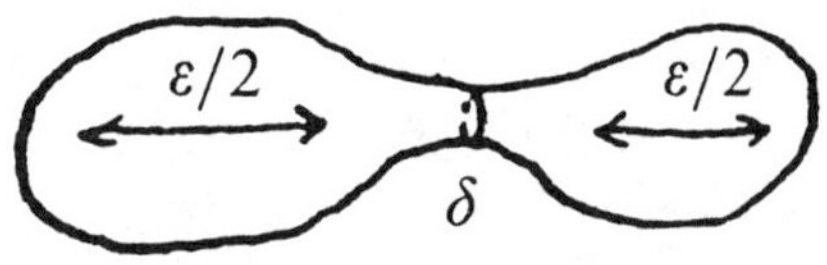

2. There is a triangulation of E^3 of small mesh (depending on δ) so that the 0-skeleton R_0 of the triangulation does not intersect S.

3. Modify S slightly to get a simple surface S' such that the 1-skeleton R_1 of the triangulation intersects S' in only a finite number of points. S' is not obtained by deforming S but as follows:

a. Put a spherical neighborhood about each point of R_0 so that the sphere contains no point of S.

b. Consider cylindrical tubes covering R_1 so that the ends of the tubes are in the spheres.

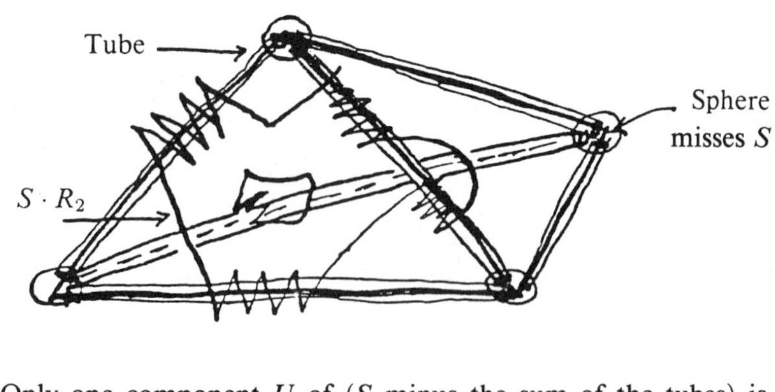

c. Only one component U of (S minus the sum of the tubes) is of large diameter.

d. If G is the decomposition of $\overline{U}$ whose elements are the points of U and the components of $\overline{U} - U$, the decomposition space is S^2.

e. If T is the surface of a tube, only a finite number of elements of G separate the two ends of the tube. Suppose the tubes were adjusted slightly (shrunk) so that no component of $T \cdot S$ divides S into as many as three pieces.

f. Each of the elements of G mentioned in (e) are shrunk to points as shown causing the tube to collapse as shown. The other elements of G on T which do not separate the ends of T are also shrunk so that a cactoid results (*Some monotone decompositions of a cube*, Annals, vol. 61, 1955, pp. 279–288).

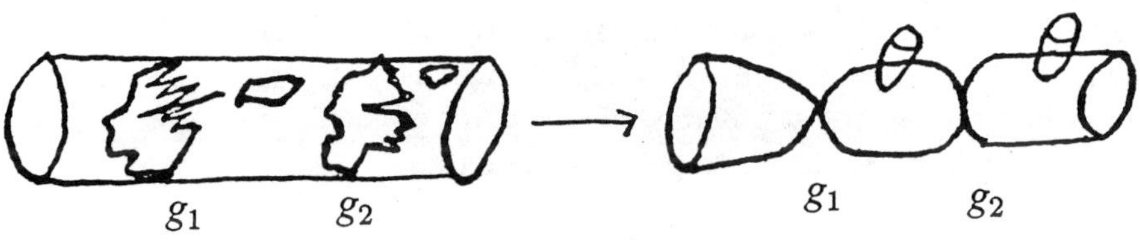

g. The nondegenerate elements of G have been shrunk to points so $\overline{U}$ has been changed to a simple surface S'. There is a homeomorphism of S onto S' so that a point and its image are close together. $S' \cdot R_1$ is finite.

4. Consider the intersection of a face F of the triangulation and S'. Modify F slightly (keeping its boundary fixed and pulling at its center) so that if X is a component of $S' \cdot$ Interior F, then $\overline{X}$ does not contain three points of R_1. Modify each such face.

5. For each such continuum $\overline{X}$ with two points on R_1, let A_X be a polygonal arc in (Interior F) $+\overline{X}$ which has its end points on R_1. If $\overline{X}_1, \overline{X}_2$ are two such continua in F with $\overline{X}_1 \cdot R_1 = \overline{X}_2 \cdot R_1$, then $A_{X_1} = A_{X_2}$; otherwise A_{X_1} does not intersect A_{X_2} except possibly at its endpoints.

6. If $\overline{X}_1, \overline{X}_2$, are two continua in the same face F as mentioned above so that $\overline{X}_1 \cdot R_1 = \overline{X}_2 \cdot R_1$, then only one component of $S' - (\overline{X}_1 + \overline{X}_2)$ is of large diameter. Add $\overline{X}_1, \overline{X}_2$, and the small components together to get a maximum continuum $\overline{X}$. The boundary (with respect to S') of $\overline{X}$ will be a continuum in F

and $\overline{X}$ will not separate S'. There will be only a finite number of such maximal continua $\overline{X}$ and their sum K will divide S' as shown. Each component of $S'-K$ will be of small diameter.

7. For each such continuum $\overline{X}$, place an arc B_X in S' near $\overline{X}$ so that the B's divide S' essentially as do the $\overline{X}$'s and B_X has the same ends as A_X. The B's do not intersect except at possibly an end point of each.

8. Throw away inside B's in positions like these.

or

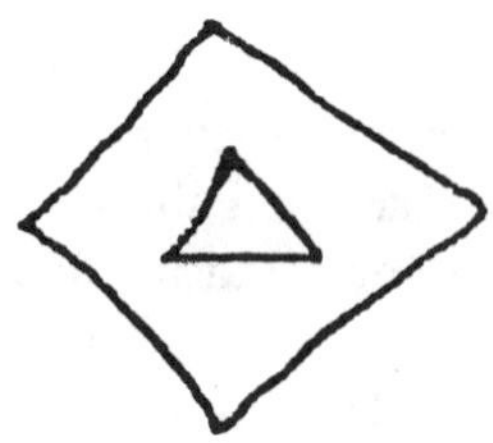

9. Let h be a homeomorphism carrying the remaining B's onto the corresponding A's. It may be noted that each end point of a B is fixed under h.

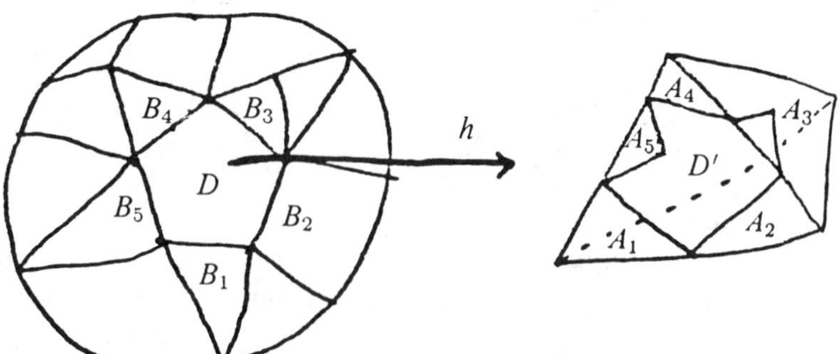

10. If a set of B's bound a disk D in S' and Interior D contains none of the remaining B's, this disk is associated with a tetrahedron C of the subdivision that contains the X's associated with the B's on the boundary of the disk as

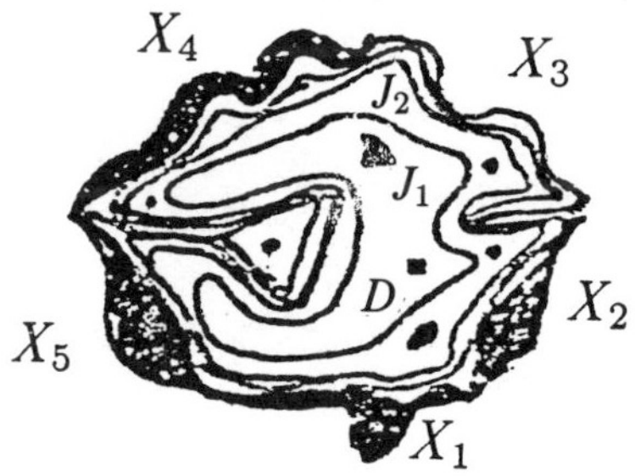

well as much of D. To get one such tetrahedron C (there may be two but we use only one) we consider the $\overline{X}$'s that are associated with the B's bounding D. Suppose the X's bound D''.

There is a sequence of simple closed curves $J_1, J_2, \ldots$ in D'' as shown such that the boundary of D'' belongs to the limiting set of these J's and no J_i intersects the 2-skeleton of our triangulation. Then there is a C containing infinitely many of the J's and we use it.

11. There is a polyhedral disk D' in C which lies, except for its boundary, in Interior C and whose boundary is the sum of the A's associated with the B's on the boundary of D. These D's may be chosen so that no two of them intersect except along a boundary of each.

12. The homeomorphism h of the B's into the A's may be extended to take the D's onto the D''s.

The sum of the D's is the required polyhedral surface S'' and the homeomorphism is h.

The theorem we have just discussed treated a simple surface (set homeomorphic to S^2). However, the methods we employed can be used to get the following extension of it.

THEOREM. *Suppose M_2 is a 2-manifold (with or without boundary) in a triangulated 3-manifold M_3 and f is a nonnegative continuous function defined on M_2. Then there is a 2-manifold M_2' (with or without boundary) in M_3 and a homeomorphism h of M_2 onto M_2' such that (1) M_2' is locally polyhedral at each point $h(x)$ for which $f(x)$ is positive, and (2) the distance between x and $h(x)$ is not more than $f(x)$.*

Reference

1. R. H. Bing, *Approximating surfaces with polyhedral ones*, Ann. of Math., vol. 65 (1957) pp. 456–483.

ANNALS OF MATHEMATICS
Vol. 65, No. 3, May, 1957
Printed in U.S.A.

APPROXIMATING SURFACES WITH POLYHEDRAL ONES

BY R. H. BING*

(Received May 29, 1956)

1. Introduction

If J is a simple closed curve in the plane E^2, the Schoenflies Theorem says that there is a homeomorphism of E^2 onto itself that takes J onto a circle. The theorem does not generalize directly to E^3 because there is a simple surface S in E^3 such that there is no homeomorphism of E^3 onto itself that takes S onto the surface of a sphere. However, if S is polyhedral, there is a homeomorphism of E^3 onto itself that takes S onto the surface of a sphere [1]. Hence, polyhedral surfaces are imbedded in E^3 in a simpler fashion than are some other surfaces.

The intersection of two polyhedral surfaces is much less complicated than the intersection of some other surfaces. For this reason, mathematicians sometimes impose the additional condition of being polyhedral on some of the surfaces they use. Applications of the results of this paper show that this extra condition is not always necessary.

Harrold gives several applications of the approximation theorem for surfaces (Theorem 7) of this paper in [7]. I make extensive use of it in showing that there is a simple closed curve in E^3 which pierces no disk [4]. In a subsequent paper I shall show that an extension of this theorem to topological 2-complexes can be used to give alternate proofs of the theorems due to Moise to the effect that each 3-manifold can be triangulated [9] (also a 3-manifold with boundary can be triangulated [2]) and that each homeomorphism of one 3-manifold with boundary into another can be approximated arbitrarily closely by a piecewise linear homeomorphism [8].

In this paper we consider the extent to which arbitrary surfaces can be approximated by polyhedral ones. We show that each surface can be homeomorphically approximated arbitrarily closely by one that is locally polyhedral. Therefore, it can be approximated by a polyhedron if it is compact [Lemma 1 of 2].

A *simple surface* is a continuum topologically equivalent to the surface of a sphere. We use the term "*surface*" synonomously with "2-manifold with boundary."

An *n-manifold* is a separable metric space K such that each point of K lies in an open set that is topologically equivalent to E^n (Euclidean n-space). An *n-manifold with boundary* is a separable metric space K such that each point of K lies in an open set N such that $\bar{N}$ is topologically equivalent to I^n (cube in E^n). We note that an n-manifold is an n-manifold with boundary but not conversely.

The term *boundary* is used in two senses. In the point set sense we say that the *boundary of a set* X is the intersection of $\bar{X}$ (the closure of X) and the closure

* Work on this paper supported in part by Contract NSF-G479.

of the complement of X. However, if K is an n-manifold with boundary, we use Bd K to denote the set of points of K which do not have neighborhoods topologically equivalent to E^n. Hence if K is respectively an arc, an annulus, a disk, or a tetrahedron, Bd K is respectively two points, the sum of two simple closed curves, a simple closed curve, or a simple surface. We use Int K to denote $K -$ Bd K.

A surface is *polyhedral* if it is the sum of a finite number of triangular disks. It is *locally polyhedral* at a point if there is an open set N containing the point such that $\bar{N}$ is the sum of a finite number of triangular disks.

In this paper theorems will be listed in a chronological, rather than a logical, order. Instead of basing each theorem on the preceding ones, we first list the theorem (Theorem 1) that engendered the paper, then give the proof of this theorem based on certain lemmas and supporting theorems, then prove the lemmas and auxiliary theorems, and finally give two generalizations (Theorems 7, 8) of the main theorem. With certain exceptions, this is the chronological order in which the theory was developed although it must be admitted that some of the ideas came in bunches rather than linearly. Much of mathematics seems to come forth in this order.

2. Approximating simple surfaces in E^3: Simple case

In this section, we consider the closeness with which a simple surface in E^3 can be approximated by a polyhedral one. A diamond cutter faces this problem when he undertakes to polish a rough stone until its surface is polyhedral. Perhaps the stones with which he works are not as complicated as some wild solids that topologists can describe. In this section we treat surfaces that have a certain property of niceness and leave until Section 4 the completion of the proof of Theorem 1 for the arbitrary case.

Suppose a person has a large rough lump of glass that he wants to have polished and cut to resemble a diamond. What instructions can he give to have the work done? If he gives instructions to have the rough spots filed off, a workman, blindly following his instructions, may file into tunnels from the other side so that the surface of the lump is no longer like the surface of a sphere. (See Figure 1.) If he orders that holes and crevices be filled in, handles may result as shown in Figure 2.

Perhaps it would be easier to use the large rough lump as a model and make a polyhedral solid approximating it. To get a suitable approximation, it might be well to get a ruling or lattice subdivision of E^3 so as to see where to put our approximation. A person who has used lines of latitude and longitude on a map in helping him to copy it has used such a scheme. Reducing the mesh of the ruling is likely to improve the approximation.

The following proof is the one outlined in [3]. The reader may find it useful to go over the short summary of proof in [3] before going through the more detailed proof in the present paper.

THEOREM 1. *If S is a simple surface in E^3 and ε is a positive number, there is a*

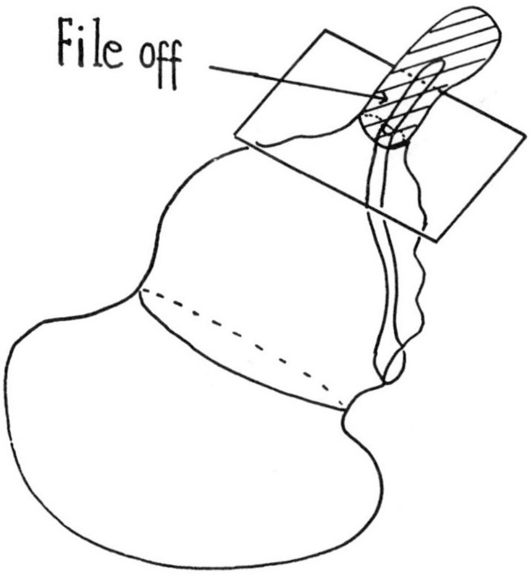

FIG. 1

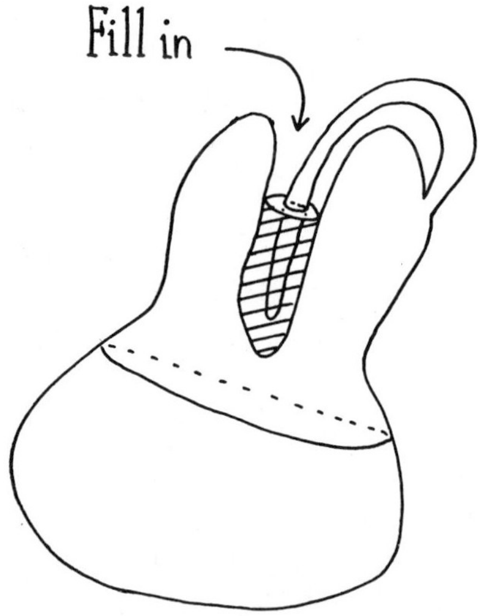

FIG. 2

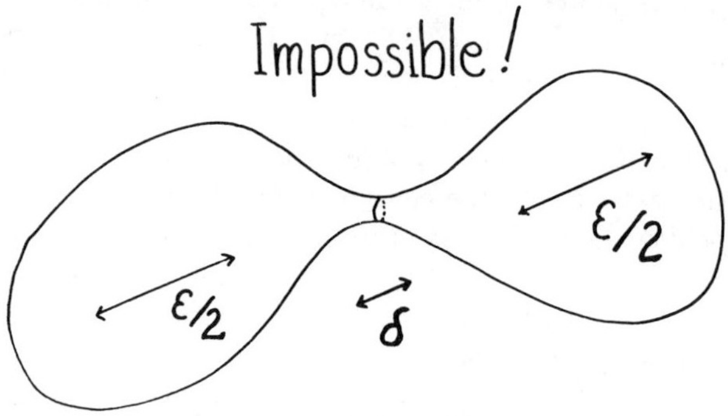

FIG. 3

polyhedral simple surface S' in E^3 and a homeomorphism h of S onto S' such that h moves no point by more than ε.

PROOF. There is a positive number δ such that no continuum in S of diameter less than δ has two complementary domains in S each of diameter more than $\varepsilon/2$. We do not have a situation as depicted in Figure 3. We suppose that $\delta < \varepsilon/2$.

There is a triangulation R of E^3 of mesh less than δ such that no vertex of R lies on S. We shall use R_i $(i = 0, 1, 2)$ to denote the i-skeleton of R. Then R_0 is the sum of the vertices of R, R_1 is the sum of the edges of the tetrahedrons in R, and R_2 is the sum of the faces of the tetrahedrons in R.

We first prove the theorem for a special case. This case is much simpler than the general case but it reveals the method of attack to be used in the general case. In the special case, we suppose that the following condition holds:

Assumption for special case. $S \cdot R_2$ is a finite graph G such that $S \cdot R_1$ may be regarded as the sum of the vertices of G.

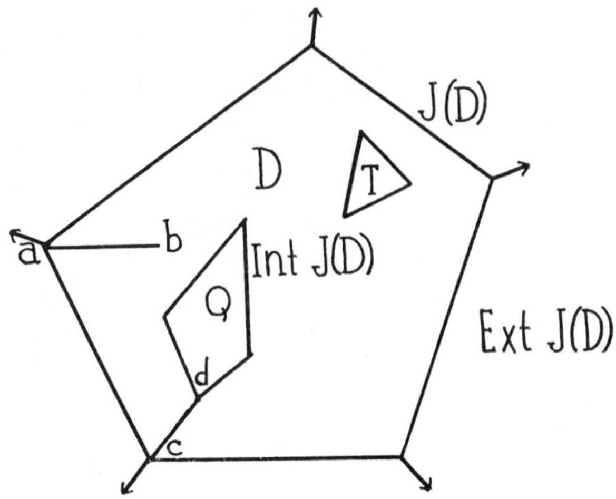

FIG. 4

A finite graph is the sum of a finite number of arcs such that no one of the arcs intersects the interior of another one of them. The ends of the arcs are regarded as the vertices of G. For each finite graph, there are many finite collections of points each of which may be regarded as an admissible set of vertices for G, since G may be expressed as the sum of arcs in many ways.

We shall enumerate steps in the consideration of the special case.

1. It would be convenient for our purposes if each component of $S - G$ were the interior of a topological disk. Therefore, in this step we eliminate certain unneeded portions of the finite graph G. If a part of G looked as shown in Figure 4, we would throw away the sect $(ab]$, the sect $(cd]$ + the quadrilateral Q, and the triangle T.

If D is a component of $S - G$, $S - \bar{D}$ is the sum of a collection of open disks. Only one component of $S - \bar{D}$ is of diameter greater than ε and we denote the boundary of this component by $J(D)$. We note that $J(D)$ is a simple closed curve even though Bd D may not be one. We use Ext $J(D)$ and Int $J(D)$ to denote the components of $S - J(D)$ of diameters greater than and less than ε. Then $D \subset \text{Int } J(D)$.

Form a finite graph G' by eliminating from G each point p of G such that for some component D of $S - G$, $p \in \text{Int } J(D)$. Then G' is a finite connected graph without cut points and for each component D' of $S - G'$, there is a component D of $S - G$ such that $D' = \text{Int } J(D)$.

Let $E_1, E_2, \cdots, E_n$ be the components of $S - G'$ and $D_1, D_2, \cdots, D_n$ be the components of $S - G$ such that $E_i = \text{Int } J(D_i)$. Each E_i is of diameter less than $\varepsilon/2$ because its boundary $J(D_i)$ is of diameter less than δ.

2. In this step we polygonize G'.

If F is a face of a tetrahedron of the triangulation R and $(a_1x_1b_1)$, $(a_2x_2b_2)$, $\cdots$, $(a_nx_nb_n)$ are the components of $G' \cdot \text{Int } F$, replace each of the open arcs $(a_ix_ib_i)$

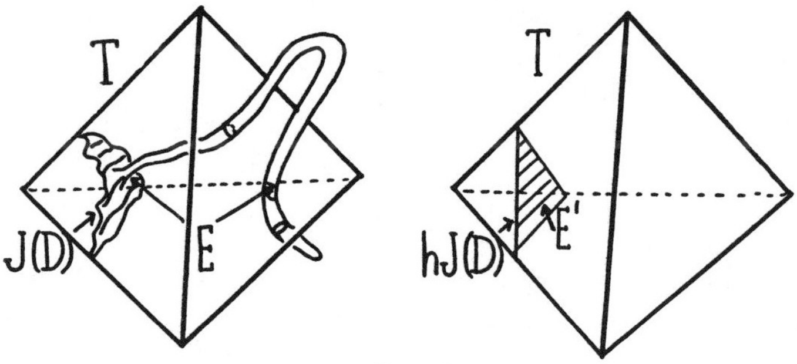

FIG. 5

by an open arc $(a_iy_ib_i)$ in Int F such that $a_iy_ib_i$ is a polygonal arc with the same ends as $a_ix_ib_i$ and $(a_iy_ib_i)\cdot(a_jy_jb_j) = 0$ if $i \neq j$.

3. In this step we start our homeomorphism h. Let h be a homeomorphism which takes each such $a_ix_ib_i$ of Step 2 onto the corresponding $a_iy_ib_i$ and which is fixed at a_i and b_i. Then h takes G' homeomorphically onto a finite polygonal graph and h moves no point of G' by as much as δ.

4. In this step we define the surface S'. For each E_i there is a tetrahedron T_i of the triangulation R such that $D_i \subset$ Int T_i. Then $hJ(D_i)$ is a polygon on Bd T_i. Let E_i' be a polygonal disk such that Bd $E_i' = hJ(D_i)$, Int $E_i' \subset$ Int T_i, and Int $E_i'\cdot$Int $E_j' = 0$ if $i \neq j$. It follows from Theorem 2 that we can pick the E_i''s in this fashion if no $hJ(D_i)$ intersects two components of any Bd $T_i - hJ(D_j)$. Theorem 3 reveals that $J(D_i)$ does not intersect two components of Bd $T_i - J(D_j)$ if D_i, D_j are components of $S\cdot$Int T_i. Since the homeomorphism h of $J(D_i) + J(D_j)$ can be extended to a homeomorphism of Bd T_i onto itself, $hJ(D_i)$ does not intersect two components of Bd $T_i - hJ(D_j)$.

Then $E_1' + E_2' + \cdots + E_n'$ is the required simple surface S'. Figure 5 shows how an E is replaced by an E'.

5. Finally, we finish the homeomorphism h. For each E_i, we have already defined a homeomorphism h of $J(D_i) =$ Bd E_i onto Bd E_i'. This homeomorphism is extended to take E_i onto E_i'. The homeomorphism h moves no point by more than $\delta + \varepsilon/2 < \varepsilon$ because $E_i' \subset T_i$, diameter $T_i < \delta$, E_i intersects T_i, and diameter $E_i < \varepsilon/2$.

Now h takes S homeomorphically onto $S' = \sum E_i'$ and moves no point by as much as ε. This disposes of the special case. Before proceeding to the general case, we pause to prove Theorems 2 and 3 which were used in Step 4.

3. Building disks in a tetrahedron

In the proof of Theorem 1 for the special case, we replaced topological disks in S with polyhedral disks in tetrahedrons. The following theorem shows how polyhedral disks may be found in a tetrahedron.

THEOREM 2. *Suppose $J_1, J_2, \cdots, J_n$ is a finite collection of simple polygons on the boundary of a tetrahedron T such that no J_i intersects both components of*

Bd $T - J_j$. *Then there is a finite collection of polygonal disks* $D_1, D_2, \cdots, D_n$ *such that* D_i *has* J_i *for a boundary,* $\text{Int}\, D_i \subset \text{Int}\, T$, *and* $\text{Int}\, D_i \cdot \text{Int}\, D_j = 0$ $(i \neq j)$.

PROOF. Let w be a point of Bd $T - \sum J_i$. Use $I(J_i)$ to denote the component of Bd $T - J_i$ which does not contain w. The $I(J_i)$'s are partially ordered by inclusion. This partial ordering may be extended to a simple ordering. We suppose that $J_1, J_2, \cdots, J_n$ are ordered with the small elements first in the sense that $I(J_i) \subset I(J_{i+k})$ if $I(J_i) \cdot I(J_{i+k}) \neq 0$.

Let R be a triangulation of Bd T such that each J_i is a subset of the 1-skeleton of R, no J_i lies on the boundary of a single 2-simplex of R, and if two vertices of a 2-simplex of R lie on J_i, the edge joining them does also.

We now describe D_i. Suppose p_0 is a point of Int T and $x_1x_2x_3$ is a 2-simplex of R that lies in $J_i + I(J_i)$. Then $y_1y_2y_3$ is a triangle in D_i where $y_j = x_j$ if $x_j \in J_i$ and y_j divides the interval from x_j to p_0 is the ratio i to n if $x_j \in I(J_i)$. The sum of all such triangles $y_1y_2y_3$ is the polyhedral disk D_i.

In Theorem 2 it was necessary to suppose that no J_i intersected two components of Bd $T - J_j$. To show that this condition was satisfied in the proof of the special case of Theorem 1, the following result was used.

THEOREM 3. *Suppose S is a simple surface, T is a tetrahedron,* $S' \cdot$ Bd T *is a finite graph,* D_1, D_2 *are two components of* $S \cdot$ Int T, *and* J_i $(i = 1, 2)$ *is a simple closed curve in* Bd D_i. *Then* J_1 *does not intersect each component of* Bd $T - J_2$.

PROOF. Suppose that J_1 does intersect both components of Bd $T - J_2$. We shall show by the use of linking simple closed curves that this is impossible. A treatment of such linking is given in [5] and last section of this paper. We enumerate steps in our proof.

1. Let axb be an arc in J_1 such that $axb \cdot J_2$ is connected and a, b belong to different components of Bd $T - J_2$. Let cxd be an arc in J_2 such that $cxd \cdot J_1 = cxd \cdot axb$ and neither c nor d belongs to axb.

Denote the closures of $J_1 - axb$ and $J_2 - cxd$ by aub and cvd respectively.

Let arb be an arc from a to b which lies except for its end points in D_1 and ctd be an arc from c to d which lies except for its end points in D_2.

Let asb be an arc from a to b which lies except for its end points in the complement of T. Then the simple closed curve $arb + bsa$ links $J_2 = cxd + dvc$. See Figure 6.

2. Since $arb + bsa$ links $cxd + dvc$, it links either $cxd + dtc$ or $ctd + dvc$. (This is because if θ is a theta-curve and J is a simple closed curve in its complement, then J links exactly two simple closed curves in θ if it links one. See last section of this paper.) We show that neither of these alternatives is possible.

3. There is a homotopy pulling bsa to bua in the complement of $cxd + dtc$. It is impossible that $arb + bsa$ link $cxd + dtc$ because if they did link, $arb + bua$ would link $cxd + dtc$. (If J_1' is homotopic to J_2' in the complement of J_3', then J_2' links J_3' if and only if J_1' does. See Theorem 10.) But $arb + bua$ cannot link $cxd + dtc$ because both lie on the same simple surface S and either can be shrunk to a point in the complement of the other.

4. There is a homotopy pulling bsa to bxa in the complement $ctd + dvc$. Hence

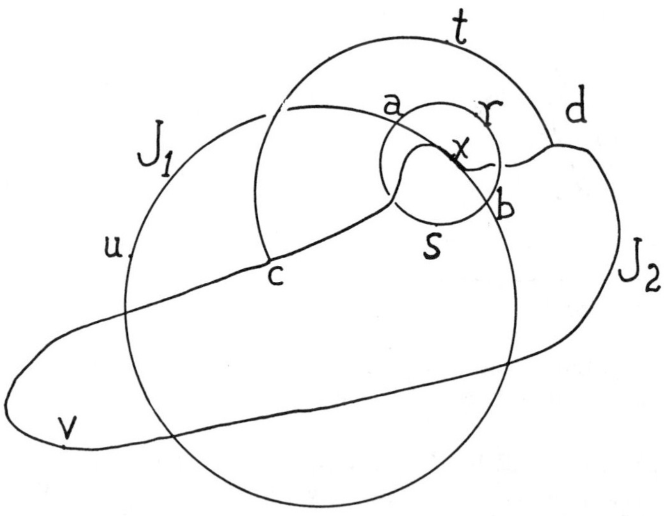

FIG. 6

arb + *bsa* cannot link *ctd* + *dvc*. The contradiction between Steps 2, 3, and 4 results from the false assumption that J_1 intersects both components of Bd $T - J_2$.

4. Approximating simple surfaces in E^3: General case

Having proved Theorems 2 and 3, we return to consider the proof of the general case of Theorem 1.

The intersection of R_2 and S is no longer supposed to be a finite graph. It may be a very bad sort of closed set. However, since it lies on S, it is not skew. Again we enumerate the steps of our proof.

1. We suppose that $S \cdot R_1$ is finite. Theorem 4 shows that there is no loss of generality in supposing this.

2. We suppose that if F is a face of a tetrahedron in R and X is a component of $S \cdot \text{Int } F$, then $\bar{X}$ does not contain three points of Bd F. If this is not already the case we modify S slightly to make it so. We want to prevent a situation as depicted in Figure 7a and have a situation as shown in Figure 7b instead.

Before including some details as to why we may suppose that $\bar{X}$ does not contain three points of Bd F, we mention the usefulness of this condition. In the general case, we use continua $\bar{X}$ in somewhat the same way that we used 1-simplexes of G in the special case. Our assumption for the special case prevented any 1-simplex in G from intersecting Bd F in three points and also prevented any two 1-simplexes of G from intersecting in Int F. When we suppose that no $\bar{X}$ has three points on Bd F we are a little closer to the situation as it existed in the special case.

Briefly, the way we modify S is to pull at the center of F (leaving Bd F fixed) without moving S so that the adjusted F has the desired property—then the

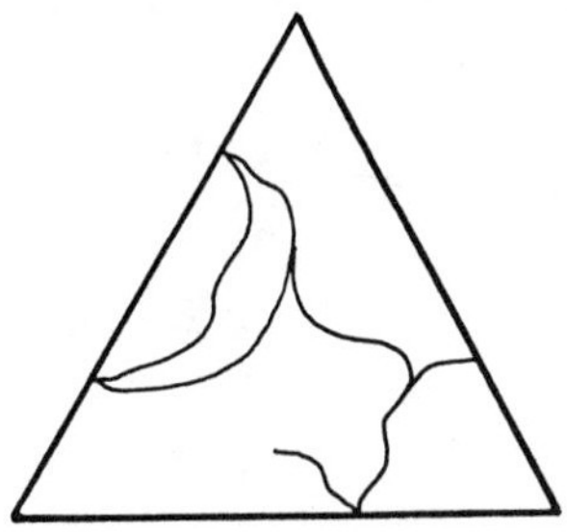

FIG. 7a

FIG. 7b

adjusted F is pushed back into the original F with a homeomorphism that moves S. The adjusted S has the required property. We include some more details.

Suppose a is a point of Int F and ab is a short interval perpendicular to F. For each point x of ab let F_x be the cone with vertex x and base Bd F; also, for each interval I of ab let F_I be the sum of all such cones F_x, $x \in I$. If p, q, r are three points of $S \cdot \text{Bd } F$, there are not three mutually exclusive intervals I_1, I_2, I_3 on ab such that there is a component X_i $(i = 1, 2, 3)$ of $S \cdot (F_{I_i} - \text{Bd } F)$ with each of p, q, r as a limit point. If there were three such points, then in S and near X_i there would be a triod with vertices p, q, r. The three triods could be chosen so that no two of them intersect except at their end points. The sum of the triods would be skew and could not lie in S. Since there cannot be three such triods, there cannot be three X_i's, and there cannot be three mutually exclusive intervals I_1, I_2, I_3 on ab such that each $S \cdot (F_{I_i} - \text{Bd } F)$ contains a component whose closure intersects each of p, q, r.

Since $S \cdot \text{Bd } F$ has only a finite number of triples of points, for each positive number γ there is an interval I of ab each point of which is within γ of a such that $S \cdot (F_I - \text{Bd } F)$ does not contain a component with three limit points on Bd F. Let x be a nonend point of I such that F_x contains no open subset of S. A homeomorphism of E^3 onto itself that moves F_x onto F would adjust S so that no component of Int F and the adjusted S would have three limit points on Bd F nor contain any open subset of S.

3. Let γ be a positive number less than the distance between any two points of $S \cdot R_1$. We suppose that if F is a face of a tetrahedron of R, then $S \cdot \text{Int } F$ does not contain infinitely many components with diameters greater than γ. If this is not already the case, we take a triangulation of F of mesh less than γ such that no vertex of the triangulation is on S. Then we adjust S near the 1-skeleton of this triangulation by applying Theorem 4. We do not wish Step 3 to destroy Step 2 so by taking this adjustment to be the identity except near this 1-skeleton, we can prevent the intersection of Int F and the adjusted S from having any component with three limit point on Bd F since in Step 2, $S \cdot (F_I - \text{Bd } F)$ did not have a component with three limit points on Bd F. The intersection of Int F and the adjusted S would not have infinitely many

components with diameters greater than γ because each such component would contain a point of the 1-skeleton of the triangulation of F and there are only a finite number of such points.

If after modifications of S are made as suggested in Steps 1, 2, and 3, one looks at the resulting set $S - R_2$ in the large rather than in the small, it looks something like a finite graph—we are a little closer to the situation as it existed in the special case.

4. In the special case, each component of $S - G$ was of diameter less than δ because it was in one of the tetrahedrons of the triangulation R. In this step we show that if $\bar{X}_1, \bar{X}_2, \cdots, \bar{X}_x$ are the continua of $S \cdot R_2$ such that for some F, X_i is a component of $S \cdot \text{Int } F$ with two limit points on Bd F, then no component of $S - \sum \bar{X}_i$ is of diameter more than $\varepsilon + 3\delta$. (It follows from Step 8 that each such component is actually of diameter less than $\varepsilon/2$.)

Let p_0, q_0 be two points of $S - \sum \bar{X}_i$ whose distance apart is more than $\varepsilon + 3\delta$. Let $F_1, F_2, \cdots, F_f$ be a collection of faces of tetrahedrons of R such that $\sum F_i$ separates p_0 from q_0 in E^3 and $\rho(F_i, p_0 + q_0) > \varepsilon/2$. (We use $\rho(a, b)$ to denote the distance between a and b.) It follows from the unicoherence and local connectedness of S that some component Z of $S \cdot \sum F_i$ separates p_0 from q_0 in S. Then Z is of diameter greater than δ since $S - Z$ has two components with diameters greater than $\varepsilon/2$.

If X is a component of $S \cdot \text{Int } F_i$ with no limit point on Bd F_i, then $Z \cdot X = 0$. Let X_Z be the sum of all elements of $\bar{X}_1, \bar{X}_2, \cdots, \bar{X}_x$ in Z.

For each point r of $Z \cdot \text{Bd } F_i$, let $Y(r, F_i)$ be the sum of the closures of all components of $S \cdot \text{Int } F_i$ with r as the only limit point on Bd F_i. Since $S \cdot \text{Int } F_i$ has only a finite number of components with diameters more than γ, $\bar{Y}(r, F_i) - Y(r, F_i)$ is a connected subset of X_Z. Then Z is the sum of X_Z and a finite number of such $Y(r, F_i)$'s. Suppose $Z = X_Z + Y_1 + Y_2 + \cdots + Y_y$. Since each Y_i is of diameter less than δ, no one of the Y_i's separates p_0 from q_0.

If the sum of two closed sets separates p_0 from q_0 in S and the intersection of these two sets is connected, then one of them separates p_0 from q_0. Since the intersection of Y_1 and $X_Z + Y_2 + Y_3 + \cdots + Y_y$ is a connected subset of X_Z, $X_Z + Y_2 + Y_3 + \cdots + Y_y$ separates p_0 from q_0 in S. Also, $X_Z + Y_3 + Y_4 + \cdots + Y_y$ does. Continuing in this fashion, we find that X_Z separates p_0 from q_0. This shows that no component of $S - \sum \bar{X}_i$ is of diameter more than $\varepsilon + 3\delta$.

5. In the special case, $S - G$ had only a finite number of components and each of these components was of diameter less than δ. In this step we define a number of sets in S, the last of which is a closed set $\sum K_i$ which divides S into a finite number of connected domains with small diameters.

As in Step 4, let X be a component of $S \cdot \text{Int } F$ (where F is a face of a tetrahedron of R) such that $\bar{X}$ contains two points a, b of Bd F. There are but a finite number of such components of $S \cdot \text{Int } F$ whose closures contain $a + b$ and the closure of their sum is a continuum $C(X)$ in $F \cdot S$ such that $C(X) \cdot \text{Bd } F = a + b$. If X_1, X_2 are two components of $S \cdot \text{Int } F$ such that each $\bar{X}_i$ contains $a + b$, $C(X_1) = C(X_2)$.

Now $C(X)$ is a continuum in S of diameter less than δ and hence has only one complementary domain in S of diameter greater than $\varepsilon/2$. Let $K(X)$ be the sum of $C(X)$ and all the components of $S - C(X)$ but the large one. The two points of $C(X)\cdot R_1$ are called the end points of $K(X)$. We look at each face F of each tetrahedron of R and define such continua $K(X)$ for each of them.

If K', K'' are two such continua, it may be that one contains the other. Otherwise they do not intersect except at possibly their end points. Let $K_1, K_2, \cdots, K_k$ be the collection of all such $K(X)$'s that are not contained in a larger $K(X)$.

No one of the K_i's separates S and the intersection of any two of them is at most two points. Hence $S - \sum K_i$ has only a finite number (less than the number of K_i's) of components $D_1, D_2, \cdots, D_d$. It follows from Step 4 that each of these D_i's is of diameter less than $\varepsilon + 3\delta$. We show in Step 8 that each is of diameter less than $\varepsilon/2$.

FIG. 8

Figure 8 shows how $\sum K_i$ divides S. The K_i's do not necessarily lie in faces of tetrahedrons of R but the boundary of each K_i (with respect to S) lies in such a face. Hence each K_i is of diameter less than $\varepsilon/2$. The D_i's do not necessarily lie on interiors of tetrahedrons of R but in Steps 8 and 11 we shall associate certain of the D_i's kept at that stage with certain of the tetrahedrons.

6. In Step 1 of the proof of the special case of this theorem, we threw away a part of the finite graph $S\cdot R_2$ under consideration and got some larger domains on S. In a similar fashion, we now throw away some of the K_i's and replace each D_i with a possibly larger domain E_i. A domain of D_i such as shown in the center of Figure 8 would be enlarged.

For each D_i, let $B(D_i)$ be the sum of all K_i's that have a non end point on Bd D_i. Let p_i be a point of S whose distance from $D_i + B(D_i)$ is more than ε. Let $B(E_i)$ be the sum of a collection of K_i's in $B(D_i)$ such that this collection but no proper subcollection of K_i's in $B(D_i)$ has a sum which separates p_i from D_i. Our reason for calling this continuum $B(E_i)$ is that we shall call E_i the component of $S - B(E_i)$ that contains D_i.

The continuum $B(E_i)$ is much like a circle but instead of being the sum of arcs, it is the sum of K_i's. In Figure 8, $B(E_i) = K_1 + K_2 + K_3 + K_4 + K_5$. This continuum $B(E_i)$ has exactly two complementary domains in $S \cdots E_i$ and the one containing p_i. We show that E_i (and hence D_i) is of diameter less than $\varepsilon/2$ by showing in Step 8 that Bd (E_i) is of diameter less than δ since it lies on the sum of the faces of some tetrahedron in R.

7. We shall need to associate each E_i with a tetrahedron T_i. This is a bit more involved here than in the special case. There D_i was a subset of a tetrahedron T_i of R; in the general case this is not true. To handle the general case, we show that there is a component E_i' of $E_i - R_2$ with a boundary in S that contains Bd E_i. As a step toward finding an E_i', we show that no component of $R_2 \cdot E_i$ has two limit points on Bd E_i.

In this step we show that if Z is a component of $R_2 \cdot E_i$, then Z does not have two limit points on Bd E_i. In order to do this, we consider three kinds of components of $S \cdot \text{Int } F$ where F is a face of a tetrahedron of R. Here are the three kinds of such components.

1. Those with no limit point on Bd F.
2. Those with two limit points on Bd F.
3. Those with exactly one limit point on Bd F.

If Z contains a point p of a component of the first kind, Z is this component and Z does not have any limit point of Bd E_i. To see that this is so, assume that $\bar{Z}$ does not lie in Int F. There is a subcontinuum Z' of $\bar{Z}$ irreducible from p to Bd F and it follows from Theorem 37, page 22 of [11] that $Z' -$ Bd F is connected and contains a limit point on Bd F. But then the component of $S \cdot \text{Int } F$ containing p contains $Z' -$ Bd F, has a limit point on Bd F, and is not of the first kind.

We use $Z(p, F)$ to denote the sum of all components of $Z \cdot \text{Int } F$ that have the point p of R_1 as a limit point. We now show that if $Z(p, F)$ contains a component of the third kind but p does not belong to Z, then Z is this component and Z has only one limit point on Bd E_i.

Assume $Z(p, F)$ contains a component of the third kind but p does not belong to Z. Since there are not infinitely many components of the second kind, one can show inductively by "peeling off" the elements of the second kind that $Z(p, F)$ contains a subset A such that A contains a component of the third kind but none of the second, and neither A nor $Z(p, F) - A$ contains a point or a limit point of the other. But then Z is the sum of A and $Z - A$ and neither of these sets contains a limit point of the other.

Assume Z has two limit points on Bd E_i. The Z can be expressed as $p_1 + \sum_j Z(p_1, F_j) + p_2 + \sum_j Z(p_2, F_j) + \cdots + p_n + \sum_j Z(p_n, F_j) + X_1 + X_2 + \cdots + X_m$ where $p_1, p_2, \cdots, p_n$ are points of R_1 in Z; $X_1, X_2, \cdots, X_m$ are sets of the second kind in Z; and the $Z(p, F)$'s are as indicated in the preceding paragraph.

Now $Z(p_k, F_j)$ does not have a limit point on Bd E_i unless there is an X_r in it that has the same limit point, so $Z_0 = p_1 + p_2 + \cdots + p_n + X_1 + X_2 + \cdots +$

X_m has two limit points on Bd E_i if Z does. We find by removing the part of the $Z(p, F)$'s not in Z_0 one at a time that Z_0 is connected.

If Z_0 contains two limit points on Bd E_i, there is a continuum Z' in $\bar{Z}_0$ that intersects Bd E_i in exactly two points p and q. Also, $B(E_i)$ is the sum of two continua Z_1, Z_2 such that $Z_1 \cdot Z_2 = p + q$, each $Z_j - (p + q)$ is connected, and each $Z_j - (p + q)$ contains points of Bd E_i. It follows from Theorem 101, page 287 of [11] that Z' separates $Z_1 - (p + q)$ from $Z_2 - (p + q)$ in $E_i + B(E_i)$. This is impossible since D_i lies in $E_i - Z'$, Z' separates $Z_1 - (p + q)$ from $Z_2 - (p + q)$ in $E_i + B(E_i)$, and Bd D_i intersects both $Z_1 - (p + q)$ and $Z_2 - (p + q)$. This completes the proof that Z does not have two limit points on Bd E_i.

8. It follows from Theorem 5 that there is a component E_i' of $E_i - R_2$ such that Bd E_i' contains Bd E_i. Although Theorem 5 is stated for the plane, a version of it works for S. In applying Theorem 5, we let E_i and $S \cdot R_2$ be the D and K of that theorem. We showed in Step 7 that if Z is a component of $E_i \cdot R_2$, Z does not have two limit points on Bd E_i. Also, $\bar{Z}$ does not intersect Bd E_i except possibly at a point of R_1 and such a point would not separate Bd E_i. Hence $\bar{Z}$ does not separate Bd E_i. Since $E_i \cdot R_2$ has only a finite number of components with diameters more than γ, there is an ε' such that not each point of Bd E_i is within ε' of any Z.

Now E_i' would lie on the interior of some tetrahedron T_i of R. Therefore Bd $E_i \subset$ Bd T_i and Bd E_i is of diameter less than δ. Hence, the diameter of E_i is less than $\varepsilon/2$.

9. The E_i's described in Step 8 are not dependent on the points p_i selected. Two E_i's do not intersect unless one contains the other.

Let $E_1, E_2, \cdots, E_n$ be the collection of maximal E_i's — no one of them is contained in another one. Then

$$S = E_1 + E_2 + \cdots + E_n + B(E_1) + B(E_2) + \cdots + B(E_n).$$

The E's discussed in Steps 6, 7, 8 differ from the E's used in the special case in that their boundaries are not simple closed curves. In this step we replace the E's with H's which do have simple closed curves for boundaries.

Let $K_1, K_2, \cdots, K_m$ be the elements of $K_1, K_2, \cdots, K_k$ in $\sum B(E_i)(i = 1, 2, \cdots, n)$ and $G_1, G_2, \cdots, G_m$ be topological disks in S such that G_i contains K_i, diameter $G_i < \varepsilon/2$, $G_i \cdot G_j = K_i \cdot K_j$. Let A_i be an arc in G_i from one end of K_i to the other. Then $S - \sum A_i$ is the sum of a finite number of topological open disks $H_1, H_2, \cdots, H_n$ such that $H_i + \sum G_i = E_i + \sum G_i$.

Figure 9 shows how the continua $K_1, K_2, \cdots, K_5$ in $B(E_i)$ are replaced by arcs $A_1, A_2, \cdots, A_5$ (shown dotted). Then H_i is the interior of the simple closed curve $A_1 + A_2 + \cdots + A_5$.

Each H_i is of diameter less than $3\varepsilon/2$ because it is a subset of the sum of E_i (which is of diameter less than $\varepsilon/2$) and the various G_j's (each of diameter less than $\varepsilon/2$) intersecting E_i. What is more significant for our purposes is that each

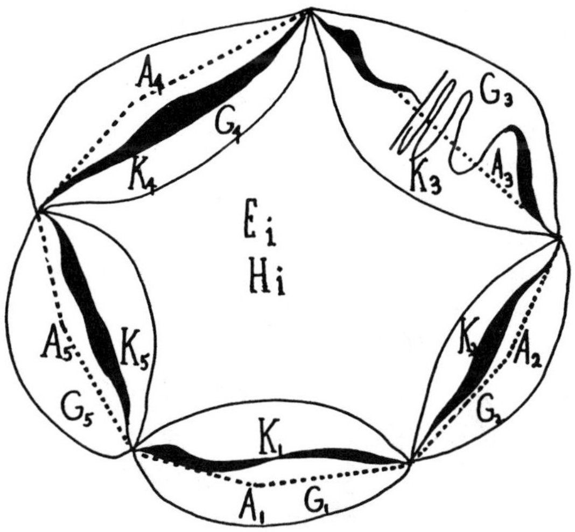

FIG. 9

point of H_i is within $\varepsilon/2$ of Bd E_i. As in the proof of the special case, we now describe a collection of polygonal arcs in R_2 and a homeomorphism of the A_i's onto them.

10. In Step 2 of the special case we replaced arcs $a_ix_ib_i$ by polygonal arcs $a_iy_ib_i$. In this step we now describe some polygonal arcs $A_1', A_2', \cdots, A_n'$ in R_2 to match the arcs $A_1, A_2, \cdots, A_n$ in S.

The arc A_i is in a disk of G_i in S; G_i contains a continuum K_i; the boundary of K_i in S is the sum of one or two continua $\bar{X}$; $\bar{X}$ is in the face $F(A_i)$ of a tetrahedron of R such that $\bar{X}\cdot$Bd $F(A_i)$ is the sum of the ends of K_i and the sum of the ends of A_i.

In $F(A_i)$ there is a polygonal arc A_i' such that A_i' lies except for its ends in Int $F(A_i)$ and these ends are the ends of A_i. For each A_i we pick an A_i' such that no two A_i''s intersect except possibly at their end points. This is possible because there cannot be two mutually exclusive continua $\bar{X}_1, \bar{X}_2$ in $F(A_i)$ such that two points of $\bar{X}_1\cdot$Bd $F(A_i)$ separate two points of $\bar{X}_2\cdot$Bd $F(A_i)$ on Bd $F(A_i)$.

Let h be a homeomorphism of $\sum A_i$ onto $\sum A_i'$ such that A_i goes onto A_i' and each end point of A_i goes onto itself. We note that since A_i is of diameter less than $\epsilon/2$ and A_i' is of diameter less than δ, h moves no point by as much as $\varepsilon/2 + \delta < \varepsilon$.

11. Now we describe polyhedral disks H_i' to replace the topological disks $\bar{H}_i$ in S.

We recall that H_i was formed from E_i by replacing continua K_j on $B(E_i)$ by arcs A_j. There was a component E_i' of $E_i - R_2$ such that Bd E_i' contained Bd E_i.

Let T_i be the tetrahedron of R such that E_i' lies in Int T_i. Then the boundary of each K_j (in S) lies in Bd T_i and each A_j' defined in Step 10 lies in Bd T_i. The polyhedral disk H_i' will be in T_i.

The boundary of $\bar{H}_i$ in S is the sum of a finite number of arcs $A_1, A_2, \cdots, A_n$ and $h(\sum A_j)$ is a polygon on Bd T_i. Then H_i' is a polyhedral disk which lies except for its boundary $h(\sum A_j)$ on Int T_i. The H_i''s are selected so that Int $H_i' \cdot$ Int $H_j' = 0$ if $i \neq j$. Theorem 2 shows that there are such H_i''s if it is shown that $T_i = T_j$ implies that $h(\text{Bd } \bar{H}_i)$ does not intersect two components of Bd $T_i - h(\text{Bd } \bar{H}_j)$. This is shown by Theorem 6.

The homeomorphism h is already defined to take Bd $\bar{H}_i$ onto Bd H_i'. The homeomorphism h is extended to H_i so as to take $\bar{H}_i$ onto H_i' and to agree with h on Bd $\bar{H}_i$. Since each point of H_i is within $\varepsilon/2$ of T_i and $H_i' \subset T_i$, h does not move any point by more than $\varepsilon/2 + \delta$.

5. Intersection of a line and a surface

The following result should not be remembered as a theorem because it follows as a corollary of Theorem 7. However, it is a powerful tool used in proving Theorems 1 and 7 so the proof is not avoided.

THEOREM 4. *Suppose in E^3 that S is a simple surface and L is a straight line. Then for each positive number ε there is a simple surface S' and a homeomorphism h of S onto S' such that* (a) *$S' \cdot L$ does not contain infinitely many points,* (b) *h does not move any point by as much as ε, and* (c) *h is fixed at each point of S whose distance from L is more than ε.*

QUESTION. The continuum S' that we use is not obtained from deforming S and we do not show that the homeomorphism h of S onto S' can be extended to a homeomorphism of E^3 onto itself. In fact, it seems likely that there is a wild simple surface S, a straight line L, and a positive number ε such that $h(S)$ intersects L is infinitely many points if h is any homeomorphism whatsoever of E^3 onto itself that moves no point of S by more than ε. It would be interesting to know if there is an S so wild.

PROOF OF THEOREM 4. We suppose that $S \cdot L$ is totally disconnected. If this is not already the case, a slight adjustment in S near L will give us a simple surface where this is the case.

Let δ be a positive number less than $\varepsilon/2$ such that no subset of S of diameter less than δ separates S into two pieces each of diameter greater than $\varepsilon/2$.

Cover $S \cdot L$ with a finite number of mutually exclusive solid right circular cylinders $C_1, C_2, \cdots, C_c$ such that each C_i is of diameter less than δ and has L for an axis, while S intersects neither end of C_i.

It will be convenient to have $S \cdot$Bd C_i as decent as possible since we shall be shrinking certain continua in it to points. For example, it would be nice if for no continuum g in $S \cdot$Bd C_i does $S - g$ have more than two components. By shrinking the radius of C_i slightly and leaving its ends in the same planes, we can arrive at a solid cylinder C'' such that for no continuum g in $S \cdot$Bd C'' does $S - g$ have more than two components. This is because S does not have uncountably many mutually exclusive continua such that each separates S into more than two pieces and C'' can be chosen in uncountably many ways.

We shall suppose that if g is a continuum in $S \cdot$Bd C_i, then $S - g$ does not

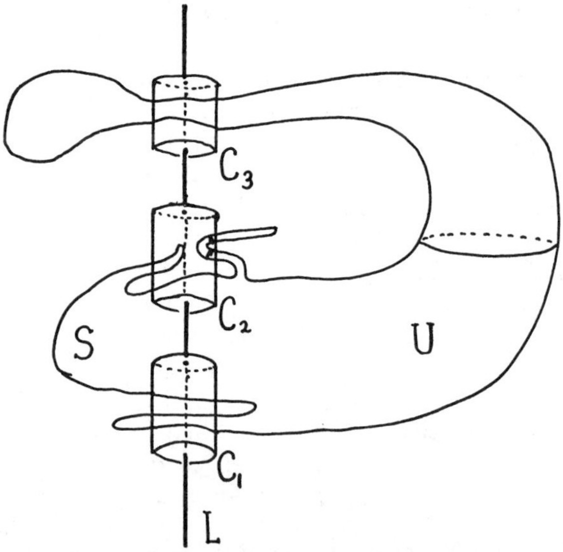

FIG. 10

have three components. Since lying on a sphere and separating it into more than two pieces is a topological property for continua, Bd $C_i - g$ does not have three components.

No components of $S \cdot \sum C_i$ is of diameter as much as δ. Therefore $S - \sum C_i$ has only one component U of diameter more than $\varepsilon/2$. See Figure 10.

If G is the upper semicontinuous decomposition of $\bar{U}$ whose elements are the points of U and the components of $\bar{U} - U$, the decomposition space is topologically equivalent to S. This may be seen by applying Moore's theorem [10] to the effect that if G' is an upper semicontinuous decomposition of S such that each element of G' is a proper subcontinuum of S that does not separate S, then the decomposition space is topologically S. We apply Moore's theorem to the decomposition space G' formed by adding to each element g of G that separates S the sum of all components of $S - g$ of diameters less than $\varepsilon/2$.

Roughly speaking, we shall obtain S' from $\bar{U}$ by shrinking each nondegenerate element of G in $\bar{U}$ to a point.

Let C_i' be a solid cylinder slightly larger than C_i but still of diameter less than δ such that C_i' has L as an axis, has ends in the same plane as C_i, and neither of these ends intersects S.

There is a homeomorphism f' of G onto S that does not move any point of $U - \sum C_i'$. If f' is considered as a map on $\bar{U}$ (rather than a homeomorphism on G), f' is fixed on $U - \sum C_i'$ and does not move any point by as much as $\varepsilon/ 2 + \delta < \varepsilon$. To see that there is such an f', note that $\bar{U} - U$ corresponds to a closed 0-dimensional set in the decomposition space G and there is a finite collection of mutually exclusive open disks $D_1, D_2, \cdots, D_n$ in G covering the 0-dimensional set such that the set associated with each D_j lies in $\sum C_i$. Then

f' is the identity on $\bar{U} - \sum D_i$ and takes D_j onto the small disk in S bounded by Bd D_j.

The lemma for Theorem 4 reveals that there is a map f_i of T_i = (closure of $(C_i' - C_i)$) into C_i' such that f is fixed on $T_i \cdot \text{Bd } C_i'$, $f(T_i) \cdot L$ is finite, f takes each element of G in C_i into a point but no two such elements into the same point. The homeomorphism h is defined as follows. If $p \in U - \Sigma C_i'$, $h(p) = p$— if $f'^{-1}(p) \in C_i'$, $h(p) = f_i f'^{-1}(p)$. Then $S' = h(S)$.

In order to apply the lemma we need to show that if 0_1, 0_2 are two open subsets of $M_i = T_i \cdot \text{Bd } C_i$, there are not infinitely many components of $\bar{U} \cdot M_i$ that separate 0_1 from 0_2 in M_i. We do this with homotopy.

Let J be a simple closed curve which is the sum of arcs axb and ayb where $a \in 0_1$, $b \in 0_2$, $(axb) \subset E^3 - C_i$, $(ayb) \subset \text{Int } C_i$. There is a positive number γ such that $\rho(J, \bar{U} \cdot M_i - (0_1 + 0_2) > \gamma$.

It follows from the local connectedness of S that for only a finite number of components g of $\bar{U} \cdot M_i$ is it true that neither components of $S - g$ is within $\gamma/2$ of g. If one component on $S - g$ is within $\gamma/2$ of g, g can be shrunk to a point on S so that each point of g stays within γ of g during the shrinking. However, no component of $\bar{U} \cdot M_i$ that separates a from b can be shrunk to a point without hitting J. Therefore, $\bar{U} \cdot M_i$ does not have infinitely many components that separate 0_1 from 0_2 in M_i.

Decomposition of a torus. The following lemma is not given in the chronological order in which it occurred since its truth is immediately evident to one who understands the methods of [6]. It belongs to the following type of problem: Describe the decomposition space resulting from decomposing a familiar object (solid torus in this case) into continua. There is an interesting and important theory involved in this sort of problem but we do not pursue it here—we follow it only far enough to get the result for which we have immediate need. Hence, we label our result a lemma.

In the lemma, we suppose that C, C' are two right circular solid cylinders with the same axis L, with bases in the same planes, and with C smaller than C'. We denote the closure of $C' - C$ by T and $T \cdot C$ by M. A closed set X on M has the following properties:

1. X intersects neither base of M.
2. No two open subsets of M are separated by infinitely many components of X.
3. No component of X separates M into more than two pieces.

LEMMA FOR THEOREM 4. *There is a map f of T into C' such that f is fixed on $T \cdot \text{Bd } C'$, $f(T)$ does not intersect L in infinitely many points, and for each point p of $f(T)$, $f^{-1}(p)$ is either a point of $T - X$ or a component of X.*

We let G be the upper semicontinuous decomposition of T whose elements are the components of X and the points of $T - X$. The lemma says that *there is a homeomorphism of the decomposition space G into C' such that f is fixed* (in the obvious way) *on $T \cdot \text{Bd } C'$ and does not take infinitely many elements of G into L.*

PROOF. We break the proof of this lemma into two steps.

1. In this step we shrink to points the elements $g_1, g_2, \cdots, g_n$ of G that separate the two bases of M from each other in M. Conditions 1 and 2 satisfied by X implies that there are only a finite number of such elements.

Let $A_{i1}, A_{i2}, A_{i3}, \cdots$ be a sequence of topological annuli in M such that $A_{i(j+1)} \subset \text{Int } A_{ij}$, $A_{i1} \cdot A_{i2} \cdot \cdots = g_i$, $A_{i1} \cdot A_{j1} = 0$ if $i \neq j$, and Bd A_{ij} does not intersect X.

We now define some surfaces and solids in T. If A is an annulus or disk in M, use $F(A, \varepsilon)$ to denote the set of all points t of T such that the perpendicular projection $p(t)$ of t on M lies on A and $\rho(t, p(t)) = \varepsilon \cdot \rho(p(t), \text{Bd } T - A)$—also $T(A, \varepsilon)$ is the set of all points t of T such that $\rho(t, p(t)) \leqq \varepsilon \cdot \rho(p(t), \text{Bd } T - A)$. Then $F(A, \varepsilon)$ is an annulus or disk and $T(A, \varepsilon)$ is a torus or cube according as A is an annulus or disk.

For brevity, we denote $F(A_{ij}, 1/j)$ by F_{ij}.

We shall suppose that M is the cylinder $(x^2 + y^2 = 1, 0 \leqq z \leqq 1)$ and that $g_1, g_2, \cdots, g_n$ are ordered so that g_{i+1} separates the upper base of M from g_i. Let K_i be the double napped cone generated by revolving the closed segment from $(0, -1, (i-1)/n)$ to $(0, 1, i/n)$ about the z axis.

Let $M' = \sum K_i$, T' be the sum of T, M' and all points between T and M', A'_{ij} be the set of all points of K_i whose distances from the vertex of K_i is no more than $1/(2jn)$; and F'_{ij} be the cylinder whose bases are the extreme circles in A'_{ij}. See Figure 11.

We use T_0 and T'_0 to denote the topological tori bounded by Bd $T - \sum A_{i1} + \sum F_{i1}$ and Bd $T' - \sum A'_{i1} + \Sigma F'_{i1}$ respectively. Also, T_{ij} and T'_{ij} denotes the topological tori bounded by $F_{ij} + A_{ij} - A_{i(j+1)} + F_{i(j+1)}$ and $F'_{ij} + A'_{ij} - A'_{i(j+1)} + F'_{i(j+1)}$ respectively.

There is a homeomorphism f' of T_0 onto T'_0 that is fixed on Bd $T - M$ and takes $\sum F_{i1}$ onto $\sum F'_{i1}$. This homeomorphism may be extended to take T_{i1} onto T'_{i1} in such a way that F_{i2} goes onto F'_{i2}. The homeomorphism may be

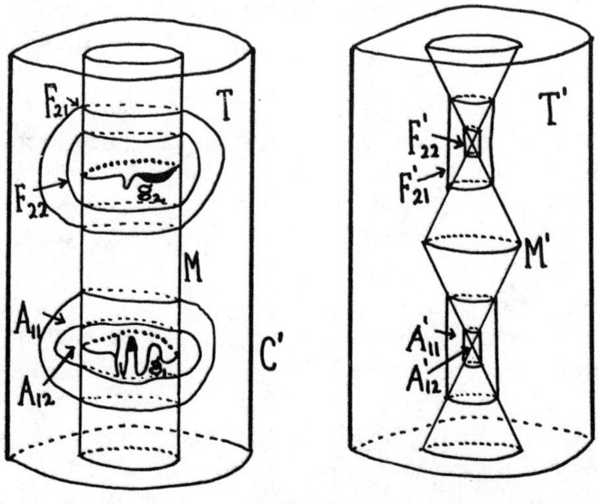

FIG. 11

extended to $\sum\sum T_{ij}$ so that $f'(T_{ij}) = T'_{ij}$. If we define $f'(g_i)$ to be the vertex of K_i, f' is a map that takes T onto T', takes $g'_i, g_2, \cdots, g_n$ into points on L and is a homeomorphism on $T - \sum g_i$.

2. We now modify f' to get a map f that shrinks each element of G to a point. The map f' is an approximation to f in that f agrees with f' on Bd $T - M + \sum\sum F_{ij}$ and $f(T_0) \subset f'(T_0)$, $f(T_{ij}) \subset f'(T_{ij})$. This insures that $f(T)\cdot L$ is finite. We only describe the modification of f' to get f on T_{ij} since the modification is similar on T_0.

The closure of $A_{ij} - A_{i(j+1)}$ is the sum of two annuli A and A'. There is a number ε such that $T(A, \varepsilon) + T(A', \varepsilon)$ is contained in T_{ij}. We define f to be f' on $T_{ij} - (T(A, \varepsilon) + T(A', \varepsilon))$. In fact, if either $T(A, \varepsilon)$ or $T(A', \varepsilon)$ failed to contain a nondegenerate element of G, we would define f to equal f' on it. We now define f on $T(A, \varepsilon)$ for the case where A contains nondegenerate elements of G.

No element of G separates the two components of Bd A from each other in A. If there were such an element, it would separate the two bases of M from each other in M, would have been shrunk to a point at Step 1, and would not lie in A. The elements of G in A that separate A may be ordered $g_1, g_2, \cdots$ so that g_{i+k} does not separate g_i from Bd A in A. This is because X satisfies the second condition listed for it.

Let $\delta_1, \delta_2, \cdots$ be a sequence of positive numbers converging to zero. To describe f on $T(A, \varepsilon)$, we use a sequence of maps $f_1, f_2, \cdots$ such that f_{2i-1} shrinks each element of G in A to a set of diameter less than δ_i and f_{2i} shrinks g_i to a point. The map f will be the limit of $f_1, f_2, \cdots$.

Description of f_1. Let X_A denote the sum of $X\cdot A$ and the set of points that $X\cdot A$ separates from Bd A in A. There is a map f'' of A onto itself that is fixed on Bd A such that for each point p of A, $f''^{-1}(p)$ is either a component of X_A or a point of $A - X_A$. Since $f''(X_A)$ is totally disconnected, there is a finite graph H' in $A - f''(X_A)$ such that each component of $A - H'$ is an open disk U such that $f'(U)$ is of diameter less than δ_1 and for some point p of U, each point of $f''^{-1}(U)$ is within δ_1 of $f''^{-1}(p)$. We use H to denote $f''^{-1}(H')$ and note that it has the following property:

Property of H. If V is a component of $A - H$, it is an open disk and if g, g' are two elements of G in V, there is an element g'' of G in V such that either g'' separates one of g, g' from Bd V in V or both g and g' lie within δ_1 of g''.

There is a homeomorphism f'_1 of A onto itself that agrees with f'' on H. The homeomorphism f'_1 may be extended to $T(A, \varepsilon)$ so as to be fixed on $F(A, \varepsilon)$. The homeomorphism $f_1 = f'f'_1$ takes $T(A, \varepsilon)$ onto $f'(T(A, \varepsilon))$, agrees with f' on $F(A, \varepsilon)$, and shrinks each element of G in A to a set of diameter less than δ_1.

The map f_1 is a closer approximation to f than f'. In fact, if $V_1, V_2, \cdots, V_n$ are the closures of the open disks in $A - H$, we pick a positive number $\varepsilon' \leqq \varepsilon$ such that the diameter of each $f_1T(V_i, \varepsilon') < \delta_1$ and prescribe that f agree with f_1 at each point of $T(A, \varepsilon) - \sum T(V_i, \varepsilon')$. For convenience we suppose that $\varepsilon = \varepsilon'$.

Description of f_2. We only define f_2 on $T(V_1, \varepsilon)$ since each f_i agrees with f_1 on

$T(A, \varepsilon) - \sum T(V_i, \varepsilon)$. Our purpose at this stage is to shrink g_1 to a point. If g_1 is not in V_1, $f_2 = f_1$ in $T(V_1, \varepsilon)$.

If g_1 is in V_1, the description of f_2 on $T(V_1, \varepsilon)$ is patterned after the description of f' on T. Using the same approach as used in Step 1, we get a map f_2' of $T(V_1, \varepsilon)$ onto $T'(V_1, \varepsilon)$ where $T'(V_1, \varepsilon)$ is $T(V_1, \varepsilon)$ minus the interior of a sphere Z tangent to M at a point p of V_1 and such that $f_2'(g_1) = p$, f_2' is fixed on $F(V_1, \varepsilon)$, f_2' is a homeomorphism except on $T(V_1, \varepsilon) - g_1$. Then $f_2 = f_1 f_2'$.

We give just a few of the details in the description of f_2' since it is obtained in much the same way that f' was gotten in Step 1. There is a sequence of annuli $A_1, A_2, \cdots$ in V_i such that $A_{i+1} \subset \text{Int } A_i$, $A_1 \cdot A_2 \cdot \cdots = g_1$, Bd A_i does not intersect X, and no element of G other than g_1 separates the two simple closed curves on Bd A_1 from each other in A_1. For each A_i we define a surface $F(A_i, \varepsilon_i)$ in $T(V_1, \varepsilon)$ such that $F(A_{i+1}, \varepsilon_{i+1}) \subset T(A_i, \varepsilon_i)$, $T(A_1, \varepsilon_1) \cdot T(A_2, \varepsilon_2) \cdot \cdots = g_1$. See Figure 12. The map f_2' is fixed on $F(V_1, \varepsilon)$, $f'(g_1) = p$, f_2' (Ext g_1) $= V_1 - p$, $f_2'(\text{Int } g_1) = \text{Bd } Z - p$, and $f_2'(F(A_i, \varepsilon_i))$ is a right circular cylinder with one base on Bd Z and the other on V_1.

The map f will agree with $f_2 = f_1 f_2'$ on each $F(A_i, \varepsilon_i)$.

Description of f_3. Suppose A is the closure of a maximal piece of A_{ij} on which f has not already been prescribed. It is either a disk left over from the description of f_1 or an annulus left from the description of f_2. In either case, we pick an ε such that f has not been prescribed on $T(A, \varepsilon) - F(A, \varepsilon)$, get a finite graph H in A that satisfies the condition satisfied by H in the description of f_1 with δ_2 substituted for δ_1, and get a map f_3 of $T(A, \varepsilon)$ onto $f_2 T(A, \varepsilon)$ such that f_3 agrees with f_2 on $F(A, \varepsilon)$ and takes each component of $A - H$ into a set of diameter less than δ_2. If $V_1, V_2, \cdots, V_n$ are the closures of the components of $A - H$, there is an ε' such that each $f_3 T(V_i, \varepsilon')$ is of diameter less than δ_2. Then we define f to equal f_3 on $T(A, \varepsilon) - \sum T(V_i, \varepsilon')$.

Definition of f. Using methods employed in the description of f_2, we modify f_3 and get a map f_4 that shrinks g_2 to a point. We prescribe that $f = f_4$ except on some $T(A, \varepsilon)$'s where the A's are either annuli or disks. Then we modify f_4 on these $T(A, \varepsilon)$'s and get a map f_5 that shrinks each element of G to a set of diameter less than δ_3 and satisfies conditions similar to those satisfied by f_1 and

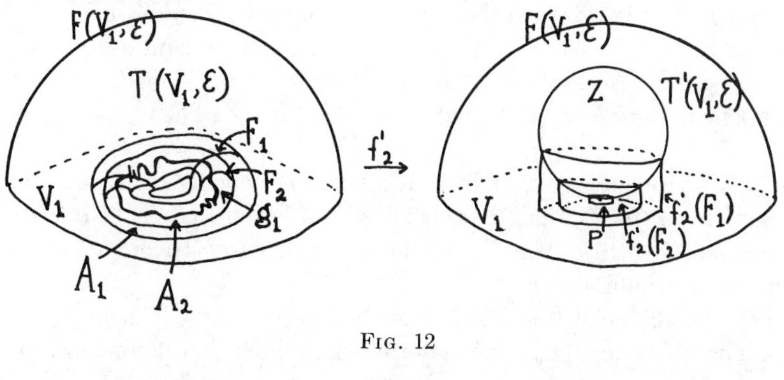

FIG. 12

f_3. We prescribe that $f = f_5$ at many places. Continuing in this fashion we define $f_6, f_7, \cdots$.

The map $f = \lim f_i$ is continuous because each f_i is continuous and $\rho(f_{2i-1}(p), f_{2i+k}(p)) < \delta_i$. If g is an element of G, $f(g)$ is a point because $f_{2i}(g)$ is of diameter less d_i. That finite graphs H used in the description of $f_1, f_3, \cdots$ have the *Property of* H displayed in the description of f_1 may be used to show that no two elements of G are taken into the same point.

6. The boundary of a plane domain

We used the following result from plane topology in Step 8 of the proof of the general case for Theorem 1.

THEOREM 5. *Suppose D is a simply connected, connected, bounded open set in E^2, ε is a positive number, and K is a closed set containing* Bd D *such that for no component Z of $D \cdot K$ does Z have two limit points on* Bd D, *does $\bar{Z}$ separate* Bd D, *and is each point of* Bd D *within ε of Z. Then there is a component D' of $D - K$ such that* Bd $D \subset$ Bd D'.

PROOF. For each component Z of $D \cdot K$, use Int $\bar{Z}$ to denote the sum of all bounded components of $E^2 - \bar{Z}$.

Case 1. First, we consider the case where no $\bar{Z}$ separates E^2—that is each Int Z is null. We show in this case that $D - K$ is connected.

Suppose $D - K$ is the sum of two mutually separated open sets A, B. Let a, b be points of A, B. Then $K \cdot D$ separates a from b in D and it follows from the unicoherence and local connectedness of D that some component Z' of $K \cdot D$ separates a from b in D. Then Bd $D + \bar{Z}'$ would separate a from b. This is impossible since neither $\bar{Z}'$ nor Bd D separates a from b in E^2 and $Z' \cdot$ Bd D is connected or null.

We now show that Bd $D \subset$ Bd D'. Since D is simply connected, no point of Bd D is isolated and in each neighborhood of a point of Bd D there is an arc which has its end points on Bd D and which lies except for these end points in D. This arc does not belong to K since no component of $K \cdot D$ has two limit points on Bd D. Hence the arc contains a point of D'.

Case 2. Now we consider the case where some $\bar{Z}$ does separate E^2 but that if p is a point in any Int $\bar{Z}$, then there is a Z_p such that $p \in$ Int $\bar{Z}_p$ but for no other Z' does Int $\bar{Z}'$ contain Z_p. In this case we add to K the sum of all Int $\bar{Z}$'s and get an enlarged set K'. Now $K \subset K'$ and for each component Z' of $K' \cdot D$ there is a component Z of $K \cdot D$ such that $Z' = Z +$ Int $\bar{Z}$. Then Case 2 reduces to Case 1.

Case 3. Now we suppose that there is a point p_0 in some Int $\bar{Z}$ such that if Z' is any Z with $p \in$ Int $\bar{Z}'$, then there is another Z'' such that $Z' \subset$ Int $\bar{Z}''$. There is no maximal Z with $p_0 \in$ Int $\bar{Z}$. We finish the proof of Theorem 5 by showing that this case cannot occur.

Let $\{Z_\alpha\}$ be the collection of all Z such that $p_0 \in$ Int $\bar{Z}$. Then $\sum Z_\alpha \subset \sum$ Int $\bar{Z}_\alpha$. The collection $\{Z_\alpha\}$ can be ordered by inclusion of the corresponding interiors, so if for no one Z is each point of Bd D within ε of Z, then Bd D does not belong to the closure of $\sum$ Int Z_α.

Let abc be an arc with end points on Bd D, interior in D, and which does not intersect the closure of $\sum$ Int Z_α. Then some component Z_0 of $K \cdot D$ in Bd ($\sum$ Int Z_α) separates p_0 from b in D. But Int $\bar{Z}_0$ does not contain p_0 since if $p \in$ Int $\bar{Z}_0$, $Z_0 \in \{Z_\alpha\}$, $Z_0 \subset \sum$ Int $\bar{Z}_\alpha$, and $Z_0 \not\subset$ Bd $\sum$ Int $\bar{Z}_\alpha$. Also, Int $\bar{Z}_0$ does not contain b since Z_0 does not intersect the arc abc. As in Case 1, we find that it is impossible for $\bar{Z}_0 +$ Bd D to separate p_0 from b.

7. Intersections of surfaces

Suppose T is a tetrahedron, S is a simple surface, and E_1, E_2 are two components of $S \cdot$ Int T. Then Bd E_1 does not intersect two components of $S -$ Bd E_2 since E_1 is connected. However, examples can be constructed to show that Bd E_1 may intersect two components of Bd $T -$ Bd E_2.

QUESTION. If B_1 is a continuum in Bd E_1 that is irreducible with respect to separating two points of Bd T (or S), can B_1 intersect two components of Bd $T -$ Bd E_2?

We do not answer this question but do extend Theorem 3 enough to get a result we can use in Step 11 of the proof of the general case of Theorem 1.

It must be admitted that the proof of the following theorem seems a little artificial. It was other considerations that led the author to suspect that the theorem was true and the notion of linking was seized upon to give the proof since the other notions were difficult to describe.

THEOREM 6. *Suppose T is a tetrahedron; J_1, J_2 are two simple closed curves on* Bd T; *$J_1 + J_2$ is the sum of a finite number of arcs $A_1, A_2, \cdots, A_n$ such that $A_i \cdot$* Int *$A_j = 0$ if $i \neq j$; $K_1, K_2, \cdots, K_n$ are topological disks on* Bd T *such that* Int *$A_i \subset$* Int *K_i and $K_i \cdot K_j = A_i \cdot A_j$ if $i \neq j$. Suppose S is a simple surface and E_1, E_2 are two components of $S \cdot$* Int *T such that a component of $K_i \cdot$* Bd *E_j intersects both ends of A_i if $A_i \subset J_j$. Then J_1 does not intersect both components of* Bd $T - J_2$.

PROOF. The proof of this theorem is modeled after the proof of Theorem 3.

Assume J_1 intersects two components of Bd $T - J_2$. Let axb be an arc of J_1 such that a, b belong to different components of Bd $T - J_2$ and $axb \cdot J_2$ is connected. Also, let cxd be an arc in J_2 such that c, d do not belong to J_1 but $cxd \cdot J_1 = cxd \cdot axb$. See Figure 13.

Let (x_1ay_1) and (x_2by_2) be the components of $J_1 - J_2$ containing a and b respectively such that x_i is between a and b on axb. Also, let (x_1cz_1) and (x_2dz_2) be the components of $J_2 - J_1$ containing c and d respectively and x_1, x_2 are as before. It may be that some of the y's are also z's.

For each arc rs, let $K(rs)$ be the sum of the K_i's with an interior point on rs and $X_i(rs)$ be a continuum (if any) in $K(rs) \cdot$ Bd E_i which is irreducible from r to s. Also, if M is any set, $K(M)$ is the sum of the K_i's whose interiors intersect M.

For the arcs x_1cy_1 and x_1dy_1 of J_2, there are such continua $X_2(x_1cy_1)$ and

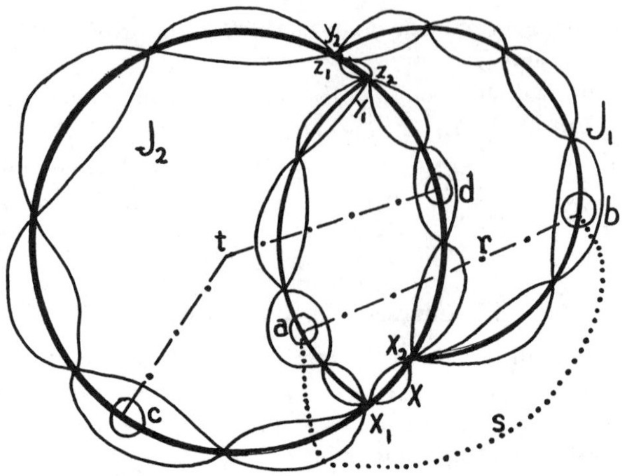

FIG. 13

$X_2(x_1dy_1)$. The intersection of these continua is $x_1 + y_1$. Therefore, there is no continuum $X_2(x_1ay_1)$ because no connected open set in S can have a boundary containing three continua such that the intersection of any two of them is two points. Hence, we suppose that the point a belongs to $X_1(x_1ay_1)$ but not to Bd $E_2 + K(J_2)$. (To get a so situated, it may be necessary to shift a and x_1ay_1 slightly.)

Furthermore, we suppose that b belongs to $X_1(x_2by_2)$ but not to Bd $E_2 + K(J_2)$; c belongs to $X_2(x_1cz_1)$ but not to Bd $E_1 + K(J_1)$; d belongs to $X_2(x_2dz_2)$ but not to Bd $E_1 + K(J_1)$ We suppose furthermore that a, b, c, d belong to the interiors of K_i's.

Let D_a, D_b, D_c, D_d be topological disks in S containing a, b, c, d on their respective interiors such that each of the disks is of diameter less than the minimum of $\rho(a + b, E_2 + K(J_2))$, $\rho(c + d, E_1 + K(J_1))$, and such that $D_a \cdot \text{Bd } T \subset \text{Int } K(a)$, $D_b \cdot \text{Bd } T \subset \text{Int } K(b)$, $D_c \cdot \text{Bd } T \subset \text{Int } K(c)$, $Dd \cdot \text{Bd } T \subset \text{Int } K(d)$.

Let *arb* be an arc from a to b in $D_a + E_1 + D_b$ and *asb* be an arc from a to b in $a + b + (E^3 - T)$ such that the projection of *asb* on Bd T from the center p_0 of the center of T does not intersect the projection of $D_c + D_d$ on Bd T from p_0.

Let *ctd* be an arc from c to d in $D_c + E_2 + D_d$.

Since $arb + asb$ links $cx_1d + cz_1d$, it links either $cx_1d + ctd$ or $cz_1d + ctd$. (These may be loops or images of simple closed curves rather than simple closed curves themselves, but since slight modifications in E^3 will take them into simple closed curves, modifications of the linking theorems used in the proof of Theorem 3 may still be used. See Theorems 9′ and 10′ of Section 9.) We shall show that $arb + asb$ links neither $cx_1d + ctd$ nor $cz_1d + ctd$, thereby arriving at a contradiction that refutes the assumption that J_1 intersects both components of Bd $T - J_2$.

Let $X_2(z_1, D_c)$ be the sum of D_c and a continuum in $X_2(z_1\ cx_1)$ irreducible from z_1 to D_c. Similarly we define $X_2(x_1, D_c)$, $X_2(x_2, D_d)$, $X_2(z_2, D_d)$, $X_1(x_1, D_a)$, $X_1(y_1, D_a)$, $X_1(x_2, D_b)$, and $X_1(y_2, D_b)$.

Let $f(cz_1d)$ be an image of cz_1d such that

(1) $f(cz_1d) \subset S$,

(2) there is a homotopy pulling cz_1d to $f(cz_1d)$ in the complement of $arb + asb$, and

(3) $f(cz_1d)$ is so near $X_2(z_1, D_c) + X_2(z_1 z_2) + X_2(z_2, D_d)$ that there is a homotopy in the complement of $f(cz_1d) + ctd$ pulling asb onto a set $f'(asb)$ in S near $X_1(x_1, D_a) + X_1(x_1 x_2) + X_1(x_2, D_b)$.

Similarly, let $g(cxd)$ be an image of cxd such that

(1) $g(cxd) \subset S$,

(2) there is homotopy pulling cxd to $g(cxd)$ in the complement of $arb + asb$, and

(3) $g(cxd)$ is so near $X_2(x_1, D_c) + X_2(x_1 x_2) + X_2(x_2, D_d)$ that there is a homotopy in the complement of $g(cxd) + ctd$ pulling asb onto a set $g'(asb)$ in S and near $X_1(y_1, D_a) + X_1(y_1 y_2) + X_1(y_2, D_b)$.

Finally we arrive at a contradiction resulting from the false assumption that J_1 intersects two components of Bd $T - J_2$. Since $arb + asb$ links J_2, it also links $f(cz_1d) + g(cxd)$. Then $arb + asb$ either links $f(cz_1d) + ctd$ or $g(cxd) + ctd$. If it links $f(cz_1d) + ctd$, $arb + f'(asb)$ links $f(cz_1d) + ctd$; if it links $f(cxd) + ctd$, $arb + g'(asb)$ links $g(cxd) + ctd$. Both of these alternatives are impossible since all sets concerned lie on S.

8. Approximation theorems for surfaces

Here we give two extensions of Theorem 1.

THEOREM 7. *If in a triangulated 3-manifold S, M is a 2-manifold with boundary and f is a nonnegative continuous function defined on M, there is a manifold M' and a homeomorphism h of M onto M' such that M' is locally polyhedral at $h(x)$ if $f(x) > 0$ and $\rho(x, h(x)) \leqq f(x)$.*

PROOF. First we consider the case where M is a topological disk. In working on this case, we need not fear cutting off a handle.

Case where M is a disk. Our proof will be a modification of the proof of the general case of Theorem 1 so we omit many details. Temporarily we suppose that $f = 0$ on Bd M. Let V be the open subset of M consisting of those points of M at which f is positive. Then $V \subset$ Int M.

For each point p of V let $\delta(p)$ be the least upper bound of all numbers ε with the property that if G is an ε-continuum in M containing p, and $K(G)$ is the sum of G and all components of $M - G$ in Int M, then the diameter of $K(G)$ is less than the minimum value of $f/2$ on $K(G)$.

Let U be an open set in S containing V but no point of $M - V$. There is a triangulation R of U (compatible with prescribed triangulation of S) such that no vertex of R intersects V and the diameter of any tetrahedron in R intersecting V is less than the minimum value of $\delta(p)$ for all points p of V on the tetrahedron.

Following Steps 1, 2, 3 of the proof of the general case of Theorem 1 we adjust V so that if F is a face of a tetrahedron of R, the following conditions are satisfied: (a) $M \cdot \text{Bd } F$ is not infinite, (b) if X is a component of $M \cdot \text{Int } F$, X does not have more than two limit points on Bd F, and (c) all but a finite number of components of $M \cdot F$ are of diameter less than γ_F where γ_F is less than the distance between any two points of $M \cdot \text{Bd } F$.

Following Steps 4, 5, 6, 7, 8 we get a collection of compact continua $K_1, K_2, \cdots$ in V such that for each K_i there is a face F_i of a tetrahedron of R and two points p_i, q_i of Bd F_i such that K_i is the sum of $C_i = (p_i + q_i +$ all components X of $M \cdot \text{Int } F_i$ with both p_i and q_i as limit points) and all components of $M - C_i$ in Int M.

Furthermore the K_i's are such that the components of $V - \sum K_i$ are open disks $E_1, E_2, \cdots$ such that for each E_i there is a tetrahedron T_i of T such that for some component E_i' of $E \cdot \text{Int } T_i$, the boundary of E_i' in M contains the boundary of E_i in M. Since the boundary of E_i lies in Bd T_i, each E_i is of diameter less than the minimum value of $f/2$ on $\bar{E}_i$.

As in Steps 9, 10, 11, we replace each K_i by an arc A_i in V and each E_i by an open disk H_i in V. Then we get polygonal arcs A_i' in F_i from p_i to q_i such that $A_i' \cdot \text{Int } A_j' = 0$ if $i \neq j$. Next comes a homeomorphism h of A_i onto A_i' that is fixed at p_i and q_i. Then we build polygonal disks H_i' in T_i such that Bd H_i' is the image of Bd H_i under h and $H_i' \cdot \text{Int } H_j' = 0$ if $i \neq j$. The homeomorphism h is extended to take H_i onto H_i'. If $M'' = M - V + \sum H_i'$ and h is the identity on $M - V$, h is a homeomorphism of M onto M'' that does not move any point x of M by more than $f(x)$.

Now M'' as yet has not been defined so as to be locally polyhedral at points of Bd M''. At points of Bd M where f is positive, we can trim back Bd M'' slightly at the appropriate place to get the desired manifold M'. The homeomorphism h can be adjusted accordingly.

General case. The preceding argument fails for the general case because the K_i's and E_i's described might have handles. However, this is avoided if we made each element of R of diameter less than the minimum value of $\delta/3$ on it and place the additional restriction on $\delta(x)$ that $K(G)$ is a subset of the interior of a disk in M. With this alteration, the proof goes through.

THEOREM 8. *If in Theorem 7 we had supposed that M is locally polyhedral at each point of a closed point set N, we could have chosen M' and h so that h is the identity on N.*

PROOF. With no loss of generality we suppose that $f = 0$ on N. We point out how the proof of Theorem 7 could be altered so as to get a proof of Theorem 8.

Suppose that in the paragraph where we duplicated Steps 9, 10, 11 of the proof of Theorem 1, we had used the following additional precautions. If K_i is a polygonal arc, then we make $K_i = A_i = A_i'$ and h the identity on K_i; if E_i is a polygonal disk which lies except for its boundary in Int T_i, we let $E_i = H_i = H_i'$ and h be the identity on E_i. Then it would be true that if p is a point of the boundary of V and M was locally polyhedral at p, there is a neighborhood of p such that h would not move any point in this neighborhood. Hence, M' is locally polyhedral at each point of N.

9. Linking curves

The proofs of Theorems 3 and 5 made use of linking simple closed curves. We include a treatment of such linking in this section. The notion of linking used here resembles the notion of linking used in combinatorial topology shorn of some of its algebraic notation.

A 2-complex (for our present purposes) is a finite collection of triangular disks called 2-simplexes. (On many occasions it is more convenient to think of a 2-complex as the sum of a finite number of triangular disks such that if two of them intersect, their intersection is either a vertex or an edge of each. These extra suppositions act more as a hindrance than a help in discussing linking so we do not use them here. For example, with the extra condition it is more difficult to show the existence of 2-complexes X and Y as mentioned in the fifth and seventh paragraphs of this section.)

The collection of all edges of 2-simplexes of a 2-complex X that are edges of exactly an odd number of 2-simplexes of X is called Bd X. We note that Bd X is a collection of edges rather than a point set.

If J is a polygonal simple closed curve, it may be regarded either as a point set or as a collection of edges. As convenient, we oscillate back and forth between these two points of view. Except in the following sentence, we do not point out which of these senses we are using since the meaning is generally clear.

If J_1, J_2 are mutually exclusive (in the point set sense) polygonal simple closed curves, there is a 2-complex X such that Bd $X = J_1$ (in the collection of edges sense), J_2 (in the point set sense) intersects no edge of any 2-simplex in X, an no 2-simplex in X contains a vertex of J_2. To get one such X, one can take a cone over J_1 from a point that is not coplanar with any three noncollinear vertices of $J_1 + J_2$. Had we used the alternate definition of a 2-complex as suggested in the second paragraph of this section, we could take for the vertex of X a point that is not on any line which intersects three edges of $J_1 + J_2$. That there is such a point follows from the result in solid analytics that if L_1, L_2, L_3 are three nonconcurrent lines, the sum of all lines intersecting each of L_1, L_2, L_3 is a quadric surface.

We say that J_1 *links* J_2 if $N(X, J_2) = \sum N(T_i, J_2)$ is odd where $T_1, T_2, \cdots, T_n$ are the 2-simplexes of a 2-complex X as described in the preceding paragraph and $N(A, B)$ denotes the number of points in $A \cdot B$ in case A, B are point sets. It may be convenient to think of X as a 2-manifold with boundary and $N(X, J_2)$ as the number of times J_2 passes through X.

We show that whether or not J_1 links J_2 is not dependent on the 2-complex X chosen and that J_2 links J_1 if J_1 links J_2. To this end, we let X_1, X_2 be two 2-complexes such that Bd $X_i = J_1$, J_2 intersects no edge of any 2-simplex in X_i, and no 2-simplex in X_i contains a vertex of J_2. Let Y be a 2-complex such that Bd $Y = J_2$, no edge of any element of Y intersects any edge of any element of $X_1 + X_2$, no element of $X_1 + X_2$ contains any vertex of any element of Y, and no element of Y contains any vertex of any element of $X_1 + X_2$. To get one such Y, one could take a cone over J_2 from a point that is not coplanar with any three noncollinear vertices of elements of $X_1 + X_2$ or J_2.

Denote the 2-simplexes of X_1 and Y by $T_1, T_2, \cdots, T_n$ and $S_1, S_2, \cdots, S_m$ respectively. We note that $N(T_i, \text{Bd } S_j) = N(S_j, \text{Bd } T_i) \bmod 2$. (Here we regarded Bd S_j and Bd T_i as simple closed curves.) Using this result and the fact that some edges belong to an even number of triangles, we find that the following relations hold mod 2:

$$\begin{aligned} N(X_1, J_2) &= \sum N(T_i, J_2) \\ &= \sum \sum N(T_i, \text{Bd } S_j) \\ &= \sum \sum N(S_j, \text{Bd } T_i) \\ &= \sum N(S_j, J_1) \\ &= N(Y, J_1). \end{aligned}$$

Also, $N(X_2, J_2) = N(Y, J_1) \bmod 2$ so $N(X_1, J_2)$ is odd if and only if $N(X_2, J_2)$ is odd. This shows that whether or not J_1 links J_2 is not a function of the 2-complex X chosen.

Since $N(X_1, J_2) = N(Y, J_1) \bmod 2$, J_2 links J_1 if and only if J_1 links J_2.

Suppose J_1, J_2 are two arbitrary simple closed curves (not necessarily polygonal) and $\rho(J_1, J_2) > \varepsilon$. Let K_j^i $(i, j = 1, 2)$ be polygonal simple closed curves such that there is a homeomorphism h_j^i of J_j onto K_j^i such that for each x in J_j, $\rho(x, h_j^i(x)) < \varepsilon/2$. Then it follows from Theorem 10 that K_1^1 links K_2^1 if and only if K_1^2 links K_2^2. Hence, to determine whether or not two arbitrary disjoint simple closed curves link, we investigate the linking of nearby polygonal simple closed curves—that is, J_1 *links* J_2 if and only if K_1^1 links K_2^1.

Similarly, suppose f_1, f_2 are two maps of simple closed curves C_1, C_2 into E^3 such that $\rho(f_1(C_1), f_2(C_2)) > \varepsilon$. Consider two homeomorphisms h_1, h_2 of C_1, C_2 into E^3 such that for each point x of C_i, $\rho(f_i(x), h_i(x)) < \varepsilon/2$. The map f_1 *links* the map f_2 if and only if $h_1(C_1)$ links $h_2(C_2)$.

THEOREM 9. *If a polygonal simple closed curve J does not intersect a certain polygonal θ curve, then J links exactly two simple closed curves in the θ curve if it links one of them.*

PROOF. Suppose the θ curve is the sum of the polygonal arcs pxq, pyq, pzq and J links $pxq + pyq$. Let X be a finite 2-complex bounded by J such that no edge of any element of X intersects $pxq + pyq + pzq$ and no vertex in $pxq + pyq + pzq$ lies on any element of X.

We use $N(X, pwq)$ to denote $\sum N(T_i, pwq)$ where $T_1, T_2, \cdots, T_n$ are the 2-simplexes in X. Since $pxq + pyq$ links J, one of $N(X, pxq)$, $N(X, pyq)$ is odd (say $N(X, pxq)$) and the other is even. Then J links $pxq + pzq$ but not $pyq + pzq$ if $N(X, pzq)$ is even; J links $pyq + pzq$ but not $pxq + pzq$ if $N(X, pzq)$ is odd.

The following result is proved by repeated applications of Theorem 9.

THEOREM 10. *If J_1, J_2, J_3 are three simple closed curves in E^3 such that J_1 links J_2 and J_2 is homotopic to J_3 in the complement of J_1, then J_1 links J_3.*

PROOF. First we consider the case where J_1, J_2, J_3 are polygonal.

Case 1. J_1, J_2, J_3 are polygonal. Since J_2, J_3 are homotopic in the complement of J_1 there is a sequence of polygonal simple closed curves $C_1, C_2, \cdots, C_n$ such that

1. $C_1 = J_2$, $C_n = J_3$.
2. The closure of $C_i + C_{i+1} - C_i \cdot C_{i+1}$ is a triangle T_i such that T_i plus its interior does not intersect J_1.

No T_i links J_1 since the triangular disk bounded by it does not intersect J_1. Then since $J_2 = C_1$ links J_1, it follows from Theorem 9 that each of C_2, C_3, $\cdots$, $C_n = J_3$ does also.

Case 2. If J_1, J_2, J_3 are not polygons, there is a polygonal simple closed curve K_i $(i = 1, 2, 3)$ and a homeomorphism h_i of J_i onto K_i such that for each x of J_i, $\rho(x, h_i(x))$ is less than one half the distance between J_1 and the path swept out as J_2 was homotopically pulled to J_3. Then the proof of Case 2 follows from Case 1 and the definition of linking of arbitrary simple closed curves. From Case 1, we find that it does not matter what K_i's we pick so long as they are close enough to the J_i's.

Theorems 9 and 10 have their analogues in combinatorial topology with various coefficient groups which are rings. In the case of Theorem 9 where the coefficient group has more than two elements, we conclude merely that J links "at least" two of the simple closed curves in the θ curve instead of "exactly" two of them.

We now state Theorems 9 and 10 in terms of mappings.

THEOREM 9′. *Suppose f is the map of a θ curve T into E^3 where f_1, f_2, f_3 are the parts of f operating on the simple closed curves in T. If f_4 is a map of a circle J into E^3 such that $f_4(J) \cdot f(T) = 0$, then f_4 links exactly two of f_1, f_2, f_3 if it links one of them.*

THEOREM 10′. *Suppose f_1, f_2, f_3 are maps of circles C_1, C_2, C_3 into E^3 such that f_1 links f_2 and f_2 is homotopic to f_3 in the complement of $f_1(C_1)$. Then f_1 links f_3.*

UNIVERSITY OF WISCONSIN

REFERENCES

[1] J. W. ALEXANDER, *On the subdivision of 3-space by a polyhedron*, Proc. Nat. Acad. Sci. U.S.A., vol. 10 (1924), pp. 6–8.

[2] R. H. BING, *Locally tame sets are tame*, Ann. of Math., vol. 59 (1954), pp. 145–158.

[3] ——, *Approximating surfaces with polyhedral ones*, Summary of Lectures and Seminars, Summer Institute on Set Theoretic Topology, Madison, 1955, pp. 47–52.

[4] ——, *A simple closed curve that pierces no disk*, J. Math. Pures Appl., vol. 35 (1956), pp. 337–343.

[5] ——, *A decomposition of E^3 into points and tame arcs such that the decomposition space is topologically different from E^3*, Ann. of Math., vol. 65 (1957), pp. 484–500.

[6] ———, *Some monotone decompositions of a cube*, Ann. of Math., vol. 61 (1955), pp. 279–288.

[7] O. G. HARROLD, JR., *Some consequences of the approximation theorem of Bing*, Summary of Lectures and Seminars, Summer Institute on Set Theoretic Topology, Madison, 1955, pp. 53–56.

[8] E. E. MOISE, *Affine structures in 3-manifolds*, IV. *Piecewise linear approximations of homeomorphisms*, Ann. of Math., vol. 55 (1952), pp. 215–222.

[9] ———, *Affine structures in 3-manifolds*, V. *The triangulation theorem and Hauptvermutung*, Ann. of Math., vol. 56 (1952), pp. 96–114.

[10] R. L. MOORE, *Concerning upper semi-continuous collections*, Monatscheften für Mathematik und Physik, vol. 36 (1929), pp. 81–88.

[11] ———, Foundations of point set theory, American Mathematical Society Colloquium Publication, vol. 13, New York, 1932.

APPLICATIONS OF THE SIDE APPROXIMATION THEOREM FOR SURFACES

R. H. BING

Madison

By a surface we mean a closed set in a 3-manifold which is locally like the plane – each point of the surface lies in an open subset of the surface which is topologically equivalent to the interior of a circle.

Two surfaces may intersect in a peculiar set in Euclidean 3-space. Even two smooth surfaces may intersect in a set locally like the closure of the graph of $y = \sin(1/x)$. In fact, if K is any closed subset of a round 2-sphere S, there is a second two sphere S' (not necessarily round) such that S' contains K and lies except for K on the interior of S. On the other hand, the intersection of two polyhedral surfaces is always a polyhedron. (A surface is polyhedral if it is a geometric complex – the sum of a locally finite collection of triangular planar disks.) To bring about a reasonable intersection of surfaces, the Approximation Theorem for surfaces was proved. It is proved in [2] and may be stated as follows.

Theorem 1. Approximation Theorem for Surfaces. *Suppose S is a surface in a triangulated 3-manifold M and f is a non negative continuous function defined on S. Then there is a surface S' and a homeomorphism h of S onto S' such that*

a) $p(x, h(x)) \leqq f(x)$, $x \in S$ *and*

b) S' *is locally polyhedral at* $h(x)$ *if* $f(x) > 0$.

This Approximation Theorem for Surfaces has had several applications. O. G. Harrold used it [10] to show that if P is polygon in E^3 which bounds a disk, then P bounds a polyhedral disk. He also used it to show that in various theorems where the hypothesis has previously required that certain 2-spheres be locally polyhedral on certain parts, these extra conditions could be dropped.

E. E. Moise has shown [11] that if h is a homeomorphism of an open subset U of E^3 into E^3 and f is a non negative continuous function defined on U, then there is a homeomorphism h' of U onto $h(U)$ such that

a) h' is locally polyhedral at $x \in U$ if $f(x) > 0$ and

b) $p(h(x), h'(x)) \leqq f(x)$.

This is a very useful theorem and was used in [12] by Moise to prove that any 3-manifold can be triangulated.

The Approximation Theorem for Surfaces was extended to an Approximation

Theorem for 2-complexes in [6] to give alternative proofs of these two results of MOISE.

In hearing papers given about mappings on 3-manifolds, I frequently hear the restrictions, "We restrict ourselves to piecewise linear homeomorphisms". While this restriction is sometimes necessary (as in the case of certain periodic homeomorphisms) it frequently is not and shows that the author is not proving as strong a theorem as he might have proved if he were acquainted with the results mentioned in the preceding two paragraphs.

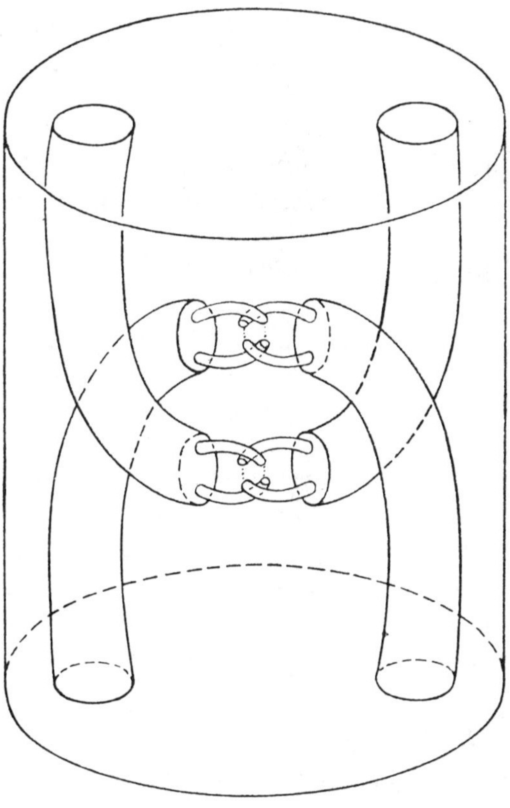

Fig. 1.

If S is a 2-sphere and T is a tetrahedron in E^3, there is a homeomorphism of S onto BdT (the boundary of T). If this homeomorphism can be extended to take E^3 onto itself, we say that S is tame. If the homeomorphism cannot be extended, we say it is wild. The 2-sphere shown in Figure 1 was obtained by starting with the boundary of a cylindrical can, replacing disks in the ends of the can by tubes that almost link, replacing disk in the ends of tubes with tubes, etc.

It is a wild 2-sphere since its interior (bounded complementary domain) is not simply connected (1-connected). Note that the 2-sphere can be approximated by a polyhedral 2-sphere which lies on the exterior of the wild 2-sphere. However, it cannot be approximated from the interior of the wild 2-sphere since an approximating polyhedral 2-sphere which was on the interior would cut off some of the small tubes. Hence, at best some parts of the approximating 2-sphere would fail to lie on the interior of the wild 2-sphere.

In a seminar BOB WILLIAMS asked if it might be possible to get a polyhedral approximation to any 2-sphere which would lie except for small disks on a prescribed side of the 2-sphere. It turns out that the answer is in the affirmative as given by the following result.

Theorem 2. Side Approximation Theorem for 2-Spheres. *Suppose S is a 2-sphere in E^3, $\varepsilon > 0$, and U is a component of $E^3 - S$. Then there is a polyhedral 2-sphere S' and a homeomorphism h of S onto S' such that*

a) *h moves no point more than ε and*

b) *S' contains a finite family of mutually exclusive disks, each of diameter less than ε, such that S' minus these disks lies in U.*

The proof of this theorem has been distributed in ditto form and will appear in the Annals of Mathematics [3]. The proof is modeled (with certain simplifications) after the proof the Approximation Theorem for Surfaces given in [2]. The construction is as might be expected in that S is used as a model and a polyhedral surface S' is constructed beside S with S' not being permitted to intersect S except in some feelers.

Without using the full strength of the Side Approximation Theorem but using merely the Approximation Theorem for Surfaces the following result had been proved [4]:

Theorem 3. *A 2-sphere S in E^3 is tame if for each $\varepsilon > 0$ and each component U of $E^3 - S$ there is a 2-sphere S in U, and a homeomorphism of S onto S' that moves no point more than ε.*

By using this result and the Side Approximation Theorem the following result may be established:

Theorem 4. *A 2-sphere S in E^3 is tame if $E^3 - S$ is 1-ulc.*

We say that a set is 1-*ulc* if for each $\varepsilon > 0$ there is a $\delta > 0$ such that if D is a disk and f is a map of BdD into a δ-subset of X, then f can be extended to map D into an ε-subset of X.

Outline of proof of Theorem 4. Theorem 4 is proved in [5] but the outline of the proof is as follows. We shall apply Theorem 3 so we only need to show that for each component U of $E^3 - S$, there is a 2-sphere in U that is a close homeomorphic approximation to S. It follows from the Side Approximation Theorem that there is a polyhedral 2-sphere S' which lies except for small polyhedral disks $D_1, D_2, \ldots, D_n$ in U. Since U is 1-*ulc*, D_i can be shrunk to a point on a small polyhedral singular disk E_i in U. Then $(S' - \sum D_i) + \sum E_i$ is a singular 2-sphere in U. By using results of C. D. Papakyriakopoulos regarding Dehn's lemma [3] the singular 2-sphere may be changed to a 2-sphere in U which is a close approximation to S.

Theorem 4 may be extended to apply to surfaces as well as 2-spheres. How this may be done and the details of the proof of Theorem 4 are to be found in [5].

A Sierpinski curve is a 1-dimensional subset of a disk D obtained by deleting from D a null sequence of interiors of disks such that each of these small disks lies on the interior of D and no two of them intersect each other. We say that a Sierpinski curve is tame if it lies on a tame 2-sphere. The following is another application of the Side Approximation Theorem which gives a strong affirmative answer to the question as to whether or not each 2-sphere in E^3 contains a tame arc.

Theorem 5. *For each 2-sphere S in E^3 and each $\varepsilon > 0$ there is a tame Sierpinski curve X on S such that each component of $S - X$ is of diameter less than ε.*

Outline of proof. It follows from an extension of Theorem 2 as shown in Figure 2 that there is a polyhedral 2-sphere S' such that S' lies except for small feelers (like F_1) on Int S. Cut off feelers like F_1 and replace them with a smooth disk so that the resulting 2-sphere S_1 lies on Int S'. An approximation to S_1 almost from the other side enables us to replace F_2 by a smooth disk so that the resulting 2-sphere S_2 can be approximated closely from both sides. In a similar fashion we replace S_2 by S_3 so as to remove small feelers like F_3 and replace S_3 by S_4 to remove feelers like F_4. If great care is exercised as done in [8], S_1, $S_2, \ldots$ converges to a 2-sphere S_∞ which shares a Sierpinski curve X with S so that each component of $S - X$ is of diameter less than ε. Furthermore S_∞ was chosen so that it could be homeomorphically approximated arbitrarily closely from both sides. It follows from Theorem 3 that S_∞ is tame.

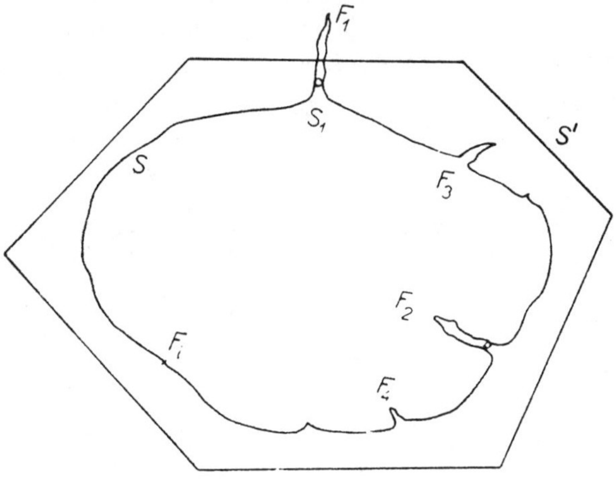

Fig. 2.

Some peculiar wild 2-spheres have been described. Some have the property that although their complements are simply connected, no disk in them is tame but each arc in them is tame. Some of them are even such that each Sierpinski curve in them is tame. David Gillman and W. R. Alford have obtained such examples as modification of the ones given in [7]. However, Theorem 5 reveals that any 2-sphere whatsoever in E^3 can be obtained from a tame 2-sphere by removing small disks and putting on feelers. Also, the reversal of this operation changes every 2-sphere into a tame 2-sphere. This adds to our understanding of wild 2-spheres. Alford has shown [1] that there are uncountably many different kinds of wild 2-spheres.

Theorem 5 is used to prove the following result:

Theorem 6. *Each 2-sphere in each 3-manifold can be pierced by a tame arc.*

Details of the proof of the above result are found in [9] where it is shown that the 2-sphere is pierced at each point of the tame Sierpinski curve. D. Gillman has extended Theorem 6 to show that the places at which the 2-sphere can be pierced by a tame arc are actually the points of the 2-sphere which lie on tame arcs in the 2-sphere.

The following questions suggest themselves:

Question 1. What is the category of the points of a 2-sphere in E^3 at which it cannot be pierced by a tame arc? Conjecture – a 0-dimensional F_σ set.

Question 2. Can Dehn's lemma be extended to imply that if f is a map of a disk D into E^3 such that f^{-1} is $1-1$ on $f(\mathrm{Bd}D)$, then $f(\mathrm{Bd}D)$ bounds a disk in E^3.

Question 3. Can a homotopy 3-cell in E^3 be approximated by a real cell in the following sense.

Suppose U is a bounded, contractible open subset of E^3 and V is a neighborhood of $\mathrm{Bd}U$. Is there a 3 cell C in E^3 such that $U - V \subset C \subset U + V$.

Question 4. (Raised by Borsuk). If X is a contractible continuous curve in E^3 and $\varepsilon > 0$, is there a map f of X onto a contractible polyhedron such that f moves no point more than ε?

References

[1] *W. R. Alford*: Some „nice" wild two spheres in three space. Notices Amer. Math. Soc., vol. *8* (1961), 348.

[2] *R. H. Bing*: Approximating surfaces with polyhedral ones. Ann. Math., vol. *65* (1957), 456–483.

[3] *R. H. Bing*: Approximating surfaces from the side. Ann. Math. to appear.

[4] *R. H. Bing*: Conditions under which a surface in E^3 is tame. Fund. Math., vol. *47* (1959), 105–139.

[5] *R. H. Bing*: A surface is tame if its complement is 1-*ulc*. Trans. Amer. Math. Soc., vol. *101* (1961), 294—305.

[6] *R. H. Bing*: An alternative proof that 3-manifolds can be triangulated. Ann. Math., vol. *69* (1959), 37–65.

[7] *R. H. Bing*: A wild surface each of whose arcs is tame. Duke Journal, *28* (1961), 1–15.

[8] *R. H. Bing*: Each disk in E^3 contains a tame arc, Am. J. Math., to appear.

[9] *R. H. Bing*: Each disk in E^3 is pierced by a tame arc, Am. J. Math., to appear.

[10] *O. G. Harrold Jr.*: Some consequences of the approximation theorem of Bing. Proc. Amer. Math. Soc., vol. *8* (1961), 204–206.

[11] *Edwin E. Moise*: Affine structures in 3-manifolds, IV, Piecewise linear approximations of homeomorphisms. Ann. Math., vol. *66* (1959), 215–222.

[12] *Edwin E. Moise*: Affine structures in 3-manifolds, V, The triangulation theorem and Hauptvermutung. Ann. Math., vol. *56* (1952), 96–114.

[13] *C. D. Papakyriakopoulos*: On Dehn's lemma and the asphericity of knots. Ann. Math., vol. *66* (1951), 1–26.

ANNALS OF MATHEMATICS
Vol. 77, No. 1, January, 1963
Printed in Japan

APPROXIMATING SURFACES FROM THE SIDE*

BY R. H. BING

(Received January 8, 1962)

1. Introduction

The following theorem was proved in [2].

APPROXIMATION THEOREM FOR SURFACES. *For each 2-manifold M in a triangulated 3-manifold-with-boundary and each non-negative continuous function f defined on M, there is a 2-manifold M' and a homeomorphism h of M onto M' such that*

$$\rho(x, h(x)) \leqq f(x) \qquad (x \in M)$$

and M' is locally polyhedral at $h(x)$ if $f(x) > 0$.

In the above theorem we used ρ to denote the distance function.

Numerous applications of the *approximation theorem for surfaces* have been made. For example, it has been used to show the following:

(i) A surface is tame if it can be homeomorphically approximated from both sides [6].

(ii) A compact connected 3-manifold is topologically S^3 if each simple closed curve in it lies in a topological cube in it [14].

(iii) There is a simple closed curve that pierces no disk [5].

(iv) E^3 does not contain uncountably many mutually exclusive wild 2-spheres [12].

(v) Each disk in a 3-manifold can be cut down to size so as to lie in a small 2-sphere—that is, for each disk D in a 3-manifold, each point p of $\operatorname{Int} D$, and each neighborhood N of p in the 3-manifold, there is a 2-sphere S in N and a subdisk D' of D such that $p \in \operatorname{Int} D' \subset S$, [11].

(vi) Each homeomorphism of a polyhedron into a triangulated 3-manifold with boundary can be approximated by a piecewise linear homeomorphism [7].

The result was used in [7] to give alternative proofs of the following two results.

(vii) Each 3-manifold can be triangulated. (Also see [20].)

(viii) Each homeomorphism of a triangulated 3-manifold with boundary into a triangulated 3-manifold with boundary can be approximated by a piecewise linear homeomorphism. (Also see [19].)

* Work on this paper is supported by contract NSF-G11665.

This last result in turn is critical in proving the following.

(ix) Each locally tame set in a triangulated 3-manifold is tame [4, 21]. Harrold has used the result to give characterizations of tame arcs [18].

Let $[-1, 1]$ denote the closed real interval from -1 to 1 and $S \times [-1, 1]$ the cartesian product of S with this interval. In the present paper we extend the *approximation theorem for surfaces* as applied to 2-spheres and prove the following result.

THEOREM 1. *For each 2-sphere S in E^3 and each positive number ε there is a homeomorphism h of $S \times [-1, 1]$ into E^3 such that:*

(1) $\rho(s, h(s \times t)) < \varepsilon\ (s \in S,\ -1 \leqq t \leqq 1)$,

(2) *S contains a finite collection of mutually exclusive ε-disks such that S minus the disks lies between $h(S \times -1)$ and $h(S \times 1)$, and*

(3) *for each t, $h(S \times t)$ is polyhedral.*

Theorem 1 is proved by reconstructing the proof of the *approximation theorem for surfaces* and making alterations as needed to extend it. Several simplifications are made in the proof of Theorem 1 over that given for the *approximation theorem for surfaces* as given in [2]. Also, as a result of comments made by readers of [2], we give additional explanations for certain difficult spots. It is hoped that the present paper is easier for the reader to follow than [2]. The author is grateful to David Gillman, John Hempel, C. E. Burgess, and Bert Casler for their readings of a preliminary version of this paper and comments which resulted in improvements.

The proof of Theorem 1 is completed in § 6. However, here we indicate how the proof would go in the two dimensional case. If one understands this lower dimensional case, he is in better condition to approach the proof of Theorem 1. Consider a simple closed curve J as shown in Figure 1a.

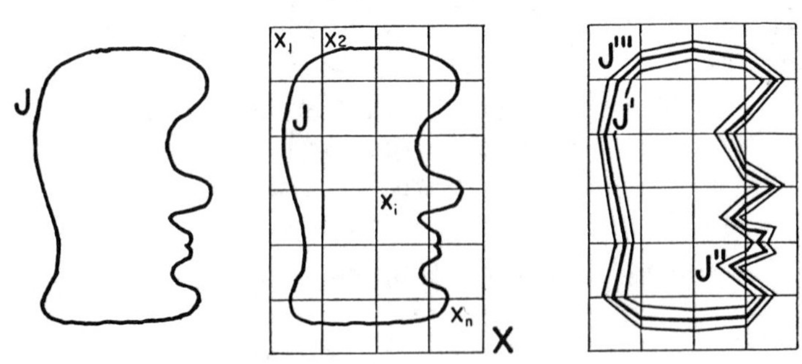

a b c

Figure 1

If a person considers how J may be approximated by a polyhedral simple closed curve, he may get a clue as to how to prove Theorem 1. One such method is to consider a rectangle X containing J and subdivide X into smaller rectangles $X_1, X_2, \cdots, X_n$ as shown in Figure 1b. If J intersects the boundaries of the X_i's in only a finite number of points, the continuation is relatively easy. Consider one of the small rectangles X_i and replace each non-degenerate component of $X_i \cdot J$ by a polygonal arc which spans X_i and has the same end points as the arc it is replacing. This is done for each non-degenerate component of $X_i \cdot J$ and care is taken that no two of the replacements intersect except possibly at their end points. If this replacement is done for each of the small rectangles into which X_i is divided, there results an approximation J' to J. If the X_i's are small, the approximation is close. If J' is shoved slightly inward onto a polygonal simple closed curve J'' and outward onto a simple closed curve J''' as indicated in Figure 1c, then J will lie except for small pieces between J'' and J'''.

If J intersects the boundaries of certain X_i's in infinitely many points, it is more difficult to use the X_i's to approximate J with a polygonal simple closed curve. Reflection on how to overcome such difficulties may help in seeing why certain steps are used in the proof of Theorem 1. Of course, Theorem 1 is a 3-dimensional version of the plane case we have been considering. The proof of Theorem 1 is started in the next section and completed in § 6. Along the way, various properties of E^3 are developed.

In § 7 we show that Theorem 1 implies the following result.

THEOREM 16 (Side approximation theorem for 2-spheres). *Each 2-sphere S in E^3 can be polyhedrally approximated almost from either side—that is, for each $\varepsilon > 0$ and each component U of $E^3 - S$ there is a homeomorphism h of S onto a polyhedral 2-sphere such that:*

(1) *h moves no point more than ε, and*

(2) *$h(S)$ contains a finite collection of mutually exclusive disks each of diameter less than ε such that $h(S)$ minus the sum of these disks lies in U.*

The use of the word "almost" is essential in the statement of Theorem 16, since no wild spheres (see [1, 9, 10, 16] for examples of wild spheres) can be approximated from both sides. Various extensions of Theorems 1 and 16 are found in § 8.

A proof of Theorem 16 was obtained in 1958, and various applications of this theorem have been made subsequently. For example, the theorem was used to show each of the following.

(x) Each 2-sphere in E^3 contains a tame arc [15].
(xi) Each disk in each 3-manifold is pierced by a tame arc [16].
(xii) A surface is tame if its complement is 1-*ULC* [11].
(xiii) Each 2-sphere in E^3 can be slightly pushed except for a Cantor set to either side.

The preceding statement means that for each 2-sphere S in E^3, each positive number ε, and each component U of $E^3 - S$, there is a Cantor set C on S and an ε-map f of S into $U + C$ such that f is a homeomorphism on $S - C$. Wilder has used a converse of this result to get a positional characterization of the 2-sphere in E^3 [24].

We finish § 1 by recalling certain definitions that we shall use.

An *n-manifold* is a metric space locally like euclidean n-space E^n. An *n-manifold-with-boundary* is a metric space each of whose points has a neighborhood whose closure is homeomorphic with an n-cell. The set of points which have neighborhoods homeomorphic with E^n is the *interior* of the n-manifold-with-boundary. The complement (in the n-manifold with boundary) of the interior is the *boundary*.

We used boundary in several senses. It is used in the combinatorial sense as above where if D is a disk, Bd D denotes the rim of the disk and Int D denotes the point of the disk not on the rim. If $A \subset B$, the *boundary of A in B* (denoted by Bd A in B or by Bd A if no confusion results) is the intersection of the closure of A and the closure of $B - A$.

A 2-sphere is a 2-manifold homeomorphic with the boundary of a tetrahedron. An arc is said to *pierce* a 2-sphere in E^3 if the arc intersects the 2-sphere at only one point and the ends of the arc lie in different components of the complement of the 2-sphere.

Alternative definitions of piercing are used. If D is a disk in E^3 and apb is an arc that intersects D only at the point p of Int D, then *apb pierces D* at p if and only if no small neighborhood of p contains an arc from ap to bp that misses D. This definition is compatible with the preceding one since, as previously mentioned, a disk can be cut down to size to lie in a 2-sphere and each component of the complement of a 2-sphere is uniformly locally arcwise connected (0-*ULC*). See [25, Th. 5.35]. Another equivalent definition of piercing is that *apb pierces D* if apb lies in a simple closed curve J such that each neighborhood of p contains a simple closed curve in D that links J. Note that piercing is a local property and depends on the local two-sidedness of a surface in a 3-manifold. We could say that *apb pierces a surface* at p if ap and bp abut on the surface from opposite sides of the surface.

A *spanning arc* of a disk is an arc whose end points belong to the boundary of the disk and whose non-end points belong to the interior of the disk. A *punctured disk* is a 2-manifold-with-boundary homeomorphic with a closed connected subset of a disk. Its boundary is the sum of a finite number of mutually exclusive simple closed curves.

A set is *simply connected* if each map of the boundary of a disk into the set can be extended to a map of the disk into the set.

A set of diameter less than or equal to ε is called an *ε-set*. We speak of *ε-disks* and *ε-components*. Sets with diameters less than ε are sometimes called *($<\varepsilon$)-sets* and those with diameters more than ε are sometimes called *($>\varepsilon$)-sets*.

An *n-simplex* in $E^m (m \geqq n)$ is the convex hull of $n+1$ points in E^m such that these points do not lie in an $n-1$ plane in E^m. Each convex hull of a subset of the $n+1$ points is a simplex which is a *face* of the original n-simplex. In an abstract metric set, an *n-simplex* is a set isometric with an n-simplex in E^n. The *faces* of this abstract n-simplex are the images of the faces of its isometric image in E^n.

A *geometric complex* is a metric space which is the sum of a locally finite collection of simplexes such that if two of these simplexes intersect, the intersection is a face of each. Suppose this locally finite collection R of simplexes contains each face of each element of R. Then R is called a *triangulation* of the geometric complex. The *mesh* of R is the least upper bound of the diameters of the elements of R.

A *triangulated manifold* is a geometric complex which is a manifold together with some particular triangulation of the geometric complex. To triangulate a manifold, one may need to re-metrize it so that it will be the sum of simplexes of the right sort. However, when we speak of a triangulation of E^n, unless we indicate to the contrary, we mean a rectilinear triangulation of E^n where the simplexes of the triangulation are simplexes under the ordinary euclidean metric.

The *i-skeleton* of a triangulation is the sum of all i-simplexes in it.

Note that a geometric complex may have several triangulations. A set X in a geometric complex is called a *polyhedron* if the geometric complex has a triangulation (without re-metrization of the complex) such that X is the sum of simplexes of the triangulation. A set X is *tame* in a geometric complex C if there is a homeomorphism h of C onto itself such that $h(X)$ is a polyhedron. A closed set X in C which is not tame but is homeomorphic to a polyhedron is called *wild*. A set is *locally tame* at a point p of an n-manifold if there is a neighborhood N of p and a homeomorphism h of $\bar{N}$ into E^n such that $h(\bar{N} \cdot X)$ is a polyhedron in E^n.

A map f of a geometric complex C is called *piecewise linear* if there is a triangulation R of C such that f is linear on each simplex of R. The map f is *locally piecewise linear* at a point p if for some neighborhood N of p there is a triangulation R' of the closure of this neighborhood such that f is linear on each element of R'. If f is locally piecewise linear except possibly at points of a set X, we say that it is *locally piecewise linear* mod X. We speak of *piecewise linear homeomorphisms* and *locally piecewise linear homeomorphisms*.

2. Triangulating E^3

To make the approximation of J suggested in Figure 1 close, one should take the mesh of the subdivision of R small. By taking this mesh very small, a very good approximation may be obtained. A mosaic may appear very life-like if it is made with small tiles.

To approximate the 2-sphere S of Theorem 1 by a polyhedral 2-sphere, we could subdivide a rectangular solid containing S. It is irrelevant for our purposes whether we triangulate the solid or subdivide it into small rectangular solids. Since triangulations are more fashionable, we use a triangulation of E^3.

If T is a 3-simplex of a triangulation and a feeler of S reaches through Bd T as shown in Figure 2, we wish to be able to cut off this feeler and replace it by something near Bd T so that the resulting 2-sphere will be a close approximation to the original. If the mesh of the triangulation is small, the part of the feeler sticking through Bd T will not be long. We suppose the diameter of S is large compared to ε.

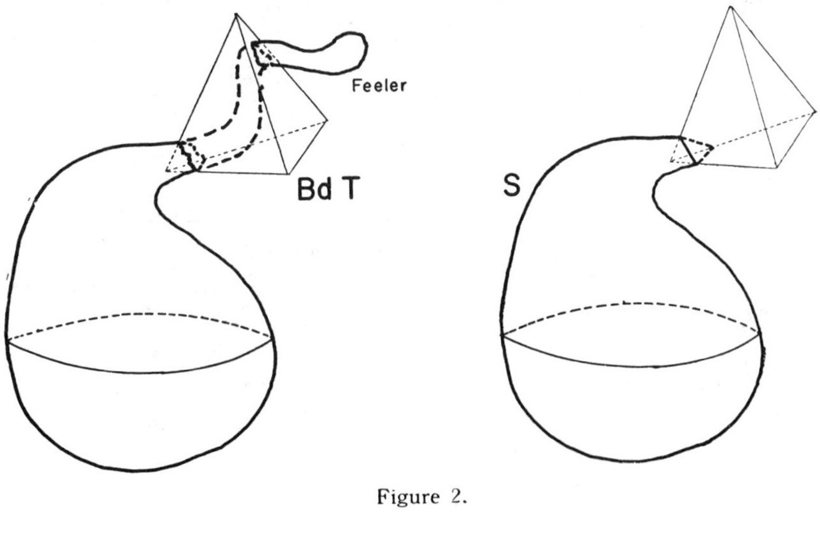

Figure 2.

Let δ be a positive number such that each 7δ-subset of S lies in a disk of S of diameter less than $\varepsilon/3$. The reason for using 7δ and $\varepsilon/3$ is not apparent until we treat the general case of Theorem 1 in § 6. In the special cases we can put on less stringent conditions.

Let R be a triangulation of E^3 of mesh less than δ and such that no vertex of R belongs to S.

We denote the i-skeleton of R by R_i. We would like for R_0, R_1, R_2 to meet S in as nice a set as possible. We achieved this for R_0 but not for R_1 and R_2. We find a substitute for a nice intersection of R_1 and S in § 6. We simplify the intersection of R_2 with S in §§ 4 and 5.

3. Very, very special case

Suppose ε is the positive number in the statement of Theorem 1, δ is the positive number defined in the last section, and R is the triangulation defined in that section. As we mentioned, we would like for S to meet R_1 and R_2 in very nice sets. If one understands how to prove Theorem 1 for the *very, very special case*, he has insight into how to proceed in the general case, Hence, consider the following.

Assumption of the very, very special case. For each 2-simplex F of R, $F \cdot S$ has only a finite number of components and each of these components is either a point of Bd F or a spanning arc of F.

The above assumption is a bit different from the one given in [2] for there we contended with the case where $F \cdot S$ might contain several spanning arcs of F with a common end point.

We show in seven steps how to complete the proof of Theorem 1 in this *very, very special case*.

1. Let $D_1, D_2, \cdots, D_n$ be the components of $S - R_2$. Since each D_i lies on the interior of a 3-simplex of R, each D_i is of diameter less than δ. It would be convenient if each D_i were the interior of a disk. However, these components may appear as in Figure 3 where $\bar{D}_1$ is a disk with two holes, $\bar{D}_2$ is an annulus ring, $\bar{D}_3$ is a disk (although D_3 is not the interior of any disk on S), and D_4 is an open disk. In any case, $\bar{D}_i$ is a disk with or without holes and $S - \bar{D}_i$ is the sum of a finite number of open disks. Only one of these open disks is of diameter more than $\varepsilon/2$ and we denote the complement of this big open disk by E_i. We note that $\bar{D}_i \subset E_i$, Bd $E_i \subset \bar{D}_i$, and diameter $E_i < \varepsilon/2$.

It may happen that certain E's are subsets of other E's. For example, in Figure 3, $E_4 \subset E_2 \subset E_1$.

Suppose $D_1, D_2, \cdots, D_m$ are the elements of $D_1, D_2, \cdots, D_n$ such that

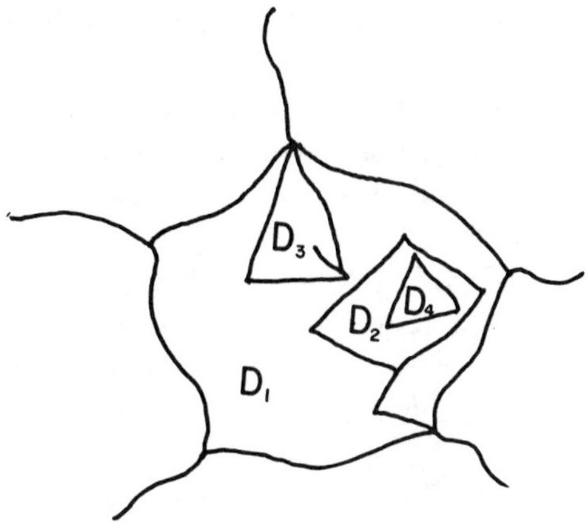

Figure 3.

$E_i (i = 1, 2, \cdots, m)$ is not a subset of any other E. Then $E_1, E_2, \cdots, E_m$ provides a cellular subdivision of S in that $E_1 + E_2 + \cdots + E_m = S$ and if two of these E's intersect, the intersection lies on the boundary of each.

2. In this step we consider polyhedral annuli about certain simple closed curves in $G = \text{Bd}\, E_1 + \text{Bd}\, E_2 + \cdots + \text{Bd}\, E_m$. We find from the last step that G is a connected finite graph without cut points and each component of $S - G$ is the interior of an E_i.

Consider the collection $\{A_j\}$ of all the arcs in G which are spanning arcs of a 2-simplex of R. Let θ be a positive number less than the distance between any two elements of $\{A_j\}$ without an end point in common and less than the distance between any pair of points of $R_0 + S \cdot R_1$.

Suppose the arc A_j spans the 2-simplex F of R. Let C_j be a polyhedral disk in F such that A_j spans C_j, each point of C_j is within $\theta/2$ of A_j, and $C_j \cdot \text{Bd}\, F$ is the sum of two segments each of length θ and centered at an end point of A_j. There is a C_j for each A_j. See Figure 4.

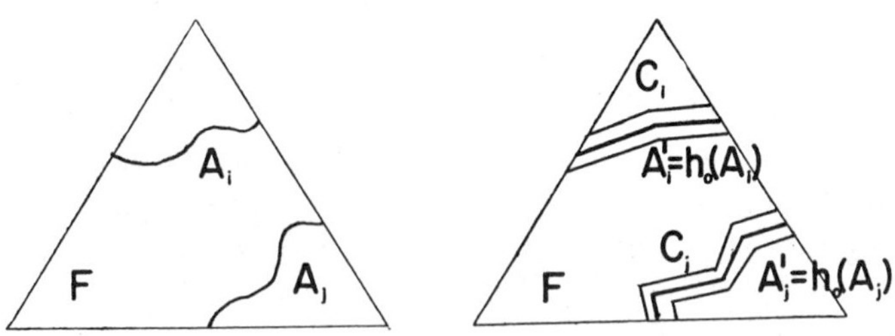

Figure 4.

Note that $\text{Bd}\, E_i (i = 1, 2, \cdots, m)$ is a simple closed curve which is the sum of certain A_j's all of which lie on the boundary of one 3-simplex of

T (namely the one which contains D_i). If we consider the sum of the C_j's whose A_j's lie on Bd E_i, this sum is a polyhedral annulus within $\theta/2$ of Bd E_j and the interior of this annulus contains Bd E_i.

3. In this step we start a homeomorphism h_0. Let A'_j be a polyhedral spanning arc of C_j that has the same end points as A_j. See Figure 4. Let h_0 be a homeomorphism that leaves the end points of A_j fixed and takes A_j onto A'_j. Note that h_0 takes G onto $\sum A'_j$.

4. In this step we define a polyhedral 2-sphere S' and a homeomorphism h_0 of S onto S'. The homeomorphism h_0 is an extension of the homeomorphism h_0 defined on G in Step 3.

Suppose T is a 3-simplex of R and Bd E_i $(i = 1, 2, \cdots, m)$ lies on Bd T. Let E'_i be a polyhedral disk in T such that Int $E'_i \subset$ Int T and Bd $E'_i = h'(\text{Bd } E_i)$. We suppose that if E'_i, E'_j are two such disks in T, they do not intersect. See Figure 5. This is possible since Bd E_i, Bd E_j are mutually

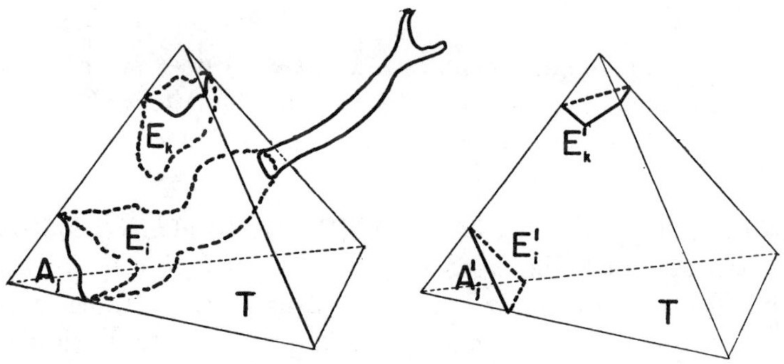

Figure 5.

exclusive simple closed curves on Bd T. (The proof that there are E''s such that $E'_i \cdot E'_j = 0$ is easier here than the corresponding proof that Int $E'_i \cdot$ Int $E'_j = 0$ was in Step 4 on page 460 of [2] because there we permitted Bd E_i to intersect Bd E_j and we had to go to considerable trouble to show that Bd E_i and Bd E_j did not cross on Bd T.) Extend h_0 to take E_i onto E'_i. When this is done for each E_i $(i = 1, 2, \cdots, m)$ we have a homeomorphism h_0 of S onto a polyhedral 2-sphere

$$h_0(S) = S' = E'_1 + E'_2 + \cdots + E'_m .$$

5. In this step we thicken S' slightly so that the thickened 2-sphere will contain G. We define a homeomorphism h of $S \times [-1, 1]$ into E^3. We use $h_t(s)$ to denote $h(s \times t)$. Then $h_t(-1 \leqq t \leqq 1)$ is an extension of the homeomorphism h_0 we defined in Steps 3 and 4.

Consider a point p of $G \cdot R_1$. Then h takes $p \times [-1, 1]$ linearly onto the interval of length θ centered at p and lying on the 1-simplex of R with $h(p \times -1)$ lying in the bounded component of $E^3 - S'$ and $h(p \times 1)$ in the unbounded component.

Now we define h on $G \times [-1, 1]$. We have already defined h_t on the ends of an arc A_j and defined h_0 on A_j so we extend this map so that h takes $A_j \times [-1, 1]$ onto C_j. This extension is done for each A_j so that each $h_t(A_j)$ is a polygonal arc.

Suppose E_i' lies in 3-simplex T of R. We have already defined $h(\text{Bd}\, E_i \times [-1,1])$ to be a closed polyhedral annulus on Bd T and $h(E_i \times 0) = E_i'$ to lie in T. Thicken E_i' to a polyhedral cube K_i in T such that $\text{Bd}\, K_i \cdot \text{Bd}\, T = h(\text{Bd}\, E_i \times [-1,1])$, E_i' lies except for its boundary on $\text{Int}\, K_i$, and no two K_i's in T intersect each other. Then h is extended to take $E_i \times [-1, 1]$ homeomorphically onto K_i. This is done for each E_i so that each $h_t(E_i)$ is polyhedral.

We have now defined h to take $S \times [-1, 1]$ homeomorphically into E^3 and $G \subset \text{Int}\, h(S \times [-1, 1])$.

6. In this step we show that $h_t(-1 \leqq t \leqq 1)$ does not move any point of S by as much as $\delta + \varepsilon/2 < \varepsilon$. This is because diameter $h(E_i \times [-1,1]) < \delta$ since it lies in a 3-simplex T of R, E_i intersects this same T, and diameter $E_i \leqq \varepsilon/2$.

7. Note that $\sum \text{Bd} E_i$ lies between $h(S \times -1)$ and $h(S \times 1)$. Any subset of E_i that does not lie between $h(S \times -1)$ and $h(S \times 1)$ lies in a disk in E_i. This disk is of diameter less than $\varepsilon/2$ since E_i is. Hence S contains a finite collection of mutually exclusive $\varepsilon/2$-disks such that S lies except for these disks between $h(S \times -1)$ and $h(S \times 1)$.

This completes the *very, very special case*. We find that the methods used in constructing $h(S \times [-1, 1])$ in this *very, very special case* are essentially the same as those used in the less special cases and indeed for the general case. We use the proof of this *very, very special case* for a model in what is to follow.

Note that in the *very, very special case* we did not need as stringent definition of δ as that given in § 2. We merely needed that each δ-subset of S lies in an $\varepsilon/2$-disk in S. Also, we proved more, namely, that h_t moves no point more than $\varepsilon/2 + \delta$ and the small disks of S that do not lie between $h(S \times -1)$ and $h(S \times 1)$ have diameters less than $\varepsilon/2$.

4. Very special case

In this section we put less restrictions on $S \cdot R_2$ but find that $f(S \times [-1, 1])$ can be constructed in much the same way as in the preceding section.

Assumption for the very special case. In this case we suppose:

$S \cdot R_1$ is finite and for each 2-simplex F of R, no component of $S \cdot F$ contains three points of Bd F.

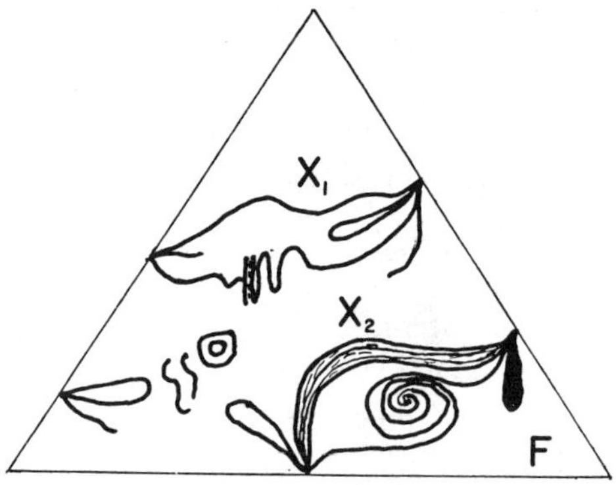

Figure 6.

The intersection of S and F might appear as in Figure 6. Note that the components of $S \cdot F$ are not assumed to be either points or spanning arcs as in the *very, very special case,* but any such component that has two boundary points on Bd F is a spanning continuum of F.

In this *very special case* we will not need as stringent definition of δ as that given in § 3, but a more stringent one than the one needed in the *very, very special case.* We need that each 3δ subset of S lies in a ($< \varepsilon/2$)-disk in S. As in the *very, very special case* we show that h_t moves no point more than $\varepsilon/2 + \delta$ and the small disks of S that do not lie between $h(S \times -1)$ and $h(S \times 1)$ have diameters less than $\varepsilon/2$.

Let $X_1, X_2, \cdots, X_x$ be the continua of $S \cdot R_2$ such that for some 2-simplex F of R and some pair of points of Bd F that belong to the same component of $S \cdot F$, X_i is the closure of the sum of all components of $S \cdot \text{Int } F$ whose closures contain these two points. Note that X_i may be a proper subcontinuum of a component of $S \cdot F$. We did not define X_i to be the whole component, for we wish to avoid complications in defining K in Step 2 and $B(D_i)$ in Step 3. In the *very special case* the X's play somewhat the same role as do the spanning arcs in the *very, very special case.*

We give the proof of Theorem 1 in the *very special case* in twelve steps. The first six of these steps closely parallel Steps 4–9 on pages 464–469 of [2] and the last six of the steps closely parallel Steps 2–7 of the *very, very special case.*

1. In this step we show that each component of $S - \sum X_i$ is of diameter

less than $\varepsilon + 3\delta$. (It follows from Step 5 that each is actually of diameter less than $\varepsilon/2$.)

Suppose p_0, q_0 are two points of $S - \sum X_i$ such that $\rho(p_0, q_0) \geqq \varepsilon + 3\delta$. We show that $\sum X_i$ separates p_0 from q_0 in S. Let M be the sum of the 2-simplexes of R whose distances from each of p_0 and q_0 is more than $\varepsilon/2$. Then M separates p_0 from q_0 in E^3 and $M \cdot S$ separates p_0 from q_0 in S. Let Z be a subset of $M \cdot S$ irreducible with respect to separating p_0 from q_0 in S. It follows from the unicoherence of S that Z is connected and from the definition of δ that diameter $Z > \delta$. We finish Step 1 by showing that $Z \subset \sum X_i$.

Suppose $Z \not\subset \sum X_i$ and p is a point of $Z - \sum X_i$. We suppose that $p \notin R_1$. Let F be the 2-simplex of R containing p, and A the component of $Z \cdot F$ containing p. We note that A does not separate p_0 from q_0 in S because it is of diameter less than δ.

If $A \cdot \text{Bd}\, F$ is one point, then A and the closure of $Z - A$ are continua neither of which separates p_0 from q_0. However in this case the sum would not separate p_0 from q_0 since their intersection is connected.

If $A \cdot \text{Bd}\, F$ is two points, let A' be a subcontinuum of A irreducible between these two points. Then $A \neq A'$ since $A' \subset \sum X_i$ [22, p. 22, Th. 37]. But the intersection of the continua A and $(Z - A) + A'$ is connected, neither separates p_0 from q_0, so their sum Z could not separate p_0 from q_0 in S. The assumption that $Z \not\subset \sum X_i$ led to this contradiction. An alternative proof of this step is given in Step 4 on page 464 of [2].

2. In the *very, very special case*, $S - G$ had only a finite number of components. In this step we fill in the holes of the X's to bring about an analogous situation.

For each X_i, let $K(X_i)$ be the sum of X_i and all components of $S - X_i$ of diameter less than $\varepsilon/2$. The points of $X_i \cdot R_1$ are called the end points of $K(X_i)$.

If K', K'' are two such continua, it may be that one contains the other. Otherwise, they do not intersect at all except possibly in their end points. Let $K_1, K_2, \cdots, K_k$ be the collection of all such $K(X)$'s that are not contained in a larger $K(X)$.

None of the K's separate S and the intersection of any two of them is at most two points. Hence $S - \sum K_i$ has only a finite number (no more than the number of K's) of components $D_1, D_2, \cdots, D_d$. It follows from Step 1 that each of these D's is of diameter less than $\varepsilon + 3\delta$. (It will follow from Step 5 that each is actually of diameter less than $\varepsilon/2$.)

Figure 7 shows how $\sum K_i$ divides S. The K's do not necessarily lie in

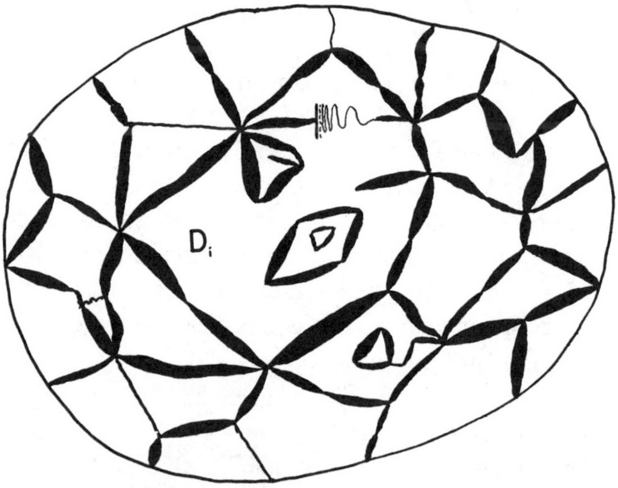

Figure 7.

2-simplexes of R but the boundary of each K (with respect to S) lies in such a 2-simplex. Hence, each K is of diameter less than $\varepsilon/2$. The D's do not necessarily lie in the interior of a 3-simplex of R, but in Step 5 we shall associate certain of the D's that are kept with 3-simplexes of R. The precaution we took in defining the X's insures that no end point of K separates it, and K minus its end points intersects the boundary of at most two D's.

3. In Step 1 of the proof of the *very, very special case* of Theorem 1, we threw away a part of the finite graph $S \cdot R_2$ under consideration and

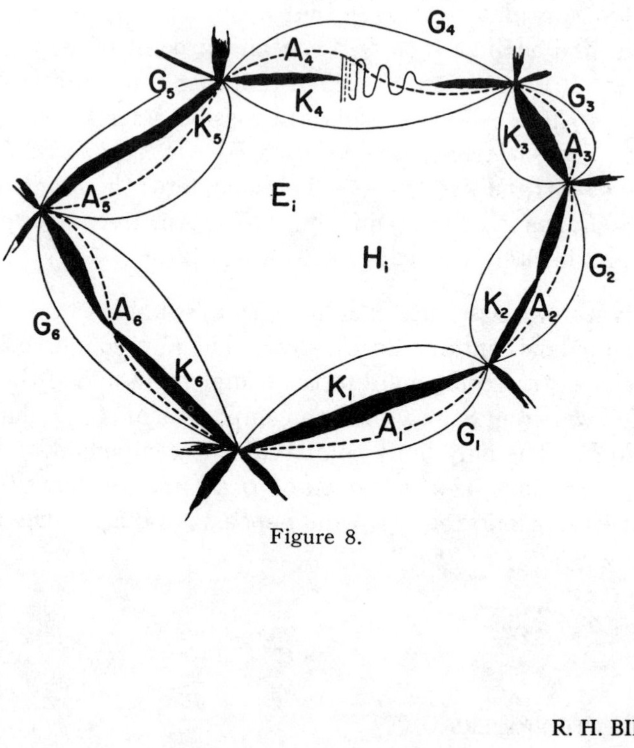

Figure 8.

got some larger domains on S. In a similar fashion, we now throw away some of the K's and replace each D_i with a possibly larger domain E_i. A domain D_i such as shown in the center of Figure 7 is enlarged. See Figure 8.

We used Bd D_i to denote the boundary of D_i on S. For each D_i, let $B(D_i)$ be the sum of all K's that have a non-end point on Bd D_i. Let p_i be a point of S whose distance from $D_i + B(D_i)$ is more than ε. Let $B(E_i)$ be the sum of the collection of K's in $B(D_i)$ such that this collection, but no proper subcollection of K's in $B(D_i)$, has a sum that separates p_i from D_i. Our reason for calling this continuum $B(E_i)$ is that we shall call E_i the component of $S - B(E_i)$ that contains D_i.

The continuum $B(E_i)$ is much like a circle but instead of being the sum of arcs, it is the sum of K's. See Figures 7 and 8. In Figure 7, $B(E_i)$ has exactly two complementary domains in S, namely, E_i and the one containing p_i. We show that E_i (and hence D_i) is of diameter less than $\varepsilon/2$ by showing in Step 5 that Bd E_i is of diameter less than δ since it lies on the boundary of some 3-simplex of R.

4. We shall need to associate with each E_i a 3-simplex (tetrahedron) of R. In order to do this in Step 5, we show in the present step that no component of $E_i \cdot R_2$ has two limit points on Bd E_i. If some component A of $E_i \cdot R_2$ has two such limit points on Bd E_i, let A' be a subcontinuum of $\bar{A}$ irreducible with respect to having two points on Bd E_i. By an argument similar to that used in Step 1 it follows that $A' \subset \sum X_i$. Hence $A' \subset \sum K_i$. Also, A' has exactly two points p and q on $B(E_i)$. Furthermore, $B(E_i)$ is the sum of two continua Z_1, Z_2 such that $Z_1 \cdot Z_2 = p + q$, each of $Z_j - (p+q)$ is connected, and each $Z_j - (p+q)$ contains a point of Bd E_i. It follows from Theorem 101 on pages 287–288 of [22] that A' separates $Z_1 - (p+q)$ from $Z_2 - (p+q)$ in $E_i + Z_1 + Z_2$. This is impossible since D_i lies in $E_i - A'$, A' separates $Z_1 - (p+q)$ from $Z_2 - (p+q)$ in $E_i + B(E_i)$, and Bd D_i intersects each of $Z_1 - (p+q)$ and $Z_2 - (p+q)$. This completes the proof that no component of $E_i \cdot R_2$ has two limit points on Bd E_i. An alternative proof of this result is found in Step 7 on pages 466–467 of [2].

5. In this step we associate each E_i with a 3-simplex of R, namely, the one whose boundary contains Bd E_i. Note that E_i is connected and simply connected since it is a component of the complement in S of the continuum $B(E_i)$. It follows from Step 4 that no component of $E_i \cdot R_2$ has two limit points on Bd E_i. The only limit point that any component of $E_i \cdot R_2$ could possibly have on Bd E_i is at a vertex of $B(E_i)$ and as pointed out in Step 3, no such vertex separates $B(E_i)$ and hence not Bd E_i. Since Bd E_i inter-

sects two open 2-simplexes of R, no sequence of components of $E_i \cdot R_2$ converges to Bd E_i. It follows from Theorem 2 cited at the end of this section that there is a component E_i' of $E_i - R_2$ such that Bd $E_i \subset$ Bd E_i'. Since E_i' lies in a 3-simplex of R, so does Bd E_i', and hence so does Bd E_i. Therefore, diameter Bd $E_i < \delta$. Hence, diameter $E_i < \varepsilon/2$.

6. The E's discussed in Steps 3, 4, 5 differ from the E's used in the *very, very special case* in that their boundaries are not simple closed curves. In this step we replace the E's with H's which have simple closed curves for boundaries.

Two E's do not intersect each other unless one contains the other. Let $E_1, E_2, \cdots, E_n$ be the collection of maximal E's no one of them contained in another of them. Then

$$S = E_1 + E_2 + \cdots + E_n + B(E_1) + B(E_2) + \cdots + B(E_n) .$$

Let $K_1, K_2, \cdots, K_m$ be the elements of $K_1, K_2, \cdots, K_k$ in $B(E_1) + B(E_2) + \cdots + B(E_n)$. Let G_i $(i = 1, 2, \cdots, m)$ be a topological disk in S of diameter less than $\varepsilon/2$ such that G_i contains K_i and $G_i \cdot G_j = K_i \cdot K_j$ $(i \neq j)$. One way to get the G_i's is to expand the K_i's slightly to make the results locally connected and then fill in holes to make the results disks, taking care that the K_i's are not expanded enough to make any resulting disk intersect any other resulting disk unnecessarily. Another approach is to get the G_i's one at a time. To get G_1, let M_1 be a continuum on S such that M_1 contains each K_i other than K_1, diameter $S - M_1 < \varepsilon/2$, and K_1 intersects M_1 only at the end points p, q of K_1. It follows from Theorem 42 on page 217 of [23] that there is a simple closed curve J in S which separates $K_1 - (\{p\} + \{q\})$ from $M_1 - (\{p\} + \{q\})$. The disk bounded by J_1 and containing K_1 is G_1. The other G_i's may be obtained in a similar manner.

Let A_i be an arc in G_i from one end of K_i to the other such that each point of A_i is within δ of the X_i determining K_i. Then $S - \sum A_j$ is the sum of a finite number of topological open disks $H_1, H_2, \cdots, H_n$ such that $H_j + \sum G_i = E_j + \sum G_i$.

Figure 8 shows how the continua $K_1, K_2, \cdots, K_5$ in $B(E_i)$ are replaced by arcs $A_1, A_2, \cdots, A_5$ (shown dotted). Then H_i is the interior of the simple closed curve $A_1 + A_2 + \cdots + A_5$.

Each Bd H_i is of diameter less than 3δ since it lies within δ of a 3-simplex of R; namely, the one containing Bd E_i. Hence, diameter $H_i < \varepsilon/2$ and each point of H_i lies within $\varepsilon/2$ of Bd E_i.

7. This step is the analogue of Step 2 for the *very, very special case*. Let θ be a positive number less than the distance between any two of the X_j's without an end point in common and less than the distance between

any two points of $R_0 + S \cdot R_1$. We shall build a polyhedral disk about each X_j which determines a K_i $(i = 1, 2, \cdots, m)$.

Suppose X_j lies in the 2-simplex F of R. Let C_j be a polyhedral disk in F such that $X_j \cdot \mathrm{Int} F \subset \mathrm{Int} C_j$, each point of $\mathrm{Bd}\, C_j$ is within $\theta/2$ of X_j, and $C_j \cdot \mathrm{Bd}\, F$ is the sum of two segments each of length θ and centered at an end point of X_j. There is a C_j for each X_j. For any E_i, the sum of the C's associated with X's determined by the K's in $B(E_i)$ is an annulus ring on the boundary of some 3-simplex of R.

8. In this analogue of Step 3 of the *very, very special case* we start a homeomorphism h_0. Let A'_j be a polygonal spanning arc of C_j that has the same end points as A_j. Let h_0 be a homeomorphism that leaves the end points of the A_j of Step 6 fixed and which takes A_i onto A'_i. Note that h_0 takes $\sum A_i$ onto $\sum A'_i$.

9. In this analogue of Step 4 of the *very, very special case* we define a polyhedral 2-sphere S' and a homeomorphism h_0 (which is an extension of the h_0 of the preceding step) of S onto S'. In fact, there is essentially no change from Step 4 of the *very, very special case* except that h_0 maps H_i rather than E_i. Let the description of Step 4 for the *very, very special case* suffice for the description of h_0 in the present case.

10. In this case we thicken S' slightly and complete the description of h_t. The details are as given in Step 5 of the *very, very special case* except that we use H_i instead of E_i.

11. In this analogue of Step 6 of the *very, very special case* we show that $h_t(-1 \leqq t \leqq 1)$ defined in the preceding step does not move any point of S by as much as $\delta + \varepsilon/2 < \varepsilon$. Recall that $h_t(H_i)$ lies in the 3-simplex of R whose boundary contains $\mathrm{Bd}\, E_i$. As pointed out in Step 6, each point of H_i lies within $\varepsilon/2$ of this 3-simplex. Since the 3-simplex is of diameter less than δ, h moves no point more than $\varepsilon/2 + \delta$.

12. We complete the proof of Theorem 1 for the *very special case* by showing that S contains a finite collection of mutually exclusive $(< \varepsilon/2)$-disks such that S minus these disks lie in $h(S \times [-1, 1])$. Note that each X_i associated with a K_i $(i = 1, 2, \cdots, m)$ lies in $h(S \times [-1, 1])$ and only a finite number of components of S minus the sum of these X's could fail to lie in $h(S \times [-1, 1])$. Components of S minus the sum of these X's are either E_i's or small components of S minus a single X. Each such component is simply connected and of diameter less than $\varepsilon/2$ so the result follows.

The following theorem is a restatement of Theorem 5 of [2]. We state it here since it was used in Step 5 of this section and will be used again in the next section.

THEOREM 2. *Suppose D is a simply connected, connected open set on a 2-sphere and K is a closed set containing* Bd D *such that no sequence of components of $D \cdot K$ converges to* Bd D, *no component of $K \cdot D$ has two limit points of* Bd D, *and no component of $K \cdot D$ has a closure which separates* BdD. *Then there is a unique component D' of $D - K$ such that* Bd$D \subset$ BdD'.

5. Special case

In this section we weaken the *assumption for the very special case* and merely assume the following.

Assumption for the special case. $S \cdot R_1$ is finite.

We need a more stringent definition of δ in the *special case* than we did in the *very special case*; namely, $\delta < \varepsilon/8$ and each 5δ-subset of S lies in a disk in S of diameter less than $\varepsilon/4$. Note that this implies that each 5δ-subset of S lies in a disk of diameter less than $\varepsilon/2 - 2\delta$.

We treat the *special case* by showing in Theorem 8 that if the *assumption for the special case* is satisfied, then there is a piecewise linear homeomorphism g of E^3 onto itself such that g moves no point more than δ, g leaves the points of R_1 fixed, and $g(S)$ satisfies the *assumptions for the very special case*. (The g mentioned here is the g^{-1} of Theorem 8.) Note that each 3δ-subset X of $g(S)$ lies in a disk D of $g(S)$ of diameter less than $\varepsilon/2$. This is because $g^{-1}(X)$ is of diameter less than 5δ and hence lies in a disk D' of diameter less than $\varepsilon/2 - 2\delta$. Then $D = g(D')$.

Once Theorem 8 is established, one merely applies the *assumptions for the very special case* to $g(S)$ to get a homeomorphism h' of $g(S) \times [-1, 1]$ into E^3 such that h'_t moves no point more than $\varepsilon/2 + \delta$, $g(S)$ lies except for a collection of mutually exclusive $\varepsilon/2$-disks between $h'(g(S) \times -1)$ and $h'(g(S) \times 1)$, and for each t, $h'_t(S)$ is polyhedral. Then $h_t = g^{-1}h'_t g$ takes S into E^3, h_t moves no point more than $\varepsilon/2 + 3\delta$ and S lies except for a finite collection of mutually exclusive $(\varepsilon/2 + 2\delta)$-disks between $h(S \times -1)$ and $h(S \times 1)$.

We prove Theorem 8 by using two essentially different types of homeomorphisms of E^3 onto itself. The first of these is the result of pushing at the center of a triangular disk while leaving its boundary fixed (used in Theorems 4, 5) and the second is obtained by unwinding an arc around a cylinder (used in Theorems 3, 6). We used the first of these types of homeomorphisms in [2] but the second type was not discovered until later. It was the discovery of this unwinding type of homeomorphism that permitted us to make certain simplifications in this paper over that given in [2].

Before proving Theorem 8, we establish Theorems 3–7 along the way to help us. The positive number ε mentioned in the statements of these theorems need not be the same as the one used in previous sections.

A spanning arc of a disk is an arc whose ends lie on the boundary of the disk and which lies except for its ends on the interior of the disk. An effective way of showing that no component of $S \cdot F$ contains each of two points of Bd F is to show that there is a spanning arc of F that misses S whose ends separate the two points from each other on Bd F.

THEOREM 3. *Suppose R is a triangulation of E^3, F is a 2-simplex of R such that $S \cdot \mathrm{Bd}\, F$ is finite, $p \in S \cdot \mathrm{Bd}\, F$, and apb is a straight line interval on* Bd F *which intersects S only at p. If a, b belong to the same component of $E^3 - S$, then for each $\varepsilon > 0$, there is a piecewise linear homeomorphism g of E^3 onto itself such that g is fixed outside an ε-neighborhood of p, g is fixed on a neighborhood of* Bd F, *and some spanning arc of $g(F)$ from a to b misses S.*

PROOF. Let axb be a polygonal arc from a to b in $E^3 - S$ such that $axb + apb$ is a simple closed curve. Let C be a triangular cylinder with apb as axis as shown in Figure 9 such that $p \in \mathrm{Int}\, C$, S does not intersect

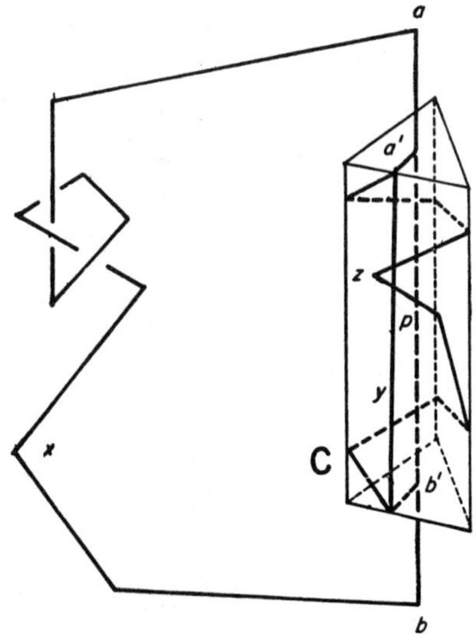

Figure 9.

either base of C, diameter C is less than ε, $C \cdot (apb + axb + \mathrm{Bd}\, F)$ is a subinterval $a'pb'$ of apb. Then $F \cdot \mathrm{Bd}\, C$ is a polygonal arc $a'yb'$.

No component of $S \cdot \mathrm{Bd}\, C$ separates a' from b' on Bd C. If one such component did separate a' from b' on Bd C, there would be a simple closed curve J on Bd C near it that links the simple closed curve $apb + axb$ and a singular closed curve J' on S near J that links $apb + axb$. (See [6, p. 119] for details for getting such a J'.) However, no singular closed curve J' on S links $apb + axb$ since such a J' can be shrunk to a point on $S - p$ and hence in $E^3 - (apb + axb)$. (See [2, Th. 10].)

Let $a'zb'$ be a polygonal arc on Bd C that misses S. We suppose that $a'zb'$ agrees with $a'yb'$ on the bases of C and unwinds enough near one of these bases so that there is a homeomorphism of E^3 onto itself which is fixed on the bases of C and, except near the lateral sides of C, which takes Bd C onto itself and $a'yb'$ onto $a'zb'$. The homeomorphsim g promised by the theorem is such a homeomorphism which is piecewise linear.

The following lemma shows how $a'zb'$ could be taken to unwind near a base of C.

LEMMA. *Suppose D_1, D_2 are two polyhedral disks such that $D_1 + D_2$ is an annulus A and $D_1 \cdot D_2$ is the sum of two arcs ayb and cxd; azb is a polygonal arc which lies except for a, b in* Int A *such that $azb \cdot cxd$ is a finite set of points, azb crosses cxd at each of these points, and azb (in the order from a to b) crosses cxd in the direction from D_1 to D_2 the same number of times it crosses from D_2 to D_1. Then there is a piecewise linear isotopy of A onto itself that is fixed on* Bd A *and which pulls azb onto ayb.*

We do not give details of the proof of the lemma except to say that the required isotopy may be obtained by pulling azb to $az'b$ so that $az'b \cdot cxd = 0$, then pulling $az'b$ to $az''b$ so that $az''b$ intersects ayb only at its ends, and finally using the disk bounded by $ayb + az''b$ to pull $az''b$ to ayb.

THEOREM 4. *Suppose S is a 2-sphere in E^3, F is a triangular disk such that $S \cdot$* Bd F *is finite, and I is a line interval perpendicular to F through a point of* Int F. *Then I contains a subinterval I' such that if $V(I', \text{Bd}\, F)$ denote the join of I' and* Bd F *(the sum of all intervals with one end point on I' and the other on* Bd F) *then no component of $S \cdot (V(I', \text{Bd}\, F) - \text{Bd}\, F)$ has three limit points on* Bd F.

PROOF. Consider n mutually exclusive subintervals $I_1, I_2, \cdots, I_n$ of I where n is more than twice the number of triples of points of $S \cdot \text{Bd}\, F$. It will follow that one of the I_j's will suffice for I' when we show that if p, q, r are three points of $S \cdot \text{Bd}\, F$, then it is impossible that three of the $S \cdot (V(I_i, \text{Bd}\, F) - \text{Bd}\, F)$'s have components such that each of these components has each of p, q, r as a limit point.

Assume C_1, C_2, C_3 are three continua in different $V(I_i, \text{Bd}\, F)$'s such that each is the closure of a component of an $S\cdot(V(I_i, \text{Bd } F) - \text{Bd } F)$ that has each of p, q, r as a limit point. Then $C_j \cdot C_k = p + q + r$ $(j \neq k)$. Suppose each C_i is expanded to a locally connected continuum A_i on S such that $A_j \cdot A_k = p + q + r$ $(j \neq k)$. Let T_i be a triod in A_i with end points at p, q, r. However, $T_1 + T_2 + T_3$ is a skew set and cannot lie in S. Our assumption that there are three different $S\cdot(V(I_i, \text{Bd } F) - \text{Bd } F)$'s with components having each of p, q, r as a limit point is false.

THEOREM 5. *Suppose S is a 2-sphere, R is a triangulation of E^3. F is a 2-simplex of R such that $S\cdot\text{Bd } F$ is finite, and $\varepsilon > 0$. Then there is a piecewise linear homeomorphism g of E^3 onto itself such that:*

(1) *g moves no point more than ε,*

(2) *g is fixed on the 1-skeleton of R,*

(3) *no component of $S\cdot g(\text{Int } F)$ has three limit points on $\text{Bd } F$, and*

(4) *if E is a component of $g(\text{Int } F) - S$ with three limit points on $\text{Bd } F$, then each such limit point is accessible from E.*

PROOF. Consider a short interval centered at a point of $\text{Int } F$ and perpendicular to F. Apply Theorem 4 to get a subinterval I' of I such that $S\cdot(V(I', \text{Bd } F) - \text{Bd } F)$ has no component with three limit points on $\text{Bd } F$. The homeomorphism g results from keeping $\text{Bd } F$ fixed and pushing the center of I onto the center of I'. The disk F goes onto the sum of three disks—the cone over $\text{Bd } F$ from the center of I'. This homeomorphism g is fixed outside $V(I, \text{Bd } F)$ and moves points only in a direction perpendicular to F.

Let us note two things that are outlawed by the homeomorphism g. No component of $S\cdot g(\text{Int } F)$ has three limit points on $\text{Bd } F$ as shown in Figure 10a. (We do not outlaw some component of $S\cdot g(F)$ having three points on $\text{Bd } F$ as shown in Figure 10c until Theorem 6.)

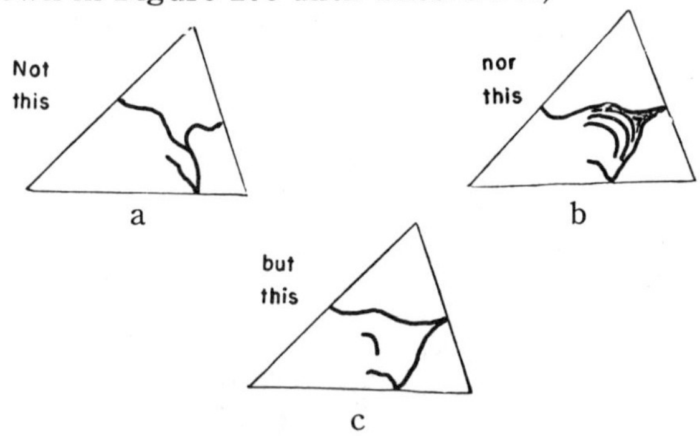

Figure 10.

It does not even have two components as shown in Figure 10b each with two limit points on $\text{Bd } F$ such that each of these components intersects the inferior limit of a sequence of other components. (If there were such a sequence, $S\cdot(V(I', \mathrm{Bd}\, F) - \mathrm{Bd}\, F)$ would have a component with three limit points on Bd F.) The situation is more as depicted in Figure 10c.

One simplification of the present paper over the treatments in [2] and [3] results from the simple way in which we avoided situations as shown in Figure 10b. We used much more complicated procedures in Step 10 on page 53 of [3] and in Step 3 on page 463 of [2] to bypass difficulties we have so easily overcome here by avoiding situations as depicted in Figure 10b.

For each pair of points a, b of Bd F such that each is a limit point of the same component of $S\cdot g(\mathrm{Int}\, F)$, let $X(a, b)$ denote the closure of the sum of all such components each of which has both a and b as limit points. Let $Y(a, b)$ denote the sum of $X(a, b)$ and all components of $g(F) - X(a, b)$ that do not contain a point of Bd F. (We obtain $Y(a, b)$ by filling in holes in $X(a, b)$.) Since $S\cdot \mathrm{Bd}\, F$ has only a finite number of pairs of points, there are at most a finite number of Y's. We denote them by $Y_1, Y_2, \cdots, Y_y$ as shown in Figure 11. We note that each Y_i has exactly two points on Bd F. Y_i separates $g(F)$ into exactly two pieces, and $g(\mathrm{Int}\, F) - \sum Y_i$ has one more component than there are Y_i's. Let $D_1, D_2, \cdots, D_{y+1}$ be these components.

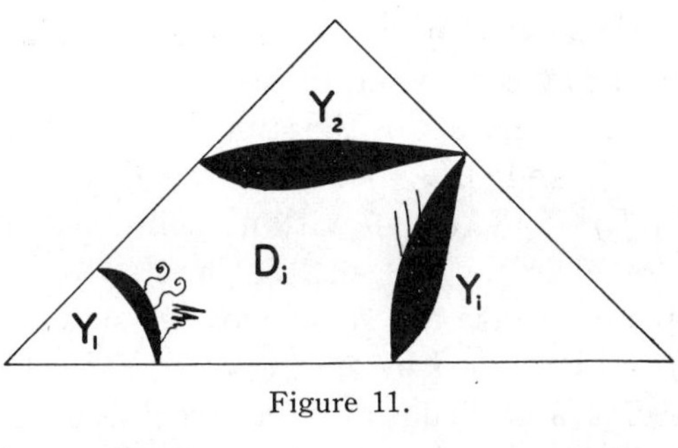

Figure 11.

We show that D_i and Bd $F + S\cdot g(F)$ satisfy the conditions for D and K of Theorem 2. Since D_i is a component of the complement in $g(F)$ of the continuum Bd $F + \sum Y_j$, it is connected and simply connected. There is no sequence of components of $D_i\cdot S$ converging to Bd D_i since either Bd D_i contains a subarc of Bd F or Bd $D_i\cdot g(\mathrm{Int}\, F)$ contains points of two X's, and Theorem 4 tells us that no continuum in $S\cdot(V(I', \mathrm{Bd}\, F) - \mathrm{Bd}\, F)$ intersects two X's. No component of $D_i\cdot(\mathrm{Bd}\, F + S\cdot g(F))$ has a limit point on Bd D_i except possibly at a point of Bd F, and such a point would not separate Bd D_i. No component of $D_i\cdot(\mathrm{Bd}\, F + S\cdot g(F))$ has two limit points on BdD_i or else the component would be in an X. Hence D_i and

$\mathrm{Bd}F+S\cdot g(F)$ satisfy the conditions for D and K of Theorem 2 on any 2-sphere containing $g(F)$.

It follows from Theorem 2 that there is a unique component E_j of $g(F)-S$ such that $\bar{D}_j - D_j \subset \bar{E}_j - E_j$. In the next two paragraphs we show that points of $\bar{E}_j\cdot \mathrm{Bd}\, F$ are accessible from E_j.

In this paragraph we enlarge $S\cdot g(F)$ to a closed set M such that:

(1) $M \subset g(F)$,

(2) $M\cdot \mathrm{Bd}\, F = S\cdot \mathrm{Bd}\, F$,

(3) $S\cdot g(\mathrm{Int}\, F)$ lies in $\mathrm{Int}\, M$ in $g(F)$,

(4) M is locally connected except possibly at points of $\mathrm{Bd}\, F$, and

(5) if two points of $S\cdot g(\mathrm{Int}\, F)$ are connected by an arc in $M\cdot g(\mathrm{Int}\, F)$, these same two points are connected by an arc in $S\cdot(V(I', \mathrm{Bd}\, F) - \mathrm{Bd}\, F)$.

The purpose of enlarging $S\cdot g(F)$ to M is to get $X(a, b)$ enclosed in a disk so as to show that each point of $\mathrm{Bd}\, F\cdot \mathrm{Bd}\, E_j$ is accessible from E_j. For each point p of $S\cdot g(\mathrm{Int}\, F)$, let $\theta(p)$ be a positive number so small that any point of $S\cdot g(\mathrm{Int}\, F)$ within $\theta(p)$ of p lies on an arc with p in $S\cdot(V(I', \mathrm{Bd}\, F) - \mathrm{Bd}\, F)$. Let D_p be a disk in $g(\mathrm{Int}\, F)$ such that $p \in \mathrm{Int}\, D_p$ and diameter $D_p < \theta(p)/2$. We obtain M by covering $S\cdot g(\mathrm{Int}\, F)$ by a family of such disks such that the family is locally finite except at points of $S\cdot \mathrm{Bd}\, F$ and letting M be the sum of $S\cdot \mathrm{Bd}\, F$ and these closed disks. We suppose that the disks were adjusted so that no point of $M - \mathrm{Bd}\, F$ is a separating point of any component of M.

Let $M(X(a, b))$ be the closure of the sum of the components of $M - \mathrm{Bd}\, F$ intersecting $X(a, b)$, and $M(a, b)$ denote the sum of $M(X(a, b))$ and all components of $g(F) - M(X(a, b))$ with no points on $\mathrm{Bd}\, F$. Then $M(a, b)$ is a disk (see, for example, [23, p. 216, Th. 4]). Also $M(a, b)$ intersects $\mathrm{Bd}\, F$ only at a and b because the $\theta(p)$'s were so small that any two points of $S\cdot \mathrm{Int}\, M(a, b)$ can be joined by an arc in $S\cdot(V(I', \mathrm{Bd}\, F) - \mathrm{Bd}\, F)$. The boundary of the disk $M(a, b)$ does not intersect S except at a and b since points of $S\cdot \mathrm{Int} g(F)$ lie on the interior of M. The disk $M(a, b)$ contains $Y(a, b)$ and arcs on the boundary of this disk show that each point of $Y(a, b)\cdot \mathrm{Bd}\, F$ is accessible from each component of $g(F) - S$ that has $Y(a, b)$ as a "side".

If p is a point of $\mathrm{Bd}\, F\cdot \mathrm{Bd}\, E_j$ that does not belong to any $M(a, b)$, there is an interval pq in $\mathrm{Bd}\, F\cdot \mathrm{Bd}\, E_j$ that does not intersect S except possibly at p. Some arc tangent to pq at p shows that p is accessible from E_j at p.

THEOREM 6. *Suppose S is a 2-sphere in E^3, R is a triangulation of E^3, F is a 2-simplex of R such that $S\cdot \mathrm{Bd}\, F$ is finite, p is a point of $S\cdot \mathrm{Bd}\, F$*

at which Bd F *pierces* S, pxq *is an arc in* F *that intersects* $S + \text{Bd}\,F$ *only at* p, $\varepsilon > 0$, *and* rp *is a segment of* Bd F *which lies except for* p *in the component of* $E^3 - S$ *containing* q. *Then there is a piecewise linear homeomorphism* g *of* E^3 *onto itself such that:*

(1) g *is fixed on the* 1-*skeleton of* R,

(2) g *is fixed outside an* ε-*neighborhood of* p, *and*

(3) *there is a polygonal arc in* $g(F)$ *from* q *to* rp *which misses* S *and lies in an* ε-*neighborhood of* pxq.

PROOF. The proof of this theorem is much like that of Theorem 3. With no loss of generality we suppose that pxq is locally polyhedral mod p, and p is not a vertex of F. See Figure 12. Let qyr be a polygonal arc from

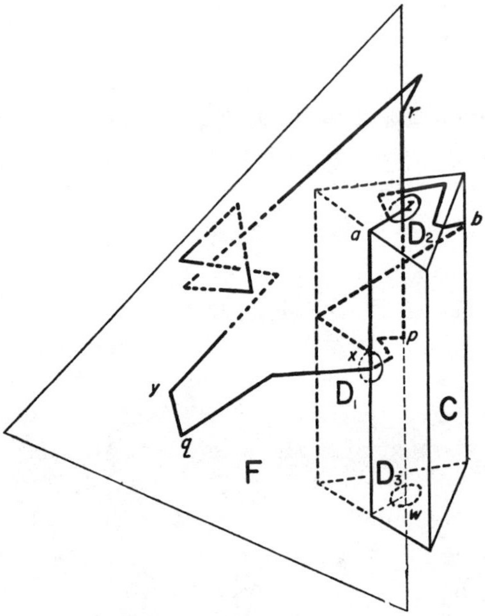

Figure 12.

q to r in $E^3 - S$ such that $qyr + rp$ (of Bd F) $+ pxq$ is a simple closed curve.

Let C be a small polyhedral cube of diameter less than ε such that $C \cdot \text{Bd}\,F$ is a straight line interval wpz missing r, $pxq \cdot \text{Bd}\,C$ is the point x, $F \cdot \text{Bd}\,C$ is an arc wxz, and $C \cdot qyr = 0$. Let z be $\text{Bd}\,C \cdot rp$.

Let D_1, D_2, D_3 be mutually exclusive disks on Bd C such that each misses S, $x \in \text{Int}\,D_1$, $z \in \text{Int}\,D_2$, $w \in \text{Int}\,D_3$, and D_3 misses the subarc xaz of $F \cdot \text{Bd}\,C = wxz$ from x to z. Let xbz be a polygonal arc on Bd C such that

xbz agrees with xaz on D_1 and on D_2, xbz misses $D_3 + S$, and xbz unspirals around Bd D_1 and Bd D_2 just enough so that there is an isotopy of Bd C onto itself that leaves D_1, D_2, D_3 fixed and takes xaz onto xbz. That there is such an arc xbz follows from the fact as demonstrated in the proof of Theorem 3 that no component of $S \cdot \text{Bd}\, C$ separates D_1 from D_2 on Bd C. (If some component X of $S \cdot \text{Bd}\, C$ separated D_1 from D_2 on Bd C, then a simple closed curve near X and on S links the simple closed curve $pxq + qyr + rzp$. But each simple closed curve on $S - \{p\}$ can be shrunk to a point in $S - \{p\} = S - (pxq + qyr + rzp)$.) The homeomorphism g promised by Theorem 6 is fixed except near Bd $C - (D_1 + D_2 + D_3)$ and takes xaz onto xbz.

THEOREM 7. *Suppose S is a 2-sphere, R is a triangulation of E^3, F is a 2-simplex of R such that $S \cdot \text{Bd}\, F$ is finite, $\varepsilon > 0$, and a, b, c are three points of $S \cdot \text{Bd}\, F$. Then there is a piecewise linear homeomorphism g of E^3 onto itself such that:*

(1) *g is fixed on the 1-skeleton of R,*

(2) *g moves no point more than ε, and*

(3) *$g(F)$ contains a spanning arc which misses S and separates some two of a, b, c on $g(F)$; that is, no component of $S \cdot g(F)$ contains each of a, b, c.*

PROOF. We find from repeated applications of Theorem 3 that there is a homeomorphism g_1 of E^3 onto itself such that g_1 is fixed on R_1, g_1 moves no point more than $\varepsilon/3$, and if p is a point of $S \cdot \text{Bd}\, F$ at which Bd F does not pierce S, then the component of $S \cdot g_1(F)$ that contains p does not intersect any other component of Bd F. Note that this implies that if a component of $S \cdot g(F)$ has two points on Bd F, then Bd F pierces S at each of these points. For convenience we suppose that g_1 is the identity.

Let δ be a positive number less than $\varepsilon/3$ such that for each point p of $S \cdot \text{Bd}\, F$ at which Bd F does not pierce S, there is a spanning arc of F which is at a distance of more than δ from S and which separates p from each other point of $S \cdot \text{Bd}\, F$ on F. It follows from Theorem 5 that there is a homeomorphism g_2 of E^3 onto itself such that g_2 moves no point more than δ, g_2 is fixed on R_1, no component of $S \cdot g_2(\text{Int}\, F)$ has three limit points on Bd F, and if E is a component of $g_2(\text{Int}\, F) - S$ with three limit points on Bd F, each of these limit points is accessible from E. Note that if a component of $S \cdot g_2(F)$ contains two points of Bd F, then Bd F pierces S at each of these points. (If we had not supposed that g_1 is the identity we would have used $g_1^{-1}(S)$ instead of S in applying Theorem 5 and shown that $g_2 g_1$ had the above properties.) For convenience in notation we suppose that g_2 is also the identity.

First, consider the case where some $X(r, s)$ separates some pairs of a, b, c (say a, b) on F. Then $M(r, s)$ separates a from b on F and an arc rxs on Bd $M(r, s)$ is a spanning arc of F that intersects S only at its end points. We may apply Theorem 6 since it follows from the preceding paragraph that Bd F pierces S at r and s. Hence there is a piecewise linear homeomorphism g of E^3 onto itself fixed on the 1-skeleton of R, fixed except near r and s, and such that there is a spanning arc of $g(F)$ near rxs that misses S and has end points near r and s. This spanning arc would separate a from b on $g(F)$.

If no $X(r, s)$ separates any two of a, b, c on F, then a, b, c all are limit points of the same component E of $F - S$. Let d be a point of Bd F near a that belongs to the same component of $E^3 - S$ as E. There is such a point d near a since Bd F pierces S at a. With no loss of generality we suppose that $b + d$ separates a and c on Bd F. It follows from Theorem 5 that there is a spanning arc axb of F which lies except for its endpoints in E. Applying Theorem 6 we find that there is a piecewise linear homeomorphism g of E^3 onto itself that is fixed on the 1-skeleton of R, is fixed except near a and b, and is such that $g(F)$ has a spanning arc near axb that misses S, lies in the component of $E^3 - S$ containing E, and has end points near a and near b. Furthermore we may assume that this spanning arc hits at d since d is in the same component of $E^3 - S$ as E whereas points of Bd F slightly to the other side of a do not. This spanning arc would separate a from c on $g(F)$.

Note that no component of $S \cdot g(F)$ contains each of a, b, c.

THEOREM 8. *Suppose S is a 2-sphere in E^3, R is a triangulation of E^3 such that S intersects the* 1*-skeleton R_1 of R in only a finite number of points. Then for each positive number ε there is a piecewise linear homeomorphism g of E^3 onto itself such that:*

(1) *g moves no point more than ε,*

(2) *g is fixed on R_1, and*

(3) *for each* 2*-simplex F of R, no component of $S \cdot g(F)$ has three points on* Bd F.

PROOF. The theorem is proved by repeated applications of Theorem 7. Let $(T_1, F_1), (T_2, F_2), \cdots, (T_n, F_n)$ be the set of all pairs where F_i is a 2-simplex of R and T_i is a triple of points of $R_1 \cdot S$ on Bd F_i. (It may be that $F_i = F_j$ or $T_r = T_s$ even though $i \neq j$ or $r \neq s$.)

It follows from Theorem 7 that there is a piecewise linear homeomorphism g_1 of E^3 onto itself such that:

(1) g_1 moves no point more than ε/n,

(2) g_1 is fixed on R_1, and

(3) there is a spanning arc A_1 in F_1 such that $g_1(A_1)$ misses S and $g_1(A_1)$ separates some pair of points of T_1 on $g_1(F_1)$.

Note that A_1 misses $g_1^{-1}(S)$.

Let δ be a positive number less that ε/n or the distance from A_1 to $g_1^{-1}(S)$. It follows from another application of Theorem 7 that there a piecewise linear homeomorphism g_2 of E^3 onto itself such that:

(1) g_2 moves no point more than δ,

(2) g_2 is fixed on R_1, and

(3) there is a spanning arc A_2 in F_2 such that $g_2(A_2)$ misses $g_1^{-1}(S)$ and $g_2(A_2)$ separates some pair of points of T_2 on $g_2(F_2)$.

Note that A_2 misses $g_2^{-1}g_1^{-1}(S)$. Also A_1 misses $g_2^{-1}g_1^{-1}(S)$ since g_2^{-1} moves no point by as much as the distance from A_1 to $g_1^{-1}(S)$.

The process is repeated to get a sequence of homeomorphisms $g_1, g_2, \cdots, g_n$ such that:

(1) $g_1g_2\cdots g_n$ moves no point more than ε,

(2) $g_1g_2\cdots g_n$ is fixed on R_1, and

(3) for each pair (T_i, F_i), there is a spanning arc A_i in F_i such that $g_1g_2\cdots g_i(A_i)$ misses S and A_i separates some pair of points of T_i on F_i.

We let $g^{-1} = g_1g_2\cdots g_n$ and note that for each 2-simplex F of R, no component of $S\cdot g^{-1}(F)$ has three points on Bd F.

6. General case

In this section we suppose that $S, R, R_0, R_1, R_2, \varepsilon, \delta$ are as in § 3 but with no restriction on the number of components of $S\cdot R_1$. We do suppose however that $S\cdot R_0 = 0$.

In the proof of Theorem 1 for the *very, very special case* we used a certain non-polyhedral finite graph in S which was the 1-skeleton of a curvilinear cellular subdivision of S and showed that S was almost between $h(S\times -1)$ and $h(S\times 1)$ by showing that the finite graph was between. In the *very special case* and *special case* we did not have a finite graph at our disposal but used a "fake 1-skeleton" which was the sum of X's instead of being the sum of arcs. In Part A of § 6 we pick an open subset U of S such that $\bar{U}$ will contain a "fake 1-skeleton" of S with both "fake arcs" and "fake vertices". The purpose of using $\bar{U}$ instead of S is that S may intersect R_1 in infinitely many points (contrary to the assumption for the *special case*).

Section 6 on the general case is broken into four parts. In Part A we describe U and give some of its properties.

A. *An open subset U of S.* About each vertex of R, put a small ball which misses S. Also, about each 1-simplex of R, consider a narrow tube joining the balls about the ends of the simplex. See Figure 13. The tubes have diameters less than δ. The sum of all such balls and tubes is the closure of a tubular neighborhood of R_1. Denote the closure of this tubular neighborhood by N.

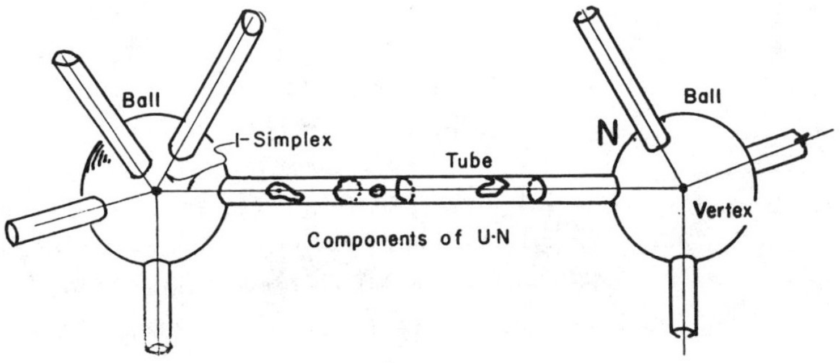

Figure 13.

Each component of $S \cdot N$ is of diameter less than δ since it lies in a narrow tube about a 1-simplex of R. It follows from Theorem 9 that $S - N$ has a component U such that each component of $S - U$ is of diameter less than $\varepsilon/3$. We use Bd U to denote the boundary of U in S; that is, Bd $U = \bar{U} \cdot (S - U)$. If follows from Theorem 9 that one obtains a 2-sphere from $\bar{U}$ if each boundary component of $\bar{U}$ is shrunk to a point. In [2] we actually shrank all these boundary components to points, but we will not do this in the present treatment. However, in Part B of this section we do shrink a finite number of them to points; namely, those that link R_1.

Theorem 10 shows that there are only a finite number of the components of Bd U that link R_1. It is these components which correspond in the general case to the points of $S \cdot R_1$ of the *special case*. When these components are shrunk to points, the points are the "fake vertices" of the "fake 1-skeleton" we shall use.

The ε used in the next theorem is arbitrary and need not be the same as the one mentioned in the statement of Theorem 1.

THEOREM 9. *Suppose $\varepsilon > 0$, S is a 2-sphere of diameter more than 2ε, and X is a closed subset of S such that each component of X lies in a disk in S of diameter less than ε. Then $S - X$ contains a component U such that $S - U$ lies in the sum of a collection of mutually exclusive disks* *on S each of diameter less than ε. Also, the decomposition of $\bar{U}$ whose elements are the points of U and the components of $\bar{U}\cdot(S-U)$ is a 2-sphere. In fact, there is a map g of $\bar{U}$ onto S such that g moves no point as much as ε and for each point s of S, $g^{-1}(s)$ is either a point of U or a component of* Bd U *in S.*

PROOF. For each point s of S that belongs to X or to a component of $S-X$ that lies in an ε-disk of S, let V_s denote the sum of all connected ε-subsets of S containing s such that the boundary (in S) of the ε-subset lies in a component of X. Note that V_s is a continuum lying in a disk of S of diameter less than ε, V_s does not separate S, the boundary of V_s is connected and lies in X, and if p is a point of V_s then $V_p = V_s$.

The sum of the V_s's is closed. This is because each point of the boundary (in S) of this sum is a point of X and hence belongs to a V_s. Let V be a component of the sum of the V_s's. We show that V is a V_s. To do this, subtract from V the interior of each V_s in it. The remainder is connected because if it were not, some simple closed curve J in S would separate this remainder in S. However, there is no place for J to be. It would intersect the interior of some V_s in V so it would lie in the interior of this V_s. But then one component of $S-J$ would belong to the interior of this V_s and would not intersect the remainder. Hence this remainder is connected and lies in X. Hence it lies in a V_s and V is a V_s.

The open subset U of S promised by Theorem 9 is S minus the sum of the V_s's. We showed in the last paragraph that each component of $S-U$ lies in a disk in S of diameter less than ε. We now show that U is connected. If it were not connected, it would follow from the unicoherence of S that some component of $S-U$ separates S between two points of U. This is impossible since each component of $S-U$ is a V_s which does not separate S.

Consider the decomposition of S whose elements are the points of U and the components (V_s's) of $S-U$. It follows from Moore's *sphere decomposition theorem* [22] that the decomposition space is a 2-sphere. Another way of proving that it is a 2-sphere is to apply the Kline *sphere characterization* [8] after noting that the decomposition space is a locally connected metric continuum which is separated by each simple closed curve in it but by no pair of its points.

Let t be a map of S onto itself such that for each point q of S, $t^{-1}(q)$ is either a point of U or a V_s. The map t applied to $\bar{U}$ shows that the decomposition space of $\bar{U}$ whose elements are the points of U and the components of Bd U is a 2-sphere. We merely replace the V_s's of the preceding paragraph by their boundaries.

Note that $t(S - U)$ is a closed 0-dimensional subset of S. Each point q of $t(S - U)$ lies in an open disk D_q in S such that diameter $t^{-1}(D_q) < \varepsilon$ and $\text{Bd}\, D_q \cdot t(S - U) = 0$. Cover $t(S - U)$ with a finite collection $D_1, D_2, \cdots, D_n$ of such mutually exclusive open disks. The map g promised by the theorem is the identity except on $\sum t^{-1}(D_i)$. To get g on $\bar{U} \cdot t^{-1}(D_i)$, let h_i be any homeomorphism of $\bar{D}_i$ onto $t^{-1}(\bar{D}_i)$ that agrees with t^{-1} on $\text{Bd}\, D_i$ and let $g = h_i t$ on $\bar{U} \cdot t^{-1}(D_i)$. The reason that g moves no point by as much as ε is that for each point which is moved at all, both it and its image under g lie in a disk $t^{-1}(D_i)$ of diameter less than ε.

THEOREM 10. *Suppose as shown in Figure* 13 *that C is a tube joining two balls and making up part of N. Then* Bd U *has at most a finite number of components which separate the two ends of* Bd C *from each other in* Bd C.

PROOF. This result is given on page 471 of [2] and by [6, Th. 5.3], but we repeat its proof for completeness.

Let J be a simple closed curve in E^3 which intersects the solid cylinder C only in its axis. Each simple closed curve on Bd C that separates the two ends of Bd C from each other links J.

Let $V_1, V_2, \cdots$ be the components of Bd U that separate the ends of Bd C from each other on Bd C and $M(V_i)$ be the sum of V_i and all components of $S - V_i$ other than the one containing U. The $M(V_i)$'s are mutually exclusive so not more than a finite number of them contain a point which is as far away from Bd C as J is from the lateral boundary of C. However, if an $M(V_i)$ misses J, it lies in a disk D_i which misses J. This is impossible because, as shown in the proofs of Theorems 3 and 6, there would be a simple closed curve J_i on Bd C that links J and is so near V_i that J_i is homotopic in $E^3 - J$ to a closed curve J_i' on D_i. This is impossible since J_i links J but J_i' does not. See [2, Th. 10']. Thus there are at most a finite number of V_i's.

THEOREM 11. *If $S, R, \varepsilon, \delta$ are as described in* § 3, *and U is the open set described at the beginning of Part* A *of* § 6, *then for no* 5δ*-subset X of $\bar{U}$ does $\bar{U} - X$ have two components each with diameters more than* $\varepsilon/3$.

PROOF. Assume B_1, B_2 are two components of $\bar{U} - X$ with diameters more than $\varepsilon/3$. Then $X + (S - \bar{U})$ separates B_1 from B_2 in S. In fact, if W is an open subset of S containing Bd U, then $(X + W) \cdot (S - \bar{U})$ separates B_1 from B_2 in S. Pick such a W each of whose components is of diameter less than δ.

A closed subset Z in $X + W \cdot (S - \bar{U})$ is irreducible with respect to separating B_1 from B_2 in S. It follows from the unicoherence of S that Z is connected. We show that it is of diameter less than 7δ by showing that each point of it lies within δ of X. Suppose $p \in Z - X$ and W_1 is the component of W containing p. Let Z' be a subcontinuum of Z irreducible from p to Bd W_1. Then Z' is of diameter less than δ and intersects $\bar{U}$ and hence X.

The assumption that X separates B_1 from B_2 in S led to the contradiction that a 7δ-subset of S separates S into two pieces each of diameter more than $\varepsilon/3$. This violates the definition of δ.

B. *Shrinking "fake vertices" to points.* Theorem 10 shows that there are only a finite number of components of Bd U that link R_1. We use these components as substitutes for vertices in a "fake 1-skeleton" of S. In this part of § 6 we shrink these "fake vertices" to points so that they will be more like real vertices. The map k that we describe is a simplified version of the part of the map described on pages 471–475 of [2]. Perhaps Figure 14 describes this map k better than words.

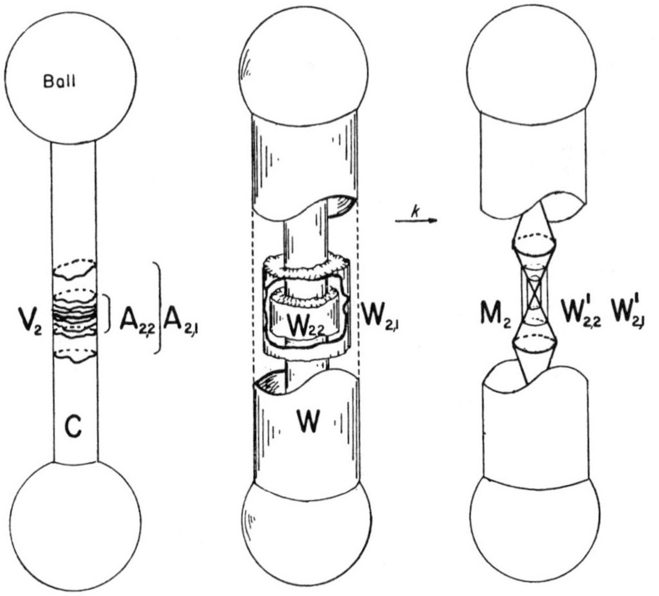

Figure 14.

Let C be a cylinder as shown in Figures 13 and 14 making up part of N and $V_1, V_2, \cdots, V_n$ be the components of Bd U that link R_1 as described in Theorem 10. In picking N, we could pick the tubes so that no component of Bd U separates S into more than two pieces. This is because we can pick the radii of the tubes in uncountably many ways but S contains only a countable number of mutually exclusive closed sets V such that $S - V$ has three components. Since separating a 2-sphere into a certain number of pieces is a topological invariant for closed subsets of the 2-sphere, we suppose that $\operatorname{Bd} C - V_i$ has exactly two components.

Let $A_{i1}, A_{i2}, \cdots$ be a sequence of topological annuli on the lateral boundary of C such that $A_{ij+1} \subset \operatorname{Int} A_{ij}$, $A_{i1} \cdot A_{i2} \cdots = V_i$, and $A_{i1} \cdot A_{j1} = 0$ if $i \neq j$. Let W be a cylinder slightly larger than C such that C and W have their ends in the same planes and C lies except for its ends in Int W. We suppose diameter $W < \delta$. Let $W_{i1}, W_{i2}, \cdots$ be a decreasing sequence of tame cubes in Int W such that $W_{ij+1} \subset \operatorname{Int} W_{ij}$, $W_{ij} \cdot \operatorname{Bd} C = A_{ij}$, $W_{i1} \cdot W_{j1} = 0$ if $i \neq j$, and $W_{ij} - C$ lies within $1/j$ of V_i.

For convenience in describing the map k we suppose that the lateral boundary of C is the cylinder $(x^2 + y^2 = 1, 0 \leqq z \leqq 1)$ and that $V_1, V_2, \cdots, V_n$ are ordered so that V_{i+1} separates the upper base of $\operatorname{Bd} C$ from V_i on $\operatorname{Bd} C$. Let M_i be the double napped cone generated by revolving the closed segment from $(0, -1, (i-1)/n)$ to $(0, 1, i/n)$ about the z axis. Let W'_{ij} be the cylinder such that the boundary of its bases are the circles which are the intersection of M_i and the 2-sphere with center at $(0, 0, (2i-1)/2n)$ and radius $1/4jn$.

Let k be a map of W onto itself such that:

(1) k is fixed on $\operatorname{Bd} W$,

(2) $k(V_i) = k(W_{i1} \cdot W_{i2} \cdots)$ is the point $(0, 0, (2i-1)/2n)$,

(3) k takes the lateral boundary of C onto $\sum M_i$,

(4) $k(W_{ij}) = W'_{ij}$, and

(5) k is a homeomorphism except on $k^{-1}k(\sum V_i) = \sum (W_{i1} \cdot W_{i2} \cdots)$.

Note that $k(C \cdot \operatorname{Bd} U)$ lies on $\sum M_i$. No component of $k(\operatorname{Bd} U)$ other than $k(V_1), k(V_2), \cdots, k(V_n)$ separates the top of $\sum M_i$ from the bottom in $\sum M_i$. Hence we suppose that there are planes containing the axis of C that intersect $k(C \cdot \bar{U})$ only in the separating points of $\sum M_i$. In fact we suppose that for each 2-simplex F of R, the barycenter of F bisects a segment I perpendicular to F such that *the join* $V(I, \operatorname{Bd} F)$ *intersects* $k\,(C \cdot \bar{U})$ *only at the separating points of* $\sum M_i$.

We place one last restriction on k. For convenience we supposed that cylinders were round rather than prismatic. If we had made N polyhedral we could have supposed that k is locally piecewise linear except at points of $k^{-1}k(\sum V_i)$. We suppose that N was polyhedral and that k *is locally piecewise linear* mod $k^{-1}k(\sum V_i)$.

So far we have only defined k near one tube C of N. Consider mutually exclusive W's about the various tubes, define k on each such W, and extend k to be the identity on the rest of E^3. Note that k is a map of E^3 onto itself and that *k moves no point more than δ, and $k(\bar{U})\cdot R_1$ is finite.*

The thing we have accomplished by the map k is to put ourselves in a position to employ the techniques used in the *special case*. What we have done here is easier than the corresponding step in [2] for there we pulled all components of Bd U to points rather than just those that linked R_1.

The V_i's are the "fake vertices" of a "fake 1-skeleton" (which we have not yet described) of $\bar{U}$ and the images of the V_i's under k are the "fake vertices" of a "fake 1-skeleton" of $k(\bar{U})$ that we describe in Part C. Note that the "fake vertices" of $k(\bar{U})$ are points.

C. *The "fake 1-skeleton" of $k(\bar{U})$.* In this section we use $k(\bar{U})$ in place of a 2-sphere. In applying the extensions of Theorems 3, 6, 7, 8 we need to decide when an arc pierces $k(\bar{U})$. Recall that $k(\bar{U})$ is a 2-sphere with holes and the boundary of each of these holes lies on a double napped cone M_i such that this boundary does not separate a vertex of M_i from either of its bases. Let Z denote the set of all points z such that either z is a point of $k(\bar{U})$ on some M_i or z is separated on some M_i from both bases of M_i by a component of $k(\bar{U})\cdot M_i$. We show in the next paragraph that $Z + k(\bar{U})$ separates E^3. Since $k(U)$ locally has only two sides and for each point p of $E^3 - (Z + k(\bar{U}))$ there is an arc from p to $k(U)$ that misses Z, $E^3 - (Z + k(\bar{U}))$ has exactly two components.

Let apb be an arc that misses Z and pierces $k(U)$ at p. Assume that $Z + k(\bar{U})$ does not separate E^3. Then apb lies on a simple closed curve J that intersects $Z + k(\bar{U})$ only at p. Let D be a punctured disk in $k(\bar{U})$ such that $p \in \operatorname{Int} D$ and each boundary component of D is so close to Z that it does not link J. Then no simple closed curve in $D - \{p\}$ links J. This contradicts the assumption that J pierces $k(U)$ at p. The contradiction arose from the false assumption that $Z + k(\bar{U})$ did not separate E^3.

The arc apb pierces $k(\bar{U})$ at p if $p \in k(\bar{U})$, p is the only point common to apb and $Z + k(\bar{U})$, and $Z + k(\bar{U})$ separates a from b in E^3. We only needed this extra definition of piercing for the case where p is a point of the boundary of $k(\bar{U})$.

As pointed out on pages 471–475 of [2] it is possible to shrink the boundary components of $k(\bar{U})$ to points so as to change $k(\bar{U})$ into a real 2-sphere. For each open set O in E^3 which contains the Z minus the sum of the vertices of the M_i's we may take the shrinking to be fixed outside O. However, we shall not use the shrinking in the present treatment.

The argument of Theorems 4 and 5 shows that there is no loss of generality in supposing that for each 2-simplex F of R, no component of $k(\bar{U})\cdot \operatorname{Int} F$ has three limit points on Bd F. Applying the techniques of

Theorems 3, 6, 7, 8 we find further that we can adjust k so that if a, b, c are three points of $k(\bar{U})\cdot \mathrm{Bd}\, F$, some spanning arc of F misses $k(\bar{U})$ and separates some two of a, b, c from each other in F. Hence we suppose that *no component of* $F\cdot k(\bar{U})$ *has three limit points on* Bd F. In making these alterations in k, we maintain the properties that:

(1) k moves no point more than δ, and

(2) $F\cdot k(\mathrm{Bd}\, U) = k(\bar{U})\cdot \mathrm{Bd}\, F$ is finite and is the image of V_i's.

We can treat $k(\bar{U})$ much like a 2-sphere since the decomposition of $k(\bar{U})$ whose non-degenerate elements are the images under k of the non-degenerate boundary components of $\bar{U}$ which did not link R is a 2-sphere. Let g be a map of $\bar{U}$ onto S as described in Theorem 9 such that g^{-1} is fixed except on a finite collection of mutually exclusive $\varepsilon/3$-disks in S such that the boundary of each of the disks is of diameter less than δ and for each point s of S, $g^{-1}(s)$ is either a point of U or a component of Bd U. Then gk^{-1} takes $k(\bar{U})$ continuously onto S and moves no point more than $\varepsilon/3 + \delta$. See Figure 15.

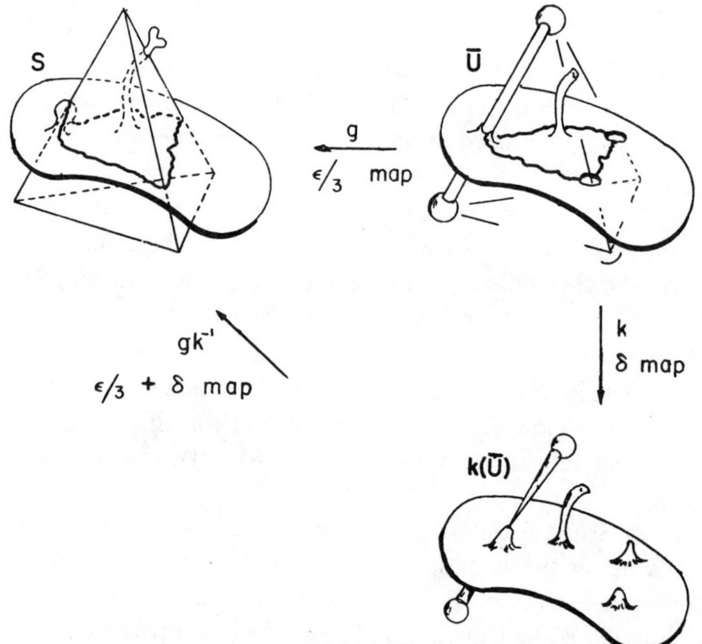

Figure 15.

If X is a subset of $k(\bar{U})$ of diameter less than 3δ, its image under k^{-1} has diameter less than 5δ. If a subset of $k(\bar{U})$ is of diameter more than $\varepsilon/3 + 2\delta$, its image under k^{-1} has diameter more than $\varepsilon/3$. Hence *for no*

3δ-subset X of $k(\bar{U})$ does $k(\bar{U}) - X$ have two components with diameters more than $\varepsilon/3 + 2\delta$.

We are now in a position to duplicate the argument used in the *very special case*. Let $X_1, X_2, \cdots, X_x$ be the continua of $k(\bar{U}) \cdot R_2$ such that for some 2-simplex F of R and some pair of points of Bd F that belong to some component of $k(\bar{U}) \cdot F$, X_i is the closure of the sum of all components of $k(\bar{U}) \cdot \text{Int}\, F$ that contain these two points. Certain of these X's are the 1-simplexes of the "fake 1-skeleton" of $k(\bar{U})$ we shall describe.

Duplicating the argument of Step 1 for the *very special case* we find that each component of $k(\bar{U}) - \sum X_i$ is of diameter no more than $2\varepsilon/3 + 7\delta$. We do this by showing that if p_0, q_0 are two points of $k(\bar{U}) - \sum X_i$ whose distance apart is greater than $2\varepsilon/3 + 7\delta$, then $\sum X_i$ separates p_0 from q_0 in $k(\bar{U})$. Let M be the sum of the 2-simplexes of R whose distances from each of p_0 and q_0 is more than $\varepsilon/3 + 2\delta$. Then M separates p_0 from q_0 in E^3 and $M \cdot k(\bar{U})$ separates p_0 from q_0 in $k(\bar{U})$. Let Z be a subset of $M \cdot k(\bar{U})$ irreducible with respect to separating p_0 from q_0 in $k(\bar{U})$. Since Z intersects no non-degenerate boundary component of $k(\bar{U})$, gk^{-1} is a homeomorphism on Z. Then $gk^{-1}(Z)$ is irreducible with respect to separating $gk^{-1}(p_0)$ from $gk^{-1}(q_0)$ in the 2-sphere $gk^{-1}(k(\bar{U}))$, and it follows from the unicoherence of a 2-sphere that $gk^{-1}(Z)$ is connected. Hence Z is connected. It follows from the paragraph before last that diameter $Z > 3\delta$. The argument used in Step 1 of the *very special case* then shows that $Z \subset \sum X_i$.

As in Step 2 of the *very special case* we let $K(X_i)$ be the sum of X_i and all components of $k(\bar{U}) - X_i$ with diameters less than $\varepsilon/3 + 2\delta$. We let $K_1, K_2, \cdots, K_k$ be the collection of all such $K(X)$'s not contained in a larger $K(X)$. Let $D_1, D_2, \cdots, D_d$ be the components of $k(\bar{U}) - \sum K_i$.

As in Step 3 of the *very special case*, we throw away certain of the K's and enlarge certain of the D's to E's. The K's that have a non-end point on the boundary (in $k(\bar{U})$) of an E_i form something like a circle as shown in Figure 8. The sum of these K's associated with the boundary of E_i is called $B(E_i)$.

As shown in Steps 4 and 5 of the *very special case*, each Bd E_i (boundary in $k(\bar{U})$) was on the boundary of a 3-simplex of R. Since the simplex is of diameter less than δ, diameter $E_i < \varepsilon/3 + 2\delta$.

Let $E_1, E_2, \cdots, E_n$ be the maximal E's. As in Step 6 of the *very special case* we associate a finite graph with $B(E_1) + B(E_2) + \cdots + B(E_n)$. To get the finite graph we replace each K_j in $B(E_i)$ with an arc A_j in $k(\bar{U})$ such that:

(1) A_j has the same ends as K_j,

(2) A_j lies except for its ends in $k(U)$,

(3) each point of A_j lies within δ of the X_j determining K_j,

(4) no two A_i's intersect except possibly at an end point of each, and

(5) $k(\bar{U}) - \sum A_i$ is the sum of the components $H_1, H_2, \cdots, H_n$ such that $\mathrm{Bd}\, H_i$ is the sum of the A's associated with the K's in $B(E_i)$.

Note that $gk^{-1}(A_j)$ is an arc in S and $gk^{-1}(H_i)$ is a disk in S. We know that diameter $A_j < 3\delta$, diameter $k^{-1}(A_j) < 5\delta$ and diameter $gk^{-1}(A_j) < 2\varepsilon/3 + 5\delta$. The last inequality is not good enough for our purposes so we now suppose that g is modified to improve it. Recall that g^{-1} was fixed except on a family of mutually exclusive disks such that the boundary of each disk was of diameter less than δ. For each 1-dimensional subset of any of the disks there is a homeomorphism of the disk onto itself that leaves the boundary of the disk fixed and moves the 1-dimensional subset to near the boundary. We suppose that g is adjusted (and for convenience we call the adjusted map g also) such that if $gk^{-1}(A_j)$ intersects one of these disks D, its intersection with D lies so near the boundary of this disk that diameter $D \cdot gk^{-1}(A_j) < \delta$. Then diameter $gk^{-1}(A_j) < 7\delta$.

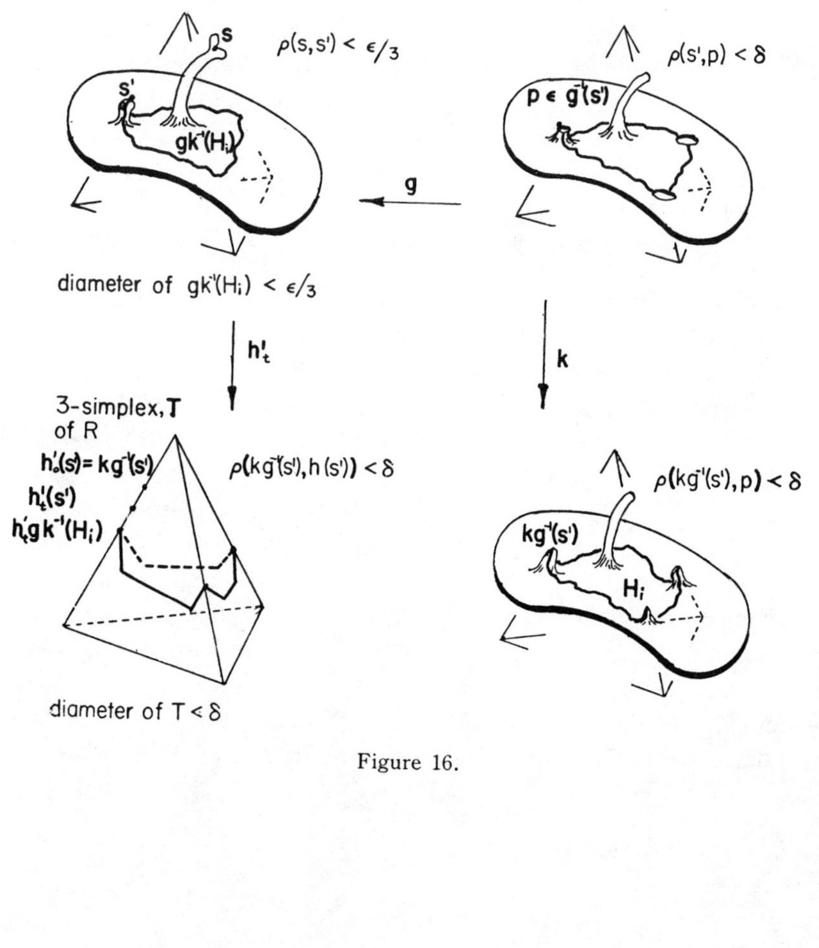

Figure 16.

The argument of the preceding paragraph shows that diameter $\operatorname{Bd} H_i < 3\delta$, diameter $k^{-1}(\operatorname{Bd} H_i) < 5\delta$, and diameter $gk^{-1}(\operatorname{Bd} H_i) < 7\delta$. Hence diameter $gk^{-1}(H_i) < \varepsilon/3$.

Aping Steps 7, 8, 9, 10 of the *very special case* we get a homeomorphism h' of $S \times [-1, 1]$ into E^3 such that $h'_t(gk^{-1}(H_i))$ is a polyhedral disk in the 3-simplex T of R whose boundary contains $\operatorname{Bd} E_i$ (in $k(\bar{U})$) and each X_i associated with a K_j in a $B(E_k)$ lies between $h'(S \times -1)$ and $h'(S \times 1)$. The sum of such X's is the "fake 1-skeleton" of $k(\bar{U})$. We suppose that the only points of $k(\bar{U}) \cdot R_1$ in $h'(S \times [-1, 1])$ are the "fake vertices" of this "fake 1-skeleton." We wish to avoid blowing up points of $k(\bar{U}) \cdot R_1$ that are not end points of any K and would find it particularly inconvenient to have one of these points be on the boundary of $h'(S \times [-1, 1])$.

We now show that h'_t moves no point more than $\varepsilon/3 + 3\delta$. Let s be a point of $gk^{-1}(H_i)$ and s' be a vertex of $gk^{-1}(\operatorname{Bd} H_i)$. Since diameter $gk^{-1}(\operatorname{Bd} H_i) < 7\delta$, diameter $gk^{-1}(H_i) < \varepsilon/3$. We read from Figure 16 that $\rho(s, s') < \varepsilon/3$, $\rho(s', kg^{-1}(s')) < 2\delta$, and $\rho(kg^{-1}(s'), h'_t(s)) < \delta$. Hence h'_t moves no point more than $\varepsilon/3 + 3\delta$.

D. *The homeomorphism* h_t. We obtain the h_t required by Theorem 1 by modifying $k^{-1}h'_t$ where h'_t is the homeomorphism described in Part C of this section. About the "fake vertices" in $k(\bar{U})$ put mutually exclusive polyhedral cubes in $h'(S \times [-1, 1])$ such that the images under k^{-1} of these cubes have diameters less than δ. Let k' be a piecewise linear homeomorphism of $k^{-1}h'(S \times [-1, 1])$ onto $h'(S \times [-1, 1])$ such that $k' = k$ except on the images of the cubes under k^{-1}. Then define h_t as follows:

$$h_t = k'^{-1}h'_t \, .$$

Note that k moves no point more than δ and k' moves no point more than 2δ. Since h'_t moves no point more than $\varepsilon/3 + 3\delta$, h_t moves no point more than $\varepsilon/3 + 5\delta < \varepsilon$.

Let W be a component of $S - h(S \times [-1, 1])$. We show that W lies in a disk of diameter less than ε. We suppose W intersects U since each component of $S - U$ lies in a disk of diameter less than $\varepsilon/3$.

Note that the image under k^{-1} of the "fake 1-skeleton" of $k(\bar{U})$ lies between $h(S \times -1)$ and $h(S \times 1)$. If X_i is a "1-simplex" of this "fake 1-skeleton", then $k^{-1}(X_i)$ is of diameter less than 3δ and lies in an $\varepsilon/3$-disk in S. This disk contains W if $k(W)$ intersects a K_i obtained by expanding X_i. Each $\operatorname{Bd} E_i$ (in $k(U)$) is of diameter less than δ so $k^{-1}(E_i)$ lies in an $\varepsilon/3$-disk in S. If $k(W)$ intersects E_i, W lies in this disk since $k^{-1}(\operatorname{Bd} E_i)$ lies between $h(S \times -1)$ and $h(S \times 1)$. We have considered all cases because each point of S either lies in $S - \bar{U}$, in a $k^{-1}(K_i)$, or in a $k^{-1}(E_i)$.

We have actually proved the following stronger version of Theorem 1.

THEOREM 1′. *Suppose* S *is a polyhedral 2-sphere,* h' *is a homeomorphism of* S *into* E^3*, and* $\varepsilon > 0$. *Then there is a piecewise linear homeomorphism* h *of* $S \times [-1, 1]$ *into* E^3 *such that* $\rho(h'(s), h(s \times t)) < \varepsilon (s \in S, -1 \leqq t \leqq 1)$ *and* $h'(S)$ *contains a family of mutually exclusive* ε*-disks such that* $h'(S)$ *minus the disks lies between* $h(S \times -1)$ *and* $h(S \times 1)$.

7. Side approximation of 2-spheres

Theorem 16 stated in § 1 follows easily from Theorem 1 by means of the following four results.

THEOREM 12. *For each 2-sphere* S *in* E^3 *and each* $\varepsilon > 0$ *there is a* $\delta > 0$ *such that if* h *is a* δ*-homeomorphism of* S *into* E^3 *and* X *is a closed subset of* $h(S)$ *each of whose components is a* δ*-set, then* $h(S)$ *contains a collection of mutually exclusive disks with diameters less than* ε *such that* X *lies in their sum.*

PROOF. Let δ be a positive number so small that each 3δ-subset of S lies in an $\varepsilon/2$-disk in S. This is the δ promised by the theorem.

Suppose h and S are as given in the statement of the theorem. Let X_1 be a component of X. Then

(1) diameter $X_1 \leqq \delta$,
(2) diameter $h^{-1}(X_1) \leqq 3\delta$,
(3) $h^{-1}(X_1)$ lies in an $\varepsilon/2$-disk D in S, and
(4) X_1 lies in the $(\varepsilon/2 + 2\delta)$-disk $h(D)$ of $h(S)$.

Theorem 12 then follows from Theorem 9.

THEOREM 13. *Suppose* D_1, D_2 *are two disks in* E^3, J *is a simple closed curve that misses* D_2, X *is a component of* $D_1 - D_2$ *which misses* Bd D_1, *and* $J \cdot X$ *contains only one point* p. *Then* J *does not pierce* X *at* p; *that is, there is a neighborhood* N *of* p *such that each simple closed curve in* $N \cdot X$ *that misses* p *can be shrunk to a point in* $E^3 - J$.

PROOF. Let U be a neighborhood of D_2 such that each simple closed curve in U can be shrunk to a point in $E^3 - J$. Let K be a punctured disk in X such that $p \in \operatorname{Int} K$ and $\operatorname{Bd} K \subset U$. Each simple closed curve in K that misses p can be shrunk to a point in $E^3 - J$.

THEOREM 14. *Suppose* $A_1, A_2, \cdots, A_n$ *is a finite collection of mutually exclusive disks in* E^3 *and* $B_1, B_2, \cdots, B_m$ *is a collection of mutually exclusive disks in* E^3. *Then if* p, q *are two points in* $E^3 - (\sum A_i + \sum B_j)$ *separated in* E^3 *by* $\sum A_i + \sum B_j$ *then there is an* r *and an* s *such that* $A_r + A_s$ *separates* p *from* q *in* E^3.

PROOF. Assume the theorem false. Suppose the collections $A_1, A_2, \cdots, A_n$ and $B_1, B_2, \cdots, B_m$ are irreducible with respect to making the theorem false in the sense that $m + n > 2$ and if any A_i or B_j is eliminated, then the sum of the rest does not separate p from q.

It is known that for each 2-sphere S in E^3, each component of $E^3 - S$ is uniformly locally arcwise connected (0-ULC). See [25, p. 66]. Hence, if ab is an arc intersecting S and U is an open set containing $ab \cdot S$, then there is a homeomorphism h of ab into $ab + U$ such that h is the identity outside U, $h(ab)$ is homotopic to ab in $ab + U$, and $S \cdot h(ab)$ is finite. We first pull parts of ab onto S so that the adjusted ab intersects S in only a finite number of components. We then pull the adjusted ab to the side of S so that it intersects only where it locally pierces S. Since a disk can be cut down to size to lie in a 2-sphere, for any given arc and any given surface, there is an adjustment of the arc which intersects the surface in at most a finite number of points.

Let pq be an arc from p to q that misses $A_2 + A_3 + \cdots + A_n + B_1 + \cdots + B_m$ and intersects A_1 in as few points as possible.

A 2-sphere in E^3 separates E^3 into two pieces and each of the pieces is uniformly locally arcwise connected. Theorem 5 of [11] tells us that each interior point of a disk is the interior point of a subdisk which lies on a 2-sphere. Hence a disk in E^3 locally has two sides at interior points of the disk and these sides are locally arcwise connected. Thus any arc on the interior of a disk can be approximated by arcs on either side of the disk.

Let x be a point of $A_1 \cdot pq$. Each arc xy in A_1 from x to Bd A_1 intersects $\sum B_i$ or else a gap in pq could be cut out of pq near x and this gap replaced by an arc that goes around A_1 near xy and misses A_1—thereby reducing the number of points of $A_1 \cdot pq$. A similar argument shows that no component of $A_1 - \sum B_i$ contains two points of pq. Hence there is a B_i (say B_1) such that a component of $A_1 - B_1$ misses Bd A_1 and contains only one point of pq.

Since $A_1 + B_1$ does not separate p from q, there is an arc pzq from p to q that misses $A_1 + B_1$ such that $pq + pzq$ is a simple closed curve J. Theorem 13 shows that J does not pierce A_1 at x. Hence, in a neighborhood of x, J lies on one side of A_1 and pq can be adjusted near x so as to reduce the number of components of $A_1 \cdot pq$.

Question. Would Theorem 14 remain true if we supposed that the A_i's and B_i's were contractible continua instead of disks?

THEOREM 15. *Suppose S is a 2-sphere, ε is a positive number such that* *the bounded component of $E^3 - S$ contains a point whose distance from S is more than ε, and h is a homeomorphism of $S \times [-1, 1]$ into E^3 such that:*

(1) *$h(S \times [-1, 1])$ lies in an ε-neiphborhood of S,*

(2) *S contains a finite collection of mutually exclusive ε-disks $A_1, A_2, \cdots, A_n$ such that $S - \sum A_i$ lies between $h(S \times -1)$ and $h(S \times 1)$,*

(3) *$h(S \times -1)$ contains a finite collection of mutually exclusive ε-disks $D_1, D_2, \cdots, D_d$ such that $h(S \times -1) - \sum D_i$ misses S, and*

(4) *$h(S \times 1)$ contains a finite collection of mutually exclusive ε-disks $E_1, E_2, \cdots, E_e$ such that $h(S \times 1) - \sum E_i$ misses S.*

Then $h(S \times -1) - \sum D_i$ lies in one component of $E^3 - S$ and $h(S \times 1) - \sum E_i$ lies in the other.

PROOF. Let p, q be points in different components of $E^3 - S$ each of whose distances from S is more than ε. It follows from Theorem 14 that there is an arc pq from p to q that misses $\sum A_i + (\sum D_i + \sum E_i)$.

Let x be the first point of $S + h(S \times -1) + h(S \times 1)$ on pq in the order from p to q. Then $x \notin S - \sum A_i$ since p does not lie between $h(S \times -1)$ and $h(S \times 1)$ but $S - \sum A_i$ does. The one of $h(S \times -1) - \sum D_i$, $h(S \times 1) - \sum E_i$ that contains x lies in the component of $E^3 - S$ containing p.

Theorem 16 follows immediately from Theorems 1, 12, and 15. In fact, we have proved the following stronger version of Theorem 16.

THEOREM 16′. *Suppose S is a polyhedral 2-sphere, h' is a homeomorphism of S into E^3, and $\varepsilon > 0$. Then there is a piecewise linear homeomorphism h of $S \times [-1, 1]$ into E^3 such that:*

(1) *$\rho(h'(s), h(s \times t)) < \varepsilon$ $(s \in S, -1 \leqq t \leqq 1)$,*

(2) *$h(S)$ lies except for a finite collection of mutually exclusive ε-disks between $h(S \times -1)$ and $h(S \times 1)$, and*

(3) *$h(S \times -1)$ lies except for a finite number of mutually exclusive ε-disks in one component of $E^3 - h'(S)$ while $h(S \times 1)$ lies except for a finite number of mutually exclusive ε-disks in the other component of $E^3 - h'(S)$.*

8. Side approximations of open subsets of 2-spheres

In this section we show that open subsets of a 2-sphere in E^3 can be almost approximated from either side.

THEOREM 17. *Suppose*

(1) *S is a polyhedral 2-sphere,*

(2) *h' is a homeomorphism of S into E^3,*

(3) *V is an open subset of S, and*

(4) *f is a non-negative continuous function defined on S such that f is positive on each point of V.*

Then there is a locally piecewise linear homeomorphism h of $V \times [-1, 1]$ into E^3 and a collection of mutually exclusive disks $D_1, D_2, \cdots$ in V such that:

(1) $\rho(h'(p), h(p \times t)) < f(p)$ $(p \in V, -1 \leqq t \leqq 1)$,

(2) $h'(V - \sum D_i) \subset h(V \times (-1, 1))$,

(3) *diameter $h'(D_i) <$ minimum value of f on D_i, and*

(4) *$\{D_i\}$ is locally finite at each point of V.*

PROOF. The proof of Theorem 17 is a modification of the proof of Theorem 1. The first task is to get a triangulation R of an open subset of E^3 such that this open set contains $h'(V)$.

Let f' be a continuous real valued function defined on E^3 such that f' is 0 on $h'(S - V)$ and positive elsewhere and $f' \leqq fh'^{-1}$ on $h'(S)$. The function f' plays the role of the ε of Theorem 1.

Let f'' be a continuous positive function defined on $E^3 - h'(S - V)$ such that for each point p of $h'(V)$ and each subset of $h'(S)$ within $10f''(p)$ of p lies in a disk D in $h'(V)$ such that diameter $D <$ (minimum value of f' on D) $- 16f''(p)$. Note that $f''(x)$ approaches 0 as x approaches $h'(S - V)$ on $h'(V)$. The function f'' plays the role of δ in Theorem 1. Epsilontics are more complicated in the proof of this theorem than in Theorem 1 since f', f'' vary while ε, δ did not.

Let R be a triangulation of $E^3 - h'(S - V)$ such that no vertex of R is on $h'(S)$, each simplex of R is of diameter less than the minimum value of f'' on the simplex, and the ratio of the maximum value of f'' on a simplex of R to the minimum value of f'' on this simplex is less than 2. The simplexes of R are rectilinear. Of course, the triangulation is not locally finite at points of $h'(S - V)$ but R is a triangulation of $E^3 - h'(S - V)$ rather than a triangulation of E^3.

Build a closed tubular neighborhood N about the 1-skeleton of R as illustrated in Figure 13. The balls about the vertices are so small that they miss $h'(S)$ and the cylinders joining the balls are so thin that each set which is the sum of a 3-simplex of R and the balls and cylinders intersecting it is such that it is of diameter less than the maximum value of f'' on it and the ratio of the maximum value of f'' on this sum to the minimum value of f'' there is less than 2.

We suppose with no loss of generality that V is connected. If the boundary components of $h'(V)$ were shrunk to points, $h'(\bar{V})$ would become a 2-sphere. Let U be the big component of $h'(V) - N$. Each component of $\mathrm{Bd}\, U - \mathrm{Bd}\, h'(V)$ lies on N and is of diameter less than the minimum value of f'' on this component. Hence the component lies in a disk in $h'(V)$ of diameter less than the minimum value of f' on the disk minus the maximum value of $16f''$ on the component. We suppose that N was adjusted so that each boundary component C of U which was not a boundary component of $h'(V)$ and which separates $h'(S)$ is such that $h'(S) - C$ has exactly two components and C is the boundary of each. The collection of these boundary components $V_1, V_2, \cdots$ which separate two ends of cylinders of N from each other on the boundary of the cylinders is locally finite at each point of $h'(V)$.

Let g be a map of $\bar{U}$ into $h'(\bar{V})$ such that the inverse of each point of $h'(\bar{V})$ under g is either a point of $U + h'(\text{Bd}\, V)$ or a component of $\text{Bd}\, U - h'(\text{Bd}\, V)$ and g^{-1} is fixed except on a collection of mutually exclusive disks in $h'(V)$ where the boundary of each of these disks is of diameter less than the minimum value of f'' on this boundary and the ratio of the maximum value of f'' to the minimum value of f'' on this boundary is less than 2.

We let k be a map similar to the map k defined in Part B of the general case which shrinks the V_i's to points. We note that $k(\bar{U})$ now satisfies conditions for the *special case* in that $k(\bar{U})$ does not intersect any 1-simplex of R in more than a finite number of points. As shown by the argument given in the *special case*, we suppose with no loss of generality that if F is a 2-simplex of R, no component of $k(\bar{U})\cdot F$ has three points on $\text{Bd}\, F$. Also, k is such that for each point q of $\bar{U}$,

$$\rho(q, k(q)) \leqq f''(q)$$

and

$$f''k(q)/f''(q) < 2\,.$$

As a result of gk^{-1} we may treat $k(\bar{U})$ much like $h'(\bar{V})$.

Consider all continua X_i such that for some 2-simplex F of R and some two points a, b of $k(\bar{U})\cdot \text{Bd}\, F$, X_i is the closure of all components of $k(\bar{U})\cdot \text{Int}\, F$ with both a and b as limit points. No F contains more than a finite number of X_i's. Certain of these X_i's are "fake 1-simplexes" of a "fake 1-skeleton" of $k(\bar{U})$. If X is a closed subset of $k(\bar{U})$ such that diameter $X <$ maximum $3f''$ on $k^{-1}(X)$, diameter $k^{-1}(X) <$ maximum $5f''$ on $k^{-1}(X)$. We find as done in Part C of § 6 that $k(\bar{U}) - X$ does not have two components with diameters more than the maximum f' on $k(\bar{U})$ plus the maximum $2f''$ on $k(\bar{U})$. Using this fact and duplicating the argument of Step 1 for the *very special case* we find that no component of $(k(U) - h'(\text{Bd}\, V)) - \sum X_i$ is of diameter as much as the maximum f' on $h'(\bar{V})$ plus the maximum $7f''$ on $h'(\bar{V})$. These restrictions are not good enough for our purposes but point the direction in which we are going.

In the next part of the proof of Theorem 17 we avoid details as given in the *very special case* and summarize even more concisely than we did in Part C of § 6. Following this Part C we define $K(X_i)$'s, D's, E's, and $B(E)$'s. Let B_i be the sum of the X's determining $B(E_i)$. Since B_i lies on the boundary of a 3-simplex of R,

diameter $B_i <$ minimum f'' on $B_i <$ minimum $2f''$ on $k^{-1}(B_i)$,
diameter $k^{-1}(B_i) <$ maximum $4f''$ on $k^{-1}(B_i)$, and
diameter $k^{-1}(E_i + B(E_i))$
$< \text{minimum } f' \text{ on } k^{-1}(E_i + B(E_i)) - \text{maximum } 16f'' \text{ on } k^{-1}(B_i)$.

The arc A_j replacing K_j is so near the X_j determining K_j that

diameter Bd $H_i <$ minimum f'' on Bd H_i
$<$ minimum $2f''$ on $k^{-1}(\text{Bd } H_i)$, and
diameter $k^{-1}(\text{Bd } H_i) <$ diameter Bd $H_i +$ maximum $2f''$ on $k^{-1}(\text{Bd } H_i)$
$<$ maximum $4f''$ on $k^{-1}(\text{Bd } H_i)$.

The map g is adjusted in the disks on which it is not fixed (and for convenience we call the adjusted map g) to take the part of A_j in the disks to so near the boundary of the disks that

$f''k^{-1}(\text{end of } A_j)/f''gk^{-1}(\text{same end of } A_j) < 2$,
diameter $k^{-1}(\text{Bd } H_i) <$ maximum $8f''$ on $gk^{-1}(\text{Bd } H_i)$, and
diameter $gk^{-1}(\text{Bd } H_i) <$ maximum $10f''$ on $gk^{-1}(\text{Bd } H_i)$.

The last of the above inequalities implies that

diameter $gk^{-1}(H_i) <$ minimum $f'gk^{-1}(H_i) - 16$ maximum $f''gk^{-1}(\text{Bd } H_i)$.

Copying Steps 7, 8, 9, 10 of the *very special case* as we did in Part C of § 6, we get a homeomorphism h'' of $h'(V) \times [-1, 1]$ into E^3 such that $h''_t gk^{-1}(H_i)$ is a disk in the 3-simplex of R whose boundary contains Bd E_i in $k(\bar{U})$ and each X_i associated with a K_j in a $B(E_n)$ lies in

$$h''(h'(V) \times (-1, 1)) .$$

As in Part D of § 6 we take polyhedral cubes C_i in $h''(h'(V) \times (-1, 1))$ about the "fake vertices" of $k(U)$ (where k^{-1} is not a homeomorphism) such that

diameter $k^{-1}(C_i) <$ minimum f'' on $k^{-1}(C_i)$, and
maximum f'' on $k^{-1}(C_i)$/minimum f'' on $k^{-1}(C_i) < 2$.

Let k' be a homeomorphism of E^3 onto itself such that $k' = k$ outside the $k^{-1}(C_i)$'s and k' is piecewise linear on each $k^{-1}(C_i)$.

The h_t promised by Theorem 17 is

$$h_t = k'^{-1} h''_t h' .$$

We now show that for each point p of V, $\rho(h'(p), h_t(p)) < f(p)$. Suppose

$h'(p)$ is a point of $gk^{-1}(H_i)$ and s is a vertex of $gk^{-1}(\mathrm{Bd}\, H_i)$. We read the following from Figure 17:

$$\rho(h'(p), s) < f'(h'(p)) - 16f''(s) ,$$
$$\rho(s, h_0''(s)) < 2f''(q) ,$$
$$\rho(h_0''(s), h_t''h'(p)) < f''k(q) ,$$
$$\rho(h_t''h'(p), h_t(p)) < f''h_t''h'(p) .$$

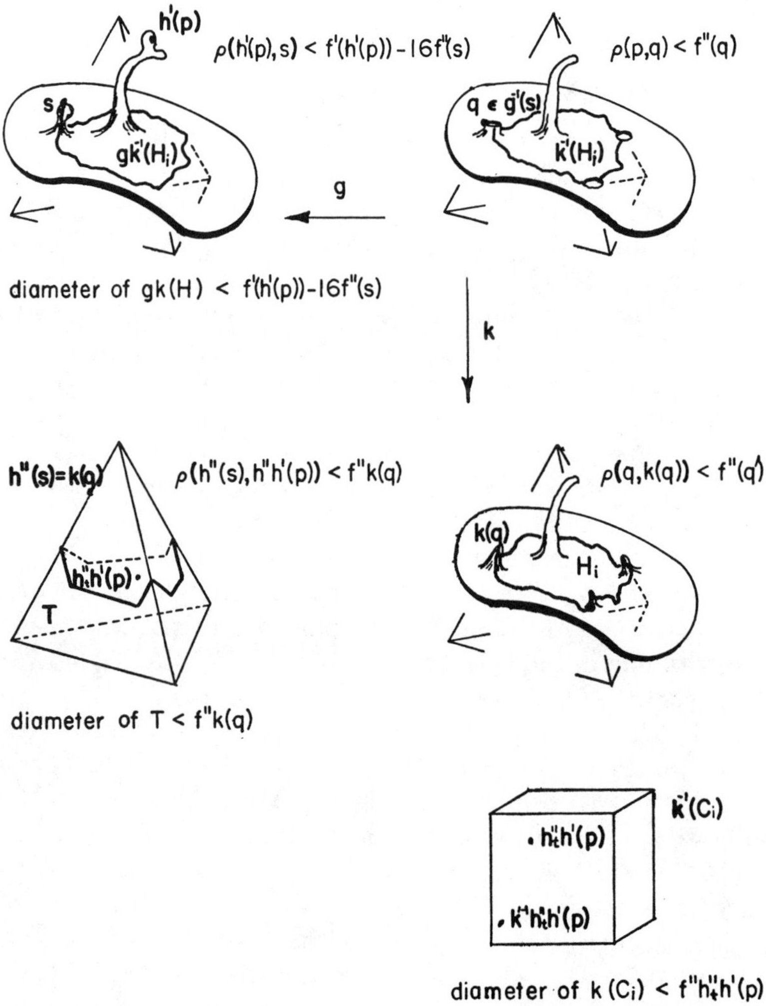

Figure 17.

Hence

$$\rho'(h'(p), h_t(p)) < f'h'(p) - 16f''(s) + 2f''(q) + f''k(q) + f''h_t''h'(p) .$$

However,

$$2f''(q) < 4f''(s)$$

since $f''k^{-1}(\text{end of } A_j)/f''gk^{-1}(\text{same end of } A_j) < 2$. Also

$$f''k(q) < 2f''(q) < 4f''(s) ,$$

since $f''k(q)/f''(q) < 2$. Finally

$$f''h_t''h'(p) < 2f''k(q) < 8f''(s) ,$$

since $h_t''h'(p)$, $k(q)$ belong to the same 3-simplex of R. Hence

$$\rho(h'(p), h_t(p)) < f'(h'(p)) < f(p) .$$

We now show that each component W of $h'(V) - (h(V) \times (-1,1))$ lies in a disk in $h'(V)$ of diameter less than the minimum value of f' on this disk. Since $\sum X_i \subset h''(h'(V) \times (-1,1)) = k'h(V \times (-1,1))$ and $k'^{-1} = k^{-1}$ except on $\sum C_i$, which lies in $h''(h'(V) \times (-1,1))$, then $k^{-1}(\sum X_i) \subset h(V \times (-1,1))$. Hence $W \subset h'(V) - k^{-1}(\sum X_i)$.

We suppose that W intersects U since each component of $h'(V) - U$ lies in a disk in V of diameter less than the minimum value of f'' on this disk. This is because the components of $\operatorname{Bd} U - \operatorname{Bd} V$ lie on N and each has a diameter less than minimum value of f'' on it.

If X_i is a "fake 1-simplex" of the "fake skeleton" of $k(U)$.

diameter $X_i <$ minimum f'' on $X_i <$ minimum $2f''$ on $k^{-1}(X_i)$, and

diameter $k^{-1}(X_i) <$ diameter $X_i +$ maximum $2f''$ on $k^{-1}(X_i)$

$<$ maximum $4f''$ on $k^{-1}(X_i)$.

Hence $k^{-1}(X_i)$ lies in a disk in $h'(V)$ of diameter less than the minimum value of f' on the disk. This disk contains W if $k(W)$ intersects a K_i obtained by filling in the holes in an X_i.

If W intersects U but does not lie in a $k^{-1}(K_i)$, then $k(W)$ intersects an E_i. Since $\operatorname{Bd} E_i$ (in $k(U)$) is of diameter less than the minimum value of f'' on $\operatorname{Bd} E_i$, diameter $k^{-1}(\operatorname{Bd} E_i) <$ maximum $4f''$ on $k^{-1}(\operatorname{Bd} E_i)$ as shown above and $k^{-1}(\operatorname{Bd} E_i)$ lies in a disk in $h'(V)$ of diameter less than the minimum value of f' on this disk. This disk contains W.

The disk in $h'(V)$ which contained W could be adjusted so that its boundary lies in $h(V \times (-1, 1))$. Let $G_1, G_2, \cdots$ be a collection of such disks which covers $h'(V) - h(V \times (-1,1))$ such that the collection is locally finite at each point of $h'(V)$. The G's are adjusted (some are thrown away and others chopped up) so that no two of the resulting disks intersect each other. The D's promised by Theorem 17 are the images under h'^{-1} of these adjusted G's.

THEOREM 18. *Under the hypotheses of Theorem* 17 *there is a locally* *piecewise linear homeomorphism h of $V \times [-1, 1]$ into E^3 such that:*

(1) $\rho(h'(p), h_t(p)) < f(p)$ $(p \in V, -1 \leqq t \leqq 1)$,

(2) *$h'(V)$ lies except for a family of mutually exclusive disks in $h(V \times (-1, 1))$ such that each of these disks has diameter less than the minimum value of f on its inverse in V under h'^{-1}, and*

(3) *$h(V \times -1)$ lies except for a family of mutually exclusive disks in one component of $E^3 - h'(S)$ while $h(V \times 1)$ lies except for a family of mutually exclusive disks in the other component of $E^3 - h'(S)$ where each of the disks is of diameter less than the minimum value of f on its inverse in V.*

In fact, if K is a closed subset of V on which h' is locally piecewise linear, we may choose h so that $h_0 = h'$ on K.

PROOF. At first we ignore K and later show how to adjust our argument to take care of it.

We show that Theorem 18 follows from Theorem 17 in much the same way that we showed in § 7 that Theorem 16 follows from Theorem 1. We define a new function f_2 on V so that if Theorem 17 is satisfied with respect to this new function, then $h(V \times -1)$ and $h(V \times 1)$ automatically lie except for small disks in different components of $E^3 - h'(S)$.

Let a, b be points in different components of $E^3 - h'(S)$ and ab be an arc from a to b in $E^3 - h'(S - V)$. Let f_1 be a continuous positive function on V such that $f_1 < f/6$, $f_1 < \rho(ab, h'(S - V))$, $f_1 < \rho(a, h'(S))/2$, $f_1 < \rho(b, h'(S))/2$, and each point p of V lies in a disk D in V such that:

(1) diameter $h'(D) < f(p)/6$,

(2) maximum f on D/minimum f on $D < 2$, and

(3) $h'(D)$ contains each point of $h'(S)$ within $5f_1(p)$ of $h'(p)$.

Let f_2 be a positive continuous function on V such that

$$f_2 < f_1$$

and

$$1/2 < f_1(p)/f_1(q) < 2 \qquad \text{if } \rho(h'(p), h'(q)) < f_2(p) .$$

Suppose h is a homeomorphism of $V \times [-1, 1]$ into E^3 such that $\rho(h'(p), h(p \times t)) < f_2(p)$ and $h'(V)$ contains a family of mutually exclusive disks $D_1, D_2, \cdots$ such that $\{D_i\}$ is locally finite at each point of $h'(V)$, diameter $D_i <$ minimum value of f_2 on $h'^{-1}(D_i)$ and $h'(V) - \sum D_i \subset h(V \times (-1, 1))$. We show that $h(V \times -1) = h_{-1}(V)$ and $h(V \times 1) = h_1(V)$ lie except for small disks in different components of $E^3 - h'(S)$.

Let X be a component of $h'(V) \cdot h_1(V)$ and p be a point of $h'^{-1}(X)$. We show that there is a disk E in $h_1(V)$ such that $X \subset E$ and diameter $E <$ minimum f on $h_1^{-1}(E)$. We note that diameter $X < f_2(p) < f_1(p)$.

It follows from the definition of f_1 that there is a disk D in V such that:
(1) diameter $h'(D) < f(p)/6$,
(2) maximum f on D/minimum f on $D < 2$, and
(3) $h'(D)$ contains each point of $h'(S)$ within $5f_1(p)$ of $h'(p)$.
The disk E promised is $h_1(D)$.

We note that

$$\begin{aligned} \text{diameter } h_1(D) &< \text{diameter } h'(D) + 2 \text{ maximum } f_2 \text{ on } D \\ &< f(p)/6 + 2 \text{ maximum } f/6 \text{ on } D \\ &< \text{maximum } f/2 \text{ on } D < \text{minimum } f \text{ on } D. \end{aligned}$$

Hence diameter $E <$ minimum f on $h_1^{-1}(E)$.

We now show that each point $h_1(q)$ of X is contained in E. Let $q' = h'^{-1}h_1(q)$. Then $h_1(q) = h'(q')$ and

$$\rho(h'(q), h'(q')) < f_2(q)$$

by definition of h_1. Hence

$$f_1(q)/f_1(q') < 2 .$$

Since $h'(q')$ and $h'(p)$ both belong to X,

$$\rho(h'(p), h'(q')) < f_2(p)$$

and

$$f_1(q')/f_1(p) < 2 .$$

Note that

$$\begin{aligned} \rho(h'(p), h'(q)) &< f_2(p) + f_2(q) < f_1(p) + f_1(q) \\ &< f_1(p) + 2f_1(q') < f_1(p) + 4f_1(p) = 5f_1(p) . \end{aligned}$$

The last inequality insures that q lies in D. Hence $h_1(q)$ lies in $h_1(D) = E$. Since q was an arbitrary point of $h_1^{-1}(X)$, this shows that $X \subset E$.

Since each component of $h_1(V) \cdot h'(V)$ lies in such a disk E in $h_1(V)$, it follows that there is a family of mutually exclusive disks $E_1, E_2, \cdots$ in $h_1(V)$ such that:
(1) $\{E_i\}$ covers $h_1(V) \cdot h'(V)$,
(2) $\{E_i\}$ is locally finite at each point of V, and
(3) diameter $E_i <$ minimum value of f on $h_1^{-1}(E_i)$.
Similarly $h_{-1}(V)$ contains a family of mutually exclusive disks $F_1, F_2, \cdots$ satisfying similar conditions with h_{-1} replacing h_1.

By an argument similar to that given in Theorem 15 it follows that there is an arc axb from a to b that misses $\sum D_i + \sum E_i + \sum F_i + h'(S - V)$. Notice that neither $h_1(V)$ nor $h_{-1}(V)$ intersects $h'(S - V)$ because of the definition of f_1 and f_2. Let a' be the first point of $h(V \times [-1, 1])$ on axb

in the order from a to b and b' be the last. One of a', b' belongs to $h_1(V) - \sum E_i$ and the other belongs to $h_{-1}(V) - \sum F_i$. Since each of $h_1(V) - \sum E_i$, $h_{-1}(V) - \sum F_i$ is connected, they belong to different components of $E^3 - h'(S)$.

Finally we return to a consideration of K and show that we could have defined h so that $h_0 = h'$ on K. Note that if h' is locally piecewise linear at a point, it is locally piecewise linear in a neighborhood of the point. Hence, there is no loss of generality in supposing that K is locally polyhedral at each point of K. In the proof of Theorem 17 we used a triangulation R of $E^3 - h'(S - V)$ so that the simplexes of R are small. Suppose we had used a triangulation R' with just as stringent restrictions on the sizes of its simplexes and adjusted so that if T is a 3-simplex of R' which intersects $h'(K)$, then $h'(K)\cdot T$ is the sum of a finite collection of mutually exclusive polyhedral disks $D_1, D_2, \cdots, D_n$ such that $\operatorname{Int} D_i \subset \operatorname{Int} T$, $\operatorname{Bd} D_i \subset \operatorname{Bd} T$, and for each face F of T the components of $F \cdot \operatorname{Bd} D_i$ are spanning arcs A_j of F. Then when we came to replace X_j by an arc A_j (as done in Step 6 of the *very special case* of the proof of Theorem 1), we would replace A_j by itself and when we defined h_0 on D_i (as in Step 9 of the *very special case* of the proof of Theorem 1) we would take it to be the identity.

There is one final precaution to take to insure that $h_0 = h'$ on K. There might be a component of K which is on a feeler as shown in Figure 2 or Figure 5 if we did not take precautions to prevent this. We avoid this by taking a subdivision R'' of R' such that no 3-simplex of R' that intersects K is subdivided but others are subdivided so that no 3-simplex of R'' which misses F together with a narrow tubular neighborhood N of the 1-skeleton of R'' separates any point of K from $h'(\operatorname{Bd} V)$ on $h'(\bar{V})$. If we use the triangulation R'' to define h as done in the proof of Theorem 17 using the precaution on h_0 given in the preceding paragraph, we obtain a homeomorphism h of $V \times [-1, 1]$ into E^3 satisfying the conditions of Theorem 18.

UNIVERSITY OF WISCONSIN

REFERENCES

1. J. W. ALEXANDER, *An example of a simply connected surface bounding a region which is not simply connected*, Proc. Nat. Acad. Sci., U. S. A., 19 (1924), 8-10.
2. R. H. BING, *Approximating surfaces with polyhedral ones*, Ann. of Math., 65 (1957), 456-483.
3. ———, Approximating surfaces with polyhedral ones, Summary of Lectures and Seminars, Summer Institute on Set Theoretic Topology, Madison, 1955, pp. 47-53.

4. ———, *Locally tame sets are tame*, Ann. of Math., 59 (1954), 145–158.
5. ———, *A simple closed curve that pierces no disk*. J. Math. Pures Appl., 35 (1956), 337–343.
6. ———, *Conditions under which a surface in E^3 is tame*, Fund. Math., 47 (1959), 105–139.
7. ———, *An alternative proof that 3-manifolds can be triangulated*, Ann. of Math., 69 (1959), 37–65.
8. ———, *The Kline sphere characterization problem*, Bull. Amer. Math. Soc., 52 (1946), 644–653.
9. ———, *A homeomorphism between the 3-sphere and the sum of two solid horned spheres*, Ann. of Math., 56 (1952), 354–362.
10. ———, *A wild surface each of whose arcs is tame*, Duke J., 28 (1961), 1–16.
11. ———, *A surface is tame if its complement is 1-ULC*, Trans. Amer. Math. Soc., 101 (1961), 294–305.
12. ———, *E^3 does not contain uncountably many mutually exclusive wild surfaces*, Abstract 63-801t, Bull. Amer. Math. Soc., 63 (1957), 404.
13. ———, *Pushing a 2-sphere into its complement*, Abstract 564-165, Amer. Math. Soc. Notices, 6 (1959), 838.
14. ———, *Necessary and sufficient conditions that a 3-manifold be S^3*, Ann. of Math., 68 (1958), 17–37.
15. ———, *Each 2-sphere in E^3 contains a tame arc*, Abstract 559-19 Amer. Math. Soc. Notices, 6 (1959), 405. (Submitted to Amer. J. Math.)
16. ———, *Each disk in each 3-manifold is pierced by a tame arc*, Abstract 559-114, Amer. Math. Soc. Notices, 6 (1959), 510. (Submitted to Amer. J. Math.)
17. R. H. FOX and E. ARTIN, *Some wild cells and spheres in three-dimensinal space*, Ann. of Math., 49 (1948), 979–990.
18. O. G. HARROLD, JR. *Some consequences of the approximation theorem of Bing*, Proc. Amer. Math. Soc., 8 (1957), 204–206.
19. E. E. MOISE, *Affine structures in 3-manifolds*, IV. *Piecewise linear approximations of homeomorphisms*, Ann. of Math., 55 (1952), 215–222.
20. ———, ———, V. *The triangulation theorem and Hauptvermutung*, Ann. of Math., 56 (1952), 96–114.
21. ———, ———, VIII. *Invariance of the knot type; local tame embedding*, Ann. of Math., 59 (1954), 159–170.
22. R. L. MOORE, *Concerning upper semicontinuous collections of continua*, Trans. Amer. Math. Soc., 27 (1925), 416–428.
23. ———, Foundations of Point Set Theory, Amer. Math. Soc. Colloquium Publications, vol. 13, New York, 1932.
24. R. L. WILDER, *A converse of a theorem of R. H. Bing and its generalization*, Fund. Math., to appear.
25. ———, Topology of Manifolds, Amer. Math. Soc. Colloquium Publications, vol. 32, New York, 1949.

EMBEDDING SURFACES IN 3-MANIFOLDS

By R. H. BING

Is a 2-sphere in E^3 tame if it can be pierced at each point by a straight line interval?(1) Is it tame if it can be pierced along each arc in it by a tame disk?

Is a 2-sphere in E^3 tame if it is homogeneous under a space homeomorphism?

Is a 2-sphere in E^3 tame if each Sierpinski curve in it can be pushed to either side? Is it tame if it can be approximated from either side by a singular 2-sphere?

Does one obtain E^3 if the interior of a wild 2-sphere is replaced by the interior of a tame one?

What restrictions imply that a surface is tamely embedded in a 3-manifold?

These are samples of some of the interesting unsolved problems about the embedding of surfaces in 3-manifolds.

In addition to raising questions about the embedding of surfaces in 3-manifolds, recent discoveries were discussed. A surface is known to be tame if its complement is 1-ULC. It is tame if it can be pushed to either side with a map. It is tame if it can be homeomorphically approximated from either side. The fact that surfaces can be approximated by polyheral surfaces has been useful in proving some of these results and may be the means for learning more about the embedding of surfaces in 3-manifolds.

One advantage of working with polyhedral surfaces instead of arbitrary ones is that they intersect in decent sets. However, it has been shown recently that if in E^3, L is a straight line, S is a 2-sphere, and $\varepsilon > 0$, then there is a homeomorphism h of E^3 onto itself such that h does not move any point more than ε, h is fixed outside an ε-neighborhood of L, and $S \cdot h(L)$ does not have infinitely many points. A simple proof of this result would improve the proof of the approximation theorems. Suppose S_1, S_2 are two arbitrary surfaces in E^3. One might wonder if it is possible to move S_1 slightly with a space homeomorphism so that the resulting 2-sphere intersects S_2 "nicely".

Hand in hand with proofs go counterexamples. Examples have recently been given of wild spheres all of whose arcs are tame. It has been shown however that all wild surfaces contain tame arcs and can be pierced by tame arcs. The set of points that belong to tame arcs in the surfaces is precisely the set of points of the surface at which it can be pierced by tame arcs.

Does a simple closed curve bound a disk in a 3-manifold if the curve can be shrunk to a point in its own complement? This is an unsolved version of Dehn's lemma.

If a singular plane P in E^3 links two points in E^3, does each ε-neighbor-

(1) Following the lecture, M. K. Fort, Jr. answered the first question raised in this lecture in the negative by giving an example of a wild 2-sphere (which might be called a "wild porcupine") which could be pierced at each point by a straight line interval. The example is described in a paper submitted to the *Proc. Amer. Math. Soc.*

hood of P contain a non-singular plane that links the two points? This version of the sphere theorem remains unanswered partly because in Papakyriakopoklos' proof of Dehn's lemma, the splitting and resewing along double lines introduced long feelers. Suppose D is a disk on a polyhedral 2-sphere S and $\varepsilon > 0$. Is there a positive number $\delta > 0$ such that if g is a piecewise linear map of S into E^3 that is fixed on $S - D$ and moves no point more than δ, then each neighborhood of $g(S)$ contains a 2-sphere whose homeomorphic distance from S is less than ε?

A contemplation of E^3 and a study of recent discoveries about 3-manifolds reveals that although we have learned much about 3-space in recent years, there is much yet to learn.

PUSHING A 2-SPHERE INTO ITS COMPLEMENT

R. H. Bing

1. INTRODUCTION

We investigate the extent to which a closed set in E^3 can be slightly pushed to one side of a surface. One of the main results (Theorem 2.1) states that if U is a complementary domain of a 2-sphere S (possibly wild) in E^3, then $\overline{U}$ can be slightly shoved into U plus a Cantor set. We show in Theorem 5.1 that for each $\varepsilon > 0$ there exists a Cantor set C on S and an ε-map $f: S \to U + C$ that takes S - C homeomorphically into U. This and other related results are given for surfaces in 3-manifolds as well as for 2-spheres in E^3.

Wilder has given a converse for Theorem 5.1 in which he shows [19] that a compact locally connected continuum S in E^3 is a topological 2-sphere if its first homology is trivial, if S separates E^3, and if for each $\varepsilon > 0$ and each component U of E^3 - S there exists a Cantor set C on S and an ε-map $f: S \to U + C$. He gives a corresponding characterization for 2-manifolds, where instead of supposing that the first homology of S is trivial, he supposes it is finitely generated, and he insists furthermore that C does not locally separate S.

For a 2-sphere S in E^3 we use Int S and Ext S to denote respectively the bounded and the unbounded components of E^3 - S. In case D is a cell, we use Bd D to denote the combinatorial boundary of D, and Int D to denote D - Bd D.

The distance function is denoted by D. In case f, g are maps of a set A into a metric space, we use D(f, g) to denote the least upper bound of D(f(a), g(a)), $a \in A$. We call a map f an *ε-map* if $D(f, I) \leq \varepsilon$, where I is the identity map. A *null sequence* of sets is a sequence of sets whose diameters converge to zero. We use $V(X, \varepsilon)$ to denote the set of all points q whose distance from X is less than ε. We call $V(X, \varepsilon)$ the *ε-neighborhood of* X, but we note that this is only a special sort of neighborhood of X. In general, we call any open set containing X a *neighborhood* of X.

A subset X of E^3 (or of a triangulated manifold M) is called *tame* if there exists a homeomorphism $h: E^3 \to E^3$ (or $M \to M$) such that h(X) is a polyhedron. A closed set which is a topological complex but for which there exists no such homeomorphism is called *wild.* They say that X is *locally tame* at a point p of X if there exists a neighborhood N of p and a homeomorphism h of $\overline{N}$ onto a combinatorial cell that takes $\overline{N} \cdot X$ onto a polyhedron. We say that X is *locally tame* mod K if X is locally tame at each point of X - K. A *finite graph* is the sum of a finite number of arcs (topological segments) such that if two of the segments intersect each other, the intersection is an end point of each.

There are several criteria for determining whether or not a 2-sphere in E^3 is tame [1, 3, 4, 8, 9, 11, 13, 14, 16]. One of the most useful of these [4] says that a 2-sphere is tame if its complement is 1-ULC. Using I^m to denote an m-cell, we say that a set Y is n-ULC if for each $\varepsilon > 0$ there exists a $\delta > 0$ such that each map of Bd I^{n+1} into a δ-subset of Y can be extended to map I^{n+1} into an ε-subset of Y.

Received June 10, 1963.

Research on this paper was supported by contracts NSF G21514 and NSF G11665.

We say that a set Y is *locally simply connected* at a point p of $\overline{Y}$ if for each neighborhood N_1 of p there exists a neighborhood N_2 of p such that each map of Bd I^2 into $N_2 \cdot Y$ can be extended to map I^2 into $N_1 \cdot Y$.

Suppose S is a 2-sphere and $D_1, D_2, \cdots$ is a null sequence of mutually exclusive disks in S such that ΣD_i is dense in S. Any set homeomorphic with $S - \Sigma \text{Int } D_i$ is called a *Sierpiński curve* or *universal plane curve.* A Sierpiński curve X in E^3 is called *tame* if there exists a homeomorphism $h: E^3 \to E^3$ such that $h(X)$ lies in a horizontal plane. A point in the image of one of the Bd D_i's is called an *accessible point* of the Sierpiński curve, and an image point of $S - \Sigma D_i$ is called an *inaccessible point.* Accessible points have the property that they lie on arcs in the Sierpiński curve that do not locally separate it, but inaccessible points do not have this property. Hence, the property of being an accessible point is a topological property (Theorem 3.2 of [6]). It has been shown [18] that any two Sierpiński curves are homeomorphic, and that if X_1, X_2 are two Sierpiński curves in the same plane P, then there exists a homeomorphism of P onto itself taking X_1 onto X_2.

In the last section of this paper we prove some theorems about tame Sierpiński curves. It was shown in [5] that for each 2-sphere S (possibly wild) in E^3 there exists a tame Sierpiński curve X in S such that each component of $S - X$ is small. We show in Section 9 that certain sets in S can be buried in such tame Sierpiński curves. This enables one to construct tame arcs in S with great abundance and some precision. Tame arcs may be extended to bigger tame arcs. Tame arcs are accessible at interior points from either side with other tame arcs. Triangulations of S with tame 1-skeletons can be refined by other triangulations with tame 1-skeletons. David Gillman [10] and Joseph Martin have obtained results of this type.

C. E. Burgess read an early version of this paper, and the author is indebted to him for some helpful comments.

2. FREEING ALL BUT A CANTOR SET

THEOREM 2.1. *Suppose* S *is a 2-sphere in* E^3 *and* U *is a component of* $E^3 - S$. *Then for each* $\varepsilon > 0$ *there exists a Cantor set* C *on* S *and a map* $f: \overline{U} \to U + C$ *such that*

$$D(f, I) < \varepsilon,$$

$$f = I \text{ on } U - V(S, \varepsilon), \text{ and}$$

$$f \text{ is a homeomorphism on } \overline{U} - f^{-1}(C).$$

Proof. We shove $\overline{U}$ "almost" into U with a sequence of shoves. At intermediate stages it is permitted that parts of $\overline{U}$ be moved into $E^3 - \overline{U}$ provided they go into certain controlled neighborhoods of certain open 2-simplexes on S. The theorem will follow from one application of Theorem 3.3 of the next section and then repeated applications of Theorem 3.2.

Let T_1 be a curvilinear triangulation of S of mesh less than $\varepsilon/4$ such that the 1-skeleton K_1 of T_1 is tame. That there exists such a triangulation follows from [5]. Let U_1 be an open set containing $S - K_1$ such that each component of U_1 is of diameter less than $\varepsilon/4$, and such that for each component U_{1i} of U_1 there exists a 2-simplex D_{1i} of T_1 with Int $D_{1i} \subset U_{1i}$ and $S \cdot \overline{U}_{1i} = D_{1i}$. Note that $U_1 \cdot K_1 = 0$.

We find from Theorem 3.3 that there exists a homeomorphism

$$f_1: U_1 + U + K_1 \to U_1 + U$$

such that $D(f_1, I) < \varepsilon/2$. In applying Theorem 3.3, we let T_1 be both the T_1 and T_2 of the statement of that theorem, and U_1 both the U_1 and the U_2, and we let U be $\text{Ext } S$.

Let T_2' be a curvilinear triangulation of S of mesh less than $\varepsilon/8$ such that T_2' refines T_1 and the 1-skeleton K_2' of T_2' is tame. That there exists such a refinement follows from Theorem 9.2 of the last section of this paper. Let U_2' be an open set containing $S - K_2'$ such that $U_2' \subset U_1$, each component of U_2' is of diameter less than $\varepsilon/8$, and for each component U_{2i}' of U_2' there exists a 2-simplex D_{2i}' of T_2' with $\text{Int } D_{2i}' \subset U_{2i}'$ and $S \cdot \overline{U}_{2i}' = D_{2i}'$.

It follows from Theorem 3.2 that there exists a homeomorphism

$$f_2: U_1 + U + K_1 \to U_2' + U + K_1$$

such that $f_2 = I$ on $(U + K_1) - U_1$ and f_2 takes each component of U_1 into itself. Since $f_1(\overline{U}) \subset U_1 + U$, $f_2 f_1(\overline{U}) \subset U_2' + U$. Since each component of U_1 is of diameter less than $\varepsilon/4$, $D(f_2, I) < \varepsilon/4$.

The homeomorphisms f_1, f_2 we have defined are the first two terms of a sequence of homeomorphisms $f_1, f_2, f_3, \cdots$ such that $D(f_i, I) < \varepsilon/2^i$. Hence the sequence $f_1, f_2 f_1, f_3 f_2 f_1, \cdots$ converges to a continuous map f that moves no point as much as ε. We note that since $f_2 f_1(\overline{U}) \subset U_2' + U$, $K_1 \cdot f_2 f_1(\overline{U}) = 0$. We wish to preserve this property for the limit map f, and with this end in mind we subdivide T_2' before reapplying Theorem 3.2. If we were to define f_3 before subdividing, we might pull points of $f_2 f_1(\overline{U})$ near K_1.

Let T_2 be a curvilinear triangulation of S such that T_2 refines T_2', the 1-skeleton K_2 of T_2 is tame, and for each 2-simplex D_{2i}' of T_2' there is a core 2-simplex D_{2i} of T_2 such that

$$D_{2i}' \cdot f_2 f_1(\overline{U}) \subset \text{Int } D_{2i} \subset D_{2i} \subset \text{Int } D_{2i}'.$$

We pick such a core D_{2i} for D_{2i}' even if $D_{2i}' \cdot f_2 f_1(\overline{U}) = 0$. Let U_2 be an open set containing $S - K_2$ such that $U_2 \subset U_2'$, such that $U_{2i}' \cdot f_2 f_1(\overline{U}) \subset U_{2i}$ (where U_{2i} is the component of U_2 associated with the core 2-simplex D_{2i} of T_2 in $\text{Int } D_{2i}'$), and such that for each component U_{2j} of U_2 there exists a 2-simplex D_{2j} of T_2 with $\text{Int } D_{2j} \subset U_{2j}$ and $S \cdot \overline{U}_{2j} = D_{2j}$. Note that $f_2 f_1(\overline{U}) \subset U_2 + U$. When we apply f_3, we shall not pull points of $f_2 f_1(\overline{U})$ near K_1 since f_3 reduces to I on $f_2 f_1(U) - U_2$ and takes each component of U_2 onto itself.

We are now ready to define f_3. Skipping details of how we go from T_3' to T_3, we find the following: a triangulation T_3 of S that refines T_2 so that the 1-skeleton K_3 of T_3 is tame; an open set U_3 containing $S - K_3$ such that $U_3 \subset U_2$, each component of U_3 is of diameter less than $\varepsilon/16$, and for each such component U_{3i} there exists a 2-simplex D_{3i} of T_3 with $\text{Int } D_{3i} \subset U_{3i}$ and $S \cdot \overline{U}_{3i} = D_{3i}$; a homeomorphism $f_3: f_2 f_1(\overline{U}) \to U_3 + U$ such that

$f_3 = I$ on $f_2 f_1(\overline{U}) - U_2$,

$f_3(U_{2i} \cdot f_2 f_1(\overline{U})) \subset U_{2i}$ for each component U_{2i} of U_2, and

$f_3 f_2 f_1(\overline{U}) \cdot U_{3i} = 0$ if $D_{3i} \cdot K_2 \neq 0$.

Continuing in this fashion, we define T_4, K_4, U_4, f_4, T_5, K_5, U_5, f_5, $\cdots$. As we noted previously, f_1, $f_2 f_1$, $f_3 f_2 f_1$, $\cdots$ converges to a continuous map f. It takes $\overline{U}$ into $\overline{U}$. Since $f(\overline{U})$ misses each K_i, $f(\overline{U}) \cdot S$ is at most a closed 0-dimensional set C' (which in turn lies in a Cantor set C). Since f_i is a homeomorphism and is the identity except very near S, f is a homeomorphism on $\overline{U} - f^{-1}(C)$.

3. FREEING 1-SKELETONS

In this section we take care of some housecleaning details and justify some of the steps used in the proof of Theorem 2.1. For concreteness of treatment we deal with Ext S rather than the arbitrary set U of the previous section.

THEOREM 3.1. *Suppose*

S *is a 2-sphere in* E^3,

T_1 *is a curvilinear triangulation of* S *whose 1-skeleton* K_1 *is tame, and*

U_1 *is an open set containing* $S - K_1$ *such that for each component* U_{1i} *of* U_1 *there exists a 2-simplex* D_{1i} *of* T *so that* $\text{Int } D_{1i} \subset U_{1i}$ *and* $S \cdot \overline{U}_{1i} = D_{1i}$.

Then for each open set V *containing* $S - K_1$, *there exists a homeomorphism*

$$f: U_1 + \text{Ext } S + K_1 \rightarrow V + \text{Ext } S + K_1$$

such that $f = I$ *on* $\text{Ext } S + K_1 - U_1$.

Proof. Let g be the continuous real-valued function defined on S by the relation

$$g(p) = \text{minimum } (D(p, E^3 - U_1), D(p, E^3 - V)).$$

It follows from Theorem 7 of [2] that there exists a homeomorphism h of S into $K_1 + V$ such that for each point p of $S - K_1$,

$$D(p, h(p)) < g(p) \text{ and}$$

$$h(S) \text{ is locally tame at } h(p).$$

It follows from [9] that h(S) is tame, and from Theorem VI 10 of [15] that

$$V + \text{Ext } S = V + \text{Ext } h(S) \supset \text{Ext } h(S).$$

We now describe a homeomorphism

$$f: U_1 + \text{Ext } S + K_1 \rightarrow \text{Ext } h(S) + K_1$$

satisfying the conclusion of Theorem 3.1. The homeomorphism f is fixed except on the U_{1i} and takes each of these into itself, so we define only the part f_i of f restricted to U_{1i}.

Let abcd be a solid tetrahedron in E^3 and p_i a point of $h(S) - (K_1 + D_{1i})$. Since h(S) is tame, there exists a homeomorphism $h_i: E^3 \rightarrow E^3$ such that

$$h_i h(S) = \text{Bd abcd}, \quad h_i h(D_{1i}) = \text{abc}, \quad h_i h(p_i) = d.$$

By pushing some points radially away from d, one finds that there exists a homeomorphism $f_i': E^3 \to E^3$ such that

$$f_i' = I \text{ outside Int abcd} + h_i(V \cdot U_{1i}) \text{ and}$$

$$f_i' h_i(U_{1i}) \subset E^3 - \text{abcd} = \text{Ext Bd abcd}.$$

Then

$$f_i = h_i^{-1} f_i' h_i \text{ on } U_{1i}.$$

THEOREM 3.2. *Suppose* S, T_1, K_1, U_1 *are as in Theorem* 3.1, *and* T_2 *is a curvilinear triangulation of* S *such that* T_2 *refines* T_1 *and has a tame* 1-*skeleton* K_2.

Then for each open set U_2 *containing* $S - K_2$ *there exists a homeomorphism*

$$f: U_1 + \text{Ext } S + K_1 \to U_2 + \text{Ext } S + K_1$$

such that

$$f = I \textit{ on } K_1 + \text{Ext } S - U_1 \textit{ and}$$

$$f \textit{ takes each component of } U_1 \textit{ into itself.}$$

Proof. This result follows from the proof of Theorem 3.1 if we insist that $K_2 \subset h(S) \subset U_2 + K_2$; so we suppose this. The homeomorphism f is the homeomorphism f of Theorem 3.1.

The theorem is true without the hypothesis that T_2 refines T_1, but we made the simplifying assumption because it is part of the context in which we used the theorem.

THEOREM 3.3. *Suppose* $S, T_1, K_1, U_1, T_2, K_2, U_2$ *are as in Theorem* 3.2. *Then for each* $\varepsilon > 0$ *there exists a homeomorphism*

$$f': U_1 + \text{Ext } S + K_1 \to U_2 + \text{Ext } S$$

such that

$$f' = I \textit{ on } \text{Ext } S - V(S, \varepsilon),$$

$$f'(U_{1i}) \subset V(U_{1i}, \varepsilon) \textit{ for each component } U_{1i} \textit{ of } U_1, \textit{ and}$$

$$f' \textit{ moves no point of } (K_1 + \text{Ext } S) - U_1 \textit{ more than } \varepsilon.$$

Proof. The homeomorphism f' is the homeomorphism f of Theorem 2 followed by a homeomorphism that shoves h(S) + Ext h(S) into Ext h(S).

Of course, Theorems 3.1, 3.2, 3.3 hold as well for Int S as for Ext S.

4. SIMPLE CONNECTIVITY OF THE COMPLEMENT OF A 2-SPHERE

The following is an application of Theorem 2.1.

THEOREM 4.1. *Suppose* S *is a 2-sphere in* E^3. *For each* $\varepsilon > 0$ *there exists a* $\delta > 0$ *and a sequence of Cantor sets* $C_1, C_2, \cdots$ *on* S *such that if* J *is a simple closed curve in* $E^3 - S$ *of diameter less than* δ *such that* $D(J, S) > 1/i$, *then* J *can be shrunk to a point on an* ε*-subset of* $(E^3 - S) + C_i$.

In addition the following theorem holds.

THEOREM 4.2. *Suppose* S *is a 2-sphere in* E^3 *and* U *is a component of* $E^3 - S$. *Then there exists a 0-dimensional* F_σ*-set* F *on* S *such that* U + F *is* 1-ULC.

Proof. It follows from repeated applications of Theorem 4.1 that an F_σ-set F exists such that to each $\varepsilon > 0$ there corresponds a $\delta > 0$ such that, for each disk D, each map of Bd D into a δ-subset of U can be extended to map D into an ε-subset of U + F. We show that U + F is 1-ULC by showing that each map f of Bd D into a $(< \delta)$-subset of U + F can be extended to map D into a 3ε-subset of U + F.

We suppose that D is a round planar disk and that the closed set $f^{-1}(F)$ is 0-dimensional. Let $p_1 q_1, p_2 q_2, \cdots$ be a null sequence of mutually exclusive secants for D such that the closure of each component of $D - \Sigma p_i q_i$ is a disk which misses $f^{-1}(F)$, and such that f can be extended to map $\text{Bd } D + \Sigma p_i q_i$ into a subset of U + F with each $f(p_i q_i)$ contained in U. That the $p_i q_i$'s can be chosen so that f can be so extended follows from the fact (Theorem 5.35 in Chapter II of [20]) that U is 0-ULC. Only a few of the $f(p_i q_i)$ will have large diameters, and we discard them so that the diameter of the image of Bd D and the remaining $p_i q_i$'s is less than δ. For convenience in notation we suppose none were thrown away. Since F can be extended to map each component of $D - \Sigma p_i q_i$ into a small subset of U + F (in no case requiring an image with a diameter as much as ε but in most cases requiring a much smaller image), f can be extended to map D into a subset of U + F that lies in an ε-neighborhood of $f(\text{Bd } D + \Sigma p_i q_i)$.

5. PUSHING A 2-SPHERE "ALMOST" TO THE SIDE

THEOREM 5.1. *Suppose* S *is a 2-sphere in* E^3 *and* U *is a component of* $E^3 - S$. *Then for each* $\varepsilon > 0$ *there exists a Cantor set* C *on* S *and a map* g *of* S *into* U + C *such that* $D(g, I) < \varepsilon$ *and* g *takes* S - C *homeomorphically into* U.

Proof. Let δ be a positive number so small that each δ-subset of S lies in an $\varepsilon/2$-disk in S.

It follows from Theorem 2.1 that there exists a Cantor set C_1 on S and a map f of S into $U + C_1$ such that $D(f, I) < \delta/2$ and f is a homeomorphism on $S - f^{-1}(C_1)$. Each component of $f^{-1}(C_1)$ is of diameter less than δ and lies in an $\varepsilon/2$-disk on S.

As pointed out in Theorem 9 of [7], there exists a component Y of $S - f^{-1}(C_1)$ such that each component of S - Y is of diameter less than $\varepsilon/2$. Also, as pointed out there, there exists a map f_1 of $\overline{Y}$ onto S such that $D(f_1, I) < \varepsilon/2$, and such that for each point p of S, $f_1^{-1}(p)$ is either a point of Y or a component of $\overline{Y} - Y$.

Let C_2 be a Cantor set in S that contains $f_1(\overline{Y} - Y)$. Then a C and a g satisfying the conclusion of Theorem 5.1 are

$$C = C_1 + C_2 \quad \text{and} \quad g(p) = ff_1^{-1}(p).$$

Questions. Is a 2-sphere S in E^3 tame if for each $\varepsilon > 0$ and each component U of E^3 - S there exists an ε-map f: S → U? Hempel has shown [14] that S is tame if there exists a homotopy $H_t(s)$ $(0 \leq t \leq 1,\ s \in S)$ of S into S + U such that $H_0 = I$ and $H_t(S) \subset U$ for $t > 0$. Is S tame if for each Sierpiński curve X on S there exists a homotopy H_t of X into X + U such that $H_0 = I$ and $H_t(X) \subset U$ for $t > 0$? Burgess has shown [8] that S is tame if there exists such an H_t and if in addition each H_t is a homeomorphism and $H_{t_1}(X) \cdot H_{t_2}(X) = 0$ if $t_1 \neq t_2$.

6. PUSHING TO THE SIDES OF SURFACES

In the preceding sections we have dealt with 2-spheres in E^3, but since we were dealing with them only locally, our results apply equally well to arbitrary two-sided 2-manifolds embedded in 3-manifolds.

THEOREM 6.1. *Suppose*

M^3 *is a connected* 3-*manifold,*

M^2 *is a connected* 2-*manifold in* M^3 *that separates* M^3,

U *is a component of* $M^3 - M^2$, *and*

g *is a nonnegative continuous real function defined on* M^3 *such that* g *is positive on an open subset* V *of* M^2.

Then there exists a 0-*dimensional subset* C *of* V *and a map* $f: \overline{U} \to (\overline{U} - V) + C$ *such that*

$$D(x, f(x)) \leq g(x),$$

$$f \text{ is a homeomorphism on } \overline{U} - f^{-1}(C).$$

In fact, for each neighborhood N *of* V *there exists such an* f *that is the identity on* $\overline{U}$ - N.

Proof. We suppose g = 0 on M^2 - V.

The only change of approach used here over that used in the proof of Theorem 2.1 is to start with a triangulation T_1 of V such that each 1-simplex of T_1 is tame, each 2-simplex D_i of T_1 lies in an open 3-cell O_i of diameter less than the minimum value of g on O_i, and such that D_i lies in an open subset of M^2 which in turn lies on a 2-sphere in O_i. This is possible by Theorem 5 of [4]. We take small mutually exclusive open subsets U_i in the O_i's about the Int D_i's and shove as before—first freeing the 1-skeleton of T_1 and then in a countable number of steps freeing all but at most a Cantor set on each D_i.

The following result follows from the methods used in the proof of Theorem 4.2.

THEOREM 6.2. *Suppose* M^3, M^2, U *are as in Theorem* 6.1. *Then there exists a* 0-*dimensional* F_σ-*set* F *on* M^2 *such that* U + F *is locally simply connected at each point of* M^2.

Duplicating the arguments used in proving Theorem 5.1, we obtain the following extension of it.

THEOREM 6.3. *Suppose* M^3, M^2, U *are as in Theorem* 6.1. *Then for each* $\varepsilon > 0$ *there exists a closed* 0-*dimensional set* C *on* M^2 *and an* ε-*map* $f: M^2 \to U + C$ *such that* f *takes* M^2 - C *homeomorphically into* U.

7 FREE GROUPS

We now set forth some known results from algebra [12, Chapter 7], one of which we use to prove a topological theorem that will help us in the next section.

Consider a group expressed with generators and relations. An element of the group is an equivalence class of words where the letters of the words are generators and their inverses. If the group is free, two words belong to the same equivalence class if and only if one can be changed to the other with a finite number of operations, where each operation consists either of the insertion of two adjacent letters xx^{-1} somewhere in the word or the cancellation of two such letters. The following result shows that any word that can be reduced to the trivial word can be so reduced with cancellations alone.

THEOREM 7.1. *For a free group, any word that belongs to the equivalence class of the identity element contains two adjacent letters, one of which is the inverse of the other.*

Proof. Suppose w_1 is a word in the equivalence class of the identity. Then there exists a sequence $w_1, w_2, \cdots, w_n$ such that w_n has no letters and w_{i+1} is obtained from w_i by either adjacently inserting two letters xx^{-1} somewhere in the word or cancelling two such letters. We suppose that the operations of cancellation and insertion have been performed so as to minimize the number of steps in going from w_1 to w_n (the word with no letters). We show that in this case, each operation was a cancellation. This will establish the theorem.

Suppose not each operation was a cancellation. There exists a j such that the operation of going from w_j to w_{j+1} was not a cancellation but all the following operations were cancellations. With no loss of generality, we suppose that w_j is the shortest word in the equivalence class of the identity element with the property that

1) w_j cannot be reduced to w_n by cancellation alone, but
2) w_j can be changed, with one insertion, to a word that can be reduced to w_n by cancellation alone.

Suppose w_{j+1} results from w_j by inserting xx^{-1} somewhere, and w_{j+2} results from w_{j+1} by cancelling yy^{-1}. Since w_{j+2} differs as a word from w_j, neither of the letters inserted is one of the letters cancelled. Let w' be the word obtained from w_j by cancelling the same yy^{-1} as was cancelled in going from w_{j+1} to w_{j+2}. This leads to a contradiction because w' is shorter than w_j and has properties (1) and (2) mentioned above. Hence the theorem follows.

We do not need the following two results in this paper, but they are listed since they follow from the same methods and seem of interest.

THEOREM 7.2. *If* w *and* w' *are words in the same equivalence class of a free group, then they can be reduced to the same word by cancellation alone.*

Proof. Theorem 7.2 is an immediate consequence of Theorem 7.1 if we consider that ww'^{-1} is equivalent to the trivial word $w'w'^{-1}$.

Suppose $G = \{a_1, a_2, \cdots / r_1 = r_2 = \cdots = 1\}$ is a group and two words belong to the same equivalence class if one can be changed to the other with a finite sequence of operations of the following types:

1) Insert an $a_i a_i^{-1}$ or an $a_i^{-1} a_i$ somewhere.
2) Insert r_i or r_i^{-1} somewhere.

3) Cancel r_i or r_i^{-1} somewhere.

4) Cancel $a_i a_i^{-1}$ or $a_i^{-1} a_i$ somewhere.

There is no need of operations of type 3, since any such operation can be replaced by one of type 2 followed by a sequence of those of type 4. The following theorem shows that these can be done in a prescribed order.

THEOREM 7.3. *If two words are equivalent, it is possible to change the first to the second by first performing a finite number of operations of type* 1, *then a finite number of type* 2, *and finally a finite number of type* 4.

Proof. We prove the theorem by showing that if w_1 can be changed to w_2 by one operation of type 1, 2, or 4 (say of type j) and w_2 can be changed to w_3 by taking the operations in the prescribed order, then w_1 can be changed to w_3 by operations in the prescribed order.

Case 1. $j = 1$. This order of operations is already as prescribed.

Case 2. $j = 2$. Let $w_1 = v_1 v_2$ and $w_2 = v_1 r v_2$. Change $w_1 = v_1 v_2$ to $v_1 r^{-1} v_1^{-1} v_1 r v_2$ by a sequence of operations of type 1. Since $v_1 r^{-1} v_1^{-1}$ can be changed to 1 by one operation of type 2 and a sequence of type 4 (and by hypothesis $v_1 r v_2$ can be reduced to w_3 by operations in the prescribed order), interspersion of these operations reduces $v_1 r^{-1} v_1^{-1} v_1 r v_2$ to w_3 by a sequence of operations in the prescribed order.

Case 3. $j = 4$. Let $w_1 = v_1 x x^{-1} v_2$ and $w_2 = v_1 v_2$. Change $w_1 = v_1 x x^{-1} v_2$ to $v_1 x x^{-1} v_1^{-1} v_1 v_2$ by operations of type 1. Since each of $v_1 x x^{-1} v_1^{-1}$ and $v_1 v_2$ can be reduced by operations in the prescribed order, their product can also be reduced.

Finally we come to the topological theorem responsible for our excursion into group theory.

THEOREM 7.4. *Suppose* $D_1, D_2, \cdots, D_n$ *is a finite collection of mutually exclusive polyhedral disks in* E^3 *and* J *is a polygonal simple closed curve in general position with respect to* ΣD_i *that can be shrunk to a point in* $E^3 - \Sigma \operatorname{Bd} D_i$. *Then, if* J *intersects* ΣD_i, *there exists an arc* axb *in* J *that intersects* ΣD_i *only at its end points and such that* xa *and* xb *abut on the same* D_j *from the same side.*

Proof. We suppose, without loss of generality, that each D_i lies in the same horizontal plane. Theorem 7.4 then follows from Theorem 7.1, since the fundamental group of $E^3 - \Sigma \operatorname{Bd} D_i$ is a free group on n generators. A loop corresponds to a word, where a letter x of the word is represented by the loop crossing a D_i in one direction and the inverse x^{-1} of the letter is a crossing in the other direction. The letters in the word are ordered as the crossings of the D's by the loop. We note in the following paragraph why a trivial loop corresponds to a word in the fundamental group of $E^3 - \Sigma \operatorname{Bd} D_i$, which is in the equivalence class of the identity element.

Suppose E is an oriented 2-simplex, p_0 is a vertex of E, and f is a map of Bd E into $E^3 - \operatorname{Bd} D_i$ such that $f(p_0)$ goes to the starting point used in computing the fundamental group of $E^3 - \Sigma \operatorname{Bd} D_i$. The map under f of the positively oriented path around Bd E starting and ending at p_0 represents a loop. For each loop there corresponds a word whose letters are ordered in the order that the loop crosses the D_i's and the letter a_i or a_i^{-1} is assigned to the crossing according to whether the crossing is from below or above. If the loop can be shrunk to a point in $E^3 - \Sigma \operatorname{Bd} D_i$, then the map f can be extended to map E into $E^3 - \Sigma \operatorname{Bd} D_i$. We call this extended map f also, and we suppose that it is so nice that there exists a triangulation T of E such that f takes each 2-simplex of T homeomorphically onto a 2-simplex in E^3 without vertices on any of the D_i's. By shelling T, we get a sequence of disks

$E = E_1, E_2, E_3, \cdots, E_m$ such that E_m is a single 2-simplex of T and each E_{i-1} is the sum of E_i and such a single 2-simplex. We suppose that E_m misses each D_i. The word corresponding to f(Bd E_i) belongs to the same equivalence class as the word corresponding to f(Bd E_{i+1}), since each the words are the same or one word can be changed to the other by the single insertion or deletion of two letters of the sort xx^{-1}. This is because the boundary of each 2-simplex of T either misses a D_i or crosses it twice in opposite directions. Since f(Bd E_m) corresponds to the word with no letters, f(E) corresponds to a word in the equivalence class of the identity.

8. SPHERES LOCALLY TAME MODULO SIERPIŃSKI CURVES

In a certain sense the theorems of this section are generalizations of the result of Doyle and Hocking [9] that a 2-sphere is tame if it is tame modulo a tame finite graph. We use the results of this section only mildly in the rest of this paper (to help establish Theorem 9.2, which in turn was used in the proofs of Theorems 2.1 and 6.1), but the results have outside interest and will be used extensively in another paper to give a simplified proof of the result that any open subset of any two-sided surface in any 3-manifold can be "almost" approximated from either side.

THEOREM 8.1. *Suppose* X *is a tame Sierpiński curve that lies in a* 2*-sphere* S *(possibly wild) in* E^3. *Then for each* $\varepsilon > 0$ *there exists a* $\delta > 0$ *such that each simple closed curve of diameter less than* δ *in* E^3 - S *can be shrunk to a point on an* ε*-subset of* E^3 - X.

Proof. We suppose with no loss of generality that X lies in a horizontal plane P.

Since each component of P - X is 0-ULC and P - X contains only a finite number of such components with diameters greater than $\varepsilon/3$, there exists a $\delta > 0$ such that each pair of points of the same component of P - X whose distance apart is less than δ can be joined by an arc in P - X of diameter less than $\varepsilon/3$. We show that this is the δ promised by the theorem. Note that $\delta \leq \varepsilon/3$.

Suppose J_1 is a simple closed curve in E^3 - S of diameter less than δ. Without loss of generality we suppose that J_1 is polygonal and contains no vertex on P. We show that if N is the convex hull of the $\varepsilon/3$-neighborhood of J_1, then J_1 can be shrunk to a point in $N - N \cdot X$. In the easy case that J_1 misses P, we can shrink J_1 to a point in $N - N \cdot P$ and hence in $N - N \cdot X$. We show in any case that J_1 can be shrunk by reducing the number of points in $J_1 \cdot P$.

Let $D_1, D_2, \cdots, D_m$ be the closures of the components of P - X intersecting J_1. It follows from Theorem 7 of [2] that there exists a 2-sphere S' in $E^3 - J_1$ that contains Bd D_1 + Bd D_2 + ⋯ + Bd D_m and is locally polyhedral

$$\text{mod Bd } D_1 + \text{Bd } D_2 + \cdots + \text{Bd } D_m.$$

It follows from [9] that S' is tame. Hence, J_1 can be shrunk to a point in E^3 - S'. In particular, J_1 can be shrunk to a point in E^3 - (Bd D_1 + Bd D_2 + ⋯ + Bd D_m).

For convenience we suppose each D_i is a disk. (If some D_i were unbounded, we would replace it with a disk D_i' such that Bd D_i' = Bd D_i and $D_i' \subset D_i$ + Bd C, where C is a large oriented cube whose interior contains $X + J_1$.) Hence we suppose the D_i's are mutually exclusive polyhedral disks.

It follows from Theorem 7.4 that there exists an arc $p_1 x_1 q_1$ of J_1 that intersects P only in its end points, and p_1, q_1 belong to the same component of P - X. Let $p_1 y_1 q_1$ be an arc in P - X of diameter less than $\varepsilon/3$ such that

$$J_2 = (J_1 - p_1 x_1 q_1) + p_1 y_1 q_1$$

is a simple closed curve. Shrink J_1 to J_2 in N - X by leaving the points of $J_1 - p_1 x_1 q_1$ fixed and moving points of $p_1 x_1 q_1$ along rectilinear segments to points of $p_1 y_1 q_1$. In a neighborhood of $p_1 y_1 q_1$ we shove J_2 slightly to one side of P and onto a polygonal simple closed curve J_2' that intersects P in two fewer points than J_1 does. It follows again from Theorem 7.4 that, unless J_2' misses P, some arc $p_2 x_2 q_2$ of J_2' intersects P only at its end points and has its end points on the same component of P - X. Let $p_2 y_2 q_2$ be a polygonal arc in P - X of diameter less than $\varepsilon/3$ such that $J_3 = (J_2' - p_2 x_2 q_2) + p_2 y_2 q_2$ is a polygonal simple closed curve. Push J_2' straight to J_3 in N, and adjust J_3 to J_3' by pushing J_3 in a neighborhood of $p_2 y_2 q_2$ slightly to one side of P.

We continue this procedure and get a sequence of polygonal simple closed curves $J_1, J_2, J_2', J_3, J_3', \cdots, J_n, J_n'$ such that each can be pushed to the next in $N - N \cdot X$ and J_n' misses P. Since J_n' can be shrunk to a point in $N - N \cdot P$, J_1 can be shrunk to a point in an ε-subset of $E^3 - X$.

Although Theorem 8.1 was stated only for simple closed curves, each map of a simple closed curve into E^3 can be approximated by a homeomorphism. Hence, we have shown that each map of Bd I^2 into an ε-subset of $E^3 - S$ can be extended to map I^2 into an ε-subset of $E^3 - X$. It is in this form that we use Theorem 8.1 in the proof of the following result.

THEOREM 8.2. *A 2-sphere* S *in* E^3 *is tame if it is locally tame modulo a tame Sierpiński curve* X *in* S.

Proof. We show that S is tame by showing that $E^3 - S$ is 1-ULC. See Theorem 2 of [4].

It follows from [9] that the closure of each component of S - X is a tame disk. Let δ_1 be a positive number so small that each δ_1-subset of one of these disks lies in a subdisk of diameter less than $\varepsilon/3$. Let δ_2 be a number promised by the previous theorem such that each simple closed curve of diameter less than δ_2 in $E^3 - S$ can be shrunk to a point on a δ_1-subset of $E^3 - X$.

Let f be a map of Bd I^2 into a δ_2-subset of $E^3 - S$. Then f can be extended to map I^2 into a δ_1-subset of $E^3 - X$. Let Y be the component of $I^2 - f^{-1}(S \cdot f(I^2))$ containing Bd I^2. Let g be a map of I^2 into f(Y) + (S - X) such that each component of $I^2 - Y$ is sent into a disk in S - X of diameter less than $\varepsilon/3$. Since each such disk in S - X is tame, we can adjust f slightly in a neighborhood of $I^2 - Y$ by shoving to one side of these disks so that the adjusted $g(I^2)$ misses S.

It may be convenient to use Theorem 8.2 in the following more general form.

THEOREM 8.3. *Suppose* S *is a 2-sphere in* E^3, X *is a time Sierpiński curve in* S, *and* U *is an open subset of* S *such that* S *is locally tame at each point of* U - X. *Then* S *is locally tame on* U.

Proof. To show that S is locally tame at a point p of U, use Theorem 8 of [2] to adjust S to a 2-sphere S' such that S' contains X, S' is locally tame mod X, and S' agrees with S in a neighborhood of p. Since S' is tame by Theorem 8.2, S is locally tame at p.

We can extend Theorem 8.2 as follows.

THEOREM 8.4. *A 2-sphere* S *in* E^3 *is tame if it is locally tame modulo the sum of a finite number of tame Sierpiński curves* $X_1, X_2, \cdots, X_n$.

Proof. It follows from repeated applications of Theorem 8.3 that S is locally tame mod $X_2 + X_3 + \cdots + X_n$, locally tame mod $X_3 + X_4 + \cdots + X_n, \cdots$, locally tame mod X_n, and finally locally tame everywhere. Hence it is tame.

Our arguments apply to arcs as well as to Sierpiński curves, so we have the following generalization.

THEOREM 8.5. *Suppose* S *is a 2-sphere in* E^3; $X_1, X_2, \cdots, X_n$ *is a finite collection of subsets of* S *such that each is either a tame finite graph or a tame Sierpiński curve;* U *is an open subset of* S *such that* S *is locally tame at each point of* $U - U \cdot (X_1 + X_2 + \cdots + X_n)$. *Then* S *is locally tame on* U.

9. BURYING SETS IN SIERPINSKI CURVES

THEOREM 9.1. *Suppose* S *is a 2-sphere in* E^3 *and* K *is a tame finite graph (or tame Sierpiński curve) in* S. *Then for each* $\varepsilon > 0$ *there exists a tame Sierpiński curve* X *on* S *such that each point of* K *is an inaccessible point of* X *and each component of* S - X *is of diameter less than* ε.

Proof. We deal only with the case where K is connected, since the general case can be treated in an analogous manner. We suppose diameter $K > \varepsilon$.

It follows from Theorem 1 of [5] that there exists a tame Sierpiński curve X_1 on S such that each component of $S - X_1$ has diameter less than ε. For each integer $i > 1$, let X_i be a tame Sierpiński curve on S such that $X_i \subset V(K, 1/i)$, and such that each component of $S - X_i$ of diameter more than ε/i is at a positive distance from K. The existence of such X_i's follows from the facts [5] that we can get tame Sierpiński curves in S with arbitrarily small holes and that we can cut out parts of these curves that are far from K. The required Sierpiński curve X will lie in $K + \Sigma X_i$.

Let G_i be the decomposition of S such that an element of G_i is either an inaccessible point of X_i or the closure of a component of $S - X_i$. Let G be the decomposition of S such that the element of G containing p is also a component of the intersection of the various elements of the G_i that contain p. No element of G separates S. Also, G has only a null sequence of nondegenerate elements, and none of these intersects K. The sum of these nondegenerate elements is dense in S. Using the fact that the decomposition space of G is a 2-sphere [17], we find that there exists a null sequence of mutually exclusive disks $D_1, D_2, \cdots$ in S such that each of these disks has diameter less than ε, none of them intersects K, and each nondegenerate element of G lies on the interior of one of the disks. Then $S - \Sigma \operatorname{Int} D_i = X$ is the required Sierpiński curve.

We show that X is tame by showing that $K + \Sigma X_i$ lies on a tame 2-sphere. Adjust S to a 2-sphere S' such that S' contains $K + \Sigma X_i$ and S' is locally tame mod $K + \Sigma X_i$. It follows from Theorem 8.5 that S' is locally tame mod K, and from another application of Theorem 8.5 that S' is locally tame everywhere—and hence tame.

The following theorem is an application of Theorem 9.1.

THEOREM 9.2. *Suppose* S *is a 2-sphere in* E^3 *and* T_1 *is a triangulation of* S *such that the* 1*-skeleton of* T_1 *is tame. Then for each* $\varepsilon > 0$, *there exists a triangulation* T_2 *of mesh less than* ε *such that* T_2 *refines* T_1 *and the* 1*-skeleton of* T_2 *is tame.*

REFERENCES

1. R. H. Bing, *Locally tame sets are tame*, Ann. of Math. (2), 59 (1954), 145-158.
2. ———, *Approximating surfaces with polyhedral ones*, Ann. of Math. (2) 65 (1957), 456-483.
3. ———, *Conditions under which a surface in* E^3 *is tame*, Fund. Math. 47 (1959), 105-139.
4. ———, *A surface is tame if its complement is* 1-ULC, Trans. Amer. Math. Soc. 101 (1961), 294-305.
5. ———, *Each disk in* E^3 *contains a tame arc*, Amer. J. Math. 84 (1962), 583-590.
6. ———, *Each disk in* E^3 *is pierced by a tame arc*, Amer. J. Math. 84 (1962), 591-599.
7. ———, *Approximating surfaces from the side*, Ann. of Math. (2) 77 (1963), 145-192.
8. C. E. Burgess, *A characterization of tame surfaces in* E^3, Abstract 518-12, Amer. Math. Soc. Notices 10 (1963), 185-186.
9. P. H. Doyle and J. G. Hocking, *Some results on tame disks and spheres in* E^3, Proc. Amer. Math. Soc. 11 (1960), 832-836.
10. D. Gillman, *Tame subsets of* 2-*spheres in* E^3, Topology of 3-manifolds and related topics, Prentice Hall, Englewood Cliffs, N.J., (1962), 26-28.
11. H. C. Griffith, *A characterization of tame surfaces in three space*, Ann. of Math. (2) 69 (1959), 291-308.
12. M. Hall, Jr., *The theory of groups*, The Macmillan Co., N.Y., 1959.
13. O. G. Harrold, Jr., *Locally peripherally unknotted surfaces in* E^3, Ann. of Math. (2) 69 (1959), 276-290.
14. J. Hempel, *A surface in* E^3 *is tame if it can be deformed into each complementary domain* (to appear).
15. W. Hurewicz and H. Wallman, *Dimension theory*, Princeton University Press (1959).
16. E. E. Moise, *Affine structures in* 3-*manifolds*, VIII. *Invariance of the knot-types; local tame imbedding*, Ann. of Math. (2) 59 (1954), 159-170.
17. R. L. Moore, *Concerning upper semicontinuous collections of continua*, Trans. Amer. Math. Soc. 27 (1925), 416-428.
18. G. T. Whyburn, *Topological characterization of the Sierpiński curve*, Fund. Math. 45 (1958), 320-324.
19. R. L. Wilder, *A converse of a theorem of R. H. Bing and its generalization*, Fund. Math. 50 (1961/62), 119-122.
20. ———, *Topology of manifolds*, Amer. Math. Soc. Colloquium Publications 32 (1949).

The Institute for Advanced Study
and
The University of Wisconsin

SPHERES IN E^3

R. H. BING, University of Wisconsin and Institute for Advanced Study

1. Introduction. In this paper we shall be concerned with two dimensional spheres embedded in Euclidean three space E^3.

A 2-sphere is a set homeomorphic to the boundary of a ball. It might have an equation like $x^2+y^2+z^2=1$, but in general it is only homeomorphic with something having such an equation. Figure 1 shows some 2-spheres. The first picture shown in Figure 1 is that of a piece of lava. This particular piece of lava was not examined with enough care to determine whether or not its boundary is actually a 2-sphere. Such a 2-sphere (if indeed it is one) seems much more representative of a general 2-sphere, however, than the rather special ones shown in the other parts of Figure 1.

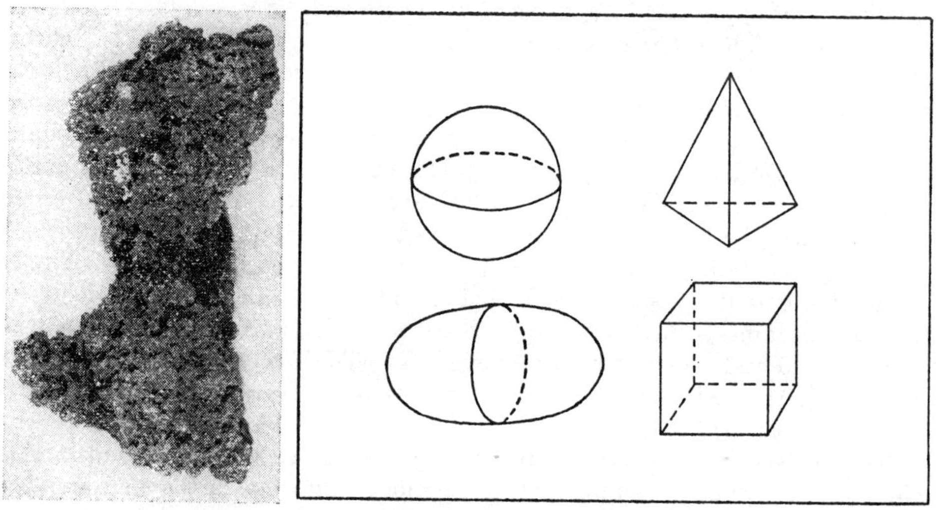

FIG. 1

There are certain properties possessed by all 2-spheres in E^3 irrespective of their shapes. For example, topologists have shown (see for example Theorem 8-38 of [17]) that the Jordan Curve Theorem for the plane E^2 extends to Euclidean spaces of higher dimensions. Hence, if S is a 2-sphere in E^3, S separates E^3 into precisely two pieces. These two pieces are called the interior and exterior of S as follows:

$$\text{Int}\, S = \text{bounded component of } E^3 - S,$$

$$\text{Ext}\, S = \text{unbounded component of } E^3 - S.$$

There are certain properties possessed by certain special 2-spheres that are not possessed by 2-spheres in general. The boundaries of the ball and the ellipsoid shown in Figure 1 are smooth and differentiable whereas the other 2-spheres shown are not. The boundaries of the tetrahedron and the cube are polyhedral while the others are not. The boundary of the rock stands a chance of having infinite area but the others do not. Some of the 2-spheres can be pierced by a straight line interval at each point but one cannot.

It may come as a surprise to some that the solid bounded by one 2-sphere may be topologically different from the solid bounded by another. Those who think the Schoenflies Theorem needs no proof in E^2 because it is intuitively obvious, might also think that the extension to E^3 is true—where indeed it is false. [The Schoenflies Theorem says that if J_1, J_2 are two 1-spheres in E^2 (topological circles, frequently called simple closed curves), then there is a homeomorphism of E^2 onto itself taking J_1 onto J_2. See Corollary 9.25 of [15]. The generalization to E^3 (which is false) would say that if S_1, S_2 are two 2-spheres in E^3, then there is a homeomorphism of E^3 onto itself taking S_1 onto S_2.] Once one realizes that 2-spheres can be so exotic, one may be surprised that they behave as well as they do. We mentioned in a preceding paragraph that the Jordan Curve Theorem extends. In Section 5 we discuss some results about retractions that apply to all 2-spheres in E^3. These and other results of this paper extend to Euclidean spaces of higher dimensions and even to abstract spaces of certain sorts but since this is an expository paper, we restrict ourselves to the simplest cases of interest.

A 2-sphere S in E^3 is called *tame* if there is a homeomorphism of E^3 onto itself that takes S onto a polyhedral 2-sphere. Note that for each 2-sphere S' there is a homeomorphism of S' onto a polyhedral 2-sphere but in order that S' be tame we insist that the homeomorphism can be extended to E^3. It is known that, if a 2-sphere S is tame, then there are homeomorphisms of E^3 onto E^3 that take S onto the boundary of a round ball. Examples are known of 2-spheres in E^3 that are not tame. Such 2-spheres are called *wild.* We describe two wild 2-spheres in Sections 3 and 6 of this paper. Sketches of wild 2-spheres are found in [2], [6], [7], [10], [13], and [17] while descriptions of these and others are also found in [3] and [12].

Topologists are interested in the problem of determining conditions under which a 2-sphere in E^3 is tame. At Stockholm in 1962 one of the addresses [9] was devoted to reviewing known sets of conditions that imply that a 2-sphere in E^3 is tame. Some questions were raised, one of which is as follows: *Is a 2-sphere S in E^3 tame if S can be pierced by a straight line segment at each point?*

The question is not wholly topological since the notion of straightness is geometric rather than topological. However, the question seemed of interest even for topologists. A segment axb is said to pierce S at x if $S \cdot axb = \{x\}$ and a, b belong to different components of $E^3 - S$. M. K. Fort, Jr. found [12] a negative answer to the question. An example giving this answer is given in Section 3. The example relates to the game of pick-up-sticks which is discussed in the next section.

2. Pick-up-sticks. In pick-up-sticks one is confronted with a tangled pile of sticks and challenged to remove them one at a time in such a way that none of the remaining ones is moved. One finds that there are piles in which each stick has another resting on it so that it is impossible to proceed.

Suppose P_1, P_2 are two horizontal planes with P_2 above P_1. Let X be a collection of mutually exclusive straight line segments such that each has one end in P_1 and the other in P_2. See Figure 2. Is it possible to adjust the segments gradually so that each assumes a vertical position, each keeps its bottom end fixed, and during the movement the segments remain mutually exclusive?

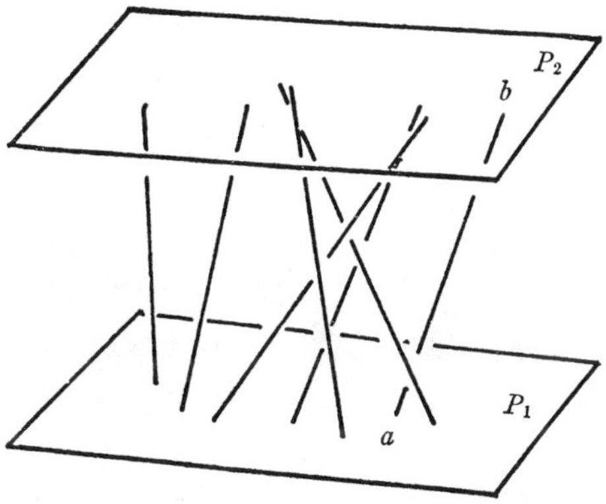

FIG. 2

An affirmative answer to the above question is provided by the following method suggested by R. H. Fox. Consider a segment ab of X with $\{a\}$ in P_1. The lower end of the segment ab is left fixed but at time t the modified segment intersects P_2 at the point which is the vertical projection of the point of ab which divides ab in the ratio $1-t$ to t. Note that at time $t=1$ the modified segment is vertical. In showing that the modified segments do not intersect each other we suppose that a and b have coordinates $(x_a, y_a, 0)$ and $(x_b, y_b, 1)$ respectively. The modified segment at time $t(0 \leqq t < 1)$ is the lower portion of the elongated segment from $(x_a, y_a, 0)$ to $(x_b, y_b, 1/(1-t))$. The elongated segments do not intersect each other since they are the same as the originals under a change of z coordinates. Hence the lower portions of these segments do not intersect each other either.

Let us consider a collection X of segments in a certain instructive position. Let V_1 be the sum of two segments each with ends in P_1, P_2 so that the segments intersect in their upper end points. The segments in V_1 are not elements of X but are merely used as a first approximation to the elements of X. Let N_1 be a tubular neighborhood on V_1. See Figure 3.

Let V_2 be the sum of two inverted V's in N_1 such that the V's hook as shown in Figure 3. Let N_2 be a thin tubular neighborhood of V_2. Let V_3 be the sum of four inverted V's in N_2 so that the part of V_3 in a component of N_2 consists of two hooked V's linked like V_2 in N_1. We continue defining N_3, V_4, N_4, V_5, $\cdots$. Elements of the set X are the components of the intersection of $\overline{N}_1$, $\overline{N}_2$, $\overline{N}_3$, $\cdots$. Note that X is a Cantor set of segments.

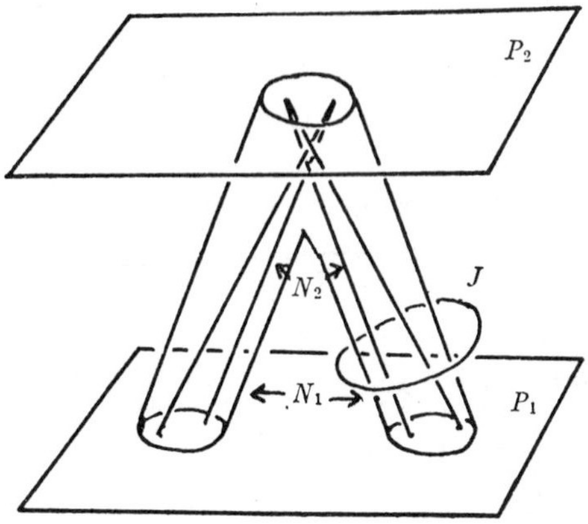

FIG. 3

To see that X is an unusual collection of segments, consider a simple closed curve J as shown in Figure 3 which circles one leg of N_1. In the halfplane above P_1, J cannot be lifted above P_2 without intersecting $\overline{N}_1$. Neither can it be so lifted without intersecting $\overline{N}_2$, nor $\overline{N}_3$ nor . . . . A rigorous proof that it cannot be lifted can be based on Theorem 9 of [7]. Hence it cannot be lifted without intersecting some element of X. If the elements of X were vertical, however, it could be so lifted. Is there a paradox here? Was the argument sound that the elements of X could be lifted into a vertical position?

It was not the verticalizing argument that was in error. The family of homeomorphisms making the segments vertical was defined on the segments only and not on other points of E^3. The family of homeomorphisms could not be extended in a continuous fashion to the rest of space.

3. A wild porcupine. We describe a wild 2-sphere S_0 which can be pierced at each point with a straight line segment. The example is a modification of that given by M. K. Fort, Jr. in [12].

Suppose that S' is the boundary of a cube in E^3 such that the upper face of the cube lies in P_1 of Section 2 and contains the lower portion of N_1 as shown in Figure 4.

Suppose two holes are cut in S' where N_1 intersects P_1 and the holes are replaced by cylinders which run halfway along the vertical sides on N_1 and capped with horizontal disks halfway between P_1 and P_2. Suppose two holes are cut in each cap and cylinders along the sides on N_2 are run up half the remaining distance to P_2. Caps are put on these cylinders but two holes are cut in each cap and cylinders along the sides of N_3 are run up half the remaining distance to P_2. The procedure is continued so as to get a 2-sphere S_0 as shown in Figure 4 so that all intervals of X lie, except for their upper ends, in Int S_0.

Each tame 2-sphere S has the property that each simple closed curve in E^3-S can be shrunk to a point in E^3-S. The simple closed curve J shown in Figure 3 cannot be shrunk to a point in E^3-S_0; hence S_0 is not tame. It can be shown that the embedding of S_0 is the same as the embedding of Alexander's Horned Sphere [2].

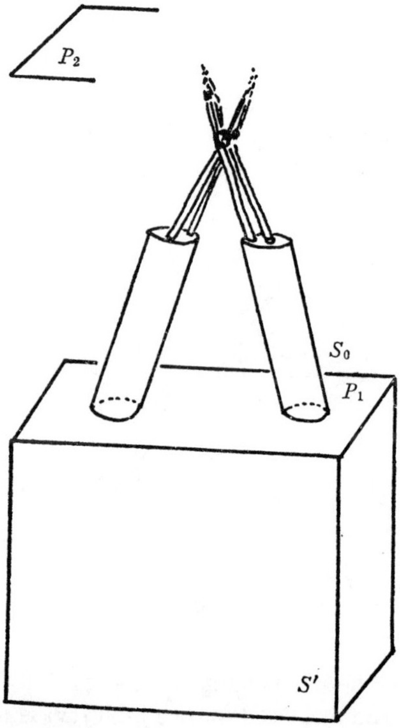

FIG. 4

At each point of S_0, S_0 can be pierced with a straight line segment. The only questionable points are the points of $P_2 \cdot S_0$ and the segments in X show that S_0 can be pierced there.

We can define the segments piercing S_0 so that the directions in which the segments run is continuous. If we do this, however, we will have to permit some of the segments that pierce near $P_2 \cdot S_0$ to be very short. Also, we could have defined the segments so that all of them reach the same distance into Int S_0 and Ext S_0, but under this definition we would not make the directions continuous. That we cannot do both at the same time follows from a result of John Hempel [16].

THEOREM 1. *Any 2-sphere in E^3 for which there is a continuous family of piercing segments is tame.*

The quills of S_0 are tangentially sharp since the cylinders get progressively smaller very fast. If one sharpened a pencil, it would not be possible to touch S_0 from the inside with the tip of the pencil.

Questions. The above considerations suggest the question as to whether or not a 2-sphere S is tame if at each point it can be wedged between two tangent round balls which lie, except for the point of contact, in different components of E^3-S. The left part of Figure 5 suggests the question. The answer is unknown. Similarly we do not know the answer to the corresponding question suggested by the right part of Figure 5 where we show an arbitrary point of S wedged between two cones. These questions go beyond the notions of Topology and involve the idea of straightness from Geometry.

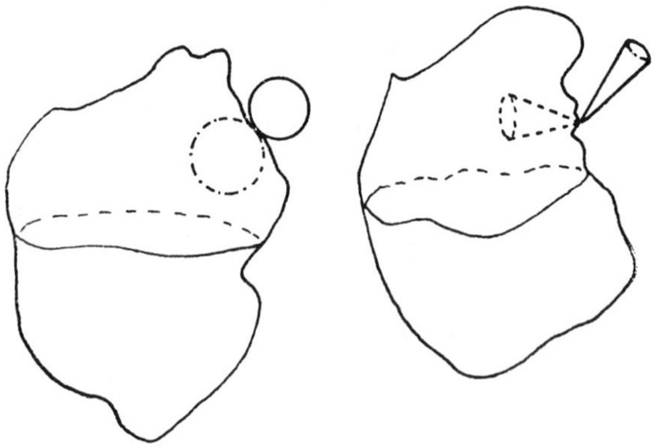

FIG. 5

4. Tietze Extension Theorem. Suppose that one is given an integrable function $y=f(x)$ defined on the line E^1. Even though $f(x)$ is not continuous, for each $\epsilon>0$ the function

$$g_\epsilon(x) = \int_{x-\epsilon}^{x+\epsilon} f(t)dt$$

is continuous. We get a corresponding result if we replace the constant ϵ by a continuous positive function $\epsilon(x)$.

We use the above notion in proving a version of the Tietze Extension Theorem which in turn will be used in the next section. We use calculus (something novel in point set topology) as an averaging tool to give the proof.

Although the Tietze Extension Theorem is normally proved for more abstract spaces (see for example pages 59–61 of [**17**]), we prove it here only for metric spaces since the proof is more elementary in this case and this is the context in which we use it. Furthermore, we get to exhibit an interesting application of calculus. Another elementary proof is found in [**18**].

THEOREM 2. (Tietze Extension Theorem for metric spaces.) *Each map f of a closed subset X of a metric space Y into the segment $[0, 1]$ can be extended to take Y onto $[0, 1]$.*

Proof. Our job is to consider a point p of $Y-X$ and decide what value of $[0, 1]$ to assign to it. If points of X near p are assigned a particular value of $[0, 1]$, we wish to assign p a value nearby. If D is the distance function for Y and

$$V(p, r) = \{q \mid q \in Y,\ D(p, q) \leqq r\},$$

it does no good to look on the interior of the ball $V(p, D(p, X))$ since there are no points of X there. However, there will be points of X in the hollow ball $V(p, 2D(p, X)) - \text{Int } V(p, D(p, X))$, and we seek a certain average over the f values of X in this annulus or hollow ball. See Figure 6.

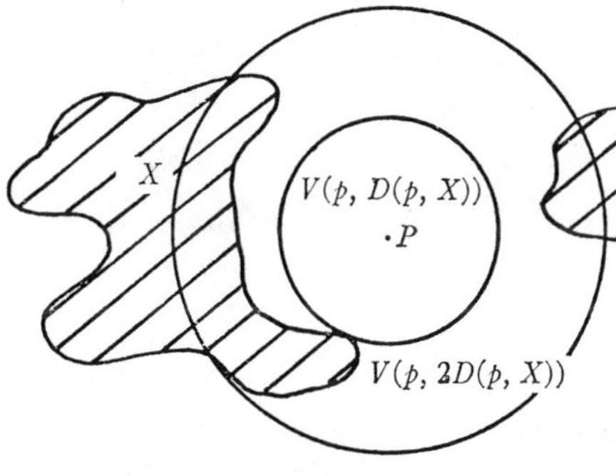

FIG. 6

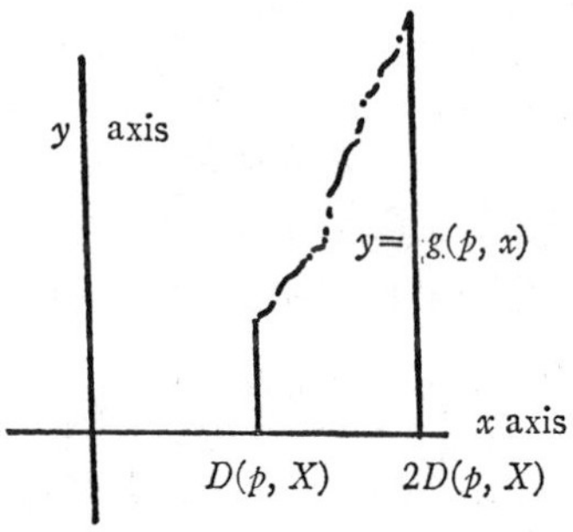

FIG. 7

Consider the function

$$g(p, r) = \text{least upper bound } f(x), \qquad x \in X \cdot V(p, r).$$

There is no reason to suppose that g is continuous in either p or r. It is not defined for $r < D(p, X)$, but for p fixed it is a bounded monotone nondecreasing function for $r > D(p, X)$ and hence integrable. The graph of $y = g(p, x)$ with p fixed and $D(p, X) < x \leqq 2D(p, X)$ might look somewhat like that shown in Figure 7. If we take the area under the curve and divide by the length of the base line, we find an average value of $g(p, x)$. It is immaterial as to whether or not g is defined on the left end of the base line.

For values of p not in X we define

$$f(p) = \frac{1}{D(p, X)} \int_{D(p,X)}^{2D(p,X)} g(p, t)\, dt.$$

It is an exercise in calculus to show that this function is continuous at each point of $Y-X$. It is a continuous extension of the given function f since for each point q of X and each $\epsilon > 0$, there is a positive number $\delta(q)$ such that if $q' \in X \cdot V(q, \delta)$, $|f(q) - f(q')| < \epsilon$. Then for each $p \in V(q, \delta/3)$, $|f(q) - f(p)| \leqq \epsilon$.

It is sometimes convenient to generalize the Tietze Extension Theorem as follows.

THEOREM 3. *Each map of a closed subset of a metric space Y into a disk can be extended to all of Y.*

Proof. We regard the disk as a square in E^2 with opposite vertices at (0, 0) and (1, 1). For each point p on which f is defined let $(f_x(p), f_y(p))$ be the coordinates of $f(p)$. It follows from the preceding theorem that f_x and f_y can be extended to all of Y. If f_x, f_y are used to denote these extensions, the extended f may be defined so that $f(p)$ has coordinates $(f_x(p), f_y(p))$.

5. Retractions onto spheres. A *retraction* of a set Y onto a set X is a map f of Y onto X such that f is fixed on each point of X.

A round 2-sphere has the property that if p is a point of Int S, then there is a retraction of $E^3-\{p\}$ onto S. As pointed out in Theorem 4, each wild 2-sphere has this property. For the round 2-sphere S, one can take the half lines emanating from p onto the points where these half lines pierce S. For arbitrary 2-spheres more ingenuity needs to be used in getting the retraction.

THEOREM 4. *If S is an arbitrary 2-sphere in E^3 and p is a point of* Int S, *then there is a retraction of $E^3-\{p\}$ onto S.*

This theorem was proved in [8]. An effort was made to avoid the existence type of proof, but rather to exhibit a method for getting a retraction of $E^3-\{p\}$ onto S. The proof given in [8] was somewhat constructive but failed to indicate what the retraction of a neighborhood of S into S would look like. The purpose of giving a proof of the Tietze Extension Theorem in the last section by defining the extension rather than by existence techniques was to enable us to show in Theorem 5 what the retraction of the neighborhood of S is.

A set X is an ANR (*absolute neighborhood retract*) if for each embedding of X into a metric space Y there is a neighborhood N of X in Y and a retraction of N onto X. It is known that each 2-sphere is an ANR. See for example Theorem 2-36 of [17]. Hence, no matter how wildly a 2-sphere S is embedded in E^3, there is a neighborhood N of S in E^3 and a retraction of N onto S.

THEOREM 5. *If S is a 2-sphere in E^3, then there is a neighborhood N of S and a retraction of N onto S.*

Proof. Let D_1, D_2, D_3 be three disks in S with $D_1\subset \text{Int } D_2\subset D_2\subset \text{Int } D_3$. Let f be the identity map of D_3 onto itself and use Theorem 3 to extend f so as to obtain a retraction of E^3 onto D^3. Let N be a neighborhood of Bd D_2 such that $\overline{N}\cdot S+f(\overline{N})\subset D_3-D_1$. Let N_1, N_2 be neighborhoods of D_2 and $S-\text{Int } D_2$ respectively such that $\overline{N}_1\cdot\overline{N}_2\subset N$. Let g be a map of $(S-\text{Int } D_1)+\overline{N}$ onto $S-\text{Int } D_1$ that is the identity on $S-\text{Int } D_1$ and f on $\overline{N}$. It follows from Theorem 3 that the map g can be extended to a retraction of $(S-\text{Int } D_1)+\overline{N}+\overline{N}_2$ onto $S-\text{Int } D_1$. Then the retraction of N_1+N+N_2 given by f on N_1+N and by the extended g on N_2+N satisfies the requirements of Theorem 5.

Fixed point properties of cubes imply that if p is not removed in Theorem 4, there is no retraction. Again, this theorem holds as well for arbitrary 2-spheres in E^3 as for nice ones.

THEOREM 6. *There is no 2-sphere in E^3 which is a retract of E^3.*

Proof. Assume there is a retraction r of E^3 onto a particular 2-sphere S. Consider a homeomorphism h of S onto itself that moves every point of S. That here is such a homeomorphism follows from the fact that the antipodal homeomorphism of a round 2-sphere onto itself throws each point into its diametrically opposite point and hence moves every point. The map hr would be a fixed point free map of E^3 onto a compact subset of E^3. Theorems like Theorem 6-39 of [17] say that this is impossible.

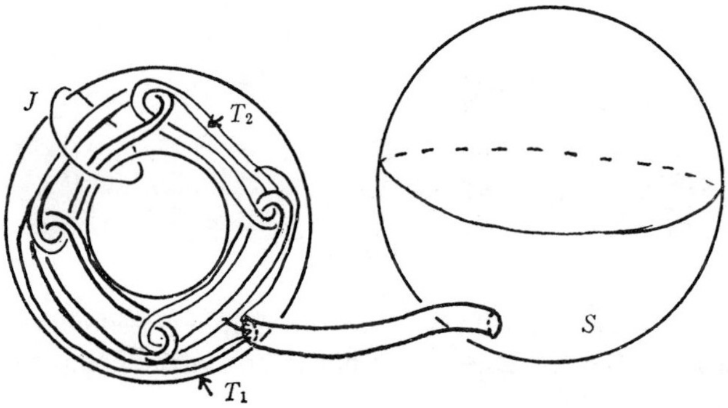

FIG. 8

6. **Antoine's necklace.** An Antoine's necklace is illustrated in the left portion of Figure 8. It is described in [3, 4, 5, 6, 10, 11, 17] and is the intersection of a solid torus T_1, a set T_2 which is the sum of several small solid tori which form a chain circling T_1, a set T_3 which is the sum of some very small solid tori which intersect each component of T_2 in the same fashion that T_2 lies in T_1, $\cdots$. There are countably many layers of T's, and the set of points belonging to their intersection is Antoine's necklace. We denote it by A. It may be shown that A is topologically equivalent to an ordinary Cantor set. Its complement, however, is topologically different from the ordinary Cantor set, because the simple closed curve J shown in Figure 8 cannot be shrunk to a point without hitting A. See [11] for proof of this.

Consider a dendron running down to a Cantor set as shown in Figure 9. It looks something like a form used to record results of an elimination tournament in which only winners remain at each stage to engage other winners. There is no starting round, however, since at each stage there was a preceding one. We use a combination of the notion of a dendron of this type and of Antoine's necklace to describe what was perhaps the first known wild 2-sphere [3].

Let S be a 2-sphere missing T_1 as shown in Figure 8. Alter S by removing a small disk and replacing the hole with a cylinder leading over to T_1 with a cap on Bd T_1. Holes are cut in this cap and tubes are run in T_1 over to the components of T_2 where the tubes are capped. Holes are cut in these caps and tubes are run in T_2 over to the components of T_3. The process is continued. Note that the tubes get shorter since they lie in the T's which at the later stages get arbitrarily small.

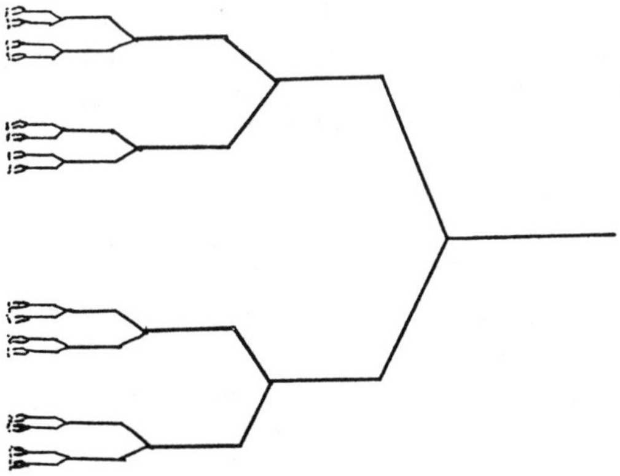

FIG. 9

The resulting 2-sphere is wild because the simple closed curve J shown cannot be shrunk to a point in $E^3 - A$ and hence not in the complement of the resulting 2-sphere.

7. The grooved ball. There are many criteria known for determining whether or not a 2-sphere is tame. One of the earliest of these [1] shows that a 2-sphere is tame if it is polyhedral. More understandable proofs of this are found in [**14**] and [**19**].

A condition that is somewhat in doubt is involved in the following question: *Is a 2-sphere S in E^3 tame if each horizontal cross section of S is either a point or a simple closed curve?*

An affirmative answer is suggested in the literature but the proof is very sketchy, saying merely that the result follows by "well-known theorems on logarithmic potential in 2-dimensions."

We give an example to show that even if the above condition does imply that a 2-sphere is tame, it is not enough to ensure that there is a homeomorphism of E^3 onto itself which is invariant on horizontal planes and which takes S onto a round 2-sphere.

Suppose B is a croquet ball to be left out in the rain. One wishes to groove the ball so that as water runs down the ball, one section of the equator does not have water run over it. See Figure 10. The grooves are very shallow so that each horizontal cross section of the ball is a disk. The cross section through the

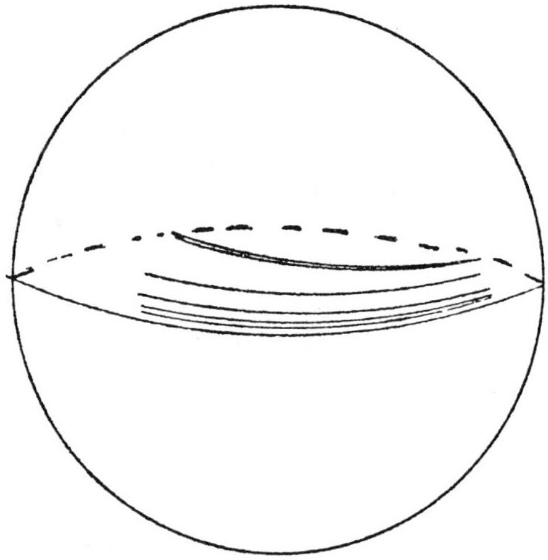

FIG. 10

equator is round but some close to it are disks with long feelers. These disks converge, but not homeomorphically, to the disk containing the equator. If we chose to make the groves spiral we could get even longer feelers.

We have learned much about E^3 but there is much we do not know. For the person with ingenuity and ability this is a fruitful area in which to work.

This Research was supported by grants NSF-G21514 and NSF-G11665. This paper contains material from several different lectures at the Southwestern, Allegheny Mountain, and Minnesota, Sections which the author visited as President of the MAA in the Spring of 1963.

References

1. J. W. Alexander, On the subdivision of 3-space by a polyhedron, Proc. Nat. Acad. Sci. U.S.A., 10 (1924) 6–8.

2. ———, An example of a simply connected surface bounding a region which is not simply connected, Proc. Nat. Acad. Sci. U.S.A., 10 (1924) 8–10.

3. ———, Remarks on a point set constructed by Antoine, Proc. Nat. Acad. Sci. U.S.A., 10 (1924) 11–12.

4. M. L. Antoine, Sur l'homéomorphie de deux figures et de leurs voisinages, J. Math. Pures Appl., 86 (1921) 221–325.

5. ———, Sur la possibilité d'étendre l'homéomorphie de deux figures à leurs voisinages, C. R. Acad. Sci. Paris, 171 (1920) 661–663.

6. R. H. Bing, Examples and counter examples, Pi Mu Epsilon J., 1 (1953) 311–317.

7. ———, A wild surface each of whose arcs is tame, Duke Math. J., 28 (1961) 1–16.

8, ———, Retractions onto spheres, this MONTHLY, to appear in May issue.

9. ———, Embedding surfaces in 3-manifolds, Internat. Congress of Math., Outlines of one-hour and half-hour addresses, Stockholm, 1962, 2.11–2.12.

10. W. A. Blankenship and R. H. Fox, Remarks on certain pathological open subsets of 3-space and their fundamental groups, Proc. Amer. Math. Soc., 1 (1950) 618–624.

11. R. P. Coelbo, On the groups of certain linkages, Portugal. Math., 6 (1947) 57–65.

12. M. K. Fort, Jr., A wild sphere which can be pierced at each point by a straight line segment. Proc. Amer. Math. Soc., 14 (1963) 994–995.

13. R. H. Fox and E. Artin, Some wild cells and spheres in three-dimensional space, Ann. of Math., 49 (1948) 979–990.

14. W. Graeub, Die semilinearen Abbildungen, S.-B. Heidelberger Akad. Wiss., part 4 (1950) 205–272.

15. D. W. Hall and G. L. Spencer II, Elementary Topology, Wiley, New York, 1955.

16. J. P. Hempel, A surface in E^3 is tame if it can be deformed into each complementary domain, to appear.

17. J. G. Hocking and G. S. Young, Topology, Addison-Wesley, Reading, Mass., 1961.

18. K. Kuratowski, Topologie, vol. 1, Warsaw, 1948.

19. E. E. Moise, Affine structures in 3-manifolds II, Positional properties of 2-spheres, Ann. of Math., 55 (1952) 172–176.

IMPROVING THE SIDE APPROXIMATION THEOREM(1,2)

BY

R. H. BING

1. **Introduction.** The following version of the Side Approximation Theorem for 2-spheres is proved as Theorem 16′ in [5].

THEOREM. *For each 2-sphere S in E^3 and each $\epsilon > 0$ there is a homeomorphism h: $S \times [-1,1] \to E^3$ such that*

each $h_t(S)$ is a polyhedral 2-sphere,

$D(s, h_t(s)) < \epsilon \quad (s \in S, -1 \leqq t \leqq 1)$,

S lies except for a finite collection of mutually exclusive ϵ-disks in $h(S \times (-1,1))$,

$h_{-1}(S)$ lies except for a finite collection of mutually exclusive ϵ-disks in the interior of S, and

$h_1(S)$ lies except for a finite collection of mutually exclusive ϵ-disks in the exterior of S.

In the above we use D to represent the distance function and interior of S, exterior of S respectively to denote the bounded, unbounded components of $E^3 - S$. We say that $h_1(S) = h(S \times 1)$ *approximates S "almost" from the exterior of S and $h_{-1}(S)$ approximates S "almost" from the interior of S.*

In the last section of [5] the above result was generalized to open subsets of 2-spheres but the proof is complicated and should not be studied. Instead, we recommend the relatively easy treatment of the following result.

THEOREM 1.1. SIDE APPROXIMATION THEOREM FOR 2-MANIFOLDS. *Suppose M^2 is a connected 2-manifold (perhaps noncompact) in a connected 3-manifold M^3 such that*

$$M^3 - M^2 = U_1 + U_2 \text{ (mutually separated)}$$

and f is a positive continuous function defined on M^2. Then there is a homeomorphism h: $M^2 \times [-1,1] \to M^3$ such that

(1) *each $h_t(M^2)$ is tame,*

(2) $D(m, h_t(m)) < f(m) \ (m \in M^2, -1 \leqq t \leqq 1)$,

(3) *$M^2 - h(M^2 \times (-1,1))$ is covered by the interiors of a locally finite*

Presented to the Society, August 20, 1965; received by the editors March 30, 1964.

(1) Work on this paper supported by National Science Foundation under grant GP-2244.

(2) This paper is one of a collection of articles prepared by friends of A. D. Wallace and to be presented to him in celebration of his sixtieth birthday.

collection of mutually exclusive disks in M^2 *such that the diameter of each is less than the minimum value of* f *on it,*

(4) *for* $0 < t \leqq 1$, $\bar{U}_1 \cdot h_{-t}(M^2)$ *is covered by the interiors of a locally finite collection of mutually exclusive disks in* $h_t(M^2)$ *each of diameter less than* $f(x)$ *if* $h_t(x)$ *lies in the disk, and*

(5) *for* $0 < t \leqq 1$, $\bar{U}_2 \cdot h_{-t}(M^2)$ *is covered by the interiors of a locally finite collection of mutually exclusive disks in* $h_{-t}(M^2)$ *each of diameter less than* $f(x)$ if $h_{-t}(x)$ *lies in the disk.*

In fact, if $M^{2\prime}$ *is a closed subset of* M^2 *which is a tame 2-manifold with boundary,* h *can be chosen so that* h_0 *is the identity on* $M^{2\prime}$.

The details of the proof of Theorem 1.1 are delayed until the next section. However, the idea of the description of the homeomorphism h: $M^2 \times [-1,1] \to M^3$ is so simple that we outline it here. The homeomorphism h is obtained as follows.

(1) Get a locally finite collection of tame simple closed curves $J_1, J_2, \cdots$ in M^2 so that each component of $M^2 - \sum J_i$ is small.

(2) Use the approximation theorem for open subsets of 2-spheres (Theorem 8 of [2]) to get a homeomorphism h_0: $M^2 \to M^3$ such that $h_0 = I$ (identity) on $M^{2\prime} + \sum J_i$, $h_0(M^2)$ is locally tame mod $h_0(\sum J_i)$, and h_0 does not move points far.

(3) Use a variation of Theorem 8.5 of [7] to show that $h_0(M^2)$ is tame (and hence bicollared). Get a homeomorphism h: $M^2 \times [-1,1] \to M^3$ so that h: $M^2 \times 0$ is the h_0 we have already described. If the "normals" $h(m \times [-1,1])$ are taken small, the h satisfies the conclusion of Theorem 1.1.

An *n-manifold* is a metric space each of whose points has a neighborhood homeomorphic with Euclidean n-space E^n. Each point of an *n-manifold with boundary* has a neighborhood whose closure is a topological n-cell. Hence, an n-manifold is an n-manifold with boundary but not conversely. Since each component of an n-manifold is separable, we frequently work only with their components to have the added security of separability. If M is an n-manifold with boundary, we denote the set of its points which have neighborhoods homeomorphic with E^n by $\operatorname{Int} M$ (called interior M) and denote $M - \operatorname{Int} M$ by $\operatorname{Bd} M$ (called boundary M). For a disk D in the plane, the point set boundary is $\operatorname{Bd} D$ but if D is embedded in E^3, the point set boundary is more.

A 2-manifold with boundary $M^{2\prime}$ is *tame* in a 3-manifold M^3 if there is a triangulation of M^3 such that $M^{2\prime}$ is the closed sum of elements of the triangulation. Being such a tame manifold is a local property since a closed subset of a 3-manifold is tame if it is locally tame [1], [11]. A set X is *locally tame at a point* p of X if there is a neighborhood N of p and a homeomorphism of $\bar{N}$ into a combinatorial cube that takes $\bar{N} \cdot X$ onto a polyhedron. A set is *locally tame* if it is locally tame at each of its points.

In general, a set is *tame* in a manifold if the set is a geometric complex of some combinatorial triangulation (perhaps curvilinear) of the manifold. If the manifold already has a combinatorial structure we insist that the combinatorial triangulation that makes the tame set a geometric complex is isomorphic to a subdivision of the original combinatorial triangulation. Hence, a set X in a combinatorial manifold is tame if there is a homeomorphism of the manifold onto itself that takes X onto a polyhedron.

A *Sierpiński curve* is a set homeomorphic to the set obtained by removing from a 2-sphere the interiors of a null sequence of mutually exclusive subdisks whose sum is dense in the 2-sphere. A *null* sequence is one which for each positive number ϵ has at most a finite number of elements with diameters more than ϵ. A Sierpiński curve in a 3-manifold is called *tame* if it lies on some tame disk in the 3-manifold.

2. **Proof of Theorem 1.1.** Let p_0, q_0 be points in U_1, U_2 respectively and ϵ be a positive number such that no 2ϵ-subset of $V(M^2, \epsilon)$ separates p_0 from q_0 in M^3. We use $V(A, \epsilon)$ to denote the set of all points whose distance from A is less than ϵ. We suppose that $f < \epsilon$ and so small that for each point p of M^2, $V(p, 2f(p))$ lies in an open 3-cell in M^3. The reason for making f so small is to make it easy to show that $h_1(M^2)$ and $h_{-1}(M^2)$ lie except for small holes in different ones of U_1, U_2. See Theorem 5.2 of §5.

We use $\min_f(X)$ to denote the minimum value of f on X.

The proof is broken into four pieces—the first three following our outline in the introduction and the fourth showing that the homeomorphism h we describe satisfies the conditions of the theorem.

1. *Chopping M^2 into small pieces.* It is known that M^2 can be triangulated into small simplexes. It follows further from Theorem 5 of [4] that there is a locally finite collection $\{D_i\}$ of disks in M^2 such that

$\{\operatorname{Int} D_i\}$ covers M^2 but no subcollection does,

diameter $D_i < \min_f(D_i)/2$,

for each D_i there is an open 3-cell O_i in M^3 and a 2-sphere S_i so that $D_i \subset S_i \subset O_i$.

Let $\{D_i''\}$ be a collection of disks such that

$$D_i'' \subset \operatorname{Int} D_i \quad \text{and} \quad \{\operatorname{Int} D_i''\} \text{ covers } M^2.$$

It follows from Theorem 1 of [6] that for each positive integer i there is a tame Sierpiński curve X_i in S_i such that each component of $S_i - X_i$ is of diameter less than $D(\operatorname{Bd} D_i, \operatorname{Bd} D_i'')/2$. Let J_i be a simple closed curve in $X_i \cdot (\operatorname{Int} D_i - D_i'')$ that separates D_i'' from $\operatorname{Bd} D_i$ in D_i. To find such a J_i, let g_i be a map of S_i onto a 2-sphere $g_i(S_i)$ such that the inverse of each point of $g_i(S_i)$ is either the closure of a component of $S_i - X_i$ or a point of X_i not on such a closure. Let J_i' be a simple closed curve on $g_i(S_i)$ that separates $g_i(D_i'')$ from $g_i(S_i - \operatorname{Int} D_i)$ and which misses each of the countable set of points of $g_i(S_i)$ which have nondegenerate inverses. Then $J_i = g_i^{-1}(J_i')$. Denote the disk in D_i bounded by J_i by D_i'. Then $D_i'' \subset \text{Int}\, D_i' \subset D_i' \subset \text{Int}\, D_i$. Also, $J_i = \text{Bd}\, D_i'$ is tame.

2. *Defining* h_0. We first get a function f' which tells us how close to approximate M^2. Let f' be a continuous function defined on M^2 such that for each element D_i of $\{D_i\}$,

$$p \in D_i \Rightarrow f'(p) < \min_f(D_i)/4,$$

$$p \notin \text{Int}\, D_i \Rightarrow D(p, D_i') > f'(p), \text{ and}$$

$$D(p,q) < f'(q) \Rightarrow f(p) < 2f(q).$$

Let U be an open set containing $M^2 - (M^{2\prime} + \sum J_i)$ such that each component of U lies in a 3-cell in M^3 and no component of U contains two components of $M^2 - (M^{2\prime} + \sum J_i)$. It follows from Theorem 7 of [2] that there is a homeomorphism h_0 of M^2 into M^3 such that

$$h_0(M^2) \text{ is locally tame at } h_0(p) \text{ if } p \in M^2 - (\text{Bd}\, M^{2\prime} + \textstyle\sum J_i),$$

$$D(m, h_0(m) < f'(m)/2 < f(m)/8,$$

$$h_0 = I \text{ (identity) on } M^{2\prime} + \textstyle\sum J_i,$$

and h_0 is so close to the identity that there is a homotopy g: $M^2 \times [0,1] \rightarrow M^3$ such that

$$g_0 = h_0,$$

$$g_1 = I,$$

$$g_t = I \text{ on } M^{2\prime} + \textstyle\sum J_i, \text{ and}$$

$$g_t(M^2 \cdot U) \subset U.$$

Although Theorem 7 of [2] says nothing about a homotopy g, it does imply that h_0 can be taken close to the identity. To see that taking h_0 near the identity implies a g, consider a locally finite collection of topological cubes such that each component u of U lies in one of them (say C_u). Let h_u be a homeomorphism of C_u onto a canonical cube. Then there is a homotopy g if we take h_0 close enough to the identity that for each u and for each point $p \in M^2 \cap u$, the point $h_0(p)$ lies in C_u and the segment from $h_u(p)$ to $h_u h_0(p)$ lies in $h_u(u)$.

Let h_i' be the homeomorphism that is h_0 on $\sum_1^i D_j'$ and the identity on $M^2 - \sum_1^i D_j'$. It follows from Corollary 7.2 of §7 that each $h_i'(M^2)$ separates p from q in M^3. Hence $h_0(M^2)$ is two sided in M^3.

3. *Defining* h. It follows from a modification of Theorem 8.5 of [7] that $h_0(M^2)$ is locally tame. This modification is spelled out in Theorem 3.1 of the next section. Hence $h_0(M^2)$ locally has a cartesian product neighborhood and it follows from [8] that there is a homeomorphism h: M^2

$\times [-1,1] \to M^3$ such that h_0 is the homeomorphism previously defined and each $h_t(M^2)$ is locally tame. Suppose that for each point p of M^2, diameter $h(p \times [-1,1]) < f'(p)/2$. This implies that

$$D(p, h_t(p)) < f'(p) < f(p)/4.$$

Except for a possible exchange of t and $-t$ this homeomorphism h satisfies the conclusion of Theorem 1.1.

4. *Showing that h satisfies Conditions* 2, 3, 4, 5. Condition 2 of the conclusion of Theorem 1 is satisfied because

$$D(m, h_0(m)) < f(m)/8 \text{ and}$$

$$\text{diameter } h(m \times [-1,1]) < f(m)/8.$$

Each component of $M^2 - h(M^2 \times (-1,1))$ lies on the interior of an element of $\{D_i\}$ since

$$\sum J_i \subset h_0(M^2) \subset h(M^2 \times (-1,1)) \text{ and}$$

$$\text{each component of } M^2 - \sum J_i \text{ lies in a disk } D_i' \text{ of } \{D_i'\}.$$

It follows from Theorem 4.1 that there is a locally finite collection of mutually exclusive disks $G_1, G_2, \cdots$ in M^2 such that the interiors of the G_i's cover $M^2 - h(M^2 \times (-1,1))$. The G_i's are small since if $G_i \subset D_j$,

$$\text{diameter } G_i \leqq \text{diameter } D_j < \min_f(D_j)/2 \leqq \min_f(G_i)/2.$$

Hence Condition 3 is satisfied.

As a step toward showing that Conditions 4 and 5 are satisfied we show that each component A of $M^2 \cdot h_1(M^2)$ lies in an $h_1(\operatorname{Int} D_i)$ for some element D_i of $\{D_i\}$. Since $\sum J_i \cdot h_1(M^2) = 0$, A lies in a component of $M^2 - \sum J_i$ and hence in a D_i'. Then A lies in the corresponding $h_1(\operatorname{Int} D_i)$ because $\operatorname{Cl}(h_1(M^2 - D_i))$ misses D_i' as can be seen from the facts that $D(p, h_1(p)) < f'(p)$ and $f'(q) > D(q, D_i')$ if $q \notin \operatorname{Int} D_i$.

The $h_1(D_i)$'s give a locally finite collection of disks whose interiors cover $M^2 \cdot h_1(M^2)$. It follows from Theorem 4.1 that there is a locally finite collection of mutually exclusive disks $E_1^1, E_2^1, \cdots$ in $h_1(M^2)$ whose interiors cover $M^2 \cdot h_1(M^2)$ such that each of these disks lies in an $h_1(D_i)$. If $E_i^1 \subset h_1(D_j)$, then

$$\text{diameter } E_i^1 < \text{diameter } h_1(D_j) \leqq \text{diameter } D_j + 2\min_f(D_j)/4$$
$$\leqq \min_f(D_j) \leqq \min_f h^{-1}(E_i^1).$$

Since $h_1(M^2) - \sum E_i^1$ is connected, it lies in one of U_1, U_2—so with a possible exchange of U_1, U_2, Condition 4 is satisfied for $t = 1$.

Similarly, it follows that for $-1 \leqq t < 0$ or $0 < t \leqq 1$ there is a collection of mutually exclusive small disks $E_1^t E_2^t, \cdots$ in $h_t(M^2)$ such that $h_t(M^2) - \sum E_i^t$ lies in one of U_1, U_2. If t, t' are of the same sign, $h_t(M^2) - \sum E_i^t$

and $h_{t'}(M^2) - \sum E_i^{t'}$ lie in the same one of U_1, U_2 or else there is a $t_0 \neq 0$ and a sequence $t_1, t_2, \cdots$ converging to t_0 such that $h_{t_0}(M^2) - \sum E_i^{t_0}$ and $h_{t_j}(M^2) - \sum E_i^{t_j} (j > 0)$ lie in different ones of U_1, U_2. This is impossible since for j sufficiently large, $h_{t_j}\{h_{t_0}^{-1}[h_{t_0}(M^2) - \sum E_i^{t_0}]\}$ lies in the one of U_1, U_2 containing $h_{t_0}(M^2) - \sum E_i^{t_0}$ and there would not be a finite collection of small disks in $h_{t_j}(M^2)$ covering $h_{t_j}\{h_{t_0}^{-1}[h_{t_0}(M^2) - \sum E_i^{t_0}]\}$.

Condition 5 is established when we show that it is not the same one of U_1, U_2 containing $h_1(M^2) - \sum E_i^1$ and $h_{-1}(M^2) - \sum E_i^{-1}$. This follows from Theorem 5.2 of §5. The only difficult condition to check in applying Theorem 5.2 is to see that Condition 5 of the hypothesis of that theorem is satisfied. We do this in the next paragraph.

Suppose G_k is one of the disks of Condition 3 that intersects the disk E_i of Condition 4 in a point p and $h_1(q) = p$. We show that $G_k + E_i$ lies in an open 3-cell in M^3 by showing that each lies in $V(q, 2f(q))$. First, there is a D_j such that $E_i \subset h_1(D_j)$. Then since h_1 does not move x farther than $f'(x)$,

$$E_i \subset h_1(D_j) \subset V(D_j, \min_f(D_j)/4) \subset V(q, \text{diameter}\, D_j + \min_f(D_j)/4)$$
$$\subset V(q, \min_f(D_j)/2 + \min_f(D_j)/4) \subset V(q, f(q)).$$

Also, since $D(p,q) < f'(q), f(p) < 2f(q)$ and

$$G_k \subset V(p, f(p)/2) \subset V(p, f(q)) \subset V(q, D(p,q) + f(q))$$
$$\subset V(q, f'(q) + f(q)) \subset V(q, f(q)/4 + f(q)) \subset V(q, 2f(q)).$$

Using the fact that any homeomorphism of a triangulated 3-manifold with boundary into a triangulated 3-manifold can be approximated with a piecewise linear homeomorphism as shown by Theorem 9 of [3] and Theorem 2 of [10], we obtain the following variation of Theorem 1.1.

THEOREM 1.1′. *Suppose M^2 is a connected polyhedral 2-manifold, h' is a homeomorphism of M^2 into a triangulated 3-manifold M^3 such that $h'(M^2)$ separates M^3, and f is a positive continuous function defined on M^2. Then there are three locally finite collections $\{G_i\}$, $\{E_i\}$, $\{F_i\}$ of mutually exclusive disks in M^2 and a piecewise linear homeomorphism h: $M^2 \times [-1,1] \to M^3$ such that*

$$D(h'(m), h_t(m)) < f(m) \qquad (m \in M^2, -1 \leqq t \leqq 1),$$

$$\text{diameter}\, h'(G_i) < \min_f(G_i),$$

$$\text{diameter}\, h_{-1}(E_i) < \min_f(E_i),$$

$$\text{diameter}\, h_1(F_i) < \min_f(F_i),$$

$$h'(M^2 - \sum \text{Int}\, G_i) \subset h(M^2 \times (-1,1)),$$

and

$$h'(M^2) \text{ separates } h_{-1}(M^2 - \sum \operatorname{Int} E_i) \text{ from } h_1(M^2 - \sum \operatorname{Int} F_i) \text{ in } M^3.$$

In fact, if P is a polyhedron in M^2 on which h' is piecewise linear, h can be chosen so that $h_0 = h'$ on P.

3. **Tameness** mod **tame sets.** It was shown in [7] that a 2-sphere in E^3 is tame if it is locally tame mod the sum of a finite number of sets each of which is either a tame arc or a tame Sierpiński Curve. (This result was used in showing that the 2-manifold $h_0(M^2)$ considered in the preceding two sections was locally tame.) The following is a mild extension of Theorem 8.5 of [7].

THEOREM 3.1. *Suppose M^2 is a 2-manifold embedded in a 3-manifold M^3, $\{X_i\}$ is a countable collection each of whose elements is either a tame arc or a tame Sierpiński curve in M^2, U is an open subset of M^2 such that M^2 is locally tame at each point of $U - \sum X_i$. Then M^2 is locally tame at each point of U.*

Proof. We first prove the theorem in the case where there is only one element X_1 in $\{X_i\}$. Let p be a point of $X \cdot U$ under consideration, D a disk in U, O^3 an open 3-cell, and S a 2-sphere, such that $p \in \operatorname{Int} D \subset S \subset O^3 \subset M^3$. Then there is an arc or Sierpiński curve X_1' in $X_1 \cdot D$ such that for some open subset N of $\operatorname{Int} D$ $p \in N \cdot X_1 \subset X_1'$. It follows from Theorem 8.5 of [7] S is locally tame at each point of N. Since $p \in N \subset M^2$, M^2 is locally tame at p.

Now that we have disposed of the special case, let Y be the set of all points of U at which M^2 is not locally tame. Then Y is a relatively closed subset of M^2 which lies in $\sum X_i$. It follows from the Baire Category Theorem that there is a point q of Y and an open subset N of U such that $q \in N$ and a single X_i contains $Y \cdot N$. It follows from the special case treated in the preceding paragraph that M^2 is locally tame at each point of N. But then M^2 is locally tame at q.

4. **Building disks about sets.** In the proof of Theorem 1.1 we showed that $h_1(M^2)$ lay except for a set with small components in $M^3 - M^2$. In this section we point out why this implies that $h_1(M^2)$ lies except for a locally finite collection of mutually exclusive small disks on one side of M^2.

THEOREM 4.1. *Suppose M^2 is a connected 2-manifold, X is a closed subset of M^2, and $\{D_i\}$ is a locally finite collection of disks in M^2 such that each component of X lies on the interior of one of the disks. Then there is a locally finite collection of mutually exclusive disks $E_1, E_2, \cdots$ in M^2 such that each component of X lies on the interior of an E_i and each E_i lies in some* $\operatorname{Int} D_j$.

Proof. If C is a component of X in $\operatorname{Int} D_1$, there is a disk D_C such that $C \subset \operatorname{Int} D_C \subset D_1$ and $X \cdot \operatorname{Bd} D_C = 0$. By so covering each component of X not in any $\operatorname{Int} D_2, \operatorname{Int} D_3, \cdots$, we find that there is a finite collection of disks $D_{11}, D_{12}, \cdots, D_{1n_1}$ in D_1 such that each $X \cdot \mathrm{Bd}\, D_{1i} = 0$, and each component of X lies in the interior of one of $D_{11}, D_{12}, \cdots, D_{1n_1}, D_2, D_3, \cdots$.

We next replace D_2 with a finite collection of disks $D_{21}, D_{22}, \cdots, D_{2n_2}$ in D_2 such that each $X \cdot \mathrm{Bd}\, D_{2i} = 0$ and each component of X lies in the interior of one of $D_{11}, D_{12}, \cdots, D_{1n_1}, D_{21}, D_{22}, \cdots, D_{2n_2}, D_3, D_4, \cdots$. Continuing in this fashion we get a locally finite collection of D_{ij}'s such that each D_{ij} lies in D_i, each $X \cdot \mathrm{Bd}\, D_{ij} = 0$, and each component of X lies in some $\mathrm{Int}\, D_{ij}$.

Throw away each D_{ij} that lies in a larger one. This causes the closure of each component of $D_{ij} - \sum \mathrm{Bd}\, D_{rs}$ to be a disk rather than an annulus or something worse. The collection of closures of such components may not be a locally finite collection but the subcollection of those which intersect X is a locally finite collection. If two of these disks intersect, the intersection is on the boundary so if each is shrunk slightly but not enough to uncover any point of X or even bring it to the boundary, we obtain a collection of disks satisfying the conditions of the theorem.

The following result follows from Theorem 4.1 and the fact that a connected 2-manifold minus the sum of a locally finite collection of mutually exclusive disks is connected.

COROLLARY 4.2. *Under the hypotheses of Theorem* 4.1 *there is a component* U *of* $M^2 - X$ *and a locally finite collection of mutually exclusive disks* E_1, $E_2, \cdots$ *in* M^2 *such that each* E_i *lies in a* D_j *and the* E_i*'s cover* $M^2 - U$.

EXAMPLE. Corollary 4.2 is not true if we weaken the hypothesis by supposing that each component of X lies in some D_i rather than in some $\mathrm{Int}\, D_i$. One could let M^2 be a 2-sphere, D_1 and D_2 be two disks in M^2 with a common boundary, and X be the sum of D_1 and a sequence of mutually exclusive simple closed curves in $\mathrm{Int}\, D_2$ converging homeomorphically to $\mathrm{Bd}\, D_1 = \mathrm{Bd}\, D_2$.

The above is essentially the only counterexample to the modified Corollary 4.2 as may be seen by the following theorems which are not used elsewhere in this paper but included since they show an extent to which Theorem 4.1 and Corollary 4.2 can be strengthened.

THEOREM 4.3. *Suppose* M^2 *is a 2-manifold,* $\{D_i\}$ *is a locally finite collection of disks in* M^2, *and* X *is a closed subset of* M^2 *such that each component of* X *lies in a* D_i. *Then* $M - X$ *has at least one component which does not lie in any* D_i *unless possibly* M^2 *is a 2-sphere which is the sum of two* D_i*'s.*

Proof. Suppose each component of $M^2 - X$ lies in some D_i. We show that under this condition there are two D_i's whose sum is M^2.

Let U_0 be a component of $M^2 - X$ and $D_1, D_2, \cdots, D_n$ be the D_i's containing U_0. Let Y_i be the set of all points p of D_i such that each arc in

M^2 from p to $\operatorname{Bd} D_i$ intersects X, and C_i be the component of Y_i containing U_0. Denote the point set boundary in M^2 of C_i by F_i—that is, $F_i = C_i \cdot \overline{(M^2 - C_i)}$. Then F_i is connected and lies in X.

Consider the case where there is a D_j containing an F_i such that $C_i \not\subset D_j$. We show that under this condition that $D_i + D_j = M^2$. Suppose not. Let $q_0 \in C_i - D_j$. There would be a disk E_j in M^2 slightly larger than D_j such that $D_j \subset \operatorname{Int} E_j$, $q_0 \notin E_j$, and $D_i + E_j \neq M^2$. Since $F_i \subset \operatorname{Int} E_j$, there is a disk $E_i \subset C_i$ such that $q_0 \in \operatorname{Int} E_i$ and $\operatorname{Bd} E_i \subset E_j$. Let E_j' be the disk in E_j bounded by $\operatorname{Bd} E_i$. The disks E_i, E_j' have the same boundary but not the same interiors since $q_0 \in \operatorname{Int} E_i$, $q_0 \notin \operatorname{Int} E_j'$. Hence $E_i + E_j'$ is a 2-sphere. This is contrary to the condition that $D_i + E_j \neq M^2$ implied by the false assumption that $D_i + D_j \neq M^2$. We suppose henceforth that $F_i \subset D_j$ implies that $C_i \subset D_j$.

Since there are only a finite number of C_i's, we pick one of these (say C_1) such that it (C_1) is not properly contained in any other C_i. Then C_1 contains the component of X containing F_1 or else there is a larger C_i and C_1 is a component of $C_1 + X$. Let $G_1, G_2, \cdots$ be a decreasing sequence of disks in M^2 such that $G_{i+1} \subset \operatorname{Int} G_i$, $G_1 \cdot G_2 \cdot \cdots = C_1$, and $X \cdot \operatorname{Bd} G_i = 0$. Let U_i be the component of $M^2 - X$ containing $\operatorname{Bd} G_i$. It follows from the local finiteness of $\{D_i\}$ that there is one of the D_i's (say D_j) such that D_j contains infinitely many of the U_i's. For convenience we suppose D_j contains all the U_i's. Unless $D_1 + D_j$ covers M^2, there is an integer k such that $G_k + D_j$ does not cover M^2. Since $\operatorname{Bd} G_k$ bounds a disk in D_j, this implies that $G_k \subset D_j$. But since U_k lies in D_j, each arc in M^2 from $\operatorname{Bd} G_k$ to $\operatorname{Bd} D_j$ intersects X. Also, each arc in M^2 from G_k to $\operatorname{Bd} D_j$ intersects X and C_1 is not as large as supposed—it should have contained G_k.

Example. Theorem 4.3 is not true if we do not insist that $\{D\}$ is locally finite as can be seen by letting M^2 be the plane, D_i be the round disk with center at the origin and radius i, and $X = \sum \operatorname{Bd} D_i$. However, the following result shows that even without local finiteness on $\{D_i\}$, there cannot be two large components of $M^2 - X$.

Theorem 4.4. *Suppose M^2 is a connected 2-manifold, $\{D_i\}$ is a collection of disks in M^2, X is a closed subset of M^2 such that each component of X lies in some D_i, and U is a component of $M^2 - X$ that does not lie in any D_i. Then each component of $M^2 - U$ lies in some $\{D_i\}$.*

Proof. Suppose the component C of $M^2 - U$ fails to lie in any D_i. Let p be a point of C such that each neighborhood of p intersects U. Let C' be the component of X containing p. Then C' lies in some D_i (say D_1). Since D_1 does not contain C, there is a disk E in M_2 such that $C' \subset D_1 \subset \operatorname{Int} E$ but $C \not\subset E$ and $U \not\subset E$. Since the component of $X \cdot E$ containing C' does not intersect $\operatorname{Bd} E$, there is a disk E' in E such that $C' \subset \operatorname{Int} E'$

and $X \cdot \operatorname{Bd} E' = 0$. Then $\operatorname{Bd} E' \subset U$ and $C \subset E' \subset E$. The assumption that the component C of $M^2 - U$ fails to lie in any D_i led to the contradiction that $C \subset E$ and $C \not\subset E$.

5. **Separating 3-manifolds with disks.** If S^2 is a 2-sphere in E^3, h a homeomorphism of $S^2 \times [-1,1] \to E^3$ such that $S^2 \subset h(S^2 \times (-1,1))$, then one of $h_{-1}(S^2)$, $h_1(S^2)$ lies in the bounded component of $E^3 - S^2$ and the other lies in the unbounded component unless $h(S^2 \times (-1,1))$ contains the bounded component of $E^3 - S^2$. This result follows from the Invariance of Domain Theorem. Something can be said about the matter in the case $h(S^2 \times (-1,1))$ contains all of S^2 except for some small holes as evidenced by Theorem 15 of [5]. In this section we generalize Theorems 14, 15 of [5] to see what can be concluded about M^2 in M^3.

THEOREM 5.1. *Suppose $\{A_i\}$, $\{B_i\}$ are locally finite collections of mutually exclusive disks in a 3-manifold M^3 such that if an element A_r of $\{A_i\}$ intersects an element B_s of $\{B_i\}$, then $A_r + B_s$ lie in an open 3-cell in M^3. Then if $\sum A_i + \sum B_i$ separates two points p,q in M^3, there are elements A_u, B_v in $\{A_i\}$, $\{B_i\}$ respectively such that $A_u \cdot B_v \neq 0$ and $A_u + B_v$ separates p from q in M^3.*

Proof. This theorem is an extension of Theorem 14 of [5] and is proved the same way. We suppose that $\sum B_i + \sum_{i \geq 2} A_i$ does not separate p from q but $\sum A_i + \sum B_i$ does. Let pq be an arc from p to q in $\sum B_i + \sum_{i \geq 2} A_i$ which intersects A_1 as few times as possible. As pointed out in the proof of Theorem 14 of [5], pq intersects A_1 in only a finite number of points and pierces it at each point at which it intersects it. Also, as pointed out in that same proof, there is an element B_t of $\{B_i\}$ and a component U of $A_1 - B_t$ such that $U \cdot \operatorname{Bd} A_1 = 0$ and $U \cdot pq$ is precisely one point.

Let O^3 be an open 3-cell in M^3 containing $A_1 + B_t$. If $A_1 + B_t$ does not separate p from q, there is a simple closed curve J' in M^3 such that $pq \subset J'$ and $(J' - pq) \cdot (A_1 + B_t) = 0$. Adjust J' near $M^3 - O^3$ to get a simple closed curve J in O^3 such that for some neighborhood N of $A_1 + B_t$, $J' \cdot N = J \cdot N$. However, this violates Theorem 13 of [5]. Hence, the assumption that $A_1 + B_t$ does not separate p from q is false.

THEOREM 5.2. *Suppose*

M^2 is a connected 2-manifold (perhaps noncompact) in a connected 3-manifold M^3,

p,q are points in different components of $M^3 - M^2$,

ϵ is a positive number such that no 2ϵ-subset of $V(M^2,\epsilon)$ separates p from q in M^3, and

h is a homeomorphism of $M^2 \times [-1,1] \to M^3$ such that

(1) $h(M^2 \times [-1,1]) \subset V(M^2,\epsilon)$,

(2) M^2 contains a locally finite collection $\{D_i\}$ of mutually exclusive ϵ-disks

such that $M^2 - \sum D_i \subset h(M^2 \times (-1,1))$,

(3) $h_{-1}(M^2)$ *contains a locally finite collection* $\{E_i\}$ *of mutually exclusive* ϵ*-disks such that* $M^2 \cdot h_{-1}(M^2) \subset \sum E_i$,

(4) $h_1(M^2)$ *contains a locally finite collection* $\{F_i\}$ *of mutually exclusive* ϵ*-disks such that* $M^2 \cdot h_1(M^2) \subset \sum F_i$, *and*

(5) *if an element* D *of* $\{D_i\}$ *intersects an element* G *of* $\{E_i\} + \{F_i\}$, *then* $D + G$ *lies in an open* 3*-cell in* M^3.

Then $h_{-1}(M^2) - \sum E_i$ *lies in one component of* $M^3 - M^2$ *and* $h_1(M^2) - \sum F_i$ *lies in the other.*

Proof. This theorem is an analogue of Theorem 15 of [5] and is proved in the same way. We note that $\{E_i\} + \{F_i\}$ is a locally finite collection of mutually exclusive disks and $\{D_i\}$ is another such collection. It follows from Theorem 5.1 that there is an arc from p to q in $M^3 - (\sum D_i + \sum E_i + \sum F_i)$. The first point of $M^2 + h(M^2 \times [-1,1])$ on this arc in the order from p to q belongs to one of $h_{-1}(M^2) - \sum E_i$, $h_1(M^2) - \sum F_i$ and the last such point belongs to the other.

6. **Triangulations with tame skeletons.** One might consider our proof of Theorem 1.1 neater if instead of chopping M^2 up helter-skelter by the J_i's we had taken a triangulation of it with a tame 1-skeleton. In this section we show that any 2-manifold M^2 embedded in a 3-manifold can be given a fine triangulation with a tame 1-skeleton. The 1-skeleton is picked in a certain tame Sierpiński-like set to ensure that it is tame.

THEOREM 6.1. *Suppose* M^2 *is a connected 2-manifold embedded in a* 3*-manifold* M^3 *and* f *is a positive continuous function defined on* M^2. *Then there is a null sequence of mutually exclusive disks* $E_1, E_2, \cdots$ *in* M^2 *such that*

$$\text{diameter } E_i < \min_f E_i,$$

$$\textstyle\sum E_i \textit{ is dense in } M_2, \textit{ and}$$

$$M^2 - \textstyle\sum \operatorname{Int} E_i \textit{ lies in a tame 2-manifold in } M^3.$$

Proof. As in the proof of Theorem 1.1 we let $\{D_i'\}$ be a locally finite collection of disks in M^2 such that

$$\{\operatorname{Int} D_i'\} \text{ covers } M^2 \text{ but no subcollection does,}$$

$$\operatorname{Bd} D_i' \text{ is tame,}$$

$$\text{diameter } D_i' < \min_f D_i', \text{ and}$$

$$\text{each } D_i' \text{ lies on a 2-sphere in an open 3-cell in } M^3.$$

The closure of each component of $M^2 - \sum \operatorname{Bd} D_i'$ is a disk with a tame boundary. Let $F_1, F_2, \cdots$ be the collection of these closures and S_i, O_i be a 2-sphere and open 3-cell respectively such that $F_i \subset S_i \subset O_i \subset M^3$.

It follows from Theorem 9.1 of [7] that there is a tame Sierpiński curve X_i' in S_i such that each component of $S_i - X_i'$ is of diameter less than $1/i$ and $\mathrm{Bd}\, F_i$ belongs to the set of inaccessible points of X_i'. Then $X_i = X_i' \cdot F_i$ is a tame Sierpiński curve in F_i that contains $\mathrm{Bd}\, F_i$.

The E_i's are chosen so that $M^2 - \sum \mathrm{Int}\, E_i = \sum \mathrm{Bd}\, F_i + \sum X_i$. It follows from the Approximation Theorem for Surfaces (Theorem 7 of [2]) that there is a homeomorphism h of M^2 into M^3 such that h is the identity on $\sum \mathrm{Bd}\, F_i + \sum X_i$ and $h(M^2)$ is locally tame off $\sum \mathrm{Bd}\, F_i + \sum X_i$. It follows from Theorem 3.1 that $h(M^2)$ is tame.

THEOREM 6.2. *The E_i's of Theorem* 6.1 *may be chosen so that there is a monotone map g of M^2 onto itself such that for each point m of M^2*

$$D(m,g(m)) < f(m) \text{ and}$$

$$g^{-1}(m) \text{ is either an } E_i \text{ or a point of } M^2 - \textstyle\sum E_i.$$

Proof. It follows from [12] that there is a monotone map g_i of F_i onto itself that is the identity on $\mathrm{Bd}\, F_i$ and such that for each point x of F_i, $g_i^{-1}(x)$ is either a point of $F_i - \sum E_j$ or an E_j in F_i. The map g required by Theorem 6.2 is g_i on each F_i and the identity elsewhere.

THEOREM 6.3. *Suppose M^2 is a 2-manifold embedded in a 3-manifold and f is a positive continuous function defined on M^2. Then there is a triangulation T of M^2 such that the* 1-*skeleton of T is tame and for each simplex σ of T,* diameter $\sigma < \min_f(\sigma)$.

Proof. Since we can operate on the components of M^2 one at a time, we suppose with no loss of generality that M^2 is connected.

Let f_1 be a positive continuous function defined on M^2 such that

$$D(p,q) < f_1(p) \Rightarrow f(q) < 2f(p).$$

It follows from Theorem 6.2 that there is a null sequence of mutually exclusive disks $E_1, E_2, \cdots$ and a monotone map g of M^2 onto itself such that

$$M^2 - \textstyle\sum \mathrm{Int}\, E_i \text{ lies on a tame 2-manifold in } M^3,$$

$$\text{diameter } E_i < \min_{f_1} E_i,$$

$$D(m,g(m)) < f(m)/5,$$

$$g^{-1} \text{ is either an } E_i \text{ or a point of } M^2 - \textstyle\sum E_i.$$

Let f_2 be a continuous function defined on M^2 such that

$$D(g(p),g(a)) < f_2 g(a) \Rightarrow f(p) < 2f(a).$$

Also, let f_3 be a continuous positive function defined on M^2 such that

$$f_3 g(p) < f(p)/5.$$

Consider a triangulation T' of M^2 such that the 1-skeleton of T' misses each $g(E_i)$ and for each simplex s of T'

$$\text{diameter of } s \text{ is less than either } \min_{f_2}(s) \text{ or } \min_{f_3}(s).$$

Let T be the triangulation of M^2 such that for each simplex s of T', $g^{-1}(s)$ is a simplex of T. To see that diameter $g^{-1}(s) < \min_f g^{-1}(s)$, consider three points p, q, a of $g^{-1}(s)$ such that $D(p,q) = \text{diameter } g^{-1}(s), f(a) = \min_f g^{-1}(s)$. Then

$$\begin{aligned}\text{diameter } g^{-1}(s) &= D(p,q) < f(p)/5 + f(q)/5 + D(g(p),g(q)) \\ &< f(p)/5 + f(q)/5 + f_3 g(a) \leqq 2f(a)/5 + 2f(a)/5 + f(a)/5 \\ &= f(a) = \min_f g^{-1}(s).\end{aligned}$$

THEOREM 6.4. *Suppose T_1 is a triangulation of a 2-manifold M^2 embedded in a 3-manifold M^3 so that the 1-skeleton of T_1 is tame. Then for each positive continuous function f defined on M^2 there is a triangulation T_2 of M^2 refining T_1 such that the 1-skeleton of T_2 is tame and for each simplex σ of T_2,* diameter $\sigma < \min_f(\sigma)$.

Proof. We pick f_1, $\{E_i\}$, g, f_2, f_3 as in the proof of Theorem 6.3 but with the added precaution that the E_i's miss the 1-skeleton of T_1 (as we picked the E_i's to miss $\sum \text{Bd}\, F_i$ in the proof of Theorem 6.1) and in defining g we pick it to be the identity on the 1-skeleton of T_1 (as we picked g to be the identity on $\sum \text{Bd}\, F_i$ in the proof of Theorem 6.2). The triangulation T' is taken to refine T_1 so that the 1-skeleton of T' misses the $g(E_i)$'s and each simplex s of T' is of diameter less than the minimum value of $\min_{f_2}(s)$ or $\min_{f_3}(s)$. The simplexes of T_2 are the $g^{-1}(s)$'s.

7. Maintaining separation.

We finally give the theorem which shows that the tame 2-manifold $h_0(M)$ constructed in the proof of Theorem 1.1 is two-sided.

THEOREM 7.1. *Suppose X is a subset of E^n that separates the point p from the point q, U is a bounded subset of X, and h_t $(0 \leqq t \leqq 1)$ is a homotopy of $X \to E^n - (\{p\} + \{q\})$ such that h_0 is the identity and h_t is fixed on $X - U$. Then $h_1(X)$ separates p from q.*

Proof. We suppose that X is closed since if a set separates two points of E^n, some closed set in it does.

We alter E^n and X by adding the point at infinity to each. This makes X compact. We then recover E^n by deleting the point q. We now have that p belongs to a bounded component of $E^n - X$ and wish to show that p belongs to a bounded component of $E^n - h_1(X)$.

Let S^{n-1} be the unit $(n-1)$-sphere with center at p and let r be the retraction of $E^n - \{p\}$ onto S^{n-1} that takes open rays from p onto the point where they pierce S^{n-1}. It follows from Theorem VI 10 on page 97 of [9] that $r: X \to S^{n-1}$ is not homotopic to a constant map. It also follows from this same Theorem VI 10 that p belongs to a bounded component of $E^n - h_1(X)$ unless $r: h_1(X) \to S^{n-1}$ is homotopic to a constant map.

Assume that p does not belong to a bounded component of $E^n - h_1(X)$ and f_t $(0 \leqq t \leqq 1)$ is a homotopy of $h_1(X) \to S^{n-1}$ such that $f_0 = r$ and $f_1 =$ constant map. Then

$$g_t = \begin{cases} rh_{2t} & (0 \leqq t \leqq 1/2), \\ f_{2t-1}h_1 & (1/2 \leqq t \leqq 1) \end{cases}$$

shows that $r: X \to S^{n-1}$ is homotopic to a constant map since $g_0 = r$ and $g_1 =$ constant map. The assumption that p does not belong to a bounded component of $E^n - h_1(X)$ led to this contradiction.

COROLLARY 7.2. *Suppose M^2 is a 2-manifold embedded as a closed set in a 3-manifold M^3, U is an open subset of M^3 homeomorphic to E^3, X is a compact subset of $M^2 \cdot U$ and h is a homeomorphism of M^2 into M^3 such that h is fixed on $M^2 - X$ and $H(X) \subset U$. Then $h(M^2)$ separates M^3 if and only if M^2 does. If p, q are two points of $M^3 - U$ separated in M^3 by M^2, they are also separated in M^3 by $h(M^2)$.*

QUESTION. Can Theorem 7.1 be generalized by replacing E^n by an n-manifold where it is understood that U has a compact closure?

8. **Corrections to [5].** While studying [5], Stephen Slack pointed out some minor errors in it. Although the proof given in [5] contains some simplifications over the proof given in [2], three of the figures used in [5] do not reflect these simplifications and are more applicable to [2]. For example, the assumption for the *very, very special* case (pp. 151-154 of [5]) is so strong that no component of $S - R_2$ has a triod in its boundary. Hence, the situation could not be as bad as depicted in Figure 3 of [5]. Since each component of $S \cdot R_2$ intersects R_1, Figure 5 is inapplicable. Also, the assumption for the *very special case* is enough to prevent anything as bad as that shown in Figure 7.

The proof of the *very special case* (pp. 154-161 of [5]) overlooked the fact that two X_i's as defined on page 155 might intersect in two points and the corresponding $K(X_i)$'s as defined on page 156 might intersect in a nondegenerate subcontinuum of each. This is remedied if one replaces the definition of $K(X_i)$ on page 156 with the following.

"For each X_i that does not share two points with any preceding X_i, let $K(X_i)$ be the sum of X_i and all components of $S - X_i$ with diameters less than $\epsilon/2$. Other K's are defined inductively. If X_i shares two points with some preceding X's and the K's on these preceding X's have been defined, let X_i' be the closure of X_i minus the K's associated with these preceding X's and let $K(X_i)$ be the sum of X_i' and all components of $S - X_i'$ with diameters less than $\epsilon/2$."

References

1. R. H. Bing, *Locally tame sets are tame*, Ann. of Math. **59** (1954), 145-158.

2. ———, *Approximating surfaces with polyhedral ones*, Ann. of Math. **65** (1957), 456-483.

3. ———, *An alternative proof that 3-manifolds can be triangulated*, Ann. of Math. **69** (1959), 37-65.

4. ———, *A surface is tame if its complement is* 1-ULC, Trans. Amer. Math. Soc. **101** (1961), 294-305.

5. ———, *Approximating surfaces from the side*, Ann. of Math. **77** (1963), 145-192.

6. ———, *Each disk in* E^3 *contains a tame arc*, Amer. J. Math. **84** (1962), 583-590.

7. ———, *Pushing a 2-sphere into its complement*, Michigan Math. J. **11** (1964), 33-45.

8. Morton Brown, *Locally flat imbeddings of topological manifolds*, Ann. of Math. **75** (1962), 331-341.

9. W. Hurewicz and H. Wallman, *Dimension theory*, Princeton Univ. Press, Princeton, N. J., 1948.

10. E. E. Moise, *Affine structures in 3-manifolds*. IV. *Piecewise linear approximations of homeomorphisms*, Ann. of Math. **55** (1952), 215-222.

11. ———, *Affine structures in 3-manifolds*. VIII. *Invariance of the knot type; local tame embedding*, Ann. of Math. **59** (1954), 159-170.

12. R. L. Moore, *Concerning upper semicontinuous collections of continua*, Trans. Amer. Math. Soc. **27** (1925), 416-428.

The University of Wisconsin,
Madison, Wisconsin

WILD SURFACES HAVE SOME NICE PROPERTIES

R. H. Bing

This paper was stimulated by hearing in lectures such remarks as the following: "We consider only smoothable surfaces in 3-manifolds so that we can suppose that their intersections with the 2-skeletons of triangulations are nice." This led me to wonder if we might not make the intersections nice even if the surfaces were not smoothable but wild. See Theorem V.3.

I would prefer to give complete proofs of the results in this paper so that the results would be believed even by mathematical agnostics—those who doubt things for which they do not have complete proofs. Complete proofs are given here for the theorems in the main section (§V) of this paper but the proofs of the preliminary theorems come mostly from the literature. Section V deals with the intersection of wild surfaces in triangulated 3-manifolds with other objects in these 3-manifolds, and the preceding sections deal with theorems related to this treatment.

Throughout this paper we deal with 3-manifolds without boundary. *Surfaces* are 2-manifolds imbedded as closed sets in 3-manifolds. Although there are related results about 3-manifolds with boundaries and surfaces without boundaries, they are not treated here. We do not suppose that manifolds and surfaces are compact. Throughout this paper we use M^3 to denote a triangulated 3-manifold with metric ρ.

I. Pushing tame sets to polyhedral ones. One of the important results of the fifties shows that homeomorphisms of 3-manifolds are not wild. A result [10, Theorem 2; 2, Theorem 9] may be stated as follows. We use M_0^3 to denote a triangulated 3-manifold (perhaps different from M^3).

THEOREM I.1. *Suppose*
U is an open subset of M_0^3,
h is a homeomorphism of U into M^3, and
$\epsilon(x)$ is a positive continuous function defined on U.
Then there is a homeomorphism $g: U \to h(U)$ such that
g is PL and
$\rho(g(x), h(x)) < \epsilon(x)$ for each $x \in U$.

At the expense of complicating the statement of Theorem I.1 we could have added that if C is a closed subset of U on which h is locally PL, there is such a g that agrees with h on C.

QUESTION. Suppose U, h are as given in Theorem I.1. Is there an isotopy H_t $(0 \le t \le 1)$ of U onto $h(U)$ such that $h = H_0$ and each H_s is polyhedral for $s \in (0,1]$? It would be especially nice to get an H_t that connects h with a nearby g in

Received July 2, 1985. Revision received October 7, 1985.
This work supported by NSF Grant MCS-810425.
Michigan Math. J. 33 (1986).

some canonical way—or even if g_1, g_2 are two close approximations to h, to isotope g_1 to g_2 in some canonical way.

Suppose T is a triangulation of M^3. A surface S in M^3 is *polyhedral* if it is the union of 2-simplexes of some subdivision of T. It is *tame* if there is a homeomorphism h of M^3 onto itself such that $h(S)$ is polyhedral. We say that *S has a Cartesian product neighborhood* if for some neighborhood V of S there is a homeomorphism $h: (S \times R^1, S \times 0) \to (V, S)$.

A surface S in M^3 is called *2-sided* if it separates some connected open subset of M^3 containing S. It is known that any 2-sphere in M^3 is 2-sided and any surface S in R^3 is 2-sided in R^3.

THEOREM I.2. *Each 2-sided tame surface in M^3 has a Cartesian product neighborhood.*

Theorem I.2 has the following converse.

THEOREM I.3. *A surface S in M^3 is tame if it has a Cartesian product neighborhood.*

The proof comes from giving S a triangulation $T(S)$ and extending $T(S)$ to a prismatic triangulation T_0 of $S \times R^1 = M_0$. Use Theorem I.1 to get a homeomorphism g of $S \times R^1$ onto itself that is PL with respect to T_0 on $S \times (0,1) = U$ and the identity on $M_0 - U$. Let f be a homeomorphism of $S \times R^1$ onto itself that is fixed outside $S \times (-1, 1)$ and shoves $S \times 0$ onto a surface in $S \times (0,1)$ that is polyhedral with respect to T_0. The homeomorphism that is gfg^{-1} on $S \times R^1$ and the identity on $M^3 - [S \times R^1]$ shows that S is tame. We have pushed S to the side. □

A surface S is *locally polyhedral* at a point $s \in S$ if there is a polyhedral disk D in S with $s \in \text{Int } D$. Also S is *locally tame* at a point $s \in S$ if there is a homeomorphism $H: M^3 \to M^3$ such that $h(S)$ is locally polyhedral at $h(s)$.

In our discussion of the proof of Theorem I.3 we could have considered pushing only a part of S. The next result follows.

THEOREM I.4. *Suppose*

S is a 2-sided surface in M^3,

U is an open subset of S on which S is locally tame, and

$\epsilon(x)$ is a continuous non-negative function defined on M^3 that is positive on U.

Then there is an isotopy H_t $(0 \le t \le 1)$ of M^3 onto itself such that

H_0 is the identity,

each $\rho(x, H_t(x)) \le \epsilon(x)$,

$H_t(S)$ is locally polyhedral at $H_t(s)$ if $t > 0$, $s \in U$, and

$H_{t_1}(s_1) \ne H_{t_2}(s_2)$ for $t_1 \ne t_2$, $s_1 \in U$, and $s_2 \in S$.

We say that a surface S in M^3 is *bicollared* if it has a Cartesian product neighborhood. It is *locally bicollared* if for each point $p \in S$ there is a neighborhood N

of p in M^3 such that $S \cap N$ is bicollared in N. Theorems I.2 and I.3 can be extended as follows.

THEOREM I.5. *A surface in M^3 is tame if and only if it is locally bicollared.*

QUESTION. If one omits the requirement in Theorem I.4 that S be 2-sided, can one get an H_t satisfying all but the last of the four restrictions?

II. Recognizing tame surfaces. Theorem I.5 gives a criteria for recognizing tame surfaces but we consider others. We do not give the proof of Theorem II.1 here but mention that one is found in [3, Theorem 2.2].

THEOREM II.1. *A 2-sphere S in R^3 is tame if for each component U of $R^3 - S$ there is a sequence $S_1, S_2, \ldots$ of 2-spheres in U and homeomorphisms $h_1: S \to S_1$, $h_2: S \to S_2, \ldots$ such that each $\rho(x, h_i(x)) < 1/i$.*

We say that the S above can be homeomorphically approximated from both sides.

QUESTION. The following question [9, p. 280] has received considerable attention. Is a 2-sphere S in R^3 tame if for each component U of $R^3 - S$ and each positive integer i there is a map g_i of S into U such that $\rho(s, g_i(s)) < 1/i$?

Note that in the above we hypothesized maps rather than homeomorphisms as hypothesized in Theorem II.1. However, it is known that S is tame if it can be homotoped to either side as described in Theorem II.2. See [9, Theorem 1].

THEOREM II.2. *A 2-sphere S in R^3 is tame if for each component U of $R^3 - S$ there is a map H of $S \times [0, 1]$ into R^3 such that for each $s \in S$*
$H(s \times 0) = s$, and
$H(s \times t) \subset U$ for $t > 0$.

A set X is 1-ULC if for each $\epsilon > 0$ there is a $\delta > 0$ such that each map of Bd D^2 into a subset of X of diameter less than δ can be extended to send D^2 into a subset of X of diameter less than ϵ. It is 1-LC at a point $p \in \bar{X}$ if for each neighborhood N_1 of p there is a neighborhood N_2 of p such that each map of Bd D^2 into $X \cap N_2$ can be extended to map D^2 into $X \cap N_1$. A useful result that is proved in [4, Theorem 7] but not here is the following.

THEOREM II.3. *A surface S in M^3 is tame if $M^3 - S$ is* 1-LC *at each point of S.*

THEOREM II.4. *A 2-sphere S^2 in R^3 is tame if each component of $R^3 - S$ is* 1-ULC [4, Theorem 2].

III. Approximating surfaces. The following is a much-used theorem about approximating surfaces. See [1, Theorem 1].

THEOREM III.1. *Suppose $\epsilon(x)$ is a non-negative continuous function defined on a surface S in M^3. Then there is a homeomorphism h of S into M^3 such that*
$h(S)$ is locally polyhedral at $h(s)$ if $\epsilon(s) > 0$, and
$\rho(s, h(s)) \leq \epsilon(s)$.

In working with the local structure of a surface S it sometimes suffices to deal with 2-spheres as shown by the following application of Theorem III.1. See [4, Theorem 5].

THEOREM III.2. *If p is a point of a surface S in M^3 and $\epsilon > 0$, then there is a disk D^2 in S such that*

$P \in \text{Int } S$, and

D^2 lies in a 2-sphere K^2 in M^3 where diameter $K^2 < \epsilon$ and K^2 is locally polyhedral on $K^2 - D^2$.

An extension of Theorem III.1 that has had far-reaching applications is the following.

THEOREM III.3 (Side Approximation Theorem). *Suppose S is a 2-sphere in R^3 and $\epsilon > 0$. Then there is a homeomorphism h of $S \times [-1, 1]$ into R^3 such that*

each component of $S - h(S \times [-1, 1])$ has diameter less than ϵ,

each $h(S \times t)$ is polyhedral, and

each $\rho(s, h(s \times t)) < \epsilon$.

We say that $h(S \times (-1))$ and $h(S \times 1)$ are approximating S almost from the side — in fact from different sides.

The known proofs of Theorem IV.3 are difficult and will not be given here. One such proof is given in Chapter XIII of [7].

IV. Tame Sierpinski curves. A sequence of sets $X_1, X_2, \ldots$ is called a *null sequence* if for each $\epsilon > 0$ there are at most finitely many X_i's with diameters more than ϵ. Let $D_1, D_2, \ldots$ be a null sequence of mutually disjoint disks on a 2-sphere S^2 such that their union is dense in S^2. A *Sierpinski curve* is any set homeomorphic to $S^2 - \bigcup \text{Int } D_i$. Some people regard it as looking like a thin slice of cheese. A disk D can be changed to a Sierpinski curve by removing from it the interiors of a null sequence of mutually disjoint disks whose union is dense in D.

It may be shown that the topology of a Sierpinski curve does not depend on the null sequence of disks whose interiors are removed from S^2. A point of a Sierpinski curve Y is called an inaccessible point if it is the image of a point of $S^2 - \bigcup D_i$ under the homeomorphism of $S^2 - \bigcup \text{Int } D_i$ onto Y. It follows from Moore's decomposition theorem [11] for a 2-sphere that the set of inaccessible points of Y is homeomorphic to $S^2 - Z$ where Z is a countable dense set of points in S^2. We use $I(Y)$ and $A(Y)$ to denote (respectively) the set of inaccessible points and the set of accessible points of Y.

A Sierpinski curve Y in M^3 is called tame if it lies on a tame 2-sphere in M^3. This is not to say that the 2-sphere S^2 used to define Y is tame or even that it lies in M^3. However, if Y is tame in M^3, it lies on a tame 2-sphere S_0^2 in M^3 and there is a null sequence of mutually disjoint disks $E_1, E_2, \ldots$ on S_0^2 such that $S_0^2 - \bigcup \text{Int } E_i = Y$.

An isotopy H_t $(0 \le t \le 1)$ is a ϵ-isotopy if H_0 is the identity and each

$$\rho(H_0, H_t) \le \epsilon.$$

The following result is known.

THEOREM IV.1. *Suppose* Y_1, Y_2 *are Sierpinski curves in the same disk* D *such that each* Y_i *contains* Bd D. *Then there is an isotopy* H_t $(0 \leq t \leq 1)$ *of* D *onto itself such that*

H_0 *is the identity,*

each H_t *is fixed on* Bd D, *and*

$H_1(Y_1) = Y_2$.

If ϵ *is a positive number such that each component of* $D - Y_1$ *or of* $D - Y_2$ *has diameter less than* $\epsilon/2$, *then we can pick such an* H *that is an* ϵ*-isotopy.*

If C is a closed set on the interior of a disk D which does not contain an open subset of D, there is a Sierpinski curve Y in D such that Bd $D \subset Y$ and C belongs to the inaccessible part of Y. Hence Theorem IV.1 provides us with an engulfing theorem that permits a Sierpinski curve to gobble up certain closed sets. See the Lemma in [6, p. 156].

THEOREM IV.2. *Suppose*

D *is a disk,*

$\epsilon > 0$,

Y *is a Sierpinski curve in* D *containing* Bd D *such that each component of* $D - Y$ *has diameter less than* ϵ, *and*

C *is a nowhere dense closed set in* D.

Then there is an ϵ*-isotopy* H_t $(0 \leq t \leq 1)$ *of* D *onto itself such that*

each H_t *is fixed on* Bd D, *and*

$C \cap \text{Int}\, D$ *lies in the inaccessible part of* $H_1(Y)$.

We shall apply Theorem IV.2 in the case where D is a tame disk in M^3. If N is an open subset of M^3 containing Int D we can use the Cartesian product structure on Int D to feather H_t $(0 \leq t \leq 1)$ out into M^3 so that each of the extended H_t's is fixed off N.

We make use of the following result, whose proof is given in [5, Theorem 8.2].

THEOREM IV.3. *A 2-sphere in* M^3 *is tame if it is locally tame mod a tame Sierpinski curve.*

Suppose Y is a tame Sierpinski curve in M^3 in a possibly wild 2-sphere S^2 in M^3 and $\epsilon(x)$ is a continuous function on S^2 that is 0 on Y and positive on $S^2 - Y$. A combination of Theorems IV.3 and III.1 enables us to know that there is a homeomorphism h of S^2 into M^3 such that h is fixed on Y, $h(S^2)$ is locally polyhedral mod Y, and $\rho(x, h(x)) \leq \epsilon(x)$. The holes in $S - Y$ correspond to those in $h(S) - Y$, and corresponding holes have the same boundaries and are close.

The Side Approximation Theorem (Theorem III.3) can be used to show that even wild 2-spheres in M^3 contain tame Sierpinski curves. The result is stated as follows and the proof is given in [5, Theorem 9.1].

THEOREM IV.4. *If* S^2 *is a 2-sphere (possibly wild) in* M^3, *then there is a sequence* $Y_1, Y_2, \ldots$ *of tame Sierpinski curves in* S^2 *such that*

each component of $S^2 - Y_i$ *has diameter less than* $1/i$, *and*

each Y_i *lies in the inaccessible part of* Y_{i+1}.

Theorems IV.4, IV.3, and III.1 imply that any wild 2-sphere in M^3 can be obtained by starting with a tame 2-sphere, getting a Sierpinski curve in it with small holes, and filling in the small holes with small disks (possibly wild).

V. Moving wild spheres. A theorem is now given to introduce methods we shall use to ambiently isotope wild surfaces in M^3 so as to simplify the intersection of the moved surfaces with other objects.

THEOREM V.1. *Suppose*
S^2 is a 2-sphere (possibly wild) in R^3,
P is a plane in R^3, and
$\epsilon > 0$.
Then there is an ϵ-isotopy H_t $(0 \leq t \leq 1)$ of R^3 onto itself such that
$P \cap H_t(S^2)$ has only a finite number of components with diameters more than ϵ, and
each component of $P \cap H_t(S^2)$ with diameter more than ϵ is a polygonal simply closed curve.

Proof. It follows from Theorem IV.4 that there is a tame Sierpinski curve Y in S^2 such that each component of $S^2 - Y$ has diameter less than ϵ/n, where n is to be selected later. Change S^2 to a tame 2-sphere S_0^2 by using Theorems III.1 and IV.3 to replace the components of $S^2 - Y$ so that the components of $S_0^2 - Y$ have diameters less than ϵ/n.

Since S_0^2 is tame, it follows from Theorem I.4 that there is an (ϵ/n)-isotopy F_t $(0 \leq t \leq 1)$ such that $F_1(S_0^2)$ is polyhedral and in general position with respect to P. The components of $F_1(S_0^2 - Y)$ have diameters less than $3\epsilon/n$. The components of $P \cap F_1(S_0^2)$ are polygonal simple closed curves.

It follows from Theorem IV.2 that there is a $(3\epsilon/n)$-isotopy G_t $(0 \leq t \leq 1)$ of $F_1(S_0^2)$ onto itself such that $P \cap G_1 F_1(S_0^2)$ lies in the inaccessible part of $G_1 F_1(Y)$. Without changing its name, we suppose G_t $(0 \leq t \leq 1)$ is extended to a $(3\epsilon/n)$-isotopy on R^3 that is invariant on $F_1(S_0^2)$. The isotopy H obtained by following F with G satisfies the conditions of the theorem for a suitably chosen n.

Now to pick n. Since F is an (ϵ/n)-isotopy and G is a $(3\epsilon/n)$-isotopy, H is a $(4\epsilon/n)$-isotopy. Since components of $S^2 - Y$ have diameters less than ϵ/n, the diameters of the images of these components under H_1 are less than $9\epsilon/n$. Hence we pick $n = 9$.

The 2-sphere S is the union of the accessible part of Y, the inaccessible part of Y, and $S - Y$. The image under H_1 of the accessible part of Y misses P since $P \cap H_1(S^2)$ lies in the inaccessible part of $H_1(Y)$. The images under H_1 of the components of $S - Y$ have diameters less than ϵ. □

We now extend Theorem V.1 by making repeated applications of the techniques of its proof. Theorem V.2 is a variation of Theorem 1 of [6].

THEOREM V.2. *Under the hypotheses of Theorem* V.1 *there is an ϵ-isotopy H_t $(0 \leq t \leq 1)$ of R^3 onto itself such that*

each H_t is fixed outside the ϵ-neighborhood of $P \cap S^2$;
each nondegenerate component of $P \cap H_1(S^2)$ is a polygonal simple closed curve;
for each $\delta > 0$, at most a finite number of components of $P \cap H_1(S^2)$ have diameters of more than δ;
each arc in P that pierces $P \cap H_1(S^2)$ also pierces $H_1(S^2)$; and
each point of $P \cap H_1(S^2)$ is a limit point of the union of the nondegenerate components of $P \cap H_1(S^2)$.

Proof. We shall pick a sequence $\epsilon_1, Y_1, g_1 h_1, \epsilon_2, Y_2, g_2 h_2, \epsilon_3, \ldots$, but an element of the sequence is not selected until the previous ones have been used. First we stipulate that $\epsilon_1 = \epsilon/2$ and although we do not pick the other ϵ_i's now, we remark that they will be selected so that $\sum \epsilon_i < \epsilon$.

Let Y_1 be a tame Sierpinski curve S^2 such that each component of $S^2 - Y_1$ has diameter less than $\epsilon_1/9$.

Pick h_1 and g_1 as we picked F_1 and G_1 in the proof of Theorem V.1, subject to the restrictions that the isotopies leading to h_1 and g_1 are fixed outside the ϵ_1-neighborhood of $P \cap S^2$. Rather than $g_1 h_1(S_0^2)$ being polyhedral everywhere we only need that it be polyhedral near P.

In selecting ϵ_2, we want to start taking precautions to make the limit of the $g_1 h_1$, $g_2 h_2 g_1 h_1$, $g_3 h_3 g_2 h_2 g_1 h_1$, ... be a homeomorphism rather than merely a map. We want to prevent two points from being moved to the same point. We pick ϵ_2 to be less than $\epsilon_1/2$ and also subject to the further restriction that if p, q are points of R^3 with $\rho(p,q) > \epsilon_1$, then $\rho(g_1 h_1(p), g_1 h_1(q)) > \epsilon_2$.

Pick Y_2 so that the components of $g_1 h_1(S^2 - Y_2)$ have diameters less than $\epsilon_2/9$ and so that Y_1 lies in the inaccessible part of Y_2. In defining $g_2 h_2$ we use $g_1 h_1(S^2)$ and $g_1 h_1(Y_2)$ instead of S^2 and Y_1. The accessible part of Y_1 was sent by $g_1 h_1$ into the complement of P so we pick $g_2 h_2$ to be the identity on $g_1 h_1(Y_1)$.

We follow this pattern to define $\epsilon_1, Y_1, g_1 h_1, \epsilon_2, Y_2, g_2 h_2, \epsilon_3, \ldots$ and note that $H_1 = \text{limit}(g_1 h_1, g_2 h_2 g_1 h_1, \ldots)$ satisfies most of the requirements of the theorem.

We now examine the condition on piercing. If an arc A in P pierces (in P) one of the polygonal arcs in $P \cap g_1 h_1(Y_1)$ at a point p, it pierces $g_1 h_1(S_0^2)$ at p, where S_0^2 is a replacement for S^2 that is locally tame near P. Since p is an inaccessible point of $g_1 h_1(Y_1)$, there are small simple closed curves in $g_1 h_1(Y_1)$ that circle p in $g_1 h_1(S_0^2)$. Hence they circle A. These simple closed curves are not moved further since $H_1 = g_1 h_1$ on Y_1. Hence, if A intersects $H_1(S^2)$ only at p, A pierces $H_1(S^2)$. Similarly, if an arc in P pierces (in P) one of the polygonal arcs in $P \cap g_i h_i \cdots g_2 h_2 g_1 h_1(Y_i)$ at a point p, then the arc pierces $H_1(S^2)$ at p if it does not intersect $H_1(S^2)$ at any other point.

If there are points of $P \cap H_1(S^2)$ that are not limit points of the union of the nondegenerate components of $P \cap H_1(S^2)$, $H_1(S)$ may be pushed to the side of P near these points. This push does not alter the piercing. □

We now consider how to move a 2-sphere S^2 in R^3 to improve its intersection with the 2-skeleton T^2 of a triangulation T of R^3. To begin with, S^2 might inter-

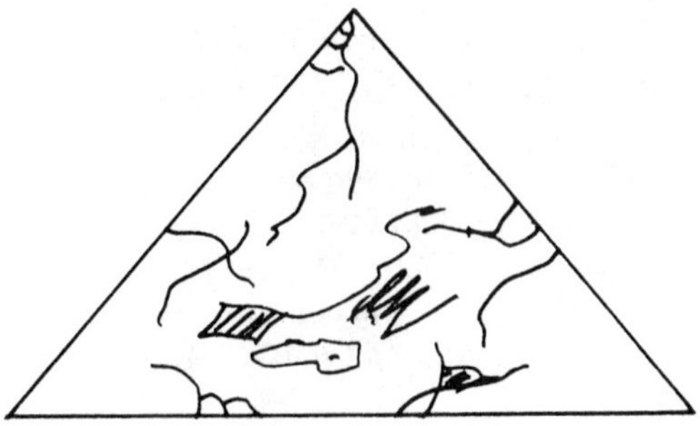

Figure 1

sect T^2 in a terribly looking set. In fact, if Δ^2 is a 2-simplex of T, $T \cap \Delta^2$ might look even worse than shown in Figure 1.

For each $\epsilon > 0$, S^2 can be ϵ-approximated by a polyhedral 2-sphere. Hence, there is an ϵ-homeomorphism $g: S^2 \to R^3$ such that $g(S^2)$ is polyhedral. Then $\Delta^2 \cap g(S^2)$ might look like Figure 2.

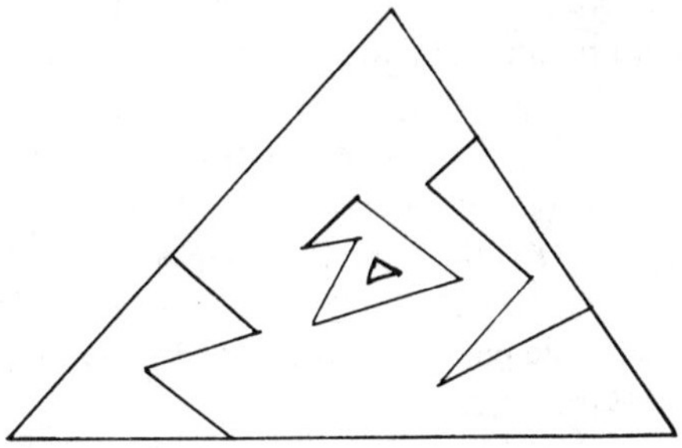

Figure 2

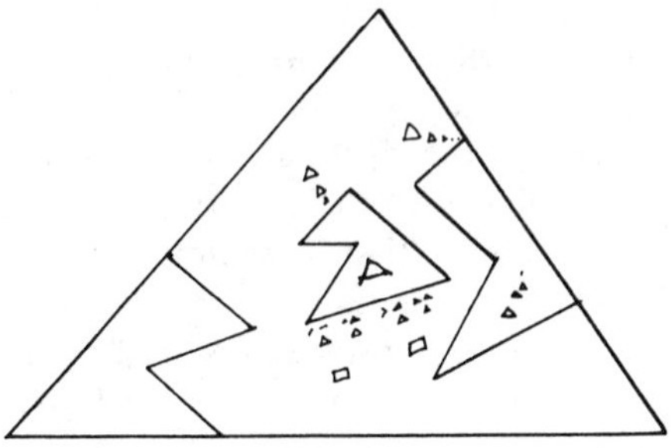

Figure 3

However, g is not an ambient homeomorphism and its domain is S^2 rather than R^3. The following theorem shows that there is an ϵ-homeomorphism $H_1: R^3 \to R^3$ such that $\Delta^2 \cap H_1(S^2)$ might look like Figure 3.

THEOREM V.3. *Suppose*

T is a triangulation of R^3,

S^2 is a 2-sphere (possibly wild) in R^3, and

$\epsilon > 0$.

Then there is an ϵ-isotopy H_t $(0 \le t \le 1)$ of R^3 onto itself such that:

each H_t is fixed outside the ϵ-neighborhood of $S^2 \cap T^2$;

$H_1(S^2)$ misses each vertex of T;

if Δ^1 is a 1-simplex of T, then

$\Delta^1 \cap H_1(S^2)$ is finite and

Δ^1 pierces $H_1(S^2)$ at each point of $\Delta^1 \cap H_1(S^2)$; and

if Δ^2 is a 2-simplex of T, then

each nondegenerate component of $\Delta^2 \cap H_1(S^2)$ is either a polygonal spanning arc of Δ^2 or a polygonal simple closed curve in Int Δ^2,

any arc in Δ^2 that pierces $\Delta^2 \cap H_1(S^2)$ in Δ^2 also pierces $H_1(S^2)$,

for each $\delta > 0$, at most finitely many components of $\Delta^2 \cap H_1(S^2)$ have diameters more than δ; and

each degenerate component of $\Delta^2 \cap H_1(S^2)$ lies in Int Δ^2 *and is a limit point of the union of the nondegenerate components.*

Proof. It is easy to ambiently move S^2 so that the adjusted S^2 misses T^0, but getting an adjustment that is nice near T^1 is harder. If one blindly follows the techniques of the proof of Theorem V.2 and uses T^2 instead of P, one gets an H such that $T^2 \cap H_1(S^2)$ is somewhat nice except near T^1—but there is a problem in making it nice there. Even if one follows [6, Theorem 2] and first gets an H_t $(0 \leq t \leq 1/2)$ so that $T^0 \cap H_{1/2}(S^2) = 0$ and $T^1 \cap H_{1/2}(S^2)$ is finite and then continues the isotopy following techniques of the proof of Theorem V.2 (but without moving T_1 further), one gets an H such that $T^1 \cap H_1(S^2)$ is finite; but there is no assurance that, for some 2-simplex Δ of T, $\Delta \cap H_1(S^2)$ does not contain several arcs whose intersection is a point of Bd Δ.

To avoid such difficulties we proceed as follows. First, suppose without loss of generality that $T^0 \cap S^2 = 0$ and $T^1 \cap S^2$ is of dimension 0. We then model a proof after that of Theorem V.2 by picking $\epsilon_1, Y_1, g_1 h_1, \epsilon_2, \ldots$ but with restrictions. We pick ϵ_1 even smaller than $\epsilon/2$ but leave that detail until later. Then we select Y_1, S_0^2, and $g_1 h_1$ so that $h_1(S_0^2)$ is locally polyhedral near T^2 and in general position with respect to T^2. An additional restriction that we impose is that $T^1 \cap h_1(S^2) \subset h_1(I(Y_1))$. Before showing why this restriction is possible, we show how to proceed if it is possible.

Pick g_1 to be invariant on $h_1(S_0^2)$ and so that $T_2 \cap g_1 h_1(S_0^2)$ lies in the image under $g_1 h_1$ of the inaccessible part of Y_1 while $g_1 h_1(S^2 - Y_1)$ misses T^1.

The homeomorphism $g_1 h_1$ defines $H_1 = \text{limit}(g_1 h_1, g_2 h_2 g_1 h_1, \ldots)$ on Y_1, and it only remains to define H_1 near the components of $g_1 h_1(S^2 - Y_1)$. The closures of each of these components is a disk D such that $T^1 \cap D = \emptyset$ and $T^2 \cap \text{Bd}\, D = \emptyset$. We pick an open set $N(D)$ about Int D so that $T^1 \cap \bar{N}(D) = \emptyset$ and pick the $g_i h_i$'s so that $H_1(D) \subset \bar{N}(D)$.

Finally, we show how to get h_1 so that $h_1(S^2 - Y_1)$ does not intersect T^1. Let $R_1, R_2, \ldots, R_n$ be a finite number of mutually disjoint rectangular planar disks each of diameter less than $\epsilon/2$, and such that no vertex of T lies on any R_i, $T^1 \cap S^2 \subset \bigcup \text{Int}\, R_i$, and $T^1 \cap R_i = p_i q_i$ is a bisector of R_i perpendicular to two edges E_{i_1}, E_{i_2} of R_i (where $S^2 \cap (E_{i_1} \cup E_{i_2}) = \emptyset$). See Figure 4. Let δ_i be the minimum of the length of E_{i_1} and $\rho(S^2, E_{i_1} \cup E_{i_2})$. Let ϵ_1 be less than any δ_i.

Now pick a preliminary h_1 (which we call k) so that $k(S_0^2)$ is locally polyhedral and in general position not only near T^2 but also near each R_i. We pick it in much the same way that we picked $g_1 h_1$ in the proof of Theorem V.2, but the components of $k(S^2 - Y_1)$ have diameters less than ϵ_1 rather than less than $\epsilon/4$. We do not yet suppose that $k(S^2 - Y_1)$ misses T^1. When we modify k to get this, we will designate the modified k by h_1.

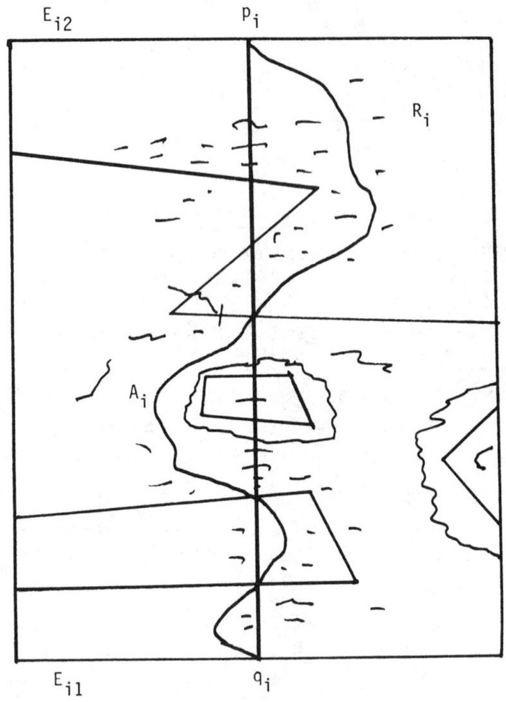

Figure 4

Figure 4 shows $R_i \cap k(S_0^2)$ as polygonal and $R_i \cap k(S^2 - Y_1)$ as the union of squiggles. Since these squiggles lie in $k(S^2 - Y_1)$ and $R_i \cap k(S^2)$ misses $k(A(Y_1))$, then if $p_1, p_2, \ldots$ is a sequence of points of $R_i \cap k(S^2 - Y_1)$ converging to a point of $R_i \cap k(S_0^2)$ and Z_i is a squiggle containing p_i, the diameters of the Z_i's converge to 0. Consider the decomposition of R_i whose nondegenerate elements are squiggles with holes filled. It follows from [10] that there is a spanning arc A_i of R_i from p_i to q_i that misses each squiggle and each vertex of $R_i \cap k(S_0^2)$ and is such that $A_i \cap k(S_0^2)$ is finite and pierces $k(S_0^2)$ at each point of $A_i \cap k(S_0^2)$. Since there is an isotopy of R_i onto itself that is fixed on $R_i \cap k(S_0^2)$ and makes A_i polygonal, and this isotopy can be feathered into R^3 without moving $k(S_0^2)$, we suppose without loss of generality that A_i is polygonal and normal to $R_i \cap k(S_0^2)$ at each point where they intersect.

Let f_i be a PL homeomorphism of R_i onto itself that is fixed on Bd R_i, takes A_i to the straight segment from p_i to q_i, and is an isometry near each point of $A_i \cap k(S_0^2)$. Let C_i be a small rectangular solid having R_i as a spanning disk. Extend f_i to a PL homeomorphism of C_i onto itself that is fixed on Bd C_i, f_i on R_i and an isometry near each point of $A_i \cap k(S_0^2)$. We suppose T^2 and the extended $f_i(S_0^2)$ are in general position. The required h_1 is k modified by the various extended f_i's. Note that $h_1(S_0^2)$ is locally polyhedral near T^2 and $h_1(S^2 - Y_1)$ misses T^1. Craggs gave an alternative description of such an h_1 in [8, Theorem 6.1]. □

In a similar fashion we could treat a noncompact S and a variable ϵ to get the following variation of Theorem V.3.

THEOREM V.4. *Suppose*
T is a triangulation of M^3,
S is a surface (possibly wild) in M^3, and
$\epsilon(x)$ is a non-negative continuous function defined on R^3 and positive on S.
Then there is an isotopy H_t $(0 \leq t \leq 1)$ of M^3 onto itself such that
H_0 is the identity,
each $\rho(x, H_t(x)) \leq \epsilon(x)$, and
$H_1(S)$ satisfies the same restrictions about its intersections with the skeletons of T as did $H_1(S^2)$ in Theorem V.3.

The following result was stated as Theorem 4 in [6] with the statement that the proof would be given in a later paper. This is that later paper.

THEOREM V.5. *Suppose S_1, S_2 are compact surfaces in M^3. Then for each $\epsilon > 0$ there is an ϵ-isotopy H_t $(0 \leq t \leq 1)$ of M^3 onto itself such that*
each H_t is fixed outside the ϵ-neighborhood of $S_1 \cap S_2$, and
the collection of nondegenerate elements of $S_1 \cap H_1(S_2)$ is a null sequence of tame simple closed curves whose union is dense in $S_1 \cap H_1(S_2)$.

Proof. So that we can speak of Sierpinski curves rather than sets which are locally like Sierpinski curves, we suppose S_1 and S_2 are 2-spheres. Our proof will be a modification of the preceding three proofs but we avoid epsilonics. We use Y_i $(i=1,2)$ to denote a tame Sierpinski curve in S_i with small holes and $I(Y_i)$, $A(Y_i)$ to denote (respectively) the inaccessible and accessible parts of Y_i. Also, S_{i0} denotes a tame 2-sphere obtained by replacing closures of components of $S_i - Y_i$ by nearby closed disks. Let $h_1, h_2 h_1$ be homeomorphisms of M^3 onto itself that are the ends of short isotopies starting at the identity and carrying S_{10}, S_{20} (respectively) to polyhedral sets, so that $h_1(S_{10}) \cap h_2 h_1(S_{20})$ is the union of a finite number of polyhedral simple closed curves. Then

(a) $$h_1(S_{10}) \cap h_2 h_1(S_{20}) = \bigcup J_i.$$

Let g_1 be the end of a short isotopy invariant on $h_1(S_{10})$ that makes $h_1(I(Y_1))$ engulf $h_1(S_{10}) \cap h_2 h_1(S_{20})$. Then

(b) $$g_1 h_1(I(Y_1)) \supset \bigcup J_i$$

and

(c) $$g_1 h_1(A(Y_1)) \cap h_2 h_1(S_{20}) = \varnothing.$$

Also, g_2 is the end of a short isotopy invariant on $h_2 h_1(S_{20})$ that makes $h_2 h_1(I(Y_2))$ engulf $h_2 h_1(S_{20}) \cap g_1 h_1(S_1)$. Then

(d) $$g_2 h_2 h_1(I(Y_2)) \supset \bigcup J_i$$

and

(e) $$g_2 h_2 h_1(A(Y_2)) \cap g_1 h_1(S_1) = \varnothing.$$

Then

$$\begin{aligned}\bigcup(J_i) &= h_1(S_{10})\cap h_2h_1(S_{20}) \quad \text{[from (a)]}\\ &= g_1h_1(S_{10})\cap g_2h_2h_1(S_{20}) \quad \text{[from definitions of } g_1, g_2]\\ &\subset g_1h_1(I(Y_1))\cap g_2h_2h_1(I(Y_2)) \quad \text{[from (b) and (d)]}\\ &\subset g_1h_1(S_{10})\cap g_2h_2h_1(S_{20}) = \bigcup J_i.\end{aligned}$$

Hence,

(f) $$g_1h_1(I(Y_1))\cap g_2h_2h_1(I(Y_2)) = \bigcup J_i.$$

The required isotopy H_t $(0\le t\le 1)$ is defined on $0\le t\le 1/2$, so that

(g) $$H_{1/2} = h_1^{-1}g_1^{-1}g_2h_2h_1 = k.$$

The image of $S_1\cap k(S_2)$ under g_1h_1 is

$$g_1h_1(S_1)\cap g_2h_2h_1(I(Y_2)\cup A(Y_2)\cup(S_2-Y_2)).$$

However,

$$g_1h_1(S_1)\cap g_2h_2h_1(A(Y_2)) = \varnothing \quad \text{[by (e)]},$$

so $g_1h_1(S\cap kS_2)$ is the union of the two closed sets $g_1h_1(S_1)\cap g_2h_2h_1(I(Y_2))$ and $g_1h_1(S_1)\cap g_2h_2h_1(S_2-Y_2)$. The components of the second of these sets are small, so we examine the first set:

$$g_1h_1(I(Y_1)\cup A(Y_1)\cup(S_1-Y_1))\cap g_2h_2h_1(I(Y_2)).$$

Note that

$$g_1h_1(A(Y_1))\cap g_2h_2h_1(I(Y_2)) = \varnothing \quad \text{[by (c)]},$$

and components of $g_1h_1(S-Y_1)\cap g_2h_2h_1(I(Y_2))$ are small. We look at

$$g_1h_1(I(Y_1))\cap g_2h_2h_1(I(Y_2)).$$

However, this is $\bigcup(J_i)$ by condition (f). Then the components of $S_1\cap k(S_2)$ that are not small are simple closed curves in $h_1^{-1}g_1^{-1}(\bigcup J_i)$.

We continue the description of H_t $(0\le t\le 1)$ as in the proof of Theorem V.2, but with one exception. Recall in the proof of Theorem V.2 that as a final move we pushed isolated points of $P\cap H_1(S^2)$ to one side of P. We might encounter difficulty in pushing isolated points off $S_1\cap H_1(S_2)$ at the end so if the possibility of an isolated point developing as t approaches 1 occurs, we change the S_{10} and S_{20} at later stages to introduce small simple closed curves in the intersection of their respective images so that each point of $S_1\cap H_1(S_2)$ is a limit point of the union of the images of the J_i's. □

Just as we varied Theorem V.3 to consider noncompact surfaces and variable ϵ's, a variation of Theorem V.5 could be given.

REFERENCES

1. R. H. Bing, *Approximating surfaces with polyhedral ones,* Ann. of Math. (2) 65 (1957), 456–483.

2. ———, *An alternative proof that* 3-*manifolds can be triangulated,* Ann. of Math. (2) 69 (1959), 37–65.
3. ———, *Conditions under which a surface in* E^3 *is tame,* Fund. Math. 47 (1959), 105–139.
4. ———, *A surface is tame if its complement is* 1-ULC, Trans. Amer. Math. Soc. 101 (1961), 294–305.
5. ———, *Pushing a* 2-*sphere into its complement,* Michigan Math. J. 11 (1964), 33–45.
6. ———, *Improving the intersections of lines and surfaces,* Michigan Math. J. 14 (1967), 155–159.
7. ———, *The geometric topology of* 3-*manifolds,* Amer. Math. Soc., Providence, R.I., 1983.
8. R. Craggs, *Improving the intersection of polyhedra in* 3-*manifolds,* Illinois J. Math. 12 (1968), 567–586.
9. J. Hempel, *A surface in* S^3 *is tame if it can be deformed into each complementary domain,* Trans. Amer. Math. Soc. 111 (1964), 273–287.
10. E. E. Moise, *Affine structures in* 3-*manifolds,* V. *The triangulation theorem and Hauptvermutung,* Ann. of Math. (2) 56 (1952), 96–114.
11. R. L. Moore, *Concerning upper semicontinuous collections of continua,* Trans. Amer. Math. Soc. 27 (1925), 416–429.

Department of Mathematics
University of Texas
Austin, Texas

D. Recognizing Tameness and Related Topics

ANNALS OF MATHEMATICS
Vol. 59, No. 1, January, 1954
Printed in U.S.A.

LOCALLY TAME SETS ARE TAME*

R. H. BING

(Received April 20, 1953)

1. Introduction

In this paper we prove some theorems concerning 3-manifolds with boundaries. Any such set can be triangulated (Theorem 5). Furthermore, any locally tame closed subset of such a set is tame (Theorem 9). Working independently, Moise has subsequently obtained these results and is including his proofs along with other results in [5]. Much of the terminology used in this paper resembles that used in [4].

An *n-manifold* is a separable metric space K such that each point p of K lies in an open set that is topologically equivalent to E^n (Euclidean n-space). An *n-manifold with boundary* is a separable metric space K such that each point p of K lies in an open set N such that $\bar{N}$ is topologically equivalent to I^n (a cube plus its interior in E^n). We note that an n-manifold is an n-manifold with boundary but not conversely.

The term "*boundary*" is used in two senses. In the point set sense we say that the *boundary* of a set X is the intersection of the closure of X and the closure of the complement of X. However, if K is an n-manifold with boundary, we use Bd K to denote the points of K which do not have a neighborhood topologically equivalent to E^n. For example, if in a triangulated 3-manifold M with boundary, St(v) represents the star of a vertex v (the sum of all simplexes having a vertex at v) then Bd $\overline{\text{St}}(v)$ will be larger than the point set boundary of $\overline{\text{St}}(v)$ if v belongs to Bd M. In this paper we always use Bd(K) to denote the set of points of K which do not have a neighborhood topologically equivalent to E^n.

The *interior* of an n-manifold K with boundary is K − Bd K and is denoted by Int K.

A set K can be *triangulated* if there is a homeomorphism h carrying a geometric complex C onto K. (We suppose that complexes are locally finite.) Let $T(K)$ denote the collection of all subsets σ of K such that σ is the image under h of a simplex of C. If $(a_1x_1, a_2x_2, \cdots, a_ix_i)$ represents a point p of C

$$(a_j \geqq 0, \textstyle\sum a_j = 1, x_j = \text{vertex of } C),$$

we use $(a_1h(x_1), a_2h(x_2), \cdots, a_ih(x_i))$ to represent the point $h(p)$ of K. Hence $T(K)$ imposes an affine structure on K. Then $T(K)$ is called a *triangulation* of K. Its elements are called *simplexes*.

If $T(K)$ is a triangulation of a set K, it is convenient to think of assigning a metric to K compatible with its original topology so that each n-simplex in $T(K)$ is isometric to a geometric n-simplex in E^n. To get such a metric we may take

* This work was supported in part by funds from the National Science Foundation under Contract NSF-G238

the distance between two points to be the square root of the sum of the squares of the differences of the coordinates of the points, where it is understood that if v is a vertex and p is a point not in $\mathrm{St}(v)$, the coordinate of p corresponding to v is 0. This metric will help us in considering subdivisions of $T(K)$ and in deciding what sorts of things are polyhedrons in K under $T(K)$.

A set P is called a *polyhedron* in a geometric complex C if it is equal to a subcomplex of some subdivision of C. If the subcomplex contains only a finite number of simplexes of the subdivision, P is called a *finite polyhedron*. (Some authors [3] use the term *polytope*.) For example, in E^2 a finite polyhedron is always the sum of a finite number of points, closed intervals, and triangles plus their interiors. Conversely, the sum of such sets is always a polyhedron. In E^3, a polyhedron might also contain some solid tetrahedrons.

When we are associating a particular triangulation $T(K)$ with a set K, we regard K as a complex. However, if $T'(K)$ is a different triangulation of K, a set P might be a polyhedron under one of the triangulations and not under the other. Two triangulations are called *compatible* if they have a common subdivision.

Let K be a set in a geometric complex C. Then (1) K is *tamely imbedded* in C if there is a homeomorphism of C onto itself that throws K onto a polyhedron; (2) K is *locally tamely imbedded* if for each point p of K there is a neighborhood N of p and a homeomorphism h_p of $\bar{N}$ onto a polyhedron in C such that $h_p(\bar{N} \cdot K)$ is a polyhedron. (The condition that $h_p(\bar{N})$ is a polyhedron prevents such things as the closure of the graph of $y = | x \sin 1/x |$ from being locally tame at the origin in the upper half plane $y \geqq 0$.) Here (1) is a generalization of a definition due to Artin and Fox [1].

A continuous transformation t of a geometric complex C into a geometric complex G is *piecewise linear* if there are subdivisions C' and G' of C and G respectively such that t carries each simplex of C' linearly onto a simplex of G'. Such a transformation would carry polyhedrons onto polyhedrons. However, a transformation of a 1-complex onto itself might carry polyhedrons onto polyhedrons and still not be piecewise linear. Whether or not a transformation is piecewise linear depends on the triangulations of the sets involved. Even the identity map of an n-manifold onto itself will not be piecewise linear with respect to two incompatible triangulations.

We call t *locally piecewise linear* at a point p of C if some neighborhood of p lies in a subcomplex of C on which t is piecewise linear. Similarly, a point set K in C is *locally polyhedral* at a point p of K if there is an open set U containing p such that $\bar{U} \cdot K$ is a polyhedron in C. Here is a result concerning closed sets which are locally polyhedral.

LEMMA 1. *Suppose K is a closed set in a triangulated 3-manifold with boundary M such that K is locally polyhedral at each of its points. Then K is a polyhedron—that is, there is a subdivision $T'(M)$ of the original triangulation $T(M)$ of M such that K is the sum of simplexes of M under $T'(M)$.*

PROOF. The first step is to consider each 3-simplex s^3 of M under $T(M)$. We show that there is a subdivision $T(\bar{s}^3)$ of $\bar{s}^3$ so that $K\cdot\bar{s}^3$ is the sum of simplexes under $T(\bar{s}^3)$.

For each point p of $\bar{s}^3\cdot K$ there is an open set U_p containing p and a subdivision T_p of $\bar{s}^3$ such that $\bar{U}\cdot K$ is the sum of the simplexes of T_p. Let $U_1, U_2, \cdots, U_n$ be a finite number of such open sets covering $K\cdot\bar{s}^3$ and $T_1, T_2, \cdots, T_n$ be the corresponding triangulations of $\bar{s}^3$.

There are only a finite number of planes such that each of these planes contains 3 noncollinear points each of which is a vertex of some T_i. These planes subdivide $\bar{s}^3$ into convex polyhedra. Without dividing the edges of the polyhedra, the faces may be triangulated. The triangulation of the faces may be extended to the interiors by radial lines drawn to a point on the interior. This gives $T(\bar{s}^3)$.

The next step is to consider the 2-simplexes s^2 of M under $T(M)$. Triangulate $\bar{s}^2$ by a subdivision so that each vertex and edge introduced into $\bar{s}^2$ at the first step is the sum of vertices and edges in this triangulation of $\bar{s}^2$.

Finally, we consider a 3-simplex introduced at the first step. If its boundary was altered at the second step, we put a vertex on its interior and subdivide it in a natural fashion so as to introduce no new edges on its boundary.

Similarly, the following results may be demonstrated.

LEMMA 2. *A homeomorphism between two triangulated 3-manifolds with boundaries is piecewise linear if it is locally piecewise linear at each point.*

LEMMA 3. *Two triangulations of a 3-manifold with boundary are compatible if they are locally so.*

Two triangulations T_1 and T_2 are *locally compatible* at a point p if there is a neighborhood N of p and subdivisions T_1' and T_2' of T_1 and T_2 respectively such that each simplex of T_1' intersecting N is a simplex of T_2'. Even though two simplexes are the same as point sets, we do not call them equal if their affine structures differ.

LEMMA 4. *If U is an open subset of a triangulated 3-manifold with boundary M, then there is a triangulation of U which is locally compatible with that of M. In fact, if ϕ is a continuous positive function on U, there is such a triangulation with the additional property that each of its simplexes has a diameter that is less than the minimum value of ϕ on its closure.*

2. Piecewise Linear Approximations on Manifolds

If each of f and g is a homeomorphism of K onto a part of M (not necessarily the same part), we call f an *ε-approximation* of g if for each point k of K, $\rho(f(k), g(k)) \leqq \varepsilon$. (We use $\rho(p, q)$ to denote the distance between p and q.)

In this paper we lean heavily on two results due to Moise. The first [Lemma 4 of 4] deals with the fitting together of two homeomorphisms and may be stated as follows.

THEOREM 1. *Suppose K is a compact triangulated 3-manifold with boundary and K_1, K_2 are subcomplexes of K such that K_1 and K_2 are 3-manifolds with boundaries, $K = K_1 + K_2$, and $K_1\cdot K_2$ is a 2-manifold L with boundary. Let f be a homeomorphism throwing K into a triangulated 3-manifold M, U be a neighborhood* *of L in K, and ε be a positive number. Then there is a positive number δ such that if f_1' and f_2' are piecewise linear δ-approximations of $f_1 = f \mid K_1$ and $f \mid (K_2 + U)$ respectively, then there is a piecewise linear ε-approximation f' of f such that* (1) *$f' \mid K_1 = f_1'$ and* (2) *$f' \mid (K_2 - U) = f_2' \mid (K_2 - U)$.*

In view of Theorem 2′, Theorem 1 can also be stated in the following interesting form.

THEOREM 1′. *Let K, K_1, K_2, M, and f be as in Theorem 1. Then for each positive number ε there is a positive number γ such that each piecewise linear γ-approximation of f on K_1 can be extended to get an ε-approximation of f on K.*

The second theorem due to Moise [Theorem 2 of 4] that we shall use extensively deals with piecewise linear approximations of homeomorphisms.

THEOREM 2. *Let K and K' be triangulated 3-manifolds, U be an open subset of K, f be a homeomorphism throwing U into K', and ϕ be a positive continuous function defined on U. Then there is a homeomorphism f' throwing U into K', such that* (1) *f' is locally piecewise linear on U, and* (2) $\rho(f(p), f'(p)) < \phi(p)$.

The following modification of Theorem 2 is the result shown in the proof of Lemma 5 of [4].

THEOREM 2′. *If f is a homeomorphism of a triangulated 3-manifold with boundary into a triangulated 3-manifold and ε is a positive number, then there is a piecewise linear ε-approximation of f.*

3. Piecewise Linear Approximations on 3-Manifolds with Boundaries

In order to cope with 3-manifolds with boundaries we need to strengthen Theorem 2 to get the following result.

THEOREM 3. *Suppose M, M' are triangulated 3-manifolds with boundaries, U is an open subset of M, f is a homeomorphism throwing U into M' such that f is locally piecewise linear in carrying $U \cdot \mathrm{Bd}\, M$ onto $f(U \cdot \mathrm{Bd}\, M)$, and ϕ is a positive continuous function defined on U. Then there is a homeomorphism f' of U onto $f(U)$ such that* (1) *f' is locally piecewise linear on U,* (2) *$\rho(f(p), f'(p)) < \phi(p)$, and* (3) *$f = f'$ on $U \cdot \mathrm{Bd}\, M$.*

PROOF. We find from Lemma 4 that there is no loss of generality in supposing the $U = M$, so we do this.

Since $f(M)$ is open in M' except possibly at points of $f(\mathrm{Bd}\, M)$ and it is closed at these points, there is an open subset U' of M' such that $f(M)$ is a closed subset of U'. Using Lemma 4, we suppose that $M' = U'$. Since f is locally piecewise linear on $\mathrm{Bd}\, M$, $f(M)$ is locally polyhedral in M' and by Lemma 1, $f(M)$ has a triangulation compatible with that of M'. Hence we suppose that $f(M) = M'$.

Since we can operate on each component of M, we also suppose that M is connected.

Consider the cartesian product $L \times H$ where $L = \mathrm{Bd}\, M$ and H is the collection of non-negative real numbers. Suppose $L \times H$ is tied to M so that for each point p of L, $p \times 0 = p$, and $M \cdot (L \times H) = L$. Tie $f(L) \times H$ to M' in a similar fashion.

The triangulations $T(M)$ and $T(M')$ of M and M' are extended to $L \times H$ and

$f(L) \times H$ in a natural fashion. Extend the homeomorphism f of M onto M' so that $f(p \times a) = f(p) \times a$ $(p \in L,\ a \in H)$. Then f is locally piecewise linear in carrying $L \times H$ onto $f(L \times H)$.

Let $M_1, M_2, \cdots$ be a sequence of compact 3-manifolds with boundaries satisfying the following conditions:

(a) M_i is polyhedral under $T(M)$.

(b) $\sum M_i = M$.

(c) M_i intersects M_j if and only if $| i - j | \leqq 1$.

(d) $M_i \cdot L = L_i$ is either null or a 2-manifold with boundary.

(e) $M_i + M_{i+2} + (L_i + L_{i+1} + L_{i+2}) \times (0, 1) = K_{i+1}$ is a 3-manifold with boundary and $M_{i+1} \cdot K_{i+1} = \text{Bd}\, M_{i+1}$.

Let ε_i be a positive number less than either $1/i$, the minimum value of ϕ on M_i, or one-half the distance between $f(M_i)$ and any point not in $f(M_i + K_i)$. (We suppose the metric on M' has been extended to $M' + f(L \times H)$.)

By Theorem 1′ there is a positive number γ_{2i} such that if f' is a piecewise linear γ_{2i}-approximation of f on K_{2i}, then f' may be extended to get a piecewise linear ε_{2i}-approximation to f on $K_{2i} + M_{2i}$. Also by Theorem 1′, the homeomorphism f on $(L_{2i} + L_{2i+1} + L_{2i+2}) \times (0, 1)$ may be extended to M_{2i+1} so as to get a piecewise linear δ_{2i+1}-approximation g_{2i+1} of f on $M_{2i+1} + (L_{2i} + L_{2i+1} + L_{2i+2}) \times (0, 1)$ where δ_{2i+1} is the minimum of $\gamma_{2i}, \gamma_{2i+2}, \varepsilon_{2i+1}$. (Ignore γ_0.) Then $f' = g_{2i+1}$ on M_{2i+1}. After f' has been defined on $L \times H$ and all the odd M_i's, we use Theorem 1′ again to define it on the even M_i's as suggested in the first sentence of this paragraph.

Since f' is piecewise linear on each M_i, it is locally piecewise linear on U. Since $f'(U)$ is both open and closed in $f(U)$, $f'(U) = f(U)$.

We shall use the following two dimensional version of Theorem 3 as a lemma in the proof of Theorem 4.

Lemma 5. *Suppose f is a homeomorphism of a triangulated closed disk D into a triangulated 2-manifold, L is an open subset of* Bd D, *and f is locally piecewise linear on* Bd D *at each point of L. Then for each positive number ε there is a homeomorphism f' of D onto $f(D)$ such that (1) f' is locally piecewise linear at each point of L +* Int D, *(2) $\rho(f, f') < \varepsilon$, and (3) $f = f'$ on* Bd D.

In order to get both conditions (1) and (3) of the conclusion of Theorem 3 it was necessary to suppose that f was locally piecewise linear when operating on $U \cdot \text{Bd}\, M$. However, this could be altered to get the following result.

Theorem 4. *Suppose U and U' are open subsets of triangulated 3-manifolds with boundaries and f is a homeomorphism of U onto U'. Then there is a locally piecewise linear homeomorphism f' of U onto U'.*

Furthermore, if K is a closed subset of U such that f is locally piecewise linear at each point of K and ϕ is a positive continuous function on $U - K$, we can pick f' so that $f = f'$ on K and $\rho(f(p), f'(p)) < \phi(p)$ on $U - K$.

Proof. Let V be the open set consisting of all points at which f is locally piecewise linear. As noted in Lemma 4, there is no loss of generality in supposing that U and U' are triangulated 3-manifolds with boundaries. By taking $\phi(p) <$

$(\rho(p, K) + \rho(p, U - V))/2$ in Lemma 4, we suppose further that the triangulation of U is so fine that for each vertex v of its triangulation, f is piecewise linear on $\overline{\text{St}}(v)$ if $\overline{\text{St}}(v) \cdot K = 0$. Let $v_1, v_2, \cdots$ be the vertices of U for which $\overline{\text{St}}(v_i) \cdot K = 0$. Then $V + \text{St}(v_1) + \text{St}(v_2) + \cdots = U$.

We shall describe a sequence of homeomorphism $f_1, f_2, \cdots$ of U onto U' such that $f_i = f_{i-1}$ on $U - \overline{\text{St}}(v_i)$, f_i is locally piecewise linear at each point of $V + \text{St}(v_1) + \text{St}(v_2) + \cdots + \text{St}(v_i)$, and the distance from $f_i(x)$ to $f_{i-1}(x)$ is less than $\phi(x)/2^i$. Since V, $\text{St}(v_1)$, $\text{St}(v_2)$, $\cdots$ is a locally finite covering of U, $f' = \lim f_i$ is the required homeomorphism.

Given f_i, we describe f_{i+1}. This is easier to do if v_{i+1} is an element of Int U. In this case we find from Theorem 3 that there is a homeomorphism g of $\overline{\text{St}}(v_{i+1})$ onto $f_i(\overline{\text{St}}(v_{i+1}))$ such that (a) $g = f_i$ on Bd $\overline{\text{St}}(v_{i+1})$, (b) the distance between g and f_i is less than the minimum value of $\phi/2^{i+1}$ on $\overline{\text{St}}(v_{i+1})$, and (c) g is locally piecewise linear at each point of $\text{St}(v_{i+1})$ and also at each point of Bd $\overline{\text{St}}(v_{i+1})$ at which f_i is locally piecewise linear. In using Theorem 3, we let $M = \overline{\text{St}}(v_{i+1})$, $U =$ Int M plus all points on Bd M at which f_i is locally piecewise linear, and $\phi(p)$ approach 0 and p approaches $M - U$. Then f_{i+1} is g on $\text{St}(v_{i+1})$ and f_i elsewhere.

Now we consider the case where v_{i+1} is a point of Bd U. Let L be the set of points of Bd $\overline{\text{St}}(v_{i+1})$ at which f_i is locally piecewise linear and D be the disk whose interior is $\text{St}(v_{i+1}) \cdot \text{Bd } U$. We find from Lemma 5 that for each positive number ε there is a homeomorphism h of D onto $f_i(D)$ such that (a) $h = f_i$ on Bd D and (b) h is locally piecewise linear on $L \cdot \text{Bd } D + \text{Int } D$. Since $\overline{\text{St}}(v_{i+1})$ is a polyhedral 3-cell, for ε sufficiently small the homeomorphism h may be extended to carry $\overline{\text{St}}(v_{i+1})$ onto $f_i(\overline{\text{St}}(v_{i+1}))$ so that $h = f_i$ on Bd $\overline{\text{St}}(v_{i+1}) -$ Int D and the distance between h and f_i is less than the minimum value of $\phi/2^{i+2}$ on $\overline{\text{St}}(v_{i+1})$. Then it follows from Theorem 3 that there is a homeomorphism g of $\overline{\text{St}}(v_{i+1})$ onto $f_i(\overline{\text{St}}(v_{i+1}))$ such that (a) $g = f_i$ on Bd $\overline{\text{St}}(v_{i+1}) -$ Int D, (b) the distance from g to f_i is less than the minimum value of $\phi/2^{i+1}$ on $\overline{\text{St}}(v_{i+1})$, and (c) g is locally piecewise linear at each point of $\text{St}(v_{i+1})$ and also each point of L. Then $f_{i+1} = g$ on $\text{St}(v_{i+1})$ and $f_{i+1} = f_i$ elsewhere.

COROLLARY 1. *Suppose M, M' are homeomorphic 3-manifolds with boundaries and with triangulations $T(M)$ and $T(M')$. Then there are subdivisions $T'(M)$ and $T'(M')$ of $T(M)$ and $T(M')$ respectively and an isomorphism of $T'(M)$ onto $T'(M')$.*

COROLLARY 2. *Suppose M is a triangulated 3-manifold with boundary and K is a subset that is locally tame at a point p of K. Then for each neighborhood U of p there is a homeomorphism h of M onto itself such that h is the identity on $M - U$ and $h(K)$ is locally polyhedral at $h(p)$.*

PROOF. First suppose p belongs to Bd M. Since K is locally tame at p there is a neighborhood N of p and a homeomorphism f of $\bar{N}$ onto I^3 such that $f(\bar{N} \cdot K)$ is a polyhedron in I^3, $\bar{N} \cdot \text{Bd } M$ is a disk whose interior lies in N, and (c) $N \subset U$.

By Theorem 4 there is a homeomorphism g of I^3 onto $\bar{N}$ such that $g = f^{-1}$ on each point of $f(\bar{N} - N)$ and g is locally piecewise linear in N. In applying

Theorem 4 we ignore K, let $U = f(N)$, and pick $\phi(p)$ so that it approaches 0 as p approaches $f(\bar{N} - N)$. The required homeomorphism h is $g \cdot f$ on $\bar{N}$ and the identity elsewhere.

The case is the same if $p \notin \text{Bd } M$ except that we choose N so that $\bar{N} \cdot \text{Bd } M = 0$.

4. Triangulation of 3-Manifolds with Boundaries

Moise showed in [4] that any 3-manifold could be triangulated. We extend the result to 3-manifolds with boundaries.

THEOREM 5. *Each 3-manifold with boundary M can be triangulated. In fact, if $T(\text{Bd } M)$ is any trangulation of* Bd M, *there is a triangulation of M compatible* with $T(\text{Bd } M)$.

PROOF. Since M is locally compact and metric, it has a locally finite countable open covering $\{D_i\}$ such that $\bar{D}_i$ is compact. There is an open covering $\{D_i'\}$ of M such that D_i contains $\bar{D}_i'$. For each point p of D_i' there is a neighborhood N of p such that $\bar{N}$ is topologically equivalent to I^3, D_i contains N, and $\bar{N} \cdot \text{Bd } M$ is either null or a polyhedral disk under $T(\text{Bd } M)$ whose interior lies in N. Since a finite number of such N's cover $\bar{D}_i'$, there is a locally finite collection $N_1, N_2, \cdots$ of such N's covering M.

We now choose a triangulation $T(\bar{N}_i)$ for each $\bar{N}_i$. Since any triangulation of the surface of a cube may be extended to the interior by considering radial lines running from the center of the cube to points on the surface, we may suppose that if $\bar{N}_i$ intersects Bd M, then $T(\bar{N}_i)$ is compatible with $T(\text{Bd } M)$ on $\bar{N}_i \cdot \text{Bd } M$. Let $T(N_i)$ be a triangulation of N_i locally compatible with $T(\bar{N}_i)$ on N_i. Since N_i is open in M, $T(N_i)$ has infinitely many simplexes.

Use M_i to denote $N_1 + N_2 + \cdots + N_i$. We shall show a sequence of triangulations $T(M_1), T(M_2), \cdots$ of $M_1, M_2, \cdots$ respectively such that:

(1) $T(M_i)$ is compatible with $T(\text{Bd } M)$ on $M_i \cdot \text{Bd } M$.

(2) Each simplex in $T(M_i)$ that intersects $M - \sum_{j>i} N_j$ is a simplex in $T(M_{i+1})$.

The sequence of triangulations $T(M_1)$, $T(M_2)$, $\cdots$ provides a triangulation $T(M)$ of M. Since $\{N_i\}$ is locally finite, any compact set will lie in some $M - \sum_{j>i} N_j$. Therefore for each point p, there is an integer i_p such that p does not lie in $\sum_{j>i_p} \bar{N}_j$. We find from condition (2) that we can take the simplex of $T(M_{i_p})$ that contains p to be a simplex of $T(M)$. We find from condition (1) that such a $T(M)$ is locally compatible with $T(\text{Bd } M)$ on Bd M and from a 2-dimensional version of Lemma 3 that it is compatible.

For $T(M_1)$ we take $T(N_1)$. If $T(M_i)$ has been obtained, we now show how to get $T(M_{i+1})$.

Now both $T(M_i)$ and $T(N_{i+1})$ prescribe triangulations for $M_i \cdot N_{i+1}$, but the triangulations they prescribe may not be locally compatible. However, they are locally compatible on $M_i \cdot N_{i+1} \cdot \text{Bd } M$. We find from Theorem 3 that there is a homeomorphism g of $\bar{N}_{i+1}$ onto itself such that g is the identity on $\text{Bd } N_{i+1} + (N_{i+1} - M_i)$ and is locally piecewise linear on $M_i \cdot N_{i+1}$ in interchanging $T(M_i)$ and $T(N_{i+1})$. We apply Theorem 3 to $M_i \cdot N_{i+1}$ using for the M, M', U, f, and

$\phi(p)$ of that theorem respectively, $M_i \cdot N_{i+1}$ under $T(M_i)$, $M_i \cdot N_{i+1}$ under $T(N_{i+1})$, $M_i \cdot N_{i+1}$, the identity transformation, and a positive continuous function that approaches 0 as p approaches $\bar{M}_i \cdot \bar{N}_{i+1} - M_i \cdot N_{i+1}$. Let $T'(N_{i+1})$ be the triangulation of N_{i+1} whose simplexes are the images of simplexes of $T(N_{i+1})$ under g. Then $T(M_i)$ and $T'(N_{i+1})$ are locally compatible on $M_i \cdot M_{i+1}$.

We can describe $T(M_{i+1})$ by telling what its 3-simplexes are. Let W be the set of all 3-simplexes of $T(M_i)$ whose closures intersect $M_i - N_{i+1}$. The elements of W are 3-simplexes in $T(M_{i+1})$. This insures that condition (2) is satisfied. We now describe the other 3-simplexes in $T(M_{i+1})$.

Let V be the set of all 3-simplexes of $T(M_i)$ which do not belong to W but which have a vertex in common with an element of W. Let Z be the set of all 2-simplexes z such that z is the face of an element of V but does not have a vertex in common with an element of W. Each z lies in $M_i \cdot N_{i+1}$ and since $T(M_i)$ and $T'(N_{i+1})$ are locally compatible there, it follows from Lemma 1 that there is a subdivision $T''(N_{i+1})$ of $T'(N_{i+1})$ such that each z in Z is the sum of simplexes of $T''(N_{i+1})$. The 3-simplexes of $T''(N_{i+1})$ which do not intersect any element of $W + V$ are elements of $T(M_{i+1})$.

If v is an element of V, its boundary is triangulated as follows: each face which is a face of an element of W is not subdivided at all; each face in Z is triangulated by $T''(N_{i+1})$; other faces are triangulated only if a new vertex was introduced at the last step. The triangulation of the boundary of v imposes a triangulation on v and the 3-simplexes in this triangulation are elements of $T(M_{i+1})$. We note that $T(M_{i+1})$ satisfies conditions (1) and (2).

5. Taming Surfaces

The following result provides a clue as to how we shall proceed to show in Theorem 9 that any locally tame closed subset of a triangulated 3-manifold with boundary is tame.

THEOREM 6. *Suppose L is a 2-manifold with boundary which is locally tamely imbedded in a 3-manifold M with triangulation $T(M)$. Then L is tamely imbedded in M.*

In fact, if $T(L)$ is a triangulation of L, there is a homeomorphism g of M onto itself such that g is piecewise linear in carrying L under $T(L)$ onto $g(L)$ under $T(M)$.

PROOF. We show that there is a triangulation $T'(M)$ of M such that L is a polyhedron under $T'(M)$ with a triangulation compatible with $T(L)$; then the theorem will follow from Theorem 4 for the following reason. Since there is a piecewise linear transformation g of M under $T'(M)$ onto M under $T(M)$, this transformation carries M onto itself, L onto a polyhedron, and L under $T(L)$ piecewise linearly onto $g(L)$ under $T(M)$.

To see that there is such a triangulation $T'(M)$ of M, split M along L to get a 3-manifold with boundary M'. (Each point of Int L becomes two points of M'.) Then $T(L)$ induces a triangulation $T(\text{Bd } M')$ on Bd M'. By Theorem 5 there is a triangulation $T(M')$ of M' compatible with $T(\text{Bd } M')$. Suppose M' is now sewn back together to form M again. The triangulation $T(M')$ induces a partition $S(L)$ of L for which there is a triangulation $T'(L)$ that refines both $T(L)$ and $S(L)$. Then there is a triangulation $T'(M)$ of M such that each 3-simplex in $T(M')$ which is not adjacent to Bd M' is a 3-simplex in $T'(M)$, and each simplex in $T'(L)$ is a simplex in $T'(M)$.

6. Taming an Arc

After considering surfaces, we turn our attention to arcs. As a preliminary to showing that a locally tame arc is tame, we consider first an arc which is already locally polyhedral except at one point.

LEMMA 6. *Suppose ab is an arc in E^3, p is a nonend point of ab, ab is locally tame at p, and each arc in ab-p is polygonal. Then for each neighborhood U of p there is a homeomorphism of E^3 onto itself that is the identity outside U and which carries ab onto a polygonal line.*

PROOF. There are essentially two steps to the proof. The first is to shift the part of ab near p into the surface of a cube. The second step is to straighten out the part of ab shifted into the surface so that the resulting arc is polygonal.

STEP 1. Since ab is locally tame at p there is a neighborhood N of p in U and a homeomorphism f of $\bar{N}$ onto I^3 that carries $\bar{N}\cdot ab$ onto a straight line interval. By Theorem 2 there is a homeomorphism f' of $\bar{N}$ onto I^3 such that $f = f'$ on $(\bar{N} - N) + \bar{N}\cdot ab$ and is locally piecewise linear elsewhere on $N - ab$. In applying Theorem 2, we let $N - ab$ be the U of that theorem and $\phi(p)$ be a positive continuous function that approaches 0 as p approaches $E^3 - U$. We suppose that the ends a', b' of $f'(\bar{N}\cdot ab)$ lie in Bd I^3.

Let M be a cube plus its interior such that Int M contains p, N contains M, $ab\cdot$Bd M is a finite set of points, and ab does not intersect an edge or vertex of Bd M.

Let D be a plane disk in I^3 such that D contains $f'(\bar{N}\cdot ab)$, Bd I^3 contains Bd D, and f'(Bd M) contains no open subset of D.

There is a simple closed curve J in $D\cdot f'$(Bd M) which has $f'(p)$ on its interior in the disk D. To see that this is true, note that $D\cdot f'$(Bd M) separates $f'(p)$ from Bd D in D. This is true because Bd M separates p from $\bar{N} - N$. Hence, a component of $D\cdot f'$(Bd M) separates $f'(p)$ from Bd D in D. This component is locally connected since $D\cdot f'$(Bd M) is locally polyhedral except possibly at $f'(ab\cdot$Bd $M)$. Therefore the component contains a simple closed curve J which separates $f'(p)$ from Bd D in D.

There is an arc $a'xb'$ from a' to b' such that $a'xb'$ does not contain p and $a'xb' - a'b'$ is an open arc in J. There is a homeomorphism h of D onto itself that carries $a'b'$ onto $a'xb'$ and leaves each point of Bd D fixed. The homeomorphism h of D onto itself may be extended to a homeomorphism h' of I^3 onto itself that leaves each point of Bd I^3 fixed.

Consider the following homeomorphism t of E^3 onto itself. $t(x) = x$ if x is an element of $E^3 - N$; $t(x) = f'^{-1}h'f'(x)$ if x is an element of N. Under the homeomorphism t of E^3 onto itself, a subarc of ab containing p is moved onto an arc in Bd M and the rest of ab is left fixed.

Step 2. Now $t(ab)$ is the sum of an arc rs on Bd M and two polygonal lines that intersect Bd M in only a finite number of points. In fact, any subarc of rs not intersecting $r + s$ is polygonal. Let Δ be the interior of a triangle in Bd M which contains r but not s nor any point of $t(ab) - rs$. Then there is a homeomorphism t' of Bd M onto itself that leaves r and each point of Bd $M - \Delta$ fixed and carries each arc in $rs - s$ onto a polygonal line. The homeomorphism t' may be extended to a homeomorphism of E^3 onto itself so as not to move any point of $t(ab) - rs$. Similarly, there is a homeomorphism t'' that straightens rs out near s and does not disturb $t(ab) - rs$. Then $t''t't$ is a homeomorphism of E^3 onto itself that takes ab onto a polygonal line.

We now use this lemma to tame a locally tame arc.

Theorem 7. *If A is a locally tame arc in a triangulated 3-manifold with boundary M, then A is tame.*

In fact, if U is an open subset of M containing A and ε is a positive number, there is a homeomorphism g of M onto itself such that g is an ε-approximation of the identity, is the identity on $M - U$, and carries A onto a polygonal line.

Proof. For each point p of A there is a neighborhood N_p of p in the space M and a homeomorphism f_p of $\bar{N}_p$ onto I^3 such that $f_p(\bar{N} \cdot A)$ is a polygonal line, U contains N_p, $\bar{N} \cdot \text{Bd}\, M$ is either null or a disk whose interior is $N \cdot \text{Bd}\, M$, $A \cdot (\bar{N}_p - N_p)$ is at most two points, and neither of these points is either an end point of A or a point of $A \cdot \text{Bd}\, M$ which is a limit point of $A \cdot \text{Int}\, M$. Let $N_1, N_2, \cdots, N_j$ be a finite number of such neighborhoods covering A and f_i be the homeomorphism associated with $\bar{N}_i$.

By Theorem 4 there is a homeomorphism f_i' of $\bar{N}_i$ onto I^3 such that $f_i' = f_i$ on $(\bar{N}_1 + \bar{N}_2 + \cdots + \bar{N}_{i-1}) \cdot \bar{N}_i + (\bar{N}_i - N_i)$, f_i' is locally piecewise linear elsewhere on N_i, and $f_i'^{-1}$ is an $\varepsilon/2$-approximation of f_i^{-1}. Let g_i be the homeomorphism of M onto itself that is the identity of $M - N_i$ and is $f_i'^{-1} f_i$ on $\bar{N}_i$.

Now $g = g_1 g_2 \cdots g_j$ is an $\varepsilon/2$-approximation of the identity that carries each closed arc in $A - \sum(\bar{N}_i - N_i)$ into a polygonal line. Therefore $g(A)$ is locally polygonal except at a finite number of points. We can straighten out $g(A)$ at each of these in Int M by Lemma 6. At the bad points of $g(A)$ on Bd M, we can use the methods of Step 2 of Lemma 6 alone.

7. Taming Closed Sets

Any closed set in E^2 which is topologically equivalent to a polyhedron is tame. This is not true in E^3 because of knotted arcs and horned surfaces [1]. For example, there is a homeomorphism of period 2 of E^3 onto itself that has for its fixed point set a wild set which is topologically equivalent to a plane [2]. However, we find that for closed sets in E^3, local tameness implies tameness.

Theorem 8. *If K is a locally tame closed subset of a 3-manifold with boundary, then M has a triangulation under which K is a polyhedron.*

Proof. There is no loss of generality in supposing that K is two dimensional, so we do this. We break the proof into several steps.

STEP 1. First, we triangulate M so that K behaves well on a set of isolated points of K.

Let K_1 be the set of all points p of $K + \text{Bd}\, M$ such that $K + \text{Bd}\, M$ is not locally a 2-manifold at p. (No open subset of $K + \text{Bd}\, M$ containing p is topologically equivalent to an open disk.) Let K_0 be the set of all points p of K_1 at which K_1 is not locally a 1-manifold. Then K_0 is a closed set of isolated points (isolated in K_0).

We find from Theorem 5 that M has a triangulation $T(M)$. We now show by using Theorem 4 that there is no loss of generality in supposing that K is locally polyhedral at each point of K_0.

In applying Theorem 4, take a neighborhood N of a point p of K_0 and a homeomorphism f of $\bar{N}$ onto I^3 such that $f(\bar{N} \cdot K)$ is a polyhedron under the natural triangulation of I^3. Letting $f(N)$ be the U mentioned in Theorem 4 and $\phi(x)$ be a continuous positive function on U that approaches 0 as x approaches $f(\bar{N} - N)$, we find from Theorem 4 that there is a homeomorphism f' of I^3 onto $\bar{N}$ such that $f' = f^{-1}$ on $f(\bar{N} - N)$ and f' is piecewise linear on $f(N)$ in taking I^3 under its natural triangulation into M under $T(M)$. The homeomorphism $f'f$ is extended to be the identity on $M - N$. Then K will be locally polyhedral at p under the triangulation whose simplexes are the images of simplexes of $T(M)$ and under $f'f$. Since this procedure can be carried out near each point of K_0, we suppose without loss of generality that under $T(M)$, K is locally polyhedral at each point of K_0.

STEP 2. Our second step is to get a triangulation $T(K + \text{Bd}\, M)$ of $K + \text{Bd}\, M$ so that for each point p of K_0, $T(K + \text{Bd}\, M)$ is compatible near p with $T(M)$. To get $T(K + \text{Bd}\, M)$, we first triangulate K_1 so that the identity transformation of K_1 under this triangulation onto K_1 under $T(M)$ is piecewise linear near each point of K_0. This triangulation of K_1 may be extended to get a triangulation $T(K + \text{Bd}\, M)$ so that K is a polyhedron under $T(K + \text{Bd}\, M)$ and for each point p of K_0, there is a neighborhood N in M of p such that the identity transformation of $\bar{N} \cdot K$ under $T(K + \text{Bd}\, M)$ onto $\bar{N} \cdot K$ under $T(M)$ is piecewise linear.

STEP 3. The third step is to divide M into 3-manifolds with boundaries so that with the exception of points of K near K_0 or in K_1, K lies in the sum of the boundaries of these 3-manifolds.

Let $T'(M)$ be a subdivision of $T(M)$ so that (a) each point k_i of K_0 is a vertex of M under $T'(M)$, (b) $K \cdot (\overline{\text{St}} k_i)$ is the sum of simplexes of M under $T'(M)$ and is polyhedral under $T(K + \text{Bd}\, M)$, (c) $\overline{\text{St}}(k_i) \cdot \text{Bd}\, M$ is either a disk or null, and (d) $\overline{\text{St}}(k_i) \cdot (\overline{\text{St}} k_j) = 0$ if $k_i \neq k_j$. Each $\overline{\text{St}}(k_i)$ is one of the 3-manifolds with boundaries into which we divide M. Slit $M - \sum \text{St}(k_i)$ along $\overline{K - K_1}$ to get the others. In this division, some points of K go into more than one point.

STEP 4. We now triangulate the 3-manifolds with boundaries into which we divided M at the third step. Each $\overline{\text{St}}(k_i)$ is left with the triangulation it had under $T'(M)$. If M_n is another 3-manifold with boundary, Bd M_n is triangulated so that the triangulation of Bd(M_n) locally is compatible with $T(K + \text{Bd } M)$ on Bd $M_n \cdot (K + \text{Bd } M)$ and with $T'(M)$ on Bd $M_n \cdot \sum$ Bd $\overline{\text{St}}(k_i)$. By Theorem 5 this triangulation of Bd M_n may be extended to get a triangulation $T(M_n)$ of M_n.

STEP 5. Now we join the 3-manifolds with boundaries back together to get M. By the methods used in the proof of Theorem 6, the triangulations we have imposed on the various 3-manifolds with boundaries may be refined to get a triangulation $T''(M)$ of M. Under $T''(M)$, $\overline{K - K_1}$ is a polyhedron. Since $T''(M)$ is compatible with $T'(M)$ on $\overline{\text{St}}(k_i)$, K under $T''(M)$ is locally polyhedral at each point of $\overline{K - K_1}$.

STEP 6. K is now locally polyhedral except on a set of mutually exclusive arcs. Without moving any point of $K - K_1$, each of these arcs can be straightened by the methods used in Theorem 7. Hence, there is a homeomorphism h of M onto itself such that $h(K)$ is a polyhedron under $T''(M)$. Then K is a polyhedron under the triangulation of M whose simplexes are images of simplexes of $T''(M)$ under h^{-1}.

THEOREM 8′. *If in Theorem* 8, $T(K + \text{Bd } M)$ *were a given triangulation of* $K + \text{Bd } M$ *which makes* K *a polyhedron, a triangulation of* M *can be selected so as to be compatible with* $T(K + \text{Bd } M)$ *on* $K + \text{Bd } M$.

PROOF. If we get a triangulation of M compatible with $T(K + \text{Bd } M)$ on Bd M + the point set boundary of K, it will follow from the method of Theorem 3 that we can get one compatible with $K + \text{Bd } M$. Hence, we suppose that $K + \text{Bd } M$ is not 3-dimensional.

We shall show that in Step 1 of the proof of Theorem 8, we can adjust $T(M)$ so that near each point of K_0, $T(M)$ and $T(K + \text{Bd } M)$ are compatible on $K + \text{Bd } M$. Once this is done, the rest of the proof needs little alteration. We do some pushing and pulling along K_1 to insure that the final triangulation is compatible with $T(K + \text{Bd } M)$ along the part of K_1 not in the closure of $K - K_1$.

We finish consideration of Theorem 8′ by showing that in Step 1, $T(M)$ can be suitably adjusted near points of K_0.

As in Step 1 we note that since K is locally tame at points of K_0, we may suppose from Theorem 4 that K is locally polyhedral at each point of K_0 under $T(M)$. Further, we suppose that points of K_0 are vertices of $T(M)$ and that for each such vertex k_i, $\text{St}(k_i) \cdot K$ is the sum of simplexes of $T(M)$. By pulling and pushing points towards and away from k_i, we may adjust $T(M)$ so that the identity transformation is piecewise linear in taking $\text{St}(k_i) \cdot K_1$ under $T(M)$ onto itself under $T(K + \text{Bd } M)$.

If V is a component of $K + \text{Bd } M \cdot \text{St}(k_i) - K_1$, it is the sum of 2-simplexes and 1-simplexes of $T(M)$. Using the notions involved in Lemma 4, we find that there is a homeomorphism h of $\bar{V}$ onto itself that is the identity on $\bar{V} - V$ and is locally piecewise linear on $\bar{V} \cdot \text{St}(k_i)$ in taking $\bar{V}$ under $T(K + \text{Bd } M)$ onto $\bar{V}$

under $T(M)$. We show that h can be extended to M so as not to move any point of $K + \text{Bd}\, M - V$.

There is no loss of generality in supposing that $\overline{\text{St}}(k_i)$ is a convex polygon in E^3 and that V lies in a plane. Then V is either an open disk, an open disk minus a point, or an open disk minus an interval. If it is an open disk, by considering pyramids with $\bar{V}$ as bases, one can see how to extend h to M so as not to move any point of $K + \text{Bd}\, M - V$. If V is an open disk minus an interval, the procedure is not much different. If it is a disk minus a point, we employ the following lemma.

LEMMA 7. *Suppose that in E^3, K is the point set obtained by subtracting the open cone $z > r$ (cylindrical coordinates) from the cylinder $r \leqq 1$, $0 \leqq z \leqq 1$. Suppose that f is a homeomorphism of the base of the cylinder onto itself that leaves its center and each point of its circumference fixed. Then f may be extended to K so as to remain fixed at each nonbase point of the boundary of K.*

PROOF. For each $0 \leqq \theta < 2\pi$, $0 < a \leqq 1$, let $P(a, \theta)$ be the sum of the interval from the origin to (a, θ, a^2) and the interval from (a, θ, a^2) to $(a, \theta, 0)$. For each point p on the broken line $P(a, \theta)$ let $f(p)$ be the point that divides $P(a', \theta')$ in the same ratio that p divides $P(a, \theta)$ where $(a', \theta', 0) = f(a, \theta, 0)$.

THEOREM 9. *Each locally tame closed subset K of a triangulated 3-manifold with boundary is tame.*

Furthermore, if C is a closed subset of M such that K is locally polyhedral at each point of $K \cdot C$, and ϕ is a positive continuous function on $M - C$, there is a homeomorphism f of M onto itself such that $x = f(x)$ on C, $\rho(x, f(x)) < \phi(x)$ on $M - C$, and $f(K)$ is a polyhedron.

PROOF. Let $T(M)$ be the given triangulation of M and $T'(M)$ be a triangulation of it under which K is a polyhedron. If $T'(M)$ is such that the identity transformation is locally piecewise linear at points of C in interchanging $T(M)$ and $T'(M)$, Theorem 9 follows from Theorem 4. We complete the proof by showing that there is such a $T'(M)$ compatible with $T(M)$ near C.

Since K is locally polyhedral at each point of C, there is an open set U containing C such that $\bar{U} \cdot K$ is a polyhedron. Using Lemma 1 we shall suppose that $T(M)$ is such that $\bar{U} \cdot K$ is the sum of simplexes of $T(M)$. Furthermore, we suppose $T(M)$ is so fine that $\overline{\text{St}}(\overline{\text{St}}(C)) \subset U$ where $\text{St}(A)$ denotes the sum of all simplexes of $T(M)$ whose closures intersect A.

Use K' to denote the sum of $K - \text{St}\,\overline{\text{St}}(C)$ and all the vertices, edges, and faces in $\text{St}\,\overline{\text{St}}(C)$ under $T(M)$. Since K' is locally tame, it follows from Theorem 8 and Lemma 1 that there is a triangulation $T''(M)$ of M such that K' is the sum of simplexes of $T''(M)$. Then the 3-simplexes in $\text{St}\,\overline{\text{St}}(C)$ which are adjacent to $M - \text{St}\,\overline{\text{St}}(C)$ may be divided so as to get a triangulation $T'(M)$ of M such that each simplex under $T''(M)$ in $M - \text{St}\,\overline{\text{St}}(C)$ is a simplex in $T'(M)$ and each simplex under $T(M)$ in $\text{St}\,\overline{\text{St}}(C)$ is a simplex of $T'(M)$ except those 3-simplexes which were adjacent to $M - \text{St}\,\overline{\text{St}}(C)$. Then K is a polyhedron under $T'(M)$ and $T'(M)$ is compatible with $T(M)$ near C.

THE UNIVERSITY OF WISCONSIN

References

1. E. Artin and R. H. Fox, *Some wild cells and spheres in three-dimensional space*, Ann. of Math., vol. 49 (1948), pp. 979–990.
2. R. H. Bing, *A homeomorphism between the 3-sphere and the sum of two solid horned spheres.* Ann. of Math., vol. 56 (1952), pp. 354–362.
3. W. Hurewicz and H. Wallman, Dimension Theory, Princeton University Press.
4. E. E. Moise, *Affine structures in 3-manifolds, V. The triangulation theorem and Hauptvermutung*, Ann. of Math., vol. 56 (1952), pp. 96–114.
5. E. E. Moise, *Affine structures in 3-manifolds*, VIII. *Invariance of the knot-type; local tame imbedding*, Ann. of Math., vol. 59 (1954), pp. 159–170.

A simple closed curve that pierces no disk [1];

BY **R. H. BING**.

1. Introduction.

A simple closed curve J is said to pierce a disk D if J links BdD (boundary of D) and J.D is a single point.

We shall describe a simple closed curve J in E^3 such that J pierces no disk in E^3. This description came as a result of questions posed by O. G. Harrold during stimulating discussions of the properties of E^3 at the 1955 Summer Institute on Set Theoretic Topology sponsored by the American Mathematical Society.

O. G. Harrold has defined [6] a simple closed curve J to have property P at a point x of J provided that for each neighbourhood U of x there is in U a neighbourhood V of x such that (1) BdV is a 2-sphere, (2) J.BdV is precisely two points and (3) BdV is locally polyhedral except at these two points. A simple closed curve with property P at one of its points would pierce a disk.

Harrold pointed out in [7] that condition (3) can be omitted as a result of an approximation theorem of Bing [4]. We shall use this approximation theorem so we state it here.

APPROXIMATION THEOREM FOR SURFACES. — *If* M *is a 2-manifold (with or without boundary) in* E^3 *and* f *is a continuous non-negative function*

([1]) Work on this paper was supported in part by funds from the National Science Foundation under Contract NSF-G 479.

defined on M, *there is a 2-manifold* M′ *in* E^3 *and a homeomorphism* h *of* M *onto* M′ *such that* (1) M′ *is locally polyhedral at all points* y *such that* $fh^{-1}(y) > 0$ *and* (2) *the distance between* x *and* $h(x)$ *is no more than* $f(x)$.

An abstract covering this result appears in [3], some of the details are found in [4], and a more precise treatment will be contained in a paper to be submitted to the *Annals of Mathematics*.

If two simple closed curves have the property that one cannot be shrunk to a point in the complement of the other, it does not follow that they link. Let C be a Cantor set as described by Antoine [2] and J_1 a simple closed curve in $E^3 - C$ that cannot be shrunk to a point in $E^3 - C$. There is a wild 2-sphere in $E^3 - J_1$ containing C and a simple closed curve J_2 on this wild 2-sphere containing C. Then J_1, J_2 do not link as J_2 can be shrunk to a point in the complement of J_1 but J_1 can not be shrunk to a point in the complement of J_2. Hence, the following statement about the simple closed curve we shall define is stronger than the statement that it pierces no disk.

Property of J. — *The simple closed curve* J *to be described has the property that if* D *is a disk such that* J . Bd D = 0 *but* Bd D *bounds no disk which misses* J *then* D . J *contains a Cantor set.*

2. Description of J.

This simple closed curve J is the intersection of a decreasing sequence of topological tori T_1, T_2, ... The first of these, T_1, is shown in the figure and may be regarded as the sum of a finite number of solid cylinders C_1, C_2, ..., C_n as shown. We suppose each of these cylinders is polyhedral.

The torus T_2 winds about in T_1 as shown. Except in part of C_1, the figure represents T_2 as linear, but it is actually the sum of polyhedral solid cylinders also. One may observe the resemblance between the way in which T_2 is imbedded in T_1 and the description of an arc in [5].

The torus T_3 is the sum of small polyhedral solid cylinders and is imbedded in T_2 as T_2 was imbetted in T_1. The figure shows only the

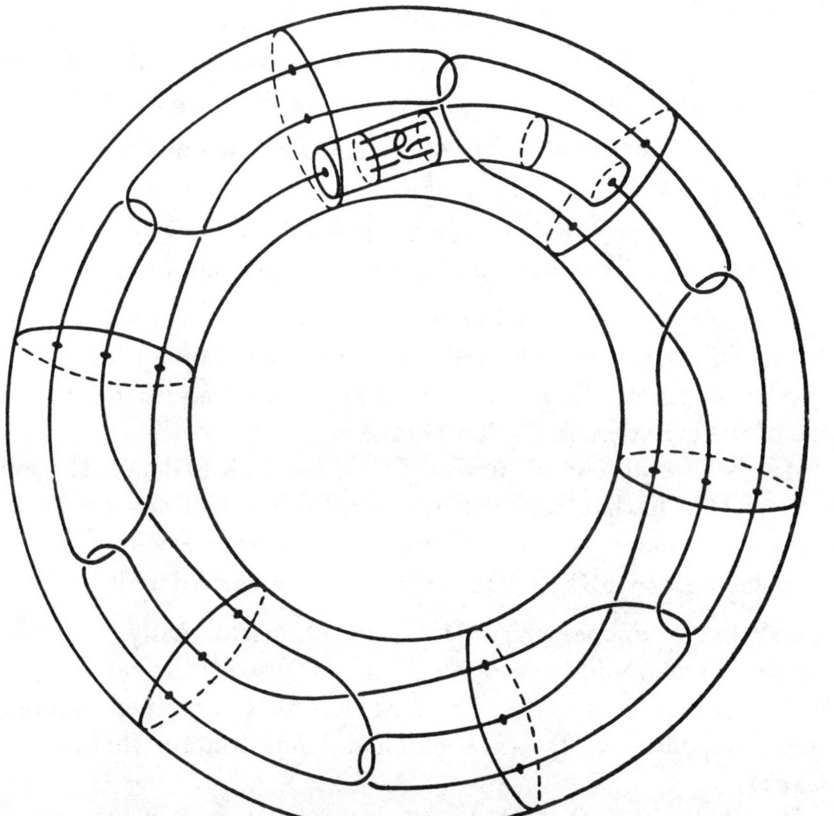

part of T_3 in one of the cylinders making up T_2. Similarly, the tori T_4, T_5, ... are defined. The intersection of T_1, T_2, ... is the simple closed curve J.

3. The intersection of J with disks.

Suppose D is a disk such that J.BdD = 0 and BdD bounds no disk which misses J. We shall show that J.D contains a Cantor set. In doing this we make certain adjustments and simplifications in the structure of D but none of these will add any points to J.D. Instead of describing a new disk at each stage, we for simplicity suppose that the original disk D had the simpler properties.

1. We suppose that D is locally polyhedral except at J.D + BdD. The approximation theorem assures us that this is permissible.

2. Some T_i does not intersect BdD, so for convenience, we suppose that T_1 does not.

3. We suppose that D and each BdT_i are in general position. Then D.BdT_i is the sum of a finite number of simple closed curves.

4. One of the solid cylinders making up one of the T_i's misses D so we suppose that one of the cylinders making up T_1 misses D. Use O_i to denote the open annulus ring that is equal to BdT_i minus one of the cylinders in T_i that D misses.

If K is a simple closed curve on D, we use IntK to denote the component of D — K that does not intersect BdD. If K is a simple closed curve in O_i that separates the two components of the boundary of O_i from each other in $\overline{O}_i$, we say that K cuts O_i meridianally.

5. Some component K_1 of D.O_1 cuts O_1 meridianally. If this were not so, there would be a topological disk D′ with the same boundary of D such that D′.J = 0. The disk D′ could be obtained by replacing each component of D — O_1 which did not contain BdD by a disk near O_1.

We shall show that D.J contains a Cantor set by showing that there are two simple closed curves K_{11}, K_{12} in O_2 IntK_1 such that IntK_{11} IntK_{12} = 0 and each of K_{11}, K_{12} cuts O_2 meridianally; there are two simple closed curves K_{1ij} $(i, j = 1, 2)$ in O_3 IntK_{1i} such that IntK_{1i1} IntK_{1i2} = 0, and each K_{1ij} cuts O_3 meridianally, etc. The intersection of IntK_1, Σ IntK_{1i}, Σ IntK_{1ij}, ... belongs to J.D and contains a Cantor set. We end our argument by showing the existence to the two curves K_{11}, K_{12} since the existence of the other K's is established in a similar fashion.

6. We suppose that IntK_1 does not contain any component of D.O_1 that cuts O_1 meridianally. If it did, we could use it instead of K_1.

7. We shall suppose that IntK_1 does not intersect O_1. If it intersected O_1 in some simple closed curves that did not cut O_1 meridianally, these simple closed curves would bound disks in O_1. If K′ were such a simple closed curve which bounded a disk D′ in O_1 such that $D'.D = K'$, we could modify D by replacing IntK′ by D′ and shoving this new disk slightly to one side of O_1. Since these unwanted simple closed curves in $K.O_1$ could be removed one by one, we suppose that they are not there.

8. Also, we suppose that O_2 IntK_1 does not contain any simple closed curves that fail to cut O_2 meridianally. If there were any such curves, they could be removed.

9. Since K_1 links J, IntK_1 intersects BdT_2. Each component of O_2 IntK_1 is a simple closed curve that cuts O_2 meridianally. Either there are two such simple closed curves with mutually exclusive interiors (what we are proving) or there is one K′ such that IntK_1 — IntK′ does not intersect T_2. We shall assume that there is such a K′ and show that this leads to a contradiction.

10. Modify IntK_1 by replacing IntK′ by a polyhedral disk which lies except for its boundary in IntT_2 and whose boundary is K′. Use F to denote this new disk. If L is a simple closed curve on F, henceforth we use IntL to denote the component of F — L that does not intersect K′. We have no further need to refer to the interiors of simple closed curves in D.

11. We shall suppose that K_1 does not intersect two of the solid cylinders making up T_1 and K′ does not intersect two of the solid cylinders making up T_2. If this is not already the case, we could adjust F by sliding it near O_1 and O_2.

12. Let E_i be the disk which is $C_iC_j (j = i + 1, \mod n)$ and A_i be the closure of $E_i - T_2$. We shall suppose that F and each E are in general position so that $F.A_1$ will be a collection of simple closed curves.

13. We shall suppose that each simple closed curve in $F.A_i$ separates two components of BdA_i from each other in A_i. If there were some components without this property, they could be eliminated one by one as indicated in steps 7 and 8.

14. Suppose L is a component of $A_i.F$. We show that $\operatorname{Int} L$ intersects T_2. Suppose that $\operatorname{Int} L$ contains no point of ΣA_j. Then $\operatorname{Int} L - T_2$ lies in either C_i or C_{i+1} (mod n). We suppose that $\operatorname{Int} L$ lies in C_i.

Denote the components of $T_2.C_i$ by M_1, M_2, M_3, where M_1 intersects both E_i and E_{i-1} (mod n), $M_2.E_{i-1}$ (mod n) $= 0$, and $M_3.E_i = 0$. Also, denote the components of $\operatorname{Bd} A_i$ by B_0, B_1, B_2, B_3 where B_0 belongs to T_1 and B_1 belongs to M_1. If L separates B_0 from B_1 in A_i, $\operatorname{Int} L$ intersects M_1. To see that this is so, we note that there is a 3-cell bounded by $\operatorname{Int} L$ and a disk in E_i [1]. Both the interior and the exterior of this 3-cell woul dintersect M_1 so $\operatorname{Int} L$ would intersect M_1.

A similar argument shows that if L separates B_2 from B_3 in A_i, then $\operatorname{Int} L$ intersects M_2.

15. If L separates B_0 from each of B_2 and B_3 in A_i, $\operatorname{Int} L$ intersects either M_2 or M_3. If this were not true, a 3-cell bounded by the sum of $\operatorname{Int} L$ and a disk D' in E_i would contain M_2 and fail to intersect M_3. This would contradict the fact that some simple closed curve in $M_2 + D'$ links a simple closed curve in $M_3 + E_{i-1}$ (mod n).

If L separates B_0 from each of B_1, B_2, B_3, then $F.\operatorname{Bd} T_2$ would not be connected and this would contradict the assumption that $F.\operatorname{Bd} T_2$ is the simple closed curve K' of steps 9 and 10.

16. We shall suppose that the components of $F.\Sigma A_i$ are simple closed curves $L_1, L_2, \ldots, L_m$ such that $L_{i+1} \subset \operatorname{Int} L_i$, $L_i \subset A_{x_i}$, and L_1 separates in A_{x_1} the component of $\operatorname{Bd} A_{x_1}$ on T_1 from each of the other three components of $\operatorname{Bd} A_{x_1}$. The case where there are no L's is similar to the case where there is exactly one so we do not consider it.

If $F.\operatorname{Bd} T_2$ is the assumed simple closed curve K', it would have to belong to $\operatorname{Int} L_m$ by the results of steps 15, 16 and $F - \operatorname{Int} L_m$ would not intersect T_2. It could be shown by the methods used in steps 14, 15 that L_2 separates $T_1.\operatorname{Bd} A_{x_2}$ from each other component of $\operatorname{Bd} A_{x_2}$ in A_{x_2}, L_3 separates $T_1.\operatorname{Bd} A_{x_3}$ from each other component of $\operatorname{Bd} A_{x_3}$ in A_{x_3}, ..., and L_m separates $T_1.\operatorname{Bd} A_{x_m}$ from each other component of $\operatorname{Bd} A_{x_m}$ in A_{x_m}. But this is a contradiction as pointed out in the last sentence of step 15.

QUESTIONS. — Is there a topological disk in E^3 which is pierced by no tame arc? Is there a surface in E^3 so wild that it contains no tame arc? Does some topological disk in E^3 contain an arc of the simple closed curve J we described, that is, does J have property Q?

BIBLIOGRAPHY.

[1] J. W. ALEXANDER, *On the subdivision of 3-space by a polyhedron* (*Proc. Nat. Acad. Sc.*, U. S. A., vol. **10**, 1924, p. 6-8).

[2] Louis ANTOINE, *Sur l'homéomorphie de deux figures et de leurs voisinages* (*J. Math. pures et appl.*, vol. **86**, 1921, p. 221-325).

[3] R. H. BING, *Approximating surfaces with polyhedral ones* (*Bull. Amer. Math. Soc.*, vol. **61**, 1955, Abstract).

[4] R. H. BING, *Approximating surfaces with polyhedral ones*, Summary of Lectures and Seminars, Summer Institute on Set Theoretic Topology, Madison, Wisconsin, 1955, p. 47-52.

[5] R. H. FOX and Emil ARTIN, *Some wild cells and spheres in three-dimensional space* (*Ann. Math.*, vol. **49**, 1948, p. 979-989).

[6] O. G. HARROLD, *The enclosing of simple arcs and curves by polyhedra* (*Duke Math. J.*, vol. **21**, 1954 p. 615-622).

[7] O. G. HARROLD, *Some consequences of the approximation theorem of Bing*, Summary of Lectures and Seminars, Summer Institute on Set Theoretic Topology, Madison, Wisconsin, 1955, p. 53-56.

FUNDAMENTA
MATHEMATICAE
XLVII (1959)

Conditions under which a surface in E^3 is tame *

by

R. H. Bing (Wisconsin)

1. Introduction. A surface (closed set that is a 2-manifold) M in E^3 is *tame* if there is a homeomorphism of E^3 onto itself that takes M onto a polyhedron (finite or infinite). If there is no such homeomorphism, M is called *wild.* The main purpose of this paper is to give a condition under which surfaces are tame. This paper deals with surfaces in E^3, so if a surface is mentioned, it is to be understood that this surface lies in E^3.

If A, B are two homeomorphic sets,

$$H(A, B) \leqslant \varepsilon$$

is used to denote the fact that there is a homeomorphism of A onto B that moves no point by more than ε.

If S is a 2-sphere, we use $\operatorname{Int} S$ and $\operatorname{Ext} S$ to denote the bounded and unbounded components respectively of $E^3 - S$. Theorem 2.2 of the next section states that a 2-sphere S is tame if for each positive number ε there are 2-spheres S', S'' in $\operatorname{Int} S$, $\operatorname{Ext} S$ respectively such that

$$H(S, S') \leqslant \varepsilon, \quad H(S, S'') \leqslant \varepsilon .$$

This theorem (Theorem 2.2) is not proved directly but is reduced by way of one theorem (Theorem 2.1) to a simpler one (Theorem 3.1) which in turn is proved on the basis of a still easier one (Theorem 3.2). The proof given in Section 4 of this last theorem is based on an assortment of results developed in Sections 5-9.

While this treatment is not the logical one where simpler results are proved first and the big ones are proved in terms of those already proved, it does reveal the chronological order of a logical approach in trying to prove a big result where the big result is reduced to a simpler result, the simpler result is reduced, ..., and finally a result is proved on which the sequence of theorems leading to the main result depends.

* Work was supported by the National Science Foundation under N. S. F. Grant G-3248 at the Institute for Advanced Study, Princeton, New Jersey.

Such an approach shows the purpose of the subsidiary results which might seem unexciting in themselves.

The person preferring the logical approach may start by studying Sections 5, 6, 7, 8, 9, 10 which are independent except for 6 which depends on 5, then 4 which depends on 6-9, then 3 which depends on 4, and then 2 which depends on 3 and 10.

Some of the results developed in Sections 5-10 are pursued for their own interest rather than for their contributions alone to the study of tame embedding.

In Section 11 we discuss the possibility of extending the result (Theorem 2.2) about tame 2-spheres to apply to tame surfaces. We give conditions under which a surface is locally tame at a point p. A surface S is locally tame at a point p if there is a neighborhood N of p in E^3 and a homeomorphism of $\overline{N}$ into E^3 that takes $\overline{N} \cdot S$ onto a polyhedron.

Previous papers have given conditions under which surfaces are tame. Bing [4] and Moise [18] showed that a surface is tame if it is locally tame. Griffiths [11] and Harrold [14] have given other sets of conditions under which a surface is tame.

In another paper [7] we shall apply the results of the present paper to show that E^3 does not contain uncountably many mutually exclusive wild surfaces. This result will be used to give still another proof that each 3-manifold can be triangulated. Moise gave a proof of the triangulation theorem in [17] and Bing gave an alternate proof in [6].

Another future paper [8] will extend the results of the present paper to show that a surface in E^3 is tame if its complement is uniformly locally simply connected. A set X is *uniformly locally simply connected* if for each positive number ε there is a positive number δ such that each closed curve in X of diameter less than δ can be shrunk to a point on a subset of X of diameter less than ε.

The above results do not all extend to surfaces with boundaries. Stallings has given an example of an uncountable collection of mutually exclusive wild disks in E^3. In another paper [9] we shall show that a disk in E^3 is tame if its complement is uniformly locally simply connected.

In showing in [3] that each surface M in E^3 could be approximated by a polyhedral surface, we used the 2-skeleton of a triangulation to chop up E^3, and hence M, into small pieces. We could use such an approach for the present paper but decided instead to try a slightly different approach, – namely, adjust M so that it contains no vertical interval and then use a fence to chop up the adjusted M. If a person chose to redo the work in [3], he might choose to use fences as defined below rather than triangulations.

A *fence* is defined to be the sum of all vertical lines intersecting the 1-skeleton of a rectilinear triangulation of a horizontal plane. The mesh of the triangulation is called the *mesh* of the fence. Each vertical line intersecting a vertex of the triangulation is called a *corner* of the fence and the sum of all vertical lines intersecting an edge of the triangulation is called a *section* of the fence. The term "fence" was used to define a somewhat similar set in [5] but there the fences were not infinite and extended in only one direction.

Closely related to a fence is a vertical triangular cylinder. A *vertical triangular cylinder* C is the sum of a collection of vertical lines such that a horizontal cross section of C is a triangle. The sum of the lines intersecting the interior of the triangle is denoted by $\operatorname{Int} C$ and $\operatorname{Ext} C = E^3 - (C + \operatorname{Int} C)$. If L is a vertical line in $\operatorname{Int} C$, we call C a vertical triangular cylinder about L. We use vertical triangular cylinders about the corners of a fence to study a 2-sphere S near these corners.

If M is a manifold with boundary, we use $\operatorname{Bd} M$ to denote this boundary and $\operatorname{Int} M$ to denote $M - \operatorname{Bd} M$. There is an inconsistency here in the way that we defined $\operatorname{Int} S$ for a 2-sphere S in E^3 and $\operatorname{Int} C$ for a cylinder C, but it seems unlikely that this double meaning of the symbol Int will lead to confusion.

A finite graph is used to subdivide a 2-sphere. A *finite graph* is the sum of a finite collection of arcs such that if two of these arcs have a point in common, this point is an end point of each. A finite graph is called *planar* if it can be embedded in a 2-sphere. A connected finite graph is called *stable* if it is planar and if each homeomorphism between two of its images in a 2-sphere S can be extended to a homeomorphism of S onto itself.

The distance function is denoted by ϱ.

An isotopy $H_t (0 \leqslant t \leqslant 1)$ on X is a one parameter (denoted by t) family of homeomorphisms of X onto itself. All of the isotopies we use start at the identity map — that is $H_0(x) = x$. Hence, it is to be understood without the condition being imposed further that if we use an isotopy $H_t (0 \leqslant t \leqslant 1)$, then

$$H_0 = I \text{ (the identity)}.$$

A map f defined on E^3 is called piecewise linear if there is a rectilinear triangulation of E^3 such that f is linear on each tetrahedron of the triangulation. An isotopy H_t $(0 \leqslant t \leqslant 1)$ is called piecewise linear if each of the homeomorphisms H_t in the one parameter family is piecewise linear.

We say that H_t moves no point by more than ε if $\varrho\big(H_t(x), x\big) \leqslant \varepsilon$ for each point x and write

$$\varrho(H_t, I) \leqslant \varepsilon .$$

In general, if f and g are two maps defined on X

$$\varrho(f, g) \leqslant \varepsilon \quad \text{means} \quad \varrho(f(x), g(x)) \leqslant \varepsilon \quad \text{for each} \quad x \in X.$$

2. A condition under which 2-spheres in E^3 are tame. In this section we give the main results of the paper — a condition under which a 2-sphere in E^3 is tame. The possibility of extending this result is discussed in Section 11. The intervening sections are used to verify some of the facts used in the present section.

THEOREM 2.1. *Suppose S is a 2-sphere in E^3 such that for each positive number ε there is a 2-sphere S' on* Int S *such that $H(S, S') \leqslant \varepsilon$. Then $S + \operatorname{Int} S$ is a topological 3-cell.*

Proof. We suppose that S contains no vertical interval. That there is no loss of generality in supposing this follows from Theorem 10.1.

Let $S_1, S_2, \ldots$ be a sequence of 2-spheres on Int S such that $H(S, S_i) \leqslant 1/i$. Since each 2-sphere can be approximated by a polyhedral 2-sphere [3], we suppose with no loss of generality that each S_i is polyhedral. We also suppose that $S_i \subset \operatorname{Int} S_{i+1}$.

Let $T(x)$ $(0 < x \leqslant 1)$ be the 2-sphere in E^3 with center at the origin and radius x. We shall prove Theorem 2.1 by showing that there is a homeomorphism h taking $S + \operatorname{Int} S$ onto $T(1) + \operatorname{Int} T(1)$.

Let $X(i, j)$ denote $S_j + \operatorname{Int} S_j - \operatorname{Int} S_i$ and $Y(i, j)$ denote $T(j/(j+1)) + \operatorname{Int} T(j/(j+1)) - \operatorname{Int} T(i/(i+1))$. See Figure 1.

We prove Theorem 2.1 by sewing together homeomorphisms between certain of the $X(i, j)$'s and the corresponding $Y(i, j)$'s so as to get a homeomorphism of $S + \operatorname{Int} S$ onto $T(1) + \operatorname{Int} T(1)$. We must be able to control the homeomorphisms of the $X(i, j)$'s onto the $Y(i, j)$'s so that the homeomorphisms on the individual parts combine to give a homeomorphism on the sum.

Let h be a homeomorphism of S onto $T(1)$ and g_i a homeomorphism of S_i onto S such that g_i moves no point by more than $1/i$. Let h_i be the homeomorphism of S_i onto $T(i/(i+1))$ defined so that the interval from the origin to $hg_i(x)$ goes through $h_i(x)$. See Figure 1. The h_i's give an approximation of the homeomorphism h of $S + \operatorname{Int} S$ onto $T(1) + \operatorname{Int} T(1)$ but rather than agreeing with h on each S_i, they only agree on a sequence of S_i's converging to S.

It follows from Theorem 3.1 that for each positive number ε there is an integer k so large that if $k < i < j$, then there is a homeomorphism $f(i, j; x)$ of $X(i, j)$ onto $Y(i, j)$ such that

$f(i, j) = h_i$ on S_i, and h_j on S_j while diameter $f^{-1}(i, j; A) < \varepsilon$ for each straight line interval A in $Y(i, j)$ on a ray through the origin.

Let $k_1, k_2, \ldots$ be a monotone increasing sequence of positive integers such that if $k_n \leqslant i < j$, then there is such a homeomorphism $f(i, j)$ for $\varepsilon < 1/n$. Hence we suppose that $f(k_n, k_{n+1})$ is a homeomorphism of $X(k_n, k_{n+1})$ onto $Y(k_n, k_{n+1})$ such that

$f(k_n, k_{n+1}) = h_{k_n}$ on S_{k_n} and $h_{k_{n+1}}$ on $S_{k_{n+1}}$, while diameter $f^{-1}(k_n, k_{n+1}; A) < 1/n$ for each straight line interval A in $Y(k_n, k_{n+1})$ on a line through the origin.

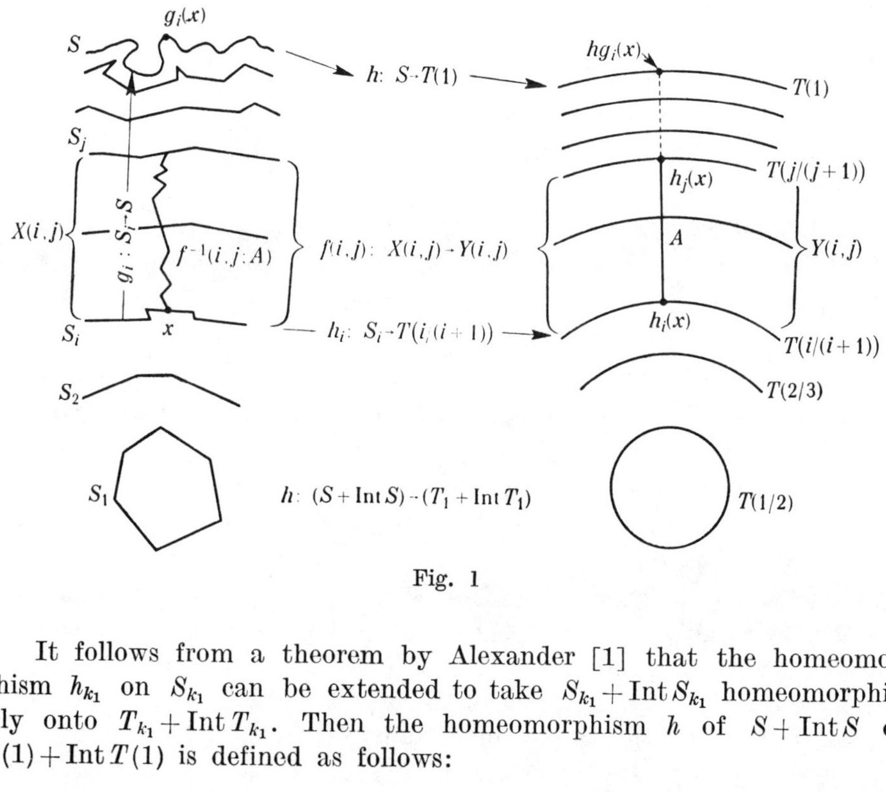

Fig. 1

It follows from a theorem by Alexander [1] that the homeomorphism h_{k_1} on S_{k_1} can be extended to take $S_{k_1} + \operatorname{Int} S_{k_1}$ homeomorphically onto $T_{k_1} + \operatorname{Int} T_{k_1}$. Then the homeomorphism h of $S + \operatorname{Int} S$ on $T(1) + \operatorname{Int} T(1)$ is defined as follows:

$$\begin{aligned} h &= h_{k_1} && \text{on} \quad S_{k_1} + \operatorname{Int} S_{k_1}, \\ h &= f(k_n, k_{n+1}) && \text{on} \quad X(k_n, k_{n+1}), \\ h &= h && \text{on} \quad S. \end{aligned}$$

By applying Theorem 2.1 to both the interior and exterior of a 2-sphere in E^3, one can obtain the following result.

THEOREM 2.2. *A 2-sphere S in E^3 is tame if for each positive number ε there are 2-spheres S', S'' in* $\operatorname{Int} S, \operatorname{Ext} S$ *respectively such that $H(S, S') < \varepsilon$, $H(S, S'') < \varepsilon$.*

3. A set bounded by two polyhedral 2-spheres. As in the proof of Theorem 2.1 we keep the following notation which will be used throughout Sections 2-9.

S is a 2-sphere in E^3 that contains no vertical interval.

$S_1, S_2, \ldots$ is a sequence of polyhedral 2-spheres in $\operatorname{Int} S$ such that $S_i \subset \operatorname{Int} S_{i+1}$.

g_i is a homeomorphism of S_i onto S that moves no point by more than $1/i$.

$T(r)$ $(0 < r \leqslant 1)$ is the 2-sphere with center at the origin and radius r.

h is a homeomorphism of S onto $T(1)$.

h_i is a homeomorphism of S_i onto $T\big(i/(i+1)\big)$ such that $h_i(x)$ is between $hg_i(x)$ and the origin.

$X(i, j) = S_j + \operatorname{Int} S_j - \operatorname{Int} S_i$.

$Y(i, j) = T\big(j/(j+1)\big) + \operatorname{Int} T\big(j/(j+1)\big) - \operatorname{Int} T\big(i/(i+1)\big)$.

We use fences to chop up images of certain $X(i, j)$'s so as to get a homeomorphism $f(i, j)$ of such an $X(i, j)$ onto the corresponding $Y(i, j)$ satisfying the conditions of Theorem 3.1.

THEOREM 3.1. *For each positive number* ε *there is an integer* k *such that if* $k < i < j$, *there is a homeomorphism* $f(i, j; x)$ *of* $X(i, j)$ *onto* $Y(i, j)$ *such that*

1. $f(i, j; x)$ *agrees with* $h_i(x)$ *on* S_i *and* $h_j(x)$ *on* S_j *and*
2. *diameter* $f^{-1}(i, j; A) < \varepsilon$ *for each straight line interval* A *in* $Y(i, j)$ *on a line through the origin.*

We now state Theorem 3.2 on which the proof of Theorem 3.1 is based. The proof of Theorem 3.2 is outlined in Section 4. Much tedious pushing and pulling, discussed in Section 7, is a basis for the proof of Theorem 3.2. It is in the proof of Theorem 3.2 that we start the use of fences to chop up the images of certain $X(i, j)$'s.

THEOREM 3.2. *For each positive number* ε *there are an integer* k *and a positive number* δ *such that if* F *is a fence of mesh less than* δ *and* $k < i < j$, *then there is a homeomorphism* H_3 *of* E^3 *onto itself, a stable finite graph* G *on* S, *and a homeomorphism* g *of* $G \times [0, 1]$ *into* F *such that*

1. $\varrho(H_3, I) < \varepsilon$,
2. $H_3(S_i + S_j)$ *is a polyhedron*,
3. *each component of* $S - G$ *is of diameter less than* ε,
4. $g(G \times 0) \subset H_3(S_i)$ *with* $g_i H_3^{-1} g(a \times 0) = a$,
5. $g(G \times 1) \subset H_3(S_j)$ *with* $g_j H_3^{-1} g(a \times 1) = a$,
6. *each* $g(a \times [0, 1])$ *is of diameter less than* ε,
7. $H_3(S_i + S_j) \cdot g(G \times [0, 1]) = g(G \times 0) + g(G \times 1)$.

See Figure 2. We note that H_3 removes certain feelers from S_i and S_j so that these feelers will not run in and out of the fence F and prevent Condition 7 from being satisfied.

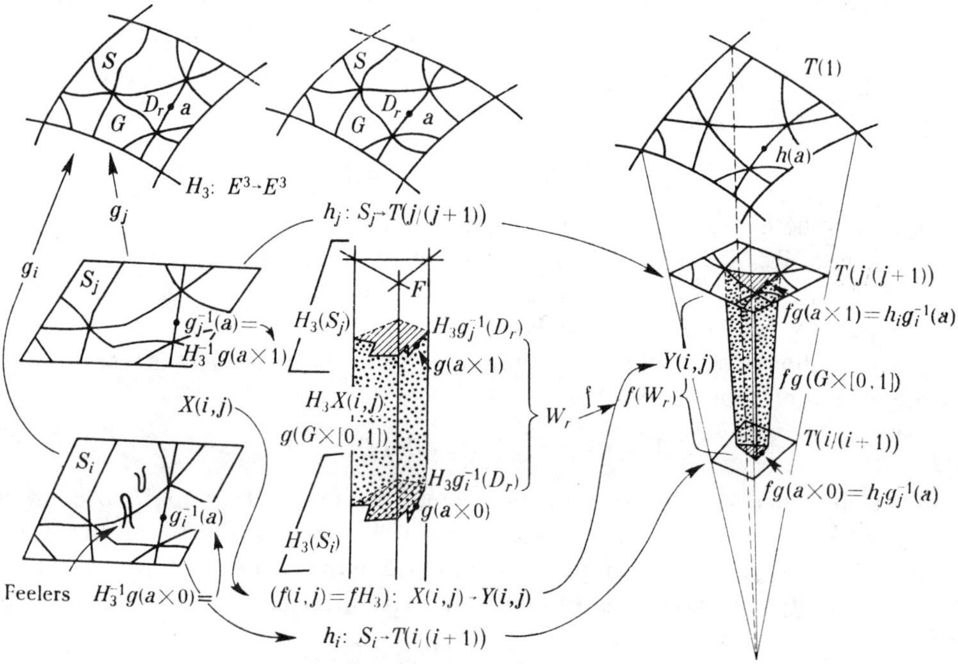

Fig. 2

Proof of Theorem 3.1. Let $\varepsilon_1 < \frac{1}{5}\varepsilon$ and k, F, G, H_3, and g be as promised in Theorem 3.2 for the positive number ε_1. We suppose $1/k < \varepsilon_1 <$ (diameter S)/11.

We describe the homeomorphism $f(i, j)$ in terms of a homeomorphism f of $H_3(X(i, j))$ onto $Y(i, j)$ where

$$f(i, j) = fH_3 .$$

Since the conclusion of Theorem 3.1 requires that

$$f(i, j) = h_i \text{ on } S_i \text{ and } h_j \text{ on } S_j ,$$

it follows from Conditions 4 and 5 of the statement of Theorem 3.2 that for each point a of G

$$fg(a \times 0) = h_i g_i^{-1}(a) = fH_3 g_i^{-1}(a)$$

and

$$fg(a \times 1) = h_j g_j^{-1}(a) = fH_3 g_j^{-1}(a) .$$

See Figure 2. The homeomorphism f is extended to the rest of $g(G\times[0,1])$ so that

$$fg(a\times t) \text{ is the point of } T\big((1-t)i/(i+1)+jt/(j+1)\big) \text{ between the origin and } h(a).$$

Let $D_1, D_2, \ldots, D_n$ be the disks which are the closures of the components of $S-G$ and

$$W_r = H_3 g_i^{-1}(D_r)+H_3 g_j^{-1}(D_r)+g(\mathrm{Bd}\,D_r\times[0,1]).$$

The 2-sphere W_r is shown in Figure 2.

Then W_m is a polyhedral 2-sphere of diameter less than $5\varepsilon_1$ since $H_3 g_i^{-1}(D_m)+H_3 g_j^{-1}(D_m)$ lies in a $2\varepsilon_1$ neighborhood of D_m and W_m lies in the convex hull of this neighborhood.

It follows from the fact that $G-\mathrm{Bd}\,D_r$ is a connected set of diameter more than $9\varepsilon_1$ that

$$\text{diameter } H_3 g_i^{-1}(G-\mathrm{Bd}\,D_r) > 5\varepsilon_1 .$$

Then W_r does not intersect $\mathrm{Int}\,W_m$ since $\mathrm{Int}\,W_m$ is too small to contain $H_3 g_i^{-1}(G-\mathrm{Bd}\,D_r)$. Hence no two of the W's have interiors that intersect and $H_3\big(X(i,j)\big)$ is the sum of polyhedral cubes bounded by the W's.

We suppose that f is extended to the interiors of the W's so that

$$f(W_r+\mathrm{Int}\,W_r) = f(W_r)+\mathrm{Int}\,f(W_r).$$

Then f takes $H_3\big(X(i,j)\big)$ homeomorphically onto $Y(i,j)$. Since $f(i,j) = fH_3$ and $H_3^{-1}(W_r)$ is of diameter less than $5\varepsilon_1$, for each arc A in $Y(i,j)$ on a line through the origin

$$\text{diameter } f^{-1}(i,j;A) < 5\varepsilon_1 < \varepsilon .$$

4. An isotopy that simplifies 2-spheres near fences. In this section we give the proof of Theorem 3.2. The proof depends on some somewhat unrelated theorems developed in Sections 5-9. As the proof of Theorem 3.2 is somewhat long, we break it into 8 steps.

Step 1. *Building cylinders about fence corners.* Consider any fence F. Let $L_1, L_2, \ldots, L_n$ be the corners of the fence that intersect S. It follows from the fact that S contains no vertical interval that for each positive number ε_1, we can find about each L_r a vertical triangular cylinder C_r with small horizontal cross section and such that

1. no corner of C_r lies on F,
2. each C_r is in general position with respect to each S_j,
3. for i sufficiently large, $S_i-\sum C_r$ has a component U_i such that each component of S_i-U_i is of diameter less than ε_1.

Let $K(r, j)$ denote the collection of components of $\mathrm{Bd}\, U_j$ on C_r that separate C_r into two unbounded pieces. It follows from Theorem 6.3 that we can find such C's and an integer k so that for i, j larger than k, there is a 1-1 correspondence between the elements of $K(r, i)$ and $K(r, j)$ such that

1. the distance between corresponding elements is less than ε_1,
2. no element of $K(r, i)+K(r, j)$ separates two corresponding elements from each other on C_r.

We suppose that k is so large that S_i+S_j misses each corner of F that S misses and that $1/k < \varepsilon_1$. If we take ε_1 to be very small, k must be large. As a substitute for saying "for k sufficiently large" we can say "for ε_1 sufficiently small".

This ε_1 we have introduced is the first of a sequence $\varepsilon_1, \varepsilon_2, \ldots, \varepsilon_7$ that we shall use. In general we suppose that each ε_r is very small but this can only be accomplished by making the preceding ones very very small. We are interested in knowing that such ε_r's exist rather than in knowing how big they are. If one wants a workout in epsilontics, he can start with ε_7 and work back to ε_1 but this is not recommended.

Step 2. *An isotopy that simplifies near fence corners.* Since a solid cylinder can be chopped up into topological cubes, it follows from Theorems 8.3 and 7.5 that there is a piecewise linear isotopy H_t $(0 \leqslant t \leqslant 1)$ of E^3 onto itself such that

1. H_1 is fixed on U_i+U_j,
2. $\varrho(H_1, I) < 9\varepsilon_1$,
3. $H_1(S_i+S_j)$ is in general position with respect to F,
4. each component of $H_1(S_i+S_j)-(U_i+U_j)$ is a disk E of diameter less than ε_1 such that $\mathrm{Int}\, E \subset \sum \mathrm{Int}\, C_r$, $E \cdot L_r$ contains only one point if $\mathrm{Bd}\, E$ separates C_r into two unbounded pieces and E misses L_r otherwise.

Figure 3 shows that H_1 removes bulges (shown in S_i-U_i) having extraneous intersections with C_r and also removes feelers (shown in S_j-U_j) winding through C_r.

The correspondence between the elements of $K(r, i)$ and the elements of $K(r, j)$ sets up a 1-1 correspondence between points of $L_r \cdot H_1(S_i)$ and $L_r \cdot H_1(S_j)$. If π takes a point of $L_r \cdot H_1(S_i)$ onto the corresponding point of $L_r \cdot H_1(S_j)$, then

$$\varrho(\pi, I) < 3\varepsilon_1 \quad \text{on} \quad L_r \cdot H_1(S_i)\,.$$

There is no point of $H_1(S_i+S_j)$ between a point and its image under π.

It follows from Theorem 8.4 that if k is sufficiently large and the disks E are properly chosen, then for each section V of F and arc pq in

$V \cdot H_1(S_i)$ joining two corner points of F, there is an arc in $V \cdot H_1(S_j)$ joining the points corresponding to p and q. See Figure 4.

Hence for each section V of F, we have a 1-1 correspondence between the components of $V \cdot H_1(S_i)$ that are arcs and the components of $V \cdot H_1(S_j)$ that are arcs.

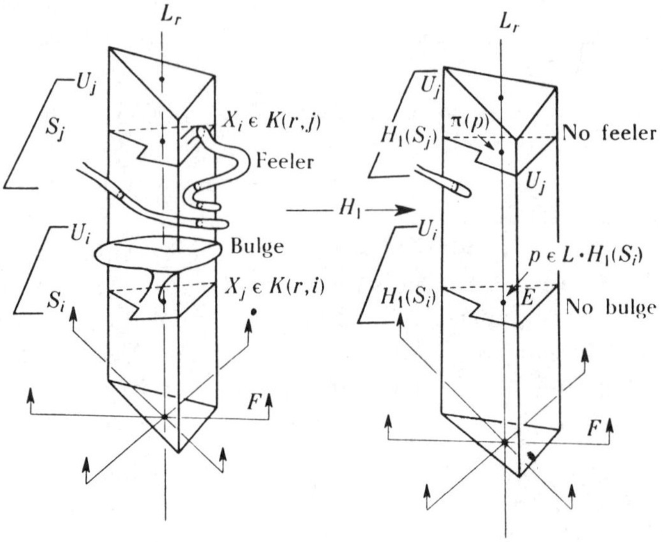

Fig. 3

Let G_i be the sum of all such arcs in $F \cdot H_1(S_i)$ and G_j be the sum of all such arcs in $F \cdot H_1(S_j)$. The aforementioned map π can be extended to a homeomorphism π of G_i onto G_j such that an arc in $V \cdot H_1(S_i)$ goes into the corresponding arc in $V \cdot H_1(S_j)$. Then $F \cdot H_1(S_i)$ is the sum of G_i and various simple closed curves in the sections of F. Also, $F \cdot H_1(S_j)$ is the sum of $G_j = \pi(G_i)$ and various simple closed curves in the sections of F.

Step 3. *The finite graph G.* The aforementioned G_i is a finite graph on $H_1(S_i)$ and $g_i H_1^{-1}(G_i)$ is a finite graph on S. We use these to get G.

Let ε_2 be the mesh of F.

Let ε_3 be a positive number such that each component of $H_1(S_i + S_j) - (G_i + G_j)$ is of diameter less than ε_3. Since S contains no vertical interval we can make ε_3 small by making ε_1 and ε_2 sufficiently small.

For each section V of F and each component A of $V \cdot (G_i + G_j)$

$$\text{diameter}\, A < \varepsilon_3 .$$

Hence

$$\varrho(\pi, I) < 3\varepsilon_1 + 2\varepsilon_3 \text{ on } G_i .$$

Since $\varrho(g_i, I) < \varepsilon_1$ and $\varrho(H_1, I) < 9\varepsilon_1$, each component of $S - g_i H_1^{-1}(G_i)$ is of diameter less than $\varepsilon_3 + 20\varepsilon_1$. It follows from Theorems 9.2 and 9.6 that for each positive number ε_4 and $\varepsilon_3 + 20\varepsilon_1$ sufficiently small one can cause $g_i H_1^{-1}(G_i)$ to contain a stable graph G such that

3. each component of $S - G$ is of diameter less than ε_4.

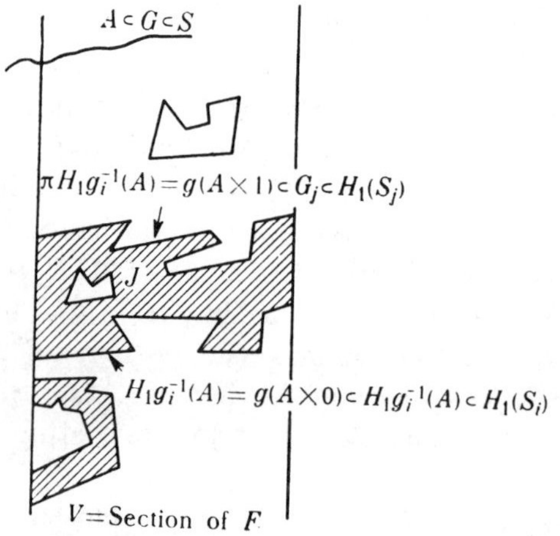

Fig. 4

For $\varepsilon_1, \varepsilon_2, \varepsilon_3, \varepsilon_4$, sufficiently small, this finite graph G is the one mentioned in the statement of Theorem 3.2. We have just shown that Condition 3 in the statement of Theorem 3.2 can be realized.

Step 4. *The homeomorphism g.* We recall that G_i is the sum of arcs in the sections of F and that G_j is the sum of corresponding arcs under π. Also $H_1 g_i^{-1}(G)$ is the sum of a subcollection of the arcs in G_i.

For each point a of G we define

$$g(a \times 0) = H_1 g_i^{-1}(a) \quad \text{and} \quad g(a \times 1) = \pi H_1 g_i^{-1}(a).$$

Since H_3 will be defined to be H_1 on $g_i^{-1}(G)$, we find that Condition 4 of the statement of Theorem 3.2 can be realized.

If V is a section of F, A is an arc in G such that $H_1 g_i^{-1}(A)$ is a component of $V \cdot H_1 g_i^{-1}(G)$, then g takes $A \times [0, 1]$ onto the disk in V bounded by the arcs $H_1 g_i^{-1}(A)$, $\pi H_1 g_i^{-1}(A)$, and two vertical intervals. See Figure 4. Since the diameter of this disk is less than $3\varepsilon_1 + 2\varepsilon_3$, it follows that

$$\text{diameter}\, g(a \times [0, 1]) < 3\varepsilon_1 + 2\varepsilon_3.$$

This suggests that Condition 6 of the statement of Theorem 3.2 can be realized.

Step 5. *Diameters of* $g(G\times 0)$ *and* $g(G\times 1)$. We have the following inequalities:

$$\text{diameter}\, G > \text{diameter}\, S - 2\varepsilon_4,$$

$$\text{diameter}\left(H_1 g_i^{-1}(G) = g(G\times 0)\right) > \text{diameter}\, S - 2\varepsilon_4 - 20\varepsilon_1,$$

$$\text{diameter}\left(\pi H_1 g_i^{-1}(G) = g(G\times 1)\right) > \text{diameter}\, S - 2\varepsilon_4 - 4\varepsilon_3 - 26\varepsilon_1.$$

Hence we suppose that the diameters of $g(G\times 0)$ and $g(G\times 1)$ approximate the diameter of S.

Step 6. *Pulling closed curves off sections.* There is no assurance that $H_1(S_i+S_j)\cdot g(G\times[0,1]) = g(G\times 0)+g(G\times 1)$ because of the simple closed curves (such as J shown in Figure 4) that lie in the intersections of the sections of F with $H_1(S_i+S_j)$. We extend the isotopy H_t $(0\leqslant t\leqslant 1)$ to a piecewise linear isotopy H_t $(0\leqslant t\leqslant 2)$ so that $F\cdot H_2(S_i+S_j)\subset G_i+G_j$.

Let ε_5 be a positive number such that if J is a simple closed curve in $F\cdot H_1(S_i)$ or $F\cdot H_1(S_j)$ that misses the corners of F, then J bounds a disk in $H_1(S_i)$ or $H_1(S_j)$ of diameter less than ε_5. Since S contains no vertical interval, we can make ε_5 small by restricting ε_1 and ε_4. In fact we can set $\varepsilon_5 = \varepsilon_4 + 20\varepsilon_1$.

By putting 2-spheres about these simple closed curves in $F\cdot H_1(S_i)$ and $F\cdot H_1(S_j)$ that miss corners of F, we find from Theorem 7.5 that we can extend H_t $(0\leqslant t\leqslant 1)$ to a piecewise linear isotopy H_t $(0\leqslant t\leqslant 2)$ so that

1. $\varrho(H_1, H_2) < 9\varepsilon_5$,
2. $F\cdot H_2(S_i)\subset G_i$, $F\cdot H_2(S_j)\subset G_j$, and
3. $H_t(1\leqslant t\leqslant 2) = H_1$ on components of $H_1^{-1}(G_i)$ and $H_1^{-1}(G_j)$ containing

$$H_1^{-1}\left(g(G\times 0)+g(G\times 1)\right).$$

We note that

$$\varrho(H_2, I) < 9\varepsilon_1 + 9\varepsilon_5.$$

One reason we did not suppose that $H_t(1\leqslant t\leqslant 2)$ equals H_1 on $H_1^{-1}(G_i)$ (or $H_1^{-1}(G_j)$) is that G_i (or G_j) may not be connected and one of the small simple closed curves (such as J shown in Figure 4) in a section of F may separate two points of it from each other in $H_1(S_i)$ (or $H_1(S_j)$).

We now have

$$H_2(S_i+S_j)\cdot g(G\times[0,1]) = g(G\times 0)+g(G\times 1).$$

Since we shall define H_3 so that $H_3(S_i+S_j) = H_2(S_i+S_j)$, we find that Conditions 2 and 7 of the statement of Theorem 3.2 can be realized.

Step 7. *The homeomorphism* H_3. The reason for not using H_2 for the H_3 promised by Theorem 3.2 is that we are not sure that $g_j H_2^{-1} g(a \times 1) = a$ for each point a of G. Other important requirements other than Condition 5 can be met. The difficulty is that even though

$$\pi H_2 g_i^{-1} \text{ is close to } H_2 g_j^{-1} \text{ on } G,$$

they are not necessarily equal.

We shall extend H_t $(0 \leqslant t \leqslant 2)$ to an isotopy H_t $(0 \leqslant t \leqslant 3)$ so that

1. $H_t(2 \leqslant t \leqslant 3) = H_2$ on S_i,
2. $H_t(S_j)$ $(2 \leqslant t \leqslant 3)$ is the same point set as $H_2(S_j)$, and
3. $\pi H_3 g_i^{-1} = H_3 g_j^{-1}$ on G.

This last condition will insure that Condition 5 of the statement of Theorem 3.2 can be attained.

Consider an arbitrary positive number ε_6. We show in the next step that for $\varepsilon_1, \varepsilon_2, \varepsilon_3, \varepsilon_4, \varepsilon_5$, suitably chosen, there is an isotopy F_t $(0 \leqslant t \leqslant 1)$ on S such that

1. F_t moves no point more than ε_6, and
2. $F_1 = g_j H_2^{-1} \pi H_2 g_i^{-1}$ on G.

Then $H_2 g_j^{-1} F_t g_j H_2^{-1}$ is an isotopy on $H_2(S_j)$ such that

$$\varrho(H_2 g_j^{-1} F_t g_j H_2^{-1}, I) < \varepsilon_6 + 18\varepsilon_5 + 20\varepsilon_1 .$$

Since $H_2(S_j)$ is a polyhedral 2-sphere, for each open set U in E^3 containing it, the isotopy $H_2 g_j^{-1} F_t g_j H_2^{-1} = F'_t$ can be extended to an isotopy on E^3 so that

1. $\varrho(F'_t, I) < \varepsilon_6 + 18\varepsilon_5 + 20\varepsilon_1$ and
2. $F'_t = I$ outside U.

In particular, we suppose $F'_t = I$ on $H_2(S_i)$.

We define

$$H_{t+2} = F'_t H_2$$

and note that

$$\varrho(H_3, I) < \varepsilon_6 + 27\varepsilon_5 + 29\varepsilon_1 .$$

Hence, we see that Condition 1 of the statement of Theorem 3.2 can be attained.

Furthermore

$$\begin{aligned} H_3 g_j^{-1} &= H_2 g_j^{-1} F_t g_j H_2^{-1} H_2 g_j^{-1} \\ &= H_2 g_j^{-1} g_j H_2^{-1} \pi H_2 g_i^{-1} \\ &= \pi H_2 g_i^{-1} = \pi H_3 g_i^{-1} \text{ on } G . \end{aligned}$$

This shows that for each point a of G

$$g(a \times 1) = \pi H_1 g_i^{-1}(a) = \pi H_3 g_i^{-1}(a) = H_3 g_j^{-1}(a) .$$

Hence Condition 5 of the statement of Theorem 3.2 can be attained.

Step 8. *An isotopy on S.* A 2-sphere S has the property that for each positive number ε_6 there is a positive number ε_7 such that if h is a homeomorphism of S onto itself that moves no point by more than ε_7, there is an isotopy F_t $(0 \leqslant t \leqslant 1)$ on S such that $F_0 = I$, $F_1 = h$, and $\varrho(F_t, I) < \varepsilon_6$. See [10].

We have that G is a stable graph on S, each component of $S-G$ is of diameter less than ε_4, and for each point a of G,

$$\varrho\big(g_j H_2^{-1}\pi H_2 g_i^{-1}(a), a\big) < \varepsilon_1 + (9\varepsilon_1 + 9\varepsilon_5) + (3\varepsilon_1 + 2\varepsilon_3) + (9\varepsilon_1 + 9\varepsilon_5) + \varepsilon_1 = 18\varepsilon_5 + 2\varepsilon_3 + 23\varepsilon_1 .$$

It follows from Theorem 9.5 that for ε_1 and $18\varepsilon_5 + 2\varepsilon_3 + 23\varepsilon_1$ sufficiently small, $g_j H_2^{-1} f H_2 g_i^{-1}$ on G may be extended to a homeomorphism h of S onto itself such that $\varrho(h, I) < \varepsilon_7$. Then

$$F_1 = g_j H_2^{-1}\pi H_2 g_i^{-1} \text{ on } G .$$

5. The intersection of 2-spheres with cylinders. Let L be a vertical line. It may be that $S\cdot L$ contains infinitely many points. In this case, the number of components of $S_i\cdot L$ may increase with i. In this and the next section we seek to control these intersections.

Question. There is no known method for slightly adjusting an arbitrary 2-sphere with an isotopy of E^3 onto itself so as to make the intersection of the adjusted surface with a fixed line finite. The lack of such a method led to complications in [3] and [13]. It would be convenient to know if each 2-sphere is pierced by a tame arc.

Let L be a vertical line and Δ be a triangle in a horizontal plane P such that $P\cdot L$ is the center of Δ. Let $\Delta(x)$ denote the triangle with center at $P\cdot L$, diameter x, edges parallel to the edges of Δ, and vertices on the rays from $P\cdot L$ through the vertices of Δ. We use $C(x)$ to denote the vertical triangular cylinder such that $P\cdot C(x) = \Delta(x)$.

We suppose that c is so small that only one component of $S - C(c)$ has a large diameter and this component $U(c)$ lies in $\operatorname{Ext} C(c)$. For each $0 < x \leqslant c$, we use $U(x)$ to denote the component of $S - C(x)$ containing $U(c)$ and $K(x)$ to denote the collection of components X of $\overline{U}(x)\cdot C(x)$ such that $C(x) - X$ has two unbounded components.

THEOREM 5.1. *If $0 < x \leqslant c$, each element of $K(x)$ separates $U(x)$ from a point of $S\cdot L$ in S.*

Proof. Assume X is an element of $K(x)$ that separates no point of $S\cdot L$ from $U(x)$ in S. We show in six steps that the assumption that such an X exists leads to a contradiction.

1. If X does not separate any point of $S\cdot L$ from $U(x)$ in S, there is a continuum Y in $S - X$ that contains $S\cdot L$. Let X^+ be the component

of $S-U(x)$ containing X. Then X^+ is a component of $(S-U(x))+Y$ and there is a simple closed curve J in S separating X^+ from Y in $S-[(S-U(x))+Y]$. This simple closed curve J bounds a disk D in $S-Y$.

2. Let x_1 be a positive number less than one half the distance from L to $C(x)$ and x_2 be a positive number so small that if p_1, p_2 are two points of D such that $\varrho(p_1, p_2) < x_2$, then $p_1 + p_2$ lies on an arc in D of diameter less than x_1. The existence of such a number x_2 follows from the uniform local connectedness of disk D.

3. Since each unbounded component of $C(x)-X$ is topologically equivalent to the interior of a circle minus its center, there is a simple closed curve J in $C(x)$ such that $C(x)-J$ has two unbounded components and $J \subset V(X, \frac{1}{3}x_2)$ where $V(A, \varepsilon)$ denotes the set of all points whose distances from A are less than ε. We call $V(A, \varepsilon)$ the ε neighborhood of A.

4. The simple closed curve J cannot be shrunk to a point without hitting L — that is, there is no map of a disk E into E^3-L that takes $\mathrm{Bd}E$ homeomorphically onto J. If there were such a map, one could find by using a projection from L that there is a map f_1 of E into $C(x)$ that takes $\mathrm{Bd}E$ homeomorphically onto J; there is a retraction f_2 of $C(x)$ onto J; then $f_1^{-1}f_2f_1$ is a map of E onto $\mathrm{Bd}E$ that leaves each point of $\mathrm{Bd}E$ fixed. Rather than using the set of impossible maps to arrive at a contradiction to the assumption that J can be shrunk to a point without hitting L, certain linking arguments could have been used instead.

5. Since $J \subset V(X, \frac{1}{3}x_2)$, there is a map g of J into D such that $\varrho(p, g(p)) < x_1 + x_2 < \varrho(L, C(x))$. To get such a g one could consider points $p_1, p_2, \ldots, p_n = p_1$ on J such that $\varrho(p_i, p_{i+1}) < \frac{1}{3}x_2$; points $g(p_1)$, $g(p_2), \ldots, g(p_n) = g(p_1)$ on X such that $\varrho(p_i, g(p_i)) < \frac{1}{3}x_2$; and extend g to an arc p_ip_{i+1} of J onto an arc in D of diameter less than x_1.

6. Finally we show that the false assumption that X separates no point of $S\cdot L$ from $U(x)$ in S leads to the contradiction that J can be shrunk to a point without hitting L (violating Step 4). Let E be the disk $x^2+y^2 \leqslant 4$ in the x, y plane and f be a homeomorphism of $\mathrm{Bd}E$ onto J. Then f can be extended to $1 \leqslant x^2+y^2 \leqslant 4$ so that for each point (x, y) of $\mathrm{Bd}E$, f takes the interval from (x, y) to $(\frac{1}{2}x, \frac{1}{2}y)$ linearly onto the interval from $f(x, y)$ to $gf(x, y)$. Since $g(J) \subset D$, the map f can be extended to map $x^2+y^2 \leqslant 1$ into D. The map f we have described shrinks J to a point in E^3-L and contradicts Step 4.

THEOREM 5.2. *If* $0 < x \leqslant c$, $X \in K(x)$, *and* $\varepsilon > 0$, *there are simple closed curves* J_1, J_2 *on* S *in an* ε *neighborhood of* X *such that* X *separates* J_1 *from* J_2 *on* S *and neither* J_1 *nor* J_2 *can be shrunk to a point in* E^3-L.

Proof. We suppose $\varepsilon < \varrho(L, C(x))$. Let D be an open subset of S containing X such that $\bar{D} \subset V(X, \varepsilon)$ and $\bar{D} - D$ is the sum of a finite collection of simple closed curves $J_1, J_2, \ldots, J_n$ in different components of $S - X$. We show that the assumption that each J_i with the possible exception of J_1 can be shrunk to a point on a set R_i in $E^3 - L$ leads to a contradiction.

It follows from Steps 3 and 4 of the proof of Theorem 5.1 that there is a simple closed curve J on $C(x)$ such that J cannot be shrunk to a point in $E^3 - L$ but J can be shrunk in $E^3 - L$ to a closed set Y in D. But the set Y in turn can be shrunk to a point in $\bar{D} + R_2 + \ldots + R_n$. This shrinking violates the condition that J cannot be shrunk to a point in $E^3 - L$.

THEOREM 5.3. *For each fixed positive number $x \leqslant c$, $K(x)$ has only a finite number of elements.*

Proof. Cover $S \cdot L$ with a finite collection $V_1, V_2, \ldots, V_n$ of connected open subsets of S such that each V is of diameter less than $\varrho(L, C(x))$. Then each element of $K(x)$ separates a V_i from $U(x)$ in S but no two elements of $K(x)$ separates the same V_i from $U(x)$ in S. Hence, there are not more than n elements in $K(x)$.

THEOREM 5.4. *For each element X of $K(c)$ and each $0 < x < c$, X separates some element of $K(x)$ from $U(c)$.*

Proof. Assume otherwise. Then there is a continuum Y in $S - X$ containing all elements of $K(x)$ and a disk D in $S - Y$ such that $X \subset D$, $\operatorname{Bd} D \subset U(c)$. Let D' be the component of $D - C(x)$ containing $\operatorname{Bd} D$. It may be that $D \cdot C(x)$ contains a component that separates $C(x)$ into two unbounded sets but since D contains no element of $K(x)$, $\bar{D}' \cdot C(x)$ contains no component that separates $C(x)$ into two unbounded sets.

Following the proof of Theorem 5.3, we find that there is a simple closed curve J in $C(c)$ that cannot be shrunk to a point in $E^3 - L$ but J can be shrunk in $E^3 - C(x)$ to a closed set on $\bar{D}'$. Then there is a map g of the disk $x^2 + y^2 \leqslant 4$ into E^3 such that g takes the circle $x^2 + y^2 = 4$ homeomorphically onto J,

$$g(1 \leqslant x^2 + y^2 \leqslant 4) \subset \operatorname{Ext} C(x), \quad g(x^2 + y^2 = 1) \subset \bar{D}', \quad g(x^2 + y^2 \leqslant 1) \subset D.$$

Let Z be the set of all points p of the disk $x^2 + y^2 \leqslant 4$ such that there is an arc A in $x^2 + y^2 \leqslant 4$ from p to $\operatorname{Bd}(x^2 + y^2 \leqslant 4)$ such that $g(A)$ misses $C(x)$. Let $f = g$ on $\bar{Z}$. The map f of $\bar{Z}$ into $E^3 - \operatorname{Int} C(x)$ can be extended to take $x^2 + y^2 \leqslant 4$ into $E^3 - \operatorname{Int} C(x)$ since $C(x) \cdot f(\bar{Z})$ contains no component that separates $C(x)$ into two unbounded sets. Then f shrinks J to a point in $E^3 - L$ and this contradicts the definition of J.

COROLLARY 5.5. *If $0 < x \leqslant y \leqslant c$, the number of elements in $K(y)$ is less than or equal to the number of elements in $K(x)$.*

THEOREM 5.6. *Suppose that f is a homeomorphism of the plane set $(1 \leqslant x^2+y^2 \leqslant 4)$ into E^3-L such that $f(x^2+y^2=4)$ cannot be shrunk to a point in E^3-L. If a component of $C(x)\cdot f(1 \leqslant x^2+y^2 \leqslant 4)$ separates $f(x^2+y^2=1)$ from $f(x^2+y^2=4)$ in $f(1 \leqslant x^2+y^2 \leqslant 4)$, this component also separates $C(x)$ into two unbounded sets.*

Proof. The proof is a modification of the last paragraph of the proof of Theorem 5.4. Let D' be the component of $f(1 \leqslant x^2+y^2 \leqslant 4)-C(x)$ containing $f(x^2+y^2=4)$. If no component of $\bar{D}'\cdot C(x)$ separates $C(x)$ into two unbounded sets, there is a map f' of $(x^2+y^2 \leqslant 4)$ into $D'+C(x)$ such that f' agrees with f on $(x^2+y^2=4)$. This contradicts the condition that $f(x^2+y^2=4)$ cannot be shrunk to a point in E^3-L.

The preceding argument also gives the following result.

THEOREM 5.7. *If $0<x<c$ and J is a simple closed curve in $U(x)-U(c)$ that cannot be shrunk to a point in E^3-L, then J separates an element of $K(x)$ from $U(c)$.*

THEOREM 5.8. *Each subinterval $[a, b]$ of $[0, c]$ contains a subinterval $[a_1, b_1]$ such that if x, y are elements of $[a_1, b_1]$, then $K(x), K(y)$ have the same number of elements.*

Proof. Suppose $K(a), K(b)$ have m, n elements respectively. If we consider $m-n+1$ mutually exclusive subintervals of $[a, b]$, it follows from Corollary 5.5 that one of them, $[a_1, b_1]$ satisfies the conclusions of Theorem 5.8.

THEOREM 5.9. *The interval $[a_1, b_1]$ of Theorem 5.8 contains a subinterval $[a_2, b_2]$ such that if x, y are elements of $[a_2, b_2]$, the elements of $K(x)$ and $K(y)$ may be ordered $X(x)_1, X(x)_2, \dots, X(x)_r$ and $X(y)_1, X(y)_1, \dots, X(y)_r$ so that $X(x)_i, X(y)_i$ separate the same subset of $S\cdot L$ from $U(c)$.*

Proof. If $a_1 \leqslant x \leqslant y \leqslant b_1$, it follows from Theorems 5.4 and 5.8 that if $X(x)$ is an element of $K(x)$, then there is one and only one element $X(y)$ of $K(y)$ that separates $X(x)$ from $U(c)$ in S. These are corresponding elements and are given the same subscripts.

Let $V_1, V_2, \dots, V_r$ be connected subsets of $S\cdot \operatorname{Int} C(a_1)$ covering $S\cdot L$. If an element of $K(x)$ separates V_1 from $U(c)$ in S, so does the corresponding element of $K(y)$ if $a_1 \leqslant x \leqslant y \leqslant b_1$. Hence some subinterval $[a', b']$ of $[a_1, b_1]$ has the property that if x, y are elements of $[a', b']$, an element of $K(x)$ separates V_1 from $U(c)$ in S if, and only if, the corresponding elements of $K(y)$ does. A subinterval of $[a', b']$ has this property with respect to V_2. By taking subintervals for as many times as there are elements in $V_1, V_2, \dots, V_r$, we arrive at a subinterval $[a_2, b_2]$ of $[a_1, b_1]$ such that if x, y are elements of $[a_2, b_2]$, an element of $K(x)$ separates a V_i from $U(c)$ in S if, and only if, the corresponding element of $K(y)$ does.

THEOREM 5.10. *If* X', X'' *are corresponding elements of* $K(a_2)$, $K(b_2)$ *respectively of Theorem* 5.9 *and* J *is a simple closed curve in* S *that separates* X' *from* X'' *in* S, *then* J *cannot be shrunk to a point in* E^3-L.

Proof. Since X', X'' separates the same subset of $S \cdot L$ from $U(c)$ in S, the component U' of $S-(X'+X'')$ between X' and X'' contains no point of L.

Let X''' be the element of $K\big((a_2+b_2)/2\big)$ corresponding to X', X''. It follows from Steps 3 and 4 of the proof of Theorem 5.1 that there is a simple closed curve J' on $C\big((a_2+b_2)/2\big)$ such that J' cannot be shrunk to a point in E^3-L but it can be shrunk in E^3-L to a closed set Y in U' very near X'''. Since U' is topologically equivalent to the plane set $(1 \leqslant x^2+y^2 < 4)$, Y can be shrunk into J in U'. But J cannot be shrunk to a point in E^3-L or else J' could be shrunk to a point by way of Y and J.

THEOREM 5.11. *For each positive number* ε *there is a subinterval* $[a_3, b_3]$ *of* $[a_2, b_2]$ *of Theorem* 5.9 *such that if* $x \in [a_3, b_3]$ *and* $X(x), X(a_3), X(b_3)$ *are corresponding elements of* $K(x), K(a_3), K(b_3)$, *then the component of* $S-\big(X(a_3)+X(b_3)\big)$ *between* $X(a_3)$ *and* $X(b_3)$ *is in an* ε *neighborhood of* $X(x)$ *and is between* $C(a_2)$ *and* $C(b_2)$.

Proof. Pick an element $X(a_2)$ of $K(a_2)$ for preliminary consideration. For each element x of $[a_2, b_2]$ we let $X(x)$ denote the element of $K(x)$ corresponding to $X(a_2)$. The $X(x)$'s are linearly ordered on S in the sense that if $x<y<z$, $X(y)$ separates $X(x)$ from $X(z)$ on S. Using this linear ordering and the fact that there are uncountably many x's between a_2 and b_2 we find a subinterval $[a_2', b_2']$ of $[a_2, b_2]$ so that if C is a continuum in S separating $X(a_2')$ from $X(b_2')$, C lies between $C(a_2)$ and $C(b_2)$ while the component of $S-\big(X(a_2')+X(b_2')\big)$ between $X(a_2)$ and $X(b_2)$ is in an ε neighborhood of C.

If $[a_2', b_2']$ is successively shortened by an iteration of the above process in considering the various other elements of $K(a_2)$, we arrive at an interval $[a_3, b_3]$ satisfying the conclusion of the theorem.

6. The intersections of cylinders and approximating 2-spheres. In the preceding section we considered the intersection of S with certain vertical triangular cylinders $C(x)$. The $C(x)$'s we use in this section are the same as those used in Section 5. In the present section we find that if S' is a close approximation to S, its intersection with certain $C(x)$'s has some properties in common with the intersection of S with these $C(x)$'s.

Suppose $0<x<c$ and h is a homeomorphism of S onto a 2-sphere S' such that h moves no point as far as $\varrho\big(U(c), C(x)\big)$. We use $U'(x)$ to denote the component of $S'-C(x)$ containing $h\big(U(c)\big)$ and $K(x, S')$ to

denote the collection of all components of $\bar{U}'(x)\cdot C(x)$ that separate $C(x)$ into two unbounded pieces.

THEOREM 6.1. *Suppose $[a_3, b_3]$ is the subinterval promised by Theorem 5.11, $a_3 < a_4 < b_4 < b_3$, and h is a homeomorphism of S onto a 2-sphere S' such that*

$$\varrho(h, I) < \min\big(\varrho(C(a_3), C(a_4)), \varrho(C(b_4), C(b_3))\big).$$

Then if $x \in [a_4, b_4]$, $K(x, S')$ has the same number of elements as $K(x)$ and there is a correspondence between the elements of $K(x)$ and $K(x, S')$ such that an element of $K(x)$ separates a point p of $S\cdot L$ from $U(c)$ if and only if the corresponding element of $K(x, S')$ separates $h(p)$ from $h\big(U(c)\big)$.

Proof. Since h is so near I,

$$h\big(U(b_3)\big) \subset U'(b_4) \subset h\big(U(a_3)\big).$$

Let X, X'' be corresponding elements of $K(b_3), K(a_3)$. Applying Theorems 5.2 and 5.7 to an annulus on S' between $h(X)$ and $h(X'')$, we find that for each element x of $[a_4, b_4]$, an element X' of $K(x, S')$ separates $h(X)$ from $h(X'')$ in S. Then X separates a point p of $S\cdot L$ from $U(c)$ on S if and only if X' separates $h(p)$ from $h\big(U(c)\big)$ on S.

We finish the proof of Theorem 6.1 by showing that all elements of $K(x, S')$ are of the sort described above. Suppose there is an element X' of $K(x, S')$ such that $h^{-1}(X')$ does not separate from $U(c)$ any element of $K(a_3)$. It follows from Theorem 5.2 that there is a simple closed curve J in S' such that J cannot be shrunk to a point in E^3-L, X separates J from $h\big(U(c)\big)$, and J is so near X that $h^{-1}(J) \subset U(a_3)-U(c)$ and J can be shrunk to $h^{-1}(J)$ in E^3-L. But Theorem 5.7 gives the contradiction that $h^{-1}(J)$ separates an element of $K(a_3)$ from $U(c)$ in S.

THEOREM 6.2. *If X, X' are corresponding elements of $K(x)$, $K(x, S')$ of Theorem 6.1, then each lies in a 2ε neighborhood of the other where ε is as given in Theorem 5.11. If ε is taken to be sufficiently small, no element of $K(x)+K(x, S')$ separates X from X' on $C(x)$.*

Proof. Let $X(a_3), X(b_3)$ be elements of $K(a_3), K(b_3)$ corresponding to X of $K(x)$. Since $h^{-1}(X')$ separates $X(a_3)$ from $X(b_3)$ as shown in proof of Theorem 6.1, each of $X, h^{-1}(X')$ lies in an ε neighborhood of the other as shown in Theorem 5.11, and $\varrho(h, I) < \frac{1}{2}\varepsilon$, each of X, X' lies in a $\frac{3}{2}\varepsilon$ neighborhood of the other.

If X_1, X_2 are two elements of $K(b_3)$, the subsets of S separated from $U(c)$ by X_1, X_2 are a finite distance apart and for $x \in [a_4, b_4]$, the corresponding elements of $K(x)$ are at least this far apart. Decreasing the size of ε of Theorem 5.11 does not make noncorresponding elements close together. Hence, for ε sufficiently small, the last sentence in the statement of Theorem 6.2 is satisfied.

Applications of the theorems of Sections 5 and 6 to vertical triangular cylinders about the corners of a fence give the following result.

THEOREM 6.3. *Suppose S is a 2-sphere in E^3 that contains no vertical interval; $S_1, S_2, \ldots$ is a sequence of polyhedral 2-spheres in* $\operatorname{Int} S$ *such that $S_i \subset \operatorname{Int} S_{i+1}$ and $H(S, S_i) < 1/i$; F is a fence; and ε is a positive number. Then there are mutually exclusive vertical triangular cylinders $C_1, C_2, \ldots, C_n$ about the corners of F that intersect S such that*

1. *no corner of any C_r lies in F,*
2. *each C_r is in general position with respect to each S_i,*
3. *for i sufficiently large, $S_i - \sum C_r$ has a component U_i such that each component of $S_i - U_i$ is of diameter less than ε,*

Furthermore, if $K(r, j)$ denotes the collection of components of $C_r \cdot \operatorname{Bd} U_j$ that separate C_r into two unbounded pieces, there is an integer k such that for i, j greater than k, there is a 1-1 *correspondence between the elements of $K(r, i)$ and $K(r, j)$ such that*

1. *the distance between corresponding elements is less than ε,*
2. *and no element of $K(r, i) + K(r, j)$ separates two corresponding elements from each other on C_r.*

7. Isotopies near 2-spheres. A polyhedral 2-sphere S may have feelers that wander in and out of other 2-spheres $M_1, M_2, \ldots, M_n$. In this section we learn how to pull these feelers back without moving points too far.

THEOREM 7.1. *Suppose S, M are polyhedral 2-spheres in E^3 in relative general position; D is a disk in M such that $D \cdot S = \operatorname{Bd} D$; E is a disk in S bounded by* $\operatorname{Bd} D$; *V_D, V_E are interiors of polyhedral 3-cells such that these interiors contain* $\operatorname{Int} D$, $\operatorname{Int} E$ *respectively; C is the polyhedral cube bounded by $D + E$; and $\varepsilon > 0$. Then there is a piecewise linear isotopy H_t $(0 \leqslant t \leqslant 1)$ of E^3 onto itself such that*

1. $H_1(E) = D$,
2. $H_1 = I$ *(Identity) outside* $C + V_D + V_E$,
3. $H_1(C) \subset D + V_D$,
4. *H_t moves no point which is outside $V_E + C$ by as much as ε,*
5. *$\varrho(H_s(x), H_t(x)) < \varepsilon$ if s, t are two values of $[0, 1]$ and x is a point such that neither $H_s(x), H_t(x)$ belong to $C + V_E$.*

Proof. Figure 5 shows how we use H_t to pull back a feeler E of S.

It follows from the extension of Alexander's theorem [1] due to Moise [16] and Graeub [12] that there is a piecewise linear homeomorphism h of E^3 onto itself that takes $D + E$ onto the surface of a tetrahedron $abcp$ whose base abc is $h(D)$. Let q be the center of abc and r, s

be points such that $rpqs$ is a straight line interval (with points r, p, q, s in the order indicated) such that

$$abcr \subset h(C+V_E),$$

$$\operatorname{Int} abcs \subset h(V_D), \quad \text{and} \quad s \text{ is very close to } q.$$

Let x_t be the point of pq whose distance from p is t times the length of pq. Let F_t be the homeomorphism of E^3 onto itself fixed outside $abcr + abcs$, that takes p to x_t and is linear on the tetrahedra $abpr$, $bcpr$, $acpr$, $abps$, $bcps$, $acps$. Then $H_t = h^{-1}F_t h$.

Conditions 1, 2, 3 of the hypothesis of the theorem are met without restricting s to be very close to q but a sufficiently stringent enforcement of this restriction causes Conditions 4 and 5 to be satisfied.

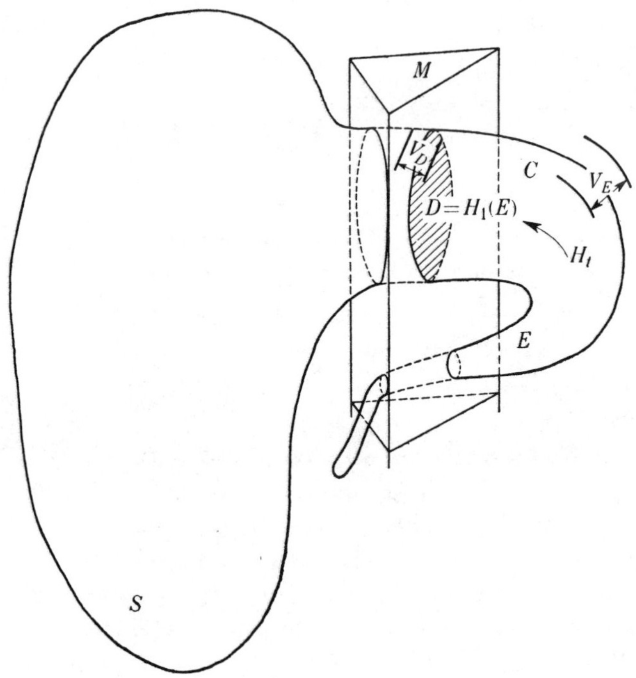

Fig. 5

THEOREM 7.2. *Suppose K is a polyhedral cube in E^3 and D, E are two polyhedral disks such that $D \cdot \operatorname{Bd} K = \operatorname{Bd} D = \operatorname{Bd} E = E \cdot \operatorname{Bd} K$, $\operatorname{Int} D \subset \operatorname{Int} K$, $\operatorname{Int} E \subset \operatorname{Int} K$. Then there is a piecewise linear isotopy H_t $(0 \leqslant t \leqslant 1)$ of E^3 onto itself that is the identity except on K and such that $H_1(E) = D$.*

Proof. With the exception of the piecewise linear part of the conclusion, this theorem was essentially proved by Alexander in [2]. It has been used as a lemma since but we include a short proof for completeness.

Let E' be a polyhedral disk such that $\operatorname{Bd} E' = \operatorname{Bd} D$, $\operatorname{Int} E' \subset \operatorname{Int} K$, $E' \cdot (D+E) = \operatorname{Bd} E'$. It follows from Theorem 7.1 that there is an isotopy $H_t (0 \leqslant t \leqslant \frac{1}{2})$ of E^3 onto itself so that H_t is the identity except on K and $H_{1/2}(E) = E'$. Another application of Theorem 7.1 shows that $H_t (0 \leqslant t \leqslant \frac{1}{2})$ can be extended to $H_t (0 \leqslant t \leqslant 1)$ so that H_t is the identity except on K and $H_1(E) = D$.

THEOREM 7.3. *Suppose $M_1, M_2, \ldots, M_n$ are mutually exclusive polyhedral 2-spheres in E^3 each of diameter less than ε.*

Suppose S is a polyhedral 2-sphere in general position with respect to each of the M_i's and U is a component of $S - \sum M_i$ of diameter more than 2ε such that each component of $S-U$ is of diameter less than ε.

Then for each continuum X in $E^3 - \big(\sum M_i + (S-U)\big)$ that intersects U there is an isotopy $H_t (0 \leqslant t \leqslant 1)$ of E^3 onto itself such that

1. *H_t moves no point of $\overline{U} + X$,*
2. *for each component E_j of $S-U$, $H_1(\operatorname{Int} E_j)$ lies in some $\operatorname{Int} M_i$,*
3. *H_1 moves no point of S by as much as 3ε,*
4. *H_t moves no point of S by as much as 6ε,*
5. *H_1 moves no point of E^3 by as much as 9ε,*
6. *H_t moves no point of E^3 by as much as 12ε.*

Proof. Some of the coefficients ε given in Conditions 3, 4, 5, 6 may be a bit extravagant but they are good enough for our purposes when we apply Theorem 7.3 in Theorem 3.2. We divide the proof into seven steps.

Step 1. *Preliminary simplifications.* Let $J_1, J_2, \ldots, J_n$ be the simple closed curves in $\operatorname{Bd} \overline{U}$ and E_i be the disk in $S-U$ bounded by J_i.

With no loss of generality we suppose that each M_j contains one of the J_i's since we can discard from consideration any that does not.

We suppose that no M_j contains two J_i's because if M_1 contains $J_1 + J_2$, we can split it and get two mutually exclusive polyhedral 2-spheres M', M'' such that $J_1 \subset M'$, $J_2 \subset M''$, $M_1 \cdot S \subset M' + M'' \subset M_1 + \operatorname{Int} M_1$. Hence we suppose that $J_i \subset M_i$.

If $\operatorname{Int} E_i \subset \operatorname{Int} M_i$, we consider E_i as joined to U and ignore both E_i and M_i. There is no need for the isotopy to move E_i.

Step 2. *Description of H_t.* The isotopy promised by Theorem 7.3 is given by iterated applications of Theorem 7.1 used to reduce the number of components of $S \cdot \sum M_i$. Suppose there are $n+k$ such components.

If $\operatorname{Int} E_j \not\subset \operatorname{Int} M_j$, consider a disk D in M_j such that $D \cdot S = \operatorname{Bd} D \subset S - \overline{U}$. Let E be the disk in $S - \overline{U}$ bounded by $\operatorname{Bd} D$ and C be the polyhedral cube bounded by $D+E$.

We take V_D and V_E to be open sets containing D and E respectively such that each point of V_D is very near D and each point of V_E is very near E. We shall discuss how close this should be later when we use this restriction to show that an isotopy we describe does not move points too far.

We apply Theorem 7.1 to get an isotopy H_t $(0 \leqslant t \leqslant 1/2k)$ pulling D to E and fixed at each point of $S-E$ and at each point outside $C+V_D+V_E$. The V's we described in the preceding paragraph are slightly larger than those mentioned in Theorem 7.1 but we use these larger ones since we shall need them in the next paragraph when we shove points across D. We suppose the isotopy H_t $(0 \leqslant t \leqslant 1/2k)$ satisfies Conditions 1, 2, 3, 4, 5 of Theorem 7.1 (with $1/2k$ substituted for 1 in determining range of t) where the size of the ε mentioned in Conditions 4 and 5 is to be given later. It is not the same as the ε mentioned in the statement of Theorem 7.3.

Moving nothing except in V_D, we extend H_t $(0 \leqslant t \leqslant 1/2k)$ to H_t $(0 \leqslant t \leqslant 1/k)$ by shoving D to one side of M_j so that nothing moves far (far is explained later) and $H_{1/k}(S)$ is a polyhedral 2-sphere in general position with respect to the M_i's, $H_{1/k}(S) \subset (S-E)+V_D$, and $H_{1/k}(S)\cdot \sum M_i$ is a proper subset of $S \cdot \sum M_i$.

Using $H_{1/k}(S)$ instead of S, the process is continued to reduce the number of components of the intersection of $\sum M_i$ and the resulting image of S. Taking a disk D in some M_j such that $D \cdot H_{1/k}(S) = \mathrm{Bd}D \subset \big(H_{1/k}(S)-\overline{U}\big)$ and the disk E in $H_{1/k}(S)-\overline{U}$ bounded by $\mathrm{Bd}D$, we consider neighborhoods V_D, V_E about D and $H_{1/k}(E)$ and extend H_t $(0 \leqslant t \leqslant 1/k)$ to H_t $(0 \leqslant t \leqslant 2/k)$ so as to shove E across D as before. In general we suppose that H_t is extended so that $H_{j/k}(S)\cdot \sum M_i$ is the sum of at most $n+k-j$ components of $S\cdot \sum M_i$.

It is clear that if the V's are chosen so as not to intersect $\overline{U}+X$, then H_t $(0 \leqslant t \leqslant 1)$ satisfies Condition 1 of the conclusion of the theorem and $H_1(S)$ satisfies Condition 2. We need to place restrictions on the V's in order to show that Conditions 3, 4, 5, 6 are satisfied.

Step 3. *Distance H_t moves points of S.* Let us consider the points of the E_i's and see how far they move under H_t. Let E_i^+ be the sum of E_i and all the M_j's intersecting it. If we suppose that each point of V_D is within δ of D, then each point of $H_{j/k}(E_i)$ is within δ of E_i^+. Since diameter E_i^+ is less than 3ε, we may restrict the V_D's and suppose that each $H_{j/k}(E_i)$ is of diameter less than 3ε and $\varrho\big(p, H_{j/k}(p)\big) < 3\varepsilon$ if $p \in S$. Hence Condition 3 is satisfied.

The cube C used at any stage is near the appropriate E_i^+ so we may suppose that for each point x of E^3,

$$\big(H_{j/k}(x), H_t(x)\big) \leqslant 3\varepsilon \quad \text{if} \quad j/k \leqslant t \leqslant (j+1)/k\,.$$

The preceding two paragraphs show that by restricting the V_D's and V_E's, we can keep H_t from moving any point of S by more than 6ε. Hence Condition 4 is satisfied.

Step 4. *Traps to be avoided.* Now let us turn our attention to Conditions 5 and 6 as applied to points of E^3-S. We must control our isotopies so as not to move a point too far. A trap to be avoided is illustrated by Figure 6. The point p_1 is near the leftmost sticker and is pulled into the left cube when the sticker is pulled in. If it goes into p_2 near another sticker, this point may in turn be pulled to p_3 when this sticker is removed. Care must be taken so that the point p_1 does not move to p_2, then to p_3, then to p_4, then to p_5, etc. and hence move more than 12ε. This is accomplished by restricting the sizes of the V_D's and V_E's used.

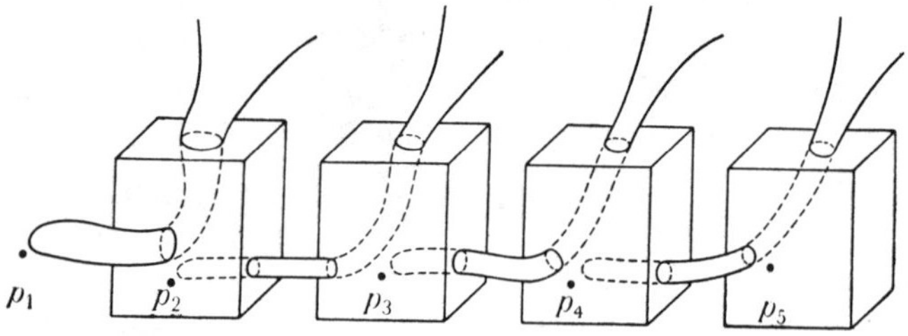

Fig. 6

Points of the V_E's offer special difficulties as suggested in the preceding paragraph so we first consider a point p of E^3-S such that for $j=r, r+1, \ldots, \text{or}, k-1$, the point $H_{j/k}(p)$ is not a point of the V_E about the disk E in $H_{j/k}(S)$ used in extending H_t past $H_{j/k}$. We let $s=r/k$ and show that with suitable restrictions on the V's, $\varrho\big(H_s(p), H_t(p)\big) < 6\varepsilon$ $(s \leqslant t \leqslant 1)$.

Step 5. *Diameter of* $H_s(p)^+$. Let $H_s(p)^+$ denote the sum of $H_s(p)$ and all M_j's such that each arc from $H_s(p)$ to U in $E^3-H_s\big(\sum E_i\big)$ intersects M_j. As a help in showing that $\varrho\big(H_s(p), H_t(p)\big) < 6\varepsilon$, we show in the next four paragraphs that diameter $H_s(p)^+ < 3\varepsilon$.

If no M_j separates $H_s(p)$ from U in $E^3-H_s\big(\sum E_i\big)$, then $H_s(p)^+$ is the point $H_s(p)$ and its diameter is less than 3ε. Hence we suppose that some M_j separates $H_s(p)$ from U in $E^3-H_s\big(\sum E_i\big)$.

We use $H_s(E_i)^+$ to denote the sum of $H_s(E_i)$ and all M_j's that intersect it. We suppose $H_s(E_i)^+$ is so close to E_i^+ that it is of diameter less than 3ε.

If $H_s(p) \in M_j$, let rq be an arc on M_j irreducible from $r=H_s(p)$ to $H_s\big(\sum E_i\big)$. Suppose $q \in H_s(E_j)$. Then $H_s(p)^+ \subset H_s(E_j)^+$ and is therefore of diameter less than 3ε. A similar argument shows that $H_s(p)^+$ is of diameter less than 3ε if $H_s(p) \subset \operatorname{Int} M_i$.

If $H_s(p)$ lies in the exterior of each M_i, it follows from the unicoherence of $E^3-\sum M_i$ that there is an $H_s(E_j)$ accessible from $H_s(p)$ in $E^3-\left(\sum M_i+H_s\left(\sum E_i\right)\right)$ such that $H_s(E_j)+\sum M_i$ separates $H_s(p)$ from U in E^3. It follows from the unicoherence of $E^3-H_s(E_j)$ that $H_s(E_j)^+$ separates $H_s(p)$ from U in E^3. Since $H_s(p)^+\subset H_s(p)+H_s(E_j)^+$, diameter $H_s(p)^+\leqslant$ diameter $H_s(p)+H_s(E_j)^+=$ diameter $H_s(E^+)<3\varepsilon$. This completes the proof that for each point p of E^3-S, diameter $H_s(p)^+<3\varepsilon$.

Step 6. *Distance moved by a point p whose image is never in V_E again.* Again we suppose that if $j\geqslant r$, $H_{j/k}(p)$ does not lie in the V_E used in extending $H_{j/k}$. We show that if $s=r/k$, then $\varrho\left(H_s(p), H_t(p)\right)<6\varepsilon$ if $s\leqslant t\leqslant 1$. We can accomplish this by showing that $\varrho\left(H_s(p), H_{j/k}(p)\right)<3\varepsilon$ if $s=r/k\leqslant j/k$ since it has already been demonstrated that $\varrho(H_{j/k}, H_t)<3\varepsilon$ $(j/k\leqslant t\leqslant (j+1)/k)$.

Since $H_s(p)$ does not belong to the V_E used in extending H_s, as t moves from $s=r/k$ to $(r+1)/k$, $H_s(p)$ does not move far unless it is in the C used at this stage. If it is in the C, $H_s(p)$ is moved close to D which lies in $H_s(p)^+$. In any case, by a proper choice of the ε of Theorem 7.1, we can cause each $H_{(r+1)/k}(p)$ to be very close to its $H_s(p)^+$.

If M_j belongs to $M_{(r+1)/k}(p)^+$, it belongs to $H_{r/k}(p)^+$. Hence, as t takes on the values $r/k, (r+1)/k, (r+2)/k, \dots, 1$, $H_t(p)$ remains close to $H_s(p)^+$. We may suppose that $\varrho\left(H_s(p), H_{j/k}(p)\right)<3\varepsilon$ $(s=r/k\leqslant j/k)$ and $\varrho\left(H_s(p), H_t(p)\right)<6\varepsilon$ $(s\leqslant t\leqslant 1)$ if $H_{j/k}(p)$ is never in the V_E used in extending $H_{j/k}$ $(j=r, r+1, \dots, k-1)$.

The preceding argument shows that H_t does not move p by as much as 6ε if for no j does $H_{j/k}(p)$ belong to the V_E used in extending $H_{j/k}$.

Step 7. *Distances moved by a point one of whose images is in a V_E.* At any stage we have disk E and can select a V_E each of whose points is very close to E. No point of E has been moved by as much as 6ε and $H_{j/k}$ moves no point of E by as much as 3ε. Hence we may select V_E so that for each point p of V_E, $\varrho\left(p, H_{j/k}(p)\right)<3\varepsilon$ and $\varrho\left(p, H_t(p)\right)<6\varepsilon$ if $0\leqslant t\leqslant j/k$. We now show that this restriction on V_E insures that Conditions 5 and 6 are satisfied.

Suppose $H_{r/k}(p)$ lies in the V_E used in extending $H_{r/k}$ but if $r<j$, $H_{j/k}(p)$ does not belong to the V_E used in extending $H_{j/k}$. It follows from the restriction placed on V_E in the preceding paragraph that $\varrho\left(p, H_{r/k}(p)\right)<3\varepsilon$ and $\varrho\left(p, H_t(p)\right)<6\varepsilon$ for $0\leqslant t\leqslant r/k$. We found in Step 3 that $\varrho(H_{r/k}, H_t)<3\varepsilon$ if $r/k\leqslant t\leqslant (r+1)/k$ and in Step 6 that $\varrho\left(H_{(r+1)/k}(p), H_1(p)\right)<3\varepsilon$ and $\varrho\left(H_{(r+1)/k}(p), H_1(p)\right)<6\varepsilon$ if $(r+1)/k\leqslant t\leqslant 1$. These relations imply Conditions 5 and 6.

THEOREM 7.4. *If $E_1, E_2, \ldots, E_n$ are the disks of $S-U$ of Theorem 7.3 and $D_1, D_2, \ldots, D_n$ are mutually exclusive polyhedral disks such that $\mathrm{Bd}\,E_i = \mathrm{Bd}\,D_i$ and $\mathrm{Int}\,D_i \subset \mathrm{Int}\,M_i$, then the isotopy H_t $(0 \leqslant t \leqslant 1)$ may be chosen satisfying the conclusion of Theorem 7.3 and such that $H_1(E_i) = D_i$.*

Proof. The proof of this theorem follows the pattern of the proof of Theorem 7.3 except that we use $n+k$ rather than k steps in describing H_t, the extra steps being used to apply Theorem 7.2 and pull the E_i's onto the D_i's.

Theorem 7.4 may be extended by replacing S by a finite collection of mutually exclusive 2-spheres. We need the following result where there are two S's,

THEOREM 7.5. *Suppose $M_1, M_2, \ldots, M_n$ are mutually exclusive polyhedral 2-spheres in E^3 such that each is of diameter less than ε.*

Suppose S', S'' are mutually exclusive polyhedral 2-spheres each in general position with respect to the M_i's and U', U'' are components of $S'-\sum M_i$, $S''-\sum M_i$ respectively each of diameter more than 2ε and such that each component of $S'-U'+S''-U''$ is of diameter less than ε.

Suppose $D_1, D_2, \ldots, D_m$ are mutually exclusive disks such that $U'+U''+ +\sum D_i$ is the sum of two polyhedral 2-spheres and each $\mathrm{Int}\,D_j$ lies in $\sum \mathrm{Int}\,M_i$.

Then for each continuum X in $E^3-\left(\sum M_i + S'-U'+S''-U''\right)$ that intersects $U'+U''$ there is an isotopy H_t $(0 \leqslant t \leqslant 1)$ of E^3 onto itself such that

1. *H_t moves no point of $\bar{U}'+\bar{U}''+X$,*
2. *$H_1(S'+S'') = U'+U''+\sum D_i$,*
3. *H_1 moves no point of $S'+S''$ by as much as 3ε,*
4. *H_t moves no point of $S'+S''$ by as much as 6ε,*
5. *H_1 moves no point of E^3 by as much as 9ε,*
6. *H_t moves no point of E^3 by as much as 12ε.*

Proof. The proof of Theorem 7.5 follows the pattern of the proof of Theorem 7.3 so we only give a broad outline of it.

As in the proof of Theorem 7.3 we subdivide the M_i's so that each of those that needs to be considered contains one and only one simple closed curve in $\mathrm{Bd}\,\bar{U}'+\mathrm{Bd}\,\bar{U}''$, and the interior of this M_i contains the appropriate $\mathrm{Int}\,D_j$.

In a finite sequence of applications of Theorem 7.1 each followed by pushing a disk E to one side of a disk D, we obtain an isotopy of E^3 onto itself that reduces the number of components of $(S'+S'')\cdot\sum M_i$ to the number of components of $\mathrm{Bd}\,\bar{U}'+\mathrm{Bd}\,\bar{U}''$. Then Theorem 7.2 is applied to push $S'+S''$ onto $U'+U''+\sum D_i$.

For each component F of $S'-U'+S''-U''$, the image of F at each stage is near the sum of F and the M_i's that F intersects. This insures that Condition 3 is satisfied.

At no stage is any point moved more than 3ε. This insures that Condition 4 is satisfied.

If at a certain stage the image of a point p fails to belong to an open set V_E about the disk E used at this certain stage, the image is not moved much unless it moves near an M_i that separates p from $\bar{U}'+\bar{U}''$ in $E^3-(S'-U'+S''-U'')$. This insures that H_t does not move too far those points whose images fail to belong to V_E's.

By choosing the V_E's so that points in them have not been moved too far already, we insure that the isotopy we describe satisfies Conditions 5 and 6.

THEOREM 7.6. *Suppose S; $S_1, S_2, \ldots$; F; $C_1, C_2, \ldots, C_n$; and the $K(r,j)$'s are taken as in Theorem 6.3, and ε is taken so small and k so large that if $k<i<j$, noncorresponding elements of $K(r,i)$, $K(r,j)$ are farther apart than 2ε. If J_s is an element of $K(r,s)$ that separates on C_r two corresponding elements J_i, J_j of $K(r,i)$, $K(r,j)$ and $i,j,s>k$, then $i<s<j$ and J_i, J_s, J_j are corresponding elements.*

Proof. We let U_i, U_s, U_j be the large components of $S_i-\sum C_r$, $S_s-\sum C_r$, $S_j-\sum C_r$ and find from the techniques of Theorems 7.3 and 7.5 (we are not interested here in preventing points from moving far) that there is an isotopy on E^3 that is fixed on $U_i+U_s+U_j$ that takes S_i, S_s, S_j onto polyhedral 2-spheres S'_i, S'_s, S'_j such that each of $S'_i-\bar{U}_i$, $S'_s-\bar{U}_s$, $S'_j-\bar{U}_j$ misses $\sum\text{-}C_r$.

Since J_s separates J_i from J_j on C_r, then J_i, J_s, J_j are corresponding elements because noncorresponding elements are not within 2ε of each other.

Since neither J_i nor J_j separates the other from J_s on C_r, neither S'_i nor S'_j separates the other from S'_s in E^3. Therefore S'_s separates S'_i from S'_j and $i<s<j$.

The preceding argument also gives the following result.

THEOREM 7.7. *If in Theorem 7.6, J_t is an element of $K(r,t)$ corresponding to J_i, J_j and $k<i<t<j$, then J_t is between J_i and J_j on C_r.*

8. Special disks with respect to cylinders. Each polygonal simple closed curve J on C bounds a polygonal disk D which lies except for $J=\operatorname{Bd} D$ in $\operatorname{Int} C$. Suppose L is a vertical line in $\operatorname{Int} C$. If J bounds a disk in C, we may choose such a D which misses L; if $C-J$ has two unbounded components, we can pick D so that D intersects L in just one point.

Suppose D', D'' are two such disks bounded by J and P is a plane containing L such that P contains no vertex of some rectilinear triangulation of $D'+D''$. Each component of $P\cdot D'$ with a point p on J is an arc with only its end points on J. The same goes for components of $P\cdot D''$. However, the arc in $P\cdot D'$ containing p may not have the same other end point as the arc in $P\cdot D''$ containing p. We wish to define a special property such that if D', D'' have this special property, then if an arc which is a component of $P\cdot D'$ shares an end with a component of $P\cdot D''$, then their other ends are identical also.

Special property for a disk. If J is a polygonal simple closed curve on a vertical triangular cylinder C and L is a vertical line in $\mathrm{Int}\, C$, then a polygonal disk D is said to have the *special property with respect to J, C, L* provided:

1. $\mathrm{Bd} D = J$,
2. $\mathrm{Int} D \subset \mathrm{Int}\, C$,
3. $\mathrm{Diameter} D = \mathrm{diameter} J$,

4a. $D\cdot L = 0$ if J bounds a disk in C,

4b. $D\cdot L$ contains only one point if $C-J$ has two unbounded components,

5. if P is a plane containing L but no vertex of a rectilinear triangulation of D and ab is an arc in $P\cdot C$ with end points on J and interior in the component of $C-J$ that does not reach below some horizontal plane, then some component of $D\cdot P$ is an arc with end points at a, b.

THEOREM 8.1. *Suppose J is a polygonal simple closed curve on a vertical triangular cylinder C, L is a vertical line in* $\mathrm{Int}\, C$, *and D', D'' are two polyhedral disks each with the special property with respect to J, C, L. If P is a plane containing L but no vertex of some triangulation of $D'+D''$ and ab is a component of $P\cdot D'$ intersecting J, then there is a component of $P\cdot D''$ which is an arc with the same end points as ab.*

Proof. Since P contains no vertex of some triangulation of $D'+D''$, each component of $P\cdot D'$ (of $P\cdot D''$ also) that intersects J is an arc with only its end points on J.

For the case where J bounds a disk in C, Condition 5 of the definition of the special property of a disk causes two points of $J\cdot P$ to belong to an arc in $P\cdot D$ if and only if they belong to an arc in $P\cdot D''$.

If $C-J$ has two unbounded components, $J\cdot P$ has two points p, q such that each pair of points of $(J\cdot P)-(p+q)$ belong to an arc in $P\cdot D'$ if and only if they belong to an arc in $P\cdot D''$ according as they do or do not belong to an arc in $P\cdot C$ whose interior belongs to the upper component of $C-J$. Since p, q are the only two points of $J\cdot P$ left to consider, there is an arc from p to q in $P\cdot D'$ and also an arc from p to q in $P\cdot D''$.

THEOREM 8.2. *If J is a polygonal simple closed curve on a vertical triangular cylinder C, and L is a vertical line in* $\operatorname{Int} C$, *then there is a polygonal disk D that has the special disk property with respect to J, C, L.*

If $P_1, P_2, \ldots, P_n$ is a finite collection of planes each containing L but no point at which J is broken, such a disk D may be selected with a rectilinear triangulation with no vertex on $P_1+P_2+\ldots+P_n$ and no edge intersecting L.

Proof. The proof of the case where J bounds a disk E in C is easy. Here we obtain D by pushing $\operatorname{Int} E$ in slightly toward L.

Suppose $C-J$ has two unbounded components U^+ and U^- where U^+ is above some horizontal plane. The selection of D is immediate if J lies in a plane so we suppose that the convex hull of J is a polyhedral 3-cell K.

Let E be the disk on $\operatorname{Bd} K$ bounded by J and missing U^-. Although E satisfies conditions 1, 3, 4, 5 of the definition of the special property, $\operatorname{Int} E$ may not lie in $\operatorname{Int} C$. Also $L \cdot E$ may be on an edge of every triangulation of E. This latter difficulty may be taken care of by adjusting E slightly near $L \cdot E$. The former exception may be removed by shoving certain points of $\operatorname{Int} E$ near C slightly toward L. Care is taken to see that no "corner" of the resulting disk D lies on any P_i.

THEOREM 8.3. *Suppose $J_1, J_2, \ldots, J_m$ are m mutually exclusive polygonal simple closed curves on a vertical triangular cylinder C, L is a vertical line in* $\operatorname{Int} C$, *and $P_1, P_2, \ldots, P_n$ are n planes each containing L but none containing a point where any J is broken. Then there are m mutually exclusive polyhedral disks $D_1, D_2, \ldots, D_m$ such that each D_i has the special property with respect to J_i, C, L and D_i has a triangulation with no vertex on $P_1+P_2+\ldots+P_n$ and no edge intersecting L.*

Proof. Suppose $J_1, J_2, \ldots, J_m$ are ordered so that if $C-J_i$ has two unbounded components U^+, U^- with U^- the lower one, then $i \leqslant j$ if J_j either bounds a disk in U^- or fails to intersect U^-. The disks $D_1, D_2, \ldots, D_m$ may be obtained by an iteration of the process described in the proof of Theorem 8.2 where we define D_1, then $D_2, \ldots$, and finally D_n.

THEOREM 8.4. *Suppose S; $S_1, S_2, \ldots$; F; $C_1, C_2, \ldots, C_n$; and the $K(r,j)$'s are as in Theorem 6.3 and 7.6 and ε, k are as in Theorem 7.6.*

Suppose U_i $(i=1,2,\ldots)$ is the large component of $S-\sum C_r$ and S_i' is a 2-sphere in general position with respect to F formed by replacing each component D of S_i-U_i by a disk E which lies except for its boundary in some $\operatorname{Int} C_r$ *and has the special property with respect to* $\operatorname{Bd} D$, C_r, L_r *where L_r is the center of C_r.*

For each point p of $S_i' \cdot L_r$, let $\pi_j(p_i)$ be the corresponding point of $S_j' \cdot L_r$ as determined by the correspondence between $K(r,i)$ and $K(r,j)$.

There is an integer k' such that if $i, j > k'$, V is a section of F, and $p_i q_i$ is an arc of $V \cdot S'_i$ between two corner points of F, then there is an arc in $V \cdot S'_j$ between $\pi_j(p_i)$ and $\pi_j(q_i)$.

Proof. Suppose $p \in S'_1 \cdot L_r$ such that $\pi_i(p) = p_i$. Since for each integer j there is an arc in $V \cdot S'_j$ from $\pi_j(p)$ to a corner point of F, there is a point q of $S'_1 \cdot L_r$ such that for infinitely many j's, there is an arc in $V \cdot S'_j$ from $\pi_j(p)$ to $\pi_j(q)$. The truth of Theorem 8.4 will follow if we show that if $k < i < s < j$ and arcs in $V \cdot S'_i$ and $V \cdot S'_j$ from $\pi_i(p)$ to $\pi_i(q)$ and $\pi_j(p)$ to $\pi_j(q)$ imply that there is an arc in $V \cdot S'_s$ from $\pi_s(p)$ to $\pi_s(q)$.

We find from Theorem 8.1 that whether or not there is an arc in $V \cdot S'_s$ from $\pi_s(p)$ to $\pi_s(q)$ is not determined by what E's we use to replace the D's as long as they have the special property. Hence we suppose that the E's do not intersect. Theorem 8.3 shows that we can pick these E's so they do not intersect and the techniques used in Theorems 7.3 and 7.5 show that there is an isotopy on E^3 pulling $S_i + S_s + S_j$ onto $S'_i + S'_s + S'_j$. Here we do not need the full strength of Theorems 7.3 and 7.5 since, although we do not want to move points of $U_i + U_s + U_j$, we are not interested in preventing other points from moving far. It follows from Theorem 7.7 that $\pi_s(p)$ is between $\pi_i(p)$ and $\pi_j(p)$ while $\pi_s(q)$ is between $\pi_i(q)$ and $\pi_j(q)$.

The arc in $V \cdot S'_s$ from $\pi_s(p)$ is trapped on V between the arcs in $V \cdot S'_i$ from $\pi_i(p)$ to $\pi_i(q)$ and the arc in $V \cdot S'_j$ from $\pi_j(p)$ to $\pi_j(q)$ so it can lead only to $\pi_s(q)$.

9. Finite graphs on a 2-sphere. Recall that $T(1)$ is the 2-sphere with center at the origin and radius 1. We prove some theorems about stable graphs on $T(1)$ and extend these to theorems about graphs on arbitrary 2-spheres.

THEOREM 9.1. *Suppose G' is a finite graph on $T(1)$ such that each component of $T(1) - G'$ is of diameter less than $\varepsilon < \frac{1}{3}$. Then G' contains a finite graph G such that each component of $T(1) - G$ is an open 2-cell of diameter less than 3ε and no two of the closures of these open 2-cells meet in a disconnected set.*

Proof. An advantage of working on $T(1)$ rather than on an arbitrary 2-sphere is that the diameter of a small set is equal to the diameter of its boundary.

For each component U of $S - G'$, let $J(U)$ denote the simple closed curve in $\overline{U}$ bounding the large component of $S - \overline{U}$ and $D(U)$ denote the disk in S containing U and bounded by $J(U)$. The diameter of $D(U)$ is the same as the diameter of U.

We note that if two $D(U)$'s have an interior point in common, one contains the other. Hence there is a collection of $D(U)$'s covering S such that the interiors of these $D(U)$'s are mutually exclusive. If no two of these $D(U)$'s intersected in a disconnected set, we could use the sum of their boundaries for G.

We wish to adjust the above mentioned $D(U)$'s so that the intersection of two of the adjusted disks that have a point in common meet in a connected set. We are willing for the diameters of the adjusted disks to be more than ε.

The finite graph shown in Figure 7 is not stable because a homeomorphism of the graph onto itself interchanging arcs axb and ayb or czd and cwd could not be extended to the plane. However, if D_1 and D_5 are combined (by removing $\operatorname{Int} axb$) and if D_6 is added to D_3 (by removing $\operatorname{Int} czd$) the resulting graph is stable. We prove the theorem by considering a scheme to combine such disks.

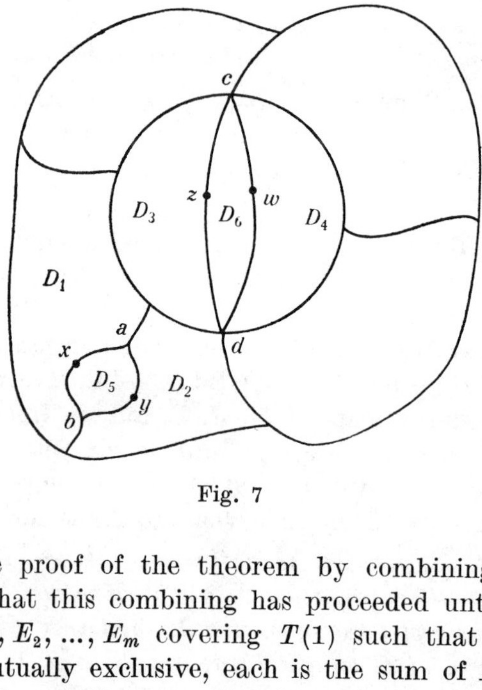

Fig. 7

We finish the proof of the theorem by combining certain of the $D(U)$'s. Suppose that this combining has proceeded until we have disks $D_1, D_2, \ldots, D_n, E_1, E_2, \ldots, E_m$ covering $T(1)$ such that the interiors of these disks are mutually exclusive, each is the sum of $D(U)$'s described earlier, diameter $D_i < \varepsilon$, diameter $E_i < 3\varepsilon$, and no one of the E's meets one of the other disks in a disconnected set. If we can eliminate the D's and make all the disks E's, the sum of the boundaries of the E's will be the required finite graph G.

If J is a simple closed curve on $T(1)$ of diameter less than 1, we use $D(J)$ to denote the smaller disk in $T(1)$ bounded by J. Let E_{m+1} be the sum of all disks R of $D_1, D_2, \ldots, D_n, E_1, \ldots, E_m$ such that there is a simple closed curve J in $T(1)$ such that $R \subset D(J)$ and

$J \cdot (\mathrm{Bd}\,D_1 + \mathrm{Bd}\,D_2 + \ldots + \mathrm{Bd}\,E_m) \subset \mathrm{Bd}\,D_n$. In general, J need not lie in $\mathrm{Bd}\,D_1 + \mathrm{Bd}\,D_2 + \ldots + \mathrm{Bd}\,E_m$. For example, if $D_n = D_4$ of Figure 4, the simple closed curve J that shows that D_6 (which has been renamed an E) is to be combined with D_4 lies in $D_4 + \mathrm{Ind}\,D_3$. Then E_{m+1} is a disk of diameter less than 3ε, and it has swallowed up some of the D's and perhaps some of the E's, but it does not intersect any of the remaining disks in a disconnected set.

A continuation of this process changes all the D's into E's.

The following extension of Theorem 9.1 follows from use of the fact that each 2-sphere is the image of $T(1)$ unter a uniformly continuous homeomorphism.

THEOREM 9.2. *If S is a 2-sphere and $\varepsilon > 0$, there is a positive number δ such that if G' is a finite graph on S such that each component of $S - G'$ is of diameter less than δ, then G' has a finite subgraph G such that each component of $S - G$ is an open 2-cell of diameter less than ε and no two of the closures of these open 2-cells meet in a disconnected set.*

THEOREM 9.3. *Suppose $D_1, D_2, \ldots, D_n$ are disks whose sum is a 2-sphere S and the intersection of two of these disks with a point in common is either a point or an arc. Then $\sum \mathrm{Bd}\,D_i$ is connected and no $\mathrm{Bd}\,D_j$ separates it.*

Proof. If $\sum \mathrm{Bd}\,D_i$ were not connected, a simple closed curve in $S - \sum \mathrm{Bd}\,D_i$ would separate $\sum \mathrm{Bd}\,D_i$ in S. But this simple closed curve would not lie in any D_j.

If $\mathrm{Bd}\,D_j$ separates $\sum \mathrm{Bd}\,D_i$, there is a simple closed curve J in $S - (\sum \mathrm{Bd}\,D_i - \mathrm{Bd}\,D_j)$ such that $(\sum \mathrm{Bd}\,D_i) - \mathrm{Bd}\,D_j$ contains points p, q lying in different components of $S - J$. It follows from the unicoherence of $S - \mathrm{Int}\,D_j$ that there is an arc A in J irreducible with respect to separating p from q in $S - \mathrm{Int}\,D_j$. But A would intersect D_j in only two points and would lie in a D_i such that $D_i \cdot D_j$ is not connected.

THEOREM 9.4. *Suppose G is a finite graph on $T(1)$ such that each component of $T(1) - G$ is an open 2-cell of diameter less than $\varepsilon < \frac{1}{3}$ and no two of the closures of these open 2-cells meet in a disconnected set. If h is a homeomorphism of G onto a finite graph in $T(1)$ such that h moves no point by more than ε, h can be extended to a homeomorphism of $T(1)$ onto itself that does not move any point by as much as 3ε.*

Proof. Let $D_1, D_2, \ldots, D_n$ be the disks which are the closures of the components of $T(1) - G$. Then for each i, h can be extended from $\mathrm{Bd}\,D_i$ to map D_i onto the smaller disk in $T(1)$ bounded by $h(\mathrm{Bd}\,D_i)$. This extension moves no point by as much as 3ε. We now show that the combined map of h on all the D_i's is a homeomorphism of $T(1)$ onto itself.

If $h(\mathrm{Int}\,D_i)$ intersects $h(\mathrm{Int}\,D_j)$, $i \neq j$, then one of $h(D_i)$, $h(D_j)$ contains the other. Assume $h(D_j) \subset h(D_i)$. Since $h(G - \mathrm{Bd}\,D_i)$ is connected, $h(D_i)$ contains $h(G)$. This contradicts the facts that diameter $h(D_i) < 3\varepsilon$, diameter $G > 5\varepsilon$, diameter $h(G) > 3\varepsilon$.

The preceding paragraph showes that h sends no two points into the same point. Since $T(1)$ is compact and h is continuous, h is a homeomorphism. Since no proper subset of $T(1)$ is homeomorphic with $T(1)$, h takes $T(1)$ onto itself.

THEOREM 9.5. *For each 2-sphere S and each positive number ε, there is a positive number δ such that if G is a finite graph on S such that each component of $S - G$ is an open 2-cell of diameter less than δ but the closures of no two of the open 2-cells meet in a disconnected set and h is a homeomorphism of G into S that moves no point by more than ε, then h can be extended to a homeomorphism of S onto itself that does not move any point by as much as 3ε.*

Theorem 9.5 follows from Theorem 9.4 and the fact that S is the image of $T(1)$ under a uniformly continuous homeomorphism.

The following theorem shows that we have been considering stable graphs in Theorem 9.1 to 9.5.

THEOREM 9.6. *A finite graph G on a 2-sphere S is stable if there is a finite collection of disks $D_1, D_2, \ldots, D_n$ covering S such that $\mathrm{Int}\,D_i \cdot D_j = 0$ if $i \neq j$, $\sum \mathrm{Bd}\,D_i = G$, and if two of the D's have a point in common, their intersection is connected.*

Proof. We denote that $\mathrm{Bd}\,D_i$ does not separate G or else some D_j intersects D_i in a disconnected set. A homeomorphism h of G into a 2-sphere S' may be extended to a homeomorphism of S onto S' by taking D_i onto the disk on S' bounded by $h(\mathrm{Bd}\,D_i)$ and containing no point of $h(G - \mathrm{Bd}\,D_i)$.

10. Freeing surfaces of intervals. In the proof of Theorem 2.1 we supposed that there is no loss of generality in supposing that a 2-sphere S contains no vertical interval. We justify this supposition with Theorem 10.1.

A different proof of Theorem 10.1 was originally employed which made use of tame Cantor sets. It was decided to include this different proof in another paper devoted to properties of tame Cantor sets and substitute here a shorter proof suggested by M. K. Fort, Jr. and modeled after the proof of Theorem 4 in [10].

THEOREM 10.1. *If X is a closed 2-dimensional set in E^3 and ε is a positive number, there is a homeomorphism h of E^3 onto itself such that h moves no point by more than ε and $h(X)$ contains no straight line interval.*

Proof. Let H be the set of all homeomorphisms h_a of E^3 onto itself and such that each of $\varrho(h_a, I)$, $\varrho(h_a^{-1}, I)$ is finite. If H is metrized with the metric

$$D(h_1, h_2) = \varrho(h_1, h_2) + \varrho(h_1^{-1}, h_2^{-1}),$$

the resulting space is a complete metric space.

Let H_n be the set of all elements h_n such that $h_n(X)$ contains a straight line interval of length no less than $1/n$ and at a distance from the origin of no more than n. Then H_n is closed and $H_1 + H_2 + \ldots$ is an F_σ set.

We now show that no H_i contains an open subset of H by showing that for each element h_i of H_i, there is an element h_0 of $H - H_i$ very close to h_i. Let T be a triangulation of E^3 of very small mesh and such that the 2-skeleton of T contains no straight line interval of length more than $1/2i$ and no center of a 3-simplex of T intersects $h_i(X)$. Then h_0 can be taken as fh_i where f is a homeomorphism fixed on the 2-skeleton of T which moves parts of the 3-simplexes of T in a straight line directly away from the centers of these 3-simplexes. In fact, f can be taken so that $fh_i(X)$ is so close to the 2-skeleton of T as to contain no straight line interval of length $1/i$.

Since no H_i contains an open subset of H, it follows from the Baire category theorem that H contains an element h near I that does not belong to any H_i.

11. Extension of preceding results to surfaces other than 2-spheres. Theorem 2.2 gave a condition under which a 2-sphere in E^3 is tame. This result may be extended by the methods we have used to show the following results.

THEOREM 11.1. *A surface S in E^3 is tame if for each positive number ε there are surfaces S', S'' on different sides of S such that*

$$H(S, S') \leqslant \varepsilon, \quad H(S, S'') \leqslant \varepsilon.$$

THEOREM 11.2. *A surface S in E^3 is locally tame at a point p of S if there is a disk D with $p \in \operatorname{Int} D \subset S$ such that for each positive number ε, there are disks D', D'' on opposite sides of S such that*

$$H(D, D') \leqslant \varepsilon, \quad H(D, D'') \leqslant \varepsilon.$$

The original intention was to prove Theorems 11.1 and 11.2 in the present paper but the paper seems long enough already and their proofs will follow briefly from extensions of Theorem 2.2 to be given in another paper. This will be accomplished as follows.

It will be shown in [8] that a 2-sphere in E^3 is tame if its complement is uniformly locally simply connected. Theorem 11.2 will be established by showing that if p is a point of a surface S satisfying the hypothesis of Theorem 11.2, then there is a disk E and a 2-sphere K such that $p \in \mathrm{Int} E \subset K$, $E \subset S$, and $E^3 - K$ is uniformly locally simply connected. Theorem 11.1 will follow from Theorem 11.2 and the fact that locally tame surfaces are tame.

The above theorems are true in 3-manifolds as well as in E^3.

References

[1] J. W. Alexander, *On the subdivision of 3-space by a polyhedron*, Proc. Nat. Acad. Sci., U.S.A., 10 (1924), p. 6-8.

[2] — *On the deformation of an n-cell*, Proc. Acad. Sci., U.S.A., 9 (1923), p. 406-407.

[3] R. H. Bing, *Approximating surfaces with polyhedral ones*, Ann. of Math. 65 (1957), p. 456-483.

[4] — *Locally tame sets are tame*, Ann. of Math. 59 (1954), p. 145-158.

[5] — *Some monotone decompositions of a cube*, Ann. of Math. 61 (1955), p. 279-288.

[6] — *An alternative proof that 3-manifolds can be triangulated*, Ann. of Math. 69 (1959), p. 37-65.

[7] — *E^3 does not contain uncountably many mutually exclusive wild surfaces*, Abstract 63-801t, Bull. Amer. Math. Soc. 63 (1957), p. 404.

[8] — *A surface S is tame in E^3 if $E^3 - S$ is locally simply connected at each point of S*, Abstract, 542-74, Amer. Math. Soc. Notices 5 (1958), p. 180-181.

[9] — *A 2-manifold-with-boundary in E^3 is tame if its complement is 1-ULC*, Abstract 546-547, Amer. Math. Soc. Notices 5 (1958), p. 365.

[10] M. K. Fort, Jr., *A proof that the group of all homeomorphisms of the plane onto itself is locally arcwise connected*, Proc. Amer. Math. Soc. 1 (1950), p. 59-62.

[11] H. C. Griffiths, *A characterization of tame surfaces in three space*, Ann. of Math., to appear.

[12] W. Graeub, *Die Semilinear Abbildungen*, Sitzungsberichte der Heiderlberg Akad. der Wiss., part 4 (1950), p. 205-272.

[13] R. P. Goblirsch, *An area for simple surfaces*, Ann. of Math. 68 (1958), p. 231-246.

[14] O. G. Harrold, Jr., *Locally peripherally unknotted surfaces in E^3*, to appear.

[15] John Jewitt, *Differentiable approximations to interior functions*, Duke Math. J. 24 (1947), p. 227-232.

[16] E. E. Moise, *Affine structures in 3-manifolds, II. Positional properties of 2-spheres*, Ann. of Math. 55 (1952), p. 172-176.

[17] — *Affine structure in 3-manifolds, V. The triangulation theorem and Hauptvermutung*, Ann. of Math. 56 (1952), p. 96-114.

[18] — *Affine structure in 3-manifolds, VII. Invariance of the knot type; local tame imbedding*, Ann. of Math. 59 (1954), p. 159-170.

THE INSTITUTE FOR ADVANCED STUDY
THE UNIVERSITY OF WISCONSIN

Reçu par la Rédaction le 15. 5. 1958

A WILD SURFACE EACH OF WHOSE ARCS IS TAME

By R. H. Bing

1. Introduction. A closed set X in E^3 is *tame* if there is a homeomorphism h of E^3 onto itself such that $h(X)$ is a polyhedron (the sum of a locally finite collection of tetrahedra, triangles, segments, and points). A closed set in E^3 that is not tame is called *wild.* Examples of wild surfaces are found in [1], [9]. Since a tame surface divides space in the same fashion as does a polyhedral surface, it is convenient to have criteria for determining which surfaces are tame and which are wild.

Consider the following question.

Question. Is a 2-sphere in E^3 tame if each arc in it is tame?

In this paper we give a negative answer. Had the answer been in the affirmative, it would have provided us with another characterization of tame 2-spheres. This would have been fortunate since such characterizations are few. We know that a 2-sphere is tame if any of the following conditions are met.

1. It is locally tame [2], [13].
2. Its complement is uniformly locally simply connected [5].
3. It can be homeomorphically approximated from both sides [4].

It is known [6] that each surface (either tame or wild) in E^3 contains many tame arcs. The examples of wild surfaces described in the literature contain wild arcs. These examples are locally tame except at a 0-dimensional set and as pointed out in Theorem 8, any simple closed curve on the 2-sphere that contains this 0-dimensional set is wild.

Suppose S is a 2-sphere in E^3 and J is a simple closed curve in $E^3 - S$ that cannot be shrunk to a point in $E^3 - S$. The following plan was devised for getting a partial affirmative solution to the above question by showing that S contains a wild simple closed curve. Instead of leading to an affirmative solution, the plan backfired and led to a counterexample and a negative answer.

Faulty Plan

1. Consider a decreasing sequence of curvilinear triangulations $T_1, T_2, \cdots$ of S such that the 1-skelton of each triangulation is tame. That there is such a decreasing sequence of triangulations follows from the fact that S contains a sequence of Sierpinski curves $S_1, S_2, \cdots$ such that $S_i \subset S_{i+1}$, each S_i lies on some tame 2-sphere, and each component of $S - S_i$ is of diameter less than $1/i$. See [6].

2. Let D be a 2-simplex of T_1 such that J cannot be shrunk to a point in $E^3 - (S - \text{Int } D)$. Replace D by a disk D' such that Int D' is locally tame

Received May 7, 1960. Work on this contract was supported by contract NSF-G11665.

and $(S - D) + D'$ is a 2-sphere very close to S. That D can be replaced by such a D' very close to D follows from the fact that if f is a continuous non-negative function on S, there is a homeomorphism h of S into E^3 such that for each point s of S, the distance between s and $h(s)$ is no more than $f(s)$ and $h(s)$ is locally tame at $h(s)$ if $f(s) > 0$ [3]. Since D' is locally tame at each of its interior points and Bd D' is tame by Step 1, it follows from [13] that D' is tame.

3. Iterate the procedure started in Step 2 by replacing as many 2-simplexes of T_1 as can be done so that J cannot be shrunk to a point in the complement of the adjusted S. If 2-simplexes D_1, D_2 with a common edge are adjusted, then it follows from [8] that their adjusted sum is tame.

4. Now move to the 2-simplexes of T_2 that have not been adjusted and replace them by tame disks retaining the property that J cannot be shrunk to a point in the complement of the adjusted S.

5. It was hoped to continue this process to get an adjusted 2-sphere S' such that S' is locally tame except at a 0-dimensional set X.

6. The last step was to get a simple closed curve J' in S' such that J' contains X and lies in the sum of X and the various 1-skeletons of the triangulations $T_1, T_2, \cdots$. Then J' would be wild by Theorem 8 and would lie in S as well as in S'. This would have shown that S contains a wild arc.

Just where does the faulty plan break down? It was in Step 5. Not only was it impossible to continue the process but in trying to prove that the process could be carried out, an example was found which showed that it could not even be started as suggested in Step 2.

THEOREM 1. *There is a 2-sphere S in E^3 and a simple closed curve J in $E^3 - S$ such that J cannot be shrunk to a point in $E^3 - S$, but for each disk Y of S, J can be shrunk to a point in $E^3 - Y$.*

Theorem 1 is shown to be true by Example 2 described in §3 of this paper.

An examination of Example 2 also shows that it has the remarkable property that each arc in it is tame. That each arc in 2-sphere S of Example 2 is tame is shown in §5 of this paper.

THEOREM 2. *There is a wild 2-sphere S in E^3 such that each arc in S is tame.*

2. An approximation to the example. Before describing Example 2, we give a simpler example.

EXAMPLE 1. *A 2-sphere S and a simple closed curve J in $E^3 - S$ such that J cannot be shrunk to a point in $E^3 - S$ but for certain large disks Y_i in S, J can be shrunk to a point in $E^3 - Y_i$.*

Suppose S_0 is a tame 2-sphere in E^3 and S_0 has a cellular subdivision such that the disks of the subdivision may be ordered $E_1, E_2, \cdots, E_n$ such that each of $E_1 \cdot E_2, E_2 \cdot E_3, \cdots, E_{n-1} \cdot E_n, E_n \cdot E_1$ is an edge. If S_0 is the boundary of a cube, we could take $n = 6$ and let E_1, E_4 be the top and bottom faces and E_2, E_5 be the front and back faces of the cube.

From the center of each face E_i erect a solid feeler with a solid loop on the far end of it. The feeler and loop are three-dimensional as shown in Figure 1

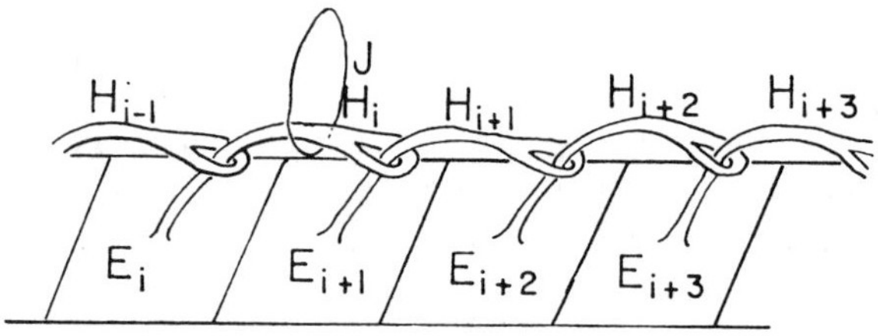

Figure 1

and resemble a bent eye bolt. We shall denote the "eye bolt" from E_i as H_i. Although it is topologically a solid torus, it more nearly resembles a solid torus on a curved stem. The loop in H_1 circles around the stem of H_2, the loop of H_2 circles around the stem of H_3, $\cdots$, and the loop of H_n circles around the stem of H_1 as shown in Figure 1.

Suppose J is a simple closed curve that circles around one of the solid feelers as shown in Figure 1. By using Theorem 11 it can be shown that J cannot be shrunk to a point in the complement of $S_0 + \sum H_i$. To show this, let C^i be a cube which contains the loop of H_i such that Bd $C^i \cdot H_i = B_3^i$ is a disk which is a cross section of the stem of H_i, Bd $C^i \cdot H_{i+1} = B_1^i + B_2^i$ is the sum of two disks which are cross sections of the stem of H_{i+1}, and C_i intersects $S_0 + \sum H_i$ only in a torus of H_i and a cylinder of H_{i+1}. The C^i's do not intersect each other. Then C^i, B_1^i, B_2^i, B_3^i, $C^i \cdot H_i$, $C^i \cdot H_{i+1}$ correspond to the C, B_1, B_2, B_3, T, C_1 of Theorem 11 as shown in Figure 7. If S_0' denotes the sum of S_0 and the component of $E^3 - S_0$ missing the H_i's, $S_0' + \sum H_i + \sum C^i$ is a cube with handles such that J circles one of these handles as does each B_1^i and each B_2^i. We use Theorem 11 to replace C^1 by $C^1 \cdot (H_1 + H_2)$ in $S_0 + \sum H_i + \sum C^i$, then replace C^2 by $C^2 \cdot (H_2 + H_3)$, $\cdots$, and finally C^n by $C_n \cdot (H_n + H_1)$. At each stage we find that J cannot be shrunk to a point in the complement of what is left. Neither can any B_j^i $(j = 1, 2)$ be shrunk to a point in the sum of B_j^i and the complement of what is left so we can continue to apply Theorem 11 at the next step. After the n-th step we are left only with $S_0' + H_i$ and J cannot be shrunk to a point in its complement.

If H_{i+1} is deleted from $S_0 + \sum H_i$, the simple closed curve J shown in Figure 1 can be shrunk to a point over the right end of H_i.

If instead of deleting H_{i+1}, we omitted H_{i+2}, it is a bit trickier to shrink J to a point without hitting S_0 or any remaining H. However, this may be accomplished as follows. Let J bound a disk which extends over the right end of H_i and which intersects the stem of H_{i+1} in two disks, one above the loop of H_i and one below. The two disks are adjusted so that they reach over the right end of H_{i+1}. The adjusted lower disk goes through the hole in H_i.

Continuing in this fashion it may be shown that if H_j is any eye bolt whatever, J may be shrunk to a point in

$$E^3 - (S_0 + \sum_{k \neq j} H_k).$$

Although $\sum (E_i + H_i)$ is not a 2-sphere, it is an approximation to the 2-sphere of Example 1. Of course a 2-sphere does not contain loops as shown in Figure 1 but Alexander's horned sphere does contain loops of a certain sort. If each $E_i + H_i$ is replaced by a wild disk D_i as shown in Figure 2, there would result

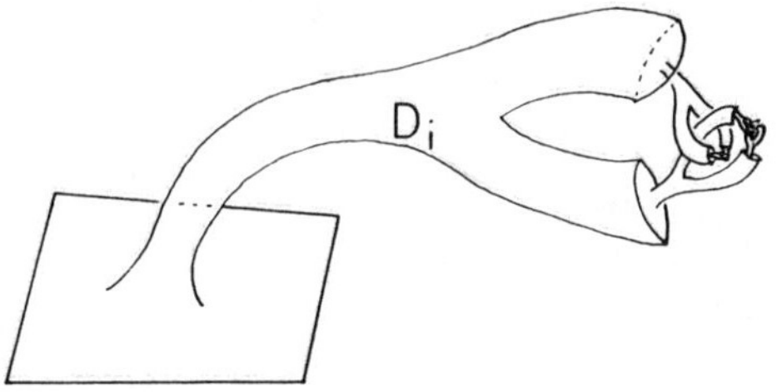

Figure 2

a 2-sphere $S = \sum D_i$ such that J cannot be shrunk to a point in $E^3 - S$. However if $Y_j = \sum_{i \neq j} D_i$, J can be shrunk to a point in $E^3 - Y_j$.

The 2-sphere S of Example 1 is $\sum D_i$. Although it has the property that J can be shrunk to a point in the complement of the sum of any proper subcollections of the D_i's, it is not a good enough example to establish Theorem 1. However, it points the way.

Although J can be shrunk to a point in each $E^3 - Y_i$, we could get a disk Y by deleting a small tame subset of some D_i and we could not shrink J to a point in $E^3 - Y$. Although we could make the mesh of the subdivision of S_0 very small, and the corresponding D_i's also small, this would not help since one is permitted to choose the Y last and our example cannot be made to have the property claimed by Theorem 1. However, the example does provide a clue as to how to proceed. It is an approximation to a better example. The better example is obtained by choosing different D_i's.

3. The example. In this section we describe a 2-sphere with the following properties.

EXAMPLE 2. *A 2-sphere S and a simple closed curve J in $E^3 - S$ such that J cannot be shrunk to a point in $E^3 - S$ but for each disk Y in S, J can be shrunk to a point in $E^3 - Y$.*

This 2-sphere is a modification of the one given in Example 1. As before we let $E_1, E_2, \cdots, E_n$ be a cellular decomposition of a tame 2-sphere S_0 such that adjacent E_i's in the sequence have an edge in common (the first and the last also share a common edge). Before adding to the "eye bolt" H_i as in

Example 1, we thicken E_i first so that the thickened E_i plus the eye bolt H_i is topologically a solid torus T_i. We call T_i a cube-with-eye-bolt.

The 2-sphere S of Example 2 lies in $\sum T_i = W_1$ and intersects $\sum \text{Bd } T_i$ only in $\sum \text{Bd } E_i$. It follows from Theorem 11 that the simple closed curve J of Example 1 cannot be shrunk to a point in $E^3 - W_1$.

A slice is removed from the loop of T_i to form a topological cube K_i. Then Bd E_i separates Bd K_i into two disks one of which has a "hook" in it. The interior of this hooked disk is pushed slightly into Int T_i so as to form a disk D_i as shown in Figure 3. The disk D_i lies except for its boundary Bd D_i =

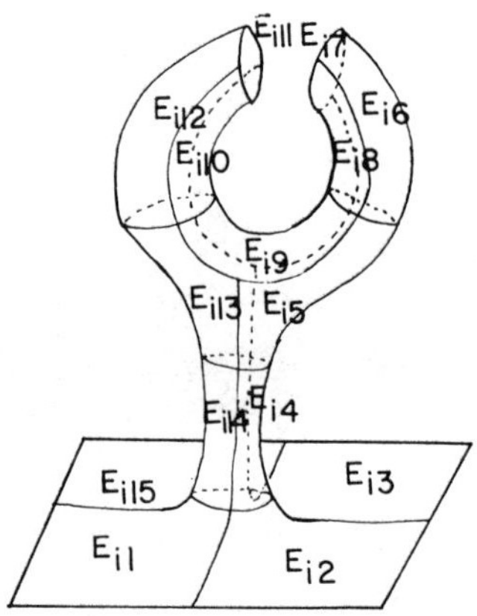

Figure 3

Bd E_i in Int T_i. The 2-sphere $S_1 = \sum D_i$ is an approximation to the 2-sphere S of our example.

Each D_i is given a cellular subdivision into small disks $E_{i1}, E_{i2}, \cdots, E_{i15}$ as shown in Figure 3 such that adjacent D_{ij}'s in the sequence share an edge. It is not essential that there be exactly 15 elements in the subdivision and the scheme suggested is only for convenience.

Each E_{ij} is thickened and an eye bolt added to it to form a cube-with-eye-bolt T_{ij}. The loop in the eye bolt T_{ij} goes around the stem of the eye bolt H_{ij+1} where $j + 1$ is computed mod 15. Also, the stems of T_{i7} and T_{i11} intertwine as shown in Figure 4 so that a simple closed curve linking T_i cannot be shrunk to a point in the complement of $\sum_{j=1}^{15} T_{ij}$. It follows from Theorem 9 followed by Theorem 11 that no such shrinking is possible. The eye bolts run so close to D_i that $\sum_{j=1}^{15} T_{ij} \subset T_i$.

Just as a disk D_i is placed in each T_i, a disk D_{ij} is placed in each T_{ij}. Let $S_2 = \sum D_{ij}$ and $W_2 = \sum T_{ij}$.

Perhaps we have gone far enough for the reader to anticipate the description of S. It is to be the limit of a sequence of 2 spheres $S_0, S_1, S_2, S_3 \cdots$ of which we have already described S_0, S_1, S_2. Also, it is to be the intersection of a

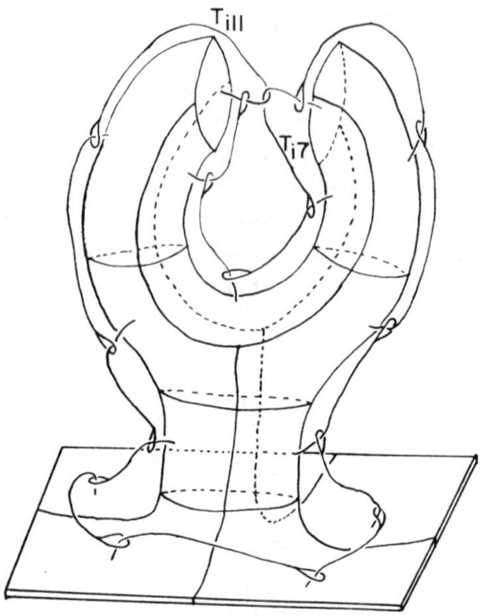

Figure 4

decreasing sequence of 3 manifolds with boundaries W_1, W_2, W_3, $\cdots$ of which we have already described W_1, W_2.

To get S_{n+1} from S_n, each of the special disks D in S_n is subdivided into 15 small special subdisks, each of the special subdisks is thickened into a topological cube and a set topologically equivalent to an eye bolt is added as suggested by Figure 4. The cube-with-eye-bolts T shown in Figure 4 are solid even though the figure shows the eye bolts as linear and the thickened subdisks as thin. The 3 manifold with boundary W_{n+1} is the sum of the T's at the $(n + 1)$-st stage and lies in W_n. A slice is cut in the loop of each T at the $(n + 1)$-st stage and a disk D like that shown in Figure 3 is put in each sliced T. The sum of these D's is the 2-sphere S_{n+1}.

The limit of the S_n's is a 2-sphere because each boundary of a special disk D at the n-th stages lies in each of the following S's (and hence in the limit) and if two D's at the n-th stage fail to intersect each other, their replacement at following stages lies in the W's at the n-th stages containing the D's so the limits of the replacements do not intersect each other.

J cannot be shrunk to a point in $E^3 - S$.

Let J be the simple closed curve shown in Figure 1. If J could be shrunk to a point in $E^3 - S$, it could be shrunk to a point in some $E^3 - W_n$. We show that this cannot occur.

Let W_i' be the set obtained from W_i by filling each hole in each eyebolt in W_i. Let W_i'' be the set obtained by removing slices from the eyebolts of W_i

and then adding back to each of these eye bolts the intertwining stems of the 7-th and 11-th eye bolts at the next stage.

Since W_1' is a hollow ball with handles and J circles one of these handles exactly once, J cannot be shrunk to a point in $E^3 - W_1'$. It follows from Theorem 11 that J cannot be shrunk to a point in $E^3 - W_1$. It follows from Theorem 9 that J cannot be shrunk to a point in $E^3 - W_1''$. Since W_2' is like W_1'' (after a slight isotopy shrinking W_1'') with more handles added, J cannot be shrunk to a point in $E^3 - W'$. It follows from Theorem 9 that J cannot be shrunk to a point in $E^3 = W_2$.

Suppose it has been shown that J cannot be shrunk to a point in $E^3 - W_i$. It follows from Theorem 9 that J cannot be shrunk to a point in $E^3 - W_i''$, by a slight isotopy of E^3 with the adding of handles that J cannot be shrunk to a point in $E^3 - W_{i+1}'$, and from Theorem 11 that J cannot be shrunk to a point in $E^3 - W_{i+1}$.

It is shown in Theorem 6 that for each disk Y in S. J can be shrunk to a point in $E^3 - Y$.

4. Cubes about subsets of the 2-sphere S of Example 2. We introduce the notion of associated annuli and associated cubes to prove some additional things about the 2-sphere S of Example 2. In this section we use D's, E's, T's and S's with the same meaning that they had in the preceding section.

Associated annulus. Recall that S_n is the sum of a collection of special disks D_i and that each D_i lies except for its boundary on the interior of a cube-with-eye-bolt T_i. Each T_i intersects the sum of the other T_i's in an annulus ring on Bd T_i. This annulus is called the annulus associated with T_i. The D's are defined at each succeeding stage so that each S_{n+k} intersects the annulus in exactly Bd D_i. In fact, the regularity of the annuli A_i associated with T_i is maintained so that A_i may be regarded as the Cartesian product $S^1 \times [0, 1]$ with Bd D_i corresponding to $S^1 \times [1/2]$ and if A' is an annulus at a later stage intersecting A_i, then there is an $\epsilon > 0$ and a subset X of S^1 such that $A' \cdot A_i$ corresponds to $X \times [1/2 - \epsilon, 1/2 + \epsilon]$.

Associated cube. Suppose C_i is a tame cube in the cube-with-eye-bolt T_i such that Bd $C_i \cdot$Bd T_i is the annulus associated with T_i. This cube C_i is called a cube associated with T_i. If $T_{i1}, T_{i2}, \cdots, T_{i15}$ are the cubes-with-eye-bolts in T_i, there is no cube associated with T_i that contains $\sum_{j=1}^{15} T_{ij}$ since one of the meridianal simple closed curves linking the eye of T_i can be shrunk to a point in the complement of any cube associated with T_i but not in the complement of $\sum_{j=1}^{15} T_{ij}$. However, we have the following result.

THEOREM 3. *Suppose $T_{i1}, T_{i2}, \cdots, T_{i15}$ are the cubes-with-eye-bolts in T_i and C_{i1} is a cube associated with T_{i1}. Then there is a cube C_i associated with T_i that contains $C_{i1} + \sum_{j=2}^{15} T_{ij}$.*

Proof. Let X_{ij} $(j > 1)$ be the solid cylinder in the eye of T_{ij}. Then $X_{ij} + T_{ij}$ is a tame 3-cell in T_i, X_{ij} misses T_{ik} $(k \neq j, j + 1)$ and $X_{ij} \cdot T_{ij+1}$ is a solid cylinder whose boundary consists of two disks on Bd X_{ij} and an annulus on Bd T_{ij+1}. With no loss of generality we suppose that C_{i1} and each T_{ij}, X_{ij} are polyhedral and Bd $X_{i15}\cdot$Bd C_{i1} is the sum of a collection of mutually exclusive simple closed curves.

In a fashion reminiscent of the manner used in §2 to show that J can be shrunk to a point if we omitted H_{i+2} instead of H_{i+1}, we describe a cube in T_i containing $C_{i1} + T_{i15}$. However, we obtain this cube by pushing things over the far end of C_{i1} rather than over the far end of H_{i+1} as was done in §2.

Let J be a component of Bd $X_{i15}\cdot$Bd C_{i1} that is an inside simple closed curve on Bd C_{i1} in that it bounds a disk D_J on Bd C_{i1} such that Int D_J does not intersect Bd X_{i15} + Bd D_{i1}. Let E_J be the disk in Bd X_{i15} which is bounded by J and misses T_{i15}. Let X'_{i15} be the 3-cell whose boundary is (Bd $X_{i15} - E_J) + D_J$. Suppose X''_{i15} is obtained from X'_{i15} by shoving it slightly to one side of Bd C_{i1} near D_J so that Bd $X''_{i15}\cdot$ Bd C_{i1} has fewer components than does Bd $X_{i15}\cdot$Bd C_{i1}.

An iteration of the procedure described in the preceding paragraph yields a polyhedral 3-cell Y_{i15} in Int T_i such that

$$Y_{i15}\cdot T_{i15} = X_{i15}\cdot T_{i15} \subset \text{Bd } Y_{i15}.$$

$$Y_{i15}\cdot C_{i1} = 0, \quad \text{and}$$

$$Y_{i15}\cdot \sum_{j=2}^{14} (T_{ij} + X_{ij}) = 0.$$

Next we replace X_{i14} by a polyhedral 3-cell Y_{i14} which misses $C_{i1} + T_{i15} + Y_{i15} + \sum_{j=2}^{13} (T_{ij} + X_{ij})$. Only two steps are needed in making this replacement since Bd $X_{i14}\cdot$(Bd $T_{i15} + Y_{i15} + C_{i1}$) is the sum of only two simple closed curves on Bd T_{i15}.

We continue to replace the X_{ij}'s with Y_{ij}'s. Now $C_{i1} + \sum_{j=2}^{15} (T_{ij} + Y_{ij})$ is a polyhedral cube in T_i that contains $C_{i1} + \sum_{j=2}^{15} T_{ij}$. It contains a middle part of the annulus associated with T_i and it is a simple matter to expand $C_{i1} + \sum_{j=2}^{15} (T_{ij} + Y_{ij})$ until it contains all of this associated annulus.

THEOREM 4. *Suppose T is a cube-with-eye-bolt at some stage in the description of S and T' is a cube-with-eye-bolt at a later stage which lies in T. For each cube C' associated with T' there is a cube associated with T such that C contains C' and each cube-with-eye-bolt other than T' at the stage of T' which lies in T.*

Proof. We suppose that $T = T_1, T_2, \cdots, T_n = T'$ is a sequence such that the cube-with-eye-bolt T_i at one stage contains the cube-with-eye-bolt T_{i+1} at the next stage. For the stage containing T_i, let Z_i denote the sum of the T's at this stage other than T_i.

Theorem 3 shows that there is a cube C_{n-1} associated with T_{n-1} that contains $C' + Z_n\cdot T_{n-1}$. Another application of Theorem 3 shows that there is a cube C_{n-2} associated with T_{n-2} which contains $C_{n-1} + Z_{n-1}\cdot T_{n-2}$. We note that C_{n-2} also contains $C' + Z_n\cdot T_{n-2}$.

A succession of $n - 1$ applications of Theorem 3 shows that there is a cube $C_1 = C$ associated with $T_1 = H$ which contains $C' + Z_n \cdot T_1$.

The following theorem is an immediate consequence of Theorem 4.

THEOREM 5. *For each cube-with-eye-bolt H at some stage and each closed proper subset Y of $H \cdot S$ there is a tame cube C in H which contains Y.*

THEOREM 6. *For each closed proper subset Y of S of Example 2, the simple closed curve J shown in Figure 1 can be shrunk to a point in $E^3 - Y$.*

Proof. There is an integer n and a cube-with-eye-bolt H_n at the n-th stage that misses Y. Let H_1 be the cube-with-eye-bolt at the first stage containing H_n. It follows from Theorem 5 that there is a cube C_1 associated with H_1 that contains every cube-with-eye-bolt at the n-th stage in H_1 other than H_n.

It follows from the methods used in the proof of Theorem 3 that there is a tame hollow ball W (Cartesian product of a 2-sphere and an arc) in $E^3 - J$ that contains C_1 plus the cubes-with-eye-bolts at the first stage other than H_1. Since each component of $E^3 - W$ is simple connected, J can be shrunk to a point in $E^3 - W$ and hence in $E^3 - Y$.

5. Each simple closed curve in S is tame. Harrold, Griffith, and Posey have shown [10] that a simple closed curve J in E^3 is tame if it has the following two properties at each point p.

Property P. For each $\epsilon > 0$ there is a 2-sphere K of diameter less than ϵ such that p lies in the bounded component of $E^3 - K$ and $J \cdot K$ is a set of two points.

Property Q. An open subset of J containing p lies on a disk in E^3.

Originally, the conditions were imposed that the 2-sphere K mentioned in the definition of Property P was locally polyhedral mod the two points of $J \cdot K$ and the disk mentioned in the definition of Property Q were locally polyhedral mod the intersection of J and the disk. However, as pointed out by Harrold in [11], it follows as a consequence of [3] that these extra conditions are not necessary. Harrold [12] has called a simple closed curve locally peripherally unknotted if it has Property P and locally unknotted if it has Property Q.

Each simple closed curve J in S has Property Q at each point since the required disks can be found on S. Hence, we only need to show that Property P is satisfied at each point p of J.

THEOREM 7. *For each point p of each simple closed curve J' in 2-sphere S of Example 2, J' has Property* P *at p.*

Proof. Let D be an $\epsilon/2$ disk in S whose interior contains p and whose boundary meets J' in exactly two points. We use Theorem 3 or 4 to adjust D to get the required 2-sphere K.

First we consider the case where D lies in the cube-with-eye-bolt H_i, Bd D lies on Bd H_i, diameter $H_i < \epsilon$, and J' misses H_{i1}. In this case Theorem 3 gives the result immediately. We let C_{i1} be any cube whatsoever associated with H_{i1} and find from Theorem 3 that there is a cube C_i associated with H_i that contains $C_{i1} + \sum_{j=2}^{15} H_{ij}$. Then $J' \cdot D \sum_{j=2}^{15} H_{ij} \subset C_i$ and $J' \cdot \text{Bd } C_i$ is exactly two points. In this case we let $K = \text{Bd } C_i$.

If we omit the assumption that J' misses H_{i1}, it is not much more difficult to show that there is a required 2-sphere K since we only use Theorem 4 instead of Theorem 3. As J' does not fill the part of S in H_i and since the H's at later stages become very small, there is an H' at a later stage which lies in H_i but which misses J'. Theorem 4 shows that there is a cube C_i associated with H_i which contains all of the cubes-with-eye-bolts at this later stage which lie in H_i and which intersect J'. Again we let $K = \text{Bd } C_i$.

We now consider the general case where D is not necessarily the intersection of S and a small H. Consider the cubes-with-eye-bolts at the various stages that intersect S only in Int D. Let $H_1, H_2, \cdots$ be an infinite collection of these which cover Int D and are such that each H_i is of diameter less than $\epsilon/4$, no one of the H's is a subset of any other H, and the collection of H's is locally finite at each point of Int D. It follows from Theorem 4 that there is a cube C_i associated with H_i that covers all of J' in H_i. Then Bd $D + \sum C_i$ is a 3-cell C of diameter less than ϵ which contains $D \cdot J'$. Furthermore, Bd C intersects J' only in the end points of $D \cdot J'$.

THEOREM 8. *If a wild 2-sphere is locally tame except at each point of a set X, each simple closed curve containing X in the 2-sphere is wild.*

Proof. Let D_1, D_2, be the two disks on the 2-sphere bounded by the simple closed curve. Each is locally tame except possibly on its boundary.

If Bd D_1 = Bd D_2 were tame, it would follow from [13] that each D_i is tame. In turn this implies [13] that the 2-sphere $D_1 + D_2$ is tame.

6. Another example. Suppose Example 2 is modified so that T_{i+15} loops around the stem of $T_{i+1,1}$ instead of around the stem of T_{i+1} where it is understood if T_n is the last T, $T_{n+1} = T_1$. If this change is made at each stage, there results a wild 2-sphere S such that for each disk D in S and each simple closed curve J in $E^3 - S$, J can be shrunk to a point $E^3 - D$.

Question. It is not evident that the arcs in the two sphere of the above example are tame. Is a 2-sphere S in E^3 tame if it has the following two properties?

1) Each arc in S is tame.

2) For each disk D in S and each simple closed curve J in $E^3 - S$, J can be shrunk to a point in $E^3 - D$.

In fact, I know of no wild 2-sphere in E^3 such that each of its arcs is tame and each of its complementary domains is simply connected.

7. Reducing sets without simplifying their complements. In this section we show that certain reductions may be made in a set without trivializing any of the elements of the fundamental group of its complement.

THEOREM 9. *Suppose C is a cylinder as shown in Figure 5 with bases B_1, B_2; T_1, T_2 are two linked topological tori imbedded in C as shown such that $T_i \cdot$ Bd $C = B_i$; and X is a closed subset of E^3 such that $X \cdot C = B_1 + B_2$. If J is a simple closed curve in $E^3 - (X + C)$ that cannot be shrunk to a point in $E^3 - (X + C)$, it cannot be shrunk to a point in $E^3 - (X + T_1 + T_2)$.*

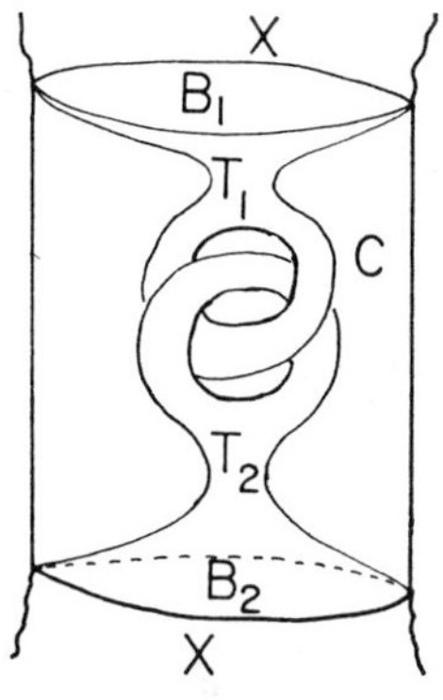

Figure 5

Proof. Assume J can be shrunk to a point in $E^3 - (X + T_1 + T_2)$. Then there is a map f of a disk D into $E^3 - (X + T_1 + T_2)$ that takes Bd D homeomorphically onto J.

If f^{-1} (Bd C) $= 0$, $f(D)$ misses C and J can be shrunk to a point in $E^3 - (X + C)$. This is contrary to hypothesis.

We suppose that f^{-1} (Bd C) is the sum of a finite number of simple closed curves. Suppose furthermore that n is the least integer for which there is a map f of D into $E^3 - (X + T_1 + T_2)$ that takes Bd D homeomorphically onto J and such that f^{-1} (Bd C) is the sum of n simple closed curves. We finish the proof of Theorem 9 by showing that we can reduce n if it is not already 0.

Let J_1 be a simple closed curve in f^{-1} (Bd C) such that J_1 bounds a disk D_1 in D whose interior misses f^{-1} (Bd C). If we show that $f(J_1)$ can be shrunk to a point on the lateral boundary of C (that is, the map f of J_1 can be extended to map D_1 into the lateral boundary of C), then we can adjust f on a disk slightly larger than D_1 to reduce the number of components of f^{-1} (Bd C). We finish the proof of Theorem 9 by showing that $f(J_1)$ can be shrunk to a point on the lateral boundary of C. Either $f(\text{Int } D_1)$ belongs to Int C or Ext C and we treat each case separately.

Case 1. $f(\text{Int } D_1) \subset \text{Int } C$. Now $M_1 = C - (T_1 + T_2)$ is a 3-manifold with boundary. The fundamental group of M_1 is equivalent to the fundamental group of the complement of the finite graph G shown in Figure 6. This group has five generators a, b, c, d, e where a generator is represented by a path that starts at the eye (starting point), goes under G in the direction indicated by the arrow, and returns to the eye. The only relations are given at the crossing points and the branch points and these (reading from top to bottom of Figure 6)

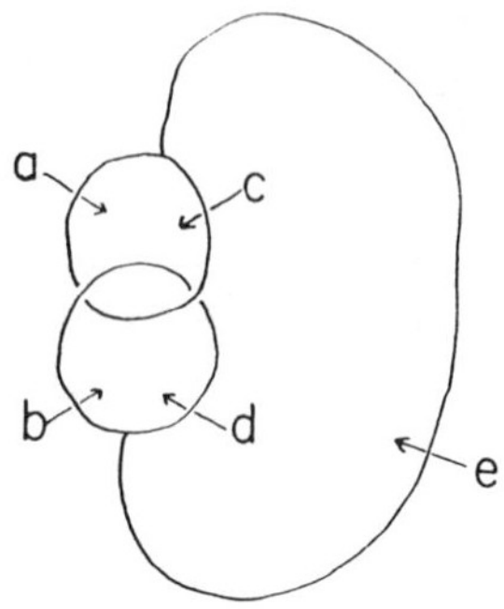

Figure 6

are $a = ec$, $cb = dc$, $ab = bc$, and $b = ed$. The four relations tend to eliminate the generators c, d, e and show that the fundamental group of M_1 is a free group on two generators a and b. Solving for e in terms of a and b we find that $e = ab^{-1}a^{-1}b$ which is nontrivial in the free group. Any loop that circles Bd C an integral number of times different from zero cannot be shrunk to a point in M_1. Any loop on Bd M_1 that can be shrunk to a point in M_1 can be shrunk to a point on Bd M_1.

Case 2. $f(\text{Int } D_1) \subset \text{Ext } C$. Let M_2 be the 3-manifold with boundary $E^3 - (X + \text{Int } C)$. If there is a singular closed curve on Bd M_2 that can be shrunk to a point in M_2 but not in Bd M_2, it follows from [14] that there is a simple closed curve with the same property. Such a simple closed curve could only be one which circles the lateral boundary of C once. The existence of such a simple closed curve would show that any loop on Bd M_2 could be shrunk to a point in M_2.

If any loop on Bd M_2 can be shrunk to a point in M_2, we arrive at the contradiction that f can be adjusted to shrink J to a point in $E^3 - (X + C)$. The first approximation to the adjusted f is unchanged on the component E of $D - f^{-1}(C)$ containing Bd D. This first approximation of f takes each component of $D - E$ into M_2. This first approximation to f takes certain points of D into Bd C but a pushing out gives an adjusted f which shrinks J to a point in $E^3 - (X + C)$.

Blankinship and Fox showed [7] that the exterior of a solid Alexander horned sphere [1] is not simple connected by computing its fundamental group. However, it is not necessary to compute this fundamental group to show the complement is not simply connected. We offer an alternate proof.

Theorem 10. *The complement of a solid Alexander horned sphere is not simply connected.*

Proof. The solid Alexander horned sphere may be described as the intersection of a decreasing sequence of 3-manifolds with boundaries H_1, H_2, $\cdots$. H_1 is a torus. H_2 is obtained by cutting a slice in H_1 and replacing the slice with two solid tori as shown in Figure 5. H_3 is obtained by cutting slices in each of these two solid tori and replacing each slice with two solid tori as indicated in Figure 5. The process is continued. If J is a simple closed curve in $E^3 - H_1$ that cannot be shrunk to a point in $E^3 - H_1$, it follows from repeated use of Theorem 9 that J cannot be shrunk to a point in any $E^3 - H_i$. Hence it cannot be shrunk to a point in the complement of the solid Alexander horned sphere.

THEOREM 11. *Suppose C, B_1, B_2, B_3, C_1, T are as shown in Figure 7 where C is a topological cube; B_1, B_2, B_3 are three mutually exclusive disks on* Bd C;

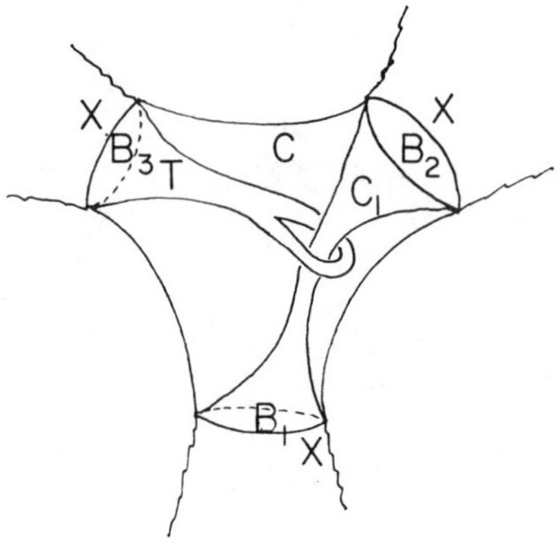

Figure 7

C_1 is a cube which lies except for disks B_1, B_2 on Int C; *and T is a solid torus which intersects* Bd C *in B_3 and circles C_1 as shown. Suppose X is a closed set which intersects C only in $B_1 + B_2 + B_3$ and is such that for $i = 1, 2$,* Bd B_i *cannot be shrunk to a point in $(E^3 - (X + C))$ +* Bd B_i. *Then if J is a simple closed curve in $E^3 - (X + C)$ that cannot be shrunk to a point in $E^3 - (X + C)$, then J cannot be shrunk to a point in $E^3 - (X + C_1 + T)$.*

Proof. If Bd B_3 can be shrunk to a point in $(E^3 - (X + C))$ + Bd B_3, J cannot even be shrunk to a point in $E^3 - (X + C_1)$ so we consider only the case in which Bd B_3 cannot be shrunk to a point in $(E^3 - (X + C))$ + Bd B_3. To see that this is true, consider a thin cylinder C' obtained by slicing off C slightly to one side of B_3. Then Bd C' intersects Bd C − Int B_3 in an annulus A. If J can be shrunk to a point in $E^3 - (X + C_1)$, it can be shrunk to a point in $E^3 - (X$ + Int C' + Int $A)$. However, if Bd B_3 can be shrunk to a point in $(E^3 - (X + C))$ + Bd B_3, each loop in Int A can be shrunk to a point in $(E^3 - (X + C))$ + Int A so there is no need to intersect Int A in shrinking J

to a point in $E^3 - (X + \text{Int } C' + \text{Int } A)$. If Int A is avoided, J is shrunk in $E^3 - (X + C)$.

Let $M_1 = C - (T + C_1)$. It is a 3-manifold with boundary. We compute its fundamental group is a fashion similar to that used in computing the fundamental group of M_1 of Theorem 9 and find that it is a free group on two generators.

Each simple closed curve in $A = \text{Bd } C - (\text{Int } B_1 + \text{Int } B_2 + \text{Int } B_3)$ either bounds a disk in A or is homotopic in A to either Bd B_1, Bd B_2, or Bd B_3. Since each Bd B_i corresponds to a nontrivial element of the fundamental group of $M_1 + \sum \text{Bd } B_i$, it follows from a result of Papakyriakopoulos [13] that each loop on Bd M_1 that can be shrunk to a point in M_1 can be shrunk to a point in Bd M_1.

Let $M_2 = E^3 - (X + \text{Int } C)$. Using Papakyriakopoulos' result again, the fact that each simple closed curve in A either bounds a disk in A or is homotopic in A to some Bd B_i ($i = 1, 2, 3$), and the conditions that no Bd B_i ($i = 1, 2, 3$) can be shrunk to a point in $M_2 + \text{Bd } B_i$, it follows that a loop on Bd M_2 can be shrunk to a point on Bd M_2 if it can be shrunk to a point in M_2.

The proof is completed in a manner analogous to the way in which the proof of Theorem 9 was finished. We suppose that there is a map f of a disk D into $E^3 - (X + C_1 + T)$ that maps Bd D homeomorphically onto J. We consider the number of components of f^{-1} (Bd C) and show that f can be adjusted to reduce this number to zero. But this adjusted f shows that J could be shrunk to a point in $E^3 - (X + C)$.

References

1. J. W. Alexander, *An example of a simply connected surface bounding a region which is not simply connected*, Proceedings of the National Academy of Sciences, vol. 10(1924), pp. 8–10.
2. R. H. Bing, *Locally tame sets are tame*, Annals of Mathematics, vol. 59(1954), pp. 145–158.
3. R. H. Bing, *Approximating surfaces with polyhedral ones*, Annals of Mathematics, vol. 65(1957), pp. 456–483.
4. R. H. Bing, *Conditions under which a surface in E^3 is tame*, Fundamenta Mathematicae, vol. 47(1959), pp. 105–139.
5. R. H. Bing, *A 2-manifold-with-boundary in E^3 is tame if its complement is 1-ULC*, Abstract 546–547, American Mathematical Society Notices, vol. 5(1958), p. 365.
6. R. H. Bing, *Each 2-sphere in E^3 contains a tame arc*, Abstract 559–19, American Mathematical Society Notices, vol. 6(1959), p. 405.
7. W. A. Blankinship and R. H. Fox, *Remarks on certain pathological open subsets of 3-space and their fundamental groups*, Proceedings of the American Mathematical Society, vol. 1(1950), pp. 618–624.
8. P. H. Doyle, *Union of cell pairs in E^3*, Pacific Journal of Mathematics, vol. 10(1960), pp. 521–524.
9. R. H. Fox and E. Artin, *Some wild cells and spheres in three-dimensional space*, Annals Mathematics, vol. 49(1948), pp. 979–990.
10. O. G. Harrold, Jr., H. C. Griffith and E. E. Posey, *A characterization of tame curves in 3-space*, Transactions of the American Mathematical Society, vol. 79(1955), pp. 12–34.

11. O. G. HARROLD, JR., *Some consequences of the approximation theorem of Bing*, Proceedings of the American Mathematical Society, vol. 8(1957), pp. 204–206.
12. O. G. HARROLD, JR., *Locally tame curves and surfaces in three-dimensional manifolds*, Bulletin of the American Mathematical Society, vol. 63(1957), pp. 293–305.
13. E. E. MOISE, *Affine structures in 3-manifolds*, VIII. *Invariance of the knot-types; local tame imbedding*, Annals of Mathematics, vol. 59(1954), pp. 159–170.
14. C. D. PAPAKYRIAKOPOULOS, *On solid tori*, Proceedings of the London Mathematical Society, vol. 7(1957), pp. 281–299.

UNIVERSITY OF WISCONSIN

TAME CANTOR SETS IN E^3

R. H. BING

1. Introduction. Let M_1 be the straight line interval $[0, 1]$, M_2 be the sum of $[0, 1/3]$ and $[2/3, 1]$ obtained by deleting the open middle third of M_1, M_3 be the sum of the four intervals obtained by deleting from M_2 the open middle thirds of the components of $M_2, \cdots$. The intersection M of $M_1, M_2, \cdots$ is the familiar Cantor set. Any homeomorphic image of M is called a *Cantor set.*

It may be shown that if C_1 is a Cantor set on the real line, there is a homeomorphism of the line onto itself taking C_1 onto M. Also, if C_2 is a Cantor set in a plane, there is a homeomorphism of the plane onto itself taking C_2 into a line. For these reasons, we say that each Cantor set in a line or a plane is tame. A Cantor set C_3 in E^3 is *tame* if there is a homeomorphism of E^3 onto itself taking C_3 into a line. Not each Cantor set in E^3 is tame. Antoine gives an example of such a wild Cantor set in [1]. A diagram of one is found in [3, 8]. Blankenship has described [7] wild Cantor sets in Euclidean spaces of all dimensions greater than 2.

Characterizations of tame Cantor sets are provided by Theorems 1.1, 3.1, 4.1 and 5.1. In § 6 we prove theorems about the sums of tame Cantor sets and apply these results in § 7 to show that for each closed 2-dimensional set X in E^3, there is a homeomorphism h of E^3 onto itself that is close to the identity and such that $h(X)$ contains no straight line interval. An example is given of a disk containing intervals pointing in all directions showing that such a homeomorphism h may need to be something more than a rigid motion.

A *simple neighborhood* in E^3 is an open set topologically equivalent to the interior of a sphere. By using simple neighborhoods, we give the following characterization of tame Cantor sets. The proof is obtained in a straight forward fashion.

THEOREM 1.1. *A necessary and sufficient condition that a Cantor set C in E^3 be tame is that for each positive number ε, C can be covered by a finite collection of mutually exclusive simple neighborhoods of mesh less than ε.*

Proof. That the condition is necessary follows from the facts that a homeomorphism of a closed and bounded set in E^3 is uniformly continuous and if the homeomorphic image of C lies on a line, this image

Received March 30, 1960. Work was supported by the National Science Foundation under N.S.F. Grant G-3248 at the Institute for Advanced Study, Princeton, New Jersey.

can be covered with a finite collection of mutually exclusive small interiors of spheres.

The sufficiency condition is a bit more involved but its proof may be obtained as outlined below.

1. Using results on piecewise linear approximations to homeomorphisms [12, 5] show that for each $\varepsilon > 0$, there is a finite collection of mutually exclusive polyhedral cubes of mesh less than ε such that the interiors of these polyhedral cubes cover C.

2. Let $X_1, X_2, \cdots$ be a sequence such that each X_i is a collection of mutually exclusive polyhedral cubes of mesh less than $1/i$, each element of X_{i+1} lies on the interior of an element of X_i, and the interiors of the elements of X_i cover C.

3. Using Moise's extension of Alexander's theorem on polyhedral 2-spheres [11], show that there is a piecewise linear homeomorphism h_1 of E^3 onto itself that takes each element of X_1 onto a cube of diameter less than 1 and with center on the x-axis.

4. Reapplying Moise's result, show that there is a piecewise linear homeomorphism h_2 of E^3 onto itself such that $h_1 = h_2$ except in elements of X_1, h_2 takes each element of X_2 onto a cube of diameter less than 1/2 with center on the x-axis. Continuing in this fashion, we obtain $h_3, h_4, \cdots$. The limit of $h_1, h_2, h_3, \cdots$ is a homeomorphism of E^3 onto itself that takes C into the x-axis.

COROLLARY 1.2. *A Cantor set in E^3 is tame if its complement is topologically equivalent to the complement of a tame Cantor set.*

Theorem 1.1 reveals how to find a tame Cantor set in any Cantor set in E^3. Hence, we have the following theorem.

THEOREM 1.3. *Each Cantor set in E^3 contains a tame Cantor set.*

2. Making 2-spheres mutually exclusive. The Cantor set C mentioned in the statement of Theorem 1.1 did not intersect the boundaries of the simple neighborhoods mentioned there. We shall see in Theorem 3.1 of the next section that we can dispense with the condition that the simple neighborhoods be mutually exclusive if we retain this property. As a start in this direction, we consider the following two properties of a finite collection $(N_1, N_2, \cdots, N_n)$ of simple neighborhoods that cover a closed set X.

Property 1. X misses the boundary of each N_i.

Property 2. The N_i's are mutually exclusive.

In this section we see how to replace a collection satisfying Property 1 with a collection satisfying Property 2.

As noted in the proof of Theorem 1.1, a simple neighborhood can be reduced slightly so that the boundary of the resulting neighborhood is a polyhedral 2-sphere. We now prove some theorems about polyhedral 2-spheres. We use $V(A, \varepsilon)$ to denote the set of all points whose distances from A is less than ε.

THEOREM 2.1. *Suppose* $(K_1, K_2, \cdots, K_n)$ *is a collection of mutually exclusive polyhedral 2-spheres;* X *is a closed set that lies in each* $\operatorname{Ext} K_i$; $(S_1, S_2, \cdots, S_m)$ *is a collection of mutually exclusive polyhedral 2-spheres such that* $X \subset (\operatorname{Int} S_1 + \operatorname{Int} S_2 + \cdots + \operatorname{Int} S_m)$; *and* $\varepsilon > 0$. *Then there is a collection of mutually exclusive polyhedral 2-spheres* $(S_1', S_2', \cdots, S_r')$ *such that*

$$X \subset \operatorname{Int} S_1' + \operatorname{Int} S_2' + \cdots + \operatorname{Int} S_r' ,$$
$$\Sigma S_j' \cdot \Sigma K_i = 0 , \quad \text{and}$$
$$S_j' \subset V(\Sigma S_i + \Sigma K_i, \varepsilon) .$$

Proof. It would be convenient to have each S_i and each K_j in general position in the sense that if p is a point of $S_i \cdot K_j$, there is a neighborhood N of p and a homeomorphism h of N onto E^3 such that $h(N \cdot S_i)$ and $h(N \cdot K_j)$ are perpendicular planes. Such a convenient position would insure that each component of $S_i \cdot K_j$ is a simple closed curve.

We show how a slight adjustment of S_i brings about such a convenient position. Suppose S_i and each K_j is triangulated. Let $v_1, v_2, \cdots, v_n$ be the vertices of the triangulation of S_i and X be the collection of the vertices of the triangulation of ΣK_j. Adjust v_1 slightly to v_1' so that there results a 2-sphere close to S_i but v_1' is not coplanar with any three noncollinear points of X. Next adjust v_2 to v_2' so that v_2' is not coplanar with any three noncollinear points of $X + v_1'$. If we continue this adjusting of the vertices of S_i one at a time, we obtain a 2-sphere S_i' close to S_i so that no vertex of S_i' lies on any K_j, no vertex of an K_j lies on S_i', and no edge of S_i' intersects any edge of any K_j.

In view of remarks in the preceding paragraph we suppose with no loss of generality that each S_i and each K_j are in general position. (If the S's are not conveniently placed, we replace them by some that are.) The proof of Theorem 2.1 is by induction on the number of components of $\Sigma K_i \cdot \Sigma S_i$.

Let J be such a component that bounds a disk D in K_i such that $D \cdot \Sigma S_j = J$. Assume $J \subset S_m$ and E_1, E_2 are the two disks in S_m bounded by J. Let S_{m_1}, S_{m_2} be two polyhedral 2-spheres obtained by separate deformations of $E_1 + D$ and $E_2 + D$ in a neighborhood of D such that

$$X \cdot \text{Int}\, S_m \subset \text{Int}\, S_{m_1} + \text{Int}\, S_{m_2} ,$$
$$(S_{m_1} + S_{m_2}) \cdot \Sigma K_i = S_m \cdot \Sigma K_i - J ,$$
$$S_{m_1} \cdot S_j \quad (j \neq m) = 0 .$$

The truth of the theorem now follows by induction because $(S_1 + S_2 + \cdots + S_{m-1} + S_{m_1} + S_{m_2}) \cdot \Sigma K_i$ has one less component than $\Sigma S_i \cdot \Sigma K_i$.

THEOREM 2.2. *Suppose* $(K_1, K_2, \cdots, K_n)$ *is a collection of mutually exclusive polyhedral 2-spheres,* X *is a closed set which lies in each* Ext K_i, S *is a polyhedral 2-sphere such that* $X \subset \text{Int}\, S$, *and* $\varepsilon > 0$. *Then there is a polyhedral 2-sphere* S' *such that*

$$X \subset \text{Int}\, S' ,$$
$$S' \cdot \Sigma K_i = 0 ,$$
$$S' \subset V(S + \Sigma K_i, \varepsilon) .$$

Proof. It follows from the preceding theorem that there is a collection of mutually exclusive polyhedral 2-spheres $(S_1, S_2, \cdots, S_m)$ such that

$$X \subset \text{Int}\, S_1 + \text{Int}\, S_2 + \cdots + \text{Int}\, S_m ,$$
$$S_j \cdot \Sigma K_i = 0 , \quad \text{and}$$
$$S_i \subset V(S + \Sigma K_{,}, \varepsilon) .$$

The polyhedral 2-sphere S' required by the theorem is obtained by cutting holes in some of the S_j's and joining the boundaries of these holes with tubes near $S + \Sigma K_i$. The expression "some of" is to suggest that we ignore the S's on the interior of any K_j or any other S_i.

THEOREM 2.3. *Suppose* X *is a closed, bounded subset of* E^3, $(K_1, K_2, \cdots, K_n)$ *is a finite collection of topological 2-spheres such that* $X \subset \Sigma\, \text{Int}\, K_i$ *but* $X \cdot \Sigma K_i = 0$; *and* $\varepsilon > 0$. *Then there is a collection of mutually exclusive polyhedral 2-spheres* $(K_1', K_2', \cdots, K_n')$ *such that*

$$X \subset \Sigma\, \text{Int}\, K_i'$$
$$\Sigma K_i' \subset V(\Sigma K_i, \varepsilon) .$$

Proof. Using the approximation theorem for surfaces [4] we suppose with no loss of generality that the 2-spheres $(K_1, K_2, \cdots, K_n)$ are polyhedral. We then let $K_1' = K_1$.

Next we use the preceding theorem to pick K_2' so that

$$X\cdot(\text{Int } K_1 + \text{Int } K_2) \subset X\cdot(\text{Int } K_1' + \text{Int } K_2') ,$$
$$K_2' \text{ is very close to } K_1 + K_2 \text{ and misses } X + K_1 .$$

Repeated applications of Theorem 2.2 gives the collection $(K_1', K_2', \cdots, K_n')$ promised by the theorem.

The following result is an immediate consequence of Theorem 2.3.

THEOREM 2.4. *Suppose X is a closed bounded subset of E^3; $(N_1, N_2, \cdots, N_n)$ is a finite collection of simple neighborhoods covering X and satisfying Property 1; and $\varepsilon > 0$. Then there is a collection of mutually exclusive simple neighborhoods $(N_1', N_2', \cdots, N_n')$ covering X such that*

$$\Sigma \text{ Bd } N_i' \subset V(\Sigma BdN_i, \varepsilon)$$

and each Bd N_i' *is a polyhedral 2-sphere.*

3. Separating Cantor sets with small 2-spheres. We use the results of the preceding section to refine the characterization of a tame Cantor set given by Theorem 1.1.

An example is known [2] of a wild Cantor set C such that if p, q are two points of C, there is a 2-sphere in $E^3 - C$ that separates p from q. However, in this particular example, there is no such 2-sphere of small size. This example along with Theorem 1.1 suggest the following theorem. I am indebted to O. G. Harrold, Jr. for suggesting its proof.

THEOREM 3.1. *A Cantor set C in E^3 is tame if for each positive number ε and each pair of points p, q of C there is an ε 2-sphere in $E^3 - C$ separating p from q.*

Proof. The theorem is established by way of Theorem 1.1 since we show that for each positive number ε there is an ε collection of mutually exclusive simple neighborhoods covering C.

Let U be an open set containing C such that $E^3 - U$ is connected and each component of U is of diameter less than ε.

The hypothesis of the theorem tells us that for each point p of C with the possible exception of one point p_0, there is a 2-sphere in U that misses C and contains p on its interior. The approximation theorem for surfaces [4] reveals that there is a polyhedral cube C_p in U such that $p \in C_p$, and $C\cdot\text{Bd } C_p = 0$.

Let $(C_{p_0}, C_1, C_2, \cdots, C_n)$ be a finite collection of polyhedral cubes in U whose interiors cover C and such that $C\cdot(\text{Bd } C_1 + \text{Bd } C_2 + \cdots + \text{Bd } C_n) = 0$.

It follows from Theorem 2.4 that the collection of C_i's may be replaced by mutually exclusive C_i''s in U so that

$$C\cdot(C_1 + C_2 + \cdots + C_2) \subset C\cdot(C'_1 + C'_2 + \cdots + C'_n)$$

and

$$C\cdot(\text{Bd } C'_1 + \text{Bd } C'_2 + \cdots + \text{Bd } C'_n) = 0 .$$

Letting X be the closed set consisting of the part of C not in $(C'_1 + C'_2 + \cdots + C'_n)$ we find from Theorem 2.2 that there is a polyhedral cube C'_0 in U such that

$$C'_0, C'_1, \cdots, C'_2 \text{ are mutually exclusive,}$$
$$X \subset \text{Int } C'_0 .$$

Then $(C'_0, C'_1, \cdots, C'_2)$ is an ε collection of mutually exclusive polyhedral cubes whose interiors covers C and it follows from Theorem 1.1 that C is tame.

COROLLARY 3.2. *A Cantor set C in E^3 is tame if for each neighborhood N of a point p of C there is a simple neighborhood N' of p such that $N' \subset N$ and* Bd $N'\cdot C = 0$.

4. Local tameness.

DEFINITION. A Cantor set C is locally tame at a point p of C if there is a neighborhood N of p such that $p\cdot N$ is a tame Cantor set. The following result follows from Corollary 3.2.

THEOREM 4.1. *A Cantor set in E^3 is tame if it is locally tame at each of its points.*

The proof of Theorem 3.1 reveals even more—namely, C is tame if it is locally tame except possibly on a discrete set of points. Theorem 6.1 of a following section shows that a Cantor set is tame if it is locally tame except possibly at the points of a tame Cantor set.

THEOREM 4.2. *Suppose C is a Cantor set in E^3, $p \in C$, N is a neighborhood of p, and C is locally tame at each point of $N\cdot(C - p)$. Then C is locally tame at p.*

THEOREM 4.3. *Each wild Cantor set in E^3 contains a Cantor set that is not locally tame at any point.*

Proof. Let C be a wild Cantor set and C' be the set of all points of C at which C is not locally tame. It follows from Theorem 4.2 that C' contains no isolated point and is therefore a Cantor set. We show that C' is not locally tame at any point.

If C' were locally tame at a point p_0, there would be a neighborhood N' of p_0 such that $N'\cdot C'$ is a tame Cantor set. Let N be a second neighborhood of p_0 such that $N \cdot C$ is a Cantor set and $N \cdot C' = N' \cdot C'$. We show that $N \cdot C$ is tame, thereby contradicting the assumption that $p_0 \in C'$.

We show that $N \cdot C$ is tame by showing that if p, q are two points of $N \cdot C$, then there is an ε 2-sphere in $E^3 - N \cdot C$ separating p from q. We suppose the distance from p to q is more than ε and consider only the more difficult case where $p \in N \cdot C'$.

Let C_0 be a polyhedral cube of diameter less than $\varepsilon/3$ and such that $p \in C_0$, $\operatorname{Bd} C_0 \cdot (N \cdot C') = 0$.

Let $(C_1, C_2, \cdots, C_n)$ be a finite collection of polyhedral $\varepsilon/3$ cubes covering $C \cdot \operatorname{Bd} C_0$ such that $C \cdot (\operatorname{Bd} C_1 + \operatorname{Bd} C_2 + \cdots + \operatorname{Bd} C_n) = 0$ and let X be the part of $C \cdot C_0$ not contained in $\operatorname{Int} C_1 + \operatorname{Int} C_2 + \cdots + \operatorname{Int} C_n$. By using Theorem 2.3 we suppose with no loss of generality that the C_i's are mutually exclusive. It follows from Theorem 2.2 that there is a 2-sphere K in $\operatorname{Ext} C_1 + \operatorname{Ext} C_2 + \cdots + \operatorname{Ext} C_n$ such that $X \subset \operatorname{Int} K$ and K is so close to $\operatorname{Bd} C_0 + \operatorname{Bd} C_1 + \cdots + \operatorname{Bd} C_n$ that it misses C and is of diameter less than ε.

Question. Is there a universally wild Cantor set C in E^3 in the sense that for each Cantor set C' in E^3 there is a homeomorphism h of E^3 onto itself taking C' into C?

5. The fundamental group of the complement. The complement of each tame Cantor set C in E^3 is 1-*ULC*—that is, for each positive number ε there is a positive number δ such that each closed curve of diameter less than δ in $E^3 - C$ can be shrunk to a point on subset of $E^3 - C$ of diameter less than ε. We find that this property of the complement characterized a tame Cantor set. Kirkor has given [10] an example of a wild Cantor set whose complement is simply connected.

THEOREM 5.1. *A Cantor set C in E^3 is tame if and only if $E^3 - C$ is* 1-*ULC*.

Proof. Let p be a point of C and K a small polyhedral 2-sphere whose interior contains p. Let $D_1, D_2, \cdots, D_n$ be a finite number of mutually exclusive disks in K whose interiors cover $K \cdot C$. It follows from that fact that $E^3 - C$ is 1-*ULC* that these D_i's can be replaced by singular disks E_i that miss C and from Dehn's lemma as proved by Papakyriakopoulos in [13] that these singular disks E_i can be replaced by disks F_i so that $(K - \Sigma D_i) + \Sigma F_i$ is a small polyhedral 2-sphere which misses C and has p on its interior. Then Theorem 5.1 follows from Theorem 3.1.

6. The sum of Cantor sets. In this section we prove some theorems that will be used in the next sections about the sums of Cantor sets.

THEOREM 6.1. *Any Cantor set which is the sum of a countable number of tame Cantor sets is tame.*

Proof. Suppose the Cantor set C is the sum of the tame Cantor sets $C_1, C_2, \cdots$. If C is wild, it follows from Theorem 4.3 that C contains a wild Cantor set C' that is not locally tame at any point. The Baire category theorem assures us that such a C' would contain a Cantor set C'' which is open in C' and which lies in one C_i. The C'' is not locally tame at any point since C' is not. This contradicts the fact that C'' is a tame Cantor set since it lies in a tame C_i.

Another proof of Theorem 6.1 also follows from Theorem 1.1 and Theorem 7.2 of the next section.

Examples of Cantor sets that cast dense shadows are known [9] but we include the example described in Theorem 6.2 for completeness.

THEOREM 6.2. *If R is a rectangular solid in E^3, there is a tame Cantor set C in R such that each straight line intersecting two opposite faces of R also intersects C.*

Proof. Let F, F' be two opposite faces of R. First, we construct a tame Cantor set C' in R such that each line intersecting each of F, F' also intersects C'.

1. Let $R = M_1$.

2a. Let P be the plane halfway between F and F'. For convenience we suppose that P is horizontal, F lies above P, and F' lies below. Let X_1, X_2, X_3 be three collections of rectangles in $P \cdot R$ such that the elements of X_i $(i = 1, 2, 3)$ are mutually exclusive and of diameter less than 1/2 while any line intersecting each of F, F', Int R also intersects the interior of an element of $X_1 + X_2 + X_3$.

2b. Push the elements of X_1 up and the elements of X_2 down but so slightly that if X_2', X_2' denotes the collection of adjusted rectangles, any line that intersects each of F, F', Int R also intersects the interior of a rectangle of $X_1' + X_2' + X_3$.

2c. Expand each rectangle in $X_2' + X_2' + X_3$ slightly so as to form a collection of mutually exclusive horizontally based rectangular solids in R of mesh less than 1/2 such that any line that intersects each of F_1, F_2 intersects both the upper and lower face of some one of the solids. Denote the sum of these solids by M_2.

3. Let M_3 be the sum of a collection of horizontally based rectangular solids in M_2 of mesh less than 1/3 such that any line intersecting each of F, F' intersects both the top and bottom face of one of these solids. Continuing in this fashion we obtain $M_4, M_5, \cdots$. The intersection of $M_1, M_2, \cdots$ is the Cantor set C'. It follows from Theorem 1.1 that C' is tame.

For each pair of opposite faces of R there is a tame Cantor set in R such that any line intersecting each of these opposite faces also intersects the Cantor set. Since R has only three pairs of opposite faces and it follows from Theorem 6.1 that the sum of three tame Cantor sets is a tame Cantor set, there is a Cantor set C in R such that any line intersecting two opposite faces of R also intersects C.

THEOREM 6.3. *There is a countable collection of tame Cantor sets in E^3 such that each straight line interval in E^3 intersects one of these Cantor sets.*

Proof. Let $R_1, R_2, \cdots$ be the set of all rectangular solids in E^3 with rational vertices and C_i be a tame Cantor set in R_i such that each line intersecting two opposite faces of R_i also intersects C_i. Then ΣC_i intersects each straight line interval in E^3.

7. Avoiding Cantor sets. In [6] it is shown that a 2-sphere S in E^3 is tame provided that for each positive number ε there are 2-spheres S', S'' in different complementary domains of S and homeomorphisms h', h'' of S onto S', S'' respectively that move no point by as much as ε. As a first step in this proof, a homeomorphism h of E^3 onto itself is used such that $h(S)$ contains no vertical interval. Theorem 7.3, which is proved by means of tame Cantor sets, assures us that there is such a homeomorphism h. An alternate proof of this result is found in Theorem 10.1 of [6].

The following result is used as a lemma in getting Theorem 7.3. We use ρ to denote the distance function and if f, g are two maps of a metric space X onto a metric space Y, $\rho(f, g)$ is the least upper bound of $\rho(f(x), g(x))$ for all points x of X.

THEOREM 7.1. *Suppose $h_1, h_2, \cdots$ is a sequence of homeomorphisms of E^3 onto itself such that*

1. $\rho(h_{i+1}, h_i) < \varepsilon/2^i$ *and* h_1 *is the identity,*
2. $\rho(h_{i+1}, h_i) < \rho(h_i(x_1), h_i(x_2))/2^i$ *if* $\rho(x_1, x_2) > 1/i$.

Since this paper was written, a proof of Theorem 7.1 by T. Homma has appeared in the Yokohama Mathematics Journal, vol. 7, pp. 191-195.

Then the limit of $h_1, h_2, \cdots$ is a homeomorphism of E^3 onto itself that moves no point by as much as ε.

Proof. That the limit of $h_1, h_2, \cdots$ is continuous and moves no point by as much as ε follows from Condition 1 of the hypothesis. That no two points go into the same point under this limit follows from Condition 2. The limit is onto because each continuous map of E^3 onto itself that moves no point by more than ε is an onto map. It is a homeomorphism because it is a homeomorphism on each closed bounded set.

THEOREM 7.2. *If X is a closed 2 dimensional set in E^3, ε is a positive number, and $C_1, C_2, C_3, \cdots$ is a sequence of tame Cantor sets, there is a homeomorphism h of E^3 onto itself that moves no point by as much as ε and $h(X)\cdot\Sigma C_i = 0$.*

Proof. The homeomorphism h is the inverse of the limit of a sequence of homeomorphism as considered in Theorem 7.1. The identity homeomorphism is denoted by h_1.

Let $y_1, y_2, \cdots, y_n$ be finite collection of mutually exclusive simple neighborhoods covering C_1 of mesh less than $\varepsilon/2$. Let h_2 be a homeomorphism of E^3 onto itself that is fixed outside Σy_i and which takes each $C_1\cdot y_i$ into $y_i - X$.

We now describe h_3. Suppose δ is a positive number such that $\delta < \rho(X, h_2(C_1))$, $\delta < \varepsilon/4$, and $\delta < \rho(h_2(x_1), h_2(x_2))/4$ if $p(x_1, x_2) > 1/2$. Let $z_1, z_2, \cdots, z_m$ be a finite collection of mutually exclusive simple neighborhoods covering $h_2(C_2)$ of mesh less than δ. Let g be a homeomorphism of E^3 onto itself that is fixed outside Σz_i, g is fixed on each z_i missing X, and $gh_2(C_2)$ misses X. Then $h_3 = gh_2$ is fixed on $h_2(C_1)$ and $h_3(C_1 + C_2)$ misses X.

Similarly we obtain $h_4, h_5, \cdots$ such that the sequence $h_1, h_2, \cdots$ satisfies the hypothesis of Theorem 7.1, $h_{i+1} = h_i$ on $C_1 + C_2 + \cdots + C_{i-1}$, and $h_{i+1}(C_1 + C_2 + \cdots + C_i)$ misses X.

It follows from Theorem 7.1 that the limit h_∞ of $h_1, h_2, \cdots$ is a homeomorphism of E^3 onto itself such that h_∞ moves no point by as much as ε and $X\cdot h(\Sigma C_i) = 0$. Then $h = h_\infty^{-1}$ satisfies the conclusion of Theorem 7.2.

EXAMPLE. It is necessary in the hypothesis of Theorem 7.2 to assume that the Cantor sets are tame. The wild Cantor set C described by Antoine [1] has a complement which is not simply connected. There is a disk D in E^3 such that $C\cdot \mathrm{Bd}\, D = 0$ but $\mathrm{Bd}\, D$ can not be shrunk to a point in $E^3 - C$. If $e = \rho(C, \mathrm{Bd}\, D)$, there is no homeomorphism h of E^3 onto itself that moves no point by as much as ε and such that $h(D)\cdot C = 0$.

THEOREM 7.3. *If X is a closed 2 dimensional set in E^3 and ε is a positive number, there is a homeomorphism h of E^3 onto itself such that h moves no point by as much as ε and $h(X)$ contains no straight line interval.*

Proof. The result follows from Theorems 6.3 and 7.2.

EXAMPLE. Theorem 7.3 was used in [6] to adjust a 2-sphere S in E^3 so that the adjusted S will contain no vertical interval. One might hope to get such an adjustment by tilting S with a rigid motion. However, we give an example of a tame disk D such that for each direction there is a unit interval in D pointing in that direction.

Let X be the Cantor set obtained by starting with the interval $[-1, 1]$ and removing middle thirds. Let f be the map of $[-1, 1]$ onto $[-1, 1]$ such that $f(-1) = -1$, $f(1) = 1$, $f[-1/3, 1/3] = 0$, $f[-7/9, -5/9] = -1/2$, $f[5/9, 7/9] = 1/2$, $f[-25/27, 23/27] = -3/4, \cdots$. Then f is a function that is continuous and has a derivative equal to 0 except on X.

Let K be the cube with edges parallel to the axes and opposite vertices at $(-1, -1, -1)$, $(1, 1, 1)$. Let X' be the set of all points (x, y, z) on Bd K such that each coordinate of (x, y, z) is a point of X. See Figure 1.

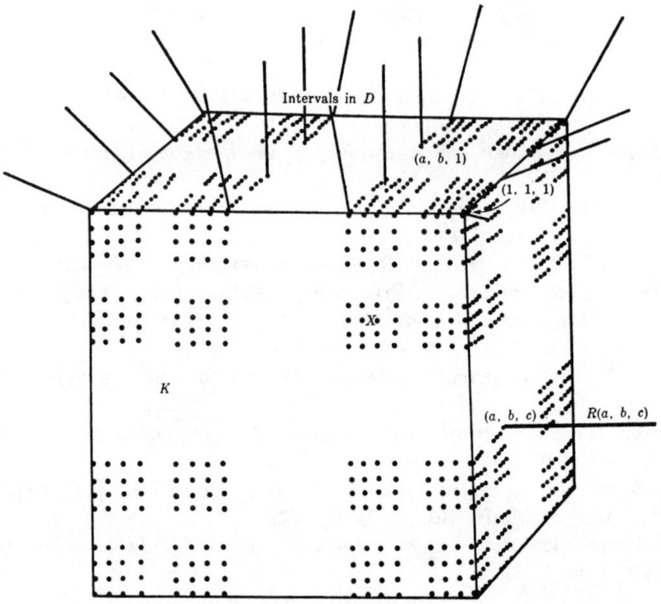

From each point (x, y, z) of Bd K let $R(x, y, z)$ be the ray from the point (x, y, z) in the direction $f(x) : f(y) : f(z)$. We shall show in the next two paragraphs that no two of these rays intersect. For each direction there is one of these rays from a point of X' pointing in that direction. There is an arc A on Bd K that contains X'. A disk D containing a unit interval from each of the rays $R(x, y, z)$ intersecting A contains intervals pointing in all directions.

The ray $R(a_i, b_i, 1)$ has the parametric equation

$$x = a_i + f(a_i)t , \quad y = b_i + f(b_i)t , \quad z = 1 + t \qquad (t \geqq 0) .$$

If $a_1 \neq a_2$, $a_1 + f(a_1)t \neq a_2 + f(a_2)t$. Since the same thing is true for b's, $R(a_1, b_1, 1)$ does not intersect $R(a_2, b_2, 1)$ if $(a_1, b_1, 1) \neq (a_2, b_2, 1)$. Hence, no two of the rays starting on the same face of K intersect.

If $c_3 \neq 1$, $R(a_1, b, 1)$ does not intersect $R(a_3, b_3, c_3)$ because if (x, y, z) is a point of $R(a_1, b_1, 1)$

$$z \geqq \max (| x |, | y |)$$

but this inequality does not hold for points of $R(a_3, b_3, c_3)$. Hence, no two of the rays emanating from different faces of K intersect.

References

1. Louis Antoine, *Sur l'homeomorphie de deux figures et de leurs voisanages*, J. Math. Pures. Appl., **86** (1921), 221-325.

2. R. H. Bing, *A homeomorphism between the 3-sphere and the sum of two solid horned spheres*, Ann. of Math., **56** (1952), 354-362.

3. ———, *Examples and counter examples*, Pi Mu Epsilon Journal, **1** (1953), 310-317.

4. ———, *Approximating surfaces with polyhedral ones*, Ann. of Math., **65** (1957), 456-483.

5. ———, *An alternative proof that 3-manifolds can be triangulated*, Ann. of Math., **69** (1959), 37-65.

6. ———, *Conditions under which a surface in E^3 is tame*, Fund. Math., **47** (1959), 105-139.

7. W. A. Blankenship, *Generalization of a construction by Antoine*, Ann. of Math., **53** (1951), 276-297.

8. W. A. Blankenship and R. H. Fox, *Remarks on certain pathological open subsets of 3-space and their fundamental groups*, Proc. Amer. Math. Soc., **1** (1950), 618-624.

9. Karol Borsuk, *An example of a simple arc in space whose projection in every plane has interior points*, Fund. Math., **34** (1946), 272-277.

10. A. Kirkor, *Wild 0-dimensional sets and the fundamental group*, Fund. Math., **45** (1958), 228-236.

11. E. E. Moise, *Affine structures in 3-manifolds, II. Positional properties of 2-spheres*, Ann. of Math., **55** (1952), 172-176.

12. ———, *Affine structures in 3-manifolds, IV. Piecewise linear approximations of homeomorphism*, Ann. of Math., **55** (1952), 215-222.

13. C. D. Papakyriakopoulos, *On Dehn's lemma and the asphericity of knots*, Ann. of Math., **66** (1957), 1-26.

INSTITUTE FOR ADVANCED STUDY
UNIVERSITY OF WISCONSIN

A SURFACE IS TAME IF ITS COMPLEMENT IS 1-ULC(1)

BY

R. H. BING

1. **Introduction.** In this paper we show that a 2-manifold M in Euclidean 3-space E^3 is tame if E^3-M is uniformly locally simply connected.

A closed subset X of a triangulated manifold Y is *tame* if there is a homeomorphism of Y onto itself taking X onto a polyhedron (geometric complex) in Y. If there is no such homeomorphism, X is called *wild*. Examples of 2-manifolds wildly embedded in E^3 are found in [1; 8; 6].

An *n-manifold* is a separable metric space each of whose points lies in a neighborhood homeomorphic to Euclidean n-space. An *n-manifold-with-boundary* is a separable metric space each of whose points lies in a neighborhood whose closure is a topological n-cell. If M is an n-manifold-with-boundary, we use Int M to denote the set of points of M with neighborhoods homeomorphic to Euclidean n-space and Bd D to denote $M-\text{Int } M$. For example, if D is a disk, Bd D is a simple closed curve which is the rim of the disk. If we have a manifold embedded in a larger space and treat the manifold as a subset rather than a space, we insist that it be closed. If S is a 2-sphere embedded in E^3, we use Int S and Ext S to denote the bounded and unbounded components of E^3-S. The double meaning of the symbol Int should not lead to confusion.

A subset X of a manifold-with-boundary is *locally tame* at a point p of X if there is a neighborhood N of p and a homeomorphism of $\overline{N}$ (the closure of N) onto a cell that takes $X\cdot\overline{N}$ onto a polyhedron. If the manifold-with-boundary is triangulated, we say that X is locally polyhedral at p if there is a neighborhood N of p such that $X\cdot\overline{N}$ is a polyhedron.

Suppose D is a disk. We say that *a map of* Bd D *into a set* Y *can be shrunk to a constant in* Y if the map can be extended to take D into Y. If each map of Bd D into Y can be shrunk to a constant in Y, we say that Y is *simply connected*. Also, Y is *locally simply connected* at a point p of $\overline{Y}$ if for each neighborhood U of p there is a neighborhood V of p such that each map of Bd D into $V\cdot Y$ can be shrunk to a point in $U\cdot Y$. A metric space Y is *uniformly locally simply connected* (or 1-ULC) if for each $\epsilon>0$ there is a $\delta>0$ such that each map of Bd D into a δ subset of Y can be shrunk to a point on an ϵ subset of Y. If $\overline{Y}$ is a compact subset of a metric space, it can be shown that Y is 1-ULC if it is locally simply connected at each point of $\overline{Y}$.

Presented to the Society, January 28, 1958 under the title *A surface S is tame in E^3 if E^3-S is locally simply connected at each point of S*; received by the editors March 1, 1961.

(1) Work on this paper was supported by contract NSF-G 11665.

The surfaces studied in solid geometry are tame. It is useful to have criteria for determining which surfaces are tame. We mention three such criteria.

1. A closed subset of a triangulated 3-manifold-with-boundary is tame if and only if it is locally tame at each of its points.

2. A 2-sphere in E^3 is tame if and only if it can be homeomorphically approximated from both sides.

3. A 2-manifold M in a triangulated 3-manifold M_3 is tame if and only if $M_3 - M$ is locally simply connected at each point of M.

Criterion 1 is proved in [2] and [10]. Criterion 2 is proved in [5] and restated below as Theorem 0. Criterion 3 is Theorem 7 of the present paper.

The distance function is denoted by ρ. We shall make use of the following approximation theorem proved in [3].

APPROXIMATION THEOREM. *For each 2-manifold M in a triangulated 3-manifold-with-boundary and each non-negative continuous function f defined on M, there is a 2-manifold M' and a homeomorphism h of M onto M' such that*

$$\rho(x, h(x)) \leqq f(x) \qquad (x \in M)$$

and M' is locally polyhedral at $h(x)$ if $f(x) > 0$.

Another theorem that we shall use is the Side Approximation Theorem for 2-Spheres. It is proved by the same methods as the Approximation Theorem and its proof will be given in another paper [7].

SIDE APPROXIMATION THEOREM FOR 2-SPHERES. *Each 2-sphere S in E^3 can be polyhedrally approximated almost from either side—that is for each $\epsilon > 0$ and each component U of $E^3 - S$ there is a homeomorphism h of S onto a polyhedral 2-sphere such that*

h moves no point more than ϵ and

$h(S)$ contains a finite collection of mutually exclusive disks each of diameter less than ϵ such that $h(S)$ minus the sum of the disks lies in U.

If A and B are two sets and there is a homeomorphism of A onto B that moves no point by more than ϵ, we write

$$H(A, B) \leqq \epsilon.$$

The following theorem is proved in [5].

THEOREM 0. *A 2-sphere S in E^3 is tame if it can be homeomorphically approximated from both sides—that is, for each $\epsilon > 0$ and each component U of $E^3 - S$, there is a 2-sphere S' in U such that*

$$H(S, S') < \epsilon.$$

To show that a 2-sphere is tame, a first step might be to show that the hypothesis of Theorem 0 is met. As a step toward proving that a 2-sphere in E^3 is tame if its complement is 1-ULC, we prove the following.

THEOREM 1. *If S is a 2-sphere in E^3 such that* Int S *is* 1-ULC, *then for each* $\epsilon>0$, S *can be homeomorphically approximated from* Int S—*that is, for each* $\epsilon>0$ *there is a 2-sphere S' in* Int S *such that*

$$H(S, S') < \epsilon.$$

Our plan for proving Theorem 1 is to get a special cellular decomposition T of S, get a homeomorphism of S onto itself that pulls the boundaries of the cells of the decomposition T into Int S, and finally pull the cells themselves into Int S.

It is hoped that the serious reader will understand the why of the attack as well as the details. Hence, we give our over-all plan of attack first and reserve epsilontics to last so that the reader can see why these particular ϵ's are used. When we need a close approximation, we let it be an ϵ_i approximation and decide later how small the ϵ_i would need to be to make the details work.

2. **Proof of Theorem** 1. a. **Special cellular decomposition** T **of** S. We need a cellular decomposition T of S with the following properties.

The mesh of T is less than ϵ_1 (ϵ_1 is a small number whose size is to be described later).

The collection of 2-cells of T is the sum of three subcollections A_1, A_2, A_3 such that no two elements of A_i ($i=1, 2, 3$) intersect each other.

That for each $\epsilon_1>0$ there is such a cellular decomposition T of S follows from a consideration of a triangulation T' of S of mesh less than $\epsilon_1/2$. The vertices of T' are swelled into 2-cells and become the elements of A_1. The parts of the 1-simplexes of T' not in elements of A_1 are expanded into the elements of A_2. The closures of the parts of the 2-simplexes of T' not in elements of A_1 or A_2 are the elements of A_3.

b. **Pulling elements of** T **partially into** S. Suppose T is a fixed special cellular decomposition of S such as mentioned in the preceding section. The 1-skeleton of T is the sum of the boundaries of the 2-cells in T and is denoted by K_1. We select a small number ϵ_2 whose size is described later. Then there is a polyhedral 2-sphere S_1 and a homeomorphism h_1 of S onto S_1 such that

h_1 moves no point more than ϵ_2,
$h_1(K_1) \subset \text{Int } S$, and
S_1 contains a finite collection of mutually exclusive ϵ_2 disks such that S_1 minus the sum of the interiors of these disks lies in Int S.

That for each ϵ_2 there are such an S_1 and an h_1 follows from the Side Approximation Theorem of 2-Spheres. We let $\epsilon_2/2$ be the ϵ in the statement of that theorem and S_1 be the S' guaranteed by the conclusion of that theorem. The homeomorphism h_1 is the homeomorphism h guaranteed by that theorem followed by a homeomorphism of S_1 onto itself that moves no point by more than $\epsilon_2/2$ but pulls the image of K_1 off of the disks.

c. **The next approximations to elements of** T. For each 2-cell D of T, $h_1(D)$ is a first approximation to D. We note that $h_1(D)$ is homeomorphically close to D and $h_1(\text{Int } D)$ contains a finite collection $E_1, E_2, \cdots, E_n$ of mutually exclusive ϵ_2 disks such that $h_1(D) - \sum \text{Int } E_i \subset \text{Int } S$. We suppose that $h_1(\text{Bd } D)$ is a polygon.

The second approximation $h_2(D)$ to D may not be quite as close homeomorphically to D as is $h_1(D)$ but it will still be close. However, it will have the advantage that the components of $S \cdot h_2(D)$ will have diameters much smaller than ϵ_2 (which is an upper bound on the diameters of $E_1, E_2, \cdots, E_n$) and simple closed curves in Int S near the components of $S \cdot h_2(D)$ can be shrunk to points in Int S without hitting $h_1(K_1)$.

Let ϵ_3 be a very small positive number selected in a fashion to be described later and S' be a polyhedral 2-sphere which is homeomorphically within ϵ_3 of S and which contains a finite collection of mutually exclusive ϵ_3 disks such that S' minus the sum of the interiors of these disks is contained in Int S. We select ϵ_3 so that $S' \cdot h_1(D) \subset \sum \text{Int } E_i$ and suppose that $S' \cdot h_1(D)$ is the sum of a finite collection of mutually exclusive simple closed curves $J_1, J_2, \cdots, J_r$.

The ϵ_2 and ϵ_3 were selected so that each J_i bounds a disk F_i on S' of small diameter. We suppose that these disks F_i are ordered by size with the small ones first so that no F_i contains an F_{i+j}.

The disk in $h_1(D)$ bounded by J_1 is first replaced by F_1 and then pushed slightly to one side of S' so as to reduce the number of components with the intersection with S'. The process is continued by replacing disks in the adjusted $h_1(D)$ by F_i's and then pushing slightly so as to get a polyhedral disk $h_2(D)$ which is close to D homeomorphically and which lies on Int S'. We select h_2 so that it agrees with h_1 in a neighborhood of Bd D and such that the components of $S \cdot h_2(D)$ are not much bigger than those of $S' \cdot h_1(D)$. By selecting ϵ_3 very small, we can insure that the components of $h_2(D) \cdot S$ are very small—in fact of diameter less than some preselected positive number ϵ_4.

Although we could have chosen the sum of the $h_2(D)$'s to be a 2-sphere, we did not insist on this since at the third approximation of D, there seems to be no easy way to prevent the approximating disks from intersecting at interior points of each.

d. **Third approximation to** D. Since each component of $S \cdot h_2(D)$ is of diameter less than ϵ_4, $h_2(\text{Int } D)$ contains a collection of mutually exclusive disks $E_1', E_2', \cdots, E_m'$ which cover $S \cdot h_2(D)$ such that each Bd E_i' lies close to S and is of diameter less than ϵ_4. Each Bd E_i' lies in Int S since Bd D does.

Each Bd E_i' is of such small diameter that it can be shrunk to a point on a small subset of Int S where ϵ_4 and ϵ_3 have been selected so that this subset will not intersect $h_1(K_1) = h_2(K_1)$. Hence there is a map g of $h_2(D)$ into Int S

such that $gh_2(D)$ lies close to $h_2(D)$, and g is the identity in a neighborhood of $h_2(\text{Bd } D)$, $gh_2(D)$ intersects $h_2(K_1)$ only in $h_2(\text{Bd } D)$, and $gh_2(D)$ has no singular points near $gh_2(\text{Bd } D)$. It follows from Dehn's lemma as proved by Papakyriokopoulos [11] that for each neighborhood U of the set of singular points of $gh_2(D)$, there is a homeomorphism h_3 of D onto a polyhedral disk $h_3(D)$ in $gh_2(D)+U$ that agrees with h_2 in a neighborhood of Bd D. The third approximation to D is $h_3(D)$. The advantage it has over $h_2(D)$ is that it lies in Int S. The only thing that makes $h_3(D)$ and D homeomorphically close is that each is of small diameter and their boundaries are homeomorphically close.

e. **The fourth approximation to D.** Our task is now to untangle the $h_3(D)$'s so that their sum forms a 2-sphere in Int S. We recall that the collection of 2-cells of T is the sum of three subcollections A_1, A_2, A_3 so that no two elements of any A_i have a point in common. We will have enlarged the $h_1(D)$'s so little as we changed them to $h_2(D)$'s and then to $h_3(D)$'s that two $h_3(D)$'s will not intersect if the corresponding D's do not intersect.

For each element D of A_1, the fourth approximation $h(D)$ to D is $h_3(D)$.

We may suppose that for each element D of A_1, the intersection of $h_3(\text{Int } D)=h(\text{Int } D)$ and the sum of the images under h_3 of the elements of A_2 is the sum of a finite collection of mutually exclusive simple closed curves $J_1, J_2, \cdots, J_s$ such that J_i bounds a disk G_i in Int $h(D)$ and the J_i's are ordered according to the size of the disks G_i they bound in $h(D)$, with the small ones coming first.

If J_1 lies in an $h_3(D')$ for an element D' of A_2, we replace the disk in $h_3(D')$ by G_1 and shove this replaced disk slightly to one side of $h_3(D)$. This process is continued until for each element D' of A_2, there is an h defined on D' so that $h(D)$ and the $h(D')$'s fit together only along their boundaries as they should.

Finally we turn to the elements of A_3. We suppose that for each element D of A_1+A_2, $h(\text{Int } D)$ intersects the sum of the images under h_3 of the elements of A_3 in the sum of a collection of mutually exclusive simple closed curves. These simple closed curves are eliminated one by one, starting at the inside in a manner already described. For each element D' of A_3, the resulting adjustment of $h_3(D')$ is called $h(D')$.

The sum of the $h(D)$'s is a 2-sphere in Int S. We note that $h=h_1=h_2=h_3$ on K_1. We will cause h to be near the identity by picking the ϵ_i's so that the D's and the $h(D)$'s are small and h is near the identity on K_1.

f. **Epsilontics.** In this section we explain the sizes for ϵ_1, ϵ_2, ϵ_3, ϵ_4 and the reasons for these selections. We recall that

ϵ_1 limits the mesh of T,

ϵ_2 limits h_1 to be near the identity and limits the sizes of the E_i's,

ϵ_3 limits S' to be near S and limits the sizes of the disks in S' such that S' minus these disks lies in Int S, and

ϵ_4 limits the sizes of the components of $h_2(D)\cdot S$.

We choose

$$\epsilon_1 < \epsilon/8,$$

where ϵ is the number mentioned in the statement of Theorem 1. This is the only restriction we place on ϵ_1.

Let δ_1 be a positive number so small that the distance between two elements of T without a common point is more than δ_1.

Our goal is to choose ϵ_2, ϵ_3, ϵ_4 so that for each 2-cell D of T, the following conditions are satisfied.

$$h_1(D) \text{ lies in a } \delta_1/6 \text{ neighborhood of } D.$$

$$h_2(D) \text{ lies in a } \delta_1/6 \text{ neighborhood of } h_1(D).$$

$$h_3(D) \text{ lies in a } \delta_1/6 \text{ neighborhood of } h_2(D).$$

This will insure that if D_1, D_2 are two 2-cells of T without a common point, then $h_3(D_1)\cdot h_3(D_2)=0$.

Since each point of $h_3(D)$ lies within $3\delta_1/6$ of a point of D and $\delta_1<\epsilon_1$,

$$\text{diameter } h_3(D) < \epsilon_1 + \delta_1 < 2\epsilon_1.$$

Since $h=h_3$ on elements of A_1,

$$\text{diameter } h(D) < 2\epsilon_1 \text{ if } D \text{ is an element of } A_1.$$

In changing from h_3 to h on elements D of A_2, the adjustment was made so that $h(D)$ lies within $2\epsilon_1$ of $h_3(D)$. Hence,

$$\text{diameter } h(D) < 6\epsilon_1 \text{ if } D \text{ is an element of } A_2.$$

For each element D of A_3,

$$h(D) \text{ lies within } 6\epsilon_1 \text{ of } h_3(D).$$

Since each point of $h(D)$ lies within $6\epsilon_1$ of a point of $h_3(D)$, this point in turn lies within $3\delta_1/6$ of a point of D, and diameter $D<\epsilon_1$, we find that h moves no point as much as $8\epsilon_1$. This is the reason we selected $\epsilon_1<\epsilon/8$.

Let ϵ_2 be a positive number so small that each subset of S of diameter $3\epsilon_2$ lies in a disk on S of diameter less than $\delta_1/10$. We note that

$$\epsilon_2 < \delta_1/30 < \delta_1/6.$$

Since $\epsilon_2<\delta_1/6$, for each 2-cell D of T, $h_1(D)$ lies in a $\delta_1/6$ neighborhood of D. The more stringent condition that $\epsilon_2<\delta_1/30$ is used later to help insure that $h_2(D)$ lies in a $\delta_1/6$ neighborhood of $h_1(D)$.

Let δ_2 be a positive number so small that for each 2-cell D of T, δ_2 is less than the distance between S and $h_1(D)-\sum E_i$. We note that

$$\delta_2 < \epsilon_2.$$

Although we shall place more stringent conditions on ϵ_3, we first consider the sizes of the components of $S'\cdot h_1(D)$ if we merely suppose

$$\epsilon_3 < \delta_2.$$

Let X be a component of $S'\cdot h_1(D)$ and h' be a homeomorphism of S onto S' that moves no point more than ϵ_3. Then X lies in an E_i (which is of diameter less than ϵ_2), $h'^{-1}(X)$ is of diameter less than $\epsilon_2+2\epsilon_3<3\epsilon_2$, $h'^{-1}(X)$ lies in a disk of diameter less than $\delta_1/10$, and the image of this disk under h' is a disk in S' of diameter less than $\delta_1/10+2\epsilon_3<\delta_1/10+2\delta_1/30=\delta_1/6$. Hence, the restrictions we have placed on ϵ_2, ϵ_3 are enough to insure that we can select h_2 so that $h_2(D)$ lies in a $\delta_1/6$ neighborhood of $h_1(D)$.

Let ϵ_4 be a positive number so small that each simple closed curve in E^3-S of diameter less than ϵ_4 can be shrunk to a point on a subset of E^3-S of diameter less than the minimum of $\delta_1/6$ and $\delta_2/2$. Note that ϵ_4 does not depend on ϵ_3. The Bd E_i''s are selected to lie within $\epsilon_4<\delta_2/2$ of S. Also, the Bd E_i''s have diameters less than ϵ_4 so that $gh_2(D)$ intersects $h_2(K_1)$ only in $h_2(\text{Bd } D)$. Furthermore, $h_3(D)$ lies in a $\delta_1/6$ neighborhood of $h_2(D)$.

The final restriction we place on ϵ_3 is to insure that each component of $S\cdot h_2(D)$ is of diameter less than ϵ_4. Let δ_3 be a number so small that each δ_3 subset of S lies in a disk in S of diameter less than ϵ_4. We suppose

$$\epsilon_3 < \delta_3.$$

Each component X of $S\cdot h_2(D)$ lies on Int S' and is separated from the big component of $S-S'$ by a disk in S' of diameter less than $\epsilon_3<\delta_3$. Hence, diameter $X<\epsilon_4$.

3. **Conditions under which a 2-sphere is tame.** Tame 2-spheres in E^3 have uniformly locally simply connected complements. A converse is as follows.

THEOREM 2. *A 2-sphere S in E^3 is tame if each component of E^3-S is* 1-ULC.

Proof. It follows from Theorem 1 that for each positive number ϵ there is a 2-sphere S' in Int S such that

$$H(S, S') < \epsilon.$$

Similarly, it follows that there is a 2-sphere S'' in Ext S such that

$$H(S, S'') < \epsilon.$$

That S is tame then follows from Theorem 0.

COROLLARY. *A 2-sphere S in E^3 is tame if E^3-S is locally simply connected at each point of S.*

4. **Retaining simple connectivity.** We shall be applying the Approximation Theorem to a 2-manifold to make part of the 2-manifold locally poly-

hedral. We are interested in seeing what this does to the local simple connectivity of the complement. First we examine the effect of throwing away part of the 2-manifold.

THEOREM 3. *Suppose M is a 2-manifold in E^3, D is a disk in M, and p is a point of D at which E^3-M is locally simply connected. Then E^3-D is locally simply connected at p.*

Proof. We only consider the case where $p \in \text{Bd } D$. Suppose U is a given neighborhood of p. Let V' be a neighborhood of p such that each map of a circle into $V' \cdot (E^3-M)$ can be shrunk to a point in $U \cdot (E^3-M)$ and V be a neighborhood of p such that each pair of points of $V \cdot (M-D)$ lie in an arc in $V' \cdot (M-D)$. We show that if E is a plane circular disk and f is a map of Bd E into $V \cdot (E^3-D)$, then there is a map of E into $U \cdot (E^3-D)$ that agrees with f on Bd E.

We want to simplify f so that $M \cdot f(\text{Bd } E)$ does not have infinitely many components. Suppose aa' is an arc on Bd E and bb' is an arc on $V \cdot (M-D)$ such that $f(a)=b$, $f(a')=b'$, and $bb'+f(aa')$ lies in a convex subset of $V-D$. Then there is a homotopy F_t on aa' such that $F_0=f$, F_t is constant on a and a', $F_1(aa')=bb'$, and each $F_t(aa')$ lies in $V-D$. Hence, we suppose with no loss of generality that $f^{-1}(M \cdot f(\text{Bd } E))$ is the sum of a finite number of arcs $a_1a_2, a_3a_4, \cdots, a_{2n-1}a_{2n}$ on Bd E ordered so that there is no a_j between any a_i and a_{i+1}.

Let q be the center of E and $f(q)$ be any point of $V \cdot (M-D)$. Extend f to map the radius qa_i of E onto an arc in $V' \cdot (M-D)$ from $f(q)$ to $f(a_i)$. Since each component of M divides E^3 into two pieces and any arc in M can be approximated by arcs on either side of M, the map f on the boundary of the sector a_iqa_{i+1} of E can be extended to map an annulus ring in the sector one of whose boundary components is the boundary of the sector into $V' \cdot (E^3-D)$ so that the image of the other boundary component of the annulus misses M. The map f can be further extended to take the rest of the sector into $U \cdot (E^3-M)$.

THEOREM 4. *Suppose M is a 2-manifold in E^3, D is a disk in M, p is a point of M at which E^3-M is locally simply connected, and h is a homeomorphism of M into E^3 such that h is the identity on D and $h(M)$ is locally polyhedral at each point of $h(M)-D$. Then $E^3-h(M)$ is locally simply connected at p.*

Proof. The only case we consider is the one in which p is a point of Bd D. Suppose U is a given neighborhood of p. Let V' be a neighborhood of p such that each closed set in $V' \cdot (h(M)-D)$ lies in a disk in $U \cdot (h(M)-D)$. It follows from Theorem 3 that there is a neighborhood V of p such that each map of a circle into $V \cdot (E^3-D)$ can be shrunk to a point in $V' \cdot (E^3-D)$. We show that if E is a disk and f is a map of Bd E into $V\cdot(E^3-h(M))$, then f can be extended to map E into $U\cdot(E^3-h(M))$.

Let f' be a map of E into $V'\cdot(E^3-D)$ that is an extension of f on Bd E. Let E' be the component of $E-f'^{-1}(h(M)\cdot f(E))$ containing Bd E. With no loss of generality we suppose that E' is E minus a finite collection of mutually exclusive subdisks $E_1, E_2, \cdots, E_n$ of E.

Since $f'(\text{Bd } E_i)$ lies on the interior of a polyhedral disk in $U\cdot(h(M)-D)$, it is possible to adjust f' on disks in E slightly larger than the E_i's so as to take the larger disks slightly to one side of $h(M)$. The adjusted f' is f and takes E into $U\cdot(E^3-h(M))$.

QUESTION. Would Theorem 4 be true if we supposed that D were merely a closed subset of M with only nondegenerate components rather than actually a disk in M?

5. **Enlarging a disk to a 2-sphere.** Not each disk in E^3 lies on a 2-sphere. An example of such a disk is obtained by taking a horizontal disk D in E^3; removing two circular holes from Int D; adding tubes from the holes, one tube going up and the other down and around; and finally adding hooked disks as shown in [1]. The disk does not lie on a 2-sphere since its boundary cannot be shrunk to a point in the complement of the disk. Although it does not lie on a 2-sphere, it does lie on a torus as was pointed out to me by David Gillman. If instead of removing a pair of holes from the horizontal disk and replacing the holes with hooked wild disks, one had removed an infinite collection of pairs of holes converging to a boundary point of D and replaced each pair of holes with wild disks hooked over the boundary of D, there would have resulted a wild disk in E^3 that does not lie on any 2-manifold in E^3.

The following result shows that each disk contains many small disks each of which lies on a 2-sphere.

THEOREM 5. *Suppose M is a 2-manifold in E^3, p is a point of M, and U is a neighborhood of p. Then there is a disk D in $M\cdot U$ and a 2-sphere S in U such that $p\in\text{Int } D\subset S$ and S is locally polyhedral at each point of $S-D$.*

Proof. Let E be a disk in M such that $p\in\text{Int } E$ and C be a cube in U whose interior contains p and whose exterior contains Bd E. Let D be a disk in $M\cdot\text{Int } C$ such that $p\in\text{Int } D$. It follows from the Approximation Theorem that there is a homeomorphism h of E into E^3 such that h is the identity on D, h takes Bd E into Ext C, and $h(E)$ is locally polyhedral at each point of $h(E)-D$. We suppose with no loss of generality that Bd $C\cdot h(E)$ is the sum of a finite collection of mutually exclusive polygons.

Let E' be the component of $h(E)-\text{Bd } C$ containing D. It is topologically equivalent to a 2-sphere minus the sum of a finite collection of mutually exclusive disks. By adding polygonal disks in $U\cdot(C+\text{Ext } C)$ to E', one obtains the required 2-sphere S.

6. **Locally tame subsets of 2-manifolds.** A 2-sphere S in E^3 may not be

locally tame at a point p even if E^3-S is locally simply connected at p as shown by the following example.

EXAMPLE. Consider a spherical 2-sphere S' and a sequence of mutually exclusive spherical disks $E_1, E_2, \cdots$ in S' converging to a point p of S'. Fox and Artin have described [8] a wild arc which is locally polyhedral except at one end point. For each E_i let A_i be such an arc reaching out from the center of E_i such that the arc is of diameter less than the radius of E_i and such that A_i intersects S only at the polyhedral end of A_i. By replacing each E_i in S by a disk obtained by swelling up A_i as done in [7] one can obtain a 2-sphere S such that E^3-S is locally simply connected at p even though S is not locally tame at p.

THEOREM 6. *Suppose M_2 is a 2-manifold embedded in a 3-manifold M_3 and U is an open subset of M_2 such that M_3-M_2 is locally simply connected at each point of U. Then M_2 is locally tame at each point of U.*

Proof. Since local tameness is only a local property, we suppose that M_3 is E^3 and U is all of M_2. If this were not the case already we would take a homeomorphism h of a neighborhood of p in M_3 into E^3 such that this neighborhood did not intersect M_2-U.

It follows from Theorem 5 that there is a disk D in M_2 and a 2-sphere S such that $p\in \text{Int } D\subset S$ and S is locally polyhedral at each point of $S-D$. It follows from Theorem 4 that E^3-S is locally simply connected at each point of S. Since S is compact, E^3-S is 1-ULC. Theorem 2 implies that S is tame. Since S is tame, M_2 is locally tame at p.

Since a closed set in a 3-manifold is tame if it is locally tame [2; 10] we have the following result.

THEOREM 7. *A 2-manifold M_2 in a triangulated 3-manifold M_3 is tame if and only if M_3-M_2 is locally simply connected at each point of M_2.*

COROLLARY. *A 2-manifold M_2 in a 3-manifold M_3 is tame if M_3-M_2 is* 1-ULC.

7. Tame 2-manifolds-with-boundaries.

In this section we extend our results about 2-manifolds to 2-manifolds-with-boundaries.

THEOREM 8. *Suppose M_2 is a 2-manifold-with-boundary embedded in a 3-manifold M_3 and U is an open subset of M_2 such that M_3-M_2 is locally simply connected at each point of U. Then M_2 is locally tame at each point of U.*

Proof. Since we only look at M_3 locally, we suppose that M_3 is E^3 and U is M_2. Since Theorem 6 takes care of points of Int M_2, we only show that M_2 is locally tame at a point p of Bd M_2.

Let D be a disk in M_2 such that Bd $D\cdot$Bd M_2 is an arc containing p as a non end point. An argument like that used in the proof of Theorem 3 shows that E^3-D is locally simply connected at each point of D. We finish the proof of Theorem 8 by showing that D is tame.

Since Theorem 6 shows that D is locally tame at each point of Int D, it follows from [4; 9] that there is a homeomorphism h of E^3 onto itself such that $h(D)$ is locally polyhedral at each point of $h(\text{Int } D)$. Hence, we suppose with no loss of generality that D is locally polyhedral at each point of Int D.

By pushing D to one side at points of Int D, one obtains a disk D' such that Bd $D=$ Bd D', D' is locally polyhedral at points of Int D', and $D+D'$ bounds a topological cube C. Since E^3-Bd C is locally simply connected at each point of Bd C, it follows from either Theorem 2 or 6 that Bd C is tame. Hence D is tame and M_2 is locally tame at p.

8. **Extensions of Theorem** 0. As pointed out in [5], we can use Theorem 6 to extend Theorem 0 as follows.

THEOREM 9. *A 2-manifold M_2 in a 3-manifold M_3 is locally tame at a point p of M_2 if there is a disk D with $p \in \text{Int } D \subset M_2$ such that for each positive number ϵ, there are disks D', D'' on opposite sides of M_2 such that*

$$H(D, D') < \epsilon, \qquad H(D, D'') < \epsilon.$$

Proof. Theorem 9 follows from Theorem 6 when we show that M_3-M_2 is locally simply connected at each point of Int D.

Suppose E is a disk, q is a point of Int D, and f is a map of Bd E into a small subset of M_3-M_2 near q. Suppose f is extended to map E into a small subset of M_3. Let D', D'' be disks on opposite sides of M_2 which are homeomorphically close to D and whose sum separates $f(\text{Bd } E)$ from $M_2 \cdot f(E)$ in $f(E)$. If E' is the component of $E-f^{-1}(D'+D'')$ containing Bd E, f on E' can be extended to map E into a small subset of $f(E')+D'+D''$. Hence, M_3-M_2 is locally simply connected at q.

Since locally tame closed subsets are tame in triangulated 3-manifolds [2; 10], we have the following as a corollary of Theorem 9.

COROLLARY. *A 2-manifold M_2 in a triangulated 3-manifold M_3 is tame if and only if for each positive number ϵ the appropriate one of the following conditions is satisfied:*

CASE 1. *If M_2 is two sided in some neighborhood N of M_2 there are 2-manifolds M', M'' on opposite sides of M_2 in N such that*

$$H(M_2, M') < \epsilon, \qquad H(M_2, M'') < \epsilon.$$

CASE 2. *If M_2 is one sided in each neighborhood of M_2 there is a connected double covering M' of M_2 with projection map π and a homeomorphism h' of M' into M_3-M_2 such that*

$$p(h(x), \pi(x)) < \epsilon, \qquad x \in M'.$$

Neither Theorem 9 nor its corollary can be extended to 2-manifolds-with-boundaries. Besides having to speak with care about the two sides of a 2-manifold-with-boundary in a 3-manifold, one would have to contend with the example of Stallings [12] in which he describes an uncountable family of mutually exclusive wild disks in E^3. It would follow from an application of Theorem 9 that most of these disks are locally tame except on their boundaries.

References

1. J. W. Alexander, *An example of a simply connected surface bounding a region which is not simply connected*, Proc. Nat. Acad. Sci. U.S.A. vol. 10 (1924) pp. 8–10.

2. R. H. Bing, *Locally tame sets are tame*, Ann. of Math. vol. 59 (1954) pp. 145–158.

3. ———, *Approximating surfaces with polyhedral ones*, Ann. of Math. vol. 61 (1957) pp. 456–483.

4. ———, *An alternative proof that 3-manifolds can be triangulated*, Ann. of Math. vol. 69 (1959) pp. 37–65.

5. ———, *Conditions under which a surface in E^3 is tame*, Fund. Math. vol. 47 (1959) pp. 105–139.

6. ———, *A wild sphere each of whose arcs is tame*, Duke Math. J.

7. ———, *Side approximations of 2-spheres*, submitted to Annals of Math.

8. R. H. Fox and E. Artin, *Some wild cells and spheres in three-dimensional space*, Ann. of Math. vol. 49 (1948) pp. 979–990.

9. E. E. Moise, *Affine structures in 3-manifolds*. IV. *Piecewise linear approximations of homeomorphisms*, Ann. of Math. vol. 55 (1952) pp. 215–222.

10. ———, *Affine structures in 3-manifolds*, VIII. *Invariance of the knot-type; local tame imbedding*, Ann. of Math. vol. 59 (1954) pp. 159–170.

11. C. D. Papakyriokopoulos, *On Dehn's lemma and the asphericity of knots*, Ann. of Math. vol. 66 (1957) pp. 1–26.

12. J. R. Stallings, *Uncountably many wild disks*, Ann. of Math. vol. 71 (1960) pp. 185–186.

University of Wisconsin,
Madison, Wisconsin

EACH DISK IN E^3 CONTAINS A TAME ARC.*

By R. H. BING.[1]

An arc ab in E^3 is tame if there is a homeomorphim h of E^3 onto itself such that $h(ab)$ is a polygonal arc. The purpose of this note is to show that each topological disk in E^3 contains many tame arcs.

A 2-sphere S is tame if there is a homeomorphism of E^3 onto itself taking S onto a polyhedral 2-sphere. It is known that each arc on a tame 2-sphere is tame. We show that each disk contains tame arcs by showing that it shares arcs with a tame 2-sphere.

We recall that a Sierpinski curve is obtained by removing from a 2-sphere the interiors of a null sequence of mutually exclusive disks such that the sum of the disks is dense in the 2-sphere. (A null sequence is one the diameter of whose elements converge to zero.) Any set topologically equivalent to such a Sierpinski curve is called a Sierpinski curve. It may be shown that any two Sierpinski curves are homeomorphic.

One of the principal theorems of this note is the following.

THEOREM 1. *For each 2-sphere S in E^3 and each positive number ϵ there is a Sierpinski curve X in S such that*

1. *X lies in a tame 2-sphere, and*
2. *each component of $S - X$ is of diameter less than ϵ.*

Listing the above theorem first gives direction to this note. We digress to give some definitions and other theorems. These other theorems point to the proof of Theorem 1 and that proof is given after Theorem 7.

If A, B are homeomorphic sets we say

$$H(A,B) \leqq \epsilon$$

if there is a homeomorphism of A onto B that moves no point by more than ϵ.

We denote the distance function for a metric space by ρ. If f_1, f_2 are two mappings of a set A into a bounded metric space B, we use

$$\rho(f_1, f_2) = \text{least upper bound } \rho(f_1(x), f_2(x)), \quad x \in A.$$

An ϵ set is a set of diameter no more than ϵ.

---

* Received May 8, 1962.

[1] Work on this paper was supported by contract NSF-G11665.

We use interior in two senses but this seems unlikely to cause confusion. If S is a 2-sphere in E^3, we use $\operatorname{Int} S$ and $\operatorname{Ext} S$ to denote the bounded and unbounded components respectively of $E^3 - S$. If D is a disk, we let $\operatorname{Int} D$ be the difference of D and its combinatorial boundary.

If S is a 2-sphere in E^3 and ϵ is a positive number, we say that S can be ϵ approximated from both sides if there are 2-spheres S', S'' in $\operatorname{Int} S$, $\operatorname{Ext} S$ respectively such that $H(S, S') < \epsilon$, $H(S, S'') < \epsilon$.

Theorem 2 provides a clue as to how we show the 2-sphere mentioned in Condition 1 of Theorem 1 is tame. Theorem 2 is lifted from [2].

THEOREM 2. *A 2-sphere S in E^3 is tame if for each positive number ϵ, S can be ϵ approximated from both sides.*

If S is a 2-sphere, U is a component of $E^3 - S$, and $\epsilon > 0$, we say that *S can be ϵ almost approximated from U* provided there is a 2-sphere S' such that

1. $H(S, S') < \epsilon$ and
2. each component of $S' - U$ is an ϵ set.

We say that S can be almost approximated from both sides if for each positive number ϵ, it can be ϵ almost approximated from each of its complementary domains.

We make use of the following theorem lifted from [3]. It is proved by methods similar to those employed in [1].

THEOREM 3. *Each 2-sphere in E^3 can be almost approximated from both sides.*

The tame 2-sphere mentioned in Condition 1 of Theorem 1 is obtained by a sequence of adjustments of the 2-sphere S of Theorem 1. Iterated use of an extension of the following result is made in getting these adjustments.

THEOREM 4. *For each 2-sphere S in E^3 and each positive number ϵ there is a 2-sphere S' such that*

1. *S' is obtained by removing a finite collection of mutually exclusive ϵ disks from S and replacing them with other ϵ disks, and*
2. *S' can be ϵ approximated from its interior.*

Proof of Theorem 4. We give the proof of Theorem 4 in seven steps.

1. Let δ be a positive number such that each δ set in S lies in a disk in S of diameter less than $\epsilon/6$.

2. It follows from Theorem 3 that there is a polyhedral 2-sphere S'' such that S'' contains a finite collection of mutually exclusive δ disks $D_1, D_2, \cdots, D_n$ such that

$$S'' - \sum \operatorname{Int} D_i \subset \operatorname{Int} S \text{ and } H(S, S'') \leqq \epsilon/6.$$

We note that each component of $S \cdot S''$ is a δ set.

3. We suppose that diameter $S > 2\epsilon$ and that there is a point p of $\operatorname{Int} S$ whose distance from S is more than ϵ. Let $E_1, E_2, \cdots, E_m$ be a finite collection of mutually exclusive $\epsilon/6$ disks in S such that

$$S \cdot S'' \subset \sum \operatorname{Int} E_j.$$

To see that there are such disks, consider the decomposition space whose nondegenerate elements are components of $S \cdot S''$. Then consider the big 2-sphere of the resulting cactoid [5]. A 0-dimensional set on this 2-sphere comes from $S \cdot S''$ and this closed 0-dimensional set can be covered with a finite collection of interiors of mutually exclusive very small disks on the 2-sphere. The boundaries of these disks in the 2-sphere of the cactoid correspond to the boundaries of the E_i's in S. The above reasoning is given in more expanded form in Theorem 9 of [3].

We note that

$$S - \sum E_i \subset \operatorname{Ext} S''$$

To see that this true, let qp be an arc which misses $\sum D_i + \sum E_i$ such that $q \in \operatorname{Ext} S$, $p \in \operatorname{Int} S$, and $\rho(\{p\} + \{q\}, S) > \epsilon$. It follows from Theorem 14 of [3] that there is such an arc. The first point of $(S - \sum E_i) + (S'' - \sum D_i)$ on qp in the order from q to p is a point of S since $S'' - \sum D_i \subset \operatorname{Int} S$.

4. It follows from Theorem 7 of [1] that each E_i can be replaced by a disk F_i such that

F_i is locally polyhedral at each point of $\operatorname{Int} F_i$,
diameter $F_i < \epsilon/4$, and
$(S - \sum E_i) + \sum F_i = S_1$ is a 2-sphere that contains $S'' - \sum \operatorname{Int} D_i$ on its interior.

5. With no loss of generality we suppose that each D_i and each F_j are in general position in the sense that each component of $D_i \cdot F_j$ is a polygonal simple closed curve. Such general position may be achieved by insuring that no four vertices of the triangulations of the D_i's and the F_j's are coplanar. Let U be the component of $S_1 \cdot \operatorname{Ext} S''$ containing $S - \sum E_i$. We note that

it is obtained by subtracting a finite collection of mutually exclusive disks from S_1. Each of these subtracted disks has its boundary on a D_i.

6. The 2-sphere S' is obtained as follows. Each component of $S_1 - U$ is a disk with its boundary on a D_i. Enlarge this disk slightly so that its boundary no longer intersects this D_i and replace the enlarged disk by an $\epsilon/6$ disk near this D_i and in $\operatorname{Ext} S''$. The holes in U may thus be enlarged and filled in one at a time. So that the resulting "fill-ins" will not intersect each other it may be convenient to plug up the small holes first by starting with a component J of $\operatorname{Bd} U$ on a D_i such that $\operatorname{Int} J$ on D_i does not intersect $\bar{U}$. Each hole in U may be thus treated so as to obtain a 2-sphere S' so that

$$S'' \subset \operatorname{Int} S', \text{ and } S - \sum E_i \subset S'.$$

Each component of $S' - (S - \sum E_i)$ is of diameter less than $\epsilon/4 + 2\epsilon/6 = 7\epsilon/12$.

7. Since the disks E_i have diameter less than $\epsilon/5$ and the disks of $S' - (S - \sum \operatorname{Int} E_i)$ have diameter less than $7\epsilon/12$,

$$H(S, S') < \epsilon/5 + 7\epsilon/12 < 5\epsilon/6.$$

This result with the inequality $H(S, S'') < \epsilon/6$ implies that

$$H(S', S'') < \epsilon.$$

The following is a slight extension of Theorem 4.

THEOREM 5. *For each 2-sphere S and each positive number ϵ there is a 2-sphere S' such that*

1. *S' is obtained by subtracting a finite collection of mutually exclusive ϵ disks from S and replacing them with ϵ disks, and*

2. *S' can be ϵ approximated from its interior and its exterior.*

Proof of Theorem 5. Apply Theorem 4 to get a 2-sphere S_1 such that

1. S_1 is obtained by subtracting a finite collection G of $\epsilon/4$ disks from S and replacing them with $\epsilon/4$ disks, and

2. there is a 2-sphere S'' on the interior of S_1 such that $H(S_1, S'') < \epsilon/4$.

Pick a small δ. (We indicate how small after the following paragraph.)

Reapply Theorem 4 (replacing $\operatorname{Int} S$ of the statement of that theorem by $\operatorname{Ext} S_1$) to get a 2-sphere S' such that

1. S' is obtained by subtracting a finite number of mutually exclusive δ disks from S_1 and replacing them with δ disks, and

2. there is a 2-sphere S''' on Ext S' such that $H(S', S''') < \delta$.

If we have picked δ small enough, S' is the required 2-sphere.

Finally, we indicate how small to take δ. We put on 4 restrictions.

Restriction 1. We take $\delta < \epsilon/4$. (We make more stringent restrictions later but this makes for an easy start.) This insures that $H(S_1, S') < \epsilon/2$ and $H(S', S'') < \epsilon$.

Restriction 2. We take δ so small that S'' lies on the interior of any 2δ approximation of S_1. This insures that $S'' \subset \operatorname{Int} S'$.

Restriction 3. We take δ so small that if A is a closed set in S each component of which is a δ set, then there is a finite collection H of mutually exclusive $\epsilon/3$ disks in S such that each point of A lies on the interior of one of these disks as does each element of G. We let A be the intersection of S and the δ disks of S_i which are given by the second application of Theorem 4. The resulting collection H is the collection of disks in S to be replaced as suggested in the first condition of the statement of Theorem 5.

Restriction 4. Finally we take $\delta < \epsilon/12$. The purpose of this restriction is to insure that the disks in S' replacing the elements of H have diameters less than ϵ. Since each element of H is of diameter less than $\epsilon/3$ and the disks in S_1 replacing elements of G have diameters less than $\epsilon/4$, the small disks in S_1 sharing boundaries with disks of H have diameters less than $\epsilon/3 + \epsilon/4 + \epsilon/4 = 5\epsilon/6$. Since the disks in S' replacing disks in S_1 have diameters less than δ, the small disks of S' sharing boundaries with disks of H have diameters less than $5\epsilon/6 + 2\delta$, so if $\delta < \epsilon/12$, the disks in S' replacing the disks of H have diameters less than ϵ.

Another theorem we mention before proving Theorem 1 is the following.

THEOREM 6. *Suppose $\epsilon > 0$ and $X_1, X_2, \cdots$ is a sequence of compact sets in a metric space such that*

1. *each X_i has only a finite number of components,*
2. *each component of X_1 is an $\epsilon/2$ set, and*
3. *each component of X_{i+1} $(i \geqq 1)$ is of diameter less than $e/2^{i+1}$ and $1/2^{i+1}$ times the distance between any two components of*

$$X_1 + X_2 + \cdots + X_i.$$

Then there is a countable collection of mutually exclusive closed connected sets $A_1, A_2, \cdots$ such that

1. *each* A_k *is of diameter less than* ϵ,

2. $\lim (i\to\infty)$ *diameter* $A_i = 0$, *and*

3. *each component of each* X_i *lies in some* A_j.

For each point p in X_i, let $C(j,p)$ $(j > i)$ be the component of $X_1 + X_2 + \cdots + X_j$ containing p. The A_k containing p is the closure of $C(i,p) + C(i+1,p) + \cdots$. The proof that these A's satisfy the conclusion of the theorem is straightforward and we do not include it in this note.

We need the following result giving conditions under which the limit of a sequence of homeomorphisms is a homeomorphism

THEOREM 7. *Suppose* $f_1, f_2, \cdots$ *is a sequence of homeomorphisms of a compact metric space* X *into a complete metric space* Y *and* $\epsilon_1, \epsilon_2 \cdots$ *is a sequence of positive numbers such that* $\rho(f_i, f_{i+1}) < \epsilon_i$. *Then* $\lim f_i(X)$ *is a homeomorphism if* $\sum_j^\infty \epsilon_i$ *is a positive number so small that if* $\rho(x_1, x_2) > 1/j$ $(x_1, x_2 \in X)$, *then* $\rho(f_j(x_1), f_j(x_2)) > 2\sum_j^\infty \epsilon_i$.

Again the proof is straightforward and we do not include it.

We finally give the proof of Theorem 1.

Proof of Theorem 1. The tame 2-sphere that is to share a Sierpinski curve with S is the limit of a sequence of 2-spheres $S_1, S_2, \cdots$. We indicate restrictions to be placed on the S_i's to insure that

1. $\lim S_i$ is a 2-sphere,

2. $\lim S_i$ is tame, and

3. $\lim S_i$ shares a Sierpinski curve X with S.

Step 1. *Restrictions to make* $\lim S_i$ *a 2-sphere.* Each S_i is the image of S under a homeomorphism h_i where $\rho(h_{i+1}, h_i) < \epsilon_i$ and $\epsilon_1, \epsilon_2, \cdots$ is a sequence of positive numbers chosen according to the criteria of Theorem 7 to make $\lim h_i$ a homeomorphism. Although in later steps we place more restrictions on $\epsilon_1, \epsilon_2, \cdots$, it follows from Theorem 7 that $\lim h_i$ is a homeomorphism if the ϵ_i's satisfy the following conditions. Insist (along with other restrictions to be imposed later) that $\epsilon_1 < 1/6$. After h_1 is defined (so that it moves no point more than ϵ_1), pick a positive number δ_1 so small that if x_1, x_2 are two points of S_1 such that $\rho(x_1, x_2) > 1/2$, then $\rho(h_1(x_1), h_1(x_2) > \delta_1$. Insist (among other things) that ϵ_2 is less than either

$\epsilon_1/2$ or $\delta_1/6$. After h_2 is defined (so that $\rho(h_1, h_2) < \epsilon_2$), pick a positive number δ_2 such that if x_1, x_2 are points of S_1 such that $\rho(x_1, x_2) > 1/3$, then $\rho(h_2(x_1), h_2(x_2)) > \delta_2$. Insist that $\epsilon_3 < \epsilon_2/2$, $\epsilon_3 < \delta_2/6$. This procedure is continued. We note that $\sum \epsilon_i < 1/3$ and $\sum_{j+1}^{\infty} \epsilon_i < \delta_j/3$ where δ_j is so small that if $\rho(x_1, x_2) > 1/j$, then $\rho(h_j(x_1), h_j(x_2)) > \delta_j$. Hence

$$\rho(h_{j+1}(x_1), h_{j+1}(x_2)) > \delta_j - 2\epsilon_{j+1} > 2\delta_{j/3} > 2 \sum_{j+1}^{\infty} \epsilon_i.$$

Step 2. Restrictions to make $\lim S_i$ *tame.* Theorem 2 and Condition 2 of Theorem 5 provide the clues. We note from Theorem 5 that there is a 2-sphere S_1 and a homeomorphism h_1 of S onto S_1 such that h_1 moves no point by more than ϵ_1 and S_1 can be ϵ_1 approximated from $\operatorname{Int} S_1$ and $\operatorname{Ext} S_1$ by 2-spheres S_1', S_1'' respectively We pick $\sum_2^{\infty} \epsilon_i$ to be so small that S_1', S_1'' lie in the interior and exterior respectively of $\lim S_i = \lim h_i(S)$. In general, S_j is chosen so that it can be ϵ_j approximated by 2-spheres S_j', S_j'' in $\operatorname{Int} S_j$, $\operatorname{Ext} S_j$ respectively and $\sum_j^{\infty} \epsilon_i$ is chosen so small that S_j', S_j'' lie respectively in the interior and exterior of $\lim S_i = \lim h_i(S)$. Theorem 2 insures that $\lim S_i = \lim h_i(S)$ is tame.

Step 3. Restrictions to make $S \cdot \lim S_i$ *contain a Sierpinski curve* X. We use Theorem 6 and Condition 1 of Theorem 5. Let δ be a positive number such that if K is a δ set in S, $S - K$ has only one component of diameter as much as $\epsilon/2$. The number δ is to correspond to the ϵ of Theorem 6 so we pick $\epsilon_1 < \delta/2$. We find from Condition 1 of Theorem 5 that we can pick S_1 so that it is obtained by removing a finite collection H_1 of mutually exclusive ϵ_1 disks from S and replacing them with other ϵ_1 disks. The sum of the disks of H_1 corresponds to the set X_1 of Theorem 6.

We pick ϵ_2 less than either $\epsilon_1/4$ or $1/4$ the distance between any two elements of H_1 and choose S_2 so that it is obtained by removing a finite collection H_2 of mutually exclusive ϵ_2 disks from S_1 and replacing them with ϵ_2 disks. The set X_2 is a subset of S with only a finite number of components (each of diameter less than ϵ_2) such that X_2 contains the intersection of S with the sum of the elements of H_2. If this pattern is continued to get $S_3, S_4, \cdots$, Theorem 6 assures us that there is a null sequence $A_1, A_2, \cdots$, of mutually exclusive continua in S such that each A_j is a δ set and $S - \sum A_i$ lies in each S_i and hence in $S \cdot \lim S_i$.

The sum B_i of A_i and all small components of $S - A_i$ is an ϵ set. If one considers the upper semicontinuous decomposition G of S whose nondegenerate elements are the maximal B_i's, one finds from a theorem due to R. L. Moore [5] that the corresponding decomposition space is a 2-sphere. Since there are only a countable number of nondegenerate elements in G, one can find in this decomposition space a Sierpinski curve missing the points corresponding to the nondegenerate elements of G such that the corresponding Sierpinski curve X in S satisfies Condition 2 of Theorem 1.

The following result was announced at the International Congress of Mathematicians in Edinburgh in 1958. See Theorem 5 of [4].

THEOREM 8. *For each interior point p of each disk D in E^3 there is a 2-sphere S that contains an open subset of D containing p.*

A combination of Theorems 1 and 8 gives the following result.

THEOREM 9. *Each disk in E^3 contains many tame arcs.*

UNIVERSITY OF WISCONSIN.

REFERENCES.

[1] R. H. Bing, "Approximating surfaces with polyhedral ones," *Annals of Mathematics*, vol. 65 (1957), pp. 456-483.

[2] ———, "Conditions under which a surface in E^3 is tame," *Fundamenta Mathematicae*, vol. 47 (1959), pp. 105-139.

[3] ———, "Approximating surfaces from the side," *Annals of Mathematics*, vol. 77 (1963), in press.

[4] ———, "A surface is tame if its complement is 1-ULC," *Transactions of the American Mathematical Society*, vol. 101 (1961), pp. 456-483.

[5] R. L. Moore, "Concerning upper semicontinuous collections of continua," *Transactions of American Mathematical Society*, vol. 27 (1925), pp. 416-428.

[Reprinted from *American Journal of Mathematics*, Vol. LXXXIV, No. 4, October, 1962.]

EACH DISK IN E^3 IS PIERCED BY A TAME ARC.*

By R. H. Bing.[1]

1. Introduction. An arc ab in E^3 is *tame* if there is a homeomorphism of E^3 onto itself that takes ab onto a straight line interval. If ab lies in a plane in E^3, it is tame since if ab is an arc in a plane, there is a homeomorphism of the plane onto itself taking ab onto a straight line interval; this homeomorphism may be extended to E^3.

A disk is *pierced by* an arc ab at a point p if $p \in \operatorname{Int} D$, $p \in \operatorname{Int} ab$, p is the only point of $D \cdot ab$, and the arcs ap, pb abut on D from opposite sides in that there is an $\epsilon > 0$ such that no connected ϵ-set in $E^3 - D$ intersects both ap and pb. (An ϵ-set is a set of diameter less than or equal to ϵ.) If D lies on a 2-sphere which intersects ab only at p, the 2-sphere separates a from b in E^3 since each complementary domain of a 2-sphere in E^3 is uniformly locally connected. For convenience we state this result as a theorem.

THEOREM 1.1. *Suppose in E^3 that D is a disk that lies on a 2-sphere S and ab is an arc which intersects S only at a point p of* $\operatorname{Int} D$. *Then ab pierces D if and only if S separtes a from b.*

It is sometimes more convenient to study the interior and exterior of a 2-sphere in E^3 than to study the sides of disk. We show in Theorem 2.1 that this approach may be used since for each interior point p of a disk D in E^3, there is a disk D' in D such that $\operatorname{Int} D'$ contains p and lies on a 2-sphere.

Bing has given an example [1] of an arc in E^3 that pierces no disk. Fox and Artin have given an example [5] of a 2-sphere S with a point p such that no tame arc in E^3 pierces S at p. This 2-sphere S has something of a knotted feeler. By using a null sequence of knotted feelers one can construct a 2-sphere that cannot be pierced at any point of a countable dense set.

One might wonder if there is a disk so wild that it is pierced by no time arc. See the question on page 343 of [1]. One of the principal results

* Received May 8, 1962.

[1] Work on this paper was supported by contract NSF-G11665.

of this paper is to show in Theorem 5.3 that each disk in E^3 can be pierced by a tame arc.

In showing that each disk in E^3 can be pierced by a tame arc, we make extensive use of the fact (Theorem 1 of [4]) that each 2-sphere contains a Sierpinski curve X that lies on a tame 2-sphere. Section 3 of our paper is devoted to a study of Sierpinski curves.

2. Extending disks to 2-spheres. An arc A in the plane locally has two sides and we speak of arcs abutting on A either from the same side or from opposite sides. To aid us in studying the sides of A we may use either of two approaches. (1) We may use the fact that there is a homeomorphism of the plane onto itself taking A onto a horizontal segment and use the elementary notions of "above" and "below." (2) We may use the fact that A lies on a simple closed curve J and use the notion of the interior and exterior of J.

Neither of these approaches works with great ease in studying the sides of a disk in E^3. That the first fails may be seen by considering a wild disk D in E^3 such that there is no homeomorphism of E^3 onto itself taking D onto a polyhedral disk. The second is not completely successful since there are disks in E^3 which are not a subset of any 2-sphere. The disk shown in Figure 1 has a boundary which cannot be shrunk to a point in the complement of the disk. However, the second approach is partially successful as a result of the following theorem.

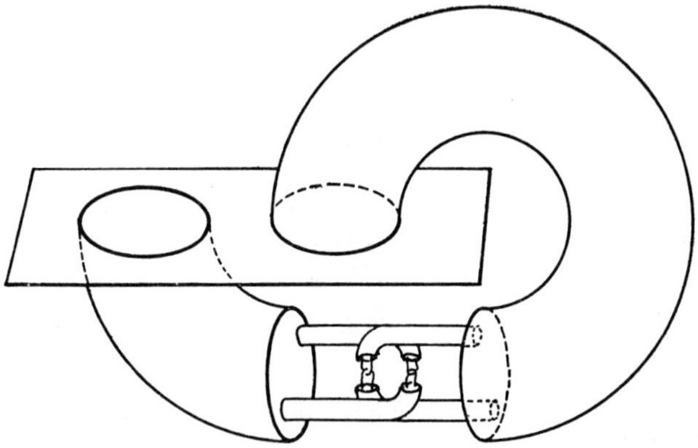

FIGURE 1.

THEOREM 2.1. *Suppose D is a disk in E^3, $p \in \operatorname{Int} D$, and $\epsilon > 0$. Then there is an ϵ 2-sphere S in E^3 and a disk D' in D such that $p \in \operatorname{Int} D' \subset S$.*

Theorem 2.1 is Theorem 5 of [3] but it is included here for completeness.

Proof. Let C be a cube in E of diameter less than ϵ and such that $p \in \operatorname{Int} C$, $\operatorname{Bd} D \subset \operatorname{Ext} C$.

Let f be a nonnegative continuous function defined on E^3 such that $f = 0$ in a neighborhood of p, f is positive on $\operatorname{Bd} C$, and $f < \rho(C, \operatorname{Bd} D)$.

It follows from Theorem 7 of [2] that there is a homeomorphism h of D into E^3 such that for each point x of D, $\rho(x, h(x)) \leqq f(x)$ and $h(D)$ is locally polyhedral at $h(x)$ if $f(x) > 0$. With no loss of generality we suppose $h(D)$ and $\operatorname{Bd} C$ are in general position.

Let K be the component of $h(D) \cdot C$ that contains p. It is a punctured disk and has as a boundary the sum of a finite collection of mutually exclusive simple closed curves $J_1, J_2, \cdots, J_n$. Let $E_1, E_2, \cdots, E_n$ be a finite collection of mutually exclusive disks in $\operatorname{Ext} C + \operatorname{Bd} C$ such that $\operatorname{Bd} E_i = J_i$ and diameter $(C + \sum E_i) < \epsilon$. The 2-sphere S promised by the theorem is $K + \sum E_i$.

3. Properties of Sierpinski curves. In this section we review some properties of Sierpinski curves. We recall that a Sierpinski curve is the set obtained by removing from a 2-sphere S the interiors of a null sequence of mutually exclusive disks $\{D_i\}$ in S such that $\sum D_i$ is dense in S. The Sierpinski curve is

$$X = S - \sum \operatorname{Int} D_i.$$

Any two Sierpinski curves described in such a fashion are homeomorphic and any continuum homeomorphic to such a Sierpinski is called a Sierpinski curve irrespective of whether or not it is imbedded in a 2-sphere.

A convenient way to learn the properties of $X = S - \sum \operatorname{Int} D_i$ is to note that the decomposition G of $S - \operatorname{Int} D_1$ whose elements are the disks of $\{D_i\}$; $i = 2, 3, \cdots$ and the points of $(S - \sum D_i) + \operatorname{Bd} D_1$ is upper semicontinuous and use the result of R. L. Moore [7] that there is a monotone map f of $S - \operatorname{Int} D_1$ onto a disk D taking elements of the decomposition G onto unique points of D.

By considering such a decomposition of S one may obtain the following result.

THEOREM 3.1. *Suppose X is a Sierpinski curve on a 2-sphere S. For* *each* $\epsilon > 0$ *and each point* p *of* X *that is not a boundary point of a component of* $S - X$ *there is an* ϵ *disk* D *in* S *such that*

$$p \in \operatorname{Int} D, \qquad \operatorname{Bd} D \subset X.$$

A point of $X = S - \sum \operatorname{Int} D_i$ is an *accessible point* of X if it is a boundary point of some D_i; otherwise it is called *inaccessible.* Each such accessible point lies on an arc in X whose complement in X is uniformly locally connected; no inaccessible point lies on such an arc. Since lying on such an arc is a topological property, we have the following result.

THEOREM 3.2. *Suppose* X *is a Sierpinski curve on a 2-sphere* S *and* h *is a homeomorphism of* X *into a 2-sphere* S'. *The point* p *of* X *is a boundary point of a component of* $S - X$ *if and only if* $h(p)$ *is a boundary point of a component of* $S' - h(X)$.

A simple closed curve J is said to pierce a 2-manifold M at a point p if J contains an arc ab and M a disk D such that ab pierces D at p. The following definition of linking is a modification of the one given in [2]. The simple closed curve J_1 links the simple closed curve J_2 if J_1 bounds a 2-manifold M such that J_2 intersects M in an odd number of points and J_2 pierces M at each point that it intersects it. Furthermore, if J_1 links J_2 in this fashion and J_1', J_2' are closed curves (simple or singular) such that J_1' is homotopic to J_1 in $E^3 - J_2'$ and J_2' is homotopic to J_2 in $E^3 - J_1$, then J_1' is said to link J_2'. It was shown in [2] that if J_1' links J_2', then J_2' links J_1'. Hence, if J_1' links J_2', we say that J_1' and J_2' link.

Although the following theorem does not treat Sierpinski curves, it is used in Theorem 3.4 which does.

THEOREM 3.3. *Suppose* S *is a 2-sphere in* E^3, J *is a simple closed curve in* S *that bounds a disk* D *in* S, *and* apb, aqb *are arcs in* E^3 *such that*

$$\begin{aligned}&a, b \text{ belong to different components of } E^3 - S,\\ &apb \cdot (S - D) = 0,\\ &aqb \cdot D = 0.\end{aligned}$$

Then $apb + aqb$ *and* J *link.*

Proof. Let h be a homeomorphism of S onto a polyhedral 2-sphere in E^3 such that

$h(S)$ separates a from b,
$apb \cdot h(\overline{S - D}) = 0$,
$aqb \cdot h(D) = 0$,
there is a homotopy taking J to $h(J)$ in $E^3 - (apb + aqb)$.

That there is such a homeomorphism close to the identity follows from Theorem 1 of [2] and that $h(S)$ separates a from b under such a homeomorphism follows from Theorem VI 10 on page 97 of [6].

Let f be a homeomorphism of $apb + aqb$ onto a polygon in E^3 such that

f is the identity on a and b.
$f(apb) \cdot h(\overline{S - D}) = 0$,
$f(aqb) \cdot h(d) = 0$,
there is a homotopy taking $apb + aqb$ to $f(apb + aqb)$ in $E^3 - h(J)$.

With no loss of generality we suppose $f(apb)$ and $h(D)$ are in general position.

Then $f(apb + aqb)$ links $h(J)$ since $f(apb)$ crosses $h(D)$ an odd number of times and $f(aqb) \cdot h(D) = 0$. See page 480 of [2]. Since there is a homotopy taking $(apb + aqb)$ to $f(apb + aqb)$ in $E^3 - h(J)$, it follows from Theorem 10 of [2] that $apb + aqb$ links $h(J)$. Since there is a homotopy taking J to $h(J)$ in $E^3 - (apb + aqb)$, it follows from this same theorem that $apq + aqb$ links J.

THEOREM 3.4. *Suppose X is a Sierpinski curve that lies in each of two disks D_1, D_2 in E^3 and apb is an arc that intersects $D_1 + D_2$ only at an inaccessible point p of X. Then apb pierces D_1 if and only if it pierces D_2.*

Proof. Let D_i' be a subdisk of D_i such that D_i' lies in a 2-sphere S_i and p is in interior point of D_i'. That there is such a disk D_i' and a 2-sphere S_i follows from Theorem 2.1. Then apb pierces D_i if and only if it pierces D_i'. With no loss of generality we suppose that apb is so short that it intersects $S_1 + S_2$ only at p.

It follows from Theorem 1.1 that apb pierces D_i if and only if a, b belong to different components of $E^3 - S_i$. Suppose a, b belong to one component of $E^3 - S_1$. We show in the next two paragraphs that in this case they also belong to the same component of $E^3 - S_2$.

Let aqb be an arc from a to b in $E^3 - S_1$ such that $apb + aqb$ is a simple closed curve. Since $apb + aqb$ intersects S_1 only at p, any simple closed curve in $(X - p) \cdot D_1'$ can be shrunk to a point in $S_1 - (apb + aqb)$. Hence no simple closed curve in $(X - p) \cdot D_1'$ links $apb + aqb$. See Theorem 10 of [2].

Let J be a simple closed curve in $X \cdot D_2'$ which bounds a disk E in D_2' such that $p \in \operatorname{Int} E$ and diameter E is less than the distance between p and $aqb + (D_1 - D_1')$. That there is such a disk E follows from Theorem 3.1. Then a, b belong to the same component of $E^3 - S_2$ because if they did not, it would follow from Theorem 3.3 that J would link $apb + aqb$. Thus would contradict the conclusion of the preceding paragraph.

We have shown that if apb does not pierce D_1, then it does not pierce D_2. Similarly, it may be shown if it does not pierce D_2, it does not pierce D_1.

4. Arcs in disks. The tame arcs that we will show pierce an arbitrary 2-sphere in E^3 lie in tame disks. In this section we study how arcs in disks may miss certain closed sets. An arc xy is said to span a disk D if the ends of xy lie on $\operatorname{Bd} D$ and xy lies except for its ends in $\operatorname{Int} D$.

THEOREM 4.1. *Suppose D is a disk, xy is a spanning arc of D, and Y is a closed subset of D containing xy such that for each $\epsilon > 0$ and each point q of* $\operatorname{Int} xy$ *there is a neighborhood N of q that intersects no component of $Y - xy$ of diameter greater than ϵ. Then there are points a, b of D on opposite sides of xy such that for each point p of xy there is an arc apb in $(D - Y) + \{p\}$ from a to b.*

Proof. Let E be a subdisk of D such that xy spans E and no component of $(Y - xy) \cdot E$ has either x or y as a limit point. The disk E is obtained by removing parts of D on both sides of xy near x and y without removing any point of xy. With no loss of generality we suppose that $D = E$.

Let C be the collection of components of Y other than xy and C' be the collection of all continua which are the sum of an element c of C and all components of $D - c$ which fail to contain xy. Ig G is the decomposition of D whose nondegenerate elements are the maximal elements of C', G is upper semicontinuous. Since no element of G separates D, it follows from a result of R. L. Moore [7] that there is a map f of D onto itself that takes each element of G onto a point of D and no two elements of G onto the same point. We suppose with no loss of generality that f is the identity on xy.

Since the set $f(Y)-xy$ is 0-dimensional, there is a null sequence Z of mutually exclusive disks in $D-xy$ which cover $f(Y)-xy$ such that each point of $f(Y)-(xy+\operatorname{Bd} D)$ lies on the interior of one of these disks and each point of $f(Y)-xy$ on $\operatorname{Bd} D$ lies on an open arc on $\operatorname{Bd} D$ which is on the boundary of an element of Z. We suppose that if a disk of Z intersects $\operatorname{Bd} D$, the intersection is connected. We note that the inverses under f of the elements of Z give a null sequence Z' of mutually exclusive disks in $D-xy$.

Let a, b be points in different components of $\operatorname{Int} D-xy$ such that neither a nor b lies in an element of Z'. Let $(apb)'$ be an arc from a to b in $(D-xy)+\{p\}$. By replacing certain arcs in $(apb)'$ by arcs on the boundaries of elements of Z', one may obtain an arc apb in D that intersects Y only at p.

THEOREM 4.2. *Under the hypothesis of Theorem 4.1 there is a family of arcs from a to b such that no two of these arcs intersect except in their end points and for each point p of xy there is one of these arcs in $(D-Y)+\{p\}$.*

Proof. If $Y-xy$ were only a countable number of points it would be easy to get the prescribed family of arcs.

If g is a monotone map of D onto itself such that the elements of Z' (of Theorem 4.1) are the nondegenerate inverses under g of points of D, $g(Y-xy)$ is countable. We suppose g is the identity on xy. Let F be a family of arcs from $g(a)$ to $g(b)$ such that no two of these arcs intersect except in their end points and for each point p of xy there is an element of F in $(D-g(Y))+\{p\}$. The inverses of the elements of F are members of a family satisfying the conclusion of the theorem.

5. Piercing disks with tame arcs. In this section we use the results of the preceding sections to show that each disk in E^3 is pierced by a tame arc.

THEOREM 5.1. *Suppose S is a 2-sphere, E is a tame disk, and X is a Sierpinski curve in $S\cdot E$. Then S can be pierced by a tame arc at each point of X.*

Proof. Let x be a given point of X. With no loss of generality we suppose that X lies in a horizontal plane P and xy is a straight line interval in X that does not intersect the boundary of a component of $P-X$ except possibly at x.

Let P' be a vertical plane containing xy and D be a circular disk in P'

with xy as diameter. We shall show that there is a tame arc in D that pierces S at x.

For each ϵ, $S-X$ has only a finite number of components of diameter greater than ϵ and no one of these components has any point of $xy-x$ as a limit point. Hence, for each $\epsilon>0$ and each point p of $\mathrm{Int}\, xy$ there is a neighborhood N of p such that N intersects no component of $D\cdot S-xy$ of diameter greater than ϵ.

It follows from Theorem 4.1 that there are points a, b of D on opposite sides of xy such that for each point p of xy there is an arc apb in D that intersects S only at p. The arc apb pierces the plane P at p and it follows from Theorem 3.4 that if $p\neq x$, apb pierces S at p. Hence a and b belong to different components of E^3-S. Hence, even in the case where $p=x$, apb pierces S at p.

Question. Theorem 4.2 can be used to show that there is a family of mutually exclusive tame arsc such that for each point p of xy there is a member of the family that pierces S at xy. One might wonder if there is such a family with xy replaced by X. A more general question—is there a family F of mutually exclusive tame arcs such that for each point of S at which S can be pierced by a tame arc, there is a member of F piercing S there?

THEOREM 5.2. *Each 2-sphere S in E^3 contains a 0 dimensional G_δ set Y such that S can be pierced by a tame arc at each point of $S-Y$.*

Proof. It follows from [4] that for each integer i, S contains a Sierpinski curve X_i such that X_i lies in a tame 2-sphere and each component of $S-X_i$ is of diameter less than $1/i$. It follows from Theorem 5.1 that S can be pierced by a tame arc at each point of the F_σ set $X_1+X_2+\cdots$. The set Y of Theorem 5.2 is $S-(X_1+X_2+\cdots)$.

The following theorem is a consequence of Theorem 5.2 and Theorem 2.1.

THEOREM 5.3. *Each disk D in E^3 contains a 0 dimensional G_δ set Y such that D can be pierced by a tame arc at each point of* $\mathrm{Int}\, D-Y$.

Questions. Suppose S is a 2-sphere and Y is the set of points of S at which S cannot be pierced by a tame arc. What can be said about the category of Y? Is it a 0 dimensional F_σ set?

If xy is a tame arc in a 2-sphere S, can S be pierced by a tame arc at each point of xy? (Note. Since this paper was written, David Gillman has answered this question in the affirmative.)

Suppose S is a 2-sphere in E^3, L is a straight line, and $\epsilon>0$. Is there a homeomorphism h of E^3 onto itself such that h moves no point more than ϵ, h is the identity outside an ϵ neighborhood of L, and $H(S)-L$ is finite. An easy positive answer to this question would have simplified the approximation work in [2]. (Note. Since this paper was written, an affirmative answer to the question was announced in the 1962 Notices, Abstract 589-13. The methods used however were not easy.)

UNIVERSITY OF WISCONSIN.

REFERENCES.

[1] R. H. Bing, "A simple closed curve that pierces no disk," *Journal de Mathematiques Pures Appliquées*, vol. 35 (1956), pp. 337-343.

[2] ———, "Approximating surfaces with polyhedral ones," *Annals of Mathematics*, vol. 65 (1957), pp. 456-483.

[3] ———, "A surface is tame if its complement is 1-ULC," *Transactions of the American Mathematical Society*, vol. 101 (1961), pp. 284-305.

[4] ———, "Each disk in E^3 contains a tame arc," *American Journal of Mathematics*, vol. 84 (1962), pp. 583-590.

[5] Ralph H. Fox and Emil Artin, "Some wild cells and spheres in three-dimensional space," *Annals of Mathematics*, vol. 49 (1948), pp. 979-990.

[6] W. Hurewicz and H. Wallman, *Dimension Theory*, Princeton University Press, 1948.

[7] R. L. Moore, "Concerning upper semicontinuous collections of continua," *Transactions of the American Mathematical Society*, vol. 27 (1925), pp. 416-428.

FUNDAMENTA
MATHEMATICAE
LV (1964)

An arc is tame in 3-space if and only if it is strongly cellular *

by

R. H. Bing (Wisconsin, USA) and **A. Kirkor** (Warszawa)

A set Z in Euclidean n-space E^n is *tame* if there exists such a homeomorphism f of E^n onto itself that $f(Z)$ is a polyhedron. There are known some necessary and sufficient conditions of tameness of an arc in E^3, e.g. [4] and [5]. We shall give here another one based on the reinforced notion of a cellular set [2]. A set Z in E^n is *strongly cellular* if there is an n-cell C in E^n and a homotopy $h\colon\ C\times I\to C$ such that, if $h_t(x) = h(x, t)$ and $S = \operatorname{Bd} C$ is the boundary of C, then

(1) $h_0 =$ identity mapping and $h_t|Z =$ identity for all t,

(2) $h_t|S =$ homeomorphism for $t<1$,

(3) $h_t(S)\cap h_{t'}(S) = 0$ for $t\neq t'$,

(4) $h_1(C) = Z$.

The set Z will be said to be a *strong deformation retract* of the cell C. By M. Brown's generalization of the Schoenflies Theorem [2] there will be no loss of generality if the cell C is assumed to be a ball.

THEOREM 1. *An arc is tame in E^3 if and only if it is strongly cellular.*

COROLLARY. *An arc A in E^3 is tame if and only if there are two concentric balls B_0 and B, $B_0\subset \operatorname{Int} B$, and a mapping f of B into E^3 such that*

(1) $f|B-B_0 =$ homeomorphism,

(2) $f(B_0) = A$.

Proof of the Corollary. It is obvious that the conditions of the Corollary are necessary. In order to prove that they are also sufficient let us observe first that $f(B)$ is a 3-cell by M. Brown's Theorem 1 of [2]. Then assume $C = f(B)$ and next consider the homotopy $r\colon\ B\times I\to B$ retracting B to B_0. Now define the homotopy $h\colon\ C\times I\to C$ in the following way:

$$h(x, t) = \operatorname{fr}[f^{-1}(x), t] \quad \text{for} \quad x\in C-A\,,$$
$$h(x, t) = x \quad \text{for} \quad x\in A\,.$$

* Work on this paper is supported by contract NSF-611665.

It is easy to check that A is a strong deformation retract of C and so the proof has been reduced to the Theorem 1.

QUESTION. Is the Corollary still valid if the set B_0 is assumed to be an acyclic continuum?

The Theorem 1 is an easy corollary of the following one:

THEOREM 2. *Let Q be the ball $x^2+y^2+z^2 \leqslant 4$, I_0 be the segment $-1 \leqslant x \leqslant 1$, $y = z = 0$ and let A be an arc which is a strong deformation retract of Q. Then any homeomorphism of I_0 onto A can be extended to a homeomorphism of Q onto itself.*

Proof of the Theorem 2 will be preceded by some lemmas. Let the symbols Z, C, S and h keep the meaning they got in the definition of strong cellularity.

Remark. If X and Y are disjoint subsets of S and $h_1(X) \cap h_1(Y) = 0$ then

$$h(X \times I) \cap h(Y \times I) = 0 .$$

Let henceforth $Z = A$ and $f = h_1|S$.

LEMMA 1. *If B is a continuum in A then $f^{-1}(B)$ is a continuum in S.*

Proof. Suppose $f^{-1}(B) = B' = X \cup Y$ and $\overline{X} \cap \overline{Y} = 0$. Since the 0-dimensional homology group $H_0(B') \neq 0$, then by a standard duality theorem $H_{n-1}(S-B') \neq 0$ and there exist two cycles γ^0 in B' and γ_2^{n-1} in $S-B'$ which are linked in S. They can be chosen so that the carrier of γ^0 is the set $\{x, y\}$ and the carrier of γ_2^{n-1} is a continuum M. Then $F = f(M)$ is an arc or a point and does not intersect B.

Denote by $f(x)f(y)$ the subarc of B with the end-points $f(x)$ and $f(y)$. Let J_1 be the arc $h[\{x, y\} \times I] \cup f(x)f(y)$ and J_2 be an arc with the end points x and y which lies in $(E^n - C) \cup \{x, y\}$. The simple closed curve $J = J_1 \cup J_2$ carries a cycle γ_1^1 which links γ_2^{n-1} in E^n. On the other hand $h(M \times I)$ and J are disjoint by our Remark and therefore γ_2^{n-1} is homotopic in $E^n - J$ to a cycle lying in F; but every cycle bounds in F and so $\gamma_2^{n-1} \sim 0$ in $E^n - J$. This contradiction was a result of our false assumption that $f^{-1}(B)$ is not connected.

Further considerations will be restricted to the 3-space E^3.

LEMMA 2. *If B is a connected subset of A containing one end-point of A, then $f^{-1}(B)$ is a simply connected subset of S.*

Proof. Let $B \neq A$ for otherwise the Lemma 2 is obvious. If B is open in A then its complement B' is an arc or a point. Therefore $f^{-1}(B') = S - f^{-1}(B)$ is a connected set by Lemma 1. Hence $f^{-1}(B)$ is simply connected. If B is closed in A then B' is open and $f^{-1}(B')$ is connected; hence $f^{-1}(B)$ is simply connected.

Let $S_t = h_t(S)$ and C_t be the 3-cell bounded by S_t in E^3.

LEMMA 3. *For any interior point x of the arc A there exists a topological disk D_x in C such that the following conditions are satisfied:*

(1) *$D_x \cap S_t$ is a simple closed curve for $t < 1$, and the point x for $t = 1$.*

(2) *D_x cuts the cell C between the end-points of A.*

(3) *D_x is locally tame at every point except perhaps at x.*

(4) *For any two disks D_x and D_y, $x \neq y$, there is a number t_0 such that $0 \leqslant t_0 < 1$ and $C_{t_0} \cap D_x \cap D_y = 0$.*

(5) *If $0 \leqslant t_0 < t_1 < 1$ and W is the closure of $C_{t_0} - C_{t_1}$, Q' is the closure of any component of $W - D_x$ and P is the closure of that component of $C_{t_0} - (D_x \cup D_y)$ which contains no end-points of A, then Q' is a tame 3-cell and $T = W \cap P$ is a tame solid torus.*

Proof. Construction of D_x. We adopt the following notation: Let a, b be the end-points of A; $B_x = [a, x)$ be a closed-open subarc of A; $G_x = f^{-1}(B_x)$ is a simply connected domain in S with the boundary $\mathrm{Bd}\, G_x$ contained in $f^{-1}(x)$; $F = f^{-1}(a)$ is a closed, simply connected set in G_x; $R_x = G_x - F$ is an open annulus.

Define a sequence $\{J_m\}$ of simple closed curves in R_x as follows: Let $J_1 = J_x$ be a simple closed curve which cuts G_x between F and $\mathrm{Bd}\, G_x$ and such that the distance $\varrho(J_1, \mathrm{Bd}\, G_x) < 1$. Suppose that J_m has been already defined for some $m \geqslant 1$. Let J_{m+1} be a simple closed curve which cuts G_x between J_m and $\mathrm{Bd}\, B_x$ and such that $\varrho(J_{m+1}, \mathrm{Bd}\, B_x) < 1/(m+1)$.

Denote the annulus in G_x bounded by J_m and J_{m+1} by R_m and construct the following deformations:

a. A sequence of isotopies $g^{(x)m}\colon J_m \times [(m-1)/m, m/(m+1)] \to R_m$, $m = 1, 2, 3, \ldots$, simply denoted by g^m and such that if $g_t^m(p) = g^m(p, t)$, then

$$g^m_{(m-1)/m}(p) = p\,, \qquad g^m_{m/(m+1)}(J_m) = J_{m+1}\,.$$

b. An isotopic deformation $g^{(x)}\colon J_x \times [0, 1) \to R_x$ such that

$$g_t^{(x)}(p) = g_t^m\, g^{m-1}_{(m-1)/m} \ldots g^1_{1/2}(p) \qquad \text{for} \qquad \frac{m-1}{m} \leqslant t \leqslant \frac{m}{m+1} \text{ and } m = 1, 2, 3, \ldots$$

c. A pseudoisotopy $\psi_x\colon J_x \times I \to C$ such that

$$\psi_x(p, t) = h_t g_t^{(x)}(p) \qquad \text{for} \qquad 0 \leqslant t < 1\,,$$
$$\psi_x(p, 1) = x\,.$$

Then $D_x = \psi_x(J_x \times I)$ is the desired disk. It is evident that D_x satisfies condition (1). Now we check the other ones.

Condition (2). Let Γ be that component of $C - D_x$ which contains the point a. It follows from the construction that $\overline{\Gamma} \subset \{x\} \cup h[f^{-1}(B_x) \times I]$, but $b \notin \{x\} \cup h[f^{-1}(B_x) \times I]$ and so there is no continuum in $C - D_x$ which could contain both a and b.

Condition (3). Let $p \in D_x - \{x\}$. There is a number s such that $0 < s < 1$ and $p \in D_x - C_s$. Let D be the closure of a component of $S - D_x$, D' be the closure of a component of $S_s - D_x$ and $R = \overline{D_x - C_s}$. Then

$$p \in R \subset D \cup R \cup D' = S'$$

and S' is a sphere. It can be easily shown that S' is bicollared and hence tame by M. Brown's Theorem 1 [2]. Therefore D_x is locally tame at p.

Condition (4). By construction $D_x = \psi_x(J_x \times I)$ and similarly $D_y = \psi_y(J_y \times I)$. Notice that $\psi_z(J_z \times 1) = z$ for $z = x, y$. Given disjoint neighbourhoods U_x and U_y of x and y respectively we can choose a number t_0 such that $0 \leqslant t_0 < 1$ and $\psi_z(J_z \times [t_0, 1]) \subset U_z$ for $z = x, y$. Then t_0 is the right number.

Condition (5). It is easy to see that Q' is a tame 3-cell because its boundary is a tame sphere by much the same reason that S' is one.

In order to prove that T is a tame solid torus it will be sufficient to show that its boundary is a tame torus and that there are two tame disjoint disks in T which span $\operatorname{Bd} T$.

a. $\operatorname{Bd} T$ *is a tame torus*: It is clear that $\operatorname{Bd} T$ is a torus for

$$\operatorname{Bd} T = T \cap (D_x \cup D_y \cup S_{t_0} \cup S_{t_1}) .$$

Also it is quite obvious that $\operatorname{Bd} T$ is locally tame at the interior points of the sets $T \cap D_z$, $z = x, y$, and $T \cap S_{t_i}$, $i = 0, 1$. And at the boundary points of these sets $\operatorname{Bd} T$ is locally tame by Doyle's theorem concerning the unions of cell pairs in E^3 [3]. Hence, by the theorem of Bing and Moise [1], $\operatorname{Bd} T$ is tame.

b. *There are two disjoint, tame disks D_1, D_2 in T such that*

$$D_i \cap \operatorname{Bd} T = \operatorname{Bd} D_i, \quad i = 1, 2 .$$

By condition (4) the simple closed curves $J_{z,t_0} = g_{t_0}^{(z)}(J_z)$, $z = x, y$, are disjoint subsets of S. In the annulus bounded by these curves choose two disjoint arcs A_1, A_2 having with each curve only one end-point in common. The mappings $g_t^{(x)}, g_t^{(y)}$ define an isotopy of the set $J_{x,t_0} \cup J_{y,t_0}$ which can be extended to an isotopy

$$g^*\colon (J_{x,t_0} \cup J_{y,t_0} \cup A_1 \cup A_2) \times [t_0, t_1] \to S .$$

Let $hg^*(p, t)$ stand for $h[g^*(p, t), t]$. Then the sets $D_i = hg^*(A_i \times [t_0, t_1])$, $i = 1, 2$, are the desired disks.

Proof of the Theorem 2. Let $\chi\colon I_0 \to A$ be a homeomorphism of the segment I_0 onto the arc A. In order to produce an extension of this homeomorphism to a homeomorphism of Q onto itself there will be given two isomorphic decompositions P_1 and P_2 of Q into tame 3-cells, tame solid tori and points of I_0 or A respectively.

Standard decomposition P_1. Let K_m, $m = 0, 1, 2, \ldots$, be the set

$$\left(\frac{m+1}{m+2}x\right)^2 + \left(\frac{m+1}{2}y\right)^2 + \left(\frac{m+1}{2}z\right)^2 \leqslant 1\,.$$

Notice that $K_0 = Q$, $K_{m+1} \subset \operatorname{Int} K_m$ and $\bigcap_{m=0}^{\infty} K_m = I_0$.

For each fraction of the form $p/2^q$ in lowest terms and with $-1 < p/2^q < 1$, let

$$E_{p/2^q} = \{(x, y, z) \in K_q |\, x = p/2^q\}\,.$$

Note that $E_{p/2^q}$ is a disk which spans K_q and $E_0 \cup E_{-1/2} \cup E_{1/2} \cup E_{-3/4} \cup \cup \ldots \cup E_{1-2^{-q}}$ separates K_q into 2^{1+q} pieces. The closure of each component of $K_q - K_{q+1}$ minus the sum of the E's is either a tame 3-cell or a tame solid torus. The elements of P_1 are the points of I_0, these 3-cells, and these solid tori.

Curvilinear decomposition P_2. Adopt the following notation. The number s will stand for both the point $(s, 0, 0)$ of I_0 and for its image in A under the homeomorphism χ. This should cause no confusion. With the aid of Lemma 3 we construct disks $D'_{p/2^q}$ which meet the arc A at the points $p/2^q$ and are the counterparts of the disks $E_{p/2^q}$.

Let D'_0 be the disk D_0 promised by Lemma 3. Let t_1 be the number suggested by Condition 4 of that lemma such that no pair of $D_{-1/2} \cap C_{t_1}$, $D_0 \cap C_{t_1}$, $D_{1/2} \cap C_{t_1}$ have a point in common. Then let $D'_{-1/2} = D_{-1/2} \cap C_{t_1}$ and $D'_{1/2} = D_{1/2} \cap C_{t_1}$.

Let t_2 be a number such that $t_1 < t_2 < 1$, $1/2 < t_2$, and no pair of $D_{-3/4}$, $D_{-1/2}$, $D_{-1/4}$, D_0, $D_{1/4}$, $D_{1/2}$, $D_{3/4}$ of Lemma 3 intersect in C_{t_2}. Then $D'_{-3/4}$, $D'_{-1/4}$, $D'_{1/4}$, $D'_{3/4}$ are the intersections of the appropriate D's with C_{t_2}.

Continuing in this fashion we describe a monotone increasing sequence of numbers $t_0 = 0, t_1, t_2, \ldots$ converging to 1 such that $D_{a/2^b} \cap \cap D_{c/2^d} \cap C_{t_i} = 0$ if $a/2^b \neq c/2^d$, $2^b \leqslant 2^i$, and $2^d \leqslant 2^i$. Then $D'_{p/2^q} = D_{p/2^q} \cap C_{t_q}$.

The elements of P_2 are the points of A and the closure of the components of the $(C_{t_i} - C_{t_{i+1}})$'s minus the D''s. Note that these elements are either points, tame 3-cells, or tame solid tori. The decomposition P_1 is isomorphic to the decomposition P_2.

Extending the homeomorphism χ. Let χ'_0: $I_0 \cup \operatorname{Bd} Q \to A \cup \operatorname{Bd} Q$ be a homeomorphism such that χ'_0 is an extension of the homeomorphism mentioned in the statement of Theorem 2, χ'_0: $\operatorname{Bd} E_0 \to \operatorname{Bd} D'_0$, and χ'_0 takes the right half of $\operatorname{Bd} Q = \operatorname{Bd} K_0$ into the closure of the component of $Q - D'_0$ containing the point $(1, 0, 0)$ of A. Extend χ'_0 to a homeomorphism χ_0: $I_0 \cup \operatorname{Bd} Q \cup E_0 \to A \cup \operatorname{Bd} Q \cup D'_0$.

Now extend χ_0 to a homeomorphism χ_1: $I_0 \cup \overline{(K_0 - K_1)} \cup E_0 \cup E_{-1/2} \cup \cup E_{1/2} \to A \cup \overline{(Q - C_{t_1})} \cup D'_0 \cup D'_{-1/2} \cup D'_{1/2}$. We continue extending χ so that in general χ_i takes $(\overline{K_0 - K_{i+1}})$ onto $(\overline{Q - C_{t_{i+1}}})$, I_0 onto A, and $E_{p/2^q}$ onto $D'_{p/2^q}$ if $2^q \leqslant 2^i$. The limit of the χ_i's is the required extension taking Q onto itself.

References

[1] R. H. Bing, *Locally tame sets are tame*, Ann. of Math. 59 (1954), pp. 145-158.

[2] M. Brown, *A proof of generalized Schoenflies theorem*, Bull. Amer. Math. Soc. 66 (1960), p. 74.

[3] P. H. Doyle, *Unions of cell pairs in* E^3, Pacific J. Math. 10 (1960), pp. 521-524.

[4] Harrold, Griffith and Posey, *A characterization of tame curves in 3-space*, Trans. Amer. Math. Soc. 79 (1955), pp. 12-35.

[5] A. B. Sosinskii, *About the embedding of a k-cell into* E^n, Dokl. Akad. Nauk SSSR 139 (1961), pp. 1311-1313.

Reçu par la Rédaction le 12. 5. 1963

ERRATA

Page 175 – in condition (2) of definition insert "into $E^n - Z$" after "homeomorphism".

Page 176 – line 3. A for B_0

Page 1?7 – lines 18, 21 $\overline{G}_x$ for G_x

line 21 $\mathrm{Bd}G_x$ for $\mathrm{Bd}B_x$

IMPROVING THE INTERSECTIONS OF LINES AND SURFACES

R. H. Bing

Dedicated to R. L. Wilder on his seventieth birthday.

Suppose S^2 is a topological 2-sphere and L^1 is a straight line in Euclidean 3-space E^3. If one had an easy proof of the following corollary, one could use the corollary to simplify the proof of the Approximation Theorem (Theorem 7 of [2]) and the Side Approximation Theorem (Theorem 16 of [6]).

COROLLARY TO THEOREM 2. *For each* $\varepsilon > 0$ *there exists a homeomorphism* $h: E^3 \to E^3$ *such that* $L^1 \cap h(S^2)$ *is finite and* $h = I$ *(identity) outside an* ε*-neighborhood of* $S^2 \cap L^1$.

We do not get a proof of the corollary that makes it useful in proving these theorems, since we use consequences of the Side Approximation Theorem to the hilt in proving Theorem 2, the basis for the corollary. There seems to be good reason for giving a proof of Theorem 2 even if it is complicated, since aside from its intrinsic interest, Theorem 2 is needed in proving some other interesting theorems, such as the result by McMillan (Theorem 2 of [8]) that each 2-sphere topologically embedded in an arbitrary 3-manifold has a neighborhood that can be embedded in E^3.

We denote the distance function by ρ. If f_1, f_2 are two maps of the space X into the metric space Y, we use $\rho(f_1, f_2)$ to denote the least upper bound $\rho(f_1(x), f_2(x))$ $(x \in X)$.

THEOREM 1. *Suppose* S_1, S_2 *are two 2-spheres in* E^3, S_2 *is tame,* $\varepsilon > 0$, *and* X *is a tame Sierpinski curve in* S_1 *such that each component of* $S_1 - X$ *has diameter less than* ε. *Then there exists an isotopy* h_t $(0 \le t \le 1)$ *of* E^3 *onto itself such that* $h_0 = I$, *the set* $S_2 \cap h_1(X)$ *is the union of a finite number of mutually exclusive simple closed curves each in the inaccessible part of* $h_1(X)$, $h_1(X)$ *locally lies on different sides of* S_2 *near these simple closed curves,* $h_t = I$ *outside the* ε*-neighborhood of* $S_1 \cap S_2$, *and* $\rho(h_t, I) < \varepsilon$.

Proof. Let δ be a positive number such that each component of $S_1 - X$ has diameter less than $\varepsilon - 4\delta$.

Let S_1' be a tame 2-sphere obtained by replacing each component of $S_1 - X$ by the interior of a tame disk homeomorphically so close to the component that the tame disk has diameter less than $\varepsilon - 4\delta$ and each point of $S_1' \cap S_2$ lies within δ of $S_1 \cap S_2$. The Approximation Theorem (Theorem 7 of [2]) tells us that this replacement is possible, and Theorem 8.2 of [7] assures us that S_1' is tame.

Since S_2 is tame, we suppose it is polyhedral.

Since S_1' is tame, there exists a homeomorphism $g: K^2 \times [-1, 1] \to E^3$, where K^2 is a polyhedral 2-sphere and $g(K^2 \times 0) = S_1'$. Using Theorem 9 of [3] or Theorem 2 of [9], we learn that g can be chosen to be locally piecewise linear on $K^2 \times (0, 1]$. Let f_t $(0 \le t \le 1)$ be an isotopy of $K^2 \times [-1, 1]$ onto itself that is fixed outside a small neighborhood of $g^{-1}(S_1' \cap S_2)$ and pushes a small neighborhood of $g^{-1}(S_1' \cap S_2)$ in $K^2 \times 0$ onto a polyhedron on the positive side of $K^2 \times 0$. By considering $g f_t g^{-1}$

Received October 27, 1966.
Work on this paper was supported by NSF grant GP-3857.

and putting the adjusted S_1' into general position with respect to S_2 near S_2, one finds that there exists an isotopy g_t $(0 \leq t \leq 1)$ of E^3 onto itself such that $S_2 \cap g_1(S_1')$ is the union of a finite collection $\{J_i\}$ of mutually exclusive simple closed curves along which S_2 and $g_1(S_1')$ cross, and such that $\rho(g_t, I) < \delta$, $g_0 = I$, and $g_t = I$ outside the δ-neighborhood of $S_2 \cap S_1'$. Note that each point of each J_i lies within δ of $S_1' \cap S_2$ and within 2δ of $S_1 \cap S_2$.

We would be finished with the proof of Theorem 1 if the J_i were in the inaccessible part of $h_1(X)$ rather than merely in $g_1(S_1')$. Since this is not necessarily the case, we modify g_1. Note that $g_1(S_1')$ is tame and the components of $g_1(S_1' - X)$ have diameters less than $\varepsilon - 2\delta$. The following lemma implies that there exists an isotopy h_t': $E^3 \to E^3$ such that $h_0' = I$, $h_t'(g_1(S_1')) = g_1(S_1')$, the J_i lie in the inaccessible part of $h_1' g_1(X)$, $\rho(h_t', I) < \varepsilon - 2\delta$, and $h_t' = I$ outside the $(\varepsilon - 2\delta)$-neighborhood of the union of the J_i. An isotopy satisfying the requirements of Theorem 1 is given by $h_t = h_t' g_t$.

The following is an engulfing theorem for Sierpinski curves, and it shows how Sierpinski curves can be made to swallow certain 1-dimensional sets.

LEMMA. *Suppose* S *is a 2-sphere,* $\varepsilon > 0$, X *is a Sierpinski curve in* S *such that each component of* $S - X$ *has diameter less than* ε, *and* Y *is a closed* 1-*dimensional subset of* S. *Then there exists an isotopy* h_t: $S \to S$ *such that* $h_0 = I$, Y *lies in the inaccessible part of* $h_1(X)$, *each orbit* $h(x, t)$ $(0 \leq t \leq 1)$ *has diameter less than* ε, *and* $h_t = I$ *outside an* ε-*neighborhood of* Y. *In fact, if* S *is a tame sphere in* E^3, h_t *may be extended to an isotopy of* E^3 *onto itself with the same properties.*

Proof. Let X' be a Sierpinski curve in the inaccessible part of X such that each component of $S - X'$ has diameter less than ε. To see that such an X' exists, use [10] to obtain a map f: $S \to S$ such that each $f^{-1}(x)$ is either an inaccessible point of X or the closure of a component of $S - X$. If one lets X'' be a Sierpinski curve in S that has small holes and misses the countably many points of S with nondegenerate inverses under f^{-1}, one can use $f^{-1}(X'')$ for X'. We let $h_t = I$ on X'.

If D is a component of $S - X'$ that misses Y, we permit h_t to be the identity on D. In the next four paragraphs we describe h_t on each component D that intersects Y.

It follows from [10] that there exists a map g_D of $\overline{D}$ onto itself such that $g_D | \operatorname{Bd} \overline{D} = I$ and each $g_D^{-1}(x)$ is either an inaccessible point of X or the closure of a component of $D \cap (S - X)$. Let W_D be the countable set of points of D with nondegenerate inverses under g_D^{-1}, and let h_D be a homeomorphism of $\overline{D}$ onto itself such that $h_D | \operatorname{Bd} \overline{D} = I$ and $W_D \cap h_D(D \cap Y) = 0$.

Let X_D be a Sierpinski curve in $\overline{D}$ that contains $\operatorname{Bd} \overline{D} \cup h_D(D \cap Y)$ but no point of W_D. The curve X_D is obtained by removing from $\overline{D}$ some small holes about the points of W_D, the holes being so small that they do not reach $h_D(D \cap Y)$. Note that $h_D^{-1}(X_D)$ contains $D \cap Y$, $g_D^{-1} | X_D$ is a homeomorphism, and $g_D^{-1}(X_D) \subset X$.

Let $h_1 = h_D^{-1} g_D$ on $g_D^{-1}(X_D)$. This h_1 is extended to take $\overline{D}$ homeomorphically onto itself. Note that $h_1(X) \supset D \cap Y$, since

$$h_1(X) \supset h_1(g_D^{-1}(X_D)) = h_D^{-1}(X_D) \supset D \cap Y.$$

It follows from [1] that the homeomorphisms $h_0 | \overline{D}$ and $h_1 | \overline{D}$ can be extended to an isotopy $h_t | \overline{D}$ that is fixed on $\operatorname{Bd} \overline{D}$. The collection of such $h_t | \overline{D}$ defines an isotopy h_t: $S \to S$.

If S is tamely embedded in E^3, we let δ be a positive number such that each $h(x \times [0, 1])$ has diameter less than $\varepsilon - 2\delta$, and we let $f: (S \times [-1, 1]) \to E^3$ be a homeomorphism such that $f(x \times 0) = x$ and $\operatorname{diam} f(x \times [-1, 1]) < \delta$ for each x. To extend h_t to E^3, we let

$$h_t = I \text{ on } E^3 - f(S \times [-1, 1]) \quad \text{and} \quad h_t f(x \times s) = f(h_{(1-s^2)t}(x) \times s).$$

THEOREM 2. *Suppose* S *is a 2-sphere and* A *is a tame arc in* E^3. *Then for each* $\varepsilon > 0$ *there is an isotopy* $h_t: E^3 \to E^3$ $(0 \leq t \leq 1)$ *such that* $h_0 = I$, A *intersects* $h_1(S)$ *at only a finite number of points and pierces it at each of these points,* $\rho(h_t, I) < \varepsilon$, *and* $h_t = I$ *outside the* ε*-neighborhood of* $S \cap A$.

Proof. We suppose without loss of generality that A is straight, $A \cap S$ is 0-dimensional, and S misses the ends of A.

Let $\{C_i\}$ be a finite collection of mutually exclusive solid right circular cylinders, each of diameter less than $\varepsilon/2$, and each with a subarc of A as an axis, such that no base of any of these cylinders intersects S but their interiors cover $S \cap A$. We suppose that neither end of A lies in a C_i.

Let δ be a positive number less than the radius of any C_i, the distance between any pair of C_i, the distance from S to the base of any C_i, and the distance from S to $A - \bigcup C_i$.

Let X be a tame Sierpinski curve in S such that each component of $S - X$ has diameter less than $\delta/4$. That such a curve exists follows from [4].

Theorem 1 implies that there exists an isotopy h_t': $E^3 \to E^3$ $(0 \leq t \leq 1)$ such that $h_0' = I$, $\rho(h_t', I) < \delta/4$, $h_t' = I$ outside the $\delta/4$-neighborhood of $\bigcup \operatorname{Bd} C_i$, and $h_1'(X) \cap \bigcup \operatorname{Bd} C_i$ is the union of a finite collection $\{J_{ij}\}$ of mutually exclusive simple closed curves such that each J_{ij} lies in the inaccessible part of $h_1'(X)$ and the parts of $h_1'(X)$ on different sides of J_{ij} in S near J_{ij} lie on different sides of $\operatorname{Bd} C_i$. Note that $h_t'(S)$ misses the bases of the C_i and also $A - \bigcup C_i$. Each component of $h_1'(S - X)$ has diameter less than $3\delta/4$; therefore the only components of $\operatorname{Bd} C_i \cap h_1'(S)$ that could separate the bases of C_i from each other on $\operatorname{Bd} C_i$ are the J_{ij}.

In this paragraph we point out why there is an arc $p_i q_i$ on $\operatorname{Bd} C_i$ joining the points p_i, q_i of $A \cap \operatorname{Bd} C_i$ such that $p_i q_i \cap h_1'(S)$ is a finite set of points and each of these points is a point of a J_{ij} at which $p_i q_i$ crosses J_{ij}. A similar argument is given in the proof of Theorem 4.1 of [5], but for completeness we include the argument. Since J_{ij} is in the inaccessible part of $h_1'(X)$, we see that if $R_1, R_2, \cdots$ is a sequence of different components of $\operatorname{Bd} C_i \cap h_1'(S)$ converging to a set intersecting a J_{ij}, then the diameters of these components converge to 0. It follows from [10] that there exists a map f_i: $\operatorname{Bd} C_i \to \operatorname{Bd} C_i$ such that each $f_i^{-1}(x)$ $(x \in \operatorname{Bd} C_i)$ is either a continuum which is the union of a component of $\operatorname{Bd} C_i \cap h_1(S)$ other than a J_{ij} and all points that this component separates from p_i, or a point of $\operatorname{Bd} C_i - h_1'(S)$, or a point of a J_{ij}. Since $f(\operatorname{Bd} C_i \cap h_1'(S))$ is the union of the sets $f(J_{ij})$ and a 0-dimensional set, there exists an arc A_i on $\operatorname{Bd} C_i$ from $f(p_i)$ to $f(q_i)$ that intersects $f_i h_1'(S)$ at only a finite set of points, at each of which A_i crosses an $f_i(J_{ij})$. Consequently $p_i q_i = f_i^{-1}(A_i)$. Note that $p_i q_i$ pierces $h_1'(S)$ at each point of $p_i q_i \cap h_1'(S)$, since $h_1'(X)$ lies on different sides of $\operatorname{Bd} C_i$ near J_{ij}. (See Theorem 3.4 of [5].)

Let A' be the arc obtained from A by replacing each $A \cap C_i$ by $p_i q_i$. Note that A' intersects $h_1'(S)$ at only a finite number of points and pierces it at each of these points. Let $h_t''\colon E^3 \to E^3$ $(0 \le t \le 1)$ be an isotopy such that $h_0'' = I$, $h_t'' = I$ outside the $\delta/2$-neighborhood of $\bigcup C_i$, and $h_1''(A') = A$. Then $h_t''h_t' = h_t$ is the required isotopy.

THEOREM 3. *Suppose* M^3 *is a* 3-*manifold (not necessarily compact, but without boundary),* K *is a locally tame finite graph in* M^3, C^2 *is a* 2-*complex (not necessarily compact),* h *is a homeomorphism of* C^2 *onto a closed subset of* M^3, *and* $\varepsilon > 0$. *Then there exists an isotopy* $h_t\colon M^3 \to M^3$ $(0 \le t \le 1)$ *such that* $h_0 = I$, $\rho(h_t, I) < \varepsilon$, $h_t = I$ *outside the* ε-*neighborhood of* $K \cap h(C^2)$, *and* $K \cap h_1 h(C^2)$ *is a finite set of points such that for each such point there is a* 2-*simplex* σ *of* C^2 *such that* K *pierces* $h_1 h(\sigma)$ *at the point.*

Proof. We suppose $h(C^2)$ is adjusted so that $K \cap h(C^2)$ is 0-dimensional and misses the image under h of the 1-skeleton of C^2, while K has a triangulation such that the set of the vertices of this triangulation misses $h(C^2)$. Since h_t will only move points of M^3 that are near the 1-simplexes of K under this triangulation but not near the 0-skeleton, we suppose with no loss of generality that K is a tame arc. Let U be an open subset of M^3 such that $K \subset U$ and U is homeomorphic to E^3.

It follows from Theorem 2.1 of [5] that for each point p of $K \cap h(C^2)$ there is a disk D_p and a 2-sphere S_p such that $p \in \operatorname{Int} D_p \subset S_p \subset U$ and D_p lies in the image under h of a 2-simplex of C^2. Let $D_1, D_2, \cdots, D_n$ be a finite collection of such disks such that each point $K \cap h(C^2)$ lies in the interior of one of the disks and no two of the disks intersect each other. Let S_i be the 2-sphere in U containing D_i.

Let $A_1, A_2, \cdots, A_m$ be a finite collection of mutually exclusive subarcs of K such that each point of $K \cap h(C^2)$ is an interior point of one of these arcs, no one of the arcs intersects two of the D_i, and the A_j intersecting D_i does not intersect $S_i - D_i$. If we apply Theorem 2 to get a suitable isotopy h_t^i that is the identity except near A_i, we obtain the required isotopy $h_t = h_t^1 h_t^2 \cdots h_t^m$.

In another paper, we use the techniques of the present paper to get the following general position theorem for arbitrary 2-manifolds embedded in a 3-manifold.

THEOREM 4. *Suppose* M^3 *is a* 3-*manifold and* M_1^2, M_2^2 *are compact* 2-*manifolds embedded in* M^3. *Then, for each positive* ε, *there exists an isotopy* $h_t\colon M^3 \to M^3$ $(0 \le t \le 1)$ *such that* $h_0 = I$, $\rho(h_t, I) < \varepsilon$, $h_t = I$ *outside the* ε-*neighborhood of* $M_1^2 \cap M_2^2$, *and the collection of nondegenerate elements of* $M_1^2 \cap h_1(M_2^2)$ *is a null-sequence of tame, simple closed curves whose union is dense in* $M_1^2 \cap h_1(M_1^2)$.

REFERENCES

1. J. W. Alexander II, *On the deformation of an* n-*cell*, Proc. Nat. Acad. Sci. U.S.A. 9 (1923), 406-407.

2. R. H. Bing, *Approximating surfaces with polyhedral ones*, Ann. of Math. (2) 65 (1957), 456-483.

3. ———, *An alternative proof that* 3-*manifolds can be triangulated*, Ann. of Math. (2) 69 (1959), 37-65.

4. ———, *Each disk in* E^3 *contains a tame arc*, Amer. J. Math. 84 (1962), 583-590.

5. R. H. Bing, *Each disk in* E *is pierced by a tame arc*, Amer. J. Math. 84 (1962), 591-599.

6. ———, *Approximating surfaces from the side*, Ann. of Math. (2) 77 (1963), 145-192.

7. ———, *Pushing a 2-sphere into its complement*, Michigan Math. J. 11 (1964), 33-45.

8. D. R. McMillan, Jr., *Neighborhoods of surfaces in 3-manifolds*, Michigan Math. J. 14 (1967), 161-170.

9. E. E. Moise, *Affine structure in 3-manifolds, V. The triangulation theorem and Hauptvermutung*, Ann. of Math. (2) 56 (1952), 96-114.

10. R. L. Moore, *Concerning upper semi-continuous collections of continua*, Trans. Amer. Math. Soc. 27 (1925), 416-428.

The University of Wisconsin

Istituto Nazionale di Alta Matematica
Symposia Mathematica
Volume XVI (1975)

PULLING BACK FEELERS (*) (**)

R. H. BING

1. Pull-back theorems.

The purpose of this paper is to give some extensions and applications of the following result.

THEOREM 1.1: Suppose in Euclidean 3-space R^3 that

D is a tame disk,

V is an open set containing Int D,

X is a compact locally connected subset of $R^3 - \mathrm{Bd}\, D$ such that each loop in X can be shrunk to a point in $R^3 - \mathrm{Bd}\, D$, and

M is a continuum in $X - D$.

Then there is a homeomorphism h of X into $(X \cup V) - D$ such that

h is fixed on M,

any point that is moved at all is moved into $V - D$, and

if $p \in X - V$ lies in the component of X with M and h moves p into V, then Int D separates p from M in X.

Recall that a geometric object in R^3 is called *tame* if there is a homeomorphism of R^3 onto itself that takes the geometric object onto a polyhedron.

Theorem 1.1 has applications in the case where X is a topological 2-sphere S. Feelers of S may make it difficult to compute S's area or approximate it with either a PL (piecewise linear) or smooth (differentiable) 2-sphere. The homeomorphism h may be used to pull back feelers of S as shown in Figure 1. The feelers may intertwine and even

(*) I risultati conseguiti in questo lavoro sono stati esposti nella conferenza tenuta il 10 marzo 1973.

(**) Research in connection with this paper supported by U. S. Army Research Office under Grant AR-ARO-D-31-124-72-G201.

be hooked together as in the Alexander horned sphere [1] but the homeomorphism pulls them back anyway. The pull-back operation is a homotopy operation in S^3, rather than an isotopy. The end homeomorphism h is defined on S and perhaps cannot be extended to all of R^3.

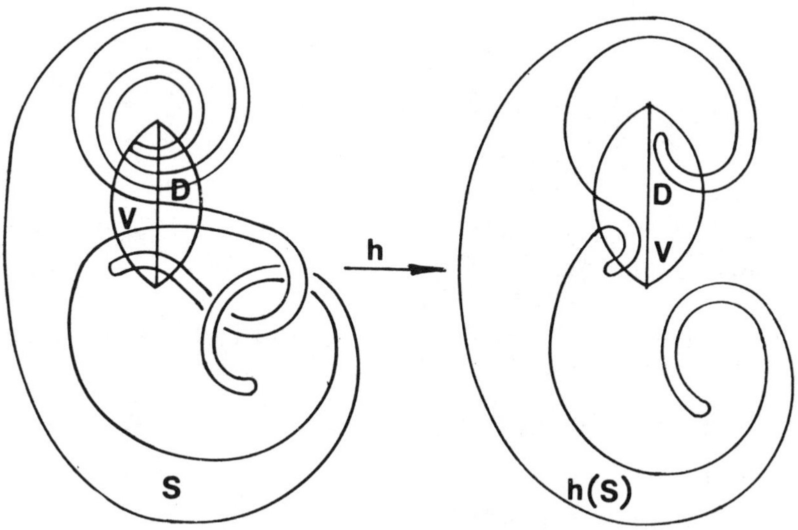

Figure 1

If X were a tame 2-sphere which intersected D in a finite number of simple closed curves, we would be tempted to prove the theorem by cutting holes in X near D and capping off these holes with disks in V to one side of D. This is a technique that has been used previously [7, 8]. However, we shall use the theorem in a more general setting.

We regard X as the union of M and some feelers reaching out from M and poking through Int D. We even permit loops in the feelers to encircle other feelers but do not allow loops in the feelers to link Bd D.

Proof of Theorem 1.1: The proof of Theorem 1.1 is not difficult. One adds the point at infinity to R^3 to change it to a 3-sphere S^3. Note that $S^3 - \text{Bd}\, D$ is an open solid torus and there is a homeomorphism f of $S^3 - \text{Bd}\, D$ onto $S^1 \times \text{Int}\, D$ such that for $(1, 0) = p_0 \in S^1$, $f^{-1}(p_0 \times y) = y$ for each $y \in \text{Int}\, D$. Let $R^1 \times \text{Int}\, D$ be a universal covering space of $S^1 \times \text{Int}\, D$ with the natural projection map $\varrho[\varrho(x, y) = ((\cos x, \sin x), y)]$.

Since each loop in X can be shrunk to a point in $R^3 - \text{Bd}\, D$, $f(X)$ can be lifted—there is a map g of $f(X)$ into $R^1 \times \text{Int}\, D$ such that ϱg

is the identity on $f(X)$. We suppose that $gf(M) \subset (0, 2\pi) \times \operatorname{Int} D$. See Figure 2.

It is convenient to regard V as a slight thickening of $\operatorname{Int} D$ that misses M and $\varrho^{-1}f(V) \cap [0, 2\pi] \times \operatorname{Int} D$ as the union of thin blisters on the ends of $[0, 2\pi] \times \operatorname{Int} D$. Let r be a homeomorphism of $R^1 \times \times \operatorname{Int} D$ onto $(0, 2\pi) \times \operatorname{Int} D$ that shoves $(-\infty, 0] \times \operatorname{Int} D$ into the left blister and $[2\pi, \infty] \times \operatorname{Int} D$ into the right blister. We require that r be the identity on $gf(M)$ and that it send any point that moves at all into $\varrho^{-1}f(V)$. The required homeomorphism h is $f^{-1}\varrho r g f$. The lower part of Figure 2 suggests how loops can be unhooked.

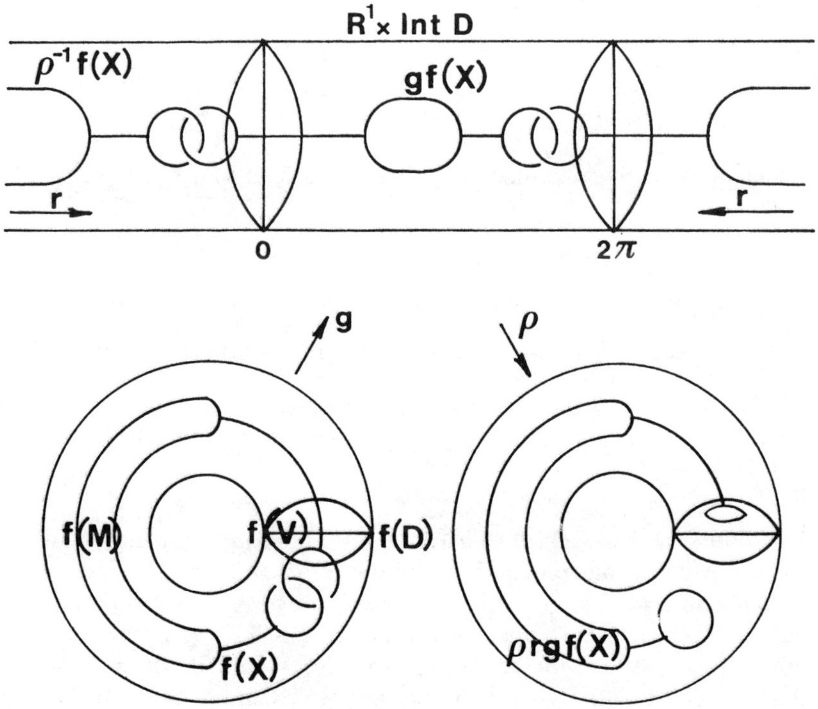

Figure 2

The homeomorphism h we have described satisfies a condition a bit more restrictive than we gave in the statement of Theorem 1.1. If $p \in X - V$, $h(p) \in V$, and A is an arc from p to M in X, then not only is it true that A intersects D but it must indeed cross D algebraically a nonzero number of times.

The distance function is denoted by d. A homeomorphism h is a δ-homeomorphism if both the domain and its image lie in the same space with metric d and for each point p of the domain of definition, $d(p, h(p)) \leqslant \delta$.

Repeated applications of Theorem 1.1 extend it as follows.

THEOREM 1.2: Suppose in R^3 that

X is a compact locally connected continuum,

$D_1, D_2, \ldots, D_n$ is a finite collection of disjoint tame disks such that each $X \cap \operatorname{Bd} D_i = 0$ and no $\operatorname{Bd} D_i$ links any simple closed curve in X,

$V_1, V_2, \ldots, V_n$ is a finite collection of disjoint open sets with $\operatorname{Int} D_i \subset V_i$, and

M is a continuum in $X - \bigcup D_i$.

Then there is a homeomorphism h of X into $(X \cup \bigcup V_i) - \bigcup D_i$ such that

h is fixed on M,

any point that is moved at all is moved into some V_i, and

if h moves $p \in X - V_i$ into V_i, then D_i separates p from M in X.

Furthermore, h is a 2ε-homeomorphism if $\varepsilon > 0$ such that each diameter $V_i < \varepsilon$ and if D_i separates a point p from M in X, than $d(p, D_i) < \varepsilon$.

PROOF: One uses Theorem 1.1 to get a homeomorphism h_1 pulling X off D_1, a homeomorphism $h_2 h_1$ pulling X off $D_1 \cup D_2, \ldots$, and a homeomorphism $h = h_n h_{n-1} \ldots h_2 h_1$ pulling X off $\bigcup D_i$. Separation is preserved in that if D_{i+j} does not separate $q \in X - D_{i+j}$ from M in X, then D_{i+j} does not separate $h_i \ldots h_2 h_1(q)$ from M in $h_i \ldots h_2 h_1(X)$. One can check this by considering an arc A from q to M in $X - D_{i+j}$ and observing that $h_i \ldots h_2 h_1(A)$ misses D_{i+j}.

A point $p \in X$ might have been moved many different places by the various h_1, $h_2 h_1 \ldots h = h_n \ldots h_2 h_1$ but its image does not stray far from p. To see this, note that if $h(p) \in X - \bigcup V_i$, then $h(p) = p$; if $h(p) \in V_i$ and $p \in V_i$, then $d(p, h(p)) < \varepsilon$ since diameter $V_i < \varepsilon$; and if $h(p) \in V_i$ and $p \notin V_i$, D_i separates p from M in X. In this latter case, p lies in the ε-neighborhood of D_i and $d(p, h(p)) < 2\varepsilon$.

The ordering of the D_i's makes a difference in the homeomorphism h finally obtained as can be seen from Figure 3. If in that figure one pulls the feelers off $D_1 \cup D_2$ by considering D_1 first and then D_2, the point p is moved. However, if one considers D_2 first and then D_1, p is not moved.

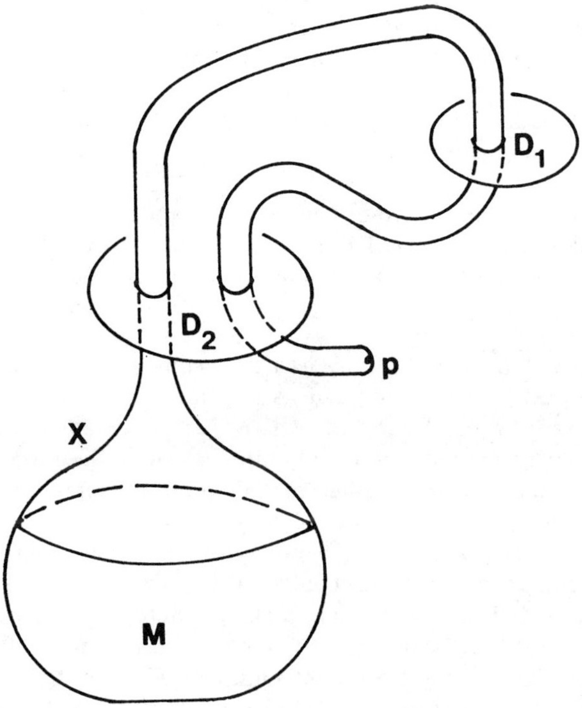

Figure 3

We give two applications (Theorems 2.2 and 3.1) of Theorem 1.2 in the next two sections.

2. Unions of crumpled cubes and 3-balls.

A *crumpled cube* is a set homeomorphic to the union of a topological 2-sphere in R^3 and its interior (bounded complementary domain). The wildness of the embedded 2-sphere determines the topology of the crumpled cube. The union of two 3-balls sewed together with a homeomorphism between their boundaries provides us with a model of a 3-sphere. The union of two crumpled cubes united in this fashion might be a dogbone space (or worse). See [2, 10, 13, 14, 16, 20]. It does not seem immediately apparent what the union of a crumpled cube and a 3-ball might be but the following result was discovered independently by Norman Hosay [17] and L. L. Lininger [18].

THEOREM 2.1 (Hosay-Lininger): If a crumpled cube K^3 is sewed to a 3-ball B^3 with a homeomorphism between their boundaries, their union is a topological 3-sphere.

Lininger gave his treatment of the above result in [18] but Hosay did not write up his proof for publication. In proving Theorem 2.1, both Hosay and Lininger proved the following.

THEOREM 2.2: For each crumpled cube K^3 in R^3 and each $\varepsilon > 0$ there is an ε-homeomorphism h of K^3 into R^3 such that $R^3 - h(K^3)$ is 1-ULC.

Theorem 2.1 follows from Theorem 2.2 and Bing's result [9] that a crumpled cube is a 3-ball if its interior is 1-ULC. (The crumpled cube shown to be a 3-ball is the one point compactification of the closure of $R^3 - h(K^3)$.)

PROOF OF THEOREM 2.2: A homeomorphism h such as promised by Theorem 2.2 can be described as the limit of a sequence of homeomorphisms h_1, h_2h_1, $h_3h_2h_1, \ldots$ where for some small ε_i, h_i is an ε_i-homeomorphism of $h_{i-1} \ldots h_1(K^3)$ into R^3 such that $h_ih_{i-1} \ldots h_1(K^3)$ lies in the bounded complementary domain of some polyhedral 2-sphere S_i such that there is an ε_i-homeomorphism of $h_{i-1} \ldots h_1(\text{Bd } K^3)$ onto S_i. To make $h_1, h_2h_1, \ldots,$ converge to a homeomorphism rather than to a map requires careful control of the size of the ε_i's since these cannot be selected in advance but each ε_i is determined by $h_{i-1}, \ldots, h_1$. To make $h_1, h_2h_1, \ldots,$ converge to a continuous ε-map it is enough to make $\varepsilon_i \leqslant \varepsilon/2^i$. To keep different points of K^3 from having the same image we could take the ε_i's so small that if p, q are two points of X with $d(p, q) > \frac{1}{2}^i$, then

$$\varepsilon_i + \varepsilon_{i+1} + \ldots < d(h_{i-1} \ldots h_1(p), h_{i-1} \ldots h_1(q))/2 .$$

See Theorem 7 of [7]. However, the crucial result is the following.

THEOREM 2.3: For each crumpled cube K^3 in R^3 and each $\varepsilon_i > 0$ there is an ε_i-homeomorphism h_i from K^3 into R^3 and a polyhedral 2-sphere S_i homeomorphically within ε_i of Bd K such that $h_i(K^3)$ lies in the bounded complementary domain of S_i.

PROOF: Recall that a *punctured disk* is a compact connected 2-manifold with boundary which can be embedded in a disk. Its boundary consists of a finite number of simple closed curves.

It follows from the Side Approximation Theorem [4, 5, 11] that there is an ε_i-homeomorphism of Bd K^3 onto a polyhedral 2-sphere S_i in R^3, a punctured disk M in Bd $K^3 \cap$ (interior of S_i) and a punctured disk M' in $S_i - K^3$ such that the components of Bd $K^3 - M$, $S_i - M'$ have diameters less than $\varepsilon_i/4$. If D is disk which is a component of $S_i -$ Int M' that separates a point p of K^3 from M in K^3 it may be shown that p lies in the $\varepsilon_i/2$ neighborhood of D. Theorem 2.3 follows directly from using Theorem 1.2 to get a homeomorphism h_i that pulls the feelers of K^3 off of the components of $S_i -$ Int M'.

Both Daverman [15] and Lininger [18] gave proofs of Theorem 2.3 but the above proof may be shorter than either of theirs (once Theorem 1.2 is known). Daverman's proof was rather easy but instead of using the Side Approximation Theorem, he used a more sophisticated version of the Side Appoximation Theorem due to Lister [19].

Instead of applying Theorem 1.2 globally, one could have applied it locally on open balls to obtain the following variation of Theorem 2.2.

THEOREM 2.4: Suppose

M^3 is a metric 3-manifold without boundary,

K^3 is an object in M^3,

S is a subset of the boundary of K^3 in M^3 which is open in this boundary and is locally a 2-manifold, and

$\varepsilon(x)$ is a map of K^3 into $[0, 1)$ that is positive on S.

Then there is a homeomorphism h of K^3 into M^3 such that

$d(x, h(x)) \leqslant \varepsilon(x)$, and

$M^3 - h(K^3)$ is 1-LC at each point of $h(S)$.

3. Tame Sierpinski curves in wild spheres.

A sequence of sets is called a *null sequence* if for each positive number ε, not more than a finite number of the sets in the sequence have diameters more than ε. A *Sierpinski curve* is a set homeomorphic to $S - \bigcup \text{Int } D_i$ where S is a 2-sphere and $D_1, D_2, \ldots$ is a null sequence of mutually exclusive disks on S whose union is dense in S. A Sierpinski curve in R^3 is *tame* if it lies on a tame 2-sphere. In this section we apply Theorem 1.2 to get a new proof of the following known result [7, 8].

THEOREM 3.1: Suppose S is a 2-sphere (possibly wild) in R^3 and $\varepsilon > 0$. Then there is a Sierpinski curve K in S and an ε-homeomorphism h of S onto a tame 2-sphere in R^3 such that

each component of $S - K$ is of diameter less than ε and

h is the identity on K.

Theorem 3.1 follows from repeated applications of the following result.

THEOREM 3.2: If S is a 2-sphere in R^3 and $\varepsilon > 0$, then there are disjoint polyhedral 2-spheres S', S'' in R^3 each homeomorphically within ε of S, a punctured disk M in S such that each component of $S - M$ is of diameter less than ε, and an ε-homeomorphism h of S into R^3 such that

$h(S)$ lies interior to S' and exterior to S'' while

h is fixed on M.

At the time that Theorem 3.2 was originally proved, Theorem 1.2 had not been formulated. Theorem 3.2 was established by using the Approximation Theorem for Surfaces [3] to replace the feelers with piecewise linear feelers, then cutting holes in the feelers, and then capping off the holes. See proof of Theorem 5 in [7]. However, Theorem 1.2 provides a shorter approach.

Theorem 3.2 can be proved by approximating S almost from the exterior by S', then using Theorem 1.2 to get a homeomorphism g_1 that pulls S to the interior of S', then approximating $g_1(S)$ even closer almost from its interior, and finally using Theorem 1.2 again to get a homeomorphism g_2 that pulls $g_1(S)$ between S' and S''. The required homeomorphism is $g_2 g_1$.

To use Theorem 3.2 to get Theorem 3.1, one would apply Theorem 3.2 infinitely often to get homeomorphisms $h_1, h_2 h_1, h_3 h_2 h_1, \ldots$. We would let $f_n = h_n h_{n-1} \ldots h_1$ and $h = \lim f_n$. The h_i's are ε_i-homeomorphisms and the ε_i's are picked so small that the following three conditions prevail.

$\lim f_n$ is a homeomorphism.

Each component of $R^3 - h(S)$ is 1-ULC.

$M_1 \cap M_2 \cap \ldots$ contains the required Sierpinski curve.

In the above, M_i is the punctured disk in S such that h_i is the identity on $h_{i-1} \ldots h_i(M_i)$. The following result insures that $M_i \cap \cap M_2 \cap \ldots$ contains the required Sierpinski curve.

THEOREM 3.3: Suppose S is a 2-sphere, $\varepsilon > 0$, δ is a positive number such that each δ-subset of S lies in an ε-disk, and $M_1, M_2, \ldots$ is a sequence of punctured disks in S such that

each component of $S - M_i$ is of diameter less than $\delta/2^i$ and

each component of $S - M_{i+1}$ is of diameter less than (distance between any two components of the closure of $S - (M_1 \cap \cap M_2 \cap \ldots \cap M_i))/2^{i+2}$.

Then $\bigcap M_i$ contains a Sierpinski curve K such that each component of $S-K$ is of diameter less than ε.

PROOF: Let X_i be the closure of $S-M_i$. It is the union of a finite number of small disks.

For each point $p\in\bigcup X_i$, let $Y(p,i)$ be the component of $X_1\cup X_2\cup\ldots\cup X_i$ containing p and $Y(p)$ be the closure of $\bigcup Y(p,i)$. If p,q belong to different components of $X_1\cup X_2\cup\ldots\cup X_i$, the second restriction on the sizes of the components of the X_i's implies that $d(Y(p),Y(q))>d(Y(p,i),Y(q,i))/2$. Hence, two $Y(p)$'s are either identical as closed sets or do not intersect.

Each $Y(p)$ is of diameter less than δ and hence lies in a disk in S of diameter less than ε. There are only a countable number of distinct $Y(p)$'s and they form a null sequence $Y_1, Y_2, \ldots$.

It follows from Moore's decomposition theorem [21] that Y_1 lies in a disk D_1 such that

diameter $D_1<\varepsilon$,

$\operatorname{Bd} D_1\cap\bigcup Y_i=0$, and

no disk in S of less than one-half the diameter of D_1 contains Y_1.

Similarly, we find a sequence of disks $D_2, D_3, \ldots$ satisfying conditions like those above with the additional restrictions that $Y_i\subset D_i$ and $\operatorname{Bd} D_i\cap(\operatorname{Bd} D_{i-1}\cup\ldots\cup\operatorname{Bd} D_1)=0$.

If the union of the above D_i's is dense in S, the required Sierpinski curve K is $S-\bigcup\operatorname{Int} D_i$. If the union is not dense, we obtain K by also removing the interior of additional disks.

By applying Theorem 1.2 locally one could have obtained the following variation of Theorem 3.1. This theorem is used to show that surfaces in 3-manifolds contain many tame arcs.

THEOREM 3.4: Suppose

M^3 is a piecewise linear 3-manifold without boundary,

S is a closed subset of M^3 which is a 2-manifold without boundary, and

$\varepsilon(x)$ is a positive continuous function on S.

Then there is a homeomorphism h of S into M^3 and a null sequence of disjoint disks $D_1, D_2, \ldots$ on S such that

$d(x,h(x))<\varepsilon(x)$,

$h(S)$ is tame in M^3,

diameter $D_i<\max\limits_{x\in D_i}\varepsilon(x)$, and

h is the identity on $S-\bigcup\operatorname{Int} D_i$.

4. **Pulling feelers off a null sequence of disks.**

In this section we consider how we may pull back feelers from a null sequence of disks.

THEOREM 4.1: Suppose in Euclidean 3-space R^3 that

X is a compact locally connected continuum,

$D_1, D_2, \dots$ is a null sequence of disjoint tame disks such that each $X \cap \operatorname{Bd} D_i = 0$ and no $\operatorname{Bd} D_i$ links any simple closed curve in X,

$V_1, V_2, \dots$ is a sequence of disjoint open sets with $\operatorname{Int} D_i \subset V_i$, and

M is a continuum in $X - \bigcup D_i$.

Then there is a map f of X into $(X \cup \bigcup V_i) - \bigcup D_i$ such that

f is fixed on M,

any point that is moved at all is moved into $\overline{\bigcup V_i}$, and

if $p \in X - \bigcup V_i$ and p is moved by f, then some D_i separates p from M in X.

Furthermore, if ε is a positive number such that each D_i is of diameter less than ε and any point that D_i separates from M in X lies in the ε-neighborhood of D_i, then there is such a map f with $d(f, \text{identity}) \leqslant 2\varepsilon$.

PROOF: We suppose that V_i's are thin neighborhoods of the Int D_i's and form a null sequence. If this is not already the case, we reduce the V_i's to bring this about. As in the proof of Theorem 1.2 we use Theorem 1.1 to get a sequence of homeomorphisms $h_1, h_2h_1, h_3h_2h_1, \dots$ of X into R^3. We use f_n to denote $h_n h_{n-1} \dots h_1$ and define $f(p)$ to be $\lim f_n(p)$.

We now check that $f(p)$ is well defined for each $p \in X$. It is clearly well defined if there is an integer n_0 such that for each $j > n_0$, $f_j(p) = f_{n_0}(p)$. Let $n_1, n_2, \dots$ be the collection of all integers n_i such that $f_{n_i}(p) \neq f_{n_i-1}(p)$. Consider the case where this is an infinite monotone increasing sequence. Then $f_{n_i}(p) \in V_{n_i}$. Let pq be an arc irreducible from p to M and p_i be the first point of Bd V_{n_i} on $f_{n_i}(pq)$ in the order from $f_{n_i}(p)$ to q. Then $p_i \notin \bigcup V_j$ and $f_j(p_i) = p_i$ for $j \leqslant n_i$. Hence $p_{i+1} \in p_i q$ of pq. Hence $\lim p_i$ is a unique point p_0 of pq. Since $\{V_i\}$ is a null sequence, $\lim f_n(p)$ is this same point p_0 and $\lim f_n(p)$ is well defined.

We now consider the continuity of f at an arbitrary point $p \in X$. We show that f is continuous by showing that if $V(f(p), \delta)$ is the open δ-neighborhood of $f(p)$ in R^3, then there is an open subset U_p of X containing p such that $f(U_p) \subset V(f(p), \delta)$. The proof breaks into three cases.

CASE 1: If for infinitely many integers j, $f_j(p) \neq f_{j-1}(p)$, we pick the monotone increasing sequence of integers $n_1, n_2, \ldots$ such that $f_{n_i}(p) \neq f_{n_i-1}(p)$ and let U_p be a connected open subset of X containing p such that $f_{n_1}(U_p) \subset V_{n_1}$. One finds that for each i, $f_{n_i}(U_p) \subset V_{n_i}$ and $f(U_p) = f(p) \subset V(f(p), \delta)$.

CASE 2: If for some n_0, $f_{n_0}(p) \in V_{n_0}$ and for $j \geqslant n_0$, $f_j(p) = f_{n_0}(p)$, we let U_p be a connected open subset of X containing p such that $f_{n_0}(U_p) \subset V(f(p), \delta) \cap V_{n_0}$ and note that $f(U_p) = f_{n_0}(U_p) \subset V(f(p), \delta)$.

CASE 3: If for each j, $f_j(p) \in X - \bigcup V_i$, then each $f_i(p) = p$ and $f(p) = p$. Let n_0 be an integer so large that for $j \geqslant n_0$, diameter $V_{n_0} < \delta/2$ and U_p be an open subset of X containing p such that each point t of U_p there is an arc pt in X such that $f_{n_0}(pt)$ is of diameter less than $\delta/2$. For $j \geqslant n_0$, $f_j(t)$ lies either on $f_{n_0}(pt)$ or in a $V_m (m \geqslant n_0)$ whose boundary contains a point of $f_{n_0}(pt)$. Hence $f_j(t) \in V(f(p), \delta)$ and $f(t) \in V(f(p), \delta)$.

Since each $d(f_i, \text{identity}) \leqslant 2\varepsilon$, $d(f, \text{identity}) \leqslant 2\varepsilon$.

We can put additional restrictions on X and M to insure that the map f of Theorem 4.1 is a homeomorphism.

THEOREM 4.2: Suppose in Euclidean 3-space R^3 that

X is a compact locally connected continuum with no separating points,

$D_1, D_2, \ldots$ is a null sequence of disjoint tame disks such that each $X \cap \text{Bd}\, D_i = 0$ and no $\text{Bd}\, D_i$ links any simple closed curve in X,

$V_1, V_2, \ldots$ is a sequence of disjoint open sets with $\text{Int}\, D_i \subset V_i$, and

M is a nondegenerate continuum in $X - \bigcup D_i$.

Then there is a homeomorphism h of X into $(X \cup \bigcup V_i) - \bigcup D_i$ such that

h is fixed on M,

any point that is moved at all is moved into $\bigcup V_i$, and

if $p \in X - \bigcup V_i$ and p is moved at all, then some D_i separates p from M in X.

Furthermore, if ε is a positive number such that each V_i is of diameter less than ε and any point that D_i separates from M in X lies in the ε-neighborhood of D_i, then there is such a homeomorphism h such that $d(h, \text{identity}) \leqslant 2\varepsilon$.

PROOF: The required homeomorphism h is $\lim f_n$ as given in the proof of Theorem 4.1. The reason that the limit is a homeomorphism rather than merely a map is that for each point p, then except for a finite number of integers, $f_{i+1}(p) = f_i(p)$. To see that this is true, we would note from the proof of Theorem 4.1 that if for infinitely many $n, f_n(p) \neq f_{n-1}(p)$, then for each arc A from p to M, then $\lim f_n(p) = h(p) \in A$. But there is an arc A' from p to M that misses $h(p)$ so it is impossible that for infinitely many $n, f_n(p) \neq f_{n-1}(p)$. Since no two points have the same image under any f_n, no two have the same image under h.

THEOREM 4.3: Theorem 4.2 holds if instead of supposing that X has no separating points and M is nondegenerate, we suppose instead that if $p \in X - M$, not more than a finite number of D_i's separate p from M in X.

In Section 7 we give another result about pulling feelers of X off of a sequence (not necessarily a null sequence) of disks.

5. Easing proof of approximation theorem.

One version of the Approximation Theorem of Surfaces is as follows.

APPROXIMATION THEOREM OF 2-SPHERES IN R^3: Suppose S is a topological 2-sphere in R^3. Then for each $\varepsilon > 0$ there is an ε-homeomorphism h of S onto a polyhedral 2-sphere $h(S)$ in R^3.

The above theorem has had many applications but its proof is still not widely known. Bing [3] gave the original proof of this result and then gave [4, 5] a proof of a deeper result called the Side Approximation Theorem. Cannon [11] has given an alternative proof of the Side Approximation Theorem which implies the Approximation Theorem for 2-spheres in R^3 above.

We outline Bing's proof so that we can show how pulling back feelers can ease the most difficult step in the proof. We give the outline in seven steps.

Step 1: One considers a triangulation T of R^3 of small mesh so that the 0-skeleton T^0 of T misses S. We suppress details here as to how small the mesh of T must be but once this size is determined, it is easy to obtain a T.

Step 2: The next step is to adjust S so that the adjustment S_1 intersects the 1-skeleton T^1 of T in only a finite number of points. Perhaps this was the most difficult part of Bing's proof. The original proof made use of cactoids to verify the construction and some readers did not understand the procedure. The use of Theorem 6.1 given in Section 6 provides an alternative approach which we believe is simpler.

Step 3: Adjust S_1 to get a 2-sphere S_2 such that $S_2 \cap T_0 = 0$, $S_2 \cap T_1$ is finite, and if A is a 1-simplex of T with $p \in S_2 \cap A$, then A crosses from the interior of S_2 to the exterior of S_2 at p. The 2-sphere S_2 is obtained as the image of S_1 under a space homeomorphism of R^3 that is fixed except near certain points of $S_1 \cap A$ where S_1 is moved to eliminate superfluous intersections. The procedure is an easy one and involves unwinding arcs around cylinders as described in Theorem 3 on page 162 of [4].

Step 4: Adjust S_2 with a space homeomorphism to get a 2-sphere S_3 such that $S_3 \cap T^1 = S_2 \cap T^1$ but if F is any 2-simplex of T, then each component of $S_3 \cap F$ that intersects Bd F intersects it in precisely two points. Bd F crosses S_3 at each point of $S_3 \cap$ Bd F. The procedure of getting S_3 is given by Theorem 8 on page 169 of [4].

Special case: We digress from listing the steps in the proof to examine a special case whose consideration may make the outline easier to follow. Suppose each component of $F \cap S_3$ that intersects T^1 is a spanning arc A of F and the union of all such arcs (considering all F's) is a finite graph G which is the union of the 1-cells of a cellular subdivision of S_3. In this special case we would replace each component A of $G \cap F$ by a polygonal spanning arc A' of F with the same ends as A and get a homeomorphism h of G onto a polygonal 1-complex. Note that if Δ^3 is a 3-simplex of T, then Bd $\Delta^3 \cap h(G)$ is the union of a collection of disjoint polygonal simple closed curves. By building disjoint polygonal disks in Δ^3 whose boundaries are the simple closed curves and extending h to take the 2 cells of the cellular subdivision of S_3 into such disks, one can get a homeomorphism of S_3 onto a polygonal 2-sphere.

Step 5: We return to the case where a component of $S_3 \cap F$ that intersects Bd F is not necessarily a spanning arc A of F but is rather a continuum A'' in F with exactly two points on Bd F. Let $A_1'', A_2'', \ldots, A_k''$ be all such continua (considering all F's) and call each point of $A_i'' \cap$ Bd F an end point of A_i''. Let $A_1, A_2, \ldots, A_k$ be arcs in S_3 approximating the A_i'''s and having the same end points as the A_i'''s. Two A's intersect if and only if they share an end point. No component of $\bigcup A_i$

has a separating point (Figure 3 of [4] is in error) and each component of $S_3 - \bigcup A_i$ is small. Hence there is a component G of $\bigcup A_i$ such that G is the 1-skeleton of a cellular curvilinear subdivision of S_3 and the closure of each component of $S_3 - G$ is a small disk D. The A_i's could been selected so that the A_i'''s associated with the A_i's on any such Bd D all lie on the boundary of the same 3-simplex of T so we suppose they were selected in this way.

Step 6: If A_i is an arc as above in G and F is the 2-simplex of T containing A_i'', let A_i' be a polygonal arc in F so that $A_i' \cap A_j' = A_i \cap A_j$. Let h be a homeomorphism of G into T^2 so that $h(A_i) = A_i'$.

Step 7: For each component D_i of $S - G$ and 3-simplex Δ^3 of T containing the A_j'''s associated with Bd D_i, let D_i' be a polygonal disk in Δ^3 so that $D_i' = h(\text{Bd}\, D_i)$. Pick the D_i''s so that no two intersect needlessly and extend h to take each D_i onto the corresponding D_i'. This extended homeomorphism takes S_2 onto the polyhedral 2-sphere $\bigcup D_i'$.

The reader who is interested in the details of the above steps is referred to the original treatments [3, 4, 5]. However, in Section 6 we give a complete treatment of the difficult Step 2.

A 2-manifold embedded in a 3-manifold as a closed set is sometimes called a *surface*. We deal with surfaces (such as disks) that have boundaries and others (such as 2-spheres, tori, planes) that do not. The proof of the following theorem follows the proof of the Approximation Theorem of 2-spheres in R^3.

Approximation Theorem for Surfaces: Suppose S is a surface in a piecewise linear 3-manifold M^3 and $\varepsilon(x)$ is a map of S into the nonnegative reals. Then there is a homeomorphism h of S into M^3 such that

$d(x, h(x)) \leqslant \varepsilon(x)$, and
$h(S)$ is locally polyhedral at x if $\varepsilon(x) > 0$.

The following is a version of a theorem we have mentioned.

Side Approximation Theorem for 2-Spheres in R^3: Suppose S is a 2-sphere in R^3 and $\varepsilon > 0$. Then there are disjoint polyhedral 2-spheres S', S'' in R^3 each homeomorphically within ε of S and a punctured disk M in S such that each component of $S - M$ is of diameter less than ε and M lies interior to S' and exterior to S''.

The steps in the proof of the Side Approximation Theorem are similar to those we have given except that in Step 5, instead of taking A_i' to have the same ends as A_i, we consider two spanning arcs B_i, C_i as shown in Figure 4 so that A_i'' lies between them. The B_i's corresponding to the A_i's in G lie in one of S' and the C_i's corresponding to these A_i's in G lie in S''. The reason that there is an M is that the union of the A_i'''s corresponding to A_i's in G lie between S' and S''.

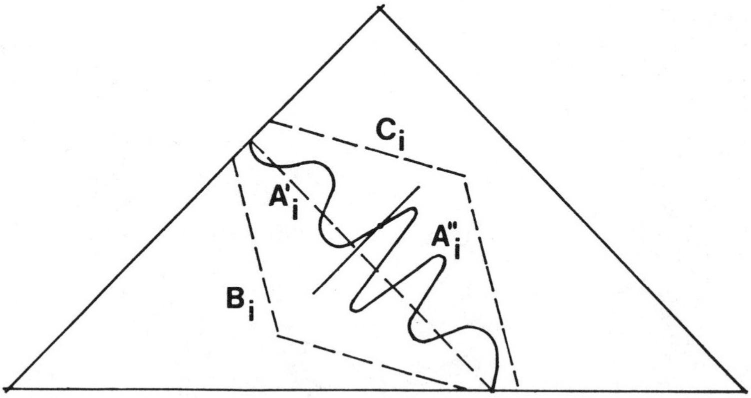

Figure 4

Of course, this only builds side approximations to the S_3 of Step 4. To get a side approximation of the original S, one would need to reverse the homeomorphisms that took S to S_3.

6. Adjusting surfaces with respect to intervals.

Theorem 6.1 given below guarantees that Step 2 of the proof of the approximation theorems given in the preceding section is valid. There is a stronger version of Theorem 6.1 which states that the homeomorphism h mentioned there can be extended to an ε-homeomorphism of R^3 onto itself so as to be the identity on $R^3 - U$. This stronger version is given in [6] and [2] but we do not find it useful to use the stronger version to substantiate Step 2 since the proofs of the stronger version start with either the assumption that certain approximation theorems hold or that surfaces contain tame arcs.

While Theorem 6.1 holds equally well for surfaces in piecewise linear 3-manifolds, for simplicity, we state it for 2-spheres in R^3.

THEOREM 6.1: Suppose in R^3 that

S is a topological 2-sphere,

A is a straight line interval,

U is a neighborhood of $A \cap S$, and

$\varepsilon > 0$.

Then there is an ε-homeomorphism h of S into $S \cup U$ such that

h is the identity on $S - U$, and

$A \cap h(S)$ is finite.

PROOF: We suppose with no loss of generality that S misses the ends of A and dimension $(A \cap S) \leqslant 0$. Theorem 6.2 to follow shows that this is permissible.

Since no 0-dimensional set separates S, there is a punctured disk M in $S - A$ such that each disk of $S - \text{Int}\, M$ is of diameter less than $\varepsilon/4$. We suppose that ε is so small that the ε-neighborhood of $S - M$ lies in U. Let C be a solid right circular cylinder with A as axis such that

S misses the ends of C,

$S \cap M = 0$,

diameter (cross section of C) $< \varepsilon/8$, and

each component of the perpendicular projection of $S \cap C$ onto A is of diameter less than $\varepsilon/8$.

The above conditions can be forced by making the radius of the cross section of C small.

Our aim is to adjust S so as to reduce the number of components of $S \cap \text{Bd}\, C$. Let Y be such a component and let us consider how and if we can adjust S so as to remove it.

If Y lies in a disk on the lateral side of C, it lies on the interior of such a disk D whose boundary misses S. If we let V be a thin neighborhood of $\text{Int}\, D$ and apply Theorem 1.1 we find that we can adjust S so that the adjustment misses $\text{Int}\, D$. This adjustment may remove many more components of $S \cap \text{Bd}\, C$ and if V were sufficiently thin, it would not add any.

If Y does not lie in a disk on the lateral side of C, it separates the ends of C from each other on $\text{Bd}\, C$. If there are only a finite number of such Y's, we might throw the part of S away that these Y's separate from M and then shrink these Y's to points by pinching C as shown in Figure 11 on page 472 of [3]. A problem that arises is that there may be feelers reaching through the Y's so that when the Y's are pinched, singularities with the feelers are introduced. This suggests that the feelers should be removed before the circling Y's are pinched. This is where the concept of pulling back feelers is useful.

Let $\{Y_a\}$ be the collection of components of $S \cap \text{Bd}\, C$ that separate the ends of C from each other on $\text{Bd}\, C$. Lemma 6.3 provides us with a sequence of disjoint $\varepsilon/4$ disks $D_1, D_2, \ldots$ on the lateral side of C such that

> each $\text{Bd}\, D_1 \cap S = 0$,
>
> each component of $S \cap \text{Bd}\, C$ not an element of $\{Y_a\}$ lies in some D_i, and
>
> the collection $\{D_i\}$ is locally finite except possibly at $\bigcup Y_a$.

Note that each D_i lies in U.

If this sequence of disks were a null sequence, one could apply Theorem 4.2 to get a homeomorphism f of S into $S \cup U$ such that each component of $\text{Bd}\, C \cap f(S)$ is an element of $\{Y_a\}$. However an examination of the geometry of topological 2-spheres shows that the noncircling components of $S \cap \text{Bd}\, C$ may get long and thin as the components get close to $\bigcup Y_a$. Hence it may be impossible to select suitable disks with diameters converging to 0. To pull feelers off many long disks we develop Theorem 7.1 in the next section. We need to know that if $p \in S - \bigcup Y_a$, then not more than a finite number of D_i's separate p from M in S. This can be verified by the use of Lemma 6.4 which is listed.

Let $V_1, V_2, \ldots$ be a sequence of disjoint thin neighborhoods of the $\text{Int}\, D_i$'s such that each V_i lies in U and intersects $\text{Bd}\, C$ only in $\text{Int}\, D_i$.

We apply Theorem 7.1 letting S, $\bigcup Y_a$, $\{D_i\}$, $\{V_i\}$, M, $\bigcup Y'_\beta$ be the S, Y, $\{D_i\}$, $\{V_i\}$, M, Y' of the statement of that theorem. It does not bother us that the limiting function f we obtain in Theorem 7.1 is not continuous on $\bigcup Y'_\beta$ because we are going to shrink certain elements of Y'_β to points later and thereby recover continuity. It follows from Theorem 7.1 that there is a subcollection $\{Y'_\beta\}$ of $\{Y_\alpha\}$ and a one-to-one function f (perhaps not continuous) of S into $S \cup U$ such that

> f is fixed on $\bigcup Y'_\beta \cup M$,
>
> if f moves a point $p \in S$, $f(p) \in U - \text{Bd}\, C$,
>
> $f(S) \cap \text{Bd}\, C = \bigcup Y'_\beta$,
>
> $d(f, \text{identity}) \leqslant \varepsilon/2$,
>
> f is continuous on $S - \bigcup Y'_\beta$, and
>
> if a sequence of points $p_1, p_2, \ldots$ of S converges to a point of any element Y'_γ of $\{Y'_\beta\}$, each open set containing Y'_γ contains all but a finite number of $f(p_1), f(p_2), \ldots$.

The function f pulls S off of $\text{Bd}\, C$ except for some circling components.

Let W be the component of $S - \bigcup Y'_\beta$ containing M. If L_A is the line containing A, each Y'_β separates M from a point of $L_A \cap (S - M)$ in S. Hence $\overline{W}$ is the union of W and a finite number of elements $Y_1, Y_2, \ldots, Y_j$ of $\{Y'_\beta\}$. Let Y_i^+ be the union of Y_i and the set of all points that it separates from M in S. It follows from Moore's decomposition theorem [21] that there is a map g of S onto S such that g is fixed on M and each nondegenerate point inverse is one of $Y_1^+, Y_2^+, \dots, Y_j^+$. Let $g_1 = g|\overline{W}$. Note that $f(W) \cap C = 0$, $fg_1^{-1}(S) \cap \text{Bd}\, C = \bigcup_{i=1}^{j} Y_i$, and $d(g_1, \text{identity}) < \varepsilon/4$.

Since each component of the perpendicular projection of $S \cap C$ onto A is of diameter less than $\varepsilon/8$, we can subdivide C by cross sections into cylinders with diameters less than $\varepsilon/4$ such that each Y_i lies in one of the cylinders. Applying Lemma 6.5 to each small cylinder we find that there is a map g_2 of R^3 onto itself such that

g_2 is fixed outside U,
$d(g_2(p), p) < \varepsilon$ for each $p \in R^3$,
g_2 takes $R^3 - C$ homeomorphically onto $R^3 - A$, and
$g_2(Y_1), g_2(Y_2), \dots, g_2(Y_j)$ are different points of A.

The homeomorphism h promised by Theorem 6.1 is given by $h = g_2 f g_1^{-1}$. Since g_2, f, g_1 are within $\varepsilon/4$, $\varepsilon/2$, $\varepsilon/4$ respectively of the identity, $d(h, \text{identity}) \leqslant \varepsilon$.

THEOREM 6.2: Suppose in R^3 that S is a set which contains no open set and A is a straight line interval. Then for each $\delta > 0$ there is a homeomorphism $h: R^3 \to R^3$ such that

dimension $(A \cap h(S)) \leqslant 0$,
$h(S)$ misses the ends of A,
$d(h, \text{identity}) < \delta$, and
h is fixed outside the δ neighborhood of A.

The homeomorphism is obtained by a sequence of homeomorphisms the first of which pulls S off one end of A, the second of which leaves the adjusted S just as far off this end as before but also pulls it off the other end, the third also pulls S off the center of A, ... The homeomorphisms are selected so as to converge to a homeomorphism with the required properties.

LEMMA 6.3: Suppose in R^3 that S is a topological 2-sphere and C is a right circular cylinder such that

S misses the ends of C and
each component of $S \cap \text{Bd}\, C$ is of diameter less than $\varepsilon/4$.

Let $\{Y_a\}$ be the collection of components of $S \cap \text{Bd}\, C$ that separates the ends of C from each other in Bd C. Then there is a sequence of mutually exclusive disks $D_1, D_2, \ldots$ on the lateral side of C such that

each $S \cap \operatorname{Bd} D_i = 0$,

each diameter $D_i < \varepsilon/4$,

each component of $S \cap \operatorname{Bd} C$ not in $\{Y_a\}$ lies in some D_i, and

the collection $\{D_i\}$ is locally finite except possibly at $\bigcup Y_a$.

PROOF: Let Z be a component of $\operatorname{Bd} C - (\bigcup Y_a \cup \text{ends of } C)$. We show how to find the D's in Z.

For each component Y of $Z \cap S$, let Y^+ denote the union of Y and all points of $\operatorname{Bd} C$ that Y separates from the ends of C on $\operatorname{Bd} C$. Then Y^+ does not separate $\operatorname{Bd} C$. For each point $p \in Z$, let Y_p denote

$\{p\}$ if p is not in any Y^+, and

the union of all Y^+'s containing p if p lies in some Y^+.

Let G be the decomposition of $\operatorname{Bd} C$ whose elements are either Y_p's or components of $\operatorname{Bd} C - Z$.

It follows from Moore's decomposition theorem [21] that there is a map g of Z onto (R^2-origin) such that each point inverse is an element of G in Z. Then $g(S \cap Z)$ is a 0-dimensional set closed relative to (R^2-origin). Hence there is a sequence of mutually exclusive disks $E_1, E_2, \ldots$ in R^2-origin such that the boundary of each of these disks misses $g(S \cap Z)$, each point of $g(S \cap Z)$ lies in one of the disks, $\{E_i\}$ is locally finite except possibly at the origin, and each $g^{-1}(E_i)$ is of diameter less than $\varepsilon/4$. The D's in Z are the inverses of the E's.

The following two results are not deep but they are quoted here since we made use of them in the proof of Theorem 6.1.

LEMMA 6.4: Suppose in R^3 that S is a 2-sphere and E_1, E_2 are two disjoint disks in S. Then there is a positive number δ such that no subset of $S - (E_1 \cup E_2)$ that separates E_1 from E_2 in S can be shrunk to a point in the δ-neighborhood of $S - (E_1 \cup E_2)$.

LEMMA 6.5: Suppose in R^3 that L is a straight line, C is a solid right circular cylinder with axis A on L, U is an open set containing C, and $Y_1, Y_2, \ldots, Y_n$ are disjoint continua on the lateral side of C each of which separates the ends of C from each other on $\operatorname{Bd} C$. Then there is a map g of R^3 onto itself such that

g is fixed on $(R^3 - U) \cup L$,

g takes $R^3 - C$ homeomorphically onto $R^3 - A$, and

$g(Y_1), g(Y_2), \ldots, g(Y_n)$ are different points of A.

7. Pulling feelers off many long disks.

In Theorems 4.1, 4.2, 4.3 we pulled X off a null sequence of disks but in the proof of Theorem 6.1 we needed to consider a sequence of disks which was not a null sequence. The needed theorem follows.

THEOREM 7.1: Suppose in R^3 that

S is a compact locally connected continuum,

Y is a closed subset of S,

$D_1, D_2, \ldots$ is a sequence of disjoint tame disks in $R^3 - Y$ such that each $S \cap \operatorname{Bd} D_i = 0$, no $\operatorname{Bd} D_i$ links any simple closed curve in S, and $\{D_i\}$ is locally finite mod Y,

$V_1, V_2, \ldots$ is a sequence of disjoint open sets with $\operatorname{Int} D_i \subset V_i$, and

M is continuum in $S - \bigcup D_i$ such that no point p of $S - (M \cup Y)$ is separated from M by infinitely many D_i's.

Then there is a closed set Y' which is the union of components of Y and a one-to-one function f (not necessarily continuous) of S into $(S \cup \bigcup V_i) - \bigcup D_i$ such that

f is fixed on $Y' \cup M$,

$f(S) \cap Y = Y'$,

f is continuous on $S - Y'$,

if $p \in S - \bigcup V_i$ and p is moved by f, then some D_i separates p from C in S and p is moved into $\bigcup V_i$, and

if a sequence of points $p_1, p_2, \ldots$ of S converge to a point of a component of Y', each open set containing this component contains all but a finite number of $f(p_1), f(p_2), \ldots$.

Furthermore, if ε is a positive number such that each D_i is of diameter less than ε and any point that D_i separates from M in S lies in the ε-neighborhood of D_i, then there is such a function f such that $d(f, \text{identity}) \leqslant 2\varepsilon$.

PROOF: We suppose that V_i is a thin connected neighborhood of $\operatorname{Int} D_i$ of diameter less than ε which lies in the $\frac{1}{2}^i$-neighborhood of D_i and misses Y. The function f can be defined as the limit of a sequence of homeomorphisms $h_1, h_2h_1, \ldots$ such as used to define the f's of Theorem 4.1, 4.2. We let $f_j = h_j \ldots h_2 h_1$ and let $f = \lim f_j$. We recall that h_i is a homeomorphism of $f_{i-1}(S)$ into $(f_{i-1}(S) \cup V_i) - D_i$ such that h_i is the identity on M, any point that is moved at all is moved into $V_i - D_i$, and if a point $f_{i-1}(p)$ outside V_i is moved into V_i, then D_i separates p from M in S.

A component of Y is in Y' if and only if for each i, f_i is fixed on the component. Note that $\{Y \cap f_i(Y),\ i = 1, 2, ...\}$ is a decreasing sequence of closed sets each of which is the union of components of Y. Hence $f = \lim f_i$ is defined and fixed on Y'.

We now show that f is defined and continuous at each point $p \in \in S - Y$. Since $\{D_i\}$ is locally finite at p and at most a finite number of D_i's separate p from C in S, there is a connected neighborhood N of p in S and an integer j such that for $n > j$, h_n is the identity on $f_j(N)$. Then f agrees with f_j on N.

We now show that f is defined and continuous at each point $p \in \in Y - Y'$. There is an integer i and a connected neighborhood N of p in S such that $f_i(N) \subset V_i$ and $N \not\subset Y$. If q is a point of $N - Y$, at most a finite number of D_i's separate q from C in S. Hence there is an integer $j \geqslant i$ such that for $n > j$, $f_j(q) \in V_j$ and h_n is the identity on $f_j(q)$. It follows that $f_j(N) \subset V_j$ also and f agrees with f_j on N.

Finally we consider the condition about a sequence $p_1, p_2, ...$ of points converging to a point $p' \in Y'$. Some subsequence of $f(p_1), f(p_2), ...$ converges to a point p'' so we suppose the whole sequence does. Assume $p' \neq p''$ and $f(p_i) \in V_{n_i}$. Let N be a small connected neighborhood of p' in S containing p_j: Then V_{n_j} intersects N or else $f_{n_j}(N) \subset V_{n_j}$ since $f_{n_j}(p_j) \in V_{n_j}$. This violates the fact that $f_{n_j}(p') \notin V_{n_j}$. Hence, each neighborhood of p' intersects infinitely many V_{n_i}'s and each neighborhood of p'' intersects infinitely many V_{n_i}'s. It follows from Theorem 7.2 that p', p'' belong to the same component of Y.

Recall that if $X_1, X_2, ...$ is a sequence of sets, then *Sup* $\{X_i\}$ denotes the set of all points p such that each neighborhood of p intersects infinitely many X_i's and *Inf*$\{X_i\}$ denotes the set of all points p such that each neighborhood of p intersects all but finitely many X_i's. Since the $\{V_i\}$ of Theorem 7.2 is locally finite mod Y, $\mathrm{Sup}\{V_i\} \subset Y$.

The following result is standard and is only listed because it was used in the proof of Theorem 7.1.

LEMMA 7.2: If $X_1, X_2, ...$ are connected subsets of a compact set and if $\mathrm{Inf}\{X_i\}$ contains a point, then $\mathrm{Sup}\{X_i\}$ is connected.

Testo pervenuto il 9 marzo 1973.
Bozze licenziate il 23 settembre 1975.

REFERENCES

[1] J. W. ALEXANDER, *An example of a simply connected surface bounding a region which is not simply connected*, Proc. Nat. Acad. Sci., U.S.A., 19 (1924), 8-10.

[2] B. J. BALL, *The sum of two solid horned spheres*, Ann. of Math. (2) 69 (1959), 253-257.

[3] R. H. BING, *Approximating surfaces with polyhedral ones*, Ann. of Math. (2) 65 (1957), 456-483.

[4] R. H. BING, *Approximating surfaces from the side*, Ann. of Math. (2) 77 (1963), 145-192.

[5] R. H. BING, *Improving the side approximation theorem*, Trans. Amer. Math. Soc. 116 (1965), 511-525.

[6] R. H. BING, *Improving the intersections of lines and surfaces*, Mich. Math. J., 14 (1967), 155-159.

[7] R. H. BING, *Each disk in E^3 contains a tame arc*, Amer. J. Math., 84 (1962), 583-590.

[8] R. H. BING, *Applications of the side approximation theorem for surfaces*, Proc. Sympos. General Topology and its Relations to Modern Analysis and Algebra (Prague, 1961), 91-95.

[9] R. H. BING, *A surface is tame if its complement is $1 - ULC$*, Trans. Amer. Math. Soc., 101 (1961), 294-305.

[10] R. H. BING, *A homeomorphism between the 3-sphere and the sum of two solid horned spheres*, Ann. of Math. (2) 56 (1952), 354-362.

[11] J. W. CANNON, *New proofs of Bing's approximation theorem for surfaces*, Pac. J. Math., accepted.

[12] J. W. CANNON, *Piercing disks with tame arcs*, Mich. Math. J., submitted.

[13] L. O. CANNON, *Sums of solid horned spheres*, Trans. Amer. Math. Soc., 122 (1966), 203-228.

[14] B. G. CASLER, *On the sum of two solid Alexander horned spheres*, Trans. Amer. Math. Soc., 116 (1965), 135-150.

[15] R. J. DAVERMAN, *A new proof of the Hosay-Lininger theorem about crumpled cubes*, Proc. Amer. Math. Soc., 23 (1969), 52-54.

[16] R. J. DAVERMAN and W. T. EATON, *A dense set of sewings of two crumpled cubes yields S^3*, Fund. Math., 65 (1969), 51-60.

[17] N. HOSAY, *The sum of a real cube and a crumpled cube is S^3*, Notices Amer. Math. Soc., 10 (1963), 666. Abstract 607-17.

[18] L. L. LININGER, *Some results on crumpled cubes*, Trans. Amer. Math. Soc., 118 (1965), 544-549.

[19] F. M. LISTER, *Simplifying intersections of disks in Bing's side approximation theorem*, Pac. J. Math., 22 (1967), 281-295.

[20] J. M. MARTIN, *The sum of two crumpled cubes*, Mich. Math. J., 19 (1966), 147-151.

[21] R. L. MOORE, *Concerning supper semi-continuous collections of continua*, Trans. Amer. Math. Soc., 27 (1925), 416-428.

E. Topics Related to the Poincaré Conjecture

3

Some Aspects of the Topology of 3-Manifolds Related to the Poincaré Conjecture

R. H. Bing

1. INTRODUCTION

One of the most interesting questions in the study of 3-manifolds asks if the following conjecture is true.

Poincaré Conjecture. A compact connected 3-manifold is topologically a 3-sphere if it is simple connected.

Our discussion will center around some of the aspects of 3-manifolds as they relate to this conjecture.

The last decade has seen the solutions of many interesting and important problems about 3-manifolds. When we consider some of the remarkable results obtained, it is a bit surprising that the Poincaré conjecture remains unconquered. Before reviewing some definitions at the end of this section, we mention several of these problems.

Moise showed [15] that for open subsets U of Euclidean 3-space E^3, homeomorphisms can be approximated by piecewise linear homeomorphisms. If U is an open subset of E^3, f is a continuous positive function defined on U, and $h: U \to E^3$ is a homeomorphism, there is a piecewise linear homeomorphism $g: U \to E^3$ such that distance $[h(x), g(x)] < f(x)$ for each point x of U (see Figure 1). This result was used to show [16] that 3-manifolds can be triangulated. The result also yields the Hauptvermutung. If T_1, T_2 are two triangulations of the same 3-manifold, then T_1, T_2 have isomorphic rectilinear subdivisions. Hence, in attacking the Poincaré conjecture, it suffices to limit our consideration to piecewise linear 3-manifolds.

Papakyriakopoulos [18] made a remarkable breakthrough when he proved Dehn's lemma, which states that if D is a 2-simplex, M^3 is a combinatorial 3-manifold, and $f: D \to M^3$ is a piecewise linear map such that for some neighborhood N of $Bd\ D$ in $D, f^{-1}: f(N) \to D$ is a homeomorphism, then $f(Bd\ D)$ bounds a polyhedral disk in M^3 (see Figure 2). The theorem was proved by scissors and paste techniques. Using similar techniques, Papa established [17] the loop theorem—if M^3 is a triangulated 3-manifold with boundary and L is a polyhedral singular closed curve on $Bd\ M^3$ that can be shrunk to a point in M^3 but not on $Bd\ M^3$, then there is a polyhedral simple closed curve with the same property.

Another noteworthy theorem by Papa [18] was the sphere theorem, which states that if S^2 is a 2-sphere, M^3 a compact triangulated 3-manifold, and $f: S^2 \to M^3$ a map of S^2 into M^3 that is not homotopic to a constant, there is a polyhedral 2-sphere in M^3

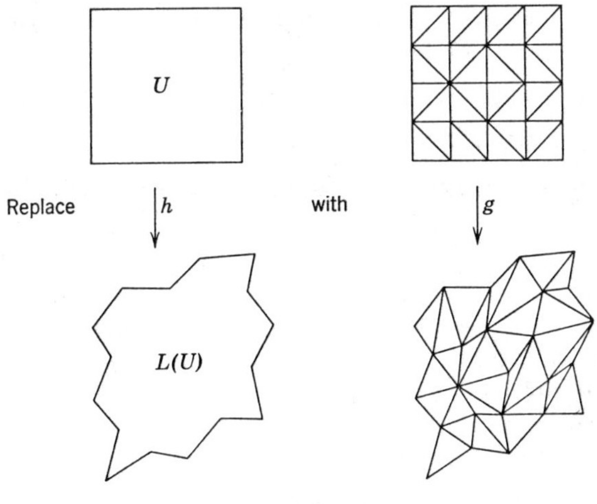

Figure 1

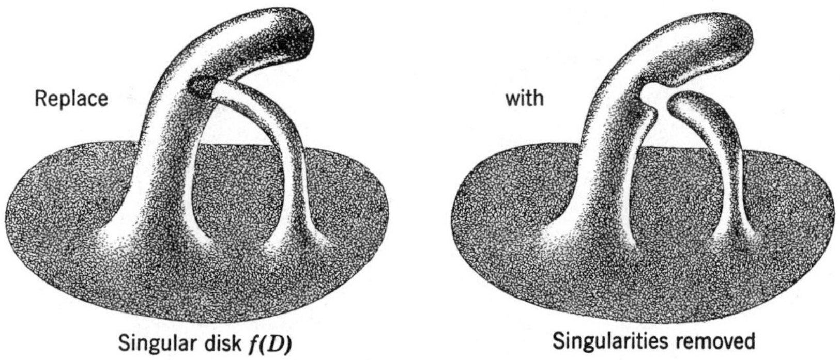

Figure 2

which cannot be shrunk to a point in M^3. We might hope that this Sphere theorem would be of some help in classifying compact 3-manifolds, but again the Poincaré conjecture gets in the way [19, 14]. Papa made an attack [20] on the Poincaré conjecture, but the victory was only partial.

Bing partially bridged the gap between the study of the topological embeddings of objects in triangulated 3-manifolds and the piecewise linear embeddings of these objects when he showed [4] that any surface in E^3 can be approximated by a polyhedral surface. If M^2 is a 2-manifold, f is a continuous nonnegative function defined on M^2, and $h: M^2 \to E^3$ is a homeomorphism of M^2 onto a closed subset of E^3, there is a piecewise linear homeomorphism $g: M^2 \to E^3$ such that for each point $x \in M^2$, distance $[h(x), g(x)] < f(x)$ and g is locally piecewise linear at $x \in M^2$ if $f(x) > 0$. Figure 3 indicates how to start building an approximation to $h(M^2)$. This theorem (with later refinements of it) were used to give [6] an alternative proof of Moise's results on the triangulation of 3-manifolds and of his results that homeomorphisms of open subsets of E^3 can be approximated by piecewise linear homeomorphisms.

Bing made an attack on the Poincaré conjecture in [5] but achieved only a partial solution. Other partial solutions not previously appearing in the literature are provided by Theorems 7, 17, and 18 of this chapter.

An *n-manifold* is a separable metric space, each of whose points has a neighborhood homeomorphic with Euclidean n-space E^n. Some mathematicians denote E^n by R^n since it is the cartesian

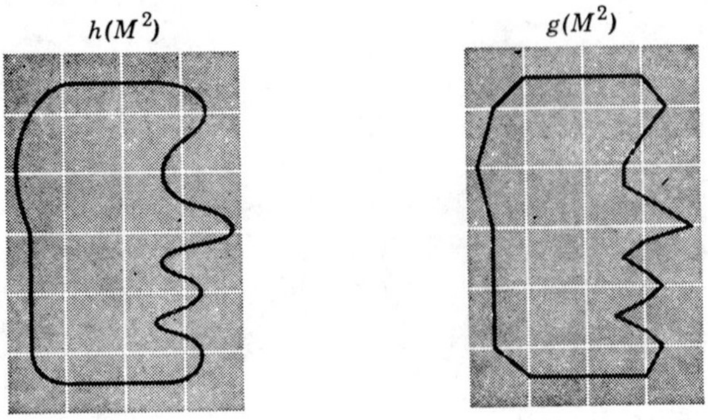

Figure 3

product of n copies of the real line. An *n-manifold with boundary* is a separable metric space, each of whose points has a neighborhood whose closure is topologically equivalent to an n-cell I^n. Hence, an n-manifold is an n-manifold with boundary but not conversely. If M^n is an n-manifold with boundary, we use Int M^n to denote the set of points of M^n which have neighborhoods homeomorphic to E^n and $Bd\ M^n$ to denote M^n − Int M^n. If $Bd\ M^n \neq 0$, we say that M^n is an n-manifold with nonnull boundary. It may be shown that here $Bd\ M^n$ is an $(n - 1)$-manifold.

An *n-simplex* in $E^m (m \geq n)$ is the convex hull of $n + 1$ points in E^m such that these points do not lie in any $n - 1$ plane in E^m. Each convex hull of a subset of the $n + 1$ points is a simplex which is a *face* of the original n-simplex. In an abstract metric set, an *n-simplex* is a set isometric with an n-simplex in E^n. The faces of this abstract n-simplex are the images of the faces of its isometric image in E^n.

A *geometric complex* is a metric space which is the sum of a locally finite collection of simplexes such that if two of these simplexes intersect, the intersection is a face of each. Suppose this locally finite collection T of simplexes contains each face of each element of T. Then T is called a *triangulation* of the geometric complex. For a triangulation of a 3-manifold, the 3-simplexes are tetrahedra. We sometimes refer to the 2-simplexes and 1-simplexes as faces and edges of the tetrahedra, although strictly speaking from the definitions given previously, a face need not be 2-dimensional.

A *triangulated manifold* is a geometric complex which is a manifold together with some particular triangulation of the geometric complex. To triangulate a manifold, we may need to re-metrize it so that we have the sum of simplexes of the right sort. When we speak of a triangulation of E^n, however, unless we indicate to the contrary, we mean a rectilinear triangulation of E^n where the simplexes of the triangulation are simplexes under the ordinary Euclidean metric.

Suppose T_1 is a triangulation of a geometric complex C. A *subdivision* T_2 of T_1 is a triangulation of C (keeping the same metric for C) such that each simplex of T_2 lies in a simplex of T_1.

The *i-skeleton* of a triangulation is the sum of all i-simplexes in it.

A geometric complex may have several triangulations. A set Y in a geometric complex is called a *polyhedron* if the geometric complex has a triangulation (without re-metrization of the complex) such that Y is the sum of simplexes of the triangulation. A set Z is *tame* in a geometric complex C if there is a homeomorphism h of C onto itself such that $h(Z)$ is a polyhedron. A closed set X in C which is not tame but is homeomorphic with a polyhedron is called *wild.* A set X is *locally tame* at a point p of an n-manifold if there is a neighborhood N of p and a homeomorphism $h:\bar{N} \rightarrow E^n$ such that $h(\bar{N} \cdot X)$ is a polyhedron in E^n.

A map f of a geometric complex C into a geometric complex C' is *piecewise linear* if there is a triangulation T of C such that f is linear on each simplex of T. The map is called *locally piecewise linear* at a point p if for some neighborhood N of p there is a triangulation T' of $\bar{N}$ such that f is linear on each element of T'. If f is locally piecewise linear except possibly at points of a set X, we say that it is *locally piecewise linear mod X.*

A triangulated n-manifold M^n is a *piecewise linear manifold* if for each point p of M^n there is a piecewise linear homeomorphism of E^n onto a neighborhood of p in M^n. Although it is known that any triangulation of a 3-manifold yields a piecewise linear manifold, the corresponding result is not known for triangulated 4-manifolds (see Section 8).

If s is a simplex of a triangulation T of a geometric complex, *Star* (s, T) is the sum of all simplexes of T that contain s. If s, s' are mutually exclusive simplexes that are faces of the same simplex s'', the *join* of s and s' is the sum of all segments in s'' from a point of s to a point of s'. *Link* (s, T) is the sum of all simplexes s' of T such that the join of s and s' is a simplex of T. Some authors refer to the link of s as the *complement* of s. When T is a triangulation of a manifold and s is a vertex of T, Link $(s, T) = Bd$ Star (s, T).

We use I^n to denote an n-cell. A topological set X is *n-connected* if each map of $Bd\ I^{n+1}$ into X can be extended to map I^{n+1} into X. This is a bit more delicate than terminology used by some authors who say that a set is i-connected if and only if for each $j \leq i$, the set is j-connected as we have defined it. A set is called *simply connected* if it is 1-connected. A set U is *locally n-connected* at a point p of $\bar{U}$ if for each neighborhood N_1 of p there is a neighborhood N_2 of p such that each map of $Bd\ I^{n+1}$ into $N_2 \cdot U$ can be extended to map I^{n+1} into $N_1 \cdot U$. Note that p need not be a point of U.

A compact connected simply connected 3-manifold is called a *homotopy 3-sphere*. The Poincaré conjecture claims that a homotopy 3-sphere is a 3-sphere, but the issue is still in doubt.

A *homotopy* is a continuous 1-parameter family of maps $f_t(0 \leq t \leq 1)$. If $f:X \times [0, 1] \rightarrow Y$, we find it convenient to denote $f(x \times t)$ by $f_t(x)$. We say that the maps $g_1:X \rightarrow Y$, $g_2:X \rightarrow Y$ *are homotopic in* Y provided there is a map $f:X \times [0, 1] \rightarrow Y$ such that $f_0 = g_1, f_1 = g_2$. The homotopy is called an *isotopy* if each f_t is a homeomorphism.

A constant map of X is a map that sends all of X into a single point. We say that a set X can be shrunk to a point in a set Y if there is a map $f:X \times [0, 1] \rightarrow Y$ such that $f_0 = I$ (identity map) and f_1 is a constant map.

2. EXAMPLES OF COMPACT 3-MANIFOLDS

Perhaps the best known example of a 3-manifold is Euclidean 3-space E^3 itself. Space E^3 is not compact, however, since it has an infinite set of points with no limit point. Examples of compact 3-manifolds are somewhat artificial since there are no physical objects in E^3, the world in which we presume we live, which are examples of compact 3-manifolds. We do have objects in E^3 which are examples of compact 3-manifolds with boundaries such as a cube, a solid torus, a cube with handles, a cube with a knotted hole, and a punctured cube. We also have examples of compact 2-manifolds. In this section we use these real examples to describe examples of compact 3-manifolds which are abstract in the sense that in E^3, there are no physical objects topologically equivalent to them.

Sewing 3-manifolds with boundaries together is a convenient way of constructing examples of 3-manifolds. Suppose X, Y are two topological sets and h is a homeomorphism of a closed subset A of X onto a closed subset $h(A)$ of Y. Then $X +_h Y$ denotes the topological set obtained by sewing X and Y together along A and $h(A)$ so that a point of A is identified with its image under h. It is convenient to think of X and Y as intersecting only in $A = h(A)$ so that a coincides with $h(a)$. An open set in $X +_h Y$ is either an open set in $X - A$, an open set in $Y - h(A)$, or a set of the form $U_1 + U_1$ where U_1 is an open set in X, U_2 is an open set in Y, and $h(U_1 \cdot A) = U_2 \cdot h(A)$.

1. S^n. An n-sphere S^n is a set homeomorphic to the boundary of a unit $(n + 1)$-ball in E^{n+1}. It might have the equation $x_1{}^2 + x_2{}^2 + \cdots + x_{n+1}^2 = 1$. Similarly, S^n is a set homeomorphic with $Bd\ I^{n+1}$.

2. S^3. Since E^4 does not seem as real to many of us as E^3, it may be more convenient to regard S^3 as the one-point-compactification of E^3. Here, a deleted neighborhood of the ideal point at infinity is the exterior of a closed 3-ball—the larger the 3-ball, the smaller the neighborhood.

3. S^3 as a toroidal manifold. Suppose T is the solid torus obtained by revolving in E^3 the disk $(x - 2)^2 + y^2 \leq 1$ in the xy-plane about the y-axis. The simple closed curve J_M defined by $z = 0$, $(x - 2)^2 + y^2 = 1$ bounds a disk in T but not a disk on $Bd\ T$ and is called a meridianal simple closed curve of $Bd\ T$. The simple closed curve J_L defined by $y = 0$, $x^2 + z^2 = 1$ bounds a 2-manifold in $E^3 - \text{Int}\ T$ but none on $Bd\ T$ and is called a longitudinal simple closed curve of $Bd\ T$. Suppose T_1, T_2 are two copies of T with J_{M_1}, J_{L_1}, J_{M_1}, J_{M_2} being the corresponding meridianal and longitudinal simple closed curves. If h is a homeomorphism of $Bd\ T_1$ onto $Bd\ T_2$ that takes J_{M_1} onto J_{L_2}, then $T_1 +_h T_2$ is topologically S^3. It may be convenient to regard T_1, T_2 as being two linked inner tubes plus their interiors (see Figure 4). The point at infinity is moved to be on the interior of T_2. Then the two inner tubes are blown up so that $Bd\ T_1$ and $Bd\ T_2$ coincide and $T_1 + T_2$ fills up E^3 plus the point added at infinity.

4. Toroidal manifold $T(m, n)$. Suppose that m, n are relatively prime positive integers; T_1, T_2 are as in the preceding example; and $p_1, p_2, \ldots, p_n$ are n points on J_{M_2} cyclicly ordered on J_{M_2} as

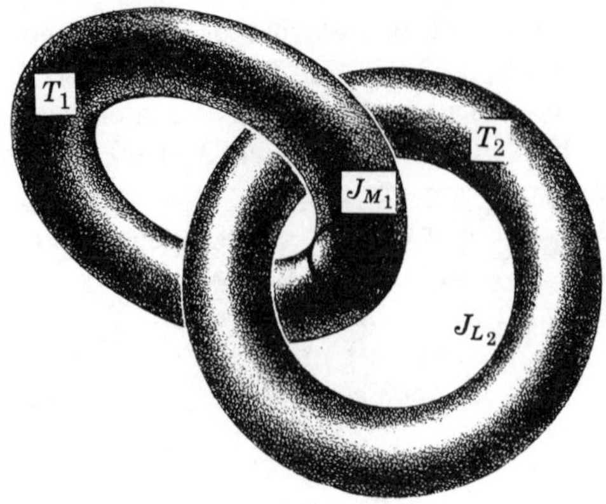

$T_1 +_h T_2 = S^3$

Figure 4

indicated by their subscripts. Let $J(m, n)$ be a simple closed curve on $Bd\ T_2$ that intersects J_{M_2} only at $p_1, p_2, \ldots, p_n$; crosses J_{M_2} at each such point of intersection, with the p_i's cyclicly ordered on $J(m, n)$ as $p_1, p_{m+1}, p_{2m+1}, \ldots, p_{(n-1)m+1}$ (see Figure 5). Let h be a homeomorphism of $Bd\ T_1$ onto $Bd\ T_2$ taking J_{M_1} onto $J(m, n)$. Then $T_1 +_h T_2 = T(m, n)$. It is known that $T(1, 1)$ is the only simply connected toroidal manifold. As pointed out in the previous example, $T(1, 1)$ is topologically S^3. Hence toroidal manifolds do not furnish a clue as to whether or not the Poincaré conjecture is true. Had $T(0, 1)$ been defined, it would represent the cartesian product $S^2 \times S^1$.

5. Double toroidal manifolds. Suppose that in E^3 that X is a tubular neighborhood of a figure eight. It may be regarded as a cube with two handles. Suppose X_1, X_2 are two copies of X and h is a homeomorphism of $Bd\ X_1$ onto $Bd\ X_2$. Then $X_1 +_h X_2$ is a double toroidal manifold. It is not known whether or not each simply connected double toroidal manifold is topologically S^3.

6. n-Tuple toroidal manifolds. Suppose X be a cube with n handles and X_1, X_2 are two copies of X. For each homeomorphism $h: Bd\ X_1 \rightarrow X_2$, $X_1 +_h X_2$ is an n-tuple toroidal manifold. It is known that each compact connected orientable (does not contain a solid Kline bottle) 3-manifold can be regarded as an n-tuple toroidal

manifold. To regard it as such, consider a triangulation of the manifold and let X_1 be a tubular neighborhood of its 1-skeleton. If there is a counterexample to the Poincaré conjecture, it lies among the n-tuple toroidal manifolds. It would be interesting to know for what integers n there is no counterexample. This attack seems difficult since we have only found the answer for $n = 1$.

7. Homology spheres. Let C_1, C_2 be cubes with knotted holes (see Figure 6). Let J_{M_i} be a simple closed curve on $Bd\ C_i$ such that J_{M_i} bounds an orientable 2-manifold in C_i but not on $Bd\ C_i$ and J_{L_i} be another simple closed curve on $Bd\ C_i$ that intersects J_{M_i} at just one point and crosses it there. If $h{:}Bd\ C_1 \rightarrow Bd\ C_2$ is a homeomorphism that takes J_{M_1}, J_{L_1} onto J_{L_2}, J_{M_2} respectively, then $C_1 +_h C_2$ is a 3-manifold with the same homology as S^3, but which is not a homotopy sphere. The reason that $C_1 +_h C_2$ is not a homotopy sphere is as follows. If a tame 2-dimension torus separates a homotopy 3-sphere, it follows from the Loop theorem and Dehn's Lemma that one of the pieces into which the torus separates the homotopy 3-sphere has the same fundamental group as a circle. These homology spheres we have described give counterexamples to the Poincaré conjecture in its original form (see Section 8).

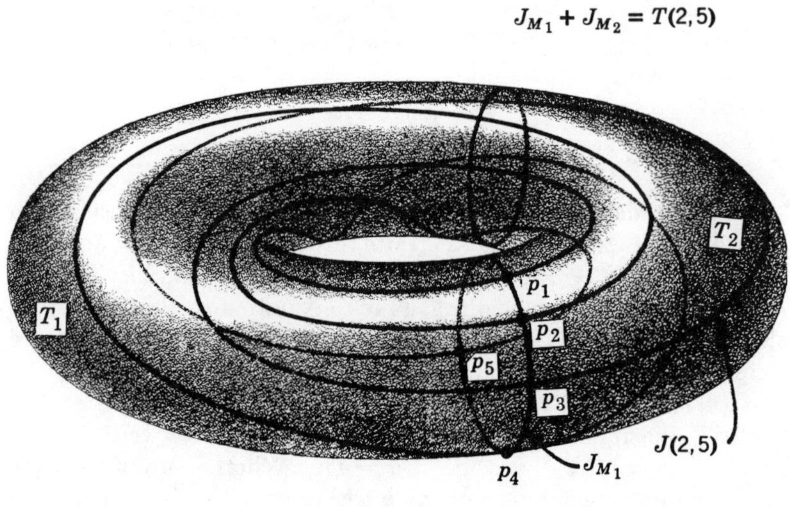

Figure 5

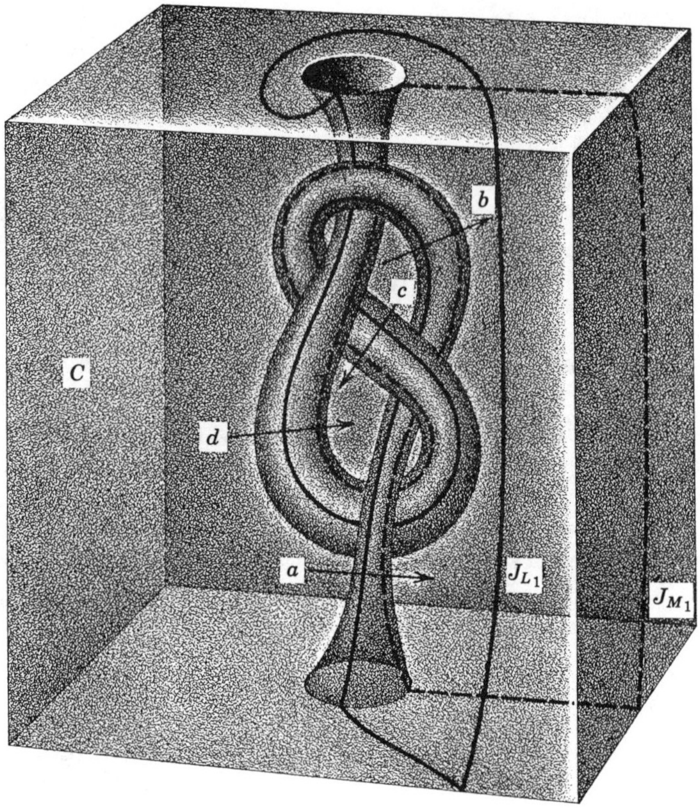

Figure 6

8. A candidate for a counterexample. Suppose J_{L_1} is selected in the $Bd\ C_1$ of Figure 6 so that J_{L_1} goes through the hole once and does not bound in $E^3 - \text{Int}\ C_1$. Let T be a solid torus and $h{:}Bd\ C_1 \to Bd\ T$ be a homeomorphism that takes J_{L_1} onto a meridianal simple closed curve on $Bd\ T$. Is $C_1 +_h T$ a counterexample to the Poincaré conjecture? The fundamental group of C_1 can be found to be $G = \{a, b, c, d/cac^{-1}b^{-1} = bdc^{-1}d^{-1} = aca^{-1}d^{-1} = 1\}$. After T is sewn in, the fundamental group of the sum is $G' = \{cac^{-1}b^{-1} = bdc^{-1}d^{-1} = aca^{-1}d^{-1} = c^{-1}da^{-1}ba = 1\}$. Whether or not $C_1 +_h T$ is a homotopy sphere depends on whether or not G' is trivial.

Perhaps by drilling knotted holes (possibly different from the one shown in Figure 6) and sewing back in solid tori, we obtain a group whose relations, through some quirk, makes the group trivial but obtains a sum that is not S^3. Two difficulties serve as blocks to this approach. First, the algebra is fierce and it seems only by chance that we can show a group to be trivial. Second, even if the algebra succeeds, we still need to show that the resulting space is topologically different from S^3. Bing has shown [5] that no counterexample results if the knot in C_1 is a trefoil knot, and Hempel [11] has extended this result by showing that no counterexample results for certain other knots. The four-crossing knot shown in Figure 6 still baffles us.

9. Projective 3-space P^3. Suppose B is a unit ball in E^3 and P^3 is the space obtained by sewing each point p of $Bd\,B$ onto its diametrically opposite point (antipodal point). Note that under the identification, $Bd\,D$ goes onto a projective plane which is a one sided surface in P^3. It follows readily that P^3 is not simply connected since this projective plane does not separate P^3. In a simply connected 3-manifold, each 2-manifold separates it.

10. Lens space $L(m, n)$. Let B be the unit ball in E^3 described by the equation $x^2 + y^2 + z^2 \leq 1$; m, n be relatively prime positive integers; and $L(m, n)$ be the decomposition space of B obtained by identifying the upper half of $Bd\,B$ with the lower half under a rotation through $2m\pi/n$ radians and a reflection through the $z = 0$ plane. Each point on $z = 0$, $x^2 + y^2 = 1$ is identified with $n - 1$ other points; each point of $x^2 + y^2 + z^2 = 1$, $z > 0$ is identified with one of $x^2 + y^2 + z^2 = 1$, $z < 0$; and points of $x^2 + y^2 + z^2 < 1$ are left alone. It may be shown that each lens space is a toroidal manifold. The tori correspond to tubular neighborhoods of the axis and the equator. The upper and lower halves of $Bd\,B$ shown in Figure 7 are shown making a 72° angle with each other so that when sewing is done the tubular neighborhood of the equator looks like a torus without unnecessary distortion. Note that $L(1, 1)$ is topologically equivalent to S^3. In fact $L(m, n)$ is topologically equivalent to $T(m, n)$. Each lens space is a toroidal manifold and $S^2 \times S^1$ is the only toroidal manifold which is not a lens space.

11. Cartesian products. Taking the cartesian product of S^1 with various compact 2-manifolds provides us with an abundant supply of compact 3-manifolds. None of these are homotopy spheres.

12. Replacing tori. It is possible to obtain a set topologically equivalent to $S^2 \times S^1$ by removing an unknotted solid torus from

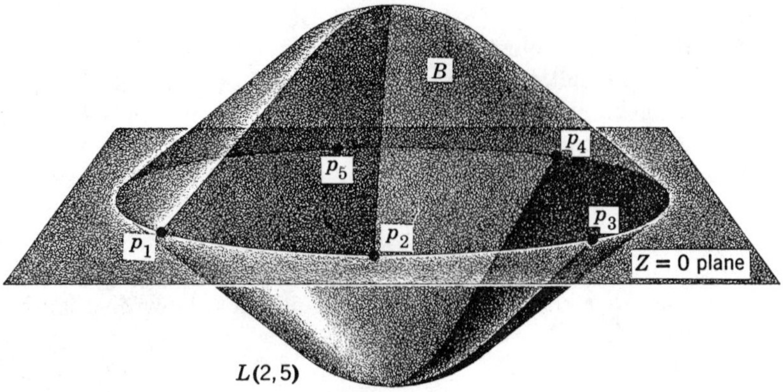

Figure 7

S^3 and sewing it back in so that the meridianal and longitudinal simple closed curves on the boundary are interchanged. In fact, A. H. Wallace [28] and W. B. R. Lickorish [10] have shown that any compact connected orientable 3-manifold whatever can be obtained by removing a finite collection of mutually exclusive solid tori and sewing them in differently.

13. R. H. Fox has suggested [8] branched covering spaces of S^3 as a possible place to look for counterexamples to the Poincaré conjecture.

3. IMPROVING TRIANGULATIONS

Consider a 2-dimensional torus with a small hole. If the hole is large, we can easily see how to triangulate the 2-manifold with boundary so that all the vertices lie on the boundary. If the hole is small, the result remains true but is more surprising.

There are many different ways of triangulating the same 3-manifold with boundary. If we triangulate a simply connected 3-manifold and remove the interior of one 3-simplex, what is left is called a *fake cube*. Like a 3-cell, its boundary is a 2-sphere and it is connected as well as simply connected. If it is a 3-cell, the original 3-manifold was a 3-sphere, but if it is not a 3-cell, the original 3-manifold was a counterexample to the Poincaré conjecture.

It might conceivably be of some help in deciding whether a fake cube is a real cube if the fake cube has a "nice" triangulation. Theorem 1 shows that each fake cube has a triangulation all of whose vertices lies on the boundary. The theorem can be proved for arbitrary compact, connected, combinatorial n-manifolds, but for simplicity we only discuss the case $n = 3$.

Theorem 1. Each compact connected 3-manifold M^3 with non-null boundary has a triangulation all of whose vertices are on $Bd\ M^3$.

PROOF. It follows from [3] that M^3 has a triangulation T'. If a vertex of T' lies in Int M^3, we use two steps to reduce the number of vertices in Int M^3. Let vv' be an edge of T' with $v \in Bd\ M^3$ and $v' \in \text{Int}\ M^3$. If Link (v', T') has a 2-simplex on $Bd\ M^3$, we can skip the first step.

1. Let U be the open half-star of v—that is, the set of all points of Star (v, T') which lie closer to v than to the other end of the segment through them from v to Link (v, T'). Note that $M^3 - U$ is topologically M^3. However, T' is not a triangulation of $M^3 - U$ since some of the original 3-simplexes of T' intersect $M^3 - U$ in a tetrahedron with one end blunted. Suppose T' is a triangulation of $M^3 - U$ obtained by leaving the elements of T' in $M^3 -$ Star (v, T') alone and subdividing the blunted tetrahedra so as not to introduce new corners and so that some 3-simplex of T' with a face on $Bd(M^3 - U)$ has what is left of vv' as an edge (see Figure 8).
2. Let s^3 be a 3-simplex of T' with v' as vertex and with a face on $Bd(M^3 - U)$. Denote the closure of $(M^3 - U) - s^3$ by N^3. Then T' has one fewer vertex in Int N^3 than T' has in Int M^3. Since N^3 is topologically equivalent to M^3, there is a triangulation of M^3 that has fewer vertices on Int M^3 than T' does. An iteration of this procedure yields T.

Theorem 2. Suppose T is a triangulation of a compact 3-manifold with boundary M such that each vertex of T lies on $Bd\ M$ and each 3-simplex of M has at least two edges on $Bd\ M$. Then M has a triangulation T' such that each vertex of T' lies on $Bd\ M$ and each 2-simplex of T' has an odd number of edges on $Bd\ M$.

PROOF. Suppose T is chosen so as to have a minimal number of 2-simplexes not on $Bd\ M$ while maintaining the property that all vertices of T lie on $Bd\ M$ and each 3-simplex of T has at least two edges on $Bd\ M$. We show that this minimal triangulation of M suffices for T'.

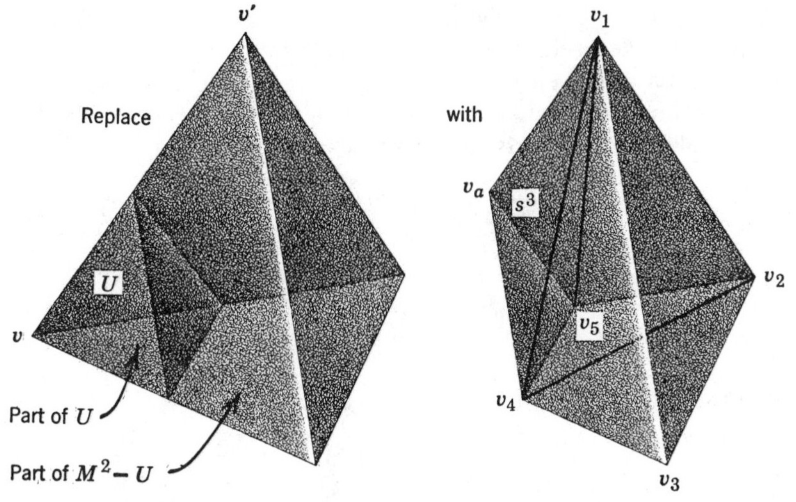

Figure 8

Assume that T has a 2-simplex $v_1v_2v_3$ with precisely two edges v_1v_2, v_1v_3 on $Bd\ M$. We split M along this 2-simplex (see Figure 9). Let X be the star of v_1 with respect to T (sum of all simplexes of T that contain v_1). Then X is a cube which is the sum of two cubes X_1, X_2 such that $X_1 \cdot X_2 = v_1v_2v_3$. Split M along $v_1v_2v_3$ to get a 3-manifold with boundary homeomorphic with M with a triangula-

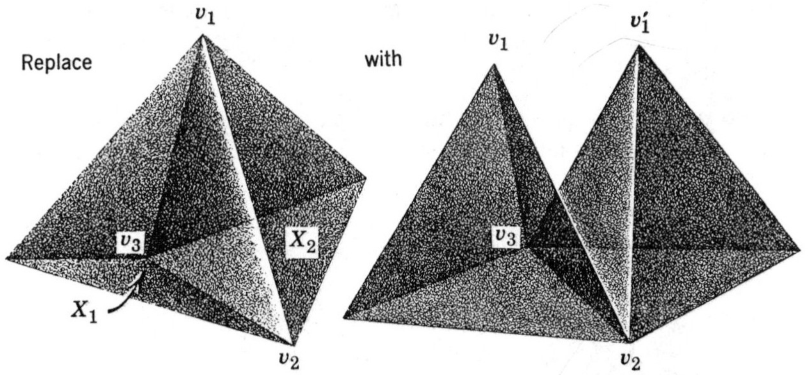

Figure 9

tion T_1 such that a simplex of T_1 is either a simplex of T that misses v_1, a simplex of T in X_1, or a simplex of T in X_2 with the vertex v_1 replaced with a new vertex v_1'. Note that T' has an extra vertex v_1', two extra edges $v_1'v_2$, $v_1'v_3$, and an extra face $v_1'v_2v_3$, but it has one fewer 2-simplex not on *Bd M*. Since T was already minimal in this respect, there is no $v_1v_2v_3$ as assumed at the beginning of this paragraph.

We now show that under the triangulation T we have chosen, each 2-simplex of T has at least one edge on *Bd M*. Suppose $v_1v_2v_3$ is a 2-simplex of T with no edge on T; then if $v_1v_2v_3v_4$ is a 3-simplex of T, $v_1v_2v_3v_4$ has at least two edges on *Bd M* by hypothesis and one of the three 2-simplexes $v_2v_3v_4$, $v_1v_3v_4$, $v_1v_2v_4$ has precisely two edges on *Bd M*. This was ruled out in the preceding paragraph.

4. A FAKE CUBE IS REAL IF IT HAS A "NICE" TRIANGULATION

A natural way of attempting to prove the Poincaré conjecture is to try to show that a fake cube has a triangulation that can be shelled. We say that a triangulation T of an n-cell can be *shelled* if the n-simplexes of T can be ordered $s_1, s_2, \ldots, s_n$ so that for each integer $k \leq n$, $s_k + s_{k+1} + \cdots + s_n$ is an n-cell. A triangulation of a fake cube can be *shelled* if the 3-simplexes of the triangulation can be ordered $s_1, s_2, \ldots, s_n$, such that for each $k \leq n$, $s_k + s_{k+1} + \cdots + s_n$ is a fake cube.

Theorem 3. Suppose T is a triangulation of a 2-cell D and D' is a disk in D which is the sum of elements of T. Then the 2-simplexes of T not contained in D' can be ordered $s_1, s_2, \ldots, s_n$ so that for each integer $k \leq n$, $D' + s_k + s_{k+1} + \cdots + s_n$ is a disk.

PROOF. The theorem is proved by induction on n. Let s be a 2-simplex of T which does not lie in D' but which has an edge on *Bd D*. We can either let s be s_1 and shell it or else some edge s' of s not on D' is a spanning arc of D. For the latter we consider the two disks into which s' separates D and expand D' to be one of them. We have thereby reduced n and our inductive process is started (see Figure 10).

Theorem 4. A fake cube is real if it has a triangulation that can be shelled.

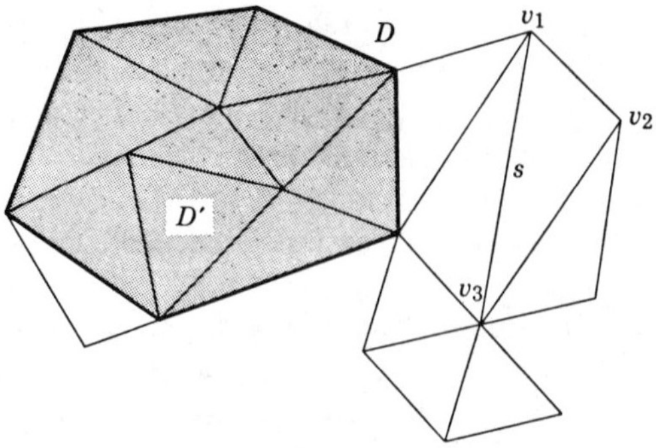

Figure 10

PROOF. It can be shown inductively that each of s_n, $(s_{n-1} + s_n)$, $(s_{n-2} + s_{n-1} + s_n)$, . . . , $(s_1 + s_2 + \cdots + s_n)$ is a real cube.

Contrasted with the easiness with which any triangulation of a 2-cell can be shelled, there is difficulty in shelling a cube. It sometimes comes as a shock to the beginner that even real cubes have triangulations which cannot be shelled. Mary E. Rudin described [23] a rectilinear triangulation of a real tetrahedron that could not be shelled. Earlier unshellable triangulations of topological cubes were given in [9] and [27]. We describe two triangulations of a topological cube that cannot be shelled.

Example 1. Let C be the house with two rooms shown in Figure 11. The walls are made out of one layer of brick (without mortar) so that each brick touches air in two components, and if two bricks intersect, the intersection is a face, edge, or corner of each. The object is a house with two floors so that to enter the lower floor, one goes through a tunnel from the roof and to get to the upper floor one goes through a tunnel from below. There is a partition in each room from the wall to the tunnel leading to the other room. This causes the room to be simply connected.

We see that C is a topological cube if we regard it as a real cube with indentations from above and below.

Suppose C is triangulated as follows. Order the corners of the bricks as follows. First put in any vertex which is a component of the intersection of a brick with *Bd C*. There is one of these on each inside corner of a room. Next, list the vertices not of this sort but which lie on an edge of a brick that is a component of the intersection of the brick with *Bd C*. Finally list the other vertices. The bricks are not triangulated as follows. A face of a brick is divided with a diagonal that contains the first vertex on the face. Once the faces of a brick are triangulated, the brick is triangulated by taking the cone from the first vertex of the brick to the triangles on the boundary of the brick that do not contain this vertex.

Note that each tetrahedra of the triangulation intersects the boundary of the cube in a disconnected set. Hence the triangulation cannot be shelled. It may be noted that each vertex of the triangulation lies on *Bd C*.

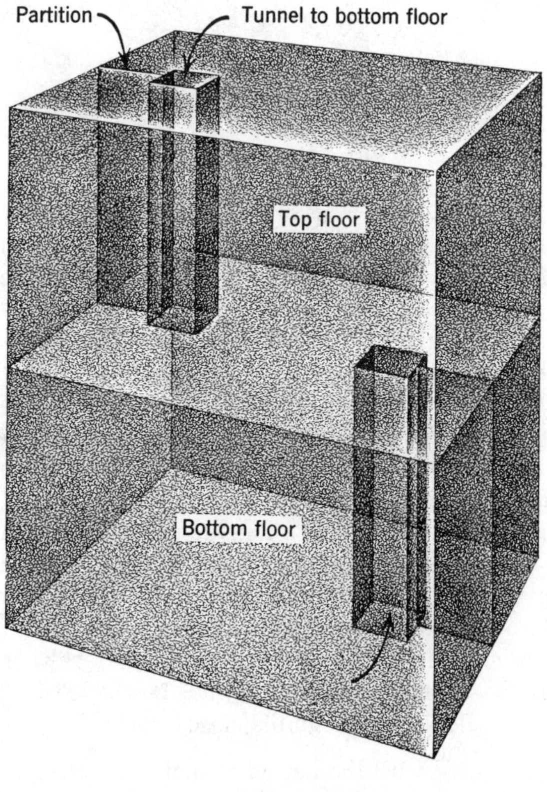

Figure 11

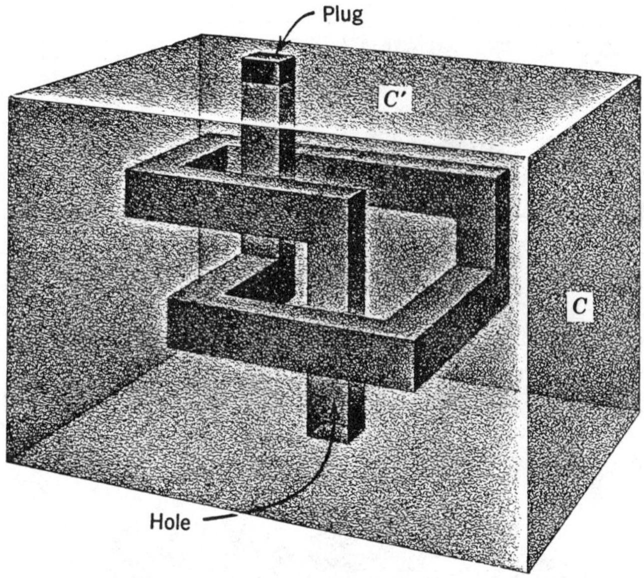

Figure 12

Example 2. Consider the cube with the plugged knotted hole shown in Figure 12. The object is topologically a cube since the hole is plugged at the upper end with a small cube C'. The resulting topological cube C is triangulated so that the edges of C' are l-simplexes of the triangulation. This may be accomplished by first subdividing C into small cubes of the same size of C', subdividing each cube into two prisms, and then subdividing each prism into three tetrahedra—taking care that the resulting tetrahedra fit together right.

Consider a simple closed curve J made up of a spanning segment of C which is an edge of C' and an arc of $Bd\ C$. Then J is knotted. If we were to start shelling the triangulation of C, at each stage there would be a knotted simple closed curve which lies, except for one spanning simplex, on the boundary of the resulting cube. It follows that the triangulation of C cannot be shelled for at the last stage there could be no such simple closed curve.

Closely connected with the notion of shelling is that of sequential connectedness. A triangulation of a finite n-complex all of whose simplexes are faces of an n-simplex is *sequentially connected* if its n-simplexes can be ordered $s_1, s_2, \ldots, s_n$ so that for each $i < n$, s_i intersects $(s_{i+1} + s_{i+2} + \cdots + s_n)$ in a connected set. The notion of sequential connectedness is stronger than either that of connectedness or unicoherence since a projective plane is connected and unicoherent. It has no sequentially connected triangulation or else it would be simply connected.

The following theorem comes from [2].

Theorem 5. A fake cube is real if it has a sequentially connected triangulation.

A still more general sort of shelling is frequently used where simplexes of all dimensions are considered. A complex C is *collapsible* if it has a decreasing sequence of subcomplexes $C = C_1, C_2, \ldots, C_n$ such that C_n has only one vertex and for each $i < n$, C_{i+1} is obtained from C_i by eliminating from C_i exactly two simplexes such that one is a major face of the other. If s, s' are the simplexes eliminated from C_i, with s' a face of s, it follows that s is not a face of any other simplex of C_i, and s' is not a face of any simplex except s and s' itself.

Theorem 6. A fake cube is real if it has a collapsible triangulation.

PROOF. The proof is found in works of H. C. Whitehead [32], but for completeness we indicate a proof in the following paragraph.

For each simplex s of the triangulation, let $S(s)$ be the star of the baricenter of s under the second baricentric subdivision of the collapsible triangulation. Then these $S(s)$'s provide the fake cube with a cellular subdivision. By following the way that the original triangulation can be collapsed, we find that this cellular subdivision can be shelled. Then the proof follows that of Theorem 3.

It is not immediately obvious whether or not the triangulation given by Mary E. Rudin in [23] or that of a house with two rooms or that of a cube with a plugged knotted hole is either sequentially connected or collapsible. We give an example that is not.

Example 3. This example is the same as Example 2 except that the knot is more complicated and consists of a double trefoil knot instead of a single one (see Figure 13). The proof that the resulting triangulation is not collapsible is given elsewhere. By getting more complicated holes we can show that for each integer n there is a

Figure 13

triangulation of a cube whose nth baricentric subdivision is not contractible. We now consider another condition on the triangulation of a fake cube that will insure that it is real cube.

Theorem 7. A fake cube K is a cube if and only if it has a triangulation T such that each vertex of T lies on $Bd\,K$ and each 3-simplex of T has at least two edges on $Bd\,K$.

PROOF. The necessity of the condition follows from the fact that a cube has a triangulation with only one 3-simplex. We now show that a fake cube is real if it has a triangulation satisfying the prescribed condition.

Following Theorem 2, we suppose with no loss of generality that each 2-simplex of T has precisely one or three edges on $Bd\,K$.

We show here that each 1-simplex of T lies on $Bd\,K$. Assume T has 1-simplexes $s_1^1, s_2^1, \ldots, s_m^1$ not on $Bd\,K$. The fundamental group of $Bd\,K + \Sigma s_1^1$ is the free group on m generators

$$G = \{a_1, a_2, \ldots, a_m\}$$

Adding a 2-simplex s^2 of T with three edges on $Bd\,K$ to $Bd\,K + \Sigma s_i^1$ does not change the fundamental group of the resulting set since $Bd\,s^2$ could already be shrunk to a point on $Bd\,K$. However, adding a 2-simplex s^2 of T with one edge on $Bd\,K$ to $Bd\,K + \Sigma s_i^1$ results in a set whose fundamental group is a group with m generators and one relation $w = 1$ where w is a two letter word (since s^2 has precisely two edges among the s_i^1's). Accordingly, the fundamental group of the 2-skeleton of T is

$$G' = \{a_1, a_2, \ldots, a_n | w_1 = w_2 = \cdots = w_n = 1\}$$

where each of the w_i's is a two letter word. This group is nontrivial because the homomorphism $f(a_i) = b$ sends G' onto the group $H = \{b/b^2 = 1\}$. This contradicts the fact that K is simply connected. This contradiction arises from assuming that T has a 1-simplex not on $Bd\,K$.

Since each 1-simplex of T lies on $Bd\,K$, each 2-simplex of T either lies on $Bd\,K$ or separates K. Hence T can be shelled. Theorem 7 then follows from Theorem 4.

5. EULER CHARACTERISTICS?

For a triangulation of a complex, let V_i denote the number of i-simplexes in the triangulation. For triangulations of a compact connected orientable 2-manifold, the number

$$V_2 - V_1 + V_0$$

is important since it determines the topology of the 2-manifold. We might hope we could tell whether or not a fake cube is real by considering a triangulation and computing the V_i's. We find the following, however.

Theorem 8. For each triangulation T of a compact 3-manifold, $V_2 = 2V_3$.

PROOF. We merely note that each 3-simplex of T has four faces and each face belongs to precisely two 3-simplexes.

Theorem 8 shows us that we do not need to concern ourselves with the number of 3-simplexes. This might seem temporarily encouraging, for whether or not a triangulated homotopy sphere is topologically S^3 depends not on the number of 3-simplexes, but rather on whether or not the 2-skeleton of the triangulation can be embedded in E^3. We then find the following discouraging result, however.

Theorem 9. For each triangulation T of a compact 3-manifold, $V_1 = V_3 + V_0$.

PROOF. For each vertex v of T, let $V_i(v)$ be the number of i-simplexes of T containing v. It follows from Euler's formula applied to the 2-sphere $L(v, T) = Bd\,\text{Star}\,(v, T)$ that

$$V_3(v) - V_2(v) + V_1(v) = 2$$

Summing with respect to all the vertices of T, we find that

$$\Sigma[V_3(v) - V_2(v) + V_1(v)] = 2V_0$$

Each i-simplex contributes to $i + 1$ of the $V_i(v)$'s so

$$\Sigma V_i(v) = (i + 1)V_i$$

Hence,

$$4V_3 - 3V_2 + 2V_1 = 2V_0$$

Theorem 9 follows by substituting the equation from Theorem 8 in the foregoing equation.

As a final discouraging factor, we have the following theorem.

Theorem 10. If M_1, M_2 are two (possibly topologically different) compact 3-manifolds, then there are triangulations T_1, T_2 of M_1, M_2 respectively such that T_1, T_2 have precisely the same number of simplexes in each dimension.

PROOF. I had a rather cumbersome proof of this result for orientable manifolds using an interesting theorem by Wallace [28] and Lickorish [10], but Leslie Glaser pointed out the following easy proof.

Let T_1', T_2' be arbitrary triangulations of M_1, M_2. By subdividing one of them we get triangulations with the same number of vertices. If the triangulations also have the same number of 3-simplexes, Theorems 8 and 9 imply that they have the same number of simplexes of each dimension.

For the one of the triangulations with the larger number of 3-simplexes, subdivide the triangulation by adding a vertex at the baricenter of a 3-simplex. This increases the number of vertices by one and the number of 3-simplexes by three. To the other, add a vertex at the baricenter of a 2-simplex. This increases the number of vertices by one and the number of 3-simplexes by four. Hence, in a finite number of such subdivisions, we bring T_1' and T_2' into balance with respect to the number of simplexes in each dimension (see Figure 14).

This result suggests that knowing the V_i's does not give enough information. Although T_1 and T_2 have the same number of simplexes in each dimension, they may differ in how these simplexes fit together. Something might be accomplished by putting coefficients on the simplexes and weighting them according to how other simplexes hang on to them—but this would be more difficult than counting.

Theorem 11. For each triangulation of a compact 3-manifold, $V_3 - V_2 + V_1 - V_0 = 0$.

PROOF. The result is an immediate consequence of Theorems 8 and 9.

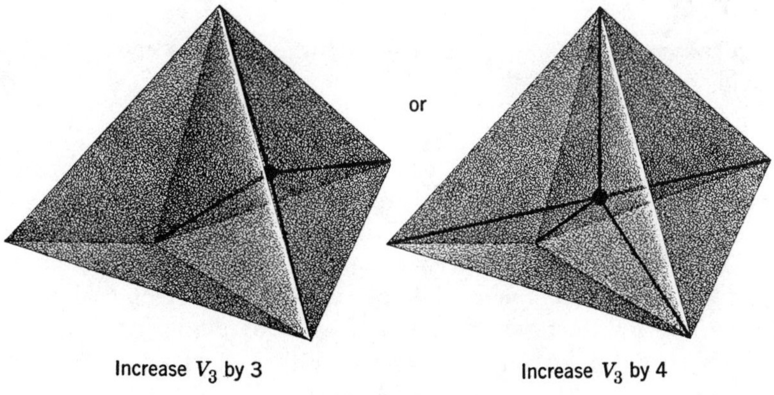

Figure 14

6. CURVES ENCIRCLED BY TORI

An affirmative answer to the following problems would provide a partial solution to the Poincaré conjecture.

Question 1. Is a compact connected 3-manifold M topologically S^3 if each tame simple closed curve in M can be isotopically shrunk to small size in M?

A simple closed curve J in M can be isotopically shrunk to small size in M if for each $\epsilon > 0$ there is an isotopy $h: J \times [0, 1] \rightarrow M$ such that $h_0 = I$ (identity), and diameter $h_1(J) < \epsilon$.

Theorem 19 of Section 7 would provide an affirmative answer to Question 1 if we have an affirmative answer to the following question.

Question 2. Can each simple closed curve in a homotopy 3-manifold be encircled by a solid torus?

A *solid torus* is an object topologically equivalent to the cartesian product of a circle S^1 with a disk D. If $p_0 \in \text{Int } D$, the image of $S^1 \times p_0$ is called a *core* of the solid torus. A solid torus T *encircles* a simple closed curve J if $J \subset \text{Int } T$ and J is homotopic in Int T to a core of T. In fact, for map $f: J \rightarrow \text{Int } T$ we say that T encircles $f(J)$ if there is a homeomorphism $h: J \rightarrow \text{Int } T$ such that T encircles $h(J)$ and f is homotopic to h in Int T.

Not every map of a simple closed curve into E^3 can be encircled by a solid torus but it can if the map is a homeomorphism.

Theorem 12. Each simple closed curve in E^3 is encircled by a solid torus.

PROOF. Let a, b be two points of a simple closed curve J and P be a plane separating a from b. Let axb be one arc from a to b in J and ayb be the other. There is a polygonal simple closed curve K in P separating $axb \cdot P$ from $ayb \cdot P$ in P. Let T be a tubular neighborhood of K so small that it misses J. If the point at infinity is added to E^3, the complement of T is a solid torus which encircles J.

To get a solid torus encircling J without adding the point at infinity, we could consider a large round ball B whose interior contains $K + J$. Let A be a polygonal arc in $P - J$ from K to $Bd\ B$. Then the solid torus is B minus a tubular neighborhood of $A + K$.

Each polygonal simple closed curve J is encircled by a solid torus each point of which lies near J. We should not conclude that the torus promised by Theorem 12 lies close to the simple closed curve it encircles because the simple closed curve was not assumed to be polyhedral. Neither should we conclude that Theorem 12 holds in 3-manifolds other than E^3 and S^3.

Example 4. There is a simple closed curve J in E^3 such that each solid torus encircling it extends far away. Consider a wild 2-sphere S in E^3 with only a 0-dimensional set of wild points such that the complement of S is not simply connected (see Figure 15). Let J be

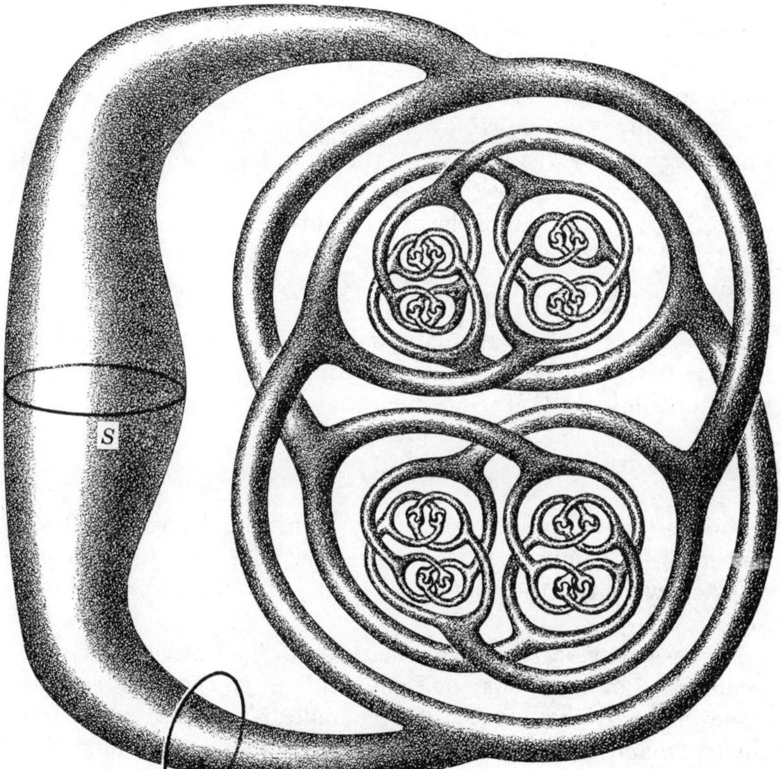

Figure 15

a simple closed curve on S which contains all the wild points of S, and K be a polygonal simple closed curve in $E^3 - S$ which cannot be shrunk to a point in $E^3 - S$. Then K cannot be shrunk to a point in $E^3 - J$ even though J bounds a disk in $S \subset E^3 - K$. We indicate in the following paragraph why each solid torus encircling J must intersect K.

If T were a solid torus in $E^3 - K$ encircling J, it would have a tame core C. This tame core C could be shrunk to a point in $E^3 - K$ so C would not homologically link K. Then K would not homologically link C. It follows from Theorem 15 that C is unknotted so K could be shrunk to a point in $E^3 - C$. If K could be shrunk to a point in $E^3 - C$, however, it could be shrunk in $E^3 - \text{Int } T$ and hence in $E^3 - J$.

Suppose simple closed curve J is oriented as is the longitudinal direction around a solid torus T. In the next two theorems, for each map f of J into T, we let $L(f)$ be the number of times f circles T longitudinally. To find $L(f)$ we could let C be a core of T with length oriented as T and r be a retraction of T onto C. Then subdivide J into n small positively oriented arcs $x_1x_2, x_2x_3, \ldots, x_nx_1$ and let $L(f)$ be the sum of the oriented distances along C from $rf(x_i)$ to $rf(x_{i+1})$ divided by the length of C where it is understood that we denote x_{n+1} by x_1.

Theorem 13. If f_1, f_2 are maps of J into $Bd\ T$ such that $L(f_1) \geq 2$, $L(f_2) \neq 0$, then either $L(f_2)$ is a multiple of $L(f_1)$ or else there is a map f of J into $f_2(J) + f_1(J)$ such that

$$0 < L(f) < L(f_1)$$

PROOF. We suppose $L(f_2) > 0$. Assume the theorem is false and that $L(f_2)$ is minimal in the sense that if f_1, f_3 are two other maps showing the theorem false; then $L(f_2) \leq L(f_3)$. Note that $L(f_1) < L(f_2)$.

First we consider the case where $f_1(J)$, $f_2(J)$ have a point p in common. Let f_3 be a map of J onto $f_1(J) + f_2(J)$ so that as a point moves half way around J, its image under f_3 starts at p and traces out $f_2(J)$, and as the point goes the other half way around J, the image under f_3 traces out $f_1(J)$ in the negative direction so that

$$L(f_3) = L(f_2) - L(f_1)$$

But $L(f_3)$ is not a multiple of $L(f_1)$ unless $L(f_2)$ is and $L(f_3) < L(f_2)$. This contradicts the minimality we had assumed for $L(f_2)$.

Finally we consider the case where $f_1(J) \cdot f_2(J) = 0$. Let f_i' $(i = 1, 2)$ be a homeomorphism of J into $f_i(J)$ such that $L(f_i') > 0$. Since $f_1'(J) \cdot f_2'(J) = 0$,

$$L(f_1') = L(f_2')$$

Since $f_1(J) \cdot f_2'(J) = 0$, $L(f_1)$ is a multiple of $L(f_2')$ and hence a multiple of $L(f_1')$. Then $L(f_1') \leq L(f_1)$, and under the assumption that there is no f with $0 < L(f) < L(f_1)$ we have

$$L(f_1) = L(f_1')$$

Since $f_2(J) \cdot f_1'(J) = 0$, $L(f_2)$ is a multiple of $L(f_1')$ and hence a multiple of $L(f_1)$.

A *punctured disk* is a set homeomorphic to $D_0 - (\text{Int } D_1 + \text{Int } D_2 + \cdots + \text{Int } D_n)$ where D_0 is a disk and $D_1, D_2, \ldots, D_n$ is a collection of mutually exclusive disks each lying in Int D_0.

Theorem 14. Suppose f is a map of a punctured disk $D_0 - (\text{Int } D_1 + \text{Int } D_2 + \cdots + \text{Int } D_n)$ into a solid torus T such that $L(f/Bd\ D_0) = 1$ and $f(Bd\ D_1 + Bd\ D_2 + \cdots + Bd\ D_n) \subset Bd\ T$. Then each open subset of $Bd\ T$ containing $f(Bd\ D_1 + Bd\ D_2 + \cdots + Bd\ D_n)$ also contains a simple closed curve isotopic in T to a core of T.

PROOF. Suppose each $Bd\ D_i$ $(i = 1, 2, \ldots, n)$ is ordered opposite that of $Bd\ D_0$. By joining the $Bd\ D_i$'s with cross cuts in the punctured disk, we have,

$$L(f|Bd\ D_0) = L(f|Bd\ D_1) + L(f|Bd\ D_2) + \cdots + L(f|Bd\ D_n) = 1$$

Let n_0 be a positive integer minimal with respect to there being a map g of S^1 into $f(Bd\ D_1 + Bd\ D_2 + \cdots + Bd\ D_n)$ such that $L(g) = n_0$. Either $n_0 = 1$ or else by the foregoing relation there is an $L(f/Bd\ D_i)$ which is not a multiple of n_0. Theorem 13 shows that this second alternative leads to the contradiction that n_0 is not minimal. Hence $n_0 = 1$.

Let U be an open subset of $Bd\ T$ containing $g(S^1)$ and such that each boundary component of U is a simple closed curve. We show that U contains a simple closed curve that is isotopic in T to a core of T.

If each boundary component of U bounds a disk in T, Theorem 14 follows since we can get a simple closed curve C in $Bd\ T$ which circles T longitudinally once and then push this C off the disks of $Bd\ T - U$. If J is a boundary component of U which does not bound a disk, J must circle T longitudinally, for if it circles it meridianally but not longitudinally, $g(S^1)$ could not circle longitudinally, since $g(S^1)$ misses J. In fact, $L(g)$ is a multiple of the number of times J circles longitudinally. Since $L(g) = 1$, J circles T once longitudinally. A simple closed curve J' in U runs "parallel" to J and circles T once longitudinally also. Hence J' is isotopic in T to a core of T.

It may be noted that we could not conclude that there is a simple closed curve in $f(Bd\ D_1 + Bd\ D_2 + \cdots + Bd\ D_n)$ isotopic to a core of T because it might be that $n = 2$, $f(Bd\ D_1)$ is a simple closed curve circling T twice and $f(Bd\ D_2)$ is a simple closed curve circling T three times with $f(Bd\ D_1) \cdot f(Bd\ D_2)$ being one point.

Theorem 15. If D is a disk in E^3 and T is a solid torus in E^3 encircling $Bd\ D$, each core of T is unknotted.

PROOF. With no loss of generality, we suppose $Bd\ T$ is polyhedral. To see that there is no loss of generality in supposing this we argue as follows. Suppose T is the image of a polyhedral torus T' under a homeomorphism h. It follows from [6] or [15] that we can adjust h slightly to get a homeomorphism h' of T' onto T such that h' is locally piecewise linear mod $Bd\ T'$. Reduce T' slightly to a polyhedral torus T'' so that T'' has the same core as T', $T' \subset$ Int T', and $Bd\ D \subset h'$ (Int T''). Then $h'(T'')$ is a polyhedral torus with the same core as T.

We suppose that D is locally polyhedral mod $Bd\ D$, and from [4] we can support this. We suppose furthermore that $D \cdot Bd\ T$ is the sum of a finite number of mutually exclusive simple closed curves. Such a desirable condition results from a slight adjustment of either the vertices of $Bd\ T$ or of certain vertices on $Bd\ D$.

If a component of $D \cdot Bd\ T$ bounds a disk on $Bd\ T$, we take the inside of such a disk, replace a disk in D with it, and shove it to one side of T so as to eliminate a component of $D \cdot Bd\ T$. For convenience we call the new disk D also. This elimination of components of $D \cdot Bd\ T$ is continued until no component of $D \cdot Bd\ T$ bounds a disk on $Bd\ T$.

It follows from Theorem 14 that each simple closed curve in

$D \cdot Bd\ T$ circles T longitudinally exactly once. Let D' be a subdisk of D such that $D' \cdot Bd\ T = Bd\ D'$. A core C of T is unknotted since it can be homotopically pulled in its own complement over to $Bd\ D'$ and can then be shrunk to a point in D'. A tame simple closed curve is unknotted if it can be shrunk to a point in its own complement.

7. NICE SHRINKING

If J is a simple closed curve in E^3, then J can be shrunk to a point so that at each stage of the shrinking, except the last, J has been shrunk to a simple closed curve. A natural way to get this shrinking is to pick a point p of E^3 and decide that at time t the point x of J has been moved to the point of the segment xp that divides it in the ratio t to $(1 - t)$. Note that if J is a polygonal simple closed curve, then at each stage of the homotopy except the last, J is a polygonal simple closed curve. It is not known whether or not this property holds in an arbitrary simply connected compact 3-manifold, but Theorem 17 shows that unless the Poincaré conjecture is true, it does not.

In this section we lean heavily on the following result proved in [5]. McMillan [13] has an even stronger version of this result.

Theorem 16. A compact, connected 3-manifold M is topologically S^3 if each tame simple closed curve in M lies in a topological cube in M.

PROOF. Although we do not repeat the proof given in detail in [5], we list the steps used in proving the theorem.

1. Let T be a triangulation of M. We suppose each polygonal simple closed curve in this triangulation lies in a topological cube in M.
2. Show that step 1 implies that each polygonal simple closed curve lies on the interior of a polyhedral cube.
3. Show that step 2 implies that the 1-skeleton of T lies on the interior of a polyhedral cube.
4. Show that step 3 implies that the 2-skeleton of T lies in a polyhedral punctured cube.
5. Fill in the holes of the punctured cube of step 4 to show that M is topologically S^3.

The following result is new and its proof is included.

Theorem 17. A compact connected 3-manifold M is topologically S^3 if for each tame simple closed curve J in M there is a homotopy $f_t: J \to M (0 \leq t \leq 1)$ such that f_0 is the identity, each $f_t(J)$ is encircled by a solid torus in M, and $f_1(J)$ lies on the interior of a topological cube in M.

PROOF. We prove Theorem 17 by showing that each simple closed curve in M lies on the interior of a topological cube in M. Then Theorem 17 follows from Theorem 16.

Let X be the set of all points t of $[0, 1]$ such that $f_t(J)$ lies on the interior of a cube in M. The proof is finished if we show that $0 \in X$. It follows easily that X is an open subset of $[0, 1]$ that contains 1. It follows that $0 \in X$ when we show that X is closed. Toward this end we let t_0 be a point of the closure of X and show that $t_0 \in X$.

Let T_0 be a solid torus in M encircling $f_{t_0}(J)$. We select such a T_0 whose boundary is polyhedral. That we can do this follows from an argument used in the proof of Theorem 15.

If there is a cube C in M whose interior contains a simple closed curve on $Bd\ T_0$ that circles T_0 once longitudinally, there is a cube C' whose interior contains a core K of T_0. Expanding a small tubular neighborhood of K to contain $f_{t_0}(J)$, we see how to adjust C' so that its interior contains $f_{t_0}(J)$. Hence we finish the proof of Theorem 17 by showing that there is a cube C in M such that Int C contains a simple closed curve on $Bd\ T_0$ that circles T_0 once longitudinally.

Let t_1 be a point of X so near t_0 that T_0 encircles $f_{t_1}(J)$ and C be a topological cube in M whose interior contains $f_{t_1}(J)$. Let D be a polyhedral disk and $h: J \to Bd\ D$ be a homeomorphism. Let $g: D \to \text{Int}\ C$ be a map such that $g = f_{t_1}h^{-1}$ on $Bd\ D$ and g is piecewise linear in a neighborhood of $g^{-1}[Bd\ T_0 \cdot g(D)]$. We suppose that g is in a general position with respect to $Bd\ T_0$ in the sense that $g^{-1}[Bd\ T_0 \cdot g(D)]$ is the sum of a finite collection of mutually exclusive simple closed curves. Let E be the component of $D - g^{-1}[Bd\ T_0 \cdot g(D)]$ containing $Bd\ D$. Then $\bar{E}$ is a punctured disk, and it follows from Theorem 14 that there is a simple closed curve in $Bd\ T_0 \cdot \text{Int}\ C$ that is isotopic in T_0 to a core of T_0.

We have the following as a corollary to Theorem 17.

Theorem 18. A compact connected 3-manifold M is topologically S^3 if each tame simple closed curve J in M can be isotopically shrunk to small size in M so that the image of J at each stage of the isotopy is a tame simple closed curve.

Although it is not known that each tame simple closed curve J in a homotopy 3-sphere can be shrunk to a point so that at each intermediate stage the image is a simple closed curve, it is known that J can be shrunk so that at each intermediate stage the image of J is either a simple closed curve or a figure eight.

Theorem 19. If J is a polygonal simple closed curve which can be shrunk to a point in a triangulated 3-manifold M, there is a homotopy $f_t:J \to M$ such that f_0 is the identity, f_1 is a constant map, and each $f_t (0 \leq t < 1)$ is a piecewise linear map which is a homeomorphism except at a finite number of t's and has precisely two singularities at each of these.

PROOF. Let D be a 2-simplex and $g:D \to M$ be a piecewise linear map such that $g: Bd\ D \to J$ is a homeomorphism and g is a general position map in the sense that there is a triangulation T of D such that g is a piecewise linear homeomorphism on each simplex of T; and if s', s'' are two simplexes of T,

$$\text{dimension } [g(s') \cdot g(s'') - g(s' \cdot s'')] \leq \text{dimension } s' + \text{dimension } s'' - 3$$

Since T can be shelled (Theorem 3), there is an ordering $s_1, s_2, \ldots, s_n$ of the 2-simplexes of T such that for each $i \leq n$, $s_i + s_{i+1} + \cdots + s_n$ is a disk D_i. Note that g is a homeomorphism on each $Bd\ D_i$ and g^{-1} is locally a homeomorphism modulo a finite set of points X_i on $g(s_i + Bd\ D_{i+1})$ where each point of X_i has precisely two preimages under g.

The homotopy $f_t:J \to g(D)$ is described as follows.

$$f_0 = I$$
$$f_{i/n}(J) = g(Bd\ D_{i+1})$$
$$f_t = f_{i/n} \text{ on } f_{i/n}^{-1} g(Bd\ D_{i+1} \cdot Bd\ D_{i+2})$$
$$f_t = g h_t g^{-1} f_{i/n} \text{ on } f_{i/n}^{-1} g(Bd\ D_{i+1} - Bd\ D_{i+2})$$

for $i/n \leq t \leq (i+1)/n$ where h_t is a piecewise linear isotopy that shoves $Bd\ D_{i+1} - Bd\ D_{i+2}$ straight to $Bd\ D_{i+2} - Bd\ D_{i+1}$ through s_{i+1} in such a way that for no t does $gh_t(Bd\ D_{i+2} - Bd\ D_{i+1})$ contain more than one point of X_{i+1} and for each point x of $s_{i+1} \cdot X_{i+1}$ there is exactly one h_t with $x \in gh_t(Bd\ D_{i+1} - Bd\ D_{i+2})$. For the values of t for which $gh_t(Bd\ D_{i+1} - Bd\ D_{i+2})$ misses X_{i+1}, f_t is a homeomorphism and for the finite number of others, $f_t(J)$ is a figure eight.

8. WHY?

Why are mathematicians interested in finding a solution to the Poincaré conjecture? For one thing, a solution would show the power of the algebraic technique. For compact 3-manifolds, it would show at least in one case that the algebraic properties determine the topology. It would also provide us with a handy characterization of S^3.

Aside from this, there is a historical interest in the Poincaré conjecture. Poincaré announced in [21] that two compact combinatorial n-manifolds were topologically equivalent if they had the same homological characters, but he disproved this in [22] by constructing a 3-manifold which is not simply connected, even though all its 0, 1, and 2 cycles bounded. The homology spheres given in Section 2 were such examples. Alexander showed [1] that there were combinatorially different toroidal 3-manifolds which had the same homotopy properties. It follows as a result of Moise's proof of the Hauptvermutung [16] that these homotopically equivalent toroidal manifolds were topologically different. We might still hope, however, that if the first homotopy group is trivial, this would imply that the manifolds were topologically equivalent also. J. H. C. Whitehead gave a faulty proof of the Poincaré conjecture in [30] but pointed out his error shortly thereafter in [31].

There were two competing solutions for the Poincaré conjecture being rumored at the Institute for Advanced Study in Princeton in 1957. Although both proofs fell flat, the rumor was so strong for awhile that the author considered withdrawing his partial solution [5] which had already been accepted by the Annals.

A "proof" of over 100 pages appeared [12] in 1958. The last half of the paper gives the true result that a compact connected triangulated 3-manifold is topologically S^3 if its 2-skeleton can be embedded in E^3. The first half asserts that for any triangulation of a homotopy 3-sphere, the 2-skeleton can be embedded in E^3. The plan of the proof is to embed the 2-simplexes one at a time, taking care that the images fan out around edges and vertices in the way the original simplexes do. Most of the first half is clear where it is shown how to embed the first few 2-simplexes. It is then claimed that if an obstruction arises, it would have been possible to start over and by a different approach embed more of the 2-simplexes. The argument here is not clear.

It was shown by Stallings, Zeeman, Smale, and Wallace that in high dimensions certain other versions of the Poincaré conjecture hold [25, 33, 34, 24, 29]. Theorem 20 was shown.

Theorem 20. If M^n $(n \geq 5)$ is a compact combinatorial n-manifold that has the same connectedness properties as S^n, then M^n is topologically equivalent to S^n.

PROOF. Although we do not include details, the proof may be given with the following steps.

1. Using an engulfing lemma [26], show that any $(n - 3)$-dimensional polyhedron in M^n lies on the interior of a polyhedral n-cell in M^n.
2. Put the $(n - 3)$-skeleton K^{n-3} of the triangulation T of M^n on the interior of a topological n-cell C_1.
3. Get the dual of K^{n-3} by taking the first baricentric subdivision of T and letting the dual H^2 be the sum of all simplexes of this subdivision that miss K^{n-3}. It is to be noted that H^2 is of dimension 2.
4. If $2 \leq n - 3$ (and this occurs only when $n \geq 5$), there is a topological n-cell C_2 whose interior covers H^2.
5. Swell up C_1 and C_2 until the sum of their interiors cover M^n.
6. It follows from the Schoenflies theorem proved by Brown [7] that any compact n-manifold is topologically S^n if it is the sum of two open n-cells.

One of the reasons we have been able to prove so many interesting things about 3-manifolds is that we know so much about S^2. Not knowing more about S^3 is a bottleneck in learning more about higher dimensional manifolds. The following theorem gives one reason for wanting to get an answer to the Poincaré conjecture.

Theorem 21. Any triangulated 4-manifold is a combinatorial 4-manifold if the Poincaré conjecture holds.

This result follows from the following theorem.

Theorem 22. If T is a triangulation of a 4-manifold M^4 and v is a vertex of T, then $L(v, T) = Bd$ Star (v, t) is a simply connected 3-manifold.

PROOF. Note that $L(v, T)$ consists of the sum of all 3-simplexes s^3 of T such that the join of v and s^3 is a 4-simplex of T. Let T' be the set of simplexes of T in $L(v, T)$. We note that $L(v, T)$ is simply connected. If it were not, some simple closed curve near v in $M^4 - v$ could not be shrunk to a point on a small subset of $M^4 - v$.

To show that $L(v, T)$ is a manifold, it suffices to show that the link $L(v', T')$ of each vertex v' of T' is a 2-sphere. This link $L(v', T')$ is the sum of all 2-simplexes s^2 of T' such that the join of s^2 and v' is a 3-simplex of T'. We note that $L(v', T')$ is a 2-complex. Also, $L(v', T')$ is connected or else v' locally separates $L(v', T')$ and vv' locally separates M^4. Each 1-simplex s^1 of T' in $L(v', T')$ is a face of precisely two 2-simplexes of T' in $L(v', T')$ or else the invariance of domain is violated in M^4 near the join of s^1 and v. No point s of $L(v', T')$ locally separates $L(v', T')$ or else $v's$ locally separates $L(v, T)$ and $vv's$ locally separates M^4. Hence $L(v', T')$ is a connected 2-manifold.

To show that the connected 2-manifold $L(v', T')$ is a 2-sphere, it suffices to show that each simple closed curve in $L(v', T')$ homologically bounds in $L(v', T')$. If such a simple closed curve J_1 did not bound, there would be a small simple closed curve J_2 in $L(v, T) - v'$ near v' such that J_2 does not bound on a small subset of $L(v, T) - v'$. Then there would be a small simple closed curve J_3 on the join of v and J_2 near the center of vv' such that J_3 does not bound on a small set in $M^4 - vv'$. This violates the fact that the first homology of a 4-cell remains trivial even on the deletion of a closed set homeomorphic to a subset of an arc.

REFERENCES

1. J. W. Alexander, "Note on two three-dimensional manifolds with the same group," *Trans. Amer. Math. Soc.* **20** (1919) 339–342.
2. R. H. Bing, "A characterization of 3-space by partitioning," *Trans. Amer. Math. Soc.* **70** (1951) 15–27.
3. ———, "Locally tame sets are tame," *Ann. of Math.* **59** (1954) 145–158.
4. ———, "Approximating surfaces with polyhedral ones," *Ann. of Math.* **65** (1957) 454–483.
5. ———, "Necessary and sufficient conditions that a 3-manifold be S^3," *Ann. of Math.* **68** (1958) 17–37.

6. ———, "An alternative proof that 3-manifolds can be triangulated," *Ann. of Math.* **69** (1959) 37–65.
7. M. Brown, "A proof of the generalized Schoenflies theorem," *Bull. Amer. Math. Soc.* **66** (1960) 74–76.
8. R. H. Fox, *Construction of Simply Connected 3-Manifolds, Topology of 3-Manifolds and Related Topics*, Englewood Cliffs, N.J. (1962) Prentice-Hall, 213–216.
9. F. Frankl, "Zur Topologie des dreidimensionalen Raumes," *Monatsh. Math. Phys.* **38** (1931) 357–364.
10. J. Hempel, *Construction of Orientable 3-Manifolds, Topology of 3-Manifolds and Related Topics*, Englewood Cliffs, N.J. (1962) Prentice-Hall, 207–212.
11. ———, "A simply connected 3-manifold is S^3 if it is the sum of a solid torus and the complement of a torus knot," *Bull. Amer. Math. Soc.* **15** (1964) 154–158.
12. K. Koseki, "Poincarésche Vermutung in Topologie," *Math. J. Okayama Univ.* **8** (1958) 1–106.
13. D. R. McMillan, Jr., "On homologically trivial 3-manifolds," *Trans. Amer. Math. Soc.* **98** (1961) 350–367.
14. J. Milnor, "A unique decomposition of 3-manifolds," *Amer. J. Math.* **84** (1962) 1–7.
15. E. E. Moise, "Affine structures in 3-manifolds, IV. Piecewise linear approximations of homeomorphisms," *Ann. of Math.* **55** (1952) 215–222.
16. ———, "Affine structures in 3-manifolds, V. The triangulation theorem and Hauptvermutung," *Ann. of Math.* **56** (1952) 96–114.
17. C. D. Papakyriakopoulos, "On solid tori," *Proc. London Math. Soc.* (3) **7** (1957) 281–299.
18. ———, "On Dehn's lemma and the asphericity of knots," *Ann. of Math.* **66** (1957) 1–26.
19. ———, "Some problems on 3-dimensional manifolds," *Bull. Amer. Math. Soc.* **64** (1958) 311–335.
20. ———, "Reduction of the Poincaré conjecture to other conjectures," *Bull. Amer. Math. Soc.* **68** (1962) 360–366.
21. H. Poincaré, "Second complément a l' analysis situs," *Proc. London Math. Soc.* **32** (1900) 277–308.
22. ———, "Cinquième complément a l' analysis situs," *Rend. Circ. Mat. Palermo* **18** (1904) 45–110.
23. Mary E. Rudin, "An unshellable triangulation of a tetrahedron," *Bull. Amer. Math. Soc.* **64** (1958) 90–91.
24. S. Smale, "Generalized Poincaré's conjecture in dimensions greater than four," *Ann. of Math.* (2) **74** (1961) 391–406.
25. J. R. Stallings, "Polyhedral homotopy spheres," *Bull. Amer. Math. Soc.* **66** (1960) 485–488.
26. ———, "The piecewise-linear structure of euclidean space," *Proc. Cambridge Philos. Soc.* **58** (1962) 481–488.
27. E. R. Van Kampen, Remark on the address of S. S. Cairns, Michigan Lectures in Topology (1941).
28. A. H. Wallace, "Modifications and cobounding manifolds," *Canad. J. Math.* **12** (1960) 503–528.

29. A. H. Wallace, "Modifications and cobounding manifolds," II, *J. Math. Mech.* **10** (1961) 773–809.
30. J. H. C. Whitehead, "Certain theorems about three-dimensional manifolds," I, *Quart. J. Math. Oxford* **5** (1934) 308–320.
31. ———, "Three-dimensional manifolds (corrigendum)," *Quart. J. Math. Oxford* **6** (1935) 80.
32. ———, "Simplicial spaces, nuclei, and m-groups," *Proc. London Math. Soc.* **45** (1939) 243–327.
33. E. C. Zeeman, "The generalized Poincaré conjecture," *Bull. Amer. Math. Soc.* **67** (1961) 270.
34. ———, *The Poincaré conjecture for $n \geq 5$, Topology of 3-manifolds and related topics,* Englewood Cliffs, N.J. (1962) Prentice-Hall, 198–204.

MAPPING A 3-SPHERE ONTO A HOMOTOPY 3-SPHERE

by

R. H. Bing*

1. Introduction

We shall use S^3 to denote a 3-sphere—that is a set homeomorphic to the one point compactification of Euclidean 3-space. It may be convenient to regard S^3 as the union of two cubes sewed together with a homeomorphism on their boundaries. We suppose that S^3 is triangulated.

We shall use M^3 to denote a homotopy 3-sphere—that is a compact connected simply connected 3-manifold. We also suppose it is triangulated.

The Poincaré conjecture is that M^3 is topologically equivalent to S^3. See [1] for a discussion of attempts before 1964 to solve the conjecture. There were serious attempts made during 1964 to prove the Poincaré conjecture in the affirmative. For a while during the summer of 1964 it was felt by some that Wolfgang Haken or (and) Valentin Poénaru would be able to construct an airtight proof. Their plans of attack were essentially the same although Haken used a piecewise linear approach and Poénaru a differentiable approach.

This common approach consisted of removing the interior of a 3-simplex from M^3 so as to have left a compact connected, simply connected 3-manifold with boundary whose boundary is a 2-sphere. We denote this remaining homotopy 3-cell by K^3 and call it a fake cube. It follows from a theorem of Hurewicz that since $\pi_1(K^3) = 0$, $H_2(K^3) = 0$, then $\pi_2(K^3) = 0$. Hence, there is a homotopy pulling the boundary of K^3 to a point in K^3. Both Haken and Poénaru try to make this homotopy "nice" so as to conclude that K^3 is a real cube. Neither succeeded.

Suppose H_t: Bd $K^3 \to K^3$ $(0 \leq t \leq 1)$ is a homotopy of Bd K^3 into K^3 such that H_0 is the identity map and H_1 is a constant map. It can be shown that if each H_t $(0 \leq t \leq 1)$ takes Bd K^3 homeomorphically into a tame 2-sphere in Int K^3, then K^3 is a topological cube.

As far as I know, the proof has not appeared in print but the approach is to show that if $0 < t_0 \leq 1$, then Bd K^3 and H_{t_0}(Bd K^3) co-

* Work on this paper was supported by the National Science Foundation under Grant GP-3857.

bound a hollow ball (3-dimensional annulus) in K^3.

Question 1. Can one prove that K^3 is a real cube if one merely assumes that each H_t $(0 \leq t \leq 1)$ takes Bd K^3 homeomorphically into a 2-sphere (perhaps wild) in Int K^3?

(After this talk was given, Robert F. Craggs answered the above question in the affirmative by using the result of McMillan [4] that if a 2-sphere is embedded in a 3-manifold M^3 then there is a hollow ball with handles in M^3 such that the two sphere separates in M^3 the two boundary components of the hollow ball with handles.)

Instead of pushing the boundary of a fake cube, one might try pushing spanning arcs to near the boundary. It follows from techniques used in [1] that a fake cube K^3 is a topological 3-cell if for each polyhedral spanning arc A^1 of K^3 there is an isotopy H_t: $A^1 \to K^3$ $(0 \leq t \leq 1)$ such that H_0 is the identity, H_t is the identity on the ends of A^1, $H_1(A^1)$ lies in a Cartesian product neighborhood of Bd K^3, and each $H_t(A^1)$ is a tame arc.

Question 2. Can one prove that K^3 is a real cube if one drops the assumption that each $H_t(A^1)$ is tame?

Although neither Haken nor Poénaru got a solution to the Poincaré conjecture, each was able to salvage some good results out of his effort. An interesting result obtained by Haken was the following:

THEOREM 1. There is a map f of S^3 onto M^3 such that the image of the singularities of f lie on the interior of a cube with handles X^3.

A point $p \in S^3$ is a nonsingular point of f if there is a neighborhood U of $f(p)$ in $f(S^3)$ such that f^{-1} is a homeomorphism on U. Other points of S^3 are singular points of f. A cube with handles is a set homeomorphic to the union of a finite number of 3-balls C^3, $H_1{}^3$, $H_2{}^3$, ..., $H_n{}^3$ such that $C^3 \cap H_i{}^3 = \text{Bd } C^3 \cap \text{Bd } H_i{}^3$ is the union of two mutually exclusive disks and no two of the handles $H_i{}^3$ intersect each other.

A stronger form of Theorem 1 was stated by Moise in [3]. However, we do not have Moise's proof so we will model ours after Haken's given in [2].

2. Three Lemmas

We shall list three lemmas that can be used to help prove Theorem 1.

Consider the Cartesian product of a disk with an arc. If the disk is round and the arc is short straight segment, the resulting topological cube resembles a pill box. Figure 1 shows two pill boxes P^3, Q^3 sewed together along a disk S^2 on the lateral side of each.

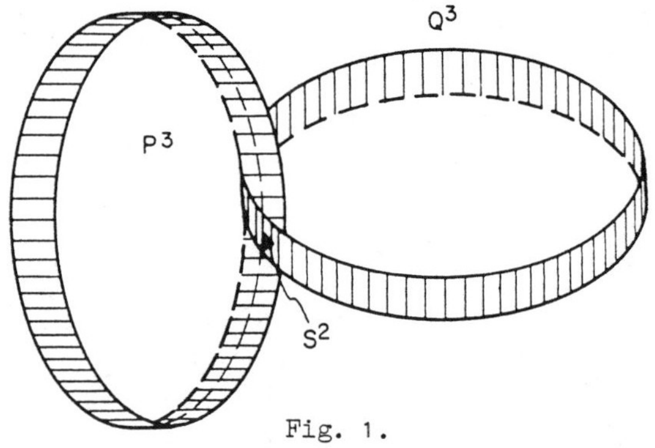

Fig. 1.

Pill Box Lemma 1. There is a homeomorphism of $P^3 \cup Q^3$ of Fig. 1 onto a ball that takes the two bases of P^3 and the exposed part of the lateral side of Q^3 onto the upper hemisphere of the ball and the two bases of Q^3 and the exposed part of the lateral side of P^3 onto the lower hemisphere.

We will not include a proof of the Pill Box Lemma. We note that the simple closed curve on $P^3 \cup Q^3$ corresponding to the equator of the ball is the union of four circles each with a gap cut in it where P^3 intersects Q^3.

We mention where the Pill Box Lemma is used. Suppose one is using the approach of Haken or Poénaru and pushing a two sphere in toward a 2-complex in M^3. A regular neighborhood of this 2-complex resembles the union of a tubular neighborhood T^3 of the 1-skeleton of the 2-complex together with some pill boxes $P_i{}^3$ sewed onto T^3 to cover the middle portion of the 2-simplexes of the 2-complex as shown in Fig. 2. If the 2-sphere has been pushed in to cover the bases of the $P_i{}^3$'s and the part Bd T^3 not on the lateral edge of a $P_i{}^3$ then one can push in further across the double pill box shown in Fig. 2 so that the part of the sphere that once went onto the bases of $P_i{}^3$ and the lateral edge of the right pill box now go onto the lateral edges of $P_i{}^3$ and the bases of the right pill box. Such a push is cometimes called a baseball move since the part of the double pill box corresponding to the equator of a ball is in somewhat the same position as the seams on a baseball. We have essentially replaced one of the two halves of a baseball cover with the other.

If the $P_i{}^3$ of Fig. 2 is added to T^3, then the new T^3 and the new S^3 - Int T^3 each has one fewer handle.

The second lemma we shall find useful is the following:

LEMMA 2. Suppose G^1 is a finite polygonal graph that can be pulled to a point in a triangulated 3-manifold M^3 in which it lies. Then there is a polyhedral cube with handles T^3 in M^3 containing G^1 on its interior such that G^1 can be shrunk to a point in T^3.

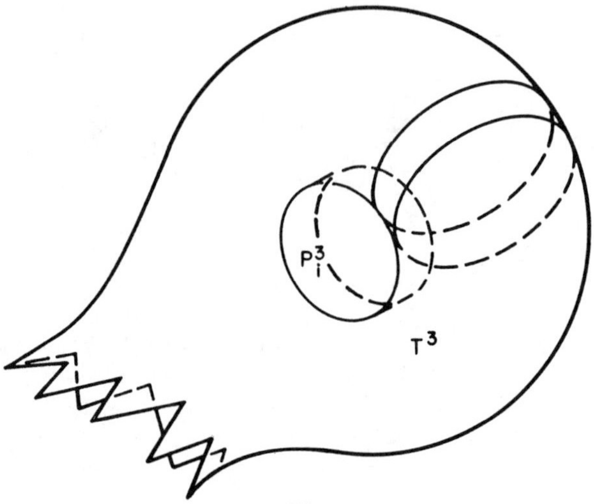

Fig. 2

With no loss of generality we suppose that G^1 is connected. In this connected G^1 a maximal tree can be shrunk to a point such that G^1 becomes a bouquet of simple closed curves. In Section 6 of [2] Haken shows this bouquet of simple closed curves bounds a singular bouquet of disks whose regular neighborhood is a cube with handles.

LEMMA 3. Any orientable compact, connected, 3-manifold N^3 with non-null boundary is the union of a cube with handles T^3 and a finite number of 3-cells $P_1^3, P_2^3, \ldots, P_n^3$ such that each $P_i^3 \cap T^3$ is an annulus on $\mathrm{Bd}\, P_i^3 \cap \mathrm{Bd}\, T^3$.

This is a lemma used by McMillan in [4].

We suppose N^3 is triangulated and T^1 is the 1-skeleton of the triangulation. A tubular neighborhood of T^1 is a cube with handles. One can obtain this tubular neighborhood of T^1 as a remainder if one bores holes in M^3 along the dual 1-skeleton of the triangulation. These holes correspond to the P_i^3's. Although as originally drilled, a hole may have its ends on the lateral sides of previous holes rather than on $\mathrm{Bd}\, N^3$, an isotopy remedies this situation.

Another way of proving Lemma 3 involves collapsing N^3 to a 2-spine. A tubular neighborhood of the 1-skeleton of the spine provides us with T^3 and a thickening of the middle part of the 2-simplexes of the spine provides us with the P_i^3's. In this case the P_i^3's resemble pill boxes. In case N^3 is a fake cube, the 2-spine is a contractible 2-complex such as shown in Fig. 3. It shows a house-with-two rooms, an igloo shaped figure, and a dunce's cap strung together.

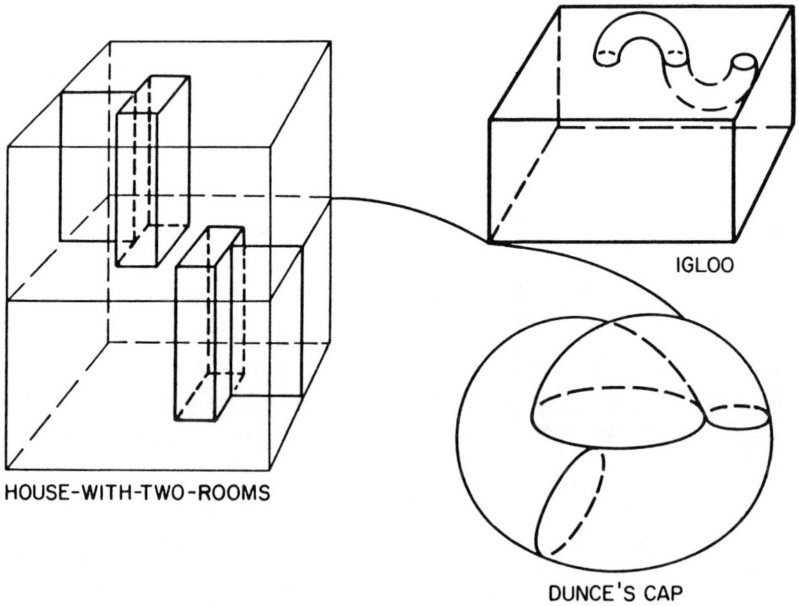

Fig. 3

A fake cube is the contractible union of a cube with handles and a finite number of mutually exclusive pill boxes each attached to the cube with handles along an annulus. The fundamental group of the cube with handles is a free group on n generators where n is the number of handles. Each pill box provides a relation so the fundamental group of the fake cube is

$$G = \{a_1, a_2, \ldots, a_n / r_1 = r_2 = \cdots = r_n = 1\} .$$

Since a fake cube is simply connected, the fundamental group G for a fake cube is trivial.

Question 3. What are some exotic presentations of the trivial group of the following sort

$$G = \{a_1, a_2, \ldots, a_n / r_1 = r_2 = \cdots = r_n = 1\}?$$

An example of such a presentation is

$$G = \{a, b / a^i b^j = a^r b^s = 1 \quad \text{where} \quad |is - rj| = 1\} .$$

This one does not lead to a counterexample to the Poincaré conjecture because it is possible to obtain a real cube by adding two pill boxes to a cube with handles so that the pill boxes yield the suggested relations. Perhaps some other group does lead to a counterexample.

3. Proof of Theorem 1

The map f is defined in three parts. First we define it outside a 3-ball B^3. Then it is defined on some pits in B^3. Finally, it is defined on the part of B^3 not in these pits. See Fig. 4.

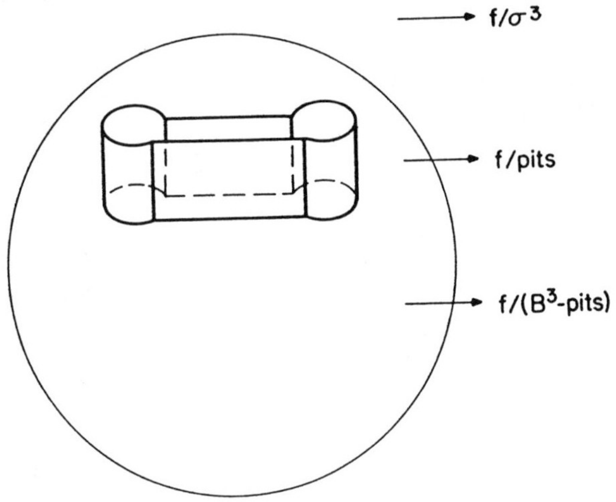

Fig. 4

Part 1. Let f be the map of a 3-simplex σ^3 of S^3 onto a 3-simplex of M^3. We let B^3 be the 3-cell $S^3 - (\text{Int } \sigma^3)$ and K^3 be the fake cube $M^3 - f(\text{Int } \sigma^3)$. Our task now is to extend f to B^3.

Part 2. It follows from Lemma 3 that K^3 is the union of a cube with handles T^3 and a finite number of mutually exclusive pill boxes $P_1^3, P_2^3, \ldots, P_n^3$ sewed onto T^3 along their lateral sides. It follows from Lemma 2 that T^3 lies in a cube with handles X^3 in M^3 such that T^3 can be shrunk to a point in $\text{Int } X^3$ and some open subset U^3 of $f(\sigma^3)$ misses X^3.

Let A_i^2 be the lateral side of P_i^3 and S_i^2 be a disk on A_i^2 intersecting each end of P_i^3 in an arc. Let R_i^2 be a disk on $\text{Bd } T^3 - \cup \text{ Int } A_j$ such that $R_i^2 \cap (P_j^3 \cup R_j^2) = 0$ $(i \neq j)$ and $R_i^2 \cap P_i^3 = S_i^2 - \text{Int } A_i$. See Fig. 5.

If $R_i^2 + S_i^2$ were the lateral edge of a pill box Q_i^3 in T^3, $P_i^3 \cup Q_i^3$ would be a double pill gox as mentioned in Lemma 1. There is no assurance that this is the case. However, since $R_i^2 \cup S_i^2$ lies in T^3, which in turn can be shrunk to a point in $\text{Int } X^3$ there is a map g_i of

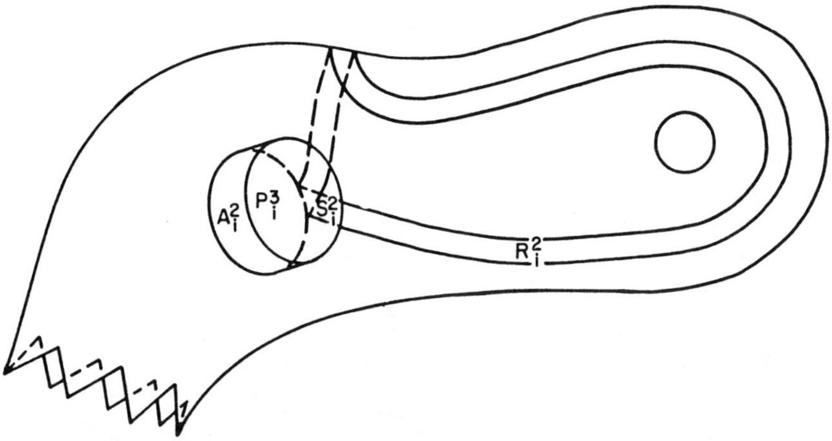

Fig. 5

the double pill box $P^3 \cup Q^3$ of Lemma 1 such that g_i takes P^3 homeomorphically onto P_i^3, the later sides of P^3 onto A_i^2, the lateral side of Q^3 homeomorphically onto $S_i^2 \cup R_i^2$ and Q^3 into $\text{Int } X^3$ (perhaps singularly).

Let $B_i^2 = f^{-1}(R_i^2 \cup \text{ends of } P_i^3)$. It is a disk on $\text{Bd } B^3$ resembling two disks joined by a narrow band. Let $C_1^3, C_2^3, \ldots, C_n^3$ be mutually exclusive 3-cells in B^3 such that $C_i^3 \cap \text{Bd } B^3 = B_i^2$. Let h_i be a homeomorphism of C_i^3 onto $P^3 \cup Q^3$ of Lemma 1 such that $g_i h_i / B_i^2 = f/B_i^2$. Note that $f(\text{Bd } C_i^3 - B_i^2) \subset X^3$. We define f on the pits C_i^3 by $f/C_i^3 = g_i h_i$.

<u>Part</u> 3. Let $B^{3'}$ denote the 3-ball which is the closure of $B^3 - \cup C_i^3$. Note that $f(\text{Bd } B^{3'}) \subset X^3$. Since X^3 is a cube with handles, $\pi_2(X^3) = 0$. Hence $f/\text{Bd } B^{3'}$ can be extended to map of $B^{3'}$ into $\text{Int } X^3$.

It may be shown that $f: S^3 \to K^3$ is of degree 1 since f^{-1} is a homeomorphism on U^3. The only points of M^3 that have more than one inverse lie on a compact subset of $\text{Int } X^3$. Hence the image of the singularities of f lie in $\text{Int } X^3$.

4. Replacing cubes with holes by cubes with handles

Suppose C^3 is a polyhedral 3-cell and $C_1^3, C_2^3, \ldots, C_n^3$ is a collection of mutually exclusive polyhedral 3-cells in C^3 such that $\text{Bd } C^3 \cap \text{Bd } C_i^3$ is the union of two mutually exclusive disks $A_i^2, A_i^{2'}$. We regard C_i^3 as $A_i^2 \times [0,1]$ where $A_i^2 \times 0 = A_i^2$ and $A_i^2 \times 1 = A_i^{2'}$.

Then $C^3 - \cup (\text{Int } A_i \times [0,1])$ is a <u>cube with holes</u> and each $\text{Int } A_i \times [0,1]$ is a <u>hole</u>.

Let p_i be a point of A_i^2. The hole $\text{Int } A_i^2 \times [0,1]$ is <u>knotted</u> or <u>unknotted</u> according as $p_i \times [0,1]$ is a knotted or unknotted spanning arc of C^3. A tame spanning arc of C^3 is <u>unknotted</u> or <u>knotted</u> according as it, together with an arc on $\text{Bd } C^3$, does or does not bound a disk in C^3.

It is known that each polyhedron in S^3 bounded by a connected 2-manifold is topologically a cube with holes. Hence $f^{-1}(X^3)$ of Theorem 1 is a cube with holes and M^3 was obtained from S^3 by removing a cube with holes and replacing it with a cube with handles. Theorem 1 advises us that any homotopy 3-sphere can be obtained as such a simplification of S^3. Theorem 1 tells us a way of getting a counterexample to the Poincaré conjecture if there is one.

The following result is rather easy to verify.

THEOREM 2. Suppose $Y^{3'}$ is a tame cube with holes in S^3, q is a map $Y^{3'}$ onto a cube with handles Y^3 such that $g/\text{Bd } Y^{3'}$ takes $\text{Bd } Y^{3'}$ homeomorphically onto $\text{Bd } Y^3$. If one replaces $Y^{3'}$ by Y^3 in S^3 (the sewing being done by a g along $\text{Bd } Y^{3'}$), one obtains a homotopy 3-sphere.

Theorem 1 tells us that any homotopy 3-sphere can be obtained from S^3 by removing some cube with holes and replacing it with a cube with handles while Theorem 2 tells us that anytime this replacement is done in a certain way, one obtains a homotopy 3-sphere.

Hempel showed that if X is a cube with one hole, then there is a map f of X onto a cube with a handle such that $f/\text{Bd } X$ takes $\text{Bd } X$ homeomorphically onto $\text{Bd } f(X)$. Whether or not each cube with holes in S^3 can be replaced in the "certain way" depends on the answers to the following questions.

<u>Question</u> 4. If X is a cube with two holes, is there a map f of X onto a cube with two handles such that $f/\text{Bd } X$ takes $\text{Bd } X$ homeomorphically onto the boundary of the cube with handles?

<u>Question</u> 5. If X is a cube with holes, under what conditions is there a map f of X onto a cube with handles such that f takes $\text{Bd } X$ homeomorphically onto the boundary of the cube with handles?

A cube with a knotted hole is topologically different from a cube with an unknotted hole since one has a non-Abelian fundamental group and the other has a fundamental group which is infinite cyclic.

However, a cube with two holes one of which is knotted and the other of which is unknotted may be topologically equivalent to a cube with two straight holes. Figure 6 shows how to start with a cube with two straight holes, gradually move the bottom base of the right hole along parallel to the dotted line until the right hole becomes knotted.

Question 6. Suppose $C^3 - (\text{Int } A_1 \times [0,1]) - (\text{Int } A_2 \times [0,1])$ is a cube with two holes tamely embedded in E^3 and topologically equivalent to a cube with handles such that $\text{Int } A_1 \times [0,1]$ is unknotted but $\text{Int } A_2 \times [0,1]$ is knotted. In Fig. 6 the knotted hole is a trefoil knot. Other toroidal knots could have been similarly obtained. What kind of knots could have been obtained?

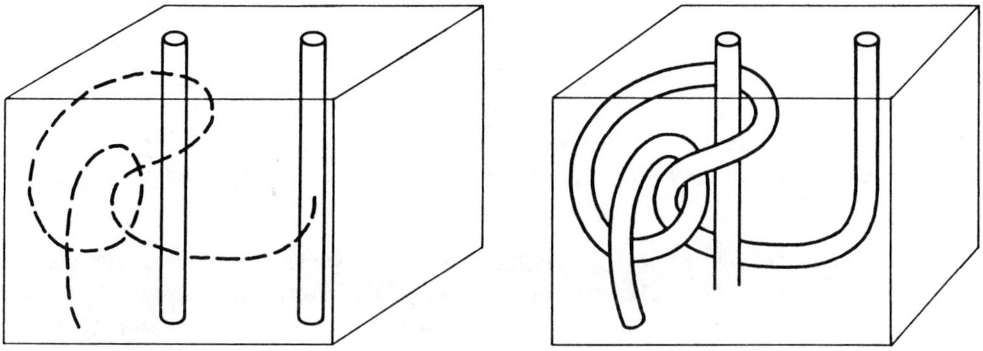

Fig. 6

One cannot tell at a glance whether or not a cube with holes is topologically equivalent to a cube with handles. Even if both holes are knotted, one cannot be sure it is topologically inequivalent to a cube with handles. For example if one winds the straight hole of Fig. 6 about the knotted hole, one obtains two holes as shown in the bottom part of Fig. 7.

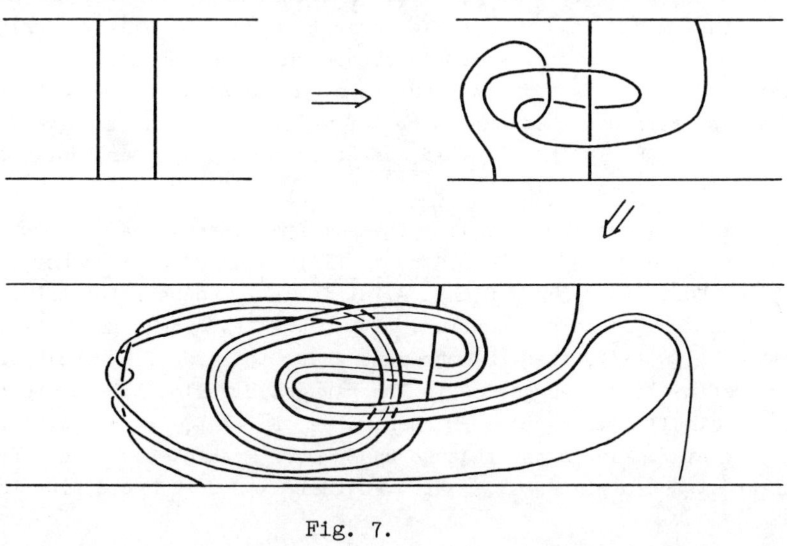

Fig. 7.

5. Simplifying handles

One might try to prove the Poincaré conjecture by simplifying either f or X^3 of Theorem 1.

THEOREM 3. In order for M^3 of Theorem 1 to be a counterexample to the Poincaré conjecture it is necessary that X^3 have more than one handle.

<u>Proof</u>: If X^3 has no handles, each of X^3 and M^3 - Int X^3 is a 3-cell. Hence, in this case, M^3 is topologically equivalent to S^3.

If X^3 has one handle, one of $f^{-1}(X^3)$, $S^3 - f^{-1}(\text{Int } X^3)$ is a solid torus. If $S^3 - f^{-1}(\text{Int } X^3)$ is a solid torus, each of X^3, M^3 - Int X^3 is a solid torus and no counterexample results. If $f^{-1}(X^3)$ is a solid torus, there is a homeomorphism h of $f^{-1}(X^3)$ onto X^3 that agrees with f on Bd $f^{-1}(X^3)$ and M^3 is homeomorphic to S^3.

One might try proving the Poincaré conjecture by reducing the number of handles in X^3. If one can find a handle in $f^{-1}(X^3)$ one can cut off this handle and redefine f so that the images of the singularities of the modified f lie in a cube with fewer handles than X^3 has.

<u>Question</u> 7. What special can be said about a cube with handles tamely embedded in S^3 so that some simple closed curve in Bd X bounds a disk in S^3 - Int X but none on Bd X? It might be of help in reducing the handles on X^3 of Theorem 1 if one knew, for example, that such a Bd X contained two transverse simple closed curves one of which bounded a disk in X and the other of which bounded a disk in E^3 - X. This may be difficult to show since I do not believe it is even known to be true in case S^3 - Int X is a cube with handles. As noted in the following paragraph, if we start modifying f and X^3 so as to reduce handles we arrive at a situation where there is a simple closed curve on the modified Bd X^3 which bounds a disk in M^3 - Int X^3 but none on Bd X^3. If Bd X^3 contained two transverse simple closed curves one of which bounded a disk in X^3 and the other a disk D in M^3 - Int X^3, we could remove another handle from X^3 by thickening D and adding it to X^3.

It follows from the loop theorem that there is a disk D in S^3 such that $D \cap f^{-1}(X^3) = \text{Bd } D \cap \text{Bd } f^{-1}(X^3)$ and Bd D does not bound a disk on Bd $f^{-1}(X^3)$. If $D \subset S^3 - f^{-1}(\text{Int } X^3)$, we have the situation mentioned in the preceding question. If $D \subset f^{-1}(X^3)$ and Bd D fails to separate Bd $f^{-1}(X^3)$, then D shows us where to cut off a handle from X^3 and redefine f as suggested in the paragraph following the proof of Theorem 3. Finally, we suppose D separates $f^{-1}(X^3)$. We suppose f is a homeomorphism near D so that we can remove a thickened $f(D)$ from X^3 and be left with two pieces X_1^3, X_2^3 so that $X_1^3 \cup X_2^3$ contains the images of the singularities of f. By continuing splitting we finally arrive at a finite set of X_i^3's whose union contains the images of the singularities of f and the sum of the handles in the X_i^3's is the same as the number of handles in X^3. Then we find a disk D such that $D \cap f^{-1}(\cup X_i^3) = \text{Bd } D \cap \text{Bd } f^{-1}(\cup X_i^3)$ and Bd D does not separate the Bd $f^{-1}(X_i^3)$ on which it lies. We join the $f^{-1}(X_i^3)$'s by tubes which miss D to obtain a new $f^{-1}(X^3)$ such that the new X^3 has the same number of handles as the old but also, either f(D) enables us to move a handle from X^3 or there is a simple closed curve on Bd X^3 which does not bound a disk on Bd X^3 but does bound one in M^3 - Int X^3.

Instead of reducing the number of handles in X^3, one instead might try to simplify f^{-1} on the handles. If $D_i \times [0,1]$ is a handle of X^3, it is possible to define f so that each $f^{-1}(D_i \times t)$ is a disk with handles such that the set of singularities of f on $f^{-1}(D_i \times t)$ is a finite number of figure eights. By drilling holes in each $f^{-1}(D_i \times [0,1])$ so as to separate the figure eights in the middle level one can change X^3 (by increasing handles) so that the inverse of each handle is the Cartesian product of an arc and a disk-with-a-handle. Hence, we obtain the following result:

THEOREM 4. For each homotopy 3-sphere M^3 there is a map f of S^3 onto M^3 such that the image of the singular set of f is a bouquet of simple closed curves, the inverse of the vertex point of the bouquet is a cube with holes and the inverse of each other point of the bouquet is a figure eight.

REFERENCES

[1] R. H. Bing, "Some aspects of the topology of 3-manifolds related to the Poincaré conjecture," Lectures on Modern Mathematics, John Wiley and Sons (1964), vol. 2, pp. 93-128.

[2] Wolfgang Haken, "On homotopy 3-spheres," Illinois Journal of Mathematics, to appear.

[3] E. E. Moise, "Simply connected 3-manifolds," Topology of 3-Manifolds and Related Topics, Prentice Hall (1962), pp. 196-197.

[4] Russel McMillan, these notes.

TRANSACTIONS OF THE
AMERICAN MATHEMATICAL SOCIETY
Volume 155, Number 1, March 1971

CUBES WITH KNOTTED HOLES

BY
R. H. BING AND J. M. MARTIN[1]

Abstract. The statement that a knot K has *Property* P means that (1) if C is a cube with a K-knotted hole, and M^3 is a simply connected 3-manifold obtained by sewing a solid torus onto C, then $M^3 \cong S^3$, and (2) any piecewise linear homeomorphism of a cube with a K-knotted hole into S^3 can be extended to a piecewise linear homeomorphism of S^3 onto S^3. In this paper it is shown that many familiar classes of knots have Property P.

1. **Introduction.** The 3-dimensional Poincaré conjecture is that a compact, connected, simply connected 3-manifold without boundary is topologically a 3-sphere S^3. Despite efforts to prove the conjecture, it has withstood attack [4].

Since no solution has been found, some believe that the conjecture is false, and that there is a counterexample.

In attempting to construct a counterexample to the Poincaré conjecture, the first hurdle seems to be that of showing that the proposed example has a trivial fundamental group. This is difficult in that the information about the fundamental group usually appears as an exotic presentation of the group. Having somehow verified that this group is indeed trivial, one encounters the second hurdle of somehow showing that the example is topologically different from S^3.

It is known that every orientable 3-manifold may be obtained by removing a collection of disjoint solid tori from S^3 and sewing them back differently [14], [22], [12]. In this paper we examine some of the possibilities for constructing a counterexample to the Poincaré conjecture by removing a single solid torus from S^3 and sewing it back differently. Actually, we examine not only this process but one analogous to it which we call "attaching a pillbox to a cube with a knotted hole".

We find it useful to consider two sorts of surgery which are inverses of each other. If M^3 is a 3-manifold with boundary, A^2 is an annulus on Bd M^3 and P^3 is a 3-cell such that $M^3 \cap P^3 = A^2 = \text{Bd } P^3 \cap \text{Bd } M^3$, then we say that $M^3 \cup P^3$ was obtained by *sewing a pillbox* P^3 *to* M^3 *along* A^2. Suppose that H^3 is a 3-cell in M^3 such that $H^3 \cap \text{Bd } M^3$ is the union of two disjoint disks D_1^2 and D_2^2, and $N^3 = \text{Cl}(M^3 - H^3)$ is a 3-manifold with boundary. We say that N^3 was obtained by *boring a hole in* M^3.

Received by the editors September 29, 1969.

AMS 1970 *subject classifications*. Primary 55A05, 55A25, 55A40.

Key words and phrases. Knots, cubes with knotted holes, Poincaré conjecture, 3-manifolds, Property P.

[1] The first author was partially supported by ONR grant NR 043-374, and the second by NSF grant GP-3857.

If M^3 is a 3-cell, the manifold N^3 described in the preceding paragraph is called a *cube with a hole.* We name the hole after a polygonal simple closed curve $h(J)$ in Euclidean 3-space, E^3, where h is a homeomorphism of M^3 into E^3, and J is a simple closed curve on Bd N^3 which is the union of an arc from D_1^2 to D_2^2 on Bd H^3 and an arc from D_1^2 to D_2^2 on Bd M^3. Thus, we speak of a cube with a trefoil (overhand) knotted hole, a cube with a figure eight knotted hole, etc. If K is a knot, then a cube with this type of hole is called a *cube with a K-knotted hole* and is denoted by $C^3(K)$. A cube with an unknotted hole is topologically a *solid torus*, that is, the Cartesian product of a disk and a circle. We reserve the terminology "cube with a knotted hole", for a manifold of this type different from a solid torus. One obtains the fundamental group of N^3 by computing the fundamental group of $E^3-h(J)$ since $\pi_1(N^3)=\pi_1(E^3-h(J))$.

Let M^3 be a compact, connected 3-manifold with boundary whose boundary is a 2-sphere. Then M^3 is a *homology* 3-*cell* if its 1st homology group is trivial and a *homotopy* 3-*cell* if its fundamental group is trivial. The Poincaré conjecture is equivalent to the conjecture that each homotopy 3-cell is homeomorphic with a 3-cell.

Poincaré's original conjecture was [19] that the homology of a compact combinatorial n-manifold determines its topology. This would have implied that each homology 3-cell is a 3-cell, but Poincaré himself obtained [20] a counterexample to this shortly after he made the original conjecture. Peculiar homology 3-cells can be built by sewing pillboxes onto cubes with knotted holes so that the connecting annuli run through the holes.

One way to try for a counterexample to the Poincaré conjecture is to seek a homotopy 3-cell obtained by sewing a pillbox onto a cube with a knotted hole so that the connecting annulus runs through the hole in the cube (perhaps several times). It is known that no homotopy 3-cell results by using a trefoil knot [2] or even a torus knot [11], [13]. We show in this paper that none results for many other sorts of knots. We do not know the answer for all knots despite an assertion made on p. 36 of [2]. See [3].

The preceding paragraph implies that if K is a torus knot and $C^3(K)$ is a polyhedron in E^3 which is a cube with a K knotted hole, then each piecewise linear homeomorphism of $C^3(K)$ into E^3 can be extended to a piecewise linear homeomorphism of E^3 onto itself. We show in this paper that many other knots have this property. A counterexample to the Poincaré conjecture would be at hand if one could find a knot J with this property and such that $C^3(J)$ could be changed to a homotopy cube by sewing on a pillbox along an annulus that runs through the hole.

We say that a knot K has *Property* P if (1) each homotopy 3-cell obtained by sewing a pillbox onto $C^3(K)$ is a 3-cell, and (2) if $C^3(K)$ is a polyhedral cube with a K-knotted hole in E^3, then any homeomorphism of $C^3(K)$ into E^3 can be extended to a piecewise linear homeomorphism of E^3 onto itself. The trivial knot satisfies the first restriction but not the second. In §§2–5, we extend the class of knots known to have Property P. We are advised by Ralph Fox that some of the results in this paper have been obtained independently by F. Gonzales in his Princeton thesis.

2. **Cube with a figure eight knotted hole.** The figure eight knot, T_2, shown in Figure 1 is a special kind of a twist knot, T_n to be considered later. However, the study of $C^3(T_2)$ has an interesting history, and so is treated independently.

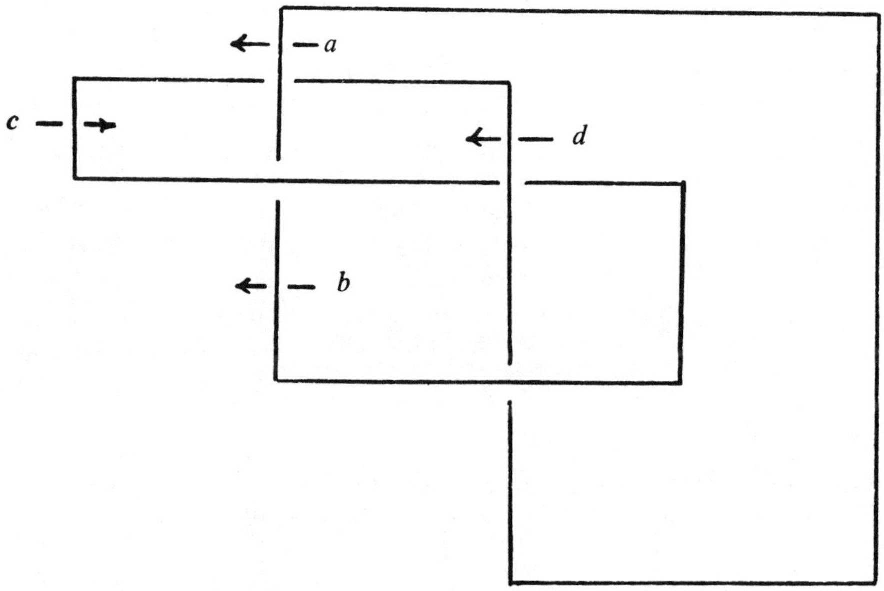

FIGURE 1

If one reads the fundamental group of E^3-T_2 from Figure 1 as suggested in [5] or [10], it is found to have a presentation

$$\pi_1(E^3-T_2) = \{a, b, c, d; aca^{-1}d^{-1} = ac^{-1}b^{-1}c = dc^{-1}d^{-1}b = dba^{-1}b^{-1} = 1\}.$$

Since $d=aca^{-1}$ and $b=cac^{-1}$, the presentation simplifies to

$$\pi_1(E^3-T_2) = \{a, c : aca^{-1}c^{-1}ac^{-1}a^{-1}cac^{-1} = 1\}.$$

This is also the fundamental group of $C^3(T_2)$.

If one forms a 3-manifold $C^3(T_2) \cup P^3$ by sewing a pillbox P^3 to $C^3(T_2)$ along an annulus that runs through the hole n-times, one finds that the relation $(b^{-1}ad^{-1}c)^n a^i = 1$ is added. The abelianization of the resulting group yields the integers mod $|i|$, so we find that in order to construct a homology cube we must have $i = \pm 1$. Hence

$$\begin{aligned}\pi_1(C^3(T_2) \cup P^3) &= \{a, c; aca^{-1}c^{-1}ac^{-1}a^{-1}cac^{-1} \\ &= (ca^{-1}c^{-1}a^2c^{-1}a^{-1}c)^n a^{\pm 1} = 1\}.\end{aligned}$$

Letting n take on negative values as well as positive ones, we drop the $\pm$ sign from the final exponent on a.

The question was raised in [3] as to whether or not, for some $n\neq 0$, $C^3(T_2)\cup P^3$ might be a homotopy 3-cell. In the hopes of getting something started, Bing raised the question in [4] as to whether or not it might be a homotopy 3-cell in the special case where $n=-1$. Independent answers were received in letters to Bing for the question about the special case. In 1964 J. Mennicke wrote Bing that in the special case no homotopy 3-cell results since for $n=-1$ there is a homomorphism of $\pi_1(C^3(T_2)\cup P^3)$ onto the simple group of order 168. Later J. M. Martin reported a similar result. In 1967 W. R. Alford and D. E. Penny also reported this result and said that they felt they were on the verge of solving the general case. (However, at last report, they had not succeeded.) In 1968, John McKay reported from the Atlas Computer Laboratory in England that with the aid of two colleagues, J. Leech and M. S. Patterson, and a computer that he had shown that Klein's 168 group is a factor group of the special group. The homomorphisms differed in detail, but a typical one would send a to (12)(3456)(7) and c to (1736)(25)(4). In 1968 J. M. Van Buskirk reported that $\pi_1(C^3(T_2)\cup P^3)$ is nontrivial for all positive values of n if and only if it is nontrivial for all negative values.

THEOREM 2.1. *For each integer $n\neq 0$, no homotopy 3-cell results from sewing a pillbox onto a cube with a figure eight knotted hole along an annulus that runs through the hole n times.*

Proof. For $n\neq 0$, we show the nontriviality of the group

$$\{a, c : aca^{-1}c^{-1}ac^{-1}a^{-1}cac^{-1} = (ca^{-1}c^{-1}a^2c^{-1}a^{-1}c)^n a = 1\}.$$

First we simplify the presentation. If we let $w=ca^{-1}$ and eliminate c by replacing it with wa, the first relation is shortened to $awa^{-1}w^{-2}a^{-1}waw^{-1}=1$. If we let $k=w^{-1}a$ and replace a by wk, the relation becomes $wk^2w=kw^3k$. The presentation of the group is changed to

$$\{w, k : kw^3k = wk^2w, (wk^{-1}w^{-1}kw^{-1}k^{-1}wk)^n(wk) = 1\}.$$

Now we simplify the second relation. The first relation $wk^2w=kw^3k$ implies that the factors $wk^{-1}w^{-1}kw^{-1}k^{-1}wk$ and wk commute. (This is also implied by the fact that $wk^{-1}w^{-1}kw^{-1}k^{-1}wk$ and wk are represented by curves on the boundary of a tubular neighborhood of T_2 and the fundamental group of a 2-dimensional torus is abelian.) Hence the second relations becomes $(k^{-2}w^{-1}kw^{-1}k^{-2}w^{-1})^n(wk)^{3n+1}=1$. The first relation simplifies this so we obtain

$$\pi_1(C^3(T_2)\cup P^3) = \{w, k; kw^3k = wk^2w, (kw^4k^2)^n = (wk)^{3n+1}\}.$$

We complete the proof of Theorem 2.1 by considering two cases.

Case 1. $n=-1$. In this case the presentation reduces to

$$\{x, y : xy^2x = yx^3y, yx^4y = xyx\}.$$

The substitution $z=xy^{-1}$ changes it to $\{x, z; x^4z^{-1}x^{-1}zx^{-1}=zxz^{-1}=x^{-1}z\}$, which simplifies to $\{x, z; x^7 = z^3 = (xz)^2\}$. This group is nontrivial since there is a homomorphism of it onto the group $G = \{u, v; u^7 = v^3 = (uv)^2 = 1\}$. It was noted on p. 35 of [2] that G is nontrivial since there is a homomorphism of G onto the permutation group generated by $u = (1234567)$, $v = (721)(643)(5)$.

Case 2. $n \neq 0, -1$. In this case $|3n+1| > 2$. By adding the relations $w^4 = 1$, $k^3 = 1$, we find that there is a homomorphism of $\pi_1(C^3(T_2) \cup P^3)$ onto $\{x, y : x^4 = y^3 = (yx^{-1})^3 = (xy)^{3n+1} = 1\}$. We show in §4 that this group is nontrivial if $|3n+1| > 2$.

THEOREM 2.2. *The figure eight knot has Property* P.

Proof. Theorem 2.1 shows that T_2 satisfies the first half of Property P. To see that the second condition is satisfied, regard $C^3(T_2)$ as standardly embedded in E^3. Let $h\colon C^3(T_2) \to E^3$ be a piecewise linear homeomorphism. Let J_1 be a polygonal simple closed curve on Bd $C^3(T_2)$ that goes around, but not through the hole. Since the closure (in S^3) of one complementary domain of $h(\text{Bd }(C^3(T_2)))$ is a solid torus, there is a polygonal simple closed curve J_2 on Bd $(C^3(T_2))$ such that $h(J_2)$ bounds a disk D^2 in $h(J_2) \cup (E^3 - h(C^3(T_2)))$. Notice that $\pi_1(D^2 \cup h(C^2(T_2)))$ is trivial.

Now if J_1 is isotopic to J_2 on Bd $(C^3(T_2))$ then there is a polyhedral 3-cell P^3 in Cl $(E^3 - C^3(T_2))$ whose intersection with $C^3(T_2)$ is an annulus having J_2 for a centerline. Then h can be extended to take P^3 onto a regular neighborhood of D^2 and then to take E^3 onto E^3.

If J_2 is not isotopic to J_1 on Bd $(C^3(T_2))$, it runs through the hole. Since $\pi_1(D^2 \cup h(C^3(T_2)))$ is trivial, a homotopy 3-cell results from sewing a pillbox onto $C^3(T_1)$ along an annulus whose centerline is J_2. This violates Theorem 2.1.

3. **Cubes with twist knotted holes.** A *twist knot* is a simple closed curve in E^3 which is doubled [23], with twists about an unknotted circle. See Figures 2 and 3. Twist knots are alternating knots and are determined by the number of half twists. This class includes the trefoil knot (1 half twist), the figure eight (2 half twists) and the stevedore's knot (4 half twists). In this section we show that cubes with twist knotted holes have Property P. We let T_j denote a twist knot with j half twists.

Figure 2 shows a twist knot T_{2m} with an even number of half twists. In computing $\pi_1(E^3 - T_{2m})$ we use $a_1, a_2, \ldots, a_{m+1}, b_1, b_2, \ldots, b_{m+1}$ as generators. Since all the generators are functions of a_1 and b_1 we drop their subscripts in writing the relation as follows:

$$\begin{aligned}
b_2 &= a_1^{-1}b_1a_1 = a^{-1}ba\\
a_2 &= b_2a_1b_2^{-1} = (a^{-1}b)a(a^{-1}b)^{-1}\\
b_3 &= a_2^{-1}b_2a_2 = (a^{-1}b)^2b(a^{-1}b)^{-2}\\
&\;\vdots\\
b_{m+1} &= a_m^{-1}b_ma_m = (a^{-1}b)^mb(a^{-1}b)^{-m}\\
a_{m+1} &= b_{m+1}a_mb_{m+1}^{-1} = (a^{-1}b)^ma(a^{-1}b)^{-m}\\
b_1 = b &= a_{m+1}a_1a_{m+1}^{-1} = (a^{-1}b)^ma(a^{-1}b)^{-m}a(a^{-1}b)^ma^{-1}(a^{-1}b)^{-m}.
\end{aligned}$$

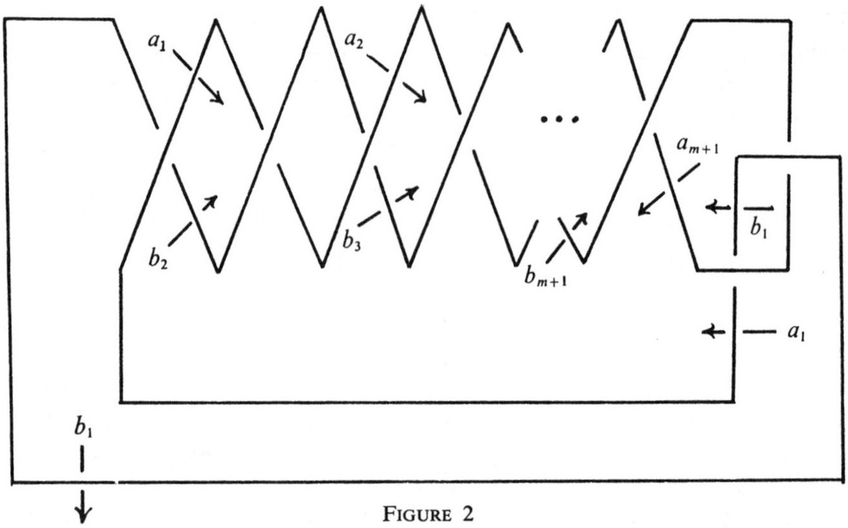

FIGURE 2

There is another relation given by the rightmost crossing but we ignore it since any one relation is a consequence of the others.

Letting $w=a^{-1}b$ and replacing b with aw we find that

$$\pi_1(E^3-T_{2m}) = \{a, w : aw = w^m aw^{-m} aw^m a^{-1} w^{-m}\}.$$

Substituting $k=aw^{-m}$ and replacing a with kw^m leads to the further simplification

$$\pi_1(E^3-T_{2m}) = \{w, k; kw^{2m+1}k = w^m k^2 w^m\}.$$

Figure 3 shows a twist knot T_{2m-1} with an odd number of half twists. Again, letting $a_1=a$ and $b_1=b$ we find that the crucial relation is

$$b = b_{m+1}^{-1} a b_{m+1} = (a^{-1}b)^m b^{-1}(a^{-1}b)^{-m} a (a^{-1}b)^m b (a^{-1}b)^{-m}.$$

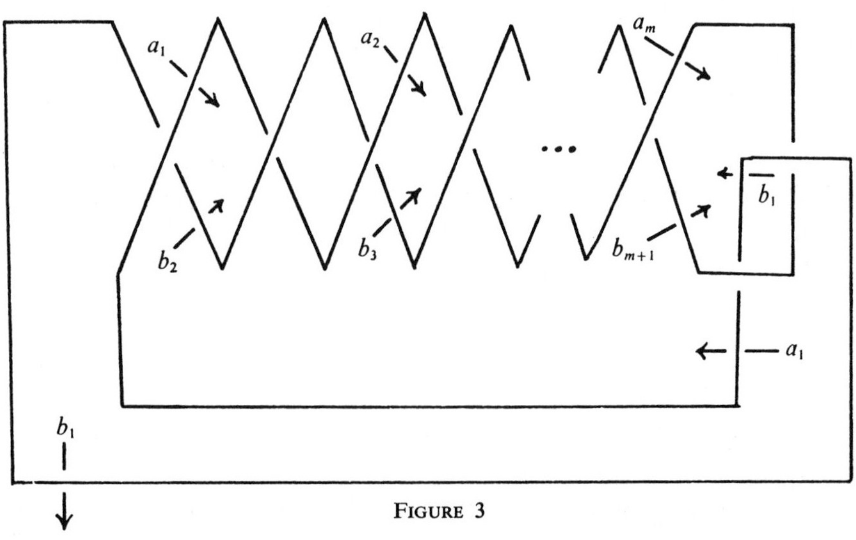

FIGURE 3

By letting $w=a^{-1}b$ and eliminating a we obtain $b=w^mb^{-1}w^{-m}bw^{-1}w^mbw^{-m}$. Substituting $k=bw^{-m}$ and eliminating b we obtain

$$\pi_1(E^3-T_{2m-1}) = \{w, k : kw^{2m-1}k = w^mk^2w^m\}.$$

In the last section we converted a cube with a figure eight knotted hole into a homology 3-cell by sewing on a pillbox along an annulus that runs through the hole n times. If we do the same with $C^3(T_{2m})$ the added relation is

$$(a_{m+1}^{-1}a_1a_2\cdots a_mb_1^{-1}b_{m+1}\cdots b_2a^{2-2m})^na = 1$$

where $a=a_1$, and a and the expression in the parentheses commute. Introducing $w=a^{-1}b$ to eliminate b we obtain

$$(w^ma^{-1}w^{-m}(aw)^mw^{-1-m}a^{-1}w^ma^{2-m})^na = 1.$$

Since $aw=w^maw^{-m}a^{-1}w^{-m}$, then $(aw)^m=w^maw^{-m}a^mw^ma^{-1}w^{-m}$, and we get the simplification $(w^ma^{-1}w^{-2m-1}a^{-1}w^ma^2)^na=1$. By letting $k=aw^{-m}$ and using the commuting relation, we find that

$$\pi_1(C^3(T_{2m})\cup P^3) = \{w, k : kw^{2m+1}k = w^mk^2w^m, (k^2w^{3m+1}k)^n = (kw^m)^{3n+1}\}.$$

For $\pi_1(C^3T_{2m-1})\cup P^3)$, the extra relation is

$$(a_1a_2\cdots a_mb_1b_mb_{m-1}\cdots b_2b_{m+1}b_1^{-1-2m})^nb = 1.$$

A simplification after letting $w=a^{-1}b$ yields $(w^{-m}bw^{2m-1}bw^{-m}b^{-2})^nb=1$, and the substitution $k=bw^{-m}$ shows that

$$\pi_1(C^3(T_{2m-1})\cup P^3) = \{w, k; kw^{2m-1}k = w^mk^2w^m, (k^2w^{3m-1}k)^n = (kw^m)^{3n-1}\}.$$

THEOREM 3.1. *The twist knots T_j, $j\neq 0$, have Property* P.

Proof. In order to establish Theorem 3.1, it is sufficient to show that $\pi_1(C^3(T_j)\cup P^3)$ is nontrivial unless $n=0$.

We first consider the case where $j=2m$. In this case we are interested in the group with generators w, k and relations (1) $kw^{2m+1}k=w^mk^2w^m$ and (2) $(k^2w^{3m+1}k)^n=(kw^m)^{3n+1}$. Now if we adjoin the relations (3) $w^{3m+1}=1$ and (4) $k^3=1$, then (1) becomes (1) $(w^mk^{-1})^3=1$ and (2) becomes (2) $(kw^m)^{3n+1}=1$. Now since $w^{3m+1}=1$ this group is generated by w^m and k and hence is isomorphic with

$$\{a, b : a^{3m+1} = b^3 = (ab^{-1})^3 = (ab)^{3n+1} = 1\}.$$

In the next section we show that this group is nontrivial if $|3n+1|>2$.

We make a special argument for the case $n=-1$. In this case we do not adjoin the relations $w^{3m+1}=1=k^3$ but instead consider the group generated by w and k with relations (1) $kw^{2m+1}k=w^mk^2w^m$ and (2) $(k^2w^{3m+1}k)^{-1}=(kw^m)^{-2}$, or (2) $kw^{3m+1}k=w^mkw^m$. We may write out (1) as $(w^mkw^m)w^{-2m}(w^mkw^m)=kw^{2m+1}k$ and then substitute (2) in (1) arriving at (1) $k^{-1}w^{2m}k^{-1}=w^{4m+1}$. Making the substitution

$k=w^m x^{-1}$ we have (1) $xw^m xw^m = w^{6m+1}$, and (2) $w^{6m+1}=w^m xw^m x^{-1} w^m xw^m$. Substituting (1) in (2), we have

(1) $w^{6m+1}=(xw^m)^2$ and

(2) $(xw^m)^2=x^3$, and so our group is $\{x, w; w^{6m+1}=x^3=(xw^m)^2\}$.

There is a homomorphism of this group onto $\{x, w; w^{6m+1}=x^3=(xw^m)^2=1\}$ and since this group is generated by x and w^m it is isomorphic with $\{a, b; a^{6m+1}=b^3 =(ab)^2=1\}$. In [2], it is shown that this group is nontrivial, since there is a homomorphism of it onto the permutation group generated by $a=(1, 2, \ldots, 6m+1)$ and $b=(6m+1, 2, 1)(6m, 4, 3)\cdots(4m+2, 4m, 4m-1)(4m+1)$.

We now turn our attention to the case where $j=2m-1$. In this case we are interested in the group with generators w, k, and relations (1) $kw^{2m-1}k=w^m k^2 w^m$ and (2) $(k^2 w^{3m-1}k)^n=(kw^m)^{3n-1}$. If we adjoin the relations (3) $w^{3m-1}=1$ and (4) $k^3=1$ then we have (1) $(kw^{-m})^3=1$ and (2) $(kw^m)^{3n-1}=1$. Now since $w^{3m-1}=1$ this group is generated by w^m and k and is isomorphic with $\{a, b; a^{3m-1}=b^3=(ab^{-1})^3 =(ab)^{3n-1}=1\}$. In the next section we show that this group is nontrivial unless $3m-1=2$ or $3n-1=2$. In case $m=1$, the knot under consideration is the trefoil, which is known to have Property P [2]. We handle the case $n=1$ by a special method. We do not add the relations $w^{3m-1}=1=k^3$ but consider the group generated by w and k with relations (1) $kw^{2m-1}k=w^m k^2 w^m$ and (2) $k^2 w^{3m-1}k =(kw^m)^2$ or (2) $kw^{3m-1}k=w^m kw^m$. We rewrite (1) as $(w^m kw^m)w^{-2m}(w^m kw^m) =kw^{2m-1}k$ and then substitute (2) in (1) to obtain

(1) $kw^{-2m}k=w^{1-4m}$ and

(2) $kw^{3m-1}k=w^m kw^m$.

We now make the substitution $k=xw^m$ and arrive at (1) $xw^{-m}xw^{-m}=w^{1-6m}$ and (2) $w^{1-6m}=w^{-m}xw^{-m}x^{-1}w^{-m}xw^{-m}$. Now substituting (1) in (2) we have (1) $w^{1-6m}=(xw^{-m})^2$ and (2) $x^3=(xw^{-m})^2$. This gives the group $\{x, w; w^{1-6m}=x^3 =(xw^{-m})^2\}$. Now there is a homomorphism of this group onto $\{x, w; w^{1-6m}=x^3 =(xw^{-m})^2=1\}$ which in turn is isomorphic with $\{a, b; a^{6m-1}=b^3=(ab)^2=1\}$. In [2] it is shown that this group is nontrivial, since there is a homomorphism of it onto the permutation group generated by

$$a = (1, 2, \ldots, 6m-1),$$

and

$$b = (6m-1, 2, 1)(6m-2, 4, 3)\cdots(4m+1, 4m-1, 4m-3)(4m)(4m-2).$$

This concludes the proof of Theorem 3.1.

4. **The nontriviality of certain groups.** The groups $(l, p|q, r)=\{a, b; a^l=b^p =(ab)^q=(a^{-1}b)^r=1\}$ have been studied by Coxeter [8], [9] and others. In particular, in [9] H. Coxeter establishes finiteness theorems for $(3, p|q, r)$ and shows that $(3, p|3, r)$ is nontrivial if $p, r \geqq 3$. For completeness we include a proof of the nontriviality of the groups $\{a, b : a^3=b^p=(ab)^3=(a^{-1}b)^r\}$ if $p, r \geqq 3$. We first notice that this group is isomorphic with $\{A, B : A^3=B^3=(AB)^r=(A^{-1}B)^p$

$=1\}$ by sending a to A and ab to B. This group in turn is isomorphic with $\{C, D : C^3=D^3=(CD)^p=(C^{-1}D)^r=1\}$ and hence in $\{A, B : A^3=B^3=(AB)^r=(A^{-1}B)^p=1\}$ we may assume that $r \geqq p \geqq 3$.

We now choose a complex number c such that $c\bar{c}=4\cos^2(\pi/r)$ and $c+\bar{c}=4\cos^2(\pi/p)-4\cos^2(\pi/r)-1$. This choice is always possible if $r\geqq p\geqq 3$. Let A be the 3×3 matrix

$$\begin{bmatrix} 1 & c & c+1 \\ 0 & -1 & -1 \\ 0 & 1 & 0 \end{bmatrix}$$

and B be the matrix

$$\begin{bmatrix} 0 & 0 & 1 \\ 1+\bar{c} & 1 & 1 \\ -1 & 0 & -1 \end{bmatrix}.$$

Then

$$AB = \begin{bmatrix} c\bar{c}-1 & c & 0 \\ -\bar{c} & -1 & 0 \\ 1+\bar{c} & 1 & 1 \end{bmatrix} \quad \text{and} \quad A^{-1}B = \begin{bmatrix} c+\bar{c}+c\bar{c} & c+1 & c+1 \\ -1 & 0 & -1 \\ -\bar{c} & -1 & 0 \end{bmatrix}.$$

Now computing characteristic polynomials we have

$$P_A(x) = 1-x^3, \qquad P_B(x) = 1-x^3,$$

$$P_{AB}(x) = 1-(c\bar{c}-1)x+(c\bar{c}-1)x^2-x^3 = -(x-1)(x^2-2\cos(2\pi/r)x+1),$$

and

$$P_{A^{-1}B}(x) = 1-(c+\bar{c}+c\bar{c})x+(c+\bar{c}+c\bar{c})x^2-x^3 = -(x-1)(x^2-2\cos(2\pi/p)x+1).$$

Now P_{AB} divides x^r-1 and $P_{A^{-1}B}$ divides x^p-1.

Since a matrix satisfies its characteristic polynomial [6, p. 299], this shows that A and B satisfy the relations $A^3=B^3=(AB)^r=(A^{-1}B)^p=1$.

5. **Other knots.** In this section we will show that many other knots have Property P. First, we need some additional definitions.

Let M be a homotopy 3-sphere and let S be a polyhedral subset of M which is homeomorphic with $S^1\times S^1$. Then it follows from the loop theorem and Dehn's lemma [17] that there is a disk D such that $D\cap S=\text{Bd } D$ and Bd B is not homotopic to 0 on S. The closure of the complementary domain of S containing Int D is then a *homotopy solid torus*, that is, homeomorphic with a homotopy 3-cell with a pair of disks on its boundary identified.

If K_1 and K_2 are knots in S^3, then the sum of K_1 and K_2, denoted K_1+K_2 is described in [10]. For example, the square knot is a composite.

Suppose that M is a 3-manifold, and T is a solid torus in Int M. If we remove Int T from M and then sew back a solid torus T' so that a meridianal curve of T' is not isotopic on Bd T with a meridianal curve of T, then we obtain a 3-manifold M'. In this case we will say that M' is obtained from M by *removing T and sewing it back differently*.

THEOREM 5.1. *If K is a composite knot, then K has Property* P.

Theorem 5.1 is only a slight extension of the result by D. Noga [16, Satz 1], that one cannot recover S^3 by removing a solid torus about K and sewing it back differently. Also Theorem 5.1 is a special case of a more general theorem of Conner [7] in which he shows that "splittable" knots have Property P.

Before proceeding with the proof of Theorem 5.1, we first prove two lemmas.

LEMMA 1. *Suppose that T is a homotopy solid torus, and K is a cube with a knotted hole in the interior of T. Then there is a homotopy 3-cell B in T such that $K \subset B$.*

Proof. Let D be a disk in T such that $D \cap \text{Bd } T = \text{Bd } D$, and Bd D is not homotopic to 0 in Bd T. Putting D in general position with respect to Bd K we find that $D \cap \text{Bd } K$ is a finite collection of mutually disjoint simple closed curves. We may assume that each curve of intersection is not homotopic to 0 on Bd K, for these may be removed by a simple disk replacement. Let J be an innermost curve of intersection on D and let H be the subdisk of D bounded by J. It is impossible that $H \subset K$, since K is a cube with a knotted hole and $i_*\colon \pi_1(\text{Bd } K) \to \pi_1(K)$ is one-to-one. By adding a regular neighborhood of H to K we obtain a homotopy 3-cell which contains K. If there are no curves of intersection then the removal of the interior of a regular neighborhood of D from T results in a homotopy 3-cell which contains K. This establishes Lemma 1.

LEMMA 2. *Let T_1 be a solid torus and let T_2 be a solid torus in the interior of T_1 such that* (a) *there is a meridianal disk of T_1 whose intersection with T_2 is a meridianal disk of T_2, and* (b) *T_2 is not parallel to T_1. See Figure* 4. *Then the result of removing T_2 from T_1 and sewing it back differently is not a homotopy solid torus.*

Proof. First, let F be a meridianal disk of T_1 which intersects T_2 in a meridianal disk of T_2. Let N be a regular neighborhood in T_1 of F and let K_1 be the closure of $T_1 - (N \cup T_2)$. Then K_1 is a cube with a knotted hole since T_2 is not parallel to T_1. Now $K_1 \cap \text{Bd } T_1$ is an annulus and we push this annulus slightly to the interior of T_1 and continue to call the resulting cube with a knotted hole K_1. Suppose that we remove T_2 from T_1 and sew it back differently. Call the resulting manifold T_1', and the replaced T_2, T_2'. Suppose that T_1' is a homotopy solid torus. Then there is a disk D in T_1' such that $D \cap \text{Bd } T_1' = \text{Bd } D$ and Bd D is not homotopic to 0 on Bd T_1'. Now since, by Lemma 1, K_1 lies in a homotopy 3-cell in T_1', we may assume that $D \cap K_1 = \varnothing$. We now put D and Bd T_2 in general position so that $D \cap \text{Bd } T_2$ is a

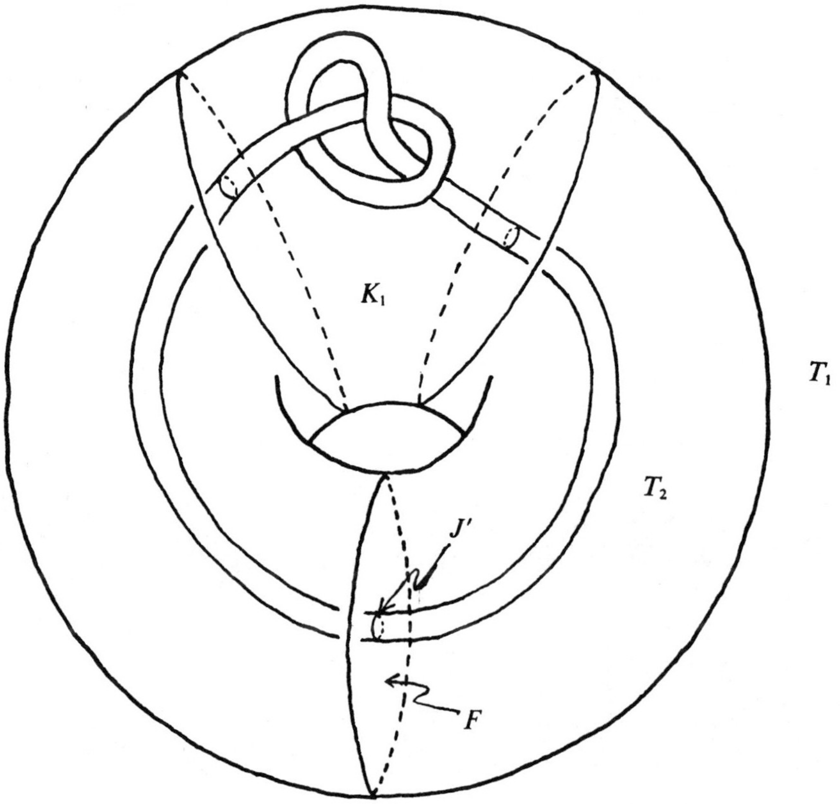

FIGURE 4

finite collection of mutually disjoint simple closed curves. We may assume that none of these simple closed curves is homotopic to 0 on Bd T_2. Since $D \cap K_1 = \varnothing$, it follows that each curve of intersection of $D \cap$ Bd T_2 is homotopic to the simple closed curve $J' = F \cap$ Bd T_2 that was originally meridianal on T_2. Let J be an innermost curve of intersection on D and let H be the subdisk of D bounded by J. Now H is not contained in T_2' for otherwise T_2 would not have been sewn in differently. But H is not contained in Cl $(T_1' - T_2')$ since J' is not homotopic to 0 in Cl $(T_1' - T_2')$. Consequently there are no curves of intersection. But then Bd D is homotopic to 0 in Cl $(T_1' - T_2')$ and hence in Bd T_1'. This contradicts the choice of D. Hence T_1' is not a homotopy solid torus. This establishes Lemma 2.

Proof of Theorem 5.1. Let $K = K_1 + K_2$ be a composite knot in S^3. We will show that the result of removing a regular neighborhood of K and sewing it back differently is not a simply connected manifold. It is easy to see that this process is equivalent to attaching a pillbox to a cube with a K-knotted hole in a nonstandard way. Let S by a polyhedrally embedded $S^1 \times S^1$ which envelopes the knot K_1 and follows the knot K_2. See Figure 5. Notice that (S^3 − interior S) is a cube with a knotted hole and hence is not a homotopy solid torus.

Now let N be a regular neighborhood of K in interior S and let M be a 3-manifold which results from removing N from S^3 and sewing it back differently. Now it follows from Lemma 2 that the torus S in M does not bound a homotopy solid torus in M. Hence M is not simply connected. This establishes Theorem 5.1.

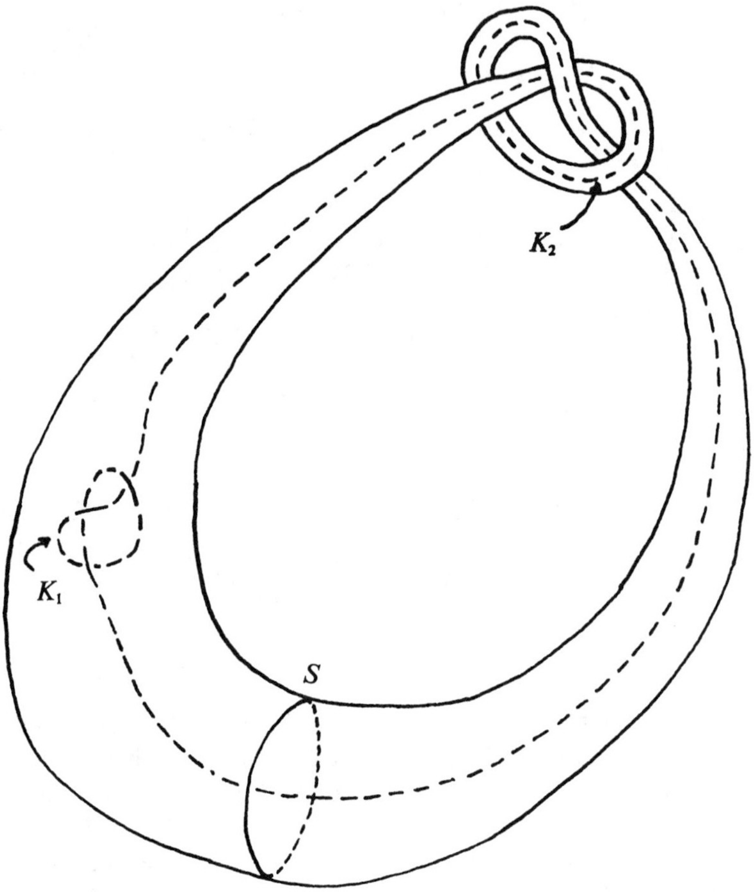

FIGURE 5

THEOREM 5.2. *Suppose that K is a knot in S^3 with Property* P *and that T is an unknotted polyhedral solid torus in S^3 which contains K in its interior. Let L be a simple closed curve on* Bd T *which bounds a disk in S^3-*Int T *but not on* Bd T, *and let $h: T \to S^3$ be a piecewise linear homeomorphism such that $h(L)$ is homologous to* 0 *in S^3-*Int $[h(T)]$. *Then $h(K)$ has a Property* P.

Proof. Let N be a regular neighborhood of K in Int T and let M be a meridianal simple closed curve on Bd T which is transverse to L.

Suppose that we remove N from T and sew it back differently, arriving at the manifold T'. Since K has Property P, there are two possibilities. First, T' may not be a homotopy solid torus, or, secondly, T' is a homotopy solid torus whose meridianal curve is $pM+qL$ where p and q are relatively prime integers and $|p|\neq 1$.

Now, suppose that we remove $h(N)$ from S^3 and sew it back differently, obtaining the manifold M^3. If $h(T)$ is unknotted, then h may be extended to a homeomorphism of S^3 onto S^3. Since K has Property P, it follows that in this case M^3 is not simply connected.

If $h(T)$ is knotted and the result of sewing back $h(N)$ in $h(T)$ is not a homotopy solid torus, then M^3 is not simply connected since $h(\text{Bd } T)$ does not bound a homotopy solid torus in M^3.

Finally, suppose that $h(T)$ is knotted and the result of sewing back $h(N)$ in $h(T)$ is a homotopy solid torus. Then a meridianal curve for this homotopy solid torus is $ph(M)+qh(L)$, $|p|\neq 1$, and it follows that the first homology group of M^3 is either Z or Z_p. In any case M^3 is not simply connected. This establishes Theorem 5.2.

COROLLARY 5.2. *Doubled knots* [21] *with* ρ *twists* $\rho\neq 0$ *have Property* P.

Proof. This follows directly from the definition of doubled knots, Theorem 3.1 and Theorem 5.2. Actually, it is true that doubled knots with no twists also have Property P if they are knotted. We will show this in a later paper.

6. **Questions.** In this section we raise some questions related to Property P.

(1) In examining the knot table of Alexander and Briggs [1], one finds that the first two knots which are not known to have Property P are 6_2 and 6_3. These knots are shown in Figures 6 and 7, respectively. Do these knots have Property P?

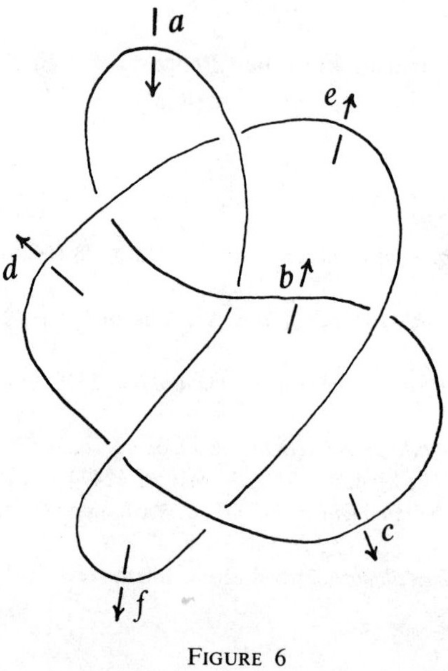

FIGURE 6

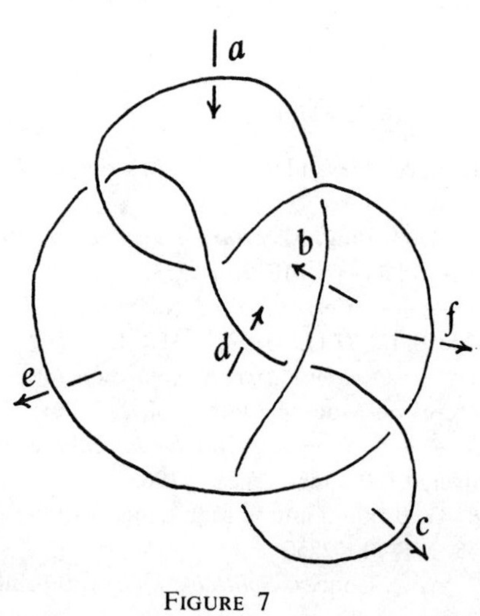

FIGURE 7

If one converts the cube with a knotted hole $C^3(6_2)$ to a homology cube by sewing on a pillbox P^3, along an annulus which runs through the hole $|n|$ times, one presentation of $\pi_1(C^3(6_2) \cup P^3)$ has generators a, b, c, d, e and f and relations $b = dad^{-1}$, $c = e^{-1}be$, $d = f^{-1}cf$, $e = ada^{-1}$, $f = c^{-1}ec$, and $(b^{-1}c^{-1}af^{-1}e^{-1}da^2)^h a = 1$.

If one converts the cube with a knotted hole $C^3(6_3)$ to a homology cube by sewing on a pillbox P^3, along an annulus which runs through the hole $|n|$ times, one presentation of $\pi_1(C^3(6_3) \cup P^3)$ has generators a, b, c, d, e and f, and relations $b = faf^{-1}$, $c = e^{-1}be$, $d = b^{-1}cb$, $e = ada^{-1}$, $f = c^{-1}ec$, and $(f^{-1}eba^{-1}cd^{-1})^n a = 1$.

(2) Do knots whose groups have finitely generated commutator subgroups [15] have Property P?

(3) Do pretzel knots [21] have Property P?

(4) If the Poincaré conjecture is false, can a homotopy 3-cell which is not a topological 3-cell be obtained by attaching a pillbox to a cube with a knotted hole?

(5) If k_1 and k_2 are polygonal knots in E^3 with homeomorphic complements, is there a homeomorphism of E^3 onto itself which carries k_1 onto k_2?

(6) In showing that the figure eight knot (and other twist knots) have Property P, we used the fact that certain matrix groups are nontrivial. In dealing with the trefoil (and other torus knots), finite permutation groups sufficed for the same purpose. Is there a collection of nontrivial finite permutation groups such that for each positive integer n, each of the groups

$$\{w, k : kw^3k = wk^2w, (kw^4k^2)^n = (wk)^{3n\pm 1}\}$$

maps homomorphically onto one of them? In Case 1 of the proof of Theorem 2.1, we noted that, if $3n \pm 1 = 2$, the group maps onto the permutation group generated by (1, 2, 3, 4, 5, 6, 7) and (7, 2, 1)(6, 4, 3)(5). Such permutation groups can also be found if $3n \pm 1$ is divisible by 4, 5 or 7.

(7) Is there a geometric proof that the trefoil knot has Property P? Such an argument might provide a clue for showing that all nontrivial knots have Property P.

References

1. J. W. Alexander and G. B. Briggs, *On types of knotted curves*, Ann. of Math. **28** (1926), 562–586.

2. R. H. Bing, *Necessary and sufficient conditions that a 3-manifold be S^3*, Ann. of Math. (2) **68** (1958), 17–37. MR **20** #1973.

3. ———, *Correction to "Necessary and sufficient conditions that a 3-manifold be S^3"*, Ann. of Math. (2) **77** (1963), 210. MR **25** #5508.

4. ———, *Some aspects of the topology of 3-manifolds related to the Poincaré conjecture*, Lectures on Modern Math., vol. 2, Wiley, New York, 1964, pp. 93–128. MR **30** #2474.

5. ———, *Computing the fundamental group of the complement of curves*, Washington State University, Pullman, Wash., 1965.

6. G. Birkhoff and S. Mac Lane, *A survey of modern algebra*, 3rd ed., Macmillan, New York, 1965. MR **31** #2250.

7. A. C. Conner, *Splittable knots* (preprint).

8. H. S. M. Coxeter, *The abstract groups* $G^{m,n,p}$, Trans. Amer. Math. Soc. **45** (1939), 73–150.

9. ———, *The abstract group* $G^{3,7,16}$, Proc. Edinburgh Math. Soc. (2) **13** (1962), 47–61, 189. MR **26** #190; MR **26** #6267.

10. R. H. Fox, *A quick trip through knot theory*, Topology of 3-Manifolds and Related Topics (Proc. Univ. of Georgia Inst., 1961), Prentice-Hall, Englewood Cliffs, N. J., 1962, pp. 120–167. MR **25** #3522.

11. H. Gluck, Ph.D. Thesis, Princeton University, Princeton, N. J., 1961.

12. J. Hempel, *Construction of orientable* 3-*manifolds*, Topology of 3-Manifolds and Related Topics (Proc. Univ. of Geogria Inst., 1961), Prentice-Hall, Englewood Cliffs, N. J., 1962, pp. 207–212. MR **25** #3538.

13. ———, *A simply connected* 3-*manifold is* S^3 *if it is the sum of a solid torus and the complement of a torus knot*, Proc. Amer. Math. Soc. **15** (1964), 154–158. MR **28** #599.

14. W. B. R. Lickorish, *A representation of orientable combinatorial* 3-*manifolds*, Ann. of Math. (2) **76** (1962), 531–540. MR **27** #1929.

15. L. Neuwirth, *The algebraic determination of the genus of knots*, Amer. J. Math. **82** (1960), 791–798. MR **22** #11397.

16. D. Noga, *Über den Aussenraum von Produktknoten und die Bedeutung der Fixgruppen*, Math. Z. **101** (1967), 131–141. MR **36** #2137.

17. C. D. Papakyriakopoulos, *On solid tori*, Proc. London Math. Soc. (3) **7** (1957), 281–299. MR **19**, 441.

18. ———, *On Dehn's lemma and the asphericity of knots*, Ann. of Math. (2) **66** (1957), 1–26. MR **19**, 761.

19. H. Poincaré, *Second complément a l'analysis situs*, Proc. London Math. Soc. (2) **32** (1900), 277–308.

20. ———, *Cinquieme complément a l'analysis situs*, Rend. Circ. Mat. Palermo **18** (1904), 45–110.

21. H. F. Trotter, *Non-invertible knots exist*, Topology **2** (1963), 275–280. MR **28** #1618; errata MR **30** p. 1203.

22. A. H. Wallace, *Modifications and cobounding manifolds*, Canad. J. Math. **12** (1960), 503–528. MR **23** #A2887.

23. J. H. C. Whitehead, *On doubled knots*, J. London Math. Soc. **12** (1937), 63–71.

UNIVERSITY OF WISCONSIN,
MADISON, WISCONSIN 53706

F. Triangulations

ANNALS OF MATHEMATICS
Vol. 69, No. 1, January, 1959
Printed in Japan

AN ALTERNATIVE PROOF THAT 3-MANIFOLDS CAN BE TRIANGULATED

BY R. H. BING*

(Received October 30, 1957)

1. Introduction

It was shown in Theorem 1 of [3] that for each topolological 2-sphere S in E^3 and each positive number ε there is a polyhedral 2-sphere S' and a homeomorphism h of S onto S' such that h moves no point by more than ε. This result was extended [3, Theorems 7, 8] to show that topological surfaces in triangulated 3-manifolds can be approximated by polyhedral surfaces.

The main purpose of this paper is to extend the above mentioned results and show (Theorem 5) that if in E^3 (or any other triangulated 3-manifold with boundary), P is a closed set which is the topological image of a 2-complex (finite or infinite) and $f(x)$ is a positive continuous function defined on P, then there is a homeomorphism h of P onto a polyhedron P' such that h does not move any point x of P by as much as $f(x)$. Three applications of this *approximation theorem for 2-complexes* are then given. Each of the applications is treated independently and any one of them may be studied without reference to the other two. Perhaps the most difficult application is given first rather than last but this is because it may be of the most interest.

This paper was studied in a seminar at the University of Wisconsin in the spring of 1957 and members of that seminar offered many suggestions for improving the original draft. The author is particularly indebted to C. E. Burgess, L. F. McAuley, E. R. Fadell, J. B. Kurskal for helpful suggestions.

TRIANGULATING 3-MANIFOLDS. Moise showed in [10] that each 3-manifold can be triangulated. This result was extended in [2] and [11] to show that 3-manifolds with boundaries can be triangulated. We give an alternative proof in this paper based on the approximation theorem for 2-complexes. See Theorems 6 and 7.

In another paper we shall give another proof that each 3-manifold can be triangulated by using the result that E^3 does not contain uncountably many mutually exclusive wild surfaces.

Just what does it mean to say that a 3-manifold M can be triangulated? For one thing it means that M is topologically equivalent to some geomet-

* Work on this paper was supported by contract NSF-G479

ric 3-complex. However, a more useful notion for our present purposes is that there exists a partitioning of M of a certain sort. If g is an element of the partitioning, $\bar{g}$ is topologically equivalent to a tetrahedron. A homeomorphism of a tetrahedron onto $\bar{g}$ is used to designate certain parts of $\bar{g}$ as edges, vertices, and faces. The partitioning is such that if the closures of two elements of it have a point in common, their intersection is either a face of each, an edge of each, or a vertex of each. The sum of the faces forms a 2-skeleton or grid that cuts M into small pieces. We use the approximation theorem for 2-complexes to build such a grid to triangulate an arbitrary 3-manifold M.

It follows from the definition of a 3-manifold that it can be triangulated locally. However, if U_1, U_2 are two open subsets of a 3-manifold M and T_i is a triangulation of U_i, the triangulation T_1 of U_1 may appear curvilinear rather than polyhedral when viewed from U_2 under the triangulation T_2. See Figure 1. An extension of the approximation theorem for

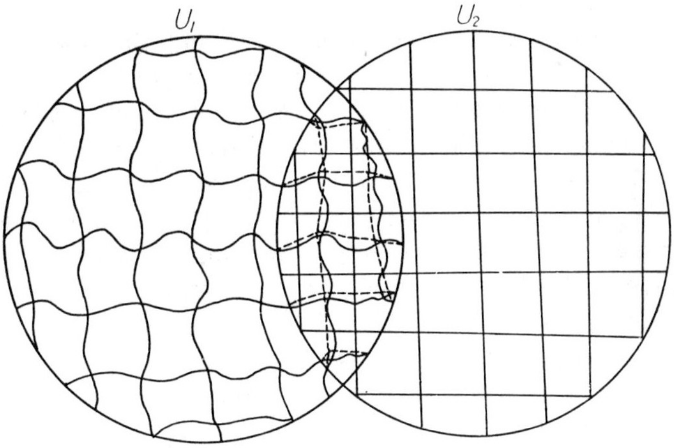

Figure 1

2-complexes may be used to modify the part of the 2-skeleton of T_1 in $U_1 \cdot U_2$ so as to make it appear locally polyhedral when viewed from U_2 under T_2. This adjustment makes it possible to fit together the triangulations of U_1 and U_2 to get a triangulation of $U_1 + U_2$. In turn, the local triangulations of open subsets of M are fitted together to get a triangulation of all of M.

The results of this paper could be given in short concise form, or they could be given in long expanded form. It is hoped that the short preview given in this introduction will acquaint the reader with the overall plan of attack, and the details that follow will make it possible for the serious reader to understand each step.

APPROXIMATING HOMEOMORPHISMS. Another application of the approximation theorem for 2-complexes shows that if f is a homeomorphism of a tetrahedron X (we suppose that a tetrahedron is solid) into E^3 and ε is a positive number, then there is a piecewise linear homeomorphism h of X into E^3 such that for each point x of X, $\rho(f(x), h(x)) < \varepsilon$. (We use $\rho(x, y)$ to denote the distance between x and y.) Extensive use of Vietoris homology theory was used by Moise in getting the preliminary theorem on which to base the original proof of this theorem [9, Theorem 1]. The present paper gets a proof of this approximation theorem for homeomorphisms (Theorem 9) by using the approximation theorem for 2-complexes (Theorem 5) which, in turn, was proved by using methods of plane topology. The above approximation theorem for homeomorphisms is frequently stated in more general forms where X is a 3-manifold with boundary rather than a tetrahedron, ε is a positive function on X rather than a positive number, and the imbedding space is a 3-manifold with boundary rather than E^3, but these are technicalities that we shall not concern ourselves with in this introduction. The simple case is the important one and extensions of various sorts may be made. See [9, Theorem 1], [10, Lemma 5 and Theorem 2], [2, Theorems 3 and 4], and Theorems 8 and 9 of this paper.

The general idea of attack for proving the approximation theorem for homeomorphisms of the tetrahedron X is to triangulate X and let R_2 be

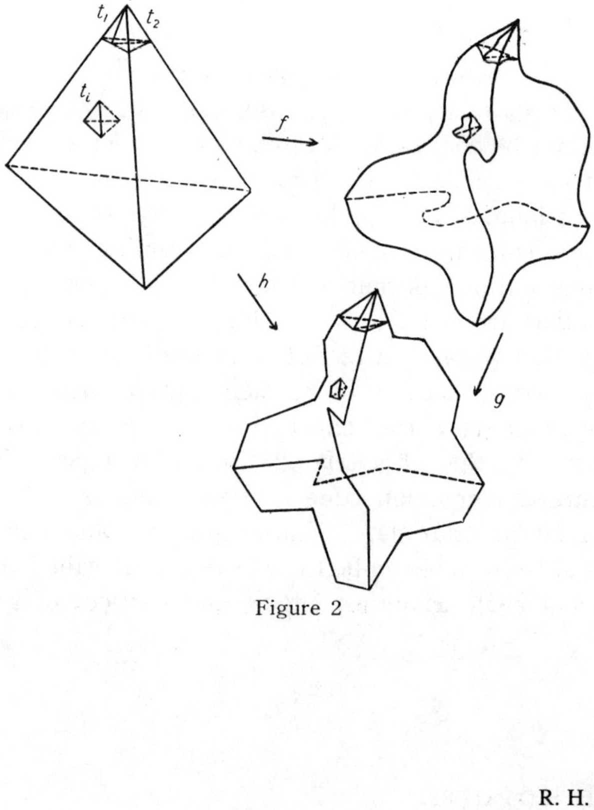

Figure 2

the 2-skeleton of the triangulation. Note that R_2 is a 2-complex and use the approximation theorem for 2-complexes to show that there is a homeomorphism g of $f(R_2)$ onto a polyhedron in E^3. Let h be a piecewise linear homeomorphism of R_2 onto $gf(R_2)$, such that for each face F of a tetrahedron t of the triangulation of X, h takes F piecewise linearly onto $gf(F)$. See Figure 2. The piecewise linear homeomorphism h of Bd t into E^3 is extended to take t piecewise linearly onto the polyhedral cube bounded by $h(\text{Bd } t)$. If the triangulation of X is sufficiently fine and g is sufficiently close to the identity, $\rho(f(x), h(x)) < \varepsilon$. The details of proof are found in the proofs of Theorems 8 and 9 and the corollary to Theorem 5.

APPROXIMATING POLYHEDRA. Whereas the preceding two applications of the approximation theorem for 2-complexes were merely new proofs of known results ; the third application (Theorem 10), to the effect that in a triangulated 3-manifold with boundary, each closed set P that is homeomorphic to a polyhedron can be homeomorphically approximated by a polyhedron, is new. In some respects this result might be considered the highlight of the paper except that it comes somewhat as an anticlimax. It is not surprising that P can be approximated by a polyhedron after the approximation theorem for 2-complexes is known, since it is relatively easy to fit in isolated vertices, dangling edges, and inclosed tetrahedra after faces have been located properly.

DEFINITIONS. The definitions we use are similar to those used in [3], [9], and [10]. We include some of these for completeness.

A *geometric 2-complex* (we use the abbreviated term 2-*complex* instead, since we shall not be dealing with combinatorial complexes in this paper) is a subset of a separable metric space which is the sum of a locally finite collection of triangles (triangular disks), such that if two of these triangles have a point in common, their intersection is a vertex or edge of each. Some authors permit a 2-complex to contain isolated vertices and intervals that are not edges of triangles, but we do not find that convenient in this paper. (A collection is *locally finite* if, for each point p, there is a neighborhood N of p, such that N does not intersect infinitely many elements of the collection.) Each of these triangles has a linear structure like that of a triangle in the euclidean plane, and the linear structure on a common edge of two triangles is the same when viewed from either triangle. These triangles and edges are called 2-*simplexes* and 1-*simplexes* while their interiors are called *open simplexes*. The collection of such triangles, edges, and vertices of a 2-complex is called a *trangulation* of the 2-complex.

A 3-*manifold* is a separable metric space which is locally like E^3 (euclidean 3-space). A 3-*manifold with boundary* is a separable metric space M, such that each point of M has a neighborhood whose closure is topologically equivalent to a cube. We note that a 3-manifold is a 3-manifold with boundary. The set of points at which M is locally like E^3 is called Int M and Bd $M = M -$ Int M. In general, if X is an n-manifold with boundary, we use Bd X to denote the set of points of X at which it is not locally like E^n.

Consider a locally finite collection of tetrahedra filling M, such that if two of these tetrahedra have a point in common, their intersection is either a vertex, an edge, or a face of each. The collection of such tetrahedra (3-simplexes), faces (2-simplexes), edges (1-simplexes), and vertices (0-simplexes) is called a *triangulation* T of M. Each of the tetrahedra in T has a linear structure, and if two tetrahedra have an edge or face in common, the linear structure on this edge or face is the same when viewed from either of the tetrahedra.

The sum of the vertices, sum of the edges, and sum of the faces of a triangulation are called respectively, the 0-*skeleton*, 1-*skeleton*, and 2-*skeleton* of a triangulation.

A triangulation T' of M is called a *subdivision* of a triangulation T of M, if each tetrahedron in T' is a tetrahedron in some element of T. Also, if $T(U)$ is a triangulation of an open subset U of M, we say that $T(U)$ is a subdivision of T, if each tetrahedron of $T(U)$ is a tetrahedron in some element of T. ($T(U)$ may have infinitely many elements even though T may be finite.) We say that two triangulations T and T'' are *compatible* if they have a common subdivision. It is known that each open subset U of M has a triangulation compatible with that of M [2, Lemma 4].

A closed set X is called a *polyhedron* in triangulated M if there is a subdivision T' of the triangulation of M, such that X is the sum of some of the tetrahedra, faces, edges, and vertices of T'. It is a *finite* or *infinite polyhedron* according as X is the sum of a finite or an infinite number of such objects. A *polyhedral cube* in triangulated M is a polyhedron which is topologically equivalent to a cube.

A closed set Y is *locally polyhedral* at a point y of Y in triangulated M, if there is a neighborhood N of y, such that $\bar{N} \cdot Y$ is a finite polyhedron in M. If a set Y is locally polyhedral except possibly at those points of a set K, we say that Y is *locally polyhedral* mod K. A closed set is known to be a polyhedron if it is locally polyhedral at each of its points [2, Lemma 1].

An open set U is called *polyhedral* in triangulated M if it is the sum of vertices and open simplexes of some subdivision of M. (See statement of Lemma 3 for an alternate definition for bounded subsets of E^3.) An open set U is said to be *locally polyhedral* at a point p of $\overline{U}$ if there is a neighborhood N of p such that $N \cdot U$ is polyhedral. (See statements of Lemmas 4, 5 and Theorem 3.)

The euclidean metric on E^3 determines what is a tetrahedron in E^3, although we may at times want to ignore this metric and get a curvilinear triangulation of E^3, rather than an ordinary rectilinear triangulation. If M is a 3-manifold with boundary where the tetrahedrons are not given in advance, we are free to break M into topological tetrahedra, give each of these tetrahedra a linear structure, and let this triangulation determine the locally linear structure of M. We may thus obtain different triangulations of M which are not compatible. Even though two simplexes are the same as point sets, we do not call them equal as simplexes if their linear structures differ.

If M_i $(i = 1, 2)$ is a triangulated 3-manifold with boundary, and h is a homeomorphism of M_1 onto M_2, h is called *piecewise linear* if there are subdivisions T_1, T_2 of the triangulations of M_1, M_2 respectively, such that h takes each tetrahedron of T_1 linearly onto a tetrahedron of T_2. A homeomorphism h of an open subset of M_1 into M_2 is said to be *locally piecewise linear* at the point p of M_1, if there is a neighborhood U of p and subdivisions T_1, T_2 of M_1, M_2 respectively (U, T_1, T_2 depend on p), such that each element of T_1 that intersects U is taken linearly by h onto an element of T_2. It is known that a homeomorphism of M_1 onto M_2 is piecewise linear if it is locally piecewise linear at each point of M_1 [2, Lemma 2].

We have only defined triangulations for 3-manifolds with boundaries and piecewise linear homeomorphisms on such triangulated manifolds. These definitions may be extended in a natural way to polyhedra.

A homeomorphism of one tetrahedron (or simplex) onto another is *linear*, if straight line intervals go onto straight line intervals in a ratio preserving manner. Actually, the map is affine because ratios are preserved in addition to lines. A projection in E^3 of one triangle onto another gives a map that preserves lines and cross ratios rather than ratios. However, it is just as easy (if not easier) to get a map that preserves ratios. For those acquainted with barycentric coordinates, it may be convenient to think of the point with barycentric coordinates $(a_1, a_2, \cdots, a_n)$ with respect to vertices $x_1, \cdots, x_n$, going into the point $(a_1, \cdots, a_n)$ whose barycentric coordinates are taken with respect to the vertices $h(x_1), \cdots, h(x_n)$.

In a euclidean figure, certain triples of points may be collinear, and one point of such a triple divides the other two in a certain ratio. A *linear structure* of a figure is the set of all quadruples (p_1, p_2, p_3, r) where p_1, p_2, p_3 are three points, such that $\rho(p_1, p_2) + \rho(p_2, p_3) = \rho(p_1, p_3)$ and $\rho(p_1, p_2)/\rho(p_2, p_3) = r$. The linear structure of a topological simplex may be determined by the inverse of a homeomorphism of a euclidean simplex onto the topological simplex.

Suppose M is a 3-manifold with boundary. Two triangulations T_1, T_2 of M are *locally compatible* at a point p of M if there is a neghborhood U of p, such that T_1 and T_2 have a common subdivision on U. If two triangulations are locally compatible at each point of a closed set K, they are locally compatible on an open set containing K. It is known that two triangulations of M are compatible if they are locally so at each point of M [3, Lemma 3].

2. Approximation theorem for 2-complexes

In this section we prove the approximation theorem for 2-complexes. We proceed toward this proof as follows. Consider a 2-complex C with

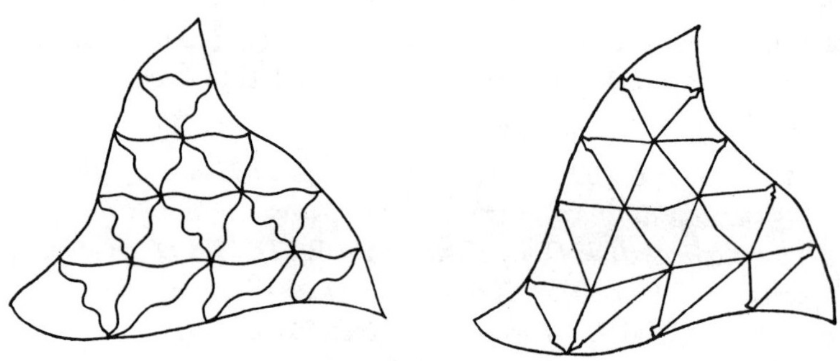

Figure 3

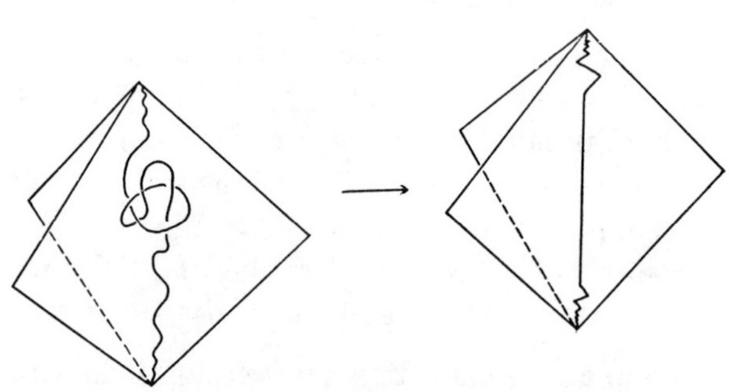

Figure 4

a triangulation $T(C)$ and a homeomorphism of C into a closed set P in E^3. It is shown in Theorem 1 (which is borrowed from [3]) that P can be adjusted so as to be locally polyhedral on the image of each open 2-simplex

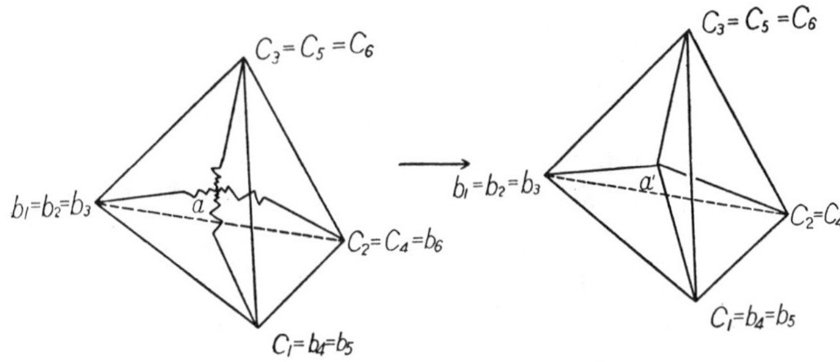

Figure 5

of $T(C)$. See Figure 3. It is shown in Theorem 3 that a further adjustment is possible to make the result locally polyhedral at the images of the open 1-simplexes of $T(C)$. See Figure 4. Finally, it is shown in Theorem 4 that the result may be made polyhedral at the images of vertices of $T(C)$. See Figure 5. A combination of Theorems 1, 3, and 4 yields the approximation theorem for 2-complexes (Theorem 5).

In a certain sense, this whole paper may be regarded as applications of extensions of Theorem 1. Other applications of this theorem are given in [4], [5], and [7].

THEOREM 1. *Suppose D is a topological disk in E^3 and f is a positive continuous function defined on* Int D. *Then there is a homeomorphism h of D onto a topological disk D' in E^3 such that* Bd $D=$ Bd D', $\rho(x, h(x)) \leqq f(x)$ *for each point x of D, and D' is locally polyhedral at each point of* Int D' *and at each point of* Bd D' *at which D is locally polyhedral.*

If K is a closed subset of D at which D is locally polyhedral, h may be chosen so as to be the identity on K.

PROOF. Let f' be a continuous function defined on D such that f' is positive on Int D, $f' = 0$ on Bd D and $f' \leqq f$. Let X be the set of points of Bd D at which D is not locally polyhedral. Let Y be the set of all points y of $D - X$ such that D is locally polyhedral in a $2\rho(x, y)$ neighborhood of some point x of $K +$ Bd $D - X$. The truth of Theorem 1 follows as a corollary of Theorems 7 and 8 of [3], if we let $E^3 - X$, $D - X$, Y, and f' be the S, M, Y, and f of those theorems.

QUESTION. One might wonder to what extent disks can be approximated from either side by polyhedral disks. Suppose S is a topological

2-sphere in E^3, C is a component of $E^3 - S$ and $\varepsilon > 0$. Is there a homeomorphism h of S onto a polyhedral 2-sphere S' in E^3, such that $\rho(x,h(x))<\varepsilon$ and each component of $S' - C$ is of diameter less than ε?

THEOREM 2. *Suppose ab is an arc and $D_1, D_2, \cdots, D_n$ are topological disks in E^3, such that* Bd D_i *contains ab, $D_i \cdot D_j = ab$ ($i \neq j$), and D_i is locally polyhedal* mod ab. *Then for each open set U containing* Int ab, *there is an arc axb in D_1 and a collection of topological disks $D_1', D_2', \cdots, D_n'$, such that* Bd $D_i' =$ (Bd $D_i - ab) + axb$, $D_i' - U = D_i - U$, $D_i' \cdot D_j' = axb$, *and D_i' is locally polyhedral* mod a, b.

If each D_i is locally polyhedral at a (or b), then D_i' may be chosen so as to be locally polyhedral there also.

If $D_1 + D_2$ is already locally polyhedral mod (a, b), *D_1', D_2' may be chosen so that $D_1' + D_2' = D_1 + D_2$.*

The proof of this theorem involves little more than applications of Theorem 1 and building new disks from old ones by "scissors and paste" methods. We give a sequence of lemmas that may help to make clearer the methods used in constructing these new disks. The first two of these lemmas are very elementary and are in general use, but are included for completeness.

LEMMA 1. *If D is a polyhedral disk in the plane E^2, there is a piecewise linear homeomorphism h of E^2 onto itself that takes D onto a triangular disk.*

PROOF. Suppose D is triangulated into n 2-simplexes. Since a triangulated disk can be shelled, these 2-simplexes may be ordered S_1, $S_2, \cdots, S_n$ so that for each positive integer i less than n, the closure of $D - (S_1 + S_2 + \cdots + S_i)$ is a topological disk. We now exhibit a piecewise linear homeomorphism h_1 of E^2 onto itself that takes D onto the closure of $(D - S_1)$.

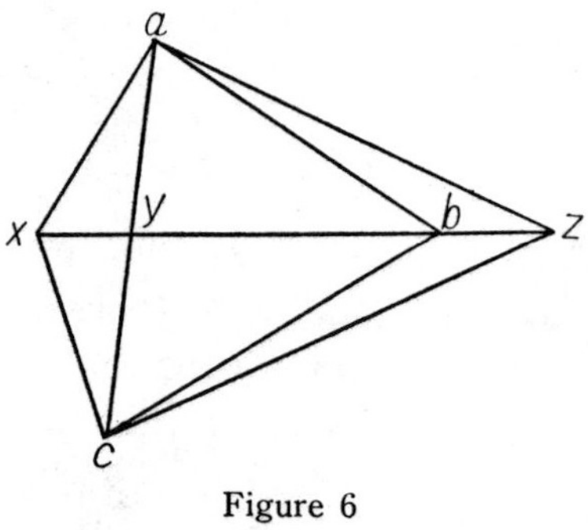

Figure 6

CASE 1. *Two edges of* S_1 *lie on* Bd D. Denote the vertices of S_1 by a, b, c and suppose $ab + bc \subset$ Bd D. Let y be the midpoint of ac. The median yb of S_1 is extended to form the interval $xybz$ such that the quadrilateral $xazc$ intersects D in the quadrilateral $xabc$. See Figure 6. Then the homeomorphism h_1 that is fixed outside the quadrilateral $xazc$, and takes the triangles abz, cbz, axb, cxb linearly onto the triangles ayz, cyz, axy, cxy respectively, is a piecewise linear homeomorphism of E^2 onto itself taking D onto the closure of $D - S_1$.

CASE 2. *Only one edge of* S_1 *lies on* Bd D. Let this edge be ac. We take the interval $xybz$ as in Case 1, except that this time the intersection of the quadrilateral $xazc$ and D is the triangle azc. The homeomorphism h_1 in this case is the inverse of the one described in Case 1.

It follows from the preceding argument that there is a piecewise linear homeomorphism h_i of E^2 onto itself that takes $D - (S_1 + S_2 + \cdots + S_{i-1})$ onto $D - (S_1 + S_2 + \cdots + S_i)$. The homeomorphism h promised by the lemma is $h_{n-1} \cdots h_2 h_1$.

The preceding lemma holds on a polyhedral 2-sphere as well as in the plane. We state a corollary to it in this setting.

COROLLARY. *If* D_1, D_2 *are two polyhedral disks on the boundary* S^2 *of a tetrahedron in* E^3, *there is a piecewise linear homeomorphism* h *of* S^2 *onto itself that takes* D_1 *onto* D_2.

The corollary follows directly from the methods of Lemma 1.

LEMMA 2. *If* D *is a polyhedral disk in* E^3, *there is a piecewise linear homeomorphism* h *of* E^3 *onto itself that takes* D *into a horizontal plane.*

Lemma 2 is a weak version of Theorem 1 on page 916 of [12] but we shall give its proof for completeness. Before proving Lemma 2, we give the following example which shows that a certain generalization of Lemma 2 does not extend to E^4.

EXAMPLE. There is a polyhedral disk D in E^4 such that there is no piecewise linear homeomorphism of E^4 onto itself that takes D onto a triangle. Consider a knotted polyhedral simple closed curve in E^3 and the disk D in E^4 obtained by taking the cone over J from a point outside E^3. Professor R. H. Fox called my attention to this example several years ago.

PROOF OF LEMMA 2. Suppose T is a rectilinear triangulation of D. The proof parallels that of Lemma 1, so we only show how to get a piecewise linear homeomorphism h_1 of E^3 onto itself that takes D onto the closure of $D - S_1$, where $S_1 = abc$ is a 2-simplex that can be shelled from D.

CASE 1. $ab + bc \subset$ Bd D. Let acd denote the 2-simplex in D sharing the edge ac with S_1, y be the midpoint of ac, x be the midpoint of dy, and ybz be an interval such that $azc \cdot D = abc$. See Figure 7. Let p_1yp_2 be an interval which is perpendicular to ac, pierces D at y, and is so short

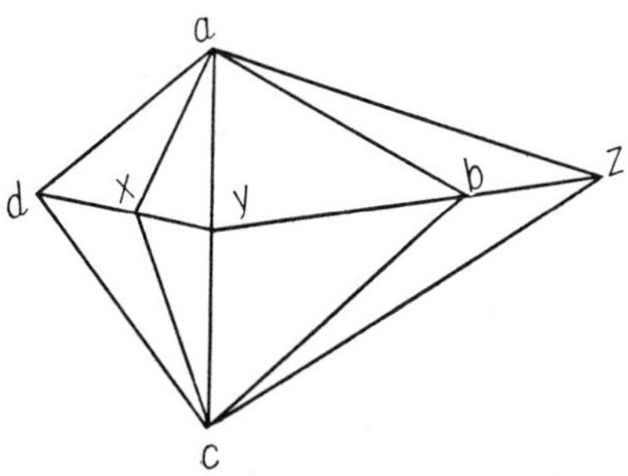

Figure 7

that the four tetrahedra with apexes at p_1 and p_2 and bases acd and acz intersect D only in the triangles acd and abc. Then h_1 is fixed outside the sum of the four tetrahedra and takes p_iabz, p_iayb, p_iady, p_icbz, p_icyb, p_icdy linearly onto the tetrahedra p_iayz, p_iaxy, p_iadx, p_icyz, p_icxy, p_icdx.

CASE 2. *S_1 has only one edge on* Bd D. Let b be the vertex of S_1 in Int D. Let D_0 be the sum of all the 2-simplexes of T in D with vertices at b. First we get a piecewise linear homeomorphism h' that takes D_0 into a horizontal plane.

Let W be a tetrahedron whose interior contains b, and J be the simple closed curve that is the projection of Bd D_0 onto Bd W from b. It follows from the preceding corollary that there is a piecewise linear homeomorphism g of Bd W onto itself that takes J onto the simple closed curve of Bd W in the horizontal plane through b. Let h' be a piecewise linear homeomorphism of E^3 onto itself, such $h'(b) = b$, $h' = g$ on Bd W and for each point p of Bd W, h' takes the ray from b through p linearly onto the ray from b through $g(p)$. Then h' is a piecewise linear homeomorphism of E^3 onto itself that takes D_0 into a horizontal plane. Now $h'(abc)$ is the sum of a finite number of triangles with vertices at b, and these triangles can be successively shelled from $h'(D)$. For convenience we suppose that $h(abc) = abc$.

We let $xybz$ be as in Case 2 of the proof of Lemma 1 where $xazc \cdot h'(D) = azc$. Let p_1yp_2 be a vertical interval through y, and so short that the pyramids with apexes at p_1 and p_2 and base $xazc$ intersects $h'(D)$ only in azc. The result is a homeomorphism h'' of E^3 onto itself that is fixed outside these pyramids, and takes p_iaxy, p_iayz, p_icxy, p_icyz linearly onto the tetrahedra p_iaxb, p_iabz, p_icxb, p_icbz. The piecewise linear homeomorphism h_1 is $h'^{-1}h''h'$.

COROLLARY. *If D_1, D_2 are two polyhedral disks on a polyhedral 2-sphere S^2 in E^3, there is a piecewise linear homeomorphism of E^3 onto itself that takes S^2 onto itself and D_1 onto D_2.*

PROOF OF COROLLARY. If $S^2 \neq D_1 + D_2$, we let D_3 be a polyhedral disk in S^2 containing $D_1 + D_2$ and apply Lemma 2 to get a homeomorphism h of E^3 onto itself, taking D_3 into a horizontal plane where $h(D_1)$, $h(D_2)$ can be dealt with more easily. Next we get a piecewise linear homeomorphism g of E^3 onto itself that takes $h(S^2)$ onto itself, and $h(D_1)$ onto $h(D_2)$.

If $D_1 + D_2 = S^2$, we consider a small polyhedral disk D_0 in S^2 such that $D_1 + D_0 \neq S^2 \neq D_2 + D_0$. It follows from the preceding paragraph that there is a piecewise linear homeomorphism h_i $(i = 1, 2)$ of E^3 onto itself that takes S^2 onto itself and D_i onto D_0. Then $h_2^{-1}h_1$ is the required homeomorphism.

LEMMA 3. *Suppose D_1, D_2, D_0 are three polyhedral disks in E^3 such that* Bd D_1 = Bd D_2, $D_1 \cdot D_2 = D_0 +$ Bd D_1, *and* Bd $D_0 \cdot$ Bd D_1 *is a point or an arc. Then for each bounded open set U which is polyhedral (sum of a finite number of interiors of tetrahedra) and contains $D_0 -$* Bd D_1, *there is a polyhedral disk D_2' in $D_2 + U$, such that $D_1 \cdot D_2' =$* Bd D_1 = Bd D_2'.

PROOF. Using Lemmas 2 and 1 we suppose, with no loss of generality, that D_1 lies in a horizontal plane, and D_0 is a triangle abc in this plane.

CASE 1. $D_0 \cdot$ Bd D_1 *is a point.* We suppose that this point is c. Let $a'b'c$ be a triangle such that $abc - c \subset$ Int $a'b'c \subset a'b'c - c \subset$ Int $D_1 \cdot U$. See left half of Figure 8. Let y be a point of Int abc and p_1yp_2 be an

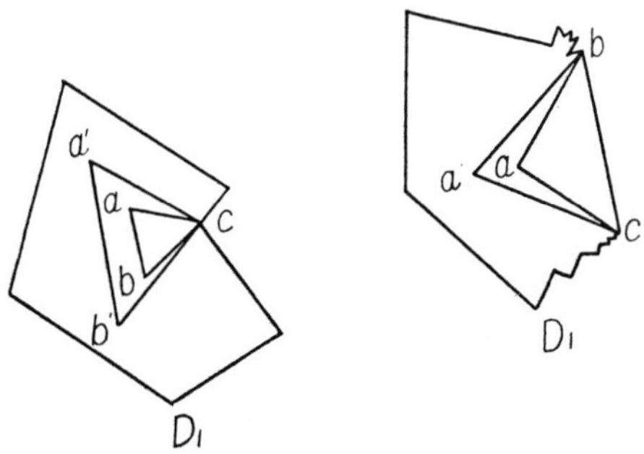

Figure 8

interval perpendicular to abc at y, and so short that if X is the sum of the two tetrahedra $p_1a'b'c$ and $p_2a'b'c$, $X - c \subset U$ and $D_2 \cdot$ Bd X is a simple closed curve. Then D_2' is formed by replacing the disk $D_2 \cdot X$ by a disk on Bd X.

CASE 2. *$D_0 \cdot$ Bd D_1 is an arc.* We suppose that this arc is bc and let $a'bc$ be a triangle, such that $abc - bc \subset \text{Int } a'bc \subset a'bc - bc \subset \text{Int } D_1 \cdot U$. See right half of Figure 8. Again D_2' is formed by replacing a disk in a pyramid with base $a'bc$ by a disk on the surface of this pyramid.

LEMMA 4. *Suppose D is a topological disk in E^3 which is locally polyhedral* mod *n points $q_1, q_2, \cdots, q_n$ on* Bd *D and U is an open set containing* Int *D that is locally polyhedral* mod $\sum q_i$ *(if $p \in \overline{U} - \sum q_i$, there is a neighborhood N of p such that $N \cdot U$ is the sum of a finite collection of interiors of tetrahedra). Then there is a topological 2-sphere S, such that $S \cdot D =$* Bd *D,* Int *D lies in the bounded component of $E^3 - S$, $S \subset U +$* Bd *D, and S is locally polyhedral* mod $\sum q_i$.

PROOF. The proof follows from a sequence of applications of Lemma 3. Let $E_1, E_2, \cdots$ be a sequence of polyhedral disks in D such that $\sum E_j = D - \sum q_i$, $E_j \subset \text{Int } E_{j+1} + \text{Bd } D$, and $D - E_j$ is the sum of n components such that the closure of each is a disk of diameter less than $1/j$ which contains a point of $\sum q_i$, and has for a boundary the sum of an arc on Bd D and an arc whose interior lies in Int D.

Using Lemmas 2 and 1 we suppose, with no loss of generality, that E_2 is a triangle abc in a horizontal plane. Let y be a point in Int abc, and p_1yp_2 be a vertical interval through y, and so short that the tetrahedron p_iabc $(i = 1, 2)$ lies in $U + \text{Bd } E_2$ and intersects D only in abc. Let S_2 be the polyhedral 2-sphere Bd $(p_1abc + p_2abc)$. (The purpose in starting with E_2 rather than E_1 is to facilitate the definition of U_j below.)

Let U_j be a polyhedral neighborhood of Int $E_j - E_{j-2}$, such that U_j has n components, each is of diameter less than $1/(j-2)$ and lies in U, and no one of these components intersects $D - E_j$.

It follows from $2n$ applications of Lemma 3 that there is a polyhedral 2-sphere S_3 in $S_2 + U_3 + \text{Bd } E_3$ that contains Bd E_3 and has a bounded complementary domain containing Int E_3. In the first two of these applications, the two disks on S_2 bounded by Bd E_2 are replaced by larger disks so that their sum is a polyhedral 2-sphere whose interior contains not only Int E_2, but also one component of Int $E_3 - E_2$. In successive similar steps, Lemma 3 is applied to the remaining components of $E_3 - E_2$.

We continue this procedure to get polyhedral 2-spheres $S_4, S_5, \cdots$ where $S_i \subset S_{i-1} + U + \text{Bd } E_i$, $E_i \cdot S_i = \text{Bd } E_i$, and Int E_i lies in the bounded component of $E^3 - S_i$. Then $S = \lim S_i$.

LEMMA 5. *Suppose ab is an arc ; D_1, D_2 are two topological disks in E^3 such that $D_1 \cdot \mathrm{Bd}\, D_2 = D_2 \cdot \mathrm{Bd}\, D_1 = ab$ and D_i $(i = 1, 2)$ is locally polyhedral* mod *points a, b ; and U is an open set containing* Int D_1 *which is locally polyhedral* mod *a, b. Then there is a topological disk D_2' in $D_2 + U$ such that* Bd D_2' = Bd D_2, *$D_1 \cdot D_2' = ab$, and D_2' is locally polyhedral* mod *a, b.*

If D_1, D_2, U are locally polyhedral at a (or b), D_2' may be chosen so as to be locally polyhedral there also.

PROOF. It follows from Lemma 4 that there is a 2-sphere S in $U + \mathrm{Bd}\, D_1$, such that $\mathrm{Bd}\, D_1 \subset S$, Int D_1 and $\mathrm{Bd}\, D_2 - ab$ lie respectively in the bounded and unbounded components of $E^3 - S$, and S is locally polyhedral mod a, b. We suppose that the vertices of S are altered slightly to put it in general position with respect to $D_2 - ab$ in the sense that $S \cdot (D_2 - ab)$ is locally topologically like a line at each of its points.

The closure of each component of $S \cdot (D_2 - ab)$ is either an arc with end points on ab or a simple closed curve with at most one point on ab. Since both D_2, S are locally polyhedral mod a, b, for each positive number ε, all but a finite number of such componets lie in an ε neighborhood of $a + b$.

Let E be the component of $D_2 - S$ which contains $\mathrm{Bd}\, D_2 - ab$. We form D_2' by replacing disks in $D_2 - E$ by disks near S.

Consider the closure of a component of $D_2 - \bar{E}$. It is a disk, and all such disks may be ordered $E_1, E_2, \cdots$ where for each positive number ε, all but a finite number of these disks lie in an ε neighborhood of $a + b$, and Bd E_{i+1} does not separate in S any point of Bd E_i from any point of Bd D_1.

It follows from Lemma 4 applied to a disk in S bounded by Bd E_1 that there is a disk E_1' such that Bd E_1 = Bd E_1', E_1' is locally polyhedral mod a, b, Int E_1' lies in the bounded component of $E^3 - S$, misses Int D_1, and lies within 1 of a disk in S bounded by Bd E_1.

Disks $E_2', E_3', \cdots$ are obtained in a similar fashion. In general, E_i' is a topological disk such that Bd E_i = Bd E_i', E_i' is locally polyhedral mod a, b, Int E_i' lies in the bounded component of $E^3 - S$, misses Int D_1 + Int $E_1' + \cdots +$ Int E_{i-1}', and lies within $1/i$ of a disk in S bounded by Bd E_i.

The disk D_2' is obtained by replacing the disks E_i in D_2 by the disks E_i'.

PROOF OF THEOREM 2. Since the disks $D_1, D_2, \cdots, D_n$ may be whittled down so as to lie (except possibly for a, b) in U, we suppose that $D_i - (a + b) \subset U$.

Let axb be an arc in Int $D_1 + a + b$ which is locally polyhedral mod a, b. Then D_1 is the sum of two disks D_0, D_1' such that $D_0 \cdot D_1' = axb$, and

Bd $D_0 = ab + axb$. We use E_i $(i = 2, 3, \cdots, n)$ to denote the disk $D_i + D_0$.

It follows from Theorem 1 that there is a disk E_i' in $U +$ Bd E_i, such that Bd $E_1' =$ Bd E_i, $E_i' \cdot D_1' = axb$, and E_i' is locally polyhedral mod a, b. If $D_1 + D_2$ is locally polyhedral mod (a, b), E_2' is taken to be $D_2 + D_0$.

We define D_2' to be E_2'.

Let U_3 be an open set such that $U_3 \subset U$, $U_3 \cdot D_1' = 0$, Int $D_2' \subset U_3$, and U_3 is locally polyhedral mod a, b. It follows from Lemma 5 that E_3' may be adjusted to get a disk D_3' in $E_3' + U_3$ such that Bd $D_3' =$ Bd E_3', $D_3' \cdot D_2' = axb = D_3' \cdot D_1'$, and D_3' is locally polyhedral mod a, b.

It requires two steps to get D_4' from E_4'—first to keep it from intersecting D_2' except along axb, and then to keep it from intersecting D_3' except along axb. To insure that the second adjustment does not cause it to intersect D_2' again, we insist that the open set U_i covering Int D_3' at this stage not intersect $D_2' + D_1'$.

The other D_i''s are obtained one by one from the E_i''s in a similar fashion.

If $D_1 + D_2$ is locally polyhedral mod (a, b), we pick our adjustment at each stage so that the adjusted $D_1 + D_2$ is the same as the original.

THEOREM 3. *Suppose ab is an arc; $D_1, D_2, \cdots, D_n$ are n topological disks in E^3 such that* Bd D_i *contains ab, $D_i \cdot D_j = ab$ $(i \neq j)$, and D_i is locally polyhedral* mod ab; *f is a nonnegative continus function defined on $\sum D_i$ that is positive on* Int ab. *Then there is a homeomorphism h of $\sum D_i$ into E^3 such that $\rho(x, h(x)) \leqq f(x)$ for each point x of $\sum D_i$, $h(a) = a$, $h(b) = b$, and $h(D_i)$ is locally polyhedral* mod a, b.

If $D_1 + D_2$ is locally polyhedral mod a, b, *h may be chosen so that $h(D_1 + D_2) = D_1 + D_2$.*

If K is a closed set on which D_i is locally polyhedral, h may be chosen so that h is the identity on K.

PROOF. We suppose with no loss of generality that $f(a) = f(b) = 0$. Let $a_1, a_2, \cdots$ be a sequence of points of ab converging to a such that a_{i+1} separates a_i from a on ab; the diameter of a_ia_{i+1} is less than one fourth the minimum value of f on the arc a_ia_{i+1}, and $\sum D_i$ is locally polyhedral at each point of a_ia_{i+1} if $K \cdot a_ia_{i+1} \neq 0$.

Let $U_1, U_2, \cdots$ be a sequence of mutually exclusive open sets such that Int $a_{2i}a_{2i-1} \subset U_i$ and the diameter of U_i is less than one fourth the minimum value of f on U_i. We note that the diameters of the U_i's converge to 0.

Using Theorem 2, we adjust the part of the D_i's in the U_j's so that the new D_i's are locally polyhedral mod $(ba_1 + a_2a_3 + a_4a_5 + \cdots + a)$, no point x is moved more than $f(x)/4$, and no point of K is moved. Another sequence of adjustments causes the resulting D_i's to be locally polyhedral mod $(ba_1 + a)$ — now no point x has been moved more than $f(x)/2$.

In two more sequences of adjustments, using a sequence of points converging to b, the D_i's are changed to disks that are locally polyhedral mod (a, b).

THEOREM 4. *Suppose $ab_1c_1, ab_2c_2, \cdots, ab_nc_n$ are n topological disks in E^3, such that each ab_ic_i is locally polyhedral* mod *the point a and $ab_ic_i \cdot ab_jc_j$ $(i \neq j)$ is either the point a or one of the arcs ab_i, ac_i of* Bd *ab_ic_i (if it is one of these arcs, it is also one of the arcs ab_j, ac_j of* Bd *ab_jc_j). Then for each neighborhood U of the point a, there are n polyhedral disks $a'b_1c_1$, $a'b_2c_2, \cdots, a'b_nc_n$, such that that $a'b_ic_i + U = ab_ic_i + U$ and $a'b_ic_i \cdot a'b_jc_j$ $(i \neq j)$ is either the point a' or one of the arcs $a'b_i$, $a'c_i$ of* Bd *$a'b_ic_i$.*

Also if $ab_1c_1 + ab_2c_2 + \cdots + ab_mc_m$ is already a polyhedral disk whose interior contains a, the $a'b_ic_i$'s, may be chosen so that $ab_1c_1 + ab_2c_2 + \cdots + ab_mc_m = a'b_1c_1 + a'b_2c_2 + \cdots + a'b_mc_m$.

The reason for complicating Theorem 4 by adding the statement about the possibility of a disk $ab_1c_1 + ab_2c_2 + \cdots + ab_mc_m$ is to make it possible to apply Theorem 4 in triangulating a 3-manifold with boundary.

PROOF. We suppose that U is the interior of a sphere and $U \cdot \sum b_ic_i = 0$. Let C be a polyhedral cube whose interior contains a, and which is so small that $C \subset U$ and only one component of each ab_ic_i − Bd C intersects $E^3 - U$.

By decreasing C slightly, uncountably many polyhedral cubes can be obtained all satisfying the above conditions, and such that no two of their boundaries intersect. With a countable number of exceptions, the boundaries of these cubes intersect the ab_ic_i's in a convenient position so we suppose that Bd C and each ab_ic_i are in general position in the sense that if p is a point of ab_ic_i · Bd C, then there is a neighborhood V of p and a homeomorphism of V onto E^3 that takes V · Bd C onto the plane $z = 0$, and $V \cdot ab_ic_i$ onto either the plane $x = 0$ or the half plane $x = 0$, $y \geqq 0$. Then ab_ic_i · Bd C contains no arc rs such that Int rs · Bd $ab_ic_i \neq 0$. Each component of ab_ic_i · Bd C is either an arc with end points on Bd ab_ic_i, or a simple closed curve in Int ab_ic_i. Also, if $ab_1c_1 + ab_2c_2 + \cdots + ab_mc_m$ is a polyhedral disk with a on its interior, C is chosen so that each of $C \cdot ab_1c_1$, $C \cdot ab_2c_2, \cdots$, $C \cdot ab_mc_m$ is a disk and Bd $C \cdot ab_1c_1$, Bd $C \cdot ab_2c_2, \cdots$, Bd $C \cdot ab_mc_m$ is an arc.

SPECIAL CASE. First we consider the special case where the ab_ic_i's and Bd C are so fortunately located that for each integer i, ab_ic_i · Bd C is the sum of a finite number of mutually exclusive arcs each irreducible from ab_i to ac_i. Let $b_i'c_i'$ be the one of these arcs that is farthest from a in the sense that no other of the arcs is separated from a by $b_i'c_i'$ in ab_ic_i.

If we do not have the condition that $ab_1c_1 + ab_2c_2 + \cdots + ab_mc_m$ is a polyhedral disk with a on its interior, we proceed as follows. Using the corollary to Lemma 2, we suppose, with no loss of generality, that there is a triangular disk F in Bd C such that $\sum ab_ic_i \cdot \text{Bd } C \subset \text{Int } F$. Let a' be a point in Int C and near F, so that the cone over F from a' lies in C. Denote the cone over $b_i'c_i'$ from a' by $a'b_i'c_i'$ and $(ab_ic_i - ab_i'c_i') + a'b_i'c_i'$, by $a'b_ic_i$. The polyhedral disks $a'b_1c_1, a'b_2c_2, \cdots, a'b_nc_n$ show that the theorem holds in this case.

If $ab_1c_1 + ab_2c_2 + \cdots + ab_mc_m$ is a polyhedral disk D whose interior contains a, $D \cdot C$ is a disk, and C is the sum of two polyhedral cubes C_1, C_2, such that $C_1 \cdot C_2 = D \cdot C$. We divide the polyhedral disk $D \cdot C$ into polyhedral disks $a'b_1'c_1', a'b_2'c_2', \cdots, a'b_m'c_m'$. We now describe $a'b_i'c_i'$ $(i > m)$ if $b_i'c_i' \subset \text{Bd } C_1$. (A similar description holds if $b_i'c_i' \subset \text{Bd } C_2$, so we do not give it.) Using the corollary to Lemma 2, we suppose, with no loss of generality, that $\sum ab_ic_i \cdot \text{Bd } C_1$ lies in a triangular disk F of Bd C_i and each of $a'b_1'c_1', a'b_2'c_2', \cdots, a'b_m'c_m'$ is an isosceles triangle in F with vertex at a'. Let a'' be a point of Int C_1 near a', and $a''b_i'c_i'$ be the cone over $b_i'c_i'$ from a''. Then $a'b_i'c_i'$ is the polyhedral disk that $a''b_i'c_i'$ goes into when a'' is shoved to a', so that $a''b_j'c_j'$ $(j \leqq m)$ goes onto $a'b_j'c_j'$. In this shoving no points are moved except in a double cone over $D \cdot C$. Again $a'b_ic_i = (ab_ic_i - ab_i'c_i') + a'b_i'c_i'$.

General case. We finish the proof of Theorem 4 by showing that C and $\sum ab_ic_i$ can be adjusted in U so as to bring about such a special position as that treated in the special case.

Suppose that C and the ab_ic_i's have been adjusted so that no further adjustment reduces $N =$(number of components of $ab_1c_1 \cdot \text{Bd } C$) + (number of components of $ab_2c_2 \cdot \text{Bd } C$) + $\cdots$ + (number of components of $ab_nc_n \cdot \text{Bd } C$). We show that in this case Bd C and the ab_ic_i's are situated as suggested in the special case. We do this in two steps by considering what a component of $ab_ic_i \cdot$ Bd might be — (1) a simple closed curve or (2) an arc.

(1) *$ab_ic_i \cdot$ Bd C does not contain a simple closed curve.* Assume it does contain a simple closed curve, and D is a disk in ab_ic_i such that $D \cdot \text{Bd } C = \text{Bd } D$. Then $D \subset \text{Int } ab_ic_i$ because $ab_ic_i \cdot \text{Bd } C$ contains no arc rs such that $\text{Int } rs \cdot \text{Bd } ab_ic_i \neq 0$. Using Lemma 2, we suppose, with no loss of generality, that a neighborhood of D relative to ab_ic_i lies in a horizontal plane P. Let C' be the polyhedral cube containing a such that $D \subset \text{Bd } C' \subset \text{Bd } C + D$. If C' is adjusted by shoving it to one side of P near D there results a polyhedral cube C'' such that the N associated with C'' is less than the N which we had supposed could not

be reduced by adjusting C.

(2) *Each component of* $ab_ic_i \cdot \text{Bd } C$ *is an arc irreducible from* ab_i *to* ac_i. Since such a component is a nondegenerate finite graph that contains neither a triod nor simple closed curve, it is an arc rxs. Both r, s belong to Bd ab_ic_i because Bd C, $(ab_ic_i + ab_jc_j)$ are in general position; also Int $rxs \cdot \text{Bd } ab_ic_i = 0$. We finish the proof by showing that not both r, s belong to ab_i. This is certainly the case if $ab_i \subset D$, since Bd $C \cdot ab_i$ contains only one point.

Assume rxs is an arc with only one bend at x; the points r, s are located on ab_i in the order a, r, s, b_i; the arc rs of ab_i is straight and intersects Bd C only at r, s; $rxs + rs$ bounds a triangular disk in ab_ic_i; and a neighborhood of this disk relative to ab_ic_i lies in a horizontal plane P. In fact we suppose that, except in a small neighborhood of a that misses rxs, ab_ic_i lies in P.

Let $r'x's'$ be an arc in ab_ic_i with only one bend at x' such that r', s' lie on ab_i in the order a, r', r, s, s', b_i, the arcs $r'x's'$, rxs do not intersect, and Bd C intersects the disk in ab_ic_i bounded by $r'xs' + r's'$ of ab_i only in the arc rxs. We suppose that this disk is a triangle in P with vertices r', x', s'. If no other disk ab_jc_j shares ab_i with ab_ic_i, we adjust ab_ic_i by removing Int (disk $r'x's'$) + Int $r's'$ from it to form $ar'x'b_ic_i$ and thereby reduce N.

For convenience we suppose $ab_{i+1}c_{i+1}, ab_{i+2}c_{i+2}, \cdots, ab_kc_k$ are the disks which share ab_i with ab_ic_i. We show how to reduce N in this case by replacing ab_ic_i by $ar'x's'b_i'c_i = (ab_ic_i)'$ and ab_jc_j $(i < j \leqq k)$ by a disk $(ab_jc_j)'$ near ab_jc_j + (disk $r'x's'$). We note that (number of components of $(ab_ic_i)' \cdot \text{Bd } C$) < (number of components of $ab_ic_i \cdot \text{Bd } C$).

We suppose that $ab_ic_i, ab_{i+1}c_{i+1}, \cdots, ab_kc_k$ are ordered so that, in any neighborhood of r, there is an arc from Int ab_ic_i to Int ab_jc_j that misses each of the disks $ab_{j+1}c_{j+1}, \cdots, ab_kc_k$.

Assume a neighborhood of $r's'$ with respect to $ab_ic_i + ab_{i+1}c_{i+1}$ lies in P, and $r'ys'$ is a polygonal arc in $r' + s' + \text{Int } ab_{i+1}c_{i+1}$, such that $r'ys'$ is broken only at y, the triangle $r'ys'$ lies in $ab_{i+1}c_{i+1}$, and Bd $C \cdot$ (disk $r'ys'$) is the sum of two arcs ru and sv. See Figure 9. Let z be a point on the side of P opposite from points of Int $ab_{i+2}c_{i+2} + \cdots +$ Int ab_kc_k so near r that the cone $(z\,r'x's'y)$ from z over the sum of the arcs $r'x's'$, $r'ys'$ meets Bd C only in an arc uv, and $(zr'x's'y) \cdot \sum ab_jc_j$ is the simple closed curve $r'x's'y$. The new disk $(ab_{i+1}c_{i+1})'$ is $ab_{i+1}c_{i+1}$ − (disk $r'ys'$) + $(zr'x's'y)$. We note that (number of components of Bd $C \cdot (ab_{i+1}c_{i+i})'$) $\leqq$ (number of components of Bd $C \cdot ab_{i+1}c_{i+1}$).

Next we obtain a new disk $(ab_{i+2}c_{i+2})'$ near $ab_{i+2}c_{i+2}$ + (disk $r'x's'$) in the same fashion. The only extra precaution we take is to see that the cone used in getting it is so near $ab_{i+2}c_{i+2}$ + (disk $r'x's'$) that it does not intersect $(ab_{i+1}c_{i+1})'$, except along the arc $ar'x's'b_i$. Similarly $(ab_{i+3}c_{i+3})'$, $\cdots$, $(ab_kc_k)'$ are obtained.

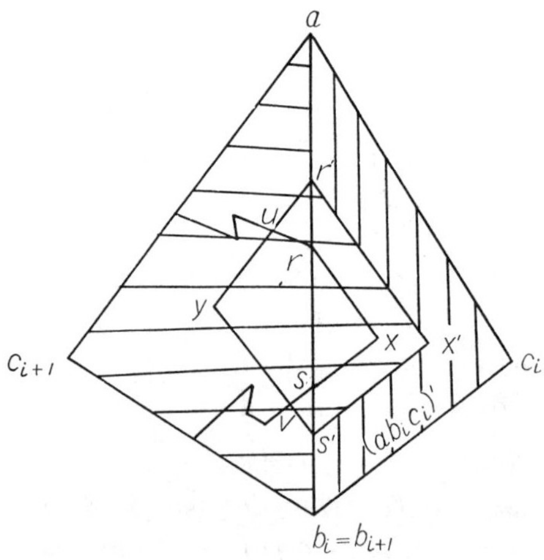

Figure 9

THEOREM 5. (Approximation theorem for 2-complexes.) *Suppose M is a triangulated* 3-*manifold with boundary, P is a closed set in M which is the image of a* 2-*complex C under a homeomorphism g, and f is a positive continuous function defined on P. Then there is a homeomorphism h of P onto a polyhedron P′ in M such that* $\rho(x, h(x)) < f(x)$.

If K is a closed set on which P is locally polyhedral, h may be chosen so as to be the identity on K.

For each triangulation of C, h may be chosen so that $hg(t) \subset \text{Bd } M$ *for each* 2-*simplex t in this triangulation for which* $g(t) \subset \text{Bd } M$.

PROOF. Suppose C is triangulated into such small triangles $t_1, t_2, \cdots$ that $g(t_i)$ lies in a neighborhood N_i in M such that there is a piecewise linear map p_i of $\overline{N_i}$ onto a cube in E^3, and the diameter of N_i is less than the minimum value of f on $g(t_i)$. We use R_0 and R_1 to denote the sum of the vertices and the sum of the edges respectively, of these triangles $t_1, t_2, \cdots$ of the triangulation of C.

It follows from Theorem 1 that there is a homeomorphism h_1 of P into M such that h_1 is the identity on $g(R_1)$, $h_1(P)$ is locally polyhedral mod $h_1g(R_1)$, and $h_1g(t_i) \subset N_i$. The details of getting h_1 on $g(t_i)$ are as follows: About each $g(\text{Int } t_i)$ put a neighborhood U_i so that no two of these neighborhoods intersect. By Theorem 1 there is a homeomorphism r_i of $p_ig(t_i)$ onto a disk in $p_i(U_i \cdot N_i) + p_ig(\text{Bd } t_i)$ such that r_i is fixed on

$p_ig(\text{Bd } t_i+K)$, $r_ip_ig(t_i)$ is locally polyhedral mod $p_ig(\text{Bd } t_i)$, and $p_i^{-1}r_ip_i$ moves no point of t_i by more than $1/i$. Then h_1 on $g(t_i)$ is defined by $p_i^{-1}r_ip_i$ if $g(t_i) \not\subset \text{Bd } M$; if $g(t_i) \subset \text{Bd } M$, h_1 is the identity on $g(t_i)$.

Similarly we find from Theorem 3 that there is a homeomorphism h_2 of $h_1(P)$ into M such that h_2 is the identity on both K and $g(R_0) = h_1g(R_0)$, $h_2h_1(P)$ is locally polyhedral mod $h_1g(R_0)$, $h_2h_1g(t_i) \subset \text{Bd } M$ if $h_1g(t_i) \subset \text{Bd } M$, and h_2h_1 moves no point x of P by as much as $f(x)$.

Finally we find from Theorem 4 that there is a homeomorphism h_3 of $h_2h_1(P)$ into M such that $h_3h_2h_1(P) = h(P) = P'$ is locally polyhedral, $h = h_3h_2h_1$ moves no point x by as much as $f(x)$, $hg(t_i) \subset \text{Bd } M$ if $h_2h_1g(t_i) \subset \text{Bd } M$, and h is the identity on K.

We shall make applications of the following corollary to Theorem 5 where it applies to noncompact 3-manifolds with boundaries with infinitely many tetrahedrons in their triangulations.

COROLLARY TO THEOREM 5. *Suppose M is a 3-manifold with boundary, f is a continuous positive function defined on M, and T_1, T_2 are two triangulations (perhaps not compatible) of M. Then there is a homeomorphism h of M onto itself such that for each point x of M, $\rho(x, h(x)) < f(x)$, and for each face F of a tetrahedron of T_1, $h(F)$ is a polyhedron under T_2.*

In fact if T_1, T_2 are locally compatible at each point of a closed set K, h may be chosen so as to be the identity on K.

PROOF. With no loss of generality we suppose that M is connected and the triangulation T_1 is so fine that if t_i is a tetrahedron of T_1, t_i lies in a neighborhood N_i such that there is a piecewise linear (under T_2) homeomorphism of $\overline{N}_i$ onto a cube in E^3, diameter N_i is less than the minimum value of f on t_i and $\overline{N}_i$ does not contain the 2-skeleton of T_1. Furthermore we suppose that if t_i intersects K, T_1 and T_2 are locally compatible at each point of t_i. Also, we suppose that no Bd t_i separates the 2-skeleton R_2 of T_1. Let f' be a continuous positive function defined on M such that for each point p of t_i, each point within $f'(p)$ of p lies in N_i but some point of the 2-skeleton of T_1 is at a distance from N_i of more than $f'(p)$.

The 2-skeleton R_2 of T_1 is a 2-complex in M under T_1. However, as viewed under T_2, R_2 is merely the homeomorphic image of a 2-complex. It follows from Theorem 5 that there is a homeomorphism h of R_2 onto a polyhedron P' (under T_2) such that $h(\text{Bd } M) = \text{Bd } M$, h does not move p by more than $f'(p)$, and h is the identity on Bd t_i if t_i intersects K. If t_i intersects K, h is extended to t_i to be the identity there. Since $h(\text{Bd } t_i)$ bounds a topological cube in N_i it follows from a theorem of Alexander

[1], that h may be extended to all of M.

The map h does not move the point p of t_i by more than $f(p)$ because $h(t_i) \subset N_i$.

Since Bd t_i does not separate R_2 and $h(\text{Bd } t_i)$ lies in N_i but $h(R_2)$ does not, $h(\text{Int } t_i) \cdot h(\text{Int } t_j) = 0$ if $i \neq j$. Hence, no two points go into the same point and h is a homeomorphism of M into itself.

To see that h is an onto map, consider a point x of $M - h(R_2)$ and an arc xy irreducible from x to $h(R_2)$ where y is a point of $h(R_2 - R_1) -$ Bd M and R_1 is the 1-skeleton of T_1. Then if t_i, t_j are the tetrahedra of T_1 containing $h^{-1}(y)$, one of $h(t_i)$, $h(t_j)$ contains x.

3. Triangulating manifolds

In this section we give details of the argument ontlined in the introduction to show that a 3-manifold with boundary can be triangulated.

THEOREM 6. *Each 3-manifold with boundary can be triangulated.*

PROOF. Suppose M is a 3-manifold with boundary. Since each point of M lies in a neighborhood O which is topologically equivalent to either E^3 or the points of E^3 on or above the xy-plane, there is a sequence of open sets $O_1, O_2, \cdots$ covering M such that each O_i can be triangulated.

Let $U_1, U_2, \cdots$ be a monotone increasing sequence of open sets covering M such that $\overline{U}_i$ is a compact subset of U_{i+1}. Since $\overline{U}_i$ can be covered by a finite number of the O_i's, it follows from Theorem 7 that U_i lies in a triangulated open subset of M and can therefore be triangulated [2, Lemma 4]. Since each $U_{i+2}-\overline{U}_i$ can be triangulated, so can each of the open sets $U_2 + (U_4-\overline{U}_2) + (U_6-\overline{U}_4) + \cdots$ and $(U_3-\overline{U}_1)+(U_5-\overline{U}_3)+\cdots$. Again it follows from Theorem 7 that their sum can be triangulated.

THEOREM 7. *Suppose U_1, U_2 are two open subsets of a 3-manifold with boundary and T_1, T_2 are triangulations (perhaps not locally compatible on $U_1 \cdot U_2$) of U_1, U_2 respectively. Then there is a triangulation T_3 of $U_1 + U_2$ such that each tetrahedron of T_1 in $U_1 - \overline{U}_2$ and each tetrahedron of T_2 in $U_2 - \overline{U}_1$ is an element of T_3.*

PROOF. Denote by M the 3-manifold with boundary containing U_1+U_2.

Before giving a detailed description of the triangulation T_3 of M, we consider the following schematic diagram (Figure 10) to show how we use the triangulations T_1, T_2 of U_1, U_2 respectively, to get the triangulation T_3 of $M = U_1 + U_2$. A tetrahedron of T_1 (represented by w_1) in $U_1 - U_2$ is left alone and is an element of T_3. A tetrahedron of T_1 only partly in $U_1 - U_2$ is adjusted so that its faces in $U_1 \cdot U_2$ are polyhedra under T_2.

The adjusted tetrahedron (represented by w_2') is subdivided so that its left faces are not divided, but its right faces (those entirely in $U_1 \cdot U_2$) are subdivided into triangles each in a tetrahedron of the triangulation

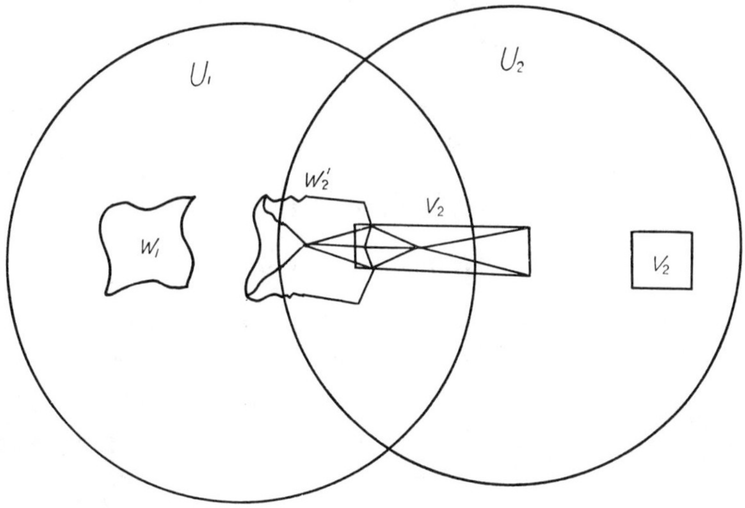

Figure 10

T_3. An element of T_2 (represented by v_1) which misses the adjusted elements of T_2 is left alone and is an element of T_3. If an element of T_2 (represented by v_2) lies only partly in the adjusted elements of T_1, the part of this element which does not lie in the adjusted elements is subdivided so that faces on the right agree with the faces of the type of v_1, and those on the left agree with the faces of the subdivisions of the tetrahedra of the type of w_2'. Now for some details.

We suppose that if K is a compact subset of U_2, all but a finite number of elements of T_1 that intersect K lie in $U_1 \cdot U_2$. What we are outlawing here is having a point of U_2 in the superior limit of a sequence of long thin tetrahedra of T_1 each intersecting $U_1 - U_2$. Lemma 7 assures us that we can bring about this situation by subdividing the elements of T_1 that intersect $U_1 \cdot \overline{U}_2$ without disturbing those in $U_1 - \overline{U}_2$.

Let f be a continuous function defined on $\overline{U}_1 \cdot \overline{U}_2$ such that f is positive on $U_1 \cdot U_2$ and zero on $\overline{U}_1 \cdot \overline{U}_2 - U_1 \cdot U_2$. ($\rho(x, \overline{U}_1 \cdot \overline{U}_2 - U_1 \cdot U_2)$ is such a function.) Let T be a triangulation of $U_1 \cdot U_2$ such that T is a subdivision of T_1. We find from the corollary of Theorem 5 that there is a homeomorphism h of $U_1 \cdot U_2$ onto itself such that each face of a tetrahedron of T goes onto a polyhedron under T_2 and $\rho(x, h(x)) < f(x)$ for each point x of $U_1 \cdot U_2$. Suppose h is extended to U_1 so as to be the identity on $U_1 - U_2$.

Let W_1 be the collection of elements of T_1 which lie in $U_1 - \overline{U}_2$, W_2 be the collection of elements of T_1 that intersect $U_1 - U_2$ but are not in W_1, and W_2' be the set of images of elements of W_2 under h. Also, let V_2 be the collection of elements of T_2 which intersect elements of W_2', and V_1 be the collection of other elements of T_2. We note that W_2' and V_2 form a buffer zone between W_1 and V_1. The elements of W_1 and V_1 are elements of T_3 but the elements of W_2', V_2 are subdivided further to bring about a fit of the triangulation in the buffer zone.

The critical part of this buffer zone is the sum B of all faces F_i of tetrahedra of W_2' such that F_i is a face of only one element of $W_1 + W_2'$ and F_i is not on Bd M.

Fortunately, in addition to being a face of an element of W_2', F_i is also a polyhedron under T_2 because of the nature of h. Let $T'(F_i)$ be a triangulation of F_i such that each triangle of $T'(F_i)$ is a triangle in some element of T_2.

For each 1-simplex E_j of W_2' in B, let $T(E_j)$ be a triangulation of E_j such that if $E_j \subset \text{Bd } F_i$ ($F_i \in W_2'$ in B), then $T(E_j)$ contains each vertex of $T'(F_i)$ in E_j. Let $T(F_i)$ be a subdivision of $T'(F_i)$ such that for each edge E_j of F_i, $T(F_i)$ contains $T(E_j)$. The triangulation of the various F_i's give a triangulation $T(B)$ of B.

We now proceed to triangulate the edges and faces of elements of W_2'. If E is an edge that does not lie in B, it is not divided and keeps its original linear structure given by h. If $E \subset B$, the triangulation of E is given by $T(B)$. If F is a face of an element of W_2' which does not have an edge in B, it is not subdivided but keeps its linear structure as a face of an element of W_2'. If $F \subset B$, its triangulation is given by $T(B)$. If F is a face not in B but with an edge in B, a vertex is placed in Int F and F is triangulated by running edges to the vertices of F and the other vertices on Bd F put there by $T(B)$. The linear structure of this triangulation $T(F)$ is made to agree with the linear structure already imposed on Bd F.

Now we triangulate the tetrahedra of W_2' to form elements of T_3. If w is such an element, Bd w has already been triangulated and the triangulation $T(w)$ of w is completed by taking a point of Int w and (regarding w as a tetrahedron with a linear structure given by h) taking the cone from this point over the triangles of the triangulation of Bd w. The linear structure of $T(w)$ is chosen to agree with the linear structure already imposed on the base of the cones. The elements of $T(w)$ are elements of T_3. We note that these elements fit the tetrahedrons of W_1

and also fit the triangles of $T(B)$ in B.

All that remains to be done is to triangulate the part of the elements of V_2 not in elements of W_2', and do this so as to get a fit with the other elements of T_3 previously defined. As a step toward this, we consider a face F of a tetrahedron of V_2 and triangulate it. Now F already has a linear structure under T_2, and if F does not intersect B, we leave F as it is. Otherwise F is given a triangulation $T(F)$ into triangles under T_2 so that if an element of $T(B)$ lies in F, this element is an element of $T(F)$ — furthermore $T(F)$ is chosen so that if F_i, F_j have an edge in common $T(F_i)$, $T(F_j)$ agree on this edge and no edge of V_1 is subdivided.

Suppose v is a tetrahedron of V_2 not covered by elements of W_2' and $T(\text{Bd}\ v)$ is a triangulation of Bd v imposed by the four triangulations of the faces of v. It follows from Lemma 6 that there is a triangulation $T(v)$ of v which is rectilinear with respect to the structure of v under T_2 and such that each element of $T(B)$ in v and each element of $T(\text{Bd}\ v)$ is an element of $T(v)$. For each tetrahedron v in V_2 not covered by elements of W_2', the elements of $T(v)$ that do not lie in the sum of elements of W_2' are the remaining elements of T_3. We note that these elements agree with $T(B)$ on B and with T_2 on the faces of the tetrahedra of V_1.

LEMMA 6. *If P is a finite polyhedron in E^3 and $T(P)$ is a rectilinear triangulation of P, there is a rectilinear triangulation T of E^3 such that each simplex of $T(P)$ is a simplex of T.*

PROOF. First we build protective cushions about certain exposed parts of P. First we cover exposed faces of $T(P)$ and then we cover exposed edges of $T(P)$.

COVERING EXPOSED FACES. If abc is an exposed face of $T(P)$ (some interval intersects P only at a point, and this point belongs to Int abc), consider a point p to one side of abc and and so near the center of abc that the tetrahedron $abcp$ intersects P only in abc. We regard $abcp$ as a pyramid with base abc and apex p. Such pyramids are placed over each exposed face in $T(P)$, on both sides of a face if both sides are exposed, and such that no two of these added pyramids intersect except possibly in a subset of their bases. Denote the sum of P and all such pyramids by P_1.

COVERING EXPOSED EDGES. Let ab be an exposed edge of $T(P)$ and xyz be a small equilateral triangle such that ab is perpendicular to xyz at the center of xyz and xyz bisects ab. We suppose that the triangles xyz are taken so small that the double pyramids $xyza + xyzb$ about the various edges ab do not intersect, except possibly at the ends of the edges, and no one of these double pyramids needlessly intersects any face, edge, vertex of $T(P)$, or pyramid added in previous step. Denote the sum of P_1 and all such double pyramids by P_2.

THE TRIANGULATION T. Let T' be a triangulation of E^3 such that each isolated vertex of $T(P)$ is a vertex of T' and the 2-skeleton of T' contains each triangle abp, bcp, acp described at the first step, and each triangle xya, xyb, yza, yzb, xza, xzb described at the second stage. Each tetrahedron of T' not in P_2 is an element of T as is each tetrahedron in $T(P)$.

We now triangulate a triangle abp introduced at the first step. The intersection of it with the boundary of the double pyramid $xyza + xyzb$ is a broken line aqb broken only at q. This broken line lies in the 1-skeleton of T' and contains certain vertices of T'. Let $T(abq)$ be the triangulation of triangle abq formed by drawing edges from the center of triangle abq to the various vertices of T' on broken line aqb. Then $T(abp)$ is a triangulation of abp containing the triangles in $T(abq)$ and the triangles of T' in the closure of $abp - abq$. Triangulations $T(bcp)$ and $T(acp)$ are given to bcp and acp in a similar fashion.

We now triangulate the tetrahedron $abcp$. Suppose the triangulations $T(abp)$, $T(bcp)$, $T(abp)$ have been described as above. The tetrahedrons of $T(abcp)$ consists of cones from the center of $abcp$ to the triangle abc, and the triangles of the triangulations of the other faces of $abcp$. Each tetrahedron in each $T(abcp)$ is an element of T. Now we have described all elements of T that lie in either P_1 or $E^3 - P_2$. We note that ab is an edge and abc is a face in this triangulation.

Now consider a tetrahedron $abxy$ in the double pyramid described at the second step. If it does not lie in P_1 (elements of T have already been defined covering it if it lies in P_1) the closure of each component of $abxy - P_1$ is a pyramid $abrs$. Then abr and abs are triangulated as abq was triangulated previously (it may be that abr or abs is an abq) and ars and brs are triangulated by T'. Then $abrs$ is triangulated by taking cones from the center of $abrs$ to the various triangles on its triangulated boundary. The tetrahedrons $bcxy$ and $acxy$ are treated similarly.

LEMMA 7. *Suppose M is a 3-manifold with boundary, T is a triangulation of M, K is a closed subset of M, and f is a positive continuous function defined on M. Then there is a subdivision T' of T such that each element of T that misses K is an element of T' and each element t of T' that intersects K is of diameter less than the minimum value of f on t.*

PROOF. Let W be the set of elements of T that miss K and f' be a positive continuous function on M such that $f' < f$ and for each point x

of K, $f'(x)$ is less than the distance from x to any element of W. Let T'' be a subdivision of T such that each element of T'' is of diameter less than the minimum value of f' on this element. [3, Lemma 4] assures us that there is such a T''.) Let V be the elements of T'' that do not intersect any element of W. Both the elements of W and the elements of V are elements of T'. (If $W = 0$, $T'' = T'$).

If F_i is a 2-simplex of T that intersects both K and $M - K$, it follows from a 2-dimensional version of Lemma 6 that there is a triangulation $T(F_i)$ of F_i compatible with linear structure of F_i such that each element of V in F_i is an element of $T(F_i)$, and each vertex of $T(F_i)$ on Bd F_i is either a vertex of F_i or belongs to V. If t_j is a tetrahedron of T that intersects both K and $M - K$, Bd t_j has already been triangulated by W, V, and the various $T(F_i)$'s. It follows from Lemma 6 that there is a subdivision $T(t_j)$ of t_j such that $T(t_j)$ contains each element of W, V, or $T(F_i)$ in t_j. Then T' is the sum of W, V, and the $T(t_j)$'s.

4. Approximating homeomorphisms

The result that a homeomorphism of a cube into E^3 can be approximated by a piecewise linear homeomorphism is very useful. This result seems to have even more applications than the result that 3-manifolds can be triangulated. Moise first proved this result by using the methods of Vietoris homology theory. As indicated in the introduction, we prove the result by using the approximation theorem for 2-complexes. Moise used extensions of this result in [10] to show that each 3-manifold can be triangulated. I used other extensions in [2] to show that each 3-manifold with boundary can be triangulated. Sanderson used such extensions [12] in showing that any orientation preserving homeomorphism of E^3 onto itself is isotopic to the identity.

The following theorem is only a brief extension of the corollary to Theorem 5.

THEOREM 8. *Suppose M is a 3-manifold with boundary, f is a continuous positive function defined on M, and T_1, T_2 are two triangulations (perhaps not compatible) of M. Then there is a piecewise linear homeomorphism h of M under triangulation T_1 onto M under triangulation T_2 such that $\rho(x, h(x)) < f(x)$.*

Since M under T_1 and under T_2 may be different as triangulated manifolds, Theorem 8 may be stated another way as follows.

If M_1, M_2 are two triangulated 3-manifolds with boundaries, g is a homeomorphism of M_1 onto M_2, and f is a continuous positive function *defined on* M_1, *then there is a piecewise linear homeomorphism* h *of* M_1 *onto* M_2 *such that* $\rho(g(x), h(x)) < f(x)$.

PROOF. This result follows from the approximation theorem for 2-complexes in the same way that the corollary to Theorem 5 follows from that approximation theorem with the following refinements. The homeomorphism h is made piecewise linear on the 1-skeleton of T_1, then on the 2-skeleton, and finally on the rest of M. Instead of using the result of Alexander [1] that each polyhedral 2-sphere in E^3 bounds a polyhedral cube, use the result due to Moise [8] and Graeub [6] to the effect that each polyhedral 2-sphere in E^3 bounds a polyhedral cube, that is, the image of a tetrahedron under a piecewise linear homeomorphism. Moise gives a short proof of this result in [8] but a slight adjustment should be made in the third paragraph of page 174 of that proof. (The fact that σ is free in $K_{im} \cdot L_i$ does not imply that $S_{i,1} \cdot S_{im}$ is a disk.)

THEOREM 9. *Suppose* M_1, M_2 *are triangulated 3-manifolds with boundaries,* f *is a continuous positive function defined on an open subset* U *of* M_1, *and* h *is a homeomorphism of* U *into* M_2. *Then there is a piecewise linear homeomorphism* h' *of* U *onto* $h(U)$, *such that* $\rho(h(x), h'(x)) < f(x)$ *for each point* x *of* U.

If K *is a closed subset of* U *on which* h *is already locally piecewise linear, the homeomorphism* h' *may be chosen so that* $h = h'$ *on* K.

PROOF. We suppose that U is connected since each component of it can be dealt with separately. Suppose T is a triangulation of U such that for each tetrahedron t_i of T, $h(t_i)$ lies in a neighborhood N_i in M_2, such that $\bar{N}_i$ is a polyhedral cube of diameter less than the minimum value of f on t_i. Also, suppose that for no tetrahedron t_i of T does Bd t_i separate the 2-skeleton R_2 of T.

Using Theorem 5 as suggested in proof Theorem 8, we get a homeomorphism g of $h(R_2)$ onto a relatively closed subset of $h(U)$, such that for each element t_i of T, $gh(t_i) \subset N_i$ and $gh = h'$ is piecewise linear on R_2. Extend h' in a piecewise linear fashion to all of U.

That Theorem 9 may be extended further may be seen by examining [2, Theorem 3].

5. Approximation theorem for polyhedra

In this last section we extend the approximation theorem for 2-complexes to apply to polyhedra.

THEOREM 10. *Suppose* M *is a triangulated 3-manifold with boundary,* P *is a closed subset of* M *that is the homeomorphic image of a polyhedron,* *and f is a continuous positive function defined on P. Then there is a homeomorphism h of P onto a polyhedron P' in M such that h moves no point x by as much as $f(x)$.*

PROOF. Suppose P is the image of a polyhedron Q under a homeomorphism g, T is a triangulation of Q, and f' is a positive continuous function defined on P that is so small that $f' < f$, and if t_i, t_j are two elements of T such that t_j does not contain t_i, then t_i contains a point p_i such that $\rho(g(p_i), g(t_j))/2 >$ maximum value of f' on $g(t_i + t_j)$.

Let T' be a subdivision of T such that for no tetrahedron t_i of T does Bd t_i separate the 2-skeleton of the triangulation of t_i under T' (we wish to prevent a 2-simplex of T' from separating a tetrahedron of T), and T' is of mesh so small that for each simplex t'_i of T', $g(t'_i)$ lies in a neighborhood N_i such that $\overline{N}_i$ is a polyhedral cube in M of diameter less than the minimum value of f' on t'_i, and $\overline{N}_i$ does not contain element of t_i containing t'_i (unless t_i is a vertex). Denote the i-skeleton of T' by R_i. It follows from application of Theorems 1, 3, and 4 that there is a piecewise linear homeomorphism h_1 of $R_0 + R_1 + R_2$ into M such that $\rho(h_1(x), g(x)) < f'g(x)$ and $h_1(\text{Bd } t'_i) \subset N_i$ if t'_i is a tetrahedron in T'. Suppose h_1 is extended to take Q into M, so that h_1 takes the tetrahedron t'_i piecewise linearly onto the polyhedral cube in N_i bounded by h_1 (Bd t'_i). If t_i is a tetrahedron of T, h_1 is a homeomorphism on t_i since for no tetrahedron t'_i of T' does Bd t'_i separate the 2-skeleton of the triangulation of t_i under T'. Also, for each point x of Q, $\rho(h_1(x), g(x)) < f'g(x)$.

Since h_1 is a homeomorphism on $R_0 + R_1 + R_2$ and on each tetrahedron of T, it follows that h_1 is a homeomorphism on Q unless there exist two elements t_i and t_j of T, such that neither of them contains the other and some two distinct points of $t_i + t_j$ go into the same point under h_1. This implies that one of these (say t_j) is a tetrahedron and that $h_1(t_i)$ is a subset of $h_1(t_j)$, since $t_i \cdot (R_0 + R_1 + R_2) -$ Bd t_j is connected. But f' was chosen so that for some point $p_i \in t_i$,

$$\rho(g(p_i), g(t_j)) > 2 \max f' \text{ on } g(t_i + t_j) .$$

Let p_j be a point of t_j such that

$$h_1(p_j) = h_1(p_i)$$

Then

$$\rho(g(p_i), g(p_j)) > 2 \max f' \text{ on } g(t_i + t_j) .$$

Since

$$\rho(h_1(p_i), g(p_i)) < f'g(p_i) < \max f' \text{ on } g(t_i + t_j) ,$$
$$\rho(h_1(p_j), g(p_j)) < f'g(p_j) < \max f' \text{ on } g(t_i + t_j) ,$$

and

$$h_1(p_i) = h_1(p_i) ,$$

we get the contradiction

$$\rho(g(p_i), g(p_j)) < 2 \max f' \text{ on } g(t_i + t_j) .$$

The homeomorphism h is given by $h_1 g^{-1}$.

If g had already been locally piecewise linear on a closed subset K of Q, we could have chosen h to be the identity on $g(K)$.

THE UNIVERSITY OF WISCONSIN AND
THE INSTITUTE FOR ADVANCED STUDY.

REFERENCES

1. J. W. ALEXANDER, *On the subdivision of 3-space by a polyhedron*, Proc. Nat. Acad. Sci. U. S. A. 10 (1924), 6–8.
2. R. H. BING, *Locally tame sets are tame*, Ann. of Math. 59 (1954), 145–158.
3. ———, *Approximating surfaces with polyhedral ones*, Ann. of Math. 65 (1957), 456–483.
4. ———, *A simple closed curve that pierces no disk*, Math. Pures. Appl. 35 (1956), 337–343.
5. R. P. GOBLIRSCH, *An area for simple surfaces*, thesis (1956) University of Wisconsin. See also Ann. of Math., 68 (1958), 231–246.
6. W. GRAEUB, *Die Semilineare Abbildungen*, Sitzungsberichte der Heidelberg Akad. der Wiss., part 4, pp. 205-272, 1950.
7. O. G. HARROLD, Jr., *Some consequences of the approximation theorem of Bing*, Summary of Lectures and Seminars, Summer Institute on Set Theoretic Topology, Madison, 1955, 53–56. See also Proc. Amer. Math. Soc. 8 (1957), 204–206.
8. E. E. MOISE, *Affine structures in 3-manifolds*, II. *Positional properties of 2-spheres*, Ann. of Math. 55 (1952), 172–176.
9. ———, IV. *Piecewise linear approximations of homeomorphisms*, Ann. of Math. 55 (1952), 215–222.
10. ———, V. *The triangulation theorem and Hauptvermutung*, Ann. of Math. 56 (1952), 96–114.
11. ———, VIII. *Invariance of the knot-type*; *local tame imbedding*, Ann. of Math. 59 (1954), pp. 159–170.
12. D. E. SANDERSON, Isotopic deformations in 3-manifolds, thesis (1953) University of Wisconsin. See also Bull. Amer. Math. Soc., Abstract 60–6–545 p. 385 and Abstract 59-6-717 p. 567. See also I. *Isotopic deformations of 2-cells and 3-cells.* Proc. Amer. Math. Soc, 8 (1957), 912-922.

G. Cartesian Factors

THE CARTESIAN PRODUCT OF A CERTAIN NONMANIFOLD AND A LINE IS E^4

BY R. H. BING[1]

Communicated February 25, 1958

An upper semicontinuous decomposition G of E^3 into points and tame arcs is defined in [1] such that the decomposition space B is topologically different from E^3. Interesting properties of this space have also been given by Fort [4], Curtis [2; 3], and Wilder [3]. We show that the cartesian product of the space B and a line E^1 is topologically E^4. Perhaps the argument used is related to that employed by Arnold Shapiro to show that the cartesian product of a manifold described by Whitehead in [5] and a line is topologically E^4.

The arcs of the decomposition G are intersections of double tori as shown in the figure. The solid double torus contains four double tori T_1, T_2, T_3, T_4 as shown; each T_i in turn contains four double tori T_{i1}, T_{i2}, T_{i3}, T_{i4} (not shown) imbedded in T_i as T_1, T_2, T_3, T_4 were imbedded in T; more double tori are imbedded in the T_{ij}'s; etc. The tame arcs of the decomposition G are the components of

$$T\cdot\Sigma T_i\cdot\Sigma T_{ij}\cdot\Sigma T_{ijk}\cdot \cdots .$$

Although these tame arcs are mutually exclusive, it is not possible to get a 2-sphere in E^3 that misses their sum and separates two of them. No topological cube in T contains $T_1+T_2+T_3+T_4$.

When the cartesian product is taken, the extra dimension enables one to unravel certain linking handles in the sense that if $[a, b]$ is

[1] Work supported by the National Science Foundation under NSF Grant G3248 at the Institute for Advanced Study, Princeton, New Jersey.

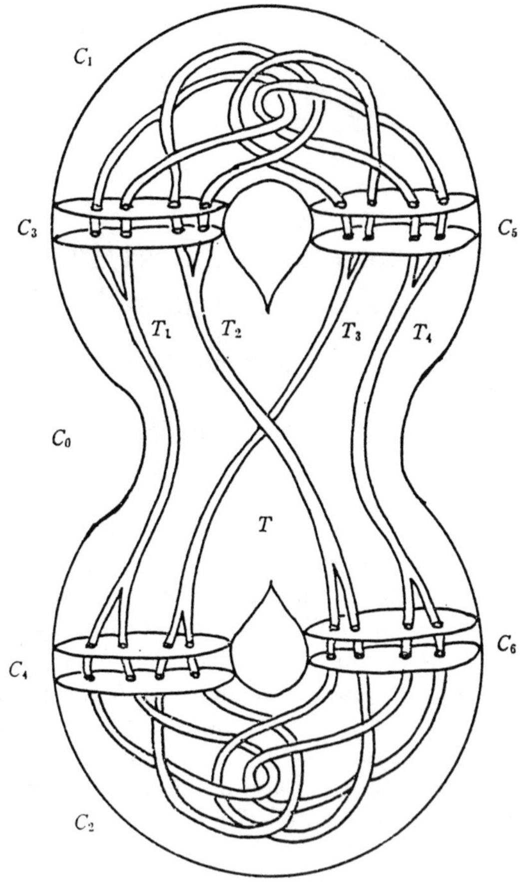

an interval and $\epsilon > 0$, then there is a topological 4-cell in

$$T \times [a - \epsilon, b + \epsilon]$$

that contains $(T_1 + T_2 + T_3 + T_4) \times [a, b]$.

In order to describe this topological 4-cell, we use eight disks to chop T into seven topological cubes C_0, C_1, C_2, C_3, C_4, C_5, C_6 as shown in the figure. The topological 4-cell in $T \times [a - \epsilon, b + \epsilon]$ that contains $(T_1 + T_2 + T_3 + T_4) \times [a, b]$ is the sum of the following thirteen 4-cells:

$C_0 \times [a - \epsilon, b + \epsilon]$,

$(C_1 + C_3) \times [a - \epsilon, a - \epsilon/2]$, $\quad (C_2 + C_4) \times [a - \epsilon, a - \epsilon/2]$,

$(C_1 + C_5) \times [b + \epsilon/2, b + \epsilon]$, $\quad (C_2 + C_6) \times [b + \epsilon/2, b + \epsilon]$,

$(T_1 \cdot (C_1 + C_3)) \times [a - \epsilon/2, b]$, $\quad (T_2 \cdot (C_1 + C_3)) \times [a - \epsilon/2, b]$,

$$(T_3\cdot(C_1+C_5))\times[a,b+\epsilon/2], \qquad (T_4\cdot(C_1+C_5))\times[a,b+\epsilon/2],$$
$$(T_1\cdot(C_2+C_4))\times[a-\epsilon/2,b], \qquad (T_3\cdot(C_2+C_4))\times[a-\epsilon/2,b],$$
$$(T_2\cdot(C_2+C_6))\times[a,b+\epsilon/2], \qquad (T_4\cdot(C_2+C_6))\times[a,b+\epsilon/2].$$

The first of these thirteen 4-cells forms the main body of the sum, while the next four serve as wings onto which are fastened the last eight.

The fact that the cartesian products of the sum of double tori and intervals lie in such 4-cells enables one to describe in isotopy on $E^3\times E^1=E^4$ that shrinks sets of the sort $A\times w$ to small size where A is a tame arc of G in E^3 and $w\in E^1$. This shrinking is continued to a pseudo isotopy that shows that $B\times E^1$ is topologically E^4. Hence we have the following result in which the equality implies topological equivalence.

THEOREM. $B\times E^1=E^4$.

The noncompactness of B is not a critical issue for if we let $\overline{B}$ denote the one point compactification of B we have the following results.

THEOREM. $\overline{B}\times E^1=S^3\times E^1$.

THEOREM. $\overline{B}\times S^1=S^3\times S^1$.

THEOREM. *The sum of two cones over the common base $\overline{B}$ is S^4.*

The preceding results show that there are homeomorphisms of period 2 of E^4 and S^4 onto themselves whose fixed point sets are B and $\overline{B}$ respectively.

REFERENCES

1. R. H. Bing, *A decomposition of E^3 into points and tame arcs such that the decomposition space is topologically different from E^3*, Ann. of Math. vol. 65 (1957) pp. 484–500.

2. M. L. Curtis, *An imbedding theorem*, Duke Math. J. vol. 24 (1957) pp. 349–352.

3. M. L. Curtis and R. L. Wilder, *The existence of certain types of manifolds*, to appear.

4. M. K. Fort, Jr., *A note concerning a decomposition space defined by Bing*, Ann. of Math. vol. 65 (1957) pp. 501–504.

5. J. H. C. Whitehead, *A certain open manifold whose group is unity*, Quart. J. Math. vol. 6 (1935) pp. 268-279.

UNIVERSITY OF WISCONSIN AND
INSTITUTE FOR ADVANCED STUDY

ANNALS OF MATHEMATICS
Vol. 70, No. 3, November, 1959
Printed in Japan

THE CARTESIAN PRODUCT OF A CERTAIN NONMANIFOLD AND A LINE IS E^4

BY R. H. BING*

(Received February 15, 1959)

1. Introduction

An upper semicontinuous decomposition G of E^3 into points and tame arcs is defined in [2] such that the decomposition space B is topologically different from E^3. We show in the present paper that the cartesian product of B and a line E^1 is topologically E^4. Stimulating discussions with Deane Montgomery helped the author in getting this result and some of its consequences. Consideration of a result by Arnold Shapiro mentioned in Section 10 of this paper was another contributing factor. Furthermore, a study of the paper [7] by Curtis suggested that B might be imbedded in E^4 as a "plane".

The space B has some interesting properties. Bing showed [2] that it is topologically different from E^3. Fort showed [9] that it could be represented as an upper semicontinuous decomposition of E^3 such that the collection of nondegenerate elements of the decomposition were broken lines with exactly two breaks. Curtis and Wilder showed [8] that B was a homotopy manifold and Curtis showed [7] that it could be imbedded in E^4.

The results of this paper are summarized in [6] and it might be useful to examine that summary before going through the details of this paper.

2. The decomposition G

The example G is described in both [1] and [2] but we repeat this description here for, in showing that $B \times E^1$ is topologically E^4, we shall be making use of the double tori used in describing the nondegenerate elements of G.

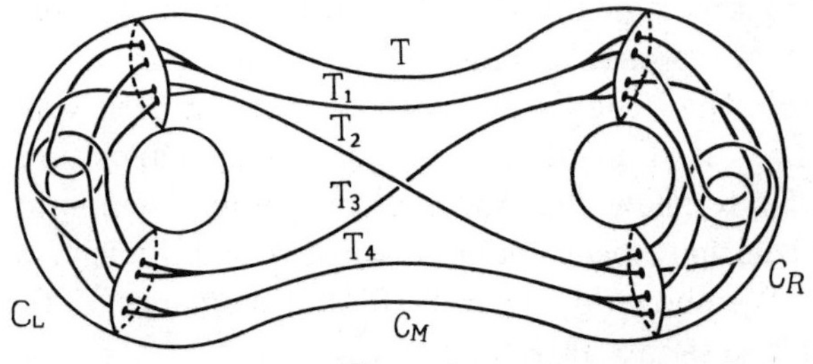

Figure 1

1. Consider a double torus T in E^3 as shown in Figure 1. It is a cube with two handles. It may be regarded as an elongated figure 8 plus all points whose distances from it are no more than some positive number ε.

* Work was supported by the National Science Foundation under NSF Grant G3248 at the Institute for Advanced Study, Princeton, New Jersey.

2. Inside T are four solid double tori T_1, T_2, T_3, T_4. The T_i's are shown as finite graphs although they are solid. Each pair of left loops of these double tori link each other as do the right loops. We denote the collection of these four double tori by X_1 and use X_1^* to denote the sum of the elements of X_1.

3. Four double tori $T_{i1}, T_{i2}, T_{i3}, T_{i4}$ are imbedded in each T_i as T_1, T_2, T_3, T_4 were imbedded in T. There are $4^2 T_{ij}$'s defined at this second stage.

4. Four double tori $T_{ij1}, T_{ij2}, T_{ij3}, T_{ij4}$ are imbedded in each T_{ij} in a similar manner. Similarly, for each integer n we define 4 double tori in each of the 4^n double tori already defined at the nth stage. The collection of 4^n double tori at the nth stage is called X_n and the sum of the elements of X_n is X_n^*.

5. We suppose that the T's are all long and thin without doubling back so that each component $X_\infty^* = T \cdot X_1^* \cdot X_2^* \cdot X_3^* \cdot \cdots$ is an arc that does not intersect in two points any plane normal to the major diameter of T. The collection of such arcs is called X_∞ and their sum is X_∞^*.

Let G be the collection of all elements g such that either g is an element of X_∞ or a point of $E^3 - X_\infty^*$. Then G is an upper semicontinuous decomposition of E^3 each of whose nondegenerate elements is a tame arc.

3. The space B

Let B be the decomposition space whose points are the elements of the upper semicontinuous decomposition G of E^3. Each nondegenerate element of G is a tame arc in X_∞ and the collection of such tame arcs corresponds to a Cantor set in B.

The cartesian product $E^3 \times E^1$ is topologically E^4. Consider the upper semicontinuous decomposition G' of E^4 whose elements are of the form $g \times w$ where g is an element of G and w is a point of E^1. The upper semicontinuous decomposition G' of E^4 has for elements tame arcs and points. The corresponding decomposition space B' is topologically $B \times E^1$. We show that $B \times E^1$ is topologically E^4 by showing that B' is.

The method we use in showing that $B \times E^1$ is topologically E^4 resembles the attack used in [4] to show that the sum of two horned spheres sewed together in a certain way was topologically S^3. It was shown in [4] that the sum of the two horned spheres was topologically equivalent to a certain decomposition space and that the decomposition space in turn was topologically S^3. The method used in showing that the decomposition space was topologically S^3 was to show that the nondegenerate elements of the corresponding decomposition space could be shrunk small without disturbing points far away. We show in Theorem 3 that elements of G' can be simultaneously shrunk to points in E^4.

4. A 4-cell in $E^3 \times E^1$

It was shown in [2] that there was no isotopy in E^3 fixed outside any X_n^* that shrinks elements of X_∞ to small size. This fact was used in showing that B is topologically different from E^3.

In showing that B' is topologically equivalent to E^4, we shall want to shrink nondegenerate elements of G' to points without disturbing points far away. In order to do this shrinking we must adjust them in a w direction. In $E^3 \times E^1$, we call the coordinate determined by E^1 the w coordinate.

As a preliminary step to shrinking an element of G', it is convenient to enclose it in a 4-cell. The existence of useful 4-cells is revealed by the following theorem. We use $[a, b]$ to denote the closed interval from a to b.

THEOREM 1. *Suppose* T, T_1, T_2, T_3, T_4 *are as shown in Figure* 1 *and* $a - \varepsilon < a < b < b + \varepsilon$. *Then* $(T_1 + T_2 + T_3 + T_4) \times [a, b]$ *lies in a* 4-*cell* K *in* $T \times [a - \varepsilon, b + \varepsilon]$.

PROOF. Consider two disks cutting off the left loop of T as shown in Figure 1 and two disks cutting off the right loop. Denote by C_L, C_M, C_R the three cubes into which these four disks divide T.

Let C_L^+ be the cube formed from C_L by removing a slightly thickened disk having as base the lower left disk and C_L^- be the cube formed from C_L by removing a thickened disk having as base the upper left disk. Figure 7 shows sets resembling the C_L^+ and C_L^- considered here. We suppose that the parts removed from C_L are so thin that

$$C_L \cdot (T_1 + T_2) \subset C_L^+ ,$$
$$C_L \cdot (T_3 + T_4) \subset C_L^- .$$

Consider the following seven 4-cells:

$$C_M \times [a - \varepsilon, b + \varepsilon] ,$$
$$C_L^+ \times [a - \varepsilon, a - \varepsilon/2] ,$$
$$C_L^- \times [b + \varepsilon/2, b + \varepsilon] ,$$
$$(C_L \cdot T_1) \times [a - \varepsilon/2, b] ,$$
$$(C_L \cdot T_2) \times [a - \varepsilon/2, b] ,$$
$$(C_L \cdot T_3) \times [a, b + \varepsilon/2] ,$$
$$(C_L \cdot T_4) \times [a, b + \varepsilon/2] .$$

The second and third are mutually exclusive and each of these two intersects the first in a 3-cell. The sum of all three is a 4-cell.

The fourth, fifth, sixth, and seventh are mutually exclusive and each intersects the sum of the first three in a 3-cell. For example, the fourth intersects the sum of the first three in

$$(C_M \cdot C_L \cdot T_1) \times [a - \varepsilon/2, b] + (C_L \cdot T_1) \times (a - \varepsilon/2)\,.$$

The sum of all seven of these 4-cells is a 4-cell.

The first of the above seven 4-cells contains the central part of $(T_1 + T_2 + T_3 + T_4) \times [a, b]$; that is, the part in $C_M \times [a, b]$. The second

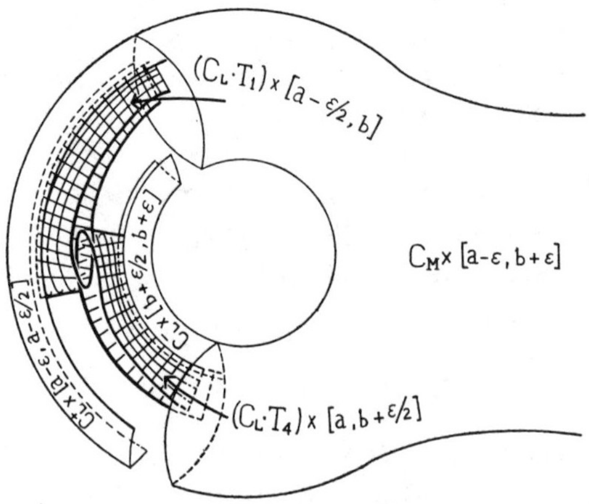

Figure 2a

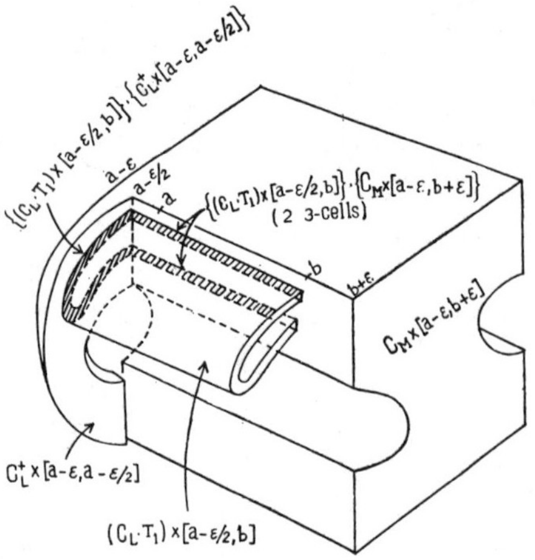

Figure 2b

and third do not contain any of $(T_1 + T_2 + T_3 + T_4) \times [a, b]$ but serve as wings onto which to attach the next four 4-cells which contain the left portion of $(T_1 + T_2 + T_3 + T_4) \times [a, b]$. See Figure 2a. The figure is schematic because of lack of dimensions but shows two apparently linking pieces being joined to different parts of

$$C_L^+ \times [a - \varepsilon, a - \varepsilon/2] + C_M \times [a - \varepsilon, b + \varepsilon] + C_L^- \times [b + \varepsilon/2, b + \varepsilon]$$

so that a 4-cell is formed. Figure 2b shows why the intersection of the fourth 4-cell intersects the sum of the first two in a 3-cell. The figure was suggested by J. G. Hocking who discussed this paper at a seminar at the University of Michigan.

In a similar fashion, we may add six 4-cells in $C_R \times [a - \varepsilon, b + \varepsilon]$ to the seven we already have, to get a 4-cell K in $T \times [a - \varepsilon, b + \varepsilon]$ containing all of $(T_1 + T_2 + T_3 + T_4) \times [a, b]$. All thirteen of these 4-cells are listed in [6].

COROLLARY TO THEOREM 1. *For each element T of X_k and each positive integer $m > 1$, there is a sequence $K_1, K_2, \cdots, K_{m-2}$ such that K_i is the sum of 4^{i-1} mutually exclusive 4-cells and*

$$\begin{aligned} T \times [0, 2m - 3] \supset K_1 \supset \operatorname{Int} K_1 \supset (T \cdot X_{k+1}^*) \times [1, 2m - 4] \\ \supset K_2 \supset \operatorname{Int} K_2 \supset (T \cdot X_{k+2}^*) \times [2, 2m - 5] \supset K_3 \supset \cdots \\ \supset K_{m-2} \supset \operatorname{Int} K_{m-2} \supset (T \cdot X_{k+m-2}^*) \times [m - 2, m - 1]\,. \end{aligned}$$

In fact the K_i*'s may be chosen so as to contain the central part of* $(T \cdot X^*_{k+i-1}) \times [i-1, 2m-i-2]$ *in the sense that*

$$(T \cdot X^*_{k+i-1} \cdot C_M) \times [i-1, 2m-i-2] = K_i \cdot (C_M \times [i-1, 2m-i-2]).$$

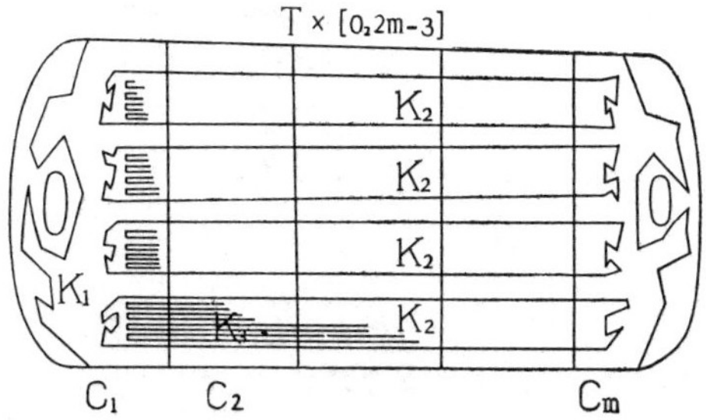

Figure 3

Figure 3 gives a schematic picture of the K's. They run smoothly through the central portion (with respect to T) of $T \times [0, 2m-3]$ but they may twist near the ends.

5. Shrinking elements of G' to small size

An element of G' on the interior of a 4-cell may be shrunk small without disturbing points outside the 4-cell. However, during this procedure, other elements of G' near the boundary of the 4-cell may be stretched. We seek a method of shrinking elements of G' without stretching others.

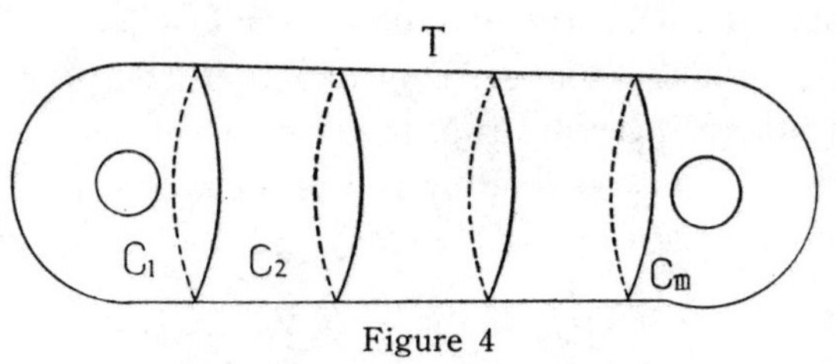

Figure 4

LEMMA 1. *Suppose a double torus* $T \in X_k$ *is broken into chambers* $C_1, C_2, \cdots, C_m$ *as shown in Figure 4. Then there is a homeomorphism* h *of* $E^3 \times E^1$ *onto itself such that*

$h = I$ *(the identity) except on parts of* $T \times [0, 2m-3]$,
$h = I$ *on* $(C_3 + C_4 + \cdots + C_m) \times ([0, 1] + [2m-4, 2m-3])$,
$(C_4 + C_5 + \cdots + C_m) \times ([1, 2] + [2m-5, 2m-4])$,
$\cdots$
and $C_m \times [m-3, m]$.

$$h((X_k^* \cdot (C_1 + C_2)) \times ([0, 1] + [2m - 4, 2m - 3])) \subset (C_1 + C_2) \times [0, 2m - 3] ,$$
$$h((X_{k+1}^* \cdot (C_1 + C_2 + C_3)) \times ([1, 2] + [2m - 5, 2m - 4])) \subset (C_2 + C_3) \times [0, 2m - 3] ,$$
$$h((X_{k+2}^* \cdot (C_1 + C_2 + C_3 + C_4)) \times ([2, 3] + [2m - 6, 2m - 5])) \subset (C_3 + C_4) \times [0, 2m - 3] ,$$
$$\cdots$$
$$h((X_{k+m-2}^* \cdot (C_1 + C_2 + \cdots + C_m)) \times [m - 2, m - 1] \subset (C_{m-1} + C_m) \times [0, 2m - 3] .$$

Diagrammatic interpretation of Lemma 1. Perhaps Figure 5 shows better than the equations where a point moves under h.

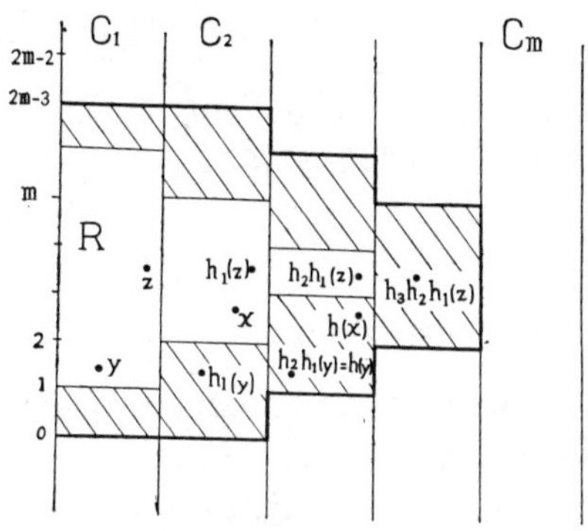

Figure 5

The columns in the diagram represents the chambers into which T is divided while a row (y value) shows the w coordinate of a point of $E^3 \times E^1$. For example the point x of $E^3 \times E^1$ shown lies in $C_2 \times 2.3$.

Figure 5 may be interpreted to show that h does not move any point unless it lies in the polygon R bounded by the vertical line from 0 to $2m - 3$ and the heavy broken line to the right of this line. This is also implied by the condition in the statement of the lemma showing where $h = 1$.

If x lies in R, consider the horizontal line through x and the shaded rectangles in R which intersect the line. Then the first coordinate of $h(x)$ lies in one of the C_i's corresponding to these shaded rectangles if $x \in X_\infty^*$. In the particular x shown in the diagram, $x \in C_2 \times 2.3$ and

$$h(x) \in (C_3 + C_4) \times [0, 2m - 3] .$$

The diagram does not reveal the w coordinate of $h(x)$. However, if x belongs to R, the w coordinate of $h(x)$ lies between 0 and $2m - 3$. Roughly speaking, h moves certain points to the right but it does not move them up and down by more than $2m - 3$.

Proof of Lemma 1. The lemma is proved by making use of the K_i's of the Corollary to Theorem 1. See Figure 3. Without moving its boundary, part of K_1 is shoved to the right, then part of the adjusted K_2 is shoved farther to the right, etc. Now for more details.

There is a homeomorphism h_1 of $E^3 \times E^1$ onto itself such that

$h_1 = I$ outside K_1,
$h_1 = I$ on $(C_3 + C_4 + \cdots + C_m) \times [0, 2m - 3]$,
$h_1((X^*_{k+1} \cdot (C_1 + C_2)) \times [1, 2m - 4]) \subset C_2 \times [0, 2m - 3]$.

This is the homeomorphism we mentioned in the preceding paragraph that moved part of K_1 to the right. Figure 5 shows that since points x, z belong to $C_1 \times [1, 2m - 4]$, then $h_1(y)$, $h_1(z)$ belong to $C_2 \times [0, 2m - 3]$.

The homeomorphism that moves part of the adjusted K_2 is called h_2.

$h_2 = I$ outside $h_1(K_2)$,
$h_2 = I$ on $(C_4 + C_5 + \cdots + C_m) \times [0, 2m - 3]$,
$h_2h_1((X^*_{k+2} \cdot (C_1 + C_2 + C_3)) \times [2, 2m - 5]) \subset C_3 \times [0, 2m - 3]$.

Figure 5 shows $h_2h_1(y)$ and $h_2h_1(z)$ belonging to $C_3 \times [0, 2m - 3]$. However $h_2h_1(y)$ might belong to $C_2 \times [0, 2m - 3]$ instead of $C_3 \times [0, 2m - 3]$ because $y \notin C_1 \times [2, 2m - 5]$. Since $h_2h_1 = h_1$ on $(E^3 \times E^1) - \operatorname{Int} K_2$, and $\operatorname{Int} K_2 \cdot (E^3 \times ([0, 1] + [2m - 4, 2m - 3])) = 0$,

$$\begin{aligned} h_2h_1((X^*_k \cdot (C_1 + C_2)) \times ([0, 1] + [2m - 4, 2m - 3])) \\ = h_1((X^*_k \cdot (C_1 + C_2)) \times ([0, 1] + [2m - 4, 2m - 3])) \\ \subset (C_1 + C_2) \times [0, 2m - 3] . \end{aligned}$$

Next we get h_3 that moves part of the adjusted K_3 still farther to the right.

$h_3 = I$ outside $h_2h_1(K_3)$,
$h_3 = I$ on $(C_5 + C_4 + \cdots + C_m) \times [0, 2m - 3]$,
$h_3h_2h_1((X^*_{k+3} \cdot (C_1 + C_2 + C_3 + C_4)) \times [3, 2m - 6]) \subset C_4 \times [0, 2m - 3]$.

We note that h_3 does not move $h_2h_1(y)$ since $y \notin K_3$ but $h_3h_2h_1(z)$ lies in $C_4 \times [0, 2m - 3]$.

Continuing in this fashion we get $h_4, h_5, \cdots, h_{m-2}$. Then

$$h = h_{m-2}h_{m-3} \cdots h_2h_1 .$$

LEMMA 2. *Suppose a double torus* $T \in X_k$ *is broken into chambers* $C_1, C_2, \cdots, C_m$ *as shown in Figure 4. Then there is a uniformly continuous homeomorphism* h *of* $E^3 \times E^1$ *onto itself such that* $h = I$ *except on* $T \times E'$ *and for each* $w \in E^1$ *there is an* i *(which is a function of* w*) such that*

$$\begin{aligned} h((T \cdot X^*_{k+m-2}) \times w) \\ \subset (C_i + C_{i+1} + C_{i+2} + C_{i+3}) \times [w - 2m + 3, w + 2m - 3] . \end{aligned}$$

PROOF. Figure 6 shows how to apply Lemma 1 to get Lemma 2.

A point x of $(T \cdot X^*_{k+m-2}) \times E^1$ represented in the regions $R_j (j = 0, 1, -1, 2, -2, \cdots)$ is moved to the right so that its image is in

$(C_i + C_{i+1}) \times E^1$ where C_i, C_{i+1} correspond to the columns containing the shaded rectangles in R_j on the same horizontal line with x. The w coordinate of x is not changed by more than $2m - 3$ in this adjustment. Similarly, points in the regions $S_j(j = 0, 1, -1, 2, -2, \cdots)$ are moved to the left.

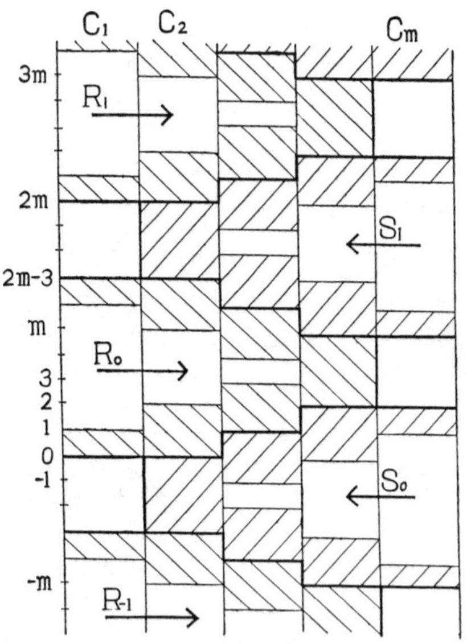

Figure 6

THEOREM 2. *For each double torus T of X_k and each positive number ε there is an integer n and a uniformly continuous homeomorphism h of $E^3 \times E^1$ onto itself such that*

$h = I$ outside $T \times E^1$

h moves no point in w direction by more than ε,

diameter $h(T' \times w) < \varepsilon$ for each double torus $T' \in X_n$ in T and each $w \in E^1$.

PROOF. Let T_1, T_2, T_3, T_4 be the four double tori of X_{k+1} in T. We suppose that each T_i is long, skinny, and divided into m chambers $C_1, C_2, \cdots, C_m$ as shown in Figure 4 such that each chamber is of diameter less than $\varepsilon/8$ and end chambers contain loops of T_i as shown. If the T_i's are not already in this shape, use a homeomorphism of E^3 fixed outside T to modify them. For convenience we suppose they are already so located. Since h will be the identity outside $(T_1 + T_2 + T_3 + T_4) \times E^1$, we only describe it on $T_1 \times E^1$.

Theorem 2 follows immediately from Lemma 2 if we modify the scale of E^1 so that $2m - 3 < \varepsilon/2$. Then the h of Lemma 2 does not move any point in the w direction by more than $\varepsilon/2$. We let $n = (k + 1) + m - 2$.

If $T' \in X_n$ in T_1, then $h(T' \times w)$ lies in $(C_i + C_{i+1} + C_{i+2} + C_{i+3}) \times E^1$ for some i. Since each C_i is of diameter less than $\varepsilon/8$, the projection of $h(T' \times w)$ into E^3 is of diameter less than $\varepsilon/2$. Hence $h(T' \times w)$ is of diameter less than ε.

6. A pseudo-isotopy on E^4

The elements of G' in E^4 may be shrunk to points in the following sense:

THEOREM 3. *There is a pseudo-isotopy* $f(x, t)(x \in E^4, 0 \leqq t \leqq 1)$ *of* E^4 *onto itself such that*

$f(x, 0)$ *is the identity—that is,* $f(x, 0) = x$,

for each fixed t_0 *less than* 1, $f(x, t_0)$ *is a homeomorphism of* E^4 *onto* E^4,

$f(x, 1)$ *takes* E^4 *onto itself and each element of* G' *onto a distinct point of* E^4.

PROOF. The homeomorphisms mentioned in Lemmas 1, 2, and Theorem 2 resulted from shrinking compact sets inside 4-cells so that they can be attained by isotopies that start at the identity, are the identity outside the 4-cells, and end at the required homeomorphisms. We obtain the isotopy promised in Theorem 3 by a sequence of applications of Theorem 2.

Let $\varepsilon_1, \varepsilon_2, \cdots$ be a sequence of positive numbers with a finite sum. We define a monotone increasing sequence $n_1 = 1, n_2, n_3, \cdots$ of integers and a sequence of isotopies

$$f(x, t)(x \in E^4, 0 \leqq t \leqq 1/2), f(x, t)(x \in E^4, 1/2 \leqq t \leqq 2/3), \cdots$$

such that

$f(x, 0) = x$,

two adjacent $f(x, t)$'s agree on their common end,

each $f(x, i/(i+1))$ is uniformly continuous,

(1) $f(x, (i-1)/i) = f(x, t)((i-1)/i \leqq t \leqq i/(i+1))$ except possibly on $X^*_{n_i} \times E^1$,

(2) diameter $f(T \times w, i/(i+1)) < \varepsilon_i$ if $T \in X_{n_{i+1}}, w \in E^1$,

(3) no point moves more than $2\varepsilon_{i-1}$ during $f(x, t)$ $((i-1)/i \leqq t \leqq i/(i+1))$,

(4) $f(E^3 \times [w - \varepsilon_i, w + \varepsilon_i], (i - 1/i)) \supset f(E^3 \times w, i/(i+1))$.

Before defining the sequence of $f(x, t)$'s we show that the existence of such a sequence is enough to guarantee the truth of Theorem 3.

Condition (3) implies that $f(x, 1) = \lim (t \to 1) f(x, t)$ is a map of E^4 onto itself,

Condition (2) insures that $f(g, 1)$ is a point for each element g of G'.

Condition (1) implies that, if $f(g_1, 1) = f(g_2, 1)(g_1, g_2 \in G')$, then one of

g_1, g_2 is above the other in the w direction. The reasons are as follows. If g_1 is a point, there is an integer i so large that $f(x, (i-1)/i) = f(x, 1)$ in a neighborhood of g_1. If $g_1 = g' \times w_1$, $g_2 = g'' \times w_2$, where g', g'' are different nondegenerate elements of G, then there is an integer i such that g', g'' are in different double tori T', T'' of X_i. Then $f(g_1, 1), f(g_2, 1)$ belong respectively to the mutually exclusive sets

$$f(T' \times E^1, (i - 1/i)), f(T'' \times E^1, (i-1)/i) .$$

Condition (4) implies that no two points with different w coordinates go into the same point under $f(x, 1)$ because if $w_1 \neq w_2$ there is an i such that $\varepsilon_i + \varepsilon_{i+1} + \cdots < |w_2 - w_1|/2$ and Condition 4 implies that

$$\begin{aligned} &f(E^3 \times [w - \varepsilon_i - \varepsilon_{i+1} - \cdots, w + \varepsilon_i + \varepsilon_{i+1} + \cdots], (i-1)/i) \\ &\quad \supset f(E^3 \times [w - \varepsilon_{i+1} - \varepsilon_{i+2} - \cdots, w + \varepsilon_{i+1} + \varepsilon_{i+2} + \cdots], i/(i+1)) \\ &\quad \supset \cdots \supset f(E^3 \times w, 1) . \end{aligned}$$

Then $f(E^3 \times w_1, 1)$ and $f(E^3 \times w_2, 1)$ lie respectively in the mutually exclusive curved "strips"

$$f(E^3 \times [w_1 - |w_2 - w_1|/2, w_1 + |w_2 - w_1|/2], (i-1)/i)$$

and $$f(E^3 \times [w_2 - |w_2 - w_1|/2, w_2 + |w_2 - w_1|/2], (i-1)/i) .$$

The existence of $f(x, t)(x \in E^4, 0 \leqq t \leqq 1/2)$ and n_2 follow directly from Theorem 2 (we ignore Condition (3) since ε_0 is not defined) so we proceed to define $f(x, t)((i-1)/i \leqq t \leqq i/(i+1))$ and n_{i+1}.

Let γ be a positive number so small that

$$\text{diameter } f(T \times [a, b], (i-1)/i) < 2\varepsilon_{i-1} \qquad \text{if } T \in X_{n_i}, |b - a| < \gamma .$$

The existence of such a γ follows from Condition (2) and the uniform continuity of $f(x, (i-1)/i)$.

Let δ be a positive number so small that for each set S of diameter less than δ, diameter $f(S, (i-1)/i) < \varepsilon_i$.

It follows from Theorem 2 that there is an isotopy $h(x, t)(x \in E^4$, $(i-1)/i \leqq t \leqq i/(i+1))$ and an integer n_{i+1} such that

$h(x, (i-1)/i) = x$,
$h(x, t) = x$ unless $x \in X^*_{n_i} \times E^1$,
$h(x, t)$ moves no point in w direction by more than minimum of (γ, ε_i),
diameter $h(T' \times w, i/(i+1)) < \delta$ if $T' \in X_{n_{i+1}}$,
$h(x, i/(i+1))$ is uniformly continuous.

Then

$$f(h(x, t), (i-1)/i) = f(x, t)((i - 1/i \leqq t \leqq i/(i+1)) .$$

The $f(x, t)(x \in E^4, (i-1)/i \leqq t \leqq i/(i+1))$ we have defined satisfies Condition (1) because $h(x, t) = x$ except possibly on $X^*_{n_i} \times E^1$. It satisfies Condition (2) because diameter $h(T \times w, i/(i+1)) < \delta$. It satisfies Condition (3) because

$$\text{diameter } f(T \times [a, b], (i-1)/i) < 2\varepsilon_{i-1}(T \in X_{n_i}, |b-a| < \gamma) ,$$

and $h(x, t)$ moves no point in the w direction by more than γ. It satisfies Condition (4) because $h(E^3 \times w, (i+1)/i) \subset E^3 \times [w - \varepsilon_i, w + \varepsilon_i]$ and

$$f(E^3 \times w, (i+1)/i) = f(h(E^3 \times w, (i+1)/i), (i-1)/i) \subset f(E^3 \times [w - \varepsilon_i, w + \varepsilon_i], (i-1)/i) .$$

7. The cartesian product of B with a line

An immediate consequence of Theorem 3 is the following result:

THEOREM 4. *The cartesian product of the decompositson space B with a line is topologically E^4.*

It is of interest that there is no space A different from E^2 such that $A \times E^1$ is topologically E^3. If $A \times E^1 = E^3$, then A is connected, locally connected, and locally compact since E^3 is. That A is topologically E^2 follows from the Kline sphere characterization [3]. If $\bar{A}$ is the one point compactification of A, it is a 2-sphere because

(1) no finite set of points separates $\bar{A}$ since no finite set of lines separates E^3, and

(2) each simple closed curve separates $\bar{A}$ since each closed set in E^3 homeomorphic to either a plane or an infinite circular cylinder separate E^3.

Theorem 4 shows that there are fiberings of E^4 of peculiar sorts.

8. The compact case

Let $\bar{B}$ be the one point compactification of B. It is the decomposition space corresponding to the upper semicontinuous decomposition of S^3 each of whose elements is either a tame arc, or a point of G, or the ideal point at infinity added to E^3 to change it into S^3.

Our argument to show that $B \times E^1$ is topologically E^4 gives the following:

THEOREM 5. *The cartesian product of $\bar{B}$ and a line is topologically equivalent to the cartesian product of S^3 and a line.*

A two point compactification of $\bar{B} \times E^1$ gives the sum of two cones over $\bar{B}$ intersecting only in their common base. Hence we have the following:

THEOREM 6. *The sum of two cones over the common base $\bar{B}$ is topologically S^4. A finite cone over $\bar{B}$ is locally euclidean except at its base (it is euclidean at its vertex). An infinite cone over $\bar{B}$ is topologically E^4.*

The ordering $\cdots, 0, 1, \cdots, 2m-3, \cdots$ we did on a line to get Lemma 2 could have been done just as effectively cyclically on a circle so we have the following result:

THEOREM 7. *The cartesian product of $\bar{B}$ and a circle is topologically equivalent to the cartesian product of S^3 and a circle.*

9. Homeomorphisms of period 2

Since there is a periodic transformation of period 2 of a line onto itself with only one point fixed, we have the following result:

THEOREM 8. *There is a homeomorphism of period 2 of E^4 onto itself that has as its fixed point set the homeomorphic image of B.*

Theorem 8 is in contrast with the result of [10, Theorem 4] that each periodic homeomorphism of E^3 onto itself has either a point, a topological line, a topological plane, or E^3 as its fixed point set.

Since S^4 is the one point compactification of E^4 we also have the following:

THEOREM 9. *There is a homeomorphism of period 2 of S^4 onto itself that has as its fixed point set the homeomorphic image of $\bar{B}$.*

The fixed point set F mentioned in Theorem 9 is locally a 3-manifold except on a Cantor set. The periodic homeomorphism h may be adjusted so that S^4 contains a polyhedral 4-cell K such that $h(K) = K$, $F \cdot K$ is a polyhedral 3-cell, and the closure of $S^4 - K$ is a 4-cell. Consideration of $S^4 - K$ suggests the following result:

THEOREM 10. *For each polyhedral 4-cell K and each tame 2-sphere S on* Bd K *there is a homeomorphism of period 2 of K onto itself with fixed point set F such that $F \cdot \text{Bd}\, K = S$ and F fails to be a 3-manifold on a Cantor set in F.*

Suppose h is a homeomorphism of period 2 of a polyhedral 4-cell K onto itself such that the fixed point set F is a polyhedral 3-cell. By application of Theorem 10 to a small neighborhood N of a point of F, h can be adjusted to get a homeomorphism h' of period 2 of K onto itself such that $h = h'$ on $K - N$ and h' has for its fixed point set a continuum that is not a 3-cell. If such an adjustment is iterated a countable number of times, a periodic homeomorphism can be obtained with a peculiar fixed point set. In fact, if one could control the iterated complications so that the limit of a sequence of bad fixed point sets did not converge to something better in the limit, one might get a fixed point set that is not locally euclidean at any point.

10. A result by Shapiro

Whitehead described [11] an example of a connected simply connected open subset U of E^3 with a connected boundary such that U is topologically different from E^3. The example may be described as follows:

Consider a solid torus T_1 imbedded in a solid torus T_2 as shown in Figure 7, and let h be a homeomorphism of E^3 onto itself that takes T_1 onto T_2. Then $U = T_1 + h(T_1) + h^2(T_1) + \cdots$.

Arnold Shapiro showed that the cartesian product of the 3-manifold U and a line is topologically E^4. This result can be used to show the existence of unusual periodic transformations of E^4 onto itself as considered in the preceding section.

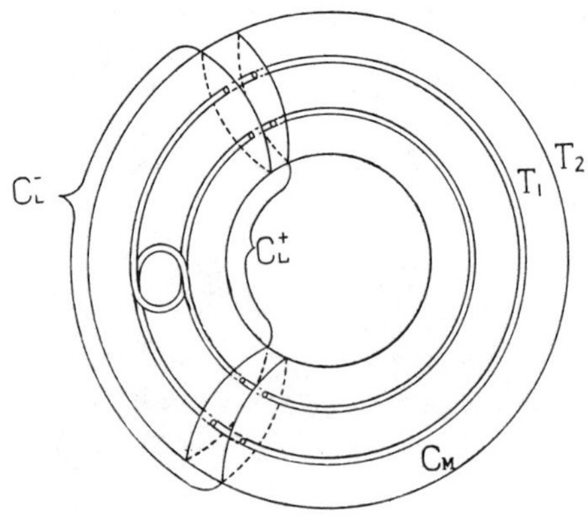

Figure 7

One method of showing that U is topologically different from E^3 is to show, as is done in [5], that there is no topological cube in U that contains T_1. One might wonder how a 4-cell in $U \times E^1$ could possibly contain $T_1 \times [a, b]$ if no cube in U contains T_1. How can one get around the handles? How far out would one have to go in $U \times E^1$ to get a 4-cell containing $T_1 \times [a, b]$? The following result gives an answer.

THEOREM 11. *Suppose* T_1, T_2 *are as shown in Figure* 7 *and* $a - \varepsilon < a < b < b + \varepsilon$. *Then* $T_2 \times [a - \varepsilon, b + \varepsilon]$ *contains a* 4*-cell which contains* $T_1 \times [a, b]$.

I discovered the proof of Theorem 11 while wondering how Shapiro obtained his result. The proof of Theorem 11 is a simplified version of the proof of Theorem 1. Recall Figure 2. Hence, the discovery of Theorem 11 led to Theorem 1 which in turn led to Theorem 4.

REFERENCES

1. R. H. BING, *Decomposition of E^3 into points and tame arcs*, Summary of Lectures and Seminars, Summer Institute on Set Theoretic Topology, Madison, 1955, pp. 40–47.

2. ———, *A decomposition of E^3 into points and tame arcs such that the decomposition space is topologically different from E^3*, Ann of Math. 65 (1957), 484–500.

3. ———, *The Kline sphere characterization problem*, Bull. Amer. Math. Soc. 52 (1946), 644-653.
4. ———, *A homeomorphism between the 3-sphere and the sum of two solid horned spheres*, Ann. of Math. 56 (1952), 354-362.
5. ———, *Necessary and sufficient conditions that a 3-manifold be* S^3, Ann. of Math. 68 (1958), 17-37.
6. ———, *The cartesian product of a certain nonmanifold and a line is* E^4, Bull. Amer. Math. Soc., 64 (1958), 82-84.
7. M. L. CURTIS, *An imbedding theorem*, Duke Math. J. 24 (1957), 349-352.
8. ——— and R. L. Wilder, *The existence of certain types of manifolds*, to appear.
9. M. K. Fort, Jr., *A note concerning a decomposition space defined by Bing*, Ann. of Math. 65 (1957), 501-504.
10. P. A. SMITH, *Fixed-point theorems for periodic transformations*, Amer. J. Math. 63 (1941), 1-8.
11. J. H. C. WHITEHEAD, *A certain open manifold whose group is unity*, Quart. J. of Math. (Oxford), 6 (1935), 268-279.

A SET IS A 3 CELL IF ITS CARTESIAN PRODUCT WITH AN ARC IS A 4 CELL[1]

R. H. BING

We are learning strange things about cartesian products.

Shapiro has shown that the cartesian product of a line and a certain example described by Whitehead [16] is topologically E^4. The example is a bounded open subset of E^3 that is simply connected but which contains a simple closed curve which does not lie in any topological cube in it [3]. Glimm has given a proof of this result in [9] and a different proof is suggested in [5].

Artin and Fox have given [8] an example of a wild arc in S^3 whose complement is simply connected though topologically different from E^3. Kirkor has shown [11] that the cartesian product of this complement and a line is E^4.

Bing has described a nonmanifold [2] whose cartesian product with a line is E^4 [4; 5]. A modification of this example gives a compact nonmanifold whose cartesian product with a circle is topologically the same as the cartesian product of a 3 sphere and a circle. The modification is a monotone decomposition of S^3 whose set of nondegenerate elements is a Cantor set of tame arcs.

Rosen has complicated the example of Bing to show [14] that there is a set which is not locally a manifold at any point but whose cartesian product with a line is E^4. A proof that the cartesian product of a line and such a decomposition is E^4 has also been given by K. W. Kwun.

Fox and Artin described [8] an arc in E^3 whose complement is not simply connected. Curtis has shown [7] that the cartesian product of a line and the decomposition of E^3 whose only nondegenerate element is this arc is E^4. He later genralized this result to show that any decomposition of E^n whose only nondegenerate element is an arc gives E^{n+1} under a cartesian product with a line. One might wonder about any monotone decomposition of E^3 each of whose nondegenerate elements is an arc.

Poénaru gives [12] an example of a 4 manifold with boundary different from a cell whose cartesian product with an arc is a 5 cell. One might wonder just how far these unexpected things go.

Presented to the Society, January 29, 1960; received by the editors March 26, 1960.

[1] Work on this paper was supported by National Science Foundation contract NSF G11665.

Question. Is there a complex different from a manifold whose cartesian product with a line is a manifold? The existence of such a complex would show that there is a triangulation of a manifold such that the star of a vertex is not a cell.

Raymond has shown [**13**] that $A \times B = M$ is a generalized manifold with possible boundary over a coefficient domain L if and only if A and B are generalized manifolds over L.

The following theorem shows another direction in which the pathologies are limited.

THEOREM. *A set is a 3 cell if its cartesian product with an interval is a 4 cell.*

PROOF. Suppose $X \times [0, 1] = I^4$. We give a sequence of lemmas exhibiting enough properties of X to insure that it is a 3 cell.

It follows from Lemma 1 that X can be embedded in E^3. Hence, we suppose that X lies in E^3 and use Bd X and Int X to denote the point set boundary and the interior respectively of the set X in E^3.

We show that Bd X is a 2 sphere. This follows from the Kline sphere characterization [**1**] since Bd X is connected (Lemma 9), locally connected (Lemma 10) and is not separated by any pair of its points (Lemma 12) but is separated by each simple closed curve in it (Lemma 13). Dyer has pointed out to me that instead of using the Kline sphere characterization, one can use results of Wilder [**17**; **18**] instead.

Since Int X is uniformly locally simply connected (Lemma 11), Bd X is tame from the inside [**6**]. This insures that X is a 3 cell.

Suppose M is a set whose cartesian product with an interval $[0, 1]$ is an n cell I^n. In the following lemmas we note certain properties of M.

LEMMA 1. *M can be embedded in E^{n-1}.*

PROOF. We show that each of $M \times 0$ and $M \times 1$ is contained in the $(n-1)$ sphere Bd I^n. Suppose $p \in M$ and $0 < \epsilon < 1$. If $p \times \epsilon$ were a point of the interior of the n cell $M \times [0, \epsilon]$, then $(p \times [\epsilon, 1]) + (M \times [0, \epsilon])$ would be an n cell with a feeler sticking out of its interior. The invariance of domain theorem [**10**, p. 96] implies that such a set cannot be imbedded in I^n. Hence $p \times \epsilon$ is a point of Bd$(M \times [0, \epsilon])$. Since there is a homeomorphism of $M \times [0, \epsilon]$ onto $M \times [0, 1]$ taking $p \times \epsilon$ onto $p \times 1$, $p \times 1$ is a point of Bd I^n. Similarly $p \times 0$ is a point of Bd I^n.

Since M can be imbedded in E^{n-1}, we suppose that it lies there. In the following lemmas we used Bd M and Int M to denote the point set boundary and interior respectively of M in E^{n-1}.

LEMMA 2. *M is of dimension $n-1$ at each of its points.*

PROOF. It is not of dimension more than $n-1$ at any point since it lies in E^{n-1}. It is not of dimension less than $n-1$ at any point or else $I^n = M\times[0, 1]$ would be of dimension less than n somewhere.

LEMMA 3. *M is the closure of an open subset of E^{n-1}.*

PROOF. If it were not such a subset, it would be of dimension less than $n-1$ at some point.

LEMMA 4. Bd $M\times(0, 1)\subset$Bd I^n.

PROOF. Suppose $p\in$Bd M and N is a spherical neighborhood of p in E^{n-1}. Then $p \times \epsilon$ $(0 < \epsilon < 1)$ is a limit point of $(N \times (0, 1)) - (M\times(0, 1))$. It follows from the invariance of domain that $p\times\epsilon$ does not lie in any open n manifold in $M\times(0, 1)$. Hence $p\times\epsilon$ is not a point of Int I^n.

LEMMA 5. Int $M\times(0, 1)\subset$Int I^n.

This result follows from the invariance of domain theorem. We summarize Lemmas 1, 4, 5 as follows:

$$\text{Bd}\, I^n = (M \times 0) + (M \times 1) + (\text{Bd}M \times (0, 1)),$$

$$\text{Int}\, I^n = \text{Int}\, M \times (0, 1).$$

We recall certain definitions. Let F_t $(0\leqq t\leqq 1)$ be a one parameter (the parameter is t) family of maps of a set A into B such that if F_t is regarded as taking $A\times t$ into B, F_t $(0\leqq t\leqq 1)$ takes $A\times[0, 1]$ continuously into B. Then F_t $(0\leqq t\leqq 1)$ is called a *homotopy pulling F_0 to F_1*. We are particularly interested in the case where $F_1(A)$ is a point. In this case we say F_1 is a *constant map*.

A map f of A into B is *inessential* if there is a homotopy of A into B pulling f to a constant map. If there is no such homotopy, f is called an *essential map*.

A subset A of B *can be shrunk to a point in B* if the cone over A can be mapped into B in such a way that the map is the identity on the base of the cone. (This cone lies in an abstract space and is not supposed to intersect B except at its base.) Hence, A can be shrunk to a point in B if the identity map of A into B is inessential.

A set is *contractible* if the identity map of the set into itself is inessential.

A set A is *uniformly locally connected in dimension n* if for each $\epsilon>0$, there is a $\delta>0$ such that each map of an n sphere S^n into a δ subset of A can be extended to map an $(n+1)$-ball bounded by S^n

into an ϵ subset of A. We say that the map of S^n into a δ subset of A can be shrunk to a constant on an ϵ subset of A.

In determining whether or not M is uniformly locally connected it is irrelevant which metric we use for M since M is compact. However, there is more of a problem when we deal with Int M. We suppose that M is given a prescribed metric and that this metric is the same on Int M. We suppose that the metric of $M\times[0, 1]=I^n$ is given by the metric of a cube in E^n rather than by the cartesian metric.

LEMMA 6. *M is uniformly locally connected in all dimensions.*

PROOF. This is because $M\times[0, 1]$ has these properties. To see that each map f of an m sphere S^m into a small subset of M can be shrunk to a constant map on a small subset of M, consider the map f' of S^m into $M\times 1/2$ given by $f'(p)=f(p)\times 1/2$. Consider a homotopy F_t $(0\leqq t\leqq 1)$ shrinking f' to a constant in a small subset of I^n. Let g be the projection of I^n onto M that takes $p\times\epsilon$ to p. Then gF_t $(0\leqq t\leqq 1)$ is a homotopy shrinking the map f to a constant on a small subset of M.

LEMMA 7. *M is contractible.*

The proof of Lemma 7 is similar to the proof of Lemma 6.

LEMMA 8. Bd $M\times 1/2$ *irreducibly separates* $M\times 0$ *from* $M\times 1$ *on* Bd I^n.

PROOF. Each point $p\times 1/2$ of Bd $M\times 1/2$ is needed to separate $M\times 0$ from $M\times 1$ on Bd I^n since $p\times[0, 1]$ is an arc on Bd I^n from $M\times 0$ to $M\times 1$.

LEMMA 9. Bd *M is connected.*

PROOF. This follows from Lemma 8 and the unicoherence of the $(n-1)$ sphere Bd I^n.

LEMMA 10. Bd *M is uniformly locally connected in all dimensions.*

PROOF. The proof is the same as the proof of Lemma 6 except that we define F_t to be a homotopy of S^m into Bd $I^n-((M\times 0)+(M\times 1))$.

REMARK. We do not prove that Bd M is homotopically trivial even in low dimensions. In fact Poénaru's example [12] shows that Bd M need not even be simply connected in the case where $n=5$.

LEMMA 11. Int *M is uniformly locally connected in all dimensions.*

PROOF. We are to prove that for each $\epsilon>0$ there is a $\delta>0$ such that each map of an m sphere S^m into a δ subset of Int M can be shrunk to a constant map on an ϵ subset of Int M.

Let h be the homeomorphism of M onto $M\times 1/2$ given by $h(p) = p\times 1/2$ and g the projection of I^n onto $M\times 1/2$ given by $g(p\times t) = p\times 1/2$. We suppose I^n has its ordinary Euclidean metric. Let $\epsilon_1>0$ be so small that each ϵ_1 subset of I^n has an image under $h^{-1}g$ of diameter less than ϵ. Let $\delta>0$ be so small that each δ subset of M has a diameter of less than ϵ_1 under h.

Let f be a map of S^m into a δ subset of Int M. Then hf takes S^m into an ϵ_1 subset of Int I^n. There is a homotopy F_t $(0\leqq t\leqq 1)$ shrinking hf to a constant map on an ϵ_1 subset of Int I^n. Then $h^{-1}gF_thf$ shrinks f to a constant map on an ϵ subset of Int M.

LEMMA 12. *No $(n-4)$ sphere in* Bd M *separates it.*

PROOF. Suppose T is a topological $(n-4)$ sphere in Bd M. Then $T\times(0, 1)$ is an $(n-3)$ manifold in Bd $M\times(0, 1)$. However, Bd $M\times(0, 1)$ is a connected open subset of Bd I^n and is hence a connected $(n-1)$ manifold. Since no $(n-3)$ manifold even locally separates an $(n-1)$ manifold, T does not separate Bd M.

LEMMA 13. *Each topological $(n-3)$ sphere T in* Bd M *separates it.*

PROOF. Consider $(M\times 0)+(M\times 1)+(T\times(0, 1))=X$. We show in Lemma 14 that there is an essential map of X onto an $(n-2)$ sphere. This shows [10] that X separates the $(n-1)$ sphere Bd I^n and that Bd $I^n-X=(\text{Bd } M-T)\times(0, 1)$ is not connected. Hence, Bd $M-T$ is not connected.

Lemma 13 can also be proved by using the methods of homology theory by showing that $(M\times 0)+(M\times 1)+(T\times(0, 1))$ contains an $(n-2)$ Čech cycle that does not bound in $(M\times 0)+(M\times 1)+(T\times(0, 1))$.

The proof of the following lemma was supplied by a conversation with E. E. Floyd and Walter Rudin.

LEMMA 14. *Suppose A_1, A_2 are two closed sets whose intersection is an n sphere S^n such that S^n can be shrunk to a point in each A_i. Then there is an essential map of A_1+A_2 onto an $(n+1)$ sphere.*

PROOF. Let C_1, C_2 be two $(n+1)$ cells sewed together along their boundaries to form an $(n+1)$ sphere. Let f be a map of A_1+A_2 into C_1+C_2 that takes the n sphere $A_1\cdot A_2$ homeomorphically onto Bd C_1 = Bd $C_2=C_1\cdot C_2$, A_1 into C_1, and A_2 into C_2.

We show that f is essential by showing that there is no homotopy F_t $(0\leqq t\leqq 1)$ of A_1+A_2 into C_1+C_2 that shrinks f to a constant map.

We show that the assumption that there is such a homotopy F_t leads to the contradiction that the identity map of C_1+C_2 onto itself is inessential.

Since $A_1 \cdot A_2$ can be shrunk to a point in A_1, there is a map g of C_1 into A_1 that is f^{-1} in taking $C_1 \cdot C_2$ onto $A_1 \cdot A_2$. Similarly, the map g can be extended to take C_2 into A_2. We call the extended map g.

Consider the map fg. It is homotopic to the identity since it takes each C_1 $(i=1, 2)$ onto itself and is the identity on Bd $C_1=$ Bd C_2. The assumption that f is inessential leads to the contradiction that F_tg $(0 \leqq t \leqq 1)$ is a homotopy pulling fg to a constant map.

It may be noted that the proof of Lemma 14 did not use the fact that $A_1 \cdot A_2$ is an n sphere but merely that $A_1 \cdot A_2$ contains an n sphere that is a retract of $A_1 \cdot A_2$.

A homological analogue of Lemma 14 gives that if A_1, A_2 are compact and $A_1 \cdot A_2$ contains an n cycle which bounds in each of A_1, A_2 but not in $A_1 \cdot A_2$, then A_1+A_2 contains an $(n+1)$ cycle that does not bound in A_1+A_2.

THEOREM. *Each factor of an n cell I^n, $n \leqq 4$, is a cell.*

PROOF. Szumbarski proved [**15**] this result for $n \leqq 3$.

Szumbarski also proved [**15**] that each 1 dimensional factor of an n cell is a 1 cell.

Young showed [**19**] that each 2 dimensional factor of an n cell is a 2 cell.

Young's proof of Theorem 1 in [**19**] showed that if one factor of an n cell is of dimension $n-1$, the other factor contains no triod. Hence, if a factor of I^4 is 3 dimensional, the other factor is an arc and it follows from our first theorem that the 3 dimensional factor is a 3 cell.

REFERENCES

1. R. H. Bing, *The Kline sphere characterization problem*, Bull. Amer. Math. Soc. vol. 52 (1946) pp. 644–653.

2. ———, *A decomposition of E^3 into points and tame arcs such that the decomposition space is topologically different from E^3*, Ann. of Math. vol. 65 (1957) pp. 484–500.

3. ———, *Necessary and sufficient conditions that a 3-manifold be S^3*, Ann. of Math. vol. 68 (1958) pp. 17–37.

4. ———, *The cartesian product of a certain nonmanifold and a line is E^4*, Bull. Amer. Math. Soc. vol. 64 (1958) pp. 82–84.

5. ———, *The cartesian product of a certain nonmanifold and a line is E^4*, Ann. of Math. vol. 70 (1959) pp. 399–412.

6. ———, *A surface S is tame in E^3 if E^3-S is locally simply connected at each point of S*, Abstract 546–74, Notices Amer. Math. Soc. vol. 5 (1958) pp. 180–181.

7. M. L. Curtis, *Corollary to a proof due to Bing*, Abstract 571–13, Notices Amer. Math. Soc. vol. 7 (1960) p. 842.

8. R. H. Fox and E. Artin, *Some wild cells and spheres in three-dimensional space*, Ann. of Math. vol. 49 (1948) pp. 979–990.

9. J. Glimm, *Two cartesian products which are Euclidean spaces*, Bull. Soc. Math. France vol. 88 (1960) pp. 131–135.

10. W. Hurewicz and H. Wallman, *Dimension theory*, Princeton, N. J., Princeton University Press, 1948.

11. A. Kirkor, *A remark about Cartesian division by a segment*, Bull. Acad. Polon. Sci. Sér. Sci. Math. Astr. Phys. vol. 6 (1958) pp. 379–381.

12. V. Poénaru, *Les décompositions de l'hypercube en produit topologique*, Bull. Soc. Math. France, vol. 88 (1960) pp. 113–129.

13. F. A. Raymond, *Separation and union theorems for generalized manifolds with boundary*, Michigan Math. J. vol. 7 (1960) pp. 7–21.

14. R. H. Rosen, *E^4 is the cartesian product of a totally non-Euclidean space and E^1*, Abstract 563-1, Notices Amer. Math. Soc. vol. 6 (1959) p. 641.

15. M. Szumbarski, *Sur la décomposition des éléments euclideans en produits cartésiens*, Mat. Sb. vol. 16 (1945) pp. 39–42.

16. J. H. C. Whitehead, *A certain open manifold whose group is unity*, Quart. J. Math. Oxford vol. 6 (1935) pp. 268–279.

17. R. L. Wilder, *A converse of the Jordan-Brouwer separation theorem in three dimensions*, Trans. Amer. Math. Soc. vol. 32 (1930) pp. 632–657.

18. ———, *On the properties of domains and their boundaries in E_n*, Math. Ann. vol. 109 (1934) pp. 273–306.

19. G. S. Young, Jr., *On the factors and fiberings of manifolds*, Proc. Amer. Math. Soc. vol. 1 (1950) pp. 215–223.

UNIVERSITY OF WISCONSIN

H. Others

Topology of 3-Manifolds

by R. H. Bing

§1. The monotone mapping problem

Euclidean 3-space R^3 provides more room for movement without collision than does the plane R^2. This partially explains why the Monotone Mapping Theorem holds for R^2 but not for R^3.

Consider how one may erode a cave into a unit 3-cell I^3. Suppose a hole is dug into the cube from one of its bases I^2. We let I_t^2 $(0 \leq t \leq 1)$ be the cross section of I^3 parallel to I^2 and at a distance t from I^2. Then I_1^2 is the base of I^3 opposite to $I_0^2 = I^2$. A disk D in Int I^2 (interior of I^2) serves as the mouth of the cave.

The cave is dug in stages. Figure 1.1 shows the result at an early stage where theleft side of the figure shows the portion removed and the right side shows the remainder. We regard the hole as an open subset of I^3 so that the remainder will be closed. The hole is the image of Int $I^3 \cup$ Int I^2 under a uniformly continuous homeomorphism. Also, Figure 1.1 is supposed to suggest that if X is a component of the intersection of the boundary (with respect to I^3) of the hole with any I_t^2,

Research in connection with this report supported by ONR.

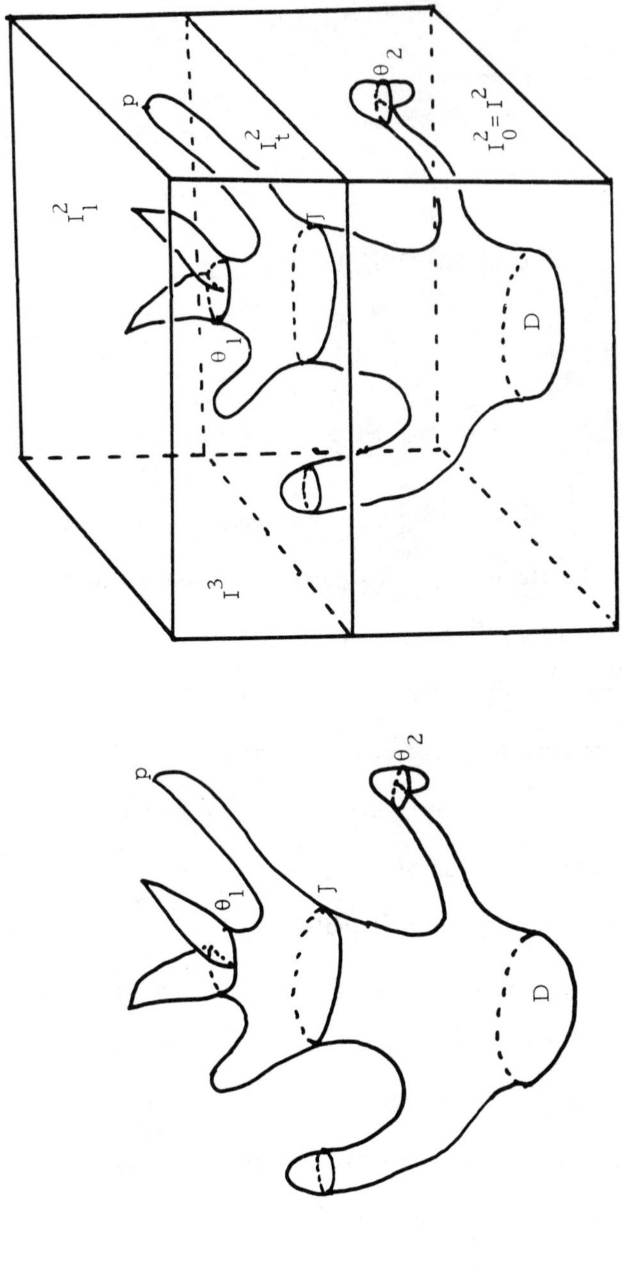

Figure 1.1

-2-

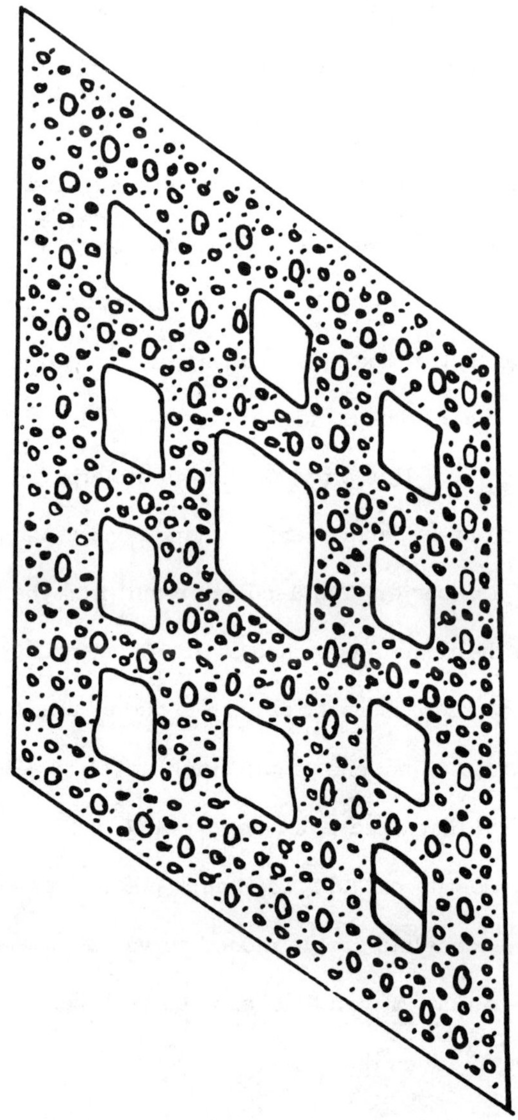

Figure 1. 2

then either X is a point p, a simple closed curve J, or a θ-curve. The θ-curve θ_1 resulted where the hole forked, and θ_2 resulted from a vertical turn.

Suppose the hole is extended by adding feelers until the cave becomes dense in I^3. Call the final cave or hole H. It can be chosen so that H is the image of Int $I^3 \cup$ Int I^2 under a uniformly continuous homeomorphism h. Note that h extends to a map of I^3 onto I^3.

One can obtain such a hole H so that for $0 < t < 1$, $I_t^2 \cap (I^3 - H)$ looks as shown in Figure 1.2. This cross section is a Sierpinski curve with the possible addition of a spanning arc to change one of the accessible simple closed curves to a θ-curve. The θ-curves result from the θ-curves mentioned earlier where the tunnels branched or turned vertically. There will not be more than a countable number of such θ-curves at all levels together and need not be more than one at any single level.

It follows from [25] that there is a map g_t of $I_t^2 \cap (I^3 - H)$ onto I_t^2 such that g_t is the identity on Bd I_t^2 and each point inverse is either the θ-curve, one of the accessible simple closed curves other than Bd I_t^2, or a point. This suggests that H can be chosen so that there is a map g of all of $I^3 - H$ onto I^3 such that

-4-

g is the identity on $\text{Bd } I^3 - \text{Int } I^2$,

$g[I_t^2 \cap (I^3 - H)] = I_t^2$ for each $0 \leq t \leq 1$, and

each nondegenerate point inverse is either one of the θ-curves, one of the accessible simple closed curves other than $\text{Bd } I_t^2$ in one of the $[I_t^2 - (I^3 - H)]$'s, or Bd D.

Results of Dyer and Hamstrom [13] are used in getting the map g.

The hole H has been called [9] a _drainage system_ for I^3. In particular, if f is a map of I^3 onto itself which is a homeomorphism on $\text{Int } I^3$ and takes I^2 homeomorphically into I^2, we call $f(\text{Int } I^2 \cup \text{Int } I^3)$ a drainage system for I^3. The hole H we have dug in I^3 establishes the following result. Details are found in [9].

Theorem 1.1. _There is a drainage system_ $H = f(\text{Int } I^2 \cup \text{Int } I^3)$ _and a monotone map_ g _of_ $I^3 - H$ _onto_ I^3 _which is the identity on_ $\text{Bd } I^3 - \text{Int } I^2$.

Recall that a map is _monotone_ if each point inverse is compact and connected. Our map g is monotone because each point inverse is either a point, a simple closed curve, or a θ-curve.

The hole we have dug in I^3 enables us to describe an interesting monotone map of I^3 onto itself.

Theorem 1.2. _There is a monotone map_ g_1 _of_ I^3 _onto itself such that_ g_1 _is the identity on_ $\text{Bd } I^3 - \text{Int } I^2$ _and each point inverse intersects_ I^2.

The proof of this result is found in [9]. We shall content ourselves here with describing the point inverses under g_1.

Let $H = f(\text{Int } I^2 \cup \text{Int } I^3)$ be the drainage system for I^3 promised by Theorem 1.1 and g be the associated monotone map of $I^3 - H$ onto I^3. Then g_1 agrees with g on $I^3 - H$. This causes g_1 to be the identity on $\text{Bd } I^3 - \text{Int } I_0^2$. Also, if $p \in I^3$, $g^{-1}(p) \subset g_1^{-1}(p)$. We shall now describe the point inverses further.

Let r be a retraction of I^3 onto $\text{Bd } I^3 - \text{Int } I^2$ such that r sends I^2 homeomorphically onto $\text{Bd } I^3 - \text{Int } I^2$, and each nondegenerate point inverse is a spanning segment of I^3. Then for each point $q \in \text{Bd } I^3 - I^2$, $fr^{-1}(q)$ is an arc in I^3 with one end on $\text{Int } D = f(\text{Int } I^2)$ and the other on $I^3 - H$. Then $g_1^{-1}(p)$ is the union of $g^{-1}(p)$ and all such arcs that intersect $g^{-1}(p)$.

It was shown in [9] that H, f, g, r could have been picked so that each $g_1^{-1}(p)$ is one of the following: $\text{Bd } D = f(\text{Bd } I^2)$, an annulus which lies in $\text{Int } I^3$ except for one of its bases which is on $\text{Int } D$, sets homeomorphic to the Cartesian product of an arc and a θ-curve where each of these sets lies in $\text{Int } I^3$ except for one of its bases which is in $\text{Int } D$, topological cones over one, two, three, or four points of $\text{Int } D$ where the rest of the cone lies in $\text{Int } I^3$, and topological cones over one, two, or three points of $\text{Int } D$ where the vertex of the cone lies on $\text{Bd } I^3 - I^2$ and the rest of the cone lies in $\text{Int } I^3$.

-6-

We use R^3_+ to denote 3-dimensional Euclidean half space--namely, the set of points of R^3 whose last coordinate is nonnegative. Since one can obtain a topological copy of R^3_+ by adding an infinite row of copies of I^3, each placed by the side of the preceding, the following result follows from Theorem 1.2.

Theorem 1.3. *There is a compact monotone map of* R^3_+ *onto itself such that each point inverse intersects* Bd R^3_+.

By letting $(x, y, -z)$ have the same image as (x, y, z), one obtains the following extension of Theorem 1.3.

Theorem 1.4. *There is a compact monotone map of* R^3 *onto* R^3_+.

The map could have been chosen so that each point inverse is either a finite graph, an annulus, or the Cartesian product of an arc and a θ-curve. The only finite graphs that are needed are arcs, simple closed curves, θ-curves, bouquets of two circles, bouquets of three circles, or the suspension of four point sets.

Questions. Is there a monotone map of R^3 onto R^3_+ with fewer sorts of point inverses? With simpler point inverses? With 1-dimensional point inverses?

While the hole H we dug in I^3 for the proof of Theorem 1.1 is not complicated, it leads to the interesting maps mentioned in

Theorems 1.2, 1.3, 1.4 and these theorems in turn lead to the final link in the solution of the monotone mapping problem. The <u>monotone mapping problem</u> asks if each monotone map of R^n onto itself is compact. Recall that a map is <u>compact</u> (or proper) if the inverse of each compact set is compact. The answer to the monotone mapping problem is trivial if $n = 1$ and G. T. Whyburn gave an affirmative solution [26] for $n = 2$. Using unpublished results of R. D. Anderson, L. C. Glaser gave a negative answer [15] for $n \geq 4$. Partial solutions have been given by E. H. Connell, Michael Olinick, and others. See [9].

Theorem 1.5. <u>There is a monotone map of</u> R^3 <u>onto itself that is not compact</u>.

A composition map $f \circ g$ will establish Theorem 1.5 where g is the monotone map of R^3 onto R^3_+ and f is a 1-1 map of R^3_+ onto R^3.

L. C. Glaser described a noncompact 1-1 map (continuous) of R^3_+ onto R^3 in [14]. Also, see [9]. While we shall not repeat the description here, we allude to how it is obtained. The set $\text{Int } I^3 \cup \text{Int } I^2$ serves as a model for R^3_+. One can obtain a 1-1 map taking $\text{Int } I^3 \cup \text{Int } I^2$ onto an open solid torus (open 3-cell with a handle) by sewing the top I^2_1 of I^3 onto the bottom I^2_0 with a homeomorphism. If the homeomorphism sews the top onto only a part of the bottom, some boundary is left. This left over boundary can be used to help cap off the handle. Glaser described a sewing of a part of $\text{Bd } I^3 - I^2$ onto I^2 so as to obtain a topological copy of R^3.

-8-

§2. R^3 versus R^n

Perhaps R^3 sheds more light on R^n than R^n does on R^3. Of course, if one knows the general properties of R^n, he knows a great deal about R^3. However, one frequently learns about high dimensional spaces by first discovering low dimensional analogues. In developing a proof, by the time three is reached, the complete pattern is frequently disclosed.

After gaining an understanding of the general case, one is inclined to seek a short proof. There is a possibility that the paring of the proof to bare essentials will result in simplification but there is also the danger that the proof will become obscure and unmotivative. The gradual unfolding of the proof may be more enlightening than a quick treatment of the end product. Many gain a better insight and appreciation of the final result by having first treated it in the more easily understood cases. In Section 3 we introduce some 3-dimensional proofs by first describing their easier 2-dimensional counterparts.

There are results which hold in certain dimensions and are either known not to hold in others or are too difficult to ascertain there. Let us consider two fundamental properties of R^n: the Schoenflies property and the approximation of homeomorphisms.

The traditional Schoenflies Theorem says that each topological 1-sphere in R^2 bounds a topological 2-cell in R^2. This theorem does not extend to R^3 to say that topological 2-spheres there bound topological 3-cells. Various wild spheres (as described by Antoine, Alexander, Fox, Artin, Bing, and others) show that this extension is not possible. Morton Brown showed [10] that a 2-sphere in R^3 does bound a topological 3-cell if it is collared from the inside. Brown proved the result for an (n-1)-sphere in R^n but the proof for $n = 3$ reveals the story. Although Brown's proof works for $n = 2$, this does not seem the proper setting for it since there are traditional proofs for this dimension that do not require the strengths of Brown's collared assumptions.

It follows from [6] that a 2-sphere in R^3 bounds a 3-cell if its bounded complementary domain is 1-ULC. It is unknown if this result extends to R^n.

In recent investigations by R. C. Kirby, L. C. Siebenmann, R. Edwards and others, it has been useful to get a canonical version of the Schoenflies Theorem as suggested in the following two paragraphs.

Suppose B^n is an n-cell with boundary Bd B^n. Let $\{h_\alpha\}$ be the set of homeomorphisms of B^n into R^n metrized by $d(h_1, h_2) = \max d(h_1(x), h_2(x))$, $x \in B^n$. Each h_α sends Bd B^n onto a topological (n-1)-sphere in R^n. Although each h_α determines a unique homeomorphism of Bd B^n into R^n, there are many different h_α's that agree on Bd B^n. It would be interesting to know if one could give an algorithm for continuously selecting from among the various h_α's that agree on Bd B^n. If $h_\beta|$ Bd B^n is close to $h_\gamma|$ Bd B^n, we would want

-10-

the element selected from the class of h_α's agreeing with h_β on Bd B^n to be close to the element selected from the class of h_α's agreeing with h_γ on Bd B^n. If this selection could be made, it could provide us with a means of canonically filling in certain holes in R^n.

Although we have not found a canonical way of continuously selecting a homeomorphism of B^n into R^n just by knowing the homeomorphism restricted to Bd B^n, the following weakened version of the question has been affirmatively answered by using the techniques of Morton Brown given in [10]. Let N be a neighborhood of B^n. There is a map $g : \{h_\alpha\} \to \{h_\alpha\}$ such that

$g(h_\alpha)$ agrees with h_α on Bd B^n, and

$g(h_\alpha) = g(h_\beta)$ if h_α agrees with h_β on N.

The technique used to prove this result for $n = 3$ works equally well in other dimensions. As we mentioned earlier, it is not known if we can get a map g which satisfies the further restriction that

$$g(h_\alpha) = g(h_\beta) \text{ if } h_\alpha \text{ agrees with } h_\beta \text{ on Bd } B^n.$$

Let us now consider a polyhedral version of the Schoenflies Theorem. There are several versions but we consider the following PL (piecewise linear) version. Suppose Δ^n is an n-simplex and h is a PL homeomorphism of Bd Δ^n into R^n. Can h be extended to a PL homeomorphism of Δ^n into R^n?

There is a relatively easy proof of the affirmative answer to the above question if $n = 2$. For $n = 3$ the answer is still affirmative [1, 16, 21] but the problem is a bit rickier. In fact some of the earlier proofs of the result contained slips--even for the case where each horizontal cross section of $h(\mathrm{Bd}\,\Delta^n)$ is either a closed curve or a point. For $n = 4$, the problem is so elusive that neither a proof nor a counterexample is at hand.

Let us consider the fundamental problem of approximating a homeomorphism by a PL homeomorphism. Let U be a bounded open subset of R^n and $\varepsilon : U \to (0,\infty)$ be a continuous positive function on U. For each homeomorphism $h : U \to R^n$, is there a homeomorphism $h' : U \to R^n$ such that

$d(h'(x), h(x)) \leq \varepsilon(x)$ for each $x \in U$ and

h' is locally PL at each point of U?

The solution for $n = 2$ is relatively easy. One can consider a triangulation T of U such that the simplexes of T have small mesh (relative to ε), images of simplexes of T have small mesh, and the diameters of simplexes approach 0 as simplexes approach $\mathrm{Bd}\,U$. (Here Bd denotes the point set boundary rather than the combinatorial boundary.) Then h' is defined on the 1-skeleton of T so that h' is PL on this 1-skeleton and h' is close to h. Finally the polyhedral version of the Schoenflies Theorem we mentioned earlier is used to extend h' to all of U.

-12-

The proof for $n = 3$ follows in a similar fashion. It is more difficult to find a homeomorphism which is a PL approximation to h on the 2-skeleton of the triangulation of U but this is possible as shown in [5]. This is considered again in the next section.

The proof eludes us for $n \geq 4$. We do not know how to get a PL homeomorphic approximation to h on the $(n-1)$-skeleton of the triangulation of U. Even if we could get such an approximation, we might be unable to get a version of the Schoenflies Theorem we need.

Consideration of a relative version of this approximation theorem has been fruitful [18]. Let B^k be a canonical k-cell. A homeomorphism h of $B^k \times R^{n-k}$ onto a combinatorial n-manifold with boundary V^n that is PL on $\text{Bd}\, B^k \times R^{n-1}$ is called a <u>handle</u>. We say that this <u>handle</u> can be straightened if there is a homeomorphism h' of $B^k \times R^{n-k}$ onto V^n such that

$h' = h$ on $\text{Bd}\, B^k \times R^{n-k}$,

h' has compact support, and

h' is PL on $B^k \times N$ for some neighborhood N of the origin in R^n.

The straightening of handles has been used [18] in the study of such fundamental problems as the triangulation of manifolds, PL approximation of homeomorphism, and the annulus conjecture. Not all handles can be straightened but for $n = 3$, all can. The study of handle straightening can be quite difficult for the uninitiated. Perhaps a study of handle straightening for $n = 3$ might form an understandable introduction.

§3. Approximating surfaces in R^3

The following theorem (along with various refinements of it) has been very useful in unlocking some of the secrets of 3-manifolds.

Theorem 3.1. (Approximation Theorem for 2-Spheres) [2, 4, 7, 8, 12]. *Suppose* S *is a topological* 2-*sphere in* R^3. *Then for each positive number* ε *there is an* ε-*homeomorphism* h *of* S *into a polyhedral* 2-*sphere in* R^3.

The outline of the original proof of the above approximation theorem is not difficult even though the details are hard. In the last section we noted that the study of low dimensional results sometimes leads to a clearer understanding of high dimensional phenomena. An understanding of the proof of the Approximation Theorem for 2-Spheres may be approached by a proof of the following theorem for the plane. We shall give two proofs of this planar theorem since there are proofs of the Approximation Theorem for 2-Spheres modeled after each.

Theorem 3.2. *Suppose* J *is a topological* 1-*sphere (simple closed curve) in* R^2. *Then for each positive number* ε *there is an* ε-*homeomorphism of* J *onto a polygon in* R^2.

First Proof. Let T be a triangulation of R^2 of mesh less than $\varepsilon/2$. We suppose T is chosen so that J misses the geometric 0-skeleton $|T^0|$ of T.

Our first step is to adjust J so that the adjusted J intersects the geometric 1-skeleton $|T^1|$ of T in only a finite number of components. To this end we select a point p_0 in $J - |T^1|$ and let δ be a number so small that any pair of points of $J \cap |T^1|$ within δ of each other lie on an $\varepsilon/2$ arc in $J - \{p_0\}$. Let $A_1, A_2, \ldots, A_j$ be a finite number of mutually exclusive δ-arcs in $|T^1| - |T^0|$ that cover $J \cap |T^1|$. See Figure 3.1.

If $J \cap A_1$ does not have more than one component, we leave J alone and let $J_1 = J$. It it has more than one component, we let p_1q_1 be the maximal arc in $J - \{p_0\}$ with its ends on $J \cap A_1$ and form J_1 by replacing the arc p_1q_1 in J by an arc from p_1 to q_1 in A_1. If $J_1 \cap A_2$ does not have more than one component, we leave it along and let $J_2 = J_1$. Otherwise, we let p_2q_2 be the maximal arc in $J_1 - \{p_0\}$ with its ends on $J_1 \cap A_2$ and form J_2 by replacing the arc p_2q_2 in J_1 by an arc from p_2 to q_2 in A_2. A continuation of this procedure leads to a simple closed curve J_j which agrees with J except on a finite number of $\varepsilon/2$ arcs. We note that $J_j \cap |T^0| = 0$, $J_j \cap |T^1|$ has only a finite number of components, and there is an $\varepsilon/2$ homeomorphism h_1 of J onto J_j.

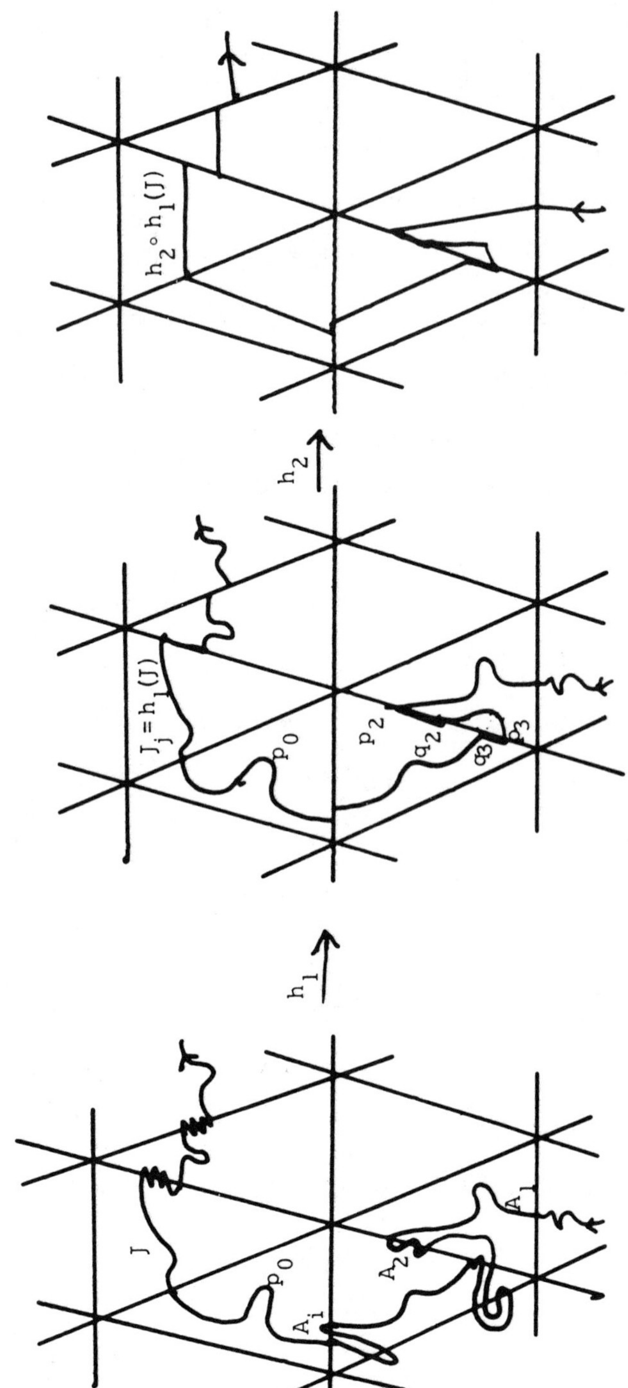

Figure 3.1

-16-

The second and final step is to get a homeomorphism h_2 of J_j onto a polygonal simple closed curve. We shall choose an h_2 that is fixed on $J_j \cap |T^1|$ so will only need to describe it on the finite set of components of $J_j - |T^1|$. If pq is the closure of such a component, it is a spanning arc of some 2-simplex Δ^2 of T. If p, q belong to different sides of Δ^2, we define $h_2(p, q)$ to be the straight segment in Δ^2 from p to q. After h_2 is defined on all such components with ends on different sides of Δ^2, it is defined on the pq's with ends on the same side of Δ^2. In this case $h_2(p, q)$ is a spanning broken line in Δ^2 from p to q. Care must be taken in defining these $h_2(p, q)$ so that h_2 is a homeomorphism on J_j.

Since each of h_1, h_2 is an $\varepsilon/2$-homeomorphism, $h_2 \circ h_1$ is an ε-homeomorphism. It may be noted that we made little use of the fact that J lies in R^2. The proof would have gone equally well if it had lain in a triangulated 2-manifold.

Second Proof. Rather than using a triangulation T of R^2 and using it to build a model for J, one could have used a curvilinear triangulation of J instead. Let T' be a curvilinear triangulation of J of mesh less than $\varepsilon/2$. We denote the 1-simplexes of T' by $A_1, A_2, \ldots, A_k$ and its 0-simplexes by $p_1, p_2, \ldots, p_k$. The required homeomorphism is fixed on the p_i's.

Let δ be a positive number less than the distance between any two nonadjacent A_i's. Let D_i be a round disk with center at p_i and radius $\delta/2$. Then A_i intersects only the two D's about its ends. Let A_i' be a subarc of A_i irreducible between these two D's.

Let γ be a positive number less than the distance between any two A_j''s. Let B_j' be a polygonal arc irreducible between the two D's about the ends of A_i and within $\gamma/2$ of A_i'. Then the required homeomorphism h sends A_i' onto B_i'. This homeomorphism h sends the closure of each component of $A_i - A_i'$ onto a radius of the appropriate D.

<u>Outline of first proof</u> [2, 4, 7] <u>of the Approximation Theorem for 2-Spheres</u>. Following the plan of the first proof of Theorem 3.2, we select a triangulation T of R^3 of small mesh and such that $S \cap |T^0| = 0$.

In the first proof of Theorem 3.2 we adjusted J so that the adjustment intersected $|T^1|$ in a nice set. In the present proof, we must select an approximation that is nice with respect to each of $|T^1|, |T^2|$.

As the first half of this adjustment we obtain a homeomorphism h_1 such that h_1 is near the identity, $|T^0| \cap h_1(S) = 0$, $|T^1| \cap h_1(S)$ is finite, and $|T^1|$ pierces $h_1(S)$ at each point of $|T^1| \cap h(S)$. It is not easy to get h_1 but getting it is not a wasted exercise since in obtaining it, one develops some fundamental properties of the intersection surfaces with cylinders in R^3. In a colloquium volume to follow these lectures a chapter will be devoted to considering such intersections. We shall not give the details for obtaining h_1 here.

-18-

After h_1 is obtained, one gets a homeomorphism h_2' of $h_1(S)$ into R^3 such that $h_2 = h_2' \circ h_1$ maintains the properties that

h_2 is near the identity,

$|T^0| \cap h_2(S) = 0$,

$|T^1| \cap h_2(S)$ is finite, and

$|T^1|$ pierces $h_2(S)$ at each point of $|T^1| \cap h_2(S)$.

It has the additional properties that if Δ^2 is a 2-simplex of T and X is a continuum in S such that $h_2(X)$ is a component of $\Delta^2 \cap h_2(S)$ that intersects $\mathrm{Bd}\, \Delta^2$, then

$\mathrm{Bd}\, \Delta^2 \cap h_2(X)$ is a set of two points and some point of $h_2(X)$ separates these two points in $h_2(X)$.

The inverses under h_2 of these two points of $\mathrm{Bd}\, \Delta^2 \cap h_2(X)$ are called the ends of X. We shall not describe h_2 since its description involves many subtle properties of R^3. We shall let $X_1, X_2, \ldots, X_k$ be the set of all such X's. The left side of Figure 3. 2 shows a typical intersection of $h_2(X)$ with some Δ^2.

We recall that some point p_i of X_i separates the two ends a_i, b_i of X_i from each other in X_i. Let X_i' be the closure of the component of $X_i - (\{a_i\} \cup \{b_i\})$ containing $\{p_i\}$. We say that X_i' was obtained from X_i by cutting off feelers. See Figure 3. 3.

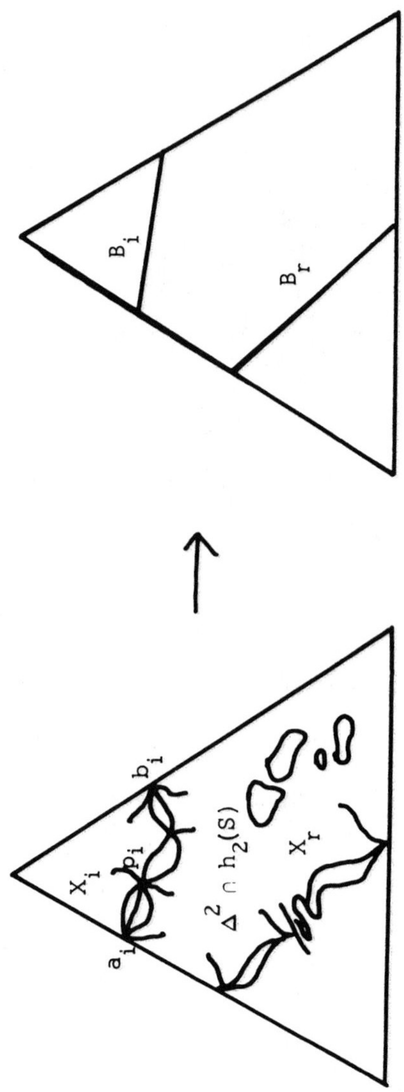

Figure 3.2

-20-

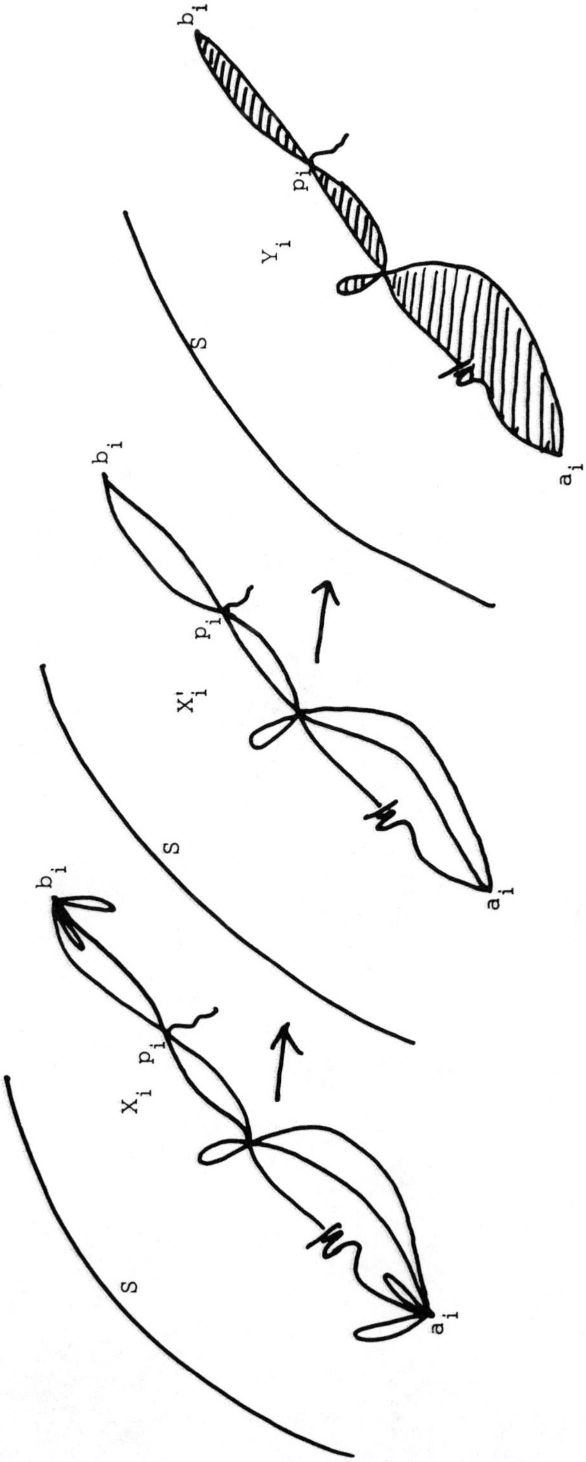

Figure 3.3

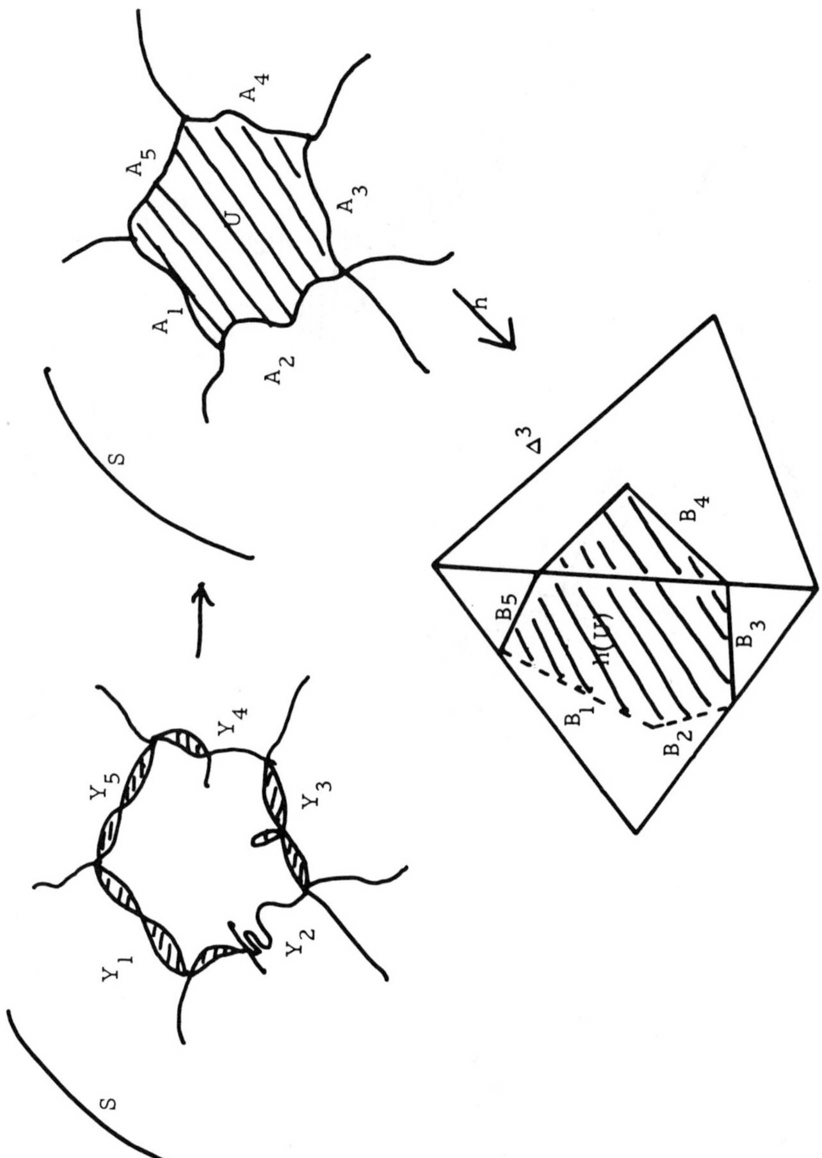

Figure 3. 4

-22-

Since X'_i is a small subset of S, $S - X'_i$ has only one large component. Let Y_i be the continuum obtained by filling in all the small holes in X'_i. The ends of X_i are called the ends of Y_i. As noted in Figure 3.3 the change from X_i to X'_i to Y_i occurs on S rather than on Δ^2.

Some component of $\cup Y_i$ chops up S into small pieces much like the 1-skeleton of a cellular subdivision of S subdivides S but with the exception that the Y_i's are continua rather than arcs. See Figure 3.4. Let $Y_1, Y_2, \ldots, Y_j$ be the Y's in this component that are not subsets of larger Y's.

We wish to replace the Y_i's with arcs A_i's. See center part of Figure 3.4. To facilitate this, we put each Y_i $(i = 1, 2, \ldots, j)$ in a disk D_i in S such that Y_i lies except for its ends in $\operatorname{Int} D_i$ and no two D_i's intersect each other except in a common end of the corresponding Y_i's. Let A_i be a spanning arc of D_i with the same ends as Y_i and which lies very close to X'_i.

If $X_i \subset \Delta^2$, let B_i be a polygonal spanning arc of Δ^2 such that no two B's in the same Δ^2 intersect each other. The homeomorphism h required by the Approximation Theorem for 2-Spheres is fixed on the ends of each A_i and sends each A_i onto the corresponding B_i. See right part of Figure 3.4.

It may be shown that if U is a component of $S - \cup A_i$, then U is an open disk in S such that $\operatorname{Bd} U$ is the union of A_i's all of whose corresponding B_i's lie on the boundary of one 3-simplex Δ^3 of T. The homeomorphism h is extended to take $\overline{U}$ into a spanning polygonal disk of Δ^3. Care is taken that no two of the h(U)'s intersect each other. Then the extended h is a homeomorphism of S onto a polyhedron.

<u>Outline of Alternative Proof</u>. James Cannon [12] can be credited for a second proof. (He actually gave a proof of the Side Approximation Theorem quoted in the next section but the essential ideas are related.) His proof might be regarded as patterned after the Second Proof of Theorem 3.2. He uses a curvilinear triangulation of S rather than a rectilinear triangulation of R^3. His proof avoids some of the complexities of the justifications of the homeomorphisms h_1, h_2 of the preceding proof but is still not available to the ordinary mathematician since it assumes knowledge of such concepts as linking, Dehn's Lemma, and the Loop Theorem. However, these are fundamental concepts for the modern 3-dimensional topologist so in this sense his treatment is elementary.

One first seeks a curvilinear triangulation T' of S whose 1-skeleton is nicely embedded in R^3. With this immediate goal in mind one uses the notion of linking to show that there is a 0-dimensional F_σ set X in S such that $R^3 - S$ is 1-ULC in $(R^3 - S) \cup X$. This means that for each $\varepsilon > 0$, there is a $\delta > 0$ such that each map of Bd Δ^2 into a δ-subset of $R^3 - S$ can be extended to send Δ^2 into an ε-subset of $(R^3 - S) \cup X$.

A curvilinear triangulation T' of S is obtained such that the mesh of T' is small and $|T'^1| \cap X = 0$. It is shown that there is a rectangular tubular neighborhood N of $|T'^1|$. Then h is extended to take $|T'^1|$ into a polyhedral core of this tubular neighborhood. Dehn's Lemma and the Loop Theorem are then used to find polyhedral disks in R^3 onto which h sends the 2-simplexes of T'.

§4. More approximation results

Some variations of the Approximation Theorem for 2-Spheres have had useful applications. The proofs of these variations are usually extensions and refinements of the proofs outlined in the preceding section. We give two such variations in this section.

Theorem 4.1. (Approximation Theorem of 2-Complexes) [5]. _Suppose_

> C _is a locally finite_ 2-_complex (not necessarily compact)_,
>
> $\varepsilon : C \to (0, \infty)$ _is a continuous positive function on_ C,
>
> M^3 _is a combinatorial_ 3-_manifold with boundary_,
>
> h _is a homeomorphism of_ C _into_ M^3.

Then there is a PL homeomorphism g _of_ C _into_ M^3 _such that_

$$d(g(x), h(x)) < \varepsilon(x) \text{ for each } x \in C.$$

Furthermore, if C' *is a subpolyhedron of* C *and* h *is already locally PL in a neighborhood of* C', *one can pick such a* g *that agrees with* h *on* C'.

The preceding theorem has many applications. We mention three.

1. If U is an open subset of a combinatorial 3-manifold with boundary and h is a homeomorphism of U into a combinatorial 3-manifold with boundary, h can be approximated by a PL homeomorphism [2, 5, 22, 24]. We mentioned a weak version of this result in Section 2. Moise gave the first proof of the result in [22].

2. Any separable metric space that is a topological 3-manifold with boundary can be triangulated [3, 5, 23, 24].

3. The Hauptvermutuang holds for 3-manifolds with boundaries. This means that if T, T' are triangulations (perhaps curvilinear) of the same topological 3-manifold with boundary, then T, T' have isomorphic subdivisions [23].

Theorem 4.2. (Side Approximation Theorem) [7, 8, 12]. *Suppose*

S *is a 2-sphere in* R^3,

ε *is a positive number, and*

U *is a component of* $R^3 - S$.

Then there are

a finite collection $D_1, D_2, \ldots, D_j$ *of mutually exclusive*

ε-*disks on* S *and*

an ε-*homeomorphism* h *of* S *onto a polyhedral 2-sphere in* R^3

such that

$$h(S - \cup D_i) \subset U.$$

One interesting application of Theorem 4.2 shows [6] that a 2-sphere in a 3-manifold is tame if its complement is 1-ULC. Another shows that if K is a crumpled cube, then there is a homeomorphism of K into R^3 so that S is a 3-sphere S^3 so that the closure of $S^3 - K$ is a *topological 3-cell* [17, 19].

The following extension of the Side Approximation Theorem is due to F.M. Lister. It is sometimes easier to apply than the original version of Theorem 4.2.

Theorem 4.3. (Lister's Side Approximation Theorem) [20]. *The homeomorphism* h *of Theorem* 4.2 *can be chosen so that* $S \cap h(D_i) \subset D_i$.

There are versions of the Side Approximation Theorem that apply to open subsets of 2-manifolds embedded in combinatorial 3-manifolds.

These and many other applications of related theorems will appear in the colloquium volume to follow these lectures. The paper by C. E. Burgess and James W. Cannon to appear in the Rocky Mountain Mathematical

Journal [11] perhaps contain the best compendium of such results at present.

University of Wisconsin
Madison, Wisconsin

References

1. J. W. Alexander, On the subdivision of 3-space by a polyhedron, Proc. Nat. Acad. Sci. U.S.A. 10 (1924), 6-8.

2. R. H. Bing, Approximating surfaces with polyhedral ones, Summary of Lectures and Seminars, Summer Institute on Set Theoretic Topology, Madison, 1955, 47-53.

3. __________, Locally tame sets are tame, Ann. of Math. (2) 59 (1954), 145-158.

4. __________, Approximating surfaces with polyhedral ones, Ann. of Math. (2) 65 (1957), 456-483.

5. __________, An alternative proof that 3-manifolds can be triangulated, Ann. of Math. (2) 69 (1959), 37-65.

6. __________, A surface is tame if its complement is 1-ULC, Trans. Amer. Math. Soc. 101 (1961), 294-305.

7. __________, Approximating surfaces from the side, Ann. of Math. (2) 77 (1963), 154-192.

8. __________, Improving the side approximation theorem, Trans. Amer. Math. Soc. 116 (1965), 511-525.

9. __________, The monotone mapping problem, Georgia Topology of Manifolds Conference, (1969), to appear.

10. Morton Brown, A proof of the generalized Schoenflies Theorem, Bull. Amer. Math. Soc. 66 (1960), 74-76.

11. C.E. Burgess and J.W. Cannon, Embeddings of surfaces in E^3, Rocky Mountain J. of Math., to appear.

12. J.W. Cannon, New proofs of Bing's approximation theorem for surfaces, Notices Amer. Math. Soc. 17 (1970), 544-445. Abstract 674-81.

13. E. Dyer and M.E. Hamstrom, Completely regular mappings, Fund. Math. 45 (1957), 103-118.

14. L.C. Glaser, Dimension lowering monotone non-compact mappings of E^n, Fund. Math., 58 (1966), 177-181.

15. __________, Monotone non-compact mappings of E^r onto E^k for $r \geq 4$ and $k \geq 3$, to appear.

16. W. Graeub, Semilinear Abbildungen, Sitz.-Ber. d. Wissensch. Heidelberg, 1950, 205-272.

17. Norman Hosay, The sum of a real cube and a crumpled cube is S^3, Notices Amer. Math. Soc. 10 (1963), 666. Abstract 607-17.

18. R.C. Kirby and L.C. Siebenmann, On the triangulation of manifolds and the Hauptvermutung, Bull. Amer. Math. Soc. 75 (1969), 742-749.

19. L.L. Lininger, Some results on crumpled cubes, Trans. Amer. Math. Soc. 118 (1965), 534-549.

20. F.M. Lister, Simplifying intersections of disks in Bing's side approximation theorem, Pacific J. Math. 22 (1967), 281-295.

21. E.E. Moise, Affine structures in 3-manifolds, II. Positional properties of 2-spheres, Ann. of Math. (2) 55 (1952), 172-176.

22. __________, Affine structures in 3-manifolds, IV. Piecewise linear approximations of homeomorphisms, Ann. of Math. (2) 55 (1952), 215-222.

23. __________, Affine structures in 3-manifolds, V. The triangulation theorem and Hauptvermutung, Ann. of Math. (2) 56 (1952), 96-114.

24. __________, Affine structures in 3-manifolds, VIII. Invariance of the knot type, local tame embedding, Ann. of Math. (2) 59 (1954), 159-170.

25. R. L. Moore, Concerning upper semi-continuous collections of continua, Trans. Amer. Math. Soc. 27 (1925), 416-428.

26. G. T. Whyburn, Compactness of certain mappings, Amer. J. Math. 81 (1959), 306-314.

PROCEEDINGS OF THE INTERNATIONAL
CONFERENCE ON GEOMETRIC TOPOLOGY
PWN – POLISH SCIENTIFIC PUBLISHERS
WARSZAWA 1980

Sets in R^3 with the two-disk property

by

R. H. Bing

Suppose T^3 is a solid torus in R^3 and $\{C_i\}$ is a finite collection of disjoint closed sets in T^3. We say that $\{C_i\}$ has the two-disk property if for each pair of diagonal meridional disks D_i, D_2 of T^3 there is a $C_j \in \{C_i\}$ such that C_j intersects both D_1 and D_2.

Suppose $X_1, X_2, \ldots$ is a sequence of closed sets in T^3 such that for each positive integer i, $X_{i+1} \subset X_i$ and the set of components of X_i has the two-disk property. Let G be the monotone decomposition of R^3 whose elements are the components of $\bigcap X_i$ and the points of $R^3 - \bigcap X_i$. Then the monotone decomposition R^3/G is topologically different from R^3. If each component of each X_{i+1} lies in a tame 3-cell in X_i, the collection G is cellular. For example, Bing and Starbird constructed such a decomposition whose collection of non-degenerate elements is a null sequence of cellular arcs. Starbird has examples where the elements are other things.

The challenging problem is to learn how one can take a 3-manifold-with-boundary M_1^3 in T^3 whose collection of components have the two-disk property and replace M_1^3 with another polyhedral 3-manifold-with-boundary M_2^3 such that $M_2^3 \subset M_1^3$, each component of M_2^3 lies in a 3-cell in M_1^3, and M_2^3 has the two-disk property. Several methods are discussed for obtaining such an M_2^3. It is shown how certain handles can be replaced.

PERMISSIONS

The American Mathematical Society gratefully acknowledges the kindness of these institutions in granting the following permissions:

Academia, Publishing House of the Czechoslovak Academy of Sciences:

Applications of the side approximation theorem for surfaces. *General Topology and its Relation to Modern Analysis and Algebra* (Proceedings of the Symposium, Prague, 1961), 91–95, Academic Press, New York-London, Publ. House Czech. Acad. Sci., Prague, 1962.

A hereditarily infinite dimensional space. *General Topology and its Relations to Modern Analysis and Algebra* II (Proceedings of the Second Prague Topological Symposium, 1966), 56–62, Academia, Prague, 1967.

Academic Press, Inc.:

A decomposition of S^3 with a null sequence of cellular arcs (with M. Starbird). *Geometric Topology* (J. C. Cantrell, editor), 3–21, Academic Press, New York-London, 1979.

Metrization problems. *General Topology and Modern Analysis* (L. F. McAuley and M. M. Rao, editors), 3–16, Academic Press, New York-London, 1981.

American Journal of Mathematics:

Converse linearity conditions. *Amer. J. Math.* **68** (1946), 309–318.

Skew sets. *Amer. J. Math.* **69** (1947), 493–498.

Solution of a problem of R. L. Wilder. *Amer. J. Math.* **70** (1948), 95–98.

Some characterizations of arcs and simple closed curves. *Amer. J. Math.* **70** (1948), 497–506.

Each disk in E^3 contains a tame arc. *Amer. J. Math.* **84** (1962), 583–590.

Embedding circle-like continua in the plane. *Canad. J. Math.* **14** (1962), 113–128.

Duke Mathematical Journal:

Conformal minimal varieties (with E. F. Beckenbach). *Duke Math. J.* **10** (1943), 637–640.

Extending a metric. *Duke Math. J.* **14** (1947), 511–519.

A homogeneous indecomposable plane continuum. *Duke Math. J.* **15** (1948), 729–742.

Snake-like continua. *Duke Math. J.* **18** (1951), 653–663.

A wild surface each of whose arcs is tame. *Duke Math. J.* **28** (1961), 1–15.

Taming complexes in hyperplanes (with J. M. Kister). *Duke Math. J.* **31** (1964), 491–511.

Encyclopaedia Britannica, Inc.:

Topology, General. *Encyclopaedia Britannica*, 15th edition, **18**, 509–514. Copyright ©1974 by Encyclopaedia Britannica, Inc.

Fundamenta Mathematicae:

Complementary domains of continuous curves. *Fund. Math.* **36** (1949), 303–318.

Conditions under which a surface in E^3 is tame. *Fund. Math.* **47** (1959), 105–139.

Point-like decompositions of E^3. *Fund. Math.* **50** (1962), 431–453.

A 3-dimensional absolute retract which does not contain any disk (with K. Borsuk). *Fund. Math.* **54** (1964), 159–175.

An arc is tame in 3-space if and only if it is strongly cellular (with A. Kirkor). *Fund. Math.* **55** (1964), 175–180.

A toroidal decomposition of E^3 (with S. Armentrout). *Fund. Math.* **60** (1967), 81–87.

Illinois Journal of Mathematics:

Decompositions that destroy simple connectivity. *Illinois J. Math.* **30** (1986), 527–535.

Institut Mittag-Leffler, The Royal Swedish Academy of Sciences:

Embedding surfaces in 3-manifolds. *Proceedings International Congr. Mathematicians* (Stockholm, 1962), 457–458, Inst. Mittag-Leffler, Djursholm, 1963.

Istituto Nazionale di Alta Matematica, Rome:

Pulling back feelers. *Symposia Mathematica XVI*, 245–266, Academic Press, New York London, 1974.

Iwanami Shoten, Publishers, Tokyo:

The elusive fixed point property. *Sugaku* **21** (1969), 203–210.

John Wiley & Sons, Inc.:

Some aspects of the topology of 3-manifolds related to the Poincaré conjecture. *Lectures in Modern Mathematics* **II** (T. L. Saaty, editor), 93–128, John Wiley & Sons, New York, 1964.

Journal de Mathematiques Pures et Appliquees:

A simple closed curve that pierces no disk. *J. Math. Pures et Appl.* **35** (1956), 337–343.

Markham Publishing Company through the courtesy of James C. Cantrell:

The monotone mapping problem. *Topology of Manifolds* (*Proceedings of the University of Georgia Topology of Manifolds Institute, 1969*) (J. C. Cantrell and C. H. Edwards, Jr., editors), 99–115, Markham Publ. Co., Chicago, 1970.

Mathematical Association of America:

An equilateral distance. *Amer. Math. Monthly* **58** (1951), 380–383.

Retractions onto spheres. *Amer. Math. Monthly* **71** (1964), 481–484.

Spheres in E^3. *Amer. Math. Monthly* **71** (1964), 353–364.

Elementary point set topology. Number 8 of the Herbert Ellsworth Slaught Memorial Papers, *Amer. Math. Monthly* **67** (1960), Part II, i–iv, 1–58.

Challenging conjectures. *Amer. Math. Monthly* **74** (1967), 56–64.

The elusive fixed point property. *Amer. Math. Monthly* **76** (1969), 119–132.

Award for distinguished service to Professor Raymond L. Wilder. *Amer. Math. Monthly* **80** (1973), 117–119.

Award for distinguished service to Professor R. D. Anderson. *Amer. Math. Monthly* **85** (1978), 73–74.

McGraw-Hill Book Company:

Set theory. *McGraw-Hill Encyclopedia of Science and Technology*, 1960, 205–206, Copyright ©1960 McGraw-Hill, New York (reprinted 1977).

Michigan Mathematical Journal:

Pushing a 2-sphere into its complement. *Michigan Math. J.* **11** (1964), 33–45.

Improving the intersections of lines and surfaces. *Michigan Math. J.* **14** (1967), 155–159.

Wild surfaces have some nice properties. *Michigan Math. J.* **33** (1986), 403–415.

MIT Press:

Point-set topology. *The Mathematical Sciences Essays for COSRIMS*, (COSRIMS, editors) 1969, 209–216. M.I.T. Press, Cambridge, Massachusetts

Oxford University Press:

Every simple closed curve in E^3 is unknotted in E^4 (with V. L. Klee). *J. London Math. Soc.* **39** (1964), 86–94.

Pacific Journal of Mathematics:

Concerning hereditarily indecomposable continua. *Pacific J. Math.* **1** (1951), 43–51.

Tame Cantor sets in E^3. *Pacific J. Math.* **11** (1961), 435–446.

The simple connectivity of the sum of two disks. *Pacific J. Math.* **14** (1964), 439–455.

Sets generated by rectangles (with W. W. Bledsoe and R. D. Mauldin). *Pacific J. Math.* **51** (1974), 27–36.

Super triangulations (with M. Starbird). *Pacific J. Math.* **74** (1978), 307–325.

Pergamon Press:

Shrinking without lengthening. *Topology*, to appear.

Pi Mu Epsilon, Inc., National Honorary Mathematics Society:

Examples and counterexamples. *Pi Mu Epsilon J.* **1** (1953), 311–317.

Prentice-Hall, Inc.:

Decompositions of E^3. *Topology of 3-Manifolds and Related Topics*, (M. K. Fort, Jr., editor), Prentice-Hall, Inc., 1962, 5–21.

Princeton University Press:

Mapping a 3-sphere onto a homotopy 3-sphere. In "Topology Seminar Wisconsin 1965," (R. H. Bing and R. J. Bean, editors), *Ann. of Math. Studies*, No. 60, 89–99, Copyright ©1966, Princeton University Press, Princeton, N.J., 1966.

PWN-Polish Scientific Publishers:

Sets in R^3 with the two-disk property. *Proceedings of the International Conference on Geometric Topology*, **43**, Polish Scientific Publishers, Warszawa, 1980.

PWS-KENT Publishing Company:

Radial engulfing. *Conference on the Topology of Manifolds 1967* (J. G. Hocking, editor), 1–18, Prindle, Weber & Schmidt, Boston, 1968.

Springer-Verlag New York Inc.:

An unusual map of a 3-cell onto itself. *Topology Conference* (R. F. Dickman, Jr., and P. Fletcher, editors), V.P.I. (1973), Lecture Notes in Math. No. 375, 15–33, Springer, Berlin-New York, 1974.

Vertical general position. In *General Topology* (L. C. Glaser and T. B. Rushing, editors), Lecture Notes in Math. No. 438, 16–41, Springer, Berlin-New York, 1975.

SUNY at Binghamton:

Monotone images of E^3 (with J. M. Martin). *Proceedings of Conference on Monotone Mappings and Open Mappings*, 1970 (L. F. McAuley, editor), 55–77, SUNY at Binghamton, N.Y., 1971.

Texas Tech University:

Models for S^3. Visiting Scholars' Lectures, Texas Tech. University Mathematics Series No. 9, 1–31, Texas Tech. Press, Lubbock, 1970–1971.

Washington State University:

Computing the fundamental group of the complements of curves. Technical Report 2, Washington State Univ., 1965, 28 pages.

The following articles were reprinted from publications of the American Mathematical Society:

On generalized convex functions (with E. F. Beckenbach). *Trans. Amer. Math. Soc.* **58** (1945), 220–230.

Generalizations of two theorems of Janiszewski. *Bull. Amer. Math. Soc.* **51** (1945), 954–960.

Collections filling up a simple plane web. *Bull. Amer. Math. Soc.* **51** (1945), 674–679.

Generalizations of two theorems of Janiszewski. II. *Bull. Amer. Math. Soc.* **52** (1946), 478–480.

Concerning simple plane webs. *Trans. Amer. Math. Soc.* **60** (1946), 133–148.

The Kline sphere characterization problem. *Bull. Amer. Math. Soc.* **52** (1946), 644–653.

Partitioning a set. *Bull. Amer. Math. Soc.* **55** (1949), 1101–1110.

Each disk in E^3 is pierced by a tame arc. *Amer. J. Math.* **84** (1962), 591–599.

Annals of Mathematics:

Sets cutting the plane. *Ann. of Math.* **47** (1946), 476–479.

A homeomorphism between the 3-sphere and the sum of two solid horned spheres. *Ann. of Math.* **56** (1952), 354–362.

Locally tame sets are tame. *Ann. of Math.* **59** (1954), 145–158.

Some monotone decompositions of a cube. *Ann. of Math.* **61** (1955), 279–288.

Upper semicontinuous decompositions of E^3. *Ann. of Math.* **65** (1957), 363–374.

Approximating surfaces with polyhedral ones. *Ann. of Math.* **65** (1957), 456–483.

A decomposition of E^3 into points and tame arcs such that the decomposition space is topologically different from E^3. *Ann. of Math.* **65** (1957), 484–500.

Necessary and sufficient conditions that a 3-manifold be S^3. *Ann. of Math.* **68** (1958), 17–37. (See also correction, *Ann. of Math.* **77** (1963), 210.)

An alternative proof that 3-manifolds can be triangulated. *Ann. of Math.* **69** (1959), 37–65.

The cartesian product of a certain nonmanifold and a line is E^4. *Ann. of Math.* **70** (1959), 399–412.

Approximating surfaces from the side. *Ann. of Math.* **77** (1963), 145–192.

Inequivalent families of periodic homeomorphisms of E^3. *Ann. of Math.* **80** (1964), 78–93.

Some remarks concerning topologically homogeneous spaces (with K. Borsuk). *Ann. of Math.* **81** (1965), 100–111.

Auburn University:

Extending monotone decompositions of 3-manifolds. *Proceedings of Auburn Topology Conference* (W. R. R. Transue, editor), 7–35, Auburn, Alabama, 1969.

Canadian Journal of Mathematics:

Metrization of topological spaces. *Canad. J. Math.* **3** (1951), 175–186.

A simple closed curve is the only homogeneous bounded plane continuum that contains an arc. *Canad. J. Math.* **12** (1960), 209–230.

A set is a 3 cell if its cartesian product with an arc is a 4 cell. *Proc. Amer. Math. Soc.* **12** (1961), 13–19.

Improving the side approximation theorem. *Trans. Amer. Math. Soc.* **116** (1965), 511–525.

A translation of the normal Moore space conjecture. *Proc. Amer. Math. Soc.* **16** (1965), 612–619.

A complete elementary proof that Hilbert space is homeomorphic to the countable infinite product of lines (with R. D. Anderson). *Bull. Amer. Math. Soc.* **74** (1968), 771–792.

Retractions onto ANR's. *Proc. Amer. Math. Soc.* **21** (1969), 618–620.

Topology of 3-manifolds. Notes distributed by American Mathematical Society in conjunction with Colloquium Lectures at the University of Wyoming, Laramie, Wyoming, August 24–28, 1970, Seventy-fifth summer meeting of the American Mathematical Society, 30 pages.

Extending monotone decompositions of 3-manifolds. *Trans. Amer. Math. Soc.* **149** (1970), 351–369.

Cubes with knotted holes (with J. M. Martin). *Trans. Amer. Math. Soc.* **155** (1971), 217–231.

Linear isotopies in E^2 (with M. Starbird). *Trans. Amer. Math. Soc.* **237** (1978), 205–222.

R H Bing, October 20, 1914–April 28, 1986. *Notices Amer. Math. Soc.* **33** (1986), 595–596.

ABCDEFGHIJ — 898

A convex metric for a locally connected continuum. *Bull. Amer. Math. Soc.* **55** (1949), 812–819.

Coverings with connected intersections (with E. E. Floyd). *Trans. Amer. Math. Soc.* **69** (1950), 387–391.

Higher-dimensional hereditarily indecomposable continua. *Trans. Amer. Math. Soc.* **71** (1951), 267–273.

A characterization of 3-space by partitionings. *Trans. Amer. Math. Soc.* **70** (1951), 15–27.

Partitioning continuous curves. *Bull. Amer. Math. Soc.* **58** (1952), 536–556.

Review of *General Topology* by W. Sierpiński. *Bull. Amer. Math. Soc.* **59** (1953), 410.

A convex metric with unique segments. *Proc. Amer. Math. Soc.* **4** (1953), 167–174.

A connected countable Hausdorff space. *Proc. Amer. Math. Soc.* **4** (1953), 474.

Partially continuous decompositions. *Proc. Amer. Math. Soc.* **6** (1955), 124–133.

What topology is here to stay? In "Summary of Lectures and Seminars," Summer Institute on Set Theoretic Topology, Madison (1955), Revised (1958), 25–27, University of Wisconsin.

Decompositions of E^3 into points and tame arc? "Summary of Lectures and Seminars, Summer Institute on Set Theoretic Topology, Madison (1955), Revised (1958), 41–48.

Approximating surfaces with polyhedral ones. In "Summary of Lectures and Seminars," Summer Institute on Set Theoretic Topology, Madison (1955), Revised (1958), 49–53, University of Wisconsin.

The pseudo-arc. In "Summary of Lectures and Seminars," Summer Institute on Set Theoretic Topology, Madison (1955), Revised (1958), 72–75, University of Wisconsin.

The cartesian product of a certain nonmanifold and a line is E^4. *Bull. Amer. Math. Soc.* **64** (1958), 82–84.

Another homogeneous plane continuum (with F. B. Jones). *Trans. Amer. Math. Soc.* **90** (1959), 171–192.

Each homogeneous nondegenerate chainable continuum is a pseudo-arc. *Proc. Amer. Math. Soc.* **10** (1959), 345–346.

Imbedding decompositions of E^3 in E^4 (with M. L. Curtis). *Proc. Amer. Math. Soc.* **11** (1960), 149–155.

A surface is tame if its complement is 1–ULC. *Trans. Amer. Math. Soc.* **101** (1961), 294–305.